TABLES OF THE HYPERGEOMETRIC PROBABILITY DISTRIBUTION

TABLES OF THE HYPERGEOMETRIC PROBABILITY DISTRIBUTION

GERALD J. LIEBERMAN *Stanford University*
DONALD B. OWEN *Sandia Corporation*

STANFORD UNIVERSITY PRESS
Stanford, California
1961

STANFORD UNIVERSITY PRESS

STANFORD, CALIFORNIA

Library of Congress Catalog Card Number: 61-6879

Printed in the United States of America

PREFACE

The hypergeometric probability distribution is not an easy function to compute, since reasonably accurate computation involves extensive factorial expansion with laborious calculations. For this reason, the function has usually been approximated by means of binomial, Poisson, and normal distributions.

In connection with another project, it was decided to undertake the programming and calculation of the hypergeometric distribution function for various sample and lot sizes. The sample and lot values were chosen to provide exact (six-decimal-place) point and cumulative probability values in the ranges where most sampling is done and where the usual approximations are poor.

The material presented here should be useful in many different disciplines. Research workers in the physical sciences and engineering, in industrial management, and in the social sciences, in particular, should find many applications of the distribution-free statistics whose probabilities can now be evaluated from these tables. Several of these statistics, including those based upon 2×2 tables and the number of exceedances, are discussed in the Introduction.

The mathematician and the statistician will find the equivalence of various sums of combinatorials of special interest. The results are similar to the well-known equivalence of the cumulative distribution of the negative binomial to the cumulative distribution of an ordinary binomial distribution. Shown for the first time, we believe, is the equivalence of the work on the hypergeometric distribution, per se, and the work on exceedance theory.

One author (G.J.L.) was partially supported in this study by the Office of Naval Research under Contract No. Nonr 225(53). The other author (D.B.O.) is at Sandia Corporation, a prime contractor to the United States Atomic Energy Commission; many of his contributions to this book first appeared in the form of technical memoranda issued by Sandia Corporation. These memoranda are listed in the Bibliography.

Drs. G. P. Steck and E. J. Gilbert contributed significantly to the Sandia Corporation memoranda as co-authors on two of them. Mr. C. M. Williams and Miss M. K. Weston, both of Sandia Corporation, programmed the computations of the logarithms of factorials and of the hypergeometric probability distribution, respectively. Mrs. Marjorie E. Endres of Sandia Corporation made spot checks of the tables for accuracy and completeness. Miss Anna Glinski of Stanford University made some calculations on the approximations to the hypergeometric distribution. The authors are indebted to all of these persons for their help, although full responsibility for the accuracy of the results presented here rests with the authors.

G. J. Lieberman
D. B. Owen

Stanford, California
Albuquerque, New Mexico
December 1960

CONTENTS

PART I

THE HYPERGEOMETRIC PROBABILITY DISTRIBUTION

$$\sum_{x=0}^{\infty} p(x) z^x = \frac{(N-k)^{[n]}}{N^{[n]}} F(-n, -k; N-k-n+1; z)$$
if $N-k \geqslant n$

THE HYPERGEOMETRIC PROBABILITY DISTRIBUTION

1. INTRODUCTION

1.1 The Hypergeometric Probability Distribution

Tabulations of the hypergeometric probability distribution have many potentially useful applications, some of which are not generally recognized as having any connection with the distribution. In Part I of this book we have attempted to present the theory and rationale of the hypergeometric probability distribution and to indicate as many of its specific applications as possible.

1.2 Definitions

The nomenclature of sampling inspection (one of the applications) is convenient to describe the parameters of the hypergeometric probability distribution and will be used here. The following symbols are defined:

$N =$ number of items in a lot,
$n =$ number of items in a sample taken from the lot,
$k =$ number of defective items in the lot,
$x =$ number of defective items observed in the sample.

Then the probability

Pr {Exactly x defectives are found in the sample}

$$= p(x) = p(N, n, k, x)$$

$$= \binom{k}{x}\binom{N-k}{n-x}\Big/\binom{N}{n} = \frac{k!\,n!}{(k-x)!\,(n-x)!\,x!}\,\frac{(N-k)!\,(N-n)!}{N!\,(N-k-n+x)!},$$

where x is an integer such that $\max[0, n+k-N] \leqq x \leqq \min[n, k]$, and

Pr {x defectives or fewer are found in the sample}

$$= P(x) = P(N, n, k, x)$$

$$= \sum_{i=\max[0,n+k-N]}^{x} \frac{k!\,n!}{(k-i)!\,(n-i)!\,i!}\,\frac{(N-k)!\,(N-n)!}{N!\,(N-k-n+i)!}.$$

1.3 Calculation of the Tables

Included at the end of the hypergeometric tables is a table of $\log N!$ for $N = 1(1)2000$ taken from [50]. This table was put on magnetic tape (all 15 decimal places for each N) and a program was prepared for an IBM 704 computer to sum the proper logarithms to give the logarithms of the point probabilities to 15 decimal places. The point probabilities were obtained by taking antilogarithms correct to at least eight decimal places; usually nine decimal places were obtained. The cumulative probabilities were calculated by summing the point probabilities. The results were rounded off to six decimal places within the IBM 704 computer and printed. The tables given here were produced by photographic means from this output.

As a check on the accuracy of the tables we made calculations on desk calculators of randomly selected values from each set of 200. We found no discrepancies. The cumulative probabilities were checked to see that for each set of values N, n, k there was an entry equal to 1 for $x = k$ and $N \leqq 100$. For $N > 100$, a check was made for

$$P(N, \tfrac{1}{2}N, k, \tfrac{1}{2}[k-1]) = \tfrac{1}{2}\,,$$

where N is even and k is odd.

1.4 Symmetries and Check

Since n and k may be interchanged in either of the probabilities $P(x)$ and $p(x)$ without changing the values of the probabilities, it is necessary to tabulate only for $k \leqq n$. *If $n < k$, it is necessary only to interchange n and k to read the probabilities directly from the tables.*

This volume tabulates the hypergeometric distribution for $N = 1(1)49$, $50(10)100$, and 1000. The values for $N = 1000$ are given only for $n = 500$. Some values for $N = 100(100)2000$ are also given.

All possible hypergeometric probabilitity distributions with $N \leqq 25$ are tabulated below, and the only symmetry that may have to be used in entering the table is the one on n and k mentioned above.

For $N > 25$, three additional symmetries were taken into account in order to keep the table to a reasonable size. These symmetries are given by the following equations. For point probabilities,

$$\begin{aligned} p(N, n, k, x) &= p(N, n, N-k, n-x) \\ &= p(N, N-n, k, k-x) \\ &= p(N, N-n, N-k, N-n-k+x)\,; \end{aligned}$$

and for cumulative probabilities,

$$\begin{aligned} P(N, n, k, x) &= P(N, N-n, N-k, N-n-k+x) \\ &= 1 - P(N, n, N-k, n-x-1) \\ &= 1 - P(N, N-n, k, k-x-1)\,, \end{aligned}$$

where the value of $P(N, n, k, x)$ is 1 if either $n-x-1$ or $k-x-1$ becomes negative. These three symmetries are immediately obvious if one considers the effect of interchanging the roles of defective and nondefective, and if one keeps in mind also that n and k can always be interchanged. They may be proved formally by substituting in the defining equations for $p(N, n, k, x)$ and $P(N, n, k, x)$ and showing that the resulting factorials and

TABLE I

N	n	k	x	$P(x)$	$p(x)$
			Set I		
16	6	4	0	0.115385	0.115385
16	6	4	1	0.510989	0.395604
16	6	4	2	0.881868	0.370879
16	6	4	3	0.991758	0.109890
16	6	4	4	1.000000	0.008242
			Set II		
16	10	4	0	0.008242	0.008242
16	10	4	1	0.118132	0.109890
16	10	4	2	0.489011	0.370879
16	10	4	3	0.884615	0.395604
16	10	4	4	1.000000	0.115385
			Set III		
16	12	6	2	0.008242	0.008242
16	12	6	3	0.118132	0.109890
16	12	6	4	0.489011	0.370879
16	12	6	5	0.884615	0.395604
16	12	6	6	1.000000	0.115385
			Set IV		
16	12	10	6	0.115385	0.115385
16	12	10	7	0.510989	0.395604
16	12	10	8	0.881868	0.370879
16	12	10	9	0.991758	0.109890
16	12	10	10	1.000000	0.008242

sums of factorials are equal. The symmetry involving n and k may be written

$$p(N, n, k, x) = p(N, k, n, x) \quad \text{and} \quad P(N, n, k, x) = P(N, k, n, x)\ .$$

The usefulness of this symmetry cannot be overemphasized and should always be kept in mind when dealing with the hypergeometric probability distribution.

As an example of the usefulness of the three symmetries mentioned above, consider the four sets of equivalent entries given in Table I. Clearly, when one of the sets of values is given, it will be easy to obtain the other three. These symmetries were used in producing the table for $N > 25$ in order to reduce the number of entries. This involved restricting the parameters so that $x \leqq k \leqq n \leqq \frac{1}{2}N$. These restrictions eliminated practically all of the duplications, the exceptions being values such that N is even and $n = \frac{1}{2}N$, where the distributions themselves are symmetric about $x = \frac{1}{2}k$. In this connection, for N even, $n = \frac{1}{2}N$, k odd, and $x = \frac{1}{2}(k-1)$, the value of $P(N, n, k, x)$ is $\frac{1}{2}$. This computation was found to be a useful and quick check on parts of the tables. Another useful and quick check on the tables is given by

$$\sum_{x=\max[0,n+k-N]}^{k} P(x) = 1 + k\left(1 - \frac{n}{N}\right), \qquad \text{for } n \geqq k .$$

The tables give all possible hypergeometric probability distributions for $N \leqq 25$. For $25 < N \leqq 50$ all possible distributions are given except that one of the set of three symmetries may have to be used to find the proper table entry. For $N = 60(10)100$ all values also are given except that one of the set of three symmetries may have to be used. For $N = 1000$, $n = 500$, all distributions are given except that point probabilities equal to 0 to six decimal places (0.000000) are not explicitly given and one of the set of three symmetries may have to be used. A table of probabilities is also given for $N = 100(100)2000$, $n = \frac{1}{2}N$ and $k = n - 1$, n, and all values of x.

As an example of the use of the symmetries in looking up probabilities in the table, consider the problem of finding $P(50, 40, 30, 20)$. The three equivalent values of the parameters are

$$P(N, N-n, N-k, N-n-k+x) = P(50, 10, 20, 0) ,$$
$$1 - P(N, n, N-k, n-x-1) = 1 - P(50, 40, 20, 19) ,$$
$$1 - P(N, N-n, k, k-x-1) = 1 - P(50, 10, 30, 9) .$$

The value that can be found in the table is the one with $n \leqq \frac{1}{2}N$ and $k \leqq n$. Therefore the appropriate value given in the table is $P(50, 20, 10, 0) =$ 0.002925. Note the interchange of n and k. Hence we find $P(50, 40, 30, 20) =$ 0.002925.

As a second example, suppose the value of $P(1000, 565, 500, 287)$ is needed. Note that this can be obtained from the tables, since n and k can be interchanged and the tables include all values of $N = 1000$, $n = 500$ which are greater than 0 to six decimal places and less than 1. The equivalent values are $P(1000, 435, 500, 222)$, $1 - P(1000, 565, 500, 277)$, and $1 - P(1000, 435, 500, 212)$. Now the value sought may be obtained from $1 - P(1000, 500, 435, 212) =$ $1 - 0.261791$. Hence we have $P(1000, 565, 500, 287) = 0.738209$.

1.5 Number of Entries in the Tables

If only the symmetry given by the interchangeability of n and k is taken into account, then the number of entries in a table of the hypergeometric distribution beginning with $N = 2$ is given by

$$S = \frac{N^4 + 12N^3 + 2N^2 - 12N - \begin{cases} 0 \text{ if } N \text{ is even} \\ 3 \text{ if } N \text{ is odd} \end{cases}}{48} .$$

If all entries such that $0 \leqq x \leqq k \leqq n \leqq \frac{1}{2}N$ are tabulated beginning with $N = 2$ up to $N = N$, the number of entries for even values of N is

$$S^* = \frac{1}{192}[N(N+2)(N^2 + 14N + 16)] ,$$

and for odd values of N,

$$S^* = \frac{1}{192}[(N-1)(N+1)(N+3)(N+13)] .$$

Thus for the tables given here, the number of entries for $N \leqq 25$ is $S = 12{,}064$; the number of entries from $26 \leqq N \leqq 50$ is $S^* = 43{,}550 - 3{,}458 = 40{,}092$. Similarly, the number of entries for $N = 60, 70, 80, 90$, and 100 can be computed, and for these values of N the number of entries is 66,750. Through $N = 100$, the total number of entries in the tables given here is then 118,906 entries. Since zero entries were eliminated for $N = 1000$, $n = 500$, and for $N = 100(100)2000$, $n = \frac{1}{2}N$ and $k = n-1$, n, it was not possible to predict the number of entries for these cases. There were, however, 15,433 entries for $N = 1000$, $n = 500$ and 1535 entries for the cases covered with $N = 100(100)2000$, giving a grand total of 135,874 entries in the hypergeometric tables given here.

2. APPLICATIONS

2.1 Applications to a Sequential Procedure

Given a lot of N items containing k defectives, a question that frequently arises is "How many items must be sampled from the lot to produce n nondefectives?" The solution to this problem may be obtained as follows:

Pr $\{x + n$ trials or fewer will be required to produce n nondefectives$\}$

$$= \frac{(N-k)!\,(N-n)!}{(N-k-n)!\,N!}\Big[1 + n\frac{k}{N-n} + \frac{n(n+1)}{2}\,\frac{k(k-1)}{(N-n)(N-n-1)} + \cdots$$
$$+ \frac{n(n+1)\cdots(n+x-1)}{x!}\,\frac{k(k-1)\cdots(k-x+1)}{(N-n)(N-n-1)\cdots(N-n-x+1)}\Big]$$
$$= \left[\binom{N-k}{n}\Big/\binom{N}{n}\right]\sum_{i=0}^{x}\binom{n+i-1}{i}\binom{k}{i}\Big/\binom{N-n}{i}$$
$$= \left[1\Big/\binom{N}{k}\right]\sum_{i=0}^{x}\binom{n+i-1}{n-1}\binom{N-n-i}{N-n-k},$$

where $0 \leqq x \leqq k$ and $N \geqq k + n$.

It can be shown that this probability reduces to

$$1 - P(N, x+n, N-k, n-1) = P(N, n+x, k, x)$$

by the following argument. Let N_n be the number of trials until n nondefectives first appear. Let y_h be the number of nondefectives in h trials. Suppose there are fewer than n nondefectives in the first h trials. Then more than h trials will be required to obtain n nondefectives. Also, if the number of trials needed to obtain n nondefectives is larger than h, then in the first h trials there will be fewer than n nondefectives. Since $N_n > h$ for exactly the same sequences for which $y_h < n$, it follows that $N_n \leqq h$ for exactly the same sequences for which $y_h \geqq n$.

Thus, we obtain

$$\begin{aligned}\Pr\{N_n \leqq h\} &= \Pr\{y_h \geqq n\} \\ &= 1 - P(N, x+n, N-k, n-1) \\ &= P(N, n+x, k, x)\,.\end{aligned}$$

This proves the identity

$$\sum_{j=0}^{N-n-k} \binom{k+j}{x}\binom{N-k-1-j}{n-1} \equiv \sum_{j=0}^{x} \binom{N-k}{n+x-j}\binom{k}{j},$$

which is a generalization of the identity (12.16) in Feller [16].

For example, suppose a lot of 50 items contains 10 defectives and it is necessary to obtain 20 nondefective items from the lot. The sampling will stop when the 20 nondefectives are obtained. What is the probability that the 20 nondefective items can be obtained with a sample of 25 or fewer? The answer is $P(50, 25, 10, 5) = 0.637399$.

The entire probability distribution for the possible sample sizes may be obtained from the hypergeometric table also. The distribution is given in Table II. Hence, for example, about 66 times out of 1000 it will be necessary to take a sample of 28 or more to obtain 20 nondefectives from the lot.

TABLE II

Sample size	Look up
20	$P(50, 20, 10, 0) = 0.002925$
21	$P(50, 21, 10, 1) = 0.022424$
22	$P(50, 22, 10, 2) = 0.085964$
23	$P(50, 23, 10, 3) = 0.219096$
24	$P(50, 24, 10, 4) = 0.417561$
25	$P(50, 25, 10, 5) = 0.637399$
26	$P(50, 26, 10, 6) = 0.820598$
27	$P(50, 27, 10, 7) = 0.934006$
28	$P(50, 28, 10, 8) = 0.983930$
29	$P(50, 29, 10, 9) = 0.998051$
30	$P(50, 30, 10, 10) = 1.000000$

As a second example, suppose a lot of 35 items is at hand, and it is necessary to obtain 20 nondefectives from this lot before sampling can cease. When sampling is stopped, it is necessary to make a statement about the number of defectives in the original lot, e.g., that the number of defectives in the lot is no more than k with 90 per cent assurance.

This problem may be solved by solving the inequality $P(35, x+20, k, x) \leqq 0.10$ for k. The results are shown in Table III. Now, for example, it can be said that if sampling stopped with a sample of 25 items to produce 20 nondefective items, one is at least 90 per cent sure that there were no more than 10 defective items in the lot before sampling; or equivalently, one is 90 per cent sure that there are no more than 5 defective items in the remaining 10 items.

TABLE III

Sample size taken	x	k	Actual probability
20	0	3	0.070
21	1	5	0.071
22	2	7	0.050
23	3	8	0.070
24	4	9	0.084
25	5	10	0.089
26	6	11	0.084
27	7	12	0.070
28	8	13	0.050
29	9	14	0.028
30	10	14	0.070
31	11	15	0.026
32	12	15	0.070

It might be instructive to compare this case with the case of an infinite lot. Suppose a sample of 25 is taken from a continuous production process to produce 20 nondefective items. With 90 per cent assurance, what is the upper bound on the proportion of defective items coming from this process?

The answer is obtained by computing the following upper confidence limit on a proportion where sampling is from a negative binomial distribution:

$$1 - \frac{n+1}{n+1+xF_{\gamma,2x,2n+2}},$$

where F_γ is an upper γ percentage point of the F distribution based on $2x$ degrees of freedom for the numerator and $2n+2$ degrees of freedom for the denominator. Here $n = 20$ and $x = 5$; hence $F_{0.90} = 1.75$ and the upper bound on the proportion of defectives produced by the process is $1 - 0.706 = 0.294$ with 90 per cent assurance. For the finite lot with $N = 35$, as pointed out above, the proportion of defectives in the original lot is less than $10/35 = 0.286$ with at least 90 per cent assurance, or the proportion of defectives in the remaining part of the lot is less than $5/10 = 0.500$ with at least 90 per cent assurance.

2.2 Applications to Tests of the Equality of Two Proportions (2 x 2 Tables)

A 2×2 contingency table is represented below.

Characteristic II	Characteristic I: Has	Characteristic I: Does not have	Totals
Has	x	$k - x$	k
Does not have	$n - x$	$N - n - k + x$	$N - k$
	n	$N - n$	N

The analysis considered here of this type of table is due to Fisher [19], p. 85; various probability tables have been prepared for easy application of the method [17], [18], [33], [37], and [62].

An example will make clear the usefulness of the hypergeometric table in testing a 2×2 table. We wish to evaluate two methods of encapsulating small batteries in plastic. The data are presented as follows:

Treatment	Performance		Totals
	Failure	Success	
Encapsulation Method I	9	6	15
Encapsulation Method II	3	11	14
	12	17	29

Hence, take $n = 12$, $k = 15$, $x = 9$, and $N = 29$. But in order to read this from the hypergeometric table, it is necessary to interchange k and n and to make use of the symmetry $P(N, n, k, x) = 1 - P(N, N - n, k, k - x - 1)$. Entries are now taken from the main table as given in Table IV.

TABLE IV

N	n	k	x	$P(x)$	$p(x)$
29	15	12	0	0.000002	0.000002
29	15	12	1	0.000107	0.000105
29	15	12	2	0.002132	0.002025
29	15	12	3	0.019685	0.017553
29	15	12	4	0.098672	0.078987
29	15	12	5	0.297267	0.198595
29	15	12	6	0.586885	0.289618
29	15	12	7	0.835130	0.248244
29	15	12	8	0.959252	0.124122
29	15	12	9	0.994357	0.035105
29	15	12	10	0.999623	0.005266
29	15	12	11	0.999991	0.000368
29	15	12	12	1.000000	0.000009

The question to be answered is "Does Method II have a better effect on performance than Method I?" This is a one-sided test since the outcome is interesting only if Method II is better than Method I. The relative proportion of successes using Method II is $11/14 = 0.79$ and for Method I is $6/15 = 0.40$. Hence, in this sample Method II shows a better performance than Method I. The question now becomes "Is this due to chance or is Method II really better than Method I?" Next, the statistical test will be performed to determine if Method II is indeed better than Method I. If Method II were worse than Method I in the sample, then no further statistical test would be performed because the hypothesis that Method II is no better than Method I is automatically accepted. Note that the procedure outlined above enables one to make a one-tailed test. For a two-tailed test no preliminary look at the proportions is necessary.

The probability of observing exactly nine failures is then $p(x) = 0.035105$. But it is necessary to find the probability of nine or more failures (a deviation as extreme as, or more extreme than, the one observed), and this can be obtained from the table by taking

$$1 - \Pr\{x \leqq 8\} = 1 - 0.959252 = 0.040748 .$$

Since this probability is less than 0.05, there is a significant difference between Method I and Method II at the 95 per cent level of significance.

If in the problem solved above there were no prior understanding that the outcome would be interesting only if Method II were better than Method I, then a two-sided test should be run. Looking at Table IV, one should say there is a significant difference at the 95 per cent level of significance only if $x \leqq 3$ or if $x \geqq 10$, since

$$\begin{aligned}\Pr\{x \leqq 3\} + \Pr\{x \geqq 10\} &= 0.019685 + 1 - 0.994357 \\ &= 0.025328 < 0.05 .\end{aligned}$$

Table IV shows that x cannot be raised to 4, since then $P(x) = 0.098672$. But x could be lowered to 9 at the upper end if x were lowered to 2 at the lower end. That is, another two-sided rule for determining significance at the 95 per cent significance level is to conclude that Method I differs from Method II if $x \leqq 2$ or if $x \geqq 9$, since

$$\begin{aligned}\Pr\{x \leqq 2\} + \Pr\{x \geqq 9\} &= 0.002132 + 1 - 0.959252 \\ &= 0.042881 < 0.05 .\end{aligned}$$

Both rules, $x \leqq 3$ or $x \geqq 10$ and $x \leqq 2$ or $x \geqq 9$, are equally good. The second might be preferred to the first on the ground that the actual test probability is closer to 0.05. The first might be preferred to the second on the ground that both tails of the distribution are below 0.025. The difficulty, of course, arises because of the discreteness of x. For further discussion of this point as applied to the above problem see [62].

2.3 Applications to the Distribution of the Number of Exceedances

Consider a random sample of size n taken from a continuous distribution. Let another random sample of size m, independent of the first sample, be drawn from the same population. The probability that x observations among the observations of the second sample will exceed the rth-largest observation in the first sample is given by

$$\begin{aligned}&\Pr\{x \text{ among } m \text{ future trials will exceed the } r\text{th-largest observation in a sample of } n\} \\ &\quad = \binom{m+n-r-x}{n-r}\binom{x+r-1}{r-1}\Big/\binom{m+n}{n} \\ &\quad = \frac{n}{m+n}p(m+n-1, m, x+r-1, x) \\ &\quad = \frac{r}{x+r}p(m+n, m, x+r, x) ,\end{aligned}$$

where $p(m+n, m, x+r, x)$ is the quantity defined in Section 1.2 and is equal to $p(x)$ in the hypergeometric table.

The probability that the largest among n past observations will be exceeded at most x times in m future trials is given by

$$\Pr\{x \text{ or fewer among } m \text{ future trials will exceed the largest among } n \text{ observations}\}$$
$$= \sum_{y=0}^{x} \binom{m+n-y-1}{n-1} \Big/ \binom{m+n}{n} = 1 - \binom{m+n-1-x}{n} \Big/ \binom{m+n}{n} .$$

The summation of the binomial coefficients was accomplished by means of Equation (12.6) of [16]. Hence,

$$\Pr\{x \text{ or fewer among } m \text{ future trials will exceed the largest among } n \text{ observations}\}$$
$$= 1 - P(m+n, n, x+1, 0), \qquad \text{for } 0 \leqq x \leqq m ,$$

and

$$\Pr\{x \text{ or more among } m \text{ future trials will exceed the largest among } n \text{ observations}\}$$
$$= P(m+n, n, x, 0), \qquad \text{for } 0 \leqq x \leqq m .$$

Also,

$$\Pr\{x \text{ or fewer among } m \text{ future trials will exceed the smallest observation in a sample of } n\}$$
$$= \sum_{y=0}^{x} \binom{y+n-1}{n-1} \Big/ \binom{m+n}{n} = \binom{n+x}{n} \Big/ \binom{m+n}{n} .$$

The summation of the binomial coefficients was accomplished by means of Equation (12.8) of [16]. Then,

$$\Pr\{x \text{ or fewer among } m \text{ future trials will exceed the smallest observation in a sample of } n\}$$
$$= P(m+n, n, m-x, 0), \qquad \text{for } 0 \leqq x \leqq m .$$

In general,

$$\Pr\{x \text{ or more among } m \text{ future trials will exceed the } r\text{th-largest among } n \text{ observations}\}$$
$$= \sum_{y=x}^{m} \binom{m+n-r-y}{n-r} \binom{y+r-1}{r-1} \Big/ \binom{m+n}{n}$$
$$= P(m+n, n, x+r-1, r-1), \qquad \text{for } 1 \leqq r \leqq n .$$

(See Section 2.1 for a proof of the equivalence of these sums.)

If $r = 1$, the formula for the probability that x or more among m future trials will exceed the largest among n observations is obtained.

Also, if $r = n$,

$$\Pr\{x \text{ or more among } m \text{ future trials will exceed the smallest among } n \text{ observations}\}$$
$$= P(m+n, n, x+n-1, n-1) ;$$

or

$$\Pr\{x \text{ or fewer among } m \text{ future trials will exceed the smallest among } n \text{ observations}\}$$
$$= 1 - P(m + n, n, x + n, n - 1)$$
$$= P(m + n, n, m - x, 0) .$$

The equivalence of the last two expressions follows from the second symmetry given in Section 1.4.

Consider the following example of the use of the hypergeometric table. What is the probability that the largest flood in the past 20 years will be exceeded at least once during the next 25 years? The answer (assuming no major change in weather patterns) is

$$\Pr\{\text{one or more among 25 future years will have a flood which exceeds the largest flood in the past 20 years}\}$$
$$= P(45, 20, 1, 0) = 0.55556 .$$

It is also possible to list the probability of x or more exceedances for all values of x such that $0 \leqq x \leqq 25$. These are

$P(45, 20, 0, 0) = 1.00000$	$P(45, 20, 8, 0) = 0.00502$
$P(45, 20, 1, 0) = 0.55556$	$P(45, 20, 9, 0) = 0.00231$
$P(45, 20, 2, 0) = 0.30303$	$P(45, 20, 10, 0) = 0.00102$
$P(45, 20, 3, 0) = 0.16209$	$P(45, 20, 11, 0) = 0.00044$
$P(45, 20, 4, 0) = 0.08490$	$P(45, 20, 12, 0) = 0.00018$
$P(45, 20, 5, 0) = 0.04349$	$P(45, 20, 13, 0) = 0.00007$
$P(45, 20, 6, 0) = 0.02174$	$P(45, 20, 14, 0) = 0.00003$
$P(45, 20, 7, 0) = 0.01059$	$P(45, 20, 15, 0) = 0.00001$

and the rest of the probabilities are 0 to five decimal places (0.00000). The use of this table makes it possible to pick a value of x such that the probability of x or more exceedances is less than some set probability.

As a second example, consider the following. In a sample of 15 capacitors, all met a certain criterion on voltage; i.e., all were operative below a stated voltage. What is the probability that in an additional sample of 20 all will also be operative below the stated voltage?

The method given here gives the following bound on the required probability:

$$\Pr\{\text{none among 20 future trials will exceed the largest among 15 observations}\}$$
$$= 1 - P(35, 15, 1, 0) = 1 - 0.571428 = 0.428572 .$$

The probabilitity requested is more than this value; i.e.,

$$\Pr\{\text{all observations in the second sample will be below the stated voltage}\} \geqq 0.428572 .$$

For a probability like this to be large, the first sample must be large compared with the second sample.

Wilks [63] gave a formula for the probability that x among m future trials will exceed the smallest in an initial sample of n values. Rosenbaum [56] tabulated critical values of x at the 5 per cent and 1 per cent levels and called the test a location test. In terms of the hypergeometric tables given here, Rosenbaum's table for the location test gives values of x such that

$$P(m+n, n, x, 0) \leqq 0.05, \text{ or } 0.01 ,$$

for $m, n = 1(1)50$. Wilks' result was extended by Gumbel and von Schelling [22] to the rth-smallest and moments of the distribution were given. Epstein [15] prepared some tables of these distributions for the case $m = n$. Wilks [63] also gave the probability distribution

$$\begin{aligned}&\Pr\{x \text{ or more among } m \text{ future observations will lie between}\\ &\quad \text{the largest and smallest values in a sample of } n\}\\ &\qquad = \sum_{y=x}^{m} n(n-1)(m-y+1)\frac{m!(n-2+y)!}{y!\,(n+m)!}\\ &\qquad = 1 - P(m+n, n, m-x+2, 1) .\end{aligned}$$

Rosenbaum [55] tabulated critical values of x for this problem also at the 5 per cent and 1 per cent levels and called the test a dispersion test. Rosenbaum's table for the dispersion test gives values of r such that $P(m+n, n, r+1, 1) \leqq 0.05$, or 0.01, where $r = m - x$, and $m, n = 1(1)50$.

Mosteller [42] derived the probability that one of c samples (of size n each) has r or more observations larger than all of the observations of the other $(c-1)$ samples. He tabulated this probability [42, p. 62]. The required probability may be obtained from the tables given here by computing $cP(cn, [c-1]n, r, 0)$. Mosteller and Tukey [43] have extended these results to unequal size samples, and have given some approximations to the probability.

As an additional example, what is the probability that one additional observation from a population will be between the extremes of a sample of size n? The answer is obtained by determining

$$\begin{aligned}&\Pr\{x \text{ or more among } m \text{ future observations will be between}\\ &\quad \text{the largest and smallest values in a sample of } n\}\\ &\qquad = 1 - P(m+n, n, m-x+2, 1) .\end{aligned}$$

Or for the problem at hand, find

$$1 - P(n+1, n, 2, 1) = 1 - \frac{2}{n+1} = \frac{n-1}{n+1} .$$

Another question can now be answered from this last equation. That is, how large a sample is needed to be 95 per cent sure that the next observation is between the extremes of the sample? The answer is obtained from $(n-1)/(n+1) = 0.95$, or $n = 39$.

It is interesting to compare this result with distribution-free tolerance limits for an infinite population. If one uses Table IV of [47], it is possible to say with 95 per cent confidence that at least 88.37 per cent of the future observations will lie between the extremes of the sample.

Wilks [63] also gives the limiting values of x/m as m tends to infinity and n remains fixed and the probability is fixed at γ. For $P(m+n, n, x, 0) = \gamma$,

$$\lim_{m\to\infty} \frac{x}{m} = 1 - \frac{n}{n + F_{\gamma,2,2n}},$$

where $F_{\gamma,2,2n}$ is the upper 100γ percentage point of the F distribution with 2 degrees of freedom for the numerator and $2n$ degrees of freedom for the denominator.

For example, if $\gamma = 0.05$ and $n = 10$,

$$\lim_{m\to\infty} \frac{x}{m} = 1 - \frac{10}{10 + 3.49} = 0.259 .$$

For $P(m + n, n, x + 1, 1) = \gamma$

$$\lim_{m\to\infty} \frac{x}{m} = 1 - \frac{(n-1)}{(n-1) + 2F_{\gamma,4,2n-2}} .$$

For example, if $\gamma = 0.05$ and $n = 10$,

$$\lim_{m\to\infty} \frac{x}{m} = 1 - \frac{9}{9 + 2(2.93)} = 0.394 .$$

2.4 Applications to the Binomial Distribution (Bayesian Prediction)

Fisher [19] gives the distribution of c successes in $c + d$ trials where a successes in $a + b$ trials have been observed and a uniform a priori distribution between 0 and 1 on p, the probability of success, is assumed. The distribution is

$$\Pr\{c|a\} = \frac{(a+b+1)!}{a!\,b!} \frac{(c+d)!}{c!\,d!} \frac{(a+c)!\,(b+d)!}{(a+b+c+d+1)!},$$

or written another way,

$$\begin{aligned}\Pr\{&c \text{ successes in } m \text{ trials} \mid a \text{ successes in } n \text{ trials} \\ &\text{and } p \text{ uniformly distributed between 0 and 1}\} \\ &= \frac{(n+1)!}{a!\,(n-a)!} \frac{m!}{c!\,(m-c)!} \frac{(a+c)!\,(m+n-c-a)!}{(m+n+1)!} \\ &= \frac{n+1}{m+n+1} p(m+n, a+c, m, c) \\ &= \frac{a+1}{a+c+1} p(m+n+1, a+c+1, m, c) ,\end{aligned}$$

and

$$\begin{aligned}\Pr\{c \leqq x \mid a, n, m, p \text{ uniform } (0, 1)\} \\ &= \sum_{j=0}^{x} \binom{a+j}{a} \binom{m+n-a-j}{n-a} \Big/ \binom{m+n+1}{m} \\ &= 1 - P(m+n+1, x+a+1, n+1, a) \\ &= P(m+n+1, x+a+1, m, x) \,.\end{aligned}$$

Hence these probabilities can be found in the hypergeometric table.

2.5 Applications to Sampling Inspection

One of the most important applications of the hypergeometric probability distribution is in the field of sampling inspection. A lot of size N containing k defectives is submitted for inspection. A sample of size n is drawn and the lot is accepted if the number of defectives found in the sample is less than or equal to x. The probability of accepting the lot (the operating characteristic curve) is given by $P(N, n, k, x)$. Because of the difficulty of obtaining $P(N, n, k, x)$ prior to the construction of these tables, most of the existing sets of sampling inspection plans used the binomial or Poisson approximations to calculate the operating characteristic curves.

By way of comparison, let us suppose that Military Standard 105B, whose operating characteristic curves are based upon the binomial approximation, is to be used with normal inspection at Level II, AQL $= 10.0\%$, and $N = 8$. The appropriate sampling plan calls for a sample size of 2 and an acceptance number of 0. The exact probability of acceptance for various fraction defectives in the lot can be read from the hypergeometric tables:

$$\begin{aligned}P(8, 2, 0, 0) &= 1.000000 & P(8, 2, 4, 0) &= 0.214286 \\ P(8, 2, 1, 0) &= 0.750000 & P(8, 2, 5, 0) &= 0.107143 \\ P(8, 2, 2, 0) &= 0.535714 & P(8, 2, 6, 0) &= 0.035714 \\ P(8, 2, 3, 0) &= 0.357143\end{aligned}$$

(If k exceeds $6 = N - n + x$, the probability of acceptance is zero since it becomes impossible to have 2 nondefective items in the sample.) It is interesting to note that had the binomial approximation been used, the entries corresponding to $k = 1$, 2, and 5 would have been 0.7657, 0.5625, and 0.1407, respectively (rounded to 4 decimal places). Obviously, the operating characteristic curves based upon the binomial are poor approximations to the true curves in certain regions of Military Standard 105B. (See Section 3.1.)

3. APPROXIMATIONS TO THE HYPERGEOMETRIC PROBABILITY DISTRIBUTION

3.1 Binomial Approximations

When dealing with a large lot and a small sample from the lot, it is common practice to use the binomial distribution to obtain relevant probabilities. That is, let n be the number of repetitions of a simple alternative and $p = k/N$ be the probability of obtaining a defective on a single trial.

Then the probability that the number of defectives in a sample of n repetitions of the simple alternative is less than or equal to x is given by

$$\sum_{i=0}^{x}\binom{n}{i}p^i(1-p)^{n-i} = 1 - E(n, x+1, p) ,$$

where the E notation is that used by the Harvard Computation Laboratory [27], which extensively tabulates this distribution.

The various symmetries of the hypergeometric probability distribution then lead to the following possible approximations:

$$P(N, n, k, x) \cong 1 - E(n, x+1, k/N) ,$$

or

$$P(N, n, k, x) \cong 1 - E(k, x+1, n/N) ,$$

or

$$P(N, n, k, x) \cong E(N-n, k-x, k/N) ,$$

or

$$P(N, n, k, x) \cong E(N-k, n-x, n/N) .$$

Examination of these various approximations shows that for any given problem one should use that approximation $E(a, b, p)$ which has the smallest value of a. For example, suppose $P(50, 20, 5, 2)$ is to be approximated. The exact value is 0.690029, and the approximate values are

$$P(50, 20, 5, 2) \cong 1 - E(20, 3, 0.1) = 0.67693 ,$$
$$P(50, 20, 5, 2) \cong 1 - E(5, 3, 0.4) = 0.68256 ,$$
$$P(50, 20, 5, 2) \cong E(30, 3, 0.1) = 0.58865 ,$$
$$P(50, 20, 5, 2) \cong E(45, 18, 0.4) = 0.55639 .$$

Obviously $1 - E(5, 3, 0.4)$ gives the best approximation and is the one with the smallest sample size for the binomial. If the rule of using that binomial approximation with the smallest sample size is followed, then the approximation is exact for $k = 1$ and $x = k$. The first approximation is usually good if n/N is small, say $n \leqq 0.1N$ [26], [16]. For the other approximations, k, $N - n$, or $N - k$ should be substituted for n in $n \leqq 0.1N$.

Intuitively, as the sample becomes a larger and larger part of the lot, the binomial approximation $1 - E(n, x+1, k/N)$ becomes poorer and poorer as an approximation for $P(N, n, k, x)$, since the finiteness of the lot plays a bigger and bigger role. However, the worst possible case is that for $n = \frac{1}{2}N$, since, as pointed out above, if $n > \frac{1}{2}N$, then an equivalent hypergeometric probability with $n < \frac{1}{2}N$ can be found. We must also remember that the roles of n and k may be interchanged so that all of the above statements concerning n can also be made with k substituted for n.

Wise [64] refined the binomial approximation and gave the following formulas for sums of hypergeometric probabilities. If N, n, k, and x are known, then

$$1 - \gamma = \sum_{r=0}^{x} \frac{k!\,n!\,(N-k)!\,(N-n)!}{(k-r)!\,(n-r)!\,r!\,N!\,(N-k-n+r)!}$$
$$\cong I_h(n-x, x+1) + \frac{n(n-1)}{24M^2} I_2 \,,$$

where

$$M = N - \tfrac{1}{2}n + \tfrac{1}{2}\,, \qquad h = (N - k - \tfrac{1}{2}n + \tfrac{1}{2}x + \tfrac{1}{2})/M\,,$$

and

$$I_2 = (n+1)\,I_h(n-x, x+1) - (n-x+1)\,I_h(n-x-2, x+1)$$
$$- (x+2)\,I_h(n-x, x-1) + 2I_h(n-x-1, x)\,,$$

and

$$I_h(n-x, x+1) = \frac{n!\displaystyle\int_0^h t^{n-x-1}(1-t)^x\,dt}{(n-x-1)!\,x!}$$

is the incomplete beta function tabulated by Karl Pearson [51]. Note that

$$I_h(n-x, x+1) = \sum_{y=n-x}^{n} \binom{n}{y} h^y (1-h)^{n-y} = E(n, n-x, h)\,.$$

The sums of the first two terms of the expansion seem accurate enough for almost all practical purposes. One term is usually sufficient when the sample size is smaller than 0.4 of the lot size, or when both sample and lot are very large. The accuracy of the first term is seen in the following examples for $N = 50$ and $n = 7$ (Table V). The values of $P(50, 7, k, x)$ are exact for the stated k and x, whereas Wise's approximation uses the first term only.

TABLE V

k	x	$P(50, 7, k, x)$	Wise's approximation
5	4	0.999990	0.999979
10	4	0.997947	0.997802
15	4	0.980293	0.979648
20	4	0.920062	0.919208
25	4	0.791287	0.791054
30	4	0.589652	0.588730
35	4	0.347663	0.348476
40	4	0.132620	0.132944
45	4	0.015637	0.016674
25	0	0.004813	0.004925
25	2	0.208713	0.208946
25	4	0.791287	0.791054
25	6	0.995187	0.995075
25	7	1.000000	1.000000

If N, n, x and γ are known, Wise's approximation can be used to find k. First solve for h_0:

$$I_{h_0}(n-x, x+1) = 1-\gamma .$$

The first approximation is then

$$k = (N - \tfrac{1}{2}n + \tfrac{1}{2})(1-h_0) + \tfrac{1}{2}x .$$

The second approximation makes use of h_0 and is

$$h = h_0 + \frac{\delta}{24M^2}, \qquad \text{where } M = N - \tfrac{1}{2}n + \tfrac{1}{2} ,$$

and

$$\delta = \left(\frac{1}{h_0} - h_0\right)(n-x)^2 + \left(1 - h_0 - \frac{1}{1-h_0}\right)(x+1)^2 + (1-2h_0)[x(n-x-1)-1] + \frac{1}{1-h_0} - \frac{1}{h_0},$$

and then

$$k = (N - \tfrac{1}{2}n + \tfrac{1}{2})(1-h) + \tfrac{1}{2}x .$$

Wise's approximation using the first term only may be rewritten

$$P(N, n, k, x) \cong 1 - E(n, x+1, p) ,$$

where

$$p = \frac{k - \dfrac{x}{2}}{N - \dfrac{n-1}{2}} .$$

In [49] it is shown that

$$P(N, n, k, 0) \leqq \left(1 - \frac{n}{N - \dfrac{k-1}{2}}\right)^k ,$$

which means that the first approximation given by Wise is in fact an upper bound on the probability when $x = 0$. Suppose that it is necessary to find a sample size such that if no defectives were found in the sample it is possible to say with 100γ per cent confidence that there are no more than k defectives in the sample. This is done by solving the above inequality for n:

$$n \leqq [1 - (1-\gamma)^{1/k}]\left(N - \frac{k-1}{2}\right) ,$$

where $P(N, n, k, 0) = \gamma$. An extensive table of $1 - (1-\gamma)^{1/k}$ is given in [49].

For $x \geqq 1$, $k \leqq n \leqq \frac{1}{2}N$, the binomial approximation becomes a lower bound on the hypergeometric probabilities; i.e.,

$$P(N, n, k, x) \geqq 1 - E(k, x+1, n/N) ,$$

when $1 \leqq x \leqq k \leqq n \leqq \frac{1}{2}N$. When $x = 0$, the binomial approximation is an upper bound but Wise's first approximation as noted above is a better upper bound. Since n and k can be interchanged, this same inequality can give a 100γ per cent upper confidence limit on k; i.e.,

$$k \leqq [1 - (1-\gamma)^{1/n}]\left(N - \frac{n-1}{2}\right),$$

when $x = 0$ and $P(N, n, k, 0) = \gamma$. Katz [29] gives methods for constructing confidence intervals on k for any x. Reference [11] gives charts from which certain of these confidence limits on k may be read. (The lower limits given in the charts are in error and need to be corrected. See the correction note in the Introduction for details.)

Burr [6] gives the following approximation for the point probabilities of the hypergeometric distribution based on the binomial distribution:

$$P(N, n, k, x) \cong \binom{n}{x}\left(\frac{k}{N}\right)^x\left(1 - \frac{k}{N}\right)^{n-x}\left\{1 + \frac{N}{2k(N-k)}\left[x - \left(x - \frac{nk}{N}\right)^2\right]\right\}.$$

Feller [16] presents bounds to individual hypergeometric probability terms. If $k/N = p$ and $q = 1 - p$, then the probability of exactly x defectives in the sample is bounded as follows:

$$\binom{n}{x}\left(p - \frac{x}{N}\right)^x\left(q - \frac{n-x}{N}\right)^{n-x} < \binom{k}{x}\binom{N-k}{n-x}\Big/\binom{N}{n}$$
$$< \binom{n}{x}p^x q^{n-x}\left(1 - \frac{n}{N}\right)^{-n}.$$

Note that as N increases both the upper and lower bounds approach the binomial term $\binom{n}{x}p^x q^{n-x}$.

Another approximation of the hypergeometric distribution based upon the binomial is due to Yamauti [65]. He shows that

$$P(N, n, k, x)$$
$$\cong \sum_{i=0}^{x}\binom{n}{i}\left(\frac{k}{N}\right)^i\left(1 - \frac{k}{N}\right)^{n-i} - n\binom{n-1}{x}\left(\frac{k}{N}\right)^x\left(1 - \frac{k}{N}\right)^{n-x-1}$$
$$\cdot\left[\frac{A_1}{N} + \frac{A_2}{N^2} - \frac{A_1^3}{N^2\left(\frac{k}{N}\right)\left(1 - \frac{k}{N}\right)}\right],$$

where

$$A_1 = -\frac{1}{2}\left(x - (n-1)\frac{k}{N}\right),$$

$$A_2 = A_1^2\left[\frac{1}{6\left(\frac{k}{N}\right)\left(1 - \frac{k}{N}\right)} + \frac{7}{\left(6\frac{k}{N}\right)}\right] + A_1\left[\frac{3x}{4\left(\frac{k}{N}\right)} + \frac{1}{6\left(1 - \frac{k}{N}\right)} + \frac{1}{6}\right]$$
$$+ \frac{x}{6} - \frac{x}{12\left(\frac{k}{N}\right)}.$$

Further, Yamauti indicates that if only the binomial term is used as an approximation,

$$\frac{nA_1}{N}\binom{n-1}{x}\left(\frac{k}{N}\right)^x\left(1-\frac{k}{N}\right)^{n-x-1}$$

is generally bounded by $0.2(n/N)$ so that a bound on the error is available.

3.2 Poisson Approximations

If $p = k/N$ is small, say $k \leqq 0.1N$, and n is large but $n \leqq 0.1N$, then the hypergeometic distribution can be approximated by the Poisson distribution [16], [26]; i.e.,

$$\Pr\{\text{Exactly } x \text{ defectives are found in the sample}\} = \frac{e^{-\lambda}\lambda^x}{x!},$$

where $\lambda = nk/N$. Kittagawa [31] and Molina [40] have given tables which are useful in this case.

Burr [6] gives the following approximation for the point probabilities of the hypergeometric distribution based on the Poisson distribution:

$$p(N, n, k, x) = \frac{e^{-\lambda}\lambda^x}{x!}\left\{1 + \left(\frac{1}{2n} + \frac{N}{2k(n-k)}\right)\left[x - \left(x - \frac{nk}{N}\right)^2\right]\right\},$$

where $\lambda = nk/N$. Note that in applying these approximation formulas, one should use the symmetries given in Section 1.4 and the set of parameters such that $0 \leqq x \leqq k \leqq n \leqq \frac{1}{2}N$ in order to get the best results.

3.3 Normal Approximations

Riordan [52] gave recurrence relations for the moments about the origin and about the mean for the hypergeometric distribution. Larsen [32] also found the moments about the mean by a different method. No doubt there are also other earlier references to these moments. The results used here are

$$\text{Mean of } x = \frac{nk}{N},$$

$$\text{Variance of } x = \frac{nk(N-k)(N-n)}{N^2(N-1)},$$

$$\text{Third moment of } x \text{ about the mean} = \frac{nk(N-k)(N-n)(N-2k)(N-2n)}{N^3(N-1)(N-2)}.$$

The classical normal approximation to the hypergeometric distribution is found [26] by setting

$$\mu = \frac{nk}{N}, \qquad \sigma^2 = \frac{n(N-n)k(N-k)}{N^2(N-1)}, \qquad Z_x = \frac{x-\mu}{\sigma},$$

and then if $\sigma^2 > 9$, computing

$$\Pr\{\lambda \leqq x \leqq \nu\} = P(Z_\nu + \tfrac{1}{2}\sigma) - P(Z_\lambda - \tfrac{1}{2}\sigma) ,$$

where

$$P(Z) = \frac{1}{\sqrt{2\pi}} \int_{-\infty}^{Z} \exp(-\tfrac{1}{2}t^2)\, dt .$$

W. L. Nicholson [44] studied another normal approximation. He defined

$$p = \frac{k+1}{N+2}, \quad q = 1 - p ; \qquad s = \frac{n+1}{N+2}, \quad t = 1 - s ;$$

$$a = \frac{(p-q)(t-s)}{6} ; \qquad \sigma^2 = (n+1)pqt .$$

Now if $\sigma > 3$ and if the required probability is $\Pr\{\lambda \leqq x \leqq \upsilon\}$, where $\lambda \geqq (n+1)p$, $\upsilon + \frac{1}{2} \leqq (n+1)p + \frac{2}{3}\sigma^2$, $n - \upsilon \geqq 4$, and $k - \upsilon \geqq 4$, then

$$\Pr\{\lambda \leqq x \leqq \upsilon\} \leqq \frac{N+1}{N+2} e^R \{P(Z_{\upsilon+1}) - P(Z_\lambda)\} ,$$

where

$$R = \frac{5(1-pq)(1-st)}{36\sigma^2} + \frac{2}{3(N+2)} ,$$

and

$$Z_x = \frac{x-(n+1)p}{\sigma} + \frac{a}{\sigma}\left\{\frac{x-(n+1)p}{\sigma}\right\}^2 + \frac{2a}{\sigma} - \frac{1}{2\sigma^2} ;$$

whereas the inequality is reversed if

$$Z_x = \frac{x-(n+1)p}{\sigma} + \frac{a}{\sigma}\left\{\frac{x-(n+1)p}{\sigma}\right\}^2 + \frac{2a}{\sigma} + \frac{M}{6\sigma} + \frac{1}{7\sigma} ,$$

where

$$M = \{\upsilon + \tfrac{1}{2} - (n+1)p\}^3 \sigma^{-4} .$$

As a point estimate for $\Pr\{\lambda \leqq x \leqq \upsilon\}$, Nicholson suggests

$$\Pr\{\lambda \leqq x \leqq \upsilon\} \cong \frac{N+1}{N+2} e^R \{P(Z_{\upsilon+1}) - P(Z_\lambda)\} ,$$

where R is defined as before, but

$$Z_x = \frac{x-(n+1)p}{\sigma} + \frac{a}{\sigma}\left\{\frac{x-(n+1)p}{\sigma}\right\}^2 + \frac{2a}{\sigma} .$$

Designating this estimate, the lower bound, and the upper bound by $\hat{P}$, L, and U, respectively, an upper bound on the relative absolute error is given by the expression $(1/L)\max(\hat{P} - L, U - \hat{P})$.

It must be pointed out that the condition $\sigma > 3$ implies $N > 142$, which limits the usefulness of this approximation for small N. The same condition on σ for the classical normal approximation also requires that N be essentially greater than this value. Hence, both the classical and Nicholson approximations have limited usefulness when N is less than 142.

4. INTERPOLATION IN THE HYPERGEOMETRIC TABLES

Since all possible hypergeometric probability distributions are given for $N \leqq 50$, interpolation is unnecessary in this region. However, since N is tabulated in steps of 10 between 50 and 100, interpolation for some intermediate value may be required. If linear interpolation is deemed adequate, the method recommended for finding a value of $P(N, n, k, x)$ not in the table $(50 < N \leqq 100)$ is first to use the symmetry relationships and find what

TABLE VI

k	x	$P(50, 25, k, x)$	$1 - E(k, x+1, \frac{1}{2})$	Interpolated value	$P(1000, 500, k, x)$
2	0	0.244898	0.250000	0.249745	0.249750
2	1	0.755102	0.750000	0.750255	0.750250
2	2	1.000000	1.000000	1.000000	1.000000
3	0	0.117347	0.125000	0.124617	0.124625
3	1	0.500000	0.500000	0.500000	0.500000
3	2	0.882653	0.875000	0.875383	0.875375
3	3	1.000000	1.000000	1.000000	1.000000
12	0	0.000043	0.000244	0.000234	0.000228
12	1	0.000961	0.003174	0.003063	0.003030
12	2	0.009038	0.019287	0.018775	0.018725
12	3	0.047744	0.072998	0.071735	0.071787
12	4	0.160446	0.193848	0.192178	0.192388
12	5	0.370822	0.387207	0.386388	0.386524
12	6	0.629179	0.612793	0.613612	0.613475
12	7	0.839555	0.806151	0.807821	0.807612
12	8	0.952257	0.927002	0.928265	0.928213
12	9	0.990962	0.980713	0.981225	0.981275
12	10	0.999040	0.996826	0.996937	0.996969
12	11	0.999958	0.999756	0.999766	0.999772
12	12	1.000000	1.000000	1.000000	1.000000
25	0	0.000000	0.000000	0.000000	0.000000
25	1	0.000000	0.000001	0.000001	0.000001
25	2	0.000000	0.000010	0.000010	0.000008
25	3	0.000000	0.000078	0.000074	0.000066
25	4	0.000001	0.000455	0.000432	0.000395
25	5	0.000024	0.002039	0.001938	0.001824
25	6	0.000272	0.007317	0.006965	0.006726
25	7	0.002100	0.021643	0.020666	0.020362
25	8	0.011354	0.053876	0.051750	0.051679
25	9	0.044371	0.114761	0.111242	0.111811
25	10	0.128896	0.212178	0.208014	0.209213
25	11	0.286070	0.345019	0.342072	0.343126
25	12	0.500000	0.500000	0.500000	0.500000
25	13	0.713931	0.654981	0.657929	0.656874
25	14	0.871105	0.787822	0.791986	0.790787
25	15	0.955629	0.885239	0.888759	0.888189
25	16	0.988647	0.946124	0.948250	0.948321
25	17	0.997901	0.978357	0.979334	0.979638
25	18	0.999729	0.992683	0.993035	0.993274
25	19	0.999977	0.997961	0.998062	0.998176
25	20	0.999999	0.999545	0.999568	0.999605
25	21	1.000000	0.999922	0.999926	0.999934
25	22	1.000000	0.999990	0.999991	0.999992
25	23	1.000000	0.999999	0.999999	0.999999
25	24	1.000000	1.000000	1.000000	1.000000
25	25	1.000000	1.000000	1.000000	1.000000

values of the parameters correspond to $0 \leqq x \leqq k \leqq n \leqq \frac{1}{2}N$. Then locate in the table the adjacent values of N, say N_1 and N_2, between which the desired value of N lies. Next, choose appropriate values of n, say n_1 and n_2, such that the ratios n_1/N_1 and n_2/N_2 are equal and the same as that for the required probability. Look up the required probabilities $P(N_1, n_1, k, x)$ and $P(N_2, n_2, k, x)$ and interpolate on N. For example, suppose that $P(76, 38, 26, 12)$ is desired. N_1 and N_2 are 70 and 80, respectively, and $n/N = n_1/N_1 = n_2/N_2 = \frac{1}{2}$, so that $n_1 = 35$ and $n_2 = 40$. From the tables $P(70, 35, 26, 12) = 0.402466$ and $P(80, 40, 26, 12) = 0.405823$. The desired value for $P(76, 38, 26, 12)$ is 0.404480.

The tables may also be used to obtain interpolated values of $N > 100$ by inverse linear interpolation between $N = 100$ (or 50, 60, 70, 80, 90 if more convenient) and $N = \infty$, with $N = \infty$ corresponding to the binomial distribution. As an example, Table VI compares values for $P(1000, 500, k, x)$ obtained by linear interpolation with the exact values.

The column headed "interpolated value" is obtained by linear interpolation with respect to $1/N$ between $N = 50$ and $N = \infty$. Interpolation is most easily done by using $50/N$ as the argument. Then 50 corresponds to 1, infinity corresponds to 0, and 1000, for example, corresponds to 0.05. Compare the "interpolated value" column with the $P(1000, 500, k, x)$ column. Where applicable, this method can be used to approximate to three or more decimal places.

Specifically, the method recommended for finding a value of $P(N, n, k, x)$ not in the table ($N > 100$) is first to use the symmetry relationships and find what values of the parameters correspond to $0 \leqq x \leqq k \leqq n \leqq \frac{1}{2}N$. Then locate the values of n and N in the table with $N \leqq 50$ or $N = 60, 70, 80, 90, 100$ such that the ratio n/N is the same as that for the required probability. Look up the required k and x in the table if they appear, and interpolate as above.

If k and x are not in the table, as could happen if, say, we wanted $P(2000, 500, 126, 12)$, then it is recommended that one of the aforementioned approximations be used directly.

5. SUMMARY OF SOME USEFUL FORMULAS ON SUMS OF COMBINATORIALS

Formulas which have proved most useful in dealing with sums of combinatorials are given below. Reference to the source from which it was obtained is given after each formula.

(1) $$\sum_{j=0}^{x} \binom{a}{b-j}\binom{c}{j} \equiv \sum_{j=0}^{a-b+x} \binom{c+j}{x}\binom{a-1-j}{b-x-1}, \qquad [46, \text{ p. } 17]$$

for a, b, c, x positive integers, $0 \leqq x \leqq \min(c, b-1)$, and $0 \leqq b \leqq a$. If $x = c$, then the above identity (1) reduces to Equation (12.16) of Feller [16]. If $b = c = x + 1$, then (1) reduces to Equation (12.8) of Feller [16].

(2) $$\sum_{j=0}^{x}\binom{a}{b-j}\binom{c}{j} \equiv \sum_{j=0}^{x}\binom{c-x-1+j}{c-x-1}\binom{a+x-j}{a+x-b}, \qquad [46, p. 17]$$

where $0 \leqq b \leqq a$, $0 \leqq x \leqq \min(c-1, b)$, and a, b, c, x are positive integers. If $x = c-1$, the above identity reduces to Equation (12.6) of Feller [16] and if $x = b$, then (2) reduces to (12.16) of Feller [16].

(3) $$\sum_{j=0}^{x}\binom{a}{b-j}\binom{c}{j} \equiv \sum_{j=0}^{x}\binom{a}{b-x+j}\binom{c}{x-j}.$$

This identity (3) reverses the order of the summation. It follows from the symmetry relationships of the hypergeometric distribution.

(4) $$\sum_{i=r}^{n}\binom{n}{i}p^i(1-p)^{n-i} \equiv p^r\sum_{i=0}^{n-r}\binom{r+i-1}{r-1}(1-p)^i \equiv p^r\sum_{i=r}^{n}\binom{i-1}{r-1}(1-p)^{i-r}. \qquad [48, p. 11]$$

(5) $$\sum_{i=0}^{k}(-1)^i\binom{x+n}{k-i}\binom{x-k+i}{i} \equiv \binom{n-1+k}{k}$$

for $k = 0, 1, \cdots, x$. [48, p. 22]

(6) $$\sum_{i=r}^{n} i\binom{n}{i}p^i(1-p)^{n-i} \equiv npE(n, r, p) + r\binom{n}{r}p^r(1-p)^{n-r+1}. \qquad [28]$$

In identities (4), (5), and (6), $0 \leqq p \leqq 1$, and n and r are positive integers.

The following five formulas are taken from Birnbaum and Pyke [4], who quoted the first from Abel.

(7) $$\sum_{i=0}^{n}\binom{n}{i}(a+i)^i(b-i)^{n-i-1} \equiv \frac{(a+b)^n}{(b-n)},$$

for all real a, b, and integer $n \geqq 0$ (for $b = n$, the left hand term is defined as the limit for $b \to n$).

(8) $$\sum_{i=0}^{n}\binom{n}{i}(a+i)^i(b-i)^{n-1} \equiv n!\sum_{i=0}^{n}\frac{(a+b)^i}{i!},$$

for all real a, b, and integer $n \geqq 0$.

(9) $$\sum_{i=0}^{n-1}\binom{n}{i}(a+i)^i(b-i)^{n-i-1} \equiv \sum_{i=0}^{n-1}(a+b)^i(a+n)^{n-i-1},$$

for all real a, b, and integer $n \geqq 0$.

(10) $$(a-1)(b-n)\sum_{i=0}^{n}\frac{1}{i+1}\binom{n}{i}(a+i)^i(b-i)^{n-i-1} \equiv \frac{1}{n+1}[(a+b)^n(a+b-n-1)-(b+1)^n(b-n)],$$

for all real a, b, and integer $n \geqq 0$.

(11) $$\sum_{i=0}^{n-1} \frac{1}{i+1} \binom{n}{i} i^i (n-i)^{n-i-1} \equiv (n+1)^{n-1} ,$$

for all integers $n \geqq 0$.

Additional relationships for positive integral values of a and b, $b \leqq a$, are given as follows:

(12) $$\sum_{j=0}^{b} \binom{2a-2b-1}{a-j} \binom{2b+1}{j} \equiv \frac{1}{2} \binom{2a}{a}$$

(13) $$\sum_{j=0}^{a} \binom{2a+1}{j} \equiv 2^{2a}$$

(14) $$\sum_{j=1}^{a+1} \binom{2a+1-j}{a} \equiv \binom{2a+1}{a} .$$

BIBLIOGRAPHY

[1] ARMSEN, P., "Tables for significance tests of 2 × 2 contingency tables," *Biometrika*, Vol. 42 (1955), pp. 494–511.

[2] BARKER, B. E., "Attributes Sampling Plans for Assuring High Lot Reliability," Sandia Corporation Technical Memorandum SCTM 352-58-51, Albuquerque, N.M., December 19, 1958. (Available from the Office of Technical Services, Department of Commerce, Washington, D.C.)

[3] BARNARD, G. A., "Significance tests for 2 × 2 tables," *Biometrika*, Vol. 34 (1947), pp. 123–138.

[4] BIRNBAUM, Z. W., and R. PYKE, "On some distributions related to the statistic D_n^+," *Annals of Mathematical Statistics*, Vol. 29 (1958), pp. 179–187.

[5] BROSS, I. D., and E. L. KASTON, "Rapid analysis of 2 × 2 tables," *Journal of the American Statistical Association*, Vol. 52 (1957), pp. 18–28.

[6] BURR, I. W., *Engineering Statistics and Quality Control*, McGraw-Hill Book Co., Inc., New York, 1953, p. 209.

[7] CHAPMAN, D. G., "Some properties of the hypergeometric distribution with applications to zoological sample censuses," *University of California Publications in Statistics*, Vol. 1, No. 7 (1951), pp. 131–160.

[8] CHAPMAN, D. G., "A mathematical study of confidence limits of salmon populations calculated from sample tag ratios," *International Pacific Salmon Fisheries Commission, Bulletin II* (1948), pp. 69–85.

[9] CHAPMAN, D. G., "Inverse, multiple and sequential sample censuses," *Biometrics*, Vol. 8 (1952), pp. 286–306.

[10] CHAPMAN, D. G., "The estimation of biological populations," *Annals of Mathematical Statistics*, Vol. 25 (1954), pp. 1–15.

[11] CHUNG, J. H. and D. B. DELURY, *Confidence Limits for the Hypergeometric Distribution*, University of Toronto Press, Toronto, 1950.

[12] CLARK, C. R. and L. H. KOOPMANS, "Graphs of the Hypergeometric O.C. and A.O.Q. Functions for Lot Sizes 10 to 225," Sandia Corporation Monograph SCR-121, Albuquerque, N.M., September 1959. (Available from the Office of Technical Services, Department of Commerce, Washington, D.C.)

[13] COCHRAN, W. G., "The χ^2 test of goodness of fit," *Annals of Mathematical Statistics*, Vol. 23 (1952), pp. 315–345 (especially pp. 324–327).

[14] DODGE, H. F. and H. G. ROMIG, *Sampling Inspection Tables*, John Wiley and Sons, Inc., New York, 2nd ed., 1959.

[15] EPSTEIN, B., "Tables for the distribution of the number of exceedances," *Annals of Mathematical Statistics*, Vol. 25 (1954), pp. 762-768.

[16] FELLER, W., *An Introduction to Probability Theory and Its Applications*, Vol. 1, John Wiley and Sons, Inc., New York, 2nd ed., 1957, pp. 57, 161, 162, and 180.

[17] FINNEY, D. J., "Tests of significance in a 2 × 2 contingency table," *Biometrika*, Vol. 35 (1948), pp. 145-156. (Tables reproduced in *Biometrika Tables for Statisticians*, Vol. 1, by Hartley and Pearson, Cambridge University Press, Cambridge, 1954, pp. 188-193).

[18] FISHER, R. A. and F. YATES, *Statistical Tables for Biological, Agricultural and Medical Research*, Oliver and Boyd, Edinburgh, 3rd ed., 1948, pp. 4-6 and 47.

[19] FISHER, R. A., *Statistical Methods and Scientific Inference*, Hafner Publishing Company, New York, 1956, p. 111.

[20] FRY, T. C., *Probability and Its Engineering Uses*, D. Van Nostrand Co., Inc., New York, 1928, pp. 429-438.

[21] GUMBEL, E. J., *Statistics of Extremes*, Columbia University Press, New York, 1958, pp. 58-67.

[22] GUMBEL, E. J. and H. VON SCHELLING, "The distribution of the number of exceedances," *Annals of Mathematical Statistics*, Vol. 21 (1950), pp. 247-262.

[23] HAIGHT, F. A., *Index to the Distribution of Mathematical Statistics*, Auckland University College, Auckland, 1955, p. 51.

[24] HALD, A., "The compound hypergeometric distribution and a system of single sampling inspection plans based on prior distributions and costs," *Technometrics*, Vol. 2 (1960), pp. 275-340.

[25] HALD, A., *Statistical Tables and Formulas*, John Wiley and Sons, Inc., New York, 1952, pp. 72-75.

[26] HALD, A., *Statistical Theory with Engineering Applications*, John Wiley and Sons, Inc., New York, 1952, p. 691.

[27] HARVARD UNIVERSITY COMPUTATION LABORATORY, *Tables of the Cumulative Binomial Probability Distribution*, Harvard University Press, Cambridge, Mass., 1955.

[28] JOHNSON, N. L., "A note on the mean deviation of the binomial distribution," *Biometrika*, Vol. 44 (1957), pp. 532-533.

[29] KATZ, L., "Confidence intervals for the number showing a certain characteristic in a population when sampling is without replacement," *Journal of the American Statistical Association*, Vol. 48 (1953), pp. 256-261.

[30] KEMP, C. D. and A. W. KEMP, "Generalized hypergeometric distribution," *Journal of the Royal Statistical Society*, Series B, Vol. 18 (1956), pp. 202-211.

[31] KITTAGAWA, T., *Tables of Poisson Distribution*, Baifukan, Tokyo, 1952.

[32] LARSEN, H. D., "Moments about the arithmetic mean of a hypergeometric frequency distribution," *Annals of Mathematical Statistics*, Vol. 10 (1939), pp. 198-201.

[33] LATSCHA, R., "Tests of significance in a 2 × 2 contingency table: Extension of Finney's tables," *Biometrika*, Vol. 40 (1953), pp. 74-86.

[34] LEHMANN, E. L., *Theory of Testing Hypotheses, Lecture Notes*, Associated Students Store, Berkeley, Calif., 1949, pp. 5-23 (duplicated).

[35] LESLIE, P. H., "A simple method of calculating the exact probability in 2 × 2 contingency tables with small marginal totals," *Biometrika*, Vol. 42 (1955), pp. 522-523.

[36] MAINLAND, D., L. HERRERA, and M. I. SUTCLIFFE, *Tables for Use with Binomial Samples*, New York University College of Medicine, New York, 1956.

[37] MAINLAND, D. and I. M. MURRAY, "Tables for use in fourfold contingency tests," *Science*, Vol. 116 (November 1952), pp. 591-594.

[38] MANN, H. B. and D. R. WHITNEY, "On a test of whether one of two random variables is stochastically larger than the other," *Annals of Mathematical Statistics*, Vol. 18 (1947), pp. 50-60.

[39] *Military Standard Sampling Procedures and Tables for Inspection by Attributes*, MIL-STD-105B, U.S. Government Printing Office, Washington, D.C., December 31, 1958.

[40] MOLINA, E. C., *Poisson's Exponential Limit*, D. Van Nostrand, Inc., New York, 1942.

[41] MOOD, A. M., "Tests of independence in contingency tables as unconditional tests," *Annals of Mathematical Statistics*, Vol. 20 (1949), pp. 114-116.

[42] MOSTELLER, F., "A k-sample slippage test for an extreme population," *Annals of Mathematical Statistics*, Vol. 19 (1948), pp. 58-65.

[43] MOSTELLER, F. and J. W. TUKEY, "Significance levels for a k-sample slippage test," *Annals of Mathematical Statistics*, Vol. 21 (1950), pp. 120-123.

[44] NICHOLSON, W. L., "On the normal approximation to the hypergeometric distribution," *Annals of Mathematical Statistics*, Vol. 27 (1956), pp. 471-483.

[45] ODELL, P. L., "Tables and Graphs for Determining an Upper Confidence Bound on the Number of Defectives in a Finite Population," NavOrd Report 7123, U.S. Naval Ordnance Evaluation Unit, Albuquerque, N.M. (July 15, 1960).

[46] OWEN, D. B., "Tabulation of the Hypergeometric Probability Distribution for Lot Sizes Less than or Equal to 50," Sandia Corporation Technical Memorandum SCTM 215-59(51), Albuquerque, N.M., August 13, 1959. (Available from the Office of Technical Services, Department of Commerce, Washington, D.C.)

[47] OWEN, D. B., "Distribution-Free Tolerance Limits," Sandia Corporation Technical Memorandum SCTM 66A-57(51), Albuquerque, N.M., June 24, 1957. (Available from the Office of Technical Services, Department of Commerce, Washington, D.C.)

[48] OWEN, D. B. and E. J. GILBERT, "The Relationship of the Binomial Probability Distribution to Other Probability Distributions with a Selected Bibliography on the Subject," Sandia Corporation Technical Memorandum SCTM 1-59(51), Albuquerque, N.M., June 22, 1959. (Available from the Office of Technical Services, Department of Commerce, Washington, D.C.)

[49] OWEN, D. B., E. J. GILBERT, G. P. STECK, and D. A. YOUNG, "A Formula for Determining Sample Size in Hypergeometric Sampling When Zero Defectives are Observed in the Sample," Sandia Corporation Technical Memorandum SCTM 178-59(51), Albuquerque, N.M., (Available from the Office of Technical Services, Department of Commerce, Washington, D.C.)

[50] OWEN, D. B. and C. M. WILLIAMS, "Logarithms of Factorials from 1 to 2000," Sandia Corporation Monograph SCR-158, Albuquerque, N.M., December 1959. (Available from the Office of Technical Services, Department of Commerce, Washington, D.C.)

[51] PEARSON, K., *Tables of the Incomplete Beta-Function*, The University Press, Cambridge, England (1934).

[52] RIORDAN, J., "Moment recurrence relations for binomial, Poisson and hypergeometric frequency distributions," *Annals of Mathematical Statistics*, Vol. 8 (1937), pp. 103-111.

[53] ROKHSAR, A. E. and G. E. KANE, "The effect of hypergeometric probability

distribution on the design of sampling plans for small lot sizes," *Journal of Industrial Engineering*, Vol. 10 (1959), pp. 467-469.

[54] ROMANOWSKI, V. I., "Duality theorems for the hypergeometric distribution," *Akademiya Nauk Uzbek, SSR, Trudy Inst. Math. Meh.*, Vol. 11 (1953), pp. 22-28 (Russian).

[55] ROSENBAUM, S., "Tables for a nonparametric test of dispersion," *Annals of Mathematical Statistics*, Vol. 24 (1953), pp. 663-668.

[56] ROSENBAUM, S., "Tables for a nonparametric test of location," *Annals of Mathematical Statistics*, Vol. 25 (1954), pp. 146-150.

[57] SANDELIUS, M., "Unbiased estimation based on inverse hypergeometric sampling," *Kungl. Lantbrukshögskolas Annaler*, Vol. 18 (1951), pp. 123-127.

[58] SIYA, R., "Multidimensional hypergeometric distribution," *Sankhyā*, Vol. 15 (1955), pp. 391-398.

[59] TOCHER, K. D., "Extension of the Neyman-Pearson Theory of tests to discontinuous variates," *Biometrika*, Vol. 37 (1950), pp. 130-144.

[60] WALLACE, D. L., "Asymptotic approximations to distributions," *Annals of Mathematical Statistics*, Vol. 29 (1958), pp. 635-654.

[61] WIESEN, J. M., "Extension of Existing Tables of Attributes Sampling Plans," Sandia Corporation Technical Memorandum SCTM 42-58(12), Albuquerque, N.M., February 4, 1958. (Available from the Office of Technical Services, Department of Commerce, Washington, D.C.)

[62] WIESEN, J. M., D. B. OWEN, and G. P. STECK, "A Discussion of the Analysis of 2 × 2 Tables," Sandia Corporation Technical Memorandum SCTM 170-56(51), Albuquerque, N.M., October 24, 1956. (Available from the Office of Technical Services, Department of Commerce, Washington, D.C.)

[63] WILKS, S. S., "Statistical prediction with special reference to the problem of tolerance limits," *Annals of Mathematical Statistics*, Vol. 13 (1942), pp. 400-409.

[64] WISE, M. E., "A quickly convergent expansion for cumulative hypergeometric probabilities, direct and inverse," *Biometrika*, Vol. 41 (1954), pp. 317-329.

[65] YAMAUTI, Z., "Sampling inspection," Chapter 6 in *Statistical Quality Control*, Denki Gakkai, Tokyo, 1953 (Japanese).

[66] YATES, F., "A note on the application of probabilities test to a set of 2 × 2 tables," *Biometrika*, Vol. 42 (1955), pp. 404-411.

PART II

TABLES OF THE HYPERGEOMETRIC PROBABILITY DISTRIBUTION

Table for $N = 2$, $n = 1$, through $N = 100$, $n = 50$

N	n	k	x	P(x)	p(x)
2	1	1	0	0.500000	0.500000
2	1	1	1	1.000000	0.500000
3	1	1	0	0.666667	0.666667
3	1	1	1	1.000000	0.333333
3	2	1	0	0.333333	0.333333
3	2	1	1	1.000000	0.666667
3	2	2	1	0.666667	0.666667
3	2	2	2	1.000000	0.333333
4	1	1	0	0.750000	0.750000
4	1	1	1	1.000000	0.250000
4	2	1	0	0.500000	0.500000
4	2	1	1	1.000000	0.500000
4	2	2	0	0.166667	0.166667
4	2	2	1	0.833333	0.666667
4	2	2	2	1.000000	0.166667
4	3	1	0	0.250000	0.250000
4	3	1	1	1.000000	0.750000
4	3	2	1	0.500000	0.500000
4	3	2	2	1.000000	0.500000
4	3	3	2	0.750000	0.750000
4	3	3	3	1.000000	0.250000
5	1	1	0	0.800000	0.800000
5	1	1	1	1.000000	0.200000
5	2	1	0	0.600000	0.600000
5	2	1	1	1.000000	0.400000
5	2	2	0	0.300000	0.300000
5	2	2	1	0.900000	0.600000
5	2	2	2	1.000000	0.100000
5	3	1	0	0.400000	0.400000
5	3	1	1	1.000000	0.600000
5	3	2	0	0.100000	0.100000
5	3	2	1	0.700000	0.600000
5	3	2	2	1.000000	0.300000
5	3	3	1	0.300000	0.300000
5	3	3	2	0.900000	0.600000
5	3	3	3	1.000000	0.100000
5	4	1	0	0.200000	0.200000
5	4	1	1	1.000000	0.800000
5	4	2	1	0.400000	0.400000
5	4	2	2	1.000000	0.600000
5	4	3	2	0.600000	0.600000
5	4	3	3	1.000000	0.400000
5	4	4	3	0.800000	0.800000
5	4	4	4	1.000000	0.200000
6	1	1	0	0.833333	0.833333
6	1	1	1	1.000000	0.166667
6	2	1	0	0.666667	0.666667
6	2	1	1	1.000000	0.333333
6	2	2	0	0.400000	0.400000
6	2	2	1	0.933333	0.533333
6	2	2	2	1.000000	0.066667
6	3	1	0	0.500000	0.500000
6	3	1	1	1.000000	0.500000
6	3	2	0	0.200000	0.200000
6	3	2	1	0.800000	0.600000
6	3	2	2	1.000000	0.200000
6	3	3	0	0.050000	0.050000
6	3	3	1	0.500000	0.450000
6	3	3	2	0.950000	0.450000
6	3	3	3	1.000000	0.050000
6	4	1	0	0.333333	0.333333
6	4	1	1	1.000000	0.666667
6	4	2	0	0.066667	0.066667
6	4	2	1	0.600000	0.533333
6	4	2	2	1.000000	0.400000
6	4	3	1	0.200000	0.200000
6	4	3	2	0.800000	0.600000
6	4	3	3	1.000000	0.200000
6	4	4	2	0.400000	0.400000
6	4	4	3	0.933333	0.533333
6	4	4	4	1.000000	0.066667
6	5	1	0	0.166667	0.166667
6	5	1	1	1.000000	0.833333
6	5	2	1	0.333333	0.333333
6	5	2	2	1.000000	0.666667
6	5	3	2	0.500000	0.500000
6	5	3	3	1.000000	0.500000
6	5	4	3	0.666667	0.666667
6	5	4	4	1.000000	0.333333
6	5	5	4	0.833333	0.833333
6	5	5	5	1.000000	0.166667
7	1	1	0	0.857143	0.857143
7	1	1	1	1.000000	0.142857
7	2	1	0	0.714286	0.714286
7	2	1	1	1.000000	0.285714
7	2	2	0	0.476190	0.476190
7	2	2	1	0.952381	0.476190
7	2	2	2	1.000000	0.047619
7	3	1	0	0.571429	0.571429
7	3	1	1	1.000000	0.428571
7	3	2	0	0.285714	0.285714
7	3	2	1	0.857143	0.571429
7	3	2	2	1.000000	0.142857
7	3	3	0	0.114286	0.114286
7	3	3	1	0.628571	0.514286
7	3	3	2	0.971428	0.342857
7	3	3	3	1.000000	0.028571
7	4	1	0	0.428571	0.428571
7	4	1	1	1.000000	0.571429
7	4	2	0	0.142857	0.142857
7	4	2	1	0.714286	0.571429
7	4	2	2	1.000000	0.285714
7	4	3	0	0.028571	0.028571
7	4	3	1	0.371429	0.342857
7	4	3	2	0.885714	0.514286
7	4	3	3	1.000000	0.114286
7	4	4	1	0.114286	0.114286
7	4	4	2	0.628571	0.514286
7	4	4	3	0.971428	0.342857
7	4	4	4	1.000000	0.028571
7	5	1	0	0.285714	0.285714
7	5	1	1	1.000000	0.714286
7	5	2	0	0.047619	0.047619
7	5	2	1	0.523809	0.476190
7	5	2	2	1.000000	0.476190
7	5	3	1	0.142857	0.142857
7	5	3	2	0.714286	0.571429
7	5	3	3	1.000000	0.285714
7	5	4	2	0.285714	0.285714
7	5	4	3	0.857143	0.571429
7	5	4	4	1.000000	0.142857
7	5	5	3	0.476190	0.476190
7	5	5	4	0.952381	0.476190
7	5	5	5	1.000000	0.047619
7	6	1	0	0.142857	0.142857
7	6	1	1	1.000000	0.857143
7	6	2	1	0.285714	0.285714
7	6	2	2	1.000000	0.714286
7	6	3	2	0.428571	0.428571
7	6	3	3	1.000000	0.571429
7	6	4	3	0.571429	0.571429
7	6	4	4	1.000000	0.428571
7	6	5	4	0.714286	0.714286
7	6	5	5	1.000000	0.285714
7	6	6	5	0.857143	0.857143
7	6	6	6	1.000000	0.142857
8	1	1	0	0.875000	0.875000
8	1	1	1	1.000000	0.125000
8	2	1	0	0.750000	0.750000
8	2	1	1	1.000000	0.250000
8	2	2	0	0.535714	0.535714
8	2	2	1	0.964286	0.428571
8	2	2	2	1.000000	0.035714
8	3	1	0	0.625000	0.625000
8	3	1	1	1.000000	0.375000
8	3	2	0	0.357143	0.357143
8	3	2	1	0.892857	0.535714
8	3	2	2	1.000000	0.107143
8	3	3	0	0.178571	0.178571
8	3	3	1	0.714286	0.535714
8	3	3	2	0.982143	0.267857
8	3	3	3	1.000000	0.017857
8	4	1	0	0.500000	0.500000
8	4	1	1	1.000000	0.500000
8	4	2	0	0.214286	0.214286
8	4	2	1	0.785714	0.571429
8	4	2	2	1.000000	0.214286
8	4	3	0	0.071429	0.071429
8	4	3	1	0.500000	0.428571
8	4	3	2	0.928571	0.428571
8	4	3	3	1.000000	0.071429
8	4	4	0	0.014286	0.014286
8	4	4	1	0.242857	0.228571
8	4	4	2	0.757143	0.514286
8	4	4	3	0.985714	0.228571
8	4	4	4	1.000000	0.014286
8	5	1	0	0.375000	0.375000
8	5	1	1	1.000000	0.625000
8	5	2	0	0.107143	0.107143
8	5	2	1	0.642857	0.535714
8	5	2	2	1.000000	0.357143
8	5	3	0	0.017857	0.017857
8	5	3	1	0.285714	0.267857
8	5	3	2	0.821429	0.535714
8	5	3	3	1.000000	0.178571
8	5	4	1	0.071429	0.071429
8	5	4	2	0.500000	0.428571
8	5	4	3	0.928571	0.428571
8	5	4	4	1.000000	0.071429
8	5	5	2	0.178571	0.178571
8	5	5	3	0.714286	0.535714
8	5	5	4	0.982143	0.267857
8	5	5	5	1.000000	0.017857
8	6	1	0	0.250000	0.250000
8	6	1	1	1.000000	0.750000
8	6	2	0	0.035714	0.035714
8	6	2	1	0.464286	0.428571
8	6	2	2	1.000000	0.535714
8	6	3	1	0.107143	0.107143
8	6	3	2	0.642857	0.535714
8	6	3	3	1.000000	0.357143
8	6	4	2	0.214286	0.214286
8	6	4	3	0.785714	0.571429
8	6	4	4	1.000000	0.214286
8	6	5	3	0.357143	0.357143
8	6	5	4	0.892857	0.535714
8	6	5	5	1.000000	0.107143
8	6	6	4	0.535714	0.535714
8	6	6	5	0.964286	0.428571
8	6	6	6	1.000000	0.035714

Table for $N = 2$, $n = 1$, through $N = 100$, $n = 50$

N	n	k	x	P(x)	p(x)
8	7	1	0	0.125000	0.125000
8	7	1	1	1.000000	0.875000
8	7	2	1	0.250000	0.250000
8	7	2	2	1.000000	0.750000
8	7	3	2	0.375000	0.375000
8	7	3	3	1.000000	0.625000
8	7	4	3	0.500000	0.500000
8	7	4	4	1.000000	0.500000
8	7	5	4	0.625000	0.625000
8	7	5	5	1.000000	0.375000
8	7	6	5	0.750000	0.750000
8	7	6	6	1.000000	0.250000
8	7	7	6	0.875000	0.875000
8	7	7	7	1.000000	0.125000
9	1	1	0	0.888889	0.888889
9	1	1	1	1.000000	0.111111
9	2	1	0	0.777778	0.777778
9	2	1	1	1.000000	0.222222
9	2	2	0	0.583333	0.583333
9	2	2	1	0.972222	0.388889
9	2	2	2	1.000000	0.027778
9	3	1	0	0.666667	0.666667
9	3	1	1	1.000000	0.333333
9	3	2	0	0.416667	0.416667
9	3	2	1	0.916667	0.500000
9	3	2	2	1.000000	0.083333
9	3	3	0	0.238095	0.238095
9	3	3	1	0.773809	0.535714
9	3	3	2	0.988095	0.214286
9	3	3	3	1.000000	0.011905
9	4	1	0	0.555556	0.555556
9	4	1	1	1.000000	0.444444
9	4	2	0	0.277778	0.277778
9	4	2	1	0.833333	0.555556
9	4	2	2	1.000000	0.166667
9	4	3	0	0.119048	0.119048
9	4	3	1	0.595238	0.476190
9	4	3	2	0.952381	0.357143
9	4	3	3	1.000000	0.047619
9	4	4	0	0.039683	0.039683
9	4	4	1	0.357143	0.317460
9	4	4	2	0.833333	0.476190
9	4	4	3	0.992063	0.158730
9	4	4	4	1.000000	0.007936
9	5	1	0	0.444444	0.444444
9	5	1	1	1.000000	0.555556
9	5	2	0	0.166667	0.166667
9	5	2	1	0.722222	0.555556
9	5	2	2	1.000000	0.277778
9	5	3	0	0.047619	0.047619

N	n	k	x	P(x)	p(x)
9	5	3	1	0.404762	0.357143
9	5	3	2	0.880952	0.476190
9	5	3	3	1.000000	0.119048
9	5	4	0	0.007936	0.007936
9	5	4	1	0.166667	0.158730
9	5	4	2	0.642857	0.476190
9	5	4	3	0.960317	0.317460
9	5	4	4	1.000000	0.039683
9	5	5	1	0.039683	0.039683
9	5	5	2	0.357143	0.317460
9	5	5	3	0.833333	0.476190
9	5	5	4	0.992063	0.158730
9	5	5	5	1.000000	0.007936
9	6	1	0	0.333333	0.333333
9	6	1	1	1.000000	0.666667
9	6	2	0	0.083333	0.083333
9	6	2	1	0.583333	0.500000
9	6	2	2	1.000000	0.416667
9	6	3	0	0.011905	0.011905
9	6	3	1	0.226190	0.214286
9	6	3	2	0.761905	0.535714
9	6	3	3	1.000000	0.238095
9	6	4	1	0.047619	0.047619
9	6	4	2	0.404762	0.357143
9	6	4	3	0.880952	0.476190
9	6	4	4	1.000000	0.119048
9	6	5	2	0.119048	0.119048
9	6	5	3	0.595238	0.476190
9	6	5	4	0.952381	0.357143
9	6	5	5	1.000000	0.047619
9	6	6	3	0.238095	0.238095
9	6	6	4	0.773809	0.535714
9	6	6	5	0.988095	0.214286
9	6	6	6	1.000000	0.011905
9	7	1	0	0.222222	0.222222
9	7	1	1	1.000000	0.777778
9	7	2	0	0.027778	0.027778
9	7	2	1	0.416667	0.388889
9	7	2	2	1.000000	0.583333
9	7	3	1	0.083333	0.083333
9	7	3	2	0.583333	0.500000
9	7	3	3	1.000000	0.416667
9	7	4	2	0.166667	0.166667
9	7	4	3	0.722222	0.555556
9	7	4	4	1.000000	0.277778
9	7	5	3	0.277778	0.277778
9	7	5	4	0.833333	0.555556
9	7	5	5	1.000000	0.166667
9	7	6	4	0.416667	0.416667
9	7	6	5	0.916667	0.500000

N	n	k	x	P(x)	p(x)
9	7	6	6	1.000000	0.083333
9	7	7	5	0.583333	0.583333
9	7	7	6	0.972222	0.388889
9	7	7	7	1.000000	0.027778
9	8	1	0	0.111111	0.111111
9	8	1	1	1.000000	0.888889
9	8	2	1	0.222222	0.222222
9	8	2	2	1.000000	0.777778
9	8	3	2	0.333333	0.333333
9	8	3	3	1.000000	0.666667
9	8	4	3	0.444444	0.444444
9	8	4	4	1.000000	0.555556
9	8	5	4	0.555556	0.555556
9	8	5	5	1.000000	0.444444
9	8	6	5	0.666667	0.666667
9	8	6	6	1.000000	0.333333
9	8	7	6	0.777778	0.777778
9	8	7	7	1.000000	0.222222
9	8	8	7	0.888889	0.888889
9	8	8	8	1.000000	0.111111
10	1	1	0	0.900000	0.900000
10	1	1	1	1.000000	0.100000
10	2	1	0	0.800000	0.800000
10	2	1	1	1.000000	0.200000
10	2	2	0	0.622222	0.622222
10	2	2	1	0.977778	0.355556
10	2	2	2	1.000000	0.022222
10	3	1	0	0.700000	0.700000
10	3	1	1	1.000000	0.300000
10	3	2	0	0.466667	0.466667
10	3	2	1	0.933333	0.466667
10	3	2	2	1.000000	0.066667
10	3	3	0	0.291667	0.291667
10	3	3	1	0.816667	0.525000
10	3	3	2	0.991667	0.175000
10	3	3	3	1.000000	0.008333
10	4	1	0	0.600000	0.600000
10	4	1	1	1.000000	0.400000
10	4	2	0	0.333333	0.333333
10	4	2	1	0.866667	0.533333
10	4	2	2	1.000000	0.133333
10	4	3	0	0.166667	0.166667
10	4	3	1	0.666667	0.500000
10	4	3	2	0.966667	0.300000
10	4	3	3	1.000000	0.033333
10	4	4	0	0.071429	0.071429
10	4	4	1	0.452381	0.380952
10	4	4	2	0.880952	0.428571
10	4	4	3	0.995238	0.114286
10	4	4	4	1.000000	0.004762

N	n	k	x	P(x)	p(x)
10	5	1	0	0.500000	0.500000
10	5	1	1	1.000000	0.500000
10	5	2	0	0.222222	0.222222
10	5	2	1	0.777778	0.555556
10	5	2	2	1.000000	0.222222
10	5	3	0	0.083333	0.083333
10	5	3	1	0.500000	0.416667
10	5	3	2	0.916667	0.416667
10	5	3	3	1.000000	0.083333
10	5	4	0	0.023810	0.023810
10	5	4	1	0.261905	0.238095
10	5	4	2	0.738095	0.476190
10	5	4	3	0.976190	0.238095
10	5	4	4	1.000000	0.023810
10	5	5	0	0.003968	0.003968
10	5	5	1	0.103175	0.099206
10	5	5	2	0.500000	0.396825
10	5	5	3	0.896825	0.396825
10	5	5	4	0.996032	0.099206
10	5	5	5	1.000000	0.003968
10	6	1	0	0.400000	0.400000
10	6	1	1	1.000000	0.600000
10	6	2	0	0.133333	0.133333
10	6	2	1	0.666667	0.533333
10	6	2	2	1.000000	0.333333
10	6	3	0	0.033333	0.033333
10	6	3	1	0.333333	0.300000
10	6	3	2	0.833333	0.500000
10	6	3	3	1.000000	0.166667
10	6	4	0	0.004762	0.004762
10	6	4	1	0.119048	0.114286
10	6	4	2	0.547619	0.428571
10	6	4	3	0.928571	0.380952
10	6	4	4	1.000000	0.071429
10	6	5	1	0.023810	0.023810
10	6	5	2	0.261905	0.238095
10	6	5	3	0.738095	0.476190
10	6	5	4	0.976190	0.238095
10	6	5	5	1.000000	0.023810
10	6	6	2	0.071429	0.071429
10	6	6	3	0.452381	0.380952
10	6	6	4	0.880952	0.428571
10	6	6	5	0.995238	0.114286
10	6	6	6	1.000000	0.004762
10	7	1	0	0.300000	0.300000
10	7	1	1	1.000000	0.700000
10	7	2	0	0.066667	0.066667
10	7	2	1	0.533333	0.466667
10	7	2	2	1.000000	0.466667
10	7	3	0	0.008333	0.008333

Table for $N = 2$, $n = 1$, through $N = 100$, $n = 50$

N	n	k	x	P(x)	p(x)
10	7	3	1	0.183333	0.175000
10	7	3	2	0.708333	0.525000
10	7	3	3	1.000000	0.291667
10	7	4	1	0.033333	0.033333
10	7	4	2	0.333333	0.300000
10	7	4	3	0.833333	0.500000
10	7	4	4	1.000000	0.166667
10	7	5	2	0.083333	0.083333
10	7	5	3	0.500000	0.416667
10	7	5	4	0.916667	0.416667
10	7	5	5	1.000000	0.083333
10	7	6	3	0.166667	0.166667
10	7	6	4	0.666667	0.500000
10	7	6	5	0.966667	0.300000
10	7	6	6	1.000000	0.033333
10	7	7	4	0.291667	0.291667
10	7	7	5	0.816667	0.525000
10	7	7	6	0.991667	0.175000
10	7	7	7	1.000000	0.008333
10	8	1	0	0.200000	0.200000
10	8	1	1	1.000000	0.800000
10	8	2	0	0.022222	0.022222
10	8	2	1	0.377778	0.355556
10	8	2	2	1.000000	0.622222
10	8	3	1	0.066667	0.066667
10	8	3	2	0.533333	0.466667
10	8	3	3	1.000000	0.466667
10	8	4	2	0.133333	0.133333
10	8	4	3	0.666667	0.533333
10	8	4	4	1.000000	0.333333
10	8	5	3	0.222222	0.222222
10	8	5	4	0.777778	0.555556
10	8	5	5	1.000000	0.222222
10	8	6	4	0.333333	0.333333
10	8	6	5	0.866667	0.533333
10	8	6	6	1.000000	0.133333
10	8	7	5	0.466667	0.466667
10	8	7	6	0.933333	0.466667
10	8	7	7	1.000000	0.066667
10	8	8	6	0.622222	0.622222
10	8	8	7	0.977778	0.355556
10	8	8	8	1.000000	0.022222
10	9	1	0	0.100000	0.100000
10	9	1	1	1.000000	0.900000
10	9	2	1	0.200000	0.200000
10	9	2	2	1.000000	0.800000
10	9	3	2	0.300000	0.300000
10	9	3	3	1.000000	0.700000
10	9	4	3	0.400000	0.400000
10	9	4	4	1.000000	0.600000

N	n	k	x	P(x)	p(x)
10	9	5	4	0.500000	0.500000
10	9	5	5	1.000000	0.500000
10	9	6	5	0.600000	0.600000
10	9	6	6	1.000000	0.400000
10	9	7	6	0.700000	0.700000
10	9	7	7	1.000000	0.300000
10	9	8	7	0.800000	0.800000
10	9	8	8	1.000000	0.200000
10	9	9	8	0.900000	0.900000
10	9	9	9	1.000000	0.100000
11	1	1	0	0.909091	0.909091
11	1	1	1	1.000000	0.090909
11	2	1	0	0.818182	0.818182
11	2	1	1	1.000000	0.181818
11	2	2	0	0.654545	0.654545
11	2	2	1	0.981818	0.327273
11	2	2	2	1.000000	0.018182
11	3	1	0	0.727273	0.727273
11	3	1	1	1.000000	0.272727
11	3	2	0	0.509091	0.509091
11	3	2	1	0.945455	0.436364
11	3	2	2	1.000000	0.054545
11	3	3	0	0.339394	0.339394
11	3	3	1	0.848485	0.509091
11	3	3	2	0.993939	0.145455
11	3	3	3	1.000000	0.006061
11	4	1	0	0.636364	0.636364
11	4	1	1	1.000000	0.363636
11	4	2	0	0.381818	0.381818
11	4	2	1	0.890909	0.509091
11	4	2	2	1.000000	0.109091
11	4	3	0	0.212121	0.212121
11	4	3	1	0.721212	0.509091
11	4	3	2	0.975758	0.254545
11	4	3	3	1.000000	0.024242
11	4	4	0	0.106061	0.106061
11	4	4	1	0.530303	0.424242
11	4	4	2	0.912121	0.381818
11	4	4	3	0.996970	0.084848
11	4	4	4	1.000000	0.003030
11	5	1	0	0.545455	0.545455
11	5	1	1	1.000000	0.454545
11	5	2	0	0.272727	0.272727
11	5	2	1	0.818182	0.545455
11	5	2	2	1.000000	0.181818
11	5	3	0	0.121212	0.121212
11	5	3	1	0.575758	0.454545
11	5	3	2	0.939394	0.363636
11	5	3	3	1.000000	0.060606
11	5	4	0	0.045455	0.045455

N	n	k	x	P(x)	p(x)
11	5	4	1	0.348485	0.303030
11	5	4	2	0.803030	0.454545
11	5	4	3	0.984848	0.181818
11	5	4	4	1.000000	0.015152
11	5	5	0	0.012987	0.012987
11	5	5	1	0.175325	0.162338
11	5	5	2	0.608225	0.432900
11	5	5	3	0.932900	0.324675
11	5	5	4	0.997835	0.064935
11	5	5	5	1.000000	0.002164
11	6	1	0	0.454545	0.454545
11	6	1	1	1.000000	0.545455
11	6	2	0	0.181818	0.181818
11	6	2	1	0.727273	0.545455
11	6	2	2	1.000000	0.272727
11	6	3	0	0.060606	0.060606
11	6	3	1	0.424242	0.363636
11	6	3	2	0.878788	0.454545
11	6	3	3	1.000000	0.121212
11	6	4	0	0.015152	0.015152
11	6	4	1	0.196970	0.181818
11	6	4	2	0.651515	0.454545
11	6	4	3	0.954545	0.303030
11	6	4	4	1.000000	0.045455
11	6	5	0	0.002164	0.002164
11	6	5	1	0.067100	0.064935
11	6	5	2	0.391775	0.324675
11	6	5	3	0.824675	0.432900
11	6	5	4	0.987013	0.162338
11	6	5	5	1.000000	0.012987
11	6	6	1	0.012987	0.012987
11	6	6	2	0.175325	0.162338
11	6	6	3	0.608225	0.432900
11	6	6	4	0.932900	0.324675
11	6	6	5	0.997835	0.064935
11	6	6	6	1.000000	0.002164
11	7	1	0	0.363636	0.363636
11	7	1	1	1.000000	0.636364
11	7	2	0	0.109091	0.109091
11	7	2	1	0.618182	0.509091
11	7	2	2	1.000000	0.381818
11	7	3	0	0.024242	0.024242
11	7	3	1	0.278788	0.254545
11	7	3	2	0.787879	0.509091
11	7	3	3	1.000000	0.212121
11	7	4	0	0.003030	0.003030
11	7	4	1	0.087879	0.084848
11	7	4	2	0.469697	0.381818
11	7	4	3	0.893939	0.424242
11	7	4	4	1.000000	0.106061

N	n	k	x	P(x)	p(x)
11	7	5	1	0.015152	0.015152
11	7	5	2	0.196970	0.181818
11	7	5	3	0.651515	0.454545
11	7	5	4	0.954545	0.303030
11	7	5	5	1.000000	0.045455
11	7	6	2	0.045455	0.045455
11	7	6	3	0.348485	0.303030
11	7	6	4	0.803030	0.454545
11	7	6	5	0.984848	0.181818
11	7	6	6	1.000000	0.015152
11	7	7	3	0.106061	0.106061
11	7	7	4	0.530303	0.424242
11	7	7	5	0.912121	0.381818
11	7	7	6	0.996970	0.084848
11	7	7	7	1.000000	0.003030
11	8	1	0	0.272727	0.272727
11	8	1	1	1.000000	0.727273
11	8	2	0	0.054545	0.054545
11	8	2	1	0.490909	0.436364
11	8	2	2	1.000000	0.509091
11	8	3	0	0.006061	0.006061
11	8	3	1	0.151515	0.145455
11	8	3	2	0.660606	0.509091
11	8	3	3	1.000000	0.339394
11	8	4	1	0.024242	0.024242
11	8	4	2	0.278788	0.254545
11	8	4	3	0.787879	0.509091
11	8	4	4	1.000000	0.212121
11	8	5	2	0.060606	0.060606
11	8	5	3	0.424242	0.363636
11	8	5	4	0.878788	0.454545
11	8	5	5	1.000000	0.121212
11	8	6	3	0.121212	0.121212
11	8	6	4	0.575758	0.454545
11	8	6	5	0.939394	0.363636
11	8	6	6	1.000000	0.060606
11	8	7	4	0.212121	0.212121
11	8	7	5	0.721212	0.509091
11	8	7	6	0.975758	0.254545
11	8	7	7	1.000000	0.024242
11	8	8	5	0.339394	0.339394
11	8	8	6	0.848485	0.509091
11	8	8	7	0.993939	0.145455
11	8	8	8	1.000000	0.006061
11	9	1	0	0.181818	0.181818
11	9	1	1	1.000000	0.818182
11	9	2	0	0.018182	0.018182
11	9	2	1	0.345454	0.327273
11	9	2	2	1.000000	0.654545
11	9	3	1	0.054545	0.054545

Table for $N = 2$, $n = 1$, through $N = 100$, $n = 50$

N	n	k	x	P(x)	p(x)
11	9	3	2	0.490909	0.436364
11	9	3	3	1.000000	0.509091
11	9	4	2	0.109091	0.109091
11	9	4	3	0.618182	0.509091
11	9	4	4	1.000000	0.381818
11	9	5	3	0.181818	0.181818
11	9	5	4	0.727273	0.545455
11	9	5	5	1.000000	0.272727
11	9	6	4	0.272727	0.272727
11	9	6	5	0.818182	0.545455
11	9	6	6	1.000000	0.181818
11	9	7	5	0.381818	0.381818
11	9	7	6	0.890909	0.509091
11	9	7	7	1.000000	0.109091
11	9	8	6	0.509091	0.509091
11	9	8	7	0.945455	0.436364
11	9	8	8	1.000000	0.054545
11	9	9	7	0.654545	0.654545
11	9	9	8	0.981818	0.327273
11	9	9	9	1.000000	0.018182
11	10	1	0	0.090909	0.090909
11	10	1	1	1.000000	0.909091
11	10	2	1	0.181818	0.181818
11	10	2	2	1.000000	0.818182
11	10	3	2	0.272727	0.272727
11	10	3	3	1.000000	0.727273
11	10	4	3	0.363636	0.363636
11	10	4	4	1.000000	0.636364
11	10	5	4	0.454545	0.454545
11	10	5	5	1.000000	0.545455
11	10	6	5	0.545455	0.545455
11	10	6	6	1.000000	0.454545
11	10	7	6	0.636364	0.636364
11	10	7	7	1.000000	0.363636
11	10	8	7	0.727273	0.727273
11	10	8	8	1.000000	0.272727
11	10	9	8	0.818182	0.818182
11	10	9	9	1.000000	0.181818
11	10	10	9	0.909091	0.909091
11	10	10	10	1.000000	0.090909
12	1	1	0	0.916667	0.916667
12	1	1	1	1.000000	0.083333
12	2	1	0	0.833333	0.833333
12	2	1	1	1.000000	0.166667
12	2	2	0	0.681818	0.681818
12	2	2	1	0.984848	0.303030
12	2	2	2	1.000000	0.015152
12	3	1	0	0.750000	0.750000
12	3	1	1	1.000000	0.250000
12	3	2	0	0.545455	0.545455

N	n	k	x	P(x)	p(x)
12	3	2	1	0.954545	0.409091
12	3	2	2	1.000000	0.045455
12	3	3	0	0.381818	0.381818
12	3	3	1	0.872727	0.490909
12	3	3	2	0.995454	0.122727
12	3	3	3	1.000000	0.004545
12	4	1	0	0.666667	0.666667
12	4	1	1	1.000000	0.333333
12	4	2	0	0.424242	0.424242
12	4	2	1	0.909091	0.484848
12	4	2	2	1.000000	0.090909
12	4	3	0	0.254545	0.254545
12	4	3	1	0.763636	0.509091
12	4	3	2	0.981818	0.218182
12	4	3	3	1.000000	0.018182
12	4	4	0	0.141414	0.141414
12	4	4	1	0.593939	0.452525
12	4	4	2	0.933333	0.339394
12	4	4	3	0.997980	0.064646
12	4	4	4	1.000000	0.002020
12	5	1	0	0.583333	0.583333
12	5	1	1	1.000000	0.416667
12	5	2	0	0.318182	0.318182
12	5	2	1	0.848485	0.530303
12	5	2	2	1.000000	0.151515
12	5	3	0	0.159091	0.159091
12	5	3	1	0.636364	0.477273
12	5	3	2	0.954545	0.318182
12	5	3	3	1.000000	0.045455
12	5	4	0	0.070707	0.070707
12	5	4	1	0.424242	0.353535
12	5	4	2	0.848485	0.424242
12	5	4	3	0.989899	0.141414
12	5	4	4	1.000000	0.010101
12	5	5	0	0.026515	0.026515
12	5	5	1	0.247475	0.220960
12	5	5	2	0.689394	0.441919
12	5	5	3	0.954545	0.265152
12	5	5	4	0.998737	0.044192
12	5	5	5	1.000000	0.001263
12	6	1	0	0.500000	0.500000
12	6	1	1	1.000000	0.500000
12	6	2	0	0.227273	0.227273
12	6	2	1	0.772727	0.545455
12	6	2	2	1.000000	0.227273
12	6	3	0	0.090909	0.090909
12	6	3	1	0.500000	0.409091
12	6	3	2	0.909091	0.409091
12	6	3	3	1.000000	0.090909
12	6	4	0	0.030303	0.030303

N	n	k	x	P(x)	p(x)
12	6	4	1	0.272727	0.242424
12	6	4	2	0.727273	0.454545
12	6	4	3	0.969697	0.242424
12	6	4	4	1.000000	0.030303
12	6	5	0	0.007576	0.007576
12	6	5	1	0.121212	0.113636
12	6	5	2	0.500000	0.378788
12	6	5	3	0.878788	0.378788
12	6	5	4	0.992424	0.113636
12	6	5	5	1.000000	0.007576
12	6	6	0	0.001082	0.001082
12	6	6	1	0.040043	0.038961
12	6	6	2	0.283550	0.243506
12	6	6	3	0.716450	0.432900
12	6	6	4	0.959957	0.243506
12	6	6	5	0.998918	0.038961
12	6	6	6	1.000000	0.001082
12	7	1	0	0.416667	0.416667
12	7	1	1	1.000000	0.583333
12	7	2	0	0.151515	0.151515
12	7	2	1	0.681818	0.530303
12	7	2	2	1.000000	0.318182
12	7	3	0	0.045455	0.045455
12	7	3	1	0.363636	0.318182
12	7	3	2	0.840909	0.477273
12	7	3	3	1.000000	0.159091
12	7	4	0	0.010101	0.010101
12	7	4	1	0.151515	0.141414
12	7	4	2	0.575758	0.424242
12	7	4	3	0.929293	0.353535
12	7	4	4	1.000000	0.070707
12	7	5	0	0.001263	0.001263
12	7	5	1	0.045455	0.044192
12	7	5	2	0.310606	0.265152
12	7	5	3	0.752525	0.441919
12	7	5	4	0.973485	0.220960
12	7	5	5	1.000000	0.026515
12	7	6	1	0.007576	0.007576
12	7	6	2	0.121212	0.113636
12	7	6	3	0.500000	0.378788
12	7	6	4	0.878788	0.378788
12	7	6	5	0.992424	0.113636
12	7	6	6	1.000000	0.007576
12	7	7	2	0.026515	0.026515
12	7	7	3	0.247475	0.220960
12	7	7	4	0.689394	0.441919
12	7	7	5	0.954545	0.265152
12	7	7	6	0.998737	0.044192
12	7	7	7	1.000000	0.001263
12	8	1	0	0.333333	0.333333

N	n	k	x	P(x)	p(x)
12	8	1	1	1.000000	0.666667
12	8	2	0	0.090909	0.090909
12	8	2	1	0.575758	0.484848
12	8	2	2	1.000000	0.424242
12	8	3	0	0.018182	0.018182
12	8	3	1	0.236364	0.218182
12	8	3	2	0.745455	0.509091
12	8	3	3	1.000000	0.254545
12	8	4	0	0.002020	0.002020
12	8	4	1	0.066667	0.064646
12	8	4	2	0.406061	0.339394
12	8	4	3	0.858586	0.452525
12	8	4	4	1.000000	0.141414
12	8	5	1	0.010101	0.010101
12	8	5	2	0.151515	0.141414
12	8	5	3	0.575758	0.424242
12	8	5	4	0.929293	0.353535
12	8	5	5	1.000000	0.070707
12	8	6	2	0.030303	0.030303
12	8	6	3	0.272727	0.242424
12	8	6	4	0.727273	0.454545
12	8	6	5	0.969697	0.242424
12	8	6	6	1.000000	0.030303
12	8	7	3	0.070707	0.070707
12	8	7	4	0.424242	0.353535
12	8	7	5	0.848485	0.424242
12	8	7	6	0.989899	0.141414
12	8	7	7	1.000000	0.010101
12	8	8	4	0.141414	0.141414
12	8	8	5	0.593939	0.452525
12	8	8	6	0.933333	0.339394
12	8	8	7	0.997980	0.064646
12	8	8	8	1.000000	0.002020
12	9	1	0	0.250000	0.250000
12	9	1	1	1.000000	0.750000
12	9	2	0	0.045455	0.045455
12	9	2	1	0.454545	0.409091
12	9	2	2	1.000000	0.545455
12	9	3	0	0.004545	0.004545
12	9	3	1	0.127273	0.122727
12	9	3	2	0.618182	0.490909
12	9	3	3	1.000000	0.381818
12	9	4	1	0.018182	0.018182
12	9	4	2	0.236364	0.218182
12	9	4	3	0.745455	0.509091
12	9	4	4	1.000000	0.254545
12	9	5	2	0.045455	0.045455
12	9	5	3	0.363636	0.318182
12	9	5	4	0.840909	0.477273
12	9	5	5	1.000000	0.159091

Table for $N = 2$, $n = 1$, through $N = 100$, $n = 50$

N	n	k	x	P(x)	p(x)
12	9	6	3	0.090909	0.090909
12	9	6	4	0.500000	0.409091
12	9	6	5	0.909091	0.409091
12	9	6	6	1.000000	0.090909
12	9	7	4	0.159091	0.159091
12	9	7	5	0.636364	0.477273
12	9	7	6	0.954545	0.318182
12	9	7	7	1.000000	0.045455
12	9	8	5	0.254545	0.254545
12	9	8	6	0.763636	0.509091
12	9	8	7	0.981818	0.218182
12	9	8	8	1.000000	0.018182
12	9	9	6	0.381818	0.381818
12	9	9	7	0.872727	0.490909
12	9	9	8	0.995454	0.122727
12	9	9	9	1.000000	0.004545
12	10	1	0	0.166667	0.166667
12	10	1	1	1.000000	0.833333
12	10	2	0	0.015152	0.015152
12	10	2	1	0.318182	0.303030
12	10	2	2	1.000000	0.681818
12	10	3	1	0.045455	0.045455
12	10	3	2	0.454545	0.409091
12	10	3	3	1.000000	0.545455
12	10	4	2	0.090909	0.090909
12	10	4	3	0.575758	0.484848
12	10	4	4	1.000000	0.424242
12	10	5	3	0.151515	0.151515
12	10	5	4	0.681818	0.530303
12	10	5	5	1.000000	0.318182
12	10	6	4	0.227273	0.227273
12	10	6	5	0.772727	0.545455
12	10	6	6	1.000000	0.227273
12	10	7	5	0.318182	0.318182
12	10	7	6	0.848485	0.530303
12	10	7	7	1.000000	0.151515
12	10	8	6	0.424242	0.424242
12	10	8	7	0.909091	0.484848
12	10	8	8	1.000000	0.090909
12	10	9	7	0.545455	0.545455
12	10	9	8	0.954545	0.409091
12	10	9	9	1.000000	0.045455
12	10	10	8	0.681818	0.681818
12	10	10	9	0.984848	0.303030
12	10	10	10	1.000000	0.015152
12	11	1	0	0.083333	0.083333
12	11	1	1	1.000000	0.916667
12	11	2	1	0.166667	0.166667
12	11	2	2	1.000000	0.833333
12	11	3	2	0.250000	0.250000

N	n	k	x	P(x)	p(x)
12	11	3	3	1.000000	0.750000
12	11	4	3	0.333333	0.333333
12	11	4	4	1.000000	0.666667
12	11	5	4	0.416667	0.416667
12	11	5	5	1.000000	0.583333
12	11	6	5	0.500000	0.500000
12	11	6	6	1.000000	0.500000
12	11	7	6	0.583333	0.583333
12	11	7	7	1.000000	0.416667
12	11	8	7	0.666667	0.666667
12	11	8	8	1.000000	0.333333
12	11	9	8	0.750000	0.750000
12	11	9	9	1.000000	0.250000
12	11	10	9	0.833333	0.833333
12	11	10	10	1.000000	0.166667
12	11	11	10	0.916667	0.916667
12	11	11	11	1.000000	0.083333
13	1	1	0	0.923077	0.923077
13	1	1	1	1.000000	0.076923
13	2	1	0	0.846154	0.846154
13	2	1	1	1.000000	0.153846
13	2	2	0	0.705128	0.705128
13	2	2	1	0.987179	0.282051
13	2	2	2	1.000000	0.012821
13	3	1	0	0.769231	0.769231
13	3	1	1	1.000000	0.230769
13	3	2	0	0.576923	0.576923
13	3	2	1	0.961538	0.384615
13	3	2	2	1.000000	0.038462
13	3	3	0	0.419580	0.419580
13	3	3	1	0.891608	0.472028
13	3	3	2	0.996503	0.104895
13	3	3	3	1.000000	0.003496
13	4	1	0	0.692308	0.692308
13	4	1	1	1.000000	0.307692
13	4	2	0	0.461538	0.461538
13	4	2	1	0.923077	0.461538
13	4	2	2	1.000000	0.076923
13	4	3	0	0.293706	0.293706
13	4	3	1	0.797203	0.503497
13	4	3	2	0.986014	0.188811
13	4	3	3	1.000000	0.013986
13	4	4	0	0.176224	0.176224
13	4	4	1	0.646154	0.469930
13	4	4	2	0.948252	0.302098
13	4	4	3	0.998601	0.050350
13	4	4	4	1.000000	0.001399
13	5	1	0	0.615385	0.615385
13	5	1	1	1.000000	0.384615
13	5	2	0	0.358974	0.358974

N	n	k	x	P(x)	p(x)
13	5	2	1	0.871795	0.512821
13	5	2	2	1.000000	0.128205
13	5	3	0	0.195804	0.195804
13	5	3	1	0.685315	0.489510
13	5	3	2	0.965035	0.279720
13	5	3	3	1.000000	0.034965
13	5	4	0	0.097902	0.097902
13	5	4	1	0.489510	0.391608
13	5	4	2	0.881119	0.391608
13	5	4	3	0.993007	0.111888
13	5	4	4	1.000000	0.006993
13	5	5	0	0.043512	0.043512
13	5	5	1	0.315462	0.271950
13	5	5	2	0.750583	0.435120
13	5	5	3	0.968143	0.217560
13	5	5	4	0.999223	0.031080
13	5	5	5	1.000000	0.000777
13	6	1	0	0.538462	0.538462
13	6	1	1	1.000000	0.461538
13	6	2	0	0.269231	0.269231
13	6	2	1	0.807692	0.538462
13	6	2	2	1.000000	0.192308
13	6	3	0	0.122378	0.122378
13	6	3	1	0.562937	0.440559
13	6	3	2	0.930070	0.367133
13	6	3	3	1.000000	0.069930
13	6	4	0	0.048951	0.048951
13	6	4	1	0.342657	0.293706
13	6	4	2	0.783217	0.440559
13	6	4	3	0.979021	0.195804
13	6	4	4	1.000000	0.020979
13	6	5	0	0.016317	0.016317
13	6	5	1	0.179487	0.163170
13	6	5	2	0.587413	0.407925
13	6	5	3	0.913753	0.326340
13	6	5	4	0.995338	0.081585
13	6	5	5	1.000000	0.004662
13	6	6	0	0.004079	0.004079
13	6	6	1	0.077506	0.073427
13	6	6	2	0.383450	0.305944
13	6	6	3	0.791375	0.407925
13	6	6	4	0.974942	0.183566
13	6	6	5	0.999417	0.024476
13	6	6	6	1.000000	0.000583
13	7	1	0	0.461538	0.461538
13	7	1	1	1.000000	0.538462
13	7	2	0	0.192308	0.192308
13	7	2	1	0.730769	0.538462
13	7	2	2	1.000000	0.269231
13	7	3	0	0.069930	0.069930

N	n	k	x	P(x)	p(x)
13	7	3	1	0.437063	0.367133
13	7	3	2	0.877622	0.440559
13	7	3	3	1.000000	0.122378
13	7	4	0	0.020979	0.020979
13	7	4	1	0.216783	0.195804
13	7	4	2	0.657343	0.440559
13	7	4	3	0.951049	0.293706
13	7	4	4	1.000000	0.048951
13	7	5	0	0.004662	0.004662
13	7	5	1	0.086247	0.081585
13	7	5	2	0.412587	0.326340
13	7	5	3	0.820513	0.407925
13	7	5	4	0.983683	0.163170
13	7	5	5	1.000000	0.016317
13	7	6	0	0.000583	0.000583
13	7	6	1	0.025058	0.024476
13	7	6	2	0.208625	0.183566
13	7	6	3	0.616550	0.407925
13	7	6	4	0.922494	0.305944
13	7	6	5	0.995921	0.073427
13	7	6	6	1.000000	0.004079
13	7	7	1	0.004079	0.004079
13	7	7	2	0.077506	0.073427
13	7	7	3	0.383450	0.305944
13	7	7	4	0.791375	0.407925
13	7	7	5	0.974942	0.183566
13	7	7	6	0.999417	0.024476
13	7	7	7	1.000000	0.000583
13	8	1	0	0.384615	0.384615
13	8	1	1	1.000000	0.615385
13	8	2	0	0.128205	0.128205
13	8	2	1	0.641026	0.512821
13	8	2	2	1.000000	0.358974
13	8	3	0	0.034965	0.034965
13	8	3	1	0.314685	0.279720
13	8	3	2	0.804196	0.489510
13	8	3	3	1.000000	0.195804
13	8	4	0	0.006993	0.006993
13	8	4	1	0.118881	0.111888
13	8	4	2	0.510489	0.391608
13	8	4	3	0.902098	0.391608
13	8	4	4	1.000000	0.097902
13	8	5	0	0.000777	0.000777
13	8	5	1	0.031857	0.031080
13	8	5	2	0.249417	0.217560
13	8	5	3	0.684538	0.435120
13	8	5	4	0.956488	0.271950
13	8	5	5	1.000000	0.043512
13	8	6	1	0.004662	0.004662
13	8	6	2	0.086247	0.081585

Table for $N = 2$, $n = 1$, through $N = 100$, $n = 50$

N	n	k	x	P(x)	p(x)
13	8	6	3	0.412587	0.326340
13	8	6	4	0.820513	0.407925
13	8	6	5	0.983683	0.163170
13	8	6	6	1.000000	0.016317
13	8	7	2	0.016317	0.016317
13	8	7	3	0.179487	0.163170
13	8	7	4	0.587413	0.407925
13	8	7	5	0.913753	0.326340
13	8	7	6	0.995338	0.081585
13	8	7	7	1.000000	0.004662
13	8	8	3	0.043512	0.043512
13	8	8	4	0.315462	0.271950
13	8	8	5	0.750583	0.435120
13	8	8	6	0.968143	0.217560
13	8	8	7	0.999223	0.031080
13	8	8	8	1.000000	0.000777
13	9	1	0	0.307692	0.307692
13	9	1	1	1.000000	0.692308
13	9	2	0	0.076923	0.076923
13	9	2	1	0.538462	0.461538
13	9	2	2	1.000000	0.461538
13	9	3	0	0.013986	0.013986
13	9	3	1	0.202797	0.188811
13	9	3	2	0.706294	0.503497
13	9	3	3	1.000000	0.293706
13	9	4	0	0.001399	0.001399
13	9	4	1	0.051748	0.050350
13	9	4	2	0.353846	0.302098
13	9	4	3	0.823776	0.469930
13	9	4	4	1.000000	0.176224
13	9	5	1	0.006993	0.006993
13	9	5	2	0.118881	0.111888
13	9	5	3	0.510489	0.391608
13	9	5	4	0.902098	0.391608
13	9	5	5	1.000000	0.097902
13	9	6	2	0.020979	0.020979
13	9	6	3	0.216783	0.195804
13	9	6	4	0.657343	0.440559
13	9	6	5	0.951049	0.293706
13	9	6	6	1.000000	0.048951
13	9	7	3	0.048951	0.048951
13	9	7	4	0.342657	0.293706
13	9	7	5	0.783217	0.440559
13	9	7	6	0.979021	0.195804
13	9	7	7	1.000000	0.020979
13	9	8	4	0.097902	0.097902
13	9	8	5	0.489510	0.391608
13	9	8	6	0.881119	0.391608
13	9	8	7	0.993007	0.111888
13	9	8	8	1.000000	0.006993

N	n	k	x	P(x)	p(x)
13	9	9	5	0.176224	0.176224
13	9	9	6	0.646154	0.469930
13	9	9	7	0.948252	0.302098
13	9	9	8	0.998601	0.050350
13	9	9	9	1.000000	0.001399
13	10	1	0	0.230769	0.230769
13	10	1	1	1.000000	0.769231
13	10	2	0	0.038462	0.038462
13	10	2	1	0.423077	0.384615
13	10	2	2	1.000000	0.576923
13	10	3	0	0.003496	0.003496
13	10	3	1	0.108392	0.104895
13	10	3	2	0.580420	0.472028
13	10	3	3	1.000000	0.419580
13	10	4	1	0.013986	0.013986
13	10	4	2	0.202797	0.188811
13	10	4	3	0.706294	0.503497
13	10	4	4	1.000000	0.293706
13	10	5	2	0.034965	0.034965
13	10	5	3	0.314685	0.279720
13	10	5	4	0.804196	0.489510
13	10	5	5	1.000000	0.195804
13	10	6	3	0.069930	0.069930
13	10	6	4	0.437063	0.367133
13	10	6	5	0.877622	0.440559
13	10	6	6	1.000000	0.122378
13	10	7	4	0.122378	0.122378
13	10	7	5	0.562937	0.440559
13	10	7	6	0.930070	0.367133
13	10	7	7	1.000000	0.069930
13	10	8	5	0.195804	0.195804
13	10	8	6	0.685315	0.489510
13	10	8	7	0.965035	0.279720
13	10	8	8	1.000000	0.034965
13	10	9	6	0.293706	0.293706
13	10	9	7	0.797203	0.503497
13	10	9	8	0.986014	0.188811
13	10	9	9	1.000000	0.013986
13	10	10	7	0.419580	0.419580
13	10	10	8	0.891608	0.472028
13	10	10	9	0.996503	0.104895
13	10	10	10	1.000000	0.003496
13	11	1	0	0.153846	0.153846
13	11	1	1	1.000000	0.846154
13	11	2	0	0.012821	0.012821
13	11	2	1	0.294872	0.282051
13	11	2	2	1.000000	0.705128
13	11	3	1	0.038462	0.038462
13	11	3	2	0.423077	0.384615
13	11	3	3	1.000000	0.576923

N	n	k	x	P(x)	p(x)
13	11	4	2	0.076923	0.076923
13	11	4	3	0.538462	0.461538
13	11	4	4	1.000000	0.461538
13	11	5	3	0.128205	0.128205
13	11	5	4	0.641026	0.512821
13	11	5	5	1.000000	0.358974
13	11	6	4	0.192308	0.192308
13	11	6	5	0.730769	0.538462
13	11	6	6	1.000000	0.269231
13	11	7	5	0.269231	0.269231
13	11	7	6	0.807692	0.538462
13	11	7	7	1.000000	0.192308
13	11	8	6	0.358974	0.358974
13	11	8	7	0.871795	0.512821
13	11	8	8	1.000000	0.128205
13	11	9	7	0.461538	0.461538
13	11	9	8	0.923077	0.461538
13	11	9	9	1.000000	0.076923
13	11	10	8	0.576923	0.576923
13	11	10	9	0.961538	0.384615
13	11	10	10	1.000000	0.038462
13	11	11	9	0.705128	0.705128
13	11	11	10	0.987179	0.282051
13	11	11	11	1.000000	0.012821
13	12	1	0	0.076923	0.076923
13	12	1	1	1.000000	0.923077
13	12	2	1	0.153846	0.153846
13	12	2	2	1.000000	0.846154
13	12	3	2	0.230769	0.230769
13	12	3	3	1.000000	0.769231
13	12	4	3	0.307692	0.307692
13	12	4	4	1.000000	0.692308
13	12	5	4	0.384615	0.384615
13	12	5	5	1.000000	0.615385
13	12	6	5	0.461538	0.461538
13	12	6	6	1.000000	0.538462
13	12	7	6	0.538462	0.538462
13	12	7	7	1.000000	0.461538
13	12	8	7	0.615385	0.615385
13	12	8	8	1.000000	0.384615
13	12	9	8	0.692308	0.692308
13	12	9	9	1.000000	0.307692
13	12	10	9	0.769231	0.769231
13	12	10	10	1.000000	0.230769
13	12	11	10	0.846154	0.846154
13	12	11	11	1.000000	0.153846
13	12	12	11	0.923077	0.923077
13	12	12	12	1.000000	0.076923
14	1	1	0	0.928571	0.928571
14	1	1	1	1.000000	0.071429

N	n	k	x	P(x)	p(x)
14	2	1	0	0.857143	0.857143
14	2	1	1	1.000000	0.142857
14	2	2	0	0.725275	0.725275
14	2	2	1	0.989011	0.263736
14	2	2	2	1.000000	0.010989
14	3	1	0	0.785714	0.785714
14	3	1	1	1.000000	0.214286
14	3	2	0	0.604396	0.604396
14	3	2	1	0.967033	0.362637
14	3	2	2	1.000000	0.032967
14	3	3	0	0.453297	0.453297
14	3	3	1	0.906593	0.453297
14	3	3	2	0.997253	0.090659
14	3	3	3	1.000000	0.002747
14	4	1	0	0.714286	0.714286
14	4	1	1	1.000000	0.285714
14	4	2	0	0.494505	0.494505
14	4	2	1	0.934066	0.439560
14	4	2	2	1.000000	0.065934
14	4	3	0	0.329670	0.329670
14	4	3	1	0.824176	0.494505
14	4	3	2	0.989011	0.164835
14	4	3	3	1.000000	0.010989
14	4	4	0	0.209790	0.209790
14	4	4	1	0.689311	0.479520
14	4	4	2	0.959041	0.269730
14	4	4	3	0.999001	0.039960
14	4	4	4	1.000000	0.000999
14	5	1	0	0.642857	0.642857
14	5	1	1	1.000000	0.357143
14	5	2	0	0.395604	0.395604
14	5	2	1	0.890110	0.494505
14	5	2	2	1.000000	0.109890
14	5	3	0	0.230769	0.230769
14	5	3	1	0.725275	0.494505
14	5	3	2	0.972527	0.247253
14	5	3	3	1.000000	0.027473
14	5	4	0	0.125874	0.125874
14	5	4	1	0.545454	0.419580
14	5	4	2	0.905095	0.359640
14	5	4	3	0.995005	0.089910
14	5	4	4	1.000000	0.004995
14	5	5	0	0.062937	0.062937
14	5	5	1	0.377622	0.314685
14	5	5	2	0.797203	0.419580
14	5	5	3	0.977023	0.179820
14	5	5	4	0.999500	0.022478
14	5	5	5	1.000000	0.000499
14	6	1	0	0.571429	0.571429
14	6	1	1	1.000000	0.428571

Table for $N = 2,\ n = 1,$ through $N = 100,\ n = 50$

N	n	k	x	P(x)	p(x)
14	6	2	0	0.307692	0.307692
14	6	2	1	0.835165	0.527473
14	6	2	2	1.000000	0.164835
14	6	3	0	0.153846	0.153846
14	6	3	1	0.615385	0.461538
14	6	3	2	0.945055	0.329670
14	6	3	3	1.000000	0.054945
14	6	4	0	0.069930	0.069930
14	6	4	1	0.405594	0.335664
14	6	4	2	0.825175	0.419580
14	6	4	3	0.985015	0.159840
14	6	4	4	1.000000	0.014985
14	6	5	0	0.027972	0.027972
14	6	5	1	0.237762	0.209790
14	6	5	2	0.657343	0.419580
14	6	5	3	0.937063	0.279720
14	6	5	4	0.997003	0.059940
14	6	5	5	1.000000	0.002997
14	6	6	0	0.009324	0.009324
14	6	6	1	0.121212	0.111888
14	6	6	2	0.470862	0.349650
14	6	6	3	0.843823	0.372960
14	6	6	4	0.983683	0.139860
14	6	6	5	0.999667	0.015984
14	6	6	6	1.000000	0.000333
14	7	1	0	0.500000	0.500000
14	7	1	1	1.000000	0.500000
14	7	2	0	0.230769	0.230769
14	7	2	1	0.769231	0.538462
14	7	2	2	1.000000	0.230769
14	7	3	0	0.096154	0.096154
14	7	3	1	0.500000	0.403846
14	7	3	2	0.903846	0.403846
14	7	3	3	1.000000	0.096154
14	7	4	0	0.034965	0.034965
14	7	4	1	0.279720	0.244755
14	7	4	2	0.720280	0.440559
14	7	4	3	0.965035	0.244755
14	7	4	4	1.000000	0.034965
14	7	5	0	0.010490	0.010490
14	7	5	1	0.132867	0.122378
14	7	5	2	0.500000	0.367133
14	7	5	3	0.867133	0.367133
14	7	5	4	0.989510	0.122378
14	7	5	5	1.000000	0.010490
14	7	6	0	0.002331	0.002331
14	7	6	1	0.051282	0.048951
14	7	6	2	0.296037	0.244755
14	7	6	3	0.703963	0.407925
14	7	6	4	0.948718	0.244755

N	n	k	x	P(x)	p(x)
14	7	6	5	0.997669	0.048951
14	7	6	6	1.000000	0.002331
14	7	7	0	0.000291	0.000291
14	7	7	1	0.014569	0.014277
14	7	7	2	0.143065	0.128496
14	7	7	3	0.500000	0.356935
14	7	7	4	0.856935	0.356935
14	7	7	5	0.985431	0.128496
14	7	7	6	0.999709	0.014277
14	7	7	7	1.000000	0.000291
14	8	1	0	0.428571	0.428571
14	8	1	1	1.000000	0.571429
14	8	2	0	0.164835	0.164835
14	8	2	1	0.692308	0.527473
14	8	2	2	1.000000	0.307692
14	8	3	0	0.054945	0.054945
14	8	3	1	0.384615	0.329670
14	8	3	2	0.846154	0.461538
14	8	3	3	1.000000	0.153846
14	8	4	0	0.014985	0.014985
14	8	4	1	0.174825	0.159840
14	8	4	2	0.594406	0.419580
14	8	4	3	0.930070	0.335664
14	8	4	4	1.000000	0.069930
14	8	5	0	0.002997	0.002997
14	8	5	1	0.062937	0.059940
14	8	5	2	0.342657	0.279720
14	8	5	3	0.762238	0.419580
14	8	5	4	0.972028	0.209790
14	8	5	5	1.000000	0.027972
14	8	6	0	0.000333	0.000333
14	8	6	1	0.016317	0.015984
14	8	6	2	0.156177	0.139860
14	8	6	3	0.529138	0.372960
14	8	6	4	0.878788	0.349650
14	8	6	5	0.990676	0.111888
14	8	6	6	1.000000	0.009324
14	8	7	1	0.002331	0.002331
14	8	7	2	0.051282	0.048951
14	8	7	3	0.296037	0.244755
14	8	7	4	0.703963	0.407925
14	8	7	5	0.948718	0.244755
14	8	7	6	0.997669	0.048951
14	8	7	7	1.000000	0.002331
14	8	8	2	0.009324	0.009324
14	8	8	3	0.121212	0.111888
14	8	8	4	0.470862	0.349650
14	8	8	5	0.843823	0.372960
14	8	8	6	0.983683	0.139860
14	8	8	7	0.999667	0.015984

N	n	k	x	P(x)	p(x)
14	8	8	8	1.000000	0.000333
14	9	1	0	0.357143	0.357143
14	9	1	1	1.000000	0.642857
14	9	2	0	0.109890	0.109890
14	9	2	1	0.604396	0.494505
14	9	2	2	1.000000	0.395604
14	9	3	0	0.027473	0.027473
14	9	3	1	0.274725	0.247253
14	9	3	2	0.769231	0.494505
14	9	3	3	1.000000	0.230769
14	9	4	0	0.004995	0.004995
14	9	4	1	0.094905	0.089910
14	9	4	2	0.454545	0.359640
14	9	4	3	0.874126	0.419580
14	9	4	4	1.000000	0.125874
14	9	5	0	0.000499	0.000499
14	9	5	1	0.022977	0.022478
14	9	5	2	0.202797	0.179820
14	9	5	3	0.622378	0.419580
14	9	5	4	0.937063	0.314685
14	9	5	5	1.000000	0.062937
14	9	6	1	0.002997	0.002997
14	9	6	2	0.062937	0.059940
14	9	6	3	0.342657	0.279720
14	9	6	4	0.762238	0.419580
14	9	6	5	0.972028	0.209790
14	9	6	6	1.000000	0.027972
14	9	7	2	0.010490	0.010490
14	9	7	3	0.132867	0.122378
14	9	7	4	0.500000	0.367133
14	9	7	5	0.867133	0.367133
14	9	7	6	0.989510	0.122378
14	9	7	7	1.000000	0.010490
14	9	8	3	0.027972	0.027972
14	9	8	4	0.237762	0.209790
14	9	8	5	0.657343	0.419580
14	9	8	6	0.937063	0.279720
14	9	8	7	0.997003	0.059940
14	9	8	8	1.000000	0.002997
14	9	9	4	0.062937	0.062937
14	9	9	5	0.377622	0.314685
14	9	9	6	0.797203	0.419580
14	9	9	7	0.977023	0.179820
14	9	9	8	0.999500	0.022478
14	9	9	9	1.000000	0.000499
14	10	1	0	0.285714	0.285714
14	10	1	1	1.000000	0.714286
14	10	2	0	0.065934	0.065934
14	10	2	1	0.505494	0.439560
14	10	2	2	1.000000	0.494505

N	n	k	x	P(x)	p(x)
14	10	3	0	0.010989	0.010989
14	10	3	1	0.175824	0.164835
14	10	3	2	0.670330	0.494505
14	10	3	3	1.000000	0.329670
14	10	4	0	0.000999	0.000999
14	10	4	1	0.040959	0.039960
14	10	4	2	0.310689	0.269730
14	10	4	3	0.790210	0.479520
14	10	4	4	1.000000	0.209790
14	10	5	1	0.004995	0.004995
14	10	5	2	0.094905	0.089910
14	10	5	3	0.454545	0.359640
14	10	5	4	0.874126	0.419580
14	10	5	5	1.000000	0.125874
14	10	6	2	0.014985	0.014985
14	10	6	3	0.174825	0.159840
14	10	6	4	0.594406	0.419580
14	10	6	5	0.930070	0.335664
14	10	6	6	1.000000	0.069930
14	10	7	3	0.034965	0.034965
14	10	7	4	0.279720	0.244755
14	10	7	5	0.720280	0.440559
14	10	7	6	0.965035	0.244755
14	10	7	7	1.000000	0.034965
14	10	8	4	0.069930	0.069930
14	10	8	5	0.405594	0.335664
14	10	8	6	0.825175	0.419580
14	10	8	7	0.985015	0.159840
14	10	8	8	1.000000	0.014985
14	10	9	5	0.125874	0.125874
14	10	9	6	0.545454	0.419580
14	10	9	7	0.905095	0.359640
14	10	9	8	0.995005	0.089910
14	10	9	9	1.000000	0.004995
14	10	10	6	0.209790	0.209790
14	10	10	7	0.689311	0.479520
14	10	10	8	0.959041	0.269730
14	10	10	9	0.999001	0.039960
14	10	10	10	1.000000	0.000999
14	11	1	0	0.214286	0.214286
14	11	1	1	1.000000	0.785714
14	11	2	0	0.032967	0.032967
14	11	2	1	0.395604	0.362637
14	11	2	2	1.000000	0.604396
14	11	3	0	0.002747	0.002747
14	11	3	1	0.093407	0.090659
14	11	3	2	0.546703	0.453297
14	11	3	3	1.000000	0.453297
14	11	4	1	0.010989	0.010989
14	11	4	2	0.175824	0.164835

Table for $N = 2$, $n = 1$, through $N = 100$, $n = 50$

N	n	k	x	P(x)	p(x)
14	11	4	3	0.670330	0.494505
14	11	4	4	1.000000	0.329670
14	11	5	2	0.027473	0.027473
14	11	5	3	0.274725	0.247253
14	11	5	4	0.769231	0.494505
14	11	5	5	1.000000	0.230769
14	11	6	3	0.054945	0.054945
14	11	6	4	0.384615	0.329670
14	11	6	5	0.846154	0.461538
14	11	6	6	1.000000	0.153846
14	11	7	4	0.096154	0.096154
14	11	7	5	0.500000	0.403846
14	11	7	6	0.903846	0.403846
14	11	7	7	1.000000	0.096154
14	11	8	5	0.153846	0.153846
14	11	8	6	0.615385	0.461538
14	11	8	7	0.945055	0.329670
14	11	8	8	1.000000	0.054945
14	11	9	6	0.230769	0.230769
14	11	9	7	0.725275	0.494505
14	11	9	8	0.972527	0.247253
14	11	9	9	1.000000	0.027473
14	11	10	7	0.329670	0.329670
14	11	10	8	0.824176	0.494505
14	11	10	9	0.989011	0.164835
14	11	10	10	1.000000	0.010989
14	11	11	8	0.453297	0.453297
14	11	11	9	0.906593	0.453297
14	11	11	10	0.997253	0.090659
14	11	11	11	1.000000	0.002747
14	12	1	0	0.142857	0.142857
14	12	1	1	1.000000	0.857143
14	12	2	0	0.010989	0.010989
14	12	2	1	0.274725	0.263736
14	12	2	2	1.000000	0.725275
14	12	3	1	0.032967	0.032967
14	12	3	2	0.395604	0.362637
14	12	3	3	1.000000	0.604396
14	12	4	2	0.065934	0.065934
14	12	4	3	0.505494	0.439560
14	12	4	4	1.000000	0.494505
14	12	5	3	0.109890	0.109890
14	12	5	4	0.604396	0.494505
14	12	5	5	1.000000	0.395604
14	12	6	4	0.164835	0.164835
14	12	6	5	0.692308	0.527473
14	12	6	6	1.000000	0.307692
14	12	7	5	0.230769	0.230769
14	12	7	6	0.769231	0.538462
14	12	7	7	1.000000	0.230769

N	n	k	x	P(x)	p(x)
14	12	8	6	0.307692	0.307692
14	12	8	7	0.835165	0.527473
14	12	8	8	1.000000	0.164835
14	12	9	7	0.395604	0.395604
14	12	9	8	0.890110	0.494505
14	12	9	9	1.000000	0.109890
14	12	10	8	0.494505	0.494505
14	12	10	9	0.934066	0.439560
14	12	10	10	1.000000	0.065934
14	12	11	9	0.604396	0.604396
14	12	11	10	0.967033	0.362637
14	12	11	11	1.000000	0.032967
14	12	12	10	0.725275	0.725275
14	12	12	11	0.989011	0.263736
14	12	12	12	1.000000	0.010989
14	13	1	0	0.071429	0.071429
14	13	1	1	1.000000	0.928571
14	13	2	1	0.142857	0.142857
14	13	2	2	1.000000	0.857143
14	13	3	2	0.214286	0.214286
14	13	3	3	1.000000	0.785714
14	13	4	3	0.285714	0.285714
14	13	4	4	1.000000	0.714286
14	13	5	4	0.357143	0.357143
14	13	5	5	1.000000	0.642857
14	13	6	5	0.428571	0.428571
14	13	6	6	1.000000	0.571429
14	13	7	6	0.500000	0.500000
14	13	7	7	1.000000	0.500000
14	13	8	7	0.571429	0.571429
14	13	8	8	1.000000	0.428571
14	13	9	8	0.642857	0.642857
14	13	9	9	1.000000	0.357143
14	13	10	9	0.714286	0.714286
14	13	10	10	1.000000	0.285714
14	13	11	10	0.785714	0.785714
14	13	11	11	1.000000	0.214286
14	13	12	11	0.857143	0.857143
14	13	12	12	1.000000	0.142857
14	13	13	12	0.928571	0.928571
14	13	13	13	1.000000	0.071429
15	1	1	0	0.933333	0.933333
15	1	1	1	1.000000	0.066667
15	2	1	0	0.866667	0.866667
15	2	1	1	1.000000	0.133333
15	2	2	0	0.742857	0.742857
15	2	2	1	0.990476	0.247619
15	2	2	2	1.000000	0.009524
15	3	1	0	0.800000	0.800000
15	3	1	1	1.000000	0.200000

N	n	k	x	P(x)	p(x)
15	3	2	0	0.628571	0.628571
15	3	2	1	0.971429	0.342857
15	3	2	2	1.000000	0.028571
15	3	3	0	0.483516	0.483516
15	3	3	1	0.918681	0.435165
15	3	3	2	0.997802	0.079121
15	3	3	3	1.000000	0.002198
15	4	1	0	0.733333	0.733333
15	4	1	1	1.000000	0.266667
15	4	2	0	0.523810	0.523810
15	4	2	1	0.942857	0.419048
15	4	2	2	1.000000	0.057143
15	4	3	0	0.362637	0.362637
15	4	3	1	0.846154	0.483516
15	4	3	2	0.991209	0.145055
15	4	3	3	1.000000	0.008791
15	4	4	0	0.241758	0.241758
15	4	4	1	0.725275	0.483516
15	4	4	2	0.967033	0.241758
15	4	4	3	0.999267	0.032234
15	4	4	4	1.000000	0.000733
15	5	1	0	0.666667	0.666667
15	5	1	1	1.000000	0.333333
15	5	2	0	0.428571	0.428571
15	5	2	1	0.904762	0.476190
15	5	2	2	1.000000	0.095238
15	5	3	0	0.263736	0.263736
15	5	3	1	0.758242	0.494505
15	5	3	2	0.978022	0.219780
15	5	3	3	1.000000	0.021978
15	5	4	0	0.153846	0.153846
15	5	4	1	0.593407	0.439560
15	5	4	2	0.923077	0.329670
15	5	4	3	0.996337	0.073260
15	5	4	4	1.000000	0.003663
15	5	5	0	0.083916	0.083916
15	5	5	1	0.433566	0.349650
15	5	5	2	0.833167	0.399600
15	5	5	3	0.983017	0.149850
15	5	5	4	0.999667	0.016650
15	5	5	5	1.000000	0.000333
15	6	1	0	0.600000	0.600000
15	6	1	1	1.000000	0.400000
15	6	2	0	0.342857	0.342857
15	6	2	1	0.857143	0.514286
15	6	2	2	1.000000	0.142857
15	6	3	0	0.184615	0.184615
15	6	3	1	0.659341	0.474725
15	6	3	2	0.956044	0.296703
15	6	3	3	1.000000	0.043956

N	n	k	x	P(x)	p(x)
15	6	4	0	0.092308	0.092308
15	6	4	1	0.461538	0.369231
15	6	4	2	0.857143	0.395604
15	6	4	3	0.989011	0.131868
15	6	4	4	1.000000	0.010989
15	6	5	0	0.041958	0.041958
15	6	5	1	0.293706	0.251748
15	6	5	2	0.713287	0.419580
15	6	5	3	0.953047	0.239760
15	6	5	4	0.998002	0.044955
15	6	5	5	1.000000	0.001998
15	6	6	0	0.016783	0.016783
15	6	6	1	0.167832	0.151049
15	6	6	2	0.545455	0.377622
15	6	6	3	0.881119	0.335664
15	6	6	4	0.989011	0.107892
15	6	6	5	0.999800	0.010789
15	6	6	6	1.000000	0.000200
15	7	1	0	0.533333	0.533333
15	7	1	1	1.000000	0.466667
15	7	2	0	0.266667	0.266667
15	7	2	1	0.800000	0.533333
15	7	2	2	1.000000	0.200000
15	7	3	0	0.123077	0.123077
15	7	3	1	0.553846	0.430769
15	7	3	2	0.923077	0.369231
15	7	3	3	1.000000	0.076923
15	7	4	0	0.051282	0.051282
15	7	4	1	0.338462	0.287179
15	7	4	2	0.769231	0.430769
15	7	4	3	0.974359	0.205128
15	7	4	4	1.000000	0.025641
15	7	5	0	0.018648	0.018648
15	7	5	1	0.181818	0.163170
15	7	5	2	0.573427	0.391608
15	7	5	3	0.899767	0.326340
15	7	5	4	0.993007	0.093240
15	7	5	5	1.000000	0.006993
15	7	6	0	0.005594	0.005594
15	7	6	1	0.083916	0.078322
15	7	6	2	0.377622	0.293706
15	7	6	3	0.769231	0.391608
15	7	6	4	0.965035	0.195804
15	7	6	5	0.998601	0.033566
15	7	6	6	1.000000	0.001399
15	7	7	0	0.001243	0.001243
15	7	7	1	0.031702	0.030458
15	7	7	2	0.214452	0.182751
15	7	7	3	0.595183	0.380730
15	7	7	4	0.899767	0.304584

Table for $N = 2$, $n = 1$, through $N = 100$, $n = 50$

N	n	k	x	P(x)	p(x)
15	7	7	5	0.991142	0.091375
15	7	7	6	0.999845	0.008702
15	7	7	7	1.000000	0.000155
15	8	1	0	0.466667	0.466667
15	8	1	1	1.000000	0.533333
15	8	2	0	0.200000	0.200000
15	8	2	1	0.733333	0.533333
15	8	2	2	1.000000	0.266667
15	8	3	0	0.076923	0.076923
15	8	3	1	0.446154	0.369231
15	8	3	2	0.876923	0.430769
15	8	3	3	1.000000	0.123077
15	8	4	0	0.025641	0.025641
15	8	4	1	0.230769	0.205128
15	8	4	2	0.661538	0.430769
15	8	4	3	0.948718	0.287179
15	8	4	4	1.000000	0.051282
15	8	5	0	0.006993	0.006993
15	8	5	1	0.100233	0.093240
15	8	5	2	0.426573	0.326340
15	8	5	3	0.818182	0.391608
15	8	5	4	0.981352	0.163170
15	8	5	5	1.000000	0.018648
15	8	6	0	0.001399	0.001399
15	8	6	1	0.034965	0.033566
15	8	6	2	0.230769	0.195804
15	8	6	3	0.622378	0.391608
15	8	6	4	0.916084	0.293706
15	8	6	5	0.994406	0.078322
15	8	6	6	1.000000	0.005594
15	8	7	0	0.000155	0.000155
15	8	7	1	0.008858	0.008702
15	8	7	2	0.100233	0.091375
15	8	7	3	0.404817	0.304584
15	8	7	4	0.785548	0.380730
15	8	7	5	0.968298	0.182751
15	8	7	6	0.998757	0.030458
15	8	7	7	1.000000	0.001243
15	8	8	1	0.001243	0.001243
15	8	8	2	0.031702	0.030458
15	8	8	3	0.214452	0.182751
15	8	8	4	0.595183	0.380730
15	8	8	5	0.899767	0.304584
15	8	8	6	0.991142	0.091375
15	8	8	7	0.999845	0.008702
15	8	8	8	1.000000	0.000155
15	9	1	0	0.400000	0.400000
15	9	1	1	1.000000	0.600000
15	9	2	0	0.142857	0.142857
15	9	2	1	0.657143	0.514286

N	n	k	x	P(x)	p(x)
15	9	2	2	1.000000	0.342857
15	9	3	0	0.043956	0.043956
15	9	3	1	0.340659	0.296703
15	9	3	2	0.815385	0.474725
15	9	3	3	1.000000	0.184615
15	9	4	0	0.010989	0.010989
15	9	4	1	0.142857	0.131868
15	9	4	2	0.538461	0.395604
15	9	4	3	0.907692	0.369231
15	9	4	4	1.000000	0.092308
15	9	5	0	0.001998	0.001998
15	9	5	1	0.046953	0.044955
15	9	5	2	0.286713	0.239760
15	9	5	3	0.706294	0.419580
15	9	5	4	0.958042	0.251748
15	9	5	5	1.000000	0.041958
15	9	6	0	0.000200	0.000200
15	9	6	1	0.010989	0.010789
15	9	6	2	0.118881	0.107892
15	9	6	3	0.454545	0.335664
15	9	6	4	0.832168	0.377622
15	9	6	5	0.983217	0.151049
15	9	6	6	1.000000	0.016783
15	9	7	1	0.001399	0.001399
15	9	7	2	0.034965	0.033566
15	9	7	3	0.230769	0.195804
15	9	7	4	0.622378	0.391608
15	9	7	5	0.916084	0.293706
15	9	7	6	0.994406	0.078322
15	9	7	7	1.000000	0.005594
15	9	8	2	0.005594	0.005594
15	9	8	3	0.083916	0.078322
15	9	8	4	0.377622	0.293706
15	9	8	5	0.769231	0.391608
15	9	8	6	0.965035	0.195804
15	9	8	7	0.998601	0.033566
15	9	8	8	1.000000	0.001399
15	9	9	3	0.016783	0.016783
15	9	9	4	0.167832	0.151049
15	9	9	5	0.545455	0.377622
15	9	9	6	0.881119	0.335664
15	9	9	7	0.989011	0.107892
15	9	9	8	0.999800	0.010789
15	9	9	9	1.000000	0.000200
15	10	1	0	0.333333	0.333333
15	10	1	1	1.000000	0.666667
15	10	2	0	0.095238	0.095238
15	10	2	1	0.571429	0.476190
15	10	2	2	1.000000	0.428571
15	10	3	0	0.021978	0.021978

N	n	k	x	P(x)	p(x)
15	10	3	1	0.241758	0.219780
15	10	3	2	0.736264	0.494505
15	10	3	3	1.000000	0.263736
15	10	4	0	0.003663	0.003663
15	10	4	1	0.076923	0.073260
15	10	4	2	0.406593	0.329670
15	10	4	3	0.846154	0.439560
15	10	4	4	1.000000	0.153846
15	10	5	0	0.000333	0.000333
15	10	5	1	0.016983	0.016650
15	10	5	2	0.166833	0.149850
15	10	5	3	0.566434	0.399600
15	10	5	4	0.916084	0.349650
15	10	5	5	1.000000	0.083916
15	10	6	1	0.001998	0.001998
15	10	6	2	0.046953	0.044955
15	10	6	3	0.286713	0.239760
15	10	6	4	0.706294	0.419580
15	10	6	5	0.958042	0.251748
15	10	6	6	1.000000	0.041958
15	10	7	2	0.006993	0.006993
15	10	7	3	0.100233	0.093240
15	10	7	4	0.426573	0.326340
15	10	7	5	0.818182	0.391608
15	10	7	6	0.981352	0.163170
15	10	7	7	1.000000	0.018648
15	10	8	3	0.018648	0.018648
15	10	8	4	0.181818	0.163170
15	10	8	5	0.573427	0.391608
15	10	8	6	0.899767	0.326340
15	10	8	7	0.993007	0.093240
15	10	8	8	1.000000	0.006993
15	10	9	4	0.041958	0.041958
15	10	9	5	0.293706	0.251748
15	10	9	6	0.713287	0.419580
15	10	9	7	0.953047	0.239760
15	10	9	8	0.998002	0.044955
15	10	9	9	1.000000	0.001998
15	10	10	5	0.083916	0.083916
15	10	10	6	0.433566	0.349650
15	10	10	7	0.833167	0.399600
15	10	10	8	0.983017	0.149850
15	10	10	9	0.999667	0.016650
15	10	10	10	1.000000	0.000333
15	11	1	0	0.266667	0.266667
15	11	1	1	1.000000	0.733333
15	11	2	0	0.057143	0.057143
15	11	2	1	0.476190	0.419048
15	11	2	2	1.000000	0.523810
15	11	3	0	0.008791	0.008791

N	n	k	x	P(x)	p(x)
15	11	3	1	0.153846	0.145055
15	11	3	2	0.637363	0.483516
15	11	3	3	1.000000	0.362637
15	11	4	0	0.000733	0.000733
15	11	4	1	0.032967	0.032234
15	11	4	2	0.274725	0.241758
15	11	4	3	0.758242	0.483516
15	11	4	4	1.000000	0.241758
15	11	5	1	0.003663	0.003663
15	11	5	2	0.076923	0.073260
15	11	5	3	0.406593	0.329670
15	11	5	4	0.846154	0.439560
15	11	5	5	1.000000	0.153846
15	11	6	2	0.010989	0.010989
15	11	6	3	0.142857	0.131868
15	11	6	4	0.538461	0.395604
15	11	6	5	0.907692	0.369231
15	11	6	6	1.000000	0.092308
15	11	7	3	0.025641	0.025641
15	11	7	4	0.230769	0.205128
15	11	7	5	0.661538	0.430769
15	11	7	6	0.948718	0.287179
15	11	7	7	1.000000	0.051282
15	11	8	4	0.051282	0.051282
15	11	8	5	0.338462	0.287179
15	11	8	6	0.769231	0.430769
15	11	8	7	0.974359	0.205128
15	11	8	8	1.000000	0.025641
15	11	9	5	0.092308	0.092308
15	11	9	6	0.461538	0.369231
15	11	9	7	0.857143	0.395604
15	11	9	8	0.989011	0.131868
15	11	9	9	1.000000	0.010989
15	11	10	6	0.153846	0.153846
15	11	10	7	0.593407	0.439560
15	11	10	8	0.923077	0.329670
15	11	10	9	0.996337	0.073260
15	11	10	10	1.000000	0.003663
15	11	11	7	0.241758	0.241758
15	11	11	8	0.725275	0.483516
15	11	11	9	0.967033	0.241758
15	11	11	10	0.999267	0.032234
15	11	11	11	1.000000	0.000733
15	12	1	0	0.200000	0.200000
15	12	1	1	1.000000	0.800000
15	12	2	0	0.028571	0.028571
15	12	2	1	0.371429	0.342857
15	12	2	2	1.000000	0.628571
15	12	3	0	0.002198	0.002198
15	12	3	1	0.081319	0.079121

Table for $N = 2$, $n = 1$, through $N = 100$, $n = 50$

N	n	k	x	P(x)	p(x)
15	12	3	2	0.516483	0.435165
15	12	3	3	1.000000	0.483516
15	12	4	1	0.008791	0.008791
15	12	4	2	0.153846	0.145055
15	12	4	3	0.637363	0.483516
15	12	4	4	1.000000	0.362637
15	12	5	2	0.021978	0.021978
15	12	5	3	0.241758	0.219780
15	12	5	4	0.736264	0.494505
15	12	5	5	1.000000	0.263736
15	12	6	3	0.043956	0.043956
15	12	6	4	0.340659	0.296703
15	12	6	5	0.815385	0.474725
15	12	6	6	1.000000	0.184615
15	12	7	4	0.076923	0.076923
15	12	7	5	0.446154	0.369231
15	12	7	6	0.876923	0.430769
15	12	7	7	1.000000	0.123077
15	12	8	5	0.123077	0.123077
15	12	8	6	0.553846	0.430769
15	12	8	7	0.923077	0.369231
15	12	8	8	1.000000	0.076923
15	12	9	6	0.184615	0.184615
15	12	9	7	0.659341	0.474725
15	12	9	8	0.956044	0.296703
15	12	9	9	1.000000	0.043956
15	12	10	7	0.263736	0.263736
15	12	10	8	0.758242	0.494505
15	12	10	9	0.978022	0.219780
15	12	10	10	1.000000	0.021978
15	12	11	8	0.362637	0.362637
15	12	11	9	0.846154	0.483516
15	12	11	10	0.991209	0.145055
15	12	11	11	1.000000	0.008791
15	12	12	9	0.483516	0.483516
15	12	12	10	0.918681	0.435165
15	12	12	11	0.997802	0.079121
15	12	12	12	1.000000	0.002198
15	13	1	0	0.133333	0.133333
15	13	1	1	1.000000	0.866667
15	13	2	0	0.009524	0.009524
15	13	2	1	0.257143	0.247619
15	13	2	2	1.000000	0.742857
15	13	3	1	0.028571	0.028571
15	13	3	2	0.371429	0.342857
15	13	3	3	1.000000	0.628571
15	13	4	2	0.057143	0.057143
15	13	4	3	0.476190	0.419048
15	13	4	4	1.000000	0.523810
15	13	5	3	0.095238	0.095238

N	n	k	x	P(x)	p(x)
15	13	5	4	0.571429	0.476190
15	13	5	5	1.000000	0.428571
15	13	6	4	0.142857	0.142857
15	13	6	5	0.657143	0.514286
15	13	6	6	1.000000	0.342857
15	13	7	5	0.200000	0.200000
15	13	7	6	0.733333	0.533333
15	13	7	7	1.000000	0.266667
15	13	8	6	0.266667	0.266667
15	13	8	7	0.800000	0.533333
15	13	8	8	1.000000	0.200000
15	13	9	7	0.342857	0.342857
15	13	9	8	0.857143	0.514286
15	13	9	9	1.000000	0.142857
15	13	10	8	0.428571	0.428571
15	13	10	9	0.904762	0.476190
15	13	10	10	1.000000	0.095238
15	13	11	9	0.523810	0.523810
15	13	11	10	0.942857	0.419048
15	13	11	11	1.000000	0.057143
15	13	12	10	0.628571	0.628571
15	13	12	11	0.971429	0.342857
15	13	12	12	1.000000	0.028571
15	13	13	11	0.742857	0.742857
15	13	13	12	0.990476	0.247619
15	13	13	13	1.000000	0.009524
15	14	1	0	0.066667	0.066667
15	14	1	1	1.000000	0.933333
15	14	2	1	0.133333	0.133333
15	14	2	2	1.000000	0.866667
15	14	3	2	0.200000	0.200000
15	14	3	3	1.000000	0.800000
15	14	4	3	0.266667	0.266667
15	14	4	4	1.000000	0.733333
15	14	5	4	0.333333	0.333333
15	14	5	5	1.000000	0.666667
15	14	6	5	0.400000	0.400000
15	14	6	6	1.000000	0.600000
15	14	7	6	0.466667	0.466667
15	14	7	7	1.000000	0.533333
15	14	8	7	0.533333	0.533333
15	14	8	8	1.000000	0.466667
15	14	9	8	0.600000	0.600000
15	14	9	9	1.000000	0.400000
15	14	10	9	0.666667	0.666667
15	14	10	10	1.000000	0.333333
15	14	11	10	0.733333	0.733333
15	14	11	11	1.000000	0.266667
15	14	12	11	0.800000	0.800000
15	14	12	12	1.000000	0.200000

N	n	k	x	P(x)	p(x)
15	14	13	12	0.866667	0.866667
15	14	13	13	1.000000	0.133333
15	14	14	13	0.933333	0.933333
15	14	14	14	1.000000	0.066667
16	1	1	0	0.937500	0.937500
16	1	1	1	1.000000	0.062500
16	2	1	0	0.875000	0.875000
16	2	1	1	1.000000	0.125000
16	2	2	0	0.758333	0.758333
16	2	2	1	0.991667	0.233333
16	2	2	2	1.000000	0.008333
16	3	1	0	0.812500	0.812500
16	3	1	1	1.000000	0.187500
16	3	2	0	0.650000	0.650000
16	3	2	1	0.975000	0.325000
16	3	2	2	1.000000	0.025000
16	3	3	0	0.510714	0.510714
16	3	3	1	0.928571	0.417857
16	3	3	2	0.998214	0.069643
16	3	3	3	1.000000	0.001786
16	4	1	0	0.750000	0.750000
16	4	1	1	1.000000	0.250000
16	4	2	0	0.550000	0.550000
16	4	2	1	0.950000	0.400000
16	4	2	2	1.000000	0.050000
16	4	3	0	0.392857	0.392857
16	4	3	1	0.864286	0.471429
16	4	3	2	0.992857	0.128571
16	4	3	3	1.000000	0.007143
16	4	4	0	0.271978	0.271978
16	4	4	1	0.755494	0.483516
16	4	4	2	0.973077	0.217582
16	4	4	3	0.999451	0.026374
16	4	4	4	1.000000	0.000549
16	5	1	0	0.687500	0.687500
16	5	1	1	1.000000	0.312500
16	5	2	0	0.458333	0.458333
16	5	2	1	0.916667	0.458333
16	5	2	2	1.000000	0.083333
16	5	3	0	0.294643	0.294643
16	5	3	1	0.785714	0.491071
16	5	3	2	0.982143	0.196429
16	5	3	3	1.000000	0.017857
16	5	4	0	0.181319	0.181319
16	5	4	1	0.634615	0.453297
16	5	4	2	0.936813	0.302198
16	5	4	3	0.997253	0.060440
16	5	4	4	1.000000	0.002747
16	5	5	0	0.105769	0.105769
16	5	5	1	0.483516	0.377747

N	n	k	x	P(x)	p(x)
16	5	5	2	0.861264	0.377747
16	5	5	3	0.987179	0.125916
16	5	5	4	0.999771	0.012592
16	5	5	5	1.000000	0.000229
16	6	1	0	0.625000	0.625000
16	6	1	1	1.000000	0.375000
16	6	2	0	0.375000	0.375000
16	6	2	1	0.875000	0.500000
16	6	2	2	1.000000	0.125000
16	6	3	0	0.214286	0.214286
16	6	3	1	0.696429	0.482143
16	6	3	2	0.964286	0.267857
16	6	3	3	1.000000	0.035714
16	6	4	0	0.115385	0.115385
16	6	4	1	0.510989	0.395604
16	6	4	2	0.881868	0.370879
16	6	4	3	0.991758	0.109890
16	6	4	4	1.000000	0.008242
16	6	5	0	0.057692	0.057692
16	6	5	1	0.346154	0.288462
16	6	5	2	0.758242	0.412088
16	6	5	3	0.964286	0.206044
16	6	5	4	0.998626	0.034341
16	6	5	5	1.000000	0.001374
16	6	6	0	0.026224	0.026224
16	6	6	1	0.215035	0.188811
16	6	6	2	0.608392	0.393357
16	6	6	3	0.908092	0.299700
16	6	6	4	0.992383	0.084291
16	6	6	5	0.999875	0.007493
16	6	6	6	1.000000	0.000125
16	7	1	0	0.562500	0.562500
16	7	1	1	1.000000	0.437500
16	7	2	0	0.300000	0.300000
16	7	2	1	0.825000	0.525000
16	7	2	2	1.000000	0.175000
16	7	3	0	0.150000	0.150000
16	7	3	1	0.600000	0.450000
16	7	3	2	0.937500	0.337500
16	7	3	3	1.000000	0.062500
16	7	4	0	0.069231	0.069231
16	7	4	1	0.392308	0.323077
16	7	4	2	0.807692	0.415385
16	7	4	3	0.980769	0.173077
16	7	4	4	1.000000	0.019231
16	7	5	0	0.028846	0.028846
16	7	5	1	0.230769	0.201923
16	7	5	2	0.634615	0.403846
16	7	5	3	0.923077	0.288462
16	7	5	4	0.995192	0.072115

Table for $N = 2$, $n = 1$, through $N = 100$, $n = 50$

N	n	k	x	P(x)	p(x)
16	7	5	5	1.000000	0.004808
16	7	6	0	0.010490	0.010490
16	7	6	1	0.120629	0.110140
16	7	6	2	0.451049	0.330420
16	7	6	3	0.818182	0.367133
16	7	6	4	0.975524	0.157343
16	7	6	5	0.999126	0.023601
16	7	6	6	1.000000	0.000874
16	7	7	0	0.003147	0.003147
16	7	7	1	0.054545	0.051399
16	7	7	2	0.285839	0.231294
16	7	7	3	0.671329	0.385489
16	7	7	4	0.928322	0.256993
16	7	7	5	0.994405	0.066084
16	7	7	6	0.999912	0.005507
16	7	7	7	1.000000	0.000087
16	8	1	0	0.500000	0.500000
16	8	1	1	1.000000	0.500000
16	8	2	0	0.233333	0.233333
16	8	2	1	0.766667	0.533333
16	8	2	2	1.000000	0.233333
16	8	3	0	0.100000	0.100000
16	8	3	1	0.500000	0.400000
16	8	3	2	0.900000	0.400000
16	8	3	3	1.000000	0.100000
16	8	4	0	0.038462	0.038462
16	8	4	1	0.284615	0.246154
16	8	4	2	0.715385	0.430769
16	8	4	3	0.961538	0.246154
16	8	4	4	1.000000	0.038462
16	8	5	0	0.012821	0.012821
16	8	5	1	0.141026	0.128205
16	8	5	2	0.500000	0.358974
16	8	5	3	0.858974	0.358974
16	8	5	4	0.987179	0.128205
16	8	5	5	1.000000	0.012821
16	8	6	0	0.003496	0.003496
16	8	6	1	0.059441	0.055944
16	8	6	2	0.304196	0.244755
16	8	6	3	0.695804	0.391608
16	8	6	4	0.940559	0.244755
16	8	6	5	0.996503	0.055944
16	8	6	6	1.000000	0.003496
16	8	7	0	0.000699	0.000699
16	8	7	1	0.020280	0.019580
16	8	7	2	0.157343	0.137063
16	8	7	3	0.500000	0.342657
16	8	7	4	0.842657	0.342657
16	8	7	5	0.979720	0.137063
16	8	7	6	0.999301	0.019580

N	n	k	x	P(x)	p(x)
16	8	7	7	1.000000	0.000699
16	8	8	0	0.000078	0.000078
16	8	8	1	0.005050	0.004973
16	8	8	2	0.065967	0.060917
16	8	8	3	0.309635	0.243667
16	8	8	4	0.690365	0.380730
16	8	8	5	0.934033	0.243667
16	8	8	6	0.994949	0.060917
16	8	8	7	0.999922	0.004973
16	8	8	8	1.000000	0.000078
16	9	1	0	0.437500	0.437500
16	9	1	1	1.000000	0.562500
16	9	2	0	0.175000	0.175000
16	9	2	1	0.700000	0.525000
16	9	2	2	1.000000	0.300000
16	9	3	0	0.062500	0.062500
16	9	3	1	0.400000	0.337500
16	9	3	2	0.850000	0.450000
16	9	3	3	1.000000	0.150000
16	9	4	0	0.019231	0.019231
16	9	4	1	0.192308	0.173077
16	9	4	2	0.607692	0.415385
16	9	4	3	0.930769	0.323077
16	9	4	4	1.000000	0.069231
16	9	5	0	0.004808	0.004808
16	9	5	1	0.076923	0.072115
16	9	5	2	0.365385	0.288462
16	9	5	3	0.769231	0.403846
16	9	5	4	0.971154	0.201923
16	9	5	5	1.000000	0.028846
16	9	6	0	0.000874	0.000874
16	9	6	1	0.024476	0.023601
16	9	6	2	0.181818	0.157343
16	9	6	3	0.548951	0.367133
16	9	6	4	0.879371	0.330420
16	9	6	5	0.989510	0.110140
16	9	6	6	1.000000	0.010490
16	9	7	0	0.000087	0.000087
16	9	7	1	0.005594	0.005507
16	9	7	2	0.071678	0.066084
16	9	7	3	0.328671	0.256993
16	9	7	4	0.714161	0.385489
16	9	7	5	0.945454	0.231294
16	9	7	6	0.996853	0.051399
16	9	7	7	1.000000	0.003147
16	9	8	1	0.000699	0.000699
16	9	8	2	0.020280	0.019580
16	9	8	3	0.157343	0.137063
16	9	8	4	0.500000	0.342657
16	9	8	5	0.842657	0.342657

N	n	k	x	P(x)	p(x)
16	9	8	6	0.979720	0.137063
16	9	8	7	0.999301	0.019580
16	9	8	8	1.000000	0.000699
16	9	9	2	0.003147	0.003147
16	9	9	3	0.054545	0.051399
16	9	9	4	0.285839	0.231294
16	9	9	5	0.671329	0.385489
16	9	9	6	0.928322	0.256993
16	9	9	7	0.994405	0.066084
16	9	9	8	0.999912	0.005507
16	9	9	9	1.000000	0.000087
16	10	1	0	0.375000	0.375000
16	10	1	1	1.000000	0.625000
16	10	2	0	0.125000	0.125000
16	10	2	1	0.625000	0.500000
16	10	2	2	1.000000	0.375000
16	10	3	0	0.035714	0.035714
16	10	3	1	0.303571	0.267857
16	10	3	2	0.785714	0.482143
16	10	3	3	1.000000	0.214286
16	10	4	0	0.008242	0.008242
16	10	4	1	0.118132	0.109890
16	10	4	2	0.489011	0.370879
16	10	4	3	0.884615	0.395604
16	10	4	4	1.000000	0.115385
16	10	5	0	0.001374	0.001374
16	10	5	1	0.035714	0.034341
16	10	5	2	0.241758	0.206044
16	10	5	3	0.653846	0.412088
16	10	5	4	0.942308	0.288462
16	10	5	5	1.000000	0.057692
16	10	6	0	0.000125	0.000125
16	10	6	1	0.007617	0.007493
16	10	6	2	0.091908	0.084291
16	10	6	3	0.391608	0.299700
16	10	6	4	0.784965	0.393357
16	10	6	5	0.973776	0.188811
16	10	6	6	1.000000	0.026224
16	10	7	1	0.000874	0.000874
16	10	7	2	0.024476	0.023601
16	10	7	3	0.181818	0.157343
16	10	7	4	0.548951	0.367133
16	10	7	5	0.879371	0.330420
16	10	7	6	0.989510	0.110140
16	10	7	7	1.000000	0.010490
16	10	8	2	0.003496	0.003496
16	10	8	3	0.059441	0.055944
16	10	8	4	0.304196	0.244755
16	10	8	5	0.695804	0.391608
16	10	8	6	0.940559	0.244755

N	n	k	x	P(x)	p(x)
16	10	8	7	0.996503	0.055944
16	10	8	8	1.000000	0.003496
16	10	9	3	0.010490	0.010490
16	10	9	4	0.120629	0.110140
16	10	9	5	0.451049	0.330420
16	10	9	6	0.818182	0.367133
16	10	9	7	0.975524	0.157343
16	10	9	8	0.999126	0.023601
16	10	9	9	1.000000	0.000874
16	10	10	4	0.026224	0.026224
16	10	10	5	0.215035	0.188811
16	10	10	6	0.608392	0.393357
16	10	10	7	0.908092	0.299700
16	10	10	8	0.992383	0.084291
16	10	10	9	0.999875	0.007493
16	10	10	10	1.000000	0.000125
16	11	1	0	0.312500	0.312500
16	11	1	1	1.000000	0.687500
16	11	2	0	0.083333	0.083333
16	11	2	1	0.541667	0.458333
16	11	2	2	1.000000	0.458333
16	11	3	0	0.017857	0.017857
16	11	3	1	0.214286	0.196429
16	11	3	2	0.705357	0.491071
16	11	3	3	1.000000	0.294643
16	11	4	0	0.002747	0.002747
16	11	4	1	0.063187	0.060440
16	11	4	2	0.365385	0.302198
16	11	4	3	0.818681	0.453297
16	11	4	4	1.000000	0.181319
16	11	5	0	0.000229	0.000229
16	11	5	1	0.012821	0.012592
16	11	5	2	0.138736	0.125916
16	11	5	3	0.516483	0.377747
16	11	5	4	0.894231	0.377747
16	11	5	5	1.000000	0.105769
16	11	6	1	0.001374	0.001374
16	11	6	2	0.035714	0.034341
16	11	6	3	0.241758	0.206044
16	11	6	4	0.653846	0.412088
16	11	6	5	0.942308	0.288462
16	11	6	6	1.000000	0.057692
16	11	7	2	0.004808	0.004808
16	11	7	3	0.076923	0.072115
16	11	7	4	0.365385	0.288462
16	11	7	5	0.769231	0.403846
16	11	7	6	0.971154	0.201923
16	11	7	7	1.000000	0.028846
16	11	8	3	0.012821	0.012821
16	11	8	4	0.141026	0.128205

Table for $N = 2$, $n = 1$, through $N = 100$, $n = 50$

N	n	k	x	P(x)	p(x)
16	11	8	5	0.500000	0.358974
16	11	8	6	0.858974	0.358974
16	11	8	7	0.987179	0.128205
16	11	8	8	1.000000	0.012821
16	11	9	4	0.028846	0.028846
16	11	9	5	0.230769	0.201923
16	11	9	6	0.634615	0.403846
16	11	9	7	0.923077	0.288462
16	11	9	8	0.995192	0.072115
16	11	9	9	1.000000	0.004808
16	11	10	5	0.057692	0.057692
16	11	10	6	0.346154	0.288462
16	11	10	7	0.758242	0.412088
16	11	10	8	0.964286	0.206044
16	11	10	9	0.998626	0.034341
16	11	10	10	1.000000	0.001374
16	11	11	6	0.105769	0.105769
16	11	11	7	0.483516	0.377747
16	11	11	8	0.861264	0.377747
16	11	11	9	0.987179	0.125916
16	11	11	10	0.999771	0.012592
16	11	11	11	1.000000	0.000229
16	12	1	0	0.250000	0.250000
16	12	1	1	1.000000	0.750000
16	12	2	0	0.050000	0.050000
16	12	2	1	0.450000	0.400000
16	12	2	2	1.000000	0.550000
16	12	3	0	0.007143	0.007143
16	12	3	1	0.135714	0.128571
16	12	3	2	0.607143	0.471429
16	12	3	3	1.000000	0.392857
16	12	4	0	0.000549	0.000549
16	12	4	1	0.026923	0.026374
16	12	4	2	0.244505	0.217582
16	12	4	3	0.728022	0.483516
16	12	4	4	1.000000	0.271978
16	12	5	1	0.002747	0.002747
16	12	5	2	0.063187	0.060440
16	12	5	3	0.365385	0.302198
16	12	5	4	0.818681	0.453297
16	12	5	5	1.000000	0.181319
16	12	6	2	0.008242	0.008242
16	12	6	3	0.118132	0.109890
16	12	6	4	0.489011	0.370879
16	12	6	5	0.884615	0.395604
16	12	6	6	1.000000	0.115385
16	12	7	3	0.019231	0.019231
16	12	7	4	0.192308	0.173077
16	12	7	5	0.607692	0.415385
16	12	7	6	0.930769	0.323077

N	n	k	x	P(x)	p(x)
16	12	7	7	1.000000	0.069231
16	12	8	4	0.038462	0.038462
16	12	8	5	0.284615	0.246154
16	12	8	6	0.715385	0.430769
16	12	8	7	0.961538	0.246154
16	12	8	8	1.000000	0.038462
16	12	9	5	0.069231	0.069231
16	12	9	6	0.392308	0.323077
16	12	9	7	0.807692	0.415385
16	12	9	8	0.980769	0.173077
16	12	9	9	1.000000	0.019231
16	12	10	6	0.115385	0.115385
16	12	10	7	0.510989	0.395604
16	12	10	8	0.881868	0.370879
16	12	10	9	0.991758	0.109890
16	12	10	10	1.000000	0.008242
16	12	11	7	0.181319	0.181319
16	12	11	8	0.634615	0.453297
16	12	11	9	0.936813	0.302198
16	12	11	10	0.997253	0.060440
16	12	11	11	1.000000	0.002747
16	12	12	8	0.271978	0.271978
16	12	12	9	0.755494	0.483516
16	12	12	10	0.973077	0.217582
16	12	12	11	0.999451	0.026374
16	12	12	12	1.000000	0.000549
16	13	1	0	0.187500	0.187500
16	13	1	1	1.000000	0.812500
16	13	2	0	0.025000	0.025000
16	13	2	1	0.350000	0.325000
16	13	2	2	1.000000	0.650000
16	13	3	0	0.001786	0.001786
16	13	3	1	0.071429	0.069643
16	13	3	2	0.489286	0.417857
16	13	3	3	1.000000	0.510714
16	13	4	1	0.007143	0.007143
16	13	4	2	0.135714	0.128571
16	13	4	3	0.607143	0.471429
16	13	4	4	1.000000	0.392857
16	13	5	2	0.017857	0.017857
16	13	5	3	0.214286	0.196429
16	13	5	4	0.705357	0.491071
16	13	5	5	1.000000	0.294643
16	13	6	3	0.035714	0.035714
16	13	6	4	0.303571	0.267857
16	13	6	5	0.785714	0.482143
16	13	6	6	1.000000	0.214286
16	13	7	4	0.062500	0.062500
16	13	7	5	0.400000	0.337500
16	13	7	6	0.850000	0.450000

N	n	k	x	P(x)	p(x)
16	13	7	7	1.000000	0.150000
16	13	8	5	0.100000	0.100000
16	13	8	6	0.500000	0.400000
16	13	8	7	0.900000	0.400000
16	13	8	8	1.000000	0.100000
16	13	9	6	0.150000	0.150000
16	13	9	7	0.600000	0.450000
16	13	9	8	0.937500	0.337500
16	13	9	9	1.000000	0.062500
16	13	10	7	0.214286	0.214286
16	13	10	8	0.696429	0.482143
16	13	10	9	0.964286	0.267857
16	13	10	10	1.000000	0.035714
16	13	11	8	0.294643	0.294643
16	13	11	9	0.785714	0.491071
16	13	11	10	0.982143	0.196429
16	13	11	11	1.000000	0.017857
16	13	12	9	0.392857	0.392857
16	13	12	10	0.864286	0.471429
16	13	12	11	0.992857	0.128571
16	13	12	12	1.000000	0.007143
16	13	13	10	0.510714	0.510714
16	13	13	11	0.928571	0.417857
16	13	13	12	0.998214	0.069643
16	13	13	13	1.000000	0.001786
16	14	1	0	0.125000	0.125000
16	14	1	1	1.000000	0.875000
16	14	2	0	0.008333	0.008333
16	14	2	1	0.241667	0.233333
16	14	2	2	1.000000	0.758333
16	14	3	1	0.025000	0.025000
16	14	3	2	0.350000	0.325000
16	14	3	3	1.000000	0.650000
16	14	4	2	0.050000	0.050000
16	14	4	3	0.450000	0.400000
16	14	4	4	1.000000	0.550000
16	14	5	3	0.083333	0.083333
16	14	5	4	0.541667	0.458333
16	14	5	5	1.000000	0.458333
16	14	6	4	0.125000	0.125000
16	14	6	5	0.625000	0.500000
16	14	6	6	1.000000	0.375000
16	14	7	5	0.175000	0.175000
16	14	7	6	0.700000	0.525000
16	14	7	7	1.000000	0.300000
16	14	8	6	0.233333	0.233333
16	14	8	7	0.766667	0.533333
16	14	8	8	1.000000	0.233333
16	14	9	7	0.300000	0.300000
16	14	9	8	0.825000	0.525000

N	n	k	x	P(x)	p(x)
16	14	9	9	1.000000	0.175000
16	14	10	8	0.375000	0.375000
16	14	10	9	0.875000	0.500000
16	14	10	10	1.000000	0.125000
16	14	11	9	0.458333	0.458333
16	14	11	10	0.916667	0.458333
16	14	11	11	1.000000	0.083333
16	14	12	10	0.550000	0.550000
16	14	12	11	0.950000	0.400000
16	14	12	12	1.000000	0.050000
16	14	13	11	0.650000	0.650000
16	14	13	12	0.975000	0.325000
16	14	13	13	1.000000	0.025000
16	14	14	12	0.758333	0.758333
16	14	14	13	0.991667	0.233333
16	14	14	14	1.000000	0.008333
16	15	1	0	0.062500	0.062500
16	15	1	1	1.000000	0.937500
16	15	2	1	0.125000	0.125000
16	15	2	2	1.000000	0.875000
16	15	3	2	0.187500	0.187500
16	15	3	3	1.000000	0.812500
16	15	4	3	0.250000	0.250000
16	15	4	4	1.000000	0.750000
16	15	5	4	0.312500	0.312500
16	15	5	5	1.000000	0.687500
16	15	6	5	0.375000	0.375000
16	15	6	6	1.000000	0.625000
16	15	7	6	0.437500	0.437500
16	15	7	7	1.000000	0.562500
16	15	8	7	0.500000	0.500000
16	15	8	8	1.000000	0.500000
16	15	9	8	0.562500	0.562500
16	15	9	9	1.000000	0.437500
16	15	10	9	0.625000	0.625000
16	15	10	10	1.000000	0.375000
16	15	11	10	0.687500	0.687500
16	15	11	11	1.000000	0.312500
16	15	12	11	0.750000	0.750000
16	15	12	12	1.000000	0.250000
16	15	13	12	0.812500	0.812500
16	15	13	13	1.000000	0.187500
16	15	14	13	0.875000	0.875000
16	15	14	14	1.000000	0.125000
16	15	15	14	0.937500	0.937500
16	15	15	15	1.000000	0.062500
17	1	1	0	0.941176	0.941176
17	1	1	1	1.000000	0.058824
17	2	1	0	0.882353	0.882353
17	2	1	1	1.000000	0.117647

Table for $N = 2$, $n = 1$, through $N = 100$, $n = 50$

N	n	k	x	P(x)	p(x)
17	2	2	0	0.772059	0.772059
17	2	2	1	0.992647	0.220588
17	2	2	2	1.000000	0.007353
17	3	1	0	0.823529	0.823529
17	3	1	1	1.000000	0.176471
17	3	2	0	0.669118	0.669118
17	3	2	1	0.977941	0.308824
17	3	2	2	1.000000	0.022059
17	3	3	0	0.535294	0.535294
17	3	3	1	0.936765	0.401471
17	3	3	2	0.998529	0.061765
17	3	3	3	1.000000	0.001471
17	4	1	0	0.764706	0.764706
17	4	1	1	1.000000	0.235294
17	4	2	0	0.573529	0.573529
17	4	2	1	0.955882	0.382353
17	4	2	2	1.000000	0.044118
17	4	3	0	0.420588	0.420588
17	4	3	1	0.879412	0.458824
17	4	3	2	0.994118	0.114706
17	4	3	3	1.000000	0.005882
17	4	4	0	0.300420	0.300420
17	4	4	1	0.781092	0.480672
17	4	4	2	0.977731	0.196639
17	4	4	3	0.999580	0.021849
17	4	4	4	1.000000	0.000420
17	5	1	0	0.705882	0.705882
17	5	1	1	1.000000	0.294118
17	5	2	0	0.485294	0.485294
17	5	2	1	0.926471	0.441176
17	5	2	2	1.000000	0.073529
17	5	3	0	0.323529	0.323529
17	5	3	1	0.808823	0.485294
17	5	3	2	0.985294	0.176471
17	5	3	3	1.000000	0.014706
17	5	4	0	0.207983	0.207983
17	5	4	1	0.670168	0.462185
17	5	4	2	0.947479	0.277311
17	5	4	3	0.997899	0.050420
17	5	4	4	1.000000	0.002101
17	5	5	0	0.127990	0.127990
17	5	5	1	0.527957	0.399968
17	5	5	2	0.883484	0.355527
17	5	5	3	0.990142	0.106658
17	5	5	4	0.999838	0.009696
17	5	5	5	1.000000	0.000162
17	6	1	0	0.647059	0.647059
17	6	1	1	1.000000	0.352941
17	6	2	0	0.404412	0.404412
17	6	2	1	0.889706	0.485294

N	n	k	x	P(x)	p(x)
17	6	2	2	1.000000	0.110294
17	6	3	0	0.242647	0.242647
17	6	3	1	0.727941	0.485294
17	6	3	2	0.970588	0.242647
17	6	3	3	1.000000	0.029412
17	6	4	0	0.138655	0.138655
17	6	4	1	0.554622	0.415966
17	6	4	2	0.901260	0.346639
17	6	4	3	0.993697	0.092437
17	6	4	4	1.000000	0.006303
17	6	5	0	0.074661	0.074661
17	6	5	1	0.394635	0.319974
17	6	5	2	0.794602	0.399968
17	6	5	3	0.972366	0.177763
17	6	5	4	0.999030	0.026665
17	6	5	5	1.000000	0.000970
17	6	6	0	0.037330	0.037330
17	6	6	1	0.261312	0.223982
17	6	6	2	0.661280	0.399968
17	6	6	3	0.927925	0.266645
17	6	6	4	0.994586	0.066661
17	6	6	5	0.999919	0.005333
17	6	6	6	1.000000	0.000081
17	7	1	0	0.588235	0.588235
17	7	1	1	1.000000	0.411765
17	7	2	0	0.330882	0.330882
17	7	2	1	0.845588	0.514706
17	7	2	2	1.000000	0.154412
17	7	3	0	0.176471	0.176471
17	7	3	1	0.639706	0.463235
17	7	3	2	0.948529	0.308824
17	7	3	3	1.000000	0.051471
17	7	4	0	0.088235	0.088235
17	7	4	1	0.441176	0.352941
17	7	4	2	0.838235	0.397059
17	7	4	3	0.985294	0.147059
17	7	4	4	1.000000	0.014706
17	7	5	0	0.040724	0.040724
17	7	5	1	0.278281	0.237557
17	7	5	2	0.685520	0.407240
17	7	5	3	0.940045	0.254525
17	7	5	4	0.996606	0.056561
17	7	5	5	1.000000	0.003394
17	7	6	0	0.016968	0.016968
17	7	6	1	0.159502	0.142534
17	7	6	2	0.515837	0.356335
17	7	6	3	0.855204	0.339367
17	7	6	4	0.982466	0.127262
17	7	6	5	0.999434	0.016968
17	7	6	6	1.000000	0.000566

N	n	k	x	P(x)	p(x)
17	7	7	0	0.006170	0.006170
17	7	7	1	0.081756	0.075586
17	7	7	2	0.353867	0.272110
17	7	7	3	0.731798	0.377931
17	7	7	4	0.947758	0.215960
17	7	7	5	0.996349	0.048591
17	7	7	6	0.999948	0.003599
17	7	7	7	1.000000	0.000051
17	8	1	0	0.529412	0.529412
17	8	1	1	1.000000	0.470588
17	8	2	0	0.264706	0.264706
17	8	2	1	0.794118	0.529412
17	8	2	2	1.000000	0.205882
17	8	3	0	0.123529	0.123529
17	8	3	1	0.547059	0.423529
17	8	3	2	0.917647	0.370588
17	8	3	3	1.000000	0.082353
17	8	4	0	0.052941	0.052941
17	8	4	1	0.335294	0.282353
17	8	4	2	0.758824	0.423529
17	8	4	3	0.970588	0.211765
17	8	4	4	1.000000	0.029412
17	8	5	0	0.020362	0.020362
17	8	5	1	0.183258	0.162896
17	8	5	2	0.563348	0.380090
17	8	5	3	0.889140	0.325792
17	8	5	4	0.990950	0.101810
17	8	5	5	1.000000	0.009050
17	8	6	0	0.006787	0.006787
17	8	6	1	0.088235	0.081448
17	8	6	2	0.373303	0.285068
17	8	6	3	0.753394	0.380090
17	8	6	4	0.957014	0.203620
17	8	6	5	0.997738	0.040724
17	8	6	6	1.000000	0.002262
17	8	7	0	0.001851	0.001851
17	8	7	1	0.036405	0.034554
17	8	7	2	0.217812	0.181407
17	8	7	3	0.580625	0.362814
17	8	7	4	0.882970	0.302345
17	8	7	5	0.986631	0.103661
17	8	7	6	0.999589	0.012958
17	8	7	7	1.000000	0.000411
17	8	8	0	0.000370	0.000370
17	8	8	1	0.012217	0.011847
17	8	8	2	0.108967	0.096750
17	8	8	3	0.399218	0.290251
17	8	8	4	0.762032	0.362814
17	8	8	5	0.955533	0.193501
17	8	8	6	0.996997	0.041464

N	n	k	x	P(x)	p(x)
17	8	8	7	0.999959	0.002962
17	8	8	8	1.000000	0.000041
17	9	1	0	0.470588	0.470588
17	9	1	1	1.000000	0.529412
17	9	2	0	0.205882	0.205882
17	9	2	1	0.735294	0.529412
17	9	2	2	1.000000	0.264706
17	9	3	0	0.082353	0.082353
17	9	3	1	0.452941	0.370588
17	9	3	2	0.876471	0.423529
17	9	3	3	1.000000	0.123529
17	9	4	0	0.029412	0.029412
17	9	4	1	0.241176	0.211765
17	9	4	2	0.664706	0.423529
17	9	4	3	0.947059	0.282353
17	9	4	4	1.000000	0.052941
17	9	5	0	0.009050	0.009050
17	9	5	1	0.110860	0.101810
17	9	5	2	0.436652	0.325792
17	9	5	3	0.816742	0.380090
17	9	5	4	0.979638	0.162896
17	9	5	5	1.000000	0.020362
17	9	6	0	0.002262	0.002262
17	9	6	1	0.042986	0.040724
17	9	6	2	0.246606	0.203620
17	9	6	3	0.626697	0.380090
17	9	6	4	0.911765	0.285068
17	9	6	5	0.993213	0.081448
17	9	6	6	1.000000	0.006787
17	9	7	0	0.000411	0.000411
17	9	7	1	0.013369	0.012958
17	9	7	2	0.117030	0.103661
17	9	7	3	0.419375	0.302345
17	9	7	4	0.782188	0.362814
17	9	7	5	0.963595	0.181407
17	9	7	6	0.998149	0.034554
17	9	7	7	1.000000	0.001851
17	9	8	0	0.000041	0.000041
17	9	8	1	0.003003	0.002962
17	9	8	2	0.044467	0.041464
17	9	8	3	0.237968	0.193501
17	9	8	4	0.600782	0.362814
17	9	8	5	0.891032	0.290251
17	9	8	6	0.987783	0.096750
17	9	8	7	0.999630	0.011847
17	9	8	8	1.000000	0.000370
17	9	9	1	0.000370	0.000370
17	9	9	2	0.012217	0.011847
17	9	9	3	0.108967	0.096750
17	9	9	4	0.399218	0.290251

N	n	k	x	P(x)	p(x)
17	9	9	5	0.762032	0.362814
17	9	9	6	0.955533	0.193501
17	9	9	7	0.996997	0.041464
17	9	9	8	0.999959	0.002962
17	9	9	9	1.000000	0.000041
17	10	1	0	0.411765	0.411765
17	10	1	1	1.000000	0.588235
17	10	2	0	0.154412	0.154412
17	10	2	1	0.669118	0.514706
17	10	2	2	1.000000	0.330882
17	10	3	0	0.051471	0.051471
17	10	3	1	0.360294	0.308824
17	10	3	2	0.823529	0.463235
17	10	3	3	1.000000	0.176471
17	10	4	0	0.014706	0.014706
17	10	4	1	0.161765	0.147059
17	10	4	2	0.558823	0.397059
17	10	4	3	0.911765	0.352941
17	10	4	4	1.000000	0.088235
17	10	5	0	0.003394	0.003394
17	10	5	1	0.059955	0.056561
17	10	5	2	0.314480	0.254525
17	10	5	3	0.721719	0.407240
17	10	5	4	0.959276	0.237557
17	10	5	5	1.000000	0.040724
17	10	6	0	0.000566	0.000566
17	10	6	1	0.017534	0.016968
17	10	6	2	0.144796	0.127262
17	10	6	3	0.484163	0.339367
17	10	6	4	0.840498	0.356335
17	10	6	5	0.983032	0.142534
17	10	6	6	1.000000	0.016968
17	10	7	0	0.000051	0.000051
17	10	7	1	0.003651	0.003599
17	10	7	2	0.052242	0.048591
17	10	7	3	0.268202	0.215960
17	10	7	4	0.646133	0.377931
17	10	7	5	0.918243	0.272110
17	10	7	6	0.993830	0.075586
17	10	7	7	1.000000	0.006170
17	10	8	1	0.000411	0.000411
17	10	8	2	0.013369	0.012958
17	10	8	3	0.117030	0.103661
17	10	8	4	0.419375	0.302345
17	10	8	5	0.782188	0.362814
17	10	8	6	0.963595	0.181407
17	10	8	7	0.998149	0.034554
17	10	8	8	1.000000	0.001851
17	10	9	2	0.001851	0.001851
17	10	9	3	0.036405	0.034554
17	10	9	4	0.217812	0.181407
17	10	9	5	0.580625	0.362814
17	10	9	6	0.882970	0.302345
17	10	9	7	0.986631	0.103661
17	10	9	8	0.999589	0.012958
17	10	9	9	1.000000	0.000411
17	10	10	3	0.006170	0.006170
17	10	10	4	0.081756	0.075586
17	10	10	5	0.353867	0.272110
17	10	10	6	0.731798	0.377931
17	10	10	7	0.947758	0.215960
17	10	10	8	0.996349	0.048591
17	10	10	9	0.999948	0.003599
17	10	10	10	1.000000	0.000051
17	11	1	0	0.352941	0.352941
17	11	1	1	1.000000	0.647059
17	11	2	0	0.110294	0.110294
17	11	2	1	0.595588	0.485294
17	11	2	2	1.000000	0.404412
17	11	3	0	0.029412	0.029412
17	11	3	1	0.272059	0.242647
17	11	3	2	0.757353	0.485294
17	11	3	3	1.000000	0.242647
17	11	4	0	0.006303	0.006303
17	11	4	1	0.098739	0.092437
17	11	4	2	0.445378	0.346639
17	11	4	3	0.861345	0.415966
17	11	4	4	1.000000	0.138655
17	11	5	0	0.000970	0.000970
17	11	5	1	0.027634	0.026665
17	11	5	2	0.205398	0.177763
17	11	5	3	0.605365	0.399968
17	11	5	4	0.925339	0.319974
17	11	5	5	1.000000	0.074661
17	11	6	0	0.000081	0.000081
17	11	6	1	0.005414	0.005333
17	11	6	2	0.072075	0.066661
17	11	6	3	0.338720	0.266645
17	11	6	4	0.738688	0.399968
17	11	6	5	0.962670	0.223982
17	11	6	6	1.000000	0.037330
17	11	7	1	0.000566	0.000566
17	11	7	2	0.017534	0.016968
17	11	7	3	0.144796	0.127262
17	11	7	4	0.484163	0.339367
17	11	7	5	0.840498	0.356335
17	11	7	6	0.983032	0.142534
17	11	7	7	1.000000	0.016968
17	11	8	2	0.002262	0.002262
17	11	8	3	0.042986	0.040724
17	11	8	4	0.246606	0.203620
17	11	8	5	0.626697	0.380090
17	11	8	6	0.911765	0.285068
17	11	8	7	0.993213	0.081448
17	11	8	8	1.000000	0.006787
17	11	9	3	0.006787	0.006787
17	11	9	4	0.088235	0.081448
17	11	9	5	0.373303	0.285068
17	11	9	6	0.753394	0.380090
17	11	9	7	0.957014	0.203620
17	11	9	8	0.997738	0.040724
17	11	9	9	1.000000	0.002262
17	11	10	4	0.016968	0.016968
17	11	10	5	0.159502	0.142534
17	11	10	6	0.515837	0.356335
17	11	10	7	0.855204	0.339367
17	11	10	8	0.982466	0.127262
17	11	10	9	0.999434	0.016968
17	11	10	10	1.000000	0.000566
17	11	11	5	0.037330	0.037330
17	11	11	6	0.261312	0.223982
17	11	11	7	0.661280	0.399968
17	11	11	8	0.927925	0.266645
17	11	11	9	0.994586	0.066661
17	11	11	10	0.999919	0.005333
17	11	11	11	1.000000	0.000081
17	12	1	0	0.294118	0.294118
17	12	1	1	1.000000	0.705882
17	12	2	0	0.073529	0.073529
17	12	2	1	0.514706	0.441176
17	12	2	2	1.000000	0.485294
17	12	3	0	0.014706	0.014706
17	12	3	1	0.191176	0.176471
17	12	3	2	0.676471	0.485294
17	12	3	3	1.000000	0.323529
17	12	4	0	0.002101	0.002101
17	12	4	1	0.052521	0.050420
17	12	4	2	0.329832	0.277311
17	12	4	3	0.792017	0.462185
17	12	4	4	1.000000	0.207983
17	12	5	0	0.000162	0.000162
17	12	5	1	0.009858	0.009696
17	12	5	2	0.116516	0.106658
17	12	5	3	0.472043	0.355527
17	12	5	4	0.872010	0.399968
17	12	5	5	1.000000	0.127990
17	12	6	1	0.000970	0.000970
17	12	6	2	0.027634	0.026665
17	12	6	3	0.205398	0.177763
17	12	6	4	0.605365	0.399968
17	12	6	5	0.925339	0.319974
17	12	6	6	1.000000	0.074661
17	12	7	2	0.003394	0.003394
17	12	7	3	0.059955	0.056561
17	12	7	4	0.314480	0.254525
17	12	7	5	0.721719	0.407240
17	12	7	6	0.959276	0.237557
17	12	7	7	1.000000	0.040724
17	12	8	3	0.009050	0.009050
17	12	8	4	0.110860	0.101810
17	12	8	5	0.436652	0.325792
17	12	8	6	0.816742	0.380090
17	12	8	7	0.979638	0.162896
17	12	8	8	1.000000	0.020362
17	12	9	4	0.020362	0.020362
17	12	9	5	0.183258	0.162896
17	12	9	6	0.563348	0.380090
17	12	9	7	0.889140	0.325792
17	12	9	8	0.990950	0.101810
17	12	9	9	1.000000	0.009050
17	12	10	5	0.040724	0.040724
17	12	10	6	0.278281	0.237557
17	12	10	7	0.685520	0.407240
17	12	10	8	0.940045	0.254525
17	12	10	9	0.996606	0.056561
17	12	10	10	1.000000	0.003394
17	12	11	6	0.074661	0.074661
17	12	11	7	0.394635	0.319974
17	12	11	8	0.794602	0.399968
17	12	11	9	0.972366	0.177763
17	12	11	10	0.999030	0.026665
17	12	11	11	1.000000	0.000970
17	12	12	7	0.127990	0.127990
17	12	12	8	0.527957	0.399968
17	12	12	9	0.883484	0.355527
17	12	12	10	0.990142	0.106658
17	12	12	11	0.999838	0.009696
17	12	12	12	1.000000	0.000162
17	13	1	0	0.235294	0.235294
17	13	1	1	1.000000	0.764706
17	13	2	0	0.044118	0.044118
17	13	2	1	0.426471	0.382353
17	13	2	2	1.000000	0.573529
17	13	3	0	0.005882	0.005882
17	13	3	1	0.120588	0.114706
17	13	3	2	0.579412	0.458824
17	13	3	3	1.000000	0.420588
17	13	4	0	0.000420	0.000420
17	13	4	1	0.022269	0.021849
17	13	4	2	0.218908	0.196639

Table for $N = 2$, $n = 1$, through $N = 100$, $n = 50$

N	n	k	x	P(x)	p(x)
17	13	4	3	0.699580	0.480672
17	13	4	4	1.000000	0.300420
17	13	5	1	0.002101	0.002101
17	13	5	2	0.052521	0.050420
17	13	5	3	0.329832	0.277311
17	13	5	4	0.792017	0.462185
17	13	5	5	1.000000	0.207983
17	13	6	2	0.006303	0.006303
17	13	6	3	0.098739	0.092437
17	13	6	4	0.445378	0.346639
17	13	6	5	0.861345	0.415966
17	13	6	6	1.000000	0.138655
17	13	7	3	0.014706	0.014706
17	13	7	4	0.161765	0.147059
17	13	7	5	0.558823	0.397059
17	13	7	6	0.911765	0.352941
17	13	7	7	1.000000	0.088235
17	13	8	4	0.029412	0.029412
17	13	8	5	0.241176	0.211765
17	13	8	6	0.664706	0.423529
17	13	8	7	0.947059	0.282353
17	13	8	8	1.000000	0.052941
17	13	9	5	0.052941	0.052941
17	13	9	6	0.335294	0.282353
17	13	9	7	0.758824	0.423529
17	13	9	8	0.970588	0.211765
17	13	9	9	1.000000	0.029412
17	13	10	6	0.088235	0.088235
17	13	10	7	0.441176	0.352941
17	13	10	8	0.838235	0.397059
17	13	10	9	0.985294	0.147059
17	13	10	10	1.000000	0.014706
17	13	11	7	0.138655	0.138655
17	13	11	8	0.554622	0.415966
17	13	11	9	0.901260	0.346639
17	13	11	10	0.993697	0.092437
17	13	11	11	1.000000	0.006303
17	13	12	8	0.207983	0.207983
17	13	12	9	0.670168	0.462185
17	13	12	10	0.947479	0.277311
17	13	12	11	0.997899	0.050420
17	13	12	12	1.000000	0.002101
17	13	13	9	0.300420	0.300420
17	13	13	10	0.781092	0.480672
17	13	13	11	0.977731	0.196639
17	13	13	12	0.999580	0.021849
17	13	13	13	1.000000	0.000420
17	14	1	0	0.176471	0.176471
17	14	1	1	1.000000	0.823529
17	14	2	0	0.022059	0.022059

N	n	k	x	P(x)	p(x)
17	14	2	1	0.330882	0.308824
17	14	2	2	1.000000	0.669118
17	14	3	0	0.001471	0.001471
17	14	3	1	0.063235	0.061765
17	14	3	2	0.464706	0.401471
17	14	3	3	1.000000	0.535294
17	14	4	1	0.005882	0.005882
17	14	4	2	0.120588	0.114706
17	14	4	3	0.579412	0.458824
17	14	4	4	1.000000	0.420588
17	14	5	2	0.014706	0.014706
17	14	5	3	0.191176	0.176471
17	14	5	4	0.676471	0.485294
17	14	5	5	1.000000	0.323529
17	14	6	3	0.029412	0.029412
17	14	6	4	0.272059	0.242647
17	14	6	5	0.757353	0.485294
17	14	6	6	1.000000	0.242647
17	14	7	4	0.051471	0.051471
17	14	7	5	0.360294	0.308824
17	14	7	6	0.823529	0.463235
17	14	7	7	1.000000	0.176471
17	14	8	5	0.082353	0.082353
17	14	8	6	0.452941	0.370588
17	14	8	7	0.876471	0.423529
17	14	8	8	1.000000	0.123529
17	14	9	6	0.123529	0.123529
17	14	9	7	0.547059	0.423529
17	14	9	8	0.917647	0.370588
17	14	9	9	1.000000	0.082353
17	14	10	7	0.176471	0.176471
17	14	10	8	0.639706	0.463235
17	14	10	9	0.948529	0.308824
17	14	10	10	1.000000	0.051471
17	14	11	8	0.242647	0.242647
17	14	11	9	0.727941	0.485294
17	14	11	10	0.970588	0.242647
17	14	11	11	1.000000	0.029412
17	14	12	9	0.323529	0.323529
17	14	12	10	0.808823	0.485294
17	14	12	11	0.985294	0.176471
17	14	12	12	1.000000	0.014706
17	14	13	10	0.420588	0.420588
17	14	13	11	0.879412	0.458824
17	14	13	12	0.994118	0.114706
17	14	13	13	1.000000	0.005882
17	14	14	11	0.535294	0.535294
17	14	14	12	0.936765	0.401471
17	14	14	13	0.998529	0.061765
17	14	14	14	1.000000	0.001471

N	n	k	x	P(x)	p(x)
17	15	1	0	0.117647	0.117647
17	15	1	1	1.000000	0.882353
17	15	2	0	0.007353	0.007353
17	15	2	1	0.227941	0.220588
17	15	2	2	1.000000	0.772059
17	15	3	1	0.022059	0.022059
17	15	3	2	0.330882	0.308824
17	15	3	3	1.000000	0.669118
17	15	4	2	0.044118	0.044118
17	15	4	3	0.426471	0.382353
17	15	4	4	1.000000	0.573529
17	15	5	3	0.073529	0.073529
17	15	5	4	0.514706	0.441176
17	15	5	5	1.000000	0.485294
17	15	6	4	0.110294	0.110294
17	15	6	5	0.595588	0.485294
17	15	6	6	1.000000	0.404412
17	15	7	5	0.154412	0.154412
17	15	7	6	0.669118	0.514706
17	15	7	7	1.000000	0.330882
17	15	8	6	0.205882	0.205882
17	15	8	7	0.735294	0.529412
17	15	8	8	1.000000	0.264706
17	15	9	7	0.264706	0.264706
17	15	9	8	0.794118	0.529412
17	15	9	9	1.000000	0.205882
17	15	10	8	0.330882	0.330882
17	15	10	9	0.845588	0.514706
17	15	10	10	1.000000	0.154412
17	15	11	9	0.404412	0.404412
17	15	11	10	0.889706	0.485294
17	15	11	11	1.000000	0.110294
17	15	12	10	0.485294	0.485294
17	15	12	11	0.926471	0.441176
17	15	12	12	1.000000	0.073529
17	15	13	11	0.573529	0.573529
17	15	13	12	0.955882	0.382353
17	15	13	13	1.000000	0.044118
17	15	14	12	0.669118	0.669118
17	15	14	13	0.977941	0.308824
17	15	14	14	1.000000	0.022059
17	15	15	13	0.772059	0.772059
17	15	15	14	0.992647	0.220588
17	15	15	15	1.000000	0.007353
17	16	1	0	0.058824	0.058824
17	16	1	1	1.000000	0.941176
17	16	2	1	0.117647	0.117647
17	16	2	2	1.000000	0.882353
17	16	3	2	0.176471	0.176471
17	16	3	3	1.000000	0.823529

N	n	k	x	P(x)	p(x)
17	16	4	3	0.235294	0.235294
17	16	4	4	1.000000	0.764706
17	16	5	4	0.294118	0.294118
17	16	5	5	1.000000	0.705882
17	16	6	5	0.352941	0.352941
17	16	6	6	1.000000	0.647059
17	16	7	6	0.411765	0.411765
17	16	7	7	1.000000	0.588235
17	16	8	7	0.470588	0.470588
17	16	8	8	1.000000	0.529412
17	16	9	8	0.529412	0.529412
17	16	9	9	1.000000	0.470588
17	16	10	9	0.588235	0.588235
17	16	10	10	1.000000	0.411765
17	16	11	10	0.647059	0.647059
17	16	11	11	1.000000	0.352941
17	16	12	11	0.705882	0.705882
17	16	12	12	1.000000	0.294118
17	16	13	12	0.764706	0.764706
17	16	13	13	1.000000	0.235294
17	16	14	13	0.823529	0.823529
17	16	14	14	1.000000	0.176471
17	16	15	14	0.882353	0.882353
17	16	15	15	1.000000	0.117647
17	16	16	15	0.941176	0.941176
17	16	16	16	1.000000	0.058824
18	1	1	0	0.944444	0.944444
18	1	1	1	1.000000	0.055556
18	2	1	0	0.888889	0.888889
18	2	1	1	1.000000	0.111111
18	2	2	0	0.784314	0.784314
18	2	2	1	0.993464	0.209150
18	2	2	2	1.000000	0.006536
18	3	1	0	0.833333	0.833333
18	3	1	1	1.000000	0.166667
18	3	2	0	0.686275	0.686275
18	3	2	1	0.980392	0.294118
18	3	2	2	1.000000	0.019608
18	3	3	0	0.557598	0.557598
18	3	3	1	0.943627	0.386029
18	3	3	2	0.998774	0.055147
18	3	3	3	1.000000	0.001225
18	4	1	0	0.777778	0.777778
18	4	1	1	1.000000	0.222222
18	4	2	0	0.594771	0.594771
18	4	2	1	0.960784	0.366013
18	4	2	2	1.000000	0.039216
18	4	3	0	0.446078	0.446078
18	4	3	1	0.892157	0.446078
18	4	3	2	0.995098	0.102941

$N = 17–18$ $n = 13–4$

Table for $N = 2$, $n = 1$, through $N = 100$, $n = 50$

N	n	k	x	P(x)	p(x)
18	4	3	3	1.000000	0.004902
18	4	4	0	0.327124	0.327124
18	4	4	1	0.802941	0.475817
18	4	4	2	0.981373	0.178431
18	4	4	3	0.999673	0.018301
18	4	4	4	1.000000	0.000327
18	5	1	0	0.722222	0.722222
18	5	1	1	1.000000	0.277778
18	5	2	0	0.509804	0.509804
18	5	2	1	0.934640	0.424837
18	5	2	2	1.000000	0.065359
18	5	3	0	0.350490	0.350490
18	5	3	1	0.828431	0.477941
18	5	3	2	0.987745	0.159314
18	5	3	3	1.000000	0.012255
18	5	4	0	0.233660	0.233660
18	5	4	1	0.700980	0.467320
18	5	4	2	0.955882	0.254902
18	5	4	3	0.998366	0.042484
18	5	4	4	1.000000	0.001634
18	5	5	0	0.150210	0.150210
18	5	5	1	0.567460	0.417250
18	5	5	2	0.901260	0.333800
18	5	5	3	0.992297	0.091036
18	5	5	4	0.999883	0.007586
18	5	5	5	1.000000	0.000117
18	6	1	0	0.666667	0.666667
18	6	1	1	1.000000	0.333333
18	6	2	0	0.431373	0.431373
18	6	2	1	0.901961	0.470588
18	6	2	2	1.000000	0.098039
18	6	3	0	0.269608	0.269608
18	6	3	1	0.754902	0.485294
18	6	3	2	0.975490	0.220588
18	6	3	3	1.000000	0.024510
18	6	4	0	0.161765	0.161765
18	6	4	1	0.593137	0.431373
18	6	4	2	0.916667	0.323529
18	6	4	3	0.995098	0.078431
18	6	4	4	1.000000	0.004902
18	6	5	0	0.092437	0.092437
18	6	5	1	0.439076	0.346639
18	6	5	2	0.824230	0.385154
18	6	5	3	0.978291	0.154062
18	6	5	4	0.999300	0.021008
18	6	5	5	1.000000	0.000700
18	6	6	0	0.049774	0.049774
18	6	6	1	0.305753	0.255979
18	6	6	2	0.705721	0.399968
18	6	6	3	0.942739	0.237018
18	6	6	4	0.996068	0.053329
18	6	6	5	0.999946	0.003878
18	6	6	6	1.000000	0.000054
18	7	1	0	0.611111	0.611111
18	7	1	1	1.000000	0.388889
18	7	2	0	0.359477	0.359477
18	7	2	1	0.862745	0.503268
18	7	2	2	1.000000	0.137255
18	7	3	0	0.202206	0.202206
18	7	3	1	0.674020	0.471814
18	7	3	2	0.957108	0.283088
18	7	3	3	1.000000	0.042892
18	7	4	0	0.107843	0.107843
18	7	4	1	0.485294	0.377451
18	7	4	2	0.862745	0.377451
18	7	4	3	0.988562	0.125817
18	7	4	4	1.000000	0.011438
18	7	5	0	0.053922	0.053922
18	7	5	1	0.323529	0.269608
18	7	5	2	0.727941	0.404412
18	7	5	3	0.952614	0.224673
18	7	5	4	0.997549	0.044935
18	7	5	5	1.000000	0.002451
18	7	6	0	0.024887	0.024887
18	7	6	1	0.199095	0.174208
18	7	6	2	0.572398	0.373303
18	7	6	3	0.883484	0.311086
18	7	6	4	0.987179	0.103695
18	7	6	5	0.999623	0.012443
18	7	6	6	1.000000	0.000377
18	7	7	0	0.010370	0.010370
18	7	7	1	0.111991	0.101621
18	7	7	2	0.416855	0.304864
18	7	7	3	0.779789	0.362934
18	7	7	4	0.961256	0.181467
18	7	7	5	0.997549	0.036293
18	7	7	6	0.999968	0.002420
18	7	7	7	1.000000	0.000031
18	8	1	0	0.555556	0.555556
18	8	1	1	1.000000	0.444444
18	8	2	0	0.294118	0.294118
18	8	2	1	0.816993	0.522876
18	8	2	2	1.000000	0.183007
18	8	3	0	0.147059	0.147059
18	8	3	1	0.588235	0.441176
18	8	3	2	0.931372	0.343137
18	8	3	3	1.000000	0.068627
18	8	4	0	0.068627	0.068627
18	8	4	1	0.382353	0.313726
18	8	4	2	0.794118	0.411765
18	8	4	3	0.977124	0.183007
18	8	4	4	1.000000	0.022876
18	8	5	0	0.029412	0.029412
18	8	5	1	0.225490	0.196078
18	8	5	2	0.617647	0.392157
18	8	5	3	0.911765	0.294118
18	8	5	4	0.993464	0.081699
18	8	5	5	1.000000	0.006536
18	8	6	0	0.011312	0.011312
18	8	6	1	0.119909	0.108597
18	8	6	2	0.436652	0.316742
18	8	6	3	0.798642	0.361991
18	8	6	4	0.968326	0.169683
18	8	6	5	0.998492	0.030166
18	8	6	6	1.000000	0.001508
18	8	7	0	0.003771	0.003771
18	8	7	1	0.056561	0.052790
18	8	7	2	0.278281	0.221719
18	8	7	3	0.647813	0.369532
18	8	7	4	0.911765	0.263952
18	8	7	5	0.990950	0.079186
18	8	7	6	0.999749	0.008798
18	8	7	7	1.000000	0.000251
18	8	8	0	0.001028	0.001028
18	8	8	1	0.022967	0.021939
18	8	8	2	0.157343	0.134375
18	8	8	3	0.479844	0.322501
18	8	8	4	0.815782	0.335939
18	8	8	5	0.969354	0.153572
18	8	8	6	0.998149	0.028795
18	8	8	7	0.999977	0.001828
18	8	8	8	1.000000	0.000023
18	9	1	0	0.500000	0.500000
18	9	1	1	1.000000	0.500000
18	9	2	0	0.235294	0.235294
18	9	2	1	0.764706	0.529412
18	9	2	2	1.000000	0.235294
18	9	3	0	0.102941	0.102941
18	9	3	1	0.500000	0.397059
18	9	3	2	0.897059	0.397059
18	9	3	3	1.000000	0.102941
18	9	4	0	0.041176	0.041176
18	9	4	1	0.288235	0.247059
18	9	4	2	0.711765	0.423529
18	9	4	3	0.958824	0.247059
18	9	4	4	1.000000	0.041176
18	9	5	0	0.014706	0.014706
18	9	5	1	0.147059	0.132353
18	9	5	2	0.500000	0.352941
18	9	5	3	0.852941	0.352941
18	9	5	4	0.985294	0.132353
18	9	5	5	1.000000	0.014706
18	9	6	0	0.004525	0.004525
18	9	6	1	0.065611	0.061086
18	9	6	2	0.309955	0.244344
18	9	6	3	0.690045	0.380090
18	9	6	4	0.934389	0.244344
18	9	6	5	0.995475	0.061086
18	9	6	6	1.000000	0.004525
18	9	7	0	0.001131	0.001131
18	9	7	1	0.024887	0.023756
18	9	7	2	0.167421	0.142534
18	9	7	3	0.500000	0.332579
18	9	7	4	0.832579	0.332579
18	9	7	5	0.975113	0.142534
18	9	7	6	0.998869	0.023756
18	9	7	7	1.000000	0.001131
18	9	8	0	0.000206	0.000206
18	9	8	1	0.007610	0.007404
18	9	8	2	0.076717	0.069107
18	9	8	3	0.318593	0.241876
18	9	8	4	0.681407	0.362814
18	9	8	5	0.923283	0.241876
18	9	8	6	0.992390	0.069107
18	9	8	7	0.999794	0.007404
18	9	8	8	1.000000	0.000206
18	9	9	0	0.000021	0.000021
18	9	9	1	0.001687	0.001666
18	9	9	2	0.028342	0.026656
18	9	9	3	0.173468	0.145125
18	9	9	4	0.500000	0.326532
18	9	9	5	0.826532	0.326532
18	9	9	6	0.971658	0.145125
18	9	9	7	0.998313	0.026656
18	9	9	8	0.999979	0.001666
18	9	9	9	1.000000	0.000021
18	10	1	0	0.444444	0.444444
18	10	1	1	1.000000	0.555556
18	10	2	0	0.183007	0.183007
18	10	2	1	0.705882	0.522876
18	10	2	2	1.000000	0.294118
18	10	3	0	0.068627	0.068627
18	10	3	1	0.411765	0.343137
18	10	3	2	0.852941	0.441176
18	10	3	3	1.000000	0.147059
18	10	4	0	0.022876	0.022876
18	10	4	1	0.205882	0.183007
18	10	4	2	0.617647	0.411765
18	10	4	3	0.931373	0.313726
18	10	4	4	1.000000	0.068627

Table for $N = 2$, $n = 1$, through $N = 100$, $n = 50$

N	n	k	x	P(x)	p(x)
18	10	5	0	0.006536	0.006536
18	10	5	1	0.088235	0.081699
18	10	5	2	0.382353	0.294118
18	10	5	3	0.774510	0.392157
18	10	5	4	0.970588	0.196078
18	10	5	5	1.000000	0.029412
18	10	6	0	0.001508	0.001508
18	10	6	1	0.031674	0.030166
18	10	6	2	0.201357	0.169683
18	10	6	3	0.563348	0.361991
18	10	6	4	0.880090	0.316742
18	10	6	5	0.988688	0.108597
18	10	6	6	1.000000	0.011312
18	10	7	0	0.000251	0.000251
18	10	7	1	0.009050	0.008798
18	10	7	2	0.088235	0.079186
18	10	7	3	0.352187	0.263952
18	10	7	4	0.721719	0.369532
18	10	7	5	0.943439	0.221719
18	10	7	6	0.996229	0.052790
18	10	7	7	1.000000	0.003771
18	10	8	0	0.000023	0.000023
18	10	8	1	0.001851	0.001828
18	10	8	2	0.030646	0.028795
18	10	8	3	0.184218	0.153572
18	10	8	4	0.520156	0.335939
18	10	8	5	0.842657	0.322501
18	10	8	6	0.977033	0.134375
18	10	8	7	0.998972	0.021939
18	10	8	8	1.000000	0.001028
18	10	9	1	0.000206	0.000206
18	10	9	2	0.007610	0.007404
18	10	9	3	0.076717	0.069107
18	10	9	4	0.318593	0.241876
18	10	9	5	0.681407	0.362814
18	10	9	6	0.923283	0.241876
18	10	9	7	0.992390	0.069107
18	10	9	8	0.999794	0.007404
18	10	9	9	1.000000	0.000206
18	10	10	2	0.001028	0.001028
18	10	10	3	0.022967	0.021939
18	10	10	4	0.157343	0.134375
18	10	10	5	0.479844	0.322501
18	10	10	6	0.815782	0.335939
18	10	10	7	0.969354	0.153572
18	10	10	8	0.998149	0.028795
18	10	10	9	0.999977	0.001828
18	10	10	10	1.000000	0.000023
18	11	1	0	0.388889	0.388889
18	11	1	1	1.000000	0.611111

N	n	k	x	P(x)	p(x)
18	11	2	0	0.137255	0.137255
18	11	2	1	0.640523	0.503268
18	11	2	2	1.000000	0.359477
18	11	3	0	0.042892	0.042892
18	11	3	1	0.325980	0.283088
18	11	3	2	0.797794	0.471814
18	11	3	3	1.000000	0.202206
18	11	4	0	0.011438	0.011438
18	11	4	1	0.137255	0.125817
18	11	4	2	0.514706	0.377451
18	11	4	3	0.892157	0.377451
18	11	4	4	1.000000	0.107843
18	11	5	0	0.002451	0.002451
18	11	5	1	0.047386	0.044935
18	11	5	2	0.272059	0.224673
18	11	5	3	0.676471	0.404412
18	11	5	4	0.946078	0.269608
18	11	5	5	1.000000	0.053922
18	11	6	0	0.000377	0.000377
18	11	6	1	0.012821	0.012443
18	11	6	2	0.116516	0.103695
18	11	6	3	0.427602	0.311086
18	11	6	4	0.800905	0.373303
18	11	6	5	0.975113	0.174208
18	11	6	6	1.000000	0.024887
18	11	7	0	0.000031	0.000031
18	11	7	1	0.002451	0.002420
18	11	7	2	0.038744	0.036293
18	11	7	3	0.220211	0.181467
18	11	7	4	0.583145	0.362934
18	11	7	5	0.888009	0.304864
18	11	7	6	0.989630	0.101621
18	11	7	7	1.000000	0.010370
18	11	8	1	0.000251	0.000251
18	11	8	2	0.009050	0.008798
18	11	8	3	0.088235	0.079186
18	11	8	4	0.352187	0.263952
18	11	8	5	0.721719	0.369532
18	11	8	6	0.943439	0.221719
18	11	8	7	0.996229	0.052790
18	11	8	8	1.000000	0.003771
18	11	9	2	0.001131	0.001131
18	11	9	3	0.024887	0.023756
18	11	9	4	0.167421	0.142534
18	11	9	5	0.500000	0.332579
18	11	9	6	0.832579	0.332579
18	11	9	7	0.975113	0.142534
18	11	9	8	0.998869	0.023756
18	11	9	9	1.000000	0.001131
18	11	10	3	0.003771	0.003771

N	n	k	x	P(x)	p(x)
18	11	10	4	0.056561	0.052790
18	11	10	5	0.278281	0.221719
18	11	10	6	0.647813	0.369532
18	11	10	7	0.911765	0.263952
18	11	10	8	0.990950	0.079186
18	11	10	9	0.999749	0.008798
18	11	10	10	1.000000	0.000251
18	11	11	4	0.010370	0.010370
18	11	11	5	0.111991	0.101621
18	11	11	6	0.416855	0.304864
18	11	11	7	0.779789	0.362934
18	11	11	8	0.961256	0.181467
18	11	11	9	0.997549	0.036293
18	11	11	10	0.999968	0.002420
18	11	11	11	1.000000	0.000031
18	12	1	0	0.333333	0.333333
18	12	1	1	1.000000	0.666667
18	12	2	0	0.098039	0.098039
18	12	2	1	0.568627	0.470588
18	12	2	2	1.000000	0.431373
18	12	3	0	0.024510	0.024510
18	12	3	1	0.245098	0.220588
18	12	3	2	0.730392	0.485294
18	12	3	3	1.000000	0.269608
18	12	4	0	0.004902	0.004902
18	12	4	1	0.083333	0.078431
18	12	4	2	0.406863	0.323529
18	12	4	3	0.838235	0.431373
18	12	4	4	1.000000	0.161765
18	12	5	0	0.000700	0.000700
18	12	5	1	0.021709	0.021008
18	12	5	2	0.175770	0.154062
18	12	5	3	0.560924	0.385154
18	12	5	4	0.907563	0.346639
18	12	5	5	1.000000	0.092437
18	12	6	0	0.000054	0.000054
18	12	6	1	0.003932	0.003878
18	12	6	2	0.057261	0.053329
18	12	6	3	0.294279	0.237018
18	12	6	4	0.694247	0.399968
18	12	6	5	0.950226	0.255979
18	12	6	6	1.000000	0.049774
18	12	7	1	0.000377	0.000377
18	12	7	2	0.012821	0.012443
18	12	7	3	0.116516	0.103695
18	12	7	4	0.427602	0.311086
18	12	7	5	0.800905	0.373303
18	12	7	6	0.975113	0.174208
18	12	7	7	1.000000	0.024887
18	12	8	2	0.001508	0.001508

N	n	k	x	P(x)	p(x)
18	12	8	3	0.031674	0.030166
18	12	8	4	0.201357	0.169683
18	12	8	5	0.563348	0.361991
18	12	8	6	0.880090	0.316742
18	12	8	7	0.988688	0.108597
18	12	8	8	1.000000	0.011312
18	12	9	3	0.004525	0.004525
18	12	9	4	0.065611	0.061086
18	12	9	5	0.309955	0.244344
18	12	9	6	0.690045	0.380090
18	12	9	7	0.934389	0.244344
18	12	9	8	0.995475	0.061086
18	12	9	9	1.000000	0.004525
18	12	10	4	0.011312	0.011312
18	12	10	5	0.119909	0.108597
18	12	10	6	0.436652	0.316742
18	12	10	7	0.798642	0.361991
18	12	10	8	0.968326	0.169683
18	12	10	9	0.998492	0.030166
18	12	10	10	1.000000	0.001508
18	12	11	5	0.024887	0.024887
18	12	11	6	0.199095	0.174208
18	12	11	7	0.572398	0.373303
18	12	11	8	0.883484	0.311086
18	12	11	9	0.987179	0.103695
18	12	11	10	0.999623	0.012443
18	12	11	11	1.000000	0.000377
18	12	12	6	0.049774	0.049774
18	12	12	7	0.305753	0.255979
18	12	12	8	0.705721	0.399968
18	12	12	9	0.942739	0.237018
18	12	12	10	0.996068	0.053329
18	12	12	11	0.999946	0.003878
18	12	12	12	1.000000	0.000054
18	13	1	0	0.277778	0.277778
18	13	1	1	1.000000	0.722222
18	13	2	0	0.065359	0.065359
18	13	2	1	0.490196	0.424837
18	13	2	2	1.000000	0.509804
18	13	3	0	0.012255	0.012255
18	13	3	1	0.171569	0.159314
18	13	3	2	0.649510	0.477941
18	13	3	3	1.000000	0.350490
18	13	4	0	0.001634	0.001634
18	13	4	1	0.044118	0.042484
18	13	4	2	0.299020	0.254902
18	13	4	3	0.766340	0.467320
18	13	4	4	1.000000	0.233660
18	13	5	0	0.000117	0.000117
18	13	5	1	0.007703	0.007586

N	n	k	x	P(x)	p(x)
18	13	5	2	0.098739	0.091036
18	13	5	3	0.432540	0.333800
18	13	5	4	0.849790	0.417250
18	13	5	5	1.000000	0.150210
18	13	6	1	0.000700	0.000700
18	13	6	2	0.021709	0.021008
18	13	6	3	0.175770	0.154062
18	13	6	4	0.560924	0.385154
18	13	6	5	0.907563	0.346639
18	13	6	6	1.000000	0.092437
18	13	7	2	0.002451	0.002451
18	13	7	3	0.047386	0.044935
18	13	7	4	0.272059	0.224673
18	13	7	5	0.676471	0.404412
18	13	7	6	0.946078	0.269608
18	13	7	7	1.000000	0.053922
18	13	8	3	0.006536	0.006536
18	13	8	4	0.088235	0.081699
18	13	8	5	0.382353	0.294118
18	13	8	6	0.774510	0.392157
18	13	8	7	0.970588	0.196078
18	13	8	8	1.000000	0.029412
18	13	9	4	0.014706	0.014706
18	13	9	5	0.147059	0.132353
18	13	9	6	0.500000	0.352941
18	13	9	7	0.852941	0.352941
18	13	9	8	0.985294	0.132353
18	13	9	9	1.000000	0.014706
18	13	10	5	0.029412	0.029412
18	13	10	6	0.225490	0.196078
18	13	10	7	0.617647	0.392157
18	13	10	8	0.911765	0.294118
18	13	10	9	0.993464	0.081699
18	13	10	10	1.000000	0.006536
18	13	11	6	0.053922	0.053922
18	13	11	7	0.323529	0.269608
18	13	11	8	0.727941	0.404412
18	13	11	9	0.952614	0.224673
18	13	11	10	0.997549	0.044935
18	13	11	11	1.000000	0.002451
18	13	12	7	0.092437	0.092437
18	13	12	8	0.439076	0.346639
18	13	12	9	0.824230	0.385154
18	13	12	10	0.978291	0.154062
18	13	12	11	0.999300	0.021008
18	13	12	12	1.000000	0.000700
18	13	13	8	0.150210	0.150210
18	13	13	9	0.567460	0.417250
18	13	13	10	0.901260	0.333800
18	13	13	11	0.992297	0.091036

N	n	k	x	P(x)	p(x)
18	13	13	12	0.999883	0.007586
18	13	13	13	1.000000	0.000117
18	14	1	0	0.222222	0.222222
18	14	1	1	1.000000	0.777778
18	14	2	0	0.039216	0.039216
18	14	2	1	0.405229	0.366013
18	14	2	2	1.000000	0.594771
18	14	3	0	0.004902	0.004902
18	14	3	1	0.107843	0.102941
18	14	3	2	0.553921	0.446078
18	14	3	3	1.000000	0.446078
18	14	4	0	0.000327	0.000327
18	14	4	1	0.018627	0.018301
18	14	4	2	0.197059	0.178431
18	14	4	3	0.672876	0.475817
18	14	4	4	1.000000	0.327124
18	14	5	1	0.001634	0.001634
18	14	5	2	0.044118	0.042484
18	14	5	3	0.299020	0.254902
18	14	5	4	0.766340	0.467320
18	14	5	5	1.000000	0.233660
18	14	6	2	0.004902	0.004902
18	14	6	3	0.083333	0.078431
18	14	6	4	0.406863	0.323529
18	14	6	5	0.838235	0.431373
18	14	6	6	1.000000	0.161765
18	14	7	3	0.011438	0.011438
18	14	7	4	0.137255	0.125817
18	14	7	5	0.514706	0.377451
18	14	7	6	0.892157	0.377451
18	14	7	7	1.000000	0.107843
18	14	8	4	0.022876	0.022876
18	14	8	5	0.205882	0.183007
18	14	8	6	0.617647	0.411765
18	14	8	7	0.931373	0.313726
18	14	8	8	1.000000	0.068627
18	14	9	5	0.041176	0.041176
18	14	9	6	0.288235	0.247059
18	14	9	7	0.711765	0.423529
18	14	9	8	0.958824	0.247059
18	14	9	9	1.000000	0.041176
18	14	10	6	0.068627	0.068627
18	14	10	7	0.382353	0.313726
18	14	10	8	0.794118	0.411765
18	14	10	9	0.977124	0.183007
18	14	10	10	1.000000	0.022876
18	14	11	7	0.107843	0.107843
18	14	11	8	0.485294	0.377451
18	14	11	9	0.862745	0.377451
18	14	11	10	0.988562	0.125817

N	n	k	x	P(x)	p(x)
18	14	11	11	1.000000	0.011438
18	14	12	8	0.161765	0.161765
18	14	12	9	0.593137	0.431373
18	14	12	10	0.916667	0.323529
18	14	12	11	0.995098	0.078431
18	14	12	12	1.000000	0.004902
18	14	13	9	0.233660	0.233660
18	14	13	10	0.700980	0.467320
18	14	13	11	0.955882	0.254902
18	14	13	12	0.998366	0.042484
18	14	13	13	1.000000	0.001634
18	14	14	10	0.327124	0.327124
18	14	14	11	0.802941	0.475817
18	14	14	12	0.981373	0.178431
18	14	14	13	0.999673	0.018301
18	14	14	14	1.000000	0.000327
18	15	1	0	0.166667	0.166667
18	15	1	1	1.000000	0.833333
18	15	2	0	0.019608	0.019608
18	15	2	1	0.313725	0.294118
18	15	2	2	1.000000	0.686275
18	15	3	0	0.001225	0.001225
18	15	3	1	0.056373	0.055147
18	15	3	2	0.442402	0.386029
18	15	3	3	1.000000	0.557598
18	15	4	1	0.004902	0.004902
18	15	4	2	0.107843	0.102941
18	15	4	3	0.553921	0.446078
18	15	4	4	1.000000	0.446078
18	15	5	2	0.012255	0.012255
18	15	5	3	0.171569	0.159314
18	15	5	4	0.649510	0.477941
18	15	5	5	1.000000	0.350490
18	15	6	3	0.024510	0.024510
18	15	6	4	0.245098	0.220588
18	15	6	5	0.730392	0.485294
18	15	6	6	1.000000	0.269608
18	15	7	4	0.042892	0.042892
18	15	7	5	0.325980	0.283088
18	15	7	6	0.797794	0.471814
18	15	7	7	1.000000	0.202206
18	15	8	5	0.068627	0.068627
18	15	8	6	0.411765	0.343137
18	15	8	7	0.852941	0.441176
18	15	8	8	1.000000	0.147059
18	15	9	6	0.102941	0.102941
18	15	9	7	0.500000	0.397059
18	15	9	8	0.897059	0.397059
18	15	9	9	1.000000	0.102941
18	15	10	7	0.147059	0.147059

N	n	k	x	P(x)	p(x)
18	15	10	8	0.588235	0.441176
18	15	10	9	0.931372	0.343137
18	15	10	10	1.000000	0.068627
18	15	11	8	0.202206	0.202206
18	15	11	9	0.674020	0.471814
18	15	11	10	0.957108	0.283088
18	15	11	11	1.000000	0.042892
18	15	12	9	0.269608	0.269608
18	15	12	10	0.754902	0.485294
18	15	12	11	0.975490	0.220588
18	15	12	12	1.000000	0.024510
18	15	13	10	0.350490	0.350490
18	15	13	11	0.828431	0.477941
18	15	13	12	0.987745	0.159314
18	15	13	13	1.000000	0.012255
18	15	14	11	0.446078	0.446078
18	15	14	12	0.892157	0.446078
18	15	14	13	0.995098	0.102941
18	15	14	14	1.000000	0.004902
18	15	15	12	0.557598	0.557598
18	15	15	13	0.943627	0.386029
18	15	15	14	0.998774	0.055147
18	15	15	15	1.000000	0.001225
18	16	1	0	0.111111	0.111111
18	16	1	1	1.000000	0.888889
18	16	2	0	0.006536	0.006536
18	16	2	1	0.215686	0.209150
18	16	2	2	1.000000	0.784314
18	16	3	1	0.019608	0.019608
18	16	3	2	0.313725	0.294118
18	16	3	3	1.000000	0.686275
18	16	4	2	0.039216	0.039216
18	16	4	3	0.405229	0.366013
18	16	4	4	1.000000	0.594771
18	16	5	3	0.065359	0.065359
18	16	5	4	0.490196	0.424837
18	16	5	5	1.000000	0.509804
18	16	6	4	0.098039	0.098039
18	16	6	5	0.568627	0.470588
18	16	6	6	1.000000	0.431373
18	16	7	5	0.137255	0.137255
18	16	7	6	0.640523	0.503268
18	16	7	7	1.000000	0.359477
18	16	8	6	0.183007	0.183007
18	16	8	7	0.705882	0.522876
18	16	8	8	1.000000	0.294118
18	16	9	7	0.235294	0.235294
18	16	9	8	0.764706	0.529412
18	16	9	9	1.000000	0.235294
18	16	10	8	0.294118	0.294118

N	n	k	x	P(x)	p(x)
18	16	10	9	0.816993	0.522876
18	16	10	10	1.000000	0.183007
18	16	11	9	0.359477	0.359477
18	16	11	10	0.862745	0.503268
18	16	11	11	1.000000	0.137255
18	16	12	10	0.431373	0.431373
18	16	12	11	0.901961	0.470588
18	16	12	12	1.000000	0.098039
18	16	13	11	0.509804	0.509804
18	16	13	12	0.934640	0.424837
18	16	13	13	1.000000	0.065359
18	16	14	12	0.594771	0.594771
18	16	14	13	0.960784	0.366013
18	16	14	14	1.000000	0.039216
18	16	15	13	0.686275	0.686275
18	16	15	14	0.980392	0.294118
18	16	15	15	1.000000	0.019608
18	16	16	14	0.784314	0.784314
18	16	16	15	0.993464	0.209150
18	16	16	16	1.000000	0.006536
18	17	1	0	0.055556	0.055556
18	17	1	1	1.000000	0.944444
18	17	2	1	0.111111	0.111111
18	17	2	2	1.000000	0.888889
18	17	3	2	0.166667	0.166667
18	17	3	3	1.000000	0.833333
18	17	4	3	0.222222	0.222222
18	17	4	4	1.000000	0.777778
18	17	5	4	0.277778	0.277778
18	17	5	5	1.000000	0.722222
18	17	6	5	0.333333	0.333333
18	17	6	6	1.000000	0.666667
18	17	7	6	0.388889	0.388889
18	17	7	7	1.000000	0.611111
18	17	8	7	0.444444	0.444444
18	17	8	8	1.000000	0.555556
18	17	9	8	0.500000	0.500000
18	17	9	9	1.000000	0.500000
18	17	10	9	0.555556	0.555556
18	17	10	10	1.000000	0.444444
18	17	11	10	0.611111	0.611111
18	17	11	11	1.000000	0.388889
18	17	12	11	0.666667	0.666667
18	17	12	12	1.000000	0.333333
18	17	13	12	0.722222	0.722222
18	17	13	13	1.000000	0.277778
18	17	14	13	0.777778	0.777778
18	17	14	14	1.000000	0.222222
18	17	15	14	0.833333	0.833333
18	17	15	15	1.000000	0.166667

N	n	k	x	P(x)	p(x)
18	17	16	15	0.888889	0.888889
18	17	16	16	1.000000	0.111111
18	17	17	16	0.944444	0.944444
18	17	17	17	1.000000	0.055556
19	1	1	0	0.947368	0.947368
19	1	1	1	1.000000	0.052632
19	2	1	0	0.894737	0.894737
19	2	1	1	1.000000	0.105263
19	2	2	0	0.795322	0.795322
19	2	2	1	0.994152	0.198830
19	2	2	2	1.000000	0.005848
19	3	1	0	0.842105	0.842105
19	3	1	1	1.000000	0.157895
19	3	2	0	0.701754	0.701754
19	3	2	1	0.982456	0.280702
19	3	2	2	1.000000	0.017544
19	3	3	0	0.577915	0.577915
19	3	3	1	0.949432	0.371517
19	3	3	2	0.998968	0.049536
19	3	3	3	1.000000	0.001032
19	4	1	0	0.789474	0.789474
19	4	1	1	1.000000	0.210526
19	4	2	0	0.614035	0.614035
19	4	2	1	0.964912	0.350877
19	4	2	2	1.000000	0.035088
19	4	3	0	0.469556	0.469556
19	4	3	1	0.902993	0.433437
19	4	3	2	0.995872	0.092879
19	4	3	3	1.000000	0.004128
19	4	4	0	0.352167	0.352167
19	4	4	1	0.821723	0.469556
19	4	4	2	0.984262	0.162539
19	4	4	3	0.999742	0.015480
19	4	4	4	1.000000	0.000258
19	5	1	0	0.736842	0.736842
19	5	1	1	1.000000	0.263158
19	5	2	0	0.532164	0.532164
19	5	2	1	0.941520	0.409357
19	5	2	2	1.000000	0.058480
19	5	3	0	0.375645	0.375645
19	5	3	1	0.845201	0.469556
19	5	3	2	0.989680	0.144479
19	5	3	3	1.000000	0.010320
19	5	4	0	0.258256	0.258256
19	5	4	1	0.727812	0.469556
19	5	4	2	0.962590	0.234778
19	5	4	3	0.998710	0.036120
19	5	4	4	1.000000	0.001290
19	5	5	0	0.172171	0.172171
19	5	5	1	0.602597	0.430427

N	n	k	x	P(x)	p(x)
19	5	5	2	0.915635	0.313037
19	5	5	3	0.993894	0.078259
19	5	5	4	0.999914	0.006020
19	5	5	5	1.000000	0.000086
19	6	1	0	0.684211	0.684211
19	6	1	1	1.000000	0.315789
19	6	2	0	0.456140	0.456140
19	6	2	1	0.912281	0.456140
19	6	2	2	1.000000	0.087719
19	6	3	0	0.295150	0.295150
19	6	3	1	0.778122	0.482972
19	6	3	2	0.979360	0.201238
19	6	3	3	1.000000	0.020640
19	6	4	0	0.184469	0.184469
19	6	4	1	0.627193	0.442724
19	6	4	2	0.929051	0.301858
19	6	4	3	0.996130	0.067079
19	6	4	4	1.000000	0.003870
19	6	5	0	0.110681	0.110681
19	6	5	1	0.479618	0.368937
19	6	5	2	0.848555	0.368937
19	6	5	3	0.982714	0.134159
19	6	5	4	0.999484	0.016770
19	6	5	5	1.000000	0.000516
19	6	6	0	0.063246	0.063246
19	6	6	1	0.347855	0.284609
19	6	6	2	0.743145	0.395290
19	6	6	3	0.953966	0.210821
19	6	6	4	0.997088	0.043123
19	6	6	5	0.999963	0.002875
19	6	6	6	1.000000	0.000037
19	7	1	0	0.631579	0.631579
19	7	1	1	1.000000	0.368421
19	7	2	0	0.385965	0.385965
19	7	2	1	0.877193	0.491228
19	7	2	2	1.000000	0.122807
19	7	3	0	0.227038	0.227038
19	7	3	1	0.703818	0.476780
19	7	3	2	0.963880	0.260062
19	7	3	3	1.000000	0.036120
19	7	4	0	0.127709	0.127709
19	7	4	1	0.525026	0.397317
19	7	4	2	0.882611	0.357585
19	7	4	3	0.990970	0.108359
19	7	4	4	1.000000	0.009030
19	7	5	0	0.068111	0.068111
19	7	5	1	0.366099	0.297988
19	7	5	2	0.763416	0.397317
19	7	5	3	0.962074	0.198658
19	7	5	4	0.998194	0.036120

N	n	k	x	P(x)	p(x)
19	7	5	5	1.000000	0.001806
19	7	6	0	0.034056	0.034056
19	7	6	1	0.238390	0.204334
19	7	6	2	0.621517	0.383127
19	7	6	3	0.905315	0.283798
19	7	6	4	0.990454	0.085139
19	7	6	5	0.999742	0.009288
19	7	6	6	1.000000	0.000258
19	7	7	0	0.015718	0.015718
19	7	7	1	0.144082	0.128364
19	7	7	2	0.474160	0.330079
19	7	7	3	0.817992	0.343832
19	7	7	4	0.970806	0.152814
19	7	7	5	0.998313	0.027507
19	7	7	6	0.999980	0.001667
19	7	7	7	1.000000	0.000020
19	8	1	0	0.578947	0.578947
19	8	1	1	1.000000	0.421053
19	8	2	0	0.321637	0.321637
19	8	2	1	0.836257	0.514620
19	8	2	2	1.000000	0.163743
19	8	3	0	0.170279	0.170279
19	8	3	1	0.624355	0.454076
19	8	3	2	0.942208	0.317853
19	8	3	3	1.000000	0.057792
19	8	4	0	0.085139	0.085139
19	8	4	1	0.425697	0.340557
19	8	4	2	0.823013	0.397317
19	8	4	3	0.981940	0.158927
19	8	4	4	1.000000	0.018060
19	8	5	0	0.039732	0.039732
19	8	5	1	0.266770	0.227038
19	8	5	2	0.664087	0.397317
19	8	5	3	0.928965	0.264878
19	8	5	4	0.995184	0.066219
19	8	5	5	1.000000	0.004816
19	8	6	0	0.017028	0.017028
19	8	6	1	0.153251	0.136223
19	8	6	2	0.493808	0.340557
19	8	6	3	0.834365	0.340557
19	8	6	4	0.976264	0.141899
19	8	6	5	0.998968	0.022704
19	8	6	6	1.000000	0.001032
19	8	7	0	0.006549	0.006549
19	8	7	1	0.079900	0.073351
19	8	7	2	0.336628	0.256728
19	8	7	3	0.703382	0.366754
19	8	7	4	0.932603	0.229221
19	8	7	5	0.993729	0.061126
19	8	7	6	0.999841	0.006113

N	n	k	x	P(x)	p(x)
19	8	7	7	1.000000	0.000159
19	8	8	0	0.002183	0.002183
19	8	8	1	0.037112	0.034929
19	8	8	2	0.208264	0.171152
19	8	8	3	0.550568	0.342304
19	8	8	4	0.856196	0.305628
19	8	8	5	0.978447	0.122251
19	8	8	6	0.998822	0.020375
19	8	8	7	0.999987	0.001164
19	8	8	8	1.000000	0.000013
19	9	1	0	0.526316	0.526316
19	9	1	1	1.000000	0.473684
19	9	2	0	0.263158	0.263158
19	9	2	1	0.789474	0.526316
19	9	2	2	1.000000	0.210526
19	9	3	0	0.123839	0.123839
19	9	3	1	0.541796	0.417957
19	9	3	2	0.913313	0.371517
19	9	3	3	1.000000	0.086687
19	9	4	0	0.054180	0.054180
19	9	4	1	0.332817	0.278638
19	9	4	2	0.750774	0.417957
19	9	4	3	0.967492	0.216718
19	9	4	4	1.000000	0.032508
19	9	5	0	0.021672	0.021672
19	9	5	1	0.184210	0.162539
19	9	5	2	0.555728	0.371517
19	9	5	3	0.880805	0.325077
19	9	5	4	0.989164	0.108359
19	9	5	5	1.000000	0.010836
19	9	6	0	0.007740	0.007740
19	9	6	1	0.091331	0.083591
19	9	6	2	0.369969	0.278638
19	9	6	3	0.741486	0.371517
19	9	6	4	0.950464	0.208978
19	9	6	5	0.996904	0.046440
19	9	6	6	1.000000	0.003096
19	9	7	0	0.002382	0.002382
19	9	7	1	0.039890	0.037509
19	9	7	2	0.219933	0.180043
19	9	7	3	0.570017	0.350083
19	9	7	4	0.870088	0.300071
19	9	7	5	0.982615	0.112527
19	9	7	6	0.999285	0.016671
19	9	7	7	1.000000	0.000714
19	9	8	0	0.000595	0.000595
19	9	8	1	0.014884	0.014289
19	9	8	2	0.114908	0.100024
19	9	8	3	0.394975	0.280067
19	9	8	4	0.745058	0.350083
19	9	8	5	0.945106	0.200048
19	9	8	6	0.995118	0.050012
19	9	8	7	0.999881	0.004763
19	9	8	8	1.000000	0.000119
19	9	9	0	0.000108	0.000108
19	9	9	1	0.004492	0.004384
19	9	9	2	0.051257	0.046764
19	9	9	3	0.242211	0.190955
19	9	9	4	0.585929	0.343718
19	9	9	5	0.872361	0.286432
19	9	9	6	0.981478	0.109117
19	9	9	7	0.999015	0.017537
19	9	9	8	0.999989	0.000974
19	9	9	9	1.000000	0.000011
19	10	1	0	0.473684	0.473684
19	10	1	1	1.000000	0.526316
19	10	2	0	0.210526	0.210526
19	10	2	1	0.736842	0.526316
19	10	2	2	1.000000	0.263158
19	10	3	0	0.086687	0.086687
19	10	3	1	0.458204	0.371517
19	10	3	2	0.876161	0.417957
19	10	3	3	1.000000	0.123839
19	10	4	0	0.032508	0.032508
19	10	4	1	0.249226	0.216718
19	10	4	2	0.667183	0.417957
19	10	4	3	0.945820	0.278638
19	10	4	4	1.000000	0.054180
19	10	5	0	0.010836	0.010836
19	10	5	1	0.119195	0.108359
19	10	5	2	0.444272	0.325077
19	10	5	3	0.815789	0.371517
19	10	5	4	0.978328	0.162539
19	10	5	5	1.000000	0.021672
19	10	6	0	0.003096	0.003096
19	10	6	1	0.049536	0.046440
19	10	6	2	0.258514	0.208978
19	10	6	3	0.630031	0.371517
19	10	6	4	0.908669	0.278638
19	10	6	5	0.992260	0.083591
19	10	6	6	1.000000	0.007740
19	10	7	0	0.000714	0.000714
19	10	7	1	0.017385	0.016671
19	10	7	2	0.129912	0.112527
19	10	7	3	0.429983	0.300071
19	10	7	4	0.780067	0.350083
19	10	7	5	0.960109	0.180043
19	10	7	6	0.997618	0.037509
19	10	7	7	1.000000	0.002382
19	10	8	0	0.000119	0.000119
19	10	8	1	0.004882	0.004763
19	10	8	2	0.054894	0.050012
19	10	8	3	0.254942	0.200048
19	10	8	4	0.605025	0.350083
19	10	8	5	0.885092	0.280067
19	10	8	6	0.985115	0.100024
19	10	8	7	0.999405	0.014289
19	10	8	8	1.000000	0.000595
19	10	9	0	0.000011	0.000011
19	10	9	1	0.000985	0.000974
19	10	9	2	0.018522	0.017537
19	10	9	3	0.127639	0.109117
19	10	9	4	0.414070	0.286432
19	10	9	5	0.757789	0.343718
19	10	9	6	0.948743	0.190955
19	10	9	7	0.995507	0.046764
19	10	9	8	0.999892	0.004384
19	10	9	9	1.000000	0.000108
19	10	10	1	0.000108	0.000108
19	10	10	2	0.004492	0.004384
19	10	10	3	0.051257	0.046764
19	10	10	4	0.242211	0.190955
19	10	10	5	0.585929	0.343718
19	10	10	6	0.872361	0.286432
19	10	10	7	0.981478	0.109117
19	10	10	8	0.999015	0.017537
19	10	10	9	0.999989	0.000974
19	10	10	10	1.000000	0.000011
19	11	1	0	0.421053	0.421053
19	11	1	1	1.000000	0.578947
19	11	2	0	0.163743	0.163743
19	11	2	1	0.678363	0.514620
19	11	2	2	1.000000	0.321637
19	11	3	0	0.057792	0.057792
19	11	3	1	0.375645	0.317853
19	11	3	2	0.829721	0.454076
19	11	3	3	1.000000	0.170279
19	11	4	0	0.018060	0.018060
19	11	4	1	0.176987	0.158927
19	11	4	2	0.574303	0.397317
19	11	4	3	0.914861	0.340557
19	11	4	4	1.000000	0.085139
19	11	5	0	0.004816	0.004816
19	11	5	1	0.071035	0.066219
19	11	5	2	0.335913	0.264878
19	11	5	3	0.733230	0.397317
19	11	5	4	0.960268	0.227038
19	11	5	5	1.000000	0.039732
19	11	6	0	0.001032	0.001032
19	11	6	1	0.023736	0.022704
19	11	6	2	0.165635	0.141899
19	11	6	3	0.506192	0.340557
19	11	6	4	0.846749	0.340557
19	11	6	5	0.982972	0.136223
19	11	6	6	1.000000	0.017028
19	11	7	0	0.000159	0.000159
19	11	7	1	0.006271	0.006113
19	11	7	2	0.067397	0.061126
19	11	7	3	0.296618	0.229221
19	11	7	4	0.663372	0.366754
19	11	7	5	0.920100	0.256728
19	11	7	6	0.993451	0.073351
19	11	7	7	1.000000	0.006549
19	11	8	0	0.000013	0.000013
19	11	8	1	0.001178	0.001164
19	11	8	2	0.021553	0.020375
19	11	8	3	0.143804	0.122251
19	11	8	4	0.449432	0.305628
19	11	8	5	0.791736	0.342304
19	11	8	6	0.962888	0.171152
19	11	8	7	0.997817	0.034929
19	11	8	8	1.000000	0.002183
19	11	9	1	0.000119	0.000119
19	11	9	2	0.004882	0.004763
19	11	9	3	0.054894	0.050012
19	11	9	4	0.254942	0.200048
19	11	9	5	0.605025	0.350083
19	11	9	6	0.885092	0.280067
19	11	9	7	0.985115	0.100024
19	11	9	8	0.999405	0.014289
19	11	9	9	1.000000	0.000595
19	11	10	2	0.000595	0.000595
19	11	10	3	0.014884	0.014289
19	11	10	4	0.114908	0.100024
19	11	10	5	0.394975	0.280067
19	11	10	6	0.745058	0.350083
19	11	10	7	0.945106	0.200048
19	11	10	8	0.995118	0.050012
19	11	10	9	0.999881	0.004763
19	11	10	10	1.000000	0.000119
19	11	11	3	0.002183	0.002183
19	11	11	4	0.037112	0.034929
19	11	11	5	0.208264	0.171152
19	11	11	6	0.550568	0.342304
19	11	11	7	0.856196	0.305628
19	11	11	8	0.978447	0.122251
19	11	11	9	0.998822	0.020375
19	11	11	10	0.999987	0.001164
19	11	11	11	1.000000	0.000013
19	12	1	0	0.368421	0.368421

Table for $N = 2$, $n = 1$, through $N = 100$, $n = 50$

N	n	k	x	P(x)	p(x)
19	12	1	1	1.000000	0.631579
19	12	2	0	0.122807	0.122807
19	12	2	1	0.614035	0.491228
19	12	2	2	1.000000	0.385965
19	12	3	0	0.036120	0.036120
19	12	3	1	0.296182	0.260062
19	12	3	2	0.772962	0.476780
19	12	3	3	1.000000	0.227038
19	12	4	0	0.009030	0.009030
19	12	4	1	0.117389	0.108359
19	12	4	2	0.474974	0.357585
19	12	4	3	0.872291	0.397317
19	12	4	4	1.000000	0.127709
19	12	5	0	0.001806	0.001806
19	12	5	1	0.037926	0.036120
19	12	5	2	0.236584	0.198658
19	12	5	3	0.633901	0.397317
19	12	5	4	0.931888	0.297988
19	12	5	5	1.000000	0.068111
19	12	6	0	0.000258	0.000258
19	12	6	1	0.009546	0.009288
19	12	6	2	0.094685	0.085139
19	12	6	3	0.378483	0.283798
19	12	6	4	0.761610	0.383127
19	12	6	5	0.965944	0.204334
19	12	6	6	1.000000	0.034056
19	12	7	0	0.000020	0.000020
19	12	7	1	0.001687	0.001667
19	12	7	2	0.029193	0.027507
19	12	7	3	0.182008	0.152814
19	12	7	4	0.525839	0.343832
19	12	7	5	0.855918	0.330079
19	12	7	6	0.984282	0.128364
19	12	7	7	1.000000	0.015718
19	12	8	1	0.000159	0.000159
19	12	8	2	0.006271	0.006113
19	12	8	3	0.067397	0.061126
19	12	8	4	0.296618	0.229221
19	12	8	5	0.663372	0.366754
19	12	8	6	0.920100	0.256728
19	12	8	7	0.993451	0.073351
19	12	8	8	1.000000	0.006549
19	12	9	2	0.000714	0.000714
19	12	9	3	0.017385	0.016671
19	12	9	4	0.129912	0.112527
19	12	9	5	0.429983	0.300071
19	12	9	6	0.780067	0.350083
19	12	9	7	0.960109	0.180043
19	12	9	8	0.997618	0.037509
19	12	9	9	1.000000	0.002382

N	n	k	x	P(x)	p(x)
19	12	10	3	0.002382	0.002382
19	12	10	4	0.039890	0.037509
19	12	10	5	0.219933	0.180043
19	12	10	6	0.570017	0.350083
19	12	10	7	0.870088	0.300071
19	12	10	8	0.982615	0.112527
19	12	10	9	0.999285	0.016671
19	12	10	10	1.000000	0.000714
19	12	11	4	0.006549	0.006549
19	12	11	5	0.079900	0.073351
19	12	11	6	0.336628	0.256728
19	12	11	7	0.703382	0.366754
19	12	11	8	0.932603	0.229221
19	12	11	9	0.993729	0.061126
19	12	11	10	0.999841	0.006113
19	12	11	11	1.000000	0.000159
19	12	12	5	0.015718	0.015718
19	12	12	6	0.144082	0.128364
19	12	12	7	0.474160	0.330079
19	12	12	8	0.817992	0.343832
19	12	12	9	0.970806	0.152814
19	12	12	10	0.998313	0.027507
19	12	12	11	0.999980	0.001667
19	12	12	12	1.000000	0.000020
19	13	1	0	0.315789	0.315789
19	13	1	1	1.000000	0.684211
19	13	2	0	0.087719	0.087719
19	13	2	1	0.543860	0.456140
19	13	2	2	1.000000	0.456140
19	13	3	0	0.020640	0.020640
19	13	3	1	0.221878	0.201238
19	13	3	2	0.704850	0.482972
19	13	3	3	1.000000	0.295150
19	13	4	0	0.003870	0.003870
19	13	4	1	0.070949	0.067079
19	13	4	2	0.372807	0.301858
19	13	4	3	0.815531	0.442724
19	13	4	4	1.000000	0.184469
19	13	5	0	0.000516	0.000516
19	13	5	1	0.017286	0.016770
19	13	5	2	0.151445	0.134159
19	13	5	3	0.520382	0.368937
19	13	5	4	0.889319	0.368937
19	13	5	5	1.000000	0.110681
19	13	6	0	0.000037	0.000037
19	13	6	1	0.002912	0.002875
19	13	6	2	0.046034	0.043123
19	13	6	3	0.256855	0.210821
19	13	6	4	0.652145	0.395290
19	13	6	5	0.936754	0.284609

N	n	k	x	P(x)	p(x)
19	13	6	6	1.000000	0.063246
19	13	7	1	0.000258	0.000258
19	13	7	2	0.009546	0.009288
19	13	7	3	0.094685	0.085139
19	13	7	4	0.378483	0.283798
19	13	7	5	0.761610	0.383127
19	13	7	6	0.965944	0.204334
19	13	7	7	1.000000	0.034056
19	13	8	2	0.001032	0.001032
19	13	8	3	0.023736	0.022704
19	13	8	4	0.165635	0.141899
19	13	8	5	0.506192	0.340557
19	13	8	6	0.846749	0.340557
19	13	8	7	0.982972	0.136223
19	13	8	8	1.000000	0.017028
19	13	9	3	0.003096	0.003096
19	13	9	4	0.049536	0.046440
19	13	9	5	0.258514	0.208978
19	13	9	6	0.630031	0.371517
19	13	9	7	0.908669	0.278638
19	13	9	8	0.992260	0.083591
19	13	9	9	1.000000	0.007740
19	13	10	4	0.007740	0.007740
19	13	10	5	0.091331	0.083591
19	13	10	6	0.369969	0.278638
19	13	10	7	0.741486	0.371517
19	13	10	8	0.950464	0.208978
19	13	10	9	0.996904	0.046440
19	13	10	10	1.000000	0.003096
19	13	11	5	0.017028	0.017028
19	13	11	6	0.153251	0.136223
19	13	11	7	0.493808	0.340557
19	13	11	8	0.834365	0.340557
19	13	11	9	0.976264	0.141899
19	13	11	10	0.998968	0.022704
19	13	11	11	1.000000	0.001032
19	13	12	6	0.034056	0.034056
19	13	12	7	0.238390	0.204334
19	13	12	8	0.621517	0.383127
19	13	12	9	0.905315	0.283798
19	13	12	10	0.990454	0.085139
19	13	12	11	0.999742	0.009288
19	13	12	12	1.000000	0.000258
19	13	13	7	0.063246	0.063246
19	13	13	8	0.347855	0.284609
19	13	13	9	0.743145	0.395290
19	13	13	10	0.953966	0.210821
19	13	13	11	0.997088	0.043123
19	13	13	12	0.999963	0.002875
19	13	13	13	1.000000	0.000037

N	n	k	x	P(x)	p(x)
19	14	1	0	0.263158	0.263158
19	14	1	1	1.000000	0.736842
19	14	2	0	0.058480	0.058480
19	14	2	1	0.467836	0.409357
19	14	2	2	1.000000	0.532164
19	14	3	0	0.010320	0.010320
19	14	3	1	0.154799	0.144479
19	14	3	2	0.624355	0.469556
19	14	3	3	1.000000	0.375645
19	14	4	0	0.001290	0.001290
19	14	4	1	0.037410	0.036120
19	14	4	2	0.272188	0.234778
19	14	4	3	0.741744	0.469556
19	14	4	4	1.000000	0.258256
19	14	5	0	0.000086	0.000086
19	14	5	1	0.006106	0.006020
19	14	5	2	0.084365	0.078259
19	14	5	3	0.397403	0.313037
19	14	5	4	0.827829	0.430427
19	14	5	5	1.000000	0.172171
19	14	6	1	0.000516	0.000516
19	14	6	2	0.017286	0.016770
19	14	6	3	0.151445	0.134159
19	14	6	4	0.520382	0.368937
19	14	6	5	0.889319	0.368937
19	14	6	6	1.000000	0.110681
19	14	7	2	0.001806	0.001806
19	14	7	3	0.037926	0.036120
19	14	7	4	0.236584	0.198658
19	14	7	5	0.633901	0.397317
19	14	7	6	0.931888	0.297988
19	14	7	7	1.000000	0.068111
19	14	8	3	0.004816	0.004816
19	14	8	4	0.071035	0.066219
19	14	8	5	0.335913	0.264878
19	14	8	6	0.733230	0.397317
19	14	8	7	0.960268	0.227038
19	14	8	8	1.000000	0.039732
19	14	9	4	0.010836	0.010836
19	14	9	5	0.119195	0.108359
19	14	9	6	0.444272	0.325077
19	14	9	7	0.815789	0.371517
19	14	9	8	0.978328	0.162539
19	14	9	9	1.000000	0.021672
19	14	10	5	0.021672	0.021672
19	14	10	6	0.184210	0.162539
19	14	10	7	0.555728	0.371517
19	14	10	8	0.880805	0.325077
19	14	10	9	0.989164	0.108359
19	14	10	10	1.000000	0.010836

$N = 19$ $n = 12$–14

Table for $N = 2$, $n = 1$, through $N = 100$, $n = 50$

N	n	k	x	P(x)	p(x)
19	14	11	6	0.039732	0.039732
19	14	11	7	0.266770	0.227038
19	14	11	8	0.664087	0.397317
19	14	11	9	0.928965	0.264878
19	14	11	10	0.995184	0.066219
19	14	11	11	1.000000	0.004816
19	14	12	7	0.068111	0.068111
19	14	12	8	0.366099	0.297988
19	14	12	9	0.763416	0.397317
19	14	12	10	0.962074	0.198658
19	14	12	11	0.998194	0.036120
19	14	12	12	1.000000	0.001806
19	14	13	8	0.110681	0.110681
19	14	13	9	0.479618	0.368937
19	14	13	10	0.848555	0.368937
19	14	13	11	0.982714	0.134159
19	14	13	12	0.999484	0.016770
19	14	13	13	1.000000	0.000516
19	14	14	9	0.172171	0.172171
19	14	14	10	0.602597	0.430427
19	14	14	11	0.915635	0.313037
19	14	14	12	0.993894	0.078259
19	14	14	13	0.999914	0.006020
19	14	14	14	1.000000	0.000086
19	15	1	0	0.210526	0.210526
19	15	1	1	1.000000	0.789474
19	15	2	0	0.035088	0.035088
19	15	2	1	0.385965	0.350877
19	15	2	2	1.000000	0.614035
19	15	3	0	0.004128	0.004128
19	15	3	1	0.097007	0.092879
19	15	3	2	0.530444	0.433437
19	15	3	3	1.000000	0.469556
19	15	4	0	0.000258	0.000258
19	15	4	1	0.015738	0.015480
19	15	4	2	0.178277	0.162539
19	15	4	3	0.647833	0.469556
19	15	4	4	1.000000	0.352167
19	15	5	1	0.001290	0.001290
19	15	5	2	0.037410	0.036120
19	15	5	3	0.272188	0.234778
19	15	5	4	0.741744	0.469556
19	15	5	5	1.000000	0.258256
19	15	6	2	0.003870	0.003870
19	15	6	3	0.070949	0.067079
19	15	6	4	0.372807	0.301858
19	15	6	5	0.815531	0.442724
19	15	6	6	1.000000	0.184469
19	15	7	3	0.009030	0.009030
19	15	7	4	0.117389	0.108359

N	n	k	x	P(x)	p(x)
19	15	7	5	0.474974	0.357585
19	15	7	6	0.872291	0.397317
19	15	7	7	1.000000	0.127709
19	15	8	4	0.018060	0.018060
19	15	8	5	0.176987	0.158927
19	15	8	6	0.574303	0.397317
19	15	8	7	0.914861	0.340557
19	15	8	8	1.000000	0.085139
19	15	9	5	0.032508	0.032508
19	15	9	6	0.249226	0.216718
19	15	9	7	0.667183	0.417957
19	15	9	8	0.945820	0.278638
19	15	9	9	1.000000	0.054180
19	15	10	6	0.054180	0.054180
19	15	10	7	0.332817	0.278638
19	15	10	8	0.750774	0.417957
19	15	10	9	0.967492	0.216718
19	15	10	10	1.000000	0.032508
19	15	11	7	0.085139	0.085139
19	15	11	8	0.425697	0.340557
19	15	11	9	0.823013	0.397317
19	15	11	10	0.981940	0.158927
19	15	11	11	1.000000	0.018060
19	15	12	8	0.127709	0.127709
19	15	12	9	0.525026	0.397317
19	15	12	10	0.882611	0.357585
19	15	12	11	0.990970	0.108359
19	15	12	12	1.000000	0.009030
19	15	13	9	0.184469	0.184469
19	15	13	10	0.627193	0.442724
19	15	13	11	0.929051	0.301858
19	15	13	12	0.996130	0.067079
19	15	13	13	1.000000	0.003870
19	15	14	10	0.258256	0.258256
19	15	14	11	0.727812	0.469556
19	15	14	12	0.962590	0.234778
19	15	14	13	0.998710	0.036120
19	15	14	14	1.000000	0.001290
19	15	15	11	0.352167	0.352167
19	15	15	12	0.821723	0.469556
19	15	15	13	0.984262	0.162539
19	15	15	14	0.999742	0.015480
19	15	15	15	1.000000	0.000258
19	16	1	0	0.157895	0.157895
19	16	1	1	1.000000	0.842105
19	16	2	0	0.017544	0.017544
19	16	2	1	0.298246	0.280702
19	16	2	2	1.000000	0.701754
19	16	3	0	0.001032	0.001032
19	16	3	1	0.050568	0.049536

N	n	k	x	P(x)	p(x)
19	16	3	2	0.422085	0.371517
19	16	3	3	1.000000	0.577915
19	16	4	1	0.004128	0.004128
19	16	4	2	0.097007	0.092879
19	16	4	3	0.530444	0.433437
19	16	4	4	1.000000	0.469556
19	16	5	2	0.010320	0.010320
19	16	5	3	0.154799	0.144479
19	16	5	4	0.624355	0.469556
19	16	5	5	1.000000	0.375645
19	16	6	3	0.020640	0.020640
19	16	6	4	0.221878	0.201238
19	16	6	5	0.704850	0.482972
19	16	6	6	1.000000	0.295150
19	16	7	4	0.036120	0.036120
19	16	7	5	0.296182	0.260062
19	16	7	6	0.772962	0.476780
19	16	7	7	1.000000	0.227038
19	16	8	5	0.057792	0.057792
19	16	8	6	0.375645	0.317853
19	16	8	7	0.829721	0.454076
19	16	8	8	1.000000	0.170279
19	16	9	6	0.086687	0.086687
19	16	9	7	0.458204	0.371517
19	16	9	8	0.876161	0.417957
19	16	9	9	1.000000	0.123839
19	16	10	7	0.123839	0.123839
19	16	10	8	0.541796	0.417957
19	16	10	9	0.913313	0.371517
19	16	10	10	1.000000	0.086687
19	16	11	8	0.170279	0.170279
19	16	11	9	0.624355	0.454076
19	16	11	10	0.942208	0.317853
19	16	11	11	1.000000	0.057792
19	16	12	9	0.227038	0.227038
19	16	12	10	0.703818	0.476780
19	16	12	11	0.963880	0.260062
19	16	12	12	1.000000	0.036120
19	16	13	10	0.295150	0.295150
19	16	13	11	0.778122	0.482972
19	16	13	12	0.979360	0.201238
19	16	13	13	1.000000	0.020640
19	16	14	11	0.375645	0.375645
19	16	14	12	0.845201	0.469556
19	16	14	13	0.989680	0.144479
19	16	14	14	1.000000	0.010320
19	16	15	12	0.469556	0.469556
19	16	15	13	0.902993	0.433437
19	16	15	14	0.995872	0.092879
19	16	15	15	1.000000	0.004128

N	n	k	x	P(x)	p(x)
19	16	16	13	0.577915	0.577915
19	16	16	14	0.949432	0.371517
19	16	16	15	0.998968	0.049536
19	16	16	16	1.000000	0.001032
19	17	1	0	0.105263	0.105263
19	17	1	1	1.000000	0.894737
19	17	2	0	0.005848	0.005848
19	17	2	1	0.204678	0.198830
19	17	2	2	1.000000	0.795322
19	17	3	1	0.017544	0.017544
19	17	3	2	0.298246	0.280702
19	17	3	3	1.000000	0.701754
19	17	4	2	0.035088	0.035088
19	17	4	3	0.385965	0.350877
19	17	4	4	1.000000	0.614035
19	17	5	3	0.058480	0.058480
19	17	5	4	0.467836	0.409357
19	17	5	5	1.000000	0.532164
19	17	6	4	0.087719	0.087719
19	17	6	5	0.543860	0.456140
19	17	6	6	1.000000	0.456140
19	17	7	5	0.122807	0.122807
19	17	7	6	0.614035	0.491228
19	17	7	7	1.000000	0.385965
19	17	8	6	0.163743	0.163743
19	17	8	7	0.678363	0.514620
19	17	8	8	1.000000	0.321637
19	17	9	7	0.210526	0.210526
19	17	9	8	0.736842	0.526316
19	17	9	9	1.000000	0.263158
19	17	10	8	0.263158	0.263158
19	17	10	9	0.789474	0.526316
19	17	10	10	1.000000	0.210526
19	17	11	9	0.321637	0.321637
19	17	11	10	0.836257	0.514620
19	17	11	11	1.000000	0.163743
19	17	12	10	0.385965	0.385965
19	17	12	11	0.877193	0.491228
19	17	12	12	1.000000	0.122807
19	17	13	11	0.456140	0.456140
19	17	13	12	0.912281	0.456140
19	17	13	13	1.000000	0.087719
19	17	14	12	0.532164	0.532164
19	17	14	13	0.941520	0.409357
19	17	14	14	1.000000	0.058480
19	17	15	13	0.614035	0.614035
19	17	15	14	0.964912	0.350877
19	17	15	15	1.000000	0.035088
19	17	16	14	0.701754	0.701754
19	17	16	15	0.982456	0.280702

Table for $N = 2$, $n = 1$, through $N = 100$, $n = 50$

N	n	k	x	P(x)	p(x)
19	17	16	16	1.000000	0.017544
19	17	17	15	0.795322	0.795322
19	17	17	16	0.994152	0.198830
19	17	17	17	1.000000	0.005848
19	18	1	0	0.052632	0.052632
19	18	1	1	1.000000	0.947368
19	18	2	1	0.105263	0.105263
19	18	2	2	1.000000	0.894737
19	18	3	2	0.157895	0.157895
19	18	3	3	1.000000	0.842105
19	18	4	3	0.210526	0.210526
19	18	4	4	1.000000	0.789474
19	18	5	4	0.263158	0.263158
19	18	5	5	1.000000	0.736842
19	18	6	5	0.315789	0.315789
19	18	6	6	1.000000	0.684211
19	18	7	6	0.368421	0.368421
19	18	7	7	1.000000	0.631579
19	18	8	7	0.421053	0.421053
19	18	8	8	1.000000	0.578947
19	18	9	8	0.473684	0.473684
19	18	9	9	1.000000	0.526316
19	18	10	9	0.526316	0.526316
19	18	10	10	1.000000	0.473684
19	18	11	10	0.578947	0.578947
19	18	11	11	1.000000	0.421053
19	18	12	11	0.631579	0.631579
19	18	12	12	1.000000	0.368421
19	18	13	12	0.684211	0.684211
19	18	13	13	1.000000	0.315789
19	18	14	13	0.736842	0.736842
19	18	14	14	1.000000	0.263158
19	18	15	14	0.789474	0.789474
19	18	15	15	1.000000	0.210526
19	18	16	15	0.842105	0.842105
19	18	16	16	1.000000	0.157895
19	18	17	16	0.894737	0.894737
19	18	17	17	1.000000	0.105263
19	18	18	17	0.947368	0.947368
19	18	18	18	1.000000	0.052632
20	1	1	0	0.950000	0.950000
20	1	1	1	1.000000	0.050000
20	2	1	0	0.900000	0.900000
20	2	1	1	1.000000	0.100000
20	2	2	0	0.805263	0.805263
20	2	2	1	0.994737	0.189474
20	2	2	2	1.000000	0.005263
20	3	1	0	0.850000	0.850000
20	3	1	1	1.000000	0.150000
20	3	2	0	0.715789	0.715789
20	3	2	1	0.984210	0.268421
20	3	2	2	1.000000	0.015789
20	3	3	0	0.596491	0.596491
20	3	3	1	0.954386	0.357895
20	3	3	2	0.999123	0.044737
20	3	3	3	1.000000	0.000877
20	4	1	0	0.800000	0.800000
20	4	1	1	1.000000	0.200000
20	4	2	0	0.631579	0.631579
20	4	2	1	0.968421	0.336842
20	4	2	2	1.000000	0.031579
20	4	3	0	0.491228	0.491228
20	4	3	1	0.912281	0.421053
20	4	3	2	0.996491	0.084211
20	4	3	3	1.000000	0.003509
20	4	4	0	0.375645	0.375645
20	4	4	1	0.837977	0.462332
20	4	4	2	0.986584	0.148607
20	4	4	3	0.999794	0.013209
20	4	4	4	1.000000	0.000206
20	5	1	0	0.750000	0.750000
20	5	1	1	1.000000	0.250000
20	5	2	0	0.552632	0.552632
20	5	2	1	0.947368	0.394737
20	5	2	2	1.000000	0.052632
20	5	3	0	0.399123	0.399123
20	5	3	1	0.859649	0.460526
20	5	3	2	0.991228	0.131579
20	5	3	3	1.000000	0.008772
20	5	4	0	0.281734	0.281734
20	5	4	1	0.751290	0.469556
20	5	4	2	0.968008	0.216718
20	5	4	3	0.998968	0.030960
20	5	4	4	1.000000	0.001032
20	5	5	0	0.193692	0.193692
20	5	5	1	0.633901	0.440209
20	5	5	2	0.927374	0.293473
20	5	5	3	0.995098	0.067724
20	5	5	4	0.999935	0.004837
20	5	5	5	1.000000	0.000064
20	6	1	0	0.700000	0.700000
20	6	1	1	1.000000	0.300000
20	6	2	0	0.478947	0.478947
20	6	2	1	0.921053	0.442105
20	6	2	2	1.000000	0.078947
20	6	3	0	0.319298	0.319298
20	6	3	1	0.798246	0.478947
20	6	3	2	0.982456	0.184211
20	6	3	3	1.000000	0.017544
20	6	4	0	0.206605	0.206605
20	6	4	1	0.657379	0.450774
20	6	4	2	0.939112	0.281734
20	6	4	3	0.996904	0.057792
20	6	4	4	1.000000	0.003096
20	6	5	0	0.129128	0.129128
20	6	5	1	0.516512	0.387384
20	6	5	2	0.868679	0.352167
20	6	5	3	0.986068	0.117389
20	6	5	4	0.999613	0.013545
20	6	5	5	1.000000	0.000387
20	6	6	0	0.077477	0.077477
20	6	6	1	0.387384	0.309907
20	6	6	2	0.774768	0.387384
20	6	6	3	0.962590	0.187822
20	6	6	4	0.997807	0.035217
20	6	6	5	0.999974	0.002167
20	6	6	6	1.000000	0.000026
20	7	1	0	0.650000	0.650000
20	7	1	1	1.000000	0.350000
20	7	2	0	0.410526	0.410526
20	7	2	1	0.889474	0.478947
20	7	2	2	1.000000	0.110526
20	7	3	0	0.250877	0.250877
20	7	3	1	0.729825	0.478947
20	7	3	2	0.969298	0.239474
20	7	3	3	1.000000	0.030702
20	7	4	0	0.147575	0.147575
20	7	4	1	0.560784	0.413209
20	7	4	2	0.898865	0.338080
20	7	4	3	0.992776	0.093911
20	7	4	4	1.000000	0.007224
20	7	5	0	0.083011	0.083011
20	7	5	1	0.405831	0.322820
20	7	5	2	0.793215	0.387384
20	7	5	3	0.969298	0.176084
20	7	5	4	0.998645	0.029347
20	7	5	5	1.000000	0.001354
20	7	6	0	0.044272	0.044272
20	7	6	1	0.276703	0.232430
20	7	6	2	0.664087	0.387384
20	7	6	3	0.922343	0.258256
20	7	6	4	0.992776	0.070433
20	7	6	5	0.999819	0.007043
20	7	6	6	1.000000	0.000181
20	7	7	0	0.022136	0.022136
20	7	7	1	0.177090	0.154954
20	7	7	2	0.525735	0.348645
20	7	7	3	0.848555	0.322820
20	7	7	4	0.977683	0.129128
20	7	7	5	0.998813	0.021130
20	7	7	6	0.999987	0.001174
20	7	7	7	1.000000	0.000013
20	8	1	0	0.600000	0.600000
20	8	1	1	1.000000	0.400000
20	8	2	0	0.347368	0.347368
20	8	2	1	0.852632	0.505263
20	8	2	2	1.000000	0.147368
20	8	3	0	0.192982	0.192982
20	8	3	1	0.656140	0.463158
20	8	3	2	0.950877	0.294737
20	8	3	3	1.000000	0.049123
20	8	4	0	0.102167	0.102167
20	8	4	1	0.465428	0.363261
20	8	4	2	0.846852	0.381424
20	8	4	3	0.985552	0.138700
20	8	4	4	1.000000	0.014448
20	8	5	0	0.051084	0.051084
20	8	5	1	0.306502	0.255418
20	8	5	2	0.703818	0.397317
20	8	5	3	0.942208	0.238390
20	8	5	4	0.996388	0.054180
20	8	5	5	1.000000	0.003612
20	8	6	0	0.023839	0.023839
20	8	6	1	0.187306	0.163467
20	8	6	2	0.544892	0.357585
20	8	6	3	0.862745	0.317853
20	8	6	4	0.981940	0.119195
20	8	6	5	0.999278	0.017337
20	8	6	6	1.000000	0.000722
20	8	7	0	0.010217	0.010217
20	8	7	1	0.105573	0.095356
20	8	7	2	0.391641	0.286068
20	8	7	3	0.749226	0.357585
20	8	7	4	0.947884	0.198658
20	8	7	5	0.995562	0.047678
20	8	7	6	0.999897	0.004334
20	8	7	7	1.000000	0.000103
20	8	8	0	0.003930	0.003930
20	8	8	1	0.054227	0.050298
20	8	8	2	0.259609	0.205382
20	8	8	3	0.611693	0.352084
20	8	8	4	0.886759	0.275065
20	8	8	5	0.984560	0.097801
20	8	8	6	0.999230	0.014670
20	8	8	7	0.999992	0.000762
20	8	8	8	1.000000	0.000008
20	9	1	0	0.550000	0.550000
20	9	1	1	1.000000	0.450000
20	9	2	0	0.289474	0.289474
20	9	2	1	0.810526	0.521053

Table for $N=2$, $n=1$, through $N=100$, $n=50$

N	n	k	x	P(x)	p(x)
20	9	2	2	1.000000	0.189474
20	9	3	0	0.144737	0.144737
20	9	3	1	0.578947	0.434210
20	9	3	2	0.926316	0.347368
20	9	3	3	1.000000	0.073684
20	9	4	0	0.068111	0.068111
20	9	4	1	0.374613	0.306502
20	9	4	2	0.783282	0.408669
20	9	4	3	0.973994	0.190712
20	9	4	4	1.000000	0.026006
20	9	5	0	0.029799	0.029799
20	9	5	1	0.221362	0.191563
20	9	5	2	0.604489	0.383127
20	9	5	3	0.902477	0.297988
20	9	5	4	0.991873	0.089396
20	9	5	5	1.000000	0.008127
20	9	6	0	0.011919	0.011919
20	9	6	1	0.119195	0.107276
20	9	6	2	0.425697	0.306502
20	9	6	3	0.783282	0.357585
20	9	6	4	0.962074	0.178793
20	9	6	5	0.997833	0.035759
20	9	6	6	1.000000	0.002167
20	9	7	0	0.004257	0.004257
20	9	7	1	0.057895	0.053638
20	9	7	2	0.272446	0.214551
20	9	7	3	0.630031	0.357585
20	9	7	4	0.898220	0.268189
20	9	7	5	0.987616	0.089396
20	9	7	6	0.999536	0.011919
20	9	7	7	1.000000	0.000464
20	9	8	0	0.001310	0.001310
20	9	8	1	0.024887	0.023577
20	9	8	2	0.156918	0.132031
20	9	8	3	0.464992	0.308073
20	9	8	4	0.795070	0.330079
20	9	8	5	0.960109	0.165039
20	9	8	6	0.996785	0.036675
20	9	8	7	0.999928	0.003144
20	9	8	8	1.000000	0.000071
20	9	9	0	0.000327	0.000327
20	9	9	1	0.009169	0.008841
20	9	9	2	0.079900	0.070731
20	9	9	3	0.310955	0.231055
20	9	9	4	0.657537	0.346583
20	9	9	5	0.905096	0.247559
20	9	9	6	0.987616	0.082520
20	9	9	7	0.999405	0.011789
20	9	9	8	0.999994	0.000589
20	9	9	9	1.000000	0.000006

N	n	k	x	P(x)	p(x)
20	10	1	0	0.500000	0.500000
20	10	1	1	1.000000	0.500000
20	10	2	0	0.236842	0.236842
20	10	2	1	0.763158	0.526316
20	10	2	2	1.000000	0.236842
20	10	3	0	0.105263	0.105263
20	10	3	1	0.500000	0.394737
20	10	3	2	0.894737	0.394737
20	10	3	3	1.000000	0.105263
20	10	4	0	0.043344	0.043344
20	10	4	1	0.291022	0.247678
20	10	4	2	0.708978	0.417957
20	10	4	3	0.956656	0.247678
20	10	4	4	1.000000	0.043344
20	10	5	0	0.016254	0.016254
20	10	5	1	0.151703	0.135449
20	10	5	2	0.500000	0.348297
20	10	5	3	0.848297	0.348297
20	10	5	4	0.983746	0.135449
20	10	5	5	1.000000	0.016254
20	10	6	0	0.005418	0.005418
20	10	6	1	0.070433	0.065015
20	10	6	2	0.314241	0.243808
20	10	6	3	0.685758	0.371517
20	10	6	4	0.929567	0.243808
20	10	6	5	0.994582	0.065015
20	10	6	6	1.000000	0.005418
20	10	7	0	0.001548	0.001548
20	10	7	1	0.028638	0.027090
20	10	7	2	0.174923	0.146285
20	10	7	3	0.500000	0.325077
20	10	7	4	0.825077	0.325077
20	10	7	5	0.971362	0.146285
20	10	7	6	0.998452	0.027090
20	10	7	7	1.000000	0.001548
20	10	8	0	0.000357	0.000357
20	10	8	1	0.009883	0.009526
20	10	8	2	0.084901	0.075018
20	10	8	3	0.324958	0.240057
20	10	8	4	0.675042	0.350083
20	10	8	5	0.915099	0.240057
20	10	8	6	0.990117	0.075018
20	10	8	7	0.999643	0.009526
20	10	8	8	1.000000	0.000357
20	10	9	0	0.000060	0.000060
20	10	9	1	0.002739	0.002679
20	10	9	2	0.034889	0.032150
20	10	9	3	0.184925	0.150036
20	10	9	4	0.500000	0.315075
20	10	9	5	0.815075	0.315075

N	n	k	x	P(x)	p(x)
20	10	9	6	0.965111	0.150036
20	10	9	7	0.997261	0.032150
20	10	9	8	0.999940	0.002679
20	10	9	9	1.000000	0.000060
20	10	10	0	0.000005	0.000005
20	10	10	1	0.000547	0.000541
20	10	10	2	0.011507	0.010960
20	10	10	3	0.089448	0.077941
20	10	10	4	0.328141	0.238693
20	10	10	5	0.671859	0.343718
20	10	10	6	0.910552	0.238693
20	10	10	7	0.988493	0.077941
20	10	10	8	0.999453	0.010960
20	10	10	9	0.999994	0.000541
20	10	10	10	1.000000	0.000005
20	11	1	0	0.450000	0.450000
20	11	1	1	1.000000	0.550000
20	11	2	0	0.189474	0.189474
20	11	2	1	0.710526	0.521053
20	11	2	2	1.000000	0.289474
20	11	3	0	0.073684	0.073684
20	11	3	1	0.421053	0.347368
20	11	3	2	0.855263	0.434210
20	11	3	3	1.000000	0.144737
20	11	4	0	0.026006	0.026006
20	11	4	1	0.216718	0.190712
20	11	4	2	0.625387	0.408669
20	11	4	3	0.931888	0.306502
20	11	4	4	1.000000	0.068111
20	11	5	0	0.008127	0.008127
20	11	5	1	0.097523	0.089396
20	11	5	2	0.395511	0.297988
20	11	5	3	0.778638	0.383127
20	11	5	4	0.970201	0.191563
20	11	5	5	1.000000	0.029799
20	11	6	0	0.002167	0.002167
20	11	6	1	0.037926	0.035759
20	11	6	2	0.216718	0.178793
20	11	6	3	0.574303	0.357585
20	11	6	4	0.880805	0.306502
20	11	6	5	0.988080	0.107276
20	11	6	6	1.000000	0.011919
20	11	7	0	0.000464	0.000464
20	11	7	1	0.012384	0.011919
20	11	7	2	0.101780	0.089396
20	11	7	3	0.369969	0.268189
20	11	7	4	0.727554	0.357585
20	11	7	5	0.942105	0.214551
20	11	7	6	0.995743	0.053638
20	11	7	7	1.000000	0.004257

N	n	k	x	P(x)	p(x)
20	11	8	0	0.000071	0.000071
20	11	8	1	0.003215	0.003144
20	11	8	2	0.039890	0.036675
20	11	8	3	0.204930	0.165039
20	11	8	4	0.535008	0.330079
20	11	8	5	0.843082	0.308073
20	11	8	6	0.975113	0.132031
20	11	8	7	0.998690	0.023577
20	11	8	8	1.000000	0.001310
20	11	9	0	0.000006	0.000006
20	11	9	1	0.000595	0.000589
20	11	9	2	0.012384	0.011789
20	11	9	3	0.094904	0.082520
20	11	9	4	0.342462	0.247559
20	11	9	5	0.689045	0.346583
20	11	9	6	0.920100	0.231055
20	11	9	7	0.990831	0.070731
20	11	9	8	0.999672	0.008841
20	11	9	9	1.000000	0.000327
20	11	10	1	0.000060	0.000060
20	11	10	2	0.002739	0.002679
20	11	10	3	0.034889	0.032150
20	11	10	4	0.184925	0.150036
20	11	10	5	0.500000	0.315075
20	11	10	6	0.815075	0.315075
20	11	10	7	0.965111	0.150036
20	11	10	8	0.997261	0.032150
20	11	10	9	0.999940	0.002679
20	11	10	10	1.000000	0.000060
20	11	11	2	0.000327	0.000327
20	11	11	3	0.009169	0.008841
20	11	11	4	0.079900	0.070731
20	11	11	5	0.310955	0.231055
20	11	11	6	0.657537	0.346583
20	11	11	7	0.905096	0.247559
20	11	11	8	0.987616	0.082520
20	11	11	9	0.999405	0.011789
20	11	11	10	0.999994	0.000589
20	11	11	11	1.000000	0.000006
20	12	1	0	0.400000	0.400000
20	12	1	1	1.000000	0.600000
20	12	2	0	0.147368	0.147368
20	12	2	1	0.652632	0.505263
20	12	2	2	1.000000	0.347368
20	12	3	0	0.049123	0.049123
20	12	3	1	0.343860	0.294737
20	12	3	2	0.807017	0.463158
20	12	3	3	1.000000	0.192982
20	12	4	0	0.014448	0.014448
20	12	4	1	0.153148	0.138700

Table for $N = 2$, $n = 1$, through $N = 100$, $n = 50$

N	n	k	x	P(x)	p(x)
20	12	4	2	0.534572	0.381424
20	12	4	3	0.897833	0.363261
20	12	4	4	1.000000	0.102167
20	12	5	0	0.003612	0.003612
20	12	5	1	0.057792	0.054180
20	12	5	2	0.296182	0.238390
20	12	5	3	0.693498	0.397317
20	12	5	4	0.948916	0.255418
20	12	5	5	1.000000	0.051084
20	12	6	0	0.000722	0.000722
20	12	6	1	0.018060	0.017337
20	12	6	2	0.137255	0.119195
20	12	6	3	0.455108	0.317853
20	12	6	4	0.812693	0.357585
20	12	6	5	0.976161	0.163467
20	12	6	6	1.000000	0.023839
20	12	7	0	0.000103	0.000103
20	12	7	1	0.004438	0.004334
20	12	7	2	0.052116	0.047678
20	12	7	3	0.250774	0.198658
20	12	7	4	0.608359	0.357585
20	12	7	5	0.894427	0.286068
20	12	7	6	0.989783	0.095356
20	12	7	7	1.000000	0.010217
20	12	8	0	0.000008	0.000008
20	12	8	1	0.000770	0.000762
20	12	8	2	0.015440	0.014670
20	12	8	3	0.113241	0.097801
20	12	8	4	0.388307	0.275065
20	12	8	5	0.740391	0.352084
20	12	8	6	0.945773	0.205382
20	12	8	7	0.996070	0.050298
20	12	8	8	1.000000	0.003930
20	12	9	1	0.000071	0.000071
20	12	9	2	0.003215	0.003144
20	12	9	3	0.039890	0.036675
20	12	9	4	0.204930	0.165039
20	12	9	5	0.535008	0.330079
20	12	9	6	0.843082	0.308073
20	12	9	7	0.975113	0.132031
20	12	9	8	0.998690	0.023577
20	12	9	9	1.000000	0.001310
20	12	10	2	0.000357	0.000357
20	12	10	3	0.009883	0.009526
20	12	10	4	0.084901	0.075018
20	12	10	5	0.324958	0.240057
20	12	10	6	0.675042	0.350083
20	12	10	7	0.915099	0.240057
20	12	10	8	0.990117	0.075018
20	12	10	9	0.999643	0.009526
20	12	10	10	1.000000	0.000357
20	12	11	3	0.001310	0.001310
20	12	11	4	0.024887	0.023577
20	12	11	5	0.156918	0.132031
20	12	11	6	0.464992	0.308073
20	12	11	7	0.795070	0.330079
20	12	11	8	0.960109	0.165039
20	12	11	9	0.996785	0.036675
20	12	11	10	0.999928	0.003144
20	12	11	11	1.000000	0.000071
20	12	12	4	0.003930	0.003930
20	12	12	5	0.054227	0.050298
20	12	12	6	0.259609	0.205382
20	12	12	7	0.611693	0.352084
20	12	12	8	0.886759	0.275065
20	12	12	9	0.984560	0.097801
20	12	12	10	0.999230	0.014670
20	12	12	11	0.999992	0.000762
20	12	12	12	1.000000	0.000008
20	13	1	0	0.350000	0.350000
20	13	1	1	1.000000	0.650000
20	13	2	0	0.110526	0.110526
20	13	2	1	0.589474	0.478947
20	13	2	2	1.000000	0.410526
20	13	3	0	0.030702	0.030702
20	13	3	1	0.270175	0.239474
20	13	3	2	0.749123	0.478947
20	13	3	3	1.000000	0.250877
20	13	4	0	0.007224	0.007224
20	13	4	1	0.101135	0.093911
20	13	4	2	0.439216	0.338080
20	13	4	3	0.852425	0.413209
20	13	4	4	1.000000	0.147575
20	13	5	0	0.001354	0.001354
20	13	5	1	0.030702	0.029347
20	13	5	2	0.206785	0.176084
20	13	5	3	0.594169	0.387384
20	13	5	4	0.916989	0.322820
20	13	5	5	1.000000	0.083011
20	13	6	0	0.000181	0.000181
20	13	6	1	0.007224	0.007043
20	13	6	2	0.077657	0.070433
20	13	6	3	0.335913	0.258256
20	13	6	4	0.723297	0.387384
20	13	6	5	0.955727	0.232430
20	13	6	6	1.000000	0.044272
20	13	7	0	0.000013	0.000013
20	13	7	1	0.001187	0.001174
20	13	7	2	0.022317	0.021130
20	13	7	3	0.151445	0.129128
20	13	7	4	0.474265	0.322820
20	13	7	5	0.822910	0.348645
20	13	7	6	0.977864	0.154954
20	13	7	7	1.000000	0.022136
20	13	8	1	0.000103	0.000103
20	13	8	2	0.004438	0.004334
20	13	8	3	0.052116	0.047678
20	13	8	4	0.250774	0.198658
20	13	8	5	0.608359	0.357585
20	13	8	6	0.894427	0.286068
20	13	8	7	0.989783	0.095356
20	13	8	8	1.000000	0.010217
20	13	9	2	0.000464	0.000464
20	13	9	3	0.012384	0.011919
20	13	9	4	0.101780	0.089396
20	13	9	5	0.369969	0.268189
20	13	9	6	0.727554	0.357585
20	13	9	7	0.942105	0.214551
20	13	9	8	0.995743	0.053638
20	13	9	9	1.000000	0.004257
20	13	10	3	0.001548	0.001548
20	13	10	4	0.028638	0.027090
20	13	10	5	0.174923	0.146285
20	13	10	6	0.500000	0.325077
20	13	10	7	0.825077	0.325077
20	13	10	8	0.971362	0.146285
20	13	10	9	0.998452	0.027090
20	13	10	10	1.000000	0.001548
20	13	11	4	0.004257	0.004257
20	13	11	5	0.057895	0.053638
20	13	11	6	0.272446	0.214551
20	13	11	7	0.630031	0.357585
20	13	11	8	0.898220	0.268189
20	13	11	9	0.987616	0.089396
20	13	11	10	0.999536	0.011919
20	13	11	11	1.000000	0.000464
20	13	12	5	0.010217	0.010217
20	13	12	6	0.105573	0.095356
20	13	12	7	0.391641	0.286068
20	13	12	8	0.749226	0.357585
20	13	12	9	0.947884	0.198658
20	13	12	10	0.995562	0.047678
20	13	12	11	0.999897	0.004334
20	13	12	12	1.000000	0.000103
20	13	13	6	0.022136	0.022136
20	13	13	7	0.177090	0.154954
20	13	13	8	0.525735	0.348645
20	13	13	9	0.848555	0.322820
20	13	13	10	0.977683	0.129128
20	13	13	11	0.998813	0.021130
20	13	13	12	0.999987	0.001174
20	13	13	13	1.000000	0.000013
20	14	1	0	0.300000	0.300000
20	14	1	1	1.000000	0.700000
20	14	2	0	0.078947	0.078947
20	14	2	1	0.521053	0.442105
20	14	2	2	1.000000	0.478947
20	14	3	0	0.017544	0.017544
20	14	3	1	0.201754	0.184211
20	14	3	2	0.680702	0.478947
20	14	3	3	1.000000	0.319298
20	14	4	0	0.003096	0.003096
20	14	4	1	0.060888	0.057792
20	14	4	2	0.342621	0.281734
20	14	4	3	0.793395	0.450774
20	14	4	4	1.000000	0.206605
20	14	5	0	0.000387	0.000387
20	14	5	1	0.013932	0.013545
20	14	5	2	0.131321	0.117389
20	14	5	3	0.483488	0.352167
20	14	5	4	0.870872	0.387384
20	14	5	5	1.000000	0.129128
20	14	6	0	0.000026	0.000026
20	14	6	1	0.002193	0.002167
20	14	6	2	0.037410	0.035217
20	14	6	3	0.225232	0.187822
20	14	6	4	0.612616	0.387384
20	14	6	5	0.922523	0.309907
20	14	6	6	1.000000	0.077477
20	14	7	1	0.000181	0.000181
20	14	7	2	0.007224	0.007043
20	14	7	3	0.077657	0.070433
20	14	7	4	0.335913	0.258256
20	14	7	5	0.723297	0.387384
20	14	7	6	0.955727	0.232430
20	14	7	7	1.000000	0.044272
20	14	8	2	0.000722	0.000722
20	14	8	3	0.018060	0.017337
20	14	8	4	0.137255	0.119195
20	14	8	5	0.455108	0.317853
20	14	8	6	0.812693	0.357585
20	14	8	7	0.976161	0.163467
20	14	8	8	1.000000	0.023839
20	14	9	3	0.002167	0.002167
20	14	9	4	0.037926	0.035759
20	14	9	5	0.216718	0.178793
20	14	9	6	0.574303	0.357585
20	14	9	7	0.880805	0.306502
20	14	9	8	0.988080	0.107276
20	14	9	9	1.000000	0.011919

Table for $N=2$, $n=1$, through $N=100$, $n=50$

N	n	k	x	P(x)	p(x)
20	14	10	4	0.005418	0.005418
20	14	10	5	0.070433	0.065015
20	14	10	6	0.314241	0.243808
20	14	10	7	0.685758	0.371517
20	14	10	8	0.929567	0.243808
20	14	10	9	0.994582	0.065015
20	14	10	10	1.000000	0.005418
20	14	11	5	0.011919	0.011919
20	14	11	6	0.119195	0.107276
20	14	11	7	0.425697	0.306502
20	14	11	8	0.783282	0.357585
20	14	11	9	0.962074	0.178793
20	14	11	10	0.997833	0.035759
20	14	11	11	1.000000	0.002167
20	14	12	6	0.023839	0.023839
20	14	12	7	0.187306	0.163467
20	14	12	8	0.544892	0.357585
20	14	12	9	0.862745	0.317853
20	14	12	10	0.981940	0.119195
20	14	12	11	0.999278	0.017337
20	14	12	12	1.000000	0.000722
20	14	13	7	0.044272	0.044272
20	14	13	8	0.276703	0.232430
20	14	13	9	0.664087	0.387384
20	14	13	10	0.922343	0.258256
20	14	13	11	0.992776	0.070433
20	14	13	12	0.999819	0.007043
20	14	13	13	1.000000	0.000181
20	14	14	8	0.077477	0.077477
20	14	14	9	0.387384	0.309907
20	14	14	10	0.774768	0.387384
20	14	14	11	0.962590	0.187822
20	14	14	12	0.997807	0.035217
20	14	14	13	0.999974	0.002167
20	14	14	14	1.000000	0.000026
20	15	1	0	0.250000	0.250000
20	15	1	1	1.000000	0.750000
20	15	2	0	0.052632	0.052632
20	15	2	1	0.447368	0.394737
20	15	2	2	1.000000	0.552632
20	15	3	0	0.008772	0.008772
20	15	3	1	0.140351	0.131579
20	15	3	2	0.600877	0.460526
20	15	3	3	1.000000	0.399123
20	15	4	0	0.001032	0.001032
20	15	4	1	0.031992	0.030960
20	15	4	2	0.248710	0.216718
20	15	4	3	0.718266	0.469556
20	15	4	4	1.000000	0.281734
20	15	5	0	0.000064	0.000064

N	n	k	x	P(x)	p(x)
20	15	5	1	0.004902	0.004837
20	15	5	2	0.072626	0.067724
20	15	5	3	0.366099	0.293473
20	15	5	4	0.806308	0.440209
20	15	5	5	1.000000	0.193692
20	15	6	1	0.000387	0.000387
20	15	6	2	0.013932	0.013545
20	15	6	3	0.131321	0.117389
20	15	6	4	0.483488	0.352167
20	15	6	5	0.870872	0.387384
20	15	6	6	1.000000	0.129128
20	15	7	2	0.001354	0.001354
20	15	7	3	0.030702	0.029347
20	15	7	4	0.206785	0.176084
20	15	7	5	0.594169	0.387384
20	15	7	6	0.916989	0.322820
20	15	7	7	1.000000	0.083011
20	15	8	3	0.003612	0.003612
20	15	8	4	0.057792	0.054180
20	15	8	5	0.296182	0.238390
20	15	8	6	0.693498	0.397317
20	15	8	7	0.948916	0.255418
20	15	8	8	1.000000	0.051084
20	15	9	4	0.008127	0.008127
20	15	9	5	0.097523	0.089396
20	15	9	6	0.395511	0.297988
20	15	9	7	0.778638	0.383127
20	15	9	8	0.970201	0.191563
20	15	9	9	1.000000	0.029799
20	15	10	5	0.016254	0.016254
20	15	10	6	0.151703	0.135449
20	15	10	7	0.500000	0.348297
20	15	10	8	0.848297	0.348297
20	15	10	9	0.983746	0.135449
20	15	10	10	1.000000	0.016254
20	15	11	6	0.029799	0.029799
20	15	11	7	0.221362	0.191563
20	15	11	8	0.604489	0.383127
20	15	11	9	0.902477	0.297988
20	15	11	10	0.991873	0.089396
20	15	11	11	1.000000	0.008127
20	15	12	7	0.051084	0.051084
20	15	12	8	0.306502	0.255418
20	15	12	9	0.703818	0.397317
20	15	12	10	0.942208	0.238390
20	15	12	11	0.996388	0.054180
20	15	12	12	1.000000	0.003612
20	15	13	8	0.083011	0.083011
20	15	13	9	0.405831	0.322820
20	15	13	10	0.793215	0.387384

N	n	k	x	P(x)	p(x)
20	15	13	11	0.969298	0.176084
20	15	13	12	0.998645	0.029347
20	15	13	13	1.000000	0.001354
20	15	14	9	0.129128	0.129128
20	15	14	10	0.516512	0.387384
20	15	14	11	0.868679	0.352167
20	15	14	12	0.986068	0.117389
20	15	14	13	0.999613	0.013545
20	15	14	14	1.000000	0.000387
20	15	15	10	0.193692	0.193692
20	15	15	11	0.633901	0.440209
20	15	15	12	0.927374	0.293473
20	15	15	13	0.995098	0.067724
20	15	15	14	0.999935	0.004837
20	15	15	15	1.000000	0.000064
20	16	1	0	0.200000	0.200000
20	16	1	1	1.000000	0.800000
20	16	2	0	0.031579	0.031579
20	16	2	1	0.368421	0.336842
20	16	2	2	1.000000	0.631579
20	16	3	0	0.003509	0.003509
20	16	3	1	0.087719	0.084211
20	16	3	2	0.508772	0.421053
20	16	3	3	1.000000	0.491228
20	16	4	0	0.000206	0.000206
20	16	4	1	0.013416	0.013209
20	16	4	2	0.162023	0.148607
20	16	4	3	0.624355	0.462332
20	16	4	4	1.000000	0.375645
20	16	5	1	0.001032	0.001032
20	16	5	2	0.031992	0.030960
20	16	5	3	0.248710	0.216718
20	16	5	4	0.718266	0.469556
20	16	5	5	1.000000	0.281734
20	16	6	2	0.003096	0.003096
20	16	6	3	0.060888	0.057792
20	16	6	4	0.342621	0.281734
20	16	6	5	0.793395	0.450774
20	16	6	6	1.000000	0.206605
20	16	7	3	0.007224	0.007224
20	16	7	4	0.101135	0.093911
20	16	7	5	0.439216	0.338080
20	16	7	6	0.852425	0.413209
20	16	7	7	1.000000	0.147575
20	16	8	4	0.014448	0.014448
20	16	8	5	0.153148	0.138700
20	16	8	6	0.534572	0.381424
20	16	8	7	0.897833	0.363261
20	16	8	8	1.000000	0.102167
20	16	9	5	0.026006	0.026006

N	n	k	x	P(x)	p(x)
20	16	9	6	0.216718	0.190712
20	16	9	7	0.625387	0.408669
20	16	9	8	0.931888	0.306502
20	16	9	9	1.000000	0.068111
20	16	10	6	0.043344	0.043344
20	16	10	7	0.291022	0.247678
20	16	10	8	0.708978	0.417957
20	16	10	9	0.956656	0.247678
20	16	10	10	1.000000	0.043344
20	16	11	7	0.068111	0.068111
20	16	11	8	0.374613	0.306502
20	16	11	9	0.783282	0.408669
20	16	11	10	0.973994	0.190712
20	16	11	11	1.000000	0.026006
20	16	12	8	0.102167	0.102167
20	16	12	9	0.465428	0.363261
20	16	12	10	0.846852	0.381424
20	16	12	11	0.985552	0.138700
20	16	12	12	1.000000	0.014448
20	16	13	9	0.147575	0.147575
20	16	13	10	0.560784	0.413209
20	16	13	11	0.898865	0.338080
20	16	13	12	0.992776	0.093911
20	16	13	13	1.000000	0.007224
20	16	14	10	0.206605	0.206605
20	16	14	11	0.657379	0.450774
20	16	14	12	0.939112	0.281734
20	16	14	13	0.996904	0.057792
20	16	14	14	1.000000	0.003096
20	16	15	11	0.281734	0.281734
20	16	15	12	0.751290	0.469556
20	16	15	13	0.968008	0.216718
20	16	15	14	0.998968	0.030960
20	16	15	15	1.000000	0.001032
20	16	16	12	0.375645	0.375645
20	16	16	13	0.837977	0.462332
20	16	16	14	0.986584	0.148607
20	16	16	15	0.999794	0.013209
20	16	16	16	1.000000	0.000206
20	17	1	0	0.150000	0.150000
20	17	1	1	1.000000	0.850000
20	17	2	0	0.015789	0.015789
20	17	2	1	0.284210	0.268421
20	17	2	2	1.000000	0.715789
20	17	3	0	0.000877	0.000877
20	17	3	1	0.045614	0.044737
20	17	3	2	0.403509	0.357895
20	17	3	3	1.000000	0.596491
20	17	4	1	0.003509	0.003509
20	17	4	2	0.087719	0.084211

Table for $N = 2$, $n = 1$, through $N = 100$, $n = 50$

N	n	k	x	P(x)	p(x)
20	17	4	3	0.508772	0.421053
20	17	4	4	1.000000	0.491228
20	17	5	2	0.008772	0.008772
20	17	5	3	0.140351	0.131579
20	17	5	4	0.600877	0.460526
20	17	5	5	1.000000	0.399123
20	17	6	3	0.017544	0.017544
20	17	6	4	0.201754	0.184211
20	17	6	5	0.680702	0.478947
20	17	6	6	1.000000	0.319298
20	17	7	4	0.030702	0.030702
20	17	7	5	0.270175	0.239474
20	17	7	6	0.749123	0.478947
20	17	7	7	1.000000	0.250877
20	17	8	5	0.049123	0.049123
20	17	8	6	0.343860	0.294737
20	17	8	7	0.807017	0.463158
20	17	8	8	1.000000	0.192982
20	17	9	6	0.073684	0.073684
20	17	9	7	0.421053	0.347368
20	17	9	8	0.855263	0.434210
20	17	9	9	1.000000	0.144737
20	17	10	7	0.105263	0.105263
20	17	10	8	0.500000	0.394737
20	17	10	9	0.894737	0.394737
20	17	10	10	1.000000	0.105263
20	17	11	8	0.144737	0.144737
20	17	11	9	0.578947	0.434210
20	17	11	10	0.926316	0.347368
20	17	11	11	1.000000	0.073684
20	17	12	9	0.192982	0.192982
20	17	12	10	0.656140	0.463158
20	17	12	11	0.950877	0.294737
20	17	12	12	1.000000	0.049123
20	17	13	10	0.250877	0.250877
20	17	13	11	0.729825	0.478947
20	17	13	12	0.969298	0.239474
20	17	13	13	1.000000	0.030702
20	17	14	11	0.319298	0.319298
20	17	14	12	0.798246	0.478947
20	17	14	13	0.982456	0.184211
20	17	14	14	1.000000	0.017544
20	17	15	12	0.399123	0.399123
20	17	15	13	0.859649	0.460526
20	17	15	14	0.991228	0.131579
20	17	15	15	1.000000	0.008772
20	17	16	13	0.491228	0.491228
20	17	16	14	0.912281	0.421053
20	17	16	15	0.996491	0.084211
20	17	16	16	1.000000	0.003509

N	n	k	x	P(x)	p(x)
20	17	17	14	0.596491	0.596491
20	17	17	15	0.954386	0.357895
20	17	17	16	0.999123	0.044737
20	17	17	17	1.000000	0.000877
20	18	1	0	0.100000	0.100000
20	18	1	1	1.000000	0.900000
20	18	2	0	0.005263	0.005263
20	18	2	1	0.194737	0.189474
20	18	2	2	1.000000	0.805263
20	18	3	1	0.015789	0.015789
20	18	3	2	0.284210	0.268421
20	18	3	3	1.000000	0.715789
20	18	4	2	0.031579	0.031579
20	18	4	3	0.368421	0.336842
20	18	4	4	1.000000	0.631579
20	18	5	3	0.052632	0.052632
20	18	5	4	0.447368	0.394737
20	18	5	5	1.000000	0.552632
20	18	6	4	0.078947	0.078947
20	18	6	5	0.521053	0.442105
20	18	6	6	1.000000	0.478947
20	18	7	5	0.110526	0.110526
20	18	7	6	0.589474	0.478947
20	18	7	7	1.000000	0.410526
20	18	8	6	0.147368	0.147368
20	18	8	7	0.652632	0.505263
20	18	8	8	1.000000	0.347368
20	18	9	7	0.189474	0.189474
20	18	9	8	0.710526	0.521053
20	18	9	9	1.000000	0.289474
20	18	10	8	0.236842	0.236842
20	18	10	9	0.763158	0.526316
20	18	10	10	1.000000	0.236842
20	18	11	9	0.289474	0.289474
20	18	11	10	0.810526	0.521053
20	18	11	11	1.000000	0.189474
20	18	12	10	0.347368	0.347368
20	18	12	11	0.852632	0.505263
20	18	12	12	1.000000	0.147368
20	18	13	11	0.410526	0.410526
20	18	13	12	0.889474	0.478947
20	18	13	13	1.000000	0.110526
20	18	14	12	0.478947	0.478947
20	18	14	13	0.921053	0.442105
20	18	14	14	1.000000	0.078947
20	18	15	13	0.552632	0.552632
20	18	15	14	0.947368	0.394737
20	18	15	15	1.000000	0.052632
20	18	16	14	0.631579	0.631579
20	18	16	15	0.968421	0.336842

N	n	k	x	P(x)	p(x)
20	18	16	16	1.000000	0.031579
20	18	17	15	0.715789	0.715789
20	18	17	16	0.984210	0.268421
20	18	17	17	1.000000	0.015789
20	18	18	16	0.805263	0.805263
20	18	18	17	0.994737	0.189474
20	18	18	18	1.000000	0.005263
20	19	1	0	0.050000	0.050000
20	19	1	1	1.000000	0.950000
20	19	2	1	0.100000	0.100000
20	19	2	2	1.000000	0.900000
20	19	3	2	0.150000	0.150000
20	19	3	3	1.000000	0.850000
20	19	4	3	0.200000	0.200000
20	19	4	4	1.000000	0.800000
20	19	5	4	0.250000	0.250000
20	19	5	5	1.000000	0.750000
20	19	6	5	0.300000	0.300000
20	19	6	6	1.000000	0.700000
20	19	7	6	0.350000	0.350000
20	19	7	7	1.000000	0.650000
20	19	8	7	0.400000	0.400000
20	19	8	8	1.000000	0.600000
20	19	9	8	0.450000	0.450000
20	19	9	9	1.000000	0.550000
20	19	10	9	0.500000	0.500000
20	19	10	10	1.000000	0.500000
20	19	11	10	0.550000	0.550000
20	19	11	11	1.000000	0.450000
20	19	12	11	0.600000	0.600000
20	19	12	12	1.000000	0.400000
20	19	13	12	0.650000	0.650000
20	19	13	13	1.000000	0.350000
20	19	14	13	0.700000	0.700000
20	19	14	14	1.000000	0.300000
20	19	15	14	0.750000	0.750000
20	19	15	15	1.000000	0.250000
20	19	16	15	0.800000	0.800000
20	19	16	16	1.000000	0.200000
20	19	17	16	0.850000	0.850000
20	19	17	17	1.000000	0.150000
20	19	18	17	0.900000	0.900000
20	19	18	18	1.000000	0.100000
20	19	19	18	0.950000	0.950000
20	19	19	19	1.000000	0.050000
21	1	1	0	0.952381	0.952381
21	1	1	1	1.000000	0.047619
21	2	1	0	0.904762	0.904762
21	2	1	1	1.000000	0.095238
21	2	2	0	0.814286	0.814286

N	n	k	x	P(x)	p(x)
21	2	2	1	0.995238	0.180952
21	2	2	2	1.000000	0.004762
21	3	1	0	0.857143	0.857143
21	3	1	1	1.000000	0.142857
21	3	2	0	0.728571	0.728571
21	3	2	1	0.985714	0.257143
21	3	2	2	1.000000	0.014286
21	3	3	0	0.613534	0.613534
21	3	3	1	0.958647	0.345113
21	3	3	2	0.999248	0.040601
21	3	3	3	1.000000	0.000752
21	4	1	0	0.809524	0.809524
21	4	1	1	1.000000	0.190476
21	4	2	0	0.647619	0.647619
21	4	2	1	0.971429	0.323810
21	4	2	2	1.000000	0.028571
21	4	3	0	0.511278	0.511278
21	4	3	1	0.920301	0.409023
21	4	3	2	0.996992	0.076692
21	4	3	3	1.000000	0.003008
21	4	4	0	0.397661	0.397661
21	4	4	1	0.852130	0.454470
21	4	4	2	0.988471	0.136341
21	4	4	3	0.999833	0.011362
21	4	4	4	1.000000	0.000167
21	5	1	0	0.761905	0.761905
21	5	1	1	1.000000	0.238095
21	5	2	0	0.571429	0.571429
21	5	2	1	0.952381	0.380952
21	5	2	2	1.000000	0.047619
21	5	3	0	0.421053	0.421053
21	5	3	1	0.872180	0.451128
21	5	3	2	0.992481	0.120301
21	5	3	3	1.000000	0.007519
21	5	4	0	0.304094	0.304094
21	5	4	1	0.771930	0.467836
21	5	4	2	0.972431	0.200501
21	5	4	3	0.999165	0.026733
21	5	4	4	1.000000	0.000835
21	5	5	0	0.214654	0.214654
21	5	5	1	0.661851	0.447196
21	5	5	2	0.937048	0.275198
21	5	5	3	0.996019	0.058971
21	5	5	4	0.999951	0.003931
21	5	5	5	1.000000	0.000049
21	6	1	0	0.714286	0.714286
21	6	1	1	1.000000	0.285714
21	6	2	0	0.500000	0.500000
21	6	2	1	0.928571	0.428571
21	6	2	2	1.000000	0.071429

N	n	k	x	P(x)	p(x)
21	6	3	0	0.342105	0.342105
21	6	3	1	0.815789	0.473684
21	6	3	2	0.984962	0.169173
21	6	3	3	1.000000	0.015038
21	6	4	0	0.228070	0.228070
21	6	4	1	0.684211	0.456140
21	6	4	2	0.947368	0.263158
21	6	4	3	0.997494	0.050125
21	6	4	4	1.000000	0.002506
21	6	5	0	0.147575	0.147575
21	6	5	1	0.550052	0.402477
21	6	5	2	0.885449	0.335397
21	6	5	3	0.988648	0.103199
21	6	5	4	0.999705	0.011057
21	6	5	5	1.000000	0.000295
21	6	6	0	0.092234	0.092234
21	6	6	1	0.424278	0.332043
21	6	6	2	0.801600	0.377322
21	6	6	3	0.969298	0.167699
21	6	6	4	0.998323	0.029025
21	6	6	5	0.999981	0.001659
21	6	6	6	1.000000	0.000018
21	7	1	0	0.666667	0.666667
21	7	1	1	1.000000	0.333333
21	7	2	0	0.433333	0.433333
21	7	2	1	0.900000	0.466667
21	7	2	2	1.000000	0.100000
21	7	3	0	0.273684	0.273684
21	7	3	1	0.752632	0.478947
21	7	3	2	0.973684	0.221053
21	7	3	3	1.000000	0.026316
21	7	4	0	0.167251	0.167251
21	7	4	1	0.592982	0.425731
21	7	4	2	0.912281	0.319298
21	7	4	3	0.994152	0.081871
21	7	4	4	1.000000	0.005848
21	7	5	0	0.098383	0.098383
21	7	5	1	0.442724	0.344341
21	7	5	2	0.818369	0.375645
21	7	5	3	0.974888	0.156519
21	7	5	4	0.998968	0.024080
21	7	5	5	1.000000	0.001032
21	7	6	0	0.055341	0.055341
21	7	6	1	0.313596	0.258256
21	7	6	2	0.700980	0.387384
21	7	6	3	0.935758	0.234778
21	7	6	4	0.994453	0.058695
21	7	6	5	0.999871	0.005418
21	7	6	6	1.000000	0.000129
21	7	7	0	0.029515	0.029515

N	n	k	x	P(x)	p(x)
21	7	7	1	0.210294	0.180779
21	7	7	2	0.571852	0.361558
21	7	7	3	0.873151	0.301299
21	7	7	4	0.982714	0.109563
21	7	7	5	0.999149	0.016434
21	7	7	6	0.999991	0.000843
21	7	7	7	1.000000	0.000009
21	8	1	0	0.619048	0.619048
21	8	1	1	1.000000	0.380952
21	8	2	0	0.371429	0.371429
21	8	2	1	0.866667	0.495238
21	8	2	2	1.000000	0.133333
21	8	3	0	0.215038	0.215038
21	8	3	1	0.684210	0.469173
21	8	3	2	0.957895	0.273684
21	8	3	3	1.000000	0.042105
21	8	4	0	0.119465	0.119465
21	8	4	1	0.501754	0.382289
21	8	4	2	0.866667	0.364912
21	8	4	3	0.988304	0.121637
21	8	4	4	1.000000	0.011696
21	8	5	0	0.063246	0.063246
21	8	5	1	0.344341	0.281095
21	8	5	2	0.737874	0.393533
21	8	5	3	0.952528	0.214654
21	8	5	4	0.997248	0.044720
21	8	5	5	1.000000	0.002752
21	8	6	0	0.031623	0.031623
21	8	6	1	0.221362	0.189739
21	8	6	2	0.590299	0.368937
21	8	6	3	0.885449	0.295150
21	8	6	4	0.986068	0.100619
21	8	6	5	0.999484	0.013416
21	8	6	6	1.000000	0.000516
21	8	7	0	0.014757	0.014757
21	8	7	1	0.132817	0.118060
21	8	7	2	0.442724	0.309907
21	8	7	3	0.787066	0.344341
21	8	7	4	0.959236	0.172171
21	8	7	5	0.996801	0.037564
21	8	7	6	0.999931	0.003130
21	8	7	7	1.000000	0.000069
21	8	8	0	0.006325	0.006325
21	8	8	1	0.073787	0.067463
21	8	8	2	0.309907	0.236120
21	8	8	3	0.664087	0.354180
21	8	8	4	0.910045	0.245958
21	8	8	5	0.988751	0.078707
21	8	8	6	0.999484	0.010733
21	8	8	7	0.999995	0.000511

N	n	k	x	P(x)	p(x)
21	8	8	8	1.000000	0.000005
21	9	1	0	0.571429	0.571429
21	9	1	1	1.000000	0.428571
21	9	2	0	0.314286	0.314286
21	9	2	1	0.828571	0.514286
21	9	2	2	1.000000	0.171429
21	9	3	0	0.165414	0.165414
21	9	3	1	0.612030	0.446617
21	9	3	2	0.936842	0.324812
21	9	3	3	1.000000	0.063158
21	9	4	0	0.082707	0.082707
21	9	4	1	0.413534	0.330827
21	9	4	2	0.810526	0.396992
21	9	4	3	0.978947	0.168421
21	9	4	4	1.000000	0.021053
21	9	5	0	0.038921	0.038921
21	9	5	1	0.257850	0.218930
21	9	5	2	0.647059	0.389208
21	9	5	3	0.919505	0.272446
21	9	5	4	0.993808	0.074303
21	9	5	5	1.000000	0.006192
21	9	6	0	0.017028	0.017028
21	9	6	1	0.148386	0.131358
21	9	6	2	0.476780	0.328394
21	9	6	3	0.817337	0.340557
21	9	6	4	0.970588	0.153251
21	9	6	5	0.998452	0.027864
21	9	6	6	1.000000	0.001548
21	9	7	0	0.006811	0.006811
21	9	7	1	0.078328	0.071517
21	9	7	2	0.323529	0.245201
21	9	7	3	0.681115	0.357585
21	9	7	4	0.919505	0.238390
21	9	7	5	0.991022	0.071517
21	9	7	6	0.999690	0.008669
21	9	7	7	1.000000	0.000310
21	9	8	0	0.002433	0.002433
21	9	8	1	0.037461	0.035029
21	9	8	2	0.200929	0.163467
21	9	8	3	0.527864	0.326935
21	9	8	4	0.834365	0.306502
21	9	8	5	0.970588	0.136223
21	9	8	6	0.997833	0.027245
21	9	8	7	0.999956	0.002123
21	9	8	8	1.000000	0.000044
21	9	9	0	0.000748	0.000748
21	9	9	1	0.015905	0.015157
21	9	9	2	0.112908	0.097003
21	9	9	3	0.376971	0.264063
21	9	9	4	0.716480	0.339509

N	n	k	x	P(x)	p(x)
21	9	9	5	0.928673	0.212193
21	9	9	6	0.991545	0.062872
21	9	9	7	0.999629	0.008084
21	9	9	8	0.999996	0.000367
21	9	9	9	1.000000	0.000003
21	10	1	0	0.523810	0.523810
21	10	1	1	1.000000	0.476190
21	10	2	0	0.261905	0.261905
21	10	2	1	0.785714	0.523810
21	10	2	2	1.000000	0.214286
21	10	3	0	0.124060	0.124060
21	10	3	1	0.537594	0.413534
21	10	3	2	0.909774	0.372180
21	10	3	3	1.000000	0.090226
21	10	4	0	0.055138	0.055138
21	10	4	1	0.330827	0.275689
21	10	4	2	0.744361	0.413534
21	10	4	3	0.964912	0.220551
21	10	4	4	1.000000	0.035088
21	10	5	0	0.022704	0.022704
21	10	5	1	0.184874	0.162170
21	10	5	2	0.549757	0.364883
21	10	5	3	0.874097	0.324340
21	10	5	4	0.987616	0.113519
21	10	5	5	1.000000	0.012384
21	10	6	0	0.008514	0.008514
21	10	6	1	0.093653	0.085139
21	10	6	2	0.367315	0.273662
21	10	6	3	0.732198	0.364883
21	10	6	4	0.945046	0.212848
21	10	6	5	0.996130	0.051084
21	10	6	6	1.000000	0.003870
21	10	7	0	0.002838	0.002838
21	10	7	1	0.042570	0.039732
21	10	7	2	0.221362	0.178793
21	10	7	3	0.561919	0.340557
21	10	7	4	0.859907	0.297988
21	10	7	5	0.979102	0.119195
21	10	7	6	0.998968	0.019866
21	10	7	7	1.000000	0.001032
21	10	8	0	0.000811	0.000811
21	10	8	1	0.017028	0.016217
21	10	8	2	0.119195	0.102167
21	10	8	3	0.391641	0.272446
21	10	8	4	0.732198	0.340557
21	10	8	5	0.936532	0.204334
21	10	8	6	0.993292	0.056760
21	10	8	7	0.999779	0.006487
21	10	8	8	1.000000	0.000221
21	10	9	0	0.000187	0.000187

N	n	k	x	P(x)	p(x)
21	10	9	1	0.005801	0.005614
21	10	9	2	0.056323	0.050522
21	10	9	3	0.244939	0.188616
21	10	9	4	0.575018	0.330079
21	10	9	5	0.857942	0.282924
21	10	9	6	0.975827	0.117885
21	10	9	7	0.998282	0.022454
21	10	9	8	0.999966	0.001684
21	10	9	9	1.000000	0.000034
21	10	10	0	0.000031	0.000031
21	10	10	1	0.001591	0.001559
21	10	10	2	0.022641	0.021051
21	10	10	3	0.134913	0.112272
21	10	10	4	0.409979	0.275065
21	10	10	5	0.740057	0.330079
21	10	10	6	0.936532	0.196475
21	10	10	7	0.992668	0.056136
21	10	10	8	0.999685	0.007017
21	10	10	9	0.999997	0.000312
21	10	10	10	1.000000	0.000003
21	11	1	0	0.476190	0.476190
21	11	1	1	1.000000	0.523810
21	11	2	0	0.214286	0.214286
21	11	2	1	0.738095	0.523810
21	11	2	2	1.000000	0.261905
21	11	3	0	0.090226	0.090226
21	11	3	1	0.462406	0.372180
21	11	3	2	0.875940	0.413534
21	11	3	3	1.000000	0.124060
21	11	4	0	0.035088	0.035088
21	11	4	1	0.255639	0.220551
21	11	4	2	0.669173	0.413534
21	11	4	3	0.944862	0.275689
21	11	4	4	1.000000	0.055138
21	11	5	0	0.012384	0.012384
21	11	5	1	0.125903	0.113519
21	11	5	2	0.450243	0.324340
21	11	5	3	0.815126	0.364883
21	11	5	4	0.977296	0.162170
21	11	5	5	1.000000	0.022704
21	11	6	0	0.003870	0.003870
21	11	6	1	0.054954	0.051084
21	11	6	2	0.267802	0.212848
21	11	6	3	0.632685	0.364883
21	11	6	4	0.906347	0.273662
21	11	6	5	0.991486	0.085139
21	11	6	6	1.000000	0.008514
21	11	7	0	0.001032	0.001032
21	11	7	1	0.020898	0.019866
21	11	7	2	0.140093	0.119195
21	11	7	3	0.438080	0.297988
21	11	7	4	0.778638	0.340557
21	11	7	5	0.957430	0.178793
21	11	7	6	0.997162	0.039732
21	11	7	7	1.000000	0.002838
21	11	8	0	0.000221	0.000221
21	11	8	1	0.006708	0.006487
21	11	8	2	0.063467	0.056760
21	11	8	3	0.267802	0.204334
21	11	8	4	0.608359	0.340557
21	11	8	5	0.880805	0.272446
21	11	8	6	0.982972	0.102167
21	11	8	7	0.999189	0.016217
21	11	8	8	1.000000	0.000811
21	11	9	0	0.000034	0.000034
21	11	9	1	0.001718	0.001684
21	11	9	2	0.024172	0.022454
21	11	9	3	0.142058	0.117885
21	11	9	4	0.424982	0.282924
21	11	9	5	0.755061	0.330079
21	11	9	6	0.943677	0.188616
21	11	9	7	0.994199	0.050522
21	11	9	8	0.999813	0.005614
21	11	9	9	1.000000	0.000187
21	11	10	0	0.000003	0.000003
21	11	10	1	0.000315	0.000312
21	11	10	2	0.007332	0.007017
21	11	10	3	0.063467	0.056136
21	11	10	4	0.259943	0.196475
21	11	10	5	0.590021	0.330079
21	11	10	6	0.865087	0.275065
21	11	10	7	0.977358	0.112272
21	11	10	8	0.998409	0.021051
21	11	10	9	0.999969	0.001559
21	11	10	10	1.000000	0.000031
21	11	11	1	0.000031	0.000031
21	11	11	2	0.001591	0.001559
21	11	11	3	0.022641	0.021051
21	11	11	4	0.134913	0.112272
21	11	11	5	0.409979	0.275065
21	11	11	6	0.740057	0.330079
21	11	11	7	0.936532	0.196475
21	11	11	8	0.992668	0.056136
21	11	11	9	0.999685	0.007017
21	11	11	10	0.999997	0.000312
21	11	11	11	1.000000	0.000003
21	12	1	0	0.428571	0.428571
21	12	1	1	1.000000	0.571429
21	12	2	0	0.171429	0.171429
21	12	2	1	0.685714	0.514286
21	12	2	2	1.000000	0.314286
21	12	3	0	0.063158	0.063158
21	12	3	1	0.387970	0.324812
21	12	3	2	0.834586	0.446617
21	12	3	3	1.000000	0.165414
21	12	4	0	0.021053	0.021053
21	12	4	1	0.189474	0.168421
21	12	4	2	0.586466	0.396992
21	12	4	3	0.917293	0.330827
21	12	4	4	1.000000	0.082707
21	12	5	0	0.006192	0.006192
21	12	5	1	0.080495	0.074303
21	12	5	2	0.352941	0.272446
21	12	5	3	0.742149	0.389208
21	12	5	4	0.961079	0.218930
21	12	5	5	1.000000	0.038921
21	12	6	0	0.001548	0.001548
21	12	6	1	0.029412	0.027864
21	12	6	2	0.182662	0.153251
21	12	6	3	0.523220	0.340557
21	12	6	4	0.851614	0.328394
21	12	6	5	0.982972	0.131358
21	12	6	6	1.000000	0.017028
21	12	7	0	0.000310	0.000310
21	12	7	1	0.008978	0.008669
21	12	7	2	0.080495	0.071517
21	12	7	3	0.318885	0.238390
21	12	7	4	0.676471	0.357585
21	12	7	5	0.921672	0.245201
21	12	7	6	0.993189	0.071517
21	12	7	7	1.000000	0.006811
21	12	8	0	0.000044	0.000044
21	12	8	1	0.002167	0.002123
21	12	8	2	0.029412	0.027245
21	12	8	3	0.165635	0.136223
21	12	8	4	0.472136	0.306502
21	12	8	5	0.799071	0.326935
21	12	8	6	0.962539	0.163467
21	12	8	7	0.997567	0.035029
21	12	8	8	1.000000	0.002433
21	12	9	0	0.000003	0.000003
21	12	9	1	0.000371	0.000367
21	12	9	2	0.008454	0.008084
21	12	9	3	0.071326	0.062872
21	12	9	4	0.283520	0.212193
21	12	9	5	0.623029	0.339509
21	12	9	6	0.887092	0.264063
21	12	9	7	0.984095	0.097003
21	12	9	8	0.999251	0.015157
21	12	9	9	1.000000	0.000748
21	12	10	1	0.000034	0.000034
21	12	10	2	0.001718	0.001684
21	12	10	3	0.024172	0.022454
21	12	10	4	0.142058	0.117885
21	12	10	5	0.424982	0.282924
21	12	10	6	0.755061	0.330079
21	12	10	7	0.943677	0.188616
21	12	10	8	0.994199	0.050522
21	12	10	9	0.999813	0.005614
21	12	10	10	1.000000	0.000187
21	12	11	2	0.000187	0.000187
21	12	11	3	0.005801	0.005614
21	12	11	4	0.056323	0.050522
21	12	11	5	0.244939	0.188616
21	12	11	6	0.575018	0.330079
21	12	11	7	0.857942	0.282924
21	12	11	8	0.975827	0.117885
21	12	11	9	0.998282	0.022454
21	12	11	10	0.999966	0.001684
21	12	11	11	1.000000	0.000034
21	12	12	3	0.000748	0.000748
21	12	12	4	0.015905	0.015157
21	12	12	5	0.112908	0.097003
21	12	12	6	0.376971	0.264063
21	12	12	7	0.716480	0.339509
21	12	12	8	0.928673	0.212193
21	12	12	9	0.991545	0.062872
21	12	12	10	0.999629	0.008084
21	12	12	11	0.999996	0.000367
21	12	12	12	1.000000	0.000003
21	13	1	0	0.380952	0.380952
21	13	1	1	1.000000	0.619048
21	13	2	0	0.133333	0.133333
21	13	2	1	0.628571	0.495238
21	13	2	2	1.000000	0.371429
21	13	3	0	0.042105	0.042105
21	13	3	1	0.315789	0.273684
21	13	3	2	0.784962	0.469173
21	13	3	3	1.000000	0.215038
21	13	4	0	0.011696	0.011696
21	13	4	1	0.133333	0.121637
21	13	4	2	0.498246	0.364912
21	13	4	3	0.880535	0.382289
21	13	4	4	1.000000	0.119465
21	13	5	0	0.002752	0.002752
21	13	5	1	0.047472	0.044720
21	13	5	2	0.262126	0.214654
21	13	5	3	0.655659	0.393533
21	13	5	4	0.936754	0.281095
21	13	5	5	1.000000	0.063246

N	n	k	x	P(x)	p(x)
21	13	6	0	0.000516	0.000516
21	13	6	1	0.013932	0.013416
21	13	6	2	0.114551	0.100619
21	13	6	3	0.409701	0.295150
21	13	6	4	0.778638	0.368937
21	13	6	5	0.968377	0.189739
21	13	6	6	1.000000	0.031623
21	13	7	0	0.000069	0.000069
21	13	7	1	0.003199	0.003130
21	13	7	2	0.040764	0.037564
21	13	7	3	0.212934	0.172171
21	13	7	4	0.557276	0.344341
21	13	7	5	0.867183	0.309907
21	13	7	6	0.985242	0.118060
21	13	7	7	1.000000	0.014757
21	13	8	0	0.000005	0.000005
21	13	8	1	0.000516	0.000511
21	13	8	2	0.011249	0.010733
21	13	8	3	0.089955	0.078707
21	13	8	4	0.335913	0.245958
21	13	8	5	0.690093	0.354180
21	13	8	6	0.926213	0.236120
21	13	8	7	0.993675	0.067463
21	13	8	8	1.000000	0.006325
21	13	9	1	0.000044	0.000044
21	13	9	2	0.002167	0.002123
21	13	9	3	0.029412	0.027245
21	13	9	4	0.165635	0.136223
21	13	9	5	0.472136	0.306502
21	13	9	6	0.799071	0.326935
21	13	9	7	0.962539	0.163467
21	13	9	8	0.997567	0.035029
21	13	9	9	1.000000	0.002433
21	13	10	2	0.000221	0.000221
21	13	10	3	0.006708	0.006487
21	13	10	4	0.063467	0.056760
21	13	10	5	0.267802	0.204334
21	13	10	6	0.608359	0.340557
21	13	10	7	0.880805	0.272446
21	13	10	8	0.982972	0.102167
21	13	10	9	0.999189	0.016217
21	13	10	10	1.000000	0.000811
21	13	11	3	0.000811	0.000811
21	13	11	4	0.017028	0.016217
21	13	11	5	0.119195	0.102167
21	13	11	6	0.391641	0.272446
21	13	11	7	0.732198	0.340557
21	13	11	8	0.936532	0.204334
21	13	11	9	0.993292	0.056760
21	13	11	10	0.999779	0.006487

N	n	k	x	P(x)	p(x)
21	13	11	11	1.000000	0.000221
21	13	12	4	0.002433	0.002433
21	13	12	5	0.037461	0.035029
21	13	12	6	0.200929	0.163467
21	13	12	7	0.527864	0.326935
21	13	12	8	0.834365	0.306502
21	13	12	9	0.970588	0.136223
21	13	12	10	0.997833	0.027245
21	13	12	11	0.999956	0.002123
21	13	12	12	1.000000	0.000044
21	13	13	5	0.006325	0.006325
21	13	13	6	0.073787	0.067463
21	13	13	7	0.309907	0.236120
21	13	13	8	0.664087	0.354180
21	13	13	9	0.910045	0.245958
21	13	13	10	0.988751	0.078707
21	13	13	11	0.999484	0.010733
21	13	13	12	0.999995	0.000511
21	13	13	13	1.000000	0.000005
21	14	1	0	0.333333	0.333333
21	14	1	1	1.000000	0.666667
21	14	2	0	0.100000	0.100000
21	14	2	1	0.566667	0.466667
21	14	2	2	1.000000	0.433333
21	14	3	0	0.026316	0.026316
21	14	3	1	0.247368	0.221053
21	14	3	2	0.726316	0.478947
21	14	3	3	1.000000	0.273684
21	14	4	0	0.005848	0.005848
21	14	4	1	0.087719	0.081871
21	14	4	2	0.407017	0.319298
21	14	4	3	0.832748	0.425731
21	14	4	4	1.000000	0.167251
21	14	5	0	0.001032	0.001032
21	14	5	1	0.025112	0.024080
21	14	5	2	0.181631	0.156519
21	14	5	3	0.557275	0.375645
21	14	5	4	0.901617	0.344341
21	14	5	5	1.000000	0.098383
21	14	6	0	0.000129	0.000129
21	14	6	1	0.005547	0.005418
21	14	6	2	0.064241	0.058695
21	14	6	3	0.299020	0.234778
21	14	6	4	0.686403	0.387384
21	14	6	5	0.944659	0.258256
21	14	6	6	1.000000	0.055341
21	14	7	0	0.000009	0.000009
21	14	7	1	0.000851	0.000843
21	14	7	2	0.017286	0.016434
21	14	7	3	0.126849	0.109563

N	n	k	x	P(x)	p(x)
21	14	7	4	0.428148	0.301299
21	14	7	5	0.789706	0.361558
21	14	7	6	0.970485	0.180779
21	14	7	7	1.000000	0.029515
21	14	8	1	0.000069	0.000069
21	14	8	2	0.003199	0.003130
21	14	8	3	0.040764	0.037564
21	14	8	4	0.212934	0.172171
21	14	8	5	0.557276	0.344341
21	14	8	6	0.867183	0.309907
21	14	8	7	0.985242	0.118060
21	14	8	8	1.000000	0.014757
21	14	9	2	0.000310	0.000310
21	14	9	3	0.008978	0.008669
21	14	9	4	0.080495	0.071517
21	14	9	5	0.318885	0.238390
21	14	9	6	0.676471	0.357585
21	14	9	7	0.921672	0.245201
21	14	9	8	0.993189	0.071517
21	14	9	9	1.000000	0.006811
21	14	10	3	0.001032	0.001032
21	14	10	4	0.020898	0.019866
21	14	10	5	0.140093	0.119195
21	14	10	6	0.438080	0.297988
21	14	10	7	0.778638	0.340557
21	14	10	8	0.957430	0.178793
21	14	10	9	0.997162	0.039732
21	14	10	10	1.000000	0.002838
21	14	11	4	0.002838	0.002838
21	14	11	5	0.042570	0.039732
21	14	11	6	0.221362	0.178793
21	14	11	7	0.561919	0.340557
21	14	11	8	0.859907	0.297988
21	14	11	9	0.979102	0.119195
21	14	11	10	0.998968	0.019866
21	14	11	11	1.000000	0.001032
21	14	12	5	0.006811	0.006811
21	14	12	6	0.078328	0.071517
21	14	12	7	0.323529	0.245201
21	14	12	8	0.681115	0.357585
21	14	12	9	0.919505	0.238390
21	14	12	10	0.991022	0.071517
21	14	12	11	0.999690	0.008669
21	14	12	12	1.000000	0.000310
21	14	13	6	0.014757	0.014757
21	14	13	7	0.132817	0.118060
21	14	13	8	0.442724	0.309907
21	14	13	9	0.787066	0.344341
21	14	13	10	0.959236	0.172171
21	14	13	11	0.996801	0.037564

N	n	k	x	P(x)	p(x)
21	14	13	12	0.999931	0.003130
21	14	13	13	1.000000	0.000069
21	14	14	7	0.029515	0.029515
21	14	14	8	0.210294	0.180779
21	14	14	9	0.571852	0.361558
21	14	14	10	0.873151	0.301299
21	14	14	11	0.982714	0.109563
21	14	14	12	0.999149	0.016434
21	14	14	13	0.999991	0.000843
21	14	14	14	1.000000	0.000009
21	15	1	0	0.285714	0.285714
21	15	1	1	1.000000	0.714286
21	15	2	0	0.071429	0.071429
21	15	2	1	0.500000	0.428571
21	15	2	2	1.000000	0.500000
21	15	3	0	0.015038	0.015038
21	15	3	1	0.184211	0.169173
21	15	3	2	0.657895	0.473684
21	15	3	3	1.000000	0.342105
21	15	4	0	0.002506	0.002506
21	15	4	1	0.052632	0.050125
21	15	4	2	0.315789	0.263158
21	15	4	3	0.771930	0.456140
21	15	4	4	1.000000	0.228070
21	15	5	0	0.000295	0.000295
21	15	5	1	0.011352	0.011057
21	15	5	2	0.114551	0.103199
21	15	5	3	0.449948	0.335397
21	15	5	4	0.852425	0.402477
21	15	5	5	1.000000	0.147575
21	15	6	0	0.000018	0.000018
21	15	6	1	0.001677	0.001659
21	15	6	2	0.030702	0.029025
21	15	6	3	0.198400	0.167699
21	15	6	4	0.575722	0.377322
21	15	6	5	0.907766	0.332043
21	15	6	6	1.000000	0.092234
21	15	7	1	0.000129	0.000129
21	15	7	2	0.005547	0.005418
21	15	7	3	0.064241	0.058695
21	15	7	4	0.299020	0.234778
21	15	7	5	0.686403	0.387384
21	15	7	6	0.944659	0.258256
21	15	7	7	1.000000	0.055341
21	15	8	2	0.000516	0.000516
21	15	8	3	0.013932	0.013416
21	15	8	4	0.114551	0.100619
21	15	8	5	0.409701	0.295150
21	15	8	6	0.778638	0.368937
21	15	8	7	0.968377	0.189739

Table for $N = 2$, $n = 1$, through $N = 100$, $n = 50$

N	n	k	x	P(x)	p(x)
21	15	8	8	1.000000	0.031623
21	15	9	3	0.001548	0.001548
21	15	9	4	0.029412	0.027864
21	15	9	5	0.182662	0.153251
21	15	9	6	0.523220	0.340557
21	15	9	7	0.851614	0.328394
21	15	9	8	0.982972	0.131358
21	15	9	9	1.000000	0.017028
21	15	10	4	0.003870	0.003870
21	15	10	5	0.054954	0.051084
21	15	10	6	0.267802	0.212848
21	15	10	7	0.632685	0.364883
21	15	10	8	0.906347	0.273662
21	15	10	9	0.991486	0.085139
21	15	10	10	1.000000	0.008514
21	15	11	5	0.008514	0.008514
21	15	11	6	0.093653	0.085139
21	15	11	7	0.367315	0.273662
21	15	11	8	0.732198	0.364883
21	15	11	9	0.945046	0.212848
21	15	11	10	0.996130	0.051084
21	15	11	11	1.000000	0.003870
21	15	12	6	0.017028	0.017028
21	15	12	7	0.148386	0.131358
21	15	12	8	0.476780	0.328394
21	15	12	9	0.817337	0.340557
21	15	12	10	0.970588	0.153251
21	15	12	11	0.998452	0.027864
21	15	12	12	1.000000	0.001548
21	15	13	7	0.031623	0.031623
21	15	13	8	0.221362	0.189739
21	15	13	9	0.590299	0.368937
21	15	13	10	0.885449	0.295150
21	15	13	11	0.986068	0.100619
21	15	13	12	0.999484	0.013416
21	15	13	13	1.000000	0.000516
21	15	14	8	0.055341	0.055341
21	15	14	9	0.313596	0.258256
21	15	14	10	0.700980	0.387384
21	15	14	11	0.935758	0.234778
21	15	14	12	0.994453	0.058695
21	15	14	13	0.999871	0.005418
21	15	14	14	1.000000	0.000129
21	15	15	9	0.092234	0.092234
21	15	15	10	0.424278	0.332043
21	15	15	11	0.801600	0.377322
21	15	15	12	0.969298	0.167699
21	15	15	13	0.998323	0.029025
21	15	15	14	0.999981	0.001659
21	15	15	15	1.000000	0.000018

N	n	k	x	P(x)	p(x)
21	16	1	0	0.238095	0.238095
21	16	1	1	1.000000	0.761905
21	16	2	0	0.047619	0.047619
21	16	2	1	0.428571	0.380952
21	16	2	2	1.000000	0.571429
21	16	3	0	0.007519	0.007519
21	16	3	1	0.127820	0.120301
21	16	3	2	0.578947	0.451128
21	16	3	3	1.000000	0.421053
21	16	4	0	0.000835	0.000835
21	16	4	1	0.027569	0.026733
21	16	4	2	0.228070	0.200501
21	16	4	3	0.695906	0.467836
21	16	4	4	1.000000	0.304094
21	16	5	0	0.000049	0.000049
21	16	5	1	0.003981	0.003931
21	16	5	2	0.062951	0.058971
21	16	5	3	0.338149	0.275198
21	16	5	4	0.785346	0.447196
21	16	5	5	1.000000	0.214654
21	16	6	1	0.000295	0.000295
21	16	6	2	0.011352	0.011057
21	16	6	3	0.114551	0.103199
21	16	6	4	0.449948	0.335397
21	16	6	5	0.852425	0.402477
21	16	6	6	1.000000	0.147575
21	16	7	2	0.001032	0.001032
21	16	7	3	0.025112	0.024080
21	16	7	4	0.181631	0.156519
21	16	7	5	0.557275	0.375645
21	16	7	6	0.901617	0.344341
21	16	7	7	1.000000	0.098383
21	16	8	3	0.002752	0.002752
21	16	8	4	0.047472	0.044720
21	16	8	5	0.262126	0.214654
21	16	8	6	0.655659	0.393533
21	16	8	7	0.936754	0.281095
21	16	8	8	1.000000	0.063246
21	16	9	4	0.006192	0.006192
21	16	9	5	0.080495	0.074303
21	16	9	6	0.352941	0.272446
21	16	9	7	0.742149	0.389208
21	16	9	8	0.961079	0.218930
21	16	9	9	1.000000	0.038921
21	16	10	5	0.012384	0.012384
21	16	10	6	0.125903	0.113519
21	16	10	7	0.450243	0.324340
21	16	10	8	0.815126	0.364883
21	16	10	9	0.977296	0.162170
21	16	10	10	1.000000	0.022704

N	n	k	x	P(x)	p(x)
21	16	11	6	0.022704	0.022704
21	16	11	7	0.184874	0.162170
21	16	11	8	0.549757	0.364883
21	16	11	9	0.874097	0.324340
21	16	11	10	0.987616	0.113519
21	16	11	11	1.000000	0.012384
21	16	12	7	0.038921	0.038921
21	16	12	8	0.257850	0.218930
21	16	12	9	0.647059	0.389208
21	16	12	10	0.919505	0.272446
21	16	12	11	0.993808	0.074303
21	16	12	12	1.000000	0.006192
21	16	13	8	0.063246	0.063246
21	16	13	9	0.344341	0.281095
21	16	13	10	0.737874	0.393533
21	16	13	11	0.952528	0.214654
21	16	13	12	0.997248	0.044720
21	16	13	13	1.000000	0.002752
21	16	14	9	0.098383	0.098383
21	16	14	10	0.442724	0.344341
21	16	14	11	0.818369	0.375645
21	16	14	12	0.974888	0.156519
21	16	14	13	0.998968	0.024080
21	16	14	14	1.000000	0.001032
21	16	15	10	0.147575	0.147575
21	16	15	11	0.550052	0.402477
21	16	15	12	0.885449	0.335397
21	16	15	13	0.988648	0.103199
21	16	15	14	0.999705	0.011057
21	16	15	15	1.000000	0.000295
21	16	16	11	0.214654	0.214654
21	16	16	12	0.661851	0.447196
21	16	16	13	0.937048	0.275198
21	16	16	14	0.996019	0.058971
21	16	16	15	0.999951	0.003931
21	16	16	16	1.000000	0.000049
21	17	1	0	0.190476	0.190476
21	17	1	1	1.000000	0.809524
21	17	2	0	0.028571	0.028571
21	17	2	1	0.352381	0.323810
21	17	2	2	1.000000	0.647619
21	17	3	0	0.003008	0.003008
21	17	3	1	0.079699	0.076692
21	17	3	2	0.488722	0.409023
21	17	3	3	1.000000	0.511278
21	17	4	0	0.000167	0.000167
21	17	4	1	0.011529	0.011362
21	17	4	2	0.147870	0.136341
21	17	4	3	0.602339	0.454470
21	17	4	4	1.000000	0.397661

N	n	k	x	P(x)	p(x)
21	17	5	1	0.000835	0.000835
21	17	5	2	0.027569	0.026733
21	17	5	3	0.228070	0.200501
21	17	5	4	0.695906	0.467836
21	17	5	5	1.000000	0.304094
21	17	6	2	0.002506	0.002506
21	17	6	3	0.052632	0.050125
21	17	6	4	0.315789	0.263158
21	17	6	5	0.771930	0.456140
21	17	6	6	1.000000	0.228070
21	17	7	3	0.005848	0.005848
21	17	7	4	0.087719	0.081871
21	17	7	5	0.407017	0.319298
21	17	7	6	0.832748	0.425731
21	17	7	7	1.000000	0.167251
21	17	8	4	0.011696	0.011696
21	17	8	5	0.133333	0.121637
21	17	8	6	0.498246	0.364912
21	17	8	7	0.880535	0.382289
21	17	8	8	1.000000	0.119465
21	17	9	5	0.021053	0.021053
21	17	9	6	0.189474	0.168421
21	17	9	7	0.586466	0.396992
21	17	9	8	0.917293	0.330827
21	17	9	9	1.000000	0.082707
21	17	10	6	0.035088	0.035088
21	17	10	7	0.255639	0.220551
21	17	10	8	0.669173	0.413534
21	17	10	9	0.944862	0.275689
21	17	10	10	1.000000	0.055138
21	17	11	7	0.055138	0.055138
21	17	11	8	0.330827	0.275689
21	17	11	9	0.744361	0.413534
21	17	11	10	0.964912	0.220551
21	17	11	11	1.000000	0.035088
21	17	12	8	0.082707	0.082707
21	17	12	9	0.413534	0.330827
21	17	12	10	0.810526	0.396992
21	17	12	11	0.978947	0.168421
21	17	12	12	1.000000	0.021053
21	17	13	9	0.119465	0.119465
21	17	13	10	0.501754	0.382289
21	17	13	11	0.866667	0.364912
21	17	13	12	0.988304	0.121637
21	17	13	13	1.000000	0.011696
21	17	14	10	0.167251	0.167251
21	17	14	11	0.592982	0.425731
21	17	14	12	0.912281	0.319298
21	17	14	13	0.994152	0.081871
21	17	14	14	1.000000	0.005848

Table for $N = 2$, $n = 1$, through $N = 100$, $n = 50$

N	n	k	x	P(x)	p(x)
21	17	15	11	0.228070	0.228070
21	17	15	12	0.684211	0.456140
21	17	15	13	0.947368	0.263158
21	17	15	14	0.997494	0.050125
21	17	15	15	1.000000	0.002506
21	17	16	12	0.304094	0.304094
21	17	16	13	0.771930	0.467836
21	17	16	14	0.972431	0.200501
21	17	16	15	0.999165	0.026733
21	17	16	16	1.000000	0.000835
21	17	17	13	0.397661	0.397661
21	17	17	14	0.852130	0.454470
21	17	17	15	0.988471	0.136341
21	17	17	16	0.999833	0.011362
21	17	17	17	1.000000	0.000167
21	18	1	0	0.142857	0.142857
21	18	1	1	1.000000	0.857143
21	18	2	0	0.014286	0.014286
21	18	2	1	0.271429	0.257143
21	18	2	2	1.000000	0.728571
21	18	3	0	0.000752	0.000752
21	18	3	1	0.041353	0.040601
21	18	3	2	0.386466	0.345113
21	18	3	3	1.000000	0.613534
21	18	4	1	0.003008	0.003008
21	18	4	2	0.079699	0.076692
21	18	4	3	0.488722	0.409023
21	18	4	4	1.000000	0.511278
21	18	5	2	0.007519	0.007519
21	18	5	3	0.127820	0.120301
21	18	5	4	0.578947	0.451128
21	18	5	5	1.000000	0.421053
21	18	6	3	0.015038	0.015038
21	18	6	4	0.184211	0.169173
21	18	6	5	0.657895	0.473684
21	18	6	6	1.000000	0.342105
21	18	7	4	0.026316	0.026316
21	18	7	5	0.247368	0.221053
21	18	7	6	0.726316	0.478947
21	18	7	7	1.000000	0.273684
21	18	8	5	0.042105	0.042105
21	18	8	6	0.315789	0.273684
21	18	8	7	0.784962	0.469173
21	18	8	8	1.000000	0.215038
21	18	9	6	0.063158	0.063158
21	18	9	7	0.387970	0.324812
21	18	9	8	0.834586	0.446617
21	18	9	9	1.000000	0.165414
21	18	10	7	0.090226	0.090226
21	18	10	8	0.462406	0.372180

N	n	k	x	P(x)	p(x)
21	18	10	9	0.875940	0.413534
21	18	10	10	1.000000	0.124060
21	18	11	8	0.124060	0.124060
21	18	11	9	0.537594	0.413534
21	18	11	10	0.909774	0.372180
21	18	11	11	1.000000	0.090226
21	18	12	9	0.165414	0.165414
21	18	12	10	0.612030	0.446617
21	18	12	11	0.936842	0.324812
21	18	12	12	1.000000	0.063158
21	18	13	10	0.215038	0.215038
21	18	13	11	0.684210	0.469173
21	18	13	12	0.957895	0.273684
21	18	13	13	1.000000	0.042105
21	18	14	11	0.273684	0.273684
21	18	14	12	0.752632	0.478947
21	18	14	13	0.973684	0.221053
21	18	14	14	1.000000	0.026316
21	18	15	12	0.342105	0.342105
21	18	15	13	0.815789	0.473684
21	18	15	14	0.984962	0.169173
21	18	15	15	1.000000	0.015038
21	18	16	13	0.421053	0.421053
21	18	16	14	0.872180	0.451128
21	18	16	15	0.992481	0.120301
21	18	16	16	1.000000	0.007519
21	18	17	14	0.511278	0.511278
21	18	17	15	0.920301	0.409023
21	18	17	16	0.996992	0.076692
21	18	17	17	1.000000	0.003008
21	18	18	15	0.613534	0.613534
21	18	18	16	0.958647	0.345113
21	18	18	17	0.999248	0.040601
21	18	18	18	1.000000	0.000752
21	19	1	0	0.095238	0.095238
21	19	1	1	1.000000	0.904762
21	19	2	0	0.004762	0.004762
21	19	2	1	0.185714	0.180952
21	19	2	2	1.000000	0.814286
21	19	3	1	0.014286	0.014286
21	19	3	2	0.271429	0.257143
21	19	3	3	1.000000	0.728571
21	19	4	2	0.028571	0.028571
21	19	4	3	0.352381	0.323810
21	19	4	4	1.000000	0.647619
21	19	5	3	0.047619	0.047619
21	19	5	4	0.428571	0.380952
21	19	5	5	1.000000	0.571429
21	19	6	4	0.071429	0.071429
21	19	6	5	0.500000	0.428571

N	n	k	x	P(x)	p(x)
21	19	6	6	1.000000	0.500000
21	19	7	5	0.100000	0.100000
21	19	7	6	0.566667	0.466667
21	19	7	7	1.000000	0.433333
21	19	8	6	0.133333	0.133333
21	19	8	7	0.628571	0.495238
21	19	8	8	1.000000	0.371429
21	19	9	7	0.171429	0.171429
21	19	9	8	0.685714	0.514286
21	19	9	9	1.000000	0.314286
21	19	10	8	0.214286	0.214286
21	19	10	9	0.738095	0.523810
21	19	10	10	1.000000	0.261905
21	19	11	9	0.261905	0.261905
21	19	11	10	0.785714	0.523810
21	19	11	11	1.000000	0.214286
21	19	12	10	0.314286	0.314286
21	19	12	11	0.828571	0.514286
21	19	12	12	1.000000	0.171429
21	19	13	11	0.371429	0.371429
21	19	13	12	0.866667	0.495238
21	19	13	13	1.000000	0.133333
21	19	14	12	0.433333	0.433333
21	19	14	13	0.900000	0.466667
21	19	14	14	1.000000	0.100000
21	19	15	13	0.500000	0.500000
21	19	15	14	0.928571	0.428571
21	19	15	15	1.000000	0.071429
21	19	16	14	0.571429	0.571429
21	19	16	15	0.952381	0.380952
21	19	16	16	1.000000	0.047619
21	19	17	15	0.647619	0.647619
21	19	17	16	0.971429	0.323810
21	19	17	17	1.000000	0.028571
21	19	18	16	0.728571	0.728571
21	19	18	17	0.985714	0.257143
21	19	18	18	1.000000	0.014286
21	19	19	17	0.814286	0.814286
21	19	19	18	0.995238	0.180952
21	19	19	19	1.000000	0.004762
21	20	1	0	0.047619	0.047619
21	20	1	1	1.000000	0.952381
21	20	2	1	0.095238	0.095238
21	20	2	2	1.000000	0.904762
21	20	3	2	0.142857	0.142857
21	20	3	3	1.000000	0.857143
21	20	4	3	0.190476	0.190476
21	20	4	4	1.000000	0.809524
21	20	5	4	0.238095	0.238095
21	20	5	5	1.000000	0.761905

N	n	k	x	P(x)	p(x)
21	20	6	5	0.285714	0.285714
21	20	6	6	1.000000	0.714286
21	20	7	6	0.333333	0.333333
21	20	7	7	1.000000	0.666667
21	20	8	7	0.380952	0.380952
21	20	8	8	1.000000	0.619048
21	20	9	8	0.428571	0.428571
21	20	9	9	1.000000	0.571429
21	20	10	9	0.476190	0.476190
21	20	10	10	1.000000	0.523810
21	20	11	10	0.523810	0.523810
21	20	11	11	1.000000	0.476190
21	20	12	11	0.571429	0.571429
21	20	12	12	1.000000	0.428571
21	20	13	12	0.619048	0.619048
21	20	13	13	1.000000	0.380952
21	20	14	13	0.666667	0.666667
21	20	14	14	1.000000	0.333333
21	20	15	14	0.714286	0.714286
21	20	15	15	1.000000	0.285714
21	20	16	15	0.761905	0.761905
21	20	16	16	1.000000	0.238095
21	20	17	16	0.809524	0.809524
21	20	17	17	1.000000	0.190476
21	20	18	17	0.857143	0.857143
21	20	18	18	1.000000	0.142857
21	20	19	18	0.904762	0.904762
21	20	19	19	1.000000	0.095238
21	20	20	19	0.952381	0.952381
21	20	20	20	1.000000	0.047619
22	1	1	0	0.954545	0.954545
22	1	1	1	1.000000	0.045455
22	2	1	0	0.909091	0.909091
22	2	1	1	1.000000	0.090909
22	2	2	0	0.822511	0.822511
22	2	2	1	0.995671	0.173160
22	2	2	2	1.000000	0.004329
22	3	1	0	0.863636	0.863636
22	3	1	1	1.000000	0.136364
22	3	2	0	0.740260	0.740260
22	3	2	1	0.987013	0.246753
22	3	2	2	1.000000	0.012987
22	3	3	0	0.629221	0.629221
22	3	3	1	0.962338	0.333117
22	3	3	2	0.999351	0.037013
22	3	3	3	1.000000	0.000649
22	4	1	0	0.818182	0.818182
22	4	1	1	1.000000	0.181818
22	4	2	0	0.662338	0.662338
22	4	2	1	0.974026	0.311688

Table for $N = 2$, $n = 1$, through $N = 100$, $n = 50$

N	n	k	x	P(x)	p(x)
22	4	2	2	1.000000	0.025974
22	4	3	0	0.529870	0.529870
22	4	3	1	0.927273	0.397403
22	4	3	2	0.997403	0.070130
22	4	3	3	1.000000	0.002597
22	4	4	0	0.418319	0.418319
22	4	4	1	0.864525	0.446206
22	4	4	2	0.990020	0.125496
22	4	4	3	0.999863	0.009843
22	4	4	4	1.000000	0.000137
22	5	1	0	0.772727	0.772727
22	5	1	1	1.000000	0.227273
22	5	2	0	0.588745	0.588745
22	5	2	1	0.956710	0.367965
22	5	2	2	1.000000	0.043290
22	5	3	0	0.441558	0.441558
22	5	3	1	0.883117	0.441558
22	5	3	2	0.993506	0.110390
22	5	3	3	1.000000	0.006494
22	5	4	0	0.325359	0.325359
22	5	4	1	0.790157	0.464798
22	5	4	2	0.976076	0.185919
22	5	4	3	0.999316	0.023240
22	5	4	4	1.000000	0.000684
22	5	5	0	0.234981	0.234981
22	5	5	1	0.686869	0.451887
22	5	5	2	0.945090	0.258221
22	5	5	3	0.996734	0.051644
22	5	5	4	0.999962	0.003228
22	5	5	5	1.000000	0.000038
22	6	1	0	0.727273	0.727273
22	6	1	1	1.000000	0.272727
22	6	2	0	0.519480	0.519480
22	6	2	1	0.935065	0.415584
22	6	2	2	1.000000	0.064935
22	6	3	0	0.363636	0.363636
22	6	3	1	0.831169	0.467532
22	6	3	2	0.987013	0.155844
22	6	3	3	1.000000	0.012987
22	6	4	0	0.248804	0.248804
22	6	4	1	0.708134	0.459330
22	6	4	2	0.954204	0.246070
22	6	4	3	0.997949	0.043746
22	6	4	4	1.000000	0.002051
22	6	5	0	0.165869	0.165869
22	6	5	1	0.580542	0.414673
22	6	5	2	0.899521	0.318979
22	6	5	3	0.990658	0.091137
22	6	5	4	0.999772	0.009114
22	6	5	5	1.000000	0.000228
22	6	6	0	0.107327	0.107327
22	6	6	1	0.458580	0.351252
22	6	6	2	0.824468	0.365888
22	6	6	3	0.974575	0.150108
22	6	6	4	0.998700	0.024124
22	6	6	5	0.999986	0.001287
22	6	6	6	1.000000	0.000013
22	7	1	0	0.681818	0.681818
22	7	1	1	1.000000	0.318182
22	7	2	0	0.454545	0.454545
22	7	2	1	0.909091	0.454545
22	7	2	2	1.000000	0.090909
22	7	3	0	0.295455	0.295455
22	7	3	1	0.772727	0.477273
22	7	3	2	0.977273	0.204545
22	7	3	3	1.000000	0.022727
22	7	4	0	0.186603	0.186603
22	7	4	1	0.622010	0.435407
22	7	4	2	0.923445	0.301435
22	7	4	3	0.995215	0.071770
22	7	4	4	1.000000	0.004785
22	7	5	0	0.114035	0.114035
22	7	5	1	0.476874	0.362839
22	7	5	2	0.839713	0.362839
22	7	5	3	0.979266	0.139553
22	7	5	4	0.999202	0.019936
22	7	5	5	1.000000	0.000797
22	7	6	0	0.067079	0.067079
22	7	6	1	0.348813	0.281734
22	7	6	2	0.732996	0.384182
22	7	6	3	0.946430	0.213435
22	7	6	4	0.995684	0.049254
22	7	6	5	0.999906	0.004222
22	7	6	6	1.000000	0.000094
22	7	7	0	0.037732	0.037732
22	7	7	1	0.243163	0.205431
22	7	7	2	0.612939	0.369776
22	7	7	3	0.893072	0.280133
22	7	7	4	0.986449	0.093378
22	7	7	5	0.999378	0.012929
22	7	7	6	0.999994	0.000616
22	7	7	7	1.000000	0.000006
22	8	1	0	0.636364	0.636364
22	8	1	1	1.000000	0.363636
22	8	2	0	0.393939	0.393939
22	8	2	1	0.878788	0.484848
22	8	2	2	1.000000	0.121212
22	8	3	0	0.236364	0.236364
22	8	3	1	0.709091	0.472727
22	8	3	2	0.963636	0.254545
22	8	3	3	1.000000	0.036364
22	8	4	0	0.136842	0.136842
22	8	4	1	0.534928	0.398086
22	8	4	2	0.883253	0.348325
22	8	4	3	0.990431	0.107177
22	8	4	4	1.000000	0.009569
22	8	5	0	0.076023	0.076023
22	8	5	1	0.380117	0.304094
22	8	5	2	0.767145	0.387028
22	8	5	3	0.960659	0.193514
22	8	5	4	0.997873	0.037214
22	8	5	5	1.000000	0.002127
22	8	6	0	0.040248	0.040248
22	8	6	1	0.254902	0.214654
22	8	6	2	0.630547	0.375645
22	8	6	3	0.903743	0.273196
22	8	6	4	0.989117	0.085374
22	8	6	5	0.999625	0.010508
22	8	6	6	1.000000	0.000375
22	8	7	0	0.020124	0.020124
22	8	7	1	0.160991	0.140867
22	8	7	2	0.489680	0.328689
22	8	7	3	0.818369	0.328689
22	8	7	4	0.967774	0.149404
22	8	7	5	0.997655	0.029881
22	8	7	6	0.999953	0.002299
22	8	7	7	1.000000	0.000047
22	8	8	0	0.009391	0.009391
22	8	8	1	0.095253	0.085862
22	8	8	2	0.358204	0.262951
22	8	8	3	0.708806	0.350602
22	8	8	4	0.927933	0.219126
22	8	8	5	0.991678	0.063746
22	8	8	6	0.999647	0.007968
22	8	8	7	0.999997	0.000350
22	8	8	8	1.000000	0.000003
22	9	1	0	0.590909	0.590909
22	9	1	1	1.000000	0.409091
22	9	2	0	0.337662	0.337662
22	9	2	1	0.844156	0.506494
22	9	2	2	1.000000	0.155844
22	9	3	0	0.185714	0.185714
22	9	3	1	0.641558	0.455844
22	9	3	2	0.945454	0.303896
22	9	3	3	1.000000	0.054545
22	9	4	0	0.097744	0.097744
22	9	4	1	0.449624	0.351880
22	9	4	2	0.833493	0.383869
22	9	4	3	0.982775	0.149282
22	9	4	4	1.000000	0.017225
22	9	5	0	0.048872	0.048872
22	9	5	1	0.293233	0.244361
22	9	5	2	0.684210	0.390977
22	9	5	3	0.933014	0.248804
22	9	5	4	0.995215	0.062201
22	9	5	5	1.000000	0.004785
22	9	6	0	0.022999	0.022999
22	9	6	1	0.178240	0.155241
22	9	6	2	0.523220	0.344980
22	9	6	3	0.845201	0.321981
22	9	6	4	0.976921	0.131720
22	9	6	5	0.998874	0.021953
22	9	6	6	1.000000	0.001126
22	9	7	0	0.010062	0.010062
22	9	7	1	0.100619	0.090557
22	9	7	2	0.372291	0.271672
22	9	7	3	0.724458	0.352167
22	9	7	4	0.935758	0.211300
22	9	7	5	0.993386	0.057627
22	9	7	6	0.999789	0.006403
22	9	7	7	1.000000	0.000211
22	9	8	0	0.004025	0.004025
22	9	8	1	0.052322	0.048297
22	9	8	2	0.245511	0.193189
22	9	8	3	0.583591	0.338080
22	9	8	4	0.865325	0.281734
22	9	8	5	0.978018	0.112693
22	9	8	6	0.998508	0.020490
22	9	8	7	0.999972	0.001464
22	9	8	8	1.000000	0.000028
22	9	9	0	0.001437	0.001437
22	9	9	1	0.024724	0.023286
22	9	9	2	0.148916	0.124193
22	9	9	3	0.438700	0.289783
22	9	9	4	0.764706	0.326006
22	9	9	5	0.945820	0.181115
22	9	9	6	0.994118	0.048297
22	9	9	7	0.999763	0.005645
22	9	9	8	0.999998	0.000235
22	9	9	9	1.000000	0.000002
22	10	1	0	0.545455	0.545455
22	10	1	1	1.000000	0.454545
22	10	2	0	0.285714	0.285714
22	10	2	1	0.805195	0.519480
22	10	2	2	1.000000	0.194805
22	10	3	0	0.142857	0.142857
22	10	3	1	0.571429	0.428571
22	10	3	2	0.922078	0.350649
22	10	3	3	1.000000	0.077922
22	10	4	0	0.067669	0.067669

Table for $N = 2$, $n = 1$, through $N = 100$, $n = 50$

N	n	k	x	P(x)	p(x)
22	10	4	1	0.368421	0.300752
22	10	4	2	0.774436	0.406015
22	10	4	3	0.971292	0.196856
22	10	4	4	1.000000	0.028708
22	10	5	0	0.030075	0.030075
22	10	5	1	0.218045	0.187970
22	10	5	2	0.593985	0.375940
22	10	5	3	0.894737	0.300752
22	10	5	4	0.990430	0.095694
22	10	5	5	1.000000	0.009569
22	10	6	0	0.012384	0.012384
22	10	6	1	0.118532	0.106148
22	10	6	2	0.417072	0.298540
22	10	6	3	0.770898	0.353826
22	10	6	4	0.956656	0.185759
22	10	6	5	0.997185	0.040529
22	10	6	6	1.000000	0.002815
22	10	7	0	0.004644	0.004644
22	10	7	1	0.058824	0.054180
22	10	7	2	0.267802	0.208978
22	10	7	3	0.616099	0.348297
22	10	7	4	0.886997	0.270898
22	10	7	5	0.984520	0.097523
22	10	7	6	0.999296	0.014776
22	10	7	7	1.000000	0.000704
22	10	8	0	0.001548	0.001548
22	10	8	1	0.026316	0.024768
22	10	8	2	0.156347	0.130031
22	10	8	3	0.453560	0.297214
22	10	8	4	0.778638	0.325077
22	10	8	5	0.952012	0.173375
22	10	8	6	0.995356	0.043344
22	10	8	7	0.999859	0.004503
22	10	8	8	1.000000	0.000141
22	10	9	0	0.000442	0.000442
22	10	9	1	0.010394	0.009951
22	10	9	2	0.082043	0.071650
22	10	9	3	0.304954	0.222910
22	10	9	4	0.639319	0.334365
22	10	9	5	0.890093	0.250774
22	10	9	6	0.982972	0.092879
22	10	9	7	0.998894	0.015922
22	10	9	8	0.999980	0.001086
22	10	9	9	1.000000	0.000020
22	10	10	0	0.000102	0.000102
22	10	10	1	0.003504	0.003402
22	10	10	2	0.037951	0.034447
22	10	10	3	0.184925	0.146974
22	10	10	4	0.484996	0.300071
22	10	10	5	0.793641	0.308645

N	n	k	x	P(x)	p(x)
22	10	10	6	0.954394	0.160753
22	10	10	7	0.995220	0.040826
22	10	10	8	0.999813	0.004593
22	10	10	9	0.999998	0.000186
22	10	10	10	1.000000	0.000002
22	11	1	0	0.500000	0.500000
22	11	1	1	1.000000	0.500000
22	11	2	0	0.238095	0.238095
22	11	2	1	0.761905	0.523810
22	11	2	2	1.000000	0.238095
22	11	3	0	0.107143	0.107143
22	11	3	1	0.500000	0.392857
22	11	3	2	0.892857	0.392857
22	11	3	3	1.000000	0.107143
22	11	4	0	0.045113	0.045113
22	11	4	1	0.293233	0.248120
22	11	4	2	0.706767	0.413534
22	11	4	3	0.954887	0.248120
22	11	4	4	1.000000	0.045113
22	11	5	0	0.017544	0.017544
22	11	5	1	0.155388	0.137845
22	11	5	2	0.500000	0.344611
22	11	5	3	0.844611	0.344611
22	11	5	4	0.982456	0.137845
22	11	5	5	1.000000	0.017544
22	11	6	0	0.006192	0.006192
22	11	6	1	0.074303	0.068111
22	11	6	2	0.317559	0.243255
22	11	6	3	0.682441	0.364883
22	11	6	4	0.925697	0.243255
22	11	6	5	0.993808	0.068111
22	11	6	6	1.000000	0.006192
22	11	7	0	0.001935	0.001935
22	11	7	1	0.031734	0.029799
22	11	7	2	0.180728	0.148994
22	11	7	3	0.500000	0.319272
22	11	7	4	0.819272	0.319272
22	11	7	5	0.968266	0.148994
22	11	7	6	0.998065	0.029799
22	11	7	7	1.000000	0.001935
22	11	8	0	0.000516	0.000516
22	11	8	1	0.011868	0.011352
22	11	8	2	0.091331	0.079463
22	11	8	3	0.329721	0.238390
22	11	8	4	0.670279	0.340557
22	11	8	5	0.908669	0.238390
22	11	8	6	0.988132	0.079463
22	11	8	7	0.999484	0.011352
22	11	8	8	1.000000	0.000516
22	11	9	0	0.000111	0.000111

N	n	k	x	P(x)	p(x)
22	11	9	1	0.003759	0.003649
22	11	9	2	0.040248	0.036488
22	11	9	3	0.193498	0.153251
22	11	9	4	0.500000	0.306502
22	11	9	5	0.806501	0.306502
22	11	9	6	0.959752	0.153251
22	11	9	7	0.996240	0.036488
22	11	9	8	0.999889	0.003649
22	11	9	9	1.000000	0.000111
22	11	10	0	0.000017	0.000017
22	11	10	1	0.000953	0.000936
22	11	10	2	0.014987	0.014034
22	11	10	3	0.099190	0.084204
22	11	10	4	0.334961	0.235770
22	11	10	5	0.665039	0.330079
22	11	10	6	0.900810	0.235770
22	11	10	7	0.985013	0.084204
22	11	10	8	0.999047	0.014034
22	11	10	9	0.999983	0.000936
22	11	10	10	1.000000	0.000017
22	11	11	0	0.000001	0.000001
22	11	11	1	0.000173	0.000172
22	11	11	2	0.004461	0.004288
22	11	11	3	0.043054	0.038593
22	11	11	4	0.197428	0.154373
22	11	11	5	0.500000	0.302572
22	11	11	6	0.802572	0.302572
22	11	11	7	0.956945	0.154373
22	11	11	8	0.995539	0.038593
22	11	11	9	0.999827	0.004288
22	11	11	10	0.999999	0.000172
22	11	11	11	1.000000	0.000001
22	12	1	0	0.454545	0.454545
22	12	1	1	1.000000	0.545455
22	12	2	0	0.194805	0.194805
22	12	2	1	0.714286	0.519480
22	12	2	2	1.000000	0.285714
22	12	3	0	0.077922	0.077922
22	12	3	1	0.428571	0.350649
22	12	3	2	0.857143	0.428571
22	12	3	3	1.000000	0.142857
22	12	4	0	0.028708	0.028708
22	12	4	1	0.225564	0.196856
22	12	4	2	0.631579	0.406015
22	12	4	3	0.932331	0.300752
22	12	4	4	1.000000	0.067669
22	12	5	0	0.009569	0.009569
22	12	5	1	0.105263	0.095694
22	12	5	2	0.406015	0.300752
22	12	5	3	0.781955	0.375940

N	n	k	x	P(x)	p(x)
22	12	5	4	0.969925	0.187970
22	12	5	5	1.000000	0.030075
22	12	6	0	0.002815	0.002815
22	12	6	1	0.043344	0.040529
22	12	6	2	0.229102	0.185759
22	12	6	3	0.582928	0.353826
22	12	6	4	0.881468	0.298540
22	12	6	5	0.987616	0.106148
22	12	6	6	1.000000	0.012384
22	12	7	0	0.000704	0.000704
22	12	7	1	0.015480	0.014776
22	12	7	2	0.113003	0.097523
22	12	7	3	0.383901	0.270898
22	12	7	4	0.732198	0.348297
22	12	7	5	0.941176	0.208978
22	12	7	6	0.995356	0.054180
22	12	7	7	1.000000	0.004644
22	12	8	0	0.000141	0.000141
22	12	8	1	0.004644	0.004503
22	12	8	2	0.047988	0.043344
22	12	8	3	0.221362	0.173375
22	12	8	4	0.546440	0.325077
22	12	8	5	0.843653	0.297214
22	12	8	6	0.973684	0.130031
22	12	8	7	0.998452	0.024768
22	12	8	8	1.000000	0.001548
22	12	9	0	0.000020	0.000020
22	12	9	1	0.001106	0.001086
22	12	9	2	0.017028	0.015922
22	12	9	3	0.109907	0.092879
22	12	9	4	0.360681	0.250774
22	12	9	5	0.695046	0.334365
22	12	9	6	0.917957	0.222910
22	12	9	7	0.989606	0.071650
22	12	9	8	0.999558	0.009951
22	12	9	9	1.000000	0.000442
22	12	10	0	0.000002	0.000002
22	12	10	1	0.000187	0.000186
22	12	10	2	0.004780	0.004593
22	12	10	3	0.045606	0.040826
22	12	10	4	0.206359	0.160753
22	12	10	5	0.515004	0.308645
22	12	10	6	0.815075	0.300071
22	12	10	7	0.962049	0.146974
22	12	10	8	0.996496	0.034447
22	12	10	9	0.999898	0.003402
22	12	10	10	1.000000	0.000102
22	12	11	1	0.000017	0.000017
22	12	11	2	0.000953	0.000936
22	12	11	3	0.014987	0.014034

Table for $N = 2$, $n = 1$, through $N = 100$, $n = 50$

N	n	k	x	P(x)	p(x)
22	12	11	4	0.099190	0.084204
22	12	11	5	0.334961	0.235770
22	12	11	6	0.665039	0.330079
22	12	11	7	0.900810	0.235770
22	12	11	8	0.985013	0.084204
22	12	11	9	0.999047	0.014034
22	12	11	10	0.999983	0.000936
22	12	11	11	1.000000	0.000017
22	12	12	2	0.000102	0.000102
22	12	12	3	0.003504	0.003402
22	12	12	4	0.037951	0.034447
22	12	12	5	0.184925	0.146974
22	12	12	6	0.484996	0.300071
22	12	12	7	0.793641	0.308645
22	12	12	8	0.954394	0.160753
22	12	12	9	0.995220	0.040826
22	12	12	10	0.999813	0.004593
22	12	12	11	0.999998	0.000186
22	12	12	12	1.000000	0.000002
22	13	1	0	0.409091	0.409091
22	13	1	1	1.000000	0.590909
22	13	2	0	0.155844	0.155844
22	13	2	1	0.662338	0.506494
22	13	2	2	1.000000	0.337662
22	13	3	0	0.054545	0.054545
22	13	3	1	0.358442	0.303896
22	13	3	2	0.814286	0.455844
22	13	3	3	1.000000	0.185714
22	13	4	0	0.017225	0.017225
22	13	4	1	0.166507	0.149282
22	13	4	2	0.550376	0.383869
22	13	4	3	0.902256	0.351880
22	13	4	4	1.000000	0.097744
22	13	5	0	0.004785	0.004785
22	13	5	1	0.066986	0.062201
22	13	5	2	0.315789	0.248804
22	13	5	3	0.706767	0.390977
22	13	5	4	0.951128	0.244361
22	13	5	5	1.000000	0.048872
22	13	6	0	0.001126	0.001126
22	13	6	1	0.023079	0.021953
22	13	6	2	0.154799	0.131720
22	13	6	3	0.476780	0.321981
22	13	6	4	0.821760	0.344980
22	13	6	5	0.977001	0.155241
22	13	6	6	1.000000	0.022999
22	13	7	0	0.000211	0.000211
22	13	7	1	0.006614	0.006403
22	13	7	2	0.064241	0.057627
22	13	7	3	0.275542	0.211300

N	n	k	x	P(x)	p(x)
22	13	7	4	0.627709	0.352167
22	13	7	5	0.899381	0.271672
22	13	7	6	0.989938	0.090557
22	13	7	7	1.000000	0.010062
22	13	8	0	0.000028	0.000028
22	13	8	1	0.001492	0.001464
22	13	8	2	0.021981	0.020490
22	13	8	3	0.134675	0.112693
22	13	8	4	0.416409	0.281734
22	13	8	5	0.754489	0.338080
22	13	8	6	0.947678	0.193189
22	13	8	7	0.995975	0.048297
22	13	8	8	1.000000	0.004025
22	13	9	0	0.000002	0.000002
22	13	9	1	0.000237	0.000235
22	13	9	2	0.005882	0.005645
22	13	9	3	0.054180	0.048297
22	13	9	4	0.235294	0.181115
22	13	9	5	0.561300	0.326006
22	13	9	6	0.851083	0.289783
22	13	9	7	0.975276	0.124193
22	13	9	8	0.998562	0.023286
22	13	9	9	1.000000	0.001437
22	13	10	1	0.000020	0.000020
22	13	10	2	0.001106	0.001086
22	13	10	3	0.017028	0.015922
22	13	10	4	0.109907	0.092879
22	13	10	5	0.360681	0.250774
22	13	10	6	0.695046	0.334365
22	13	10	7	0.917957	0.222910
22	13	10	8	0.989606	0.071650
22	13	10	9	0.999558	0.009951
22	13	10	10	1.000000	0.000442
22	13	11	2	0.000111	0.000111
22	13	11	3	0.003759	0.003649
22	13	11	4	0.040248	0.036488
22	13	11	5	0.193498	0.153251
22	13	11	6	0.500000	0.306502
22	13	11	7	0.806501	0.306502
22	13	11	8	0.959752	0.153251
22	13	11	9	0.996240	0.036488
22	13	11	10	0.999889	0.003649
22	13	11	11	1.000000	0.000111
22	13	12	3	0.000442	0.000442
22	13	12	4	0.010394	0.009951
22	13	12	5	0.082043	0.071650
22	13	12	6	0.304954	0.222910
22	13	12	7	0.639319	0.334365
22	13	12	8	0.890093	0.250774
22	13	12	9	0.982972	0.092879

N	n	k	x	P(x)	p(x)
22	13	12	10	0.998894	0.015922
22	13	12	11	0.999980	0.001086
22	13	12	12	1.000000	0.000020
22	13	13	4	0.001437	0.001437
22	13	13	5	0.024724	0.023286
22	13	13	6	0.148916	0.124193
22	13	13	7	0.438700	0.289783
22	13	13	8	0.764706	0.326006
22	13	13	9	0.945820	0.181115
22	13	13	10	0.994118	0.048297
22	13	13	11	0.999763	0.005645
22	13	13	12	0.999998	0.000235
22	13	13	13	1.000000	0.000002
22	14	1	0	0.363636	0.363636
22	14	1	1	1.000000	0.636364
22	14	2	0	0.121212	0.121212
22	14	2	1	0.606061	0.484848
22	14	2	2	1.000000	0.393939
22	14	3	0	0.036364	0.036364
22	14	3	1	0.290909	0.254545
22	14	3	2	0.763636	0.472727
22	14	3	3	1.000000	0.236364
22	14	4	0	0.009569	0.009569
22	14	4	1	0.116746	0.107177
22	14	4.	2	0.465072	0.348325
22	14	4	3	0.863158	0.398086
22	14	4	4	1.000000	0.136842
22	14	5	0	0.002127	0.002127
22	14	5	1	0.039341	0.037214
22	14	5	2	0.232855	0.193514
22	14	5	3	0.619883	0.387028
22	14	5	4	0.923977	0.304094
22	14	5	5	1.000000	0.076023
22	14	6	0	0.000375	0.000375
22	14	6	1	0.010883	0.010508
22	14	6	2	0.096257	0.085374
22	14	6	3	0.369453	0.273196
22	14	6	4	0.745098	0.375645
22	14	6	5	0.959752	0.214654
22	14	6	6	1.000000	0.040248
22	14	7	0	0.000047	0.000047
22	14	7	1	0.002345	0.002299
22	14	7	2	0.032226	0.029881
22	14	7	3	0.181631	0.149404
22	14	7	4	0.510320	0.328689
22	14	7	5	0.839009	0.328689
22	14	7	6	0.979876	0.140867
22	14	7	7	1.000000	0.020124
22	14	8	0	0.000003	0.000003
22	14	8	1	0.000353	0.000350

N	n	k	x	P(x)	p(x)
22	14	8	2	0.008322	0.007968
22	14	8	3	0.072067	0.063746
22	14	8	4	0.291194	0.219126
22	14	8	5	0.641796	0.350602
22	14	8	6	0.904747	0.262951
22	14	8	7	0.990609	0.085862
22	14	8	8	1.000000	0.009391
22	14	9	1	0.000028	0.000028
22	14	9	2	0.001492	0.001464
22	14	9	3	0.021981	0.020490
22	14	9	4	0.134675	0.112693
22	14	9	5	0.416409	0.281734
22	14	9	6	0.754489	0.338080
22	14	9	7	0.947678	0.193189
22	14	9	8	0.995975	0.048297
22	14	9	9	1.000000	0.004025
22	14	10	2	0.000141	0.000141
22	14	10	3	0.004644	0.004503
22	14	10	4	0.047988	0.043344
22	14	10	5	0.221362	0.173375
22	14	10	6	0.546440	0.325077
22	14	10	7	0.843653	0.297214
22	14	10	8	0.973684	0.130031
22	14	10	9	0.998452	0.024768
22	14	10	10	1.000000	0.001548
22	14	11	3	0.000516	0.000516
22	14	11	4	0.011868	0.011352
22	14	11	5	0.091331	0.079463
22	14	11	6	0.329721	0.238390
22	14	11	7	0.670279	0.340557
22	14	11	8	0.908669	0.238390
22	14	11	9	0.988132	0.079463
22	14	11	10	0.999484	0.011352
22	14	11	11	1.000000	0.000516
22	14	12	4	0.001548	0.001548
22	14	12	5	0.026316	0.024768
22	14	12	6	0.156347	0.130031
22	14	12	7	0.453560	0.297214
22	14	12	8	0.778638	0.325077
22	14	12	9	0.952012	0.173375
22	14	12	10	0.995356	0.043344
22	14	12	11	0.999859	0.004503
22	14	12	12	1.000000	0.000141
22	14	13	5	0.004025	0.004025
22	14	13	6	0.052322	0.048297
22	14	13	7	0.245511	0.193189
22	14	13	8	0.583591	0.338080
22	14	13	9	0.865325	0.281734
22	14	13	10	0.978018	0.112693
22	14	13	11	0.998508	0.020490

Table for $N = 2$, $n = 1$, through $N = 100$, $n = 50$

N	n	k	x	P(x)	p(x)
22	14	13	12	0.999972	0.001464
22	14	13	13	1.000000	0.000028
22	14	14	6	0.009391	0.009391
22	14	14	7	0.095253	0.085862
22	14	14	8	0.358204	0.262951
22	14	14	9	0.708806	0.350602
22	14	14	10	0.927933	0.219126
22	14	14	11	0.991678	0.063746
22	14	14	12	0.999647	0.007968
22	14	14	13	0.999997	0.000350
22	14	14	14	1.000000	0.000003
22	15	1	0	0.318182	0.318182
22	15	1	1	1.000000	0.681818
22	15	2	0	0.090909	0.090909
22	15	2	1	0.545454	0.454545
22	15	2	2	1.000000	0.454545
22	15	3	0	0.022727	0.022727
22	15	3	1	0.227273	0.204545
22	15	3	2	0.704545	0.477273
22	15	3	3	1.000000	0.295455
22	15	4	0	0.004785	0.004785
22	15	4	1	0.076555	0.071770
22	15	4	2	0.377990	0.301435
22	15	4	3	0.813397	0.435407
22	15	4	4	1.000000	0.186603
22	15	5	0	0.000797	0.000797
22	15	5	1	0.020734	0.019936
22	15	5	2	0.160287	0.139553
22	15	5	3	0.523126	0.362839
22	15	5	4	0.885965	0.362839
22	15	5	5	1.000000	0.114035
22	15	6	0	0.000094	0.000094
22	15	6	1	0.004316	0.004222
22	15	6	2	0.053570	0.049254
22	15	6	3	0.267004	0.213435
22	15	6	4	0.651187	0.384182
22	15	6	5	0.932921	0.281734
22	15	6	6	1.000000	0.067079
22	15	7	0	0.000006	0.000006
22	15	7	1	0.000622	0.000616
22	15	7	2	0.013551	0.012929
22	15	7	3	0.106928	0.093378
22	15	7	4	0.387061	0.280133
22	15	7	5	0.756837	0.369776
22	15	7	6	0.962268	0.205431
22	15	7	7	1.000000	0.037732
22	15	8	1	0.000047	0.000047
22	15	8	2	0.002345	0.002299
22	15	8	3	0.032226	0.029881
22	15	8	4	0.181631	0.149404

N	n	k	x	P(x)	p(x)
22	15	8	5	0.510320	0.328689
22	15	8	6	0.839009	0.328689
22	15	8	7	0.979876	0.140867
22	15	8	8	1.000000	0.020124
22	15	9	2	0.000211	0.000211
22	15	9	3	0.006614	0.006403
22	15	9	4	0.064241	0.057627
22	15	9	5	0.275542	0.211300
22	15	9	6	0.627709	0.352167
22	15	9	7	0.899381	0.271672
22	15	9	8	0.989938	0.090557
22	15	9	9	1.000000	0.010062
22	15	10	3	0.000704	0.000704
22	15	10	4	0.015480	0.014776
22	15	10	5	0.113003	0.097523
22	15	10	6	0.383901	0.270898
22	15	10	7	0.732198	0.348297
22	15	10	8	0.941176	0.208978
22	15	10	9	0.995356	0.054180
22	15	10	10	1.000000	0.004644
22	15	11	4	0.001935	0.001935
22	15	11	5	0.031734	0.029799
22	15	11	6	0.180728	0.148994
22	15	11	7	0.500000	0.319272
22	15	11	8	0.819272	0.319272
22	15	11	9	0.968266	0.148994
22	15	11	10	0.998065	0.029799
22	15	11	11	1.000000	0.001935
22	15	12	5	0.004644	0.004644
22	15	12	6	0.058824	0.054180
22	15	12	7	0.267802	0.208978
22	15	12	8	0.616099	0.348297
22	15	12	9	0.886997	0.270898
22	15	12	10	0.984520	0.097523
22	15	12	11	0.999296	0.014776
22	15	12	12	1.000000	0.000704
22	15	13	6	0.010062	0.010062
22	15	13	7	0.100619	0.090557
22	15	13	8	0.372291	0.271672
22	15	13	9	0.724458	0.352167
22	15	13	10	0.935758	0.211300
22	15	13	11	0.993386	0.057627
22	15	13	12	0.999789	0.006403
22	15	13	13	1.000000	0.000211
22	15	14	7	0.020124	0.020124
22	15	14	8	0.160991	0.140867
22	15	14	9	0.489680	0.328689
22	15	14	10	0.818369	0.328689
22	15	14	11	0.967774	0.149404
22	15	14	12	0.997655	0.029881

N	n	k	x	P(x)	p(x)
22	15	14	13	0.999953	0.002299
22	15	14	14	1.000000	0.000047
22	15	15	8	0.037732	0.037732
22	15	15	9	0.243163	0.205431
22	15	15	10	0.612939	0.369776
22	15	15	11	0.893072	0.280133
22	15	15	12	0.986449	0.093378
22	15	15	13	0.999378	0.012929
22	15	15	14	0.999994	0.000616
22	15	15	15	1.000000	0.000006
22	16	1	0	0.272727	0.272727
22	16	1	1	1.000000	0.727273
22	16	2	0	0.064935	0.064935
22	16	2	1	0.480519	0.415584
22	16	2	2	1.000000	0.519480
22	16	3	0	0.012987	0.012987
22	16	3	1	0.168831	0.155844
22	16	3	2	0.636364	0.467532
22	16	3	3	1.000000	0.363636
22	16	4	0	0.002051	0.002051
22	16	4	1	0.045796	0.043746
22	16	4	2	0.291866	0.246070
22	16	4	3	0.751196	0.459330
22	16	4	4	1.000000	0.248804
22	16	5	0	0.000228	0.000228
22	16	5	1	0.009342	0.009114
22	16	5	2	0.100478	0.091137
22	16	5	3	0.419458	0.318979
22	16	5	4	0.834131	0.414673
22	16	5	5	1.000000	0.165869
22	16	6	0	0.000013	0.000013
22	16	6	1	0.001300	0.001287
22	16	6	2	0.025425	0.024124
22	16	6	3	0.175532	0.150108
22	16	6	4	0.541420	0.365888
22	16	6	5	0.892673	0.351252
22	16	6	6	1.000000	0.107327
22	16	7	1	0.000094	0.000094
22	16	7	2	0.004316	0.004222
22	16	7	3	0.053570	0.049254
22	16	7	4	0.267004	0.213435
22	16	7	5	0.651187	0.384182
22	16	7	6	0.932921	0.281734
22	16	7	7	1.000000	0.067079
22	16	8	2	0.000375	0.000375
22	16	8	3	0.010883	0.010508
22	16	8	4	0.096257	0.085374
22	16	8	5	0.369453	0.273196
22	16	8	6	0.745098	0.375645
22	16	8	7	0.959752	0.214654

N	n	k	x	P(x)	p(x)
22	16	8	8	1.000000	0.040248
22	16	9	3	0.001126	0.001126
22	16	9	4	0.023079	0.021953
22	16	9	5	0.154799	0.131720
22	16	9	6	0.476780	0.321981
22	16	9	7	0.821760	0.344980
22	16	9	8	0.977001	0.155241
22	16	9	9	1.000000	0.022999
22	16	10	4	0.002815	0.002815
22	16	10	5	0.043344	0.040529
22	16	10	6	0.229102	0.185759
22	16	10	7	0.582928	0.353826
22	16	10	8	0.881468	0.298540
22	16	10	9	0.987616	0.106148
22	16	10	10	1.000000	0.012384
22	16	11	5	0.006192	0.006192
22	16	11	6	0.074303	0.068111
22	16	11	7	0.317559	0.243255
22	16	11	8	0.682441	0.364883
22	16	11	9	0.925697	0.243255
22	16	11	10	0.993808	0.068111
22	16	11	11	1.000000	0.006192
22	16	12	6	0.012384	0.012384
22	16	12	7	0.118532	0.106148
22	16	12	8	0.417072	0.298540
22	16	12	9	0.770898	0.353826
22	16	12	10	0.956656	0.185759
22	16	12	11	0.997185	0.040529
22	16	12	12	1.000000	0.002815
22	16	13	7	0.022999	0.022999
22	16	13	8	0.178240	0.155241
22	16	13	9	0.523220	0.344980
22	16	13	10	0.845201	0.321981
22	16	13	11	0.976921	0.131720
22	16	13	12	0.998874	0.021953
22	16	13	13	1.000000	0.001126
22	16	14	8	0.040248	0.040248
22	16	14	9	0.254902	0.214654
22	16	14	10	0.630547	0.375645
22	16	14	11	0.903743	0.273196
22	16	14	12	0.989117	0.085374
22	16	14	13	0.999625	0.010508
22	16	14	14	1.000000	0.000375
22	16	15	9	0.067079	0.067079
22	16	15	10	0.348813	0.281734
22	16	15	11	0.732996	0.384182
22	16	15	12	0.946430	0.213435
22	16	15	13	0.995684	0.049254
22	16	15	14	0.999906	0.004222
22	16	15	15	1.000000	0.000094

Table for $N = 2$, $n = 1$, through $N = 100$, $n = 50$

N	n	k	x	P(x)	p(x)
22	16	16	10	0.107327	0.107327
22	16	16	11	0.458580	0.351252
22	16	16	12	0.824468	0.365888
22	16	16	13	0.974575	0.150108
22	16	16	14	0.998700	0.024124
22	16	16	15	0.999986	0.001287
22	16	16	16	1.000000	0.000013
22	17	1	0	0.227273	0.227273
22	17	1	1	1.000000	0.772727
22	17	2	0	0.043290	0.043290
22	17	2	1	0.411255	0.367965
22	17	2	2	1.000000	0.588745
22	17	3	0	0.006494	0.006494
22	17	3	1	0.116883	0.110390
22	17	3	2	0.558442	0.441558
22	17	3	3	1.000000	0.441558
22	17	4	0	0.000684	0.000684
22	17	4	1	0.023923	0.023240
22	17	4	2	0.209843	0.185919
22	17	4	3	0.674641	0.464798
22	17	4	4	1.000000	0.325359
22	17	5	0	0.000038	0.000038
22	17	5	1	0.003266	0.003228
22	17	5	2	0.054910	0.051644
22	17	5	3	0.313131	0.258221
22	17	5	4	0.765019	0.451887
22	17	5	5	1.000000	0.234981
22	17	6	1	0.000228	0.000228
22	17	6	2	0.009342	0.009114
22	17	6	3	0.100478	0.091137
22	17	6	4	0.419458	0.318979
22	17	6	5	0.834131	0.414673
22	17	6	6	1.000000	0.165869
22	17	7	2	0.000797	0.000797
22	17	7	3	0.020734	0.019936
22	17	7	4	0.160287	0.139553
22	17	7	5	0.523126	0.362839
22	17	7	6	0.885965	0.362839
22	17	7	7	1.000000	0.114035
22	17	8	3	0.002127	0.002127
22	17	8	4	0.039341	0.037214
22	17	8	5	0.232855	0.193514
22	17	8	6	0.619883	0.387028
22	17	8	7	0.923977	0.304094
22	17	8	8	1.000000	0.076023
22	17	9	4	0.004785	0.004785
22	17	9	5	0.066986	0.062201
22	17	9	6	0.315789	0.248804
22	17	9	7	0.706767	0.390977
22	17	9	8	0.951128	0.244361

N	n	k	x	P(x)	p(x)
22	17	9	9	1.000000	0.048872
22	17	10	5	0.009569	0.009569
22	17	10	6	0.105263	0.095694
22	17	10	7	0.406015	0.300752
22	17	10	8	0.781955	0.375940
22	17	10	9	0.969925	0.187970
22	17	10	10	1.000000	0.030075
22	17	11	6	0.017544	0.017544
22	17	11	7	0.155388	0.137845
22	17	11	8	0.500000	0.344611
22	17	11	9	0.844611	0.344611
22	17	11	10	0.982456	0.137845
22	17	11	11	1.000000	0.017544
22	17	12	7	0.030075	0.030075
22	17	12	8	0.218045	0.187970
22	17	12	9	0.593985	0.375940
22	17	12	10	0.894737	0.300752
22	17	12	11	0.990430	0.095694
22	17	12	12	1.000000	0.009569
22	17	13	8	0.048872	0.048872
22	17	13	9	0.293233	0.244361
22	17	13	10	0.684210	0.390977
22	17	13	11	0.933014	0.248804
22	17	13	12	0.995215	0.062201
22	17	13	13	1.000000	0.004785
22	17	14	9	0.076023	0.076023
22	17	14	10	0.380117	0.304094
22	17	14	11	0.767145	0.387028
22	17	14	12	0.960659	0.193514
22	17	14	13	0.997873	0.037214
22	17	14	14	1.000000	0.002127
22	17	15	10	0.114035	0.114035
22	17	15	11	0.476874	0.362839
22	17	15	12	0.839713	0.362839
22	17	15	13	0.979266	0.139553
22	17	15	14	0.999202	0.019936
22	17	15	15	1.000000	0.000797
22	17	16	11	0.165869	0.165869
22	17	16	12	0.580542	0.414673
22	17	16	13	0.899521	0.318979
22	17	16	14	0.990658	0.091137
22	17	16	15	0.999772	0.009114
22	17	16	16	1.000000	0.000228
22	17	17	12	0.234981	0.234981
22	17	17	13	0.686869	0.451887
22	17	17	14	0.945090	0.258221
22	17	17	15	0.996734	0.051644
22	17	17	16	0.999962	0.003228
22	17	17	17	1.000000	0.000038
22	18	1	0	0.181818	0.181818

N	n	k	x	P(x)	p(x)
22	18	1	1	1.000000	0.818182
22	18	2	0	0.025974	0.025974
22	18	2	1	0.337662	0.311688
22	18	2	2	1.000000	0.662338
22	18	3	0	0.002597	0.002597
22	18	3	1	0.072727	0.070130
22	18	3	2	0.470130	0.397403
22	18	3	3	1.000000	0.529870
22	18	4	0	0.000137	0.000137
22	18	4	1	0.009979	0.009843
22	18	4	2	0.135475	0.125496
22	18	4	3	0.581681	0.446206
22	18	4	4	1.000000	0.418319
22	18	5	1	0.000684	0.000684
22	18	5	2	0.023923	0.023240
22	18	5	3	0.209843	0.185919
22	18	5	4	0.674641	0.464798
22	18	5	5	1.000000	0.325359
22	18	6	2	0.002051	0.002051
22	18	6	3	0.045796	0.043746
22	18	6	4	0.291866	0.246070
22	18	6	5	0.751196	0.459330
22	18	6	6	1.000000	0.248804
22	18	7	3	0.004785	0.004785
22	18	7	4	0.076555	0.071770
22	18	7	5	0.377990	0.301435
22	18	7	6	0.813397	0.435407
22	18	7	7	1.000000	0.186603
22	18	8	4	0.009569	0.009569
22	18	8	5	0.116746	0.107177
22	18	8	6	0.465072	0.348325
22	18	8	7	0.863158	0.398086
22	18	8	8	1.000000	0.136842
22	18	9	5	0.017225	0.017225
22	18	9	6	0.166507	0.149282
22	18	9	7	0.550376	0.383869
22	18	9	8	0.902256	0.351880
22	18	9	9	1.000000	0.097744
22	18	10	6	0.028708	0.028708
22	18	10	7	0.225564	0.196856
22	18	10	8	0.631579	0.406015
22	18	10	9	0.932331	0.300752
22	18	10	10	1.000000	0.067669
22	18	11	7	0.045113	0.045113
22	18	11	8	0.293233	0.248120
22	18	11	9	0.706767	0.413534
22	18	11	10	0.954887	0.248120
22	18	11	11	1.000000	0.045113
22	18	12	8	0.067669	0.067669
22	18	12	9	0.368421	0.300752

N	n	k	x	P(x)	p(x)
22	18	12	10	0.774436	0.406015
22	18	12	11	0.971292	0.196856
22	18	12	12	1.000000	0.028708
22	18	13	9	0.097744	0.097744
22	18	13	10	0.449624	0.351880
22	18	13	11	0.833493	0.383869
22	18	13	12	0.982775	0.149282
22	18	13	13	1.000000	0.017225
22	18	14	10	0.136842	0.136842
22	18	14	11	0.534928	0.398086
22	18	14	12	0.883253	0.348325
22	18	14	13	0.990431	0.107177
22	18	14	14	1.000000	0.009569
22	18	15	11	0.186603	0.186603
22	18	15	12	0.622010	0.435407
22	18	15	13	0.923445	0.301435
22	18	15	14	0.995215	0.071770
22	18	15	15	1.000000	0.004785
22	18	16	12	0.248804	0.248804
22	18	16	13	0.708134	0.459330
22	18	16	14	0.954204	0.246070
22	18	16	15	0.997949	0.043746
22	18	16	16	1.000000	0.002051
22	18	17	13	0.325359	0.325359
22	18	17	14	0.790157	0.464798
22	18	17	15	0.976076	0.185919
22	18	17	16	0.999316	0.023240
22	18	17	17	1.000000	0.000684
22	18	18	14	0.418319	0.418319
22	18	18	15	0.864525	0.446206
22	18	18	16	0.990020	0.125496
22	18	18	17	0.999863	0.009843
22	18	18	18	1.000000	0.000137
22	19	1	0	0.136364	0.136364
22	19	1	1	1.000000	0.863636
22	19	2	0	0.012987	0.012987
22	19	2	1	0.259740	0.246753
22	19	2	2	1.000000	0.740260
22	19	3	0	0.000649	0.000649
22	19	3	1	0.037662	0.037013
22	19	3	2	0.370779	0.333117
22	19	3	3	1.000000	0.629221
22	19	4	1	0.002597	0.002597
22	19	4	2	0.072727	0.070130
22	19	4	3	0.470130	0.397403
22	19	4	4	1.000000	0.529870
22	19	5	2	0.006494	0.006494
22	19	5	3	0.116883	0.110390
22	19	5	4	0.558442	0.441558
22	19	5	5	1.000000	0.441558

N	n	k	x	P(x)	p(x)
22	19	6	3	0.012987	0.012987
22	19	6	4	0.168831	0.155844
22	19	6	5	0.636364	0.467532
22	19	6	6	1.000000	0.363636
22	19	7	4	0.022727	0.022727
22	19	7	5	0.227273	0.204545
22	19	7	6	0.704545	0.477273
22	19	7	7	1.000000	0.295455
22	19	8	5	0.036364	0.036364
22	19	8	6	0.290909	0.254545
22	19	8	7	0.763636	0.472727
22	19	8	8	1.000000	0.236364
22	19	9	6	0.054545	0.054545
22	19	9	7	0.358442	0.303896
22	19	9	8	0.814286	0.455844
22	19	9	9	1.000000	0.185714
22	19	10	7	0.077922	0.077922
22	19	10	8	0.428571	0.350649
22	19	10	9	0.857143	0.428571
22	19	10	10	1.000000	0.142857
22	19	11	8	0.107143	0.107143
22	19	11	9	0.500000	0.392857
22	19	11	10	0.892857	0.392857
22	19	11	11	1.000000	0.107143
22	19	12	9	0.142857	0.142857
22	19	12	10	0.571429	0.428571
22	19	12	11	0.922078	0.350649
22	19	12	12	1.000000	0.077922
22	19	13	10	0.185714	0.185714
22	19	13	11	0.641558	0.455844
22	19	13	12	0.945454	0.303896
22	19	13	13	1.000000	0.054545
22	19	14	11	0.236364	0.236364
22	19	14	12	0.709091	0.472727
22	19	14	13	0.963636	0.254545
22	19	14	14	1.000000	0.036364
22	19	15	12	0.295455	0.295455
22	19	15	13	0.772727	0.477273
22	19	15	14	0.977273	0.204545
22	19	15	15	1.000000	0.022727
22	19	16	13	0.363636	0.363636
22	19	16	14	0.831169	0.467532
22	19	16	15	0.987013	0.155844
22	19	16	16	1.000000	0.012987
22	19	17	14	0.441558	0.441558
22	19	17	15	0.883117	0.441558
22	19	17	16	0.993506	0.110390
22	19	17	17	1.000000	0.006494
22	19	18	15	0.529870	0.529870
22	19	18	16	0.927273	0.397403

N	n	k	x	P(x)	p(x)
22	19	18	17	0.997403	0.070130
22	19	18	18	1.000000	0.002597
22	19	19	16	0.629221	0.629221
22	19	19	17	0.962338	0.333117
22	19	19	18	0.999351	0.037013
22	19	19	19	1.000000	0.000649
22	20	1	0	0.090909	0.090909
22	20	1	1	1.000000	0.909091
22	20	2	0	0.004329	0.004329
22	20	2	1	0.177489	0.173160
22	20	2	2	1.000000	0.822511
22	20	3	1	0.012987	0.012987
22	20	3	2	0.259740	0.246753
22	20	3	3	1.000000	0.740260
22	20	4	2	0.025974	0.025974
22	20	4	3	0.337662	0.311688
22	20	4	4	1.000000	0.662338
22	20	5	3	0.043290	0.043290
22	20	5	4	0.411255	0.367965
22	20	5	5	1.000000	0.588745
22	20	6	4	0.064935	0.064935
22	20	6	5	0.480519	0.415584
22	20	6	6	1.000000	0.519480
22	20	7	5	0.090909	0.090909
22	20	7	6	0.545454	0.454545
22	20	7	7	1.000000	0.454545
22	20	8	6	0.121212	0.121212
22	20	8	7	0.606061	0.484848
22	20	8	8	1.000000	0.393939
22	20	9	7	0.155844	0.155844
22	20	9	8	0.662338	0.506494
22	20	9	9	1.000000	0.337662
22	20	10	8	0.194805	0.194805
22	20	10	9	0.714286	0.519480
22	20	10	10	1.000000	0.285714
22	20	11	9	0.238095	0.238095
22	20	11	10	0.761905	0.523810
22	20	11	11	1.000000	0.238095
22	20	12	10	0.285714	0.285714
22	20	12	11	0.805195	0.519480
22	20	12	12	1.000000	0.194805
22	20	13	11	0.337662	0.337662
22	20	13	12	0.844156	0.506494
22	20	13	13	1.000000	0.155844
22	20	14	12	0.393939	0.393939
22	20	14	13	0.878788	0.484848
22	20	14	14	1.000000	0.121212
22	20	15	13	0.454545	0.454545
22	20	15	14	0.909091	0.454545
22	20	15	15	1.000000	0.090909

N	n	k	x	P(x)	p(x)
22	20	16	14	0.519480	0.519480
22	20	16	15	0.935065	0.415584
22	20	16	16	1.000000	0.064935
22	20	17	15	0.588745	0.588745
22	20	17	16	0.956710	0.367965
22	20	17	17	1.000000	0.043290
22	20	18	16	0.662338	0.662338
22	20	18	17	0.974026	0.311688
22	20	18	18	1.000000	0.025974
22	20	19	17	0.740260	0.740260
22	20	19	18	0.987013	0.246753
22	20	19	19	1.000000	0.012987
22	20	20	18	0.822511	0.822511
22	20	20	19	0.995671	0.173160
22	20	20	20	1.000000	0.004329
22	21	1	0	0.045455	0.045455
22	21	1	1	1.000000	0.954545
22	21	2	1	0.090909	0.090909
22	21	2	2	1.000000	0.909091
22	21	3	2	0.136364	0.136364
22	21	3	3	1.000000	0.863636
22	21	4	3	0.181818	0.181818
22	21	4	4	1.000000	0.818182
22	21	5	4	0.227273	0.227273
22	21	5	5	1.000000	0.772727
22	21	6	5	0.272727	0.272727
22	21	6	6	1.000000	0.727273
22	21	7	6	0.318182	0.318182
22	21	7	7	1.000000	0.681818
22	21	8	7	0.363636	0.363636
22	21	8	8	1.000000	0.636364
22	21	9	8	0.409091	0.409091
22	21	9	9	1.000000	0.590909
22	21	10	9	0.454545	0.454545
22	21	10	10	1.000000	0.545455
22	21	11	10	0.500000	0.500000
22	21	11	11	1.000000	0.500000
22	21	12	11	0.545455	0.545455
22	21	12	12	1.000000	0.454545
22	21	13	12	0.590909	0.590909
22	21	13	13	1.000000	0.409091
22	21	14	13	0.636364	0.636364
22	21	14	14	1.000000	0.363636
22	21	15	14	0.681818	0.681818
22	21	15	15	1.000000	0.318182
22	21	16	15	0.727273	0.727273
22	21	16	16	1.000000	0.272727
22	21	17	16	0.772727	0.772727
22	21	17	17	1.000000	0.227273
22	21	18	17	0.818182	0.818182

N	n	k	x	P(x)	p(x)
22	21	18	18	1.000000	0.181818
22	21	19	18	0.863636	0.863636
22	21	19	19	1.000000	0.136364
22	21	20	19	0.909091	0.909091
22	21	20	20	1.000000	0.090909
22	21	21	20	0.954545	0.954545
22	21	21	21	1.000000	0.045455
23	1	1	0	0.956522	0.956522
23	1	1	1	1.000000	0.043478
23	2	1	0	0.913043	0.913043
23	2	1	1	1.000000	0.086957
23	2	2	0	0.830040	0.830040
23	2	2	1	0.996047	0.166008
23	2	2	2	1.000000	0.003953
23	3	1	0	0.869565	0.869565
23	3	1	1	1.000000	0.130435
23	3	2	0	0.750988	0.750988
23	3	2	1	0.988142	0.237154
23	3	2	2	1.000000	0.011858
23	3	3	0	0.643704	0.643704
23	3	3	1	0.965556	0.321852
23	3	3	2	0.999435	0.033879
23	3	3	3	1.000000	0.000565
23	4	1	0	0.826087	0.826087
23	4	1	1	1.000000	0.173913
23	4	2	0	0.675889	0.675889
23	4	2	1	0.976285	0.300395
23	4	2	2	1.000000	0.023715
23	4	3	0	0.547148	0.547148
23	4	3	1	0.933371	0.386222
23	4	3	2	0.997741	0.064370
23	4	3	3	1.000000	0.002259
23	4	4	0	0.437719	0.437719
23	4	4	1	0.875438	0.437719
23	4	4	2	0.991304	0.115867
23	4	4	3	0.999887	0.008583
23	4	4	4	1.000000	0.000113
23	5	1	0	0.782609	0.782609
23	5	1	1	1.000000	0.217391
23	5	2	0	0.604743	0.604743
23	5	2	1	0.960474	0.355731
23	5	2	2	1.000000	0.039526
23	5	3	0	0.460757	0.460757
23	5	3	1	0.892716	0.431959
23	5	3	2	0.994353	0.101637
23	5	3	3	1.000000	0.005647
23	5	4	0	0.345567	0.345567
23	5	4	1	0.806324	0.460757
23	5	4	2	0.979108	0.172784
23	5	4	3	0.999435	0.020327

Table for $N=2$, $n=1$, through $N=100$, $n=50$

N	n	k	x	P(x)	p(x)
23	5	4	4	1.000000	0.000565
23	5	5	0	0.254629	0.254629
23	5	5	1	0.709323	0.454694
23	5	5	2	0.951826	0.242503
23	5	5	3	0.997296	0.045469
23	5	5	4	0.999970	0.002675
23	5	5	5	1.000000	0.000030
23	6	1	0	0.739130	0.739130
23	6	1	1	1.000000	0.260870
23	6	2	0	0.537549	0.537549
23	6	2	1	0.940711	0.403162
23	6	2	2	1.000000	0.059289
23	6	3	0	0.383964	0.383964
23	6	3	1	0.844720	0.460757
23	6	3	2	0.988707	0.143986
23	6	3	3	1.000000	0.011293
23	6	4	0	0.268775	0.268775
23	6	4	1	0.729531	0.460757
23	6	4	2	0.959910	0.230378
23	6	4	3	0.998306	0.038396
23	6	4	4	1.000000	0.001694
23	6	5	0	0.183898	0.183898
23	6	5	1	0.608280	0.424381
23	6	5	2	0.911409	0.303129
23	6	5	3	0.992243	0.080834
23	6	5	4	0.999822	0.007578
23	6	5	5	1.000000	0.000178
23	6	6	0	0.122599	0.122599
23	6	6	1	0.490396	0.367797
23	6	6	2	0.844047	0.353651
23	6	6	3	0.978771	0.134724
23	6	6	4	0.998980	0.020209
23	6	6	5	0.999990	0.001010
23	6	6	6	1.000000	0.000010
23	7	1	0	0.695652	0.695652
23	7	1	1	1.000000	0.304348
23	7	2	0	0.474308	0.474308
23	7	2	1	0.916996	0.442688
23	7	2	2	1.000000	0.083004
23	7	3	0	0.316206	0.316206
23	7	3	1	0.790514	0.474308
23	7	3	2	0.980237	0.189723
23	7	3	3	1.000000	0.019763
23	7	4	0	0.205534	0.205534
23	7	4	1	0.648221	0.442688
23	7	4	2	0.932806	0.284585
23	7	4	3	0.996047	0.063241
23	7	4	4	1.000000	0.003953
23	7	5	0	0.129811	0.129811
23	7	5	1	0.508425	0.378615
23	7	5	2	0.857916	0.349490
23	7	5	3	0.982733	0.124818
23	7	5	4	0.999376	0.016642
23	7	5	5	1.000000	0.000624
23	7	6	0	0.079329	0.079329
23	7	6	1	0.382220	0.302892
23	7	6	2	0.760835	0.378615
23	7	6	3	0.954996	0.194161
23	7	6	4	0.996602	0.041606
23	7	6	5	0.999931	0.003328
23	7	6	6	1.000000	0.000069
23	7	7	0	0.046664	0.046664
23	7	7	1	0.275317	0.228653
23	7	7	2	0.649478	0.374160
23	7	7	3	0.909311	0.259833
23	7	7	4	0.989260	0.079949
23	7	7	5	0.999539	0.010279
23	7	7	6	0.999996	0.000457
23	7	7	7	1.000000	0.000004
23	8	1	0	0.652174	0.652174
23	8	1	1	1.000000	0.347826
23	8	2	0	0.415020	0.415020
23	8	2	1	0.889328	0.474308
23	8	2	2	1.000000	0.110672
23	8	3	0	0.256917	0.256917
23	8	3	1	0.731225	0.474308
23	8	3	2	0.968379	0.237154
23	8	3	3	1.000000	0.031621
23	8	4	0	0.154150	0.154150
23	8	4	1	0.565217	0.411067
23	8	4	2	0.897233	0.332016
23	8	4	3	0.992095	0.094862
23	8	4	4	1.000000	0.007905
23	8	5	0	0.089245	0.089245
23	8	5	1	0.413772	0.324527
23	8	5	2	0.792386	0.378615
23	8	5	3	0.967131	0.174745
23	8	5	4	0.998336	0.031204
23	8	5	5	1.000000	0.001664
23	8	6	0	0.049580	0.049580
23	8	6	1	0.287567	0.237986
23	8	6	2	0.666181	0.378615
23	8	6	3	0.918591	0.252410
23	8	6	4	0.991401	0.072810
23	8	6	5	0.999723	0.008321
23	8	6	6	1.000000	0.000277
23	8	7	0	0.026248	0.026248
23	8	7	1	0.189572	0.163324
23	8	7	2	0.532553	0.342980
23	8	7	3	0.844353	0.311800
23	8	7	4	0.974269	0.129917
23	8	7	5	0.998254	0.023985
23	8	7	6	0.999967	0.001713
23	8	7	7	1.000000	0.000033
23	8	8	0	0.013124	0.013124
23	8	8	1	0.118118	0.104994
23	8	8	2	0.403935	0.285817
23	8	8	3	0.746915	0.342980
23	8	8	4	0.941790	0.194875
23	8	8	5	0.993757	0.051967
23	8	8	6	0.999753	0.005996
23	8	8	7	0.999998	0.000245
23	8	8	8	1.000000	0.000002
23	9	1	0	0.608696	0.608696
23	9	1	1	1.000000	0.391304
23	9	2	0	0.359684	0.359684
23	9	2	1	0.857707	0.498024
23	9	2	2	1.000000	0.142292
23	9	3	0	0.205534	0.205534
23	9	3	1	0.667984	0.462451
23	9	3	2	0.952569	0.284585
23	9	3	3	1.000000	0.047431
23	9	4	0	0.113043	0.113043
23	9	4	1	0.483004	0.369960
23	9	4	2	0.852964	0.369960
23	9	4	3	0.985771	0.132806
23	9	4	4	1.000000	0.014229
23	9	5	0	0.059497	0.059497
23	9	5	1	0.327231	0.267735
23	9	5	2	0.716663	0.389432
23	9	5	3	0.943832	0.227169
23	9	5	4	0.996255	0.052424
23	9	5	5	1.000000	0.003745
23	9	6	0	0.029748	0.029748
23	9	6	1	0.208238	0.178490
23	9	6	2	0.565217	0.356979
23	9	6	3	0.868109	0.302892
23	9	6	4	0.981693	0.113584
23	9	6	5	0.999168	0.017475
23	9	6	6	1.000000	0.000832
23	9	7	0	0.013999	0.013999
23	9	7	1	0.124243	0.110244
23	9	7	2	0.418226	0.293983
23	9	7	3	0.761206	0.342980
23	9	7	4	0.948286	0.187080
23	9	7	5	0.995056	0.046770
23	9	7	6	0.999853	0.004797
23	9	7	7	1.000000	0.000147
23	9	8	0	0.006125	0.006125
23	9	8	1	0.069121	0.062996
23	9	8	2	0.289608	0.220487
23	9	8	3	0.632588	0.342980
23	9	8	4	0.889824	0.257235
23	9	8	5	0.983364	0.093540
23	9	8	6	0.998954	0.015590
23	9	8	7	0.999982	0.001028
23	9	8	8	1.000000	0.000018
23	9	9	0	0.002450	0.002450
23	9	9	1	0.035523	0.033073
23	9	9	2	0.186714	0.151191
23	9	9	3	0.495396	0.308682
23	9	9	4	0.804079	0.308682
23	9	9	5	0.958420	0.154341
23	9	9	6	0.995836	0.037416
23	9	9	7	0.999845	0.004009
23	9	9	8	0.999999	0.000154
23	9	9	9	1.000000	0.000001
23	10	1	0	0.565217	0.565217
23	10	1	1	1.000000	0.434783
23	10	2	0	0.308300	0.308300
23	10	2	1	0.822134	0.513834
23	10	2	2	1.000000	0.177866
23	10	3	0	0.161491	0.161491
23	10	3	1	0.601920	0.440429
23	10	3	2	0.932242	0.330322
23	10	3	3	1.000000	0.067758
23	10	4	0	0.080745	0.080745
23	10	4	1	0.403727	0.322981
23	10	4	2	0.800113	0.396386
23	10	4	3	0.976284	0.176172
23	10	4	4	1.000000	0.023715
23	10	5	0	0.038248	0.038248
23	10	5	1	0.250736	0.212488
23	10	5	2	0.633213	0.382478
23	10	5	3	0.911379	0.278166
23	10	5	4	0.992511	0.081132
23	10	5	5	1.000000	0.007489
23	10	6	0	0.016999	0.016999
23	10	6	1	0.144492	0.127493
23	10	6	2	0.463223	0.318732
23	10	6	3	0.803204	0.339980
23	10	6	4	0.965467	0.162263
23	10	6	5	0.997920	0.032453
23	10	6	6	1.000000	0.002080
23	10	7	0	0.007000	0.007000
23	10	7	1	0.076996	0.069996
23	10	7	2	0.313232	0.236236
23	10	7	3	0.663212	0.349980
23	10	7	4	0.908197	0.244986
23	10	7	5	0.988375	0.080177

N	n	k	x	P(x)	p(x)
23	10	7	6	0.999510	0.011136
23	10	7	7	1.000000	0.000489
23	10	8	0	0.002625	0.002625
23	10	8	1	0.037623	0.034998
23	10	8	2	0.195114	0.157491
23	10	8	3	0.510095	0.314982
23	10	8	4	0.816328	0.306232
23	10	8	5	0.963319	0.146991
23	10	8	6	0.996726	0.033407
23	10	8	7	0.999908	0.003182
23	10	8	8	1.000000	0.000092
23	10	9	0	0.000875	0.000875
23	10	9	1	0.016624	0.015749
23	10	9	2	0.111119	0.094495
23	10	9	3	0.363104	0.251985
23	10	9	4	0.693835	0.330731
23	10	9	5	0.914322	0.220487
23	10	9	6	0.987818	0.073496
23	10	9	7	0.999272	0.011454
23	10	9	8	0.999988	0.000716
23	10	9	9	1.000000	0.000012
23	10	10	0	0.000250	0.000250
23	10	10	1	0.006500	0.006250
23	10	10	2	0.057122	0.050622
23	10	10	3	0.237111	0.179990
23	10	10	4	0.552093	0.314982
23	10	10	5	0.835577	0.283484
23	10	10	6	0.966819	0.131242
23	10	10	7	0.996817	0.029998
23	10	10	8	0.999885	0.003068
23	10	10	9	0.999999	0.000114
23	10	10	10	1.000000	0.000001
23	11	1	0	0.521739	0.521739
23	11	1	1	1.000000	0.478261
23	11	2	0	0.260870	0.260870
23	11	2	1	0.782609	0.521739
23	11	2	2	1.000000	0.217391
23	11	3	0	0.124224	0.124224
23	11	3	1	0.534161	0.409938
23	11	3	2	0.906832	0.372671
23	11	3	3	1.000000	0.093168
23	11	4	0	0.055901	0.055901
23	11	4	1	0.329193	0.273292
23	11	4	2	0.739130	0.409938
23	11	4	3	0.962733	0.223602
23	11	4	4	1.000000	0.037267
23	11	5	0	0.023537	0.023537
23	11	5	1	0.185355	0.161818
23	11	5	2	0.544949	0.359595
23	11	5	3	0.868584	0.323635

N	n	k	x	P(x)	p(x)
23	11	5	4	0.986270	0.117685
23	11	5	5	1.000000	0.013730
23	11	6	0	0.009153	0.009153
23	11	6	1	0.095456	0.086303
23	11	6	2	0.365152	0.269696
23	11	6	3	0.724747	0.359595
23	11	6	4	0.940503	0.215757
23	11	6	5	0.995423	0.054920
23	11	6	6	1.000000	0.004577
23	11	7	0	0.003231	0.003231
23	11	7	1	0.044690	0.041459
23	11	7	2	0.222372	0.177682
23	11	7	3	0.555526	0.333154
23	11	7	4	0.851662	0.296137
23	11	7	5	0.976040	0.124377
23	11	7	6	0.998654	0.022614
23	11	7	7	1.000000	0.001346
23	11	8	0	0.001010	0.001010
23	11	8	1	0.018778	0.017768
23	11	8	2	0.122426	0.103648
23	11	8	3	0.388949	0.266523
23	11	8	4	0.722102	0.333154
23	11	8	5	0.929398	0.207296
23	11	8	6	0.991587	0.062189
23	11	8	7	0.999663	0.008076
23	11	8	8	1.000000	0.000337
23	11	9	0	0.000269	0.000269
23	11	9	1	0.006932	0.006663
23	11	9	2	0.060237	0.053305
23	11	9	3	0.246803	0.186566
23	11	9	4	0.566631	0.319828
23	11	9	5	0.846480	0.279849
23	11	9	6	0.970857	0.124377
23	11	9	7	0.997510	0.026652
23	11	9	8	0.999933	0.002423
23	11	9	9	1.000000	0.000067
23	11	10	0	0.000058	0.000058
23	11	10	1	0.002173	0.002115
23	11	10	2	0.025970	0.023797
23	11	10	3	0.140194	0.114224
23	11	10	4	0.406717	0.266523
23	11	10	5	0.726545	0.319828
23	11	10	6	0.926437	0.199892
23	11	10	7	0.989895	0.063458
23	11	10	8	0.999413	0.009519
23	11	10	9	0.999990	0.000577
23	11	10	10	1.000000	0.000010
23	11	11	0	0.000009	0.000009
23	11	11	1	0.000546	0.000537
23	11	11	2	0.009495	0.008949

N	n	k	x	P(x)	p(x)
23	11	11	3	0.069902	0.060407
23	11	11	4	0.263204	0.193302
23	11	11	5	0.578932	0.315727
23	11	11	6	0.849555	0.270623
23	11	11	7	0.970369	0.120814
23	11	11	8	0.997217	0.026848
23	11	11	9	0.999902	0.002685
23	11	11	10	0.999999	0.000098
23	11	11	11	1.000000	0.000001
23	12	1	0	0.478261	0.478261
23	12	1	1	1.000000	0.521739
23	12	2	0	0.217391	0.217391
23	12	2	1	0.739130	0.521739
23	12	2	2	1.000000	0.260870
23	12	3	0	0.093168	0.093168
23	12	3	1	0.465838	0.372671
23	12	3	2	0.875776	0.409938
23	12	3	3	1.000000	0.124224
23	12	4	0	0.037267	0.037267
23	12	4	1	0.260870	0.223602
23	12	4	2	0.670807	0.409938
23	12	4	3	0.944099	0.273292
23	12	4	4	1.000000	0.055901
23	12	5	0	0.013730	0.013730
23	12	5	1	0.131415	0.117685
23	12	5	2	0.455051	0.323635
23	12	5	3	0.814645	0.359595
23	12	5	4	0.976463	0.161818
23	12	5	5	1.000000	0.023537
23	12	6	0	0.004577	0.004577
23	12	6	1	0.059497	0.054920
23	12	6	2	0.275253	0.215757
23	12	6	3	0.634848	0.359595
23	12	6	4	0.904544	0.269696
23	12	6	5	0.990847	0.086303
23	12	6	6	1.000000	0.009153
23	12	7	0	0.001346	0.001346
23	12	7	1	0.023960	0.022614
23	12	7	2	0.148338	0.124377
23	12	7	3	0.444474	0.296137
23	12	7	4	0.777628	0.333154
23	12	7	5	0.955310	0.177682
23	12	7	6	0.996769	0.041459
23	12	7	7	1.000000	0.003231
23	12	8	0	0.000337	0.000337
23	12	8	1	0.008413	0.008076
23	12	8	2	0.070602	0.062189
23	12	8	3	0.277897	0.207296
23	12	8	4	0.611051	0.333154
23	12	8	5	0.877574	0.266523

N	n	k	x	P(x)	p(x)
23	12	8	6	0.981222	0.103648
23	12	8	7	0.998990	0.017768
23	12	8	8	1.000000	0.001010
23	12	9	0	0.000067	0.000067
23	12	9	1	0.002490	0.002423
23	12	9	2	0.029143	0.026652
23	12	9	3	0.153520	0.124377
23	12	9	4	0.433369	0.279849
23	12	9	5	0.753197	0.319828
23	12	9	6	0.939763	0.186566
23	12	9	7	0.993068	0.053305
23	12	9	8	0.999731	0.006663
23	12	9	9	1.000000	0.000269
23	12	10	0	0.000010	0.000010
23	12	10	1	0.000587	0.000577
23	12	10	2	0.010105	0.009519
23	12	10	3	0.073563	0.063458
23	12	10	4	0.273455	0.199892
23	12	10	5	0.593283	0.319828
23	12	10	6	0.859806	0.266523
23	12	10	7	0.974030	0.114224
23	12	10	8	0.997827	0.023797
23	12	10	9	0.999942	0.002115
23	12	10	10	1.000000	0.000058
23	12	11	0	0.000001	0.000001
23	12	11	1	0.000098	0.000098
23	12	11	2	0.002783	0.002685
23	12	11	3	0.029631	0.026848
23	12	11	4	0.150445	0.120814
23	12	11	5	0.421068	0.270623
23	12	11	6	0.736795	0.315727
23	12	11	7	0.930098	0.193302
23	12	11	8	0.990505	0.060407
23	12	11	9	0.999454	0.008949
23	12	11	10	0.999991	0.000537
23	12	11	11	1.000000	0.000009
23	12	12	1	0.000009	0.000009
23	12	12	2	0.000546	0.000537
23	12	12	3	0.009495	0.008949
23	12	12	4	0.069902	0.060407
23	12	12	5	0.263204	0.193302
23	12	12	6	0.578932	0.315727
23	12	12	7	0.849555	0.270623
23	12	12	8	0.970369	0.120814
23	12	12	9	0.997217	0.026848
23	12	12	10	0.999902	0.002685
23	12	12	11	0.999999	0.000098
23	12	12	12	1.000000	0.000001
23	13	1	0	0.434783	0.434783
23	13	1	1	1.000000	0.565217

Table for $N = 2$, $n = 1$, through $N = 100$, $n = 50$

N	n	k	x	P(x)	p(x)
23	13	2	0	0.177866	0.177866
23	13	2	1	0.691700	0.513834
23	13	2	2	1.000000	0.308300
23	13	3	0	0.067758	0.067758
23	13	3	1	0.398080	0.330322
23	13	3	2	0.838509	0.440429
23	13	3	3	1.000000	0.161491
23	13	4	0	0.023715	0.023715
23	13	4	1	0.199887	0.176172
23	13	4	2	0.596273	0.396386
23	13	4	3	0.919255	0.322981
23	13	4	4	1.000000	0.080745
23	13	5	0	0.007489	0.007489
23	13	5	1	0.088621	0.081132
23	13	5	2	0.366786	0.278166
23	13	5	3	0.749264	0.382478
23	13	5	4	0.961752	0.212488
23	13	5	5	1.000000	0.038248
23	13	6	0	0.002080	0.002080
23	13	6	1	0.034533	0.032453
23	13	6	2	0.196796	0.162263
23	13	6	3	0.536777	0.339980
23	13	6	4	0.855508	0.318732
23	13	6	5	0.983001	0.127493
23	13	6	6	1.000000	0.016999
23	13	7	0	0.000489	0.000489
23	13	7	1	0.011625	0.011136
23	13	7	2	0.091802	0.080177
23	13	7	3	0.336788	0.244986
23	13	7	4	0.686768	0.349980
23	13	7	5	0.923004	0.236236
23	13	7	6	0.993000	0.069996
23	13	7	7	1.000000	0.007000
23	13	8	0	0.000092	0.000092
23	13	8	1	0.003273	0.003182
23	13	8	2	0.036681	0.033407
23	13	8	3	0.183672	0.146991
23	13	8	4	0.489904	0.306232
23	13	8	5	0.804886	0.314982
23	13	8	6	0.962377	0.157491
23	13	8	7	0.997375	0.034998
23	13	8	8	1.000000	0.002625
23	13	9	0	0.000012	0.000012
23	13	9	1	0.000728	0.000716
23	13	9	2	0.012182	0.011454
23	13	9	3	0.085678	0.073496
23	13	9	4	0.306165	0.220487
23	13	9	5	0.636896	0.330731
23	13	9	6	0.888881	0.251985
23	13	9	7	0.983376	0.094495
23	13	9	8	0.999125	0.015749
23	13	9	9	1.000000	0.000875
23	13	10	0	0.000001	0.000001
23	13	10	1	0.000115	0.000114
23	13	10	2	0.003183	0.003068
23	13	10	3	0.033181	0.029998
23	13	10	4	0.164423	0.131242
23	13	10	5	0.447907	0.283484
23	13	10	6	0.762889	0.314982
23	13	10	7	0.942878	0.179990
23	13	10	8	0.993500	0.050622
23	13	10	9	0.999750	0.006250
23	13	10	10	1.000000	0.000250
23	13	11	1	0.000010	0.000010
23	13	11	2	0.000587	0.000577
23	13	11	3	0.010105	0.009519
23	13	11	4	0.073563	0.063458
23	13	11	5	0.273455	0.199892
23	13	11	6	0.593283	0.319828
23	13	11	7	0.859806	0.266523
23	13	11	8	0.974030	0.114224
23	13	11	9	0.997827	0.023797
23	13	11	10	0.999942	0.002115
23	13	11	11	1.000000	0.000058
23	13	12	2	0.000058	0.000058
23	13	12	3	0.002173	0.002115
23	13	12	4	0.025970	0.023797
23	13	12	5	0.140194	0.114224
23	13	12	6	0.406717	0.266523
23	13	12	7	0.726545	0.319828
23	13	12	8	0.926437	0.199892
23	13	12	9	0.989895	0.063458
23	13	12	10	0.999413	0.009519
23	13	12	11	0.999990	0.000577
23	13	12	12	1.000000	0.000010
23	13	13	3	0.000250	0.000250
23	13	13	4	0.006500	0.006250
23	13	13	5	0.057122	0.050622
23	13	13	6	0.237111	0.179990
23	13	13	7	0.552093	0.314982
23	13	13	8	0.835577	0.283484
23	13	13	9	0.966819	0.131242
23	13	13	10	0.996817	0.029998
23	13	13	11	0.999885	0.003068
23	13	13	12	0.999999	0.000114
23	13	13	13	1.000000	0.000001
23	14	1	0	0.391304	0.391304
23	14	1	1	1.000000	0.608696
23	14	2	0	0.142292	0.142292
23	14	2	1	0.640316	0.498024
23	14	2	2	1.000000	0.359684
23	14	3	0	0.047431	0.047431
23	14	3	1	0.332016	0.284585
23	14	3	2	0.794466	0.462451
23	14	3	3	1.000000	0.205534
23	14	4	0	0.014229	0.014229
23	14	4	1	0.147036	0.132806
23	14	4	2	0.516996	0.369960
23	14	4	3	0.886956	0.369960
23	14	4	4	1.000000	0.113043
23	14	5	0	0.003745	0.003745
23	14	5	1	0.056168	0.052424
23	14	5	2	0.283337	0.227169
23	14	5	3	0.672769	0.389432
23	14	5	4	0.940503	0.267735
23	14	5	5	1.000000	0.059497
23	14	6	0	0.000832	0.000832
23	14	6	1	0.018307	0.017475
23	14	6	2	0.131891	0.113584
23	14	6	3	0.434783	0.302892
23	14	6	4	0.791762	0.356979
23	14	6	5	0.970252	0.178490
23	14	6	6	1.000000	0.029748
23	14	7	0	0.000147	0.000147
23	14	7	1	0.004944	0.004797
23	14	7	2	0.051714	0.046770
23	14	7	3	0.238794	0.187080
23	14	7	4	0.581774	0.342980
23	14	7	5	0.875757	0.293983
23	14	7	6	0.986001	0.110244
23	14	7	7	1.000000	0.013999
23	14	8	0	0.000018	0.000018
23	14	8	1	0.001046	0.001028
23	14	8	2	0.016636	0.015590
23	14	8	3	0.110176	0.093540
23	14	8	4	0.367411	0.257235
23	14	8	5	0.710392	0.342980
23	14	8	6	0.930879	0.220487
23	14	8	7	0.993875	0.062996
23	14	8	8	1.000000	0.006125
23	14	9	0	0.000001	0.000001
23	14	9	1	0.000155	0.000154
23	14	9	2	0.004164	0.004009
23	14	9	3	0.041580	0.037416
23	14	9	4	0.195921	0.154341
23	14	9	5	0.504604	0.308682
23	14	9	6	0.813286	0.308682
23	14	9	7	0.964477	0.151191
23	14	9	8	0.997550	0.033073
23	14	9	9	1.000000	0.002450
23	14	10	1	0.000012	0.000012
23	14	10	2	0.000728	0.000716
23	14	10	3	0.012182	0.011454
23	14	10	4	0.085678	0.073496
23	14	10	5	0.306165	0.220487
23	14	10	6	0.636896	0.330731
23	14	10	7	0.888881	0.251985
23	14	10	8	0.983376	0.094495
23	14	10	9	0.999125	0.015749
23	14	10	10	1.000000	0.000875
23	14	11	2	0.000067	0.000067
23	14	11	3	0.002490	0.002423
23	14	11	4	0.029143	0.026652
23	14	11	5	0.153520	0.124377
23	14	11	6	0.433369	0.279849
23	14	11	7	0.753197	0.319828
23	14	11	8	0.939763	0.186566
23	14	11	9	0.993068	0.053305
23	14	11	10	0.999731	0.006663
23	14	11	11	1.000000	0.000269
23	14	12	3	0.000269	0.000269
23	14	12	4	0.006932	0.006663
23	14	12	5	0.060237	0.053305
23	14	12	6	0.246803	0.186566
23	14	12	7	0.566631	0.319828
23	14	12	8	0.846480	0.279849
23	14	12	9	0.970857	0.124377
23	14	12	10	0.997510	0.026652
23	14	12	11	0.999933	0.002423
23	14	12	12	1.000000	0.000067
23	14	13	4	0.000875	0.000875
23	14	13	5	0.016624	0.015749
23	14	13	6	0.111119	0.094495
23	14	13	7	0.363104	0.251985
23	14	13	8	0.693835	0.330731
23	14	13	9	0.914322	0.220487
23	14	13	10	0.987818	0.073496
23	14	13	11	0.999272	0.011454
23	14	13	12	0.999988	0.000716
23	14	13	13	1.000000	0.000012
23	14	14	5	0.002450	0.002450
23	14	14	6	0.035523	0.033073
23	14	14	7	0.186714	0.151191
23	14	14	8	0.495396	0.308682
23	14	14	9	0.804079	0.308682
23	14	14	10	0.958420	0.154341
23	14	14	11	0.995836	0.037416
23	14	14	12	0.999845	0.004009
23	14	14	13	0.999999	0.000154
23	14	14	14	1.000000	0.000001

Table for $N = 2$, $n = 1$, through $N = 100$, $n = 50$

N	n	k	x	P(x)	p(x)
23	15	1	0	0.347826	0.347826
23	15	1	1	1.000000	0.652174
23	15	2	0	0.110672	0.110672
23	15	2	1	0.584980	0.474308
23	15	2	2	1.000000	0.415020
23	15	3	0	0.031621	0.031621
23	15	3	1	0.268775	0.237154
23	15	3	2	0.743083	0.474308
23	15	3	3	1.000000	0.256917
23	15	4	0	0.007905	0.007905
23	15	4	1	0.102767	0.094862
23	15	4	2	0.434783	0.332016
23	15	4	3	0.845850	0.411067
23	15	4	4	1.000000	0.154150
23	15	5	0	0.001664	0.001664
23	15	5	1	0.032869	0.031204
23	15	5	2	0.207614	0.174745
23	15	5	3	0.586228	0.378615
23	15	5	4	0.910755	0.324527
23	15	5	5	1.000000	0.089245
23	15	6	0	0.000277	0.000277
23	15	6	1	0.008599	0.008321
23	15	6	2	0.081409	0.072810
23	15	6	3	0.333819	0.252410
23	15	6	4	0.712433	0.378615
23	15	6	5	0.950419	0.237986
23	15	6	6	1.000000	0.049580
23	15	7	0	0.000033	0.000033
23	15	7	1	0.001746	0.001713
23	15	7	2	0.025730	0.023985
23	15	7	3	0.155647	0.129917
23	15	7	4	0.467447	0.311800
23	15	7	5	0.810428	0.342980
23	15	7	6	0.973751	0.163324
23	15	7	7	1.000000	0.026248
23	15	8	0	0.000002	0.000002
23	15	8	1	0.000247	0.000245
23	15	8	2	0.006243	0.005996
23	15	8	3	0.058210	0.051967
23	15	8	4	0.253085	0.194875
23	15	8	5	0.596065	0.342980
23	15	8	6	0.881882	0.285817
23	15	8	7	0.986876	0.104994
23	15	8	8	1.000000	0.013124
23	15	9	1	0.000018	0.000018
23	15	9	2	0.001046	0.001028
23	15	9	3	0.016636	0.015590
23	15	9	4	0.110176	0.093540
23	15	9	5	0.367411	0.257235
23	15	9	6	0.710392	0.342980

N	n	k	x	P(x)	p(x)
23	15	9	7	0.930879	0.220487
23	15	9	8	0.993875	0.062996
23	15	9	9	1.000000	0.006125
23	15	10	2	0.000092	0.000092
23	15	10	3	0.003273	0.003182
23	15	10	4	0.036681	0.033407
23	15	10	5	0.183672	0.146991
23	15	10	6	0.489904	0.306232
23	15	10	7	0.804886	0.314982
23	15	10	8	0.962377	0.157491
23	15	10	9	0.997375	0.034998
23	15	10	10	1.000000	0.002625
23	15	11	3	0.000337	0.000337
23	15	11	4	0.008413	0.008076
23	15	11	5	0.070602	0.062189
23	15	11	6	0.277897	0.207296
23	15	11	7	0.611051	0.333154
23	15	11	8	0.877574	0.266523
23	15	11	9	0.981222	0.103648
23	15	11	10	0.998990	0.017768
23	15	11	11	1.000000	0.001010
23	15	12	4	0.001010	0.001010
23	15	12	5	0.018778	0.017768
23	15	12	6	0.122426	0.103648
23	15	12	7	0.388949	0.266523
23	15	12	8	0.722102	0.333154
23	15	12	9	0.929398	0.207296
23	15	12	10	0.991587	0.062189
23	15	12	11	0.999663	0.008076
23	15	12	12	1.000000	0.000337
23	15	13	5	0.002625	0.002625
23	15	13	6	0.037623	0.034998
23	15	13	7	0.195114	0.157491
23	15	13	8	0.510095	0.314982
23	15	13	9	0.816328	0.306232
23	15	13	10	0.963319	0.146991
23	15	13	11	0.996726	0.033407
23	15	13	12	0.999908	0.003182
23	15	13	13	1.000000	0.000092
23	15	14	6	0.006125	0.006125
23	15	14	7	0.069121	0.062996
23	15	14	8	0.289608	0.220487
23	15	14	9	0.632588	0.342980
23	15	14	10	0.889824	0.257235
23	15	14	11	0.983364	0.093540
23	15	14	12	0.998954	0.015590
23	15	14	13	0.999982	0.001028
23	15	14	14	1.000000	0.000018
23	15	15	7	0.013124	0.013124
23	15	15	8	0.118118	0.104994

N	n	k	x	P(x)	p(x)
23	15	15	9	0.403935	0.285817
23	15	15	10	0.746915	0.342980
23	15	15	11	0.941790	0.194875
23	15	15	12	0.993757	0.051967
23	15	15	13	0.999753	0.005996
23	15	15	14	0.999998	0.000245
23	15	15	15	1.000000	0.000002
23	16	1	0	0.304348	0.304348
23	16	1	1	1.000000	0.695652
23	16	2	0	0.083004	0.083004
23	16	2	1	0.525692	0.442688
23	16	2	2	1.000000	0.474308
23	16	3	0	0.019763	0.019763
23	16	3	1	0.209486	0.189723
23	16	3	2	0.683794	0.474308
23	16	3	3	1.000000	0.316206
23	16	4	0	0.003953	0.003953
23	16	4	1	0.067194	0.063241
23	16	4	2	0.351779	0.284585
23	16	4	3	0.794466	0.442688
23	16	4	4	1.000000	0.205534
23	16	5	0	0.000624	0.000624
23	16	5	1	0.017266	0.016642
23	16	5	2	0.142084	0.124818
23	16	5	3	0.491575	0.349490
23	16	5	4	0.870189	0.378615
23	16	5	5	1.000000	0.129811
23	16	6	0	0.000069	0.000069
23	16	6	1	0.003398	0.003328
23	16	6	2	0.045004	0.041606
23	16	6	3	0.239165	0.194161
23	16	6	4	0.617780	0.378615
23	16	6	5	0.920671	0.302892
23	16	6	6	1.000000	0.079329
23	16	7	0	0.000004	0.000004
23	16	7	1	0.000461	0.000457
23	16	7	2	0.010740	0.010279
23	16	7	3	0.090689	0.079949
23	16	7	4	0.350522	0.259833
23	16	7	5	0.724683	0.374160
23	16	7	6	0.953336	0.228653
23	16	7	7	1.000000	0.046664
23	16	8	1	0.000033	0.000033
23	16	8	2	0.001746	0.001713
23	16	8	3	0.025730	0.023985
23	16	8	4	0.155647	0.129917
23	16	8	5	0.467447	0.311800
23	16	8	6	0.810428	0.342980
23	16	8	7	0.973751	0.163324
23	16	8	8	1.000000	0.026248

N	n	k	x	P(x)	p(x)
23	16	9	2	0.000147	0.000147
23	16	9	3	0.004944	0.004797
23	16	9	4	0.051714	0.046770
23	16	9	5	0.238794	0.187080
23	16	9	6	0.581774	0.342980
23	16	9	7	0.875757	0.293983
23	16	9	8	0.986001	0.110244
23	16	9	9	1.000000	0.013999
23	16	10	3	0.000489	0.000489
23	16	10	4	0.011625	0.011136
23	16	10	5	0.091802	0.080177
23	16	10	6	0.336788	0.244986
23	16	10	7	0.686768	0.349980
23	16	10	8	0.923004	0.236236
23	16	10	9	0.993000	0.069996
23	16	10	10	1.000000	0.007000
23	16	11	4	0.001346	0.001346
23	16	11	5	0.023960	0.022614
23	16	11	6	0.148338	0.124377
23	16	11	7	0.444474	0.296137
23	16	11	8	0.777628	0.333154
23	16	11	9	0.955310	0.177682
23	16	11	10	0.996769	0.041459
23	16	11	11	1.000000	0.003231
23	16	12	5	0.003231	0.003231
23	16	12	6	0.044690	0.041459
23	16	12	7	0.222372	0.177682
23	16	12	8	0.555526	0.333154
23	16	12	9	0.851662	0.296137
23	16	12	10	0.976040	0.124377
23	16	12	11	0.998654	0.022614
23	16	12	12	1.000000	0.001346
23	16	13	6	0.007000	0.007000
23	16	13	7	0.076996	0.069996
23	16	13	8	0.313232	0.236236
23	16	13	9	0.663212	0.349980
23	16	13	10	0.908197	0.244986
23	16	13	11	0.988375	0.080177
23	16	13	12	0.999510	0.011136
23	16	13	13	1.000000	0.000489
23	16	14	7	0.013999	0.013999
23	16	14	8	0.124243	0.110244
23	16	14	9	0.418226	0.293983
23	16	14	10	0.761206	0.342980
23	16	14	11	0.948286	0.187080
23	16	14	12	0.995056	0.046770
23	16	14	13	0.999853	0.004797
23	16	14	14	1.000000	0.000147
23	16	15	8	0.026248	0.026248
23	16	15	9	0.189572	0.163324

Table for $N = 2$, $n = 1$, through $N = 100$, $n = 50$

N	n	k	x	P(x)	p(x)
23	16	15	10	0.532553	0.342980
23	16	15	11	0.844353	0.311800
23	16	15	12	0.974269	0.129917
23	16	15	13	0.998254	0.023985
23	16	15	14	0.999967	0.001713
23	16	15	15	1.000000	0.000033
23	16	16	9	0.046664	0.046664
23	16	16	10	0.275317	0.228653
23	16	16	11	0.649478	0.374160
23	16	16	12	0.909311	0.259833
23	16	16	13	0.989260	0.079949
23	16	16	14	0.999539	0.010279
23	16	16	15	0.999996	0.000457
23	16	16	16	1.000000	0.000004
23	17	1	0	0.260870	0.260870
23	17	1	1	1.000000	0.739130
23	17	2	0	0.059289	0.059289
23	17	2	1	0.462451	0.403162
23	17	2	2	1.000000	0.537549
23	17	3	0	0.011293	0.011293
23	17	3	1	0.155279	0.143986
23	17	3	2	0.616036	0.460757
23	17	3	3	1.000000	0.383964
23	17	4	0	0.001694	0.001694
23	17	4	1	0.040090	0.038396
23	17	4	2	0.270469	0.230378
23	17	4	3	0.731225	0.460757
23	17	4	4	1.000000	0.268775
23	17	5	0	0.000178	0.000178
23	17	5	1	0.007757	0.007578
23	17	5	2	0.088591	0.080834
23	17	5	3	0.391720	0.303129
23	17	5	4	0.816101	0.424381
23	17	5	5	1.000000	0.183898
23	17	6	0	0.000010	0.000010
23	17	6	1	0.001020	0.001010
23	17	6	2	0.021229	0.020209
23	17	6	3	0.155953	0.134724
23	17	6	4	0.509604	0.353651
23	17	6	5	0.877401	0.367797
23	17	6	6	1.000000	0.122599
23	17	7	1	0.000069	0.000069
23	17	7	2	0.003398	0.003328
23	17	7	3	0.045004	0.041606
23	17	7	4	0.239165	0.194161
23	17	7	5	0.617780	0.378615
23	17	7	6	0.920671	0.302892
23	17	7	7	1.000000	0.079329
23	17	8	2	0.000277	0.000277
23	17	8	3	0.008599	0.008321

N	n	k	x	P(x)	p(x)
23	17	8	4	0.081409	0.072810
23	17	8	5	0.333819	0.252410
23	17	8	6	0.712433	0.378615
23	17	8	7	0.950419	0.237986
23	17	8	8	1.000000	0.049580
23	17	9	3	0.000832	0.000832
23	17	9	4	0.018307	0.017475
23	17	9	5	0.131891	0.113584
23	17	9	6	0.434783	0.302892
23	17	9	7	0.791762	0.356979
23	17	9	8	0.970252	0.178490
23	17	9	9	1.000000	0.029748
23	17	10	4	0.002080	0.002080
23	17	10	5	0.034533	0.032453
23	17	10	6	0.196796	0.162263
23	17	10	7	0.536777	0.339980
23	17	10	8	0.855508	0.318732
23	17	10	9	0.983001	0.127493
23	17	10	10	1.000000	0.016999
23	17	11	5	0.004577	0.004577
23	17	11	6	0.059497	0.054920
23	17	11	7	0.275253	0.215757
23	17	11	8	0.634848	0.359595
23	17	11	9	0.904544	0.269696
23	17	11	10	0.990847	0.086303
23	17	11	11	1.000000	0.009153
23	17	12	6	0.009153	0.009153
23	17	12	7	0.095456	0.086303
23	17	12	8	0.365152	0.269696
23	17	12	9	0.724747	0.359595
23	17	12	10	0.940503	0.215757
23	17	12	11	0.995423	0.054920
23	17	12	12	1.000000	0.004577
23	17	13	7	0.016999	0.016999
23	17	13	8	0.144492	0.127493
23	17	13	9	0.463223	0.318732
23	17	13	10	0.803204	0.339980
23	17	13	11	0.965467	0.162263
23	17	13	12	0.997920	0.032453
23	17	13	13	1.000000	0.002080
23	17	14	8	0.029748	0.029748
23	17	14	9	0.208238	0.178490
23	17	14	10	0.565217	0.356979
23	17	14	11	0.868109	0.302892
23	17	14	12	0.981693	0.113584
23	17	14	13	0.999168	0.017475
23	17	14	14	1.000000	0.000832
23	17	15	9	0.049580	0.049580
23	17	15	10	0.287567	0.237986
23	17	15	11	0.666181	0.378615

N	n	k	x	P(x)	p(x)
23	17	15	12	0.918591	0.252410
23	17	15	13	0.991401	0.072810
23	17	15	14	0.999723	0.008321
23	17	15	15	1.000000	0.000277
23	17	16	10	0.079329	0.079329
23	17	16	11	0.382220	0.302892
23	17	16	12	0.760835	0.378615
23	17	16	13	0.954996	0.194161
23	17	16	14	0.996602	0.041606
23	17	16	15	0.999931	0.003328
23	17	16	16	1.000000	0.000069
23	17	17	11	0.122599	0.122599
23	17	17	12	0.490396	0.367797
23	17	17	13	0.844047	0.353651
23	17	17	14	0.978771	0.134724
23	17	17	15	0.998980	0.020209
23	17	17	16	0.999990	0.001010
23	17	17	17	1.000000	0.000010
23	18	1	0	0.217391	0.217391
23	18	1	1	1.000000	0.782609
23	18	2	0	0.039526	0.039526
23	18	2	1	0.395257	0.355731
23	18	2	2	1.000000	0.604743
23	18	3	0	0.005647	0.005647
23	18	3	1	0.107284	0.101637
23	18	3	2	0.539243	0.431959
23	18	3	3	1.000000	0.460757
23	18	4	0	0.000565	0.000565
23	18	4	1	0.020892	0.020327
23	18	4	2	0.193676	0.172784
23	18	4	3	0.654432	0.460757
23	18	4	4	1.000000	0.345567
23	18	5	0	0.000030	0.000030
23	18	5	1	0.002704	0.002675
23	18	5	2	0.048174	0.045469
23	18	5	3	0.290677	0.242503
23	18	5	4	0.745371	0.454694
23	18	5	5	1.000000	0.254629
23	18	6	1	0.000178	0.000178
23	18	6	2	0.007757	0.007578
23	18	6	3	0.088591	0.080834
23	18	6	4	0.391720	0.303129
23	18	6	5	0.816101	0.424381
23	18	6	6	1.000000	0.183898
23	18	7	2	0.000624	0.000624
23	18	7	3	0.017266	0.016642
23	18	7	4	0.142084	0.124818
23	18	7	5	0.491575	0.349490
23	18	7	6	0.870189	0.378615
23	18	7	7	1.000000	0.129811

N	n	k	x	P(x)	p(x)
23	18	8	3	0.001664	0.001664
23	18	8	4	0.032869	0.031204
23	18	8	5	0.207614	0.174745
23	18	8	6	0.586228	0.378615
23	18	8	7	0.910755	0.324527
23	18	8	8	1.000000	0.089245
23	18	9	4	0.003745	0.003745
23	18	9	5	0.056168	0.052424
23	18	9	6	0.283337	0.227169
23	18	9	7	0.672769	0.389432
23	18	9	8	0.940503	0.267735
23	18	9	9	1.000000	0.059497
23	18	10	5	0.007489	0.007489
23	18	10	6	0.088621	0.081132
23	18	10	7	0.366786	0.278166
23	18	10	8	0.749264	0.382478
23	18	10	9	0.961752	0.212488
23	18	10	10	1.000000	0.038248
23	18	11	6	0.013730	0.013730
23	18	11	7	0.131415	0.117685
23	18	11	8	0.455051	0.323635
23	18	11	9	0.814645	0.359595
23	18	11	10	0.976463	0.161818
23	18	11	11	1.000000	0.023537
23	18	12	7	0.023537	0.023537
23	18	12	8	0.185355	0.161818
23	18	12	9	0.544949	0.359595
23	18	12	10	0.868584	0.323635
23	18	12	11	0.986270	0.117685
23	18	12	12	1.000000	0.013730
23	18	13	8	0.038248	0.038248
23	18	13	9	0.250736	0.212488
23	18	13	10	0.633213	0.382478
23	18	13	11	0.911379	0.278166
23	18	13	12	0.992511	0.081132
23	18	13	13	1.000000	0.007489
23	18	14	9	0.059497	0.059497
23	18	14	10	0.327231	0.267735
23	18	14	11	0.716663	0.389432
23	18	14	12	0.943832	0.227169
23	18	14	13	0.996255	0.052424
23	18	14	14	1.000000	0.003745
23	18	15	10	0.089245	0.089245
23	18	15	11	0.413772	0.324527
23	18	15	12	0.792386	0.378615
23	18	15	13	0.967131	0.174745
23	18	15	14	0.998336	0.031204
23	18	15	15	1.000000	0.001664
23	18	16	11	0.129811	0.129811
23	18	16	12	0.508425	0.378615

Table for $N = 2$, $n = 1$, through $N = 100$, $n = 50$

N	n	k	x	P(x)	p(x)
23	18	16	13	0.857916	0.349490
23	18	16	14	0.982733	0.124818
23	18	16	15	0.999376	0.016642
23	18	16	16	1.000000	0.000624
23	18	17	12	0.183898	0.183898
23	18	17	13	0.608280	0.424381
23	18	17	14	0.911409	0.303129
23	18	17	15	0.992243	0.080834
23	18	17	16	0.999822	0.007578
23	18	17	17	1.000000	0.000178
23	18	18	13	0.254629	0.254629
23	18	18	14	0.709323	0.454694
23	18	18	15	0.951826	0.242503
23	18	18	16	0.997296	0.045469
23	18	18	17	0.999970	0.002675
23	18	18	18	1.000000	0.000030
23	19	1	0	0.173913	0.173913
23	19	1	1	1.000000	0.826087
23	19	2	0	0.023715	0.023715
23	19	2	1	0.324111	0.300395
23	19	2	2	1.000000	0.675889
23	19	3	0	0.002259	0.002259
23	19	3	1	0.066629	0.064370
23	19	3	2	0.452851	0.386222
23	19	3	3	1.000000	0.547148
23	19	4	0	0.000113	0.000113
23	19	4	1	0.008696	0.008583
23	19	4	2	0.124562	0.115867
23	19	4	3	0.562281	0.437719
23	19	4	4	1.000000	0.437719
23	19	5	1	0.000565	0.000565
23	19	5	2	0.020892	0.020327
23	19	5	3	0.193676	0.172784
23	19	5	4	0.654432	0.460757
23	19	5	5	1.000000	0.345567
23	19	6	2	0.001694	0.001694
23	19	6	3	0.040090	0.038396
23	19	6	4	0.270469	0.230378
23	19	6	5	0.731225	0.460757
23	19	6	6	1.000000	0.268775
23	19	7	3	0.003953	0.003953
23	19	7	4	0.067194	0.063241
23	19	7	5	0.351779	0.284585
23	19	7	6	0.794466	0.442688
23	19	7	7	1.000000	0.205534
23	19	8	4	0.007905	0.007905
23	19	8	5	0.102767	0.094862
23	19	8	6	0.434783	0.332016
23	19	8	7	0.845850	0.411067
23	19	8	8	1.000000	0.154150

N	n	k	x	P(x)	p(x)
23	19	9	5	0.014229	0.014229
23	19	9	6	0.147036	0.132806
23	19	9	7	0.516996	0.369960
23	19	9	8	0.886956	0.369960
23	19	9	9	1.000000	0.113043
23	19	10	6	0.023715	0.023715
23	19	10	7	0.199887	0.176172
23	19	10	8	0.596273	0.396386
23	19	10	9	0.919255	0.322981
23	19	10	10	1.000000	0.080745
23	19	11	7	0.037267	0.037267
23	19	11	8	0.260870	0.223602
23	19	11	9	0.670807	0.409938
23	19	11	10	0.944099	0.273292
23	19	11	11	1.000000	0.055901
23	19	12	8	0.055901	0.055901
23	19	12	9	0.329193	0.273292
23	19	12	10	0.739130	0.409938
23	19	12	11	0.962733	0.223602
23	19	12	12	1.000000	0.037267
23	19	13	9	0.080745	0.080745
23	19	13	10	0.403727	0.322981
23	19	13	11	0.800113	0.396386
23	19	13	12	0.976284	0.176172
23	19	13	13	1.000000	0.023715
23	19	14	10	0.113043	0.113043
23	19	14	11	0.483004	0.369960
23	19	14	12	0.852964	0.369960
23	19	14	13	0.985771	0.132806
23	19	14	14	1.000000	0.014229
23	19	15	11	0.154150	0.154150
23	19	15	12	0.565217	0.411067
23	19	15	13	0.897233	0.332016
23	19	15	14	0.992095	0.094862
23	19	15	15	1.000000	0.007905
23	19	16	12	0.205534	0.205534
23	19	16	13	0.648221	0.442688
23	19	16	14	0.932806	0.284585
23	19	16	15	0.996047	0.063241
23	19	16	16	1.000000	0.003953
23	19	17	13	0.268775	0.268775
23	19	17	14	0.729531	0.460757
23	19	17	15	0.959910	0.230378
23	19	17	16	0.998306	0.038396
23	19	17	17	1.000000	0.001694
23	19	18	14	0.345567	0.345567
23	19	18	15	0.806324	0.460757
23	19	18	16	0.979108	0.172784
23	19	18	17	0.999435	0.020327
23	19	18	18	1.000000	0.000565

N	n	k	x	P(x)	p(x)
23	19	19	15	0.437719	0.437719
23	19	19	16	0.875438	0.437719
23	19	19	17	0.991304	0.115867
23	19	19	18	0.999887	0.008583
23	19	19	19	1.000000	0.000113
23	20	1	0	0.130435	0.130435
23	20	1	1	1.000000	0.869565
23	20	2	0	0.011858	0.011858
23	20	2	1	0.249012	0.237154
23	20	2	2	1.000000	0.750988
23	20	3	0	0.000565	0.000565
23	20	3	1	0.034444	0.033879
23	20	3	2	0.356296	0.321852
23	20	3	3	1.000000	0.643704
23	20	4	1	0.002259	0.002259
23	20	4	2	0.066629	0.064370
23	20	4	3	0.452851	0.386222
23	20	4	4	1.000000	0.547148
23	20	5	2	0.005647	0.005647
23	20	5	3	0.107284	0.101637
23	20	5	4	0.539243	0.431959
23	20	5	5	1.000000	0.460757
23	20	6	3	0.011293	0.011293
23	20	6	4	0.155279	0.143986
23	20	6	5	0.616036	0.460757
23	20	6	6	1.000000	0.383964
23	20	7	4	0.019763	0.019763
23	20	7	5	0.209486	0.189723
23	20	7	6	0.683794	0.474308
23	20	7	7	1.000000	0.316206
23	20	8	5	0.031621	0.031621
23	20	8	6	0.268775	0.237154
23	20	8	7	0.743083	0.474308
23	20	8	8	1.000000	0.256917
23	20	9	6	0.047431	0.047431
23	20	9	7	0.332016	0.284585
23	20	9	8	0.794466	0.462451
23	20	9	9	1.000000	0.205534
23	20	10	7	0.067758	0.067758
23	20	10	8	0.398080	0.330322
23	20	10	9	0.838509	0.440429
23	20	10	10	1.000000	0.161491
23	20	11	8	0.093168	0.093168
23	20	11	9	0.465838	0.372671
23	20	11	10	0.875776	0.409938
23	20	11	11	1.000000	0.124224
23	20	12	9	0.124224	0.124224
23	20	12	10	0.534161	0.409938
23	20	12	11	0.906832	0.372671
23	20	12	12	1.000000	0.093168

N	n	k	x	P(x)	p(x)
23	20	13	10	0.161491	0.161491
23	20	13	11	0.601920	0.440429
23	20	13	12	0.932242	0.330322
23	20	13	13	1.000000	0.067758
23	20	14	11	0.205534	0.205534
23	20	14	12	0.667984	0.462451
23	20	14	13	0.952569	0.284585
23	20	14	14	1.000000	0.047431
23	20	15	12	0.256917	0.256917
23	20	15	13	0.731225	0.474308
23	20	15	14	0.968379	0.237154
23	20	15	15	1.000000	0.031621
23	20	16	13	0.316206	0.316206
23	20	16	14	0.790514	0.474308
23	20	16	15	0.980237	0.189723
23	20	16	16	1.000000	0.019763
23	20	17	14	0.383964	0.383964
23	20	17	15	0.844720	0.460757
23	20	17	16	0.988707	0.143986
23	20	17	17	1.000000	0.011293
23	20	18	15	0.460757	0.460757
23	20	18	16	0.892716	0.431959
23	20	18	17	0.994353	0.101637
23	20	18	18	1.000000	0.005647
23	20	19	16	0.547148	0.547148
23	20	19	17	0.933371	0.386222
23	20	19	18	0.997741	0.064370
23	20	19	19	1.000000	0.002259
23	20	20	17	0.643704	0.643704
23	20	20	18	0.965556	0.321852
23	20	20	19	0.999435	0.033879
23	20	20	20	1.000000	0.000565
23	21	1	0	0.086957	0.086957
23	21	1	1	1.000000	0.913043
23	21	2	0	0.003953	0.003953
23	21	2	1	0.169960	0.166008
23	21	2	2	1.000000	0.830040
23	21	3	1	0.011858	0.011858
23	21	3	2	0.249012	0.237154
23	21	3	3	1.000000	0.750988
23	21	4	2	0.023715	0.023715
23	21	4	3	0.324111	0.300395
23	21	4	4	1.000000	0.675889
23	21	5	3	0.039526	0.039526
23	21	5	4	0.395257	0.355731
23	21	5	5	1.000000	0.604743
23	21	6	4	0.059289	0.059289
23	21	6	5	0.462451	0.403162
23	21	6	6	1.000000	0.537549
23	21	7	5	0.083004	0.083004

Table for $N = 2$, $n = 1$, through $N = 100$, $n = 50$

N	n	k	x	P(x)	p(x)
23	21	7	6	0.525692	0.442688
23	21	7	7	1.000000	0.474308
23	21	8	6	0.110672	0.110672
23	21	8	7	0.584980	0.474308
23	21	8	8	1.000000	0.415020
23	21	9	7	0.142292	0.142292
23	21	9	8	0.640316	0.498024
23	21	9	9	1.000000	0.359684
23	21	10	8	0.177866	0.177866
23	21	10	9	0.691700	0.513834
23	21	10	10	1.000000	0.308300
23	21	11	9	0.217391	0.217391
23	21	11	10	0.739130	0.521739
23	21	11	11	1.000000	0.260870
23	21	12	10	0.260870	0.260870
23	21	12	11	0.782609	0.521739
23	21	12	12	1.000000	0.217391
23	21	13	11	0.308300	0.308300
23	21	13	12	0.822134	0.513834
23	21	13	13	1.000000	0.177866
23	21	14	12	0.359684	0.359684
23	21	14	13	0.857707	0.498024
23	21	14	14	1.000000	0.142292
23	21	15	13	0.415020	0.415020
23	21	15	14	0.889328	0.474308
23	21	15	15	1.000000	0.110672
23	21	16	14	0.474308	0.474308
23	21	16	15	0.916996	0.442688
23	21	16	16	1.000000	0.083004
23	21	17	15	0.537549	0.537549
23	21	17	16	0.940711	0.403162
23	21	17	17	1.000000	0.059289
23	21	18	16	0.604743	0.604743
23	21	18	17	0.960474	0.355731
23	21	18	18	1.000000	0.039526
23	21	19	17	0.675889	0.675889
23	21	19	18	0.976285	0.300395
23	21	19	19	1.000000	0.023715
23	21	20	18	0.750988	0.750988
23	21	20	19	0.988142	0.237154
23	21	20	20	1.000000	0.011858
23	21	21	19	0.830040	0.830040
23	21	21	20	0.996047	0.166008
23	21	21	21	1.000000	0.003953
23	22	1	0	0.043478	0.043478
23	22	1	1	1.000000	0.956522
23	22	2	1	0.086957	0.086957
23	22	2	2	1.000000	0.913043
23	22	3	2	0.130435	0.130435
23	22	3	3	1.000000	0.869565
23	22	4	3	0.173913	0.173913
23	22	4	4	1.000000	0.826087
23	22	5	4	0.217391	0.217391
23	22	5	5	1.000000	0.782609
23	22	6	5	0.260870	0.260870
23	22	6	6	1.000000	0.739130
23	22	7	6	0.304348	0.304348
23	22	7	7	1.000000	0.695652
23	22	8	7	0.347826	0.347826
23	22	8	8	1.000000	0.652174
23	22	9	8	0.391304	0.391304
23	22	9	9	1.000000	0.608696
23	22	10	9	0.434783	0.434783
23	22	10	10	1.000000	0.565217
23	22	11	10	0.478261	0.478261
23	22	11	11	1.000000	0.521739
23	22	12	11	0.521739	0.521739
23	22	12	12	1.000000	0.478261
23	22	13	12	0.565217	0.565217
23	22	13	13	1.000000	0.434783
23	22	14	13	0.608696	0.608696
23	22	14	14	1.000000	0.391304
23	22	15	14	0.652174	0.652174
23	22	15	15	1.000000	0.347826
23	22	16	15	0.695652	0.695652
23	22	16	16	1.000000	0.304348
23	22	17	16	0.739130	0.739130
23	22	17	17	1.000000	0.260870
23	22	18	17	0.782609	0.782609
23	22	18	18	1.000000	0.217391
23	22	19	18	0.826087	0.826087
23	22	19	19	1.000000	0.173913
23	22	20	19	0.869565	0.869565
23	22	20	20	1.000000	0.130435
23	22	21	20	0.913043	0.913043
23	22	21	21	1.000000	0.086957
23	22	22	21	0.956522	0.956522
23	22	22	22	1.000000	0.043478
24	1	1	0	0.958333	0.958333
24	1	1	1	1.000000	0.041667
24	2	1	0	0.916667	0.916667
24	2	1	1	1.000000	0.083333
24	2	2	0	0.836957	0.836957
24	2	2	1	0.996377	0.159420
24	2	2	2	1.000000	0.003623
24	3	1	0	0.875000	0.875000
24	3	1	1	1.000000	0.125000
24	3	2	0	0.760870	0.760870
24	3	2	1	0.989130	0.228261
24	3	2	2	1.000000	0.010870
24	3	3	0	0.657115	0.657115
24	3	3	1	0.968379	0.311265
24	3	3	2	0.999506	0.031126
24	3	3	3	1.000000	0.000494
24	4	1	0	0.833333	0.833333
24	4	1	1	1.000000	0.166667
24	4	2	0	0.688406	0.688406
24	4	2	1	0.978261	0.289855
24	4	2	2	1.000000	0.021739
24	4	3	0	0.563241	0.563241
24	4	3	1	0.938735	0.375494
24	4	3	2	0.998024	0.059289
24	4	3	3	1.000000	0.001976
24	4	4	0	0.455957	0.455957
24	4	4	1	0.885093	0.429136
24	4	4	2	0.992377	0.107284
24	4	4	3	0.999906	0.007529
24	4	4	4	1.000000	0.000094
24	5	1	0	0.791667	0.791667
24	5	1	1	1.000000	0.208333
24	5	2	0	0.619565	0.619565
24	5	2	1	0.963768	0.344203
24	5	2	2	1.000000	0.036232
24	5	3	0	0.478755	0.478755
24	5	3	1	0.901186	0.422431
24	5	3	2	0.995059	0.093874
24	5	3	3	1.000000	0.004941
24	5	4	0	0.364766	0.364766
24	5	4	1	0.820723	0.455957
24	5	4	2	0.981649	0.160926
24	5	4	3	0.999529	0.017881
24	5	4	4	1.000000	0.000471
24	5	5	0	0.273574	0.273574
24	5	5	1	0.729531	0.455957
24	5	5	2	0.957510	0.227979
24	5	5	3	0.997741	0.040231
24	5	5	4	0.999976	0.002235
24	5	5	5	1.000000	0.000024
24	6	1	0	0.750000	0.750000
24	6	1	1	1.000000	0.250000
24	6	2	0	0.554348	0.554348
24	6	2	1	0.945652	0.391304
24	6	2	2	1.000000	0.054348
24	6	3	0	0.403162	0.403162
24	6	3	1	0.856719	0.453557
24	6	3	2	0.990119	0.133399
24	6	3	3	1.000000	0.009881
24	6	4	0	0.287973	0.287973
24	6	4	1	0.748729	0.460757
24	6	4	2	0.964709	0.215980
24	6	4	3	0.998588	0.033879
24	6	4	4	1.000000	0.001412
24	6	5	0	0.201581	0.201581
24	6	5	1	0.633540	0.431959
24	6	5	2	0.921513	0.287973
24	6	5	3	0.993506	0.071993
24	6	5	4	0.999859	0.006352
24	6	5	5	1.000000	0.000141
24	6	6	0	0.137924	0.137924
24	6	6	1	0.519867	0.381943
24	6	6	2	0.860887	0.341021
24	6	6	3	0.982139	0.121252
24	6	6	4	0.999190	0.017051
24	6	6	5	0.999993	0.000802
24	6	6	6	1.000000	0.000007
24	7	1	0	0.708333	0.708333
24	7	1	1	1.000000	0.291667
24	7	2	0	0.492754	0.492754
24	7	2	1	0.923913	0.431159
24	7	2	2	1.000000	0.076087
24	7	3	0	0.335968	0.335968
24	7	3	1	0.806324	0.470356
24	7	3	2	0.982707	0.176383
24	7	3	3	1.000000	0.017292
24	7	4	0	0.223979	0.223979
24	7	4	1	0.671937	0.447958
24	7	4	2	0.940711	0.268775
24	7	4	3	0.996706	0.055995
24	7	4	4	1.000000	0.003294
24	7	5	0	0.145586	0.145586
24	7	5	1	0.537549	0.391963
24	7	5	2	0.873518	0.335968
24	7	5	3	0.985507	0.111989
24	7	5	4	0.999506	0.013999
24	7	5	5	1.000000	0.000494
24	7	6	0	0.091949	0.091949
24	7	6	1	0.413772	0.321822
24	7	6	2	0.785105	0.371333
24	7	6	3	0.961930	0.176825
24	7	6	4	0.997296	0.035365
24	7	6	5	0.999948	0.002652
24	7	6	6	1.000000	0.000052
24	7	7	0	0.056191	0.056191
24	7	7	1	0.306497	0.250306
24	7	7	2	0.681957	0.375459
24	7	7	3	0.922636	0.240679
24	7	7	4	0.991401	0.068765
24	7	7	5	0.999653	0.008252
24	7	7	6	0.999997	0.000344
24	7	7	7	1.000000	0.000003

N	n	k	x	P(x)	p(x)
24	8	1	0	0.666667	0.666667
24	8	1	1	1.000000	0.333333
24	8	2	0	0.434783	0.434783
24	8	2	1	0.898551	0.463768
24	8	2	2	1.000000	0.101449
24	8	3	0	0.276680	0.276680
24	8	3	1	0.750988	0.474308
24	8	3	2	0.972332	0.221344
24	8	3	3	1.000000	0.027668
24	8	4	0	0.171278	0.171278
24	8	4	1	0.592885	0.421607
24	8	4	2	0.909091	0.316206
24	8	4	3	0.993412	0.084321
24	8	4	4	1.000000	0.006588
24	8	5	0	0.102767	0.102767
24	8	5	1	0.445323	0.342556
24	8	5	2	0.814229	0.368906
24	8	5	3	0.972332	0.158103
24	8	5	4	0.998682	0.026350
24	8	5	5	1.000000	0.001318
24	8	6	0	0.059497	0.059497
24	8	6	1	0.319118	0.259621
24	8	6	2	0.697732	0.378615
24	8	6	3	0.930726	0.232994
24	8	6	4	0.993135	0.062409
24	8	6	5	0.999792	0.006657
24	8	6	6	1.000000	0.000208
24	8	7	0	0.033054	0.033054
24	8	7	1	0.218154	0.185100
24	8	7	2	0.571528	0.353374
24	8	7	3	0.866006	0.294478
24	8	7	4	0.979266	0.113261
24	8	7	5	0.998682	0.019416
24	8	7	6	0.999977	0.001294
24	8	7	7	1.000000	0.000023
24	8	8	0	0.017499	0.017499
24	8	8	1	0.141936	0.124437
24	8	8	2	0.446808	0.304871
24	8	8	3	0.779394	0.332587
24	8	8	4	0.952617	0.173222
24	8	8	5	0.995256	0.042639
24	8	8	6	0.999825	0.004568
24	8	8	7	0.999999	0.000174
24	8	8	8	1.000000	0.000001
24	9	1	0	0.625000	0.625000
24	9	1	1	1.000000	0.375000
24	9	2	0	0.380435	0.380435
24	9	2	1	0.869565	0.489130
24	9	2	2	1.000000	0.130435
24	9	3	0	0.224802	0.224802

N	n	k	x	P(x)	p(x)
24	9	3	1	0.691700	0.466897
24	9	3	2	0.958498	0.266798
24	9	3	3	1.000000	0.041502
24	9	4	0	0.128458	0.128458
24	9	4	1	0.513834	0.385375
24	9	4	2	0.869565	0.355731
24	9	4	3	0.988142	0.118577
24	9	4	4	1.000000	0.011858
24	9	5	0	0.070652	0.070652
24	9	5	1	0.359684	0.289032
24	9	5	2	0.745059	0.385375
24	9	5	3	0.952569	0.207510
24	9	5	4	0.997035	0.044466
24	9	5	5	1.000000	0.002964
24	9	6	0	0.037185	0.037185
24	9	6	1	0.237986	0.200801
24	9	6	2	0.603079	0.365093
24	9	6	3	0.887040	0.283961
24	9	6	4	0.985334	0.098294
24	9	6	5	0.999376	0.014042
24	9	6	6	1.000000	0.000624
24	9	7	0	0.018593	0.018593
24	9	7	1	0.148741	0.130149
24	9	7	2	0.461098	0.312357
24	9	7	3	0.792386	0.331288
24	9	7	4	0.958030	0.165644
24	9	7	5	0.996255	0.038225
24	9	7	6	0.999896	0.003641
24	9	7	7	1.000000	0.000104
24	9	8	0	0.008749	0.008749
24	9	8	1	0.087495	0.078745
24	9	8	2	0.332481	0.244986
24	9	8	3	0.675461	0.342980
24	9	8	4	0.909311	0.233850
24	9	8	5	0.987261	0.077950
24	9	8	6	0.999253	0.011992
24	9	8	7	0.999988	0.000734
24	9	8	8	1.000000	0.000012
24	9	9	0	0.003828	0.003828
24	9	9	1	0.048122	0.044294
24	9	9	2	0.225299	0.177177
24	9	9	3	0.546843	0.321544
24	9	9	4	0.836233	0.289389
24	9	9	5	0.967774	0.131541
24	9	9	6	0.997005	0.029231
24	9	9	7	0.999896	0.002891
24	9	9	8	0.999999	0.000103
24	9	9	9	1.000000	0.000001
24	10	1	0	0.583333	0.583333
24	10	1	1	1.000000	0.416667

N	n	k	x	P(x)	p(x)
24	10	2	0	0.329710	0.329710
24	10	2	1	0.836956	0.507246
24	10	2	2	1.000000	0.163043
24	10	3	0	0.179842	0.179842
24	10	3	1	0.629447	0.449605
24	10	3	2	0.940711	0.311265
24	10	3	3	1.000000	0.059289
24	10	4	0	0.094203	0.094203
24	10	4	1	0.436759	0.342556
24	10	4	2	0.822134	0.385375
24	10	4	3	0.980237	0.158103
24	10	4	4	1.000000	0.019763
24	10	5	0	0.047101	0.047101
24	10	5	1	0.282609	0.235507
24	10	5	2	0.667984	0.385375
24	10	5	3	0.924901	0.256917
24	10	5	4	0.994071	0.069170
24	10	5	5	1.000000	0.005929
24	10	6	0	0.022311	0.022311
24	10	6	1	0.171053	0.148741
24	10	6	2	0.505721	0.334668
24	10	6	3	0.830247	0.324527
24	10	6	4	0.972228	0.141980
24	10	6	5	0.998440	0.026212
24	10	6	6	1.000000	0.001560
24	10	7	0	0.009916	0.009916
24	10	7	1	0.096682	0.086766
24	10	7	2	0.356979	0.260297
24	10	7	3	0.704043	0.347063
24	10	7	4	0.924901	0.220858
24	10	7	5	0.991159	0.066258
24	10	7	6	0.999653	0.008495
24	10	7	7	1.000000	0.000347
24	10	8	0	0.004083	0.004083
24	10	8	1	0.050747	0.046664
24	10	8	2	0.234486	0.183739
24	10	8	3	0.561134	0.326648
24	10	8	4	0.846951	0.285817
24	10	8	5	0.971671	0.124720
24	10	8	6	0.997654	0.025983
24	10	8	7	0.999939	0.002284
24	10	8	8	1.000000	0.000061
24	10	9	0	0.001531	0.001531
24	10	9	1	0.024499	0.022967
24	10	9	2	0.142617	0.118118
24	10	9	3	0.418226	0.275609
24	10	9	4	0.739770	0.321544
24	10	9	5	0.932696	0.192926
24	10	9	6	0.991159	0.058463
24	10	9	7	0.999510	0.008352

N	n	k	x	P(x)	p(x)
24	10	9	8	0.999992	0.000482
24	10	9	9	1.000000	0.000008
24	10	10	0	0.000510	0.000510
24	10	10	1	0.010718	0.010208
24	10	10	2	0.079620	0.068902
24	10	10	3	0.289608	0.209988
24	10	10	4	0.611152	0.321544
24	10	10	5	0.868387	0.257235
24	10	10	6	0.975569	0.107181
24	10	10	7	0.997840	0.022271
24	10	10	8	0.999928	0.002088
24	10	10	9	0.999999	0.000071
24	10	10	10	1.000000	0.000001
24	11	1	0	0.541667	0.541667
24	11	1	1	1.000000	0.458333
24	11	2	0	0.282609	0.282609
24	11	2	1	0.800725	0.518116
24	11	2	2	1.000000	0.199275
24	11	3	0	0.141304	0.141304
24	11	3	1	0.565217	0.423913
24	11	3	2	0.918478	0.353261
24	11	3	3	1.000000	0.081522
24	11	4	0	0.067288	0.067288
24	11	4	1	0.363354	0.296066
24	11	4	2	0.767081	0.403727
24	11	4	3	0.968944	0.201863
24	11	4	4	1.000000	0.031056
24	11	5	0	0.030279	0.030279
24	11	5	1	0.215321	0.185041
24	11	5	2	0.585404	0.370083
24	11	5	3	0.888199	0.302795
24	11	5	4	0.989130	0.100932
24	11	5	5	1.000000	0.010870
24	11	6	0	0.012749	0.012749
24	11	6	1	0.117931	0.105181
24	11	6	2	0.410101	0.292171
24	11	6	3	0.760706	0.350605
24	11	6	4	0.951945	0.191239
24	11	6	5	0.996567	0.044622
24	11	6	6	1.000000	0.003432
24	11	7	0	0.004958	0.004958
24	11	7	1	0.059497	0.054539
24	11	7	2	0.264016	0.204519
24	11	7	3	0.604882	0.340866
24	11	7	4	0.877574	0.272693
24	11	7	5	0.981693	0.104119
24	11	7	6	0.999046	0.017353
24	11	7	7	1.000000	0.000953
24	11	8	0	0.001750	0.001750
24	11	8	1	0.027415	0.025665

N	n	k	x	P(x)	p(x)
24	11	8	2	0.155741	0.128326
24	11	8	3	0.444474	0.288733
24	11	8	4	0.765289	0.320815
24	11	8	5	0.944945	0.179656
24	11	8	6	0.993943	0.048997
24	11	8	7	0.999776	0.005833
24	11	8	8	1.000000	0.000224
24	11	9	0	0.000547	0.000547
24	11	9	1	0.011374	0.010827
24	11	9	2	0.083558	0.072183
24	11	9	3	0.300108	0.216550
24	11	9	4	0.624933	0.324825
24	11	9	5	0.877574	0.252642
24	11	9	6	0.978631	0.101057
24	11	9	7	0.998317	0.019686
24	11	9	8	0.999958	0.001641
24	11	9	9	1.000000	0.000042
24	11	10	0	0.000146	0.000146
24	11	10	1	0.004156	0.004010
24	11	10	2	0.040248	0.036092
24	11	10	3	0.184614	0.144367
24	11	10	4	0.473348	0.288733
24	11	10	5	0.776518	0.303170
24	11	10	6	0.944945	0.168428
24	11	10	7	0.993068	0.048122
24	11	10	8	0.999630	0.006562
24	11	10	9	0.999994	0.000365
24	11	10	10	1.000000	0.000006
24	11	11	0	0.000031	0.000031
24	11	11	1	0.001292	0.001260
24	11	11	2	0.017046	0.015754
24	11	11	3	0.102119	0.085073
24	11	11	4	0.328981	0.226862
24	11	11	5	0.646588	0.317607
24	11	11	6	0.884793	0.238205
24	11	11	7	0.979318	0.094526
24	11	11	8	0.998224	0.018905
24	11	11	9	0.999942	0.001719
24	11	11	10	0.999999	0.000057
24	11	11	11	1.000000	0.000000
24	12	1	0	0.500000	0.500000
24	12	1	1	1.000000	0.500000
24	12	2	0	0.239130	0.239130
24	12	2	1	0.760870	0.521739
24	12	2	2	1.000000	0.239130
24	12	3	0	0.108696	0.108696
24	12	3	1	0.500000	0.391304
24	12	3	2	0.891304	0.391304
24	12	3	3	1.000000	0.108696
24	12	4	0	0.046584	0.046584

N	n	k	x	P(x)	p(x)
24	12	4	1	0.295031	0.248447
24	12	4	2	0.704969	0.409938
24	12	4	3	0.953416	0.248447
24	12	4	4	1.000000	0.046584
24	12	5	0	0.018634	0.018634
24	12	5	1	0.158385	0.139752
24	12	5	2	0.500000	0.341615
24	12	5	3	0.841615	0.341615
24	12	5	4	0.981366	0.139752
24	12	5	5	1.000000	0.018634
24	12	6	0	0.006865	0.006865
24	12	6	1	0.077476	0.070611
24	12	6	2	0.320203	0.242726
24	12	6	3	0.679797	0.359595
24	12	6	4	0.922524	0.242726
24	12	6	5	0.993135	0.070611
24	12	6	6	1.000000	0.006865
24	12	7	0	0.002288	0.002288
24	12	7	1	0.034325	0.032037
24	12	7	2	0.185355	0.151030
24	12	7	3	0.500000	0.314645
24	12	7	4	0.814645	0.314645
24	12	7	5	0.965675	0.151030
24	12	7	6	0.997712	0.032037
24	12	7	7	1.000000	0.002288
24	12	8	0	0.000673	0.000673
24	12	8	1	0.013595	0.012922
24	12	8	2	0.096514	0.082918
24	12	8	3	0.333423	0.236909
24	12	8	4	0.666577	0.333154
24	12	8	5	0.903486	0.236909
24	12	8	6	0.986405	0.082918
24	12	8	7	0.999327	0.012922
24	12	8	8	1.000000	0.000673
24	12	9	0	0.000168	0.000168
24	12	9	1	0.004711	0.004543
24	12	9	2	0.044690	0.039978
24	12	9	3	0.200162	0.155472
24	12	9	4	0.500000	0.299838
24	12	9	5	0.799838	0.299838
24	12	9	6	0.955310	0.155472
24	12	9	7	0.995289	0.039978
24	12	9	8	0.999832	0.004543
24	12	9	9	1.000000	0.000168
24	12	10	0	0.000034	0.000034
24	12	10	1	0.001380	0.001346
24	12	10	2	0.018037	0.016658
24	12	10	3	0.106878	0.088841
24	12	10	4	0.340086	0.233208
24	12	10	5	0.659914	0.319828

N	n	k	x	P(x)	p(x)
24	12	10	6	0.893121	0.233208
24	12	10	7	0.981963	0.088841
24	12	10	8	0.998620	0.016658
24	12	10	9	0.999966	0.001346
24	12	10	10	1.000000	0.000034
24	12	11	0	0.000005	0.000005
24	12	11	1	0.000322	0.000317
24	12	11	2	0.006139	0.005817
24	12	11	3	0.049766	0.043627
24	12	11	4	0.206825	0.157058
24	12	11	5	0.500000	0.293175
24	12	11	6	0.793175	0.293175
24	12	11	7	0.950234	0.157058
24	12	11	8	0.993861	0.043627
24	12	11	9	0.999678	0.005817
24	12	11	10	0.999995	0.000317
24	12	11	11	1.000000	0.000005
24	12	12	0	0.000000	0.000000
24	12	12	1	0.000054	0.000053
24	12	12	2	0.001664	0.001611
24	12	12	3	0.019563	0.017898
24	12	12	4	0.110173	0.090610
24	12	12	5	0.342136	0.231963
24	12	12	6	0.657864	0.315727
24	12	12	7	0.889826	0.231963
24	12	12	8	0.980437	0.090610
24	12	12	9	0.998335	0.017898
24	12	12	10	0.999946	0.001611
24	12	12	11	0.999999	0.000053
24	12	12	12	1.000000	0.000000
24	13	1	0	0.458333	0.458333
24	13	1	1	1.000000	0.541667
24	13	2	0	0.199275	0.199275
24	13	2	1	0.717391	0.518116
24	13	2	2	1.000000	0.282609
24	13	3	0	0.081522	0.081522
24	13	3	1	0.434783	0.353261
24	13	3	2	0.858696	0.423913
24	13	3	3	1.000000	0.141304
24	13	4	0	0.031056	0.031056
24	13	4	1	0.232919	0.201863
24	13	4	2	0.636646	0.403727
24	13	4	3	0.932712	0.296066
24	13	4	4	1.000000	0.067288
24	13	5	0	0.010870	0.010870
24	13	5	1	0.111801	0.100932
24	13	5	2	0.414596	0.302795
24	13	5	3	0.784679	0.370083
24	13	5	4	0.969720	0.185041
24	13	5	5	1.000000	0.030279

N	n	k	x	P(x)	p(x)
24	13	6	0	0.003432	0.003432
24	13	6	1	0.048055	0.044622
24	13	6	2	0.239294	0.191239
24	13	6	3	0.589899	0.350605
24	13	6	4	0.882069	0.292171
24	13	6	5	0.987251	0.105181
24	13	6	6	1.000000	0.012749
24	13	7	0	0.000953	0.000953
24	13	7	1	0.018307	0.017353
24	13	7	2	0.122426	0.104119
24	13	7	3	0.395118	0.272693
24	13	7	4	0.735984	0.340866
24	13	7	5	0.940503	0.204519
24	13	7	6	0.995042	0.054539
24	13	7	7	1.000000	0.004958
24	13	8	0	0.000224	0.000224
24	13	8	1	0.006057	0.005833
24	13	8	2	0.055055	0.048997
24	13	8	3	0.234711	0.179656
24	13	8	4	0.555526	0.320815
24	13	8	5	0.844259	0.288733
24	13	8	6	0.972585	0.128326
24	13	8	7	0.998250	0.025665
24	13	8	8	1.000000	0.001750
24	13	9	0	0.000042	0.000042
24	13	9	1	0.001683	0.001641
24	13	9	2	0.021369	0.019686
24	13	9	3	0.122426	0.101057
24	13	9	4	0.375067	0.252642
24	13	9	5	0.699892	0.324825
24	13	9	6	0.916442	0.216550
24	13	9	7	0.988626	0.072183
24	13	9	8	0.999453	0.010827
24	13	9	9	1.000000	0.000547
24	13	10	0	0.000006	0.000006
24	13	10	1	0.000370	0.000365
24	13	10	2	0.006932	0.006562
24	13	10	3	0.055055	0.048122
24	13	10	4	0.223482	0.168428
24	13	10	5	0.526652	0.303170
24	13	10	6	0.815386	0.288733
24	13	10	7	0.959752	0.144367
24	13	10	8	0.995844	0.036092
24	13	10	9	0.999854	0.004010
24	13	10	10	1.000000	0.000146
24	13	11	0	0.000000	0.000000
24	13	11	1	0.000058	0.000057
24	13	11	2	0.001776	0.001719
24	13	11	3	0.020681	0.018905
24	13	11	4	0.115207	0.094526

Table for $N = 2$, $n = 1$, through $N = 100$, $n = 50$

N	n	k	x	P(x)	p(x)
24	13	11	5	0.353412	0.238205
24	13	11	6	0.671019	0.317607
24	13	11	7	0.897881	0.226862
24	13	11	8	0.982954	0.085073
24	13	11	9	0.998708	0.015754
24	13	11	10	0.999969	0.001260
24	13	11	11	1.000000	0.000031
24	13	12	1	0.000005	0.000005
24	13	12	2	0.000322	0.000317
24	13	12	3	0.006139	0.005817
24	13	12	4	0.049766	0.043627
24	13	12	5	0.206825	0.157058
24	13	12	6	0.500000	0.293175
24	13	12	7	0.793175	0.293175
24	13	12	8	0.950234	0.157058
24	13	12	9	0.993861	0.043627
24	13	12	10	0.999678	0.005817
24	13	12	11	0.999995	0.000317
24	13	12	12	1.000000	0.000005
24	13	13	2	0.000031	0.000031
24	13	13	3	0.001292	0.001260
24	13	13	4	0.017046	0.015754
24	13	13	5	0.102119	0.085073
24	13	13	6	0.328981	0.226862
24	13	13	7	0.646588	0.317607
24	13	13	8	0.884793	0.238205
24	13	13	9	0.979318	0.094526
24	13	13	10	0.998224	0.018905
24	13	13	11	0.999942	0.001719
24	13	13	12	0.999999	0.000057
24	13	13	13	1.000000	0.000000
24	14	1	0	0.416667	0.416667
24	14	1	1	1.000000	0.583333
24	14	2	0	0.163043	0.163043
24	14	2	1	0.670290	0.507246
24	14	2	2	1.000000	0.329710
24	14	3	0	0.059289	0.059289
24	14	3	1	0.370553	0.311265
24	14	3	2	0.820158	0.449605
24	14	3	3	1.000000	0.179842
24	14	4	0	0.019763	0.019763
24	14	4	1	0.177866	0.158103
24	14	4	2	0.563241	0.385375
24	14	4	3	0.905797	0.342556
24	14	4	4	1.000000	0.094203
24	14	5	0	0.005929	0.005929
24	14	5	1	0.075099	0.069170
24	14	5	2	0.332016	0.256917
24	14	5	3	0.717391	0.385375
24	14	5	4	0.952898	0.235507
24	14	5	5	1.000000	0.047101
24	14	6	0	0.001560	0.001560
24	14	6	1	0.027772	0.026212
24	14	6	2	0.169752	0.141980
24	14	6	3	0.494279	0.324527
24	14	6	4	0.828947	0.334668
24	14	6	5	0.977689	0.148741
24	14	6	6	1.000000	0.022311
24	14	7	0	0.000347	0.000347
24	14	7	1	0.008841	0.008495
24	14	7	2	0.075099	0.066258
24	14	7	3	0.295957	0.220858
24	14	7	4	0.643021	0.347063
24	14	7	5	0.903318	0.260297
24	14	7	6	0.990084	0.086766
24	14	7	7	1.000000	0.009916
24	14	8	0	0.000061	0.000061
24	14	8	1	0.002345	0.002284
24	14	8	2	0.028329	0.025983
24	14	8	3	0.153049	0.124720
24	14	8	4	0.438866	0.285817
24	14	8	5	0.765513	0.326648
24	14	8	6	0.949253	0.183739
24	14	8	7	0.995917	0.046664
24	14	8	8	1.000000	0.004083
24	14	9	0	0.000008	0.000008
24	14	9	1	0.000489	0.000482
24	14	9	2	0.008841	0.008352
24	14	9	3	0.067304	0.058463
24	14	9	4	0.260230	0.192926
24	14	9	5	0.581774	0.321544
24	14	9	6	0.857383	0.275609
24	14	9	7	0.975501	0.118118
24	14	9	8	0.998469	0.022967
24	14	9	9	1.000000	0.001531
24	14	10	0	0.000001	0.000001
24	14	10	1	0.000072	0.000071
24	14	10	2	0.002160	0.002088
24	14	10	3	0.024431	0.022271
24	14	10	4	0.131613	0.107181
24	14	10	5	0.388848	0.257235
24	14	10	6	0.710392	0.321544
24	14	10	7	0.920380	0.209988
24	14	10	8	0.989282	0.068902
24	14	10	9	0.999490	0.010208
24	14	10	10	1.000000	0.000510
24	14	11	1	0.000006	0.000006
24	14	11	2	0.000370	0.000365
24	14	11	3	0.006932	0.006562
24	14	11	4	0.055055	0.048122
24	14	11	5	0.223482	0.168428
24	14	11	6	0.526652	0.303170
24	14	11	7	0.815386	0.288733
24	14	11	8	0.959752	0.144367
24	14	11	9	0.995844	0.036092
24	14	11	10	0.999854	0.004010
24	14	11	11	1.000000	0.000146
24	14	12	2	0.000034	0.000034
24	14	12	3	0.001380	0.001346
24	14	12	4	0.018037	0.016658
24	14	12	5	0.106878	0.088841
24	14	12	6	0.340086	0.233208
24	14	12	7	0.659914	0.319828
24	14	12	8	0.893121	0.233208
24	14	12	9	0.981963	0.088841
24	14	12	10	0.998620	0.016658
24	14	12	11	0.999966	0.001346
24	14	12	12	1.000000	0.000034
24	14	13	3	0.000146	0.000146
24	14	13	4	0.004156	0.004010
24	14	13	5	0.040248	0.036092
24	14	13	6	0.184614	0.144367
24	14	13	7	0.473348	0.288733
24	14	13	8	0.776518	0.303170
24	14	13	9	0.944945	0.168428
24	14	13	10	0.993068	0.048122
24	14	13	11	0.999630	0.006562
24	14	13	12	0.999994	0.000365
24	14	13	13	1.000000	0.000006
24	14	14	4	0.000510	0.000510
24	14	14	5	0.010718	0.010208
24	14	14	6	0.079620	0.068902
24	14	14	7	0.289608	0.209988
24	14	14	8	0.611152	0.321544
24	14	14	9	0.868387	0.257235
24	14	14	10	0.975569	0.107181
24	14	14	11	0.997840	0.022271
24	14	14	12	0.999928	0.002088
24	14	14	13	0.999999	0.000071
24	14	14	14	1.000000	0.000001
24	15	1	0	0.375000	0.375000
24	15	1	1	1.000000	0.625000
24	15	2	0	0.130435	0.130435
24	15	2	1	0.619565	0.489130
24	15	2	2	1.000000	0.380435
24	15	3	0	0.041502	0.041502
24	15	3	1	0.308300	0.266798
24	15	3	2	0.775198	0.466897
24	15	3	3	1.000000	0.224802
24	15	4	0	0.011858	0.011858
24	15	4	1	0.130435	0.118577
24	15	4	2	0.486166	0.355731
24	15	4	3	0.871541	0.385375
24	15	4	4	1.000000	0.128458
24	15	5	0	0.002964	0.002964
24	15	5	1	0.047431	0.044466
24	15	5	2	0.254941	0.207510
24	15	5	3	0.640316	0.385375
24	15	5	4	0.929348	0.289032
24	15	5	5	1.000000	0.070652
24	15	6	0	0.000624	0.000624
24	15	6	1	0.014666	0.014042
24	15	6	2	0.112960	0.098294
24	15	6	3	0.396921	0.283961
24	15	6	4	0.762014	0.365093
24	15	6	5	0.962815	0.200801
24	15	6	6	1.000000	0.037185
24	15	7	0	0.000104	0.000104
24	15	7	1	0.003745	0.003641
24	15	7	2	0.041970	0.038225
24	15	7	3	0.207614	0.165644
24	15	7	4	0.538902	0.331288
24	15	7	5	0.851258	0.312357
24	15	7	6	0.981407	0.130149
24	15	7	7	1.000000	0.018593
24	15	8	0	0.000012	0.000012
24	15	8	1	0.000746	0.000734
24	15	8	2	0.012739	0.011992
24	15	8	3	0.090689	0.077950
24	15	8	4	0.324539	0.233850
24	15	8	5	0.667519	0.342980
24	15	8	6	0.912505	0.244986
24	15	8	7	0.991250	0.078745
24	15	8	8	1.000000	0.008749
24	15	9	0	0.000001	0.000001
24	15	9	1	0.000104	0.000103
24	15	9	2	0.002995	0.002891
24	15	9	3	0.032226	0.029231
24	15	9	4	0.163767	0.131541
24	15	9	5	0.453156	0.289389
24	15	9	6	0.774700	0.321544
24	15	9	7	0.951878	0.177177
24	15	9	8	0.996172	0.044294
24	15	9	9	1.000000	0.003828
24	15	10	1	0.000008	0.000008
24	15	10	2	0.000489	0.000482
24	15	10	3	0.008841	0.008352
24	15	10	4	0.067304	0.058463
24	15	10	5	0.260230	0.192926
24	15	10	6	0.581774	0.321544

N	n	k	x	P(x)	p(x)
24	15	10	7	0.857383	0.275609
24	15	10	8	0.975501	0.118118
24	15	10	9	0.998469	0.022967
24	15	10	10	1.000000	0.001531
24	15	11	2	0.000042	0.000042
24	15	11	3	0.001683	0.001641
24	15	11	4	0.021369	0.019686
24	15	11	5	0.122426	0.101057
24	15	11	6	0.375067	0.252642
24	15	11	7	0.699892	0.324825
24	15	11	8	0.916442	0.216550
24	15	11	9	0.988626	0.072183
24	15	11	10	0.999453	0.010827
24	15	11	11	1.000000	0.000547
24	15	12	3	0.000168	0.000168
24	15	12	4	0.004711	0.004543
24	15	12	5	0.044690	0.039978
24	15	12	6	0.200162	0.155472
24	15	12	7	0.500000	0.299838
24	15	12	8	0.799838	0.299838
24	15	12	9	0.955310	0.155472
24	15	12	10	0.995289	0.039978
24	15	12	11	0.999832	0.004543
24	15	12	12	1.000000	0.000168
24	15	13	4	0.000547	0.000547
24	15	13	5	0.011374	0.010827
24	15	13	6	0.083558	0.072183
24	15	13	7	0.300108	0.216550
24	15	13	8	0.624933	0.324825
24	15	13	9	0.877574	0.252642
24	15	13	10	0.978631	0.101057
24	15	13	11	0.998317	0.019686
24	15	13	12	0.999958	0.001641
24	15	13	13	1.000000	0.000042
24	15	14	5	0.001531	0.001531
24	15	14	6	0.024499	0.022967
24	15	14	7	0.142617	0.118118
24	15	14	8	0.418226	0.275609
24	15	14	9	0.739770	0.321544
24	15	14	10	0.932696	0.192926
24	15	14	11	0.991159	0.058463
24	15	14	12	0.999510	0.008352
24	15	14	13	0.999992	0.000482
24	15	14	14	1.000000	0.000008
24	15	15	6	0.003828	0.003828
24	15	15	7	0.048122	0.044294
24	15	15	8	0.225299	0.177177
24	15	15	9	0.546843	0.321544
24	15	15	10	0.836233	0.289389
24	15	15	11	0.967774	0.131541

N	n	k	x	P(x)	p(x)
24	15	15	12	0.997005	0.029231
24	15	15	13	0.999896	0.002891
24	15	15	14	0.999999	0.000103
24	15	15	15	1.000000	0.000001
24	16	1	0	0.333333	0.333333
24	16	1	1	1.000000	0.666667
24	16	2	0	0.101449	0.101449
24	16	2	1	0.565217	0.463768
24	16	2	2	1.000000	0.434783
24	16	3	0	0.027668	0.027668
24	16	3	1	0.249012	0.221344
24	16	3	2	0.723320	0.474308
24	16	3	3	1.000000	0.276680
24	16	4	0	0.006588	0.006588
24	16	4	1	0.090909	0.084321
24	16	4	2	0.407115	0.316206
24	16	4	3	0.828722	0.421607
24	16	4	4	1.000000	0.171278
24	16	5	0	0.001318	0.001318
24	16	5	1	0.027668	0.026350
24	16	5	2	0.185771	0.158103
24	16	5	3	0.554677	0.368906
24	16	5	4	0.897233	0.342556
24	16	5	5	1.000000	0.102767
24	16	6	0	0.000208	0.000208
24	16	6	1	0.006865	0.006657
24	16	6	2	0.069274	0.062409
24	16	6	3	0.302267	0.232994
24	16	6	4	0.680882	0.378615
24	16	6	5	0.940503	0.259621
24	16	6	6	1.000000	0.059497
24	16	7	0	0.000023	0.000023
24	16	7	1	0.001318	0.001294
24	16	7	2	0.020734	0.019416
24	16	7	3	0.133994	0.113261
24	16	7	4	0.428472	0.294478
24	16	7	5	0.781846	0.353374
24	16	7	6	0.966946	0.185100
24	16	7	7	1.000000	0.033054
24	16	8	0	0.000001	0.000001
24	16	8	1	0.000175	0.000174
24	16	8	2	0.004744	0.004568
24	16	8	3	0.047383	0.042639
24	16	8	4	0.220606	0.173222
24	16	8	5	0.553192	0.332587
24	16	8	6	0.858064	0.304871
24	16	8	7	0.982501	0.124437
24	16	8	8	1.000000	0.017499
24	16	9	1	0.000012	0.000012
24	16	9	2	0.000746	0.000734

N	n	k	x	P(x)	p(x)
24	16	9	3	0.012739	0.011992
24	16	9	4	0.090689	0.077950
24	16	9	5	0.324539	0.233850
24	16	9	6	0.667519	0.342980
24	16	9	7	0.912505	0.244986
24	16	9	8	0.991250	0.078745
24	16	9	9	1.000000	0.008749
24	16	10	2	0.000061	0.000061
24	16	10	3	0.002345	0.002284
24	16	10	4	0.028329	0.025983
24	16	10	5	0.153049	0.124720
24	16	10	6	0.438866	0.285817
24	16	10	7	0.765513	0.326648
24	16	10	8	0.949253	0.183739
24	16	10	9	0.995917	0.046664
24	16	10	10	1.000000	0.004083
24	16	11	3	0.000224	0.000224
24	16	11	4	0.006057	0.005833
24	16	11	5	0.055055	0.048997
24	16	11	6	0.234711	0.179656
24	16	11	7	0.555526	0.320815
24	16	11	8	0.844259	0.288733
24	16	11	9	0.972585	0.128326
24	16	11	10	0.998250	0.025665
24	16	11	11	1.000000	0.001750
24	16	12	4	0.000673	0.000673
24	16	12	5	0.013595	0.012922
24	16	12	6	0.096514	0.082918
24	16	12	7	0.333423	0.236909
24	16	12	8	0.666577	0.333154
24	16	12	9	0.903486	0.236909
24	16	12	10	0.986405	0.082918
24	16	12	11	0.999327	0.012922
24	16	12	12	1.000000	0.000673
24	16	13	5	0.001750	0.001750
24	16	13	6	0.027415	0.025665
24	16	13	7	0.155741	0.128326
24	16	13	8	0.444474	0.288733
24	16	13	9	0.765289	0.320815
24	16	13	10	0.944945	0.179656
24	16	13	11	0.993943	0.048997
24	16	13	12	0.999776	0.005833
24	16	13	13	1.000000	0.000224
24	16	14	6	0.004083	0.004083
24	16	14	7	0.050747	0.046664
24	16	14	8	0.234486	0.183739
24	16	14	9	0.561134	0.326648
24	16	14	10	0.846951	0.285817
24	16	14	11	0.971671	0.124720
24	16	14	12	0.997654	0.025983

N	n	k	x	P(x)	p(x)
24	16	14	13	0.999939	0.002284
24	16	14	14	1.000000	0.000061
24	16	15	7	0.008749	0.008749
24	16	15	8	0.087495	0.078745
24	16	15	9	0.332481	0.244986
24	16	15	10	0.675461	0.342980
24	16	15	11	0.909311	0.233850
24	16	15	12	0.987261	0.077950
24	16	15	13	0.999253	0.011992
24	16	15	14	0.999988	0.000734
24	16	15	15	1.000000	0.000012
24	16	16	8	0.017499	0.017499
24	16	16	9	0.141936	0.124437
24	16	16	10	0.446808	0.304871
24	16	16	11	0.779394	0.332587
24	16	16	12	0.952617	0.173222
24	16	16	13	0.995256	0.042639
24	16	16	14	0.999825	0.004568
24	16	16	15	0.999999	0.000174
24	16	16	16	1.000000	0.000001
24	17	1	0	0.291667	0.291667
24	17	1	1	1.000000	0.708333
24	17	2	0	0.076087	0.076087
24	17	2	1	0.507246	0.431159
24	17	2	2	1.000000	0.492754
24	17	3	0	0.017292	0.017292
24	17	3	1	0.193676	0.176383
24	17	3	2	0.664032	0.470356
24	17	3	3	1.000000	0.335968
24	17	4	0	0.003294	0.003294
24	17	4	1	0.059289	0.055995
24	17	4	2	0.328063	0.268775
24	17	4	3	0.776021	0.447958
24	17	4	4	1.000000	0.223979
24	17	5	0	0.000494	0.000494
24	17	5	1	0.014493	0.013999
24	17	5	2	0.126482	0.111989
24	17	5	3	0.462451	0.335968
24	17	5	4	0.854414	0.391963
24	17	5	5	1.000000	0.145586
24	17	6	0	0.000052	0.000052
24	17	6	1	0.002704	0.002652
24	17	6	2	0.038069	0.035365
24	17	6	3	0.214895	0.176825
24	17	6	4	0.586228	0.371333
24	17	6	5	0.908051	0.321822
24	17	6	6	1.000000	0.091949
24	17	7	0	0.000003	0.000003
24	17	7	1	0.000347	0.000344
24	17	7	2	0.008599	0.008252

Table for $N = 2$, $n = 1$, through $N = 100$, $n = 50$

N	n	k	x	P(x)	p(x)
24	17	7	3	0.077364	0.068765
24	17	7	4	0.318043	0.240679
24	17	7	5	0.693502	0.375459
24	17	7	6	0.943809	0.250306
24	17	7	7	1.000000	0.056191
24	17	8	1	0.000023	0.000023
24	17	8	2	0.001318	0.001294
24	17	8	3	0.020734	0.019416
24	17	8	4	0.133994	0.113261
24	17	8	5	0.428472	0.294478
24	17	8	6	0.781846	0.353374
24	17	8	7	0.966946	0.185100
24	17	8	8	1.000000	0.033054
24	17	9	2	0.000104	0.000104
24	17	9	3	0.003745	0.003641
24	17	9	4	0.041970	0.038225
24	17	9	5	0.207614	0.165644
24	17	9	6	0.538902	0.331288
24	17	9	7	0.851258	0.312357
24	17	9	8	0.981407	0.130149
24	17	9	9	1.000000	0.018593
24	17	10	3	0.000347	0.000347
24	17	10	4	0.008841	0.008495
24	17	10	5	0.075099	0.066258
24	17	10	6	0.295957	0.220858
24	17	10	7	0.643021	0.347063
24	17	10	8	0.903318	0.260297
24	17	10	9	0.990084	0.086766
24	17	10	10	1.000000	0.009916
24	17	11	4	0.000953	0.000953
24	17	11	5	0.018307	0.017353
24	17	11	6	0.122426	0.104119
24	17	11	7	0.395118	0.272693
24	17	11	8	0.735984	0.340866
24	17	11	9	0.940503	0.204519
24	17	11	10	0.995042	0.054539
24	17	11	11	1.000000	0.004958
24	17	12	5	0.002288	0.002288
24	17	12	6	0.034325	0.032037
24	17	12	7	0.185355	0.151030
24	17	12	8	0.500000	0.314645
24	17	12	9	0.814645	0.314645
24	17	12	10	0.965675	0.151030
24	17	12	11	0.997712	0.032037
24	17	12	12	1.000000	0.002288
24	17	13	6	0.004958	0.004958
24	17	13	7	0.059497	0.054539
24	17	13	8	0.264016	0.204519
24	17	13	9	0.604882	0.340866
24	17	13	10	0.877574	0.272693
24	17	13	11	0.981693	0.104119
24	17	13	12	0.999046	0.017353
24	17	13	13	1.000000	0.000953
24	17	14	7	0.009916	0.009916
24	17	14	8	0.096682	0.086766
24	17	14	9	0.356979	0.260297
24	17	14	10	0.704043	0.347063
24	17	14	11	0.924901	0.220858
24	17	14	12	0.991159	0.066258
24	17	14	13	0.999653	0.008495
24	17	14	14	1.000000	0.000347
24	17	15	8	0.018593	0.018593
24	17	15	9	0.148741	0.130149
24	17	15	10	0.461098	0.312357
24	17	15	11	0.792386	0.331288
24	17	15	12	0.958030	0.165644
24	17	15	13	0.996255	0.038225
24	17	15	14	0.999896	0.003641
24	17	15	15	1.000000	0.000104
24	17	16	9	0.033054	0.033054
24	17	16	10	0.218154	0.185100
24	17	16	11	0.571528	0.353374
24	17	16	12	0.866006	0.294478
24	17	16	13	0.979266	0.113261
24	17	16	14	0.998682	0.019416
24	17	16	15	0.999977	0.001294
24	17	16	16	1.000000	0.000023
24	17	17	10	0.056191	0.056191
24	17	17	11	0.306497	0.250306
24	17	17	12	0.681957	0.375459
24	17	17	13	0.922636	0.240679
24	17	17	14	0.991401	0.068765
24	17	17	15	0.999653	0.008252
24	17	17	16	0.999997	0.000344
24	17	17	17	1.000000	0.000003
24	18	1	0	0.250000	0.250000
24	18	1	1	1.000000	0.750000
24	18	2	0	0.054348	0.054348
24	18	2	1	0.445652	0.391304
24	18	2	2	1.000000	0.554348
24	18	3	0	0.009881	0.009881
24	18	3	1	0.143281	0.133399
24	18	3	2	0.596838	0.453557
24	18	3	3	1.000000	0.403162
24	18	4	0	0.001412	0.001412
24	18	4	1	0.035291	0.033879
24	18	4	2	0.251270	0.215980
24	18	4	3	0.712027	0.460757
24	18	4	4	1.000000	0.287973
24	18	5	0	0.000141	0.000141
24	18	5	1	0.006494	0.006352
24	18	5	2	0.078487	0.071993
24	18	5	3	0.366460	0.287973
24	18	5	4	0.798419	0.431959
24	18	5	5	1.000000	0.201581
24	18	6	0	0.000007	0.000007
24	18	6	1	0.000810	0.000802
24	18	6	2	0.017861	0.017051
24	18	6	3	0.139113	0.121252
24	18	6	4	0.480133	0.341021
24	18	6	5	0.862076	0.381943
24	18	6	6	1.000000	0.137924
24	18	7	1	0.000052	0.000052
24	18	7	2	0.002704	0.002652
24	18	7	3	0.038069	0.035365
24	18	7	4	0.214895	0.176825
24	18	7	5	0.586228	0.371333
24	18	7	6	0.908051	0.321822
24	18	7	7	1.000000	0.091949
24	18	8	2	0.000208	0.000208
24	18	8	3	0.006865	0.006657
24	18	8	4	0.069274	0.062409
24	18	8	5	0.302267	0.232994
24	18	8	6	0.680882	0.378615
24	18	8	7	0.940503	0.259621
24	18	8	8	1.000000	0.059497
24	18	9	3	0.000624	0.000624
24	18	9	4	0.014666	0.014042
24	18	9	5	0.112960	0.098294
24	18	9	6	0.396921	0.283961
24	18	9	7	0.762014	0.365093
24	18	9	8	0.962815	0.200801
24	18	9	9	1.000000	0.037185
24	18	10	4	0.001560	0.001560
24	18	10	5	0.027772	0.026212
24	18	10	6	0.169752	0.141980
24	18	10	7	0.494279	0.324527
24	18	10	8	0.828947	0.334668
24	18	10	9	0.977689	0.148741
24	18	10	10	1.000000	0.022311
24	18	11	5	0.003432	0.003432
24	18	11	6	0.048055	0.044622
24	18	11	7	0.239294	0.191239
24	18	11	8	0.589899	0.350605
24	18	11	9	0.882069	0.292171
24	18	11	10	0.987251	0.105181
24	18	11	11	1.000000	0.012749
24	18	12	6	0.006865	0.006865
24	18	12	7	0.077476	0.070611
24	18	12	8	0.320203	0.242726
24	18	12	9	0.679797	0.359595
24	18	12	10	0.922524	0.242726
24	18	12	11	0.993135	0.070611
24	18	12	12	1.000000	0.006865
24	18	13	7	0.012749	0.012749
24	18	13	8	0.117931	0.105181
24	18	13	9	0.410101	0.292171
24	18	13	10	0.760706	0.350605
24	18	13	11	0.951945	0.191239
24	18	13	12	0.996567	0.044622
24	18	13	13	1.000000	0.003432
24	18	14	8	0.022311	0.022311
24	18	14	9	0.171053	0.148741
24	18	14	10	0.505721	0.334668
24	18	14	11	0.830247	0.324527
24	18	14	12	0.972228	0.141980
24	18	14	13	0.998440	0.026212
24	18	14	14	1.000000	0.001560
24	18	15	9	0.037185	0.037185
24	18	15	10	0.237986	0.200801
24	18	15	11	0.603079	0.365093
24	18	15	12	0.887040	0.283961
24	18	15	13	0.985334	0.098294
24	18	15	14	0.999376	0.014042
24	18	15	15	1.000000	0.000624
24	18	16	10	0.059497	0.059497
24	18	16	11	0.319118	0.259621
24	18	16	12	0.697732	0.378615
24	18	16	13	0.930726	0.232994
24	18	16	14	0.993135	0.062409
24	18	16	15	0.999792	0.006657
24	18	16	16	1.000000	0.000208
24	18	17	11	0.091949	0.091949
24	18	17	12	0.413772	0.321822
24	18	17	13	0.785105	0.371333
24	18	17	14	0.961930	0.176825
24	18	17	15	0.997296	0.035365
24	18	17	16	0.999948	0.002652
24	18	17	17	1.000000	0.000052
24	18	18	12	0.137924	0.137924
24	18	18	13	0.519867	0.381943
24	18	18	14	0.860887	0.341021
24	18	18	15	0.982139	0.121252
24	18	18	16	0.999190	0.017051
24	18	18	17	0.999993	0.000802
24	18	18	18	1.000000	0.000007
24	19	1	0	0.208333	0.208333
24	19	1	1	1.000000	0.791667
24	19	2	0	0.036232	0.036232
24	19	2	1	0.380435	0.344203

N	n	k	x	P(x)	p(x)
24	19	2	2	1.000000	0.619565
24	19	3	0	0.004941	0.004941
24	19	3	1	0.098814	0.093874
24	19	3	2	0.521245	0.422431
24	19	3	3	1.000000	0.478755
24	19	4	0	0.000471	0.000471
24	19	4	1	0.018351	0.017881
24	19	4	2	0.179277	0.160926
24	19	4	3	0.635234	0.455957
24	19	4	4	1.000000	0.364766
24	19	5	0	0.000024	0.000024
24	19	5	1	0.002259	0.002235
24	19	5	2	0.042490	0.040231
24	19	5	3	0.270469	0.227979
24	19	5	4	0.726426	0.455957
24	19	5	5	1.000000	0.273574
24	19	6	1	0.000141	0.000141
24	19	6	2	0.006494	0.006352
24	19	6	3	0.078487	0.071993
24	19	6	4	0.366460	0.287973
24	19	6	5	0.798419	0.431959
24	19	6	6	1.000000	0.201581
24	19	7	2	0.000494	0.000494
24	19	7	3	0.014493	0.013999
24	19	7	4	0.126482	0.111989
24	19	7	5	0.462451	0.335968
24	19	7	6	0.854414	0.391963
24	19	7	7	1.000000	0.145586
24	19	8	3	0.001318	0.001318
24	19	8	4	0.027668	0.026350
24	19	8	5	0.185771	0.158103
24	19	8	6	0.554677	0.368906
24	19	8	7	0.897233	0.342556
24	19	8	8	1.000000	0.102767
24	19	9	4	0.002964	0.002964
24	19	9	5	0.047431	0.044466
24	19	9	6	0.254941	0.207510
24	19	9	7	0.640316	0.385375
24	19	9	8	0.929348	0.289032
24	19	9	9	1.000000	0.070652
24	19	10	5	0.005929	0.005929
24	19	10	6	0.075099	0.069170
24	19	10	7	0.332016	0.256917
24	19	10	8	0.717391	0.385375
24	19	10	9	0.952898	0.235507
24	19	10	10	1.000000	0.047101
24	19	11	6	0.010870	0.010870
24	19	11	7	0.111801	0.100932
24	19	11	8	0.414596	0.302795
24	19	11	9	0.784679	0.370083

N	n	k	x	P(x)	p(x)
24	19	11	10	0.969720	0.185041
24	19	11	11	1.000000	0.030279
24	19	12	7	0.018634	0.018634
24	19	12	8	0.158385	0.139752
24	19	12	9	0.500000	0.341615
24	19	12	10	0.841615	0.341615
24	19	12	11	0.981366	0.139752
24	19	12	12	1.000000	0.018634
24	19	13	8	0.030279	0.030279
24	19	13	9	0.215321	0.185041
24	19	13	10	0.585404	0.370083
24	19	13	11	0.888199	0.302795
24	19	13	12	0.989130	0.100932
24	19	13	13	1.000000	0.010870
24	19	14	9	0.047101	0.047101
24	19	14	10	0.282609	0.235507
24	19	14	11	0.667984	0.385375
24	19	14	12	0.924901	0.256917
24	19	14	13	0.994071	0.069170
24	19	14	14	1.000000	0.005929
24	19	15	10	0.070652	0.070652
24	19	15	11	0.359684	0.289032
24	19	15	12	0.745059	0.385375
24	19	15	13	0.952569	0.207510
24	19	15	14	0.997035	0.044466
24	19	15	15	1.000000	0.002964
24	19	16	11	0.102767	0.102767
24	19	16	12	0.445323	0.342556
24	19	16	13	0.814229	0.368906
24	19	16	14	0.972332	0.158103
24	19	16	15	0.998682	0.026350
24	19	16	16	1.000000	0.001318
24	19	17	12	0.145586	0.145586
24	19	17	13	0.537549	0.391963
24	19	17	14	0.873518	0.335968
24	19	17	15	0.985507	0.111989
24	19	17	16	0.999506	0.013999
24	19	17	17	1.000000	0.000494
24	19	18	13	0.201581	0.201581
24	19	18	14	0.633540	0.431959
24	19	18	15	0.921513	0.287973
24	19	18	16	0.993506	0.071993
24	19	18	17	0.999859	0.006352
24	19	18	18	1.000000	0.000141
24	19	19	14	0.273574	0.273574
24	19	19	15	0.729531	0.455957
24	19	19	16	0.957510	0.227979
24	19	19	17	0.997741	0.040231
24	19	19	18	0.999976	0.002235
24	19	19	19	1.000000	0.000024

N	n	k	x	P(x)	p(x)
24	20	1	0	0.166667	0.166667
24	20	1	1	1.000000	0.833333
24	20	2	0	0.021739	0.021739
24	20	2	1	0.311594	0.289855
24	20	2	2	1.000000	0.688406
24	20	3	0	0.001976	0.001976
24	20	3	1	0.061265	0.059289
24	20	3	2	0.436759	0.375494
24	20	3	3	1.000000	0.563241
24	20	4	0	0.000094	0.000094
24	20	4	1	0.007623	0.007529
24	20	4	2	0.114907	0.107284
24	20	4	3	0.544043	0.429136
24	20	4	4	1.000000	0.455957
24	20	5	1	0.000471	0.000471
24	20	5	2	0.018351	0.017881
24	20	5	3	0.179277	0.160926
24	20	5	4	0.635234	0.455957
24	20	5	5	1.000000	0.364766
24	20	6	2	0.001412	0.001412
24	20	6	3	0.035291	0.033879
24	20	6	4	0.251270	0.215980
24	20	6	5	0.712027	0.460757
24	20	6	6	1.000000	0.287973
24	20	7	3	0.003294	0.003294
24	20	7	4	0.059289	0.055995
24	20	7	5	0.328063	0.268775
24	20	7	6	0.776021	0.447958
24	20	7	7	1.000000	0.223979
24	20	8	4	0.006588	0.006588
24	20	8	5	0.090909	0.084321
24	20	8	6	0.407115	0.316206
24	20	8	7	0.828722	0.421607
24	20	8	8	1.000000	0.171278
24	20	9	5	0.011858	0.011858
24	20	9	6	0.130435	0.118577
24	20	9	7	0.486166	0.355731
24	20	9	8	0.871541	0.385375
24	20	9	9	1.000000	0.128458
24	20	10	6	0.019763	0.019763
24	20	10	7	0.177866	0.158103
24	20	10	8	0.563241	0.385375
24	20	10	9	0.905797	0.342556
24	20	10	10	1.000000	0.094203
24	20	11	7	0.031056	0.031056
24	20	11	8	0.232919	0.201863
24	20	11	9	0.636646	0.403727
24	20	11	10	0.932712	0.296066
24	20	11	11	1.000000	0.067288
24	20	12	8	0.046584	0.046584

N	n	k	x	P(x)	p(x)
24	20	12	9	0.295031	0.248447
24	20	12	10	0.704969	0.409938
24	20	12	11	0.953416	0.248447
24	20	12	12	1.000000	0.046584
24	20	13	9	0.067288	0.067288
24	20	13	10	0.363354	0.296066
24	20	13	11	0.767081	0.403727
24	20	13	12	0.968944	0.201863
24	20	13	13	1.000000	0.031056
24	20	14	10	0.094203	0.094203
24	20	14	11	0.436759	0.342556
24	20	14	12	0.822134	0.385375
24	20	14	13	0.980237	0.158103
24	20	14	14	1.000000	0.019763
24	20	15	11	0.128458	0.128458
24	20	15	12	0.513834	0.385375
24	20	15	13	0.869565	0.355731
24	20	15	14	0.988142	0.118577
24	20	15	15	1.000000	0.011858
24	20	16	12	0.171278	0.171278
24	20	16	13	0.592885	0.421607
24	20	16	14	0.909091	0.316206
24	20	16	15	0.993412	0.084321
24	20	16	16	1.000000	0.006588
24	20	17	13	0.223979	0.223979
24	20	17	14	0.671937	0.447958
24	20	17	15	0.940711	0.268775
24	20	17	16	0.996706	0.055995
24	20	17	17	1.000000	0.003294
24	20	18	14	0.287973	0.287973
24	20	18	15	0.748729	0.460757
24	20	18	16	0.964709	0.215980
24	20	18	17	0.998588	0.033879
24	20	18	18	1.000000	0.001412
24	20	19	15	0.364766	0.364766
24	20	19	16	0.820723	0.455957
24	20	19	17	0.981649	0.160926
24	20	19	18	0.999529	0.017881
24	20	19	19	1.000000	0.000471
24	20	20	16	0.455957	0.455957
24	20	20	17	0.885093	0.429136
24	20	20	18	0.992377	0.107284
24	20	20	19	0.999906	0.007529
24	20	20	20	1.000000	0.000094
24	21	1	0	0.125000	0.125000
24	21	1	1	1.000000	0.875000
24	21	2	0	0.010870	0.010870
24	21	2	1	0.239130	0.228261
24	21	2	2	1.000000	0.760870
24	21	3	0	0.000494	0.000494

N	n	k	x	P(x)	p(x)
24	21	3	1	0.031621	0.031126
24	21	3	2	0.342885	0.311265
24	21	3	3	1.000000	0.657115
24	21	4	1	0.001976	0.001976
24	21	4	2	0.061265	0.059289
24	21	4	3	0.436759	0.375494
24	21	4	4	1.000000	0.563241
24	21	5	2	0.004941	0.004941
24	21	5	3	0.098814	0.093874
24	21	5	4	0.521245	0.422431
24	21	5	5	1.000000	0.478755
24	21	6	3	0.009881	0.009881
24	21	6	4	0.143281	0.133399
24	21	6	5	0.596838	0.453557
24	21	6	6	1.000000	0.403162
24	21	7	4	0.017292	0.017292
24	21	7	5	0.193676	0.176383
24	21	7	6	0.664032	0.470356
24	21	7	7	1.000000	0.335968
24	21	8	5	0.027668	0.027668
24	21	8	6	0.249012	0.221344
24	21	8	7	0.723320	0.474308
24	21	8	8	1.000000	0.276680
24	21	9	6	0.041502	0.041502
24	21	9	7	0.308300	0.266798
24	21	9	8	0.775198	0.466897
24	21	9	9	1.000000	0.224802
24	21	10	7	0.059289	0.059289
24	21	10	8	0.370553	0.311265
24	21	10	9	0.820158	0.449605
24	21	10	10	1.000000	0.179842
24	21	11	8	0.081522	0.081522
24	21	11	9	0.434783	0.353261
24	21	11	10	0.858696	0.423913
24	21	11	11	1.000000	0.141304
24	21	12	9	0.108696	0.108696
24	21	12	10	0.500000	0.391304
24	21	12	11	0.891304	0.391304
24	21	12	12	1.000000	0.108696
24	21	13	10	0.141304	0.141304
24	21	13	11	0.565217	0.423913
24	21	13	12	0.918478	0.353261
24	21	13	13	1.000000	0.081522
24	21	14	11	0.179842	0.179842
24	21	14	12	0.629447	0.449605
24	21	14	13	0.940711	0.311265
24	21	14	14	1.000000	0.059289
24	21	15	12	0.224802	0.224802
24	21	15	13	0.691700	0.466897
24	21	15	14	0.958498	0.266798

N	n	k	x	P(x)	p(x)
24	21	15	15	1.000000	0.041502
24	21	16	13	0.276680	0.276680
24	21	16	14	0.750988	0.474308
24	21	16	15	0.972332	0.221344
24	21	16	16	1.000000	0.027668
24	21	17	14	0.335968	0.335968
24	21	17	15	0.806324	0.470356
24	21	17	16	0.982707	0.176383
24	21	17	17	1.000000	0.017292
24	21	18	15	0.403162	0.403162
24	21	18	16	0.856719	0.453557
24	21	18	17	0.990119	0.133399
24	21	18	18	1.000000	0.009881
24	21	19	16	0.478755	0.478755
24	21	19	17	0.901186	0.422431
24	21	19	18	0.995059	0.093874
24	21	19	19	1.000000	0.004941
24	21	20	17	0.563241	0.563241
24	21	20	18	0.938735	0.375494
24	21	20	19	0.998024	0.059289
24	21	20	20	1.000000	0.001976
24	21	21	18	0.657115	0.657115
24	21	21	19	0.968379	0.311265
24	21	21	20	0.999506	0.031126
24	21	21	21	1.000000	0.000494
24	22	1	0	0.083333	0.083333
24	22	1	1	1.000000	0.916667
24	22	2	0	0.003623	0.003623
24	22	2	1	0.163043	0.159420
24	22	2	2	1.000000	0.836957
24	22	3	1	0.010870	0.010870
24	22	3	2	0.239130	0.228261
24	22	3	3	1.000000	0.760870
24	22	4	2	0.021739	0.021739
24	22	4	3	0.311594	0.289855
24	22	4	4	1.000000	0.688406
24	22	5	3	0.036232	0.036232
24	22	5	4	0.380435	0.344203
24	22	5	5	1.000000	0.619565
24	22	6	4	0.054348	0.054348
24	22	6	5	0.445652	0.391304
24	22	6	6	1.000000	0.554348
24	22	7	5	0.076087	0.076087
24	22	7	6	0.507246	0.431159
24	22	7	7	1.000000	0.492754
24	22	8	6	0.101449	0.101449
24	22	8	7	0.565217	0.463768
24	22	8	8	1.000000	0.434783
24	22	9	7	0.130435	0.130435
24	22	9	8	0.619565	0.489130

N	n	k	x	P(x)	p(x)
24	22	9	9	1.000000	0.380435
24	22	10	8	0.163043	0.163043
24	22	10	9	0.670290	0.507246
24	22	10	10	1.000000	0.329710
24	22	11	9	0.199275	0.199275
24	22	11	10	0.717391	0.518116
24	22	11	11	1.000000	0.282609
24	22	12	10	0.239130	0.239130
24	22	12	11	0.760870	0.521739
24	22	12	12	1.000000	0.239130
24	22	13	11	0.282609	0.282609
24	22	13	12	0.800725	0.518116
24	22	13	13	1.000000	0.199275
24	22	14	12	0.329710	0.329710
24	22	14	13	0.836956	0.507246
24	22	14	14	1.000000	0.163043
24	22	15	13	0.380435	0.380435
24	22	15	14	0.869565	0.489130
24	22	15	15	1.000000	0.130435
24	22	16	14	0.434783	0.434783
24	22	16	15	0.898551	0.463768
24	22	16	16	1.000000	0.101449
24	22	17	15	0.492754	0.492754
24	22	17	16	0.923913	0.431159
24	22	17	17	1.000000	0.076087
24	22	18	16	0.554348	0.554348
24	22	18	17	0.945652	0.391304
24	22	18	18	1.000000	0.054348
24	22	19	17	0.619565	0.619565
24	22	19	18	0.963768	0.344203
24	22	19	19	1.000000	0.036232
24	22	20	18	0.688406	0.688406
24	22	20	19	0.978261	0.289855
24	22	20	20	1.000000	0.021739
24	22	21	19	0.760870	0.760870
24	22	21	20	0.989130	0.228261
24	22	21	21	1.000000	0.010870
24	22	22	20	0.836957	0.836957
24	22	22	21	0.996377	0.159420
24	22	22	22	1.000000	0.003623
24	23	1	0	0.041667	0.041667
24	23	1	1	1.000000	0.958333
24	23	2	1	0.083333	0.083333
24	23	2	2	1.000000	0.916667
24	23	3	2	0.125000	0.125000
24	23	3	3	1.000000	0.875000
24	23	4	3	0.166667	0.166667
24	23	4	4	1.000000	0.833333
24	23	5	4	0.208333	0.208333
24	23	5	5	1.000000	0.791667

N	n	k	x	P(x)	p(x)
24	23	6	5	0.250000	0.250000
24	23	6	6	1.000000	0.750000
24	23	7	6	0.291667	0.291667
24	23	7	7	1.000000	0.708333
24	23	8	7	0.333333	0.333333
24	23	8	8	1.000000	0.666667
24	23	9	8	0.375000	0.375000
24	23	9	9	1.000000	0.625000
24	23	10	9	0.416667	0.416667
24	23	10	10	1.000000	0.583333
24	23	11	10	0.458333	0.458333
24	23	11	11	1.000000	0.541667
24	23	12	11	0.500000	0.500000
24	23	12	12	1.000000	0.500000
24	23	13	12	0.541667	0.541667
24	23	13	13	1.000000	0.458333
24	23	14	13	0.583333	0.583333
24	23	14	14	1.000000	0.416667
24	23	15	14	0.625000	0.625000
24	23	15	15	1.000000	0.375000
24	23	16	15	0.666667	0.666667
24	23	16	16	1.000000	0.333333
24	23	17	16	0.708333	0.708333
24	23	17	17	1.000000	0.291667
24	23	18	17	0.750000	0.750000
24	23	18	18	1.000000	0.250000
24	23	19	18	0.791667	0.791667
24	23	19	19	1.000000	0.208333
24	23	20	19	0.833333	0.833333
24	23	20	20	1.000000	0.166667
24	23	21	20	0.875000	0.875000
24	23	21	21	1.000000	0.125000
24	23	22	21	0.916667	0.916667
24	23	22	22	1.000000	0.083333
24	23	23	22	0.958333	0.958333
24	23	23	23	1.000000	0.041667
25	1	1	0	0.960000	0.960000
25	1	1	1	1.000000	0.040000
25	2	1	0	0.920000	0.920000
25	2	1	1	1.000000	0.080000
25	2	2	0	0.843333	0.843333
25	2	2	1	0.996667	0.153333
25	2	2	2	1.000000	0.003333
25	3	1	0	0.880000	0.880000
25	3	1	1	1.000000	0.120000
25	3	2	0	0.770000	0.770000
25	3	2	1	0.990000	0.220000
25	3	2	2	1.000000	0.010000
25	3	3	0	0.669565	0.669565
25	3	3	1	0.970870	0.301304

Table for $N = 2$, $n = 1$, through $N = 100$, $n = 50$

N	n	k	x	P(x)	p(x)
25	3	3	2	0.999565	0.028696
25	3	3	3	1.000000	0.000435
25	4	1	0	0.840000	0.840000
25	4	1	1	1.000000	0.160000
25	4	2	0	0.700000	0.700000
25	4	2	1	0.980000	0.280000
25	4	2	2	1.000000	0.020000
25	4	3	0	0.578261	0.578261
25	4	3	1	0.943478	0.365217
25	4	3	2	0.998261	0.054783
25	4	3	3	1.000000	0.001739
25	4	4	0	0.473122	0.473122
25	4	4	1	0.893676	0.420553
25	4	4	2	0.993281	0.099605
25	4	4	3	0.999921	0.006640
25	4	4	4	1.000000	0.000079
25	5	1	0	0.800000	0.800000
25	5	1	1	1.000000	0.200000
25	5	2	0	0.633333	0.633333
25	5	2	1	0.966667	0.333333
25	5	2	2	1.000000	0.033333
25	5	3	0	0.495652	0.495652
25	5	3	1	0.908696	0.413043
25	5	3	2	0.995652	0.086957
25	5	3	3	1.000000	0.004348
25	5	4	0	0.383004	0.383004
25	5	4	1	0.833597	0.450593
25	5	4	2	0.983794	0.150198
25	5	4	3	0.999605	0.015810
25	5	4	4	1.000000	0.000395
25	5	5	0	0.291812	0.291812
25	5	5	1	0.747770	0.455957
25	5	5	2	0.962338	0.214568
25	5	5	3	0.998099	0.035761
25	5	5	4	0.999981	0.001882
25	5	5	5	1.000000	0.000019
25	6	1	0	0.760000	0.760000
25	6	1	1	1.000000	0.240000
25	6	2	0	0.570000	0.570000
25	6	2	1	0.950000	0.380000
25	6	2	2	1.000000	0.050000
25	6	3	0	0.421304	0.421304
25	6	3	1	0.867391	0.446087
25	6	3	2	0.991304	0.123913
25	6	3	3	1.000000	0.008696
25	6	4	0	0.306403	0.306403
25	6	4	1	0.766008	0.459605
25	6	4	2	0.968775	0.202767
25	6	4	3	0.998814	0.030040
25	6	4	4	1.000000	0.001186

N	n	k	x	P(x)	p(x)
25	6	5	0	0.218859	0.218859
25	6	5	1	0.656578	0.437719
25	6	5	2	0.930152	0.273574
25	6	5	3	0.994523	0.064370
25	6	5	4	0.999887	0.005364
25	6	5	5	1.000000	0.000113
25	6	6	0	0.153202	0.153202
25	6	6	1	0.547148	0.393947
25	6	6	2	0.875438	0.328289
25	6	6	3	0.984867	0.109430
25	6	6	4	0.999351	0.014483
25	6	6	5	0.999994	0.000644
25	6	6	6	1.000000	0.000006
25	7	1	0	0.720000	0.720000
25	7	1	1	1.000000	0.280000
25	7	2	0	0.510000	0.510000
25	7	2	1	0.930000	0.420000
25	7	2	2	1.000000	0.070000
25	7	3	0	0.354783	0.354783
25	7	3	1	0.820435	0.465652
25	7	3	2	0.984782	0.164348
25	7	3	3	1.000000	0.015217
25	7	4	0	0.241897	0.241897
25	7	4	1	0.693439	0.451541
25	7	4	2	0.947431	0.253992
25	7	4	3	0.997233	0.049802
25	7	4	4	1.000000	0.002767
25	7	5	0	0.161265	0.161265
25	7	5	1	0.564427	0.403162
25	7	5	2	0.886956	0.322530
25	7	5	3	0.987747	0.100790
25	7	5	4	0.999605	0.011858
25	7	5	5	1.000000	0.000395
25	7	6	0	0.104822	0.104822
25	7	6	1	0.443478	0.338656
25	7	6	2	0.806324	0.362846
25	7	6	3	0.967589	0.161265
25	7	6	4	0.997826	0.030237
25	7	6	5	0.999960	0.002134
25	7	6	6	1.000000	0.000040
25	7	7	0	0.066203	0.066203
25	7	7	1	0.336534	0.270331
25	7	7	2	0.710838	0.374304
25	7	7	3	0.933638	0.222800
25	7	7	4	0.993052	0.059413
25	7	7	5	0.999736	0.006684
25	7	7	6	0.999998	0.000262
25	7	7	7	1.000000	0.000002
25	8	1	0	0.680000	0.680000
25	8	1	1	1.000000	0.320000

N	n	k	x	P(x)	p(x)
25	8	2	0	0.453333	0.453333
25	8	2	1	0.906667	0.453333
25	8	2	2	1.000000	0.093333
25	8	3	0	0.295652	0.295652
25	8	3	1	0.768696	0.473043
25	8	3	2	0.975652	0.206957
25	8	3	3	1.000000	0.024348
25	8	4	0	0.188142	0.188142
25	8	4	1	0.618182	0.430040
25	8	4	2	0.919209	0.301028
25	8	4	3	0.994466	0.075257
25	8	4	4	1.000000	0.005534
25	8	5	0	0.116469	0.116469
25	8	5	1	0.474835	0.358366
25	8	5	2	0.833201	0.358366
25	8	5	3	0.976548	0.143346
25	8	5	4	0.998946	0.022398
25	8	5	5	1.000000	0.001054
25	8	6	0	0.069881	0.069881
25	8	6	1	0.349407	0.279526
25	8	6	2	0.725692	0.376285
25	8	6	3	0.940711	0.215020
25	8	6	4	0.994466	0.053755
25	8	6	5	0.999842	0.005375
25	8	6	6	1.000000	0.000158
25	8	7	0	0.040458	0.040458
25	8	7	1	0.246424	0.205966
25	8	7	2	0.606865	0.360441
25	8	7	3	0.884127	0.277262
25	8	7	4	0.983149	0.099022
25	8	7	5	0.998993	0.015844
25	8	7	6	0.999983	0.000990
25	8	7	7	1.000000	0.000017
25	8	8	0	0.022476	0.022476
25	8	8	1	0.166326	0.143849
25	8	8	2	0.486718	0.320392
25	8	8	3	0.807110	0.320392
25	8	8	4	0.961145	0.154035
25	8	8	5	0.996352	0.035208
25	8	8	6	0.999873	0.003521
25	8	8	7	0.999999	0.000126
25	8	8	8	1.000000	0.000001
25	9	1	0	0.640000	0.640000
25	9	1	1	1.000000	0.360000
25	9	2	0	0.400000	0.400000
25	9	2	1	0.880000	0.480000
25	9	2	2	1.000000	0.120000
25	9	3	0	0.243478	0.243478
25	9	3	1	0.713043	0.469565
25	9	3	2	0.963478	0.250435

N	n	k	x	P(x)	p(x)
25	9	3	3	1.000000	0.036522
25	9	4	0	0.143874	0.143874
25	9	4	1	0.542292	0.398419
25	9	4	2	0.883794	0.341502
25	9	4	3	0.990039	0.106245
25	9	4	4	1.000000	0.009960
25	9	5	0	0.082213	0.082213
25	9	5	1	0.390514	0.308300
25	9	5	2	0.769960	0.379447
25	9	5	3	0.959684	0.189723
25	9	5	4	0.997628	0.037945
25	9	5	5	1.000000	0.002372
25	9	6	0	0.045217	0.045217
25	9	6	1	0.267194	0.221976
25	9	6	2	0.637154	0.369960
25	9	6	3	0.902767	0.265613
25	9	6	4	0.988142	0.085375
25	9	6	5	0.999526	0.011383
25	9	6	6	1.000000	0.000474
25	9	7	0	0.023799	0.023799
25	9	7	1	0.173730	0.149931
25	9	7	2	0.500853	0.327123
25	9	7	3	0.818889	0.318036
25	9	7	4	0.965675	0.146786
25	9	7	5	0.997129	0.031454
25	9	7	6	0.999925	0.002796
25	9	7	7	1.000000	0.000075
25	9	8	0	0.011899	0.011899
25	9	8	1	0.107094	0.095194
25	9	8	2	0.373638	0.266545
25	9	8	3	0.712877	0.339239
25	9	8	4	0.924901	0.212024
25	9	8	5	0.990139	0.065238
25	9	8	6	0.999459	0.009320
25	9	8	7	0.999992	0.000533
25	9	8	8	1.000000	0.000008
25	9	9	0	0.005600	0.005600
25	9	9	1	0.062296	0.056697
25	9	9	2	0.263885	0.201588
25	9	9	3	0.593146	0.329261
25	9	9	4	0.862541	0.269395
25	9	9	5	0.974789	0.112248
25	9	9	6	0.997814	0.023025
25	9	9	7	0.999929	0.002115
25	9	9	8	0.999999	0.000070
25	9	9	9	1.000000	0.000000
25	10	1	0	0.600000	0.600000
25	10	1	1	1.000000	0.400000
25	10	2	0	0.350000	0.350000
25	10	2	1	0.850000	0.500000

Table for $N = 2$, $n = 1$, through $N = 100$, $n = 50$

N	n	k	x	P(x)	p(x)
25	10	2	2	1.000000	0.150000
25	10	3	0	0.197826	0.197826
25	10	3	1	0.654348	0.456522
25	10	3	2	0.947826	0.293478
25	10	3	3	1.000000	0.052174
25	10	4	0	0.107905	0.107905
25	10	4	1	0.467589	0.359684
25	10	4	2	0.841107	0.373518
25	10	4	3	0.983399	0.142292
25	10	4	4	1.000000	0.016601
25	10	5	0	0.056522	0.056522
25	10	5	1	0.313439	0.256917
25	10	5	2	0.698814	0.385375
25	10	5	3	0.935968	0.237154
25	10	5	4	0.995257	0.059289
25	10	5	5	1.000000	0.004743
25	10	6	0	0.028261	0.028261
25	10	6	1	0.197826	0.169565
25	10	6	2	0.544664	0.346838
25	10	6	3	0.852964	0.308300
25	10	6	4	0.977470	0.124506
25	10	6	5	0.998814	0.021344
25	10	6	6	1.000000	0.001186
25	10	7	0	0.013387	0.013387
25	10	7	1	0.117506	0.104119
25	10	7	2	0.398627	0.281121
25	10	7	3	0.739380	0.340753
25	10	7	4	0.938153	0.198773
25	10	7	5	0.993197	0.055045
25	10	7	6	0.999750	0.006553
25	10	7	7	1.000000	0.000250
25	10	8	0	0.005950	0.005950
25	10	8	1	0.065446	0.059497
25	10	8	2	0.273684	0.208238
25	10	8	3	0.606865	0.333181
25	10	8	4	0.871895	0.265030
25	10	8	5	0.977907	0.106012
25	10	8	6	0.998294	0.020387
25	10	8	7	0.999958	0.001664
25	10	8	8	1.000000	0.000042
25	10	9	0	0.002450	0.002450
25	10	9	1	0.033948	0.031498
25	10	9	2	0.175690	0.141742
25	10	9	3	0.469673	0.293983
25	10	9	4	0.778355	0.308682
25	10	9	5	0.946727	0.168372
25	10	9	6	0.993497	0.046770
25	10	9	7	0.999665	0.006167
25	10	9	8	0.999995	0.000330
25	10	9	9	1.000000	0.000005

N	n	k	x	P(x)	p(x)
25	10	10	0	0.000919	0.000919
25	10	10	1	0.016230	0.015312
25	10	10	2	0.104819	0.088589
25	10	10	3	0.341055	0.236236
25	10	10	4	0.662599	0.321544
25	10	10	5	0.894111	0.231512
25	10	10	6	0.981805	0.087694
25	10	10	7	0.998508	0.016704
25	10	10	8	0.999954	0.001445
25	10	10	9	1.000000	0.000046
25	10	10	10	1.000000	0.000000
25	11	1	0	0.560000	0.560000
25	11	1	1	1.000000	0.440000
25	11	2	0	0.303333	0.303333
25	11	2	1	0.816667	0.513333
25	11	2	2	1.000000	0.183333
25	11	3	0	0.158261	0.158261
25	11	3	1	0.593478	0.435217
25	11	3	2	0.928261	0.334783
25	11	3	3	1.000000	0.071739
25	11	4	0	0.079130	0.079130
25	11	4	1	0.395652	0.316522
25	11	4	2	0.791304	0.395652
25	11	4	3	0.973913	0.182609
25	11	4	4	1.000000	0.026087
25	11	5	0	0.037681	0.037681
25	11	5	1	0.244928	0.207246
25	11	5	2	0.621739	0.376812
25	11	5	3	0.904348	0.282609
25	11	5	4	0.991304	0.086957
25	11	5	5	1.000000	0.008696
25	11	6	0	0.016957	0.016957
25	11	6	1	0.141304	0.124348
25	11	6	2	0.452174	0.310870
25	11	6	3	0.791304	0.339130
25	11	6	4	0.960870	0.169565
25	11	6	5	0.997391	0.036522
25	11	6	6	1.000000	0.002609
25	11	7	0	0.007140	0.007140
25	11	7	1	0.075858	0.068719
25	11	7	2	0.304920	0.229062
25	11	7	3	0.648513	0.343593
25	11	7	4	0.898398	0.249886
25	11	7	5	0.985858	0.087460
25	11	7	6	0.999313	0.013455
25	11	7	7	1.000000	0.000686
25	11	8	0	0.002777	0.002777
25	11	8	1	0.037681	0.034905
25	11	8	2	0.190389	0.152708
25	11	8	3	0.495805	0.305416

N	n	k	x	P(x)	p(x)
25	11	8	4	0.801220	0.305416
25	11	8	5	0.956705	0.155484
25	11	8	6	0.995576	0.038871
25	11	8	7	0.999847	0.004272
25	11	8	8	1.000000	0.000153
25	11	9	0	0.000980	0.000980
25	11	9	1	0.017149	0.016169
25	11	9	2	0.109544	0.092395
25	11	9	3	0.352080	0.242536
25	11	9	4	0.675461	0.323381
25	11	9	5	0.901828	0.226367
25	11	9	6	0.984143	0.082315
25	11	9	7	0.998842	0.014699
25	11	9	8	0.999973	0.001131
25	11	9	9	1.000000	0.000027
25	11	10	0	0.000306	0.000306
25	11	10	1	0.007043	0.006737
25	11	10	2	0.057572	0.050528
25	11	10	3	0.230812	0.173240
25	11	10	4	0.533982	0.303170
25	11	10	5	0.816940	0.282959
25	11	10	6	0.958420	0.141479
25	11	10	7	0.995167	0.036748
25	11	10	8	0.999761	0.004593
25	11	10	9	0.999997	0.000236
25	11	10	10	1.000000	0.000003
25	11	11	0	0.000082	0.000082
25	11	11	1	0.002552	0.002470
25	11	11	2	0.027255	0.024703
25	11	11	3	0.138417	0.111162
25	11	11	4	0.392502	0.254085
25	11	11	5	0.703757	0.311255
25	11	11	6	0.911260	0.207503
25	11	11	7	0.985368	0.074108
25	11	11	8	0.998842	0.013474
25	11	11	9	0.999965	0.001123
25	11	11	10	1.000000	0.000035
25	11	11	11	1.000000	0.000000
25	12	1	0	0.520000	0.520000
25	12	1	1	1.000000	0.480000
25	12	2	0	0.260000	0.260000
25	12	2	1	0.780000	0.520000
25	12	2	2	1.000000	0.220000
25	12	3	0	0.124348	0.124348
25	12	3	1	0.531304	0.406957
25	12	3	2	0.904348	0.373043
25	12	3	3	1.000000	0.095652
25	12	4	0	0.056522	0.056522
25	12	4	1	0.327826	0.271304
25	12	4	2	0.734783	0.406957

N	n	k	x	P(x)	p(x)
25	12	4	3	0.960870	0.226087
25	12	4	4	1.000000	0.039130
25	12	5	0	0.024224	0.024224
25	12	5	1	0.185714	0.161491
25	12	5	2	0.540994	0.355280
25	12	5	3	0.863975	0.322981
25	12	5	4	0.985093	0.121118
25	12	5	5	1.000000	0.014907
25	12	6	0	0.009689	0.009689
25	12	6	1	0.096894	0.087205
25	12	6	2	0.363354	0.266460
25	12	6	3	0.718634	0.355280
25	12	6	4	0.936646	0.218012
25	12	6	5	0.994783	0.058137
25	12	6	6	1.000000	0.005217
25	12	7	0	0.003570	0.003570
25	12	7	1	0.046407	0.042838
25	12	7	2	0.223112	0.176705
25	12	7	3	0.550343	0.327231
25	12	7	4	0.844851	0.294508
25	12	7	5	0.973364	0.128513
25	12	7	6	0.998352	0.024989
25	12	7	7	1.000000	0.001648
25	12	8	0	0.001190	0.001190
25	12	8	1	0.020229	0.019039
25	12	8	2	0.124943	0.104714
25	12	8	3	0.386728	0.261785
25	12	8	4	0.713959	0.327231
25	12	8	5	0.923387	0.209428
25	12	8	6	0.990023	0.066636
25	12	8	7	0.999542	0.009519
25	12	8	8	1.000000	0.000458
25	12	9	0	0.000350	0.000350
25	12	9	1	0.007910	0.007560
25	12	9	2	0.063346	0.055437
25	12	9	3	0.248136	0.184789
25	12	9	4	0.559968	0.311832
25	12	9	5	0.837152	0.277184
25	12	9	6	0.966504	0.129353
25	12	9	7	0.996742	0.030238
25	12	9	8	0.999892	0.003150
25	12	9	9	1.000000	0.000108
25	12	10	0	0.000087	0.000087
25	12	10	1	0.002712	0.002625
25	12	10	2	0.028698	0.025986
25	12	10	3	0.144192	0.115493
25	12	10	4	0.404052	0.259860
25	12	10	5	0.715884	0.311832
25	12	10	6	0.917997	0.202113
25	12	10	7	0.987293	0.069296

N	n	k	x	P(x)	p(x)
25	12	10	8	0.999105	0.011812
25	12	10	9	0.999980	0.000875
25	12	10	10	1.000000	0.000020
25	12	11	0	0.000017	0.000017
25	12	11	1	0.000787	0.000770
25	12	11	2	0.011374	0.010587
25	12	11	3	0.074896	0.063521
25	12	11	4	0.265460	0.190564
25	12	11	5	0.570362	0.304902
25	12	11	6	0.837152	0.266790
25	12	11	7	0.964194	0.127043
25	12	11	8	0.995955	0.031761
25	12	11	9	0.999805	0.003850
25	12	11	10	0.999997	0.000192
25	12	11	11	1.000000	0.000003
25	12	12	0	0.000002	0.000002
25	12	12	1	0.000182	0.000180
25	12	12	2	0.003812	0.003630
25	12	12	3	0.034061	0.030248
25	12	12	4	0.156566	0.122505
25	12	12	5	0.417911	0.261345
25	12	12	6	0.722813	0.304902
25	12	12	7	0.918822	0.196009
25	12	12	8	0.986880	0.068059
25	12	12	9	0.998980	0.012099
25	12	12	10	0.999970	0.000990
25	12	12	11	1.000000	0.000030
25	12	12	12	1.000000	0.000000
25	13	1	0	0.480000	0.480000
25	13	1	1	1.000000	0.520000
25	13	2	0	0.220000	0.220000
25	13	2	1	0.740000	0.520000
25	13	2	2	1.000000	0.260000
25	13	3	0	0.095652	0.095652
25	13	3	1	0.468696	0.373043
25	13	3	2	0.875652	0.406957
25	13	3	3	1.000000	0.124348
25	13	4	0	0.039130	0.039130
25	13	4	1	0.265217	0.226087
25	13	4	2	0.672174	0.406957
25	13	4	3	0.943478	0.271304
25	13	4	4	1.000000	0.056522
25	13	5	0	0.014907	0.014907
25	13	5	1	0.136025	0.121118
25	13	5	2	0.459006	0.322981
25	13	5	3	0.814286	0.355280
25	13	5	4	0.975776	0.161491
25	13	5	5	1.000000	0.024224
25	13	6	0	0.005217	0.005217
25	13	6	1	0.063354	0.058137

N	n	k	x	P(x)	p(x)
25	13	6	2	0.281366	0.218012
25	13	6	3	0.636646	0.355280
25	13	6	4	0.903106	0.266460
25	13	6	5	0.990311	0.087205
25	13	6	6	1.000000	0.009689
25	13	7	0	0.001648	0.001648
25	13	7	1	0.026636	0.024989
25	13	7	2	0.155149	0.128513
25	13	7	3	0.449657	0.294508
25	13	7	4	0.776888	0.327231
25	13	7	5	0.953593	0.176705
25	13	7	6	0.996430	0.042838
25	13	7	7	1.000000	0.003570
25	13	8	0	0.000458	0.000458
25	13	8	1	0.009977	0.009519
25	13	8	2	0.076613	0.066636
25	13	8	3	0.286041	0.209428
25	13	8	4	0.613272	0.327231
25	13	8	5	0.875057	0.261785
25	13	8	6	0.979771	0.104714
25	13	8	7	0.998810	0.019039
25	13	8	8	1.000000	0.001190
25	13	9	0	0.000108	0.000108
25	13	9	1	0.003257	0.003150
25	13	9	2	0.033496	0.030238
25	13	9	3	0.162848	0.129353
25	13	9	4	0.440032	0.277184
25	13	9	5	0.751864	0.311832
25	13	9	6	0.936654	0.184789
25	13	9	7	0.992090	0.055437
25	13	9	8	0.999650	0.007560
25	13	9	9	1.000000	0.000350
25	13	10	0	0.000020	0.000020
25	13	10	1	0.000895	0.000875
25	13	10	2	0.012707	0.011812
25	13	10	3	0.082003	0.069296
25	13	10	4	0.284116	0.202113
25	13	10	5	0.595948	0.311832
25	13	10	6	0.855808	0.259860
25	13	10	7	0.971302	0.115493
25	13	10	8	0.997288	0.025986
25	13	10	9	0.999912	0.002625
25	13	10	10	1.000000	0.000087
25	13	11	0	0.000003	0.000003
25	13	11	1	0.000195	0.000192
25	13	11	2	0.004045	0.003850
25	13	11	3	0.035806	0.031761
25	13	11	4	0.162848	0.127043
25	13	11	5	0.429638	0.266790
25	13	11	6	0.734540	0.304902

N	n	k	x	P(x)	p(x)
25	13	11	7	0.925104	0.190564
25	13	11	8	0.988626	0.063521
25	13	11	9	0.999212	0.010587
25	13	11	10	0.999982	0.000770
25	13	11	11	1.000000	0.000017
25	13	12	0	0.000000	0.000000
25	13	12	1	0.000030	0.000030
25	13	12	2	0.001020	0.000990
25	13	12	3	0.013119	0.012099
25	13	12	4	0.081178	0.068059
25	13	12	5	0.277187	0.196009
25	13	12	6	0.582089	0.304902
25	13	12	7	0.843434	0.261345
25	13	12	8	0.965939	0.122505
25	13	12	9	0.996188	0.030248
25	13	12	10	0.999817	0.003630
25	13	12	11	0.999997	0.000180
25	13	12	12	1.000000	0.000002
25	13	13	1	0.000002	0.000002
25	13	13	2	0.000182	0.000180
25	13	13	3	0.003812	0.003630
25	13	13	4	0.034061	0.030248
25	13	13	5	0.156566	0.122505
25	13	13	6	0.417911	0.261345
25	13	13	7	0.722813	0.304902
25	13	13	8	0.918822	0.196009
25	13	13	9	0.986880	0.068059
25	13	13	10	0.998980	0.012099
25	13	13	11	0.999970	0.000990
25	13	13	12	1.000000	0.000030
25	13	13	13	1.000000	0.000000
25	14	1	0	0.440000	0.440000
25	14	1	1	1.000000	0.560000
25	14	2	0	0.183333	0.183333
25	14	2	1	0.696667	0.513333
25	14	2	2	1.000000	0.303333
25	14	3	0	0.071739	0.071739
25	14	3	1	0.406522	0.334783
25	14	3	2	0.841739	0.435217
25	14	3	3	1.000000	0.158261
25	14	4	0	0.026087	0.026087
25	14	4	1	0.208696	0.182609
25	14	4	2	0.604348	0.395652
25	14	4	3	0.920869	0.316522
25	14	4	4	1.000000	0.079130
25	14	5	0	0.008696	0.008696
25	14	5	1	0.095652	0.086957
25	14	5	2	0.378261	0.282609
25	14	5	3	0.755072	0.376812
25	14	5	4	0.962319	0.207246

N	n	k	x	P(x)	p(x)
25	14	5	5	1.000000	0.037681
25	14	6	0	0.002609	0.002609
25	14	6	1	0.039130	0.036522
25	14	6	2	0.208696	0.169565
25	14	6	3	0.547826	0.339130
25	14	6	4	0.858696	0.310870
25	14	6	5	0.983043	0.124348
25	14	6	6	1.000000	0.016957
25	14	7	0	0.000686	0.000686
25	14	7	1	0.014142	0.013455
25	14	7	2	0.101602	0.087460
25	14	7	3	0.351487	0.249886
25	14	7	4	0.695080	0.343593
25	14	7	5	0.924142	0.229062
25	14	7	6	0.992860	0.068719
25	14	7	7	1.000000	0.007140
25	14	8	0	0.000153	0.000153
25	14	8	1	0.004424	0.004272
25	14	8	2	0.043295	0.038871
25	14	8	3	0.198780	0.155484
25	14	8	4	0.504195	0.305416
25	14	8	5	0.809611	0.305416
25	14	8	6	0.962319	0.152708
25	14	8	7	0.997223	0.034905
25	14	8	8	1.000000	0.002777
25	14	9	0	0.000027	0.000027
25	14	9	1	0.001158	0.001131
25	14	9	2	0.015857	0.014699
25	14	9	3	0.098172	0.082315
25	14	9	4	0.324539	0.226367
25	14	9	5	0.647920	0.323381
25	14	9	6	0.890456	0.242536
25	14	9	7	0.982851	0.092395
25	14	9	8	0.999020	0.016169
25	14	9	9	1.000000	0.000980
25	14	10	0	0.000003	0.000003
25	14	10	1	0.000239	0.000236
25	14	10	2	0.004832	0.004593
25	14	10	3	0.041580	0.036748
25	14	10	4	0.183060	0.141479
25	14	10	5	0.466018	0.282959
25	14	10	6	0.769188	0.303170
25	14	10	7	0.942428	0.173240
25	14	10	8	0.992957	0.050528
25	14	10	9	0.999694	0.006737
25	14	10	10	1.000000	0.000306
25	14	11	0	0.000000	0.000000
25	14	11	1	0.000035	0.000035
25	14	11	2	0.001158	0.001123
25	14	11	3	0.014632	0.013474

N	n	k	x	P(x)	p(x)
25	14	11	4	0.088740	0.074108
25	14	11	5	0.296243	0.207503
25	14	11	6	0.607498	0.311255
25	14	11	7	0.861583	0.254085
25	14	11	8	0.972745	0.111162
25	14	11	9	0.997448	0.024703
25	14	11	10	0.999918	0.002470
25	14	11	11	1.000000	0.000082
25	14	12	1	0.000003	0.000003
25	14	12	2	0.000195	0.000192
25	14	12	3	0.004045	0.003850
25	14	12	4	0.035806	0.031761
25	14	12	5	0.162848	0.127043
25	14	12	6	0.429638	0.266790
25	14	12	7	0.734540	0.304902
25	14	12	8	0.925104	0.190564
25	14	12	9	0.988626	0.063521
25	14	12	10	0.999212	0.010587
25	14	12	11	0.999982	0.000770
25	14	12	12	1.000000	0.000017
25	14	13	2	0.000017	0.000017
25	14	13	3	0.000787	0.000770
25	14	13	4	0.011374	0.010587
25	14	13	5	0.074896	0.063521
25	14	13	6	0.265460	0.190564
25	14	13	7	0.570362	0.304902
25	14	13	8	0.837152	0.266790
25	14	13	9	0.964194	0.127043
25	14	13	10	0.995955	0.031761
25	14	13	11	0.999805	0.003850
25	14	13	12	0.999997	0.000192
25	14	13	13	1.000000	0.000003
25	14	14	3	0.000082	0.000082
25	14	14	4	0.002552	0.002470
25	14	14	5	0.027255	0.024703
25	14	14	6	0.138417	0.111162
25	14	14	7	0.392502	0.254085
25	14	14	8	0.703757	0.311255
25	14	14	9	0.911260	0.207503
25	14	14	10	0.985368	0.074108
25	14	14	11	0.998842	0.013474
25	14	14	12	0.999965	0.001123
25	14	14	13	1.000000	0.000035
25	14	14	14	1.000000	0.000000
25	15	1	0	0.400000	0.400000
25	15	1	1	1.000000	0.600000
25	15	2	0	0.150000	0.150000
25	15	2	1	0.650000	0.500000
25	15	2	2	1.000000	0.350000
25	15	3	0	0.052174	0.052174

N	n	k	x	P(x)	p(x)
25	15	3	1	0.345652	0.293478
25	15	3	2	0.802174	0.456522
25	15	3	3	1.000000	0.197826
25	15	4	0	0.016601	0.016601
25	15	4	1	0.158893	0.142292
25	15	4	2	0.532411	0.373518
25	15	4	3	0.892095	0.359684
25	15	4	4	1.000000	0.107905
25	15	5	0	0.004743	0.004743
25	15	5	1	0.064032	0.059289
25	15	5	2	0.301186	0.237154
25	15	5	3	0.686561	0.385375
25	15	5	4	0.943478	0.256917
25	15	5	5	1.000000	0.056522
25	15	6	0	0.001186	0.001186
25	15	6	1	0.022530	0.021344
25	15	6	2	0.147036	0.124506
25	15	6	3	0.455336	0.308300
25	15	6	4	0.802174	0.346838
25	15	6	5	0.971739	0.169565
25	15	6	6	1.000000	0.028261
25	15	7	0	0.000250	0.000250
25	15	7	1	0.006803	0.006553
25	15	7	2	0.061847	0.055045
25	15	7	3	0.260620	0.198773
25	15	7	4	0.601373	0.340753
25	15	7	5	0.882494	0.281121
25	15	7	6	0.986613	0.104119
25	15	7	7	1.000000	0.013387
25	15	8	0	0.000042	0.000042
25	15	8	1	0.001706	0.001664
25	15	8	2	0.022093	0.020387
25	15	8	3	0.128105	0.106012
25	15	8	4	0.393135	0.265030
25	15	8	5	0.726316	0.333181
25	15	8	6	0.934554	0.208238
25	15	8	7	0.994050	0.059497
25	15	8	8	1.000000	0.005950
25	15	9	0	0.000005	0.000005
25	15	9	1	0.000335	0.000330
25	15	9	2	0.006503	0.006167
25	15	9	3	0.053273	0.046770
25	15	9	4	0.221645	0.168372
25	15	9	5	0.530327	0.308682
25	15	9	6	0.824310	0.293983
25	15	9	7	0.966052	0.141742
25	15	9	8	0.997550	0.031498
25	15	9	9	1.000000	0.002450
25	15	10	0	0.000000	0.000000
25	15	10	1	0.000046	0.000046

N	n	k	x	P(x)	p(x)
25	15	10	2	0.001492	0.001445
25	15	10	3	0.018195	0.016704
25	15	10	4	0.105889	0.087694
25	15	10	5	0.337401	0.231512
25	15	10	6	0.658945	0.321544
25	15	10	7	0.895181	0.236236
25	15	10	8	0.983770	0.088589
25	15	10	9	0.999081	0.015312
25	15	10	10	1.000000	0.000919
25	15	11	1	0.000003	0.000003
25	15	11	2	0.000239	0.000236
25	15	11	3	0.004832	0.004593
25	15	11	4	0.041580	0.036748
25	15	11	5	0.183060	0.141479
25	15	11	6	0.466018	0.282959
25	15	11	7	0.769188	0.303170
25	15	11	8	0.942428	0.173240
25	15	11	9	0.992957	0.050528
25	15	11	10	0.999694	0.006737
25	15	11	11	1.000000	0.000306
25	15	12	2	0.000020	0.000020
25	15	12	3	0.000895	0.000875
25	15	12	4	0.012707	0.011812
25	15	12	5	0.082003	0.069296
25	15	12	6	0.284116	0.202113
25	15	12	7	0.595948	0.311832
25	15	12	8	0.855808	0.259860
25	15	12	9	0.971302	0.115493
25	15	12	10	0.997288	0.025986
25	15	12	11	0.999912	0.002625
25	15	12	12	1.000000	0.000087
25	15	13	3	0.000087	0.000087
25	15	13	4	0.002712	0.002625
25	15	13	5	0.028698	0.025986
25	15	13	6	0.144192	0.115493
25	15	13	7	0.404052	0.259860
25	15	13	8	0.715884	0.311832
25	15	13	9	0.917997	0.202113
25	15	13	10	0.987293	0.069296
25	15	13	11	0.999105	0.011812
25	15	13	12	0.999980	0.000875
25	15	13	13	1.000000	0.000020
25	15	14	4	0.000306	0.000306
25	15	14	5	0.007043	0.006737
25	15	14	6	0.057572	0.050528
25	15	14	7	0.230812	0.173240
25	15	14	8	0.533982	0.303170
25	15	14	9	0.816940	0.282959
25	15	14	10	0.958420	0.141479
25	15	14	11	0.995167	0.036748

N	n	k	x	P(x)	p(x)
25	15	14	12	0.999761	0.004593
25	15	14	13	0.999997	0.000236
25	15	14	14	1.000000	0.000003
25	15	15	5	0.000919	0.000919
25	15	15	6	0.016230	0.015312
25	15	15	7	0.104819	0.088589
25	15	15	8	0.341055	0.236236
25	15	15	9	0.662599	0.321544
25	15	15	10	0.894111	0.231512
25	15	15	11	0.981805	0.087694
25	15	15	12	0.998508	0.016704
25	15	15	13	0.999954	0.001445
25	15	15	14	1.000000	0.000046
25	15	15	15	1.000000	0.000000
25	16	1	0	0.360000	0.360000
25	16	1	1	1.000000	0.640000
25	16	2	0	0.120000	0.120000
25	16	2	1	0.600000	0.480000
25	16	2	2	1.000000	0.400000
25	16	3	0	0.036522	0.036522
25	16	3	1	0.286957	0.250435
25	16	3	2	0.756522	0.469565
25	16	3	3	1.000000	0.243478
25	16	4	0	0.009960	0.009960
25	16	4	1	0.116206	0.106245
25	16	4	2	0.457707	0.341502
25	16	4	3	0.856126	0.398419
25	16	4	4	1.000000	0.143874
25	16	5	0	0.002372	0.002372
25	16	5	1	0.040316	0.037945
25	16	5	2	0.230040	0.189723
25	16	5	3	0.609486	0.379447
25	16	5	4	0.917787	0.308300
25	16	5	5	1.000000	0.082213
25	16	6	0	0.000474	0.000474
25	16	6	1	0.011858	0.011383
25	16	6	2	0.097233	0.085375
25	16	6	3	0.362846	0.265613
25	16	6	4	0.732806	0.369960
25	16	6	5	0.954783	0.221976
25	16	6	6	1.000000	0.045217
25	16	7	0	0.000075	0.000075
25	16	7	1	0.002871	0.002796
25	16	7	2	0.034325	0.031454
25	16	7	3	0.181111	0.146786
25	16	7	4	0.499147	0.318036
25	16	7	5	0.826270	0.327123
25	16	7	6	0.976201	0.149931
25	16	7	7	1.000000	0.023799
25	16	8	0	0.000008	0.000008

Table for $N = 2$, $n = 1$, through $N = 100$, $n = 50$

N	n	k	x	P(x)	p(x)
25	16	8	1	0.000541	0.000533
25	16	8	2	0.009861	0.009320
25	16	8	3	0.075099	0.065238
25	16	8	4	0.287123	0.212024
25	16	8	5	0.626362	0.339239
25	16	8	6	0.892906	0.266545
25	16	8	7	0.988101	0.095194
25	16	8	8	1.000000	0.011899
25	16	9	0	0.000000	0.000000
25	16	9	1	0.000071	0.000070
25	16	9	2	0.002186	0.002115
25	16	9	3	0.025211	0.023025
25	16	9	4	0.137459	0.112248
25	16	9	5	0.406854	0.269395
25	16	9	6	0.736115	0.329261
25	16	9	7	0.937703	0.201588
25	16	9	8	0.994400	0.056697
25	16	9	9	1.000000	0.005600
25	16	10	1	0.000005	0.000005
25	16	10	2	0.000335	0.000330
25	16	10	3	0.006503	0.006167
25	16	10	4	0.053273	0.046770
25	16	10	5	0.221645	0.168372
25	16	10	6	0.530327	0.308682
25	16	10	7	0.824310	0.293983
25	16	10	8	0.966052	0.141742
25	16	10	9	0.997550	0.031498
25	16	10	10	1.000000	0.002450
25	16	11	2	0.000027	0.000027
25	16	11	3	0.001158	0.001131
25	16	11	4	0.015857	0.014699
25	16	11	5	0.098172	0.082315
25	16	11	6	0.324539	0.226367
25	16	11	7	0.647920	0.323381
25	16	11	8	0.890456	0.242536
25	16	11	9	0.982851	0.092395
25	16	11	10	0.999020	0.016169
25	16	11	11	1.000000	0.000980
25	16	12	3	0.000108	0.000108
25	16	12	4	0.003257	0.003150
25	16	12	5	0.033496	0.030238
25	16	12	6	0.162848	0.129353
25	16	12	7	0.440032	0.277184
25	16	12	8	0.751864	0.311832
25	16	12	9	0.936654	0.184789
25	16	12	10	0.992090	0.055437
25	16	12	11	0.999650	0.007560
25	16	12	12	1.000000	0.000350
25	16	13	4	0.000350	0.000350
25	16	13	5	0.007910	0.007560

N	n	k	x	P(x)	p(x)
25	16	13	6	0.063346	0.055437
25	16	13	7	0.248136	0.184789
25	16	13	8	0.559968	0.311832
25	16	13	9	0.837152	0.277184
25	16	13	10	0.966504	0.129353
25	16	13	11	0.996742	0.030238
25	16	13	12	0.999892	0.003150
25	16	13	13	1.000000	0.000108
25	16	14	5	0.000980	0.000980
25	16	14	6	0.017149	0.016169
25	16	14	7	0.109544	0.092395
25	16	14	8	0.352080	0.242536
25	16	14	9	0.675461	0.323381
25	16	14	10	0.901828	0.226367
25	16	14	11	0.984143	0.082315
25	16	14	12	0.998842	0.014699
25	16	14	13	0.999973	0.001131
25	16	14	14	1.000000	0.000027
25	16	15	6	0.002450	0.002450
25	16	15	7	0.033948	0.031498
25	16	15	8	0.175690	0.141742
25	16	15	9	0.469673	0.293983
25	16	15	10	0.778355	0.308682
25	16	15	11	0.946727	0.168372
25	16	15	12	0.993497	0.046770
25	16	15	13	0.999665	0.006167
25	16	15	14	0.999995	0.000330
25	16	15	15	1.000000	0.000005
25	16	16	7	0.005600	0.005600
25	16	16	8	0.062296	0.056697
25	16	16	9	0.263885	0.201588
25	16	16	10	0.593146	0.329261
25	16	16	11	0.862541	0.269395
25	16	16	12	0.974789	0.112248
25	16	16	13	0.997814	0.023025
25	16	16	14	0.999929	0.002115
25	16	16	15	0.999999	0.000070
25	16	16	16	1.000000	0.000000
25	17	1	0	0.320000	0.320000
25	17	1	1	1.000000	0.680000
25	17	2	0	0.093333	0.093333
25	17	2	1	0.546667	0.453333
25	17	2	2	1.000000	0.453333
25	17	3	0	0.024348	0.024348
25	17	3	1	0.231304	0.206957
25	17	3	2	0.704348	0.473043
25	17	3	3	1.000000	0.295652
25	17	4	0	0.005534	0.005534
25	17	4	1	0.080791	0.075257
25	17	4	2	0.381818	0.301028

N	n	k	x	P(x)	p(x)
25	17	4	3	0.811858	0.430040
25	17	4	4	1.000000	0.188142
25	17	5	0	0.001054	0.001054
25	17	5	1	0.023452	0.022398
25	17	5	2	0.166798	0.143346
25	17	5	3	0.525165	0.358366
25	17	5	4	0.883531	0.358366
25	17	5	5	1.000000	0.116469
25	17	6	0	0.000158	0.000158
25	17	6	1	0.005534	0.005375
25	17	6	2	0.059289	0.053755
25	17	6	3	0.274308	0.215020
25	17	6	4	0.650593	0.376285
25	17	6	5	0.930119	0.279526
25	17	6	6	1.000000	0.069881
25	17	7	0	0.000017	0.000017
25	17	7	1	0.001007	0.000990
25	17	7	2	0.016850	0.015844
25	17	7	3	0.115873	0.099022
25	17	7	4	0.393135	0.277262
25	17	7	5	0.753576	0.360441
25	17	7	6	0.959542	0.205966
25	17	7	7	1.000000	0.040458
25	17	8	0	0.000001	0.000001
25	17	8	1	0.000127	0.000126
25	17	8	2	0.003647	0.003521
25	17	8	3	0.038855	0.035208
25	17	8	4	0.192890	0.154035
25	17	8	5	0.513282	0.320392
25	17	8	6	0.833674	0.320392
25	17	8	7	0.977523	0.143849
25	17	8	8	1.000000	0.022476
25	17	9	1	0.000008	0.000008
25	17	9	2	0.000541	0.000533
25	17	9	3	0.009861	0.009320
25	17	9	4	0.075099	0.065238
25	17	9	5	0.287123	0.212024
25	17	9	6	0.626362	0.339239
25	17	9	7	0.892906	0.266545
25	17	9	8	0.988101	0.095194
25	17	9	9	1.000000	0.011899
25	17	10	2	0.000042	0.000042
25	17	10	3	0.001706	0.001664
25	17	10	4	0.022093	0.020387
25	17	10	5	0.128105	0.106012
25	17	10	6	0.393135	0.265030
25	17	10	7	0.726316	0.333181
25	17	10	8	0.934554	0.208238
25	17	10	9	0.994050	0.059497
25	17	10	10	1.000000	0.005950

N	n	k	x	P(x)	p(x)
25	17	11	3	0.000153	0.000153
25	17	11	4	0.004424	0.004272
25	17	11	5	0.043295	0.038871
25	17	11	6	0.198780	0.155484
25	17	11	7	0.504195	0.305416
25	17	11	8	0.809611	0.305416
25	17	11	9	0.962319	0.152708
25	17	11	10	0.997223	0.034905
25	17	11	11	1.000000	0.002777
25	17	12	4	0.000458	0.000458
25	17	12	5	0.009977	0.009519
25	17	12	6	0.076613	0.066636
25	17	12	7	0.286041	0.209428
25	17	12	8	0.613272	0.327231
25	17	12	9	0.875057	0.261785
25	17	12	10	0.979771	0.104714
25	17	12	11	0.998810	0.019039
25	17	12	12	1.000000	0.001190
25	17	13	5	0.001190	0.001190
25	17	13	6	0.020229	0.019039
25	17	13	7	0.124943	0.104714
25	17	13	8	0.386728	0.261785
25	17	13	9	0.713959	0.327231
25	17	13	10	0.923387	0.209428
25	17	13	11	0.990023	0.066636
25	17	13	12	0.999542	0.009519
25	17	13	13	1.000000	0.000458
25	17	14	6	0.002777	0.002777
25	17	14	7	0.037681	0.034905
25	17	14	8	0.190389	0.152708
25	17	14	9	0.495805	0.305416
25	17	14	10	0.801220	0.305416
25	17	14	11	0.956705	0.155484
25	17	14	12	0.995576	0.038871
25	17	14	13	0.999847	0.004272
25	17	14	14	1.000000	0.000153
25	17	15	7	0.005950	0.005950
25	17	15	8	0.065446	0.059497
25	17	15	9	0.273684	0.208238
25	17	15	10	0.606865	0.333181
25	17	15	11	0.871895	0.265030
25	17	15	12	0.977907	0.106012
25	17	15	13	0.998294	0.020387
25	17	15	14	0.999958	0.001664
25	17	15	15	1.000000	0.000042
25	17	16	8	0.011899	0.011899
25	17	16	9	0.107094	0.095194
25	17	16	10	0.373638	0.266545
25	17	16	11	0.712877	0.339239
25	17	16	12	0.924901	0.212024

$N = 25$ $n = 16–17$

N	n	k	x	P(x)	p(x)
25	17	16	13	0.990139	0.065238
25	17	16	14	0.999459	0.009320
25	17	16	15	0.999992	0.000533
25	17	16	16	1.000000	0.000008
25	17	17	9	0.022476	0.022476
25	17	17	10	0.166326	0.143849
25	17	17	11	0.486718	0.320392
25	17	17	12	0.807110	0.320392
25	17	17	13	0.961145	0.154035
25	17	17	14	0.996352	0.035208
25	17	17	15	0.999873	0.003521
25	17	17	16	0.999999	0.000126
25	17	17	17	1.000000	0.000001
25	18	1	0	0.280000	0.280000
25	18	1	1	1.000000	0.720000
25	18	2	0	0.070000	0.070000
25	18	2	1	0.490000	0.420000
25	18	2	2	1.000000	0.510000
25	18	3	0	0.015217	0.015217
25	18	3	1	0.179565	0.164348
25	18	3	2	0.645217	0.465652
25	18	3	3	1.000000	0.354783
25	18	4	0	0.002767	0.002767
25	18	4	1	0.052569	0.049802
25	18	4	2	0.306561	0.253992
25	18	4	3	0.758103	0.451541
25	18	4	4	1.000000	0.241897
25	18	5	0	0.000395	0.000395
25	18	5	1	0.012253	0.011858
25	18	5	2	0.113043	0.100790
25	18	5	3	0.435573	0.322530
25	18	5	4	0.838735	0.403162
25	18	5	5	1.000000	0.161265
25	18	6	0	0.000040	0.000040
25	18	6	1	0.002174	0.002134
25	18	6	2	0.032411	0.030237
25	18	6	3	0.193676	0.161265
25	18	6	4	0.556522	0.362846
25	18	6	5	0.895178	0.338656
25	18	6	6	1.000000	0.104822
25	18	7	0	0.000002	0.000002
25	18	7	1	0.000264	0.000262
25	18	7	2	0.006948	0.006684
25	18	7	3	0.066362	0.059413
25	18	7	4	0.289162	0.222800
25	18	7	5	0.663466	0.374304
25	18	7	6	0.933796	0.270331
25	18	7	7	1.000000	0.066203
25	18	8	1	0.000017	0.000017
25	18	8	2	0.001007	0.000990
25	18	8	3	0.016850	0.015844
25	18	8	4	0.115873	0.099022
25	18	8	5	0.393135	0.277262
25	18	8	6	0.753576	0.360441
25	18	8	7	0.959542	0.205966
25	18	8	8	1.000000	0.040458
25	18	9	2	0.000075	0.000075
25	18	9	3	0.002871	0.002796
25	18	9	4	0.034325	0.031454
25	18	9	5	0.181111	0.146786
25	18	9	6	0.499147	0.318036
25	18	9	7	0.826270	0.327123
25	18	9	8	0.976201	0.149931
25	18	9	9	1.000000	0.023799
25	18	10	3	0.000250	0.000250
25	18	10	4	0.006803	0.006553
25	18	10	5	0.061847	0.055045
25	18	10	6	0.260620	0.198773
25	18	10	7	0.601373	0.340753
25	18	10	8	0.882494	0.281121
25	18	10	9	0.986613	0.104119
25	18	10	10	1.000000	0.013387
25	18	11	4	0.000686	0.000686
25	18	11	5	0.014142	0.013455
25	18	11	6	0.101602	0.087460
25	18	11	7	0.351487	0.249886
25	18	11	8	0.695080	0.343593
25	18	11	9	0.924142	0.229062
25	18	11	10	0.992860	0.068719
25	18	11	11	1.000000	0.007140
25	18	12	5	0.001648	0.001648
25	18	12	6	0.026636	0.024989
25	18	12	7	0.155149	0.128513
25	18	12	8	0.449657	0.294508
25	18	12	9	0.776888	0.327231
25	18	12	10	0.953593	0.176705
25	18	12	11	0.996430	0.042838
25	18	12	12	1.000000	0.003570
25	18	13	6	0.003570	0.003570
25	18	13	7	0.046407	0.042838
25	18	13	8	0.223112	0.176705
25	18	13	9	0.550343	0.327231
25	18	13	10	0.844851	0.294508
25	18	13	11	0.973364	0.128513
25	18	13	12	0.998352	0.024989
25	18	13	13	1.000000	0.001648
25	18	14	7	0.007140	0.007140
25	18	14	8	0.075858	0.068719
25	18	14	9	0.304920	0.229062
25	18	14	10	0.648513	0.343593
25	18	14	11	0.898398	0.249886
25	18	14	12	0.985858	0.087460
25	18	14	13	0.999313	0.013455
25	18	14	14	1.000000	0.000686
25	18	15	8	0.013387	0.013387
25	18	15	9	0.117506	0.104119
25	18	15	10	0.398627	0.281121
25	18	15	11	0.739380	0.340753
25	18	15	12	0.938153	0.198773
25	18	15	13	0.993197	0.055045
25	18	15	14	0.999750	0.006553
25	18	15	15	1.000000	0.000250
25	18	16	9	0.023799	0.023799
25	18	16	10	0.173730	0.149931
25	18	16	11	0.500853	0.327123
25	18	16	12	0.818889	0.318036
25	18	16	13	0.965675	0.146786
25	18	16	14	0.997129	0.031454
25	18	16	15	0.999925	0.002796
25	18	16	16	1.000000	0.000075
25	18	17	10	0.040458	0.040458
25	18	17	11	0.246424	0.205966
25	18	17	12	0.606865	0.360441
25	18	17	13	0.884127	0.277262
25	18	17	14	0.983149	0.099022
25	18	17	15	0.998993	0.015844
25	18	17	16	0.999983	0.000990
25	18	17	17	1.000000	0.000017
25	18	18	11	0.066203	0.066203
25	18	18	12	0.336534	0.270331
25	18	18	13	0.710838	0.374304
25	18	18	14	0.933638	0.222800
25	18	18	15	0.993052	0.059413
25	18	18	16	0.999736	0.006684
25	18	18	17	0.999998	0.000262
25	18	18	18	1.000000	0.000002
25	19	1	0	0.240000	0.240000
25	19	1	1	1.000000	0.760000
25	19	2	0	0.050000	0.050000
25	19	2	1	0.430000	0.380000
25	19	2	2	1.000000	0.570000
25	19	3	0	0.008696	0.008696
25	19	3	1	0.132609	0.123913
25	19	3	2	0.578696	0.446087
25	19	3	3	1.000000	0.421304
25	19	4	0	0.001186	0.001186
25	19	4	1	0.031225	0.030040
25	19	4	2	0.233992	0.202767
25	19	4	3	0.693597	0.459605
25	19	4	4	1.000000	0.306403
25	19	5	0	0.000113	0.000113
25	19	5	1	0.005477	0.005364
25	19	5	2	0.069848	0.064370
25	19	5	3	0.343422	0.273574
25	19	5	4	0.781141	0.437719
25	19	5	5	1.000000	0.218859
25	19	6	0	0.000006	0.000006
25	19	6	1	0.000649	0.000644
25	19	6	2	0.015133	0.014483
25	19	6	3	0.124562	0.109430
25	19	6	4	0.452851	0.328289
25	19	6	5	0.846798	0.393947
25	19	6	6	1.000000	0.153202
25	19	7	1	0.000040	0.000040
25	19	7	2	0.002174	0.002134
25	19	7	3	0.032411	0.030237
25	19	7	4	0.193676	0.161265
25	19	7	5	0.556522	0.362846
25	19	7	6	0.895178	0.338656
25	19	7	7	1.000000	0.104822
25	19	8	2	0.000158	0.000158
25	19	8	3	0.005534	0.005375
25	19	8	4	0.059289	0.053755
25	19	8	5	0.274308	0.215020
25	19	8	6	0.650593	0.376285
25	19	8	7	0.930119	0.279526
25	19	8	8	1.000000	0.069881
25	19	9	3	0.000474	0.000474
25	19	9	4	0.011858	0.011383
25	19	9	5	0.097233	0.085375
25	19	9	6	0.362846	0.265613
25	19	9	7	0.732806	0.369960
25	19	9	8	0.954783	0.221976
25	19	9	9	1.000000	0.045217
25	19	10	4	0.001186	0.001186
25	19	10	5	0.022530	0.021344
25	19	10	6	0.147036	0.124506
25	19	10	7	0.455336	0.308300
25	19	10	8	0.802174	0.346838
25	19	10	9	0.971739	0.169565
25	19	10	10	1.000000	0.028261
25	19	11	5	0.002609	0.002609
25	19	11	6	0.039130	0.036522
25	19	11	7	0.208696	0.169565
25	19	11	8	0.547826	0.339130
25	19	11	9	0.858696	0.310870
25	19	11	10	0.983043	0.124348
25	19	11	11	1.000000	0.016957
25	19	12	6	0.005217	0.005217
25	19	12	7	0.063354	0.058137

Table for $N = 2$, $n = 1$, through $N = 100$, $n = 50$

N	n	k	x	P(x)	p(x)
25	19	12	8	0.281366	0.218012
25	19	12	9	0.636646	0.355280
25	19	12	10	0.903106	0.266460
25	19	12	11	0.990311	0.087205
25	19	12	12	1.000000	0.009689
25	19	13	7	0.009689	0.009689
25	19	13	8	0.096894	0.087205
25	19	13	9	0.363354	0.266460
25	19	13	10	0.718634	0.355280
25	19	13	11	0.936646	0.218012
25	19	13	12	0.994783	0.058137
25	19	13	13	1.000000	0.005217
25	19	14	8	0.016957	0.016957
25	19	14	9	0.141304	0.124348
25	19	14	10	0.452174	0.310870
25	19	14	11	0.791304	0.339130
25	19	14	12	0.960870	0.169565
25	19	14	13	0.997391	0.036522
25	19	14	14	1.000000	0.002609
25	19	15	9	0.028261	0.028261
25	19	15	10	0.197826	0.169565
25	19	15	11	0.544664	0.346838
25	19	15	12	0.852964	0.308300
25	19	15	13	0.977470	0.124506
25	19	15	14	0.998814	0.021344
25	19	15	15	1.000000	0.001186
25	19	16	10	0.045217	0.045217
25	19	16	11	0.267194	0.221976
25	19	16	12	0.637154	0.369960
25	19	16	13	0.902767	0.265613
25	19	16	14	0.988142	0.085375
25	19	16	15	0.999526	0.011383
25	19	16	16	1.000000	0.000474
25	19	17	11	0.069881	0.069881
25	19	17	12	0.349407	0.279526
25	19	17	13	0.725692	0.376285
25	19	17	14	0.940711	0.215020
25	19	17	15	0.994466	0.053755
25	19	17	16	0.999842	0.005375
25	19	17	17	1.000000	0.000158
25	19	18	12	0.104822	0.104822
25	19	18	13	0.443478	0.338656
25	19	18	14	0.806324	0.362846
25	19	18	15	0.967589	0.161265
25	19	18	16	0.997826	0.030237
25	19	18	17	0.999960	0.002134
25	19	18	18	1.000000	0.000040
25	19	19	13	0.153202	0.153202
25	19	19	14	0.547148	0.393947
25	19	19	15	0.875438	0.328289

N	n	k	x	P(x)	p(x)
25	19	19	16	0.984867	0.109430
25	19	19	17	0.999351	0.014483
25	19	19	18	0.999994	0.000644
25	19	19	19	1.000000	0.000006
25	20	1	0	0.200000	0.200000
25	20	1	1	1.000000	0.800000
25	20	2	0	0.033333	0.033333
25	20	2	1	0.366667	0.333333
25	20	2	2	1.000000	0.633333
25	20	3	0	0.004348	0.004348
25	20	3	1	0.091304	0.086957
25	20	3	2	0.504348	0.413043
25	20	3	3	1.000000	0.495652
25	20	4	0	0.000395	0.000395
25	20	4	1	0.016206	0.015810
25	20	4	2	0.166403	0.150198
25	20	4	3	0.616996	0.450593
25	20	4	4	1.000000	0.383004
25	20	5	0	0.000019	0.000019
25	20	5	1	0.001901	0.001882
25	20	5	2	0.037662	0.035761
25	20	5	3	0.252230	0.214568
25	20	5	4	0.708187	0.455957
25	20	5	5	1.000000	0.291812
25	20	6	1	0.000113	0.000113
25	20	6	2	0.005477	0.005364
25	20	6	3	0.069848	0.064370
25	20	6	4	0.343422	0.273574
25	20	6	5	0.781141	0.437719
25	20	6	6	1.000000	0.218859
25	20	7	2	0.000395	0.000395
25	20	7	3	0.012253	0.011858
25	20	7	4	0.113043	0.100790
25	20	7	5	0.435573	0.322530
25	20	7	6	0.838735	0.403162
25	20	7	7	1.000000	0.161265
25	20	8	3	0.001054	0.001054
25	20	8	4	0.023452	0.022398
25	20	8	5	0.166798	0.143346
25	20	8	6	0.525165	0.358366
25	20	8	7	0.883531	0.358366
25	20	8	8	1.000000	0.116469
25	20	9	4	0.002372	0.002372
25	20	9	5	0.040316	0.037945
25	20	9	6	0.230040	0.189723
25	20	9	7	0.609486	0.379447
25	20	9	8	0.917787	0.308300
25	20	9	9	1.000000	0.082213
25	20	10	5	0.004743	0.004743
25	20	10	6	0.064032	0.059289

N	n	k	x	P(x)	p(x)
25	20	10	7	0.301186	0.237154
25	20	10	8	0.686561	0.385375
25	20	10	9	0.943478	0.256917
25	20	10	10	1.000000	0.056522
25	20	11	6	0.008696	0.008696
25	20	11	7	0.095652	0.086957
25	20	11	8	0.378261	0.282609
25	20	11	9	0.755072	0.376812
25	20	11	10	0.962319	0.207246
25	20	11	11	1.000000	0.037681
25	20	12	7	0.014907	0.014907
25	20	12	8	0.136025	0.121118
25	20	12	9	0.459006	0.322981
25	20	12	10	0.814286	0.355280
25	20	12	11	0.975776	0.161491
25	20	12	12	1.000000	0.024224
25	20	13	8	0.024224	0.024224
25	20	13	9	0.185714	0.161491
25	20	13	10	0.540994	0.355280
25	20	13	11	0.863975	0.322981
25	20	13	12	0.985093	0.121118
25	20	13	13	1.000000	0.014907
25	20	14	9	0.037681	0.037681
25	20	14	10	0.244928	0.207246
25	20	14	11	0.621739	0.376812
25	20	14	12	0.904348	0.282609
25	20	14	13	0.991304	0.086957
25	20	14	14	1.000000	0.008696
25	20	15	10	0.056522	0.056522
25	20	15	11	0.313439	0.256917
25	20	15	12	0.698814	0.385375
25	20	15	13	0.935968	0.237154
25	20	15	14	0.995257	0.059289
25	20	15	15	1.000000	0.004743
25	20	16	11	0.082213	0.082213
25	20	16	12	0.390514	0.308300
25	20	16	13	0.769960	0.379447
25	20	16	14	0.959684	0.189723
25	20	16	15	0.997628	0.037945
25	20	16	16	1.000000	0.002372
25	20	17	12	0.116469	0.116469
25	20	17	13	0.474835	0.358366
25	20	17	14	0.833201	0.358366
25	20	17	15	0.976548	0.143346
25	20	17	16	0.998946	0.022398
25	20	17	17	1.000000	0.001054
25	20	18	13	0.161265	0.161265
25	20	18	14	0.564427	0.403162
25	20	18	15	0.886956	0.322530
25	20	18	16	0.987747	0.100790

N	n	k	x	P(x)	p(x)
25	20	18	17	0.999605	0.011858
25	20	18	18	1.000000	0.000395
25	20	19	14	0.218859	0.218859
25	20	19	15	0.656578	0.437719
25	20	19	16	0.930152	0.273574
25	20	19	17	0.994523	0.064370
25	20	19	18	0.999887	0.005364
25	20	19	19	1.000000	0.000113
25	20	20	15	0.291812	0.291812
25	20	20	16	0.747770	0.455957
25	20	20	17	0.962338	0.214568
25	20	20	18	0.998099	0.035761
25	20	20	19	0.999981	0.001882
25	20	20	20	1.000000	0.000019
25	21	1	0	0.160000	0.160000
25	21	1	1	1.000000	0.840000
25	21	2	0	0.020000	0.020000
25	21	2	1	0.300000	0.280000
25	21	2	2	1.000000	0.700000
25	21	3	0	0.001739	0.001739
25	21	3	1	0.056522	0.054783
25	21	3	2	0.421739	0.365217
25	21	3	3	1.000000	0.578261
25	21	4	0	0.000079	0.000079
25	21	4	1	0.006719	0.006640
25	21	4	2	0.106324	0.099605
25	21	4	3	0.526877	0.420553
25	21	4	4	1.000000	0.473122
25	21	5	1	0.000395	0.000395
25	21	5	2	0.016206	0.015810
25	21	5	3	0.166403	0.150198
25	21	5	4	0.616996	0.450593
25	21	5	5	1.000000	0.383004
25	21	6	2	0.001186	0.001186
25	21	6	3	0.031225	0.030040
25	21	6	4	0.233992	0.202767
25	21	6	5	0.693597	0.459605
25	21	6	6	1.000000	0.306403
25	21	7	3	0.002767	0.002767
25	21	7	4	0.052569	0.049802
25	21	7	5	0.306561	0.253992
25	21	7	6	0.758103	0.451541
25	21	7	7	1.000000	0.241897
25	21	8	4	0.005534	0.005534
25	21	8	5	0.080791	0.075257
25	21	8	6	0.381818	0.301028
25	21	8	7	0.811858	0.430040
25	21	8	8	1.000000	0.188142
25	21	9	5	0.009960	0.009960
25	21	9	6	0.116206	0.106245

N	n	k	x	P(x)	p(x)
25	21	9	7	0.457707	0.341502
25	21	9	8	0.856126	0.398419
25	21	9	9	1.000000	0.143874
25	21	10	6	0.016601	0.016601
25	21	10	7	0.158893	0.142292
25	21	10	8	0.532411	0.373518
25	21	10	9	0.892095	0.359684
25	21	10	10	1.000000	0.107905
25	21	11	7	0.026087	0.026087
25	21	11	8	0.208696	0.182609
25	21	11	9	0.604348	0.395652
25	21	11	10	0.920869	0.316522
25	21	11	11	1.000000	0.079130
25	21	12	8	0.039130	0.039130
25	21	12	9	0.265217	0.226087
25	21	12	10	0.672174	0.406957
25	21	12	11	0.943478	0.271304
25	21	12	12	1.000000	0.056522
25	21	13	9	0.056522	0.056522
25	21	13	10	0.327826	0.271304
25	21	13	11	0.734783	0.406957
25	21	13	12	0.960870	0.226087
25	21	13	13	1.000000	0.039130
25	21	14	10	0.079130	0.079130
25	21	14	11	0.395652	0.316522
25	21	14	12	0.791304	0.395652
25	21	14	13	0.973913	0.182609
25	21	14	14	1.000000	0.026087
25	21	15	11	0.107905	0.107905
25	21	15	12	0.467589	0.359684
25	21	15	13	0.841107	0.373518
25	21	15	14	0.983399	0.142292
25	21	15	15	1.000000	0.016601
25	21	16	12	0.143874	0.143874
25	21	16	13	0.542292	0.398419
25	21	16	14	0.883794	0.341502
25	21	16	15	0.990039	0.106245
25	21	16	16	1.000000	0.009960
25	21	17	13	0.188142	0.188142
25	21	17	14	0.618182	0.430040
25	21	17	15	0.919209	0.301028
25	21	17	16	0.994466	0.075257
25	21	17	17	1.000000	0.005534
25	21	18	14	0.241897	0.241897
25	21	18	15	0.693439	0.451541
25	21	18	16	0.947431	0.253992
25	21	18	17	0.997233	0.049802
25	21	18	18	1.000000	0.002767
25	21	19	15	0.306403	0.306403
25	21	19	16	0.766008	0.459605

N	n	k	x	P(x)	p(x)
25	21	19	17	0.968775	0.202767
25	21	19	18	0.998814	0.030040
25	21	19	19	1.000000	0.001186
25	21	20	16	0.383004	0.383004
25	21	20	17	0.833597	0.450593
25	21	20	18	0.983794	0.150198
25	21	20	19	0.999605	0.015810
25	21	20	20	1.000000	0.000395
25	21	21	17	0.473122	0.473122
25	21	21	18	0.893676	0.420553
25	21	21	19	0.993281	0.099605
25	21	21	20	0.999921	0.006640
25	21	21	21	1.000000	0.000079
25	22	1	0	0.120000	0.120000
25	22	1	1	1.000000	0.880000
25	22	2	0	0.010000	0.010000
25	22	2	1	0.230000	0.220000
25	22	2	2	1.000000	0.770000
25	22	3	0	0.000435	0.000435
25	22	3	1	0.029130	0.028696
25	22	3	2	0.330435	0.301304
25	22	3	3	1.000000	0.669565
25	22	4	1	0.001739	0.001739
25	22	4	2	0.056522	0.054783
25	22	4	3	0.421739	0.365217
25	22	4	4	1.000000	0.578261
25	22	5	2	0.004348	0.004348
25	22	5	3	0.091304	0.086957
25	22	5	4	0.504348	0.413043
25	22	5	5	1.000000	0.495652
25	22	6	3	0.008696	0.008696
25	22	6	4	0.132609	0.123913
25	22	6	5	0.578696	0.446087
25	22	6	6	1.000000	0.421304
25	22	7	4	0.015217	0.015217
25	22	7	5	0.179565	0.164348
25	22	7	6	0.645217	0.465652
25	22	7	7	1.000000	0.354783
25	22	8	5	0.024348	0.024348
25	22	8	6	0.231304	0.206957
25	22	8	7	0.704348	0.473043
25	22	8	8	1.000000	0.295652
25	22	9	6	0.036522	0.036522
25	22	9	7	0.286957	0.250435
25	22	9	8	0.756522	0.469565
25	22	9	9	1.000000	0.243478
25	22	10	7	0.052174	0.052174
25	22	10	8	0.345652	0.293478
25	22	10	9	0.802174	0.456522
25	22	10	10	1.000000	0.197826

N	n	k	x	P(x)	p(x)
25	22	11	8	0.071739	0.071739
25	22	11	9	0.406522	0.334783
25	22	11	10	0.841739	0.435217
25	22	11	11	1.000000	0.158261
25	22	12	9	0.095652	0.095652
25	22	12	10	0.468696	0.373043
25	22	12	11	0.875652	0.406957
25	22	12	12	1.000000	0.124348
25	22	13	10	0.124348	0.124348
25	22	13	11	0.531304	0.406957
25	22	13	12	0.904348	0.373043
25	22	13	13	1.000000	0.095652
25	22	14	11	0.158261	0.158261
25	22	14	12	0.593478	0.435217
25	22	14	13	0.928261	0.334783
25	22	14	14	1.000000	0.071739
25	22	15	12	0.197826	0.197826
25	22	15	13	0.654348	0.456522
25	22	15	14	0.947826	0.293478
25	22	15	15	1.000000	0.052174
25	22	16	13	0.243478	0.243478
25	22	16	14	0.713043	0.469565
25	22	16	15	0.963478	0.250435
25	22	16	16	1.000000	0.036522
25	22	17	14	0.295652	0.295652
25	22	17	15	0.768696	0.473043
25	22	17	16	0.975652	0.206957
25	22	17	17	1.000000	0.024348
25	22	18	15	0.354783	0.354783
25	22	18	16	0.820435	0.465652
25	22	18	17	0.984782	0.164348
25	22	18	18	1.000000	0.015217
25	22	19	16	0.421304	0.421304
25	22	19	17	0.867391	0.446087
25	22	19	18	0.991304	0.123913
25	22	19	19	1.000000	0.008696
25	22	20	17	0.495652	0.495652
25	22	20	18	0.908696	0.413043
25	22	20	19	0.995652	0.086957
25	22	20	20	1.000000	0.004348
25	22	21	18	0.578261	0.578261
25	22	21	19	0.943478	0.365217
25	22	21	20	0.998261	0.054783
25	22	21	21	1.000000	0.001739
25	22	22	19	0.669565	0.669565
25	22	22	20	0.970870	0.301304
25	22	22	21	0.999565	0.028696
25	22	22	22	1.000000	0.000435
25	23	1	0	0.080000	0.080000
25	23	1	1	1.000000	0.920000

N	n	k	x	P(x)	p(x)
25	23	2	0	0.003333	0.003333
25	23	2	1	0.156667	0.153333
25	23	2	2	1.000000	0.843333
25	23	3	1	0.010000	0.010000
25	23	3	2	0.230000	0.220000
25	23	3	3	1.000000	0.770000
25	23	4	2	0.020000	0.020000
25	23	4	3	0.300000	0.280000
25	23	4	4	1.000000	0.700000
25	23	5	3	0.033333	0.033333
25	23	5	4	0.366667	0.333333
25	23	5	5	1.000000	0.633333
25	23	6	4	0.050000	0.050000
25	23	6	5	0.430000	0.380000
25	23	6	6	1.000000	0.570000
25	23	7	5	0.070000	0.070000
25	23	7	6	0.490000	0.420000
25	23	7	7	1.000000	0.510000
25	23	8	6	0.093333	0.093333
25	23	8	7	0.546667	0.453333
25	23	8	8	1.000000	0.453333
25	23	9	7	0.120000	0.120000
25	23	9	8	0.600000	0.480000
25	23	9	9	1.000000	0.400000
25	23	10	8	0.150000	0.150000
25	23	10	9	0.650000	0.500000
25	23	10	10	1.000000	0.350000
25	23	11	9	0.183333	0.183333
25	23	11	10	0.696667	0.513333
25	23	11	11	1.000000	0.303333
25	23	12	10	0.220000	0.220000
25	23	12	11	0.740000	0.520000
25	23	12	12	1.000000	0.260000
25	23	13	11	0.260000	0.260000
25	23	13	12	0.780000	0.520000
25	23	13	13	1.000000	0.220000
25	23	14	12	0.303333	0.303333
25	23	14	13	0.816667	0.513333
25	23	14	14	1.000000	0.183333
25	23	15	13	0.350000	0.350000
25	23	15	14	0.850000	0.500000
25	23	15	15	1.000000	0.150000
25	23	16	14	0.400000	0.400000
25	23	16	15	0.880000	0.480000
25	23	16	16	1.000000	0.120000
25	23	17	15	0.453333	0.453333
25	23	17	16	0.906667	0.453333
25	23	17	17	1.000000	0.093333
25	23	18	16	0.510000	0.510000
25	23	18	17	0.930000	0.420000

N	n	k	x	P(x)	p(x)
25	23	18	18	1.000000	0.070000
25	23	19	17	0.570000	0.570000
25	23	19	18	0.950000	0.380000
25	23	19	19	1.000000	0.050000
25	23	20	18	0.633333	0.633333
25	23	20	19	0.966667	0.333333
25	23	20	20	1.000000	0.033333
25	23	21	19	0.700000	0.700000
25	23	21	20	0.980000	0.280000
25	23	21	21	1.000000	0.020000
25	23	22	20	0.770000	0.770000
25	23	22	21	0.990000	0.220000
25	23	22	22	1.000000	0.010000
25	23	23	21	0.843333	0.843333
25	23	23	22	0.996667	0.153333
25	23	23	23	1.000000	0.003333
25	24	1	0	0.040000	0.040000
25	24	1	1	1.000000	0.960000
25	24	2	1	0.080000	0.080000
25	24	2	2	1.000000	0.920000
25	24	3	2	0.120000	0.120000
25	24	3	3	1.000000	0.880000
25	24	4	3	0.160000	0.160000
25	24	4	4	1.000000	0.840000
25	24	5	4	0.200000	0.200000
25	24	5	5	1.000000	0.800000
25	24	6	5	0.240000	0.240000
25	24	6	6	1.000000	0.760000
25	24	7	6	0.280000	0.280000
25	24	7	7	1.000000	0.720000
25	24	8	7	0.320000	0.320000
25	24	8	8	1.000000	0.680000
25	24	9	8	0.360000	0.360000
25	24	9	9	1.000000	0.640000
25	24	10	9	0.400000	0.400000
25	24	10	10	1.000000	0.600000
25	24	11	10	0.440000	0.440000
25	24	11	11	1.000000	0.560000
25	24	12	11	0.480000	0.480000
25	24	12	12	1.000000	0.520000
25	24	13	12	0.520000	0.520000
25	24	13	13	1.000000	0.480000
25	24	14	13	0.560000	0.560000
25	24	14	14	1.000000	0.440000
25	24	15	14	0.600000	0.600000
25	24	15	15	1.000000	0.400000
25	24	16	15	0.640000	0.640000
25	24	16	16	1.000000	0.360000
25	24	17	16	0.680000	0.680000
25	24	17	17	1.000000	0.320000

N	n	k	x	P(x)	p(x)
25	24	18	17	0.720000	0.720000
25	24	18	18	1.000000	0.280000
25	24	19	18	0.760000	0.760000
25	24	19	19	1.000000	0.240000
25	24	20	19	0.800000	0.800000
25	24	20	20	1.000000	0.200000
25	24	21	20	0.840000	0.840000
25	24	21	21	1.000000	0.160000
25	24	22	21	0.880000	0.880000
25	24	22	22	1.000000	0.120000
25	24	23	22	0.920000	0.920000
25	24	23	23	1.000000	0.080000
25	24	24	23	0.960000	0.960000
25	24	24	24	1.000000	0.040000
26	1	1	0	0.961538	0.961538
26	1	1	1	1.000000	0.038462
26	2	1	0	0.923077	0.923077
26	2	1	1	1.000000	0.076923
26	2	2	0	0.849231	0.849231
26	2	2	1	0.996923	0.147692
26	2	2	2	1.000000	0.003077
26	3	1	0	0.884615	0.884615
26	3	1	1	1.000000	0.115385
26	3	2	0	0.778462	0.778462
26	3	2	1	0.990769	0.212308
26	3	2	2	1.000000	0.009231
26	3	3	0	0.681154	0.681154
26	3	3	1	0.973077	0.291923
26	3	3	2	0.999615	0.026538
26	3	3	3	1.000000	0.000385
26	4	1	0	0.846154	0.846154
26	4	1	1	1.000000	0.153846
26	4	2	0	0.710769	0.710769
26	4	2	1	0.981538	0.270769
26	4	2	2	1.000000	0.018462
26	4	3	0	0.592308	0.592308
26	4	3	1	0.947692	0.355385
26	4	3	2	0.998462	0.050769
26	4	3	3	1.000000	0.001538
26	4	4	0	0.489298	0.489298
26	4	4	1	0.901338	0.412040
26	4	4	2	0.994047	0.092709
26	4	4	3	0.999933	0.005886
26	4	4	4	1.000000	0.000067
26	5	1	0	0.807692	0.807692
26	5	1	1	1.000000	0.192308
26	5	2	0	0.646154	0.646154
26	5	2	1	0.969231	0.323077
26	5	2	2	1.000000	0.030769
26	5	3	0	0.511538	0.511538

N	n	k	x	P(x)	p(x)
26	5	3	1	0.915385	0.403846
26	5	3	2	0.996154	0.080769
26	5	3	3	1.000000	0.003846
26	5	4	0	0.400334	0.400334
26	5	4	1	0.845150	0.444816
26	5	4	2	0.985619	0.140468
26	5	4	3	0.999665	0.014047
26	5	4	4	1.000000	0.000334
26	5	5	0	0.309349	0.309349
26	5	5	1	0.764275	0.454925
26	5	5	2	0.966464	0.202189
26	5	5	3	0.998388	0.031925
26	5	5	4	0.999985	0.001596
26	5	5	5	1.000000	0.000015
26	6	1	0	0.769231	0.769231
26	6	1	1	1.000000	0.230769
26	6	2	0	0.584615	0.584615
26	6	2	1	0.953846	0.369231
26	6	2	2	1.000000	0.046154
26	6	3	0	0.438462	0.438462
26	6	3	1	0.876923	0.438462
26	6	3	2	0.992308	0.115385
26	6	3	3	1.000000	0.007692
26	6	4	0	0.324080	0.324080
26	6	4	1	0.781605	0.457525
26	6	4	2	0.972241	0.190635
26	6	4	3	0.998997	0.026756
26	6	4	4	1.000000	0.001003
26	6	5	0	0.235695	0.235695
26	6	5	1	0.677622	0.441928
26	6	5	2	0.937580	0.259957
26	6	5	3	0.995348	0.057768
26	6	5	4	0.999909	0.004561
26	6	5	5	1.000000	0.000091
26	6	6	0	0.168353	0.168353
26	6	6	1	0.572401	0.404048
26	6	6	2	0.888064	0.315663
26	6	6	3	0.987095	0.099031
26	6	6	4	0.999474	0.012379
26	6	6	5	0.999996	0.000521
26	6	6	6	1.000000	0.000004
26	7	1	0	0.730769	0.730769
26	7	1	1	1.000000	0.269231
26	7	2	0	0.526154	0.526154
26	7	2	1	0.935385	0.409231
26	7	2	2	1.000000	0.064615
26	7	3	0	0.372692	0.372692
26	7	3	1	0.833077	0.460385
26	7	3	2	0.986538	0.153462
26	7	3	3	1.000000	0.013462

N	n	k	x	P(x)	p(x)
26	7	4	0	0.259264	0.259264
26	7	4	1	0.712977	0.453712
26	7	4	2	0.953177	0.240201
26	7	4	3	0.997659	0.044482
26	7	4	4	1.000000	0.002341
26	7	5	0	0.176771	0.176771
26	7	5	1	0.589237	0.412466
26	7	5	2	0.898586	0.309349
26	7	5	3	0.989571	0.090985
26	7	5	4	0.999681	0.010109
26	7	5	5	1.000000	0.000319
26	7	6	0	0.117847	0.117847
26	7	6	1	0.471389	0.353542
26	7	6	2	0.824932	0.353542
26	7	6	3	0.972241	0.147309
26	7	6	4	0.998236	0.025996
26	7	6	5	0.999970	0.001733
26	7	6	6	1.000000	0.000030
26	7	7	0	0.076601	0.076601
26	7	7	1	0.365327	0.288726
26	7	7	2	0.736546	0.371219
26	7	7	3	0.942779	0.206233
26	7	7	4	0.994337	0.051558
26	7	7	5	0.999796	0.005459
26	7	7	6	0.999998	0.000202
26	7	7	7	1.000000	0.000002
26	8	1	0	0.692308	0.692308
26	8	1	1	1.000000	0.307692
26	8	2	0	0.470769	0.470769
26	8	2	1	0.913846	0.443077
26	8	2	2	1.000000	0.086154
26	8	3	0	0.313846	0.313846
26	8	3	1	0.784615	0.470769
26	8	3	2	0.978461	0.193846
26	8	3	3	1.000000	0.021538
26	8	4	0	0.204682	0.204682
26	8	4	1	0.641338	0.436655
26	8	4	2	0.927893	0.286555
26	8	4	3	0.995318	0.067425
26	8	4	4	1.000000	0.004682
26	8	5	0	0.130252	0.130252
26	8	5	1	0.502402	0.372150
26	8	5	2	0.849742	0.347340
26	8	5	3	0.979994	0.130252
26	8	5	4	0.999149	0.019155
26	8	5	5	1.000000	0.000851
26	8	6	0	0.080632	0.080632
26	8	6	1	0.378352	0.297720
26	8	6	2	0.750502	0.372150
26	8	6	3	0.948981	0.198480

N	n	k	x	P(x)	p(x)
26	8	6	4	0.995500	0.046519
26	8	6	5	0.999878	0.004378
26	8	6	6	1.000000	0.000122
26	8	7	0	0.048379	0.048379
26	8	7	1	0.274150	0.225771
26	8	7	2	0.638857	0.364707
26	8	7	3	0.899361	0.260505
26	8	7	4	0.986196	0.086835
26	8	7	5	0.999222	0.013025
26	8	7	6	0.999988	0.000766
26	8	7	7	1.000000	0.000012
26	8	8	0	0.028009	0.028009
26	8	8	1	0.190971	0.162962
26	8	8	2	0.523686	0.332715
26	8	8	3	0.830808	0.307121
26	8	8	4	0.967915	0.137108
26	8	8	5	0.997165	0.029250
26	8	8	6	0.999907	0.002742
26	8	8	7	0.999999	0.000092
26	8	8	8	1.000000	0.000001
26	9	1	0	0.653846	0.653846
26	9	1	1	1.000000	0.346154
26	9	2	0	0.418462	0.418462
26	9	2	1	0.889231	0.470769
26	9	2	2	1.000000	0.110769
26	9	3	0	0.261538	0.261538
26	9	3	1	0.732308	0.470769
26	9	3	2	0.967692	0.235385
26	9	3	3	1.000000	0.032308
26	9	4	0	0.159197	0.159197
26	9	4	1	0.568562	0.409365
26	9	4	2	0.896053	0.327492
26	9	4	3	0.991572	0.095518
26	9	4	4	1.000000	0.008428
26	9	5	0	0.094071	0.094071
26	9	5	1	0.419702	0.325631
26	9	5	2	0.791852	0.372150
26	9	5	3	0.965521	0.173670
26	9	5	4	0.998084	0.032563
26	9	5	5	1.000000	0.001915
26	9	6	0	0.053755	0.053755
26	9	6	1	0.295652	0.241897
26	9	6	2	0.667802	0.372150
26	9	6	3	0.915901	0.248100
26	9	6	4	0.990331	0.074430
26	9	6	5	0.999635	0.009304
26	9	6	6	1.000000	0.000365
26	9	7	0	0.029565	0.029565
26	9	7	1	0.198893	0.169328
26	9	7	2	0.537549	0.338656
26	9	7	3	0.841472	0.303922
26	9	7	4	0.971724	0.130252
26	9	7	5	0.997774	0.026050
26	9	7	6	0.999945	0.002171
26	9	7	7	1.000000	0.000055
26	9	8	0	0.015561	0.015561
26	9	8	1	0.127597	0.112037
26	9	8	2	0.412781	0.285184
26	9	8	3	0.745496	0.332715
26	9	8	4	0.937447	0.191951
26	9	8	5	0.992290	0.054843
26	9	8	6	0.999602	0.007312
26	9	8	7	0.999994	0.000392
26	9	8	8	1.000000	0.000006
26	9	9	0	0.007780	0.007780
26	9	9	1	0.077803	0.070023
26	9	9	2	0.301876	0.224073
26	9	9	3	0.634591	0.332715
26	9	9	4	0.884127	0.249536
26	9	9	5	0.980103	0.095975
26	9	9	6	0.998384	0.018281
26	9	9	7	0.999951	0.001567
26	9	9	8	1.000000	0.000049
26	9	9	9	1.000000	0.000000
26	10	1	0	0.615385	0.615385
26	10	1	1	1.000000	0.384615
26	10	2	0	0.369231	0.369231
26	10	2	1	0.861538	0.492308
26	10	2	2	1.000000	0.138462
26	10	3	0	0.215385	0.215385
26	10	3	1	0.676923	0.461538
26	10	3	2	0.953846	0.276923
26	10	3	3	1.000000	0.046154
26	10	4	0	0.121739	0.121739
26	10	4	1	0.496321	0.374582
26	10	4	2	0.857525	0.361204
26	10	4	3	0.985953	0.128428
26	10	4	4	1.000000	0.014047
26	10	5	0	0.066403	0.066403
26	10	5	1	0.343083	0.276680
26	10	5	2	0.726178	0.383095
26	10	5	3	0.945090	0.218911
26	10	5	4	0.996169	0.051079
26	10	5	5	1.000000	0.003831
26	10	6	0	0.034783	0.034783
26	10	6	1	0.224506	0.189723
26	10	6	2	0.580237	0.355731
26	10	6	3	0.872119	0.291882
26	10	6	4	0.981575	0.109456
26	10	6	5	0.999088	0.017513
26	10	6	6	1.000000	0.000912
26	10	7	0	0.017391	0.017391
26	10	7	1	0.139130	0.121739
26	10	7	2	0.437945	0.298814
26	10	7	3	0.769960	0.332016
26	10	7	4	0.948738	0.178778
26	10	7	5	0.994709	0.045971
26	10	7	6	0.999817	0.005108
26	10	7	7	1.000000	0.000182
26	10	8	0	0.008238	0.008238
26	10	8	1	0.081464	0.073227
26	10	8	2	0.312128	0.230664
26	10	8	3	0.647639	0.335511
26	10	8	4	0.892282	0.244643
26	10	8	5	0.982612	0.090330
26	10	8	6	0.998742	0.016130
26	10	8	7	0.999971	0.001229
26	10	8	8	1.000000	0.000029
26	10	9	0	0.003661	0.003661
26	10	9	1	0.044851	0.041190
26	10	9	2	0.209611	0.164760
26	10	9	3	0.517162	0.307551
26	10	9	4	0.810734	0.293572
26	10	9	5	0.957520	0.146786
26	10	9	6	0.995158	0.037637
26	10	9	7	0.999766	0.004609
26	10	9	8	0.999997	0.000230
26	10	9	9	1.000000	0.000003
26	10	10	0	0.001508	0.001508
26	10	10	1	0.023045	0.021537
26	10	10	2	0.132077	0.109032
26	10	10	3	0.390524	0.258447
26	10	10	4	0.707121	0.316597
26	10	10	5	0.914348	0.207227
26	10	10	6	0.986302	0.071954
26	10	10	7	0.998953	0.012651
26	10	10	8	0.999970	0.001017
26	10	10	9	1.000000	0.000030
26	10	10	10	1.000000	0.000000
26	11	1	0	0.576923	0.576923
26	11	1	1	1.000000	0.423077
26	11	2	0	0.323077	0.323077
26	11	2	1	0.830769	0.507692
26	11	2	2	1.000000	0.169231
26	11	3	0	0.175000	0.175000
26	11	3	1	0.619231	0.444231
26	11	3	2	0.936538	0.317308
26	11	3	3	1.000000	0.063462
26	11	4	0	0.091304	0.091304
26	11	4	1	0.426087	0.334783
26	11	4	2	0.812375	0.386288
26	11	4	3	0.977926	0.165552
26	11	4	4	1.000000	0.022074
26	11	5	0	0.045652	0.045652
26	11	5	1	0.273913	0.228261
26	11	5	2	0.654348	0.380435
26	11	5	3	0.917726	0.263378
26	11	5	4	0.992977	0.075251
26	11	5	5	1.000000	0.007023
26	11	6	0	0.021739	0.021739
26	11	6	1	0.165217	0.143478
26	11	6	2	0.491304	0.326087
26	11	6	3	0.817391	0.326087
26	11	6	4	0.967893	0.150502
26	11	6	5	0.997993	0.030100
26	11	6	6	1.000000	0.002007
26	11	7	0	0.009783	0.009783
26	11	7	1	0.093478	0.083696
26	11	7	2	0.344565	0.251087
26	11	7	3	0.686956	0.342391
26	11	7	4	0.915217	0.228261
26	11	7	5	0.988963	0.073746
26	11	7	6	0.999498	0.010535
26	11	7	7	1.000000	0.000502
26	11	8	0	0.004119	0.004119
26	11	8	1	0.049428	0.045309
26	11	8	2	0.225629	0.176201
26	11	8	3	0.542792	0.317162
26	11	8	4	0.831121	0.288329
26	11	8	5	0.965675	0.134554
26	11	8	6	0.996726	0.031051
26	11	8	7	0.999894	0.003168
26	11	8	8	1.000000	0.000106
26	11	9	0	0.001602	0.001602
26	11	9	1	0.024256	0.022654
26	11	9	2	0.137529	0.113272
26	11	9	3	0.401831	0.264302
26	11	9	4	0.718993	0.317162
26	11	9	5	0.920824	0.201831
26	11	9	6	0.988101	0.067277
26	11	9	7	0.999190	0.011090
26	11	9	8	0.999982	0.000792
26	11	9	9	1.000000	0.000018
26	11	10	0	0.000565	0.000565
26	11	10	1	0.010930	0.010365
26	11	10	2	0.077561	0.066631
26	11	10	3	0.277453	0.199892
26	11	10	4	0.588397	0.310944
26	11	10	5	0.849589	0.261193
26	11	10	6	0.968313	0.118724

Table for $N = 2$, $n = 1$, through $N = 100$, $n = 50$

N	n	k	x	P(x)	p(x)
26	11	10	7	0.996581	0.028268
26	11	10	8	0.999843	0.003262
26	11	10	9	0.999998	0.000155
26	11	10	10	1.000000	0.000002
26	11	11	0	0.000177	0.000177
26	11	11	1	0.004452	0.004275
26	11	11	2	0.040081	0.035629
26	11	11	3	0.177507	0.137426
26	11	11	4	0.452359	0.274852
26	11	11	5	0.751642	0.299283
26	11	11	6	0.931212	0.179570
26	11	11	7	0.989514	0.058302
26	11	11	8	0.999231	0.009717
26	11	11	9	0.999978	0.000747
26	11	11	10	1.000000	0.000021
26	11	11	11	1.000000	0.000000
26	12	1	0	0.538462	0.538462
26	12	1	1	1.000000	0.461538
26	12	2	0	0.280000	0.280000
26	12	2	1	0.796923	0.516923
26	12	2	2	1.000000	0.203077
26	12	3	0	0.140000	0.140000
26	12	3	1	0.560000	0.420000
26	12	3	2	0.915385	0.355385
26	12	3	3	1.000000	0.084615
26	12	4	0	0.066956	0.066956
26	12	4	1	0.359130	0.292174
26	12	4	2	0.760870	0.401739
26	12	4	3	0.966890	0.206020
26	12	4	4	1.000000	0.033110
26	12	5	0	0.030435	0.030435
26	12	5	1	0.213043	0.182609
26	12	5	2	0.578261	0.365217
26	12	5	3	0.882609	0.304348
26	12	5	4	0.987960	0.105351
26	12	5	5	1.000000	0.012040
26	12	6	0	0.013043	0.013043
26	12	6	1	0.117391	0.104348
26	12	6	2	0.404348	0.286956
26	12	6	3	0.752174	0.347826
26	12	6	4	0.947826	0.195652
26	12	6	5	0.995986	0.048161
26	12	6	6	1.000000	0.004013
26	12	7	0	0.005217	0.005217
26	12	7	1	0.060000	0.054783
26	12	7	2	0.260870	0.200870
26	12	7	3	0.595652	0.334783
26	12	7	4	0.869565	0.273913
26	12	7	5	0.979130	0.109565
26	12	7	6	0.998796	0.019666

N	n	k	x	P(x)	p(x)
26	12	7	7	1.000000	0.001204
26	12	8	0	0.001922	0.001922
26	12	8	1	0.028284	0.026362
26	12	8	2	0.155149	0.126865
26	12	8	3	0.437071	0.281922
26	12	8	4	0.754233	0.317162
26	12	8	5	0.938764	0.184531
26	12	8	6	0.992586	0.053822
26	12	8	7	0.999683	0.007097
26	12	8	8	1.000000	0.000317
26	12	9	0	0.000641	0.000641
26	12	9	1	0.012174	0.011533
26	12	9	2	0.084668	0.072494
26	12	9	3	0.296110	0.211442
26	12	9	4	0.613272	0.317162
26	12	9	5	0.867002	0.253730
26	12	9	6	0.974645	0.107643
26	12	9	7	0.997712	0.023066
26	12	9	8	0.999929	0.002218
26	12	9	9	1.000000	0.000070
26	12	10	0	0.000188	0.000188
26	12	10	1	0.004711	0.004523
26	12	10	2	0.042024	0.037313
26	12	10	3	0.184170	0.142146
26	12	10	4	0.464019	0.279849
26	12	10	5	0.762525	0.298506
26	12	10	6	0.936654	0.174128
26	12	10	7	0.990927	0.054274
26	12	10	8	0.999408	0.008480
26	12	10	9	0.999988	0.000580
26	12	10	10	1.000000	0.000012
26	12	11	0	0.000047	0.000047
26	12	11	1	0.001602	0.001555
26	12	11	2	0.018704	0.017102
26	12	11	3	0.104213	0.085509
26	12	11	4	0.324095	0.219882
26	12	11	5	0.631929	0.307834
26	12	11	6	0.871355	0.239427
26	12	11	7	0.973967	0.102611
26	12	11	8	0.997288	0.023321
26	12	11	9	0.999879	0.002591
26	12	11	10	0.999998	0.000120
26	12	11	11	1.000000	0.000002
26	12	12	0	0.000009	0.000009
26	12	12	1	0.000462	0.000452
26	12	12	2	0.007302	0.006841
26	12	12	3	0.052908	0.045605
26	12	12	4	0.206825	0.153917
26	12	12	5	0.488273	0.281448
26	12	12	6	0.775585	0.287312

N	n	k	x	P(x)	p(x)
26	12	12	7	0.939763	0.164178
26	12	12	8	0.991069	0.051306
26	12	12	9	0.999361	0.008292
26	12	12	10	0.999982	0.000622
26	12	12	11	1.000000	0.000017
26	12	12	12	1.000000	0.000000
26	13	1	0	0.500000	0.500000
26	13	1	1	1.000000	0.500000
26	13	2	0	0.240000	0.240000
26	13	2	1	0.760000	0.520000
26	13	2	2	1.000000	0.240000
26	13	3	0	0.110000	0.110000
26	13	3	1	0.500000	0.390000
26	13	3	2	0.890000	0.390000
26	13	3	3	1.000000	0.110000
26	13	4	0	0.047826	0.047826
26	13	4	1	0.296522	0.248696
26	13	4	2	0.703478	0.406957
26	13	4	3	0.952174	0.248696
26	13	4	4	1.000000	0.047826
26	13	5	0	0.019565	0.019565
26	13	5	1	0.160870	0.141304
26	13	5	2	0.500000	0.339130
26	13	5	3	0.839130	0.339130
26	13	5	4	0.980435	0.141304
26	13	5	5	1.000000	0.019565
26	13	6	0	0.007453	0.007453
26	13	6	1	0.080124	0.072671
26	13	6	2	0.322360	0.242236
26	13	6	3	0.677640	0.355280
26	13	6	4	0.919876	0.242236
26	13	6	5	0.992547	0.072671
26	13	6	6	1.000000	0.007453
26	13	7	0	0.002609	0.002609
26	13	7	1	0.036522	0.033913
26	13	7	2	0.189130	0.152609
26	13	7	3	0.500000	0.310870
26	13	7	4	0.810870	0.310870
26	13	7	5	0.963478	0.152609
26	13	7	6	0.997391	0.033913
26	13	7	7	1.000000	0.002609
26	13	8	0	0.000824	0.000824
26	13	8	1	0.015103	0.014279
26	13	8	2	0.100778	0.085675
26	13	8	3	0.336384	0.235606
26	13	8	4	0.663615	0.327231
26	13	8	5	0.899222	0.235606
26	13	8	6	0.984897	0.085675
26	13	8	7	0.999176	0.014279
26	13	8	8	1.000000	0.000824

N	n	k	x	P(x)	p(x)
26	13	9	0	0.000229	0.000229
26	13	9	1	0.005584	0.005355
26	13	9	2	0.048421	0.042838
26	13	9	3	0.205492	0.157071
26	13	9	4	0.500000	0.294508
26	13	9	5	0.794508	0.294508
26	13	9	6	0.951579	0.157071
26	13	9	7	0.994416	0.042838
26	13	9	8	0.999771	0.005355
26	13	9	9	1.000000	0.000229
26	13	10	0	0.000054	0.000054
26	13	10	1	0.001804	0.001750
26	13	10	2	0.020703	0.018899
26	13	10	3	0.113097	0.092395
26	13	10	4	0.344084	0.230987
26	13	10	5	0.655916	0.311832
26	13	10	6	0.886903	0.230987
26	13	10	7	0.979297	0.092395
26	13	10	8	0.998196	0.018899
26	13	10	9	0.999946	0.001750
26	13	10	10	1.000000	0.000054
26	13	11	0	0.000010	0.000010
26	13	11	1	0.000491	0.000481
26	13	11	2	0.007710	0.007218
26	13	11	3	0.055351	0.047641
26	13	11	4	0.214154	0.158803
26	13	11	5	0.500000	0.285846
26	13	11	6	0.785846	0.285846
26	13	11	7	0.944649	0.158803
26	13	11	8	0.992290	0.047641
26	13	11	9	0.999509	0.007218
26	13	11	10	0.999990	0.000481
26	13	11	11	1.000000	0.000010
26	13	12	0	0.000001	0.000001
26	13	12	1	0.000106	0.000105
26	13	12	2	0.002416	0.002310
26	13	12	3	0.023590	0.021174
26	13	12	4	0.118872	0.095282
26	13	12	5	0.347549	0.228677
26	13	12	6	0.652451	0.304902
26	13	12	7	0.881128	0.228677
26	13	12	8	0.976410	0.095282
26	13	12	9	0.997584	0.021174
26	13	12	10	0.999893	0.002310
26	13	12	11	0.999998	0.000105
26	13	12	12	1.000000	0.000001
26	13	13	0	0.000000	0.000000
26	13	13	1	0.000016	0.000016
26	13	13	2	0.000601	0.000585
26	13	13	3	0.008466	0.007865

Table for $N = 2$, $n = 1$, through $N = 100$, $n = 50$

N	n	k	x	P(x)	p(x)
26	13	13	4	0.057619	0.049153
26	13	13	5	0.216876	0.159257
26	13	13	6	0.500000	0.283124
26	13	13	7	0.783124	0.283124
26	13	13	8	0.942381	0.159257
26	13	13	9	0.991534	0.049153
26	13	13	10	0.999399	0.007865
26	13	13	11	0.999983	0.000585
26	13	13	12	1.000000	0.000016
26	13	13	13	1.000000	0.000000
27	1	1	0	0.962963	0.962963
27	1	1	1	1.000000	0.037037
27	2	1	0	0.925926	0.925926
27	2	1	1	1.000000	0.074074
27	2	2	0	0.854701	0.854701
27	2	2	1	0.997151	0.142450
27	2	2	2	1.000000	0.002849
27	3	1	0	0.888889	0.888889
27	3	1	1	1.000000	0.111111
27	3	2	0	0.786325	0.786325
27	3	2	1	0.991453	0.205128
27	3	2	2	1.000000	0.008547
27	3	3	0	0.691966	0.691966
27	3	3	1	0.975043	0.283077
27	3	3	2	0.999658	0.024615
27	3	3	3	1.000000	0.000342
27	4	1	0	0.851852	0.851852
27	4	1	1	1.000000	0.148148
27	4	2	0	0.720798	0.720798
27	4	2	1	0.982906	0.262108
27	4	2	2	1.000000	0.017094
27	4	3	0	0.605470	0.605470
27	4	3	1	0.951453	0.345983
27	4	3	2	0.998632	0.047179
27	4	3	3	1.000000	0.001368
27	4	4	0	0.504558	0.504558
27	4	4	1	0.908205	0.403647
27	4	4	2	0.994701	0.086496
27	4	4	3	0.999943	0.005242
27	4	4	4	1.000000	0.000057
27	5	1	0	0.814815	0.814815
27	5	1	1	1.000000	0.185185
27	5	2	0	0.658120	0.658120
27	5	2	1	0.971510	0.313390
27	5	2	2	1.000000	0.028490
27	5	3	0	0.526496	0.526496
27	5	3	1	0.921368	0.394872
27	5	3	2	0.996581	0.075214
27	5	3	3	1.000000	0.003419
27	5	4	0	0.416809	0.416809

N	n	k	x	P(x)	p(x)
27	5	4	1	0.855556	0.438746
27	5	4	2	0.987179	0.131624
27	5	4	3	0.999715	0.012536
27	5	4	4	1.000000	0.000285
27	5	5	0	0.326198	0.326198
27	5	5	1	0.779252	0.453053
27	5	5	2	0.970011	0.190759
27	5	5	3	0.998625	0.028614
27	5	5	4	0.999988	0.001363
27	5	5	5	1.000000	0.000012
27	6	1	0	0.777778	0.777778
27	6	1	1	1.000000	0.222222
27	6	2	0	0.598291	0.598291
27	6	2	1	0.957265	0.358974
27	6	2	2	1.000000	0.042735
27	6	3	0	0.454701	0.454701
27	6	3	1	0.885470	0.430769
27	6	3	2	0.993162	0.107692
27	6	3	3	1.000000	0.006838
27	6	4	0	0.341026	0.341026
27	6	4	1	0.795726	0.454701
27	6	4	2	0.975214	0.179487
27	6	4	3	0.999145	0.023932
27	6	4	4	1.000000	0.000855
27	6	5	0	0.252062	0.252062
27	6	5	1	0.696878	0.444816
27	6	5	2	0.943998	0.247120
27	6	5	3	0.996024	0.052025
27	6	5	4	0.999926	0.003902
27	6	5	5	1.000000	0.000074
27	6	6	0	0.183318	0.183318
27	6	6	1	0.595784	0.412466
27	6	6	2	0.899068	0.303284
27	6	6	3	0.988929	0.089862
27	6	6	4	0.999571	0.010642
27	6	6	5	0.999997	0.000426
27	6	6	6	1.000000	0.000003
27	7	1	0	0.740741	0.740741
27	7	1	1	1.000000	0.259259
27	7	2	0	0.541311	0.541311
27	7	2	1	0.940171	0.398860
27	7	2	2	1.000000	0.059829
27	7	3	0	0.389744	0.389744
27	7	3	1	0.844444	0.454701
27	7	3	2	0.988034	0.143590
27	7	3	3	1.000000	0.011966
27	7	4	0	0.276068	0.276068
27	7	4	1	0.730769	0.454701
27	7	4	2	0.958120	0.227350
27	7	4	3	0.998006	0.039886

N	n	k	x	P(x)	p(x)
27	7	4	4	1.000000	0.001994
27	7	5	0	0.192048	0.192048
27	7	5	1	0.612152	0.420104
27	7	5	2	0.908696	0.296544
27	7	5	3	0.991069	0.082373
27	7	5	4	0.999740	0.008671
27	7	5	5	1.000000	0.000260
27	7	6	0	0.130942	0.130942
27	7	6	1	0.497578	0.366636
27	7	6	2	0.841299	0.343721
27	7	6	3	0.976092	0.134793
27	7	6	4	0.998557	0.022465
27	7	6	5	0.999976	0.001419
27	7	6	6	1.000000	0.000024
27	7	7	0	0.087294	0.087294
27	7	7	1	0.392825	0.305530
27	7	7	2	0.759461	0.366636
27	7	7	3	0.950417	0.190956
27	7	7	4	0.995348	0.044931
27	7	7	5	0.999841	0.004493
27	7	7	6	0.999999	0.000158
27	7	7	7	1.000000	0.000001
27	8	1	0	0.703704	0.703704
27	8	1	1	1.000000	0.296296
27	8	2	0	0.487179	0.487179
27	8	2	1	0.920228	0.433048
27	8	2	2	1.000000	0.079772
27	8	3	0	0.331282	0.331282
27	8	3	1	0.798974	0.467692
27	8	3	2	0.980855	0.181880
27	8	3	3	1.000000	0.019145
27	8	4	0	0.220855	0.220855
27	8	4	1	0.662564	0.441709
27	8	4	2	0.935385	0.272820
27	8	4	3	0.996011	0.060627
27	8	4	4	1.000000	0.003989
27	8	5	0	0.144036	0.144036
27	8	5	1	0.528131	0.384095
27	8	5	2	0.864214	0.336083
27	8	5	3	0.982832	0.118618
27	8	5	4	0.999306	0.016475
27	8	5	5	1.000000	0.000694
27	8	6	0	0.091659	0.091659
27	8	6	1	0.405919	0.314260
27	8	6	2	0.772555	0.366636
27	8	6	3	0.955873	0.183318
27	8	6	4	0.996311	0.040438
27	8	6	5	0.999905	0.003594
27	8	6	6	1.000000	0.000095
27	8	7	0	0.056741	0.056741

N	n	k	x	P(x)	p(x)
27	8	7	1	0.301165	0.244424
27	8	7	2	0.667802	0.366636
27	8	7	3	0.912226	0.244424
27	8	7	4	0.988608	0.076383
27	8	7	5	0.999392	0.010783
27	8	7	6	0.999991	0.000599
27	8	7	7	1.000000	0.000009
27	8	8	0	0.034045	0.034045
27	8	8	1	0.215617	0.181572
27	8	8	2	0.557811	0.342194
27	8	8	3	0.851120	0.293309
27	8	8	4	0.973332	0.122212
27	8	8	5	0.997774	0.024442
27	8	8	6	0.999931	0.002157
27	8	8	7	0.999999	0.000068
27	8	8	8	1.000000	0.000000
27	9	1	0	0.666667	0.666667
27	9	1	1	1.000000	0.333333
27	9	2	0	0.435897	0.435897
27	9	2	1	0.897436	0.461538
27	9	2	2	1.000000	0.102564
27	9	3	0	0.278974	0.278974
27	9	3	1	0.749744	0.470769
27	9	3	2	0.971282	0.221538
27	9	3	3	1.000000	0.028718
27	9	4	0	0.174359	0.174359
27	9	4	1	0.592820	0.418462
27	9	4	2	0.906667	0.313846
27	9	4	3	0.992820	0.086154
27	9	4	4	1.000000	0.007179
27	9	5	0	0.106132	0.106132
27	9	5	1	0.447269	0.341137
27	9	5	2	0.811148	0.363880
27	9	5	3	0.970345	0.159197
27	9	5	4	0.998439	0.028094
27	9	5	5	1.000000	0.001561
27	9	6	0	0.062714	0.062714
27	9	6	1	0.323219	0.260505
27	9	6	2	0.695368	0.372150
27	9	6	3	0.926928	0.231560
27	9	6	4	0.992054	0.065126
27	9	6	5	0.999716	0.007662
27	9	6	6	1.000000	0.000284
27	9	7	0	0.035837	0.035837
27	9	7	1	0.223979	0.188142
27	9	7	2	0.571318	0.347340
27	9	7	3	0.860768	0.289450
27	9	7	4	0.976548	0.115780
27	9	7	5	0.998257	0.021709
27	9	7	6	0.999959	0.001703

Table for $N = 2$, $n = 1$, through $N = 100$, $n = 50$

N	n	k	x	P(x)	p(x)
27	9	7	7	1.000000	0.000041
27	9	8	0	0.019710	0.019710
27	9	8	1	0.148722	0.129012
27	9	8	2	0.449750	0.301028
27	9	8	3	0.773933	0.324184
27	9	8	4	0.947603	0.173670
27	9	8	5	0.993915	0.046312
27	9	8	6	0.999704	0.005789
27	9	8	7	0.999996	0.000292
27	9	8	8	1.000000	0.000004
27	9	9	0	0.010374	0.010374
27	9	9	1	0.094401	0.084027
27	9	9	2	0.338845	0.244444
27	9	9	3	0.671559	0.332715
27	9	9	4	0.901900	0.230341
27	9	9	5	0.984165	0.082265
27	9	9	6	0.998790	0.014625
27	9	9	7	0.999965	0.001175
27	9	9	8	1.000000	0.000035
27	9	9	9	1.000000	0.000000
27	10	1	0	0.629630	0.629630
27	10	1	1	1.000000	0.370370
27	10	2	0	0.387464	0.387464
27	10	2	1	0.871795	0.484330
27	10	2	2	1.000000	0.128205
27	10	3	0	0.232479	0.232479
27	10	3	1	0.697436	0.464957
27	10	3	2	0.958974	0.261538
27	10	3	3	1.000000	0.041026
27	10	4	0	0.135613	0.135613
27	10	4	1	0.523077	0.387464
27	10	4	2	0.871795	0.348718
27	10	4	3	0.988034	0.116239
27	10	4	4	1.000000	0.011966
27	10	5	0	0.076651	0.076651
27	10	5	1	0.371460	0.294810
27	10	5	2	0.750502	0.379041
27	10	5	3	0.952657	0.202155
27	10	5	4	0.996878	0.044221
27	10	5	5	1.000000	0.003122
27	10	6	0	0.041809	0.041809
27	10	6	1	0.250856	0.209047
27	10	6	2	0.612668	0.361812
27	10	6	3	0.888335	0.275666
27	10	6	4	0.984818	0.096483
27	10	6	5	0.999291	0.014472
27	10	6	6	1.000000	0.000709
27	10	7	0	0.021900	0.021900
27	10	7	1	0.161265	0.139365
27	10	7	2	0.474835	0.313570
27	10	7	3	0.796446	0.321611
27	10	7	4	0.957251	0.160805
27	10	7	5	0.995845	0.038593
27	10	7	6	0.999865	0.004020
27	10	7	7	1.000000	0.000135
27	10	8	0	0.010950	0.010950
27	10	8	1	0.098551	0.087601
27	10	8	2	0.349407	0.250856
27	10	8	3	0.683882	0.334475
27	10	8	4	0.909010	0.225128
27	10	8	5	0.986196	0.077187
27	10	8	6	0.999061	0.012864
27	10	8	7	0.999980	0.000919
27	10	8	8	1.000000	0.000020
27	10	9	0	0.005187	0.005187
27	10	9	1	0.057056	0.051869
27	10	9	2	0.243783	0.186728
27	10	9	3	0.560655	0.316871
27	10	9	4	0.837917	0.277262
27	10	9	5	0.965884	0.127967
27	10	9	6	0.996352	0.030468
27	10	9	7	0.999834	0.003482
27	10	9	8	0.999998	0.000163
27	10	9	9	1.000000	0.000002
27	10	10	0	0.002305	0.002305
27	10	10	1	0.031121	0.028816
27	10	10	2	0.160793	0.129672
27	10	10	3	0.437427	0.276634
27	10	10	4	0.745496	0.308069
27	10	10	5	0.930338	0.184842
27	10	10	6	0.989582	0.059244
27	10	10	7	0.999254	0.009672
27	10	10	8	0.999980	0.000725
27	10	10	9	1.000000	0.000020
27	10	10	10	1.000000	0.000000
27	11	1	0	0.592593	0.592593
27	11	1	1	1.000000	0.407407
27	11	2	0	0.341880	0.341880
27	11	2	1	0.843305	0.501424
27	11	2	2	1.000000	0.156695
27	11	3	0	0.191453	0.191453
27	11	3	1	0.642735	0.451282
27	11	3	2	0.943590	0.300855
27	11	3	3	1.000000	0.056410
27	11	4	0	0.103704	0.103704
27	11	4	1	0.454701	0.350997
27	11	4	2	0.830769	0.376068
27	11	4	3	0.981197	0.150427
27	11	4	4	1.000000	0.018803
27	11	5	0	0.054106	0.054106
27	11	5	1	0.302093	0.247987
27	11	5	2	0.683612	0.381519
27	11	5	3	0.928874	0.245262
27	11	5	4	0.994277	0.065403
27	11	5	5	1.000000	0.005723
27	11	6	0	0.027053	0.027053
27	11	6	1	0.189372	0.162319
27	11	6	2	0.527536	0.338164
27	11	6	3	0.839688	0.312152
27	11	6	4	0.973467	0.133779
27	11	6	5	0.998439	0.024972
27	11	6	6	1.000000	0.001561
27	11	7	0	0.012882	0.012882
27	11	7	1	0.112077	0.099195
27	11	7	2	0.382609	0.270531
27	11	7	3	0.720773	0.338164
27	11	7	4	0.928874	0.208101
27	11	7	5	0.991304	0.062430
27	11	7	6	0.999628	0.008324
27	11	7	7	1.000000	0.000372
27	11	8	0	0.005797	0.005797
27	11	8	1	0.062480	0.056683
27	11	8	2	0.260870	0.198390
27	11	8	3	0.585507	0.324638
27	11	8	4	0.856039	0.270531
27	11	8	5	0.972575	0.116537
27	11	8	6	0.997547	0.024972
27	11	8	7	0.999926	0.002378
27	11	8	8	1.000000	0.000074
27	11	9	0	0.002441	0.002441
27	11	9	1	0.032647	0.030206
27	11	9	2	0.166895	0.134249
27	11	9	3	0.448818	0.281922
27	11	9	4	0.756369	0.307551
27	11	9	5	0.935774	0.179405
27	11	9	6	0.990976	0.055202
27	11	9	7	0.999425	0.008449
27	11	9	8	0.999988	0.000563
27	11	9	9	1.000000	0.000012
27	11	10	0	0.000949	0.000949
27	11	10	1	0.015866	0.014917
27	11	10	2	0.099771	0.083905
27	11	10	3	0.323519	0.223748
27	11	10	4	0.636766	0.313247
27	11	10	5	0.875973	0.239207
27	11	10	6	0.975642	0.099669
27	11	10	7	0.997547	0.021905
27	11	10	8	0.999894	0.002347
27	11	10	9	0.999999	0.000104
27	11	10	10	1.000000	0.000001
27	11	11	0	0.000335	0.000335
27	11	11	1	0.007091	0.006756
27	11	11	2	0.055351	0.048259
27	11	11	3	0.218226	0.162875
27	11	11	4	0.507782	0.289556
27	11	11	5	0.791547	0.283765
27	11	11	6	0.946327	0.154781
27	11	11	7	0.992393	0.046066
27	11	11	8	0.999480	0.007087
27	11	11	9	0.999986	0.000506
27	11	11	10	1.000000	0.000013
27	11	11	11	1.000000	0.000000
27	12	1	0	0.555556	0.555556
27	12	1	1	1.000000	0.444444
27	12	2	0	0.299145	0.299145
27	12	2	1	0.811966	0.512821
27	12	2	2	1.000000	0.188034
27	12	3	0	0.155556	0.155556
27	12	3	1	0.586325	0.430769
27	12	3	2	0.924786	0.338462
27	12	3	3	1.000000	0.075214
27	12	4	0	0.077778	0.077778
27	12	4	1	0.388889	0.311111
27	12	4	2	0.783761	0.394872
27	12	4	3	0.971795	0.188034
27	12	4	4	1.000000	0.028205
27	12	5	0	0.037198	0.037198
27	12	5	1	0.240097	0.202899
27	12	5	2	0.612077	0.371981
27	12	5	3	0.898216	0.286139
27	12	5	4	0.990189	0.091973
27	12	5	5	1.000000	0.009810
27	12	6	0	0.016908	0.016908
27	12	6	1	0.138647	0.121739
27	12	6	2	0.442995	0.304348
27	12	6	3	0.781159	0.338164
27	12	6	4	0.956745	0.175585
27	12	6	5	0.996878	0.040134
27	12	6	6	1.000000	0.003122
27	12	7	0	0.007246	0.007246
27	12	7	1	0.074879	0.067633
27	12	7	2	0.298068	0.223188
27	12	7	3	0.636232	0.338164
27	12	7	4	0.889855	0.253623
27	12	7	5	0.983500	0.093645
27	12	7	6	0.999108	0.015608
27	12	7	7	1.000000	0.000892
27	12	8	0	0.002899	0.002899
27	12	8	1	0.037681	0.034783
27	12	8	2	0.186473	0.148792

Table for $N = 2$, $n = 1$, through $N = 100$, $n = 50$

N	n	k	x	P(x)	p(x)
27	12	8	3	0.484058	0.297584
27	12	8	4	0.788406	0.304348
27	12	8	5	0.950725	0.162319
27	12	8	6	0.994426	0.043701
27	12	8	7	0.999777	0.005351
27	12	8	8	1.000000	0.000223
27	12	9	0	0.001068	0.001068
27	12	9	1	0.017544	0.016476
27	12	9	2	0.108162	0.090618
27	12	9	3	0.343097	0.234935
27	12	9	4	0.660259	0.317162
27	12	9	5	0.890923	0.230664
27	12	9	6	0.980625	0.089703
27	12	9	7	0.998369	0.017743
27	12	9	8	0.999953	0.001584
27	12	9	9	1.000000	0.000047
27	12	10	0	0.000356	0.000356
27	12	10	1	0.007475	0.007119
27	12	10	2	0.057818	0.050343
27	12	10	3	0.225629	0.167811
27	12	10	4	0.519298	0.293669
27	12	10	5	0.801220	0.281922
27	12	10	6	0.950725	0.149504
27	12	10	7	0.993440	0.042715
27	12	10	8	0.999601	0.006161
27	12	10	9	0.999992	0.000391
27	12	10	10	1.000000	0.000008
27	12	11	0	0.000105	0.000105
27	12	11	1	0.002869	0.002764
27	12	11	2	0.028205	0.025336
27	12	11	3	0.136788	0.108583
27	12	11	4	0.381101	0.244313
27	12	11	5	0.685135	0.304034
27	12	11	6	0.897958	0.212824
27	12	11	7	0.980877	0.082918
27	12	11	8	0.998151	0.017275
27	12	11	9	0.999923	0.001772
27	12	11	10	0.999999	0.000076
27	12	11	11	1.000000	0.000001
27	12	12	0	0.000026	0.000026
27	12	12	1	0.000968	0.000942
27	12	12	2	0.012370	0.011401
27	12	12	3	0.075710	0.063340
27	12	12	4	0.258945	0.183235
27	12	12	5	0.552120	0.293175
27	12	12	6	0.818150	0.266029
27	12	12	7	0.954965	0.136815
27	12	12	8	0.993833	0.038868
27	12	12	9	0.999591	0.005758
27	12	12	10	0.999990	0.000399

N	n	k	x	P(x)	p(x)
27	12	12	11	1.000000	0.000010
27	12	12	12	1.000000	0.000000
27	13	1	0	0.518519	0.518519
27	13	1	1	1.000000	0.481481
27	13	2	0	0.259259	0.259259
27	13	2	1	0.777778	0.518519
27	13	2	2	1.000000	0.222222
27	13	3	0	0.124444	0.124444
27	13	3	1	0.528889	0.404444
27	13	3	2	0.902222	0.373333
27	13	3	3	1.000000	0.097778
27	13	4	0	0.057037	0.057037
27	13	4	1	0.326667	0.269630
27	13	4	2	0.731111	0.404444
27	13	4	3	0.959259	0.228148
27	13	4	4	1.000000	0.040741
27	13	5	0	0.024799	0.024799
27	13	5	1	0.185990	0.161192
27	13	5	2	0.537681	0.351691
27	13	5	3	0.860064	0.322383
27	13	5	4	0.984058	0.123994
27	13	5	5	1.000000	0.015942
27	13	6	0	0.010145	0.010145
27	13	6	1	0.098068	0.087923
27	13	6	2	0.361836	0.263768
27	13	6	3	0.713526	0.351691
27	13	6	4	0.933333	0.219807
27	13	6	5	0.994203	0.060870
27	13	6	6	1.000000	0.005797
27	13	7	0	0.003865	0.003865
27	13	7	1	0.047826	0.043961
27	13	7	2	0.223671	0.175845
27	13	7	3	0.546055	0.322383
27	13	7	4	0.839130	0.293076
27	13	7	5	0.971014	0.131884
27	13	7	6	0.998068	0.027053
27	13	7	7	1.000000	0.001932
27	13	8	0	0.001353	0.001353
27	13	8	1	0.021449	0.020097
27	13	8	2	0.126957	0.105507
27	13	8	3	0.384863	0.257907
27	13	8	4	0.707246	0.322383
27	13	8	5	0.918261	0.211014
27	13	8	6	0.988599	0.070338
27	13	8	7	0.999420	0.010821
27	13	8	8	1.000000	0.000580
27	13	9	0	0.000427	0.000427
27	13	9	1	0.008757	0.008330
27	13	9	2	0.065873	0.057117
27	13	9	3	0.249123	0.183249

N	n	k	x	P(x)	p(x)
27	13	9	4	0.554539	0.305416
27	13	9	5	0.829413	0.274874
27	13	9	6	0.962685	0.133272
27	13	9	7	0.996003	0.033318
27	13	9	8	0.999847	0.003844
27	13	9	9	1.000000	0.000153
27	13	10	0	0.000119	0.000119
27	13	10	1	0.003204	0.003085
27	13	10	2	0.030969	0.027765
27	13	10	3	0.147318	0.116349
27	13	10	4	0.401831	0.254513
27	13	10	5	0.707246	0.305416
27	13	10	6	0.910857	0.203610
27	13	10	7	0.984897	0.074040
27	13	10	8	0.998779	0.013883
27	13	10	9	0.999966	0.001187
27	13	10	10	1.000000	0.000034
27	13	11	0	0.000028	0.000028
27	13	11	1	0.001026	0.000998
27	13	11	2	0.013003	0.011977
27	13	11	3	0.078877	0.065874
27	13	11	4	0.267088	0.188211
27	13	11	5	0.563521	0.296433
27	13	11	6	0.827017	0.263496
27	13	11	7	0.958765	0.131748
27	13	11	8	0.994696	0.035931
27	13	11	9	0.999687	0.004990
27	13	11	10	0.999994	0.000307
27	13	11	11	1.000000	0.000006
27	13	12	0	0.000005	0.000005
27	13	12	1	0.000277	0.000272
27	13	12	2	0.004769	0.004491
27	13	12	3	0.037706	0.032937
27	13	12	4	0.161220	0.123514
27	13	12	5	0.415305	0.254085
27	13	12	6	0.711738	0.296433
27	13	12	7	0.909360	0.197622
27	13	12	8	0.983468	0.074108
27	13	12	9	0.998439	0.014971
27	13	12	10	0.999936	0.001497
27	13	12	11	0.999999	0.000063
27	13	12	12	1.000000	0.000001
27	13	13	0	0.000001	0.000001
27	13	13	1	0.000060	0.000059
27	13	13	2	0.001475	0.001415
27	13	13	3	0.015748	0.014273
27	13	13	4	0.087111	0.071363
27	13	13	5	0.279793	0.192681
27	13	13	6	0.573402	0.293610
27	13	13	7	0.830311	0.256908

N	n	k	x	P(x)	p(x)
27	13	13	8	0.958765	0.128454
27	13	13	9	0.994447	0.035682
27	13	13	10	0.999637	0.005190
27	13	13	11	0.999991	0.000354
27	13	13	12	1.000000	0.000009
27	13	13	13	1.000000	0.000000
28	1	1	0	0.964286	0.964286
28	1	1	1	1.000000	0.035714
28	2	1	0	0.928571	0.928571
28	2	1	1	1.000000	0.071429
28	2	2	0	0.859788	0.859788
28	2	2	1	0.997354	0.137566
28	2	2	2	1.000000	0.002645
28	3	1	0	0.892857	0.892857
28	3	1	1	1.000000	0.107143
28	3	2	0	0.793651	0.793651
28	3	2	1	0.992063	0.198413
28	3	2	2	1.000000	0.007936
28	3	3	0	0.702076	0.702076
28	3	3	1	0.976801	0.274725
28	3	3	2	0.999695	0.022894
28	3	3	3	1.000000	0.000305
28	4	1	0	0.857143	0.857143
28	4	1	1	1.000000	0.142857
28	4	2	0	0.730159	0.730159
28	4	2	1	0.984127	0.253968
28	4	2	2	1.000000	0.015873
28	4	3	0	0.617827	0.617827
28	4	3	1	0.954823	0.336996
28	4	3	2	0.998779	0.043956
28	4	3	3	1.000000	0.001221
28	4	4	0	0.518974	0.518974
28	4	4	1	0.914383	0.395409
28	4	4	2	0.995262	0.080879
28	4	4	3	0.999951	0.004689
28	4	4	4	1.000000	0.000049
28	5	1	0	0.821429	0.821429
28	5	1	1	1.000000	0.178571
28	5	2	0	0.669312	0.669312
28	5	2	1	0.973545	0.304233
28	5	2	2	1.000000	0.026455
28	5	3	0	0.540598	0.540598
28	5	3	1	0.926740	0.386142
28	5	3	2	0.996947	0.070208
28	5	3	3	1.000000	0.003053
28	5	4	0	0.432479	0.432479
28	5	4	1	0.864957	0.432479
28	5	4	2	0.988523	0.123565
28	5	4	3	0.999756	0.011233
28	5	4	4	1.000000	0.000244

Table for $N = 2$, $n = 1$, through $N = 100$, $n = 50$

N	n	k	x	P(x)	p(x)
28	5	5	0	0.342379	0.342379
28	5	5	1	0.792877	0.450499
28	5	5	2	0.973077	0.180199
28	5	5	3	0.998820	0.025743
28	5	5	4	0.999990	0.001170
28	5	5	5	1.000000	0.000010
28	6	1	0	0.785714	0.785714
28	6	1	1	1.000000	0.214286
28	6	2	0	0.611111	0.611111
28	6	2	1	0.960317	0.349206
28	6	2	2	1.000000	0.039683
28	6	3	0	0.470085	0.470085
28	6	3	1	0.893162	0.423077
28	6	3	2	0.993895	0.100733
28	6	3	3	1.000000	0.006105
28	6	4	0	0.357265	0.357265
28	6	4	1	0.808547	0.451282
28	6	4	2	0.977778	0.169231
28	6	4	3	0.999267	0.021490
28	6	4	4	1.000000	0.000733
28	6	5	0	0.267949	0.267949
28	6	5	1	0.714530	0.446581
28	6	5	2	0.949573	0.235043
28	6	5	3	0.996581	0.047009
28	6	5	4	0.999939	0.003358
28	6	5	5	1.000000	0.000061
28	6	6	0	0.198049	0.198049
28	6	6	1	0.617447	0.419398
28	6	6	2	0.908696	0.291249
28	6	6	3	0.990450	0.081754
28	6	6	4	0.999647	0.009197
28	6	6	5	0.999997	0.000350
28	6	6	6	1.000000	0.000003
28	7	1	0	0.750000	0.750000
28	7	1	1	1.000000	0.250000
28	7	2	0	0.555556	0.555556
28	7	2	1	0.944444	0.388889
28	7	2	2	1.000000	0.055556
28	7	3	0	0.405983	0.405983
28	7	3	1	0.854701	0.448718
28	7	3	2	0.989316	0.134615
28	7	3	3	1.000000	0.010684
28	7	4	0	0.292308	0.292308
28	7	4	1	0.747009	0.454701
28	7	4	2	0.962393	0.215385
28	7	4	3	0.998291	0.035897
28	7	4	4	1.000000	0.001709
28	7	5	0	0.207051	0.207051
28	7	5	1	0.633333	0.426282
28	7	5	2	0.917521	0.284188
28	7	5	3	0.992308	0.074786
28	7	5	4	0.999786	0.007479
28	7	5	5	1.000000	0.000214
28	7	6	0	0.144036	0.144036
28	7	6	1	0.522129	0.378094
28	7	6	2	0.855741	0.333612
28	7	6	3	0.979301	0.123560
28	7	6	4	0.998811	0.019509
28	7	6	5	0.999981	0.001171
28	7	6	6	1.000000	0.000019
28	7	7	0	0.098206	0.098206
28	7	7	1	0.419013	0.320807
28	7	7	2	0.779920	0.360908
28	7	7	3	0.956836	0.176915
28	7	7	4	0.996150	0.039315
28	7	7	5	0.999875	0.003725
28	7	7	6	0.999999	0.000124
28	7	7	7	1.000000	0.000001
28	8	1	0	0.714286	0.714286
28	8	1	1	1.000000	0.285714
28	8	2	0	0.502645	0.502645
28	8	2	1	0.925926	0.423280
28	8	2	2	1.000000	0.074074
28	8	3	0	0.347985	0.347985
28	8	3	1	0.811966	0.463980
28	8	3	2	0.982906	0.170940
28	8	3	3	1.000000	0.017094
28	8	4	0	0.236630	0.236630
28	8	4	1	0.682051	0.445421
28	8	4	2	0.941880	0.259829
28	8	4	3	0.996581	0.054701
28	8	4	4	1.000000	0.003419
28	8	5	0	0.157753	0.157753
28	8	5	1	0.552137	0.394383
28	8	5	2	0.876923	0.324786
28	8	5	3	0.985185	0.108262
28	8	5	4	0.999430	0.014245
28	8	5	5	1.000000	0.000570
28	8	6	0	0.102883	0.102883
28	8	6	1	0.432107	0.329224
28	8	6	2	0.792196	0.360089
28	8	6	3	0.961650	0.169454
28	8	6	4	0.996953	0.035303
28	8	6	5	0.999926	0.002973
28	8	6	6	1.000000	0.000074
28	8	7	0	0.065471	0.065471
28	8	7	1	0.327354	0.261883
28	8	7	2	0.693990	0.366636
28	8	7	3	0.923138	0.229148
28	8	7	4	0.990534	0.067396
28	8	7	5	0.999520	0.008986
28	8	7	6	0.999993	0.000473
28	8	7	7	1.000000	0.000007
28	8	8	0	0.040530	0.040530
28	8	8	1	0.240059	0.199530
28	8	8	2	0.589237	0.349177
28	8	8	3	0.868579	0.279342
28	8	8	4	0.977697	0.109118
28	8	8	5	0.998236	0.020540
28	8	8	6	0.999948	0.001712
28	8	8	7	1.000000	0.000051
28	8	8	8	1.000000	0.000000
28	9	1	0	0.678571	0.678571
28	9	1	1	1.000000	0.321429
28	9	2	0	0.452381	0.452381
28	9	2	1	0.904762	0.452381
28	9	2	2	1.000000	0.095238
28	9	3	0	0.295788	0.295788
28	9	3	1	0.765568	0.469780
28	9	3	2	0.974359	0.208791
28	9	3	3	1.000000	0.025641
28	9	4	0	0.189304	0.189304
28	9	4	1	0.615238	0.425934
28	9	4	2	0.915897	0.300659
28	9	4	3	0.993846	0.077949
28	9	4	4	1.000000	0.006154
28	9	5	0	0.118315	0.118315
28	9	5	1	0.473260	0.354945
28	9	5	2	0.828205	0.354945
28	9	5	3	0.974359	0.146154
28	9	5	4	0.998718	0.024359
28	9	5	5	1.000000	0.001282
28	9	6	0	0.072018	0.072018
28	9	6	1	0.349801	0.277783
28	9	6	2	0.720178	0.370377
28	9	6	3	0.936232	0.216053
28	9	6	4	0.993422	0.057191
28	9	6	5	0.999777	0.006355
28	9	6	6	1.000000	0.000223
28	9	7	0	0.042556	0.042556
28	9	7	1	0.248789	0.206233
28	9	7	2	0.602331	0.353542
28	9	7	3	0.877308	0.274977
28	9	7	4	0.980425	0.103116
28	9	7	5	0.998622	0.018197
28	9	7	6	0.999970	0.001348
28	9	7	7	1.000000	0.000030
28	9	8	0	0.024318	0.024318
28	9	8	1	0.170224	0.145906
28	9	8	2	0.484484	0.314260
28	9	8	3	0.798743	0.314260
28	9	8	4	0.955873	0.157130
28	9	8	5	0.995155	0.039282
28	9	8	6	0.999777	0.004621
28	9	8	7	0.999997	0.000220
28	9	8	8	1.000000	0.000003
28	9	9	0	0.013375	0.013375
28	9	9	1	0.111861	0.098487
28	9	9	2	0.374493	0.262631
28	9	9	3	0.704465	0.329973
28	9	9	4	0.916591	0.212125
28	9	9	5	0.987299	0.070708
28	9	9	6	0.999084	0.011785
28	9	9	7	0.999975	0.000891
28	9	9	8	1.000000	0.000025
28	9	9	9	1.000000	0.000000
28	10	1	0	0.642857	0.642857
28	10	1	1	1.000000	0.357143
28	10	2	0	0.404762	0.404762
28	10	2	1	0.880952	0.476190
28	10	2	2	1.000000	0.119048
28	10	3	0	0.249084	0.249084
28	10	3	1	0.716117	0.467033
28	10	3	2	0.963370	0.247253
28	10	3	3	1.000000	0.036630
28	10	4	0	0.149451	0.149451
28	10	4	1	0.547985	0.398535
28	10	4	2	0.884249	0.336264
28	10	4	3	0.989744	0.105494
28	10	4	4	1.000000	0.010256
28	10	5	0	0.087179	0.087179
28	10	5	1	0.398535	0.311355
28	10	5	2	0.772161	0.373626
28	10	5	3	0.958974	0.186813
28	10	5	4	0.997436	0.038462
28	10	5	5	1.000000	0.002564
28	10	6	0	0.049275	0.049275
28	10	6	1	0.276700	0.227425
28	10	6	2	0.642204	0.365504
28	10	6	3	0.902118	0.259914
28	10	6	4	0.987402	0.085284
28	10	6	5	0.999443	0.012040
28	10	6	6	1.000000	0.000557
28	10	7	0	0.026877	0.026877
28	10	7	1	0.183663	0.156785
28	10	7	2	0.509294	0.325631
28	10	7	3	0.819418	0.310125
28	10	7	4	0.964143	0.144725
28	10	7	5	0.996706	0.032563
28	10	7	6	0.999899	0.003192

Table for $N = 2$, $n = 1$, through $N = 100$, $n = 50$

N	n	k	x	P(x)	p(x)
28	10	7	7	1.000000	0.000101
28	10	8	0	0.014079	0.014079
28	10	8	1	0.116469	0.102390
28	10	8	2	0.385244	0.268775
28	10	8	3	0.716043	0.330800
28	10	8	4	0.922793	0.206750
28	10	8	5	0.988953	0.066160
28	10	8	6	0.999290	0.010337
28	10	8	7	0.999985	0.000695
28	10	8	8	1.000000	0.000014
28	10	9	0	0.007039	0.007039
28	10	9	1	0.070393	0.063354
28	10	9	2	0.277734	0.207340
28	10	9	3	0.600263	0.322530
28	10	9	4	0.860768	0.260505
28	10	9	5	0.972413	0.111645
28	10	9	6	0.997223	0.024810
28	10	9	7	0.999881	0.002658
28	10	9	8	0.999998	0.000117
28	10	9	9	1.000000	0.000001
28	10	10	0	0.003334	0.003334
28	10	10	1	0.040384	0.037049
28	10	10	2	0.190433	0.150049
28	10	10	3	0.481437	0.291004
28	10	10	4	0.778503	0.297067
28	10	10	5	0.943033	0.164529
28	10	10	6	0.992000	0.048967
28	10	10	7	0.999461	0.007462
28	10	10	8	0.999986	0.000525
28	10	10	9	1.000000	0.000014
28	10	10	10	1.000000	0.000000
28	11	1	0	0.607143	0.607143
28	11	1	1	1.000000	0.392857
28	11	2	0	0.359788	0.359788
28	11	2	1	0.854497	0.494709
28	11	2	2	1.000000	0.145503
28	11	3	0	0.207570	0.207570
28	11	3	1	0.664225	0.456654
28	11	3	2	0.949634	0.285409
28	11	3	3	1.000000	0.050366
28	11	4	0	0.116239	0.116239
28	11	4	1	0.481563	0.365324
28	11	4	2	0.846886	0.365324
28	11	4	3	0.983883	0.136996
28	11	4	4	1.000000	0.016117
28	11	5	0	0.062963	0.062963
28	11	5	1	0.329345	0.266382
28	11	5	2	0.709890	0.380545
28	11	5	3	0.938217	0.228327
28	11	5	4	0.995299	0.057082

N	n	k	x	P(x)	p(x)
28	11	5	5	1.000000	0.004701
28	11	6	0	0.032850	0.032850
28	11	6	1	0.213527	0.180676
28	11	6	2	0.560981	0.347454
28	11	6	3	0.858799	0.297818
28	11	6	4	0.977926	0.119127
28	11	6	5	0.998774	0.020847
28	11	6	6	1.000000	0.001226
28	11	7	0	0.016425	0.016425
28	11	7	1	0.131401	0.114976
28	11	7	2	0.418841	0.287440
28	11	7	3	0.750502	0.331661
28	11	7	4	0.940022	0.189521
28	11	7	5	0.993088	0.053066
28	11	7	6	0.999721	0.006633
28	11	7	7	1.000000	0.000279
28	11	8	0	0.007821	0.007821
28	11	8	1	0.076651	0.068829
28	11	8	2	0.295652	0.219002
28	11	8	3	0.624155	0.328502
28	11	8	4	0.876849	0.252694
28	11	8	5	0.977926	0.101078
28	11	8	6	0.998142	0.020216
28	11	8	7	0.999947	0.001805
28	11	8	8	1.000000	0.000053
28	11	9	0	0.003520	0.003520
28	11	9	1	0.042236	0.038716
28	11	9	2	0.197101	0.154865
28	11	9	3	0.492754	0.295652
28	11	9	4	0.788406	0.295652
28	11	9	5	0.947603	0.159197
28	11	9	6	0.993088	0.045485
28	11	9	7	0.999586	0.006498
28	11	9	8	0.999992	0.000406
28	11	9	9	1.000000	0.000008
28	11	10	0	0.001482	0.001482
28	11	10	1	0.021859	0.020377
28	11	10	2	0.123744	0.101885
28	11	10	3	0.368268	0.244524
28	11	10	4	0.679481	0.311213
28	11	10	5	0.897330	0.217849
28	11	10	6	0.981118	0.083788
28	11	10	7	0.998218	0.017100
28	11	10	8	0.999928	0.001710
28	11	10	9	0.999999	0.000071
28	11	10	10	1.000000	0.000001
28	11	11	0	0.000576	0.000576
28	11	11	1	0.010538	0.009962
28	11	11	2	0.072802	0.062263
28	11	11	3	0.259591	0.186789

N	n	k	x	P(x)	p(x)
28	11	11	4	0.558454	0.298863
28	11	11	5	0.824714	0.266260
28	11	11	6	0.957844	0.133130
28	11	11	7	0.994418	0.036574
28	11	11	8	0.999643	0.005225
28	11	11	9	0.999991	0.000348
28	11	11	10	1.000000	0.000009
28	11	11	11	1.000000	0.000000
28	12	1	0	0.571429	0.571429
28	12	1	1	1.000000	0.428571
28	12	2	0	0.317460	0.317460
28	12	2	1	0.825397	0.507936
28	12	2	2	1.000000	0.174603
28	12	3	0	0.170940	0.170940
28	12	3	1	0.610501	0.439560
28	12	3	2	0.932845	0.322344
28	12	3	3	1.000000	0.067155
28	12	4	0	0.088889	0.088889
28	12	4	1	0.417094	0.328205
28	12	4	2	0.803907	0.386813
28	12	4	3	0.975824	0.171917
28	12	4	4	1.000000	0.024176
28	12	5	0	0.044444	0.044444
28	12	5	1	0.266667	0.222222
28	12	5	2	0.642735	0.376068
28	12	5	3	0.911355	0.268620
28	12	5	4	0.991941	0.080586
28	12	5	5	1.000000	0.008059
28	12	6	0	0.021256	0.021256
28	12	6	1	0.160386	0.139130
28	12	6	2	0.479227	0.318841
28	12	6	3	0.806243	0.327016
28	12	6	4	0.963911	0.157668
28	12	6	5	0.997547	0.033636
28	12	6	6	1.000000	0.002453
28	12	7	0	0.009662	0.009662
28	12	7	1	0.090821	0.081159
28	12	7	2	0.334299	0.243478
28	12	7	3	0.672464	0.338164
28	12	7	4	0.906577	0.234114
28	12	7	5	0.986845	0.080268
28	12	7	6	0.999331	0.012486
28	12	7	7	1.000000	0.000669
28	12	8	0	0.004141	0.004141
28	12	8	1	0.048309	0.044168
28	12	8	2	0.218357	0.170048
28	12	8	3	0.527536	0.309179
28	12	8	4	0.817391	0.289855
28	12	8	5	0.960089	0.142698
28	12	8	6	0.995764	0.035674

N	n	k	x	P(x)	p(x)
28	12	8	7	0.999841	0.004077
28	12	8	8	1.000000	0.000159
28	12	9	0	0.001656	0.001656
28	12	9	1	0.024017	0.022360
28	12	9	2	0.133333	0.109317
28	12	9	3	0.388406	0.255072
28	12	9	4	0.701449	0.313043
28	12	9	5	0.910145	0.208696
28	12	9	6	0.985061	0.074916
28	12	9	7	0.998821	0.013760
28	12	9	8	0.999968	0.001147
28	12	9	9	1.000000	0.000032
28	12	10	0	0.000610	0.000610
28	12	10	1	0.011071	0.010461
28	12	10	2	0.075798	0.064727
28	12	10	3	0.267582	0.191784
28	12	10	4	0.569641	0.302059
28	12	10	5	0.833257	0.263616
28	12	10	6	0.961403	0.128146
28	12	10	7	0.995200	0.033797
28	12	10	8	0.999727	0.004526
28	12	10	9	0.999995	0.000268
28	12	10	10	1.000000	0.000005
28	12	11	0	0.000203	0.000203
28	12	11	1	0.004678	0.004475
28	12	11	2	0.039839	0.035160
28	12	11	3	0.171690	0.131851
28	12	11	4	0.435393	0.263703
28	12	11	5	0.730740	0.295347
28	12	11	6	0.918688	0.187948
28	12	11	7	0.985812	0.067124
28	12	11	8	0.998721	0.012909
28	12	11	9	0.999950	0.001229
28	12	11	10	0.999999	0.000049
28	12	11	11	1.000000	0.000001
28	12	12	0	0.000060	0.000060
28	12	12	1	0.001783	0.001723
28	12	12	2	0.019156	0.017373
28	12	12	3	0.101886	0.082730
28	12	12	4	0.311297	0.209411
28	12	12	5	0.609126	0.297829
28	12	12	6	0.852353	0.243227
28	12	12	7	0.966070	0.113716
28	12	12	8	0.995683	0.029614
28	12	12	9	0.999733	0.004050
28	12	12	10	0.999994	0.000260
28	12	12	11	1.000000	0.000006
28	12	12	12	1.000000	0.000000
28	13	1	0	0.535714	0.535714
28	13	1	1	1.000000	0.464286

Table for $N = 2$, $n = 1$, through $N = 100$, $n = 50$

N	n	k	x	P(x)	p(x)
28	13	2	0	0.277778	0.277778
28	13	2	1	0.793651	0.515873
28	13	2	2	1.000000	0.206349
28	13	3	0	0.138889	0.138889
28	13	3	1	0.555556	0.416667
28	13	3	2	0.912698	0.357143
28	13	3	3	1.000000	0.087302
28	13	4	0	0.066667	0.066667
28	13	4	1	0.355556	0.288889
28	13	4	2	0.755555	0.400000
28	13	4	3	0.965079	0.209524
28	13	4	4	1.000000	0.034921
28	13	5	0	0.030556	0.030556
28	13	5	1	0.211111	0.180556
28	13	5	2	0.572222	0.361111
28	13	5	3	0.877778	0.305555
28	13	5	4	0.986905	0.109127
28	13	5	5	1.000000	0.013095
28	13	6	0	0.013285	0.013285
28	13	6	1	0.116908	0.103623
28	13	6	2	0.399517	0.282609
28	13	6	3	0.744927	0.345411
28	13	6	4	0.944203	0.199275
28	13	6	5	0.995445	0.051242
28	13	6	6	1.000000	0.004555
28	13	7	0	0.005435	0.005435
28	13	7	1	0.060386	0.054952
28	13	7	2	0.258213	0.197826
28	13	7	3	0.587923	0.329710
28	13	7	4	0.862681	0.274758
28	13	7	5	0.976811	0.114130
28	13	7	6	0.998551	0.021739
28	13	7	7	1.000000	0.001449
28	13	8	0	0.002070	0.002070
28	13	8	1	0.028985	0.026915
28	13	8	2	0.154589	0.125604
28	13	8	3	0.430918	0.276328
28	13	8	4	0.744927	0.314010
28	13	8	5	0.933333	0.188406
28	13	8	6	0.991304	0.057971
28	13	8	7	0.999586	0.008282
28	13	8	8	1.000000	0.000414
28	13	9	0	0.000725	0.000725
28	13	9	1	0.012836	0.012112
28	13	9	2	0.085507	0.072671
28	13	9	3	0.292754	0.207246
28	13	9	4	0.603623	0.310870
28	13	9	5	0.857971	0.254348
28	13	9	6	0.971014	0.113043
28	13	9	7	0.997101	0.026087

N	n	k	x	P(x)	p(x)
28	13	9	8	0.999896	0.002795
28	13	9	9	1.000000	0.000104
28	13	10	0	0.000229	0.000229
28	13	10	1	0.005187	0.004958
28	13	10	2	0.043435	0.038248
28	13	10	3	0.183677	0.140242
28	13	10	4	0.456369	0.272693
28	13	10	5	0.750877	0.294508
28	13	10	6	0.929367	0.178490
28	13	10	7	0.988863	0.059497
28	13	10	8	0.999161	0.010297
28	13	10	9	0.999978	0.000817
28	13	10	10	1.000000	0.000022
28	13	11	0	0.000064	0.000064
28	13	11	1	0.001882	0.001818
28	13	11	2	0.020061	0.018179
28	13	11	3	0.105764	0.085703
28	13	11	4	0.320023	0.214258
28	13	11	5	0.619985	0.299962
28	13	11	6	0.859954	0.239969
28	13	11	7	0.969031	0.109077
28	13	11	8	0.996300	0.027269
28	13	11	9	0.999796	0.003496
28	13	11	10	0.999996	0.000200
28	13	11	11	1.000000	0.000004
28	13	12	0	0.000015	0.000015
28	13	12	1	0.000598	0.000583
28	13	12	2	0.008298	0.007700
28	13	12	3	0.055351	0.047053
28	13	12	4	0.206592	0.151241
28	13	12	5	0.478826	0.272234
28	13	12	6	0.761143	0.282317
28	13	12	7	0.930533	0.169390
28	13	12	8	0.988280	0.057747
28	13	12	9	0.998974	0.010694
28	13	12	10	0.999961	0.000987
28	13	12	11	0.999999	0.000038
28	13	12	12	1.000000	0.000000
28	13	13	0	0.000003	0.000003
28	13	13	1	0.000161	0.000158
28	13	13	2	0.003004	0.002844
28	13	13	3	0.025943	0.022938
28	13	13	4	0.121519	0.095576
28	13	13	5	0.342709	0.221190
28	13	13	6	0.637629	0.294920
28	13	13	7	0.867012	0.229383
28	13	13	8	0.970234	0.103222
28	13	13	9	0.996300	0.026066
28	13	13	10	0.999776	0.003475
28	13	13	11	0.999995	0.000219

N	n	k	x	P(x)	p(x)
28	13	13	12	1.000000	0.000005
28	13	13	13	1.000000	0.000000
28	14	1	0	0.500000	0.500000
28	14	1	1	1.000000	0.500000
28	14	2	0	0.240741	0.240741
28	14	2	1	0.759259	0.518519
28	14	2	2	1.000000	0.240741
28	14	3	0	0.111111	0.111111
28	14	3	1	0.500000	0.388889
28	14	3	2	0.888889	0.388889
28	14	3	3	1.000000	0.111111
28	14	4	0	0.048889	0.048889
28	14	4	1	0.297778	0.248889
28	14	4	2	0.702222	0.404444
28	14	4	3	0.951111	0.248889
28	14	4	4	1.000000	0.048889
28	14	5	0	0.020370	0.020370
28	14	5	1	0.162963	0.142593
28	14	5	2	0.500000	0.337037
28	14	5	3	0.837037	0.337037
28	14	5	4	0.979630	0.142593
28	14	5	5	1.000000	0.020370
28	14	6	0	0.007971	0.007971
28	14	6	1	0.082367	0.074396
28	14	6	2	0.324155	0.241787
28	14	6	3	0.675845	0.351691
28	14	6	4	0.917633	0.241787
28	14	6	5	0.992029	0.074396
28	14	6	6	1.000000	0.007971
28	14	7	0	0.002899	0.002899
28	14	7	1	0.038406	0.035507
28	14	7	2	0.192271	0.153865
28	14	7	3	0.500000	0.307729
28	14	7	4	0.807729	0.307729
28	14	7	5	0.961594	0.153865
28	14	7	6	0.997101	0.035507
28	14	7	7	1.000000	0.002899
28	14	8	0	0.000966	0.000966
28	14	8	1	0.016425	0.015459
28	14	8	2	0.104348	0.087923
28	14	8	3	0.338808	0.234461
28	14	8	4	0.661192	0.322383
28	14	8	5	0.895652	0.234461
28	14	8	6	0.983575	0.087923
28	14	8	7	0.999034	0.015459
28	14	8	8	1.000000	0.000966
28	14	9	0	0.000290	0.000290
28	14	9	1	0.006377	0.006087
28	14	9	2	0.051594	0.045217
28	14	9	3	0.209855	0.158261

N	n	k	x	P(x)	p(x)
28	14	9	4	0.500000	0.290145
28	14	9	5	0.790145	0.290145
28	14	9	6	0.948406	0.158261
28	14	9	7	0.993623	0.045217
28	14	9	8	0.999710	0.006087
28	14	9	9	1.000000	0.000290
28	14	10	0	0.000076	0.000076
28	14	10	1	0.002212	0.002136
28	14	10	2	0.023036	0.020824
28	14	10	3	0.118230	0.095194
28	14	10	4	0.347292	0.229062
28	14	10	5	0.652708	0.305416
28	14	10	6	0.881770	0.229062
28	14	10	7	0.976964	0.095194
28	14	10	8	0.997788	0.020824
28	14	10	9	0.999924	0.002136
28	14	10	10	1.000000	0.000076
28	14	11	0	0.000017	0.000017
28	14	11	1	0.000670	0.000653
28	14	11	2	0.009153	0.008484
28	14	11	3	0.060056	0.050903
28	14	11	4	0.220036	0.159980
28	14	11	5	0.500000	0.279964
28	14	11	6	0.779964	0.279964
28	14	11	7	0.939944	0.159980
28	14	11	8	0.990847	0.050903
28	14	11	9	0.999330	0.008484
28	14	11	10	0.999983	0.000653
28	14	11	11	1.000000	0.000017
28	14	12	0	0.000003	0.000003
28	14	12	1	0.000170	0.000168
28	14	12	2	0.003165	0.002994
28	14	12	3	0.027119	0.023954
28	14	12	4	0.125930	0.098811
28	14	12	5	0.351783	0.225854
28	14	12	6	0.648216	0.296433
28	14	12	7	0.874070	0.225854
28	14	12	8	0.972881	0.098811
28	14	12	9	0.996835	0.023954
28	14	12	10	0.999829	0.002994
28	14	12	11	0.999997	0.000168
28	14	12	12	1.000000	0.000003
28	14	13	0	0.000000	0.000000
28	14	13	1	0.000034	0.000034
28	14	13	2	0.000919	0.000885
28	14	13	3	0.010650	0.009731
28	14	13	4	0.064173	0.053523
28	14	13	5	0.224741	0.160568
28	14	13	6	0.500000	0.275259
28	14	13	7	0.775259	0.275259

Table for $N = 2$, $n = 1$, through $N = 100$, $n = 50$

N	n	k	x	P(x)	p(x)
28	14	13	8	0.935827	0.160568
28	14	13	9	0.989350	0.053523
28	14	13	10	0.999081	0.009731
28	14	13	11	0.999966	0.000885
28	14	13	12	1.000000	0.000034
28	14	13	13	1.000000	0.000000
28	14	14	0	0.000000	0.000000
28	14	14	1	0.000005	0.000005
28	14	14	2	0.000211	0.000206
28	14	14	3	0.003514	0.003303
28	14	14	4	0.028491	0.024977
28	14	14	5	0.128400	0.099909
28	14	14	6	0.353195	0.224795
28	14	14	7	0.646805	0.293610
28	14	14	8	0.871600	0.224795
28	14	14	9	0.971509	0.099909
28	14	14	10	0.996486	0.024977
28	14	14	11	0.999789	0.003303
28	14	14	12	0.999995	0.000206
28	14	14	13	1.000000	0.000005
28	14	14	14	1.000000	0.000000
29	1	1	0	0.965517	0.965517
29	1	1	1	1.000000	0.034483
29	2	1	0	0.931034	0.931034
29	2	1	1	1.000000	0.068966
29	2	2	0	0.864532	0.864532
29	2	2	1	0.997537	0.133005
29	2	2	2	1.000000	0.002463
29	3	1	0	0.896552	0.896552
29	3	1	1	1.000000	0.103448
29	3	2	0	0.800493	0.800493
29	3	2	1	0.992611	0.192118
29	3	2	2	1.000000	0.007389
29	3	3	0	0.711549	0.711549
29	3	3	1	0.978380	0.266831
29	3	3	2	0.999726	0.021346
29	3	3	3	1.000000	0.000274
29	4	1	0	0.862069	0.862069
29	4	1	1	1.000000	0.137931
29	4	2	0	0.738916	0.738916
29	4	2	1	0.985222	0.246305
29	4	2	2	1.000000	0.014778
29	4	3	0	0.629447	0.629447
29	4	3	1	0.957854	0.328407
29	4	3	2	0.998905	0.041051
29	4	3	3	1.000000	0.001095
29	4	4	0	0.532609	0.532609
29	4	4	1	0.919961	0.387352
29	4	4	2	0.995747	0.075786
29	4	4	3	0.999958	0.004210

N	n	k	x	P(x)	p(x)
29	4	4	4	1.000000	0.000042
29	5	1	0	0.827586	0.827586
29	5	1	1	1.000000	0.172414
29	5	2	0	0.679803	0.679803
29	5	2	1	0.975369	0.295566
29	5	2	2	1.000000	0.024631
29	5	3	0	0.553914	0.553914
29	5	3	1	0.931582	0.377668
29	5	3	2	0.997263	0.065681
29	5	3	3	1.000000	0.002737
29	5	4	0	0.447392	0.447392
29	5	4	1	0.873479	0.426087
29	5	4	2	0.989685	0.116206
29	5	4	3	0.999789	0.010105
29	5	4	4	1.000000	0.000211
29	5	5	0	0.357913	0.357913
29	5	5	1	0.805305	0.447392
29	5	5	2	0.975740	0.170435
29	5	5	3	0.998981	0.023241
29	5	5	4	0.999992	0.001010
29	5	5	5	1.000000	0.000008
29	6	1	0	0.793103	0.793103
29	6	1	1	1.000000	0.206897
29	6	2	0	0.623153	0.623153
29	6	2	1	0.963054	0.339901
29	6	2	2	1.000000	0.036946
29	6	3	0	0.484674	0.484674
29	6	3	1	0.900109	0.415435
29	6	3	2	0.994526	0.094417
29	6	3	3	1.000000	0.005473
29	6	4	0	0.372826	0.372826
29	6	4	1	0.820218	0.447392
29	6	4	2	0.980001	0.159783
29	6	4	3	0.999368	0.019368
29	6	4	4	1.000000	0.000632
29	6	5	0	0.283348	0.283348
29	6	5	1	0.730740	0.447392
29	6	5	2	0.954436	0.223696
29	6	5	3	0.997044	0.042609
29	6	5	4	0.999949	0.002905
29	6	5	5	1.000000	0.000051
29	6	6	0	0.212511	0.212511
29	6	6	1	0.637533	0.425022
29	6	6	2	0.917153	0.279620
29	6	6	3	0.991718	0.074565
29	6	6	4	0.999707	0.007989
29	6	6	5	0.999998	0.000291
29	6	6	6	1.000000	0.000002
29	7	1	0	0.758621	0.758621
29	7	1	1	1.000000	0.241379

N	n	k	x	P(x)	p(x)
29	7	2	0	0.568966	0.568966
29	7	2	1	0.948276	0.379310
29	7	2	2	1.000000	0.051724
29	7	3	0	0.421456	0.421456
29	7	3	1	0.863985	0.442529
29	7	3	2	0.990421	0.126437
29	7	3	3	1.000000	0.009579
29	7	4	0	0.307987	0.307987
29	7	4	1	0.761863	0.453876
29	7	4	2	0.966107	0.204244
29	7	4	3	0.998526	0.032420
29	7	4	4	1.000000	0.001474
29	7	5	0	0.221751	0.221751
29	7	5	1	0.652932	0.431182
29	7	5	2	0.925258	0.272325
29	7	5	3	0.993339	0.068081
29	7	5	4	0.999823	0.006484
29	7	5	5	1.000000	0.000177
29	7	6	0	0.157073	0.157073
29	7	6	1	0.545137	0.388064
29	7	6	2	0.868523	0.323386
29	7	6	3	0.981992	0.113469
29	7	6	4	0.999013	0.017020
29	7	6	5	0.999985	0.000973
29	7	6	6	1.000000	0.000015
29	7	7	0	0.109268	0.109268
29	7	7	1	0.443903	0.334635
29	7	7	2	0.798222	0.354319
29	7	7	3	0.962259	0.164037
29	7	7	4	0.996793	0.034534
29	7	7	5	0.999901	0.003108
29	7	7	6	0.999999	0.000099
29	7	7	7	1.000000	0.000001
29	8	1	0	0.724138	0.724138
29	8	1	1	1.000000	0.275862
29	8	2	0	0.517241	0.517241
29	8	2	1	0.931034	0.413793
29	8	2	2	1.000000	0.068966
29	8	3	0	0.363985	0.363985
29	8	3	1	0.823755	0.459770
29	8	3	2	0.984674	0.160920
29	8	3	3	1.000000	0.015326
29	8	4	0	0.251989	0.251989
29	8	4	1	0.699970	0.447981
29	8	4	2	0.947539	0.247569
29	8	4	3	0.997053	0.049514
29	8	4	4	1.000000	0.002947
29	8	5	0	0.171353	0.171353
29	8	5	1	0.574536	0.403183
29	8	5	2	0.888123	0.313587

N	n	k	x	P(x)	p(x)
29	8	5	3	0.987150	0.099027
29	8	5	4	0.999528	0.012378
29	8	5	5	1.000000	0.000472
29	8	6	0	0.114235	0.114235
29	8	6	1	0.456941	0.342706
29	8	6	2	0.809726	0.352785
29	8	6	3	0.966519	0.156793
29	8	6	4	0.997465	0.030946
29	8	6	5	0.999941	0.002476
29	8	6	6	1.000000	0.000059
29	8	7	0	0.074501	0.074501
29	8	7	1	0.352639	0.278138
29	8	7	2	0.717695	0.365056
29	8	7	3	0.932434	0.214739
29	8	7	4	0.992083	0.059650
29	8	7	5	0.999618	0.007535
29	8	7	6	0.999995	0.000377
29	8	7	7	1.000000	0.000005
29	8	8	0	0.047410	0.047410
29	8	8	1	0.264141	0.216731
29	8	8	2	0.618134	0.353994
29	8	8	3	0.883629	0.265495
29	8	8	4	0.981238	0.097609
29	8	8	5	0.998591	0.017353
29	8	8	6	0.999961	0.001370
29	8	8	7	1.000000	0.000039
29	8	8	8	1.000000	0.000000
29	9	1	0	0.689655	0.689655
29	9	1	1	1.000000	0.310345
29	9	2	0	0.467980	0.467980
29	9	2	1	0.911330	0.443350
29	9	2	2	1.000000	0.088670
29	9	3	0	0.311987	0.311987
29	9	3	1	0.779967	0.467980
29	9	3	2	0.977011	0.197044
29	9	3	3	1.000000	0.022988
29	9	4	0	0.203991	0.203991
29	9	4	1	0.635973	0.431982
29	9	4	2	0.923961	0.287988
29	9	4	3	0.994695	0.070734
29	9	4	4	1.000000	0.005305
29	9	5	0	0.130554	0.130554
29	9	5	1	0.497739	0.367185
29	9	5	2	0.843324	0.345585
29	9	5	3	0.977719	0.134394
29	9	5	4	0.998939	0.021220
29	9	5	5	1.000000	0.001061
29	9	6	0	0.081597	0.081597
29	9	6	1	0.375344	0.293748
29	9	6	2	0.742529	0.367185

N	n	k	x	P(x)	p(x)
29	9	6	3	0.944120	0.201592
29	9	6	4	0.994518	0.050398
29	9	6	5	0.999823	0.005305
29	9	6	6	1.000000	0.000177
29	9	7	0	0.049667	0.049667
29	9	7	1	0.273171	0.223504
29	9	7	2	0.630777	0.357606
29	9	7	3	0.891531	0.260754
29	9	7	4	0.983562	0.092031
29	9	7	5	0.998900	0.015338
29	9	7	6	0.999977	0.001076
29	9	7	7	1.000000	0.000023
29	9	8	0	0.029349	0.029349
29	9	8	1	0.191897	0.162548
29	9	8	2	0.516993	0.325096
29	9	8	3	0.820416	0.303423
29	9	8	4	0.962646	0.142230
29	9	8	5	0.996112	0.033466
29	9	8	6	0.999830	0.003718
29	9	8	7	0.999998	0.000168
29	9	8	8	1.000000	0.000002
29	9	9	0	0.016771	0.016771
29	9	9	1	0.129974	0.113203
29	9	9	2	0.408628	0.278654
29	9	9	3	0.733724	0.325096
29	9	9	4	0.928782	0.195058
29	9	9	5	0.989737	0.060956
29	9	9	6	0.999299	0.009562
29	9	9	7	0.999982	0.000683
29	9	9	8	1.000000	0.000018
29	9	9	9	1.000000	0.000000
29	10	1	0	0.655172	0.655172
29	10	1	1	1.000000	0.344828
29	10	2	0	0.421182	0.421182
29	10	2	1	0.889162	0.467980
29	10	2	2	1.000000	0.110837
29	10	3	0	0.265189	0.265189
29	10	3	1	0.733169	0.467980
29	10	3	2	0.967159	0.233990
29	10	3	3	1.000000	0.032841
29	10	4	0	0.163193	0.163193
29	10	4	1	0.571176	0.407983
29	10	4	2	0.895162	0.323986
29	10	4	3	0.991158	0.095996
29	10	4	4	1.000000	0.008842
29	10	5	0	0.097916	0.097916
29	10	5	1	0.424302	0.326386
29	10	5	2	0.791487	0.367185
29	10	5	3	0.964279	0.172793
29	10	5	4	0.997878	0.033599

N	n	k	x	P(x)	p(x)
29	10	5	5	1.000000	0.002122
29	10	6	0	0.057118	0.057118
29	10	6	1	0.301907	0.244790
29	10	6	2	0.669092	0.367185
29	10	6	3	0.913881	0.244790
29	10	6	4	0.989478	0.075597
29	10	6	5	0.999558	0.010080
29	10	6	6	1.000000	0.000442
29	10	7	0	0.032284	0.032284
29	10	7	1	0.206120	0.173836
29	10	7	2	0.541375	0.335255
29	10	7	3	0.839380	0.298005
29	10	7	4	0.969757	0.130377
29	10	7	5	0.997367	0.027609
29	10	7	6	0.999923	0.002556
29	10	7	7	1.000000	0.000077
29	10	8	0	0.017609	0.017609
29	10	8	1	0.135005	0.117396
29	10	8	2	0.419464	0.284459
29	10	8	3	0.744561	0.325096
29	10	8	4	0.934200	0.189639
29	10	8	5	0.991092	0.056892
29	10	8	6	0.999458	0.008366
29	10	8	7	0.999989	0.000531
29	10	8	8	1.000000	0.000010
29	10	9	0	0.009224	0.009224
29	10	9	1	0.084693	0.075469
29	10	9	2	0.311099	0.226406
29	10	9	3	0.636195	0.325096
29	10	9	4	0.880017	0.243822
29	10	9	5	0.977546	0.097529
29	10	9	6	0.997865	0.020319
29	10	9	7	0.999913	0.002049
29	10	9	8	0.999999	0.000085
29	10	9	9	1.000000	0.000001
29	10	10	0	0.004612	0.004612
29	10	10	1	0.050732	0.046120
29	10	10	2	0.220536	0.169805
29	10	10	3	0.522411	0.301875
29	10	10	4	0.806871	0.284459
29	10	10	5	0.953164	0.146293
29	10	10	6	0.993801	0.040637
29	10	10	7	0.999606	0.005805
29	10	10	8	0.999990	0.000384
29	10	10	9	1.000000	0.000009
29	10	10	10	1.000000	0.000000
29	11	1	0	0.620690	0.620690
29	11	1	1	1.000000	0.379310
29	11	2	0	0.376847	0.376847
29	11	2	1	0.864532	0.487685

N	n	k	x	P(x)	p(x)
29	11	2	2	1.000000	0.135468
29	11	3	0	0.223317	0.223317
29	11	3	1	0.683908	0.460591
29	11	3	2	0.954844	0.270936
29	11	3	3	1.000000	0.045156
29	11	4	0	0.128837	0.128837
29	11	4	1	0.506758	0.377921
29	11	4	2	0.861058	0.354301
29	11	4	3	0.986106	0.125047
29	11	4	4	1.000000	0.013894
29	11	5	0	0.072149	0.072149
29	11	5	1	0.355589	0.283441
29	11	5	2	0.733510	0.377921
29	11	5	3	0.946091	0.212580
29	11	5	4	0.996110	0.050019
29	11	5	5	1.000000	0.003890
29	11	6	0	0.039080	0.039080
29	11	6	1	0.237489	0.198408
29	11	6	2	0.591790	0.354301
29	11	6	3	0.875230	0.283441
29	11	6	4	0.981521	0.106290
29	11	6	5	0.999027	0.017507
29	11	6	6	1.000000	0.000973
29	11	7	0	0.020390	0.020390
29	11	7	1	0.151224	0.130835
29	11	7	2	0.453150	0.301926
29	11	7	3	0.776642	0.323492
29	11	7	4	0.949171	0.172529
29	11	7	5	0.994460	0.045289
29	11	7	6	0.999788	0.005328
29	11	7	7	1.000000	0.000211
29	11	8	0	0.010195	0.010195
29	11	8	1	0.091754	0.081559
29	11	8	2	0.329635	0.237881
29	11	8	3	0.659009	0.329374
29	11	8	4	0.894276	0.235267
29	11	8	5	0.982109	0.087833
29	11	8	6	0.998578	0.016469
29	11	8	7	0.999962	0.001384
29	11	8	8	1.000000	0.000038
29	11	9	0	0.004855	0.004855
29	11	9	1	0.052916	0.048062
29	11	9	2	0.227686	0.174770
29	11	9	3	0.533533	0.305847
29	11	9	4	0.815854	0.282320
29	11	9	5	0.957014	0.141160
29	11	9	6	0.994656	0.037643
29	11	9	7	0.999698	0.005041
29	11	9	8	0.999994	0.000297
29	11	9	9	1.000000	0.000005

N	n	k	x	P(x)	p(x)
29	11	10	0	0.002185	0.002185
29	11	10	1	0.028886	0.026701
29	11	10	2	0.149040	0.120154
29	11	10	3	0.411194	0.262155
29	11	10	4	0.717041	0.305847
29	11	10	5	0.914666	0.197624
29	11	10	6	0.985246	0.070580
29	11	10	7	0.998690	0.013444
29	11	10	8	0.999950	0.001260
29	11	10	9	0.999999	0.000049
29	11	10	10	1.000000	0.000001
29	11	11	0	0.000920	0.000920
29	11	11	1	0.014832	0.013913
29	11	11	2	0.092125	0.077292
29	11	11	3	0.300813	0.208689
29	11	11	4	0.604361	0.303547
29	11	11	5	0.852258	0.247897
29	11	11	6	0.966672	0.114414
29	11	11	7	0.995859	0.029187
29	11	11	8	0.999751	0.003892
29	11	11	9	0.999994	0.000243
29	11	11	10	1.000000	0.000006
29	11	11	11	1.000000	0.000000
29	12	1	0	0.586207	0.586207
29	12	1	1	1.000000	0.413793
29	12	2	0	0.334975	0.334975
29	12	2	1	0.837438	0.502463
29	12	2	2	1.000000	0.162562
29	12	3	0	0.186097	0.186097
29	12	3	1	0.632731	0.446634
29	12	3	2	0.939792	0.307061
29	12	3	3	1.000000	0.060208
29	12	4	0	0.100206	0.100206
29	12	4	1	0.443771	0.343564
29	12	4	2	0.821692	0.377921
29	12	4	3	0.979159	0.157467
29	12	4	4	1.000000	0.020841
29	12	5	0	0.052107	0.052107
29	12	5	1	0.292602	0.240495
29	12	5	2	0.670523	0.377921
29	12	5	3	0.922471	0.251947
29	12	5	4	0.993331	0.070860
29	12	5	5	1.000000	0.006669
29	12	6	0	0.026054	0.026054
29	12	6	1	0.182375	0.156322
29	12	6	2	0.513056	0.330681
29	12	6	3	0.827990	0.314934
29	12	6	4	0.969711	0.141720
29	12	6	5	0.998055	0.028344
29	12	6	6	1.000000	0.001945

Table for $N = 2$, $n = 1$, through $N = 100$, $n = 50$

N	n	k	x	P(x)	p(x)
29	12	7	0	0.012460	0.012460
29	12	7	1	0.107613	0.095152
29	12	7	2	0.369282	0.261669
29	12	7	3	0.704755	0.335473
29	12	7	4	0.920417	0.215661
29	12	7	5	0.989428	0.069012
29	12	7	6	0.999492	0.010064
29	12	7	7	1.000000	0.000507
29	12	8	0	0.005664	0.005664
29	12	8	1	0.060037	0.054373
29	12	8	2	0.250341	0.190305
29	12	8	3	0.567516	0.317175
29	12	8	4	0.841994	0.274478
29	12	8	5	0.967470	0.125476
29	12	8	6	0.996748	0.029278
29	12	8	7	0.999885	0.003137
29	12	8	8	1.000000	0.000115
29	12	9	0	0.002427	0.002427
29	12	9	1	0.031556	0.029128
29	12	9	2	0.159720	0.128164
29	12	9	3	0.431584	0.271864
29	12	9	4	0.737431	0.305847
29	12	9	5	0.925645	0.188214
29	12	9	6	0.988383	0.062738
29	12	9	7	0.999138	0.010755
29	12	9	8	0.999978	0.000840
29	12	9	9	1.000000	0.000022
29	12	10	0	0.000971	0.000971
29	12	10	1	0.015535	0.014564
29	12	10	2	0.095638	0.080103
29	12	10	3	0.309245	0.213607
29	12	10	4	0.615092	0.305847
29	12	10	5	0.859770	0.244678
29	12	10	6	0.969561	0.109791
29	12	10	7	0.996449	0.026888
29	12	10	8	0.999810	0.003361
29	12	10	9	0.999997	0.000187
29	12	10	10	1.000000	0.000003
29	12	11	0	0.000358	0.000358
29	12	11	1	0.007103	0.006745
29	12	11	2	0.053479	0.046375
29	12	11	3	0.208063	0.154584
29	12	11	4	0.486315	0.278252
29	12	11	5	0.769626	0.283311
29	12	11	6	0.934890	0.165265
29	12	11	7	0.989373	0.054483
29	12	11	8	0.999102	0.009729
29	12	11	9	0.999967	0.000865
29	12	11	10	1.000000	0.000032
29	12	11	11	1.000000	0.000000

N	n	k	x	P(x)	p(x)
29	12	12	0	0.000119	0.000119
29	12	12	1	0.002981	0.002862
29	12	12	2	0.027714	0.024733
29	12	12	3	0.130771	0.103056
29	12	12	4	0.362647	0.231877
29	12	12	5	0.659449	0.296802
29	12	12	6	0.879802	0.220353
29	12	12	7	0.974239	0.094437
29	12	12	8	0.996940	0.022701
29	12	12	9	0.999823	0.002883
29	12	12	10	0.999996	0.000173
29	12	12	11	1.000000	0.000004
29	12	12	12	1.000000	0.000000
29	13	1	0	0.551724	0.551724
29	13	1	1	1.000000	0.448276
29	13	2	0	0.295566	0.295566
29	13	2	1	0.807882	0.512315
29	13	2	2	1.000000	0.192118
29	13	3	0	0.153257	0.153257
29	13	3	1	0.580186	0.426929
29	13	3	2	0.921730	0.341543
29	13	3	3	1.000000	0.078270
29	13	4	0	0.076628	0.076628
29	13	4	1	0.383142	0.306513
29	13	4	2	0.777230	0.394089
29	13	4	3	0.969896	0.192666
29	13	4	4	1.000000	0.030104
29	13	5	0	0.036782	0.036782
29	13	5	1	0.236015	0.199234
29	13	5	2	0.603831	0.367816
29	13	5	3	0.892830	0.288998
29	13	5	4	0.989162	0.096333
29	13	5	5	1.000000	0.010837
29	13	6	0	0.016858	0.016858
29	13	6	1	0.136398	0.119540
29	13	6	2	0.435249	0.298851
29	13	6	3	0.772414	0.337165
29	13	6	4	0.953038	0.180624
29	13	6	5	0.996387	0.043350
29	13	6	6	1.000000	0.003612
29	13	7	0	0.007330	0.007330
29	13	7	1	0.074030	0.066700
29	13	7	2	0.292320	0.218291
29	13	7	3	0.625820	0.333500
29	13	7	4	0.882359	0.256538
29	13	7	5	0.981309	0.098951
29	13	7	6	0.998900	0.017591
29	13	7	7	1.000000	0.001099
29	13	8	0	0.002998	0.002998
29	13	8	1	0.037648	0.034649

N	n	k	x	P(x)	p(x)
29	13	8	2	0.183175	0.145527
29	13	8	3	0.474230	0.291054
29	13	8	4	0.777411	0.303182
29	13	8	5	0.945327	0.167916
29	13	8	6	0.993303	0.047976
29	13	8	7	0.999700	0.006397
29	13	8	8	1.000000	0.000300
29	13	9	0	0.001142	0.001142
29	13	9	1	0.017848	0.016706
29	13	9	2	0.106946	0.089098
29	13	9	3	0.335632	0.228686
29	13	9	4	0.647476	0.311844
29	13	9	5	0.881359	0.233883
29	13	9	6	0.977311	0.095952
29	13	9	7	0.997872	0.020561
29	13	9	8	0.999929	0.002056
29	13	9	9	1.000000	0.000071
29	13	10	0	0.000400	0.000400
29	13	10	1	0.007825	0.007425
29	13	10	2	0.057942	0.050118
29	13	10	3	0.221289	0.163347
29	13	10	4	0.507146	0.285857
29	13	10	5	0.787806	0.280660
29	13	10	6	0.943728	0.155922
29	13	10	7	0.991704	0.047976
29	13	10	8	0.999414	0.007710
29	13	10	9	0.999986	0.000571
29	13	10	10	1.000000	0.000014
29	13	11	0	0.000126	0.000126
29	13	11	1	0.003135	0.003009
29	13	11	2	0.028927	0.025792
29	13	11	3	0.135317	0.106390
29	13	11	4	0.371740	0.236423
29	13	11	5	0.669634	0.297893
29	13	11	6	0.886283	0.216650
29	13	11	7	0.976554	0.090271
29	13	11	8	0.997385	0.020832
29	13	11	9	0.999865	0.002480
29	13	11	10	0.999998	0.000132
29	13	11	11	1.000000	0.000002
29	13	12	0	0.000035	0.000035
29	13	12	1	0.001129	0.001094
29	13	12	2	0.013165	0.012036
29	13	12	3	0.076211	0.063046
29	13	12	4	0.253529	0.177317
29	13	12	5	0.537237	0.283708
29	13	12	6	0.802030	0.264794
29	13	12	7	0.946464	0.144433
29	13	12	8	0.991599	0.045135
29	13	12	9	0.999314	0.007715

N	n	k	x	P(x)	p(x)
29	13	12	10	0.999976	0.000661
29	13	12	11	1.000000	0.000024
29	13	12	12	1.000000	0.000000
29	13	13	0	0.000008	0.000008
29	13	13	1	0.000357	0.000349
29	13	13	2	0.005377	0.005020
29	13	13	3	0.039126	0.033748
29	13	13	4	0.159655	0.120529
29	13	13	5	0.403727	0.244072
29	13	13	6	0.692998	0.289271
29	13	13	7	0.895487	0.202489
29	13	13	8	0.978324	0.082837
29	13	13	9	0.997499	0.019175
29	13	13	10	0.999859	0.002360
29	13	13	11	0.999997	0.000138
29	13	13	12	1.000000	0.000003
29	13	13	13	1.000000	0.000000
29	14	1	0	0.517241	0.517241
29	14	1	1	1.000000	0.482759
29	14	2	0	0.258621	0.258621
29	14	2	1	0.775862	0.517241
29	14	2	2	1.000000	0.224138
29	14	3	0	0.124521	0.124521
29	14	3	1	0.526820	0.402299
29	14	3	2	0.900383	0.373563
29	14	3	3	1.000000	0.099617
29	14	4	0	0.057471	0.057471
29	14	4	1	0.325670	0.268199
29	14	4	2	0.727969	0.402299
29	14	4	3	0.957854	0.229885
29	14	4	4	1.000000	0.042146
29	14	5	0	0.025287	0.025287
29	14	5	1	0.186207	0.160920
29	14	5	2	0.534866	0.348659
29	14	5	3	0.856705	0.321839
29	14	5	4	0.983142	0.126437
29	14	5	5	1.000000	0.016858
29	14	6	0	0.010536	0.010536
29	14	6	1	0.099042	0.088506
29	14	6	2	0.360536	0.261494
29	14	6	3	0.709195	0.348659
29	14	6	4	0.930460	0.221264
29	14	6	5	0.993678	0.063218
29	14	6	6	1.000000	0.006322
29	14	7	0	0.004123	0.004123
29	14	7	1	0.049017	0.044894
29	14	7	2	0.224105	0.175087
29	14	7	3	0.542445	0.318341
29	14	7	4	0.834258	0.291812
29	14	7	5	0.968940	0.134683

N	n	k	x	P(x)	p(x)
29	14	7	6	0.997801	0.028861
29	14	7	7	1.000000	0.002199
29	14	8	0	0.001499	0.001499
29	14	8	1	0.022489	0.020989
29	14	8	2	0.128602	0.106114
29	14	8	3	0.383275	0.254673
29	14	8	4	0.701616	0.318341
29	14	8	5	0.913843	0.212227
29	14	8	6	0.987306	0.073463
29	14	8	7	0.999300	0.011994
29	14	8	8	1.000000	0.000700
29	14	9	0	0.000500	0.000500
29	14	9	1	0.009495	0.008995
29	14	9	2	0.067966	0.058471
29	14	9	3	0.249875	0.181909
29	14	9	4	0.550025	0.300150
29	14	9	5	0.822888	0.272864
29	14	9	6	0.959320	0.136432
29	14	9	7	0.995302	0.035982
29	14	9	8	0.999800	0.004498
29	14	9	9	1.000000	0.000200
29	14	10	0	0.000150	0.000150
29	14	10	1	0.003648	0.003498
29	14	10	2	0.032884	0.029235
29	14	10	3	0.149825	0.116941
29	14	10	4	0.399950	0.250125
29	14	10	5	0.700100	0.300150
29	14	10	6	0.904748	0.204648
29	14	10	7	0.982709	0.077961
29	14	10	8	0.998451	0.015742
29	14	10	9	0.999950	0.001499
29	14	10	10	1.000000	0.000050
29	14	11	0	0.000039	0.000039
29	14	11	1	0.001255	0.001215
29	14	11	2	0.014419	0.013164
29	14	11	3	0.082122	0.067703
29	14	11	4	0.268305	0.186183
29	14	11	5	0.557924	0.289618
29	14	11	6	0.818580	0.260657
29	14	11	7	0.953986	0.135406
29	14	11	8	0.993479	0.039493
29	14	11	9	0.999555	0.006076
29	14	11	10	0.999989	0.000434
29	14	11	11	1.000000	0.000011
29	14	12	0	0.000009	0.000009
29	14	12	1	0.000377	0.000368
29	14	12	2	0.005643	0.005266
29	14	12	3	0.040748	0.035105
29	14	12	4	0.164870	0.124122
29	14	12	5	0.413114	0.248244

N	n	k	x	P(x)	p(x)
29	14	12	6	0.702733	0.289618
29	14	12	7	0.901328	0.198595
29	14	12	8	0.980315	0.078987
29	14	12	9	0.997868	0.017553
29	14	12	10	0.999893	0.002025
29	14	12	11	0.999998	0.000105
29	14	12	12	1.000000	0.000002
29	14	13	0	0.000002	0.000002
29	14	13	1	0.000095	0.000094
29	14	13	2	0.001926	0.001830
29	14	13	3	0.018033	0.016107
29	14	13	4	0.091857	0.073824
29	14	13	5	0.281691	0.189834
29	14	13	6	0.566442	0.284751
29	14	13	7	0.819554	0.253112
29	14	13	8	0.952437	0.132884
29	14	13	9	0.992705	0.040268
29	14	13	10	0.999416	0.006711
29	14	13	11	0.999980	0.000563
29	14	13	12	1.000000	0.000020
29	14	13	13	1.000000	0.000000
29	14	14	0	0.000000	0.000000
29	14	14	1	0.000019	0.000019
29	14	14	2	0.000553	0.000534
29	14	14	3	0.006959	0.006406
29	14	14	4	0.045717	0.038758
29	14	14	5	0.174909	0.129192
29	14	14	6	0.424066	0.249157
29	14	14	7	0.708817	0.284751
29	14	14	8	0.902606	0.193789
29	14	14	9	0.980121	0.077515
29	14	14	10	0.997739	0.017617
29	14	14	11	0.999874	0.002135
29	14	14	12	0.999997	0.000123
29	14	14	13	1.000000	0.000003
29	14	14	14	1.000000	0.000000
30	1	1	0	0.966667	0.966667
30	1	1	1	1.000000	0.033333
30	2	1	0	0.933333	0.933333
30	2	1	1	1.000000	0.066667
30	2	2	0	0.868966	0.868966
30	2	2	1	0.997701	0.128736
30	2	2	2	1.000000	0.002299
30	3	1	0	0.900000	0.900000
30	3	1	1	1.000000	0.100000
30	3	2	0	0.806897	0.806897
30	3	2	1	0.993103	0.186207
30	3	2	2	1.000000	0.006897
30	3	3	0	0.720443	0.720443
30	3	3	1	0.979803	0.259360

N	n	k	x	P(x)	p(x)
30	3	3	2	0.999754	0.019951
30	3	3	3	1.000000	0.000246
30	4	1	0	0.866667	0.866667
30	4	1	1	1.000000	0.133333
30	4	2	0	0.747126	0.747126
30	4	2	1	0.986207	0.239080
30	4	2	2	1.000000	0.013793
30	4	3	0	0.640394	0.640394
30	4	3	1	0.960591	0.320197
30	4	3	2	0.999015	0.038424
30	4	3	3	1.000000	0.000985
30	4	4	0	0.545521	0.545521
30	4	4	1	0.925014	0.379493
30	4	4	2	0.996169	0.071155
30	4	4	3	0.999963	0.003795
30	4	4	4	1.000000	0.000036
30	5	1	0	0.833333	0.833333
30	5	1	1	1.000000	0.166667
30	5	2	0	0.689655	0.689655
30	5	2	1	0.977011	0.287356
30	5	2	2	1.000000	0.022988
30	5	3	0	0.566502	0.566502
30	5	3	1	0.935961	0.369458
30	5	3	2	0.997537	0.061576
30	5	3	3	1.000000	0.002463
30	5	4	0	0.461595	0.461595
30	5	4	1	0.881226	0.419631
30	5	4	2	0.990695	0.109469
30	5	4	3	0.999817	0.009122
30	5	4	4	1.000000	0.000182
30	5	5	0	0.372826	0.372826
30	5	5	1	0.816667	0.443841
30	5	5	2	0.978064	0.161397
30	5	5	3	0.999116	0.021052
30	5	5	4	0.999993	0.000877
30	5	5	5	1.000000	0.000007
30	6	1	0	0.800000	0.800000
30	6	1	1	1.000000	0.200000
30	6	2	0	0.634483	0.634483
30	6	2	1	0.965517	0.331034
30	6	2	2	1.000000	0.034483
30	6	3	0	0.498522	0.498522
30	6	3	1	0.906404	0.407882
30	6	3	2	0.995074	0.088670
30	6	3	3	1.000000	0.004926
30	6	4	0	0.387739	0.387739
30	6	4	1	0.830870	0.443131
30	6	4	2	0.981938	0.151067
30	6	4	3	0.999453	0.017515
30	6	4	4	1.000000	0.000547

N	n	k	x	P(x)	p(x)
30	6	5	0	0.298261	0.298261
30	6	5	1	0.745653	0.447392
30	6	5	2	0.958696	0.213044
30	6	5	3	0.997432	0.038735
30	6	5	4	0.999958	0.002526
30	6	5	5	1.000000	0.000042
30	6	6	0	0.226678	0.226678
30	6	6	1	0.656174	0.429496
30	6	6	2	0.924609	0.268435
30	6	6	3	0.992783	0.068174
30	6	6	4	0.999756	0.006972
30	6	6	5	0.999998	0.000243
30	6	6	6	1.000000	0.000002
30	7	1	0	0.766667	0.766667
30	7	1	1	1.000000	0.233333
30	7	2	0	0.581609	0.581609
30	7	2	1	0.951724	0.370115
30	7	2	2	1.000000	0.048276
30	7	3	0	0.436207	0.436207
30	7	3	1	0.872414	0.436207
30	7	3	2	0.991379	0.118965
30	7	3	3	1.000000	0.008621
30	7	4	0	0.323116	0.323116
30	7	4	1	0.775479	0.452363
30	7	4	2	0.969349	0.193870
30	7	4	3	0.998723	0.029374
30	7	4	4	1.000000	0.001277
30	7	5	0	0.236123	0.236123
30	7	5	1	0.671087	0.434964
30	7	5	2	0.932066	0.260978
30	7	5	3	0.994204	0.062138
30	7	5	4	0.999853	0.005649
30	7	5	5	1.000000	0.000147
30	7	6	0	0.170009	0.170009
30	7	6	1	0.566696	0.396687
30	7	6	2	0.879870	0.313174
30	7	6	3	0.984262	0.104391
30	7	6	4	0.999175	0.014913
30	7	6	5	0.999988	0.000813
30	7	6	6	1.000000	0.000012
30	7	7	0	0.120423	0.120423
30	7	7	1	0.467524	0.347101
30	7	7	2	0.814626	0.347101
30	7	7	3	0.966863	0.152237
30	7	7	4	0.997311	0.030447
30	7	7	5	0.999920	0.002610
30	7	7	6	0.999999	0.000079
30	7	7	7	1.000000	0.000000
30	8	1	0	0.733333	0.733333
30	8	1	1	1.000000	0.266667

N	n	k	x	P(x)	p(x)
30	8	2	0	0.531034	0.531034
30	8	2	1	0.935632	0.404598
30	8	2	2	1.000000	0.064368
30	8	3	0	0.379310	0.379310
30	8	3	1	0.834483	0.455172
30	8	3	2	0.986207	0.151724
30	8	3	3	1.000000	0.013793
30	8	4	0	0.266922	0.266922
30	8	4	1	0.716475	0.449553
30	8	4	2	0.952490	0.236015
30	8	4	3	0.997446	0.044955
30	8	4	4	1.000000	0.002554
30	8	5	0	0.184792	0.184792
30	8	5	1	0.595442	0.410649
30	8	5	2	0.898025	0.302584
30	8	5	3	0.988800	0.090775
30	8	5	4	0.999607	0.010807
30	8	5	5	1.000000	0.000393
30	8	6	0	0.125659	0.125659
30	8	6	1	0.480460	0.354801
30	8	6	2	0.825405	0.344945
30	8	6	3	0.970645	0.145240
30	8	6	4	0.997878	0.027233
30	8	6	5	0.999953	0.002075
30	8	6	6	1.000000	0.000047
30	8	7	0	0.083772	0.083772
30	8	7	1	0.376976	0.293204
30	8	7	2	0.739169	0.362193
30	8	7	3	0.940387	0.201218
30	8	7	4	0.993339	0.052952
30	8	7	5	0.999693	0.006354
30	8	7	6	0.999996	0.000303
30	8	7	7	1.000000	0.000004
30	8	8	0	0.054634	0.054634
30	8	8	1	0.287740	0.233106
30	8	8	2	0.644684	0.356944
30	8	8	3	0.896644	0.251960
30	8	8	4	0.984130	0.087486
30	8	8	5	0.998865	0.014735
30	8	8	6	0.999970	0.001105
30	8	8	7	1.000000	0.000030
30	8	8	8	1.000000	0.000000
30	9	1	0	0.700000	0.700000
30	9	1	1	1.000000	0.300000
30	9	2	0	0.482759	0.482759
30	9	2	1	0.917241	0.434483
30	9	2	2	1.000000	0.082759
30	9	3	0	0.327586	0.327586
30	9	3	1	0.793103	0.465517
30	9	3	2	0.979310	0.186207
30	9	3	3	1.000000	0.020690
30	9	4	0	0.218391	0.218391
30	9	4	1	0.655172	0.436782
30	9	4	2	0.931034	0.275862
30	9	4	3	0.995402	0.064368
30	9	4	4	1.000000	0.004598
30	9	5	0	0.142794	0.142794
30	9	5	1	0.520778	0.377984
30	9	5	2	0.856764	0.335986
30	9	5	3	0.980548	0.123784
30	9	5	4	0.999116	0.018568
30	9	5	5	1.000000	0.000884
30	9	6	0	0.091388	0.091388
30	9	6	1	0.399823	0.308435
30	9	6	2	0.762688	0.362865
30	9	6	3	0.950840	0.188152
30	9	6	4	0.995402	0.044562
30	9	6	5	0.999858	0.004456
30	9	6	6	1.000000	0.000141
30	9	7	0	0.057118	0.057118
30	9	7	1	0.297011	0.239894
30	9	7	2	0.656852	0.359841
30	9	7	3	0.903802	0.246950
30	9	7	4	0.986118	0.082317
30	9	7	5	0.999116	0.012997
30	9	7	6	0.999982	0.000866
30	9	7	7	1.000000	0.000018
30	9	8	0	0.034767	0.034767
30	9	8	1	0.213570	0.178803
30	9	8	2	0.547336	0.333765
30	9	8	3	0.839380	0.292045
30	9	8	4	0.968223	0.128843
30	9	8	5	0.996855	0.028632
30	9	8	6	0.999869	0.003014
30	9	8	7	0.999998	0.000129
30	9	8	8	1.000000	0.000002
30	9	9	0	0.020544	0.020544
30	9	9	1	0.148551	0.128007
30	9	9	2	0.441137	0.292587
30	9	9	3	0.759732	0.318594
30	9	9	4	0.938941	0.179209
30	9	9	5	0.991650	0.052709
30	9	9	6	0.999458	0.007809
30	9	9	7	0.999987	0.000528
30	9	9	8	1.000000	0.000013
30	9	9	9	1.000000	0.000000
30	10	1	0	0.666667	0.666667
30	10	1	1	1.000000	0.333333
30	10	2	0	0.436782	0.436782
30	10	2	1	0.896552	0.459770
30	10	2	2	1.000000	0.103448
30	10	3	0	0.280788	0.280788
30	10	3	1	0.748768	0.467980
30	10	3	2	0.970443	0.221675
30	10	3	3	1.000000	0.029557
30	10	4	0	0.176793	0.176793
30	10	4	1	0.592775	0.415982
30	10	4	2	0.904762	0.311987
30	10	4	3	0.992337	0.087575
30	10	4	4	1.000000	0.007663
30	10	5	0	0.108795	0.108795
30	10	5	1	0.448781	0.339986
30	10	5	2	0.808766	0.359985
30	10	5	3	0.968759	0.159993
30	10	5	4	0.998232	0.029472
30	10	5	5	1.000000	0.001768
30	10	6	0	0.065277	0.065277
30	10	6	1	0.326386	0.261109
30	10	6	2	0.693571	0.367185
30	10	6	3	0.923961	0.230390
30	10	6	4	0.991158	0.067197
30	10	6	5	0.999646	0.008488
30	10	6	6	1.000000	0.000354
30	10	7	0	0.038078	0.038078
30	10	7	1	0.228470	0.190392
30	10	7	2	0.571176	0.342706
30	10	7	3	0.856764	0.285588
30	10	7	4	0.974359	0.117595
30	10	7	5	0.997878	0.023519
30	10	7	6	0.999941	0.002063
30	10	7	7	1.000000	0.000059
30	10	8	0	0.021523	0.021523
30	10	8	1	0.153969	0.132447
30	10	8	2	0.451974	0.298005
30	10	8	3	0.769846	0.317872
30	10	8	4	0.943682	0.173836
30	10	8	5	0.992765	0.049083
30	10	8	6	0.999582	0.006817
30	10	8	7	0.999992	0.000410
30	10	8	8	1.000000	0.000008
30	10	9	0	0.011740	0.011740
30	10	9	1	0.099786	0.088047
30	10	9	2	0.343609	0.243822
30	10	9	3	0.668705	0.325096
30	10	9	4	0.896272	0.227567
30	10	9	5	0.981610	0.085338
30	10	9	6	0.998343	0.016733
30	10	9	7	0.999936	0.001594
30	10	9	8	0.999999	0.000063
30	10	9	9	1.000000	0.000001
30	10	10	0	0.006149	0.006149
30	10	10	1	0.062052	0.055903
30	10	10	2	0.250724	0.188672
30	10	10	3	0.560339	0.309615
30	10	10	4	0.831253	0.270913
30	10	10	5	0.961291	0.130038
30	10	10	6	0.995155	0.033864
30	10	10	7	0.999709	0.004553
30	10	10	8	0.999993	0.000285
30	10	10	9	1.000000	0.000007
30	10	10	10	1.000000	0.000000
30	11	1	0	0.633333	0.633333
30	11	1	1	1.000000	0.366667
30	11	2	0	0.393103	0.393103
30	11	2	1	0.873563	0.480460
30	11	2	2	1.000000	0.126437
30	11	3	0	0.238670	0.238670
30	11	3	1	0.701970	0.463300
30	11	3	2	0.959360	0.257389
30	11	3	3	1.000000	0.040640
30	11	4	0	0.141434	0.141434
30	11	4	1	0.530378	0.388944
30	11	4	2	0.873563	0.343185
30	11	4	3	0.987958	0.114395
30	11	4	4	1.000000	0.012042
30	11	5	0	0.081597	0.081597
30	11	5	1	0.380784	0.299187
30	11	5	2	0.754768	0.373984
30	11	5	3	0.952760	0.197992
30	11	5	4	0.996758	0.043998
30	11	5	5	1.000000	0.003242
30	11	6	0	0.045694	0.045694
30	11	6	1	0.261109	0.215415
30	11	6	2	0.620134	0.359025
30	11	6	3	0.889402	0.269269
30	11	6	4	0.984438	0.095036
30	11	6	5	0.999222	0.014783
30	11	6	6	1.000000	0.000778
30	11	7	0	0.024751	0.024751
30	11	7	1	0.171353	0.146602
30	11	7	2	0.485499	0.314147
30	11	7	3	0.799646	0.314147
30	11	7	4	0.956720	0.157073
30	11	7	5	0.995526	0.038806
30	11	7	6	0.999838	0.004312
30	11	7	7	1.000000	0.000162
30	11	8	0	0.012914	0.012914
30	11	8	1	0.107613	0.094699
30	11	8	2	0.362573	0.254960
30	11	8	3	0.690378	0.327805

N	n	k	x	P(x)	p(x)
30	11	8	4	0.908915	0.218537
30	11	8	5	0.985403	0.076488
30	11	8	6	0.998900	0.013498
30	11	8	7	0.999972	0.001071
30	11	8	8	1.000000	0.000028
30	11	9	0	0.006457	0.006457
30	11	9	1	0.064568	0.058111
30	11	9	2	0.258271	0.193703
30	11	9	3	0.571176	0.312905
30	11	9	4	0.839380	0.268204
30	11	9	5	0.964542	0.125162
30	11	9	6	0.995833	0.031291
30	11	9	7	0.999777	0.003944
30	11	9	8	0.999996	0.000219
30	11	9	9	1.000000	0.000004
30	11	10	0	0.003075	0.003075
30	11	10	1	0.036896	0.033821
30	11	10	2	0.175255	0.138359
30	11	10	3	0.451974	0.276719
30	11	10	4	0.749979	0.298005
30	11	10	5	0.928782	0.178803
30	11	10	6	0.988383	0.059601
30	11	10	7	0.999026	0.010643
30	11	10	8	0.999965	0.000939
30	11	10	9	1.000000	0.000035
30	11	10	10	1.000000	0.000000
30	11	11	0	0.001384	0.001384
30	11	11	1	0.019985	0.018602
30	11	11	2	0.112993	0.093008
30	11	11	3	0.341286	0.228293
30	11	11	4	0.645677	0.304391
30	11	11	5	0.875141	0.229464
30	11	11	6	0.973482	0.098342
30	11	11	7	0.996897	0.023415
30	11	11	8	0.999824	0.002927
30	11	11	9	0.999996	0.000172
30	11	11	10	1.000000	0.000004
30	11	11	11	1.000000	0.000000
30	12	1	0	0.600000	0.600000
30	12	1	1	1.000000	0.400000
30	12	2	0	0.351724	0.351724
30	12	2	1	0.848276	0.496552
30	12	2	2	1.000000	0.151724
30	12	3	0	0.200985	0.200985
30	12	3	1	0.653202	0.452217
30	12	3	2	0.945813	0.292611
30	12	3	3	1.000000	0.054187
30	12	4	0	0.111658	0.111658
30	12	4	1	0.468966	0.357307
30	12	4	2	0.837438	0.368473

N	n	k	x	P(x)	p(x)
30	12	4	3	0.981938	0.144499
30	12	4	4	1.000000	0.018062
30	12	5	0	0.060124	0.060124
30	12	5	1	0.317797	0.257673
30	12	5	2	0.695718	0.377921
30	12	5	3	0.931919	0.236201
30	12	5	4	0.994442	0.062524
30	12	5	5	1.000000	0.005558
30	12	6	0	0.031264	0.031264
30	12	6	1	0.204421	0.173156
30	12	6	2	0.544550	0.340129
30	12	6	3	0.846886	0.302337
30	12	6	4	0.974435	0.127548
30	12	6	5	0.998444	0.024009
30	12	6	6	1.000000	0.001556
30	12	7	0	0.015632	0.015632
30	12	7	1	0.125057	0.109425
30	12	7	2	0.402829	0.277772
30	12	7	3	0.733510	0.330681
30	12	7	4	0.931919	0.198408
30	12	7	5	0.991441	0.059523
30	12	7	6	0.999611	0.008170
30	12	7	7	1.000000	0.000389
30	12	8	0	0.007476	0.007476
30	12	8	1	0.072724	0.065247
30	12	8	2	0.282059	0.209335
30	12	8	3	0.604113	0.322054
30	12	8	4	0.862907	0.258794
30	12	8	5	0.973326	0.110419
30	12	8	6	0.997480	0.024154
30	12	8	7	0.999915	0.002436
30	12	8	8	1.000000	0.000085
30	12	9	0	0.003398	0.003398
30	12	9	1	0.040100	0.036702
30	12	9	2	0.186907	0.146807
30	12	9	3	0.472364	0.285457
30	12	9	4	0.768800	0.296436
30	12	9	5	0.938192	0.169392
30	12	9	6	0.990892	0.052700
30	12	9	7	0.999362	0.008470
30	12	9	8	0.999985	0.000623
30	12	9	9	1.000000	0.000015
30	12	10	0	0.001456	0.001456
30	12	10	1	0.020875	0.019419
30	12	10	2	0.116999	0.096123
30	12	10	3	0.350025	0.233026
30	12	10	4	0.655872	0.305847
30	12	10	5	0.881728	0.225856
30	12	10	6	0.975835	0.094107
30	12	10	7	0.997345	0.021510

N	n	k	x	P(x)	p(x)
30	12	10	8	0.999866	0.002521
30	12	10	9	0.999998	0.000132
30	12	10	10	1.000000	0.000002
30	12	11	0	0.000583	0.000583
30	12	11	1	0.010195	0.009612
30	12	11	2	0.068937	0.058742
30	12	11	3	0.245163	0.176226
30	12	11	4	0.533533	0.288370
30	12	11	5	0.802679	0.269145
30	12	11	6	0.947603	0.144924
30	12	11	7	0.991968	0.044365
30	12	11	8	0.999362	0.007394
30	12	11	9	0.999978	0.000616
30	12	11	10	1.000000	0.000022
30	12	11	11	1.000000	0.000000
30	12	12	0	0.000215	0.000215
30	12	12	1	0.004630	0.004415
30	12	12	2	0.038020	0.033390
30	12	12	3	0.161688	0.123667
30	12	12	4	0.412114	0.250427
30	12	12	5	0.703520	0.291406
30	12	12	6	0.901837	0.198318
30	12	12	7	0.980293	0.078455
30	12	12	8	0.997805	0.017512
30	12	12	9	0.999881	0.002076
30	12	12	10	0.999997	0.000117
30	12	12	11	1.000000	0.000002
30	12	12	12	1.000000	0.000000
30	13	1	0	0.566667	0.566667
30	13	1	1	1.000000	0.433333
30	13	2	0	0.312644	0.312644
30	13	2	1	0.820690	0.508046
30	13	2	2	1.000000	0.179310
30	13	3	0	0.167488	0.167488
30	13	3	1	0.602956	0.435468
30	13	3	2	0.929557	0.326601
30	13	3	3	1.000000	0.070443
30	13	4	0	0.086845	0.086845
30	13	4	1	0.409414	0.322569
30	13	4	2	0.796497	0.387083
30	13	4	3	0.973910	0.177413
30	13	4	4	1.000000	0.026090
30	13	5	0	0.043423	0.043423
30	13	5	1	0.260536	0.217114
30	13	5	2	0.632731	0.372195
30	13	5	3	0.905674	0.272943
30	13	5	4	0.990969	0.085295
30	13	5	5	1.000000	0.009031
30	13	6	0	0.020843	0.020843
30	13	6	1	0.156322	0.135479

N	n	k	x	P(x)	p(x)
30	13	6	2	0.468965	0.312644
30	13	6	3	0.796497	0.327531
30	13	6	4	0.960263	0.163766
30	13	6	5	0.997110	0.036847
30	13	6	6	1.000000	0.002890
30	13	7	0	0.009553	0.009553
30	13	7	1	0.088582	0.079029
30	13	7	2	0.325670	0.237088
30	13	7	3	0.660026	0.334355
30	13	7	4	0.898851	0.238825
30	13	7	5	0.984828	0.085977
30	13	7	6	0.999157	0.014329
30	13	7	7	1.000000	0.000843
30	13	8	0	0.004153	0.004153
30	13	8	1	0.047350	0.043196
30	13	8	2	0.212280	0.164931
30	13	8	3	0.514654	0.302373
30	13	8	4	0.805397	0.290743
30	13	8	5	0.954922	0.149525
30	13	8	6	0.994796	0.039873
30	13	8	7	0.999780	0.004984
30	13	8	8	1.000000	0.000220
30	13	9	0	0.001699	0.001699
30	13	9	1	0.023788	0.022089
30	13	9	2	0.129815	0.106027
30	13	9	3	0.377211	0.247396
30	13	9	4	0.686457	0.309245
30	13	9	5	0.900550	0.214093
30	13	9	6	0.982109	0.081559
30	13	9	7	0.998421	0.016312
30	13	9	8	0.999950	0.001529
30	13	9	9	1.000000	0.000050
30	13	10	0	0.000647	0.000647
30	13	10	1	0.011166	0.010519
30	13	10	2	0.074277	0.063111
30	13	10	3	0.259404	0.185126
30	13	10	4	0.553923	0.294519
30	13	10	5	0.818990	0.265067
30	13	10	6	0.954922	0.135932
30	13	10	7	0.993760	0.038838
30	13	10	8	0.999586	0.005826
30	13	10	9	0.999990	0.000405
30	13	10	10	1.000000	0.000010
30	13	11	0	0.000227	0.000227
30	13	11	1	0.004855	0.004628
30	13	11	2	0.039566	0.034711
30	13	11	3	0.166840	0.127274
30	13	11	4	0.421389	0.254549
30	13	11	5	0.712963	0.291574
30	13	11	6	0.907346	0.194383

Table for $N = 2$, $n = 1$, through $N = 100$, $n = 50$

N	n	k	x	P(x)	p(x)
30	13	11	7	0.982109	0.074763
30	13	11	8	0.998129	0.016021
30	13	11	9	0.999909	0.001780
30	13	11	10	0.999998	0.000089
30	13	11	11	1.000000	0.000001
30	13	12	0	0.000072	0.000072
30	13	12	1	0.001932	0.001860
30	13	12	2	0.019470	0.017538
30	13	12	3	0.099854	0.080384
30	13	12	4	0.300813	0.200960
30	13	12	5	0.590195	0.289382
30	13	12	6	0.835732	0.245536
30	13	12	7	0.958500	0.122768
30	13	12	8	0.993913	0.035414
30	13	12	9	0.999535	0.005621
30	13	12	10	0.999984	0.000450
30	13	12	11	1.000000	0.000015
30	13	12	12	1.000000	0.000000
30	13	13	0	0.000020	0.000020
30	13	13	1	0.000692	0.000672
30	13	13	2	0.008752	0.008061
30	13	13	3	0.055196	0.046444
30	13	13	4	0.200334	0.145138
30	13	13	5	0.461581	0.261248
30	13	13	6	0.740245	0.278664
30	13	13	7	0.917577	0.177332
30	13	13	8	0.984076	0.066499
30	13	13	9	0.998285	0.014209
30	13	13	10	0.999909	0.001624
30	13	13	11	0.999998	0.000089
30	13	13	12	1.000000	0.000002
30	13	13	13	1.000000	0.000000
30	14	1	0	0.533333	0.533333
30	14	1	1	1.000000	0.466667
30	14	2	0	0.275862	0.275862
30	14	2	1	0.790805	0.514943
30	14	2	2	1.000000	0.209195
30	14	3	0	0.137931	0.137931
30	14	3	1	0.551724	0.413793
30	14	3	2	0.910345	0.358621
30	14	3	3	1.000000	0.089655
30	14	4	0	0.066411	0.066411
30	14	4	1	0.352490	0.286079
30	14	4	2	0.750958	0.398467
30	14	4	3	0.963474	0.212516
30	14	4	4	1.000000	0.036526
30	14	5	0	0.030651	0.030651
30	14	5	1	0.209451	0.178799
30	14	5	2	0.567050	0.357599
30	14	5	3	0.873563	0.306513

N	n	k	x	P(x)	p(x)
30	14	5	4	0.985951	0.112388
30	14	5	5	1.000000	0.014049
30	14	6	0	0.013487	0.013487
30	14	6	1	0.116475	0.102988
30	14	6	2	0.395402	0.278927
30	14	6	3	0.738697	0.343295
30	14	6	4	0.940996	0.202299
30	14	6	5	0.994942	0.053946
30	14	6	6	1.000000	0.005057
30	14	7	0	0.005619	0.005619
30	14	7	1	0.060690	0.055070
30	14	7	2	0.255939	0.195249
30	14	7	3	0.581354	0.325415
30	14	7	4	0.856705	0.275351
30	14	7	5	0.974713	0.118008
30	14	7	6	0.998314	0.023602
30	14	7	7	1.000000	0.001686
30	14	8	0	0.002199	0.002199
30	14	8	1	0.029563	0.027364
30	14	8	2	0.154070	0.124507
30	14	8	3	0.425720	0.271651
30	14	8	4	0.736987	0.311267
30	14	8	5	0.928536	0.191549
30	14	8	6	0.990105	0.061569
30	14	8	7	0.999487	0.009382
30	14	8	8	1.000000	0.000513
30	14	9	0	0.000800	0.000800
30	14	9	1	0.013393	0.012594
30	14	9	2	0.086157	0.072764
30	14	9	3	0.289895	0.203738
30	14	9	4	0.595502	0.305607
30	14	9	5	0.850175	0.254673
30	14	9	6	0.967716	0.117541
30	14	9	7	0.996502	0.028786
30	14	9	8	0.999860	0.003358
30	14	9	9	1.000000	0.000140
30	14	10	0	0.000267	0.000267
30	14	10	1	0.005597	0.005331
30	14	10	2	0.044578	0.038980
30	14	10	3	0.183175	0.138597
30	14	10	4	0.449975	0.266800
30	14	10	5	0.741029	0.291054
30	14	10	6	0.922938	0.181909
30	14	10	7	0.986906	0.063968
30	14	10	8	0.998900	0.011994
30	14	10	9	0.999967	0.001066
30	14	10	10	1.000000	0.000033
30	14	11	0	0.000080	0.000080
30	14	11	1	0.002132	0.002052
30	14	11	2	0.021189	0.019057

N	n	k	x	P(x)	p(x)
30	14	11	3	0.106947	0.085757
30	14	11	4	0.316575	0.209628
30	14	11	5	0.610055	0.293480
30	14	11	6	0.850175	0.240120
30	14	11	7	0.964518	0.114343
30	14	11	8	0.995302	0.030785
30	14	11	9	0.999700	0.004398
30	14	11	10	0.999993	0.000293
30	14	11	11	1.000000	0.000007
30	14	12	0	0.000021	0.000021
30	14	12	1	0.000728	0.000707
30	14	12	2	0.009153	0.008425
30	14	12	3	0.057298	0.048144
30	14	12	4	0.206244	0.148947
30	14	12	5	0.471038	0.264794
30	14	12	6	0.749072	0.278034
30	14	12	7	0.922391	0.173320
30	14	12	8	0.985581	0.063189
30	14	12	9	0.998543	0.012962
30	14	12	10	0.999932	0.001389
30	14	12	11	0.999999	0.000067
30	14	12	12	1.000000	0.000001
30	14	13	0	0.000005	0.000005
30	14	13	1	0.000217	0.000213
30	14	13	2	0.003536	0.003319
30	14	13	3	0.027876	0.024340
30	14	13	4	0.123496	0.095620
30	14	13	5	0.338641	0.215145
30	14	13	6	0.625501	0.286860
30	14	13	7	0.854989	0.229488
30	14	13	8	0.964518	0.109528
30	14	13	9	0.994942	0.030425
30	14	13	10	0.999623	0.004681
30	14	13	11	0.999988	0.000365
30	14	13	12	1.000000	0.000012
30	14	13	13	1.000000	0.000000
30	14	14	0	0.000001	0.000001
30	14	14	1	0.000055	0.000054
30	14	14	2	0.001194	0.001139
30	14	14	3	0.012127	0.010933
30	14	14	4	0.067249	0.055122
30	14	14	5	0.224741	0.157492
30	14	14	6	0.490508	0.265767
30	14	14	7	0.760494	0.269986
30	14	14	8	0.925861	0.165366
30	14	14	9	0.985994	0.060133
30	14	14	10	0.998522	0.012528
30	14	14	11	0.999923	0.001402
30	14	14	12	0.999998	0.000075
30	14	14	13	1.000000	0.000002

N	n	k	x	P(x)	p(x)
30	14	14	14	1.000000	0.000000
30	15	1	0	0.500000	0.500000
30	15	1	1	1.000000	0.500000
30	15	2	0	0.241379	0.241379
30	15	2	1	0.758621	0.517241
30	15	2	2	1.000000	0.241379
30	15	3	0	0.112069	0.112069
30	15	3	1	0.500000	0.387931
30	15	3	2	0.887931	0.387931
30	15	3	3	1.000000	0.112069
30	15	4	0	0.049808	0.049808
30	15	4	1	0.298851	0.249042
30	15	4	2	0.701149	0.402299
30	15	4	3	0.950192	0.249042
30	15	4	4	1.000000	0.049808
30	15	5	0	0.021073	0.021073
30	15	5	1	0.164751	0.143678
30	15	5	2	0.500000	0.335249
30	15	5	3	0.835249	0.335249
30	15	5	4	0.978927	0.143678
30	15	5	5	1.000000	0.021073
30	15	6	0	0.008429	0.008429
30	15	6	1	0.084291	0.075862
30	15	6	2	0.325670	0.241379
30	15	6	3	0.674329	0.348659
30	15	6	4	0.915709	0.241379
30	15	6	5	0.991571	0.075862
30	15	6	6	1.000000	0.008429
30	15	7	0	0.003161	0.003161
30	15	7	1	0.040038	0.036877
30	15	7	2	0.194923	0.154885
30	15	7	3	0.500000	0.305077
30	15	7	4	0.805077	0.305077
30	15	7	5	0.959962	0.154885
30	15	7	6	0.996839	0.036877
30	15	7	7	1.000000	0.003161
30	15	8	0	0.001099	0.001099
30	15	8	1	0.017591	0.016492
30	15	8	2	0.107380	0.089788
30	15	8	3	0.340830	0.233450
30	15	8	4	0.659170	0.318341
30	15	8	5	0.892620	0.233450
30	15	8	6	0.982409	0.089788
30	15	8	7	0.998900	0.016492
30	15	8	8	1.000000	0.001099
30	15	9	0	0.000350	0.000350
30	15	9	1	0.007096	0.006747
30	15	9	2	0.054323	0.047226
30	15	9	3	0.213493	0.159170
30	15	9	4	0.500000	0.286507

Table for $N = 2$, $n = 1$, through $N = 100$, $n = 50$

N	n	k	x	P(x)	p(x)
30	15	9	5	0.786507	0.286507
30	15	9	6	0.945677	0.159170
30	15	9	7	0.992904	0.047226
30	15	9	8	0.999650	0.006747
30	15	9	9	1.000000	0.000350
30	15	10	0	0.000100	0.000100
30	15	10	1	0.002599	0.002499
30	15	10	2	0.025087	0.022489
30	15	10	3	0.122539	0.097451
30	15	10	4	0.349925	0.227386
30	15	10	5	0.650075	0.300150
30	15	10	6	0.877461	0.227386
30	15	10	7	0.974912	0.097451
30	15	10	8	0.997401	0.022489
30	15	10	9	0.999900	0.002499
30	15	10	10	1.000000	0.000100
30	15	11	0	0.000025	0.000025
30	15	11	1	0.000850	0.000825
30	15	11	2	0.010470	0.009620
30	15	11	3	0.064068	0.053598
30	15	11	4	0.224863	0.160795
30	15	11	5	0.500000	0.275137
30	15	11	6	0.775137	0.275137
30	15	11	7	0.935932	0.160795
30	15	11	8	0.989530	0.053598
30	15	11	9	0.999150	0.009620
30	15	11	10	0.999975	0.000825
30	15	11	11	1.000000	0.000025
30	15	12	0	0.000005	0.000005
30	15	12	1	0.000242	0.000237
30	15	12	2	0.003888	0.003646
30	15	12	3	0.030216	0.026329
30	15	12	4	0.131771	0.101554
30	15	12	5	0.355191	0.223420
30	15	12	6	0.644809	0.289618
30	15	12	7	0.868229	0.223420
30	15	12	8	0.969783	0.101554
30	15	12	9	0.996112	0.026329
30	15	12	10	0.999758	0.003646
30	15	12	11	0.999995	0.000237
30	15	12	12	1.000000	0.000005
30	15	13	0	0.000001	0.000001
30	15	13	1	0.000058	0.000057
30	15	13	2	0.001255	0.001197
30	15	13	3	0.012664	0.011409
30	15	13	4	0.069710	0.057046
30	15	13	5	0.231069	0.161359
30	15	13	6	0.500000	0.268931
30	15	13	7	0.768931	0.268931
30	15	13	8	0.930290	0.161359

N	n	k	x	P(x)	p(x)
30	15	13	9	0.987336	0.057046
30	15	13	10	0.998745	0.011409
30	15	13	11	0.999942	0.001197
30	15	13	12	0.999999	0.000057
30	15	13	13	1.000000	0.000001
30	15	14	0	0.000000	0.000000
30	15	14	1	0.000011	0.000011
30	15	14	2	0.000339	0.000329
30	15	14	3	0.004610	0.004271
30	15	14	4	0.032798	0.028187
30	15	14	5	0.136152	0.103354
30	15	14	6	0.357625	0.221473
30	15	14	7	0.642375	0.284751
30	15	14	8	0.863848	0.221473
30	15	14	9	0.967202	0.103354
30	15	14	10	0.995390	0.028187
30	15	14	11	0.999660	0.004271
30	15	14	12	0.999989	0.000329
30	15	14	13	1.000000	0.000011
30	15	14	14	1.000000	0.000000
30	15	15	0	0.000000	0.000000
30	15	15	1	0.000001	0.000001
30	15	15	2	0.000073	0.000071
30	15	15	3	0.001407	0.001335
30	15	15	4	0.013419	0.012012
30	15	15	5	0.071555	0.058137
30	15	15	6	0.233046	0.161491
30	15	15	7	0.500000	0.266954
30	15	15	8	0.766954	0.266954
30	15	15	9	0.928444	0.161491
30	15	15	10	0.986581	0.058137
30	15	15	11	0.998593	0.012012
30	15	15	12	0.999927	0.001335
30	15	15	13	0.999998	0.000071
30	15	15	14	1.000000	0.000001
30	15	15	15	1.000000	0.000000
31	1	1	0	0.967742	0.967742
31	1	1	1	1.000000	0.032258
31	2	1	0	0.935484	0.935484
31	2	1	1	1.000000	0.064516
31	2	2	0	0.873118	0.873118
31	2	2	1	0.997849	0.124731
31	2	2	2	1.000000	0.002151
31	3	1	0	0.903226	0.903226
31	3	1	1	1.000000	0.096774
31	3	2	0	0.812903	0.812903
31	3	2	1	0.993548	0.180645
31	3	2	2	1.000000	0.006452
31	3	3	0	0.728810	0.728810
31	3	3	1	0.981090	0.252280

N	n	k	x	P(x)	p(x)
31	3	3	2	0.999778	0.018687
31	3	3	3	1.000000	0.000222
31	4	1	0	0.870968	0.870968
31	4	1	1	1.000000	0.129032
31	4	2	0	0.754839	0.754839
31	4	2	1	0.987097	0.232258
31	4	2	2	1.000000	0.012903
31	4	3	0	0.650723	0.650723
31	4	3	1	0.963070	0.312347
31	4	3	2	0.999110	0.036040
31	4	3	3	1.000000	0.000890
31	4	4	0	0.557763	0.557763
31	4	4	1	0.929604	0.371842
31	4	4	2	0.996536	0.066931
31	4	4	3	0.999968	0.003432
31	4	4	4	1.000000	0.000032
31	5	1	0	0.838710	0.838710
31	5	1	1	1.000000	0.161290
31	5	2	0	0.698925	0.698925
31	5	2	1	0.978495	0.279570
31	5	2	2	1.000000	0.021505
31	5	3	0	0.578420	0.578420
31	5	3	1	0.939933	0.361513
31	5	3	2	0.997775	0.057842
31	5	3	3	1.000000	0.002225
31	5	4	0	0.475131	0.475131
31	5	4	1	0.888289	0.413157
31	5	4	2	0.991578	0.103289
31	5	4	3	0.999841	0.008263
31	5	4	4	1.000000	0.000159
31	5	5	0	0.387144	0.387144
31	5	5	1	0.827080	0.439936
31	5	5	2	0.980101	0.153021
31	5	5	3	0.999229	0.019128
31	5	5	4	0.999994	0.000765
31	5	5	5	1.000000	0.000006
31	6	1	0	0.806452	0.806452
31	6	1	1	1.000000	0.193548
31	6	2	0	0.645161	0.645161
31	6	2	1	0.967742	0.322581
31	6	2	2	1.000000	0.032258
31	6	3	0	0.511680	0.511680
31	6	3	1	0.912125	0.400445
31	6	3	2	0.995551	0.083426
31	6	3	3	1.000000	0.004449
31	6	4	0	0.402034	0.402034
31	6	4	1	0.840617	0.438583
31	6	4	2	0.983633	0.143016
31	6	4	3	0.999523	0.015891
31	6	4	4	1.000000	0.000477

N	n	k	x	P(x)	p(x)
31	6	5	0	0.312693	0.312693
31	6	5	1	0.759398	0.446704
31	6	5	2	0.962445	0.203047
31	6	5	3	0.997758	0.035313
31	6	5	4	0.999965	0.002207
31	6	5	5	1.000000	0.000035
31	6	6	0	0.240533	0.240533
31	6	6	1	0.673493	0.432960
31	6	6	2	0.931207	0.257714
31	6	6	3	0.993683	0.062476
31	6	6	4	0.999795	0.006112
31	6	6	5	0.999999	0.000204
31	6	6	6	1.000000	0.000001
31	7	1	0	0.774194	0.774194
31	7	1	1	1.000000	0.225806
31	7	2	0	0.593548	0.593548
31	7	2	1	0.954839	0.361290
31	7	2	2	1.000000	0.045161
31	7	3	0	0.450278	0.450278
31	7	3	1	0.880089	0.429811
31	7	3	2	0.992214	0.112125
31	7	3	3	1.000000	0.007786
31	7	4	0	0.337709	0.337709
31	7	4	1	0.787987	0.450278
31	7	4	2	0.972191	0.184205
31	7	4	3	0.998888	0.026696
31	7	4	4	1.000000	0.001112
31	7	5	0	0.250154	0.250154
31	7	5	1	0.687925	0.437770
31	7	5	2	0.938079	0.250154
31	7	5	3	0.994933	0.056853
31	7	5	4	0.999876	0.004944
31	7	5	5	1.000000	0.000124
31	7	6	0	0.182805	0.182805
31	7	6	1	0.586901	0.404096
31	7	6	2	0.889973	0.303072
31	7	6	3	0.986186	0.096213
31	7	6	4	0.999306	0.013120
31	7	6	5	0.999990	0.000685
31	7	6	6	1.000000	0.000010
31	7	7	0	0.131620	0.131620
31	7	7	1	0.489918	0.358298
31	7	7	2	0.829358	0.339440
31	7	7	3	0.970792	0.141433
31	7	7	4	0.997731	0.026940
31	7	7	5	0.999936	0.002204
31	7	7	6	0.999999	0.000064
31	7	7	7	1.000000	0.000000
31	8	1	0	0.741935	0.741935
31	8	1	1	1.000000	0.258064

N	n	k	x	P(x)	p(x)
31	8	2	0	0.544086	0.544086
31	8	2	1	0.939785	0.395699
31	8	2	2	1.000000	0.060215
31	8	3	0	0.393993	0.393993
31	8	3	1	0.844271	0.450278
31	8	3	2	0.987542	0.143270
31	8	3	3	1.000000	0.012458
31	8	4	0	0.281424	0.281424
31	8	4	1	0.731702	0.450278
31	8	4	2	0.956841	0.225139
31	8	4	3	0.997775	0.040934
31	8	4	4	1.000000	0.002225
31	8	5	0	0.198039	0.198039
31	8	5	1	0.614963	0.416924
31	8	5	2	0.906810	0.291847
31	8	5	3	0.990195	0.083385
31	8	5	4	0.999670	0.009476
31	8	5	5	1.000000	0.000330
31	8	6	0	0.137104	0.137104
31	8	6	1	0.502714	0.365610
31	8	6	2	0.839461	0.336746
31	8	6	3	0.974159	0.134699
31	8	6	4	0.998213	0.024053
31	8	6	5	0.999962	0.001749
31	8	6	6	1.000000	0.000038
31	8	7	0	0.093231	0.093231
31	8	7	1	0.400343	0.307113
31	8	7	2	0.758642	0.358298
31	8	7	3	0.947219	0.188578
31	8	7	4	0.994364	0.047144
31	8	7	5	0.999752	0.005388
31	8	7	6	0.999997	0.000245
31	8	7	7	1.000000	0.000003
31	8	8	0	0.062154	0.062154
31	8	8	1	0.310769	0.248615
31	8	8	2	0.669067	0.358298
31	8	8	3	0.907932	0.238865
31	8	8	4	0.986507	0.078574
31	8	8	5	0.999078	0.012572
31	8	8	6	0.999976	0.000898
31	8	8	7	1.000000	0.000023
31	8	8	8	1.000000	0.000000
31	9	1	0	0.709677	0.709677
31	9	1	1	1.000000	0.290323
31	9	2	0	0.496774	0.496774
31	9	2	1	0.922581	0.425806
31	9	2	2	1.000000	0.077419
31	9	3	0	0.342603	0.342603
31	9	3	1	0.805117	0.462514
31	9	3	2	0.981312	0.176196

N	n	k	x	P(x)	p(x)
31	9	3	3	1.000000	0.018687
31	9	4	0	0.232480	0.232480
31	9	4	1	0.672970	0.440489
31	9	4	2	0.937264	0.264294
31	9	4	3	0.995995	0.058732
31	9	4	4	1.000000	0.004004
31	9	5	0	0.154987	0.154987
31	9	5	1	0.542455	0.387468
31	9	5	2	0.868743	0.326288
31	9	5	3	0.982944	0.114201
31	9	5	4	0.999258	0.016314
31	9	5	5	1.000000	0.000742
31	9	6	0	0.101338	0.101338
31	9	6	1	0.423234	0.321896
31	9	6	2	0.780896	0.357662
31	9	6	3	0.956590	0.175694
31	9	6	4	0.996121	0.039531
31	9	6	5	0.999886	0.003765
31	9	6	6	1.000000	0.000114
31	9	7	0	0.064856	0.064856
31	9	7	1	0.320227	0.255371
31	9	7	2	0.680751	0.360524
31	9	7	3	0.914423	0.233673
31	9	7	4	0.988215	0.073791
31	9	7	5	0.999283	0.011069
31	9	7	6	0.999986	0.000703
31	9	7	7	1.000000	0.000014
31	9	8	0	0.040535	0.040535
31	9	8	1	0.235103	0.194568
31	9	8	2	0.575598	0.340495
31	9	8	3	0.856005	0.280407
31	9	8	4	0.972841	0.116836
31	9	8	5	0.997439	0.024597
31	9	8	6	0.999898	0.002460
31	9	8	7	0.999999	0.000100
31	9	8	8	1.000000	0.000001
31	9	9	0	0.024674	0.024674
31	9	9	1	0.167427	0.142754
31	9	9	2	0.471969	0.304542
31	9	9	3	0.782855	0.310886
31	9	9	4	0.947442	0.164587
31	9	9	5	0.993161	0.045719
31	9	9	6	0.999578	0.006417
31	9	9	7	0.999990	0.000412
31	9	9	8	1.000000	0.000010
31	9	9	9	1.000000	0.000000
31	10	1	0	0.677419	0.677419
31	10	1	1	1.000000	0.322581
31	10	2	0	0.451613	0.451613
31	10	2	1	0.903226	0.451613

N	n	k	x	P(x)	p(x)
31	10	2	2	1.000000	0.096774
31	10	3	0	0.295884	0.295884
31	10	3	1	0.763070	0.467186
31	10	3	2	0.973304	0.210234
31	10	3	3	1.000000	0.026696
31	10	4	0	0.190211	0.190211
31	10	4	1	0.612903	0.422692
31	10	4	2	0.913237	0.300334
31	10	4	3	0.993326	0.080089
31	10	4	4	1.000000	0.006674
31	10	5	0	0.119763	0.119763
31	10	5	1	0.472006	0.352243
31	10	5	2	0.824249	0.352243
31	10	5	3	0.972562	0.148313
31	10	5	4	0.998517	0.025955
31	10	5	5	1.000000	0.001483
31	10	6	0	0.073700	0.073700
31	10	6	1	0.350076	0.276375
31	10	6	2	0.715867	0.365791
31	10	6	3	0.932632	0.216765
31	10	6	4	0.992527	0.059896
31	10	6	5	0.999715	0.007187
31	10	6	6	1.000000	0.000285
31	10	7	0	0.044220	0.044220
31	10	7	1	0.250580	0.206360
31	10	7	2	0.598813	0.348233
31	10	7	3	0.871937	0.273124
31	10	7	4	0.978152	0.106215
31	10	7	5	0.998277	0.020125
31	10	7	6	0.999954	0.001677
31	10	7	7	1.000000	0.000046
31	10	8	0	0.025795	0.025795
31	10	8	1	0.173195	0.147400
31	10	8	2	0.482736	0.309541
31	10	8	3	0.792276	0.309541
31	10	8	4	0.951599	0.159322
31	10	8	5	0.994085	0.042486
31	10	8	6	0.999675	0.005590
31	10	8	7	0.999994	0.000319
31	10	8	8	1.000000	0.000006
31	10	9	0	0.014580	0.014580
31	10	9	1	0.115517	0.100937
31	10	9	2	0.375070	0.259553
31	10	9	3	0.698068	0.322999
31	10	9	4	0.910036	0.211968
31	10	9	5	0.984848	0.074812
31	10	9	6	0.998703	0.013854
31	10	9	7	0.999953	0.001250
31	10	9	8	0.999999	0.000047
31	10	9	9	1.000000	0.000000

N	n	k	x	P(x)	p(x)
31	10	10	0	0.007953	0.007953
31	10	10	1	0.074224	0.066272
31	10	10	2	0.280687	0.206462
31	10	10	3	0.595296	0.314609
31	10	10	4	0.852227	0.256931
31	10	10	5	0.967846	0.115619
31	10	10	6	0.996184	0.028338
31	10	10	7	0.999782	0.003598
31	10	10	8	0.999995	0.000213
31	10	10	9	1.000000	0.000005
31	10	10	10	1.000000	0.000000
31	11	1	0	0.645161	0.645161
31	11	1	1	1.000000	0.354839
31	11	2	0	0.408602	0.408602
31	11	2	1	0.881720	0.473118
31	11	2	2	1.000000	0.118280
31	11	3	0	0.253615	0.253615
31	11	3	1	0.718576	0.464961
31	11	3	2	0.963293	0.244716
31	11	3	3	1.000000	0.036707
31	11	4	0	0.153981	0.153981
31	11	4	1	0.552519	0.398538
31	11	4	2	0.884634	0.332115
31	11	4	3	0.989512	0.104878
31	11	4	4	1.000000	0.010488
31	11	5	0	0.091248	0.091248
31	11	5	1	0.404912	0.313664
31	11	5	2	0.773929	0.369017
31	11	5	3	0.958437	0.184508
31	11	5	4	0.997281	0.038844
31	11	5	5	1.000000	0.002719
31	11	6	0	0.052643	0.052643
31	11	6	1	0.284272	0.231629
31	11	6	2	0.646192	0.361920
31	11	6	3	0.901665	0.255473
31	11	6	4	0.986823	0.085158
31	11	6	5	0.999372	0.012550
31	11	6	6	1.000000	0.000627
31	11	7	0	0.029480	0.029480
31	11	7	1	0.191620	0.162140
31	11	7	2	0.515901	0.324281
31	11	7	3	0.819914	0.304013
31	11	7	4	0.962979	0.143065
31	11	7	5	0.996361	0.033382
31	11	7	6	0.999874	0.003514
31	11	7	7	1.000000	0.000125
31	11	8	0	0.015968	0.015968
31	11	8	1	0.124062	0.108094
31	11	8	2	0.394296	0.270234
31	11	8	3	0.718576	0.324281

Table for $N = 2$, $n = 1$, through $N = 100$, $n = 50$

N	n	k	x	P(x)	p(x)
31	11	8	4	0.921251	0.202675
31	11	8	5	0.988015	0.066764
31	11	8	6	0.999142	0.011127
31	11	8	7	0.999979	0.000837
31	11	8	8	1.000000	0.000021
31	11	9	0	0.008331	0.008331
31	11	9	1	0.077065	0.068733
31	11	9	2	0.288552	0.211487
31	11	9	3	0.605783	0.317231
31	11	9	4	0.859568	0.253785
31	11	9	5	0.970598	0.111031
31	11	9	6	0.996723	0.026125
31	11	9	7	0.999833	0.003110
31	11	9	8	0.999997	0.000164
31	11	9	9	1.000000	0.000003
31	11	10	0	0.004166	0.004166
31	11	10	1	0.045822	0.041657
31	11	10	2	0.202034	0.156212
31	11	10	3	0.490426	0.288392
31	11	10	4	0.778818	0.288392
31	11	10	5	0.940317	0.161499
31	11	10	6	0.990786	0.050469
31	11	10	7	0.999268	0.008482
31	11	10	8	0.999975	0.000707
31	11	10	9	1.000000	0.000025
31	11	10	10	1.000000	0.000000
31	11	11	0	0.001984	0.001984
31	11	11	1	0.025986	0.024002
31	11	11	2	0.135086	0.109101
31	11	11	3	0.380563	0.245476
31	11	11	4	0.682687	0.302125
31	11	11	5	0.894175	0.211487
31	11	11	6	0.978770	0.084595
31	11	11	7	0.997652	0.018883
31	11	11	8	0.999874	0.002222
31	11	11	9	0.999997	0.000123
31	11	11	10	1.000000	0.000003
31	11	11	11	1.000000	0.000000
31	12	1	0	0.612903	0.612903
31	12	1	1	1.000000	0.387097
31	12	2	0	0.367742	0.367742
31	12	2	1	0.858064	0.490323
31	12	2	2	1.000000	0.141935
31	12	3	0	0.215573	0.215573
31	12	3	1	0.672080	0.456507
31	12	3	2	0.951057	0.278977
31	12	3	3	1.000000	0.048943
31	12	4	0	0.123184	0.123184
31	12	4	1	0.492738	0.369553
31	12	4	2	0.851422	0.358684

N	n	k	x	P(x)	p(x)
31	12	4	3	0.984268	0.132846
31	12	4	4	1.000000	0.015732
31	12	5	0	0.068436	0.068436
31	12	5	1	0.342179	0.273743
31	12	5	2	0.718576	0.376397
31	12	5	3	0.939986	0.221410
31	12	5	4	0.995339	0.055353
31	12	5	5	1.000000	0.004661
31	12	6	0	0.036850	0.036850
31	12	6	1	0.226365	0.189515
31	12	6	2	0.573808	0.347443
31	12	6	3	0.863344	0.289536
31	12	6	4	0.978307	0.114963
31	12	6	5	0.998745	0.020438
31	12	6	6	1.000000	0.001255
31	12	7	0	0.019162	0.019162
31	12	7	1	0.142978	0.123816
31	12	7	2	0.434831	0.291852
31	12	7	3	0.759111	0.324281
31	12	7	4	0.941519	0.182408
31	12	7	5	0.993022	0.051503
31	12	7	6	0.999699	0.006676
31	12	7	7	1.000000	0.000301
31	12	8	0	0.009581	0.009581
31	12	8	1	0.086229	0.076648
31	12	8	2	0.313225	0.226996
31	12	8	3	0.637506	0.324281
31	12	8	4	0.880716	0.243210
31	12	8	5	0.978001	0.097284
31	12	8	6	0.998030	0.020029
31	12	8	7	0.999937	0.001908
31	12	8	8	1.000000	0.000063
31	12	9	0	0.004582	0.004582
31	12	9	1	0.049571	0.044989
31	12	9	2	0.214531	0.164960
31	12	9	3	0.510614	0.296082
31	12	9	4	0.796121	0.285508
31	12	9	5	0.948392	0.152271
31	12	9	6	0.992805	0.044412
31	12	9	7	0.999522	0.006718
31	12	9	8	0.999989	0.000467
31	12	9	9	1.000000	0.000011
31	12	10	0	0.002083	0.002083
31	12	10	1	0.027077	0.024994
31	12	10	2	0.139550	0.112473
31	12	10	3	0.389489	0.249940
31	12	10	4	0.692300	0.302811
31	12	10	5	0.899942	0.207642
31	12	10	6	0.980692	0.080750
31	12	10	7	0.997996	0.017304

N	n	k	x	P(x)	p(x)
31	12	10	8	0.999904	0.001908
31	12	10	9	0.999998	0.000094
31	12	10	10	1.000000	0.000001
31	12	11	0	0.000893	0.000893
31	12	11	1	0.013985	0.013092
31	12	11	2	0.085991	0.072006
31	12	11	3	0.282372	0.196381
31	12	11	4	0.576944	0.294572
31	12	11	5	0.830728	0.253785
31	12	11	6	0.957621	0.126892
31	12	11	7	0.993876	0.036255
31	12	11	8	0.999541	0.005665
31	12	11	9	0.999985	0.000444
31	12	11	10	1.000000	0.000015
31	12	11	11	1.000000	0.000000
31	12	12	0	0.000357	0.000357
31	12	12	1	0.006784	0.006427
31	12	12	2	0.049988	0.043204
31	12	12	3	0.194001	0.144013
31	12	12	4	0.459115	0.265114
31	12	12	5	0.741904	0.282789
31	12	12	6	0.919553	0.177649
31	12	12	7	0.984812	0.065259
31	12	12	8	0.998408	0.013596
31	12	12	9	0.999918	0.001511
31	12	12	10	0.999998	0.000080
31	12	12	11	1.000000	0.000002
31	12	12	12	1.000000	0.000000
31	13	1	0	0.580645	0.580645
31	13	1	1	1.000000	0.419355
31	13	2	0	0.329032	0.329032
31	13	2	1	0.832258	0.503226
31	13	2	2	1.000000	0.167742
31	13	3	0	0.181535	0.181535
31	13	3	1	0.624027	0.442492
31	13	3	2	0.936374	0.312347
31	13	3	3	1.000000	0.063626
31	13	4	0	0.097251	0.097251
31	13	4	1	0.434387	0.337136
31	13	4	2	0.813666	0.379279
31	13	4	3	0.977276	0.163610
31	13	4	4	1.000000	0.022724
31	13	5	0	0.050426	0.050426
31	13	5	1	0.284549	0.234123
31	13	5	2	0.659145	0.374596
31	13	5	3	0.916680	0.257535
31	13	5	4	0.992425	0.075746
31	13	5	5	1.000000	0.007575
31	13	6	0	0.025213	0.025213
31	13	6	1	0.176492	0.151279

N	n	k	x	P(x)	p(x)
31	13	6	2	0.500662	0.324170
31	13	6	3	0.817628	0.316966
31	13	6	4	0.966206	0.148578
31	13	6	5	0.997669	0.031464
31	13	6	6	1.000000	0.002331
31	13	7	0	0.012102	0.012102
31	13	7	1	0.103878	0.091776
31	13	7	2	0.358027	0.254149
31	13	7	3	0.690842	0.332814
31	13	7	4	0.912718	0.221876
31	13	7	5	0.987601	0.074883
31	13	7	6	0.999347	0.011746
31	13	7	7	1.000000	0.000653
31	13	8	0	0.005547	0.005547
31	13	8	1	0.057990	0.052443
31	13	8	2	0.241542	0.183552
31	13	8	3	0.552169	0.310627
31	13	8	4	0.829514	0.277345
31	13	8	5	0.962640	0.133126
31	13	8	6	0.995921	0.033281
31	13	8	7	0.999837	0.003915
31	13	8	8	1.000000	0.000163
31	13	9	0	0.002412	0.002412
31	13	9	1	0.030629	0.028217
31	13	9	2	0.153757	0.123128
31	13	9	3	0.417114	0.263357
31	13	9	4	0.720988	0.303874
31	13	9	5	0.916335	0.195347
31	13	9	6	0.985792	0.069457
31	13	9	7	0.998815	0.013023
31	13	9	8	0.999964	0.001149
31	13	9	9	1.000000	0.000035
31	13	10	0	0.000987	0.000987
31	13	10	1	0.015238	0.014251
31	13	10	2	0.092193	0.076955
31	13	10	3	0.297406	0.205213
31	13	10	4	0.596676	0.299270
31	13	10	5	0.845300	0.248624
31	13	10	6	0.963692	0.118392
31	13	10	7	0.995264	0.031571
31	13	10	8	0.999703	0.004440
31	13	10	9	0.999993	0.000290
31	13	10	10	1.000000	0.000006
31	13	11	0	0.000376	0.000376
31	13	11	1	0.007094	0.006718
31	13	11	2	0.051883	0.044789
31	13	11	3	0.199685	0.147803
31	13	11	4	0.468417	0.268732
31	13	11	5	0.750586	0.282169
31	13	11	6	0.924228	0.173642

Table for $N = 2$, $n = 1$, through $N = 100$, $n = 50$

N	n	k	x	P(x)	p(x)
31	13	11	7	0.986243	0.062015
31	13	11	8	0.998646	0.012403
31	13	11	9	0.999938	0.001292
31	13	11	10	0.999999	0.000061
31	13	11	11	1.000000	0.000001
31	13	12	0	0.000132	0.000132
31	13	12	1	0.003063	0.002932
31	13	12	2	0.027249	0.024186
31	13	12	3	0.125784	0.098535
31	13	12	4	0.347488	0.221704
31	13	12	5	0.637718	0.290231
31	13	12	6	0.863453	0.225735
31	13	12	7	0.967639	0.104185
31	13	12	8	0.995545	0.027907
31	13	12	9	0.999680	0.004134
31	13	12	10	0.999990	0.000310
31	13	12	11	1.000000	0.000010
31	13	12	12	1.000000	0.000000
31	13	13	0	0.000042	0.000042
31	13	13	1	0.001212	0.001170
31	13	13	2	0.013247	0.012035
31	13	13	3	0.073924	0.060677
31	13	13	4	0.242470	0.168547
31	13	13	5	0.515516	0.273046
31	13	13	6	0.780288	0.264772
31	13	13	7	0.934738	0.154450
31	13	13	8	0.988202	0.053464
31	13	13	9	0.998809	0.010608
31	13	13	10	0.999941	0.001131
31	13	13	11	0.999999	0.000058
31	13	13	12	1.000000	0.000001
31	13	13	13	1.000000	0.000000
31	14	1	0	0.548387	0.548387
31	14	1	1	1.000000	0.451613
31	14	2	0	0.292473	0.292473
31	14	2	1	0.804301	0.511828
31	14	2	2	1.000000	0.195699
31	14	3	0	0.151279	0.151279
31	14	3	1	0.574861	0.423582
31	14	3	2	0.919021	0.344160
31	14	3	3	1.000000	0.080979
31	14	4	0	0.075640	0.075640
31	14	4	1	0.378198	0.302558
31	14	4	2	0.771524	0.393326
31	14	4	3	0.968187	0.196663
31	14	4	4	1.000000	0.031813
31	14	5	0	0.036419	0.036419
31	14	5	1	0.232522	0.196103
31	14	5	2	0.596712	0.364191
31	14	5	3	0.888065	0.291353

N	n	k	x	P(x)	p(x)
31	14	5	4	0.988217	0.100152
31	14	5	5	1.000000	0.011783
31	14	6	0	0.016809	0.016809
31	14	6	1	0.134470	0.117662
31	14	6	2	0.428624	0.294154
31	14	6	3	0.764800	0.336176
31	14	6	4	0.949697	0.184897
31	14	6	5	0.995921	0.046224
31	14	6	6	1.000000	0.004079
31	14	7	0	0.007396	0.007396
31	14	7	1	0.073286	0.065890
31	14	7	2	0.287430	0.214144
31	14	7	3	0.616883	0.329452
31	14	7	4	0.875738	0.258855
31	14	7	5	0.979281	0.103542
31	14	7	6	0.998695	0.019414
31	14	7	7	1.000000	0.001305
31	14	8	0	0.003082	0.003082
31	14	8	1	0.037596	0.034514
31	14	8	2	0.180358	0.142763
31	14	8	3	0.465884	0.285525
31	14	8	4	0.767882	0.301998
31	14	8	5	0.940452	0.172570
31	14	8	6	0.992223	0.051771
31	14	8	7	0.999619	0.007396
31	14	8	8	1.000000	0.000381
31	14	9	0	0.001206	0.001206
31	14	9	1	0.018088	0.016882
31	14	9	2	0.105873	0.087786
31	14	9	3	0.329328	0.223455
31	14	9	4	0.636578	0.307250
31	14	9	5	0.872925	0.236346
31	14	9	6	0.974216	0.101291
31	14	9	7	0.997368	0.023152
31	14	9	8	0.999901	0.002532
31	14	9	9	1.000000	0.000099
31	14	10	0	0.000438	0.000438
31	14	10	1	0.008112	0.007674
31	14	10	2	0.057990	0.049878
31	14	10	3	0.217601	0.159611
31	14	10	4	0.496919	0.279318
31	14	10	5	0.776238	0.279318
31	14	10	6	0.937383	0.161145
31	14	10	7	0.990002	0.052619
31	14	10	8	0.999210	0.009208
31	14	10	9	0.999977	0.000767
31	14	10	10	1.000000	0.000023
31	14	11	0	0.000146	0.000146
31	14	11	1	0.003362	0.003216
31	14	11	2	0.029488	0.026127

N	n	k	x	P(x)	p(x)
31	14	11	3	0.133995	0.104507
31	14	11	4	0.363910	0.229915
31	14	11	5	0.656530	0.292619
31	14	11	6	0.875994	0.219464
31	14	11	7	0.972462	0.096468
31	14	11	8	0.996579	0.024117
31	14	11	9	0.999795	0.003216
31	14	11	10	0.999996	0.000201
31	14	11	11	1.000000	0.000004
31	14	12	0	0.000044	0.000044
31	14	12	1	0.001272	0.001228
31	14	12	2	0.013812	0.012541
31	14	12	3	0.076517	0.062704
31	14	12	4	0.248953	0.172436
31	14	12	5	0.524851	0.275898
31	14	12	6	0.788208	0.263357
31	14	12	7	0.938698	0.150490
31	14	12	8	0.989344	0.050646
31	14	12	9	0.998991	0.009647
31	14	12	10	0.999955	0.000965
31	14	12	11	0.999999	0.000044
31	14	12	12	1.000000	0.000001
31	14	13	0	0.000012	0.000012
31	14	13	1	0.000432	0.000420
31	14	13	2	0.005892	0.005460
31	14	13	3	0.040214	0.034322
31	14	13	4	0.158197	0.117983
31	14	13	5	0.394162	0.235966
31	14	13	6	0.677321	0.283159
31	14	13	7	0.883255	0.205934
31	14	13	8	0.973351	0.090096
31	14	13	9	0.996452	0.023102
31	14	13	10	0.999752	0.003300
31	14	13	11	0.999992	0.000240
31	14	13	12	1.000000	0.000007
31	14	13	13	1.000000	0.000000
31	14	14	0	0.000003	0.000003
31	14	14	1	0.000128	0.000126
31	14	14	2	0.002252	0.002123
31	14	14	3	0.019239	0.016988
31	14	14	4	0.092651	0.073411
31	14	14	5	0.276180	0.183529
31	14	14	6	0.551473	0.275293
31	14	14	7	0.803169	0.251697
31	14	14	8	0.943319	0.140149
31	14	14	9	0.990035	0.046716
31	14	14	10	0.999019	0.008984
31	14	14	11	0.999952	0.000933
31	14	14	12	0.999999	0.000047
31	14	14	13	1.000000	0.000001

N	n	k	x	P(x)	p(x)
31	14	14	14	1.000000	0.000000
31	15	1	0	0.516129	0.516129
31	15	1	1	1.000000	0.483871
31	15	2	0	0.258064	0.258064
31	15	2	1	0.774193	0.516129
31	15	2	2	1.000000	0.225806
31	15	3	0	0.124583	0.124583
31	15	3	1	0.525028	0.400445
31	15	3	2	0.898776	0.373749
31	15	3	3	1.000000	0.101224
31	15	4	0	0.057842	0.057842
31	15	4	1	0.324805	0.266963
31	15	4	2	0.725250	0.400445
31	15	4	3	0.956618	0.231368
31	15	4	4	1.000000	0.043382
31	15	5	0	0.025708	0.025708
31	15	5	1	0.186380	0.160672
31	15	5	2	0.532443	0.346064
31	15	5	3	0.853788	0.321345
31	15	5	4	0.982326	0.128538
31	15	5	5	1.000000	0.017674
31	15	6	0	0.010876	0.010876
31	15	6	1	0.099864	0.088988
31	15	6	2	0.359412	0.259548
31	15	6	3	0.705475	0.346064
31	15	6	4	0.927945	0.222469
31	15	6	5	0.993202	0.065258
31	15	6	6	1.000000	0.006798
31	15	7	0	0.004351	0.004351
31	15	7	1	0.050031	0.045680
31	15	7	2	0.224447	0.174416
31	15	7	3	0.539365	0.314918
31	15	7	4	0.830058	0.290693
31	15	7	5	0.967099	0.137041
31	15	7	6	0.997553	0.030454
31	15	7	7	1.000000	0.002447
31	15	8	0	0.001631	0.001631
31	15	8	1	0.023384	0.021753
31	15	8	2	0.129972	0.106588
31	15	8	3	0.381906	0.251934
31	15	8	4	0.696824	0.314918
31	15	8	5	0.909999	0.213175
31	15	8	6	0.986133	0.076134
31	15	8	7	0.999184	0.013052
31	15	8	8	1.000000	0.000816
31	15	9	0	0.000567	0.000567
31	15	9	1	0.010143	0.009576
31	15	9	2	0.069726	0.059583
31	15	9	3	0.250462	0.180735
31	15	9	4	0.546211	0.295749

N	n	k	x	P(x)	p(x)
31	15	9	5	0.817314	0.271103
31	15	9	6	0.956341	0.139027
31	15	9	7	0.994644	0.038303
31	15	9	8	0.999752	0.005107
31	15	9	9	1.000000	0.000248
31	15	10	0	0.000181	0.000181
31	15	10	1	0.004050	0.003869
31	15	10	2	0.034518	0.030469
31	15	10	3	0.151879	0.117361
31	15	10	4	0.398336	0.246457
31	15	10	5	0.694085	0.295749
31	15	10	6	0.899466	0.205381
31	15	10	7	0.980716	0.081250
31	15	10	8	0.998127	0.017411
31	15	10	9	0.999932	0.001806
31	15	10	10	1.000000	0.000068
31	15	11	0	0.000052	0.000052
31	15	11	1	0.001470	0.001419
31	15	11	2	0.015657	0.014186
31	15	11	3	0.084816	0.069159
31	15	11	4	0.269240	0.184424
31	15	11	5	0.553252	0.284013
31	15	11	6	0.811446	0.258193
31	15	11	7	0.949764	0.138318
31	15	11	8	0.992323	0.042559
31	15	11	9	0.999416	0.0C7093
31	15	11	10	0.999984	0.000567
31	15	11	11	1.000000	0.000016
31	15	12	0	0.000013	0.000013
31	15	12	1	0.000477	0.000464
31	15	12	2	0.006435	0.005958
31	15	12	3	0.043320	0.036885
31	15	12	4	0.167806	0.124486
31	15	12	5	0.411246	0.243440
31	15	12	6	0.695259	0.284013
31	15	12	7	0.894437	0.199178
31	15	12	8	0.977427	0.082991
31	15	12	9	0.997288	0.019861
31	15	12	10	0.999842	0.002554
31	15	12	11	0.999997	0.000155
31	15	12	12	1.000000	0.000003
31	15	13	0	0.000003	0.000003
31	15	13	1	0.000135	0.000132
31	15	13	2	0.002359	0.002224
31	15	13	3	0.020025	0.017666
31	15	13	4	0.095735	0.075711
31	15	13	5	0.283120	0.187384
31	15	13	6	0.560726	0.277606
31	15	13	7	0.810572	0.249846
31	15	13	8	0.946852	0.136280

N	n	k	x	P(x)	p(x)
31	15	13	9	0.991016	0.044165
31	15	13	10	0.999170	0.008153
31	15	13	11	0.999964	0.000794
31	15	13	12	0.999999	0.000035
31	15	13	13	1.000000	0.000001
31	15	14	0	0.000000	0.000000
31	15	14	1	0.000032	0.000032
31	15	14	2	0.000753	0.000721
31	15	14	3	0.008247	0.007495
31	15	14	4	0.049468	0.041220
31	15	14	5	0.179017	0.129550
31	15	14	6	0.421923	0.242906
31	15	14	7	0.699530	0.277606
31	15	14	8	0.893854	0.194325
31	15	14	9	0.976295	0.082441
31	15	14	10	0.996905	0.020610
31	15	14	11	0.999788	0.002883
31	15	14	12	0.999993	0.000206
31	15	14	13	1.000000	0.000006
31	15	14	14	1.000000	0.000000
31	15	15	0	0.000000	0.000000
31	15	15	1	0.000006	0.000006
31	15	15	2	0.000202	0.000196
31	15	15	3	0.002957	0.002755
31	15	15	4	0.022796	0.019839
31	15	15	5	0.102812	0.080016
31	15	15	6	0.293326	0.190514
31	15	15	7	0.568891	0.275565
31	15	15	8	0.813838	0.244947
31	15	15	9	0.947198	0.133360
31	15	15	10	0.990843	0.043645
31	15	15	11	0.999109	0.008266
31	15	15	12	0.999957	0.000848
31	15	15	13	0.999999	0.000042
31	15	15	14	1.000000	0.000001
31	15	15	15	1.000000	0.000000
32	1	1	0	0.968750	0.968750
32	1	1	1	1.000000	0.031250
32	2	1	0	0.937500	0.937500
32	2	1	1	1.000000	0.062500
32	2	2	0	0.877016	0.877016
32	2	2	1	0.997984	0.120968
32	2	2	2	1.000000	0.002016
32	3	1	0	0.906250	0.906250
32	3	1	1	1.000000	0.093750
32	3	2	0	0.818548	0.818548
32	3	2	1	0.993952	0.175403
32	3	2	2	1.000000	0.006048
32	3	3	0	0.736694	0.736694
32	3	3	1	0.982258	0.245564

N	n	k	x	P(x)	p(x)
32	3	3	2	0.999798	0.017540
32	3	3	3	1.000000	0.000202
32	4	1	0	0.875000	0.875000
32	4	1	1	1.000000	0.125000
32	4	2	0	0.762097	0.762097
32	4	2	1	0.987903	0.225806
32	4	2	2	1.000000	0.012097
32	4	3	0	0.660484	0.660484
32	4	3	1	0.965323	0.304839
32	4	3	2	0.999194	0.033871
32	4	3	3	1.000000	0.000806
32	4	4	0	0.569383	0.569383
32	4	4	1	0.933787	0.364405
32	4	4	2	0.996858	0.063070
32	4	4	3	0.999972	0.003115
32	4	4	4	1.000000	0.000028
32	5	1	0	0.843750	0.843750
32	5	1	1	1.000000	0.156250
32	5	2	0	0.707661	0.707661
32	5	2	1	0.979839	0.272177
32	5	2	2	1.000000	0.020161
32	5	3	0	0.589718	0.589718
32	5	3	1	0.943548	0.353831
32	5	3	2	0.997984	0.054435
32	5	3	3	1.000000	0.002016
32	5	4	0	0.488042	0.488042
32	5	4	1	0.894744	0.406702
32	5	4	2	0.992353	0.097608
32	5	4	3	0.999861	0.007508
32	5	4	4	1.000000	0.000139
32	5	5	0	0.400892	0.400892
32	5	5	1	0.836644	0.435752
32	5	5	2	0.981895	0.145251
32	5	5	3	0.999325	0.017430
32	5	5	4	0.999995	0.000670
32	5	5	5	1.000000	0.000005
32	6	1	0	0.812500	0.812500
32	6	1	1	1.000000	0.187500
32	6	2	0	0.655242	0.655242
32	6	2	1	0.969758	0.314516
32	6	2	2	1.000000	0.030242
32	6	3	0	0.524194	0.524194
32	6	3	1	0.917339	0.393145
32	6	3	2	0.995968	0.078629
32	6	3	3	1.000000	0.004032
32	6	4	0	0.415740	0.415740
32	6	4	1	0.849555	0.433815
32	6	4	2	0.985122	0.135567
32	6	4	3	0.999583	0.014461
32	6	4	4	1.000000	0.000417

N	n	k	x	P(x)	p(x)
32	6	5	0	0.326653	0.326653
32	6	5	1	0.772088	0.445435
32	6	5	2	0.965756	0.193668
32	6	5	3	0.998033	0.032278
32	6	5	4	0.999970	0.001937
32	6	5	5	1.000000	0.000030
32	6	6	0	0.254063	0.254063
32	6	6	1	0.689600	0.435537
32	6	6	2	0.937064	0.247464
32	6	6	3	0.994447	0.057383
32	6	6	4	0.999827	0.005380
32	6	6	5	0.999999	0.000172
32	6	6	6	1.000000	0.000001
32	7	1	0	0.781250	0.781250
32	7	1	1	1.000000	0.218750
32	7	2	0	0.604839	0.604839
32	7	2	1	0.957661	0.352823
32	7	2	2	1.000000	0.042339
32	7	3	0	0.463710	0.463710
32	7	3	1	0.887097	0.423387
32	7	3	2	0.992944	0.105847
32	7	3	3	1.000000	0.007056
32	7	4	0	0.351780	0.351780
32	7	4	1	0.799499	0.447720
32	7	4	2	0.974694	0.175195
32	7	4	3	0.999027	0.024333
32	7	4	4	1.000000	0.000973
32	7	5	0	0.263835	0.263835
32	7	5	1	0.703559	0.439725
32	7	5	2	0.943409	0.239850
32	7	5	3	0.995551	0.052141
32	7	5	4	0.999896	0.004345
32	7	5	5	1.000000	0.000104
32	7	6	0	0.195433	0.195433
32	7	6	1	0.605843	0.410410
32	7	6	2	0.898993	0.293150
32	7	6	3	0.987826	0.088833
32	7	6	4	0.999413	0.011587
32	7	6	5	0.999992	0.000579
32	7	6	6	1.000000	0.000008
32	7	7	0	0.142817	0.142817
32	7	7	1	0.511133	0.368316
32	7	7	2	0.842618	0.331485
32	7	7	3	0.974159	0.131542
32	7	7	4	0.998076	0.023917
32	7	7	5	0.999948	0.001872
32	7	7	6	1.000000	0.000052
32	7	7	7	1.000000	0.000000
32	8	1	0	0.750000	0.750000
32	8	1	1	1.000000	0.250000

Table for $N = 2$, $n = 1$, through $N = 100$, $n = 50$

N	n	k	x	P(x)	p(x)
32	8	2	0	0.556452	0.556452
32	8	2	1	0.943548	0.387097
32	8	2	2	1.000000	0.056452
32	8	3	0	0.408065	0.408065
32	8	3	1	0.853226	0.445161
32	8	3	2	0.988710	0.135484
32	8	3	3	1.000000	0.011290
32	8	4	0	0.295495	0.295495
32	8	4	1	0.745773	0.450278
32	8	4	2	0.960678	0.214905
32	8	4	3	0.998053	0.037375
32	8	4	4	1.000000	0.001947
32	8	5	0	0.211068	0.211068
32	8	5	1	0.633204	0.422136
32	8	5	2	0.914627	0.281424
32	8	5	3	0.991379	0.076752
32	8	5	4	0.999722	0.008343
32	8	5	5	1.000000	0.000278
32	8	6	0	0.148529	0.148529
32	8	6	1	0.523761	0.375232
32	8	6	2	0.852089	0.328328
32	8	6	3	0.977166	0.125077
32	8	6	4	0.998486	0.021320
32	8	6	5	0.999969	0.001483
32	8	6	6	1.000000	0.000031
32	8	7	0	0.102828	0.102828
32	8	7	1	0.422737	0.319909
32	8	7	2	0.776321	0.353584
32	8	7	3	0.953113	0.176792
32	8	7	4	0.995206	0.042093
32	8	7	5	0.999798	0.004592
32	8	7	6	0.999998	0.000200
32	8	7	7	1.000000	0.000002
32	8	8	0	0.069923	0.069923
32	8	8	1	0.333162	0.263239
32	8	8	2	0.691461	0.358298
32	8	8	3	0.917754	0.226294
32	8	8	4	0.988471	0.070717
32	8	8	5	0.999247	0.010776
32	8	8	6	0.999982	0.000735
32	8	8	7	1.000000	0.000018
32	8	8	8	1.000000	0.000000
32	9	1	0	0.718750	0.718750
32	9	1	1	1.000000	0.281250
32	9	2	0	0.510081	0.510081
32	9	2	1	0.927419	0.417339
32	9	2	2	1.000000	0.072581
32	9	3	0	0.357056	0.357056
32	9	3	1	0.816129	0.459073
32	9	3	2	0.983064	0.166935

N	n	k	x	P(x)	p(x)
32	9	3	3	1.000000	0.016935
32	9	4	0	0.246246	0.246246
32	9	4	1	0.689488	0.443242
32	9	4	2	0.942770	0.253281
32	9	4	3	0.996496	0.053726
32	9	4	4	1.000000	0.003504
32	9	5	0	0.167095	0.167095
32	9	5	1	0.562848	0.395752
32	9	5	2	0.879449	0.316602
32	9	5	3	0.984983	0.105534
32	9	5	4	0.999374	0.014391
32	9	5	5	1.000000	0.000626
32	9	6	0	0.111397	0.111397
32	9	6	1	0.445588	0.334191
32	9	6	2	0.797367	0.351780
32	9	6	3	0.961531	0.164164
32	9	6	4	0.996709	0.035178
32	9	6	5	0.999907	0.003198
32	9	6	6	1.000000	0.000093
32	9	7	0	0.072836	0.072836
32	9	7	1	0.342760	0.269923
32	9	7	2	0.702657	0.359898
32	9	7	3	0.923647	0.220990
32	9	7	4	0.989944	0.066297
32	9	7	5	0.999415	0.009471
32	9	7	6	0.999989	0.000574
32	9	7	7	1.000000	0.000011
32	9	8	0	0.046615	0.046615
32	9	8	1	0.256384	0.209769
32	9	8	2	0.601886	0.345502
32	9	8	3	0.870610	0.268724
32	9	8	4	0.976685	0.106075
32	9	8	5	0.997900	0.021215
32	9	8	6	0.999920	0.002020
32	9	8	7	0.999999	0.000079
32	9	8	8	1.000000	0.000001
32	9	9	0	0.029135	0.029135
32	9	9	1	0.186461	0.157327
32	9	9	2	0.501115	0.314653
32	9	9	3	0.803429	0.302314
32	9	9	4	0.954586	0.151157
32	9	9	5	0.994364	0.039778
32	9	9	6	0.999668	0.005304
32	9	9	7	0.999993	0.000325
32	9	9	8	1.000000	0.000007
32	9	9	9	1.000000	0.000000
32	10	1	0	0.687500	0.687500
32	10	1	1	1.000000	0.312500
32	10	2	0	0.465726	0.465726
32	10	2	1	0.909274	0.443548

N	n	k	x	P(x)	p(x)
32	10	2	2	1.000000	0.090726
32	10	3	0	0.310484	0.310484
32	10	3	1	0.776210	0.465726
32	10	3	2	0.975806	0.199597
32	10	3	3	1.000000	0.024194
32	10	4	0	0.203420	0.203420
32	10	4	1	0.631674	0.428254
32	10	4	2	0.920745	0.289071
32	10	4	3	0.994160	0.073415
32	10	4	4	1.000000	0.005840
32	10	5	0	0.130770	0.130770
32	10	5	1	0.494021	0.363251
32	10	5	2	0.838153	0.344132
32	10	5	3	0.975806	0.137653
32	10	5	4	0.998749	0.022942
32	10	5	5	1.000000	0.001251
32	10	6	0	0.082337	0.082337
32	10	6	1	0.372937	0.290601
32	10	6	2	0.736188	0.363251
32	10	6	3	0.940119	0.203930
32	10	6	4	0.993650	0.053532
32	10	6	5	0.999768	0.006118
32	10	6	6	1.000000	0.000232
32	10	7	0	0.050669	0.050669
32	10	7	1	0.272345	0.221676
32	10	7	2	0.624419	0.352074
32	10	7	3	0.885214	0.260795
32	10	7	4	0.981297	0.096083
32	10	7	5	0.998592	0.017295
32	10	7	6	0.999964	0.001373
32	10	7	7	1.000000	0.000036
32	10	8	0	0.030401	0.030401
32	10	8	1	0.192542	0.162140
32	10	8	2	0.511755	0.319214
32	10	8	3	0.812192	0.300436
32	10	8	4	0.958237	0.146045
32	10	8	5	0.995133	0.036896
32	10	8	6	0.999745	0.004612
32	10	8	7	0.999996	0.000251
32	10	8	8	1.000000	0.000004
32	10	9	0	0.017734	0.017734
32	10	9	1	0.131739	0.114005
32	10	9	2	0.405351	0.273612
32	10	9	3	0.724564	0.319214
32	10	9	4	0.921726	0.197161
32	10	9	5	0.987446	0.065720
32	10	9	6	0.998976	0.011530
32	10	9	7	0.999964	0.000988
32	10	9	8	0.999999	0.000035
32	10	9	9	1.000000	0.000000

N	n	k	x	P(x)	p(x)
32	10	10	0	0.010024	0.010024
32	10	10	1	0.087128	0.077105
32	10	10	2	0.310181	0.223053
32	10	10	3	0.627412	0.317231
32	10	10	4	0.870292	0.242880
32	10	10	5	0.973159	0.102867
32	10	10	6	0.996971	0.023812
32	10	10	7	0.999835	0.002865
32	10	10	8	0.999996	0.000161
32	10	10	9	1.000000	0.000003
32	10	10	10	1.000000	0.000000
32	11	1	0	0.656250	0.656250
32	11	1	1	1.000000	0.343750
32	11	2	0	0.423387	0.423387
32	11	2	1	0.889113	0.465726
32	11	2	2	1.000000	0.110887
32	11	3	0	0.268145	0.268145
32	11	3	1	0.733871	0.465726
32	11	3	2	0.966734	0.232863
32	11	3	3	1.000000	0.033266
32	11	4	0	0.166435	0.166435
32	11	4	1	0.573276	0.406841
32	11	4	2	0.894466	0.321190
32	11	4	3	0.990823	0.096357
32	11	4	4	1.000000	0.009177
32	11	5	0	0.101050	0.101050
32	11	5	1	0.427975	0.326926
32	11	5	2	0.791226	0.363251
32	11	5	3	0.963292	0.172066
32	11	5	4	0.997706	0.034413
32	11	5	5	1.000000	0.002294
32	11	6	0	0.059881	0.059881
32	11	6	1	0.306892	0.247011
32	11	6	2	0.670143	0.363251
32	11	6	3	0.912310	0.242167
32	11	6	4	0.988784	0.076474
32	11	6	5	0.999490	0.010706
32	11	6	6	1.000000	0.000510
32	11	7	0	0.034547	0.034547
32	11	7	1	0.211888	0.177341
32	11	7	2	0.544402	0.332514
32	11	7	3	0.837797	0.293395
32	11	7	4	0.968195	0.130398
32	11	7	5	0.997019	0.028825
32	11	7	6	0.999902	0.002882
32	11	7	7	1.000000	0.000098
32	11	8	0	0.019346	0.019346
32	11	8	1	0.140951	0.121605
32	11	8	2	0.424697	0.283745
32	11	8	3	0.743911	0.319214

N	n	k	x	P(x)	p(x)
32	11	8	4	0.931683	0.187773
32	11	8	5	0.990101	0.058418
32	11	8	6	0.999325	0.009224
32	11	8	7	0.999984	0.000659
32	11	8	8	1.000000	0.000016
32	11	9	0	0.010479	0.010479
32	11	9	1	0.090283	0.079803
32	11	9	2	0.318292	0.228010
32	11	9	3	0.637506	0.319214
32	11	9	4	0.876916	0.239410
32	11	9	5	0.975497	0.098581
32	11	9	6	0.997404	0.021907
32	11	9	7	0.999874	0.002471
32	11	9	8	0.999998	0.000124
32	11	9	9	1.000000	0.000002
32	11	10	0	0.005467	0.005467
32	11	10	1	0.055586	0.050118
32	11	10	2	0.229071	0.173486
32	11	10	3	0.526475	0.297404
32	11	10	4	0.804052	0.277577
32	11	10	5	0.949780	0.145728
32	11	10	6	0.992641	0.042861
32	11	10	7	0.999445	0.006803
32	11	10	8	0.999982	0.000537
32	11	10	9	1.000000	0.000018
32	11	10	10	1.000000	0.000000
32	11	11	0	0.002734	0.002734
32	11	11	1	0.032805	0.030071
32	11	11	2	0.158100	0.125295
32	11	11	3	0.418328	0.260229
32	11	11	4	0.715732	0.297404
32	11	11	5	0.910036	0.194304
32	11	11	6	0.982900	0.072864
32	11	11	7	0.998208	0.015308
32	11	11	8	0.999909	0.001701
32	11	11	9	0.999998	0.000090
32	11	11	10	1.000000	0.000002
32	11	11	11	1.000000	0.000000
32	12	1	0	0.625000	0.625000
32	12	1	1	1.000000	0.375000
32	12	2	0	0.383065	0.383065
32	12	2	1	0.866935	0.483871
32	12	2	2	1.000000	0.133065
32	12	3	0	0.229839	0.229839
32	12	3	1	0.689516	0.459677
32	12	3	2	0.955645	0.266129
32	12	3	3	1.000000	0.044355
32	12	4	0	0.134733	0.134733
32	12	4	1	0.515156	0.380423
32	12	4	2	0.863876	0.348721

N	n	k	x	P(x)	p(x)
32	12	4	3	0.986235	0.122358
32	12	4	4	1.000000	0.013765
32	12	5	0	0.076990	0.076990
32	12	5	1	0.365704	0.288714
32	12	5	2	0.739333	0.373629
32	12	5	3	0.946905	0.207572
32	12	5	4	0.996067	0.049162
32	12	5	5	1.000000	0.003933
32	12	6	0	0.042772	0.042772
32	12	6	1	0.248080	0.205307
32	12	6	2	0.600952	0.352872
32	12	6	3	0.877715	0.276762
32	12	6	4	0.981500	0.103786
32	12	6	5	0.998980	0.017480
32	12	6	6	1.000000	0.001020
32	12	7	0	0.023031	0.023031
32	12	7	1	0.161219	0.138188
32	12	7	2	0.465232	0.304013
32	12	7	3	0.781912	0.316680
32	12	7	4	0.949566	0.167654
32	12	7	5	0.994274	0.044708
32	12	7	6	0.999765	0.005490
32	12	7	7	1.000000	0.000235
32	12	8	0	0.011976	0.011976
32	12	8	1	0.100416	0.088440
32	12	8	2	0.343627	0.243210
32	12	8	3	0.667907	0.324281
32	12	8	4	0.895917	0.228010
32	12	8	5	0.981756	0.085839
32	12	8	6	0.998447	0.016691
32	12	8	7	0.999953	0.001506
32	12	8	8	1.000000	0.000047
32	12	9	0	0.005988	0.005988
32	12	9	1	0.059881	0.053893
32	12	9	2	0.242289	0.182408
32	12	9	3	0.546302	0.304013
32	12	9	4	0.819914	0.273612
32	12	9	5	0.956720	0.136806
32	12	9	6	0.994274	0.037555
32	12	9	7	0.999639	0.005365
32	12	9	8	0.999992	0.000353
32	12	9	9	1.000000	0.000008
32	12	10	0	0.002864	0.002864
32	12	10	1	0.034106	0.031242
32	12	10	2	0.162981	0.128875
32	12	10	3	0.427340	0.264359
32	12	10	4	0.724744	0.297404
32	12	10	5	0.915083	0.190339
32	12	10	6	0.984477	0.069394
32	12	10	7	0.998473	0.013995

N	n	k	x	P(x)	p(x)
32	12	10	8	0.999931	0.001458
32	12	10	9	0.999999	0.000068
32	12	10	10	1.000000	0.000001
32	12	11	0	0.001302	0.001302
32	12	11	1	0.018485	0.017183
32	12	11	2	0.104402	0.085917
32	12	11	3	0.319194	0.214792
32	12	11	4	0.616598	0.297404
32	12	11	5	0.854521	0.237923
32	12	11	6	0.965552	0.111031
32	12	11	7	0.995292	0.029740
32	12	11	8	0.999666	0.004374
32	12	11	9	0.999990	0.000324
32	12	11	10	1.000000	0.000010
32	12	11	11	1.000000	0.000000
32	12	12	0	0.000558	0.000558
32	12	12	1	0.009484	0.008926
32	12	12	2	0.063489	0.054005
32	12	12	3	0.227140	0.163651
32	12	12	4	0.503301	0.276161
32	12	12	5	0.775213	0.271912
32	12	12	6	0.933829	0.158615
32	12	12	7	0.988211	0.054382
32	12	12	8	0.998833	0.010622
32	12	12	9	0.999943	0.001111
32	12	12	10	0.999999	0.000056
32	12	12	11	1.000000	0.000001
32	12	12	12	1.000000	0.000000
32	13	1	0	0.593750	0.593750
32	13	1	1	1.000000	0.406250
32	13	2	0	0.344758	0.344758
32	13	2	1	0.842742	0.497984
32	13	2	2	1.000000	0.157258
32	13	3	0	0.195363	0.195363
32	13	3	1	0.643548	0.448185
32	13	3	2	0.942339	0.298790
32	13	3	3	1.000000	0.057661
32	13	4	0	0.107786	0.107786
32	13	4	1	0.458092	0.350306
32	13	4	2	0.829004	0.370912
32	13	4	3	0.980117	0.151112
32	13	4	4	1.000000	0.019883
32	13	5	0	0.057743	0.057743
32	13	5	1	0.307961	0.250218
32	13	5	2	0.683289	0.375328
32	13	5	3	0.926148	0.242859
32	13	5	4	0.993609	0.067461
32	13	5	5	1.000000	0.006391
32	13	6	0	0.029941	0.029941
32	13	6	1	0.196753	0.166812

N	n	k	x	P(x)	p(x)
32	13	6	2	0.530378	0.333625
32	13	6	3	0.836200	0.305823
32	13	6	4	0.971122	0.134922
32	13	6	5	0.998106	0.026984
32	13	6	6	1.000000	0.001894
32	13	7	0	0.014970	0.014970
32	13	7	1	0.119763	0.104792
32	13	7	2	0.389229	0.269466
32	13	7	3	0.718576	0.329347
32	13	7	4	0.924418	0.205842
32	13	7	5	0.989803	0.065385
32	13	7	6	0.999490	0.009687
32	13	7	7	1.000000	0.000510
32	13	8	0	0.007186	0.007186
32	13	8	1	0.069462	0.062277
32	13	8	2	0.270664	0.201201
32	13	8	3	0.586837	0.316174
32	13	8	4	0.850315	0.263478
32	13	8	5	0.968880	0.118565
32	13	8	6	0.996778	0.027898
32	13	8	7	0.999878	0.003100
32	13	8	8	1.000000	0.000122
32	13	9	0	0.003293	0.003293
32	13	9	1	0.038324	0.035031
32	13	9	2	0.178446	0.140122
32	13	9	3	0.455098	0.276652
32	13	9	4	0.751511	0.296413
32	13	9	5	0.929358	0.177848
32	13	9	6	0.988641	0.059283
32	13	9	7	0.999103	0.010462
32	13	9	8	0.999974	0.000872
32	13	9	9	1.000000	0.000025
32	13	10	0	0.001432	0.001432
32	13	10	1	0.020047	0.018615
32	13	10	2	0.111431	0.091384
32	13	10	3	0.334815	0.223383
32	13	10	4	0.635523	0.300709
32	13	10	5	0.867498	0.231975
32	13	10	6	0.970598	0.103100
32	13	10	7	0.996373	0.025775
32	13	10	8	0.999785	0.003411
32	13	10	9	0.999995	0.000211
32	13	10	10	1.000000	0.000004
32	13	11	0	0.000586	0.000586
32	13	11	1	0.009893	0.009308
32	13	11	2	0.065739	0.055846
32	13	11	3	0.233277	0.167538
32	13	11	4	0.512506	0.279229
32	13	11	5	0.783144	0.270638
32	13	11	6	0.937794	0.154650

Table for $N = 2$, $n = 1$, through $N = 100$, $n = 50$

N	n	k	x	P(x)	p(x)
32	13	11	7	0.989344	0.051550
32	13	11	8	0.999010	0.009666
32	13	11	9	0.999957	0.000948
32	13	11	10	0.999999	0.000042
32	13	11	11	1.000000	0.000001
32	13	12	0	0.000223	0.000223
32	13	12	1	0.004575	0.004352
32	13	12	2	0.036487	0.031912
32	13	12	3	0.153497	0.117010
32	13	12	4	0.392836	0.239339
32	13	12	5	0.680044	0.287207
32	13	12	6	0.886244	0.206200
32	13	12	7	0.974615	0.088371
32	13	12	8	0.996708	0.022093
32	13	12	9	0.999777	0.003068
32	13	12	10	0.999993	0.000217
32	13	12	11	1.000000	0.000007
32	13	12	12	1.000000	0.000000
32	13	13	0	0.000078	0.000078
32	13	13	1	0.001964	0.001886
32	13	13	2	0.018935	0.016971
32	13	13	3	0.094992	0.076057
32	13	13	4	0.285134	0.190142
32	13	13	5	0.565161	0.280027
32	13	13	6	0.814074	0.248913
32	13	13	7	0.948104	0.134030
32	13	13	8	0.991185	0.043081
32	13	13	9	0.999163	0.007978
32	13	13	10	0.999961	0.000798
32	13	13	11	0.999999	0.000038
32	13	13	12	1.000000	0.000001
32	13	13	13	1.000000	0.000000
32	14	1	0	0.562500	0.562500
32	14	1	1	1.000000	0.437500
32	14	2	0	0.308468	0.308468
32	14	2	1	0.816532	0.508064
32	14	2	2	1.000000	0.183468
32	14	3	0	0.164516	0.164516
32	14	3	1	0.596371	0.431855
32	14	3	2	0.926613	0.330242
32	14	3	3	1.000000	0.073387
32	14	4	0	0.085095	0.085095
32	14	4	1	0.402781	0.317686
32	14	4	2	0.789961	0.387180
32	14	4	3	0.972163	0.182202
32	14	4	4	1.000000	0.027836
32	14	5	0	0.042547	0.042547
32	14	5	1	0.255284	0.212736
32	14	5	2	0.624027	0.368743
32	14	5	3	0.900584	0.276557

N	n	k	x	P(x)	p(x)
32	14	5	4	0.990058	0.089474
32	14	5	5	1.000000	0.009942
32	14	6	0	0.020486	0.020486
32	14	6	1	0.152855	0.132369
32	14	6	2	0.460141	0.307286
32	14	6	3	0.787912	0.327772
32	14	6	4	0.956920	0.169007
32	14	6	5	0.996686	0.039766
32	14	6	6	1.000000	0.003314
32	14	7	0	0.009455	0.009455
32	14	7	1	0.086670	0.077215
32	14	7	2	0.318317	0.231646
32	14	7	3	0.649240	0.330923
32	14	7	4	0.891917	0.242677
32	14	7	5	0.982921	0.091004
32	14	7	6	0.998980	0.016060
32	14	7	7	1.000000	0.001020
32	14	8	0	0.004160	0.004160
32	14	8	1	0.046518	0.042358
32	14	8	2	0.207126	0.160608
32	14	8	3	0.503634	0.296507
32	14	8	4	0.794846	0.291212
32	14	8	5	0.950159	0.155313
32	14	8	6	0.993841	0.043682
32	14	8	7	0.999714	0.005873
32	14	8	8	1.000000	0.000285
32	14	9	0	0.001733	0.001733
32	14	9	1	0.023574	0.021841
32	14	9	2	0.126822	0.103248
32	14	9	3	0.367734	0.240912
32	14	9	4	0.673508	0.305773
32	14	9	5	0.891917	0.218409
32	14	9	6	0.979281	0.087364
32	14	9	7	0.998001	0.018721
32	14	9	8	0.999929	0.001927
32	14	9	9	1.000000	0.000071
32	14	10	0	0.000678	0.000678
32	14	10	1	0.011229	0.010551
32	14	10	2	0.072954	0.061724
32	14	10	3	0.252516	0.179562
32	14	10	4	0.540563	0.288047
32	14	10	5	0.806452	0.265890
32	14	10	6	0.948893	0.142441
32	14	10	7	0.992304	0.043411
32	14	10	8	0.999426	0.007122
32	14	10	9	0.999984	0.000559
32	14	10	10	1.000000	0.000016
32	14	11	0	0.000247	0.000247
32	14	11	1	0.004995	0.004748
32	14	11	2	0.039286	0.034291

N	n	k	x	P(x)	p(x)
32	14	11	3	0.162735	0.123449
32	14	11	4	0.409632	0.246897
32	14	11	5	0.697679	0.288047
32	14	11	6	0.897096	0.199417
32	14	11	7	0.978491	0.081395
32	14	11	8	0.997483	0.018992
32	14	11	9	0.999857	0.002374
32	14	11	10	0.999997	0.000140
32	14	11	11	1.000000	0.000003
32	14	12	0	0.000082	0.000082
32	14	12	1	0.002055	0.001973
32	14	12	2	0.019691	0.017636
32	14	12	3	0.098071	0.078380
32	14	12	4	0.292062	0.193991
32	14	12	5	0.574231	0.282169
32	14	12	6	0.821128	0.246897
32	14	12	7	0.951360	0.130232
32	14	12	8	0.992057	0.040697
32	14	12	9	0.999292	0.007235
32	14	12	10	0.999970	0.000678
32	14	12	11	0.999999	0.000029
32	14	12	12	1.000000	0.000000
32	14	13	0	0.000025	0.000025
32	14	13	1	0.000773	0.000748
32	14	13	2	0.009110	0.008337
32	14	13	3	0.054962	0.045852
32	14	13	4	0.195067	0.140105
32	14	13	5	0.447255	0.252188
32	14	13	6	0.722369	0.275114
32	14	13	7	0.905779	0.183410
32	14	13	8	0.979848	0.074069
32	14	13	9	0.997483	0.017636
32	14	13	10	0.999835	0.002351
32	14	13	11	0.999995	0.000160
32	14	13	12	1.000000	0.000005
32	14	13	13	1.000000	0.000000
32	14	14	0	0.000006	0.000006
32	14	14	1	0.000261	0.000254
32	14	14	2	0.003844	0.003583
32	14	14	3	0.028416	0.024572
32	14	14	4	0.121327	0.092911
32	14	14	5	0.327797	0.206470
32	14	14	6	0.606531	0.278734
32	14	14	7	0.838207	0.231675
32	14	14	8	0.956458	0.118251
32	14	14	9	0.992843	0.036385
32	14	14	10	0.999340	0.006497
32	14	14	11	0.999970	0.000630
32	14	14	12	0.999999	0.000030
32	14	14	13	1.000000	0.000001

N	n	k	x	P(x)	p(x)
32	14	14	14	1.000000	0.000000
32	15	1	0	0.531250	0.531250
32	15	1	1	1.000000	0.468750
32	15	2	0	0.274194	0.274194
32	15	2	1	0.788306	0.514113
32	15	2	2	1.000000	0.211694
32	15	3	0	0.137097	0.137097
32	15	3	1	0.548387	0.411290
32	15	3	2	0.908266	0.359879
32	15	3	3	1.000000	0.091734
32	15	4	0	0.066185	0.066185
32	15	4	1	0.349833	0.283648
32	15	4	2	0.746941	0.397108
32	15	4	3	0.962041	0.215100
32	15	4	4	1.000000	0.037959
32	15	5	0	0.030729	0.030729
32	15	5	1	0.208009	0.177280
32	15	5	2	0.562569	0.354561
32	15	5	3	0.869855	0.307286
32	15	5	4	0.985088	0.115232
32	15	5	5	1.000000	0.014912
32	15	6	0	0.013657	0.013657
32	15	6	1	0.116086	0.102429
32	15	6	2	0.391855	0.275769
32	15	6	3	0.733284	0.341429
32	15	6	4	0.938141	0.204857
32	15	6	5	0.994477	0.056336
32	15	6	6	1.000000	0.005523
32	15	7	0	0.005778	0.005778
32	15	7	1	0.060932	0.055154
32	15	7	2	0.253970	0.193039
32	15	7	3	0.575701	0.321731
32	15	7	4	0.851471	0.275769
32	15	7	5	0.972809	0.121339
32	15	7	6	0.998088	0.025279
32	15	7	7	1.000000	0.001912
32	15	8	0	0.002311	0.002311
32	15	8	1	0.030046	0.027735
32	15	8	2	0.153590	0.123545
32	15	8	3	0.421271	0.267680
32	15	8	4	0.730132	0.308862
32	15	8	5	0.924274	0.194142
32	15	8	6	0.988988	0.064714
32	15	8	7	0.999388	0.010400
32	15	8	8	1.000000	0.000612
32	15	9	0	0.000867	0.000867
32	15	9	1	0.013867	0.013001
32	15	9	2	0.086670	0.072803
32	15	9	3	0.287430	0.200760
32	15	9	4	0.588571	0.301140

Table for $N = 2$, $n = 1$, through $N = 100$, $n = 50$

N	n	k	x	P(x)	p(x)
32	15	9	5	0.843381	0.254811
32	15	9	6	0.964720	0.121339
32	15	9	7	0.995921	0.031201
32	15	9	8	0.999821	0.003900
32	15	9	9	1.000000	0.000178
32	15	10	0	0.000301	0.000301
32	15	10	1	0.005954	0.005652
32	15	10	2	0.045521	0.039567
32	15	10	3	0.182686	0.137165
32	15	10	4	0.444547	0.261861
32	15	10	5	0.732594	0.288047
32	15	10	6	0.917240	0.184646
32	15	10	7	0.985069	0.067829
32	15	10	8	0.998634	0.013566
32	15	10	9	0.999953	0.001319
32	15	10	10	1.000000	0.000047
32	15	11	0	0.000096	0.000096
32	15	11	1	0.002357	0.002261
32	15	11	2	0.022140	0.019783
32	15	11	3	0.107869	0.085728
32	15	11	4	0.313617	0.205748
32	15	11	5	0.601664	0.288047
32	15	11	6	0.841703	0.240039
32	15	11	7	0.960404	0.118701
32	15	11	8	0.994318	0.033914
32	15	11	9	0.999594	0.005276
32	15	11	10	0.999989	0.000396
32	15	11	11	1.000000	0.000011
32	15	12	0	0.000027	0.000027
32	15	12	1	0.000850	0.000822
32	15	12	2	0.009893	0.009044
32	15	12	3	0.058881	0.048988
32	15	12	4	0.205844	0.146963
32	15	12	5	0.464498	0.258655
32	15	12	6	0.738829	0.274331
32	15	12	7	0.915184	0.176355
32	15	12	8	0.983013	0.067829
32	15	12	9	0.998086	0.015073
32	15	12	10	0.999895	0.001809
32	15	12	11	0.999998	0.000103
32	15	12	12	1.000000	0.000002
32	15	13	0	0.000007	0.000007
32	15	13	1	0.000274	0.000267
32	15	13	2	0.004015	0.003741
32	15	13	3	0.029488	0.025474
32	15	13	4	0.125014	0.095526
32	15	13	5	0.335171	0.210157
32	15	13	6	0.615380	0.280209
32	15	13	7	0.844642	0.229262
32	15	13	8	0.959273	0.114631

N	n	k	x	P(x)	p(x)
32	15	13	9	0.993564	0.034291
32	15	13	10	0.999443	0.005879
32	15	13	11	0.999977	0.000534
32	15	13	12	1.000000	0.000022
32	15	13	13	1.000000	0.000000
32	15	14	0	0.000001	0.000001
32	15	14	1	0.000077	0.000076
32	15	14	2	0.001455	0.001378
32	15	14	3	0.013400	0.011945
32	15	14	4	0.069710	0.056310
32	15	14	5	0.224562	0.154852
32	15	14	6	0.482650	0.258087
32	15	14	7	0.748111	0.265461
32	15	14	8	0.917041	0.168930
32	15	14	9	0.982736	0.065695
32	15	14	10	0.997896	0.015160
32	15	14	11	0.999865	0.001969
32	15	14	12	0.999996	0.000131
32	15	14	13	1.000000	0.000004
32	15	14	14	1.000000	0.000000
32	15	15	0	0.000000	0.000000
32	15	15	1	0.000018	0.000018
32	15	15	2	0.000460	0.000442
32	15	15	3	0.005437	0.004977
32	15	15	4	0.035298	0.029861
32	15	15	5	0.138533	0.103235
32	15	15	6	0.353606	0.215073
32	15	15	7	0.630128	0.276522
32	15	15	8	0.851346	0.221218
32	15	15	9	0.960837	0.109492
32	15	15	10	0.993685	0.032847
32	15	15	11	0.999427	0.005743
32	15	15	12	0.999974	0.000547
32	15	15	13	0.999999	0.000025
32	15	15	14	1.000000	0.000000
32	15	15	15	1.000000	0.000000
32	16	1	0	0.500000	0.500000
32	16	1	1	1.000000	0.500000
32	16	2	0	0.241935	0.241935
32	16	2	1	0.758065	0.516129
32	16	2	2	1.000000	0.241935
32	16	3	0	0.112903	0.112903
32	16	3	1	0.500000	0.387097
32	16	3	2	0.887097	0.387097
32	16	3	3	1.000000	0.112903
32	16	4	0	0.050612	0.050612
32	16	4	1	0.299778	0.249166
32	16	4	2	0.700222	0.400445
32	16	4	3	0.949388	0.249166
32	16	4	4	1.000000	0.050612

N	n	k	x	P(x)	p(x)
32	16	5	0	0.021691	0.021691
32	16	5	1	0.166296	0.144605
32	16	5	2	0.500000	0.333704
32	16	5	3	0.833704	0.333704
32	16	5	4	0.978309	0.144605
32	16	5	5	1.000000	0.021691
32	16	6	0	0.008837	0.008837
32	16	6	1	0.085960	0.077123
32	16	6	2	0.326968	0.241009
32	16	6	3	0.673032	0.346064
32	16	6	4	0.914040	0.241009
32	16	6	5	0.991163	0.077123
32	16	6	6	1.000000	0.008837
32	16	7	0	0.003399	0.003399
32	16	7	1	0.041466	0.038067
32	16	7	2	0.197194	0.155729
32	16	7	3	0.500000	0.302806
32	16	7	4	0.802806	0.302806
32	16	7	5	0.958534	0.155729
32	16	7	6	0.996601	0.038067
32	16	7	7	1.000000	0.003399
32	16	8	0	0.001224	0.001224
32	16	8	1	0.018626	0.017402
32	16	8	2	0.109986	0.091361
32	16	8	3	0.342541	0.232555
32	16	8	4	0.657459	0.314918
32	16	8	5	0.890014	0.232555
32	16	8	6	0.981374	0.091361
32	16	8	7	0.998776	0.017402
32	16	8	8	1.000000	0.001224
32	16	9	0	0.000408	0.000408
32	16	9	1	0.007749	0.007341
32	16	9	2	0.056693	0.048943
32	16	9	3	0.216574	0.159881
32	16	9	4	0.500000	0.283426
32	16	9	5	0.783426	0.283426
32	16	9	6	0.943307	0.159881
32	16	9	7	0.992251	0.048943
32	16	9	8	0.999592	0.007341
32	16	9	9	1.000000	0.000408
32	16	10	0	0.000124	0.000124
32	16	10	1	0.002961	0.002837
32	16	10	2	0.026901	0.023940
32	16	10	3	0.126206	0.099305
32	16	10	4	0.352126	0.225919
32	16	10	5	0.647874	0.295749
32	16	10	6	0.873794	0.225919
32	16	10	7	0.973099	0.099305
32	16	10	8	0.997038	0.023940
32	16	10	9	0.999876	0.002837

N	n	k	x	P(x)	p(x)
32	16	10	10	1.000000	0.000124
32	16	11	0	0.000034	0.000034
32	16	11	1	0.001027	0.000993
32	16	11	2	0.011667	0.010640
32	16	11	3	0.067526	0.055859
32	16	11	4	0.228897	0.161371
32	16	11	5	0.500000	0.271103
32	16	11	6	0.771103	0.271103
32	16	11	7	0.932474	0.161371
32	16	11	8	0.988333	0.055859
32	16	11	9	0.998973	0.010640
32	16	11	10	0.999966	0.000993
32	16	11	11	1.000000	0.000034
32	16	12	0	0.000008	0.000008
32	16	12	1	0.000318	0.000310
32	16	12	2	0.004574	0.004256
32	16	12	3	0.032946	0.028373
32	16	12	4	0.136685	0.103738
32	16	12	5	0.357994	0.221309
32	16	12	6	0.642006	0.284013
32	16	12	7	0.863315	0.221309
32	16	12	8	0.967054	0.103738
32	16	12	9	0.995426	0.028373
32	16	12	10	0.999682	0.004256
32	16	12	11	0.999992	0.000310
32	16	12	12	1.000000	0.000008
32	16	13	0	0.000002	0.000002
32	16	13	1	0.000085	0.000084
32	16	13	2	0.001594	0.001509
32	16	13	3	0.014504	0.012910
32	16	13	4	0.074442	0.059938
32	16	13	5	0.236274	0.161832
32	16	13	6	0.500000	0.263726
32	16	13	7	0.763726	0.263726
32	16	13	8	0.925558	0.161832
32	16	13	9	0.985496	0.059938
32	16	13	10	0.998405	0.012910
32	16	13	11	0.999914	0.001509
32	16	13	12	0.999998	0.000084
32	16	13	13	1.000000	0.000002
32	16	14	0	0.000000	0.000000
32	16	14	1	0.000019	0.000019
32	16	14	2	0.000483	0.000463
32	16	14	3	0.005671	0.005189
32	16	14	4	0.036586	0.030915
32	16	14	5	0.142582	0.105995
32	16	14	6	0.361197	0.218615
32	16	14	7	0.638803	0.277607
32	16	14	8	0.857418	0.218615
32	16	14	9	0.963414	0.105995

Table for $N = 2$, $n = 1$, through $N = 100$, $n = 50$

N	n	k	x	P(x)	p(x)
32	16	14	10	0.994329	0.030915
32	16	14	11	0.999517	0.005189
32	16	14	12	0.999981	0.000463
32	16	14	13	1.000000	0.000019
32	16	14	14	1.000000	0.000000
32	16	15	0	0.000000	0.000000
32	16	15	1	0.000003	0.000003
32	16	15	2	0.000122	0.000119
32	16	15	3	0.001924	0.001802
32	16	15	4	0.015976	0.014052
32	16	15	5	0.077807	0.061831
32	16	15	6	0.239744	0.161937
32	16	15	7	0.500000	0.260256
32	16	15	8	0.760256	0.260256
32	16	15	9	0.922193	0.161937
32	16	15	10	0.984024	0.061831
32	16	15	11	0.998076	0.014052
32	16	15	12	0.999878	0.001802
32	16	15	13	0.999996	0.000119
32	16	15	14	1.000000	0.000003
32	16	15	15	1.000000	0.000000
32	16	16	0	0.000000	0.000000
32	16	16	1	0.000000	0.000000
32	16	16	2	0.000024	0.000024
32	16	16	3	0.000546	0.000522
32	16	16	4	0.006057	0.005511
32	16	16	5	0.037799	0.031742
32	16	16	6	0.144487	0.106688
32	16	16	7	0.362217	0.217731
32	16	16	8	0.637783	0.275565
32	16	16	9	0.855513	0.217731
32	16	16	10	0.962201	0.106688
32	16	16	11	0.993943	0.031742
32	16	16	12	0.999454	0.005511
32	16	16	13	0.999975	0.000522
32	16	16	14	0.999999	0.000024
32	16	16	15	1.000000	0.000000
32	16	16	16	1.000000	0.000000
33	1	1	0	0.969697	0.969697
33	1	1	1	1.000000	0.030303
33	2	1	0	0.939394	0.939394
33	2	1	1	1.000000	0.060606
33	2	2	0	0.880682	0.880682
33	2	2	1	0.998106	0.117424
33	2	2	2	1.000000	0.001894
33	3	1	0	0.909091	0.909091
33	3	1	1	1.000000	0.090909
33	3	2	0	0.823864	0.823864
33	3	2	1	0.994318	0.170455
33	3	2	2	1.000000	0.005682

N	n	k	x	P(x)	p(x)
33	3	3	0	0.744135	0.744135
33	3	3	1	0.983321	0.239186
33	3	3	2	0.999817	0.016496
33	3	3	3	1.000000	0.000183
33	4	1	0	0.878788	0.878788
33	4	1	1	1.000000	0.121212
33	4	2	0	0.768939	0.768939
33	4	2	1	0.988636	0.219697
33	4	2	2	1.000000	0.011364
33	4	3	0	0.669721	0.669721
33	4	3	1	0.967375	0.297654
33	4	3	2	0.999267	0.031891
33	4	3	3	1.000000	0.000733
33	4	4	0	0.580425	0.580425
33	4	4	1	0.937610	0.357185
33	4	4	2	0.997141	0.059531
33	4	4	3	0.999976	0.002835
33	4	4	4	1.000000	0.000024
33	5	1	0	0.848485	0.848485
33	5	1	1	1.000000	0.151515
33	5	2	0	0.715909	0.715909
33	5	2	1	0.981061	0.265152
33	5	2	2	1.000000	0.018939
33	5	3	0	0.600440	0.600440
33	5	3	1	0.946847	0.346408
33	5	3	2	0.998167	0.051320
33	5	3	3	1.000000	0.001833
33	5	4	0	0.500367	0.500367
33	5	4	1	0.900660	0.400293
33	5	4	2	0.993035	0.092375
33	5	4	3	0.999878	0.006843
33	5	4	4	1.000000	0.000122
33	5	5	0	0.414096	0.414096
33	5	5	1	0.845447	0.431350
33	5	5	2	0.983479	0.138032
33	5	5	3	0.999406	0.015927
33	5	5	4	0.999996	0.000590
33	5	5	5	1.000000	0.000004
33	6	1	0	0.818182	0.818182
33	6	1	1	1.000000	0.181818
33	6	2	0	0.664773	0.664773
33	6	2	1	0.971591	0.306818
33	6	2	2	1.000000	0.028409
33	6	3	0	0.536107	0.536107
33	6	3	1	0.922104	0.385997
33	6	3	2	0.996334	0.074230
33	6	3	3	1.000000	0.003666
33	6	4	0	0.428886	0.428886
33	6	4	1	0.857771	0.428886
33	6	4	2	0.986437	0.128666

N	n	k	x	P(x)	p(x)
33	6	4	3	0.999633	0.013196
33	6	4	4	1.000000	0.000367
33	6	5	0	0.340151	0.340151
33	6	5	1	0.783825	0.443675
33	6	5	2	0.968690	0.184864
33	6	5	3	0.998268	0.029578
33	6	5	4	0.999975	0.001706
33	6	5	5	1.000000	0.000025
33	6	6	0	0.267261	0.267261
33	6	6	1	0.704598	0.437337
33	6	6	2	0.942281	0.237683
33	6	6	3	0.995099	0.052818
33	6	6	4	0.999853	0.004754
33	6	6	5	0.999999	0.000146
33	6	6	6	1.000000	0.000001
33	7	1	0	0.787879	0.787879
33	7	1	1	1.000000	0.212121
33	7	2	0	0.615530	0.615530
33	7	2	1	0.960227	0.344697
33	7	2	2	1.000000	0.039773
33	7	3	0	0.476540	0.476540
33	7	3	1	0.893512	0.416972
33	7	3	2	0.993585	0.100073
33	7	3	3	1.000000	0.006415
33	7	4	0	0.365347	0.365347
33	7	4	1	0.810117	0.444770
33	7	4	2	0.976906	0.166789
33	7	4	3	0.999145	0.022239
33	7	4	4	1.000000	0.000855
33	7	5	0	0.277160	0.277160
33	7	5	1	0.718096	0.440936
33	7	5	2	0.948149	0.230054
33	7	5	3	0.996077	0.047928
33	7	5	4	0.999911	0.003834
33	7	5	5	1.000000	0.000088
33	7	6	0	0.207870	0.207870
33	7	6	1	0.623610	0.415740
33	7	6	2	0.907068	0.283459
33	7	6	3	0.989230	0.082162
33	7	6	4	0.999501	0.010270
33	7	6	5	0.999994	0.000493
33	7	6	6	1.000000	0.000006
33	7	7	0	0.153978	0.153978
33	7	7	1	0.531223	0.377245
33	7	7	2	0.854576	0.323353
33	7	7	3	0.977058	0.122482
33	7	7	4	0.998360	0.021301
33	7	7	5	0.999957	0.001598
33	7	7	6	1.000000	0.000043
33	7	7	7	1.000000	0.000000

N	n	k	x	P(x)	p(x)
33	8	1	0	0.757576	0.757576
33	8	1	1	1.000000	0.242424
33	8	2	0	0.568182	0.568182
33	8	2	1	0.946970	0.378788
33	8	2	2	1.000000	0.053030
33	8	3	0	0.421554	0.421554
33	8	3	1	0.861437	0.439883
33	8	3	2	0.989736	0.128299
33	8	3	3	1.000000	0.010264
33	8	4	0	0.309140	0.309140
33	8	4	1	0.758798	0.449658
33	8	4	2	0.964076	0.205279
33	8	4	3	0.998289	0.034213
33	8	4	4	1.000000	0.001711
33	8	5	0	0.223860	0.223860
33	8	5	1	0.650260	0.426400
33	8	5	2	0.921605	0.271345
33	8	5	3	0.992391	0.070786
33	8	5	4	0.999764	0.007374
33	8	5	5	1.000000	0.000236
33	8	6	0	0.159900	0.159900
33	8	6	1	0.543660	0.383760
33	8	6	2	0.863459	0.319800
33	8	6	3	0.979750	0.116291
33	8	6	4	0.998711	0.018960
33	8	6	5	0.999975	0.001264
33	8	6	6	1.000000	0.000025
33	8	7	0	0.112522	0.112522
33	8	7	1	0.444166	0.331644
33	8	7	2	0.792393	0.348226
33	8	7	3	0.958215	0.165822
33	8	7	4	0.995902	0.037687
33	8	7	5	0.999834	0.003933
33	8	7	6	0.999998	0.000164
33	8	7	7	1.000000	0.000002
33	8	8	0	0.077900	0.077900
33	8	8	1	0.354878	0.276978
33	8	8	2	0.712033	0.357155
33	8	8	3	0.926326	0.214293
33	8	8	4	0.990104	0.063778
33	8	8	5	0.999380	0.009277
33	8	8	6	0.999985	0.000605
33	8	8	7	1.000000	0.000014
33	8	8	8	1.000000	0.000000
33	9	1	0	0.727273	0.727273
33	9	1	1	1.000000	0.272727
33	9	2	0	0.522727	0.522727
33	9	2	1	0.931818	0.409091
33	9	2	2	1.000000	0.068182
33	9	3	0	0.370968	0.370968

Table for $N=2$, $n=1$, through $N=100$, $n=50$

N	n	k	x	P(x)	p(x)
33	9	3	1	0.826246	0.455279
33	9	3	2	0.984604	0.158358
33	9	3	3	1.000000	0.015396
33	9	4	0	0.259677	0.259677
33	9	4	1	0.704839	0.445161
33	9	4	2	0.947654	0.242815
33	9	4	3	0.996921	0.049267
33	9	4	4	1.000000	0.003079
33	9	5	0	0.179088	0.179088
33	9	5	1	0.582036	0.402948
33	9	5	2	0.889043	0.307008
33	9	5	3	0.986728	0.097684
33	9	5	4	0.999469	0.012741
33	9	5	5	1.000000	0.000531
33	9	6	0	0.121524	0.121524
33	9	6	1	0.466908	0.345384
33	9	6	2	0.812291	0.345384
33	9	6	3	0.965795	0.153504
33	9	6	4	0.997194	0.031399
33	9	6	5	0.999924	0.002730
33	9	6	6	1.000000	0.000076
33	9	7	0	0.081016	0.081016
33	9	7	1	0.364572	0.283556
33	9	7	2	0.722747	0.358176
33	9	7	3	0.931683	0.208936
33	9	7	4	0.991379	0.059696
33	9	7	5	0.999520	0.008140
33	9	7	6	0.999991	0.000472
33	9	7	7	1.000000	0.000008
33	9	8	0	0.052972	0.052972
33	9	8	1	0.277324	0.224352
33	9	8	2	0.626316	0.348992
33	9	8	3	0.883467	0.257152
33	9	8	4	0.979899	0.096432
33	9	8	5	0.998267	0.018368
33	9	8	6	0.999937	0.001670
33	9	8	7	0.999999	0.000062
33	9	8	8	1.000000	0.000001
33	9	9	0	0.033902	0.033902
33	9	9	1	0.205531	0.171629
33	9	9	2	0.528598	0.323067
33	9	9	3	0.821751	0.293153
33	9	9	4	0.960613	0.138862
33	9	9	5	0.995328	0.034715
33	9	9	6	0.999737	0.004408
33	9	9	7	0.999994	0.000258
33	9	9	8	1.000000	0.000006
33	9	9	9	1.000000	0.000000
33	10	1	0	0.696970	0.696970
33	10	1	1	1.000000	0.303030
33	10	2	0	0.479167	0.479167
33	10	2	1	0.914773	0.435606
33	10	2	2	1.000000	0.085227
33	10	3	0	0.324597	0.324597
33	10	3	1	0.788306	0.463710
33	10	3	2	0.978006	0.189699
33	10	3	3	1.000000	0.021994
33	10	4	0	0.216398	0.216398
33	10	4	1	0.649194	0.432796
33	10	4	2	0.927419	0.278226
33	10	4	3	0.994868	0.067449
33	10	4	4	1.000000	0.005132
33	10	5	0	0.141778	0.141778
33	10	5	1	0.514878	0.373100
33	10	5	2	0.850667	0.335790
33	10	5	3	0.978587	0.127920
33	10	5	4	0.998938	0.020351
33	10	5	5	1.000000	0.001062
33	10	6	0	0.091143	0.091143
33	10	6	1	0.394953	0.303810
33	10	6	2	0.754727	0.359775
33	10	6	3	0.946607	0.191880
33	10	6	4	0.994577	0.047970
33	10	6	5	0.999810	0.005233
33	10	6	6	1.000000	0.000190
33	10	7	0	0.057386	0.057386
33	10	7	1	0.293683	0.236296
33	10	7	2	0.648127	0.354445
33	10	7	3	0.896861	0.248733
33	10	7	4	0.983917	0.087057
33	10	7	5	0.998841	0.014924
33	10	7	6	0.999972	0.001131
33	10	7	7	1.000000	0.000028
33	10	8	0	0.035315	0.035315
33	10	8	1	0.211888	0.176573
33	10	8	2	0.539068	0.327180
33	10	8	3	0.829894	0.290826
33	10	8	4	0.963827	0.133933
33	10	8	5	0.995971	0.032144
33	10	8	6	0.999798	0.003827
33	10	8	7	0.999997	0.000199
33	10	8	8	1.000000	0.000003
33	10	9	0	0.021189	0.021189
33	10	9	1	0.148321	0.127133
33	10	9	2	0.434370	0.286049
33	10	9	3	0.748463	0.314093
33	10	9	4	0.931683	0.183221
33	10	9	5	0.989542	0.057859
33	10	9	6	0.999186	0.009643
33	10	9	7	0.999973	0.000787
33	10	9	8	1.000000	0.000027
33	10	9	9	1.000000	0.000000
33	10	10	0	0.012360	0.012360
33	10	10	1	0.100647	0.088287
33	10	10	2	0.339021	0.238374
33	10	10	3	0.656852	0.317832
33	10	10	4	0.885878	0.229026
33	10	10	5	0.977488	0.091610
33	10	10	6	0.997578	0.020090
33	10	10	7	0.999874	0.002296
33	10	10	8	0.999997	0.000123
33	10	10	9	1.000000	0.000002
33	10	10	10	1.000000	0.000000
33	11	1	0	0.666667	0.666667
33	11	1	1	1.000000	0.333333
33	11	2	0	0.437500	0.437500
33	11	2	1	0.895833	0.458333
33	11	2	2	1.000000	0.104167
33	11	3	0	0.282258	0.282258
33	11	3	1	0.747984	0.465726
33	11	3	2	0.969758	0.221774
33	11	3	3	1.000000	0.030242
33	11	4	0	0.178763	0.178763
33	11	4	1	0.592742	0.413978
33	11	4	2	0.903226	0.310484
33	11	4	3	0.991935	0.088710
33	11	4	4	1.000000	0.008065
33	11	5	0	0.110957	0.110957
33	11	5	1	0.449991	0.339034
33	11	5	2	0.806869	0.356878
33	11	5	3	0.967464	0.160595
33	11	5	4	0.998053	0.030590
33	11	5	5	1.000000	0.001947
33	11	6	0	0.067366	0.067366
33	11	6	1	0.328907	0.261541
33	11	6	2	0.692158	0.363251
33	11	6	3	0.921579	0.229422
33	11	6	4	0.990406	0.068826
33	11	6	5	0.999583	0.009177
33	11	6	6	1.000000	0.000417
33	11	7	0	0.039921	0.039921
33	11	7	1	0.232040	0.192119
33	11	7	2	0.571074	0.339034
33	11	7	3	0.853603	0.282528
33	11	7	4	0.972562	0.118959
33	11	7	5	0.997544	0.024981
33	11	7	6	0.999923	0.002379
33	11	7	7	1.000000	0.000077
33	11	8	0	0.023031	0.023031
33	11	8	1	0.158148	0.135117
33	11	8	2	0.453716	0.295568
33	11	8	3	0.766671	0.312955
33	11	8	4	0.940535	0.173864
33	11	8	5	0.991779	0.051244
33	11	8	6	0.999465	0.007687
33	11	8	7	0.999988	0.000523
33	11	8	8	1.000000	0.000012
33	11	9	0	0.012898	0.012898
33	11	9	1	0.104101	0.091204
33	11	9	2	0.347312	0.243210
33	11	9	3	0.666525	0.319214
33	11	9	4	0.891853	0.225327
33	11	9	5	0.979480	0.087627
33	11	9	6	0.997928	0.018448
33	11	9	7	0.999904	0.001977
33	11	9	8	0.999998	0.000094
33	11	9	9	1.000000	0.000001
33	11	10	0	0.006986	0.006986
33	11	10	1	0.066100	0.059114
33	11	10	2	0.256108	0.190008
33	11	10	3	0.560121	0.304013
33	11	10	4	0.826132	0.266011
33	11	10	5	0.957573	0.131441
33	11	10	6	0.994084	0.036511
33	11	10	7	0.999575	0.005490
33	11	10	8	0.999987	0.000412
33	11	10	9	1.000000	0.000013
33	11	10	10	1.000000	0.000000
33	11	11	0	0.003645	0.003645
33	11	11	1	0.040398	0.036753
33	11	11	2	0.181757	0.141359
33	11	11	3	0.454377	0.272620
33	11	11	4	0.745172	0.290795
33	11	11	5	0.923284	0.178112
33	11	11	6	0.986147	0.062863
33	11	11	7	0.998620	0.012473
33	11	11	8	0.999933	0.001313
33	11	11	9	0.999999	0.000066
33	11	11	10	1.000000	0.000001
33	11	11	11	1.000000	0.000000
33	12	1	0	0.636364	0.636364
33	12	1	1	1.000000	0.363636
33	12	2	0	0.397727	0.397727
33	12	2	1	0.875000	0.477273
33	12	2	2	1.000000	0.125000
33	12	3	0	0.243768	0.243768
33	12	3	1	0.705645	0.461877
33	12	3	2	0.959677	0.254032
33	12	3	3	1.000000	0.040323
33	12	4	0	0.146261	0.146261

Table for $N = 2$, $n = 1$, through $N = 100$, $n = 50$

N	n	k	x	P(x)	p(x)
33	12	4	1	0.536290	0.390029
33	12	4	2	0.875000	0.338710
33	12	4	3	0.987903	0.112903
33	12	4	4	1.000000	0.012097
33	12	5	0	0.085739	0.085739
33	12	5	1	0.388348	0.302609
33	12	5	2	0.758204	0.369855
33	12	5	3	0.952864	0.194661
33	12	5	4	0.996663	0.043799
33	12	5	5	1.000000	0.003337
33	12	6	0	0.048994	0.048994
33	12	6	1	0.269466	0.220472
33	12	6	2	0.626112	0.356646
33	12	6	3	0.890295	0.264182
33	12	6	4	0.984149	0.093854
33	12	6	5	0.999166	0.015017
33	12	6	6	1.000000	0.000834
33	12	7	0	0.027219	0.027219
33	12	7	1	0.179644	0.152425
33	12	7	2	0.494021	0.314377
33	12	7	3	0.802234	0.308213
33	12	7	4	0.956340	0.154106
33	12	7	5	0.995272	0.038932
33	12	7	6	0.999815	0.004542
33	12	7	7	1.000000	0.000185
33	12	8	0	0.014656	0.014656
33	12	8	1	0.115156	0.100500
33	12	8	2	0.373107	0.257950
33	12	8	3	0.695545	0.322438
33	12	8	4	0.908923	0.213378
33	12	8	5	0.984791	0.075868
33	12	8	6	0.998766	0.013976
33	12	8	7	0.999964	0.001198
33	12	8	8	1.000000	0.000036
33	12	9	0	0.007621	0.007621
33	12	9	1	0.070936	0.063315
33	12	9	2	0.269927	0.198990
33	12	9	3	0.579467	0.309541
33	12	9	4	0.840642	0.261175
33	12	9	5	0.963548	0.122906
33	12	9	6	0.995412	0.031864
33	12	9	7	0.999725	0.004312
33	12	9	8	0.999994	0.000270
33	12	9	9	1.000000	0.000006
33	12	10	0	0.003811	0.003811
33	12	10	1	0.041917	0.038106
33	12	10	2	0.187014	0.145097
33	12	10	3	0.463389	0.276375
33	12	10	4	0.753584	0.290194
33	12	10	5	0.927700	0.174117

N	n	k	x	P(x)	p(x)
33	12	10	6	0.987446	0.059746
33	12	10	7	0.998826	0.011380
33	12	10	8	0.999949	0.001123
33	12	10	9	0.999999	0.000050
33	12	10	10	1.000000	0.000001
33	12	11	0	0.001822	0.001822
33	12	11	1	0.023692	0.021870
33	12	11	2	0.123928	0.100236
33	12	11	3	0.355243	0.231314
33	12	11	4	0.652647	0.297404
33	12	11	5	0.874708	0.222062
33	12	11	6	0.971860	0.097152
33	12	11	7	0.996352	0.024492
33	12	11	8	0.999754	0.003402
33	12	11	9	0.999993	0.000239
33	12	11	10	1.000000	0.000007
33	12	11	11	1.000000	0.000000
33	12	12	0	0.000828	0.000828
33	12	12	1	0.012757	0.011929
33	12	12	2	0.078366	0.065609
33	12	12	3	0.260614	0.182248
33	12	12	4	0.544500	0.283886
33	12	12	5	0.804052	0.259553
33	12	12	6	0.945364	0.141312
33	12	12	7	0.990786	0.045422
33	12	12	8	0.999135	0.008350
33	12	12	9	0.999960	0.000825
33	12	12	10	0.999999	0.000039
33	12	12	11	1.000000	0.000001
33	12	12	12	1.000000	0.000000
33	13	1	0	0.606061	0.606061
33	13	1	1	1.000000	0.393939
33	13	2	0	0.359848	0.359848
33	13	2	1	0.852273	0.492424
33	13	2	2	1.000000	0.147727
33	13	3	0	0.208944	0.208944
33	13	3	1	0.661657	0.452713
33	13	3	2	0.947581	0.285924
33	13	3	3	1.000000	0.052419
33	13	4	0	0.118402	0.118402
33	13	4	1	0.480572	0.362170
33	13	4	2	0.842742	0.362170
33	13	4	3	0.982527	0.139785
33	13	4	4	1.000000	0.017473
33	13	5	0	0.065325	0.065325
33	13	5	1	0.330708	0.265383
33	13	5	2	0.705367	0.374659
33	13	5	3	0.934325	0.228958
33	13	5	4	0.994577	0.060252
33	13	5	5	1.000000	0.005423

N	n	k	x	P(x)	p(x)
33	13	6	0	0.034996	0.034996
33	13	6	1	0.216973	0.181977
33	13	6	2	0.558180	0.341207
33	13	6	3	0.852554	0.294375
33	13	6	4	0.975210	0.122656
33	13	6	5	0.998451	0.023240
33	13	6	6	1.000000	0.001549
33	13	7	0	0.018146	0.018146
33	13	7	1	0.136094	0.117948
33	13	7	2	0.419169	0.283075
33	13	7	3	0.743527	0.324357
33	13	7	4	0.934325	0.190798
33	13	7	5	0.991565	0.057240
33	13	7	6	0.999598	0.008034
33	13	7	7	1.000000	0.000402
33	13	8	0	0.009073	0.009073
33	13	8	1	0.081656	0.072583
33	13	8	2	0.299407	0.217750
33	13	8	3	0.618774	0.319367
33	13	8	4	0.868280	0.249506
33	13	8	5	0.973952	0.105673
33	13	8	6	0.997435	0.023483
33	13	8	7	0.999907	0.002472
33	13	8	8	1.000000	0.000093
33	13	9	0	0.004355	0.004355
33	13	9	1	0.046816	0.042461
33	13	9	2	0.203597	0.156780
33	13	9	3	0.491027	0.287430
33	13	9	4	0.778457	0.287430
33	13	9	5	0.940137	0.161680
33	13	9	6	0.990860	0.050723
33	13	9	7	0.999314	0.008454
33	13	9	8	0.999981	0.000667
33	13	9	9	1.000000	0.000019
33	13	10	0	0.001996	0.001996
33	13	10	1	0.025586	0.023590
33	13	10	2	0.131739	0.106153
33	13	10	3	0.371264	0.239525
33	13	10	4	0.670671	0.299407
33	13	10	5	0.886244	0.215573
33	13	10	6	0.976066	0.089822
33	13	10	7	0.997200	0.021135
33	13	10	8	0.999842	0.002642
33	13	10	9	0.999997	0.000154
33	13	10	10	1.000000	0.000003
33	13	11	0	0.000868	0.000868
33	13	11	1	0.013278	0.012410
33	13	11	2	0.080970	0.067692
33	13	11	3	0.267123	0.186153
33	13	11	4	0.553512	0.286389

N	n	k	x	P(x)	p(x)
33	13	11	5	0.811262	0.257750
33	13	11	6	0.948729	0.137467
33	13	11	7	0.991687	0.042958
33	13	11	8	0.999268	0.007581
33	13	11	9	0.999970	0.000702
33	13	11	10	0.999999	0.000030
33	13	11	11	1.000000	0.000000
33	13	12	0	0.000355	0.000355
33	13	12	1	0.006509	0.006154
33	13	12	2	0.047124	0.040615
33	13	12	3	0.182508	0.135384
33	13	12	4	0.436353	0.253845
33	13	12	5	0.717535	0.281182
33	13	12	6	0.904989	0.187455
33	13	12	7	0.979971	0.074982
33	13	12	8	0.997545	0.017574
33	13	12	9	0.999842	0.002297
33	13	12	10	0.999995	0.000153
33	13	12	11	1.000000	0.000004
33	13	12	12	1.000000	0.000000
33	13	13	0	0.000135	0.000135
33	13	13	1	0.002992	0.002857
33	13	13	2	0.025849	0.022857
33	13	13	3	0.118039	0.092190
33	13	13	4	0.327562	0.209523
33	13	13	5	0.610418	0.282856
33	13	13	6	0.842504	0.232087
33	13	13	7	0.958548	0.116043
33	13	13	8	0.993361	0.034813
33	13	13	9	0.999405	0.006044
33	13	13	10	0.999974	0.000569
33	13	13	11	0.999999	0.000026
33	13	13	12	1.000000	0.000000
33	13	13	13	1.000000	0.000000
33	14	1	0	0.575758	0.575758
33	14	1	1	1.000000	0.424242
33	14	2	0	0.323864	0.323864
33	14	2	1	0.827651	0.503788
33	14	2	2	1.000000	0.172348
33	14	3	0	0.177603	0.177603
33	14	3	1	0.616386	0.438783
33	14	3	2	0.933284	0.316899
33	14	3	3	1.000000	0.066716
33	14	4	0	0.094721	0.094721
33	14	4	1	0.426246	0.331525
33	14	4	2	0.806525	0.380279
33	14	4	3	0.975538	0.169013
33	14	4	4	1.000000	0.024462
33	14	5	0	0.048994	0.048994
33	14	5	1	0.277632	0.228638

Table for $N = 2$, $n = 1$, through $N = 100$, $n = 50$

N	n	k	x	P(x)	p(x)
33	14	5	2	0.649168	0.371537
33	14	5	3	0.911429	0.262261
33	14	5	4	0.991565	0.080135
33	14	5	5	1.000000	0.008435
33	14	6	0	0.024497	0.024497
33	14	6	1	0.171478	0.146981
33	14	6	2	0.489938	0.318460
33	14	6	3	0.808398	0.318460
33	14	6	4	0.962945	0.154547
33	14	6	5	0.997289	0.034344
33	14	6	6	1.000000	0.002711
33	14	7	0	0.011795	0.011795
33	14	7	1	0.100710	0.088915
33	14	7	2	0.348401	0.247691
33	14	7	3	0.678655	0.330255
33	14	7	4	0.905705	0.227050
33	14	7	5	0.985841	0.080135
33	14	7	6	0.999197	0.013356
33	14	7	7	1.000000	0.000803
33	14	8	0	0.005444	0.005444
33	14	8	1	0.056252	0.050808
33	14	8	2	0.234082	0.177829
33	14	8	3	0.538932	0.304851
33	14	8	4	0.818378	0.279446
33	14	8	5	0.958102	0.139723
33	14	8	6	0.995087	0.036986
33	14	8	7	0.999784	0.004697
33	14	8	8	1.000000	0.000216
33	14	9	0	0.002395	0.002395
33	14	9	1	0.029832	0.027437
33	14	9	2	0.148723	0.118892
33	14	9	3	0.404798	0.256074
33	14	9	4	0.706600	0.301802
33	14	9	5	0.907801	0.201201
33	14	9	6	0.983252	0.075450
33	14	9	7	0.998469	0.015217
33	14	9	8	0.999948	0.001479
33	14	9	9	1.000000	0.000052
33	14	10	0	0.000998	0.000998
33	14	10	1	0.014970	0.013972
33	14	10	2	0.089278	0.074307
33	14	10	3	0.287430	0.198153
33	14	10	4	0.580849	0.293419
33	14	10	5	0.832351	0.251502
33	14	10	6	0.958102	0.125751
33	14	10	7	0.994030	0.035929
33	14	10	8	0.999578	0.005548
33	14	10	9	0.999989	0.000411
33	14	10	10	1.000000	0.000011
33	14	11	0	0.000391	0.000391
33	14	11	1	0.007073	0.006682
33	14	11	2	0.050509	0.043436
33	14	11	3	0.192662	0.142153
33	14	11	4	0.453276	0.260614
33	14	11	5	0.733937	0.280661
33	14	11	6	0.914362	0.180425
33	14	11	7	0.983095	0.068733
33	14	11	8	0.998131	0.015035
33	14	11	9	0.999900	0.001769
33	14	11	10	0.999998	0.000098
33	14	11	11	1.000000	0.000002
33	14	12	0	0.000142	0.000142
33	14	12	1	0.003124	0.002982
33	14	12	2	0.026816	0.023692
33	14	12	3	0.121585	0.094769
33	14	12	4	0.334815	0.213230
33	14	12	5	0.619121	0.284306
33	14	12	6	0.848753	0.229632
33	14	12	7	0.961226	0.112473
33	14	12	8	0.994030	0.032805
33	14	12	9	0.999498	0.005467
33	14	12	10	0.999980	0.000482
33	14	12	11	1.000000	0.000019
33	14	12	12	1.000000	0.000000
33	14	13	0	0.000047	0.000047
33	14	13	1	0.001278	0.001231
33	14	13	2	0.013278	0.012000
33	14	13	3	0.071944	0.058666
33	14	13	4	0.233277	0.161333
33	14	13	5	0.497276	0.263999
33	14	13	6	0.761274	0.263999
33	14	13	7	0.923735	0.162461
33	14	13	8	0.984658	0.060923
33	14	13	9	0.998196	0.013538
33	14	13	10	0.999888	0.001692
33	14	13	11	0.999997	0.000109
33	14	13	12	1.000000	0.000003
33	14	13	13	1.000000	0.000000
33	14	14	0	0.000014	0.000014
33	14	14	1	0.000478	0.000464
33	14	14	2	0.006078	0.005600
33	14	14	3	0.039678	0.033600
33	14	14	4	0.152611	0.112933
33	14	14	5	0.378476	0.225865
33	14	14	6	0.655675	0.277199
33	14	14	7	0.866874	0.211199
33	14	14	8	0.966381	0.099507
33	14	14	9	0.994811	0.028431
33	14	14	10	0.999550	0.004738
33	14	14	11	0.999981	0.000431
33	14	14	12	1.000000	0.000019
33	14	14	13	1.000000	0.000000
33	14	14	14	1.000000	0.000000
33	15	1	0	0.545455	0.545455
33	15	1	1	1.000000	0.454545
33	15	2	0	0.289773	0.289773
33	15	2	1	0.801136	0.511364
33	15	2	2	1.000000	0.198864
33	15	3	0	0.149560	0.149560
33	15	3	1	0.570198	0.420638
33	15	3	2	0.916605	0.346408
33	15	3	3	1.000000	0.083394
33	15	4	0	0.074780	0.074780
33	15	4	1	0.373900	0.299120
33	15	4	2	0.766496	0.392595
33	15	4	3	0.966642	0.200147
33	15	4	4	1.000000	0.033358
33	15	5	0	0.036101	0.036101
33	15	5	1	0.229497	0.193397
33	15	5	2	0.590505	0.361007
33	15	5	3	0.883823	0.293318
33	15	5	4	0.987347	0.103524
33	15	5	5	1.000000	0.012653
33	15	6	0	0.016761	0.016761
33	15	6	1	0.132799	0.116038
33	15	6	2	0.422894	0.290095
33	15	6	3	0.758115	0.335221
33	15	6	4	0.946677	0.188562
33	15	6	5	0.995481	0.048804
33	15	6	6	1.000000	0.004519
33	15	7	0	0.007449	0.007449
33	15	7	1	0.072631	0.065182
33	15	7	2	0.283219	0.210588
33	15	7	3	0.609128	0.325909
33	15	7	4	0.869855	0.260727
33	15	7	5	0.977405	0.107550
33	15	7	6	0.998494	0.021088
33	15	7	7	1.000000	0.001506
33	15	8	0	0.003152	0.003152
33	15	8	1	0.037533	0.034382
33	15	8	2	0.177925	0.140392
33	15	8	3	0.458708	0.280783
33	15	8	4	0.759548	0.300839
33	15	8	5	0.936040	0.176492
33	15	8	6	0.991194	0.055154
33	15	8	7	0.999536	0.008343
33	15	8	8	1.000000	0.000463
33	15	9	0	0.001261	0.001261
33	15	9	1	0.018280	0.017019
33	15	9	2	0.104921	0.086642
33	15	9	3	0.323932	0.219011
33	15	9	4	0.627178	0.303246
33	15	9	5	0.865443	0.238265
33	15	9	6	0.971338	0.105895
33	15	9	7	0.996867	0.025528
33	15	9	8	0.999870	0.003003
33	15	9	9	1.000000	0.000130
33	15	10	0	0.000473	0.000473
33	15	10	1	0.008352	0.007879
33	15	10	2	0.057990	0.049638
33	15	10	3	0.214427	0.156436
33	15	10	4	0.488191	0.273764
33	15	10	5	0.766166	0.277976
33	15	10	6	0.931628	0.165462
33	15	10	7	0.988357	0.056730
33	15	10	8	0.998994	0.010637
33	15	10	9	0.999967	0.000973
33	15	10	10	1.000000	0.000032
33	15	11	0	0.000164	0.000164
33	15	11	1	0.003556	0.003391
33	15	11	2	0.029934	0.026378
33	15	11	3	0.132808	0.102874
33	15	11	4	0.357260	0.224452
33	15	11	5	0.645307	0.288047
33	15	11	6	0.866882	0.221575
33	15	11	7	0.968625	0.101743
33	15	11	8	0.995757	0.027132
33	15	11	9	0.999714	0.003957
33	15	11	10	0.999993	0.000279
33	15	11	11	1.000000	0.000007
33	15	12	0	0.000052	0.000052
33	15	12	1	0.001398	0.001345
33	15	12	2	0.014347	0.012949
33	15	12	3	0.076695	0.062348
33	15	12	4	0.245034	0.168339
33	15	12	5	0.514377	0.269343
33	15	12	6	0.776238	0.261861
33	15	12	7	0.931628	0.155390
33	15	12	8	0.987124	0.055496
33	15	12	9	0.998634	0.011510
33	15	12	10	0.999929	0.001295
33	15	12	11	0.999999	0.000069
33	15	12	12	1.000000	0.000001
33	15	13	0	0.000015	0.000015
33	15	13	1	0.000501	0.000486
33	15	13	2	0.006331	0.005830
33	15	13	3	0.041067	0.034737
33	15	13	4	0.156856	0.115789
33	15	13	5	0.386118	0.229262
33	15	13	6	0.664011	0.277893

Table for $N=2$, $n=1$, through $N=100$, $n=50$

N	n	k	x	P(x)	p(x)
33	15	13	7	0.872431	0.208420
33	15	13	8	0.968625	0.096194
33	15	13	9	0.995346	0.026720
33	15	13	10	0.999621	0.004275
33	15	13	11	0.999985	0.000364
33	15	13	12	1.000000	0.000014
33	15	13	13	1.000000	0.000000
33	15	14	0	0.000004	0.000004
33	15	14	1	0.000161	0.000157
33	15	14	2	0.002541	0.002381
33	15	14	3	0.020225	0.017684
33	15	14	4	0.093172	0.072947
33	15	14	5	0.271487	0.178315
33	15	14	6	0.538959	0.267472
33	15	14	7	0.789063	0.250104
33	15	14	8	0.934957	0.145894
33	15	14	9	0.987330	0.052372
33	15	14	10	0.998552	0.011223
33	15	14	11	0.999913	0.001360
33	15	14	12	0.999998	0.000085
33	15	14	13	1.000000	0.000002
33	15	14	14	1.000000	0.000000
33	15	15	0	0.000001	0.000001
33	15	15	1	0.000045	0.000044
33	15	15	2	0.000912	0.000867
33	15	15	3	0.009056	0.008144
33	15	15	4	0.050940	0.041883
33	15	15	5	0.177637	0.126697
33	15	15	6	0.412262	0.234625
33	15	15	7	0.683757	0.271494
33	15	15	8	0.881207	0.197450
33	15	15	9	0.970791	0.089584
33	15	15	10	0.995599	0.024808
33	15	15	11	0.999626	0.004027
33	15	15	12	0.999984	0.000358
33	15	15	13	1.000000	0.000015
33	15	15	14	1.000000	0.000000
33	15	15	15	1.000000	0.000000
33	16	1	0	0.515151	0.515151
33	16	1	1	1.000000	0.484848
33	16	2	0	0.257576	0.257576
33	16	2	1	0.772727	0.515151
33	16	2	2	1.000000	0.227273
33	16	3	0	0.124633	0.124633
33	16	3	1	0.523460	0.398827
33	16	3	2	0.897361	0.373900
33	16	3	3	1.000000	0.102639
33	16	4	0	0.058162	0.058162
33	16	4	1	0.324047	0.265885
33	16	4	2	0.722874	0.398827

N	n	k	x	P(x)	p(x)
33	16	4	3	0.955523	0.232649
33	16	4	4	1.000000	0.044477
33	16	5	0	0.026073	0.026073
33	16	5	1	0.186520	0.160448
33	16	5	2	0.530337	0.343816
33	16	5	3	0.851232	0.320895
33	16	5	4	0.981596	0.130364
33	16	5	5	1.000000	0.018404
33	16	6	0	0.011174	0.011174
33	16	6	1	0.100566	0.089392
33	16	6	2	0.358429	0.257862
33	16	6	3	0.702245	0.343816
33	16	6	4	0.925725	0.223481
33	16	6	5	0.992770	0.067044
33	16	6	6	1.000000	0.007230
33	16	7	0	0.004552	0.004552
33	16	7	1	0.050904	0.046352
33	16	7	2	0.224722	0.173818
33	16	7	3	0.536704	0.311981
33	16	7	4	0.826401	0.289697
33	16	7	5	0.965455	0.139055
33	16	7	6	0.997322	0.031867
33	16	7	7	1.000000	0.002678
33	16	8	0	0.001751	0.001751
33	16	8	1	0.024163	0.022412
33	16	8	2	0.131128	0.106965
33	16	8	3	0.380713	0.249585
33	16	8	4	0.692694	0.311981
33	16	8	5	0.906625	0.213930
33	16	8	6	0.985066	0.078441
33	16	8	7	0.999073	0.014007
33	16	8	8	1.000000	0.000927
33	16	9	0	0.000630	0.000630
33	16	9	1	0.010716	0.010085
33	16	9	2	0.071227	0.060512
33	16	9	3	0.250929	0.179701
33	16	9	4	0.542943	0.292015
33	16	9	5	0.812495	0.269552
33	16	9	6	0.953689	0.141194
33	16	9	7	0.994030	0.040341
33	16	9	8	0.999703	0.005673
33	16	9	9	1.000000	0.000297
33	16	10	0	0.000210	0.000210
33	16	10	1	0.004412	0.004202
33	16	10	2	0.035929	0.031516
33	16	10	3	0.153590	0.117662
33	16	10	4	0.396936	0.243346
33	16	10	5	0.688951	0.292015
33	16	10	6	0.894858	0.205908
33	16	10	7	0.978902	0.084044

N	n	k	x	P(x)	p(x)
33	16	10	8	0.997812	0.018910
33	16	10	9	0.999913	0.002101
33	16	10	10	1.000000	0.000087
33	16	11	0	0.000064	0.000064
33	16	11	1	0.001672	0.001608
33	16	11	2	0.016745	0.015073
33	16	11	3	0.087086	0.070341
33	16	11	4	0.269973	0.182887
33	16	11	5	0.549291	0.279318
33	16	11	6	0.805333	0.256042
33	16	11	7	0.946016	0.140682
33	16	11	8	0.991235	0.045219
33	16	11	9	0.999274	0.008039
33	16	11	10	0.999977	0.000703
33	16	11	11	1.000000	0.000023
33	16	12	0	0.000017	0.000017
33	16	12	1	0.000576	0.000558
33	16	12	2	0.007153	0.006577
33	16	12	3	0.045521	0.038368
33	16	12	4	0.170216	0.124696
33	16	12	5	0.409632	0.239416
33	16	12	6	0.688951	0.279318
33	16	12	7	0.888464	0.199513
33	16	12	8	0.974792	0.086328
33	16	12	9	0.996716	0.021925
33	16	12	10	0.999786	0.003069
33	16	12	11	0.999995	0.000209
33	16	12	12	1.000000	0.000005
33	16	13	0	0.000004	0.000004
33	16	13	1	0.000177	0.000173
33	16	13	2	0.002768	0.002591
33	16	13	3	0.021769	0.019001
33	16	13	4	0.098962	0.077193
33	16	13	5	0.284224	0.185262
33	16	13	6	0.555942	0.271718
33	16	13	7	0.802958	0.247016
33	16	13	8	0.941905	0.138947
33	16	13	9	0.989408	0.047503
33	16	13	10	0.998908	0.009501
33	16	13	11	0.999945	0.001036
33	16	13	12	0.999999	0.000054
33	16	13	13	1.000000	0.000001
33	16	14	0	0.000001	0.000001
33	16	14	1	0.000047	0.000046
33	16	14	2	0.000954	0.000907
33	16	14	3	0.009418	0.008464
33	16	14	4	0.052646	0.043228
33	16	14	5	0.182330	0.129684
33	16	14	6	0.420083	0.237753
33	16	14	7	0.691801	0.271718

N	n	k	x	P(x)	p(x)
33	16	14	8	0.886326	0.194525
33	16	14	9	0.972782	0.086456
33	16	14	10	0.996058	0.023277
33	16	14	11	0.999686	0.003628
33	16	14	12	0.999988	0.000302
33	16	14	13	1.000000	0.000012
33	16	14	14	1.000000	0.000000
33	16	15	0	0.000000	0.000000
33	16	15	1	0.000011	0.000010
33	16	15	2	0.000286	0.000275
33	16	15	3	0.003627	0.003341
33	16	15	4	0.025344	0.021717
33	16	15	5	0.107250	0.081905
33	16	15	6	0.294950	0.187700
33	16	15	7	0.563092	0.268143
33	16	15	8	0.804421	0.241328
33	16	15	9	0.940930	0.136509
33	16	15	10	0.988708	0.047778
33	16	15	11	0.998731	0.010023
33	16	15	12	0.999924	0.001193
33	16	15	13	0.999998	0.000073
33	16	15	14	1.000000	0.000002
33	16	15	15	1.000000	0.000000
33	16	16	0	0.000000	0.000000
33	16	16	1	0.000002	0.000002
33	16	16	2	0.000072	0.000070
33	16	16	3	0.001214	0.001142
33	16	16	4	0.010866	0.009652
33	16	16	5	0.057197	0.046330
33	16	16	6	0.190672	0.133475
33	16	16	7	0.429021	0.238349
33	16	16	8	0.697164	0.268143
33	16	16	9	0.887843	0.190679
33	16	16	10	0.972782	0.084939
33	16	16	11	0.995947	0.023165
33	16	16	12	0.999659	0.003712
33	16	16	13	0.999986	0.000326
33	16	16	14	1.000000	0.000014
33	16	16	15	1.000000	0.000000
33	16	16	16	1.000000	0.000000
34	1	1	0	0.970588	0.970588
34	1	1	1	1.000000	0.029412
34	2	1	0	0.941176	0.941176
34	2	1	1	1.000000	0.058824
34	2	2	0	0.884135	0.884135
34	2	2	1	0.998217	0.114082
34	2	2	2	1.000000	0.001783
34	3	1	0	0.911765	0.911765
34	3	1	1	1.000000	0.088235
34	3	2	0	0.828877	0.828877

Table for $N = 2$, $n = 1$, through $N = 100$, $n = 50$

N	n	k	x	P(x)	p(x)
34	3	2	1	0.994652	0.165775
34	3	2	2	1.000000	0.005348
34	3	3	0	0.751170	0.751170
34	3	3	1	0.984291	0.233122
34	3	3	2	0.999833	0.015541
34	3	3	3	1.000000	0.000167
34	4	1	0	0.882353	0.882353
34	4	1	1	1.000000	0.117647
34	4	2	0	0.775401	0.775401
34	4	2	1	0.989305	0.213904
34	4	2	2	1.000000	0.010695
34	4	3	0	0.678476	0.678476
34	4	3	1	0.969251	0.290775
34	4	3	2	0.999331	0.030080
34	4	3	3	1.000000	0.000668
34	4	4	0	0.590931	0.590931
34	4	4	1	0.941112	0.350181
34	4	4	2	0.997391	0.056279
34	4	4	3	0.999978	0.002588
34	4	4	4	1.000000	0.000022
34	5	1	0	0.852941	0.852941
34	5	1	1	1.000000	0.147059
34	5	2	0	0.723708	0.723708
34	5	2	1	0.982175	0.258467
34	5	2	2	1.000000	0.017825
34	5	3	0	0.610628	0.610628
34	5	3	1	0.949866	0.339238
34	5	3	2	0.998329	0.048463
34	5	3	3	1.000000	0.001671
34	5	4	0	0.512140	0.512140
34	5	4	1	0.906094	0.393954
34	5	4	2	0.993639	0.087545
34	5	4	3	0.999892	0.006253
34	5	4	4	1.000000	0.000108
34	5	5	0	0.426783	0.426783
34	5	5	1	0.853566	0.426783
34	5	5	2	0.984884	0.131318
34	5	5	3	0.999475	0.014591
34	5	5	4	0.999996	0.000521
34	5	5	5	1.000000	0.000004
34	6	1	0	0.823529	0.823529
34	6	1	1	1.000000	0.176471
34	6	2	0	0.673797	0.673797
34	6	2	1	0.973262	0.299465
34	6	2	2	1.000000	0.026738
34	6	3	0	0.547460	0.547460
34	6	3	1	0.926471	0.379011
34	6	3	2	0.996658	0.070187
34	6	3	3	1.000000	0.003342
34	6	4	0	0.441500	0.441500

N	n	k	x	P(x)	p(x)
34	6	4	1	0.865340	0.423840
34	6	4	2	0.987601	0.122262
34	6	4	3	0.999677	0.012075
34	6	4	4	1.000000	0.000323
34	6	5	0	0.353200	0.353200
34	6	5	1	0.794700	0.441500
34	6	5	2	0.971300	0.176600
34	6	5	3	0.998469	0.027169
34	6	5	4	0.999978	0.001509
34	6	5	5	1.000000	0.000022
34	6	6	0	0.280124	0.280124
34	6	6	1	0.718579	0.438455
34	6	6	2	0.946941	0.228362
34	6	6	3	0.995658	0.048717
34	6	6	4	0.999874	0.004216
34	6	6	5	0.999999	0.000125
34	6	6	6	1.000000	0.000001
34	7	1	0	0.794118	0.794118
34	7	1	1	1.000000	0.205882
34	7	2	0	0.625668	0.625668
34	7	2	1	0.962567	0.336898
34	7	2	2	1.000000	0.037433
34	7	3	0	0.488803	0.488803
34	7	3	1	0.899398	0.410595
34	7	3	2	0.994151	0.094753
34	7	3	3	1.000000	0.005849
34	7	4	0	0.378428	0.378428
34	7	4	1	0.819928	0.441500
34	7	4	2	0.978868	0.158940
34	7	4	3	0.999245	0.020377
34	7	4	4	1.000000	0.000755
34	7	5	0	0.290129	0.290129
34	7	5	1	0.731628	0.441500
34	7	5	2	0.952378	0.220750
34	7	5	3	0.996528	0.044150
34	7	5	4	0.999924	0.003396
34	7	5	5	1.000000	0.000075
34	7	6	0	0.220097	0.220097
34	7	6	1	0.640284	0.420186
34	7	6	2	0.914318	0.274034
34	7	6	3	0.990439	0.076121
34	7	6	4	0.999573	0.009134
34	7	6	5	0.999995	0.000422
34	7	6	6	1.000000	0.000005
34	7	7	0	0.165073	0.165073
34	7	7	1	0.550244	0.385171
34	7	7	2	0.865383	0.315140
34	7	7	3	0.979564	0.114181
34	7	7	4	0.998594	0.019030
34	7	7	5	0.999965	0.001370

N	n	k	x	P(x)	p(x)
34	7	7	6	1.000000	0.000035
34	7	7	7	1.000000	0.000000
34	8	1	0	0.764706	0.764706
34	8	1	1	1.000000	0.235294
34	8	2	0	0.579323	0.579323
34	8	2	1	0.950089	0.370766
34	8	2	2	1.000000	0.049911
34	8	3	0	0.434492	0.434492
34	8	3	1	0.868984	0.434492
34	8	3	2	0.990642	0.121658
34	8	3	3	1.000000	0.009358
34	8	4	0	0.322365	0.322365
34	8	4	1	0.770873	0.448508
34	8	4	2	0.967095	0.196222
34	8	4	3	0.998491	0.031396
34	8	4	4	1.000000	0.001509
34	8	5	0	0.236401	0.236401
34	8	5	1	0.666221	0.429820
34	8	5	2	0.927851	0.261630
34	8	5	3	0.993258	0.065407
34	8	5	4	0.999799	0.006541
34	8	5	5	1.000000	0.000201
34	8	6	0	0.171187	0.171187
34	8	6	1	0.562471	0.391284
34	8	6	2	0.873720	0.311249
34	8	6	3	0.981981	0.108261
34	8	6	4	0.998896	0.016916
34	8	6	5	0.999979	0.001083
34	8	6	6	1.000000	0.000021
34	8	7	0	0.122276	0.122276
34	8	7	1	0.464650	0.342374
34	8	7	2	0.807024	0.342374
34	8	7	3	0.962649	0.155624
34	8	7	4	0.996480	0.033831
34	8	7	5	0.999863	0.003383
34	8	7	6	0.999998	0.000135
34	8	7	7	1.000000	0.000001
34	8	8	0	0.086046	0.086046
34	8	8	1	0.375887	0.289840
34	8	8	2	0.730941	0.355054
34	8	8	3	0.933829	0.202888
34	8	8	4	0.991468	0.057639
34	8	8	5	0.999487	0.008019
34	8	8	6	0.999988	0.000501
34	8	8	7	1.000000	0.000011
34	8	8	8	1.000000	0.000000
34	9	1	0	0.735294	0.735294
34	9	1	1	1.000000	0.264706
34	9	2	0	0.534759	0.534759
34	9	2	1	0.935829	0.401070

N	n	k	x	P(x)	p(x)
34	9	2	2	1.000000	0.064171
34	9	3	0	0.384358	0.384358
34	9	3	1	0.835561	0.451203
34	9	3	2	0.985963	0.150401
34	9	3	3	1.000000	0.014037
34	9	4	0	0.272770	0.272770
34	9	4	1	0.719122	0.446352
34	9	4	2	0.952001	0.232879
34	9	4	3	0.997283	0.045282
34	9	4	4	1.000000	0.002717
34	9	5	0	0.190939	0.190939
34	9	5	1	0.600095	0.409156
34	9	5	2	0.897663	0.297568
34	9	5	3	0.988227	0.090564
34	9	5	4	0.999547	0.011321
34	9	5	5	1.000000	0.000453
34	9	6	0	0.131682	0.131682
34	9	6	1	0.487224	0.355542
34	9	6	2	0.825836	0.338611
34	9	6	3	0.969489	0.143653
34	9	6	4	0.997595	0.028106
34	9	6	5	0.999937	0.002342
34	9	6	6	1.000000	0.000062
34	9	7	0	0.089356	0.089356
34	9	7	1	0.385641	0.296285
34	9	7	2	0.741183	0.355542
34	9	7	3	0.938706	0.197523
34	9	7	4	0.992576	0.053870
34	9	7	5	0.999603	0.007027
34	9	7	6	0.999993	0.000390
34	9	7	7	1.000000	0.000007
34	9	8	0	0.059571	0.059571
34	9	8	1	0.297853	0.238282
34	9	8	2	0.649005	0.351153
34	9	8	3	0.894812	0.245807
34	9	8	4	0.982600	0.087788
34	9	8	5	0.998562	0.015961
34	9	8	6	0.999950	0.001388
34	9	8	7	0.999999	0.000050
34	9	8	8	1.000000	0.000000
34	9	9	0	0.038950	0.038950
34	9	9	1	0.224535	0.185585
34	9	9	2	0.554464	0.329929
34	9	9	3	0.838088	0.283623
34	9	9	4	0.965718	0.127630
34	9	9	5	0.996106	0.030388
34	9	9	6	0.999790	0.003683
34	9	9	7	0.999996	0.000206
34	9	9	8	1.000000	0.000004
34	9	9	9	1.000000	0.000000

Table for $N = 2$, $n = 1$, through $N = 100$, $n = 50$

N	n	k	x	P(x)	p(x)
34	10	1	0	0.705882	0.705882
34	10	1	1	1.000000	0.294118
34	10	2	0	0.491979	0.491979
34	10	2	1	0.919786	0.427807
34	10	2	2	1.000000	0.080214
34	10	3	0	0.338235	0.338235
34	10	3	1	0.799465	0.461230
34	10	3	2	0.979946	0.180481
34	10	3	3	1.000000	0.020053
34	10	4	0	0.229127	0.229127
34	10	4	1	0.665560	0.436433
34	10	4	2	0.933371	0.267811
34	10	4	3	0.995472	0.062101
34	10	4	4	1.000000	0.004528
34	10	5	0	0.152751	0.152751
34	10	5	1	0.534630	0.381879
34	10	5	2	0.861954	0.327324
34	10	5	3	0.980981	0.119027
34	10	5	4	0.999094	0.018113
34	10	5	5	1.000000	0.000906
34	10	6	0	0.100079	0.100079
34	10	6	1	0.416116	0.316037
34	10	6	2	0.771658	0.355542
34	10	6	3	0.952251	0.180593
34	10	6	4	0.995347	0.043096
34	10	6	5	0.999844	0.004497
34	10	6	6	1.000000	0.000156
34	10	7	0	0.064336	0.064336
34	10	7	1	0.314532	0.250196
34	10	7	2	0.670075	0.355542
34	10	7	3	0.907103	0.237028
34	10	7	4	0.986112	0.079009
34	10	7	5	0.999041	0.012929
34	10	7	6	0.999978	0.000937
34	10	7	7	1.000000	0.000022
34	10	8	0	0.040508	0.040508
34	10	8	1	0.231134	0.190626
34	10	8	2	0.564729	0.333595
34	10	8	3	0.845651	0.280922
34	10	8	4	0.968554	0.122903
34	10	8	5	0.996646	0.028092
34	10	8	6	0.999839	0.003192
34	10	8	7	0.999997	0.000159
34	10	8	8	1.000000	0.000002
34	10	9	0	0.024928	0.024928
34	10	9	1	0.165148	0.140220
34	10	9	2	0.462084	0.296936
34	10	9	3	0.770018	0.307934
34	10	9	4	0.940192	0.170174
34	10	9	5	0.991244	0.051052
34	10	9	6	0.999348	0.008104
34	10	9	7	0.999979	0.000631
34	10	9	8	1.000000	0.000021
34	10	9	9	1.000000	0.000000
34	10	10	0	0.014957	0.014957
34	10	10	1	0.114669	0.099712
34	10	10	2	0.367065	0.252396
34	10	10	3	0.683796	0.316732
34	10	10	4	0.899350	0.215554
34	10	10	5	0.981034	0.081684
34	10	10	6	0.998051	0.017017
34	10	10	7	0.999903	0.001852
34	10	10	8	0.999998	0.000095
34	10	10	9	1.000000	0.000002
34	10	10	10	1.000000	0.000000
34	11	1	0	0.676471	0.676471
34	11	1	1	1.000000	0.323529
34	11	2	0	0.450980	0.450980
34	11	2	1	0.901961	0.450980
34	11	2	2	1.000000	0.098039
34	11	3	0	0.295956	0.295956
34	11	3	1	0.761029	0.465074
34	11	3	2	0.972426	0.211397
34	11	3	3	1.000000	0.027574
34	11	4	0	0.190939	0.190939
34	11	4	1	0.611006	0.420066
34	11	4	2	0.911053	0.300047
34	11	4	3	0.992884	0.081831
34	11	4	4	1.000000	0.007116
34	11	5	0	0.120928	0.120928
34	11	5	1	0.470983	0.350055
34	11	5	2	0.821039	0.350055
34	11	5	3	0.971063	0.150024
34	11	5	4	0.998340	0.027277
34	11	5	5	1.000000	0.001660
34	11	6	0	0.075059	0.075059
34	11	6	1	0.350275	0.275216
34	11	6	2	0.712401	0.362126
34	11	6	3	0.929677	0.217276
34	11	6	4	0.991756	0.062079
34	11	6	5	0.999656	0.007901
34	11	6	6	1.000000	0.000344
34	11	7	0	0.045571	0.045571
34	11	7	1	0.251983	0.206412
34	11	7	2	0.596003	0.344020
34	11	7	3	0.867598	0.271595
34	11	7	4	0.976236	0.108638
34	11	7	5	0.997963	0.021728
34	11	7	6	0.999939	0.001975
34	11	7	7	1.000000	0.000061
34	11	8	0	0.027005	0.027005
34	11	8	1	0.175535	0.148529
34	11	8	2	0.481330	0.305795
34	11	8	3	0.787125	0.305795
34	11	8	4	0.948070	0.160945
34	11	8	5	0.993135	0.045065
34	11	8	6	0.999573	0.006438
34	11	8	7	0.999991	0.000418
34	11	8	8	1.000000	0.000009
34	11	9	0	0.015580	0.015580
34	11	9	1	0.118408	0.102828
34	11	9	2	0.375478	0.257070
34	11	9	3	0.693034	0.317557
34	11	9	4	0.904739	0.211705
34	11	9	5	0.982735	0.077996
34	11	9	6	0.998335	0.015599
34	11	9	7	0.999926	0.001592
34	11	9	8	0.999999	0.000072
34	11	9	9	1.000000	0.000001
34	11	10	0	0.008725	0.008725
34	11	10	1	0.077277	0.068552
34	11	10	2	0.282933	0.205656
34	11	10	3	0.591416	0.308484
34	11	10	4	0.845462	0.254045
34	11	10	5	0.964016	0.118555
34	11	10	6	0.995215	0.031199
34	11	10	7	0.999672	0.004457
34	11	10	8	0.999990	0.000318
34	11	10	9	1.000000	0.000010
34	11	10	10	1.000000	0.000000
34	11	11	0	0.004726	0.004726
34	11	11	1	0.048713	0.043987
34	11	11	2	0.205812	0.157098
34	11	11	3	0.488588	0.282777
34	11	11	4	0.771365	0.282777
34	11	11	5	0.934378	0.163013
34	11	11	6	0.988715	0.054338
34	11	11	7	0.998929	0.010214
34	11	11	8	0.999950	0.001021
34	11	11	9	0.999999	0.000049
34	11	11	10	1.000000	0.000001
34	11	11	11	1.000000	0.000000
34	12	1	0	0.647059	0.647059
34	12	1	1	1.000000	0.352941
34	12	2	0	0.411765	0.411765
34	12	2	1	0.882353	0.470588
34	12	2	2	1.000000	0.117647
34	12	3	0	0.257353	0.257353
34	12	3	1	0.720588	0.463235
34	12	3	2	0.963235	0.242647
34	12	3	3	1.000000	0.036765
34	12	4	0	0.157732	0.157732
34	12	4	1	0.556214	0.398482
34	12	4	2	0.884962	0.328748
34	12	4	3	0.989326	0.104364
34	12	4	4	1.000000	0.010674
34	12	5	0	0.094639	0.094639
34	12	5	1	0.410104	0.315465
34	12	5	2	0.775379	0.365275
34	12	5	3	0.958017	0.182638
34	12	5	4	0.997154	0.039137
34	12	5	5	1.000000	0.002846
34	12	6	0	0.055478	0.055478
34	12	6	1	0.290445	0.234967
34	12	6	2	0.649423	0.358977
34	12	6	3	0.901336	0.251914
34	12	6	4	0.986357	0.085021
34	12	6	5	0.999313	0.012956
34	12	6	6	1.000000	0.000687
34	12	7	0	0.031702	0.031702
34	12	7	1	0.198137	0.166435
34	12	7	2	0.521216	0.323080
34	12	7	3	0.820364	0.299148
34	12	7	4	0.962066	0.141702
34	12	7	5	0.996074	0.034008
34	12	7	6	0.999853	0.003779
34	12	7	7	1.000000	0.000147
34	12	8	0	0.017612	0.017612
34	12	8	1	0.130330	0.112718
34	12	8	2	0.401557	0.271227
34	12	8	3	0.720648	0.319091
34	12	8	4	0.920080	0.199432
34	12	8	5	0.987257	0.067177
34	12	8	6	0.999013	0.011756
34	12	8	7	0.999973	0.000960
34	12	8	8	1.000000	0.000027
34	12	9	0	0.009483	0.009483
34	12	9	1	0.082642	0.073158
34	12	9	2	0.297239	0.214597
34	12	9	3	0.610194	0.312955
34	12	9	4	0.858716	0.248523
34	12	9	5	0.969171	0.110455
34	12	9	6	0.996300	0.027129
34	12	9	7	0.999788	0.003488
34	12	9	8	0.999996	0.000208
34	12	9	9	1.000000	0.000004
34	12	10	0	0.004931	0.004931
34	12	10	1	0.050452	0.045521
34	12	10	2	0.211400	0.160948
34	12	10	3	0.497530	0.286130

Table for $N = 2$, $n = 1$, through $N = 100$, $n = 50$

N	n	k	x	P(x)	p(x)
34	12	10	4	0.779189	0.281659
34	12	10	5	0.938244	0.159055
34	12	10	6	0.989789	0.051545
34	12	10	7	0.999090	0.009301
34	12	10	8	0.999963	0.000872
34	12	10	9	0.999999	0.000037
34	12	10	10	1.000000	0.000000
34	12	11	0	0.002466	0.002466
34	12	11	1	0.029588	0.027123
34	12	11	2	0.144338	0.114750
34	12	11	3	0.390231	0.245893
34	12	11	4	0.685303	0.295071
34	12	11	5	0.891853	0.206550
34	12	11	6	0.976903	0.085050
34	12	11	7	0.997153	0.020250
34	12	11	8	0.999817	0.002664
34	12	11	9	0.999995	0.000178
34	12	11	10	1.000000	0.000005
34	12	11	11	1.000000	0.000000
34	12	12	0	0.001179	0.001179
34	12	12	1	0.016617	0.015437
34	12	12	2	0.094447	0.077830
34	12	12	3	0.294012	0.199565
34	12	12	4	0.582669	0.288657
34	12	12	5	0.828990	0.246321
34	12	12	6	0.954716	0.125726
34	12	12	7	0.992751	0.038035
34	12	12	8	0.999354	0.006603
34	12	12	9	0.999972	0.000618
34	12	12	10	0.999999	0.000028
34	12	12	11	1.000000	0.000000
34	12	12	12	1.000000	0.000000
34	13	1	0	0.617647	0.617647
34	13	1	1	1.000000	0.382353
34	13	2	0	0.374332	0.374332
34	13	2	1	0.860963	0.486631
34	13	2	2	1.000000	0.139037
34	13	3	0	0.222259	0.222259
34	13	3	1	0.678476	0.456217
34	13	3	2	0.952206	0.273730
34	13	3	3	1.000000	0.047794
34	13	4	0	0.129054	0.129054
34	13	4	1	0.501876	0.372822
34	13	4	2	0.855076	0.353200
34	13	4	3	0.984582	0.129507
34	13	4	4	1.000000	0.015417
34	13	5	0	0.073130	0.073130
34	13	5	1	0.352747	0.279617
34	13	5	2	0.725569	0.372822
34	13	5	3	0.941414	0.215844

N	n	k	x	P(x)	p(x)
34	13	5	4	0.995375	0.053961
34	13	5	5	1.000000	0.004625
34	13	6	0	0.040348	0.040348
34	13	6	1	0.237044	0.196696
34	13	6	2	0.584154	0.347110
34	13	6	3	0.866985	0.282831
34	13	6	4	0.978628	0.111644
34	13	6	5	0.998724	0.020096
34	13	6	6	1.000000	0.001276
34	13	7	0	0.021615	0.021615
34	13	7	1	0.152745	0.131131
34	13	7	2	0.447789	0.295044
34	13	7	3	0.765974	0.318184
34	13	7	4	0.942743	0.176769
34	13	7	5	0.992982	0.050240
34	13	7	6	0.999681	0.006699
34	13	7	7	1.000000	0.000319
34	13	8	0	0.011208	0.011208
34	13	8	1	0.094465	0.083257
34	13	8	2	0.327586	0.233121
34	13	8	3	0.648127	0.320541
34	13	8	4	0.883820	0.235692
34	13	8	5	0.978097	0.094277
34	13	8	6	0.997944	0.019848
34	13	8	7	0.999929	0.001985
34	13	8	8	1.000000	0.000071
34	13	9	0	0.005604	0.005604
34	13	9	1	0.056039	0.050435
34	13	9	2	0.228958	0.172919
34	13	9	3	0.524842	0.295884
34	13	9	4	0.802234	0.277392
34	13	9	5	0.949088	0.146854
34	13	9	6	0.992601	0.043512
34	13	9	7	0.999471	0.006870
34	13	9	8	0.999986	0.000515
34	13	9	9	1.000000	0.000014
34	13	10	0	0.002690	0.002690
34	13	10	1	0.031830	0.029140
34	13	10	2	0.152874	0.121044
34	13	10	3	0.406489	0.253615
34	13	10	4	0.702373	0.295884
34	13	10	5	0.902095	0.199722
34	13	10	6	0.980417	0.078322
34	13	10	7	0.997822	0.017405
34	13	10	8	0.999883	0.002061
34	13	10	9	0.999998	0.000115
34	13	10	10	1.000000	0.000002
34	13	11	0	0.001233	0.001233
34	13	11	1	0.017260	0.016027
34	13	11	2	0.097395	0.080135

N	n	k	x	P(x)	p(x)
34	13	11	3	0.300816	0.203420
34	13	11	4	0.591416	0.290601
34	13	11	5	0.835521	0.244105
34	13	11	6	0.957573	0.122052
34	13	11	7	0.993471	0.035898
34	13	11	8	0.999454	0.005983
34	13	11	9	0.999979	0.000525
34	13	11	10	1.000000	0.000021
34	13	11	11	1.000000	0.000000
34	13	12	0	0.000536	0.000536
34	13	12	1	0.008898	0.008362
34	13	12	2	0.059070	0.050172
34	13	12	3	0.212372	0.153302
34	13	12	4	0.477703	0.265331
34	13	12	5	0.750615	0.272912
34	13	12	6	0.920427	0.169812
34	13	12	7	0.984106	0.063679
34	13	12	8	0.998153	0.014047
34	13	12	9	0.999887	0.001734
34	13	12	10	0.999997	0.000110
34	13	12	11	1.000000	0.000003
34	13	12	12	1.000000	0.000000
34	13	13	0	0.000219	0.000219
34	13	13	1	0.004337	0.004118
34	13	13	2	0.033984	0.029647
34	13	13	3	0.142689	0.108705
34	13	13	4	0.369159	0.226469
34	13	13	5	0.651374	0.282216
34	13	13	6	0.866396	0.215021
34	13	13	7	0.966739	0.100343
34	13	13	8	0.994961	0.028222
34	13	13	9	0.999572	0.004611
34	13	13	10	0.999982	0.000410
34	13	13	11	1.000000	0.000018
34	13	13	12	1.000000	0.000000
34	13	13	13	1.000000	0.000000
34	14	1	0	0.588235	0.588235
34	14	1	1	1.000000	0.411765
34	14	2	0	0.338681	0.338681
34	14	2	1	0.837790	0.499109
34	14	2	2	1.000000	0.162210
34	14	3	0	0.190508	0.190508
34	14	3	1	0.635027	0.444519
34	14	3	2	0.939171	0.304144
34	14	3	3	1.000000	0.060829
34	14	4	0	0.104472	0.104472
34	14	4	1	0.448616	0.344143
34	14	4	2	0.821438	0.372822
34	14	4	3	0.978415	0.156978
34	14	4	4	1.000000	0.021584

N	n	k	x	P(x)	p(x)
34	14	5	0	0.055718	0.055718
34	14	5	1	0.299487	0.243768
34	14	5	2	0.672309	0.372822
34	14	5	3	0.920857	0.248548
34	14	5	4	0.992805	0.071948
34	14	5	5	1.000000	0.007195
34	14	6	0	0.028820	0.028820
34	14	6	1	0.190211	0.161391
34	14	6	2	0.518038	0.327826
34	14	6	3	0.826580	0.308542
34	14	6	4	0.967995	0.141415
34	14	6	5	0.997767	0.029772
34	14	6	6	1.000000	0.002233
34	14	7	0	0.014410	0.014410
34	14	7	1	0.115280	0.100870
34	14	7	2	0.377541	0.262261
34	14	7	3	0.705367	0.327826
34	14	7	4	0.917490	0.212123
34	14	7	5	0.988198	0.070708
34	14	7	6	0.999362	0.011164
34	14	7	7	1.000000	0.000638
34	14	8	0	0.006938	0.006938
34	14	8	1	0.066713	0.059775
34	14	8	2	0.260980	0.194267
34	14	8	3	0.571808	0.310828
34	14	8	4	0.838926	0.267118
34	14	8	5	0.964628	0.125702
34	14	8	6	0.996054	0.031426
34	14	8	7	0.999835	0.003781
34	14	8	8	1.000000	0.000165
34	14	9	0	0.003202	0.003202
34	14	9	1	0.036825	0.033623
34	14	9	2	0.171318	0.134493
34	14	9	3	0.440304	0.268986
34	14	9	4	0.736188	0.295884
34	14	9	5	0.921116	0.184928
34	14	9	6	0.986385	0.065269
34	14	9	7	0.998817	0.012432
34	14	9	8	0.999962	0.001145
34	14	9	9	1.000000	0.000038
34	14	10	0	0.001409	0.001409
34	14	10	1	0.019341	0.017932
34	14	10	2	0.106762	0.087420
34	14	10	3	0.321950	0.215189
34	14	10	4	0.617835	0.295884
34	14	10	5	0.854542	0.236707
34	14	10	6	0.965499	0.110957
34	14	10	7	0.995336	0.029837
34	14	10	8	0.999687	0.004351
34	14	10	9	0.999992	0.000305

Table for $N = 2$, $n = 1$, through $N = 100$, $n = 50$

N	n	k	x	P(x)	p(x)
34	14	10	10	1.000000	0.000008
34	14	11	0	0.000587	0.000587
34	14	11	1	0.009628	0.009041
34	14	11	2	0.063052	0.053424
34	14	11	3	0.223322	0.160271
34	14	11	4	0.494549	0.271227
34	14	11	5	0.765777	0.271227
34	14	11	6	0.928513	0.162736
34	14	11	7	0.986633	0.058120
34	14	11	8	0.998599	0.011966
34	14	11	9	0.999929	0.001330
34	14	11	10	0.999999	0.000070
34	14	11	11	1.000000	0.000001
34	14	12	0	0.000230	0.000230
34	14	12	1	0.004518	0.004288
34	14	12	2	0.035178	0.030660
34	14	12	3	0.146671	0.111493
34	14	12	4	0.376625	0.229954
34	14	12	5	0.659644	0.283020
34	14	12	6	0.871909	0.212265
34	14	12	7	0.968944	0.097035
34	14	12	8	0.995478	0.026533
34	14	12	9	0.999640	0.004162
34	14	12	10	0.999986	0.000347
34	14	12	11	1.000000	0.000013
34	14	12	12	1.000000	0.000000
34	14	13	0	0.000084	0.000084
34	14	13	1	0.001984	0.001900
34	14	13	2	0.018454	0.016470
34	14	13	3	0.090925	0.072470
34	14	13	4	0.272100	0.181176
34	14	13	5	0.543863	0.271763
34	14	13	6	0.794722	0.250858
34	14	13	7	0.938070	0.143348
34	14	13	8	0.988241	0.050172
34	14	13	9	0.998694	0.010452
34	14	13	10	0.999923	0.001230
34	14	13	11	0.999998	0.000075
34	14	13	12	1.000000	0.000002
34	14	13	13	1.000000	0.000000
34	14	14	0	0.000028	0.000028
34	14	14	1	0.000808	0.000780
34	14	14	2	0.009043	0.008235
34	14	14	3	0.052964	0.043921
34	14	14	4	0.185826	0.132862
34	14	14	5	0.427394	0.241567
34	14	14	6	0.699157	0.271763
34	14	14	7	0.890287	0.191130
34	14	14	8	0.973906	0.083619
34	14	14	9	0.996205	0.022299

N	n	k	x	P(x)	p(x)
34	14	14	10	0.999689	0.003484
34	14	14	11	0.999987	0.000298
34	14	14	12	1.000000	0.000012
34	14	14	13	1.000000	0.000000
34	14	14	14	1.000000	0.000000
34	15	1	0	0.558824	0.558824
34	15	1	1	1.000000	0.441176
34	15	2	0	0.304813	0.304813
34	15	2	1	0.812834	0.508021
34	15	2	2	1.000000	0.187166
34	15	3	0	0.161932	0.161932
34	15	3	1	0.590575	0.428643
34	15	3	2	0.923964	0.333389
34	15	3	3	1.000000	0.076036
34	15	4	0	0.083578	0.083578
34	15	4	1	0.396994	0.313416
34	15	4	2	0.784156	0.387161
34	15	4	3	0.970567	0.186411
34	15	4	4	1.000000	0.029433
34	15	5	0	0.041789	0.041789
34	15	5	1	0.250733	0.208944
34	15	5	2	0.616386	0.365652
34	15	5	3	0.896002	0.279617
34	15	5	4	0.989208	0.093206
34	15	5	5	1.000000	0.010792
34	15	6	0	0.020174	0.020174
34	15	6	1	0.149863	0.129690
34	15	6	2	0.452472	0.302609
34	15	6	3	0.780299	0.327826
34	15	6	4	0.953854	0.173555
34	15	6	5	0.996278	0.042425
34	15	6	6	1.000000	0.003721
34	15	7	0	0.009366	0.009366
34	15	7	1	0.085019	0.075652
34	15	7	2	0.311975	0.226957
34	15	7	3	0.639802	0.327826
34	15	7	4	0.885672	0.245870
34	15	7	5	0.981127	0.095455
34	15	7	6	0.998804	0.017677
34	15	7	7	1.000000	0.001196
34	15	8	0	0.004163	0.004163
34	15	8	1	0.045792	0.041629
34	15	8	2	0.202700	0.156908
34	15	8	3	0.494101	0.291401
34	15	8	4	0.785502	0.291401
34	15	8	5	0.945773	0.160271
34	15	8	6	0.992911	0.047138
34	15	8	7	0.999645	0.006734
34	15	8	8	1.000000	0.000354
34	15	9	0	0.001761	0.001761

N	n	k	x	P(x)	p(x)
34	15	9	1	0.023376	0.021615
34	15	9	2	0.124246	0.100870
34	15	9	3	0.359608	0.235362
34	15	9	4	0.662217	0.302609
34	15	9	5	0.884130	0.221913
34	15	9	6	0.976594	0.092464
34	15	9	7	0.997573	0.020979
34	15	9	8	0.999904	0.002331
34	15	9	9	1.000000	0.000095
34	15	10	0	0.000704	0.000704
34	15	10	1	0.011272	0.010567
34	15	10	2	0.071794	0.060522
34	15	10	3	0.246634	0.174841
34	15	10	4	0.529069	0.282435
34	15	10	5	0.795365	0.266296
34	15	10	6	0.943307	0.147942
34	15	10	7	0.990860	0.047553
34	15	10	8	0.999252	0.008392
34	15	10	9	0.999977	0.000725
34	15	10	10	1.000000	0.000023
34	15	11	0	0.000264	0.000264
34	15	11	1	0.005108	0.004843
34	15	11	2	0.039011	0.033903
34	15	11	3	0.159214	0.120203
34	15	11	4	0.399620	0.240406
34	15	11	5	0.684409	0.284789
34	15	11	6	0.887829	0.203420
34	15	11	7	0.975009	0.087180
34	15	11	8	0.996804	0.021795
34	15	11	9	0.999796	0.002991
34	15	11	10	0.999995	0.000199
34	15	11	11	1.000000	0.000005
34	15	12	0	0.000092	0.000092
34	15	12	1	0.002159	0.002068
34	15	12	2	0.019848	0.017689
34	15	12	3	0.096499	0.076651
34	15	12	4	0.284643	0.188144
34	15	12	5	0.560587	0.275944
34	15	12	6	0.808230	0.247642
34	15	12	7	0.944686	0.136456
34	15	12	8	0.990171	0.045485
34	15	12	9	0.999015	0.008844
34	15	12	10	0.999952	0.000936
34	15	12	11	0.999999	0.000047
34	15	12	12	1.000000	0.000001
34	15	13	0	0.000029	0.000029
34	15	13	1	0.000844	0.000814
34	15	13	2	0.009396	0.008552
34	15	13	3	0.054690	0.045294
34	15	13	4	0.190571	0.135882

N	n	k	x	P(x)	p(x)
34	15	13	5	0.435158	0.244587
34	15	13	6	0.706921	0.271763
34	15	13	7	0.895065	0.188144
34	15	13	8	0.975698	0.080633
34	15	13	9	0.996603	0.020905
34	15	13	10	0.999739	0.003136
34	15	13	11	0.999990	0.000252
34	15	13	12	1.000000	0.000009
34	15	13	13	1.000000	0.000000
34	15	14	0	0.000008	0.000008
34	15	14	1	0.000301	0.000292
34	15	14	2	0.004102	0.003801
34	15	14	3	0.028807	0.024706
34	15	14	4	0.119395	0.090588
34	15	14	5	0.318688	0.199293
34	15	14	6	0.590451	0.271763
34	15	14	7	0.823391	0.232940
34	15	14	8	0.948821	0.125429
34	15	14	9	0.990630	0.041810
34	15	14	10	0.998992	0.008362
34	15	14	11	0.999943	0.000950
34	15	14	12	0.999998	0.000056
34	15	14	13	1.000000	0.000001
34	15	14	14	1.000000	0.000000
34	15	15	0	0.000002	0.000002
34	15	15	1	0.000096	0.000094
34	15	15	2	0.001631	0.001535
34	15	15	3	0.013984	0.012353
34	15	15	4	0.069572	0.055588
34	15	15	5	0.219042	0.149470
34	15	15	6	0.468158	0.249116
34	15	15	7	0.730215	0.262057
34	15	15	8	0.904920	0.174705
34	15	15	9	0.978088	0.073167
34	15	15	10	0.996902	0.018814
34	15	15	11	0.999753	0.002851
34	15	15	12	0.999990	0.000238
34	15	15	13	1.000000	0.000010
34	15	15	14	1.000000	0.000000
34	15	15	15	1.000000	0.000000
34	16	1	0	0.529412	0.529412
34	16	1	1	1.000000	0.470588
34	16	2	0	0.272727	0.272727
34	16	2	1	0.786096	0.513369
34	16	2	2	1.000000	0.213904
34	16	3	0	0.136364	0.136364
34	16	3	1	0.545455	0.409091
34	16	3	2	0.906417	0.360963
34	16	3	3	1.000000	0.093583
34	16	4	0	0.065982	0.065982

N	n	k	x	P(x)	p(x)
34	16	4	1	0.347507	0.281525
34	16	4	2	0.743402	0.395894
34	16	4	3	0.960755	0.217354
34	16	4	4	1.000000	0.039244
34	16	5	0	0.030792	0.030792
34	16	5	1	0.206745	0.175953
34	16	5	2	0.558651	0.351906
34	16	5	3	0.866569	0.307918
34	16	5	4	0.984302	0.117733
34	16	5	5	1.000000	0.015698
34	16	6	0	0.013803	0.013803
34	16	6	1	0.115735	0.101931
34	16	6	2	0.388765	0.273031
34	16	6	3	0.728537	0.339771
34	16	6	4	0.935585	0.207048
34	16	6	5	0.994046	0.058461
34	16	6	6	1.000000	0.005954
34	16	7	0	0.005916	0.005916
34	16	7	1	0.061129	0.055213
34	16	7	2	0.252250	0.191121
34	16	7	3	0.570786	0.318536
34	16	7	4	0.846850	0.276064
34	16	7	5	0.971079	0.124229
34	16	7	6	0.997873	0.026794
34	16	7	7	1.000000	0.002127
34	16	8	0	0.002410	0.002410
34	16	8	1	0.030455	0.028045
34	16	8	2	0.153150	0.122695
34	16	8	3	0.417417	0.264267
34	16	8	4	0.724155	0.306738
34	16	8	5	0.920467	0.196312
34	16	8	6	0.987949	0.067482
34	16	8	7	0.999291	0.011342
34	16	8	8	1.000000	0.000709
34	16	9	0	0.000927	0.000927
34	16	9	1	0.014275	0.013348
34	16	9	2	0.087083	0.072808
34	16	9	3	0.285283	0.198200
34	16	9	4	0.582583	0.297300
34	16	9	5	0.837412	0.254829
34	16	9	6	0.961995	0.124583
34	16	9	7	0.995365	0.033370
34	16	9	8	0.999782	0.004417
34	16	9	9	1.000000	0.000218
34	16	10	0	0.000334	0.000334
34	16	10	1	0.006266	0.005933
34	16	10	2	0.046311	0.040044
34	16	10	3	0.182219	0.135909
34	16	10	4	0.439879	0.257660
34	16	10	5	0.725287	0.285408

N	n	k	x	P(x)	p(x)
34	16	10	6	0.912162	0.186874
34	16	10	7	0.983352	0.071190
34	16	10	8	0.998369	0.015017
34	16	10	9	0.999939	0.001570
34	16	10	10	1.000000	0.000061
34	16	11	0	0.000111	0.000111
34	16	11	1	0.002558	0.002447
34	16	11	2	0.022951	0.020393
34	16	11	3	0.108602	0.085651
34	16	11	4	0.311049	0.202447
34	16	11	5	0.594475	0.283426
34	16	11	6	0.834297	0.239822
34	16	11	7	0.956655	0.122358
34	16	11	8	0.993363	0.036707
34	16	11	9	0.999481	0.006118
34	16	11	10	0.999985	0.000504
34	16	11	11	1.000000	0.000015
34	16	12	0	0.000034	0.000034
34	16	12	1	0.000962	0.000929
34	16	12	2	0.010538	0.009576
34	16	12	3	0.060191	0.049653
34	16	12	4	0.205425	0.145234
34	16	12	5	0.458924	0.253499
34	16	12	6	0.730027	0.271103
34	16	12	7	0.908776	0.178749
34	16	12	8	0.980595	0.071819
34	16	12	9	0.997619	0.017024
34	16	12	10	0.999853	0.002234
34	16	12	11	0.999997	0.000143
34	16	12	12	1.000000	0.000003
34	16	13	0	0.000009	0.000009
34	16	13	1	0.000329	0.000320
34	16	13	2	0.004445	0.004115
34	16	13	3	0.030851	0.026406
34	16	13	4	0.126206	0.095356
34	16	13	5	0.332174	0.205968
34	16	13	6	0.606798	0.274624
34	16	13	7	0.835651	0.228853
34	16	13	8	0.954479	0.118828
34	16	13	9	0.992202	0.037723
34	16	13	10	0.999244	0.007042
34	16	13	11	0.999964	0.000720
34	16	13	12	0.999999	0.000035
34	16	13	13	1.000000	0.000001
34	16	14	0	0.000002	0.000002
34	16	14	1	0.000101	0.000098
34	16	14	2	0.001701	0.001600
34	16	14	3	0.014504	0.012803
34	16	14	4	0.071717	0.057213
34	16	14	5	0.224286	0.152569

N	n	k	x	P(x)	p(x)
34	16	14	6	0.476025	0.251739
34	16	14	7	0.737571	0.261547
34	16	14	8	0.909211	0.171640
34	16	14	9	0.979628	0.070416
34	16	14	10	0.997232	0.017604
34	16	14	11	0.999793	0.002561
34	16	14	12	0.999993	0.000200
34	16	14	13	1.000000	0.000007
34	16	14	14	1.000000	0.000000
34	16	15	0	0.000000	0.000000
34	16	15	1	0.000027	0.000026
34	16	15	2	0.000581	0.000554
34	16	15	3	0.006182	0.005601
34	16	15	4	0.037389	0.031207
34	16	15	5	0.140373	0.102984
34	16	15	6	0.350156	0.209782
34	16	15	7	0.619875	0.269720
34	16	15	8	0.840555	0.220680
34	16	15	9	0.954982	0.114427
34	16	15	10	0.991951	0.036969
34	16	15	11	0.999152	0.007202
34	16	15	12	0.999953	0.000800
34	16	15	13	0.999999	0.000046
34	16	15	14	1.000000	0.000001
34	16	15	15	1.000000	0.000000
34	16	16	0	0.000000	0.000000
34	16	16	1	0.000006	0.000006
34	16	16	2	0.000173	0.000167
34	16	16	3	0.002350	0.002177
34	16	16	4	0.017680	0.015330
34	16	16	5	0.080751	0.063072
34	16	16	6	0.239744	0.158993
34	16	16	7	0.492113	0.252370
34	16	16	8	0.747638	0.255524
34	16	16	9	0.912825	0.165187
34	16	16	10	0.980276	0.067451
34	16	16	11	0.997257	0.016981
34	16	16	12	0.999784	0.002527
34	16	16	13	0.999991	0.000207
34	16	16	14	1.000000	0.000008
34	16	16	15	1.000000	0.000000
34	16	16	16	1.000000	0.000000
34	17	1	0	0.500000	0.500000
34	17	1	1	1.000000	0.500000
34	17	2	0	0.242424	0.242424
34	17	2	1	0.757576	0.515151
34	17	2	2	1.000000	0.242424
34	17	3	0	0.113636	0.113636
34	17	3	1	0.500000	0.386364
34	17	3	2	0.886364	0.386364

N	n	k	x	P(x)	p(x)
34	17	3	3	1.000000	0.113636
34	17	4	0	0.051320	0.051320
34	17	4	1	0.300586	0.249267
34	17	4	2	0.699413	0.398827
34	17	4	3	0.948680	0.249267
34	17	4	4	1.000000	0.051320
34	17	5	0	0.022239	0.022239
34	17	5	1	0.167644	0.145406
34	17	5	2	0.500000	0.332356
34	17	5	3	0.832356	0.332356
34	17	5	4	0.977761	0.145406
34	17	5	5	1.000000	0.022239
34	17	6	0	0.009202	0.009202
34	17	6	1	0.087420	0.078218
34	17	6	2	0.328092	0.240671
34	17	6	3	0.671908	0.343816
34	17	6	4	0.912579	0.240671
34	17	6	5	0.990798	0.078218
34	17	6	6	1.000000	0.009202
34	17	7	0	0.003615	0.003615
34	17	7	1	0.042724	0.039109
34	17	7	2	0.199161	0.156436
34	17	7	3	0.500000	0.300839
34	17	7	4	0.800839	0.300839
34	17	7	5	0.957276	0.156436
34	17	7	6	0.996385	0.039109
34	17	7	7	1.000000	0.003615
34	17	8	0	0.001339	0.001339
34	17	8	1	0.019548	0.018210
34	17	8	2	0.112252	0.092703
34	17	8	3	0.344009	0.231758
34	17	8	4	0.655991	0.311981
34	17	8	5	0.887748	0.231758
34	17	8	6	0.980451	0.092703
34	17	8	7	0.998661	0.018210
34	17	8	8	1.000000	0.001339
34	17	9	0	0.000463	0.000463
34	17	9	1	0.008343	0.007879
34	17	9	2	0.058769	0.050426
34	17	9	3	0.219217	0.160448
34	17	9	4	0.500000	0.280783
34	17	9	5	0.780783	0.280783
34	17	9	6	0.941231	0.160448
34	17	9	7	0.991657	0.050426
34	17	9	8	0.999536	0.007879
34	17	9	9	1.000000	0.000463
34	17	10	0	0.000148	0.000148
34	17	10	1	0.003300	0.003152
34	17	10	2	0.028513	0.025213
34	17	10	3	0.129366	0.100853

Table for $N = 2$, $n = 1$, through $N = 100$, $n = 50$

N	n	k	x	P(x)	p(x)
34	17	10	4	0.353993	0.224627
34	17	10	5	0.646007	0.292015
34	17	10	6	0.870634	0.224627
34	17	10	7	0.971487	0.100853
34	17	10	8	0.996700	0.025213
34	17	10	9	0.999852	0.003152
34	17	10	10	1.000000	0.000148
34	17	11	0	0.000043	0.000043
34	17	11	1	0.001199	0.001156
34	17	11	2	0.012755	0.011556
34	17	11	3	0.070535	0.057780
34	17	11	4	0.232320	0.161785
34	17	11	5	0.500000	0.267680
34	17	11	6	0.767680	0.267680
34	17	11	7	0.929465	0.161785
34	17	11	8	0.987245	0.057780
34	17	11	9	0.998801	0.011556
34	17	11	10	0.999957	0.001156
34	17	11	11	1.000000	0.000043
34	17	12	0	0.000011	0.000011
34	17	12	1	0.000395	0.000384
34	17	12	2	0.005218	0.004823
34	17	12	3	0.035365	0.030146
34	17	12	4	0.140876	0.105512
34	17	12	5	0.360341	0.219464
34	17	12	6	0.639659	0.279318
34	17	12	7	0.859124	0.219464
34	17	12	8	0.964635	0.105512
34	17	12	9	0.994782	0.030146
34	17	12	10	0.999605	0.004823
34	17	12	11	0.999989	0.000384
34	17	12	12	1.000000	0.000011
34	17	13	0	0.000003	0.000003
34	17	13	1	0.000116	0.000113
34	17	13	2	0.001930	0.001814
34	17	13	3	0.016181	0.014251
34	17	13	4	0.078528	0.062348
34	17	13	5	0.240633	0.162104
34	17	13	6	0.500000	0.259367
34	17	13	7	0.759367	0.259367
34	17	13	8	0.921471	0.162104
34	17	13	9	0.983819	0.062348
34	17	13	10	0.998070	0.014251
34	17	13	11	0.999884	0.001814
34	17	13	12	0.999997	0.000113
34	17	13	13	1.000000	0.000003
34	17	14	0	0.000000	0.000000
34	17	14	1	0.000030	0.000029
34	17	14	2	0.000634	0.000605
34	17	14	3	0.006680	0.006046

N	n	k	x	P(x)	p(x)
34	17	14	4	0.039932	0.033252
34	17	14	5	0.148002	0.108070
34	17	14	6	0.364141	0.216139
34	17	14	7	0.635859	0.271718
34	17	14	8	0.851998	0.216139
34	17	14	9	0.960068	0.108070
34	17	14	10	0.993320	0.033252
34	17	14	11	0.999366	0.006046
34	17	14	12	0.999970	0.000605
34	17	14	13	0.999999	0.000029
34	17	14	14	1.000000	0.000000
34	17	15	0	0.000000	0.000000
34	17	15	1	0.000006	0.000006
34	17	15	2	0.000181	0.000174
34	17	15	3	0.002448	0.002267
34	17	15	4	0.018318	0.015870
34	17	15	5	0.083160	0.064842
34	17	15	6	0.245264	0.162104
34	17	15	7	0.500000	0.254736
34	17	15	8	0.754736	0.254736
34	17	15	9	0.916840	0.162104
34	17	15	10	0.981682	0.064842
34	17	15	11	0.997552	0.015870
34	17	15	12	0.999819	0.002267
34	17	15	13	0.999994	0.000174
34	17	15	14	1.000000	0.000006
34	17	15	15	1.000000	0.000000
34	17	16	0	0.000000	0.000000
34	17	16	1	0.000001	0.000001
34	17	16	2	0.000043	0.000042
34	17	16	3	0.000777	0.000734
34	17	16	4	0.007460	0.006682
34	17	16	5	0.042207	0.034748
34	17	16	6	0.151415	0.109207
34	17	16	7	0.365929	0.214514
34	17	16	8	0.634071	0.268143
34	17	16	9	0.848585	0.214514
34	17	16	10	0.957793	0.109207
34	17	16	11	0.992540	0.034748
34	17	16	12	0.999223	0.006682
34	17	16	13	0.999957	0.000734
34	17	16	14	0.999999	0.000042
34	17	16	15	1.000000	0.000001
34	17	16	16	1.000000	0.000000
34	17	17	0	0.000000	0.000000
34	17	17	1	0.000000	0.000000
34	17	17	2	0.000008	0.000008
34	17	17	3	0.000206	0.000198
34	17	17	4	0.002634	0.002427
34	17	17	5	0.019042	0.016409

N	n	k	x	P(x)	p(x)
34	17	17	6	0.084677	0.065635
34	17	17	7	0.246754	0.162077
34	17	17	8	0.500000	0.253246
34	17	17	9	0.753246	0.253246
34	17	17	10	0.915323	0.162077
34	17	17	11	0.980958	0.065635
34	17	17	12	0.997366	0.016409
34	17	17	13	0.999794	0.002427
34	17	17	14	0.999992	0.000198
34	17	17	15	1.000000	0.000008
34	17	17	16	1.000000	0.000000
34	17	17	17	1.000000	0.000000
35	1	1	0	0.971429	0.971429
35	1	1	1	1.000000	0.028571
35	2	1	0	0.942857	0.942857
35	2	1	1	1.000000	0.057143
35	2	2	0	0.887395	0.887395
35	2	2	1	0.998319	0.110924
35	2	2	2	1.000000	0.001681
35	3	1	0	0.914286	0.914286
35	3	1	1	1.000000	0.085714
35	3	2	0	0.833613	0.833613
35	3	2	1	0.994958	0.161345
35	3	2	2	1.000000	0.005042
35	3	3	0	0.757830	0.757830
35	3	3	1	0.985179	0.227349
35	3	3	2	0.999847	0.014668
35	3	3	3	1.000000	0.000153
35	4	1	0	0.885714	0.885714
35	4	1	1	1.000000	0.114286
35	4	2	0	0.781513	0.781513
35	4	2	1	0.989916	0.208403
35	4	2	2	1.000000	0.010084
35	4	3	0	0.686784	0.686784
35	4	3	1	0.970970	0.284186
35	4	3	2	0.999389	0.028419
35	4	3	3	1.000000	0.000611
35	4	4	0	0.600936	0.600936
35	4	4	1	0.944328	0.343392
35	4	4	2	0.997613	0.053285
35	4	4	3	0.999981	0.002368
35	4	4	4	1.000000	0.000019
35	5	1	0	0.857143	0.857143
35	5	1	1	1.000000	0.142857
35	5	2	0	0.731092	0.731092
35	5	2	1	0.983193	0.252101
35	5	2	2	1.000000	0.016807
35	5	3	0	0.620321	0.620321
35	5	3	1	0.952636	0.332315
35	5	3	2	0.998472	0.045837

N	n	k	x	P(x)	p(x)
35	5	3	3	1.000000	0.001528
35	5	4	0	0.523396	0.523396
35	5	4	1	0.911096	0.387701
35	5	4	2	0.994175	0.083079
35	5	4	3	0.999904	0.005730
35	5	4	4	1.000000	0.000095
35	5	5	0	0.438977	0.438977
35	5	5	1	0.861070	0.422093
35	5	5	2	0.986135	0.125065
35	5	5	3	0.999535	0.013400
35	5	5	4	0.999997	0.000462
35	5	5	5	1.000000	0.000003
35	6	1	0	0.828571	0.828571
35	6	1	1	1.000000	0.171429
35	6	2	0	0.682353	0.682353
35	6	2	1	0.974790	0.292437
35	6	2	2	1.000000	0.025210
35	6	3	0	0.558289	0.558289
35	6	3	1	0.930481	0.372192
35	6	3	2	0.996944	0.066463
35	6	3	3	1.000000	0.003056
35	6	4	0	0.453610	0.453610
35	6	4	1	0.872326	0.418717
35	6	4	2	0.988636	0.116310
35	6	4	3	0.999713	0.011077
35	6	4	4	1.000000	0.000286
35	6	5	0	0.365814	0.365814
35	6	5	1	0.804791	0.438977
35	6	5	2	0.973628	0.168837
35	6	5	3	0.998641	0.025013
35	6	5	4	0.999981	0.001340
35	6	5	5	1.000000	0.000018
35	6	6	0	0.292651	0.292651
35	6	6	1	0.731628	0.438977
35	6	6	2	0.951117	0.219489
35	6	6	3	0.996140	0.045023
35	6	6	4	0.999892	0.003752
35	6	6	5	0.999999	0.000107
35	6	6	6	1.000000	0.000001
35	7	1	0	0.800000	0.800000
35	7	1	1	1.000000	0.200000
35	7	2	0	0.635294	0.635294
35	7	2	1	0.964706	0.329412
35	7	2	2	1.000000	0.035294
35	7	3	0	0.500535	0.500535
35	7	3	1	0.904813	0.404278
35	7	3	2	0.994652	0.089840
35	7	3	3	1.000000	0.005348
35	7	4	0	0.391043	0.391043
35	7	4	1	0.829011	0.437968

N	n	k	x	P(x)	p(x)
35	7	4	2	0.980615	0.151604
35	7	4	3	0.999332	0.018717
35	7	4	4	1.000000	0.000668
35	7	5	0	0.302743	0.302743
35	7	5	1	0.744243	0.441500
35	7	5	2	0.956163	0.211920
35	7	5	3	0.996916	0.040754
35	7	5	4	0.999935	0.003019
35	7	5	5	1.000000	0.000065
35	7	6	0	0.232103	0.232103
35	7	6	1	0.655943	0.423840
35	7	6	2	0.920843	0.264900
35	7	6	3	0.991483	0.070640
35	7	6	4	0.999633	0.008151
35	7	6	5	0.999996	0.000362
35	7	6	6	1.000000	0.000004
35	7	7	0	0.176078	0.176078
35	7	7	1	0.568252	0.392174
35	7	7	2	0.875170	0.306919
35	7	7	3	0.981739	0.106569
35	7	7	4	0.998790	0.017051
35	7	7	5	0.999971	0.001180
35	7	7	6	1.000000	0.000029
35	7	7	7	1.000000	0.000000
35	8	1	0	0.771429	0.771429
35	8	1	1	1.000000	0.228571
35	8	2	0	0.589916	0.589916
35	8	2	1	0.952941	0.363025
35	8	2	2	1.000000	0.047059
35	8	3	0	0.446906	0.446906
35	8	3	1	0.875936	0.429030
35	8	3	2	0.991444	0.115508
35	8	3	3	1.000000	0.008556
35	8	4	0	0.335180	0.335180
35	8	4	1	0.782086	0.446906
35	8	4	2	0.969786	0.187701
35	8	4	3	0.998663	0.028877
35	8	4	4	1.000000	0.001337
35	8	5	0	0.248682	0.248682
35	8	5	1	0.681171	0.432490
35	8	5	2	0.933457	0.252286
35	8	5	3	0.994005	0.060549
35	8	5	4	0.999827	0.005822
35	8	5	5	1.000000	0.000172
35	8	6	0	0.182366	0.182366
35	8	6	1	0.580257	0.397891
35	8	6	2	0.883000	0.302743
35	8	6	3	0.983914	0.100914
35	8	6	4	0.999051	0.015137
35	8	6	5	0.999983	0.000932
35	8	6	6	1.000000	0.000017
35	8	7	0	0.132058	0.132058
35	8	7	1	0.484214	0.352156
35	8	7	2	0.820363	0.336149
35	8	7	3	0.966515	0.146152
35	8	7	4	0.996963	0.030448
35	8	7	5	0.999886	0.002923
35	8	7	6	0.999999	0.000112
35	8	7	7	1.000000	0.000001
35	8	8	0	0.094327	0.094327
35	8	8	1	0.396175	0.301848
35	8	8	2	0.748331	0.352156
35	8	8	3	0.940416	0.192085
35	8	8	4	0.992613	0.052197
35	8	8	5	0.999573	0.006960
35	8	8	6	0.999991	0.000418
35	8	8	7	1.000000	0.000009
35	8	8	8	1.000000	0.000000
35	9	1	0	0.742857	0.742857
35	9	1	1	1.000000	0.257143
35	9	2	0	0.546218	0.546218
35	9	2	1	0.939496	0.393277
35	9	2	2	1.000000	0.060504
35	9	3	0	0.397250	0.397250
35	9	3	1	0.844156	0.446906
35	9	3	2	0.987166	0.143010
35	9	3	3	1.000000	0.012834
35	9	4	0	0.285523	0.285523
35	9	4	1	0.732429	0.446906
35	9	4	2	0.955882	0.223453
35	9	4	3	0.997594	0.041711
35	9	4	4	1.000000	0.002406
35	9	5	0	0.202629	0.202629
35	9	5	1	0.617099	0.414469
35	9	5	2	0.905425	0.288326
35	9	5	3	0.989520	0.084095
35	9	5	4	0.999612	0.010091
35	9	5	5	1.000000	0.000388
35	9	6	0	0.141841	0.141841
35	9	6	1	0.506574	0.364733
35	9	6	2	0.838149	0.331575
35	9	6	3	0.972701	0.134552
35	9	6	4	0.997930	0.025229
35	9	6	5	0.999948	0.002018
35	9	6	6	1.000000	0.000052
35	9	7	0	0.097821	0.097821
35	9	7	1	0.405958	0.308136
35	9	7	2	0.758114	0.352156
35	9	7	3	0.944863	0.186749
35	9	7	4	0.993580	0.048717
35	9	7	5	0.999670	0.006090
35	9	7	6	0.999995	0.000325
35	9	7	7	1.000000	0.000005
35	9	8	0	0.066379	0.066379
35	9	8	1	0.317919	0.251540
35	9	8	2	0.670074	0.352156
35	9	8	3	0.904845	0.234771
35	9	8	4	0.984881	0.080035
35	9	8	5	0.998800	0.013919
35	9	8	6	0.999960	0.001160
35	9	8	7	0.999999	0.000040
35	9	8	8	1.000000	0.000000
35	9	9	0	0.044252	0.044252
35	9	9	1	0.243388	0.199136
35	9	9	2	0.578775	0.335387
35	9	9	3	0.852674	0.273899
35	9	9	4	0.970059	0.117385
35	9	9	5	0.996738	0.026678
35	9	9	6	0.999831	0.003093
35	9	9	7	0.999997	0.000166
35	9	9	8	1.000000	0.000003
35	9	9	9	1.000000	0.000000
35	10	1	0	0.714286	0.714286
35	10	1	1	1.000000	0.285714
35	10	2	0	0.504202	0.504202
35	10	2	1	0.924370	0.420168
35	10	2	2	1.000000	0.075630
35	10	3	0	0.351413	0.351413
35	10	3	1	0.809778	0.458365
35	10	3	2	0.981665	0.171887
35	10	3	3	1.000000	0.018335
35	10	4	0	0.241597	0.241597
35	10	4	1	0.680863	0.439267
35	10	4	2	0.938694	0.257830
35	10	4	3	0.995989	0.057296
35	10	4	4	1.000000	0.004011
35	10	5	0	0.163662	0.163662
35	10	5	1	0.553334	0.389672
35	10	5	2	0.872157	0.318823
35	10	5	3	0.983052	0.110895
35	10	5	4	0.999224	0.016172
35	10	5	5	1.000000	0.000776
35	10	6	0	0.109108	0.109108
35	10	6	1	0.436433	0.327324
35	10	6	2	0.787137	0.350705
35	10	6	3	0.957176	0.170039
35	10	6	4	0.995989	0.038813
35	10	6	5	0.999870	0.003881
35	10	6	6	1.000000	0.000129
35	10	7	0	0.071485	0.071485
35	10	7	1	0.334849	0.263364
35	10	7	2	0.690391	0.355542
35	10	7	3	0.916132	0.225741
35	10	7	4	0.987959	0.071827
35	10	7	5	0.999201	0.011242
35	10	7	6	0.999982	0.000781
35	10	7	7	1.000000	0.000018
35	10	8	0	0.045954	0.045954
35	10	8	1	0.250196	0.204242
35	10	8	2	0.588808	0.338611
35	10	8	3	0.859697	0.270889
35	10	8	4	0.972567	0.112870
35	10	8	5	0.997194	0.024626
35	10	8	6	0.999870	0.002677
35	10	8	7	0.999998	0.000127
35	10	8	8	1.000000	0.000002
35	10	9	0	0.028934	0.028934
35	10	9	1	0.182116	0.153181
35	10	9	2	0.488478	0.306363
35	10	9	3	0.789466	0.300988
35	10	9	4	0.947485	0.158019
35	10	9	5	0.992633	0.045148
35	10	9	6	0.999474	0.006841
35	10	9	7	0.999984	0.000510
35	10	9	8	1.000000	0.000016
35	10	9	9	1.000000	0.000000
35	10	10	0	0.017806	0.017806
35	10	10	1	0.129091	0.111286
35	10	10	2	0.394213	0.265122
35	10	10	3	0.708431	0.314218
35	10	10	4	0.911019	0.202588
35	10	10	5	0.983951	0.072932
35	10	10	6	0.998422	0.014471
35	10	10	7	0.999925	0.001503
35	10	10	8	0.999998	0.000074
35	10	10	9	1.000000	0.000001
35	10	10	10	1.000000	0.000000
35	11	1	0	0.685714	0.685714
35	11	1	1	1.000000	0.314286
35	11	2	0	0.463866	0.463866
35	11	2	1	0.907563	0.443697
35	11	2	2	1.000000	0.092437
35	11	3	0	0.309244	0.309244
35	11	3	1	0.773109	0.463866
35	11	3	2	0.974790	0.201681
35	11	3	3	1.000000	0.025210
35	11	4	0	0.202941	0.202941
35	11	4	1	0.628151	0.425210
35	11	4	2	0.918067	0.289916
35	11	4	3	0.993697	0.075630

Table for $N = 2$, $n = 1$, through $N = 100$, $n = 50$

N	n	k	x	P(x)	p(x)
35	11	4	4	1.000000	0.006303
35	11	5	0	0.130930	0.130930
35	11	5	1	0.490987	0.360057
35	11	5	2	0.833898	0.342911
35	11	5	3	0.974180	0.140282
35	11	5	4	0.998577	0.024397
35	11	5	5	1.000000	0.001423
35	11	6	0	0.082922	0.082922
35	11	6	1	0.370968	0.288046
35	11	6	2	0.731025	0.360057
35	11	6	3	0.936771	0.205747
35	11	6	4	0.992884	0.056113
35	11	6	5	0.999715	0.006831
35	11	6	6	1.000000	0.000285
35	11	7	0	0.051469	0.051469
35	11	7	1	0.271642	0.220173
35	11	7	2	0.619283	0.347641
35	11	7	3	0.880014	0.260731
35	11	7	4	0.979340	0.099326
35	11	7	5	0.998302	0.018962
35	11	7	6	0.999951	0.001649
35	11	7	7	1.000000	0.000049
35	11	8	0	0.031249	0.031249
35	11	8	1	0.193009	0.161760
35	11	8	2	0.507541	0.314532
35	11	8	3	0.805519	0.297978
35	11	8	4	0.954508	0.148989
35	11	8	5	0.994239	0.039730
35	11	8	6	0.999656	0.005418
35	11	8	7	0.999993	0.000337
35	11	8	8	1.000000	0.000007
35	11	9	0	0.018518	0.018518
35	11	9	1	0.133098	0.114580
35	11	9	2	0.402697	0.269599
35	11	9	3	0.717229	0.314532
35	11	9	4	0.915881	0.198652
35	11	9	5	0.985410	0.069528
35	11	9	6	0.998653	0.013243
35	11	9	7	0.999943	0.001290
35	11	9	8	0.999999	0.000056
35	11	9	9	1.000000	0.000001
35	11	10	0	0.010683	0.010683
35	11	10	1	0.089028	0.078345
35	11	10	2	0.309374	0.220346
35	11	10	3	0.620450	0.311076
35	11	10	4	0.862398	0.241948
35	11	10	5	0.969365	0.106967
35	11	10	6	0.996106	0.026742
35	11	10	7	0.999745	0.003638
35	11	10	8	0.999993	0.000248
35	11	10	9	1.000000	0.000007
35	11	10	10	1.000000	0.000000
35	11	11	0	0.005983	0.005983
35	11	11	1	0.057690	0.051708
35	11	11	2	0.230050	0.172359
35	11	11	3	0.520906	0.290856
35	11	11	4	0.794653	0.273747
35	11	11	5	0.943693	0.149040
35	11	11	6	0.990758	0.047065
35	11	11	7	0.999162	0.008405
35	11	11	8	0.999963	0.000800
35	11	11	9	0.999999	0.000036
35	11	11	10	1.000000	0.000001
35	11	11	11	1.000000	0.000000
35	12	1	0	0.657143	0.657143
35	12	1	1	1.000000	0.342857
35	12	2	0	0.425210	0.425210
35	12	2	1	0.889076	0.463866
35	12	2	2	1.000000	0.110924
35	12	3	0	0.270588	0.270588
35	12	3	1	0.734454	0.463866
35	12	3	2	0.966386	0.231933
35	12	3	3	1.000000	0.033613
35	12	4	0	0.169118	0.169118
35	12	4	1	0.575000	0.405882
35	12	4	2	0.893907	0.318908
35	12	4	3	0.990546	0.096639
35	12	4	4	1.000000	0.009454
35	12	5	0	0.103653	0.103653
35	12	5	1	0.430977	0.327324
35	12	5	2	0.791034	0.360057
35	12	5	3	0.962490	0.171456
35	12	5	4	0.997560	0.035070
35	12	5	5	1.000000	0.002440
35	12	6	0	0.062192	0.062192
35	12	6	1	0.310958	0.248767
35	12	6	2	0.671015	0.360057
35	12	6	3	0.911053	0.240038
35	12	6	4	0.988208	0.077155
35	12	6	5	0.999431	0.011223
35	12	6	6	1.000000	0.000569
35	12	7	0	0.036457	0.036457
35	12	7	1	0.216598	0.180141
35	12	7	2	0.546858	0.330259
35	12	7	3	0.836559	0.289701
35	12	7	4	0.966924	0.130365
35	12	7	5	0.996722	0.029798
35	12	7	6	0.999882	0.003160
35	12	7	7	1.000000	0.000118
35	12	8	0	0.020833	0.020833
35	12	8	1	0.145829	0.124996
35	12	8	2	0.428908	0.283079
35	12	8	3	0.743440	0.314532
35	12	8	4	0.929677	0.186236
35	12	8	5	0.989272	0.059596
35	12	8	6	0.999205	0.009933
35	12	8	7	0.999979	0.000774
35	12	8	8	1.000000	0.000021
35	12	9	0	0.011574	0.011574
35	12	9	1	0.094904	0.083331
35	12	9	2	0.324064	0.229159
35	12	9	3	0.638596	0.314532
35	12	9	4	0.874496	0.235899
35	12	9	5	0.973822	0.099326
35	12	9	6	0.996998	0.023176
35	12	9	7	0.999836	0.002838
35	12	9	8	0.999997	0.000161
35	12	9	9	1.000000	0.000003
35	12	10	0	0.006232	0.006232
35	12	10	1	0.059649	0.053417
35	12	10	2	0.235926	0.176276
35	12	10	3	0.529720	0.293794
35	12	10	4	0.801911	0.272192
35	12	10	5	0.947080	0.145169
35	12	10	6	0.991649	0.044569
35	12	10	7	0.999290	0.007640
35	12	10	8	0.999972	0.000682
35	12	10	9	1.000000	0.000028
35	12	10	10	1.000000	0.000000
35	12	11	0	0.003241	0.003241
35	12	11	1	0.036146	0.032905
35	12	11	2	0.165415	0.129269
35	12	11	3	0.423954	0.258539
35	12	11	4	0.714810	0.290856
35	12	11	5	0.906433	0.191623
35	12	11	6	0.980953	0.074520
35	12	11	7	0.997762	0.016809
35	12	11	8	0.999863	0.002101
35	12	11	9	0.999996	0.000133
35	12	11	10	1.000000	0.000004
35	12	11	11	1.000000	0.000000
35	12	12	0	0.001620	0.001620
35	12	12	1	0.021064	0.019444
35	12	12	2	0.111553	0.090489
35	12	12	3	0.327002	0.215449
35	12	12	4	0.617858	0.290856
35	12	12	5	0.850543	0.232685
35	12	12	6	0.962323	0.111780
35	12	12	7	0.994260	0.031937
35	12	12	8	0.999513	0.005253
35	12	12	9	0.999980	0.000467
35	12	12	10	1.000000	0.000020
35	12	12	11	1.000000	0.000000
35	12	12	12	1.000000	0.000000
35	13	1	0	0.628571	0.628571
35	13	1	1	1.000000	0.371429
35	13	2	0	0.388235	0.388235
35	13	2	1	0.868908	0.480672
35	13	2	2	1.000000	0.131092
35	13	3	0	0.235294	0.235294
35	13	3	1	0.694118	0.458824
35	13	3	2	0.956302	0.262185
35	13	3	3	1.000000	0.043697
35	13	4	0	0.139706	0.139706
35	13	4	1	0.522059	0.382353
35	13	4	2	0.866176	0.344118
35	13	4	3	0.986344	0.120168
35	13	4	4	1.000000	0.013655
35	13	5	0	0.081120	0.081120
35	13	5	1	0.374051	0.292932
35	13	5	2	0.744070	0.370019
35	13	5	3	0.947581	0.203510
35	13	5	4	0.996035	0.048455
35	13	5	5	1.000000	0.003964
35	13	6	0	0.045968	0.045968
35	13	6	1	0.256879	0.210911
35	13	6	2	0.608397	0.351518
35	13	6	3	0.879744	0.271347
35	13	6	4	0.981499	0.101755
35	13	6	5	0.998943	0.017444
35	13	6	6	1.000000	0.001057
35	13	7	0	0.025362	0.025362
35	13	7	1	0.169605	0.144244
35	13	7	2	0.475062	0.305457
35	13	7	3	0.786176	0.311114
35	13	7	4	0.949920	0.163744
35	13	7	5	0.994131	0.044211
35	13	7	6	0.999745	0.005614
35	13	7	7	1.000000	0.000255
35	13	8	0	0.013587	0.013587
35	13	8	1	0.107786	0.094200
35	13	8	2	0.355061	0.247275
35	13	8	3	0.675064	0.320003
35	13	8	4	0.897288	0.222224
35	13	8	5	0.981499	0.084211
35	13	8	6	0.998341	0.016842
35	13	8	7	0.999945	0.001604
35	13	8	8	1.000000	0.000055
35	13	9	0	0.007045	0.007045
35	13	9	1	0.065920	0.058875

Table for $N = 2$, $n = 1$, through $N = 100$, $n = 50$

N	n	k	x	P(x)	p(x)
35	13	9	2	0.254320	0.188400
35	13	9	3	0.556544	0.302225
35	13	9	4	0.823213	0.266669
35	13	9	5	0.956548	0.133334
35	13	9	6	0.993975	0.037427
35	13	9	7	0.999589	0.005614
35	13	9	8	0.999990	0.000401
35	13	9	9	1.000000	0.000010
35	13	10	0	0.003522	0.003522
35	13	10	1	0.038747	0.035224
35	13	10	2	0.174612	0.135865
35	13	10	3	0.440304	0.265692
35	13	10	4	0.730905	0.290601
35	13	10	5	0.915522	0.184617
35	13	10	6	0.983898	0.068377
35	13	10	7	0.998293	0.014395
35	13	10	8	0.999913	0.001619
35	13	10	9	0.999998	0.000086
35	13	10	10	1.000000	0.000002
35	13	11	0	0.001691	0.001691
35	13	11	1	0.021839	0.020148
35	13	11	2	0.114831	0.092992
35	13	11	3	0.334027	0.219196
35	13	11	4	0.626288	0.292261
35	13	11	5	0.856444	0.230156
35	13	11	6	0.964753	0.108309
35	13	11	7	0.994838	0.030086
35	13	11	8	0.999589	0.004750
35	13	11	9	0.999985	0.000396
35	13	11	10	1.000000	0.000015
35	13	11	11	1.000000	0.000000
35	13	12	0	0.000775	0.000775
35	13	12	1	0.011765	0.010990
35	13	12	2	0.072210	0.060445
35	13	12	3	0.242696	0.170486
35	13	12	4	0.516690	0.273995
35	13	12	5	0.779726	0.263035
35	13	12	6	0.933163	0.153437
35	13	12	7	0.987317	0.054154
35	13	12	8	0.998599	0.011282
35	13	12	9	0.999919	0.001320
35	13	12	10	0.999998	0.000079
35	13	12	11	1.000000	0.000002
35	13	12	12	1.000000	0.000000
35	13	13	0	0.000337	0.000337
35	13	13	1	0.006031	0.005694
35	13	13	2	0.043301	0.037270
35	13	13	3	0.168571	0.125270
35	13	13	4	0.409475	0.240904
35	13	13	5	0.688235	0.278760

N	n	k	x	P(x)	p(x)
35	13	13	6	0.886464	0.198229
35	13	13	7	0.973190	0.086725
35	13	13	8	0.996146	0.022957
35	13	13	9	0.999689	0.003543
35	13	13	10	0.999988	0.000298
35	13	13	11	1.000000	0.000012
35	13	13	12	1.000000	0.000000
35	13	13	13	1.000000	0.000000
35	14	1	0	0.600000	0.600000
35	14	1	1	1.000000	0.400000
35	14	2	0	0.352941	0.352941
35	14	2	1	0.847059	0.494118
35	14	2	2	1.000000	0.152941
35	14	3	0	0.203209	0.203209
35	14	3	1	0.652406	0.449198
35	14	3	2	0.944385	0.291979
35	14	3	3	1.000000	0.055615
35	14	4	0	0.114305	0.114305
35	14	4	1	0.469920	0.355615
35	14	4	2	0.834893	0.364973
35	14	4	3	0.980882	0.145989
35	14	4	4	1.000000	0.019118
35	14	5	0	0.062683	0.062683
35	14	5	1	0.320791	0.258108
35	14	5	2	0.693613	0.372822
35	14	5	3	0.929080	0.235467
35	14	5	4	0.993833	0.064753
35	14	5	5	1.000000	0.006167
35	14	6	0	0.033431	0.033431
35	14	6	1	0.208944	0.175513
35	14	6	2	0.544484	0.335540
35	14	6	3	0.842742	0.298258
35	14	6	4	0.972248	0.129507
35	14	6	5	0.998150	0.025901
35	14	6	6	1.000000	0.001850
35	14	7	0	0.017292	0.017292
35	14	7	1	0.130266	0.112974
35	14	7	2	0.405640	0.275374
35	14	7	3	0.729610	0.323970
35	14	7	4	0.927591	0.197981
35	14	7	5	0.990111	0.062520
35	14	7	6	0.999490	0.009378
35	14	7	7	1.000000	0.000510
35	14	8	0	0.008646	0.008646
35	14	8	1	0.077814	0.069168
35	14	8	2	0.287623	0.209809
35	14	8	3	0.602336	0.314713
35	14	8	4	0.856883	0.254548
35	14	8	5	0.970016	0.113132
35	14	8	6	0.996810	0.026794

N	n	k	x	P(x)	p(x)
35	14	8	7	0.999872	0.003062
35	14	8	8	1.000000	0.000128
35	14	9	0	0.004163	0.004163
35	14	9	1	0.044511	0.040348
35	14	9	2	0.194374	0.149863
35	14	9	3	0.474119	0.279745
35	14	9	4	0.762607	0.288487
35	14	9	5	0.932305	0.169698
35	14	9	6	0.988871	0.056566
35	14	9	7	0.999078	0.010207
35	14	9	8	0.999972	0.000893
35	14	9	9	1.000000	0.000028
35	14	10	0	0.001921	0.001921
35	14	10	1	0.024337	0.022415
35	14	10	2	0.125206	0.100870
35	14	10	3	0.355766	0.230559
35	14	10	4	0.651650	0.295884
35	14	10	5	0.873563	0.221913
35	14	10	6	0.971466	0.097903
35	14	10	7	0.996330	0.024864
35	14	10	8	0.999765	0.003435
35	14	10	9	0.999994	0.000229
35	14	10	10	1.000000	0.000005
35	14	11	0	0.000845	0.000845
35	14	11	1	0.012681	0.011835
35	14	11	2	0.076789	0.064108
35	14	11	3	0.254320	0.177531
35	14	11	4	0.533296	0.278977
35	14	11	5	0.793674	0.260378
35	14	11	6	0.940137	0.146463
35	14	11	7	0.989368	0.049231
35	14	11	8	0.998941	0.009573
35	14	11	9	0.999949	0.001008
35	14	11	10	0.999999	0.000050
35	14	11	11	1.000000	0.000001
35	14	12	0	0.000352	0.000352
35	14	12	1	0.006270	0.005918
35	14	12	2	0.044735	0.038465
35	14	12	3	0.172951	0.128216
35	14	12	4	0.417056	0.244105
35	14	12	5	0.696033	0.278977
35	14	12	6	0.891316	0.195284
35	14	12	7	0.975009	0.083693
35	14	12	8	0.996548	0.021539
35	14	12	9	0.999739	0.003191
35	14	12	10	0.999991	0.000252
35	14	12	11	1.000000	0.000009
35	14	12	12	1.000000	0.000000
35	14	13	0	0.000138	0.000138
35	14	13	1	0.002925	0.002787

N	n	k	x	P(x)	p(x)
35	14	13	2	0.024666	0.021741
35	14	13	3	0.111630	0.086964
35	14	13	4	0.310924	0.199293
35	14	13	5	0.586868	0.275944
35	14	13	6	0.823391	0.236524
35	14	13	7	0.949537	0.126146
35	14	13	8	0.990929	0.041392
35	14	13	9	0.999045	0.008116
35	14	13	10	0.999947	0.000902
35	14	13	11	0.999999	0.000052
35	14	13	12	1.000000	0.000001
35	14	13	13	1.000000	0.000000
35	14	14	0	0.000050	0.000050
35	14	14	1	0.001278	0.001228
35	14	14	2	0.012807	0.011529
35	14	14	3	0.068148	0.055341
35	14	14	4	0.220336	0.152187
35	14	14	5	0.473982	0.253646
35	14	14	6	0.737383	0.263401
35	14	14	7	0.909400	0.172017
35	14	14	8	0.979640	0.070240
35	14	14	9	0.997200	0.017560
35	14	14	10	0.999783	0.002582
35	14	14	11	0.999991	0.000209
35	14	14	12	1.000000	0.000008
35	14	14	13	1.000000	0.000000
35	14	14	14	1.000000	0.000000
35	15	1	0	0.571429	0.571429
35	15	1	1	1.000000	0.428571
35	15	2	0	0.319328	0.319328
35	15	2	1	0.823529	0.504202
35	15	2	2	1.000000	0.176471
35	15	3	0	0.174179	0.174179
35	15	3	1	0.609626	0.435447
35	15	3	2	0.930481	0.320856
35	15	3	3	1.000000	0.069519
35	15	4	0	0.092532	0.092532
35	15	4	1	0.419118	0.326585
35	15	4	2	0.800134	0.381016
35	15	4	3	0.973930	0.173797
35	15	4	4	1.000000	0.026070
35	15	5	0	0.047759	0.047759
35	15	5	1	0.271628	0.223869
35	15	5	2	0.640353	0.368725
35	15	5	3	0.906654	0.266302
35	15	5	4	0.990749	0.084095
35	15	5	5	1.000000	0.009250
35	15	6	0	0.023879	0.023879
35	15	6	1	0.167155	0.143276
35	15	6	2	0.480572	0.313416

N	n	k	x	P(x)	p(x)
35	15	6	3	0.800134	0.319562
35	15	6	4	0.959915	0.159781
35	15	6	5	0.996916	0.037002
35	15	6	6	1.000000	0.003083
35	15	7	0	0.011528	0.011528
35	15	7	1	0.097988	0.086460
35	15	7	2	0.340075	0.242087
35	15	7	3	0.667901	0.327826
35	15	7	4	0.899308	0.231407
35	15	7	5	0.984157	0.084849
35	15	7	6	0.999043	0.014886
35	15	7	7	1.000000	0.000957
35	15	8	0	0.005352	0.005352
35	15	8	1	0.054758	0.049406
35	15	8	2	0.227677	0.172919
35	15	8	3	0.527404	0.299727
35	15	8	4	0.808398	0.280994
35	15	8	5	0.953854	0.145456
35	15	8	6	0.994258	0.040404
35	15	8	7	0.999726	0.005468
35	15	8	8	1.000000	0.000273
35	15	9	0	0.002379	0.002379
35	15	9	1	0.029140	0.026761
35	15	9	2	0.144420	0.115280
35	15	9	3	0.394192	0.249772
35	15	9	4	0.693919	0.299727
35	15	9	5	0.899981	0.206062
35	15	9	6	0.980790	0.080809
35	15	9	7	0.998106	0.017316
35	15	9	8	0.999929	0.001823
35	15	9	9	1.000000	0.000071
35	15	10	0	0.001006	0.001006
35	15	10	1	0.014730	0.013724
35	15	10	2	0.086780	0.072050
35	15	10	3	0.278913	0.192133
35	15	10	4	0.567112	0.288199
35	15	10	5	0.820727	0.253615
35	15	10	6	0.952818	0.132091
35	15	10	7	0.992778	0.039960
35	15	10	8	0.999438	0.006660
35	15	10	9	0.999984	0.000545
35	15	10	10	1.000000	0.000016
35	15	11	0	0.000403	0.000403
35	15	11	1	0.007045	0.006642
35	15	11	2	0.049314	0.042269
35	15	11	3	0.186689	0.137375
35	15	11	4	0.440304	0.253615
35	15	11	5	0.719281	0.278977
35	15	11	6	0.905265	0.185984
35	15	11	7	0.979991	0.074726
35	15	11	8	0.997573	0.017583
35	15	11	9	0.999853	0.002279
35	15	11	10	0.999997	0.000144
35	15	11	11	1.000000	0.000003
35	15	12	0	0.000151	0.000151
35	15	12	1	0.003170	0.003019
35	15	12	2	0.026418	0.023248
35	15	12	3	0.118001	0.091583
35	15	12	4	0.324064	0.206062
35	15	12	5	0.603040	0.278977
35	15	12	6	0.835521	0.232480
35	15	12	7	0.955082	0.119561
35	15	12	8	0.992445	0.037363
35	15	12	9	0.999283	0.006838
35	15	12	10	0.999967	0.000684
35	15	12	11	0.999999	0.000033
35	15	12	12	1.000000	0.000001
35	15	13	0	0.000053	0.000053
35	15	13	1	0.001332	0.001280
35	15	13	2	0.013278	0.011946
35	15	13	3	0.070219	0.056941
35	15	13	4	0.225512	0.155293
35	15	13	5	0.481746	0.256234
35	15	13	6	0.744550	0.262804
35	15	13	7	0.913496	0.168945
35	15	13	8	0.981074	0.067578
35	15	13	9	0.997499	0.016425
35	15	13	10	0.999818	0.002319
35	15	13	11	0.999994	0.000176
35	15	13	12	1.000000	0.000006
35	15	13	13	1.000000	0.000000
35	15	14	0	0.000017	0.000017
35	15	14	1	0.000518	0.000501
35	15	14	2	0.006219	0.005701
35	15	14	3	0.039160	0.032941
35	15	14	4	0.147866	0.108705
35	15	14	5	0.365276	0.217411
35	15	14	6	0.637039	0.271763
35	15	14	7	0.852061	0.215021
35	15	14	8	0.959572	0.107511
35	15	14	9	0.993019	0.033448
35	15	14	10	0.999291	0.006271
35	15	14	11	0.999962	0.000671
35	15	14	12	0.999999	0.000037
35	15	14	13	1.000000	0.000001
35	15	14	14	1.000000	0.000000
35	15	15	0	0.000005	0.000005
35	15	15	1	0.000184	0.000179
35	15	15	2	0.002690	0.002506
35	15	15	3	0.020337	0.017647
35	15	15	4	0.090925	0.070588
35	15	15	5	0.261747	0.170823
35	15	15	6	0.520569	0.258822
35	15	15	7	0.770148	0.249579
35	15	15	8	0.923735	0.153587
35	15	15	9	0.983463	0.059728
35	15	15	10	0.997798	0.014335
35	15	15	11	0.999834	0.002036
35	15	15	12	0.999994	0.000160
35	15	15	13	1.000000	0.000006
35	15	15	14	1.000000	0.000000
35	15	15	15	1.000000	0.000000
35	16	1	0	0.542857	0.542857
35	16	1	1	1.000000	0.457143
35	16	2	0	0.287395	0.287395
35	16	2	1	0.798319	0.510924
35	16	2	2	1.000000	0.201681
35	16	3	0	0.148052	0.148052
35	16	3	1	0.566081	0.418029
35	16	3	2	0.914438	0.348358
35	16	3	3	1.000000	0.085561
35	16	4	0	0.074026	0.074026
35	16	4	1	0.370130	0.296104
35	16	4	2	0.762032	0.391902
35	16	4	3	0.965241	0.203209
35	16	4	4	1.000000	0.034759
35	16	5	0	0.035819	0.035819
35	16	5	1	0.226854	0.191035
35	16	5	2	0.585044	0.358190
35	16	5	3	0.880024	0.294980
35	16	5	4	0.986545	0.106521
35	16	5	5	1.000000	0.013455
35	16	6	0	0.016716	0.016716
35	16	6	1	0.131336	0.114621
35	16	6	2	0.417889	0.286552
35	16	6	3	0.752199	0.334311
35	16	6	4	0.943936	0.191737
35	16	6	5	0.995066	0.051130
35	16	6	6	1.000000	0.004934
35	16	7	0	0.007493	0.007493
35	16	7	1	0.072050	0.064557
35	16	7	2	0.279553	0.207503
35	16	7	3	0.602336	0.322783
35	16	7	4	0.864597	0.262261
35	16	7	5	0.975672	0.111075
35	16	7	6	0.998299	0.022626
35	16	7	7	1.000000	0.001701
35	16	8	0	0.003211	0.003211
35	16	8	1	0.037466	0.034255
35	16	8	2	0.175801	0.138336
35	16	8	3	0.452472	0.276671
35	16	8	4	0.752199	0.299727
35	16	8	5	0.932036	0.179836
35	16	8	6	0.990218	0.058182
35	16	8	7	0.999453	0.009235
35	16	8	8	1.000000	0.000547
35	16	9	0	0.001308	0.001308
35	16	9	1	0.018436	0.017127
35	16	9	2	0.104072	0.085636
35	16	9	3	0.319260	0.215189
35	16	9	4	0.618987	0.299727
35	16	9	5	0.858769	0.239782
35	16	9	6	0.968669	0.109900
35	16	9	7	0.996375	0.027706
35	16	9	8	0.999838	0.003463
35	16	9	9	1.000000	0.000162
35	16	10	0	0.000503	0.000503
35	16	10	1	0.008554	0.008051
35	16	10	2	0.057960	0.049406
35	16	10	3	0.211666	0.153706
35	16	10	4	0.480652	0.268986
35	16	10	5	0.757323	0.276671
35	16	10	6	0.926400	0.169077
35	16	10	7	0.986784	0.060385
35	16	10	8	0.998772	0.011988
35	16	10	9	0.999956	0.001184
35	16	10	10	1.000000	0.000044
35	16	11	0	0.000181	0.000181
35	16	11	1	0.003724	0.003543
35	16	11	2	0.030293	0.026569
35	16	11	3	0.131739	0.101446
35	16	11	4	0.351539	0.219800
35	16	11	5	0.635588	0.284049
35	16	11	6	0.858769	0.223181
35	16	11	7	0.965046	0.106277
35	16	11	8	0.994936	0.029890
35	16	11	9	0.999625	0.004689
35	16	11	10	0.999989	0.000365
35	16	11	11	1.000000	0.000010
35	16	12	0	0.000060	0.000060
35	16	12	1	0.001510	0.001449
35	16	12	2	0.014794	0.013285
35	16	12	3	0.076789	0.061995
35	16	12	4	0.241639	0.164850
35	16	12	5	0.505399	0.263760
35	16	12	6	0.765777	0.260378
35	16	12	7	0.925192	0.159415
35	16	12	8	0.984973	0.059781
35	16	12	9	0.998257	0.013285
35	16	12	10	0.999898	0.001641

Table for $N = 2$, $n = 1$, through $N = 100$, $n = 50$

N	n	k	x	P(x)	p(x)
35	16	12	11	0.999998	0.000099
35	16	12	12	1.000000	0.000002
35	16	13	0	0.000018	0.000018
35	16	13	1	0.000564	0.000546
35	16	13	2	0.006708	0.006143
35	16	13	3	0.041748	0.035041
35	16	13	4	0.155630	0.113882
35	16	13	5	0.379253	0.223622
35	16	13	6	0.652569	0.273316
35	16	13	7	0.862812	0.210243
35	16	13	8	0.964179	0.101367
35	16	13	9	0.994214	0.030035
35	16	13	10	0.999470	0.005256
35	16	13	11	0.999976	0.000506
35	16	13	12	0.999999	0.000023
35	16	13	13	1.000000	0.000000
35	16	14	0	0.000005	0.000005
35	16	14	1	0.000192	0.000187
35	16	14	2	0.002798	0.002606
35	16	14	3	0.021043	0.018244
35	16	14	4	0.093513	0.072470
35	16	14	5	0.267441	0.173929
35	16	14	6	0.528334	0.260893
35	16	14	7	0.776803	0.248469
35	16	14	8	0.927318	0.150515
35	16	14	9	0.984658	0.057339
35	16	14	10	0.998037	0.013379
35	16	14	11	0.999861	0.001824
35	16	14	12	0.999995	0.000134
35	16	14	13	1.000000	0.000005
35	16	14	14	1.000000	0.000000
35	16	15	0	0.000001	0.000001
35	16	15	1	0.000058	0.000057
35	16	15	2	0.001061	0.001002
35	16	15	3	0.009749	0.008688
35	16	15	4	0.052101	0.042353
35	16	15	5	0.176336	0.124235
35	16	15	6	0.404100	0.227764
35	16	15	7	0.670317	0.266217
35	16	15	8	0.869979	0.199663
35	16	15	9	0.965545	0.095565
35	16	15	10	0.994214	0.028670
35	16	15	11	0.999427	0.005213
35	16	15	12	0.999970	0.000543
35	16	15	13	0.999999	0.000029
35	16	15	14	1.000000	0.000001
35	16	15	15	1.000000	0.000000
35	16	16	0	0.000000	0.000000
35	16	16	1	0.000016	0.000015
35	16	16	2	0.000359	0.000344

N	n	k	x	P(x)	p(x)
35	16	16	3	0.004102	0.003742
35	16	16	4	0.026690	0.022588
35	16	16	5	0.108007	0.081317
35	16	16	6	0.290218	0.182211
35	16	16	7	0.550519	0.260301
35	16	16	8	0.790114	0.239595
35	16	16	9	0.932097	0.141982
35	16	16	10	0.985613	0.053516
35	16	16	11	0.998124	0.012510
35	16	16	12	0.999861	0.001738
35	16	16	13	0.999995	0.000134
35	16	16	14	1.000000	0.000005
35	16	16	15	1.000000	0.000000
35	16	16	16	1.000000	0.000000
35	17	1	0	0.514286	0.514286
35	17	1	1	1.000000	0.485714
35	17	2	0	0.257143	0.257143
35	17	2	1	0.771428	0.514286
35	17	2	2	1.000000	0.228571
35	17	3	0	0.124675	0.124675
35	17	3	1	0.522078	0.397403
35	17	3	2	0.896104	0.374026
35	17	3	3	1.000000	0.103896
35	17	4	0	0.058442	0.058442
35	17	4	1	0.323377	0.264935
35	17	4	2	0.720779	0.397403
35	17	4	3	0.954545	0.233766
35	17	4	4	1.000000	0.045455
35	17	5	0	0.026393	0.026393
35	17	5	1	0.186636	0.160243
35	17	5	2	0.528488	0.341852
35	17	5	3	0.848974	0.320486
35	17	5	4	0.980938	0.131965
35	17	5	5	1.000000	0.019062
35	17	6	0	0.011437	0.011437
35	17	6	1	0.101173	0.089736
35	17	6	2	0.357562	0.256389
35	17	6	3	0.699413	0.341852
35	17	6	4	0.923754	0.224340
35	17	6	5	0.992375	0.068622
35	17	6	6	1.000000	0.007625
35	17	7	0	0.004733	0.004733
35	17	7	1	0.051663	0.046931
35	17	7	2	0.224947	0.173283
35	17	7	3	0.534382	0.309435
35	17	7	4	0.823187	0.288806
35	17	7	5	0.963980	0.140793
35	17	7	6	0.997108	0.033128
35	17	7	7	1.000000	0.002892
35	17	8	0	0.001859	0.001859

N	n	k	x	P(x)	p(x)
35	17	8	1	0.024846	0.022987
35	17	8	2	0.132116	0.107271
35	17	8	3	0.379664	0.247548
35	17	8	4	0.689099	0.309435
35	17	8	5	0.903640	0.214541
35	17	8	6	0.984093	0.080453
35	17	8	7	0.998967	0.014874
35	17	8	8	1.000000	0.001033
35	17	9	0	0.000689	0.000689
35	17	9	1	0.011224	0.010536
35	17	9	2	0.072522	0.061298
35	17	9	3	0.251306	0.178784
35	17	9	4	0.540112	0.288806
35	17	9	5	0.808289	0.268177
35	17	9	6	0.951316	0.143028
35	17	9	7	0.993458	0.042142
35	17	9	8	0.999656	0.006197
35	17	9	9	1.000000	0.000344
35	17	10	0	0.000238	0.000238
35	17	10	1	0.004741	0.004502
35	17	10	2	0.037158	0.032417
35	17	10	3	0.155038	0.117880
35	17	10	4	0.395709	0.240671
35	17	10	5	0.684515	0.288806
35	17	10	6	0.890804	0.206290
35	17	10	7	0.977250	0.086445
35	17	10	8	0.997510	0.020261
35	17	10	9	0.999894	0.002384
35	17	10	10	1.000000	0.000106
35	17	11	0	0.000076	0.000076
35	17	11	1	0.001859	0.001783
35	17	11	2	0.017707	0.015848
35	17	11	3	0.089025	0.071317
35	17	11	4	0.270560	0.181535
35	17	11	5	0.545888	0.275328
35	17	11	6	0.800037	0.254149
35	17	11	7	0.942672	0.142635
35	17	11	8	0.990217	0.047545
35	17	11	9	0.999131	0.008915
35	17	11	10	0.999970	0.000839
35	17	11	11	1.000000	0.000030
35	17	12	0	0.000022	0.000022
35	17	12	1	0.000671	0.000648
35	17	12	2	0.007802	0.007132
35	17	12	3	0.047423	0.039621
35	17	12	4	0.172228	0.124805
35	17	12	5	0.408224	0.235996
35	17	12	6	0.683552	0.275328
35	17	12	7	0.883241	0.199689
35	17	12	8	0.972387	0.089147

N	n	k	x	P(x)	p(x)
35	17	12	9	0.996160	0.023772
35	17	12	10	0.999726	0.003566
35	17	12	11	0.999992	0.000267
35	17	12	12	1.000000	0.000007
35	17	13	0	0.000006	0.000006
35	17	13	1	0.000220	0.000214
35	17	13	2	0.003151	0.002932
35	17	13	3	0.023306	0.020155
35	17	13	4	0.101686	0.078380
35	17	13	5	0.285096	0.183410
35	17	13	6	0.551873	0.266778
35	17	13	7	0.796419	0.244546
35	17	13	8	0.937504	0.141084
35	17	13	9	0.987891	0.050387
35	17	13	10	0.998640	0.010749
35	17	13	11	0.999923	0.001283
35	17	13	12	0.999998	0.000075
35	17	13	13	1.000000	0.000002
35	17	14	0	0.000001	0.000001
35	17	14	1	0.000064	0.000063
35	17	14	2	0.001152	0.001088
35	17	14	3	0.010480	0.009328
35	17	14	4	0.055371	0.044890
35	17	14	5	0.185054	0.129684
35	17	14	6	0.418485	0.233430
35	17	14	7	0.685262	0.266778
35	17	14	8	0.879787	0.194525
35	17	14	9	0.969568	0.089781
35	17	14	10	0.995220	0.025652
35	17	14	11	0.999573	0.004353
35	17	14	12	0.999981	0.000408
35	17	14	13	1.000000	0.000018
35	17	14	14	1.000000	0.000000
35	17	15	0	0.000000	0.000000
35	17	15	1	0.000016	0.000016
35	17	15	2	0.000375	0.000359
35	17	15	3	0.004262	0.003887
35	17	15	4	0.027581	0.023320
35	17	15	5	0.110949	0.083368
35	17	15	6	0.296212	0.185262
35	17	15	7	0.558225	0.262014
35	17	15	8	0.796419	0.238194
35	17	15	9	0.935366	0.138947
35	17	15	10	0.986669	0.051303
35	17	15	11	0.998329	0.011660
35	17	15	12	0.999884	0.001555
35	17	15	13	0.999996	0.000112
35	17	15	14	1.000000	0.000004
35	17	15	15	1.000000	0.000000
35	17	16	0	0.000000	0.000000

Table for $N = 2$, $n = 1$, through $N = 100$, $n = 50$

N	n	k	x	P(x)	p(x)
35	17	16	1	0.000003	0.000003
35	17	16	2	0.000106	0.000102
35	17	16	3	0.001541	0.001435
35	17	16	4	0.012424	0.010883
35	17	16	5	0.060929	0.048505
35	17	16	6	0.194317	0.133389
35	17	16	7	0.427218	0.232901
35	17	16	8	0.689232	0.262014
35	17	16	9	0.879787	0.190555
35	17	16	10	0.968713	0.088926
35	17	16	11	0.994831	0.026118
35	17	16	12	0.999495	0.004664
35	17	16	13	0.999974	0.000478
35	17	16	14	0.999999	0.000026
35	17	16	15	1.000000	0.000001
35	17	16	16	1.000000	0.000000
35	17	17	0	0.000000	0.000000
35	17	17	1	0.000001	0.000001
35	17	17	2	0.000025	0.000024
35	17	17	3	0.000484	0.000459
35	17	17	4	0.004978	0.004494
35	17	17	5	0.030294	0.025316
35	17	17	6	0.117092	0.086798
35	17	17	7	0.304639	0.187547
35	17	17	8	0.565120	0.260481
35	17	17	9	0.799554	0.234433
35	17	17	10	0.935951	0.136398
35	17	17	11	0.986584	0.050632
35	17	17	12	0.998268	0.011684
35	17	17	13	0.999873	0.001605
35	17	17	14	0.999995	0.000122
35	17	17	15	1.000000	0.000005
35	17	17	16	1.000000	0.000000
35	17	17	17	1.000000	0.000000
36	1	1	0	0.972222	0.972222
36	1	1	1	1.000000	0.027778
36	2	1	0	0.944444	0.944444
36	2	1	1	1.000000	0.055556
36	2	2	0	0.890476	0.890476
36	2	2	1	0.998413	0.107936
36	2	2	2	1.000000	0.001587
36	3	1	0	0.916667	0.916667
36	3	1	1	1.000000	0.083333
36	3	2	0	0.838095	0.838095
36	3	2	1	0.995238	0.157143
36	3	2	2	1.000000	0.004762
36	3	3	0	0.764146	0.764146
36	3	3	1	0.985994	0.221849
36	3	3	2	0.999860	0.013866
36	3	3	3	1.000000	0.000140

N	n	k	x	P(x)	p(x)
36	4	1	0	0.888889	0.888889
36	4	1	1	1.000000	0.111111
36	4	2	0	0.787302	0.787302
36	4	2	1	0.990476	0.203175
36	4	2	2	1.000000	0.009524
36	4	3	0	0.694678	0.694678
36	4	3	1	0.972549	0.277871
36	4	3	2	0.999440	0.026891
36	4	3	3	1.000000	0.000560
36	4	4	0	0.610474	0.610474
36	4	4	1	0.947288	0.336813
36	4	4	2	0.997810	0.050522
36	4	4	3	0.999983	0.002173
36	4	4	4	1.000000	0.000017
36	5	1	0	0.861111	0.861111
36	5	1	1	1.000000	0.138889
36	5	2	0	0.738095	0.738095
36	5	2	1	0.984127	0.246032
36	5	2	2	1.000000	0.015873
36	5	3	0	0.629552	0.629552
36	5	3	1	0.955182	0.325630
36	5	3	2	0.998599	0.043417
36	5	3	3	1.000000	0.001401
36	5	4	0	0.534165	0.534165
36	5	4	1	0.915712	0.381547
36	5	4	2	0.994652	0.078941
36	5	4	3	0.999915	0.005263
36	5	4	4	1.000000	0.000085
36	5	5	0	0.450702	0.450702
36	5	5	1	0.868018	0.417317
36	5	5	2	0.987252	0.119233
36	5	5	3	0.999586	0.012334
36	5	5	4	0.999997	0.000411
36	5	5	5	1.000000	0.000003
36	6	1	0	0.833333	0.833333
36	6	1	1	1.000000	0.166667
36	6	2	0	0.690476	0.690476
36	6	2	1	0.976190	0.285714
36	6	2	2	1.000000	0.023810
36	6	3	0	0.568627	0.568627
36	6	3	1	0.934174	0.365546
36	6	3	2	0.997199	0.063025
36	6	3	3	1.000000	0.002801
36	6	4	0	0.465241	0.465241
36	6	4	1	0.878788	0.413547
36	6	4	2	0.989559	0.110772
36	6	4	3	0.999745	0.010186
36	6	4	4	1.000000	0.000255
36	6	5	0	0.378008	0.378008
36	6	5	1	0.814171	0.436163

N	n	k	x	P(x)	p(x)
36	6	5	2	0.975713	0.161542
36	6	5	3	0.998790	0.023077
36	6	5	4	0.999984	0.001194
36	6	5	5	1.000000	0.000016
36	6	6	0	0.304845	0.304845
36	6	6	1	0.743822	0.438977
36	6	6	2	0.954869	0.211047
36	6	6	3	0.996557	0.041688
36	6	6	4	0.999907	0.003350
36	6	6	5	0.999999	0.000092
36	6	6	6	1.000000	0.000001
36	7	1	0	0.805556	0.805556
36	7	1	1	1.000000	0.194444
36	7	2	0	0.644444	0.644444
36	7	2	1	0.966667	0.322222
36	7	2	2	1.000000	0.033333
36	7	3	0	0.511765	0.511765
36	7	3	1	0.909804	0.398039
36	7	3	2	0.995098	0.085294
36	7	3	3	1.000000	0.004902
36	7	4	0	0.403209	0.403209
36	7	4	1	0.837433	0.434225
36	7	4	2	0.982175	0.144742
36	7	4	3	0.999406	0.017231
36	7	4	4	1.000000	0.000594
36	7	5	0	0.315007	0.315007
36	7	5	1	0.756016	0.441009
36	7	5	2	0.959559	0.203543
36	7	5	3	0.997252	0.037693
36	7	5	4	0.999944	0.002692
36	7	5	5	1.000000	0.000056
36	7	6	0	0.243876	0.243876
36	7	6	1	0.670659	0.426783
36	7	6	2	0.926729	0.256070
36	7	6	3	0.992388	0.065659
36	7	6	4	0.999684	0.007295
36	7	6	5	0.999996	0.000313
36	7	6	6	1.000000	0.000004
36	7	7	0	0.186972	0.186972
36	7	7	1	0.585303	0.398331
36	7	7	2	0.884051	0.298748
36	7	7	3	0.983634	0.099583
36	7	7	4	0.998954	0.015320
36	7	7	5	0.999975	0.001021
36	7	7	6	1.000000	0.000024
36	7	7	7	1.000000	0.000000
36	8	1	0	0.777778	0.777778
36	8	1	1	1.000000	0.222222
36	8	2	0	0.600000	0.600000
36	8	2	1	0.955556	0.355556

N	n	k	x	P(x)	p(x)
36	8	2	2	1.000000	0.044444
36	8	3	0	0.458824	0.458824
36	8	3	1	0.882353	0.423529
36	8	3	2	0.992157	0.109804
36	8	3	3	1.000000	0.007843
36	8	4	0	0.347594	0.347594
36	8	4	1	0.792513	0.444920
36	8	4	2	0.972192	0.179679
36	8	4	3	0.998812	0.026619
36	8	4	4	1.000000	0.001188
36	8	5	0	0.260695	0.260695
36	8	5	1	0.695187	0.434492
36	8	5	2	0.938503	0.243315
36	8	5	3	0.994652	0.056150
36	8	5	4	0.999851	0.005199
36	8	5	5	1.000000	0.000149
36	8	6	0	0.193419	0.193419
36	8	6	1	0.597076	0.403657
36	8	6	2	0.891409	0.294333
36	8	6	3	0.985596	0.094187
36	8	6	4	0.999181	0.013585
36	8	6	5	0.999986	0.000805
36	8	6	6	1.000000	0.000014
36	8	7	0	0.141841	0.141841
36	8	7	1	0.502889	0.361049
36	8	7	2	0.832543	0.329653
36	8	7	3	0.969898	0.137356
36	8	7	4	0.997369	0.027471
36	8	7	5	0.999905	0.002536
36	8	7	6	0.999999	0.000094
36	8	7	7	1.000000	0.000001
36	8	8	0	0.102712	0.102712
36	8	8	1	0.415740	0.313028
36	8	8	2	0.764339	0.348599
36	8	8	3	0.946216	0.181878
36	8	8	4	0.993580	0.047364
36	8	8	5	0.999643	0.006063
36	8	8	6	0.999992	0.000350
36	8	8	7	1.000000	0.000007
36	8	8	8	1.000000	0.000000
36	9	1	0	0.750000	0.750000
36	9	1	1	1.000000	0.250000
36	9	2	0	0.557143	0.557143
36	9	2	1	0.942857	0.385714
36	9	2	2	1.000000	0.057143
36	9	3	0	0.409664	0.409664
36	9	3	1	0.852101	0.442437
36	9	3	2	0.988235	0.136134
36	9	3	3	1.000000	0.011765
36	9	4	0	0.297937	0.297937

Table for $N = 2$, $n = 1$, through $N = 100$, $n = 50$

N	n	k	x	P(x)	p(x)
36	9	4	1	0.744843	0.446906
36	9	4	2	0.959358	0.214515
36	9	4	3	0.997861	0.038503
36	9	4	4	1.000000	0.002139
36	9	5	0	0.214142	0.214142
36	9	5	1	0.633117	0.418974
36	9	5	2	0.912433	0.279316
36	9	5	3	0.990642	0.078209
36	9	5	4	0.999666	0.009024
36	9	5	5	1.000000	0.000334
36	9	6	0	0.151972	0.151972
36	9	6	1	0.524994	0.373022
36	9	6	2	0.849362	0.324367
36	9	6	3	0.975505	0.126143
36	9	6	4	0.998210	0.022706
36	9	6	5	0.999957	0.001747
36	9	6	6	1.000000	0.000043
36	9	7	0	0.106380	0.106380
36	9	7	1	0.425522	0.319141
36	9	7	2	0.773676	0.348154
36	9	7	3	0.950276	0.176600
36	9	7	4	0.994426	0.044150
36	9	7	5	0.999724	0.005298
36	9	7	6	0.999996	0.000272
36	9	7	7	1.000000	0.000004
36	9	8	0	0.073366	0.073366
36	9	8	1	0.337483	0.264117
36	9	8	2	0.689639	0.352156
36	9	8	3	0.913738	0.224099
36	9	8	4	0.986814	0.073076
36	9	8	5	0.998993	0.012179
36	9	8	6	0.999967	0.000974
36	9	8	7	1.000000	0.000032
36	9	8	8	1.000000	0.000000
36	9	9	0	0.049784	0.049784
36	9	9	1	0.262021	0.212237
36	9	9	2	0.601600	0.339579
36	9	9	3	0.865717	0.264117
36	9	9	4	0.973765	0.108048
36	9	9	5	0.997253	0.023489
36	9	9	6	0.999863	0.002610
36	9	9	7	0.999997	0.000134
36	9	9	8	1.000000	0.000003
36	9	9	9	1.000000	0.000000
36	10	1	0	0.722222	0.722222
36	10	1	1	1.000000	0.277778
36	10	2	0	0.515873	0.515873
36	10	2	1	0.928571	0.412698
36	10	2	2	1.000000	0.071429
36	10	3	0	0.364146	0.364146

N	n	k	x	P(x)	p(x)
36	10	3	1	0.819328	0.455182
36	10	3	2	0.983193	0.163866
36	10	3	3	1.000000	0.016807
36	10	4	0	0.253798	0.253798
36	10	4	1	0.695187	0.441389
36	10	4	2	0.943468	0.248281
36	10	4	3	0.996435	0.052967
36	10	4	4	1.000000	0.003565
36	10	5	0	0.174486	0.174486
36	10	5	1	0.571047	0.396560
36	10	5	2	0.881398	0.310351
36	10	5	3	0.984848	0.103450
36	10	5	4	0.999332	0.014483
36	10	5	5	1.000000	0.000668
36	10	6	0	0.118200	0.118200
36	10	6	1	0.455916	0.337716
36	10	6	2	0.801307	0.345391
36	10	6	3	0.961489	0.160181
36	10	6	4	0.996528	0.035040
36	10	6	5	0.999892	0.003364
36	10	6	6	1.000000	0.000108
36	10	7	0	0.078800	0.078800
36	10	7	1	0.354601	0.275801
36	10	7	2	0.709203	0.354601
36	10	7	3	0.924113	0.214910
36	10	7	4	0.989520	0.065407
36	10	7	5	0.999331	0.009811
36	10	7	6	0.999986	0.000654
36	10	7	7	1.000000	0.000014
36	10	8	0	0.051628	0.051628
36	10	8	1	0.269008	0.217380
36	10	8	2	0.611382	0.342374
36	10	8	3	0.872238	0.260856
36	10	8	4	0.975988	0.103750
36	10	8	5	0.997640	0.021652
36	10	8	6	0.999895	0.002255
36	10	8	7	0.999998	0.000103
36	10	8	8	1.000000	0.000001
36	10	9	0	0.033189	0.033189
36	10	9	1	0.199136	0.165947
36	10	9	2	0.513561	0.314425
36	10	9	3	0.807024	0.293463
36	10	9	4	0.953756	0.146732
36	10	9	5	0.993773	0.040018
36	10	9	6	0.999573	0.005800
36	10	9	7	0.999987	0.000414
36	10	9	8	1.000000	0.000012
36	10	9	9	1.000000	0.000000
36	10	10	0	0.020897	0.020897
36	10	10	1	0.143820	0.122923

N	n	k	x	P(x)	p(x)
36	10	10	2	0.420398	0.276578
36	10	10	3	0.730941	0.310543
36	10	10	4	0.921149	0.190208
36	10	10	5	0.986363	0.065214
36	10	10	6	0.998714	0.012351
36	10	10	7	0.999941	0.001227
36	10	10	8	0.999999	0.000058
36	10	10	9	1.000000	0.000001
36	10	10	10	1.000000	0.000000
36	11	1	0	0.694444	0.694444
36	11	1	1	1.000000	0.305555
36	11	2	0	0.476190	0.476190
36	11	2	1	0.912698	0.436508
36	11	2	2	1.000000	0.087302
36	11	3	0	0.322129	0.322129
36	11	3	1	0.784314	0.462185
36	11	3	2	0.976891	0.192577
36	11	3	3	1.000000	0.023109
36	11	4	0	0.214753	0.214753
36	11	4	1	0.644258	0.429505
36	11	4	2	0.924370	0.280112
36	11	4	3	0.994398	0.070028
36	11	4	4	1.000000	0.005602
36	11	5	0	0.140931	0.140931
36	11	5	1	0.510037	0.369106
36	11	5	2	0.845588	0.335551
36	11	5	3	0.976891	0.131303
36	11	5	4	0.998774	0.021884
36	11	5	5	1.000000	0.001225
36	11	6	0	0.090923	0.090923
36	11	6	1	0.390971	0.300047
36	11	6	2	0.748170	0.357199
36	11	6	3	0.943006	0.194836
36	11	6	4	0.993833	0.050827
36	11	6	5	0.999763	0.005930
36	11	6	6	1.000000	0.000237
36	11	7	0	0.057585	0.057585
36	11	7	1	0.290955	0.233370
36	11	7	2	0.641010	0.350055
36	11	7	3	0.891050	0.250040
36	11	7	4	0.981973	0.090923
36	11	7	5	0.998577	0.016603
36	11	7	6	0.999960	0.001384
36	11	7	7	1.000000	0.000040
36	11	8	0	0.035742	0.035742
36	11	8	1	0.210483	0.174740
36	11	8	2	0.532373	0.321890
36	11	8	3	0.822073	0.289701
36	11	8	4	0.960026	0.137953
36	11	8	5	0.995142	0.035115

N	n	k	x	P(x)	p(x)
36	11	8	6	0.999722	0.004580
36	11	8	7	0.999994	0.000273
36	11	8	8	1.000000	0.000005
36	11	9	0	0.021701	0.021701
36	11	9	1	0.148075	0.126375
36	11	9	2	0.428908	0.280833
36	11	9	3	0.739302	0.310394
36	11	9	4	0.925538	0.186236
36	11	9	5	0.987617	0.062079
36	11	9	6	0.998904	0.011287
36	11	9	7	0.999955	0.001052
36	11	9	8	0.999999	0.000044
36	11	9	9	1.000000	0.000001
36	11	10	0	0.012860	0.012860
36	11	10	1	0.101270	0.088410
36	11	10	2	0.335297	0.234027
36	11	10	3	0.647333	0.312036
36	11	10	4	0.877255	0.229921
36	11	10	5	0.973822	0.096567
36	11	10	6	0.996814	0.022992
36	11	10	7	0.999800	0.002986
36	11	10	8	0.999994	0.000195
36	11	10	9	1.000000	0.000005
36	11	10	10	1.000000	0.000000
36	11	11	0	0.007419	0.007419
36	11	11	1	0.067266	0.059847
36	11	11	2	0.254288	0.187022
36	11	11	3	0.551322	0.297034
36	11	11	4	0.815353	0.264031
36	11	11	5	0.951537	0.136184
36	11	11	6	0.992392	0.040855
36	11	11	7	0.999340	0.006948
36	11	11	8	0.999972	0.000632
36	11	11	9	0.999999	0.000027
36	11	11	10	1.000000	0.000000
36	11	11	11	1.000000	0.000000
36	12	1	0	0.666667	0.666667
36	12	1	1	1.000000	0.333333
36	12	2	0	0.438095	0.438095
36	12	2	1	0.895238	0.457143
36	12	2	2	1.000000	0.104762
36	12	3	0	0.283473	0.283473
36	12	3	1	0.747339	0.463866
36	12	3	2	0.969188	0.221849
36	12	3	3	1.000000	0.030812
36	12	4	0	0.180392	0.180392
36	12	4	1	0.592717	0.412325
36	12	4	2	0.901961	0.309244
36	12	4	3	0.991597	0.089636
36	12	4	4	1.000000	0.008403

Table for $N = 2$, $n = 1$, through $N = 100$, $n = 50$

N	n	k	x	P(x)	p(x)
36	12	5	0	0.112745	0.112745
36	12	5	1	0.450980	0.338235
36	12	5	2	0.805322	0.354342
36	12	5	3	0.966387	0.161064
36	12	5	4	0.997899	0.031513
36	12	5	5	1.000000	0.002101
36	12	6	0	0.069102	0.069102
36	12	6	1	0.330961	0.261860
36	12	6	2	0.691018	0.360057
36	12	6	3	0.919626	0.228608
36	12	6	4	0.989767	0.070141
36	12	6	5	0.999526	0.009759
36	12	6	6	1.000000	0.000474
36	12	7	0	0.041461	0.041461
36	12	7	1	0.234946	0.193485
36	12	7	2	0.570999	0.336053
36	12	7	3	0.851044	0.280044
36	12	7	4	0.971063	0.120019
36	12	7	5	0.997248	0.026186
36	12	7	6	0.999905	0.002657
36	12	7	7	1.000000	0.000095
36	12	8	0	0.024305	0.024305
36	12	8	1	0.161555	0.137251
36	12	8	2	0.455119	0.293564
36	12	8	3	0.764133	0.309014
36	12	8	4	0.937954	0.173821
36	12	8	5	0.990928	0.052974
36	12	8	6	0.999355	0.008428
36	12	8	7	0.999984	0.000628
36	12	8	8	1.000000	0.000016
36	12	9	0	0.013888	0.013888
36	12	9	1	0.107635	0.093747
36	12	9	2	0.350275	0.242639
36	12	9	3	0.664807	0.314532
36	12	9	4	0.888291	0.223484
36	12	9	5	0.977684	0.089393
36	12	9	6	0.997549	0.019865
36	12	9	7	0.999871	0.002322
36	12	9	8	0.999998	0.000126
36	12	9	9	1.000000	0.000002
36	12	10	0	0.007716	0.007716
36	12	10	1	0.069442	0.061726
36	12	10	2	0.260408	0.190966
36	12	10	3	0.559963	0.299555
36	12	10	4	0.822073	0.262110
36	12	10	5	0.954508	0.132435
36	12	10	6	0.993135	0.038627
36	12	10	7	0.999441	0.006306
36	12	10	8	0.999979	0.000537
36	12	10	9	1.000000	0.000021

N	n	k	x	P(x)	p(x)
36	12	10	10	1.000000	0.000000
36	12	11	0	0.004155	0.004155
36	12	11	1	0.043327	0.039173
36	12	11	2	0.186960	0.143633
36	12	11	3	0.456271	0.269311
36	12	11	4	0.741424	0.285153
36	12	11	5	0.918853	0.177429
36	12	11	6	0.984221	0.065368
36	12	11	7	0.998229	0.014008
36	12	11	8	0.999896	0.001668
36	12	11	9	0.999997	0.000101
36	12	11	10	1.000000	0.000003
36	12	11	11	1.000000	0.000000
36	12	12	0	0.002160	0.002160
36	12	12	1	0.026091	0.023931
36	12	12	2	0.129507	0.103415
36	12	12	3	0.359319	0.229812
36	12	12	4	0.650175	0.290856
36	12	12	5	0.869173	0.218998
36	12	12	6	0.968533	0.099360
36	12	12	7	0.995427	0.026894
36	12	12	8	0.999629	0.004202
36	12	12	9	0.999985	0.000356
36	12	12	10	1.000000	0.000015
36	12	12	11	1.000000	0.000000
36	12	12	12	1.000000	0.000000
36	13	1	0	0.638889	0.638889
36	13	1	1	1.000000	0.361111
36	13	2	0	0.401587	0.401587
36	13	2	1	0.876190	0.474603
36	13	2	2	1.000000	0.123809
36	13	3	0	0.248039	0.248039
36	13	3	1	0.708683	0.460644
36	13	3	2	0.959944	0.251260
36	13	3	3	1.000000	0.040056
36	13	4	0	0.150327	0.150327
36	13	4	1	0.541176	0.390850
36	13	4	2	0.876190	0.335014
36	13	4	3	0.987862	0.111671
36	13	4	4	1.000000	0.012138
36	13	5	0	0.089257	0.089257
36	13	5	1	0.394608	0.305351
36	13	5	2	0.761029	0.366422
36	13	5	3	0.952964	0.191935
36	13	5	4	0.996586	0.043622
36	13	5	5	1.000000	0.003414
36	13	6	0	0.051826	0.051826
36	13	6	1	0.276407	0.224581
36	13	6	2	0.631009	0.354601
36	13	6	3	0.891050	0.260041

N	n	k	x	P(x)	p(x)
36	13	6	4	0.983922	0.092872
36	13	6	5	0.999119	0.015197
36	13	6	6	1.000000	0.000881
36	13	7	0	0.029368	0.029368
36	13	7	1	0.186575	0.157207
36	13	7	2	0.500988	0.314413
36	13	7	3	0.804370	0.303381
36	13	7	4	0.956060	0.151691
36	13	7	5	0.995066	0.039006
36	13	7	6	0.999794	0.004728
36	13	7	7	1.000000	0.000206
36	13	8	0	0.016203	0.016203
36	13	8	1	0.121524	0.105321
36	13	8	2	0.381728	0.260204
36	13	8	3	0.699755	0.318027
36	13	8	4	0.908984	0.209228
36	13	8	5	0.984306	0.075322
36	13	8	6	0.998653	0.014347
36	13	8	7	0.999957	0.001304
36	13	8	8	1.000000	0.000043
36	13	9	0	0.008680	0.008680
36	13	9	1	0.076386	0.067706
36	13	9	2	0.279505	0.203119
36	13	9	3	0.586174	0.306669
36	13	9	4	0.841732	0.255558
36	13	9	5	0.962785	0.121054
36	13	9	6	0.995066	0.032281
36	13	9	7	0.999678	0.004612
36	13	9	8	0.999992	0.000314
36	13	9	9	1.000000	0.000008
36	13	10	0	0.004501	0.004501
36	13	10	1	0.046295	0.041794
36	13	10	2	0.196753	0.150458
36	13	10	3	0.472593	0.275840
36	13	10	4	0.756546	0.283953
36	13	10	5	0.926918	0.170372
36	13	10	6	0.986697	0.059780
36	13	10	7	0.998653	0.011956
36	13	10	8	0.999934	0.001281
36	13	10	9	0.999999	0.000065
36	13	10	10	1.000000	0.000001
36	13	11	0	0.002250	0.002250
36	13	11	1	0.027005	0.024755
36	13	11	2	0.133098	0.106092
36	13	11	3	0.366501	0.233403
36	13	11	4	0.658254	0.291754
36	13	11	5	0.874496	0.216241
36	13	11	6	0.970603	0.096107
36	13	11	7	0.995894	0.025291
36	13	11	8	0.999688	0.003794

N	n	k	x	P(x)	p(x)
36	13	11	9	0.999989	0.000301
36	13	11	10	1.000000	0.000011
36	13	11	11	1.000000	0.000000
36	13	12	0	0.001080	0.001080
36	13	12	1	0.015123	0.014043
36	13	12	2	0.086417	0.071294
36	13	12	3	0.273139	0.186722
36	13	12	4	0.553223	0.280084
36	13	12	5	0.805298	0.252075
36	13	12	6	0.943693	0.138394
36	13	12	7	0.989824	0.046131
36	13	12	8	0.998929	0.009105
36	13	12	9	0.999941	0.001012
36	13	12	10	0.999998	0.000058
36	13	12	11	1.000000	0.000001
36	13	12	12	1.000000	0.000000
36	13	13	0	0.000495	0.000495
36	13	13	1	0.008102	0.007606
36	13	13	2	0.053741	0.045639
36	13	13	3	0.195338	0.141598
36	13	13	4	0.448192	0.252853
36	13	13	5	0.721273	0.273082
36	13	13	6	0.903328	0.182054
36	13	13	7	0.978291	0.074964
36	13	13	8	0.997032	0.018741
36	13	13	9	0.999772	0.002740
36	13	13	10	0.999991	0.000219
36	13	13	11	1.000000	0.000009
36	13	13	12	1.000000	0.000000
36	13	13	13	1.000000	0.000000
36	14	1	0	0.611111	0.611111
36	14	1	1	1.000000	0.388889
36	14	2	0	0.366667	0.366667
36	14	2	1	0.855555	0.488889
36	14	2	2	1.000000	0.144444
36	14	3	0	0.215686	0.215686
36	14	3	1	0.668627	0.452941
36	14	3	2	0.949020	0.280392
36	14	3	3	1.000000	0.050980
36	14	4	0	0.124183	0.124183
36	14	4	1	0.490196	0.366013
36	14	4	2	0.847059	0.356863
36	14	4	3	0.983006	0.135948
36	14	4	4	1.000000	0.016993
36	14	5	0	0.069853	0.069853
36	14	5	1	0.341503	0.271650
36	14	5	2	0.713235	0.371732
36	14	5	3	0.936274	0.223039
36	14	5	4	0.994689	0.058415
36	14	5	5	1.000000	0.005310

Table for $N = 2$, $n = 1$, through $N = 100$, $n = 50$

N	n	k	x	P(x)	p(x)
36	14	6	0	0.038306	0.038306
36	14	6	1	0.227585	0.189279
36	14	6	2	0.569339	0.341754
36	14	6	3	0.857132	0.287793
36	14	6	4	0.975846	0.118714
36	14	6	5	0.998458	0.022612
36	14	6	6	1.000000	0.001542
36	14	7	0	0.020430	0.020430
36	14	7	1	0.145564	0.125134
36	14	7	2	0.432638	0.287073
36	14	7	3	0.751608	0.318970
36	14	7	4	0.936274	0.184667
36	14	7	5	0.991674	0.055400
36	14	7	6	0.999589	0.007914
36	14	7	7	1.000000	0.000411
36	14	8	0	0.010567	0.010567
36	14	8	1	0.089470	0.078902
36	14	8	2	0.313849	0.224379
36	14	8	3	0.630619	0.316770
36	14	8	4	0.872596	0.241977
36	14	8	5	0.974481	0.101885
36	14	8	6	0.997406	0.022924
36	14	8	7	0.999901	0.002495
36	14	8	8	1.000000	0.000099
36	14	9	0	0.005284	0.005284
36	14	9	1	0.052836	0.047553
36	14	9	2	0.217686	0.164850
36	14	9	3	0.506173	0.288487
36	14	9	4	0.786176	0.280002
36	14	9	5	0.941733	0.155557
36	14	9	6	0.990856	0.049123
36	14	9	7	0.999277	0.008421
36	14	9	8	0.999979	0.000702
36	14	9	9	1.000000	0.000021
36	14	10	0	0.002544	0.002544
36	14	10	1	0.029941	0.027397
36	14	10	2	0.144420	0.114479
36	14	10	3	0.388642	0.244222
36	14	10	4	0.682471	0.293830
36	14	10	5	0.889880	0.207409
36	14	10	6	0.976301	0.086420
36	14	10	7	0.997094	0.020793
36	14	10	8	0.999823	0.002729
36	14	10	9	0.999996	0.000173
36	14	10	10	1.000000	0.000004
36	14	11	0	0.001174	0.001174
36	14	11	1	0.016242	0.015068
36	14	11	2	0.091583	0.075341
36	14	11	3	0.285317	0.193734
36	14	11	4	0.569460	0.284143

N	n	k	x	P(x)	p(x)
36	14	11	5	0.818085	0.248625
36	14	11	6	0.949710	0.131625
36	14	11	7	0.991496	0.041786
36	14	11	8	0.999193	0.007697
36	14	11	9	0.999963	0.000770
36	14	11	10	0.999999	0.000037
36	14	11	11	1.000000	0.000001
36	14	12	0	0.000517	0.000517
36	14	12	1	0.008407	0.007890
36	14	12	2	0.055420	0.047013
36	14	12	3	0.200074	0.144655
36	14	12	4	0.455803	0.255729
36	14	12	5	0.728580	0.272777
36	14	12	6	0.907590	0.179010
36	14	12	7	0.979796	0.072206
36	14	12	8	0.997346	0.017550
36	14	12	9	0.999809	0.002463
36	14	12	10	0.999993	0.000185
36	14	12	11	1.000000	0.000006
36	14	12	12	1.000000	0.000000
36	14	13	0	0.000215	0.000215
36	14	13	1	0.004133	0.003918
36	14	13	2	0.031913	0.027780
36	14	13	3	0.133774	0.101861
36	14	13	4	0.349249	0.215475
36	14	13	5	0.626288	0.277039
36	14	13	6	0.847920	0.221631
36	14	13	7	0.958736	0.110816
36	14	13	8	0.992958	0.034222
36	14	13	9	0.999296	0.006337
36	14	13	10	0.999963	0.000667
36	14	13	11	0.999999	0.000036
36	14	13	12	1.000000	0.000001
36	14	13	13	1.000000	0.000000
36	14	14	0	0.000084	0.000084
36	14	14	1	0.001919	0.001834
36	14	14	2	0.017419	0.015501
36	14	14	3	0.085058	0.067639
36	14	14	4	0.255564	0.170506
36	14	14	5	0.517882	0.262317
36	14	14	6	0.770831	0.252949
36	14	14	7	0.925009	0.154178
36	14	14	8	0.984030	0.059021
36	14	14	9	0.997918	0.013887
36	14	14	10	0.999847	0.001929
36	14	14	11	0.999994	0.000148
36	14	14	12	1.000000	0.000006
36	14	14	13	1.000000	0.000000
36	14	14	14	1.000000	0.000000
36	15	1	0	0.583333	0.583333

N	n	k	x	P(x)	p(x)
36	15	1	1	1.000000	0.416667
36	15	2	0	0.333333	0.333333
36	15	2	1	0.833333	0.500000
36	15	2	2	1.000000	0.166667
36	15	3	0	0.186274	0.186274
36	15	3	1	0.627451	0.441176
36	15	3	2	0.936274	0.308824
36	15	3	3	1.000000	0.063725
36	15	4	0	0.101604	0.101604
36	15	4	1	0.440285	0.338681
36	15	4	2	0.814617	0.374332
36	15	4	3	0.976827	0.162210
36	15	4	4	1.000000	0.023173
36	15	5	0	0.053977	0.053977
36	15	5	1	0.292112	0.238135
36	15	5	2	0.662545	0.370432
36	15	5	3	0.915998	0.253454
36	15	5	4	0.992034	0.076036
36	15	5	5	1.000000	0.007966
36	15	6	0	0.027859	0.027859
36	15	6	1	0.184567	0.156708
36	15	6	2	0.507202	0.322635
36	15	6	3	0.817887	0.310685
36	15	6	4	0.965054	0.147167
36	15	6	5	0.997430	0.032377
36	15	6	6	1.000000	0.002570
36	15	7	0	0.013930	0.013930
36	15	7	1	0.111437	0.097507
36	15	7	2	0.367394	0.255957
36	15	7	3	0.693613	0.326219
36	15	7	4	0.911093	0.217480
36	15	7	5	0.986638	0.075546
36	15	7	6	0.999229	0.012591
36	15	7	7	1.000000	0.000771
36	15	8	0	0.006725	0.006725
36	15	8	1	0.064364	0.057640
36	15	8	2	0.252654	0.188290
36	15	8	3	0.558626	0.305971
36	15	8	4	0.828600	0.269975
36	15	8	5	0.960588	0.131988
36	15	8	6	0.995321	0.034734
36	15	8	7	0.999787	0.004466
36	15	8	8	1.000000	0.000213
36	15	9	0	0.003122	0.003122
36	15	9	1	0.035545	0.032422
36	15	9	2	0.165234	0.129690
36	15	9	3	0.427495	0.262261
36	15	9	4	0.722539	0.295044
36	15	9	5	0.913449	0.190911
36	15	9	6	0.984157	0.070708

N	n	k	x	P(x)	p(x)
36	15	9	7	0.998511	0.014354
36	15	9	8	0.999947	0.001435
36	15	9	9	1.000000	0.000053
36	15	10	0	0.001388	0.001388
36	15	10	1	0.018733	0.017345
36	15	10	2	0.102791	0.084058
36	15	10	3	0.310935	0.208144
36	15	10	4	0.602336	0.291401
36	15	10	5	0.842742	0.240406
36	15	10	6	0.960588	0.117846
36	15	10	7	0.994258	0.033670
36	15	10	8	0.999575	0.005316
36	15	10	9	0.999988	0.000413
36	15	10	10	1.000000	0.000012
36	15	11	0	0.000587	0.000587
36	15	11	1	0.009393	0.008806
36	15	11	2	0.060762	0.051369
36	15	11	3	0.214868	0.154106
36	15	11	4	0.479051	0.264182
36	15	11	5	0.750278	0.271227
36	15	11	6	0.919795	0.169517
36	15	11	7	0.983898	0.064103
36	15	11	8	0.998143	0.014245
36	15	11	9	0.999893	0.001749
36	15	11	10	0.999998	0.000105
36	15	11	11	1.000000	0.000002
36	15	12	0	0.000235	0.000235
36	15	12	1	0.004462	0.004227
36	15	12	2	0.034050	0.029588
36	15	12	3	0.140897	0.106847
36	15	12	4	0.362810	0.221913
36	15	12	5	0.641787	0.278977
36	15	12	6	0.858769	0.216982
36	15	12	7	0.963385	0.104616
36	15	12	8	0.994155	0.030769
36	15	12	9	0.999473	0.005318
36	15	12	10	0.999977	0.000504
36	15	12	11	0.999999	0.000023
36	15	12	12	1.000000	0.000000
36	15	13	0	0.000088	0.000088
36	15	13	1	0.001996	0.001908
36	15	13	2	0.018023	0.016027
36	15	13	3	0.087474	0.069451
36	15	13	4	0.261100	0.173627
36	15	13	5	0.525547	0.264447
36	15	13	6	0.777401	0.251854
36	15	13	7	0.928513	0.151112
36	15	13	8	0.985180	0.056667
36	15	13	9	0.998143	0.012963
36	15	13	10	0.999872	0.001728

N	n	k	x	P(x)	p(x)
36	15	13	11	0.999996	0.000124
36	15	13	12	1.000000	0.000004
36	15	13	13	1.000000	0.000000
36	15	14	0	0.000031	0.000031
36	15	14	1	0.000835	0.000804
36	15	14	2	0.008964	0.008130
36	15	14	3	0.051239	0.042274
36	15	14	4	0.178062	0.126823
36	15	14	5	0.410570	0.232509
36	15	14	6	0.678849	0.268279
36	15	14	7	0.875952	0.197103
36	15	14	8	0.967934	0.091981
36	15	14	9	0.994762	0.026828
36	15	14	10	0.999496	0.004734
36	15	14	11	0.999974	0.000478
36	15	14	12	0.999999	0.000025
36	15	14	13	1.000000	0.000001
36	15	14	14	1.000000	0.000000
36	15	15	0	0.000010	0.000010
36	15	15	1	0.000323	0.000313
36	15	15	2	0.004160	0.003837
36	15	15	3	0.028180	0.024019
36	15	15	4	0.114650	0.086470
36	15	15	5	0.304884	0.190234
36	15	15	6	0.569099	0.264214
36	15	15	7	0.804278	0.235180
36	15	15	8	0.938667	0.134388
36	15	15	9	0.987445	0.048778
36	15	15	10	0.998420	0.010975
36	15	15	11	0.999887	0.001467
36	15	15	12	0.999996	0.000109
36	15	15	13	1.000000	0.000004
36	15	15	14	1.000000	0.000000
36	15	15	15	1.000000	0.000000
36	16	1	0	0.555556	0.555556
36	16	1	1	1.000000	0.444444
36	16	2	0	0.301587	0.301587
36	16	2	1	0.809524	0.507936
36	16	2	2	1.000000	0.190476
36	16	3	0	0.159664	0.159664
36	16	3	1	0.585434	0.425770
36	16	3	2	0.921569	0.336134
36	16	3	3	1.000000	0.078431
36	16	4	0	0.082251	0.082251
36	16	4	1	0.391902	0.309651
36	16	4	2	0.778966	0.387064
36	16	4	3	0.969103	0.190137
36	16	4	4	1.000000	0.030897
36	16	5	0	0.041126	0.041126
36	16	5	1	0.246753	0.205628
36	16	5	2	0.609626	0.362872
36	16	5	3	0.891860	0.282234
36	16	5	4	0.988413	0.096554
36	16	5	5	1.000000	0.011586
36	16	6	0	0.019899	0.019899
36	16	6	1	0.147256	0.127356
36	16	6	2	0.445748	0.298492
36	16	6	3	0.773503	0.327756
36	16	6	4	0.951038	0.177534
36	16	6	5	0.995889	0.044851
36	16	6	6	1.000000	0.004111
36	16	7	0	0.009286	0.009286
36	16	7	1	0.083578	0.074291
36	16	7	2	0.306452	0.222874
36	16	7	3	0.631476	0.325024
36	16	7	4	0.880024	0.248548
36	16	7	5	0.979443	0.099419
36	16	7	6	0.998629	0.019186
36	16	7	7	1.000000	0.001370
36	16	8	0	0.004163	0.004163
36	16	8	1	0.045151	0.040988
36	16	8	2	0.198857	0.153706
36	16	8	3	0.485775	0.286918
36	16	8	4	0.777177	0.291401
36	16	8	5	0.941733	0.164556
36	16	8	6	0.992014	0.050281
36	16	8	7	0.999575	0.007561
36	16	8	8	1.000000	0.000425
36	16	9	0	0.001784	0.001784
36	16	9	1	0.023193	0.021409
36	16	9	2	0.122004	0.098811
36	16	9	3	0.352563	0.230559
36	16	9	4	0.652290	0.299727
36	16	9	5	0.877086	0.224795
36	16	9	6	0.974056	0.096970
36	16	9	7	0.997144	0.023088
36	16	9	8	0.999878	0.002734
36	16	9	9	1.000000	0.000122
36	16	10	0	0.000727	0.000727
36	16	10	1	0.011299	0.010572
36	16	10	2	0.070769	0.059470
36	16	10	3	0.241553	0.170785
36	16	10	4	0.519078	0.277525
36	16	10	5	0.785502	0.266424
36	16	10	6	0.938141	0.152639
36	16	10	7	0.989448	0.051307
36	16	10	8	0.999068	0.009620
36	16	10	9	0.999968	0.000900
36	16	10	10	1.000000	0.000032
36	16	11	0	0.000280	0.000280
36	16	11	1	0.005200	0.004920
36	16	11	2	0.038747	0.033547
36	16	11	3	0.156161	0.117414
36	16	11	4	0.390990	0.234829
36	16	11	5	0.672785	0.281795
36	16	11	6	0.879434	0.206649
36	16	11	7	0.971688	0.092254
36	16	11	8	0.996108	0.024420
36	16	11	9	0.999726	0.003618
36	16	11	10	0.999993	0.000267
36	16	11	11	1.000000	0.000007
36	16	12	0	0.000101	0.000101
36	16	12	1	0.002248	0.002147
36	16	12	2	0.019960	0.017713
36	16	12	3	0.095106	0.075145
36	16	12	4	0.278272	0.183166
36	16	12	5	0.548795	0.270523
36	16	12	6	0.796774	0.247979
36	16	12	7	0.938477	0.141702
36	16	12	8	0.988294	0.049817
36	16	12	9	0.998713	0.010419
36	16	12	10	0.999929	0.001216
36	16	12	11	0.999998	0.000070
36	16	12	12	1.000000	0.000001
36	16	13	0	0.000034	0.000034
36	16	13	1	0.000906	0.000872
36	16	13	2	0.009628	0.008722
36	16	13	3	0.054402	0.044774
36	16	13	4	0.186689	0.132287
36	16	13	5	0.424805	0.238116
36	16	13	6	0.693449	0.268644
36	16	13	7	0.885338	0.191889
36	16	13	8	0.971688	0.086350
36	16	13	9	0.995674	0.023986
36	16	13	10	0.999625	0.003951
36	16	13	11	0.999984	0.000359
36	16	13	12	1.000000	0.000016
36	16	13	13	1.000000	0.000000
36	16	14	0	0.000010	0.000010
36	16	14	1	0.000337	0.000327
36	16	14	2	0.004319	0.003982
36	16	14	3	0.029095	0.024776
36	16	14	4	0.117670	0.088575
36	16	14	5	0.310924	0.193254
36	16	14	6	0.576648	0.265724
36	16	14	7	0.810251	0.233604
36	16	14	8	0.941653	0.131402
36	16	14	9	0.988374	0.046721
36	16	14	10	0.998594	0.010220
36	16	14	11	0.999906	0.001312
36	16	14	12	0.999997	0.000091
36	16	14	13	1.000000	0.000003
36	16	14	14	1.000000	0.000000
36	16	15	0	0.000003	0.000003
36	16	15	1	0.000114	0.000111
36	16	15	2	0.001785	0.001671
36	16	15	3	0.014454	0.012670
36	16	15	4	0.069356	0.054902
36	16	15	5	0.214297	0.144940
36	16	15	6	0.455864	0.241567
36	16	15	7	0.714686	0.258822
36	16	15	8	0.893871	0.179185
36	16	15	9	0.973508	0.079638
36	16	15	10	0.995807	0.022299
36	16	15	11	0.999608	0.003801
36	16	15	12	0.999980	0.000373
36	16	15	13	0.999999	0.000019
36	16	15	14	1.000000	0.000000
36	16	15	15	1.000000	0.000000
36	16	16	0	0.000001	0.000001
36	16	16	1	0.000035	0.000034
36	16	16	2	0.000671	0.000636
36	16	16	3	0.006611	0.005940
36	16	16	4	0.037984	0.031372
36	16	16	5	0.138375	0.100392
36	16	16	6	0.340832	0.202456
36	16	16	7	0.603762	0.262930
36	16	16	8	0.825610	0.221848
36	16	16	9	0.946962	0.121353
36	16	16	10	0.989436	0.042473
36	16	16	11	0.998703	0.009267
36	16	16	12	0.999909	0.001207
36	16	16	13	0.999997	0.000087
36	16	16	14	1.000000	0.000003
36	16	16	15	1.000000	0.000000
36	16	16	16	1.000000	0.000000
36	17	1	0	0.527778	0.527778
36	17	1	1	1.000000	0.472222
36	17	2	0	0.271429	0.271429
36	17	2	1	0.784127	0.512698
36	17	2	2	1.000000	0.215873
36	17	3	0	0.135714	0.135714
36	17	3	1	0.542857	0.407143
36	17	3	2	0.904762	0.361905
36	17	3	3	1.000000	0.095238
36	17	4	0	0.065801	0.065801
36	17	4	1	0.345454	0.279654
36	17	4	2	0.740260	0.394805
36	17	4	3	0.959596	0.219336
36	17	4	4	1.000000	0.040404

N	n	k	x	P(x)	p(x)
36	17	5	0	0.030844	0.030844
36	17	5	1	0.205628	0.174784
36	17	5	2	0.555195	0.349567
36	17	5	3	0.863636	0.308442
36	17	5	4	0.983586	0.119949
36	17	5	5	1.000000	0.016414
36	17	6	0	0.013930	0.013930
36	17	6	1	0.115417	0.101487
36	17	6	2	0.386049	0.270633
36	17	6	3	0.724340	0.338291
36	17	6	4	0.933284	0.208944
36	17	6	5	0.993646	0.060362
36	17	6	6	1.000000	0.006354
36	17	7	0	0.006036	0.006036
36	17	7	1	0.061290	0.055254
36	17	7	2	0.250733	0.189443
36	17	7	3	0.566471	0.315738
36	17	7	4	0.842742	0.276271
36	17	7	5	0.969501	0.126760
36	17	7	6	0.997670	0.028169
36	17	7	7	1.000000	0.002330
36	17	8	0	0.002498	0.002498
36	17	8	1	0.030805	0.028308
36	17	8	2	0.152745	0.121940
36	17	8	3	0.414046	0.261300
36	17	8	4	0.718896	0.304851
36	17	8	5	0.917049	0.198153
36	17	8	6	0.986985	0.069936
36	17	8	7	0.999197	0.012211
36	17	8	8	1.000000	0.000803
36	17	9	0	0.000981	0.000981
36	17	9	1	0.014630	0.013648
36	17	9	2	0.087420	0.072791
36	17	9	3	0.283396	0.195975
36	17	9	4	0.577359	0.293963
36	17	9	5	0.832126	0.254768
36	17	9	6	0.959510	0.127384
36	17	9	7	0.994835	0.035325
36	17	9	8	0.999742	0.004906
36	17	9	9	1.000000	0.000258
36	17	10	0	0.000363	0.000363
36	17	10	1	0.006542	0.006178
36	17	10	2	0.046981	0.040439
36	17	10	3	0.181779	0.134798
36	17	10	4	0.435821	0.254042
36	17	10	5	0.718896	0.283075
36	17	10	6	0.907613	0.188717
36	17	10	7	0.981752	0.074139
36	17	10	8	0.998106	0.016354
36	17	10	9	0.999923	0.001817
36	17	10	10	1.000000	0.000077
36	17	11	0	0.000126	0.000126
36	17	11	1	0.002740	0.002614
36	17	11	2	0.023651	0.020911
36	17	11	3	0.109195	0.085545
36	17	11	4	0.308800	0.199604
36	17	11	5	0.588246	0.279446
36	17	11	6	0.827772	0.239525
36	17	11	7	0.953237	0.125466
36	17	11	8	0.992445	0.039208
36	17	11	9	0.999364	0.006919
36	17	11	10	0.999979	0.000615
36	17	11	11	1.000000	0.000021
36	17	12	0	0.000040	0.000040
36	17	12	1	0.001067	0.001027
36	17	12	2	0.011104	0.010037
36	17	12	3	0.061290	0.050186
36	17	12	4	0.205006	0.143715
36	17	12	5	0.454112	0.249106
36	17	12	6	0.722380	0.268268
36	17	12	7	0.903051	0.180671
36	17	12	8	0.978330	0.075279
36	17	12	9	0.997150	0.018820
36	17	12	10	0.999807	0.002657
36	17	12	11	0.999995	0.000188
36	17	12	12	1.000000	0.000005
36	17	13	0	0.000012	0.000012
36	17	13	1	0.000382	0.000371
36	17	13	2	0.004831	0.004448
36	17	13	3	0.032015	0.027184
36	17	13	4	0.127160	0.095145
36	17	13	5	0.329559	0.202399
36	17	13	6	0.599424	0.269865
36	17	13	7	0.827772	0.228348
36	17	13	8	0.950101	0.122329
36	17	13	9	0.990877	0.040776
36	17	13	10	0.999032	0.008155
36	17	13	11	0.999948	0.000916
36	17	13	12	0.999999	0.000051
36	17	13	13	1.000000	0.000001
36	17	14	0	0.000003	0.000003
36	17	14	1	0.000125	0.000121
36	17	14	2	0.001930	0.001805
36	17	14	3	0.015468	0.013538
36	17	14	4	0.073382	0.057914
36	17	14	5	0.223959	0.150577
36	17	14	6	0.470358	0.246399
36	17	14	7	0.728490	0.258132
36	17	14	8	0.902233	0.173743
36	17	14	9	0.976694	0.074461
36	17	14	10	0.996550	0.019856
36	17	14	11	0.999709	0.003159
36	17	14	12	0.999988	0.000279
36	17	14	13	1.000000	0.000012
36	17	14	14	1.000000	0.000000
36	17	15	0	0.000001	0.000001
36	17	15	1	0.000036	0.000036
36	17	15	2	0.000699	0.000663
36	17	15	3	0.006853	0.006154
36	17	15	4	0.039160	0.032308
36	17	15	5	0.141826	0.102666
36	17	15	6	0.347159	0.205332
36	17	15	7	0.611157	0.263999
36	17	15	8	0.831156	0.219999
36	17	15	9	0.949617	0.118461
36	17	15	10	0.990232	0.040615
36	17	15	11	0.998848	0.008615
36	17	15	12	0.999924	0.001077
36	17	15	13	0.999998	0.000073
36	17	15	14	1.000000	0.000002
36	17	15	15	1.000000	0.000000
36	17	16	0	0.000000	0.000000
36	17	16	1	0.000009	0.000009
36	17	16	2	0.000226	0.000216
36	17	16	3	0.002750	0.002525
36	17	16	4	0.019160	0.016410
36	17	16	5	0.083160	0.064000
36	17	16	6	0.239604	0.156444
36	17	16	7	0.485444	0.245840
36	17	16	8	0.736871	0.251427
36	17	16	9	0.904489	0.167618
36	17	16	10	0.976694	0.072205
36	17	16	11	0.996386	0.019692
36	17	16	12	0.999668	0.003282
36	17	16	13	0.999984	0.000316
36	17	16	14	0.999999	0.000016
36	17	16	15	1.000000	0.000000
36	17	16	16	1.000000	0.000000
36	17	17	0	0.000000	0.000000
36	17	17	1	0.000002	0.000002
36	17	17	2	0.000063	0.000061
36	17	17	3	0.000983	0.000920
36	17	17	4	0.008494	0.007511
36	17	17	5	0.044760	0.036266
36	17	17	6	0.153560	0.108799
36	17	17	7	0.362524	0.208964
36	17	17	8	0.623729	0.261205
36	17	17	9	0.837442	0.213713
36	17	17	10	0.951422	0.113980
36	17	17	11	0.990478	0.039056
36	17	17	12	0.998848	0.008369
36	17	17	13	0.999920	0.001073
36	17	17	14	0.999997	0.000077
36	17	17	15	1.000000	0.000003
36	17	17	16	1.000000	0.000000
36	17	17	17	1.000000	0.000000
36	18	1	0	0.500000	0.500000
36	18	1	1	1.000000	0.500000
36	18	2	0	0.242857	0.242857
36	18	2	1	0.757143	0.514286
36	18	2	2	1.000000	0.242857
36	18	3	0	0.114286	0.114286
36	18	3	1	0.500000	0.385714
36	18	3	2	0.885714	0.385714
36	18	3	3	1.000000	0.114286
36	18	4	0	0.051948	0.051948
36	18	4	1	0.301299	0.249351
36	18	4	2	0.698701	0.397403
36	18	4	3	0.948052	0.249351
36	18	4	4	1.000000	0.051948
36	18	5	0	0.022727	0.022727
36	18	5	1	0.168831	0.146104
36	18	5	2	0.500000	0.331169
36	18	5	3	0.831169	0.331169
36	18	5	4	0.977273	0.146104
36	18	5	5	1.000000	0.022727
36	18	6	0	0.009531	0.009531
36	18	6	1	0.088710	0.079179
36	18	6	2	0.329074	0.240364
36	18	6	3	0.670926	0.341852
36	18	6	4	0.911290	0.240364
36	18	6	5	0.990469	0.079179
36	18	6	6	1.000000	0.009531
36	18	7	0	0.003812	0.003812
36	18	7	1	0.043842	0.040029
36	18	7	2	0.200880	0.157038
36	18	7	3	0.500000	0.299120
36	18	7	4	0.799120	0.299120
36	18	7	5	0.956158	0.157038
36	18	7	6	0.996188	0.040029
36	18	7	7	1.000000	0.003812
36	18	8	0	0.001446	0.001446
36	18	8	1	0.020376	0.018930
36	18	8	2	0.114238	0.093862
36	18	8	3	0.345283	0.231045
36	18	8	4	0.654717	0.309435
36	18	8	5	0.885762	0.231045
36	18	8	6	0.979624	0.093862
36	18	8	7	0.998554	0.018930
36	18	8	8	1.000000	0.001446

N	n	k	x	P(x)	p(x)
36	18	9	0	0.000516	0.000516
36	18	9	1	0.008883	0.008366
36	18	9	2	0.060603	0.051720
36	18	9	3	0.221509	0.160906
36	18	9	4	0.500000	0.278491
36	18	9	5	0.778491	0.278491
36	18	9	6	0.939397	0.160906
36	18	9	7	0.991117	0.051720
36	18	9	8	0.999483	0.008366
36	18	9	9	1.000000	0.000516
36	18	10	0	0.000172	0.000172
36	18	10	1	0.003615	0.003443
36	18	10	2	0.029954	0.026339
36	18	10	3	0.132116	0.102163
36	18	10	4	0.355597	0.223481
36	18	10	5	0.644403	0.288806
36	18	10	6	0.867883	0.223481
36	18	10	7	0.970046	0.102163
36	18	10	8	0.996385	0.026339
36	18	10	9	0.999828	0.003443
36	18	10	10	1.000000	0.000172
36	18	11	0	0.000053	0.000053
36	18	11	1	0.001364	0.001311
36	18	11	2	0.013745	0.012381
36	18	11	3	0.073177	0.059431
36	18	11	4	0.235261	0.162085
36	18	11	5	0.500000	0.264739
36	18	11	6	0.764739	0.264739
36	18	11	7	0.926823	0.162085
36	18	11	8	0.986254	0.059431
36	18	11	9	0.998636	0.012381
36	18	11	10	0.999947	0.001311
36	18	11	11	1.000000	0.000053
36	18	12	0	0.000015	0.000015
36	18	12	1	0.000472	0.000458
36	18	12	2	0.005821	0.005349
36	18	12	3	0.037518	0.031697
36	18	12	4	0.144494	0.106976
36	18	12	5	0.362336	0.217842
36	18	12	6	0.637664	0.275328
36	18	12	7	0.855506	0.217842
36	18	12	8	0.962482	0.106976
36	18	12	9	0.994179	0.031697
36	18	12	10	0.999527	0.005349
36	18	12	11	0.999985	0.000458
36	18	12	12	1.000000	0.000015
36	18	13	0	0.000004	0.000004
36	18	13	1	0.000148	0.000145
36	18	13	2	0.002255	0.002107
36	18	13	3	0.017707	0.015452

N	n	k	x	P(x)	p(x)
36	18	13	4	0.082091	0.064384
36	18	13	5	0.244338	0.162247
36	18	13	6	0.500000	0.255662
36	18	13	7	0.755662	0.255662
36	18	13	8	0.917909	0.162247
36	18	13	9	0.982292	0.064384
36	18	13	10	0.997744	0.015452
36	18	13	11	0.999852	0.002107
36	18	13	12	0.999996	0.000145
36	18	13	13	1.000000	0.000004
36	18	14	0	0.000001	0.000001
36	18	14	1	0.000041	0.000041
36	18	14	2	0.000790	0.000748
36	18	14	3	0.007630	0.006840
36	18	14	4	0.042901	0.035271
36	18	14	5	0.152633	0.109732
36	18	14	6	0.366611	0.213978
36	18	14	7	0.633389	0.266778
36	18	14	8	0.847367	0.213978
36	18	14	9	0.957099	0.109732
36	18	14	10	0.992370	0.035271
36	18	14	11	0.999210	0.006840
36	18	14	12	0.999959	0.000748
36	18	14	13	0.999999	0.000041
36	18	14	14	1.000000	0.000001
36	18	15	0	0.000000	0.000000
36	18	15	1	0.000010	0.000010
36	18	15	2	0.000245	0.000235
36	18	15	3	0.002966	0.002721
36	18	15	4	0.020456	0.017490
36	18	15	5	0.087792	0.067336
36	18	15	6	0.249896	0.162104
36	18	15	7	0.500000	0.250104
36	18	15	8	0.750104	0.250104
36	18	15	9	0.912208	0.162104
36	18	15	10	0.979544	0.067336
36	18	15	11	0.997034	0.017490
36	18	15	12	0.999754	0.002721
36	18	15	13	0.999990	0.000235
36	18	15	14	1.000000	0.000010
36	18	15	15	1.000000	0.000000
36	18	16	0	0.000000	0.000000
36	18	16	1	0.000002	0.000002
36	18	16	2	0.000066	0.000064
36	18	16	3	0.001023	0.000957
36	18	16	4	0.008796	0.007773
36	18	16	5	0.046108	0.037312
36	18	16	6	0.157265	0.111157
36	18	16	7	0.368993	0.211728
36	18	16	8	0.631007	0.262014

N	n	k	x	P(x)	p(x)
36	18	16	9	0.842735	0.211728
36	18	16	10	0.953892	0.111157
36	18	16	11	0.991204	0.037312
36	18	16	12	0.998977	0.007773
36	18	16	13	0.999934	0.000957
36	18	16	14	0.999998	0.000064
36	18	16	15	1.000000	0.000002
36	18	16	16	1.000000	0.000000
36	18	17	0	0.000000	0.000000
36	18	17	1	0.000000	0.000000
36	18	17	2	0.000015	0.000015
36	18	17	3	0.000305	0.000290
36	18	17	4	0.003355	0.003049
36	18	17	5	0.021855	0.018500
36	18	17	6	0.090571	0.068715
36	18	17	7	0.252543	0.161972
36	18	17	8	0.500000	0.247457
36	18	17	9	0.747457	0.247457
36	18	17	10	0.909429	0.161972
36	18	17	11	0.978145	0.068715
36	18	17	12	0.996645	0.018500
36	18	17	13	0.999695	0.003049
36	18	17	14	0.999985	0.000290
36	18	17	15	1.000000	0.000015
36	18	17	16	1.000000	0.000000
36	18	17	17	1.000000	0.000000
36	18	18	0	0.000000	0.000000
36	18	18	1	0.000000	0.000000
36	18	18	2	0.000003	0.000003
36	18	18	3	0.000076	0.000073
36	18	18	4	0.001108	0.001032
36	18	18	5	0.009197	0.008089
36	18	18	6	0.047171	0.037974
36	18	18	7	0.158769	0.111598
36	18	18	8	0.369759	0.210990
36	18	18	9	0.630241	0.260481
36	18	18	10	0.841231	0.210990
36	18	18	11	0.952829	0.111598
36	18	18	12	0.990803	0.037974
36	18	18	13	0.998892	0.008089
36	18	18	14	0.999924	0.001032
36	18	18	15	0.999997	0.000073
36	18	18	16	1.000000	0.000003
36	18	18	17	1.000000	0.000000
36	18	18	18	1.000000	0.000000
37	1	1	0	0.972973	0.972973
37	1	1	1	1.000000	0.027027
37	2	1	0	0.945946	0.945946
37	2	1	1	1.000000	0.054054
37	2	2	0	0.893393	0.893393

N	n	k	x	P(x)	p(x)
37	2	2	1	0.998498	0.105105
37	2	2	2	1.000000	0.001502
37	3	1	0	0.918919	0.918919
37	3	1	1	1.000000	0.081081
37	3	2	0	0.842342	0.842342
37	3	2	1	0.995495	0.153153
37	3	2	2	1.000000	0.004505
37	3	3	0	0.770142	0.770142
37	3	3	1	0.986744	0.216602
37	3	3	2	0.999871	0.013127
37	3	3	3	1.000000	0.000129
37	4	1	0	0.891892	0.891892
37	4	1	1	1.000000	0.108108
37	4	2	0	0.792793	0.792793
37	4	2	1	0.990991	0.198198
37	4	2	2	1.000000	0.009009
37	4	3	0	0.702188	0.702188
37	4	3	1	0.974003	0.271815
37	4	3	2	0.999485	0.025483
37	4	3	3	1.000000	0.000515
37	4	4	0	0.619578	0.619578
37	4	4	1	0.950019	0.330441
37	4	4	2	0.997986	0.047967
37	4	4	3	0.999985	0.001999
37	4	4	4	1.000000	0.000015
37	5	1	0	0.864865	0.864865
37	5	1	1	1.000000	0.135135
37	5	2	0	0.744745	0.744745
37	5	2	1	0.984985	0.240240
37	5	2	2	1.000000	0.015015
37	5	3	0	0.638353	0.638353
37	5	3	1	0.957529	0.319176
37	5	3	2	0.998713	0.041184
37	5	3	3	1.000000	0.001287
37	5	4	0	0.544477	0.544477
37	5	4	1	0.919979	0.375502
37	5	4	2	0.995079	0.075100
37	5	4	3	0.999924	0.004845
37	5	4	4	1.000000	0.000076
37	5	5	0	0.461981	0.461981
37	5	5	1	0.874463	0.412483
37	5	5	2	0.988252	0.113788
37	5	5	3	0.999631	0.011379
37	5	5	4	0.999998	0.000367
37	5	5	5	1.000000	0.000002
37	6	1	0	0.837838	0.837838
37	6	1	1	1.000000	0.162162
37	6	2	0	0.698198	0.698198
37	6	2	1	0.977477	0.279279
37	6	2	2	1.000000	0.022523

N	n	k	x	P(x)	p(x)
37	6	3	0	0.578507	0.578507
37	6	3	1	0.937580	0.359073
37	6	3	2	0.997426	0.059846
37	6	3	3	1.000000	0.002574
37	6	4	0	0.476418	0.476418
37	6	4	1	0.884776	0.408358
37	6	4	2	0.990385	0.105610
37	6	4	3	0.999773	0.009388
37	6	4	4	1.000000	0.000227
37	6	5	0	0.389796	0.389796
37	6	5	1	0.822903	0.433107
37	6	5	2	0.977584	0.154681
37	6	5	3	0.998919	0.021335
37	6	5	4	0.999986	0.001067
37	6	5	5	1.000000	0.000014
37	6	6	0	0.316709	0.316709
37	6	6	1	0.755230	0.438521
37	6	6	2	0.958249	0.203019
37	6	6	3	0.996919	0.038670
37	6	6	4	0.999920	0.003000
37	6	6	5	1.000000	0.000080
37	6	6	6	1.000000	0.000000
37	7	1	0	0.810811	0.810811
37	7	1	1	1.000000	0.189189
37	7	2	0	0.653153	0.653153
37	7	2	1	0.968468	0.315315
37	7	2	2	1.000000	0.031532
37	7	3	0	0.522523	0.522523
37	7	3	1	0.914414	0.391892
37	7	3	2	0.995495	0.081081
37	7	3	3	1.000000	0.004505
37	7	4	0	0.414944	0.414944
37	7	4	1	0.845257	0.430313
37	7	4	2	0.983572	0.138315
37	7	4	3	0.999470	0.015898
37	7	4	4	1.000000	0.000530
37	7	5	0	0.326926	0.326926
37	7	5	1	0.767018	0.440092
37	7	5	2	0.962615	0.195597
37	7	5	3	0.997543	0.034928
37	7	5	4	0.999952	0.002409
37	7	5	5	1.000000	0.000048
37	7	6	0	0.255411	0.255411
37	7	6	1	0.684501	0.429090
37	7	6	2	0.932053	0.247552
37	7	6	3	0.993177	0.061124
37	7	6	4	0.999726	0.006549
37	7	6	5	0.999997	0.000271
37	7	6	6	1.000000	0.000003
37	7	7	0	0.197737	0.197737

N	n	k	x	P(x)	p(x)
37	7	7	1	0.601451	0.403714
37	7	7	2	0.892125	0.290674
37	7	7	3	0.985290	0.093165
37	7	7	4	0.999092	0.013802
37	7	7	5	0.999979	0.000887
37	7	7	6	1.000000	0.000020
37	7	7	7	1.000000	0.000000
37	8	1	0	0.783784	0.783784
37	8	1	1	1.000000	0.216216
37	8	2	0	0.609610	0.609610
37	8	2	1	0.957958	0.348348
37	8	2	2	1.000000	0.042042
37	8	3	0	0.470270	0.470270
37	8	3	1	0.888288	0.418018
37	8	3	2	0.992793	0.104504
37	8	3	3	1.000000	0.007207
37	8	4	0	0.359618	0.359618
37	8	4	1	0.802226	0.442607
37	8	4	2	0.974351	0.172125
37	8	4	3	0.998940	0.024589
37	8	4	4	1.000000	0.001060
37	8	5	0	0.272438	0.272438
37	8	5	1	0.708339	0.435901
37	8	5	2	0.943055	0.234716
37	8	5	3	0.995214	0.052159
37	8	5	4	0.999871	0.004657
37	8	5	5	1.000000	0.000128
37	8	6	0	0.204329	0.204329
37	8	6	1	0.612986	0.408657
37	8	6	2	0.899046	0.286060
37	8	6	3	0.987065	0.088018
37	8	6	4	0.999289	0.012225
37	8	6	5	0.999988	0.000699
37	8	6	6	1.000000	0.000012
37	8	7	0	0.151599	0.151599
37	8	7	1	0.520709	0.369110
37	8	7	2	0.843680	0.322971
37	8	7	3	0.972868	0.129188
37	8	7	4	0.997712	0.024844
37	8	7	5	0.999920	0.002208
37	8	7	6	0.999999	0.000079
37	8	7	7	1.000000	0.000001
37	8	8	0	0.111172	0.111172
37	8	8	1	0.434583	0.323411
37	8	8	2	0.779085	0.344502
37	8	8	3	0.951337	0.172251
37	8	8	4	0.994399	0.043063
37	8	8	5	0.999699	0.005300
37	8	8	6	0.999994	0.000294
37	8	8	7	1.000000	0.000006

N	n	k	x	P(x)	p(x)
37	8	8	8	1.000000	0.000000
37	9	1	0	0.756757	0.756757
37	9	1	1	1.000000	0.243243
37	9	2	0	0.567568	0.567568
37	9	2	1	0.945946	0.378378
37	9	2	2	1.000000	0.054054
37	9	3	0	0.421622	0.421622
37	9	3	1	0.859459	0.437838
37	9	3	2	0.989189	0.129730
37	9	3	3	1.000000	0.010811
37	9	4	0	0.310016	0.310016
37	9	4	1	0.756439	0.446423
37	9	4	2	0.962480	0.206041
37	9	4	3	0.998092	0.035612
37	9	4	4	1.000000	0.001908
37	9	5	0	0.225466	0.225466
37	9	5	1	0.648215	0.422749
37	9	5	2	0.918774	0.270559
37	9	5	3	0.991617	0.072843
37	9	5	4	0.999711	0.008094
37	9	5	5	1.000000	0.000289
37	9	6	0	0.162054	0.162054
37	9	6	1	0.542528	0.380474
37	9	6	2	0.859590	0.317062
37	9	6	3	0.977959	0.118370
37	9	6	4	0.998446	0.020487
37	9	6	5	0.999964	0.001518
37	9	6	6	1.000000	0.000036
37	9	7	0	0.115006	0.115006
37	9	7	1	0.444341	0.329335
37	9	7	2	0.787995	0.343654
37	9	7	3	0.955049	0.167054
37	9	7	4	0.995142	0.040093
37	9	7	5	0.999768	0.004626
37	9	7	6	0.999996	0.000228
37	9	7	7	1.000000	0.000003
37	9	8	0	0.080504	0.080504
37	9	8	1	0.356518	0.276014
37	9	8	2	0.707809	0.351291
37	9	8	3	0.921638	0.213829
37	9	8	4	0.988460	0.066822
37	9	8	5	0.999151	0.010691
37	9	8	6	0.999974	0.000822
37	9	8	7	1.000000	0.000026
37	9	8	8	1.000000	0.000000
37	9	9	0	0.055520	0.055520
37	9	9	1	0.280376	0.224856
37	9	9	2	0.623015	0.342638
37	9	9	3	0.877398	0.254383
37	9	9	4	0.976939	0.099541

N	n	k	x	P(x)	p(x)
37	9	9	5	0.997676	0.020738
37	9	9	6	0.999888	0.002212
37	9	9	7	0.999998	0.000109
37	9	9	8	1.000000	0.000002
37	9	9	9	1.000000	0.000000
37	10	1	0	0.729730	0.729730
37	10	1	1	1.000000	0.270270
37	10	2	0	0.527027	0.527027
37	10	2	1	0.932432	0.405405
37	10	2	2	1.000000	0.067568
37	10	3	0	0.376448	0.376448
37	10	3	1	0.828185	0.451737
37	10	3	2	0.984556	0.156371
37	10	3	3	1.000000	0.015444
37	10	4	0	0.265728	0.265728
37	10	4	1	0.708608	0.442880
37	10	4	2	0.947763	0.239155
37	10	4	3	0.996820	0.049057
37	10	4	4	1.000000	0.003180
37	10	5	0	0.185204	0.185204
37	10	5	1	0.587822	0.402618
37	10	5	2	0.889786	0.301964
37	10	5	3	0.986414	0.096628
37	10	5	4	0.999422	0.013008
37	10	5	5	1.000000	0.000578
37	10	6	0	0.127328	0.127328
37	10	6	1	0.474586	0.347258
37	10	6	2	0.814295	0.339709
37	10	6	3	0.965277	0.150982
37	10	6	4	0.996983	0.031706
37	10	6	5	0.999910	0.002927
37	10	6	6	1.000000	0.000090
37	10	7	0	0.086254	0.086254
37	10	7	1	0.373769	0.287515
37	10	7	2	0.726628	0.352859
37	10	7	3	0.931184	0.204556
37	10	7	4	0.990846	0.059662
37	10	7	5	0.999438	0.008591
37	10	7	6	0.999988	0.000551
37	10	7	7	1.000000	0.000012
37	10	8	0	0.057503	0.057503
37	10	8	1	0.287515	0.230012
37	10	8	2	0.632532	0.345018
37	10	8	3	0.883454	0.250922
37	10	8	4	0.978914	0.095459
37	10	8	5	0.998006	0.019092
37	10	8	6	0.999915	0.001909
37	10	8	7	0.999999	0.000084
37	10	8	8	1.000000	0.000001
37	10	9	0	0.037674	0.037674

Table for $N = 2$, $n = 1$, through $N = 100$, $n = 50$

N	n	k	x	P(x)	p(x)
37	10	9	1	0.216132	0.178457
37	10	9	2	0.537355	0.321223
37	10	9	3	0.822887	0.285532
37	10	9	4	0.959164	0.136277
37	10	9	5	0.994714	0.035550
37	10	9	6	0.999652	0.004938
37	10	9	7	0.999990	0.000339
37	10	9	8	1.000000	0.000010
37	10	9	9	1.000000	0.000000
37	10	10	0	0.024219	0.024219
37	10	10	1	0.158770	0.134551
37	10	10	2	0.445577	0.286807
37	10	10	3	0.751504	0.305927
37	10	10	4	0.929961	0.178457
37	10	10	5	0.988366	0.058404
37	10	10	6	0.998946	0.010580
37	10	10	7	0.999954	0.001008
37	10	10	8	0.999999	0.000045
37	10	10	9	1.000000	0.000001
37	10	10	10	1.000000	0.000000
37	11	1	0	0.702703	0.702703
37	11	1	1	1.000000	0.297297
37	11	2	0	0.487988	0.487988
37	11	2	1	0.917417	0.429429
37	11	2	2	1.000000	0.082583
37	11	3	0	0.334620	0.334620
37	11	3	1	0.794723	0.460103
37	11	3	2	0.978764	0.184041
37	11	3	3	1.000000	0.021236
37	11	4	0	0.226361	0.226361
37	11	4	1	0.659399	0.433038
37	11	4	2	0.930048	0.270649
37	11	4	3	0.995003	0.064956
37	11	4	4	1.000000	0.004997
37	11	5	0	0.150907	0.150907
37	11	5	1	0.528175	0.377268
37	11	5	2	0.856234	0.328059
37	11	5	3	0.979256	0.123022
37	11	5	4	0.998940	0.019684
37	11	5	5	1.000000	0.001060
37	11	6	0	0.099033	0.099033
37	11	6	1	0.410279	0.311246
37	11	6	2	0.763968	0.353689
37	11	6	3	0.948501	0.184533
37	11	6	4	0.994634	0.046133
37	11	6	5	0.999801	0.005167
37	11	6	6	1.000000	0.000199
37	11	7	0	0.063892	0.063892
37	11	7	1	0.309877	0.245985
37	11	7	2	0.661284	0.351407

N	n	k	x	P(x)	p(x)
37	11	7	3	0.900879	0.239596
37	11	7	4	0.984217	0.083338
37	11	7	5	0.998801	0.014584
37	11	7	6	0.999968	0.001167
37	11	7	7	1.000000	0.000032
37	11	8	0	0.040465	0.040465
37	11	8	1	0.227882	0.187417
37	11	8	2	0.555862	0.327980
37	11	8	3	0.836987	0.281126
37	11	8	4	0.964772	0.127784
37	11	8	5	0.995884	0.031113
37	11	8	6	0.999773	0.003889
37	11	8	7	0.999996	0.000222
37	11	8	8	1.000000	0.000004
37	11	9	0	0.025116	0.025116
37	11	9	1	0.163255	0.138139
37	11	9	2	0.454075	0.290819
37	11	9	3	0.759435	0.305360
37	11	9	4	0.933927	0.174492
37	11	9	5	0.989447	0.055520
37	11	9	6	0.999103	0.009656
37	11	9	7	0.999965	0.000862
37	11	9	8	0.999999	0.000034
37	11	9	9	1.000000	0.000000
37	11	10	0	0.015249	0.015249
37	11	10	1	0.113920	0.098671
37	11	10	2	0.360597	0.246677
37	11	10	3	0.672190	0.311592
37	11	10	4	0.890304	0.218115
37	11	10	5	0.977550	0.087246
37	11	10	6	0.997379	0.019829
37	11	10	7	0.999842	0.002463
37	11	10	8	0.999996	0.000154
37	11	10	9	1.000000	0.000004
37	11	10	10	1.000000	0.000000
37	11	11	0	0.009037	0.009037
37	11	11	1	0.077375	0.068339
37	11	11	2	0.278372	0.200996
37	11	11	3	0.579866	0.301494
37	11	11	4	0.833756	0.253890
37	11	11	5	0.958162	0.124406
37	11	11	6	0.993707	0.035545
37	11	11	7	0.999477	0.005770
37	11	11	8	0.999979	0.000502
37	11	11	9	1.000000	0.000021
37	11	11	10	1.000000	0.000000
37	11	11	11	1.000000	0.000000
37	12	1	0	0.675676	0.675676
37	12	1	1	1.000000	0.324324
37	12	2	0	0.450450	0.450450

N	n	k	x	P(x)	p(x)
37	12	2	1	0.900901	0.450450
37	12	2	2	1.000000	0.099099
37	12	3	0	0.296010	0.296010
37	12	3	1	0.759331	0.463320
37	12	3	2	0.971686	0.212355
37	12	3	3	1.000000	0.028314
37	12	4	0	0.191536	0.191536
37	12	4	1	0.609433	0.417897
37	12	4	2	0.909228	0.299796
37	12	4	3	0.992505	0.083277
37	12	4	4	1.000000	0.007495
37	12	5	0	0.121887	0.121887
37	12	5	1	0.470134	0.348247
37	12	5	2	0.818381	0.348247
37	12	5	3	0.969793	0.151412
37	12	5	4	0.998183	0.028390
37	12	5	5	1.000000	0.001817
37	12	6	0	0.076179	0.076179
37	12	6	1	0.350424	0.274245
37	12	6	2	0.709554	0.359130
37	12	6	3	0.927209	0.217655
37	12	6	4	0.991086	0.063877
37	12	6	5	0.999602	0.008517
37	12	6	6	1.000000	0.000397
37	12	7	0	0.046690	0.046690
37	12	7	1	0.253111	0.206421
37	12	7	2	0.593706	0.340594
37	12	7	3	0.864019	0.270313
37	12	7	4	0.974601	0.110583
37	12	7	5	0.997679	0.023078
37	12	7	6	0.999923	0.002244
37	12	7	7	1.000000	0.000077
37	12	8	0	0.028014	0.028014
37	12	8	1	0.177424	0.149409
37	12	8	2	0.480174	0.302751
37	12	8	3	0.782925	0.302751
37	12	8	4	0.945112	0.162188
37	12	8	5	0.992294	0.047182
37	12	8	6	0.999474	0.007180
37	12	8	7	0.999987	0.000513
37	12	8	8	1.000000	0.000013
37	12	9	0	0.016422	0.016422
37	12	9	1	0.120751	0.104329
37	12	9	2	0.375777	0.255026
37	12	9	3	0.688968	0.313190
37	12	9	4	0.900371	0.211403
37	12	9	5	0.980906	0.080535
37	12	9	6	0.997989	0.017083
37	12	9	7	0.999899	0.001910
37	12	9	8	0.999998	0.000099

N	n	k	x	P(x)	p(x)
37	12	9	9	1.000000	0.000002
37	12	10	0	0.009384	0.009384
37	12	10	1	0.079765	0.070381
37	12	10	2	0.284697	0.204932
37	12	10	3	0.588299	0.303603
37	12	10	4	0.839970	0.251671
37	12	10	5	0.960772	0.120802
37	12	10	6	0.994328	0.033556
37	12	10	7	0.999558	0.005230
37	12	10	8	0.999984	0.000426
37	12	10	9	1.000000	0.000016
37	12	10	10	1.000000	0.000000
37	12	11	0	0.005213	0.005213
37	12	11	1	0.051091	0.045878
37	12	11	2	0.208796	0.157705
37	12	11	3	0.487098	0.278303
37	12	11	4	0.765401	0.278303
37	12	11	5	0.929453	0.164052
37	12	11	6	0.986871	0.057418
37	12	11	7	0.998589	0.011718
37	12	11	8	0.999921	0.001332
37	12	11	9	0.999998	0.000077
37	12	11	10	1.000000	0.000002
37	12	11	11	1.000000	0.000000
37	12	12	0	0.002807	0.002807
37	12	12	1	0.031681	0.028874
37	12	12	2	0.148140	0.116459
37	12	12	3	0.390763	0.242623
37	12	12	4	0.679769	0.289006
37	12	12	5	0.885285	0.205516
37	12	12	6	0.973621	0.088336
37	12	12	7	0.996336	0.022715
37	12	12	8	0.999716	0.003380
37	12	12	9	0.999989	0.000273
37	12	12	10	1.000000	0.000011
37	12	12	11	1.000000	0.000000
37	12	12	12	1.000000	0.000000
37	13	1	0	0.648649	0.648649
37	13	1	1	1.000000	0.351351
37	13	2	0	0.414414	0.414414
37	13	2	1	0.882883	0.468468
37	13	2	2	1.000000	0.117117
37	13	3	0	0.260489	0.260489
37	13	3	1	0.722265	0.461776
37	13	3	2	0.963192	0.240927
37	13	3	3	1.000000	0.036808
37	13	4	0	0.160890	0.160890
37	13	4	1	0.559285	0.398395
37	13	4	2	0.885245	0.325960
37	13	4	3	0.989174	0.103929

Table for $N = 2$, $n = 1$, through $N = 100$, $n = 50$

N	n	k	x	P(x)	p(x)
37	13	4	4	1.000000	0.010826
37	13	5	0	0.097509	0.097509
37	13	5	1	0.414414	0.316905
37	13	5	2	0.776592	0.362177
37	13	5	3	0.957680	0.181089
37	13	5	4	0.997047	0.039367
37	13	5	5	1.000000	0.002953
37	13	6	0	0.057896	0.057896
37	13	6	1	0.295575	0.237679
37	13	6	2	0.652093	0.356518
37	13	6	3	0.901090	0.248997
37	13	6	4	0.985975	0.084885
37	13	6	5	0.999262	0.013286
37	13	6	6	1.000000	0.000738
37	13	7	0	0.033617	0.033617
37	13	7	1	0.203570	0.169953
37	13	7	2	0.525587	0.322016
37	13	7	3	0.820769	0.295182
37	13	7	4	0.961331	0.140563
37	13	7	5	0.995833	0.034502
37	13	7	6	0.999833	0.004000
37	13	7	7	1.000000	0.000167
37	13	8	0	0.019050	0.019050
37	13	8	1	0.135589	0.116539
37	13	8	2	0.407514	0.271925
37	13	8	3	0.722375	0.314861
37	13	8	4	0.919162	0.196788
37	13	8	5	0.986633	0.067470
37	13	8	6	0.998900	0.012267
37	13	8	7	0.999967	0.001067
37	13	8	8	1.000000	0.000033
37	13	9	0	0.010510	0.010510
37	13	9	1	0.087366	0.076856
37	13	9	2	0.304370	0.217004
37	13	9	3	0.613802	0.309432
37	13	9	4	0.858090	0.244288
37	13	9	5	0.968020	0.109930
37	13	9	6	0.995939	0.027919
37	13	9	7	0.999746	0.003807
37	13	9	8	0.999994	0.000248
37	13	9	9	1.000000	0.000006
37	13	10	0	0.005630	0.005630
37	13	10	1	0.054428	0.048797
37	13	10	2	0.219118	0.164691
37	13	10	3	0.503291	0.284172
37	13	10	4	0.779569	0.276279
37	13	10	5	0.936612	0.157043
37	13	10	6	0.988959	0.052347
37	13	10	7	0.998930	0.009971
37	13	10	8	0.999950	0.001020
37	13	10	9	0.999999	0.000049
37	13	10	10	1.000000	0.000001
37	13	11	0	0.002919	0.002919
37	13	11	1	0.032740	0.029821
37	13	11	2	0.152022	0.119282
37	13	11	3	0.398042	0.246019
37	13	11	4	0.687476	0.289435
37	13	11	5	0.890081	0.202604
37	13	11	6	0.975388	0.085307
37	13	11	7	0.996714	0.021327
37	13	11	8	0.999761	0.003047
37	13	11	9	0.999992	0.000231
37	13	11	10	1.000000	0.000008
37	13	11	11	1.000000	0.000000
37	13	12	0	0.001460	0.001460
37	13	12	1	0.018977	0.017517
37	13	12	2	0.101557	0.082580
37	13	12	3	0.303419	0.201862
37	13	12	4	0.587287	0.283869
37	13	12	5	0.827741	0.240453
37	13	12	6	0.952420	0.124680
37	13	12	7	0.991793	0.039372
37	13	12	8	0.999175	0.007382
37	13	12	9	0.999956	0.000781
37	13	12	10	0.999999	0.000043
37	13	12	11	1.000000	0.000001
37	13	12	12	1.000000	0.000000
37	13	13	0	0.000701	0.000701
37	13	13	1	0.010569	0.009868
37	13	13	2	0.065221	0.054653
37	13	13	3	0.222674	0.157452
37	13	13	4	0.485095	0.262421
37	13	13	5	0.750796	0.265701
37	13	13	6	0.917510	0.166714
37	13	13	7	0.982343	0.064833
37	13	13	8	0.997699	0.015355
37	13	13	9	0.999831	0.002133
37	13	13	10	0.999994	0.000162
37	13	13	11	1.000000	0.000006
37	13	13	12	1.000000	0.000000
37	13	13	13	1.000000	0.000000
37	14	1	0	0.621622	0.621622
37	14	1	1	1.000000	0.378378
37	14	2	0	0.379880	0.379880
37	14	2	1	0.863363	0.483483
37	14	2	2	1.000000	0.136637
37	14	3	0	0.227928	0.227928
37	14	3	1	0.683784	0.455856
37	14	3	2	0.953153	0.269369
37	14	3	3	1.000000	0.046847
37	14	4	0	0.134075	0.134075
37	14	4	1	0.509486	0.375411
37	14	4	2	0.858082	0.348596
37	14	4	3	0.984844	0.126762
37	14	4	4	1.000000	0.015156
37	14	5	0	0.077195	0.077195
37	14	5	1	0.361597	0.284402
37	14	5	2	0.731319	0.369723
37	14	5	3	0.942590	0.211270
37	14	5	4	0.995407	0.052818
37	14	5	5	1.000000	0.004593
37	14	6	0	0.043422	0.043422
37	14	6	1	0.246059	0.202636
37	14	6	2	0.592674	0.346615
37	14	6	3	0.869965	0.277292
37	14	6	4	0.978902	0.108936
37	14	6	5	0.998708	0.019807
37	14	6	6	1.000000	0.001292
37	14	7	0	0.023812	0.023812
37	14	7	1	0.161082	0.137270
37	14	7	2	0.458500	0.297418
37	14	7	3	0.771572	0.313072
37	14	7	4	0.943761	0.172189
37	14	7	5	0.992958	0.049197
37	14	7	6	0.999667	0.006709
37	14	7	7	1.000000	0.000333
37	14	8	0	0.012700	0.012700
37	14	8	1	0.101598	0.088899
37	14	8	2	0.339533	0.237934
37	14	8	3	0.656779	0.317246
37	14	8	4	0.886364	0.229586
37	14	8	5	0.978199	0.091834
37	14	8	6	0.997878	0.019679
37	14	8	7	0.999922	0.002045
37	14	8	8	1.000000	0.000078
37	14	9	0	0.006569	0.006569
37	14	9	1	0.061747	0.055178
37	14	9	2	0.241077	0.179330
37	14	9	3	0.536444	0.295367
37	14	9	4	0.807197	0.270753
37	14	9	5	0.949698	0.142502
37	14	9	6	0.992449	0.042750
37	14	9	7	0.999429	0.006980
37	14	9	8	0.999984	0.000555
37	14	9	9	1.000000	0.000016
37	14	10	0	0.003284	0.003284
37	14	10	1	0.036129	0.032844
37	14	10	2	0.164221	0.128093
37	14	10	3	0.420407	0.256186
37	14	10	4	0.710499	0.290092
37	14	10	5	0.903894	0.193395
37	14	10	6	0.980234	0.076340
37	14	10	7	0.997684	0.017449
37	14	10	8	0.999865	0.002181
37	14	10	9	0.999997	0.000132
37	14	10	10	1.000000	0.000003
37	14	11	0	0.001581	0.001581
37	14	11	1	0.020315	0.018733
37	14	11	2	0.107291	0.086977
37	14	11	3	0.316035	0.208744
37	14	11	4	0.603058	0.287023
37	14	11	5	0.839429	0.236372
37	14	11	6	0.957615	0.118186
37	14	11	7	0.993160	0.035545
37	14	11	8	0.999380	0.006220
37	14	11	9	0.999973	0.000592
37	14	11	10	0.999999	0.000027
37	14	11	11	1.000000	0.000000
37	14	12	0	0.000730	0.000730
37	14	12	1	0.010948	0.010218
37	14	12	2	0.067148	0.056200
37	14	12	3	0.227720	0.160572
37	14	12	4	0.492664	0.264944
37	14	12	5	0.757609	0.264944
37	14	12	6	0.921250	0.163642
37	14	12	7	0.983590	0.062340
37	14	12	8	0.997945	0.014355
37	14	12	9	0.999859	0.001914
37	14	12	10	0.999995	0.000137
37	14	12	11	1.000000	0.000005
37	14	12	12	1.000000	0.000000
37	14	13	0	0.000321	0.000321
37	14	13	1	0.005635	0.005313
37	14	13	2	0.040172	0.034538
37	14	13	3	0.157069	0.116897
37	14	13	4	0.386687	0.229618
37	14	13	5	0.662229	0.275542
37	14	13	6	0.868885	0.206656
37	14	13	7	0.966135	0.097250
37	14	13	8	0.994500	0.028365
37	14	13	9	0.999476	0.004976
37	14	13	10	0.999973	0.000498
37	14	13	11	0.999999	0.000026
37	14	13	12	1.000000	0.000001
37	14	13	13	1.000000	0.000000
37	14	14	0	0.000134	0.000134
37	14	14	1	0.002756	0.002623
37	14	14	2	0.022903	0.020147
37	14	14	3	0.103491	0.080588
37	14	14	4	0.291013	0.187521

Table for $N = 2$, $n = 1$, through $N = 100$, $n = 50$

N	n	k	x	P(x)	p(x)
37	14	14	5	0.558900	0.267888
37	14	14	6	0.800000	0.241099
37	14	14	7	0.937770	0.137771
37	14	14	8	0.987408	0.049638
37	14	14	9	0.998439	0.011031
37	14	14	10	0.999891	0.001451
37	14	14	11	0.999996	0.000106
37	14	14	12	1.000000	0.000004
37	14	14	13	1.000000	0.000000
37	14	14	14	1.000000	0.000000
37	15	1	0	0.594595	0.594595
37	15	1	1	1.000000	0.405405
37	15	2	0	0.346847	0.346847
37	15	2	1	0.842342	0.495495
37	15	2	2	1.000000	0.157658
37	15	3	0	0.198198	0.198198
37	15	3	1	0.644144	0.445946
37	15	3	2	0.941441	0.297297
37	15	3	3	1.000000	0.058559
37	15	4	0	0.110758	0.110758
37	15	4	1	0.460519	0.349762
37	15	4	2	0.827769	0.367250
37	15	4	3	0.979332	0.151563
37	15	4	4	1.000000	0.020668
37	15	5	0	0.060413	0.060413
37	15	5	1	0.312136	0.251722
37	15	5	2	0.683095	0.370959
37	15	5	3	0.924218	0.241123
37	15	5	4	0.993111	0.068892
37	15	5	5	1.000000	0.006889
37	15	6	0	0.032095	0.032095
37	15	6	1	0.202007	0.169913
37	15	6	2	0.532393	0.330386
37	15	6	3	0.833797	0.301404
37	15	6	4	0.969429	0.135632
37	15	6	5	0.997847	0.028418
37	15	6	6	1.000000	0.002153
37	15	7	0	0.016565	0.016565
37	15	7	1	0.125272	0.108707
37	15	7	2	0.393844	0.268571
37	15	7	3	0.717124	0.323280
37	15	7	4	0.921301	0.204177
37	15	7	5	0.988680	0.067378
37	15	7	6	0.999375	0.010695
37	15	7	7	1.000000	0.000625
37	15	8	0	0.008282	0.008282
37	15	8	1	0.074542	0.066260
37	15	8	2	0.277463	0.202921
37	15	8	3	0.587812	0.310349
37	15	8	4	0.846436	0.258624
37	15	8	5	0.966220	0.119784
37	15	8	6	0.996166	0.029946
37	15	8	7	0.999833	0.003667
37	15	8	8	1.000000	0.000167
37	15	9	0	0.003998	0.003998
37	15	9	1	0.042555	0.038556
37	15	9	2	0.186498	0.143944
37	15	9	3	0.459392	0.272893
37	15	9	4	0.748338	0.288946
37	15	9	5	0.924916	0.176578
37	15	9	6	0.986873	0.061957
37	15	9	7	0.998822	0.011949
37	15	9	8	0.999960	0.001138
37	15	9	9	1.000000	0.000040
37	15	10	0	0.001856	0.001856
37	15	10	1	0.023277	0.021420
37	15	10	2	0.119667	0.096391
37	15	10	3	0.342438	0.222770
37	15	10	4	0.634823	0.292386
37	15	10	5	0.861852	0.227029
37	15	10	6	0.966958	0.105106
37	15	10	7	0.995408	0.028450
37	15	10	8	0.999675	0.004267
37	15	10	9	0.999991	0.000316
37	15	10	10	1.000000	0.000009
37	15	11	0	0.000825	0.000825
37	15	11	1	0.012170	0.011345
37	15	11	2	0.073257	0.061087
37	15	11	3	0.243429	0.170172
37	15	11	4	0.515703	0.272275
37	15	11	5	0.777767	0.262064
37	15	11	6	0.931923	0.154155
37	15	11	7	0.986978	0.055055
37	15	11	8	0.998569	0.011591
37	15	11	9	0.999921	0.001352
37	15	11	10	0.999998	0.000077
37	15	11	11	1.000000	0.000002
37	15	12	0	0.000349	0.000349
37	15	12	1	0.006061	0.005712
37	15	12	2	0.042713	0.036652
37	15	12	3	0.164888	0.122174
37	15	12	4	0.400510	0.235622
37	15	12	5	0.676973	0.276463
37	15	12	6	0.878561	0.201588
37	15	12	7	0.970038	0.091477
37	15	12	8	0.995448	0.025410
37	15	12	9	0.999609	0.004161
37	15	12	10	0.999983	0.000374
37	15	12	11	1.000000	0.000016
37	15	12	12	1.000000	0.000000
37	15	13	0	0.000140	0.000140
37	15	13	1	0.002862	0.002723
37	15	13	2	0.023654	0.020792
37	15	13	3	0.106244	0.082590
37	15	13	4	0.296836	0.190592
37	15	13	5	0.566388	0.269552
37	15	13	6	0.805990	0.239602
37	15	13	7	0.940765	0.134776
37	15	13	8	0.988333	0.047568
37	15	13	9	0.998610	0.010277
37	15	13	10	0.999909	0.001298
37	15	13	11	0.999997	0.000089
37	15	13	12	1.000000	0.000003
37	15	13	13	1.000000	0.000000
37	15	14	0	0.000052	0.000052
37	15	14	1	0.001274	0.001222
37	15	14	2	0.012392	0.011118
37	15	14	3	0.064949	0.052557
37	15	14	4	0.209482	0.144532
37	15	14	5	0.454075	0.244593
37	15	14	6	0.716139	0.262064
37	15	14	7	0.895840	0.179701
37	15	14	8	0.974459	0.078619
37	15	14	9	0.996041	0.021582
37	15	14	10	0.999638	0.003597
37	15	14	11	0.999982	0.000344
37	15	14	12	0.999999	0.000017
37	15	14	13	1.000000	0.000000
37	15	14	14	1.000000	0.000000
37	15	15	0	0.000018	0.000018
37	15	15	1	0.000530	0.000512
37	15	15	2	0.006108	0.005578
37	15	15	3	0.037528	0.031420
37	15	15	4	0.140357	0.102829
37	15	15	5	0.347730	0.207373
37	15	15	6	0.613592	0.265862
37	15	15	7	0.833335	0.219743
37	15	15	8	0.950532	0.117196
37	15	15	9	0.990411	0.039879
37	15	15	10	0.998856	0.008445
37	15	15	11	0.999922	0.001066
37	15	15	12	0.999997	0.000075
37	15	15	13	1.000000	0.000003
37	15	15	14	1.000000	0.000000
37	15	15	15	1.000000	0.000000
37	16	1	0	0.567568	0.567568
37	16	1	1	1.000000	0.432432
37	16	2	0	0.315315	0.315315
37	16	2	1	0.819820	0.504505
37	16	2	2	1.000000	0.180180
37	16	3	0	0.171171	0.171171
37	16	3	1	0.603604	0.432432
37	16	3	2	0.927928	0.324324
37	16	3	3	1.000000	0.072072
37	16	4	0	0.090620	0.090620
37	16	4	1	0.412825	0.322205
37	16	4	2	0.794383	0.381558
37	16	4	3	0.972443	0.178060
37	16	4	4	1.000000	0.027557
37	16	5	0	0.046683	0.046683
37	16	5	1	0.266368	0.219685
37	16	5	2	0.632509	0.366142
37	16	5	3	0.902298	0.269788
37	16	5	4	0.989979	0.087681
37	16	5	5	1.000000	0.010021
37	16	6	0	0.023342	0.023342
37	16	6	1	0.163391	0.140049
37	16	6	2	0.472323	0.308932
37	16	6	3	0.792696	0.320374
37	16	6	4	0.957099	0.164402
37	16	6	5	0.996555	0.039457
37	16	6	6	1.000000	0.003445
37	16	7	0	0.011294	0.011294
37	16	7	1	0.095625	0.084331
37	16	7	2	0.332805	0.237180
37	16	7	3	0.658346	0.325541
37	16	7	4	0.893459	0.235113
37	16	7	5	0.982555	0.089095
37	16	7	6	0.998889	0.016334
37	16	7	7	1.000000	0.001111
37	16	8	0	0.005271	0.005271
37	16	8	1	0.053460	0.048189
37	16	8	2	0.222121	0.168661
37	16	8	3	0.517278	0.295157
37	16	8	4	0.799414	0.282136
37	16	8	5	0.949886	0.150472
37	16	8	6	0.993444	0.043558
37	16	8	7	0.999667	0.006223
37	16	8	8	1.000000	0.000333
37	16	9	0	0.002363	0.002363
37	16	9	1	0.028534	0.026172
37	16	9	2	0.140698	0.112164
37	16	9	3	0.384966	0.244268
37	16	9	4	0.682668	0.297702
37	16	9	5	0.892811	0.210142
37	16	9	6	0.978424	0.085614
37	16	9	7	0.997735	0.019311
37	16	9	8	0.999908	0.002173
37	16	9	9	1.000000	0.000092
37	16	10	0	0.001013	0.001013

N	n	k	x	P(x)	p(x)
37	16	10	1	0.014514	0.013501
37	16	10	2	0.084616	0.070102
37	16	10	3	0.271556	0.186940
37	16	10	4	0.555082	0.283525
37	16	10	5	0.810255	0.255173
37	16	10	6	0.947848	0.137593
37	16	10	7	0.991528	0.043680
37	16	10	8	0.999287	0.007759
37	16	10	9	0.999977	0.000690
37	16	10	10	1.000000	0.000023
37	16	11	0	0.000413	0.000413
37	16	11	1	0.007013	0.006601
37	16	11	2	0.048267	0.041254
37	16	11	3	0.181548	0.133281
37	16	11	4	0.429070	0.247522
37	16	11	5	0.706295	0.277225
37	16	11	6	0.896887	0.190592
37	16	11	7	0.976968	0.080081
37	16	11	8	0.996988	0.020020
37	16	11	9	0.999798	0.002810
37	16	11	10	0.999995	0.000197
37	16	11	11	1.000000	0.000005
37	16	12	0	0.000159	0.000159
37	16	12	1	0.003205	0.003046
37	16	12	2	0.026053	0.022848
37	16	12	3	0.114907	0.088854
37	16	12	4	0.314829	0.199922
37	16	12	5	0.589008	0.274178
37	16	12	6	0.823583	0.234575
37	16	12	7	0.949248	0.125665
37	16	12	8	0.990828	0.041580
37	16	12	9	0.999042	0.008213
37	16	12	10	0.999949	0.000908
37	16	12	11	0.999999	0.000050
37	16	12	12	1.000000	0.000001
37	16	13	0	0.000057	0.000057
37	16	13	1	0.001377	0.001320
37	16	13	2	0.013258	0.011881
37	16	13	3	0.068703	0.055445
37	16	13	4	0.218867	0.150163
37	16	13	5	0.468369	0.249502
37	16	13	6	0.729753	0.261383
37	16	13	7	0.904008	0.174256
37	16	13	8	0.977522	0.073514
37	16	13	9	0.996742	0.019219
37	16	13	10	0.999731	0.002990
37	16	13	11	0.999989	0.000257
37	16	13	12	1.000000	0.000011
37	16	13	13	1.000000	0.000000
37	16	14	0	0.000019	0.000019
37	16	14	1	0.000552	0.000533
37	16	14	2	0.006328	0.005776
37	16	14	3	0.038671	0.032343
37	16	14	4	0.143785	0.105114
37	16	14	5	0.354014	0.210229
37	16	14	6	0.620843	0.266829
37	16	14	7	0.838662	0.217820
37	16	14	8	0.953018	0.114355
37	16	14	9	0.991136	0.038118
37	16	14	10	0.998984	0.007848
37	16	14	11	0.999935	0.000951
37	16	14	12	0.999998	0.000063
37	16	14	13	1.000000	0.000002
37	16	14	14	1.000000	0.000000
37	16	15	0	0.000006	0.000006
37	16	15	1	0.000204	0.000199
37	16	15	2	0.002812	0.002608
37	16	15	3	0.020390	0.017578
37	16	15	4	0.088943	0.068553
37	16	15	5	0.253470	0.164527
37	16	15	6	0.504830	0.251361
37	16	15	7	0.753429	0.248598
37	16	15	8	0.913242	0.159813
37	16	15	9	0.979535	0.066293
37	16	15	10	0.996937	0.017402
37	16	15	11	0.999729	0.002792
37	16	15	12	0.999987	0.000258
37	16	15	13	1.000000	0.000013
37	16	15	14	1.000000	0.000000
37	16	15	15	1.000000	0.000000
37	16	16	0	0.000002	0.000002
37	16	16	1	0.000069	0.000067
37	16	16	2	0.001153	0.001084
37	16	16	3	0.010003	0.008850
37	16	16	4	0.051550	0.041547
37	16	16	5	0.171206	0.119656
37	16	16	6	0.390575	0.219369
37	16	16	7	0.651729	0.261154
37	16	16	8	0.855128	0.203399
37	16	16	9	0.958442	0.103314
37	16	16	10	0.992191	0.033749
37	16	16	11	0.999094	0.006903
37	16	16	12	0.999940	0.000846
37	16	16	13	0.999998	0.000058
37	16	16	14	1.000000	0.000002
37	16	16	15	1.000000	0.000000
37	16	16	16	1.000000	0.000000
37	17	1	0	0.540541	0.540541
37	17	1	1	1.000000	0.459459
37	17	2	0	0.285285	0.285285
37	17	2	1	0.795796	0.510510
37	17	2	2	1.000000	0.204204
37	17	3	0	0.146718	0.146718
37	17	3	1	0.562420	0.413701
37	17	3	2	0.912484	0.350064
37	17	3	3	1.000000	0.087516
37	17	4	0	0.073359	0.073359
37	17	4	1	0.366795	0.293436
37	17	4	2	0.758044	0.391248
37	17	4	3	0.963964	0.205920
37	17	4	4	1.000000	0.036036
37	17	5	0	0.035568	0.035568
37	17	5	1	0.224523	0.188955
37	17	5	2	0.580204	0.355680
37	17	5	3	0.876604	0.296400
37	17	5	4	0.985804	0.109200
37	17	5	5	1.000000	0.014196
37	17	6	0	0.016673	0.016673
37	17	6	1	0.130046	0.113373
37	17	6	2	0.413478	0.283433
37	17	6	3	0.746929	0.333450
37	17	6	4	0.941441	0.194513
37	17	6	5	0.994676	0.053235
37	17	6	6	1.000000	0.005323
37	17	7	0	0.007530	0.007530
37	17	7	1	0.071530	0.064001
37	17	7	2	0.276333	0.204803
37	17	7	3	0.596338	0.320005
37	17	7	4	0.859872	0.263533
37	17	7	5	0.974069	0.114198
37	17	7	6	0.998111	0.024042
37	17	7	7	1.000000	0.001889
37	17	8	0	0.003263	0.003263
37	17	8	1	0.037397	0.034134
37	17	8	2	0.173932	0.136535
37	17	8	3	0.447003	0.273071
37	17	8	4	0.745674	0.298671
37	17	8	5	0.928390	0.182716
37	17	8	6	0.989296	0.060905
37	17	8	7	0.999370	0.010075
37	17	8	8	1.000000	0.000630
37	17	9	0	0.001350	0.001350
37	17	9	1	0.018564	0.017214
37	17	9	2	0.103310	0.084746
37	17	9	3	0.315175	0.211865
37	17	9	4	0.611787	0.296611
37	17	9	5	0.852783	0.240997
37	17	9	6	0.966194	0.113410
37	17	9	7	0.995896	0.029703
37	17	9	8	0.999805	0.003908
37	17	9	9	1.000000	0.000195
37	17	10	0	0.000530	0.000530
37	17	10	1	0.008728	0.008197
37	17	10	2	0.057911	0.049183
37	17	10	3	0.209243	0.151332
37	17	10	4	0.474074	0.264832
37	17	10	5	0.749499	0.275425
37	17	10	6	0.921640	0.172140
37	17	10	7	0.985288	0.063649
37	17	10	8	0.998548	0.013260
37	17	10	9	0.999944	0.001396
37	17	10	10	1.000000	0.000056
37	17	11	0	0.000196	0.000196
37	17	11	1	0.003870	0.003674
37	17	11	2	0.030587	0.026717
37	17	11	3	0.130774	0.100188
37	17	11	4	0.346563	0.215789
37	17	11	5	0.627088	0.280525
37	17	11	6	0.851508	0.224420
37	17	11	7	0.961715	0.110206
37	17	11	8	0.994128	0.032414
37	17	11	9	0.999531	0.005402
37	17	11	10	0.999985	0.000455
37	17	11	11	1.000000	0.000014
37	17	12	0	0.000068	0.000068
37	17	12	1	0.001609	0.001541
37	17	12	2	0.015173	0.013564
37	17	12	3	0.076827	0.061654
37	17	12	4	0.238669	0.161841
37	17	12	5	0.497615	0.258946
37	17	12	6	0.756561	0.258946
37	17	12	7	0.919328	0.162766
37	17	12	8	0.982908	0.063581
37	17	12	9	0.997868	0.014960
37	17	12	10	0.999863	0.001995
37	17	12	11	0.999997	0.000134
37	17	12	12	1.000000	0.000003
37	17	13	0	0.000022	0.000022
37	17	13	1	0.000623	0.000601
37	17	13	2	0.007035	0.006412
37	17	13	3	0.042301	0.035266
37	17	13	4	0.154511	0.112210
37	17	13	5	0.373321	0.218810
37	17	13	6	0.642625	0.269304
37	17	13	7	0.854221	0.211596
37	17	13	8	0.960019	0.105798
37	17	13	9	0.993081	0.033062
37	17	13	10	0.999304	0.006223
37	17	13	11	0.999964	0.000660
37	17	13	12	0.999999	0.000035

Table for $N = 2$, $n = 1$, through $N = 100$, $n = 50$

N	n	k	x	P(x)	p(x)
37	17	13	13	1.000000	0.000001
37	17	14	0	0.000006	0.000006
37	17	14	1	0.000222	0.000216
37	17	14	2	0.003027	0.002805
37	17	14	3	0.021729	0.018702
37	17	14	4	0.093731	0.072001
37	17	14	5	0.263916	0.170185
37	17	14	6	0.519194	0.255278
37	17	14	7	0.766056	0.246862
37	17	14	8	0.920345	0.154289
37	17	14	9	0.982060	0.061716
37	17	14	10	0.997489	0.015429
37	17	14	11	0.999799	0.002310
37	17	14	12	0.999992	0.000193
37	17	14	13	1.000000	0.000008
37	17	14	14	1.000000	0.000000
37	17	15	0	0.000002	0.000002
37	17	15	1	0.000072	0.000070
37	17	15	2	0.001198	0.001126
37	17	15	3	0.010345	0.009148
37	17	15	4	0.053034	0.042689
37	17	15	5	0.175124	0.122089
37	17	15	6	0.397104	0.221981
37	17	15	7	0.658725	0.261620
37	17	15	8	0.859971	0.201246
37	17	15	9	0.960594	0.100623
37	17	15	10	0.992794	0.032199
37	17	15	11	0.999197	0.006403
37	17	15	12	0.999950	0.000753
37	17	15	13	0.999998	0.000048
37	17	15	14	1.000000	0.000001
37	17	15	15	1.000000	0.000000
37	17	16	0	0.000000	0.000000
37	17	16	1	0.000021	0.000020
37	17	16	2	0.000430	0.000409
37	17	16	3	0.004524	0.004094
37	17	16	4	0.027809	0.023285
37	17	16	5	0.108529	0.080720
37	17	16	6	0.286114	0.177585
37	17	16	7	0.539806	0.253692
37	17	16	8	0.777643	0.237837
37	17	16	9	0.924004	0.146361
37	17	16	10	0.982548	0.058544
37	17	16	11	0.997450	0.014902
37	17	16	12	0.999779	0.002328
37	17	16	13	0.999990	0.000211
37	17	16	14	1.000000	0.000010
37	17	16	15	1.000000	0.000000
37	17	16	16	1.000000	0.000000
37	17	17	0	0.000000	0.000000
37	17	17	1	0.000005	0.000005
37	17	17	2	0.000138	0.000133
37	17	17	3	0.001795	0.001657
37	17	17	4	0.013395	0.011600
37	17	17	5	0.062403	0.049009
37	17	17	6	0.193093	0.130690
37	17	17	7	0.419000	0.225907
37	17	17	8	0.675713	0.256712
37	17	17	9	0.868247	0.192534
37	17	17	10	0.963033	0.094786
37	17	17	11	0.993193	0.030159
37	17	17	12	0.999224	0.006032
37	17	17	13	0.999949	0.000725
37	17	17	14	0.999998	0.000049
37	17	17	15	1.000000	0.000002
37	17	17	16	1.000000	0.000000
37	17	17	17	1.000000	0.000000
37	18	1	0	0.513513	0.513513
37	18	1	1	1.000000	0.486486
37	18	2	0	0.256757	0.256757
37	18	2	1	0.770270	0.513513
37	18	2	2	1.000000	0.229730
37	18	3	0	0.124710	0.124710
37	18	3	1	0.520849	0.396139
37	18	3	2	0.894981	0.374131
37	18	3	3	1.000000	0.105019
37	18	4	0	0.058687	0.058687
37	18	4	1	0.322780	0.264093
37	18	4	2	0.718919	0.396139
37	18	4	3	0.953668	0.234749
37	18	4	4	1.000000	0.046332
37	18	5	0	0.026676	0.026676
37	18	5	1	0.186732	0.160056
37	18	5	2	0.526851	0.340119
37	18	5	3	0.846964	0.320112
37	18	5	4	0.980344	0.133380
37	18	5	5	1.000000	0.019656
37	18	6	0	0.011671	0.011671
37	18	6	1	0.101702	0.090032
37	18	6	2	0.356792	0.255089
37	18	6	3	0.696911	0.340119
37	18	6	4	0.921990	0.225079
37	18	6	5	0.992015	0.070025
37	18	6	6	1.000000	0.007985
37	18	7	0	0.004894	0.004894
37	18	7	1	0.052330	0.047436
37	18	7	2	0.225133	0.172803
37	18	7	3	0.532337	0.307205
37	18	7	4	0.820342	0.288004
37	18	7	5	0.962650	0.142308
37	18	7	6	0.996909	0.034259
37	18	7	7	1.000000	0.003091
37	18	8	0	0.001958	0.001958
37	18	8	1	0.025450	0.023492
37	18	8	2	0.132971	0.107522
37	18	8	3	0.378735	0.245764
37	18	8	4	0.685940	0.307205
37	18	8	5	0.900983	0.215043
37	18	8	6	0.983205	0.082222
37	18	8	7	0.998867	0.015661
37	18	8	8	1.000000	0.001133
37	18	9	0	0.000743	0.000743
37	18	9	1	0.011679	0.010936
37	18	9	2	0.073649	0.061971
37	18	9	3	0.251616	0.177967
37	18	9	4	0.537634	0.286018
37	18	9	5	0.804584	0.266950
37	18	9	6	0.949182	0.144598
37	18	9	7	0.992926	0.043744
37	18	9	8	0.999609	0.006683
37	18	9	9	1.000000	0.000391
37	18	10	0	0.000265	0.000265
37	18	10	1	0.005039	0.004774
37	18	10	2	0.038237	0.033199
37	18	10	3	0.156277	0.118039
37	18	10	4	0.394625	0.238348
37	18	10	5	0.680643	0.286018
37	18	10	6	0.887211	0.206569
37	18	10	7	0.975741	0.088529
37	18	10	8	0.997222	0.021481
37	18	10	9	0.999874	0.002652
37	18	10	10	1.000000	0.000126
37	18	11	0	0.000088	0.000088
37	18	11	1	0.002033	0.001945
37	18	11	2	0.018564	0.016531
37	18	11	3	0.090699	0.072135
37	18	11	4	0.271037	0.180338
37	18	11	5	0.542931	0.271894
37	18	11	6	0.795403	0.252473
37	18	11	7	0.939673	0.144270
37	18	11	8	0.989266	0.049593
37	18	11	9	0.998990	0.009724
37	18	11	10	0.999963	0.000972
37	18	11	11	1.000000	0.000037
37	18	12	0	0.000027	0.000027
37	18	12	1	0.000762	0.000734
37	18	12	2	0.008391	0.007630
37	18	12	3	0.049083	0.040692
37	18	12	4	0.173932	0.124849
37	18	12	5	0.406984	0.233052
37	18	12	6	0.678877	0.271894
37	18	12	7	0.878636	0.199759
37	18	12	8	0.970192	0.091556
37	18	12	9	0.995624	0.025432
37	18	12	10	0.999663	0.004039
37	18	12	11	0.999990	0.000326
37	18	12	12	1.000000	0.000010
37	18	13	0	0.000008	0.000008
37	18	13	1	0.000262	0.000255
37	18	13	2	0.003508	0.003246
37	18	13	3	0.024668	0.021160
37	18	13	4	0.104016	0.079349
37	18	13	5	0.285797	0.181780
37	18	13	6	0.548368	0.262572
37	18	13	7	0.790742	0.242374
37	18	13	8	0.933570	0.142827
37	18	13	9	0.986469	0.052899
37	18	13	10	0.998371	0.011902
37	18	13	11	0.999898	0.001528
37	18	13	12	0.999997	0.000099
37	18	13	13	1.000000	0.000002
37	18	14	0	0.000002	0.000002
37	18	14	1	0.000082	0.000080
37	18	14	2	0.001344	0.001262
37	18	14	3	0.011443	0.010099
37	18	14	4	0.057730	0.046287
37	18	14	5	0.187332	0.129603
37	18	14	6	0.417083	0.229750
37	18	14	7	0.679654	0.262572
37	18	14	8	0.874058	0.194404
37	18	14	9	0.966631	0.092573
37	18	14	10	0.994403	0.027772
37	18	14	11	0.999453	0.005049
37	18	14	12	0.999973	0.000520
37	18	14	13	0.999999	0.000027
37	18	14	14	1.000000	0.000000
37	18	15	0	0.000000	0.000000
37	18	15	1	0.000023	0.000022
37	18	15	2	0.000466	0.000443
37	18	15	3	0.004857	0.004391
37	18	15	4	0.029555	0.024698
37	18	15	5	0.114079	0.084523
37	18	15	6	0.297213	0.183134
37	18	15	7	0.554076	0.256864
37	18	15	8	0.789535	0.235458
37	18	15	9	0.930407	0.140872
37	18	15	10	0.984744	0.054337
37	18	15	11	0.997916	0.013172
37	18	15	12	0.999837	0.001921
37	18	15	13	0.999994	0.000156

N	n	k	x	P(x)	p(x)
37	18	15	14	1.000000	0.000006
37	18	15	15	1.000000	0.000000
37	18	16	0	0.000000	0.000000
37	18	16	1	0.000005	0.000005
37	18	16	2	0.000144	0.000138
37	18	16	3	0.001863	0.001719
37	18	16	4	0.013838	0.011975
37	18	16	5	0.064133	0.050295
37	18	16	6	0.197322	0.133188
37	18	16	7	0.425645	0.228323
37	18	16	8	0.682508	0.256864
37	18	16	9	0.872777	0.190269
37	18	16	10	0.964985	0.092207
37	18	16	11	0.993725	0.028740
37	18	16	12	0.999313	0.005588
37	18	16	13	0.999958	0.000645
37	18	16	14	0.999999	0.000041
37	18	16	15	1.000000	0.000001
37	18	16	16	1.000000	0.000000
37	18	17	0	0.000000	0.000000
37	18	17	1	0.000001	0.000001
37	18	17	2	0.000038	0.000037
37	18	17	3	0.000635	0.000597
37	18	17	4	0.005855	0.005220
37	18	17	5	0.032998	0.027143
37	18	17	6	0.121214	0.088216
37	18	17	7	0.306047	0.184833
37	18	17	8	0.560192	0.254145
37	18	17	9	0.791234	0.231041
37	18	17	10	0.929858	0.138625
37	18	17	11	0.984145	0.054287
37	18	17	12	0.997717	0.013572
37	18	17	13	0.999804	0.002088
37	18	17	14	0.999991	0.000186
37	18	17	15	1.000000	0.000009
37	18	17	16	1.000000	0.000000
37	18	17	17	1.000000	0.000000
37	18	18	0	0.000000	0.000000
37	18	18	1	0.000000	0.000000
37	18	18	2	0.000009	0.000008
37	18	18	3	0.000188	0.000179
37	18	18	4	0.002201	0.002013
37	18	18	5	0.015355	0.013154
37	18	18	6	0.068284	0.052929
37	18	18	7	0.204389	0.136104
37	18	18	8	0.433120	0.228731
37	18	18	9	0.687265	0.254145
37	18	18	10	0.874408	0.187143
37	18	18	11	0.965145	0.090736
37	18	18	12	0.993645	0.028500

N	n	k	x	P(x)	p(x)
37	18	18	13	0.999283	0.005637
37	18	18	14	0.999954	0.000671
37	18	18	15	0.999998	0.000045
37	18	18	16	1.000000	0.000001
37	18	18	17	1.000000	0.000000
37	18	18	18	1.000000	0.000000
38	1	1	0	0.973684	0.973684
38	1	1	1	1.000000	0.026316
38	2	1	0	0.947368	0.947368
38	2	1	1	1.000000	0.052632
38	2	2	0	0.896159	0.896159
38	2	2	1	0.998578	0.102418
38	2	2	2	1.000000	0.001422
38	3	1	0	0.921053	0.921053
38	3	1	1	1.000000	0.078947
38	3	2	0	0.846373	0.846373
38	3	2	1	0.995733	0.149360
38	3	2	2	1.000000	0.004267
38	3	3	0	0.775842	0.775842
38	3	3	1	0.987435	0.211593
38	3	3	2	0.999881	0.012447
38	3	3	3	1.000000	0.000119
38	4	1	0	0.894737	0.894737
38	4	1	1	1.000000	0.105263
38	4	2	0	0.798009	0.798009
38	4	2	1	0.991465	0.193457
38	4	2	2	1.000000	0.008535
38	4	3	0	0.709341	0.709341
38	4	3	1	0.975344	0.266003
38	4	3	2	0.999526	0.024182
38	4	3	3	1.000000	0.000474
38	4	4	0	0.628273	0.628273
38	4	4	1	0.952544	0.324270
38	4	4	2	0.998144	0.045600
38	4	4	3	0.999986	0.001842
38	4	4	4	1.000000	0.000014
38	5	1	0	0.868421	0.868421
38	5	1	1	1.000000	0.131579
38	5	2	0	0.751067	0.751067
38	5	2	1	0.985775	0.234708
38	5	2	2	1.000000	0.014225
38	5	3	0	0.646752	0.646752
38	5	3	1	0.959696	0.312944
38	5	3	2	0.998815	0.039118
38	5	3	3	1.000000	0.001185
38	5	4	0	0.554359	0.554359
38	5	4	1	0.923931	0.369573
38	5	4	2	0.995462	0.071530
38	5	4	3	0.999932	0.004471
38	5	4	4	1.000000	0.000068

N	n	k	x	P(x)	p(x)
38	5	5	0	0.472835	0.472835
38	5	5	1	0.880452	0.407617
38	5	5	2	0.989150	0.108698
38	5	5	3	0.999669	0.010519
38	5	5	4	0.999998	0.000329
38	5	5	5	1.000000	0.000002
38	6	1	0	0.842105	0.842105
38	6	1	1	1.000000	0.157895
38	6	2	0	0.705548	0.705548
38	6	2	1	0.978663	0.273115
38	6	2	2	1.000000	0.021337
38	6	3	0	0.587956	0.587956
38	6	3	1	0.940730	0.352774
38	6	3	2	0.997629	0.056899
38	6	3	3	1.000000	0.002371
38	6	4	0	0.487164	0.487164
38	6	4	1	0.890334	0.403170
38	6	4	2	0.991126	0.100792
38	6	4	3	0.999797	0.008670
38	6	4	4	1.000000	0.000203
38	6	5	0	0.401194	0.401194
38	6	5	1	0.831044	0.429850
38	6	5	2	0.979268	0.148224
38	6	5	3	0.999032	0.019763
38	6	5	4	0.999988	0.000956
38	6	5	5	1.000000	0.000012
38	6	6	0	0.328249	0.328249
38	6	6	1	0.765915	0.437666
38	6	6	2	0.961302	0.195387
38	6	6	3	0.997235	0.035933
38	6	6	4	0.999930	0.002695
38	6	6	5	1.000000	0.000070
38	6	6	6	1.000000	0.000000
38	7	1	0	0.815789	0.815789
38	7	1	1	1.000000	0.184211
38	7	2	0	0.661451	0.661451
38	7	2	1	0.970128	0.308677
38	7	2	2	1.000000	0.029872
38	7	3	0	0.532835	0.532835
38	7	3	1	0.918682	0.385846
38	7	3	2	0.995851	0.077169
38	7	3	3	1.000000	0.004149
38	7	4	0	0.426268	0.426268
38	7	4	1	0.852537	0.426268
38	7	4	2	0.984827	0.132290
38	7	4	3	0.999526	0.014699
38	7	4	4	1.000000	0.000474
38	7	5	0	0.338507	0.338507
38	7	5	1	0.777313	0.438806
38	7	5	2	0.965372	0.188060

N	n	k	x	P(x)	p(x)
38	7	5	3	0.997796	0.032424
38	7	5	4	0.999958	0.002162
38	7	5	5	1.000000	0.000042
38	7	6	0	0.266703	0.266703
38	7	6	1	0.697530	0.430827
38	7	6	2	0.936879	0.239349
38	7	6	3	0.993866	0.056988
38	7	6	4	0.999762	0.005895
38	7	6	5	0.999997	0.000236
38	7	6	6	1.000000	0.000003
38	7	7	0	0.208361	0.208361
38	7	7	1	0.616750	0.408388
38	7	7	2	0.899480	0.282730
38	7	7	3	0.986743	0.087262
38	7	7	4	0.999209	0.012466
38	7	7	5	0.999983	0.000774
38	7	7	6	1.000000	0.000017
38	7	7	7	1.000000	0.000000
38	8	1	0	0.789474	0.789474
38	8	1	1	1.000000	0.210526
38	8	2	0	0.618777	0.618777
38	8	2	1	0.960171	0.341394
38	8	2	2	1.000000	0.039829
38	8	3	0	0.481271	0.481271
38	8	3	1	0.893788	0.412518
38	8	3	2	0.993362	0.099573
38	8	3	3	1.000000	0.006638
38	8	4	0	0.371266	0.371266
38	8	4	1	0.811285	0.440019
38	8	4	2	0.976292	0.165007
38	8	4	3	0.999052	0.022760
38	8	4	4	1.000000	0.000948
38	8	5	0	0.283909	0.283909
38	8	5	1	0.720693	0.436783
38	8	5	2	0.947173	0.226480
38	8	5	3	0.995705	0.048531
38	8	5	4	0.999888	0.004184
38	8	5	5	1.000000	0.000112
38	8	6	0	0.215083	0.215083
38	8	6	1	0.628042	0.412959
38	8	6	2	0.905995	0.277953
38	8	6	3	0.988351	0.082356
38	8	6	4	0.999381	0.011030
38	8	6	5	0.999990	0.000609
38	8	6	6	1.000000	0.000010
38	8	7	0	0.161312	0.161312
38	8	7	1	0.537707	0.376395
38	8	7	2	0.853879	0.316172
38	8	7	3	0.975483	0.121604
38	8	7	4	0.998003	0.022519

Table for $N = 2$, $n = 1$, through $N = 100$, $n = 50$

N	n	k	x	P(x)	p(x)
38	8	7	5	0.999933	0.001930
38	8	7	6	0.999999	0.000067
38	8	7	7	1.000000	0.000001
38	8	8	0	0.119683	0.119683
38	8	8	1	0.452715	0.333031
38	8	8	2	0.792684	0.339970
38	8	8	3	0.955869	0.163185
38	8	8	4	0.995097	0.039227
38	8	8	5	0.999746	0.004649
38	8	8	6	0.999995	0.000249
38	8	8	7	1.000000	0.000005
38	8	8	8	1.000000	0.000000
38	9	1	0	0.763158	0.763158
38	9	1	1	1.000000	0.236842
38	9	2	0	0.577525	0.577525
38	9	2	1	0.948791	0.371266
38	9	2	2	1.000000	0.051209
38	9	3	0	0.433144	0.433144
38	9	3	1	0.866287	0.433144
38	9	3	2	0.990043	0.123755
38	9	3	3	1.000000	0.009957
38	9	4	0	0.321764	0.321764
38	9	4	1	0.767283	0.445519
38	9	4	2	0.965292	0.198009
38	9	4	3	0.998293	0.033001
38	9	4	4	1.000000	0.001707
38	9	5	0	0.236591	0.236591
38	9	5	1	0.662455	0.425864
38	9	5	2	0.924525	0.262070
38	9	5	3	0.992469	0.067944
38	9	5	4	0.999749	0.007280
38	9	5	5	1.000000	0.000251
38	9	6	0	0.172066	0.172066
38	9	6	1	0.559215	0.387149
38	9	6	2	0.868934	0.309719
38	9	6	3	0.980116	0.111181
38	9	6	4	0.998646	0.018530
38	9	6	5	0.999969	0.001324
38	9	6	6	1.000000	0.000030
38	9	7	0	0.123673	0.123673
38	9	7	1	0.462428	0.338755
38	9	7	2	0.801183	0.338755
38	9	7	3	0.959269	0.158086
38	9	7	4	0.995751	0.036481
38	9	7	5	0.999804	0.004053
38	9	7	6	0.999997	0.000193
38	9	7	7	1.000000	0.000003
38	9	8	0	0.087768	0.087768
38	9	8	1	0.375007	0.287240
38	9	8	2	0.724690	0.349683

N	n	k	x	P(x)	p(x)
38	9	8	3	0.928672	0.203982
38	9	8	4	0.989866	0.061195
38	9	8	5	0.999281	0.009415
38	9	8	6	0.999978	0.000697
38	9	8	7	1.000000	0.000021
38	9	8	8	1.000000	0.000000
38	9	9	0	0.061437	0.061437
38	9	9	1	0.298410	0.236973
38	9	9	2	0.643097	0.344687
38	9	9	3	0.887876	0.244778
38	9	9	4	0.979667	0.091792
38	9	9	5	0.998026	0.018358
38	9	9	6	0.999909	0.001883
38	9	9	7	0.999998	0.000090
38	9	9	8	1.000000	0.000002
38	9	9	9	1.000000	0.000000
38	10	1	0	0.736842	0.736842
38	10	1	1	1.000000	0.263158
38	10	2	0	0.537696	0.537696
38	10	2	1	0.935989	0.398293
38	10	2	2	1.000000	0.064011
38	10	3	0	0.388336	0.388336
38	10	3	1	0.836415	0.448080
38	10	3	2	0.985775	0.149360
38	10	3	3	1.000000	0.014225
38	10	4	0	0.277383	0.277383
38	10	4	1	0.721195	0.443812
38	10	4	2	0.951636	0.230441
38	10	4	3	0.997155	0.045519
38	10	4	4	1.000000	0.002845
38	10	5	0	0.195799	0.195799
38	10	5	1	0.603715	0.407916
38	10	5	2	0.897414	0.293699
38	10	5	3	0.987783	0.090369
38	10	5	4	0.999498	0.011714
38	10	5	5	1.000000	0.000502
38	10	6	0	0.136466	0.136466
38	10	6	1	0.492465	0.355999
38	10	6	2	0.826215	0.333749
38	10	6	3	0.968614	0.142400
38	10	6	4	0.997368	0.028754
38	10	6	5	0.999924	0.002556
38	10	6	6	1.000000	0.000076
38	10	7	0	0.093821	0.093821
38	10	7	1	0.392341	0.298520
38	10	7	2	0.742777	0.350437
38	10	7	3	0.937464	0.194687
38	10	7	4	0.991977	0.054512
38	10	7	5	0.999525	0.007548
38	10	7	6	0.999990	0.000466

N	n	k	x	P(x)	p(x)
38	10	7	7	1.000000	0.000010
38	10	8	0	0.063556	0.063556
38	10	8	1	0.305674	0.242118
38	10	8	2	0.652342	0.346668
38	10	8	3	0.893503	0.241161
38	10	8	4	0.981426	0.087923
38	10	8	5	0.998307	0.016881
38	10	8	6	0.999930	0.001623
38	10	8	7	0.999999	0.000069
38	10	8	8	1.000000	0.000001
38	10	9	0	0.042371	0.042371
38	10	9	1	0.233038	0.190668
38	10	9	2	0.559897	0.326859
38	10	9	3	0.837232	0.277335
38	10	9	4	0.963841	0.126609
38	10	9	5	0.995494	0.031652
38	10	9	6	0.999714	0.004220
38	10	9	7	0.999992	0.000278
38	10	9	8	1.000000	0.000008
38	10	9	9	1.000000	0.000000
38	10	10	0	0.027760	0.027760
38	10	10	1	0.173866	0.146105
38	10	10	2	0.469729	0.295864
38	10	10	3	0.770289	0.300560
38	10	10	4	0.937646	0.167357
38	10	10	5	0.990036	0.052390
38	10	10	6	0.999132	0.009095
38	10	10	7	0.999963	0.000832
38	10	10	8	0.999999	0.000036
38	10	10	9	1.000000	0.000001
38	10	10	10	1.000000	0.000000
38	11	1	0	0.710526	0.710526
38	11	1	1	1.000000	0.289474
38	11	2	0	0.499289	0.499289
38	11	2	1	0.921764	0.422475
38	11	2	2	1.000000	0.078236
38	11	3	0	0.346728	0.346728
38	11	3	1	0.804410	0.457681
38	11	3	2	0.980441	0.176031
38	11	3	3	1.000000	0.019559
38	11	4	0	0.237757	0.237757
38	11	4	1	0.673644	0.435887
38	11	4	2	0.935176	0.261532
38	11	4	3	0.995529	0.060354
38	11	4	4	1.000000	0.004471
38	11	5	0	0.160835	0.160835
38	11	5	1	0.545441	0.384606
38	11	5	2	0.865947	0.320505
38	11	5	3	0.981328	0.115382
38	11	5	4	0.999079	0.017751

N	n	k	x	P(x)	p(x)
38	11	5	5	1.000000	0.000920
38	11	6	0	0.107224	0.107224
38	11	6	1	0.428894	0.321671
38	11	6	2	0.778536	0.349642
38	11	6	3	0.953357	0.174821
38	11	6	4	0.995314	0.041957
38	11	6	5	0.999833	0.004518
38	11	6	6	1.000000	0.000167
38	11	7	0	0.070365	0.070365
38	11	7	1	0.328372	0.258007
38	11	7	2	0.680199	0.351827
38	11	7	3	0.909652	0.229453
38	11	7	4	0.986136	0.076484
38	11	7	5	0.998985	0.012849
38	11	7	6	0.999974	0.000988
38	11	7	7	1.000000	0.000026
38	11	8	0	0.045397	0.045397
38	11	8	1	0.245144	0.199747
38	11	8	2	0.578056	0.332912
38	11	8	3	0.850438	0.272382
38	11	8	4	0.968865	0.118427
38	11	8	5	0.996498	0.027633
38	11	8	6	0.999814	0.003316
38	11	8	7	0.999997	0.000182
38	11	8	8	1.000000	0.000003
38	11	9	0	0.028751	0.028751
38	11	9	1	0.178562	0.149810
38	11	9	2	0.478182	0.299621
38	11	9	3	0.777803	0.299621
38	11	9	4	0.941232	0.163429
38	11	9	5	0.990972	0.049739
38	11	9	6	0.999262	0.008290
38	11	9	7	0.999972	0.000711
38	11	9	8	1.000000	0.000027
38	11	9	9	1.000000	0.000000
38	11	10	0	0.017846	0.017846
38	11	10	1	0.126903	0.109057
38	11	10	2	0.385197	0.258294
38	11	10	3	0.695149	0.309952
38	11	10	4	0.901784	0.206635
38	11	10	5	0.980681	0.078897
38	11	10	6	0.997832	0.017152
38	11	10	7	0.999874	0.002042
38	11	10	8	0.999997	0.000123
38	11	10	9	1.000000	0.000003
38	11	10	10	1.000000	0.000000
38	11	11	0	0.010835	0.010835
38	11	11	1	0.087954	0.077119
38	11	11	2	0.302174	0.214220
38	11	11	3	0.606591	0.304417

N	n	k	x	P(x)	p(x)
38	11	11	4	0.850125	0.243534
38	11	11	5	0.963774	0.113649
38	11	11	6	0.994770	0.030995
38	11	11	7	0.999583	0.004813
38	11	11	8	0.999984	0.000401
38	11	11	9	1.000000	0.000016
38	11	11	10	1.000000	0.000000
38	11	11	11	1.000000	0.000000
38	12	1	0	0.684211	0.684211
38	12	1	1	1.000000	0.315789
38	12	2	0	0.462304	0.462304
38	12	2	1	0.906117	0.443812
38	12	2	2	1.000000	0.093883
38	12	3	0	0.308203	0.308203
38	12	3	1	0.770507	0.462304
38	12	3	2	0.973921	0.203414
38	12	3	3	1.000000	0.026079
38	12	4	0	0.202533	0.202533
38	12	4	1	0.625212	0.422678
38	12	4	2	0.915803	0.290591
38	12	4	3	0.993294	0.077491
38	12	4	4	1.000000	0.006706
38	12	5	0	0.131051	0.131051
38	12	5	1	0.488463	0.357412
38	12	5	2	0.830335	0.341872
38	12	5	3	0.972782	0.142447
38	12	5	4	0.998422	0.025640
38	12	5	5	1.000000	0.001578
38	12	6	0	0.083396	0.083396
38	12	6	1	0.369325	0.285929
38	12	6	2	0.726737	0.357412
38	12	6	3	0.933933	0.207195
38	12	6	4	0.992206	0.058274
38	12	6	5	0.999665	0.007459
38	12	6	6	1.000000	0.000335
38	12	7	0	0.052123	0.052123
38	12	7	1	0.271037	0.218915
38	12	7	2	0.615046	0.344009
38	12	7	3	0.875659	0.260613
38	12	7	4	0.977638	0.101979
38	12	7	5	0.998034	0.020396
38	12	7	6	0.999937	0.001904
38	12	7	7	1.000000	0.000063
38	12	8	0	0.031946	0.031946
38	12	8	1	0.193358	0.161412
38	12	8	2	0.504076	0.310718
38	12	8	3	0.799997	0.295922
38	12	8	4	0.951321	0.151324
38	12	8	5	0.993428	0.042107
38	12	8	6	0.999569	0.006141
38	12	8	7	0.999990	0.000421
38	12	8	8	1.000000	0.000010
38	12	9	0	0.019168	0.019168
38	12	9	1	0.134174	0.115006
38	12	9	2	0.400503	0.266329
38	12	9	3	0.711221	0.310718
38	12	9	4	0.910968	0.199747
38	12	9	5	0.983603	0.072635
38	12	9	6	0.998341	0.014738
38	12	9	7	0.999920	0.001579
38	12	9	8	0.999999	0.000079
38	12	9	9	1.000000	0.000001
38	12	10	0	0.011236	0.011236
38	12	10	1	0.090551	0.079314
38	12	10	2	0.308665	0.218115
38	12	10	3	0.614791	0.306126
38	12	10	4	0.855865	0.241074
38	12	10	5	0.966070	0.110205
38	12	10	6	0.995291	0.029221
38	12	10	7	0.999647	0.004356
38	12	10	8	0.999988	0.000340
38	12	10	9	1.000000	0.000012
38	12	10	10	1.000000	0.000000
38	12	11	0	0.006421	0.006421
38	12	11	1	0.059391	0.052971
38	12	11	2	0.230767	0.171376
38	12	11	3	0.516393	0.285626
38	12	11	4	0.786987	0.270593
38	12	11	5	0.938519	0.151532
38	12	11	6	0.989030	0.050511
38	12	11	7	0.998870	0.009840
38	12	11	8	0.999939	0.001070
38	12	11	9	0.999998	0.000059
38	12	11	10	1.000000	0.000001
38	12	11	11	1.000000	0.000000
38	12	12	0	0.003567	0.003567
38	12	12	1	0.037811	0.034244
38	12	12	2	0.167295	0.129484
38	12	12	3	0.421185	0.253890
38	12	12	4	0.706811	0.285626
38	12	12	5	0.899233	0.192422
38	12	12	6	0.977805	0.078572
38	12	12	7	0.997047	0.019242
38	12	12	8	0.999781	0.002733
38	12	12	9	0.999992	0.000211
38	12	12	10	1.000000	0.000008
38	12	12	11	1.000000	0.000000
38	12	12	12	1.000000	0.000000
38	13	1	0	0.657895	0.657895
38	13	1	1	1.000000	0.342105
38	13	2	0	0.426743	0.426743
38	13	2	1	0.889047	0.462304
38	13	2	2	1.000000	0.110953
38	13	3	0	0.272641	0.272641
38	13	3	1	0.734945	0.462304
38	13	3	2	0.966098	0.231152
38	13	3	3	1.000000	0.033902
38	13	4	0	0.171374	0.171374
38	13	4	1	0.576441	0.405067
38	13	4	2	0.893450	0.317009
38	13	4	3	0.990314	0.096864
38	13	4	4	1.000000	0.009686
38	13	5	0	0.105849	0.105849
38	13	5	1	0.433476	0.327627
38	13	5	2	0.790888	0.357412
38	13	5	3	0.961824	0.170936
38	13	5	4	0.997436	0.035612
38	13	5	5	1.000000	0.002564
38	13	6	0	0.064151	0.064151
38	13	6	1	0.314339	0.250188
38	13	6	2	0.671751	0.357412
38	13	6	3	0.910025	0.238275
38	13	6	4	0.987724	0.077698
38	13	6	5	0.999378	0.011655
38	13	6	6	1.000000	0.000622
38	13	7	0	0.038090	0.038090
38	13	7	1	0.220518	0.182429
38	13	7	2	0.548891	0.328372
38	13	7	3	0.835565	0.286674
38	13	7	4	0.965871	0.130306
38	13	7	5	0.996465	0.030594
38	13	7	6	0.999864	0.003399
38	13	7	7	1.000000	0.000136
38	13	8	0	0.022117	0.022117
38	13	8	1	0.149901	0.127784
38	13	8	2	0.432371	0.282471
38	13	8	3	0.743089	0.310718
38	13	8	4	0.928040	0.184951
38	13	8	5	0.988570	0.060529
38	13	8	6	0.999096	0.010527
38	13	8	7	0.999974	0.000877
38	13	8	8	1.000000	0.000026
38	13	9	0	0.012533	0.012533
38	13	9	1	0.098787	0.086254
38	13	9	2	0.328799	0.230012
38	13	9	3	0.639517	0.310718
38	13	9	4	0.872555	0.233038
38	13	9	5	0.972428	0.099874
38	13	9	6	0.996640	0.024212
38	13	9	7	0.999798	0.003158
38	13	9	8	0.999995	0.000197
38	13	9	9	1.000000	0.000004
38	13	10	0	0.006915	0.006915
38	13	10	1	0.063096	0.056181
38	13	10	2	0.241553	0.178457
38	13	10	3	0.532373	0.290819
38	13	10	4	0.800233	0.267860
38	13	10	5	0.944877	0.144644
38	13	10	6	0.990796	0.045919
38	13	10	7	0.999145	0.008349
38	13	10	8	0.999962	0.000817
38	13	10	9	0.999999	0.000038
38	13	10	10	1.000000	0.000001
38	13	11	0	0.003704	0.003704
38	13	11	1	0.039018	0.035314
38	13	11	2	0.171445	0.132427
38	13	11	3	0.428508	0.257064
38	13	11	4	0.714135	0.285626
38	13	11	5	0.903550	0.189415
38	13	11	6	0.979316	0.075766
38	13	11	7	0.997356	0.018040
38	13	11	8	0.999816	0.002460
38	13	11	9	0.999994	0.000178
38	13	11	10	1.000000	0.000006
38	13	11	11	1.000000	0.000000
38	13	12	0	0.001921	0.001921
38	13	12	1	0.023323	0.021402
38	13	12	2	0.117493	0.094170
38	13	12	3	0.333300	0.215807
38	13	12	4	0.618926	0.285626
38	13	12	5	0.847427	0.228501
38	13	12	6	0.959673	0.112246
38	13	12	7	0.993347	0.033674
38	13	12	8	0.999360	0.006013
38	13	12	9	0.999967	0.000607
38	13	12	10	0.999999	0.000032
38	13	12	11	1.000000	0.000001
38	13	12	12	1.000000	0.000000
38	13	13	0	0.000960	0.000960
38	13	13	1	0.013445	0.012485
38	13	13	2	0.077652	0.064207
38	13	13	3	0.250297	0.172645
38	13	13	4	0.520055	0.269758
38	13	13	5	0.777119	0.257064
38	13	13	6	0.929453	0.152334
38	13	13	7	0.985576	0.056123
38	13	13	8	0.998204	0.012628
38	13	13	9	0.999874	0.001670
38	13	13	10	0.999996	0.000121
38	13	13	11	1.000000	0.000004

Table for $N = 2$, $n = 1$, through $N = 100$, $n = 50$

N	n	k	x	P(x)	p(x)
38	13	13	12	1.000000	0.000000
38	13	13	13	1.000000	0.000000
38	14	1	0	0.631579	0.631579
38	14	1	1	1.000000	0.368421
38	14	2	0	0.392603	0.392603
38	14	2	1	0.870555	0.477952
38	14	2	2	1.000000	0.129445
38	14	3	0	0.239924	0.239924
38	14	3	1	0.697961	0.458037
38	14	3	2	0.956851	0.258890
38	14	3	3	1.000000	0.043148
38	14	4	0	0.143954	0.143954
38	14	4	1	0.527833	0.383879
38	14	4	2	0.868089	0.340256
38	14	4	3	0.986439	0.118350
38	14	4	4	1.000000	0.013561
38	14	5	0	0.084679	0.084679
38	14	5	1	0.381056	0.296377
38	14	5	2	0.747999	0.366943
38	14	5	3	0.948149	0.200151
38	14	5	4	0.996011	0.047862
38	14	5	5	1.000000	0.003989
38	14	6	0	0.048755	0.048755
38	14	6	1	0.264301	0.215547
38	14	6	2	0.614565	0.350264
38	14	6	3	0.881432	0.266867
38	14	6	4	0.981508	0.100075
38	14	6	5	0.998912	0.017404
38	14	6	6	1.000000	0.001088
38	14	7	0	0.027424	0.027424
38	14	7	1	0.176736	0.149311
38	14	7	2	0.483216	0.306481
38	14	7	3	0.789697	0.306481
38	14	7	4	0.950234	0.160537
38	14	7	5	0.994017	0.043783
38	14	7	6	0.999728	0.005711
38	14	7	7	1.000000	0.000272
38	14	8	0	0.015039	0.015039
38	14	8	1	0.114121	0.099082
38	14	8	2	0.364578	0.250457
38	14	8	3	0.680946	0.316367
38	14	8	4	0.898448	0.217502
38	14	8	5	0.981306	0.082858
38	14	8	6	0.998254	0.016948
38	14	8	7	0.999939	0.001684
38	14	8	8	1.000000	0.000061
38	14	9	0	0.008021	0.008021
38	14	9	1	0.071186	0.063165
38	14	9	2	0.264396	0.193210
38	14	9	3	0.564944	0.300549

N	n	k	x	P(x)	p(x)
38	14	9	4	0.825947	0.261003
38	14	9	5	0.956449	0.130501
38	14	9	6	0.993735	0.037286
38	14	9	7	0.999545	0.005811
38	14	9	8	0.999988	0.000442
38	14	9	9	1.000000	0.000012
38	14	10	0	0.004149	0.004149
38	14	10	1	0.042870	0.038722
38	14	10	2	0.184447	0.141576
38	14	10	3	0.450943	0.266496
38	14	10	4	0.735946	0.285003
38	14	10	5	0.915948	0.180002
38	14	10	6	0.983449	0.067501
38	14	10	7	0.998143	0.014694
38	14	10	8	0.999896	0.001753
38	14	10	9	0.999998	0.000102
38	14	10	10	1.000000	0.000002
38	14	11	0	0.002074	0.002074
38	14	11	1	0.024893	0.022818
38	14	11	2	0.123771	0.098879
38	14	11	3	0.346248	0.222477
38	14	11	4	0.634159	0.287911
38	14	11	5	0.858090	0.223931
38	14	11	6	0.964163	0.106073
38	14	11	7	0.994469	0.030306
38	14	11	8	0.999520	0.005051
38	14	11	9	0.999980	0.000459
38	14	11	10	1.000000	0.000020
38	14	11	11	1.000000	0.000000
38	14	12	0	0.000999	0.000999
38	14	12	1	0.013906	0.012907
38	14	12	2	0.079825	0.065919
38	14	12	3	0.255609	0.175784
38	14	12	4	0.527526	0.271916
38	14	12	5	0.783447	0.255921
38	14	12	6	0.932734	0.149287
38	14	12	7	0.986612	0.053878
38	14	12	8	0.998398	0.011786
38	14	12	9	0.999895	0.001497
38	14	12	10	0.999997	0.000102
38	14	12	11	1.000000	0.000003
38	14	12	12	1.000000	0.000000
38	14	13	0	0.000461	0.000461
38	14	13	1	0.007452	0.006991
38	14	13	2	0.049401	0.041949
38	14	13	3	0.181239	0.131838
38	14	13	4	0.422942	0.241703
38	14	13	5	0.694859	0.271916
38	14	13	6	0.886799	0.191941
38	14	13	7	0.972107	0.085307

N	n	k	x	P(x)	p(x)
38	14	13	8	0.995678	0.023572
38	14	13	9	0.999607	0.003929
38	14	13	10	0.999981	0.000374
38	14	13	11	1.000000	0.000019
38	14	13	12	1.000000	0.000000
38	14	13	13	1.000000	0.000000
38	14	14	0	0.000203	0.000203
38	14	14	1	0.003817	0.003614
38	14	14	2	0.029266	0.025449
38	14	14	3	0.123230	0.093965
38	14	14	4	0.326261	0.203031
38	14	14	5	0.596969	0.270708
38	14	14	6	0.825378	0.228410
38	14	14	7	0.948221	0.122842
38	14	14	8	0.990021	0.041800
38	14	14	9	0.998821	0.008800
38	14	14	10	0.999921	0.001100
38	14	14	11	0.999997	0.000076
38	14	14	12	1.000000	0.000003
38	14	14	13	1.000000	0.000000
38	14	14	14	1.000000	0.000000
38	15	1	0	0.605263	0.605263
38	15	1	1	1.000000	0.394737
38	15	2	0	0.359886	0.359886
38	15	2	1	0.850640	0.490754
38	15	2	2	1.000000	0.149360
38	15	3	0	0.209934	0.209934
38	15	3	1	0.659791	0.449858
38	15	3	2	0.946064	0.286273
38	15	3	3	1.000000	0.053936
38	15	4	0	0.119962	0.119962
38	15	4	1	0.479848	0.359886
38	15	4	2	0.839734	0.359886
38	15	4	3	0.981508	0.141773
38	15	4	4	1.000000	0.018492
38	15	5	0	0.067038	0.067038
38	15	5	1	0.331660	0.264622
38	15	5	2	0.702131	0.370471
38	15	5	3	0.931470	0.229339
38	15	5	4	0.994017	0.062547
38	15	5	5	1.000000	0.005983
38	15	6	0	0.036566	0.036566
38	15	6	1	0.219396	0.182830
38	15	6	2	0.556188	0.336792
38	15	6	3	0.848074	0.291886
38	15	6	4	0.973168	0.125094
38	15	6	5	0.998187	0.025019
38	15	6	6	1.000000	0.001813
38	15	7	0	0.019426	0.019426
38	15	7	1	0.139408	0.119982

N	n	k	x	P(x)	p(x)
38	15	7	2	0.419366	0.279958
38	15	7	3	0.738617	0.319251
38	15	7	4	0.930167	0.191550
38	15	7	5	0.990368	0.060202
38	15	7	6	0.999490	0.009121
38	15	7	7	1.000000	0.000510
38	15	8	0	0.010026	0.010026
38	15	8	1	0.085222	0.075196
38	15	8	2	0.301964	0.216742
38	15	8	3	0.615036	0.313072
38	15	8	4	0.862198	0.247162
38	15	8	5	0.970949	0.108751
38	15	8	6	0.996842	0.025893
38	15	8	7	0.999868	0.003026
38	15	8	8	1.000000	0.000132
38	15	9	0	0.005013	0.005013
38	15	9	1	0.050131	0.045118
38	15	9	2	0.208043	0.157912
38	15	9	3	0.489807	0.281764
38	15	9	4	0.771572	0.281764
38	15	9	5	0.934698	0.163127
38	15	9	6	0.989074	0.054376
38	15	9	7	0.999061	0.009987
38	15	9	8	0.999969	0.000908
38	15	9	9	1.000000	0.000031
38	15	10	0	0.002420	0.002420
38	15	10	1	0.028350	0.025930
38	15	10	2	0.137255	0.108905
38	15	10	3	0.373215	0.235960
38	15	10	4	0.664695	0.291480
38	15	10	5	0.878448	0.213752
38	15	10	6	0.972199	0.093751
38	15	10	7	0.996306	0.024107
38	15	10	8	0.999750	0.003444
38	15	10	9	0.999994	0.000244
38	15	10	10	1.000000	0.000006
38	15	11	0	0.001124	0.001124
38	15	11	1	0.015385	0.014261
38	15	11	2	0.086692	0.071307
38	15	11	3	0.272089	0.185397
38	15	11	4	0.550185	0.278096
38	15	11	5	0.802108	0.251922
38	15	11	6	0.942064	0.139957
38	15	11	7	0.989418	0.047354
38	15	11	8	0.998889	0.009471
38	15	11	9	0.999941	0.001052
38	15	11	10	0.999999	0.000057
38	15	11	11	1.000000	0.000001
38	15	12	0	0.000499	0.000499
38	15	12	1	0.007990	0.007491

N	n	k	x	P(x)	p(x)
38	15	12	2	0.052359	0.044369
38	15	12	3	0.189690	0.137331
38	15	12	4	0.436887	0.247197
38	15	12	5	0.708803	0.271916
38	15	12	6	0.895412	0.186609
38	15	12	7	0.975387	0.079975
38	15	12	8	0.996434	0.021046
38	15	12	9	0.999707	0.003274
38	15	12	10	0.999988	0.000281
38	15	12	11	1.000000	0.000012
38	15	12	12	1.000000	0.000000
38	15	13	0	0.000211	0.000211
38	15	13	1	0.003957	0.003745
38	15	13	2	0.030174	0.026218
38	15	13	3	0.126306	0.096132
38	15	13	4	0.332304	0.205997
38	15	13	5	0.604220	0.271916
38	15	13	6	0.830817	0.226597
38	15	13	7	0.950780	0.119963
38	15	13	8	0.990767	0.039988
38	15	13	9	0.998952	0.008185
38	15	13	10	0.999934	0.000982
38	15	13	11	0.999998	0.000064
38	15	13	12	1.000000	0.000002
38	15	13	13	1.000000	0.000000
38	15	14	0	0.000085	0.000085
38	15	14	1	0.001859	0.001775
38	15	14	2	0.016541	0.014682
38	15	14	3	0.080163	0.063622
38	15	14	4	0.241665	0.161502
38	15	14	5	0.495453	0.253788
38	15	14	6	0.749242	0.253788
38	15	14	7	0.912392	0.163150
38	15	14	8	0.979571	0.067179
38	15	14	9	0.996988	0.017417
38	15	14	10	0.999738	0.002750
38	15	14	11	0.999988	0.000250
38	15	14	12	1.000000	0.000012
38	15	14	13	1.000000	0.000000
38	15	14	14	1.000000	0.000000
38	15	15	0	0.000032	0.000032
38	15	15	1	0.000824	0.000792
38	15	15	2	0.008588	0.007765
38	15	15	3	0.048352	0.039764
38	15	15	4	0.167643	0.119291
38	15	15	5	0.389708	0.222065
38	15	15	6	0.654071	0.264363
38	15	15	7	0.858008	0.203937
38	15	15	8	0.959977	0.101969
38	15	15	9	0.992633	0.032657

N	n	k	x	P(x)	p(x)
38	15	15	10	0.999165	0.006531
38	15	15	11	0.999946	0.000781
38	15	15	12	0.999998	0.000052
38	15	15	13	1.000000	0.000002
38	15	15	14	1.000000	0.000000
38	15	15	15	1.000000	0.000000
38	16	1	0	0.578947	0.578947
38	16	1	1	1.000000	0.421053
38	16	2	0	0.328592	0.328592
38	16	2	1	0.829303	0.500711
38	16	2	2	1.000000	0.170697
38	16	3	0	0.182551	0.182551
38	16	3	1	0.620673	0.438122
38	16	3	2	0.933618	0.312944
38	16	3	3	1.000000	0.066382
38	16	4	0	0.099099	0.099099
38	16	4	1	0.432907	0.333807
38	16	4	2	0.808440	0.375533
38	16	4	3	0.975344	0.166904
38	16	4	4	1.000000	0.024656
38	16	5	0	0.052464	0.052464
38	16	5	1	0.285639	0.233174
38	16	5	2	0.653809	0.368170
38	16	5	3	0.911528	0.257719
38	16	5	4	0.991298	0.079770
38	16	5	5	1.000000	0.008702
38	16	6	0	0.027027	0.027027
38	16	6	1	0.179650	0.152623
38	16	6	2	0.497615	0.317965
38	16	6	3	0.810002	0.312387
38	16	6	4	0.962290	0.152288
38	16	6	5	0.997099	0.034809
38	16	6	6	1.000000	0.002901
38	16	7	0	0.013514	0.013514
38	16	7	1	0.108108	0.094595
38	16	7	2	0.358506	0.250397
38	16	7	3	0.683095	0.324589
38	16	7	4	0.905182	0.222087
38	16	7	5	0.985134	0.079951
38	16	7	6	0.999093	0.013960
38	16	7	7	1.000000	0.000906
38	16	8	0	0.006539	0.006539
38	16	8	1	0.062337	0.055798
38	16	8	2	0.245423	0.183086
38	16	8	3	0.546977	0.301554
38	16	8	4	0.819213	0.272236
38	16	8	5	0.956764	0.137551
38	16	8	6	0.994590	0.037826
38	16	8	7	0.999737	0.005146
38	16	8	8	1.000000	0.000263

N	n	k	x	P(x)	p(x)
38	16	9	0	0.003051	0.003051
38	16	9	1	0.034438	0.031386
38	16	9	2	0.159983	0.125545
38	16	9	3	0.416303	0.256321
38	16	9	4	0.710318	0.294015
38	16	9	5	0.906328	0.196010
38	16	9	6	0.981981	0.075653
38	16	9	7	0.998193	0.016211
38	16	9	8	0.999930	0.001737
38	16	9	9	1.000000	0.000070
38	16	10	0	0.001368	0.001368
38	16	10	1	0.018203	0.016836
38	16	10	2	0.099375	0.081171
38	16	10	3	0.301401	0.202026
38	16	10	4	0.588657	0.287256
38	16	10	5	0.831980	0.243323
38	16	10	6	0.955894	0.123914
38	16	10	7	0.993162	0.037267
38	16	10	8	0.999451	0.006289
38	16	10	9	0.999983	0.000532
38	16	10	10	1.000000	0.000017
38	16	11	0	0.000586	0.000586
38	16	11	1	0.009184	0.008598
38	16	11	2	0.058789	0.049605
38	16	11	3	0.207603	0.148814
38	16	11	4	0.465547	0.257944
38	16	11	5	0.736389	0.270841
38	16	11	6	0.911639	0.175250
38	16	11	7	0.981183	0.069544
38	16	11	8	0.997654	0.016471
38	16	11	9	0.999850	0.002196
38	16	11	10	0.999996	0.000146
38	16	11	11	1.000000	0.000004
38	16	12	0	0.000239	0.000239
38	16	12	1	0.004408	0.004169
38	16	12	2	0.033068	0.028660
38	16	12	3	0.135952	0.102884
38	16	12	4	0.350905	0.214954
38	16	12	5	0.626046	0.275141
38	16	12	6	0.846732	0.220686
38	16	12	7	0.958002	0.111270
38	16	12	8	0.992774	0.034772
38	16	12	9	0.999281	0.006507
38	16	12	10	0.999964	0.000683
38	16	12	11	0.999999	0.000035
38	16	12	12	1.000000	0.000001
38	16	13	0	0.000092	0.000092
38	16	13	1	0.002003	0.001911
38	16	13	2	0.017636	0.015633
38	16	13	3	0.084510	0.066874

N	n	k	x	P(x)	p(x)
38	16	13	4	0.251696	0.167186
38	16	13	5	0.509640	0.257944
38	16	13	6	0.761852	0.252212
38	16	13	7	0.919485	0.157633
38	16	13	8	0.982074	0.062589
38	16	13	9	0.997529	0.015454
38	16	13	10	0.999806	0.002277
38	16	13	11	0.999992	0.000186
38	16	13	12	1.000000	0.000007
38	16	13	13	1.000000	0.000000
38	16	14	0	0.000033	0.000033
38	16	14	1	0.000856	0.000823
38	16	14	2	0.008881	0.008025
38	16	14	3	0.049735	0.040854
38	16	14	4	0.171447	0.121711
38	16	14	5	0.396145	0.224698
38	16	14	6	0.660968	0.264823
38	16	14	7	0.862737	0.201770
38	16	14	8	0.962046	0.099309
38	16	14	9	0.993201	0.031156
38	16	14	10	0.999259	0.006058
38	16	14	11	0.999955	0.000696
38	16	14	12	0.999999	0.000043
38	16	14	13	1.000000	0.000001
38	16	14	14	1.000000	0.000000
38	16	15	0	0.000011	0.000011
38	16	15	1	0.000342	0.000331
38	16	15	2	0.004200	0.003858
38	16	15	3	0.027606	0.023406
38	16	15	4	0.110591	0.082985
38	16	15	5	0.293158	0.182567
38	16	15	6	0.550625	0.257467
38	16	15	7	0.787074	0.236449
38	16	15	8	0.928943	0.141869
38	16	15	9	0.984114	0.055171
38	16	15	10	0.997745	0.013631
38	16	15	11	0.999810	0.002065
38	16	15	12	0.999991	0.000181
38	16	15	13	1.000000	0.000008
38	16	15	14	1.000000	0.000000
38	16	15	15	1.000000	0.000000
38	16	16	0	0.000003	0.000003
38	16	16	1	0.000126	0.000123
38	16	16	2	0.001851	0.001725
38	16	16	3	0.014376	0.012525
38	16	16	4	0.067294	0.052918
38	16	16	5	0.205843	0.138549
38	16	16	6	0.438683	0.232839
38	16	16	7	0.694550	0.255867
38	16	16	8	0.879597	0.185047

Table for $N = 2$, $n = 1$, through $N = 100$, $n = 50$

N	n	k	x	P(x)	p(x)
38	16	16	9	0.967323	0.087726
38	16	16	10	0.994189	0.026866
38	16	16	11	0.999361	0.005172
38	16	16	12	0.999960	0.000599
38	16	16	13	0.999999	0.000039
38	16	16	14	1.000000	0.000001
38	16	16	15	1.000000	0.000000
38	16	16	16	1.000000	0.000000
38	17	1	0	0.552632	0.552632
38	17	1	1	1.000000	0.447368
38	17	2	0	0.298720	0.298720
38	17	2	1	0.806543	0.507824
38	17	2	2	1.000000	0.193457
38	17	3	0	0.157658	0.157658
38	17	3	1	0.580844	0.423186
38	17	3	2	0.919393	0.338549
38	17	3	3	1.000000	0.080607
38	17	4	0	0.081081	0.081081
38	17	4	1	0.387387	0.306306
38	17	4	2	0.774301	0.386913
38	17	4	3	0.967757	0.193457
38	17	4	4	1.000000	0.032243
38	17	5	0	0.040541	0.040541
38	17	5	1	0.243243	0.202703
38	17	5	2	0.603604	0.360360
38	17	5	3	0.888099	0.284495
38	17	5	4	0.987672	0.099573
38	17	5	5	1.000000	0.012328
38	17	6	0	0.019656	0.019656
38	17	6	1	0.144963	0.125307
38	17	6	2	0.439803	0.294840
38	17	6	3	0.767404	0.327600
38	17	6	4	0.948446	0.181042
38	17	6	5	0.995517	0.047071
38	17	6	6	1.000000	0.004483
38	17	7	0	0.009214	0.009214
38	17	7	1	0.082310	0.073096
38	17	7	2	0.301597	0.219287
38	17	7	3	0.624079	0.322482
38	17	7	4	0.874898	0.250819
38	17	7	5	0.977865	0.102968
38	17	7	6	0.998459	0.020594
38	17	7	7	1.000000	0.001541
38	17	8	0	0.004161	0.004161
38	17	8	1	0.044583	0.040422
38	17	8	2	0.195490	0.150907
38	17	8	3	0.478442	0.282952
38	17	8	4	0.769715	0.291274
38	17	8	5	0.938007	0.168291
38	17	8	6	0.991151	0.053145

N	n	k	x	P(x)	p(x)
38	17	8	7	0.999503	0.008351
38	17	8	8	1.000000	0.000497
38	17	9	0	0.001803	0.001803
38	17	9	1	0.023024	0.021221
38	17	9	2	0.120036	0.097012
38	17	9	3	0.346398	0.226361
38	17	9	4	0.643497	0.297099
38	17	9	5	0.870690	0.227193
38	17	9	6	0.971665	0.100975
38	17	9	7	0.996719	0.025054
38	17	9	8	0.999851	0.003132
38	17	9	9	1.000000	0.000149
38	17	10	0	0.000746	0.000746
38	17	10	1	0.011316	0.010570
38	17	10	2	0.069858	0.058542
38	17	10	3	0.237120	0.167262
38	17	10	4	0.510314	0.273195
38	17	10	5	0.776679	0.266365
38	17	10	6	0.933364	0.156685
38	17	10	7	0.988080	0.054715
38	17	10	8	0.998879	0.010799
38	17	10	9	0.999959	0.001080
38	17	10	10	1.000000	0.000041
38	17	11	0	0.000293	0.000293
38	17	11	1	0.005276	0.004983
38	17	11	2	0.038496	0.033220
38	17	11	3	0.153489	0.114993
38	17	11	4	0.383474	0.229985
38	17	11	5	0.662523	0.279049
38	17	11	6	0.871809	0.209287
38	17	11	7	0.968538	0.096729
38	17	11	8	0.995408	0.026869
38	17	11	9	0.999650	0.004243
38	17	11	10	0.999990	0.000339
38	17	11	11	1.000000	0.000010
38	17	12	0	0.000109	0.000109
38	17	12	1	0.002323	0.002215
38	17	12	2	0.020041	0.017717
38	17	12	3	0.093863	0.073822
38	17	12	4	0.272740	0.178877
38	17	12	5	0.538501	0.265761
38	17	12	6	0.786545	0.248043
38	17	12	7	0.932713	0.146168
38	17	12	8	0.986451	0.053738
38	17	12	9	0.998393	0.011942
38	17	12	10	0.999902	0.001508
38	17	12	11	0.999998	0.000096
38	17	12	12	1.000000	0.000002
38	17	13	0	0.000038	0.000038
38	17	13	1	0.000960	0.000923

N	n	k	x	P(x)	p(x)
38	17	13	2	0.009819	0.008859
38	17	13	3	0.054113	0.044293
38	17	13	4	0.183302	0.129189
38	17	13	5	0.415842	0.232541
38	17	13	6	0.681603	0.265761
38	17	13	7	0.876494	0.194891
38	17	13	8	0.967850	0.091355
38	17	13	9	0.994719	0.026869
38	17	13	10	0.999495	0.004777
38	17	13	11	0.999975	0.000480
38	17	13	12	0.999999	0.000024
38	17	13	13	1.000000	0.000000
38	17	14	0	0.000012	0.000012
38	17	14	1	0.000370	0.000358
38	17	14	2	0.004504	0.004134
38	17	14	3	0.029308	0.024804
38	17	14	4	0.116123	0.086815
38	17	14	5	0.304223	0.188100
38	17	14	6	0.564668	0.260445
38	17	14	7	0.798538	0.233869
38	17	14	8	0.934962	0.136424
38	17	14	9	0.986121	0.051159
38	17	14	10	0.998158	0.012037
38	17	14	11	0.999860	0.001702
38	17	14	12	0.999995	0.000134
38	17	14	13	1.000000	0.000005
38	17	14	14	1.000000	0.000000
38	17	15	0	0.000004	0.000004
38	17	15	1	0.000131	0.000128
38	17	15	2	0.001920	0.001789
38	17	15	3	0.014839	0.012919
38	17	15	4	0.069098	0.054259
38	17	15	5	0.210173	0.141075
38	17	15	6	0.445298	0.235124
38	17	15	7	0.701092	0.255795
38	17	15	8	0.883803	0.182710
38	17	15	9	0.969068	0.085265
38	17	15	10	0.994647	0.025579
38	17	15	11	0.999435	0.004788
38	17	15	12	0.999967	0.000532
38	17	15	13	0.999999	0.000032
38	17	15	14	1.000000	0.000001
38	17	15	15	1.000000	0.000000
38	17	16	0	0.000001	0.000001
38	17	16	1	0.000042	0.000041
38	17	16	2	0.000753	0.000711
38	17	16	3	0.006975	0.006222
38	17	16	4	0.038430	0.031455
38	17	16	5	0.136569	0.098139
38	17	16	6	0.332847	0.196278

N	n	k	x	P(x)	p(x)
38	17	16	7	0.589877	0.257030
38	17	16	8	0.812307	0.222430
38	17	16	9	0.939410	0.127103
38	17	16	10	0.986862	0.047452
38	17	16	11	0.998186	0.011324
38	17	16	12	0.999851	0.001665
38	17	16	13	0.999993	0.000142
38	17	16	14	1.000000	0.000006
38	17	16	15	1.000000	0.000000
38	17	16	16	1.000000	0.000000
38	17	17	0	0.000000	0.000000
38	17	17	1	0.000012	0.000012
38	17	17	2	0.000269	0.000256
38	17	17	3	0.003016	0.002747
38	17	17	4	0.019843	0.016827
38	17	17	5	0.083039	0.063195
38	17	17	6	0.234708	0.151669
38	17	17	7	0.473045	0.238337
38	17	17	8	0.721313	0.248268
38	17	17	9	0.893191	0.171878
38	17	17	10	0.971764	0.078573
38	17	17	11	0.995097	0.023334
38	17	17	12	0.999473	0.004375
38	17	17	13	0.999967	0.000495
38	17	17	14	0.999999	0.000031
38	17	17	15	1.000000	0.000001
38	17	17	16	1.000000	0.000000
38	17	17	17	1.000000	0.000000
38	18	1	0	0.526316	0.526316
38	18	1	1	1.000000	0.473684
38	18	2	0	0.270270	0.270270
38	18	2	1	0.782361	0.512091
38	18	2	2	1.000000	0.217639
38	18	3	0	0.135135	0.135135
38	18	3	1	0.540541	0.405405
38	18	3	2	0.903272	0.362731
38	18	3	3	1.000000	0.096728
38	18	4	0	0.065637	0.065637
38	18	4	1	0.343629	0.277992
38	18	4	2	0.737452	0.393822
38	18	4	3	0.958545	0.221093
38	18	4	4	1.000000	0.041455
38	18	5	0	0.030888	0.030888
38	18	5	1	0.204633	0.173745
38	18	5	2	0.552124	0.347490
38	18	5	3	0.861004	0.308880
38	18	5	4	0.982930	0.121926
38	18	5	5	1.000000	0.017070
38	18	6	0	0.014040	0.014040
38	18	6	1	0.115128	0.101088

Table for $N = 2$, $n = 1$, through $N = 100$, $n = 50$

N	n	k	x	P(x)	p(x)
38	18	6	2	0.383643	0.268515
38	18	6	3	0.720604	0.336960
38	18	6	4	0.931204	0.210600
38	18	6	5	0.993275	0.062072
38	18	6	6	1.000000	0.006724
38	18	7	0	0.006142	0.006142
38	18	7	1	0.061425	0.055283
38	18	7	2	0.249386	0.187961
38	18	7	3	0.562654	0.313268
38	18	7	4	0.839066	0.276413
38	18	7	5	0.968059	0.128993
38	18	7	6	0.997478	0.029419
38	18	7	7	1.000000	0.002522
38	18	8	0	0.002576	0.002576
38	18	8	1	0.031109	0.028533
38	18	8	2	0.152374	0.121265
38	18	8	3	0.411072	0.258699
38	18	8	4	0.714235	0.303162
38	18	8	5	0.913965	0.199731
38	18	8	6	0.986090	0.072125
38	18	8	7	0.999105	0.013015
38	18	8	8	1.000000	0.000895
38	18	9	0	0.001030	0.001030
38	18	9	1	0.014940	0.013910
38	18	9	2	0.087699	0.072759
38	18	9	3	0.281723	0.194024
38	18	9	4	0.572759	0.291036
38	18	9	5	0.827415	0.254656
38	18	9	6	0.957240	0.129825
38	18	9	7	0.994333	0.037093
38	18	9	8	0.999702	0.005369
38	18	9	9	1.000000	0.000298
38	18	10	0	0.000391	0.000391
38	18	10	1	0.006786	0.006395
38	18	10	2	0.047556	0.040770
38	18	10	3	0.181366	0.133810
38	18	10	4	0.432259	0.250893
38	18	10	5	0.713259	0.281000
38	18	10	6	0.903520	0.190261
38	18	10	7	0.980263	0.076744
38	18	10	8	0.997850	0.017587
38	18	10	9	0.999907	0.002057
38	18	10	10	1.000000	0.000093
38	18	11	0	0.000140	0.000140
38	18	11	1	0.002903	0.002764
38	18	11	2	0.024259	0.021356
38	18	11	3	0.109682	0.085423
38	18	11	4	0.306812	0.197130
38	18	11	5	0.582795	0.275982
38	18	11	6	0.821979	0.239185

N	n	k	x	P(x)	p(x)
38	18	11	7	0.950114	0.128135
38	18	11	8	0.991569	0.041455
38	18	11	9	0.999246	0.007677
38	18	11	10	0.999973	0.000727
38	18	11	11	1.000000	0.000026
38	18	12	0	0.000047	0.000047
38	18	12	1	0.001163	0.001117
38	18	12	2	0.011604	0.010441
38	18	12	3	0.062225	0.050621
38	18	12	4	0.204597	0.142372
38	18	12	5	0.449914	0.245318
38	18	12	6	0.715675	0.265761
38	18	12	7	0.897911	0.182236
38	18	12	8	0.976215	0.078305
38	18	12	9	0.996687	0.020472
38	18	12	10	0.999758	0.003071
38	18	12	11	0.999993	0.000235
38	18	12	12	1.000000	0.000007
38	18	13	0	0.000014	0.000014
38	18	13	1	0.000433	0.000419
38	18	13	2	0.005179	0.004746
38	18	13	3	0.033020	0.027842
38	18	13	4	0.127935	0.094915
38	18	13	5	0.327255	0.199321
38	18	13	6	0.593016	0.265761
38	18	13	7	0.820811	0.227795
38	18	13	8	0.946098	0.125287
38	18	13	9	0.989601	0.043502
38	18	13	10	0.998813	0.009212
38	18	13	11	0.999930	0.001117
38	18	13	12	0.999998	0.000069
38	18	13	13	1.000000	0.000002
38	18	14	0	0.000004	0.000004
38	18	14	1	0.000148	0.000144
38	18	14	2	0.002142	0.001993
38	18	14	3	0.016315	0.014174
38	18	14	4	0.074783	0.058467
38	18	14	5	0.223609	0.148826
38	18	14	6	0.465451	0.241842
38	18	14	7	0.720581	0.255130
38	18	14	8	0.895983	0.175402
38	18	14	9	0.973940	0.077956
38	18	14	10	0.995865	0.021925
38	18	14	11	0.999617	0.003752
38	18	14	12	0.999982	0.000365
38	18	14	13	1.000000	0.000018
38	18	14	14	1.000000	0.000000
38	18	15	0	0.000001	0.000001
38	18	15	1	0.000046	0.000045
38	18	15	2	0.000813	0.000767

N	n	k	x	P(x)	p(x)
38	18	15	3	0.007457	0.006644
38	18	15	4	0.040677	0.033220
38	18	15	5	0.142995	0.102318
38	18	15	6	0.344530	0.201535
38	18	15	7	0.603647	0.259117
38	18	15	8	0.822899	0.219253
38	18	15	9	0.944706	0.121807
38	18	15	10	0.988557	0.043851
38	18	15	11	0.998523	0.009966
38	18	15	12	0.999891	0.001368
38	18	15	13	0.999996	0.000105
38	18	15	14	1.000000	0.000004
38	18	15	15	1.000000	0.000000
38	18	16	0	0.000000	0.000000
38	18	16	1	0.000013	0.000013
38	18	16	2	0.000279	0.000267
38	18	16	3	0.003124	0.002844
38	18	16	4	0.020456	0.017332
38	18	16	5	0.085163	0.064707
38	18	16	6	0.239381	0.154218
38	18	16	7	0.479721	0.240340
38	18	16	8	0.727572	0.247851
38	18	16	9	0.897043	0.169471
38	18	16	10	0.973304	0.076262
38	18	16	11	0.995490	0.022185
38	18	16	12	0.999534	0.004044
38	18	16	13	0.999973	0.000439
38	18	16	14	0.999999	0.000026
38	18	16	15	1.000000	0.000001
38	18	16	16	1.000000	0.000000
38	18	17	0	0.000000	0.000000
38	18	17	1	0.000003	0.000003
38	18	17	2	0.000085	0.000082
38	18	17	3	0.001184	0.001099
38	18	17	4	0.009426	0.008242
38	18	17	5	0.046927	0.037501
38	18	17	6	0.155262	0.108335
38	18	17	7	0.359551	0.204289
38	18	17	8	0.614913	0.255361
38	18	17	9	0.827714	0.212801
38	18	17	10	0.945573	0.117859
38	18	17	11	0.988431	0.042858
38	18	17	12	0.998431	0.010000
38	18	17	13	0.999873	0.001442
38	18	17	14	0.999994	0.000121
38	18	17	15	1.000000	0.000005
38	18	17	16	1.000000	0.000000
38	18	17	17	1.000000	0.000000
38	18	18	0	0.000000	0.000000
38	18	18	1	0.000001	0.000001

N	n	k	x	P(x)	p(x)
38	18	18	2	0.000023	0.000022
38	18	18	3	0.000399	0.000377
38	18	18	4	0.003932	0.003532
38	18	18	5	0.023712	0.019781
38	18	18	6	0.093356	0.069644
38	18	18	7	0.252543	0.159186
38	18	18	8	0.493312	0.240769
38	18	18	9	0.736513	0.243201
38	18	18	10	0.900674	0.164161
38	18	18	11	0.974145	0.073471
38	18	18	12	0.995574	0.021429
38	18	18	13	0.999530	0.003956
38	18	18	14	0.999971	0.000442
38	18	18	15	0.999999	0.000028
38	18	18	16	1.000000	0.000001
38	18	18	17	1.000000	0.000000
38	18	18	18	1.000000	0.000000
38	19	1	0	0.500000	0.500000
38	19	1	1	1.000000	0.500000
38	19	2	0	0.243243	0.243243
38	19	2	1	0.756757	0.513513
38	19	2	2	1.000000	0.243243
38	19	3	0	0.114865	0.114865
38	19	3	1	0.500000	0.385135
38	19	3	2	0.885135	0.385135
38	19	3	3	1.000000	0.114865
38	19	4	0	0.052510	0.052510
38	19	4	1	0.301931	0.249421
38	19	4	2	0.698069	0.396139
38	19	4	3	0.947490	0.249421
38	19	4	4	1.000000	0.052510
38	19	5	0	0.023166	0.023166
38	19	5	1	0.169884	0.146718
38	19	5	2	0.500000	0.330116
38	19	5	3	0.830116	0.330116
38	19	5	4	0.976834	0.146718
38	19	5	5	1.000000	0.023166
38	19	6	0	0.009828	0.009828
38	19	6	1	0.089856	0.080028
38	19	6	2	0.329940	0.240084
38	19	6	3	0.670060	0.340119
38	19	6	4	0.910144	0.240084
38	19	6	5	0.990172	0.080028
38	19	6	6	1.000000	0.009828
38	19	7	0	0.003993	0.003993
38	19	7	1	0.044840	0.040848
38	19	7	2	0.202396	0.157555
38	19	7	3	0.500000	0.297604
38	19	7	4	0.797604	0.297604
38	19	7	5	0.955160	0.157555

Table for $N = 2$, $n = 1$, through $N = 100$, $n = 50$

N	n	k	x	P(x)	p(x)
38	19	7	6	0.996007	0.040848
38	19	7	7	1.000000	0.003993
38	19	8	0	0.001546	0.001546
38	19	8	1	0.021122	0.019577
38	19	8	2	0.115994	0.094872
38	19	8	3	0.346398	0.230403
38	19	8	4	0.653602	0.307205
38	19	8	5	0.884006	0.230403
38	19	8	6	0.978878	0.094872
38	19	8	7	0.998454	0.019577
38	19	8	8	1.000000	0.001546
38	19	9	0	0.000567	0.000567
38	19	9	1	0.009376	0.008810
38	19	9	2	0.062233	0.052857
38	19	9	3	0.223516	0.161282
38	19	9	4	0.500000	0.276484
38	19	9	5	0.776484	0.276484
38	19	9	6	0.937766	0.161282
38	19	9	7	0.990624	0.052857
38	19	9	8	0.999433	0.008810
38	19	9	9	1.000000	0.000567
38	19	10	0	0.000195	0.000195
38	19	10	1	0.003908	0.003713
38	19	10	2	0.031248	0.027340
38	19	10	3	0.134532	0.103284
38	19	10	4	0.356991	0.222458
38	19	10	5	0.643009	0.286018
38	19	10	6	0.865467	0.222458
38	19	10	7	0.968752	0.103284
38	19	10	8	0.996092	0.027340
38	19	10	9	0.999804	0.003713
38	19	10	10	1.000000	0.000195
38	19	11	0	0.000063	0.000063
38	19	11	1	0.001521	0.001459
38	19	11	2	0.014649	0.013128
38	19	11	3	0.075513	0.060864
38	19	11	4	0.237817	0.162304
38	19	11	5	0.500000	0.262183
38	19	11	6	0.762183	0.262183
38	19	11	7	0.924487	0.162304
38	19	11	8	0.985351	0.060864
38	19	11	9	0.998478	0.013128
38	19	11	10	0.999937	0.001459
38	19	11	11	1.000000	0.000063
38	19	12	0	0.000019	0.000019
38	19	12	1	0.000549	0.000530
38	19	12	2	0.006383	0.005834
38	19	12	3	0.039445	0.033062
38	19	12	4	0.147648	0.108203
38	19	12	5	0.364053	0.216405

N	n	k	x	P(x)	p(x)
38	19	12	6	0.635947	0.271894
38	19	12	7	0.852352	0.216405
38	19	12	8	0.960555	0.108203
38	19	12	9	0.993616	0.033062
38	19	12	10	0.999451	0.005834
38	19	12	11	0.999981	0.000530
38	19	12	12	1.000000	0.000019
38	19	13	0	0.000005	0.000005
38	19	13	1	0.000182	0.000177
38	19	13	2	0.002569	0.002387
38	19	13	3	0.019100	0.016531
38	19	13	4	0.085223	0.066124
38	19	13	5	0.247527	0.162304
38	19	13	6	0.500000	0.252473
38	19	13	7	0.752473	0.252473
38	19	13	8	0.914777	0.162304
38	19	13	9	0.980900	0.066124
38	19	13	10	0.997431	0.016531
38	19	13	11	0.999818	0.002387
38	19	13	12	0.999995	0.000177
38	19	13	13	1.000000	0.000005
38	19	14	0	0.000001	0.000001
38	19	14	1	0.000055	0.000053
38	19	14	2	0.000946	0.000891
38	19	14	3	0.008520	0.007574
38	19	14	4	0.045549	0.037029
38	19	14	5	0.156637	0.111088
38	19	14	6	0.368714	0.212077
38	19	14	7	0.631286	0.262572
38	19	14	8	0.843363	0.212077
38	19	14	9	0.954451	0.111088
38	19	14	10	0.991480	0.037029
38	19	14	11	0.999054	0.007574
38	19	14	12	0.999945	0.000891
38	19	14	13	0.999999	0.000053
38	19	14	14	1.000000	0.000001
38	19	15	0	0.000000	0.000000
38	19	15	1	0.000015	0.000014
38	19	15	2	0.000314	0.000300
38	19	15	3	0.003470	0.003156
38	19	15	4	0.022406	0.018935
38	19	15	5	0.091836	0.069430
38	19	15	6	0.253839	0.162003
38	19	15	7	0.500000	0.246161
38	19	15	8	0.746161	0.246161
38	19	15	9	0.908164	0.162003
38	19	15	10	0.977594	0.069430
38	19	15	11	0.996530	0.018935
38	19	15	12	0.999685	0.003156
38	19	15	13	0.999985	0.000300

N	n	k	x	P(x)	p(x)
38	19	15	14	1.000000	0.000014
38	19	15	15	1.000000	0.000000
38	19	16	0	0.000000	0.000000
38	19	16	1	0.000003	0.000003
38	19	16	2	0.000093	0.000089
38	19	16	3	0.001275	0.001182
38	19	16	4	0.010057	0.008782
38	19	16	5	0.049574	0.039517
38	19	16	6	0.162272	0.112698
38	19	16	7	0.371568	0.209296
38	19	16	8	0.628432	0.256864
38	19	16	9	0.837728	0.209296
38	19	16	10	0.950426	0.112698
38	19	16	11	0.989943	0.039517
38	19	16	12	0.998725	0.008782
38	19	16	13	0.999907	0.001182
38	19	16	14	0.999996	0.000089
38	19	16	15	1.000000	0.000003
38	19	16	16	1.000000	0.000000
38	19	17	0	0.000000	0.000000
38	19	17	1	0.000001	0.000001
38	19	17	2	0.000024	0.000023
38	19	17	3	0.000415	0.000391
38	19	17	4	0.004069	0.003654
38	19	17	5	0.024427	0.020357
38	19	17	6	0.095678	0.071251
38	19	17	7	0.257407	0.161729
38	19	17	8	0.500000	0.242593
38	19	17	9	0.742593	0.242593
38	19	17	10	0.904322	0.161729
38	19	17	11	0.975573	0.071251
38	19	17	12	0.995931	0.020357
38	19	17	13	0.999585	0.003654
38	19	17	14	0.999976	0.000391
38	19	17	15	0.999999	0.000023
38	19	17	16	1.000000	0.000001
38	19	17	17	1.000000	0.000000
38	19	18	0	0.000000	0.000000
38	19	18	1	0.000000	0.000000
38	19	18	2	0.000005	0.000005
38	19	18	3	0.000117	0.000112
38	19	18	4	0.001459	0.001342
38	19	18	5	0.010855	0.009396
38	19	18	6	0.051570	0.040715
38	19	18	7	0.164990	0.113420
38	19	18	8	0.372927	0.207937
38	19	18	9	0.627073	0.254145
38	19	18	10	0.835010	0.207937
38	19	18	11	0.948430	0.113420
38	19	18	12	0.989145	0.040715

N	n	k	x	P(x)	p(x)
38	19	18	13	0.998541	0.009396
38	19	18	14	0.999883	0.001342
38	19	18	15	0.999995	0.000112
38	19	18	16	1.000000	0.000005
38	19	18	17	1.000000	0.000000
38	19	18	18	1.000000	0.000000
38	19	19	0	0.000000	0.000000
38	19	19	1	0.000000	0.000000
38	19	19	2	0.000001	0.000001
38	19	19	3	0.000027	0.000027
38	19	19	4	0.000452	0.000425
38	19	19	5	0.004278	0.003825
38	19	19	6	0.025105	0.020827
38	19	19	7	0.096938	0.071833
38	19	19	8	0.258562	0.161624
38	19	19	9	0.500000	0.241438
38	19	19	10	0.741438	0.241438
38	19	19	11	0.903062	0.161624
38	19	19	12	0.974895	0.071833
38	19	19	13	0.995722	0.020827
38	19	19	14	0.999547	0.003825
38	19	19	15	0.999972	0.000425
38	19	19	16	0.999999	0.000027
38	19	19	17	1.000000	0.000001
38	19	19	18	1.000000	0.000000
38	19	19	19	1.000000	0.000000
39	1	1	0	0.974359	0.974359
39	1	1	1	1.000000	0.025641
39	2	1	0	0.948718	0.948718
39	2	1	1	1.000000	0.051282
39	2	2	0	0.898785	0.898785
39	2	2	1	0.998650	0.099865
39	2	2	2	1.000000	0.001350
39	3	1	0	0.923077	0.923077
39	3	1	1	1.000000	0.076923
39	3	2	0	0.850202	0.850202
39	3	2	1	0.995951	0.145749
39	3	2	2	1.000000	0.004049
39	3	3	0	0.781267	0.781267
39	3	3	1	0.988073	0.206806
39	3	3	2	0.999891	0.011817
39	3	3	3	1.000000	0.000109
39	4	1	0	0.897436	0.897436
39	4	1	1	1.000000	0.102564
39	4	2	0	0.802969	0.802969
39	4	2	1	0.991903	0.188934
39	4	2	2	1.000000	0.008097
39	4	3	0	0.716162	0.716162
39	4	3	1	0.976584	0.260422
39	4	3	2	0.999562	0.022978

Table for $N = 2$, $n = 1$, through $N = 100$, $n = 50$

N	n	k	x	P(x)	p(x)
39	4	3	3	1.000000	0.000438
39	4	4	0	0.636588	0.636588
39	4	4	1	0.954882	0.318294
39	4	4	2	0.998286	0.043404
39	4	4	3	0.999988	0.001702
39	4	4	4	1.000000	0.000012
39	5	1	0	0.871795	0.871795
39	5	1	1	1.000000	0.128205
39	5	2	0	0.757085	0.757085
39	5	2	1	0.986505	0.229420
39	5	2	2	1.000000	0.013495
39	5	3	0	0.654776	0.654776
39	5	3	1	0.961703	0.306926
39	5	3	2	0.998906	0.037203
39	5	3	3	1.000000	0.001094
39	5	4	0	0.563835	0.563835
39	5	4	1	0.927600	0.363765
39	5	4	2	0.995805	0.068206
39	5	4	3	0.999939	0.004134
39	5	4	4	1.000000	0.000061
39	5	5	0	0.483287	0.483287
39	5	5	1	0.886027	0.402739
39	5	5	2	0.989959	0.103933
39	5	5	3	0.999703	0.009744
39	5	5	4	0.999998	0.000295
39	5	5	5	1.000000	0.000002
39	6	1	0	0.846154	0.846154
39	6	1	1	1.000000	0.153846
39	6	2	0	0.712551	0.712551
39	6	2	1	0.979757	0.267206
39	6	2	2	1.000000	0.020243
39	6	3	0	0.597002	0.597002
39	6	3	1	0.943648	0.346646
39	6	3	2	0.997812	0.054163
39	6	3	3	1.000000	0.002188
39	6	4	0	0.497502	0.497502
39	6	4	1	0.895503	0.398001
39	6	4	2	0.991793	0.096291
39	6	4	3	0.999818	0.008024
39	6	4	4	1.000000	0.000182
39	6	5	0	0.412216	0.412216
39	6	5	1	0.838645	0.426430
39	6	5	2	0.980789	0.142143
39	6	5	3	0.999130	0.018341
39	6	5	4	0.999990	0.000860
39	6	5	5	1.000000	0.000010
39	6	6	0	0.339472	0.339472
39	6	6	1	0.775935	0.436464
39	6	6	2	0.964066	0.188131
39	6	6	3	0.997511	0.033445

N	n	k	x	P(x)	p(x)
39	6	6	4	0.999939	0.002427
39	6	6	5	1.000000	0.000061
39	6	6	6	1.000000	0.000000
39	7	1	0	0.820513	0.820513
39	7	1	1	1.000000	0.179487
39	7	2	0	0.669366	0.669366
39	7	2	1	0.971660	0.302294
39	7	2	2	1.000000	0.028340
39	7	3	0	0.542729	0.542729
39	7	3	1	0.922639	0.379910
39	7	3	2	0.996170	0.073531
39	7	3	3	1.000000	0.003830
39	7	4	0	0.437198	0.437198
39	7	4	1	0.859321	0.422123
39	7	4	2	0.985958	0.126637
39	7	4	3	0.999574	0.013617
39	7	4	4	1.000000	0.000426
39	7	5	0	0.349759	0.349759
39	7	5	1	0.786957	0.437198
39	7	5	2	0.967867	0.180910
39	7	5	3	0.998018	0.030152
39	7	5	4	0.999963	0.001945
39	7	5	5	1.000000	0.000036
39	7	6	0	0.277749	0.277749
39	7	6	1	0.709804	0.432055
39	7	6	2	0.941262	0.231458
39	7	6	3	0.994471	0.053209
39	7	6	4	0.999792	0.005321
39	7	6	5	0.999998	0.000206
39	7	6	6	1.000000	0.000002
39	7	7	0	0.218833	0.218833
39	7	7	1	0.631249	0.412416
39	7	7	2	0.906193	0.274944
39	7	7	3	0.988021	0.081829
39	7	7	4	0.999308	0.011287
39	7	7	5	0.999985	0.000677
39	7	7	6	1.000000	0.000015
39	7	7	7	1.000000	0.000000
39	8	1	0	0.794872	0.794872
39	8	1	1	1.000000	0.205128
39	8	2	0	0.627530	0.627530
39	8	2	1	0.962213	0.334683
39	8	2	2	1.000000	0.037787
39	8	3	0	0.491848	0.491848
39	8	3	1	0.898895	0.407047
39	8	3	2	0.993872	0.094978
39	8	3	3	1.000000	0.006128
39	8	4	0	0.382549	0.382549
39	8	4	1	0.819747	0.437198
39	8	4	2	0.978043	0.158296

N	n	k	x	P(x)	p(x)
39	8	4	3	0.999149	0.021106
39	8	4	4	1.000000	0.000851
39	8	5	0	0.295109	0.295109
39	8	5	1	0.732307	0.437198
39	8	5	2	0.950906	0.218599
39	8	5	3	0.996134	0.045227
39	8	5	4	0.999903	0.003769
39	8	5	5	1.000000	0.000097
39	8	6	0	0.225671	0.225671
39	8	6	1	0.642296	0.416624
39	8	6	2	0.912330	0.270034
39	8	6	3	0.989483	0.077153
39	8	6	4	0.999459	0.009977
39	8	6	5	0.999991	0.000532
39	8	6	6	1.000000	0.000009
39	8	7	0	0.170963	0.170963
39	8	7	1	0.553921	0.382958
39	8	7	2	0.863233	0.309312
39	8	7	3	0.977793	0.114560
39	8	7	4	0.998250	0.020457
39	8	7	5	0.999943	0.001693
39	8	7	6	0.999999	0.000056
39	8	7	7	1.000000	0.000001
39	8	8	0	0.128222	0.128222
39	8	8	1	0.470149	0.341927
39	8	8	2	0.805237	0.335088
39	8	8	3	0.959893	0.154656
39	8	8	4	0.995693	0.035800
39	8	8	5	0.999784	0.004091
39	8	8	6	0.999996	0.000212
39	8	8	7	1.000000	0.000004
39	8	8	8	1.000000	0.000000
39	9	1	0	0.769231	0.769231
39	9	1	1	1.000000	0.230769
39	9	2	0	0.587045	0.587045
39	9	2	1	0.951417	0.364372
39	9	2	2	1.000000	0.048583
39	9	3	0	0.444250	0.444250
39	9	3	1	0.872634	0.428384
39	9	3	2	0.990809	0.118175
39	9	3	3	1.000000	0.009191
39	9	4	0	0.333187	0.333187
39	9	4	1	0.777437	0.444250
39	9	4	2	0.967830	0.190393
39	9	4	3	0.998468	0.030638
39	9	4	4	1.000000	0.001532
39	9	5	0	0.247511	0.247511
39	9	5	1	0.675894	0.428384
39	9	5	2	0.929752	0.253857
39	9	5	3	0.993216	0.063464

N	n	k	x	P(x)	p(x)
39	9	5	4	0.999781	0.006565
39	9	5	5	1.000000	0.000219
39	9	6	0	0.181993	0.181993
39	9	6	1	0.575098	0.393105
39	9	6	2	0.877487	0.302389
39	9	6	3	0.982016	0.104529
39	9	6	4	0.998816	0.016799
39	9	6	5	0.999974	0.001159
39	9	6	6	1.000000	0.000026
39	9	7	0	0.132359	0.132359
39	9	7	1	0.479800	0.347441
39	9	7	2	0.813344	0.333544
39	9	7	3	0.963011	0.149667
39	9	7	4	0.996270	0.033259
39	9	7	5	0.999834	0.003564
39	9	7	6	0.999998	0.000164
39	9	7	7	1.000000	0.000002
39	9	8	0	0.095133	0.095133
39	9	8	1	0.392940	0.297807
39	9	8	2	0.740381	0.347441
39	9	8	3	0.934948	0.194567
39	9	8	4	0.991073	0.056125
39	9	8	5	0.999388	0.008315
39	9	8	6	0.999982	0.000594
39	9	8	7	1.000000	0.000018
39	9	8	8	1.000000	0.000000
39	9	9	0	0.067514	0.067514
39	9	9	1	0.316086	0.248573
39	9	9	2	0.661927	0.345840
39	9	9	3	0.897290	0.235364
39	9	9	4	0.982021	0.084731
39	9	9	5	0.998315	0.016294
39	9	9	6	0.999925	0.001609
39	9	9	7	0.999999	0.000074
39	9	9	8	1.000000	0.000001
39	9	9	9	1.000000	0.000000
39	10	1	0	0.743590	0.743590
39	10	1	1	1.000000	0.256410
39	10	2	0	0.547908	0.547908
39	10	2	1	0.939271	0.391363
39	10	2	2	1.000000	0.060729
39	10	3	0	0.399825	0.399825
39	10	3	1	0.844075	0.444250
39	10	3	2	0.986869	0.142795
39	10	3	3	1.000000	0.013131
39	10	4	0	0.288762	0.288762
39	10	4	1	0.733012	0.444250
39	10	4	2	0.955137	0.222125
39	10	4	3	0.997447	0.042310
39	10	4	4	1.000000	0.002553

Table for $N = 2$, $n = 1$, through $N = 100$, $n = 50$

N	n	k	x	P(x)	p(x)
39	10	5	0	0.206259	0.206259
39	10	5	1	0.618777	0.412518
39	10	5	2	0.904366	0.285589
39	10	5	3	0.988985	0.084619
39	10	5	4	0.999562	0.010577
39	10	5	5	1.000000	0.000438
39	10	6	0	0.145595	0.145595
39	10	6	1	0.509581	0.363986
39	10	6	2	0.837168	0.327588
39	10	6	3	0.971563	0.134395
39	10	6	4	0.997696	0.026132
39	10	6	5	0.999936	0.002240
39	10	6	6	1.000000	0.000064
39	10	7	0	0.101475	0.101475
39	10	7	1	0.410312	0.308837
39	10	7	2	0.757753	0.347441
39	10	7	3	0.943055	0.185302
39	10	7	4	0.992944	0.049889
39	10	7	5	0.999596	0.006652
39	10	7	6	0.999992	0.000396
39	10	7	7	1.000000	0.000008
39	10	8	0	0.069764	0.069764
39	10	8	1	0.323451	0.253687
39	10	8	2	0.670893	0.347441
39	10	8	3	0.902520	0.231628
39	10	8	4	0.983590	0.081070
39	10	8	5	0.998557	0.014967
39	10	8	6	0.999943	0.001386
39	10	8	7	0.999999	0.000057
39	10	8	8	1.000000	0.000001
39	10	9	0	0.047260	0.047260
39	10	9	1	0.249800	0.202541
39	10	9	2	0.581231	0.331430
39	10	9	3	0.850217	0.268987
39	10	9	4	0.967899	0.117682
39	10	9	5	0.996143	0.028244
39	10	9	6	0.999764	0.003621
39	10	9	7	0.999994	0.000230
39	10	9	8	1.000000	0.000006
39	10	9	9	1.000000	0.000000
39	10	10	0	0.031506	0.031506
39	10	10	1	0.189038	0.157532
39	10	10	2	0.492849	0.303811
39	10	10	3	0.787454	0.294605
39	10	10	4	0.944363	0.156909
39	10	10	5	0.991436	0.047073
39	10	10	6	0.999281	0.007845
39	10	10	7	0.999971	0.000690
39	10	10	8	0.999999	0.000029
39	10	10	9	1.000000	0.000000
39	10	10	10	1.000000	0.000000
39	11	1	0	0.717949	0.717949
39	11	1	1	1.000000	0.282051
39	11	2	0	0.510121	0.510121
39	11	2	1	0.925776	0.415655
39	11	2	2	1.000000	0.074224
39	11	3	0	0.358464	0.358464
39	11	3	1	0.813437	0.454973
39	11	3	2	0.981945	0.168509
39	11	3	3	1.000000	0.018054
39	11	4	0	0.248933	0.248933
39	11	4	1	0.687055	0.438122
39	11	4	2	0.939818	0.252763
39	11	4	3	0.995988	0.056170
39	11	4	4	1.000000	0.004012
39	11	5	0	0.170697	0.170697
39	11	5	1	0.561878	0.391181
39	11	5	2	0.874822	0.312945
39	11	5	3	0.983149	0.108327
39	11	5	4	0.999197	0.016048
39	11	5	5	1.000000	0.000802
39	11	6	0	0.115472	0.115472
39	11	6	1	0.446824	0.331353
39	11	6	2	0.791984	0.345159
39	11	6	3	0.957660	0.165677
39	11	6	4	0.995893	0.038233
39	11	6	5	0.999858	0.003965
39	11	6	6	1.000000	0.000142
39	11	7	0	0.076981	0.076981
39	11	7	1	0.346414	0.269433
39	11	7	2	0.697849	0.351435
39	11	7	3	0.917496	0.219647
39	11	7	4	0.987783	0.070287
39	11	7	5	0.999137	0.011354
39	11	7	6	0.999978	0.000841
39	11	7	7	1.000000	0.000021
39	11	8	0	0.050519	0.050519
39	11	8	1	0.262217	0.211698
39	11	8	2	0.599008	0.336792
39	11	8	3	0.862585	0.263576
39	11	8	4	0.972408	0.109823
39	11	8	5	0.997009	0.024600
39	11	8	6	0.999847	0.002839
39	11	8	7	0.999997	0.000150
39	11	8	8	1.000000	0.000003
39	11	9	0	0.032593	0.032593
39	11	9	1	0.193927	0.161334
39	11	9	2	0.501230	0.307303
39	11	9	3	0.794565	0.293335
39	11	9	4	0.947609	0.153044
39	11	9	5	0.992247	0.044638
39	11	9	6	0.999389	0.007142
39	11	9	7	0.999978	0.000589
39	11	9	8	1.000000	0.000022
39	11	9	9	1.000000	0.000000
39	11	10	0	0.020642	0.020642
39	11	10	1	0.140149	0.119507
39	11	10	2	0.409039	0.268890
39	11	10	3	0.716342	0.307303
39	11	10	4	0.911899	0.195557
39	11	10	5	0.983320	0.071421
39	11	10	6	0.998199	0.014879
39	11	10	7	0.999899	0.001700
39	11	10	8	0.999998	0.000098
39	11	10	9	1.000000	0.000002
39	11	10	10	1.000000	0.000000
39	11	11	0	0.012812	0.012812
39	11	11	1	0.098940	0.086127
39	11	11	2	0.325590	0.226651
39	11	11	3	0.631569	0.305979
39	11	11	4	0.864696	0.233127
39	11	11	5	0.968543	0.103847
39	11	11	6	0.995633	0.027091
39	11	11	7	0.999665	0.004031
39	11	11	8	0.999987	0.000323
39	11	11	9	1.000000	0.000012
39	11	11	10	1.000000	0.000000
39	11	11	11	1.000000	0.000000
39	12	1	0	0.692308	0.692308
39	12	1	1	1.000000	0.307692
39	12	2	0	0.473684	0.473684
39	12	2	1	0.910931	0.437247
39	12	2	2	1.000000	0.089069
39	12	3	0	0.320057	0.320057
39	12	3	1	0.780939	0.460882
39	12	3	2	0.975927	0.194988
39	12	3	3	1.000000	0.024073
39	12	4	0	0.213371	0.213371
39	12	4	1	0.640114	0.426743
39	12	4	2	0.921764	0.281650
39	12	4	3	0.993982	0.072218
39	12	4	4	1.000000	0.006018
39	12	5	0	0.140215	0.140215
39	12	5	1	0.505995	0.365779
39	12	5	2	0.841292	0.335298
39	12	5	3	0.975411	0.134119
39	12	5	4	0.998624	0.023213
39	12	5	5	1.000000	0.001376
39	12	6	0	0.090728	0.090728
39	12	6	1	0.387654	0.296927
39	12	6	2	0.742675	0.355021
39	12	6	3	0.939909	0.197234
39	12	6	4	0.993162	0.053253
39	12	6	5	0.999717	0.006554
39	12	6	6	1.000000	0.000283
39	12	7	0	0.057736	0.057736
39	12	7	1	0.288679	0.230943
39	12	7	2	0.635093	0.346415
39	12	7	3	0.886118	0.251025
39	12	7	4	0.980253	0.094134
39	12	7	5	0.998326	0.018074
39	12	7	6	0.999948	0.001622
39	12	7	7	1.000000	0.000051
39	12	8	0	0.036085	0.036085
39	12	8	1	0.209292	0.173207
39	12	8	2	0.526839	0.317547
39	12	8	3	0.815517	0.288679
39	12	8	4	0.956719	0.141202
39	12	8	5	0.994373	0.037654
39	12	8	6	0.999644	0.005272
39	12	8	7	0.999992	0.000348
39	12	8	8	1.000000	0.000008
39	12	9	0	0.022117	0.022117
39	12	9	1	0.147831	0.125715
39	12	9	2	0.424404	0.276573
39	12	9	3	0.731707	0.307303
39	12	9	4	0.920280	0.188572
39	12	9	5	0.985870	0.065590
39	12	9	6	0.998624	0.012754
39	12	9	7	0.999936	0.001312
39	12	9	8	0.999999	0.000063
39	12	9	9	1.000000	0.000001
39	12	10	0	0.013270	0.013270
39	12	10	1	0.101736	0.088466
39	12	10	2	0.332213	0.230477
39	12	10	3	0.639517	0.307303
39	12	10	4	0.869994	0.230477
39	12	10	5	0.970566	0.100572
39	12	10	6	0.996073	0.025507
39	12	10	7	0.999717	0.003644
39	12	10	8	0.999990	0.000273
39	12	10	9	1.000000	0.000009
39	12	10	10	1.000000	0.000000
39	12	11	0	0.007779	0.007779
39	12	11	1	0.068180	0.060401
39	12	11	2	0.252738	0.184558
39	12	11	3	0.544147	0.291408
39	12	11	4	0.806414	0.262267
39	12	11	5	0.946290	0.139876
39	12	11	6	0.990796	0.044506

Table for $N = 2,\ n = 1,$ through $N = 100,\ n = 50$

N	n	k	x	P(x)	p(x)
39	12	11	7	0.999089	0.008293
39	12	11	8	0.999953	0.000864
39	12	11	9	0.999999	0.000046
39	12	11	10	1.000000	0.000001
39	12	11	11	1.000000	0.000000
39	12	12	0	0.004445	0.004445
39	12	12	1	0.044451	0.040006
39	12	12	2	0.186825	0.142374
39	12	12	3	0.450480	0.263655
39	12	12	4	0.731480	0.281001
39	12	12	5	0.911321	0.179840
39	12	12	6	0.981259	0.069938
39	12	12	7	0.997608	0.016349
39	12	12	8	0.999829	0.002221
39	12	12	9	0.999994	0.000165
39	12	12	10	1.000000	0.000006
39	12	12	11	1.000000	0.000000
39	12	12	12	1.000000	0.000000
39	13	1	0	0.666667	0.666667
39	13	1	1	1.000000	0.333333
39	13	2	0	0.438596	0.438596
39	13	2	1	0.894737	0.456140
39	13	2	2	1.000000	0.105263
39	13	3	0	0.284495	0.284495
39	13	3	1	0.746799	0.462304
39	13	3	2	0.968705	0.221906
39	13	3	3	1.000000	0.031294
39	13	4	0	0.181761	0.181761
39	13	4	1	0.592698	0.410937
39	13	4	2	0.900901	0.308203
39	13	4	3	0.991307	0.090406
39	13	4	4	1.000000	0.008693
39	13	5	0	0.114250	0.114250
39	13	5	1	0.451805	0.337556
39	13	5	2	0.804037	0.352232
39	13	5	3	0.965477	0.161440
39	13	5	4	0.997765	0.032288
39	13	5	5	1.000000	0.002235
39	13	6	0	0.070566	0.070566
39	13	6	1	0.332668	0.262102
39	13	6	2	0.690080	0.357412
39	13	6	3	0.917994	0.227915
39	13	6	4	0.989218	0.071223
39	13	6	5	0.999474	0.010256
39	13	6	6	1.000000	0.000526
39	13	7	0	0.042767	0.042767
39	13	7	1	0.237358	0.194591
39	13	7	2	0.570942	0.333584
39	13	7	3	0.848929	0.277987
39	13	7	4	0.969793	0.120864

N	n	k	x	P(x)	p(x)
39	13	7	5	0.996988	0.027194
39	13	7	6	0.999888	0.002901
39	13	7	7	1.000000	0.000112
39	13	8	0	0.025393	0.025393
39	13	8	1	0.164387	0.138993
39	13	8	2	0.456273	0.291886
39	13	8	3	0.762058	0.305786
39	13	8	4	0.935800	0.173742
39	13	8	5	0.990189	0.054389
39	13	8	6	0.999254	0.009065
39	13	8	7	0.999979	0.000725
39	13	8	8	1.000000	0.000021
39	13	9	0	0.014744	0.014744
39	13	9	1	0.110583	0.095838
39	13	9	2	0.352700	0.242118
39	13	9	3	0.663418	0.310718
39	13	9	4	0.885359	0.221941
39	13	9	5	0.976153	0.090794
39	13	9	6	0.997207	0.021054
39	13	9	7	0.999839	0.002632
39	13	9	8	0.999997	0.000158
39	13	9	9	1.000000	0.000003
39	13	10	0	0.008355	0.008355
39	13	10	1	0.072247	0.063892
39	13	10	2	0.263924	0.191676
39	13	10	3	0.559845	0.295922
39	13	10	4	0.818777	0.258931
39	13	10	5	0.951941	0.133165
39	13	10	6	0.992294	0.040353
39	13	10	7	0.999312	0.007018
39	13	10	8	0.999970	0.000658
39	13	10	9	0.999999	0.000029
39	13	10	10	1.000000	0.000000
39	13	11	0	0.004610	0.004610
39	13	11	1	0.045809	0.041199
39	13	11	2	0.191219	0.145410
39	13	11	3	0.457803	0.266585
39	13	11	4	0.738419	0.280615
39	13	11	5	0.915206	0.176788
39	13	11	6	0.982554	0.067348
39	13	11	7	0.997860	0.015306
39	13	11	8	0.999857	0.001996
39	13	11	9	0.999995	0.000139
39	13	11	10	1.000000	0.000004
39	13	11	11	1.000000	0.000000
39	13	12	0	0.002469	0.002469
39	13	12	1	0.028152	0.025683
39	13	12	2	0.134094	0.105941
39	13	12	3	0.362595	0.228501
39	13	12	4	0.648221	0.285626

N	n	k	x	P(x)	p(x)
39	13	12	5	0.864696	0.216475
39	13	12	6	0.965717	0.101022
39	13	12	7	0.994580	0.028863
39	13	12	8	0.999500	0.004920
39	13	12	9	0.999976	0.000475
39	13	12	10	0.999999	0.000024
39	13	12	11	1.000000	0.000001
39	13	12	12	1.000000	0.000000
39	13	13	0	0.001280	0.001280
39	13	13	1	0.016738	0.015457
39	13	13	2	0.090932	0.074195
39	13	13	3	0.277965	0.187032
39	13	13	4	0.553012	0.275048
39	13	13	5	0.800555	0.247543
39	13	13	6	0.939526	0.138971
39	13	13	7	0.988166	0.048640
39	13	13	8	0.998589	0.010423
39	13	13	9	0.999905	0.001316
39	13	13	10	0.999997	0.000092
39	13	13	11	1.000000	0.000003
39	13	13	12	1.000000	0.000000
39	13	13	13	1.000000	0.000000
39	14	1	0	0.641026	0.641026
39	14	1	1	1.000000	0.358974
39	14	2	0	0.404858	0.404858
39	14	2	1	0.877193	0.472335
39	14	2	2	1.000000	0.122807
39	14	3	0	0.251669	0.251669
39	14	3	1	0.711238	0.459569
39	14	3	2	0.960171	0.248933
39	14	3	3	1.000000	0.039829
39	14	4	0	0.153798	0.153798
39	14	4	1	0.545282	0.391485
39	14	4	2	0.877193	0.331911
39	14	4	3	0.987830	0.110637
39	14	4	4	1.000000	0.012170
39	14	5	0	0.092278	0.092278
39	14	5	1	0.399874	0.307595
39	14	5	2	0.763395	0.363521
39	14	5	3	0.953058	0.189663
39	14	5	4	0.996523	0.043465
39	14	5	5	1.000000	0.003477
39	14	6	0	0.054281	0.054281
39	14	6	1	0.282264	0.227982
39	14	6	2	0.635093	0.352830
39	14	6	3	0.891697	0.256603
39	14	6	4	0.983739	0.092042
39	14	6	5	0.999079	0.015340
39	14	6	6	1.000000	0.000920
39	14	7	0	0.031253	0.031253

N	n	k	x	P(x)	p(x)
39	14	7	1	0.192453	0.161200
39	14	7	2	0.506792	0.314339
39	14	7	3	0.806162	0.299371
39	14	7	4	0.955847	0.149685
39	14	7	5	0.994896	0.039048
39	14	7	6	0.999777	0.004881
39	14	7	7	1.000000	0.000223
39	14	8	0	0.017580	0.017580
39	14	8	1	0.126965	0.109385
39	14	8	2	0.388914	0.261949
39	14	8	3	0.703254	0.314339
39	14	8	4	0.909071	0.205817
39	14	8	5	0.983913	0.074843
39	14	8	6	0.998557	0.014643
39	14	8	7	0.999951	0.001395
39	14	8	8	1.000000	0.000049
39	14	9	0	0.009641	0.009641
39	14	9	1	0.081094	0.071453
39	14	9	2	0.287515	0.206421
39	14	9	3	0.591714	0.304199
39	14	9	4	0.842678	0.250964
39	14	9	5	0.962185	0.119507
39	14	9	6	0.994778	0.032593
39	14	9	7	0.999636	0.004859
39	14	9	8	0.999991	0.000354
39	14	9	9	1.000000	0.000009
39	14	10	0	0.005142	0.005142
39	14	10	1	0.050131	0.044989
39	14	10	2	0.204946	0.154816
39	14	10	3	0.480174	0.275228
39	14	10	4	0.759023	0.278849
39	14	10	5	0.926333	0.167310
39	14	10	6	0.986086	0.059753
39	14	10	7	0.998503	0.012416
39	14	10	8	0.999920	0.001417
39	14	10	9	0.999998	0.000079
39	14	10	10	1.000000	0.000002
39	14	11	0	0.002659	0.002659
39	14	11	1	0.029963	0.027304
39	14	11	2	0.140885	0.110922
39	14	11	3	0.375777	0.234893
39	14	11	4	0.662868	0.287091
39	14	11	5	0.874409	0.211541
39	14	11	6	0.969602	0.095193
39	14	11	7	0.995505	0.025903
39	14	11	8	0.999626	0.004121
39	14	11	9	0.999985	0.000358
39	14	11	10	1.000000	0.000015
39	14	11	11	1.000000	0.000000
39	14	12	0	0.001330	0.001330

N	n	k	x	P(x)	p(x)
39	14	12	1	0.017286	0.015957
39	14	12	2	0.093347	0.076060
39	14	12	3	0.283498	0.190151
39	14	12	4	0.560336	0.276838
39	14	12	5	0.806414	0.246078
39	14	12	6	0.942404	0.135990
39	14	12	7	0.989030	0.046625
39	14	12	8	0.998743	0.009714
39	14	12	9	0.999921	0.001177
39	14	12	10	0.999998	0.000077
39	14	12	11	1.000000	0.000002
39	14	12	12	1.000000	0.000000
39	14	13	0	0.000640	0.000640
39	14	13	1	0.009604	0.008963
39	14	13	2	0.059542	0.049939
39	14	13	3	0.206029	0.146487
39	14	13	4	0.457803	0.251774
39	14	13	5	0.724388	0.266585
39	14	13	6	0.902111	0.177723
39	14	13	7	0.976942	0.074831
39	14	13	8	0.996585	0.019643
39	14	13	9	0.999703	0.003118
39	14	13	10	0.999986	0.000283
39	14	13	11	1.000000	0.000013
39	14	13	12	1.000000	0.000000
39	14	13	13	1.000000	0.000000
39	14	14	0	0.000295	0.000295
39	14	14	1	0.005122	0.004826
39	14	14	2	0.036494	0.031372
39	14	14	3	0.144054	0.107560
39	14	14	4	0.360967	0.216913
39	14	14	5	0.632109	0.271142
39	14	14	6	0.847427	0.215318
39	14	14	7	0.956795	0.109368
39	14	14	8	0.992052	0.035257
39	14	14	9	0.999103	0.007051
39	14	14	10	0.999943	0.000839
39	14	14	11	0.999998	0.000055
39	14	14	12	1.000000	0.000002
39	14	14	13	1.000000	0.000000
39	14	14	14	1.000000	0.000000
39	15	1	0	0.615385	0.615385
39	15	1	1	1.000000	0.384615
39	15	2	0	0.372470	0.372470
39	15	2	1	0.858300	0.485830
39	15	2	2	1.000000	0.141700
39	15	3	0	0.221468	0.221468
39	15	3	1	0.674472	0.453004
39	15	3	2	0.950213	0.275741
39	15	3	3	1.000000	0.049787
39	15	4	0	0.129190	0.129190
39	15	4	1	0.498304	0.369114
39	15	4	2	0.850640	0.352336
39	15	4	3	0.983404	0.132764
39	15	4	4	1.000000	0.016596
39	15	5	0	0.073823	0.073823
39	15	5	1	0.350658	0.276836
39	15	5	2	0.719772	0.369114
39	15	5	3	0.937885	0.218113
39	15	5	4	0.994784	0.056899
39	15	5	5	1.000000	0.005216
39	15	6	0	0.041254	0.041254
39	15	6	1	0.236667	0.195413
39	15	6	2	0.578641	0.341973
39	15	6	3	0.860904	0.282264
39	15	6	4	0.976376	0.115472
39	15	6	5	0.998466	0.022090
39	15	6	6	1.000000	0.001534
39	15	7	0	0.022502	0.022502
39	15	7	1	0.153765	0.131262
39	15	7	2	0.443924	0.290159
39	15	7	3	0.758263	0.314339
39	15	7	4	0.937885	0.179622
39	15	7	5	0.991772	0.053887
39	15	7	6	0.999582	0.007810
39	15	7	7	1.000000	0.000418
39	15	8	0	0.011954	0.011954
39	15	8	1	0.096337	0.084383
39	15	8	2	0.326047	0.229709
39	15	8	3	0.640386	0.314339
39	15	8	4	0.876140	0.235754
39	15	8	5	0.974932	0.098792
39	15	8	6	0.997385	0.022453
39	15	8	7	0.999895	0.002510
39	15	8	8	1.000000	0.000105
39	15	9	0	0.006170	0.006170
39	15	9	1	0.058229	0.052059
39	15	9	2	0.229717	0.171488
39	15	9	3	0.518706	0.288989
39	15	9	4	0.792485	0.273779
39	15	9	5	0.943064	0.150579
39	15	9	6	0.990867	0.047803
39	15	9	7	0.999248	0.008381
39	15	9	8	0.999976	0.000729
39	15	9	9	1.000000	0.000024
39	15	10	0	0.003085	0.003085
39	15	10	1	0.033935	0.030850
39	15	10	2	0.155405	0.121471
39	15	10	3	0.403110	0.247705
39	15	10	4	0.692100	0.288989
39	15	10	5	0.892871	0.200771
39	15	10	6	0.976526	0.083655
39	15	10	7	0.997013	0.020487
39	15	10	8	0.999806	0.002794
39	15	10	9	0.999995	0.000189
39	15	10	10	1.000000	0.000005
39	15	11	0	0.001489	0.001489
39	15	11	1	0.019042	0.017552
39	15	11	2	0.100953	0.081911
39	15	11	3	0.300612	0.199659
39	15	11	4	0.582483	0.281871
39	15	11	5	0.823639	0.241156
39	15	11	6	0.950564	0.126924
39	15	11	7	0.991361	0.040797
39	15	11	8	0.999132	0.007771
39	15	11	9	0.999956	0.000824
39	15	11	10	0.999999	0.000043
39	15	11	11	1.000000	0.000001
39	15	12	0	0.000691	0.000691
39	15	12	1	0.010265	0.009574
39	15	12	2	0.062923	0.052657
39	15	12	3	0.215044	0.152121
39	15	12	4	0.471748	0.256704
39	15	12	5	0.737512	0.265764
39	15	12	6	0.909767	0.172255
39	15	12	7	0.979705	0.069938
39	15	12	8	0.997189	0.017484
39	15	12	9	0.999779	0.002590
39	15	12	10	0.999991	0.000212
39	15	12	11	1.000000	0.000008
39	15	12	12	1.000000	0.000000
39	15	13	0	0.000307	0.000307
39	15	13	1	0.005301	0.004994
39	15	13	2	0.037569	0.032268
39	15	13	3	0.147434	0.109865
39	15	13	4	0.367165	0.219730
39	15	13	5	0.639081	0.271916
39	15	13	6	0.852349	0.213268
39	15	13	7	0.958982	0.106634
39	15	13	8	0.992656	0.033674
39	15	13	9	0.999204	0.006548
39	15	13	10	0.999952	0.000748
39	15	13	11	0.999999	0.000046
39	15	13	12	1.000000	0.000001
39	15	13	13	1.000000	0.000000
39	15	14	0	0.000130	0.000130
39	15	14	1	0.002612	0.002482
39	15	14	2	0.021435	0.018823
39	15	14	3	0.096727	0.075292
39	15	14	4	0.274202	0.177474
39	15	14	5	0.534498	0.260296
39	15	14	6	0.778525	0.244027
39	15	14	7	0.926172	0.147647
39	15	14	8	0.983590	0.057418
39	15	14	9	0.997693	0.014103
39	15	14	10	0.999808	0.002115
39	15	14	11	0.999991	0.000183
39	15	14	12	1.000000	0.000008
39	15	14	13	1.000000	0.000000
39	15	14	14	1.000000	0.000000
39	15	15	0	0.000052	0.000052
39	15	15	1	0.001222	0.001170
39	15	15	2	0.011647	0.010425
39	15	15	3	0.060587	0.048940
39	15	15	4	0.196113	0.135526
39	15	15	5	0.430379	0.234266
39	15	15	6	0.690675	0.260296
39	15	15	7	0.878925	0.188250
39	15	15	8	0.967513	0.088588
39	15	15	9	0.994308	0.026795
39	15	15	10	0.999385	0.005077
39	15	15	11	0.999962	0.000577
39	15	15	12	0.999999	0.000037
39	15	15	13	1.000000	0.000001
39	15	15	14	1.000000	0.000000
39	15	15	15	1.000000	0.000000
39	16	1	0	0.589744	0.589744
39	16	1	1	1.000000	0.410256
39	16	2	0	0.341430	0.341430
39	16	2	1	0.838057	0.496626
39	16	2	2	1.000000	0.161943
39	16	3	0	0.193785	0.193785
39	16	3	1	0.636722	0.442937
39	16	3	2	0.938724	0.302002
39	16	3	3	1.000000	0.061276
39	16	4	0	0.107658	0.107658
39	16	4	1	0.452165	0.344506
39	16	4	2	0.821279	0.369114
39	16	4	3	0.977873	0.156594
39	16	4	4	1.000000	0.022127
39	16	5	0	0.058443	0.058443
39	16	5	1	0.304519	0.246076
39	16	5	2	0.673633	0.369114
39	16	5	3	0.919709	0.246076
39	16	5	4	0.992413	0.072704
39	16	5	5	1.000000	0.007587
39	16	6	0	0.030940	0.030940
39	16	6	1	0.195956	0.165016
39	16	6	2	0.521645	0.325689
39	16	6	3	0.825621	0.303976

Table for $N = 2$, $n = 1$, through $N = 100$, $n = 50$

N	n	k	x	P(x)	p(x)
39	16	6	4	0.966753	0.141132
39	16	6	5	0.997545	0.030792
39	16	6	6	1.000000	0.002454
39	16	7	0	0.015939	0.015939
39	16	7	1	0.120949	0.105010
39	16	7	2	0.383474	0.262525
39	16	7	3	0.705873	0.322399
39	16	7	4	0.915432	0.209559
39	16	7	5	0.987281	0.071849
39	16	7	6	0.999256	0.011975
39	16	7	7	1.000000	0.000744
39	16	8	0	0.007969	0.007969
39	16	8	1	0.071726	0.063756
39	16	8	2	0.268619	0.196894
39	16	8	3	0.574898	0.306279
39	16	8	4	0.836848	0.261949
39	16	8	5	0.962583	0.125736
39	16	8	6	0.995514	0.032931
39	16	8	7	0.999791	0.004277
39	16	8	8	1.000000	0.000209
39	16	9	0	0.003856	0.003856
39	16	9	1	0.040876	0.037020
39	16	9	2	0.179700	0.138824
39	16	9	3	0.446459	0.266759
39	16	9	4	0.735448	0.288989
39	16	9	5	0.917967	0.182519
39	16	9	6	0.984891	0.066924
39	16	9	7	0.998549	0.013658
39	16	9	8	0.999946	0.001397
39	16	9	9	1.000000	0.000054
39	16	10	0	0.001800	0.001800
39	16	10	1	0.022366	0.020566
39	16	10	2	0.114915	0.092549
39	16	10	3	0.330863	0.215948
39	16	10	4	0.619852	0.288989
39	16	10	5	0.851044	0.231191
39	16	10	6	0.962583	0.111540
39	16	10	7	0.994452	0.031868
39	16	10	8	0.999573	0.005122
39	16	10	9	0.999987	0.000414
39	16	10	10	1.000000	0.000013
39	16	11	0	0.000807	0.000807
39	16	11	1	0.011728	0.010921
39	16	11	2	0.070236	0.058508
39	16	11	3	0.234059	0.163823
39	16	11	4	0.500271	0.266212
39	16	11	5	0.763350	0.263080
39	16	11	6	0.924121	0.160771
39	16	11	7	0.984562	0.060440
39	16	11	8	0.998161	0.013599

N	n	k	x	P(x)	p(x)
39	16	11	9	0.999887	0.001727
39	16	11	10	0.999997	0.000110
39	16	11	11	1.000000	0.000003
39	16	12	0	0.000346	0.000346
39	16	12	1	0.005877	0.005532
39	16	12	2	0.040982	0.035105
39	16	12	3	0.157998	0.117016
39	16	12	4	0.386180	0.228181
39	16	12	5	0.659998	0.273818
39	16	12	6	0.866703	0.206706
39	16	12	7	0.965134	0.098431
39	16	12	8	0.994275	0.029141
39	16	12	9	0.999456	0.005181
39	16	12	10	0.999974	0.000518
39	16	12	11	0.999999	0.000026
39	16	12	12	1.000000	0.000000
39	16	13	0	0.000141	0.000141
39	16	13	1	0.002804	0.002663
39	16	13	2	0.022780	0.019975
39	16	13	3	0.101657	0.078878
39	16	13	4	0.284766	0.183109
39	16	13	5	0.548442	0.263676
39	16	13	6	0.790145	0.241703
39	16	13	7	0.932324	0.142178
39	16	13	8	0.985641	0.053317
39	16	13	9	0.998113	0.012472
39	16	13	10	0.999859	0.001746
39	16	13	11	0.999995	0.000136
39	16	13	12	1.000000	0.000005
39	16	13	13	1.000000	0.000000
39	16	14	0	0.000054	0.000054
39	16	14	1	0.001268	0.001213
39	16	14	2	0.012024	0.010756
39	16	14	3	0.062218	0.050195
39	16	14	4	0.200254	0.138036
39	16	14	5	0.436887	0.236633
39	16	14	6	0.697183	0.260296
39	16	14	7	0.883108	0.185926
39	16	14	8	0.969236	0.086127
39	16	14	9	0.994755	0.025519
39	16	14	10	0.999456	0.004701
39	16	14	11	0.999968	0.000513
39	16	14	12	0.999999	0.000031
39	16	14	13	1.000000	0.000001
39	16	14	14	1.000000	0.000000
39	16	15	0	0.000019	0.000019
39	16	15	1	0.000540	0.000520
39	16	15	2	0.006000	0.005461
39	16	15	3	0.036117	0.030117
39	16	15	4	0.133997	0.097880

N	n	k	x	P(x)	p(x)
39	16	15	5	0.332768	0.198771
39	16	15	6	0.593064	0.260296
39	16	15	7	0.816175	0.223111
39	16	15	8	0.941675	0.125500
39	16	15	9	0.987609	0.045935
39	16	15	10	0.998327	0.010718
39	16	15	11	0.999866	0.001538
39	16	15	12	0.999994	0.000128
39	16	15	13	1.000000	0.000006
39	16	15	14	1.000000	0.000000
39	16	15	15	1.000000	0.000000
39	16	16	0	0.000006	0.000006
39	16	16	1	0.000215	0.000208
39	16	16	2	0.002815	0.002600
39	16	16	3	0.019804	0.016989
39	16	16	4	0.085057	0.065253
39	16	16	5	0.241665	0.156608
39	16	16	6	0.484608	0.242943
39	16	16	7	0.732509	0.247901
39	16	16	8	0.899842	0.167333
39	16	16	9	0.974212	0.074370
39	16	16	10	0.995648	0.021436
39	16	16	11	0.999545	0.003897
39	16	16	12	0.999973	0.000427
39	16	16	13	0.999999	0.000026
39	16	16	14	1.000000	0.000001
39	16	16	15	1.000000	0.000000
39	16	16	16	1.000000	0.000000
39	17	1	0	0.564103	0.564103
39	17	1	1	1.000000	0.435897
39	17	2	0	0.311741	0.311741
39	17	2	1	0.816464	0.504723
39	17	2	2	1.000000	0.183536
39	17	3	0	0.168509	0.168509
39	17	3	1	0.598205	0.429697
39	17	3	2	0.925594	0.327388
39	17	3	3	1.000000	0.074406
39	17	4	0	0.088935	0.088935
39	17	4	1	0.407229	0.318294
39	17	4	2	0.789182	0.381953
39	17	4	3	0.971064	0.181882
39	17	4	4	1.000000	0.028936
39	17	5	0	0.045738	0.045738
39	17	5	1	0.261723	0.215985
39	17	5	2	0.625488	0.363765
39	17	5	3	0.898311	0.272823
39	17	5	4	0.989252	0.090941
39	17	5	5	1.000000	0.010748
39	17	6	0	0.022869	0.022869
39	17	6	1	0.160083	0.137214

N	n	k	x	P(x)	p(x)
39	17	6	2	0.465003	0.304920
39	17	6	3	0.785972	0.320969
39	17	6	4	0.954481	0.168509
39	17	6	5	0.996207	0.041726
39	17	6	6	1.000000	0.003793
39	17	7	0	0.011088	0.011088
39	17	7	1	0.093555	0.082467
39	17	7	2	0.326403	0.232848
39	17	7	3	0.649804	0.323400
39	17	7	4	0.888099	0.238295
39	17	7	5	0.981034	0.092935
39	17	7	6	0.998735	0.017702
39	17	7	7	1.000000	0.001264
39	17	8	0	0.005198	0.005198
39	17	8	1	0.052322	0.047124
39	17	8	2	0.217256	0.164934
39	17	8	3	0.508316	0.291060
39	17	8	4	0.791291	0.282975
39	17	8	5	0.946183	0.154892
39	17	8	6	0.992650	0.046468
39	17	8	7	0.999605	0.006954
39	17	8	8	1.000000	0.000395
39	17	9	0	0.002347	0.002347
39	17	9	1	0.027999	0.025652
39	17	9	2	0.137449	0.109449
39	17	9	3	0.376869	0.239421
39	17	9	4	0.672624	0.295755
39	17	9	5	0.886225	0.213601
39	17	9	6	0.976162	0.089937
39	17	9	7	0.997361	0.021199
39	17	9	8	0.999885	0.002524
39	17	9	9	1.000000	0.000115
39	17	10	0	0.001017	0.001017
39	17	10	1	0.014318	0.013301
39	17	10	2	0.082724	0.068406
39	17	10	3	0.265140	0.182416
39	17	10	4	0.544464	0.279324
39	17	10	5	0.800785	0.256321
39	17	10	6	0.943185	0.142400
39	17	10	7	0.990295	0.047110
39	17	10	8	0.999128	0.008833
39	17	10	9	0.999969	0.000841
39	17	10	10	1.000000	0.000031
39	17	11	0	0.000421	0.000421
39	17	11	1	0.006980	0.006559
39	17	11	2	0.047342	0.040362
39	17	11	3	0.177077	0.129735
39	17	11	4	0.419250	0.242172
39	17	11	5	0.694721	0.275471
39	17	11	6	0.889171	0.194450

N	n	k	x	P(x)	p(x)
39	17	11	7	0.974050	0.084879
39	17	11	8	0.996387	0.022337
39	17	11	9	0.999737	0.003350
39	17	11	10	0.999992	0.000255
39	17	11	11	1.000000	0.000007
39	17	12	0	0.000165	0.000165
39	17	12	1	0.003232	0.003066
39	17	12	2	0.025719	0.022487
39	17	12	3	0.112209	0.086490
39	17	12	4	0.306812	0.194603
39	17	12	5	0.576662	0.269849
39	17	12	6	0.812780	0.236118
39	17	12	7	0.943736	0.130956
39	17	12	8	0.989207	0.045471
39	17	12	9	0.998780	0.009573
39	17	12	10	0.999929	0.001149
39	17	12	11	0.999998	0.000070
39	17	12	12	1.000000	0.000002
39	17	13	0	0.000061	0.000061
39	17	13	1	0.001415	0.001353
39	17	13	2	0.013226	0.011812
39	17	13	3	0.067363	0.054136
39	17	13	4	0.213115	0.145752
39	17	13	5	0.456729	0.243614
39	17	13	6	0.716584	0.259855
39	17	13	7	0.895234	0.178650
39	17	13	8	0.974050	0.078816
39	17	13	9	0.995944	0.021893
39	17	13	10	0.999631	0.003687
39	17	13	11	0.999983	0.000352
39	17	13	12	1.000000	0.000017
39	17	13	13	1.000000	0.000000
39	17	14	0	0.000021	0.000021
39	17	14	1	0.000582	0.000561
39	17	14	2	0.006412	0.005830
39	17	14	3	0.038212	0.031800
39	17	14	4	0.140239	0.102026
39	17	14	5	0.344291	0.204053
39	17	14	6	0.606645	0.262354
39	17	14	7	0.826522	0.219877
39	17	14	8	0.946768	0.120245
39	17	14	9	0.989207	0.042440
39	17	14	10	0.998638	0.009431
39	17	14	11	0.999902	0.001263
39	17	14	12	0.999996	0.000095
39	17	14	13	1.000000	0.000003
39	17	14	14	1.000000	0.000000
39	17	15	0	0.000007	0.000007
39	17	15	1	0.000223	0.000216
39	17	15	2	0.002914	0.002691

N	n	k	x	P(x)	p(x)
39	17	15	3	0.020404	0.017490
39	17	15	4	0.087185	0.066781
39	17	15	5	0.246346	0.159161
39	17	15	6	0.491209	0.244863
39	17	15	7	0.738571	0.247362
39	17	15	8	0.903479	0.164908
39	17	15	9	0.975626	0.072147
39	17	15	10	0.995997	0.020371
39	17	15	11	0.999598	0.003601
39	17	15	12	0.999977	0.000379
39	17	15	13	0.999999	0.000022
39	17	15	14	1.000000	0.000001
39	17	15	15	1.000000	0.000000
39	17	16	0	0.000002	0.000002
39	17	16	1	0.000079	0.000077
39	17	16	2	0.001232	0.001153
39	17	16	3	0.010201	0.008969
39	17	16	4	0.051012	0.040811
39	17	16	5	0.166766	0.115754
39	17	16	6	0.378980	0.212215
39	17	16	7	0.635504	0.256523
39	17	16	8	0.841639	0.206135
39	17	16	9	0.951577	0.109939
39	17	16	10	0.990056	0.038479
39	17	16	11	0.998698	0.008642
39	17	16	12	0.999898	0.001200
39	17	16	13	0.999996	0.000097
39	17	16	14	1.000000	0.000004
39	17	16	15	1.000000	0.000000
39	17	16	16	1.000000	0.000000
39	17	17	0	0.000001	0.000001
39	17	17	1	0.000025	0.000025
39	17	17	2	0.000480	0.000455
39	17	17	3	0.004742	0.004262
39	17	17	4	0.027945	0.023203
39	17	17	5	0.106372	0.078427
39	17	17	6	0.277486	0.171114
39	17	17	7	0.523972	0.246486
39	17	17	8	0.760977	0.237005
39	17	17	9	0.913338	0.152361
39	17	17	10	0.978345	0.065007
39	17	17	11	0.996444	0.018099
39	17	17	12	0.999638	0.003194
39	17	17	13	0.999979	0.000341
39	17	17	14	0.999999	0.000021
39	17	17	15	1.000000	0.000001
39	17	17	16	1.000000	0.000000
39	17	17	17	1.000000	0.000000
39	18	1	0	0.538462	0.538462
39	18	1	1	1.000000	0.461538

N	n	k	x	P(x)	p(x)
39	18	2	0	0.283401	0.283401
39	18	2	1	0.793522	0.510121
39	18	2	2	1.000000	0.206478
39	18	3	0	0.145530	0.145530
39	18	3	1	0.559142	0.413612
39	18	3	2	0.910712	0.351570
39	18	3	3	1.000000	0.089288
39	18	4	0	0.072765	0.072765
39	18	4	1	0.363825	0.291060
39	18	4	2	0.754459	0.390634
39	18	4	3	0.962797	0.208338
39	18	4	4	1.000000	0.037203
39	18	5	0	0.035343	0.035343
39	18	5	1	0.222453	0.187110
39	18	5	2	0.575884	0.353430
39	18	5	3	0.873509	0.297626
39	18	5	4	0.985119	0.111610
39	18	5	5	1.000000	0.014881
39	18	6	0	0.016632	0.016632
39	18	6	1	0.128898	0.112266
39	18	6	2	0.409563	0.280665
39	18	6	3	0.742204	0.332640
39	18	6	4	0.939162	0.196958
39	18	6	5	0.994310	0.055148
39	18	6	6	1.000000	0.005690
39	18	7	0	0.007560	0.007560
39	18	7	1	0.071064	0.063504
39	18	7	2	0.273483	0.202419
39	18	7	3	0.591004	0.317520
39	18	7	4	0.855604	0.264600
39	18	7	5	0.972585	0.116981
39	18	7	6	0.997931	0.025346
39	18	7	7	1.000000	0.002069
39	18	8	0	0.003307	0.003307
39	18	8	1	0.037328	0.034020
39	18	8	2	0.172274	0.134946
39	18	8	3	0.442166	0.269892
39	18	8	4	0.739841	0.297675
39	18	8	5	0.925061	0.185220
39	18	8	6	0.988426	0.063365
39	18	8	7	0.999289	0.010863
39	18	8	8	1.000000	0.000711
39	18	9	0	0.001387	0.001387
39	18	9	1	0.018671	0.017284
39	18	9	2	0.102624	0.083953
39	18	9	3	0.311573	0.208949
39	18	9	4	0.605407	0.293834
39	18	9	5	0.847388	0.241981
39	18	9	6	0.963898	0.116509
39	18	9	7	0.995434	0.031536

N	n	k	x	P(x)	p(x)
39	18	9	8	0.999770	0.004336
39	18	9	9	1.000000	0.000229
39	18	10	0	0.000555	0.000555
39	18	10	1	0.008877	0.008322
39	18	10	2	0.057849	0.048972
39	18	10	3	0.207098	0.149249
39	18	10	4	0.468284	0.261186
39	18	10	5	0.742530	0.274245
39	18	10	6	0.917294	0.174764
39	18	10	7	0.983871	0.066577
39	18	10	8	0.998325	0.014454
39	18	10	9	0.999931	0.001606
39	18	10	10	1.000000	0.000069
39	18	11	0	0.000210	0.000210
39	18	11	1	0.003998	0.003788
39	18	11	2	0.030830	0.026832
39	18	11	3	0.129901	0.099071
39	18	11	4	0.342195	0.212294
39	18	11	5	0.619592	0.277398
39	18	11	6	0.844978	0.225386
39	18	11	7	0.958618	0.113640
39	18	11	8	0.993341	0.034723
39	18	11	9	0.999433	0.006092
39	18	11	10	0.999981	0.000548
39	18	11	11	1.000000	0.000019
39	18	12	0	0.000075	0.000075
39	18	12	1	0.001699	0.001623
39	18	12	2	0.015498	0.013799
39	18	12	3	0.076827	0.061329
39	18	12	4	0.236048	0.159221
39	18	12	5	0.490801	0.254753
39	18	12	6	0.748384	0.257583
39	18	12	7	0.913973	0.165589
39	18	12	8	0.980940	0.066966
39	18	12	9	0.997475	0.016535
39	18	12	10	0.999824	0.002350
39	18	12	11	0.999995	0.000171
39	18	12	12	1.000000	0.000005
39	18	13	0	0.000025	0.000025
39	18	13	1	0.000676	0.000651
39	18	13	2	0.007320	0.006644
39	18	13	3	0.042755	0.035435
39	18	13	4	0.153489	0.110734
39	18	13	5	0.368142	0.214653
39	18	13	6	0.633902	0.265761
39	18	13	7	0.846511	0.212609
39	18	13	8	0.956137	0.109626
39	18	13	9	0.991963	0.035826
39	18	13	10	0.999128	0.007165
39	18	13	11	0.999951	0.000823

Table for $N = 2$, $n = 1$, through $N = 100$, $n = 50$

N	n	k	x	P(x)	p(x)
39	18	13	12	0.999999	0.000048
39	18	13	13	1.000000	0.000001
39	18	14	0	0.000008	0.000008
39	18	14	1	0.000251	0.000243
39	18	14	2	0.003232	0.002981
39	18	14	3	0.022312	0.019080
39	18	14	4	0.093863	0.071551
39	18	14	5	0.260815	0.166952
39	18	14	6	0.511244	0.250428
39	18	14	7	0.756561	0.245318
39	18	14	8	0.913973	0.157412
39	18	14	9	0.979562	0.065588
39	18	14	10	0.996923	0.017362
39	18	14	11	0.999729	0.002806
39	18	14	12	0.999988	0.000258
39	18	14	13	1.000000	0.000012
39	18	14	14	1.000000	0.000000
39	18	15	0	0.000002	0.000002
39	18	15	1	0.000085	0.000083
39	18	15	2	0.001324	0.001238
39	18	15	3	0.010864	0.009540
39	18	15	4	0.053794	0.042931
39	18	15	5	0.174000	0.120206
39	18	15	6	0.391038	0.217038
39	18	15	7	0.648622	0.257583
39	18	15	8	0.851009	0.202387
39	18	15	9	0.955950	0.104941
39	18	15	10	0.991368	0.035418
39	18	15	11	0.998944	0.007576
39	18	15	12	0.999926	0.000982
39	18	15	13	0.999997	0.000072
39	18	15	14	1.000000	0.000003
39	18	15	15	1.000000	0.000000
39	18	16	0	0.000001	0.000001
39	18	16	1	0.000026	0.000026
39	18	16	2	0.000498	0.000472
39	18	16	3	0.004901	0.004403
39	18	16	4	0.028752	0.023850
39	18	16	5	0.108889	0.080137
39	18	16	6	0.282519	0.173630
39	18	16	7	0.530562	0.248043
39	18	16	8	0.766681	0.236118
39	18	16	9	0.916597	0.149916
39	18	16	10	0.979562	0.062965
39	18	16	11	0.996734	0.017172
39	18	16	12	0.999680	0.002946
39	18	16	13	0.999982	0.000302
39	18	16	14	0.999999	0.000017
39	18	16	15	1.000000	0.000000
39	18	16	16	1.000000	0.000000

N	n	k	x	P(x)	p(x)
39	18	17	0	0.000000	0.000000
39	18	17	1	0.000007	0.000007
39	18	17	2	0.000170	0.000163
39	18	17	3	0.002030	0.001860
39	18	17	4	0.014234	0.012204
39	18	17	5	0.063594	0.049360
39	18	17	6	0.191929	0.128335
39	18	17	7	0.411933	0.220004
39	18	17	8	0.664021	0.252088
39	18	17	9	0.857934	0.193914
39	18	17	10	0.957661	0.099727
39	18	17	11	0.991508	0.033847
39	18	17	12	0.998912	0.007404
39	18	17	13	0.999917	0.001005
39	18	17	14	0.999996	0.000080
39	18	17	15	1.000000	0.000003
39	18	17	16	1.000000	0.000000
39	18	17	17	1.000000	0.000000
39	18	18	0	0.000000	0.000000
39	18	18	1	0.000002	0.000002
39	18	18	2	0.000052	0.000050
39	18	18	3	0.000762	0.000710
39	18	18	4	0.006468	0.005706
39	18	18	5	0.034427	0.027959
39	18	18	6	0.121928	0.087501
39	18	18	7	0.301931	0.180003
39	18	18	8	0.549435	0.247504
39	18	18	9	0.778606	0.229170
39	18	18	10	0.921397	0.142791
39	18	18	11	0.980738	0.059342
39	18	18	12	0.996892	0.016154
39	18	18	13	0.999688	0.002796
39	18	18	14	0.999982	0.000294
39	18	18	15	0.999999	0.000017
39	18	18	16	1.000000	0.000001
39	18	18	17	1.000000	0.000000
39	18	18	18	1.000000	0.000000
39	19	1	0	0.512821	0.512821
39	19	1	1	1.000000	0.487179
39	19	2	0	0.256410	0.256410
39	19	2	1	0.769231	0.512821
39	19	2	2	1.000000	0.230769
39	19	3	0	0.124740	0.124740
39	19	3	1	0.519751	0.395010
39	19	3	2	0.893971	0.374220
39	19	3	3	1.000000	0.106029
39	19	4	0	0.058905	0.058905
39	19	4	1	0.322245	0.263340
39	19	4	2	0.717256	0.395010
39	19	4	3	0.952876	0.235620

N	n	k	x	P(x)	p(x)
39	19	4	4	1.000000	0.047124
39	19	5	0	0.026928	0.026928
39	19	5	1	0.186813	0.159885
39	19	5	2	0.525393	0.338580
39	19	5	3	0.845164	0.319770
39	19	5	4	0.979804	0.134640
39	19	5	5	1.000000	0.020196
39	19	6	0	0.011880	0.011880
39	19	6	1	0.102168	0.090288
39	19	6	2	0.356103	0.253935
39	19	6	3	0.694684	0.338580
39	19	6	4	0.920404	0.225720
39	19	6	5	0.991684	0.071280
39	19	6	6	1.000000	0.008316
39	19	7	0	0.005040	0.005040
39	19	7	1	0.052920	0.047880
39	19	7	2	0.225288	0.172368
39	19	7	3	0.530524	0.305235
39	19	7	4	0.817804	0.287280
39	19	7	5	0.961444	0.143640
39	19	7	6	0.996724	0.035280
39	19	7	7	1.000000	0.003276
39	19	8	0	0.002048	0.002048
39	19	8	1	0.025988	0.023940
39	19	8	2	0.133718	0.107730
39	19	8	3	0.377906	0.244188
39	19	8	4	0.683141	0.305235
39	19	8	5	0.898601	0.215460
39	19	8	6	0.982391	0.083790
39	19	8	7	0.998771	0.016380
39	19	8	8	1.000000	0.001228
39	19	9	0	0.000793	0.000793
39	19	9	1	0.012087	0.011294
39	19	9	2	0.074640	0.062553
39	19	9	3	0.251873	0.177233
39	19	9	4	0.535447	0.283573
39	19	9	5	0.801297	0.265850
39	19	9	6	0.947254	0.145957
39	19	9	7	0.992431	0.045177
39	19	9	8	0.999564	0.007133
39	19	9	9	1.000000	0.000436
39	19	10	0	0.000291	0.000291
39	19	10	1	0.005310	0.005020
39	19	10	2	0.039193	0.033883
39	19	10	3	0.157349	0.118156
39	19	10	4	0.393660	0.236311
39	19	10	5	0.677233	0.283573
39	19	10	6	0.884006	0.206772
39	19	10	7	0.974360	0.090354
39	19	10	8	0.996948	0.022589

N	n	k	x	P(x)	p(x)
39	19	10	9	0.999855	0.002906
39	19	10	10	1.000000	0.000145
39	19	11	0	0.000100	0.000100
39	19	11	1	0.002195	0.002094
39	19	11	2	0.019331	0.017136
39	19	11	3	0.092159	0.072829
39	19	11	4	0.271430	0.179271
39	19	11	5	0.540336	0.268906
39	19	11	6	0.791315	0.250979
39	19	11	7	0.936972	0.145657
39	19	11	8	0.988380	0.051408
39	19	11	9	0.998853	0.010472
39	19	11	10	0.999955	0.001102
39	19	11	11	1.000000	0.000045
39	19	12	0	0.000032	0.000032
39	19	12	1	0.000848	0.000816
39	19	12	2	0.008927	0.008078
39	19	12	3	0.050543	0.041616
39	19	12	4	0.175392	0.124849
39	19	12	5	0.405883	0.230491
39	19	12	6	0.674789	0.268906
39	19	12	7	0.874547	0.199759
39	19	12	8	0.968184	0.093637
39	19	12	9	0.995112	0.026928
39	19	12	10	0.999600	0.004488
39	19	12	11	0.999987	0.000387
39	19	12	12	1.000000	0.000013
39	19	13	0	0.000010	0.000010
39	19	13	1	0.000304	0.000295
39	19	13	2	0.003840	0.003536
39	19	13	3	0.025882	0.022041
39	19	13	4	0.106032	0.080150
39	19	13	5	0.286369	0.180338
39	19	13	6	0.545316	0.258946
39	19	13	7	0.785766	0.240450
39	19	13	8	0.930036	0.144270
39	19	13	9	0.985139	0.055103
39	19	13	10	0.998104	0.012965
39	19	13	11	0.999872	0.001768
39	19	13	12	0.999997	0.000124
39	19	13	13	1.000000	0.000003
39	19	14	0	0.000003	0.000003
39	19	14	1	0.000100	0.000098
39	19	14	2	0.001528	0.001428
39	19	14	3	0.012318	0.010789
39	19	14	4	0.059791	0.047473
39	19	14	5	0.189264	0.129473
39	19	14	6	0.415842	0.226578
39	19	14	7	0.674789	0.258946
39	19	14	8	0.868999	0.194210

Table for $N = 2$, $n = 1$, through $N = 100$, $n = 50$

N	n	k	x	P(x)	p(x)
39	19	14	9	0.963946	0.094947
39	19	14	10	0.993616	0.029671
39	19	14	11	0.999328	0.005712
39	19	14	12	0.999963	0.000635
39	19	14	13	0.999999	0.000036
39	19	14	14	1.000000	0.000001
39	19	15	0	0.000001	0.000001
39	19	15	1	0.000030	0.000029
39	19	15	2	0.000557	0.000527
39	19	15	3	0.005412	0.004855
39	19	15	4	0.031307	0.025895
39	19	15	5	0.116759	0.085452
39	19	15	6	0.298022	0.181262
39	19	15	7	0.550495	0.252473
39	19	15	8	0.783546	0.233052
39	19	15	9	0.925967	0.142420
39	19	15	10	0.982935	0.056968
39	19	15	11	0.997501	0.014566
39	19	15	12	0.999785	0.002285
39	19	15	13	0.999991	0.000205
39	19	15	14	1.000000	0.000009
39	19	15	15	1.000000	0.000000
39	19	16	0	0.000000	0.000000
39	19	16	1	0.000008	0.000008
39	19	16	2	0.000184	0.000176
39	19	16	3	0.002176	0.001992
39	19	16	4	0.015123	0.012947
39	19	16	5	0.066912	0.051789
39	19	16	6	0.199838	0.132926
39	19	16	7	0.424258	0.224420
39	19	16	8	0.676731	0.252473
39	19	16	9	0.866625	0.189894
39	19	16	10	0.961572	0.094947
39	19	16	11	0.992645	0.031074
39	19	16	12	0.999119	0.006474
39	19	16	13	0.999939	0.000820
39	19	16	14	0.999998	0.000059
39	19	16	15	1.000000	0.000002
39	19	16	16	1.000000	0.000000
39	19	17	0	0.000000	0.000000
39	19	17	1	0.000002	0.000002
39	19	17	2	0.000054	0.000052
39	19	17	3	0.000790	0.000736
39	19	17	4	0.006679	0.005889
39	19	17	5	0.035388	0.028709
39	19	17	6	0.124706	0.089318
39	19	17	7	0.307169	0.182463
39	19	17	8	0.555983	0.248814
39	19	17	9	0.784062	0.228079
39	19	17	10	0.924419	0.140356

N	n	k	x	P(x)	p(x)
39	19	17	11	0.981837	0.057419
39	19	17	12	0.997149	0.015312
39	19	17	13	0.999725	0.002576
39	19	17	14	0.999985	0.000260
39	19	17	15	1.000000	0.000014
39	19	17	16	1.000000	0.000000
39	19	17	17	1.000000	0.000000
39	19	18	0	0.000000	0.000000
39	19	18	1	0.000000	0.000000
39	19	18	2	0.000014	0.000013
39	19	18	3	0.000255	0.000241
39	19	18	4	0.002664	0.002409
39	19	18	5	0.017119	0.014455
39	19	18	6	0.071927	0.054809
39	19	18	7	0.207644	0.135717
39	19	18	8	0.431576	0.223932
39	19	18	9	0.680390	0.248814
39	19	18	10	0.867000	0.186610
39	19	18	11	0.960958	0.093958
39	19	18	12	0.992277	0.031319
39	19	18	13	0.999023	0.006746
39	19	18	14	0.999926	0.000903
39	19	18	15	0.999997	0.000071
39	19	18	16	1.000000	0.000003
39	19	18	17	1.000000	0.000000
39	19	18	18	1.000000	0.000000
39	19	19	0	0.000000	0.000000
39	19	19	1	0.000000	0.000000
39	19	19	2	0.000003	0.000003
39	19	19	3	0.000071	0.000068
39	19	19	4	0.000943	0.000872
39	19	19	5	0.007482	0.006539
39	19	19	6	0.037998	0.030516
39	19	19	7	0.130092	0.092093
39	19	19	8	0.314278	0.184187
39	19	19	9	0.561907	0.247629
39	19	19	10	0.787024	0.225117
39	19	19	11	0.925164	0.138140
39	19	19	12	0.981837	0.056673
39	19	19	13	0.997095	0.015258
39	19	19	14	0.999711	0.002616
39	19	19	15	0.999983	0.000272
39	19	19	16	0.999999	0.000016
39	19	19	17	1.000000	0.000000
39	19	19	18	1.000000	0.000000
39	19	19	19	1.000000	0.000000
40	1	1	0	0.975000	0.975000
40	1	1	1	1.000000	0.025000
40	2	1	0	0.950000	0.950000
40	2	1	1	1.000000	0.050000

N	n	k	x	P(x)	p(x)
40	2	2	0	0.901282	0.901282
40	2	2	1	0.998718	0.097436
40	2	2	2	1.000000	0.001282
40	3	1	0	0.925000	0.925000
40	3	1	1	1.000000	0.075000
40	3	2	0	0.853846	0.853846
40	3	2	1	0.996154	0.142308
40	3	2	2	1.000000	0.003846
40	3	3	0	0.786437	0.786437
40	3	3	1	0.988664	0.202227
40	3	3	2	0.999899	0.011235
40	3	3	3	1.000000	0.000101
40	4	1	0	0.900000	0.900000
40	4	1	1	1.000000	0.100000
40	4	2	0	0.807692	0.807692
40	4	2	1	0.992308	0.184615
40	4	2	2	1.000000	0.007692
40	4	3	0	0.722672	0.722672
40	4	3	1	0.977733	0.255061
40	4	3	2	0.999595	0.021862
40	4	3	3	1.000000	0.000405
40	4	4	0	0.644545	0.644545
40	4	4	1	0.957052	0.312507
40	4	4	2	0.998413	0.041361
40	4	4	3	0.999989	0.001576
40	4	4	4	1.000000	0.000011
40	5	1	0	0.875000	0.875000
40	5	1	1	1.000000	0.125000
40	5	2	0	0.762820	0.762820
40	5	2	1	0.987179	0.224359
40	5	2	2	1.000000	0.012821
40	5	3	0	0.662449	0.662449
40	5	3	1	0.963563	0.301113
40	5	3	2	0.998988	0.035425
40	5	3	3	1.000000	0.001012
40	5	4	0	0.572929	0.572929
40	5	4	1	0.931010	0.358081
40	5	4	2	0.996116	0.065106
40	5	4	3	0.999945	0.003830
40	5	4	4	1.000000	0.000055
40	5	5	0	0.493356	0.493356
40	5	5	1	0.891223	0.397867
40	5	5	2	0.990690	0.099467
40	5	5	3	0.999732	0.009042
40	5	5	4	0.999998	0.000266
40	5	5	5	1.000000	0.000002
40	6	1	0	0.850000	0.850000
40	6	1	1	1.000000	0.150000
40	6	2	0	0.719231	0.719231
40	6	2	1	0.980769	0.261538

N	n	k	x	P(x)	p(x)
40	6	2	2	1.000000	0.019231
40	6	3	0	0.605668	0.605668
40	6	3	1	0.946356	0.340688
40	6	3	2	0.997976	0.051619
40	6	3	3	1.000000	0.002024
40	6	4	0	0.507452	0.507452
40	6	4	1	0.900317	0.392866
40	6	4	2	0.992395	0.092078
40	6	4	3	0.999836	0.007441
40	6	4	4	1.000000	0.000164
40	6	5	0	0.422876	0.422876
40	6	5	1	0.845753	0.422876
40	6	5	2	0.982164	0.136412
40	6	5	3	0.999216	0.017051
40	6	5	4	0.999991	0.000775
40	6	5	5	1.000000	0.000009
40	6	6	0	0.350383	0.350383
40	6	6	1	0.785342	0.434958
40	6	6	2	0.966574	0.181233
40	6	6	3	0.997754	0.031180
40	6	6	4	0.999946	0.002192
40	6	6	5	1.000000	0.000053
40	6	6	6	1.000000	0.000000
40	7	1	0	0.825000	0.825000
40	7	1	1	1.000000	0.175000
40	7	2	0	0.676923	0.676923
40	7	2	1	0.973077	0.296154
40	7	2	2	1.000000	0.026923
40	7	3	0	0.552227	0.552227
40	7	3	1	0.926316	0.374089
40	7	3	2	0.996457	0.070142
40	7	3	3	1.000000	0.003543
40	7	4	0	0.447751	0.447751
40	7	4	1	0.865653	0.417901
40	7	4	2	0.986979	0.121326
40	7	4	3	0.999617	0.012638
40	7	4	4	1.000000	0.000383
40	7	5	0	0.360689	0.360689
40	7	5	1	0.796002	0.435314
40	7	5	2	0.970128	0.174126
40	7	5	3	0.998213	0.028085
40	7	5	4	0.999968	0.001755
40	7	5	5	1.000000	0.000032
40	7	6	0	0.288551	0.288551
40	7	6	1	0.721377	0.432826
40	7	6	2	0.945253	0.223876
40	7	6	3	0.995003	0.049750
40	7	6	4	0.999818	0.004815
40	7	6	5	0.999998	0.000181
40	7	6	6	1.000000	0.000002

N	n	k	x	P(x)	p(x)
40	7	7	0	0.229143	0.229143
40	7	7	1	0.644996	0.415853
40	7	7	2	0.912330	0.267334
40	7	7	3	0.989150	0.076820
40	7	7	4	0.999393	0.010243
40	7	7	5	0.999987	0.000595
40	7	7	6	1.000000	0.000012
40	7	7	7	1.000000	0.000000
40	8	1	0	0.800000	0.800000
40	8	1	1	1.000000	0.200000
40	8	2	0	0.635897	0.635897
40	8	2	1	0.964103	0.328205
40	8	2	2	1.000000	0.035897
40	8	3	0	0.502024	0.502024
40	8	3	1	0.903644	0.401619
40	8	3	2	0.994332	0.090688
40	8	3	3	1.000000	0.005668
40	8	4	0	0.393478	0.393478
40	8	4	1	0.827662	0.434183
40	8	4	2	0.979626	0.151964
40	8	4	3	0.999234	0.019608
40	8	4	4	1.000000	0.000766
40	8	5	0	0.306039	0.306039
40	8	5	1	0.743237	0.437198
40	8	5	2	0.954298	0.211061
40	8	5	3	0.996511	0.042212
40	8	5	4	0.999915	0.003404
40	8	5	5	1.000000	0.000085
40	8	6	0	0.236087	0.236087
40	8	6	1	0.655797	0.419710
40	8	6	2	0.918116	0.262319
40	8	6	3	0.990480	0.072364
40	8	6	4	0.999526	0.009045
40	8	6	5	0.999993	0.000467
40	8	6	6	1.000000	0.000007
40	8	7	0	0.180537	0.180537
40	8	7	1	0.569387	0.388849
40	8	7	2	0.871825	0.302438
40	8	7	3	0.979839	0.108014
40	8	7	4	0.998462	0.018623
40	8	7	5	0.999951	0.001490
40	8	7	6	0.999999	0.000048
40	8	7	7	1.000000	0.000000
40	8	8	0	0.136771	0.136771
40	8	8	1	0.486903	0.350133
40	8	8	2	0.816836	0.329933
40	8	8	3	0.963473	0.146637
40	8	8	4	0.996204	0.032731
40	8	8	5	0.999816	0.003612
40	8	8	6	0.999997	0.000181

N	n	k	x	P(x)	p(x)
40	8	8	7	1.000000	0.000003
40	8	8	8	1.000000	0.000000
40	9	1	0	0.775000	0.775000
40	9	1	1	1.000000	0.225000
40	9	2	0	0.596154	0.596154
40	9	2	1	0.953846	0.357692
40	9	2	2	1.000000	0.046154
40	9	3	0	0.454959	0.454959
40	9	3	1	0.878542	0.423583
40	9	3	2	0.991498	0.112955
40	9	3	3	1.000000	0.008502
40	9	4	0	0.344294	0.344294
40	9	4	1	0.786957	0.442663
40	9	4	2	0.970128	0.183171
40	9	4	3	0.998621	0.028493
40	9	4	4	1.000000	0.001379
40	9	5	0	0.258220	0.258220
40	9	5	1	0.688587	0.430367
40	9	5	2	0.934511	0.245924
40	9	5	3	0.993872	0.059361
40	9	5	4	0.999808	0.005936
40	9	5	5	1.000000	0.000191
40	9	6	0	0.191821	0.191821
40	9	6	1	0.590218	0.398397
40	9	6	2	0.885327	0.295109
40	9	6	3	0.983696	0.098370
40	9	6	4	0.998960	0.015264
40	9	6	5	0.999978	0.001018
40	9	6	6	1.000000	0.000022
40	9	7	0	0.141045	0.141045
40	9	7	1	0.496477	0.355433
40	9	7	2	0.824569	0.328092
40	9	7	3	0.966337	0.141768
40	9	7	4	0.996716	0.030379
40	9	7	5	0.999858	0.003143
40	9	7	6	0.999998	0.000140
40	9	7	7	1.000000	0.000002
40	9	8	0	0.102578	0.102578
40	9	8	1	0.410312	0.307734
40	9	8	2	0.754974	0.344662
40	9	8	3	0.940561	0.185587
40	9	8	4	0.992113	0.051552
40	9	8	5	0.999477	0.007365
40	9	8	6	0.999985	0.000508
40	9	8	7	1.000000	0.000015
40	9	8	8	1.000000	0.000000
40	9	9	0	0.073728	0.073728
40	9	9	1	0.333378	0.259650
40	9	9	2	0.679579	0.346201
40	9	9	3	0.905763	0.226184

N	n	k	x	P(x)	p(x)
40	9	9	4	0.984058	0.078295
40	9	9	5	0.998557	0.014499
40	9	9	6	0.999938	0.001381
40	9	9	7	0.999999	0.000061
40	9	9	8	1.000000	0.000001
40	9	9	9	1.000000	0.000000
40	10	1	0	0.750000	0.750000
40	10	1	1	1.000000	0.250000
40	10	2	0	0.557692	0.557692
40	10	2	1	0.942308	0.384615
40	10	2	2	1.000000	0.057692
40	10	3	0	0.410931	0.410931
40	10	3	1	0.851215	0.440283
40	10	3	2	0.987854	0.136640
40	10	3	3	1.000000	0.012146
40	10	4	0	0.299869	0.299869
40	10	4	1	0.744119	0.444250
40	10	4	2	0.958310	0.214192
40	10	4	3	0.997702	0.039392
40	10	4	4	1.000000	0.002298
40	10	5	0	0.216572	0.216572
40	10	5	1	0.633056	0.416484
40	10	5	2	0.910712	0.277656
40	10	5	3	0.990043	0.079330
40	10	5	4	0.999617	0.009574
40	10	5	5	1.000000	0.000383
40	10	6	0	0.154694	0.154694
40	10	6	1	0.525960	0.371266
40	10	6	2	0.847248	0.321288
40	10	6	3	0.974177	0.126929
40	10	6	4	0.997976	0.023799
40	10	6	5	0.999945	0.001970
40	10	6	6	1.000000	0.000055
40	10	7	0	0.109196	0.109196
40	10	7	1	0.427684	0.318488
40	10	7	2	0.771651	0.343967
40	10	7	3	0.948044	0.176393
40	10	7	4	0.993776	0.045732
40	10	7	5	0.999656	0.005880
40	10	7	6	0.999993	0.000338
40	10	7	7	1.000000	0.000006
40	10	8	0	0.076106	0.076106
40	10	8	1	0.340823	0.264717
40	10	8	2	0.688265	0.347441
40	10	8	3	0.910627	0.222363
40	10	8	4	0.985461	0.074834
40	10	8	5	0.998765	0.013304
40	10	8	6	0.999952	0.001188
40	10	8	7	0.999999	0.000047
40	10	8	8	1.000000	0.000001

N	n	k	x	P(x)	p(x)
40	10	9	0	0.052323	0.052323
40	10	9	1	0.266372	0.214049
40	10	9	2	0.601405	0.335033
40	10	9	3	0.861986	0.260581
40	10	9	4	0.971430	0.109444
40	10	9	5	0.996686	0.025256
40	10	9	6	0.999804	0.003118
40	10	9	7	0.999995	0.000191
40	10	9	8	1.000000	0.000005
40	10	9	9	1.000000	0.000000
40	10	10	0	0.035445	0.035445
40	10	10	1	0.204229	0.168784
40	10	10	2	0.514944	0.310716
40	10	10	3	0.803145	0.288200
40	10	10	4	0.950247	0.147102
40	10	10	5	0.992612	0.042365
40	10	10	6	0.999402	0.006789
40	10	10	7	0.999976	0.000575
40	10	10	8	1.000000	0.000023
40	10	10	9	1.000000	0.000000
40	10	10	10	1.000000	0.000000
40	11	1	0	0.725000	0.725000
40	11	1	1	1.000000	0.275000
40	11	2	0	0.520513	0.520513
40	11	2	1	0.929487	0.408974
40	11	2	2	1.000000	0.070513
40	11	3	0	0.369838	0.369838
40	11	3	1	0.821862	0.452024
40	11	3	2	0.983300	0.161437
40	11	3	3	1.000000	0.016700
40	11	4	0	0.259886	0.259886
40	11	4	1	0.699694	0.439807
40	11	4	2	0.944031	0.244337
40	11	4	3	0.996389	0.052358
40	11	4	4	1.000000	0.003611
40	11	5	0	0.180476	0.180476
40	11	5	1	0.577525	0.397048
40	11	5	2	0.882947	0.305422
40	11	5	3	0.984754	0.101807
40	11	5	4	0.999298	0.014544
40	11	5	5	1.000000	0.000702
40	11	6	0	0.123755	0.123755
40	11	6	1	0.464082	0.340327
40	11	6	2	0.804410	0.340327
40	11	6	3	0.961484	0.157074
40	11	6	4	0.996389	0.034905
40	11	6	5	0.999880	0.003491
40	11	6	6	1.000000	0.000120
40	11	7	0	0.083717	0.083717
40	11	7	1	0.363986	0.280269

Table for $N = 2$, $n = 1$, through $N = 100$, $n = 50$

N	n	k	x	P(x)	p(x)
40	11	7	2	0.714323	0.350337
40	11	7	3	0.924525	0.210202
40	11	7	4	0.989203	0.064678
40	11	7	5	0.999264	0.010061
40	11	7	6	0.999982	0.000719
40	11	7	7	1.000000	0.000018
40	11	8	0	0.055811	0.055811
40	11	8	1	0.279056	0.223245
40	11	8	2	0.618777	0.339720
40	11	8	3	0.873567	0.254790
40	11	8	4	0.975483	0.101916
40	11	8	5	0.997434	0.021951
40	11	8	6	0.999873	0.002439
40	11	8	7	0.999998	0.000124
40	11	8	8	1.000000	0.000002
40	11	9	0	0.036626	0.036626
40	11	9	1	0.209292	0.172666
40	11	9	2	0.523230	0.313938
40	11	9	3	0.809869	0.286639
40	11	9	4	0.953189	0.143320
40	11	9	5	0.993318	0.040129
40	11	9	6	0.999492	0.006174
40	11	9	7	0.999982	0.000490
40	11	9	8	1.000000	0.000017
40	11	9	9	1.000000	0.000000
40	11	10	0	0.023630	0.023630
40	11	10	1	0.153593	0.129964
40	11	10	2	0.432087	0.278494
40	11	10	3	0.735898	0.303811
40	11	10	4	0.920826	0.184928
40	11	10	5	0.985551	0.064725
40	11	10	6	0.998496	0.012945
40	11	10	7	0.999919	0.001423
40	11	10	8	0.999998	0.000079
40	11	10	9	1.000000	0.000002
40	11	10	10	1.000000	0.000000
40	11	11	0	0.014966	0.014966
40	11	11	1	0.110272	0.095307
40	11	11	2	0.348539	0.238267
40	11	11	3	0.654882	0.306343
40	11	11	4	0.877676	0.222795
40	11	11	5	0.972606	0.094930
40	11	11	6	0.996339	0.023732
40	11	11	7	0.999729	0.003390
40	11	11	8	0.999990	0.000261
40	11	11	9	1.000000	0.000010
40	11	11	10	1.000000	0.000000
40	11	11	11	1.000000	0.000000
40	12	1	0	0.700000	0.700000
40	12	1	1	1.000000	0.300000
40	12	2	0	0.484615	0.484615
40	12	2	1	0.915385	0.430769
40	12	2	2	1.000000	0.084615
40	12	3	0	0.331579	0.331579
40	12	3	1	0.790688	0.459109
40	12	3	2	0.977733	0.187045
40	12	3	3	1.000000	0.022267
40	12	4	0	0.224040	0.224040
40	12	4	1	0.654196	0.430156
40	12	4	2	0.927180	0.272984
40	12	4	3	0.994584	0.067403
40	12	4	4	1.000000	0.005416
40	12	5	0	0.149360	0.149360
40	12	5	1	0.522760	0.373400
40	12	5	2	0.851351	0.328592
40	12	5	3	0.977733	0.126381
40	12	5	4	0.998796	0.021064
40	12	5	5	1.000000	0.001204
40	12	6	0	0.098151	0.098151
40	12	6	1	0.405405	0.307255
40	12	6	2	0.757468	0.352063
40	12	6	3	0.945235	0.187767
40	12	6	4	0.993982	0.048747
40	12	6	5	0.999759	0.005777
40	12	6	6	1.000000	0.000241
40	12	7	0	0.063509	0.063509
40	12	7	1	0.305999	0.242490
40	12	7	2	0.653920	0.347921
40	12	7	3	0.895532	0.241612
40	12	7	4	0.982512	0.086980
40	12	7	5	0.998570	0.016058
40	12	7	6	0.999957	0.001388
40	12	7	7	1.000000	0.000042
40	12	8	0	0.040415	0.040415
40	12	8	1	0.225169	0.184754
40	12	8	2	0.548490	0.323320
40	12	8	3	0.829638	0.281148
40	12	8	4	0.961426	0.131788
40	12	8	5	0.995163	0.033738
40	12	8	6	0.999705	0.004542
40	12	8	7	0.999993	0.000288
40	12	8	8	1.000000	0.000006
40	12	9	0	0.025259	0.025259
40	12	9	1	0.161660	0.136401
40	12	9	2	0.447452	0.285792
40	12	9	3	0.750565	0.303113
40	12	9	4	0.928479	0.177914
40	12	9	5	0.987783	0.059305
40	12	9	6	0.998854	0.011070
40	12	9	7	0.999948	0.001095
40	12	9	8	0.999999	0.000051
40	12	9	9	1.000000	0.000001
40	12	10	0	0.015482	0.015482
40	12	10	1	0.113260	0.097778
40	12	10	2	0.355261	0.242001
40	12	10	3	0.662564	0.307303
40	12	10	4	0.882565	0.220001
40	12	10	5	0.974392	0.091827
40	12	10	6	0.996711	0.022319
40	12	10	7	0.999772	0.003061
40	12	10	8	0.999993	0.000221
40	12	10	9	1.000000	0.000007
40	12	10	10	1.000000	0.000000
40	12	11	0	0.009289	0.009289
40	12	11	1	0.077408	0.068119
40	12	11	2	0.274594	0.197186
40	12	11	3	0.570373	0.295779
40	12	11	4	0.823898	0.253525
40	12	11	5	0.952966	0.129067
40	12	11	6	0.992247	0.039281
40	12	11	7	0.999262	0.007015
40	12	11	8	0.999963	0.000701
40	12	11	9	0.999999	0.000036
40	12	11	10	1.000000	0.000001
40	12	11	11	1.000000	0.000000
40	12	12	0	0.005445	0.005445
40	12	12	1	0.051570	0.046124
40	12	12	2	0.206599	0.155029
40	12	12	3	0.478580	0.271981
40	12	12	4	0.753960	0.275381
40	12	12	5	0.921812	0.167851
40	12	12	6	0.984120	0.062308
40	12	12	7	0.998052	0.013932
40	12	12	8	0.999866	0.001814
40	12	12	9	0.999995	0.000129
40	12	12	10	1.000000	0.000004
40	12	12	11	1.000000	0.000000
40	12	12	12	1.000000	0.000000
40	13	1	0	0.675000	0.675000
40	13	1	1	1.000000	0.325000
40	13	2	0	0.450000	0.450000
40	13	2	1	0.900000	0.450000
40	13	2	2	1.000000	0.100000
40	13	3	0	0.296053	0.296053
40	13	3	1	0.757895	0.461842
40	13	3	2	0.971053	0.213158
40	13	3	3	1.000000	0.028947
40	13	4	0	0.192034	0.192034
40	13	4	1	0.608108	0.416074
40	13	4	2	0.907681	0.299573
40	13	4	3	0.992176	0.084495
40	13	4	4	1.000000	0.007824
40	13	5	0	0.122688	0.122688
40	13	5	1	0.469417	0.346728
40	13	5	2	0.816145	0.346728
40	13	5	3	0.968705	0.152560
40	13	5	4	0.998044	0.029339
40	13	5	5	1.000000	0.001956
40	13	6	0	0.077118	0.077118
40	13	6	1	0.350538	0.273420
40	13	6	2	0.707173	0.356635
40	13	6	3	0.925117	0.217943
40	13	6	4	0.990500	0.065383
40	13	6	5	0.999553	0.009053
40	13	6	6	1.000000	0.000447
40	13	7	0	0.047632	0.047632
40	13	7	1	0.254037	0.206405
40	13	7	2	0.591791	0.337754
40	13	7	3	0.861016	0.269224
40	13	7	4	0.973193	0.112177
40	13	7	5	0.997423	0.024230
40	13	7	6	0.999908	0.002485
40	13	7	7	1.000000	0.000092
40	13	8	0	0.028868	0.028868
40	13	8	1	0.178981	0.150113
40	13	8	2	0.479207	0.300226
40	13	8	3	0.779433	0.300226
40	13	8	4	0.942599	0.163166
40	13	8	5	0.991549	0.048950
40	13	8	6	0.999381	0.007832
40	13	8	7	0.999983	0.000602
40	13	8	8	1.000000	0.000017
40	13	9	0	0.017140	0.017140
40	13	9	1	0.122688	0.105548
40	13	9	2	0.376004	0.253316
40	13	9	3	0.685612	0.309608
40	13	9	4	0.896708	0.211096
40	13	9	5	0.979311	0.082603
40	13	9	6	0.997667	0.018356
40	13	9	7	0.999870	0.002203
40	13	9	8	0.999997	0.000127
40	13	9	9	1.000000	0.000003
40	13	10	0	0.009952	0.009952
40	13	10	1	0.081831	0.071879
40	13	10	2	0.286118	0.204287
40	13	10	3	0.585739	0.299621
40	13	10	4	0.835422	0.249684
40	13	10	5	0.957994	0.122572
40	13	10	6	0.993523	0.035528
40	13	10	7	0.999444	0.005921

N	n	k	x	P(x)	p(x)
40	13	10	8	0.999977	0.000533
40	13	10	9	1.000000	0.000023
40	13	10	10	1.000000	0.000000
40	13	11	0	0.005640	0.005640
40	13	11	1	0.053080	0.047440
40	13	11	2	0.211213	0.158133
40	13	11	3	0.485865	0.274652
40	13	11	4	0.760517	0.274652
40	13	11	5	0.925309	0.164791
40	13	11	6	0.985233	0.059924
40	13	11	7	0.998260	0.013027
40	13	11	8	0.999888	0.001628
40	13	11	9	0.999997	0.000109
40	13	11	10	1.000000	0.000003
40	13	11	11	1.000000	0.000000
40	13	12	0	0.003112	0.003112
40	13	12	1	0.033449	0.030338
40	13	12	2	0.151231	0.117782
40	13	12	3	0.391157	0.239926
40	13	12	4	0.675280	0.284123
40	13	12	5	0.879849	0.204569
40	13	12	6	0.970768	0.090919
40	13	12	7	0.995564	0.024796
40	13	12	8	0.999607	0.004043
40	13	12	9	0.999982	0.000374
40	13	12	10	0.999999	0.000018
40	13	12	11	1.000000	0.000000
40	13	12	12	1.000000	0.000000
40	13	13	0	0.001667	0.001667
40	13	13	1	0.020447	0.018781
40	13	13	2	0.104960	0.084512
40	13	13	3	0.305469	0.200510
40	13	13	4	0.583955	0.278486
40	13	13	5	0.821401	0.237446
40	13	13	6	0.948038	0.126638
40	13	13	7	0.990251	0.042213
40	13	13	8	0.998885	0.008634
40	13	13	9	0.999928	0.001043
40	13	13	10	0.999998	0.000070
40	13	13	11	1.000000	0.000002
40	13	13	12	1.000000	0.000000
40	13	13	13	1.000000	0.000000
40	14	1	0	0.650000	0.650000
40	14	1	1	1.000000	0.350000
40	14	2	0	0.416667	0.416667
40	14	2	1	0.883333	0.466667
40	14	2	2	1.000000	0.116667
40	14	3	0	0.263158	0.263158
40	14	3	1	0.723684	0.460526
40	14	3	2	0.963158	0.239474

N	n	k	x	P(x)	p(x)
40	14	3	3	1.000000	0.036842
40	14	4	0	0.163585	0.163585
40	14	4	1	0.561878	0.398293
40	14	4	2	0.885491	0.323613
40	14	4	3	0.989047	0.103556
40	14	4	4	1.000000	0.010953
40	14	5	0	0.099968	0.099968
40	14	5	1	0.418050	0.318081
40	14	5	2	0.777620	0.359570
40	14	5	3	0.957405	0.179785
40	14	5	4	0.996957	0.039553
40	14	5	5	1.000000	0.003043
40	14	6	0	0.059981	0.059981
40	14	6	1	0.299905	0.239924
40	14	6	2	0.654339	0.354433
40	14	6	3	0.900901	0.246562
40	14	6	4	0.985657	0.084756
40	14	6	5	0.999218	0.013561
40	14	6	6	1.000000	0.000782
40	14	7	0	0.035283	0.035283
40	14	7	1	0.208169	0.172886
40	14	7	2	0.529244	0.321075
40	14	7	3	0.821131	0.291886
40	14	7	4	0.960728	0.139598
40	14	7	5	0.995628	0.034899
40	14	7	6	0.999816	0.004188
40	14	7	7	1.000000	0.000184
40	14	8	0	0.020314	0.020314
40	14	8	1	0.140063	0.119748
40	14	8	2	0.412490	0.272427
40	14	8	3	0.723835	0.311345
40	14	8	4	0.918426	0.194591
40	14	8	5	0.986110	0.067684
40	14	8	6	0.998801	0.012691
40	14	8	7	0.999961	0.001160
40	14	8	8	1.000000	0.000039
40	14	9	0	0.011427	0.011427
40	14	9	1	0.091415	0.079988
40	14	9	2	0.310330	0.218915
40	14	9	3	0.616810	0.306481
40	14	9	4	0.857616	0.240806
40	14	9	5	0.967074	0.109457
40	14	9	6	0.995628	0.028554
40	14	9	7	0.999707	0.004079
40	14	9	8	0.999993	0.000286
40	14	9	9	1.000000	0.000007
40	14	10	0	0.006266	0.006266
40	14	10	1	0.057872	0.051605
40	14	10	2	0.225588	0.167717
40	14	10	3	0.508059	0.282471

N	n	k	x	P(x)	p(x)
40	14	10	4	0.779937	0.271878
40	14	10	5	0.935296	0.155359
40	14	10	6	0.988259	0.052963
40	14	10	7	0.998786	0.010527
40	14	10	8	0.999937	0.001151
40	14	10	9	0.999999	0.000061
40	14	10	10	1.000000	0.000001
40	14	11	0	0.003342	0.003342
40	14	11	1	0.035509	0.032167
40	14	11	2	0.158502	0.122992
40	14	11	3	0.404487	0.245985
40	14	11	4	0.689311	0.284825
40	14	11	5	0.888688	0.199377
40	14	11	6	0.974136	0.085447
40	14	11	7	0.996330	0.022194
40	14	11	8	0.999707	0.003377
40	14	11	9	0.999989	0.000281
40	14	11	10	1.000000	0.000011
40	14	11	11	1.000000	0.000000
40	14	12	0	0.001729	0.001729
40	14	12	1	0.021089	0.019361
40	14	12	2	0.107608	0.086519
40	14	12	3	0.311182	0.203574
40	14	12	4	0.591096	0.279914
40	14	12	5	0.826813	0.235717
40	14	12	6	0.950564	0.123751
40	14	12	7	0.990972	0.040409
40	14	12	8	0.999008	0.008036
40	14	12	9	0.999940	0.000932
40	14	12	10	0.999998	0.000058
40	14	12	11	1.000000	0.000002
40	14	12	12	1.000000	0.000000
40	14	13	0	0.000864	0.000864
40	14	13	1	0.012101	0.011236
40	14	13	2	0.070529	0.058428
40	14	13	3	0.231207	0.160678
40	14	13	4	0.491126	0.259920
40	14	13	5	0.751046	0.259920
40	14	13	6	0.915206	0.164160
40	14	13	7	0.980870	0.065664
40	14	13	8	0.997286	0.016416
40	14	13	9	0.999774	0.002487
40	14	13	10	0.999990	0.000216
40	14	13	11	1.000000	0.000010
40	14	13	12	1.000000	0.000000
40	14	13	13	1.000000	0.000000
40	14	14	0	0.000416	0.000416
40	14	14	1	0.006691	0.006274
40	14	14	2	0.044561	0.037870
40	14	14	3	0.165745	0.121185

N	n	k	x	P(x)	p(x)
40	14	14	4	0.394860	0.229115
40	14	14	5	0.664406	0.269547
40	14	14	6	0.866566	0.202160
40	14	14	7	0.963846	0.097280
40	14	14	8	0.993638	0.029792
40	14	14	9	0.999313	0.005675
40	14	14	10	0.999958	0.000645
40	14	14	11	0.999999	0.000041
40	14	14	12	1.000000	0.000001
40	14	14	13	1.000000	0.000000
40	14	14	14	1.000000	0.000000
40	15	1	0	0.625000	0.625000
40	15	1	1	1.000000	0.375000
40	15	2	0	0.384615	0.384615
40	15	2	1	0.865385	0.480769
40	15	2	2	1.000000	0.134615
40	15	3	0	0.232793	0.232793
40	15	3	1	0.688259	0.455466
40	15	3	2	0.953947	0.265688
40	15	3	3	1.000000	0.046053
40	15	4	0	0.138418	0.138418
40	15	4	1	0.515921	0.377503
40	15	4	2	0.860597	0.344677
40	15	4	3	0.985064	0.124467
40	15	4	4	1.000000	0.014936
40	15	5	0	0.080744	0.080744
40	15	5	1	0.369114	0.288370
40	15	5	2	0.736131	0.367017
40	15	5	3	0.943575	0.207444
40	15	5	4	0.995436	0.051861
40	15	5	5	1.000000	0.004564
40	15	6	0	0.046139	0.046139
40	15	6	1	0.253766	0.207627
40	15	6	2	0.599810	0.346044
40	15	6	3	0.872451	0.272641
40	15	6	4	0.979137	0.106686
40	15	6	5	0.998696	0.019559
40	15	6	6	1.000000	0.001304
40	15	7	0	0.025784	0.025784
40	15	7	1	0.168273	0.142489
40	15	7	2	0.467499	0.299227
40	15	7	3	0.776225	0.308726
40	15	7	4	0.944621	0.168396
40	15	7	5	0.992943	0.048322
40	15	7	6	0.999655	0.006711
40	15	7	7	1.000000	0.000345
40	15	8	0	0.014064	0.014064
40	15	8	1	0.107823	0.093759
40	15	8	2	0.349622	0.241799
40	15	8	3	0.663961	0.314339

N	n	k	x	P(x)	p(x)
40	15	8	4	0.888489	0.224528
40	15	8	5	0.978300	0.089811
40	15	8	6	0.997824	0.019524
40	15	8	7	0.999916	0.002092
40	15	8	8	1.000000	0.000084
40	15	9	0	0.007471	0.007471
40	15	9	1	0.066803	0.059332
40	15	9	2	0.251391	0.184588
40	15	9	3	0.546084	0.294693
40	15	9	4	0.811308	0.265224
40	15	9	5	0.950234	0.138927
40	15	9	6	0.992333	0.042099
40	15	9	7	0.999393	0.007060
40	15	9	8	0.999982	0.000588
40	15	9	9	1.000000	0.000018
40	15	10	0	0.003856	0.003856
40	15	10	1	0.040008	0.036152
40	15	10	2	0.173983	0.133975
40	15	10	3	0.432009	0.258026
40	15	10	4	0.717196	0.285187
40	15	10	5	0.905419	0.188223
40	15	10	6	0.980111	0.074692
40	15	10	7	0.997571	0.017460
40	15	10	8	0.999849	0.002277
40	15	10	9	0.999996	0.000148
40	15	10	10	1.000000	0.000004
40	15	11	0	0.001928	0.001928
40	15	11	1	0.023137	0.021209
40	15	11	2	0.115927	0.092790
40	15	11	3	0.328799	0.212871
40	15	11	4	0.612627	0.283829
40	15	11	5	0.842678	0.230051
40	15	11	6	0.957703	0.115025
40	15	11	7	0.992915	0.035212
40	15	11	8	0.999317	0.006402
40	15	11	9	0.999967	0.000649
40	15	11	10	0.999999	0.000032
40	15	11	11	1.000000	0.000001
40	15	12	0	0.000931	0.000931
40	15	12	1	0.012898	0.011968
40	15	12	2	0.074332	0.061433
40	15	12	3	0.240714	0.166382
40	15	12	4	0.504908	0.264254
40	15	12	5	0.763350	0.258382
40	15	12	6	0.922006	0.158656
40	15	12	7	0.983202	0.061196
40	15	12	8	0.997772	0.014570
40	15	12	9	0.999832	0.002060
40	15	12	10	0.999994	0.000161
40	15	12	11	1.000000	0.000006

N	n	k	x	P(x)	p(x)
40	15	12	12	1.000000	0.000000
40	15	13	0	0.000432	0.000432
40	15	13	1	0.006915	0.006482
40	15	13	2	0.045809	0.038895
40	15	13	3	0.169407	0.123598
40	15	13	4	0.401154	0.231747
40	15	13	5	0.671071	0.269917
40	15	13	6	0.871009	0.199938
40	15	13	7	0.965717	0.094708
40	15	13	8	0.994129	0.028412
40	15	13	9	0.999391	0.005262
40	15	13	10	0.999965	0.000574
40	15	13	11	0.999999	0.000034
40	15	13	12	1.000000	0.000001
40	15	13	13	1.000000	0.000000
40	15	14	0	0.000192	0.000192
40	15	14	1	0.003553	0.003361
40	15	14	2	0.027082	0.023529
40	15	14	3	0.114475	0.087393
40	15	14	4	0.306739	0.192264
40	15	14	5	0.571102	0.264363
40	15	14	6	0.804363	0.233261
40	15	14	7	0.937656	0.133292
40	15	14	8	0.986763	0.049108
40	15	14	9	0.998222	0.011458
40	15	14	10	0.999859	0.001637
40	15	14	11	0.999994	0.000135
40	15	14	12	1.000000	0.000006
40	15	14	13	1.000000	0.000000
40	15	14	14	1.000000	0.000000
40	15	15	0	0.000081	0.000081
40	15	15	1	0.001743	0.001662
40	15	15	2	0.015318	0.013574
40	15	15	3	0.074140	0.058822
40	15	15	4	0.225396	0.151257
40	15	15	5	0.469424	0.244027
40	15	15	6	0.723619	0.254195
40	15	15	7	0.896643	0.173024
40	15	15	8	0.973542	0.076899
40	15	15	9	0.995577	0.022035
40	15	15	10	0.999544	0.003966
40	15	15	11	0.999973	0.000429
40	15	15	12	0.999999	0.000026
40	15	15	13	1.000000	0.000001
40	15	15	14	1.000000	0.000000
40	15	15	15	1.000000	0.000000
40	16	1	0	0.600000	0.600000
40	16	1	1	1.000000	0.400000
40	16	2	0	0.353846	0.353846
40	16	2	1	0.846154	0.492308

N	n	k	x	P(x)	p(x)
40	16	2	2	1.000000	0.153846
40	16	3	0	0.204858	0.204858
40	16	3	1	0.651822	0.446964
40	16	3	2	0.943320	0.291498
40	16	3	3	1.000000	0.056680
40	16	4	0	0.116271	0.116271
40	16	4	1	0.470620	0.354349
40	16	4	2	0.833023	0.362403
40	16	4	3	0.980085	0.147062
40	16	4	4	1.000000	0.019915
40	16	5	0	0.064595	0.064595
40	16	5	1	0.322975	0.258380
40	16	5	2	0.692089	0.369114
40	16	5	3	0.926980	0.234891
40	16	5	4	0.993362	0.066382
40	16	5	5	1.000000	0.006638
40	16	6	0	0.035066	0.035066
40	16	6	1	0.212241	0.177175
40	16	6	2	0.544443	0.332203
40	16	6	3	0.839734	0.295291
40	16	6	4	0.970602	0.130868
40	16	6	5	0.997914	0.027312
40	16	6	6	1.000000	0.002086
40	16	7	0	0.018564	0.018564
40	16	7	1	0.134075	0.115511
40	16	7	2	0.407654	0.273579
40	16	7	3	0.726829	0.319175
40	16	7	4	0.924413	0.197585
40	16	7	5	0.989078	0.064664
40	16	7	6	0.999386	0.010309
40	16	7	7	1.000000	0.000614
40	16	8	0	0.009563	0.009563
40	16	8	1	0.081570	0.072007
40	16	8	2	0.291590	0.210020
40	16	8	3	0.601093	0.309503
40	16	8	4	0.852565	0.251471
40	16	8	5	0.967523	0.114958
40	16	8	6	0.996262	0.028740
40	16	8	7	0.999833	0.003570
40	16	8	8	1.000000	0.000167
40	16	9	0	0.004782	0.004782
40	16	9	1	0.047817	0.043035
40	16	9	2	0.199706	0.151889
40	16	9	3	0.475358	0.275651
40	16	9	4	0.758263	0.282905
40	16	9	5	0.928006	0.169743
40	16	9	6	0.987281	0.059275
40	16	9	7	0.998828	0.011547
40	16	9	8	0.999958	0.001130
40	16	9	9	1.000000	0.000042

N	n	k	x	P(x)	p(x)
40	16	10	0	0.002314	0.002314
40	16	10	1	0.026993	0.024680
40	16	10	2	0.131111	0.104118
40	16	10	3	0.359762	0.228651
40	16	10	4	0.648751	0.288989
40	16	10	5	0.867775	0.219023
40	16	10	6	0.968160	0.100386
40	16	10	7	0.995476	0.027316
40	16	10	8	0.999667	0.004190
40	16	10	9	0.999990	0.000324
40	16	10	10	1.000000	0.000009
40	16	11	0	0.001080	0.001080
40	16	11	1	0.014654	0.013574
40	16	11	2	0.082523	0.067869
40	16	11	3	0.260680	0.178157
40	16	11	4	0.533155	0.272475
40	16	11	5	0.787466	0.254310
40	16	11	6	0.934698	0.147232
40	16	11	7	0.987281	0.052583
40	16	11	8	0.998549	0.011268
40	16	11	9	0.999915	0.001366
40	16	11	10	0.999998	0.000083
40	16	11	11	1.000000	0.000002
40	16	12	0	0.000484	0.000484
40	16	12	1	0.007633	0.007149
40	16	12	2	0.049758	0.042126
40	16	12	3	0.180817	0.131058
40	16	12	4	0.420407	0.239591
40	16	12	5	0.691003	0.270596
40	16	12	6	0.883929	0.192925
40	16	12	7	0.970962	0.087034
40	16	12	8	0.995441	0.024478
40	16	12	9	0.999585	0.004144
40	16	12	10	0.999981	0.000396
40	16	12	11	1.000000	0.000019
40	16	12	12	1.000000	0.000000
40	16	13	0	0.000207	0.000207
40	16	13	1	0.003803	0.003596
40	16	13	2	0.028696	0.024893
40	16	13	3	0.119968	0.091273
40	16	13	4	0.317725	0.197757
40	16	13	5	0.584698	0.266972
40	16	13	6	0.815027	0.230329
40	16	13	7	0.942987	0.127961
40	16	13	8	0.988447	0.045460
40	16	13	9	0.998549	0.010102
40	16	13	10	0.999896	0.001347
40	16	13	11	0.999996	0.000100
40	16	13	12	1.000000	0.000004
40	16	13	13	1.000000	0.000000

Table for $N = 2$, $n = 1$, through $N = 100$, $n = 50$

N	n	k	x	P(x)	p(x)
40	16	14	0	0.000085	0.000085
40	16	14	1	0.001805	0.001721
40	16	14	2	0.015788	0.013983
40	16	14	3	0.076022	0.060234
40	16	14	4	0.229833	0.153811
40	16	14	5	0.475931	0.246098
40	16	14	6	0.729720	0.253788
40	16	14	7	0.900334	0.170614
40	16	14	8	0.974977	0.074644
40	16	14	9	0.995930	0.020953
40	16	14	10	0.999597	0.003667
40	16	14	11	0.999978	0.000381
40	16	14	12	0.999999	0.000022
40	16	14	13	1.000000	0.000001
40	16	14	14	1.000000	0.000000
40	16	15	0	0.000032	0.000032
40	16	15	1	0.000813	0.000780
40	16	15	2	0.008259	0.007446
40	16	15	3	0.045905	0.037646
40	16	15	4	0.158843	0.112938
40	16	15	5	0.371813	0.212969
40	16	15	6	0.632109	0.260296
40	16	15	7	0.841275	0.209166
40	16	15	8	0.952010	0.110735
40	16	15	9	0.990289	0.038279
40	16	15	10	0.998751	0.008462
40	16	15	11	0.999904	0.001154
40	16	15	12	0.999996	0.000092
40	16	15	13	1.000000	0.000004
40	16	15	14	1.000000	0.000000
40	16	15	15	1.000000	0.000000
40	16	16	0	0.000012	0.000012
40	16	16	1	0.000345	0.000333
40	16	16	2	0.004089	0.003745
40	16	16	3	0.026329	0.022240
40	16	16	4	0.104633	0.078304
40	16	16	5	0.278106	0.173473
40	16	16	6	0.527990	0.249884
40	16	16	7	0.765975	0.237985
40	16	16	8	0.916575	0.150600
40	16	16	9	0.979571	0.062996
40	16	16	10	0.996720	0.017149
40	16	16	11	0.999674	0.002954
40	16	16	12	0.999981	0.000308
40	16	16	13	0.999999	0.000018
40	16	16	14	1.000000	0.000001
40	16	16	15	1.000000	0.000000
40	16	16	16	1.000000	0.000000
40	17	1	0	0.575000	0.575000
40	17	1	1	1.000000	0.425000
40	17	2	0	0.324359	0.324359
40	17	2	1	0.825641	0.501282
40	17	2	2	1.000000	0.174359
40	17	3	0	0.179251	0.179251
40	17	3	1	0.614575	0.435324
40	17	3	2	0.931174	0.316599
40	17	3	3	1.000000	0.068826
40	17	4	0	0.096892	0.096892
40	17	4	1	0.426327	0.329434
40	17	4	2	0.802823	0.376496
40	17	4	3	0.973958	0.171135
40	17	4	4	1.000000	0.026042
40	17	5	0	0.051138	0.051138
40	17	5	1	0.279911	0.228774
40	17	5	2	0.645950	0.366038
40	17	5	3	0.907405	0.261456
40	17	5	4	0.990596	0.083190
40	17	5	5	1.000000	0.009404
40	17	6	0	0.026299	0.026299
40	17	6	1	0.175329	0.149030
40	17	6	2	0.489076	0.313747
40	17	6	3	0.802823	0.313747
40	17	6	4	0.959696	0.156873
40	17	6	5	0.996776	0.037079
40	17	6	6	1.000000	0.003224
40	17	7	0	0.013150	0.013150
40	17	7	1	0.105197	0.092048
40	17	7	2	0.350658	0.245461
40	17	7	3	0.673633	0.322975
40	17	7	4	0.899715	0.226082
40	17	7	5	0.983689	0.083973
40	17	7	6	0.998957	0.015268
40	17	7	7	1.000000	0.001043
40	17	8	0	0.006376	0.006376
40	17	8	1	0.060568	0.054193
40	17	8	2	0.239085	0.178517
40	17	8	3	0.536613	0.297528
40	17	8	4	0.810653	0.274039
40	17	8	5	0.953153	0.142500
40	17	8	6	0.993867	0.040714
40	17	8	7	0.999684	0.005816
40	17	8	8	1.000000	0.000316
40	17	9	0	0.002989	0.002989
40	17	9	1	0.033472	0.030483
40	17	9	2	0.155405	0.121933
40	17	9	3	0.406445	0.251039
40	17	9	4	0.699324	0.292879
40	17	9	5	0.899715	0.200391
40	17	9	6	0.979872	0.080156
40	17	9	7	0.997866	0.017994
40	17	9	8	0.999911	0.002045
40	17	9	9	1.000000	0.000089
40	17	10	0	0.001350	0.001350
40	17	10	1	0.017739	0.016389
40	17	10	2	0.096405	0.078667
40	17	10	3	0.293072	0.196667
40	17	10	4	0.576504	0.283432
40	17	10	5	0.822145	0.245641
40	17	10	6	0.951429	0.129285
40	17	10	7	0.992062	0.040632
40	17	10	8	0.999317	0.007256
40	17	10	9	0.999977	0.000660
40	17	10	10	1.000000	0.000023
40	17	11	0	0.000585	0.000585
40	17	11	1	0.008998	0.008413
40	17	11	2	0.057072	0.048074
40	17	11	3	0.201294	0.144222
40	17	11	4	0.453683	0.252389
40	17	11	5	0.723888	0.270205
40	17	11	6	0.904025	0.180137
40	17	11	7	0.978517	0.074493
40	17	11	8	0.997141	0.018623
40	17	11	9	0.999801	0.002660
40	17	11	10	0.999995	0.000193
40	17	11	11	1.000000	0.000005
40	17	12	0	0.000242	0.000242
40	17	12	1	0.004356	0.004114
40	17	12	2	0.032206	0.027850
40	17	12	3	0.131670	0.099464
40	17	12	4	0.340544	0.208874
40	17	12	5	0.612079	0.271536
40	17	12	6	0.835697	0.223618
40	17	12	7	0.952830	0.117133
40	17	12	8	0.991361	0.038531
40	17	12	9	0.999067	0.007706
40	17	12	10	0.999948	0.000881
40	17	12	11	0.999999	0.000051
40	17	12	12	1.000000	0.000001
40	17	13	0	0.000095	0.000095
40	17	13	1	0.002005	0.001910
40	17	13	2	0.017286	0.015281
40	17	13	3	0.081938	0.064651
40	17	13	4	0.243566	0.161629
40	17	13	5	0.495707	0.252140
40	17	13	6	0.747847	0.252140
40	17	13	7	0.910997	0.163150
40	17	13	8	0.978976	0.067979
40	17	13	9	0.996865	0.017889
40	17	13	10	0.999728	0.002862
40	17	13	11	0.999988	0.000260
40	17	13	12	1.000000	0.000012
40	17	13	13	1.000000	0.000000
40	17	14	0	0.000035	0.000035
40	17	14	1	0.000873	0.000838
40	17	14	2	0.008797	0.007924
40	17	14	3	0.048415	0.039618
40	17	14	4	0.165745	0.117330
40	17	14	5	0.383644	0.217899
40	17	14	6	0.645124	0.261479
40	17	14	7	0.850571	0.205448
40	17	14	8	0.956317	0.105745
40	17	14	9	0.991565	0.035248
40	17	14	10	0.998986	0.007421
40	17	14	11	0.999930	0.000944
40	17	14	12	0.999998	0.000067
40	17	14	13	1.000000	0.000002
40	17	14	14	1.000000	0.000000
40	17	15	0	0.000012	0.000012
40	17	15	1	0.000358	0.000345
40	17	15	2	0.004226	0.003868
40	17	15	3	0.027082	0.022857
40	17	15	4	0.107080	0.079998
40	17	15	5	0.283076	0.175996
40	17	15	6	0.534498	0.251422
40	17	15	7	0.771553	0.237055
40	17	15	8	0.919712	0.148159
40	17	15	9	0.980719	0.061007
40	17	15	10	0.996988	0.016268
40	17	15	11	0.999712	0.002724
40	17	15	12	0.999984	0.000272
40	17	15	13	0.999999	0.000015
40	17	15	14	1.000000	0.000000
40	17	15	15	1.000000	0.000000
40	17	16	0	0.000004	0.000004
40	17	16	1	0.000137	0.000133
40	17	16	2	0.001905	0.001768
40	17	16	3	0.014282	0.012378
40	17	16	4	0.065481	0.051199
40	17	16	5	0.198598	0.133117
40	17	16	6	0.423872	0.225274
40	17	16	7	0.676731	0.252859
40	17	16	8	0.866375	0.189644
40	17	16	9	0.961197	0.094822
40	17	16	10	0.992433	0.031235
40	17	16	11	0.999058	0.006626
40	17	16	12	0.999930	0.000872
40	17	16	13	0.999997	0.000067
40	17	16	14	1.000000	0.000003
40	17	16	15	1.000000	0.000000
40	17	16	16	1.000000	0.000000

Table for $N = 2$, $n = 1$, through $N = 100$, $n = 50$

N	n	k	x	P(x)	p(x)
40	17	17	0	0.000001	0.000001
40	17	17	1	0.000048	0.000047
40	17	17	2	0.000800	0.000752
40	17	17	3	0.007062	0.006263
40	17	17	4	0.037749	0.030686
40	17	17	5	0.132039	0.094291
40	17	17	6	0.320621	0.188582
40	17	17	7	0.571373	0.250752
40	17	17	8	0.795258	0.223885
40	17	17	9	0.929590	0.134331
40	17	17	10	0.983322	0.053732
40	17	17	11	0.997402	0.014080
40	17	17	12	0.999748	0.002347
40	17	17	13	0.999986	0.000238
40	17	17	14	0.999999	0.000014
40	17	17	15	1.000000	0.000000
40	17	17	16	1.000000	0.000000
40	17	17	17	1.000000	0.000000
40	18	1	0	0.550000	0.550000
40	18	1	1	1.000000	0.450000
40	18	2	0	0.296154	0.296154
40	18	2	1	0.803846	0.507692
40	18	2	2	1.000000	0.196154
40	18	3	0	0.155870	0.155870
40	18	3	1	0.576721	0.420850
40	18	3	2	0.917409	0.340688
40	18	3	3	1.000000	0.082591
40	18	4	0	0.080042	0.080042
40	18	4	1	0.383357	0.303315
40	18	4	2	0.770084	0.386727
40	18	4	3	0.966517	0.196433
40	18	4	4	1.000000	0.033483
40	18	5	0	0.040021	0.040021
40	18	5	1	0.240125	0.200104
40	18	5	2	0.598205	0.358081
40	18	5	3	0.884670	0.286465
40	18	5	4	0.986979	0.102309
40	18	5	5	1.000000	0.013021
40	18	6	0	0.019439	0.019439
40	18	6	1	0.142931	0.123493
40	18	6	2	0.434511	0.291580
40	18	6	3	0.761899	0.327388
40	18	6	4	0.946055	0.184156
40	18	6	5	0.995163	0.049108
40	18	6	6	1.000000	0.004836
40	18	7	0	0.009148	0.009148
40	18	7	1	0.081185	0.072037
40	18	7	2	0.297297	0.216112
40	18	7	3	0.617464	0.320166
40	18	7	4	0.870226	0.252763
40	18	7	5	0.976387	0.106160
40	18	7	6	0.998293	0.021906
40	18	7	7	1.000000	0.001707
40	18	8	0	0.004158	0.004158
40	18	8	1	0.044075	0.039917
40	18	8	2	0.192516	0.148441
40	18	8	3	0.471933	0.279418
40	18	8	4	0.762994	0.291060
40	18	8	5	0.934566	0.171572
40	18	8	6	0.990327	0.055761
40	18	8	7	0.999431	0.009104
40	18	8	8	1.000000	0.000569
40	18	9	0	0.001819	0.001819
40	18	9	1	0.022869	0.021050
40	18	9	2	0.118295	0.095426
40	18	9	3	0.340956	0.222661
40	18	9	4	0.635655	0.294699
40	18	9	5	0.864865	0.229210
40	18	9	6	0.969417	0.104552
40	18	9	7	0.996301	0.026885
40	18	9	8	0.999822	0.003521
40	18	9	9	1.000000	0.000178
40	18	10	0	0.000763	0.000763
40	18	10	1	0.011326	0.010563
40	18	10	2	0.069043	0.057717
40	18	10	3	0.233217	0.164174
40	18	10	4	0.502565	0.269348
40	18	10	5	0.768744	0.266179
40	18	10	6	0.928945	0.160201
40	18	10	7	0.986762	0.057817
40	18	10	8	0.998686	0.011925
40	18	10	9	0.999948	0.001262
40	18	10	10	1.000000	0.000052
40	18	11	0	0.000305	0.000305
40	18	11	1	0.005340	0.005035
40	18	11	2	0.038260	0.032920
40	18	11	3	0.151130	0.112870
40	18	11	4	0.376869	0.225739
40	18	11	5	0.653400	0.276531
40	18	11	6	0.864865	0.211465
40	18	11	7	0.965562	0.100697
40	18	11	8	0.994712	0.029149
40	18	11	9	0.999570	0.004858
40	18	11	10	0.999986	0.000416
40	18	11	11	1.000000	0.000014
40	18	12	0	0.000116	0.000116
40	18	12	1	0.002389	0.002273
40	18	12	2	0.020097	0.017709
40	18	12	3	0.092749	0.072652
40	18	12	4	0.267892	0.175143
40	18	12	5	0.529438	0.261546
40	18	12	6	0.777362	0.247924
40	18	12	7	0.927367	0.150005
40	18	12	8	0.984660	0.057293
40	18	12	9	0.998062	0.013402
40	18	12	10	0.999871	0.001809
40	18	12	11	0.999997	0.000125
40	18	12	12	1.000000	0.000003
40	18	13	0	0.000041	0.000041
40	18	13	1	0.001009	0.000967
40	18	13	2	0.009978	0.008969
40	18	13	3	0.053829	0.043851
40	18	13	4	0.180320	0.126492
40	18	13	5	0.408006	0.227685
40	18	13	6	0.671109	0.263103
40	18	13	7	0.868436	0.197327
40	18	13	8	0.964198	0.095762
40	18	13	9	0.993754	0.029556
40	18	13	10	0.999354	0.005600
40	18	13	11	0.999965	0.000611
40	18	13	12	0.999999	0.000034
40	18	13	13	1.000000	0.000001
40	18	14	0	0.000014	0.000014
40	18	14	1	0.000400	0.000386
40	18	14	2	0.004663	0.004263
40	18	14	3	0.029467	0.024804
40	18	14	4	0.114732	0.085265
40	18	14	5	0.298380	0.183647
40	18	14	6	0.554174	0.255795
40	18	14	7	0.788044	0.233869
40	18	14	8	0.928731	0.140687
40	18	14	9	0.983902	0.055171
40	18	14	10	0.997695	0.013793
40	18	14	11	0.999807	0.002112
40	18	14	12	0.999992	0.000185
40	18	14	13	1.000000	0.000008
40	18	14	14	1.000000	0.000000
40	18	15	0	0.000004	0.000004
40	18	15	1	0.000147	0.000143
40	18	15	2	0.002039	0.001892
40	18	15	3	0.015157	0.013118
40	18	15	4	0.068820	0.053663
40	18	15	5	0.206556	0.137736
40	18	15	6	0.436115	0.229559
40	18	15	7	0.689099	0.252984
40	18	15	8	0.874620	0.185521
40	18	15	9	0.964804	0.090184
40	18	15	10	0.993451	0.028647
40	18	15	11	0.999238	0.005787
40	18	15	12	0.999949	0.000711
40	18	15	13	0.999998	0.000049
40	18	15	14	1.000000	0.000002
40	18	15	15	1.000000	0.000000
40	18	16	0	0.000001	0.000001
40	18	16	1	0.000050	0.000049
40	18	16	2	0.000828	0.000778
40	18	16	3	0.007286	0.006458
40	18	16	4	0.038769	0.031482
40	18	16	5	0.134933	0.096164
40	18	16	6	0.325927	0.190993
40	18	16	7	0.577786	0.251859
40	18	16	8	0.800412	0.222626
40	18	16	9	0.932338	0.131926
40	18	16	10	0.984284	0.051946
40	18	16	11	0.997618	0.013334
40	18	16	12	0.999778	0.002161
40	18	16	13	0.999988	0.000210
40	18	16	14	1.000000	0.000011
40	18	16	15	1.000000	0.000000
40	18	16	16	1.000000	0.000000
40	18	17	0	0.000000	0.000000
40	18	17	1	0.000015	0.000015
40	18	17	2	0.000309	0.000294
40	18	17	3	0.003250	0.002941
40	18	17	4	0.020404	0.017154
40	18	17	5	0.082844	0.062440
40	18	17	6	0.230430	0.147586
40	18	17	7	0.462351	0.231921
40	18	17	8	0.707651	0.245301
40	18	17	9	0.882866	0.175215
40	18	17	10	0.966969	0.084103
40	18	17	11	0.993729	0.026760
40	18	17	12	0.999238	0.005509
40	18	17	13	0.999945	0.000706
40	18	17	14	0.999998	0.000053
40	18	17	15	1.000000	0.000002
40	18	17	16	1.000000	0.000000
40	18	17	17	1.000000	0.000000
40	18	18	0	0.000000	0.000000
40	18	18	1	0.000004	0.000004
40	18	18	2	0.000105	0.000101
40	18	18	3	0.001332	0.001227
40	18	18	4	0.009963	0.008630
40	18	18	5	0.047552	0.037589
40	18	18	6	0.153429	0.105877
40	18	18	7	0.351432	0.198003
40	18	18	8	0.600999	0.249567
40	18	18	9	0.814303	0.213305
40	18	18	10	0.937716	0.123412
40	18	18	11	0.985584	0.047869

N	n	k	x	P(x)	p(x)
40	18	18	12	0.997801	0.012217
40	18	18	13	0.999791	0.001990
40	18	18	14	0.999988	0.000197
40	18	18	15	1.000000	0.000011
40	18	18	16	1.000000	0.000000
40	18	18	17	1.000000	0.000000
40	18	18	18	1.000000	0.000000
40	19	1	0	0.525000	0.525000
40	19	1	1	1.000000	0.475000
40	19	2	0	0.269231	0.269231
40	19	2	1	0.780769	0.511538
40	19	2	2	1.000000	0.219231
40	19	3	0	0.134615	0.134615
40	19	3	1	0.538462	0.403846
40	19	3	2	0.901923	0.363462
40	19	3	3	1.000000	0.098077
40	19	4	0	0.065489	0.065489
40	19	4	1	0.341996	0.276507
40	19	4	2	0.734927	0.392931
40	19	4	3	0.957588	0.222661
40	19	4	4	1.000000	0.042412
40	19	5	0	0.030925	0.030925
40	19	5	1	0.203742	0.172817
40	19	5	2	0.549376	0.345634
40	19	5	3	0.858628	0.309252
40	19	5	4	0.982328	0.123701
40	19	5	5	1.000000	0.017672
40	19	6	0	0.014137	0.014137
40	19	6	1	0.114865	0.100728
40	19	6	2	0.381497	0.266632
40	19	6	3	0.717256	0.335759
40	19	6	4	0.929314	0.212058
40	19	6	5	0.992931	0.063617
40	19	6	6	1.000000	0.007069
40	19	7	0	0.006237	0.006237
40	19	7	1	0.061538	0.055301
40	19	7	2	0.248181	0.186642
40	19	7	3	0.559251	0.311071
40	19	7	4	0.835759	0.276507
40	19	7	5	0.966736	0.130977
40	19	7	6	0.997297	0.030561
40	19	7	7	1.000000	0.002703
40	19	8	0	0.002646	0.002646
40	19	8	1	0.031374	0.028728
40	19	8	2	0.152032	0.120658
40	19	8	3	0.408429	0.256398
40	19	8	4	0.710074	0.301644
40	19	8	5	0.911170	0.201096
40	19	8	6	0.985258	0.074088
40	19	8	7	0.999017	0.013759
40	19	8	8	1.000000	0.000983
40	19	9	0	0.001075	0.001075
40	19	9	1	0.015215	0.014140
40	19	9	2	0.087932	0.072718
40	19	9	3	0.280231	0.192298
40	19	9	4	0.568678	0.288447
40	19	9	5	0.823190	0.254512
40	19	9	6	0.955160	0.131969
40	19	9	7	0.993857	0.038698
40	19	9	8	0.999662	0.005805
40	19	9	9	1.000000	0.000338
40	19	10	0	0.000416	0.000416
40	19	10	1	0.007004	0.006588
40	19	10	2	0.048055	0.041050
40	19	10	3	0.180980	0.132925
40	19	10	4	0.429107	0.248127
40	19	10	5	0.708249	0.279143
40	19	10	6	0.899818	0.191568
40	19	10	7	0.978878	0.079060
40	19	10	8	0.997602	0.018725
40	19	10	9	0.999891	0.002289
40	19	10	10	1.000000	0.000109
40	19	11	0	0.000153	0.000153
40	19	11	1	0.003051	0.002899
40	19	11	2	0.024793	0.021741
40	19	11	3	0.110086	0.085294
40	19	11	4	0.305043	0.194957
40	19	11	5	0.577983	0.272939
40	19	11	6	0.816805	0.238822
40	19	11	7	0.947254	0.130449
40	19	11	8	0.990737	0.043483
40	19	11	9	0.999128	0.008391
40	19	11	10	0.999967	0.000839
40	19	11	11	1.000000	0.000033
40	19	12	0	0.000053	0.000053
40	19	12	1	0.001252	0.001200
40	19	12	2	0.012048	0.010796
40	19	12	3	0.063028	0.050980
40	19	12	4	0.204204	0.141176
40	19	12	5	0.446219	0.242015
40	19	12	6	0.709746	0.263528
40	19	12	7	0.893275	0.183528
40	19	12	8	0.974243	0.080968
40	19	12	9	0.996234	0.021991
40	19	12	10	0.999707	0.003472
40	19	12	11	0.999991	0.000284
40	19	12	12	1.000000	0.000009
40	19	13	0	0.000017	0.000017
40	19	13	1	0.000481	0.000464
40	19	13	2	0.005493	0.005012
40	19	13	3	0.033897	0.028403
40	19	13	4	0.128574	0.094677
40	19	13	5	0.325211	0.196637
40	19	13	6	0.587394	0.262183
40	19	13	7	0.814620	0.227225
40	19	13	8	0.942434	0.127814
40	19	13	9	0.988380	0.045946
40	19	13	10	0.998591	0.010210
40	19	13	11	0.999910	0.001319
40	19	13	12	0.999998	0.000088
40	19	13	13	1.000000	0.000002
40	19	14	0	0.000005	0.000005
40	19	14	1	0.000172	0.000167
40	19	14	2	0.002337	0.002166
40	19	14	3	0.017065	0.014728
40	19	14	4	0.075975	0.058910
40	19	14	5	0.223251	0.147276
40	19	14	6	0.461158	0.237907
40	19	14	7	0.713631	0.252473
40	19	14	8	0.890362	0.176731
40	19	14	9	0.971363	0.081002
40	19	14	10	0.995187	0.023824
40	19	14	11	0.999519	0.004332
40	19	14	12	0.999975	0.000456
40	19	14	13	0.999999	0.000025
40	19	14	14	1.000000	0.000000
40	19	15	0	0.000001	0.000001
40	19	15	1	0.000056	0.000055
40	19	15	2	0.000921	0.000865
40	19	15	3	0.008002	0.007081
40	19	15	4	0.041989	0.033987
40	19	15	5	0.143949	0.101960
40	19	15	6	0.342204	0.198256
40	19	15	7	0.597105	0.254900
40	19	15	8	0.815591	0.218486
40	19	15	9	0.940209	0.124618
40	19	15	10	0.986940	0.046732
40	19	15	11	0.998186	0.011246
40	19	15	12	0.999852	0.001666
40	19	15	13	0.999994	0.000142
40	19	15	14	1.000000	0.000006
40	19	15	15	1.000000	0.000000
40	19	16	0	0.000000	0.000000
40	19	16	1	0.000017	0.000016
40	19	16	2	0.000333	0.000316
40	19	16	3	0.003470	0.003137
40	19	16	4	0.021597	0.018126
40	19	16	5	0.086851	0.065254
40	19	16	6	0.239111	0.152260
40	19	16	7	0.474753	0.235641
40	19	16	8	0.719457	0.244704
40	19	16	9	0.890362	0.170905
40	19	16	10	0.970117	0.079755
40	19	16	11	0.994588	0.024470
40	19	16	12	0.999386	0.004798
40	19	16	13	0.999960	0.000574
40	19	16	14	0.999999	0.000039
40	19	16	15	1.000000	0.000001
40	19	16	16	1.000000	0.000000
40	19	17	0	0.000000	0.000000
40	19	17	1	0.000004	0.000004
40	19	17	2	0.000109	0.000105
40	19	17	3	0.001379	0.001270
40	19	17	4	0.010268	0.008889
40	19	17	5	0.048786	0.038518
40	19	17	6	0.156637	0.107851
40	19	17	7	0.356932	0.200295
40	19	17	8	0.607301	0.250369
40	19	17	9	0.819151	0.211850
40	19	17	10	0.940209	0.121057
40	19	17	11	0.986431	0.046222
40	19	17	12	0.997986	0.011555
40	19	17	13	0.999816	0.001830
40	19	17	14	0.999991	0.000174
40	19	17	15	1.000000	0.000009
40	19	17	16	1.000000	0.000000
40	19	17	17	1.000000	0.000000
40	19	18	0	0.000000	0.000000
40	19	18	1	0.000001	0.000001
40	19	18	2	0.000032	0.000031
40	19	18	3	0.000495	0.000464
40	19	18	4	0.004471	0.003975
40	19	18	5	0.025340	0.020869
40	19	18	6	0.095678	0.070338
40	19	18	7	0.252430	0.156753
40	19	18	8	0.487559	0.235129
40	19	18	9	0.727042	0.239483
40	19	18	10	0.892838	0.165796
40	19	18	11	0.970353	0.077515
40	19	18	12	0.994469	0.024116
40	19	18	13	0.999339	0.004870
40	19	18	14	0.999953	0.000614
40	19	18	15	0.999998	0.000045
40	19	18	16	1.000000	0.000002
40	19	18	17	1.000000	0.000000
40	19	18	18	1.000000	0.000000
40	19	19	0	0.000000	0.000000
40	19	19	1	0.000000	0.000000
40	19	19	2	0.000008	0.000008
40	19	19	3	0.000158	0.000150

N	n	k	x	P(x)	p(x)
40	19	19	4	0.001760	0.001602
40	19	19	5	0.012059	0.010299
40	19	19	6	0.054115	0.042055
40	19	19	7	0.166929	0.112814
40	19	19	8	0.369995	0.203066
40	19	19	9	0.618186	0.248192
40	19	19	10	0.825013	0.206826
40	19	19	11	0.942166	0.117153
40	19	19	12	0.986796	0.044630
40	19	19	13	0.998011	0.011215
40	19	19	14	0.999813	0.001802
40	19	19	15	0.999990	0.000177
40	19	19	16	1.000000	0.000010
40	19	19	17	1.000000	0.000000
40	19	19	18	1.000000	0.000000
40	19	19	19	1.000000	0.000000
40	20	1	0	0.500000	0.500000
40	20	1	1	1.000000	0.500000
40	20	2	0	0.243590	0.243590
40	20	2	1	0.756410	0.512821
40	20	2	2	1.000000	0.243590
40	20	3	0	0.115385	0.115385
40	20	3	1	0.500000	0.384615
40	20	3	2	0.884615	0.384615
40	20	3	3	1.000000	0.115385
40	20	4	0	0.053015	0.053015
40	20	4	1	0.302495	0.249480
40	20	4	2	0.697505	0.395010
40	20	4	3	0.946985	0.249480
40	20	4	4	1.000000	0.053015
40	20	5	0	0.023562	0.023562
40	20	5	1	0.170825	0.147263
40	20	5	2	0.500000	0.329175
40	20	5	3	0.829175	0.329175
40	20	5	4	0.976438	0.147263
40	20	5	5	1.000000	0.023562
40	20	6	0	0.010098	0.010098
40	20	6	1	0.090882	0.080784
40	20	6	2	0.330710	0.239828
40	20	6	3	0.669290	0.338580
40	20	6	4	0.909118	0.239828
40	20	6	5	0.989902	0.080784
40	20	6	6	1.000000	0.010098
40	20	7	0	0.004158	0.004158
40	20	7	1	0.045738	0.041580
40	20	7	2	0.203742	0.158004
40	20	7	3	0.500000	0.296258
40	20	7	4	0.796258	0.296258
40	20	7	5	0.954262	0.158004
40	20	7	6	0.995842	0.041580
40	20	7	7	1.000000	0.004158
40	20	8	0	0.001638	0.001638
40	20	8	1	0.021798	0.020160
40	20	8	2	0.117558	0.095760
40	20	8	3	0.347382	0.229824
40	20	8	4	0.652618	0.305235
40	20	8	5	0.882442	0.229824
40	20	8	6	0.978202	0.095760
40	20	8	7	0.998362	0.020160
40	20	8	8	1.000000	0.001638
40	20	9	0	0.000614	0.000614
40	20	9	1	0.009828	0.009214
40	20	9	2	0.063693	0.053865
40	20	9	3	0.225288	0.161595
40	20	9	4	0.500000	0.274712
40	20	9	5	0.774712	0.274712
40	20	9	6	0.936307	0.161595
40	20	9	7	0.990172	0.053865
40	20	9	8	0.999386	0.009214
40	20	9	9	1.000000	0.000614
40	20	10	0	0.000218	0.000218
40	20	10	1	0.004181	0.003963
40	20	10	2	0.032417	0.028236
40	20	10	3	0.136672	0.104255
40	20	10	4	0.358213	0.221542
40	20	10	5	0.641787	0.283573
40	20	10	6	0.863328	0.221542
40	20	10	7	0.967583	0.104255
40	20	10	8	0.995819	0.028236
40	20	10	9	0.999782	0.003963
40	20	10	10	1.000000	0.000218
40	20	11	0	0.000073	0.000073
40	20	11	1	0.001671	0.001598
40	20	11	2	0.015475	0.013804
40	20	11	3	0.077594	0.062119
40	20	11	4	0.240058	0.162464
40	20	11	5	0.500000	0.259942
40	20	11	6	0.759942	0.259942
40	20	11	7	0.922406	0.162464
40	20	11	8	0.984525	0.062119
40	20	11	9	0.998329	0.013804
40	20	11	10	0.999927	0.001598
40	20	11	11	1.000000	0.000073
40	20	12	0	0.000023	0.000023
40	20	12	1	0.000624	0.000601
40	20	12	2	0.006907	0.006283
40	20	12	3	0.041179	0.034272
40	20	12	4	0.150422	0.109243
40	20	12	5	0.365547	0.215125
40	20	12	6	0.634453	0.268906
40	20	12	7	0.849578	0.215125
40	20	12	8	0.958821	0.109243
40	20	12	9	0.993093	0.034272
40	20	12	10	0.999376	0.006283
40	20	12	11	0.999977	0.000601
40	20	12	12	1.000000	0.000023
40	20	13	0	0.000006	0.000006
40	20	13	1	0.000216	0.000209
40	20	13	2	0.002868	0.002652
40	20	13	3	0.020371	0.017503
40	20	13	4	0.087998	0.067627
40	20	13	5	0.250302	0.162304
40	20	13	6	0.500000	0.249698
40	20	13	7	0.749698	0.249698
40	20	13	8	0.912002	0.162304
40	20	13	9	0.979629	0.067627
40	20	13	10	0.997132	0.017503
40	20	13	11	0.999784	0.002652
40	20	13	12	0.999993	0.000209
40	20	13	13	1.000000	0.000006
40	20	14	0	0.000002	0.000002
40	20	14	1	0.000068	0.000067
40	20	14	2	0.001100	0.001031
40	20	14	3	0.009351	0.008251
40	20	14	4	0.047923	0.038572
40	20	14	5	0.160133	0.112210
40	20	14	6	0.370527	0.210394
40	20	14	7	0.629473	0.258946
40	20	14	8	0.839867	0.210394
40	20	14	9	0.952077	0.112210
40	20	14	10	0.990649	0.038572
40	20	14	11	0.998900	0.008251
40	20	14	12	0.999931	0.001031
40	20	14	13	0.999998	0.000067
40	20	14	14	1.000000	0.000002
40	20	15	0	0.000000	0.000000
40	20	15	1	0.000020	0.000019
40	20	15	2	0.000386	0.000366
40	20	15	3	0.003956	0.003570
40	20	15	4	0.024186	0.020230
40	20	15	5	0.095396	0.071210
40	20	15	6	0.257238	0.161841
40	20	15	7	0.500000	0.242762
40	20	15	8	0.742762	0.242762
40	20	15	9	0.904604	0.161841
40	20	15	10	0.975814	0.071210
40	20	15	11	0.996044	0.020230
40	20	15	12	0.999614	0.003570
40	20	15	13	0.999980	0.000366
40	20	15	14	1.000000	0.000019
40	20	15	15	1.000000	0.000000
40	20	16	0	0.000000	0.000000
40	20	16	1	0.000005	0.000005
40	20	16	2	0.000122	0.000117
40	20	16	3	0.001528	0.001406
40	20	16	4	0.011239	0.009710
40	20	16	5	0.052670	0.041431
40	20	16	6	0.166607	0.113936
40	20	16	7	0.373764	0.207157
40	20	16	8	0.626236	0.252473
40	20	16	9	0.833393	0.207157
40	20	16	10	0.947330	0.113936
40	20	16	11	0.988761	0.041431
40	20	16	12	0.998472	0.009710
40	20	16	13	0.999878	0.001406
40	20	16	14	0.999995	0.000117
40	20	16	15	1.000000	0.000005
40	20	16	16	1.000000	0.000000
40	20	17	0	0.000000	0.000000
40	20	17	1	0.000001	0.000001
40	20	17	2	0.000034	0.000033
40	20	17	3	0.000532	0.000498
40	20	17	4	0.004765	0.004233
40	20	17	5	0.026775	0.022010
40	20	17	6	0.100144	0.073368
40	20	17	7	0.261554	0.161410
40	20	17	8	0.500000	0.238446
40	20	17	9	0.738446	0.238446
40	20	17	10	0.899856	0.161410
40	20	17	11	0.973224	0.073368
40	20	17	12	0.995235	0.022010
40	20	17	13	0.999468	0.004233
40	20	17	14	0.999966	0.000498
40	20	17	15	0.999999	0.000033
40	20	17	16	1.000000	0.000001
40	20	17	17	1.000000	0.000000
40	20	18	0	0.000000	0.000000
40	20	18	1	0.000000	0.000000
40	20	18	2	0.000008	0.000008
40	20	18	3	0.000164	0.000156
40	20	18	4	0.001821	0.001656
40	20	18	5	0.012421	0.010600
40	20	18	6	0.055485	0.043064
40	20	18	7	0.170322	0.114837
40	20	18	8	0.375593	0.205271
40	20	18	9	0.624407	0.248814
40	20	18	10	0.829678	0.205271
40	20	18	11	0.944515	0.114837
40	20	18	12	0.987579	0.043064
40	20	18	13	0.998179	0.010600

Table for $N = 2,\ n = 1,$ through $N = 100,\ n = 50$

N	n	k	x	P(x)	p(x)
40	20	18	14	0.999836	0.001656
40	20	18	15	0.999992	0.000156
40	20	18	16	1.000000	0.000008
40	20	18	17	1.000000	0.000000
40	20	18	18	1.000000	0.000000
40	20	19	0	0.000000	0.000000
40	20	19	1	0.000000	0.000000
40	20	19	2	0.000002	0.000002
40	20	19	3	0.000044	0.000042
40	20	19	4	0.000616	0.000572
40	20	19	5	0.005193	0.004577
40	20	19	6	0.028080	0.022887
40	20	19	7	0.102464	0.074383
40	20	19	8	0.263627	0.161163
40	20	19	9	0.500000	0.236373
40	20	19	10	0.736373	0.236373
40	20	19	11	0.897536	0.161163
40	20	19	12	0.971919	0.074383
40	20	19	13	0.994807	0.022887
40	20	19	14	0.999384	0.004577
40	20	19	15	0.999956	0.000572
40	20	19	16	0.999998	0.000042
40	20	19	17	1.000000	0.000002
40	20	19	18	1.000000	0.000000
40	20	19	19	1.000000	0.000000
40	20	20	0	0.000000	0.000000
40	20	20	1	0.000000	0.000000
40	20	20	2	0.000000	0.000000
40	20	20	3	0.000010	0.000009
40	20	20	4	0.000180	0.000170
40	20	20	5	0.001924	0.001744
40	20	20	6	0.012822	0.010899
40	20	20	7	0.056417	0.043594
40	20	20	8	0.171534	0.115117
40	20	20	9	0.376185	0.204652
40	20	20	10	0.623814	0.247629
40	20	20	11	0.828466	0.204652
40	20	20	12	0.943583	0.115117
40	20	20	13	0.987177	0.043594
40	20	20	14	0.998076	0.010899
40	20	20	15	0.999820	0.001744
40	20	20	16	0.999990	0.000170
40	20	20	17	1.000000	0.000009
40	20	20	18	1.000000	0.000000
40	20	20	19	1.000000	0.000000
40	20	20	20	1.000000	0.000000
41	1	1	0	0.975610	0.975610
41	1	1	1	1.000000	0.024390
41	2	1	0	0.951220	0.951220
41	2	1	1	1.000000	0.048780

N	n	k	x	P(x)	p(x)
41	2	2	0	0.903659	0.903659
41	2	2	1	0.998780	0.095122
41	2	2	2	1.000000	0.001220
41	3	1	0	0.926829	0.926829
41	3	1	1	1.000000	0.073171
41	3	2	0	0.857317	0.857317
41	3	2	1	0.996341	0.139024
41	3	2	2	1.000000	0.003659
41	3	3	0	0.791370	0.791370
41	3	3	1	0.989212	0.197842
41	3	3	2	0.999906	0.010694
41	3	3	3	1.000000	0.000094
41	4	1	0	0.902439	0.902439
41	4	1	1	1.000000	0.097561
41	4	2	0	0.812195	0.812195
41	4	2	1	0.992683	0.180488
41	4	2	2	1.000000	0.007317
41	4	3	0	0.728893	0.728893
41	4	3	1	0.978799	0.249906
41	4	3	2	0.999625	0.020826
41	4	3	3	1.000000	0.000375
41	4	4	0	0.652167	0.652167
41	4	4	1	0.959070	0.306902
41	4	4	2	0.998529	0.039459
41	4	4	3	0.999990	0.001461
41	4	4	4	1.000000	0.000010
41	5	1	0	0.878049	0.878049
41	5	1	1	1.000000	0.121951
41	5	2	0	0.768293	0.768293
41	5	2	1	0.987805	0.219512
41	5	2	2	1.000000	0.012195
41	5	3	0	0.669794	0.669794
41	5	3	1	0.965291	0.295497
41	5	3	2	0.999062	0.033771
41	5	3	3	1.000000	0.000938
41	5	4	0	0.581663	0.581663
41	5	4	1	0.934186	0.352523
41	5	4	2	0.996396	0.062210
41	5	4	3	0.999951	0.003555
41	5	4	4	1.000000	0.000049
41	5	5	0	0.503060	0.503060
41	5	5	1	0.896075	0.393015
41	5	5	2	0.991352	0.095276
41	5	5	3	0.999758	0.008407
41	5	5	4	0.999999	0.000240
41	5	5	5	1.000000	0.000001
41	6	1	0	0.853659	0.853659
41	6	1	1	1.000000	0.146341
41	6	2	0	0.725610	0.725610
41	6	2	1	0.981707	0.256098

N	n	k	x	P(x)	p(x)
41	6	2	2	1.000000	0.018293
41	6	3	0	0.613977	0.613977
41	6	3	1	0.948874	0.334897
41	6	3	2	0.998124	0.049250
41	6	3	3	1.000000	0.001876
41	6	4	0	0.517034	0.517034
41	6	4	1	0.904809	0.387775
41	6	4	2	0.992940	0.088131
41	6	4	3	0.999852	0.006912
41	6	4	4	1.000000	0.000148
41	6	5	0	0.433190	0.433190
41	6	5	1	0.852407	0.419216
41	6	5	2	0.983412	0.131005
41	6	5	3	0.999291	0.015879
41	6	5	4	0.999992	0.000701
41	6	5	5	1.000000	0.000008
41	6	6	0	0.360992	0.360992
41	6	6	1	0.794182	0.433190
41	6	6	2	0.968856	0.174674
41	6	6	3	0.997968	0.029112
41	6	6	4	0.999953	0.001985
41	6	6	5	1.000000	0.000047
41	6	6	6	1.000000	0.000000
41	7	1	0	0.829268	0.829268
41	7	1	1	1.000000	0.170732
41	7	2	0	0.684146	0.684146
41	7	2	1	0.974390	0.290244
41	7	2	2	1.000000	0.025610
41	7	3	0	0.561351	0.561351
41	7	3	1	0.929737	0.368386
41	7	3	2	0.996717	0.066979
41	7	3	3	1.000000	0.003283
41	7	4	0	0.457944	0.457944
41	7	4	1	0.871571	0.413627
41	7	4	2	0.987903	0.116333
41	7	4	3	0.999654	0.011751
41	7	4	4	1.000000	0.000346
41	7	5	0	0.371306	0.371306
41	7	5	1	0.804496	0.433190
41	7	5	2	0.972183	0.167687
41	7	5	3	0.998384	0.026201
41	7	5	4	0.999972	0.001588
41	7	5	5	1.000000	0.000028
41	7	6	0	0.299108	0.299108
41	7	6	1	0.732298	0.433190
41	7	6	2	0.948893	0.216595
41	7	6	3	0.995473	0.046580
41	7	6	4	0.999840	0.004367
41	7	6	5	0.999998	0.000159
41	7	6	6	1.000000	0.000002

N	n	k	x	P(x)	p(x)
41	7	7	0	0.239286	0.239286
41	7	7	1	0.658037	0.418751
41	7	7	2	0.917951	0.259914
41	7	7	3	0.990149	0.072198
41	7	7	4	0.999465	0.009316
41	7	7	5	0.999989	0.000524
41	7	7	6	1.000000	0.000011
41	7	7	7	1.000000	0.000000
41	8	1	0	0.804878	0.804878
41	8	1	1	1.000000	0.195122
41	8	2	0	0.643902	0.643902
41	8	2	1	0.965854	0.321951
41	8	2	2	1.000000	0.034146
41	8	3	0	0.511820	0.511820
41	8	3	1	0.908068	0.396248
41	8	3	2	0.994747	0.086679
41	8	3	3	1.000000	0.005253
41	8	4	0	0.404068	0.404068
41	8	4	1	0.835075	0.431006
41	8	4	2	0.981060	0.145986
41	8	4	3	0.999309	0.018248
41	8	4	4	1.000000	0.000691
41	8	5	0	0.316702	0.316702
41	8	5	1	0.753533	0.436831
41	8	5	2	0.957387	0.203854
41	8	5	3	0.996843	0.039456
41	8	5	4	0.999925	0.003082
41	8	5	5	1.000000	0.000075
41	8	6	0	0.246324	0.246324
41	8	6	1	0.668594	0.422270
41	8	6	2	0.923411	0.254818
41	8	6	3	0.991363	0.067951
41	8	6	4	0.999583	0.008220
41	8	6	5	0.999994	0.000411
41	8	6	6	1.000000	0.000006
41	8	7	0	0.190021	0.190021
41	8	7	1	0.584140	0.394118
41	8	7	2	0.879728	0.295589
41	8	7	3	0.981655	0.101927
41	8	7	4	0.998643	0.016988
41	8	7	5	0.999959	0.001315
41	8	7	6	1.000000	0.000041
41	8	7	7	1.000000	0.000000
41	8	8	0	0.145310	0.145310
41	8	8	1	0.502998	0.357687
41	8	8	2	0.827566	0.324568
41	8	8	3	0.966666	0.139101
41	8	8	4	0.996645	0.029979
41	8	8	5	0.999842	0.003198
41	8	8	6	0.999997	0.000155

Table for $N=2$, $n=1$, through $N=100$, $n=50$

N	n	k	x	P(x)	p(x)
41	8	8	7	1.000000	0.000003
41	8	8	8	1.000000	0.000000
41	9	1	0	0.780488	0.780488
41	9	1	1	1.000000	0.219512
41	9	2	0	0.604878	0.604878
41	9	2	1	0.956097	0.351219
41	9	2	2	1.000000	0.043902
41	9	3	0	0.465291	0.465291
41	9	3	1	0.884052	0.418762
41	9	3	2	0.992120	0.108068
41	9	3	3	1.000000	0.007880
41	9	4	0	0.355090	0.355090
41	9	4	1	0.795892	0.440802
41	9	4	2	0.972213	0.176321
41	9	4	3	0.998756	0.026543
41	9	4	4	1.000000	0.001244
41	9	5	0	0.268717	0.268717
41	9	5	1	0.700584	0.431867
41	9	5	2	0.938855	0.238271
41	9	5	3	0.994451	0.055597
41	9	5	4	0.999832	0.005380
41	9	5	5	1.000000	0.000168
41	9	6	0	0.201538	0.201538
41	9	6	1	0.604613	0.403076
41	9	6	2	0.892524	0.287911
41	9	6	3	0.985185	0.092661
41	9	6	4	0.999084	0.013899
41	9	6	5	0.999981	0.000897
41	9	6	6	1.000000	0.000019
41	9	7	0	0.149714	0.149714
41	9	7	1	0.512482	0.362768
41	9	7	2	0.834942	0.322460
41	9	7	3	0.969301	0.134358
41	9	7	4	0.997099	0.027798
41	9	7	5	0.999879	0.002780
41	9	7	6	0.999998	0.000120
41	9	7	7	1.000000	0.000002
41	9	8	0	0.110084	0.110084
41	9	8	1	0.427125	0.317041
41	9	8	2	0.768553	0.341429
41	9	8	3	0.945590	0.177037
41	9	8	4	0.993011	0.047421
41	9	8	5	0.999552	0.006541
41	9	8	6	0.999988	0.000436
41	9	8	7	1.000000	0.000012
41	9	8	8	1.000000	0.000000
41	9	9	0	0.080061	0.080061
41	9	9	1	0.350266	0.270205
41	9	9	2	0.696129	0.345863
41	9	9	3	0.913402	0.217273

N	n	k	x	P(x)	p(x)
41	9	9	4	0.985826	0.072424
41	9	9	5	0.998759	0.012933
41	9	9	6	0.999948	0.001189
41	9	9	7	0.999999	0.000051
41	9	9	8	1.000000	0.000001
41	9	9	9	1.000000	0.000000
41	10	1	0	0.756098	0.756098
41	10	1	1	1.000000	0.243902
41	10	2	0	0.567073	0.567073
41	10	2	1	0.945122	0.378049
41	10	2	2	1.000000	0.054878
41	10	3	0	0.421670	0.421670
41	10	3	1	0.857880	0.436210
41	10	3	2	0.988743	0.130863
41	10	3	3	1.000000	0.011257
41	10	4	0	0.310704	0.310704
41	10	4	1	0.754567	0.443863
41	10	4	2	0.961193	0.206626
41	10	4	3	0.997926	0.036733
41	10	4	4	1.000000	0.002074
41	10	5	0	0.226730	0.226730
41	10	5	1	0.646600	0.419870
41	10	5	2	0.916517	0.269917
41	10	5	3	0.990977	0.074460
41	10	5	4	0.999664	0.008687
41	10	5	5	1.000000	0.000336
41	10	6	0	0.163749	0.163749
41	10	6	1	0.541633	0.377883
41	10	6	2	0.856535	0.314903
41	10	6	3	0.976498	0.119963
41	10	6	4	0.998216	0.021717
41	10	6	5	0.999953	0.001737
41	10	6	6	1.000000	0.000047
41	10	7	0	0.116964	0.116964
41	10	7	1	0.444463	0.327499
41	10	7	2	0.784558	0.340095
41	10	7	3	0.952506	0.167948
41	10	7	4	0.994493	0.041987
41	10	7	5	0.999705	0.005212
41	10	7	6	0.999995	0.000290
41	10	7	7	1.000000	0.000005
41	10	8	0	0.082563	0.082563
41	10	8	1	0.357772	0.275209
41	10	8	2	0.704535	0.346763
41	10	8	3	0.917928	0.213393
41	10	8	4	0.987083	0.069155
41	10	8	5	0.998938	0.011855
41	10	8	6	0.999960	0.001022
41	10	8	7	0.999999	0.000039
41	10	8	8	1.000000	0.000000

N	n	k	x	P(x)	p(x)
41	10	9	0	0.057544	0.057544
41	10	9	1	0.282715	0.225171
41	10	9	2	0.620471	0.337757
41	10	9	3	0.872663	0.252192
41	10	9	4	0.974510	0.101847
41	10	9	5	0.997142	0.022633
41	10	9	6	0.999837	0.002694
41	10	9	7	0.999996	0.000159
41	10	9	8	1.000000	0.000004
41	10	9	9	1.000000	0.000000
41	10	10	0	0.039561	0.039561
41	10	10	1	0.219385	0.179824
41	10	10	2	0.536032	0.316647
41	10	10	3	0.817496	0.281464
41	10	10	4	0.955413	0.137917
41	10	10	5	0.993606	0.038192
41	10	10	6	0.999500	0.005894
41	10	10	7	0.999981	0.000481
41	10	10	8	1.000000	0.000019
41	10	10	9	1.000000	0.000000
41	10	10	10	1.000000	0.000000
41	11	1	0	0.731707	0.731707
41	11	1	1	1.000000	0.268293
41	11	2	0	0.530488	0.530488
41	11	2	1	0.932927	0.402439
41	11	2	2	1.000000	0.067073
41	11	3	0	0.380863	0.380863
41	11	3	1	0.829737	0.448874
41	11	3	2	0.984522	0.154784
41	11	3	3	1.000000	0.015478
41	11	4	0	0.270613	0.270613
41	11	4	1	0.711612	0.440999
41	11	4	2	0.947862	0.236250
41	11	4	3	0.996741	0.048879
41	11	4	4	1.000000	0.003259
41	11	5	0	0.190161	0.190161
41	11	5	1	0.592423	0.402263
41	11	5	2	0.890396	0.297972
41	11	5	3	0.986173	0.095777
41	11	5	4	0.999383	0.013211
41	11	5	5	1.000000	0.000616
41	11	6	0	0.132056	0.132056
41	11	6	1	0.480684	0.348628
41	11	6	2	0.815903	0.335219
41	11	6	3	0.964889	0.148986
41	11	6	4	0.996815	0.031926
41	11	6	5	0.999897	0.003082
41	11	6	6	1.000000	0.000103
41	11	7	0	0.090553	0.090553
41	11	7	1	0.381076	0.290523

N	n	k	x	P(x)	p(x)
41	11	7	2	0.729704	0.348628
41	11	7	3	0.930835	0.201131
41	11	7	4	0.990430	0.059594
41	11	7	5	0.999369	0.008939
41	11	7	6	0.999985	0.000616
41	11	7	7	1.000000	0.000015
41	11	8	0	0.061256	0.061256
41	11	8	1	0.295628	0.234372
41	11	8	2	0.637420	0.341792
41	11	8	3	0.883510	0.246090
41	11	8	4	0.978160	0.094650
41	11	8	5	0.997791	0.019631
41	11	8	6	0.999895	0.002103
41	11	8	7	0.999998	0.000104
41	11	8	8	1.000000	0.000002
41	11	9	0	0.040837	0.040837
41	11	9	1	0.224606	0.183769
41	11	9	2	0.544204	0.319598
41	11	9	3	0.823852	0.279648
41	11	9	4	0.958083	0.134231
41	11	9	5	0.994222	0.036139
41	11	9	6	0.999576	0.005354
41	11	9	7	0.999986	0.000410
41	11	9	8	1.000000	0.000014
41	11	9	9	1.000000	0.000000
41	11	10	0	0.026800	0.026800
41	11	10	1	0.167178	0.140379
41	11	10	2	0.454317	0.287139
41	11	10	3	0.753940	0.299623
41	11	10	4	0.928720	0.174780
41	11	10	5	0.987446	0.058726
41	11	10	6	0.998739	0.011293
41	11	10	7	0.999934	0.001195
41	11	10	8	0.999998	0.000064
41	11	10	9	1.000000	0.000001
41	11	10	10	1.000000	0.000000
41	11	11	0	0.017290	0.017290
41	11	11	1	0.121895	0.104605
41	11	11	2	0.370954	0.249059
41	11	11	3	0.676618	0.305664
41	11	11	4	0.889253	0.212636
41	11	11	5	0.976079	0.086826
41	11	11	6	0.996918	0.020838
41	11	11	7	0.999780	0.002862
41	11	11	8	0.999992	0.000212
41	11	11	9	1.000000	0.000008
41	11	11	10	1.000000	0.000000
41	11	11	11	1.000000	0.000000
41	12	1	0	0.707317	0.707317
41	12	1	1	1.000000	0.292683

Table for $N = 2$, $n = 1$, through $N = 100$, $n = 50$

N	n	k	x	P(x)	p(x)
41	12	2	0	0.495122	0.495122
41	12	2	1	0.919512	0.424390
41	12	2	2	1.000000	0.080488
41	12	3	0	0.342777	0.342777
41	12	3	1	0.799812	0.457036
41	12	3	2	0.979362	0.179550
41	12	3	3	1.000000	0.020638
41	12	4	0	0.234531	0.234531
41	12	4	1	0.667513	0.432981
41	12	4	2	0.932112	0.264600
41	12	4	3	0.995112	0.063000
41	12	4	4	1.000000	0.004888
41	12	5	0	0.158467	0.158467
41	12	5	1	0.538788	0.380321
41	12	5	2	0.860599	0.321810
41	12	5	3	0.979788	0.119189
41	12	5	4	0.998943	0.019155
41	12	5	5	1.000000	0.001057
41	12	6	0	0.105645	0.105645
41	12	6	1	0.422579	0.316934
41	12	6	2	0.771207	0.348628
41	12	6	3	0.949990	0.178784
41	12	6	4	0.994686	0.044696
41	12	6	5	0.999794	0.005108
41	12	6	6	1.000000	0.000205
41	12	7	0	0.069424	0.069424
41	12	7	1	0.322971	0.253547
41	12	7	2	0.671599	0.348628
41	12	7	3	0.904018	0.232419
41	12	7	4	0.984470	0.080453
41	12	7	5	0.998773	0.014303
41	12	7	6	0.999965	0.001192
41	12	7	7	1.000000	0.000035
41	12	8	0	0.044921	0.044921
41	12	8	1	0.240941	0.196020
41	12	8	2	0.569061	0.328120
41	12	8	3	0.842495	0.273434
41	12	8	4	0.965540	0.123045
41	12	8	5	0.995828	0.030288
41	12	8	6	0.999754	0.003926
41	12	8	7	0.999995	0.000240
41	12	8	8	1.000000	0.000005
41	12	9	0	0.028586	0.028586
41	12	9	1	0.175601	0.147015
41	12	9	2	0.469631	0.294030
41	12	9	3	0.767922	0.298291
41	12	9	4	0.935711	0.167789
41	12	9	5	0.989403	0.053692
41	12	9	6	0.999040	0.009637
41	12	9	7	0.999958	0.000918
41	12	9	8	0.999999	0.000041
41	12	9	9	1.000000	0.000001
41	12	10	0	0.017866	0.017866
41	12	10	1	0.125065	0.107198
41	12	10	2	0.377747	0.252682
41	12	10	3	0.684028	0.306281
41	12	10	4	0.893764	0.209736
41	12	10	5	0.977658	0.083894
41	12	10	6	0.997234	0.019575
41	12	10	7	0.999815	0.002581
41	12	10	8	0.999994	0.000179
41	12	10	9	1.000000	0.000006
41	12	10	10	1.000000	0.000000
41	12	11	0	0.010950	0.010950
41	12	11	1	0.087027	0.076076
41	12	11	2	0.296236	0.209210
41	12	11	3	0.595108	0.298871
41	12	11	4	0.839638	0.244531
41	12	11	5	0.958714	0.119076
41	12	11	6	0.993445	0.034730
41	12	11	7	0.999399	0.005954
41	12	11	8	0.999971	0.000572
41	12	11	9	0.999999	0.000028
41	12	11	10	1.000000	0.000001
41	12	11	11	1.000000	0.000000
41	12	12	0	0.006570	0.006570
41	12	12	1	0.059132	0.052562
41	12	12	2	0.226500	0.167368
41	12	12	3	0.505446	0.278946
41	12	12	4	0.774430	0.268984
41	12	12	5	0.930930	0.156500
41	12	12	6	0.986499	0.055569
41	12	12	7	0.998406	0.011908
41	12	12	8	0.999895	0.001488
41	12	12	9	0.999996	0.000102
41	12	12	10	1.000000	0.000003
41	12	12	11	1.000000	0.000000
41	12	12	12	1.000000	0.000000
41	13	1	0	0.682927	0.682927
41	13	1	1	1.000000	0.317073
41	13	2	0	0.460976	0.460976
41	13	2	1	0.904878	0.443902
41	13	2	2	1.000000	0.095122
41	13	3	0	0.307317	0.307317
41	13	3	1	0.768293	0.460976
41	13	3	2	0.973171	0.204878
41	13	3	3	1.000000	0.026829
41	13	4	0	0.202182	0.202182
41	13	4	1	0.622721	0.420539
41	13	4	2	0.913864	0.291142
41	13	4	3	0.992940	0.079076
41	13	4	4	1.000000	0.007060
41	13	5	0	0.131145	0.131145
41	13	5	1	0.486330	0.355185
41	13	5	2	0.827308	0.340978
41	13	5	3	0.971568	0.144260
41	13	5	4	0.998282	0.026715
41	13	5	5	1.000000	0.001717
41	13	6	0	0.083787	0.083787
41	13	6	1	0.367935	0.284148
41	13	6	2	0.723120	0.355185
41	13	6	3	0.931496	0.208375
41	13	6	4	0.991604	0.060108
41	13	6	5	0.999618	0.008014
41	13	6	6	1.000000	0.000382
41	13	7	0	0.052666	0.052666
41	13	7	1	0.270513	0.217847
41	13	7	2	0.611491	0.340978
41	13	7	3	0.871960	0.260469
41	13	7	4	0.976147	0.104188
41	13	7	5	0.997786	0.021639
41	13	7	6	0.999924	0.002137
41	13	7	7	1.000000	0.000076
41	13	8	0	0.032529	0.032529
41	13	8	1	0.193626	0.161097
41	13	8	2	0.501174	0.307548
41	13	8	3	0.795351	0.294177
41	13	8	4	0.948568	0.153217
41	13	8	5	0.992695	0.044127
41	13	8	6	0.999484	0.006789
41	13	8	7	0.999986	0.000503
41	13	8	8	1.000000	0.000013
41	13	9	0	0.019715	0.019715
41	13	9	1	0.135045	0.115331
41	13	9	2	0.398658	0.263613
41	13	9	3	0.706207	0.307548
41	13	9	4	0.906782	0.200575
41	13	9	5	0.981997	0.075216
41	13	9	6	0.998043	0.016046
41	13	9	7	0.999895	0.001851
41	13	9	8	0.999998	0.000103
41	13	9	9	1.000000	0.000002
41	13	10	0	0.011706	0.011706
41	13	10	1	0.091796	0.080091
41	13	10	2	0.308041	0.216245
41	13	10	3	0.610098	0.302057
41	13	10	4	0.850370	0.240272
41	13	10	5	0.963194	0.112823
41	13	10	6	0.994533	0.031340
41	13	10	7	0.999548	0.005014
41	13	10	8	0.999982	0.000434
41	13	10	9	1.000000	0.000018
41	13	10	10	1.000000	0.000000
41	13	11	0	0.006797	0.006797
41	13	11	1	0.060793	0.053997
41	13	11	2	0.231309	0.170516
41	13	11	3	0.512660	0.281351
41	13	11	4	0.780614	0.267953
41	13	11	5	0.934078	0.153464
41	13	11	6	0.987457	0.053379
41	13	11	7	0.998577	0.011121
41	13	11	8	0.999912	0.001334
41	13	11	9	0.999997	0.000086
41	13	11	10	1.000000	0.000003
41	13	11	11	1.000000	0.000000
41	13	12	0	0.003852	0.003852
41	13	12	1	0.039195	0.035343
41	13	12	2	0.168787	0.129592
41	13	12	3	0.418877	0.250090
41	13	12	4	0.700228	0.281351
41	13	12	5	0.893154	0.192926
41	13	12	6	0.975002	0.081848
41	13	12	7	0.996353	0.021352
41	13	12	8	0.999689	0.003336
41	13	12	9	0.999986	0.000297
41	13	12	10	1.000000	0.000014
41	13	12	11	1.000000	0.000000
41	13	12	12	1.000000	0.000000
41	13	13	0	0.002125	0.002125
41	13	13	1	0.024570	0.022445
41	13	13	2	0.119631	0.095061
41	13	13	3	0.332639	0.213008
41	13	13	4	0.612912	0.280273
41	13	13	5	0.839933	0.227021
41	13	13	6	0.955245	0.115312
41	13	13	7	0.991936	0.036690
41	13	13	8	0.999114	0.007179
41	13	13	9	0.999945	0.000831
41	13	13	10	0.999998	0.000053
41	13	13	11	1.000000	0.000002
41	13	13	12	1.000000	0.000000
41	13	13	13	1.000000	0.000000
41	14	1	0	0.658537	0.658537
41	14	1	1	1.000000	0.341463
41	14	2	0	0.428049	0.428049
41	14	2	1	0.889024	0.460976
41	14	2	2	1.000000	0.110976
41	14	3	0	0.274390	0.274390
41	14	3	1	0.735366	0.460976
41	14	3	2	0.965854	0.230488

Table for $N = 2$, $n = 1$, through $N = 100$, $n = 50$

N	n	k	x	P(x)	p(x)
41	14	3	3	1.000000	0.034146
41	14	4	0	0.173299	0.173299
41	14	4	1	0.577664	0.404365
41	14	4	2	0.893068	0.315404
41	14	4	3	0.990115	0.097047
41	14	4	4	1.000000	0.009884
41	14	5	0	0.107726	0.107726
41	14	5	1	0.435590	0.327863
41	14	5	2	0.790775	0.355185
41	14	5	3	0.961263	0.170489
41	14	5	4	0.997328	0.036065
41	14	5	5	1.000000	0.002671
41	14	6	0	0.065833	0.065833
41	14	6	1	0.317195	0.251362
41	14	6	2	0.672380	0.355185
41	14	6	3	0.909170	0.236790
41	14	6	4	0.987310	0.078141
41	14	6	5	0.999332	0.012022
41	14	6	6	1.000000	0.000668
41	14	7	0	0.039500	0.039500
41	14	7	1	0.223832	0.184332
41	14	7	2	0.550602	0.326770
41	14	7	3	0.834750	0.284148
41	14	7	4	0.964984	0.130235
41	14	7	5	0.996241	0.031256
41	14	7	6	0.999847	0.003606
41	14	7	7	1.000000	0.000153
41	14	8	0	0.023235	0.023235
41	14	8	1	0.153352	0.130117
41	14	8	2	0.435271	0.281919
41	14	8	3	0.742820	0.307548
41	14	8	4	0.926680	0.183861
41	14	8	5	0.987967	0.061287
41	14	8	6	0.998999	0.011032
41	14	8	7	0.999968	0.000970
41	14	8	8	1.000000	0.000031
41	14	9	0	0.013378	0.013378
41	14	9	1	0.102094	0.088716
41	14	9	2	0.332755	0.230661
41	14	9	3	0.640303	0.307548
41	14	9	4	0.870965	0.230661
41	14	9	5	0.971252	0.100288
41	14	9	6	0.996324	0.025072
41	14	9	7	0.999763	0.003438
41	14	9	8	0.999994	0.000231
41	14	9	9	1.000000	0.000006
41	14	10	0	0.007525	0.007525
41	14	10	1	0.066053	0.058528
41	14	10	2	0.246257	0.180204
41	14	10	3	0.534584	0.288327

N	n	k	x	P(x)	p(x)
41	14	10	4	0.798883	0.264299
41	14	10	5	0.943047	0.144163
41	14	10	6	0.990056	0.047010
41	14	10	7	0.999011	0.008954
41	14	10	8	0.999951	0.000940
41	14	10	9	0.999999	0.000048
41	14	10	10	1.000000	0.000001
41	14	11	0	0.004127	0.004127
41	14	11	1	0.041509	0.037382
41	14	11	2	0.176501	0.134992
41	14	11	3	0.432274	0.255774
41	14	11	4	0.713625	0.281351
41	14	11	5	0.901193	0.187567
41	14	11	6	0.977925	0.076732
41	14	11	7	0.996989	0.019064
41	14	11	8	0.999769	0.002780
41	14	11	9	0.999991	0.000222
41	14	11	10	1.000000	0.000009
41	14	11	11	1.000000	0.000000
41	14	12	0	0.002201	0.002201
41	14	12	1	0.025310	0.023109
41	14	12	2	0.122504	0.097194
41	14	12	3	0.338491	0.215987
41	14	12	4	0.619842	0.281351
41	14	12	5	0.844922	0.225081
41	14	12	6	0.957463	0.112540
41	14	12	7	0.992540	0.035078
41	14	12	8	0.999213	0.006672
41	14	12	9	0.999954	0.000741
41	14	12	10	0.999999	0.000044
41	14	12	11	1.000000	0.000001
41	14	12	12	1.000000	0.000000
41	14	13	0	0.001138	0.001138
41	14	13	1	0.014951	0.013812
41	14	13	2	0.082286	0.067335
41	14	13	3	0.256565	0.174279
41	14	13	4	0.522824	0.266259
41	14	13	5	0.775070	0.252246
41	14	13	6	0.926417	0.151347
41	14	13	7	0.984073	0.057656
41	14	13	8	0.997832	0.013759
41	14	13	9	0.999826	0.001994
41	14	13	10	0.999992	0.000166
41	14	13	11	1.000000	0.000007
41	14	13	12	1.000000	0.000000
41	14	13	13	1.000000	0.000000
41	14	14	0	0.000569	0.000569
41	14	14	1	0.008538	0.007969
41	14	14	2	0.053428	0.044890
41	14	14	3	0.188098	0.134670

N	n	k	x	P(x)	p(x)
41	14	14	4	0.427731	0.239633
41	14	14	5	0.693991	0.266259
41	14	14	6	0.883175	0.189184
41	14	14	7	0.969659	0.086484
41	14	14	8	0.994884	0.025225
41	14	14	9	0.999470	0.004586
41	14	14	10	0.999969	0.000499
41	14	14	11	0.999999	0.000030
41	14	14	12	1.000000	0.000001
41	14	14	13	1.000000	0.000000
41	14	14	14	1.000000	0.000000
41	15	1	0	0.634146	0.634146
41	15	1	1	1.000000	0.365854
41	15	2	0	0.396341	0.396341
41	15	2	1	0.871951	0.475610
41	15	2	2	1.000000	0.128049
41	15	3	0	0.243902	0.243902
41	15	3	1	0.701220	0.457317
41	15	3	2	0.957317	0.256098
41	15	3	3	1.000000	0.042683
41	15	4	0	0.147625	0.147625
41	15	4	1	0.532734	0.385109
41	15	4	2	0.869705	0.336970
41	15	4	3	0.986521	0.116816
41	15	4	4	1.000000	0.013479
41	15	5	0	0.087777	0.087777
41	15	5	1	0.387017	0.299240
41	15	5	2	0.751310	0.364292
41	15	5	3	0.948635	0.197325
41	15	5	4	0.995993	0.047358
41	15	5	5	1.000000	0.004007
41	15	6	0	0.051203	0.051203
41	15	6	1	0.270646	0.219443
41	15	6	2	0.619760	0.349114
41	15	6	3	0.882860	0.263100
41	15	6	4	0.981522	0.098663
41	15	6	5	0.998887	0.017365
41	15	6	6	1.000000	0.001113
41	15	7	0	0.029259	0.029259
41	15	7	1	0.182869	0.153610
41	15	7	2	0.490089	0.307220
41	15	7	3	0.792654	0.302565
41	15	7	4	0.950514	0.157860
41	15	7	5	0.993925	0.043411
41	15	7	6	0.999714	0.005788
41	15	7	7	1.000000	0.000286
41	15	8	0	0.016351	0.016351
41	15	8	1	0.119618	0.103267
41	15	8	2	0.372622	0.253005
41	15	8	3	0.685866	0.313244

N	n	k	x	P(x)	p(x)
41	15	8	4	0.899442	0.213575
41	15	8	5	0.981157	0.081716
41	15	8	6	0.998181	0.017024
41	15	8	7	0.999933	0.001751
41	15	8	8	1.000000	0.000067
41	15	9	0	0.008919	0.008919
41	15	9	1	0.075807	0.066889
41	15	9	2	0.272954	0.197146
41	15	9	3	0.571959	0.299005
41	15	9	4	0.828250	0.256290
41	15	9	5	0.956395	0.128145
41	15	9	6	0.993539	0.037144
41	15	9	7	0.999508	0.005969
41	15	9	8	0.999986	0.000478
41	15	9	9	1.000000	0.000014
41	15	10	0	0.004738	0.004738
41	15	10	1	0.046544	0.041806
41	15	10	2	0.192863	0.146320
41	15	10	3	0.459832	0.266969
41	15	10	4	0.740150	0.280318
41	15	10	5	0.916350	0.176200
41	15	10	6	0.983092	0.066742
41	15	10	7	0.998016	0.014924
41	15	10	8	0.999881	0.001865
41	15	10	9	0.999997	0.000116
41	15	10	10	1.000000	0.000003
41	15	11	0	0.002445	0.002445
41	15	11	1	0.027664	0.025218
41	15	11	2	0.131503	0.103840
41	15	11	3	0.356489	0.224986
41	15	11	4	0.640682	0.284193
41	15	11	5	0.859511	0.218829
41	15	11	6	0.963715	0.104204
41	15	11	7	0.994164	0.030449
41	15	11	8	0.999460	0.005296
41	15	11	9	0.999975	0.000515
41	15	11	10	0.999999	0.000025
41	15	11	11	1.000000	0.000000
41	15	12	0	0.001223	0.001223
41	15	12	1	0.015895	0.014672
41	15	12	2	0.086506	0.070611
41	15	12	3	0.266495	0.179989
41	15	12	4	0.536478	0.269983
41	15	12	5	0.786568	0.250090
41	15	12	6	0.932454	0.145886
41	15	12	7	0.986045	0.053591
41	15	12	8	0.998224	0.012180
41	15	12	9	0.999872	0.001647
41	15	12	10	0.999995	0.000124
41	15	12	11	1.000000	0.000004

N	n	k	x	P(x)	p(x)
41	15	12	12	1.000000	0.000000
41	15	13	0	0.000590	0.000590
41	15	13	1	0.008812	0.008222
41	15	13	2	0.054853	0.046041
41	15	13	3	0.192017	0.137164
41	15	13	4	0.434071	0.242054
41	15	13	5	0.700330	0.266259
41	15	13	6	0.887179	0.186849
41	15	13	7	0.971261	0.084082
41	15	13	8	0.995284	0.024023
41	15	13	9	0.999531	0.004247
41	15	13	10	0.999974	0.000443
41	15	13	11	0.999999	0.000025
41	15	13	12	1.000000	0.000001
41	15	13	13	1.000000	0.000000
41	15	14	0	0.000274	0.000274
41	15	14	1	0.004701	0.004427
41	15	14	2	0.033477	0.028776
41	15	14	3	0.133232	0.099756
41	15	14	4	0.338978	0.205746
41	15	14	5	0.605238	0.266259
41	15	14	6	0.827121	0.221883
41	15	14	7	0.947238	0.120117
41	15	14	8	0.989279	0.042041
41	15	14	9	0.998621	0.009342
41	15	14	10	0.999895	0.001274
41	15	14	11	0.999996	0.000101
41	15	14	12	1.000000	0.000004
41	15	14	13	1.000000	0.000000
41	15	14	14	1.000000	0.000000
41	15	15	0	0.000122	0.000122
41	15	15	1	0.002406	0.002284
41	15	15	2	0.019622	0.017216
41	15	15	3	0.088896	0.069275
41	15	15	4	0.255156	0.166259
41	15	15	5	0.506623	0.251467
41	15	15	6	0.753160	0.246537
41	15	15	7	0.911647	0.158488
41	15	15	8	0.978379	0.066732
41	15	15	9	0.996545	0.018166
41	15	15	10	0.999659	0.003114
41	15	15	11	0.999981	0.000322
41	15	15	12	0.999999	0.000019
41	15	15	13	1.000000	0.000001
41	15	15	14	1.000000	0.000000
41	15	15	15	1.000000	0.000000
41	16	1	0	0.609756	0.609756
41	16	1	1	1.000000	0.390244
41	16	2	0	0.365854	0.365854
41	16	2	1	0.853658	0.487805

N	n	k	x	P(x)	p(x)
41	16	2	2	1.000000	0.146341
41	16	3	0	0.215760	0.215760
41	16	3	1	0.666041	0.450281
41	16	3	2	0.947467	0.281426
41	16	3	3	1.000000	0.052533
41	16	4	0	0.124914	0.124914
41	16	4	1	0.488299	0.363385
41	16	4	2	0.843784	0.355485
41	16	4	3	0.982028	0.138244
41	16	4	4	1.000000	0.017972
41	16	5	0	0.070897	0.070897
41	16	5	1	0.340980	0.270083
41	16	5	2	0.709276	0.368296
41	16	5	3	0.933456	0.224180
41	16	5	4	0.994171	0.060715
41	16	5	5	1.000000	0.005829
41	16	6	0	0.039387	0.039387
41	16	6	1	0.228446	0.189058
41	16	6	2	0.566050	0.337604
41	16	6	3	0.852502	0.286452
41	16	6	4	0.973933	0.121431
41	16	6	5	0.998219	0.024286
41	16	6	6	1.000000	0.001781
41	16	7	0	0.021382	0.021382
41	16	7	1	0.147421	0.126039
41	16	7	2	0.431008	0.283588
41	16	7	3	0.746105	0.315097
41	16	7	4	0.932299	0.186194
41	16	7	5	0.990586	0.058287
41	16	7	6	0.999491	0.008905
41	16	7	7	1.000000	0.000509
41	16	8	0	0.011320	0.011320
41	16	8	1	0.091815	0.080495
41	16	8	2	0.314237	0.222422
41	16	8	3	0.625627	0.311390
41	16	8	4	0.866584	0.240957
41	16	8	5	0.971729	0.105145
41	16	8	6	0.996872	0.025143
41	16	8	7	0.999865	0.002993
41	16	8	8	1.000000	0.000135
41	16	9	0	0.005831	0.005831
41	16	9	1	0.055226	0.049395
41	16	9	2	0.219876	0.164650
41	16	9	3	0.502958	0.283082
41	16	9	4	0.778963	0.276005
41	16	9	5	0.936680	0.157717
41	16	9	6	0.989253	0.052572
41	16	9	7	0.999049	0.009796
41	16	9	8	0.999967	0.000918
41	16	9	9	1.000000	0.000033

N	n	k	x	P(x)	p(x)
41	16	10	0	0.002916	0.002916
41	16	10	1	0.032072	0.029157
41	16	10	2	0.147842	0.115769
41	16	10	3	0.387956	0.240114
41	16	10	4	0.675461	0.287505
41	16	10	5	0.882465	0.207004
41	16	10	6	0.972824	0.090359
41	16	10	7	0.996294	0.023470
41	16	10	8	0.999738	0.003444
41	16	10	9	0.999993	0.000255
41	16	10	10	1.000000	0.000007
41	16	11	0	0.001411	0.001411
41	16	11	1	0.017964	0.016553
41	16	11	2	0.095559	0.077595
41	16	11	3	0.287263	0.191704
41	16	11	4	0.564169	0.276906
41	16	11	5	0.809012	0.244843
41	16	11	6	0.943676	0.134664
41	16	11	7	0.989480	0.045804
41	16	11	8	0.998849	0.009369
41	16	11	9	0.999935	0.001086
41	16	11	10	0.999999	0.000063
41	16	11	11	1.000000	0.000001
41	16	12	0	0.000658	0.000658
41	16	12	1	0.009688	0.009029
41	16	12	2	0.059348	0.049680
41	16	12	3	0.204191	0.144843
41	16	12	4	0.453407	0.249215
41	16	12	5	0.719236	0.265830
41	16	12	6	0.898788	0.179552
41	16	12	7	0.975739	0.076951
41	16	12	8	0.996350	0.020612
41	16	12	9	0.999682	0.003331
41	16	12	10	0.999986	0.000304
41	16	12	11	1.000000	0.000014
41	16	12	12	1.000000	0.000000
41	16	13	0	0.000295	0.000295
41	16	13	1	0.005017	0.004722
41	16	13	2	0.035374	0.030357
41	16	13	3	0.139261	0.103887
41	16	13	4	0.350283	0.211021
41	16	13	5	0.618404	0.268121
41	16	13	6	0.836874	0.218469
41	16	13	7	0.951857	0.114984
41	16	13	8	0.990665	0.038807
41	16	13	9	0.998878	0.008213
41	16	13	10	0.999923	0.001045
41	16	13	11	0.999997	0.000074
41	16	13	12	1.000000	0.000003
41	16	13	13	1.000000	0.000000

N	n	k	x	P(x)	p(x)
41	16	14	0	0.000126	0.000126
41	16	14	1	0.002488	0.002361
41	16	14	2	0.020196	0.017708
41	16	14	3	0.091028	0.070832
41	16	14	4	0.259845	0.168817
41	16	14	5	0.513071	0.253226
41	16	14	6	0.758849	0.245778
41	16	14	7	0.914898	0.156049
41	16	14	8	0.979577	0.064678
41	16	14	9	0.996824	0.017248
41	16	14	10	0.999699	0.002875
41	16	14	11	0.999984	0.000285
41	16	14	12	0.999999	0.000015
41	16	14	13	1.000000	0.000000
41	16	14	14	1.000000	0.000000
41	16	15	0	0.000052	0.000052
41	16	15	1	0.001176	0.001124
41	16	15	2	0.011014	0.009838
41	16	15	3	0.056924	0.045910
41	16	15	4	0.184815	0.127892
41	16	15	5	0.409905	0.225090
41	16	15	6	0.667820	0.257915
41	16	15	7	0.862882	0.195062
41	16	15	8	0.960413	0.097531
41	16	15	9	0.992353	0.031940
41	16	15	10	0.999060	0.006707
41	16	15	11	0.999931	0.000871
41	16	15	12	0.999997	0.000066
41	16	15	13	1.000000	0.000003
41	16	15	14	1.000000	0.000000
41	16	15	15	1.000000	0.000000
41	16	16	0	0.000020	0.000020
41	16	16	1	0.000527	0.000507
41	16	16	2	0.005716	0.005189
41	16	16	3	0.033969	0.028252
41	16	16	4	0.125788	0.091820
41	16	16	5	0.314675	0.188886
41	16	16	6	0.568622	0.253947
41	16	16	7	0.795360	0.226739
41	16	16	8	0.930403	0.135043
41	16	16	9	0.983753	0.053350
41	16	16	10	0.997512	0.013759
41	16	16	11	0.999764	0.002251
41	16	16	12	0.999987	0.000223
41	16	16	13	0.999999	0.000012
41	16	16	14	1.000000	0.000000
41	16	16	15	1.000000	0.000000
41	16	16	16	1.000000	0.000000
41	17	1	0	0.585366	0.585366
41	17	1	1	1.000000	0.414634

Table for $N = 2$, $n = 1$, through $N = 100$, $n = 50$

N	n	k	x	P(x)	p(x)
41	17	2	0	0.336585	0.336585
41	17	2	1	0.834146	0.497561
41	17	2	2	1.000000	0.165854
41	17	3	0	0.189869	0.189869
41	17	3	1	0.630019	0.440150
41	17	3	2	0.936210	0.306191
41	17	3	3	1.000000	0.063790
41	17	4	0	0.104927	0.104927
41	17	4	1	0.444692	0.339765
41	17	4	2	0.815345	0.370653
41	17	4	3	0.976498	0.161153
41	17	4	4	1.000000	0.023502
41	17	5	0	0.056718	0.056718
41	17	5	1	0.297767	0.241049
41	17	5	2	0.665080	0.367313
41	17	5	3	0.915521	0.250441
41	17	5	4	0.991743	0.076221
41	17	5	5	1.000000	0.008257
41	17	6	0	0.029934	0.029934
41	17	6	1	0.190634	0.160700
41	17	6	2	0.512033	0.321399
41	17	6	3	0.818128	0.306095
41	17	6	4	0.964218	0.146091
41	17	6	5	0.997247	0.033029
41	17	6	6	1.000000	0.002752
41	17	7	0	0.015395	0.015395
41	17	7	1	0.117171	0.101776
41	17	7	2	0.374291	0.257119
41	17	7	3	0.695690	0.321399
41	17	7	4	0.909956	0.214266
41	17	7	5	0.985923	0.075967
41	17	7	6	0.999135	0.013212
41	17	7	7	1.000000	0.000865
41	17	8	0	0.007697	0.007697
41	17	8	1	0.069276	0.061579
41	17	8	2	0.260856	0.191579
41	17	8	3	0.563349	0.302493
41	17	8	4	0.828031	0.264682
41	17	8	5	0.959111	0.131080
41	17	8	6	0.994860	0.035749
41	17	8	7	0.999745	0.004885
41	17	8	8	1.000000	0.000254
41	17	9	0	0.003732	0.003732
41	17	9	1	0.039420	0.035688
41	17	9	2	0.173774	0.134354
41	17	9	3	0.435018	0.261244
41	17	9	4	0.723762	0.288744
41	17	9	5	0.911446	0.187683
41	17	9	6	0.982944	0.071498
41	17	9	7	0.998265	0.015321
41	17	9	8	0.999931	0.001665
41	17	9	9	1.000000	0.000069
41	17	10	0	0.001749	0.001749
41	17	10	1	0.021576	0.019827
41	17	10	2	0.110796	0.089220
41	17	10	3	0.320724	0.209929
41	17	10	4	0.606460	0.285736
41	17	10	5	0.841064	0.234604
41	17	10	6	0.958366	0.117302
41	17	10	7	0.993477	0.035111
41	17	10	8	0.999462	0.005985
41	17	10	9	0.999983	0.000520
41	17	10	10	1.000000	0.000017
41	17	11	0	0.000790	0.000790
41	17	11	1	0.011343	0.010553
41	17	11	2	0.067625	0.056282
41	17	11	3	0.225918	0.158293
41	17	11	4	0.486635	0.260718
41	17	11	5	0.750250	0.263614
41	17	11	6	0.916743	0.166493
41	17	11	7	0.982151	0.065408
41	17	11	8	0.997725	0.015573
41	17	11	9	0.999848	0.002124
41	17	11	10	0.999996	0.000148
41	17	11	11	1.000000	0.000004
41	17	12	0	0.000342	0.000342
41	17	12	1	0.005715	0.005372
41	17	12	2	0.039484	0.033769
41	17	12	3	0.152048	0.112564
41	17	12	4	0.373658	0.221610
41	17	12	5	0.644804	0.271146
41	17	12	6	0.855696	0.210892
41	17	12	7	0.960349	0.104653
41	17	12	8	0.993053	0.032704
41	17	12	9	0.999282	0.006229
41	17	12	10	0.999962	0.000680
41	17	12	11	0.999999	0.000038
41	17	12	12	1.000000	0.000001
41	17	13	0	0.000142	0.000142
41	17	13	1	0.002751	0.002609
41	17	13	2	0.022017	0.019266
41	17	13	3	0.097707	0.075689
41	17	13	4	0.274315	0.176609
41	17	13	5	0.532606	0.258290
41	17	13	6	0.775702	0.243097
41	17	13	7	0.924261	0.148559
41	17	13	8	0.982903	0.058642
41	17	13	9	0.997564	0.014660
41	17	13	10	0.999798	0.002234
41	17	13	11	0.999991	0.000194
41	17	13	12	1.000000	0.000008
41	17	13	13	1.000000	0.000000
41	17	14	0	0.000056	0.000056
41	17	14	1	0.001260	0.001204
41	17	14	2	0.011696	0.010436
41	17	14	3	0.059862	0.048166
41	17	14	4	0.192318	0.132457
41	17	14	5	0.421910	0.229591
41	17	14	6	0.680200	0.258290
41	17	14	7	0.871205	0.191005
41	17	14	8	0.964054	0.092849
41	17	14	9	0.993375	0.029321
41	17	14	10	0.999239	0.005864
41	17	14	11	0.999950	0.000711
41	17	14	12	0.999998	0.000048
41	17	14	13	1.000000	0.000002
41	17	14	14	1.000000	0.000000
41	17	15	0	0.000021	0.000021
41	17	15	1	0.000546	0.000526
41	17	15	2	0.005898	0.005352
41	17	15	3	0.034887	0.028989
41	17	15	4	0.128543	0.093656
41	17	15	5	0.319869	0.191326
41	17	15	6	0.574971	0.255101
41	17	15	7	0.800462	0.225492
41	17	15	8	0.933104	0.132642
41	17	15	9	0.984687	0.051583
41	17	15	10	0.997719	0.013031
41	17	15	11	0.999792	0.002073
41	17	15	12	0.999989	0.000197
41	17	15	13	1.000000	0.000010
41	17	15	14	1.000000	0.000000
41	17	15	15	1.000000	0.000000
41	17	16	0	0.000007	0.000007
41	17	16	1	0.000223	0.000216
41	17	16	2	0.002810	0.002588
41	17	16	3	0.019277	0.016467
41	17	16	4	0.081715	0.062437
41	17	16	5	0.231565	0.149850
41	17	16	6	0.467043	0.235478
41	17	16	7	0.713735	0.246692
41	17	16	8	0.887190	0.173455
41	17	16	9	0.968815	0.081626
41	17	16	10	0.994210	0.025395
41	17	16	11	0.999313	0.005103
41	17	16	12	0.999951	0.000638
41	17	16	13	0.999998	0.000047
41	17	16	14	1.000000	0.000002
41	17	16	15	1.000000	0.000000
41	17	16	16	1.000000	0.000000
41	17	17	0	0.000002	0.000002
41	17	17	1	0.000085	0.000082
41	17	17	2	0.001258	0.001173
41	17	17	3	0.010056	0.008798
41	17	17	4	0.049247	0.039191
41	17	17	5	0.159637	0.110389
41	17	17	6	0.363433	0.203796
41	17	17	7	0.615058	0.251625
41	17	17	8	0.824746	0.209688
41	17	17	9	0.942695	0.117949
41	17	17	10	0.987100	0.044404
41	17	17	11	0.998089	0.010989
41	17	17	12	0.999824	0.001735
41	17	17	13	0.999991	0.000167
41	17	17	14	1.000000	0.000009
41	17	17	15	1.000000	0.000000
41	17	17	16	1.000000	0.000000
41	17	17	17	1.000000	0.000000
41	18	1	0	0.560976	0.560976
41	18	1	1	1.000000	0.439024
41	18	2	0	0.308537	0.308537
41	18	2	1	0.813415	0.504878
41	18	2	2	1.000000	0.186585
41	18	3	0	0.166135	0.166135
41	18	3	1	0.593340	0.427204
41	18	3	2	0.923452	0.330113
41	18	3	3	1.000000	0.076548
41	18	4	0	0.087439	0.087439
41	18	4	1	0.402222	0.314782
41	18	4	2	0.784457	0.382236
41	18	4	3	0.969784	0.185326
41	18	4	4	1.000000	0.030216
41	18	5	0	0.044901	0.044901
41	18	5	1	0.257592	0.212691
41	18	5	2	0.619166	0.361574
41	18	5	3	0.894651	0.275485
41	18	5	4	0.988567	0.093915
41	18	5	5	1.000000	0.011433
41	18	6	0	0.022451	0.022451
41	18	6	1	0.157155	0.134704
41	18	6	2	0.458467	0.301312
41	18	6	3	0.779866	0.321399
41	18	6	4	0.952044	0.172178
41	18	6	5	0.995871	0.043827
41	18	6	6	1.000000	0.004129
41	18	7	0	0.010905	0.010905
41	18	7	1	0.091727	0.080822
41	18	7	2	0.320724	0.228997
41	18	7	3	0.642123	0.321399
41	18	7	4	0.883173	0.241049

N	n	k	x	P(x)	p(x)
41	18	7	5	0.979593	0.096420
41	18	7	6	0.998584	0.018992
41	18	7	7	1.000000	0.001416
41	18	8	0	0.005132	0.005132
41	18	8	1	0.051316	0.046184
41	18	8	2	0.212961	0.161645
41	18	8	3	0.500330	0.287369
41	18	8	4	0.783917	0.283588
41	18	8	5	0.942726	0.158809
41	18	8	6	0.991881	0.049155
41	18	8	7	0.999542	0.007661
41	18	8	8	1.000000	0.000458
41	18	9	0	0.002333	0.002333
41	18	9	1	0.027524	0.025191
41	18	9	2	0.134587	0.107063
41	18	9	3	0.369707	0.235120
41	18	9	4	0.663607	0.293900
41	18	9	5	0.880165	0.216558
41	18	9	6	0.974007	0.093842
41	18	9	7	0.996988	0.022982
41	18	9	8	0.999861	0.002873
41	18	9	9	1.000000	0.000139
41	18	10	0	0.001020	0.001020
41	18	10	1	0.014141	0.013121
41	18	10	2	0.081056	0.066915
41	18	10	3	0.259495	0.178439
41	18	10	4	0.535026	0.275531
41	18	10	5	0.792188	0.257162
41	18	10	6	0.938816	0.146628
41	18	10	7	0.989088	0.050272
41	18	10	8	0.998963	0.009875
41	18	10	9	0.999961	0.000997
41	18	10	10	1.000000	0.000039
41	18	11	0	0.000428	0.000428
41	18	11	1	0.006946	0.006518
41	18	11	2	0.046519	0.039573
41	18	11	3	0.173153	0.126634
41	18	11	4	0.410593	0.237439
41	18	11	5	0.684346	0.273754
41	18	11	6	0.882057	0.197711
41	18	11	7	0.971250	0.089193
41	18	11	8	0.995778	0.024528
41	18	11	9	0.999671	0.003893
41	18	11	10	0.999990	0.000319
41	18	11	11	1.000000	0.000010
41	18	12	0	0.000171	0.000171
41	18	12	1	0.003252	0.003081
41	18	12	2	0.025413	0.022161
41	18	12	3	0.109836	0.084423
41	18	12	4	0.299788	0.189951
41	18	12	5	0.565720	0.265932
41	18	12	6	0.802973	0.237253
41	18	12	7	0.938546	0.135573
41	18	12	8	0.987602	0.049056
41	18	12	9	0.998503	0.010901
41	18	12	10	0.999905	0.001402
41	18	12	11	0.999998	0.000093
41	18	12	12	1.000000	0.000002
41	18	13	0	0.000065	0.000065
41	18	13	1	0.001446	0.001381
41	18	13	2	0.013187	0.011740
41	18	13	3	0.066169	0.052983
41	18	13	4	0.208087	0.141918
41	18	13	5	0.446509	0.238422
41	18	13	6	0.704799	0.258290
41	18	13	7	0.887122	0.182323
41	18	13	8	0.970686	0.083564
41	18	13	9	0.995120	0.024434
41	18	13	10	0.999518	0.004398
41	18	13	11	0.999975	0.000457
41	18	13	12	0.999999	0.000024
41	18	13	13	1.000000	0.000000
41	18	14	0	0.000023	0.000023
41	18	14	1	0.000608	0.000584
41	18	14	2	0.006478	0.005870
41	18	14	3	0.037786	0.031308
41	18	14	4	0.137128	0.099342
41	18	14	5	0.335813	0.198685
41	18	14	6	0.594103	0.258290
41	18	14	7	0.815495	0.221392
41	18	14	8	0.940842	0.125347
41	18	14	9	0.987266	0.046425
41	18	14	10	0.998262	0.010995
41	18	14	11	0.999861	0.001599
41	18	14	12	0.999994	0.000133
41	18	14	13	1.000000	0.000006
41	18	14	14	1.000000	0.000000
41	18	15	0	0.000008	0.000008
41	18	15	1	0.000240	0.000232
41	18	15	2	0.002999	0.002760
41	18	15	3	0.020392	0.017393
41	18	15	4	0.085617	0.065225
41	18	15	5	0.240150	0.154533
41	18	15	6	0.479308	0.239158
41	18	15	7	0.725298	0.245991
41	18	15	8	0.894417	0.169119
41	18	15	9	0.971791	0.077375
41	18	15	10	0.995004	0.023212
41	18	15	11	0.999446	0.004443
41	18	15	12	0.999965	0.000518
41	18	15	13	0.999999	0.000034
41	18	15	14	1.000000	0.000001
41	18	15	15	1.000000	0.000000
41	18	16	0	0.000002	0.000002
41	18	16	1	0.000088	0.000086
41	18	16	2	0.001301	0.001213
41	18	16	3	0.010358	0.009057
41	18	16	4	0.050496	0.040138
41	18	16	5	0.162884	0.112387
41	18	16	6	0.368927	0.206044
41	18	16	7	0.621225	0.252298
41	18	16	8	0.829371	0.208146
41	18	16	9	0.945008	0.115637
41	18	16	10	0.987862	0.042854
41	18	16	11	0.998250	0.010389
41	18	16	12	0.999845	0.001595
41	18	16	13	0.999992	0.000147
41	18	16	14	1.000000	0.000008
41	18	16	15	1.000000	0.000000
41	18	16	16	1.000000	0.000000
41	18	17	0	0.000001	0.000001
41	18	17	1	0.000030	0.000029
41	18	17	2	0.000525	0.000495
41	18	17	3	0.004924	0.004399
41	18	17	4	0.028019	0.023095
41	18	17	5	0.104442	0.076423
41	18	17	6	0.270026	0.165584
41	18	17	7	0.510214	0.240188
41	18	17	8	0.746113	0.235899
41	18	17	9	0.903379	0.157266
41	18	17	10	0.974148	0.070770
41	18	17	11	0.995341	0.021193
41	18	17	12	0.999462	0.004121
41	18	17	13	0.999963	0.000501
41	18	17	14	0.999999	0.000036
41	18	17	15	1.000000	0.000001
41	18	17	16	1.000000	0.000000
41	18	17	17	1.000000	0.000000
41	18	18	0	0.000000	0.000000
41	18	18	1	0.000009	0.000009
41	18	18	2	0.000195	0.000186
41	18	18	3	0.002174	0.001980
41	18	18	4	0.014547	0.012372
41	18	18	5	0.063046	0.048499
41	18	18	6	0.187234	0.124188
41	18	18	7	0.400128	0.212894
41	18	18	8	0.647822	0.247694
41	18	18	9	0.844404	0.196582
41	18	18	10	0.950558	0.106154
41	18	18	11	0.989160	0.038602
41	18	18	12	0.998432	0.009272
41	18	18	13	0.999859	0.001426
41	18	18	14	0.999993	0.000134
41	18	18	15	1.000000	0.000007
41	18	18	16	1.000000	0.000000
41	18	18	17	1.000000	0.000000
41	18	18	18	1.000000	0.000000
41	19	1	0	0.536585	0.536585
41	19	1	1	1.000000	0.463415
41	19	2	0	0.281707	0.281707
41	19	2	1	0.791463	0.509756
41	19	2	2	1.000000	0.208537
41	19	3	0	0.144465	0.144465
41	19	3	1	0.556191	0.411726
41	19	3	2	0.909099	0.352908
41	19	3	3	1.000000	0.090901
41	19	4	0	0.072233	0.072233
41	19	4	1	0.361163	0.288931
41	19	4	2	0.751219	0.390056
41	19	4	3	0.961726	0.210507
41	19	4	4	1.000000	0.038274
41	19	5	0	0.035140	0.035140
41	19	5	1	0.220602	0.185462
41	19	5	2	0.572004	0.351402
41	19	5	3	0.870696	0.298692
41	19	5	4	0.984483	0.113787
41	19	5	5	1.000000	0.015516
41	19	6	0	0.016594	0.016594
41	19	6	1	0.127871	0.111277
41	19	6	2	0.406065	0.278193
41	19	6	3	0.737944	0.331880
41	19	6	4	0.937072	0.199128
41	19	6	5	0.993966	0.056894
41	19	6	6	1.000000	0.006034
41	19	7	0	0.007586	0.007586
41	19	7	1	0.070643	0.063057
41	19	7	2	0.270942	0.200299
41	19	7	3	0.586228	0.315286
41	19	7	4	0.851732	0.265504
41	19	7	5	0.971208	0.119477
41	19	7	6	0.997759	0.026550
41	19	7	7	1.000000	0.002241
41	19	8	0	0.003347	0.003347
41	19	8	1	0.037260	0.033913
41	19	8	2	0.170793	0.133533
41	19	8	3	0.437858	0.267066
41	19	8	4	0.734598	0.296740
41	19	8	5	0.922012	0.187414
41	19	8	6	0.987607	0.065595
41	19	8	7	0.999209	0.011602

Table for $N = 2$, $n = 1$, through $N = 100$, $n = 50$

N	n	k	x	P(x)	p(x)
41	19	8	8	1.000000	0.000791
41	19	9	0	0.001420	0.001420
41	19	9	1	0.018762	0.017342
41	19	9	2	0.102003	0.083241
41	19	9	3	0.308372	0.206369
41	19	9	4	0.599716	0.291344
41	19	9	5	0.842503	0.242787
41	19	9	6	0.961767	0.119264
41	19	9	7	0.994990	0.033223
41	19	9	8	0.999736	0.004746
41	19	9	9	1.000000	0.000264
41	19	10	0	0.000577	0.000577
41	19	10	1	0.009007	0.008430
41	19	10	2	0.057781	0.048774
41	19	10	3	0.205187	0.147406
41	19	10	4	0.463148	0.257961
41	19	10	5	0.736284	0.273135
41	19	10	6	0.913316	0.177032
41	19	10	7	0.982531	0.069216
41	19	10	8	0.998105	0.015573
41	19	10	9	0.999917	0.001813
41	19	10	10	1.000000	0.000082
41	19	11	0	0.000223	0.000223
41	19	11	1	0.004112	0.003889
41	19	11	2	0.031034	0.026922
41	19	11	3	0.129107	0.098073
41	19	11	4	0.338329	0.209222
41	19	11	5	0.612932	0.274604
41	19	11	6	0.839076	0.226144
41	19	11	7	0.955738	0.116662
41	19	11	8	0.992579	0.036841
41	19	11	9	0.999333	0.006754
41	19	11	10	0.999976	0.000643
41	19	11	11	1.000000	0.000024
41	19	12	0	0.000082	0.000082
41	19	12	1	0.001779	0.001697
41	19	12	2	0.015778	0.013999
41	19	12	3	0.076801	0.061023
41	19	12	4	0.233718	0.156916
41	19	12	5	0.484784	0.251066
41	19	12	6	0.741081	0.256297
41	19	12	7	0.909073	0.167993
41	19	12	8	0.979070	0.069997
41	19	12	9	0.997081	0.018011
41	19	12	10	0.999783	0.002702
41	19	12	11	0.999994	0.000211
41	19	12	12	1.000000	0.000006
41	19	13	0	0.000028	0.000028
41	19	13	1	0.000726	0.000697
41	19	13	2	0.007572	0.006846

N	n	k	x	P(x)	p(x)
41	19	13	3	0.043133	0.035562
41	19	13	4	0.152554	0.109421
41	19	13	5	0.363579	0.211025
41	19	13	6	0.626189	0.262609
41	19	13	7	0.839559	0.213370
41	19	13	8	0.952520	0.112961
41	19	13	9	0.990871	0.038351
41	19	13	10	0.998945	0.008074
41	19	13	11	0.999935	0.000991
41	19	13	12	0.999998	0.000063
41	19	13	13	1.000000	0.000002
41	19	14	0	0.000009	0.000009
41	19	14	1	0.000277	0.000268
41	19	14	2	0.003415	0.003138
41	19	14	3	0.022812	0.019397
41	19	14	4	0.093936	0.071123
41	19	14	5	0.258067	0.164131
41	19	14	6	0.504263	0.246196
41	19	14	7	0.748115	0.243852
41	19	14	8	0.908142	0.160028
41	19	14	9	0.977174	0.069032
41	19	14	10	0.996349	0.019175
41	19	14	11	0.999652	0.003303
41	19	14	12	0.999983	0.000330
41	19	14	13	1.000000	0.000017
41	19	14	14	1.000000	0.000000
41	19	15	0	0.000003	0.000003
41	19	15	1	0.000098	0.000096
41	19	15	2	0.001439	0.001341
41	19	15	3	0.011318	0.009878
41	19	15	4	0.054423	0.043105
41	19	15	5	0.172962	0.118539
41	19	15	6	0.385724	0.212762
41	19	15	7	0.639736	0.254012
41	19	15	8	0.842946	0.203210
41	19	15	9	0.951607	0.108661
41	19	15	10	0.989958	0.038351
41	19	15	11	0.998674	0.008716
41	19	15	12	0.999897	0.001223
41	19	15	13	0.999996	0.000099
41	19	15	14	1.000000	0.000004
41	19	15	15	1.000000	0.000000
41	19	16	0	0.000001	0.000001
41	19	16	1	0.000032	0.000031
41	19	16	2	0.000563	0.000530
41	19	16	3	0.005239	0.004676
41	19	16	4	0.029554	0.024316
41	19	16	5	0.109133	0.079579
41	19	16	6	0.279343	0.170210
41	19	16	7	0.522500	0.243157

N	n	k	x	P(x)	p(x)
41	19	16	8	0.756973	0.234473
41	19	16	9	0.909814	0.152842
41	19	16	10	0.976682	0.066868
41	19	16	11	0.995992	0.019310
41	19	16	12	0.999568	0.003576
41	19	16	13	0.999973	0.000405
41	19	16	14	0.999999	0.000026
41	19	16	15	1.000000	0.000001
41	19	16	16	1.000000	0.000000
41	19	17	0	0.000000	0.000000
41	19	17	1	0.000010	0.000009
41	19	17	2	0.000202	0.000192
41	19	17	3	0.002246	0.002044
41	19	17	4	0.014965	0.012719
41	19	17	5	0.064569	0.049604
41	19	17	6	0.190834	0.126265
41	19	17	7	0.405785	0.214951
41	19	17	8	0.653805	0.248020
41	19	17	9	0.848678	0.194873
41	19	17	10	0.952610	0.103932
41	19	17	11	0.989813	0.037203
41	19	17	12	0.998566	0.008754
41	19	17	13	0.999876	0.001309
41	19	17	14	0.999994	0.000118
41	19	17	15	1.000000	0.000006
41	19	17	16	1.000000	0.000000
41	19	17	17	1.000000	0.000000
41	19	18	0	0.000000	0.000000
41	19	18	1	0.000003	0.000002
41	19	18	2	0.000066	0.000063
41	19	18	3	0.000883	0.000818
41	19	18	4	0.007016	0.006132
41	19	18	5	0.035633	0.028618
41	19	18	6	0.122440	0.086807
41	19	18	7	0.298309	0.175869
41	19	18	8	0.540129	0.241820
41	19	18	9	0.767480	0.227352
41	19	18	10	0.913635	0.146155
41	19	18	11	0.977412	0.063777
41	19	18	12	0.996013	0.018601
41	19	18	13	0.999548	0.003535
41	19	18	14	0.999969	0.000421
41	19	18	15	0.999999	0.000030
41	19	18	16	1.000000	0.000001
41	19	18	17	1.000000	0.000000
41	19	18	18	1.000000	0.000000
41	19	19	0	0.000000	0.000000
41	19	19	1	0.000001	0.000001
41	19	19	2	0.000019	0.000018
41	19	19	3	0.000314	0.000296

N	n	k	x	P(x)	p(x)
41	19	19	4	0.003016	0.002702
41	19	19	5	0.018214	0.015198
41	19	19	6	0.073376	0.055162
41	19	19	7	0.206552	0.133176
41	19	19	8	0.424476	0.217924
41	19	19	9	0.668632	0.244156
41	19	19	10	0.856444	0.187812
41	19	19	11	0.955229	0.098784
41	19	19	12	0.990352	0.035123
41	19	19	13	0.998626	0.008274
41	19	19	14	0.999878	0.001252
41	19	19	15	0.999994	0.000116
41	19	19	16	1.000000	0.000006
41	19	19	17	1.000000	0.000000
41	19	19	18	1.000000	0.000000
41	19	19	19	1.000000	0.000000
41	20	1	0	0.512195	0.512195
41	20	1	1	1.000000	0.487805
41	20	2	0	0.256098	0.256098
41	20	2	1	0.768293	0.512195
41	20	2	2	1.000000	0.231707
41	20	3	0	0.124765	0.124765
41	20	3	1	0.518762	0.393996
41	20	3	2	0.893058	0.374296
41	20	3	3	1.000000	0.106942
41	20	4	0	0.059099	0.059099
41	20	4	1	0.321764	0.262664
41	20	4	2	0.715760	0.393996
41	20	4	3	0.952158	0.236398
41	20	4	4	1.000000	0.047842
41	20	5	0	0.027154	0.027154
41	20	5	1	0.186882	0.159728
41	20	5	2	0.524086	0.337204
41	20	5	3	0.843542	0.319456
41	20	5	4	0.979311	0.135769
41	20	5	5	1.000000	0.020689
41	20	6	0	0.012068	0.012068
41	20	6	1	0.102581	0.090513
41	20	6	2	0.355484	0.252903
41	20	6	3	0.692688	0.337204
41	20	6	4	0.918970	0.226282
41	20	6	5	0.991380	0.072410
41	20	6	6	1.000000	0.008620
41	20	7	0	0.005172	0.005172
41	20	7	1	0.053446	0.048273
41	20	7	2	0.225420	0.171974
41	20	7	3	0.528903	0.303484
41	20	7	4	0.815527	0.286623
41	20	7	5	0.960347	0.144820
41	20	7	6	0.996552	0.036205

N	n	k	x	P(x)	p(x)
41	20	7	7	1.000000	0.003448
41	20	8	0	0.002130	0.002130
41	20	8	1	0.026469	0.024340
41	20	8	2	0.134375	0.107905
41	20	8	3	0.377161	0.242787
41	20	8	4	0.680645	0.303484
41	20	8	5	0.896455	0.215811
41	20	8	6	0.981644	0.085188
41	20	8	7	0.998682	0.017038
41	20	8	8	1.000000	0.001318
41	20	9	0	0.000839	0.000839
41	20	9	1	0.012456	0.011617
41	20	9	2	0.075517	0.063062
41	20	9	3	0.252089	0.176572
41	20	9	4	0.533501	0.281412
41	20	9	5	0.798360	0.264858
41	20	9	6	0.945503	0.147144
41	20	9	7	0.991970	0.046466
41	20	9	8	0.999520	0.007551
41	20	9	9	1.000000	0.000479
41	20	10	0	0.000315	0.000315
41	20	10	1	0.005558	0.005244
41	20	10	2	0.040045	0.034487
41	20	10	3	0.158285	0.118240
41	20	10	4	0.392795	0.234510
41	20	10	5	0.674207	0.281412
41	20	10	6	0.881128	0.206921
41	20	10	7	0.973093	0.091965
41	20	10	8	0.996689	0.023596
41	20	10	9	0.999835	0.003146
41	20	10	10	1.000000	0.000165
41	20	11	0	0.000112	0.000112
41	20	11	1	0.002344	0.002233
41	20	11	2	0.020020	0.017676
41	20	11	3	0.093444	0.073423
41	20	11	4	0.271758	0.178314
41	20	11	5	0.538040	0.266282
41	20	11	6	0.787680	0.249640
41	20	11	7	0.934527	0.146847
41	20	11	8	0.987555	0.053028
41	20	11	9	0.998719	0.011164
41	20	11	10	0.999947	0.001228
41	20	11	11	1.000000	0.000053
41	20	12	0	0.000037	0.000037
41	20	12	1	0.000930	0.000893
41	20	12	2	0.009415	0.008484
41	20	12	3	0.051837	0.042422
41	20	12	4	0.176657	0.124820
41	20	12	5	0.404899	0.228242
41	20	12	6	0.671181	0.266282

N	n	k	x	P(x)	p(x)
41	20	12	7	0.870893	0.199712
41	20	12	8	0.966344	0.095450
41	20	12	9	0.994625	0.028282
41	20	12	10	0.999537	0.004912
41	20	12	11	0.999984	0.000447
41	20	12	12	1.000000	0.000016
41	20	13	0	0.000012	0.000012
41	20	13	1	0.000345	0.000334
41	20	13	2	0.004149	0.003803
41	20	13	3	0.026969	0.022820
41	20	13	4	0.107791	0.080822
41	20	13	5	0.286843	0.179052
41	20	13	6	0.542631	0.255788
41	20	13	7	0.781367	0.238736
41	20	13	8	0.926847	0.145480
41	20	13	9	0.983898	0.057051
41	20	13	10	0.997844	0.013946
41	20	13	11	0.999845	0.002002
41	20	13	12	0.999995	0.000150
41	20	13	13	1.000000	0.000004
41	20	14	0	0.000003	0.000003
41	20	14	1	0.000119	0.000115
41	20	14	2	0.001704	0.001585
41	20	14	3	0.013114	0.011410
41	20	14	4	0.061607	0.048493
41	20	14	5	0.190922	0.129315
41	20	14	6	0.414737	0.223815
41	20	14	7	0.670526	0.255788
41	20	14	8	0.864499	0.193973
41	20	14	9	0.961485	0.096986
41	20	14	10	0.992863	0.031378
41	20	14	11	0.999202	0.006339
41	20	14	12	0.999953	0.000751
41	20	14	13	0.999999	0.000046
41	20	14	14	1.000000	0.000001
41	20	15	0	0.000001	0.000001
41	20	15	1	0.000038	0.000037
41	20	15	2	0.000647	0.000610
41	20	15	3	0.005930	0.005282
41	20	15	4	0.032870	0.026941
41	20	15	5	0.119080	0.086210
41	20	15	6	0.298685	0.179605
41	20	15	7	0.547368	0.248683
41	20	15	8	0.778288	0.230920
41	20	15	9	0.921972	0.143684
41	20	15	10	0.981241	0.059270
41	20	15	11	0.997089	0.015847
41	20	15	12	0.999730	0.002641
41	20	15	13	0.999987	0.000257
41	20	15	14	1.000000	0.000013

N	n	k	x	P(x)	p(x)
41	20	15	15	1.000000	0.000000
41	20	16	0	0.000000	0.000000
41	20	16	1	0.000011	0.000011
41	20	16	2	0.000225	0.000214
41	20	16	3	0.002476	0.002251
41	20	16	4	0.016291	0.013816
41	20	16	5	0.069344	0.053052
41	20	16	6	0.201975	0.132631
41	20	16	7	0.423027	0.221052
41	20	16	8	0.671710	0.248683
41	20	16	9	0.861183	0.189473
41	20	16	10	0.958446	0.097263
41	20	16	11	0.991603	0.033158
41	20	16	12	0.998917	0.007314
41	20	16	13	0.999918	0.001000
41	20	16	14	0.999997	0.000079
41	20	16	15	1.000000	0.000003
41	20	16	16	1.000000	0.000000
41	20	17	0	0.000000	0.000000
41	20	17	1	0.000003	0.000003
41	20	17	2	0.000071	0.000068
41	20	17	3	0.000945	0.000874
41	20	17	4	0.007449	0.006504
41	20	17	5	0.037512	0.030063
41	20	17	6	0.127701	0.090189
41	20	17	7	0.308080	0.180378
41	20	17	8	0.552342	0.244262
41	20	17	9	0.777815	0.225473
41	20	17	10	0.919540	0.141726
41	20	17	11	0.979667	0.060126
41	20	17	12	0.996577	0.016910
41	20	17	13	0.999638	0.003061
41	20	17	14	0.999978	0.000340
41	20	17	15	0.999999	0.000021
41	20	17	16	1.000000	0.000001
41	20	17	17	1.000000	0.000000
41	20	18	0	0.000000	0.000000
41	20	18	1	0.000001	0.000001
41	20	18	2	0.000020	0.000019
41	20	18	3	0.000326	0.000306
41	20	18	4	0.003113	0.002787
41	20	18	5	0.018723	0.015610
41	20	18	6	0.075091	0.056368
41	20	18	7	0.210375	0.135284
41	20	18	8	0.430211	0.219836
41	20	18	9	0.674473	0.244262
41	20	18	10	0.860488	0.186015
41	20	18	11	0.957119	0.096631
41	20	18	12	0.990940	0.033821
41	20	18	13	0.998745	0.007805

N	n	k	x	P(x)	p(x)
41	20	18	14	0.999893	0.001148
41	20	18	15	0.999995	0.000102
41	20	18	16	1.000000	0.000005
41	20	18	17	1.000000	0.000000
41	20	18	18	1.000000	0.000000
41	20	19	0	0.000000	0.000000
41	20	19	1	0.000000	0.000000
41	20	19	2	0.000005	0.000005
41	20	19	3	0.000100	0.000095
41	20	19	4	0.001174	0.001075
41	20	19	5	0.008543	0.007369
41	20	19	6	0.040780	0.032237
41	20	19	7	0.133910	0.093130
41	20	19	8	0.315514	0.181604
41	20	19	9	0.557652	0.242138
41	20	19	10	0.779612	0.221960
41	20	19	11	0.919307	0.139695
41	20	19	12	0.979176	0.059869
41	20	19	13	0.996370	0.017193
41	20	19	14	0.999593	0.003224
41	20	19	15	0.999973	0.000379
41	20	19	16	0.999999	0.000026
41	20	19	17	1.000000	0.000001
41	20	19	18	1.000000	0.000000
41	20	19	19	1.000000	0.000000
41	20	20	0	0.000000	0.000000
41	20	20	1	0.000000	0.000000
41	20	20	2	0.000001	0.000001
41	20	20	3	0.000026	0.000025
41	20	20	4	0.000393	0.000366
41	20	20	5	0.003519	0.003126
41	20	20	6	0.020265	0.016747
41	20	20	7	0.078879	0.058613
41	20	20	8	0.216457	0.137579
41	20	20	9	0.436583	0.220126
41	20	20	10	0.678721	0.242138
41	20	20	11	0.862159	0.183438
41	20	20	12	0.957406	0.095247
41	20	20	13	0.990899	0.033493
41	20	20	14	0.998714	0.007815
41	20	20	15	0.999886	0.001172
41	20	20	16	0.999994	0.000108
41	20	20	17	1.000000	0.000006
41	20	20	18	1.000000	0.000000
41	20	20	19	1.000000	0.000000
41	20	20	20	1.000000	0.000000
42	1	1	0	0.976190	0.976190
42	1	1	1	1.000000	0.023810
42	2	1	0	0.952381	0.952381
42	2	1	1	1.000000	0.047619

Table for $N = 2$, $n = 1$, through $N = 100$, $n = 50$

N	n	k	x	P(x)	p(x)
42	2	2	0	0.905923	0.905923
42	2	2	1	0.998839	0.092915
42	2	2	2	1.000000	0.001161
42	3	1	0	0.928571	0.928571
42	3	1	1	1.000000	0.071429
42	3	2	0	0.860627	0.860627
42	3	2	1	0.996516	0.135888
42	3	2	2	1.000000	0.003484
42	3	3	0	0.796080	0.796080
42	3	3	1	0.989721	0.193641
42	3	3	2	0.999913	0.010192
42	3	3	3	1.000000	0.000087
42	4	1	0	0.904762	0.904762
42	4	1	1	1.000000	0.095238
42	4	2	0	0.816492	0.816492
42	4	2	1	0.993031	0.176539
42	4	2	2	1.000000	0.006969
42	4	3	0	0.734843	0.734843
42	4	3	1	0.979791	0.244948
42	4	3	2	0.999652	0.019861
42	4	3	3	1.000000	0.000348
42	4	4	0	0.659475	0.659475
42	4	4	1	0.960949	0.301474
42	4	4	2	0.998633	0.037684
42	4	4	3	0.999991	0.001358
42	4	4	4	1.000000	0.000009
42	5	1	0	0.880952	0.880952
42	5	1	1	1.000000	0.119048
42	5	2	0	0.773519	0.773519
42	5	2	1	0.988386	0.214866
42	5	2	2	1.000000	0.011614
42	5	3	0	0.676829	0.676829
42	5	3	1	0.966899	0.290070
42	5	3	2	0.999129	0.032230
42	5	3	3	1.000000	0.000871
42	5	4	0	0.590056	0.590056
42	5	4	1	0.937148	0.347092
42	5	4	2	0.996650	0.059501
42	5	4	3	0.999955	0.003306
42	5	4	4	1.000000	0.000045
42	5	5	0	0.512417	0.512417
42	5	5	1	0.900612	0.388195
42	5	5	2	0.991952	0.091340
42	5	5	3	0.999781	0.007829
42	5	5	4	0.999999	0.000217
42	5	5	5	1.000000	0.000001
42	6	1	0	0.857143	0.857143
42	6	1	1	1.000000	0.142857
42	6	2	0	0.731707	0.731707
42	6	2	1	0.982578	0.250871

N	n	k	x	P(x)	p(x)
42	6	2	2	1.000000	0.017422
42	6	3	0	0.621951	0.621951
42	6	3	1	0.951219	0.329268
42	6	3	2	0.998258	0.047038
42	6	3	3	1.000000	0.001742
42	6	4	0	0.526266	0.526266
42	6	4	1	0.909006	0.382739
42	6	4	2	0.993433	0.084428
42	6	4	3	0.999866	0.006433
42	6	4	4	1.000000	0.000134
42	6	5	0	0.443172	0.443172
42	6	5	1	0.858645	0.415473
42	6	5	2	0.984546	0.125901
42	6	5	3	0.999358	0.014812
42	6	5	4	0.999993	0.000635
42	6	5	5	1.000000	0.000007
42	6	6	0	0.371306	0.371306
42	6	6	1	0.802500	0.431194
42	6	6	2	0.970935	0.168435
42	6	6	3	0.998157	0.027222
42	6	6	4	0.999959	0.001801
42	6	6	5	1.000000	0.000041
42	6	6	6	1.000000	0.000000
42	7	1	0	0.833333	0.833333
42	7	1	1	1.000000	0.166667
42	7	2	0	0.691057	0.691057
42	7	2	1	0.975610	0.284553
42	7	2	2	1.000000	0.024390
42	7	3	0	0.570122	0.570122
42	7	3	1	0.932927	0.362805
42	7	3	2	0.996951	0.064024
42	7	3	3	1.000000	0.003049
42	7	4	0	0.467792	0.467792
42	7	4	1	0.877111	0.409318
42	7	4	2	0.988743	0.111632
42	7	4	3	0.999687	0.010944
42	7	4	4	1.000000	0.000313
42	7	5	0	0.381620	0.381620
42	7	5	1	0.812481	0.430861
42	7	5	2	0.974054	0.161573
42	7	5	3	0.998535	0.024481
42	7	5	4	0.999975	0.001440
42	7	5	5	1.000000	0.000025
42	7	6	0	0.309422	0.309422
42	7	6	1	0.742612	0.433190
42	7	6	2	0.952220	0.209608
42	7	6	3	0.995889	0.043668
42	7	6	4	0.999859	0.003970
42	7	6	5	0.999999	0.000140
42	7	6	6	1.000000	0.000001

N	n	k	x	P(x)	p(x)
42	7	7	0	0.249256	0.249256
42	7	7	1	0.670414	0.421157
42	7	7	2	0.923108	0.252694
42	7	7	3	0.991037	0.067929
42	7	7	4	0.999528	0.008491
42	7	7	5	0.999991	0.000463
42	7	7	6	1.000000	0.000009
42	7	7	7	1.000000	0.000000
42	8	1	0	0.809524	0.809524
42	8	1	1	1.000000	0.190476
42	8	2	0	0.651568	0.651568
42	8	2	1	0.967480	0.315912
42	8	2	2	1.000000	0.032520
42	8	3	0	0.521254	0.521254
42	8	3	1	0.912195	0.390941
42	8	3	2	0.995122	0.082927
42	8	3	3	1.000000	0.004878
42	8	4	0	0.414330	0.414330
42	8	4	1	0.842026	0.427696
42	8	4	2	0.982364	0.140338
42	8	4	3	0.999375	0.017011
42	8	4	4	1.000000	0.000625
42	8	5	0	0.327103	0.327103
42	8	5	1	0.763240	0.436137
42	8	5	2	0.960205	0.196965
42	8	5	3	0.997136	0.036931
42	8	5	4	0.999934	0.002798
42	8	5	5	1.000000	0.000066
42	8	6	0	0.256378	0.256378
42	8	6	1	0.680728	0.424350
42	8	6	2	0.928265	0.247537
42	8	6	3	0.992146	0.063881
42	8	6	4	0.999632	0.007486
42	8	6	5	0.999995	0.000363
42	8	6	6	1.000000	0.000005
42	8	7	0	0.199405	0.199405
42	8	7	1	0.598215	0.398810
42	8	7	2	0.887009	0.288794
42	8	7	3	0.983273	0.096265
42	8	7	4	0.998800	0.015527
42	8	7	5	0.999964	0.001164
42	8	7	6	1.000000	0.000035
42	8	7	7	1.000000	0.000000
42	8	8	0	0.153827	0.153827
42	8	8	1	0.518453	0.364626
42	8	8	2	0.837501	0.319048
42	8	8	3	0.969521	0.132020
42	8	8	4	0.997025	0.027504
42	8	8	5	0.999865	0.002839
42	8	8	6	0.999998	0.000133

N	n	k	x	P(x)	p(x)
42	8	8	7	1.000000	0.000002
42	8	8	8	1.000000	0.000000
42	9	1	0	0.785714	0.785714
42	9	1	1	1.000000	0.214286
42	9	2	0	0.613240	0.613240
42	9	2	1	0.958188	0.344948
42	9	2	2	1.000000	0.041812
42	9	3	0	0.475261	0.475261
42	9	3	1	0.889199	0.413937
42	9	3	2	0.992683	0.103484
42	9	3	3	1.000000	0.007317
42	9	4	0	0.365586	0.365586
42	9	4	1	0.804288	0.438703
42	9	4	2	0.974109	0.169820
42	9	4	3	0.998874	0.024765
42	9	4	4	1.000000	0.001126
42	9	5	0	0.279000	0.279000
42	9	5	1	0.711930	0.432930
42	9	5	2	0.942826	0.230896
42	9	5	3	0.994964	0.052138
42	9	5	4	0.999852	0.004888
42	9	5	5	1.000000	0.000148
42	9	6	0	0.211135	0.211135
42	9	6	1	0.618323	0.407189
42	9	6	2	0.899143	0.280820
42	9	6	3	0.986509	0.087366
42	9	6	4	0.999191	0.012682
42	9	6	5	0.999984	0.000793
42	9	6	6	1.000000	0.000016
42	9	7	0	0.158351	0.158351
42	9	7	1	0.527837	0.369486
42	9	7	2	0.844539	0.316702
42	9	7	3	0.971948	0.127409
42	9	7	4	0.997430	0.025482
42	9	7	5	0.999896	0.002466
42	9	7	6	0.999999	0.000103
42	9	7	7	1.000000	0.000001
42	9	8	0	0.117632	0.117632
42	9	8	1	0.443383	0.325751
42	9	8	2	0.781199	0.337816
42	9	8	3	0.950106	0.168908
42	9	8	4	0.993790	0.043683
42	9	8	5	0.999614	0.005824
42	9	8	6	0.999990	0.000376
42	9	8	7	1.000000	0.000010
42	9	8	8	1.000000	0.000000
42	9	9	0	0.086494	0.086494
42	9	9	1	0.366736	0.280242
42	9	9	2	0.711648	0.344913
42	9	9	3	0.920299	0.208651

Table for $N = 2$, $n = 1$, through $N = 100$, $n = 50$

N	n	k	x	P(x)	p(x)
42	9	9	4	0.987366	0.067066
42	9	9	5	0.998929	0.011563
42	9	9	6	0.999957	0.001028
42	9	9	7	0.999999	0.000043
42	9	9	8	1.000000	0.000001
42	9	9	9	1.000000	0.000000
42	10	1	0	0.761905	0.761905
42	10	1	1	1.000000	0.238095
42	10	2	0	0.576074	0.576074
42	10	2	1	0.947735	0.371661
42	10	2	2	1.000000	0.052265
42	10	3	0	0.432056	0.432056
42	10	3	1	0.864111	0.432056
42	10	3	2	0.989547	0.125436
42	10	3	3	1.000000	0.010453
42	10	4	0	0.321272	0.321272
42	10	4	1	0.764406	0.443134
42	10	4	2	0.963817	0.199410
42	10	4	3	0.998124	0.034307
42	10	4	4	1.000000	0.001876
42	10	5	0	0.236727	0.236727
42	10	5	1	0.659453	0.422727
42	10	5	2	0.921835	0.262382
42	10	5	3	0.991804	0.069969
42	10	5	4	0.999704	0.007900
42	10	5	5	1.000000	0.000296
42	10	6	0	0.172747	0.172747
42	10	6	1	0.556628	0.383881
42	10	6	2	0.865104	0.308476
42	10	6	3	0.978567	0.113462
42	10	6	4	0.998423	0.019856
42	10	6	5	0.999960	0.001537
42	10	6	6	1.000000	0.000040
42	10	7	0	0.124761	0.124761
42	10	7	1	0.460658	0.335896
42	10	7	2	0.796554	0.335896
42	10	7	3	0.956505	0.159951
42	10	7	4	0.995113	0.038609
42	10	7	5	0.999746	0.004633
42	10	7	6	0.999996	0.000249
42	10	7	7	1.000000	0.000004
42	10	8	0	0.089115	0.089115
42	10	8	1	0.374284	0.285169
42	10	8	2	0.719778	0.345493
42	10	8	3	0.924514	0.204737
42	10	8	4	0.988495	0.063980
42	10	8	5	0.999084	0.010590
42	10	8	6	0.999967	0.000882
42	10	8	7	1.000000	0.000033
42	10	8	8	1.000000	0.000000

N	n	k	x	P(x)	p(x)
42	10	9	0	0.062905	0.062905
42	10	9	1	0.298798	0.235894
42	10	9	2	0.638485	0.339687
42	10	9	3	0.882363	0.243878
42	10	9	4	0.977204	0.094841
42	10	9	5	0.997527	0.020323
42	10	9	6	0.999863	0.002336
42	10	9	7	0.999997	0.000133
42	10	9	8	1.000000	0.000003
42	10	9	9	1.000000	0.000000
42	10	10	0	0.043843	0.043843
42	10	10	1	0.234464	0.190621
42	10	10	2	0.556137	0.321673
42	10	10	3	0.830631	0.274494
42	10	10	4	0.959960	0.129329
42	10	10	5	0.994448	0.034488
42	10	10	6	0.999580	0.005132
42	10	10	7	0.999985	0.000404
42	10	10	8	1.000000	0.000015
42	10	10	9	1.000000	0.000000
42	10	10	10	1.000000	0.000000
42	11	1	0	0.738095	0.738095
42	11	1	1	1.000000	0.261905
42	11	2	0	0.540070	0.540070
42	11	2	1	0.936121	0.396051
42	11	2	2	1.000000	0.063879
42	11	3	0	0.391551	0.391551
42	11	3	1	0.837108	0.445557
42	11	3	2	0.985627	0.148519
42	11	3	3	1.000000	0.014373
42	11	4	0	0.281113	0.281113
42	11	4	1	0.722862	0.441749
42	11	4	2	0.951353	0.228491
42	11	4	3	0.997052	0.045698
42	11	4	4	1.000000	0.002948
42	11	5	0	0.199738	0.199738
42	11	5	1	0.606613	0.406874
42	11	5	2	0.897237	0.290625
42	11	5	3	0.987431	0.090194
42	11	5	4	0.999457	0.012026
42	11	5	5	1.000000	0.000543
42	11	6	0	0.140357	0.140357
42	11	6	1	0.496647	0.356290
42	11	6	2	0.826545	0.329898
42	11	6	3	0.967930	0.141385
42	11	6	4	0.997182	0.029252
42	11	6	5	0.999912	0.002730
42	11	6	6	1.000000	0.000088
42	11	7	0	0.097470	0.097470
42	11	7	1	0.397677	0.300207

N	n	k	x	P(x)	p(x)
42	11	7	2	0.744070	0.346393
42	11	7	3	0.936511	0.192441
42	11	7	4	0.991494	0.054983
42	11	7	5	0.999457	0.007963
42	11	7	6	0.999988	0.000531
42	11	7	7	1.000000	0.000012
42	11	8	0	0.066836	0.066836
42	11	8	1	0.311904	0.245067
42	11	8	2	0.654998	0.343094
42	11	8	3	0.892524	0.237527
42	11	8	4	0.980497	0.087973
42	11	8	5	0.998092	0.017595
42	11	8	6	0.999912	0.001820
42	11	8	7	0.999999	0.000087
42	11	8	8	1.000000	0.000001
42	11	9	0	0.045213	0.045213
42	11	9	1	0.239825	0.194612
42	11	9	2	0.564179	0.324354
42	11	9	3	0.836636	0.272457
42	11	9	4	0.962385	0.125749
42	11	9	5	0.994987	0.032602
42	11	9	6	0.999644	0.004657
42	11	9	7	0.999988	0.000344
42	11	9	8	1.000000	0.000011
42	11	9	9	1.000000	0.000000
42	11	10	0	0.030142	0.030142
42	11	10	1	0.180852	0.150710
42	11	10	2	0.475719	0.294867
42	11	10	3	0.770586	0.294867
42	11	10	4	0.935711	0.165125
42	11	10	5	0.989059	0.053348
42	11	10	6	0.998938	0.009879
42	11	10	7	0.999947	0.001008
42	11	10	8	0.999999	0.000052
42	11	10	9	1.000000	0.000001
42	11	10	10	1.000000	0.000000
42	11	11	0	0.019781	0.019781
42	11	11	1	0.133755	0.113974
42	11	11	2	0.392787	0.259032
42	11	11	3	0.696869	0.304081
42	11	11	4	0.899590	0.202721
42	11	11	5	0.979056	0.079467
42	11	11	6	0.997395	0.018338
42	11	11	7	0.999821	0.002426
42	11	11	8	0.999994	0.000173
42	11	11	9	1.000000	0.000006
42	11	11	10	1.000000	0.000000
42	11	11	11	1.000000	0.000000
42	12	1	0	0.714286	0.714286
42	12	1	1	1.000000	0.285714

N	n	k	x	P(x)	p(x)
42	12	2	0	0.505226	0.505226
42	12	2	1	0.923345	0.418118
42	12	2	2	1.000000	0.076655
42	12	3	0	0.353658	0.353658
42	12	3	1	0.808362	0.454704
42	12	3	2	0.980836	0.172474
42	12	3	3	1.000000	0.019164
42	12	4	0	0.244841	0.244841
42	12	4	1	0.680113	0.435272
42	12	4	2	0.936612	0.256500
42	12	4	3	0.995578	0.058965
42	12	4	4	1.000000	0.004422
42	12	5	0	0.167522	0.167522
42	12	5	1	0.554113	0.386590
42	12	5	2	0.869112	0.314999
42	12	5	3	0.981612	0.112500
42	12	5	4	0.999069	0.017457
42	12	5	5	1.000000	0.000931
42	12	6	0	0.113191	0.113191
42	12	6	1	0.439180	0.325990
42	12	6	2	0.783977	0.344797
42	12	6	3	0.954247	0.170270
42	12	6	4	0.995294	0.041047
42	12	6	5	0.999824	0.004529
42	12	6	6	1.000000	0.000176
42	12	7	0	0.075461	0.075461
42	12	7	1	0.339573	0.264112
42	12	7	2	0.688200	0.348628
42	12	7	3	0.911680	0.223479
42	12	7	4	0.986173	0.074493
42	12	7	5	0.998943	0.012770
42	12	7	6	0.999971	0.001027
42	12	7	7	1.000000	0.000029
42	12	8	0	0.049588	0.049588
42	12	8	1	0.256566	0.206978
42	12	8	2	0.588592	0.332026
42	12	8	3	0.854214	0.265621
42	12	8	4	0.969146	0.114932
42	12	8	5	0.996389	0.027243
42	12	8	6	0.999794	0.003405
42	12	8	7	0.999996	0.000201
42	12	8	8	1.000000	0.000004
42	12	9	0	0.032087	0.032087
42	12	9	1	0.189603	0.157516
42	12	9	2	0.490938	0.301335
42	12	9	3	0.783902	0.292965
42	12	9	4	0.942103	0.158201
42	12	9	5	0.990780	0.048677
42	12	9	6	0.999193	0.008413
42	12	9	7	0.999966	0.000773

Table for $N = 2$, $n = 1$, through $N = 100$, $n = 50$

N	n	k	x	P(x)	p(x)
42	12	9	8	0.999999	0.000033
42	12	9	9	1.000000	0.000000
42	12	10	0	0.020419	0.020419
42	12	10	1	0.137097	0.116678
42	12	10	2	0.399624	0.262527
42	12	10	3	0.704003	0.304379
42	12	10	4	0.903751	0.199749
42	12	10	5	0.980455	0.076703
42	12	10	6	0.997664	0.017209
42	12	10	7	0.999849	0.002185
42	12	10	8	0.999995	0.000146
42	12	10	9	1.000000	0.000004
42	12	10	10	1.000000	0.000000
42	12	11	0	0.012762	0.012762
42	12	11	1	0.096989	0.084227
42	12	11	2	0.317584	0.220595
42	12	11	3	0.618396	0.300812
42	12	11	4	0.853814	0.235418
42	12	11	5	0.963676	0.109862
42	12	11	6	0.994437	0.030761
42	12	11	7	0.999508	0.005071
42	12	11	8	0.999977	0.000469
42	12	11	9	0.999999	0.000022
42	12	11	10	1.000000	0.000000
42	12	11	11	1.000000	0.000000
42	12	12	0	0.007822	0.007822
42	12	12	1	0.067102	0.059280
42	12	12	2	0.246425	0.179323
42	12	12	3	0.531064	0.284639
42	12	12	4	0.793061	0.261997
42	12	12	5	0.938868	0.145807
42	12	12	6	0.988483	0.049615
42	12	12	7	0.998690	0.010206
42	12	12	8	0.999916	0.001227
42	12	12	9	0.999997	0.000081
42	12	12	10	1.000000	0.000003
42	12	12	11	1.000000	0.000000
42	12	12	12	1.000000	0.000000
42	13	1	0	0.690476	0.690476
42	13	1	1	1.000000	0.309524
42	13	2	0	0.471545	0.471545
42	13	2	1	0.909408	0.437863
42	13	2	2	1.000000	0.090592
42	13	3	0	0.318293	0.318293
42	13	3	1	0.778049	0.459756
42	13	3	2	0.975087	0.197038
42	13	3	3	1.000000	0.024913
42	13	4	0	0.212195	0.212195
42	13	4	1	0.636585	0.424390
42	13	4	2	0.919512	0.282927

N	n	k	x	P(x)	p(x)
42	13	4	3	0.993612	0.074100
42	13	4	4	1.000000	0.006388
42	13	5	0	0.139602	0.139602
42	13	5	1	0.502567	0.362965
42	13	5	2	0.837612	0.335045
42	13	5	3	0.974112	0.136500
42	13	5	4	0.998487	0.024375
42	13	5	5	1.000000	0.001513
42	13	6	0	0.090553	0.090553
42	13	6	1	0.384849	0.294296
42	13	6	2	0.738004	0.353155
42	13	6	3	0.937220	0.199216
42	13	6	4	0.992558	0.055338
42	13	6	5	0.999673	0.007115
42	13	6	6	1.000000	0.000327
42	13	7	0	0.057853	0.057853
42	13	7	1	0.286750	0.228897
42	13	7	2	0.630096	0.343346
42	13	7	3	0.881882	0.251787
42	13	7	4	0.978723	0.096841
42	13	7	5	0.998092	0.019368
42	13	7	6	0.999936	0.001845
42	13	7	7	1.000000	0.000064
42	13	8	0	0.036365	0.036365
42	13	8	1	0.208271	0.171906
42	13	8	2	0.522187	0.313916
42	13	8	3	0.809943	0.287756
42	13	8	4	0.953822	0.143878
42	13	8	5	0.993665	0.039843
42	13	8	6	0.999567	0.005903
42	13	8	7	0.999989	0.000422
42	13	8	8	1.000000	0.000011
42	13	9	0	0.022461	0.022461
42	13	9	1	0.147598	0.125138
42	13	9	2	0.420626	0.273028
42	13	9	3	0.725309	0.304683
42	13	9	4	0.915736	0.190427
42	13	9	5	0.984290	0.068554
42	13	9	6	0.998352	0.014062
42	13	9	7	0.999915	0.001562
42	13	9	8	0.999998	0.000084
42	13	9	9	1.000000	0.000002
42	13	10	0	0.013612	0.013612
42	13	10	1	0.102094	0.088481
42	13	10	2	0.329617	0.227523
42	13	10	3	0.632981	0.303364
42	13	10	4	0.863801	0.230821
42	13	10	5	0.967671	0.103869
42	13	10	6	0.995369	0.027698
42	13	10	7	0.999630	0.004261

N	n	k	x	P(x)	p(x)
42	13	10	8	0.999986	0.000355
42	13	10	9	1.000000	0.000014
42	13	10	10	1.000000	0.000000
42	13	11	0	0.008082	0.008082
42	13	11	1	0.068913	0.060831
42	13	11	2	0.251406	0.182492
42	13	11	3	0.538180	0.286774
42	13	11	4	0.798883	0.260704
42	13	11	5	0.941703	0.142820
42	13	11	6	0.989310	0.047607
42	13	11	7	0.998832	0.009521
42	13	11	8	0.999930	0.001099
42	13	11	9	0.999998	0.000068
42	13	11	10	1.000000	0.000002
42	13	11	11	1.000000	0.000000
42	13	12	0	0.004693	0.004693
42	13	12	1	0.045366	0.040673
42	13	12	2	0.186650	0.141285
42	13	12	3	0.445672	0.259022
42	13	12	4	0.723195	0.277523
42	13	12	5	0.904847	0.181652
42	13	12	6	0.978560	0.073714
42	13	12	7	0.996989	0.018428
42	13	12	8	0.999753	0.002764
42	13	12	9	0.999989	0.000236
42	13	12	10	1.000000	0.000010
42	13	12	11	1.000000	0.000000
42	13	12	12	1.000000	0.000000
42	13	13	0	0.002659	0.002659
42	13	13	1	0.029097	0.026437
42	13	13	2	0.134846	0.105749
42	13	13	3	0.359331	0.224485
42	13	13	4	0.639938	0.280607
42	13	13	5	0.856406	0.216468
42	13	13	6	0.961360	0.104954
42	13	13	7	0.993303	0.031943
42	13	13	8	0.999292	0.005989
42	13	13	9	0.999958	0.000665
42	13	13	10	0.999999	0.000041
42	13	13	11	1.000000	0.000001
42	13	13	12	1.000000	0.000000
42	13	13	13	1.000000	0.000000
42	14	1	0	0.666667	0.666667
42	14	1	1	1.000000	0.333333
42	14	2	0	0.439024	0.439024
42	14	2	1	0.894309	0.455285
42	14	2	2	1.000000	0.105691
42	14	3	0	0.285366	0.285366
42	14	3	1	0.746341	0.460976
42	14	3	2	0.968293	0.221951

N	n	k	x	P(x)	p(x)
42	14	3	3	1.000000	0.031707
42	14	4	0	0.182927	0.182927
42	14	4	1	0.592683	0.409756
42	14	4	2	0.900000	0.307317
42	14	4	3	0.991057	0.091057
42	14	4	4	1.000000	0.008943
42	14	5	0	0.115533	0.115533
42	14	5	1	0.452503	0.336970
42	14	5	2	0.802952	0.350449
42	14	5	3	0.964698	0.161746
42	14	5	4	0.997647	0.032948
42	14	5	5	1.000000	0.002353
42	14	6	0	0.071818	0.071818
42	14	6	1	0.334108	0.262291
42	14	6	2	0.689293	0.355185
42	14	6	3	0.916612	0.227318
42	14	6	4	0.988742	0.072130
42	14	6	5	0.999427	0.010686
42	14	6	6	1.000000	0.000572
42	14	7	0	0.043889	0.043889
42	14	7	1	0.239392	0.195504
42	14	7	2	0.570898	0.331506
42	14	7	3	0.847153	0.276255
42	14	7	4	0.968705	0.121552
42	14	7	5	0.996756	0.028051
42	14	7	6	0.999873	0.003117
42	14	7	7	1.000000	0.000127
42	14	8	0	0.026333	0.026333
42	14	8	1	0.166776	0.140443
42	14	8	2	0.457239	0.290462
42	14	8	3	0.760330	0.303091
42	14	8	4	0.933976	0.173646
42	14	8	5	0.989543	0.055567
42	14	8	6	0.999160	0.009617
42	14	8	7	0.999974	0.000814
42	14	8	8	1.000000	0.000025
42	14	9	0	0.015490	0.015490
42	14	9	1	0.113078	0.097587
42	14	9	2	0.354723	0.241645
42	14	9	3	0.662271	0.307548
42	14	9	4	0.882904	0.220633
42	14	9	5	0.974834	0.091930
42	14	9	6	0.996897	0.022063
42	14	9	7	0.999807	0.002909
42	14	9	8	0.999995	0.000189
42	14	9	9	1.000000	0.000004
42	14	10	0	0.008919	0.008919
42	14	10	1	0.074634	0.065715
42	14	10	2	0.266852	0.192218
42	14	10	3	0.559755	0.292903

N	n	k	x	P(x)	p(x)
42	14	10	4	0.816046	0.256290
42	14	10	5	0.949762	0.133717
42	14	10	6	0.991549	0.041786
42	14	10	7	0.999190	0.007641
42	14	10	8	0.999961	0.000771
42	14	10	9	0.999999	0.000038
42	14	10	10	1.000000	0.000001
42	14	11	0	0.005017	0.005017
42	14	11	1	0.047937	0.042920
42	14	11	2	0.194770	0.146833
42	14	11	3	0.459070	0.264299
42	14	11	4	0.735955	0.276885
42	14	11	5	0.912154	0.176200
42	14	11	6	0.981102	0.068948
42	14	11	7	0.997518	0.016416
42	14	11	8	0.999816	0.002298
42	14	11	9	0.999993	0.000177
42	14	11	10	1.000000	0.000007
42	14	11	11	1.000000	0.000000
42	14	12	0	0.002751	0.002751
42	14	12	1	0.029938	0.027187
42	14	12	2	0.137932	0.107993
42	14	12	3	0.365286	0.227354
42	14	12	4	0.646637	0.281351
42	14	12	5	0.861000	0.214363
42	14	12	6	0.963309	0.102309
42	14	12	7	0.993811	0.030502
42	14	12	8	0.999372	0.005560
42	14	12	9	0.999965	0.000593
42	14	12	10	0.999999	0.000034
42	14	12	11	1.000000	0.000001
42	14	12	12	1.000000	0.000000
42	14	13	0	0.001467	0.001467
42	14	13	1	0.018157	0.016690
42	14	13	2	0.094734	0.076577
42	14	13	3	0.281923	0.187188
42	14	13	4	0.552853	0.270931
42	14	13	5	0.796691	0.243838
42	14	13	6	0.936027	0.139336
42	14	13	7	0.986694	0.050668
42	14	13	8	0.998260	0.011565
42	14	13	9	0.999866	0.001606
42	14	13	10	0.999994	0.000129
42	14	13	11	1.000000	0.000005
42	14	13	12	1.000000	0.000000
42	14	13	13	1.000000	0.000000
42	14	14	0	0.000759	0.000759
42	14	14	1	0.010675	0.009917
42	14	14	2	0.063047	0.052372
42	14	14	3	0.210920	0.147873

N	n	k	x	P(x)	p(x)
42	14	14	4	0.459429	0.248509
42	14	14	5	0.721017	0.261588
42	14	14	6	0.897589	0.176572
42	14	14	7	0.974464	0.076875
42	14	14	8	0.995867	0.021403
42	14	14	9	0.999589	0.003722
42	14	14	10	0.999977	0.000388
42	14	14	11	0.999999	0.000023
42	14	14	12	1.000000	0.000001
42	14	14	13	1.000000	0.000000
42	14	14	14	1.000000	0.000000
42	15	1	0	0.642857	0.642857
42	15	1	1	1.000000	0.357143
42	15	2	0	0.407665	0.407665
42	15	2	1	0.878049	0.470383
42	15	2	2	1.000000	0.121951
42	15	3	0	0.254791	0.254791
42	15	3	1	0.713415	0.458624
42	15	3	2	0.960366	0.246951
42	15	3	3	1.000000	0.039634
42	15	4	0	0.156794	0.156794
42	15	4	1	0.548780	0.391986
42	15	4	2	0.878049	0.329268
42	15	4	3	0.987805	0.109756
42	15	4	4	1.000000	0.012195
42	15	5	0	0.094902	0.094902
42	15	5	1	0.404365	0.309463
42	15	5	2	0.765404	0.361040
42	15	5	3	0.953145	0.187741
42	15	5	4	0.996470	0.043325
42	15	5	5	1.000000	0.003530
42	15	6	0	0.056428	0.056428
42	15	6	1	0.287271	0.230842
42	15	6	2	0.638553	0.351282
42	15	6	3	0.892256	0.253704
42	15	6	4	0.983589	0.091333
42	15	6	5	0.999046	0.015456
42	15	6	6	1.000000	0.000954
42	15	7	0	0.032916	0.032916
42	15	7	1	0.197499	0.164582
42	15	7	2	0.511701	0.314202
42	15	7	3	0.807688	0.295988
42	15	7	4	0.955682	0.147994
42	15	7	5	0.994752	0.039070
42	15	7	6	0.999761	0.005009
42	15	7	7	1.000000	0.000239
42	15	8	0	0.018809	0.018809
42	15	8	1	0.131666	0.112856
42	15	8	2	0.394997	0.263331
42	15	8	3	0.706207	0.311210

N	n	k	x	P(x)	p(x)
42	15	8	4	0.909170	0.202963
42	15	8	5	0.983589	0.074420
42	15	8	6	0.998473	0.014884
42	15	8	7	0.999945	0.001472
42	15	8	8	1.000000	0.000055
42	15	9	0	0.010511	0.010511
42	15	9	1	0.085195	0.074684
42	15	9	2	0.294311	0.209116
42	15	9	3	0.596368	0.302057
42	15	9	4	0.843505	0.247137
42	15	9	5	0.961701	0.118196
42	15	9	6	0.994533	0.032832
42	15	9	7	0.999599	0.005066
42	15	9	8	0.999989	0.000390
42	15	9	9	1.000000	0.000011
42	15	10	0	0.005733	0.005733
42	15	10	1	0.053511	0.047778
42	15	10	2	0.211932	0.158421
42	15	10	3	0.486529	0.274597
42	15	10	4	0.761126	0.274597
42	15	10	5	0.925884	0.164758
42	15	10	6	0.985579	0.059695
42	15	10	7	0.998371	0.012792
42	15	10	8	0.999906	0.001535
42	15	10	9	0.999998	0.000092
42	15	10	10	1.000000	0.000002
42	15	11	0	0.003046	0.003046
42	15	11	1	0.032608	0.029563
42	15	11	2	0.147574	0.114965
42	15	11	3	0.383555	0.235982
42	15	11	4	0.666734	0.283178
42	15	11	5	0.874397	0.207664
42	15	11	6	0.968790	0.094393
42	15	11	7	0.995173	0.026383
42	15	11	8	0.999570	0.004397
42	15	11	9	0.999981	0.000410
42	15	11	10	1.000000	0.000019
42	15	11	11	1.000000	0.000000
42	15	12	0	0.001572	0.001572
42	15	12	1	0.019258	0.017686
42	15	12	2	0.099363	0.080105
42	15	12	3	0.292208	0.192845
42	15	12	4	0.566251	0.274043
42	15	12	5	0.807409	0.241158
42	15	12	6	0.941386	0.133977
42	15	12	7	0.988365	0.046979
42	15	12	8	0.998577	0.010213
42	15	12	9	0.999901	0.001324
42	15	12	10	0.999997	0.000095
42	15	12	11	1.000000	0.000003

N	n	k	x	P(x)	p(x)
42	15	12	12	1.000000	0.000000
42	15	13	0	0.000786	0.000786
42	15	13	1	0.011004	0.010218
42	15	13	2	0.064650	0.053646
42	15	13	3	0.215070	0.150419
42	15	13	4	0.465768	0.250699
42	15	13	5	0.727023	0.261255
42	15	13	6	0.901193	0.174170
42	15	13	7	0.975837	0.074644
42	15	13	8	0.996194	0.020357
42	15	13	9	0.999636	0.003442
42	15	13	10	0.999981	0.000344
42	15	13	11	0.999999	0.000019
42	15	13	12	1.000000	0.000000
42	15	13	13	1.000000	0.000000
42	15	14	0	0.000379	0.000379
42	15	14	1	0.006071	0.005692
42	15	14	2	0.040602	0.034531
42	15	14	3	0.152827	0.112225
42	15	14	4	0.370676	0.217849
42	15	14	5	0.636935	0.266259
42	15	14	6	0.847140	0.210205
42	15	14	7	0.955245	0.108105
42	15	14	8	0.991280	0.036035
42	15	14	9	0.998924	0.007644
42	15	14	10	0.999921	0.000997
42	15	14	11	0.999997	0.000076
42	15	14	12	1.000000	0.000003
42	15	14	13	1.000000	0.000000
42	15	14	14	1.000000	0.000000
42	15	15	0	0.000176	0.000176
42	15	15	1	0.003225	0.003049
42	15	15	2	0.024570	0.021345
42	15	15	3	0.104731	0.080161
42	15	15	4	0.285092	0.180362
42	15	15	5	0.541843	0.256750
42	15	15	6	0.779574	0.237732
42	15	15	7	0.924358	0.144784
42	15	15	8	0.982272	0.057914
42	15	15	9	0.997286	0.015015
42	15	15	10	0.999743	0.002457
42	15	15	11	0.999986	0.000243
42	15	15	12	0.999999	0.000013
42	15	15	13	1.000000	0.000000
42	15	15	14	1.000000	0.000000
42	15	15	15	1.000000	0.000000
42	16	1	0	0.619048	0.619048
42	16	1	1	1.000000	0.380952
42	16	2	0	0.377468	0.377468
42	16	2	1	0.860627	0.483159

N	n	k	x	P(x)	p(x)
42	16	2	2	1.000000	0.139373
42	16	3	0	0.226481	0.226481
42	16	3	1	0.679442	0.452962
42	16	3	2	0.951219	0.271777
42	16	3	3	1.000000	0.048780
42	16	4	0	0.133566	0.133566
42	16	4	1	0.505226	0.371661
42	16	4	2	0.853659	0.348432
42	16	4	3	0.983740	0.130081
42	16	4	4	1.000000	0.016260
42	16	5	0	0.077327	0.077327
42	16	5	1	0.358518	0.281191
42	16	5	2	0.725289	0.366771
42	16	5	3	0.939238	0.213949
42	16	5	4	0.994865	0.055627
42	16	5	5	1.000000	0.005135
42	16	6	0	0.043889	0.043889
42	16	6	1	0.244522	0.200633
42	16	6	2	0.586511	0.341989
42	16	6	3	0.864067	0.277556
42	16	6	4	0.976824	0.112757
42	16	6	5	0.998473	0.021649
42	16	6	6	1.000000	0.001527
42	16	7	0	0.024383	0.024383
42	16	7	1	0.160925	0.136542
42	16	7	2	0.453515	0.292590
42	16	7	3	0.763838	0.310323
42	16	7	4	0.939238	0.175400
42	16	7	5	0.991858	0.052620
42	16	7	6	0.999576	0.007718
42	16	7	7	1.000000	0.000424
42	16	8	0	0.013236	0.013236
42	16	8	1	0.102407	0.089170
42	16	8	2	0.336479	0.234072
42	16	8	3	0.648575	0.312096
42	16	8	4	0.879101	0.230526
42	16	8	5	0.975321	0.096219
42	16	8	6	0.997371	0.022050
42	16	8	7	0.999891	0.002520
42	16	8	8	1.000000	0.000109
42	16	9	0	0.007007	0.007007
42	16	9	1	0.063067	0.056059
42	16	9	2	0.240096	0.177029
42	16	9	3	0.529244	0.289148
42	16	9	4	0.797739	0.268495
42	16	9	5	0.944191	0.146452
42	16	9	6	0.990885	0.046695
42	16	9	7	0.999224	0.008338
42	16	9	8	0.999974	0.000750
42	16	9	9	1.000000	0.000026
42	16	10	0	0.003610	0.003610
42	16	10	1	0.037585	0.033975
42	16	10	2	0.164993	0.127408
42	16	10	3	0.415338	0.250345
42	16	10	4	0.700105	0.284767
42	16	10	5	0.895374	0.195269
42	16	10	6	0.976736	0.081362
42	16	10	7	0.996950	0.020214
42	16	10	8	0.999792	0.002843
42	16	10	9	0.999994	0.000202
42	16	10	10	1.000000	0.000005
42	16	11	0	0.001805	0.001805
42	16	11	1	0.021659	0.019854
42	16	11	2	0.109252	0.087593
42	16	11	3	0.313635	0.204383
42	16	11	4	0.593317	0.279682
42	16	11	5	0.828250	0.234933
42	16	11	6	0.951310	0.123060
42	16	11	7	0.991264	0.039955
42	16	11	8	0.999082	0.007817
42	16	11	9	0.999950	0.000869
42	16	11	10	0.999999	0.000049
42	16	11	11	1.000000	0.000001
42	16	12	0	0.000873	0.000873
42	16	12	1	0.012052	0.011179
42	16	12	2	0.069694	0.057642
42	16	12	3	0.227926	0.158232
42	16	12	4	0.485053	0.257127
42	16	12	5	0.744887	0.259834
42	16	12	6	0.911613	0.166727
42	16	12	7	0.979665	0.068052
42	16	12	8	0.997064	0.017400
42	16	12	9	0.999754	0.002690
42	16	12	10	0.999989	0.000235
42	16	12	11	1.000000	0.000010
42	16	12	12	1.000000	0.000000
42	16	13	0	0.000408	0.000408
42	16	13	1	0.006463	0.006055
42	16	13	2	0.042795	0.036332
42	16	13	3	0.159359	0.116564
42	16	13	4	0.382202	0.222843
42	16	13	5	0.649614	0.267412
42	16	13	6	0.856038	0.206423
42	16	13	7	0.959249	0.103212
42	16	13	8	0.992424	0.033175
42	16	13	9	0.999126	0.006702
42	16	13	10	0.999942	0.000816
42	16	13	11	0.999998	0.000056
42	16	13	12	1.000000	0.000002
42	16	13	13	1.000000	0.000000
42	16	14	0	0.000183	0.000183
42	16	14	1	0.003331	0.003148
42	16	14	2	0.025255	0.021924
42	16	14	3	0.107106	0.081851
42	16	14	4	0.289991	0.182885
42	16	14	5	0.548182	0.258191
42	16	14	6	0.784857	0.236675
42	16	14	7	0.927218	0.142361
42	16	14	8	0.983273	0.056055
42	16	14	9	0.997509	0.014236
42	16	14	10	0.999774	0.002265
42	16	14	11	0.999988	0.000215
42	16	14	12	1.000000	0.000011
42	16	14	13	1.000000	0.000000
42	16	14	14	1.000000	0.000000
42	16	15	0	0.000078	0.000078
42	16	15	1	0.001644	0.001566
42	16	15	2	0.014293	0.012649
42	16	15	3	0.069104	0.054811
42	16	15	4	0.211612	0.142508
42	16	15	5	0.446750	0.235138
42	16	15	6	0.700330	0.253580
42	16	15	7	0.881459	0.181129
42	16	15	8	0.967257	0.085798
42	16	15	9	0.993950	0.026693
42	16	15	10	0.999288	0.005339
42	16	15	11	0.999950	0.000662
42	16	15	12	0.999998	0.000048
42	16	15	13	1.000000	0.000002
42	16	15	14	1.000000	0.000000
42	16	15	15	1.000000	0.000000
42	16	16	0	0.000032	0.000032
42	16	16	1	0.000774	0.000742
42	16	16	2	0.007734	0.006960
42	16	16	3	0.042713	0.034979
42	16	16	4	0.148275	0.105561
42	16	16	5	0.350953	0.202678
42	16	16	6	0.606412	0.255459
42	16	16	7	0.821083	0.214671
42	16	16	8	0.941835	0.120753
42	16	16	9	0.987029	0.045194
42	16	16	10	0.998102	0.011073
42	16	16	11	0.999828	0.001726
42	16	16	12	0.999991	0.000163
42	16	16	13	1.000000	0.000009
42	16	16	14	1.000000	0.000000
42	16	16	15	1.000000	0.000000
42	16	16	16	1.000000	0.000000
42	17	1	0	0.595238	0.595238
42	17	1	1	1.000000	0.404762
42	17	2	0	0.348432	0.348432
42	17	2	1	0.842044	0.493612
42	17	2	2	1.000000	0.157956
42	17	3	0	0.200348	0.200348
42	17	3	1	0.644599	0.444251
42	17	3	2	0.940766	0.296167
42	17	3	3	1.000000	0.059233
42	17	4	0	0.113017	0.113017
42	17	4	1	0.462342	0.349325
42	17	4	2	0.826856	0.364514
42	17	4	3	0.978737	0.151881
42	17	4	4	1.000000	0.021263
42	17	5	0	0.062457	0.062457
42	17	5	1	0.315258	0.252801
42	17	5	2	0.682969	0.367711
42	17	5	3	0.922781	0.239811
42	17	5	4	0.992726	0.069945
42	17	5	5	1.000000	0.007274
42	17	6	0	0.033760	0.033760
42	17	6	1	0.205939	0.172178
42	17	6	2	0.533897	0.327958
42	17	6	3	0.832041	0.298144
42	17	6	4	0.968150	0.136109
42	17	6	5	0.997641	0.029490
42	17	6	6	1.000000	0.002359
42	17	7	0	0.017818	0.017818
42	17	7	1	0.129415	0.111597
42	17	7	2	0.397248	0.267833
42	17	7	3	0.716096	0.318848
42	17	7	4	0.919000	0.202904
42	17	7	5	0.987811	0.068811
42	17	7	6	0.999279	0.011468
42	17	7	7	1.000000	0.000721
42	17	8	0	0.009164	0.009164
42	17	8	1	0.078399	0.069236
42	17	8	2	0.282462	0.204063
42	17	8	3	0.588557	0.306095
42	17	8	4	0.843636	0.255079
42	17	8	5	0.964218	0.120583
42	17	8	6	0.995675	0.031456
42	17	8	7	0.999794	0.004119
42	17	8	8	1.000000	0.000206
42	17	9	0	0.004582	0.004582
42	17	9	1	0.045818	0.041236
42	17	9	2	0.192434	0.146617
42	17	9	3	0.462518	0.270083
42	17	9	4	0.746105	0.283588
42	17	9	5	0.921660	0.175554
42	17	9	6	0.985498	0.063838
42	17	9	7	0.998582	0.013085

N	n	k	x	P(x)	p(x)
42	17	9	8	0.999945	0.001363
42	17	9	9	1.000000	0.000055
42	17	10	0	0.002221	0.002221
42	17	10	1	0.025825	0.023603
42	17	10	2	0.125790	0.099966
42	17	10	3	0.347937	0.222147
42	17	10	4	0.634389	0.286452
42	17	10	5	0.857822	0.223433
42	17	10	6	0.964218	0.106396
42	17	10	7	0.994617	0.030399
42	17	10	8	0.999574	0.004956
42	17	10	9	0.999987	0.000413
42	17	10	10	1.000000	0.000013
42	17	11	0	0.001041	0.001041
42	17	11	1	0.014023	0.012982
42	17	11	2	0.078931	0.064908
42	17	11	3	0.250748	0.171816
42	17	11	4	0.518018	0.267270
42	17	11	5	0.774035	0.256017
42	17	11	6	0.927645	0.153610
42	17	11	7	0.985118	0.057473
42	17	11	8	0.998180	0.013062
42	17	11	9	0.999883	0.001704
42	17	11	10	0.999997	0.000114
42	17	11	11	1.000000	0.000003
42	17	12	0	0.000470	0.000470
42	17	12	1	0.007323	0.006853
42	17	12	2	0.047524	0.040201
42	17	12	3	0.173153	0.125629
42	17	12	4	0.405937	0.232784
42	17	12	5	0.674931	0.268994
42	17	12	6	0.873138	0.198206
42	17	12	7	0.966578	0.093440
42	17	12	8	0.994387	0.027810
42	17	12	9	0.999444	0.005056
42	17	12	10	0.999971	0.000528
42	17	12	11	0.999999	0.000028
42	17	12	12	1.000000	0.000001
42	17	13	0	0.000204	0.000204
42	17	13	1	0.003668	0.003464
42	17	13	2	0.027423	0.023755
42	17	13	3	0.114526	0.087103
42	17	13	4	0.305064	0.190538
42	17	13	5	0.567334	0.262270
42	17	13	6	0.800462	0.233128
42	17	13	7	0.935431	0.134969
42	17	13	8	0.986045	0.050613
42	17	13	9	0.998095	0.012051
42	17	13	10	0.999848	0.001753
42	17	13	11	0.999994	0.000145

N	n	k	x	P(x)	p(x)
42	17	13	12	1.000000	0.000006
42	17	13	13	1.000000	0.000000
42	17	14	0	0.000084	0.000084
42	17	14	1	0.001757	0.001672
42	17	14	2	0.015136	0.013379
42	17	14	3	0.072477	0.057340
42	17	14	4	0.219651	0.147174
42	17	14	5	0.458808	0.239158
42	17	14	6	0.712034	0.253226
42	17	14	7	0.888890	0.176856
42	17	14	8	0.970337	0.081447
42	17	14	9	0.994771	0.024434
42	17	14	10	0.999425	0.004654
42	17	14	11	0.999964	0.000538
42	17	14	12	0.999999	0.000035
42	17	14	13	1.000000	0.000001
42	17	14	14	1.000000	0.000000
42	17	15	0	0.000033	0.000033
42	17	15	1	0.000801	0.000768
42	17	15	2	0.007969	0.007168
42	17	15	3	0.043806	0.035838
42	17	15	4	0.151320	0.107513
42	17	15	5	0.356312	0.204992
42	17	15	6	0.612552	0.256240
42	17	15	7	0.825727	0.213175
42	17	15	8	0.944158	0.118430
42	17	15	9	0.987790	0.043632
42	17	15	10	0.998262	0.010472
42	17	15	11	0.999848	0.001587
42	17	15	12	0.999992	0.000144
42	17	15	13	1.000000	0.000007
42	17	15	14	1.000000	0.000000
42	17	15	15	1.000000	0.000000
42	17	16	0	0.000012	0.000012
42	17	16	1	0.000346	0.000334
42	17	16	2	0.003987	0.003641
42	17	16	3	0.025224	0.021237
42	17	16	4	0.099554	0.074330
42	17	16	5	0.265204	0.165650
42	17	16	6	0.508158	0.242954
42	17	16	7	0.746774	0.238615
42	17	16	8	0.904681	0.157907
42	17	16	9	0.974862	0.070181
42	17	16	10	0.995547	0.020685
42	17	16	11	0.999496	0.003949
42	17	16	12	0.999966	0.000470
42	17	16	13	0.999999	0.000033
42	17	16	14	1.000000	0.000001
42	17	16	15	1.000000	0.000000
42	17	16	16	1.000000	0.000000

N	n	k	x	P(x)	p(x)
42	17	17	0	0.000004	0.000004
42	17	17	1	0.000141	0.000136
42	17	17	2	0.001886	0.001746
42	17	17	3	0.013788	0.011902
42	17	17	4	0.062389	0.048601
42	17	17	5	0.188750	0.126361
42	17	17	6	0.405370	0.216620
42	17	17	7	0.654998	0.249628
42	17	17	8	0.850021	0.195022
42	17	17	9	0.953268	0.103247
42	17	17	10	0.989978	0.036710
42	17	17	11	0.998584	0.008607
42	17	17	12	0.999875	0.001291
42	17	17	13	0.999994	0.000118
42	17	17	14	1.000000	0.000006
42	17	17	15	1.000000	0.000000
42	17	17	16	1.000000	0.000000
42	17	17	17	1.000000	0.000000
42	18	1	0	0.571429	0.571429
42	18	1	1	1.000000	0.428571
42	18	2	0	0.320557	0.320557
42	18	2	1	0.822300	0.501742
42	18	2	2	1.000000	0.177700
42	18	3	0	0.176307	0.176307
42	18	3	1	0.609059	0.432753
42	18	3	2	0.928920	0.319861
42	18	3	3	1.000000	0.071080
42	18	4	0	0.094934	0.094934
42	18	4	1	0.420423	0.325489
42	18	4	2	0.797695	0.377271
42	18	4	3	0.972661	0.174966
42	18	4	4	1.000000	0.027339
42	18	5	0	0.049965	0.049965
42	18	5	1	0.274810	0.224844
42	18	5	2	0.638844	0.364034
42	18	5	3	0.903596	0.264752
42	18	5	4	0.989928	0.086332
42	18	5	5	1.000000	0.010072
42	18	6	0	0.025658	0.025658
42	18	6	1	0.171503	0.145845
42	18	6	2	0.481424	0.309921
42	18	6	3	0.796264	0.314840
42	18	6	4	0.957262	0.160998
42	18	6	5	0.996461	0.039199
42	18	6	6	1.000000	0.003539
42	18	7	0	0.012829	0.012829
42	18	7	1	0.102632	0.089803
42	18	7	2	0.343681	0.241049
42	18	7	3	0.665080	0.321399
42	18	7	4	0.894651	0.229571

N	n	k	x	P(x)	p(x)
42	18	7	5	0.982306	0.087654
42	18	7	6	0.998820	0.016515
42	18	7	7	1.000000	0.001180
42	18	8	0	0.006231	0.006231
42	18	8	1	0.059013	0.052782
42	18	8	2	0.233487	0.174474
42	18	8	3	0.527338	0.293851
42	18	8	4	0.802823	0.275485
42	18	8	5	0.949748	0.146925
42	18	8	6	0.993158	0.043410
42	18	8	7	0.999629	0.006471
42	18	8	8	1.000000	0.000371
42	18	9	0	0.002932	0.002932
42	18	9	1	0.032622	0.029690
42	18	9	2	0.151382	0.118760
42	18	9	3	0.397698	0.246316
42	18	9	4	0.689388	0.291690
42	18	9	5	0.893571	0.204183
42	18	9	6	0.977837	0.084266
42	18	9	7	0.997536	0.019699
42	18	9	8	0.999891	0.002355
42	18	9	9	1.000000	0.000109
42	18	10	0	0.001333	0.001333
42	18	10	1	0.017327	0.015995
42	18	10	2	0.093801	0.076474
42	18	10	3	0.285736	0.191935
42	18	10	4	0.565641	0.279905
42	18	10	5	0.813135	0.247495
42	18	10	6	0.947195	0.134060
42	18	10	7	0.990969	0.043775
42	18	10	8	0.999177	0.008208
42	18	10	9	0.999970	0.000793
42	18	10	10	1.000000	0.000030
42	18	11	0	0.000583	0.000583
42	18	11	1	0.008830	0.008247
42	18	11	2	0.055564	0.046734
42	18	11	3	0.195767	0.140202
42	18	11	4	0.443182	0.247416
42	18	11	5	0.712591	0.269408
42	18	11	6	0.896923	0.184332
42	18	11	7	0.975922	0.078999
42	18	11	8	0.996612	0.020690
42	18	11	9	0.999747	0.003135
42	18	11	10	0.999992	0.000245
42	18	11	11	1.000000	0.000007
42	18	12	0	0.000245	0.000245
42	18	12	1	0.004308	0.004063
42	18	12	2	0.031444	0.027136
42	18	12	3	0.127927	0.096483
42	18	12	4	0.331446	0.203519

N	n	k	x	P(x)	p(x)
42	18	12	5	0.599613	0.268167
42	18	12	6	0.825568	0.225955
42	18	12	7	0.947890	0.122322
42	18	12	8	0.989938	0.042048
42	18	12	9	0.998837	0.008899
42	18	12	10	0.999929	0.001092
42	18	12	11	0.999998	0.000069
42	18	12	12	1.000000	0.000002
42	18	13	0	0.000098	0.000098
42	18	13	1	0.002005	0.001907
42	18	13	2	0.016971	0.014966
42	18	13	3	0.079685	0.062714
42	18	13	4	0.236471	0.156785
42	18	13	5	0.483407	0.246937
42	18	13	6	0.735186	0.251779
42	18	13	7	0.903039	0.167852
42	18	13	8	0.975922	0.072883
42	18	13	9	0.996167	0.020245
42	18	13	10	0.999638	0.003471
42	18	13	11	0.999982	0.000344
42	18	13	12	1.000000	0.000017
42	18	13	13	1.000000	0.000000
42	18	14	0	0.000037	0.000037
42	18	14	1	0.000887	0.000850
42	18	14	2	0.008714	0.007827
42	18	14	3	0.047247	0.038533
42	18	14	4	0.160781	0.113534
42	18	14	5	0.372712	0.211930
42	18	14	6	0.631002	0.258290
42	18	14	7	0.839370	0.208369
42	18	14	8	0.950790	0.111419
42	18	14	9	0.989884	0.039094
42	18	14	10	0.998680	0.008796
42	18	14	11	0.999899	0.001219
42	18	14	12	0.999996	0.000097
42	18	14	13	1.000000	0.000004
42	18	14	14	1.000000	0.000000
42	18	15	0	0.000013	0.000013
42	18	15	1	0.000371	0.000358
42	18	15	2	0.004242	0.003870
42	18	15	3	0.026604	0.022363
42	18	15	4	0.104014	0.077410
42	18	15	5	0.274315	0.170301
42	18	15	6	0.520306	0.245991
42	18	15	7	0.757511	0.237205
42	18	15	8	0.910997	0.153486
42	18	15	9	0.977318	0.066321
42	18	15	10	0.996167	0.018849
42	18	15	11	0.999594	0.003427
42	18	15	12	0.999975	0.000381

N	n	k	x	P(x)	p(x)
42	18	15	13	0.999999	0.000024
42	18	15	14	1.000000	0.000001
42	18	15	15	1.000000	0.000000
42	18	16	0	0.000004	0.000004
42	18	16	1	0.000146	0.000141
42	18	16	2	0.001948	0.001802
42	18	16	3	0.014181	0.012233
42	18	16	4	0.063876	0.049695
42	18	16	5	0.192318	0.128443
42	18	16	6	0.410977	0.218658
42	18	16	7	0.660872	0.249895
42	18	16	8	0.854151	0.193278
42	18	16	9	0.955211	0.101061
42	18	16	10	0.990582	0.035371
42	18	16	11	0.998706	0.008124
42	18	16	12	0.999891	0.001185
42	18	16	13	0.999995	0.000104
42	18	16	14	1.000000	0.000005
42	18	16	15	1.000000	0.000000
42	18	16	16	1.000000	0.000000
42	18	17	0	0.000001	0.000001
42	18	17	1	0.000053	0.000052
42	18	17	2	0.000839	0.000786
42	18	17	3	0.007123	0.006284
42	18	17	4	0.037117	0.029993
42	18	17	5	0.128097	0.090980
42	18	17	6	0.310057	0.181961
42	18	17	7	0.555147	0.245090
42	18	17	8	0.779813	0.224666
42	18	17	9	0.920229	0.140416
42	18	17	10	0.979699	0.059470
42	18	17	11	0.996519	0.016820
42	18	17	12	0.999617	0.003098
42	18	17	13	0.999975	0.000358
42	18	17	14	0.999999	0.000024
42	18	17	15	1.000000	0.000001
42	18	17	16	1.000000	0.000000
42	18	17	17	1.000000	0.000000
42	18	18	0	0.000000	0.000000
42	18	18	1	0.000018	0.000018
42	18	18	2	0.000336	0.000318
42	18	18	3	0.003353	0.003016
42	18	18	4	0.020320	0.016968
42	18	18	5	0.080787	0.060467
42	18	18	6	0.222716	0.141929
42	18	18	7	0.447308	0.224591
42	18	18	8	0.689946	0.242639
42	18	18	9	0.869679	0.179732
42	18	18	10	0.960668	0.090990
42	18	18	11	0.991809	0.031141

N	n	k	x	P(x)	p(x)
42	18	18	12	0.998874	0.007064
42	18	18	13	0.999903	0.001030
42	18	18	14	0.999995	0.000092
42	18	18	15	1.000000	0.000005
42	18	18	16	1.000000	0.000000
42	18	18	17	1.000000	0.000000
42	18	18	18	1.000000	0.000000
42	19	1	0	0.547619	0.547619
42	19	1	1	1.000000	0.452381
42	19	2	0	0.293844	0.293844
42	19	2	1	0.801394	0.507549
42	19	2	2	1.000000	0.198606
42	19	3	0	0.154268	0.154268
42	19	3	1	0.572996	0.418728
42	19	3	2	0.915592	0.342596
42	19	3	3	1.000000	0.084408
42	19	4	0	0.079112	0.079112
42	19	4	1	0.379737	0.300625
42	19	4	2	0.766256	0.386518
42	19	4	3	0.965371	0.199116
42	19	4	4	1.000000	0.034629
42	19	5	0	0.039556	0.039556
42	19	5	1	0.237336	0.197780
42	19	5	2	0.593340	0.356004
42	19	5	3	0.881533	0.288193
42	19	5	4	0.986331	0.104798
42	19	5	5	1.000000	0.013669
42	19	6	0	0.019243	0.019243
42	19	6	1	0.141119	0.121875
42	19	6	2	0.429770	0.288652
42	19	6	3	0.756909	0.327139
42	19	6	4	0.943845	0.186936
42	19	6	5	0.994828	0.050983
42	19	6	6	1.000000	0.005172
42	19	7	0	0.009087	0.009087
42	19	7	1	0.080181	0.071094
42	19	7	2	0.293463	0.213282
42	19	7	3	0.611514	0.318051
42	19	7	4	0.865955	0.254441
42	19	7	5	0.975001	0.109046
42	19	7	6	0.998132	0.023131
42	19	7	7	1.000000	0.001868
42	19	8	0	0.004154	0.004154
42	19	8	1	0.043618	0.039464
42	19	8	2	0.189869	0.146250
42	19	8	3	0.466119	0.276250
42	19	8	4	0.756909	0.290790
42	19	8	5	0.931383	0.174474
42	19	8	6	0.989541	0.058158
42	19	8	7	0.999360	0.009819

N	n	k	x	P(x)	p(x)
42	19	8	8	1.000000	0.000640
42	19	9	0	0.001833	0.001833
42	19	9	1	0.022726	0.020893
42	19	9	2	0.116744	0.094018
42	19	9	3	0.336119	0.219375
42	19	9	4	0.628619	0.292500
42	19	9	5	0.859541	0.230921
42	19	9	6	0.967304	0.107763
42	19	9	7	0.995894	0.028590
42	19	9	8	0.999793	0.003899
42	19	9	9	1.000000	0.000207
42	19	10	0	0.000778	0.000778
42	19	10	1	0.011329	0.010552
42	19	10	2	0.068310	0.056981
42	19	10	3	0.229755	0.161445
42	19	10	4	0.495664	0.265909
42	19	10	5	0.761574	0.265909
42	19	10	6	0.924852	0.163278
42	19	10	7	0.985498	0.060646
42	19	10	8	0.998493	0.012996
42	19	10	9	0.999937	0.001444
42	19	10	10	1.000000	0.000063
42	19	11	0	0.000316	0.000316
42	19	11	1	0.005394	0.005078
42	19	11	2	0.038039	0.032645
42	19	11	3	0.149033	0.110993
42	19	11	4	0.371019	0.221987
42	19	11	5	0.645239	0.274219
42	19	11	6	0.858520	0.213282
42	19	11	7	0.962755	0.104235
42	19	11	8	0.994026	0.031271
42	19	11	9	0.999486	0.005460
42	19	11	10	0.999982	0.000496
42	19	11	11	1.000000	0.000018
42	19	12	0	0.000122	0.000122
42	19	12	1	0.002445	0.002323
42	19	12	2	0.020137	0.017692
42	19	12	3	0.091746	0.071609
42	19	12	4	0.263606	0.171861
42	19	12	5	0.521398	0.257791
42	19	12	6	0.769079	0.247682
42	19	12	7	0.922406	0.153327
42	19	12	8	0.982930	0.060524
42	19	12	9	0.997725	0.014795
42	19	12	10	0.999838	0.002114
42	19	12	11	0.999995	0.000157
42	19	12	12	1.000000	0.000005
42	19	13	0	0.000045	0.000045
42	19	13	1	0.001052	0.001007
42	19	13	2	0.010112	0.009060

N	n	k	x	P(x)	p(x)
42	19	13	3	0.053554	0.043443
42	19	13	4	0.177676	0.124122
42	19	13	5	0.401095	0.223419
42	19	13	6	0.661751	0.260656
42	19	13	7	0.861075	0.199325
42	19	13	8	0.960738	0.099662
42	19	13	9	0.992793	0.032055
42	19	13	10	0.999204	0.006411
42	19	13	11	0.999953	0.000749
42	19	13	12	0.999999	0.000045
42	19	13	13	1.000000	0.000001
42	19	14	0	0.000015	0.000015
42	19	14	1	0.000427	0.000411
42	19	14	2	0.004801	0.004374
42	19	14	3	0.029586	0.024785
42	19	14	4	0.113475	0.083889
42	19	14	5	0.293238	0.179762
42	19	14	6	0.544905	0.251667
42	19	14	7	0.778596	0.233691
42	19	14	8	0.922935	0.144339
42	19	14	9	0.981740	0.058805
42	19	14	10	0.997214	0.015475
42	19	14	11	0.999747	0.002532
42	19	14	12	0.999988	0.000241
42	19	14	13	1.000000	0.000012
42	19	14	14	1.000000	0.000000
42	19	15	0	0.000005	0.000005
42	19	15	1	0.000162	0.000157
42	19	15	2	0.002145	0.001983
42	19	15	3	0.015423	0.013278
42	19	15	4	0.068535	0.053112
42	19	15	5	0.203356	0.134822
42	19	15	6	0.428059	0.224703
42	19	15	7	0.678443	0.250383
42	19	15	8	0.866230	0.187788
42	19	15	9	0.960738	0.094507
42	19	15	10	0.992240	0.031502
42	19	15	11	0.999023	0.006783
42	19	15	12	0.999928	0.000904
42	19	15	13	0.999997	0.000070
42	19	15	14	1.000000	0.000003
42	19	15	15	1.000000	0.000000
42	19	16	0	0.000001	0.000001
42	19	16	1	0.000057	0.000056
42	19	16	2	0.000897	0.000839
42	19	16	3	0.007555	0.006658
42	19	16	4	0.039028	0.031474
42	19	16	5	0.133449	0.094421
42	19	16	6	0.319869	0.186420
42	19	16	7	0.567161	0.247292

N	n	k	x	P(x)	p(x)
42	19	16	8	0.789724	0.222563
42	19	16	9	0.925735	0.136011
42	19	16	10	0.981740	0.056004
42	19	16	11	0.997013	0.015274
42	19	16	12	0.999693	0.002680
42	19	16	13	0.999982	0.000289
42	19	16	14	0.999999	0.000018
42	19	16	15	1.000000	0.000001
42	19	16	16	1.000000	0.000000
42	19	17	0	0.000000	0.000000
42	19	17	1	0.000019	0.000018
42	19	17	2	0.000348	0.000329
42	19	17	3	0.003457	0.003109
42	19	17	4	0.020870	0.017413
42	19	17	5	0.082607	0.061737
42	19	17	6	0.226659	0.144052
42	19	17	7	0.453026	0.226368
42	19	17	8	0.695563	0.242537
42	19	17	9	0.873423	0.177860
42	19	17	10	0.962353	0.088930
42	19	17	11	0.992314	0.029960
42	19	17	12	0.998972	0.006658
42	19	17	13	0.999915	0.000943
42	19	17	14	0.999996	0.000081
42	19	17	15	1.000000	0.000004
42	19	17	16	1.000000	0.000000
42	19	17	17	1.000000	0.000000
42	19	18	0	0.000000	0.000000
42	19	18	1	0.000006	0.000005
42	19	18	2	0.000124	0.000119
42	19	18	3	0.001467	0.001343
42	19	18	4	0.010423	0.008955
42	19	18	5	0.048034	0.037612
42	19	18	6	0.151752	0.103718
42	19	18	7	0.344370	0.192618
42	19	18	8	0.588847	0.244477
42	19	18	9	0.802279	0.213432
42	19	18	10	0.930339	0.128059
42	19	18	11	0.982727	0.052388
42	19	18	12	0.997107	0.014381
42	19	18	13	0.999689	0.002581
42	19	18	14	0.999980	0.000291
42	19	18	15	0.999999	0.000019
42	19	18	16	1.000000	0.000001
42	19	18	17	1.000000	0.000000
42	19	18	18	1.000000	0.000000
42	19	19	0	0.000000	0.000000
42	19	19	1	0.000001	0.000001
42	19	19	2	0.000040	0.000039
42	19	19	3	0.000572	0.000532

N	n	k	x	P(x)	p(x)
42	19	19	4	0.004826	0.004254
42	19	19	5	0.026094	0.021269
42	19	19	6	0.095572	0.069477
42	19	19	7	0.248061	0.152489
42	19	19	8	0.476795	0.228734
42	19	19	9	0.713349	0.236554
42	19	19	10	0.882316	0.168967
42	19	19	11	0.965264	0.082948
42	19	19	12	0.992913	0.027649
42	19	19	13	0.999043	0.006130
42	19	19	14	0.999919	0.000876
42	19	19	15	0.999996	0.000077
42	19	19	16	1.000000	0.000004
42	19	19	17	1.000000	0.000000
42	19	19	18	1.000000	0.000000
42	19	19	19	1.000000	0.000000
42	20	1	0	0.523810	0.523810
42	20	1	1	1.000000	0.476190
42	20	2	0	0.268293	0.268293
42	20	2	1	0.779326	0.511034
42	20	2	2	1.000000	0.220674
42	20	3	0	0.134146	0.134146
42	20	3	1	0.536585	0.402439
42	20	3	2	0.900697	0.364111
42	20	3	3	1.000000	0.099303
42	20	4	0	0.065353	0.065353
42	20	4	1	0.340525	0.275172
42	20	4	2	0.732645	0.392120
42	20	4	3	0.956714	0.224069
42	20	4	4	1.000000	0.043286
42	20	5	0	0.030957	0.030957
42	20	5	1	0.202939	0.171982
42	20	5	2	0.546904	0.343965
42	20	5	3	0.856473	0.309568
42	20	5	4	0.981774	0.125302
42	20	5	5	1.000000	0.018226
42	20	6	0	0.014223	0.014223
42	20	6	1	0.114624	0.100401
42	20	6	2	0.379570	0.264946
42	20	6	3	0.714239	0.334669
42	20	6	4	0.927590	0.213351
42	20	6	5	0.992611	0.065021
42	20	6	6	1.000000	0.007389
42	20	7	0	0.006322	0.006322
42	20	7	1	0.061635	0.055313
42	20	7	2	0.247097	0.185462
42	20	7	3	0.556201	0.309104
42	20	7	4	0.832767	0.276566
42	20	7	5	0.965519	0.132752
42	20	7	6	0.997126	0.031608

N	n	k	x	P(x)	p(x)
42	20	7	7	1.000000	0.002873
42	20	8	0	0.002709	0.002709
42	20	8	1	0.031608	0.028898
42	20	8	2	0.151716	0.120109
42	20	8	3	0.406065	0.254348
42	20	8	4	0.706337	0.300272
42	20	8	5	0.908625	0.202289
42	20	8	6	0.984483	0.075858
42	20	8	7	0.998933	0.014449
42	20	8	8	1.000000	0.001067
42	20	9	0	0.001116	0.001116
42	20	9	1	0.015458	0.014343
42	20	9	2	0.088129	0.072671
42	20	9	3	0.278890	0.190761
42	20	9	4	0.565032	0.286142
42	20	9	5	0.819380	0.254348
42	20	9	6	0.953248	0.133867
42	20	9	7	0.993408	0.040160
42	20	9	8	0.999623	0.006215
42	20	9	9	1.000000	0.000377
42	20	10	0	0.000439	0.000439
42	20	10	1	0.007200	0.006761
42	20	10	2	0.048491	0.041290
42	20	10	3	0.180620	0.132129
42	20	10	4	0.426297	0.245677
42	20	10	5	0.703768	0.277471
42	20	10	6	0.896455	0.192688
42	20	10	7	0.977587	0.081132
42	20	10	8	0.997363	0.019776
42	20	10	9	0.999874	0.002511
42	20	10	10	1.000000	0.000126
42	20	11	0	0.000165	0.000165
42	20	11	1	0.003186	0.003021
42	20	11	2	0.025265	0.022079
42	20	11	3	0.110426	0.085161
42	20	11	4	0.303458	0.193032
42	20	11	5	0.573703	0.270245
42	20	11	6	0.812154	0.238451
42	20	11	7	0.944627	0.132473
42	20	11	8	0.989947	0.045320
42	20	11	9	0.999011	0.009064
42	20	11	10	0.999961	0.000950
42	20	11	11	1.000000	0.000039
42	20	12	0	0.000058	0.000058
42	20	12	1	0.001334	0.001276
42	20	12	2	0.012445	0.011111
42	20	12	3	0.063725	0.051280
42	20	12	4	0.203829	0.140104
42	20	12	5	0.442939	0.239111
42	20	12	6	0.704467	0.261527

N	n	k	x	P(x)	p(x)
42	20	12	7	0.889074	0.184608
42	20	12	8	0.972404	0.083330
42	20	12	9	0.995795	0.023391
42	20	12	10	0.999654	0.003859
42	20	12	11	0.999988	0.000334
42	20	12	12	1.000000	0.000011
42	20	13	0	0.000019	0.000019
42	20	13	1	0.000526	0.000507
42	20	13	2	0.005779	0.005252
42	20	13	3	0.034666	0.028888
42	20	13	4	0.129107	0.094440
42	20	13	5	0.323384	0.194277
42	20	13	6	0.582421	0.259037
42	20	13	7	0.809078	0.226657
42	20	13	8	0.939072	0.129994
42	20	13	9	0.987218	0.048146
42	20	13	10	0.998368	0.011150
42	20	13	11	0.999888	0.001520
42	20	13	12	0.999997	0.000109
42	20	13	13	1.000000	0.000003
42	20	14	0	0.000006	0.000006
42	20	14	1	0.000194	0.000188
42	20	14	2	0.002519	0.002324
42	20	14	3	0.017732	0.015214
42	20	14	4	0.077002	0.059270
42	20	14	5	0.222896	0.145894
42	20	14	6	0.457369	0.234473
42	20	14	7	0.707473	0.250104
42	20	14	8	0.885281	0.177809
42	20	14	9	0.968956	0.083675
42	20	14	10	0.994523	0.025567
42	20	14	11	0.999416	0.004893
42	20	14	12	0.999967	0.000550
42	20	14	13	0.999999	0.000032
42	20	14	14	1.000000	0.000001
42	20	15	0	0.000002	0.000002
42	20	15	1	0.000067	0.000065
42	20	15	2	0.001024	0.000958
42	20	15	3	0.008495	0.007471
42	20	15	4	0.043133	0.034638
42	20	15	5	0.144738	0.101605
42	20	15	6	0.340132	0.195394
42	20	15	7	0.591353	0.251221
42	20	15	8	0.809078	0.217725
42	20	15	9	0.936084	0.127006
42	20	15	10	0.985392	0.049308
42	20	15	11	0.997844	0.012452
42	20	15	12	0.999810	0.001966
42	20	15	13	0.999991	0.000181
42	20	15	14	1.000000	0.000009

N	n	k	x	P(x)	p(x)
42	20	15	15	1.000000	0.000000
42	20	16	0	0.000000	0.000000
42	20	16	1	0.000021	0.000020
42	20	16	2	0.000386	0.000365
42	20	16	3	0.003791	0.003406
42	20	16	4	0.022607	0.018816
42	20	16	5	0.088291	0.065684
42	20	16	6	0.238817	0.150526
42	20	16	7	0.470395	0.231578
42	20	16	8	0.712311	0.241916
42	20	16	9	0.884341	0.172029
42	20	16	10	0.967130	0.082789
42	20	16	11	0.993693	0.026563
42	20	16	12	0.999227	0.005534
42	20	16	13	0.999944	0.000717
42	20	16	14	0.999998	0.000054
42	20	16	15	1.000000	0.000002
42	20	16	16	1.000000	0.000000
42	20	17	0	0.000000	0.000000
42	20	17	1	0.000006	0.000006
42	20	17	2	0.000133	0.000127
42	20	17	3	0.001565	0.001431
42	20	17	4	0.011028	0.009464
42	20	17	5	0.050396	0.039368
42	20	17	6	0.157764	0.107368
42	20	17	7	0.354606	0.196841
42	20	17	8	0.600657	0.246052
42	20	17	9	0.811559	0.210901
42	20	17	10	0.935288	0.123729
42	20	17	11	0.984498	0.049210
42	20	17	12	0.997524	0.013026
42	20	17	13	0.999751	0.002227
42	20	17	14	0.999985	0.000234
42	20	17	15	0.999999	0.000014
42	20	17	16	1.000000	0.000000
42	20	17	17	1.000000	0.000000
42	20	18	0	0.000000	0.000000
42	20	18	1	0.000002	0.000001
42	20	18	2	0.000042	0.000040
42	20	18	3	0.000591	0.000550
42	20	18	4	0.004972	0.004380
42	20	18	5	0.026775	0.021804
42	20	18	6	0.097638	0.070863
42	20	18	7	0.252248	0.154610
42	20	18	8	0.482553	0.230304
42	20	18	9	0.718762	0.236210
42	20	18	10	0.885796	0.167034
42	20	18	11	0.966782	0.080986
42	20	18	12	0.993356	0.026574
42	20	18	13	0.999128	0.005772

N	n	k	x	P(x)	p(x)
42	20	18	14	0.999929	0.000802
42	20	18	15	0.999997	0.000068
42	20	18	16	1.000000	0.000003
42	20	18	17	1.000000	0.000000
42	20	18	18	1.000000	0.000000
42	20	19	0	0.000000	0.000000
42	20	19	1	0.000000	0.000000
42	20	19	2	0.000012	0.000011
42	20	19	3	0.000202	0.000190
42	20	19	4	0.002051	0.001849
42	20	19	5	0.013148	0.011097
42	20	19	6	0.056302	0.043154
42	20	19	7	0.168501	0.112200
42	20	19	8	0.367400	0.198899
42	20	19	9	0.610499	0.243099
42	20	19	10	0.816199	0.205699
42	20	19	11	0.936413	0.120214
42	20	19	12	0.984498	0.048086
42	20	19	13	0.997444	0.012946
42	20	19	14	0.999729	0.002285
42	20	19	15	0.999983	0.000254
42	20	19	16	0.999999	0.000017
42	20	19	17	1.000000	0.000001
42	20	19	18	1.000000	0.000000
42	20	19	19	1.000000	0.000000
42	20	20	0	0.000000	0.000000
42	20	20	1	0.000000	0.000000
42	20	20	2	0.000003	0.000003
42	20	20	3	0.000061	0.000058
42	20	20	4	0.000765	0.000704
42	20	20	5	0.005911	0.005146
42	20	20	6	0.030034	0.024123
42	20	20	7	0.105084	0.075050
42	20	20	8	0.263627	0.158543
42	20	20	9	0.494235	0.230608
42	20	20	10	0.726764	0.232530
42	20	20	11	0.889372	0.162608
42	20	20	12	0.967773	0.078400
42	20	20	13	0.993504	0.025731
42	20	20	14	0.999133	0.005629
42	20	20	15	0.999927	0.000795
42	20	20	16	0.999996	0.000069
42	20	20	17	1.000000	0.000003
42	20	20	18	1.000000	0.000000
42	20	20	19	1.000000	0.000000
42	20	20	20	1.000000	0.000000
42	21	1	0	0.500000	0.500000
42	21	1	1	1.000000	0.500000
42	21	2	0	0.243902	0.243902
42	21	2	1	0.756098	0.512195

N	n	k	x	P(x)	p(x)
42	21	2	2	1.000000	0.243902
42	21	3	0	0.115854	0.115854
42	21	3	1	0.500000	0.384146
42	21	3	2	0.884146	0.384146
42	21	3	3	1.000000	0.115854
42	21	4	0	0.053471	0.053471
42	21	4	1	0.303002	0.249531
42	21	4	2	0.696998	0.393996
42	21	4	3	0.946529	0.249531
42	21	4	4	1.000000	0.053471
42	21	5	0	0.023921	0.023921
42	21	5	1	0.171670	0.147749
42	21	5	2	0.500000	0.328330
42	21	5	3	0.828330	0.328330
42	21	5	4	0.976079	0.147749
42	21	5	5	1.000000	0.023921
42	21	6	0	0.010344	0.010344
42	21	6	1	0.091806	0.081461
42	21	6	2	0.331398	0.239592
42	21	6	3	0.668602	0.337204
42	21	6	4	0.908194	0.239592
42	21	6	5	0.989656	0.081461
42	21	6	6	1.000000	0.010344
42	21	7	0	0.004310	0.004310
42	21	7	1	0.046549	0.042239
42	21	7	2	0.204946	0.158397
42	21	7	3	0.500000	0.295054
42	21	7	4	0.795054	0.295054
42	21	7	5	0.953451	0.158397
42	21	7	6	0.995690	0.042239
42	21	7	7	1.000000	0.004310
42	21	8	0	0.001724	0.001724
42	21	8	1	0.022413	0.020689
42	21	8	2	0.118959	0.096547
42	21	8	3	0.348258	0.229299
42	21	8	4	0.651742	0.303484
42	21	8	5	0.881040	0.229299
42	21	8	6	0.977587	0.096547
42	21	8	7	0.998276	0.020689
42	21	8	8	1.000000	0.001724
42	21	9	0	0.000659	0.000659
42	21	9	1	0.010243	0.009584
42	21	9	2	0.065007	0.054764
42	21	9	3	0.226865	0.161858
42	21	9	4	0.500000	0.273135
42	21	9	5	0.773135	0.273135
42	21	9	6	0.934993	0.161858
42	21	9	7	0.989757	0.054764
42	21	9	8	0.999341	0.009584
42	21	9	9	1.000000	0.000659

Table for $N = 2$, $n = 1$, through $N = 100$, $n = 50$

N	n	k	x	P(x)	p(x)
42	21	10	0	0.000240	0.000240
42	21	10	1	0.004435	0.004195
42	21	10	2	0.033476	0.029041
42	21	10	3	0.138579	0.105103
42	21	10	4	0.359294	0.220715
42	21	10	5	0.640706	0.281412
42	21	10	6	0.861421	0.220715
42	21	10	7	0.966524	0.105103
42	21	10	8	0.995565	0.029041
42	21	10	9	0.999760	0.004195
42	21	10	10	1.000000	0.000240
42	21	11	0	0.000082	0.000082
42	21	11	1	0.001813	0.001730
42	21	11	2	0.016233	0.014420
42	21	11	3	0.079458	0.063226
42	21	11	4	0.242039	0.162580
42	21	11	5	0.500000	0.257961
42	21	11	6	0.757961	0.257961
42	21	11	7	0.920541	0.162580
42	21	11	8	0.983767	0.063226
42	21	11	9	0.998187	0.014420
42	21	11	10	0.999917	0.001730
42	21	11	11	1.000000	0.000082
42	21	12	0	0.000027	0.000027
42	21	12	1	0.000696	0.000670
42	21	12	2	0.007395	0.006698
42	21	12	3	0.042747	0.035352
42	21	12	4	0.152882	0.110135
42	21	12	5	0.366859	0.213977
42	21	12	6	0.633141	0.266282
42	21	12	7	0.847118	0.213977
42	21	12	8	0.957253	0.110135
42	21	12	9	0.992605	0.035352
42	21	12	10	0.999304	0.006698
42	21	12	11	0.999973	0.000670
42	21	12	12	1.000000	0.000027
42	21	13	0	0.000008	0.000008
42	21	13	1	0.000250	0.000242
42	21	13	2	0.003152	0.002903
42	21	13	3	0.021535	0.018383
42	21	13	4	0.090472	0.068936
42	21	13	5	0.252738	0.162266
42	21	13	6	0.500000	0.247262
42	21	13	7	0.747262	0.247262
42	21	13	8	0.909528	0.162266
42	21	13	9	0.978464	0.068936
42	21	13	10	0.996847	0.018383
42	21	13	11	0.999750	0.002903
42	21	13	12	0.999992	0.000242
42	21	13	13	1.000000	0.000008
42	21	14	0	0.000002	0.000002
42	21	14	1	0.000083	0.000081
42	21	14	2	0.001251	0.001168
42	21	14	3	0.010125	0.008875
42	21	14	4	0.050061	0.039936
42	21	14	5	0.163212	0.113151
42	21	14	6	0.372106	0.208894
42	21	14	7	0.627894	0.255788
42	21	14	8	0.836788	0.208894
42	21	14	9	0.949939	0.113151
42	21	14	10	0.989875	0.039936
42	21	14	11	0.998749	0.008875
42	21	14	12	0.999917	0.001168
42	21	14	13	0.999998	0.000081
42	21	14	14	1.000000	0.000002
42	21	15	0	0.000001	0.000001
42	21	15	1	0.000025	0.000025
42	21	15	2	0.000458	0.000433
42	21	15	3	0.004420	0.003962
42	21	15	4	0.025814	0.021394
42	21	15	5	0.098554	0.072740
42	21	15	6	0.260198	0.161644
42	21	15	7	0.500000	0.239802
42	21	15	8	0.739802	0.239802
42	21	15	9	0.901446	0.161644
42	21	15	10	0.974186	0.072740
42	21	15	11	0.995580	0.021394
42	21	15	12	0.999541	0.003962
42	21	15	13	0.999975	0.000433
42	21	15	14	0.999999	0.000025
42	21	15	15	1.000000	0.000001
42	21	16	0	0.000000	0.000000
42	21	16	1	0.000007	0.000007
42	21	16	2	0.000154	0.000147
42	21	16	3	0.001779	0.001625
42	21	16	4	0.012344	0.010565
42	21	16	5	0.055449	0.043105
42	21	16	6	0.170396	0.114947
42	21	16	7	0.375658	0.205262
42	21	16	8	0.624342	0.248683
42	21	16	9	0.829604	0.205262
42	21	16	10	0.944551	0.114947
42	21	16	11	0.987656	0.043105
42	21	16	12	0.998221	0.010565
42	21	16	13	0.999846	0.001625
42	21	16	14	0.999993	0.000147
42	21	16	15	1.000000	0.000007
42	21	16	16	1.000000	0.000000
42	21	17	0	0.000000	0.000000
42	21	17	1	0.000002	0.000002
42	21	17	2	0.000046	0.000045
42	21	17	3	0.000654	0.000607
42	21	17	4	0.005436	0.004782
42	21	17	5	0.028923	0.023487
42	21	17	6	0.104080	0.075158
42	21	17	7	0.265132	0.161052
42	21	17	8	0.500000	0.234868
42	21	17	9	0.734867	0.234868
42	21	17	10	0.895919	0.161052
42	21	17	11	0.971077	0.075158
42	21	17	12	0.994564	0.023487
42	21	17	13	0.999346	0.004782
42	21	17	14	0.999953	0.000607
42	21	17	15	0.999998	0.000045
42	21	17	16	1.000000	0.000002
42	21	17	17	1.000000	0.000000
42	21	18	0	0.000000	0.000000
42	21	18	1	0.000000	0.000000
42	21	18	2	0.000012	0.000012
42	21	18	3	0.000216	0.000204
42	21	18	4	0.002184	0.001968
42	21	18	5	0.013891	0.011707
42	21	18	6	0.058986	0.045095
42	21	18	7	0.174943	0.115957
42	21	18	8	0.377869	0.202926
42	21	18	9	0.622131	0.244262
42	21	18	10	0.825057	0.202926
42	21	18	11	0.941014	0.115957
42	21	18	12	0.986109	0.045095
42	21	18	13	0.997816	0.011707
42	21	18	14	0.999783	0.001968
42	21	18	15	0.999987	0.000204
42	21	18	16	1.000000	0.000012
42	21	18	17	1.000000	0.000000
42	21	18	18	1.000000	0.000000
42	21	19	0	0.000000	0.000000
42	21	19	1	0.000000	0.000000
42	21	19	2	0.000003	0.000003
42	21	19	3	0.000063	0.000061
42	21	19	4	0.000790	0.000727
42	21	19	5	0.006086	0.005296
42	21	19	6	0.030802	0.024715
42	21	19	7	0.107301	0.076500
42	21	19	8	0.267951	0.160649
42	21	19	9	0.500000	0.232049
42	21	19	10	0.732049	0.232049
42	21	19	11	0.892698	0.160649
42	21	19	12	0.969198	0.076500
42	21	19	13	0.993913	0.024715
42	21	19	14	0.999210	0.005296
42	21	19	15	0.999936	0.000727
42	21	19	16	0.999997	0.000061
42	21	19	17	1.000000	0.000003
42	21	19	18	1.000000	0.000000
42	21	19	19	1.000000	0.000000
42	21	20	0	0.000000	0.000000
42	21	20	1	0.000000	0.000000
42	21	20	2	0.000001	0.000001
42	21	20	3	0.000016	0.000015
42	21	20	4	0.000253	0.000237
42	21	20	5	0.002402	0.002149
42	21	20	6	0.014683	0.012281
42	21	20	7	0.060736	0.046053
42	21	20	8	0.177149	0.116413
42	21	20	9	0.378931	0.201782
42	21	20	10	0.621069	0.242138
42	21	20	11	0.822851	0.201782
42	21	20	12	0.939264	0.116413
42	21	20	13	0.985317	0.046053
42	21	20	14	0.997598	0.012281
42	21	20	15	0.999747	0.002149
42	21	20	16	0.999984	0.000237
42	21	20	17	0.999999	0.000015
42	21	20	18	1.000000	0.000001
42	21	20	19	1.000000	0.000000
42	21	20	20	1.000000	0.000000
42	21	21	0	0.000000	0.000000
42	21	21	1	0.000000	0.000000
42	21	21	2	0.000000	0.000000
42	21	21	3	0.000003	0.000003
42	21	21	4	0.000070	0.000067
42	21	21	5	0.000839	0.000769
42	21	21	6	0.006310	0.005471
42	21	21	7	0.031430	0.025120
42	21	21	8	0.108360	0.076930
42	21	21	9	0.268868	0.160508
42	21	21	10	0.500000	0.231132
42	21	21	11	0.731132	0.231132
42	21	21	12	0.891640	0.160508
42	21	21	13	0.968570	0.076930
42	21	21	14	0.993690	0.025120
42	21	21	15	0.999161	0.005471
42	21	21	16	0.999930	0.000769
42	21	21	17	0.999996	0.000067
42	21	21	18	1.000000	0.000003
42	21	21	19	1.000000	0.000000
42	21	21	20	1.000000	0.000000
42	21	21	21	1.000000	0.000000
43	1	1	0	0.976744	0.976744
43	1	1	1	1.000000	0.023256

Table for $N = 2$, $n = 1$, through $N = 100$, $n = 50$

N	n	k	x	P(x)	p(x)
43	2	1	0	0.953488	0.953488
43	2	1	1	1.000000	0.046512
43	2	2	0	0.908084	0.908084
43	2	2	1	0.998893	0.090808
43	2	2	2	1.000000	0.001107
43	3	1	0	0.930233	0.930233
43	3	1	1	1.000000	0.069767
43	3	2	0	0.863787	0.863787
43	3	2	1	0.996678	0.132890
43	3	2	2	1.000000	0.003322
43	3	3	0	0.800583	0.800583
43	3	3	1	0.990195	0.189612
43	3	3	2	0.999919	0.009724
43	3	3	3	1.000000	0.000081
43	4	1	0	0.906977	0.906977
43	4	1	1	1.000000	0.093023
43	4	2	0	0.820598	0.820598
43	4	2	1	0.993355	0.172757
43	4	2	2	1.000000	0.006645
43	4	3	0	0.740540	0.740540
43	4	3	1	0.980715	0.240175
43	4	3	2	0.999676	0.018961
43	4	3	3	1.000000	0.000324
43	4	4	0	0.666486	0.666486
43	4	4	1	0.962702	0.296216
43	4	4	2	0.998728	0.036026
43	4	4	3	0.999992	0.001264
43	4	4	4	1.000000	0.000008
43	5	1	0	0.883721	0.883721
43	5	1	1	1.000000	0.116279
43	5	2	0	0.778516	0.778516
43	5	2	1	0.988926	0.210410
43	5	2	2	1.000000	0.011074
43	5	3	0	0.683575	0.683575
43	5	3	1	0.968398	0.284823
43	5	3	2	0.999190	0.030792
43	5	3	3	1.000000	0.000810
43	5	4	0	0.598128	0.598128
43	5	4	1	0.939916	0.341787
43	5	4	2	0.996880	0.056965
43	5	4	3	0.999959	0.003079
43	5	4	4	1.000000	0.000041
43	5	5	0	0.521445	0.521445
43	5	5	1	0.904861	0.383415
43	5	5	2	0.992498	0.087638
43	5	5	3	0.999802	0.007303
43	5	5	4	0.999999	0.000197
43	5	5	5	1.000000	0.000001
43	6	1	0	0.860465	0.860465
43	6	1	1	1.000000	0.139535
43	6	2	0	0.737542	0.737542
43	6	2	1	0.983389	0.245847
43	6	2	2	1.000000	0.016611
43	6	3	0	0.629609	0.629609
43	6	3	1	0.953407	0.323799
43	6	3	2	0.998379	0.044972
43	6	3	3	1.000000	0.001621
43	6	4	0	0.535167	0.535167
43	6	4	1	0.912932	0.377765
43	6	4	2	0.993882	0.080950
43	6	4	3	0.999878	0.005996
43	6	4	4	1.000000	0.000122
43	6	5	0	0.452834	0.452834
43	6	5	1	0.864501	0.411667
43	6	5	2	0.985580	0.121079
43	6	5	3	0.999417	0.013838
43	6	5	4	0.999994	0.000577
43	6	5	5	1.000000	0.000006
43	6	6	0	0.381334	0.381334
43	6	6	1	0.810334	0.429001
43	6	6	2	0.972834	0.162500
43	6	6	3	0.998325	0.025490
43	6	6	4	0.999963	0.001639
43	6	6	5	1.000000	0.000036
43	6	6	6	1.000000	0.000000
43	7	1	0	0.837209	0.837209
43	7	1	1	1.000000	0.162791
43	7	2	0	0.697674	0.697674
43	7	2	1	0.976744	0.279070
43	7	2	2	1.000000	0.023256
43	7	3	0	0.578559	0.578559
43	7	3	1	0.935905	0.357345
43	7	3	2	0.997164	0.061259
43	7	3	3	1.000000	0.002836
43	7	4	0	0.477311	0.477311
43	7	4	1	0.882303	0.404991
43	7	4	2	0.989506	0.107204
43	7	4	3	0.999716	0.010210
43	7	4	4	1.000000	0.000284
43	7	5	0	0.391640	0.391640
43	7	5	1	0.819996	0.428356
43	7	5	2	0.975762	0.155766
43	7	5	3	0.998669	0.022907
43	7	5	4	0.999978	0.001309
43	7	5	5	1.000000	0.000022
43	7	6	0	0.319496	0.319496
43	7	6	1	0.752361	0.432865
43	7	6	2	0.955267	0.202906
43	7	6	3	0.996258	0.040991
43	7	6	4	0.999875	0.003617
43	7	6	5	0.999999	0.000124
43	7	6	6	1.000000	0.000001
43	7	7	0	0.259051	0.259051
43	7	7	1	0.682167	0.423116
43	7	7	2	0.927847	0.245680
43	7	7	3	0.991826	0.063979
43	7	7	4	0.999581	0.007755
43	7	7	5	0.999992	0.000411
43	7	7	6	1.000000	0.000008
43	7	7	7	1.000000	0.000000
43	8	1	0	0.813953	0.813953
43	8	1	1	1.000000	0.186046
43	8	2	0	0.658915	0.658915
43	8	2	1	0.968992	0.310077
43	8	2	2	1.000000	0.031008
43	8	3	0	0.530346	0.530346
43	8	3	1	0.916052	0.385706
43	8	3	2	0.995462	0.079410
43	8	3	3	1.000000	0.004538
43	8	4	0	0.424277	0.424277
43	8	4	1	0.848554	0.424277
43	8	4	2	0.983551	0.134997
43	8	4	3	0.999433	0.015882
43	8	4	4	1.000000	0.000567
43	8	5	0	0.337246	0.337246
43	8	5	1	0.772401	0.435156
43	8	5	2	0.962782	0.190381
43	8	5	3	0.997397	0.034615
43	8	5	4	0.999942	0.002545
43	8	5	5	1.000000	0.000058
43	8	6	0	0.266247	0.266247
43	8	6	1	0.692241	0.425995
43	8	6	2	0.932722	0.240481
43	8	6	3	0.992842	0.060120
43	8	6	4	0.999674	0.006832
43	8	6	5	0.999995	0.000321
43	8	6	6	1.000000	0.000005
43	8	7	0	0.208680	0.208680
43	8	7	1	0.611648	0.402968
43	8	7	2	0.893725	0.282077
43	8	7	3	0.984718	0.090993
43	8	7	4	0.998935	0.014218
43	8	7	5	0.999969	0.001034
43	8	7	6	1.000000	0.000030
43	8	7	7	1.000000	0.000000
43	8	8	0	0.162306	0.162306
43	8	8	1	0.533293	0.370986
43	8	8	2	0.846712	0.313419
43	8	8	3	0.972080	0.125368
43	8	8	4	0.997356	0.025276
43	8	8	5	0.999883	0.002528
43	8	8	6	0.999998	0.000115
43	8	8	7	1.000000	0.000002
43	8	8	8	1.000000	0.000000
43	9	1	0	0.790698	0.790698
43	9	1	1	1.000000	0.209302
43	9	2	0	0.621262	0.621262
43	9	2	1	0.960133	0.338870
43	9	2	2	1.000000	0.039867
43	9	3	0	0.484888	0.484888
43	9	3	1	0.894012	0.409124
43	9	3	2	0.993193	0.099182
43	9	3	3	1.000000	0.006807
43	9	4	0	0.375788	0.375788
43	9	4	1	0.812187	0.436399
43	9	4	2	0.975837	0.163650
43	9	4	3	0.998979	0.023142
43	9	4	4	1.000000	0.001021
43	9	5	0	0.289068	0.289068
43	9	5	1	0.722669	0.433602
43	9	5	2	0.946464	0.223794
43	9	5	3	0.995419	0.048955
43	9	5	4	0.999869	0.004450
43	9	5	5	1.000000	0.000131
43	9	6	0	0.220604	0.220604
43	9	6	1	0.631385	0.410780
43	9	6	2	0.905238	0.273854
43	9	6	3	0.987689	0.082451
43	9	6	4	0.999283	0.011595
43	9	6	5	0.999986	0.000703
43	9	6	6	1.000000	0.000014
43	9	7	0	0.166944	0.166944
43	9	7	1	0.542567	0.375624
43	9	7	2	0.853428	0.310861
43	9	7	3	0.974318	0.120890
43	9	7	4	0.997717	0.023398
43	9	7	5	0.999910	0.002194
43	9	7	6	0.999999	0.000089
43	9	7	7	1.000000	0.000001
43	9	8	0	0.125208	0.125208
43	9	8	1	0.459095	0.333888
43	9	8	2	0.792983	0.333888
43	9	8	3	0.954170	0.161187
43	9	8	4	0.994467	0.040297
43	9	8	5	0.999666	0.005200
43	9	8	6	0.999991	0.000325
43	9	8	7	1.000000	0.000008
43	9	8	8	1.000000	0.000000
43	9	9	0	0.093012	0.093012
43	9	9	1	0.382778	0.289767

Table for $N = 2$, $n = 1$, through $N = 100$, $n = 50$

N	n	k	x	P(x)	p(x)
43	9	9	2	0.726205	0.343427
43	9	9	3	0.926538	0.200333
43	9	9	4	0.988710	0.062172
43	9	9	5	0.999072	0.010362
43	9	9	6	0.999964	0.000891
43	9	9	7	0.999999	0.000036
43	9	9	8	1.000000	0.000001
43	9	9	9	1.000000	0.000000
43	10	1	0	0.767442	0.767442
43	10	1	1	1.000000	0.232558
43	10	2	0	0.584718	0.584718
43	10	2	1	0.950166	0.365448
43	10	2	2	1.000000	0.049834
43	10	3	0	0.442104	0.442104
43	10	3	1	0.869946	0.427842
43	10	3	2	0.990276	0.120331
43	10	3	3	1.000000	0.009724
43	10	4	0	0.331578	0.331578
43	10	4	1	0.773681	0.442104
43	10	4	2	0.966210	0.192529
43	10	4	3	0.998298	0.032088
43	10	4	4	1.000000	0.001702
43	10	5	0	0.246558	0.246558
43	10	5	1	0.671657	0.425100
43	10	5	2	0.926717	0.255060
43	10	5	3	0.992539	0.065822
43	10	5	4	0.999738	0.007199
43	10	5	5	1.000000	0.000262
43	10	6	0	0.181674	0.181674
43	10	6	1	0.570976	0.389302
43	10	6	2	0.873020	0.302044
43	10	6	3	0.980414	0.107394
43	10	6	4	0.998601	0.018188
43	10	6	5	0.999965	0.001364
43	10	6	6	1.000000	0.000034
43	10	7	0	0.132573	0.132573
43	10	7	1	0.476281	0.343708
43	10	7	2	0.807713	0.331432
43	10	7	3	0.960096	0.152383
43	10	7	4	0.995652	0.035556
43	10	7	5	0.999781	0.004129
43	10	7	6	0.999996	0.000215
43	10	7	7	1.000000	0.000004
43	10	8	0	0.095747	0.095747
43	10	8	1	0.390354	0.294607
43	10	8	2	0.734062	0.343708
43	10	8	3	0.930466	0.196404
43	10	8	4	0.989726	0.059260
43	10	8	5	0.999208	0.009482
43	10	8	6	0.999972	0.000765

N	n	k	x	P(x)	p(x)
43	10	8	7	1.000000	0.000027
43	10	8	8	1.000000	0.000000
43	10	9	0	0.068391	0.068391
43	10	9	1	0.314598	0.246207
43	10	9	2	0.655500	0.340902
43	10	9	3	0.891185	0.235685
43	10	9	4	0.979567	0.088382
43	10	9	5	0.997853	0.018286
43	10	9	6	0.999885	0.002032
43	10	9	7	0.999997	0.000112
43	10	9	8	1.000000	0.000003
43	10	9	9	1.000000	0.000000
43	10	10	0	0.048276	0.048276
43	10	10	1	0.249425	0.201150
43	10	10	2	0.575288	0.325862
43	10	10	3	0.842662	0.267374
43	10	10	4	0.963970	0.121309
43	10	10	5	0.995164	0.031194
43	10	10	6	0.999646	0.004482
43	10	10	7	0.999987	0.000341
43	10	10	8	1.000000	0.000012
43	10	10	9	1.000000	0.000000
43	10	10	10	1.000000	0.000000
43	11	1	0	0.744186	0.744186
43	11	1	1	1.000000	0.255814
43	11	2	0	0.549280	0.549280
43	11	2	1	0.939092	0.389812
43	11	2	2	1.000000	0.060908
43	11	3	0	0.401912	0.401912
43	11	3	1	0.844016	0.442104
43	11	3	2	0.986630	0.142614
43	11	3	3	1.000000	0.013370
43	11	4	0	0.291386	0.291386
43	11	4	1	0.733490	0.442104
43	11	4	2	0.954542	0.221052
43	11	4	3	0.997326	0.042784
43	11	4	4	1.000000	0.002674
43	11	5	0	0.209201	0.209201
43	11	5	1	0.620130	0.410930
43	11	5	2	0.903530	0.283400
43	11	5	3	0.988550	0.085020
43	11	5	4	0.999520	0.010970
43	11	5	5	1.000000	0.000480
43	11	6	0	0.148642	0.148642
43	11	6	1	0.511991	0.363348
43	11	6	2	0.836409	0.324418
43	11	6	3	0.970651	0.134242
43	11	6	4	0.997499	0.026848
43	11	6	5	0.999924	0.002425
43	11	6	6	1.000000	0.000076

N	n	k	x	P(x)	p(x)
43	11	7	0	0.104451	0.104451
43	11	7	1	0.413788	0.309337
43	11	7	2	0.757496	0.343708
43	11	7	3	0.941625	0.184129
43	11	7	4	0.992420	0.050794
43	11	7	5	0.999531	0.007111
43	11	7	6	0.999990	0.000459
43	11	7	7	1.000000	0.000010
43	11	8	0	0.072536	0.072536
43	11	8	1	0.327861	0.255326
43	11	8	2	0.671569	0.343708
43	11	8	3	0.900708	0.229139
43	11	8	4	0.982543	0.081835
43	11	8	5	0.998346	0.015803
43	11	8	6	0.999926	0.001580
43	11	8	7	0.999999	0.000073
43	11	8	8	1.000000	0.000001
43	11	9	0	0.049739	0.049739
43	11	9	1	0.254911	0.205173
43	11	9	2	0.583187	0.328276
43	11	9	3	0.848333	0.265146
43	11	9	4	0.966176	0.117843
43	11	9	5	0.995637	0.029461
43	11	9	6	0.999700	0.004064
43	11	9	7	0.999990	0.000290
43	11	9	8	1.000000	0.000009
43	11	9	9	1.000000	0.000000
43	11	10	0	0.033647	0.033647
43	11	10	1	0.194566	0.160920
43	11	10	2	0.496291	0.301724
43	11	10	3	0.785946	0.289655
43	11	10	4	0.941914	0.155968
43	11	10	5	0.990438	0.048523
43	11	10	6	0.999103	0.008665
43	11	10	7	0.999956	0.000854
43	11	10	8	0.999999	0.000043
43	11	10	9	1.000000	0.000001
43	11	10	10	1.000000	0.000000
43	11	11	0	0.022431	0.022431
43	11	11	1	0.145803	0.123372
43	11	11	2	0.414002	0.268199
43	11	11	3	0.715727	0.301724
43	11	11	4	0.908830	0.193104
43	11	11	5	0.981615	0.072785
43	11	11	6	0.997790	0.016174
43	11	11	7	0.999853	0.002063
43	11	11	8	0.999995	0.000142
43	11	11	9	1.000000	0.000005
43	11	11	10	1.000000	0.000000
43	11	11	11	1.000000	0.000000

N	n	k	x	P(x)	p(x)
43	12	1	0	0.720930	0.720930
43	12	1	1	1.000000	0.279070
43	12	2	0	0.514950	0.514950
43	12	2	1	0.926910	0.411960
43	12	2	2	1.000000	0.073090
43	12	3	0	0.364233	0.364233
43	12	3	1	0.816384	0.452151
43	12	3	2	0.982173	0.165789
43	12	3	3	1.000000	0.017827
43	12	4	0	0.254963	0.254963
43	12	4	1	0.692043	0.437080
43	12	4	2	0.940726	0.248683
43	12	4	3	0.995989	0.055263
43	12	4	4	1.000000	0.004011
43	12	5	0	0.176513	0.176513
43	12	5	1	0.568764	0.392251
43	12	5	2	0.876961	0.308197
43	12	5	3	0.983236	0.106275
43	12	5	4	0.999177	0.015941
43	12	5	5	1.000000	0.000823
43	12	6	0	0.120772	0.120772
43	12	6	1	0.455218	0.334446
43	12	6	2	0.795856	0.340639
43	12	6	3	0.958065	0.162209
43	12	6	4	0.995821	0.037756
43	12	6	5	0.999848	0.004027
43	12	6	6	1.000000	0.000152
43	12	7	0	0.081603	0.081603
43	12	7	1	0.355788	0.274185
43	12	7	2	0.703792	0.348004
43	12	7	3	0.918609	0.214817
43	12	7	4	0.987658	0.069048
43	12	7	5	0.999086	0.011429
43	12	7	6	0.999975	0.000889
43	12	7	7	1.000000	0.000025
43	12	8	0	0.054402	0.054402
43	12	8	1	0.272009	0.217607
43	12	8	2	0.607124	0.335115
43	12	8	3	0.864905	0.257781
43	12	8	4	0.972314	0.107409
43	12	8	5	0.996864	0.024551
43	12	8	6	0.999827	0.002963
43	12	8	7	0.999996	0.000169
43	12	8	8	1.000000	0.000003
43	12	9	0	0.035750	0.035750
43	12	9	1	0.203618	0.167868
43	12	9	2	0.511377	0.307759
43	12	9	3	0.798618	0.287241
43	12	9	4	0.947763	0.149145
43	12	9	5	0.991954	0.044191

Table for $N = 2$, $n = 1$, through $N = 100$, $n = 50$

N	n	k	x	P(x)	p(x)
43	12	9	6	0.999319	0.007365
43	12	9	7	0.999972	0.000653
43	12	9	8	0.999999	0.000027
43	12	9	9	1.000000	0.000000
43	12	10	0	0.023132	0.023132
43	12	10	1	0.149308	0.126176
43	12	10	2	0.420860	0.271552
43	12	10	3	0.722584	0.301724
43	12	10	4	0.912670	0.190086
43	12	10	5	0.982856	0.070186
43	12	10	6	0.998019	0.015164
43	12	10	7	0.999876	0.001857
43	12	10	8	0.999996	0.000120
43	12	10	9	1.000000	0.000004
43	12	10	10	1.000000	0.000000
43	12	11	0	0.014720	0.014720
43	12	11	1	0.107249	0.092529
43	12	11	2	0.338571	0.231322
43	12	11	3	0.640295	0.301724
43	12	11	4	0.866589	0.226293
43	12	11	5	0.967968	0.101379
43	12	11	6	0.995262	0.027294
43	12	11	7	0.999595	0.004332
43	12	11	8	0.999982	0.000387
43	12	11	9	1.000000	0.000018
43	12	11	10	1.000000	0.000000
43	12	11	11	1.000000	0.000000
43	12	12	0	0.009200	0.009200
43	12	12	1	0.075442	0.066242
43	12	12	2	0.266283	0.190841
43	12	12	3	0.555436	0.289152
43	12	12	4	0.810015	0.254580
43	12	12	5	0.945791	0.135776
43	12	12	6	0.990145	0.044353
43	12	12	7	0.998918	0.008773
43	12	12	8	0.999933	0.001015
43	12	12	9	0.999998	0.000064
43	12	12	10	1.000000	0.000002
43	12	12	11	1.000000	0.000000
43	12	12	12	1.000000	0.000000
43	13	1	0	0.697674	0.697674
43	13	1	1	1.000000	0.302326
43	13	2	0	0.481728	0.481728
43	13	2	1	0.913621	0.431894
43	13	2	2	1.000000	0.086379
43	13	3	0	0.328985	0.328985
43	13	3	1	0.787213	0.458229
43	13	3	2	0.976825	0.189612
43	13	3	3	1.000000	0.023175
43	13	4	0	0.222065	0.222065
43	13	4	1	0.649745	0.427680
43	13	4	2	0.924682	0.274937
43	13	4	3	0.994206	0.069524
43	13	4	4	1.000000	0.005794
43	13	5	0	0.148043	0.148043
43	13	5	1	0.518151	0.370108
43	13	5	2	0.847135	0.328985
43	13	5	3	0.976379	0.129244
43	13	5	4	0.998663	0.022283
43	13	5	5	1.000000	0.001337
43	13	6	0	0.097397	0.097397
43	13	6	1	0.401275	0.303878
43	13	6	2	0.751903	0.350628
43	13	6	3	0.942368	0.190465
43	13	6	4	0.993385	0.051017
43	13	6	5	0.999718	0.006333
43	13	6	6	1.000000	0.000281
43	13	7	0	0.063176	0.063176
43	13	7	1	0.302720	0.239543
43	13	7	2	0.647662	0.344943
43	13	7	3	0.890891	0.243229
43	13	7	4	0.980976	0.090085
43	13	7	5	0.998349	0.017373
43	13	7	6	0.999947	0.001598
43	13	7	7	1.000000	0.000053
43	13	8	0	0.040363	0.040363
43	13	8	1	0.222872	0.182509
43	13	8	2	0.542263	0.319391
43	13	8	3	0.823327	0.281064
43	13	8	4	0.958454	0.135127
43	13	8	5	0.994488	0.036034
43	13	8	6	0.999636	0.005148
43	13	8	7	0.999991	0.000355
43	13	8	8	1.000000	0.000009
43	13	9	0	0.025371	0.025371
43	13	9	1	0.160297	0.134927
43	13	9	2	0.441883	0.281586
43	13	9	3	0.743023	0.301140
43	13	9	4	0.923708	0.180684
43	13	9	5	0.986252	0.062545
43	13	9	6	0.998607	0.012354
43	13	9	7	0.999930	0.001324
43	13	9	8	0.999999	0.000068
43	13	9	9	1.000000	0.000001
43	13	10	0	0.015670	0.015670
43	13	10	1	0.112676	0.097006
43	13	10	2	0.350782	0.238106
43	13	10	3	0.654453	0.303671
43	13	10	4	0.875879	0.221427
43	13	10	5	0.971536	0.095656
43	13	10	6	0.996063	0.024527
43	13	10	7	0.999696	0.003634
43	13	10	8	0.999988	0.000292
43	13	10	9	1.000000	0.000011
43	13	10	10	1.000000	0.000000
43	13	11	0	0.009497	0.009497
43	13	11	1	0.077401	0.067904
43	13	11	2	0.271413	0.194012
43	13	11	3	0.562431	0.291018
43	13	11	4	0.815490	0.253059
43	13	11	5	0.948346	0.132856
43	13	11	6	0.990860	0.042514
43	13	11	7	0.999036	0.008176
43	13	11	8	0.999944	0.000908
43	13	11	9	0.999998	0.000054
43	13	11	10	1.000000	0.000001
43	13	11	11	1.000000	0.000000
43	13	12	0	0.005639	0.005639
43	13	12	1	0.051937	0.046298
43	13	12	2	0.204722	0.152784
43	13	12	3	0.471488	0.266766
43	13	12	4	0.744317	0.272829
43	13	12	5	0.915132	0.170815
43	13	12	6	0.981560	0.066428
43	13	12	7	0.997503	0.015943
43	13	12	8	0.999802	0.002299
43	13	12	9	0.999992	0.000189
43	13	12	10	1.000000	0.000008
43	13	12	11	1.000000	0.000000
43	13	12	12	1.000000	0.000000
43	13	13	0	0.003274	0.003274
43	13	13	1	0.034015	0.030741
43	13	13	2	0.150508	0.116492
43	13	13	3	0.385434	0.234927
43	13	13	4	0.665109	0.279674
43	13	13	5	0.871051	0.205942
43	13	13	6	0.966560	0.095509
43	13	13	7	0.994417	0.027857
43	13	13	8	0.999431	0.005014
43	13	13	9	0.999967	0.000536
43	13	13	10	0.999999	0.000032
43	13	13	11	1.000000	0.000001
43	13	13	12	1.000000	0.000000
43	13	13	13	1.000000	0.000000
43	14	1	0	0.674419	0.674419
43	14	1	1	1.000000	0.325581
43	14	2	0	0.449612	0.449612
43	14	2	1	0.899225	0.449612
43	14	2	2	1.000000	0.100775
43	14	3	0	0.296086	0.296086
43	14	3	1	0.756665	0.460579
43	14	3	2	0.970505	0.213840
43	14	3	3	1.000000	0.029495
43	14	4	0	0.192456	0.192456
43	14	4	1	0.606977	0.414521
43	14	4	2	0.906353	0.299376
43	14	4	3	0.991889	0.085536
43	14	4	4	1.000000	0.008111
43	14	5	0	0.123369	0.123369
43	14	5	1	0.468803	0.345434
43	14	5	2	0.814237	0.345434
43	14	5	3	0.967763	0.153526
43	14	5	4	0.997920	0.030157
43	14	5	5	1.000000	0.002080
43	14	6	0	0.077917	0.077917
43	14	6	1	0.350628	0.272711
43	14	6	2	0.705153	0.354524
43	14	6	3	0.923321	0.218169
43	14	6	4	0.989984	0.066663
43	14	6	5	0.999507	0.009523
43	14	6	6	1.000000	0.000493
43	14	7	0	0.048435	0.048435
43	14	7	1	0.254811	0.206376
43	14	7	2	0.590172	0.335361
43	14	7	3	0.858460	0.268289
43	14	7	4	0.971967	0.113507
43	14	7	5	0.997191	0.025224
43	14	7	6	0.999893	0.002703
43	14	7	7	1.000000	0.000106
43	14	8	0	0.029599	0.029599
43	14	8	1	0.180286	0.150687
43	14	8	2	0.478385	0.298098
43	14	8	3	0.776483	0.298098
43	14	8	4	0.940438	0.163954
43	14	8	5	0.990885	0.050447
43	14	8	6	0.999293	0.008408
43	14	8	7	0.999979	0.000686
43	14	8	8	1.000000	0.000021
43	14	9	0	0.017760	0.017760
43	14	9	1	0.124317	0.106557
43	14	9	2	0.376180	0.251863
43	14	9	3	0.682795	0.306616
43	14	9	4	0.893593	0.210798
43	14	9	5	0.977913	0.084319
43	14	9	6	0.997371	0.019458
43	14	9	7	0.999842	0.002471
43	14	9	8	0.999996	0.000154
43	14	9	9	1.000000	0.000004
43	14	10	0	0.010447	0.010447
43	14	10	1	0.083574	0.073128

N	n	k	x	P(x)	p(x)
43	14	10	2	0.287287	0.203713
43	14	10	3	0.583596	0.296309
43	14	10	4	0.831594	0.247998
43	14	10	5	0.955593	0.123999
43	14	10	6	0.992793	0.037200
43	14	10	7	0.999333	0.006541
43	14	10	8	0.999969	0.000636
43	14	10	9	0.999999	0.000030
43	14	10	10	1.000000	0.000001
43	14	11	0	0.006015	0.006015
43	14	11	1	0.054767	0.048752
43	14	11	2	0.213210	0.158443
43	14	11	3	0.484826	0.271617
43	14	11	4	0.756443	0.271617
43	14	11	5	0.921775	0.165332
43	14	11	6	0.983775	0.061999
43	14	11	7	0.997946	0.014171
43	14	11	8	0.999854	0.001908
43	14	11	9	0.999995	0.000141
43	14	11	10	1.000000	0.000005
43	14	11	11	1.000000	0.000000
43	14	12	0	0.003383	0.003383
43	14	12	1	0.034961	0.031578
43	14	12	2	0.153793	0.118832
43	14	12	3	0.391458	0.237665
43	14	12	4	0.671563	0.280105
43	14	12	5	0.875275	0.203713
43	14	12	6	0.968275	0.092999
43	14	12	7	0.994846	0.026571
43	14	12	8	0.999496	0.004650
43	14	12	9	0.999973	0.000477
43	14	12	10	0.999999	0.000026
43	14	12	11	1.000000	0.000001
43	14	12	12	1.000000	0.000000
43	14	13	0	0.001855	0.001855
43	14	13	1	0.021719	0.019863
43	14	13	2	0.107794	0.086075
43	14	13	3	0.307126	0.199332
43	14	13	4	0.581207	0.274081
43	14	13	5	0.816133	0.234927
43	14	13	6	0.944275	0.128142
43	14	13	7	0.988846	0.044571
43	14	13	8	0.998596	0.009750
43	14	13	9	0.999896	0.001300
43	14	13	10	0.999996	0.000100
43	14	13	11	1.000000	0.000004
43	14	13	12	1.000000	0.000000
43	14	13	13	1.000000	0.000000
43	14	14	0	0.000990	0.000990
43	14	14	1	0.013111	0.012122

N	n	k	x	P(x)	p(x)
43	14	14	2	0.073364	0.060253
43	14	14	3	0.234037	0.160673
43	14	14	4	0.489846	0.255809
43	14	14	5	0.745655	0.255809
43	14	14	6	0.910104	0.164449
43	14	14	7	0.978446	0.068342
43	14	14	8	0.996646	0.018200
43	14	14	9	0.999679	0.003033
43	14	14	10	0.999982	0.000303
43	14	14	11	0.999999	0.000017
43	14	14	12	1.000000	0.000000
43	14	14	13	1.000000	0.000000
43	14	14	14	1.000000	0.000000
43	15	1	0	0.651163	0.651163
43	15	1	1	1.000000	0.348837
43	15	2	0	0.418605	0.418605
43	15	2	1	0.883721	0.465116
43	15	2	2	1.000000	0.116279
43	15	3	0	0.265457	0.265457
43	15	3	1	0.724901	0.459444
43	15	3	2	0.963131	0.238230
43	15	3	3	1.000000	0.036869
43	15	4	0	0.165910	0.165910
43	15	4	1	0.564095	0.398185
43	15	4	2	0.885706	0.321611
43	15	4	3	0.988939	0.103233
43	15	4	4	1.000000	0.011061
43	15	5	0	0.102099	0.102099
43	15	5	1	0.421157	0.319058
43	15	5	2	0.778503	0.357345
43	15	5	3	0.957175	0.178673
43	15	5	4	0.996880	0.039705
43	15	5	5	1.000000	0.003120
43	15	6	0	0.061797	0.061797
43	15	6	1	0.303609	0.241813
43	15	6	2	0.656253	0.352644
43	15	6	3	0.900752	0.244499
43	15	6	4	0.985387	0.084634
43	15	6	5	0.999179	0.013792
43	15	6	6	1.000000	0.000821
43	15	7	0	0.036744	0.036744
43	15	7	1	0.212113	0.175369
43	15	7	2	0.532351	0.320238
43	15	7	3	0.821455	0.289104
43	15	7	4	0.960225	0.138770
43	15	7	5	0.995451	0.035226
43	15	7	6	0.999800	0.004349
43	15	7	7	1.000000	0.000200
43	15	8	0	0.021434	0.021434
43	15	8	1	0.143914	0.122480

N	n	k	x	P(x)	p(x)
43	15	8	2	0.416709	0.272796
43	15	8	3	0.725087	0.308378
43	15	8	4	0.917823	0.192736
43	15	8	5	0.985666	0.067843
43	15	8	6	0.998713	0.013047
43	15	8	7	0.999955	0.001243
43	15	8	8	1.000000	0.000044
43	15	9	0	0.012248	0.012248
43	15	9	1	0.094922	0.082674
43	15	9	2	0.315385	0.220463
43	15	9	3	0.619358	0.303972
43	15	9	4	0.857249	0.237891
43	15	9	5	0.966282	0.109034
43	15	9	6	0.995358	0.029076
43	15	9	7	0.999671	0.004313
43	15	9	8	0.999991	0.000320
43	15	9	9	1.000000	0.000009
43	15	10	0	0.006844	0.006844
43	15	10	1	0.060880	0.054035
43	15	10	2	0.231090	0.170211
43	15	10	3	0.512073	0.280983
43	15	10	4	0.780284	0.268211
43	15	10	5	0.934214	0.153930
43	15	10	6	0.987662	0.053448
43	15	10	7	0.998657	0.010995
43	15	10	8	0.999925	0.001269
43	15	10	9	0.999998	0.000073
43	15	10	10	1.000000	0.000002
43	15	11	0	0.003733	0.003733
43	15	11	1	0.037956	0.034222
43	15	11	2	0.164038	0.126082
43	15	11	3	0.409898	0.245860
43	15	11	4	0.690880	0.280983
43	15	11	5	0.887568	0.196688
43	15	11	6	0.973085	0.085517
43	15	11	7	0.995991	0.022906
43	15	11	8	0.999656	0.003665
43	15	11	9	0.999985	0.000329
43	15	11	10	1.000000	0.000015
43	15	11	11	1.000000	0.000000
43	15	12	0	0.001983	0.001983
43	15	12	1	0.022983	0.021000
43	15	12	2	0.112817	0.089833
43	15	12	3	0.317700	0.204883
43	15	12	4	0.594293	0.276592
43	15	12	5	0.826103	0.231811
43	15	12	6	0.949033	0.122930
43	15	12	7	0.990265	0.041231
43	15	12	8	0.998854	0.008590
43	15	12	9	0.999923	0.001069

N	n	k	x	P(x)	p(x)
43	15	12	10	0.999997	0.000074
43	15	12	11	1.000000	0.000002
43	15	12	12	1.000000	0.000000
43	15	13	0	0.001024	0.001024
43	15	13	1	0.013499	0.012476
43	15	13	2	0.075145	0.061645
43	15	13	3	0.238390	0.163246
43	15	13	4	0.496147	0.257756
43	15	13	5	0.751326	0.255179
43	15	13	6	0.913344	0.162018
43	15	13	7	0.979624	0.066280
43	15	13	8	0.996915	0.017290
43	15	13	9	0.999716	0.002802
43	15	13	10	0.999985	0.000269
43	15	13	11	1.000000	0.000014
43	15	13	12	1.000000	0.000000
43	15	13	13	1.000000	0.000000
43	15	14	0	0.000512	0.000512
43	15	14	1	0.007677	0.007166
43	15	14	2	0.048432	0.040754
43	15	14	3	0.173092	0.124660
43	15	14	4	0.401636	0.228544
43	15	14	5	0.666266	0.264630
43	15	14	6	0.864739	0.198472
43	15	14	7	0.961950	0.097211
43	15	14	8	0.992830	0.030931
43	15	14	9	0.999156	0.006276
43	15	14	10	0.999941	0.000784
43	15	14	11	0.999998	0.000057
43	15	14	12	1.000000	0.000002
43	15	14	13	1.000000	0.000000
43	15	14	14	1.000000	0.000000
43	15	15	0	0.000247	0.000247
43	15	15	1	0.004218	0.003971
43	15	15	2	0.030163	0.025944
43	15	15	3	0.121509	0.091346
43	15	15	4	0.314947	0.193439
43	15	15	5	0.575014	0.260067
43	15	15	6	0.803144	0.228129
43	15	15	7	0.935133	0.131989
43	15	15	8	0.985414	0.050282
43	15	15	9	0.997858	0.012443
43	15	15	10	0.999805	0.001948
43	15	15	11	0.999990	0.000184
43	15	15	12	1.000000	0.000010
43	15	15	13	1.000000	0.000000
43	15	15	14	1.000000	0.000000
43	15	15	15	1.000000	0.000000
43	16	1	0	0.627907	0.627907
43	16	1	1	1.000000	0.372093

Table for $N = 2$, $n = 1$, through $N = 100$, $n = 50$

N	n	k	x	P(x)	p(x)
43	16	2	0	0.388704	0.388704
43	16	2	1	0.867110	0.478405
43	16	2	2	1.000000	0.132890
43	16	3	0	0.237015	0.237015
43	16	3	1	0.692083	0.455068
43	16	3	2	0.954623	0.262539
43	16	3	3	1.000000	0.045377
43	16	4	0	0.142209	0.142209
43	16	4	1	0.521433	0.379224
43	16	4	2	0.862734	0.341301
43	16	4	3	0.985252	0.122518
43	16	4	4	1.000000	0.014748
43	16	5	0	0.083867	0.083867
43	16	5	1	0.375577	0.291711
43	16	5	2	0.740215	0.364638
43	16	5	3	0.944413	0.204197
43	16	5	4	0.995462	0.051049
43	16	5	5	1.000000	0.004538
43	16	6	0	0.048554	0.048554
43	16	6	1	0.260428	0.211874
43	16	6	2	0.605875	0.345447
43	16	6	3	0.874556	0.268681
43	16	6	4	0.979341	0.104785
43	16	6	5	0.998686	0.019345
43	16	6	6	1.000000	0.001314
43	16	7	0	0.027558	0.027558
43	16	7	1	0.174534	0.146976
43	16	7	2	0.475166	0.300632
43	16	7	3	0.780155	0.304989
43	16	7	4	0.945357	0.165202
43	16	7	5	0.992935	0.047578
43	16	7	6	0.999645	0.006710
43	16	7	7	1.000000	0.000355
43	16	8	0	0.015310	0.015310
43	16	8	1	0.113294	0.097984
43	16	8	2	0.358253	0.244959
43	16	8	3	0.670020	0.311766
43	16	8	4	0.890289	0.220270
43	16	8	5	0.978397	0.088108
43	16	8	6	0.997781	0.019384
43	16	8	7	0.999911	0.002130
43	16	8	8	1.000000	0.000089
43	16	9	0	0.008311	0.008311
43	16	9	1	0.071301	0.062990
43	16	9	2	0.260269	0.188969
43	16	9	3	0.554221	0.293951
43	16	9	4	0.814768	0.260548
43	16	9	5	0.950706	0.135938
43	16	9	6	0.992243	0.041537
43	16	9	7	0.999363	0.007121

N	n	k	x	P(x)	p(x)
43	16	9	8	0.999980	0.000616
43	16	9	9	1.000000	0.000020
43	16	10	0	0.004400	0.004400
43	16	10	1	0.043511	0.039111
43	16	10	2	0.182459	0.138948
43	16	10	3	0.441827	0.259369
43	16	10	4	0.722810	0.280983
43	16	10	5	0.906726	0.183916
43	16	10	6	0.980026	0.073300
43	16	10	7	0.997479	0.017452
43	16	10	8	0.999835	0.002356
43	16	10	9	0.999996	0.000161
43	16	10	10	1.000000	0.000004
43	16	11	0	0.002267	0.002267
43	16	11	1	0.025733	0.023467
43	16	11	2	0.123511	0.097778
43	16	11	3	0.339652	0.216141
43	16	11	4	0.620635	0.280983
43	16	11	5	0.845421	0.224786
43	16	11	6	0.957814	0.112393
43	16	11	7	0.992719	0.034905
43	16	11	8	0.999263	0.006545
43	16	11	9	0.999962	0.000698
43	16	11	10	0.999999	0.000038
43	16	11	11	1.000000	0.000001
43	16	12	0	0.001133	0.001133
43	16	12	1	0.014733	0.013600
43	16	12	2	0.080733	0.066000
43	16	12	3	0.251845	0.171111
43	16	12	4	0.515266	0.263421
43	16	12	5	0.768151	0.252885
43	16	12	6	0.922691	0.154541
43	16	12	7	0.982902	0.060211
43	16	12	8	0.997627	0.014725
43	16	12	9	0.999809	0.002182
43	16	12	10	0.999992	0.000183
43	16	12	11	1.000000	0.000008
43	16	12	12	1.000000	0.000000
43	16	13	0	0.000548	0.000548
43	16	13	1	0.008153	0.007604
43	16	13	2	0.050927	0.042774
43	16	13	3	0.180088	0.129161
43	16	13	4	0.413297	0.233208
43	16	13	5	0.678418	0.265121
43	16	13	6	0.872839	0.194422
43	16	13	7	0.965421	0.092582
43	16	13	8	0.993827	0.028406
43	16	13	9	0.999316	0.005489
43	16	13	10	0.999957	0.000640
43	16	13	11	0.999999	0.000042

N	n	k	x	P(x)	p(x)
43	16	13	12	1.000000	0.000001
43	16	13	13	1.000000	0.000000
43	16	14	0	0.000256	0.000256
43	16	14	1	0.004351	0.004095
43	16	14	2	0.030966	0.026615
43	16	14	3	0.124118	0.093153
43	16	14	4	0.320013	0.195895
43	16	14	5	0.581207	0.261193
43	16	14	6	0.808032	0.226826
43	16	14	7	0.937647	0.129615
43	16	14	8	0.986252	0.048605
43	16	14	9	0.998035	0.011783
43	16	14	10	0.999829	0.001793
43	16	14	11	0.999992	0.000163
43	16	14	12	1.000000	0.000008
43	16	14	13	1.000000	0.000000
43	16	14	14	1.000000	0.000000
43	16	15	0	0.000115	0.000115
43	16	15	1	0.002233	0.002118
43	16	15	2	0.018117	0.015884
43	16	15	3	0.082360	0.064243
43	16	15	4	0.238953	0.156593
43	16	15	5	0.482133	0.243180
43	16	15	6	0.729816	0.247683
43	16	15	7	0.897422	0.167605
43	16	15	8	0.972844	0.075422
43	16	15	9	0.995191	0.022347
43	16	15	10	0.999458	0.004266
43	16	15	11	0.999963	0.000506
43	16	15	12	0.999999	0.000035
43	16	15	13	1.000000	0.000001
43	16	15	14	1.000000	0.000000
43	16	15	15	1.000000	0.000000
43	16	16	0	0.000049	0.000049
43	16	16	1	0.001098	0.001049
43	16	16	2	0.010175	0.009077
43	16	16	3	0.052533	0.042358
43	16	16	4	0.171842	0.119309
43	16	16	5	0.386598	0.214756
43	16	16	6	0.641358	0.254760
43	16	16	7	0.843548	0.202190
43	16	16	8	0.951295	0.107746
43	16	16	9	0.989604	0.038310
43	16	16	10	0.998543	0.008939
43	16	16	11	0.999873	0.001330
43	16	16	12	0.999994	0.000120
43	16	16	13	1.000000	0.000006
43	16	16	14	1.000000	0.000000
43	16	16	15	1.000000	0.000000
43	16	16	16	1.000000	0.000000

N	n	k	x	P(x)	p(x)
43	17	1	0	0.604651	0.604651
43	17	1	1	1.000000	0.395349
43	17	2	0	0.359911	0.359911
43	17	2	1	0.849391	0.489479
43	17	2	2	1.000000	0.150609
43	17	3	0	0.210680	0.210680
43	17	3	1	0.658375	0.447695
43	17	3	2	0.944899	0.286525
43	17	3	3	1.000000	0.055101
43	17	4	0	0.121141	0.121141
43	17	4	1	0.479297	0.358156
43	17	4	2	0.837452	0.358156
43	17	4	3	0.980715	0.143262
43	17	4	4	1.000000	0.019285
43	17	5	0	0.068336	0.068336
43	17	5	1	0.332361	0.264025
43	17	5	2	0.699700	0.367339
43	17	5	3	0.929287	0.229587
43	17	5	4	0.993572	0.064284
43	17	5	5	1.000000	0.006428
43	17	6	0	0.037765	0.037765
43	17	6	1	0.221192	0.183428
43	17	6	2	0.554698	0.333505
43	17	6	3	0.844702	0.290005
43	17	6	4	0.971579	0.126877
43	17	6	5	0.997970	0.026390
43	17	6	6	1.000000	0.002030
43	17	7	0	0.020413	0.020413
43	17	7	1	0.141872	0.121459
43	17	7	2	0.419493	0.277621
43	17	7	3	0.734971	0.315478
43	17	7	4	0.927001	0.192030
43	17	7	5	0.989411	0.062410
43	17	7	6	0.999396	0.009986
43	17	7	7	1.000000	0.000604
43	17	8	0	0.010774	0.010774
43	17	8	1	0.087891	0.077117
43	17	8	2	0.303818	0.215927
43	17	8	3	0.612285	0.308467
43	17	8	4	0.857657	0.245372
43	17	8	5	0.968608	0.110951
43	17	8	6	0.996345	0.027738
43	17	8	7	0.999832	0.003487
43	17	8	8	1.000000	0.000168
43	17	9	0	0.005541	0.005541
43	17	9	1	0.052637	0.047096
43	17	9	2	0.211277	0.158640
43	17	9	3	0.488898	0.277621
43	17	9	4	0.766519	0.277621
43	17	9	5	0.930567	0.164049

Table for $N = 2$, $n = 1$, through $N = 100$, $n = 50$

N	n	k	x	P(x)	p(x)
43	17	9	6	0.987628	0.057060
43	17	9	7	0.998836	0.011208
43	17	9	8	0.999957	0.001121
43	17	9	9	1.000000	0.000043
43	17	10	0	0.002770	0.002770
43	17	10	1	0.030474	0.027704
43	17	10	2	0.141289	0.110815
43	17	10	3	0.374584	0.233295
43	17	10	4	0.660370	0.285786
43	17	10	5	0.872668	0.212298
43	17	10	6	0.969167	0.096499
43	17	10	7	0.995539	0.026372
43	17	10	8	0.999660	0.004121
43	17	10	9	0.999990	0.000330
43	17	10	10	1.000000	0.000010
43	17	11	0	0.001343	0.001343
43	17	11	1	0.017042	0.015699
43	17	11	2	0.090919	0.073877
43	17	11	3	0.275610	0.184692
43	17	11	4	0.547787	0.272177
43	17	11	5	0.795468	0.247681
43	17	11	6	0.937001	0.141532
43	17	11	7	0.987548	0.050547
43	17	11	8	0.998536	0.010989
43	17	11	9	0.999910	0.001374
43	17	11	10	0.999998	0.000088
43	17	11	11	1.000000	0.000002
43	17	12	0	0.000630	0.000630
43	17	12	1	0.009193	0.008563
43	17	12	2	0.056289	0.047096
43	17	12	3	0.194808	0.138519
43	17	12	4	0.437215	0.242408
43	17	12	5	0.702588	0.265373
43	17	12	6	0.888349	0.185761
43	17	12	7	0.971752	0.083403
43	17	12	8	0.995446	0.023694
43	17	12	9	0.999566	0.004121
43	17	12	10	0.999978	0.000412
43	17	12	11	0.999999	0.000021
43	17	12	12	1.000000	0.000000
43	17	13	0	0.000284	0.000284
43	17	13	1	0.004773	0.004489
43	17	13	2	0.033500	0.028727
43	17	13	3	0.132251	0.098750
43	17	13	4	0.335561	0.203310
43	17	13	5	0.599863	0.264303
43	17	13	6	0.822434	0.222571
43	17	13	7	0.944848	0.122414
43	17	13	8	0.988567	0.043719
43	17	13	9	0.998503	0.009936
43	17	13	10	0.999885	0.001382
43	17	13	11	0.999995	0.000110
43	17	13	12	1.000000	0.000004
43	17	13	13	1.000000	0.000000
43	17	14	0	0.000123	0.000123
43	17	14	1	0.002379	0.002256
43	17	14	2	0.019137	0.016758
43	17	14	3	0.086167	0.067031
43	17	14	4	0.247460	0.161292
43	17	14	5	0.494142	0.246682
43	17	14	6	0.740825	0.246682
43	17	14	7	0.904043	0.163218
43	17	14	8	0.975451	0.071408
43	17	14	9	0.995853	0.020402
43	17	14	10	0.999563	0.003710
43	17	14	11	0.999973	0.000411
43	17	14	12	0.999999	0.000026
43	17	14	13	1.000000	0.000001
43	17	14	14	1.000000	0.000000
43	17	15	0	0.000051	0.000051
43	17	15	1	0.001134	0.001083
43	17	15	2	0.010469	0.009334
43	17	15	3	0.053808	0.043339
43	17	15	4	0.175156	0.121348
43	17	15	5	0.392067	0.216910
43	17	15	6	0.647255	0.255189
43	17	15	7	0.847761	0.200505
43	17	15	8	0.953290	0.105529
43	17	15	9	0.990225	0.036935
43	17	15	10	0.998667	0.008442
43	17	15	11	0.999888	0.001221
43	17	15	12	0.999995	0.000106
43	17	15	13	1.000000	0.000005
43	17	15	14	1.000000	0.000000
43	17	15	15	1.000000	0.000000
43	17	16	0	0.000020	0.000020
43	17	16	1	0.000515	0.000495
43	17	16	2	0.005468	0.004953
43	17	16	3	0.032138	0.026670
43	17	16	4	0.118816	0.086677
43	17	16	5	0.299105	0.180289
43	17	16	6	0.547003	0.247898
43	17	16	7	0.776152	0.229149
43	17	16	8	0.919370	0.143218
43	17	16	9	0.979672	0.060302
43	17	16	10	0.996557	0.016885
43	17	16	11	0.999627	0.003070
43	17	16	12	0.999976	0.000349
43	17	16	13	0.999999	0.000023
43	17	16	14	1.000000	0.000001
43	17	16	15	1.000000	0.000000
43	17	16	16	1.000000	0.000000
43	17	17	0	0.000007	0.000007
43	17	17	1	0.000222	0.000214
43	17	17	2	0.002717	0.002495
43	17	17	3	0.018309	0.015593
43	17	17	4	0.077082	0.058773
43	17	17	5	0.218976	0.141894
43	17	17	6	0.446007	0.227031
43	17	17	7	0.691282	0.245274
43	17	17	8	0.871630	0.180349
43	17	17	9	0.961805	0.090174
43	17	17	10	0.992179	0.030375
43	17	17	11	0.998945	0.006765
43	17	17	12	0.999911	0.000966
43	17	17	13	0.999996	0.000084
43	17	17	14	1.000000	0.000004
43	17	17	15	1.000000	0.000000
43	17	17	16	1.000000	0.000000
43	17	17	17	1.000000	0.000000
43	18	1	0	0.581395	0.581395
43	18	1	1	1.000000	0.418605
43	18	2	0	0.332226	0.332226
43	18	2	1	0.830565	0.498339
43	18	2	2	1.000000	0.169435
43	18	3	0	0.186371	0.186371
43	18	3	1	0.623936	0.437566
43	18	3	2	0.933879	0.309942
43	18	3	3	1.000000	0.066121
43	18	4	0	0.102504	0.102504
43	18	4	1	0.437971	0.335467
43	18	4	2	0.809902	0.371931
43	18	4	3	0.975205	0.165303
43	18	4	4	1.000000	0.024795
43	18	5	0	0.055194	0.055194
43	18	5	1	0.291742	0.236547
43	18	5	2	0.657315	0.365573
43	18	5	3	0.911627	0.254312
43	18	5	4	0.991099	0.079472
43	18	5	5	1.000000	0.008901
43	18	6	0	0.029050	0.029050
43	18	6	1	0.185918	0.156868
43	18	6	2	0.503389	0.317471
43	18	6	3	0.811240	0.307851
43	18	6	4	0.961820	0.150579
43	18	6	5	0.996955	0.035135
43	18	6	6	1.000000	0.003045
43	18	7	0	0.014917	0.014917
43	18	7	1	0.113843	0.098926
43	18	7	2	0.366104	0.252261
43	18	7	3	0.686436	0.320332
43	18	7	4	0.904844	0.218408
43	18	7	5	0.984610	0.079766
43	18	7	6	0.999012	0.014402
43	18	7	7	1.000000	0.000988
43	18	8	0	0.007459	0.007459
43	18	8	1	0.067128	0.059670
43	18	8	2	0.253988	0.186860
43	18	8	3	0.552964	0.298976
43	18	8	4	0.819907	0.266943
43	18	8	5	0.955806	0.135898
43	18	8	6	0.994212	0.038406
43	18	8	7	0.999698	0.005487
43	18	8	8	1.000000	0.000302
43	18	9	0	0.003623	0.003623
43	18	9	1	0.038146	0.034523
43	18	9	2	0.168567	0.130421
43	18	9	3	0.424832	0.256265
43	18	9	4	0.713130	0.288298
43	18	9	5	0.905329	0.192199
43	18	9	6	0.981044	0.075715
43	18	9	7	0.997974	0.016930
43	18	9	8	0.999914	0.001940
43	18	9	9	1.000000	0.000086
43	18	10	0	0.001705	0.001705
43	18	10	1	0.020884	0.019180
43	18	10	2	0.107192	0.086308
43	18	10	3	0.311774	0.204581
43	18	10	4	0.594419	0.282645
43	18	10	5	0.831841	0.237422
43	18	10	6	0.954321	0.122480
43	18	10	7	0.992496	0.038175
43	18	10	8	0.999343	0.006847
43	18	10	9	0.999977	0.000634
43	18	10	10	1.000000	0.000023
43	18	11	0	0.000775	0.000775
43	18	11	1	0.011004	0.010229
43	18	11	2	0.065346	0.054342
43	18	11	3	0.218782	0.153436
43	18	11	4	0.474509	0.255727
43	18	11	5	0.738311	0.263802
43	18	11	6	0.909783	0.171472
43	18	11	7	0.979771	0.069988
43	18	11	8	0.997268	0.017497
43	18	11	9	0.999804	0.002536
43	18	11	10	0.999994	0.000190
43	18	11	11	1.000000	0.000006
43	18	12	0	0.000339	0.000339
43	18	12	1	0.005570	0.005231
43	18	12	2	0.038175	0.032605

N	n	k	x	P(x)	p(x)
43	18	12	3	0.146859	0.108684
43	18	12	4	0.362628	0.215770
43	18	12	5	0.631142	0.268513
43	18	12	6	0.845481	0.214339
43	18	12	7	0.955713	0.110232
43	18	12	8	0.991801	0.036088
43	18	12	9	0.999091	0.007290
43	18	12	10	0.999947	0.000856
43	18	12	11	0.999999	0.000052
43	18	12	12	1.000000	0.000001
43	18	13	0	0.000142	0.000142
43	18	13	1	0.002701	0.002559
43	18	13	2	0.021347	0.018645
43	18	13	3	0.094270	0.072923
43	18	13	4	0.265184	0.170914
43	18	13	5	0.518539	0.253355
43	18	13	6	0.762511	0.243972
43	18	13	7	0.916598	0.154087
43	18	13	8	0.980159	0.063561
43	18	13	9	0.996974	0.016815
43	18	13	10	0.999726	0.002752
43	18	13	11	0.999987	0.000261
43	18	13	12	1.000000	0.000013
43	18	13	13	1.000000	0.000000
43	18	14	0	0.000057	0.000057
43	18	14	1	0.001251	0.001194
43	18	14	2	0.011402	0.010151
43	18	14	3	0.057808	0.046406
43	18	14	4	0.185424	0.127616
43	18	14	5	0.408752	0.223328
43	18	14	6	0.664922	0.256170
43	18	14	7	0.860100	0.195177
43	18	14	8	0.958972	0.098873
43	18	14	9	0.991930	0.032958
43	18	14	10	0.998992	0.007062
43	18	14	11	0.999926	0.000934
43	18	14	12	0.999997	0.000071
43	18	14	13	1.000000	0.000003
43	18	14	14	1.000000	0.000000
43	18	15	0	0.000022	0.000022
43	18	15	1	0.000551	0.000529
43	18	15	2	0.005802	0.005251
43	18	15	3	0.033805	0.028003
43	18	15	4	0.123816	0.090011
43	18	15	5	0.308640	0.184823
43	18	15	6	0.558921	0.250281
43	18	15	7	0.786067	0.227146
43	18	15	8	0.924878	0.138811
43	18	15	9	0.981702	0.056823
43	18	15	10	0.997044	0.015342

N	n	k	x	P(x)	p(x)
43	18	15	11	0.999701	0.002657
43	18	15	12	0.999982	0.000282
43	18	15	13	0.999999	0.000017
43	18	15	14	1.000000	0.000000
43	18	15	15	1.000000	0.000000
43	18	16	0	0.000008	0.000008
43	18	16	1	0.000230	0.000222
43	18	16	2	0.002801	0.002572
43	18	16	3	0.018803	0.016002
43	18	16	4	0.078811	0.060007
43	18	16	5	0.222829	0.144018
43	18	16	6	0.451657	0.228829
43	18	16	7	0.696831	0.245174
43	18	16	8	0.875303	0.178472
43	18	16	9	0.963437	0.088134
43	18	16	10	0.992660	0.029223
43	18	16	11	0.999036	0.006376
43	18	16	12	0.999922	0.000886
43	18	16	13	0.999996	0.000074
43	18	16	14	1.000000	0.000003
43	18	16	15	1.000000	0.000000
43	18	16	16	1.000000	0.000000
43	18	17	0	0.000003	0.000003
43	18	17	1	0.000090	0.000087
43	18	17	2	0.001277	0.001187
43	18	17	3	0.009913	0.008636
43	18	17	4	0.047696	0.037782
43	18	17	5	0.153487	0.105791
43	18	17	6	0.349956	0.196469
43	18	17	7	0.596945	0.246990
43	18	17	8	0.809202	0.212257
43	18	17	9	0.934059	0.124857
43	18	17	10	0.984002	0.049943
43	18	17	11	0.997383	0.013382
43	18	17	12	0.999725	0.002342
43	18	17	13	0.999983	0.000257
43	18	17	14	0.999999	0.000017
43	18	17	15	1.000000	0.000001
43	18	17	16	1.000000	0.000000
43	18	17	17	1.000000	0.000000
43	18	18	0	0.000001	0.000001
43	18	18	1	0.000033	0.000032
43	18	18	2	0.000547	0.000514
43	18	18	3	0.004931	0.004384
43	18	18	4	0.027351	0.022420
43	18	18	5	0.100591	0.073240
43	18	18	6	0.259278	0.158686
43	18	18	7	0.492450	0.233172
43	18	18	8	0.727565	0.235115
43	18	18	9	0.890839	0.163274

N	n	k	x	P(x)	p(x)
43	18	18	10	0.968635	0.077795
43	18	18	11	0.993781	0.025146
43	18	18	12	0.999185	0.005404
43	18	18	13	0.999933	0.000748
43	18	18	14	0.999997	0.000064
43	18	18	15	1.000000	0.000003
43	18	18	16	1.000000	0.000000
43	18	18	17	1.000000	0.000000
43	18	18	18	1.000000	0.000000
43	19	1	0	0.558140	0.558140
43	19	1	1	1.000000	0.441860
43	19	2	0	0.305648	0.305648
43	19	2	1	0.810631	0.504983
43	19	2	2	1.000000	0.189369
43	19	3	0	0.164006	0.164006
43	19	3	1	0.588931	0.424925
43	19	3	2	0.921481	0.332550
43	19	3	3	1.000000	0.078519
43	19	4	0	0.086103	0.086103
43	19	4	1	0.397715	0.311612
43	19	4	2	0.780147	0.382433
43	19	4	3	0.968592	0.188445
43	19	4	4	1.000000	0.031407
43	19	5	0	0.044156	0.044156
43	19	5	1	0.253894	0.209739
43	19	5	2	0.613446	0.359552
43	19	5	3	0.891282	0.277836
43	19	5	4	0.987920	0.096638
43	19	5	5	1.000000	0.012080
43	19	6	0	0.022078	0.022078
43	19	6	1	0.154544	0.132466
43	19	6	2	0.452594	0.298050
43	19	6	3	0.774298	0.321704
43	19	6	4	0.949773	0.175475
43	19	6	5	0.995549	0.045776
43	19	6	6	1.000000	0.004450
43	19	7	0	0.010741	0.010741
43	19	7	1	0.090101	0.079361
43	19	7	2	0.315652	0.225551
43	19	7	3	0.635183	0.319531
43	19	7	4	0.878635	0.243452
43	19	7	5	0.978229	0.099594
43	19	7	6	0.998436	0.020207
43	19	7	7	1.000000	0.001564
43	19	8	0	0.005072	0.005072
43	19	8	1	0.050421	0.045349
43	19	8	2	0.209142	0.158721
43	19	8	3	0.493169	0.284027
43	19	8	4	0.777196	0.284027
43	19	8	5	0.939498	0.162301

N	n	k	x	P(x)	p(x)
43	19	8	6	0.991139	0.051641
43	19	8	7	0.999479	0.008340
43	19	8	8	1.000000	0.000521
43	19	9	0	0.002319	0.002319
43	19	9	1	0.027099	0.024780
43	19	9	2	0.132049	0.104950
43	19	9	3	0.363328	0.231279
43	19	9	4	0.655471	0.292142
43	19	9	5	0.874577	0.219107
43	19	9	6	0.971958	0.097381
43	19	9	7	0.996619	0.024661
43	19	9	8	0.999836	0.003217
43	19	9	9	1.000000	0.000164
43	19	10	0	0.001023	0.001023
43	19	10	1	0.013980	0.012957
43	19	10	2	0.079574	0.065594
43	19	10	3	0.254491	0.174917
43	19	10	4	0.526584	0.272093
43	19	10	5	0.784357	0.257773
43	19	10	6	0.934724	0.150367
43	19	10	7	0.987915	0.053191
43	19	10	8	0.998795	0.010880
43	19	10	9	0.999952	0.001156
43	19	10	10	1.000000	0.000048
43	19	11	0	0.000434	0.000434
43	19	11	1	0.006912	0.006478
43	19	11	2	0.045783	0.038870
43	19	11	3	0.169682	0.123900
43	19	11	4	0.402905	0.233223
43	19	11	5	0.674999	0.272093
43	19	11	6	0.875489	0.200490
43	19	11	7	0.968573	0.093085
43	19	11	8	0.995169	0.026596
43	19	11	9	0.999601	0.004433
43	19	11	10	0.999987	0.000385
43	19	11	11	1.000000	0.000013
43	19	12	0	0.000176	0.000176
43	19	12	1	0.003268	0.003092
43	19	12	2	0.025133	0.021865
43	19	12	3	0.107733	0.082600
43	19	12	4	0.293582	0.185849
43	19	12	5	0.555958	0.262376
43	19	12	6	0.794040	0.238082
43	19	12	7	0.933666	0.139627
43	19	12	8	0.986026	0.052360
43	19	12	9	0.998216	0.012190
43	19	12	10	0.999878	0.001662
43	19	12	11	0.999997	0.000118
43	19	12	12	1.000000	0.000003
43	19	13	0	0.000068	0.000068

Table for $N = 2$, $n = 1$, through $N = 100$, $n = 50$

N	n	k	x	P(x)	p(x)
43	19	13	1	0.001473	0.001405
43	19	13	2	0.013143	0.011670
43	19	13	3	0.065101	0.051958
43	19	13	4	0.203655	0.138554
43	19	13	5	0.437466	0.233811
43	19	13	6	0.694199	0.256733
43	19	13	7	0.879617	0.185418
43	19	13	8	0.967447	0.087830
43	19	13	9	0.994284	0.026837
43	19	13	10	0.999396	0.005112
43	19	13	11	0.999966	0.000570
43	19	13	12	0.999999	0.000033
43	19	13	13	1.000000	0.000001
43	19	14	0	0.000025	0.000025
43	19	14	1	0.000630	0.000605
43	19	14	2	0.006530	0.005900
43	19	14	3	0.037390	0.030860
43	19	14	4	0.134378	0.096988
43	19	14	5	0.328354	0.193976
43	19	14	6	0.582948	0.254594
43	19	14	7	0.805450	0.222502
43	19	14	8	0.935243	0.129793
43	19	14	9	0.985338	0.050096
43	19	14	10	0.997862	0.012524
43	19	14	11	0.999814	0.001952
43	19	14	12	0.999991	0.000177
43	19	14	13	1.000000	0.000008
43	19	14	14	1.000000	0.000000
43	19	15	0	0.000009	0.000009
43	19	15	1	0.000255	0.000246
43	19	15	2	0.003071	0.002817
43	19	15	3	0.020364	0.017292
43	19	15	4	0.084211	0.063848
43	19	15	5	0.234710	0.150499
43	19	15	6	0.468820	0.234109
43	19	15	7	0.713380	0.244561
43	19	15	8	0.886011	0.172631
43	19	15	9	0.968064	0.082053
43	19	15	10	0.993975	0.025911
43	19	15	11	0.999276	0.005300
43	19	15	12	0.999949	0.000673
43	19	15	13	0.999998	0.000049
43	19	15	14	1.000000	0.000002
43	19	15	15	1.000000	0.000000
43	19	16	0	0.000003	0.000003
43	19	16	1	0.000096	0.000094
43	19	16	2	0.001361	0.001265
43	19	16	3	0.010482	0.009121
43	19	16	4	0.050007	0.039525
43	19	16	5	0.159461	0.109454

N	n	k	x	P(x)	p(x)
43	19	16	6	0.360126	0.200665
43	19	16	7	0.608568	0.248442
43	19	16	8	0.818192	0.209623
43	19	16	9	0.938759	0.120568
43	19	16	10	0.985647	0.046887
43	19	16	11	0.997761	0.012114
43	19	16	12	0.999780	0.002019
43	19	16	13	0.999987	0.000207
43	19	16	14	1.000000	0.000012
43	19	16	15	1.000000	0.000000
43	19	16	16	1.000000	0.000000
43	19	17	0	0.000001	0.000001
43	19	17	1	0.000034	0.000033
43	19	17	2	0.000565	0.000531
43	19	17	3	0.005077	0.004512
43	19	17	4	0.028049	0.022972
43	19	17	5	0.102707	0.074658
43	19	17	6	0.263509	0.160802
43	19	17	7	0.498150	0.234640
43	19	17	8	0.732790	0.234640
43	19	17	9	0.894105	0.161315
43	19	17	10	0.970018	0.075913
43	19	17	11	0.994172	0.024154
43	19	17	12	0.999257	0.005085
43	19	17	13	0.999941	0.000685
43	19	17	14	0.999997	0.000056
43	19	17	15	1.000000	0.000003
43	19	17	16	1.000000	0.000000
43	19	17	17	1.000000	0.000000
43	19	18	0	0.000000	0.000000
43	19	18	1	0.000011	0.000011
43	19	18	2	0.000218	0.000207
43	19	18	3	0.002300	0.002083
43	19	18	4	0.014796	0.012496
43	19	18	5	0.062507	0.047711
43	19	18	6	0.183108	0.120602
43	19	18	7	0.389854	0.206746
43	19	18	8	0.633519	0.243665
43	19	18	9	0.832060	0.198542
43	19	18	10	0.943740	0.111680
43	19	18	11	0.986740	0.043000
43	19	18	12	0.997888	0.011148
43	19	18	13	0.999783	0.001896
43	19	18	14	0.999987	0.000203
43	19	18	15	0.999999	0.000013
43	19	18	16	1.000000	0.000000
43	19	18	17	1.000000	0.000000
43	19	18	18	1.000000	0.000000
43	19	19	0	0.000000	0.000000
43	19	19	1	0.000003	0.000003

N	n	k	x	P(x)	p(x)
43	19	19	2	0.000077	0.000074
43	19	19	3	0.000967	0.000890
43	19	19	4	0.007299	0.006331
43	19	19	5	0.035789	0.028490
43	19	19	6	0.120395	0.084607
43	19	19	7	0.290616	0.170221
43	19	19	8	0.526306	0.235690
43	19	19	9	0.752644	0.226337
43	19	19	10	0.903535	0.150892
43	19	19	11	0.972980	0.069444
43	19	19	12	0.994766	0.021786
43	19	19	13	0.999328	0.004562
43	19	19	14	0.999946	0.000617
43	19	19	15	0.999997	0.000051
43	19	19	16	1.000000	0.000002
43	19	19	17	1.000000	0.000000
43	19	19	18	1.000000	0.000000
43	19	19	19	1.000000	0.000000
43	20	1	0	0.534884	0.534884
43	20	1	1	1.000000	0.465116
43	20	2	0	0.280177	0.280177
43	20	2	1	0.789590	0.509413
43	20	2	2	1.000000	0.210410
43	20	3	0	0.143505	0.143505
43	20	3	1	0.553521	0.410015
43	20	3	2	0.907625	0.354104
43	20	3	3	1.000000	0.092375
43	20	4	0	0.071753	0.071753
43	20	4	1	0.358763	0.287011
43	20	4	2	0.748278	0.389515
43	20	4	3	0.960741	0.212463
43	20	4	4	1.000000	0.039259
43	20	5	0	0.034956	0.034956
43	20	5	1	0.218938	0.183981
43	20	5	2	0.568502	0.349564
43	20	5	3	0.868129	0.299627
43	20	5	4	0.983893	0.115765
43	20	5	5	1.000000	0.016106
43	20	6	0	0.016558	0.016558
43	20	6	1	0.126947	0.110389
43	20	6	2	0.402919	0.275972
43	20	6	3	0.734085	0.331166
43	20	6	4	0.935150	0.201065
43	20	6	5	0.993642	0.058492
43	20	6	6	1.000000	0.006358
43	20	7	0	0.007608	0.007608
43	20	7	1	0.070261	0.062653
43	20	7	2	0.268662	0.198401
43	20	7	3	0.581928	0.313265
43	20	7	4	0.848203	0.266276

N	n	k	x	P(x)	p(x)
43	20	7	5	0.969929	0.121726
43	20	7	6	0.997594	0.027665
43	20	7	7	1.000000	0.002406
43	20	8	0	0.003381	0.003381
43	20	8	1	0.037194	0.033813
43	20	8	2	0.169462	0.132268
43	20	8	3	0.433997	0.264535
43	20	8	4	0.729859	0.295862
43	20	8	5	0.919210	0.189351
43	20	8	6	0.986836	0.067626
43	20	8	7	0.999131	0.012296
43	20	8	8	1.000000	0.000869
43	20	9	0	0.001449	0.001449
43	20	9	1	0.018839	0.017389
43	20	9	2	0.101438	0.082600
43	20	9	3	0.305508	0.204070
43	20	9	4	0.594608	0.289099
43	20	9	5	0.838059	0.243452
43	20	9	6	0.959785	0.121726
43	20	9	7	0.994564	0.034779
43	20	9	8	0.999702	0.005138
43	20	9	9	1.000000	0.000298
43	20	10	0	0.000597	0.000597
43	20	10	1	0.009121	0.008524
43	20	10	2	0.057709	0.048588
43	20	10	3	0.203473	0.145764
43	20	10	4	0.458561	0.255088
43	20	10	5	0.730654	0.272093
43	20	10	6	0.909663	0.179009
43	20	10	7	0.981266	0.071603
43	20	10	8	0.997889	0.016622
43	20	10	9	0.999904	0.002015
43	20	10	10	1.000000	0.000096
43	20	11	0	0.000235	0.000235
43	20	11	1	0.004213	0.003978
43	20	11	2	0.031206	0.026993
43	20	11	3	0.128383	0.097176
43	20	11	4	0.334882	0.206499
43	20	11	5	0.606975	0.272093
43	20	11	6	0.833720	0.226744
43	20	11	7	0.953059	0.119339
43	20	11	8	0.991844	0.038785
43	20	11	9	0.999232	0.007388
43	20	11	10	0.999971	0.000739
43	20	11	11	1.000000	0.000029
43	20	12	0	0.000088	0.000088
43	20	12	1	0.001851	0.001763
43	20	12	2	0.016023	0.014172
43	20	12	3	0.076758	0.060735
43	20	12	4	0.231632	0.154875

Table for $N = 2$, $n = 1$, through $N = 100$, $n = 50$

N	n	k	x	P(x)	p(x)
43	20	12	5	0.479432	0.247799
43	20	12	6	0.734519	0.255088
43	20	12	7	0.904578	0.170058
43	20	12	8	0.977300	0.072722
43	20	12	9	0.996692	0.019393
43	20	12	10	0.999740	0.003047
43	20	12	11	0.999992	0.000252
43	20	12	12	1.000000	0.000008
43	20	13	0	0.000031	0.000031
43	20	13	1	0.000771	0.000739
43	20	13	2	0.007794	0.007023
43	20	13	3	0.043451	0.035657
43	20	13	4	0.151697	0.108246
43	20	13	5	0.359529	0.207832
43	20	13	6	0.619318	0.259790
43	20	13	7	0.833263	0.213944
43	20	13	8	0.949149	0.115887
43	20	13	9	0.989811	0.040662
43	20	13	10	0.998757	0.008946
43	20	13	11	0.999919	0.001162
43	20	13	12	0.999998	0.000079
43	20	13	13	1.000000	0.000002
43	20	14	0	0.000010	0.000010
43	20	14	1	0.000302	0.000292
43	20	14	2	0.003580	0.003278
43	20	14	3	0.023246	0.019666
43	20	14	4	0.093966	0.070721
43	20	14	5	0.255613	0.161647
43	20	14	6	0.498083	0.242470
43	20	14	7	0.740553	0.242470
43	20	14	8	0.902795	0.162241
43	20	14	9	0.974902	0.072107
43	20	14	10	0.995775	0.020873
43	20	14	11	0.999570	0.003795
43	20	14	12	0.999977	0.000407
43	20	14	13	0.999999	0.000023
43	20	14	14	1.000000	0.000000
43	20	15	0	0.000003	0.000003
43	20	15	1	0.000111	0.000108
43	20	15	2	0.001546	0.001434
43	20	15	3	0.011717	0.010172
43	20	15	4	0.054948	0.043230
43	20	15	5	0.172002	0.117055
43	20	15	6	0.381029	0.209026
43	20	15	7	0.631860	0.250831
43	20	15	8	0.835660	0.203800
43	20	15	9	0.947551	0.111890
43	20	15	10	0.988577	0.041026
43	20	15	11	0.998392	0.009815
43	20	15	12	0.999864	0.001472

N	n	k	x	P(x)	p(x)
43	20	15	13	0.999994	0.000129
43	20	15	14	1.000000	0.000006
43	20	15	15	1.000000	0.000000
43	20	16	0	0.000001	0.000001
43	20	16	1	0.000038	0.000037
43	20	16	2	0.000623	0.000586
43	20	16	3	0.005542	0.004918
43	20	16	4	0.030245	0.024703
43	20	16	5	0.109295	0.079050
43	20	16	6	0.276516	0.167221
43	20	16	7	0.515402	0.238887
43	20	16	8	0.748317	0.232915
43	20	16	9	0.903594	0.155277
43	20	16	10	0.973925	0.070331
43	20	16	11	0.995237	0.021312
43	20	16	12	0.999444	0.004206
43	20	16	13	0.999962	0.000518
43	20	16	14	0.999998	0.000037
43	20	16	15	1.000000	0.000001
43	20	16	16	1.000000	0.000000
43	20	17	0	0.000000	0.000000
43	20	17	1	0.000012	0.000012
43	20	17	2	0.000233	0.000221
43	20	17	3	0.002445	0.002212
43	20	17	4	0.015606	0.013161
43	20	17	5	0.065378	0.049772
43	20	17	6	0.189808	0.124430
43	20	17	7	0.400383	0.210574
43	20	17	8	0.644800	0.244417
43	20	17	9	0.840333	0.195533
43	20	17	10	0.947876	0.107543
43	20	17	11	0.988133	0.040257
43	20	17	12	0.998198	0.010064
43	20	17	13	0.999827	0.001630
43	20	17	14	0.999990	0.000163
43	20	17	15	1.000000	0.000009
43	20	17	16	1.000000	0.000000
43	20	17	17	1.000000	0.000000
43	20	18	0	0.000000	0.000000
43	20	18	1	0.000003	0.000003
43	20	18	2	0.000080	0.000077
43	20	18	3	0.000999	0.000919
43	20	18	4	0.007507	0.006508
43	20	18	5	0.036663	0.029156
43	20	18	6	0.122807	0.086144
43	20	18	7	0.295096	0.172288
43	20	18	8	0.531992	0.236896
43	20	18	9	0.757607	0.225615
43	20	18	10	0.906514	0.148906
43	20	18	11	0.974198	0.067685

N	n	k	x	P(x)	p(x)
43	20	18	12	0.995101	0.020903
43	20	18	13	0.999389	0.004288
43	20	18	14	0.999953	0.000564
43	20	18	15	0.999998	0.000045
43	20	18	16	1.000000	0.000002
43	20	18	17	1.000000	0.000000
43	20	18	18	1.000000	0.000000
43	20	19	0	0.000000	0.000000
43	20	19	1	0.000001	0.000001
43	20	19	2	0.000025	0.000024
43	20	19	3	0.000374	0.000349
43	20	19	4	0.003342	0.002968
43	20	19	5	0.019169	0.015828
43	20	19	6	0.074567	0.055397
43	20	19	7	0.205506	0.130939
43	20	19	8	0.418282	0.212776
43	20	19	9	0.658337	0.240055
43	20	19	10	0.846951	0.188615
43	20	19	11	0.949832	0.102881
43	20	19	12	0.988412	0.038580
43	20	19	13	0.998188	0.009776
43	20	19	14	0.999817	0.001629
43	20	19	15	0.999989	0.000172
43	20	19	16	0.999999	0.000011
43	20	19	17	1.000000	0.000000
43	20	19	18	1.000000	0.000000
43	20	19	19	1.000000	0.000000
43	20	20	0	0.000000	0.000000
43	20	20	1	0.000000	0.000000
43	20	20	2	0.000007	0.000007
43	20	20	3	0.000127	0.000120
43	20	20	4	0.001363	0.001237
43	20	20	5	0.009277	0.007914
43	20	20	6	0.042252	0.032975
43	20	20	7	0.134580	0.092329
43	20	20	8	0.311894	0.177313
43	20	20	9	0.548311	0.236418
43	20	20	10	0.768362	0.220050
43	20	20	11	0.911252	0.142890
43	20	20	12	0.975552	0.064300
43	20	20	13	0.995337	0.019785
43	20	20	14	0.999410	0.004073
43	20	20	15	0.999953	0.000543
43	20	20	16	0.999998	0.000045
43	20	20	17	1.000000	0.000002
43	20	20	18	1.000000	0.000000
43	20	20	19	1.000000	0.000000
43	20	20	20	1.000000	0.000000
43	21	1	0	0.511628	0.511628
43	21	1	1	1.000000	0.488372

N	n	k	x	P(x)	p(x)
43	21	2	0	0.255814	0.255814
43	21	2	1	0.767442	0.511628
43	21	2	2	1.000000	0.232558
43	21	3	0	0.124787	0.124787
43	21	3	1	0.517867	0.393080
43	21	3	2	0.892229	0.374362
43	21	3	3	1.000000	0.107771
43	21	4	0	0.059274	0.059274
43	21	4	1	0.321327	0.262053
43	21	4	2	0.714407	0.393080
43	21	4	3	0.951503	0.237096
43	21	4	4	1.000000	0.048497
43	21	5	0	0.027357	0.027357
43	21	5	1	0.186941	0.159584
43	21	5	2	0.522907	0.335966
43	21	5	3	0.842074	0.319167
43	21	5	4	0.978860	0.136786
43	21	5	5	1.000000	0.021140
43	21	6	0	0.012239	0.012239
43	21	6	1	0.102950	0.090711
43	21	6	2	0.354924	0.251974
43	21	6	3	0.690890	0.335966
43	21	6	4	0.917667	0.226777
43	21	6	5	0.991099	0.073433
43	21	6	6	1.000000	0.008901
43	21	7	0	0.005292	0.005292
43	21	7	1	0.053917	0.048624
43	21	7	2	0.225532	0.171615
43	21	7	3	0.527447	0.301915
43	21	7	4	0.813472	0.286025
43	21	7	5	0.959344	0.145873
43	21	7	6	0.996391	0.037047
43	21	7	7	1.000000	0.003608
43	21	8	0	0.002205	0.002205
43	21	8	1	0.026903	0.024698
43	21	8	2	0.134957	0.108054
43	21	8	3	0.376489	0.241532
43	21	8	4	0.678404	0.301915
43	21	8	5	0.894512	0.216108
43	21	8	6	0.980955	0.086443
43	21	8	7	0.998597	0.017641
43	21	8	8	1.000000	0.001403
43	21	9	0	0.000882	0.000882
43	21	9	1	0.012790	0.011908
43	21	9	2	0.076299	0.063509
43	21	9	3	0.252273	0.175973
43	21	9	4	0.531760	0.279487
43	21	9	5	0.795720	0.263960
43	21	9	6	0.943908	0.148188
43	21	9	7	0.991540	0.047632

N	n	k	x	P(x)	p(x)
43	21	9	8	0.999479	0.007939
43	21	9	9	1.000000	0.000521
43	21	10	0	0.000337	0.000337
43	21	10	1	0.005785	0.005448
43	21	10	2	0.040809	0.035023
43	21	10	3	0.159110	0.118301
43	21	10	4	0.392016	0.232906
43	21	10	5	0.671503	0.279487
43	21	10	6	0.878531	0.207028
43	21	10	7	0.971927	0.093396
43	21	10	8	0.996443	0.024516
43	21	10	9	0.999816	0.003373
43	21	10	10	1.000000	0.000184
43	21	11	0	0.000123	0.000123
43	21	11	1	0.002483	0.002361
43	21	11	2	0.020644	0.018160
43	21	11	3	0.094582	0.073938
43	21	11	4	0.272034	0.177452
43	21	11	5	0.535995	0.263960
43	21	11	6	0.784428	0.248433
43	21	11	7	0.932304	0.147877
43	21	11	8	0.986785	0.054481
43	21	11	9	0.998590	0.011804
43	21	11	10	0.999939	0.001349
43	21	11	11	1.000000	0.000061
43	21	12	0	0.000042	0.000042
43	21	12	1	0.001008	0.000966
43	21	12	2	0.009861	0.008853
43	21	12	3	0.052992	0.043131
43	21	12	4	0.177763	0.124771
43	21	12	5	0.404014	0.226252
43	21	12	6	0.667975	0.263960
43	21	12	7	0.867608	0.199634
43	21	12	8	0.964652	0.097044
43	21	12	9	0.994163	0.029511
43	21	12	10	0.999475	0.005312
43	21	12	11	0.999981	0.000506
43	21	12	12	1.000000	0.000019
43	21	13	0	0.000014	0.000014
43	21	13	1	0.000385	0.000371
43	21	13	2	0.004435	0.004050
43	21	13	3	0.027948	0.023513
43	21	13	4	0.109340	0.081392
43	21	13	5	0.287239	0.177899
43	21	13	6	0.540252	0.253012
43	21	13	7	0.777451	0.237199
43	21	13	8	0.923957	0.146505
43	21	13	9	0.982740	0.058783
43	21	13	10	0.997590	0.014850
43	21	13	11	0.999818	0.002228
43	21	13	12	0.999994	0.000177
43	21	13	13	1.000000	0.000006
43	21	14	0	0.000004	0.000004
43	21	14	1	0.000137	0.000133
43	21	14	2	0.001870	0.001733
43	21	14	3	0.013840	0.011970
43	21	14	4	0.063218	0.049378
43	21	14	5	0.192360	0.129142
43	21	14	6	0.413746	0.221386
43	21	14	7	0.666758	0.253012
43	21	14	8	0.860471	0.193713
43	21	14	9	0.959226	0.098755
43	21	14	10	0.992145	0.032918
43	21	14	11	0.999075	0.006930
43	21	14	12	0.999941	0.000866
43	21	14	13	0.999998	0.000057
43	21	14	14	1.000000	0.000001
43	21	15	0	0.000001	0.000001
43	21	15	1	0.000045	0.000044
43	21	15	2	0.000735	0.000689
43	21	15	3	0.006410	0.005676
43	21	15	4	0.034272	0.027862
43	21	15	5	0.121109	0.086837
43	21	15	6	0.299236	0.178127
43	21	15	7	0.544614	0.245378
43	21	15	8	0.773634	0.229020
43	21	15	9	0.918362	0.144728
43	21	15	10	0.979658	0.061297
43	21	15	11	0.996685	0.017027
43	21	15	12	0.999672	0.002987
43	21	15	13	0.999983	0.000310
43	21	15	14	0.999999	0.000017
43	21	15	15	1.000000	0.000000
43	21	16	0	0.000000	0.000000
43	21	16	1	0.000014	0.000014
43	21	16	2	0.000267	0.000253
43	21	16	3	0.002762	0.002495
43	21	16	4	0.017356	0.014594
43	21	16	5	0.071488	0.054132
43	21	16	6	0.203811	0.132323
43	21	16	7	0.421925	0.218114
43	21	16	8	0.667303	0.245378
43	21	16	9	0.856336	0.189032
43	21	16	10	0.955578	0.099242
43	21	16	11	0.990604	0.035027
43	21	16	12	0.998712	0.008108
43	21	16	13	0.999894	0.001182
43	21	16	14	0.999995	0.000101
43	21	16	15	1.000000	0.000005
43	21	16	16	1.000000	0.000000
43	21	17	0	0.000000	0.000000
43	21	17	1	0.000004	0.000004
43	21	17	2	0.000089	0.000085
43	21	17	3	0.001099	0.001010
43	21	17	4	0.008167	0.007069
43	21	17	5	0.039410	0.031243
43	21	17	6	0.130298	0.090888
43	21	17	7	0.308829	0.178531
43	21	17	8	0.549158	0.240330
43	21	17	9	0.772321	0.223163
43	21	17	10	0.915146	0.142824
43	21	17	11	0.977631	0.062486
43	21	17	12	0.996010	0.018378
43	21	17	13	0.999544	0.003534
43	21	17	14	0.999969	0.000425
43	21	17	15	0.999999	0.000030
43	21	17	16	1.000000	0.000001
43	21	17	17	1.000000	0.000000
43	21	18	0	0.000000	0.000000
43	21	18	1	0.000001	0.000001
43	21	18	2	0.000027	0.000026
43	21	18	3	0.000400	0.000373
43	21	18	4	0.003545	0.003146
43	21	18	5	0.020184	0.016638
43	21	18	6	0.077863	0.057679
43	21	18	7	0.212697	0.134834
43	21	18	8	0.428994	0.216297
43	21	18	9	0.669323	0.240330
43	21	18	10	0.854720	0.185397
43	21	18	11	0.953599	0.098878
43	21	18	12	0.989648	0.036049
43	21	18	13	0.998456	0.008808
43	21	18	14	0.999855	0.001398
43	21	18	15	0.999992	0.000137
43	21	18	16	1.000000	0.000008
43	21	18	17	1.000000	0.000000
43	21	18	18	1.000000	0.000000
43	21	19	0	0.000000	0.000000
43	21	19	1	0.000000	0.000000
43	21	19	2	0.000007	0.000007
43	21	19	3	0.000131	0.000124
43	21	19	4	0.001406	0.001275
43	21	19	5	0.009535	0.008129
43	21	19	6	0.043255	0.033720
43	21	19	7	0.137190	0.093935
43	21	19	8	0.316519	0.179330
43	21	19	9	0.553965	0.237446
43	21	19	10	0.773145	0.219181
43	21	19	11	0.914047	0.140902
43	21	19	12	0.976670	0.062623
43	21	19	13	0.995638	0.018968
43	21	19	14	0.999463	0.003825
43	21	19	15	0.999959	0.000496
43	21	19	16	0.999998	0.000039
43	21	19	17	1.000000	0.000002
43	21	19	18	1.000000	0.000000
43	21	19	19	1.000000	0.000000
43	21	20	0	0.000000	0.000000
43	21	20	1	0.000000	0.000000
43	21	20	2	0.000002	0.000002
43	21	20	3	0.000038	0.000036
43	21	20	4	0.000503	0.000465
43	21	20	5	0.004116	0.003613
43	21	20	6	0.022180	0.018064
43	21	20	7	0.082395	0.060214
43	21	20	8	0.219382	0.136988
43	21	20	9	0.435242	0.215860
43	21	20	10	0.672688	0.237446
43	21	20	11	0.855338	0.182650
43	21	20	12	0.953187	0.097848
43	21	20	13	0.989315	0.036129
43	21	20	14	0.998347	0.009032
43	21	20	15	0.999835	0.001488
43	21	20	16	0.999990	0.000155
43	21	20	17	1.000000	0.000010
43	21	20	18	1.000000	0.000000
43	21	20	19	1.000000	0.000000
43	21	20	20	1.000000	0.000000
43	21	21	0	0.000000	0.000000
43	21	21	1	0.000000	0.000000
43	21	21	2	0.000000	0.000000
43	21	21	3	0.000010	0.000009
43	21	21	4	0.000159	0.000150
43	21	21	5	0.001603	0.001443
43	21	21	6	0.010399	0.008797
43	21	21	7	0.045742	0.035343
43	21	21	8	0.141955	0.096212
43	21	21	9	0.322620	0.180665
43	21	21	10	0.559127	0.236507
43	21	21	11	0.775925	0.216798
43	21	21	12	0.914898	0.138973
43	21	21	13	0.976749	0.061851
43	21	21	14	0.995598	0.018850
43	21	21	15	0.999447	0.003848
43	21	21	16	0.999956	0.000509
43	21	21	17	0.999998	0.000042
43	21	21	18	1.000000	0.000002
43	21	21	19	1.000000	0.000000
43	21	21	20	1.000000	0.000000
43	21	21	21	1.000000	0.000000

N	n	k	x	P(x)	p(x)
44	1	1	0	0.977273	0.977273
44	1	1	1	1.000000	0.022727
44	2	1	0	0.954545	0.954545
44	2	1	1	1.000000	0.045455
44	2	2	0	0.910148	0.910148
44	2	2	1	0.998943	0.088795
44	2	2	2	1.000000	0.001057
44	3	1	0	0.931818	0.931818
44	3	1	1	1.000000	0.068182
44	3	2	0	0.866808	0.866808
44	3	2	1	0.996829	0.130021
44	3	2	2	1.000000	0.003171
44	3	3	0	0.804893	0.804893
44	3	3	1	0.990637	0.185744
44	3	3	2	0.999924	0.009287
44	3	3	3	1.000000	0.000076
44	4	1	0	0.909091	0.909091
44	4	1	1	1.000000	0.090909
44	4	2	0	0.824524	0.824524
44	4	2	1	0.993658	0.169133
44	4	2	2	1.000000	0.006342
44	4	3	0	0.745998	0.745998
44	4	3	1	0.981577	0.235578
44	4	3	2	0.999698	0.018121
44	4	3	3	1.000000	0.000302
44	4	4	0	0.673218	0.673218
44	4	4	1	0.964339	0.291121
44	4	4	2	0.998814	0.034475
44	4	4	3	0.999993	0.001179
44	4	4	4	1.000000	0.000007
44	5	1	0	0.886364	0.886364
44	5	1	1	1.000000	0.113636
44	5	2	0	0.783298	0.783298
44	5	2	1	0.989429	0.206131
44	5	2	2	1.000000	0.010571
44	5	3	0	0.690048	0.690048
44	5	3	1	0.969798	0.279749
44	5	3	2	0.999245	0.029447
44	5	3	3	1.000000	0.000755
44	5	4	0	0.605896	0.605896
44	5	4	1	0.942505	0.336609
44	5	4	2	0.997090	0.054585
44	5	4	3	0.999963	0.002873
44	5	4	4	1.000000	0.000037
44	5	5	0	0.530159	0.530159
44	5	5	1	0.908844	0.378685
44	5	5	2	0.992996	0.084152
44	5	5	3	0.999819	0.006823
44	5	5	4	0.999999	0.000180
44	5	5	5	1.000000	0.000001
44	6	1	0	0.863636	0.863636
44	6	1	1	1.000000	0.136364
44	6	2	0	0.743129	0.743129
44	6	2	1	0.984144	0.241015
44	6	2	2	1.000000	0.015856
44	6	3	0	0.636968	0.636968
44	6	3	1	0.955452	0.318484
44	6	3	2	0.998490	0.043038
44	6	3	3	1.000000	0.001510
44	6	4	0	0.543753	0.543753
44	6	4	1	0.916612	0.372859
44	6	4	2	0.994291	0.077679
44	6	4	3	0.999889	0.005598
44	6	4	4	1.000000	0.000110
44	6	5	0	0.462190	0.462190
44	6	5	1	0.870005	0.407815
44	6	5	2	0.986523	0.116518
44	6	5	3	0.999470	0.012946
44	6	5	4	0.999994	0.000525
44	6	5	5	1.000000	0.000006
44	6	6	0	0.391084	0.391084
44	6	6	1	0.817721	0.426637
44	6	6	2	0.974572	0.156852
44	6	6	3	0.998474	0.023901
44	6	6	4	0.999967	0.001494
44	6	6	5	1.000000	0.000032
44	6	6	6	1.000000	0.000000
44	7	1	0	0.840909	0.840909
44	7	1	1	1.000000	0.159091
44	7	2	0	0.704017	0.704017
44	7	2	1	0.977801	0.273784
44	7	2	2	1.000000	0.022199
44	7	3	0	0.586681	0.586681
44	7	3	1	0.938689	0.352008
44	7	3	2	0.997357	0.058668
44	7	3	3	1.000000	0.002643
44	7	4	0	0.486516	0.486516
44	7	4	1	0.887176	0.400660
44	7	4	2	0.990203	0.103027
44	7	4	3	0.999742	0.009540
44	7	4	4	1.000000	0.000258
44	7	5	0	0.401375	0.401375
44	7	5	1	0.827077	0.425701
44	7	5	2	0.977324	0.150247
44	7	5	3	0.998788	0.021464
44	7	5	4	0.999981	0.001192
44	7	5	5	1.000000	0.000019
44	7	6	0	0.329334	0.329334
44	7	6	1	0.761584	0.432251
44	7	6	2	0.958062	0.196477
44	7	6	3	0.996587	0.038525
44	7	6	4	0.999889	0.003302
44	7	6	5	0.999999	0.000110
44	7	6	6	1.000000	0.000001
44	7	7	0	0.268667	0.268667
44	7	7	1	0.693334	0.424667
44	7	7	2	0.932209	0.238875
44	7	7	3	0.992531	0.060322
44	7	7	4	0.999628	0.007097
44	7	7	5	0.999993	0.000365
44	7	7	6	1.000000	0.000007
44	7	7	7	1.000000	0.000000
44	8	1	0	0.818182	0.818182
44	8	1	1	1.000000	0.181818
44	8	2	0	0.665962	0.665962
44	8	2	1	0.970402	0.304440
44	8	2	2	1.000000	0.029598
44	8	3	0	0.539112	0.539112
44	8	3	1	0.919662	0.380550
44	8	3	2	0.995772	0.076110
44	8	3	3	1.000000	0.004228
44	8	4	0	0.433919	0.433919
44	8	4	1	0.854690	0.420770
44	8	4	2	0.984634	0.129944
44	8	4	3	0.999484	0.014851
44	8	4	4	1.000000	0.000516
44	8	5	0	0.347136	0.347136
44	8	5	1	0.781055	0.433919
44	8	5	2	0.965142	0.184087
44	8	5	3	0.997628	0.032486
44	8	5	4	0.999948	0.002320
44	8	5	5	1.000000	0.000052
44	8	6	0	0.275928	0.275928
44	8	6	1	0.703172	0.427244
44	8	6	2	0.936821	0.233649
44	8	6	3	0.993463	0.056642
44	8	6	4	0.999710	0.006247
44	8	6	5	0.999996	0.000286
44	8	6	6	1.000000	0.000004
44	8	7	0	0.217838	0.217838
44	8	7	1	0.624469	0.406631
44	8	7	2	0.899929	0.275460
44	8	7	3	0.986010	0.086081
44	8	7	4	0.999053	0.013043
44	8	7	5	0.999973	0.000921
44	8	7	6	1.000000	0.000026
44	8	7	7	1.000000	0.000000
44	8	8	0	0.170738	0.170738
44	8	8	1	0.547539	0.376801
44	8	8	2	0.855260	0.307721
44	8	8	3	0.974378	0.119118
44	8	8	4	0.997643	0.023265
44	8	8	5	0.999899	0.002256
44	8	8	6	0.999998	0.000100
44	8	8	7	1.000000	0.000002
44	8	8	8	1.000000	0.000000
44	9	1	0	0.795455	0.795455
44	9	1	1	1.000000	0.204545
44	9	2	0	0.628964	0.628964
44	9	2	1	0.961945	0.332981
44	9	2	2	1.000000	0.038055
44	9	3	0	0.494186	0.494186
44	9	3	1	0.898520	0.404334
44	9	3	2	0.993657	0.095137
44	9	3	3	1.000000	0.006342
44	9	4	0	0.385706	0.385706
44	9	4	1	0.819626	0.433919
44	9	4	2	0.977414	0.157789
44	9	4	3	0.999072	0.021657
44	9	4	4	1.000000	0.000928
44	9	5	0	0.298922	0.298922
44	9	5	1	0.732842	0.433919
44	9	5	2	0.949801	0.216960
44	9	5	3	0.995823	0.046022
44	9	5	4	0.999884	0.004061
44	9	5	5	1.000000	0.000116
44	9	6	0	0.229940	0.229940
44	9	6	1	0.643833	0.413892
44	9	6	2	0.910860	0.267027
44	9	6	3	0.988743	0.077883
44	9	6	4	0.999363	0.010620
44	9	6	5	0.999988	0.000625
44	9	6	6	1.000000	0.000012
44	9	7	0	0.175481	0.175481
44	9	7	1	0.556697	0.381217
44	9	7	2	0.861671	0.304973
44	9	7	3	0.976446	0.114775
44	9	7	4	0.997966	0.021520
44	9	7	5	0.999922	0.001956
44	9	7	6	0.999999	0.000077
44	9	7	7	1.000000	0.000001
44	9	8	0	0.132796	0.132796
44	9	8	1	0.474272	0.341476
44	9	8	2	0.803973	0.329701
44	9	8	3	0.957833	0.153860
44	9	8	4	0.995058	0.037224
44	9	8	5	0.999711	0.004653
44	9	8	6	0.999993	0.000282
44	9	8	7	1.000000	0.000007
44	9	8	8	1.000000	0.000000

N	n	k	x	P(x)	p(x)
44	9	9	0	0.099597	0.099597
44	9	9	1	0.398389	0.298791
44	9	9	2	0.739864	0.341476
44	9	9	3	0.932190	0.192326
44	9	9	4	0.989888	0.057698
44	9	9	5	0.999194	0.009306
44	9	9	6	0.999969	0.000776
44	9	9	7	0.999999	0.000030
44	9	9	8	1.000000	0.000000
44	9	9	9	1.000000	0.000000
44	10	1	0	0.772727	0.772727
44	10	1	1	1.000000	0.227273
44	10	2	0	0.593023	0.593023
44	10	2	1	0.952431	0.359408
44	10	2	2	1.000000	0.047569
44	10	3	0	0.451827	0.451827
44	10	3	1	0.875415	0.423588
44	10	3	2	0.990939	0.115524
44	10	3	3	1.000000	0.009061
44	10	4	0	0.341625	0.341625
44	10	4	1	0.782432	0.440807
44	10	4	2	0.968398	0.185965
44	10	4	3	0.998453	0.030055
44	10	4	4	1.000000	0.001547
44	10	5	0	0.256219	0.256219
44	10	5	1	0.683251	0.427032
44	10	5	2	0.931205	0.247954
44	10	5	3	0.993193	0.061988
44	10	5	4	0.999768	0.006575
44	10	5	5	1.000000	0.000232
44	10	6	0	0.190522	0.190522
44	10	6	1	0.584705	0.394183
44	10	6	2	0.880342	0.295637
44	10	6	3	0.982067	0.101725
44	10	6	4	0.998756	0.016689
44	10	6	5	0.999970	0.001214
44	10	6	6	1.000000	0.000030
44	10	7	0	0.140385	0.140385
44	10	7	1	0.491346	0.350961
44	10	7	2	0.818103	0.326757
44	10	7	3	0.963328	0.145225
44	10	7	4	0.996121	0.032793
44	10	7	5	0.999810	0.003689
44	10	7	6	0.999997	0.000186
44	10	7	7	1.000000	0.000003
44	10	8	0	0.102443	0.102443
44	10	8	1	0.405977	0.303534
44	10	8	2	0.747453	0.341476
44	10	8	3	0.935853	0.188400
44	10	8	4	0.990804	0.054950
44	10	8	5	0.999312	0.008508
44	10	8	6	0.999977	0.000665
44	10	8	7	1.000000	0.000023
44	10	8	8	1.000000	0.000000
44	10	9	0	0.073986	0.073986
44	10	9	1	0.330093	0.256107
44	10	9	2	0.671569	0.341476
44	10	9	3	0.899220	0.227651
44	10	9	4	0.981645	0.082425
44	10	9	5	0.998130	0.016485
44	10	9	6	0.999903	0.001773
44	10	9	7	0.999998	0.000095
44	10	9	8	1.000000	0.000002
44	10	9	9	1.000000	0.000000
44	10	10	0	0.052847	0.052847
44	10	10	1	0.264237	0.211390
44	10	10	2	0.593518	0.329280
44	10	10	3	0.853690	0.260172
44	10	10	4	0.967515	0.113825
44	10	10	5	0.995775	0.028260
44	10	10	6	0.999700	0.003925
44	10	10	7	0.999990	0.000289
44	10	10	8	1.000000	0.000010
44	10	10	9	1.000000	0.000000
44	10	10	10	1.000000	0.000000
44	11	1	0	0.750000	0.750000
44	11	1	1	1.000000	0.250000
44	11	2	0	0.558139	0.558139
44	11	2	1	0.941860	0.383721
44	11	2	2	1.000000	0.058140
44	11	3	0	0.411960	0.411960
44	11	3	1	0.850498	0.438538
44	11	3	2	0.987541	0.137043
44	11	3	3	1.000000	0.012458
44	11	4	0	0.301434	0.301434
44	11	4	1	0.743538	0.442104
44	11	4	2	0.957459	0.213921
44	11	4	3	0.997569	0.040110
44	11	4	4	1.000000	0.002431
44	11	5	0	0.218540	0.218540
44	11	5	1	0.633012	0.414472
44	11	5	2	0.909327	0.276315
44	11	5	3	0.989547	0.080220
44	11	5	4	0.999574	0.010028
44	11	5	5	1.000000	0.000425
44	11	6	0	0.156900	0.156900
44	11	6	1	0.526737	0.369837
44	11	6	2	0.845562	0.318825
44	11	6	3	0.973091	0.127530
44	11	6	4	0.997775	0.024683
44	11	6	5	0.999934	0.002160
44	11	6	6	1.000000	0.000065
44	11	7	0	0.111482	0.111482
44	11	7	1	0.429412	0.317930
44	11	7	2	0.770050	0.340639
44	11	7	3	0.946243	0.176193
44	11	7	4	0.993228	0.046985
44	11	7	5	0.999593	0.006366
44	11	7	6	0.999991	0.000398
44	11	7	7	1.000000	0.000009
44	11	8	0	0.078339	0.078339
44	11	8	1	0.343485	0.265146
44	11	8	2	0.687192	0.343708
44	11	8	3	0.908147	0.220955
44	11	8	4	0.984339	0.076191
44	11	8	5	0.998561	0.014222
44	11	8	6	0.999938	0.001376
44	11	8	7	0.999999	0.000061
44	11	8	8	1.000000	0.000001
44	11	9	0	0.054402	0.054402
44	11	9	1	0.269833	0.215431
44	11	9	2	0.601265	0.331432
44	11	9	3	0.859046	0.257781
44	11	9	4	0.969524	0.110478
44	11	9	5	0.996191	0.026667
44	11	9	6	0.999746	0.003556
44	11	9	7	0.999992	0.000246
44	11	9	8	1.000000	0.000008
44	11	9	9	1.000000	0.000000
44	11	10	0	0.037304	0.037304
44	11	10	1	0.208281	0.170977
44	11	10	2	0.516040	0.307759
44	11	10	3	0.800125	0.284085
44	11	10	4	0.947428	0.147303
44	11	10	5	0.991619	0.044191
44	11	10	6	0.999238	0.007619
44	11	10	7	0.999964	0.000726
44	11	10	8	0.999999	0.000035
44	11	10	9	1.000000	0.000001
44	11	10	10	1.000000	0.000000
44	11	11	0	0.025235	0.025235
44	11	11	1	0.157994	0.132759
44	11	11	2	0.434574	0.276581
44	11	11	3	0.733281	0.298707
44	11	11	4	0.917101	0.183820
44	11	11	5	0.983821	0.066720
44	11	11	6	0.998118	0.014297
44	11	11	7	0.999879	0.001761
44	11	11	8	0.999996	0.000117
44	11	11	9	1.000000	0.000004
44	11	11	10	1.000000	0.000000
44	11	11	11	1.000000	0.000000
44	12	1	0	0.727273	0.727273
44	12	1	1	1.000000	0.272727
44	12	2	0	0.524313	0.524313
44	12	2	1	0.930233	0.405920
44	12	2	2	1.000000	0.069767
44	12	3	0	0.374509	0.374509
44	12	3	1	0.823920	0.449411
44	12	3	2	0.983389	0.159468
44	12	3	3	1.000000	0.016611
44	12	4	0	0.264897	0.264897
44	12	4	1	0.703347	0.438450
44	12	4	2	0.944494	0.241147
44	12	4	3	0.996354	0.051860
44	12	4	4	1.000000	0.003646
44	12	5	0	0.185428	0.185428
44	12	5	1	0.582773	0.397345
44	12	5	2	0.884207	0.301434
44	12	5	3	0.984635	0.100478
44	12	5	4	0.999271	0.014586
44	12	5	5	1.000000	0.000729
44	12	6	0	0.128373	0.128373
44	12	6	1	0.470701	0.342328
44	12	6	2	0.806916	0.336215
44	12	6	3	0.961498	0.154582
44	12	6	4	0.996279	0.034781
44	12	6	5	0.999869	0.003590
44	12	6	6	1.000000	0.000131
44	12	7	0	0.087834	0.087834
44	12	7	1	0.371606	0.283772
44	12	7	2	0.718439	0.346832
44	12	7	3	0.924886	0.206448
44	12	7	4	0.988956	0.064070
44	12	7	5	0.999208	0.010251
44	12	7	6	0.999979	0.000772
44	12	7	7	1.000000	0.000021
44	12	8	0	0.059347	0.059347
44	12	8	1	0.287242	0.227894
44	12	8	2	0.624700	0.337459
44	12	8	3	0.874669	0.249969
44	12	8	4	0.975103	0.100434
44	12	8	5	0.997268	0.022165
44	12	8	6	0.999854	0.002586
44	12	8	7	0.999997	0.000143
44	12	8	8	1.000000	0.000003
44	12	9	0	0.039565	0.039565
44	12	9	1	0.217607	0.178042
44	12	9	2	0.530962	0.313354
44	12	9	3	0.812177	0.281215

Table for $N = 2$, $n = 1$, through $N = 100$, $n = 50$

N	n	k	x	P(x)	p(x)
44	12	9	4	0.952785	0.140608
44	12	9	5	0.992958	0.040174
44	12	9	6	0.999423	0.006465
44	12	9	7	0.999977	0.000554
44	12	9	8	1.000000	0.000022
44	12	9	9	1.000000	0.000000
44	12	10	0	0.026000	0.026000
44	12	10	1	0.161651	0.135651
44	12	10	2	0.441432	0.279781
44	12	10	3	0.739864	0.298433
44	12	10	4	0.920646	0.180781
44	12	10	5	0.984924	0.064278
44	12	10	6	0.998315	0.013391
44	12	10	7	0.999898	0.001583
44	12	10	8	0.999997	0.000099
44	12	10	9	1.000000	0.000003
44	12	10	10	1.000000	0.000000
44	12	11	0	0.016823	0.016823
44	12	11	1	0.117764	0.100940
44	12	11	2	0.359143	0.241379
44	12	11	3	0.660868	0.301724
44	12	11	4	0.878109	0.217241
44	12	11	5	0.971690	0.093581
44	12	11	6	0.995952	0.024262
44	12	11	7	0.999665	0.003714
44	12	11	8	0.999985	0.000320
44	12	11	9	1.000000	0.000014
44	12	11	10	1.000000	0.000000
44	12	11	11	1.000000	0.000000
44	12	12	0	0.010706	0.010706
44	12	12	1	0.084117	0.073411
44	12	12	2	0.285998	0.201881
44	12	12	3	0.578579	0.292581
44	12	12	4	0.825444	0.246865
44	12	12	5	0.951840	0.126395
44	12	12	6	0.991541	0.039701
44	12	12	7	0.999103	0.007562
44	12	12	8	0.999947	0.000844
44	12	12	9	0.999998	0.000052
44	12	12	10	1.000000	0.000002
44	12	12	11	1.000000	0.000000
44	12	12	12	1.000000	0.000000
44	13	1	0	0.704545	0.704545
44	13	1	1	1.000000	0.295455
44	13	2	0	0.491543	0.491543
44	13	2	1	0.917548	0.426004
44	13	2	2	1.000000	0.082452
44	13	3	0	0.339399	0.339399
44	13	3	1	0.795832	0.456433
44	13	3	2	0.978405	0.182573
44	13	3	3	1.000000	0.021595
44	13	4	0	0.231785	0.231785
44	13	4	1	0.662242	0.430457
44	13	4	2	0.929422	0.267180
44	13	4	3	0.994733	0.065311
44	13	4	4	1.000000	0.005267
44	13	5	0	0.156455	0.156455
44	13	5	1	0.533105	0.376650
44	13	5	2	0.855948	0.322843
44	13	5	3	0.978405	0.122458
44	13	5	4	0.998815	0.020410
44	13	5	5	1.000000	0.001185
44	13	6	0	0.104303	0.104303
44	13	6	1	0.417212	0.312909
44	13	6	2	0.764889	0.347677
44	13	6	3	0.947006	0.182117
44	13	6	4	0.994105	0.047099
44	13	6	5	0.999757	0.005652
44	13	6	6	1.000000	0.000243
44	13	7	0	0.068620	0.068620
44	13	7	1	0.318399	0.249778
44	13	7	2	0.664246	0.345847
44	13	7	3	0.899080	0.234834
44	13	7	4	0.982950	0.083869
44	13	7	5	0.998567	0.015617
44	13	7	6	0.999955	0.001388
44	13	7	7	1.000000	0.000045
44	13	8	0	0.044511	0.044511
44	13	8	1	0.237390	0.192879
44	13	8	2	0.561427	0.324037
44	13	8	3	0.835612	0.274185
44	13	8	4	0.962549	0.126938
44	13	8	5	0.995190	0.032641
44	13	8	6	0.999692	0.004502
44	13	8	7	0.999993	0.000300
44	13	8	8	1.000000	0.000007
44	13	9	0	0.028437	0.028437
44	13	9	1	0.173097	0.144659
44	13	9	2	0.462415	0.289319
44	13	9	3	0.759449	0.297034
44	13	9	4	0.930815	0.171366
44	13	9	5	0.987937	0.057122
44	13	9	6	0.998817	0.010880
44	13	9	7	0.999943	0.001126
44	13	9	8	0.999999	0.000056
44	13	9	9	1.000000	0.000001
44	13	10	0	0.017875	0.017875
44	13	10	1	0.123499	0.105624
44	13	10	2	0.371487	0.247987
44	13	10	3	0.674582	0.303096
44	13	10	4	0.886749	0.212167
44	13	10	5	0.974880	0.088131
44	13	10	6	0.996641	0.021761
44	13	10	7	0.999750	0.003109
44	13	10	8	0.999991	0.000241
44	13	10	9	1.000000	0.000009
44	13	10	10	1.000000	0.000000
44	13	11	0	0.011040	0.011040
44	13	11	1	0.086220	0.075180
44	13	11	2	0.291255	0.205035
44	13	11	3	0.585437	0.294181
44	13	11	4	0.830587	0.245151
44	13	11	5	0.954144	0.123556
44	13	11	6	0.992161	0.038017
44	13	11	7	0.999201	0.007040
44	13	11	8	0.999955	0.000754
44	13	11	9	0.999999	0.000043
44	13	11	10	1.000000	0.000001
44	13	11	11	1.000000	0.000000
44	13	12	0	0.006691	0.006691
44	13	12	1	0.058882	0.052191
44	13	12	2	0.222910	0.164028
44	13	12	3	0.496291	0.273380
44	13	12	4	0.763728	0.267437
44	13	12	5	0.924191	0.160462
44	13	12	6	0.984097	0.059906
44	13	12	7	0.997921	0.013824
44	13	12	8	0.999841	0.001920
44	13	12	9	0.999993	0.000152
44	13	12	10	1.000000	0.000006
44	13	12	11	1.000000	0.000000
44	13	12	12	1.000000	0.000000
44	13	13	0	0.003973	0.003973
44	13	13	1	0.039310	0.035338
44	13	13	2	0.166525	0.127215
44	13	13	3	0.410859	0.244334
44	13	13	4	0.688511	0.277652
44	13	13	5	0.884075	0.195564
44	13	13	6	0.970992	0.086917
44	13	13	7	0.995329	0.024337
44	13	13	8	0.999541	0.004212
44	13	13	9	0.999974	0.000433
44	13	13	10	0.999999	0.000025
44	13	13	11	1.000000	0.000001
44	13	13	12	1.000000	0.000000
44	13	13	13	1.000000	0.000000
44	14	1	0	0.681818	0.681818
44	14	1	1	1.000000	0.318182
44	14	2	0	0.459831	0.459831
44	14	2	1	0.903805	0.443975
44	14	2	2	1.000000	0.096194
44	14	3	0	0.306554	0.306554
44	14	3	1	0.766385	0.459831
44	14	3	2	0.972516	0.206131
44	14	3	3	1.000000	0.027484
44	14	4	0	0.201877	0.201877
44	14	4	1	0.620585	0.418708
44	14	4	2	0.912185	0.291600
44	14	4	3	0.992626	0.080441
44	14	4	4	1.000000	0.007374
44	14	5	0	0.131220	0.131220
44	14	5	1	0.484505	0.353285
44	14	5	2	0.824705	0.340200
44	14	5	3	0.970505	0.145800
44	14	5	4	0.998156	0.027652
44	14	5	5	1.000000	0.001843
44	14	6	0	0.084115	0.084115
44	14	6	1	0.366743	0.282628
44	14	6	2	0.720028	0.353285
44	14	6	3	0.929382	0.209354
44	14	6	4	0.991066	0.061685
44	14	6	5	0.999574	0.008508
44	14	6	6	1.000000	0.000425
44	14	7	0	0.053126	0.053126
44	14	7	1	0.270055	0.216929
44	14	7	2	0.608464	0.338410
44	14	7	3	0.868779	0.260315
44	14	7	4	0.974834	0.106054
44	14	7	5	0.997559	0.022726
44	14	7	6	0.999910	0.002351
44	14	7	7	1.000000	0.000090
44	14	8	0	0.033024	0.033024
44	14	8	1	0.193836	0.160812
44	14	8	2	0.498710	0.304873
44	14	8	3	0.791388	0.292678
44	14	8	4	0.946170	0.154782
44	14	8	5	0.992031	0.045861
44	14	8	6	0.999402	0.007371
44	14	8	7	0.999983	0.000581
44	14	8	8	1.000000	0.000017
44	14	9	0	0.020181	0.020181
44	14	9	1	0.135765	0.115584
44	14	9	2	0.397085	0.261320
44	14	9	3	0.701959	0.304873
44	14	9	4	0.903175	0.201216
44	14	9	5	0.980566	0.077391
44	14	9	6	0.997764	0.017198
44	14	9	7	0.999870	0.002106
44	14	9	8	0.999997	0.000127
44	14	9	9	1.000000	0.000003

Table for $N = 2$, $n = 1$, through $N = 100$, $n = 50$

N	n	k	x	P(x)	p(x)
44	14	10	0	0.012109	0.012109
44	14	10	1	0.092834	0.080725
44	14	10	2	0.307490	0.214656
44	14	10	3	0.606141	0.298652
44	14	10	4	0.845685	0.239543
44	14	10	5	0.960666	0.114981
44	14	10	6	0.993833	0.033168
44	14	10	7	0.999449	0.005616
44	14	10	8	0.999975	0.000526
44	14	10	9	1.000000	0.000024
44	14	10	10	1.000000	0.000000
44	14	11	0	0.007123	0.007123
44	14	11	1	0.061968	0.054846
44	14	11	2	0.231729	0.169760
44	14	11	3	0.509519	0.277790
44	14	11	4	0.775231	0.265712
44	14	11	5	0.930230	0.154999
44	14	11	6	0.986029	0.055800
44	14	11	7	0.998293	0.012264
44	14	11	8	0.999882	0.001590
44	14	11	9	0.999996	0.000114
44	14	11	10	1.000000	0.000004
44	14	11	11	1.000000	0.000000
44	14	12	0	0.004101	0.004101
44	14	12	1	0.040363	0.036262
44	14	12	2	0.169998	0.129635
44	14	12	3	0.416922	0.246924
44	14	12	4	0.694712	0.277790
44	14	12	5	0.887957	0.193245
44	14	12	6	0.972502	0.084545
44	14	12	7	0.995691	0.023189
44	14	12	8	0.999593	0.003902
44	14	12	9	0.999979	0.000385
44	14	12	10	0.999999	0.000021
44	14	12	11	1.000000	0.000001
44	14	12	12	1.000000	0.000000
44	14	13	0	0.002307	0.002307
44	14	13	1	0.025631	0.023325
44	14	13	2	0.121385	0.095753
44	14	13	3	0.332042	0.210657
44	14	13	4	0.607903	0.275861
44	14	13	5	0.833607	0.225704
44	14	13	6	0.951366	0.117759
44	14	13	7	0.990619	0.039253
44	14	13	8	0.998862	0.008243
44	14	13	9	0.999919	0.001057
44	14	13	10	0.999997	0.000078
44	14	13	11	1.000000	0.000003
44	14	13	12	1.000000	0.000000
44	14	13	13	1.000000	0.000000
44	14	14	0	0.001265	0.001265
44	14	14	1	0.015850	0.014585
44	14	14	2	0.084319	0.068469
44	14	14	3	0.257293	0.172974
44	14	14	4	0.518915	0.261623
44	14	14	5	0.768080	0.249165
44	14	14	6	0.920976	0.152896
44	14	14	7	0.981755	0.060779
44	14	14	8	0.997266	0.015511
44	14	14	9	0.999748	0.002482
44	14	14	10	0.999987	0.000239
44	14	14	11	1.000000	0.000013
44	14	14	12	1.000000	0.000000
44	14	14	13	1.000000	0.000000
44	14	14	14	1.000000	0.000000
44	15	1	0	0.659091	0.659091
44	15	1	1	1.000000	0.340909
44	15	2	0	0.429175	0.429175
44	15	2	1	0.889006	0.459831
44	15	2	2	1.000000	0.110994
44	15	3	0	0.275898	0.275898
44	15	3	1	0.735729	0.459831
44	15	3	2	0.965645	0.229915
44	15	3	3	1.000000	0.034355
44	15	4	0	0.174960	0.174960
44	15	4	1	0.578714	0.403754
44	15	4	2	0.892745	0.314031
44	15	4	3	0.989945	0.097200
44	15	4	4	1.000000	0.010055
44	15	5	0	0.109350	0.109350
44	15	5	1	0.437400	0.328050
44	15	5	2	0.790685	0.353285
44	15	5	3	0.960785	0.170100
44	15	5	4	0.997235	0.036450
44	15	5	5	1.000000	0.002765
44	15	6	0	0.067292	0.067292
44	15	6	1	0.319638	0.252346
44	15	6	2	0.672923	0.353285
44	15	6	3	0.908446	0.235523
44	15	6	4	0.986954	0.078508
44	15	6	5	0.999291	0.012337
44	15	6	6	1.000000	0.000709
44	15	7	0	0.040730	0.040730
44	15	7	1	0.226669	0.185939
44	15	7	2	0.552063	0.325394
44	15	7	3	0.834071	0.282008
44	15	7	4	0.964228	0.130157
44	15	7	5	0.996044	0.031816
44	15	7	6	0.999832	0.003788
44	15	7	7	1.000000	0.000168
44	15	8	0	0.024218	0.024218
44	15	8	1	0.156313	0.132096
44	15	8	2	0.437735	0.281422
44	15	8	3	0.742609	0.304873
44	15	8	4	0.925533	0.182924
44	15	8	5	0.987445	0.061913
44	15	8	6	0.998911	0.011465
44	15	8	7	0.999964	0.001053
44	15	8	8	1.000000	0.000036
44	15	9	0	0.014127	0.014127
44	15	9	1	0.104943	0.090816
44	15	9	2	0.336111	0.231168
44	15	9	3	0.640984	0.304873
44	15	9	4	0.869639	0.228655
44	15	9	5	0.970247	0.100608
44	15	9	6	0.996044	0.025797
44	15	9	7	0.999730	0.003685
44	15	9	8	0.999993	0.000263
44	15	9	9	1.000000	0.000007
44	15	10	0	0.008073	0.008073
44	15	10	1	0.068616	0.060544
44	15	10	2	0.250248	0.181632
44	15	10	3	0.536456	0.286208
44	15	10	4	0.797776	0.261320
44	15	10	5	0.941502	0.143726
44	15	10	6	0.989411	0.047909
44	15	10	7	0.998887	0.009476
44	15	10	8	0.999940	0.001053
44	15	10	9	0.999999	0.000058
44	15	10	10	1.000000	0.000001
44	15	11	0	0.004511	0.004511
44	15	11	1	0.043687	0.039175
44	15	11	2	0.180801	0.137114
44	15	11	3	0.435442	0.254641
44	15	11	4	0.713231	0.277790
44	15	11	5	0.899230	0.185998
44	15	11	6	0.976729	0.077499
44	15	11	7	0.996658	0.019928
44	15	11	8	0.999723	0.003066
44	15	11	9	0.999988	0.000265
44	15	11	10	1.000000	0.000011
44	15	11	11	1.000000	0.000000
44	15	12	0	0.002461	0.002461
44	15	12	1	0.027067	0.024606
44	15	12	2	0.126786	0.099719
44	15	12	3	0.342845	0.216059
44	15	12	4	0.620635	0.277790
44	15	12	5	0.842867	0.222232
44	15	12	6	0.955593	0.112726
44	15	12	7	0.991826	0.036233
44	15	12	8	0.999073	0.007247
44	15	12	9	0.999940	0.000867
44	15	12	10	0.999998	0.000058
44	15	12	11	1.000000	0.000002
44	15	12	12	1.000000	0.000000
44	15	13	0	0.001307	0.001307
44	15	13	1	0.016302	0.014994
44	15	13	2	0.086275	0.069974
44	15	13	3	0.261823	0.175548
44	15	13	4	0.525144	0.263322
44	15	13	5	0.773419	0.248275
44	15	13	6	0.923889	0.150469
44	15	13	7	0.982768	0.058879
44	15	13	8	0.997488	0.014720
44	15	13	9	0.999778	0.002290
44	15	13	10	0.999989	0.000211
44	15	13	11	1.000000	0.000011
44	15	13	12	1.000000	0.000000
44	15	13	13	1.000000	0.000000
44	15	14	0	0.000675	0.000675
44	15	14	1	0.009530	0.008855
44	15	14	2	0.056931	0.047401
44	15	14	3	0.193869	0.136938
44	15	14	4	0.431708	0.237839
44	15	14	5	0.693331	0.261623
44	15	14	6	0.880204	0.186873
44	15	14	7	0.967573	0.087369
44	15	14	8	0.994164	0.026591
44	15	14	9	0.999334	0.005170
44	15	14	10	0.999955	0.000620
44	15	14	11	0.999998	0.000043
44	15	14	12	1.000000	0.000002
44	15	14	13	1.000000	0.000000
44	15	14	14	1.000000	0.000000
44	15	15	0	0.000337	0.000337
44	15	15	1	0.005397	0.005060
44	15	15	2	0.036391	0.030993
44	15	15	3	0.139094	0.102703
44	15	15	4	0.344500	0.205406
44	15	15	5	0.606123	0.261623
44	15	15	6	0.824142	0.218019
44	15	15	7	0.944275	0.120133
44	15	15	8	0.987959	0.043685
44	15	15	9	0.998300	0.010341
44	15	15	10	0.999851	0.001551
44	15	15	11	0.999992	0.000141
44	15	15	12	1.000000	0.000007
44	15	15	13	1.000000	0.000000
44	15	15	14	1.000000	0.000000
44	15	15	15	1.000000	0.000000

Table for $N = 2$, $n = 1$, through $N = 100$, $n = 50$

N	n	k	x	P(x)	p(x)
44	16	1	0	0.636364	0.636364
44	16	1	1	1.000000	0.363636
44	16	2	0	0.399577	0.399577
44	16	2	1	0.873150	0.473573
44	16	2	2	1.000000	0.126850
44	16	3	0	0.247357	0.247357
44	16	3	1	0.704017	0.456660
44	16	3	2	0.957717	0.253700
44	16	3	3	1.000000	0.042283
44	16	4	0	0.150828	0.150828
44	16	4	1	0.536946	0.386119
44	16	4	2	0.871087	0.334141
44	16	4	3	0.986593	0.115506
44	16	4	4	1.000000	0.013407
44	16	5	0	0.090497	0.090497
44	16	5	1	0.392152	0.301655
44	16	5	2	0.754138	0.361986
44	16	5	3	0.949054	0.194916
44	16	5	4	0.995978	0.046924
44	16	5	5	1.000000	0.004022
44	16	6	0	0.053370	0.053370
44	16	6	1	0.276131	0.222761
44	16	6	2	0.624194	0.348064
44	16	6	3	0.884082	0.259888
44	16	6	4	0.981540	0.097458
44	16	6	5	0.998865	0.017326
44	16	6	6	1.000000	0.001134
44	16	7	0	0.030898	0.030898
44	16	7	1	0.188199	0.157300
44	16	7	2	0.495960	0.307762
44	16	7	3	0.795173	0.299213
44	16	7	4	0.950763	0.155591
44	16	7	5	0.993850	0.043087
44	16	7	6	0.999701	0.005851
44	16	7	7	1.000000	0.000299
44	16	8	0	0.017537	0.017537
44	16	8	1	0.124428	0.106891
44	16	8	2	0.379510	0.255082
44	16	8	3	0.690044	0.310534
44	16	8	4	0.900302	0.210258
44	16	8	5	0.981041	0.080739
44	16	8	6	0.998120	0.017079
44	16	8	7	0.999927	0.001807
44	16	8	8	1.000000	0.000073
44	16	9	0	0.009743	0.009743
44	16	9	1	0.079890	0.070147
44	16	9	2	0.280311	0.200421
44	16	9	3	0.577907	0.297595
44	16	9	4	0.830216	0.252309
44	16	9	5	0.956370	0.126155

N	n	k	x	P(x)	p(x)
44	16	9	6	0.993376	0.037005
44	16	9	7	0.999475	0.006100
44	16	9	8	0.999984	0.000508
44	16	9	9	1.000000	0.000016
44	16	10	0	0.005289	0.005289
44	16	10	1	0.049827	0.044538
44	16	10	2	0.200143	0.150316
44	16	10	3	0.467371	0.267228
44	16	10	4	0.743710	0.276338
44	16	10	5	0.916722	0.173012
44	16	10	6	0.982803	0.066081
44	16	10	7	0.997907	0.015104
44	16	10	8	0.999868	0.001961
44	16	10	9	0.999997	0.000129
44	16	10	10	1.000000	0.000003
44	16	11	0	0.002800	0.002800
44	16	11	1	0.030178	0.027378
44	16	11	2	0.138248	0.108070
44	16	11	3	0.365196	0.226948
44	16	11	4	0.646179	0.280983
44	16	11	5	0.860747	0.214569
44	16	11	6	0.963367	0.102620
44	16	11	7	0.993909	0.030542
44	16	11	8	0.999406	0.005497
44	16	11	9	0.999970	0.000564
44	16	11	10	0.999999	0.000029
44	16	11	11	1.000000	0.000001
44	16	12	0	0.001442	0.001442
44	16	12	1	0.017733	0.016291
44	16	12	2	0.092400	0.074667
44	16	12	3	0.275792	0.183392
44	16	12	4	0.544003	0.268211
44	16	12	5	0.789224	0.245221
44	16	12	6	0.932270	0.143046
44	16	12	7	0.985579	0.053309
44	16	12	8	0.998074	0.012494
44	16	12	9	0.999850	0.001777
44	16	12	10	0.999994	0.000144
44	16	12	11	1.000000	0.000006
44	16	12	12	1.000000	0.000000
44	16	13	0	0.000721	0.000721
44	16	13	1	0.010097	0.009376
44	16	13	2	0.059733	0.049636
44	16	13	3	0.201289	0.141556
44	16	13	4	0.443424	0.242135
44	16	13	5	0.704930	0.261506
44	16	13	6	0.887568	0.182639
44	16	13	7	0.970586	0.083018
44	16	13	8	0.994950	0.024364
44	16	13	9	0.999462	0.004512

N	n	k	x	P(x)	p(x)
44	16	13	10	0.999967	0.000505
44	16	13	11	0.999999	0.000032
44	16	13	12	1.000000	0.000001
44	16	13	13	1.000000	0.000000
44	16	14	0	0.000349	0.000349
44	16	14	1	0.005560	0.005211
44	16	14	2	0.037317	0.031757
44	16	14	3	0.141927	0.104610
44	16	14	4	0.349694	0.207767
44	16	14	5	0.612137	0.262443
44	16	14	6	0.828653	0.216515
44	16	14	7	0.946484	0.117831
44	16	14	8	0.988662	0.042178
44	16	14	9	0.998443	0.009780
44	16	14	10	0.999869	0.001426
44	16	14	11	0.999994	0.000124
44	16	14	12	1.000000	0.000006
44	16	14	13	1.000000	0.000000
44	16	14	14	1.000000	0.000000
44	16	15	0	0.000163	0.000163
44	16	15	1	0.002955	0.002792
44	16	15	2	0.022497	0.019543
44	16	15	3	0.096596	0.074099
44	16	15	4	0.266587	0.169991
44	16	15	5	0.515908	0.249321
44	16	15	6	0.756481	0.240573
44	16	15	7	0.911135	0.154654
44	16	15	8	0.977415	0.066280
44	16	15	9	0.996161	0.018746
44	16	15	10	0.999584	0.003423
44	16	15	11	0.999973	0.000389
44	16	15	12	0.999999	0.000026
44	16	15	13	1.000000	0.000001
44	16	15	14	1.000000	0.000000
44	16	15	15	1.000000	0.000000
44	16	16	0	0.000073	0.000073
44	16	16	1	0.001511	0.001438
44	16	16	2	0.013063	0.011552
44	16	16	3	0.063379	0.050316
44	16	16	4	0.196246	0.132867
44	16	16	5	0.421338	0.225092
44	16	16	6	0.673525	0.252187
44	16	16	7	0.863139	0.189614
44	16	16	8	0.959131	0.095992
44	16	16	9	0.991636	0.032505
44	16	16	10	0.998876	0.007240
44	16	16	11	0.999906	0.001030
44	16	16	12	0.999995	0.000089
44	16	16	13	1.000000	0.000004
44	16	16	14	1.000000	0.000000

N	n	k	x	P(x)	p(x)
44	16	16	15	1.000000	0.000000
44	16	16	16	1.000000	0.000000
44	17	1	0	0.613636	0.613636
44	17	1	1	1.000000	0.386364
44	17	2	0	0.371036	0.371036
44	17	2	1	0.856237	0.485201
44	17	2	2	1.000000	0.143763
44	17	3	0	0.220855	0.220855
44	17	3	1	0.671398	0.450544
44	17	3	2	0.948656	0.277258
44	17	3	3	1.000000	0.051344
44	17	4	0	0.129281	0.129281
44	17	4	1	0.495576	0.366296
44	17	4	2	0.847220	0.351644
44	17	4	3	0.982468	0.135248
44	17	4	4	1.000000	0.017532
44	17	5	0	0.074336	0.074336
44	17	5	1	0.349058	0.274722
44	17	5	2	0.715354	0.366296
44	17	5	3	0.935131	0.219777
44	17	5	4	0.994302	0.059171
44	17	5	5	1.000000	0.005698
44	17	6	0	0.041933	0.041933
44	17	6	1	0.236352	0.194418
44	17	6	2	0.574471	0.338119
44	17	6	3	0.856237	0.281766
44	17	6	4	0.974578	0.118342
44	17	6	5	0.998247	0.023668
44	17	6	6	1.000000	0.001753
44	17	7	0	0.023174	0.023174
44	17	7	1	0.154491	0.131318
44	17	7	2	0.441003	0.286511
44	17	7	3	0.752428	0.311425
44	17	7	4	0.934093	0.181665
44	17	7	5	0.990772	0.056679
44	17	7	6	0.999492	0.008720
44	17	7	7	1.000000	0.000508
44	17	8	0	0.012526	0.012526
44	17	8	1	0.097705	0.085179
44	17	8	2	0.324850	0.227144
44	17	8	3	0.634592	0.309742
44	17	8	4	0.870265	0.235673
44	17	8	5	0.972390	0.102125
44	17	8	6	0.996900	0.024510
44	17	8	7	0.999863	0.002963
44	17	8	8	1.000000	0.000137
44	17	9	0	0.006611	0.006611
44	17	9	1	0.059848	0.053237
44	17	9	2	0.230206	0.170358
44	17	9	3	0.514136	0.283930

Table for $N = 2$, $n = 1$, through $N = 100$, $n = 50$

N	n	k	x	P(x)	p(x)
44	17	9	4	0.785161	0.271024
44	17	9	5	0.938348	0.153188
44	17	9	6	0.989411	0.051063
44	17	9	7	0.999040	0.009629
44	17	9	8	0.999966	0.000926
44	17	9	9	1.000000	0.000034
44	17	10	0	0.003400	0.003400
44	17	10	1	0.035511	0.032111
44	17	10	2	0.157196	0.121684
44	17	10	3	0.400564	0.243369
44	17	10	4	0.684494	0.283930
44	17	10	5	0.885827	0.201332
44	17	10	6	0.973363	0.087536
44	17	10	7	0.996289	0.022926
44	17	10	8	0.999728	0.003439
44	17	10	9	0.999992	0.000265
44	17	10	10	1.000000	0.000008
44	17	11	0	0.001700	0.001700
44	17	11	1	0.020400	0.018700
44	17	11	2	0.103511	0.083111
44	17	11	3	0.300354	0.196842
44	17	11	4	0.575933	0.275579
44	17	11	5	0.814768	0.238835
44	17	11	6	0.945042	0.130274
44	17	11	7	0.989546	0.044503
44	17	11	8	0.998817	0.009272
44	17	11	9	0.999930	0.001113
44	17	11	10	0.999998	0.000068
44	17	11	11	1.000000	0.000002
44	17	12	0	0.000824	0.000824
44	17	12	1	0.011333	0.010509
44	17	12	2	0.065733	0.054400
44	17	12	3	0.216845	0.151111
44	17	12	4	0.467371	0.250527
44	17	12	5	0.727919	0.260548
44	17	12	6	0.901618	0.173698
44	17	12	7	0.976060	0.074442
44	17	12	8	0.996289	0.020229
44	17	12	9	0.999660	0.003371
44	17	12	10	0.999984	0.000324
44	17	12	11	1.000000	0.000016
44	17	12	12	1.000000	0.000000
44	17	13	0	0.000386	0.000386
44	17	13	1	0.006079	0.005692
44	17	13	2	0.040233	0.034155
44	17	13	3	0.150733	0.110500
44	17	13	4	0.365595	0.214861
44	17	13	5	0.630214	0.264619
44	17	13	6	0.841909	0.211695
44	17	13	7	0.952797	0.110888

N	n	k	x	P(x)	p(x)
44	17	13	8	0.990599	0.037803
44	17	13	9	0.998817	0.008218
44	17	13	10	0.999913	0.001096
44	17	13	11	0.999997	0.000084
44	17	13	12	1.000000	0.000003
44	17	13	13	1.000000	0.000000
44	17	14	0	0.000174	0.000174
44	17	14	1	0.003141	0.002966
44	17	14	2	0.023707	0.020566
44	17	14	3	0.100830	0.077123
44	17	14	4	0.275492	0.174662
44	17	14	5	0.527781	0.252289
44	17	14	6	0.766791	0.239011
44	17	14	7	0.917026	0.150235
44	17	14	8	0.979624	0.062598
44	17	14	9	0.996696	0.017072
44	17	14	10	0.999666	0.002969
44	17	14	11	0.999980	0.000315
44	17	14	12	0.999999	0.000019
44	17	14	13	1.000000	0.000001
44	17	14	14	1.000000	0.000000
44	17	15	0	0.000076	0.000076
44	17	15	1	0.001559	0.001483
44	17	15	2	0.013424	0.011865
44	17	15	3	0.064839	0.051416
44	17	15	4	0.199805	0.134966
44	17	15	5	0.426865	0.227060
44	17	15	6	0.679154	0.252289
44	17	15	7	0.866948	0.187794
44	17	15	8	0.960845	0.093897
44	17	15	9	0.992144	0.031299
44	17	15	10	0.998973	0.006829
44	17	15	11	0.999917	0.000945
44	17	15	12	0.999996	0.000079
44	17	15	13	1.000000	0.000004
44	17	15	14	1.000000	0.000000
44	17	15	15	1.000000	0.000000
44	17	16	0	0.000031	0.000031
44	17	16	1	0.000740	0.000709
44	17	16	2	0.007287	0.006546
44	17	16	3	0.040018	0.032731
44	17	16	4	0.139303	0.099285
44	17	16	5	0.332909	0.193606
44	17	16	6	0.583458	0.250549
44	17	16	7	0.802191	0.218733
44	17	16	8	0.931704	0.129513
44	17	16	9	0.983510	0.051805
44	17	16	10	0.997324	0.013815
44	17	16	11	0.999722	0.002398
44	17	16	12	0.999983	0.000261

N	n	k	x	P(x)	p(x)
44	17	16	13	0.999999	0.000017
44	17	16	14	1.000000	0.000001
44	17	16	15	1.000000	0.000000
44	17	16	16	1.000000	0.000000
44	17	17	0	0.000012	0.000012
44	17	17	1	0.000335	0.000323
44	17	17	2	0.003780	0.003445
44	17	17	3	0.023652	0.019873
44	17	17	4	0.093207	0.069554
44	17	17	5	0.249935	0.156729
44	17	17	6	0.485028	0.235093
44	17	17	7	0.724072	0.239044
44	17	17	8	0.890075	0.166003
44	17	17	9	0.968708	0.078633
44	17	17	10	0.993871	0.025163
44	17	17	11	0.999208	0.005337
44	17	17	12	0.999936	0.000728
44	17	17	13	0.999997	0.000061
44	17	17	14	1.000000	0.000003
44	17	17	15	1.000000	0.000000
44	17	17	16	1.000000	0.000000
44	17	17	17	1.000000	0.000000
44	18	1	0	0.590909	0.590909
44	18	1	1	1.000000	0.409091
44	18	2	0	0.343552	0.343552
44	18	2	1	0.838266	0.494715
44	18	2	2	1.000000	0.161734
44	18	3	0	0.196315	0.196315
44	18	3	1	0.638025	0.441709
44	18	3	2	0.938387	0.300362
44	18	3	3	1.000000	0.061613
44	18	4	0	0.110128	0.110128
44	18	4	1	0.454877	0.344749
44	18	4	2	0.821173	0.366296
44	18	4	3	0.977459	0.156286
44	18	4	4	1.000000	0.022541
44	18	5	0	0.060570	0.060570
44	18	5	1	0.308359	0.247788
44	18	5	2	0.674654	0.366296
44	18	5	3	0.918851	0.244197
44	18	5	4	0.992110	0.073259
44	18	5	5	1.000000	0.007889
44	18	6	0	0.032615	0.032615
44	18	6	1	0.200348	0.167734
44	18	6	2	0.524379	0.324031
44	18	6	3	0.824929	0.300550
44	18	6	4	0.965812	0.140883
44	18	6	5	0.997370	0.031558
44	18	6	6	1.000000	0.002630
44	18	7	0	0.017166	0.017166

N	n	k	x	P(x)	p(x)
44	18	7	1	0.125310	0.108144
44	18	7	2	0.387945	0.262635
44	18	7	3	0.706291	0.318346
44	18	7	4	0.913908	0.207617
44	18	7	5	0.986574	0.072666
44	18	7	6	0.999169	0.012595
44	18	7	7	1.000000	0.000830
44	18	8	0	0.008815	0.008815
44	18	8	1	0.075622	0.066807
44	18	8	2	0.274373	0.198751
44	18	8	3	0.577232	0.302859
44	18	8	4	0.835350	0.258118
44	18	8	5	0.961043	0.125692
44	18	8	6	0.995084	0.034042
44	18	8	7	0.999753	0.004669
44	18	8	8	1.000000	0.000247
44	18	9	0	0.004407	0.004407
44	18	9	1	0.044074	0.039667
44	18	9	2	0.186039	0.141965
44	18	9	3	0.451041	0.265002
44	18	9	4	0.734971	0.283930
44	18	9	5	0.915654	0.180683
44	18	9	6	0.983737	0.068083
44	18	9	7	0.998327	0.014589
44	18	9	8	0.999931	0.001605
44	18	9	9	1.000000	0.000069
44	18	10	0	0.002141	0.002141
44	18	10	1	0.024807	0.022667
44	18	10	2	0.121141	0.096333
44	18	10	3	0.337469	0.216328
44	18	10	4	0.621399	0.283930
44	18	10	5	0.848543	0.227144
44	18	10	6	0.960394	0.111851
44	18	10	7	0.993741	0.033347
44	18	10	8	0.999473	0.005732
44	18	10	9	0.999982	0.000509
44	18	10	10	1.000000	0.000018
44	18	11	0	0.001007	0.001007
44	18	11	1	0.013474	0.012467
44	18	11	2	0.075807	0.062333
44	18	11	3	0.242030	0.166222
44	18	11	4	0.504486	0.262456
44	18	11	5	0.761694	0.257207
44	18	11	6	0.920917	0.159224
44	18	11	7	0.982953	0.062035
44	18	11	8	0.997787	0.014834
44	18	11	9	0.999847	0.002060
44	18	11	10	0.999996	0.000148
44	18	11	11	1.000000	0.000004
44	18	12	0	0.000458	0.000458

$N = 44$ $n = 17$–18

N	n	k	x	P(x)	p(x)
44	18	12	1	0.007052	0.006594
44	18	12	2	0.045585	0.038533
44	18	12	3	0.166474	0.120889
44	18	12	4	0.393141	0.226667
44	18	12	5	0.660370	0.267228
44	18	12	6	0.863018	0.202648
44	18	12	7	0.962274	0.099256
44	18	12	8	0.993292	0.031018
44	18	12	9	0.999286	0.005994
44	18	12	10	0.999960	0.000674
44	18	12	11	0.999999	0.000039
44	18	12	12	1.000000	0.000001
44	18	13	0	0.000200	0.000200
44	18	13	1	0.003549	0.003348
44	18	13	2	0.026319	0.022770
44	18	13	3	0.109808	0.083489
44	18	13	4	0.293974	0.184167
44	18	13	5	0.551808	0.257834
44	18	13	6	0.787025	0.235217
44	18	13	7	0.928155	0.141130
44	18	13	8	0.983599	0.055444
44	18	13	9	0.997600	0.014001
44	18	13	10	0.999791	0.002191
44	18	13	11	0.999990	0.000199
44	18	13	12	1.000000	0.000009
44	18	13	13	1.000000	0.000000
44	18	14	0	0.000084	0.000084
44	18	14	1	0.001713	0.001629
44	18	14	2	0.014566	0.012854
44	18	14	3	0.069410	0.054843
44	18	14	4	0.210802	0.141393
44	18	14	5	0.443684	0.232882
44	18	14	6	0.695973	0.252289
44	18	14	7	0.878076	0.182103
44	18	14	8	0.965714	0.087637
44	18	14	9	0.993535	0.027821
44	18	14	10	0.999226	0.005691
44	18	14	11	0.999945	0.000720
44	18	14	12	0.999998	0.000052
44	18	14	13	1.000000	0.000002
44	18	14	14	1.000000	0.000000
44	18	15	0	0.000034	0.000034
44	18	15	1	0.000790	0.000756
44	18	15	2	0.007711	0.006921
44	18	15	3	0.041988	0.034277
44	18	15	4	0.144819	0.102831
44	18	15	5	0.342769	0.197950
44	18	15	6	0.595058	0.252289
44	18	15	7	0.811305	0.216248
44	18	15	8	0.936501	0.125196

N	n	k	x	P(x)	p(x)
44	18	15	9	0.985189	0.048687
44	18	15	10	0.997708	0.012520
44	18	15	11	0.999778	0.002069
44	18	15	12	0.999987	0.000210
44	18	15	13	1.000000	0.000012
44	18	15	14	1.000000	0.000000
44	18	15	15	1.000000	0.000000
44	18	16	0	0.000013	0.000013
44	18	16	1	0.000346	0.000334
44	18	16	2	0.003892	0.003546
44	18	16	3	0.024259	0.020366
44	18	16	4	0.095177	0.070918
44	18	16	5	0.254033	0.158856
44	18	16	6	0.490662	0.236630
44	18	16	7	0.729280	0.238618
44	18	16	8	0.893330	0.164050
44	18	16	9	0.970079	0.076748
44	18	16	10	0.994254	0.024176
44	18	16	11	0.999278	0.005024
44	18	16	12	0.999944	0.000666
44	18	16	13	0.999997	0.000053
44	18	16	14	1.000000	0.000002
44	18	16	15	1.000000	0.000000
44	18	16	16	1.000000	0.000000
44	18	17	0	0.000005	0.000005
44	18	17	1	0.000144	0.000139
44	18	17	2	0.001866	0.001722
44	18	17	3	0.013348	0.011482
44	18	17	4	0.059718	0.046369
44	18	17	5	0.180278	0.120561
44	18	17	6	0.389250	0.208972
44	18	17	7	0.635538	0.246288
44	18	17	8	0.834741	0.199203
44	18	17	9	0.945410	0.110669
44	18	17	10	0.987347	0.041938
44	18	17	11	0.998022	0.010675
44	18	17	12	0.999801	0.001779
44	18	17	13	0.999988	0.000187
44	18	17	14	0.999999	0.000012
44	18	17	15	1.000000	0.000000
44	18	17	16	1.000000	0.000000
44	18	17	17	1.000000	0.000000
44	18	18	0	0.000002	0.000002
44	18	18	1	0.000056	0.000055
44	18	18	2	0.000846	0.000789
44	18	18	3	0.006969	0.006124
44	18	18	4	0.035674	0.028705
44	18	18	5	0.122230	0.086556
44	18	18	6	0.296373	0.174143
44	18	18	7	0.535198	0.238825

N	n	k	x	P(x)	p(x)
44	18	18	8	0.760962	0.225764
44	18	18	9	0.908520	0.147558
44	18	18	10	0.974921	0.066401
44	18	18	11	0.995255	0.020333
44	18	18	12	0.999406	0.004151
44	18	18	13	0.999953	0.000547
44	18	18	14	0.999998	0.000044
44	18	18	15	1.000000	0.000002
44	18	18	16	1.000000	0.000000
44	18	18	17	1.000000	0.000000
44	18	18	18	1.000000	0.000000
44	19	1	0	0.568182	0.568182
44	19	1	1	1.000000	0.431818
44	19	2	0	0.317125	0.317125
44	19	2	1	0.819239	0.502114
44	19	2	2	1.000000	0.180761
44	19	3	0	0.173664	0.173664
44	19	3	1	0.604047	0.430384
44	19	3	2	0.926835	0.322788
44	19	3	3	1.000000	0.073165
44	19	4	0	0.093185	0.093185
44	19	4	1	0.415098	0.321913
44	19	4	2	0.792996	0.377898
44	19	4	3	0.971448	0.178452
44	19	4	4	1.000000	0.028552
44	19	5	0	0.048922	0.048922
44	19	5	1	0.270237	0.221315
44	19	5	2	0.632389	0.362152
44	19	5	3	0.900067	0.267678
44	19	5	4	0.989293	0.089226
44	19	5	5	1.000000	0.010707
44	19	6	0	0.025088	0.025088
44	19	6	1	0.168092	0.143004
44	19	6	2	0.474528	0.306436
44	19	6	3	0.790251	0.315722
44	19	6	4	0.954975	0.164725
44	19	6	5	0.996156	0.041181
44	19	6	6	1.000000	0.003844
44	19	7	0	0.012544	0.012544
44	19	7	1	0.100353	0.087809
44	19	7	2	0.337438	0.237085
44	19	7	3	0.657315	0.319877
44	19	7	4	0.889952	0.232637
44	19	7	5	0.980984	0.091032
44	19	7	6	0.998685	0.017701
44	19	7	7	1.000000	0.001315
44	19	8	0	0.006103	0.006103
44	19	8	1	0.057635	0.051533
44	19	8	2	0.228507	0.170872
44	19	8	3	0.518990	0.290482

N	n	k	x	P(x)	p(x)
44	19	8	4	0.795640	0.276650
44	19	8	5	0.946540	0.150900
44	19	8	6	0.992466	0.045926
44	19	8	7	0.999574	0.007108
44	19	8	8	1.000000	0.000426
44	19	9	0	0.002882	0.002882
44	19	9	1	0.031869	0.028987
44	19	9	2	0.147818	0.115949
44	19	9	3	0.389887	0.242069
44	19	9	4	0.680369	0.290482
44	19	9	5	0.887856	0.207487
44	19	9	6	0.975881	0.088025
44	19	9	7	0.997204	0.021323
44	19	9	8	0.999870	0.002665
44	19	9	9	1.000000	0.000130
44	19	10	0	0.001317	0.001317
44	19	10	1	0.016961	0.015644
44	19	10	2	0.091500	0.074539
44	19	10	3	0.279227	0.187727
44	19	10	4	0.555877	0.276650
44	19	10	5	0.804862	0.248985
44	19	10	6	0.943186	0.138325
44	19	10	7	0.989894	0.046707
44	19	10	8	0.999032	0.009138
44	19	10	9	0.999963	0.000931
44	19	10	10	1.000000	0.000037
44	19	11	0	0.000581	0.000581
44	19	11	1	0.008679	0.008098
44	19	11	2	0.054231	0.045551
44	19	11	3	0.190885	0.136654
44	19	11	4	0.433825	0.242941
44	19	11	5	0.702338	0.268513
44	19	11	6	0.890298	0.187959
44	19	11	7	0.973409	0.083111
44	19	11	8	0.996075	0.022667
44	19	11	9	0.999689	0.003614
44	19	11	10	0.999990	0.000301
44	19	11	11	1.000000	0.000010
44	19	12	0	0.000247	0.000247
44	19	12	1	0.004262	0.004016
44	19	12	2	0.030765	0.026503
44	19	12	3	0.124628	0.093863
44	19	12	4	0.323398	0.198769
44	19	12	5	0.588424	0.265026
44	19	12	6	0.816253	0.227829
44	19	12	7	0.943186	0.126933
44	19	12	8	0.988520	0.045333
44	19	12	9	0.998594	0.010074
44	19	12	10	0.999908	0.001314
44	19	12	11	0.999997	0.000090

N	n	k	x	P(x)	p(x)
44	19	12	12	1.000000	0.000002
44	19	13	0	0.000100	0.000100
44	19	13	1	0.002003	0.001903
44	19	13	2	0.016685	0.014682
44	19	13	3	0.077696	0.061011
44	19	13	4	0.230224	0.152528
44	19	13	5	0.472475	0.242250
44	19	13	6	0.723697	0.251223
44	19	13	7	0.895586	0.171889
44	19	13	8	0.972936	0.077350
44	19	13	9	0.995446	0.022509
44	19	13	10	0.999538	0.004093
44	19	13	11	0.999975	0.000437
44	19	13	12	0.999999	0.000024
44	19	13	13	1.000000	0.000001
44	19	14	0	0.000039	0.000039
44	19	14	1	0.000898	0.000860
44	19	14	2	0.008634	0.007736
44	19	14	3	0.046207	0.037573
44	19	14	4	0.156421	0.110214
44	19	14	5	0.363071	0.206651
44	19	14	6	0.618346	0.255275
44	19	14	7	0.829049	0.210703
44	19	14	8	0.945490	0.116441
44	19	14	9	0.988185	0.042695
44	19	14	10	0.998350	0.010165
44	19	14	11	0.999862	0.001512
44	19	14	12	0.999994	0.000131
44	19	14	13	1.000000	0.000006
44	19	14	14	1.000000	0.000000
44	19	15	0	0.000014	0.000014
44	19	15	1	0.000383	0.000368
44	19	15	2	0.004250	0.003868
44	19	15	3	0.026168	0.021917
44	19	15	4	0.101314	0.075146
44	19	15	5	0.266634	0.165321
44	19	15	6	0.507727	0.241093
44	19	15	7	0.744768	0.237041
44	19	15	8	0.902795	0.158027
44	19	15	9	0.973953	0.071158
44	19	15	10	0.995301	0.021348
44	19	15	11	0.999459	0.004159
44	19	15	12	0.999963	0.000504
44	19	15	13	0.999999	0.000035
44	19	15	14	1.000000	0.000001
44	19	15	15	1.000000	0.000000
44	19	16	0	0.000005	0.000005
44	19	16	1	0.000154	0.000149
44	19	16	2	0.001983	0.001829
44	19	16	3	0.014075	0.012092
44	19	16	4	0.062445	0.048370
44	19	16	5	0.186824	0.124379
44	19	16	6	0.399651	0.212827
44	19	16	7	0.646682	0.247031
44	19	16	8	0.842853	0.196172
44	19	16	9	0.949416	0.106562
44	19	16	10	0.988675	0.039260
44	19	16	11	0.998312	0.009636
44	19	16	12	0.999842	0.001530
44	19	16	13	0.999991	0.000150
44	19	16	14	1.000000	0.000008
44	19	16	15	1.000000	0.000000
44	19	16	16	1.000000	0.000000
44	19	17	0	0.000002	0.000002
44	19	17	1	0.000058	0.000057
44	19	17	2	0.000873	0.000814
44	19	17	3	0.007166	0.006293
44	19	17	4	0.036533	0.029367
44	19	17	5	0.124635	0.088102
44	19	17	6	0.300839	0.176204
44	19	17	7	0.540811	0.239973
44	19	17	8	0.765786	0.224975
44	19	17	9	0.911358	0.145572
44	19	17	10	0.976056	0.064699
44	19	17	11	0.995559	0.019502
44	19	17	12	0.999459	0.003900
44	19	17	13	0.999959	0.000500
44	19	17	14	0.999998	0.000039
44	19	17	15	1.000000	0.000002
44	19	17	16	1.000000	0.000000
44	19	17	17	1.000000	0.000000
44	19	18	0	0.000000	0.000000
44	19	18	1	0.000020	0.000020
44	19	18	2	0.000360	0.000339
44	19	18	3	0.003436	0.003077
44	19	18	4	0.020218	0.016781
44	19	18	5	0.078952	0.058735
44	19	18	6	0.216000	0.137047
44	19	18	7	0.434157	0.218157
44	19	18	8	0.674130	0.239973
44	19	18	9	0.857442	0.183313
44	19	18	10	0.954490	0.097048
44	19	18	11	0.989780	0.035290
44	19	18	12	0.998448	0.008668
44	19	18	13	0.999848	0.001400
44	19	18	14	0.999991	0.000143
44	19	18	15	1.000000	0.000009
44	19	18	16	1.000000	0.000000
44	19	18	17	1.000000	0.000000
44	19	18	18	1.000000	0.000000
44	19	19	0	0.000000	0.000000
44	19	19	1	0.000007	0.000006
44	19	19	2	0.000138	0.000131
44	19	19	3	0.001543	0.001405
44	19	19	4	0.010536	0.008993
44	19	19	5	0.047326	0.036790
44	19	19	6	0.147476	0.100150
44	19	19	7	0.333469	0.185993
44	19	19	8	0.572603	0.239134
44	19	19	9	0.786937	0.214335
44	19	19	10	0.920897	0.133959
44	19	19	11	0.978922	0.058025
44	19	19	12	0.996114	0.017193
44	19	19	13	0.999525	0.003411
44	19	19	14	0.999963	0.000439
44	19	19	15	0.999998	0.000035
44	19	19	16	1.000000	0.000002
44	19	19	17	1.000000	0.000000
44	19	19	18	1.000000	0.000000
44	19	19	19	1.000000	0.000000
44	20	1	0	0.545455	0.545455
44	20	1	1	1.000000	0.454545
44	20	2	0	0.291755	0.291755
44	20	2	1	0.799154	0.507400
44	20	2	2	1.000000	0.200846
44	20	3	0	0.152824	0.152824
44	20	3	1	0.569616	0.416792
44	20	3	2	0.913923	0.344307
44	20	3	3	1.000000	0.086077
44	20	4	0	0.078276	0.078276
44	20	4	1	0.376469	0.298193
44	20	4	2	0.762764	0.386295
44	20	4	3	0.964310	0.201545
44	20	4	4	1.000000	0.035690
44	20	5	0	0.039138	0.039138
44	20	5	1	0.234827	0.195689
44	20	5	2	0.588931	0.354104
44	20	5	3	0.878653	0.289722
44	20	5	4	0.985724	0.107071
44	20	5	5	1.000000	0.014276
44	20	6	0	0.019067	0.019067
44	20	6	1	0.139491	0.120424
44	20	6	2	0.425498	0.286007
44	20	6	3	0.752364	0.326865
44	20	6	4	0.941797	0.189433
44	20	6	5	0.994509	0.052712
44	20	6	6	1.000000	0.005491
44	20	7	0	0.009032	0.009032
44	20	7	1	0.079279	0.070247
44	20	7	2	0.290021	0.210742
44	20	7	3	0.606135	0.316113
44	20	7	4	0.862036	0.255901
44	20	7	5	0.973702	0.111666
44	20	7	6	0.997977	0.024275
44	20	7	7	1.000000	0.002023
44	20	8	0	0.004150	0.004150
44	20	8	1	0.043206	0.039056
44	20	8	2	0.187498	0.144292
44	20	8	3	0.460893	0.273395
44	20	8	4	0.751376	0.290482
44	20	8	5	0.928432	0.177056
44	20	8	6	0.988792	0.060360
44	20	8	7	0.999289	0.010497
44	20	8	8	1.000000	0.000711
44	20	9	0	0.001844	0.001844
44	20	9	1	0.022593	0.020749
44	20	9	2	0.115352	0.092759
44	20	9	3	0.331790	0.216438
44	20	9	4	0.622273	0.290482
44	20	9	5	0.854658	0.232386
44	20	9	6	0.965318	0.110660
44	20	9	7	0.995498	0.030180
44	20	9	8	0.999763	0.004265
44	20	9	9	1.000000	0.000237
44	20	10	0	0.000790	0.000790
44	20	10	1	0.011329	0.010539
44	20	10	2	0.067647	0.056318
44	20	10	3	0.226663	0.159016
44	20	10	4	0.489481	0.262817
44	20	10	5	0.755064	0.265584
44	20	10	6	0.921054	0.165990
44	20	10	7	0.984289	0.063234
44	20	10	8	0.998301	0.014012
44	20	10	9	0.999925	0.001625
44	20	10	10	1.000000	0.000074
44	20	11	0	0.000325	0.000325
44	20	11	1	0.005440	0.005115
44	20	11	2	0.037832	0.032392
44	20	11	3	0.147155	0.109323
44	20	11	4	0.365802	0.218646
44	20	11	5	0.637895	0.272093
44	20	11	6	0.852706	0.214811
44	20	11	7	0.960111	0.107405
44	20	11	8	0.993355	0.033244
44	20	11	9	0.999400	0.006044
44	20	11	10	0.999978	0.000578
44	20	11	11	1.000000	0.000022
44	20	12	0	0.000128	0.000128
44	20	12	1	0.002495	0.002367
44	20	12	2	0.020164	0.017668

Table for $N = 2$, $n = 1$, through $N = 100$, $n = 50$

N	n	k	x	P(x)	p(x)
44	20	12	3	0.090837	0.070674
44	20	12	4	0.259791	0.168954
44	20	12	5	0.514216	0.254425
44	20	12	6	0.761574	0.247358
44	20	12	7	0.917800	0.156226
44	20	12	8	0.981267	0.063467
44	20	12	9	0.997385	0.016119
44	20	12	10	0.999803	0.002418
44	20	12	11	0.999994	0.000191
44	20	12	12	1.000000	0.000006
44	20	13	0	0.000048	0.000048
44	20	13	1	0.001090	0.001042
44	20	13	2	0.010225	0.009135
44	20	13	3	0.053292	0.043067
44	20	13	4	0.175314	0.122022
44	20	13	5	0.394955	0.219640
44	20	13	6	0.653355	0.258400
44	20	13	7	0.854333	0.200978
44	20	13	8	0.957466	0.103133
44	20	13	9	0.991844	0.034378
44	20	13	10	0.999047	0.007203
44	20	13	11	0.999940	0.000893
44	20	13	12	0.999998	0.000058
44	20	13	13	1.000000	0.000001
44	20	14	0	0.000017	0.000017
44	20	14	1	0.000451	0.000434
44	20	14	2	0.004921	0.004469
44	20	14	3	0.029675	0.024754
44	20	14	4	0.112335	0.082660
44	20	14	5	0.288677	0.176342
44	20	14	6	0.536658	0.247981
44	20	14	7	0.770052	0.233394
44	20	14	8	0.917544	0.147492
44	20	14	9	0.979646	0.062102
44	20	14	10	0.996724	0.017078
44	20	14	11	0.999681	0.002957
44	20	14	12	0.999983	0.000302
44	20	14	13	1.000000	0.000016
44	20	14	14	1.000000	0.000000
44	20	15	0	0.000006	0.000006
44	20	15	1	0.000176	0.000171
44	20	15	2	0.002239	0.002063
44	20	15	3	0.015647	0.013408
44	20	15	4	0.068249	0.052602
44	20	15	5	0.200506	0.132257
44	20	15	6	0.420934	0.220428
44	20	15	7	0.668915	0.247981
44	20	15	8	0.858547	0.189633
44	20	15	9	0.956875	0.098328
44	20	15	10	0.991031	0.034156

N	n	k	x	P(x)	p(x)
44	20	15	11	0.998794	0.007763
44	20	15	12	0.999903	0.001109
44	20	15	13	0.999996	0.000093
44	20	15	14	1.000000	0.000004
44	20	15	15	1.000000	0.000000
44	20	16	0	0.000002	0.000002
44	20	16	1	0.000065	0.000063
44	20	16	2	0.000959	0.000894
44	20	16	3	0.007787	0.006829
44	20	16	4	0.039228	0.031440
44	20	16	5	0.132097	0.092870
44	20	16	6	0.314520	0.182423
44	20	16	7	0.557751	0.243230
44	20	16	8	0.780078	0.222328
44	20	16	9	0.919578	0.139500
44	20	16	10	0.979253	0.059675
44	20	16	11	0.996385	0.017132
44	20	16	12	0.999597	0.003212
44	20	16	13	0.999973	0.000377
44	20	16	14	0.999999	0.000026
44	20	16	15	1.000000	0.000001
44	20	16	16	1.000000	0.000000
44	20	17	0	0.000000	0.000000
44	20	17	1	0.000022	0.000021
44	20	17	2	0.000384	0.000362
44	20	17	3	0.003641	0.003258
44	20	17	4	0.021262	0.017620
44	20	17	5	0.082346	0.061084
44	20	17	6	0.223309	0.140963
44	20	17	7	0.444822	0.221513
44	20	17	8	0.684795	0.239973
44	20	17	9	0.864775	0.179980
44	20	17	10	0.957941	0.093166
44	20	17	11	0.990878	0.032937
44	20	17	12	0.998679	0.007801
44	20	17	13	0.999879	0.001200
44	20	17	14	0.999993	0.000114
44	20	17	15	1.000000	0.000006
44	20	17	16	1.000000	0.000000
44	20	17	17	1.000000	0.000000
44	20	18	0	0.000000	0.000000
44	20	18	1	0.000007	0.000007
44	20	18	2	0.000143	0.000136
44	20	18	3	0.001590	0.001448
44	20	18	4	0.010820	0.009230
44	20	18	5	0.048410	0.037590
44	20	18	6	0.150217	0.101807
44	20	18	7	0.338168	0.187951
44	20	18	8	0.578140	0.239973
44	20	18	9	0.791450	0.213309

N	n	k	x	P(x)	p(x)
44	20	18	10	0.923435	0.131985
44	20	18	11	0.979899	0.056464
44	20	18	12	0.996368	0.016469
44	20	18	13	0.999568	0.003200
44	20	18	14	0.999968	0.000400
44	20	18	15	0.999999	0.000030
44	20	18	16	1.000000	0.000001
44	20	18	17	1.000000	0.000000
44	20	18	18	1.000000	0.000000
44	20	19	0	0.000000	0.000000
44	20	19	1	0.000002	0.000002
44	20	19	2	0.000049	0.000047
44	20	19	3	0.000644	0.000595
44	20	19	4	0.005140	0.004497
44	20	19	5	0.026724	0.021583
44	20	19	6	0.095398	0.068674
44	20	19	7	0.244192	0.148794
44	20	19	8	0.467384	0.223192
44	20	19	9	0.701204	0.233820
44	20	19	10	0.872671	0.171468
44	20	19	11	0.960354	0.087682
44	20	19	12	0.991300	0.030947
44	20	19	13	0.998706	0.007406
44	20	19	14	0.999876	0.001169
44	20	19	15	0.999993	0.000117
44	20	19	16	1.000000	0.000007
44	20	19	17	1.000000	0.000000
44	20	19	18	1.000000	0.000000
44	20	19	19	1.000000	0.000000
44	20	20	0	0.000000	0.000000
44	20	20	1	0.000000	0.000000
44	20	20	2	0.000015	0.000015
44	20	20	3	0.000239	0.000224
44	20	20	4	0.002262	0.002023
44	20	20	5	0.013774	0.011511
44	20	20	6	0.056940	0.043167
44	20	20	7	0.166819	0.109879
44	20	20	8	0.360252	0.193433
44	20	20	9	0.598323	0.238071
44	20	20	10	0.804084	0.205761
44	20	20	11	0.928788	0.124704
44	20	20	12	0.981397	0.052609
44	20	20	13	0.996633	0.015235
44	20	20	14	0.999595	0.002962
44	20	20	15	0.999969	0.000374
44	20	20	16	0.999999	0.000029
44	20	20	17	1.000000	0.000001
44	20	20	18	1.000000	0.000000
44	20	20	19	1.000000	0.000000
44	20	20	20	1.000000	0.000000

N	n	k	x	P(x)	p(x)
44	21	1	0	0.522727	0.522727
44	21	1	1	1.000000	0.477273
44	21	2	0	0.267442	0.267442
44	21	2	1	0.778013	0.510571
44	21	2	2	1.000000	0.221987
44	21	3	0	0.133721	0.133721
44	21	3	1	0.534884	0.401163
44	21	3	2	0.899577	0.364693
44	21	3	3	1.000000	0.100423
44	21	4	0	0.065230	0.065230
44	21	4	1	0.339195	0.273965
44	21	4	2	0.730573	0.391378
44	21	4	3	0.955912	0.225339
44	21	4	4	1.000000	0.044088
44	21	5	0	0.030984	0.030984
44	21	5	1	0.202212	0.171228
44	21	5	2	0.544668	0.342456
44	21	5	3	0.854509	0.309841
44	21	5	4	0.981263	0.126753
44	21	5	5	1.000000	0.018737
44	21	6	0	0.014300	0.014300
44	21	6	1	0.114403	0.100103
44	21	6	2	0.377831	0.263428
44	21	6	3	0.711506	0.333675
44	21	6	4	0.926011	0.214505
44	21	6	5	0.992313	0.066302
44	21	6	6	1.000000	0.007687
44	21	7	0	0.006398	0.006398
44	21	7	1	0.061717	0.055320
44	21	7	2	0.246117	0.184399
44	21	7	3	0.553449	0.307332
44	21	7	4	0.830048	0.276599
44	21	7	5	0.964396	0.134348
44	21	7	6	0.996966	0.032569
44	21	7	7	1.000000	0.003034
44	21	8	0	0.002766	0.002766
44	21	8	1	0.031815	0.029048
44	21	8	2	0.151425	0.119610
44	21	8	3	0.403936	0.252511
44	21	8	4	0.702962	0.299026
44	21	8	5	0.906300	0.203338
44	21	8	6	0.983762	0.077462
44	21	8	7	0.998852	0.015090
44	21	8	8	1.000000	0.001148
44	21	9	0	0.001153	0.001153
44	21	9	1	0.015677	0.014524
44	21	9	2	0.088297	0.072621
44	21	9	3	0.277681	0.189383
44	21	9	4	0.561755	0.284075
44	21	9	5	0.815927	0.254172

N	n	k	x	P(x)	p(x)
44	21	9	6	0.951486	0.135558
44	21	9	7	0.992983	0.041497
44	21	9	8	0.999585	0.006602
44	21	9	9	1.000000	0.000415
44	21	10	0	0.000461	0.000461
44	21	10	1	0.007377	0.006916
44	21	10	2	0.048875	0.041497
44	21	10	3	0.180284	0.131409
44	21	10	4	0.423776	0.243493
44	21	10	5	0.699734	0.275958
44	21	10	6	0.893389	0.193655
44	21	10	7	0.976384	0.082995
44	21	10	8	0.997133	0.020749
44	21	10	9	0.999858	0.002725
44	21	10	10	1.000000	0.000142
44	21	11	0	0.000176	0.000176
44	21	11	1	0.003309	0.003133
44	21	11	2	0.025685	0.022376
44	21	11	3	0.110714	0.085029
44	21	11	4	0.302030	0.191316
44	21	11	5	0.569872	0.267842
44	21	11	6	0.807953	0.238082
44	21	11	7	0.942210	0.134257
44	21	11	8	0.989200	0.046990
44	21	11	9	0.998896	0.009696
44	21	11	10	0.999954	0.001058
44	21	11	11	1.000000	0.000046
44	21	12	0	0.000064	0.000064
44	21	12	1	0.001410	0.001346
44	21	12	2	0.012802	0.011391
44	21	12	3	0.064335	0.051533
44	21	12	4	0.203473	0.139139
44	21	12	5	0.440009	0.236536
44	21	12	6	0.699734	0.259725
44	21	12	7	0.885253	0.185518
44	21	12	8	0.970689	0.085436
44	21	12	9	0.995370	0.024682
44	21	12	10	0.999601	0.004231
44	21	12	11	0.999986	0.000385
44	21	12	12	1.000000	0.000014
44	21	13	0	0.000022	0.000022
44	21	13	1	0.000569	0.000547
44	21	13	2	0.006038	0.005469
44	21	13	3	0.035347	0.029309
44	21	13	4	0.129556	0.094208
44	21	13	5	0.321741	0.192185
44	21	13	6	0.577988	0.256247
44	21	13	7	0.804088	0.226100
44	21	13	8	0.935980	0.131892
44	21	13	9	0.986115	0.050134

N	n	k	x	P(x)	p(x)
44	21	13	10	0.998147	0.012032
44	21	13	11	0.999866	0.001719
44	21	13	12	0.999996	0.000130
44	21	13	13	1.000000	0.000004
44	21	14	0	0.000007	0.000007
44	21	14	1	0.000216	0.000209
44	21	14	2	0.002686	0.002470
44	21	14	3	0.018329	0.015643
44	21	14	4	0.077893	0.059564
44	21	14	5	0.222549	0.144656
44	21	14	6	0.453998	0.231449
44	21	14	7	0.701979	0.247981
44	21	14	8	0.880671	0.178692
44	21	14	9	0.966708	0.086037
44	21	14	10	0.993877	0.027170
44	21	14	11	0.999311	0.005434
44	21	14	12	0.999958	0.000647
44	21	14	13	0.999999	0.000041
44	21	14	14	1.000000	0.000001
44	21	15	0	0.000002	0.000002
44	21	15	1	0.000077	0.000075
44	21	15	2	0.001122	0.001045
44	21	15	3	0.008943	0.007822
44	21	15	4	0.044140	0.035197
44	21	15	5	0.145399	0.101259
44	21	15	6	0.338273	0.192874
44	21	15	7	0.586254	0.247981
44	21	15	8	0.803238	0.216983
44	21	15	9	0.932293	0.129055
44	21	15	10	0.983915	0.051622
44	21	15	11	0.997500	0.013585
44	21	15	12	0.999764	0.002264
44	21	15	13	0.999988	0.000224
44	21	15	14	1.000000	0.000012
44	21	15	15	1.000000	0.000000
44	21	16	0	0.000001	0.000001
44	21	16	1	0.000025	0.000025
44	21	16	2	0.000437	0.000412
44	21	16	3	0.004089	0.003651
44	21	16	4	0.023508	0.019419
44	21	16	5	0.089532	0.066025
44	21	16	6	0.238511	0.148979
44	21	16	7	0.466539	0.228028
44	21	16	8	0.705969	0.239430
44	21	16	9	0.878891	0.172922
44	21	16	10	0.964334	0.085444
44	21	16	11	0.992816	0.028481
44	21	16	12	0.999061	0.006246
44	21	16	13	0.999926	0.000865
44	21	16	14	0.999997	0.000071

N	n	k	x	P(x)	p(x)
44	21	16	15	1.000000	0.000003
44	21	16	16	1.000000	0.000000
44	21	17	0	0.000000	0.000000
44	21	17	1	0.000008	0.000007
44	21	17	2	0.000158	0.000150
44	21	17	3	0.001741	0.001584
44	21	17	4	0.011717	0.009976
44	21	17	5	0.051804	0.040086
44	21	17	6	0.158701	0.106897
44	21	17	7	0.352525	0.193824
44	21	17	8	0.594805	0.242280
44	21	17	9	0.804781	0.209976
44	21	17	10	0.930767	0.125986
44	21	17	11	0.982644	0.051876
44	21	17	12	0.997054	0.014410
44	21	17	13	0.999679	0.002625
44	21	17	14	0.999979	0.000300
44	21	17	15	0.999999	0.000020
44	21	17	16	1.000000	0.000001
44	21	17	17	1.000000	0.000000
44	21	18	0	0.000000	0.000000
44	21	18	1	0.000002	0.000002
44	21	18	2	0.000052	0.000050
44	21	18	3	0.000686	0.000633
44	21	18	4	0.005436	0.004751
44	21	18	5	0.028049	0.022613
44	21	18	6	0.099314	0.071265
44	21	18	7	0.252024	0.152710
44	21	18	8	0.478152	0.226128
44	21	18	9	0.711459	0.233307
44	21	18	10	0.879440	0.167981
44	21	18	11	0.963430	0.083990
44	21	18	12	0.992250	0.028820
44	21	18	13	0.998901	0.006651
44	21	18	14	0.999901	0.001000
44	21	18	15	0.999995	0.000093
44	21	18	16	1.000000	0.000005
44	21	18	17	1.000000	0.000000
44	21	18	18	1.000000	0.000000
44	21	19	0	0.000000	0.000000
44	21	19	1	0.000001	0.000000
44	21	19	2	0.000016	0.000015
44	21	19	3	0.000247	0.000231
44	21	19	4	0.002330	0.002083
44	21	19	5	0.014133	0.011803
44	21	19	6	0.058199	0.044066
44	21	19	7	0.169795	0.111596
44	21	19	8	0.365088	0.195293
44	21	19	9	0.603779	0.238691
44	21	19	10	0.808371	0.204592

N	n	k	x	P(x)	p(x)
44	21	19	11	0.931126	0.122755
44	21	19	12	0.982274	0.051148
44	21	19	13	0.996855	0.014581
44	21	19	14	0.999632	0.002777
44	21	19	15	0.999973	0.000341
44	21	19	16	0.999999	0.000026
44	21	19	17	1.000000	0.000001
44	21	19	18	1.000000	0.000000
44	21	19	19	1.000000	0.000000
44	21	20	0	0.000000	0.000000
44	21	20	1	0.000000	0.000000
44	21	20	2	0.000004	0.000004
44	21	20	3	0.000080	0.000076
44	21	20	4	0.000914	0.000833
44	21	20	5	0.006579	0.005666
44	21	20	6	0.031760	0.025181
44	21	20	7	0.107301	0.075542
44	21	20	8	0.263536	0.156234
44	21	20	9	0.489207	0.225671
44	21	20	10	0.718350	0.229143
44	21	20	11	0.882024	0.163674
44	21	20	12	0.963861	0.081837
44	21	20	13	0.992189	0.028328
44	21	20	14	0.998855	0.006665
44	21	20	15	0.999891	0.001037
44	21	20	16	0.999994	0.000102
44	21	20	17	1.000000	0.000006
44	21	20	18	1.000000	0.000000
44	21	20	19	1.000000	0.000000
44	21	20	20	1.000000	0.000000
44	21	21	0	0.000000	0.000000
44	21	21	1	0.000000	0.000000
44	21	21	2	0.000001	0.000001
44	21	21	3	0.000023	0.000022
44	21	21	4	0.000323	0.000300
44	21	21	5	0.002802	0.002479
44	21	21	6	0.016022	0.013220
44	21	21	7	0.063235	0.047214
44	21	21	8	0.178909	0.115673
44	21	21	9	0.376371	0.197462
44	21	21	10	0.613326	0.236955
44	21	21	11	0.813827	0.200500
44	21	21	12	0.933172	0.119345
44	21	21	13	0.982746	0.049574
44	21	21	14	0.996910	0.014164
44	21	21	15	0.999632	0.002722
44	21	21	16	0.999972	0.000340
44	21	21	17	0.999999	0.000026
44	21	21	18	1.000000	0.000001
44	21	21	19	1.000000	0.000000

Table for $N = 2$, $n = 1$, through $N = 100$, $n = 50$

N	n	k	x	P(x)	p(x)
44	21	21	20	1.000000	0.000000
44	21	21	21	1.000000	0.000000
44	22	1	0	0.500000	0.500000
44	22	1	1	1.000000	0.500000
44	22	2	0	0.244186	0.244186
44	22	2	1	0.755814	0.511628
44	22	2	2	1.000000	0.244186
44	22	3	0	0.116279	0.116279
44	22	3	1	0.500000	0.383721
44	22	3	2	0.883721	0.383721
44	22	3	3	1.000000	0.116279
44	22	4	0	0.053885	0.053885
44	22	4	1	0.303460	0.249575
44	22	4	2	0.696540	0.393080
44	22	4	3	0.946115	0.249575
44	22	4	4	1.000000	0.053885
44	22	5	0	0.024248	0.024248
44	22	5	1	0.172433	0.148185
44	22	5	2	0.500000	0.327567
44	22	5	3	0.827567	0.327567
44	22	5	4	0.975751	0.148185
44	22	5	5	1.000000	0.024248
44	22	6	0	0.010570	0.010570
44	22	6	1	0.092641	0.082072
44	22	6	2	0.332017	0.239376
44	22	6	3	0.667983	0.335966
44	22	6	4	0.907358	0.239376
44	22	6	5	0.989430	0.082072
44	22	6	6	1.000000	0.010570
44	22	7	0	0.004450	0.004450
44	22	7	1	0.047286	0.042836
44	22	7	2	0.206030	0.158744
44	22	7	3	0.500000	0.293970
44	22	7	4	0.793970	0.293970
44	22	7	5	0.952714	0.158744
44	22	7	6	0.995549	0.042836
44	22	7	7	1.000000	0.004450
44	22	8	0	0.001804	0.001804
44	22	8	1	0.022974	0.021170
44	22	8	2	0.120222	0.097248
44	22	8	3	0.349042	0.228820
44	22	8	4	0.650958	0.301915
44	22	8	5	0.879777	0.228820
44	22	8	6	0.977026	0.097248
44	22	8	7	0.998196	0.021170
44	22	8	8	1.000000	0.001804
44	22	9	0	0.000702	0.000702
44	22	9	1	0.010625	0.009923
44	22	9	2	0.066195	0.055571
44	22	9	3	0.228276	0.162081
44	22	9	4	0.500000	0.271724
44	22	9	5	0.771724	0.271724
44	22	9	6	0.933804	0.162081
44	22	9	7	0.989375	0.055571
44	22	9	8	0.999298	0.009923
44	22	9	9	1.000000	0.000702
44	22	10	0	0.000261	0.000261
44	22	10	1	0.004671	0.004410
44	22	10	2	0.034441	0.029770
44	22	10	3	0.140290	0.105849
44	22	10	4	0.360256	0.219967
44	22	10	5	0.639744	0.279487
44	22	10	6	0.859710	0.219967
44	22	10	7	0.965559	0.105849
44	22	10	8	0.995329	0.029770
44	22	10	9	0.999739	0.004410
44	22	10	10	1.000000	0.000261
44	22	11	0	0.000092	0.000092
44	22	11	1	0.001947	0.001855
44	22	11	2	0.016929	0.014982
44	22	11	3	0.081139	0.064210
44	22	11	4	0.243803	0.162664
44	22	11	5	0.500000	0.256197
44	22	11	6	0.756197	0.256197
44	22	11	7	0.918861	0.162664
44	22	11	8	0.983071	0.064210
44	22	11	9	0.998053	0.014982
44	22	11	10	0.999908	0.001855
44	22	11	11	1.000000	0.000092
44	22	12	0	0.000031	0.000031
44	22	12	1	0.000767	0.000736
44	22	12	2	0.007849	0.007083
44	22	12	3	0.044170	0.036321
44	22	12	4	0.155077	0.110908
44	22	12	5	0.368020	0.212943
44	22	12	6	0.631980	0.263960
44	22	12	7	0.844923	0.212943
44	22	12	8	0.955830	0.110908
44	22	12	9	0.992151	0.036321
44	22	12	10	0.999233	0.007083
44	22	12	11	0.999969	0.000736
44	22	12	12	1.000000	0.000031
44	22	13	0	0.000010	0.000010
44	22	13	1	0.000284	0.000274
44	22	13	2	0.003422	0.003139
44	22	13	3	0.022604	0.019182
44	22	13	4	0.092692	0.070087
44	22	13	5	0.254894	0.162202
44	22	13	6	0.500000	0.245106
44	22	13	7	0.745106	0.245106
44	22	13	8	0.907308	0.162202
44	22	13	9	0.977396	0.070087
44	22	13	10	0.996577	0.019182
44	22	13	11	0.999716	0.003139
44	22	13	12	0.999990	0.000274
44	22	13	13	1.000000	0.000010
44	22	14	0	0.000003	0.000003
44	22	14	1	0.000098	0.000095
44	22	14	2	0.001397	0.001299
44	22	14	3	0.010848	0.009450
44	22	14	4	0.051996	0.041148
44	22	14	5	0.165944	0.113949
44	22	14	6	0.373494	0.207549
44	22	14	7	0.626506	0.253012
44	22	14	8	0.834055	0.207549
44	22	14	9	0.948004	0.113949
44	22	14	10	0.989152	0.041148
44	22	14	11	0.998602	0.009450
44	22	14	12	0.999902	0.001299
44	22	14	13	0.999997	0.000095
44	22	14	14	1.000000	0.000003
44	22	15	0	0.000001	0.000001
44	22	15	1	0.000031	0.000031
44	22	15	2	0.000531	0.000500
44	22	15	3	0.004862	0.004331
44	22	15	4	0.027307	0.022444
44	22	15	5	0.101374	0.074067
44	22	15	6	0.262801	0.161427
44	22	15	7	0.500000	0.237199
44	22	15	8	0.737199	0.237199
44	22	15	9	0.898626	0.161427
44	22	15	10	0.972693	0.074067
44	22	15	11	0.995137	0.022444
44	22	15	12	0.999469	0.004331
44	22	15	13	0.999969	0.000500
44	22	15	14	0.999999	0.000031
44	22	15	15	1.000000	0.000001
44	22	16	0	0.000000	0.000000
44	22	16	1	0.000009	0.000009
44	22	16	2	0.000186	0.000177
44	22	16	3	0.002025	0.001838
44	22	16	4	0.013376	0.011351
44	22	16	5	0.057955	0.044579
44	22	16	6	0.173737	0.115782
44	22	16	7	0.377311	0.203573
44	22	16	8	0.622689	0.245378
44	22	16	9	0.826262	0.203573
44	22	16	10	0.942045	0.115782
44	22	16	11	0.986624	0.044579
44	22	16	12	0.997975	0.011351
44	22	16	13	0.999813	0.001838
44	22	16	14	0.999991	0.000177
44	22	16	15	1.000000	0.000009
44	22	16	16	1.000000	0.000000
44	22	17	0	0.000000	0.000000
44	22	17	1	0.000002	0.000002
44	22	17	2	0.000060	0.000057
44	22	17	3	0.000777	0.000717
44	22	17	4	0.006079	0.005301
44	22	17	5	0.030889	0.024810
44	22	17	6	0.107576	0.076687
44	22	17	7	0.268254	0.160677
44	22	17	8	0.500000	0.231746
44	22	17	9	0.731746	0.231746
44	22	17	10	0.892424	0.160677
44	22	17	11	0.969111	0.076687
44	22	17	12	0.993921	0.024810
44	22	17	13	0.999223	0.005301
44	22	17	14	0.999940	0.000717
44	22	17	15	0.999997	0.000057
44	22	17	16	1.000000	0.000002
44	22	17	17	1.000000	0.000000
44	22	18	0	0.000000	0.000000
44	22	18	1	0.000001	0.000001
44	22	18	2	0.000017	0.000017
44	22	18	3	0.000272	0.000255
44	22	18	4	0.002544	0.002272
44	22	18	5	0.015268	0.012723
44	22	18	6	0.062132	0.046864
44	22	18	7	0.178988	0.116856
44	22	18	8	0.379835	0.200847
44	22	18	9	0.620165	0.240330
44	22	18	10	0.821012	0.200847
44	22	18	11	0.937868	0.116856
44	22	18	12	0.984732	0.046864
44	22	18	13	0.997455	0.012723
44	22	18	14	0.999727	0.002272
44	22	18	15	0.999983	0.000255
44	22	18	16	0.999999	0.000017
44	22	18	17	1.000000	0.000001
44	22	18	18	1.000000	0.000000
44	22	19	0	0.000000	0.000000
44	22	19	1	0.000000	0.000000
44	22	19	2	0.000004	0.000004
44	22	19	3	0.000086	0.000082
44	22	19	4	0.000971	0.000886
44	22	19	5	0.006949	0.005977
44	22	19	6	0.033292	0.026344
44	22	19	7	0.111571	0.078279
44	22	19	8	0.271687	0.160116

Table for $N = 2$, $n = 1$, through $N = 100$, $n = 50$

N	n	k	x	P(x)	p(x)
44	22	19	9	0.500000	0.228313
44	22	19	10	0.728313	0.228313
44	22	19	11	0.888429	0.160116
44	22	19	12	0.966707	0.078279
44	22	19	13	0.993051	0.026344
44	22	19	14	0.999028	0.005977
44	22	19	15	0.999914	0.000886
44	22	19	16	0.999995	0.000082
44	22	19	17	1.000000	0.000004
44	22	19	18	1.000000	0.000000
44	22	19	19	1.000000	0.000000
44	22	20	0	0.000000	0.000000
44	22	20	1	0.000000	0.000000
44	22	20	2	0.000001	0.000001
44	22	20	3	0.000024	0.000023
44	22	20	4	0.000334	0.000310
44	22	20	5	0.002884	0.002550
44	22	20	6	0.016432	0.013548
44	22	20	7	0.064604	0.048172
44	22	20	8	0.182022	0.117418
44	22	20	9	0.381277	0.199255
44	22	20	10	0.618723	0.237446
44	22	20	11	0.817978	0.199255
44	22	20	12	0.935396	0.117418
44	22	20	13	0.983567	0.048172
44	22	20	14	0.997116	0.013548
44	22	20	15	0.999666	0.002550
44	22	20	16	0.999976	0.000310
44	22	20	17	0.999999	0.000023
44	22	20	18	1.000000	0.000001
44	22	20	19	1.000000	0.000000
44	22	20	20	1.000000	0.000000
44	22	21	0	0.000000	0.000000
44	22	21	1	0.000000	0.000000
44	22	21	2	0.000000	0.000000
44	22	21	3	0.000006	0.000006
44	22	21	4	0.000101	0.000096
44	22	21	5	0.001078	0.000976
44	22	21	6	0.007400	0.006323
44	22	21	7	0.034497	0.027096
44	22	21	8	0.113528	0.079031
44	22	21	9	0.273347	0.159819
44	22	21	10	0.500000	0.226653
44	22	21	11	0.726653	0.226653
44	22	21	12	0.886472	0.159819
44	22	21	13	0.965503	0.079031
44	22	21	14	0.992600	0.027096
44	22	21	15	0.998922	0.006323
44	22	21	16	0.999898	0.000976
44	22	21	17	0.999994	0.000096
44	22	21	18	1.000000	0.000006
44	22	21	19	1.000000	0.000000
44	22	21	20	1.000000	0.000000
44	22	21	21	1.000000	0.000000
44	22	22	0	0.000000	0.000000
44	22	22	1	0.000000	0.000000
44	22	22	2	0.000000	0.000000
44	22	22	3	0.000001	0.000001
44	22	22	4	0.000027	0.000025
44	22	22	5	0.000356	0.000330
44	22	22	6	0.003002	0.002646
44	22	22	7	0.016825	0.013823
44	22	22	8	0.065422	0.048597
44	22	22	9	0.183015	0.117593
44	22	22	10	0.381746	0.198732
44	22	22	11	0.618254	0.236507
44	22	22	12	0.816985	0.198732
44	22	22	13	0.934578	0.117593
44	22	22	14	0.983175	0.048597
44	22	22	15	0.996998	0.013823
44	22	22	16	0.999644	0.002646
44	22	22	17	0.999973	0.000330
44	22	22	18	0.999999	0.000025
44	22	22	19	1.000000	0.000001
44	22	22	20	1.000000	0.000000
44	22	22	21	1.000000	0.000000
44	22	22	22	1.000000	0.000000
45	1	1	0	0.977778	0.977778
45	1	1	1	1.000000	0.022222
45	2	1	0	0.955556	0.955556
45	2	1	1	1.000000	0.044444
45	2	2	0	0.912121	0.912121
45	2	2	1	0.998990	0.086869
45	2	2	2	1.000000	0.001010
45	3	1	0	0.933333	0.933333
45	3	1	1	1.000000	0.066667
45	3	2	0	0.869697	0.869697
45	3	2	1	0.996970	0.127273
45	3	2	2	1.000000	0.003030
45	3	3	0	0.809020	0.809020
45	3	3	1	0.991050	0.182030
45	3	3	2	0.999929	0.008879
45	3	3	3	1.000000	0.000070
45	4	1	0	0.911111	0.911111
45	4	1	1	1.000000	0.088889
45	4	2	0	0.828283	0.828283
45	4	2	1	0.993939	0.165657
45	4	2	2	1.000000	0.006061
45	4	3	0	0.751233	0.751233
45	4	3	1	0.982382	0.231149
45	4	3	2	0.999718	0.017336
45	4	3	3	1.000000	0.000282
45	4	4	0	0.679687	0.679687
45	4	4	1	0.965871	0.286184
45	4	4	2	0.998893	0.033021
45	4	4	3	0.999993	0.001101
45	4	4	4	1.000000	0.000007
45	5	1	0	0.888889	0.888889
45	5	1	1	1.000000	0.111111
45	5	2	0	0.787879	0.787879
45	5	2	1	0.989899	0.202020
45	5	2	2	1.000000	0.010101
45	5	3	0	0.696265	0.696265
45	5	3	1	0.971106	0.274841
45	5	3	2	0.999295	0.028189
45	5	3	3	1.000000	0.000705
45	5	4	0	0.613376	0.613376
45	5	4	1	0.944931	0.331555
45	5	4	2	0.997282	0.052351
45	5	4	3	0.999966	0.002685
45	5	4	4	1.000000	0.000034
45	5	5	0	0.538574	0.538574
45	5	5	1	0.912584	0.374010
45	5	5	2	0.993451	0.080867
45	5	5	3	0.999835	0.006384
45	5	5	4	0.999999	0.000164
45	5	5	5	1.000000	0.000001
45	6	1	0	0.866667	0.866667
45	6	1	1	1.000000	0.133333
45	6	2	0	0.748485	0.748485
45	6	2	1	0.984848	0.236364
45	6	2	2	1.000000	0.015152
45	6	3	0	0.644045	0.644045
45	6	3	1	0.957364	0.313319
45	6	3	2	0.998591	0.041226
45	6	3	3	1.000000	0.001409
45	6	4	0	0.552039	0.552039
45	6	4	1	0.920064	0.368026
45	6	4	2	0.994664	0.074600
45	6	4	3	0.999899	0.005235
45	6	4	4	1.000000	0.000101
45	6	5	0	0.471253	0.471253
45	6	5	1	0.875183	0.403931
45	6	5	2	0.987386	0.112203
45	6	5	3	0.999516	0.012130
45	6	5	4	0.999995	0.000479
45	6	5	5	1.000000	0.000005
45	6	6	0	0.400565	0.400565
45	6	6	1	0.824692	0.424127
45	6	6	2	0.976166	0.151474
45	6	6	3	0.998606	0.022441
45	6	6	4	0.999971	0.001365
45	6	6	5	1.000000	0.000029
45	6	6	6	1.000000	0.000000
45	7	1	0	0.844444	0.844444
45	7	1	1	1.000000	0.155556
45	7	2	0	0.710101	0.710101
45	7	2	1	0.978788	0.268687
45	7	2	2	1.000000	0.021212
45	7	3	0	0.594503	0.594503
45	7	3	1	0.941297	0.346794
45	7	3	2	0.997533	0.056237
45	7	3	3	1.000000	0.002467
45	7	4	0	0.495419	0.495419
45	7	4	1	0.891755	0.396335
45	7	4	2	0.990839	0.099084
45	7	4	3	0.999765	0.008926
45	7	4	4	1.000000	0.000235
45	7	5	0	0.410836	0.410836
45	7	5	1	0.833754	0.422919
45	7	5	2	0.978755	0.145001
45	7	5	3	0.998894	0.020139
45	7	5	4	0.999983	0.001089
45	7	5	5	1.000000	0.000017
45	7	6	0	0.338939	0.338939
45	7	6	1	0.770317	0.431377
45	7	6	2	0.960630	0.190314
45	7	6	3	0.996880	0.036250
45	7	6	4	0.999901	0.003021
45	7	6	5	0.999999	0.000098
45	7	6	6	1.000000	0.000001
45	7	7	0	0.278104	0.278104
45	7	7	1	0.703951	0.425847
45	7	7	2	0.936231	0.232280
45	7	7	3	0.993162	0.056931
45	7	7	4	0.999669	0.006506
45	7	7	5	0.999994	0.000325
45	7	7	6	1.000000	0.000006
45	7	7	7	1.000000	0.000000
45	8	1	0	0.822222	0.822222
45	8	1	1	1.000000	0.177778
45	8	2	0	0.672727	0.672727
45	8	2	1	0.971717	0.298990
45	8	2	2	1.000000	0.028283
45	8	3	0	0.547569	0.547569
45	8	3	1	0.923044	0.375476
45	8	3	2	0.996054	0.073009
45	8	3	3	1.000000	0.003946
45	8	4	0	0.443270	0.443270
45	8	4	1	0.860465	0.417195

Table for $N = 2$, $n = 1$, through $N = 100$, $n = 50$

N	n	k	x	P(x)	p(x)
45	8	4	2	0.985624	0.125159
45	8	4	3	0.999530	0.013907
45	8	4	4	1.000000	0.000470
45	8	5	0	0.356778	0.356778
45	8	5	1	0.789237	0.432458
45	8	5	2	0.967308	0.178071
45	8	5	3	0.997834	0.030526
45	8	5	4	0.999954	0.002120
45	8	5	5	1.000000	0.000046
45	8	6	0	0.285423	0.285423
45	8	6	1	0.713556	0.428134
45	8	6	2	0.940597	0.227041
45	8	6	3	0.994018	0.053421
45	8	6	4	0.999742	0.005724
45	8	6	5	0.999996	0.000254
45	8	6	6	1.000000	0.000003
45	8	7	0	0.226874	0.226874
45	8	7	1	0.636712	0.409838
45	8	7	2	0.905668	0.268956
45	8	7	3	0.987169	0.081502
45	8	7	4	0.999155	0.011986
45	8	7	5	0.999977	0.000822
45	8	7	6	1.000000	0.000023
45	8	7	7	1.000000	0.000000
45	8	8	0	0.179111	0.179111
45	8	8	1	0.561215	0.382104
45	8	8	2	0.863201	0.301986
45	8	8	3	0.976446	0.113245
45	8	8	4	0.997893	0.021448
45	8	8	5	0.999912	0.002019
45	8	8	6	0.999998	0.000087
45	8	8	7	1.000000	0.000001
45	8	8	8	1.000000	0.000000
45	9	1	0	0.800000	0.800000
45	9	1	1	1.000000	0.200000
45	9	2	0	0.636364	0.636364
45	9	2	1	0.963636	0.327273
45	9	2	2	1.000000	0.036364
45	9	3	0	0.503171	0.503171
45	9	3	1	0.902748	0.399577
45	9	3	2	0.994080	0.091332
45	9	3	3	1.000000	0.005920
45	9	4	0	0.395349	0.395349
45	9	4	1	0.826638	0.431290
45	9	4	2	0.978858	0.152220
45	9	4	3	0.999154	0.020296
45	9	4	4	1.000000	0.000846
45	9	5	0	0.308565	0.308565
45	9	5	1	0.742484	0.433919
45	9	5	2	0.952870	0.210385
45	9	5	3	0.996184	0.043315
45	9	5	4	0.999897	0.003713
45	9	5	5	1.000000	0.000103
45	9	6	0	0.239138	0.239138
45	9	6	1	0.655700	0.416563
45	9	6	2	0.916052	0.260352
45	9	6	3	0.989687	0.073635
45	9	6	4	0.999433	0.009746
45	9	6	5	0.999990	0.000557
45	9	6	6	1.000000	0.000010
45	9	7	0	0.183952	0.183952
45	9	7	1	0.570252	0.386300
45	9	7	2	0.869322	0.299071
45	9	7	3	0.978359	0.109036
45	9	7	4	0.998183	0.019825
45	9	7	5	0.999933	0.001749
45	9	7	6	0.999999	0.000067
45	9	7	7	1.000000	0.000001
45	9	8	0	0.140385	0.140385
45	9	8	1	0.488925	0.348541
45	9	8	2	0.814230	0.325305
45	9	8	3	0.961142	0.146912
45	9	8	4	0.995575	0.034432
45	9	8	5	0.999748	0.004174
45	9	8	6	0.999994	0.000246
45	9	8	7	1.000000	0.000006
45	9	8	8	1.000000	0.000000
45	9	9	0	0.106237	0.106237
45	9	9	1	0.413565	0.307328
45	9	9	2	0.752686	0.339121
45	9	9	3	0.937319	0.184632
45	9	9	4	0.990922	0.053603
45	9	9	5	0.999297	0.008375
45	9	9	6	0.999974	0.000677
45	9	9	7	1.000000	0.000026
45	9	9	8	1.000000	0.000000
45	9	9	9	1.000000	0.000000
45	10	1	0	0.777778	0.777778
45	10	1	1	1.000000	0.222222
45	10	2	0	0.601010	0.601010
45	10	2	1	0.954545	0.353535
45	10	2	2	1.000000	0.045455
45	10	3	0	0.461240	0.461240
45	10	3	1	0.880550	0.419309
45	10	3	2	0.991543	0.110994
45	10	3	3	1.000000	0.008457
45	10	4	0	0.351421	0.351421
45	10	4	1	0.790698	0.439276
45	10	4	2	0.970402	0.179704
45	10	4	3	0.998590	0.028189
45	10	4	4	1.000000	0.001409
45	10	5	0	0.265709	0.265709
45	10	5	1	0.694271	0.428562
45	10	5	2	0.935337	0.241066
45	10	5	3	0.993778	0.058440
45	10	5	4	0.999794	0.006016
45	10	5	5	1.000000	0.000206
45	10	6	0	0.199282	0.199282
45	10	6	1	0.597845	0.398563
45	10	6	2	0.887124	0.289280
45	10	6	3	0.983551	0.096427
45	10	6	4	0.998891	0.015341
45	10	6	5	0.999974	0.001083
45	10	6	6	1.000000	0.000026
45	10	7	0	0.148184	0.148184
45	10	7	1	0.505868	0.357685
45	10	7	2	0.827785	0.321916
45	10	7	3	0.966243	0.138459
45	10	7	4	0.996531	0.030288
45	10	7	5	0.999835	0.003304
45	10	7	6	0.999997	0.000162
45	10	7	7	1.000000	0.000003
45	10	8	0	0.109188	0.109188
45	10	8	1	0.421154	0.311966
45	10	8	2	0.760013	0.338859
45	10	8	3	0.940738	0.180725
45	10	8	4	0.991749	0.051011
45	10	8	5	0.999401	0.007652
45	10	8	6	0.999980	0.000580
45	10	8	7	1.000000	0.000019
45	10	8	8	1.000000	0.000000
45	10	9	0	0.079678	0.079678
45	10	9	1	0.345270	0.265592
45	10	9	2	0.686746	0.341476
45	10	9	3	0.906547	0.219801
45	10	9	4	0.983477	0.076930
45	10	9	5	0.998367	0.014890
45	10	9	6	0.999918	0.001551
45	10	9	7	0.999998	0.000081
45	10	9	8	1.000000	0.000002
45	10	9	9	1.000000	0.000000
45	10	10	0	0.057545	0.057545
45	10	10	1	0.278872	0.221327
45	10	10	2	0.610862	0.331990
45	10	10	3	0.863808	0.252945
45	10	10	4	0.970655	0.106847
45	10	10	5	0.996299	0.025643
45	10	10	6	0.999745	0.003447
45	10	10	7	0.999991	0.000246
45	10	10	8	1.000000	0.000008
45	10	10	9	1.000000	0.000000
45	10	10	10	1.000000	0.000000
45	11	1	0	0.755556	0.755556
45	11	1	1	1.000000	0.244444
45	11	2	0	0.566667	0.566667
45	11	2	1	0.944444	0.377778
45	11	2	2	1.000000	0.055556
45	11	3	0	0.421705	0.421705
45	11	3	1	0.856589	0.434884
45	11	3	2	0.988372	0.131783
45	11	3	3	1.000000	0.011628
45	11	4	0	0.311259	0.311259
45	11	4	1	0.753045	0.441787
45	11	4	2	0.960133	0.207087
45	11	4	3	0.997785	0.037652
45	11	4	4	1.000000	0.002215
45	11	5	0	0.227750	0.227750
45	11	5	1	0.645293	0.417542
45	11	5	2	0.914675	0.269382
45	11	5	3	0.990438	0.075764
45	11	5	4	0.999622	0.009183
45	11	5	5	1.000000	0.000378
45	11	6	0	0.165119	0.165119
45	11	6	1	0.540907	0.375788
45	11	6	2	0.854064	0.313157
45	11	6	3	0.975286	0.121222
45	11	6	4	0.998015	0.022729
45	11	6	5	0.999943	0.001929
45	11	6	6	1.000000	0.000057
45	11	7	0	0.118547	0.118547
45	11	7	1	0.444551	0.326004
45	11	7	2	0.781797	0.337246
45	11	7	3	0.950419	0.168623
45	11	7	4	0.993935	0.043516
45	11	7	5	0.999646	0.005711
45	11	7	6	0.999993	0.000346
45	11	7	7	1.000000	0.000007
45	11	8	0	0.084231	0.084231
45	11	8	1	0.358760	0.274530
45	11	8	2	0.701923	0.343162
45	11	8	3	0.914920	0.212997
45	11	8	4	0.985919	0.070999
45	11	8	5	0.998745	0.012826
45	11	8	6	0.999947	0.001202
45	11	8	7	0.999999	0.000052
45	11	8	8	1.000000	0.000001
45	11	9	0	0.059189	0.059189
45	11	9	1	0.284563	0.225374
45	11	9	2	0.618451	0.333888
45	11	9	3	0.868867	0.250416

Table for $N = 2$, $n = 1$, through $N = 100$, $n = 50$

N	n	k	x	P(x)	p(x)
45	11	9	4	0.972487	0.103620
45	11	9	5	0.996665	0.024178
45	11	9	6	0.999785	0.003120
45	11	9	7	0.999994	0.000209
45	11	9	8	1.000000	0.000006
45	11	9	9	1.000000	0.000000
45	11	10	0	0.041104	0.041104
45	11	10	1	0.221959	0.180856
45	11	10	2	0.534979	0.313020
45	11	10	3	0.813219	0.278240
45	11	10	4	0.952338	0.139120
45	11	10	5	0.992635	0.040297
45	11	10	6	0.999351	0.006716
45	11	10	7	0.999970	0.000619
45	11	10	8	0.999999	0.000029
45	11	10	9	1.000000	0.000001
45	11	10	10	1.000000	0.000000
45	11	11	0	0.028185	0.028185
45	11	11	1	0.170286	0.142101
45	11	11	2	0.454488	0.284202
45	11	11	3	0.749621	0.295133
45	11	11	4	0.924514	0.174893
45	11	11	5	0.985727	0.061213
45	11	11	6	0.998392	0.012665
45	11	11	7	0.999900	0.001508
45	11	11	8	0.999997	0.000097
45	11	11	9	1.000000	0.000003
45	11	11	10	1.000000	0.000000
45	11	11	11	1.000000	0.000000
45	12	1	0	0.733333	0.733333
45	12	1	1	1.000000	0.266667
45	12	2	0	0.533333	0.533333
45	12	2	1	0.933333	0.400000
45	12	2	2	1.000000	0.066667
45	12	3	0	0.384496	0.384496
45	12	3	1	0.831008	0.446512
45	12	3	2	0.984496	0.153488
45	12	3	3	1.000000	0.015504
45	12	4	0	0.274640	0.274640
45	12	4	1	0.714064	0.439424
45	12	4	2	0.947951	0.233887
45	12	4	3	0.996678	0.048726
45	12	4	4	1.000000	0.003322
45	12	5	0	0.194258	0.194258
45	12	5	1	0.596170	0.401912
45	12	5	2	0.890906	0.294736
45	12	5	3	0.985982	0.095076
45	12	5	4	0.999352	0.013370
45	12	5	5	1.000000	0.000648
45	12	6	0	0.135980	0.135980
45	12	6	1	0.485644	0.349664
45	12	6	2	0.817222	0.331578
45	12	6	3	0.964589	0.147368
45	12	6	4	0.996678	0.032088
45	12	6	5	0.999886	0.003209
45	12	6	6	1.000000	0.000113
45	12	7	0	0.094140	0.094140
45	12	7	1	0.387021	0.292881
45	12	7	2	0.732202	0.345181
45	12	7	3	0.930582	0.198380
45	12	7	4	0.990095	0.059514
45	12	7	5	0.999311	0.009215
45	12	7	6	0.999982	0.000672
45	12	7	7	1.000000	0.000017
45	12	8	0	0.064412	0.064412
45	12	8	1	0.302240	0.237828
45	12	8	2	0.641365	0.339125
45	12	8	3	0.883597	0.242232
45	12	8	4	0.977566	0.093969
45	12	8	5	0.997613	0.020047
45	12	8	6	0.999876	0.002263
45	12	8	7	0.999998	0.000121
45	12	8	8	1.000000	0.000002
45	12	9	0	0.043521	0.043521
45	12	9	1	0.231534	0.188013
45	12	9	2	0.549709	0.318175
45	12	9	3	0.824676	0.274966
45	12	9	4	0.957249	0.132573
45	12	9	5	0.993820	0.036572
45	12	9	6	0.999509	0.005689
45	12	9	7	0.999981	0.000472
45	12	9	8	1.000000	0.000018
45	12	9	9	1.000000	0.000000
45	12	10	0	0.029014	0.029014
45	12	10	1	0.174086	0.145071
45	12	10	2	0.461327	0.287241
45	12	10	3	0.755934	0.294607
45	12	10	4	0.927788	0.171854
45	12	10	5	0.986709	0.058921
45	12	10	6	0.998561	0.011852
45	12	10	7	0.999916	0.001355
45	12	10	8	0.999998	0.000082
45	12	10	9	1.000000	0.000002
45	12	10	10	1.000000	0.000000
45	12	11	0	0.019067	0.019067
45	12	11	1	0.128492	0.109425
45	12	11	2	0.379258	0.250766
45	12	11	3	0.680178	0.300920
45	12	11	4	0.888507	0.208329
45	12	11	5	0.974925	0.086418
45	12	11	6	0.996529	0.021604
45	12	11	7	0.999722	0.003193
45	12	11	8	0.999988	0.000266
45	12	11	9	1.000000	0.000011
45	12	11	10	1.000000	0.000000
45	12	11	11	1.000000	0.000000
45	12	12	0	0.012337	0.012337
45	12	12	1	0.093090	0.080752
45	12	12	2	0.305503	0.212414
45	12	12	3	0.600523	0.295019
45	12	12	4	0.839488	0.238966
45	12	12	5	0.957133	0.117645
45	12	12	6	0.992717	0.035584
45	12	12	7	0.999253	0.006536
45	12	12	8	0.999957	0.000704
45	12	12	9	0.999999	0.000042
45	12	12	10	1.000000	0.000001
45	12	12	11	1.000000	0.000000
45	12	12	12	1.000000	0.000000
45	13	1	0	0.711111	0.711111
45	13	1	1	1.000000	0.288889
45	13	2	0	0.501010	0.501010
45	13	2	1	0.921212	0.420202
45	13	2	2	1.000000	0.078788
45	13	3	0	0.349542	0.349542
45	13	3	1	0.803946	0.454404
45	13	3	2	0.979845	0.175898
45	13	3	3	1.000000	0.020155
45	13	4	0	0.241350	0.241350
45	13	4	1	0.674117	0.432766
45	13	4	2	0.933776	0.259660
45	13	4	3	0.995201	0.061425
45	13	4	4	1.000000	0.004799
45	13	5	0	0.164825	0.164825
45	13	5	1	0.547453	0.382629
45	13	5	2	0.864111	0.316658
45	13	5	3	0.980219	0.116108
45	13	5	4	0.998947	0.018727
45	13	5	5	1.000000	0.001053
45	13	6	0	0.111257	0.111257
45	13	6	1	0.432665	0.321408
45	13	6	2	0.777030	0.344366
45	13	6	3	0.951192	0.174162
45	13	6	4	0.994733	0.043541
45	13	6	5	0.999789	0.005056
45	13	6	6	1.000000	0.000211
45	13	7	0	0.074171	0.074171
45	13	7	1	0.333770	0.259599
45	13	7	2	0.679902	0.346132
45	13	7	3	0.906536	0.226634
45	13	7	4	0.984685	0.078150
45	13	7	5	0.998752	0.014067
45	13	7	6	0.999962	0.001210
45	13	7	7	1.000000	0.000038
45	13	8	0	0.048797	0.048797
45	13	8	1	0.251791	0.202995
45	13	8	2	0.579706	0.327914
45	13	8	3	0.846895	0.267189
45	13	8	4	0.966176	0.119281
45	13	8	5	0.995791	0.029615
45	13	8	6	0.999739	0.003949
45	13	8	7	0.999994	0.000255
45	13	8	8	1.000000	0.000006
45	13	9	0	0.031652	0.031652
45	13	9	1	0.185955	0.154303
45	13	9	2	0.482218	0.296262
45	13	9	3	0.774682	0.292464
45	13	9	4	0.937162	0.162480
45	13	9	5	0.989387	0.052226
45	13	9	6	0.998992	0.009605
45	13	9	7	0.999953	0.000960
45	13	9	8	0.999999	0.000046
45	13	9	9	1.000000	0.000001
45	13	10	0	0.020222	0.020222
45	13	10	1	0.134521	0.114299
45	13	10	2	0.391693	0.257172
45	13	10	3	0.693442	0.301749
45	13	10	4	0.896542	0.203100
45	13	10	5	0.977782	0.081240
45	13	10	6	0.997125	0.019343
45	13	10	7	0.999793	0.002668
45	13	10	8	0.999993	0.000200
45	13	10	9	1.000000	0.000007
45	13	10	10	1.000000	0.000000
45	13	11	0	0.012711	0.012711
45	13	11	1	0.095333	0.082622
45	13	11	2	0.310867	0.215535
45	13	11	3	0.607228	0.296360
45	13	11	4	0.844316	0.237088
45	13	11	5	0.959213	0.114897
45	13	11	6	0.993256	0.034043
45	13	11	7	0.999335	0.006079
45	13	11	8	0.999964	0.000629
45	13	11	9	0.999999	0.000035
45	13	11	10	1.000000	0.000001
45	13	11	11	1.000000	0.000000
45	13	12	0	0.007851	0.007851
45	13	12	1	0.066172	0.058321
45	13	12	2	0.241136	0.174964
45	13	12	3	0.520063	0.278927

Table for $N=2$, $n=1$, through $N=100$, $n=50$

N	n	k	x	P(x)	p(x)
45	13	12	4	0.781557	0.261494
45	13	12	5	0.932178	0.150621
45	13	12	6	0.986247	0.054069
45	13	12	7	0.998262	0.012015
45	13	12	8	0.999872	0.001609
45	13	12	9	0.999995	0.000123
45	13	12	10	1.000000	0.000005
45	13	12	11	1.000000	0.000000
45	13	12	12	1.000000	0.000000
45	13	13	0	0.004758	0.004758
45	13	13	1	0.044964	0.040206
45	13	13	2	0.182814	0.137850
45	13	13	3	0.435539	0.252725
45	13	13	4	0.710241	0.274701
45	13	13	5	0.895664	0.185423
45	13	13	6	0.974778	0.079114
45	13	13	7	0.996078	0.021300
45	13	13	8	0.999628	0.003550
45	13	13	9	0.999980	0.000352
45	13	13	10	0.999999	0.000019
45	13	13	11	1.000000	0.000001
45	13	13	12	1.000000	0.000000
45	13	13	13	1.000000	0.000000
45	14	1	0	0.688889	0.688889
45	14	1	1	1.000000	0.311111
45	14	2	0	0.469697	0.469697
45	14	2	1	0.908081	0.438384
45	14	2	2	1.000000	0.091919
45	14	3	0	0.316772	0.316772
45	14	3	1	0.775546	0.458774
45	14	3	2	0.974348	0.198802
45	14	3	3	1.000000	0.025652
45	14	4	0	0.211182	0.211182
45	14	4	1	0.633545	0.422363
45	14	4	2	0.917547	0.284003
45	14	4	3	0.993282	0.075734
45	14	4	4	1.000000	0.006718
45	14	5	0	0.139071	0.139071
45	14	5	1	0.499625	0.360554
45	14	5	2	0.834425	0.334800
45	14	5	3	0.972963	0.138538
45	14	5	4	0.998361	0.025399
45	14	5	5	1.000000	0.001639
45	14	6	0	0.090396	0.090396
45	14	6	1	0.382445	0.292049
45	14	6	2	0.733985	0.351540
45	14	6	3	0.934865	0.200880
45	14	6	4	0.992012	0.057147
45	14	6	5	0.999631	0.007620
45	14	6	6	1.000000	0.000369
45	14	7	0	0.057946	0.057946
45	14	7	1	0.285095	0.227149
45	14	7	2	0.625819	0.340723
45	14	7	3	0.878206	0.252388
45	14	7	4	0.977359	0.099152
45	14	7	5	0.997873	0.020514
45	14	7	6	0.999924	0.002051
45	14	7	7	1.000000	0.000076
45	14	8	0	0.036598	0.036598
45	14	8	1	0.207386	0.170789
45	14	8	2	0.518222	0.310835
45	14	8	3	0.805147	0.286925
45	14	8	4	0.951266	0.146119
45	14	8	5	0.993014	0.041748
45	14	8	6	0.999492	0.006478
45	14	8	7	0.999986	0.000494
45	14	8	8	1.000000	0.000014
45	14	9	0	0.022750	0.022750
45	14	9	1	0.147379	0.124630
45	14	9	2	0.417410	0.270031
45	14	9	3	0.719845	0.302434
45	14	9	4	0.911774	0.191930
45	14	9	5	0.982859	0.071085
45	14	9	6	0.998092	0.015232
45	14	9	7	0.999893	0.001801
45	14	9	8	0.999998	0.000105
45	14	9	9	1.000000	0.000002
45	14	10	0	0.013903	0.013903
45	14	10	1	0.102374	0.088472
45	14	10	2	0.327400	0.225026
45	14	10	3	0.627434	0.300034
45	14	10	4	0.858460	0.231026
45	14	10	5	0.965088	0.106628
45	14	10	6	0.994707	0.029619
45	14	10	7	0.999542	0.004836
45	14	10	8	0.999980	0.000438
45	14	10	9	1.000000	0.000019
45	14	10	10	1.000000	0.000000
45	14	11	0	0.008342	0.008342
45	14	11	1	0.069513	0.061172
45	14	11	2	0.250248	0.180735
45	14	11	3	0.533138	0.282889
45	14	11	4	0.792453	0.259315
45	14	11	5	0.937669	0.145217
45	14	11	6	0.987937	0.050267
45	14	11	7	0.998575	0.010639
45	14	11	8	0.999905	0.001330
45	14	11	9	0.999997	0.000092
45	14	11	10	1.000000	0.000003
45	14	11	11	1.000000	0.000000
45	14	12	0	0.004907	0.004907
45	14	12	1	0.046124	0.041217
45	14	12	2	0.186459	0.140335
45	14	12	3	0.441615	0.255155
45	14	12	4	0.716184	0.274569
45	14	12	5	0.899230	0.183046
45	14	12	6	0.976109	0.076879
45	14	12	7	0.996385	0.020276
45	14	12	8	0.999670	0.003285
45	14	12	9	0.999983	0.000313
45	14	12	10	0.999999	0.000016
45	14	12	11	1.000000	0.000000
45	14	12	12	1.000000	0.000000
45	14	13	0	0.002825	0.002825
45	14	13	1	0.029887	0.027062
45	14	13	2	0.135428	0.105541
45	14	13	3	0.356563	0.221134
45	14	13	4	0.632981	0.276418
45	14	13	5	0.849308	0.216327
45	14	13	6	0.957472	0.108164
45	14	13	7	0.992084	0.034612
45	14	13	8	0.999073	0.006989
45	14	13	9	0.999936	0.000863
45	14	13	10	0.999998	0.000062
45	14	13	11	1.000000	0.000002
45	14	13	12	1.000000	0.000000
45	14	13	13	1.000000	0.000000
45	14	14	0	0.001589	0.001589
45	14	14	1	0.018893	0.017304
45	14	14	2	0.095850	0.076957
45	14	14	3	0.280548	0.184698
45	14	14	4	0.546600	0.266052
45	14	14	5	0.788466	0.241866
45	14	14	6	0.930431	0.141965
45	14	14	7	0.984513	0.054082
45	14	14	8	0.997763	0.013250
45	14	14	9	0.999801	0.002038
45	14	14	10	0.999990	0.000189
45	14	14	11	1.000000	0.000010
45	14	14	12	1.000000	0.000000
45	14	14	13	1.000000	0.000000
45	14	14	14	1.000000	0.000000
45	15	1	0	0.666667	0.666667
45	15	1	1	1.000000	0.333333
45	15	2	0	0.439394	0.439394
45	15	2	1	0.893939	0.454545
45	15	2	2	1.000000	0.106061
45	15	3	0	0.286117	0.286117
45	15	3	1	0.745948	0.459831
45	15	3	2	0.967935	0.221987
45	15	3	3	1.000000	0.032065
45	15	4	0	0.183932	0.183932
45	15	4	1	0.592671	0.408739
45	15	4	2	0.899225	0.306554
45	15	4	3	0.990838	0.091614
45	15	4	4	1.000000	0.009161
45	15	5	0	0.116640	0.116640
45	15	5	1	0.453102	0.336462
45	15	5	2	0.802025	0.348923
45	15	5	3	0.964025	0.162000
45	15	5	4	0.997542	0.033517
45	15	5	5	1.000000	0.002458
45	15	6	0	0.072900	0.072900
45	15	6	1	0.335340	0.262440
45	15	6	2	0.688625	0.353285
45	15	6	3	0.915425	0.226800
45	15	6	4	0.988325	0.072900
45	15	6	5	0.999385	0.011061
45	15	6	6	1.000000	0.000614
45	15	7	0	0.044862	0.044862
45	15	7	1	0.241131	0.196269
45	15	7	2	0.570863	0.329732
45	15	7	3	0.845640	0.274777
45	15	7	4	0.967763	0.122123
45	15	7	5	0.996549	0.028786
45	15	7	6	0.999858	0.003309
45	15	7	7	1.000000	0.000142
45	15	8	0	0.027153	0.027153
45	15	8	1	0.168821	0.141668
45	15	8	2	0.458060	0.289239
45	15	8	3	0.758868	0.300808
45	15	8	4	0.932412	0.173543
45	15	8	5	0.988974	0.056562
45	15	8	6	0.999074	0.010100
45	15	8	7	0.999970	0.000896
45	15	8	8	1.000000	0.000030
45	15	9	0	0.016145	0.016145
45	15	9	1	0.115217	0.099072
45	15	9	2	0.356436	0.241219
45	15	9	3	0.661309	0.304873
45	15	9	4	0.880818	0.219509
45	15	9	5	0.973687	0.092869
45	15	9	6	0.996618	0.022931
45	15	9	7	0.999776	0.003159
45	15	9	8	0.999994	0.000218
45	15	9	9	1.000000	0.000006
45	15	10	0	0.009418	0.009418
45	15	10	1	0.076689	0.067271
45	15	10	2	0.269329	0.192640
45	15	10	3	0.559684	0.290356

N	n	k	x	$P(x)$	$p(x)$
45	15	10	4	0.813746	0.254061
45	15	10	5	0.947890	0.134144
45	15	10	6	0.990885	0.042995
45	15	10	7	0.999074	0.008190
45	15	10	8	0.999952	0.000877
45	15	10	9	0.999999	0.000047
45	15	10	10	1.000000	0.000001
45	15	11	0	0.005382	0.005382
45	15	11	1	0.049781	0.044399
45	15	11	2	0.197777	0.147996
45	15	11	3	0.460134	0.262357
45	15	11	4	0.733898	0.273764
45	15	11	5	0.909563	0.175665
45	15	11	6	0.979829	0.070266
45	15	11	7	0.997203	0.017373
45	15	11	8	0.999776	0.002574
45	15	11	9	0.999991	0.000214
45	15	11	10	1.000000	0.000009
45	15	11	11	1.000000	0.000000
45	15	12	0	0.003007	0.003007
45	15	12	1	0.031499	0.028491
45	15	12	2	0.141190	0.109691
45	15	12	3	0.367537	0.226347
45	15	12	4	0.645327	0.277790
45	15	12	5	0.857897	0.212570
45	15	12	6	0.961229	0.103332
45	15	12	7	0.993115	0.031885
45	15	12	8	0.999247	0.006132
45	15	12	9	0.999953	0.000707
45	15	12	10	0.999998	0.000045
45	15	12	11	1.000000	0.000001
45	15	12	12	1.000000	0.000000
45	15	13	0	0.001640	0.001640
45	15	13	1	0.019411	0.017771
45	15	13	2	0.097978	0.078567
45	15	13	3	0.285229	0.187251
45	15	13	4	0.552731	0.267501
45	15	13	5	0.793482	0.240751
45	15	13	6	0.933048	0.139566
45	15	13	7	0.985385	0.052337
45	15	13	8	0.997946	0.012561
45	15	13	9	0.999825	0.001879
45	15	13	10	0.999992	0.000167
45	15	13	11	1.000000	0.000008
45	15	13	12	1.000000	0.000000
45	15	13	13	1.000000	0.000000
45	15	14	0	0.000871	0.000871
45	15	14	1	0.011637	0.010765
45	15	14	2	0.066061	0.054424
45	15	14	3	0.215010	0.148950
45	15	14	4	0.460777	0.245767
45	15	14	5	0.718247	0.257470
45	15	14	6	0.893795	0.175548
45	15	14	7	0.972301	0.078506
45	15	14	8	0.995198	0.022898
45	15	14	9	0.999472	0.004274
45	15	14	10	0.999965	0.000493
45	15	14	11	0.999999	0.000033
45	15	14	12	1.000000	0.000001
45	15	14	13	1.000000	0.000000
45	15	14	14	1.000000	0.000000
45	15	15	0	0.000450	0.000450
45	15	15	1	0.006775	0.006325
45	15	15	2	0.043238	0.036463
45	15	15	3	0.157352	0.114115
45	15	15	4	0.373569	0.216217
45	15	15	5	0.635192	0.261623
45	15	15	6	0.842829	0.207637
45	15	15	7	0.952041	0.109212
45	15	15	8	0.990028	0.037987
45	15	15	9	0.998645	0.008617
45	15	15	10	0.999886	0.001241
45	15	15	11	0.999994	0.000108
45	15	15	12	1.000000	0.000005
45	15	15	13	1.000000	0.000000
45	15	15	14	1.000000	0.000000
45	15	15	15	1.000000	0.000000
45	16	1	0	0.644444	0.644444
45	16	1	1	1.000000	0.355556
45	16	2	0	0.410101	0.410101
45	16	2	1	0.878788	0.468687
45	16	2	2	1.000000	0.121212
45	16	3	0	0.257505	0.257505
45	16	3	1	0.715292	0.457787
45	16	3	2	0.960535	0.245243
45	16	3	3	1.000000	0.039464
45	16	4	0	0.159408	0.159408
45	16	4	1	0.551797	0.392389
45	16	4	2	0.878788	0.326991
45	16	4	3	0.987785	0.108997
45	16	4	4	1.000000	0.012215
45	16	5	0	0.097200	0.097200
45	16	5	1	0.408240	0.311040
45	16	5	2	0.767132	0.358892
45	16	5	3	0.953225	0.186092
45	16	5	4	0.996425	0.043200
45	16	5	5	1.000000	0.003575
45	16	6	0	0.058320	0.058320
45	16	6	1	0.291600	0.233280
45	16	6	2	0.641520	0.349920
45	16	6	3	0.892745	0.251225
45	16	6	4	0.983465	0.090720
45	16	6	5	0.999017	0.015552
45	16	6	6	1.000000	0.000983
45	16	7	0	0.034394	0.034394
45	16	7	1	0.201877	0.167483
45	16	7	2	0.515908	0.314031
45	16	7	3	0.809003	0.293095
45	16	7	4	0.955551	0.146548
45	16	7	5	0.994630	0.039079
45	16	7	6	0.999748	0.005118
45	16	7	7	1.000000	0.000252
45	16	8	0	0.019912	0.019912
45	16	8	1	0.135765	0.115853
45	16	8	2	0.400212	0.264447
45	16	8	3	0.708734	0.308521
45	16	8	4	0.909273	0.200539
45	16	8	5	0.983318	0.074045
45	16	8	6	0.998401	0.015083
45	16	8	7	0.999940	0.001539
45	16	8	8	1.000000	0.000060
45	16	9	0	0.011302	0.011302
45	16	9	1	0.088798	0.077496
45	16	9	2	0.300151	0.211353
45	16	9	3	0.600334	0.300183
45	16	9	4	0.844233	0.243899
45	16	9	5	0.961304	0.117071
45	16	9	6	0.994324	0.033020
45	16	9	7	0.999566	0.005241
45	16	9	8	0.999987	0.000421
45	16	9	9	1.000000	0.000013
45	16	10	0	0.006279	0.006279
45	16	10	1	0.056508	0.050229
45	16	10	2	0.217958	0.161450
45	16	10	3	0.491935	0.273977
45	16	10	4	0.762933	0.270999
45	16	10	5	0.925533	0.162599
45	16	10	6	0.985152	0.059620
45	16	10	7	0.998255	0.013103
45	16	10	8	0.999893	0.001638
45	16	10	9	0.999997	0.000104
45	16	10	10	1.000000	0.000003
45	16	11	0	0.003408	0.003408
45	16	11	1	0.034981	0.031573
45	16	11	2	0.153378	0.118397
45	16	11	3	0.390172	0.236794
45	16	11	4	0.670020	0.279848
45	16	11	5	0.874430	0.204410
45	16	11	6	0.968118	0.093688
45	16	11	7	0.994886	0.026768
45	16	11	8	0.999519	0.004633
45	16	11	9	0.999977	0.000458
45	16	11	10	0.999999	0.000023
45	16	11	11	1.000000	0.000000
45	16	12	0	0.001804	0.001804
45	16	12	1	0.021052	0.019247
45	16	12	2	0.104626	0.083574
45	16	12	3	0.299633	0.195007
45	16	12	4	0.571250	0.271617
45	16	12	5	0.808297	0.237047
45	16	12	6	0.940563	0.132266
45	16	12	7	0.987800	0.047238
45	16	12	8	0.998429	0.010628
45	16	12	9	0.999882	0.001453
45	16	12	10	0.999995	0.000113
45	16	12	11	1.000000	0.000004
45	16	12	12	1.000000	0.000000
45	16	13	0	0.000930	0.000930
45	16	13	1	0.012303	0.011373
45	16	13	2	0.069170	0.056867
45	16	13	3	0.222812	0.153642
45	16	13	4	0.472480	0.249668
45	16	13	5	0.729281	0.256801
45	16	13	6	0.900482	0.171201
45	16	13	7	0.974917	0.074435
45	16	13	8	0.995852	0.020935
45	16	13	9	0.999574	0.003722
45	16	13	10	0.999975	0.000401
45	16	13	11	0.999999	0.000024
45	16	13	12	1.000000	0.000001
45	16	13	13	1.000000	0.000000
45	16	14	0	0.000465	0.000465
45	16	14	1	0.006972	0.006507
45	16	14	2	0.044291	0.037319
45	16	14	3	0.160395	0.116104
45	16	14	4	0.378855	0.218459
45	16	14	5	0.641006	0.262151
45	16	14	6	0.846982	0.205976
45	16	14	7	0.953983	0.107001
45	16	14	8	0.990619	0.036636
45	16	14	9	0.998760	0.008141
45	16	14	10	0.999900	0.001140
45	16	14	11	0.999995	0.000096
45	16	14	12	1.000000	0.000004
45	16	14	13	1.000000	0.000000
45	16	14	14	1.000000	0.000000
45	16	15	0	0.000225	0.000225
45	16	15	1	0.003823	0.003598
45	16	15	2	0.027437	0.023614
45	16	15	3	0.111706	0.084269

N	n	k	x	P(x)	p(x)
45	16	15	4	0.294290	0.182583
45	16	15	5	0.547985	0.253695
45	16	15	6	0.780538	0.232554
45	16	15	7	0.922918	0.142380
45	16	15	8	0.981164	0.058246
45	16	15	9	0.996922	0.015757
45	16	15	10	0.999679	0.002758
45	16	15	11	0.999980	0.000301
45	16	15	12	0.999999	0.000019
45	16	15	13	1.000000	0.000001
45	16	15	14	1.000000	0.000000
45	16	15	15	1.000000	0.000000
45	16	16	0	0.000105	0.000105
45	16	16	1	0.002024	0.001919
45	16	16	2	0.016417	0.014393
45	16	16	3	0.075190	0.058772
45	16	16	4	0.221256	0.146067
45	16	16	5	0.454963	0.233707
45	16	16	6	0.703020	0.248057
45	16	16	7	0.880204	0.177184
45	16	16	8	0.965632	0.085428
45	16	16	9	0.993245	0.027613
45	16	16	10	0.999128	0.005883
45	16	16	11	0.999930	0.000802
45	16	16	12	0.999997	0.000067
45	16	16	13	1.000000	0.000003
45	16	16	14	1.000000	0.000000
45	16	16	15	1.000000	0.000000
45	16	16	16	1.000000	0.000000
45	17	1	0	0.622222	0.622222
45	17	1	1	1.000000	0.377778
45	17	2	0	0.381818	0.381818
45	17	2	1	0.862626	0.480808
45	17	2	2	1.000000	0.137374
45	17	3	0	0.230867	0.230867
45	17	3	1	0.683721	0.452854
45	17	3	2	0.952079	0.268358
45	17	3	3	1.000000	0.047921
45	17	4	0	0.137421	0.137421
45	17	4	1	0.511205	0.373784
45	17	4	2	0.856237	0.345032
45	17	4	3	0.984026	0.127789
45	17	4	4	1.000000	0.015974
45	17	5	0	0.080441	0.080441
45	17	5	1	0.365338	0.284897
45	17	5	2	0.730006	0.364668
45	17	5	3	0.940391	0.210385
45	17	5	4	0.994935	0.054544
45	17	5	5	1.000000	0.005065
45	17	6	0	0.046254	0.046254

N	n	k	x	P(x)	p(x)
45	17	6	1	0.251379	0.205126
45	17	6	2	0.593255	0.341876
45	17	6	3	0.866756	0.273501
45	17	6	4	0.977208	0.110452
45	17	6	5	0.998480	0.021272
45	17	6	6	1.000000	0.001519
45	17	7	0	0.026092	0.026092
45	17	7	1	0.167225	0.141133
45	17	7	2	0.461765	0.294539
45	17	7	3	0.768576	0.306812
45	17	7	4	0.940391	0.171815
45	17	7	5	0.991935	0.051544
45	17	7	6	0.999571	0.007636
45	17	7	7	1.000000	0.000429
45	17	8	0	0.014419	0.014419
45	17	8	1	0.107801	0.093381
45	17	8	2	0.345499	0.237698
45	17	8	3	0.655540	0.310041
45	17	8	4	0.881612	0.226072
45	17	8	5	0.975658	0.094046
45	17	8	6	0.997361	0.021703
45	17	8	7	0.999887	0.002526
45	17	8	8	1.000000	0.000113
45	17	9	0	0.007794	0.007794
45	17	9	1	0.067419	0.059625
45	17	9	2	0.249135	0.181715
45	17	9	3	0.538227	0.289093
45	17	9	4	0.802182	0.263954
45	17	9	5	0.945157	0.142975
45	17	9	6	0.990909	0.045752
45	17	9	7	0.999204	0.008296
45	17	9	8	0.999972	0.000768
45	17	9	9	1.000000	0.000027
45	17	10	0	0.004114	0.004114
45	17	10	1	0.040919	0.036806
45	17	10	2	0.173420	0.132501
45	17	10	3	0.425803	0.252382
45	17	10	4	0.706865	0.281062
45	17	10	5	0.897498	0.190634
45	17	10	6	0.976929	0.079431
45	17	10	7	0.996900	0.019971
45	17	10	8	0.999780	0.002880
45	17	10	9	0.999994	0.000213
45	17	10	10	1.000000	0.000006
45	17	11	0	0.002116	0.002116
45	17	11	1	0.024094	0.021978
45	17	11	2	0.116634	0.092540
45	17	11	3	0.324850	0.208215
45	17	11	4	0.602470	0.277621
45	17	11	5	0.832138	0.229668

N	n	k	x	P(x)	p(x)
45	17	11	6	0.951965	0.119827
45	17	11	7	0.991194	0.039229
45	17	11	8	0.999040	0.007846
45	17	11	9	0.999945	0.000905
45	17	11	10	0.999999	0.000054
45	17	11	11	1.000000	0.000001
45	17	12	0	0.001058	0.001058
45	17	12	1	0.013751	0.012693
45	17	12	2	0.075807	0.062056
45	17	12	3	0.239114	0.163306
45	17	12	4	0.496321	0.257207
45	17	12	5	0.751079	0.254758
45	17	12	6	0.913197	0.162119
45	17	12	7	0.979656	0.066459
45	17	12	8	0.996963	0.017307
45	17	12	9	0.999732	0.002769
45	17	12	10	0.999988	0.000256
45	17	12	11	1.000000	0.000012
45	17	12	12	1.000000	0.000000
45	17	13	0	0.000513	0.000513
45	17	13	1	0.007597	0.007084
45	17	13	2	0.047600	0.040003
45	17	13	3	0.169832	0.122232
45	17	13	4	0.394997	0.225165
45	17	13	5	0.658440	0.263443
45	17	13	6	0.859158	0.200718
45	17	13	7	0.959517	0.100359
45	17	13	8	0.992243	0.032726
45	17	13	9	0.999061	0.006818
45	17	13	10	0.999933	0.000873
45	17	13	11	0.999997	0.000064
45	17	13	12	1.000000	0.000002
45	17	13	13	1.000000	0.000000
45	17	14	0	0.000240	0.000240
45	17	14	1	0.004055	0.003814
45	17	14	2	0.028849	0.024794
45	17	14	3	0.116356	0.087507
45	17	14	4	0.303524	0.187168
45	17	14	5	0.559649	0.256125
45	17	14	6	0.790161	0.230512
45	17	14	7	0.928155	0.137994
45	17	14	8	0.983039	0.054884
45	17	14	9	0.997356	0.014318
45	17	14	10	0.999743	0.002386
45	17	14	11	0.999985	0.000243
45	17	14	12	0.999999	0.000014
45	17	14	13	1.000000	0.000000
45	17	14	14	1.000000	0.000000
45	17	15	0	0.000109	0.000109
45	17	15	1	0.002086	0.001978

N	n	k	x	P(x)	p(x)
45	17	15	2	0.016852	0.014765
45	17	15	3	0.076836	0.059985
45	17	15	4	0.225034	0.148198
45	17	15	5	0.460504	0.235470
45	17	15	6	0.708366	0.247863
45	17	15	7	0.883641	0.175274
45	17	15	8	0.967105	0.083464
45	17	15	9	0.993661	0.026557
45	17	15	10	0.999204	0.005542
45	17	15	11	0.999938	0.000735
45	17	15	12	0.999997	0.000059
45	17	15	13	1.000000	0.000003
45	17	15	14	1.000000	0.000000
45	17	15	15	1.000000	0.000000
45	17	16	0	0.000047	0.000047
45	17	16	1	0.001031	0.000984
45	17	16	2	0.009469	0.008437
45	17	16	3	0.048843	0.039375
45	17	16	4	0.160815	0.111972
45	17	16	5	0.366316	0.205501
45	17	16	6	0.617483	0.251168
45	17	16	7	0.825216	0.207733
45	17	16	8	0.942065	0.116850
45	17	16	9	0.986580	0.044514
45	17	16	10	0.997910	0.011331
45	17	16	11	0.999791	0.001881
45	17	16	12	0.999987	0.000196
45	17	16	13	0.999999	0.000012
45	17	16	14	1.000000	0.000000
45	17	16	15	1.000000	0.000000
45	17	16	16	1.000000	0.000000
45	17	17	0	0.000019	0.000019
45	17	17	1	0.000488	0.000469
45	17	17	2	0.005105	0.004616
45	17	17	3	0.029835	0.024730
45	17	17	4	0.110621	0.080786
45	17	17	5	0.281281	0.170660
45	17	17	6	0.522213	0.240932
45	17	17	7	0.753584	0.231371
45	17	17	8	0.905802	0.152218
45	17	17	9	0.974300	0.068498
45	17	17	10	0.995175	0.020876
45	17	17	11	0.999402	0.004227
45	17	17	12	0.999954	0.000551
45	17	17	13	0.999998	0.000044
45	17	17	14	1.000000	0.000002
45	17	17	15	1.000000	0.000000
45	17	17	16	1.000000	0.000000
45	17	17	17	1.000000	0.000000
45	18	1	0	0.600000	0.600000

N	n	k	x	P(x)	p(x)
45	18	1	1	1.000000	0.400000
45	18	2	0	0.354545	0.354545
45	18	2	1	0.845454	0.490909
45	18	2	2	1.000000	0.154545
45	18	3	0	0.206131	0.206131
45	18	3	1	0.651374	0.445243
45	18	3	2	0.942495	0.291120
45	18	3	3	1.000000	0.057505
45	18	4	0	0.117789	0.117789
45	18	4	1	0.471157	0.353368
45	18	4	2	0.831592	0.360435
45	18	4	3	0.979462	0.147871
45	18	4	4	1.000000	0.020538
45	18	5	0	0.066077	0.066077
45	18	5	1	0.324638	0.258562
45	18	5	2	0.690934	0.366296
45	18	5	3	0.925363	0.234429
45	18	5	4	0.992987	0.067624
45	18	5	5	1.000000	0.007013
45	18	6	0	0.036342	0.036342
45	18	6	1	0.214750	0.178407
45	18	6	2	0.544416	0.329666
45	18	6	3	0.837452	0.293036
45	18	6	4	0.969319	0.131866
45	18	6	5	0.997721	0.028402
45	18	6	6	1.000000	0.002279
45	18	7	0	0.019569	0.019569
45	18	7	1	0.136982	0.117413
45	18	7	2	0.409168	0.272186
45	18	7	3	0.724746	0.315578
45	18	7	4	0.921982	0.197236
45	18	7	5	0.988253	0.066271
45	18	7	6	0.999299	0.011045
45	18	7	7	1.000000	0.000701
45	18	8	0	0.010299	0.010299
45	18	8	1	0.084455	0.074156
45	18	8	2	0.294564	0.210108
45	18	8	3	0.600176	0.305612
45	18	8	4	0.849316	0.249140
45	18	8	5	0.965582	0.116265
45	18	8	6	0.995811	0.030229
45	18	8	7	0.999797	0.003986
45	18	8	8	1.000000	0.000203
45	18	9	0	0.005289	0.005289
45	18	9	1	0.050384	0.045095
45	18	9	2	0.203706	0.153322
45	18	9	3	0.476279	0.272573
45	18	9	4	0.755047	0.278768
45	18	9	5	0.924732	0.169685
45	18	9	6	0.986007	0.061275

N	n	k	x	P(x)	p(x)
45	18	9	7	0.998612	0.012605
45	18	9	8	0.999945	0.001333
45	18	9	9	1.000000	0.000055
45	18	10	0	0.002644	0.002644
45	18	10	1	0.029089	0.026444
45	18	10	2	0.135563	0.106474
45	18	10	3	0.362707	0.227144
45	18	10	4	0.646637	0.283930
45	18	10	5	0.863457	0.216819
45	18	10	6	0.965582	0.102125
45	18	10	7	0.994760	0.029179
45	18	10	8	0.999575	0.004814
45	18	10	9	0.999986	0.000411
45	18	10	10	1.000000	0.000014
45	18	11	0	0.001284	0.001284
45	18	11	1	0.016244	0.014960
45	18	11	2	0.086889	0.070645
45	18	11	3	0.265359	0.178470
45	18	11	4	0.533065	0.267706
45	18	11	5	0.782924	0.249859
45	18	11	6	0.930567	0.147644
45	18	11	7	0.985590	0.055022
45	18	11	8	0.998199	0.012609
45	18	11	9	0.999880	0.001681
45	18	11	10	0.999997	0.000116
45	18	11	11	1.000000	0.000003
45	18	12	0	0.000604	0.000604
45	18	12	1	0.008764	0.008160
45	18	12	2	0.053645	0.044880
45	18	12	3	0.186622	0.132978
45	18	12	4	0.422833	0.236211
45	18	12	5	0.687389	0.264556
45	18	12	6	0.878458	0.191068
45	18	12	7	0.967788	0.089331
45	18	12	8	0.994490	0.026702
45	18	12	9	0.999435	0.004945
45	18	12	10	0.999969	0.000534
45	18	12	11	0.999999	0.000030
45	18	12	12	1.000000	0.000001
45	18	13	0	0.000275	0.000275
45	18	13	1	0.004561	0.004286
45	18	13	2	0.031884	0.027324
45	18	13	3	0.126178	0.094293
45	18	13	4	0.322623	0.196445
45	18	13	5	0.583170	0.260548
45	18	13	6	0.808978	0.225808
45	18	13	7	0.938011	0.129033
45	18	13	8	0.986399	0.048387
45	18	13	9	0.998087	0.011688
45	18	13	10	0.999840	0.001753

N	n	k	x	P(x)	p(x)
45	18	13	11	0.999993	0.000153
45	18	13	12	1.000000	0.000007
45	18	13	13	1.000000	0.000000
45	18	14	0	0.000120	0.000120
45	18	14	1	0.002284	0.002164
45	18	14	2	0.018223	0.015939
45	18	14	3	0.081978	0.063755
45	18	14	4	0.236678	0.154700
45	18	14	5	0.477323	0.240645
45	18	14	6	0.724300	0.246978
45	18	14	7	0.893656	0.169356
45	18	14	8	0.971278	0.077621
45	18	14	9	0.994800	0.023522
45	18	14	10	0.999402	0.004602
45	18	14	11	0.999959	0.000558
45	18	14	12	0.999999	0.000039
45	18	14	13	1.000000	0.000001
45	18	14	14	1.000000	0.000000
45	18	15	0	0.000050	0.000050
45	18	15	1	0.001097	0.001047
45	18	15	2	0.009996	0.008899
45	18	15	3	0.051129	0.041132
45	18	15	4	0.166813	0.115685
45	18	15	5	0.376407	0.209594
45	18	15	6	0.628696	0.252289
45	18	15	7	0.833562	0.204866
45	18	15	8	0.946239	0.112676
45	18	15	9	0.987971	0.041732
45	18	15	10	0.998214	0.010243
45	18	15	11	0.999833	0.001619
45	18	15	12	0.999991	0.000157
45	18	15	13	1.000000	0.000009
45	18	15	14	1.000000	0.000000
45	18	15	15	1.000000	0.000000
45	18	16	0	0.000020	0.000020
45	18	16	1	0.000504	0.000484
45	18	16	2	0.005250	0.004746
45	18	16	3	0.030562	0.025312
45	18	16	4	0.112827	0.082265
45	18	16	5	0.285583	0.172756
45	18	16	6	0.527781	0.242197
45	18	16	7	0.758445	0.230664
45	18	16	8	0.908680	0.150235
45	18	16	9	0.975451	0.066771
45	18	16	10	0.995482	0.020031
45	18	16	11	0.999456	0.003973
45	18	16	12	0.999959	0.000504
45	18	16	13	0.999998	0.000039
45	18	16	14	1.000000	0.000002
45	18	16	15	1.000000	0.000000

N	n	k	x	P(x)	p(x)
45	18	16	16	1.000000	0.000000
45	18	17	0	0.000008	0.000008
45	18	17	1	0.000220	0.000213
45	18	17	2	0.002632	0.002411
45	18	17	3	0.017470	0.014838
45	18	17	4	0.073113	0.055643
45	18	17	5	0.208141	0.135028
45	18	17	6	0.427561	0.219420
45	18	17	7	0.670951	0.243390
45	18	17	8	0.856875	0.185923
45	18	17	9	0.954729	0.097854
45	18	17	10	0.989957	0.035228
45	18	17	11	0.998497	0.008540
45	18	17	12	0.999855	0.001359
45	18	17	13	0.999991	0.000136
45	18	17	14	1.000000	0.000008
45	18	17	15	1.000000	0.000000
45	18	17	16	1.000000	0.000000
45	18	17	17	1.000000	0.000000
45	18	18	0	0.000003	0.000003
45	18	18	1	0.000091	0.000088
45	18	18	2	0.001254	0.001163
45	18	18	3	0.009521	0.008267
45	18	18	4	0.045291	0.035771
45	18	18	5	0.145449	0.100158
45	18	18	6	0.333524	0.188074
45	18	18	7	0.575334	0.241810
45	18	18	8	0.790473	0.215140
45	18	18	9	0.923276	0.132802
45	18	18	10	0.979891	0.056616
45	18	18	11	0.996362	0.016470
45	18	18	12	0.999564	0.003202
45	18	18	13	0.999967	0.000403
45	18	18	14	0.999998	0.000031
45	18	18	15	1.000000	0.000001
45	18	18	16	1.000000	0.000000
45	18	18	17	1.000000	0.000000
45	18	18	18	1.000000	0.000000
45	19	1	0	0.577778	0.577778
45	19	1	1	1.000000	0.422222
45	19	2	0	0.328283	0.328283
45	19	2	1	0.827273	0.498990
45	19	2	2	1.000000	0.172727
45	19	3	0	0.183228	0.183228
45	19	3	1	0.618393	0.435166
45	19	3	2	0.931712	0.313319
45	19	3	3	1.000000	0.068288
45	19	4	0	0.100339	0.100339
45	19	4	1	0.431894	0.331555
45	19	4	2	0.804893	0.372999

Table for $N = 2$, $n = 1$, through $N = 100$, $n = 50$

N	n	k	x	P(x)	p(x)
45	19	4	3	0.973986	0.169093
45	19	4	4	1.000000	0.026014
45	19	5	0	0.053840	0.053840
45	19	5	1	0.286333	0.232493
45	19	5	2	0.650235	0.363902
45	19	5	3	0.907998	0.257764
45	19	5	4	0.990483	0.082484
45	19	5	5	1.000000	0.009517
45	19	6	0	0.028266	0.028266
45	19	6	1	0.181711	0.153445
45	19	6	2	0.495576	0.313865
45	19	6	3	0.804893	0.309316
45	19	6	4	0.959551	0.154658
45	19	6	5	0.996669	0.037118
45	19	6	6	1.000000	0.003331
45	19	7	0	0.014495	0.014495
45	19	7	1	0.110891	0.096395
45	19	7	2	0.358763	0.247873
45	19	7	3	0.677994	0.319230
45	19	7	4	0.900067	0.222073
45	19	7	5	0.983344	0.083277
45	19	7	6	0.998890	0.015545
45	19	7	7	1.000000	0.001110
45	19	8	0	0.007248	0.007248
45	19	8	1	0.065230	0.057982
45	19	8	2	0.247873	0.182643
45	19	8	3	0.543581	0.295708
45	19	8	4	0.812406	0.268825
45	19	8	5	0.952663	0.140257
45	19	8	6	0.993571	0.040908
45	19	8	7	0.999649	0.006078
45	19	8	8	1.000000	0.000351
45	19	9	0	0.003526	0.003526
45	19	9	1	0.037022	0.033496
45	19	9	2	0.163956	0.126933
45	19	9	3	0.415707	0.251751
45	19	9	4	0.703423	0.287716
45	19	9	5	0.899593	0.196170
45	19	9	6	0.979198	0.079605
45	19	9	7	0.997678	0.018480
45	19	9	8	0.999896	0.002218
45	19	9	9	1.000000	0.000104
45	19	10	0	0.001665	0.001665
45	19	10	1	0.020274	0.018609
45	19	10	2	0.104015	0.083741
45	19	10	3	0.303818	0.199803
45	19	10	4	0.583542	0.279724
45	19	10	5	0.823305	0.239763
45	19	10	6	0.950452	0.127147
45	19	10	7	0.991518	0.041066

N	n	k	x	P(x)	p(x)
45	19	10	8	0.999218	0.007700
45	19	10	9	0.999971	0.000753
45	19	10	10	1.000000	0.000029
45	19	11	0	0.000761	0.000761
45	19	11	1	0.010704	0.009943
45	19	11	2	0.063341	0.052637
45	19	11	3	0.212479	0.149138
45	19	11	4	0.463660	0.251181
45	19	11	5	0.727399	0.263740
45	19	11	6	0.903226	0.175826
45	19	11	7	0.977438	0.074212
45	19	11	8	0.996798	0.019360
45	19	11	9	0.999756	0.002958
45	19	11	10	0.999992	0.000237
45	19	11	11	1.000000	0.000007
45	19	12	0	0.000336	0.000336
45	19	12	1	0.005440	0.005104
45	19	12	2	0.037022	0.031582
45	19	12	3	0.142296	0.105274
45	19	12	4	0.352845	0.210548
45	19	12	5	0.618801	0.265956
45	19	12	6	0.835998	0.217197
45	19	12	7	0.951246	0.115248
45	19	12	8	0.990535	0.039289
45	19	12	9	0.998886	0.008351
45	19	12	10	0.999930	0.001044
45	19	12	11	0.999998	0.000068
45	19	12	12	1.000000	0.000002
45	19	13	0	0.000142	0.000142
45	19	13	1	0.002656	0.002513
45	19	13	2	0.020753	0.018097
45	19	13	3	0.091254	0.070502
45	19	13	4	0.257141	0.165887
45	19	13	5	0.505971	0.248830
45	19	13	6	0.750436	0.244465
45	19	13	7	0.909337	0.158902
45	19	13	8	0.977438	0.068101
45	19	13	9	0.996355	0.018917
45	19	13	10	0.999645	0.003290
45	19	13	11	0.999982	0.000336
45	19	13	12	0.999999	0.000018
45	19	13	13	1.000000	0.000000
45	19	14	0	0.000058	0.000058
45	19	14	1	0.001242	0.001184
45	19	14	2	0.011139	0.009897
45	19	14	3	0.056004	0.044865
45	19	14	4	0.179382	0.123378
45	19	14	5	0.397108	0.217726
45	19	14	6	0.651122	0.254014
45	19	14	7	0.849749	0.198627

N	n	k	x	P(x)	p(x)
45	19	14	8	0.954029	0.104279
45	19	14	9	0.990444	0.036415
45	19	14	10	0.998720	0.008276
45	19	14	11	0.999897	0.001178
45	19	14	12	0.999996	0.000098
45	19	14	13	1.000000	0.000004
45	19	14	14	1.000000	0.000000
45	19	15	0	0.000022	0.000022
45	19	15	1	0.000554	0.000532
45	19	15	2	0.005712	0.005157
45	19	15	3	0.032848	0.027136
45	19	15	4	0.119683	0.086835
45	19	15	5	0.298780	0.179097
45	19	15	6	0.544600	0.245820
45	19	15	7	0.772861	0.228261
45	19	15	8	0.917026	0.144165
45	19	15	9	0.978697	0.061671
45	19	15	10	0.996317	0.017620
45	19	15	11	0.999594	0.003276
45	19	15	12	0.999973	0.000380
45	19	15	13	0.999999	0.000026
45	19	15	14	1.000000	0.000001
45	19	15	15	1.000000	0.000000
45	19	16	0	0.000008	0.000008
45	19	16	1	0.000235	0.000227
45	19	16	2	0.002789	0.002554
45	19	16	3	0.018375	0.015586
45	19	16	4	0.076265	0.057890
45	19	16	5	0.215201	0.138936
45	19	16	6	0.438078	0.222877
45	19	16	7	0.681557	0.243479
45	19	16	8	0.864166	0.182609
45	19	16	9	0.958140	0.093974
45	19	16	10	0.991031	0.032891
45	19	16	11	0.998720	0.007689
45	19	16	12	0.999885	0.001165
45	19	16	13	0.999994	0.000109
45	19	16	14	1.000000	0.000006
45	19	16	15	1.000000	0.000000
45	19	16	16	1.000000	0.000000
45	19	17	0	0.000003	0.000003
45	19	17	1	0.000094	0.000091
45	19	17	2	0.001292	0.001198
45	19	17	3	0.009776	0.008484
45	19	17	4	0.046322	0.036546
45	19	17	5	0.148129	0.101807
45	19	17	6	0.338168	0.190039
45	19	17	7	0.580807	0.242639
45	19	17	8	0.794900	0.214093
45	19	17	9	0.925735	0.130835

N	n	k	x	P(x)	p(x)
45	19	17	10	0.980823	0.055088
45	19	17	11	0.996599	0.015775
45	19	17	12	0.999604	0.003005
45	19	17	13	0.999971	0.000368
45	19	17	14	0.999999	0.000027
45	19	17	15	1.000000	0.000001
45	19	17	16	1.000000	0.000000
45	19	17	17	1.000000	0.000000
45	19	18	0	0.000001	0.000001
45	19	18	1	0.000036	0.000035
45	19	18	2	0.000565	0.000529
45	19	18	3	0.004928	0.004363
45	19	18	4	0.026744	0.021816
45	19	18	5	0.097225	0.070482
45	19	18	6	0.249935	0.152710
45	19	18	7	0.476819	0.226883
45	19	18	8	0.710792	0.233974
45	19	18	9	0.879008	0.168216
45	19	18	10	0.963116	0.084108
45	19	18	11	0.992091	0.028975
45	19	18	12	0.998852	0.006761
45	19	18	13	0.999892	0.001040
45	19	18	14	0.999994	0.000101
45	19	18	15	1.000000	0.000006
45	19	18	16	1.000000	0.000000
45	19	18	17	1.000000	0.000000
45	19	18	18	1.000000	0.000000
45	19	19	0	0.000000	0.000000
45	19	19	1	0.000012	0.000012
45	19	19	2	0.000232	0.000219
45	19	19	3	0.002342	0.002111
45	19	19	4	0.014624	0.012281
45	19	19	5	0.060679	0.046055
45	19	19	6	0.176408	0.115729
45	19	19	7	0.375982	0.199573
45	19	19	8	0.615470	0.239488
45	19	19	9	0.816706	0.201236
45	19	19	10	0.935080	0.118374
45	19	19	11	0.983506	0.048426
45	19	19	12	0.997100	0.013593
45	19	19	13	0.999661	0.002562
45	19	19	14	0.999975	0.000314
45	19	19	15	0.999999	0.000024
45	19	19	16	1.000000	0.000001
45	19	19	17	1.000000	0.000000
45	19	19	18	1.000000	0.000000
45	19	19	19	1.000000	0.000000
45	20	1	0	0.555556	0.555556
45	20	1	1	1.000000	0.444444
45	20	2	0	0.303030	0.303030

Table for $N = 2$, $n = 1$, through $N = 100$, $n = 50$

N	n	k	x	P(x)	p(x)
45	20	2	1	0.808081	0.505050
45	20	2	2	1.000000	0.191919
45	20	3	0	0.162086	0.162086
45	20	3	1	0.584919	0.422833
45	20	3	2	0.919662	0.334743
45	20	3	3	1.000000	0.080338
45	20	4	0	0.084902	0.084902
45	20	4	1	0.393637	0.308735
45	20	4	2	0.776200	0.382563
45	20	4	3	0.967482	0.191282
45	20	4	4	1.000000	0.032518
45	20	5	0	0.043486	0.043486
45	20	5	1	0.250565	0.207078
45	20	5	2	0.608246	0.357681
45	20	5	3	0.888170	0.279924
45	20	5	4	0.987310	0.099140
45	20	5	5	1.000000	0.012690
45	20	6	0	0.021743	0.021743
45	20	6	1	0.152203	0.130459
45	20	6	2	0.447290	0.295087
45	20	6	3	0.769202	0.321913
45	20	6	4	0.947654	0.178452
45	20	6	5	0.995241	0.047587
45	20	6	6	1.000000	0.004759
45	20	7	0	0.010593	0.010593
45	20	7	1	0.088645	0.078053
45	20	7	2	0.311096	0.222450
45	20	7	3	0.628881	0.317786
45	20	7	4	0.874443	0.245562
45	20	7	5	0.976938	0.102495
45	20	7	6	0.998292	0.021353
45	20	7	7	1.000000	0.001708
45	20	8	0	0.005018	0.005018
45	20	8	1	0.049619	0.044602
45	20	8	2	0.205724	0.156105
45	20	8	3	0.486714	0.280990
45	20	8	4	0.771049	0.284335
45	20	8	5	0.936480	0.165431
45	20	8	6	0.990425	0.053945
45	20	8	7	0.999415	0.008991
45	20	8	8	1.000000	0.000584
45	20	9	0	0.002305	0.002305
45	20	9	1	0.026716	0.024410
45	20	9	2	0.129781	0.103066
45	20	9	3	0.357611	0.227829
45	20	9	4	0.648093	0.290482
45	20	9	5	0.869413	0.221320
45	20	9	6	0.970013	0.100600
45	20	9	7	0.996257	0.026243
45	20	9	8	0.999810	0.003554

N	n	k	x	P(x)	p(x)
45	20	9	9	1.000000	0.000190
45	20	10	0	0.001025	0.001025
45	20	10	1	0.013832	0.012808
45	20	10	2	0.078249	0.064416
45	20	10	3	0.250025	0.171776
45	20	10	4	0.518990	0.268965
45	20	10	5	0.777196	0.258207
45	20	10	6	0.930891	0.153694
45	20	10	7	0.986780	0.055889
45	20	10	8	0.998626	0.011846
45	20	10	9	0.999942	0.001316
45	20	10	10	1.000000	0.000058
45	20	11	0	0.000439	0.000439
45	20	11	1	0.006880	0.006441
45	20	11	2	0.045120	0.038241
45	20	11	3	0.166591	0.121470
45	20	11	4	0.396034	0.229444
45	20	11	5	0.666537	0.270502
45	20	11	6	0.869413	0.202877
45	20	11	7	0.966021	0.096608
45	20	11	8	0.994564	0.028543
45	20	11	9	0.999528	0.004964
45	20	11	10	0.999983	0.000455
45	20	11	11	1.000000	0.000017
45	20	12	0	0.000181	0.000181
45	20	12	1	0.003281	0.003100
45	20	12	2	0.024875	0.021595
45	20	12	3	0.105855	0.080980
45	20	12	4	0.288061	0.182205
45	20	12	5	0.547197	0.259137
45	20	12	6	0.785876	0.238678
45	20	12	7	0.929083	0.143207
45	20	12	8	0.984490	0.055407
45	20	12	9	0.997922	0.013432
45	20	12	10	0.999849	0.001927
45	20	12	11	0.999995	0.000146
45	20	12	12	1.000000	0.000004
45	20	13	0	0.000071	0.000071
45	20	13	1	0.001496	0.001425
45	20	13	2	0.013096	0.011600
45	20	13	3	0.064138	0.051042
45	20	13	4	0.199719	0.135580
45	20	13	5	0.429408	0.229689
45	20	13	6	0.684618	0.255210
45	20	13	7	0.872668	0.188050
45	20	13	8	0.964342	0.091674
45	20	13	9	0.993445	0.029103
45	20	13	10	0.999265	0.005821
45	20	13	11	0.999956	0.000690
45	20	13	12	0.999999	0.000043

N	n	k	x	P(x)	p(x)
45	20	13	13	1.000000	0.000001
45	20	14	0	0.000027	0.000027
45	20	14	1	0.000650	0.000623
45	20	14	2	0.006571	0.005921
45	20	14	3	0.037022	0.030451
45	20	14	4	0.131929	0.094906
45	20	14	5	0.321741	0.189813
45	20	14	6	0.572964	0.251223
45	20	14	7	0.796273	0.223309
45	20	14	8	0.929964	0.133692
45	20	14	9	0.983441	0.053477
45	20	14	10	0.997447	0.014006
45	20	14	11	0.999762	0.002315
45	20	14	12	0.999988	0.000226
45	20	14	13	1.000000	0.000012
45	20	14	14	1.000000	0.000000
45	20	15	0	0.000009	0.000009
45	20	15	1	0.000268	0.000258
45	20	15	2	0.003133	0.002865
45	20	15	3	0.020323	0.017190
45	20	15	4	0.082945	0.062621
45	20	15	5	0.229896	0.146952
45	20	15	6	0.459508	0.229612
45	20	15	7	0.702627	0.243119
45	20	15	8	0.878213	0.175586
45	20	15	9	0.964465	0.086253
45	20	15	10	0.992929	0.028463
45	20	15	11	0.999089	0.006161
45	20	15	12	0.999930	0.000840
45	20	15	13	0.999997	0.000067
45	20	15	14	1.000000	0.000003
45	20	15	15	1.000000	0.000000
45	20	16	0	0.000003	0.000003
45	20	16	1	0.000104	0.000101
45	20	16	2	0.001414	0.001310
45	20	16	3	0.010582	0.009168
45	20	16	4	0.049547	0.038964
45	20	16	5	0.156420	0.106874
45	20	16	6	0.352356	0.195936
45	20	16	7	0.597276	0.244919
45	20	16	8	0.807978	0.210703
45	20	16	9	0.932839	0.124861
45	20	16	10	0.983441	0.050602
45	20	16	11	0.997241	0.013800
45	20	16	12	0.999706	0.002464
45	20	16	13	0.999981	0.000276
45	20	16	14	0.999999	0.000018
45	20	16	15	1.000000	0.000001
45	20	16	16	1.000000	0.000000
45	20	17	0	0.000001	0.000001

N	n	k	x	P(x)	p(x)
45	20	17	1	0.000038	0.000037
45	20	17	2	0.000601	0.000563
45	20	17	3	0.005208	0.004607
45	20	17	4	0.028049	0.022841
45	20	17	5	0.101141	0.073092
45	20	17	6	0.257767	0.156626
45	20	17	7	0.487484	0.229718
45	20	17	8	0.720791	0.233307
45	20	17	9	0.885478	0.164687
45	20	17	10	0.965992	0.080514
45	20	17	11	0.992958	0.026966
45	20	17	12	0.999026	0.006067
45	20	17	13	0.999915	0.000889
45	20	17	14	0.999996	0.000081
45	30	17	15	1.000000	0.000004
45	20	17	16	1.000000	0.000000
45	20	17	17	1.000000	0.000000
45	20	18	0	0.000000	0.000000
45	20	18	1	0.000013	0.000013
45	20	18	2	0.000239	0.000226
45	20	18	3	0.002411	0.002172
45	20	18	4	0.014997	0.012586
45	20	18	5	0.061984	0.046988
45	20	18	6	0.179454	0.117469
45	20	18	7	0.380829	0.201376
45	20	18	8	0.620802	0.239973
45	20	18	9	0.820780	0.199977
45	20	18	10	0.937237	0.116457
45	20	18	11	0.984291	0.047053
45	20	18	12	0.997292	0.013002
45	20	18	13	0.999692	0.002400
45	20	18	14	0.999978	0.000286
45	20	18	15	0.999999	0.000021
45	20	18	16	1.000000	0.000001
45	20	18	17	1.000000	0.000000
45	20	18	18	1.000000	0.000000
45	20	19	0	0.000000	0.000000
45	20	19	1	0.000004	0.000004
45	20	19	2	0.000088	0.000084
45	20	19	3	0.001043	0.000955
45	20	19	4	0.007538	0.006495
45	20	19	5	0.035880	0.028342
45	20	19	6	0.118544	0.082664
45	20	19	7	0.283871	0.165327
45	20	19	8	0.514148	0.230277
45	20	19	9	0.739307	0.225160
45	20	19	10	0.894105	0.154797
45	20	19	11	0.968606	0.074501
45	20	19	12	0.993440	0.024834
45	20	19	13	0.999070	0.005630

$N = 45 \quad n = 20$

Table for $N = 2$, $n = 1$, through $N = 100$, $n = 50$

N	n	k	x	P(x)	p(x)
45	20	19	14	0.999915	0.000845
45	20	19	15	0.999995	0.000080
45	20	19	16	1.000000	0.000005
45	20	19	17	1.000000	0.000000
45	20	19	18	1.000000	0.000000
45	20	19	19	1.000000	0.000000
45	20	20	0	0.000000	0.000000
45	20	20	1	0.000001	0.000001
45	20	20	2	0.000030	0.000029
45	20	20	3	0.000419	0.000389
45	20	20	4	0.003542	0.003123
45	20	20	5	0.019529	0.015988
45	20	20	6	0.074033	0.054503
45	20	20	7	0.201207	0.127175
45	20	20	8	0.407866	0.206659
45	20	20	9	0.644048	0.236181
45	20	20	10	0.834567	0.190520
45	20	20	11	0.942817	0.108250
45	20	20	12	0.985799	0.042982
45	20	20	13	0.997554	0.011756
45	20	20	14	0.999720	0.002166
45	20	20	15	0.999980	0.000260
45	20	20	16	0.999999	0.000019
45	20	20	17	1.000000	0.000001
45	20	20	18	1.000000	0.000000
45	20	20	19	1.000000	0.000000
45	20	20	20	1.000000	0.000000
45	21	1	0	0.533333	0.533333
45	21	1	1	1.000000	0.466667
45	21	2	0	0.278788	0.278788
45	21	2	1	0.787879	0.509091
45	21	2	2	1.000000	0.212121
45	21	3	0	0.142636	0.142636
45	21	3	1	0.551092	0.408457
45	21	3	2	0.906272	0.355180
45	21	3	3	1.000000	0.093728
45	21	4	0	0.071318	0.071318
45	21	4	1	0.356589	0.285271
45	21	4	2	0.745595	0.389006
45	21	4	3	0.959831	0.214235
45	21	4	4	1.000000	0.040169
45	21	5	0	0.034789	0.034789
45	21	5	1	0.217432	0.182643
45	21	5	2	0.565324	0.347892
45	21	5	3	0.865776	0.300452
45	21	5	4	0.983344	0.117568
45	21	5	5	1.000000	0.016655
45	21	6	0	0.016525	0.016525
45	21	6	1	0.126111	0.109586
45	21	6	2	0.400076	0.273965

N	n	k	x	P(x)	p(x)
45	21	6	3	0.730573	0.330497
45	21	6	4	0.933378	0.202805
45	21	6	5	0.993338	0.059960
45	21	6	6	1.000000	0.006662
45	21	7	0	0.007627	0.007627
45	21	7	1	0.069913	0.062286
45	21	7	2	0.266606	0.196693
45	21	7	3	0.578036	0.311430
45	21	7	4	0.844976	0.266940
45	21	7	5	0.968739	0.123763
45	21	7	6	0.997438	0.028699
45	21	7	7	1.000000	0.002562
45	21	8	0	0.003412	0.003412
45	21	8	1	0.037131	0.033719
45	21	8	2	0.168259	0.131128
45	21	8	3	0.430516	0.262257
45	21	8	4	0.725555	0.295039
45	21	8	5	0.916628	0.191073
45	21	8	6	0.986109	0.069481
45	21	8	7	0.999056	0.012947
45	21	8	8	1.000000	0.000944
45	21	9	0	0.001475	0.001475
45	21	9	1	0.018904	0.017429
45	21	9	2	0.100923	0.082019
45	21	9	3	0.302932	0.202009
45	21	9	4	0.589997	0.287065
45	21	9	5	0.834002	0.244005
45	21	9	6	0.957941	0.123939
45	21	9	7	0.994157	0.036216
45	21	9	8	0.999668	0.005511
45	21	9	9	1.000000	0.000332
45	21	10	0	0.000615	0.000615
45	21	10	1	0.009222	0.008607
45	21	10	2	0.057635	0.048414
45	21	10	3	0.201927	0.144292
45	21	10	4	0.454438	0.252511
45	21	10	5	0.725555	0.271117
45	21	10	6	0.906300	0.180745
45	21	10	7	0.980073	0.073773
45	21	10	8	0.997678	0.017605
45	21	10	9	0.999889	0.002211
45	21	10	10	1.000000	0.000111
45	21	11	0	0.000246	0.000246
45	21	11	1	0.004303	0.004058
45	21	11	2	0.031354	0.027050
45	21	11	3	0.127720	0.096366
45	21	11	4	0.331790	0.204070
45	21	11	5	0.601616	0.269826
45	21	11	6	0.828838	0.227222
45	21	11	7	0.950564	0.121726

N	n	k	x	P(x)	p(x)
45	21	11	8	0.991139	0.040575
45	21	11	9	0.999131	0.007992
45	21	11	10	0.999965	0.000834
45	21	11	11	1.000000	0.000035
45	21	12	0	0.000094	0.000094
45	21	12	1	0.001917	0.001823
45	21	12	2	0.016237	0.014321
45	21	12	3	0.076703	0.060465
45	21	12	4	0.229755	0.153053
45	21	12	5	0.474639	0.244884
45	21	12	6	0.728593	0.253954
45	21	12	7	0.900441	0.171848
45	21	12	8	0.975625	0.075184
45	21	12	9	0.996310	0.020685
45	21	12	10	0.999695	0.003385
45	21	12	11	0.999990	0.000294
45	21	12	12	1.000000	0.000010
45	21	13	0	0.000034	0.000034
45	21	13	1	0.000812	0.000778
45	21	13	2	0.007992	0.007180
45	21	13	3	0.043722	0.035729
45	21	13	4	0.150910	0.107188
45	21	13	5	0.355907	0.204998
45	21	13	6	0.613159	0.257252
45	21	13	7	0.827536	0.214377
45	21	13	8	0.946007	0.118471
45	21	13	9	0.988788	0.042781
45	21	13	10	0.998567	0.009779
45	21	13	11	0.999900	0.001333
45	21	13	12	0.999997	0.000097
45	21	13	13	1.000000	0.000003
45	21	14	0	0.000012	0.000012
45	21	14	1	0.000326	0.000314
45	21	14	2	0.003729	0.003403
45	21	14	3	0.023624	0.019895
45	21	14	4	0.093966	0.070342
45	21	14	5	0.253409	0.159443
45	21	14	6	0.492572	0.239164
45	21	14	7	0.733746	0.241174
45	21	14	8	0.897878	0.164132
45	21	14	9	0.972745	0.074867
45	21	14	10	0.995206	0.022460
45	21	14	11	0.999484	0.004278
45	21	14	12	0.999970	0.000486
45	21	14	13	0.999999	0.000029
45	21	14	14	1.000000	0.000001
45	21	15	0	0.000004	0.000004
45	21	15	1	0.000123	0.000119
45	21	15	2	0.001643	0.001520
45	21	15	3	0.012072	0.010429

N	n	k	x	P(x)	p(x)
45	21	15	4	0.055391	0.043319
45	21	15	5	0.171116	0.115724
45	21	15	6	0.376848	0.205732
45	21	15	7	0.624829	0.247981
45	21	15	8	0.829049	0.204220
45	21	15	9	0.943765	0.114716
45	21	15	10	0.987236	0.043471
45	21	15	11	0.998104	0.010868
45	21	15	12	0.999829	0.001725
45	21	15	13	0.999992	0.000163
45	21	15	14	1.000000	0.000008
45	21	15	15	1.000000	0.000000
45	21	16	0	0.000001	0.000001
45	21	16	1	0.000044	0.000042
45	21	16	2	0.000681	0.000637
45	21	16	3	0.005815	0.005134
45	21	16	4	0.030844	0.025029
45	21	16	5	0.109396	0.078552
45	21	16	6	0.273982	0.164586
45	21	16	7	0.509105	0.235123
45	21	16	8	0.740554	0.231449
45	21	16	9	0.897878	0.157325
45	21	16	10	0.971296	0.073418
45	21	16	11	0.994481	0.023185
45	21	16	12	0.999311	0.004830
45	21	16	13	0.999948	0.000637
45	21	16	14	0.999998	0.000050
45	21	16	15	1.000000	0.000002
45	21	16	16	1.000000	0.000000
45	21	17	0	0.000000	0.000000
45	21	17	1	0.000014	0.000014
45	21	17	2	0.000263	0.000249
45	21	17	3	0.002628	0.002365
45	21	17	4	0.016171	0.013544
45	21	17	5	0.066057	0.049885
45	21	17	6	0.188851	0.122794
45	21	17	7	0.395597	0.206746
45	21	17	8	0.636801	0.241203
45	21	17	9	0.832778	0.195978
45	21	17	10	0.943448	0.110670
45	21	17	11	0.986486	0.043038
45	21	17	12	0.997812	0.011326
45	21	17	13	0.999772	0.001960
45	21	17	14	0.999986	0.000213
45	21	17	15	0.999999	0.000014
45	21	17	16	1.000000	0.000000
45	21	17	17	1.000000	0.000000
45	21	18	0	0.000000	0.000000
45	21	18	1	0.000004	0.000004
45	21	18	2	0.000094	0.000090

N	n	k	x	P(x)	p(x)
45	21	18	3	0.001108	0.001013
45	21	18	4	0.007949	0.006841
45	21	18	5	0.037551	0.029602
45	21	18	6	0.123068	0.085518
45	21	18	7	0.292224	0.169156
45	21	18	8	0.524813	0.232589
45	21	18	9	0.748788	0.223975
45	21	18	10	0.899971	0.151183
45	21	18	11	0.971116	0.071145
45	21	18	12	0.994172	0.023056
45	21	18	13	0.999212	0.005041
45	21	18	14	0.999933	0.000720
45	21	18	15	0.999997	0.000064
45	21	18	16	1.000000	0.000003
45	21	18	17	1.000000	0.000000
45	21	18	18	1.000000	0.000000
45	21	19	0	0.000000	0.000000
45	21	19	1	0.000001	0.000001
45	21	19	2	0.000031	0.000030
45	21	19	3	0.000432	0.000401
45	21	19	4	0.003641	0.003209
45	21	19	5	0.020009	0.016367
45	21	19	6	0.075559	0.055550
45	21	19	7	0.204514	0.128955
45	21	19	8	0.412826	0.208312
45	21	19	9	0.649244	0.236418
45	21	19	10	0.838378	0.189134
45	21	19	11	0.944766	0.106388
45	21	19	12	0.986486	0.041721
45	21	19	13	0.997719	0.011233
45	21	19	14	0.999746	0.002027
45	21	19	15	0.999982	0.000236
45	21	19	16	0.999999	0.000017
45	21	19	17	1.000000	0.000001
45	21	19	18	1.000000	0.000000
45	21	19	19	1.000000	0.000000
45	21	20	0	0.000000	0.000000
45	21	20	1	0.000000	0.000000
45	21	20	2	0.000009	0.000009
45	21	20	3	0.000154	0.000145
45	21	20	4	0.001543	0.001389
45	21	20	5	0.009937	0.008394
45	21	20	6	0.043511	0.033574
45	21	20	7	0.135076	0.091566
45	21	20	8	0.308670	0.173593
45	21	20	9	0.540128	0.231458
45	21	20	10	0.758359	0.218232
45	21	20	11	0.903847	0.145488
45	21	20	12	0.972045	0.068197
45	21	20	13	0.994263	0.022218
45	21	20	14	0.999200	0.004937
45	21	20	15	0.999928	0.000728
45	21	20	16	0.999996	0.000068
45	21	20	17	1.000000	0.000004
45	21	20	18	1.000000	0.000000
45	21	20	19	1.000000	0.000000
45	21	20	20	1.000000	0.000000
45	21	21	0	0.000000	0.000000
45	21	21	1	0.000000	0.000000
45	21	21	2	0.000002	0.000002
45	21	21	3	0.000050	0.000047
45	21	21	4	0.000599	0.000549
45	21	21	5	0.004565	0.003966
45	21	21	6	0.023366	0.018801
45	21	21	7	0.083800	0.060433
45	21	21	8	0.218401	0.134602
45	21	21	9	0.429028	0.210627
45	21	21	10	0.662337	0.233310
45	21	21	11	0.845652	0.183315
45	21	21	12	0.947494	0.101841
45	21	21	13	0.987153	0.039659
45	21	21	14	0.997818	0.010665
45	21	21	15	0.999753	0.001935
45	21	21	16	0.999982	0.000229
45	21	21	17	0.999999	0.000017
45	21	21	18	1.000000	0.000001
45	21	21	19	1.000000	0.000000
45	21	21	20	1.000000	0.000000
45	21	21	21	1.000000	0.000000
45	22	1	0	0.511111	0.511111
45	22	1	1	1.000000	0.488889
45	22	2	0	0.255556	0.255556
45	22	2	1	0.766667	0.511111
45	22	2	2	1.000000	0.233333
45	22	3	0	0.124806	0.124806
45	22	3	1	0.517054	0.392248
45	22	3	2	0.891473	0.374419
45	22	3	3	1.000000	0.108527
45	22	4	0	0.059432	0.059432
45	22	4	1	0.320930	0.261499
45	22	4	2	0.713178	0.392248
45	22	4	3	0.950904	0.237726
45	22	4	4	1.000000	0.049096
45	22	5	0	0.027541	0.027541
45	22	5	1	0.186992	0.159450
45	22	5	2	0.521838	0.334846
45	22	5	3	0.840739	0.318901
45	22	5	4	0.978446	0.137707
45	22	5	5	1.000000	0.021554
45	22	6	0	0.012394	0.012394
45	22	6	1	0.103280	0.090887
45	22	6	2	0.354415	0.251134
45	22	6	3	0.689261	0.334846
45	22	6	4	0.916478	0.227217
45	22	6	5	0.990839	0.074362
45	22	6	6	1.000000	0.009161
45	22	7	0	0.005402	0.005402
45	22	7	1	0.054341	0.048939
45	22	7	2	0.225628	0.171287
45	22	7	3	0.526131	0.300503
45	22	7	4	0.811608	0.285478
45	22	7	5	0.958425	0.146817
45	22	7	6	0.996242	0.037817
45	22	7	7	1.000000	0.003758
45	22	8	0	0.002275	0.002275
45	22	8	1	0.027296	0.025021
45	22	8	2	0.135477	0.108181
45	22	8	3	0.375879	0.240402
45	22	8	4	0.676382	0.300503
45	22	8	5	0.892744	0.216362
45	22	8	6	0.980319	0.087575
45	22	8	7	0.998516	0.018197
45	22	8	8	1.000000	0.001483
45	22	9	0	0.000922	0.000922
45	22	9	1	0.013095	0.012173
45	22	9	2	0.077001	0.063906
45	22	9	3	0.252429	0.175429
45	22	9	4	0.530191	0.277762
45	22	9	5	0.793334	0.263143
45	22	9	6	0.942449	0.149114
45	22	9	7	0.991139	0.048690
45	22	9	8	0.999439	0.008299
45	22	9	9	1.000000	0.000561
45	22	10	0	0.000359	0.000359
45	22	10	1	0.005994	0.005635
45	22	10	2	0.041497	0.035503
45	22	10	3	0.159842	0.118345
45	22	10	4	0.391310	0.231468
45	22	10	5	0.669072	0.277762
45	22	10	6	0.876176	0.207103
45	22	10	7	0.970851	0.094676
45	22	10	8	0.996211	0.025360
45	22	10	9	0.999797	0.003586
45	22	10	10	1.000000	0.000203
45	22	11	0	0.000133	0.000133
45	22	11	1	0.002613	0.002480
45	22	11	2	0.021210	0.018597
45	22	11	3	0.095598	0.074388
45	22	11	4	0.272270	0.176672
45	22	11	5	0.534159	0.261890
45	22	11	6	0.781500	0.247340
45	22	11	7	0.930276	0.148776
45	22	11	8	0.986067	0.055791
45	22	11	9	0.998465	0.012398
45	22	11	10	0.999930	0.001465
45	22	11	11	1.000000	0.000069
45	22	12	0	0.000047	0.000047
45	22	12	1	0.001081	0.001034
45	22	12	2	0.010270	0.009189
45	22	12	3	0.054028	0.043758
45	22	12	4	0.178738	0.124709
45	22	12	5	0.403215	0.224477
45	22	12	6	0.665104	0.261890
45	22	12	7	0.864640	0.199535
45	22	12	8	0.963094	0.098455
45	22	12	9	0.993725	0.030630
45	22	12	10	0.999413	0.005688
45	22	12	11	0.999977	0.000564
45	22	12	12	1.000000	0.000022
45	22	13	0	0.000016	0.000016
45	22	13	1	0.000423	0.000407
45	22	13	2	0.004701	0.004278
45	22	13	3	0.028834	0.024133
45	22	13	4	0.110714	0.081880
45	22	13	5	0.287575	0.176861
45	22	13	6	0.538128	0.250553
45	22	13	7	0.773942	0.235814
45	22	13	8	0.921326	0.147384
45	22	13	9	0.981658	0.060333
45	22	13	10	0.997345	0.015686
45	22	13	11	0.999789	0.002445
45	22	13	12	0.999993	0.000204
45	22	13	13	1.000000	0.000007
45	22	14	0	0.000005	0.000005
45	22	14	1	0.000156	0.000151
45	22	14	2	0.002027	0.001872
45	22	14	3	0.014505	0.012478
45	22	14	4	0.064657	0.050151
45	22	14	5	0.193618	0.128961
45	22	14	6	0.412851	0.219234
45	22	14	7	0.663404	0.250553
45	22	14	8	0.856845	0.193441
45	22	14	9	0.957148	0.100303
45	22	14	10	0.991462	0.034314
45	22	14	11	0.998949	0.007487
45	22	14	12	0.999929	0.000980
45	22	14	13	0.999998	0.000069
45	22	14	14	1.000000	0.000002
45	22	15	0	0.000001	0.000001
45	22	15	1	0.000054	0.000052

Table for $N = 2$, $n = 1$, through $N = 100$, $n = 50$

N	n	k	x	P(x)	p(x)
45	22	15	2	0.000820	0.000766
45	22	15	3	0.006858	0.006038
45	22	15	4	0.035537	0.028679
45	22	15	5	0.122897	0.087361
45	22	15	6	0.299698	0.176801
45	22	15	7	0.542169	0.242470
45	22	15	8	0.769485	0.227316
45	22	15	9	0.915086	0.145601
45	22	15	10	0.978179	0.063094
45	22	15	11	0.996292	0.018113
45	22	15	12	0.999613	0.003321
45	22	15	13	0.999978	0.000365
45	22	15	14	0.999999	0.000021
45	22	15	15	1.000000	0.000000
45	22	16	0	0.000000	0.000000
45	22	16	1	0.000017	0.000017
45	22	16	2	0.000309	0.000292
45	22	16	3	0.003034	0.002725
45	22	16	4	0.018329	0.015295
45	22	16	5	0.073393	0.055064
45	22	16	6	0.205404	0.132012
45	22	16	7	0.420934	0.215529
45	22	16	8	0.663404	0.242470
45	22	16	9	0.851992	0.188588
45	22	16	10	0.952942	0.100950
45	22	16	11	0.989651	0.036709
45	22	16	12	0.998506	0.008855
45	22	16	13	0.999869	0.001362
45	22	16	14	0.999994	0.000125
45	22	16	15	1.000000	0.000006
45	22	16	16	1.000000	0.000000
45	22	17	0	0.000000	0.000000
45	22	17	1	0.000005	0.000005
45	22	17	2	0.000108	0.000103
45	22	17	3	0.001249	0.001141
45	22	17	4	0.008835	0.007587
45	22	17	5	0.041114	0.032279
45	22	17	6	0.132570	0.091456
45	22	17	7	0.309453	0.176883
45	22	17	8	0.546349	0.236896
45	22	17	9	0.767452	0.221103
45	22	17	10	0.911169	0.143717
45	22	17	11	0.975727	0.064557
45	22	17	12	0.995453	0.019726
45	22	17	13	0.999446	0.003993
45	22	17	14	0.999959	0.000513
45	22	17	15	0.999998	0.000039
45	22	17	16	1.000000	0.000002
45	22	17	17	1.000000	0.000000
45	22	18	0	0.000000	0.000000

N	n	k	x	P(x)	p(x)
45	22	18	1	0.000001	0.000001
45	22	18	2	0.000034	0.000033
45	22	18	3	0.000474	0.000440
45	22	18	4	0.003958	0.003484
45	22	18	5	0.021516	0.017558
45	22	18	6	0.080310	0.058793
45	22	18	7	0.214694	0.134385
45	22	18	8	0.427901	0.213207
45	22	18	9	0.664797	0.236896
45	22	18	10	0.849576	0.184779
45	22	18	11	0.950365	0.100789
45	22	18	12	0.988408	0.038043
45	22	18	13	0.998162	0.009755
45	22	18	14	0.999812	0.001650
45	22	18	15	0.999988	0.000176
45	22	18	16	0.999999	0.000011
45	22	18	17	1.000000	0.000000
45	22	18	18	1.000000	0.000000
45	22	19	0	0.000000	0.000000
45	22	19	1	0.000000	0.000000
45	22	19	2	0.000010	0.000010
45	22	19	3	0.000165	0.000155
45	22	19	4	0.001636	0.001471
45	22	19	5	0.010461	0.008826
45	22	19	6	0.045469	0.035008
45	22	19	7	0.140036	0.094567
45	22	19	8	0.317349	0.177313
45	22	19	9	0.550736	0.233387
45	22	19	10	0.767452	0.216716
45	22	19	11	0.909303	0.141851
45	22	19	12	0.974318	0.065015
45	22	19	13	0.994911	0.020593
45	22	19	14	0.999323	0.004413
45	22	19	15	0.999943	0.000619
45	22	19	16	0.999997	0.000054
45	22	19	17	1.000000	0.000003
45	22	19	18	1.000000	0.000000
45	22	19	19	1.000000	0.000000
45	22	20	0	0.000000	0.000000
45	22	20	1	0.000000	0.000000
45	22	20	2	0.000003	0.000002
45	22	20	3	0.000052	0.000049
45	22	20	4	0.000617	0.000566
45	22	20	5	0.004691	0.004073
45	22	20	6	0.023926	0.019235
45	22	20	7	0.085478	0.061553
45	22	20	8	0.221873	0.136395
45	22	20	9	0.434043	0.212170
45	22	20	10	0.667430	0.233387
45	22	20	11	0.849289	0.181860

N	n	k	x	P(x)	p(x)
45	22	20	12	0.949312	0.100023
45	22	20	13	0.987782	0.038470
45	22	20	14	0.997966	0.010183
45	22	20	15	0.999776	0.001810
45	22	20	16	0.999985	0.000208
45	22	20	17	0.999999	0.000015
45	22	20	18	1.000000	0.000001
45	22	20	19	1.000000	0.000000
45	22	20	20	1.000000	0.000000
45	22	21	0	0.000000	0.000000
45	22	21	1	0.000000	0.000000
45	22	21	2	0.000001	0.000001
45	22	21	3	0.000014	0.000014
45	22	21	4	0.000210	0.000196
45	22	21	5	0.001921	0.001711
45	22	21	6	0.011615	0.009695
45	22	21	7	0.048547	0.036932
45	22	21	8	0.145492	0.096945
45	22	21	9	0.323715	0.178223
45	22	21	10	0.555404	0.231689
45	22	21	11	0.769271	0.213867
45	22	21	12	0.909303	0.140032
45	22	21	13	0.973933	0.064630
45	22	21	14	0.994707	0.020774
45	22	21	15	0.999269	0.004562
45	22	21	16	0.999935	0.000665
45	22	21	17	0.999996	0.000062
45	22	21	18	1.000000	0.000003
45	22	21	19	1.000000	0.000000
45	22	21	20	1.000000	0.000000
45	22	21	21	1.000000	0.000000
45	22	22	0	0.000000	0.000000
45	22	22	1	0.000000	0.000000
45	22	22	2	0.000000	0.000000
45	22	22	3	0.000003	0.000003
45	22	22	4	0.000063	0.000060
45	22	22	5	0.000709	0.000646
45	22	22	6	0.005152	0.004443
45	22	22	7	0.025465	0.020312
45	22	22	8	0.088941	0.063476
45	22	22	9	0.227177	0.138237
45	22	22	10	0.439559	0.212382
45	22	22	11	0.671249	0.231689
45	22	22	12	0.850956	0.179708
45	22	22	13	0.949697	0.098741
45	22	22	14	0.987782	0.038086
45	22	22	15	0.997939	0.010156
45	22	22	16	0.999768	0.001830
45	22	22	17	0.999983	0.000215
45	22	22	18	0.999999	0.000016

N	n	k	x	P(x)	p(x)
45	22	22	19	1.000000	0.000001
45	22	22	20	1.000000	0.000000
45	22	22	21	1.000000	0.000000
45	22	22	22	1.000000	0.000000
46	1	1	0	0.978261	0.978261
46	1	1	1	1.000000	0.021739
46	2	1	0	0.956522	0.956522
46	2	1	1	1.000000	0.043478
46	2	2	0	0.914010	0.914010
46	2	2	1	0.999034	0.085024
46	2	2	2	1.000000	0.000966
46	3	1	0	0.934783	0.934783
46	3	1	1	1.000000	0.065217
46	3	2	0	0.872464	0.872464
46	3	2	1	0.997101	0.124638
46	3	2	2	1.000000	0.002899
46	3	3	0	0.812978	0.812978
46	3	3	1	0.991436	0.178458
46	3	3	2	0.999934	0.008498
46	3	3	3	1.000000	0.000066
46	4	1	0	0.913043	0.913043
46	4	1	1	1.000000	0.086957
46	4	2	0	0.831884	0.831884
46	4	2	1	0.994203	0.162319
46	4	2	2	1.000000	0.005797
46	4	3	0	0.756258	0.756258
46	4	3	1	0.983136	0.226877
46	4	3	2	0.999736	0.016601
46	4	3	3	1.000000	0.000263
46	4	4	0	0.685909	0.685909
46	4	4	1	0.967307	0.281398
46	4	4	2	0.998964	0.031657
46	4	4	3	0.999994	0.001030
46	4	4	4	1.000000	0.000006
46	5	1	0	0.891304	0.891304
46	5	1	1	1.000000	0.108696
46	5	2	0	0.792271	0.792271
46	5	2	1	0.990338	0.198068
46	5	2	2	1.000000	0.009662
46	5	3	0	0.702240	0.702240
46	5	3	1	0.972332	0.270092
46	5	3	2	0.999341	0.027009
46	5	3	3	1.000000	0.000659
46	5	4	0	0.620584	0.620584
46	5	4	1	0.947207	0.326623
46	5	4	2	0.997457	0.050250
46	5	4	3	0.999969	0.002512
46	5	4	4	1.000000	0.000031
46	5	5	0	0.546705	0.546705
46	5	5	1	0.916100	0.369395

N	n	k	x	P(x)	p(x)
46	5	5	2	0.993868	0.077767
46	5	5	3	0.999850	0.005982
46	5	5	4	0.999999	0.000150
46	5	5	5	1.000000	0.000001
46	6	1	0	0.869565	0.869565
46	6	1	1	1.000000	0.130435
46	6	2	0	0.753623	0.753623
46	6	2	1	0.985507	0.231884
46	6	2	2	1.000000	0.014493
46	6	3	0	0.650856	0.650856
46	6	3	1	0.959157	0.308300
46	6	3	2	0.998682	0.039526
46	6	3	3	1.000000	0.001318
46	6	4	0	0.560039	0.560039
46	6	4	1	0.923308	0.363269
46	6	4	2	0.995006	0.071698
46	6	4	3	0.999908	0.004902
46	6	4	4	1.000000	0.000092
46	6	5	0	0.480034	0.480034
46	6	5	1	0.880062	0.400028
46	6	5	2	0.988177	0.108116
46	6	5	3	0.999558	0.011381
46	6	5	4	0.999996	0.000438
46	6	5	5	1.000000	0.000004
46	6	6	0	0.409785	0.409785
46	6	6	1	0.831278	0.421493
46	6	6	2	0.977629	0.146352
46	6	6	3	0.998725	0.021096
46	6	6	4	0.999974	0.001249
46	6	6	5	1.000000	0.000026
46	6	6	6	1.000000	0.000000
46	7	1	0	0.847826	0.847826
46	7	1	1	1.000000	0.152174
46	7	2	0	0.715942	0.715942
46	7	2	1	0.979710	0.263768
46	7	2	2	1.000000	0.020290
46	7	3	0	0.602042	0.602042
46	7	3	1	0.943742	0.341700
46	7	3	2	0.997694	0.053953
46	7	3	3	1.000000	0.002306
46	7	4	0	0.504035	0.504035
46	7	4	1	0.896063	0.392027
46	7	4	2	0.991421	0.095358
46	7	4	3	0.999785	0.008365
46	7	4	4	1.000000	0.000214
46	7	5	0	0.420029	0.420029
46	7	5	1	0.840059	0.420029
46	7	5	2	0.980069	0.140010
46	7	5	3	0.998989	0.018920
46	7	5	4	0.999985	0.000996
46	7	5	5	1.000000	0.000015
46	7	6	0	0.348317	0.348317
46	7	6	1	0.778591	0.430274
46	7	6	2	0.962994	0.184403
46	7	6	3	0.997143	0.034149
46	7	6	4	0.999912	0.002769
46	7	6	5	0.999999	0.000087
46	7	6	6	1.000000	0.000001
46	7	7	0	0.287362	0.287362
46	7	7	1	0.714050	0.426688
46	7	7	2	0.939944	0.225894
46	7	7	3	0.993728	0.053784
46	7	7	4	0.999704	0.005976
46	7	7	5	0.999995	0.000291
46	7	7	6	1.000000	0.000005
46	7	7	7	1.000000	0.000000
46	8	1	0	0.826087	0.826087
46	8	1	1	1.000000	0.173913
46	8	2	0	0.679227	0.679227
46	8	2	1	0.972947	0.293720
46	8	2	2	1.000000	0.027053
46	8	3	0	0.555731	0.555731
46	8	3	1	0.926219	0.370487
46	8	3	2	0.996311	0.070092
46	8	3	3	1.000000	0.003689
46	8	4	0	0.452339	0.452339
46	8	4	1	0.865907	0.413567
46	8	4	2	0.986531	0.120624
46	8	4	3	0.999571	0.013040
46	8	4	4	1.000000	0.000429
46	8	5	0	0.366179	0.366179
46	8	5	1	0.796979	0.430799
46	8	5	2	0.969299	0.172320
46	8	5	3	0.998019	0.028720
46	8	5	4	0.999959	0.001941
46	8	5	5	1.000000	0.000041
46	8	6	0	0.294730	0.294730
46	8	6	1	0.723428	0.428698
46	8	6	2	0.944081	0.220653
46	8	6	3	0.994516	0.050435
46	8	6	4	0.999770	0.005254
46	8	6	5	0.999997	0.000227
46	8	6	6	1.000000	0.000003
46	8	7	0	0.235784	0.235784
46	8	7	1	0.648406	0.412622
46	8	7	2	0.910983	0.262577
46	8	7	3	0.988212	0.077229
46	8	7	4	0.999244	0.011033
46	8	7	5	0.999980	0.000736
46	8	7	6	1.000000	0.000020
46	8	7	7	1.000000	0.000000
46	8	8	0	0.187418	0.187418
46	8	8	1	0.574345	0.386927
46	8	8	2	0.870587	0.296241
46	8	8	3	0.978311	0.107724
46	8	8	4	0.998113	0.019802
46	8	8	5	0.999923	0.001810
46	8	8	6	0.999999	0.000075
46	8	8	7	1.000000	0.000001
46	8	8	8	1.000000	0.000000
46	9	1	0	0.804348	0.804348
46	9	1	1	1.000000	0.195652
46	9	2	0	0.643478	0.643478
46	9	2	1	0.965217	0.321739
46	9	2	2	1.000000	0.034783
46	9	3	0	0.511858	0.511858
46	9	3	1	0.906719	0.394862
46	9	3	2	0.994466	0.087747
46	9	3	3	1.000000	0.005534
46	9	4	0	0.404725	0.404725
46	9	4	1	0.833257	0.428532
46	9	4	2	0.980182	0.146925
46	9	4	3	0.999228	0.019046
46	9	4	4	1.000000	0.000772
46	9	5	0	0.317998	0.317998
46	9	5	1	0.751632	0.433634
46	9	5	2	0.955694	0.204063
46	9	5	3	0.996507	0.040813
46	9	5	4	0.999908	0.003401
46	9	5	5	1.000000	0.000092
46	9	6	0	0.248194	0.248194
46	9	6	1	0.667020	0.418827
46	9	6	2	0.920854	0.253834
46	9	6	3	0.990534	0.069680
46	9	6	4	0.999493	0.008959
46	9	6	5	0.999991	0.000498
46	9	6	6	1.000000	0.000009
46	9	7	0	0.192350	0.192350
46	9	7	1	0.583255	0.390905
46	9	7	2	0.876433	0.293179
46	9	7	3	0.980082	0.103649
46	9	7	4	0.998373	0.018291
46	9	7	5	0.999941	0.001568
46	9	7	6	0.999999	0.000058
46	9	7	7	1.000000	0.000001
46	9	8	0	0.147962	0.147962
46	9	8	1	0.503069	0.355108
46	9	8	2	0.823812	0.320742
46	9	8	3	0.964136	0.140325
46	9	8	4	0.996028	0.031892
46	9	8	5	0.999780	0.003752
46	9	8	6	0.999995	0.000214
46	9	8	7	1.000000	0.000005
46	9	8	8	1.000000	0.000000
46	9	9	0	0.112918	0.112918
46	9	9	1	0.428310	0.315392
46	9	9	2	0.764727	0.336418
46	9	9	3	0.941980	0.177252
46	9	9	4	0.991832	0.049852
46	9	9	5	0.999385	0.007553
46	9	9	6	0.999978	0.000592
46	9	9	7	1.000000	0.000022
46	9	9	8	1.000000	0.000000
46	9	9	9	1.000000	0.000000
46	10	1	0	0.782609	0.782609
46	10	1	1	1.000000	0.217391
46	10	2	0	0.608696	0.608696
46	10	2	1	0.956522	0.347826
46	10	2	2	1.000000	0.043478
46	10	3	0	0.470356	0.470356
46	10	3	1	0.885375	0.415020
46	10	3	2	0.992095	0.106719
46	10	3	3	1.000000	0.007905
46	10	4	0	0.360971	0.360971
46	10	4	1	0.798511	0.437540
46	10	4	2	0.972240	0.173729
46	10	4	3	0.998713	0.026473
46	10	4	4	1.000000	0.001287
46	10	5	0	0.275025	0.275025
46	10	5	1	0.704752	0.429727
46	10	5	2	0.939149	0.234397
46	10	5	3	0.994301	0.055152
46	10	5	4	0.999816	0.005515
46	10	5	5	1.000000	0.000184
46	10	6	0	0.207946	0.207946
46	10	6	1	0.610422	0.402476
46	10	6	2	0.893413	0.282991
46	10	6	3	0.984885	0.091472
46	10	6	4	0.999009	0.014124
46	10	6	5	0.999978	0.000969
46	10	6	6	1.000000	0.000022
46	10	7	0	0.155959	0.155959
46	10	7	1	0.519865	0.363905
46	10	7	2	0.836815	0.316950
46	10	7	3	0.968877	0.132062
46	10	7	4	0.996890	0.028013
46	10	7	5	0.999856	0.002966
46	10	7	6	0.999998	0.000141
46	10	7	7	1.000000	0.000002
46	10	8	0	0.115970	0.115970

Table for $N = 2$, $n = 1$, through $N = 100$, $n = 50$

N	n	k	x	P(x)	p(x)
46	10	8	1	0.435887	0.319917
46	10	8	2	0.771799	0.335913
46	10	8	3	0.945174	0.173374
46	10	8	4	0.992581	0.047407
46	10	8	5	0.999476	0.006896
46	10	8	6	0.999983	0.000507
46	10	8	7	1.000000	0.000017
46	10	8	8	1.000000	0.000000
46	10	9	0	0.085451	0.085451
46	10	9	1	0.360117	0.274665
46	10	9	2	0.701081	0.340964
46	10	9	3	0.913236	0.212155
46	10	9	4	0.985095	0.071859
46	10	9	5	0.998569	0.013474
46	10	9	6	0.999930	0.001361
46	10	9	7	0.999998	0.000069
46	10	9	8	1.000000	0.000001
46	10	9	9	1.000000	0.000000
46	10	10	0	0.062356	0.062356
46	10	10	1	0.293306	0.230950
46	10	10	2	0.627359	0.334053
46	10	10	3	0.873099	0.245740
46	10	10	4	0.973442	0.100344
46	10	10	5	0.996748	0.023306
46	10	10	6	0.999783	0.003035
46	10	10	7	0.999993	0.000210
46	10	10	8	1.000000	0.000007
46	10	10	9	1.000000	0.000000
46	10	10	10	1.000000	0.000000
46	11	1	0	0.760870	0.760870
46	11	1	1	1.000000	0.239130
46	11	2	0	0.574879	0.574879
46	11	2	1	0.946860	0.371981
46	11	2	2	1.000000	0.053140
46	11	3	0	0.431159	0.431159
46	11	3	1	0.862319	0.431159
46	11	3	2	0.989130	0.126812
46	11	3	3	1.000000	0.010870
46	11	4	0	0.320863	0.320863
46	11	4	1	0.762049	0.441186
46	11	4	2	0.962588	0.200539
46	11	4	3	0.997978	0.035389
46	11	4	4	1.000000	0.002022
46	11	5	0	0.236827	0.236827
46	11	5	1	0.657005	0.420177
46	11	5	2	0.919616	0.262611
46	11	5	3	0.991237	0.071621
46	11	5	4	0.999663	0.008426
46	11	5	5	1.000000	0.000337
46	11	6	0	0.173288	0.173288
46	11	6	1	0.554522	0.381234
46	11	6	2	0.861969	0.307447
46	11	6	3	0.977262	0.115293
46	11	6	4	0.998224	0.020962
46	11	6	5	0.999951	0.001726
46	11	6	6	1.000000	0.000049
46	11	7	0	0.125634	0.125634
46	11	7	1	0.459214	0.333580
46	11	7	2	0.792794	0.333580
46	11	7	3	0.954203	0.161410
46	11	7	4	0.994556	0.040352
46	11	7	5	0.999692	0.005136
46	11	7	6	0.999994	0.000302
46	11	7	7	1.000000	0.000006
46	11	8	0	0.090199	0.090199
46	11	8	1	0.373681	0.283482
46	11	8	2	0.715814	0.342133
46	11	8	3	0.921094	0.205280
46	11	8	4	0.987313	0.066219
46	11	8	5	0.998902	0.011588
46	11	8	6	0.999955	0.001053
46	11	8	7	0.999999	0.000044
46	11	8	8	1.000000	0.000001
46	11	9	0	0.064089	0.064089
46	11	9	1	0.299080	0.234992
46	11	9	2	0.634782	0.335702
46	11	9	3	0.877877	0.243095
46	11	9	4	0.975115	0.097238
46	11	9	5	0.997072	0.021957
46	11	9	6	0.999816	0.002745
46	11	9	7	0.999995	0.000178
46	11	9	8	1.000000	0.000005
46	11	9	9	1.000000	0.000000
46	11	10	0	0.045035	0.045035
46	11	10	1	0.235569	0.190534
46	11	10	2	0.553125	0.317556
46	11	10	3	0.825316	0.272191
46	11	10	4	0.956718	0.131403
46	11	10	5	0.993511	0.036793
46	11	10	6	0.999445	0.005934
46	11	10	7	0.999975	0.000530
46	11	10	8	0.999999	0.000024
46	11	10	9	1.000000	0.000000
46	11	10	10	1.000000	0.000000
46	11	11	0	0.031274	0.031274
46	11	11	1	0.182643	0.151368
46	11	11	2	0.473736	0.291093
46	11	11	3	0.764829	0.291093
46	11	11	4	0.931168	0.166339
46	11	11	5	0.987379	0.056211
46	11	11	6	0.998621	0.011242
46	11	11	7	0.999916	0.001295
46	11	11	8	0.999997	0.000081
46	11	11	9	1.000000	0.000002
46	11	11	10	1.000000	0.000000
46	11	11	11	1.000000	0.000000
46	12	1	0	0.739130	0.739130
46	12	1	1	1.000000	0.260870
46	12	2	0	0.542029	0.542029
46	12	2	1	0.936232	0.394203
46	12	2	2	1.000000	0.063768
46	12	3	0	0.394203	0.394203
46	12	3	1	0.837681	0.443478
46	12	3	2	0.985507	0.147826
46	12	3	3	1.000000	0.014493
46	12	4	0	0.284193	0.284193
46	12	4	1	0.724233	0.440040
46	12	4	2	0.951129	0.226896
46	12	4	3	0.996967	0.045838
46	12	4	4	1.000000	0.003033
46	12	5	0	0.202995	0.202995
46	12	5	1	0.608985	0.405990
46	12	5	2	0.897106	0.288122
46	12	5	3	0.987144	0.090038
46	12	5	4	0.999422	0.012278
46	12	5	5	1.000000	0.000578
46	12	6	0	0.143582	0.143582
46	12	6	1	0.500060	0.356479
46	12	6	2	0.826833	0.326772
46	12	6	3	0.967380	0.140547
46	12	6	4	0.997026	0.029647
46	12	6	5	0.999901	0.002875
46	12	6	6	1.000000	0.000099
46	12	7	0	0.100507	0.100507
46	12	7	1	0.402029	0.301522
46	12	7	2	0.745140	0.343111
46	12	7	3	0.935757	0.190617
46	12	7	4	0.991097	0.055340
46	12	7	5	0.999398	0.008301
46	12	7	6	0.999985	0.000587
46	12	7	7	1.000000	0.000015
46	12	8	0	0.069582	0.069582
46	12	8	1	0.316984	0.247402
46	12	8	2	0.657162	0.340178
46	12	8	3	0.891768	0.234606
46	12	8	4	0.979745	0.087977
46	12	8	5	0.997908	0.018163
46	12	8	6	0.999895	0.001987
46	12	8	7	0.999998	0.000103
46	12	8	8	1.000000	0.000002
46	12	9	0	0.047609	0.047609
46	12	9	1	0.245368	0.197759
46	12	9	2	0.567642	0.322274
46	12	9	3	0.836204	0.268562
46	12	9	4	0.961224	0.125020
46	12	9	5	0.994562	0.033339
46	12	9	6	0.999581	0.005019
46	12	9	7	0.999984	0.000403
46	12	9	8	1.000000	0.000015
46	12	9	9	1.000000	0.000000
46	12	10	0	0.032168	0.032168
46	12	10	1	0.186575	0.154407
46	12	10	2	0.480541	0.293966
46	12	10	3	0.770878	0.290337
46	12	10	4	0.934192	0.163315
46	12	10	5	0.988255	0.054063
46	12	10	6	0.998767	0.010512
46	12	10	7	0.999930	0.001163
46	12	10	8	0.999998	0.000068
46	12	10	9	1.000000	0.000002
46	12	10	10	1.000000	0.000000
46	12	11	0	0.021445	0.021445
46	12	11	1	0.139395	0.117949
46	12	11	2	0.398883	0.259489
46	12	11	3	0.698294	0.299410
46	12	11	4	0.897900	0.199607
46	12	11	5	0.977743	0.079843
46	12	11	6	0.997015	0.019272
46	12	11	7	0.999768	0.002753
46	12	11	8	0.999990	0.000222
46	12	11	9	1.000000	0.000009
46	12	11	10	1.000000	0.000000
46	12	11	11	1.000000	0.000000
46	12	12	0	0.014093	0.014093
46	12	12	1	0.102325	0.088232
46	12	12	2	0.324744	0.222419
46	12	12	3	0.621302	0.296559
46	12	12	4	0.852276	0.230973
46	12	12	5	0.961774	0.109499
46	12	12	6	0.993711	0.031937
46	12	12	7	0.999375	0.005664
46	12	12	8	0.999965	0.000590
46	12	12	9	0.999999	0.000034
46	12	12	10	1.000000	0.000001
46	12	12	11	1.000000	0.000000
46	12	12	12	1.000000	0.000000
46	13	1	0	0.717391	0.717391
46	13	1	1	1.000000	0.282609
46	13	2	0	0.510145	0.510145
46	13	2	1	0.924638	0.414493

N	n	k	x	P(x)	p(x)
46	13	2	2	1.000000	0.075362
46	13	3	0	0.359420	0.359420
46	13	3	1	0.811594	0.452174
46	13	3	2	0.981159	0.169565
46	13	3	3	1.000000	0.018841
46	13	4	0	0.250758	0.250758
46	13	4	1	0.685406	0.434648
46	13	4	2	0.937782	0.252376
46	13	4	3	0.995618	0.057836
46	13	4	4	1.000000	0.004382
46	13	5	0	0.173143	0.173143
46	13	5	1	0.561221	0.388078
46	13	5	2	0.871684	0.310463
46	13	5	3	0.981848	0.110164
46	13	5	4	0.999061	0.017213
46	13	5	5	1.000000	0.000939
46	13	6	0	0.118244	0.118244
46	13	6	1	0.447637	0.329393
46	13	6	2	0.788389	0.340752
46	13	6	3	0.954978	0.166590
46	13	6	4	0.995282	0.040304
46	13	6	5	0.999817	0.004534
46	13	6	6	1.000000	0.000183
46	13	7	0	0.079815	0.079815
46	13	7	1	0.348819	0.269005
46	13	7	2	0.694682	0.345863
46	13	7	3	0.913331	0.218649
46	13	7	4	0.986214	0.072883
46	13	7	5	0.998910	0.012696
46	13	7	6	0.999968	0.001058
46	13	7	7	1.000000	0.000032
46	13	8	0	0.053210	0.053210
46	13	8	1	0.266048	0.212839
46	13	8	2	0.597131	0.331083
46	13	8	3	0.857267	0.260136
46	13	8	4	0.969395	0.112128
46	13	8	5	0.996306	0.026911
46	13	8	6	0.999778	0.003472
46	13	8	7	0.999995	0.000217
46	13	8	8	1.000000	0.000005
46	13	9	0	0.035006	0.035006
46	13	9	1	0.198836	0.163830
46	13	9	2	0.501291	0.302455
46	13	9	3	0.788810	0.287519
46	13	9	4	0.942838	0.154028
46	13	9	5	0.990640	0.047802
46	13	9	6	0.999138	0.008498
46	13	9	7	0.999961	0.000822
46	13	9	8	0.999999	0.000039
46	13	9	9	1.000000	0.000001

N	n	k	x	P(x)	p(x)
46	13	10	0	0.022707	0.022707
46	13	10	1	0.145702	0.122995
46	13	10	2	0.411372	0.265670
46	13	10	3	0.711103	0.299730
46	13	10	4	0.905372	0.194270
46	13	10	5	0.980305	0.074933
46	13	10	6	0.997531	0.017226
46	13	10	7	0.999827	0.002297
46	13	10	8	0.999994	0.000167
46	13	10	9	1.000000	0.000006
46	13	10	10	1.000000	0.000000
46	13	11	0	0.014507	0.014507
46	13	11	1	0.104704	0.090197
46	13	11	2	0.330195	0.225492
46	13	11	3	0.627844	0.297649
46	13	11	4	0.856805	0.228961
46	13	11	5	0.963653	0.106848
46	13	11	6	0.994181	0.030528
46	13	11	7	0.999444	0.005263
46	13	11	8	0.999971	0.000526
46	13	11	9	0.999999	0.000028
46	13	11	10	1.000000	0.000001
46	13	11	11	1.000000	0.000000
46	13	12	0	0.009119	0.009119
46	13	12	1	0.073779	0.064660
46	13	12	2	0.259326	0.185547
46	13	12	3	0.542802	0.283475
46	13	12	4	0.797929	0.255128
46	13	12	5	0.939231	0.141301
46	13	12	6	0.988075	0.048845
46	13	12	7	0.998542	0.010467
46	13	12	8	0.999896	0.001353
46	13	12	9	0.999996	0.000100
46	13	12	10	1.000000	0.000004
46	13	12	11	1.000000	0.000000
46	13	12	12	1.000000	0.000000
46	13	13	0	0.005632	0.005632
46	13	13	1	0.050958	0.045326
46	13	13	2	0.199296	0.148339
46	13	13	3	0.459427	0.260130
46	13	13	4	0.730395	0.270969
46	13	13	5	0.905983	0.175588
46	13	13	6	0.978019	0.072036
46	13	13	7	0.996695	0.018676
46	13	13	8	0.999697	0.003001
46	13	13	9	0.999984	0.000287
46	13	13	10	0.999999	0.000015
46	13	13	11	1.000000	0.000000
46	13	13	12	1.000000	0.000000
46	13	13	13	1.000000	0.000000

N	n	k	x	P(x)	p(x)
46	14	1	0	0.695652	0.695652
46	14	1	1	1.000000	0.304348
46	14	2	0	0.479227	0.479227
46	14	2	1	0.912077	0.432850
46	14	2	2	1.000000	0.087923
46	14	3	0	0.326746	0.326746
46	14	3	1	0.784190	0.457444
46	14	3	2	0.976021	0.191831
46	14	3	3	1.000000	0.023979
46	14	4	0	0.220363	0.220363
46	14	4	1	0.645893	0.425529
46	14	4	2	0.922487	0.276594
46	14	4	3	0.993866	0.071379
46	14	4	4	1.000000	0.006134
46	14	5	0	0.146909	0.146909
46	14	5	1	0.514181	0.367272
46	14	5	2	0.843460	0.329279
46	14	5	3	0.975171	0.131711
46	14	5	4	0.998539	0.023368
46	14	5	5	1.000000	0.001461
46	14	6	0	0.096745	0.096745
46	14	6	1	0.397729	0.300984
46	14	6	2	0.747086	0.349357
46	14	6	3	0.939834	0.192748
46	14	6	4	0.992840	0.053006
46	14	6	5	0.999679	0.006839
46	14	6	6	1.000000	0.000321
46	14	7	0	0.062884	0.062884
46	14	7	1	0.299909	0.237025
46	14	7	2	0.642279	0.342369
46	14	7	3	0.886828	0.244550
46	14	7	4	0.979588	0.092760
46	14	7	5	0.998140	0.018552
46	14	7	6	0.999936	0.001795
46	14	7	7	1.000000	0.000064
46	14	8	0	0.040310	0.040310
46	14	8	1	0.220901	0.180590
46	14	8	2	0.536934	0.316033
46	14	8	3	0.817853	0.280918
46	14	8	4	0.955804	0.137951
46	14	8	5	0.993859	0.038055
46	14	8	6	0.999568	0.005708
46	14	8	7	0.999988	0.000421
46	14	8	8	1.000000	0.000012
46	14	9	0	0.025459	0.025459
46	14	9	1	0.159120	0.133661
46	14	9	2	0.437134	0.278014
46	14	9	3	0.736534	0.299400
46	14	9	4	0.919501	0.182967
46	14	9	5	0.984846	0.065345

N	n	k	x	P(x)	p(x)
46	14	9	6	0.998366	0.013520
46	14	9	7	0.999911	0.001545
46	14	9	8	0.999998	0.000087
46	14	9	9	1.000000	0.000002
46	14	10	0	0.015826	0.015826
46	14	10	1	0.112158	0.096332
46	14	10	2	0.346967	0.234809
46	14	10	3	0.647523	0.300556
46	14	10	4	0.870050	0.222527
46	14	10	5	0.968951	0.098901
46	14	10	6	0.995443	0.026491
46	14	10	7	0.999619	0.004176
46	14	10	8	0.999984	0.000365
46	14	10	9	1.000000	0.000016
46	14	10	10	1.000000	0.000000
46	14	11	0	0.009671	0.009671
46	14	11	1	0.077371	0.067700
46	14	11	2	0.268698	0.191326
46	14	11	3	0.555687	0.286989
46	14	11	4	0.808237	0.252551
46	14	11	5	0.944226	0.135989
46	14	11	6	0.989556	0.045330
46	14	11	7	0.998807	0.009251
46	14	11	8	0.999923	0.001116
46	14	11	9	0.999998	0.000074
46	14	11	10	1.000000	0.000002
46	14	11	11	1.000000	0.000000
46	14	12	0	0.005803	0.005803
46	14	12	1	0.052226	0.046423
46	14	12	2	0.203100	0.150874
46	14	12	3	0.465490	0.262390
46	14	12	4	0.736080	0.270590
46	14	12	5	0.909258	0.173177
46	14	12	6	0.979195	0.069937
46	14	12	7	0.996956	0.017762
46	14	12	8	0.999732	0.002775
46	14	12	9	0.999987	0.000255
46	14	12	10	1.000000	0.000013
46	14	12	11	1.000000	0.000000
46	14	12	12	1.000000	0.000000
46	14	13	0	0.003413	0.003413
46	14	13	1	0.034476	0.031062
46	14	13	2	0.149850	0.115374
46	14	13	3	0.380599	0.230749
46	14	13	4	0.656495	0.275896
46	14	13	5	0.863416	0.206922
46	14	13	6	0.962739	0.099322
46	14	13	7	0.993299	0.030561
46	14	13	8	0.999242	0.005942
46	14	13	9	0.999949	0.000707

Table for $N = 2$, $n = 1$, through $N = 100$, $n = 50$

N	n	k	x	$P(x)$	$p(x)$
46	14	13	10	0.999998	0.000049
46	14	13	11	1.000000	0.000002
46	14	13	12	1.000000	0.000000
46	14	13	13	1.000000	0.000000
46	14	14	0	0.001965	0.001965
46	14	14	1	0.022239	0.020274
46	14	14	2	0.107896	0.085657
46	14	14	3	0.303683	0.195787
46	14	14	4	0.572890	0.269207
46	14	14	5	0.806983	0.234093
46	14	14	6	0.938661	0.131677
46	14	14	7	0.986817	0.048156
46	14	14	8	0.998161	0.011345
46	14	14	9	0.999842	0.001681
46	14	14	10	0.999992	0.000150
46	14	14	11	1.000000	0.000008
46	14	14	12	1.000000	0.000000
46	14	14	13	1.000000	0.000000
46	14	14	14	1.000000	0.000000
46	15	1	0	0.673913	0.673913
46	15	1	1	1.000000	0.326087
46	15	2	0	0.449275	0.449275
46	15	2	1	0.898551	0.449275
46	15	2	2	1.000000	0.101449
46	15	3	0	0.296113	0.296113
46	15	3	1	0.755599	0.459486
46	15	3	2	0.970026	0.214427
46	15	3	3	1.000000	0.029974
46	15	4	0	0.192818	0.192818
46	15	4	1	0.605999	0.413181
46	15	4	2	0.905200	0.299200
46	15	4	3	0.991635	0.086436
46	15	4	4	1.000000	0.008365
46	15	5	0	0.123954	0.123954
46	15	5	1	0.468272	0.344318
46	15	5	2	0.812590	0.344318
46	15	5	3	0.966939	0.154349
46	15	5	4	0.997809	0.030870
46	15	5	5	1.000000	0.002191
46	15	6	0	0.078605	0.078605
46	15	6	1	0.350700	0.272095
46	15	6	2	0.703416	0.352716
46	15	6	3	0.921764	0.218348
46	15	6	4	0.989527	0.067763
46	15	6	5	0.999466	0.009939
46	15	6	6	1.000000	0.000534
46	15	7	0	0.049128	0.049128
46	15	7	1	0.255467	0.206339
46	15	7	2	0.588783	0.333316
46	15	7	3	0.856260	0.267476
46	15	7	4	0.970892	0.114633
46	15	7	5	0.996981	0.026089
46	15	7	6	0.999880	0.002899
46	15	7	7	1.000000	0.000120
46	15	8	0	0.030233	0.030233
46	15	8	1	0.181397	0.151164
46	15	8	2	0.477678	0.296281
46	15	8	3	0.773959	0.296281
46	15	8	4	0.938560	0.164601
46	15	8	5	0.990292	0.051732
46	15	8	6	0.999211	0.008919
46	15	8	7	0.999975	0.000765
46	15	8	8	1.000000	0.000025
46	15	9	0	0.018299	0.018299
46	15	9	1	0.125705	0.107406
46	15	9	2	0.376319	0.250614
46	15	9	3	0.680397	0.304078
46	15	9	4	0.890912	0.210516
46	15	9	5	0.976678	0.085766
46	15	9	6	0.997098	0.020420
46	15	9	7	0.999814	0.002716
46	15	9	8	0.999995	0.000181
46	15	9	9	1.000000	0.000005
46	15	10	0	0.010880	0.010880
46	15	10	1	0.085065	0.074184
46	15	10	2	0.288265	0.203200
46	15	10	3	0.581777	0.293512
46	15	10	4	0.828327	0.246550
46	15	10	5	0.953498	0.125171
46	15	10	6	0.992131	0.038633
46	15	10	7	0.999227	0.007096
46	15	10	8	0.999961	0.000734
46	15	10	9	0.999999	0.000038
46	15	10	10	1.000000	0.000001
46	15	11	0	0.006347	0.006347
46	15	11	1	0.056215	0.049868
46	15	11	2	0.214887	0.158672
46	15	11	3	0.483939	0.269052
46	15	11	4	0.752992	0.269052
46	15	11	5	0.918728	0.165736
46	15	11	6	0.982473	0.063745
46	15	11	7	0.997650	0.015177
46	15	11	8	0.999818	0.002168
46	15	11	9	0.999993	0.000174
46	15	11	10	1.000000	0.000007
46	15	11	11	1.000000	0.000000
46	15	12	0	0.003627	0.003627
46	15	12	1	0.036268	0.032641
46	15	12	2	0.155952	0.119684
46	15	12	3	0.391693	0.235741
46	15	12	4	0.668433	0.276740
46	15	12	5	0.871375	0.202942
46	15	12	6	0.966081	0.094706
46	15	12	7	0.994181	0.028100
46	15	12	8	0.999385	0.005204
46	15	12	9	0.999963	0.000578
46	15	12	10	0.999999	0.000036
46	15	12	11	1.000000	0.000001
46	15	12	12	1.000000	0.000000
46	15	13	0	0.002027	0.002027
46	15	13	1	0.022827	0.020801
46	15	13	2	0.110190	0.087363
46	15	13	3	0.308490	0.198300
46	15	13	4	0.578899	0.270409
46	15	13	5	0.811686	0.232787
46	15	13	6	0.941012	0.129326
46	15	13	7	0.987569	0.046557
46	15	13	8	0.998313	0.010744
46	15	13	9	0.999861	0.001547
46	15	13	10	0.999993	0.000133
46	15	13	11	1.000000	0.000006
46	15	13	12	1.000000	0.000000
46	15	13	13	1.000000	0.000000
46	15	14	0	0.001105	0.001105
46	15	14	1	0.014003	0.012897
46	15	14	2	0.075775	0.061772
46	15	14	3	0.236381	0.160607
46	15	14	4	0.488763	0.252382
46	15	14	5	0.741145	0.252382
46	15	14	6	0.905741	0.164597
46	15	14	7	0.976283	0.070541
46	15	14	8	0.996034	0.019752
46	15	14	9	0.999580	0.003545
46	15	14	10	0.999973	0.000394
46	15	14	11	0.999999	0.000026
46	15	14	12	1.000000	0.000001
46	15	14	13	1.000000	0.000000
46	15	14	14	1.000000	0.000000
46	15	15	0	0.000587	0.000587
46	15	15	1	0.008360	0.007773
46	15	15	2	0.050680	0.042320
46	15	15	3	0.176154	0.125474
46	15	15	4	0.402007	0.225853
46	15	15	5	0.662275	0.260269
46	15	15	6	0.859448	0.197173
46	15	15	7	0.958647	0.099199
46	15	15	8	0.991714	0.033066
46	15	15	9	0.998915	0.007201
46	15	15	10	0.999912	0.000997
46	15	15	11	0.999996	0.000084
46	15	15	12	1.000000	0.000004
46	15	15	13	1.000000	0.000000
46	15	15	14	1.000000	0.000000
46	15	15	15	1.000000	0.000000
46	16	1	0	0.652174	0.652174
46	16	1	1	1.000000	0.347826
46	16	2	0	0.420290	0.420290
46	16	2	1	0.884058	0.463768
46	16	2	2	1.000000	0.115942
46	16	3	0	0.267457	0.267457
46	16	3	1	0.725955	0.458498
46	16	3	2	0.963109	0.237154
46	16	3	3	1.000000	0.036891
46	16	4	0	0.167938	0.167938
46	16	4	1	0.566014	0.398076
46	16	4	2	0.885896	0.319882
46	16	4	3	0.988847	0.102951
46	16	4	4	1.000000	0.011153
46	16	5	0	0.103962	0.103962
46	16	5	1	0.423844	0.319882
46	16	5	2	0.779269	0.355425
46	16	5	3	0.956981	0.177712
46	16	5	4	0.996813	0.039832
46	16	5	5	1.000000	0.003187
46	16	6	0	0.063391	0.063391
46	16	6	1	0.306814	0.243423
46	16	6	2	0.657904	0.351090
46	16	6	3	0.900633	0.242729
46	16	6	4	0.985155	0.084522
46	16	6	5	0.999145	0.013990
46	16	6	6	1.000000	0.000855
46	16	7	0	0.038035	0.038035
46	16	7	1	0.215530	0.177496
46	16	7	2	0.535023	0.319492
46	16	7	3	0.821746	0.286724
46	16	7	4	0.959799	0.138052
46	16	7	5	0.995298	0.035499
46	16	7	6	0.999786	0.004488
46	16	7	7	1.000000	0.000214
46	16	8	0	0.022431	0.022431
46	16	8	1	0.147263	0.124832
46	16	8	2	0.420333	0.273070
46	16	8	3	0.726172	0.305839
46	16	8	4	0.917321	0.191149
46	16	8	5	0.985285	0.067964
46	16	8	6	0.998635	0.013350
46	16	8	7	0.999951	0.001315
46	16	8	8	1.000000	0.000049
46	16	9	0	0.012986	0.012986
46	16	9	1	0.097987	0.085001

N	n	k	x	P(x)	p(x)
46	16	9	2	0.319728	0.221741
46	16	9	3	0.621543	0.301815
46	16	9	4	0.856958	0.235415
46	16	9	5	0.965611	0.108653
46	16	9	6	0.995122	0.029511
46	16	9	7	0.999639	0.004517
46	16	9	8	0.999990	0.000350
46	16	9	9	1.000000	0.000010
46	16	10	0	0.007371	0.007371
46	16	10	1	0.063527	0.056157
46	16	10	2	0.235826	0.172299
46	16	10	3	0.515500	0.279674
46	16	10	4	0.780607	0.265107
46	16	10	5	0.933309	0.152702
46	16	10	6	0.987146	0.053837
46	16	10	7	0.998540	0.011394
46	16	10	8	0.999914	0.001373
46	16	10	9	0.999998	0.000084
46	16	10	10	1.000000	0.000002
46	16	11	0	0.004095	0.004095
46	16	11	1	0.040129	0.036034
46	16	11	2	0.168821	0.128692
46	16	11	3	0.414507	0.245686
46	16	11	4	0.692238	0.277731
46	16	11	5	0.886650	0.194412
46	16	11	6	0.972191	0.085541
46	16	11	7	0.995692	0.023500
46	16	11	8	0.999609	0.003917
46	16	11	9	0.999982	0.000373
46	16	11	10	1.000000	0.000018
46	16	11	11	1.000000	0.000000
46	16	12	0	0.002223	0.002223
46	16	12	1	0.024686	0.022463
46	16	12	2	0.117344	0.092659
46	16	12	3	0.323252	0.205908
46	16	12	4	0.597016	0.273764
46	16	12	5	0.825549	0.228533
46	16	12	6	0.947751	0.122202
46	16	12	7	0.989649	0.041898
46	16	12	8	0.998713	0.009064
46	16	12	9	0.999907	0.001194
46	16	12	10	0.999996	0.000090
46	16	12	11	1.000000	0.000003
46	16	12	12	1.000000	0.000000
46	16	13	0	0.001177	0.001177
46	16	13	1	0.014776	0.013599
46	16	13	2	0.079191	0.064415
46	16	13	3	0.244522	0.165332
46	16	13	4	0.500393	0.255871
46	16	13	5	0.751612	0.251219

N	n	k	x	P(x)	p(x)
46	16	13	6	0.911809	0.160197
46	16	13	7	0.978558	0.066749
46	16	13	8	0.996581	0.018022
46	16	13	9	0.999661	0.003081
46	16	13	10	0.999981	0.000319
46	16	13	11	0.999999	0.000019
46	16	13	12	1.000000	0.000001
46	16	13	13	1.000000	0.000000
46	16	14	0	0.000606	0.000606
46	16	14	1	0.008594	0.007988
46	16	14	2	0.051863	0.043269
46	16	14	3	0.179392	0.127529
46	16	14	4	0.407349	0.227958
46	16	14	5	0.667872	0.260523
46	16	14	6	0.863265	0.195392
46	16	14	7	0.960354	0.097089
46	16	14	8	0.992211	0.031857
46	16	14	9	0.999008	0.006796
46	16	14	10	0.999923	0.000915
46	16	14	11	0.999996	0.000074
46	16	14	12	1.000000	0.000003
46	16	14	13	1.000000	0.000000
46	16	14	14	1.000000	0.000000
46	16	15	0	0.000303	0.000303
46	16	15	1	0.004850	0.004547
46	16	15	2	0.032933	0.028083
46	16	15	3	0.127583	0.094650
46	16	15	4	0.321865	0.194282
46	16	15	5	0.578318	0.256452
46	16	15	6	0.802205	0.223887
46	16	15	7	0.933048	0.130843
46	16	15	8	0.984247	0.051199
46	16	15	9	0.997521	0.013274
46	16	15	10	0.999751	0.002230
46	16	15	11	0.999985	0.000234
46	16	15	12	0.999999	0.000014
46	16	15	13	1.000000	0.000000
46	16	15	14	1.000000	0.000000
46	16	15	15	1.000000	0.000000
46	16	16	0	0.000147	0.000147
46	16	16	1	0.002650	0.002503
46	16	16	2	0.020250	0.017600
46	16	16	3	0.087891	0.067641
46	16	16	4	0.246659	0.158768
46	16	16	5	0.487318	0.240659
46	16	16	6	0.729983	0.242665
46	16	16	7	0.895061	0.165078
46	16	16	8	0.971034	0.075973
46	16	16	9	0.994524	0.023489
46	16	16	10	0.999319	0.004796

N	n	k	x	P(x)	p(x)
46	16	16	11	0.999947	0.000628
46	16	16	12	0.999998	0.000050
46	16	16	13	1.000000	0.000002
46	16	16	14	1.000000	0.000000
46	16	16	15	1.000000	0.000000
46	16	16	16	1.000000	0.000000
46	17	1	0	0.630435	0.630435
46	17	1	1	1.000000	0.369565
46	17	2	0	0.392270	0.392270
46	17	2	1	0.868599	0.476328
46	17	2	2	1.000000	0.131401
46	17	3	0	0.240711	0.240711
46	17	3	1	0.695389	0.454677
46	17	3	2	0.955204	0.259816
46	17	3	3	1.000000	0.044796
46	17	4	0	0.145546	0.145546
46	17	4	1	0.526206	0.380660
46	17	4	2	0.864571	0.338364
46	17	4	3	0.985415	0.120844
46	17	4	4	1.000000	0.014585
46	17	5	0	0.086635	0.086635
46	17	5	1	0.381193	0.294558
46	17	5	2	0.743726	0.362533
46	17	5	3	0.945134	0.201407
46	17	5	4	0.995486	0.050352
46	17	5	5	1.000000	0.004514
46	17	6	0	0.050713	0.050713
46	17	6	1	0.266244	0.215530
46	17	6	2	0.611092	0.344849
46	17	6	3	0.876361	0.265268
46	17	6	4	0.979520	0.103160
46	17	6	5	0.998679	0.019158
46	17	6	6	1.000000	0.001321
46	17	7	0	0.029160	0.029160
46	17	7	1	0.180031	0.150871
46	17	7	2	0.481774	0.301743
46	17	7	3	0.783517	0.301743
46	17	7	4	0.945993	0.162477
46	17	7	5	0.992931	0.046938
46	17	7	6	0.999637	0.006705
46	17	7	7	1.000000	0.000363
46	17	8	0	0.016449	0.016449
46	17	8	1	0.118135	0.101686
46	17	8	2	0.365719	0.247584
46	17	8	3	0.675199	0.309480
46	17	8	4	0.891834	0.216636
46	17	8	5	0.978489	0.086654
46	17	8	6	0.997745	0.019257
46	17	8	7	0.999907	0.002161
46	17	8	8	1.000000	0.000093

N	n	k	x	P(x)	p(x)
46	17	9	0	0.009090	0.009090
46	17	9	1	0.075320	0.066230
46	17	9	2	0.267989	0.192669
46	17	9	3	0.561180	0.293191
46	17	9	4	0.817722	0.256542
46	17	9	5	0.951124	0.133402
46	17	9	6	0.992171	0.041047
46	17	9	7	0.999338	0.007167
46	17	9	8	0.999978	0.000640
46	17	9	9	1.000000	0.000022
46	17	10	0	0.004914	0.004914
46	17	10	1	0.046680	0.041767
46	17	10	2	0.189880	0.143200
46	17	10	3	0.450243	0.260363
46	17	10	4	0.727586	0.277343
46	17	10	5	0.907859	0.180273
46	17	10	6	0.979968	0.072109
46	17	10	7	0.997401	0.017433
46	17	10	8	0.999822	0.002421
46	17	10	9	0.999995	0.000173
46	17	10	10	1.000000	0.000005
46	17	11	0	0.002593	0.002593
46	17	11	1	0.028117	0.025524
46	17	11	2	0.130213	0.102096
46	17	11	3	0.348990	0.218777
46	17	11	4	0.627434	0.278444
46	17	11	5	0.847768	0.220334
46	17	11	6	0.957935	0.110167
46	17	11	7	0.992558	0.034624
46	17	11	8	0.999217	0.006658
46	17	11	9	0.999957	0.000740
46	17	11	10	0.999999	0.000042
46	17	11	11	1.000000	0.000001
46	17	12	0	0.001334	0.001334
46	17	12	1	0.016449	0.015116
46	17	12	2	0.086458	0.070009
46	17	12	3	0.261480	0.175022
46	17	12	4	0.524012	0.262533
46	17	12	5	0.772225	0.248213
46	17	12	6	0.923311	0.151086
46	17	12	7	0.982666	0.059355
46	17	12	8	0.997505	0.014839
46	17	12	9	0.999788	0.002283
46	17	12	10	0.999991	0.000203
46	17	12	11	1.000000	0.000009
46	17	12	12	1.000000	0.000000
46	17	13	0	0.000667	0.000667
46	17	13	1	0.009336	0.008669
46	17	13	2	0.055572	0.046236
46	17	13	3	0.189412	0.133840

Table for $N = 2$, $n = 1$, through $N = 100$, $n = 50$

N	n	k	x	P(x)	p(x)
46	17	13	4	0.423632	0.234220
46	17	13	5	0.684620	0.260988
46	17	13	6	0.874430	0.189810
46	17	13	7	0.965208	0.090779
46	17	13	8	0.993577	0.028368
46	17	13	9	0.999250	0.005674
46	17	13	10	0.999949	0.000698
46	17	13	11	0.999998	0.000049
46	17	13	12	1.000000	0.000002
46	17	13	13	1.000000	0.000000
46	17	14	0	0.000323	0.000323
46	17	14	1	0.005133	0.004809
46	17	14	2	0.034556	0.029423
46	17	14	3	0.132631	0.098076
46	17	14	4	0.331364	0.198732
46	17	14	5	0.589716	0.258352
46	17	14	6	0.811160	0.221445
46	17	14	7	0.937700	0.126540
46	17	14	8	0.985840	0.048140
46	17	14	9	0.997875	0.012035
46	17	14	10	0.999801	0.001926
46	17	14	11	0.999989	0.000189
46	17	14	12	1.000000	0.000010
46	17	14	13	1.000000	0.000000
46	17	14	14	1.000000	0.000000
46	17	15	0	0.000152	0.000152
46	17	15	1	0.002728	0.002577
46	17	15	2	0.020764	0.018036
46	17	15	3	0.089723	0.068959
46	17	15	4	0.250628	0.160905
46	17	15	5	0.492833	0.242205
46	17	15	6	0.735038	0.242205
46	17	15	7	0.898156	0.163118
46	17	15	8	0.972301	0.074144
46	17	15	9	0.994866	0.022566
46	17	15	10	0.999379	0.004513
46	17	15	11	0.999954	0.000574
46	17	15	12	0.999998	0.000044
46	17	15	13	1.000000	0.000002
46	17	15	14	1.000000	0.000000
46	17	15	15	1.000000	0.000000
46	17	16	0	0.000068	0.000068
46	17	16	1	0.001398	0.001330
46	17	16	2	0.012037	0.010638
46	17	16	3	0.058580	0.046543
46	17	16	4	0.183152	0.124572
46	17	16	5	0.399077	0.215925
46	17	16	6	0.649095	0.250018
46	17	16	7	0.845538	0.196443
46	17	16	8	0.950775	0.105237

N	n	k	x	P(x)	p(x)
46	17	16	9	0.989043	0.038268
46	17	16	10	0.998360	0.009317
46	17	16	11	0.999843	0.001482
46	17	16	12	0.999991	0.000148
46	17	16	13	1.000000	0.000009
46	17	16	14	1.000000	0.000000
46	17	16	15	1.000000	0.000000
46	17	16	16	1.000000	0.000000
46	17	17	0	0.000030	0.000030
46	17	17	1	0.000689	0.000659
46	17	17	2	0.006717	0.006028
46	17	17	3	0.036860	0.030142
46	17	17	4	0.129171	0.092311
46	17	17	5	0.312707	0.183536
46	17	17	6	0.557421	0.244715
46	17	17	7	0.780057	0.222635
46	17	17	8	0.919204	0.139147
46	17	17	9	0.978838	0.059634
46	17	17	10	0.996186	0.017348
46	17	17	11	0.999546	0.003360
46	17	17	12	0.999966	0.000420
46	17	17	13	0.999998	0.000032
46	17	17	14	1.000000	0.000001
46	17	17	15	1.000000	0.000000
46	17	17	16	1.000000	0.000000
46	17	17	17	1.000000	0.000000
46	18	1	0	0.608696	0.608696
46	18	1	1	1.000000	0.391304
46	18	2	0	0.365217	0.365217
46	18	2	1	0.852174	0.486956
46	18	2	2	1.000000	0.147826
46	18	3	0	0.215810	0.215810
46	18	3	1	0.664032	0.448221
46	18	3	2	0.946245	0.282213
46	18	3	3	1.000000	0.053755
46	18	4	0	0.125471	0.125471
46	18	4	1	0.486828	0.361357
46	18	4	2	0.841235	0.354408
46	18	4	3	0.981248	0.140013
46	18	4	4	1.000000	0.018752
46	18	5	0	0.071698	0.071698
46	18	5	1	0.340564	0.268867
46	18	5	2	0.706223	0.365659
46	18	5	3	0.931244	0.225021
46	18	5	4	0.993749	0.062506
46	18	5	5	1.000000	0.006251
46	18	6	0	0.040221	0.040221
46	18	6	1	0.229083	0.188862
46	18	6	2	0.563527	0.334444
46	18	6	3	0.848919	0.285392

N	n	k	x	P(x)	p(x)
46	18	6	4	0.972406	0.123487
46	18	6	5	0.998018	0.025612
46	18	6	6	1.000000	0.001982
46	18	7	0	0.022121	0.022121
46	18	7	1	0.148817	0.126695
46	18	7	2	0.429749	0.280933
46	18	7	3	0.741897	0.312148
46	18	7	4	0.929185	0.187289
46	18	7	5	0.989694	0.060509
46	18	7	6	0.999405	0.009711
46	18	7	7	1.000000	0.000595
46	18	8	0	0.011912	0.011912
46	18	8	1	0.093590	0.081679
46	18	8	2	0.314495	0.220904
46	18	8	3	0.621840	0.307345
46	18	8	4	0.861954	0.240114
46	18	8	5	0.969525	0.107571
46	18	8	6	0.996417	0.026893
46	18	8	7	0.999832	0.003415
46	18	8	8	1.000000	0.000168
46	18	9	0	0.006269	0.006269
46	18	9	1	0.057050	0.050781
46	18	9	2	0.221483	0.164433
46	18	9	3	0.500520	0.279037
46	18	9	4	0.773491	0.272971
46	18	9	5	0.932724	0.159233
46	18	9	6	0.987925	0.055201
46	18	9	7	0.998844	0.010919
46	18	9	8	0.999956	0.001112
46	18	9	9	1.000000	0.000044
46	18	10	0	0.003219	0.003219
46	18	10	1	0.033718	0.030499
46	18	10	2	0.150376	0.116658
46	18	10	3	0.387396	0.237020
46	18	10	4	0.670204	0.282808
46	18	10	5	0.876777	0.206573
46	18	10	6	0.970022	0.093245
46	18	10	7	0.995597	0.025576
46	18	10	8	0.999655	0.004058
46	18	10	9	0.999989	0.000334
46	18	10	10	1.000000	0.000011
46	18	11	0	0.001610	0.001610
46	18	11	1	0.019316	0.017706
46	18	11	2	0.098528	0.079212
46	18	11	3	0.288638	0.190110
46	18	11	4	0.560224	0.271585
46	18	11	5	0.802181	0.241958
46	18	11	6	0.938940	0.136759
46	18	11	7	0.987783	0.048842
46	18	11	8	0.998528	0.010745

N	n	k	x	P(x)	p(x)
46	18	11	9	0.999906	0.001378
46	18	11	10	0.999997	0.000092
46	18	11	11	1.000000	0.000002
46	18	12	0	0.000782	0.000782
46	18	12	1	0.010716	0.009934
46	18	12	2	0.062317	0.051601
46	18	12	3	0.207163	0.144846
46	18	12	4	0.451589	0.244427
46	18	12	5	0.712311	0.260722
46	18	12	6	0.892052	0.179740
46	18	12	7	0.972432	0.080381
46	18	12	8	0.995458	0.023026
46	18	12	9	0.999551	0.004093
46	18	12	10	0.999977	0.000425
46	18	12	11	0.999999	0.000023
46	18	12	12	1.000000	0.000000
46	18	13	0	0.000368	0.000368
46	18	13	1	0.005749	0.005381
46	18	13	2	0.038034	0.032285
46	18	13	3	0.143260	0.105226
46	18	13	4	0.350943	0.207683
46	18	13	5	0.612624	0.261681
46	18	13	6	0.828614	0.215990
46	18	13	7	0.946427	0.117813
46	18	13	8	0.988686	0.042259
46	18	13	9	0.998468	0.009782
46	18	13	10	0.999877	0.001409
46	18	13	11	0.999995	0.000118
46	18	13	12	1.000000	0.000005
46	18	13	13	1.000000	0.000000
46	18	14	0	0.000167	0.000167
46	18	14	1	0.002977	0.002810
46	18	14	2	0.022381	0.019404
46	18	14	3	0.095430	0.073049
46	18	14	4	0.262835	0.167405
46	18	14	5	0.509537	0.246702
46	18	14	6	0.750072	0.240535
46	18	14	7	0.907156	0.157084
46	18	14	8	0.975880	0.068724
46	18	14	9	0.995800	0.019920
46	18	14	10	0.999535	0.003735
46	18	14	11	0.999970	0.000435
46	18	14	12	0.999999	0.000029
46	18	14	13	1.000000	0.000001
46	18	14	14	1.000000	0.000000
46	18	15	0	0.000073	0.000073
46	18	15	1	0.001484	0.001411
46	18	15	2	0.012679	0.011194
46	18	15	3	0.061188	0.048509
46	18	15	4	0.189595	0.128407

N	n	k	x	P(x)	p(x)
46	18	15	5	0.409315	0.219719
46	18	15	6	0.659871	0.250557
46	18	15	7	0.853158	0.193287
46	18	15	8	0.954404	0.101245
46	18	15	9	0.990197	0.035794
46	18	15	10	0.998601	0.008404
46	18	15	11	0.999875	0.001273
46	18	15	12	0.999993	0.000119
46	18	15	13	1.000000	0.000006
46	18	15	14	1.000000	0.000000
46	18	15	15	1.000000	0.000000
46	18	16	0	0.000031	0.000031
46	18	16	1	0.000710	0.000680
46	18	16	2	0.006901	0.006190
46	18	16	3	0.037716	0.030815
46	18	16	4	0.131605	0.093889
46	18	16	5	0.317174	0.185569
46	18	16	6	0.562882	0.245707
46	18	16	7	0.784573	0.221691
46	18	16	8	0.921744	0.137171
46	18	16	9	0.979806	0.058062
46	18	16	10	0.996433	0.016627
46	18	16	11	0.999587	0.003154
46	18	16	12	0.999970	0.000383
46	18	16	13	0.999999	0.000028
46	18	16	14	1.000000	0.000001
46	18	16	15	1.000000	0.000000
46	18	16	16	1.000000	0.000000
46	18	17	0	0.000012	0.000012
46	18	17	1	0.000325	0.000313
46	18	17	2	0.003599	0.003274
46	18	17	3	0.022308	0.018709
46	18	17	4	0.087790	0.065482
46	18	17	5	0.236761	0.148971
46	18	17	6	0.464599	0.227838
46	18	17	7	0.703286	0.238687
46	18	17	8	0.876020	0.172734
46	18	17	9	0.962387	0.086367
46	18	17	10	0.991999	0.029612
46	18	17	11	0.998851	0.006852
46	18	17	12	0.999894	0.001043
46	18	17	13	0.999994	0.000100
46	18	17	14	1.000000	0.000006
46	18	17	15	1.000000	0.000000
46	18	17	16	1.000000	0.000000
46	18	17	17	1.000000	0.000000
46	18	18	0	0.000005	0.000005
46	18	18	1	0.000142	0.000137
46	18	18	2	0.001793	0.001651
46	18	18	3	0.012631	0.010838

N	n	k	x	P(x)	p(x)
46	18	18	4	0.056178	0.043547
46	18	18	5	0.169981	0.113803
46	18	18	6	0.370321	0.200340
46	18	18	7	0.612749	0.242428
46	18	18	8	0.816457	0.203707
46	18	18	9	0.935584	0.119127
46	18	18	10	0.983830	0.048246
46	18	18	11	0.997197	0.013367
46	18	18	12	0.999678	0.002481
46	18	18	13	0.999977	0.000299
46	18	18	14	0.999999	0.000022
46	18	18	15	1.000000	0.000001
46	18	18	16	1.000000	0.000000
46	18	18	17	1.000000	0.000000
46	18	18	18	1.000000	0.000000
46	19	1	0	0.586957	0.586957
46	19	1	1	1.000000	0.413043
46	19	2	0	0.339130	0.339130
46	19	2	1	0.834783	0.495652
46	19	2	2	1.000000	0.165217
46	19	3	0	0.192688	0.192688
46	19	3	1	0.632016	0.439328
46	19	3	2	0.936166	0.304150
46	19	3	3	1.000000	0.063834
46	19	4	0	0.107547	0.107547
46	19	4	1	0.448111	0.340564
46	19	4	2	0.815920	0.367809
46	19	4	3	0.976248	0.160327
46	19	4	4	1.000000	0.023752
46	19	5	0	0.058895	0.058895
46	19	5	1	0.302155	0.243260
46	19	5	2	0.667045	0.364890
46	19	5	3	0.915171	0.248125
46	19	5	4	0.991517	0.076346
46	19	5	5	1.000000	0.008483
46	19	6	0	0.031602	0.031602
46	19	6	1	0.195358	0.163756
46	19	6	2	0.515749	0.320392
46	19	6	3	0.818341	0.302592
46	19	6	4	0.963585	0.145244
46	19	6	5	0.997103	0.033518
46	19	6	6	1.000000	0.002897
46	19	7	0	0.016591	0.016591
46	19	7	1	0.121668	0.105077
46	19	7	2	0.379583	0.257915
46	19	7	3	0.697304	0.317722
46	19	7	4	0.909119	0.211814
46	19	7	5	0.985372	0.076253
46	19	7	6	0.999059	0.013686
46	19	7	7	1.000000	0.000941

N	n	k	x	P(x)	p(x)
46	19	8	0	0.008508	0.008508
46	19	8	1	0.073171	0.064662
46	19	8	2	0.267158	0.193988
46	19	8	3	0.566957	0.299799
46	19	8	4	0.827652	0.260695
46	19	8	5	0.957999	0.130347
46	19	8	6	0.994496	0.036497
46	19	8	7	0.999710	0.005214
46	19	8	8	1.000000	0.000290
46	19	9	0	0.004254	0.004254
46	19	9	1	0.042541	0.038287
46	19	9	2	0.180374	0.137833
46	19	9	3	0.440726	0.260352
46	19	9	4	0.724746	0.284020
46	19	9	5	0.909976	0.185230
46	19	9	6	0.982010	0.072034
46	19	9	7	0.998064	0.016053
46	19	9	8	0.999916	0.001852
46	19	9	9	1.000000	0.000084
46	19	10	0	0.002070	0.002070
46	19	10	1	0.023915	0.021845
46	19	10	2	0.117046	0.093131
46	19	10	3	0.328142	0.211096
46	19	10	4	0.609603	0.281461
46	19	10	5	0.839889	0.230286
46	19	10	6	0.956701	0.116812
46	19	10	7	0.992857	0.036156
46	19	10	8	0.999365	0.006508
46	19	10	9	0.999977	0.000612
46	19	10	10	1.000000	0.000023
46	19	11	0	0.000977	0.000977
46	19	11	1	0.012992	0.012015
46	19	11	2	0.073067	0.060075
46	19	11	3	0.234321	0.161254
46	19	11	4	0.492327	0.258006
46	19	11	5	0.750333	0.258006
46	19	11	6	0.914519	0.164186
46	19	11	7	0.980805	0.066286
46	19	11	8	0.997377	0.016572
46	19	11	9	0.999807	0.002430
46	19	11	10	0.999994	0.000187
46	19	11	11	1.000000	0.000006
46	19	12	0	0.000447	0.000447
46	19	12	1	0.006813	0.006366
46	19	12	2	0.043888	0.037075
46	19	12	3	0.160605	0.116717
46	19	12	4	0.381753	0.221148
46	19	12	5	0.647131	0.265378
46	19	12	6	0.853536	0.206405
46	19	12	7	0.958079	0.104543

N	n	k	x	P(x)	p(x)
46	19	12	8	0.992169	0.034090
46	19	12	9	0.999113	0.006944
46	19	12	10	0.999946	0.000833
46	19	12	11	0.999999	0.000052
46	19	12	12	1.000000	0.000001
46	19	13	0	0.000197	0.000197
46	19	13	1	0.003443	0.003246
46	19	13	2	0.025351	0.021908
46	19	13	3	0.105679	0.080329
46	19	13	4	0.284188	0.178508
46	19	13	5	0.537858	0.253670
46	19	13	6	0.774616	0.236759
46	19	13	7	0.921181	0.146565
46	19	13	8	0.981140	0.059958
46	19	13	9	0.997070	0.015931
46	19	13	10	0.999726	0.002655
46	19	13	11	0.999986	0.000261
46	19	13	12	1.000000	0.000013
46	19	13	13	1.000000	0.000000
46	19	14	0	0.000084	0.000084
46	19	14	1	0.001672	0.001589
46	19	14	2	0.014065	0.012392
46	19	14	3	0.066732	0.052667
46	19	14	4	0.203048	0.136316
46	19	14	5	0.430240	0.227193
46	19	14	6	0.681348	0.251108
46	19	14	7	0.867885	0.186537
46	19	14	8	0.961153	0.093269
46	19	14	9	0.992243	0.031090
46	19	14	10	0.999001	0.006759
46	19	14	11	0.999923	0.000922
46	19	14	12	0.999997	0.000074
46	19	14	13	1.000000	0.000003
46	19	14	14	1.000000	0.000000
46	19	15	0	0.000034	0.000034
46	19	15	1	0.000779	0.000745
46	19	15	2	0.007481	0.006703
46	19	15	3	0.040398	0.032917
46	19	15	4	0.139150	0.098751
46	19	15	5	0.330843	0.191694
46	19	15	6	0.579335	0.248492
46	19	15	7	0.797933	0.218598
46	19	15	8	0.929092	0.131159
46	19	15	9	0.982527	0.053435
46	19	15	10	0.997101	0.014573
46	19	15	11	0.999693	0.002592
46	19	15	12	0.999981	0.000288
46	19	15	13	0.999999	0.000019
46	19	15	14	1.000000	0.000001
46	19	15	15	1.000000	0.000000

Table for $N = 2$, $n = 1$, through $N = 100$, $n = 50$

N	n	k	x	P(x)	p(x)
46	19	16	0	0.000013	0.000013
46	19	16	1	0.000346	0.000333
46	19	16	2	0.003806	0.003459
46	19	16	3	0.023409	0.019603
46	19	16	4	0.091367	0.067958
46	19	16	5	0.244272	0.152905
46	19	16	6	0.475129	0.230857
46	19	16	7	0.713315	0.238186
46	19	16	8	0.882552	0.169237
46	19	16	9	0.965290	0.082738
46	19	16	10	0.992870	0.027579
46	19	16	11	0.999024	0.006154
46	19	16	12	0.999916	0.000892
46	19	16	13	0.999996	0.000080
46	19	16	14	1.000000	0.000004
46	19	16	15	1.000000	0.000000
46	19	16	16	1.000000	0.000000
46	19	17	0	0.000005	0.000005
46	19	17	1	0.000146	0.000142
46	19	17	2	0.001845	0.001699
46	19	17	3	0.012954	0.011108
46	19	17	4	0.057388	0.044434
46	19	17	5	0.172916	0.115528
46	19	17	6	0.375091	0.202175
46	19	17	7	0.618040	0.242949
46	19	17	8	0.820498	0.202458
46	19	17	9	0.937711	0.117212
46	19	17	10	0.984596	0.046885
46	19	17	11	0.997383	0.012787
46	19	17	12	0.999707	0.002325
46	19	17	13	0.999980	0.000272
46	19	17	14	0.999999	0.000019
46	19	17	15	1.000000	0.000001
46	19	17	16	1.000000	0.000000
46	19	17	17	1.000000	0.000000
46	19	18	0	0.000002	0.000002
46	19	18	1	0.000059	0.000057
46	19	18	2	0.000849	0.000791
46	19	18	3	0.006825	0.005976
46	19	18	4	0.034405	0.027580
46	19	18	5	0.117144	0.082739
46	19	18	6	0.284461	0.167317
46	19	18	7	0.517510	0.233049
46	19	18	8	0.743704	0.226194
46	19	18	9	0.897293	0.153589
46	19	18	10	0.970045	0.072753
46	19	18	11	0.993855	0.023810
46	19	18	12	0.999146	0.005291
46	19	18	13	0.999923	0.000777
46	19	18	14	0.999996	0.000072
46	19	18	15	1.000000	0.000004
46	19	18	16	1.000000	0.000000
46	19	18	17	1.000000	0.000000
46	19	18	18	1.000000	0.000000
46	19	19	0	0.000001	0.000001
46	19	19	1	0.000022	0.000021
46	19	19	2	0.000369	0.000347
46	19	19	3	0.003410	0.003041
46	19	19	4	0.019630	0.016219
46	19	19	5	0.075774	0.056144
46	19	19	6	0.206778	0.131004
46	19	19	7	0.417632	0.210854
46	19	19	8	0.654842	0.237210
46	19	19	9	0.842440	0.187598
46	19	19	10	0.946661	0.104221
46	19	19	11	0.987052	0.040392
46	19	19	12	0.997824	0.010771
46	19	19	13	0.999757	0.001933
46	19	19	14	0.999983	0.000226
46	19	19	15	0.999999	0.000016
46	19	19	16	1.000000	0.000001
46	19	19	17	1.000000	0.000000
46	19	19	18	1.000000	0.000000
46	19	19	19	1.000000	0.000000
46	20	1	0	0.565217	0.565217
46	20	1	1	1.000000	0.434783
46	20	2	0	0.314010	0.314010
46	20	2	1	0.816425	0.502415
46	20	2	2	1.000000	0.183575
46	20	3	0	0.171278	0.171278
46	20	3	1	0.599473	0.428195
46	20	3	2	0.924901	0.325428
46	20	3	3	1.000000	0.075099
46	20	4	0	0.091614	0.091614
46	20	4	1	0.410271	0.318657
46	20	4	2	0.788675	0.378405
46	20	4	3	0.970310	0.181634
46	20	4	4	1.000000	0.029690
46	20	5	0	0.047988	0.047988
46	20	5	1	0.266116	0.218128
46	20	5	2	0.626502	0.360386
46	20	5	3	0.896791	0.270289
46	20	5	4	0.988689	0.091898
46	20	5	5	1.000000	0.011311
46	20	6	0	0.024579	0.024579
46	20	6	1	0.165033	0.140453
46	20	6	2	0.468284	0.303251
46	20	6	3	0.784720	0.316436
46	20	6	4	0.952827	0.168107
46	20	6	5	0.995862	0.043035
46	20	6	6	1.000000	0.004138
46	20	7	0	0.012290	0.012290
46	20	7	1	0.098317	0.086028
46	20	7	2	0.331821	0.233503
46	20	7	3	0.650235	0.318414
46	20	7	4	0.885584	0.235349
46	20	7	5	0.979724	0.094140
46	20	7	6	0.998552	0.018828
46	20	7	7	1.000000	0.001448
46	20	8	0	0.005987	0.005987
46	20	8	1	0.056406	0.050419
46	20	8	2	0.224050	0.167644
46	20	8	3	0.511439	0.287389
46	20	8	4	0.789030	0.277592
46	20	8	5	0.943516	0.154486
46	20	8	6	0.991793	0.048277
46	20	8	7	0.999517	0.007724
46	20	8	8	1.000000	0.000483
46	20	9	0	0.002836	0.002836
46	20	9	1	0.031197	0.028361
46	20	9	2	0.144640	0.113443
46	20	9	3	0.382870	0.238230
46	20	9	4	0.672150	0.289280
46	20	9	5	0.882535	0.210385
46	20	9	6	0.974007	0.091472
46	20	9	7	0.996875	0.022868
46	20	9	8	0.999847	0.002973
46	20	9	9	1.000000	0.000152
46	20	10	0	0.001303	0.001303
46	20	10	1	0.016633	0.015330
46	20	10	2	0.089451	0.072818
46	20	10	3	0.273413	0.183962
46	20	10	4	0.547056	0.273643
46	20	10	5	0.797244	0.250188
46	20	10	6	0.939396	0.142152
46	20	10	7	0.988840	0.049444
46	20	10	8	0.998883	0.010043
46	20	10	9	0.999955	0.001071
46	20	10	10	1.000000	0.000045
46	20	11	0	0.000579	0.000579
46	20	11	1	0.008542	0.007963
46	20	11	2	0.053042	0.044500
46	20	11	3	0.186542	0.133500
46	20	11	4	0.425437	0.238895
46	20	11	5	0.692999	0.267562
46	20	11	6	0.884114	0.191116
46	20	11	7	0.970985	0.086871
46	20	11	8	0.995536	0.024550
46	20	11	9	0.999627	0.004092
46	20	11	10	0.999987	0.000360
46	20	11	11	1.000000	0.000013
46	20	12	0	0.000248	0.000248
46	20	12	1	0.004219	0.003971
46	20	12	2	0.030157	0.025937
46	20	12	3	0.121699	0.091543
46	20	12	4	0.316228	0.194528
46	20	12	5	0.578329	0.262101
46	20	12	6	0.807668	0.229339
46	20	12	7	0.938719	0.131051
46	20	12	8	0.987118	0.048399
46	20	12	9	0.998341	0.011223
46	20	12	10	0.999884	0.001543
46	20	12	11	0.999997	0.000112
46	20	12	12	1.000000	0.000003
46	20	13	0	0.000102	0.000102
46	20	13	1	0.002000	0.001898
46	20	13	2	0.016425	0.014425
46	20	13	3	0.075928	0.059503
46	20	13	4	0.224685	0.148757
46	20	13	5	0.462696	0.238011
46	20	13	6	0.713234	0.250538
46	20	13	7	0.888611	0.175377
46	20	13	8	0.970036	0.081425
46	20	13	9	0.994710	0.024674
46	20	13	10	0.999431	0.004720
46	20	13	11	0.999967	0.000536
46	20	13	12	0.999999	0.000032
46	20	13	13	1.000000	0.000001
46	20	14	0	0.000040	0.000040
46	20	14	1	0.000907	0.000867
46	20	14	2	0.008557	0.007650
46	20	14	3	0.045275	0.036718
46	20	14	4	0.152560	0.107285
46	20	14	5	0.354509	0.201949
46	20	14	6	0.606945	0.252436
46	20	14	7	0.819523	0.212578
46	20	14	8	0.940427	0.120904
46	20	14	9	0.986485	0.046059
46	20	14	10	0.998000	0.011515
46	20	14	11	0.999821	0.001820
46	20	14	12	0.999991	0.000171
46	20	14	13	1.000000	0.000008
46	20	14	14	1.000000	0.000000
46	20	15	0	0.000015	0.000015
46	20	15	1	0.000393	0.000377
46	20	15	2	0.004254	0.003862
46	20	15	3	0.025769	0.021514
46	20	15	4	0.098918	0.073149
46	20	15	5	0.259846	0.160928
46	20	15	6	0.496505	0.236659

Table for $N = 2$, $n = 1$, through $N = 100$, $n = 50$

N	n	k	x	P(x)	p(x)
46	20	15	7	0.733164	0.236659
46	20	15	8	0.895088	0.161925
46	20	15	9	0.970653	0.075565
46	20	15	10	0.994402	0.023749
46	20	15	11	0.999309	0.004907
46	20	15	12	0.999949	0.000640
46	20	15	13	0.999998	0.000049
46	20	15	14	1.000000	0.000002
46	20	15	15	1.000000	0.000000
46	20	16	0	0.000005	0.000005
46	20	16	1	0.000161	0.000156
46	20	16	2	0.002012	0.001851
46	20	16	3	0.013970	0.011958
46	20	16	4	0.061163	0.047193
46	20	16	5	0.181977	0.120814
46	20	16	6	0.389626	0.207649
46	20	16	7	0.633920	0.244293
46	20	16	8	0.832408	0.198488
46	20	16	9	0.943840	0.111432
46	20	16	10	0.986741	0.042901
46	20	16	11	0.997884	0.011143
46	20	16	12	0.999784	0.001899
46	20	16	13	0.999987	0.000203
46	20	16	14	0.999999	0.000013
46	20	16	15	1.000000	0.000000
46	20	16	16	1.000000	0.000000
46	20	17	0	0.000002	0.000002
46	20	17	1	0.000062	0.000061
46	20	17	2	0.000901	0.000839
46	20	17	3	0.007194	0.006292
46	20	17	4	0.035994	0.028800
46	20	17	5	0.121570	0.085577
46	20	17	6	0.292724	0.171153
46	20	17	7	0.528059	0.235336
46	20	17	8	0.753012	0.224953
46	20	17	9	0.902981	0.149969
46	20	17	10	0.972440	0.069459
46	20	17	11	0.994541	0.022101
46	20	17	12	0.999277	0.004736
46	20	17	13	0.999939	0.000662
46	20	17	14	0.999997	0.000058
46	20	17	15	1.000000	0.000003
46	20	17	16	1.000000	0.000000
46	20	17	17	1.000000	0.000000
46	20	18	0	0.000001	0.000001
46	20	18	1	0.000023	0.000022
46	20	18	2	0.000381	0.000358
46	20	18	3	0.003505	0.003124
46	20	18	4	0.020104	0.016599
46	20	18	5	0.077307	0.057202

N	n	k	x	P(x)	p(x)
46	20	18	6	0.210098	0.132791
46	20	18	7	0.422564	0.212466
46	20	18	8	0.659928	0.237364
46	20	18	9	0.846096	0.186168
46	20	18	10	0.948489	0.102392
46	20	18	11	0.987682	0.039193
46	20	18	12	0.997970	0.010288
46	20	18	13	0.999779	0.001809
46	20	18	14	0.999985	0.000206
46	20	18	15	0.999999	0.000014
46	20	18	16	1.000000	0.000001
46	20	18	17	1.000000	0.000000
46	20	18	18	1.000000	0.000000
46	20	19	0	0.000000	0.000000
46	20	19	1	0.000008	0.000008
46	20	19	2	0.000151	0.000143
46	20	19	3	0.001608	0.001458
46	20	19	4	0.010619	0.009011
46	20	19	5	0.046662	0.036043
46	20	19	6	0.143702	0.097040
46	20	19	7	0.323919	0.180217
46	20	19	8	0.558201	0.234282
46	20	19	9	0.772959	0.214758
46	20	19	10	0.911920	0.138961
46	20	19	11	0.975084	0.063164
46	20	19	12	0.995031	0.019947
46	20	19	13	0.999327	0.004296
46	20	19	14	0.999941	0.000614
46	20	19	15	0.999997	0.000056
46	20	19	16	1.000000	0.000003
46	20	19	17	1.000000	0.000000
46	20	19	18	1.000000	0.000000
46	20	19	19	1.000000	0.000000
46	20	20	0	0.000000	0.000000
46	20	20	1	0.000002	0.000002
46	20	20	2	0.000055	0.000053
46	20	20	3	0.000690	0.000635
46	20	20	4	0.005279	0.004589
46	20	20	5	0.026638	0.021359
46	20	20	6	0.093385	0.066747
46	20	20	7	0.237148	0.143763
46	20	20	8	0.454075	0.216928
46	20	20	9	0.685465	0.231389
46	20	20	10	0.860453	0.174988
46	20	20	11	0.954030	0.093577
46	20	20	12	0.989121	0.035091
46	20	20	13	0.998213	0.009092
46	20	20	14	0.999805	0.001591
46	20	20	15	0.999986	0.000182
46	20	20	16	0.999999	0.000013

N	n	k	x	P(x)	p(x)
46	20	20	17	1.000000	0.000001
46	20	20	18	1.000000	0.000000
46	20	20	19	1.000000	0.000000
46	20	20	20	1.000000	0.000000
46	21	1	0	0.543478	0.543478
46	21	1	1	1.000000	0.456522
46	21	2	0	0.289855	0.289855
46	21	2	1	0.797101	0.507246
46	21	2	2	1.000000	0.202899
46	21	3	0	0.151515	0.151515
46	21	3	1	0.566535	0.415020
46	21	3	2	0.912385	0.345850
46	21	3	3	1.000000	0.087615
46	21	4	0	0.077519	0.077519
46	21	4	1	0.373502	0.295983
46	21	4	2	0.759567	0.386065
46	21	4	3	0.963324	0.203756
46	21	4	4	1.000000	0.036676
46	21	5	0	0.038760	0.038760
46	21	5	1	0.232558	0.193798
46	21	5	2	0.584919	0.352361
46	21	5	3	0.876000	0.291081
46	21	5	4	0.985155	0.109155
46	21	5	5	1.000000	0.014845
46	21	6	0	0.018907	0.018907
46	21	6	1	0.138022	0.119115
46	21	6	2	0.421630	0.283607
46	21	6	3	0.748208	0.326578
46	21	6	4	0.939895	0.191687
46	21	6	5	0.994207	0.054311
46	21	6	6	1.000000	0.005793
46	21	7	0	0.008981	0.008981
46	21	7	1	0.078465	0.069484
46	21	7	2	0.286916	0.208451
46	21	7	3	0.601248	0.314332
46	21	7	4	0.858428	0.257180
46	21	7	5	0.972482	0.114054
46	21	7	6	0.997827	0.025345
46	21	7	7	1.000000	0.002172
46	21	8	0	0.004145	0.004145
46	21	8	1	0.042832	0.038687
46	21	8	2	0.185363	0.142531
46	21	8	3	0.456172	0.270809
46	21	8	4	0.746324	0.290152
46	21	8	5	0.925691	0.179367
46	21	8	6	0.988079	0.062388
46	21	8	7	0.999220	0.011141
46	21	8	8	1.000000	0.000780
46	21	9	0	0.001854	0.001854
46	21	9	1	0.022470	0.020616

N	n	k	x	P(x)	p(x)
46	21	9	2	0.114097	0.091627
46	21	9	3	0.327894	0.213796
46	21	9	4	0.616519	0.288625
46	21	9	5	0.850168	0.233649
46	21	9	6	0.963452	0.113284
46	21	9	7	0.995116	0.031663
46	21	9	8	0.999733	0.004618
46	21	9	9	1.000000	0.000267
46	21	10	0	0.000802	0.000802
46	21	10	1	0.011327	0.010525
46	21	10	2	0.067046	0.055719
46	21	10	3	0.223885	0.156839
46	21	10	4	0.483907	0.260023
46	21	10	5	0.749131	0.265223
46	21	10	6	0.917526	0.168396
46	21	10	7	0.983135	0.065609
46	21	10	8	0.998111	0.014976
46	21	10	9	0.999913	0.001803
46	21	10	10	1.000000	0.000087
46	21	11	0	0.000334	0.000334
46	21	11	1	0.005480	0.005145
46	21	11	2	0.037638	0.032159
46	21	11	3	0.145465	0.107827
46	21	11	4	0.361119	0.215654
46	21	11	5	0.631254	0.270135
46	21	11	6	0.847361	0.216108
46	21	11	7	0.957620	0.110259
46	21	11	8	0.992703	0.035082
46	21	11	9	0.999312	0.006610
46	21	11	10	0.999973	0.000661
46	21	11	11	1.000000	0.000026
46	21	12	0	0.000134	0.000134
46	21	12	1	0.002539	0.002406
46	21	12	2	0.020181	0.017641
46	21	12	3	0.090011	0.069831
46	21	12	4	0.256373	0.166361
46	21	12	5	0.507763	0.251391
46	21	12	6	0.754744	0.246980
46	21	12	7	0.913517	0.158773
46	21	12	8	0.979672	0.066155
46	21	12	9	0.997046	0.017374
46	21	12	10	0.999766	0.002719
46	21	12	11	0.999992	0.000227
46	21	12	12	1.000000	0.000008
46	21	13	0	0.000051	0.000051
46	21	13	1	0.001124	0.001073
46	21	13	2	0.010322	0.009198
46	21	13	3	0.053042	0.042720
46	21	13	4	0.173192	0.120150
46	21	13	5	0.389462	0.216270

Table for $N = 2$, $n = 1$, through $N = 100$, $n = 50$

N	n	k	x	P(x)	p(x)
46	21	13	6	0.645782	0.256320
46	21	13	7	0.848140	0.202358
46	21	13	8	0.954377	0.106238
46	21	13	9	0.990914	0.036537
46	21	13	10	0.998886	0.007972
46	21	13	11	0.999926	0.001040
46	21	13	12	0.999998	0.000072
46	21	13	13	1.000000	0.000002
46	21	14	0	0.000019	0.000019
46	21	14	1	0.000474	0.000455
46	21	14	2	0.005026	0.004553
46	21	14	3	0.029740	0.024714
46	21	14	4	0.111297	0.081556
46	21	14	5	0.284604	0.173307
46	21	14	6	0.529273	0.244669
46	21	14	7	0.762291	0.233018
46	21	14	8	0.912526	0.150235
46	21	14	9	0.977628	0.065102
46	21	14	10	0.996229	0.018601
46	21	14	11	0.999611	0.003382
46	21	14	12	0.999978	0.000368
46	21	14	13	0.999999	0.000021
46	21	14	14	1.000000	0.000000
46	21	15	0	0.000006	0.000006
46	21	15	1	0.000189	0.000183
46	21	15	2	0.002323	0.002134
46	21	15	3	0.015839	0.013515
46	21	15	4	0.067970	0.052131
46	21	15	5	0.197950	0.129980
46	21	15	6	0.414584	0.216634
46	21	15	7	0.660345	0.245761
46	21	15	8	0.851493	0.191148
46	21	15	9	0.953215	0.101722
46	21	15	10	0.989835	0.036620
46	21	15	11	0.998554	0.008719
46	21	15	12	0.999875	0.001321
46	21	15	13	0.999994	0.000119
46	21	15	14	1.000000	0.000006
46	21	15	15	1.000000	0.000000
46	21	16	0	0.000002	0.000002
46	21	16	1	0.000071	0.000069
46	21	16	2	0.001015	0.000944
46	21	16	3	0.007991	0.006976
46	21	16	4	0.039382	0.031391
46	21	16	5	0.130864	0.091482
46	21	16	6	0.309761	0.178898
46	21	16	7	0.549357	0.239595
46	21	16	8	0.771334	0.221978
46	21	16	9	0.913839	0.142504
46	21	16	10	0.976841	0.063002

N	n	k	x	P(x)	p(x)
46	21	16	11	0.995741	0.018901
46	21	16	12	0.999491	0.003750
46	21	16	13	0.999963	0.000472
46	21	16	14	0.999998	0.000035
46	21	16	15	1.000000	0.000001
46	21	16	16	1.000000	0.000000
46	21	17	0	0.000001	0.000001
46	21	17	1	0.000025	0.000025
46	21	17	2	0.000417	0.000392
46	21	17	3	0.003806	0.003388
46	21	17	4	0.021594	0.017788
46	21	17	5	0.082073	0.060480
46	21	17	6	0.220313	0.138239
46	21	17	7	0.437545	0.217233
46	21	17	8	0.675144	0.237599
46	21	17	9	0.856837	0.181693
46	21	17	10	0.953740	0.096903
46	21	17	11	0.989441	0.035701
46	21	17	12	0.998366	0.008925
46	21	17	13	0.999837	0.001471
46	21	17	14	0.999990	0.000153
46	21	17	15	1.000000	0.000009
46	21	17	16	1.000000	0.000000
46	21	17	17	1.000000	0.000000
46	21	18	0	0.000000	0.000000
46	21	18	1	0.000008	0.000008
46	21	18	2	0.000160	0.000152
46	21	18	3	0.001703	0.001542
46	21	18	4	0.011166	0.009464
46	21	18	5	0.048705	0.037539
46	21	18	6	0.148810	0.100104
46	21	18	7	0.332674	0.183865
46	21	18	8	0.568634	0.235960
46	21	18	9	0.781654	0.213019
46	21	18	10	0.916984	0.135330
46	21	18	11	0.977130	0.060147
46	21	18	12	0.995596	0.018466
46	21	18	13	0.999432	0.003835
46	21	18	14	0.999953	0.000522
46	21	18	15	0.999998	0.000044
46	21	18	16	1.000000	0.000002
46	21	18	17	1.000000	0.000000
46	21	18	18	1.000000	0.000000
46	21	19	0	0.000000	0.000000
46	21	19	1	0.000002	0.000002
46	21	19	2	0.000057	0.000055
46	21	19	3	0.000711	0.000654
46	21	19	4	0.005420	0.004709
46	21	19	5	0.027254	0.021834
46	21	19	6	0.095182	0.067928

N	n	k	x	P(x)	p(x)
46	21	19	7	0.240742	0.145560
46	21	19	8	0.459081	0.218340
46	21	19	9	0.690360	0.231278
46	21	19	10	0.863818	0.173459
46	21	19	11	0.955649	0.091831
46	21	19	12	0.989661	0.034011
46	21	19	13	0.998336	0.008675
46	21	19	14	0.999823	0.001487
46	21	19	15	0.999988	0.000165
46	21	19	16	0.999999	0.000011
46	21	19	17	1.000000	0.000000
46	21	19	18	1.000000	0.000000
46	21	19	19	1.000000	0.000000
46	21	20	0	0.000000	0.000000
46	21	20	1	0.000001	0.000001
46	21	20	2	0.000019	0.000018
46	21	20	3	0.000275	0.000256
46	21	20	4	0.002455	0.002180
46	21	20	5	0.014316	0.011860
46	21	20	6	0.057445	0.043129
46	21	20	7	0.165267	0.107822
46	21	20	8	0.353955	0.188688
46	21	20	9	0.587569	0.233614
46	21	20	10	0.793150	0.205581
46	21	20	11	0.921638	0.128488
46	21	20	12	0.978324	0.056686
46	21	20	13	0.995765	0.017442
46	21	20	14	0.999437	0.003672
46	21	20	15	0.999951	0.000514
46	21	20	16	0.999997	0.000046
46	21	20	17	1.000000	0.000002
46	21	20	18	1.000000	0.000000
46	21	20	19	1.000000	0.000000
46	21	20	20	1.000000	0.000000
46	21	21	0	0.000000	0.000000
46	21	21	1	0.000000	0.000000
46	21	21	2	0.000006	0.000005
46	21	21	3	0.000098	0.000092
46	21	21	4	0.001030	0.000932
46	21	21	5	0.007017	0.005987
46	21	21	6	0.032563	0.025546
46	21	21	7	0.107209	0.074646
46	21	21	8	0.259611	0.152402
46	21	21	9	0.479747	0.220137
46	21	21	10	0.706174	0.226426
46	21	21	11	0.872219	0.166046
46	21	21	12	0.958702	0.086482
46	21	21	13	0.990399	0.031697
46	21	21	14	0.998449	0.008050
46	21	21	15	0.999833	0.001384

N	n	k	x	P(x)	p(x)
46	21	21	16	0.999989	0.000156
46	21	21	17	0.999999	0.000011
46	21	21	18	1.000000	0.000000
46	21	21	19	1.000000	0.000000
46	21	21	20	1.000000	0.000000
46	21	21	21	1.000000	0.000000
46	22	1	0	0.521739	0.521739
46	22	1	1	1.000000	0.478261
46	22	2	0	0.266667	0.266667
46	22	2	1	0.776812	0.510145
46	22	2	2	1.000000	0.223188
46	22	3	0	0.133333	0.133333
46	22	3	1	0.533333	0.400000
46	22	3	2	0.898551	0.365217
46	22	3	3	1.000000	0.101449
46	22	4	0	0.065116	0.065116
46	22	4	1	0.337984	0.272868
46	22	4	2	0.728682	0.390698
46	22	4	3	0.955174	0.226491
46	22	4	4	1.000000	0.044826
46	22	5	0	0.031008	0.031008
46	22	5	1	0.201550	0.170543
46	22	5	2	0.542636	0.341085
46	22	5	3	0.852713	0.310077
46	22	5	4	0.980789	0.128075
46	22	5	5	1.000000	0.019211
46	22	6	0	0.014369	0.014369
46	22	6	1	0.114199	0.099830
46	22	6	2	0.376253	0.262053
46	22	6	3	0.709019	0.332766
46	22	6	4	0.924560	0.215542
46	22	6	5	0.992034	0.067474
46	22	6	6	1.000000	0.007966
46	22	7	0	0.006466	0.006466
46	22	7	1	0.061789	0.055322
46	22	7	2	0.245226	0.183437
46	22	7	3	0.550955	0.305729
46	22	7	4	0.827567	0.276612
46	22	7	5	0.963358	0.135791
46	22	7	6	0.996814	0.033456
46	22	7	7	1.000000	0.003186
46	22	8	0	0.002819	0.002819
46	22	8	1	0.032000	0.029181
46	22	8	2	0.151155	0.119156
46	22	8	3	0.402010	0.250854
46	22	8	4	0.699900	0.297890
46	22	8	5	0.904167	0.204267
46	22	8	6	0.983088	0.078921
46	22	8	7	0.998774	0.015686
46	22	8	8	1.000000	0.001225

Table for $N=2$, $n=1$, through $N=100$, $n=50$

N	n	k	x	P(x)	p(x)
46	22	9	0	0.001187	0.001187
46	22	9	1	0.015873	0.014686
46	22	9	2	0.088442	0.072569
46	22	9	3	0.276583	0.188141
46	22	9	4	0.558794	0.282211
46	22	9	5	0.812784	0.253990
46	22	9	6	0.949858	0.137074
46	22	9	7	0.992582	0.042724
46	22	9	8	0.999548	0.006966
46	22	9	9	1.000000	0.000451
46	22	10	0	0.000481	0.000481
46	22	10	1	0.007538	0.007057
46	22	10	2	0.049216	0.041678
46	22	10	3	0.179970	0.130754
46	22	10	4	0.421502	0.241532
46	22	10	5	0.696086	0.274584
46	22	10	6	0.890583	0.194497
46	22	10	7	0.975262	0.084679
46	22	10	8	0.996913	0.021651
46	22	10	9	0.999841	0.002929
46	22	10	10	1.000000	0.000159
46	22	11	0	0.000187	0.000187
46	22	11	1	0.003421	0.003234
46	22	11	2	0.026061	0.022640
46	22	11	3	0.110961	0.084899
46	22	11	4	0.300736	0.189775
46	22	11	5	0.566421	0.265685
46	22	11	6	0.804140	0.237718
46	22	11	7	0.939979	0.135839
46	22	11	8	0.988493	0.048514
46	22	11	9	0.998784	0.010291
46	22	11	10	0.999947	0.001163
46	22	11	11	1.000000	0.000053
46	22	12	0	0.000069	0.000069
46	22	12	1	0.001481	0.001411
46	22	12	2	0.013124	0.011643
46	22	12	3	0.064872	0.051748
46	22	12	4	0.203137	0.138265
46	22	12	5	0.437374	0.234237
46	22	12	6	0.695468	0.258094
46	22	12	7	0.881762	0.186294
46	22	12	8	0.969087	0.087325
46	22	12	9	0.994961	0.025874
46	22	12	10	0.999548	0.004587
46	22	12	11	0.999983	0.000435
46	22	12	12	1.000000	0.000017
46	22	13	0	0.000025	0.000025
46	22	13	1	0.000609	0.000585
46	22	13	2	0.006275	0.005666
46	22	13	3	0.035954	0.029679
46	22	13	4	0.129938	0.093984
46	22	13	5	0.320256	0.190317
46	22	13	6	0.574012	0.253757
46	22	13	7	0.799574	0.225561
46	22	13	8	0.933130	0.133556
46	22	13	9	0.985068	0.051938
46	22	13	10	0.997929	0.012861
46	22	13	11	0.999843	0.001913
46	22	13	12	0.999995	0.000152
46	22	13	13	1.000000	0.000005
46	22	14	0	0.000008	0.000008
46	22	14	1	0.000237	0.000229
46	22	14	2	0.002841	0.002604
46	22	14	3	0.018866	0.016025
46	22	14	4	0.078674	0.059808
46	22	14	5	0.222213	0.143539
46	22	14	6	0.450979	0.228765
46	22	14	7	0.697046	0.246067
46	22	14	8	0.876470	0.179424
46	22	14	9	0.964608	0.088138
46	22	14	10	0.993253	0.028645
46	22	14	11	0.999205	0.005952
46	22	14	12	0.999949	0.000744
46	22	14	13	0.999999	0.000050
46	22	14	14	1.000000	0.000001
46	22	15	0	0.000003	0.000003
46	22	15	1	0.000087	0.000084
46	22	15	2	0.001214	0.001127
46	22	15	3	0.009351	0.008138
46	22	15	4	0.045032	0.035681
46	22	15	5	0.145958	0.100926
46	22	15	6	0.336596	0.190638
46	22	15	7	0.581702	0.245106
46	22	15	8	0.797972	0.216270
46	22	15	9	0.928802	0.130830
46	22	15	10	0.982511	0.053709
46	22	15	11	0.997159	0.014648
46	22	15	12	0.999716	0.002558
46	22	15	13	0.999985	0.000268
46	22	15	14	1.000000	0.000015
46	22	15	15	1.000000	0.000000
46	22	16	0	0.000001	0.000001
46	22	16	1	0.000030	0.000029
46	22	16	2	0.000487	0.000457
46	22	16	3	0.004364	0.003877
46	22	16	4	0.024314	0.019951
46	22	16	5	0.090612	0.066297
46	22	16	6	0.238202	0.147591
46	22	16	7	0.463102	0.224900
46	22	16	8	0.700302	0.237199
46	22	16	9	0.873938	0.173636
46	22	16	10	0.961720	0.087783
46	22	16	11	0.991961	0.030241
46	22	16	12	0.998891	0.006930
46	22	16	13	0.999907	0.001015
46	22	16	14	0.999996	0.000089
46	22	16	15	1.000000	0.000004
46	22	16	16	1.000000	0.000000
46	22	17	0	0.000000	0.000000
46	22	17	1	0.000009	0.000009
46	22	17	2	0.000182	0.000173
46	22	17	3	0.001908	0.001726
46	22	17	4	0.012344	0.010436
46	22	17	5	0.053043	0.040699
46	22	17	6	0.159487	0.106444
46	22	17	7	0.350652	0.191165
46	22	17	8	0.589609	0.238956
46	22	17	9	0.798695	0.209087
46	22	17	10	0.926607	0.127912
46	22	17	11	0.980873	0.054266
46	22	17	12	0.996581	0.015708
46	22	17	13	0.999602	0.003021
46	22	17	14	0.999972	0.000370
46	22	17	15	0.999999	0.000027
46	22	17	16	1.000000	0.000001
46	22	17	17	1.000000	0.000000
46	22	18	0	0.000000	0.000000
46	22	18	1	0.000003	0.000003
46	22	18	2	0.000063	0.000060
46	22	18	3	0.000777	0.000714
46	22	18	4	0.005867	0.005089
46	22	18	5	0.029185	0.023318
46	22	18	6	0.100760	0.071574
46	22	18	7	0.251774	0.151014
46	22	18	8	0.474250	0.222476
46	22	18	9	0.704967	0.230716
46	22	18	10	0.873678	0.168711
46	22	18	11	0.960289	0.086611
46	22	18	12	0.991164	0.030875
46	22	18	13	0.998665	0.007500
46	22	18	14	0.999870	0.001205
46	22	18	15	0.999992	0.000122
46	22	18	16	1.000000	0.000007
46	22	18	17	1.000000	0.000000
46	22	18	18	1.000000	0.000000
46	22	19	0	0.000000	0.000000
46	22	19	1	0.000001	0.000001
46	22	19	2	0.000020	0.000019
46	22	19	3	0.000293	0.000273
46	22	19	4	0.002595	0.002302
46	22	19	5	0.015027	0.012433
46	22	19	6	0.059860	0.044832
46	22	19	7	0.170873	0.111014
46	22	19	8	0.363012	0.192139
46	22	19	9	0.597848	0.234836
46	22	19	10	0.801373	0.203525
46	22	19	11	0.926263	0.124890
46	22	19	12	0.980138	0.053874
46	22	19	13	0.996254	0.016116
46	22	19	14	0.999526	0.003272
46	22	19	15	0.999962	0.000436
46	22	19	16	0.999998	0.000036
46	22	19	17	1.000000	0.000002
46	22	19	18	1.000000	0.000000
46	22	19	19	1.000000	0.000000
46	22	20	0	0.000000	0.000000
46	22	20	1	0.000000	0.000000
46	22	20	2	0.000006	0.000006
46	22	20	3	0.000101	0.000095
46	22	20	4	0.001060	0.000959
46	22	20	5	0.007200	0.006140
46	22	20	6	0.033292	0.026093
46	22	20	7	0.109199	0.075907
46	22	20	8	0.263385	0.154185
46	22	20	9	0.484779	0.221394
46	22	20	10	0.710918	0.226139
46	22	20	11	0.875382	0.164464
46	22	20	12	0.960184	0.084802
46	22	20	13	0.990882	0.030698
46	22	20	14	0.998556	0.007674
46	22	20	15	0.999849	0.001293
46	22	20	16	0.999990	0.000141
46	22	20	17	0.999999	0.000009
46	22	20	18	1.000000	0.000000
46	22	20	19	1.000000	0.000000
46	22	20	20	1.000000	0.000000
46	22	21	0	0.000000	0.000000
46	22	21	1	0.000000	0.000000
46	22	21	2	0.000001	0.000001
46	22	21	3	0.000031	0.000030
46	22	21	4	0.000396	0.000365
46	22	21	5	0.003185	0.002789
46	22	21	6	0.017235	0.014050
46	22	21	7	0.065407	0.048172
46	22	21	8	0.180362	0.114955
46	22	21	9	0.374082	0.193720
46	22	21	10	0.606546	0.232464
46	22	21	11	0.805801	0.199255
46	22	21	12	0.927568	0.121767
46	22	21	13	0.980256	0.052688

Table for $N=2$, $n=1$, through $N=100$, $n=50$

N	n	k	x	P(x)	p(x)
46	22	21	14	0.996195	0.015939
46	22	21	15	0.999501	0.003306
46	22	21	16	0.999957	0.000457
46	22	21	17	0.999998	0.000040
46	22	21	18	1.000000	0.000002
46	22	21	19	1.000000	0.000000
46	22	21	20	1.000000	0.000000
46	22	21	21	1.000000	0.000000
46	22	22	0	0.000000	0.000000
46	22	22	1	0.000000	0.000000
46	22	22	2	0.000000	0.000000
46	22	22	3	0.000009	0.000008
46	22	22	4	0.000133	0.000125
46	22	22	5	0.001289	0.001155
46	22	22	6	0.008243	0.006955
46	22	22	7	0.036504	0.028261
46	22	22	8	0.115987	0.079483
46	22	22	9	0.273347	0.157360
46	22	22	10	0.494963	0.221616
46	22	22	11	0.718129	0.223166
46	22	22	12	0.878861	0.160732
46	22	22	13	0.961288	0.082427
46	22	22	14	0.991094	0.029806
46	22	22	15	0.998575	0.007481
46	22	22	16	0.999848	0.001273
46	22	22	17	0.999990	0.000142
46	22	22	18	0.999999	0.000010
46	22	22	19	1.000000	0.000000
46	22	22	20	1.000000	0.000000
46	22	22	21	1.000000	0.000000
46	22	22	22	1.000000	0.000000
46	23	1	0	0.500000	0.500000
46	23	1	1	1.000000	0.500000
46	23	2	0	0.244444	0.244444
46	23	2	1	0.755556	0.511111
46	23	2	2	1.000000	0.244444
46	23	3	0	0.116667	0.116667
46	23	3	1	0.500000	0.383333
46	23	3	2	0.883333	0.383333
46	23	3	3	1.000000	0.116667
46	23	4	0	0.054264	0.054264
46	23	4	1	0.303876	0.249612
46	23	4	2	0.696124	0.392248
46	23	4	3	0.945736	0.249612
46	23	4	4	1.000000	0.054264
46	23	5	0	0.024548	0.024548
46	23	5	1	0.173127	0.148579
46	23	5	2	0.500000	0.326873
46	23	5	3	0.826873	0.326873
46	23	5	4	0.975452	0.148579
46	23	5	5	1.000000	0.024548
46	23	6	0	0.010777	0.010777
46	23	6	1	0.093401	0.082624
46	23	6	2	0.332577	0.239176
46	23	6	3	0.667423	0.334846
46	23	6	4	0.906599	0.239176
46	23	6	5	0.989223	0.082624
46	23	6	6	1.000000	0.010777
46	23	7	0	0.004580	0.004580
46	23	7	1	0.047958	0.043378
46	23	7	2	0.207010	0.159052
46	23	7	3	0.500000	0.292990
46	23	7	4	0.792990	0.292990
46	23	7	5	0.952042	0.159052
46	23	7	6	0.995420	0.043378
46	23	7	7	1.000000	0.004580
46	23	8	0	0.001879	0.001879
46	23	8	1	0.023489	0.021609
46	23	8	2	0.121367	0.097878
46	23	8	3	0.349749	0.228382
46	23	8	4	0.650251	0.300503
46	23	8	5	0.878633	0.228382
46	23	8	6	0.976511	0.097878
46	23	8	7	0.998121	0.021609
46	23	8	8	1.000000	0.001879
46	23	9	0	0.000742	0.000742
46	23	9	1	0.010978	0.010236
46	23	9	2	0.067276	0.056298
46	23	9	3	0.229547	0.162271
46	23	9	4	0.500000	0.270452
46	23	9	5	0.770452	0.270452
46	23	9	6	0.932724	0.162271
46	23	9	7	0.989022	0.056298
46	23	9	8	0.999258	0.010236
46	23	9	9	1.000000	0.000742
46	23	10	0	0.000281	0.000281
46	23	10	1	0.004891	0.004611
46	23	10	2	0.035323	0.030431
46	23	10	3	0.141833	0.106510
46	23	10	4	0.361119	0.219286
46	23	10	5	0.638881	0.277762
46	23	10	6	0.858167	0.219286
46	23	10	7	0.964677	0.106510
46	23	10	8	0.995108	0.030431
46	23	10	9	0.999719	0.004611
46	23	10	10	1.000000	0.000281
46	23	11	0	0.000101	0.000101
46	23	11	1	0.002074	0.001972
46	23	11	2	0.017571	0.015498
46	23	11	3	0.082661	0.065090
46	23	11	4	0.245385	0.162724
46	23	11	5	0.500000	0.254615
46	23	11	6	0.754615	0.254615
46	23	11	7	0.917339	0.162724
46	23	11	8	0.982429	0.065090
46	23	11	9	0.997926	0.015498
46	23	11	10	0.999899	0.001972
46	23	11	11	1.000000	0.000101
46	23	12	0	0.000035	0.000035
46	23	12	1	0.000834	0.000799
46	23	12	2	0.008273	0.007439
46	23	12	3	0.045467	0.037194
46	23	12	4	0.157049	0.111582
46	23	12	5	0.369055	0.212006
46	23	12	6	0.630945	0.261890
46	23	12	7	0.842951	0.212006
46	23	12	8	0.954533	0.111582
46	23	12	9	0.991727	0.037194
46	23	12	10	0.999166	0.007439
46	23	12	11	0.999965	0.000799
46	23	12	12	1.000000	0.000035
46	23	13	0	0.000011	0.000011
46	23	13	1	0.000317	0.000306
46	23	13	2	0.003678	0.003361
46	23	13	3	0.023588	0.019910
46	23	13	4	0.094694	0.071106
46	23	13	5	0.256817	0.162122
46	23	13	6	0.500000	0.243183
46	23	13	7	0.743183	0.243183
46	23	13	8	0.905306	0.162122
46	23	13	9	0.976412	0.071106
46	23	13	10	0.996322	0.019910
46	23	13	11	0.999683	0.003361
46	23	13	12	0.999989	0.000306
46	23	13	13	1.000000	0.000011
46	23	14	0	0.000003	0.000003
46	23	14	1	0.000113	0.000110
46	23	14	2	0.001539	0.001426
46	23	14	3	0.011521	0.009982
46	23	14	4	0.053754	0.042233
46	23	14	5	0.168386	0.114632
46	23	14	6	0.374724	0.206337
46	23	14	7	0.625276	0.250553
46	23	14	8	0.831614	0.206337
46	23	14	9	0.946246	0.114632
46	23	14	10	0.988478	0.042233
46	23	14	11	0.998461	0.009982
46	23	14	12	0.999887	0.001426
46	23	14	13	0.999996	0.000110
46	23	14	14	1.000000	0.000003
46	23	15	0	0.000001	0.000001
46	23	15	1	0.000038	0.000037
46	23	15	2	0.000603	0.000566
46	23	15	3	0.005283	0.004679
46	23	15	4	0.028679	0.023396
46	23	15	5	0.103906	0.075227
46	23	15	6	0.265107	0.161201
46	23	15	7	0.500000	0.234893
46	23	15	8	0.734893	0.234893
46	23	15	9	0.896094	0.161201
46	23	15	10	0.971321	0.075227
46	23	15	11	0.994717	0.023396
46	23	15	12	0.999397	0.004679
46	23	15	13	0.999962	0.000566
46	23	15	14	0.999999	0.000037
46	23	15	15	1.000000	0.000001
46	23	16	0	0.000000	0.000000
46	23	16	1	0.000012	0.000011
46	23	16	2	0.000220	0.000209
46	23	16	3	0.002264	0.002044
46	23	16	4	0.014339	0.012075
46	23	16	5	0.060225	0.045886
46	23	16	6	0.176706	0.116481
46	23	16	7	0.378765	0.202059
46	23	16	8	0.621235	0.242470
46	23	16	9	0.823294	0.202059
46	23	16	10	0.939774	0.116481
46	23	16	11	0.985661	0.045886
46	23	16	12	0.997736	0.012075
46	23	16	13	0.999780	0.002044
46	23	16	14	0.999988	0.000209
46	23	16	15	1.000000	0.000011
46	23	16	16	1.000000	0.000000
46	23	17	0	0.000000	0.000000
46	23	17	1	0.000003	0.000003
46	23	17	2	0.000074	0.000071
46	23	17	3	0.000901	0.000827
46	23	17	4	0.006691	0.005790
46	23	17	5	0.032694	0.026002
46	23	17	6	0.110700	0.078007
46	23	17	7	0.271000	0.160300
46	23	17	8	0.500000	0.229000
46	23	17	9	0.729000	0.229000
46	23	17	10	0.889299	0.160300
46	23	17	11	0.967306	0.078007
46	23	17	12	0.993309	0.026002
46	23	17	13	0.999099	0.005790
46	23	17	14	0.999926	0.000827
46	23	17	15	0.999997	0.000071
46	23	17	16	1.000000	0.000003

Table for $N = 2$, $n = 1$, through $N = 100$, $n = 50$

N	n	k	x	P(x)	p(x)
46	23	17	17	1.000000	0.000000
46	23	18	0	0.000000	0.000000
46	23	18	1	0.000001	0.000001
46	23	18	2	0.000023	0.000022
46	23	18	3	0.000331	0.000308
46	23	18	4	0.002898	0.002567
46	23	18	5	0.016554	0.013656
46	23	18	6	0.064972	0.048418
46	23	18	7	0.182559	0.117587
46	23	18	8	0.381552	0.198993
46	23	18	9	0.618448	0.236896
46	23	18	10	0.817441	0.198993
46	23	18	11	0.935028	0.117587
46	23	18	12	0.983446	0.048418
46	23	18	13	0.997102	0.013656
46	23	18	14	0.999669	0.002567
46	23	18	15	0.999977	0.000308
46	23	18	16	0.999999	0.000022
46	23	18	17	1.000000	0.000001
46	23	18	18	1.000000	0.000000
46	23	19	0	0.000000	0.000000
46	23	19	1	0.000000	0.000000
46	23	19	2	0.000006	0.000006
46	23	19	3	0.000111	0.000105
46	23	19	4	0.001156	0.001045
46	23	19	5	0.007775	0.006619
46	23	19	6	0.035576	0.027800
46	23	19	7	0.115367	0.079791
46	23	19	8	0.274948	0.159582
46	23	19	9	0.500000	0.225051
46	23	19	10	0.725051	0.225051
46	23	19	11	0.884633	0.159582
46	23	19	12	0.964424	0.079791
46	23	19	13	0.992225	0.027800
46	23	19	14	0.998844	0.006619
46	23	19	15	0.999889	0.001045
46	23	19	16	0.999994	0.000105
46	23	19	17	1.000000	0.000006
46	23	19	18	1.000000	0.000000
46	23	19	19	1.000000	0.000000
46	23	20	0	0.000000	0.000000
46	23	20	1	0.000000	0.000000
46	23	20	2	0.000002	0.000002
46	23	20	3	0.000033	0.000032
46	23	20	4	0.000421	0.000387
46	23	20	5	0.003362	0.002942
46	23	20	6	0.018072	0.014709
46	23	20	7	0.068083	0.050011
46	23	20	8	0.186292	0.118209
46	23	20	9	0.383307	0.197015
46	23	20	10	0.616693	0.233387
46	23	20	11	0.813708	0.197015
46	23	20	12	0.931917	0.118209
46	23	20	13	0.981928	0.050011
46	23	20	14	0.996638	0.014709
46	23	20	15	0.999579	0.002942
46	23	20	16	0.999967	0.000387
46	23	20	17	0.999998	0.000032
46	23	20	18	1.000000	0.000002
46	23	20	19	1.000000	0.000000
46	23	20	20	1.000000	0.000000
46	23	21	0	0.000000	0.000000
46	23	21	1	0.000000	0.000000
46	23	21	2	0.000000	0.000000
46	23	21	3	0.000009	0.000009
46	23	21	4	0.000138	0.000129
46	23	21	5	0.001326	0.001188
46	23	21	6	0.008454	0.007128
46	23	21	7	0.037307	0.028853
46	23	21	8	0.118094	0.080788
46	23	21	9	0.277222	0.159127
46	23	21	10	0.500000	0.222778
46	23	21	11	0.722778	0.222778
46	23	21	12	0.881905	0.159127
46	23	21	13	0.962693	0.080788
46	23	21	14	0.991546	0.028853
46	23	21	15	0.998674	0.007128
46	23	21	16	0.999862	0.001188
46	23	21	17	0.999991	0.000129
46	23	21	18	1.000000	0.000009
46	23	21	19	1.000000	0.000000
46	23	21	20	1.000000	0.000000
46	23	21	21	1.000000	0.000000
46	23	22	0	0.000000	0.000000
46	23	22	1	0.000000	0.000000
46	23	22	2	0.000000	0.000000
46	23	22	3	0.000002	0.000002
46	23	22	4	0.000040	0.000038
46	23	22	5	0.000470	0.000430
46	23	22	6	0.003607	0.003136
46	23	22	7	0.018841	0.015234
46	23	22	8	0.069622	0.050781
46	23	22	9	0.188110	0.118489
46	23	22	10	0.384155	0.196045
46	23	22	11	0.615845	0.231689
46	23	22	12	0.811889	0.196045
46	23	22	13	0.930378	0.118489
46	23	22	14	0.981159	0.050781
46	23	22	15	0.996393	0.015234
46	23	22	16	0.999530	0.003136
46	23	22	17	0.999960	0.000430
46	23	22	18	0.999998	0.000038
46	23	22	19	1.000000	0.000002
46	23	22	20	1.000000	0.000000
46	23	22	21	1.000000	0.000000
46	23	22	22	1.000000	0.000000
46	23	23	0	0.000000	0.000000
46	23	23	1	0.000000	0.000000
46	23	23	2	0.000000	0.000000
46	23	23	3	0.000000	0.000000
46	23	23	4	0.000010	0.000010
46	23	23	5	0.000147	0.000138
46	23	23	6	0.001385	0.001238
46	23	23	7	0.008685	0.007300
46	23	23	8	0.037884	0.029199
46	23	23	9	0.118992	0.081108
46	23	23	10	0.277964	0.158972
46	23	23	11	0.500000	0.222036
46	23	23	12	0.722036	0.222036
46	23	23	13	0.881008	0.158972
46	23	23	14	0.962116	0.081108
46	23	23	15	0.991315	0.029199
46	23	23	16	0.998615	0.007300
46	23	23	17	0.999852	0.001238
46	23	23	18	0.999990	0.000138
46	23	23	19	1.000000	0.000010
46	23	23	20	1.000000	0.000000
46	23	23	21	1.000000	0.000000
46	23	23	22	1.000000	0.000000
46	23	23	23	1.000000	0.000000
47	1	1	0	0.978723	0.978723
47	1	1	1	1.000000	0.021277
47	2	1	0	0.957447	0.957447
47	2	1	1	1.000000	0.042553
47	2	2	0	0.915819	0.915819
47	2	2	1	0.999075	0.083256
47	2	2	2	1.000000	0.000925
47	3	1	0	0.936170	0.936170
47	3	1	1	1.000000	0.063830
47	3	2	0	0.875116	0.875116
47	3	2	1	0.997225	0.122109
47	3	2	2	1.000000	0.002775
47	3	3	0	0.816775	0.816775
47	3	3	1	0.991798	0.175023
47	3	3	2	0.999938	0.008141
47	3	3	3	1.000000	0.000062
47	4	1	0	0.914894	0.914894
47	4	1	1	1.000000	0.085106
47	4	2	0	0.835338	0.835338
47	4	2	1	0.994450	0.159112
47	4	2	2	1.000000	0.005550
47	4	3	0	0.761085	0.761085
47	4	3	1	0.983842	0.222757
47	4	3	2	0.999753	0.015911
47	4	3	3	1.000000	0.000247
47	4	4	0	0.691896	0.691896
47	4	4	1	0.968654	0.276758
47	4	4	2	0.999030	0.030376
47	4	4	3	0.999994	0.000964
47	4	4	4	1.000000	0.000006
47	5	1	0	0.893617	0.893617
47	5	1	1	1.000000	0.106383
47	5	2	0	0.796485	0.796485
47	5	2	1	0.990749	0.194265
47	5	2	2	1.000000	0.009251
47	5	3	0	0.707986	0.707986
47	5	3	1	0.973481	0.265495
47	5	3	2	0.999383	0.025902
47	5	3	3	1.000000	0.000617
47	5	4	0	0.627533	0.627533
47	5	4	1	0.949345	0.321812
47	5	4	2	0.997617	0.048272
47	5	4	3	0.999972	0.002355
47	5	4	4	1.000000	0.000028
47	5	5	0	0.554564	0.554564
47	5	5	1	0.919409	0.364845
47	5	5	2	0.994249	0.074840
47	5	5	3	0.999862	0.005613
47	5	5	4	0.999999	0.000137
47	5	5	5	1.000000	0.000001
47	6	1	0	0.872340	0.872340
47	6	1	1	1.000000	0.127660
47	6	2	0	0.758557	0.758557
47	6	2	1	0.986124	0.227567
47	6	2	2	1.000000	0.013876
47	6	3	0	0.657416	0.657416
47	6	3	1	0.960839	0.303423
47	6	3	2	0.998767	0.037928
47	6	3	3	1.000000	0.001233
47	6	4	0	0.567768	0.567768
47	6	4	1	0.926359	0.358590
47	6	4	2	0.995319	0.068960
47	6	4	3	0.999916	0.004597
47	6	4	4	1.000000	0.000084
47	6	5	0	0.488545	0.488545
47	6	5	1	0.884662	0.396117
47	6	5	2	0.988904	0.104241
47	6	5	3	0.999595	0.010691
47	6	5	4	0.999996	0.000401
47	6	5	5	1.000000	0.000004

Table for $N = 2$, $n = 1$, through $N = 100$, $n = 50$

N	n	k	x	P(x)	p(x)
47	6	6	0	0.418753	0.418753
47	6	6	1	0.837505	0.418753
47	6	6	2	0.978976	0.141470
47	6	6	3	0.998831	0.019855
47	6	6	4	0.999977	0.001146
47	6	6	5	1.000000	0.000023
47	6	6	6	1.000000	0.000000
47	7	1	0	0.851064	0.851064
47	7	1	1	1.000000	0.148936
47	7	2	0	0.721554	0.721554
47	7	2	1	0.980573	0.259019
47	7	2	2	1.000000	0.019426
47	7	3	0	0.609312	0.609312
47	7	3	1	0.946038	0.336725
47	7	3	2	0.997841	0.051804
47	7	3	3	1.000000	0.002158
47	7	4	0	0.512376	0.512376
47	7	4	1	0.900120	0.387744
47	7	4	2	0.991955	0.091834
47	7	4	3	0.999804	0.007849
47	7	4	4	1.000000	0.000196
47	7	5	0	0.428966	0.428966
47	7	5	1	0.846017	0.417050
47	7	5	2	0.981276	0.135260
47	7	5	3	0.999074	0.017797
47	7	5	4	0.999986	0.000913
47	7	5	5	1.000000	0.000014
47	7	6	0	0.357472	0.357472
47	7	6	1	0.786438	0.428966
47	7	6	2	0.965174	0.178736
47	7	6	3	0.997379	0.032205
47	7	6	4	0.999921	0.002542
47	7	6	5	0.999999	0.000078
47	7	6	6	1.000000	0.000001
47	7	7	0	0.296440	0.296440
47	7	7	1	0.723662	0.427222
47	7	7	2	0.943377	0.219714
47	7	7	3	0.994237	0.050860
47	7	7	4	0.999735	0.005498
47	7	7	5	0.999995	0.000260
47	7	7	6	1.000000	0.000004
47	7	7	7	1.000000	0.000000
47	8	1	0	0.829787	0.829787
47	8	1	1	1.000000	0.170213
47	8	2	0	0.685476	0.685476
47	8	2	1	0.974098	0.288622
47	8	2	2	1.000000	0.025902
47	8	3	0	0.563614	0.563614
47	8	3	1	0.929201	0.365587
47	8	3	2	0.996546	0.067345

N	n	k	x	P(x)	p(x)
47	8	3	3	1.000000	0.003454
47	8	4	0	0.461139	0.461139
47	8	4	1	0.871040	0.409901
47	8	4	2	0.987363	0.116323
47	8	4	3	0.999607	0.012245
47	8	4	4	1.000000	0.000392
47	8	5	0	0.375345	0.375345
47	8	5	1	0.804312	0.428966
47	8	5	2	0.971132	0.166820
47	8	5	3	0.998184	0.027052
47	8	5	4	0.999963	0.001780
47	8	5	5	1.000000	0.000037
47	8	6	0	0.303851	0.303851
47	8	6	1	0.732817	0.428966
47	8	6	2	0.947300	0.214483
47	8	6	3	0.994963	0.047663
47	8	6	4	0.999794	0.004831
47	8	6	5	0.999997	0.000203
47	8	6	6	1.000000	0.000003
47	8	7	0	0.244563	0.244563
47	8	7	1	0.659579	0.415016
47	8	7	2	0.915913	0.256333
47	8	7	3	0.989151	0.073238
47	8	7	4	0.999323	0.010172
47	8	7	5	0.999982	0.000660
47	8	7	6	1.000000	0.000017
47	8	7	7	1.000000	0.000000
47	8	8	0	0.195650	0.195650
47	8	8	1	0.586951	0.391301
47	8	8	2	0.877463	0.290511
47	8	8	3	0.979996	0.102533
47	8	8	4	0.998305	0.018310
47	8	8	5	0.999933	0.001628
47	8	8	6	0.999999	0.000066
47	8	8	7	1.000000	0.000001
47	8	8	8	1.000000	0.000000
47	9	1	0	0.808511	0.808511
47	9	1	1	1.000000	0.191489
47	9	2	0	0.650324	0.650324
47	9	2	1	0.966697	0.316374
47	9	2	2	1.000000	0.033302
47	9	3	0	0.520259	0.520259
47	9	3	1	0.910453	0.390194
47	9	3	2	0.994820	0.084366
47	9	3	3	1.000000	0.005180
47	9	4	0	0.413842	0.413842
47	9	4	1	0.839509	0.425666
47	9	4	2	0.981398	0.141889
47	9	4	3	0.999294	0.017896
47	9	4	4	1.000000	0.000706

N	n	k	x	P(x)	p(x)
47	9	5	0	0.327224	0.327224
47	9	5	1	0.760315	0.433091
47	9	5	2	0.958299	0.197984
47	9	5	3	0.996796	0.038497
47	9	5	4	0.999918	0.003121
47	9	5	5	1.000000	0.000082
47	9	6	0	0.257105	0.257105
47	9	6	1	0.677822	0.420717
47	9	6	2	0.925302	0.247480
47	9	6	3	0.991297	0.065995
47	9	6	4	0.999546	0.008249
47	9	6	5	0.999992	0.000446
47	9	6	6	1.000000	0.000008
47	9	7	0	0.200667	0.200667
47	9	7	1	0.595730	0.395063
47	9	7	2	0.883049	0.287319
47	9	7	3	0.981639	0.098590
47	9	7	4	0.998540	0.016901
47	9	7	5	0.999949	0.001408
47	9	7	6	0.999999	0.000051
47	9	7	7	1.000000	0.000001
47	9	8	0	0.155517	0.155517
47	9	8	1	0.516718	0.361201
47	9	8	2	0.832769	0.316051
47	9	8	3	0.966851	0.134082
47	9	8	4	0.996428	0.029577
47	9	8	5	0.999808	0.003380
47	9	8	6	0.999996	0.000188
47	9	8	7	1.000000	0.000004
47	9	8	8	1.000000	0.000000
47	9	9	0	0.119628	0.119628
47	9	9	1	0.442625	0.322997
47	9	9	2	0.776041	0.333416
47	9	9	3	0.946223	0.170181
47	9	9	4	0.992636	0.046413
47	9	9	5	0.999461	0.006825
47	9	9	6	0.999981	0.000520
47	9	9	7	1.000000	0.000019
47	9	9	8	1.000000	0.000000
47	9	9	9	1.000000	0.000000
47	10	1	0	0.787234	0.787234
47	10	1	1	1.000000	0.212766
47	10	2	0	0.616096	0.616096
47	10	2	1	0.958372	0.342276
47	10	2	2	1.000000	0.041628
47	10	3	0	0.479186	0.479186
47	10	3	1	0.889917	0.410731
47	10	3	2	0.992599	0.102683
47	10	3	3	1.000000	0.007401
47	10	4	0	0.370280	0.370280

N	n	k	x	P(x)	p(x)
47	10	4	1	0.805904	0.435624
47	10	4	2	0.973930	0.168026
47	10	4	3	0.998823	0.024893
47	10	4	4	1.000000	0.001177
47	10	5	0	0.284168	0.284168
47	10	5	1	0.714727	0.430558
47	10	5	2	0.942669	0.227943
47	10	5	3	0.994770	0.052101
47	10	5	4	0.999836	0.005065
47	10	5	5	1.000000	0.000164
47	10	6	0	0.216509	0.216509
47	10	6	1	0.622464	0.405955
47	10	6	2	0.899251	0.276787
47	10	6	3	0.986087	0.086835
47	10	6	4	0.999112	0.013025
47	10	6	5	0.999980	0.000868
47	10	6	6	1.000000	0.000020
47	10	7	0	0.163702	0.163702
47	10	7	1	0.533352	0.369650
47	10	7	2	0.845244	0.311892
47	10	7	3	0.971261	0.126017
47	10	7	4	0.997206	0.025945
47	10	7	5	0.999874	0.002669
47	10	7	6	0.999998	0.000124
47	10	7	7	1.000000	0.000002
47	10	8	0	0.122777	0.122777
47	10	8	1	0.450181	0.327404
47	10	8	2	0.782866	0.332685
47	10	8	3	0.949208	0.166342
47	10	8	4	0.993314	0.044106
47	10	8	5	0.999541	0.006227
47	10	8	6	0.999986	0.000445
47	10	8	7	1.000000	0.000014
47	10	8	8	1.000000	0.000000
47	10	9	0	0.091295	0.091295
47	10	9	1	0.374626	0.283331
47	10	9	2	0.714623	0.339997
47	10	9	3	0.919352	0.204729
47	10	9	4	0.986529	0.067177
47	10	9	5	0.998743	0.012214
47	10	9	6	0.999940	0.001197
47	10	9	7	0.999999	0.000059
47	10	9	8	1.000000	0.000001
47	10	9	9	1.000000	0.000000
47	10	10	0	0.067270	0.067270
47	10	10	1	0.307521	0.240251
47	10	10	2	0.643044	0.335523
47	10	10	3	0.881639	0.238594
47	10	10	4	0.975922	0.094283
47	10	10	5	0.997135	0.021214

Table for $N = 2$, $n = 1$, through $N = 100$, $n = 50$

N	n	k	x	P(x)	p(x)
47	10	10	6	0.999814	0.002678
47	10	10	7	0.999994	0.000180
47	10	10	8	1.000000	0.000006
47	10	10	9	1.000000	0.000000
47	10	10	10	1.000000	0.000000
47	11	1	0	0.765957	0.765957
47	11	1	1	1.000000	0.234043
47	11	2	0	0.582794	0.582794
47	11	2	1	0.949121	0.366327
47	11	2	2	1.000000	0.050879
47	11	3	0	0.440333	0.440333
47	11	3	1	0.867715	0.427382
47	11	3	2	0.989824	0.122109
47	11	3	3	1.000000	0.010176
47	11	4	0	0.330250	0.330250
47	11	4	1	0.770583	0.440333
47	11	4	2	0.964847	0.194265
47	11	4	3	0.998150	0.033302
47	11	4	4	1.000000	0.001850
47	11	5	0	0.245767	0.245767
47	11	5	1	0.668180	0.422412
47	11	5	2	0.924187	0.256008
47	11	5	3	0.991954	0.067767
47	11	5	4	0.999699	0.007745
47	11	5	5	1.000000	0.000301
47	11	6	0	0.181400	0.181400
47	11	6	1	0.567605	0.386206
47	11	6	2	0.869329	0.301723
47	11	6	3	0.979046	0.109718
47	11	6	4	0.998408	0.019362
47	11	6	5	0.999957	0.001549
47	11	6	6	1.000000	0.000043
47	11	7	0	0.132731	0.132731
47	11	7	1	0.473409	0.340677
47	11	7	2	0.803097	0.329688
47	11	7	3	0.957638	0.154541
47	11	7	4	0.995102	0.037465
47	11	7	5	0.999730	0.004628
47	11	7	6	0.999995	0.000264
47	11	7	7	1.000000	0.000005
47	11	8	0	0.096230	0.096230
47	11	8	1	0.388239	0.292009
47	11	8	2	0.728917	0.340677
47	11	8	3	0.926730	0.197813
47	11	8	4	0.988546	0.061816
47	11	8	5	0.999036	0.010490
47	11	8	6	0.999962	0.000926
47	11	8	7	0.999999	0.000038
47	11	8	8	1.000000	0.000001
47	11	9	0	0.069088	0.069088

N	n	k	x	P(x)	p(x)
47	11	9	1	0.313365	0.244277
47	11	9	2	0.650299	0.336934
47	11	9	3	0.886153	0.235854
47	11	9	4	0.977451	0.091298
47	11	9	5	0.997422	0.019971
47	11	9	6	0.999843	0.002421
47	11	9	7	0.999996	0.000153
47	11	9	8	1.000000	0.000004
47	11	9	9	1.000000	0.000000
47	11	10	0	0.049089	0.049089
47	11	10	1	0.249082	0.199993
47	11	10	2	0.570499	0.321417
47	11	10	3	0.836499	0.266000
47	11	10	4	0.960633	0.124133
47	11	10	5	0.994269	0.033636
47	11	10	6	0.999524	0.005256
47	11	10	7	0.999979	0.000455
47	11	10	8	1.000000	0.000020
47	11	10	9	1.000000	0.000000
47	11	10	10	1.000000	0.000000
47	11	11	0	0.034495	0.034495
47	11	11	1	0.195030	0.160535
47	11	11	2	0.492316	0.297287
47	11	11	3	0.778986	0.286669
47	11	11	4	0.937148	0.158162
47	11	11	5	0.988814	0.051666
47	11	11	6	0.998814	0.010000
47	11	11	7	0.999930	0.001116
47	11	11	8	0.999998	0.000068
47	11	11	9	1.000000	0.000002
47	11	11	10	1.000000	0.000000
47	11	11	11	1.000000	0.000000
47	12	1	0	0.744681	0.744681
47	12	1	1	1.000000	0.255319
47	12	2	0	0.550416	0.550416
47	12	2	1	0.938945	0.388529
47	12	2	2	1.000000	0.061055
47	12	3	0	0.403639	0.403639
47	12	3	1	0.843972	0.440333
47	12	3	2	0.986432	0.142461
47	12	3	3	1.000000	0.013568
47	12	4	0	0.293555	0.293555
47	12	4	1	0.733888	0.440333
47	12	4	2	0.954055	0.220166
47	12	4	3	0.997225	0.043170
47	12	4	4	1.000000	0.002775
47	12	5	0	0.211633	0.211633
47	12	5	1	0.621245	0.409612
47	12	5	2	0.902853	0.281608
47	12	5	3	0.988189	0.085336

N	n	k	x	P(x)	p(x)
47	12	5	4	0.999484	0.011294
47	12	5	5	1.000000	0.000516
47	12	6	0	0.151166	0.151166
47	12	6	1	0.513966	0.362799
47	12	6	2	0.835804	0.321838
47	12	6	3	0.969903	0.134099
47	12	6	4	0.997332	0.027429
47	12	6	5	0.999914	0.002582
47	12	6	6	1.000000	0.000086
47	12	7	0	0.106923	0.106923
47	12	7	1	0.416629	0.309707
47	12	7	2	0.757307	0.340677
47	12	7	3	0.940467	0.183160
47	12	7	4	0.991980	0.051514
47	12	7	5	0.999473	0.007493
47	12	7	6	0.999987	0.000514
47	12	7	7	1.000000	0.000013
47	12	8	0	0.074846	0.074846
47	12	8	1	0.331460	0.256614
47	12	8	2	0.672137	0.340677
47	12	8	3	0.899256	0.227118
47	12	8	4	0.981677	0.082422
47	12	8	5	0.998162	0.016484
47	12	8	6	0.999910	0.001748
47	12	8	7	0.999998	0.000088
47	12	8	8	1.000000	0.000002
47	12	9	0	0.051816	0.051816
47	12	9	1	0.259082	0.207265
47	12	9	2	0.584784	0.325703
47	12	9	3	0.846844	0.262060
47	12	9	4	0.964770	0.117927
47	12	9	5	0.995203	0.030433
47	12	9	6	0.999641	0.004438
47	12	9	7	0.999987	0.000346
47	12	9	8	1.000000	0.000013
47	12	9	9	1.000000	0.000000
47	12	10	0	0.035453	0.035453
47	12	10	1	0.199084	0.163630
47	12	10	2	0.499073	0.299989
47	12	10	3	0.784777	0.285704
47	12	10	4	0.939944	0.155167
47	12	10	5	0.989597	0.049653
47	12	10	6	0.998940	0.009343
47	12	10	7	0.999942	0.001001
47	12	10	8	0.999998	0.000057
47	12	10	9	1.000000	0.000001
47	12	10	10	1.000000	0.000000
47	12	11	0	0.023955	0.023955
47	12	11	1	0.150437	0.126482
47	12	11	2	0.417995	0.267558

N	n	k	x	P(x)	p(x)
47	12	11	3	0.715281	0.297287
47	12	11	4	0.906394	0.191113
47	12	11	5	0.980203	0.073809
47	12	11	6	0.997425	0.017222
47	12	11	7	0.999806	0.002381
47	12	11	8	0.999992	0.000186
47	12	11	9	1.000000	0.000008
47	12	11	10	1.000000	0.000000
47	12	11	11	1.000000	0.000000
47	12	12	0	0.015970	0.015970
47	12	12	1	0.111790	0.095820
47	12	12	2	0.343673	0.231884
47	12	12	3	0.640960	0.297287
47	12	12	4	0.863925	0.222965
47	12	12	5	0.965851	0.101927
47	12	12	6	0.994555	0.028704
47	12	12	7	0.999476	0.004921
47	12	12	8	0.999972	0.000496
47	12	12	9	0.999999	0.000028
47	12	12	10	1.000000	0.000001
47	12	12	11	1.000000	0.000000
47	12	12	12	1.000000	0.000000
47	13	1	0	0.723404	0.723404
47	13	1	1	1.000000	0.276596
47	13	2	0	0.518964	0.518964
47	13	2	1	0.927845	0.408881
47	13	2	2	1.000000	0.072155
47	13	3	0	0.369041	0.369041
47	13	3	1	0.818810	0.449769
47	13	3	2	0.982362	0.163552
47	13	3	3	1.000000	0.017638
47	13	4	0	0.260006	0.260006
47	13	4	1	0.696145	0.436139
47	13	4	2	0.941474	0.245328
47	13	4	3	0.995991	0.054517
47	13	4	4	1.000000	0.004009
47	13	5	0	0.181400	0.181400
47	13	5	1	0.574432	0.393033
47	13	5	2	0.878715	0.304283
47	13	5	3	0.983313	0.104597
47	13	5	4	0.999161	0.015848
47	13	5	5	1.000000	0.000839
47	13	6	0	0.125252	0.125252
47	13	6	1	0.462137	0.336885
47	13	6	2	0.799022	0.336885
47	13	6	3	0.958409	0.159386
47	13	6	4	0.995765	0.037356
47	13	6	5	0.999840	0.004075
47	13	6	6	1.000000	0.000160
47	13	7	0	0.085538	0.085538

Table for $N=2$, $n=1$, through $N=100$, $n=50$

N	n	k	x	P(x)	p(x)
47	13	7	1	0.363537	0.277999
47	13	7	2	0.708638	0.345102
47	13	7	3	0.919534	0.210896
47	13	7	4	0.987565	0.068031
47	13	7	5	0.999045	0.011480
47	13	7	6	0.999973	0.000928
47	13	7	7	1.000000	0.000027
47	13	8	0	0.057738	0.057738
47	13	8	1	0.280137	0.222399
47	13	8	2	0.613735	0.333598
47	13	8	3	0.866810	0.253075
47	13	8	4	0.972258	0.105448
47	13	8	5	0.996749	0.024491
47	13	8	6	0.999810	0.003061
47	13	8	7	0.999996	0.000186
47	13	8	8	1.000000	0.000004
47	13	9	0	0.038492	0.038492
47	13	9	1	0.211707	0.173215
47	13	9	2	0.519644	0.307937
47	13	9	3	0.801919	0.282276
47	13	9	4	0.947924	0.146005
47	13	9	5	0.991725	0.043801
47	13	9	6	0.999261	0.007536
47	13	9	7	0.999967	0.000706
47	13	9	8	0.999999	0.000032
47	13	9	9	1.000000	0.000001
47	13	10	0	0.025324	0.025324
47	13	10	1	0.157007	0.131684
47	13	10	2	0.430504	0.273497
47	13	10	3	0.727636	0.297132
47	13	10	4	0.913344	0.185708
47	13	10	5	0.982504	0.069160
47	13	10	6	0.997873	0.015369
47	13	10	7	0.999856	0.001983
47	13	10	8	0.999995	0.000139
47	13	10	9	1.000000	0.000005
47	13	10	10	1.000000	0.000000
47	13	11	0	0.016426	0.016426
47	13	11	1	0.114299	0.097873
47	13	11	2	0.349194	0.234895
47	13	11	3	0.647330	0.298136
47	13	11	4	0.868172	0.220841
47	13	11	5	0.967550	0.099379
47	13	11	6	0.994965	0.027415
47	13	11	7	0.999534	0.004569
47	13	11	8	0.999976	0.000442
47	13	11	9	0.999999	0.000023
47	13	11	10	1.000000	0.000001
47	13	11	11	1.000000	0.000000
47	13	12	0	0.010495	0.010495

N	n	k	x	P(x)	p(x)
47	13	12	1	0.081675	0.071180
47	13	12	2	0.277421	0.195746
47	13	12	3	0.564515	0.287094
47	13	12	4	0.812961	0.248447
47	13	12	5	0.945466	0.132505
47	13	12	6	0.989634	0.044168
47	13	12	7	0.998773	0.009138
47	13	12	8	0.999915	0.001142
47	13	12	9	0.999997	0.000082
47	13	12	10	1.000000	0.000003
47	13	12	11	1.000000	0.000000
47	13	12	12	1.000000	0.000000
47	13	13	0	0.006597	0.006597
47	13	13	1	0.057270	0.050674
47	13	13	2	0.215901	0.158630
47	13	13	3	0.482488	0.266587
47	13	13	4	0.749075	0.266587
47	13	13	5	0.915179	0.166104
47	13	13	6	0.980801	0.065621
47	13	13	7	0.997206	0.016405
47	13	13	8	0.999752	0.002546
47	13	13	9	0.999987	0.000236
47	13	13	10	1.000000	0.000012
47	13	13	11	1.000000	0.000000
47	13	13	12	1.000000	0.000000
47	13	13	13	1.000000	0.000000
47	14	1	0	0.702128	0.702128
47	14	1	1	1.000000	0.297872
47	14	2	0	0.488437	0.488437
47	14	2	1	0.915819	0.427382
47	14	2	2	1.000000	0.084181
47	14	3	0	0.336479	0.336479
47	14	3	1	0.792353	0.455874
47	14	3	2	0.977552	0.185199
47	14	3	3	1.000000	0.022448
47	14	4	0	0.229417	0.229417
47	14	4	1	0.657663	0.428245
47	14	4	2	0.927043	0.269380
47	14	4	3	0.994388	0.067345
47	14	4	4	1.000000	0.005612
47	14	5	0	0.154723	0.154723
47	14	5	1	0.528193	0.373470
47	14	5	2	0.851867	0.323674
47	14	5	3	0.977160	0.125293
47	14	5	4	0.998695	0.021535
47	14	5	5	1.000000	0.001305
47	14	6	0	0.103149	0.103149
47	14	6	1	0.412595	0.309446
47	14	6	2	0.759389	0.346793
47	14	6	3	0.944345	0.184956

N	n	k	x	P(x)	p(x)
47	14	6	4	0.993567	0.049222
47	14	6	5	0.999720	0.006153
47	14	6	6	1.000000	0.000280
47	14	7	0	0.067927	0.067927
47	14	7	1	0.314478	0.246551
47	14	7	2	0.657888	0.343410
47	14	7	3	0.894723	0.236835
47	14	7	4	0.981562	0.086839
47	14	7	5	0.998370	0.016808
47	14	7	6	0.999945	0.001576
47	14	7	7	1.000000	0.000055
47	14	8	0	0.044153	0.044153
47	14	8	1	0.234349	0.190196
47	14	8	2	0.554865	0.320516
47	14	8	3	0.829593	0.274728
47	14	8	4	0.959852	0.130259
47	14	8	5	0.994588	0.034736
47	14	8	6	0.999630	0.005042
47	14	8	7	0.999990	0.000360
47	14	8	8	1.000000	0.000010
47	14	9	0	0.028303	0.028303
47	14	9	1	0.170950	0.142647
47	14	9	2	0.456245	0.285295
47	14	9	3	0.752106	0.295861
47	14	9	4	0.926452	0.174347
47	14	9	5	0.986572	0.060120
47	14	9	6	0.998596	0.012024
47	14	9	7	0.999926	0.001330
47	14	9	8	0.999998	0.000073
47	14	9	9	1.000000	0.000001
47	14	10	0	0.017876	0.017876
47	14	10	1	0.122150	0.104274
47	14	10	2	0.366152	0.244002
47	14	10	3	0.666462	0.300310
47	14	10	4	0.880572	0.214110
47	14	10	5	0.972333	0.091761
47	14	10	6	0.996065	0.023731
47	14	10	7	0.999681	0.003616
47	14	10	8	0.999987	0.000306
47	14	10	9	1.000000	0.000013
47	14	10	10	1.000000	0.000000
47	14	11	0	0.011112	0.011112
47	14	11	1	0.085513	0.074401
47	14	11	2	0.287016	0.201503
47	14	11	3	0.577180	0.290164
47	14	11	4	0.822704	0.245524
47	14	11	5	0.950013	0.127309
47	14	11	6	0.990933	0.040921
47	14	11	7	0.998997	0.008063
47	14	11	8	0.999937	0.000941

N	n	k	x	P(x)	p(x)
47	14	11	9	0.999998	0.000061
47	14	11	10	1.000000	0.000002
47	14	11	11	1.000000	0.000000
47	14	12	0	0.006791	0.006791
47	14	12	1	0.058646	0.051855
47	14	12	2	0.219848	0.161202
47	14	12	3	0.488519	0.268671
47	14	12	4	0.754503	0.265984
47	14	12	5	0.918186	0.163682
47	14	12	6	0.981840	0.063654
47	14	12	7	0.997429	0.015589
47	14	12	8	0.999781	0.002352
47	14	12	9	0.999990	0.000209
47	14	12	10	1.000000	0.000010
47	14	12	11	1.000000	0.000000
47	14	12	12	1.000000	0.000000
47	14	13	0	0.004074	0.004074
47	14	13	1	0.039385	0.035311
47	14	13	2	0.164579	0.125194
47	14	13	3	0.404080	0.239501
47	14	13	4	0.678508	0.274428
47	14	13	5	0.876096	0.197588
47	14	13	6	0.967290	0.091195
47	14	13	7	0.994311	0.027021
47	14	13	8	0.999377	0.005066
47	14	13	9	0.999960	0.000582
47	14	13	10	0.999999	0.000039
47	14	13	11	1.000000	0.000001
47	14	13	12	1.000000	0.000000
47	14	13	13	1.000000	0.000000
47	14	14	0	0.002397	0.002397
47	14	14	1	0.025884	0.023487
47	14	14	2	0.120393	0.094509
47	14	14	3	0.326594	0.206201
47	14	14	4	0.597794	0.271199
47	14	14	5	0.823793	0.226000
47	14	14	6	0.945833	0.122040
47	14	14	7	0.988748	0.042915
47	14	14	8	0.998483	0.009735
47	14	14	9	0.999874	0.001391
47	14	14	10	0.999994	0.000120
47	14	14	11	1.000000	0.000006
47	14	14	12	1.000000	0.000000
47	14	14	13	1.000000	0.000000
47	14	14	14	1.000000	0.000000
47	15	1	0	0.680851	0.680851
47	15	1	1	1.000000	0.319149
47	15	2	0	0.458834	0.458834
47	15	2	1	0.902868	0.444033
47	15	2	2	1.000000	0.097132

Table for $N = 2$, $n = 1$, through $N = 100$, $n = 50$

N	n	k	x	$P(x)$	$p(x)$
47	15	3	0	0.305890	0.305890
47	15	3	1	0.764724	0.458834
47	15	3	2	0.971939	0.207216
47	15	3	3	1.000000	0.028060
47	15	4	0	0.201609	0.201609
47	15	4	1	0.618731	0.417122
47	15	4	2	0.910717	0.291985
47	15	4	3	0.992347	0.081630
47	15	4	4	1.000000	0.007653
47	15	5	0	0.131280	0.131280
47	15	5	1	0.482924	0.351644
47	15	5	2	0.822442	0.339518
47	15	5	3	0.969567	0.147124
47	15	5	4	0.998042	0.028476
47	15	5	5	1.000000	0.001958
47	15	6	0	0.084394	0.084394
47	15	6	1	0.365709	0.281315
47	15	6	2	0.717353	0.351644
47	15	6	3	0.927531	0.210178
47	15	6	4	0.990584	0.063053
47	15	6	5	0.999534	0.008950
47	15	6	6	1.000000	0.000466
47	15	7	0	0.053518	0.053518
47	15	7	1	0.269651	0.216132
47	15	7	2	0.605856	0.336206
47	15	7	3	0.866015	0.260159
47	15	7	4	0.973668	0.107652
47	15	7	5	0.997351	0.023683
47	15	7	6	0.999898	0.002547
47	15	7	7	1.000000	0.000102
47	15	8	0	0.033449	0.033449
47	15	8	1	0.194004	0.160555
47	15	8	2	0.496589	0.302585
47	15	8	3	0.787968	0.291378
47	15	8	4	0.944063	0.156095
47	15	8	5	0.991430	0.047367
47	15	8	6	0.999325	0.007894
47	15	8	7	0.999979	0.000655
47	15	8	8	1.000000	0.000020
47	15	9	0	0.020584	0.020584
47	15	9	1	0.136369	0.115785
47	15	9	2	0.395728	0.259359
47	15	9	3	0.698313	0.302585
47	15	9	4	0.900036	0.201723
47	15	9	5	0.979285	0.079248
47	15	9	6	0.997503	0.018218
47	15	9	7	0.999845	0.002342
47	15	9	8	0.999996	0.000151
47	15	9	9	1.000000	0.000004
47	15	10	0	0.012459	0.012459
47	15	10	1	0.093711	0.081253
47	15	10	2	0.307000	0.213288
47	15	10	3	0.602760	0.295760
47	15	10	4	0.841643	0.238883
47	15	10	5	0.958430	0.116787
47	15	10	6	0.993188	0.034758
47	15	10	7	0.999352	0.006164
47	15	10	8	0.999968	0.000616
47	15	10	9	0.999999	0.000031
47	15	10	10	1.000000	0.000001
47	15	11	0	0.007408	0.007408
47	15	11	1	0.062967	0.055559
47	15	11	2	0.232061	0.169093
47	15	11	3	0.506838	0.274777
47	15	11	4	0.770623	0.263786
47	15	11	5	0.926866	0.156242
47	15	11	6	0.984733	0.057868
47	15	11	7	0.998019	0.013286
47	15	11	8	0.999852	0.001833
47	15	11	9	0.999994	0.000143
47	15	11	10	1.000000	0.000006
47	15	11	11	1.000000	0.000000
47	15	12	0	0.004321	0.004321
47	15	12	1	0.041361	0.037040
47	15	12	2	0.170999	0.129638
47	15	12	3	0.415245	0.244246
47	15	12	4	0.690022	0.274777
47	15	12	5	0.883465	0.193443
47	15	12	6	0.970266	0.086801
47	15	12	7	0.995067	0.024800
47	15	12	8	0.999495	0.004429
47	15	12	9	0.999971	0.000475
47	15	12	10	0.999999	0.000029
47	15	12	11	1.000000	0.000001
47	15	12	12	1.000000	0.000000
47	15	13	0	0.002469	0.002469
47	15	13	1	0.026545	0.024076
47	15	13	2	0.122848	0.096303
47	15	13	3	0.331504	0.208656
47	15	13	4	0.603664	0.272160
47	15	13	5	0.828196	0.224532
47	15	13	6	0.947946	0.119750
47	15	13	7	0.989398	0.041452
47	15	13	8	0.998610	0.009212
47	15	13	9	0.999889	0.001279
47	15	13	10	0.999995	0.000106
47	15	13	11	1.000000	0.000005
47	15	13	12	1.000000	0.000000
47	15	13	13	1.000000	0.000000
47	15	14	0	0.001380	0.001380
47	15	14	1	0.016631	0.015252
47	15	14	2	0.086026	0.069395
47	15	14	3	0.257860	0.171834
47	15	14	4	0.515612	0.257752
47	15	14	5	0.762157	0.246545
47	15	14	6	0.916247	0.154091
47	15	14	7	0.979645	0.063397
47	15	14	8	0.996713	0.017068
47	15	14	9	0.999663	0.002950
47	15	14	10	0.999979	0.000316
47	15	14	11	0.999999	0.000020
47	15	14	12	1.000000	0.000001
47	15	14	13	1.000000	0.000000
47	15	14	14	1.000000	0.000000
47	15	15	0	0.000753	0.000753
47	15	15	1	0.010161	0.009408
47	15	15	2	0.058689	0.048528
47	15	15	3	0.195375	0.136686
47	15	15	4	0.429695	0.234320
47	15	15	5	0.687446	0.257752
47	15	15	6	0.874223	0.186776
47	15	15	7	0.964276	0.090053
47	15	15	8	0.993093	0.028817
47	15	15	9	0.999127	0.006034
47	15	15	10	0.999931	0.000805
47	15	15	11	0.999997	0.000065
47	15	15	12	1.000000	0.000003
47	15	15	13	1.000000	0.000000
47	15	15	14	1.000000	0.000000
47	15	15	15	1.000000	0.000000
47	16	1	0	0.659574	0.659574
47	16	1	1	1.000000	0.340426
47	16	2	0	0.430157	0.430157
47	16	2	1	0.888992	0.458834
47	16	2	2	1.000000	0.111008
47	16	3	0	0.277212	0.277212
47	16	3	1	0.736047	0.458834
47	16	3	2	0.965464	0.229417
47	16	3	3	1.000000	0.034536
47	16	4	0	0.176408	0.176408
47	16	4	1	0.579626	0.403218
47	16	4	2	0.892468	0.312842
47	16	4	3	0.989796	0.097329
47	16	4	4	1.000000	0.010204
47	16	5	0	0.110768	0.110768
47	16	5	1	0.438969	0.328201
47	16	5	2	0.790612	0.351644
47	16	5	3	0.960371	0.169759
47	16	5	4	0.997152	0.036781
47	16	5	5	1.000000	0.002848
47	16	6	0	0.068571	0.068571
47	16	6	1	0.321754	0.253183
47	16	6	2	0.673398	0.351644
47	16	6	3	0.907827	0.234429
47	16	6	4	0.986644	0.078817
47	16	6	5	0.999254	0.012611
47	16	6	6	1.000000	0.000746
47	16	7	0	0.041811	0.041811
47	16	7	1	0.229126	0.187315
47	16	7	2	0.553324	0.324198
47	16	7	3	0.833496	0.280171
47	16	7	4	0.963575	0.130080
47	16	7	5	0.995871	0.032296
47	16	7	6	0.999818	0.003947
47	16	7	7	1.000000	0.000182
47	16	8	0	0.025087	0.025087
47	16	8	1	0.158883	0.133796
47	16	8	2	0.439855	0.280972
47	16	8	3	0.742440	0.302585
47	16	8	4	0.924551	0.182111
47	16	8	5	0.986989	0.062438
47	16	8	6	0.998831	0.011842
47	16	8	7	0.999959	0.001128
47	16	8	8	1.000000	0.000041
47	16	9	0	0.014795	0.014795
47	16	9	1	0.107423	0.092628
47	16	9	2	0.338993	0.231570
47	16	9	3	0.641578	0.302585
47	16	9	4	0.868517	0.226939
47	16	9	5	0.969379	0.100862
47	16	9	6	0.995795	0.026416
47	16	9	7	0.999699	0.003904
47	16	9	8	0.999991	0.000293
47	16	9	9	1.000000	0.000008
47	16	10	0	0.008565	0.008565
47	16	10	1	0.070859	0.062294
47	16	10	2	0.253678	0.182819
47	16	10	3	0.538062	0.284384
47	16	10	4	0.796852	0.258790
47	16	10	5	0.940182	0.143330
47	16	10	6	0.988843	0.048661
47	16	10	7	0.998774	0.009931
47	16	10	8	0.999930	0.001156
47	16	10	9	0.999998	0.000068
47	16	10	10	1.000000	0.000002
47	16	11	0	0.004861	0.004861
47	16	11	1	0.045605	0.040743
47	16	11	2	0.184503	0.138898
47	16	11	3	0.438143	0.253640
47	16	11	4	0.712920	0.274777

N	n	k	x	P(x)	p(x)
47	16	11	5	0.897570	0.184650
47	16	11	6	0.975692	0.078121
47	16	11	7	0.996359	0.020667
47	16	11	8	0.999680	0.003321
47	16	11	9	0.999985	0.000305
47	16	11	10	1.000000	0.000014
47	16	11	11	1.000000	0.000000
47	16	12	0	0.002701	0.002701
47	16	12	1	0.028628	0.025928
47	16	12	2	0.130487	0.101859
47	16	12	3	0.346551	0.216064
47	16	12	4	0.621328	0.274777
47	16	12	5	0.841150	0.219822
47	16	12	6	0.953991	0.112842
47	16	12	7	0.991192	0.037201
47	16	12	8	0.998942	0.007750
47	16	12	9	0.999926	0.000984
47	16	12	10	0.999997	0.000071
47	16	12	11	1.000000	0.000003
47	16	12	12	1.000000	0.000000
47	16	13	0	0.001466	0.001466
47	16	13	1	0.017517	0.016050
47	16	13	2	0.089744	0.072227
47	16	13	3	0.266299	0.176555
47	16	13	4	0.527119	0.260820
47	16	13	5	0.772063	0.244944
47	16	13	6	0.921751	0.149688
47	16	13	7	0.981626	0.059875
47	16	13	8	0.997170	0.015545
47	16	13	9	0.999729	0.002559
47	16	13	10	0.999985	0.000256
47	16	13	11	1.000000	0.000014
47	16	13	12	1.000000	0.000000
47	16	13	13	1.000000	0.000000
47	16	14	0	0.000776	0.000776
47	16	14	1	0.010436	0.009659
47	16	14	2	0.060003	0.049568
47	16	14	3	0.198792	0.138789
47	16	14	4	0.435065	0.236272
47	16	14	5	0.692816	0.257752
47	16	14	6	0.877725	0.184909
47	16	14	7	0.965777	0.088052
47	16	14	8	0.993513	0.027736
47	16	14	9	0.999202	0.005689
47	16	14	10	0.999940	0.000738
47	16	14	11	0.999997	0.000057
47	16	14	12	1.000000	0.000002
47	16	14	13	1.000000	0.000000
47	16	14	14	1.000000	0.000000
47	16	15	0	0.000400	0.000400

N	n	k	x	P(x)	p(x)
47	16	15	1	0.006045	0.005645
47	16	15	2	0.038974	0.032930
47	16	15	3	0.144118	0.105143
47	16	15	4	0.349147	0.205030
47	16	15	5	0.606899	0.257752
47	16	15	6	0.821692	0.214793
47	16	15	7	0.941762	0.120071
47	16	15	8	0.986789	0.045026
47	16	15	9	0.997996	0.011207
47	16	15	10	0.999806	0.001810
47	16	15	11	0.999989	0.000183
47	16	15	12	1.000000	0.000011
47	16	15	13	1.000000	0.000000
47	16	15	14	1.000000	0.000000
47	16	15	15	1.000000	0.000000
47	16	16	0	0.000200	0.000200
47	16	16	1	0.003399	0.003199
47	16	16	2	0.024568	0.021169
47	16	16	3	0.101403	0.076836
47	16	16	4	0.272261	0.170858
47	16	16	5	0.518297	0.246036
47	16	16	6	0.754569	0.236272
47	16	16	7	0.907993	0.153424
47	16	16	8	0.975532	0.067540
47	16	16	9	0.995544	0.020012
47	16	16	10	0.999466	0.003922
47	16	16	11	0.999960	0.000494
47	16	16	12	0.999998	0.000038
47	16	16	13	1.000000	0.000002
47	16	16	14	1.000000	0.000000
47	16	16	15	1.000000	0.000000
47	16	16	16	1.000000	0.000000
47	17	1	0	0.638298	0.638298
47	17	1	1	1.000000	0.361702
47	17	2	0	0.402405	0.402405
47	17	2	1	0.874191	0.471785
47	17	2	2	1.000000	0.125809
47	17	3	0	0.250385	0.250385
47	17	3	1	0.706445	0.456059
47	17	3	2	0.958063	0.251619
47	17	3	3	1.000000	0.041936
47	17	4	0	0.153646	0.153646
47	17	4	1	0.540605	0.386959
47	17	4	2	0.872284	0.331679
47	17	4	3	0.986656	0.114372
47	17	4	4	1.000000	0.013343
47	17	5	0	0.092902	0.092902
47	17	5	1	0.396620	0.303718
47	17	5	2	0.756582	0.359962
47	17	5	3	0.949419	0.192837

N	n	k	x	P(x)	p(x)
47	17	5	4	0.995966	0.046547
47	17	5	5	1.000000	0.004034
47	17	6	0	0.055299	0.055299
47	17	6	1	0.280918	0.225619
47	17	6	2	0.628024	0.347106
47	17	6	3	0.885140	0.257116
47	17	6	4	0.981558	0.096418
47	17	6	5	0.998847	0.017289
47	17	6	6	1.000000	0.001153
47	17	7	0	0.032370	0.032370
47	17	7	1	0.192871	0.160501
47	17	7	2	0.501034	0.308163
47	17	7	3	0.797344	0.296310
47	17	7	4	0.950987	0.153642
47	17	7	5	0.993787	0.042800
47	17	7	6	0.999691	0.005903
47	17	7	7	1.000000	0.000309
47	17	8	0	0.018613	0.018613
47	17	8	1	0.128671	0.110058
47	17	8	2	0.385473	0.256802
47	17	8	3	0.693636	0.308163
47	17	8	4	0.901053	0.207417
47	17	8	5	0.980947	0.079894
47	17	8	6	0.998067	0.017120
47	17	8	7	0.999923	0.001855
47	17	8	8	1.000000	0.000077
47	17	9	0	0.010500	0.010500
47	17	9	1	0.083519	0.073019
47	17	9	2	0.286703	0.203184
47	17	9	3	0.583013	0.296310
47	17	9	4	0.831914	0.248901
47	17	9	5	0.956364	0.124450
47	17	9	6	0.993238	0.036874
47	17	9	7	0.999447	0.006208
47	17	9	8	0.999982	0.000535
47	17	9	9	1.000000	0.000018
47	17	10	0	0.005802	0.005802
47	17	10	1	0.052774	0.046971
47	17	10	2	0.206499	0.153725
47	17	10	3	0.473846	0.267348
47	17	10	4	0.746764	0.272917
47	17	10	5	0.917064	0.170300
47	17	10	6	0.982564	0.065500
47	17	10	7	0.997813	0.015249
47	17	10	8	0.999855	0.002042
47	17	10	9	0.999996	0.000141
47	17	10	10	1.000000	0.000004
47	17	11	0	0.003136	0.003136
47	17	11	1	0.032462	0.029325
47	17	11	2	0.144178	0.111716

N	n	k	x	P(x)	p(x)
47	17	11	3	0.372688	0.228510
47	17	11	4	0.650874	0.278186
47	17	11	5	0.861832	0.210958
47	17	11	6	0.963091	0.101260
47	17	11	7	0.993692	0.030600
47	17	11	8	0.999359	0.005667
47	17	11	9	0.999966	0.000607
47	17	11	10	0.999999	0.000033
47	17	11	11	1.000000	0.000001
47	17	12	0	0.001655	0.001655
47	17	12	1	0.019428	0.017773
47	17	12	2	0.097629	0.078201
47	17	12	3	0.283823	0.186193
47	17	12	4	0.550418	0.266595
47	17	12	5	0.791512	0.241095
47	17	12	6	0.932151	0.140638
47	17	12	7	0.985192	0.053041
47	17	12	8	0.997942	0.012750
47	17	12	9	0.999831	0.001889
47	17	12	10	0.999993	0.000162
47	17	12	11	1.000000	0.000007
47	17	12	12	1.000000	0.000000
47	17	13	0	0.000851	0.000851
47	17	13	1	0.011304	0.010452
47	17	13	2	0.064115	0.052811
47	17	13	3	0.209345	0.145231
47	17	13	4	0.451397	0.242051
47	17	13	5	0.708851	0.257455
47	17	13	6	0.887950	0.179099
47	17	13	7	0.970037	0.082087
47	17	13	8	0.994663	0.024626
47	17	13	9	0.999399	0.004736
47	17	13	10	0.999960	0.000561
47	17	13	11	0.999999	0.000038
47	17	13	12	1.000000	0.000001
47	17	13	13	1.000000	0.000000
47	17	14	0	0.000426	0.000426
47	17	14	1	0.006385	0.005959
47	17	14	2	0.040816	0.034431
47	17	14	3	0.149545	0.108729
47	17	14	4	0.358848	0.209303
47	17	14	5	0.617985	0.259137
47	17	14	6	0.830006	0.212021
47	17	14	7	0.945894	0.115887
47	17	14	8	0.988145	0.042251
47	17	14	9	0.998285	0.010140
47	17	14	10	0.999845	0.001560
47	17	14	11	0.999992	0.000147
47	17	14	12	1.000000	0.000008
47	17	14	13	1.000000	0.000000

N	n	k	x	P(x)	p(x)
47	17	14	14	1.000000	0.000000
47	17	15	0	0.000206	0.000206
47	17	15	1	0.003496	0.003289
47	17	15	2	0.025165	0.021670
47	17	15	3	0.103417	0.078252
47	17	15	4	0.276395	0.172978
47	17	15	5	0.523753	0.247358
47	17	15	6	0.759333	0.235579
47	17	15	7	0.910777	0.151444
47	17	15	8	0.976622	0.065845
47	17	15	9	0.995827	0.019205
47	17	15	10	0.999514	0.003687
47	17	15	11	0.999965	0.000451
47	17	15	12	0.999999	0.000033
47	17	15	13	1.000000	0.000001
47	17	15	14	1.000000	0.000000
47	17	15	15	1.000000	0.000000
47	17	16	0	0.000097	0.000097
47	17	16	1	0.001851	0.001754
47	17	16	2	0.015008	0.013157
47	17	16	3	0.069182	0.054174
47	17	16	4	0.206123	0.136941
47	17	16	5	0.430994	0.224871
47	17	16	6	0.678352	0.247358
47	17	16	7	0.863450	0.185098
47	17	16	8	0.958103	0.094652
47	17	16	9	0.991025	0.032923
47	17	16	10	0.998707	0.007682
47	17	16	11	0.999880	0.001173
47	17	16	12	0.999993	0.000113
47	17	16	13	1.000000	0.000006
47	17	16	14	1.000000	0.000000
47	17	16	15	1.000000	0.000000
47	17	16	16	1.000000	0.000000
47	17	17	0	0.000044	0.000044
47	17	17	1	0.000946	0.000902
47	17	17	2	0.008641	0.007696
47	17	17	3	0.044716	0.036075
47	17	17	4	0.148696	0.103980
47	17	17	5	0.343947	0.195251
47	17	17	6	0.590580	0.246633
47	17	17	7	0.803741	0.213161
47	17	17	8	0.930623	0.126882
47	17	17	9	0.982529	0.051906
47	17	17	10	0.996973	0.014443
47	17	17	11	0.999653	0.002681
47	17	17	12	0.999975	0.000322
47	17	17	13	0.999999	0.000024
47	17	17	14	1.000000	0.000001
47	17	17	15	1.000000	0.000000
47	17	17	16	1.000000	0.000000
47	17	17	17	1.000000	0.000000
47	18	1	0	0.617021	0.617021
47	18	1	1	1.000000	0.382979
47	18	2	0	0.375578	0.375578
47	18	2	1	0.858464	0.482886
47	18	2	2	1.000000	0.141536
47	18	3	0	0.225347	0.225347
47	18	3	1	0.676041	0.450694
47	18	3	2	0.949676	0.273635
47	18	3	3	1.000000	0.050324
47	18	4	0	0.133159	0.133159
47	18	4	1	0.501909	0.368749
47	18	4	2	0.850172	0.348263
47	18	4	3	0.982844	0.132672
47	18	4	4	1.000000	0.017156
47	18	5	0	0.077418	0.077418
47	18	5	1	0.356124	0.278706
47	18	5	2	0.720586	0.364462
47	18	5	3	0.936563	0.215977
47	18	5	4	0.994414	0.057851
47	18	5	5	1.000000	0.005586
47	18	6	0	0.044239	0.044239
47	18	6	1	0.243315	0.199076
47	18	6	2	0.581743	0.338429
47	18	6	3	0.859428	0.277685
47	18	6	4	0.975131	0.115702
47	18	6	5	0.998271	0.023140
47	18	6	6	1.000000	0.001729
47	18	7	0	0.024817	0.024817
47	18	7	1	0.160771	0.135954
47	18	7	2	0.449674	0.288903
47	18	7	3	0.757836	0.308163
47	18	7	4	0.935623	0.177786
47	18	7	5	0.990934	0.055311
47	18	7	6	0.999494	0.008560
47	18	7	7	1.000000	0.000506
47	18	8	0	0.013649	0.013649
47	18	8	1	0.102991	0.089341
47	18	8	2	0.334113	0.231122
47	18	8	3	0.642275	0.308163
47	18	8	4	0.873397	0.231122
47	18	8	5	0.972958	0.099560
47	18	8	6	0.996926	0.023968
47	18	8	7	0.999861	0.002935
47	18	8	8	1.000000	0.000139
47	18	9	0	0.007350	0.007350
47	18	9	1	0.064047	0.056697
47	18	9	2	0.239293	0.175246
47	18	9	3	0.523751	0.284458
47	18	9	4	0.790431	0.266679
47	18	9	5	0.939771	0.149340
47	18	9	6	0.989551	0.049780
47	18	9	7	0.999033	0.009482
47	18	9	8	0.999964	0.000931
47	18	9	9	1.000000	0.000036
47	18	10	0	0.003868	0.003868
47	18	10	1	0.038682	0.034814
47	18	10	2	0.165505	0.126823
47	18	10	3	0.411465	0.245960
47	18	10	4	0.692180	0.280715
47	18	10	5	0.888681	0.196501
47	18	10	6	0.973831	0.085150
47	18	10	7	0.996288	0.022457
47	18	10	8	0.999719	0.003431
47	18	10	9	0.999991	0.000272
47	18	10	10	1.000000	0.000008
47	18	11	0	0.001986	0.001986
47	18	11	1	0.022687	0.020700
47	18	11	2	0.110663	0.087976
47	18	11	3	0.311752	0.201089
47	18	11	4	0.585964	0.274212
47	18	11	5	0.819640	0.233676
47	18	11	6	0.946215	0.126575
47	18	11	7	0.989612	0.043397
47	18	11	8	0.998792	0.009180
47	18	11	9	0.999925	0.001133
47	18	11	10	0.999998	0.000073
47	18	11	11	1.000000	0.000002
47	18	12	0	0.000993	0.000993
47	18	12	1	0.012912	0.011918
47	18	12	2	0.071562	0.058651
47	18	12	3	0.227965	0.156402
47	18	12	4	0.479326	0.251361
47	18	12	5	0.735257	0.255931
47	18	12	6	0.904023	0.168766
47	18	12	7	0.976352	0.072328
47	18	12	8	0.996242	0.019890
47	18	12	9	0.999642	0.003400
47	18	12	10	0.999982	0.000340
47	18	12	11	1.000000	0.000018
47	18	12	12	1.000000	0.000000
47	18	13	0	0.000482	0.000482
47	18	13	1	0.007123	0.006640
47	18	13	2	0.044751	0.037628
47	18	13	3	0.160935	0.116185
47	18	13	4	0.378781	0.217846
47	18	13	5	0.640197	0.261415
47	18	13	6	0.846161	0.205964
47	18	13	7	0.953620	0.107459
47	18	13	8	0.990559	0.036939
47	18	13	9	0.998768	0.008209
47	18	13	10	0.999904	0.001137
47	18	13	11	0.999996	0.000092
47	18	13	12	1.000000	0.000004
47	18	13	13	1.000000	0.000000
47	18	14	0	0.000227	0.000227
47	18	14	1	0.003803	0.003576
47	18	14	2	0.027043	0.023241
47	18	14	3	0.109677	0.082634
47	18	14	4	0.289080	0.179403
47	18	14	5	0.540244	0.251164
47	18	14	6	0.773467	0.233224
47	18	14	7	0.918853	0.145386
47	18	14	8	0.979694	0.060841
47	18	14	9	0.996595	0.016900
47	18	14	10	0.999637	0.003042
47	18	14	11	0.999977	0.000340
47	18	14	12	0.999999	0.000022
47	18	14	13	1.000000	0.000001
47	18	14	14	1.000000	0.000000
47	18	15	0	0.000103	0.000103
47	18	15	1	0.001961	0.001857
47	18	15	2	0.015775	0.013814
47	18	15	3	0.072116	0.056341
47	18	15	4	0.212970	0.140853
47	18	15	5	0.441301	0.228331
47	18	15	6	0.688659	0.247358
47	18	15	7	0.870391	0.181733
47	18	15	8	0.961258	0.090866
47	18	15	9	0.991985	0.030728
47	18	15	10	0.998899	0.006914
47	18	15	11	0.999905	0.001006
47	18	15	12	0.999995	0.000090
47	18	15	13	1.000000	0.000005
47	18	15	14	1.000000	0.000000
47	18	15	15	1.000000	0.000000
47	18	16	0	0.000045	0.000045
47	18	16	1	0.000974	0.000929
47	18	16	2	0.008868	0.007894
47	18	16	3	0.045706	0.036839
47	18	16	4	0.151346	0.105640
47	18	16	5	0.348541	0.197195
47	18	16	6	0.595899	0.247358
47	18	16	7	0.807921	0.212021
47	18	16	8	0.932862	0.124941
47	18	16	9	0.983343	0.050481
47	18	16	10	0.997171	0.013827
47	18	16	11	0.999685	0.002514
47	18	16	12	0.999978	0.000293

Table for $N=2$, $n=1$, through $N=100$, $n=50$

N	n	k	x	P(x)	p(x)
47	18	16	13	0.999999	0.000021
47	18	16	14	1.000000	0.000001
47	18	16	15	1.000000	0.000000
47	18	16	16	1.000000	0.000000
47	18	17	0	0.000019	0.000019
47	18	17	1	0.000465	0.000446
47	18	17	2	0.004794	0.004329
47	18	17	3	0.027881	0.023088
47	18	17	4	0.103638	0.075757
47	18	17	5	0.265847	0.162209
47	18	17	6	0.500148	0.234301
47	18	17	7	0.732687	0.232540
47	18	17	8	0.892558	0.159871
47	18	17	9	0.968688	0.076129
47	18	17	10	0.993602	0.024915
47	18	17	11	0.999117	0.005515
47	18	17	12	0.999921	0.000804
47	18	17	13	0.999996	0.000074
47	18	17	14	1.000000	0.000004
47	18	17	15	1.000000	0.000000
47	18	17	16	1.000000	0.000000
47	18	17	17	1.000000	0.000000
47	18	18	0	0.000008	0.000008
47	18	18	1	0.000212	0.000204
47	18	18	2	0.002485	0.002273
47	18	18	3	0.016337	0.013853
47	18	18	4	0.068285	0.051947
47	18	18	5	0.195556	0.127271
47	18	18	6	0.406427	0.210871
47	18	18	7	0.647423	0.240996
47	18	18	8	0.839268	0.191845
47	18	18	9	0.945849	0.106581
47	18	18	10	0.986958	0.041110
47	18	18	11	0.997830	0.010872
47	18	18	12	0.999761	0.001930
47	18	18	13	0.999983	0.000223
47	18	18	14	0.999999	0.000016
47	18	18	15	1.000000	0.000001
47	18	18	16	1.000000	0.000000
47	18	18	17	1.000000	0.000000
47	18	18	18	1.000000	0.000000
47	19	1	0	0.595745	0.595745
47	19	1	1	1.000000	0.404255
47	19	2	0	0.349676	0.349676
47	19	2	1	0.841813	0.492137
47	19	2	2	1.000000	0.158187
47	19	3	0	0.202035	0.202035
47	19	3	1	0.644958	0.442923
47	19	3	2	0.940240	0.295282
47	19	3	3	1.000000	0.059759

N	n	k	x	P(x)	p(x)
47	19	4	0	0.114793	0.114793
47	19	4	1	0.463762	0.348970
47	19	4	2	0.826154	0.362392
47	19	4	3	0.978269	0.152115
47	19	4	4	1.000000	0.021731
47	19	5	0	0.064070	0.064070
47	19	5	1	0.317682	0.253612
47	19	5	2	0.682883	0.365201
47	19	5	3	0.921668	0.238785
47	19	5	4	0.992419	0.070751
47	19	5	5	1.000000	0.007580
47	19	6	0	0.035086	0.035086
47	19	6	1	0.208991	0.173905
47	19	6	2	0.535064	0.326072
47	19	6	3	0.830702	0.295639
47	19	6	4	0.967151	0.136449
47	19	6	5	0.997473	0.030322
47	19	6	6	1.000000	0.002527
47	19	7	0	0.018827	0.018827
47	19	7	1	0.132643	0.113816
47	19	7	2	0.399863	0.267220
47	19	7	3	0.715331	0.315468
47	19	7	4	0.917231	0.201900
47	19	7	5	0.987119	0.069888
47	19	7	6	0.999199	0.012079
47	19	7	7	1.000000	0.000801
47	19	8	0	0.009884	0.009884
47	19	8	1	0.081426	0.071541
47	19	8	2	0.286294	0.204869
47	19	8	3	0.589144	0.302850
47	19	8	4	0.841519	0.252375
47	19	8	5	0.962658	0.121140
47	19	8	6	0.995273	0.032615
47	19	8	7	0.999760	0.004487
47	19	8	8	1.000000	0.000240
47	19	9	0	0.005069	0.005069
47	19	9	1	0.048406	0.043338
47	19	9	2	0.196993	0.148586
47	19	9	3	0.464898	0.267905
47	19	9	4	0.744451	0.279553
47	19	9	5	0.919172	0.174721
47	19	9	6	0.984401	0.065229
47	19	9	7	0.998379	0.013978
47	19	9	8	0.999932	0.001553
47	19	9	9	1.000000	0.000068
47	19	10	0	0.002534	0.002534
47	19	10	1	0.027878	0.025344
47	19	10	2	0.130520	0.102642
47	19	10	3	0.352096	0.221576
47	19	10	4	0.634101	0.282006

N	n	k	x	P(x)	p(x)
47	19	10	5	0.854801	0.220700
47	19	10	6	0.962086	0.107285
47	19	10	7	0.993965	0.031879
47	19	10	8	0.999483	0.005517
47	19	10	9	0.999982	0.000500
47	19	10	10	1.000000	0.000018
47	19	11	0	0.001233	0.001233
47	19	11	1	0.015549	0.014316
47	19	11	2	0.083360	0.067811
47	19	11	3	0.256279	0.172919
47	19	11	4	0.519775	0.263496
47	19	11	5	0.771293	0.251519
47	19	11	6	0.924391	0.153098
47	19	11	7	0.983626	0.059234
47	19	11	8	0.997842	0.014216
47	19	11	9	0.999847	0.002005
47	19	11	10	0.999996	0.000149
47	19	11	11	1.000000	0.000004
47	19	12	0	0.000582	0.000582
47	19	12	1	0.008391	0.007809
47	19	12	2	0.051338	0.042947
47	19	12	3	0.179426	0.128088
47	19	12	4	0.409985	0.230559
47	19	12	5	0.673480	0.263496
47	19	12	6	0.869106	0.195626
47	19	12	7	0.963881	0.094775
47	19	12	8	0.993498	0.029617
47	19	12	9	0.999290	0.005792
47	19	12	10	0.999958	0.000668
47	19	12	11	0.999999	0.000041
47	19	12	12	1.000000	0.000001
47	19	13	0	0.000266	0.000266
47	19	13	1	0.004375	0.004109
47	19	13	2	0.030478	0.026103
47	19	13	3	0.120872	0.090394
47	19	13	4	0.311174	0.190302
47	19	13	5	0.568082	0.256908
47	19	13	6	0.796445	0.228363
47	19	13	7	0.931387	0.134942
47	19	13	8	0.984190	0.052803
47	19	13	9	0.997635	0.013445
47	19	13	10	0.999787	0.002151
47	19	13	11	0.999990	0.000203
47	19	13	12	1.000000	0.000010
47	19	13	13	1.000000	0.000000
47	19	14	0	0.000117	0.000117
47	19	14	1	0.002200	0.002082
47	19	14	2	0.017426	0.015227
47	19	14	3	0.078333	0.060907
47	19	14	4	0.227217	0.148884

N	n	k	x	P(x)	p(x)
47	19	14	5	0.462296	0.235079
47	19	14	6	0.709130	0.246833
47	19	14	7	0.883760	0.174630
47	19	14	8	0.967107	0.083346
47	19	14	9	0.993681	0.026574
47	19	14	10	0.999217	0.005536
47	19	14	11	0.999942	0.000725
47	19	14	12	0.999998	0.000056
47	19	14	13	1.000000	0.000002
47	19	14	14	1.000000	0.000000
47	19	15	0	0.000050	0.000050
47	19	15	1	0.001064	0.001014
47	19	15	2	0.009582	0.008518
47	19	15	3	0.048803	0.039220
47	19	15	4	0.159543	0.110740
47	19	15	5	0.362566	0.203023
47	19	15	6	0.611892	0.249327
47	19	15	7	0.820258	0.208366
47	19	15	8	0.939325	0.119066
47	19	15	9	0.985628	0.046304
47	19	15	10	0.997707	0.012079
47	19	15	11	0.999766	0.002059
47	19	15	12	0.999986	0.000220
47	19	15	13	0.999999	0.000014
47	19	15	14	1.000000	0.000000
47	19	15	15	1.000000	0.000000
47	19	16	0	0.000020	0.000020
47	19	16	1	0.000493	0.000473
47	19	16	2	0.005057	0.004563
47	19	16	3	0.029193	0.024136
47	19	16	4	0.107633	0.078441
47	19	16	5	0.273743	0.166110
47	19	16	6	0.510604	0.236860
47	19	16	7	0.742121	0.231518
47	19	16	8	0.898396	0.156274
47	19	16	9	0.971158	0.072763
47	19	16	10	0.994310	0.023152
47	19	16	11	0.999251	0.004941
47	19	16	12	0.999938	0.000686
47	19	16	13	0.999997	0.000059
47	19	16	14	1.000000	0.000003
47	19	16	15	1.000000	0.000000
47	19	16	16	1.000000	0.000000
47	19	17	0	0.000008	0.000008
47	19	17	1	0.000219	0.000211
47	19	17	2	0.002554	0.002336
47	19	17	3	0.016735	0.014181
47	19	17	4	0.069678	0.052943
47	19	17	5	0.198726	0.129048
47	19	17	6	0.411275	0.212549

N	n	k	x	P(x)	p(x)
47	19	17	7	0.652501	0.241226
47	19	17	8	0.842943	0.190442
47	19	17	9	0.947686	0.104743
47	19	17	10	0.987588	0.039902
47	19	17	11	0.997976	0.010388
47	19	17	12	0.999783	0.001807
47	19	17	13	0.999985	0.000203
47	19	17	14	0.999999	0.000014
47	19	17	15	1.000000	0.000001
47	19	17	16	1.000000	0.000000
47	19	17	17	1.000000	0.000000
47	19	18	0	0.000003	0.000003
47	19	18	1	0.000092	0.000089
47	19	18	2	0.001231	0.001139
47	19	18	3	0.009172	0.007941
47	19	18	4	0.043207	0.034035
47	19	18	5	0.138504	0.095297
47	19	18	6	0.319170	0.180667
47	19	18	7	0.556011	0.236841
47	19	18	8	0.773115	0.217104
47	19	18	9	0.912772	0.139657
47	19	18	10	0.975618	0.062846
47	19	18	11	0.995206	0.019588
47	19	18	12	0.999361	0.004155
47	19	18	13	0.999945	0.000584
47	19	18	14	0.999997	0.000052
47	19	18	15	1.000000	0.000003
47	19	18	16	1.000000	0.000000
47	19	18	17	1.000000	0.000000
47	19	18	18	1.000000	0.000000
47	19	19	0	0.000001	0.000001
47	19	19	1	0.000037	0.000036
47	19	19	2	0.000563	0.000527
47	19	19	3	0.004791	0.004227
47	19	19	4	0.025603	0.020812
47	19	19	5	0.092498	0.066896
47	19	19	6	0.238182	0.145684
47	19	19	7	0.458008	0.219826
47	19	19	8	0.690765	0.232757
47	19	19	9	0.864614	0.173849
47	19	19	10	0.956114	0.091500
47	19	19	11	0.989803	0.033689
47	19	19	12	0.998358	0.008556
47	19	19	13	0.999824	0.001466
47	19	19	14	0.999988	0.000164
47	19	19	15	0.999999	0.000011
47	19	19	16	1.000000	0.000000
47	19	19	17	1.000000	0.000000
47	19	19	18	1.000000	0.000000
47	19	19	19	1.000000	0.000000

N	n	k	x	P(x)	p(x)
47	20	1	0	0.574468	0.574468
47	20	1	1	1.000000	0.425532
47	20	2	0	0.324699	0.324699
47	20	2	1	0.824237	0.499537
47	20	2	2	1.000000	0.175763
47	20	3	0	0.180388	0.180388
47	20	3	1	0.613321	0.432932
47	20	3	2	0.929695	0.316374
47	20	3	3	1.000000	0.070305
47	20	4	0	0.098394	0.098394
47	20	4	1	0.426373	0.327979
47	20	4	2	0.800269	0.373896
47	20	4	3	0.972837	0.172567
47	20	4	4	1.000000	0.027163
47	20	5	0	0.052629	0.052629
47	20	5	1	0.281452	0.228823
47	20	5	2	0.643754	0.362302
47	20	5	3	0.904612	0.260858
47	20	5	4	0.989893	0.085280
47	20	5	5	1.000000	0.010107
47	20	6	0	0.027568	0.027568
47	20	6	1	0.177937	0.150369
47	20	6	2	0.488482	0.310545
47	20	6	3	0.799027	0.310545
47	20	6	4	0.957405	0.158378
47	20	6	5	0.996390	0.038985
47	20	6	6	1.000000	0.003610
47	20	7	0	0.014120	0.014120
47	20	7	1	0.108254	0.094134
47	20	7	2	0.352145	0.243891
47	20	7	3	0.670264	0.318119
47	20	7	4	0.895599	0.225335
47	20	7	5	0.982127	0.086528
47	20	7	6	0.998767	0.016640
47	20	7	7	1.000000	0.001233
47	20	8	0	0.007060	0.007060
47	20	8	1	0.063540	0.056480
47	20	8	2	0.242394	0.178854
47	20	8	3	0.535064	0.292670
47	20	8	4	0.805465	0.270401
47	20	8	5	0.949679	0.144214
47	20	8	6	0.992943	0.043264
47	20	8	7	0.999599	0.006656
47	20	8	8	1.000000	0.000401
47	20	9	0	0.003439	0.003439
47	20	9	1	0.036024	0.032585
47	20	9	2	0.159846	0.123822
47	20	9	3	0.407490	0.247644
47	20	9	4	0.694531	0.287041
47	20	9	5	0.894212	0.199681

N	n	k	x	P(x)	p(x)
47	20	9	6	0.977413	0.083200
47	20	9	7	0.997381	0.019968
47	20	9	8	0.999877	0.002496
47	20	9	9	1.000000	0.000123
47	20	10	0	0.001629	0.001629
47	20	10	1	0.019732	0.018103
47	20	10	2	0.101194	0.081462
47	20	10	3	0.296702	0.195508
47	20	10	4	0.573672	0.276970
47	20	10	5	0.815391	0.241719
47	20	10	6	0.946760	0.131369
47	20	10	7	0.990549	0.043790
47	20	10	8	0.999088	0.008539
47	20	10	9	0.999964	0.000876
47	20	10	10	1.000000	0.000036
47	20	11	0	0.000749	0.000749
47	20	11	1	0.010436	0.009687
47	20	11	2	0.061564	0.051128
47	20	11	3	0.206874	0.145310
47	20	11	4	0.453901	0.247027
47	20	11	5	0.717396	0.263496
47	20	11	6	0.897053	0.179656
47	20	11	7	0.975164	0.078111
47	20	11	8	0.996319	0.021155
47	20	11	9	0.999704	0.003385
47	20	11	10	0.999990	0.000286
47	20	11	11	1.000000	0.000010
47	20	12	0	0.000333	0.000333
47	20	12	1	0.005323	0.004990
47	20	12	2	0.036000	0.030677
47	20	12	3	0.138255	0.102255
47	20	12	4	0.344111	0.205856
47	20	12	5	0.607607	0.263496
47	20	12	6	0.827186	0.219580
47	20	12	7	0.946957	0.119771
47	20	12	8	0.989267	0.042310
47	20	12	9	0.998670	0.009402
47	20	12	10	0.999911	0.001241
47	20	12	11	0.999997	0.000087
47	20	12	12	1.000000	0.000002
47	20	13	0	0.000143	0.000143
47	20	13	1	0.002614	0.002471
47	20	13	2	0.020223	0.017609
47	20	13	3	0.088588	0.068365
47	20	13	4	0.250005	0.161417
47	20	13	5	0.494680	0.244675
47	20	13	6	0.739354	0.244675
47	20	13	7	0.902471	0.163116
47	20	13	8	0.974761	0.072290
47	20	13	9	0.995715	0.020954

N	n	k	x	P(x)	p(x)
47	20	13	10	0.999556	0.003842
47	20	13	11	0.999975	0.000419
47	20	13	12	0.999999	0.000024
47	20	13	13	1.000000	0.000001
47	20	14	0	0.000059	0.000059
47	20	14	1	0.001233	0.001174
47	20	14	2	0.010901	0.009668
47	20	14	3	0.054406	0.043505
47	20	14	4	0.174044	0.119639
47	20	14	5	0.386735	0.212691
47	20	14	6	0.638606	0.251871
47	20	14	7	0.840103	0.201497
47	20	14	8	0.949247	0.109144
47	20	14	9	0.988935	0.039689
47	20	14	10	0.998426	0.009491
47	20	14	11	0.999864	0.001438
47	20	14	12	0.999994	0.000129
47	20	14	13	1.000000	0.000006
47	20	14	14	1.000000	0.000000
47	20	15	0	0.000023	0.000023
47	20	15	1	0.000557	0.000534
47	20	15	2	0.005627	0.005071
47	20	15	3	0.031994	0.026367
47	20	15	4	0.116038	0.084044
47	20	15	5	0.290058	0.174020
47	20	15	6	0.531752	0.241694
47	20	15	7	0.760725	0.228974
47	20	15	8	0.909558	0.148833
47	20	15	9	0.975706	0.066148
47	20	15	10	0.995550	0.019844
47	20	15	11	0.999472	0.003922
47	20	15	12	0.999962	0.000490
47	20	15	13	0.999998	0.000036
47	20	15	14	1.000000	0.000001
47	20	15	15	1.000000	0.000000
47	20	16	0	0.000009	0.000009
47	20	16	1	0.000240	0.000231
47	20	16	2	0.002775	0.002535
47	20	16	3	0.017987	0.015212
47	20	16	4	0.074016	0.056029
47	20	16	5	0.208486	0.134470
47	20	16	6	0.426011	0.217525
47	20	16	7	0.667705	0.241694
47	20	16	8	0.853746	0.186041
47	20	16	9	0.952967	0.099222
47	20	16	10	0.989349	0.036381
47	20	16	11	0.998369	0.009020
47	20	16	12	0.999840	0.001471
47	20	16	13	0.999991	0.000151
47	20	16	14	1.000000	0.000009

Table for $N = 2$, $n = 1$, through $N = 100$, $n = 50$

N	n	k	x	P(x)	p(x)
47	20	16	15	1.000000	0.000000
47	20	16	16	1.000000	0.000000
47	20	17	0	0.000003	0.000003
47	20	17	1	0.000098	0.000095
47	20	17	2	0.001303	0.001205
47	20	17	3	0.009645	0.008342
47	20	17	4	0.045098	0.035453
47	20	17	5	0.143420	0.098322
47	20	17	6	0.327774	0.184354
47	20	17	7	0.566349	0.238576
47	20	17	8	0.781730	0.215381
47	20	17	9	0.917760	0.136030
47	20	17	10	0.977613	0.059853
47	20	17	11	0.995750	0.018137
47	20	17	12	0.999460	0.003710
47	20	17	13	0.999956	0.000496
47	20	17	14	0.999998	0.000041
47	20	17	15	1.000000	0.000002
47	20	17	16	1.000000	0.000000
47	20	17	17	1.000000	0.000000
47	20	18	0	0.000001	0.000001
47	20	18	1	0.000038	0.000037
47	20	18	2	0.000580	0.000542
47	20	18	3	0.004918	0.004338
47	20	18	4	0.026189	0.021272
47	20	18	5	0.094259	0.068069
47	20	18	6	0.241742	0.147483
47	20	18	7	0.462966	0.221225
47	20	18	8	0.695578	0.232611
47	20	18	9	0.867882	0.172305
47	20	18	10	0.957662	0.089780
47	20	18	11	0.990309	0.032647
47	20	18	12	0.998471	0.008162
47	20	18	13	0.999841	0.001370
47	20	18	14	0.999990	0.000149
47	20	18	15	1.000000	0.000010
47	20	18	16	1.000000	0.000000
47	20	18	17	1.000000	0.000000
47	20	18	18	1.000000	0.000000
47	20	19	0	0.000000	0.000000
47	20	19	1	0.000014	0.000013
47	20	19	2	0.000244	0.000230
47	20	19	3	0.002375	0.002131
47	20	19	4	0.014453	0.012078
47	20	19	5	0.059050	0.044597
47	20	19	6	0.170543	0.111493
47	20	19	7	0.363797	0.193254
47	20	19	8	0.599325	0.235528
47	20	19	9	0.802525	0.203201
47	20	19	10	0.926703	0.124178
47	20	19	11	0.980177	0.053474
47	20	19	12	0.996219	0.016042
47	20	19	13	0.999510	0.003291
47	20	19	14	0.999959	0.000449
47	20	19	15	0.999998	0.000039
47	20	19	16	1.000000	0.000002
47	20	19	17	1.000000	0.000000
47	20	19	18	1.000000	0.000000
47	20	19	19	1.000000	0.000000
47	20	20	0	0.000000	0.000000
47	20	20	1	0.000005	0.000005
47	20	20	2	0.000096	0.000091
47	20	20	3	0.001081	0.000985
47	20	20	4	0.007552	0.006471
47	20	20	5	0.035159	0.027608
47	20	20	6	0.114797	0.079638
47	20	20	7	0.274072	0.159275
47	20	20	8	0.498384	0.224312
47	20	20	9	0.722696	0.224312
47	20	20	10	0.882354	0.159658
47	20	20	11	0.962989	0.080635
47	20	20	12	0.991636	0.028647
47	20	20	13	0.998687	0.007051
47	20	20	14	0.999863	0.001175
47	20	20	15	0.999991	0.000128
47	20	20	16	0.999999	0.000009
47	20	20	17	1.000000	0.000000
47	20	20	18	1.000000	0.000000
47	20	20	19	1.000000	0.000000
47	20	20	20	1.000000	0.000000
47	21	1	0	0.553191	0.553191
47	21	1	1	1.000000	0.446808
47	21	2	0	0.300648	0.300648
47	21	2	1	0.805735	0.505088
47	21	2	2	1.000000	0.194265
47	21	3	0	0.160345	0.160345
47	21	3	1	0.581252	0.420907
47	21	3	2	0.917977	0.336725
47	21	3	3	1.000000	0.082023
47	21	4	0	0.083817	0.083817
47	21	4	1	0.389931	0.306114
47	21	4	2	0.772573	0.382642
47	21	4	3	0.966445	0.193872
47	21	4	4	1.000000	0.033555
47	21	5	0	0.042883	0.042883
47	21	5	1	0.247552	0.204669
47	21	5	2	0.603499	0.355946
47	21	5	3	0.885289	0.281791
47	21	5	4	0.986734	0.101445
47	21	5	5	1.000000	0.013266
47	21	6	0	0.021442	0.021442
47	21	6	1	0.150091	0.128649
47	21	6	2	0.442475	0.292384
47	21	6	3	0.764522	0.322047
47	21	6	4	0.945673	0.181151
47	21	6	5	0.994946	0.049273
47	21	6	6	1.000000	0.005054
47	21	7	0	0.010459	0.010459
47	21	7	1	0.087335	0.076876
47	21	7	2	0.306980	0.219645
47	21	7	3	0.623136	0.316156
47	21	7	4	0.870562	0.247426
47	21	7	5	0.975718	0.105156
47	21	7	6	0.998151	0.022433
47	21	7	7	1.000000	0.001849
47	21	8	0	0.004968	0.004968
47	21	8	1	0.048897	0.043929
47	21	8	2	0.202649	0.153751
47	21	8	3	0.480865	0.278217
47	21	8	4	0.765405	0.284540
47	21	8	5	0.933655	0.168250
47	21	8	6	0.989738	0.056083
47	21	8	7	0.999353	0.009614
47	21	8	8	1.000000	0.000647
47	21	9	0	0.002293	0.002293
47	21	9	1	0.026369	0.024076
47	21	9	2	0.127744	0.101375
47	21	9	3	0.352458	0.224714
47	21	9	4	0.641375	0.288918
47	21	9	5	0.864630	0.223254
47	21	9	6	0.968168	0.103538
47	21	9	7	0.995902	0.027733
47	21	9	8	0.999784	0.003883
47	21	9	9	1.000000	0.000216
47	21	10	0	0.001026	0.001026
47	21	10	1	0.013698	0.012672
47	21	10	2	0.077057	0.063359
47	21	10	3	0.246014	0.168958
47	21	10	4	0.512123	0.266108
47	21	10	5	0.770628	0.258505
47	21	10	6	0.927298	0.156670
47	21	10	7	0.985684	0.058386
47	21	10	8	0.998456	0.012772
47	21	10	9	0.999932	0.001476
47	21	10	10	1.000000	0.000068
47	21	11	0	0.000444	0.000444
47	21	11	1	0.006848	0.006404
47	21	11	2	0.044521	0.037673
47	21	11	3	0.163819	0.119298
47	21	11	4	0.389857	0.226038
47	21	11	5	0.658842	0.268985
47	21	11	6	0.863783	0.204941
47	21	11	7	0.963592	0.099809
47	21	11	8	0.993968	0.030377
47	21	11	9	0.999453	0.005485
47	21	11	10	0.999980	0.000527
47	21	11	11	1.000000	0.000020
47	21	12	0	0.000185	0.000185
47	21	12	1	0.003290	0.003105
47	21	12	2	0.024638	0.021348
47	21	12	3	0.104170	0.079532
47	21	12	4	0.283117	0.178947
47	21	12	5	0.539293	0.256176
47	21	12	6	0.778391	0.239098
47	21	12	7	0.924777	0.146386
47	21	12	8	0.982999	0.058222
47	21	12	9	0.997625	0.014626
47	21	12	10	0.999819	0.002194
47	21	12	11	0.999994	0.000176
47	21	12	12	1.000000	0.000006
47	21	13	0	0.000074	0.000074
47	21	13	1	0.001516	0.001442
47	21	13	2	0.013049	0.011533
47	21	13	3	0.063268	0.050219
47	21	13	4	0.196200	0.132932
47	21	13	5	0.422184	0.225984
47	21	13	6	0.675920	0.253737
47	21	13	7	0.866223	0.190302
47	21	13	8	0.961374	0.095151
47	21	13	9	0.992610	0.031236
47	21	13	10	0.999129	0.006519
47	21	13	11	0.999944	0.000815
47	21	13	12	0.999998	0.000054
47	21	13	13	1.000000	0.000001
47	21	14	0	0.000028	0.000028
47	21	14	1	0.000668	0.000639
47	21	14	2	0.006604	0.005936
47	21	14	3	0.036681	0.030078
47	21	14	4	0.129734	0.093052
47	21	14	5	0.315838	0.186105
47	21	14	6	0.563978	0.248139
47	21	14	7	0.787863	0.223885
47	21	14	8	0.924993	0.137130
47	21	14	9	0.981586	0.056593
47	21	14	10	0.997020	0.015435
47	21	14	11	0.999705	0.002684
47	21	14	12	0.999984	0.000280
47	21	14	13	1.000000	0.000015
47	21	14	14	1.000000	0.000000
47	21	15	0	0.000010	0.000010

N	n	k	x	P(x)	p(x)
47	21	15	1	0.000280	0.000270
47	21	15	2	0.003186	0.002906
47	21	15	3	0.020275	0.017089
47	21	15	4	0.081798	0.061522
47	21	15	5	0.225606	0.143808
47	21	15	6	0.451187	0.225581
47	21	15	7	0.692881	0.241694
47	21	15	8	0.870972	0.178090
47	21	15	9	0.961006	0.090035
47	21	15	10	0.991875	0.030869
47	21	15	11	0.998891	0.007016
47	21	15	12	0.999908	0.001017
47	21	15	13	0.999996	0.000088
47	21	15	14	1.000000	0.000004
47	21	15	15	1.000000	0.000000
47	21	16	0	0.000004	0.000004
47	21	16	1	0.000111	0.000108
47	21	16	2	0.001461	0.001349
47	21	16	3	0.010663	0.009202
47	21	16	4	0.049114	0.038451
47	21	16	5	0.153702	0.104588
47	21	16	6	0.345446	0.191744
47	21	16	7	0.587140	0.241694
47	21	16	8	0.798622	0.211482
47	21	16	9	0.927243	0.128621
47	21	16	10	0.981264	0.054021
47	21	16	11	0.996699	0.015435
47	21	16	12	0.999622	0.002923
47	21	16	13	0.999974	0.000352
47	21	16	14	0.999999	0.000025
47	21	16	15	1.000000	0.000001
47	21	16	16	1.000000	0.000000
47	21	17	0	0.000001	0.000001
47	21	17	1	0.000042	0.000041
47	21	17	2	0.000634	0.000592
47	21	17	3	0.005320	0.004686
47	21	17	4	0.028028	0.022708
47	21	17	5	0.099721	0.071693
47	21	17	6	0.252666	0.152945
47	21	17	7	0.477988	0.225321
47	21	17	8	0.709936	0.231948
47	21	17	9	0.877455	0.167518
47	21	17	10	0.962095	0.084641
47	21	17	11	0.991720	0.029624
47	21	17	12	0.998773	0.007053
47	21	17	13	0.999883	0.001110
47	21	17	14	0.999993	0.000110
47	21	17	15	1.000000	0.000006
47	21	17	16	1.000000	0.000000
47	21	17	17	1.000000	0.000000

N	n	k	x	P(x)	p(x)
47	21	18	0	0.000000	0.000000
47	21	18	1	0.000015	0.000014
47	21	18	2	0.000259	0.000244
47	21	18	3	0.002508	0.002249
47	21	18	4	0.015160	0.012652
47	21	18	5	0.061485	0.046325
47	21	18	6	0.176194	0.114709
47	21	18	7	0.372838	0.196644
47	21	18	8	0.609425	0.236587
47	21	18	9	0.810447	0.201022
47	21	18	10	0.931060	0.120613
47	21	18	11	0.981845	0.050784
47	21	18	12	0.996657	0.014812
47	21	18	13	0.999587	0.002930
47	21	18	14	0.999968	0.000381
47	21	18	15	0.999998	0.000031
47	21	18	16	1.000000	0.000001
47	21	18	17	1.000000	0.000000
47	21	18	18	1.000000	0.000000
47	21	19	0	0.000000	0.000000
47	21	19	1	0.000005	0.000005
47	21	19	2	0.000099	0.000094
47	21	19	3	0.001112	0.001013
47	21	19	4	0.007743	0.006631
47	21	19	5	0.035926	0.028183
47	21	19	6	0.116861	0.080935
47	21	19	7	0.277906	0.161045
47	21	19	8	0.503369	0.225463
47	21	19	9	0.727266	0.223897
47	21	19	10	0.885311	0.158045
47	21	19	11	0.964333	0.079022
47	21	19	12	0.992060	0.027727
47	21	19	13	0.998779	0.006719
47	21	19	14	0.999876	0.001097
47	21	19	15	0.999992	0.000116
47	21	19	16	1.000000	0.000008
47	21	19	17	1.000000	0.000000
47	21	19	18	1.000000	0.000000
47	21	19	19	1.000000	0.000000
47	21	20	0	0.000000	0.000000
47	21	20	1	0.000001	0.000001
47	21	20	2	0.000035	0.000034
47	21	20	3	0.000461	0.000426
47	21	20	4	0.003717	0.003256
47	21	20	5	0.019822	0.016104
47	21	20	6	0.073503	0.053682
47	21	20	7	0.197384	0.123881
47	21	20	8	0.398690	0.201306
47	21	20	9	0.631310	0.232620
47	21	20	10	0.823222	0.191912

N	n	k	x	P(x)	p(x)
47	21	20	11	0.936111	0.112889
47	21	20	12	0.983148	0.047037
47	21	20	13	0.996859	0.013711
47	21	20	14	0.999601	0.002742
47	21	20	15	0.999967	0.000366
47	21	20	16	0.999998	0.000031
47	21	20	17	1.000000	0.000002
47	21	20	18	1.000000	0.000000
47	21	20	19	1.000000	0.000000
47	21	20	20	1.000000	0.000000
47	21	21	0	0.000000	0.000000
47	21	21	1	0.000000	0.000000
47	21	21	2	0.000011	0.000011
47	21	21	3	0.000177	0.000166
47	21	21	4	0.001667	0.001490
47	21	21	5	0.010278	0.008611
47	21	21	6	0.043680	0.033402
47	21	21	7	0.133149	0.089469
47	21	21	8	0.301765	0.168615
47	21	21	9	0.527923	0.226159
47	21	21	10	0.745035	0.217112
47	21	21	11	0.894300	0.149265
47	21	21	12	0.967469	0.073169
47	21	21	13	0.992797	0.025328
47	21	21	14	0.998891	0.006094
47	21	21	15	0.999886	0.000995
47	21	21	16	0.999993	0.000107
47	21	21	17	1.000000	0.000007
47	21	21	18	1.000000	0.000000
47	21	21	19	1.000000	0.000000
47	21	21	20	1.000000	0.000000
47	21	21	21	1.000000	0.000000
47	22	1	0	0.531915	0.531915
47	22	1	1	1.000000	0.468085
47	22	2	0	0.277521	0.277521
47	22	2	1	0.786309	0.508788
47	22	2	2	1.000000	0.213691
47	22	3	0	0.141844	0.141844
47	22	3	1	0.548874	0.407030
47	22	3	2	0.905026	0.356152
47	22	3	3	1.000000	0.094974
47	22	4	0	0.070922	0.070922
47	22	4	1	0.354610	0.283688
47	22	4	2	0.743139	0.388529
47	22	4	3	0.958989	0.215849
47	22	4	4	1.000000	0.041011
47	22	5	0	0.034636	0.034636
47	22	5	1	0.216065	0.181428
47	22	5	2	0.562428	0.346363
47	22	5	3	0.863613	0.301185

N	n	k	x	P(x)	p(x)
47	22	5	4	0.982832	0.119219
47	22	5	5	1.000000	0.017168
47	22	6	0	0.016493	0.016493
47	22	6	1	0.125350	0.108857
47	22	6	2	0.397493	0.272142
47	22	6	3	0.727363	0.329870
47	22	6	4	0.931738	0.204376
47	22	6	5	0.993051	0.061313
47	22	6	6	1.000000	0.006949
47	22	7	0	0.007643	0.007643
47	22	7	1	0.069594	0.061951
47	22	7	2	0.264741	0.195146
47	22	7	3	0.574496	0.309756
47	22	7	4	0.842012	0.267516
47	22	7	5	0.967629	0.125616
47	22	7	6	0.997288	0.029659
47	22	7	7	1.000000	0.002712
47	22	8	0	0.003439	0.003439
47	22	8	1	0.037070	0.033631
47	22	8	2	0.167167	0.130097
47	22	8	3	0.427362	0.260195
47	22	8	4	0.721630	0.294268
47	22	8	5	0.914242	0.192612
47	22	8	6	0.985424	0.071183
47	22	8	7	0.998983	0.013559
47	22	8	8	1.000000	0.001017
47	22	9	0	0.001499	0.001499
47	22	9	1	0.018961	0.017462
47	22	9	2	0.100451	0.081490
47	22	9	3	0.300601	0.200150
47	22	9	4	0.585814	0.285214
47	22	9	5	0.830283	0.244469
47	22	9	6	0.956221	0.125938
47	22	9	7	0.993768	0.037547
47	22	9	8	0.999635	0.005867
47	22	9	9	1.000000	0.000365
47	22	10	0	0.000631	0.000631
47	22	10	1	0.009311	0.008680
47	22	10	2	0.057562	0.048250
47	22	10	3	0.200526	0.142964
47	22	10	4	0.450713	0.250187
47	22	10	5	0.720915	0.270202
47	22	10	6	0.903195	0.182279
47	22	10	7	0.978947	0.075752
47	22	10	8	0.997473	0.018526
47	22	10	9	0.999875	0.002402
47	22	10	10	1.000000	0.000125
47	22	11	0	0.000256	0.000256
47	22	11	1	0.004385	0.004129
47	22	11	2	0.031480	0.027096

Table for $N = 2$, $n = 1$, through $N = 100$, $n = 50$

N	n	k	x	P(x)	p(x)
47	22	11	3	0.127112	0.095631
47	22	11	4	0.329000	0.201889
47	22	11	5	0.596768	0.267768
47	22	11	6	0.824371	0.227603
47	22	11	7	0.948237	0.123865
47	22	11	8	0.990463	0.042227
47	22	11	9	0.999031	0.008568
47	22	11	10	0.999959	0.000928
47	22	11	11	1.000000	0.000041
47	22	12	0	0.000100	0.000100
47	22	12	1	0.001976	0.001877
47	22	12	2	0.016427	0.014451
47	22	12	3	0.076640	0.060212
47	22	12	4	0.228056	0.151416
47	22	12	5	0.470322	0.242266
47	22	12	6	0.723214	0.252892
47	22	12	7	0.896626	0.173412
47	22	12	8	0.974042	0.077416
47	22	12	9	0.995937	0.021895
47	22	12	10	0.999650	0.003713
47	22	12	11	0.999987	0.000338
47	22	12	12	1.000000	0.000012
47	22	13	0	0.000037	0.000037
47	22	13	1	0.000850	0.000813
47	22	13	2	0.008170	0.007319
47	22	13	3	0.043953	0.035783
47	22	13	4	0.150185	0.106232
47	22	13	5	0.352650	0.202465
47	22	13	6	0.607607	0.254956
47	22	13	7	0.822307	0.214700
47	22	13	8	0.943075	0.120769
47	22	13	9	0.987805	0.044729
47	22	13	10	0.998377	0.010572
47	22	13	11	0.999881	0.001504
47	22	13	12	0.999996	0.000115
47	22	13	13	1.000000	0.000004
47	22	14	0	0.000013	0.000013
47	22	14	1	0.000348	0.000335
47	22	14	2	0.003864	0.003516
47	22	14	3	0.023956	0.020092
47	22	14	4	0.093944	0.069988
47	22	14	5	0.251417	0.157473
47	22	14	6	0.487627	0.236210
47	22	14	7	0.727586	0.239959
47	22	14	8	0.893347	0.165761
47	22	14	9	0.970702	0.077355
47	22	14	10	0.994646	0.023943
47	22	14	11	0.999395	0.004749
47	22	14	12	0.999963	0.000568
47	22	14	13	0.999999	0.000036

N	n	k	x	P(x)	p(x)
47	22	14	14	1.000000	0.000001
47	22	15	0	0.000004	0.000004
47	22	15	1	0.000135	0.000130
47	22	15	2	0.001733	0.001598
47	22	15	3	0.012388	0.010655
47	22	15	4	0.055769	0.043381
47	22	15	5	0.170295	0.114526
47	22	15	6	0.373101	0.202806
47	22	15	7	0.618514	0.245413
47	22	15	8	0.823024	0.204511
47	22	15	9	0.940229	0.117205
47	22	15	10	0.985939	0.045710
47	22	15	11	0.997812	0.011873
47	22	15	12	0.999790	0.001979
47	22	15	13	0.999989	0.000199
47	22	15	14	1.000000	0.000011
47	22	15	15	1.000000	0.000000
47	22	16	0	0.000001	0.000001
47	22	16	1	0.000049	0.000048
47	22	16	2	0.000734	0.000685
47	22	16	3	0.006062	0.005327
47	22	16	4	0.031367	0.025306
47	22	16	5	0.109453	0.078086
47	22	16	6	0.271698	0.162245
47	22	16	7	0.503477	0.231779
47	22	16	8	0.733551	0.230074
47	22	16	9	0.892615	0.159064
47	22	16	10	0.968798	0.076183
47	22	16	11	0.993730	0.024933
47	22	16	12	0.999172	0.005442
47	22	16	13	0.999933	0.000761
47	22	16	14	0.999997	0.000064
47	22	16	15	1.000000	0.000003
47	22	16	16	1.000000	0.000000
47	22	17	0	0.000000	0.000000
47	22	17	1	0.000017	0.000016
47	22	17	2	0.000292	0.000275
47	22	17	3	0.002796	0.002504
47	22	17	4	0.016674	0.013877
47	22	17	5	0.066632	0.049958
47	22	17	6	0.187959	0.121327
47	22	17	7	0.391326	0.203367
47	22	17	8	0.629646	0.238321
47	22	17	9	0.825911	0.196264
47	22	17	10	0.939308	0.113397
47	22	17	11	0.984883	0.045576
47	22	17	12	0.997417	0.012533
47	22	17	13	0.999712	0.002295
47	22	17	14	0.999980	0.000268
47	22	17	15	0.999999	0.000019

N	n	k	x	P(x)	p(x)
47	22	17	16	1.000000	0.000001
47	22	17	17	1.000000	0.000000
47	22	18	0	0.000000	0.000000
47	22	18	1	0.000005	0.000005
47	22	18	2	0.000109	0.000103
47	22	18	3	0.001210	0.001102
47	22	18	4	0.008347	0.007137
47	22	18	5	0.038322	0.029975
47	22	18	6	0.123251	0.084929
47	22	18	7	0.289642	0.166391
47	22	18	8	0.518430	0.228788
47	22	18	9	0.740863	0.222433
47	22	18	10	0.893949	0.153086
47	22	18	11	0.968172	0.074224
47	22	18	12	0.993239	0.025067
47	22	18	13	0.999024	0.005785
47	22	18	14	0.999909	0.000885
47	22	18	15	0.999995	0.000086
47	22	18	16	1.000000	0.000005
47	22	18	17	1.000000	0.000000
47	22	18	18	1.000000	0.000000
47	22	19	0	0.000000	0.000000
47	22	19	1	0.000002	0.000002
47	22	19	2	0.000037	0.000036
47	22	19	3	0.000489	0.000451
47	22	19	4	0.003918	0.003429
47	22	19	5	0.020751	0.016833
47	22	19	6	0.076394	0.055643
47	22	19	7	0.203578	0.127184
47	22	19	8	0.407981	0.204403
47	22	19	9	0.641152	0.233171
47	22	19	10	0.830603	0.189451
47	22	19	11	0.940019	0.109416
47	22	19	12	0.984595	0.044577
47	22	19	13	0.997228	0.012633
47	22	19	14	0.999665	0.002436
47	22	19	15	0.999974	0.000309
47	22	19	16	0.999999	0.000025
47	22	19	17	1.000000	0.000001
47	22	19	18	1.000000	0.000000
47	22	19	19	1.000000	0.000000
47	22	20	0	0.000000	0.000000
47	22	20	1	0.000000	0.000000
47	22	20	2	0.000012	0.000011
47	22	20	3	0.000182	0.000171
47	22	20	4	0.001713	0.001531
47	22	20	5	0.010531	0.008817
47	22	20	6	0.044598	0.034067
47	22	20	7	0.135443	0.090846
47	22	20	8	0.305779	0.170336

N	n	k	x	P(x)	p(x)
47	22	20	9	0.532894	0.227114
47	22	20	10	0.749409	0.216516
47	22	20	11	0.897034	0.147624
47	22	20	12	0.968675	0.071641
47	22	20	13	0.993168	0.024493
47	22	20	14	0.998969	0.005801
47	22	20	15	0.999897	0.000928
47	22	20	16	0.999993	0.000097
47	22	20	17	1.000000	0.000006
47	22	20	18	1.000000	0.000000
47	22	20	19	1.000000	0.000000
47	22	20	20	1.000000	0.000000
47	22	21	0	0.000000	0.000000
47	22	21	1	0.000000	0.000000
47	22	21	2	0.000003	0.000003
47	22	21	3	0.000062	0.000059
47	22	21	4	0.000693	0.000630
47	22	21	5	0.004979	0.004286
47	22	21	6	0.024410	0.019431
47	22	21	7	0.084974	0.060564
47	22	21	8	0.217457	0.132483
47	22	21	9	0.423542	0.206085
47	22	21	10	0.653180	0.229638
47	22	21	11	0.836891	0.183710
47	22	21	12	0.942141	0.105251
47	22	21	13	0.985003	0.042862
47	22	21	14	0.997250	0.012246
47	22	21	15	0.999656	0.002406
47	22	21	16	0.999972	0.000316
47	22	21	17	0.999998	0.000027
47	22	21	18	1.000000	0.000001
47	22	21	19	1.000000	0.000000
47	22	21	20	1.000000	0.000000
47	22	21	21	1.000000	0.000000
47	22	22	0	0.000000	0.000000
47	22	22	1	0.000000	0.000000
47	22	22	2	0.000001	0.000001
47	22	22	3	0.000019	0.000018
47	22	22	4	0.000256	0.000237
47	22	22	5	0.002176	0.001920
47	22	22	6	0.012452	0.010276
47	22	22	7	0.050033	0.037581
47	22	22	8	0.146120	0.096087
47	22	22	9	0.320500	0.174380
47	22	22	10	0.547193	0.226694
47	22	22	11	0.759167	0.211973
47	22	22	12	0.901660	0.142493
47	22	22	13	0.970167	0.068506
47	22	22	14	0.993482	0.023315
47	22	22	15	0.999008	0.005527

N	n	k	x	P(x)	p(x)
47	22	22	16	0.999899	0.000891
47	22	22	17	0.999993	0.000094
47	22	22	18	1.000000	0.000006
47	22	22	19	1.000000	0.000000
47	22	22	20	1.000000	0.000000
47	22	22	21	1.000000	0.000000
47	22	22	22	1.000000	0.000000
47	23	1	0	0.510638	0.510638
47	23	1	1	1.000000	0.489362
47	23	2	0	0.255319	0.255319
47	23	2	1	0.765957	0.510638
47	23	2	2	1.000000	0.234043
47	23	3	0	0.124823	0.124823
47	23	3	1	0.516312	0.391489
47	23	3	2	0.890780	0.374468
47	23	3	3	1.000000	0.109220
47	23	4	0	0.059574	0.059574
47	23	4	1	0.320567	0.260993
47	23	4	2	0.712057	0.391489
47	23	4	3	0.950355	0.238298
47	23	4	4	1.000000	0.049645
47	23	5	0	0.027709	0.027709
47	23	5	1	0.187036	0.159327
47	23	5	2	0.520864	0.333828
47	23	5	3	0.839518	0.318654
47	23	5	4	0.978064	0.138545
47	23	5	5	1.000000	0.021936
47	23	6	0	0.012535	0.012535
47	23	6	1	0.103579	0.091044
47	23	6	2	0.353950	0.250371
47	23	6	3	0.687778	0.333828
47	23	6	4	0.915388	0.227610
47	23	6	5	0.990599	0.075210
47	23	6	6	1.000000	0.009401
47	23	7	0	0.005503	0.005503
47	23	7	1	0.054726	0.049223
47	23	7	2	0.225711	0.170985
47	23	7	3	0.524935	0.299224
47	23	7	4	0.809911	0.284975
47	23	7	5	0.957579	0.147669
47	23	7	6	0.996102	0.038522
47	23	7	7	1.000000	0.003898
47	23	8	0	0.002339	0.002339
47	23	8	1	0.027654	0.025315
47	23	8	2	0.135944	0.108291
47	23	8	3	0.375323	0.239379
47	23	8	4	0.674547	0.299224
47	23	8	5	0.891128	0.216581
47	23	8	6	0.979730	0.088601
47	23	8	7	0.998441	0.018711

N	n	k	x	P(x)	p(x)
47	23	8	8	1.000000	0.001559
47	23	9	0	0.000960	0.000960
47	23	9	1	0.013373	0.012414
47	23	9	2	0.077634	0.064260
47	23	9	3	0.252565	0.174931
47	23	9	4	0.528771	0.276207
47	23	9	5	0.791168	0.262396
47	23	9	6	0.941109	0.149941
47	23	9	7	0.990764	0.049656
47	23	9	8	0.999400	0.008636
47	23	9	9	1.000000	0.000600
47	23	10	0	0.000379	0.000379
47	23	10	1	0.006186	0.005808
47	23	10	2	0.042121	0.035935
47	23	10	3	0.160496	0.118374
47	23	10	4	0.390668	0.230172
47	23	10	5	0.666875	0.276207
47	23	10	6	0.874030	0.207155
47	23	10	7	0.969857	0.095827
47	23	10	8	0.995991	0.026135
47	23	10	9	0.999779	0.003788
47	23	10	10	1.000000	0.000221
47	23	11	0	0.000143	0.000143
47	23	11	1	0.002733	0.002590
47	23	11	2	0.021726	0.018993
47	23	11	3	0.096510	0.074784
47	23	11	4	0.272472	0.175962
47	23	11	5	0.532504	0.260032
47	23	11	6	0.778851	0.246347
47	23	11	7	0.928418	0.149568
47	23	11	8	0.985396	0.056978
47	23	11	9	0.998346	0.012950
47	23	11	10	0.999922	0.001576
47	23	11	11	1.000000	0.000078
47	23	12	0	0.000052	0.000052
47	23	12	1	0.001150	0.001099
47	23	12	2	0.010647	0.009496
47	23	12	3	0.054963	0.044316
47	23	12	4	0.179603	0.124640
47	23	12	5	0.402488	0.222885
47	23	12	6	0.662520	0.260032
47	23	12	7	0.861944	0.199423
47	23	12	8	0.961655	0.099712
47	23	12	9	0.993310	0.031655
47	23	12	10	0.999353	0.006043
47	23	12	11	0.999974	0.000621
47	23	12	12	1.000000	0.000026
47	23	13	0	0.000018	0.000018
47	23	13	1	0.000460	0.000442
47	23	13	2	0.004949	0.004489

N	n	k	x	P(x)	p(x)
47	23	13	3	0.029640	0.024691
47	23	13	4	0.111941	0.082302
47	23	13	5	0.287861	0.175920
47	23	13	6	0.536219	0.248358
47	23	13	7	0.770779	0.234560
47	23	13	8	0.918922	0.148143
47	23	13	9	0.980648	0.061726
47	23	13	10	0.997108	0.016460
47	23	13	11	0.999761	0.002653
47	23	13	12	0.999992	0.000231
47	23	13	13	1.000000	0.000008
47	23	14	0	0.000006	0.000006
47	23	14	1	0.000174	0.000168
47	23	14	2	0.002176	0.002003
47	23	14	3	0.015116	0.012939
47	23	14	4	0.065949	0.050833
47	23	14	5	0.194727	0.128778
47	23	14	6	0.412040	0.217313
47	23	14	7	0.660398	0.248358
47	23	14	8	0.853565	0.193167
47	23	14	9	0.955231	0.101667
47	23	14	10	0.990815	0.035583
47	23	14	11	0.998825	0.008010
47	23	14	12	0.999917	0.001092
47	23	14	13	0.999997	0.000080
47	23	14	14	1.000000	0.000002
47	23	15	0	0.000002	0.000002
47	23	15	1	0.000062	0.000060
47	23	15	2	0.000902	0.000840
47	23	15	3	0.007274	0.006372
47	23	15	4	0.036681	0.029408
47	23	15	5	0.124485	0.087803
47	23	15	6	0.300091	0.175606
47	23	15	7	0.539982	0.239891
47	23	15	8	0.765761	0.225780
47	23	15	9	0.912100	0.146339
47	23	15	10	0.976797	0.064697
47	23	15	11	0.995912	0.019115
47	23	15	12	0.999553	0.003641
47	23	15	13	0.999973	0.000420
47	23	15	14	0.999999	0.000026
47	23	15	15	1.000000	0.000001
47	23	16	0	0.000000	0.000000
47	23	16	1	0.000020	0.000020
47	23	16	2	0.000351	0.000330
47	23	16	3	0.003291	0.002941
47	23	16	4	0.019221	0.015929
47	23	16	5	0.075095	0.055875
47	23	16	6	0.206800	0.131705
47	23	16	7	0.420036	0.213236

N	n	k	x	P(x)	p(x)
47	23	16	8	0.659927	0.239891
47	23	16	9	0.848077	0.188150
47	23	16	10	0.950514	0.102437
47	23	16	11	0.988744	0.038230
47	23	16	12	0.998301	0.009558
47	23	16	13	0.999842	0.001540
47	23	16	14	0.999992	0.000150
47	23	16	15	1.000000	0.000008
47	23	16	16	1.000000	0.000000
47	23	17	0	0.000000	0.000000
47	23	17	1	0.000006	0.000006
47	23	17	2	0.000127	0.000121
47	23	17	3	0.001394	0.001267
47	23	17	4	0.009457	0.008063
47	23	17	5	0.042652	0.033194
47	23	17	6	0.134575	0.091923
47	23	17	7	0.309979	0.175404
47	23	17	8	0.543851	0.233872
47	23	17	9	0.763106	0.219255
47	23	17	10	0.907556	0.144450
47	23	17	11	0.973945	0.066389
47	23	17	12	0.994910	0.020965
47	23	17	13	0.999345	0.004435
47	23	17	14	0.999948	0.000603
47	23	17	15	0.999998	0.000049
47	23	17	16	1.000000	0.000002
47	23	17	17	1.000000	0.000000
47	23	18	0	0.000000	0.000000
47	23	18	1	0.000002	0.000002
47	23	18	2	0.000042	0.000041
47	23	18	3	0.000549	0.000507
47	23	18	4	0.004351	0.003801
47	23	18	5	0.022735	0.018385
47	23	18	6	0.082485	0.059750
47	23	18	7	0.216430	0.133945
47	23	18	8	0.426915	0.210485
47	23	18	9	0.660787	0.233872
47	23	18	10	0.844961	0.184174
47	23	18	11	0.947390	0.102428
47	23	18	12	0.987223	0.039833
47	23	18	13	0.997867	0.010644
47	23	18	14	0.999767	0.001901
47	23	18	15	0.999985	0.000217
47	23	18	16	0.999999	0.000015
47	23	18	17	1.000000	0.000001
47	23	18	18	1.000000	0.000000
47	23	19	0	0.000000	0.000000
47	23	19	1	0.000000	0.000000
47	23	19	2	0.000013	0.000013
47	23	19	3	0.000200	0.000187

Table for $N = 2$, $n = 1$, through $N = 100$, $n = 50$

N	n	k	x	P(x)	p(x)
47	23	19	4	0.001860	0.001660
47	23	19	5	0.011324	0.009464
47	23	19	6	0.047459	0.036135
47	23	19	7	0.142529	0.095070
47	23	19	8	0.318043	0.175514
47	23	19	9	0.547883	0.229840
47	23	19	10	0.762400	0.214517
47	23	19	11	0.905006	0.142605
47	23	19	12	0.972114	0.067108
47	23	19	13	0.994196	0.022083
47	23	19	14	0.999177	0.004981
47	23	19	15	0.999925	0.000747
47	23	19	16	0.999996	0.000071
47	23	19	17	1.000000	0.000004
47	23	19	18	1.000000	0.000000
47	23	19	19	1.000000	0.000000
47	23	20	0	0.000000	0.000000
47	23	20	1	0.000000	0.000000
47	23	20	2	0.000004	0.000003
47	23	20	3	0.000066	0.000063
47	23	20	4	0.000733	0.000667
47	23	20	5	0.005240	0.004507
47	23	20	6	0.025520	0.020280
47	23	20	7	0.088204	0.062684
47	23	20	8	0.224018	0.135814
47	23	20	9	0.432963	0.208945
47	23	20	10	0.662803	0.229840
47	23	20	11	0.843889	0.181086
47	23	20	12	0.945750	0.101861
47	23	20	13	0.986310	0.040560
47	23	20	14	0.997576	0.011267
47	23	20	15	0.999711	0.002135
47	23	20	16	0.999978	0.000267
47	23	20	17	0.999999	0.000021
47	23	20	18	1.000000	0.000001
47	23	20	19	1.000000	0.000000
47	23	20	20	1.000000	0.000000
47	23	21	0	0.000000	0.000000
47	23	21	1	0.000000	0.000000
47	23	21	2	0.000001	0.000001
47	23	21	3	0.000020	0.000019
47	23	21	4	0.000264	0.000244
47	23	21	5	0.002236	0.001972
47	23	21	6	0.012751	0.010516
47	23	21	7	0.051058	0.038307
47	23	21	8	0.148566	0.097508
47	23	21	9	0.324621	0.176056
47	23	21	10	0.552140	0.227518
47	23	21	11	0.763406	0.211267
47	23	21	12	0.904251	0.140845

N	n	k	x	P(x)	p(x)
47	23	21	13	0.971288	0.067037
47	23	21	14	0.993821	0.022533
47	23	21	15	0.999079	0.005258
47	23	21	16	0.999909	0.000830
47	23	21	17	0.999994	0.000085
47	23	21	18	1.000000	0.000005
47	23	21	19	1.000000	0.000000
47	23	21	20	1.000000	0.000000
47	23	21	21	1.000000	0.000000
47	23	22	0	0.000000	0.000000
47	23	22	1	0.000000	0.000000
47	23	22	2	0.000000	0.000000
47	23	22	3	0.000005	0.000005
47	23	22	4	0.000086	0.000080
47	23	22	5	0.000871	0.000785
47	23	22	6	0.005876	0.005005
47	23	22	7	0.027485	0.021609
47	23	22	8	0.092311	0.064827
47	23	22	9	0.229822	0.137511
47	23	22	10	0.438380	0.208558
47	23	22	11	0.665899	0.227518
47	23	22	12	0.844663	0.178764
47	23	22	13	0.945504	0.100841
47	23	22	14	0.986021	0.040517
47	23	22	15	0.997461	0.011440
47	23	22	16	0.999685	0.002224
47	23	22	17	0.999975	0.000289
47	23	22	18	0.999999	0.000024
47	23	22	19	1.000000	0.000001
47	23	22	20	1.000000	0.000000
47	23	22	21	1.000000	0.000000
47	23	22	22	1.000000	0.000000
47	23	23	0	0.000000	0.000000
47	23	23	1	0.000000	0.000000
47	23	23	2	0.000000	0.000000
47	23	23	3	0.000001	0.000001
47	23	23	4	0.000025	0.000023
47	23	23	5	0.000305	0.000281
47	23	23	6	0.002472	0.002167
47	23	23	7	0.013655	0.011183
47	23	23	8	0.053415	0.039760
47	23	23	9	0.152816	0.099401
47	23	23	10	0.329930	0.177114
47	23	23	11	0.556690	0.226760
47	23	23	12	0.766007	0.209317
47	23	23	13	0.905168	0.139161
47	23	23	14	0.971435	0.066267
47	23	23	15	0.993800	0.022365
47	23	23	16	0.999062	0.005262
47	23	23	17	0.999905	0.000843

N	n	k	x	P(x)	p(x)
47	23	23	18	0.999994	0.000089
47	23	23	19	1.000000	0.000006
47	23	23	20	1.000000	0.000000
47	23	23	21	1.000000	0.000000
47	23	23	22	1.000000	0.000000
47	23	23	23	1.000000	0.000000
48	1	1	0	0.979167	0.979167
48	1	1	1	1.000000	0.020833
48	2	1	0	0.958333	0.958333
48	2	1	1	1.000000	0.041667
48	2	2	0	0.917553	0.917553
48	2	2	1	0.999113	0.081560
48	2	2	2	1.000000	0.000887
48	3	1	0	0.937500	0.937500
48	3	1	1	1.000000	0.062500
48	3	2	0	0.877660	0.877660
48	3	2	1	0.997340	0.119681
48	3	2	2	1.000000	0.002660
48	3	3	0	0.820421	0.820421
48	3	3	1	0.992137	0.171716
48	3	3	2	0.999942	0.007805
48	3	3	3	1.000000	0.000058
48	4	1	0	0.916667	0.916667
48	4	1	1	1.000000	0.083333
48	4	2	0	0.838652	0.838652
48	4	2	1	0.994681	0.156028
48	4	2	2	1.000000	0.005319
48	4	3	0	0.765726	0.765726
48	4	3	1	0.984505	0.218779
48	4	3	2	0.999769	0.015264
48	4	3	3	1.000000	0.000231
48	4	4	0	0.697662	0.697662
48	4	4	1	0.969920	0.272258
48	4	4	2	0.999090	0.029171
48	4	4	3	0.999995	0.000905
48	4	4	4	1.000000	0.000005
48	5	1	0	0.895833	0.895833
48	5	1	1	1.000000	0.104167
48	5	2	0	0.800532	0.800532
48	5	2	1	0.991135	0.190603
48	5	2	2	1.000000	0.008865
48	5	3	0	0.713518	0.713518
48	5	3	1	0.974561	0.261043
48	5	3	2	0.999422	0.024861
48	5	3	3	1.000000	0.000578
48	5	4	0	0.634238	0.634238
48	5	4	1	0.951357	0.317119
48	5	4	2	0.997764	0.046408
48	5	4	3	0.999974	0.002210
48	5	4	4	1.000000	0.000026

N	n	k	x	P(x)	p(x)
48	5	5	0	0.562165	0.562165
48	5	5	1	0.922528	0.360362
48	5	5	2	0.994600	0.072072
48	5	5	3	0.999874	0.005274
48	5	5	4	0.999999	0.000126
48	5	5	5	1.000000	0.000001
48	6	1	0	0.875000	0.875000
48	6	1	1	1.000000	0.125000
48	6	2	0	0.763298	0.763298
48	6	2	1	0.986702	0.223404
48	6	2	2	1.000000	0.013298
48	6	3	0	0.663737	0.663737
48	6	3	1	0.962419	0.298682
48	6	3	2	0.998844	0.036425
48	6	3	3	1.000000	0.001156
48	6	4	0	0.575239	0.575239
48	6	4	1	0.929232	0.353993
48	6	4	2	0.995606	0.066374
48	6	4	3	0.999923	0.004317
48	6	4	4	1.000000	0.000077
48	6	5	0	0.496797	0.496797
48	6	5	1	0.889006	0.392208
48	6	5	2	0.989572	0.100566
48	6	5	3	0.999628	0.010057
48	6	5	4	0.999996	0.000368
48	6	5	5	1.000000	0.000004
48	6	6	0	0.427477	0.427477
48	6	6	1	0.843400	0.415923
48	6	6	2	0.980217	0.136817
48	6	6	3	0.998927	0.018710
48	6	6	4	0.999979	0.001052
48	6	6	5	1.000000	0.000021
48	6	6	6	1.000000	0.000000
48	7	1	0	0.854167	0.854167
48	7	1	1	1.000000	0.145833
48	7	2	0	0.726950	0.726950
48	7	2	1	0.981383	0.254433
48	7	2	2	1.000000	0.018617
48	7	3	0	0.616327	0.616327
48	7	3	1	0.948196	0.331869
48	7	3	2	0.997976	0.049780
48	7	3	3	1.000000	0.002024
48	7	4	0	0.520454	0.520454
48	7	4	1	0.903947	0.383493
48	7	4	2	0.992445	0.088498
48	7	4	3	0.999820	0.007375
48	7	4	4	1.000000	0.000180
48	7	5	0	0.437655	0.437655
48	7	5	1	0.851652	0.413998
48	7	5	2	0.982389	0.130736

Table for $N = 2$, $n = 1$, through $N = 100$, $n = 50$

N	n	k	x	P(x)	p(x)
48	7	5	3	0.999150	0.016761
48	7	5	4	0.999988	0.000838
48	7	5	5	1.000000	0.000012
48	7	6	0	0.366409	0.366409
48	7	6	1	0.793885	0.427477
48	7	6	2	0.967187	0.173301
48	7	6	3	0.997590	0.030404
48	7	6	4	0.999929	0.002339
48	7	6	5	0.999999	0.000070
48	7	6	6	1.000000	0.000001
48	7	7	0	0.305340	0.305340
48	7	7	1	0.732817	0.427477
48	7	7	2	0.946556	0.213738
48	7	7	3	0.994695	0.048139
48	7	7	4	0.999762	0.005067
48	7	7	5	0.999996	0.000234
48	7	7	6	1.000000	0.000004
48	7	7	7	1.000000	0.000000
48	8	1	0	0.833333	0.833333
48	8	1	1	1.000000	0.166667
48	8	2	0	0.691489	0.691489
48	8	2	1	0.975177	0.283688
48	8	2	2	1.000000	0.024823
48	8	3	0	0.571230	0.571230
48	8	3	1	0.932007	0.360777
48	8	3	2	0.996762	0.064755
48	8	3	3	1.000000	0.003238
48	8	4	0	0.469678	0.469678
48	8	4	1	0.875886	0.406208
48	8	4	2	0.988128	0.112242
48	8	4	3	0.999640	0.011512
48	8	4	4	1.000000	0.000360
48	8	5	0	0.384282	0.384282
48	8	5	1	0.811262	0.426980
48	8	5	2	0.972823	0.161560
48	8	5	3	0.998332	0.025509
48	8	5	4	0.999967	0.001635
48	8	5	5	1.000000	0.000033
48	8	6	0	0.312788	0.312788
48	8	6	1	0.741754	0.428966
48	8	6	2	0.950279	0.208525
48	8	6	3	0.995366	0.045087
48	8	6	4	0.999815	0.004449
48	8	6	5	0.999998	0.000183
48	8	6	6	1.000000	0.000002
48	8	7	0	0.253209	0.253209
48	8	7	1	0.670260	0.417050
48	8	7	2	0.920490	0.250230
48	8	7	3	0.989998	0.069508
48	8	7	4	0.999391	0.009393
48	8	7	5	0.999985	0.000593
48	8	7	6	1.000000	0.000015
48	8	7	7	1.000000	0.000000
48	8	8	0	0.203803	0.203803
48	8	8	1	0.599056	0.395253
48	8	8	2	0.883871	0.284815
48	8	8	3	0.981522	0.097651
48	8	8	4	0.998475	0.016953
48	8	8	5	0.999941	0.001466
48	8	8	6	0.999999	0.000058
48	8	8	7	1.000000	0.000001
48	8	8	8	1.000000	0.000000
48	9	1	0	0.812500	0.812500
48	9	1	1	1.000000	0.187500
48	9	2	0	0.656915	0.656915
48	9	2	1	0.968085	0.311170
48	9	2	2	1.000000	0.031915
48	9	3	0	0.528388	0.528388
48	9	3	1	0.913969	0.385580
48	9	3	2	0.995143	0.081175
48	9	3	3	1.000000	0.004857
48	9	4	0	0.422710	0.422710
48	9	4	1	0.845421	0.422710
48	9	4	2	0.982516	0.137095
48	9	4	3	0.999352	0.016836
48	9	4	4	1.000000	0.000648
48	9	5	0	0.336247	0.336247
48	9	5	1	0.768564	0.432317
48	9	5	2	0.960706	0.192141
48	9	5	3	0.997057	0.036351
48	9	5	4	0.999926	0.002870
48	9	5	5	1.000000	0.000074
48	9	6	0	0.265870	0.265870
48	9	6	1	0.688133	0.422264
48	9	6	2	0.929427	0.241293
48	9	6	3	0.991984	0.062558
48	9	6	4	0.999593	0.007608
48	9	6	5	0.999993	0.000400
48	9	6	6	1.000000	0.000007
48	9	7	0	0.208898	0.208898
48	9	7	1	0.607702	0.398804
48	9	7	2	0.889211	0.281509
48	9	7	3	0.983047	0.093836
48	9	7	4	0.998687	0.015639
48	9	7	5	0.999955	0.001268
48	9	7	6	0.999999	0.000044
48	9	7	7	1.000000	0.000000
48	9	8	0	0.163042	0.163042
48	9	8	1	0.529887	0.366845
48	9	8	2	0.841149	0.311262
48	9	8	3	0.969315	0.128167
48	9	8	4	0.996780	0.027464
48	9	8	5	0.999831	0.003052
48	9	8	6	0.999996	0.000165
48	9	8	7	1.000000	0.000004
48	9	8	8	1.000000	0.000000
48	9	9	0	0.126358	0.126358
48	9	9	1	0.456518	0.330160
48	9	9	2	0.786678	0.330160
48	9	9	3	0.950090	0.163413
48	9	9	4	0.993347	0.043256
48	9	9	5	0.999526	0.006179
48	9	9	6	0.999984	0.000458
48	9	9	7	1.000000	0.000016
48	9	9	8	1.000000	0.000000
48	9	9	9	1.000000	0.000000
48	10	1	0	0.791667	0.791667
48	10	1	1	1.000000	0.208333
48	10	2	0	0.623227	0.623227
48	10	2	1	0.960106	0.336879
48	10	2	2	1.000000	0.039894
48	10	3	0	0.487743	0.487743
48	10	3	1	0.894195	0.406452
48	10	3	2	0.993062	0.098867
48	10	3	3	1.000000	0.006938
48	10	4	0	0.379356	0.379356
48	10	4	1	0.812905	0.433549
48	10	4	2	0.975486	0.162581
48	10	4	3	0.998921	0.023435
48	10	4	4	1.000000	0.001079
48	10	5	0	0.293138	0.293138
48	10	5	1	0.724224	0.431086
48	10	5	2	0.945925	0.221701
48	10	5	3	0.995192	0.049267
48	10	5	4	0.999853	0.004660
48	10	5	5	1.000000	0.000147
48	10	6	0	0.224967	0.224967
48	10	6	1	0.633997	0.409030
48	10	6	2	0.904679	0.270682
48	10	6	3	0.987172	0.082493
48	10	6	4	0.999202	0.012030
48	10	6	5	0.999983	0.000780
48	10	6	6	1.000000	0.000017
48	10	7	0	0.171403	0.171403
48	10	7	1	0.546348	0.374944
48	10	7	2	0.853120	0.306773
48	10	7	3	0.973423	0.120303
48	10	7	4	0.997484	0.024061
48	10	7	5	0.999890	0.002406
48	10	7	6	0.999998	0.000108
48	10	7	7	1.000000	0.000002
48	10	8	0	0.129597	0.129597
48	10	8	1	0.464043	0.334445
48	10	8	2	0.793262	0.329219
48	10	8	3	0.952884	0.159622
48	10	8	4	0.993963	0.041079
48	10	8	5	0.999596	0.005634
48	10	8	6	0.999988	0.000391
48	10	8	7	1.000000	0.000012
48	10	8	8	1.000000	0.000000
48	10	9	0	0.097198	0.097198
48	10	9	1	0.388792	0.291594
48	10	9	2	0.727418	0.338626
48	10	9	3	0.924950	0.197532
48	10	9	4	0.987801	0.062851
48	10	9	5	0.998892	0.011091
48	10	9	6	0.999949	0.001056
48	10	9	7	0.999999	0.000050
48	10	9	8	1.000000	0.000001
48	10	9	9	1.000000	0.000000
48	10	10	0	0.072276	0.072276
48	10	10	1	0.321502	0.249226
48	10	10	2	0.657957	0.336455
48	10	10	3	0.889496	0.231539
48	10	10	4	0.978132	0.088636
48	10	10	5	0.997470	0.019339
48	10	10	6	0.999840	0.002370
48	10	10	7	0.999995	0.000155
48	10	10	8	1.000000	0.000005
48	10	10	9	1.000000	0.000000
48	10	10	10	1.000000	0.000000
48	11	1	0	0.770833	0.770833
48	11	1	1	1.000000	0.229167
48	11	2	0	0.590425	0.590425
48	11	2	1	0.951241	0.360816
48	11	2	2	1.000000	0.048759
48	11	3	0	0.449237	0.449237
48	11	3	1	0.872803	0.423566
48	11	3	2	0.990460	0.117657
48	11	3	3	1.000000	0.009540
48	11	4	0	0.339423	0.339423
48	11	4	1	0.778677	0.439254
48	11	4	2	0.966929	0.188252
48	11	4	3	0.998304	0.031375
48	11	4	4	1.000000	0.001696
48	11	5	0	0.254567	0.254567
48	11	5	1	0.678847	0.424279
48	11	5	2	0.928423	0.249576
48	11	5	3	0.992599	0.064177
48	11	5	4	0.999730	0.007131

N	n	k	x	P(x)	p(x)
48	11	5	5	1.000000	0.000270
48	11	6	0	0.189446	0.189446
48	11	6	1	0.580177	0.390732
48	11	6	2	0.876186	0.296009
48	11	6	3	0.980660	0.104474
48	11	6	4	0.998569	0.017910
48	11	6	5	0.999962	0.001393
48	11	6	6	1.000000	0.000038
48	11	7	0	0.139829	0.139829
48	11	7	1	0.487146	0.347317
48	11	7	2	0.812755	0.325610
48	11	7	3	0.960760	0.148004
48	11	7	4	0.995584	0.034825
48	11	7	5	0.999763	0.004179
48	11	7	6	0.999995	0.000232
48	11	7	7	1.000000	0.000004
48	11	8	0	0.102314	0.102314
48	11	8	1	0.402434	0.300121
48	11	8	2	0.741280	0.338846
48	11	8	3	0.931881	0.190601
48	11	8	4	0.989639	0.057758
48	11	8	5	0.999152	0.009513
48	11	8	6	0.999967	0.000815
48	11	8	7	0.999999	0.000032
48	11	8	8	1.000000	0.000000
48	11	9	0	0.074178	0.074178
48	11	9	1	0.327404	0.253227
48	11	9	2	0.665040	0.337636
48	11	9	3	0.893761	0.228721
48	11	9	4	0.979531	0.085770
48	11	9	5	0.997725	0.018194
48	11	9	6	0.999865	0.002140
48	11	9	7	0.999996	0.000131
48	11	9	8	1.000000	0.000004
48	11	9	9	1.000000	0.000000
48	11	10	0	0.053256	0.053256
48	11	10	1	0.262474	0.209219
48	11	10	2	0.587124	0.324650
48	11	10	3	0.846844	0.259720
48	11	10	4	0.964136	0.117293
48	11	10	5	0.994926	0.030789
48	11	10	6	0.999591	0.004665
48	11	10	7	0.999983	0.000392
48	11	10	8	1.000000	0.000017
48	11	10	9	1.000000	0.000000
48	11	10	10	1.000000	0.000000
48	11	11	0	0.037840	0.037840
48	11	11	1	0.207417	0.169577
48	11	11	2	0.510233	0.302816
48	11	11	3	0.792166	0.281933

N	n	k	x	P(x)	p(x)
48	11	11	4	0.942530	0.150364
48	11	11	5	0.990064	0.047534
48	11	11	6	0.998977	0.008913
48	11	11	7	0.999942	0.000965
48	11	11	8	0.999998	0.000057
48	11	11	9	1.000000	0.000002
48	11	11	10	1.000000	0.000000
48	11	11	11	1.000000	0.000000
48	12	1	0	0.750000	0.750000
48	12	1	1	1.000000	0.250000
48	12	2	0	0.558511	0.558511
48	12	2	1	0.941489	0.382979
48	12	2	2	1.000000	0.058511
48	12	3	0	0.412812	0.412812
48	12	3	1	0.849907	0.437095
48	12	3	2	0.987280	0.137373
48	12	3	3	1.000000	0.012720
48	12	4	0	0.302729	0.302729
48	12	4	1	0.743062	0.440333
48	12	4	2	0.956753	0.213691
48	12	4	3	0.997456	0.040703
48	12	4	4	1.000000	0.002544
48	12	5	0	0.220166	0.220166
48	12	5	1	0.632979	0.412812
48	12	5	2	0.908187	0.275208
48	12	5	3	0.989130	0.080944
48	12	5	4	0.999537	0.010407
48	12	5	5	1.000000	0.000463
48	12	6	0	0.158725	0.158725
48	12	6	1	0.527376	0.368651
48	12	6	2	0.844185	0.316809
48	12	6	3	0.972189	0.128004
48	12	6	4	0.997601	0.025413
48	12	6	5	0.999925	0.002323
48	12	6	6	1.000000	0.000075
48	12	7	0	0.113375	0.113375
48	12	7	1	0.430824	0.317449
48	12	7	2	0.768754	0.337930
48	12	7	3	0.944759	0.176005
48	12	7	4	0.992761	0.048001
48	12	7	5	0.999537	0.006777
48	12	7	6	0.999989	0.000452
48	12	7	7	1.000000	0.000011
48	12	8	0	0.080192	0.080192
48	12	8	1	0.345655	0.265463
48	12	8	2	0.686332	0.340677
48	12	8	3	0.906124	0.219792
48	12	8	4	0.983395	0.077271
48	12	8	5	0.998380	0.014986
48	12	8	6	0.999923	0.001543

N	n	k	x	P(x)	p(x)
48	12	8	7	0.999999	0.000076
48	12	8	8	1.000000	0.000001
48	12	9	0	0.056134	0.056134
48	12	9	1	0.272653	0.216518
48	12	9	2	0.601163	0.328510
48	12	9	3	0.856671	0.255508
48	12	9	4	0.967940	0.111270
48	12	9	5	0.995758	0.027817
48	12	9	6	0.999692	0.003934
48	12	9	7	0.999989	0.000298
48	12	9	8	1.000000	0.000011
48	12	9	9	1.000000	0.000000
48	12	10	0	0.038862	0.038862
48	12	10	1	0.211583	0.172721
48	12	10	2	0.516929	0.305346
48	12	10	3	0.797707	0.280778
48	12	10	4	0.945116	0.147408
48	12	10	5	0.990765	0.045649
48	12	10	6	0.999086	0.008321
48	12	10	7	0.999951	0.000865
48	12	10	8	0.999999	0.000048
48	12	10	9	1.000000	0.000001
48	12	10	10	1.000000	0.000000
48	12	11	0	0.026590	0.026590
48	12	11	1	0.161585	0.134995
48	12	11	2	0.436575	0.274990
48	12	11	3	0.731207	0.294632
48	12	11	4	0.914083	0.182875
48	12	11	5	0.982356	0.068273
48	12	11	6	0.997772	0.015417
48	12	11	7	0.999837	0.002065
48	12	11	8	0.999994	0.000156
48	12	11	9	1.000000	0.000006
48	12	11	10	1.000000	0.000000
48	12	11	11	1.000000	0.000000
48	12	12	0	0.017966	0.017966
48	12	12	1	0.121451	0.103485
48	12	12	2	0.362254	0.240802
48	12	12	3	0.659540	0.297287
48	12	12	4	0.874542	0.215002
48	12	12	5	0.969439	0.094897
48	12	12	6	0.995273	0.025833
48	12	12	7	0.999558	0.004286
48	12	12	8	0.999977	0.000419
48	12	12	9	0.999999	0.000023
48	12	12	10	1.000000	0.000001
48	12	12	11	1.000000	0.000000
48	12	12	12	1.000000	0.000000
48	13	1	0	0.729167	0.729167
48	13	1	1	1.000000	0.270833

N	n	k	x	P(x)	p(x)
48	13	2	0	0.527482	0.527482
48	13	2	1	0.930851	0.403369
48	13	2	2	1.000000	0.069149
48	13	3	0	0.378411	0.378411
48	13	3	1	0.825624	0.447213
48	13	3	2	0.983464	0.157840
48	13	3	3	1.000000	0.016536
48	13	4	0	0.269092	0.269092
48	13	4	1	0.706368	0.437275
48	13	4	2	0.944881	0.238514
48	13	4	3	0.996325	0.051444
48	13	4	4	1.000000	0.003675
48	13	5	0	0.189588	0.189588
48	13	5	1	0.587111	0.397523
48	13	5	2	0.885253	0.298142
48	13	5	3	0.984634	0.099381
48	13	5	4	0.999248	0.014615
48	13	5	5	1.000000	0.000752
48	13	6	0	0.132271	0.132271
48	13	6	1	0.476174	0.343903
48	13	6	2	0.808984	0.332810
48	13	6	3	0.961522	0.152538
48	13	6	4	0.996189	0.034668
48	13	6	5	0.999860	0.003671
48	13	6	6	1.000000	0.000140
48	13	7	0	0.091330	0.091330
48	13	7	1	0.377916	0.286586
48	13	7	2	0.721819	0.343903
48	13	7	3	0.925203	0.203384
48	13	7	4	0.988761	0.063557
48	13	7	5	0.999161	0.010400
48	13	7	6	0.999977	0.000816
48	13	7	7	1.000000	0.000023
48	13	8	0	0.062371	0.062371
48	13	8	1	0.294037	0.231665
48	13	8	2	0.629553	0.335516
48	13	8	3	0.875597	0.246045
48	13	8	4	0.974809	0.099212
48	13	8	5	0.997132	0.022323
48	13	8	6	0.999837	0.002706
48	13	8	7	0.999997	0.000159
48	13	8	8	1.000000	0.000003
48	13	9	0	0.042101	0.042101
48	13	9	1	0.224537	0.182437
48	13	9	2	0.537286	0.312748
48	13	9	3	0.814086	0.276800
48	13	9	4	0.952486	0.138400
48	13	9	5	0.992667	0.040181
48	13	9	6	0.999364	0.006697
48	13	9	7	0.999973	0.000609

Table for $N = 2$, $n = 1$, through $N = 100$, $n = 50$

N	n	k	x	P(x)	p(x)
48	13	9	8	1.000000	0.000027
48	13	9	9	1.000000	0.000000
48	13	10	0	0.028067	0.028067
48	13	10	1	0.168403	0.140336
48	13	10	2	0.449075	0.280672
48	13	10	3	0.743112	0.294037
48	13	10	4	0.920548	0.177436
48	13	10	5	0.984425	0.063877
48	13	10	6	0.998162	0.013737
48	13	10	7	0.999879	0.001717
48	13	10	8	0.999996	0.000117
48	13	10	9	1.000000	0.000004
48	13	10	10	1.000000	0.000000
48	13	11	0	0.018465	0.018465
48	13	11	1	0.124086	0.105621
48	13	11	2	0.367828	0.243741
48	13	11	3	0.665734	0.297906
48	13	11	4	0.878523	0.212790
48	13	11	5	0.970977	0.092454
48	13	11	6	0.995631	0.024654
48	13	11	7	0.999608	0.003976
48	13	11	8	0.999981	0.000373
48	13	11	9	0.999999	0.000019
48	13	11	10	1.000000	0.000000
48	13	11	11	1.000000	0.000000
48	13	12	0	0.011977	0.011977
48	13	12	1	0.089831	0.077853
48	13	12	2	0.295364	0.205533
48	13	12	3	0.585218	0.289854
48	13	12	4	0.826764	0.241545
48	13	12	5	0.950987	0.124223
48	13	12	6	0.990967	0.039980
48	13	12	7	0.998963	0.007996
48	13	12	8	0.999930	0.000967
48	13	12	9	0.999997	0.000067
48	13	12	10	1.000000	0.000002
48	13	12	11	1.000000	0.000000
48	13	12	12	1.000000	0.000000
48	13	13	0	0.007652	0.007652
48	13	13	1	0.063880	0.056227
48	13	13	2	0.232562	0.168683
48	13	13	3	0.504703	0.272141
48	13	13	4	0.766377	0.261674
48	13	13	5	0.923382	0.157004
48	13	13	6	0.983193	0.059811
48	13	13	7	0.997630	0.014437
48	13	13	8	0.999796	0.002166
48	13	13	9	0.999990	0.000194
48	13	13	10	1.000000	0.000010
48	13	13	11	1.000000	0.000000

N	n	k	x	P(x)	p(x)
48	13	13	12	1.000000	0.000000
48	13	13	13	1.000000	0.000000
48	14	1	0	0.708333	0.708333
48	14	1	1	1.000000	0.291667
48	14	2	0	0.497340	0.497340
48	14	2	1	0.919326	0.421986
48	14	2	2	1.000000	0.080674
48	14	3	0	0.345976	0.345976
48	14	3	1	0.800069	0.454093
48	14	3	2	0.978955	0.178885
48	14	3	3	1.000000	0.021045
48	14	4	0	0.238339	0.238339
48	14	4	1	0.668887	0.430548
48	14	4	2	0.931252	0.262365
48	14	4	3	0.994856	0.063604
48	14	4	4	1.000000	0.005144
48	14	5	0	0.162504	0.162504
48	14	5	1	0.541679	0.379176
48	14	5	2	0.859698	0.318018
48	14	5	3	0.978955	0.119257
48	14	5	4	0.998831	0.019876
48	14	5	5	1.000000	0.001169
48	14	6	0	0.109596	0.109596
48	14	6	1	0.427045	0.317449
48	14	6	2	0.770948	0.343903
48	14	6	3	0.948447	0.177499
48	14	6	4	0.994208	0.045761
48	14	6	5	0.999755	0.005547
48	14	6	6	1.000000	0.000245
48	14	7	0	0.073064	0.073064
48	14	7	1	0.328787	0.255723
48	14	7	2	0.672690	0.343903
48	14	7	3	0.901959	0.229269
48	14	7	4	0.983313	0.081354
48	14	7	5	0.998567	0.015254
48	14	7	6	0.999953	0.001387
48	14	7	7	1.000000	0.000047
48	14	8	0	0.048115	0.048115
48	14	8	1	0.247704	0.199589
48	14	8	2	0.572036	0.324332
48	14	8	3	0.840448	0.268412
48	14	8	4	0.963470	0.123022
48	14	8	5	0.995218	0.031748
48	14	8	6	0.999683	0.004465
48	14	8	7	0.999992	0.000309
48	14	8	8	1.000000	0.000008
48	14	9	0	0.031275	0.031275
48	14	9	1	0.182838	0.151563
48	14	9	2	0.474736	0.291899
48	14	9	3	0.766635	0.291899

N	n	k	x	P(x)	p(x)
48	14	9	4	0.932715	0.166080
48	14	9	5	0.988075	0.055360
48	14	9	6	0.998790	0.010715
48	14	9	7	0.999938	0.001148
48	14	9	8	0.999999	0.000061
48	14	9	9	1.000000	0.000001
48	14	10	0	0.020048	0.020048
48	14	10	1	0.132317	0.112269
48	14	10	2	0.384921	0.252605
48	14	10	3	0.684304	0.299383
48	14	10	4	0.890130	0.205826
48	14	10	5	0.975300	0.085169
48	14	10	6	0.996592	0.021292
48	14	10	7	0.999732	0.003140
48	14	10	8	0.999989	0.000258
48	14	10	9	1.000000	0.000010
48	14	10	10	1.000000	0.000000
48	14	11	0	0.012662	0.012662
48	14	11	1	0.093909	0.081247
48	14	11	2	0.305151	0.211242
48	14	11	3	0.597641	0.292489
48	14	11	4	0.835966	0.238325
48	14	11	5	0.955128	0.119162
48	14	11	6	0.992109	0.036981
48	14	11	7	0.999153	0.007044
48	14	11	8	0.999949	0.000795
48	14	11	9	0.999998	0.000050
48	14	11	10	1.000000	0.000002
48	14	11	11	1.000000	0.000000
48	14	12	0	0.007871	0.007871
48	14	12	1	0.065363	0.057492
48	14	12	2	0.236640	0.171278
48	14	12	3	0.510684	0.274044
48	14	12	4	0.771553	0.260869
48	14	12	5	0.926142	0.154589
48	14	12	6	0.984113	0.057971
48	14	12	7	0.997821	0.013707
48	14	12	8	0.999820	0.001999
48	14	12	9	0.999992	0.000172
48	14	12	10	1.000000	0.000008
48	14	12	11	1.000000	0.000000
48	14	12	12	1.000000	0.000000
48	14	13	0	0.004810	0.004810
48	14	13	1	0.044602	0.039792
48	14	13	2	0.179548	0.134946
48	14	13	3	0.426949	0.247401
48	14	13	4	0.699090	0.272141
48	14	13	5	0.887495	0.188405
48	14	13	6	0.971231	0.083736
48	14	13	7	0.995155	0.023924

N	n	k	x	P(x)	p(x)
48	14	13	8	0.999487	0.004331
48	14	13	9	0.999968	0.000481
48	14	13	10	0.999999	0.000031
48	14	13	11	1.000000	0.000001
48	14	13	12	1.000000	0.000000
48	14	13	13	1.000000	0.000000
48	14	14	0	0.002886	0.002886
48	14	14	1	0.029822	0.026936
48	14	14	2	0.133281	0.103459
48	14	14	3	0.349194	0.215914
48	14	14	4	0.621335	0.272141
48	14	14	5	0.839048	0.217713
48	14	14	6	0.952091	0.113043
48	14	14	7	0.990370	0.038279
48	14	14	8	0.998744	0.008374
48	14	14	9	0.999899	0.001155
48	14	14	10	0.999995	0.000096
48	14	14	11	1.000000	0.000005
48	14	14	12	1.000000	0.000000
48	14	14	13	1.000000	0.000000
48	14	14	14	1.000000	0.000000
48	15	1	0	0.687500	0.687500
48	15	1	1	1.000000	0.312500
48	15	2	0	0.468085	0.468085
48	15	2	1	0.906915	0.438830
48	15	2	2	1.000000	0.093085
48	15	3	0	0.315449	0.315449
48	15	3	1	0.773358	0.457909
48	15	3	2	0.973693	0.200335
48	15	3	3	1.000000	0.026307
48	15	4	0	0.210299	0.210299
48	15	4	1	0.630897	0.420598
48	15	4	2	0.915819	0.284921
48	15	4	3	0.992985	0.077166
48	15	4	4	1.000000	0.007015
48	15	5	0	0.138606	0.138606
48	15	5	1	0.497071	0.358464
48	15	5	2	0.831637	0.334567
48	15	5	3	0.971939	0.140302
48	15	5	4	0.998246	0.026307
48	15	5	5	1.000000	0.001754
48	15	6	0	0.090255	0.090255
48	15	6	1	0.380361	0.290106
48	15	6	2	0.730489	0.350128
48	15	6	3	0.932785	0.202296
48	15	6	4	0.991517	0.058731
48	15	6	5	0.999592	0.008076
48	15	6	6	1.000000	0.000408
48	15	7	0	0.058021	0.058021
48	15	7	1	0.283659	0.225638

N	n	k	x	P(x)	p(x)
48	15	7	2	0.622116	0.338457
48	15	7	3	0.874986	0.252870
48	15	7	4	0.976135	0.101148
48	15	7	5	0.997669	0.021535
48	15	7	6	0.999912	0.002243
48	15	7	7	1.000000	0.000087
48	15	8	0	0.036794	0.036794
48	15	8	1	0.206612	0.169818
48	15	8	2	0.514801	0.308189
48	15	8	3	0.800976	0.286175
48	15	8	4	0.948997	0.148022
48	15	8	5	0.992417	0.043420
48	15	8	6	0.999420	0.007003
48	15	8	7	0.999983	0.000563
48	15	8	8	1.000000	0.000017
48	15	9	0	0.022996	0.022996
48	15	9	1	0.147176	0.124180
48	15	9	2	0.414639	0.267464
48	15	9	3	0.715123	0.300484
48	15	9	4	0.908291	0.193168
48	15	9	5	0.981562	0.073271
48	15	9	6	0.997844	0.016282
48	15	9	7	0.999870	0.002026
48	15	9	8	0.999997	0.000127
48	15	9	9	1.000000	0.000003
48	15	10	0	0.014152	0.014152
48	15	10	1	0.102598	0.088447
48	15	10	2	0.325485	0.222886
48	15	10	3	0.622667	0.297182
48	15	10	4	0.853808	0.231141
48	15	10	5	0.962775	0.108967
48	15	10	6	0.994087	0.031312
48	15	10	7	0.999455	0.005368
48	15	10	8	0.999974	0.000519
48	15	10	9	0.999999	0.000025
48	15	10	10	1.000000	0.000000
48	15	11	0	0.008565	0.008565
48	15	11	1	0.070013	0.061447
48	15	11	2	0.249234	0.179221
48	15	11	3	0.528820	0.279585
48	15	11	4	0.786899	0.258079
48	15	11	5	0.934099	0.147201
48	15	11	6	0.986671	0.052572
48	15	11	7	0.998325	0.011654
48	15	11	8	0.999878	0.001554
48	15	11	9	0.999995	0.000117
48	15	11	10	1.000000	0.000004
48	15	11	11	1.000000	0.000000
48	15	12	0	0.005093	0.005093
48	15	12	1	0.046762	0.041669

N	n	k	x	P(x)	p(x)
48	15	12	2	0.186265	0.139502
48	15	12	3	0.438143	0.251879
48	15	12	4	0.710172	0.272029
48	15	12	5	0.894315	0.184143
48	15	12	6	0.973883	0.079568
48	15	12	7	0.995805	0.021922
48	15	12	8	0.999585	0.003780
48	15	12	9	0.999976	0.000392
48	15	12	10	0.999999	0.000023
48	15	12	11	1.000000	0.000001
48	15	12	12	1.000000	0.000000
48	15	13	0	0.002971	0.002971
48	15	13	1	0.030558	0.027587
48	15	13	2	0.135889	0.105331
48	15	13	3	0.354184	0.218295
48	15	13	4	0.627052	0.272869
48	15	13	5	0.843165	0.216112
48	15	13	6	0.953991	0.110827
48	15	13	7	0.990933	0.036942
48	15	13	8	0.998850	0.007916
48	15	13	9	0.999911	0.001062
48	15	13	10	0.999996	0.000085
48	15	13	11	1.000000	0.000004
48	15	13	12	1.000000	0.000000
48	15	13	13	1.000000	0.000000
48	15	14	0	0.001698	0.001698
48	15	14	1	0.019523	0.017825
48	15	14	2	0.096766	0.077243
48	15	14	3	0.279340	0.182574
48	15	14	4	0.541294	0.261954
48	15	14	5	0.781418	0.240125
48	15	14	6	0.925493	0.144075
48	15	14	7	0.982490	0.056997
48	15	14	8	0.997266	0.014777
48	15	14	9	0.999729	0.002463
48	15	14	10	0.999984	0.000255
48	15	14	11	0.999999	0.000015
48	15	14	12	1.000000	0.000000
48	15	14	13	1.000000	0.000000
48	15	14	14	1.000000	0.000000
48	15	15	0	0.000949	0.000949
48	15	15	1	0.012183	0.011234
48	15	15	2	0.067232	0.055049
48	15	15	3	0.214902	0.147670
48	15	15	4	0.456544	0.241642
48	15	15	5	0.710793	0.254249
48	15	15	6	0.887355	0.176562
48	15	15	7	0.969079	0.081723
48	15	15	8	0.994224	0.025146
48	15	15	9	0.999295	0.005071

N	n	k	x	P(x)	p(x)
48	15	15	10	0.999946	0.000652
48	15	15	11	0.999998	0.000051
48	15	15	12	1.000000	0.000002
48	15	15	13	1.000000	0.000000
48	15	15	14	1.000000	0.000000
48	15	15	15	1.000000	0.000000
48	16	1	0	0.666667	0.666667
48	16	1	1	1.000000	0.333333
48	16	2	0	0.439716	0.439716
48	16	2	1	0.893617	0.453901
48	16	2	2	1.000000	0.106383
48	16	3	0	0.286771	0.286771
48	16	3	1	0.745606	0.458834
48	16	3	2	0.967622	0.222017
48	16	3	3	1.000000	0.032377
48	16	4	0	0.184808	0.184808
48	16	4	1	0.592661	0.407853
48	16	4	2	0.898551	0.305890
48	16	4	3	0.990646	0.092096
48	16	4	4	1.000000	0.009353
48	16	5	0	0.117605	0.117605
48	16	5	1	0.453620	0.336015
48	16	5	2	0.801222	0.347602
48	16	5	3	0.963436	0.162214
48	16	5	4	0.997449	0.034013
48	16	5	5	1.000000	0.002551
48	16	6	0	0.073845	0.073845
48	16	6	1	0.336406	0.262561
48	16	6	2	0.688049	0.351644
48	16	6	3	0.914395	0.226345
48	16	6	4	0.987957	0.073562
48	16	6	5	0.999347	0.011390
48	16	6	6	1.000000	0.000653
48	16	7	0	0.045714	0.045714
48	16	7	1	0.242634	0.196920
48	16	7	2	0.570835	0.328201
48	16	7	3	0.844336	0.273501
48	16	7	4	0.966939	0.122604
48	16	7	5	0.996364	0.029425
48	16	7	6	0.999845	0.003480
48	16	7	7	1.000000	0.000155
48	16	8	0	0.027874	0.027874
48	16	8	1	0.170590	0.142716
48	16	8	2	0.458766	0.288176
48	16	8	3	0.757616	0.298850
48	16	8	4	0.931055	0.173439
48	16	8	5	0.988470	0.057414
48	16	8	6	0.998996	0.010526
48	16	8	7	0.999966	0.000970
48	16	8	8	1.000000	0.000034

N	n	k	x	P(x)	p(x)
48	16	9	0	0.016725	0.016725
48	16	9	1	0.117072	0.100347
48	16	9	2	0.357905	0.240833
48	16	9	3	0.660490	0.302585
48	16	9	4	0.879023	0.218534
48	16	9	5	0.972681	0.093657
48	16	9	6	0.996364	0.023683
48	16	9	7	0.999747	0.003383
48	16	9	8	0.999993	0.000246
48	16	9	9	1.000000	0.000007
48	16	10	0	0.009863	0.009863
48	16	10	1	0.078477	0.068613
48	16	10	2	0.271452	0.192975
48	16	10	3	0.559628	0.288176
48	16	10	4	0.811782	0.252154
48	16	10	5	0.946265	0.134482
48	16	10	6	0.990292	0.044027
48	16	10	7	0.998967	0.008675
48	16	10	8	0.999943	0.000976
48	16	10	9	0.999999	0.000056
48	16	10	10	1.000000	0.000001
48	16	11	0	0.005710	0.005710
48	16	11	1	0.051392	0.045682
48	16	11	2	0.200356	0.148963
48	16	11	3	0.461041	0.260686
48	16	11	4	0.732155	0.271113
48	16	11	5	0.907335	0.175181
48	16	11	6	0.978705	0.071370
48	16	11	7	0.996912	0.018207
48	16	11	8	0.999737	0.002825
48	16	11	9	0.999988	0.000251
48	16	11	10	1.000000	0.000011
48	16	11	11	1.000000	0.000000
48	16	12	0	0.003241	0.003241
48	16	12	1	0.032873	0.029632
48	16	12	2	0.143991	0.111119
48	16	12	3	0.369449	0.225458
48	16	12	4	0.644226	0.274777
48	16	12	5	0.855255	0.211029
48	16	12	6	0.959416	0.104162
48	16	12	7	0.992483	0.033067
48	16	12	8	0.999126	0.006643
48	16	12	9	0.999941	0.000814
48	16	12	10	0.999998	0.000057
48	16	12	11	1.000000	0.000002
48	16	12	12	1.000000	0.000000
48	16	13	0	0.001801	0.001801
48	16	13	1	0.020526	0.018726
48	16	13	2	0.100778	0.080252
48	16	13	3	0.288034	0.187255

Table for $N = 2$, $n = 1$, through $N = 100$, $n = 50$

N	n	k	x	P(x)	p(x)
48	16	13	4	0.552634	0.264600
48	16	13	5	0.790774	0.238140
48	16	13	6	0.930482	0.139709
48	16	13	7	0.984217	0.053734
48	16	13	8	0.997650	0.013434
48	16	13	9	0.999782	0.002132
48	16	13	10	0.999988	0.000206
48	16	13	11	1.000000	0.000011
48	16	13	12	1.000000	0.000000
48	16	13	13	1.000000	0.000000
48	16	14	0	0.000977	0.000977
48	16	14	1	0.012501	0.011523
48	16	14	2	0.068677	0.056177
48	16	14	3	0.218482	0.149804
48	16	14	4	0.461914	0.243432
48	16	14	5	0.715930	0.254016
48	16	14	6	0.890566	0.174636
48	16	14	7	0.970399	0.079834
48	16	14	8	0.994580	0.024180
48	16	14	9	0.999356	0.004776
48	16	14	10	0.999953	0.000597
48	16	14	11	0.999998	0.000045
48	16	14	12	1.000000	0.000002
48	16	14	13	1.000000	0.000000
48	16	14	14	1.000000	0.000000
48	16	15	0	0.000517	0.000517
48	16	15	1	0.007417	0.006900
48	16	15	2	0.045546	0.038129
48	16	15	3	0.161204	0.115658
48	16	15	4	0.375997	0.214793
48	16	15	5	0.633748	0.257752
48	16	15	6	0.839202	0.205454
48	16	15	7	0.949267	0.110065
48	16	15	8	0.988890	0.039623
48	16	15	9	0.998373	0.009482
48	16	15	10	0.999848	0.001475
48	16	15	11	0.999991	0.000144
48	16	15	12	1.000000	0.000008
48	16	15	13	1.000000	0.000000
48	16	15	14	1.000000	0.000000
48	16	15	15	1.000000	0.000000
48	16	16	0	0.000267	0.000267
48	16	16	1	0.004281	0.004014
48	16	16	2	0.029370	0.025089
48	16	16	3	0.115641	0.086272
48	16	16	4	0.297890	0.182249
48	16	16	5	0.547831	0.249941
48	16	16	6	0.776943	0.229112
48	16	16	7	0.919249	0.142306
48	16	16	8	0.979285	0.060035
48	16	16	9	0.996361	0.017077
48	16	16	10	0.999580	0.003218
48	16	16	11	0.999970	0.000390
48	16	16	12	0.999999	0.000029
48	16	16	13	1.000000	0.000001
48	16	16	14	1.000000	0.000000
48	16	16	15	1.000000	0.000000
48	16	16	16	1.000000	0.000000
48	17	1	0	0.645833	0.645833
48	17	1	1	1.000000	0.354167
48	17	2	0	0.412234	0.412234
48	17	2	1	0.879433	0.467199
48	17	2	2	1.000000	0.120567
48	17	3	0	0.259887	0.259887
48	17	3	1	0.716929	0.457042
48	17	3	2	0.960685	0.243756
48	17	3	3	1.000000	0.039315
48	17	4	0	0.161707	0.161707
48	17	4	1	0.554425	0.392718
48	17	4	2	0.879433	0.325008
48	17	4	3	0.987768	0.108336
48	17	4	4	1.000000	0.012231
48	17	5	0	0.099229	0.099229
48	17	5	1	0.411618	0.312389
48	17	5	2	0.768635	0.357016
48	17	5	3	0.953298	0.184663
48	17	5	4	0.996386	0.043088
48	17	5	5	1.000000	0.003614
48	17	6	0	0.059999	0.059999
48	17	6	1	0.295381	0.235381
48	17	6	2	0.644094	0.348713
48	17	6	3	0.893175	0.249081
48	17	6	4	0.983359	0.090184
48	17	6	5	0.998991	0.015632
48	17	6	6	1.000000	0.001009
48	17	7	0	0.035714	0.035714
48	17	7	1	0.205712	0.169998
48	17	7	2	0.519554	0.313842
48	17	7	3	0.810148	0.290594
48	17	7	4	0.955445	0.145297
48	17	7	5	0.994525	0.039080
48	17	7	6	0.999736	0.005211
48	17	7	7	1.000000	0.000264
48	17	8	0	0.020906	0.020906
48	17	8	1	0.139371	0.118465
48	17	8	2	0.404733	0.265362
48	17	8	3	0.710921	0.306187
48	17	8	4	0.909375	0.198455
48	17	8	5	0.983087	0.073712
48	17	8	6	0.998338	0.015251
48	17	8	7	0.999935	0.001598
48	17	8	8	1.000000	0.000064
48	17	9	0	0.012021	0.012021
48	17	9	1	0.091985	0.079964
48	17	9	2	0.305222	0.213238
48	17	9	3	0.603755	0.298533
48	17	9	4	0.844878	0.241122
48	17	9	5	0.960974	0.116096
48	17	9	6	0.994144	0.033170
48	17	9	7	0.999536	0.005392
48	17	9	8	0.999985	0.000449
48	17	9	9	1.000000	0.000014
48	17	10	0	0.006781	0.006781
48	17	10	1	0.059179	0.052398
48	17	10	2	0.223208	0.164029
48	17	10	3	0.496589	0.273382
48	17	10	4	0.764503	0.267914
48	17	10	5	0.925252	0.160748
48	17	10	6	0.984788	0.059536
48	17	10	7	0.998153	0.013365
48	17	10	8	0.999882	0.001728
48	17	10	9	0.999997	0.000115
48	17	10	10	1.000000	0.000003
48	17	11	0	0.003747	0.003747
48	17	11	1	0.037117	0.033369
48	17	11	2	0.158460	0.121343
48	17	11	3	0.395870	0.237410
48	17	11	4	0.672849	0.276979
48	17	11	5	0.874489	0.201640
48	17	11	6	0.967554	0.093065
48	17	11	7	0.994636	0.027082
48	17	11	8	0.999472	0.004836
48	17	11	9	0.999973	0.000500
48	17	11	10	0.999999	0.000027
48	17	11	11	1.000000	0.000001
48	17	12	0	0.002026	0.002026
48	17	12	1	0.022687	0.020661
48	17	12	2	0.109267	0.086580
48	17	12	3	0.306039	0.196772
48	17	12	4	0.575532	0.269493
48	17	12	5	0.809092	0.233560
48	17	12	6	0.939886	0.130794
48	17	12	7	0.987317	0.047431
48	17	12	8	0.998296	0.010979
48	17	12	9	0.999864	0.001568
48	17	12	10	0.999994	0.000130
48	17	12	11	1.000000	0.000006
48	17	12	12	1.000000	0.000000
48	17	13	0	0.001069	0.001069
48	17	13	1	0.013504	0.012435
48	17	13	2	0.073192	0.059688
48	17	13	3	0.229516	0.156325
48	17	13	4	0.478215	0.248699
48	17	13	5	0.731239	0.253024
48	17	13	6	0.899921	0.168683
48	17	13	7	0.974141	0.074220
48	17	13	8	0.995551	0.021410
48	17	13	9	0.999516	0.003965
48	17	13	10	0.999969	0.000453
48	17	13	11	0.999999	0.000030
48	17	13	12	1.000000	0.000001
48	17	13	13	1.000000	0.000000
48	17	14	0	0.000550	0.000550
48	17	14	1	0.007819	0.007270
48	17	14	2	0.047611	0.039792
48	17	14	3	0.166986	0.119375
48	17	14	4	0.385841	0.218855
48	17	14	5	0.644488	0.258646
48	17	14	6	0.846907	0.202419
48	17	14	7	0.952936	0.106029
48	17	14	8	0.990046	0.037110
48	17	14	9	0.998610	0.008564
48	17	14	10	0.999878	0.001269
48	17	14	11	0.999994	0.000115
48	17	14	12	1.000000	0.000006
48	17	14	13	1.000000	0.000000
48	17	14	14	1.000000	0.000000
48	17	15	0	0.000275	0.000275
48	17	15	1	0.004398	0.004124
48	17	15	2	0.030056	0.025658
48	17	15	3	0.117832	0.087776
48	17	15	4	0.302161	0.184330
48	17	15	5	0.553201	0.251039
48	17	15	6	0.781418	0.228217
48	17	15	7	0.921751	0.140332
48	17	15	8	0.980223	0.058472
48	17	15	9	0.996595	0.016372
48	17	15	10	0.999617	0.003023
48	17	15	11	0.999973	0.000356
48	17	15	12	0.999999	0.000025
48	17	15	13	1.000000	0.000001
48	17	15	14	1.000000	0.000000
48	17	15	15	1.000000	0.000000
48	17	16	0	0.000133	0.000133
48	17	16	1	0.002399	0.002266
48	17	16	2	0.018393	0.015994
48	17	16	3	0.080594	0.062200
48	17	16	4	0.229547	0.148953
48	17	16	5	0.461914	0.232367
48	17	16	6	0.705346	0.243432

Table for $N = 2$, $n = 1$, through $N = 100$, $n = 50$

N	n	k	x	P(x)	p(x)
48	17	16	7	0.879226	0.173880
48	17	16	8	0.964276	0.085050
48	17	16	9	0.992626	0.028350
48	17	16	10	0.998976	0.006350
48	17	16	11	0.999909	0.000933
48	17	16	12	0.999995	0.000086
48	17	16	13	1.000000	0.000005
48	17	16	14	1.000000	0.000000
48	17	16	15	1.000000	0.000000
48	17	16	16	1.000000	0.000000
48	17	17	0	0.000062	0.000062
48	17	17	1	0.001266	0.001204
48	17	17	2	0.010896	0.009630
48	17	17	3	0.053381	0.042485
48	17	17	4	0.169035	0.115654
48	17	17	5	0.374776	0.205742
48	17	17	6	0.621666	0.246890
48	17	17	7	0.824888	0.203222
48	17	17	8	0.940355	0.115467
48	17	17	9	0.985538	0.045183
48	17	17	10	0.997587	0.012049
48	17	17	11	0.999734	0.002147
48	17	17	12	0.999982	0.000248
48	17	17	13	0.999999	0.000018
48	17	17	14	1.000000	0.000001
48	17	17	15	1.000000	0.000000
48	17	17	16	1.000000	0.000000
48	17	17	17	1.000000	0.000000
48	18	1	0	0.625000	0.625000
48	18	1	1	1.000000	0.375000
48	18	2	0	0.385638	0.385638
48	18	2	1	0.864362	0.478723
48	18	2	2	1.000000	0.135638
48	18	3	0	0.234736	0.234736
48	18	3	1	0.687442	0.452706
48	18	3	2	0.952821	0.265379
48	18	3	3	1.000000	0.047179
48	18	4	0	0.140842	0.140842
48	18	4	1	0.516420	0.375578
48	18	4	2	0.858464	0.342044
48	18	4	3	0.984274	0.125809
48	18	4	4	1.000000	0.015726
48	18	5	0	0.083225	0.083225
48	18	5	1	0.371310	0.288086
48	18	5	2	0.734085	0.362774
48	18	5	3	0.941384	0.207300
48	18	5	4	0.994996	0.053612
48	18	5	5	1.000000	0.005004
48	18	6	0	0.048386	0.048386
48	18	6	1	0.257416	0.209029
48	18	6	2	0.599099	0.341683
48	18	6	3	0.869070	0.269972
48	18	6	4	0.977541	0.108471
48	18	6	5	0.998487	0.020946
48	18	6	6	1.000000	0.001513
48	18	7	0	0.027649	0.027649
48	18	7	1	0.172809	0.145159
48	18	7	2	0.468934	0.296125
48	18	7	3	0.772652	0.303718
48	18	7	4	0.941384	0.168732
48	18	7	5	0.992004	0.050620
48	18	7	6	0.999568	0.007564
48	18	7	7	1.000000	0.000432
48	18	8	0	0.015511	0.015511
48	18	8	1	0.112621	0.097110
48	18	8	2	0.353373	0.240752
48	18	8	3	0.661536	0.308163
48	18	8	4	0.883768	0.222233
48	18	8	5	0.975954	0.092185
48	18	8	6	0.997354	0.021400
48	18	8	7	0.999884	0.002530
48	18	8	8	1.000000	0.000116
48	18	9	0	0.008531	0.008531
48	18	9	1	0.071349	0.062818
48	18	9	2	0.257072	0.185723
48	18	9	3	0.545975	0.288903
48	18	9	4	0.805987	0.260012
48	18	9	5	0.945993	0.140007
48	18	9	6	0.990934	0.044940
48	18	9	7	0.999188	0.008254
48	18	9	8	0.999971	0.000783
48	18	9	9	1.000000	0.000029
48	18	10	0	0.004594	0.004594
48	18	10	1	0.043967	0.039373
48	18	10	2	0.180878	0.136911
48	18	10	3	0.434858	0.253980
48	18	10	4	0.712649	0.277791
48	18	10	5	0.899325	0.186676
48	18	10	6	0.977106	0.077781
48	18	10	7	0.996860	0.019754
48	18	10	8	0.999770	0.002910
48	18	10	9	0.999993	0.000223
48	18	10	10	1.000000	0.000007
48	18	11	0	0.002418	0.002418
48	18	11	1	0.026352	0.023935
48	18	11	2	0.123231	0.096879
48	18	11	3	0.334603	0.211372
48	18	11	4	0.610305	0.275702
48	18	11	5	0.835462	0.225157
48	18	11	6	0.952543	0.117082
48	18	11	7	0.991142	0.038598
48	18	11	8	0.999004	0.007863
48	18	11	9	0.999940	0.000936
48	18	11	10	0.999998	0.000058
48	18	11	11	1.000000	0.000001
48	18	12	0	0.001241	0.001241
48	18	12	1	0.015355	0.014114
48	18	12	2	0.081338	0.065982
48	18	12	3	0.248912	0.167574
48	18	12	4	0.505985	0.257074
48	18	12	5	0.756353	0.250367
48	18	12	6	0.914571	0.158218
48	18	12	7	0.979667	0.065096
48	18	12	8	0.996879	0.017213
48	18	12	9	0.999713	0.002833
48	18	12	10	0.999986	0.000273
48	18	12	11	1.000000	0.000014
48	18	12	12	1.000000	0.000000
48	18	13	0	0.000621	0.000621
48	18	13	1	0.008690	0.008070
48	18	13	2	0.052012	0.043322
48	18	13	3	0.179089	0.127077
48	18	13	4	0.406012	0.226923
48	18	13	5	0.665942	0.259930
48	18	13	6	0.861832	0.195889
48	18	13	7	0.959776	0.097945
48	18	13	8	0.992098	0.032322
48	18	13	9	0.999004	0.006906
48	18	13	10	0.999925	0.000921
48	18	13	11	0.999997	0.000072
48	18	13	12	1.000000	0.000003
48	18	13	13	1.000000	0.000000
48	18	14	0	0.000302	0.000302
48	18	14	1	0.004771	0.004469
48	18	14	2	0.032208	0.027437
48	18	14	3	0.124628	0.092420
48	18	14	4	0.315243	0.190615
48	18	14	5	0.569397	0.254154
48	18	14	6	0.794670	0.225273
48	18	14	7	0.928994	0.134324
48	18	14	8	0.982863	0.053870
48	18	14	9	0.997228	0.014365
48	18	14	10	0.999715	0.002486
48	18	14	11	0.999983	0.000268
48	18	14	12	0.999999	0.000017
48	18	14	13	1.000000	0.000001
48	18	14	14	1.000000	0.000000
48	18	15	0	0.000142	0.000142
48	18	15	1	0.002536	0.002394
48	18	15	2	0.019296	0.016760
48	18	15	3	0.083854	0.064558
48	18	15	4	0.236754	0.152900
48	18	15	5	0.472220	0.235466
48	18	15	6	0.715162	0.242941
48	18	15	7	0.885536	0.170374
48	18	15	8	0.967019	0.081483
48	18	15	9	0.993426	0.026407
48	18	15	10	0.999130	0.005704
48	18	15	11	0.999927	0.000798
48	18	15	12	0.999996	0.000069
48	18	15	13	1.000000	0.000003
48	18	15	14	1.000000	0.000000
48	18	15	15	1.000000	0.000000
48	18	16	0	0.000064	0.000064
48	18	16	1	0.001303	0.001238
48	18	16	2	0.011170	0.009867
48	18	16	3	0.054510	0.043340
48	18	16	4	0.171888	0.117378
48	18	16	5	0.379461	0.207573
48	18	16	6	0.626819	0.247358
48	18	16	7	0.828744	0.201925
48	18	16	8	0.942327	0.113583
48	18	16	9	0.986224	0.043897
48	18	16	10	0.997747	0.011523
48	18	16	11	0.999758	0.002011
48	18	16	12	0.999984	0.000226
48	18	16	13	0.999999	0.000015
48	18	16	14	1.000000	0.000001
48	18	16	15	1.000000	0.000000
48	18	16	16	1.000000	0.000000
48	18	17	0	0.000028	0.000028
48	18	17	1	0.000645	0.000617
48	18	17	2	0.006237	0.005592
48	18	17	3	0.034194	0.027958
48	18	17	4	0.120535	0.086340
48	18	17	5	0.295134	0.174599
48	18	17	6	0.534060	0.238926
48	18	17	7	0.759333	0.225273
48	18	17	8	0.906833	0.147500
48	18	17	9	0.973878	0.067045
48	18	17	10	0.994866	0.020988
48	18	17	11	0.999318	0.004452
48	18	17	12	0.999942	0.000623
48	18	17	13	0.999997	0.000055
48	18	17	14	1.000000	0.000003
48	18	17	15	1.000000	0.000000
48	18	17	16	1.000000	0.000000
48	18	17	17	1.000000	0.000000
48	18	18	0	0.000012	0.000012
48	18	18	1	0.000307	0.000295

N	n	k	x	P(x)	p(x)
48	18	18	2	0.003351	0.003044
48	18	18	3	0.020666	0.017316
48	18	18	4	0.081542	0.060876
48	18	18	5	0.221915	0.140373
48	18	18	6	0.441573	0.219657
48	18	18	7	0.679397	0.237825
48	18	18	8	0.859252	0.179855
48	18	18	9	0.954413	0.095161
48	18	18	10	0.989450	0.035037
48	18	18	11	0.998313	0.008863
48	18	18	12	0.999821	0.001508
48	18	18	13	0.999988	0.000167
48	18	18	14	0.999999	0.000011
48	18	18	15	1.000000	0.000000
48	18	18	16	1.000000	0.000000
48	18	18	17	1.000000	0.000000
48	18	18	18	1.000000	0.000000
48	19	1	0	0.604167	0.604167
48	19	1	1	1.000000	0.395833
48	19	2	0	0.359929	0.359929
48	19	2	1	0.848404	0.488475
48	19	2	2	1.000000	0.151596
48	19	3	0	0.211263	0.211263
48	19	3	1	0.657262	0.445999
48	19	3	2	0.943975	0.286714
48	19	3	3	1.000000	0.056024
48	19	4	0	0.122063	0.122063
48	19	4	1	0.478862	0.356799
48	19	4	2	0.835661	0.356799
48	19	4	3	0.980080	0.144419
48	19	4	4	1.000000	0.019920
48	19	5	0	0.069354	0.069354
48	19	5	1	0.332899	0.263545
48	19	5	2	0.697807	0.364908
48	19	5	3	0.927564	0.229757
48	19	5	4	0.993209	0.065645
48	19	5	5	1.000000	0.006791
48	19	6	0	0.038709	0.038709
48	19	6	1	0.222578	0.183869
48	19	6	2	0.553541	0.330963
48	19	6	3	0.842073	0.288532
48	19	6	4	0.970310	0.128237
48	19	6	5	0.997789	0.027479
48	19	6	6	1.000000	0.002211
48	19	7	0	0.021198	0.021198
48	19	7	1	0.143777	0.122579
48	19	7	2	0.419580	0.275803
48	19	7	3	0.732156	0.312576
48	19	7	4	0.924511	0.192355
48	19	7	5	0.988629	0.064118
48	19	7	6	0.999316	0.010686
48	19	7	7	1.000000	0.000684
48	19	8	0	0.011374	0.011374
48	19	8	1	0.089962	0.078587
48	19	8	2	0.305222	0.215261
48	19	8	3	0.610175	0.304953
48	19	8	4	0.854137	0.243962
48	19	8	5	0.966735	0.112598
48	19	8	6	0.995927	0.029192
48	19	8	7	0.999800	0.003872
48	19	8	8	1.000000	0.000200
48	19	9	0	0.005972	0.005972
48	19	9	1	0.054597	0.048626
48	19	9	2	0.213737	0.159139
48	19	9	3	0.488194	0.274457
48	19	9	4	0.762651	0.274457
48	19	9	5	0.927326	0.164674
48	19	9	6	0.986440	0.059114
48	19	9	7	0.998638	0.012198
48	19	9	8	0.999945	0.001307
48	19	9	9	1.000000	0.000055
48	19	10	0	0.003062	0.003062
48	19	10	1	0.032155	0.029092
48	19	10	2	0.144368	0.112213
48	19	10	3	0.375596	0.231228
48	19	10	4	0.657091	0.281495
48	19	10	5	0.868212	0.211121
48	19	10	6	0.966735	0.098523
48	19	10	7	0.994885	0.028149
48	19	10	8	0.999576	0.004692
48	19	10	9	0.999986	0.000410
48	19	10	10	1.000000	0.000014
48	19	11	0	0.001531	0.001531
48	19	11	1	0.018374	0.016843
48	19	11	2	0.094167	0.075793
48	19	11	3	0.278237	0.184070
48	19	11	4	0.545975	0.267737
48	19	11	5	0.790431	0.244456
48	19	11	6	0.933030	0.142599
48	19	11	7	0.985995	0.052965
48	19	11	8	0.998218	0.012223
48	19	11	9	0.999878	0.001660
48	19	11	10	0.999997	0.000119
48	19	11	11	1.000000	0.000003
48	19	12	0	0.000745	0.000745
48	19	12	1	0.010180	0.009435
48	19	12	2	0.059344	0.049163
48	19	12	3	0.198639	0.139296
48	19	12	4	0.437432	0.238793
48	19	12	5	0.697934	0.260501
48	19	12	6	0.882927	0.184994
48	19	12	7	0.968817	0.085890
48	19	12	8	0.994584	0.025767
48	19	12	9	0.999429	0.004845
48	19	12	10	0.999968	0.000538
48	19	12	11	0.999999	0.000031
48	19	12	12	1.000000	0.000001
48	19	13	0	0.000352	0.000352
48	19	13	1	0.005463	0.005111
48	19	13	2	0.036128	0.030665
48	19	13	3	0.136730	0.100603
48	19	13	4	0.337935	0.201205
48	19	13	5	0.596628	0.258692
48	19	13	6	0.816124	0.219497
48	19	13	7	0.940187	0.124063
48	19	13	8	0.986711	0.046524
48	19	13	9	0.998083	0.011372
48	19	13	10	0.999833	0.001750
48	19	13	11	0.999992	0.000159
48	19	13	12	1.000000	0.000008
48	19	13	13	1.000000	0.000000
48	19	14	0	0.000161	0.000161
48	19	14	1	0.002834	0.002673
48	19	14	2	0.021233	0.018399
48	19	14	3	0.090740	0.069507
48	19	14	4	0.251704	0.160964
48	19	14	5	0.493151	0.241446
48	19	14	6	0.734597	0.241446
48	19	14	7	0.897651	0.163055
48	19	14	8	0.972089	0.074438
48	19	14	9	0.994834	0.022745
48	19	14	10	0.999383	0.004549
48	19	14	11	0.999956	0.000573
48	19	14	12	0.999998	0.000042
48	19	14	13	1.000000	0.000002
48	19	14	14	1.000000	0.000000
48	19	15	0	0.000071	0.000071
48	19	15	1	0.001419	0.001348
48	19	15	2	0.012034	0.010615
48	19	15	3	0.058031	0.045997
48	19	15	4	0.180691	0.122660
48	19	15	5	0.393732	0.213041
48	19	15	6	0.642279	0.248548
48	19	15	7	0.840103	0.197824
48	19	15	8	0.948006	0.107904
48	19	15	9	0.988144	0.040138
48	19	15	10	0.998179	0.010035
48	19	15	11	0.999821	0.001642
48	19	15	12	0.999989	0.000168
48	19	15	13	0.999999	0.000010
48	19	15	14	1.000000	0.000000
48	19	15	15	1.000000	0.000000
48	19	16	0	0.000030	0.000030
48	19	16	1	0.000684	0.000654
48	19	16	2	0.006565	0.005882
48	19	16	3	0.035729	0.029164
48	19	16	4	0.124936	0.089207
48	19	16	5	0.303351	0.178414
48	19	16	6	0.544367	0.241016
48	19	16	7	0.768167	0.223800
48	19	16	8	0.912039	0.143872
48	19	16	9	0.975982	0.063943
48	19	16	10	0.995442	0.019461
48	19	16	11	0.999423	0.003981
48	19	16	12	0.999954	0.000531
48	19	16	13	0.999998	0.000044
48	19	16	14	1.000000	0.000002
48	19	16	15	1.000000	0.000000
48	19	16	16	1.000000	0.000000
48	19	17	0	0.000012	0.000012
48	19	17	1	0.000316	0.000304
48	19	17	2	0.003441	0.003125
48	19	17	3	0.021147	0.017707
48	19	17	4	0.083121	0.061973
48	19	17	5	0.225294	0.142174
48	19	17	6	0.446454	0.221159
48	19	17	7	0.684242	0.237788
48	19	17	8	0.862583	0.178341
48	19	17	9	0.955999	0.093417
48	19	17	10	0.989969	0.033970
48	19	17	11	0.998428	0.008459
48	19	17	12	0.999838	0.001410
48	19	17	13	0.999989	0.000152
48	19	17	14	0.999999	0.000010
48	19	17	15	1.000000	0.000000
48	19	17	16	1.000000	0.000000
48	19	17	17	1.000000	0.000000
48	19	18	0	0.000005	0.000005
48	19	18	1	0.000140	0.000135
48	19	18	2	0.001727	0.001588
48	19	18	3	0.012008	0.010281
48	19	18	4	0.053134	0.041125
48	19	18	5	0.161087	0.107953
48	19	18	6	0.353710	0.192623
48	19	18	7	0.592195	0.238485
48	19	18	8	0.799300	0.207106
48	19	18	9	0.925865	0.126565
48	19	18	10	0.980107	0.054242
48	19	18	11	0.996245	0.016138
48	19	18	12	0.999519	0.003274

N	n	k	x	P(x)	p(x)
48	19	18	13	0.999960	0.000441
48	19	18	14	0.999998	0.000038
48	19	18	15	1.000000	0.000002
48	19	18	16	1.000000	0.000000
48	19	18	17	1.000000	0.000000
48	19	18	18	1.000000	0.000000
48	19	19	0	0.000002	0.000002
48	19	19	1	0.000059	0.000057
48	19	19	2	0.000828	0.000769
48	19	19	3	0.006525	0.005698
48	19	19	4	0.032571	0.026046
48	19	19	5	0.110709	0.078138
48	19	19	6	0.270240	0.159531
48	19	19	7	0.496801	0.226561
48	19	19	8	0.723362	0.226561
48	19	19	9	0.883677	0.160315
48	19	19	10	0.963834	0.080158
48	19	19	11	0.991941	0.028107
48	19	19	12	0.998755	0.006814
48	19	19	13	0.999872	0.001117
48	19	19	14	0.999992	0.000120
48	19	19	15	0.999999	0.000008
48	19	19	16	1.000000	0.000000
48	19	19	17	1.000000	0.000000
48	19	19	18	1.000000	0.000000
48	19	19	19	1.000000	0.000000
48	20	1	0	0.583333	0.583333
48	20	1	1	1.000000	0.416667
48	20	2	0	0.335106	0.335106
48	20	2	1	0.831560	0.496454
48	20	2	2	1.000000	0.168440
48	20	3	0	0.189408	0.189408
48	20	3	1	0.626503	0.437095
48	20	3	2	0.934089	0.307586
48	20	3	3	1.000000	0.065911
48	20	4	0	0.105227	0.105227
48	20	4	1	0.441952	0.336725
48	20	4	2	0.811055	0.369103
48	20	4	3	0.975100	0.164046
48	20	4	4	1.000000	0.024900
48	20	5	0	0.057396	0.057396
48	20	5	1	0.296548	0.239151
48	20	5	2	0.660058	0.363510
48	20	5	3	0.911719	0.251661
48	20	5	4	0.990945	0.079227
48	20	5	5	1.000000	0.009054
48	20	6	0	0.030700	0.030700
48	20	6	1	0.190876	0.160176
48	20	6	2	0.507891	0.317015
48	20	6	3	0.812225	0.304334

N	n	k	x	P(x)	p(x)
48	20	6	4	0.961466	0.149241
48	20	6	5	0.996841	0.035376
48	20	6	6	1.000000	0.003159
48	20	7	0	0.016081	0.016081
48	20	7	1	0.118416	0.102335
48	20	7	2	0.372028	0.253612
48	20	7	3	0.689042	0.317015
48	20	7	4	0.904612	0.215570
48	20	7	5	0.984207	0.079595
48	20	7	6	0.998947	0.014740
48	20	7	7	1.000000	0.001053
48	20	8	0	0.008237	0.008237
48	20	8	1	0.070992	0.062756
48	20	8	2	0.260686	0.189693
48	20	8	3	0.557597	0.296911
48	20	8	4	0.820487	0.262890
48	20	8	5	0.955087	0.134600
48	20	8	6	0.993914	0.038827
48	20	8	7	0.999666	0.005752
48	20	8	8	1.000000	0.000334
48	20	9	0	0.004118	0.004118
48	20	9	1	0.041183	0.037065
48	20	9	2	0.175324	0.134140
48	20	9	3	0.431410	0.256086
48	20	9	4	0.715331	0.283921
48	20	9	5	0.904612	0.189281
48	20	9	6	0.980325	0.075712
48	20	9	7	0.997797	0.017472
48	20	9	8	0.999900	0.002103
48	20	9	9	1.000000	0.000100
48	20	10	0	0.002006	0.002006
48	20	10	1	0.023126	0.021120
48	20	10	2	0.113413	0.090287
48	20	10	3	0.319782	0.206370
48	20	10	4	0.598851	0.279068
48	20	10	5	0.831812	0.232961
48	20	10	6	0.953146	0.121334
48	20	10	7	0.991973	0.038827
48	20	10	8	0.999253	0.007280
48	20	10	9	0.999972	0.000719
48	20	10	10	1.000000	0.000028
48	20	11	0	0.000950	0.000950
48	20	11	1	0.012566	0.011616
48	20	11	2	0.070645	0.058079
48	20	11	3	0.227459	0.156814
48	20	11	4	0.481348	0.253889
48	20	11	5	0.739853	0.258505
48	20	11	6	0.908444	0.168590
48	20	11	7	0.978690	0.070246
48	20	11	8	0.996954	0.018264

N	n	k	x	P(x)	p(x)
48	20	11	9	0.999764	0.002810
48	20	11	10	0.999993	0.000229
48	20	11	11	1.000000	0.000007
48	20	12	0	0.000437	0.000437
48	20	12	1	0.006601	0.006165
48	20	12	2	0.042391	0.035789
48	20	12	3	0.155410	0.113019
48	20	12	4	0.371558	0.216149
48	20	12	5	0.635054	0.263496
48	20	12	6	0.844653	0.209599
48	20	12	7	0.954009	0.109356
48	20	12	8	0.991030	0.037022
48	20	12	9	0.998928	0.007898
48	20	12	10	0.999931	0.001002
48	20	12	11	0.999998	0.000068
48	20	12	12	1.000000	0.000002
48	20	13	0	0.000194	0.000194
48	20	13	1	0.003348	0.003154
48	20	13	2	0.024496	0.021148
48	20	13	3	0.102040	0.077544
48	20	13	4	0.275492	0.173453
48	20	13	5	0.525264	0.249772
48	20	13	6	0.763142	0.237878
48	20	13	7	0.914519	0.151377
48	20	13	8	0.978690	0.064171
48	20	13	9	0.996515	0.017825
48	20	13	10	0.999652	0.003137
48	20	13	11	0.999981	0.000329
48	20	13	12	0.999999	0.000018
48	20	13	13	1.000000	0.000000
48	20	14	0	0.000083	0.000083
48	20	14	1	0.001636	0.001553
48	20	14	2	0.013620	0.011984
48	20	14	3	0.064376	0.050756
48	20	14	4	0.196200	0.131824
48	20	14	5	0.418219	0.222019
48	20	14	6	0.667991	0.249772
48	20	14	7	0.858293	0.190302
48	20	14	8	0.956688	0.098395
48	20	14	9	0.990913	0.034224
48	20	14	10	0.998756	0.007843
48	20	14	11	0.999897	0.001141
48	20	14	12	0.999995	0.000099
48	20	14	13	1.000000	0.000005
48	20	14	14	1.000000	0.000000
48	20	15	0	0.000034	0.000034
48	20	15	1	0.000768	0.000734
48	20	15	2	0.007275	0.006507
48	20	15	3	0.038998	0.031722
48	20	15	4	0.134165	0.095167

N	n	k	x	P(x)	p(x)
48	20	15	5	0.320269	0.186105
48	20	15	6	0.565144	0.244874
48	20	15	7	0.785531	0.220387
48	20	15	8	0.921961	0.136430
48	20	15	9	0.979840	0.057879
48	20	15	10	0.996449	0.016609
48	20	15	11	0.999595	0.003146
48	20	15	12	0.999972	0.000377
48	20	15	13	0.999999	0.000027
48	20	15	14	1.000000	0.000001
48	20	15	15	1.000000	0.000000
48	20	16	0	0.000013	0.000013
48	20	16	1	0.000346	0.000332
48	20	16	2	0.003726	0.003380
48	20	16	3	0.022656	0.018930
48	20	16	4	0.088023	0.065367
48	20	16	5	0.235676	0.147653
48	20	16	6	0.461258	0.225581
48	20	16	7	0.698712	0.237454
48	20	16	8	0.872350	0.173638
48	20	16	9	0.960547	0.088197
48	20	16	10	0.991416	0.030869
48	20	16	11	0.998737	0.007321
48	20	16	12	0.999881	0.001144
48	20	16	13	0.999993	0.000113
48	20	16	14	1.000000	0.000006
48	20	16	15	1.000000	0.000000
48	20	16	16	1.000000	0.000000
48	20	17	0	0.000005	0.000005
48	20	17	1	0.000148	0.000143
48	20	17	2	0.001824	0.001676
48	20	17	3	0.012599	0.010775
48	20	17	4	0.055339	0.042740
48	20	17	5	0.166464	0.111124
48	20	17	6	0.362566	0.196102
48	20	17	7	0.602246	0.239680
48	20	17	8	0.807235	0.204990
48	20	17	9	0.930229	0.122994
48	20	17	10	0.981769	0.051540
48	20	17	11	0.996678	0.014908
48	20	17	12	0.999595	0.002917
48	20	17	13	0.999968	0.000374
48	20	17	14	0.999998	0.000030
48	20	17	15	1.000000	0.000001
48	20	17	16	1.000000	0.000000
48	20	17	17	1.000000	0.000000
48	20	18	0	0.000002	0.000002
48	20	18	1	0.000061	0.000059
48	20	18	2	0.000851	0.000791
48	20	18	3	0.006691	0.005839

N	n	k	x	P(x)	p(x)
48	20	18	4	0.033280	0.026589
48	20	18	5	0.112694	0.079414
48	20	18	6	0.274004	0.161310
48	20	18	7	0.501735	0.227731
48	20	18	8	0.727885	0.226150
48	20	18	9	0.886586	0.158702
48	20	18	10	0.965144	0.078557
48	20	18	11	0.992350	0.027206
48	20	18	12	0.998842	0.006492
48	20	18	13	0.999884	0.001042
48	20	18	14	0.999993	0.000109
48	20	18	15	1.000000	0.000007
48	20	18	16	1.000000	0.000000
48	20	18	17	1.000000	0.000000
48	20	18	18	1.000000	0.000000
48	20	19	0	0.000001	0.000001
48	20	19	1	0.000023	0.000023
48	20	19	2	0.000377	0.000354
48	20	19	3	0.003382	0.003005
48	20	19	4	0.019099	0.015717
48	20	19	5	0.072987	0.053888
48	20	19	6	0.198726	0.125739
48	20	19	7	0.403051	0.204326
48	20	19	8	0.637425	0.234373
48	20	19	9	0.828396	0.190971
48	20	19	10	0.938958	0.110562
48	20	19	11	0.984188	0.045230
48	20	19	12	0.997111	0.012923
48	20	19	13	0.999641	0.002530
48	20	19	14	0.999971	0.000330
48	20	19	15	0.999998	0.000028
48	20	19	16	1.000000	0.000001
48	20	19	17	1.000000	0.000000
48	20	19	18	1.000000	0.000000
48	20	19	19	1.000000	0.000000
48	20	20	0	0.000000	0.000000
48	20	20	1	0.000008	0.000008
48	20	20	2	0.000157	0.000149
48	20	20	3	0.001620	0.001463
48	20	20	4	0.010427	0.008807
48	20	20	5	0.045114	0.034687
48	20	20	6	0.138024	0.092910
48	20	20	7	0.311457	0.173433
48	20	20	8	0.540443	0.228986
48	20	20	9	0.755958	0.215516
48	20	20	10	0.900833	0.144875
48	20	20	11	0.970151	0.069318
48	20	20	12	0.993546	0.023395
48	20	20	13	0.999030	0.005484
48	20	20	14	0.999903	0.000873
48	20	20	15	0.999994	0.000091
48	20	20	16	1.000000	0.000006
48	20	20	17	1.000000	0.000000
48	20	20	18	1.000000	0.000000
48	20	20	19	1.000000	0.000000
48	20	20	20	1.000000	0.000000
48	21	1	0	0.562500	0.562500
48	21	1	1	1.000000	0.437500
48	21	2	0	0.311170	0.311170
48	21	2	1	0.813830	0.502660
48	21	2	2	1.000000	0.186170
48	21	3	0	0.169114	0.169114
48	21	3	1	0.595282	0.426168
48	21	3	2	0.923104	0.327821
48	21	3	3	1.000000	0.076896
48	21	4	0	0.090194	0.090194
48	21	4	1	0.405874	0.315680
48	21	4	2	0.784690	0.378816
48	21	4	3	0.969241	0.184551
48	21	4	4	1.000000	0.030759
48	21	5	0	0.047147	0.047147
48	21	5	1	0.262383	0.215236
48	21	5	2	0.621110	0.358727
48	21	5	3	0.893743	0.272633
48	21	5	4	0.988116	0.094373
48	21	5	5	1.000000	0.011884
48	21	6	0	0.024122	0.024122
48	21	6	1	0.162273	0.138152
48	21	6	2	0.462603	0.300330
48	21	6	3	0.779618	0.317015
48	21	6	4	0.950806	0.171188
48	21	6	5	0.995578	0.044772
48	21	6	6	1.000000	0.004422
48	21	7	0	0.012061	0.012061
48	21	7	1	0.096487	0.084426
48	21	7	2	0.326740	0.230253
48	21	7	3	0.643754	0.317015
48	21	7	4	0.881515	0.237761
48	21	7	5	0.978522	0.097006
48	21	7	6	0.998421	0.019899
48	21	7	7	1.000000	0.001579
48	21	8	0	0.005883	0.005883
48	21	8	1	0.055303	0.049420
48	21	8	2	0.220037	0.164734
48	21	8	3	0.504577	0.284540
48	21	8	4	0.782931	0.278354
48	21	8	5	0.940666	0.157734
48	21	8	6	0.991141	0.050475
48	21	8	7	0.999461	0.008320
48	21	8	8	1.000000	0.000539
48	21	9	0	0.002795	0.002795
48	21	9	1	0.030593	0.027799
48	21	9	2	0.141789	0.111195
48	21	9	3	0.376534	0.234746
48	21	9	4	0.664631	0.288097
48	21	9	5	0.877572	0.212941
48	21	9	6	0.972212	0.094640
48	21	9	7	0.996549	0.024336
48	21	9	8	0.999825	0.003276
48	21	9	9	1.000000	0.000175
48	21	10	0	0.001290	0.001290
48	21	10	1	0.016338	0.015048
48	21	10	2	0.087617	0.071279
48	21	10	3	0.268190	0.180573
48	21	10	4	0.539050	0.270860
48	21	10	5	0.790212	0.251161
48	21	10	6	0.935812	0.145601
48	21	10	7	0.987813	0.052000
48	21	10	8	0.998733	0.010920
48	21	10	9	0.999946	0.001213
48	21	10	10	1.000000	0.000054
48	21	11	0	0.000577	0.000577
48	21	11	1	0.008418	0.007841
48	21	11	2	0.051977	0.043559
48	21	11	3	0.182655	0.130678
48	21	11	4	0.417876	0.235221
48	21	11	5	0.684459	0.266583
48	21	11	6	0.878338	0.193879
48	21	11	7	0.968655	0.090316
48	21	11	8	0.994997	0.026342
48	21	11	9	0.999563	0.004566
48	21	11	10	0.999984	0.000421
48	21	11	11	1.000000	0.000016
48	21	12	0	0.000250	0.000250
48	21	12	1	0.004179	0.003930
48	21	12	2	0.029609	0.025429
48	21	12	3	0.119082	0.089473
48	21	12	4	0.309802	0.190719
48	21	12	5	0.569180	0.259379
48	21	12	6	0.799739	0.230559
48	21	12	7	0.934481	0.134742
48	21	12	8	0.985741	0.051261
48	21	12	9	0.998082	0.012340
48	21	12	10	0.999859	0.001777
48	21	12	11	0.999996	0.000137
48	21	12	12	1.000000	0.000004
48	21	13	0	0.000104	0.000104
48	21	13	1	0.001996	0.001892
48	21	13	2	0.016188	0.014192
48	21	13	3	0.074345	0.058158
48	21	13	4	0.219740	0.145394
48	21	13	5	0.453901	0.234161
48	21	13	6	0.703673	0.249772
48	21	13	7	0.882081	0.178408
48	21	13	8	0.967231	0.085149
48	21	13	9	0.993968	0.026738
48	21	13	10	0.999316	0.005348
48	21	13	11	0.999958	0.000642
48	21	13	12	0.999999	0.000041
48	21	13	13	1.000000	0.000001
48	21	14	0	0.000042	0.000042
48	21	14	1	0.000915	0.000873
48	21	14	2	0.008484	0.007569
48	21	14	3	0.044436	0.035952
48	21	14	4	0.149120	0.104684
48	21	14	5	0.346856	0.197736
48	21	14	6	0.596628	0.249772
48	21	14	7	0.810718	0.214090
48	21	14	8	0.935604	0.124886
48	21	14	9	0.984801	0.049197
48	21	14	10	0.997635	0.012834
48	21	14	11	0.999774	0.002139
48	21	14	12	0.999988	0.000214
48	21	14	13	1.000000	0.000011
48	21	14	14	1.000000	0.000000
48	21	15	0	0.000016	0.000016
48	21	15	1	0.000401	0.000385
48	21	15	2	0.004254	0.003853
48	21	15	3	0.025402	0.021148
48	21	15	4	0.096778	0.071375
48	21	15	5	0.253803	0.157026
48	21	15	6	0.486434	0.232631
48	21	15	7	0.722563	0.236129
48	21	15	8	0.887853	0.165290
48	21	15	9	0.967437	0.079584
48	21	15	10	0.993483	0.026046
48	21	15	11	0.999145	0.005662
48	21	15	12	0.999932	0.000786
48	21	15	13	0.999997	0.000065
48	21	15	14	1.000000	0.000003
48	21	15	15	1.000000	0.000000
48	21	16	0	0.000006	0.000006
48	21	16	1	0.000168	0.000162
48	21	16	2	0.002036	0.001868
48	21	16	3	0.013867	0.011831
48	21	16	4	0.060009	0.046142
48	21	16	5	0.177670	0.117661
48	21	16	6	0.380693	0.203023
48	21	16	7	0.622387	0.241694
48	21	16	8	0.822739	0.200352

Table for $N = 2$, $n = 1$, through $N = 100$, $n = 50$

N	n	k	x	P(x)	p(x)
48	21	16	9	0.938498	0.115759
48	21	16	10	0.984801	0.046304
48	21	16	11	0.997429	0.012628
48	21	16	12	0.999717	0.002288
48	21	16	13	0.999981	0.000264
48	21	16	14	0.999999	0.000018
48	21	16	15	1.000000	0.000001
48	21	16	16	1.000000	0.000000
48	21	17	0	0.000002	0.000002
48	21	17	1	0.000066	0.000064
48	21	17	2	0.000927	0.000860
48	21	17	3	0.007212	0.006285
48	21	17	4	0.035496	0.028284
48	21	17	5	0.118839	0.083343
48	21	17	6	0.285526	0.166687
48	21	17	7	0.516646	0.231120
48	21	17	8	0.741346	0.224700
48	21	17	9	0.895088	0.153742
48	21	17	10	0.968884	0.073796
48	21	17	11	0.993483	0.024599
48	21	17	12	0.999074	0.005591
48	21	17	13	0.999915	0.000841
48	21	17	14	0.999995	0.000080
48	21	17	15	1.000000	0.000004
48	21	17	16	1.000000	0.000000
48	21	17	17	1.000000	0.000000
48	21	18	0	0.000001	0.000001
48	21	18	1	0.000025	0.000024
48	21	18	2	0.000399	0.000375
48	21	18	3	0.003562	0.003163
48	21	18	4	0.019985	0.016423
48	21	18	5	0.075823	0.055838
48	21	18	6	0.204871	0.129048
48	21	18	7	0.412269	0.207398
48	21	18	8	0.647117	0.234848
48	21	18	9	0.835575	0.188458
48	21	18	10	0.942699	0.107124
48	21	18	11	0.985548	0.042849
48	21	18	12	0.997451	0.011903
48	21	18	13	0.999698	0.002247
48	21	18	14	0.999977	0.000279
48	21	18	15	0.999999	0.000022
48	21	18	16	1.000000	0.000001
48	21	18	17	1.000000	0.000000
48	21	18	18	1.000000	0.000000
48	21	19	0	0.000000	0.000000
48	21	19	1	0.000009	0.000009
48	21	19	2	0.000162	0.000153
48	21	19	3	0.001665	0.001502
48	21	19	4	0.010679	0.009014

N	n	k	x	P(x)	p(x)
48	21	19	5	0.046043	0.035364
48	21	19	6	0.140347	0.094304
48	21	19	7	0.315483	0.175136
48	21	19	8	0.545349	0.229866
48	21	19	9	0.760192	0.214842
48	21	19	10	0.903420	0.143228
48	21	19	11	0.971265	0.067845
48	21	19	12	0.993880	0.022615
48	21	19	13	0.999099	0.005219
48	21	19	14	0.999912	0.000813
48	21	19	15	0.999995	0.000083
48	21	19	16	1.000000	0.000005
48	21	19	17	1.000000	0.000000
48	21	19	18	1.000000	0.000000
48	21	19	19	1.000000	0.000000
48	21	20	0	0.000000	0.000000
48	21	20	1	0.000003	0.000003
48	21	20	2	0.000062	0.000059
48	21	20	3	0.000732	0.000670
48	21	20	4	0.005395	0.004663
48	21	20	5	0.026532	0.021137
48	21	20	6	0.091569	0.065037
48	21	20	7	0.230935	0.139366
48	21	20	8	0.442306	0.211371
48	21	20	9	0.671292	0.228986
48	21	20	10	0.849092	0.177801
48	21	20	11	0.947870	0.098778
48	21	20	12	0.986861	0.038991
48	21	20	13	0.997659	0.010798
48	21	20	14	0.999716	0.002057
48	21	20	15	0.999977	0.000262
48	21	20	16	0.999999	0.000021
48	21	20	17	1.000000	0.000001
48	21	20	18	1.000000	0.000000
48	21	20	19	1.000000	0.000000
48	21	20	20	1.000000	0.000000
48	21	21	0	0.000000	0.000000
48	21	21	1	0.000001	0.000001
48	21	21	2	0.000022	0.000021
48	21	21	3	0.000301	0.000279
48	21	21	4	0.002564	0.002263
48	21	21	5	0.014453	0.011890
48	21	21	6	0.056728	0.042274
48	21	21	7	0.161252	0.104524
48	21	21	8	0.344169	0.182917
48	21	21	9	0.573155	0.228986
48	21	21	10	0.779242	0.206087
48	21	21	11	0.912592	0.133350
48	21	21	12	0.974329	0.061736
48	21	21	13	0.994574	0.020245

N	n	k	x	P(x)	p(x)
48	21	21	14	0.999202	0.004628
48	21	21	15	0.999921	0.000720
48	21	21	16	0.999995	0.000074
48	21	21	17	1.000000	0.000005
48	21	21	18	1.000000	0.000000
48	21	21	19	1.000000	0.000000
48	21	21	20	1.000000	0.000000
48	21	21	21	1.000000	0.000000
48	22	1	0	0.541667	0.541667
48	22	1	1	1.000000	0.458333
48	22	2	0	0.288121	0.288121
48	22	2	1	0.795213	0.507092
48	22	2	2	1.000000	0.204787
48	22	3	0	0.150324	0.150324
48	22	3	1	0.563714	0.413390
48	22	3	2	0.910962	0.347248
48	22	3	3	1.000000	0.089038
48	22	4	0	0.076832	0.076832
48	22	4	1	0.370799	0.293966
48	22	4	2	0.756630	0.385831
48	22	4	3	0.962406	0.205777
48	22	4	4	1.000000	0.037594
48	22	5	0	0.038416	0.038416
48	22	5	1	0.230496	0.192080
48	22	5	2	0.581252	0.350755
48	22	5	3	0.873548	0.292296
48	22	5	4	0.984621	0.111073
48	22	5	5	1.000000	0.015379
48	22	6	0	0.018761	0.018761
48	22	6	1	0.136690	0.117928
48	22	6	2	0.418110	0.281420
48	22	6	3	0.744394	0.326284
48	22	6	4	0.938125	0.193731
48	22	6	5	0.993920	0.055795
48	22	6	6	1.000000	0.006080
48	22	7	0	0.008934	0.008934
48	22	7	1	0.077726	0.068792
48	22	7	2	0.284100	0.206375
48	22	7	3	0.596789	0.312689
48	22	7	4	0.855097	0.258308
48	22	7	5	0.971336	0.116239
48	22	7	6	0.997684	0.026347
48	22	7	7	1.000000	0.002316
48	22	8	0	0.004140	0.004140
48	22	8	1	0.042491	0.038351
48	22	8	2	0.183430	0.140939
48	22	8	3	0.451885	0.268455
48	22	8	4	0.741694	0.289809
48	22	8	5	0.923140	0.181446
48	22	8	6	0.987402	0.064262

N	n	k	x	P(x)	p(x)
48	22	8	7	0.999152	0.011751
48	22	8	8	1.000000	0.000847
48	22	9	0	0.001863	0.001863
48	22	9	1	0.022357	0.020494
48	22	9	2	0.112960	0.090604
48	22	9	3	0.324369	0.211408
48	22	9	4	0.611280	0.286911
48	22	9	5	0.846025	0.234746
48	22	9	6	0.961697	0.115672
48	22	9	7	0.994746	0.033049
48	22	9	8	0.999703	0.004957
48	22	9	9	1.000000	0.000297
48	22	10	0	0.000812	0.000812
48	22	10	1	0.011322	0.010510
48	22	10	2	0.066497	0.055175
48	22	10	3	0.221375	0.154878
48	22	10	4	0.478859	0.257484
48	22	10	5	0.743700	0.264841
48	22	10	6	0.914242	0.170542
48	22	10	7	0.982035	0.067793
48	22	10	8	0.997924	0.015889
48	22	10	9	0.999901	0.001977
48	22	10	10	1.000000	0.000099
48	22	11	0	0.000342	0.000342
48	22	11	1	0.005514	0.005172
48	22	11	2	0.037457	0.031944
48	22	11	3	0.143936	0.106478
48	22	11	4	0.356893	0.212957
48	22	11	5	0.625219	0.268326
48	22	11	6	0.842435	0.217216
48	22	11	7	0.955274	0.112840
48	22	11	8	0.992070	0.036796
48	22	11	9	0.999225	0.007155
48	22	11	10	0.999969	0.000744
48	22	11	11	1.000000	0.000031
48	22	12	0	0.000139	0.000139
48	22	12	1	0.002578	0.002440
48	22	12	2	0.020191	0.017612
48	22	12	3	0.089258	0.069067
48	22	12	4	0.253292	0.164034
48	22	12	5	0.501934	0.248642
48	22	12	6	0.748504	0.246570
48	22	12	7	0.909529	0.161025
48	22	12	8	0.978147	0.068619
48	22	12	9	0.996711	0.018563
48	22	12	10	0.999727	0.003017
48	22	12	11	0.999991	0.000263
48	22	12	12	1.000000	0.000009
48	22	13	0	0.000054	0.000054
48	22	13	1	0.001155	0.001101

N	n	k	x	P(x)	p(x)
48	22	13	2	0.010406	0.009251
48	22	13	3	0.052806	0.042400
48	22	13	4	0.171275	0.118469
48	22	13	5	0.384520	0.213245
48	22	13	6	0.638917	0.254397
48	22	13	7	0.842435	0.203518
48	22	13	8	0.951462	0.109027
48	22	13	9	0.990007	0.038545
48	22	13	10	0.998722	0.008715
48	22	13	11	0.999910	0.001188
48	22	13	12	0.999997	0.000087
48	22	13	13	1.000000	0.000003
48	22	14	0	0.000020	0.000020
48	22	14	1	0.000494	0.000474
48	22	14	2	0.005120	0.004625
48	22	14	3	0.029789	0.024669
48	22	14	4	0.110348	0.080559
48	22	14	5	0.280944	0.170596
48	22	14	6	0.522621	0.241677
48	22	14	7	0.755213	0.232592
48	22	14	8	0.907851	0.152638
48	22	14	9	0.975691	0.067839
48	22	14	10	0.995734	0.020043
48	22	14	11	0.999537	0.003803
48	22	14	12	0.999972	0.000436
48	22	14	13	0.999999	0.000027
48	22	14	14	1.000000	0.000001
48	22	15	0	0.000007	0.000007
48	22	15	1	0.000201	0.000194
48	22	15	2	0.002399	0.002198
48	22	15	3	0.016003	0.013604
48	22	15	4	0.067699	0.051696
48	22	15	5	0.195646	0.127947
48	22	15	6	0.408891	0.213245
48	22	15	7	0.652599	0.243708
48	22	15	8	0.845000	0.192401
48	22	15	9	0.949752	0.104752
48	22	15	10	0.988660	0.038908
48	22	15	11	0.998306	0.009647
48	22	15	12	0.999844	0.001538
48	22	15	13	0.999992	0.000148
48	22	15	14	1.000000	0.000008
48	22	15	15	1.000000	0.000000
48	22	16	0	0.000002	0.000002
48	22	16	1	0.000078	0.000075
48	22	16	2	0.001067	0.000989
48	22	16	3	0.008170	0.007103
48	22	16	4	0.039501	0.031331
48	22	16	5	0.129734	0.090233
48	22	16	6	0.305499	0.175765

N	n	k	x	P(x)	p(x)
48	22	16	7	0.541822	0.236323
48	22	16	8	0.763375	0.221553
48	22	16	9	0.908486	0.145111
48	22	16	10	0.974512	0.066025
48	22	16	11	0.995091	0.020579
48	22	16	12	0.999378	0.004287
48	22	16	13	0.999952	0.000574
48	22	16	14	0.999998	0.000046
48	22	16	15	1.000000	0.000002
48	22	16	16	1.000000	0.000000
48	22	17	0	0.000001	0.000001
48	22	17	1	0.000028	0.000028
48	22	17	2	0.000449	0.000420
48	22	17	3	0.003953	0.003504
48	22	17	4	0.021878	0.017925
48	22	17	5	0.081798	0.059920
48	22	17	6	0.217616	0.135819
48	22	17	7	0.431046	0.213429
48	22	17	8	0.666446	0.235400
48	22	17	9	0.849535	0.183089
48	22	17	10	0.949752	0.100217
48	22	17	11	0.988017	0.038265
48	22	17	12	0.998038	0.010022
48	22	17	13	0.999790	0.001752
48	22	17	14	0.999986	0.000196
48	22	17	15	0.999999	0.000013
48	22	17	16	1.000000	0.000000
48	22	17	17	1.000000	0.000000
48	22	18	0	0.000000	0.000000
48	22	18	1	0.000010	0.000009
48	22	18	2	0.000177	0.000168
48	22	18	3	0.001805	0.001628
48	22	18	4	0.011470	0.009665
48	22	18	5	0.048938	0.037469
48	22	18	6	0.147516	0.098578
48	22	18	7	0.327774	0.180257
48	22	18	8	0.560136	0.232363
48	22	18	9	0.772756	0.212619
48	22	18	10	0.910958	0.138203
48	22	18	11	0.974439	0.063481
48	22	18	12	0.994806	0.020367
48	22	18	13	0.999282	0.004476
48	22	18	14	0.999936	0.000654
48	22	18	15	0.999996	0.000061
48	22	18	16	1.000000	0.000003
48	22	18	17	1.000000	0.000000
48	22	18	18	1.000000	0.000000
48	22	19	0	0.000000	0.000000
48	22	19	1	0.000003	0.000003
48	22	19	2	0.000066	0.000063

N	n	k	x	P(x)	p(x)
48	22	19	3	0.000774	0.000709
48	22	19	4	0.005671	0.004897
48	22	19	5	0.027706	0.022035
48	22	19	6	0.094941	0.067235
48	22	19	7	0.237645	0.142704
48	22	19	8	0.451700	0.214055
48	22	19	9	0.680621	0.228920
48	22	19	10	0.855677	0.175057
48	22	19	11	0.951163	0.095485
48	22	19	12	0.988017	0.036854
48	22	19	13	0.997939	0.009922
48	22	19	14	0.999761	0.001822
48	22	19	15	0.999982	0.000221
48	22	19	16	0.999999	0.000017
48	22	19	17	1.000000	0.000001
48	22	19	18	1.000000	0.000000
48	22	19	19	1.000000	0.000000
48	22	20	0	0.000000	0.000000
48	22	20	1	0.000001	0.000001
48	22	20	2	0.000022	0.000022
48	22	20	3	0.000310	0.000288
48	22	20	4	0.002632	0.002322
48	22	20	5	0.014789	0.012157
48	22	20	6	0.057846	0.043057
48	22	20	7	0.163833	0.105987
48	22	20	8	0.348363	0.184530
48	22	20	9	0.578001	0.229638
48	22	20	10	0.783240	0.205239
48	22	20	11	0.914944	0.131704
48	22	20	12	0.975308	0.060364
48	22	20	13	0.994860	0.019551
48	22	20	14	0.999259	0.004399
48	22	20	15	0.999929	0.000670
48	22	20	16	0.999996	0.000067
48	22	20	17	1.000000	0.000004
48	22	20	18	1.000000	0.000000
48	22	20	19	1.000000	0.000000
48	22	20	20	1.000000	0.000000
48	22	21	0	0.000000	0.000000
48	22	21	1	0.000000	0.000000
48	22	21	2	0.000007	0.000007
48	22	21	3	0.000115	0.000108
48	22	21	4	0.001139	0.001024
48	22	21	5	0.007408	0.006269
48	22	21	6	0.033242	0.025834
48	22	21	7	0.107054	0.073812
48	22	21	8	0.256098	0.149044
48	22	21	9	0.471384	0.215286
48	22	21	10	0.695280	0.223897
48	22	21	11	0.863203	0.167923

N	n	k	x	P(x)	p(x)
48	22	21	12	0.953750	0.090547
48	22	21	13	0.988575	0.034826
48	22	21	14	0.998002	0.009426
48	22	21	15	0.999761	0.001760
48	22	21	16	0.999981	0.000220
48	22	21	17	0.999999	0.000018
48	22	21	18	1.000000	0.000001
48	22	21	19	1.000000	0.000000
48	22	21	20	1.000000	0.000000
48	22	21	21	1.000000	0.000000
48	22	22	0	0.000000	0.000000
48	22	22	1	0.000000	0.000000
48	22	22	2	0.000002	0.000002
48	22	22	3	0.000039	0.000037
48	22	22	4	0.000456	0.000417
48	22	22	5	0.003461	0.003005
48	22	22	6	0.017933	0.014472
48	22	22	7	0.066047	0.048115
48	22	22	8	0.178816	0.112769
48	22	22	9	0.367728	0.188912
48	22	22	10	0.595771	0.228043
48	22	22	11	0.794790	0.199020
48	22	22	12	0.920214	0.125424
48	22	22	13	0.976967	0.056753
48	22	22	14	0.995209	0.018242
48	22	22	15	0.999305	0.004096
48	22	22	16	0.999932	0.000627
48	22	22	17	0.999996	0.000063
48	22	22	18	1.000000	0.000004
48	22	22	19	1.000000	0.000000
48	22	22	20	1.000000	0.000000
48	22	22	21	1.000000	0.000000
48	22	22	22	1.000000	0.000000
48	23	1	0	0.520833	0.520833
48	23	1	1	1.000000	0.479167
48	23	2	0	0.265957	0.265957
48	23	2	1	0.775709	0.509752
48	23	2	2	1.000000	0.224291
48	23	3	0	0.132979	0.132979
48	23	3	1	0.531915	0.398936
48	23	3	2	0.897606	0.365691
48	23	3	3	1.000000	0.102394
48	23	4	0	0.065012	0.065012
48	23	4	1	0.336879	0.271868
48	23	4	2	0.726950	0.390071
48	23	4	3	0.954492	0.227541
48	23	4	4	1.000000	0.045508
48	23	5	0	0.031028	0.031028
48	23	5	1	0.200946	0.169917
48	23	5	2	0.540780	0.339834

Table for $N = 2$, $n = 1$, through $N = 100$, $n = 50$

N	n	k	x	P(x)	p(x)
48	23	5	3	0.851064	0.310284
48	23	5	4	0.980349	0.129285
48	23	5	5	1.000000	0.019651
48	23	6	0	0.014432	0.014432
48	23	6	1	0.114011	0.099579
48	23	6	2	0.374814	0.260803
48	23	6	3	0.706746	0.331931
48	23	6	4	0.923223	0.216477
48	23	6	5	0.991774	0.068551
48	23	6	6	1.000000	0.008226
48	23	7	0	0.006529	0.006529
48	23	7	1	0.061851	0.055322
48	23	7	2	0.244413	0.182562
48	23	7	3	0.548683	0.304270
48	23	7	4	0.825293	0.276609
48	23	7	5	0.962395	0.137102
48	23	7	6	0.996670	0.034276
48	23	7	7	1.000000	0.003330
48	23	8	0	0.002866	0.002866
48	23	8	1	0.032166	0.029299
48	23	8	2	0.150905	0.118740
48	23	8	3	0.400259	0.249353
48	23	8	4	0.697108	0.296849
48	23	8	5	0.902204	0.205096
48	23	8	6	0.982458	0.080255
48	23	8	7	0.998700	0.016242
48	23	8	8	1.000000	0.001299
48	23	9	0	0.001218	0.001218
48	23	9	1	0.016051	0.014833
48	23	9	2	0.088567	0.072516
48	23	9	3	0.275582	0.187015
48	23	9	4	0.556104	0.280522
48	23	9	5	0.809910	0.253806
48	23	9	6	0.948350	0.138440
48	23	9	7	0.992204	0.043854
48	23	9	8	0.999513	0.007309
48	23	9	9	1.000000	0.000487
48	23	10	0	0.000500	0.000500
48	23	10	1	0.007684	0.007184
48	23	10	2	0.049520	0.041836
48	23	10	3	0.179677	0.130157
48	23	10	4	0.419440	0.239763
48	23	10	5	0.692769	0.273330
48	23	10	6	0.888005	0.195235
48	23	10	7	0.974212	0.086208
48	23	10	8	0.996701	0.022489
48	23	10	9	0.999825	0.003123
48	23	10	10	1.000000	0.000175
48	23	11	0	0.000197	0.000197
48	23	11	1	0.003525	0.003327

N	n	k	x	P(x)	p(x)
48	23	11	2	0.026400	0.022875
48	23	11	3	0.111173	0.084773
48	23	11	4	0.299558	0.188385
48	23	11	5	0.563297	0.263739
48	23	11	6	0.800663	0.237365
48	23	11	7	0.937915	0.137252
48	23	11	8	0.987824	0.049910
48	23	11	9	0.998674	0.010850
48	23	11	10	0.999940	0.001266
48	23	11	11	1.000000	0.000060
48	23	12	0	0.000075	0.000075
48	23	12	1	0.001546	0.001472
48	23	12	2	0.013417	0.011870
48	23	12	3	0.065350	0.051933
48	23	12	4	0.202820	0.137470
48	23	12	5	0.434992	0.232172
48	23	12	6	0.691603	0.256611
48	23	12	7	0.878562	0.186959
48	23	12	8	0.967591	0.089028
48	23	12	9	0.994569	0.026978
48	23	12	10	0.999495	0.004926
48	23	12	11	0.999980	0.000485
48	23	12	12	1.000000	0.000019
48	23	13	0	0.000027	0.000027
48	23	13	1	0.000647	0.000620
48	23	13	2	0.006492	0.005845
48	23	13	3	0.036498	0.030006
48	23	13	4	0.130266	0.093768
48	23	13	5	0.318906	0.188640
48	23	13	6	0.570425	0.251520
48	23	13	7	0.795469	0.225044
48	23	13	8	0.930495	0.135026
48	23	13	9	0.984077	0.053582
48	23	13	10	0.997716	0.013639
48	23	13	11	0.999819	0.002102
48	23	13	12	0.999994	0.000175
48	23	13	13	1.000000	0.000006
48	23	14	0	0.000009	0.000009
48	23	14	1	0.000257	0.000248
48	23	14	2	0.002985	0.002728
48	23	14	3	0.019352	0.016367
48	23	14	4	0.079364	0.060012
48	23	14	5	0.221891	0.142528
48	23	14	6	0.448259	0.226368
48	23	14	7	0.692592	0.244333
48	23	14	8	0.872627	0.180035
48	23	14	9	0.962645	0.090018
48	23	14	10	0.992650	0.030006
48	23	14	11	0.999098	0.006448
48	23	14	12	0.999939	0.000841

N	n	k	x	P(x)	p(x)
48	23	14	13	0.999998	0.000059
48	23	14	14	1.000000	0.000002
48	23	15	0	0.000003	0.000003
48	23	15	1	0.000097	0.000094
48	23	15	2	0.001300	0.001203
48	23	15	3	0.009724	0.008424
48	23	15	4	0.045828	0.036103
48	23	15	5	0.146435	0.100608
48	23	15	6	0.335075	0.188640
48	23	15	7	0.577612	0.242537
48	23	15	8	0.793200	0.215588
48	23	15	9	0.925578	0.132379
48	23	15	10	0.981178	0.055599
48	23	15	11	0.996822	0.015645
48	23	15	12	0.999667	0.002844
48	23	15	13	0.999981	0.000314
48	23	15	14	0.999999	0.000019
48	23	15	15	1.000000	0.000000
48	23	16	0	0.000001	0.000001
48	23	16	1	0.000034	0.000033
48	23	16	2	0.000534	0.000500
48	23	16	3	0.004619	0.004084
48	23	16	4	0.025041	0.020422
48	23	16	5	0.091558	0.066518
48	23	16	6	0.237897	0.146339
48	23	16	7	0.460018	0.222121
48	23	16	8	0.695205	0.235187
48	23	16	9	0.869418	0.174213
48	23	16	10	0.959275	0.089857
48	23	16	11	0.991133	0.031858
48	23	16	12	0.998719	0.007585
48	23	16	13	0.999886	0.001167
48	23	16	14	0.999994	0.000109
48	23	16	15	1.000000	0.000005
48	23	16	16	1.000000	0.000000
48	23	17	0	0.000000	0.000000
48	23	17	1	0.000011	0.000011
48	23	17	2	0.000206	0.000195
48	23	17	3	0.002066	0.001860
48	23	17	4	0.012915	0.010849
48	23	17	5	0.054142	0.041227
48	23	17	6	0.160155	0.106012
48	23	17	7	0.348958	0.188803
48	23	17	8	0.584961	0.236004
48	23	17	9	0.793200	0.208239
48	23	17	10	0.922770	0.129571
48	23	17	11	0.979186	0.056416
48	23	17	12	0.996111	0.016925
48	23	17	13	0.999521	0.003410
48	23	17	14	0.999964	0.000443

N	n	k	x	P(x)	p(x)
48	23	17	15	0.999998	0.000035
48	23	17	16	1.000000	0.000001
48	23	17	17	1.000000	0.000000
48	23	18	0	0.000000	0.000000
48	23	18	1	0.000003	0.000003
48	23	18	2	0.000074	0.000071
48	23	18	3	0.000866	0.000792
48	23	18	4	0.006266	0.005400
48	23	18	5	0.030204	0.023938
48	23	18	6	0.102019	0.071815
48	23	18	7	0.251511	0.149492
48	23	18	8	0.470766	0.219255
48	23	18	9	0.699157	0.228391
48	23	18	10	0.868434	0.169278
48	23	18	11	0.957348	0.088914
48	23	18	12	0.990106	0.032758
48	23	18	13	0.998421	0.008315
48	23	18	14	0.999835	0.001414
48	23	18	15	0.999989	0.000154
48	23	18	16	0.999999	0.000010
48	23	18	17	1.000000	0.000000
48	23	18	18	1.000000	0.000000
48	23	19	0	0.000000	0.000000
48	23	19	1	0.000001	0.000001
48	23	19	2	0.000025	0.000024
48	23	19	3	0.000338	0.000313
48	23	19	4	0.002846	0.002508
48	23	19	5	0.015841	0.012995
48	23	19	6	0.061324	0.045483
48	23	19	7	0.171782	0.110458
48	23	19	8	0.361138	0.189357
48	23	19	9	0.592574	0.231436
48	23	19	10	0.795081	0.202506
48	23	19	11	0.921783	0.126702
48	23	19	12	0.978094	0.056312
48	23	19	13	0.995649	0.017555
48	23	19	14	0.999411	0.003762
48	23	19	15	0.999948	0.000537
48	23	19	16	0.999997	0.000049
48	23	19	17	1.000000	0.000003
48	23	19	18	1.000000	0.000000
48	23	19	19	1.000000	0.000000
48	23	20	0	0.000000	0.000000
48	23	20	1	0.000000	0.000000
48	23	20	2	0.000008	0.000007
48	23	20	3	0.000122	0.000114
48	23	20	4	0.001203	0.001081
48	23	20	5	0.007775	0.006572
48	23	20	6	0.034661	0.026886
48	23	20	7	0.110839	0.076178

Table for $N = 2$, $n = 1$, through $N = 100$, $n = 50$

N	n	k	x	P(x)	p(x)
48	23	20	8	0.263195	0.152356
48	23	20	9	0.480847	0.217651
48	23	20	10	0.704302	0.223455
48	23	20	11	0.869354	0.165052
48	23	20	12	0.956735	0.087381
48	23	20	13	0.989596	0.032861
48	23	20	14	0.998244	0.008648
48	23	20	15	0.999800	0.001557
48	23	20	16	0.999985	0.000185
48	23	20	17	0.999999	0.000014
48	23	20	18	1.000000	0.000001
48	23	20	19	1.000000	0.000000
48	23	20	20	1.000000	0.000000
48	23	21	0	0.000000	0.000000
48	23	21	1	0.000000	0.000000
48	23	21	2	0.000002	0.000002
48	23	21	3	0.000040	0.000038
48	23	21	4	0.000469	0.000429
48	23	21	5	0.003550	0.003081
48	23	21	6	0.018338	0.014787
48	23	21	7	0.067309	0.048972
48	23	21	8	0.181576	0.114267
48	23	21	9	0.372021	0.190445
48	23	21	10	0.600555	0.228534
48	23	21	11	0.798618	0.198063
48	23	21	12	0.922407	0.123789
48	23	21	13	0.977860	0.055453
48	23	21	14	0.995464	0.017604
48	23	21	15	0.999355	0.003891
48	23	21	16	0.999939	0.000584
48	23	21	17	0.999996	0.000057
48	23	21	18	1.000000	0.000003
48	23	21	19	1.000000	0.000000
48	23	21	20	1.000000	0.000000
48	23	21	21	1.000000	0.000000
48	23	22	0	0.000000	0.000000
48	23	22	1	0.000000	0.000000
48	23	22	2	0.000000	0.000000
48	23	22	3	0.000012	0.000011
48	23	22	4	0.000167	0.000155
48	23	22	5	0.001496	0.001329
48	23	22	6	0.009027	0.007531
48	23	22	7	0.038289	0.029262
48	23	22	8	0.118094	0.079805
48	23	22	9	0.273272	0.155177
48	23	22	10	0.490520	0.217248
48	23	22	11	0.710590	0.220070
48	23	22	12	0.871974	0.161384
48	23	22	13	0.957322	0.085348
48	23	22	14	0.989596	0.032274
48	23	22	15	0.998202	0.008606
48	23	22	16	0.999788	0.001585
48	23	22	17	0.999984	0.000196
48	23	22	18	0.999999	0.000016
48	23	22	19	1.000000	0.000001
48	23	22	20	1.000000	0.000000
48	23	22	21	1.000000	0.000000
48	23	22	22	1.000000	0.000000
48	23	23	0	0.000000	0.000000
48	23	23	1	0.000000	0.000000
48	23	23	2	0.000000	0.000000
48	23	23	3	0.000003	0.000003
48	23	23	4	0.000054	0.000051
48	23	23	5	0.000576	0.000522
48	23	23	6	0.004103	0.003527
48	23	23	7	0.020282	0.016179
48	23	23	8	0.072053	0.051771
48	23	23	9	0.189715	0.117662
48	23	23	10	0.381896	0.192181
48	23	23	11	0.609019	0.227123
48	23	23	12	0.803696	0.194677
48	23	23	13	0.924496	0.120800
48	23	23	14	0.978424	0.053928
48	23	23	15	0.995554	0.017130
48	23	23	16	0.999361	0.003807
48	23	23	17	0.999938	0.000577
48	23	23	18	0.999996	0.000058
48	23	23	19	1.000000	0.000004
48	23	23	20	1.000000	0.000000
48	23	23	21	1.000000	0.000000
48	23	23	22	1.000000	0.000000
48	23	23	23	1.000000	0.000000
48	24	1	0	0.500000	0.500000
48	24	1	1	1.000000	0.500000
48	24	2	0	0.244681	0.244681
48	24	2	1	0.755319	0.510638
48	24	2	2	1.000000	0.244681
48	24	3	0	0.117021	0.117021
48	24	3	1	0.500000	0.382979
48	24	3	2	0.882979	0.382979
48	24	3	3	1.000000	0.117021
48	24	4	0	0.054610	0.054610
48	24	4	1	0.304255	0.249645
48	24	4	2	0.695745	0.391489
48	24	4	3	0.945390	0.249645
48	24	4	4	1.000000	0.054610
48	24	5	0	0.024823	0.024823
48	24	5	1	0.173759	0.148936
48	24	5	2	0.500000	0.326241
48	24	5	3	0.826241	0.326241
48	24	5	4	0.975177	0.148936
48	24	5	5	1.000000	0.024823
48	24	6	0	0.010968	0.010968
48	24	6	1	0.094095	0.083127
48	24	6	2	0.333086	0.238991
48	24	6	3	0.666914	0.333828
48	24	6	4	0.905905	0.238991
48	24	6	5	0.989032	0.083127
48	24	6	6	1.000000	0.010968
48	24	7	0	0.004701	0.004701
48	24	7	1	0.048573	0.043873
48	24	7	2	0.207900	0.159327
48	24	7	3	0.500000	0.292100
48	24	7	4	0.792100	0.292100
48	24	7	5	0.951427	0.159327
48	24	7	6	0.995299	0.043873
48	24	7	7	1.000000	0.004701
48	24	8	0	0.001949	0.001949
48	24	8	1	0.023962	0.022013
48	24	8	2	0.122408	0.098446
48	24	8	3	0.350388	0.227980
48	24	8	4	0.649612	0.299224
48	24	8	5	0.877592	0.227980
48	24	8	6	0.976038	0.098446
48	24	8	7	0.998051	0.022013
48	24	8	8	1.000000	0.001949
48	24	9	0	0.000780	0.000780
48	24	9	1	0.011304	0.010525
48	24	9	2	0.068262	0.056958
48	24	9	3	0.230698	0.162436
48	24	9	4	0.500000	0.269302
48	24	9	5	0.769302	0.269302
48	24	9	6	0.931737	0.162436
48	24	9	7	0.988695	0.056958
48	24	9	8	0.999220	0.010525
48	24	9	9	1.000000	0.000780
48	24	10	0	0.000300	0.000300
48	24	10	1	0.005098	0.004798
48	24	10	2	0.036132	0.031035
48	24	10	3	0.143233	0.107101
48	24	10	4	0.361897	0.218664
48	24	10	5	0.638103	0.276207
48	24	10	6	0.856767	0.218664
48	24	10	7	0.963868	0.107101
48	24	10	8	0.994902	0.031035
48	24	10	9	0.999700	0.004798
48	24	10	10	1.000000	0.000300
48	24	11	0	0.000110	0.000110
48	24	11	1	0.002194	0.002083
48	24	11	2	0.018165	0.015971
48	24	11	3	0.084046	0.065881
48	24	11	4	0.246810	0.162765
48	24	11	5	0.500000	0.253190
48	24	11	6	0.753189	0.253190
48	24	11	7	0.915954	0.162765
48	24	11	8	0.981835	0.065881
48	24	11	9	0.997806	0.015971
48	24	11	10	0.999889	0.002083
48	24	11	11	1.000000	0.000110
48	24	12	0	0.000039	0.000039
48	24	12	1	0.000899	0.000860
48	24	12	2	0.008668	0.007770
48	24	12	3	0.046654	0.037985
48	24	12	4	0.158830	0.112176
48	24	12	5	0.369984	0.211154
48	24	12	6	0.630016	0.260032
48	24	12	7	0.841170	0.211154
48	24	12	8	0.953346	0.112176
48	24	12	9	0.991331	0.037985
48	24	12	10	0.999101	0.007770
48	24	12	11	0.999961	0.000860
48	24	12	12	1.000000	0.000039
48	24	13	0	0.000013	0.000013
48	24	13	1	0.000349	0.000336
48	24	13	2	0.003920	0.003571
48	24	13	3	0.024496	0.020575
48	24	13	4	0.096510	0.072014
48	24	13	5	0.258541	0.162032
48	24	13	6	0.500000	0.241459
48	24	13	7	0.741459	0.241459
48	24	13	8	0.903490	0.162032
48	24	13	9	0.975504	0.072014
48	24	13	10	0.996080	0.020575
48	24	13	11	0.999651	0.003571
48	24	13	12	0.999987	0.000336
48	24	13	13	1.000000	0.000013
48	24	14	0	0.000004	0.000004
48	24	14	1	0.000128	0.000124
48	24	14	2	0.001676	0.001547
48	24	14	3	0.012150	0.010475
48	24	14	4	0.055359	0.043208
48	24	14	5	0.170581	0.115222
48	24	14	6	0.375821	0.205240
48	24	14	7	0.624179	0.248358
48	24	14	8	0.829419	0.205240
48	24	14	9	0.944641	0.115222
48	24	14	10	0.987849	0.043208
48	24	14	11	0.998324	0.010475
48	24	14	12	0.999872	0.001547
48	24	14	13	0.999996	0.000124

Table for $N = 2$, $n = 1$, through $N = 100$, $n = 50$

N	n	k	x	P(x)	p(x)
48	24	14	14	1.000000	0.000004
48	24	15	0	0.000001	0.000001
48	24	15	1	0.000044	0.000043
48	24	15	2	0.000674	0.000630
48	24	15	3	0.005681	0.005006
48	24	15	4	0.029942	0.024261
48	24	15	5	0.106192	0.076250
48	24	15	6	0.267165	0.160972
48	24	15	7	0.500000	0.232835
48	24	15	8	0.732835	0.232835
48	24	15	9	0.893808	0.160972
48	24	15	10	0.970058	0.076250
48	24	15	11	0.994319	0.024261
48	24	15	12	0.999326	0.005006
48	24	15	13	0.999956	0.000630
48	24	15	14	0.999999	0.000043
48	24	15	15	1.000000	0.000001
48	24	16	0	0.000000	0.000000
48	24	16	1	0.000014	0.000014
48	24	16	2	0.000254	0.000240
48	24	16	3	0.002495	0.002241
48	24	16	4	0.015238	0.012743
48	24	16	5	0.062291	0.047052
48	24	16	6	0.179362	0.117071
48	24	16	7	0.380055	0.200693
48	24	16	8	0.619945	0.239891
48	24	16	9	0.820638	0.200693
48	24	16	10	0.937709	0.117071
48	24	16	11	0.984762	0.047052
48	24	16	12	0.997505	0.012743
48	24	16	13	0.999746	0.002241
48	24	16	14	0.999986	0.000240
48	24	16	15	1.000000	0.000014
48	24	16	16	1.000000	0.000000
48	24	17	0	0.000000	0.000000
48	24	17	1	0.000004	0.000004
48	24	17	2	0.000089	0.000085
48	24	17	3	0.001025	0.000935
48	24	17	4	0.007274	0.006249
48	24	17	5	0.034353	0.027080
48	24	17	6	0.113509	0.079156
48	24	17	7	0.273436	0.159927
48	24	17	8	0.500000	0.226564
48	24	17	9	0.726564	0.226564
48	24	17	10	0.886491	0.159927
48	24	17	11	0.965647	0.079156
48	24	17	12	0.992726	0.027080
48	24	17	13	0.998975	0.006249
48	24	17	14	0.999911	0.000935
48	24	17	15	0.999996	0.000085
48	24	17	16	1.000000	0.000004
48	24	17	17	1.000000	0.000000
48	24	18	0	0.000000	0.000000
48	24	18	1	0.000001	0.000001
48	24	18	2	0.000029	0.000028
48	24	18	3	0.000391	0.000362
48	24	18	4	0.003242	0.002851
48	24	18	5	0.017756	0.014514
48	24	18	6	0.067548	0.049792
48	24	18	7	0.185734	0.118187
48	24	18	8	0.383064	0.197330
48	24	18	9	0.616936	0.233872
48	24	18	10	0.814266	0.197330
48	24	18	11	0.932452	0.118187
48	24	18	12	0.982244	0.049792
48	24	18	13	0.996758	0.014514
48	24	18	14	0.999609	0.002851
48	24	18	15	0.999971	0.000362
48	24	18	16	0.999999	0.000028
48	24	18	17	1.000000	0.000001
48	24	18	18	1.000000	0.000000
48	24	19	0	0.000000	0.000000
48	24	19	1	0.000000	0.000000
48	24	19	2	0.000009	0.000008
48	24	19	3	0.000138	0.000129
48	24	19	4	0.001341	0.001204
48	24	19	5	0.008564	0.007223
48	24	19	6	0.037673	0.029109
48	24	19	7	0.118762	0.081089
48	24	19	8	0.277822	0.159060
48	24	19	9	0.500000	0.222178
48	24	19	10	0.722178	0.222178
48	24	19	11	0.881238	0.159060
48	24	19	12	0.962327	0.081089
48	24	19	13	0.991436	0.029109
48	24	19	14	0.998659	0.007223
48	24	19	15	0.999862	0.001204
48	24	19	16	0.999991	0.000129
48	24	19	17	1.000000	0.000008
48	24	19	18	1.000000	0.000000
48	24	19	19	1.000000	0.000000
48	24	20	0	0.000000	0.000000
48	24	20	1	0.000000	0.000000
48	24	20	2	0.000002	0.000002
48	24	20	3	0.000044	0.000042
48	24	20	4	0.000511	0.000467
48	24	20	5	0.003832	0.003321
48	24	20	6	0.019605	0.015773
48	24	20	7	0.071227	0.051622
48	24	20	8	0.190064	0.118838
48	24	20	9	0.385080	0.195016
48	24	20	10	0.614920	0.229840
48	24	20	11	0.809935	0.195016
48	24	20	12	0.928773	0.118838
48	24	20	13	0.980395	0.051622
48	24	20	14	0.996168	0.015773
48	24	20	15	0.999489	0.003321
48	24	20	16	0.999956	0.000467
48	24	20	17	0.999998	0.000042
48	24	20	18	1.000000	0.000002
48	24	20	19	1.000000	0.000000
48	24	20	20	1.000000	0.000000
48	24	21	0	0.000000	0.000000
48	24	21	1	0.000000	0.000000
48	24	21	2	0.000001	0.000001
48	24	21	3	0.000013	0.000012
48	24	21	4	0.000178	0.000165
48	24	21	5	0.001578	0.001401
48	24	21	6	0.009465	0.007887
48	24	21	7	0.039885	0.030420
48	24	21	8	0.122157	0.082272
48	24	21	9	0.280607	0.158450
48	24	21	10	0.500000	0.219393
48	24	21	11	0.719393	0.219393
48	24	21	12	0.877843	0.158450
48	24	21	13	0.960115	0.082272
48	24	21	14	0.990535	0.030420
48	24	21	15	0.998421	0.007887
48	24	21	16	0.999822	0.001401
48	24	21	17	0.999987	0.000165
48	24	21	18	0.999999	0.000012
48	24	21	19	1.000000	0.000001
48	24	21	20	1.000000	0.000000
48	24	21	21	1.000000	0.000000
48	24	22	0	0.000000	0.000000
48	24	22	1	0.000000	0.000000
48	24	22	2	0.000000	0.000000
48	24	22	3	0.000003	0.000003
48	24	22	4	0.000055	0.000052
48	24	22	5	0.000593	0.000537
48	24	22	6	0.004207	0.003615
48	24	22	7	0.020732	0.016524
48	24	22	8	0.073403	0.052672
48	24	22	9	0.192580	0.119176
48	24	22	10	0.386241	0.193661
48	24	22	11	0.613759	0.227518
48	24	22	12	0.807420	0.193661
48	24	22	13	0.926596	0.119176
48	24	22	14	0.979268	0.052672
48	24	22	15	0.995793	0.016524
48	24	22	16	0.999407	0.003615
48	24	22	17	0.999944	0.000537
48	24	22	18	0.999997	0.000052
48	24	22	19	1.000000	0.000003
48	24	22	20	1.000000	0.000000
48	24	22	21	1.000000	0.000000
48	24	22	22	1.000000	0.000000
48	24	23	0	0.000000	0.000000
48	24	23	1	0.000000	0.000000
48	24	23	2	0.000000	0.000000
48	24	23	3	0.000001	0.000001
48	24	23	4	0.000015	0.000015
48	24	23	5	0.000200	0.000185
48	24	23	6	0.001705	0.001505
48	24	23	7	0.009927	0.008222
48	24	23	8	0.040990	0.031063
48	24	23	9	0.123824	0.082834
48	24	23	10	0.281962	0.158138
48	24	23	11	0.500000	0.218038
48	24	23	12	0.718038	0.218038
48	24	23	13	0.876176	0.158138
48	24	23	14	0.959010	0.082834
48	24	23	15	0.990073	0.031063
48	24	23	16	0.998295	0.008222
48	24	23	17	0.999800	0.001505
48	24	23	18	0.999985	0.000185
48	24	23	19	0.999999	0.000015
48	24	23	20	1.000000	0.000001
48	24	23	21	1.000000	0.000000
48	24	23	22	1.000000	0.000000
48	24	23	23	1.000000	0.000000
48	24	24	0	0.000000	0.000000
48	24	24	1	0.000000	0.000000
48	24	24	2	0.000000	0.000000
48	24	24	3	0.000000	0.000000
48	24	24	4	0.000004	0.000003
48	24	24	5	0.000060	0.000056
48	24	24	6	0.000621	0.000562
48	24	24	7	0.004336	0.003715
48	24	24	8	0.021110	0.016774
48	24	24	9	0.074124	0.053014
48	24	24	10	0.193405	0.119281
48	24	24	11	0.386620	0.193215
48	24	24	12	0.613380	0.226760
48	24	24	13	0.806595	0.193215
48	24	24	14	0.925876	0.119281
48	24	24	15	0.978890	0.053014
48	24	24	16	0.995664	0.016774
48	24	24	17	0.999378	0.003715
48	24	24	18	0.999940	0.000562

Table for $N=2$, $n=1$, through $N=100$, $n=50$

N	n	k	x	P(x)	p(x)
48	24	24	19	0.999996	0.000056
48	24	24	20	1.000000	0.000003
48	24	24	21	1.000000	0.000000
48	24	24	22	1.000000	0.000000
48	24	24	23	1.000000	0.000000
48	24	24	24	1.000000	0.000000
49	1	1	0	0.979592	0.979592
49	1	1	1	1.000000	0.020408
49	2	1	0	0.959184	0.959184
49	2	1	1	1.000000	0.040816
49	2	2	0	0.919218	0.919218
49	2	2	1	0.999150	0.079932
49	2	2	2	1.000000	0.000850
49	3	1	0	0.938776	0.938776
49	3	1	1	1.000000	0.061224
49	3	2	0	0.880102	0.880102
49	3	2	1	0.997449	0.117347
49	3	2	2	1.000000	0.002551
49	3	3	0	0.823925	0.823925
49	3	3	1	0.992455	0.168530
49	3	3	2	0.999946	0.007490
49	3	3	3	1.000000	0.000054
49	4	1	0	0.918367	0.918367
49	4	1	1	1.000000	0.081633
49	4	2	0	0.841837	0.841837
49	4	2	1	0.994898	0.153061
49	4	2	2	1.000000	0.005102
49	4	3	0	0.770191	0.770191
49	4	3	1	0.985128	0.214937
49	4	3	2	0.999783	0.014655
49	4	3	3	1.000000	0.000217
49	4	4	0	0.703218	0.703218
49	4	4	1	0.971110	0.267893
49	4	4	2	0.999146	0.028035
49	4	4	3	0.999995	0.000850
49	4	4	4	1.000000	0.000005
49	5	1	0	0.897959	0.897959
49	5	1	1	1.000000	0.102041
49	5	2	0	0.804422	0.804422
49	5	2	1	0.991497	0.187075
49	5	2	2	1.000000	0.008503
49	5	3	0	0.718845	0.718845
49	5	3	1	0.975575	0.256730
49	5	3	2	0.999457	0.023882
49	5	3	3	1.000000	0.000543
49	5	4	0	0.640710	0.640710
49	5	4	1	0.953251	0.312541
49	5	4	2	0.997900	0.044649
49	5	4	3	0.999976	0.002077
49	5	4	4	1.000000	0.000024

N	n	k	x	P(x)	p(x)
49	5	5	0	0.569520	0.569520
49	5	5	1	0.925469	0.355950
49	5	5	2	0.994923	0.069454
49	5	5	3	0.999884	0.004961
49	5	5	4	0.999999	0.000115
49	5	5	5	1.000000	0.000001
49	6	1	0	0.877551	0.877551
49	6	1	1	1.000000	0.122449
49	6	2	0	0.767857	0.767857
49	6	2	1	0.987245	0.219388
49	6	2	2	1.000000	0.012755
49	6	3	0	0.669833	0.669833
49	6	3	1	0.963906	0.294073
49	6	3	2	0.998914	0.035009
49	6	3	3	1.000000	0.001086
49	6	4	0	0.582463	0.582463
49	6	4	1	0.931941	0.349478
49	6	4	2	0.995870	0.063929
49	6	4	3	0.999929	0.004059
49	6	4	4	1.000000	0.000071
49	6	5	0	0.504802	0.504802
49	6	5	1	0.893110	0.388309
49	6	5	2	0.990188	0.097077
49	6	5	3	0.999658	0.009471
49	6	5	4	0.999997	0.000338
49	6	5	5	1.000000	0.000003
49	6	6	0	0.435965	0.435965
49	6	6	1	0.848984	0.413019
49	6	6	2	0.981362	0.132378
49	6	6	3	0.999013	0.017650
49	6	6	4	0.999981	0.000969
49	6	6	5	1.000000	0.000018
49	6	6	6	1.000000	0.000000
49	7	1	0	0.857143	0.857143
49	7	1	1	1.000000	0.142857
49	7	2	0	0.732143	0.732143
49	7	2	1	0.982143	0.250000
49	7	2	2	1.000000	0.017857
49	7	3	0	0.623100	0.623100
49	7	3	1	0.950228	0.327128
49	7	3	2	0.998100	0.047872
49	7	3	3	1.000000	0.001900
49	7	4	0	0.528281	0.528281
49	7	4	1	0.907559	0.379278
49	7	4	2	0.992897	0.085338
49	7	4	3	0.999835	0.006938
49	7	4	4	1.000000	0.000165
49	7	5	0	0.446104	0.446104
49	7	5	1	0.856989	0.410885
49	7	5	2	0.983415	0.126426

N	n	k	x	P(x)	p(x)
49	7	5	3	0.999218	0.015803
49	7	5	4	0.999989	0.000771
49	7	5	5	1.000000	0.000011
49	7	6	0	0.375133	0.375133
49	7	6	1	0.800959	0.425826
49	7	6	2	0.969048	0.168089
49	7	6	3	0.997781	0.028733
49	7	6	4	0.999936	0.002155
49	7	6	5	0.999999	0.000063
49	7	6	6	1.000000	0.000000
49	7	7	0	0.314065	0.314065
49	7	7	1	0.741541	0.427477
49	7	7	2	0.949503	0.207962
49	7	7	3	0.995109	0.045606
49	7	7	4	0.999786	0.004677
49	7	7	5	0.999997	0.000210
49	7	7	6	1.000000	0.000003
49	7	7	7	1.000000	0.000000
49	8	1	0	0.836735	0.836735
49	8	1	1	1.000000	0.163265
49	8	2	0	0.697279	0.697279
49	8	2	1	0.976190	0.278912
49	8	2	2	1.000000	0.023810
49	8	3	0	0.578593	0.578593
49	8	3	1	0.934650	0.356057
49	8	3	2	0.996960	0.062310
49	8	3	3	1.000000	0.003040
49	8	4	0	0.477968	0.477968
49	8	4	1	0.880468	0.402500
49	8	4	2	0.988833	0.108365
49	8	4	3	0.999670	0.010837
49	8	4	4	1.000000	0.000330
49	8	5	0	0.392996	0.392996
49	8	5	1	0.817857	0.424861
49	8	5	2	0.974384	0.156528
49	8	5	3	0.998465	0.024081
49	8	5	4	0.999971	0.001505
49	8	5	5	1.000000	0.000029
49	8	6	0	0.321542	0.321542
49	8	6	1	0.750265	0.428723
49	8	6	2	0.953040	0.202774
49	8	6	3	0.995729	0.042689
49	8	6	4	0.999834	0.004105
49	8	6	5	0.999998	0.000164
49	8	6	6	1.000000	0.000002
49	8	7	0	0.261720	0.261720
49	8	7	1	0.680473	0.418753
49	8	7	2	0.924746	0.244272
49	8	7	3	0.990765	0.066020
49	8	7	4	0.999452	0.008687

N	n	k	x	P(x)	p(x)
49	8	7	5	0.999986	0.000535
49	8	7	6	1.000000	0.000013
49	8	7	7	1.000000	0.000000
49	8	8	0	0.211869	0.211869
49	8	8	1	0.610681	0.398812
49	8	8	2	0.889849	0.279168
49	8	8	3	0.982906	0.093056
49	8	8	4	0.998625	0.015719
49	8	8	5	0.999948	0.001324
49	8	8	6	0.999999	0.000051
49	8	8	7	1.000000	0.000001
49	8	8	8	1.000000	0.000000
49	9	1	0	0.816327	0.816327
49	9	1	1	1.000000	0.183673
49	9	2	0	0.663265	0.663265
49	9	2	1	0.969388	0.306122
49	9	2	2	1.000000	0.030612
49	9	3	0	0.536257	0.536257
49	9	3	1	0.917282	0.381025
49	9	3	2	0.995441	0.078159
49	9	3	3	1.000000	0.004559
49	9	4	0	0.431337	0.431337
49	9	4	1	0.851017	0.419679
49	9	4	2	0.983547	0.132530
49	9	4	3	0.999405	0.015858
49	9	4	4	1.000000	0.000595
49	9	5	0	0.345070	0.345070
49	9	5	1	0.776407	0.431337
49	9	5	2	0.962931	0.186524
49	9	5	3	0.997291	0.034360
49	9	5	4	0.999934	0.002643
49	9	5	5	1.000000	0.000066
49	9	6	0	0.274487	0.274487
49	9	6	1	0.697982	0.423495
49	9	6	2	0.933257	0.235275
49	9	6	3	0.992605	0.059349
49	9	6	4	0.999634	0.007028
49	9	6	5	0.999994	0.000360
49	9	6	6	1.000000	0.000006
49	9	7	0	0.217036	0.217036
49	9	7	1	0.619192	0.402156
49	9	7	2	0.894956	0.275764
49	9	7	3	0.984324	0.089368
49	9	7	4	0.998816	0.014492
49	9	7	5	0.999960	0.001144
49	9	7	6	0.999999	0.000039
49	9	7	7	1.000000	0.000000
49	9	8	0	0.170529	0.170529
49	9	8	1	0.542591	0.372063
49	9	8	2	0.848996	0.306404

Table for $N = 2$, $n = 1$, through $N = 100$, $n = 50$

N	n	k	x	P(x)	p(x)
49	9	8	3	0.971557	0.122562
49	9	8	4	0.997091	0.025534
49	9	8	5	0.999851	0.002760
49	9	8	6	0.999997	0.000145
49	9	8	7	1.000000	0.000003
49	9	8	8	1.000000	0.000000
49	9	9	0	0.133096	0.133096
49	9	9	1	0.469994	0.336898
49	9	9	2	0.796683	0.326689
49	9	9	3	0.953621	0.156939
49	9	9	4	0.993977	0.040356
49	9	9	5	0.999582	0.005605
49	9	9	6	0.999986	0.000404
49	9	9	7	1.000000	0.000014
49	9	9	8	1.000000	0.000000
49	9	9	9	1.000000	0.000000
49	10	1	0	0.795918	0.795918
49	10	1	1	1.000000	0.204082
49	10	2	0	0.630102	0.630102
49	10	2	1	0.961735	0.331633
49	10	2	2	1.000000	0.038265
49	10	3	0	0.496038	0.496038
49	10	3	1	0.898231	0.402193
49	10	3	2	0.993487	0.095256
49	10	3	3	1.000000	0.006513
49	10	4	0	0.388203	0.388203
49	10	4	1	0.819541	0.431337
49	10	4	2	0.976920	0.157380
49	10	4	3	0.999009	0.022088
49	10	4	4	1.000000	0.000991
49	10	5	0	0.301936	0.301936
49	10	5	1	0.733273	0.431337
49	10	5	2	0.948942	0.215669
49	10	5	3	0.995573	0.046631
49	10	5	4	0.999868	0.004295
49	10	5	5	1.000000	0.000132
49	10	6	0	0.233314	0.233314
49	10	6	1	0.645045	0.411731
49	10	6	2	0.909729	0.264684
49	10	6	3	0.988154	0.078425
49	10	6	4	0.999282	0.011128
49	10	6	5	0.999985	0.000703
49	10	6	6	1.000000	0.000015
49	10	7	0	0.179055	0.179055
49	10	7	1	0.558869	0.379814
49	10	7	2	0.860486	0.301617
49	10	7	3	0.975387	0.114902
49	10	7	4	0.997729	0.022342
49	10	7	5	0.999903	0.002174
49	10	7	6	0.999999	0.000095

N	n	k	x	P(x)	p(x)
49	10	7	7	1.000000	0.000001
49	10	8	0	0.136423	0.136423
49	10	8	1	0.477480	0.341057
49	10	8	2	0.803035	0.325555
49	10	8	3	0.956237	0.153202
49	10	8	4	0.994538	0.038301
49	10	8	5	0.999644	0.005107
49	10	8	6	0.999989	0.000345
49	10	8	7	1.000000	0.000010
49	10	8	8	1.000000	0.000000
49	10	9	0	0.103149	0.103149
49	10	9	1	0.402614	0.299465
49	10	9	2	0.739512	0.336898
49	10	9	3	0.930081	0.190569
49	10	9	4	0.988933	0.058852
49	10	9	5	0.999022	0.010089
49	10	9	6	0.999956	0.000934
49	10	9	7	0.999999	0.000043
49	10	9	8	1.000000	0.000001
49	10	9	9	1.000000	0.000000
49	10	10	0	0.077362	0.077362
49	10	10	1	0.335234	0.257873
49	10	10	2	0.672132	0.336898
49	10	10	3	0.896731	0.224599
49	10	10	4	0.980105	0.083374
49	10	10	5	0.997760	0.017656
49	10	10	6	0.999862	0.002102
49	10	10	7	0.999996	0.000133
49	10	10	8	1.000000	0.000004
49	10	10	9	1.000000	0.000000
49	10	10	10	1.000000	0.000000
49	11	1	0	0.775510	0.775510
49	11	1	1	1.000000	0.224490
49	11	2	0	0.597789	0.597789
49	11	2	1	0.953231	0.355442
49	11	2	2	1.000000	0.046769
49	11	3	0	0.457881	0.457881
49	11	3	1	0.877605	0.419724
49	11	3	2	0.991044	0.113439
49	11	3	3	1.000000	0.008956
49	11	4	0	0.348388	0.348388
49	11	4	1	0.786361	0.437973
49	11	4	2	0.968850	0.182489
49	11	4	3	0.998442	0.029593
49	11	4	4	1.000000	0.001558
49	11	5	0	0.263226	0.263226
49	11	5	1	0.689033	0.425807
49	11	5	2	0.932352	0.243318
49	11	5	3	0.993181	0.060830
49	11	5	4	0.999758	0.006576

N	n	k	x	P(x)	p(x)
49	11	5	5	1.000000	0.000242
49	11	6	0	0.197420	0.197420
49	11	6	1	0.592259	0.394839
49	11	6	2	0.882582	0.290323
49	11	6	3	0.982121	0.099539
49	11	6	4	0.998711	0.016590
49	11	6	5	0.999967	0.001255
49	11	6	6	1.000000	0.000033
49	11	7	0	0.146917	0.146917
49	11	7	1	0.500436	0.353519
49	11	7	2	0.821817	0.321381
49	11	7	3	0.963603	0.141786
49	11	7	4	0.996011	0.032408
49	11	7	5	0.999792	0.003781
49	11	7	6	0.999996	0.000204
49	11	7	7	1.000000	0.000004
49	11	8	0	0.108439	0.108439
49	11	8	1	0.416265	0.307826
49	11	8	2	0.752950	0.336685
49	11	8	3	0.936596	0.183646
49	11	8	4	0.990609	0.054014
49	11	8	5	0.999252	0.008642
49	11	8	6	0.999972	0.000720
49	11	8	7	0.999999	0.000028
49	11	8	8	1.000000	0.000000
49	11	9	0	0.079345	0.079345
49	11	9	1	0.341185	0.261840
49	11	9	2	0.679043	0.337858
49	11	9	3	0.900762	0.221719
49	11	9	4	0.981387	0.080625
49	11	9	5	0.997987	0.016599
49	11	9	6	0.999884	0.001897
49	11	9	7	0.999997	0.000113
49	11	9	8	1.000000	0.000003
49	11	9	9	1.000000	0.000000
49	11	10	0	0.057525	0.057525
49	11	10	1	0.275725	0.218200
49	11	10	2	0.603025	0.327300
49	11	10	3	0.856419	0.253393
49	11	10	4	0.967278	0.110860
49	11	10	5	0.995497	0.028219
49	11	10	6	0.999647	0.004150
49	11	10	7	0.999986	0.000339
49	11	10	8	1.000000	0.000014
49	11	10	9	1.000000	0.000000
49	11	10	10	1.000000	0.000000
49	11	11	0	0.041300	0.041300
49	11	11	1	0.219777	0.178476
49	11	11	2	0.527494	0.307718
49	11	11	3	0.804440	0.276946

N	n	k	x	P(x)	p(x)
49	11	11	4	0.947380	0.142940
49	11	11	5	0.991156	0.043775
49	11	11	6	0.999115	0.007959
49	11	11	7	0.999951	0.000836
49	11	11	8	0.999999	0.000048
49	11	11	9	1.000000	0.000001
49	11	11	10	1.000000	0.000000
49	11	11	11	1.000000	0.000000
49	12	1	0	0.755102	0.755102
49	12	1	1	1.000000	0.244898
49	12	2	0	0.566326	0.566326
49	12	2	1	0.943877	0.377551
49	12	2	2	1.000000	0.056122
49	12	3	0	0.421733	0.421733
49	12	3	1	0.855515	0.433782
49	12	3	2	0.988059	0.132544
49	12	3	3	1.000000	0.011941
49	12	4	0	0.311715	0.311715
49	12	4	1	0.751784	0.440069
49	12	4	2	0.959245	0.207461
49	12	4	3	0.997664	0.038419
49	12	4	4	1.000000	0.002336
49	12	5	0	0.228591	0.228591
49	12	5	1	0.644212	0.415620
49	12	5	2	0.913143	0.268931
49	12	5	3	0.989980	0.076837
49	12	5	4	0.999585	0.009605
49	12	5	5	1.000000	0.000415
49	12	6	0	0.166248	0.166248
49	12	6	1	0.540307	0.374058
49	12	6	2	0.852022	0.311715
49	12	6	3	0.974263	0.122241
49	12	6	4	0.997838	0.023575
49	12	6	5	0.999934	0.002096
49	12	6	6	1.000000	0.000066
49	12	7	0	0.119853	0.119853
49	12	7	1	0.444617	0.324764
49	12	7	2	0.779530	0.334913
49	12	7	3	0.948678	0.169148
49	12	7	4	0.993452	0.044774
49	12	7	5	0.999593	0.006140
49	12	7	6	0.999991	0.000398
49	12	7	7	1.000000	0.000009
49	12	8	0	0.085610	0.085610
49	12	8	1	0.359560	0.273950
49	12	8	2	0.699789	0.340229
49	12	8	3	0.912432	0.212643
49	12	8	4	0.984924	0.072492
49	12	8	5	0.998569	0.013646
49	12	8	6	0.999934	0.001365

N	n	k	x	P(x)	p(x)
49	12	8	7	0.999999	0.000065
49	12	8	8	1.000000	0.000001
49	12	9	0	0.060553	0.060553
49	12	9	1	0.286061	0.225508
49	12	9	2	0.616806	0.330745
49	12	9	3	0.865754	0.248948
49	12	9	4	0.970779	0.105025
49	12	9	5	0.996240	0.025461
49	12	9	6	0.999734	0.003495
49	12	9	7	0.999991	0.000257
49	12	9	8	1.000000	0.000009
49	12	9	9	1.000000	0.000000
49	12	10	0	0.042387	0.042387
49	12	10	1	0.224046	0.181659
49	12	10	2	0.534120	0.310074
49	12	10	3	0.809741	0.275621
49	12	10	4	0.949774	0.140033
49	12	10	5	0.991784	0.042010
49	12	10	6	0.999210	0.007426
49	12	10	7	0.999959	0.000749
49	12	10	8	0.999999	0.000040
49	12	10	9	1.000000	0.000001
49	12	10	10	1.000000	0.000000
49	12	11	0	0.029345	0.029345
49	12	11	1	0.172809	0.143464
49	12	11	2	0.454614	0.281805
49	12	11	3	0.746136	0.291522
49	12	11	4	0.921049	0.174913
49	12	11	5	0.984244	0.063194
49	12	11	6	0.998068	0.013824
49	12	11	7	0.999863	0.001795
49	12	11	8	0.999995	0.000132
49	12	11	9	1.000000	0.000005
49	12	11	10	1.000000	0.000000
49	12	11	11	1.000000	0.000000
49	12	12	0	0.020078	0.020078
49	12	12	1	0.131280	0.111202
49	12	12	2	0.380455	0.249175
49	12	12	3	0.677091	0.296637
49	12	12	4	0.884225	0.207134
49	12	12	5	0.972603	0.088377
49	12	12	6	0.995885	0.023282
49	12	12	7	0.999627	0.003742
49	12	12	8	0.999981	0.000354
49	12	12	9	0.999999	0.000019
49	12	12	10	1.000000	0.000000
49	12	12	11	1.000000	0.000000
49	12	12	12	1.000000	0.000000
49	13	1	0	0.734694	0.734694
49	13	1	1	1.000000	0.265306
49	13	2	0	0.535714	0.535714
49	13	2	1	0.933673	0.397959
49	13	2	2	1.000000	0.066327
49	13	3	0	0.387538	0.387538
49	13	3	1	0.832067	0.444529
49	13	3	2	0.984477	0.152410
49	13	3	3	1.000000	0.015523
49	13	4	0	0.278016	0.278016
49	13	4	1	0.716103	0.438086
49	13	4	2	0.948031	0.231928
49	13	4	3	0.996625	0.048594
49	13	4	4	1.000000	0.003375
49	13	5	0	0.197701	0.197701
49	13	5	1	0.599280	0.401579
49	13	5	2	0.891337	0.292058
49	13	5	3	0.985827	0.094489
49	13	5	4	0.999325	0.013498
49	13	5	5	1.000000	0.000675
49	13	6	0	0.139289	0.139289
49	13	6	1	0.489758	0.350469
49	13	6	2	0.818323	0.328565
49	13	6	3	0.964352	0.146029
49	13	6	4	0.996564	0.032212
49	13	6	5	0.999877	0.003313
49	13	6	6	1.000000	0.000123
49	13	7	0	0.097178	0.097178
49	13	7	1	0.391953	0.294774
49	13	7	2	0.734271	0.342319
49	13	7	3	0.930392	0.196120
49	13	7	4	0.989822	0.059430
49	13	7	5	0.999261	0.009439
49	13	7	6	0.999980	0.000719
49	13	7	7	1.000000	0.000020
49	13	8	0	0.067099	0.067099
49	13	8	1	0.307732	0.240632
49	13	8	2	0.644617	0.336885
49	13	8	3	0.883696	0.239080
49	13	8	4	0.977087	0.093391
49	13	8	5	0.997463	0.020376
49	13	8	6	0.999860	0.002397
49	13	8	7	0.999997	0.000137
49	13	8	8	1.000000	0.000003
49	13	9	0	0.045824	0.045824
49	13	9	1	0.237303	0.191479
49	13	9	2	0.554233	0.316930
49	13	9	3	0.825384	0.271151
49	13	9	4	0.956586	0.131202
49	13	9	5	0.993487	0.036901
49	13	9	6	0.999451	0.005964
49	13	9	7	0.999977	0.000526
49	13	9	8	1.000000	0.000023
49	13	9	9	1.000000	0.000000
49	13	10	0	0.030931	0.030931
49	13	10	1	0.179859	0.148928
49	13	10	2	0.467077	0.287218
49	13	10	3	0.757596	0.290519
49	13	10	4	0.927066	0.169470
49	13	10	5	0.986107	0.059041
49	13	10	6	0.998407	0.012300
49	13	10	7	0.999898	0.001491
49	13	10	8	0.999997	0.000099
49	13	10	9	1.000000	0.000003
49	13	10	10	1.000000	0.000000
49	13	11	0	0.020621	0.020621
49	13	11	1	0.134035	0.113414
49	13	11	2	0.386067	0.252032
49	13	11	3	0.683104	0.297037
49	13	11	4	0.887957	0.204853
49	13	11	5	0.973996	0.086038
49	13	11	6	0.996199	0.022203
49	13	11	7	0.999669	0.003469
49	13	11	8	0.999984	0.000315
49	13	11	9	0.999999	0.000015
49	13	11	10	1.000000	0.000000
49	13	11	11	1.000000	0.000000
49	13	12	0	0.013566	0.013566
49	13	12	1	0.098220	0.084654
49	13	12	2	0.313110	0.214890
49	13	12	3	0.604936	0.291826
49	13	12	4	0.839440	0.234503
49	13	12	5	0.955883	0.116443
49	13	12	6	0.992109	0.036227
49	13	12	7	0.999121	0.007012
49	13	12	8	0.999943	0.000822
49	13	12	9	0.999998	0.000055
49	13	12	10	1.000000	0.000002
49	13	12	11	1.000000	0.000000
49	13	12	12	1.000000	0.000000
49	13	13	0	0.008800	0.008800
49	13	13	1	0.070765	0.061965
49	13	13	2	0.249224	0.178459
49	13	13	3	0.526064	0.276841
49	13	13	4	0.782398	0.256334
49	13	13	5	0.930706	0.148307
49	13	13	6	0.985256	0.054550
49	13	13	7	0.997984	0.012728
49	13	13	8	0.999832	0.001848
49	13	13	9	0.999992	0.000160
49	13	13	10	1.000000	0.000008
49	13	13	11	1.000000	0.000000
49	13	13	12	1.000000	0.000000
49	13	13	13	1.000000	0.000000
49	14	1	0	0.714286	0.714286
49	14	1	1	1.000000	0.285714
49	14	2	0	0.505952	0.505952
49	14	2	1	0.922619	0.416667
49	14	2	2	1.000000	0.077381
49	14	3	0	0.355243	0.355243
49	14	3	1	0.807371	0.452128
49	14	3	2	0.980243	0.172872
49	14	3	3	1.000000	0.019757
49	14	4	0	0.247126	0.247126
49	14	4	1	0.679596	0.432470
49	14	4	2	0.935146	0.255550
49	14	4	3	0.995275	0.060129
49	14	4	4	1.000000	0.004724
49	14	5	0	0.170242	0.170242
49	14	5	1	0.554660	0.384418
49	14	5	2	0.866999	0.312339
49	14	5	3	0.980577	0.113578
49	14	5	4	0.998950	0.018373
49	14	5	5	1.000000	0.001050
49	14	6	0	0.116074	0.116074
49	14	6	1	0.441082	0.325008
49	14	6	2	0.781816	0.340734
49	14	6	3	0.952183	0.170367
49	14	6	4	0.994774	0.042592
49	14	6	5	0.999785	0.005011
49	14	6	6	1.000000	0.000215
49	14	7	0	0.078283	0.078283
49	14	7	1	0.342824	0.264541
49	14	7	2	0.686727	0.343903
49	14	7	3	0.908600	0.221873
49	14	7	4	0.984869	0.076269
49	14	7	5	0.998736	0.013867
49	14	7	6	0.999960	0.001224
49	14	7	7	1.000000	0.000040
49	14	8	0	0.052188	0.052188
49	14	8	1	0.260942	0.208754
49	14	8	2	0.588469	0.327527
49	14	8	3	0.850491	0.262022
49	14	8	4	0.966710	0.116219
49	14	8	5	0.995765	0.029055
49	14	8	6	0.999727	0.003962
49	14	8	7	0.999993	0.000266
49	14	8	8	1.000000	0.000007
49	14	9	0	0.034368	0.034368
49	14	9	1	0.194752	0.160384
49	14	9	2	0.492607	0.297856
49	14	9	3	0.780192	0.287585

Table for $N = 2$, $n = 1$, through $N = 100$, $n = 50$

N	n	k	x	P(x)	p(x)
49	14	9	4	0.938364	0.158172
49	14	9	5	0.989387	0.051023
49	14	9	6	0.998954	0.009567
49	14	9	7	0.999948	0.000994
49	14	9	8	0.999999	0.000051
49	14	9	9	1.000000	0.000001
49	14	10	0	0.022339	0.022339
49	14	10	1	0.142627	0.120288
49	14	10	2	0.403251	0.260624
49	14	10	3	0.701106	0.297856
49	14	10	4	0.898821	0.197715
49	14	10	5	0.977907	0.079086
49	14	10	6	0.997040	0.019134
49	14	10	7	0.999774	0.002733
49	14	10	8	0.999991	0.000217
49	14	10	9	1.000000	0.000009
49	14	10	10	1.000000	0.000000
49	14	11	0	0.014320	0.014320
49	14	11	1	0.102531	0.088211
49	14	11	2	0.323059	0.220528
49	14	11	3	0.617096	0.294037
49	14	11	4	0.848125	0.231029
49	14	11	5	0.959656	0.111531
49	14	11	6	0.993116	0.033459
49	14	11	7	0.999283	0.006168
49	14	11	8	0.999958	0.000675
49	14	11	9	0.999999	0.000041
49	14	11	10	1.000000	0.000001
49	14	11	11	1.000000	0.000000
49	14	12	0	0.009044	0.009044
49	14	12	1	0.072354	0.063309
49	14	12	2	0.253418	0.181065
49	14	12	3	0.531980	0.278561
49	14	12	4	0.787328	0.255348
49	14	12	5	0.933241	0.145913
49	14	12	6	0.986072	0.052831
49	14	12	7	0.998147	0.012076
49	14	12	8	0.999851	0.001704
49	14	12	9	0.999993	0.000142
49	14	12	10	1.000000	0.000006
49	14	12	11	1.000000	0.000000
49	14	12	12	1.000000	0.000000
49	14	13	0	0.005622	0.005622
49	14	13	1	0.050110	0.044488
49	14	13	2	0.194695	0.144585
49	14	13	3	0.449164	0.254470
49	14	13	4	0.718315	0.269151
49	14	13	5	0.897749	0.179434
49	14	13	6	0.974649	0.076900
49	14	13	7	0.995862	0.021214

N	n	k	x	P(x)	p(x)
49	14	13	8	0.999575	0.003712
49	14	13	9	0.999974	0.000399
49	14	13	10	0.999999	0.000025
49	14	13	11	1.000000	0.000001
49	14	13	12	1.000000	0.000000
49	14	13	13	1.000000	0.000000
49	14	14	0	0.003436	0.003436
49	14	14	1	0.034045	0.030609
49	14	14	2	0.146500	0.112455
49	14	14	3	0.371410	0.224910
49	14	14	4	0.643551	0.272141
49	14	14	5	0.852890	0.209339
49	14	14	6	0.957560	0.104670
49	14	14	7	0.991738	0.034178
49	14	14	8	0.998956	0.007219
49	14	14	9	0.999919	0.000962
49	14	14	10	0.999996	0.000078
49	14	14	11	1.000000	0.000004
49	14	14	12	1.000000	0.000000
49	14	14	13	1.000000	0.000000
49	14	14	14	1.000000	0.000000
49	15	1	0	0.693878	0.693878
49	15	1	1	1.000000	0.306122
49	15	2	0	0.477041	0.477041
49	15	2	1	0.910714	0.433673
49	15	2	2	1.000000	0.089286
49	15	3	0	0.324794	0.324794
49	15	3	1	0.781535	0.456741
49	15	3	2	0.975304	0.193769
49	15	3	3	1.000000	0.024696
49	15	4	0	0.218883	0.218883
49	15	4	1	0.642527	0.423644
49	15	4	2	0.920543	0.278016
49	15	4	3	0.993557	0.073014
49	15	4	4	1.000000	0.006442
49	15	5	0	0.145922	0.145922
49	15	5	1	0.510726	0.364805
49	15	5	2	0.840227	0.329501
49	15	5	3	0.974087	0.133860
49	15	5	4	0.998425	0.024338
49	15	5	5	1.000000	0.001575
49	15	6	0	0.096176	0.096176
49	15	6	1	0.394652	0.298476
49	15	6	2	0.742875	0.348223
49	15	6	3	0.937580	0.194705
49	15	6	4	0.992341	0.054761
49	15	6	5	0.999642	0.007301
49	15	6	6	1.000000	0.000358
49	15	7	0	0.062626	0.062626
49	15	7	1	0.297474	0.234848

N	n	k	x	P(x)	p(x)
49	15	7	2	0.637598	0.340124
49	15	7	3	0.883243	0.245645
49	15	7	4	0.978332	0.095089
49	15	7	5	0.997944	0.019612
49	15	7	6	0.999925	0.001981
49	15	7	7	1.000000	0.000075
49	15	8	0	0.040260	0.040260
49	15	8	1	0.219191	0.178932
49	15	8	2	0.532322	0.313130
49	15	8	3	0.813059	0.280738
49	15	8	4	0.953428	0.140369
49	15	8	5	0.993274	0.039847
49	15	8	6	0.999500	0.006226
49	15	8	7	0.999986	0.000485
49	15	8	8	1.000000	0.000014
49	15	9	0	0.025530	0.025530
49	15	9	1	0.158093	0.132562
49	15	9	2	0.433036	0.274944
49	15	9	3	0.730892	0.297856
49	15	9	4	0.915768	0.184876
49	15	9	5	0.983556	0.067788
49	15	9	6	0.998134	0.014578
49	15	9	7	0.999891	0.001757
49	15	9	8	0.999997	0.000106
49	15	9	9	1.000000	0.000002
49	15	10	0	0.015957	0.015957
49	15	10	1	0.111696	0.095739
49	15	10	2	0.343680	0.231984
49	15	10	3	0.641535	0.297856
49	15	10	4	0.864927	0.223392
49	15	10	5	0.966609	0.101682
49	15	10	6	0.994854	0.028245
49	15	10	7	0.999540	0.004686
49	15	10	8	0.999979	0.000439
49	15	10	9	1.000000	0.000021
49	15	10	10	1.000000	0.000000
49	15	11	0	0.009819	0.009819
49	15	11	1	0.077328	0.067508
49	15	11	2	0.266352	0.189024
49	15	11	3	0.549887	0.283536
49	15	11	4	0.801919	0.252032
49	15	11	5	0.940537	0.138617
49	15	11	6	0.988336	0.047799
49	15	11	7	0.998578	0.010243
49	15	11	8	0.999900	0.001322
49	15	11	9	0.999996	0.000096
49	15	11	10	1.000000	0.000004
49	15	11	11	1.000000	0.000000
49	15	12	0	0.005943	0.005943
49	15	12	1	0.052456	0.046513

N	n	k	x	P(x)	p(x)
49	15	12	2	0.201686	0.149229
49	15	12	3	0.460350	0.258664
49	15	12	4	0.728963	0.268613
49	15	12	5	0.904058	0.175096
49	15	12	6	0.977015	0.072957
49	15	12	7	0.996422	0.019407
49	15	12	8	0.999656	0.003235
49	15	12	9	0.999981	0.000325
49	15	12	10	0.999999	0.000018
49	15	12	11	1.000000	0.000000
49	15	12	12	1.000000	0.000000
49	15	13	0	0.003534	0.003534
49	15	13	1	0.034857	0.031323
49	15	13	2	0.149254	0.114397
49	15	13	3	0.376459	0.227205
49	15	13	4	0.649105	0.272646
49	15	13	5	0.856735	0.207630
49	15	13	6	0.959269	0.102534
49	15	13	7	0.992226	0.032957
49	15	13	8	0.999045	0.006819
49	15	13	9	0.999928	0.000884
49	15	13	10	0.999997	0.000068
49	15	13	11	1.000000	0.000003
49	15	13	12	1.000000	0.000000
49	15	13	13	1.000000	0.000000
49	15	14	0	0.002061	0.002061
49	15	14	1	0.022676	0.020614
49	15	14	2	0.107944	0.085268
49	15	14	3	0.300724	0.192780
49	15	14	4	0.565796	0.265072
49	15	14	5	0.799060	0.233264
49	15	14	6	0.933635	0.134575
49	15	14	7	0.984902	0.051267
49	15	14	8	0.997719	0.012817
49	15	14	9	0.999781	0.002062
49	15	14	10	0.999987	0.000206
49	15	14	11	1.000000	0.000012
49	15	14	12	1.000000	0.000000
49	15	14	13	1.000000	0.000000
49	15	14	14	1.000000	0.000000
49	15	15	0	0.001178	0.001178
49	15	15	1	0.014430	0.013252
49	15	15	2	0.076273	0.061843
49	15	15	3	0.234628	0.158355
49	15	15	4	0.482488	0.247860
49	15	15	5	0.732413	0.249925
49	15	15	6	0.899030	0.166617
49	15	15	7	0.973184	0.074154
49	15	15	8	0.995155	0.021971
49	15	15	9	0.999428	0.004272

Table for $N = 2$, $n = 1$, through $N = 100$, $n = 50$

N	n	k	x	P(x)	p(x)
49	15	15	10	0.999958	0.000530
49	15	15	11	0.999998	0.000040
49	15	15	12	1.000000	0.000002
49	15	15	13	1.000000	0.000000
49	15	15	14	1.000000	0.000000
49	15	15	15	1.000000	0.000000
49	16	1	0	0.673469	0.673469
49	16	1	1	1.000000	0.326531
49	16	2	0	0.448980	0.448980
49	16	2	1	0.897959	0.448980
49	16	2	2	1.000000	0.102041
49	16	3	0	0.296135	0.296135
49	16	3	1	0.754668	0.458532
49	16	3	2	0.969605	0.214937
49	16	3	3	1.000000	0.030395
49	16	4	0	0.193132	0.193132
49	16	4	1	0.605146	0.412015
49	16	4	2	0.904189	0.299043
49	16	4	3	0.991410	0.087221
49	16	4	4	1.000000	0.008590
49	16	5	0	0.124463	0.124463
49	16	5	1	0.467808	0.343345
49	16	5	2	0.811154	0.343345
49	16	5	3	0.966213	0.155059
49	16	5	4	0.997709	0.031496
49	16	5	5	1.000000	0.002291
49	16	6	0	0.079204	0.079204
49	16	6	1	0.350759	0.271555
49	16	6	2	0.701907	0.351149
49	16	6	3	0.920400	0.218493
49	16	6	4	0.989119	0.068719
49	16	6	5	0.999427	0.010308
49	16	6	6	1.000000	0.000573
49	16	7	0	0.049732	0.049732
49	16	7	1	0.256030	0.206298
49	16	7	2	0.587580	0.331550
49	16	7	3	0.854344	0.266764
49	16	7	4	0.969942	0.115598
49	16	7	5	0.996790	0.026849
49	16	7	6	0.999867	0.003076
49	16	7	7	1.000000	0.000133
49	16	8	0	0.030787	0.030787
49	16	8	1	0.182352	0.151566
49	16	8	2	0.477063	0.294711
49	16	8	3	0.771774	0.294711
49	16	8	4	0.936914	0.165140
49	16	8	5	0.989759	0.052845
49	16	8	6	0.999134	0.009376
49	16	8	7	0.999971	0.000837
49	16	8	8	1.000000	0.000029

N	n	k	x	P(x)	p(x)
49	16	9	0	0.018772	0.018772
49	16	9	1	0.126902	0.108129
49	16	9	2	0.376430	0.249529
49	16	9	3	0.678329	0.301899
49	16	9	4	0.888580	0.210251
49	16	9	5	0.975581	0.087000
49	16	9	6	0.996848	0.021267
49	16	9	7	0.999788	0.002940
49	16	9	8	0.999994	0.000207
49	16	9	9	1.000000	0.000006
49	16	10	0	0.011263	0.011263
49	16	10	1	0.086353	0.075090
49	16	10	2	0.289095	0.202742
49	16	10	3	0.580212	0.291117
49	16	10	4	0.825505	0.245293
49	16	10	5	0.951656	0.126151
49	16	10	6	0.991531	0.039875
49	16	10	7	0.999126	0.007595
49	16	10	8	0.999953	0.000827
49	16	10	9	0.999999	0.000046
49	16	10	10	1.000000	0.000001
49	16	11	0	0.006643	0.006643
49	16	11	1	0.057472	0.050830
49	16	11	2	0.216316	0.158844
49	16	11	3	0.483173	0.266857
49	16	11	4	0.750030	0.266857
49	16	11	5	0.916075	0.166044
49	16	11	6	0.981306	0.065232
49	16	11	7	0.997373	0.016067
49	16	11	8	0.999783	0.002410
49	16	11	9	0.999991	0.000207
49	16	11	10	1.000000	0.000009
49	16	11	11	1.000000	0.000000
49	16	12	0	0.003846	0.003846
49	16	12	1	0.037408	0.033562
49	16	12	2	0.157795	0.120387
49	16	12	3	0.391880	0.234085
49	16	12	4	0.665760	0.273880
49	16	12	5	0.868009	0.202250
49	16	12	6	0.964140	0.096131
49	16	12	7	0.993568	0.029428
49	16	12	8	0.999276	0.005708
49	16	12	9	0.999952	0.000676
49	16	12	10	0.999998	0.000046
49	16	12	11	1.000000	0.000002
49	16	12	12	1.000000	0.000000
49	16	13	0	0.002183	0.002183
49	16	13	1	0.023802	0.021619
49	16	13	2	0.112243	0.088441
49	16	13	3	0.309634	0.197391

N	n	k	x	P(x)	p(x)
49	16	13	4	0.576934	0.267300
49	16	13	5	0.807881	0.230947
49	16	13	6	0.938159	0.130278
49	16	13	7	0.986410	0.048251
49	16	13	8	0.998042	0.011632
49	16	13	9	0.999825	0.001783
49	16	13	10	0.999991	0.000166
49	16	13	11	1.000000	0.000009
49	16	13	12	1.000000	0.000000
49	16	13	13	1.000000	0.000000
49	16	14	0	0.001213	0.001213
49	16	14	1	0.014794	0.013581
49	16	14	2	0.077849	0.063055
49	16	14	3	0.238354	0.160505
49	16	14	4	0.487834	0.249480
49	16	14	5	0.737314	0.249480
49	16	14	6	0.901970	0.164657
49	16	14	7	0.974347	0.072377
49	16	14	8	0.995457	0.021110
49	16	14	9	0.999478	0.004021
49	16	14	10	0.999963	0.000485
49	16	14	11	0.999998	0.000035
49	16	14	12	1.000000	0.000001
49	16	14	13	1.000000	0.000000
49	16	14	14	1.000000	0.000000
49	16	15	0	0.000658	0.000658
49	16	15	1	0.008973	0.008315
49	16	15	2	0.052627	0.043654
49	16	15	3	0.178738	0.126111
49	16	15	4	0.402298	0.223560
49	16	15	5	0.658906	0.256608
49	16	15	6	0.854926	0.196020
49	16	15	7	0.955736	0.100810
49	16	15	8	0.990632	0.034896
49	16	15	9	0.998674	0.008042
49	16	15	10	0.999880	0.001206
49	16	15	11	0.999993	0.000113
49	16	15	12	1.000000	0.000006
49	16	15	13	1.000000	0.000000
49	16	15	14	1.000000	0.000000
49	16	15	15	1.000000	0.000000
49	16	16	0	0.000348	0.000348
49	16	16	1	0.005305	0.004956
49	16	16	2	0.034652	0.029347
49	16	16	3	0.130519	0.095867
49	16	16	4	0.323394	0.192875
49	16	16	5	0.575885	0.252491
49	16	16	6	0.797273	0.221387
49	16	16	7	0.929051	0.131778
49	16	16	8	0.982421	0.053370

N	n	k	x	P(x)	p(x)
49	16	16	9	0.997018	0.014597
49	16	16	10	0.999667	0.002649
49	16	16	11	0.999977	0.000310
49	16	16	12	0.999999	0.000022
49	16	16	13	1.000000	0.000001
49	16	16	14	1.000000	0.000000
49	16	16	15	1.000000	0.000000
49	16	16	16	1.000000	0.000000
49	17	1	0	0.653061	0.653061
49	17	1	1	1.000000	0.346939
49	17	2	0	0.421769	0.421769
49	17	2	1	0.884354	0.462585
49	17	2	2	1.000000	0.115646
49	17	3	0	0.269214	0.269214
49	17	3	1	0.726878	0.457664
49	17	3	2	0.963092	0.236214
49	17	3	3	1.000000	0.036908
49	17	4	0	0.169722	0.169722
49	17	4	1	0.567691	0.397969
49	17	4	2	0.886065	0.318375
49	17	4	3	0.988767	0.102702
49	17	4	4	1.000000	0.011233
49	17	5	0	0.105605	0.105605
49	17	5	1	0.426190	0.320586
49	17	5	2	0.779940	0.353750
49	17	5	3	0.956815	0.176875
49	17	5	4	0.996755	0.039939
49	17	5	5	1.000000	0.003245
49	17	6	0	0.064803	0.064803
49	17	6	1	0.309614	0.244811
49	17	6	2	0.659344	0.349730
49	17	6	3	0.900537	0.241193
49	17	6	4	0.984955	0.084418
49	17	6	5	0.999115	0.014160
49	17	6	6	1.000000	0.000885
49	17	7	0	0.039183	0.039183
49	17	7	1	0.218521	0.179338
49	17	7	2	0.537345	0.318824
49	17	7	3	0.822009	0.284664
49	17	7	4	0.959433	0.137424
49	17	7	5	0.995163	0.035730
49	17	7	6	0.999773	0.004610
49	17	7	7	1.000000	0.000226
49	17	8	0	0.023323	0.023323
49	17	8	1	0.150202	0.126879
49	17	8	2	0.423479	0.273277
49	17	8	3	0.727121	0.303642
49	17	8	4	0.916897	0.189776
49	17	8	5	0.984955	0.068058
49	17	8	6	0.998566	0.013612

Table for $N = 2$, $n = 1$, through $N = 100$, $n = 50$

N	n	k	x	P(x)	p(x)
49	17	8	7	0.999946	0.001380
49	17	8	8	1.000000	0.000054
49	17	9	0	0.013653	0.013653
49	17	9	1	0.100688	0.087036
49	17	9	2	0.323500	0.222812
49	17	9	3	0.623438	0.299939
49	17	9	4	0.856724	0.233286
49	17	9	5	0.965035	0.108311
49	17	9	6	0.994914	0.029879
49	17	9	7	0.999609	0.004695
49	17	9	8	0.999988	0.000379
49	17	9	9	1.000000	0.000012
49	17	10	0	0.007850	0.007850
49	17	10	1	0.065874	0.058024
49	17	10	2	0.239946	0.174071
49	17	10	3	0.518460	0.278514
49	17	10	4	0.780906	0.262446
49	17	10	5	0.932542	0.151636
49	17	10	6	0.986697	0.054156
49	17	10	7	0.998436	0.011738
49	17	10	8	0.999903	0.001467
49	17	10	9	0.999997	0.000095
49	17	10	10	1.000000	0.000002
49	17	11	0	0.004428	0.004428
49	17	11	1	0.042069	0.037641
49	17	11	2	0.172995	0.130926
49	17	11	3	0.418480	0.245485
49	17	11	4	0.693424	0.274944
49	17	11	5	0.885885	0.192461
49	17	11	6	0.971423	0.085538
49	17	11	7	0.995426	0.024003
49	17	11	8	0.999564	0.004138
49	17	11	9	0.999978	0.000414
49	17	11	10	0.999999	0.000021
49	17	11	11	1.000000	0.000000
49	17	12	0	0.002447	0.002447
49	17	12	1	0.026221	0.023773
49	17	12	2	0.121314	0.095093
49	17	12	3	0.328038	0.206725
49	17	12	4	0.599364	0.271326
49	17	12	5	0.825108	0.225743
49	17	12	6	0.946662	0.121554
49	17	12	7	0.989109	0.042447
49	17	12	8	0.998584	0.009475
49	17	12	9	0.999891	0.001307
49	17	12	10	0.999995	0.000105
49	17	12	11	1.000000	0.000004
49	17	12	12	1.000000	0.000000
49	17	13	0	0.001323	0.001323
49	17	13	1	0.015940	0.014617
49	17	13	2	0.082763	0.066822
49	17	13	3	0.249818	0.167056
49	17	13	4	0.504034	0.254215
49	17	13	5	0.751894	0.247860
49	17	13	6	0.910524	0.158630
49	17	13	7	0.977637	0.067113
49	17	13	8	0.996279	0.018642
49	17	13	9	0.999608	0.003329
49	17	13	10	0.999976	0.000367
49	17	13	11	0.999999	0.000023
49	17	13	12	1.000000	0.000001
49	17	13	13	1.000000	0.000000
49	17	14	0	0.000698	0.000698
49	17	14	1	0.009444	0.008745
49	17	14	2	0.054920	0.045476
49	17	14	3	0.184852	0.129932
49	17	14	4	0.412234	0.227382
49	17	14	5	0.669274	0.257040
49	17	14	6	0.862054	0.192780
49	17	14	7	0.958994	0.096941
49	17	14	8	0.991619	0.032624
49	17	14	9	0.998869	0.007250
49	17	14	10	0.999904	0.001036
49	17	14	11	0.999995	0.000091
49	17	14	12	1.000000	0.000005
49	17	14	13	1.000000	0.000000
49	17	14	14	1.000000	0.000000
49	17	15	0	0.000359	0.000359
49	17	15	1	0.005446	0.005087
49	17	15	2	0.035430	0.029984
49	17	15	3	0.132879	0.097449
49	17	15	4	0.327778	0.194898
49	17	15	5	0.581146	0.253368
49	17	15	6	0.801466	0.220320
49	17	15	7	0.931297	0.129831
49	17	15	8	0.983230	0.051933
49	17	15	9	0.997212	0.013982
49	17	15	10	0.999697	0.002486
49	17	15	11	0.999980	0.000282
49	17	15	12	0.999999	0.000019
49	17	15	13	1.000000	0.000001
49	17	15	14	1.000000	0.000000
49	17	15	15	1.000000	0.000000
49	17	16	0	0.000180	0.000180
49	17	16	1	0.003052	0.002872
49	17	16	2	0.022202	0.019150
49	17	16	3	0.092753	0.070551
49	17	16	4	0.253258	0.160505
49	17	16	5	0.491722	0.238464
49	17	16	6	0.730186	0.238464
49	17	16	7	0.893111	0.162926
49	17	16	8	0.969483	0.076371
49	17	16	9	0.993922	0.024439
49	17	16	10	0.999185	0.005264
49	17	16	11	0.999930	0.000744
49	17	16	12	0.999996	0.000066
49	17	16	13	1.000000	0.000004
49	17	16	14	1.000000	0.000000
49	17	16	15	1.000000	0.000000
49	17	16	16	1.000000	0.000000
49	17	17	0	0.000087	0.000087
49	17	17	1	0.001659	0.001572
49	17	17	2	0.013497	0.011838
49	17	17	3	0.062822	0.049325
49	17	17	4	0.190029	0.127206
49	17	17	5	0.405008	0.214979
49	17	17	6	0.650698	0.245690
49	17	17	7	0.843740	0.193042
49	17	17	8	0.948654	0.104914
49	17	17	9	0.987997	0.039343
49	17	17	10	0.998069	0.010072
49	17	17	11	0.999794	0.001726
49	17	17	12	0.999986	0.000192
49	17	17	13	0.999999	0.000013
49	17	17	14	1.000000	0.000001
49	17	17	15	1.000000	0.000000
49	17	17	16	1.000000	0.000000
49	17	17	17	1.000000	0.000000
49	18	1	0	0.632653	0.632653
49	18	1	1	1.000000	0.367347
49	18	2	0	0.395408	0.395408
49	18	2	1	0.869898	0.474490
49	18	2	2	1.000000	0.130102
49	18	3	0	0.243975	0.243975
49	18	3	1	0.698274	0.454299
49	18	3	2	0.955710	0.257436
49	18	3	3	1.000000	0.044290
49	18	4	0	0.148507	0.148507
49	18	4	1	0.530381	0.381874
49	18	4	2	0.866167	0.335786
49	18	4	3	0.985558	0.119391
49	18	4	4	1.000000	0.014442
49	18	5	0	0.089104	0.089104
49	18	5	1	0.386117	0.297013
49	18	5	2	0.746776	0.360659
49	18	5	3	0.945761	0.198984
49	18	5	4	0.995507	0.049746
49	18	5	5	1.000000	0.004493
49	18	6	0	0.052652	0.052652
49	18	6	1	0.271362	0.218710
49	18	6	2	0.615628	0.344265
49	18	6	3	0.877925	0.262297
49	18	6	4	0.979678	0.101753
49	18	6	5	0.998672	0.018994
49	18	6	6	1.000000	0.001328
49	18	7	0	0.030612	0.030612
49	18	7	1	0.184896	0.154284
49	18	7	2	0.487529	0.302633
49	18	7	3	0.786426	0.298897
49	18	7	4	0.946549	0.160123
49	18	7	5	0.992930	0.046381
49	18	7	6	0.999629	0.006699
49	18	7	7	1.000000	0.000370
49	18	8	0	0.017492	0.017492
49	18	8	1	0.122447	0.104955
49	18	8	2	0.372240	0.249793
49	18	8	3	0.679677	0.307437
49	18	8	4	0.893175	0.213498
49	18	8	5	0.978574	0.085399
49	18	8	6	0.997715	0.019141
49	18	8	7	0.999903	0.002188
49	18	8	8	1.000000	0.000097
49	18	9	0	0.009813	0.009813
49	18	9	1	0.078929	0.069117
49	18	9	2	0.274760	0.195830
49	18	9	3	0.567200	0.292440
49	18	9	4	0.820273	0.253073
49	18	9	5	0.951496	0.131223
49	18	9	6	0.992113	0.040617
49	18	9	7	0.999316	0.007203
49	18	9	8	0.999976	0.000660
49	18	9	9	1.000000	0.000024
49	18	10	0	0.005397	0.005397
49	18	10	1	0.049555	0.044158
49	18	10	2	0.196428	0.146873
49	18	10	3	0.457535	0.261107
49	18	10	4	0.731698	0.274163
49	18	10	5	0.908849	0.177151
49	18	10	6	0.979928	0.071079
49	18	10	7	0.997335	0.017407
49	18	10	8	0.999811	0.002476
49	18	10	9	0.999995	0.000183
49	18	10	10	1.000000	0.000005
49	18	11	0	0.002906	0.002906
49	18	11	1	0.030307	0.027401
49	18	11	2	0.136172	0.105866
49	18	11	3	0.357109	0.220937
49	18	11	4	0.633280	0.276171
49	18	11	5	0.849798	0.216518
49	18	11	6	0.958057	0.108259

N	n	k	x	P(x)	p(x)
49	18	11	7	0.992425	0.034368
49	18	11	8	0.999176	0.006751
49	18	11	9	0.999952	0.000776
49	18	11	10	0.999999	0.000047
49	18	11	11	1.000000	0.000001
49	18	12	0	0.001530	0.001530
49	18	12	1	0.018048	0.016519
49	18	12	2	0.091597	0.073549
49	18	12	3	0.269897	0.178300
49	18	12	4	0.531533	0.261636
49	18	12	5	0.775726	0.244193
49	18	12	6	0.923870	0.148144
49	18	12	7	0.982477	0.058606
49	18	12	8	0.997400	0.014923
49	18	12	9	0.999768	0.002369
49	18	12	10	0.999989	0.000221
49	18	12	11	1.000000	0.000011
49	18	12	12	1.000000	0.000000
49	18	13	0	0.000785	0.000785
49	18	13	1	0.010459	0.009673
49	18	13	2	0.059792	0.049334
49	18	13	3	0.197613	0.137821
49	18	13	4	0.432536	0.234922
49	18	13	5	0.689929	0.257393
49	18	13	6	0.875824	0.185895
49	18	13	7	0.965053	0.089230
49	18	13	8	0.993367	0.028313
49	18	13	9	0.999192	0.005826
49	18	13	10	0.999941	0.000749
49	18	13	11	0.999998	0.000056
49	18	13	12	1.000000	0.000002
49	18	13	13	1.000000	0.000000
49	18	14	0	0.000393	0.000393
49	18	14	1	0.005891	0.005498
49	18	14	2	0.037866	0.031976
49	18	14	3	0.140188	0.102322
49	18	14	4	0.341177	0.200989
49	18	14	5	0.596981	0.255804
49	18	14	6	0.813859	0.216878
49	18	14	7	0.937789	0.123930
49	18	14	8	0.985502	0.047713
49	18	14	9	0.997736	0.012234
49	18	14	10	0.999775	0.002039
49	18	14	11	0.999987	0.000212
49	18	14	12	1.000000	0.000013
49	18	14	13	1.000000	0.000000
49	18	14	14	1.000000	0.000000
49	18	15	0	0.000191	0.000191
49	18	15	1	0.003220	0.003030
49	18	15	2	0.023249	0.020029

N	n	k	x	P(x)	p(x)
49	18	15	3	0.096336	0.073087
49	18	15	4	0.260781	0.164446
49	18	15	5	0.501968	0.241187
49	18	15	6	0.739501	0.237532
49	18	15	7	0.898839	0.159339
49	18	15	8	0.971869	0.073030
49	18	15	9	0.994590	0.022720
49	18	15	10	0.999309	0.004719
49	18	15	11	0.999944	0.000636
49	18	15	12	0.999997	0.000053
49	18	15	13	1.000000	0.000003
49	18	15	14	1.000000	0.000000
49	18	15	15	1.000000	0.000000
49	18	16	0	0.000090	0.000090
49	18	16	1	0.001706	0.001616
49	18	16	2	0.013824	0.012118
49	18	16	3	0.064092	0.050268
49	18	16	4	0.193069	0.128977
49	18	16	5	0.409750	0.216681
49	18	16	6	0.655666	0.245916
49	18	16	7	0.847289	0.191623
49	18	16	8	0.950390	0.103101
49	18	16	9	0.988576	0.038186
49	18	16	10	0.998199	0.009623
49	18	16	11	0.999814	0.001615
49	18	16	12	0.999988	0.000174
49	18	16	13	0.999999	0.000011
49	18	16	14	1.000000	0.000000
49	18	16	15	1.000000	0.000000
49	18	16	16	1.000000	0.000000
49	18	17	0	0.000041	0.000041
49	18	17	1	0.000873	0.000832
49	18	17	2	0.007948	0.007075
49	18	17	3	0.041243	0.033294
49	18	17	4	0.138351	0.097108
49	18	17	5	0.324390	0.186039
49	18	17	6	0.566242	0.241851
49	18	17	7	0.783414	0.217173
49	18	17	8	0.919147	0.135733
49	18	17	9	0.978161	0.059014
49	18	17	10	0.995866	0.017704
49	18	17	11	0.999471	0.003605
49	18	17	12	0.999956	0.000485
49	18	17	13	0.999998	0.000041
49	18	17	14	1.000000	0.000002
49	18	17	15	1.000000	0.000000
49	18	17	16	1.000000	0.000000
49	18	17	17	1.000000	0.000000
49	18	18	0	0.000018	0.000018
49	18	18	1	0.000431	0.000413

N	n	k	x	P(x)	p(x)
49	18	18	2	0.004411	0.003980
49	18	18	3	0.025636	0.021225
49	18	18	4	0.095866	0.070230
49	18	18	5	0.248812	0.152946
49	18	18	6	0.475547	0.226736
49	18	18	7	0.708761	0.233214
49	18	18	8	0.876731	0.167969
49	18	18	9	0.961564	0.084833
49	18	18	10	0.991440	0.029876
49	18	18	11	0.998682	0.007243
49	18	18	12	0.999865	0.001183
49	18	18	13	0.999991	0.000126
49	18	18	14	1.000000	0.000008
49	18	18	15	1.000000	0.000000
49	18	18	16	1.000000	0.000000
49	18	18	17	1.000000	0.000000
49	18	18	18	1.000000	0.000000
49	19	1	0	0.612245	0.612245
49	19	1	1	1.000000	0.387755
49	19	2	0	0.369898	0.369898
49	19	2	1	0.854592	0.484694
49	19	2	2	1.000000	0.145408
49	19	3	0	0.220365	0.220365
49	19	3	1	0.668964	0.448600
49	19	3	2	0.947406	0.278441
49	19	3	3	1.000000	0.052594
49	19	4	0	0.129344	0.129344
49	19	4	1	0.493425	0.364081
49	19	4	2	0.844503	0.351078
49	19	4	3	0.981706	0.137203
49	19	4	4	1.000000	0.018294
49	19	5	0	0.074732	0.074732
49	19	5	1	0.347793	0.273061
49	19	5	2	0.711874	0.364081
49	19	5	3	0.932923	0.221049
49	19	5	4	0.993902	0.060979
49	19	5	5	1.000000	0.006098
49	19	6	0	0.042462	0.042462
49	19	6	1	0.236086	0.193625
49	19	6	2	0.571206	0.335120
49	19	6	3	0.852541	0.281335
49	19	6	4	0.973114	0.120572
49	19	6	5	0.998060	0.024946
49	19	6	6	1.000000	0.001940
49	19	7	0	0.023699	0.023699
49	19	7	1	0.155034	0.131335
49	19	7	2	0.438717	0.283683
49	19	7	3	0.747859	0.309142
49	19	7	4	0.931054	0.183195
49	19	7	5	0.989938	0.058884

N	n	k	x	P(x)	p(x)
49	19	7	6	0.999413	0.009476
49	19	7	7	1.000000	0.000587
49	19	8	0	0.012978	0.012978
49	19	8	1	0.098748	0.085770
49	19	8	2	0.323893	0.225145
49	19	8	3	0.630090	0.306197
49	19	8	4	0.865627	0.235536
49	19	8	5	0.970310	0.104683
49	19	8	6	0.996480	0.026171
49	19	8	7	0.999832	0.003352
49	19	8	8	1.000000	0.000168
49	19	9	0	0.006964	0.006964
49	19	9	1	0.061093	0.054129
49	19	9	2	0.230540	0.169447
49	19	9	3	0.510599	0.280059
49	19	9	4	0.779455	0.268856
49	19	9	5	0.934564	0.155109
49	19	9	6	0.988182	0.053618
49	19	9	7	0.998851	0.010669
49	19	9	8	0.999955	0.001104
49	19	9	9	1.000000	0.000045
49	19	10	0	0.003656	0.003656
49	19	10	1	0.036735	0.033079
49	19	10	2	0.158525	0.121790
49	19	10	3	0.398575	0.240050
49	19	10	4	0.678634	0.280059
49	19	10	5	0.880276	0.201642
49	19	10	6	0.970756	0.090480
49	19	10	7	0.995651	0.024894
49	19	10	8	0.999651	0.004001
49	19	10	9	0.999989	0.000337
49	19	10	10	1.000000	0.000011
49	19	11	0	0.001875	0.001875
49	19	11	1	0.021468	0.019593
49	19	11	2	0.105437	0.083969
49	19	11	3	0.300093	0.194656
49	19	11	4	0.570919	0.270826
49	19	11	5	0.807892	0.236973
49	19	11	6	0.940596	0.132705
49	19	11	7	0.987991	0.047395
49	19	11	8	0.998523	0.010532
49	19	11	9	0.999902	0.001379
49	19	11	10	0.999997	0.000095
49	19	11	11	1.000000	0.000003
49	19	12	0	0.000937	0.000937
49	19	12	1	0.012187	0.011249
49	19	12	2	0.067872	0.055685
49	19	12	3	0.218133	0.150261
49	19	12	4	0.464014	0.245881
49	19	12	5	0.720586	0.256572

Table for $N = 2$, $n = 1$, through $N = 100$, $n = 50$

N	n	k	x	P(x)	p(x)
49	19	12	6	0.895197	0.174611
49	19	12	7	0.973024	0.077827
49	19	12	8	0.995474	0.022450
49	19	12	9	0.999539	0.004065
49	19	12	10	0.999975	0.000436
49	19	12	11	0.999999	0.000025
49	19	12	12	1.000000	0.000001
49	19	13	0	0.000456	0.000456
49	19	13	1	0.006714	0.006258
49	19	13	2	0.042287	0.035573
49	19	13	3	0.153155	0.110868
49	19	13	4	0.364332	0.211177
49	19	13	5	0.623505	0.259172
49	19	13	6	0.833847	0.210343
49	19	13	7	0.947783	0.113936
49	19	13	8	0.988800	0.041017
49	19	13	9	0.998441	0.009641
49	19	13	10	0.999869	0.001428
49	19	13	11	0.999994	0.000125
49	19	13	12	1.000000	0.000006
49	19	13	13	1.000000	0.000000
49	19	14	0	0.000215	0.000215
49	19	14	1	0.003585	0.003370
49	19	14	2	0.025489	0.021904
49	19	14	3	0.103880	0.078392
49	19	14	4	0.276342	0.172462
49	19	14	5	0.522715	0.246374
49	19	14	6	0.757890	0.235175
49	19	14	7	0.909805	0.151914
49	19	14	8	0.976267	0.066462
49	19	14	9	0.995763	0.019496
49	19	14	10	0.999512	0.003749
49	19	14	11	0.999966	0.000454
49	19	14	12	0.999999	0.000032
49	19	14	13	1.000000	0.000001
49	19	14	14	1.000000	0.000000
49	19	15	0	0.000098	0.000098
49	19	15	1	0.001852	0.001754
49	19	15	2	0.014850	0.012998
49	19	15	3	0.068044	0.053194
49	19	15	4	0.202430	0.134386
49	19	15	5	0.424166	0.221736
49	19	15	6	0.670540	0.246374
49	19	15	7	0.857720	0.187180
49	19	15	8	0.955379	0.097659
49	19	15	9	0.990192	0.034814
49	19	15	10	0.998548	0.008355
49	19	15	11	0.999862	0.001315
49	19	15	12	0.999992	0.000130
49	19	15	13	1.000000	0.000007

N	n	k	x	P(x)	p(x)
49	19	15	14	1.000000	0.000000
49	19	15	15	1.000000	0.000000
49	19	16	0	0.000043	0.000043
49	19	16	1	0.000924	0.000880
49	19	16	2	0.008351	0.007427
49	19	16	3	0.043011	0.034661
49	19	16	4	0.143142	0.100130
49	19	16	5	0.332863	0.189721
49	19	16	6	0.576338	0.243475
49	19	16	7	0.791656	0.215318
49	19	16	8	0.923783	0.132127
49	19	16	9	0.979953	0.056170
49	19	16	10	0.996336	0.016383
49	19	16	11	0.999553	0.003217
49	19	16	12	0.999965	0.000412
49	19	16	13	0.999998	0.000033
49	19	16	14	1.000000	0.000002
49	19	16	15	1.000000	0.000000
49	19	16	16	1.000000	0.000000
49	19	17	0	0.000018	0.000018
49	19	17	1	0.000444	0.000425
49	19	17	2	0.004525	0.004081
49	19	17	3	0.026206	0.021681
49	19	17	4	0.097628	0.071422
49	19	17	5	0.252375	0.154747
49	19	17	6	0.480424	0.228048
49	19	17	7	0.713359	0.232935
49	19	17	8	0.879741	0.166382
49	19	17	9	0.962932	0.083191
49	19	17	10	0.991868	0.028936
49	19	17	11	0.998773	0.006905
49	19	17	12	0.999878	0.001105
49	19	17	13	0.999992	0.000114
49	19	17	14	1.000000	0.000007
49	19	17	15	1.000000	0.000000
49	19	17	16	1.000000	0.000000
49	19	17	17	1.000000	0.000000
49	19	18	0	0.000007	0.000007
49	19	18	1	0.000204	0.000197
49	19	18	2	0.002357	0.002152
49	19	18	3	0.015366	0.013009
49	19	18	4	0.064149	0.048784
49	19	18	5	0.184673	0.120524
49	19	18	6	0.387779	0.203106
49	19	18	7	0.626008	0.238229
49	19	18	8	0.822547	0.196539
49	19	18	9	0.936935	0.114388
49	19	18	10	0.983730	0.046795
49	19	18	11	0.997047	0.013317
49	19	18	12	0.999636	0.002589

N	n	k	x	P(x)	p(x)
49	19	18	13	0.999971	0.000335
49	19	18	14	0.999998	0.000028
49	19	18	15	1.000000	0.000001
49	19	18	16	1.000000	0.000000
49	19	18	17	1.000000	0.000000
49	19	18	18	1.000000	0.000000
49	19	19	0	0.000003	0.000003
49	19	19	1	0.000090	0.000087
49	19	19	2	0.001176	0.001086
49	19	19	3	0.008651	0.007475
49	19	19	4	0.040544	0.031893
49	19	19	5	0.130243	0.089699
49	19	19	6	0.302606	0.172363
49	19	19	7	0.533790	0.231185
49	19	19	8	0.752807	0.219017
49	19	19	9	0.900035	0.147228
49	19	19	10	0.970144	0.070109
49	19	19	11	0.993610	0.023466
49	19	19	12	0.999052	0.005441
49	19	19	13	0.999906	0.000855
49	19	19	14	0.999994	0.000088
49	19	19	15	1.000000	0.000006
49	19	19	16	1.000000	0.000000
49	19	19	17	1.000000	0.000000
49	19	19	18	1.000000	0.000000
49	19	19	19	1.000000	0.000000
49	20	1	0	0.591837	0.591837
49	20	1	1	1.000000	0.408163
49	20	2	0	0.345238	0.345238
49	20	2	1	0.838435	0.493197
49	20	2	2	1.000000	0.161565
49	20	3	0	0.198328	0.198328
49	20	3	1	0.639058	0.440729
49	20	3	2	0.938124	0.299066
49	20	3	3	1.000000	0.061876
49	20	4	0	0.112099	0.112099
49	20	4	1	0.457017	0.344919
49	20	4	2	0.821098	0.364081
49	20	4	3	0.977133	0.156035
49	20	4	4	1.000000	0.022867
49	20	5	0	0.062277	0.062277
49	20	5	1	0.311385	0.249108
49	20	5	2	0.675466	0.364081
49	20	5	3	0.918186	0.242721
49	20	5	4	0.991869	0.073683
49	20	5	5	1.000000	0.008131
49	20	6	0	0.033969	0.033969
49	20	6	1	0.203816	0.169846
49	20	6	2	0.526524	0.322708
49	20	6	3	0.824408	0.297884

N	n	k	x	P(x)	p(x)
49	20	6	4	0.965076	0.140668
49	20	6	5	0.997228	0.032153
49	20	6	6	1.000000	0.002772
49	20	7	0	0.018170	0.018170
49	20	7	1	0.128767	0.110598
49	20	7	2	0.391437	0.262669
49	20	7	3	0.706640	0.315203
49	20	7	4	0.912734	0.206094
49	20	7	5	0.986012	0.073278
49	20	7	6	0.999097	0.013085
49	20	7	7	1.000000	0.000902
49	20	8	0	0.009517	0.009517
49	20	8	1	0.078735	0.069218
49	20	8	2	0.278864	0.200129
49	20	8	3	0.579058	0.300194
49	20	8	4	0.834222	0.255164
49	20	8	5	0.959841	0.125619
49	20	8	6	0.994736	0.034894
49	20	8	7	0.999721	0.004985
49	20	8	8	1.000000	0.000279
49	20	9	0	0.004875	0.004875
49	20	9	1	0.046659	0.041784
49	20	9	2	0.191002	0.144344
49	20	9	3	0.454587	0.263585
49	20	9	4	0.734646	0.280059
49	20	9	5	0.913883	0.179238
49	20	9	6	0.982821	0.068937
49	20	9	7	0.998140	0.015319
49	20	9	8	0.999918	0.001778
49	20	9	9	1.000000	0.000082
49	20	10	0	0.002437	0.002437
49	20	10	1	0.026811	0.024374
49	20	10	2	0.126048	0.099236
49	20	10	3	0.342564	0.216516
49	20	10	4	0.622622	0.280059
49	20	10	5	0.846669	0.224047
49	20	10	6	0.958692	0.112023
49	20	10	7	0.993161	0.034469
49	20	10	8	0.999385	0.006224
49	20	10	9	0.999977	0.000593
49	20	10	10	1.000000	0.000022
49	20	11	0	0.001187	0.001187
49	20	11	1	0.014937	0.013749
49	20	11	2	0.080246	0.065309
49	20	11	3	0.248185	0.167939
49	20	11	4	0.507726	0.259541
49	20	11	5	0.760497	0.252771
49	20	11	6	0.918479	0.157982
49	20	11	7	0.981672	0.063193
49	20	11	8	0.997470	0.015798

N	n	k	x	P(x)	p(x)
49	20	11	9	0.999810	0.002340
49	20	11	10	0.999994	0.000184
49	20	11	11	1.000000	0.000006
49	20	12	0	0.000562	0.000562
49	20	12	1	0.008062	0.007500
49	20	12	2	0.049310	0.041248
49	20	12	3	0.173054	0.123744
49	20	12	4	0.398446	0.225391
49	20	12	5	0.660719	0.262273
49	20	12	6	0.860275	0.199556
49	20	12	7	0.960053	0.099778
49	20	12	8	0.992481	0.032428
49	20	12	9	0.999133	0.006652
49	20	12	10	0.999946	0.000813
49	20	12	11	0.999999	0.000053
49	20	12	12	1.000000	0.000001
49	20	13	0	0.000258	0.000258
49	20	13	1	0.004211	0.003953
49	20	13	2	0.029244	0.025033
49	20	13	3	0.116199	0.086955
49	20	13	4	0.300979	0.184780
49	20	13	5	0.554392	0.253413
49	20	13	6	0.784767	0.230375
49	20	13	7	0.924996	0.140228
49	20	13	8	0.981964	0.056968
49	20	13	9	0.997155	0.015191
49	20	13	10	0.999726	0.002571
49	20	13	11	0.999986	0.000260
49	20	13	12	1.000000	0.000014
49	20	13	13	1.000000	0.000000
49	20	14	0	0.000115	0.000115
49	20	14	1	0.002125	0.002010
49	20	14	2	0.016727	0.014602
49	20	14	3	0.075137	0.058409
49	20	14	4	0.218855	0.143718
49	20	14	5	0.448803	0.229949
49	20	14	6	0.695177	0.246374
49	20	14	7	0.874358	0.179181
49	20	14	8	0.962975	0.088617
49	20	14	9	0.992513	0.029539
49	20	14	10	0.999012	0.006499
49	20	14	11	0.999921	0.000909
49	20	14	12	0.999997	0.000076
49	20	14	13	1.000000	0.000003
49	20	14	14	1.000000	0.000000
49	20	15	0	0.000049	0.000049
49	20	15	1	0.001034	0.000985
49	20	15	2	0.009217	0.008184
49	20	15	3	0.046766	0.037549
49	20	15	4	0.153155	0.106389

N	n	k	x	P(x)	p(x)
49	20	15	5	0.350254	0.197099
49	20	15	6	0.596628	0.246374
49	20	15	7	0.807805	0.211177
49	20	15	8	0.932592	0.124787
49	20	15	9	0.983230	0.050638
49	20	15	10	0.997155	0.013925
49	20	15	11	0.999687	0.002532
49	20	15	12	0.999979	0.000292
49	20	15	13	0.999999	0.000020
49	20	15	14	1.000000	0.000001
49	20	15	15	1.000000	0.000000
49	20	16	0	0.000020	0.000020
49	20	16	1	0.000484	0.000463
49	20	16	2	0.004885	0.004401
49	20	16	3	0.027992	0.023107
49	20	16	4	0.103090	0.075098
49	20	16	5	0.263299	0.160209
49	20	16	6	0.495180	0.231881
49	20	16	7	0.727061	0.231881
49	20	16	8	0.888549	0.161489
49	20	16	9	0.966847	0.078297
49	20	16	10	0.993059	0.026213
49	20	16	11	0.999017	0.005957
49	20	16	12	0.999910	0.000894
49	20	16	13	0.999995	0.000085
49	20	16	14	1.000000	0.000005
49	20	16	15	1.000000	0.000000
49	20	16	16	1.000000	0.000000
49	20	17	0	0.000008	0.000008
49	20	17	1	0.000217	0.000209
49	20	17	2	0.002484	0.002267
49	20	17	3	0.016088	0.013604
49	20	17	4	0.066679	0.050590
49	20	17	5	0.190476	0.123798
49	20	17	6	0.396806	0.206329
49	20	17	7	0.635714	0.238908
49	20	17	8	0.829826	0.194113
49	20	17	9	0.940748	0.110921
49	20	17	10	0.985116	0.044369
49	20	17	11	0.997392	0.012276
49	20	17	12	0.999694	0.002302
49	20	17	13	0.999977	0.000283
49	20	17	14	0.999999	0.000022
49	20	17	15	1.000000	0.000001
49	20	17	16	1.000000	0.000000
49	20	17	17	1.000000	0.000000
49	20	18	0	0.000003	0.000003
49	20	18	1	0.000093	0.000090
49	20	18	2	0.001209	0.001116
49	20	18	3	0.008861	0.007652

N	n	k	x	P(x)	p(x)
49	20	18	4	0.041384	0.032522
49	20	18	5	0.132446	0.091063
49	20	18	6	0.306537	0.174090
49	20	18	7	0.538657	0.232121
49	20	18	8	0.757034	0.218377
49	20	18	9	0.902618	0.145584
49	20	18	10	0.971251	0.068633
49	20	18	11	0.993939	0.022688
49	20	18	12	0.999118	0.005179
49	20	18	13	0.999915	0.000797
49	20	18	14	0.999995	0.000080
49	20	18	15	1.000000	0.000005
49	20	18	16	1.000000	0.000000
49	20	18	17	1.000000	0.000000
49	20	18	18	1.000000	0.000000
49	20	19	0	0.000001	0.000001
49	20	19	1	0.000038	0.000037
49	20	19	2	0.000561	0.000523
49	20	19	3	0.004665	0.004104
49	20	19	4	0.024598	0.019933
49	20	19	5	0.088384	0.063786
49	20	19	6	0.227915	0.139532
49	20	19	7	0.441316	0.213401
49	20	19	8	0.672501	0.231185
49	20	19	9	0.850959	0.178458
49	20	19	10	0.949111	0.098152
49	20	19	11	0.987353	0.038241
49	20	19	12	0.997782	0.010429
49	20	19	13	0.999735	0.001953
49	20	19	14	0.999979	0.000244
49	20	19	15	0.999999	0.000020
49	20	19	16	1.000000	0.000001
49	20	19	17	1.000000	0.000000
49	20	19	18	1.000000	0.000000
49	20	19	19	1.000000	0.000000
49	20	20	0	0.000000	0.000000
49	20	20	1	0.000015	0.000014
49	20	20	2	0.000247	0.000232
49	20	20	3	0.002339	0.002092
49	20	20	4	0.013967	0.011628
49	20	20	5	0.056491	0.042524
49	20	20	6	0.162800	0.106310
49	20	20	7	0.348843	0.186042
49	20	20	8	0.580027	0.231185
49	20	20	9	0.785525	0.205497
49	20	20	10	0.916394	0.130869
49	20	20	11	0.975880	0.059486
49	20	20	12	0.995001	0.019121
49	20	20	13	0.999279	0.004279
49	20	20	14	0.999931	0.000651

N	n	k	x	P(x)	p(x)
49	20	20	15	0.999996	0.000065
49	20	20	16	1.000000	0.000004
49	20	20	17	1.000000	0.000000
49	20	20	18	1.000000	0.000000
49	20	20	19	1.000000	0.000000
49	20	20	20	1.000000	0.000000
49	21	1	0	0.571429	0.571429
49	21	1	1	1.000000	0.428571
49	21	2	0	0.321429	0.321429
49	21	2	1	0.821429	0.500000
49	21	2	2	1.000000	0.178571
49	21	3	0	0.177812	0.177812
49	21	3	1	0.608663	0.430851
49	21	3	2	0.927811	0.319149
49	21	3	3	1.000000	0.072188
49	21	4	0	0.096637	0.096637
49	21	4	1	0.421336	0.324699
49	21	4	2	0.795989	0.374653
49	21	4	3	0.971752	0.175763
49	21	4	4	1.000000	0.028248
49	21	5	0	0.051540	0.051540
49	21	5	1	0.277025	0.225486
49	21	5	2	0.637802	0.360777
49	21	5	3	0.901447	0.263645
49	21	5	4	0.989329	0.087882
49	21	5	5	1.000000	0.010671
49	21	6	0	0.026941	0.026941
49	21	6	1	0.174532	0.147591
49	21	6	2	0.482012	0.307480
49	21	6	3	0.793592	0.311580
49	21	6	4	0.955374	0.161782
49	21	6	5	0.996119	0.040745
49	21	6	6	1.000000	0.003880
49	21	7	0	0.013784	0.013784
49	21	7	1	0.105885	0.092101
49	21	7	2	0.346149	0.240264
49	21	7	3	0.663163	0.317015
49	21	7	4	0.891414	0.228251
49	21	7	5	0.980958	0.089544
49	21	7	6	0.998646	0.017688
49	21	7	7	1.000000	0.001354
49	21	8	0	0.006892	0.006892
49	21	8	1	0.062027	0.055135
49	21	8	2	0.237458	0.175431
49	21	8	3	0.527300	0.289842
49	21	8	4	0.799027	0.271727
49	21	8	5	0.946846	0.147819
49	21	8	6	0.992329	0.045483
49	21	8	7	0.999549	0.007220
49	21	8	8	1.000000	0.000451

N	n	k	x	P(x)	p(x)
49	21	9	0	0.003362	0.003362
49	21	9	1	0.035132	0.031770
49	21	9	2	0.156161	0.121029
49	21	9	3	0.400052	0.243891
49	21	9	4	0.686360	0.286307
49	21	9	5	0.889161	0.202801
49	21	9	6	0.975689	0.086528
49	21	9	7	0.997083	0.021394
49	21	9	8	0.999857	0.002773
49	21	9	9	1.000000	0.000143
49	21	10	0	0.001597	0.001597
49	21	10	1	0.019247	0.017650
49	21	10	2	0.098672	0.079425
49	21	10	3	0.290301	0.191629
49	21	10	4	0.564679	0.274378
49	21	10	5	0.808040	0.243361
49	21	10	6	0.943241	0.135201
49	21	10	7	0.989595	0.046355
49	21	10	8	0.998956	0.009360
49	21	10	9	0.999957	0.001001
49	21	10	10	1.000000	0.000043
49	21	11	0	0.000737	0.000737
49	21	11	1	0.010196	0.009459
49	21	11	2	0.059978	0.049782
49	21	11	3	0.201857	0.141879
49	21	11	4	0.445078	0.243221
49	21	11	5	0.708200	0.263121
49	21	11	6	0.891241	0.183041
49	21	11	7	0.972955	0.081715
49	21	11	8	0.995835	0.022880
49	21	11	9	0.999649	0.003813
49	21	11	10	0.999988	0.000339
49	21	11	11	1.000000	0.000012
49	21	12	0	0.000330	0.000330
49	21	12	1	0.005217	0.004888
49	21	12	2	0.035087	0.029869
49	21	12	3	0.134651	0.099564
49	21	12	4	0.336269	0.201618
49	21	12	5	0.597412	0.261143
49	21	12	6	0.818988	0.221576
49	21	12	7	0.942850	0.123862
49	21	12	8	0.988008	0.045158
49	21	12	9	0.998445	0.010437
49	21	12	10	0.999890	0.001445
49	21	12	11	0.999997	0.000107
49	21	12	12	1.000000	0.000003
49	21	13	0	0.000143	0.000143
49	21	13	1	0.002575	0.002433
49	21	13	2	0.019748	0.017173
49	21	13	3	0.086214	0.066466

N	n	k	x	P(x)	p(x)
49	21	13	4	0.243634	0.157419
49	21	13	5	0.484485	0.240851
49	21	13	6	0.729160	0.244675
49	21	13	7	0.895983	0.166824
49	21	13	8	0.972142	0.076159
49	21	13	9	0.995060	0.022918
49	21	13	10	0.999460	0.004400
49	21	13	11	0.999968	0.000508
49	21	13	12	0.999999	0.000031
49	21	13	13	1.000000	0.000001
49	21	14	0	0.000059	0.000059
49	21	14	1	0.001224	0.001164
49	21	14	2	0.010685	0.009461
49	21	14	3	0.052981	0.042296
49	21	14	4	0.169297	0.116315
49	21	14	5	0.377440	0.208143
49	21	14	6	0.627212	0.249772
49	21	14	7	0.831107	0.203895
49	21	14	8	0.944640	0.113533
49	21	14	9	0.987420	0.042780
49	21	14	10	0.998116	0.010695
49	21	14	11	0.999827	0.001711
49	21	14	12	0.999991	0.000165
49	21	14	13	1.000000	0.000008
49	21	14	14	1.000000	0.000000
49	21	15	0	0.000024	0.000024
49	21	15	1	0.000558	0.000535
49	21	15	2	0.005549	0.004990
49	21	15	3	0.031229	0.025680
49	21	15	4	0.112801	0.081572
49	21	15	5	0.282289	0.169488
49	21	15	6	0.520167	0.237878
49	21	15	7	0.749549	0.229382
49	21	15	8	0.902471	0.152922
49	21	15	9	0.972753	0.070282
49	21	15	10	0.994754	0.022001
49	21	15	11	0.999338	0.004584
49	21	15	12	0.999949	0.000611
49	21	15	13	0.999998	0.000049
49	21	15	14	1.000000	0.000002
49	21	15	15	1.000000	0.000000
49	21	16	0	0.000009	0.000009
49	21	16	1	0.000244	0.000235
49	21	16	2	0.002760	0.002516
49	21	16	3	0.017634	0.014873
49	21	16	4	0.072015	0.054381
49	21	16	5	0.202530	0.130515
49	21	16	6	0.415221	0.212691
49	21	16	7	0.655098	0.239877
49	21	16	8	0.844001	0.188903

N	n	k	x	P(x)	p(x)
49	21	16	9	0.947947	0.103947
49	21	16	10	0.987636	0.039689
49	21	16	11	0.997990	0.010354
49	21	16	12	0.999787	0.001797
49	21	16	13	0.999986	0.000199
49	21	16	14	0.999999	0.000013
49	21	16	15	1.000000	0.000000
49	21	16	16	1.000000	0.000000
49	21	17	0	0.000003	0.000003
49	21	17	1	0.000102	0.000098
49	21	17	2	0.001311	0.001210
49	21	17	3	0.009521	0.008209
49	21	17	4	0.044000	0.034479
49	21	17	5	0.139250	0.095249
49	21	17	6	0.318543	0.179293
49	21	17	7	0.553332	0.234789
49	21	17	8	0.769584	0.216253
49	21	17	9	0.910149	0.140564
49	21	17	10	0.974407	0.064258
49	21	17	11	0.994852	0.020446
49	21	17	12	0.999297	0.004445
49	21	17	13	0.999938	0.000641
49	21	17	14	0.999997	0.000059
49	21	17	15	1.000000	0.000003
49	21	17	16	1.000000	0.000000
49	21	17	17	1.000000	0.000000
49	21	18	0	0.000001	0.000001
49	21	18	1	0.000040	0.000039
49	21	18	2	0.000593	0.000553
49	21	18	3	0.004903	0.004310
49	21	18	4	0.025683	0.020780
49	21	18	5	0.091625	0.065942
49	21	18	6	0.234499	0.142874
49	21	18	7	0.450612	0.216112
49	21	18	8	0.681732	0.231120
49	21	18	9	0.857437	0.175705
49	21	18	10	0.952318	0.094881
49	21	18	11	0.988463	0.036145
49	21	18	12	0.998047	0.009584
49	21	18	13	0.999778	0.001731
49	21	18	14	0.999984	0.000206
49	21	18	15	0.999999	0.000015
49	21	18	16	1.000000	0.000001
49	21	18	17	1.000000	0.000000
49	21	18	18	1.000000	0.000000
49	21	19	0	0.000000	0.000000
49	21	19	1	0.000015	0.000015
49	21	19	2	0.000254	0.000239
49	21	19	3	0.002400	0.002146
49	21	19	4	0.014288	0.011887

N	n	k	x	P(x)	p(x)
49	21	19	5	0.057590	0.043303
49	21	19	6	0.165367	0.107776
49	21	19	7	0.353012	0.187646
49	21	19	8	0.584810	0.231798
49	21	19	9	0.789422	0.204612
49	21	19	10	0.918650	0.129228
49	21	19	11	0.976803	0.058153
49	21	19	12	0.995264	0.018461
49	21	19	13	0.999331	0.004067
49	21	19	14	0.999937	0.000606
49	21	19	15	0.999996	0.000059
49	21	19	16	1.000000	0.000004
49	21	19	17	1.000000	0.000000
49	21	19	18	1.000000	0.000000
49	21	19	19	1.000000	0.000000
49	21	20	0	0.000000	0.000000
49	21	20	1	0.000005	0.000005
49	21	20	2	0.000103	0.000097
49	21	20	3	0.001113	0.001010
49	21	20	4	0.007552	0.006439
49	21	20	5	0.034496	0.026944
49	21	20	6	0.111479	0.076983
49	21	20	7	0.265444	0.153966
49	21	20	8	0.484365	0.218920
49	21	20	9	0.707577	0.223213
49	21	20	10	0.871267	0.163689
49	21	20	11	0.957419	0.086152
49	21	20	12	0.989726	0.032307
49	21	20	13	0.998247	0.008521
49	21	20	14	0.999796	0.001549
49	21	20	15	0.999984	0.000189
49	21	20	16	0.999999	0.000015
49	21	20	17	1.000000	0.000001
49	21	20	18	1.000000	0.000000
49	21	20	19	1.000000	0.000000
49	21	20	20	1.000000	0.000000
49	21	21	0	0.000000	0.000000
49	21	21	1	0.000002	0.000002
49	21	21	2	0.000039	0.000037
49	21	21	3	0.000486	0.000447
49	21	21	4	0.003777	0.003291
49	21	21	5	0.019630	0.015853
49	21	21	6	0.071660	0.052030
49	21	21	7	0.191116	0.119456
49	21	21	8	0.386228	0.195112
49	21	21	9	0.615213	0.228986
49	21	21	10	0.809178	0.193964
49	21	21	11	0.927711	0.118534
49	21	21	12	0.979700	0.051988
49	21	21	13	0.995896	0.016196

Table for $N = 2$, $n = 1$, through $N = 100$, $n = 50$

N	n	k	x	P(x)	p(x)
49	21	21	14	0.999422	0.003526
49	21	21	15	0.999945	0.000524
49	21	21	16	0.999997	0.000051
49	21	21	17	1.000000	0.000003
49	21	21	18	1.000000	0.000000
49	21	21	19	1.000000	0.000000
49	21	21	20	1.000000	0.000000
49	21	21	21	1.000000	0.000000
49	22	1	0	0.551020	0.551020
49	22	1	1	1.000000	0.448980
49	22	2	0	0.298469	0.298469
49	22	2	1	0.803571	0.505102
49	22	2	2	1.000000	0.196429
49	22	3	0	0.158760	0.158760
49	22	3	1	0.577888	0.419127
49	22	3	2	0.916413	0.338526
49	22	3	3	1.000000	0.083587
49	22	4	0	0.082831	0.082831
49	22	4	1	0.386547	0.303715
49	22	4	2	0.769228	0.382681
49	22	4	3	0.965475	0.196247
49	22	4	4	1.000000	0.034525
49	22	5	0	0.042336	0.042336
49	22	5	1	0.244813	0.202477
49	22	5	2	0.599148	0.354335
49	22	5	3	0.882615	0.283468
49	22	5	4	0.986190	0.103575
49	22	5	5	1.000000	0.013810
49	22	6	0	0.021168	0.021168
49	22	6	1	0.148176	0.127008
49	22	6	2	0.438086	0.289910
49	22	6	3	0.760209	0.322122
49	22	6	4	0.943818	0.183610
49	22	6	5	0.994664	0.050846
49	22	6	6	1.000000	0.005336
49	22	7	0	0.010338	0.010338
49	22	7	1	0.086149	0.075811
49	22	7	2	0.303244	0.217095
49	22	7	3	0.617876	0.314631
49	22	7	4	0.866959	0.249083
49	22	7	5	0.974562	0.107604
49	22	7	6	0.998015	0.023452
49	22	7	7	1.000000	0.001985
49	22	8	0	0.004923	0.004923
49	22	8	1	0.048243	0.043321
49	22	8	2	0.199866	0.151622
49	22	8	3	0.475542	0.275677
49	22	8	4	0.760209	0.284666
49	22	8	5	0.931008	0.170800
49	22	8	6	0.989080	0.058072
49	22	8	7	0.999291	0.010210
49	22	8	8	1.000000	0.000709
49	22	9	0	0.002281	0.002281
49	22	9	1	0.026055	0.023774
49	22	9	2	0.125904	0.099849
49	22	9	3	0.347790	0.221886
49	22	9	4	0.635233	0.287443
49	22	9	5	0.860189	0.224956
49	22	9	6	0.966418	0.106229
49	22	9	7	0.995555	0.029137
49	22	9	8	0.999758	0.004202
49	22	9	9	1.000000	0.000242
49	22	10	0	0.001027	0.001027
49	22	10	1	0.013574	0.012547
49	22	10	2	0.075979	0.062405
49	22	10	3	0.242394	0.166415
49	22	10	4	0.505884	0.263490
49	22	10	5	0.764583	0.258699
49	22	10	6	0.923927	0.159344
49	22	10	7	0.984629	0.060702
49	22	10	8	0.998287	0.013658
49	22	10	9	0.999921	0.001634
49	22	10	10	1.000000	0.000079
49	22	11	0	0.000447	0.000447
49	22	11	1	0.006818	0.006370
49	22	11	2	0.043976	0.037159
49	22	11	3	0.161320	0.117344
49	22	11	4	0.384273	0.222953
49	22	11	5	0.651817	0.267544
49	22	11	6	0.858555	0.206738
49	22	11	7	0.961282	0.102727
49	22	11	8	0.993384	0.032102
49	22	11	9	0.999376	0.005992
49	22	11	10	0.999976	0.000599
49	22	11	11	1.000000	0.000024
49	22	12	0	0.000188	0.000188
49	22	12	1	0.003297	0.003109
49	22	12	2	0.024419	0.021122
49	22	12	3	0.102648	0.078229
49	22	12	4	0.278664	0.176016
49	22	12	5	0.532126	0.253462
49	22	12	6	0.771507	0.239381
49	22	12	7	0.920732	0.149225
49	22	12	8	0.981557	0.060825
49	22	12	9	0.997326	0.015770
49	22	12	10	0.999786	0.002460
49	22	12	11	0.999993	0.000206
49	22	12	12	1.000000	0.000007
49	22	13	0	0.000076	0.000076
49	22	13	1	0.001533	0.001456
49	22	13	2	0.013002	0.011469
49	22	13	3	0.062477	0.049475
49	22	13	4	0.193035	0.130558
49	22	13	5	0.415670	0.222636
49	22	13	6	0.667991	0.252321
49	22	13	7	0.860235	0.192244
49	22	13	8	0.958542	0.098307
49	22	13	9	0.991786	0.033244
49	22	13	10	0.998989	0.007203
49	22	13	11	0.999931	0.000943
49	22	13	12	0.999998	0.000066
49	22	13	13	1.000000	0.000002
49	22	14	0	0.000030	0.000030
49	22	14	1	0.000683	0.000654
49	22	14	2	0.006630	0.005947
49	22	14	3	0.036365	0.029735
49	22	14	4	0.127756	0.091391
49	22	14	5	0.310537	0.182781
49	22	14	6	0.555848	0.245312
49	22	14	7	0.780133	0.224285
49	22	14	8	0.920312	0.140178
49	22	14	9	0.979781	0.059469
49	22	14	10	0.996588	0.016807
49	22	14	11	0.999643	0.003056
49	22	14	12	0.999980	0.000336
49	22	14	13	0.999999	0.000020
49	22	14	14	1.000000	0.000000
49	22	15	0	0.000011	0.000011
49	22	15	1	0.000291	0.000280
49	22	15	2	0.003232	0.002941
49	22	15	3	0.020223	0.016991
49	22	15	4	0.080755	0.060531
49	22	15	5	0.221757	0.141003
49	22	15	6	0.443706	0.221949
49	22	15	7	0.684011	0.240305
49	22	15	8	0.864240	0.180229
49	22	15	9	0.957692	0.093452
49	22	15	10	0.990825	0.033133
49	22	15	11	0.998683	0.007858
49	22	15	12	0.999884	0.001200
49	22	15	13	0.999994	0.000111
49	22	15	14	1.000000	0.000005
49	22	15	15	1.000000	0.000000
49	22	16	0	0.000004	0.000004
49	22	16	1	0.000118	0.000114
49	22	16	2	0.001502	0.001384
49	22	16	3	0.010728	0.009226
49	22	16	4	0.048709	0.037981
49	22	16	5	0.151256	0.102547
49	22	16	6	0.339260	0.188004
49	22	16	7	0.577994	0.238735
49	22	16	8	0.790028	0.212034
49	22	16	9	0.921961	0.131932
49	22	16	10	0.979131	0.057171
49	22	16	11	0.996141	0.017009
49	22	16	12	0.999530	0.003390
49	22	16	13	0.999965	0.000435
49	22	16	14	0.999998	0.000034
49	22	16	15	1.000000	0.000001
49	22	16	16	1.000000	0.000000
49	22	17	0	0.000001	0.000001
49	22	17	1	0.000045	0.000044
49	22	17	2	0.000663	0.000618
49	22	17	3	0.005416	0.004753
49	22	17	4	0.027992	0.022576
49	22	17	5	0.098429	0.070437
49	22	17	6	0.248106	0.149678
49	22	17	7	0.469478	0.221372
49	22	17	8	0.700075	0.230596
49	22	17	9	0.869987	0.169913
49	22	17	10	0.958342	0.088355
49	22	17	11	0.990471	0.032129
49	22	17	12	0.998503	0.008032
49	22	17	13	0.999846	0.001343
49	22	17	14	0.999990	0.000144
49	22	17	15	1.000000	0.000009
49	22	17	16	1.000000	0.000000
49	22	17	17	1.000000	0.000000
49	22	18	0	0.000000	0.000000
49	22	18	1	0.000016	0.000016
49	22	18	2	0.000277	0.000261
49	22	18	3	0.002594	0.002317
49	22	18	4	0.015293	0.012699
49	22	18	5	0.061009	0.045716
49	22	18	6	0.173267	0.112258
49	22	18	7	0.365710	0.192443
49	22	18	8	0.599189	0.233478
49	22	18	9	0.800960	0.201772
49	22	18	10	0.925209	0.124249
49	22	18	11	0.979427	0.054218
49	22	18	12	0.995993	0.016566
49	22	18	13	0.999469	0.003475
49	22	18	14	0.999954	0.000486
49	22	18	15	0.999997	0.000043
49	22	18	16	1.000000	0.000002
49	22	18	17	1.000000	0.000000
49	22	18	18	1.000000	0.000000
49	22	19	0	0.000000	0.000000
49	22	19	1	0.000006	0.000005
49	22	19	2	0.000109	0.000103

Table for $N = 2$, $n = 1$, through $N = 100$, $n = 50$

N	n	k	x	$P(x)$	$p(x)$
49	22	19	3	0.001174	0.001065
49	22	19	4	0.007919	0.006745
49	22	19	5	0.035939	0.028020
49	22	19	6	0.115328	0.079389
49	22	19	7	0.272593	0.157265
49	22	19	8	0.493747	0.221154
49	22	19	9	0.716346	0.222600
49	22	19	10	0.877113	0.160766
49	22	19	11	0.960188	0.083075
49	22	19	12	0.990649	0.030461
49	22	19	13	0.998460	0.007811
49	22	19	14	0.999829	0.001369
49	22	19	15	0.999988	0.000159
49	22	19	16	0.999999	0.000012
49	22	19	17	1.000000	0.000000
49	22	19	18	1.000000	0.000000
49	22	19	19	1.000000	0.000000
49	22	20	0	0.000000	0.000000
49	22	20	1	0.000002	0.000002
49	22	20	2	0.000040	0.000038
49	22	20	3	0.000499	0.000459
49	22	20	4	0.003872	0.003373
49	22	20	5	0.020061	0.016189
49	22	20	6	0.072987	0.052926
49	22	20	7	0.193960	0.120973
49	22	20	8	0.390542	0.196581
49	22	20	9	0.619887	0.229345
49	22	20	10	0.812806	0.192920
49	22	20	11	0.929727	0.116921
49	22	20	12	0.980495	0.050768
49	22	20	13	0.996116	0.015621
49	22	20	14	0.999464	0.003347
49	22	20	15	0.999951	0.000487
49	22	20	16	0.999997	0.000046
49	22	20	17	1.000000	0.000003
49	22	20	18	1.000000	0.000000
49	22	20	19	1.000000	0.000000
49	22	20	20	1.000000	0.000000
49	22	21	0	0.000000	0.000000
49	22	21	1	0.000001	0.000000
49	22	21	2	0.000014	0.000013
49	22	21	3	0.000198	0.000185
49	22	21	4	0.001779	0.001580
49	22	21	5	0.010571	0.008792
49	22	21	6	0.043787	0.033215
49	22	21	7	0.131388	0.087601
49	22	21	8	0.295640	0.164252
49	22	21	9	0.517077	0.221437
49	22	21	10	0.732977	0.215901
49	22	21	11	0.885378	0.152400

N	n	k	x	$P(x)$	$p(x)$
49	22	21	12	0.962989	0.077611
49	22	21	13	0.991269	0.028279
49	22	21	14	0.998540	0.007272
49	22	21	15	0.999833	0.001293
49	22	21	16	0.999987	0.000154
49	22	21	17	0.999999	0.000012
49	22	21	18	1.000000	0.000001
49	22	21	19	1.000000	0.000000
49	22	21	20	1.000000	0.000000
49	22	21	21	1.000000	0.000000
49	22	22	0	0.000000	0.000000
49	22	22	1	0.000000	0.000000
49	22	22	2	0.000004	0.000004
49	22	22	3	0.000073	0.000069
49	22	22	4	0.000763	0.000690
49	22	22	5	0.005233	0.004470
49	22	22	6	0.024806	0.019573
49	22	22	7	0.084459	0.059652
49	22	22	8	0.213514	0.129055
49	22	22	9	0.414267	0.200753
49	22	22	10	0.640449	0.226182
49	22	22	11	0.825506	0.185058
49	22	22	12	0.935271	0.109765
49	22	22	13	0.982179	0.046908
49	22	22	14	0.996463	0.014284
49	22	22	15	0.999510	0.003047
49	22	22	16	0.999954	0.000444
49	22	22	17	0.999997	0.000043
49	22	22	18	1.000000	0.000003
49	22	22	19	1.000000	0.000000
49	22	22	20	1.000000	0.000000
49	22	22	21	1.000000	0.000000
49	22	22	22	1.000000	0.000000
49	23	1	0	0.530612	0.530612
49	23	1	1	1.000000	0.469388
49	23	2	0	0.276360	0.276360
49	23	2	1	0.784864	0.508503
49	23	2	2	1.000000	0.215136
49	23	3	0	0.141120	0.141120
49	23	3	1	0.546841	0.405721
49	23	3	2	0.903875	0.357034
49	23	3	3	1.000000	0.096125
49	23	4	0	0.070560	0.070560
49	23	4	1	0.352801	0.282241
49	23	4	2	0.740881	0.388081
49	23	4	3	0.958207	0.217325
49	23	4	4	1.000000	0.041793
49	23	5	0	0.034496	0.034496
49	23	5	1	0.214816	0.180320
49	23	5	2	0.559777	0.344961

N	n	k	x	$P(x)$	$p(x)$
49	23	5	3	0.861618	0.301841
49	23	5	4	0.982354	0.120736
49	23	5	5	1.000000	0.017646
49	23	6	0	0.016464	0.016464
49	23	6	1	0.124656	0.108192
49	23	6	2	0.395137	0.270481
49	23	6	3	0.724417	0.329281
49	23	6	4	0.930218	0.205800
49	23	6	5	0.992781	0.062563
49	23	6	6	1.000000	0.007219
49	23	7	0	0.007658	0.007658
49	23	7	1	0.069302	0.061644
49	23	7	2	0.263042	0.193740
49	23	7	3	0.571264	0.308222
49	23	7	4	0.839283	0.268019
49	23	7	5	0.966592	0.127309
49	23	7	6	0.997146	0.030554
49	23	7	7	1.000000	0.002854
49	23	8	0	0.003464	0.003464
49	23	8	1	0.037012	0.033548
49	23	8	2	0.166172	0.129160
49	23	8	3	0.424491	0.258319
49	23	8	4	0.718036	0.293545
49	23	8	5	0.912031	0.193995
49	23	8	6	0.984779	0.072748
49	23	8	7	0.998913	0.014134
49	23	8	8	1.000000	0.001087
49	23	9	0	0.001521	0.001521
49	23	9	1	0.019011	0.017490
49	23	9	2	0.100017	0.081006
49	23	9	3	0.298482	0.198465
49	23	9	4	0.582003	0.283521
49	23	9	5	0.826862	0.244859
49	23	9	6	0.954615	0.127753
49	23	9	7	0.993397	0.038782
49	23	9	8	0.999602	0.006205
49	23	9	9	1.000000	0.000398
49	23	10	0	0.000646	0.000646
49	23	10	1	0.009391	0.008745
49	23	10	2	0.057489	0.048097
49	23	10	3	0.199249	0.141761
49	23	10	4	0.447330	0.248081
49	23	10	5	0.716676	0.269345
49	23	10	6	0.900320	0.183644
49	23	10	7	0.977884	0.077564
49	23	10	8	0.997275	0.019391
49	23	10	9	0.999861	0.002585
49	23	10	10	1.000000	0.000139
49	23	11	0	0.000265	0.000265
49	23	11	1	0.004458	0.004193

N	n	k	x	$P(x)$	$p(x)$
49	23	11	2	0.031590	0.027134
49	23	11	3	0.126552	0.094961
49	23	11	4	0.326470	0.199919
49	23	11	5	0.592362	0.265892
49	23	11	6	0.820270	0.227907
49	23	11	7	0.946063	0.125793
49	23	11	8	0.989817	0.043754
49	23	11	9	0.998933	0.009115
49	23	11	10	0.999953	0.001021
49	23	11	11	1.000000	0.000046
49	23	12	0	0.000105	0.000105
49	23	12	1	0.002031	0.001926
49	23	12	2	0.016596	0.014566
49	23	12	3	0.076572	0.059976
49	23	12	4	0.226511	0.149939
49	23	12	5	0.466414	0.239903
49	23	12	6	0.718311	0.251898
49	23	12	7	0.893098	0.174786
49	23	12	8	0.972546	0.079448
49	23	12	9	0.995574	0.023028
49	23	12	10	0.999604	0.004030
49	23	12	11	0.999985	0.000381
49	23	12	12	1.000000	0.000015
49	23	13	0	0.000040	0.000040
49	23	13	1	0.000885	0.000846
49	23	13	2	0.008329	0.007444
49	23	13	3	0.044153	0.035823
49	23	13	4	0.149515	0.105363
49	23	13	5	0.349704	0.200189
49	23	13	6	0.602574	0.252870
49	23	13	7	0.817514	0.214940
49	23	13	8	0.940337	0.122823
49	23	13	9	0.986861	0.046524
49	23	13	10	0.998188	0.011328
49	23	13	11	0.999862	0.001673
49	23	13	12	0.999995	0.000134
49	23	13	13	1.000000	0.000004
49	23	14	0	0.000014	0.000014
49	23	14	1	0.000369	0.000354
49	23	14	2	0.003987	0.003619
49	23	14	3	0.024251	0.020264
49	23	14	4	0.093907	0.069656
49	23	14	5	0.249610	0.155703
49	23	14	6	0.483164	0.233554
49	23	14	7	0.721986	0.238822
49	23	14	8	0.889161	0.167175
49	23	14	9	0.968768	0.079607
49	23	14	10	0.994098	0.025330
49	23	14	11	0.999304	0.005206
49	23	14	12	0.999955	0.000651

N	n	k	x	P(x)	p(x)
49	23	14	13	0.999999	0.000044
49	23	14	14	1.000000	0.000001
49	23	15	0	0.000005	0.000005
49	23	15	1	0.000146	0.000141
49	23	15	2	0.001816	0.001670
49	23	15	3	0.012672	0.010856
49	23	15	4	0.056094	0.043422
49	23	15	5	0.169534	0.113440
49	23	15	6	0.369723	0.200189
49	23	15	7	0.612810	0.243087
49	23	15	8	0.817514	0.204705
49	23	15	9	0.936925	0.119411
49	23	15	10	0.984690	0.047764
49	23	15	11	0.997519	0.012829
49	23	15	12	0.999750	0.002231
49	23	15	13	0.999986	0.000236
49	23	15	14	1.000000	0.000013
49	23	15	15	1.000000	0.000000
49	23	16	0	0.000002	0.000002
49	23	16	1	0.000055	0.000053
49	23	16	2	0.000784	0.000730
49	23	16	3	0.006286	0.005501
49	23	16	4	0.031828	0.025542
49	23	16	5	0.109477	0.077649
49	23	16	6	0.269629	0.160151
49	23	16	7	0.498416	0.228787
49	23	16	8	0.727203	0.228787
49	23	16	9	0.887756	0.160553
49	23	16	10	0.966427	0.078671
49	23	16	11	0.992991	0.026564
49	23	16	12	0.999028	0.006037
49	23	16	13	0.999917	0.000888
49	23	16	14	0.999996	0.000079
49	23	16	15	1.000000	0.000004
49	23	16	16	1.000000	0.000000
49	23	17	0	0.000000	0.000000
49	23	17	1	0.000019	0.000019
49	23	17	2	0.000320	0.000301
49	23	17	3	0.002952	0.002632
49	23	17	4	0.017122	0.014170
49	23	17	5	0.067123	0.050001
49	23	17	6	0.187126	0.120003
49	23	17	7	0.387489	0.200362
49	23	17	8	0.623209	0.235720
49	23	17	9	0.819643	0.196434
49	23	17	10	0.935435	0.115792
49	23	17	11	0.983331	0.047896
49	23	17	12	0.997016	0.013685
49	23	17	13	0.999647	0.002632
49	23	17	14	0.999974	0.000327

N	n	k	x	P(x)	p(x)
49	23	17	15	0.999999	0.000025
49	23	17	16	1.000000	0.000001
49	23	17	17	1.000000	0.000000
49	23	18	0	0.000000	0.000000
49	23	18	1	0.000006	0.000006
49	23	18	2	0.000123	0.000116
49	23	18	3	0.001307	0.001184
49	23	18	4	0.008708	0.007402
49	23	18	5	0.038998	0.030289
49	23	18	6	0.123375	0.084377
49	23	18	7	0.287308	0.163933
49	23	18	8	0.512715	0.225408
49	23	18	9	0.733703	0.220988
49	23	18	10	0.888395	0.154692
49	23	18	11	0.965370	0.076976
49	23	18	12	0.992312	0.026941
49	23	18	13	0.998825	0.006513
49	23	18	14	0.999882	0.001057
49	23	18	15	0.999993	0.000110
49	23	18	16	1.000000	0.000007
49	23	18	17	1.000000	0.000000
49	23	18	18	1.000000	0.000000
49	23	19	0	0.000000	0.000000
49	23	19	1	0.000002	0.000002
49	23	19	2	0.000044	0.000042
49	23	19	3	0.000543	0.000499
49	23	19	4	0.004172	0.003629
49	23	19	5	0.021410	0.017238
49	23	19	6	0.077103	0.055693
49	23	19	7	0.202697	0.125594
49	23	19	8	0.403647	0.200950
49	23	19	9	0.633902	0.230255
49	23	19	10	0.823524	0.189622
49	23	19	11	0.935573	0.112049
49	23	19	12	0.982752	0.047179
49	23	19	13	0.996724	0.013972
49	23	19	14	0.999575	0.002851
49	23	19	15	0.999964	0.000389
49	23	19	16	0.999998	0.000034
49	23	19	17	1.000000	0.000002
49	23	19	18	1.000000	0.000000
49	23	19	19	1.000000	0.000000
49	23	20	0	0.000000	0.000000
49	23	20	1	0.000001	0.000001
49	23	20	2	0.000015	0.000014
49	23	20	3	0.000210	0.000196
49	23	20	4	0.001874	0.001663
49	23	20	5	0.011067	0.009194
49	23	20	6	0.045544	0.034477
49	23	20	7	0.135714	0.090170

N	n	k	x	P(x)	p(x)
49	23	20	8	0.303172	0.167458
49	23	20	9	0.526450	0.223278
49	23	20	10	0.741355	0.214905
49	23	20	11	0.890754	0.149399
49	23	20	12	0.965453	0.074700
49	23	20	13	0.992067	0.026614
49	23	20	14	0.998720	0.006653
49	23	20	15	0.999861	0.001141
49	23	20	16	0.999990	0.000130
49	23	20	17	0.999999	0.000009
49	23	20	18	1.000000	0.000000
49	23	20	19	1.000000	0.000000
49	23	20	20	1.000000	0.000000
49	23	21	0	0.000000	0.000000
49	23	21	1	0.000000	0.000000
49	23	21	2	0.000004	0.000004
49	23	21	3	0.000075	0.000071
49	23	21	4	0.000784	0.000709
49	23	21	5	0.005361	0.004577
49	23	21	6	0.025334	0.019973
49	23	21	7	0.085965	0.060631
49	23	21	8	0.216556	0.130591
49	23	21	9	0.418661	0.202105
49	23	21	10	0.645018	0.226357
49	23	21	11	0.828933	0.183915
49	23	21	12	0.937119	0.108185
49	23	21	13	0.982890	0.045771
49	23	21	14	0.996655	0.013766
49	23	21	15	0.999546	0.002891
49	23	21	16	0.999959	0.000413
49	23	21	17	0.999998	0.000039
49	23	21	18	1.000000	0.000002
49	23	21	19	1.000000	0.000000
49	23	21	20	1.000000	0.000000
49	23	21	21	1.000000	0.000000
49	23	22	0	0.000000	0.000000
49	23	22	1	0.000000	0.000000
49	23	22	2	0.000001	0.000001
49	23	22	3	0.000025	0.000023
49	23	22	4	0.000303	0.000278
49	23	22	5	0.002418	0.002115
49	23	22	6	0.013207	0.010789
49	23	22	7	0.051318	0.038111
49	23	22	8	0.146596	0.095278
49	23	22	9	0.317608	0.171012
49	23	22	10	0.539923	0.222315
49	23	22	11	0.750112	0.210189
49	23	22	12	0.894617	0.144505
49	23	22	13	0.966543	0.071926
49	23	22	14	0.992230	0.025688

N	n	k	x	P(x)	p(x)
49	23	22	15	0.998720	0.006490
49	23	22	16	0.999856	0.001136
49	23	22	17	0.999989	0.000134
49	23	22	18	0.999999	0.000010
49	23	22	19	1.000000	0.000000
49	23	22	20	1.000000	0.000000
49	23	22	21	1.000000	0.000000
49	23	22	22	1.000000	0.000000
49	23	23	0	0.000000	0.000000
49	23	23	1	0.000000	0.000000
49	23	23	2	0.000000	0.000000
49	23	23	3	0.000007	0.000007
49	23	23	4	0.000107	0.000100
49	23	23	5	0.001008	0.000901
49	23	23	6	0.006414	0.005406
49	23	23	7	0.028734	0.022320
49	23	23	8	0.093664	0.064930
49	23	23	9	0.228935	0.135271
49	23	23	10	0.432883	0.203947
49	23	23	11	0.656695	0.223812
49	23	23	12	0.835745	0.179050
49	23	23	13	0.939904	0.104159
49	23	23	14	0.983668	0.043764
49	23	23	15	0.996797	0.013129
49	23	23	16	0.999561	0.002764
49	23	23	17	0.999960	0.000398
49	23	23	18	0.999998	0.000038
49	23	23	19	1.000000	0.000002
49	23	23	20	1.000000	0.000000
49	23	23	21	1.000000	0.000000
49	23	23	22	1.000000	0.000000
49	23	23	23	1.000000	0.000000
49	24	1	0	0.510204	0.510204
49	24	1	1	1.000000	0.489796
49	24	2	0	0.255102	0.255102
49	24	2	1	0.765306	0.510204
49	24	2	2	1.000000	0.234694
49	24	3	0	0.124837	0.124837
49	24	3	1	0.515632	0.390795
49	24	3	2	0.890143	0.374511
49	24	3	3	1.000000	0.109857
49	24	4	0	0.059705	0.059705
49	24	4	1	0.320234	0.260530
49	24	4	2	0.711029	0.390795
49	24	4	3	0.949848	0.238819
49	24	4	4	1.000000	0.050152
49	24	5	0	0.027862	0.027862
49	24	5	1	0.187075	0.159213
49	24	5	2	0.519974	0.332899
49	24	5	3	0.838399	0.318425

Table for $N = 2$, $n = 1$, through $N = 100$, $n = 50$

N	n	k	x	P(x)	p(x)
49	24	5	4	0.977710	0.139311
49	24	5	5	1.000000	0.022290
49	24	6	0	0.012665	0.012665
49	24	6	1	0.103850	0.091185
49	24	6	2	0.353524	0.249674
49	24	6	3	0.686423	0.332899
49	24	6	4	0.914387	0.227963
49	24	6	5	0.990375	0.075988
49	24	6	6	1.000000	0.009625
49	24	7	0	0.005596	0.005596
49	24	7	1	0.055076	0.049480
49	24	7	2	0.225784	0.170708
49	24	7	3	0.523845	0.298061
49	24	7	4	0.808357	0.284513
49	24	7	5	0.956799	0.148441
49	24	7	6	0.995971	0.039172
49	24	7	7	1.000000	0.004029
49	24	8	0	0.002398	0.002398
49	24	8	1	0.027980	0.025582
49	24	8	2	0.136366	0.108386
49	24	8	3	0.374814	0.238449
49	24	8	4	0.672875	0.298061
49	24	8	5	0.889647	0.216771
49	24	8	6	0.979183	0.089536
49	24	8	7	0.998369	0.019186
49	24	8	8	1.000000	0.001631
49	24	9	0	0.000994	0.000994
49	24	9	1	0.013629	0.012635
49	24	9	2	0.078208	0.064578
49	24	9	3	0.252682	0.174475
49	24	9	4	0.527480	0.274798
49	24	9	5	0.789192	0.261712
49	24	9	6	0.939874	0.150683
49	24	9	7	0.990414	0.050540
49	24	9	8	0.999363	0.008950
49	24	9	9	1.000000	0.000636
49	24	10	0	0.000398	0.000398
49	24	10	1	0.006364	0.005966
49	24	10	2	0.042690	0.036325
49	24	10	3	0.161083	0.118393
49	24	10	4	0.390081	0.228998
49	24	10	5	0.664878	0.274798
49	24	10	6	0.872067	0.207189
49	24	10	7	0.968934	0.096867
49	24	10	8	0.995784	0.026849
49	24	10	9	0.999761	0.003978
49	24	10	10	1.000000	0.000239
49	24	11	0	0.000153	0.000153
49	24	11	1	0.002846	0.002693
49	24	11	2	0.022198	0.019353

N	n	k	x	P(x)	p(x)
49	24	11	3	0.097333	0.075134
49	24	11	4	0.272646	0.175313
49	24	11	5	0.531003	0.258357
49	24	11	6	0.776442	0.245439
49	24	11	7	0.926710	0.150269
49	24	11	8	0.984769	0.058058
49	24	11	9	0.998231	0.013463
49	24	11	10	0.999914	0.001683
49	24	11	11	1.000000	0.000086
49	24	12	0	0.000056	0.000056
49	24	12	1	0.001216	0.001159
49	24	12	2	0.010994	0.009778
49	24	12	3	0.055811	0.044817
49	24	12	4	0.180376	0.124565
49	24	12	5	0.401824	0.221449
49	24	12	6	0.660181	0.258357
49	24	12	7	0.859485	0.199304
49	24	12	8	0.960323	0.100838
49	24	12	9	0.992917	0.032594
49	24	12	10	0.999294	0.006377
49	24	12	11	0.999971	0.000676
49	24	12	12	1.000000	0.000029
49	24	13	0	0.000020	0.000020
49	24	13	1	0.000495	0.000475
49	24	13	2	0.005180	0.004685
49	24	13	3	0.030374	0.025194
49	24	13	4	0.113044	0.082669
49	24	13	5	0.288108	0.175064
49	24	13	6	0.534494	0.246386
49	24	13	7	0.767913	0.233419
49	24	13	8	0.916717	0.148804
49	24	13	9	0.979703	0.062986
49	24	13	10	0.996881	0.017178
49	24	13	11	0.999733	0.002852
49	24	13	12	0.999990	0.000257
49	24	13	13	1.000000	0.000009
49	24	14	0	0.000007	0.000007
49	24	14	1	0.000191	0.000185
49	24	14	2	0.002317	0.002126
49	24	14	3	0.015678	0.013361
49	24	14	4	0.067116	0.051439
49	24	14	5	0.195713	0.128596
49	24	14	6	0.411301	0.215588
49	24	14	7	0.657687	0.246386
49	24	14	8	0.850582	0.192895
49	24	14	9	0.953459	0.102877
49	24	14	10	0.990201	0.036742
49	24	14	11	0.998703	0.008502
49	24	14	12	0.999905	0.001201
49	24	14	13	0.999997	0.000092

N	n	k	x	P(x)	p(x)
49	24	14	14	1.000000	0.000003
49	24	15	0	0.000002	0.000002
49	24	15	1	0.000070	0.000068
49	24	15	2	0.000981	0.000911
49	24	15	3	0.007661	0.006680
49	24	15	4	0.037723	0.030061
49	24	15	5	0.125903	0.088180
49	24	15	6	0.300427	0.174524
49	24	15	7	0.538014	0.237587
49	24	15	8	0.762401	0.224388
49	24	15	9	0.909369	0.146967
49	24	15	10	0.975504	0.066135
49	24	15	11	0.995545	0.020041
49	24	15	12	0.999493	0.003947
49	24	15	13	0.999968	0.000475
49	24	15	14	0.999999	0.000031
49	24	15	15	1.000000	0.000001
49	24	16	0	0.000001	0.000001
49	24	16	1	0.000024	0.000023
49	24	16	2	0.000391	0.000367
49	24	16	3	0.003535	0.003144
49	24	16	4	0.020040	0.016504
49	24	16	5	0.076626	0.056586
49	24	16	6	0.208032	0.131406
49	24	16	7	0.419220	0.211188
49	24	16	8	0.656807	0.237587
49	24	16	9	0.844530	0.187723
49	24	16	10	0.948272	0.103742
49	24	16	11	0.987882	0.039610
49	24	16	12	0.998099	0.010217
49	24	16	13	0.999814	0.001715
49	24	16	14	0.999990	0.000176
49	24	16	15	1.000000	0.000010
49	24	16	16	1.000000	0.000000
49	24	17	0	0.000000	0.000000
49	24	17	1	0.000008	0.000008
49	24	17	2	0.000147	0.000139
49	24	17	3	0.001535	0.001388
49	24	17	4	0.010037	0.008502
49	24	17	5	0.044046	0.034009
49	24	17	6	0.136356	0.092310
49	24	17	7	0.310426	0.174070
49	24	17	8	0.541614	0.231187
49	24	17	9	0.759202	0.217588
49	24	17	10	0.904260	0.145059
49	24	17	11	0.972278	0.068018
49	24	17	12	0.994384	0.022106
49	24	17	13	0.999243	0.004858
49	24	17	14	0.999937	0.000694
49	24	17	15	0.999997	0.000060

N	n	k	x	P(x)	p(x)
49	24	17	16	1.000000	0.000003
49	24	17	17	1.000000	0.000000
49	24	18	0	0.000000	0.000000
49	24	18	1	0.000002	0.000002
49	24	18	2	0.000051	0.000049
49	24	18	3	0.000624	0.000573
49	24	18	4	0.004723	0.004099
49	24	18	5	0.023853	0.019130
49	24	18	6	0.084432	0.060578
49	24	18	7	0.217951	0.133520
49	24	18	8	0.426020	0.208069
49	24	18	9	0.657207	0.231187
49	24	18	10	0.840797	0.183590
49	24	18	11	0.944646	0.103849
49	24	18	12	0.986095	0.041448
49	24	18	13	0.997573	0.011478
49	24	18	14	0.999720	0.002147
49	24	18	15	0.999980	0.000260
49	24	18	16	0.999999	0.000019
49	24	18	17	1.000000	0.000001
49	24	18	18	1.000000	0.000000
49	24	19	0	0.000000	0.000000
49	24	19	1	0.000001	0.000001
49	24	19	2	0.000016	0.000016
49	24	19	3	0.000236	0.000219
49	24	19	4	0.002078	0.001842
49	24	19	5	0.012128	0.010050
49	24	19	6	0.049257	0.037129
49	24	19	7	0.144731	0.095474
49	24	19	8	0.318630	0.173899
49	24	19	9	0.545343	0.226713
49	24	19	10	0.757886	0.212543
49	24	19	11	0.901097	0.143211
49	24	19	12	0.970050	0.068953
49	24	19	13	0.993500	0.023450
49	24	19	14	0.999027	0.005527
49	24	19	15	0.999904	0.000877
49	24	19	16	0.999994	0.000090
49	24	19	17	1.000000	0.000006
49	24	19	18	1.000000	0.000000
49	24	19	19	1.000000	0.000000
49	24	20	0	0.000000	0.000000
49	24	20	1	0.000000	0.000000
49	24	20	2	0.000005	0.000005
49	24	20	3	0.000082	0.000077
49	24	20	4	0.000850	0.000768
49	24	20	5	0.005763	0.004913
49	24	20	6	0.026980	0.021216
49	24	20	7	0.090629	0.063649
49	24	20	8	0.225884	0.135255

N	n	k	x	P(x)	p(x)
49	24	20	9	0.431986	0.206102
49	24	20	10	0.658699	0.226713
49	24	20	11	0.839038	0.180340
49	24	20	12	0.942469	0.103430
49	24	20	13	0.984901	0.042433
49	24	20	14	0.997185	0.012283
49	24	20	15	0.999641	0.002457
49	24	20	16	0.999970	0.000329
49	24	20	17	0.999998	0.000028
49	24	20	18	1.000000	0.000001
49	24	20	19	1.000000	0.000000
49	24	20	20	1.000000	0.000000
49	24	21	0	0.000000	0.000000
49	24	21	1	0.000000	0.000000
49	24	21	2	0.000001	0.000001
49	24	21	3	0.000026	0.000025
49	24	21	4	0.000321	0.000294
49	24	21	5	0.002544	0.002224
49	24	21	6	0.013811	0.011267
49	24	21	7	0.053317	0.039506
49	24	21	8	0.151260	0.097943
49	24	21	9	0.325381	0.174121
49	24	21	10	0.549251	0.223870
49	24	21	11	0.758197	0.208945
49	24	21	12	0.899670	0.141473
49	24	21	13	0.968806	0.069136
49	24	21	14	0.992949	0.024143
49	24	21	15	0.998879	0.005930
49	24	21	16	0.999879	0.001001
49	24	21	17	0.999992	0.000112
49	24	21	18	1.000000	0.000008
49	24	21	19	1.000000	0.000000
49	24	21	20	1.000000	0.000000
49	24	21	21	1.000000	0.000000
49	24	22	0	0.000000	0.000000
49	24	22	1	0.000000	0.000000
49	24	22	2	0.000000	0.000000
49	24	22	3	0.000008	0.000007
49	24	22	4	0.000110	0.000103
49	24	22	5	0.001035	0.000925
49	24	22	6	0.006568	0.005533
49	24	22	7	0.029331	0.022763
49	24	22	8	0.095293	0.065962
49	24	22	9	0.232102	0.136809
49	24	22	10	0.437316	0.205214
49	24	22	11	0.661186	0.223870
49	24	22	12	0.839038	0.177852
49	24	22	13	0.941645	0.102607
49	24	22	14	0.984327	0.042681
49	24	22	15	0.996973	0.012646

N	n	k	x	P(x)	p(x)
49	24	22	16	0.999594	0.002621
49	24	22	17	0.999964	0.000370
49	24	22	18	0.999998	0.000034
49	24	22	19	1.000000	0.000002
49	24	22	20	1.000000	0.000000
49	24	22	21	1.000000	0.000000
49	24	22	22	1.000000	0.000000
49	24	23	0	0.000000	0.000000
49	24	23	1	0.000000	0.000000
49	24	23	2	0.000000	0.000000
49	24	23	3	0.000002	0.000002
49	24	23	4	0.000034	0.000032
49	24	23	5	0.000384	0.000350
49	24	23	6	0.002880	0.002495
49	24	23	7	0.014999	0.012119
49	24	23	8	0.056205	0.041206
49	24	23	9	0.156097	0.099893
49	24	23	10	0.330909	0.174812
49	24	23	11	0.553397	0.222488
49	24	23	12	0.759993	0.206596
49	24	23	13	0.899843	0.139850
49	24	23	14	0.968519	0.068676
49	24	23	15	0.992757	0.024239
49	24	23	16	0.998817	0.006060
49	24	23	17	0.999868	0.001051
49	24	23	18	0.999990	0.000123
49	24	23	19	0.999999	0.000009
49	24	23	20	1.000000	0.000000
49	24	23	21	1.000000	0.000000
49	24	23	22	1.000000	0.000000
49	24	23	23	1.000000	0.000000
49	24	24	0	0.000000	0.000000
49	24	24	1	0.000000	0.000000
49	24	24	2	0.000000	0.000000
49	24	24	3	0.000000	0.000000
49	24	24	4	0.000009	0.000009
49	24	24	5	0.000128	0.000119
49	24	24	6	0.001152	0.001024
49	24	24	7	0.007075	0.005923
49	24	24	8	0.030847	0.023773
49	24	24	9	0.098467	0.067620
49	24	24	10	0.236780	0.138313
49	24	24	11	0.442153	0.205374
49	24	24	12	0.664641	0.222488
49	24	24	13	0.840675	0.176034
49	24	24	14	0.942105	0.101429
49	24	24	15	0.984367	0.042262
49	24	24	16	0.996953	0.012585
49	24	24	17	0.999585	0.002632
49	24	24	18	0.999962	0.000377

N	n	k	x	P(x)	p(x)
49	24	24	19	0.999998	0.000036
49	24	24	20	1.000000	0.000002
49	24	24	21	1.000000	0.000000
49	24	24	22	1.000000	0.000000
49	24	24	23	1.000000	0.000000
49	24	24	24	1.000000	0.000000
50	1	1	0	0.980000	0.980000
50	1	1	1	1.000000	0.020000
50	2	1	0	0.960000	0.960000
50	2	1	1	1.000000	0.040000
50	2	2	0	0.920816	0.920816
50	2	2	1	0.999184	0.078367
50	2	2	2	1.000000	0.000816
50	3	1	0	0.940000	0.940000
50	3	1	1	1.000000	0.060000
50	3	2	0	0.882449	0.882449
50	3	2	1	0.997551	0.115102
50	3	2	2	1.000000	0.002449
50	3	3	0	0.827296	0.827296
50	3	3	1	0.992755	0.165459
50	3	3	2	0.999949	0.007194
50	3	3	3	1.000000	0.000051
50	4	1	0	0.920000	0.920000
50	4	1	1	1.000000	0.080000
50	4	2	0	0.844898	0.844898
50	4	2	1	0.995102	0.150204
50	4	2	2	1.000000	0.004898
50	4	3	0	0.774490	0.774490
50	4	3	1	0.985714	0.211224
50	4	3	2	0.999796	0.014082
50	4	3	3	1.000000	0.000204
50	4	4	0	0.708576	0.708576
50	4	4	1	0.972232	0.263656
50	4	4	2	0.999197	0.026965
50	4	4	3	0.999996	0.000799
50	4	4	4	1.000000	0.000004
50	5	1	0	0.900000	0.900000
50	5	1	1	1.000000	0.100000
50	5	2	0	0.808163	0.808163
50	5	2	1	0.991837	0.183673
50	5	2	2	1.000000	0.008163
50	5	3	0	0.723980	0.723980
50	5	3	1	0.976531	0.252551
50	5	3	2	0.999490	0.022959
50	5	3	3	1.000000	0.000510
50	5	4	0	0.646960	0.646960
50	5	4	1	0.955037	0.308076
50	5	4	2	0.998024	0.042987
50	5	4	3	0.999978	0.001954
50	5	4	4	1.000000	0.000022

N	n	k	x	P(x)	p(x)
50	5	5	0	0.576639	0.576639
50	5	5	1	0.928248	0.351609
50	5	5	2	0.995221	0.066973
50	5	5	3	0.999893	0.004673
50	5	5	4	0.999999	0.000106
50	5	5	5	1.000000	0.000000
50	6	1	0	0.880000	0.880000
50	6	1	1	1.000000	0.120000
50	6	2	0	0.772245	0.772245
50	6	2	1	0.987755	0.215510
50	6	2	2	1.000000	0.012245
50	6	3	0	0.675714	0.675714
50	6	3	1	0.965306	0.289592
50	6	3	2	0.998980	0.033673
50	6	3	3	1.000000	0.001020
50	6	4	0	0.589453	0.589453
50	6	4	1	0.934498	0.345046
50	6	4	2	0.996114	0.061615
50	6	4	3	0.999935	0.003821
50	6	4	4	1.000000	0.000065
50	6	5	0	0.512568	0.512568
50	6	5	1	0.896994	0.384426
50	6	5	2	0.990756	0.093762
50	6	5	3	0.999686	0.008930
50	6	5	4	0.999997	0.000312
50	6	5	5	1.000000	0.000003
50	6	6	0	0.444225	0.444225
50	6	6	1	0.854279	0.410054
50	6	6	2	0.982421	0.128142
50	6	6	3	0.999090	0.016669
50	6	6	4	0.999983	0.000893
50	6	6	5	1.000000	0.000017
50	6	6	6	1.000000	0.000000
50	7	1	0	0.860000	0.860000
50	7	1	1	1.000000	0.140000
50	7	2	0	0.737143	0.737143
50	7	2	1	0.982857	0.245714
50	7	2	2	1.000000	0.017143
50	7	3	0	0.629643	0.629643
50	7	3	1	0.952143	0.322500
50	7	3	2	0.998214	0.046071
50	7	3	3	1.000000	0.001786
50	7	4	0	0.535866	0.535866
50	7	4	1	0.910973	0.375106
50	7	4	2	0.993313	0.082340
50	7	4	3	0.999848	0.006535
50	7	4	4	1.000000	0.000152
50	7	5	0	0.454321	0.454321
50	7	5	1	0.862046	0.407724
50	7	5	2	0.984363	0.122317

N	n	k	x	P(x)	p(x)
50	7	5	3	0.999280	0.014917
50	7	5	4	0.999990	0.000710
50	7	5	5	1.000000	0.000010
50	7	6	0	0.383649	0.383649
50	7	6	1	0.807682	0.424033
50	7	6	2	0.970772	0.163090
50	7	6	3	0.997954	0.027182
50	7	6	4	0.999943	0.001989
50	7	6	5	1.000000	0.000057
50	7	6	6	1.000000	0.000000
50	7	7	0	0.322614	0.322614
50	7	7	1	0.749860	0.427246
50	7	7	2	0.952239	0.202380
50	7	7	3	0.995483	0.043243
50	7	7	4	0.999807	0.004324
50	7	7	5	0.999997	0.000190
50	7	7	6	1.000000	0.000003
50	7	7	7	1.000000	0.000000
50	8	1	0	0.840000	0.840000
50	8	1	1	1.000000	0.160000
50	8	2	0	0.702857	0.702857
50	8	2	1	0.977143	0.274286
50	8	2	2	1.000000	0.022857
50	8	3	0	0.585714	0.585714
50	8	3	1	0.937143	0.351429
50	8	3	2	0.997143	0.060000
50	8	3	3	1.000000	0.002857
50	8	4	0	0.486018	0.486018
50	8	4	1	0.884802	0.398784
50	8	4	2	0.989483	0.104681
50	8	4	3	0.999696	0.010213
50	8	4	4	1.000000	0.000304
50	8	5	0	0.401493	0.401493
50	8	5	1	0.824118	0.422625
50	8	5	2	0.975829	0.151711
50	8	5	3	0.998586	0.022757
50	8	5	4	0.999974	0.001388
50	8	5	5	1.000000	0.000026
50	8	6	0	0.330117	0.330117
50	8	6	1	0.758376	0.428260
50	8	6	2	0.955601	0.197225
50	8	6	3	0.996057	0.040456
50	8	6	4	0.999850	0.003793
50	8	6	5	0.999998	0.000148
50	8	6	6	1.000000	0.000002
50	8	7	0	0.270096	0.270096
50	8	7	1	0.690244	0.420149
50	8	7	2	0.928707	0.238463
50	8	7	3	0.991460	0.062753
50	8	7	4	0.999505	0.008045

N	n	k	x	P(x)	p(x)
50	8	7	5	0.999988	0.000483
50	8	7	6	1.000000	0.000012
50	8	7	7	1.000000	0.000000
50	8	8	0	0.219845	0.219845
50	8	8	1	0.621848	0.402003
50	8	8	2	0.895433	0.273585
50	8	8	3	0.984163	0.088730
50	8	8	4	0.998757	0.014594
50	8	8	5	0.999954	0.001197
50	8	8	6	0.999999	0.000045
50	8	8	7	1.000000	0.000001
50	8	8	8	1.000000	0.000000
50	9	1	0	0.820000	0.820000
50	9	1	1	1.000000	0.180000
50	9	2	0	0.669388	0.669388
50	9	2	1	0.970612	0.301224
50	9	2	2	1.000000	0.029388
50	9	3	0	0.543878	0.543878
50	9	3	1	0.920408	0.376531
50	9	3	2	0.995714	0.075306
50	9	3	3	1.000000	0.004286
50	9	4	0	0.439731	0.439731
50	9	4	1	0.856318	0.416587
50	9	4	2	0.984498	0.128181
50	9	4	3	0.999453	0.014954
50	9	4	4	1.000000	0.000547
50	9	5	0	0.353697	0.353697
50	9	5	1	0.783868	0.430171
50	9	5	2	0.964993	0.181125
50	9	5	3	0.997502	0.032510
50	9	5	4	0.999940	0.002438
50	9	5	5	1.000000	0.000059
50	9	6	0	0.282957	0.282957
50	9	6	1	0.707393	0.424436
50	9	6	2	0.936818	0.229425
50	9	6	3	0.993168	0.056350
50	9	6	4	0.999670	0.006502
50	9	6	5	0.999995	0.000325
50	9	6	6	1.000000	0.000005
50	9	7	0	0.225080	0.225080
50	9	7	1	0.630223	0.405143
50	9	7	2	0.900318	0.270096
50	9	7	3	0.985484	0.085165
50	9	7	4	0.998931	0.013447
50	9	7	5	0.999965	0.001034
50	9	7	6	1.000000	0.000034
50	9	7	7	1.000000	0.000000
50	9	8	0	0.177970	0.177970
50	9	8	1	0.554847	0.376877
50	9	8	2	0.856349	0.301502

N	n	k	x	P(x)	p(x)
50	9	8	3	0.973600	0.117251
50	9	8	4	0.997367	0.023767
50	9	8	5	0.999869	0.002502
50	9	8	6	0.999997	0.000128
50	9	8	7	1.000000	0.000003
50	9	8	8	1.000000	0.000000
50	9	9	0	0.139833	0.139833
50	9	9	1	0.483061	0.343228
50	9	9	2	0.806099	0.323038
50	9	9	3	0.956850	0.150751
50	9	9	4	0.994538	0.037688
50	9	9	5	0.999631	0.005093
50	9	9	6	0.999988	0.000357
50	9	9	7	1.000000	0.000012
50	9	9	8	1.000000	0.000000
50	9	9	9	1.000000	0.000000
50	10	1	0	0.800000	0.800000
50	10	1	1	1.000000	0.200000
50	10	2	0	0.636735	0.636735
50	10	2	1	0.963265	0.326531
50	10	2	2	1.000000	0.036735
50	10	3	0	0.504082	0.504082
50	10	3	1	0.902041	0.397959
50	10	3	2	0.993878	0.091837
50	10	3	3	1.000000	0.006122
50	10	4	0	0.396830	0.396830
50	10	4	1	0.825836	0.429006
50	10	4	2	0.978246	0.152410
50	10	4	3	0.999088	0.020842
50	10	4	4	1.000000	0.000912
50	10	5	0	0.310563	0.310563
50	10	5	1	0.741900	0.431337
50	10	5	2	0.951740	0.209840
50	10	5	3	0.995916	0.044177
50	10	5	4	0.999881	0.003965
50	10	5	5	1.000000	0.000119
50	10	6	0	0.241549	0.241549
50	10	6	1	0.655633	0.414084
50	10	6	2	0.914435	0.258802
50	10	6	3	0.989044	0.074610
50	10	6	4	0.999352	0.010308
50	10	6	5	0.999987	0.000634
50	10	6	6	1.000000	0.000013
50	10	7	0	0.186651	0.186651
50	10	7	1	0.570934	0.384282
50	10	7	2	0.867380	0.296446
50	10	7	3	0.977175	0.109795
50	10	7	4	0.997947	0.020772
50	10	7	5	0.999915	0.001968
50	10	7	6	0.999999	0.000084

N	n	k	x	P(x)	p(x)
50	10	7	7	1.000000	0.000001
50	10	8	0	0.143244	0.143244
50	10	8	1	0.490502	0.347258
50	10	8	2	0.812227	0.321725
50	10	8	3	0.959301	0.147074
50	10	8	4	0.995048	0.035747
50	10	8	5	0.999686	0.004637
50	10	8	6	0.999991	0.000305
50	10	8	7	1.000000	0.000009
50	10	8	8	1.000000	0.000000
50	10	9	0	0.109138	0.109138
50	10	9	1	0.416090	0.306952
50	10	9	2	0.750946	0.334856
50	10	9	3	0.934789	0.183843
50	10	9	4	0.989942	0.055153
50	10	9	5	0.999134	0.009192
50	10	9	6	0.999962	0.000828
50	10	9	7	0.999999	0.000037
50	10	9	8	1.000000	0.000001
50	10	9	9	1.000000	0.000000
50	10	10	0	0.082519	0.082519
50	10	10	1	0.348710	0.266191
50	10	10	2	0.685608	0.336898
50	10	10	3	0.903401	0.217793
50	10	10	4	0.981870	0.078469
50	10	10	5	0.998013	0.016142
50	10	10	6	0.999881	0.001868
50	10	10	7	0.999996	0.000115
50	10	10	8	1.000000	0.000003
50	10	10	9	1.000000	0.000000
50	10	10	10	1.000000	0.000000
50	11	1	0	0.780000	0.780000
50	11	1	1	1.000000	0.220000
50	11	2	0	0.604898	0.604898
50	11	2	1	0.955102	0.350204
50	11	2	2	1.000000	0.044898
50	11	3	0	0.466275	0.466275
50	11	3	1	0.882143	0.415867
50	11	3	2	0.991582	0.109439
50	11	3	3	1.000000	0.008418
50	11	4	0	0.357147	0.357147
50	11	4	1	0.793660	0.436513
50	11	4	2	0.970625	0.176965
50	11	4	3	0.998567	0.027942
50	11	4	4	1.000000	0.001433
50	11	5	0	0.271742	0.271742
50	11	5	1	0.698766	0.427024
50	11	5	2	0.936002	0.237235
50	11	5	3	0.993708	0.057706
50	11	5	4	0.999782	0.006074

N	n	k	x	P(x)	p(x)
50	11	5	5	1.000000	0.000218
50	11	6	0	0.205316	0.205316
50	11	6	1	0.603872	0.398556
50	11	6	2	0.888555	0.284682
50	11	6	3	0.983449	0.094894
50	11	6	4	0.998837	0.015388
50	11	6	5	0.999971	0.001134
50	11	6	6	1.000000	0.000029
50	11	7	0	0.153987	0.153987
50	11	7	1	0.513291	0.359304
50	11	7	2	0.830324	0.317033
50	11	7	3	0.966195	0.135871
50	11	7	4	0.996389	0.030194
50	11	7	5	0.999816	0.003427
50	11	7	6	0.999997	0.000180
50	11	7	7	1.000000	0.000003
50	11	8	0	0.114595	0.114595
50	11	8	1	0.429732	0.315137
50	11	8	2	0.763968	0.334236
50	11	8	3	0.940917	0.176949
50	11	8	4	0.991474	0.050557
50	11	8	5	0.999338	0.007864
50	11	8	6	0.999976	0.000638
50	11	8	7	1.000000	0.000024
50	11	8	8	1.000000	0.000000
50	11	9	0	0.084582	0.084582
50	11	9	1	0.354700	0.270117
50	11	9	2	0.692346	0.337647
50	11	9	3	0.907212	0.214866
50	11	9	4	0.983047	0.075835
50	11	9	5	0.998214	0.015167
50	11	9	6	0.999900	0.001685
50	11	9	7	0.999997	0.000098
50	11	9	8	1.000000	0.000003
50	11	9	9	1.000000	0.000000
50	11	10	0	0.061889	0.061889
50	11	10	1	0.288817	0.226928
50	11	10	2	0.618229	0.329411
50	11	10	3	0.865287	0.247059
50	11	10	4	0.970100	0.104813
50	11	10	5	0.995995	0.025895
50	11	10	6	0.999694	0.003699
50	11	10	7	0.999988	0.000294
50	11	10	8	1.000000	0.000012
50	11	10	9	1.000000	0.000000
50	11	10	10	1.000000	0.000000
50	11	11	0	0.044870	0.044870
50	11	11	1	0.232085	0.187215
50	11	11	2	0.544111	0.312026
50	11	11	3	0.815876	0.271764

N	n	k	x	P(x)	p(x)
50	11	11	4	0.951758	0.135882
50	11	11	5	0.992111	0.040353
50	11	11	6	0.999232	0.007121
50	11	11	7	0.999958	0.000727
50	11	11	8	0.999999	0.000040
50	11	11	9	1.000000	0.000001
50	11	11	10	1.000000	0.000000
50	11	11	11	1.000000	0.000000
50	12	1	0	0.760000	0.760000
50	12	1	1	1.000000	0.240000
50	12	2	0	0.573878	0.573878
50	12	2	1	0.946122	0.372245
50	12	2	2	1.000000	0.053878
50	12	3	0	0.430408	0.430408
50	12	3	1	0.860816	0.430408
50	12	3	2	0.988775	0.127959
50	12	3	3	1.000000	0.011224
50	12	4	0	0.320517	0.320517
50	12	4	1	0.760082	0.439566
50	12	4	2	0.961550	0.201468
50	12	4	3	0.997851	0.036300
50	12	4	4	1.000000	0.002149
50	12	5	0	0.236904	0.236904
50	12	5	1	0.654969	0.418065
50	12	5	2	0.917753	0.262784
50	12	5	3	0.990748	0.072995
50	12	5	4	0.999626	0.008878
50	12	5	5	1.000000	0.000374
50	12	6	0	0.173729	0.173729
50	12	6	1	0.552775	0.379046
50	12	6	2	0.859356	0.306581
50	12	6	3	0.976149	0.116793
50	12	6	4	0.998048	0.021899
50	12	6	5	0.999942	0.001894
50	12	6	6	1.000000	0.000058
50	12	7	0	0.126349	0.126349
50	12	7	1	0.458014	0.331665
50	12	7	2	0.789679	0.331665
50	12	7	3	0.952260	0.162581
50	12	7	4	0.994066	0.041807
50	12	7	5	0.999640	0.005574
50	12	7	6	0.999992	0.000352
50	12	7	7	1.000000	0.000008
50	12	8	0	0.091089	0.091089
50	12	8	1	0.373169	0.282081
50	12	8	2	0.712547	0.339378
50	12	8	3	0.918231	0.205684
50	12	8	4	0.986288	0.068057
50	12	8	5	0.998733	0.012445
50	12	8	6	0.999943	0.001210

N	n	k	x	P(x)	p(x)
50	12	8	7	0.999999	0.000056
50	12	8	8	1.000000	0.000001
50	12	9	0	0.065063	0.065063
50	12	9	1	0.299291	0.234228
50	12	9	2	0.631743	0.332452
50	12	9	3	0.874156	0.242413
50	12	9	4	0.973325	0.099169
50	12	9	5	0.996659	0.023334
50	12	9	6	0.999770	0.003111
50	12	9	7	0.999992	0.000222
50	12	9	8	1.000000	0.000008
50	12	9	9	1.000000	0.000000
50	12	10	0	0.046020	0.046020
50	12	10	1	0.236449	0.190429
50	12	10	2	0.550657	0.314208
50	12	10	3	0.820943	0.270286
50	12	10	4	0.953975	0.133032
50	12	10	5	0.992675	0.038700
50	12	10	6	0.999315	0.006640
50	12	10	7	0.999965	0.000650
50	12	10	8	0.999999	0.000034
50	12	10	9	1.000000	0.000001
50	12	10	10	1.000000	0.000000
50	12	11	0	0.032214	0.032214
50	12	11	1	0.184081	0.151867
50	12	11	2	0.472105	0.288024
50	12	11	3	0.760129	0.288024
50	12	11	4	0.927369	0.167240
50	12	11	5	0.985903	0.058534
50	12	11	6	0.998319	0.012416
50	12	11	7	0.999884	0.001565
50	12	11	8	0.999996	0.000112
50	12	11	9	1.000000	0.000004
50	12	11	10	1.000000	0.000000
50	12	11	11	1.000000	0.000000
50	12	12	0	0.022302	0.022302
50	12	12	1	0.141247	0.118945
50	12	12	2	0.398253	0.257006
50	12	12	3	0.693662	0.295409
50	12	12	4	0.893063	0.199401
50	12	12	5	0.975396	0.082333
50	12	12	6	0.996409	0.021012
50	12	12	7	0.999683	0.003275
50	12	12	8	0.999984	0.000301
50	12	12	9	1.000000	0.000015
50	12	12	10	1.000000	0.000000
50	12	12	11	1.000000	0.000000
50	12	12	12	1.000000	0.000000
50	13	1	0	0.740000	0.740000
50	13	1	1	1.000000	0.260000

N	n	k	x	P(x)	p(x)
50	13	2	0	0.543673	0.543673
50	13	2	1	0.936327	0.392653
50	13	2	2	1.000000	0.063673
50	13	3	0	0.396429	0.396429
50	13	3	1	0.838163	0.441735
50	13	3	2	0.985408	0.147245
50	13	3	3	1.000000	0.014592
50	13	4	0	0.286778	0.286778
50	13	4	1	0.725380	0.438602
50	13	4	2	0.950947	0.225567
50	13	4	3	0.996895	0.045949
50	13	4	4	1.000000	0.003105
50	13	5	0	0.205732	0.205732
50	13	5	1	0.610962	0.405230
50	13	5	2	0.897007	0.286045
50	13	5	3	0.986906	0.089900
50	13	5	4	0.999392	0.012486
50	13	5	5	1.000000	0.000607
50	13	6	0	0.146298	0.146298
50	13	6	1	0.502901	0.356602
50	13	6	2	0.827085	0.324184
50	13	6	3	0.966929	0.139844
50	13	6	4	0.996895	0.029967
50	13	6	5	0.999892	0.002997
50	13	6	6	1.000000	0.000108
50	13	7	0	0.103074	0.103074
50	13	7	1	0.405646	0.302572
50	13	7	2	0.746039	0.340393
50	13	7	3	0.935146	0.189107
50	13	7	4	0.990766	0.055620
50	13	7	5	0.999347	0.008581
50	13	7	6	0.999983	0.000636
50	13	7	7	1.000000	0.000017
50	13	8	0	0.071912	0.071912
50	13	8	1	0.321207	0.249295
50	13	8	2	0.658961	0.337754
50	13	8	3	0.891168	0.232206
50	13	8	4	0.979124	0.087957
50	13	8	5	0.997751	0.018626
50	13	8	6	0.999879	0.002129
50	13	8	7	0.999997	0.000118
50	13	8	8	1.000000	0.000002
50	13	9	0	0.049654	0.049654
50	13	9	1	0.249980	0.200326
50	13	9	2	0.570502	0.320522
50	13	9	3	0.835880	0.265378
50	13	9	4	0.960276	0.124396
50	13	9	5	0.994203	0.033926
50	13	9	6	0.999524	0.005322
50	13	9	7	0.999981	0.000456

Table for $N = 2$, $n = 1$, through $N = 100$, $n = 50$

N	n	k	x	$P(x)$	$p(x)$
50	13	9	8	1.000000	0.000019
50	13	9	9	1.000000	0.000000
50	13	10	0	0.033910	0.033910
50	13	10	1	0.191348	0.157438
50	13	10	2	0.484508	0.293160
50	13	10	3	0.771154	0.286646
50	13	10	4	0.932970	0.161816
50	13	10	5	0.987583	0.054613
50	13	10	6	0.998616	0.011033
50	13	10	7	0.999914	0.001298
50	13	10	8	0.999997	0.000083
50	13	10	9	1.000000	0.000003
50	13	10	10	1.000000	0.000000
50	13	11	0	0.022889	0.022889
50	13	11	1	0.144116	0.121227
50	13	11	2	0.403889	0.259773
50	13	11	3	0.699492	0.295603
50	13	11	4	0.896561	0.197069
50	13	11	5	0.976660	0.080099
50	13	11	6	0.996685	0.020025
50	13	11	7	0.999719	0.003034
50	13	11	8	0.999987	0.000268
50	13	11	9	1.000000	0.000013
50	13	11	10	1.000000	0.000000
50	13	11	11	1.000000	0.000000
50	13	12	0	0.015259	0.015259
50	13	12	1	0.106816	0.091556
50	13	12	2	0.330620	0.223804
50	13	12	3	0.623697	0.293077
50	13	12	4	0.851084	0.227387
50	13	12	5	0.960230	0.109146
50	13	12	6	0.993091	0.032861
50	13	12	7	0.999252	0.006161
50	13	12	8	0.999953	0.000700
50	13	12	9	0.999998	0.000046
50	13	12	10	1.000000	0.000002
50	13	12	11	1.000000	0.000000
50	13	12	12	1.000000	0.000000
50	13	13	0	0.010039	0.010039
50	13	13	1	0.077903	0.067864
50	13	13	2	0.265834	0.187931
50	13	13	3	0.546571	0.280737
50	13	13	4	0.797229	0.250658
50	13	13	5	0.937252	0.140023
50	13	13	6	0.987038	0.049786
50	13	13	7	0.998279	0.011242
50	13	13	8	0.999860	0.001581
50	13	13	9	0.999993	0.000133
50	13	13	10	1.000000	0.000006
50	13	13	11	1.000000	0.000000
50	13	13	12	1.000000	0.000000
50	13	13	13	1.000000	0.000000
50	14	1	0	0.720000	0.720000
50	14	1	1	1.000000	0.280000
50	14	2	0	0.514286	0.514286
50	14	2	1	0.925714	0.411429
50	14	2	2	1.000000	0.074286
50	14	3	0	0.364286	0.364286
50	14	3	1	0.814286	0.450000
50	14	3	2	0.981428	0.167143
50	14	3	3	1.000000	0.018571
50	14	4	0	0.255775	0.255775
50	14	4	1	0.689818	0.434043
50	14	4	2	0.938754	0.248936
50	14	4	3	0.995653	0.056900
50	14	4	4	1.000000	0.004346
50	14	5	0	0.177930	0.177930
50	14	5	1	0.567153	0.389223
50	14	5	2	0.873814	0.306661
50	14	5	3	0.982047	0.108233
50	14	5	4	0.999055	0.017008
50	14	5	5	1.000000	0.000945
50	14	6	0	0.122574	0.122574
50	14	6	1	0.454711	0.332137
50	14	6	2	0.792038	0.337327
50	14	6	3	0.955590	0.163552
50	14	6	4	0.995275	0.039685
50	14	6	5	0.999811	0.004535
50	14	6	6	1.000000	0.000189
50	14	7	0	0.083573	0.083573
50	14	7	1	0.356580	0.273006
50	14	7	2	0.700040	0.343460
50	14	7	3	0.914702	0.214662
50	14	7	4	0.986256	0.071554
50	14	7	5	0.998883	0.012627
50	14	7	6	0.999966	0.001082
50	14	7	7	1.000000	0.000034
50	14	8	0	0.056363	0.056363
50	14	8	1	0.274043	0.217680
50	14	8	2	0.604190	0.330147
50	14	8	3	0.859788	0.255598
50	14	8	4	0.969616	0.109827
50	14	8	5	0.996240	0.026625
50	14	8	6	0.999764	0.003524
50	14	8	7	0.999994	0.000230
50	14	8	8	1.000000	0.000006
50	14	9	0	0.037576	0.037576
50	14	9	1	0.206666	0.169090
50	14	9	2	0.509863	0.303197
50	14	9	3	0.792846	0.282983
50	14	9	4	0.943466	0.150620
50	14	9	5	0.990535	0.047069
50	14	9	6	0.999093	0.008558
50	14	9	7	0.999956	0.000863
50	14	9	8	0.999999	0.000043
50	14	9	9	1.000000	0.000001
50	14	10	0	0.024745	0.024745
50	14	10	1	0.153052	0.128307
50	14	10	2	0.421122	0.268070
50	14	10	3	0.716924	0.295802
50	14	10	4	0.906730	0.189806
50	14	10	5	0.980203	0.073473
50	14	10	6	0.997423	0.017220
50	14	10	7	0.999809	0.002385
50	14	10	8	0.999993	0.000184
50	14	10	9	1.000000	0.000007
50	14	10	10	1.000000	0.000000
50	14	11	0	0.016084	0.016084
50	14	11	1	0.111352	0.095268
50	14	11	2	0.340701	0.229349
50	14	11	3	0.635578	0.294877
50	14	11	4	0.859278	0.223700
50	14	11	5	0.963671	0.104393
50	14	11	6	0.993979	0.030308
50	14	11	7	0.999391	0.005412
50	14	11	8	0.999965	0.000574
50	14	11	9	0.999999	0.000034
50	14	11	10	1.000000	0.000001
50	14	11	11	1.000000	0.000000
50	14	12	0	0.010310	0.010310
50	14	12	1	0.079596	0.069286
50	14	12	2	0.270132	0.190536
50	14	12	3	0.552408	0.282276
50	14	12	4	0.801919	0.249511
50	14	12	5	0.939581	0.137661
50	14	12	6	0.987762	0.048182
50	14	12	7	0.998420	0.010658
50	14	12	8	0.999877	0.001457
50	14	12	9	0.999995	0.000118
50	14	12	10	1.000000	0.000005
50	14	12	11	1.000000	0.000000
50	14	12	12	1.000000	0.000000
50	14	13	0	0.006512	0.006512
50	14	13	1	0.055893	0.049381
50	14	13	2	0.209963	0.154070
50	14	13	3	0.470696	0.260733
50	14	13	4	0.736258	0.265562
50	14	13	5	0.906977	0.170718
50	14	13	6	0.977619	0.070642
50	14	13	7	0.996457	0.018838
50	14	13	8	0.999647	0.003190
50	14	13	9	0.999979	0.000332
50	14	13	10	0.999999	0.000020
50	14	13	11	1.000000	0.000001
50	14	13	12	1.000000	0.000000
50	14	13	13	1.000000	0.000000
50	14	14	0	0.004048	0.004048
50	14	14	1	0.038543	0.034495
50	14	14	2	0.159994	0.121451
50	14	14	3	0.393181	0.233187
50	14	14	4	0.664485	0.271304
50	14	14	5	0.865450	0.200966
50	14	14	6	0.962345	0.096894
50	14	14	7	0.992893	0.030548
50	14	14	8	0.999129	0.006237
50	14	14	9	0.999934	0.000805
50	14	14	10	0.999997	0.000063
50	14	14	11	1.000000	0.000003
50	14	14	12	1.000000	0.000000
50	14	14	13	1.000000	0.000000
50	14	14	14	1.000000	0.000000
50	15	1	0	0.700000	0.700000
50	15	1	1	1.000000	0.300000
50	15	2	0	0.485714	0.485714
50	15	2	1	0.914286	0.428571
50	15	2	2	1.000000	0.085714
50	15	3	0	0.333929	0.333929
50	15	3	1	0.789286	0.455357
50	15	3	2	0.976786	0.187500
50	15	3	3	1.000000	0.023214
50	15	4	0	0.227356	0.227356
50	15	4	1	0.653647	0.426292
50	15	4	2	0.924924	0.271277
50	15	4	3	0.994073	0.069149
50	15	4	4	1.000000	0.005927
50	15	5	0	0.153218	0.153218
50	15	5	1	0.523906	0.370688
50	15	5	2	0.848259	0.324352
50	15	5	3	0.976034	0.127775
50	15	5	4	0.998583	0.022549
50	15	5	5	1.000000	0.001417
50	15	6	0	0.102145	0.102145
50	15	6	1	0.408581	0.306436
50	15	6	2	0.754557	0.345976
50	15	6	3	0.941961	0.187404
50	15	6	4	0.993071	0.051110
50	15	6	5	0.999685	0.006614
50	15	6	6	1.000000	0.000315
50	15	7	0	0.067323	0.067323
50	15	7	1	0.311079	0.243756

Table for $N = 2$, $n = 1$, through $N = 100$, $n = 50$

N	n	k	x	P(x)	p(x)
50	15	7	2	0.652337	0.341258
50	15	7	3	0.890851	0.238514
50	15	7	4	0.980293	0.089443
50	15	7	5	0.998182	0.017889
50	15	7	6	0.999935	0.001754
50	15	7	7	1.000000	0.000064
50	15	8	0	0.043838	0.043838
50	15	8	1	0.231716	0.187878
50	15	8	2	0.549166	0.317449
50	15	8	3	0.824289	0.275123
50	15	8	4	0.957412	0.133124
50	15	8	5	0.994022	0.036609
50	15	8	6	0.999568	0.005547
50	15	8	7	0.999988	0.000420
50	15	8	8	1.000000	0.000012
50	15	9	0	0.028182	0.028182
50	15	9	1	0.169090	0.140909
50	15	9	2	0.450908	0.281817
50	15	9	3	0.745682	0.294774
50	15	9	4	0.922547	0.176865
50	15	9	5	0.985305	0.062758
50	15	9	6	0.998380	0.013075
50	15	9	7	0.999908	0.001528
50	15	9	8	0.999998	0.000090
50	15	9	9	1.000000	0.000002
50	15	10	0	0.017871	0.017871
50	15	10	1	0.120975	0.103104
50	15	10	2	0.361551	0.240576
50	15	10	3	0.659407	0.297856
50	15	10	4	0.875095	0.215689
50	15	10	5	0.969998	0.094903
50	15	10	6	0.995510	0.025512
50	15	10	7	0.999610	0.004100
50	15	10	8	0.999983	0.000373
50	15	10	9	1.000000	0.000017
50	15	10	10	1.000000	0.000000
50	15	11	0	0.011170	0.011170
50	15	11	1	0.084889	0.073719
50	15	11	2	0.283364	0.198475
50	15	11	3	0.570050	0.286686
50	15	11	4	0.815781	0.245731
50	15	11	5	0.946272	0.130492
50	15	11	6	0.989770	0.043497
50	15	11	7	0.998790	0.009020
50	15	11	8	0.999917	0.001128
50	15	11	9	0.999997	0.000080
50	15	11	10	1.000000	0.000003
50	15	11	11	1.000000	0.000000
50	15	12	0	0.006874	0.006874
50	15	12	1	0.058426	0.051552

N	n	k	x	P(x)	p(x)
50	15	12	2	0.217206	0.158780
50	15	12	3	0.481839	0.264633
50	15	12	4	0.746472	0.264633
50	15	12	5	0.912813	0.166341
50	15	12	6	0.979732	0.066919
50	15	12	7	0.996940	0.017208
50	15	12	8	0.999715	0.002775
50	15	12	9	0.999985	0.000270
50	15	12	10	1.000000	0.000015
50	15	12	11	1.000000	0.000000
50	15	12	12	1.000000	0.000000
50	15	13	0	0.004160	0.004160
50	15	13	1	0.039433	0.035272
50	15	13	2	0.162886	0.123453
50	15	13	3	0.398270	0.235384
50	15	13	4	0.669868	0.271597
50	15	13	5	0.869039	0.199171
50	15	13	6	0.963883	0.094844
50	15	13	7	0.993317	0.029434
50	15	13	8	0.999204	0.005887
50	15	13	9	0.999942	0.000738
50	15	13	10	0.999998	0.000055
50	15	13	11	1.000000	0.000002
50	15	13	12	1.000000	0.000000
50	15	13	13	1.000000	0.000000
50	15	14	0	0.002474	0.002474
50	15	14	1	0.026086	0.023613
50	15	14	2	0.119511	0.093424
50	15	14	3	0.321930	0.202419
50	15	14	4	0.589123	0.267193
50	15	14	5	0.815209	0.226086
50	15	14	6	0.940813	0.125604
50	15	14	7	0.986953	0.046140
50	15	14	8	0.998090	0.011137
50	15	14	9	0.999822	0.001732
50	15	14	10	0.999990	0.000168
50	15	14	11	1.000000	0.000010
50	15	14	12	1.000000	0.000000
50	15	14	13	1.000000	0.000000
50	15	14	14	1.000000	0.000000
50	15	15	0	0.001443	0.001443
50	15	15	1	0.016904	0.015461
50	15	15	2	0.085774	0.068870
50	15	15	3	0.254457	0.168682
50	15	15	4	0.507480	0.253024
50	15	15	5	0.752407	0.244927
50	15	15	6	0.909412	0.157004
50	15	15	7	0.976699	0.067288
50	15	15	8	0.995924	0.019225
50	15	15	9	0.999534	0.003609

N	n	k	x	P(x)	p(x)
50	15	15	10	0.999967	0.000433
50	15	15	11	0.999998	0.000032
50	15	15	12	1.000000	0.000001
50	15	15	13	1.000000	0.000000
50	15	15	14	1.000000	0.000000
50	15	15	15	1.000000	0.000000
50	16	1	0	0.680000	0.680000
50	16	1	1	1.000000	0.320000
50	16	2	0	0.457959	0.457959
50	16	2	1	0.902041	0.444082
50	16	2	2	1.000000	0.097959
50	16	3	0	0.305306	0.305306
50	16	3	1	0.763265	0.457959
50	16	3	2	0.971429	0.208163
50	16	3	3	1.000000	0.028571
50	16	4	0	0.201372	0.201372
50	16	4	1	0.617108	0.415736
50	16	4	2	0.909422	0.292314
50	16	4	3	0.992097	0.082675
50	16	4	4	1.000000	0.007903
50	16	5	0	0.131330	0.131330
50	16	5	1	0.481542	0.350212
50	16	5	2	0.820457	0.338915
50	16	5	3	0.968733	0.148275
50	16	5	4	0.997938	0.029206
50	16	5	5	1.000000	0.002062
50	16	6	0	0.084635	0.084635
50	16	6	1	0.364805	0.280170
50	16	6	2	0.715017	0.350212
50	16	6	3	0.925897	0.210881
50	16	6	4	0.990150	0.064253
50	16	6	5	0.999496	0.009346
50	16	6	6	1.000000	0.000504
50	16	7	0	0.053858	0.053858
50	16	7	1	0.269292	0.215434
50	16	7	2	0.603586	0.334294
50	16	7	3	0.863592	0.260006
50	16	7	4	0.972627	0.109035
50	16	7	5	0.997160	0.024533
50	16	7	6	0.999885	0.002726
50	16	7	7	1.000000	0.000115
50	16	8	0	0.033818	0.033818
50	16	8	1	0.194141	0.160323
50	16	8	2	0.494746	0.300605
50	16	8	3	0.784985	0.290239
50	16	8	4	0.942198	0.157213
50	16	8	5	0.990884	0.048685
50	16	8	6	0.999251	0.008368
50	16	8	7	0.999976	0.000724
50	16	8	8	1.000000	0.000024

N	n	k	x	P(x)	p(x)
50	16	9	0	0.020935	0.020935
50	16	9	1	0.136883	0.115948
50	16	9	2	0.394544	0.257662
50	16	9	3	0.695149	0.300605
50	16	9	4	0.897280	0.202131
50	16	9	5	0.978133	0.080852
50	16	9	6	0.997259	0.019126
50	16	9	7	0.999821	0.002562
50	16	9	8	0.999995	0.000175
50	16	9	9	1.000000	0.000005
50	16	10	0	0.012765	0.012765
50	16	10	1	0.094463	0.081698
50	16	10	2	0.306562	0.212099
50	16	10	3	0.599836	0.293273
50	16	10	4	0.838120	0.238285
50	16	10	5	0.956441	0.118321
50	16	10	6	0.992594	0.036154
50	16	10	7	0.999258	0.006664
50	16	10	8	0.999961	0.000703
50	16	10	9	0.999999	0.000038
50	16	10	10	1.000000	0.000001
50	16	11	0	0.007659	0.007659
50	16	11	1	0.063826	0.056167
50	16	11	2	0.232327	0.168501
50	16	11	3	0.504522	0.272194
50	16	11	4	0.766635	0.262113
50	16	11	5	0.923902	0.157268
50	16	11	6	0.983556	0.059653
50	16	11	7	0.997759	0.014203
50	16	11	8	0.999821	0.002062
50	16	11	9	0.999992	0.000172
50	16	11	10	1.000000	0.000007
50	16	11	11	1.000000	0.000000
50	16	12	0	0.004517	0.004517
50	16	12	1	0.042223	0.037707
50	16	12	2	0.171840	0.129616
50	16	12	3	0.413790	0.241950
50	16	12	4	0.685985	0.272194
50	16	12	5	0.879545	0.193560
50	16	12	6	0.968260	0.088715
50	16	12	7	0.994481	0.026221
50	16	12	8	0.999398	0.004916
50	16	12	9	0.999962	0.000564
50	16	12	10	0.999999	0.000037
50	16	12	11	1.000000	0.000001
50	16	12	12	1.000000	0.000000
50	16	13	0	0.002615	0.002615
50	16	13	1	0.027339	0.024724
50	16	13	2	0.124086	0.096747
50	16	13	3	0.331018	0.206931

Table for $N = 2$, $n = 1$, through $N = 100$, $n = 50$

N	n	k	x	P(x)	p(x)
50	16	13	4	0.600028	0.269011
50	16	13	5	0.823514	0.223486
50	16	13	6	0.944914	0.121400
50	16	13	7	0.988271	0.043357
50	16	13	8	0.998363	0.010092
50	16	13	9	0.999858	0.001495
50	16	13	10	0.999993	0.000135
50	16	13	11	1.000000	0.000007
50	16	13	12	1.000000	0.000000
50	16	13	13	1.000000	0.000000
50	16	14	0	0.001484	0.001484
50	16	14	1	0.017316	0.015832
50	16	14	2	0.087479	0.070163
50	16	14	3	0.258312	0.170833
50	16	14	4	0.512782	0.254470
50	16	14	5	0.757073	0.244291
50	16	14	6	0.912103	0.155031
50	16	14	7	0.977725	0.065621
50	16	14	8	0.996181	0.018456
50	16	14	9	0.999575	0.003394
50	16	14	10	0.999971	0.000396
50	16	14	11	0.999999	0.000028
50	16	14	12	1.000000	0.000001
50	16	14	13	1.000000	0.000000
50	16	14	14	1.000000	0.000000
50	16	15	0	0.000825	0.000825
50	16	15	1	0.010719	0.009895
50	16	15	2	0.060194	0.049474
50	16	15	3	0.196623	0.136429
50	16	15	4	0.427959	0.231336
50	16	15	5	0.682428	0.254470
50	16	15	6	0.869039	0.186611
50	16	15	7	0.961319	0.092280
50	16	15	8	0.992079	0.030760
50	16	15	9	0.998915	0.006836
50	16	15	10	0.999905	0.000990
50	16	15	11	0.999995	0.000090
50	16	15	12	1.000000	0.000005
50	16	15	13	1.000000	0.000000
50	16	15	14	1.000000	0.000000
50	16	15	15	1.000000	0.000000
50	16	16	0	0.000448	0.000448
50	16	16	1	0.006479	0.006031
50	16	16	2	0.040404	0.033925
50	16	16	3	0.145949	0.105545
50	16	16	4	0.348643	0.202694
50	16	16	5	0.602452	0.253809
50	16	16	6	0.815722	0.213270
50	16	16	7	0.937590	0.121868
50	16	16	8	0.985048	0.047458
50	16	16	9	0.997548	0.012499
50	16	16	10	0.999735	0.002187
50	16	16	11	0.999982	0.000247
50	16	16	12	0.999999	0.000017
50	16	16	13	1.000000	0.000001
50	16	16	14	1.000000	0.000000
50	16	16	15	1.000000	0.000000
50	16	16	16	1.000000	0.000000
50	17	1	0	0.660000	0.660000
50	17	1	1	1.000000	0.340000
50	17	2	0	0.431020	0.431020
50	17	2	1	0.888980	0.457959
50	17	2	2	1.000000	0.111020
50	17	3	0	0.278367	0.278367
50	17	3	1	0.736326	0.457959
50	17	3	2	0.965306	0.228980
50	17	3	3	1.000000	0.034694
50	17	4	0	0.177681	0.177681
50	17	4	1	0.580426	0.402744
50	17	4	2	0.892227	0.311802
50	17	4	3	0.989666	0.097438
50	17	4	4	1.000000	0.010334
50	17	5	0	0.112016	0.112016
50	17	5	1	0.440341	0.328324
50	17	5	2	0.790553	0.350212
50	17	5	3	0.960010	0.169458
50	17	5	4	0.997079	0.037069
50	17	5	5	1.000000	0.002921
50	17	6	0	0.069699	0.069699
50	17	6	1	0.323603	0.253904
50	17	6	2	0.673815	0.350212
50	17	6	3	0.907290	0.233475
50	17	6	4	0.986370	0.079080
50	17	6	5	0.999221	0.012851
50	17	6	6	1.000000	0.000779
50	17	7	0	0.042770	0.042770
50	17	7	1	0.231274	0.188504
50	17	7	2	0.554425	0.323151
50	17	7	3	0.833003	0.278578
50	17	7	4	0.963006	0.130003
50	17	7	5	0.995716	0.032710
50	17	7	6	0.999805	0.004089
50	17	7	7	1.000000	0.000195
50	17	8	0	0.025861	0.025861
50	17	8	1	0.161133	0.135272
50	17	8	2	0.441698	0.280565
50	17	8	3	0.742303	0.300605
50	17	8	4	0.923703	0.181400
50	17	8	5	0.986588	0.062885
50	17	8	6	0.998759	0.012171
50	17	8	7	0.999955	0.001195
50	17	8	8	1.000000	0.000045
50	17	9	0	0.015393	0.015393
50	17	9	1	0.109601	0.094208
50	17	9	2	0.341496	0.231895
50	17	9	3	0.642101	0.300605
50	17	9	4	0.867555	0.225454
50	17	9	5	0.968621	0.101065
50	17	9	6	0.995572	0.026951
50	17	9	7	0.999670	0.004099
50	17	9	8	0.999990	0.000320
50	17	9	9	1.000000	0.000010
50	17	10	0	0.009011	0.009011
50	17	10	1	0.072837	0.063826
50	17	10	2	0.256656	0.183819
50	17	10	3	0.539456	0.282799
50	17	10	4	0.796070	0.256614
50	17	10	5	0.939041	0.142971
50	17	10	6	0.988341	0.049300
50	17	10	7	0.998670	0.010330
50	17	10	8	0.999920	0.001250
50	17	10	9	0.999998	0.000078
50	17	10	10	1.000000	0.000002
50	17	11	0	0.005181	0.005181
50	17	11	1	0.047306	0.042125
50	17	11	2	0.187724	0.140418
50	17	11	3	0.440476	0.252752
50	17	11	4	0.712670	0.272194
50	17	11	5	0.896149	0.183479
50	17	11	6	0.974783	0.078634
50	17	11	7	0.996088	0.021305
50	17	11	8	0.999639	0.003551
50	17	11	9	0.999982	0.000344
50	17	11	10	1.000000	0.000017
50	17	11	11	1.000000	0.000000
50	17	12	0	0.002923	0.002923
50	17	12	1	0.030024	0.027102
50	17	12	2	0.133717	0.103693
50	17	12	3	0.349745	0.216027
50	17	12	4	0.621939	0.272194
50	17	12	5	0.839694	0.217755
50	17	12	6	0.952604	0.112910
50	17	12	7	0.990625	0.038021
50	17	12	8	0.998819	0.008194
50	17	12	9	0.999912	0.001093
50	17	12	10	0.999996	0.000085
50	17	12	11	1.000000	0.000003
50	17	12	12	1.000000	0.000000
50	17	13	0	0.001615	0.001615
50	17	13	1	0.018613	0.016998
50	17	13	2	0.092786	0.074173
50	17	13	3	0.270156	0.177370
50	17	13	4	0.528820	0.258664
50	17	13	5	0.770929	0.242110
50	17	13	6	0.919920	0.148991
50	17	13	7	0.980620	0.060700
50	17	13	8	0.996879	0.016259
50	17	13	9	0.999682	0.002803
50	17	13	10	0.999981	0.000299
50	17	13	11	0.999999	0.000018
50	17	13	12	1.000000	0.000001
50	17	13	13	1.000000	0.000000
50	17	14	0	0.000873	0.000873
50	17	14	1	0.011263	0.010390
50	17	14	2	0.062716	0.051453
50	17	14	3	0.203043	0.140327
50	17	14	4	0.437938	0.234895
50	17	14	5	0.692407	0.254470
50	17	14	6	0.875625	0.183218
50	17	14	7	0.964214	0.088589
50	17	14	8	0.992924	0.028709
50	17	14	9	0.999076	0.006152
50	17	14	10	0.999924	0.000849
50	17	14	11	0.999996	0.000072
50	17	14	12	1.000000	0.000003
50	17	14	13	1.000000	0.000000
50	17	14	14	1.000000	0.000000
50	17	15	0	0.000461	0.000461
50	17	15	1	0.006645	0.006184
50	17	15	2	0.041277	0.034632
50	17	15	3	0.148471	0.107194
50	17	15	4	0.353115	0.204643
50	17	15	5	0.607584	0.254470
50	17	15	6	0.819642	0.212058
50	17	15	7	0.939606	0.119964
50	17	15	8	0.985746	0.046140
50	17	15	9	0.997709	0.011962
50	17	15	10	0.999759	0.002051
50	17	15	11	0.999984	0.000225
50	17	15	12	0.999999	0.000015
50	17	15	13	1.000000	0.000001
50	17	15	14	1.000000	0.000000
50	17	15	15	1.000000	0.000000
50	17	16	0	0.000237	0.000237
50	17	16	1	0.003818	0.003581
50	17	16	2	0.026435	0.022617
50	17	16	3	0.105594	0.079159
50	17	16	4	0.277104	0.171511
50	17	16	5	0.520337	0.243233
50	17	16	6	0.752995	0.232658

Table for $N = 2$, $n = 1$, through $N = 100$, $n = 50$

N	n	k	x	P(x)	p(x)
50	17	16	7	0.905331	0.152336
50	17	16	8	0.973882	0.068551
50	17	16	9	0.994974	0.021093
50	17	16	10	0.999349	0.004375
50	17	16	11	0.999946	0.000597
50	17	16	12	0.999997	0.000051
50	17	16	13	1.000000	0.000003
50	17	16	14	1.000000	0.000000
50	17	16	15	1.000000	0.000000
50	17	16	16	1.000000	0.000000
50	17	17	0	0.000118	0.000118
50	17	17	1	0.002133	0.002014
50	17	17	2	0.016457	0.014324
50	17	17	3	0.072999	0.056542
50	17	17	4	0.211527	0.138528
50	17	17	5	0.434490	0.222964
50	17	17	6	0.677724	0.243233
50	17	17	7	0.860526	0.182803
50	17	17	8	0.955736	0.095210
50	17	17	9	0.990011	0.034275
50	17	17	10	0.998449	0.008437
50	17	17	11	0.999840	0.001392
50	17	17	12	0.999990	0.000149
50	17	17	13	1.000000	0.000010
50	17	17	14	1.000000	0.000000
50	17	17	15	1.000000	0.000000
50	17	17	16	1.000000	0.000000
50	17	17	17	1.000000	0.000000
50	18	1	0	0.640000	0.640000
50	18	1	1	1.000000	0.360000
50	18	2	0	0.404898	0.404898
50	18	2	1	0.875102	0.470204
50	18	2	2	1.000000	0.124898
50	18	3	0	0.253061	0.253061
50	18	3	1	0.708571	0.455510
50	18	3	2	0.958367	0.249796
50	18	3	3	1.000000	0.041633
50	18	4	0	0.156144	0.156144
50	18	4	1	0.543812	0.387668
50	18	4	2	0.873330	0.329518
50	18	4	3	0.986713	0.113383
50	18	4	4	1.000000	0.013287
50	18	5	0	0.095044	0.095044
50	18	5	1	0.400544	0.305499
50	18	5	2	0.758715	0.358172
50	18	5	3	0.949740	0.191025
50	18	5	4	0.995956	0.046216
50	18	5	5	1.000000	0.004044
50	18	6	0	0.057027	0.057027
50	18	6	1	0.285133	0.228106

N	n	k	x	P(x)	p(x)
50	18	6	2	0.631365	0.346233
50	18	6	3	0.886065	0.254700
50	18	6	4	0.981578	0.095512
50	18	6	5	0.998832	0.017254
50	18	6	6	1.000000	0.001168
50	18	7	0	0.033698	0.033698
50	18	7	1	0.197001	0.163303
50	18	7	2	0.505463	0.308462
50	18	7	3	0.799236	0.293773
50	18	7	4	0.951188	0.151952
50	18	7	5	0.993734	0.042546
50	18	7	6	0.999681	0.005947
50	18	7	7	1.000000	0.000319
50	18	8	0	0.019592	0.019592
50	18	8	1	0.132439	0.112847
50	18	8	2	0.390686	0.258247
50	18	8	3	0.696757	0.306071
50	18	8	4	0.901715	0.204958
50	18	8	5	0.980871	0.079156
50	18	8	6	0.998022	0.017151
50	18	8	7	0.999918	0.001897
50	18	8	8	1.000000	0.000082
50	18	9	0	0.011195	0.011195
50	18	9	1	0.086763	0.075567
50	18	9	2	0.292306	0.205544
50	18	9	3	0.587446	0.295140
50	18	9	4	0.833396	0.245950
50	18	9	5	0.956370	0.122975
50	18	9	6	0.993121	0.036751
50	18	9	7	0.999422	0.006300
50	18	9	8	0.999981	0.000559
50	18	9	9	1.000000	0.000019
50	18	10	0	0.006280	0.006280
50	18	10	1	0.055430	0.049150
50	18	10	2	0.212094	0.156664
50	18	10	3	0.479468	0.267374
50	18	10	4	0.749413	0.269945
50	18	10	5	0.917378	0.167966
50	18	10	6	0.982365	0.064987
50	18	10	7	0.997731	0.015366
50	18	10	8	0.999844	0.002113
50	18	10	9	0.999996	0.000151
50	18	10	10	1.000000	0.000004
50	18	11	0	0.003454	0.003454
50	18	11	1	0.034541	0.031087
50	18	11	2	0.149428	0.114887
50	18	11	3	0.379203	0.229774
50	18	11	4	0.654932	0.275729
50	18	11	5	0.862790	0.207857
50	18	11	6	0.962869	0.100080

N	n	k	x	P(x)	p(x)
50	18	11	7	0.993506	0.030637
50	18	11	8	0.999316	0.005810
50	18	11	9	0.999962	0.000646
50	18	11	10	0.999999	0.000037
50	18	11	11	1.000000	0.000001
50	18	12	0	0.001860	0.001860
50	18	12	1	0.020990	0.019131
50	18	12	2	0.102295	0.081305
50	18	12	3	0.290828	0.188533
50	18	12	4	0.555952	0.265124
50	18	12	5	0.793504	0.237551
50	18	12	6	0.932075	0.138572
50	18	12	7	0.984864	0.052789
50	18	12	8	0.997826	0.012962
50	18	12	9	0.999813	0.001986
50	18	12	10	0.999991	0.000179
50	18	12	11	1.000000	0.000008
50	18	12	12	1.000000	0.000000
50	18	13	0	0.000979	0.000979
50	18	13	1	0.012432	0.011453
50	18	13	2	0.068062	0.055630
50	18	13	3	0.216407	0.148346
50	18	13	4	0.458275	0.241868
50	18	13	5	0.712236	0.253961
50	18	13	6	0.888316	0.176080
50	18	13	7	0.969583	0.081268
50	18	13	8	0.994415	0.024832
50	18	13	9	0.999342	0.004927
50	18	13	10	0.999954	0.000612
50	18	13	11	0.999998	0.000044
50	18	13	12	1.000000	0.000002
50	18	13	13	1.000000	0.000000
50	18	14	0	0.000503	0.000503
50	18	14	1	0.007170	0.006667
50	18	14	2	0.044006	0.036836
50	18	14	3	0.156267	0.112261
50	18	14	4	0.366757	0.210490
50	18	14	5	0.623006	0.256249
50	18	14	6	0.831209	0.208202
50	18	14	7	0.945423	0.114214
50	18	14	8	0.987704	0.042281
50	18	14	9	0.998144	0.010440
50	18	14	10	0.999821	0.001678
50	18	14	11	0.999990	0.000168
50	18	14	12	1.000000	0.000010
50	18	14	13	1.000000	0.000000
50	18	14	14	1.000000	0.000000
50	18	15	0	0.000251	0.000251
50	18	15	1	0.004021	0.003770
50	18	15	2	0.027634	0.023613

N	n	k	x	P(x)	p(x)
50	18	15	3	0.109491	0.081857
50	18	15	4	0.284900	0.175409
50	18	15	5	0.530472	0.245572
50	18	15	6	0.761808	0.231336
50	18	15	7	0.910524	0.148716
50	18	15	8	0.975959	0.065435
50	18	15	9	0.995534	0.019575
50	18	15	10	0.999449	0.003915
50	18	15	11	0.999957	0.000508
50	18	15	12	0.999998	0.000041
50	18	15	13	1.000000	0.000002
50	18	15	14	1.000000	0.000000
50	18	15	15	1.000000	0.000000
50	18	16	0	0.000122	0.000122
50	18	16	1	0.002190	0.002068
50	18	16	2	0.016840	0.014650
50	18	16	3	0.074410	0.057570
50	18	16	4	0.214737	0.140327
50	18	16	5	0.439260	0.224523
50	18	16	6	0.682493	0.243233
50	18	16	7	0.863785	0.181292
50	18	16	8	0.957263	0.093479
50	18	16	9	0.990500	0.033237
50	18	16	10	0.998554	0.008054
50	18	16	11	0.999855	0.001302
50	18	16	12	0.999991	0.000136
50	18	16	13	1.000000	0.000009
50	18	16	14	1.000000	0.000000
50	18	16	15	1.000000	0.000000
50	18	16	16	1.000000	0.000000
50	18	17	0	0.000057	0.000057
50	18	17	1	0.001156	0.001099
50	18	17	2	0.009946	0.008790
50	18	17	3	0.049011	0.039065
50	18	17	4	0.156955	0.107944
50	18	17	5	0.353413	0.196458
50	18	17	6	0.596646	0.243233
50	18	17	7	0.805132	0.208486
50	18	17	8	0.929770	0.124638
50	18	17	9	0.981702	0.051933
50	18	17	10	0.996659	0.014957
50	18	17	11	0.999587	0.002929
50	18	17	12	0.999967	0.000380
50	18	17	13	0.999998	0.000031
50	18	17	14	1.000000	0.000002
50	18	17	15	1.000000	0.000000
50	18	17	16	1.000000	0.000000
50	18	17	17	1.000000	0.000000
50	18	18	0	0.000026	0.000026
50	18	18	1	0.000590	0.000564

Table for $N = 2$, $n = 1$, through $N = 100$, $n = 50$

N	n	k	x	P(x)	p(x)
50	18	18	2	0.005684	0.005094
50	18	18	3	0.031254	0.025570
50	18	18	4	0.111161	0.079906
50	18	18	5	0.276020	0.164860
50	18	18	6	0.508197	0.232177
50	18	18	7	0.735636	0.227439
50	18	18	8	0.892001	0.156364
50	18	18	9	0.967539	0.075538
50	18	18	10	0.993033	0.025494
50	18	18	11	0.998966	0.005933
50	18	18	12	0.999898	0.000932
50	18	18	13	0.999994	0.000096
50	18	18	14	1.000000	0.000006
50	18	18	15	1.000000	0.000000
50	18	18	16	1.000000	0.000000
50	18	18	17	1.000000	0.000000
50	18	18	18	1.000000	0.000000
50	19	1	0	0.620000	0.620000
50	19	1	1	1.000000	0.380000
50	19	2	0	0.379592	0.379592
50	19	2	1	0.860408	0.480816
50	19	2	2	1.000000	0.139592
50	19	3	0	0.229337	0.229337
50	19	3	1	0.680102	0.450765
50	19	3	2	0.950561	0.270459
50	19	3	3	1.000000	0.049439
50	19	4	0	0.136626	0.136626
50	19	4	1	0.507468	0.370842
50	19	4	2	0.852735	0.345267
50	19	4	3	0.983170	0.130434
50	19	4	4	1.000000	0.016830
50	19	5	0	0.080194	0.080194
50	19	5	1	0.362356	0.282163
50	19	5	2	0.725137	0.362781
50	19	5	3	0.937801	0.212664
50	19	5	4	0.994512	0.056711
50	19	5	5	1.000000	0.005488
50	19	6	0	0.046334	0.046334
50	19	6	1	0.249491	0.203157
50	19	6	2	0.588086	0.338595
50	19	6	3	0.862187	0.274101
50	19	6	4	0.975608	0.113421
50	19	6	5	0.998292	0.022684
50	19	6	6	1.000000	0.001707
50	19	7	0	0.026326	0.026326
50	19	7	1	0.166381	0.140055
50	19	7	2	0.457266	0.290884
50	19	7	3	0.762514	0.305249
50	19	7	4	0.936942	0.174428
50	19	7	5	0.991075	0.054133

N	n	k	x	P(x)	p(x)
50	19	7	6	0.999495	0.008421
50	19	7	7	1.000000	0.000504
50	19	8	0	0.014694	0.014694
50	19	8	1	0.107754	0.093060
50	19	8	2	0.342265	0.234511
50	19	8	3	0.648933	0.306668
50	19	8	4	0.876095	0.227162
50	19	8	5	0.973450	0.097355
50	19	8	6	0.996950	0.023499
50	19	8	7	0.999859	0.002909
50	19	8	8	1.000000	0.000141
50	19	9	0	0.008047	0.008047
50	19	9	1	0.067871	0.059824
50	19	9	2	0.247344	0.179473
50	19	9	3	0.532107	0.284764
50	19	9	4	0.794966	0.262859
50	19	9	5	0.940999	0.146033
50	19	9	6	0.989676	0.048678
50	19	9	7	0.999028	0.009352
50	19	9	8	0.999963	0.000935
50	19	9	9	1.000000	0.000037
50	19	10	0	0.004318	0.004318
50	19	10	1	0.041606	0.037289
50	19	10	2	0.172928	0.131322
50	19	10	3	0.420980	0.248052
50	19	10	4	0.698798	0.277818
50	19	10	5	0.891134	0.192336
50	19	10	6	0.974242	0.083108
50	19	10	7	0.996291	0.022049
50	19	10	8	0.999712	0.003421
50	19	10	9	0.999991	0.000279
50	19	10	10	1.000000	0.000009
50	19	11	0	0.002267	0.002267
50	19	11	1	0.024827	0.022560
50	19	11	2	0.117116	0.092290
50	19	11	3	0.321759	0.204643
50	19	11	4	0.594616	0.272857
50	19	11	5	0.823816	0.229200
50	19	11	6	0.947232	0.123415
50	19	11	7	0.989676	0.042444
50	19	11	8	0.998771	0.009095
50	19	11	9	0.999921	0.001150
50	19	11	10	0.999998	0.000077
50	19	11	11	1.000000	0.000002
50	19	12	0	0.001162	0.001162
50	19	12	1	0.014414	0.013252
50	19	12	2	0.076887	0.062473
50	19	12	3	0.237803	0.160916
50	19	12	4	0.489671	0.251868
50	19	12	5	0.741539	0.251868

N	n	k	x	P(x)	p(x)
50	19	12	6	0.906093	0.164554
50	19	12	7	0.976616	0.070523
50	19	12	8	0.996206	0.019590
50	19	12	9	0.999626	0.003420
50	19	12	10	0.999980	0.000354
50	19	12	11	1.000000	0.000019
50	19	12	12	1.000000	0.000000
50	19	13	0	0.000581	0.000581
50	19	13	-1	0.008137	0.007556
50	19	13	2	0.048939	0.040802
50	19	13	3	0.170049	0.121110
50	19	13	4	0.390250	0.220200
50	19	13	5	0.648746	0.258496
50	19	13	6	0.849798	0.201053
50	19	13	7	0.954346	0.104547
50	19	13	8	0.990535	0.036189
50	19	13	9	0.998726	0.008191
50	19	13	10	0.999896	0.001170
50	19	13	11	0.999995	0.000099
50	19	13	12	1.000000	0.000004
50	19	13	13	1.000000	0.000000
50	19	14	0	0.000283	0.000283
50	19	14	1	0.004461	0.004179
50	19	14	2	0.030192	0.025731
50	19	14	3	0.117677	0.087485
50	19	14	4	0.300979	0.183302
50	19	14	5	0.550936	0.249957
50	19	14	6	0.779158	0.228222
50	19	14	7	0.920439	0.141280
50	19	14	8	0.979776	0.059338
50	19	14	9	0.996512	0.016736
50	19	14	10	0.999612	0.003099
50	19	14	11	0.999974	0.000362
50	19	14	12	0.999999	0.000025
50	19	14	13	1.000000	0.000001
50	19	14	14	1.000000	0.000000
50	19	15	0	0.000134	0.000134
50	19	15	1	0.002372	0.002238
50	19	15	2	0.018041	0.015669
50	19	15	3	0.078795	0.060754
50	19	15	4	0.224603	0.145808
50	19	15	5	0.453731	0.229127
50	19	15	6	0.696745	0.243014
50	19	15	7	0.873345	0.176600
50	19	15	8	0.961645	0.088300
50	19	15	9	0.991863	0.030218
50	19	15	10	0.998837	0.006973
50	19	15	11	0.999894	0.001057
50	19	15	12	0.999994	0.000101
50	19	15	13	1.000000	0.000006

N	n	k	x	P(x)	p(x)
50	19	15	14	1.000000	0.000000
50	19	15	15	1.000000	0.000000
50	19	16	0	0.000061	0.000061
50	19	16	1	0.001221	0.001160
50	19	16	2	0.010431	0.009210
50	19	16	3	0.051022	0.040591
50	19	16	4	0.162114	0.111092
50	19	16	5	0.362080	0.199966
50	19	16	6	0.606482	0.244403
50	19	16	7	0.812796	0.206314
50	19	16	8	0.933894	0.121097
50	19	16	9	0.983230	0.049336
50	19	16	10	0.997044	0.013814
50	19	16	11	0.999652	0.002608
50	19	16	12	0.999974	0.000322
50	19	16	13	0.999999	0.000025
50	19	16	14	1.000000	0.000001
50	19	16	15	1.000000	0.000000
50	19	16	16	1.000000	0.000000
50	19	17	0	0.000027	0.000027
50	19	17	1	0.000607	0.000580
50	19	17	2	0.005826	0.005219
50	19	17	3	0.031920	0.026094
50	19	17	4	0.113103	0.081183
50	19	17	5	0.279741	0.166638
50	19	17	6	0.513034	0.233293
50	19	17	7	0.739980	0.226945
50	19	17	8	0.894715	0.154735
50	19	17	9	0.968719	0.074004
50	19	17	10	0.993387	0.024668
50	19	17	11	0.999038	0.005651
50	19	17	12	0.999908	0.000869
50	19	17	13	0.999994	0.000087
50	19	17	14	1.000000	0.000005
50	19	17	15	1.000000	0.000000
50	19	17	16	1.000000	0.000000
50	19	17	17	1.000000	0.000000
50	19	18	0	0.000011	0.000011
50	19	18	1	0.000291	0.000279
50	19	18	2	0.003137	0.002847
50	19	18	3	0.019268	0.016131
50	19	18	4	0.076202	0.056933
50	19	18	5	0.209046	0.132844
50	19	18	6	0.421131	0.212085
50	19	18	7	0.657454	0.236323
50	19	18	8	0.843137	0.185683
50	19	18	9	0.946294	0.103157
50	19	18	10	0.986659	0.040366
50	19	18	11	0.997668	0.011009
50	19	18	12	0.999723	0.002055

N	n	k	x	P(x)	p(x)
50	19	18	13	0.999979	0.000255
50	19	18	14	0.999999	0.000020
50	19	18	15	1.000000	0.000001
50	19	18	16	1.000000	0.000000
50	19	18	17	1.000000	0.000000
50	19	18	18	1.000000	0.000000
50	19	19	0	0.000005	0.000005
50	19	19	1	0.000134	0.000129
50	19	19	2	0.001625	0.001491
50	19	19	3	0.011203	0.009578
50	19	19	4	0.049514	0.038311
50	19	19	5	0.150927	0.101412
50	19	19	6	0.334971	0.184045
50	19	19	7	0.568833	0.233862
50	19	19	8	0.779308	0.210475
50	19	19	9	0.914057	0.134749
50	19	19	10	0.975307	0.061249
50	19	19	11	0.994916	0.019610
50	19	19	12	0.999274	0.004358
50	19	19	13	0.999931	0.000657
50	19	19	14	0.999996	0.000065
50	19	19	15	1.000000	0.000004
50	19	19	16	1.000000	0.000000
50	19	19	17	1.000000	0.000000
50	19	19	18	1.000000	0.000000
50	19	19	19	1.000000	0.000000
50	20	1	0	0.600000	0.600000
50	20	1	1	1.000000	0.400000
50	20	2	0	0.355102	0.355102
50	20	2	1	0.844898	0.489796
50	20	2	2	1.000000	0.155102
50	20	3	0	0.207143	0.207143
50	20	3	1	0.651020	0.443878
50	20	3	2	0.941837	0.290816
50	20	3	3	1.000000	0.058163
50	20	4	0	0.118997	0.118997
50	20	4	1	0.471581	0.352584
50	20	4	2	0.830460	0.358880
50	20	4	3	0.978962	0.148502
50	20	4	4	1.000000	0.021038
50	20	5	0	0.067259	0.067259
50	20	5	1	0.325948	0.258689
50	20	5	2	0.690029	0.364081
50	20	5	3	0.924081	0.234052
50	20	5	4	0.992682	0.068601
50	20	5	5	1.000000	0.007317
50	20	6	0	0.037366	0.037366
50	20	6	1	0.216724	0.179358
50	20	6	2	0.544397	0.327673
50	20	6	3	0.835661	0.291265
50	20	6	4	0.968291	0.132629
50	20	6	5	0.997561	0.029270
50	20	6	6	1.000000	0.002439
50	20	7	0	0.020382	0.020382
50	20	7	1	0.139274	0.118892
50	20	7	2	0.410349	0.271075
50	20	7	3	0.723127	0.312779
50	20	7	4	0.920062	0.196935
50	20	7	5	0.987582	0.067520
50	20	7	6	0.999224	0.011641
50	20	7	7	1.000000	0.000776
50	20	8	0	0.010902	0.010902
50	20	8	1	0.086740	0.075838
50	20	8	2	0.296876	0.210135
50	20	8	3	0.599471	0.302595
50	20	8	4	0.846784	0.247313
50	20	8	5	0.964029	0.117245
50	20	8	6	0.995434	0.031405
50	20	8	7	0.999765	0.004332
50	20	8	8	1.000000	0.000235
50	20	9	0	0.005710	0.005710
50	20	9	1	0.052432	0.046722
50	20	9	2	0.206818	0.154385
50	20	9	3	0.476992	0.270174
50	20	9	4	0.752569	0.275578
50	20	9	5	0.922156	0.169586
50	20	9	6	0.984965	0.062810
50	20	9	7	0.998425	0.013459
50	20	9	8	0.999933	0.001508
50	20	9	9	1.000000	0.000067
50	20	10	0	0.002925	0.002925
50	20	10	1	0.030781	0.027856
50	20	10	2	0.139039	0.108258
50	20	10	3	0.364968	0.225930
50	20	10	4	0.645027	0.280059
50	20	10	5	0.860112	0.215085
50	20	10	6	0.963518	0.103406
50	20	10	7	0.994157	0.030639
50	20	10	8	0.999491	0.005334
50	20	10	9	0.999982	0.000491
50	20	10	10	1.000000	0.000018
50	20	11	0	0.001462	0.001462
50	20	11	1	0.017549	0.016087
50	20	11	2	0.090323	0.072773
50	20	11	3	0.268948	0.178626
50	20	11	4	0.533003	0.264055
50	20	11	5	0.779455	0.246452
50	20	11	6	0.927326	0.147871
50	20	11	7	0.984199	0.056873
50	20	11	8	0.997891	0.013692
50	20	11	9	0.999847	0.001956
50	20	11	10	0.999995	0.000148
50	20	11	11	1.000000	0.000004
50	20	12	0	0.000712	0.000712
50	20	12	1	0.009712	0.009000
50	20	12	2	0.056735	0.047023
50	20	12	3	0.191086	0.134351
50	20	12	4	0.424673	0.233587
50	20	12	5	0.684666	0.259993
50	20	12	6	0.874244	0.189578
50	20	12	7	0.965241	0.090997
50	20	12	8	0.993678	0.028437
50	20	12	9	0.999295	0.005617
50	20	12	10	0.999957	0.000662
50	20	12	11	0.999999	0.000041
50	20	12	12	1.000000	0.000001
50	20	13	0	0.000337	0.000337
50	20	13	1	0.005212	0.004875
50	20	13	2	0.034461	0.029249
50	20	13	3	0.130981	0.096521
50	20	13	4	0.326321	0.195339
50	20	13	5	0.582037	0.255717
50	20	13	6	0.804400	0.222362
50	20	13	7	0.934111	0.129711
50	20	13	8	0.984698	0.050587
50	20	13	9	0.997669	0.012971
50	20	13	10	0.999783	0.002114
50	20	13	11	0.999989	0.000206
50	20	13	12	1.000000	0.000011
50	20	13	13	1.000000	0.000000
50	20	14	0	0.000155	0.000155
50	20	14	1	0.002709	0.002554
50	20	14	2	0.020232	0.017523
50	20	14	3	0.086634	0.066402
50	20	14	4	0.241850	0.155215
50	20	14	5	0.478368	0.236519
50	20	14	6	0.720262	0.241894
50	20	14	7	0.888537	0.168274
50	20	14	8	0.968291	0.079755
50	20	14	9	0.993813	0.025522
50	20	14	10	0.999212	0.005399
50	20	14	11	0.999939	0.000727
50	20	14	12	0.999997	0.000058
50	20	14	13	1.000000	0.000002
50	20	14	14	1.000000	0.000000
50	20	15	0	0.000069	0.000069
50	20	15	1	0.001361	0.001292
50	20	15	2	0.011470	0.010109
50	20	15	3	0.055277	0.043807
50	20	15	4	0.172865	0.117587
50	20	15	5	0.379819	0.206954
50	20	15	6	0.626192	0.246374
50	20	15	7	0.827771	0.201578
50	20	15	8	0.941707	0.113936
50	20	15	9	0.986015	0.044308
50	20	15	10	0.997712	0.011697
50	20	15	11	0.999757	0.002045
50	20	15	12	0.999984	0.000227
50	20	15	13	0.999999	0.000015
50	20	15	14	1.000000	0.000001
50	20	15	15	1.000000	0.000000
50	20	16	0	0.000030	0.000030
50	20	16	1	0.000660	0.000630
50	20	16	2	0.006271	0.005612
50	20	16	3	0.034000	0.027728
50	20	16	4	0.119111	0.085111
50	20	16	5	0.291124	0.172014
50	20	16	6	0.527643	0.236519
50	20	16	7	0.752899	0.225256
50	20	16	8	0.902643	0.149744
50	20	16	9	0.972089	0.069446
50	20	16	10	0.994370	0.022281
50	20	16	11	0.999231	0.004861
50	20	16	12	0.999932	0.000701
50	20	16	13	0.999996	0.000064
50	20	16	14	1.000000	0.000003
50	20	16	15	1.000000	0.000000
50	20	16	16	1.000000	0.000000
50	20	17	0	0.000012	0.000012
50	20	17	1	0.000308	0.000295
50	20	17	2	0.003300	0.002992
50	20	17	3	0.020136	0.016835
50	20	17	4	0.079058	0.058923
50	20	17	5	0.215236	0.136177
50	20	17	6	0.430253	0.215017
50	20	17	7	0.666772	0.236519
50	20	17	8	0.849792	0.183020
50	20	17	9	0.949621	0.099829
50	20	17	10	0.987817	0.038196
50	20	17	11	0.997944	0.010128
50	20	17	12	0.999767	0.001823
50	20	17	13	0.999983	0.000216
50	20	17	14	0.999999	0.000016
50	20	17	15	1.000000	0.000001
50	20	17	16	1.000000	0.000000
50	20	17	17	1.000000	0.000000
50	20	18	0	0.000005	0.000005
50	20	18	1	0.000137	0.000133
50	20	18	2	0.001668	0.001530
50	20	18	3	0.011463	0.009795

Table for $N = 2$, $n = 1$, through $N = 100$, $n = 50$

N	n	k	x	P(x)	p(x)
50	20	18	4	0.050490	0.039027
50	20	18	5	0.153337	0.102847
50	20	18	6	0.339034	0.185696
50	20	18	7	0.573598	0.234564
50	20	18	8	0.783239	0.209642
50	20	18	9	0.916345	0.133106
50	20	18	10	0.976242	0.059898
50	20	18	11	0.995182	0.018940
50	20	18	12	0.999325	0.004143
50	20	18	13	0.999937	0.000612
50	20	18	14	0.999996	0.000059
50	20	18	15	1.000000	0.000003
50	20	18	16	1.000000	0.000000
50	20	18	17	1.000000	0.000000
50	20	18	18	1.000000	0.000000
50	20	19	0	0.000002	0.000002
50	20	19	1	0.000059	0.000057
50	20	19	2	0.000807	0.000748
50	20	19	3	0.006259	0.005452
50	20	19	4	0.030976	0.024717
50	20	19	5	0.105127	0.074151
50	20	19	6	0.257791	0.152664
50	20	19	7	0.478306	0.220515
50	20	19	8	0.704624	0.226318
50	20	19	9	0.870590	0.165966
50	20	19	10	0.957524	0.086935
50	20	19	11	0.989856	0.032331
50	20	19	12	0.998290	0.008434
50	20	19	13	0.999804	0.001514
50	20	19	14	0.999985	0.000182
50	20	19	15	0.999999	0.000014
50	20	19	16	1.000000	0.000001
50	20	19	17	1.000000	0.000000
50	20	19	18	1.000000	0.000000
50	20	19	19	1.000000	0.000000
50	20	20	0	0.000001	0.000001
50	20	20	1	0.000024	0.000023
50	20	20	2	0.000373	0.000349
50	20	20	3	0.003269	0.002897
50	20	20	4	0.018219	0.014950
50	20	20	5	0.069248	0.051029
50	20	20	6	0.188846	0.119598
50	20	20	7	0.385832	0.196986
50	20	20	8	0.617017	0.231185
50	20	20	9	0.811699	0.194682
50	20	20	10	0.929481	0.117782
50	20	20	11	0.980469	0.050988
50	20	20	12	0.996113	0.015644
50	20	20	13	0.999462	0.003349
50	20	20	14	0.999950	0.000488

N	n	k	x	P(x)	p(x)
50	20	20	15	0.999997	0.000047
50	20	20	16	1.000000	0.000003
50	20	20	17	1.000000	0.000000
50	20	20	18	1.000000	0.000000
50	20	20	19	1.000000	0.000000
50	20	20	20	1.000000	0.000000
50	21	1	0	0.580000	0.580000
50	21	1	1	1.000000	0.420000
50	21	2	0	0.331429	0.331429
50	21	2	1	0.828571	0.497143
50	21	2	2	1.000000	0.171429
50	21	3	0	0.186429	0.186429
50	21	3	1	0.621429	0.435000
50	21	3	2	0.932143	0.310714
50	21	3	3	1.000000	0.067857
50	21	4	0	0.103131	0.103131
50	21	4	1	0.436322	0.333191
50	21	4	2	0.806535	0.370213
50	21	4	3	0.974012	0.167477
50	21	4	4	1.000000	0.025988
50	21	5	0	0.056049	0.056049
50	21	5	1	0.291456	0.235407
50	21	5	2	0.653621	0.362165
50	21	5	3	0.908478	0.254857
50	21	5	4	0.990396	0.081918
50	21	5	5	1.000000	0.009604
50	21	6	0	0.029893	0.029893
50	21	6	1	0.186831	0.156938
50	21	6	2	0.500707	0.313876
50	21	6	3	0.806535	0.305828
50	21	6	4	0.959449	0.152914
50	21	6	5	0.996585	0.037136
50	21	6	6	1.000000	0.003415
50	21	7	0	0.015626	0.015626
50	21	7	1	0.115495	0.099870
50	21	7	2	0.365170	0.249674
50	21	7	3	0.681423	0.316254
50	21	7	4	0.900368	0.218945
50	21	7	5	0.983081	0.082713
50	21	7	6	0.998836	0.015755
50	21	7	7	1.000000	0.001164
50	21	8	0	0.007995	0.007995
50	21	8	1	0.069044	0.061050
50	21	8	2	0.254848	0.185804
50	21	8	3	0.549038	0.294190
50	21	8	4	0.813809	0.264771
50	21	8	5	0.952304	0.138495
50	21	8	6	0.993340	0.041036
50	21	8	7	0.999621	0.006281
50	21	8	8	1.000000	0.000379

N	n	k	x	P(x)	p(x)
50	21	9	0	0.003997	0.003997
50	21	9	1	0.039973	0.035976
50	21	9	2	0.170794	0.130821
50	21	9	3	0.422957	0.252163
50	21	9	4	0.706640	0.283683
50	21	9	5	0.899544	0.192904
50	21	9	6	0.978684	0.079140
50	21	9	7	0.997527	0.018843
50	21	9	8	0.999883	0.002355
50	21	9	9	1.000000	0.000117
50	21	10	0	0.001950	0.001950
50	21	10	1	0.022424	0.020474
50	21	10	2	0.110170	0.087746
50	21	10	3	0.312251	0.202081
50	21	10	4	0.589015	0.276764
50	21	10	5	0.824264	0.235249
50	21	10	6	0.949731	0.125466
50	21	10	7	0.991093	0.041362
50	21	10	8	0.999136	0.008043
50	21	10	9	0.999966	0.000830
50	21	10	10	1.000000	0.000034
50	21	11	0	0.000926	0.000926
50	21	11	1	0.012187	0.011261
50	21	11	2	0.068491	0.056304
50	21	11	3	0.221315	0.152824
50	21	11	4	0.471391	0.250076
50	21	11	5	0.730165	0.258774
50	21	11	6	0.902681	0.172516
50	21	11	7	0.976616	0.073935
50	21	11	8	0.996522	0.019906
50	21	11	9	0.999717	0.003195
50	21	11	10	0.999990	0.000274
50	21	11	11	1.000000	0.000009
50	21	12	0	0.000427	0.000427
50	21	12	1	0.006412	0.005985
50	21	12	2	0.041061	0.034648
50	21	12	3	0.150780	0.109720
50	21	12	4	0.362383	0.211603
50	21	12	5	0.624001	0.261618
50	21	12	6	0.836328	0.212327
50	21	12	7	0.950075	0.113747
50	21	12	8	0.989887	0.039811
50	21	12	9	0.998734	0.008847
50	21	12	10	0.999913	0.001180
50	21	12	11	0.999997	0.000084
50	21	12	12	1.000000	0.000002
50	21	13	0	0.000191	0.000191
50	21	13	1	0.003262	0.003071
50	21	13	2	0.023736	0.020474
50	21	13	3	0.098808	0.075071

N	n	k	x	P(x)	p(x)
50	21	13	4	0.267719	0.168911
50	21	13	5	0.513846	0.246127
50	21	13	6	0.752515	0.238669
50	21	13	7	0.908169	0.155654
50	21	13	8	0.976267	0.068098
50	21	13	9	0.995940	0.019673
50	21	13	10	0.999572	0.003632
50	21	13	11	0.999975	0.000404
50	21	13	12	0.999999	0.000024
50	21	13	13	1.000000	0.000001
50	21	14	0	0.000083	0.000083
50	21	14	1	0.001602	0.001520
50	21	14	2	0.013223	0.011620
50	21	14	3	0.062287	0.049064
50	21	14	4	0.190111	0.127824
50	21	14	5	0.407413	0.217302
50	21	14	6	0.655757	0.248345
50	21	14	7	0.849273	0.193515
50	21	14	8	0.952340	0.103068
50	21	14	9	0.989559	0.037219
50	21	14	10	0.998492	0.008933
50	21	14	11	0.999866	0.001374
50	21	14	12	0.999993	0.000127
50	21	14	13	1.000000	0.000006
50	21	14	14	1.000000	0.000000
50	21	15	0	0.000034	0.000034
50	21	15	1	0.000758	0.000724
50	21	15	2	0.007090	0.006332
50	21	15	3	0.037755	0.030665
50	21	15	4	0.129750	0.091995
50	21	15	5	0.310834	0.181085
50	21	15	6	0.552280	0.241446
50	21	15	7	0.774017	0.221736
50	21	15	8	0.915121	0.141105
50	21	15	9	0.977153	0.062032
50	21	15	10	0.995763	0.018609
50	21	15	11	0.999484	0.003722
50	21	15	12	0.999962	0.000477
50	21	15	13	0.999998	0.000037
50	21	15	14	1.000000	0.000001
50	21	15	15	1.000000	0.000000
50	21	16	0	0.000014	0.000014
50	21	16	1	0.000345	0.000331
50	21	16	2	0.003653	0.003308
50	21	16	3	0.021984	0.018332
50	21	16	4	0.085066	0.063082
50	21	16	5	0.228053	0.142986
50	21	16	6	0.448803	0.220751
50	21	16	7	0.685322	0.236519
50	21	16	8	0.862711	0.177389

N	n	k	x	P(x)	p(x)
50	21	16	9	0.955885	0.093174
50	21	16	10	0.989914	0.034029
50	21	16	11	0.998421	0.008507
50	21	16	12	0.999839	0.001418
50	21	16	13	0.999990	0.000151
50	21	16	14	0.999999	0.000010
50	21	16	15	1.000000	0.000000
50	21	16	16	1.000000	0.000000
50	21	17	0	0.000005	0.000005
50	21	17	1	0.000150	0.000145
50	21	17	2	0.001804	0.001654
50	21	17	3	0.012279	0.010475
50	21	17	4	0.053525	0.041246
50	21	17	5	0.160765	0.107240
50	21	17	6	0.351413	0.190648
50	21	17	7	0.587932	0.236519
50	21	17	8	0.794886	0.206954
50	21	17	9	0.923000	0.128114
50	21	17	10	0.978905	0.055904
50	21	17	11	0.995919	0.017014
50	21	17	12	0.999464	0.003545
50	21	17	13	0.999954	0.000491
50	21	17	14	0.999998	0.000043
50	21	17	15	1.000000	0.000002
50	21	17	16	1.000000	0.000000
50	21	17	17	1.000000	0.000000
50	21	18	0	0.000002	0.000002
50	21	18	1	0.000062	0.000060
50	21	18	2	0.000852	0.000789
50	21	18	3	0.006565	0.005714
50	21	18	4	0.032277	0.025712
50	21	18	5	0.108770	0.076493
50	21	18	6	0.264755	0.155985
50	21	18	7	0.487591	0.222836
50	21	18	8	0.713359	0.225768
50	21	18	9	0.876413	0.163055
50	21	18	10	0.960270	0.083857
50	21	18	11	0.990763	0.030493
50	21	18	12	0.998497	0.007734
50	21	18	13	0.999836	0.001339
50	21	18	14	0.999988	0.000153
50	21	18	15	0.999999	0.000011
50	21	18	16	1.000000	0.000000
50	21	18	17	1.000000	0.000000
50	21	18	18	1.000000	0.000000
50	21	19	0	0.000001	0.000001
50	21	19	1	0.000025	0.000024
50	21	19	2	0.000383	0.000358
50	21	19	3	0.003351	0.002968
50	21	19	4	0.018618	0.015266

N	n	k	x	P(x)	p(x)
50	21	19	5	0.070524	0.051906
50	21	19	6	0.191637	0.121113
50	21	19	7	0.390100	0.198463
50	21	19	8	0.621640	0.231540
50	21	19	9	0.815268	0.193627
50	21	19	10	0.931444	0.116176
50	21	19	11	0.981234	0.049790
50	21	19	12	0.996322	0.015088
50	21	19	13	0.999501	0.003179
50	21	19	14	0.999955	0.000454
50	21	19	15	0.999997	0.000042
50	21	19	16	1.000000	0.000002
50	21	19	17	1.000000	0.000000
50	21	19	18	1.000000	0.000000
50	21	19	19	1.000000	0.000000
50	21	20	0	0.000000	0.000000
50	21	20	1	0.000009	0.000009
50	21	20	2	0.000163	0.000154
50	21	20	3	0.001628	0.001465
50	21	20	4	0.010246	0.008618
50	21	20	5	0.043734	0.033488
50	21	20	6	0.133034	0.089300
50	21	20	7	0.300472	0.167438
50	21	20	8	0.524543	0.224071
50	21	20	9	0.740315	0.215772
50	21	20	10	0.890220	0.149905
50	21	20	11	0.965173	0.074953
50	21	20	12	0.991941	0.026769
50	21	20	13	0.998680	0.006739
50	21	20	14	0.999852	0.001172
50	21	20	15	0.999989	0.000137
50	21	20	16	0.999999	0.000010
50	21	20	17	1.000000	0.000000
50	21	20	18	1.000000	0.000000
50	21	20	19	1.000000	0.000000
50	21	20	20	1.000000	0.000000
50	21	21	0	0.000000	0.000000
50	21	21	1	0.000003	0.000003
50	21	21	2	0.000066	0.000062
50	21	21	3	0.000749	0.000683
50	21	21	4	0.005362	0.004613
50	21	21	5	0.025873	0.020511
50	21	21	6	0.088384	0.062510
50	21	21	7	0.222334	0.133950
50	21	21	8	0.427445	0.205111
50	21	21	9	0.654006	0.226561
50	21	21	10	0.835255	0.181249
50	21	21	11	0.940189	0.104934
50	21	21	12	0.983911	0.043722
50	21	21	13	0.996883	0.012973

N	n	k	x	P(x)	p(x)
50	21	21	14	0.999579	0.002696
50	21	21	15	0.999962	0.000383
50	21	21	16	0.999998	0.000036
50	21	21	17	1.000000	0.000002
50	21	21	18	1.000000	0.000000
50	21	21	19	1.000000	0.000000
50	21	21	20	1.000000	0.000000
50	21	21	21	1.000000	0.000000
50	22	1	0	0.560000	0.560000
50	22	1	1	1.000000	0.440000
50	22	2	0	0.308571	0.308571
50	22	2	1	0.811429	0.502857
50	22	2	2	1.000000	0.188571
50	22	3	0	0.167143	0.167143
50	22	3	1	0.591429	0.424286
50	22	3	2	0.921429	0.330000
50	22	3	3	1.000000	0.078571
50	22	4	0	0.088906	0.088906
50	22	4	1	0.401854	0.312948
50	22	4	2	0.781003	0.379149
50	22	4	3	0.968237	0.187234
50	22	4	4	1.000000	0.031763
50	22	5	0	0.046386	0.046386
50	22	5	1	0.258986	0.212601
50	22	5	2	0.616156	0.357169
50	22	5	3	0.890901	0.274746
50	22	5	4	0.987571	0.096670
50	22	5	5	1.000000	0.012429
50	22	6	0	0.023708	0.023708
50	22	6	1	0.159773	0.136064
50	22	6	2	0.457414	0.297641
50	22	6	3	0.774898	0.317484
50	22	6	4	0.948903	0.174006
50	22	6	5	0.995305	0.046401
50	22	6	6	1.000000	0.004695
50	22	7	0	0.011854	0.011854
50	22	7	1	0.094833	0.082979
50	22	7	2	0.322122	0.227290
50	22	7	3	0.637802	0.315680
50	22	7	4	0.877719	0.239917
50	22	7	5	0.977377	0.099658
50	22	7	6	0.998292	0.020916
50	22	7	7	1.000000	0.001707
50	22	8	0	0.005789	0.005789
50	22	8	1	0.054308	0.048519
50	22	8	2	0.216406	0.162098
50	22	8	3	0.498316	0.281909
50	22	8	4	0.777289	0.278973
50	22	8	5	0.937977	0.160688
50	22	8	6	0.990510	0.052533

N	n	k	x	P(x)	p(x)
50	22	8	7	0.999404	0.008894
50	22	8	8	1.000000	0.000596
50	22	9	0	0.002757	0.002757
50	22	9	1	0.030049	0.027292
50	22	9	2	0.139217	0.109168
50	22	9	3	0.370785	0.231569
50	22	9	4	0.657729	0.286944
50	22	9	5	0.872937	0.215208
50	22	9	6	0.970497	0.097561
50	22	9	7	0.996228	0.025730
50	22	9	8	0.999801	0.003574
50	22	9	9	1.000000	0.000199
50	22	10	0	0.001278	0.001278
50	22	10	1	0.016070	0.014792
50	22	10	2	0.085964	0.069894
50	22	10	3	0.263473	0.177509
50	22	10	4	0.531754	0.268281
50	22	10	5	0.783704	0.251950
50	22	10	6	0.932425	0.148721
50	22	10	7	0.986814	0.054389
50	22	10	8	0.998581	0.011767
50	22	10	9	0.999937	0.001356
50	22	10	10	1.000000	0.000063
50	22	11	0	0.000575	0.000575
50	22	11	1	0.008304	0.007729
50	22	11	2	0.051017	0.042713
50	22	11	3	0.179156	0.128139
50	22	11	4	0.411027	0.231871
50	22	11	5	0.676625	0.265598
50	22	11	6	0.872937	0.196311
50	22	11	7	0.966418	0.093482
50	22	11	8	0.994463	0.028044
50	22	11	9	0.999496	0.005034
50	22	11	10	0.999981	0.000485
50	22	11	11	1.000000	0.000019
50	22	12	0	0.000251	0.000251
50	22	12	1	0.004142	0.003892
50	22	12	2	0.029113	0.024971
50	22	12	3	0.116729	0.087617
50	22	12	4	0.304010	0.187280
50	22	12	5	0.560852	0.256842
50	22	12	6	0.792399	0.231547
50	22	12	7	0.930464	0.138065
50	22	12	8	0.984395	0.053932
50	22	12	9	0.997818	0.013423
50	22	12	10	0.999832	0.002013
50	22	12	11	0.999995	0.000163
50	22	12	12	1.000000	0.000005
50	22	13	0	0.000106	0.000106
50	22	13	1	0.001992	0.001886

N	n	k	x	P(x)	p(x)
50	22	13	2	0.015970	0.013979
50	22	13	3	0.072921	0.056951
50	22	13	4	0.215298	0.142377
50	22	13	5	0.445949	0.230651
50	22	13	6	0.694905	0.248956
50	22	13	7	0.875964	0.181059
50	22	13	8	0.964526	0.088562
50	22	13	9	0.993226	0.028700
50	22	13	10	0.999196	0.005970
50	22	13	11	0.999947	0.000751
50	22	13	12	0.999998	0.000051
50	22	13	13	1.000000	0.000001
50	22	14	0	0.000043	0.000043
50	22	14	1	0.000921	0.000878
50	22	14	2	0.008414	0.007493
50	22	14	3	0.043676	0.035262
50	22	14	4	0.146034	0.102358
50	22	14	5	0.339974	0.193941
50	22	14	6	0.587248	0.247274
50	22	14	7	0.802562	0.215314
50	22	14	8	0.931016	0.128454
50	22	14	9	0.983142	0.052126
50	22	14	10	0.997260	0.014118
50	22	14	11	0.999724	0.002464
50	22	14	12	0.999985	0.000261
50	22	14	13	1.000000	0.000015
50	22	14	14	1.000000	0.000000
50	22	15	0	0.000017	0.000017
50	22	15	1	0.000409	0.000392
50	22	15	2	0.004251	0.003843
50	22	15	3	0.025066	0.020814
50	22	15	4	0.094855	0.069789
50	22	15	5	0.248391	0.153536
50	22	15	6	0.477349	0.228958
50	22	15	7	0.712848	0.235499
50	22	15	8	0.881062	0.168214
50	22	15	9	0.964319	0.083257
50	22	15	10	0.992554	0.028235
50	22	15	11	0.998971	0.006417
50	22	15	12	0.999912	0.000941
50	22	15	13	0.999996	0.000084
50	22	15	14	1.000000	0.000004
50	22	15	15	1.000000	0.000000
50	22	16	0	0.000006	0.000006
50	22	16	1	0.000173	0.000167
50	22	16	2	0.002056	0.001882
50	22	16	3	0.013767	0.011711
50	22	16	4	0.058963	0.045197
50	22	16	5	0.173816	0.114853
50	22	16	6	0.372682	0.198866

N	n	k	x	P(x)	p(x)
50	22	16	7	0.611920	0.239237
50	22	16	8	0.813776	0.201856
50	22	16	9	0.933395	0.119619
50	22	16	10	0.982873	0.049479
50	22	16	11	0.996954	0.014081
50	22	16	12	0.999643	0.002689
50	22	16	13	0.999974	0.000331
50	22	16	14	0.999999	0.000025
50	22	16	15	1.000000	0.000001
50	22	16	16	1.000000	0.000000
50	22	17	0	0.000002	0.000002
50	22	17	1	0.000070	0.000068
50	22	17	2	0.000948	0.000878
50	22	17	3	0.007222	0.006274
50	22	17	4	0.035036	0.027813
50	22	17	5	0.116390	0.081354
50	22	17	6	0.279098	0.162709
50	22	17	7	0.506374	0.227275
50	22	17	8	0.730659	0.224285
50	22	17	9	0.887658	0.156999
50	22	17	10	0.965410	0.077752
50	22	17	11	0.992399	0.026988
50	22	17	12	0.998853	0.006454
50	22	17	13	0.999887	0.001034
50	22	17	14	0.999993	0.000106
50	22	17	15	1.000000	0.000007
50	22	17	16	1.000000	0.000000
50	22	17	17	1.000000	0.000000
50	22	18	0	0.000001	0.000001
50	22	18	1	0.000027	0.000026
50	22	18	2	0.000416	0.000389
50	22	18	3	0.003610	0.003194
50	22	18	4	0.019865	0.016255
50	22	18	5	0.074480	0.054615
50	22	18	6	0.200209	0.125729
50	22	18	7	0.403067	0.202857
50	22	18	8	0.635508	0.232441
50	22	18	9	0.825810	0.190302
50	22	18	10	0.937137	0.111327
50	22	18	11	0.983403	0.046266
50	22	18	12	0.996897	0.013494
50	22	18	13	0.999605	0.002708
50	22	18	14	0.999967	0.000363
50	22	18	15	0.999998	0.000031
50	22	18	16	1.000000	0.000002
50	22	18	17	1.000000	0.000000
50	22	18	18	1.000000	0.000000
50	22	19	0	0.000000	0.000000
50	22	19	1	0.000010	0.000009
50	22	19	2	0.000173	0.000163

N	n	k	x	P(x)	p(x)
50	22	19	3	0.001714	0.001541
50	22	19	4	0.010721	0.009008
50	22	19	5	0.045466	0.034744
50	22	19	6	0.137345	0.091879
50	22	19	7	0.307977	0.170633
50	22	19	8	0.533815	0.225837
50	22	19	9	0.748500	0.214685
50	22	19	10	0.895389	0.146890
50	22	19	11	0.967499	0.072109
50	22	19	12	0.992680	0.025181
50	22	19	13	0.998843	0.006163
50	22	19	14	0.999877	0.001034
50	22	19	15	0.999991	0.000115
50	22	19	16	1.000000	0.000008
50	22	19	17	1.000000	0.000000
50	22	19	18	1.000000	0.000000
50	22	19	19	1.000000	0.000000
50	22	20	0	0.000000	0.000000
50	22	20	1	0.000003	0.000003
50	22	20	2	0.000068	0.000064
50	22	20	3	0.000769	0.000702
50	22	20	4	0.005491	0.004722
50	22	20	5	0.026412	0.020921
50	22	20	6	0.089923	0.063511
50	22	20	7	0.225413	0.135490
50	22	20	8	0.431824	0.206410
50	22	20	9	0.658471	0.226647
50	22	20	10	0.838529	0.180058
50	22	20	11	0.941912	0.103383
50	22	20	12	0.984557	0.042645
50	22	20	13	0.997054	0.012497
50	22	20	14	0.999610	0.002556
50	22	20	15	0.999966	0.000356
50	22	20	16	0.999998	0.000032
50	22	20	17	1.000000	0.000002
50	22	20	18	1.000000	0.000000
50	22	20	19	1.000000	0.000000
50	22	20	20	1.000000	0.000000
50	22	21	0	0.000000	0.000000
50	22	21	1	0.000001	0.000001
50	22	21	2	0.000025	0.000024
50	22	21	3	0.000325	0.000300
50	22	21	4	0.002658	0.002333
50	22	21	5	0.014557	0.011899
50	22	21	6	0.056051	0.041494
50	22	21	7	0.157668	0.101617
50	22	21	8	0.335499	0.177831
50	22	21	9	0.560257	0.224758
50	22	21	10	0.766506	0.206249
50	22	21	11	0.904005	0.137499

N	n	k	x	P(x)	p(x)
50	22	21	12	0.970342	0.066337
50	22	21	13	0.993305	0.022963
50	22	21	14	0.998928	0.005624
50	22	21	15	0.999883	0.000954
50	22	21	16	0.999992	0.000109
50	22	21	17	1.000000	0.000008
50	22	21	18	1.000000	0.000000
50	22	21	19	1.000000	0.000000
50	22	21	20	1.000000	0.000000
50	22	21	21	1.000000	0.000000
50	22	22	0	0.000000	0.000000
50	22	22	1	0.000000	0.000000
50	22	22	2	0.000008	0.000008
50	22	22	3	0.000128	0.000120
50	22	22	4	0.001210	0.001082
50	22	22	5	0.007582	0.006372
50	22	22	6	0.033158	0.025576
50	22	22	7	0.105107	0.071950
50	22	22	8	0.249650	0.144542
50	22	22	9	0.459503	0.209854
50	22	22	10	0.681161	0.221658
50	22	22	11	0.851850	0.170689
50	22	22	12	0.947467	0.095617
50	22	22	13	0.986178	0.038711
50	22	22	14	0.997377	0.011199
50	22	22	15	0.999652	0.002275
50	22	22	16	0.999969	0.000317
50	22	22	17	0.999998	0.000029
50	22	22	18	1.000000	0.000002
50	22	22	19	1.000000	0.000000
50	22	22	20	1.000000	0.000000
50	22	22	21	1.000000	0.000000
50	22	22	22	1.000000	0.000000
50	23	1	0	0.540000	0.540000
50	23	1	1	1.000000	0.460000
50	23	2	0	0.286531	0.286531
50	23	2	1	0.793469	0.506939
50	23	2	2	1.000000	0.206531
50	23	3	0	0.149235	0.149235
50	23	3	1	0.561122	0.411888
50	23	3	2	0.909643	0.348520
50	23	3	3	1.000000	0.090357
50	23	4	0	0.076205	0.076205
50	23	4	1	0.368324	0.292119
50	23	4	2	0.753921	0.385597
50	23	4	3	0.961550	0.207629
50	23	4	4	1.000000	0.038450
50	23	5	0	0.038102	0.038102
50	23	5	1	0.228615	0.190512
50	23	5	2	0.577887	0.349273

Table for $N = 2$, $n = 1$, through $N = 100$, $n = 50$

N	n	k	x	P(x)	p(x)
50	23	5	3	0.871277	0.293389
50	23	5	4	0.984118	0.112842
50	23	5	5	1.000000	0.015881
50	23	6	0	0.018628	0.018628
50	23	6	1	0.135475	0.116848
50	23	6	2	0.414894	0.279418
50	23	6	3	0.740881	0.325988
50	23	6	4	0.936474	0.195593
50	23	6	5	0.993647	0.057173
50	23	6	6	1.000000	0.006353
50	23	7	0	0.008891	0.008891
50	23	7	1	0.077052	0.068161
50	23	7	2	0.281535	0.204483
50	23	7	3	0.592705	0.311170
50	23	7	4	0.852014	0.259308
50	23	7	5	0.970258	0.118245
50	23	7	6	0.997546	0.027287
50	23	7	7	1.000000	0.002454
50	23	8	0	0.004135	0.004135
50	23	8	1	0.042179	0.038043
50	23	8	2	0.181671	0.139492
50	23	8	3	0.447975	0.266304
50	23	8	4	0.737435	0.289461
50	23	8	5	0.920760	0.183325
50	23	8	6	0.986758	0.065997
50	23	8	7	0.999087	0.012329
50	23	8	8	1.000000	0.000913
50	23	9	0	0.001871	0.001871
50	23	9	1	0.022251	0.020380
50	23	9	2	0.111925	0.089674
50	23	9	3	0.321163	0.209239
50	23	9	4	0.606489	0.285325
50	23	9	5	0.842193	0.235704
50	23	9	6	0.960044	0.117852
50	23	9	7	0.994390	0.034345
50	23	9	8	0.999674	0.005284
50	23	9	9	1.000000	0.000326
50	23	10	0	0.000821	0.000821
50	23	10	1	0.011315	0.010494
50	23	10	2	0.065994	0.054679
50	23	10	3	0.219096	0.153101
50	23	10	4	0.474265	0.255169
50	23	10	5	0.738713	0.264448
50	23	10	6	0.911179	0.172466
50	23	10	7	0.980987	0.069808
50	23	10	8	0.997741	0.016754
50	23	10	9	0.999889	0.002148
50	23	10	10	1.000000	0.000111
50	23	11	0	0.000349	0.000349
50	23	11	1	0.005544	0.005195

N	n	k	x	P(x)	p(x)
50	23	11	2	0.037288	0.031744
50	23	11	3	0.142545	0.105257
50	23	11	4	0.353060	0.210515
50	23	11	5	0.619711	0.266652
50	23	11	6	0.837881	0.218170
50	23	11	7	0.953064	0.115183
50	23	11	8	0.991458	0.038394
50	23	11	9	0.999137	0.007679
50	23	11	10	0.999964	0.000827
50	23	11	11	1.000000	0.000036
50	23	12	0	0.000143	0.000143
50	23	12	1	0.002613	0.002470
50	23	12	2	0.020195	0.017581
50	23	12	3	0.088567	0.068372
50	23	12	4	0.250501	0.161934
50	23	12	5	0.496641	0.246140
50	23	12	6	0.742781	0.246140
50	23	12	7	0.905809	0.163028
50	23	12	8	0.976691	0.070882
50	23	12	9	0.996380	0.019689
50	23	12	10	0.999688	0.003308
50	23	12	11	0.999989	0.000301
50	23	12	12	1.000000	0.000011
50	23	13	0	0.000057	0.000057
50	23	13	1	0.001183	0.001127
50	23	13	2	0.010479	0.009295
50	23	13	3	0.052582	0.042103
50	23	13	4	0.169534	0.116953
50	23	13	5	0.380049	0.210515
50	23	13	6	0.632666	0.252617
50	23	13	7	0.837166	0.204500
50	23	13	8	0.948711	0.111545
50	23	13	9	0.989126	0.040415
50	23	13	10	0.998556	0.009430
50	23	13	11	0.999894	0.001337
50	23	13	12	0.999997	0.000103
50	23	13	13	1.000000	0.000003
50	23	14	0	0.000021	0.000021
50	23	14	1	0.000513	0.000492
50	23	14	2	0.005203	0.004690
50	23	14	3	0.029823	0.024620
50	23	14	4	0.109477	0.079654
50	23	14	5	0.277636	0.168159
50	23	14	6	0.516599	0.238962
50	23	14	7	0.748734	0.232135
50	23	14	8	0.903490	0.154757
50	23	14	9	0.973834	0.070344
50	23	14	10	0.995243	0.021409
50	23	14	11	0.999460	0.004217
50	23	14	12	0.999966	0.000506

N	n	k	x	P(x)	p(x)
50	23	14	13	0.999999	0.000033
50	23	14	14	1.000000	0.000001
50	23	15	0	0.000008	0.000008
50	23	15	1	0.000213	0.000205
50	23	15	2	0.002467	0.002255
50	23	15	3	0.016145	0.013678
50	23	15	4	0.067438	0.051292
50	23	15	5	0.193557	0.126119
50	23	15	6	0.403755	0.210198
50	23	15	7	0.645562	0.241807
50	23	15	8	0.839008	0.193446
50	23	15	9	0.946478	0.107470
50	23	15	10	0.987512	0.041034
50	23	15	11	0.998054	0.010542
50	23	15	12	0.999811	0.001757
50	23	15	13	0.999990	0.000178
50	23	15	14	1.000000	0.000010
50	23	15	15	1.000000	0.000000
50	23	16	0	0.000003	0.000003
50	23	16	1	0.000084	0.000081
50	23	16	2	0.001115	0.001031
50	23	16	3	0.008329	0.007215
50	23	16	4	0.039593	0.031264
50	23	16	5	0.128696	0.089102
50	23	16	6	0.301659	0.172963
50	23	16	7	0.535022	0.233363
50	23	16	8	0.756103	0.221081
50	23	16	9	0.903490	0.147387
50	23	16	10	0.972271	0.068781
50	23	16	11	0.994440	0.022169
50	23	16	12	0.999259	0.004819
50	23	16	13	0.999939	0.000680
50	23	16	14	0.999997	0.000058
50	23	16	15	1.000000	0.000003
50	23	16	16	1.000000	0.000000
50	23	17	0	0.000001	0.000001
50	23	17	1	0.000031	0.000030
50	23	17	2	0.000478	0.000447
50	23	17	3	0.004085	0.003607
50	23	17	4	0.022122	0.018037
50	23	17	5	0.081524	0.059402
50	23	17	6	0.215177	0.133653
50	23	17	7	0.425204	0.210027
50	23	17	8	0.658567	0.233363
50	23	17	9	0.842801	0.184234
50	23	17	10	0.945972	0.103171
50	23	17	11	0.986616	0.040643
50	23	17	12	0.997700	0.011084
50	23	17	13	0.999739	0.002039
50	23	17	14	0.999982	0.000243

N	n	k	x	P(x)	p(x)
50	23	17	15	0.999999	0.000017
50	23	17	16	1.000000	0.000001
50	23	17	17	1.000000	0.000000
50	23	18	0	0.000000	0.000000
50	23	18	1	0.000011	0.000011
50	23	18	2	0.000194	0.000183
50	23	18	3	0.001899	0.001705
50	23	18	4	0.011737	0.009838
50	23	18	5	0.049123	0.037386
50	23	18	6	0.146325	0.097203
50	23	18	7	0.323373	0.177047
50	23	18	8	0.552493	0.229120
50	23	18	9	0.764641	0.212148
50	23	18	10	0.905329	0.140688
50	23	18	11	0.971836	0.066507
50	23	18	12	0.994005	0.022169
50	23	18	13	0.999121	0.005116
50	23	18	14	0.999916	0.000794
50	23	18	15	0.999995	0.000079
50	23	18	16	1.000000	0.000005
50	23	18	17	1.000000	0.000000
50	23	18	18	1.000000	0.000000
50	23	19	0	0.000000	0.000000
50	23	19	1	0.000004	0.000004
50	23	19	2	0.000074	0.000070
50	23	19	3	0.000833	0.000759
50	23	19	4	0.005896	0.005063
50	23	19	5	0.028094	0.022198
50	23	19	6	0.094687	0.066593
50	23	19	7	0.234849	0.140163
50	23	19	8	0.445093	0.210244
50	23	19	9	0.671827	0.226734
50	23	19	10	0.848175	0.176348
50	23	19	11	0.946896	0.098721
50	23	19	12	0.986385	0.039489
50	23	19	13	0.997522	0.011138
50	23	19	14	0.999692	0.002170
50	23	19	15	0.999975	0.000283
50	23	19	16	0.999999	0.000024
50	23	19	17	1.000000	0.000001
50	23	19	18	1.000000	0.000000
50	23	19	19	1.000000	0.000000
50	23	20	0	0.000000	0.000000
50	23	20	1	0.000001	0.000001
50	23	20	2	0.000026	0.000025
50	23	20	3	0.000343	0.000317
50	23	20	4	0.002793	0.002450
50	23	20	5	0.015205	0.012412
50	23	20	6	0.058168	0.042963
50	23	20	7	0.162507	0.104339

Table for $N = 2$, $n = 1$, through $N = 100$, $n = 50$

N	n	k	x	P(x)	p(x)
50	23	20	8	0.343362	0.180855
50	23	20	9	0.569431	0.226069
50	23	20	10	0.774222	0.204792
50	23	20	11	0.908681	0.134459
50	23	20	12	0.972373	0.063691
50	23	20	13	0.993930	0.021557
50	23	20	14	0.999062	0.005133
50	23	20	15	0.999902	0.000840
50	23	20	16	0.999993	0.000091
50	23	20	17	1.000000	0.000006
50	23	20	18	1.000000	0.000000
50	23	20	19	1.000000	0.000000
50	23	20	20	1.000000	0.000000
50	23	21	0	0.000000	0.000000
50	23	21	1	0.000000	0.000000
50	23	21	2	0.000009	0.000008
50	23	21	3	0.000132	0.000123
50	23	21	4	0.001241	0.001110
50	23	21	5	0.007758	0.006516
50	23	21	6	0.033822	0.026064
50	23	21	7	0.106860	0.073038
50	23	21	8	0.252935	0.146075
50	23	21	9	0.463932	0.210997
50	23	21	10	0.685479	0.221547
50	23	21	11	0.854898	0.169418
50	23	21	12	0.949019	0.094121
50	23	21	13	0.986744	0.037725
50	23	21	14	0.997522	0.010779
50	23	21	15	0.999678	0.002156
50	23	21	16	0.999972	0.000294
50	23	21	17	0.999998	0.000026
50	23	21	18	1.000000	0.000001
50	23	21	19	1.000000	0.000000
50	23	21	20	1.000000	0.000000
50	23	21	21	1.000000	0.000000
50	23	22	0	0.000000	0.000000
50	23	22	1	0.000000	0.000000
50	23	22	2	0.000003	0.000003
50	23	22	3	0.000047	0.000044
50	23	22	4	0.000515	0.000468
50	23	22	5	0.003713	0.003199
50	23	22	6	0.018543	0.014830
50	23	22	7	0.066563	0.048020
50	23	22	8	0.177379	0.110816
50	23	22	9	0.362071	0.184693
50	23	22	10	0.586165	0.224094
50	23	22	11	0.784794	0.198629
50	23	22	12	0.913318	0.128524
50	23	22	13	0.973735	0.060417
50	23	22	14	0.994177	0.020442

N	n	k	x	P(x)	p(x)
50	23	22	15	0.999083	0.004906
50	23	22	16	0.999901	0.000818
50	23	22	17	0.999993	0.000092
50	23	22	18	1.000000	0.000007
50	23	22	19	1.000000	0.000000
50	23	22	20	1.000000	0.000000
50	23	22	21	1.000000	0.000000
50	23	22	22	1.000000	0.000000
50	23	23	0	0.000000	0.000000
50	23	23	1	0.000000	0.000000
50	23	23	2	0.000001	0.000001
50	23	23	3	0.000015	0.000015
50	23	23	4	0.000197	0.000182
50	23	23	5	0.001657	0.001460
50	23	23	6	0.009539	0.007882
50	23	23	7	0.039123	0.029584
50	23	23	8	0.118013	0.078890
50	23	23	9	0.269725	0.151712
50	23	23	10	0.482121	0.212397
50	23	23	11	0.699667	0.217546
50	23	23	12	0.862826	0.163159
50	23	23	13	0.952158	0.089331
50	23	23	14	0.987607	0.035449
50	23	23	15	0.997682	0.010075
50	23	23	16	0.999697	0.002015
50	23	23	17	0.999973	0.000277
50	23	23	18	0.999998	0.000025
50	23	23	19	1.000000	0.000001
50	23	23	20	1.000000	0.000000
50	23	23	21	1.000000	0.000000
50	23	23	22	1.000000	0.000000
50	23	23	23	1.000000	0.000000
50	24	1	0	0.520000	0.520000
50	24	1	1	1.000000	0.480000
50	24	2	0	0.265306	0.265306
50	24	2	1	0.774694	0.509388
50	24	2	2	1.000000	0.225306
50	24	3	0	0.132653	0.132653
50	24	3	1	0.530612	0.397959
50	24	3	2	0.896735	0.366122
50	24	3	3	1.000000	0.103265
50	24	4	0	0.064915	0.064915
50	24	4	1	0.335866	0.270951
50	24	4	2	0.725358	0.389492
50	24	4	3	0.953860	0.228502
50	24	4	4	1.000000	0.046140
50	24	5	0	0.031046	0.031046
50	24	5	1	0.200391	0.169344
50	24	5	2	0.539079	0.338689
50	24	5	3	0.849544	0.310465

N	n	k	x	P(x)	p(x)
50	24	5	4	0.979939	0.130395
50	24	5	5	1.000000	0.020061
50	24	6	0	0.014488	0.014488
50	24	6	1	0.113837	0.099349
50	24	6	2	0.373498	0.259661
50	24	6	3	0.704661	0.331162
50	24	6	4	0.921986	0.217325
50	24	6	5	0.991530	0.069544
50	24	6	6	1.000000	0.008470
50	24	7	0	0.006586	0.006586
50	24	7	1	0.061905	0.055319
50	24	7	2	0.243668	0.181763
50	24	7	3	0.546606	0.302938
50	24	7	4	0.823202	0.276596
50	24	7	5	0.961499	0.138298
50	24	7	6	0.996535	0.035035
50	24	7	7	1.000000	0.003465
50	24	8	0	0.002910	0.002910
50	24	8	1	0.032315	0.029406
50	24	8	2	0.150673	0.118357
50	24	8	3	0.398659	0.247987
50	24	8	4	0.694552	0.295893
50	24	8	5	0.900391	0.205839
50	24	8	6	0.981869	0.081478
50	24	8	7	0.998630	0.016761
50	24	8	8	1.000000	0.001370
50	24	9	0	0.001247	0.001247
50	24	9	1	0.016212	0.014965
50	24	9	2	0.088676	0.072464
50	24	9	3	0.274666	0.185990
50	24	9	4	0.553651	0.278985
50	24	9	5	0.807274	0.253623
50	24	9	6	0.946950	0.139676
50	24	9	7	0.991846	0.044896
50	24	9	8	0.999478	0.007632
50	24	9	9	1.000000	0.000522
50	24	10	0	0.000517	0.000517
50	24	10	1	0.007817	0.007300
50	24	10	2	0.049793	0.041976
50	24	10	3	0.179403	0.129610
50	24	10	4	0.417561	0.238158
50	24	10	5	0.689741	0.272180
50	24	10	6	0.885629	0.195887
50	24	10	7	0.973230	0.087602
50	24	10	8	0.996500	0.023269
50	24	10	9	0.999809	0.003309
50	24	10	10	1.000000	0.000191
50	24	11	0	0.000207	0.000207
50	24	11	1	0.003620	0.003413
50	24	11	2	0.026706	0.023087

N	n	k	x	P(x)	p(x)
50	24	11	3	0.111358	0.084651
50	24	11	4	0.298482	0.187124
50	24	11	5	0.560455	0.261974
50	24	11	6	0.797479	0.237024
50	24	11	7	0.936000	0.138520
50	24	11	8	0.987192	0.051192
50	24	11	9	0.998568	0.011376
50	24	11	10	0.999933	0.001365
50	24	11	11	1.000000	0.000067
50	24	12	0	0.000080	0.000080
50	24	12	1	0.001607	0.001527
50	24	12	2	0.013683	0.012076
50	24	12	3	0.065776	0.052093
50	24	12	4	0.202521	0.136744
50	24	12	5	0.432827	0.230307
50	24	12	6	0.688084	0.255256
50	24	12	7	0.875619	0.187535
50	24	12	8	0.966190	0.090571
50	24	12	9	0.994193	0.028003
50	24	12	10	0.999443	0.005250
50	24	12	11	0.999978	0.000535
50	24	12	12	1.000000	0.000022
50	24	13	0	0.000029	0.000029
50	24	13	1	0.000682	0.000653
50	24	13	2	0.006692	0.006009
50	24	13	3	0.036988	0.030296
50	24	13	4	0.130550	0.093562
50	24	13	5	0.317674	0.187124
50	24	13	6	0.567173	0.249499
50	24	13	7	0.791722	0.224549
50	24	13	8	0.928055	0.136333
50	24	13	9	0.983139	0.055084
50	24	13	10	0.997509	0.014370
50	24	13	11	0.999795	0.002286
50	24	13	12	0.999993	0.000198
50	24	13	13	1.000000	0.000007
50	24	14	0	0.000010	0.000010
50	24	14	1	0.000276	0.000266
50	24	14	2	0.003119	0.002842
50	24	14	3	0.019793	0.016674
50	24	14	4	0.079976	0.060183
50	24	14	5	0.221583	0.141607
50	24	14	6	0.445795	0.224212
50	24	14	7	0.688550	0.242755
50	24	14	8	0.869100	0.180549
50	24	14	9	0.960807	0.091708
50	24	14	10	0.992071	0.031264
50	24	14	11	0.998992	0.006920
50	24	14	12	0.999929	0.000937
50	24	14	13	0.999998	0.000069

Table for $N = 2$, $n = 1$, through $N = 100$, $n = 50$

N	n	k	x	P(x)	p(x)
50	24	14	14	1.000000	0.000002
50	24	15	0	0.000003	0.000003
50	24	15	1	0.000106	0.000103
50	24	15	2	0.001382	0.001275
50	24	15	3	0.010066	0.008684
50	24	15	4	0.046541	0.036475
50	24	15	5	0.146846	0.100305
50	24	15	6	0.333689	0.186843
50	24	15	7	0.573916	0.240227
50	24	15	8	0.788856	0.214940
50	24	15	9	0.922596	0.133740
50	24	15	10	0.979913	0.057317
50	24	15	11	0.996493	0.016579
50	24	15	12	0.999616	0.003124
50	24	15	13	0.999977	0.000360
50	24	15	14	0.999999	0.000023
50	24	15	15	1.000000	0.000001
50	24	16	0	0.000001	0.000001
50	24	16	1	0.000039	0.000038
50	24	16	2	0.000580	0.000541
50	24	16	3	0.004856	0.004275
50	24	16	4	0.025698	0.020843
50	24	16	5	0.092395	0.066696
50	24	16	6	0.237598	0.145204
50	24	16	7	0.457234	0.219636
50	24	16	8	0.690597	0.233363
50	24	16	9	0.865279	0.174681
50	24	16	10	0.956986	0.091708
50	24	16	11	0.990335	0.033348
50	24	16	12	0.998545	0.008211
50	24	16	13	0.999863	0.001318
50	24	16	14	0.999993	0.000129
50	24	16	15	1.000000	0.000007
50	24	16	16	1.000000	0.000000
50	24	17	0	0.000000	0.000000
50	24	17	1	0.000013	0.000013
50	24	17	2	0.000230	0.000217
50	24	17	3	0.002215	0.001985
50	24	17	4	0.013438	0.011223
50	24	17	5	0.055123	0.041685
50	24	17	6	0.160726	0.105603
50	24	17	7	0.347416	0.186691
50	24	17	8	0.580780	0.233363
50	24	17	9	0.788213	0.207434
50	24	17	10	0.919224	0.131011
50	24	17	11	0.977584	0.058359
50	24	17	12	0.995647	0.018064
50	24	17	13	0.999437	0.003790
50	24	17	14	0.999955	0.000518
50	24	17	15	0.999998	0.000043
50	24	17	16	1.000000	0.000002
50	24	17	17	1.000000	0.000000
50	24	18	0	0.000000	0.000000
50	24	18	1	0.000004	0.000004
50	24	18	2	0.000085	0.000081
50	24	18	3	0.000952	0.000866
50	24	18	4	0.006636	0.005684
50	24	18	5	0.031122	0.024486
50	24	18	6	0.103124	0.072002
50	24	18	7	0.251242	0.148118
50	24	18	8	0.467634	0.216391
50	24	18	9	0.693925	0.226292
50	24	18	10	0.863644	0.169719
50	24	18	11	0.954594	0.090950
50	24	18	12	0.989079	0.034485
50	24	18	13	0.998174	0.009095
50	24	18	14	0.999798	0.001624
50	24	18	15	0.999986	0.000188
50	24	18	16	0.999999	0.000013
50	24	18	17	1.000000	0.000000
50	24	18	18	1.000000	0.000000
50	24	19	0	0.000000	0.000000
50	24	19	1	0.000001	0.000001
50	24	19	2	0.000030	0.000028
50	24	19	3	0.000383	0.000354
50	24	19	4	0.003083	0.002700
50	24	19	5	0.016584	0.013500
50	24	19	6	0.062623	0.046040
50	24	19	7	0.172555	0.109931
50	24	19	8	0.359438	0.186883
50	24	19	9	0.587851	0.228413
50	24	19	10	0.789392	0.201541
50	24	19	11	0.917645	0.128253
50	24	19	12	0.976147	0.058502
50	24	19	13	0.995047	0.018900
50	24	19	14	0.999290	0.004243
50	24	19	15	0.999933	0.000643
50	24	19	16	0.999996	0.000063
50	24	19	17	1.000000	0.000004
50	24	19	18	1.000000	0.000000
50	24	19	19	1.000000	0.000000
50	24	20	0	0.000000	0.000000
50	24	20	1	0.000000	0.000000
50	24	20	2	0.000009	0.000009
50	24	20	3	0.000144	0.000134
50	24	20	4	0.001341	0.001198
50	24	20	5	0.008309	0.006968
50	24	20	6	0.035891	0.027581
50	24	20	7	0.112270	0.076379
50	24	20	8	0.262982	0.150712
50	24	20	9	0.477329	0.214347
50	24	20	10	0.698374	0.221045
50	24	20	11	0.863862	0.165488
50	24	20	12	0.953501	0.089639
50	24	20	13	0.988341	0.034840
50	24	20	14	0.997922	0.009581
50	24	20	15	0.999747	0.001825
50	24	20	16	0.999980	0.000233
50	24	20	17	0.999999	0.000019
50	24	20	18	1.000000	0.000001
50	24	20	19	1.000000	0.000000
50	24	20	20	1.000000	0.000000
50	24	21	0	0.000000	0.000000
50	24	21	1	0.000000	0.000000
50	24	21	2	0.000003	0.000003
50	24	21	3	0.000050	0.000047
50	24	21	4	0.000543	0.000493
50	24	21	5	0.003896	0.003353
50	24	21	6	0.019342	0.015446
50	24	21	7	0.068988	0.049646
50	24	21	8	0.182602	0.113614
50	24	21	9	0.370155	0.187553
50	24	21	10	0.595219	0.225064
50	24	21	11	0.792150	0.196931
50	24	21	12	0.917645	0.125495
50	24	21	13	0.975566	0.057921
50	24	21	14	0.994728	0.019162
50	24	21	15	0.999199	0.004471
50	24	21	16	0.999918	0.000719
50	24	21	17	0.999994	0.000077
50	24	21	18	1.000000	0.000005
50	24	21	19	1.000000	0.000000
50	24	21	20	1.000000	0.000000
50	24	21	21	1.000000	0.000000
50	24	22	0	0.000000	0.000000
50	24	22	1	0.000000	0.000000
50	24	22	2	0.000001	0.000001
50	24	22	3	0.000016	0.000015
50	24	22	4	0.000203	0.000187
50	24	22	5	0.001699	0.001496
50	24	22	6	0.009755	0.008056
50	24	22	7	0.039885	0.030130
50	24	22	8	0.119919	0.080034
50	24	22	9	0.273145	0.153227
50	24	22	10	0.486568	0.213423
50	24	22	11	0.703871	0.217303
50	24	22	12	0.865716	0.161845
50	24	22	13	0.953596	0.087880
50	24	22	14	0.988121	0.034524
50	24	22	15	0.997812	0.009691
50	24	22	16	0.999719	0.001908
50	24	22	17	0.999976	0.000257
50	24	22	18	0.999999	0.000023
50	24	22	19	1.000000	0.000001
50	24	22	20	1.000000	0.000000
50	24	22	21	1.000000	0.000000
50	24	22	22	1.000000	0.000000
50	24	23	0	0.000000	0.000000
50	24	23	1	0.000000	0.000000
50	24	23	2	0.000000	0.000000
50	24	23	3	0.000004	0.000004
50	24	23	4	0.000069	0.000065
50	24	23	5	0.000684	0.000615
50	24	23	6	0.004576	0.003892
50	24	23	7	0.021592	0.017016
50	24	23	8	0.074185	0.052593
50	24	23	9	0.191059	0.116874
50	24	23	10	0.379856	0.188797
50	24	23	11	0.602980	0.223124
50	24	23	12	0.796354	0.193374
50	24	23	13	0.919072	0.122718
50	24	23	14	0.975790	0.056718
50	24	23	15	0.994696	0.018906
50	24	23	16	0.999174	0.004478
50	24	23	17	0.999912	0.000738
50	24	23	18	0.999994	0.000082
50	24	23	19	1.000000	0.000006
50	24	23	20	1.000000	0.000000
50	24	23	21	1.000000	0.000000
50	24	23	22	1.000000	0.000000
50	24	23	23	1.000000	0.000000
50	24	24	0	0.000000	0.000000
50	24	24	1	0.000000	0.000000
50	24	24	2	0.000000	0.000000
50	24	24	3	0.000001	0.000001
50	24	24	4	0.000021	0.000020
50	24	24	5	0.000251	0.000230
50	24	24	6	0.001981	0.001730
50	24	24	7	0.010878	0.008897
50	24	24	8	0.043019	0.032140
50	24	24	9	0.126129	0.083111
50	24	24	10	0.281962	0.155832
50	24	24	11	0.495550	0.213588
50	24	24	12	0.710410	0.214860
50	24	24	13	0.869076	0.158666
50	24	24	14	0.954784	0.085708
50	24	24	15	0.988394	0.033611
50	24	24	16	0.997848	0.009453
50	24	24	17	0.999721	0.001873
50	24	24	18	0.999975	0.000255

Table for $N = 2$, $n = 1$, through $N = 100$, $n = 50$

N	n	k	x	P(x)	p(x)
50	24	24	19	0.999998	0.000023
50	24	24	20	1.000000	0.000001
50	24	24	21	1.000000	0.000000
50	24	24	22	1.000000	0.000000
50	24	24	23	1.000000	0.000000
50	24	24	24	1.000000	0.000000
50	25	1	0	0.500000	0.500000
50	25	1	1	1.000000	0.500000
50	25	2	0	0.244898	0.244898
50	25	2	1	0.755102	0.510204
50	25	2	2	1.000000	0.244898
50	25	3	0	0.117347	0.117347
50	25	3	1	0.500000	0.382653
50	25	3	2	0.882653	0.382653
50	25	3	3	1.000000	0.117347
50	25	4	0	0.054928	0.054928
50	25	4	1	0.304603	0.249674
50	25	4	2	0.695397	0.390795
50	25	4	3	0.945072	0.249674
50	25	4	4	1.000000	0.054928
50	25	5	0	0.025076	0.025076
50	25	5	1	0.174338	0.149262
50	25	5	2	0.500000	0.325662
50	25	5	3	0.825662	0.325662
50	25	5	4	0.974924	0.149262
50	25	5	5	1.000000	0.025076
50	25	6	0	0.011145	0.011145
50	25	6	1	0.094731	0.083587
50	25	6	2	0.333550	0.238819
50	25	6	3	0.666450	0.332899
50	25	6	4	0.905268	0.238819
50	25	6	5	0.988855	0.083587
50	25	6	6	1.000000	0.011145
50	25	7	0	0.004813	0.004813
50	25	7	1	0.049139	0.044326
50	25	7	2	0.208713	0.159574
50	25	7	3	0.500000	0.291287
50	25	7	4	0.791287	0.291287
50	25	7	5	0.950861	0.159574
50	25	7	6	0.995187	0.044326
50	25	7	7	1.000000	0.004813
50	25	8	0	0.002015	0.002015
50	25	8	1	0.024399	0.022384
50	25	8	2	0.123359	0.098961
50	25	8	3	0.350970	0.227610
50	25	8	4	0.649030	0.298061
50	25	8	5	0.876640	0.227610
50	25	8	6	0.975601	0.098961
50	25	8	7	0.997985	0.022384
50	25	8	8	1.000000	0.002015

N	n	k	x	P(x)	p(x)
50	25	9	0	0.000815	0.000815
50	25	9	1	0.011608	0.010792
50	25	9	2	0.069167	0.057559
50	25	9	3	0.231745	0.162579
50	25	9	4	0.500000	0.268255
50	25	9	5	0.768255	0.268255
50	25	9	6	0.930833	0.162579
50	25	9	7	0.988392	0.057559
50	25	9	8	0.999185	0.010792
50	25	9	9	1.000000	0.000815
50	25	10	0	0.000318	0.000318
50	25	10	1	0.005290	0.004972
50	25	10	2	0.036877	0.031587
50	25	10	3	0.144508	0.107630
50	25	10	4	0.362601	0.218093
50	25	10	5	0.637399	0.274798
50	25	10	6	0.855492	0.218093
50	25	10	7	0.963122	0.107630
50	25	10	8	0.994710	0.031587
50	25	10	9	0.999682	0.004972
50	25	10	10	1.000000	0.000318
50	25	11	0	0.000119	0.000119
50	25	11	1	0.002307	0.002188
50	25	11	2	0.018715	0.016408
50	25	11	3	0.085311	0.066596
50	25	11	4	0.248102	0.162791
50	25	11	5	0.500000	0.251898
50	25	11	6	0.751898	0.251898
50	25	11	7	0.914689	0.162791
50	25	11	8	0.981285	0.066596
50	25	11	9	0.997693	0.016408
50	25	11	10	0.999881	0.002188
50	25	11	11	1.000000	0.000119
50	25	12	0	0.000043	0.000043
50	25	12	1	0.000961	0.000918
50	25	12	2	0.009038	0.008078
50	25	12	3	0.047744	0.038706
50	25	12	4	0.160446	0.112702
50	25	12	5	0.370822	0.210376
50	25	12	6	0.629178	0.258357
50	25	12	7	0.839554	0.210376
50	25	12	8	0.952256	0.112702
50	25	12	9	0.990962	0.038706
50	25	12	10	0.999039	0.008078
50	25	12	11	0.999957	0.000918
50	25	12	12	1.000000	0.000043
50	25	13	0	0.000015	0.000015
50	25	13	1	0.000381	0.000366
50	25	13	2	0.004149	0.003768
50	25	13	3	0.025336	0.021186

N	n	k	x	P(x)	p(x)
50	25	13	4	0.098163	0.072828
50	25	13	5	0.260097	0.161934
50	25	13	6	0.500000	0.239903
50	25	13	7	0.739903	0.239903
50	25	13	8	0.901837	0.161934
50	25	13	9	0.974664	0.072828
50	25	13	10	0.995851	0.021186
50	25	13	11	0.999619	0.003768
50	25	13	12	0.999985	0.000366
50	25	13	13	1.000000	0.000015
50	25	14	0	0.000005	0.000005
50	25	14	1	0.000143	0.000139
50	25	14	2	0.001807	0.001663
50	25	14	3	0.012738	0.010931
50	25	14	4	0.056829	0.044090
50	25	14	5	0.172565	0.115737
50	25	14	6	0.376807	0.204241
50	25	14	7	0.623193	0.246386
50	25	14	8	0.827435	0.204241
50	25	14	9	0.943171	0.115737
50	25	14	10	0.987262	0.044090
50	25	14	11	0.998193	0.010931
50	25	14	12	0.999857	0.001663
50	25	14	13	0.999995	0.000139
50	25	14	14	1.000000	0.000005
50	25	15	0	0.000001	0.000001
50	25	15	1	0.000051	0.000050
50	25	15	2	0.000744	0.000693
50	25	15	3	0.006058	0.005314
50	25	15	4	0.031109	0.025051
50	25	15	5	0.108267	0.077158
50	25	15	6	0.269013	0.160746
50	25	15	7	0.500000	0.230987
50	25	15	8	0.730987	0.230987
50	25	15	9	0.891733	0.160746
50	25	15	10	0.968891	0.077158
50	25	15	11	0.993942	0.025051
50	25	15	12	0.999256	0.005314
50	25	15	13	0.999949	0.000693
50	25	15	14	0.999998	0.000050
50	25	15	15	1.000000	0.000001
50	25	16	0	0.000000	0.000000
50	25	16	1	0.000017	0.000017
50	25	16	2	0.000289	0.000272
50	25	16	3	0.002718	0.002429
50	25	16	4	0.016078	0.013361
50	25	16	5	0.064177	0.048098
50	25	16	6	0.181751	0.117574
50	25	16	7	0.381206	0.199456
50	25	16	8	0.618793	0.237587

N	n	k	x	P(x)	p(x)
50	25	16	9	0.818249	0.199456
50	25	16	10	0.935823	0.117574
50	25	16	11	0.983921	0.048098
50	25	16	12	0.997282	0.013361
50	25	16	13	0.999711	0.002429
50	25	16	14	0.999983	0.000272
50	25	16	15	0.999999	0.000017
50	25	16	16	1.000000	0.000000
50	25	17	0	0.000000	0.000000
50	25	17	1	0.000005	0.000005
50	25	17	2	0.000105	0.000100
50	25	17	3	0.001146	0.001041
50	25	17	4	0.007826	0.006680
50	25	17	5	0.035884	0.028057
50	25	17	6	0.116048	0.080164
50	25	17	7	0.275612	0.159565
50	25	17	8	0.500000	0.224388
50	25	17	9	0.724388	0.224388
50	25	17	10	0.883952	0.159565
50	25	17	11	0.964116	0.080164
50	25	17	12	0.992174	0.028057
50	25	17	13	0.998854	0.006680
50	25	17	14	0.999895	0.001041
50	25	17	15	0.999995	0.000100
50	25	17	16	1.000000	0.000005
50	25	17	17	1.000000	0.000000
50	25	18	0	0.000000	0.000000
50	25	18	1	0.000002	0.000001
50	25	18	2	0.000035	0.000034
50	25	18	3	0.000452	0.000416
50	25	18	4	0.003575	0.003123
50	25	18	5	0.018879	0.015304
50	25	18	6	0.069893	0.051013
50	25	18	7	0.188577	0.118684
50	25	18	8	0.384406	0.195829
50	25	18	9	0.615594	0.231187
50	25	18	10	0.811423	0.195829
50	25	18	11	0.930107	0.118684
50	25	18	12	0.981121	0.051013
50	25	18	13	0.996425	0.015304
50	25	18	14	0.999548	0.003123
50	25	18	15	0.999964	0.000416
50	25	18	16	0.999998	0.000034
50	25	18	17	1.000000	0.000001
50	25	18	18	1.000000	0.000000
50	25	19	0	0.000000	0.000000
50	25	19	1	0.000000	0.000000
50	25	19	2	0.000011	0.000011
50	25	19	3	0.000166	0.000155
50	25	19	4	0.001526	0.001360

Table for $N = 2$, $n = 1$, through $N = 100$, $n = 50$

N	n	k	x	P(x)	p(x)
50	25	19	5	0.009314	0.007789
50	25	19	6	0.039603	0.030289
50	25	19	7	0.121817	0.082214
50	25	19	8	0.280372	0.158555
50	25	19	9	0.500000	0.219628
50	25	19	10	0.719628	0.219628
50	25	19	11	0.878183	0.158555
50	25	19	12	0.960396	0.082214
50	25	19	13	0.990686	0.030289
50	25	19	14	0.998474	0.007789
50	25	19	15	0.999834	0.001360
50	25	19	16	0.999989	0.000155
50	25	19	17	0.999999	0.000011
50	25	19	18	1.000000	0.000000
50	25	19	19	1.000000	0.000000
50	25	20	0	0.000000	0.000000
50	25	20	1	0.000000	0.000000
50	25	20	2	0.000003	0.000003
50	25	20	3	0.000056	0.000053
50	25	20	4	0.000604	0.000548
50	25	20	5	0.004289	0.003685
50	25	20	6	0.021039	0.016750
50	25	20	7	0.074080	0.053041
50	25	20	8	0.193423	0.119342
50	25	20	9	0.386644	0.193221
50	25	20	10	0.613356	0.226713
50	25	20	11	0.806577	0.193221
50	25	20	12	0.925920	0.119342
50	25	20	13	0.978961	0.053041
50	25	20	14	0.995711	0.016750
50	25	20	15	0.999396	0.003685
50	25	20	16	0.999944	0.000548
50	25	20	17	0.999997	0.000053
50	25	20	18	1.000000	0.000003
50	25	20	19	1.000000	0.000000
50	25	20	20	1.000000	0.000000
50	25	21	0	0.000000	0.000000
50	25	21	1	0.000000	0.000000
50	25	21	2	0.000001	0.000001
50	25	21	3	0.000017	0.000016
50	25	21	4	0.000220	0.000203
50	25	21	5	0.001833	0.001612
50	25	21	6	0.010431	0.008598
50	25	21	7	0.042255	0.031825
50	25	21	8	0.125795	0.083540
50	25	21	9	0.283592	0.157797
50	25	21	10	0.500000	0.216408
50	25	21	11	0.716408	0.216408
50	25	21	12	0.874205	0.157797
50	25	21	13	0.957744	0.083540
50	25	21	14	0.989569	0.031825
50	25	21	15	0.998167	0.008598
50	25	21	16	0.999779	0.001612
50	25	21	17	0.999983	0.000203
50	25	21	18	0.999999	0.000016
50	25	21	19	1.000000	0.000001
50	25	21	20	1.000000	0.000000
50	25	21	21	1.000000	0.000000
50	25	22	0	0.000000	0.000000
50	25	22	1	0.000000	0.000000
50	25	22	2	0.000000	0.000000
50	25	22	3	0.000005	0.000005
50	25	22	4	0.000073	0.000069
50	25	22	5	0.000721	0.000647
50	25	22	6	0.004798	0.004077
50	25	22	7	0.022502	0.017705
50	25	22	8	0.076824	0.054321
50	25	22	9	0.196532	0.119708
50	25	22	10	0.388065	0.191533
50	25	22	11	0.611935	0.223870
50	25	22	12	0.803468	0.191533
50	25	22	13	0.923176	0.119708
50	25	22	14	0.977498	0.054321
50	25	22	15	0.995202	0.017705
50	25	22	16	0.999279	0.004077
50	25	22	17	0.999927	0.000647
50	25	22	18	0.999995	0.000069
50	25	22	19	1.000000	0.000005
50	25	22	20	1.000000	0.000000
50	25	22	21	1.000000	0.000000
50	25	22	22	1.000000	0.000000
50	25	23	0	0.000000	0.000000
50	25	23	1	0.000000	0.000000
50	25	23	2	0.000000	0.000000
50	25	23	3	0.000001	0.000001
50	25	23	4	0.000022	0.000021
50	25	23	5	0.000258	0.000236
50	25	23	6	0.002031	0.001773
50	25	23	7	0.011121	0.009089
50	25	23	8	0.043843	0.032722
50	25	23	9	0.128127	0.084284
50	25	23	10	0.285458	0.157331
50	25	23	11	0.500000	0.214542
50	25	23	12	0.714542	0.214542
50	25	23	13	0.871873	0.157331
50	25	23	14	0.956157	0.084284
50	25	23	15	0.988879	0.032722
50	25	23	16	0.997969	0.009089
50	25	23	17	0.999742	0.001773
50	25	23	18	0.999978	0.000236
50	25	23	19	0.999999	0.000021
50	25	23	20	1.000000	0.000001
50	25	23	21	1.000000	0.000000
50	25	23	22	1.000000	0.000000
50	25	23	23	1.000000	0.000000
50	25	24	0	0.000000	0.000000
50	25	24	1	0.000000	0.000000
50	25	24	2	0.000000	0.000000
50	25	24	3	0.000000	0.000000
50	25	24	4	0.000006	0.000006
50	25	24	5	0.000083	0.000077
50	25	24	6	0.000784	0.000700
50	25	24	7	0.005061	0.004277
50	25	24	8	0.023240	0.018179
50	25	24	9	0.078181	0.054941
50	25	24	10	0.198052	0.119871
50	25	24	11	0.388756	0.190704
50	25	24	12	0.611244	0.222488
50	25	24	13	0.801948	0.190704
50	25	24	14	0.921819	0.119871
50	25	24	15	0.976760	0.054941
50	25	24	16	0.994939	0.018179
50	25	24	17	0.999216	0.004277
50	25	24	18	0.999917	0.000700
50	25	24	19	0.999994	0.000077
50	25	24	20	1.000000	0.000006
50	25	24	21	1.000000	0.000000
50	25	24	22	1.000000	0.000000
50	25	24	23	1.000000	0.000000
50	25	24	24	1.000000	0.000000
50	25	25	0	0.000000	0.000000
50	25	25	1	0.000000	0.000000
50	25	25	2	0.000000	0.000000
50	25	25	3	0.000000	0.000000
50	25	25	4	0.000001	0.000001
50	25	25	5	0.000024	0.000022
50	25	25	6	0.000272	0.000248
50	25	25	7	0.002100	0.001828
50	25	25	8	0.011354	0.009254
50	25	25	9	0.044371	0.033017
50	25	25	10	0.128896	0.084524
50	25	25	11	0.286069	0.157174
50	25	25	12	0.500000	0.213931
50	25	25	13	0.713931	0.213931
50	25	25	14	0.871104	0.157174
50	25	25	15	0.955629	0.084524
50	25	25	16	0.988646	0.033017
50	25	25	17	0.997900	0.009254
50	25	25	18	0.999728	0.001828
50	25	25	19	0.999976	0.000248
50	25	25	20	0.999999	0.000022
50	25	25	21	1.000000	0.000001
50	25	25	22	1.000000	0.000000
50	25	25	23	1.000000	0.000000
50	25	25	24	1.000000	0.000000
50	25	25	25	1.000000	0.000000

Table for $N = 2$, $n = 1$, through $N = 100$, $n = 50$

N	n	k	x	P(x)	p(x)
60	1	1	0	0.983333	0.983333
60	1	1	1	1.000000	0.016667
60	2	1	0	0.966667	0.966667
60	2	1	1	1.000000	0.033333
60	2	2	0	0.933898	0.933898
60	2	2	1	0.999435	0.065537
60	2	2	2	1.000000	0.000565
60	3	1	0	0.950000	0.950000
60	3	1	1	1.000000	0.050000
60	3	2	0	0.901695	0.901695
60	3	2	1	0.998305	0.096610
60	3	2	2	1.000000	0.001695
60	3	3	0	0.855056	0.855056
60	3	3	1	0.994974	0.139918
60	3	3	2	0.999971	0.004997
60	3	3	3	1.000000	0.000029
60	4	1	0	0.933333	0.933333
60	4	1	1	1.000000	0.066667
60	4	2	0	0.870056	0.870056
60	4	2	1	0.996610	0.126554
60	4	2	2	1.000000	0.003390
60	4	3	0	0.810053	0.810053
60	4	3	1	0.990064	0.180012
60	4	3	2	0.999883	0.009819
60	4	3	3	1.000000	0.000117
60	4	4	0	0.753207	0.753207
60	4	4	1	0.980590	0.227383
60	4	4	2	0.999539	0.018949
60	4	4	3	0.999998	0.000459
60	4	4	4	1.000000	0.000002
60	5	1	0	0.916667	0.916667
60	5	1	1	1.000000	0.083333
60	5	2	0	0.838983	0.838983
60	5	2	1	0.994350	0.155367
60	5	2	2	1.000000	0.005650
60	5	3	0	0.766657	0.766657
60	5	3	1	0.983635	0.216978
60	5	3	2	0.999708	0.016072
60	5	3	3	1.000000	0.000292
60	5	4	0	0.699406	0.699406
60	5	4	1	0.968409	0.269002
60	5	4	2	0.998862	0.030453
60	5	4	3	0.999990	0.001128
60	5	4	4	1.000000	0.000010
60	5	5	0	0.636959	0.636959
60	5	5	1	0.949194	0.312235
60	5	5	2	0.997230	0.048036
60	5	5	3	0.999949	0.002719
60	5	5	4	1.000000	0.000050
60	5	5	5	1.000000	0.000000

N	n	k	x	P(x)	p(x)
60	6	1	0	0.900000	0.900000
60	6	1	1	1.000000	0.100000
60	6	2	0	0.808475	0.808475
60	6	2	1	0.991525	0.183051
60	6	2	2	1.000000	0.008475
60	6	3	0	0.724839	0.724839
60	6	3	1	0.975745	0.250906
60	6	3	2	0.999416	0.023670
60	6	3	3	1.000000	0.000584
60	6	4	0	0.648540	0.648540
60	6	4	1	0.953736	0.305195
60	6	4	2	0.997754	0.044019
60	6	4	3	0.999969	0.002215
60	6	4	4	1.000000	0.000031
60	6	5	0	0.579054	0.579054
60	6	5	1	0.926486	0.347432
60	6	5	2	0.994610	0.068124
60	6	5	3	0.999851	0.005240
60	6	5	4	0.999999	0.000148
60	6	5	5	1.000000	0.000001
60	6	6	0	0.515884	0.515884
60	6	6	1	0.894901	0.379017
60	6	6	2	0.989656	0.094754
60	6	6	3	0.999565	0.009909
60	6	6	4	0.999993	0.000429
60	6	6	5	1.000000	0.000006
60	6	6	6	1.000000	0.000000
60	7	1	0	0.883333	0.883333
60	7	1	1	1.000000	0.116667
60	7	2	0	0.778531	0.778531
60	7	2	1	0.988136	0.209604
60	7	2	2	1.000000	0.011864
60	7	3	0	0.684570	0.684570
60	7	3	1	0.966452	0.281882
60	7	3	2	0.998977	0.032525
60	7	3	3	1.000000	0.001023
60	7	4	0	0.600500	0.600500
60	7	4	1	0.936780	0.336280
60	7	4	2	0.996124	0.059344
60	7	4	3	0.999928	0.003804
60	7	4	4	1.000000	0.000072
60	7	5	0	0.525438	0.525438
60	7	5	1	0.900750	0.375313
60	7	5	2	0.990826	0.090075
60	7	5	3	0.999656	0.008831
60	7	5	4	0.999996	0.000340
60	7	5	5	1.000000	0.000004
60	7	6	0	0.458564	0.458564
60	7	6	1	0.859807	0.401243
60	7	6	2	0.982637	0.122830

N	n	k	x	P(x)	p(x)
60	7	6	3	0.999014	0.016377
60	7	6	4	0.999978	0.000963
60	7	6	5	1.000000	0.000022
60	7	6	6	1.000000	0.000000
60	7	7	0	0.399120	0.399120
60	7	7	1	0.815225	0.416104
60	7	7	2	0.971264	0.156039
60	7	7	3	0.997801	0.026537
60	7	7	4	0.999924	0.002123
60	7	7	5	0.999999	0.000075
60	7	7	6	1.000000	0.000001
60	7	7	7	1.000000	0.000000
60	8	1	0	0.866667	0.866667
60	8	1	1	1.000000	0.133333
60	8	2	0	0.749153	0.749153
60	8	2	1	0.984181	0.235028
60	8	2	2	1.000000	0.015819
60	8	3	0	0.645821	0.645821
60	8	3	1	0.955815	0.309994
60	8	3	2	0.998363	0.042548
60	8	3	3	1.000000	0.001636
60	8	4	0	0.555180	0.555180
60	8	4	1	0.917746	0.362566
60	8	4	2	0.993885	0.076139
60	8	4	3	0.999856	0.005972
60	8	4	4	1.000000	0.000144
60	8	5	0	0.475868	0.475868
60	8	5	1	0.872425	0.396557
60	8	5	2	0.985727	0.113302
60	8	5	3	0.999323	0.013596
60	8	5	4	0.999990	0.000666
60	8	5	5	1.000000	0.000010
60	8	6	0	0.406651	0.406651
60	8	6	1	0.821954	0.415303
60	8	6	2	0.973367	0.151413
60	8	6	3	0.998087	0.024720
60	8	6	4	0.999941	0.001854
60	8	6	5	0.999999	0.000058
60	8	6	6	1.000000	0.000001
60	8	7	0	0.346406	0.346406
60	8	7	1	0.768119	0.421712
60	8	7	2	0.956543	0.188425
60	8	7	3	0.995798	0.039255
60	8	7	4	0.999804	0.004006
60	8	7	5	0.999996	0.000192
60	8	7	6	1.000000	0.000004
60	8	7	7	1.000000	0.000000
60	8	8	0	0.294119	0.294119
60	8	8	1	0.712421	0.418302
60	8	8	2	0.935212	0.222791

N	n	k	x	P(x)	p(x)
60	8	8	3	0.992095	0.056883
60	8	8	4	0.999502	0.007407
60	8	8	5	0.999985	0.000484
60	8	8	6	1.000000	0.000015
60	8	8	7	1.000000	0.000000
60	8	8	8	1.000000	0.000000
60	9	1	0	0.850000	0.850000
60	9	1	1	1.000000	0.150000
60	9	2	0	0.720339	0.720339
60	9	2	1	0.979661	0.259322
60	9	2	2	1.000000	0.020339
60	9	3	0	0.608562	0.608562
60	9	3	1	0.943892	0.335330
60	9	3	2	0.997545	0.053653
60	9	3	3	1.000000	0.002455
60	9	4	0	0.512473	0.512473
60	9	4	1	0.896829	0.384355
60	9	4	2	0.990956	0.094128
60	9	4	3	0.999742	0.008785
60	9	4	4	1.000000	0.000258
60	9	5	0	0.430112	0.430112
60	9	5	1	0.841921	0.411809
60	9	5	2	0.979190	0.137270
60	9	5	3	0.998800	0.019610
60	9	5	4	0.999977	0.001177
60	9	5	5	1.000000	0.000023
60	9	6	0	0.359730	0.359730
60	9	6	1	0.782021	0.422291
60	9	6	2	0.961720	0.179698
60	9	6	3	0.996661	0.034941
60	9	6	4	0.999870	0.003209
60	9	6	5	0.999998	0.000128
60	9	6	6	1.000000	0.000002
60	9	7	0	0.299775	0.299775
60	9	7	1	0.719459	0.419685
60	9	7	2	0.938425	0.218966
60	9	7	3	0.992779	0.054353
60	9	7	4	0.999573	0.006794
60	9	7	5	0.999989	0.000416
60	9	7	6	1.000000	0.000011
60	9	7	7	1.000000	0.000000
60	9	8	0	0.248870	0.248870
60	9	8	1	0.656111	0.407241
60	9	8	2	0.909505	0.253395
60	9	8	3	0.986625	0.077120
60	9	8	4	0.998932	0.012306
60	9	8	5	0.999957	0.001026
60	9	8	6	0.999999	0.000042
60	9	8	7	1.000000	0.000001
60	9	8	8	1.000000	0.000000

Table for $N = 2$, $n = 1$, through $N = 100$, $n = 50$

N	n	k	x	P(x)	p(x)
60	9	9	0	0.205796	0.205796
60	9	9	1	0.593458	0.387662
60	9	9	2	0.875395	0.281936
60	9	9	3	0.977727	0.102332
60	9	9	4	0.997749	0.020022
60	9	9	5	0.999878	0.002130
60	9	9	6	0.999997	0.000118
60	9	9	7	1.000000	0.000003
60	9	9	8	1.000000	0.000000
60	9	9	9	1.000000	0.000000
60	10	1	0	0.833333	0.833333
60	10	1	1	1.000000	0.166667
60	10	2	0	0.692090	0.692090
60	10	2	1	0.974576	0.282486
60	10	2	2	1.000000	0.025424
60	10	3	0	0.572764	0.572764
60	10	3	1	0.930742	0.357978
60	10	3	2	0.996493	0.065751
60	10	3	3	1.000000	0.003507
60	10	4	0	0.472279	0.472279
60	10	4	1	0.874219	0.401940
60	10	4	2	0.987265	0.113046
60	10	4	3	0.999569	0.012304
60	10	4	4	1.000000	0.000431
60	10	5	0	0.387944	0.387944
60	10	5	1	0.809622	0.421678
60	10	5	2	0.971116	0.161494
60	10	5	3	0.998031	0.026916
60	10	5	4	0.999954	0.001923
60	10	5	5	1.000000	0.000046
60	10	6	0	0.317409	0.317409
60	10	6	1	0.740620	0.423211
60	10	6	2	0.947626	0.207006
60	10	6	3	0.994606	0.046980
60	10	6	4	0.999744	0.005138
60	10	6	5	0.999996	0.000252
60	10	6	6	1.000000	0.000004
60	10	7	0	0.258629	0.258629
60	10	7	1	0.670085	0.411456
60	10	7	2	0.916958	0.246873
60	10	7	3	0.988516	0.071558
60	10	7	4	0.999173	0.010657
60	10	7	5	0.999972	0.000799
60	10	7	6	1.000000	0.000027
60	10	7	7	1.000000	0.000000
60	10	8	0	0.209831	0.209831
60	10	8	1	0.600215	0.390384
60	10	8	2	0.879694	0.279479
60	10	8	3	0.979065	0.099370
60	10	8	4	0.997967	0.018902
60	10	8	5	0.999897	0.001930
60	10	8	6	0.999998	0.000101
60	10	8	7	1.000000	0.000002
60	10	8	8	1.000000	0.000000
60	10	9	0	0.169479	0.169479
60	10	9	1	0.532649	0.363169
60	10	9	2	0.836697	0.304049
60	10	9	3	0.965688	0.128990
60	10	9	4	0.995786	0.030098
60	10	9	5	0.999711	0.003926
60	10	9	6	0.999990	0.000278
60	10	9	7	1.000000	0.000010
60	10	9	8	1.000000	0.000000
60	10	9	9	1.000000	0.000000
60	10	10	0	0.136248	0.136248
60	10	10	1	0.468560	0.332312
60	10	10	2	0.789003	0.320444
60	10	10	3	0.947983	0.158980
60	10	10	4	0.992245	0.044261
60	10	10	5	0.999327	0.007082
60	10	10	6	0.999968	0.000641
60	10	10	7	0.999999	0.000031
60	10	10	8	1.000000	0.000001
60	10	10	9	1.000000	0.000000
60	10	10	10	1.000000	0.000000
60	11	1	0	0.816667	0.816667
60	11	1	1	1.000000	0.183333
60	11	2	0	0.664407	0.664407
60	11	2	1	0.968926	0.304520
60	11	2	2	1.000000	0.031073
60	11	3	0	0.538399	0.538399
60	11	3	1	0.916423	0.378025
60	11	3	2	0.995178	0.078755
60	11	3	3	1.000000	0.004822
60	11	4	0	0.434497	0.434497
60	11	4	1	0.850103	0.415606
60	11	4	2	0.982743	0.132640
60	11	4	3	0.999323	0.016580
60	11	4	4	1.000000	0.000677
60	11	5	0	0.349149	0.349149
60	11	5	1	0.775888	0.426738
60	11	5	2	0.961426	0.185538
60	11	5	3	0.996955	0.035529
60	11	5	4	0.999915	0.002961
60	11	5	5	1.000000	0.000085
60	11	6	0	0.279320	0.279320
60	11	6	1	0.698299	0.418979
60	11	6	2	0.931065	0.232766
60	11	6	3	0.991787	0.060722
60	11	6	4	0.999539	0.007752
60	11	6	5	0.999991	0.000452
60	11	6	6	1.000000	0.000009
60	11	7	0	0.222421	0.222421
60	11	7	1	0.620710	0.398289
60	11	7	2	0.892271	0.271561
60	11	7	3	0.982791	0.090520
60	11	7	4	0.998534	0.015743
60	11	7	5	0.999940	0.001407
60	11	7	6	0.999999	0.000059
60	11	7	7	1.000000	0.000001
60	11	8	0	0.176258	0.176258
60	11	8	1	0.545561	0.369303
60	11	8	2	0.846157	0.300595
60	11	8	3	0.969128	0.122971
60	11	8	4	0.996454	0.027327
60	11	8	5	0.999781	0.003327
60	11	8	6	0.999994	0.000212
60	11	8	7	1.000000	0.000006
60	11	8	8	1.000000	0.000000
60	11	9	0	0.138973	0.138973
60	11	9	1	0.474541	0.335569
60	11	9	2	0.794131	0.319589
60	11	9	3	0.950209	0.156078
60	11	9	4	0.992776	0.042567
60	11	9	5	0.999397	0.006622
60	11	9	6	0.999973	0.000576
60	11	9	7	0.999999	0.000026
60	11	9	8	1.000000	0.000001
60	11	9	9	1.000000	0.000000
60	11	10	0	0.108998	0.108998
60	11	10	1	0.408744	0.299745
60	11	10	2	0.737732	0.328989
60	11	10	3	0.925726	0.187994
60	11	10	4	0.986933	0.061207
60	11	10	5	0.998618	0.011685
60	11	10	6	0.999917	0.001298
60	11	10	7	0.999997	0.000081
60	11	10	8	1.000000	0.000003
60	11	10	9	1.000000	0.000000
60	11	10	10	1.000000	0.000000
60	11	11	0	0.085019	0.085019
60	11	11	1	0.348795	0.263776
60	11	11	2	0.678514	0.329720
60	11	11	3	0.895647	0.217133
60	11	11	4	0.978364	0.082717
60	11	11	5	0.997216	0.018852
60	11	11	6	0.999787	0.002571
60	11	11	7	0.999991	0.000204
60	11	11	8	1.000000	0.000009
60	11	11	9	1.000000	0.000000
60	11	11	10	1.000000	0.000000
60	11	11	11	1.000000	0.000000
60	12	1	0	0.800000	0.800000
60	12	1	1	1.000000	0.200000
60	12	2	0	0.637288	0.637288
60	12	2	1	0.962712	0.325424
60	12	2	2	1.000000	0.037288
60	12	3	0	0.505435	0.505435
60	12	3	1	0.900994	0.395558
60	12	3	2	0.993571	0.092577
60	12	3	3	1.000000	0.006429
60	12	4	0	0.399028	0.399028
60	12	4	1	0.824658	0.425630
60	12	4	2	0.977329	0.152672
60	12	4	3	0.998985	0.021656
60	12	4	4	1.000000	0.001015
60	12	5	0	0.313522	0.313522
60	12	5	1	0.741052	0.427530
60	12	5	2	0.950067	0.209015
60	12	5	3	0.995504	0.045438
60	12	5	4	0.999855	0.004350
60	12	5	5	1.000000	0.000145
60	12	6	0	0.245117	0.245117
60	12	6	1	0.655546	0.410429
60	12	6	2	0.912064	0.256518
60	12	6	3	0.988069	0.076005
60	12	6	4	0.999222	0.011153
60	12	6	5	0.999981	0.000759
60	12	6	6	1.000000	0.000018
60	12	7	0	0.190647	0.190647
60	12	7	1	0.571940	0.381293
60	12	7	2	0.864561	0.292621
60	12	7	3	0.975402	0.110841
60	12	7	4	0.997570	0.022168
60	12	7	5	0.999883	0.002313
60	12	7	6	0.999998	0.000115
60	12	7	7	1.000000	0.000002
60	12	8	0	0.147481	0.147481
60	12	8	1	0.492804	0.345322
60	12	8	2	0.809349	0.316545
60	12	8	3	0.956580	0.147230
60	12	8	4	0.994224	0.037644
60	12	8	5	0.999578	0.005354
60	12	8	6	0.999985	0.000407
60	12	8	7	1.000000	0.000015
60	12	8	8	1.000000	0.000000
60	12	9	0	0.113447	0.113447
60	12	9	1	0.419755	0.306308
60	12	9	2	0.748475	0.328720
60	12	9	3	0.931097	0.182622

$N = 60$ $n = 9$–12

N	n	k	x	P(x)	p(x)
60	12	9	4	0.988432	0.057335
60	12	9	5	0.998857	0.010425
60	12	9	6	0.999938	0.001081
60	12	9	7	0.999998	0.000060
60	12	9	8	1.000000	0.000002
60	12	9	9	1.000000	0.000000
60	12	10	0	0.086754	0.086754
60	12	10	1	0.353688	0.266935
60	12	10	2	0.684020	0.330332
60	12	10	3	0.898870	0.214850
60	12	10	4	0.979439	0.080569
60	12	10	5	0.997426	0.017987
60	12	10	6	0.999811	0.002385
60	12	10	7	0.999992	0.000182
60	12	10	8	1.000000	0.000007
60	12	10	9	1.000000	0.000000
60	12	10	10	1.000000	0.000000
60	12	11	0	0.065933	0.065933
60	12	11	1	0.294963	0.229030
60	12	11	2	0.617954	0.322991
60	12	11	3	0.860197	0.242243
60	12	11	4	0.966548	0.106351
60	12	11	5	0.994908	0.028360
60	12	11	6	0.999524	0.004617
60	12	11	7	0.999974	0.000450
60	12	11	8	0.999999	0.000025
60	12	11	9	1.000000	0.000001
60	12	11	10	1.000000	0.000000
60	12	11	11	1.000000	0.000000
60	12	12	0	0.049786	0.049786
60	12	12	1	0.243548	0.193762
60	12	12	2	0.552037	0.308489
60	12	12	3	0.815703	0.263666
60	12	12	4	0.949184	0.133481
60	12	12	5	0.990856	0.041672
60	12	12	6	0.998959	0.008103
60	12	12	7	0.999928	0.000969
60	12	12	8	0.999997	0.000069
60	12	12	9	1.000000	0.000003
60	12	12	10	1.000000	0.000000
60	12	12	11	1.000000	0.000000
60	12	12	12	1.000000	0.000000
60	13	1	0	0.783333	0.783333
60	13	1	1	1.000000	0.216667
60	13	2	0	0.610734	0.610734
60	13	2	1	0.955932	0.345198
60	13	2	2	1.000000	0.044068
60	13	3	0	0.473846	0.473846
60	13	3	1	0.884512	0.410666
60	13	3	2	0.991642	0.107130

N	n	k	x	P(x)	p(x)
60	13	3	3	1.000000	0.008358
60	13	4	0	0.365776	0.365776
60	13	4	1	0.798056	0.432280
60	13	4	2	0.970968	0.172912
60	13	4	3	0.998534	0.027566
60	13	4	4	1.000000	0.001466
60	13	5	0	0.280863	0.280863
60	13	5	1	0.705424	0.424561
60	13	5	2	0.937003	0.231579
60	13	5	3	0.993611	0.056608
60	13	5	4	0.999764	0.006153
60	13	5	5	1.000000	0.000236
60	13	6	0	0.214478	0.214478
60	13	6	1	0.612793	0.398315
60	13	6	2	0.890687	0.277894
60	13	6	3	0.983319	0.092631
60	13	6	4	0.998757	0.015439
60	13	6	5	0.999966	0.001208
60	13	6	6	1.000000	0.000034
60	13	7	0	0.162844	0.162844
60	13	7	1	0.524278	0.361434
60	13	7	2	0.834079	0.309801
60	13	7	3	0.966165	0.132086
60	13	7	4	0.996184	0.030019
60	13	7	5	0.999787	0.003602
60	13	7	6	0.999995	0.000209
60	13	7	7	1.000000	0.000004
60	13	8	0	0.122901	0.122901
60	13	8	1	0.442444	0.319543
60	13	8	2	0.769781	0.327337
60	13	8	3	0.941243	0.171462
60	13	8	4	0.991087	0.049844
60	13	8	5	0.999243	0.008156
60	13	8	6	0.999968	0.000725
60	13	8	7	0.999999	0.000032
60	13	8	8	1.000000	0.000000
60	13	9	0	0.092176	0.092176
60	13	9	1	0.368703	0.276528
60	13	9	2	0.700537	0.331833
60	13	9	3	0.908270	0.207733
60	13	9	4	0.982460	0.074190
60	13	9	5	0.997988	0.015528
60	13	9	6	0.999870	0.001882
60	13	9	7	0.999996	0.000125
60	13	9	8	1.000000	0.000004
60	13	9	9	1.000000	0.000000
60	13	10	0	0.068680	0.068680
60	13	10	1	0.303638	0.234958
60	13	10	2	0.628965	0.325327
60	13	10	3	0.867538	0.238573

N	n	k	x	P(x)	p(x)
60	13	10	4	0.969368	0.101830
60	13	10	5	0.995552	0.026185
60	13	10	6	0.999612	0.004060
60	13	10	7	0.999981	0.000369
60	13	10	8	0.999999	0.000018
60	13	10	9	1.000000	0.000000
60	13	10	10	1.000000	0.000000
60	13	11	0	0.050823	0.050823
60	13	11	1	0.247248	0.196425
60	13	11	2	0.557393	0.310145
60	13	11	3	0.819823	0.262430
60	13	11	4	0.951038	0.131215
60	13	11	5	0.991363	0.040325
60	13	11	6	0.999044	0.007681
60	13	11	7	0.999937	0.000893
60	13	11	8	0.999998	0.000061
60	13	11	9	1.000000	0.000002
60	13	11	10	1.000000	0.000000
60	13	11	11	1.000000	0.000000
60	13	12	0	0.037340	0.037340
60	13	12	1	0.199144	0.161805
60	13	12	2	0.487769	0.288624
60	13	12	3	0.766266	0.278497
60	13	12	4	0.926937	0.160671
60	13	12	5	0.984779	0.057842
60	13	12	6	0.997946	0.013167
60	13	12	7	0.999827	0.001881
60	13	12	8	0.999991	0.000164
60	13	12	9	1.000000	0.000008
60	13	12	10	1.000000	0.000000
60	13	12	11	1.000000	0.000000
60	13	12	12	1.000000	0.000000
60	13	13	0	0.027227	0.027227
60	13	13	1	0.158693	0.131466
60	13	13	2	0.421626	0.262933
60	13	13	3	0.708246	0.286620
60	13	13	4	0.896811	0.188566
60	13	13	5	0.975139	0.078327
60	13	13	6	0.996026	0.020887
60	13	13	7	0.999592	0.003566
60	13	13	8	0.999974	0.000382
60	13	13	9	0.999999	0.000025
60	13	13	10	1.000000	0.000001
60	13	13	11	1.000000	0.000000
60	13	13	12	1.000000	0.000000
60	13	13	13	1.000000	0.000000
60	14	1	0	0.766667	0.766667
60	14	1	1	1.000000	0.233333
60	14	2	0	0.584746	0.584746
60	14	2	1	0.948588	0.363842

N	n	k	x	P(x)	p(x)
60	14	2	2	1.000000	0.051412
60	14	3	0	0.443600	0.443600
60	14	3	1	0.867037	0.423437
60	14	3	2	0.989363	0.122326
60	14	3	3	1.000000	0.010637
60	14	4	0	0.334646	0.334646
60	14	4	1	0.770464	0.435818
60	14	4	2	0.963610	0.193146
60	14	4	3	0.997947	0.034337
60	14	4	4	1.000000	0.002053
60	14	5	0	0.250984	0.250984
60	14	5	1	0.669292	0.418307
60	14	5	2	0.922221	0.252930
60	14	5	3	0.991202	0.068981
60	14	5	4	0.999633	0.008431
60	14	5	5	1.000000	0.000367
60	14	6	0	0.187097	0.187097
60	14	6	1	0.570419	0.383322
60	14	6	2	0.867037	0.296618
60	14	6	3	0.977406	0.110369
60	14	6	4	0.998100	0.020694
60	14	6	5	0.999940	0.001839
60	14	6	6	1.000000	0.000060
60	14	7	0	0.138591	0.138591
60	14	7	1	0.478138	0.339547
60	14	7	2	0.801122	0.322984
60	14	7	3	0.954924	0.153802
60	14	7	4	0.994268	0.039345
60	14	7	5	0.999633	0.005365
60	14	7	6	0.999991	0.000358
60	14	7	7	1.000000	0.000009
60	14	8	0	0.101982	0.101982
60	14	8	1	0.394853	0.292871
60	14	8	2	0.727993	0.333141
60	14	8	3	0.923002	0.195009
60	14	8	4	0.986845	0.063842
60	14	8	5	0.998722	0.011878
60	14	8	6	0.999937	0.001215
60	14	8	7	0.999999	0.000062
60	14	8	8	1.000000	0.000001
60	14	9	0	0.074525	0.074525
60	14	9	1	0.321635	0.247110
60	14	9	2	0.651115	0.329480
60	14	9	3	0.881750	0.230636
60	14	9	4	0.974567	0.092817
60	14	9	5	0.996667	0.022099
60	14	9	6	0.999750	0.003084
60	14	9	7	0.999990	0.000240
60	14	9	8	1.000000	0.000009
60	14	9	9	1.000000	0.000000

N	n	k	x	P(x)	p(x)
60	14	10	0	0.054067	0.054067
60	14	10	1	0.258646	0.204579
60	14	10	2	0.573590	0.314944
60	14	10	3	0.832005	0.258415
60	14	10	4	0.956368	0.124362
60	14	10	5	0.992767	0.036399
60	14	10	6	0.999266	0.006500
60	14	10	7	0.999957	0.000691
60	14	10	8	0.999999	0.000041
60	14	10	9	1.000000	0.000001
60	14	10	10	1.000000	0.000000
60	14	11	0	0.038928	0.038928
60	14	11	1	0.205456	0.166527
60	14	11	2	0.498003	0.292548
60	14	11	3	0.775154	0.277151
60	14	11	4	0.931495	0.156341
60	14	11	5	0.986215	0.054719
60	14	11	6	0.998226	0.012012
60	14	11	7	0.999861	0.001634
60	14	11	8	0.999994	0.000133
60	14	11	9	1.000000	0.000006
60	14	11	10	1.000000	0.000000
60	14	11	11	1.000000	0.000000
60	14	12	0	0.027806	0.027806
60	14	12	1	0.161275	0.133469
60	14	12	2	0.426359	0.265084
60	14	12	3	0.712937	0.286577
60	14	12	4	0.899589	0.186652
60	14	12	5	0.976164	0.076575
60	14	12	6	0.996265	0.020101
60	14	12	7	0.999627	0.003362
60	14	12	8	0.999977	0.000350
60	14	12	9	0.999999	0.000022
60	14	12	10	1.000000	0.000001
60	14	12	11	1.000000	0.000000
60	14	12	12	1.000000	0.000000
60	14	13	0	0.019696	0.019696
60	14	13	1	0.125127	0.105431
60	14	13	2	0.360088	0.234961
60	14	13	3	0.647263	0.287175
60	14	13	4	0.860703	0.213441
60	14	13	5	0.961806	0.101103
60	14	13	6	0.992915	0.031109
60	14	13	7	0.999137	0.006222
60	14	13	8	0.999934	0.000797
60	14	13	9	0.999997	0.000063
60	14	13	10	1.000000	0.000003
60	14	13	11	1.000000	0.000000
60	14	13	12	1.000000	0.000000
60	14	13	13	1.000000	0.000000

N	n	k	x	P(x)	p(x)
60	14	14	0	0.013829	0.013829
60	14	14	1	0.095965	0.082136
60	14	14	2	0.300098	0.204133
60	14	14	3	0.580052	0.279953
60	14	14	4	0.815290	0.235239
60	14	14	5	0.942446	0.127156
60	14	14	6	0.987620	0.045174
60	14	14	7	0.998210	0.010590
60	14	14	8	0.999832	0.001622
60	14	14	9	0.999990	0.000158
60	14	14	10	1.000000	0.000009
60	14	14	11	1.000000	0.000000
60	14	14	12	1.000000	0.000000
60	14	14	13	1.000000	0.000000
60	14	14	14	1.000000	0.000000
60	15	1	0	0.750000	0.750000
60	15	1	1	1.000000	0.250000
60	15	2	0	0.559322	0.559322
60	15	2	1	0.940678	0.381356
60	15	2	2	1.000000	0.059322
60	15	3	0	0.414670	0.414670
60	15	3	1	0.848626	0.433957
60	15	3	2	0.986704	0.138077
60	15	3	3	1.000000	0.013296
60	15	4	0	0.305546	0.305546
60	15	4	1	0.742041	0.436494
60	15	4	2	0.955212	0.213172
60	15	4	3	0.997201	0.041988
60	15	4	4	1.000000	0.002799
60	15	5	0	0.223703	0.223703
60	15	5	1	0.632917	0.409214
60	15	5	2	0.905726	0.272809
60	15	5	3	0.988203	0.082477
60	15	5	4	0.999450	0.011247
60	15	5	5	1.000000	0.000550
60	15	6	0	0.162693	0.162693
60	15	6	1	0.528754	0.366060
60	15	6	2	0.841244	0.312490
60	15	6	3	0.970208	0.128964
60	15	6	4	0.997201	0.026993
60	15	6	5	0.999900	0.002699
60	15	6	6	1.000000	0.000100
60	15	7	0	0.117501	0.117501
60	15	7	1	0.433849	0.316348
60	15	7	2	0.766015	0.332166
60	15	7	3	0.941549	0.175535
60	15	7	4	0.991702	0.050153
60	15	7	5	0.999400	0.007698
60	15	7	6	0.999983	0.000583
60	15	7	7	1.000000	0.000017

N	n	k	x	P(x)	p(x)
60	15	8	0	0.084246	0.084246
60	15	8	1	0.350285	0.266039
60	15	8	2	0.684540	0.334255
60	15	8	3	0.901806	0.217266
60	15	8	4	0.981293	0.079487
60	15	8	5	0.997948	0.016654
60	15	8	6	0.999884	0.001937
60	15	8	7	0.999997	0.000113
60	15	8	8	1.000000	0.000003
60	15	9	0	0.059944	0.059944
60	15	9	1	0.278659	0.218715
60	15	9	2	0.600976	0.322317
60	15	9	3	0.851667	0.250691
60	15	9	4	0.964478	0.112811
60	15	9	5	0.994745	0.030266
60	15	9	6	0.999549	0.004804
60	15	9	7	0.999980	0.000431
60	15	9	8	0.999999	0.000020
60	15	9	9	1.000000	0.000000
60	15	10	0	0.042314	0.042314
60	15	10	1	0.218620	0.176306
60	15	10	2	0.518817	0.300197
60	15	10	3	0.792681	0.273864
60	15	10	4	0.940147	0.147465
60	15	10	5	0.988810	0.048664
60	15	10	6	0.998701	0.009891
60	15	10	7	0.999912	0.001211
60	15	10	8	0.999997	0.000084
60	15	10	9	1.000000	0.000003
60	15	10	10	1.000000	0.000000
60	15	11	0	0.029619	0.029619
60	15	11	1	0.169254	0.139635
60	15	11	2	0.440766	0.271512
60	15	11	3	0.726954	0.286188
60	15	11	4	0.907704	0.180750
60	15	11	5	0.979078	0.071373
60	15	11	6	0.996921	0.017843
60	15	11	7	0.999719	0.002798
60	15	11	8	0.999985	0.000266
60	15	11	9	0.999999	0.000014
60	15	11	10	1.000000	0.000000
60	15	11	11	1.000000	0.000000
60	15	12	0	0.020552	0.020552
60	15	12	1	0.129358	0.108806
60	15	12	2	0.368732	0.239374
60	15	12	3	0.656867	0.288135
60	15	12	4	0.867128	0.210261
60	15	12	5	0.964512	0.097384
60	15	12	6	0.993644	0.029132
60	15	12	7	0.999262	0.005618

N	n	k	x	P(x)	p(x)
60	15	12	8	0.999947	0.000685
60	15	12	9	0.999998	0.000051
60	15	12	10	1.000000	0.000002
60	15	12	11	1.000000	0.000000
60	15	12	12	1.000000	0.000000
60	15	13	0	0.014130	0.014130
60	15	13	1	0.097623	0.083494
60	15	13	2	0.303902	0.206278
60	15	13	3	0.584833	0.280932
60	15	13	4	0.818943	0.234110
60	15	13	5	0.944223	0.125280
60	15	13	6	0.988181	0.043958
60	15	13	7	0.998325	0.010144
60	15	13	8	0.999847	0.001522
60	15	13	9	0.999991	0.000144
60	15	13	10	1.000000	0.000008
60	15	13	11	1.000000	0.000000
60	15	13	12	1.000000	0.000000
60	15	13	13	1.000000	0.000000
60	15	14	0	0.009620	0.009620
60	15	14	1	0.072753	0.063133
60	15	14	2	0.246846	0.174093
60	15	14	3	0.513106	0.266260
60	15	14	4	0.764151	0.251045
60	15	14	5	0.917568	0.153417
60	15	14	6	0.979764	0.062196
60	15	14	7	0.996599	0.016835
60	15	14	8	0.999620	0.003022
60	15	14	9	0.999973	0.000353
60	15	14	10	0.999999	0.000026
60	15	14	11	1.000000	0.000001
60	15	14	12	1.000000	0.000000
60	15	14	13	1.000000	0.000000
60	15	14	14	1.000000	0.000000
60	15	15	0	0.006483	0.006483
60	15	15	1	0.053539	0.047055
60	15	15	2	0.197646	0.144107
60	15	15	3	0.443647	0.246001
60	15	15	4	0.704119	0.260472
60	15	15	5	0.884216	0.180098
60	15	15	6	0.967595	0.083379
60	15	15	7	0.993671	0.026076
60	15	15	8	0.999161	0.005490
60	15	15	9	0.999927	0.000766
60	15	15	10	0.999996	0.000069
60	15	15	11	1.000000	0.000004
60	15	15	12	1.000000	0.000000
60	15	15	13	1.000000	0.000000
60	15	15	14	1.000000	0.000000
60	15	15	15	1.000000	0.000000

Table for $N = 2$, $n = 1$, through $N = 100$, $n = 50$

N	n	k	x	P(x)	p(x)
60	16	1	0	0.733333	0.733333
60	16	1	1	1.000000	0.266667
60	16	2	0	0.534463	0.534463
60	16	2	1	0.932203	0.397740
60	16	2	2	1.000000	0.067797
60	16	3	0	0.387025	0.387025
60	16	3	1	0.829340	0.442314
60	16	3	2	0.983635	0.154296
60	16	3	3	1.000000	0.016365
60	16	4	0	0.278386	0.278386
60	16	4	1	0.712941	0.434555
60	16	4	2	0.945738	0.232797
60	16	4	3	0.996268	0.050530
60	16	4	4	1.000000	0.003732
60	16	5	0	0.198847	0.198847
60	16	5	1	0.596542	0.397695
60	16	5	2	0.887539	0.290996
60	16	5	3	0.984538	0.096999
60	16	5	4	0.999200	0.014663
60	16	5	5	1.000000	0.000800
60	16	6	0	0.141001	0.141001
60	16	6	1	0.488080	0.347079
60	16	6	2	0.813467	0.325387
60	16	6	3	0.961611	0.148144
60	16	6	4	0.996001	0.034390
60	16	6	5	0.999840	0.003839
60	16	6	6	1.000000	0.000160
60	16	7	0	0.099223	0.099223
60	16	7	1	0.391669	0.292446
60	16	7	2	0.729107	0.337438
60	16	7	3	0.925946	0.196839
60	16	7	4	0.988359	0.062412
60	16	7	5	0.999058	0.010699
60	16	7	6	0.999970	0.000912
60	16	7	7	1.000000	0.000030
60	16	8	0	0.069269	0.069269
60	16	8	1	0.308901	0.239633
60	16	8	2	0.639973	0.331071
60	16	8	3	0.877665	0.237692
60	16	8	4	0.974228	0.096562
60	16	8	5	0.996837	0.022610
60	16	8	6	0.999798	0.002961
60	16	8	7	0.999995	0.000197
60	16	8	8	1.000000	0.000005
60	16	9	0	0.047955	0.047955
60	16	9	1	0.239777	0.191821
60	16	9	2	0.550838	0.311062
60	16	9	3	0.818242	0.267404
60	16	9	4	0.951944	0.133702
60	16	9	5	0.992055	0.040111

N	n	k	x	P(x)	p(x)
60	16	9	6	0.999229	0.007174
60	16	9	7	0.999961	0.000732
60	16	9	8	0.999999	0.000038
60	16	9	9	1.000000	0.000001
60	16	10	0	0.032911	0.032911
60	16	10	1	0.183359	0.150448
60	16	10	2	0.465449	0.282090
60	16	10	3	0.750080	0.284632
60	16	10	4	0.920485	0.170404
60	16	10	5	0.983403	0.062919
60	16	10	6	0.997822	0.014419
60	16	10	7	0.999832	0.002010
60	16	10	8	0.999993	0.000161
60	16	10	9	1.000000	0.000007
60	16	10	10	1.000000	0.000000
60	16	11	0	0.022379	0.022379
60	16	11	1	0.138224	0.115845
60	16	11	2	0.386464	0.248239
60	16	11	3	0.676076	0.289613
60	16	11	4	0.879588	0.203512
60	16	11	5	0.969561	0.089974
60	16	11	6	0.994938	0.025377
60	16	11	7	0.999470	0.004532
60	16	11	8	0.999967	0.000497
60	16	11	9	0.999999	0.000032
60	16	11	10	1.000000	0.000001
60	16	11	11	1.000000	0.000000
60	16	12	0	0.015072	0.015072
60	16	12	1	0.102761	0.087690
60	16	12	2	0.315538	0.212777
60	16	12	3	0.599240	0.283702
60	16	12	4	0.829748	0.230508
60	16	12	5	0.949363	0.119615
60	16	12	6	0.989759	0.040396
60	16	12	7	0.998638	0.008878
60	16	12	8	0.999886	0.001249
60	16	12	9	0.999994	0.000108
60	16	12	10	1.000000	0.000005
60	16	12	11	1.000000	0.000000
60	16	12	12	1.000000	0.000000
60	16	13	0	0.010048	0.010048
60	16	13	1	0.075358	0.065311
60	16	13	2	0.253478	0.178120
60	16	13	3	0.522404	0.268926
60	16	13	4	0.772121	0.249717
60	16	13	5	0.921951	0.149830
60	16	13	6	0.981343	0.059392
60	16	13	7	0.996973	0.015630
60	16	13	8	0.999678	0.002705
60	16	13	9	0.999979	0.000301

N	n	k	x	P(x)	p(x)
60	16	13	10	0.999999	0.000021
60	16	13	11	1.000000	0.000001
60	16	13	12	1.000000	0.000000
60	16	13	13	1.000000	0.000000
60	16	14	0	0.006627	0.006627
60	16	14	1	0.054515	0.047887
60	16	14	2	0.200421	0.145907
60	16	14	3	0.448020	0.247599
60	16	14	4	0.708364	0.260343
60	16	14	5	0.886885	0.178521
60	16	14	6	0.968707	0.081822
60	16	14	7	0.993980	0.025273
60	16	14	8	0.999218	0.005238
60	16	14	9	0.999934	0.000716
60	16	14	10	0.999996	0.000063
60	16	14	11	1.000000	0.000003
60	16	14	12	1.000000	0.000000
60	16	14	13	1.000000	0.000000
60	16	14	14	1.000000	0.000000
60	16	15	0	0.004322	0.004322
60	16	15	1	0.038899	0.034577
60	16	15	2	0.156015	0.117116
60	16	15	3	0.378047	0.222032
60	16	15	4	0.640448	0.262401
60	16	15	5	0.844195	0.203747
60	16	15	6	0.950919	0.106725
60	16	15	7	0.989035	0.038116
60	16	15	8	0.998307	0.009271
60	16	15	9	0.999825	0.001518
60	16	15	10	0.999988	0.000163
60	16	15	11	0.999999	0.000011
60	16	15	12	1.000000	0.000000
60	16	15	13	1.000000	0.000000
60	16	15	14	1.000000	0.000000
60	16	15	15	1.000000	0.000000
60	16	16	0	0.002785	0.002785
60	16	16	1	0.027373	0.024588
60	16	16	2	0.119579	0.092205
60	16	16	3	0.313904	0.194325
60	16	16	4	0.570474	0.256570
60	16	16	5	0.794390	0.223916
60	16	16	6	0.927203	0.132813
60	16	16	7	0.981412	0.054209
60	16	16	8	0.996658	0.015246
60	16	16	9	0.999589	0.002930
60	16	16	10	0.999967	0.000378
60	16	16	11	0.999998	0.000032
60	16	16	12	1.000000	0.000002
60	16	16	13	1.000000	0.000000
60	16	16	14	1.000000	0.000000

N	n	k	x	P(x)	p(x)
60	16	16	15	1.000000	0.000000
60	16	16	16	1.000000	0.000000
60	17	1	0	0.716667	0.716667
60	17	1	1	1.000000	0.283333
60	17	2	0	0.510169	0.510169
60	17	2	1	0.923164	0.412994
60	17	2	2	1.000000	0.076836
60	17	3	0	0.360637	0.360637
60	17	3	1	0.809234	0.448597
60	17	3	2	0.980128	0.170894
60	17	3	3	1.000000	0.019871
60	17	4	0	0.253079	0.253079
60	17	4	1	0.683312	0.430234
60	17	4	2	0.935156	0.251844
60	17	4	3	0.995119	0.059963
60	17	4	4	1.000000	0.004881
60	17	5	0	0.176251	0.176251
60	17	5	1	0.560388	0.384137
60	17	5	2	0.867698	0.307310
60	17	5	3	0.980128	0.112430
60	17	5	4	0.998867	0.018738
60	17	5	5	1.000000	0.001133
60	17	6	0	0.121774	0.121774
60	17	6	1	0.448639	0.326866
60	17	6	2	0.783886	0.335247
60	17	6	3	0.951510	0.167624
60	17	6	4	0.994438	0.042928
60	17	6	5	0.999753	0.005315
60	17	6	6	1.000000	0.000247
60	17	7	0	0.083437	0.083437
60	17	7	1	0.351790	0.268353
60	17	7	2	0.690762	0.338972
60	17	7	3	0.908052	0.217290
60	17	7	4	0.984103	0.076051
60	17	7	5	0.998572	0.014468
60	17	7	6	0.999949	0.001378
60	17	7	7	1.000000	0.000050
60	17	8	0	0.056674	0.056674
60	17	8	1	0.270778	0.214104
60	17	8	2	0.594827	0.324049
60	17	8	3	0.850655	0.255828
60	17	8	4	0.965449	0.114795
60	17	8	5	0.995296	0.029847
60	17	8	6	0.999664	0.004368
60	17	8	7	0.999990	0.000327
60	17	8	8	1.000000	0.000009
60	17	9	0	0.038146	0.038146
60	17	9	1	0.204900	0.166754
60	17	9	2	0.501351	0.296451
60	17	9	3	0.781778	0.280427

Table for $N=2$, $n=1$, through $N=100$, $n=50$

N	n	k	x	P(x)	p(x)
60	17	9	4	0.936751	0.154973
60	17	9	5	0.988408	0.051658
60	17	9	6	0.998740	0.010332
60	17	9	7	0.999928	0.001188
60	17	9	8	0.999998	0.000071
60	17	9	9	1.000000	0.000002
60	17	10	0	0.025431	0.025431
60	17	10	1	0.152585	0.127154
60	17	10	2	0.414160	0.261575
60	17	10	3	0.704798	0.290638
60	17	10	4	0.897248	0.192450
60	17	10	5	0.976253	0.079006
60	17	10	6	0.996511	0.020258
60	17	10	7	0.999695	0.003183
60	17	10	8	0.999986	0.000291
60	17	10	9	1.000000	0.000014
60	17	10	10	1.000000	0.000000
60	17	11	0	0.016784	0.016784
60	17	11	1	0.111896	0.095111
60	17	11	2	0.335687	0.223792
60	17	11	3	0.623419	0.287732
60	17	11	4	0.847211	0.223792
60	17	11	5	0.957292	0.110081
60	17	11	6	0.992055	0.034762
60	17	11	7	0.999058	0.007003
60	17	11	8	0.999933	0.000875
60	17	11	9	0.999997	0.000064
60	17	11	10	1.000000	0.000002
60	17	11	11	1.000000	0.000000
60	17	12	0	0.010961	0.010961
60	17	12	1	0.080839	0.069878
60	17	12	2	0.267180	0.186341
60	17	12	3	0.541210	0.274030
60	17	12	4	0.787837	0.246627
60	17	12	5	0.930333	0.142496
60	17	12	6	0.984251	0.053917
60	17	12	7	0.997629	0.013378
60	17	12	8	0.999773	0.002144
60	17	12	9	0.999987	0.000214
60	17	12	10	0.999999	0.000013
60	17	12	11	1.000000	0.000000
60	17	12	12	1.000000	0.000000
60	17	13	0	0.007079	0.007079
60	17	13	1	0.057546	0.050467
60	17	13	2	0.208948	0.151402
60	17	13	3	0.461285	0.252336
60	17	13	4	0.721043	0.259758
60	17	13	5	0.894709	0.173667
60	17	13	6	0.971895	0.077185
60	17	13	7	0.994842	0.022947

N	n	k	x	P(x)	p(x)
60	17	13	8	0.999371	0.004529
60	17	13	9	0.999951	0.000581
60	17	13	10	0.999998	0.000046
60	17	13	11	1.000000	0.000002
60	17	13	12	1.000000	0.000000
60	17	13	13	1.000000	0.000000
60	17	14	0	0.004519	0.004519
60	17	14	1	0.040366	0.035847
60	17	14	2	0.160628	0.120262
60	17	14	3	0.386121	0.225492
60	17	14	4	0.649195	0.263074
60	17	14	5	0.850369	0.201174
60	17	14	6	0.953830	0.103461
60	17	14	7	0.989959	0.036129
60	17	14	8	0.998503	0.008544
60	17	14	9	0.999852	0.001349
60	17	14	10	0.999991	0.000138
60	17	14	11	1.000000	0.000009
60	17	14	12	1.000000	0.000000
60	17	14	13	1.000000	0.000000
60	17	14	14	1.000000	0.000000
60	17	15	0	0.002849	0.002849
60	17	15	1	0.027897	0.025049
60	17	15	2	0.121412	0.093515
60	17	15	3	0.317493	0.196080
60	17	15	4	0.574848	0.257355
60	17	15	5	0.797889	0.223041
60	17	15	6	0.929089	0.131201
60	17	15	7	0.982105	0.053016
60	17	15	8	0.996832	0.014727
60	17	15	9	0.999618	0.002786
60	17	15	10	0.999970	0.000352
60	17	15	11	0.999998	0.000029
60	17	15	12	1.000000	0.000001
60	17	15	13	1.000000	0.000000
60	17	15	14	1.000000	0.000000
60	17	15	15	1.000000	0.000000
60	17	16	0	0.001773	0.001773
60	17	16	1	0.018991	0.017219
60	17	16	2	0.090241	0.071250
60	17	16	3	0.256490	0.166249
60	17	16	4	0.500501	0.244011
60	17	16	5	0.738411	0.237910
60	17	16	6	0.897018	0.158607
60	17	16	7	0.970324	0.073306
60	17	16	8	0.993886	0.023563
60	17	16	9	0.999122	0.005236
60	17	16	10	0.999915	0.000792
60	17	16	11	0.999995	0.000080
60	17	16	12	1.000000	0.000005

N	n	k	x	P(x)	p(x)
60	17	16	13	1.000000	0.000000
60	17	16	14	1.000000	0.000000
60	17	16	15	1.000000	0.000000
60	17	16	16	1.000000	0.000000
60	17	17	0	0.001088	0.001088
60	17	17	1	0.012730	0.011642
60	17	17	2	0.065951	0.053221
60	17	17	3	0.203592	0.137641
60	17	17	4	0.428406	0.224814
60	17	17	5	0.673526	0.245120
60	17	17	6	0.857366	0.183840
60	17	17	7	0.953663	0.096297
60	17	17	8	0.989067	0.035403
60	17	17	9	0.998170	0.009104
60	17	17	10	0.999789	0.001618
60	17	17	11	0.999984	0.000195
60	17	17	12	0.999999	0.000015
60	17	17	13	1.000000	0.000001
60	17	17	14	1.000000	0.000000
60	17	17	15	1.000000	0.000000
60	17	17	16	1.000000	0.000000
60	17	17	17	1.000000	0.000000
60	18	1	0	0.700000	0.700000
60	18	1	1	1.000000	0.300000
60	18	2	0	0.486441	0.486441
60	18	2	1	0.913559	0.427119
60	18	2	2	1.000000	0.086441
60	18	3	0	0.335476	0.335476
60	18	3	1	0.788369	0.452893
60	18	3	2	0.976154	0.187785
60	18	3	3	1.000000	0.023846
60	18	4	0	0.229536	0.229536
60	18	4	1	0.653296	0.423760
60	18	4	2	0.923443	0.270147
60	18	4	3	0.993725	0.070282
60	18	4	4	1.000000	0.006275
60	18	5	0	0.155757	0.155757
60	18	5	1	0.524655	0.368898
60	18	5	2	0.846258	0.321603
60	18	5	3	0.974899	0.128641
60	18	5	4	0.998431	0.023532
60	18	5	5	1.000000	0.001569
60	18	6	0	0.104782	0.104782
60	18	6	1	0.410632	0.305850
60	18	6	2	0.752701	0.342069
60	18	6	3	0.939815	0.187115
60	18	6	4	0.992441	0.052626
60	18	6	5	0.999629	0.007188
60	18	6	6	1.000000	0.000371
60	18	7	0	0.069855	0.069855

N	n	k	x	P(x)	p(x)
60	18	7	1	0.314346	0.244491
60	18	7	2	0.651347	0.337001
60	18	7	3	0.887839	0.236492
60	18	7	4	0.978797	0.090958
60	18	7	5	0.997899	0.019101
60	18	7	6	0.999917	0.002019
60	18	7	7	1.000000	0.000082
60	18	8	0	0.046130	0.046130
60	18	8	1	0.235924	0.189794
60	18	8	2	0.549611	0.313687
60	18	8	3	0.820907	0.271297
60	18	8	4	0.954771	0.133863
60	18	8	5	0.993213	0.038443
60	18	8	6	0.999460	0.006247
60	18	8	7	0.999983	0.000522
60	18	8	8	1.000000	0.000017
60	18	9	0	0.030162	0.030162
60	18	9	1	0.173876	0.143714
60	18	9	2	0.453092	0.279216
60	18	9	3	0.742649	0.289557
60	18	9	4	0.918730	0.176082
60	18	9	5	0.983603	0.064872
60	18	9	6	0.998019	0.014416
60	18	9	7	0.999872	0.001853
60	18	9	8	0.999997	0.000124
60	18	9	9	1.000000	0.000003
60	18	10	0	0.019517	0.019517
60	18	10	1	0.125971	0.106455
60	18	10	2	0.365495	0.239523
60	18	10	3	0.657485	0.291990
60	18	10	4	0.870394	0.212909
60	18	10	5	0.967067	0.096672
60	18	10	6	0.994627	0.027560
60	18	10	7	0.999472	0.004846
60	18	10	8	0.999972	0.000500
60	18	10	9	0.999999	0.000027
60	18	10	10	1.000000	0.000001
60	18	11	0	0.012491	0.012491
60	18	11	1	0.089777	0.077286
60	18	11	2	0.288847	0.199070
60	18	11	3	0.569888	0.281041
60	18	11	4	0.810780	0.240892
60	18	11	5	0.941932	0.131152
60	18	11	6	0.988012	0.046081
60	18	11	7	0.998406	0.010394
60	18	11	8	0.999872	0.001466
60	18	11	9	0.999994	0.000122
60	18	11	10	1.000000	0.000005
60	18	11	11	1.000000	0.000000
60	18	12	0	0.007902	0.007902

Table for $N = 2$, $n = 1$, through $N = 100$, $n = 50$

N	n	k	x	P(x)	p(x)
60	18	12	1	0.062963	0.055061
60	18	12	2	0.223845	0.160881
60	18	12	3	0.483855	0.260010
60	18	12	4	0.741953	0.258098
60	18	12	5	0.907136	0.165183
60	18	12	6	0.976727	0.069591
60	18	12	7	0.996073	0.019346
60	18	12	8	0.999573	0.003500
60	18	12	9	0.999972	0.000399
60	18	12	10	0.999999	0.000027
60	18	12	11	1.000000	0.000001
60	18	12	12	1.000000	0.000000
60	18	13	0	0.004939	0.004939
60	18	13	1	0.043462	0.038524
60	18	13	2	0.170217	0.126755
60	18	13	3	0.402602	0.232384
60	18	13	4	0.666675	0.264073
60	18	13	5	0.862399	0.195725
60	18	13	6	0.959330	0.096930
60	18	13	7	0.991640	0.032310
60	18	13	8	0.998844	0.007204
60	18	13	9	0.999897	0.001053
60	18	13	10	0.999994	0.000097
60	18	13	11	1.000000	0.000005
60	18	13	12	1.000000	0.000000
60	18	13	13	1.000000	0.000000
60	18	14	0	0.003047	0.003047
60	18	14	1	0.029528	0.026481
60	18	14	2	0.127067	0.097538
60	18	14	3	0.328437	0.201370
60	18	14	4	0.588015	0.259578
60	18	14	5	0.808263	0.220248
60	18	14	6	0.934581	0.126319
60	18	14	7	0.984078	0.049496
60	18	14	8	0.997311	0.013233
60	18	14	9	0.999696	0.002384
60	18	14	10	0.999978	0.000282
60	18	14	11	0.999999	0.000021
60	18	14	12	1.000000	0.000001
60	18	14	13	1.000000	0.000000
60	18	14	14	1.000000	0.000000
60	18	15	0	0.001855	0.001855
60	18	15	1	0.019742	0.017887
60	18	15	2	0.093140	0.073398
60	18	15	3	0.262773	0.169632
60	18	15	4	0.509013	0.246240
60	18	15	5	0.746019	0.237006
60	18	15	6	0.901629	0.155610
60	18	15	7	0.972242	0.070613
60	18	15	8	0.994434	0.022193

N	n	k	x	P(x)	p(x)
60	18	15	9	0.999229	0.004795
60	18	15	10	0.999929	0.000700
60	18	15	11	0.999996	0.000067
60	18	15	12	1.000000	0.000004
60	18	15	13	1.000000	0.000000
60	18	15	14	1.000000	0.000000
60	18	15	15	1.000000	0.000000
60	18	16	0	0.001113	0.001113
60	18	16	1	0.012985	0.011872
60	18	16	2	0.067043	0.054059
60	18	16	3	0.206228	0.139185
60	18	16	4	0.432405	0.226176
60	18	16	5	0.677550	0.245146
60	18	16	6	0.860133	0.182582
60	18	16	7	0.954981	0.094848
60	18	16	8	0.989503	0.034522
60	18	16	9	0.998270	0.008767
60	18	16	10	0.999804	0.001534
60	18	16	11	0.999985	0.000181
60	18	16	12	0.999999	0.000014
60	18	16	13	1.000000	0.000001
60	18	16	14	1.000000	0.000000
60	18	16	15	1.000000	0.000000
60	18	16	16	1.000000	0.000000
60	18	17	0	0.000658	0.000658
60	18	17	1	0.008398	0.007740
60	18	17	2	0.047386	0.038988
60	18	17	3	0.158779	0.111393
60	18	17	4	0.360439	0.201660
60	18	17	5	0.605121	0.244681
60	18	17	6	0.810338	0.205217
60	18	17	7	0.931269	0.120931
60	18	17	8	0.981657	0.050388
60	18	17	9	0.996477	0.014820
60	18	17	10	0.999525	0.003049
60	18	17	11	0.999957	0.000431
60	18	17	12	0.999997	0.000041
60	18	17	13	1.000000	0.000002
60	18	17	14	1.000000	0.000000
60	18	17	15	1.000000	0.000000
60	18	17	16	1.000000	0.000000
60	18	17	17	1.000000	0.000000
60	18	18	0	0.000382	0.000382
60	18	18	1	0.005338	0.004955
60	18	18	2	0.032879	0.027541
60	18	18	3	0.119921	0.087042
60	18	18	4	0.294783	0.174862
60	18	18	5	0.531147	0.236365
60	18	18	6	0.753068	0.221920
60	18	18	7	0.900333	0.147265

N	n	k	x	P(x)	p(x)
60	18	18	8	0.969939	0.069606
60	18	18	9	0.993375	0.023436
60	18	18	10	0.998958	0.005583
60	18	18	11	0.999886	0.000928
60	18	18	12	0.999992	0.000105
60	18	18	13	1.000000	0.000008
60	18	18	14	1.000000	0.000000
60	18	18	15	1.000000	0.000000
60	18	18	16	1.000000	0.000000
60	18	18	17	1.000000	0.000000
60	18	18	18	1.000000	0.000000
60	19	1	0	0.683333	0.683333
60	19	1	1	1.000000	0.316667
60	19	2	0	0.463277	0.463277
60	19	2	1	0.903390	0.440113
60	19	2	2	1.000000	0.096610
60	19	3	0	0.311514	0.311514
60	19	3	1	0.766803	0.455289
60	19	3	2	0.971683	0.204880
60	19	3	3	1.000000	0.028317
60	19	4	0	0.207676	0.207676
60	19	4	1	0.623027	0.415352
60	19	4	2	0.910579	0.287551
60	19	4	3	0.992051	0.081473
60	19	4	4	1.000000	0.007949
60	19	5	0	0.137214	0.137214
60	19	5	1	0.489522	0.352307
60	19	5	2	0.823286	0.333765
60	19	5	3	0.968773	0.145487
60	19	5	4	0.997871	0.029097
60	19	5	5	1.000000	0.002129
60	19	6	0	0.089813	0.089813
60	19	6	1	0.374221	0.284408
60	19	6	2	0.720123	0.345902
60	19	6	3	0.926450	0.206327
60	19	6	4	0.989935	0.063485
60	19	6	5	0.999458	0.009523
60	19	6	6	1.000000	0.000542
60	19	7	0	0.058212	0.058212
60	19	7	1	0.279418	0.221206
60	19	7	2	0.611228	0.331809
60	19	7	3	0.865316	0.254088
60	19	7	4	0.972300	0.106985
60	19	7	5	0.996989	0.024689
60	19	7	6	0.999869	0.002880
60	19	7	7	1.000000	0.000130
60	19	8	0	0.037344	0.037344
60	19	8	1	0.204292	0.166948
60	19	8	2	0.504798	0.300507
60	19	8	3	0.788610	0.283812

N	n	k	x	P(x)	p(x)
60	19	8	4	0.942022	0.153412
60	19	8	5	0.990468	0.048446
60	19	8	6	0.999163	0.008695
60	19	8	7	0.999970	0.000807
60	19	8	8	1.000000	0.000030
60	19	9	0	0.023699	0.023699
60	19	9	1	0.146502	0.122803
60	19	9	2	0.406556	0.260054
60	19	9	3	0.701283	0.294728
60	19	9	4	0.897768	0.196485
60	19	9	5	0.977424	0.079656
60	19	9	6	0.996989	0.019565
60	19	9	7	0.999784	0.002795
60	19	9	8	0.999994	0.000210
60	19	9	9	1.000000	0.000006
60	19	10	0	0.014870	0.014870
60	19	10	1	0.103160	0.088290
60	19	10	2	0.319871	0.216711
60	19	10	3	0.608820	0.288949
60	19	10	4	0.839979	0.231159
60	19	10	5	0.955558	0.115579
60	19	10	6	0.992002	0.036444
60	19	10	7	0.999126	0.007124
60	19	10	8	0.999948	0.000822
60	19	10	9	0.999999	0.000050
60	19	10	10	1.000000	0.000001
60	19	11	0	0.009219	0.009219
60	19	11	1	0.071375	0.062156
60	19	11	2	0.246189	0.174814
60	19	11	3	0.516356	0.270167
60	19	11	4	0.770631	0.254275
60	19	11	5	0.923196	0.152565
60	19	11	6	0.982527	0.059331
60	19	11	7	0.997417	0.014890
60	19	11	8	0.999768	0.002351
60	19	11	9	0.999989	0.000221
60	19	11	10	1.000000	0.000011
60	19	11	11	1.000000	0.000000
60	19	12	0	0.005644	0.005644
60	19	12	1	0.048543	0.042898
60	19	12	2	0.185540	0.136997
60	19	12	3	0.428138	0.242599
60	19	12	4	0.692792	0.264653
60	19	12	5	0.879606	0.186814
60	19	12	6	0.966786	0.087180
60	19	12	7	0.993770	0.026984
60	19	12	8	0.999240	0.005470
60	19	12	9	0.999944	0.000704
60	19	12	10	0.999998	0.000054
60	19	12	11	1.000000	0.000002

Table for $N = 2$, $n = 1$, through $N = 100$, $n = 50$

N	n	k	x	P(x)	p(x)
60	19	12	12	1.000000	0.000000
60	19	13	0	0.003410	0.003410
60	19	13	1	0.032456	0.029046
60	19	13	2	0.137020	0.104564
60	19	13	3	0.347272	0.210252
60	19	13	4	0.610088	0.262815
60	19	13	5	0.825118	0.215031
60	19	13	6	0.943175	0.118056
60	19	13	7	0.987024	0.043849
60	19	13	8	0.997986	0.010962
60	19	13	9	0.999797	0.001811
60	19	13	10	0.999988	0.000191
60	19	13	11	1.000000	0.000012
60	19	13	12	1.000000	0.000000
60	19	13	13	1.000000	0.000000
60	19	14	0	0.002032	0.002032
60	19	14	1	0.021332	0.019300
60	19	14	2	0.099199	0.077867
60	19	14	3	0.275697	0.176498
60	19	14	4	0.526210	0.250513
60	19	14	5	0.761067	0.234856
60	19	14	6	0.910521	0.149454
60	19	14	7	0.975828	0.065308
60	19	14	8	0.995421	0.019592
60	19	14	9	0.999412	0.003991
60	19	14	10	0.999951	0.000539
60	19	14	11	0.999997	0.000046
60	19	14	12	1.000000	0.000002
60	19	14	13	1.000000	0.000000
60	19	14	14	1.000000	0.000000
60	19	15	0	0.001192	0.001192
60	19	15	1	0.013780	0.012587
60	19	15	2	0.070422	0.056642
60	19	15	3	0.214306	0.143884
60	19	15	4	0.444521	0.230215
60	19	15	5	0.689589	0.245068
60	19	15	6	0.868284	0.178695
60	19	15	7	0.958792	0.090508
60	19	15	8	0.990736	0.031944
60	19	15	9	0.998544	0.007809
60	19	15	10	0.999846	0.001301
60	19	15	11	0.999989	0.000144
60	19	15	12	1.000000	0.000010
60	19	15	13	1.000000	0.000000
60	19	15	14	1.000000	0.000000
60	19	15	15	1.000000	0.000000
60	19	16	0	0.000689	0.000689
60	19	16	1	0.008745	0.008056
60	19	16	2	0.049024	0.040279
60	19	16	3	0.163147	0.114124

N	n	k	x	P(x)	p(x)
60	19	16	4	0.367783	0.204635
60	19	16	5	0.613345	0.245563
60	19	16	6	0.816661	0.203315
60	19	16	7	0.934656	0.117995
60	19	16	8	0.982927	0.048271
60	19	16	9	0.996809	0.013882
60	19	16	10	0.999585	0.002776
60	19	16	11	0.999964	0.000379
60	19	16	12	0.999998	0.000034
60	19	16	13	1.000000	0.000002
60	19	16	14	1.000000	0.000000
60	19	16	15	1.000000	0.000000
60	19	16	16	1.000000	0.000000
60	19	17	0	0.000391	0.000391
60	19	17	1	0.005449	0.005058
60	19	17	2	0.033461	0.028012
60	19	17	3	0.121648	0.088186
60	19	17	4	0.298021	0.176373
60	19	17	5	0.535212	0.237191
60	19	17	6	0.756590	0.221378
60	19	17	7	0.902476	0.145885
60	19	17	8	0.970859	0.068384
60	19	17	9	0.993654	0.022795
60	19	17	10	0.999017	0.005363
60	19	17	11	0.999895	0.000878
60	19	17	12	0.999992	0.000098
60	19	17	13	1.000000	0.000007
60	19	17	14	1.000000	0.000000
60	19	17	15	1.000000	0.000000
60	19	17	16	1.000000	0.000000
60	19	17	17	1.000000	0.000000
60	19	18	0	0.000218	0.000218
60	19	18	1	0.003332	0.003114
60	19	18	2	0.022387	0.019055
60	19	18	3	0.088834	0.066447
60	19	18	4	0.236495	0.147661
60	19	18	5	0.457987	0.221492
60	19	18	6	0.689662	0.231675
60	19	18	7	0.861763	0.172101
60	19	18	8	0.953366	0.091602
60	19	18	9	0.988353	0.034987
60	19	18	10	0.997895	0.009542
60	19	18	11	0.999732	0.001837
60	19	18	12	0.999977	0.000245
60	19	18	13	0.999999	0.000022
60	19	18	14	1.000000	0.000001
60	19	18	15	1.000000	0.000000
60	19	18	16	1.000000	0.000000
60	19	18	17	1.000000	0.000000
60	19	18	18	1.000000	0.000000

N	n	k	x	P(x)	p(x)
60	19	19	0	0.000120	0.000120
60	19	19	1	0.001998	0.001878
60	19	19	2	0.014674	0.012676
60	19	19	3	0.063521	0.048847
60	19	19	4	0.183759	0.120238
60	19	19	5	0.384156	0.200397
60	19	19	6	0.617953	0.233797
60	19	19	7	0.812592	0.194639
60	19	19	8	0.929375	0.116783
60	19	19	9	0.980023	0.050648
60	19	19	10	0.995850	0.015827
60	19	19	11	0.999382	0.003532
60	19	19	12	0.999936	0.000554
60	19	19	13	0.999995	0.000060
60	19	19	14	1.000000	0.000004
60	19	19	15	1.000000	0.000000
60	19	19	16	1.000000	0.000000
60	19	19	17	1.000000	0.000000
60	19	19	18	1.000000	0.000000
60	19	19	19	1.000000	0.000000
60	20	1	0	0.666667	0.666667
60	20	1	1	1.000000	0.333333
60	20	2	0	0.440678	0.440678
60	20	2	1	0.892655	0.451977
60	20	2	2	1.000000	0.107345
60	20	3	0	0.288720	0.288720
60	20	3	1	0.744594	0.455874
60	20	3	2	0.966686	0.222092
60	20	3	3	1.000000	0.033314
60	20	4	0	0.187415	0.187415
60	20	4	1	0.592636	0.405221
60	20	4	2	0.896552	0.303916
60	20	4	3	0.990064	0.093513
60	20	4	4	1.000000	0.009936
60	20	5	0	0.120481	0.120481
60	20	5	1	0.455150	0.334669
60	20	5	2	0.798864	0.343714
60	20	5	3	0.961676	0.162812
60	20	5	4	0.997161	0.035485
60	20	5	5	1.000000	0.002839
60	20	6	0	0.076670	0.076670
60	20	6	1	0.339537	0.262867
60	20	6	2	0.686376	0.346839
60	20	6	3	0.911353	0.224977
60	20	6	4	0.986838	0.075486
60	20	6	5	0.999226	0.012387
60	20	6	6	1.000000	0.000774
60	20	7	0	0.048273	0.048273
60	20	7	1	0.247047	0.198773
60	20	7	2	0.570763	0.323716

N	n	k	x	P(x)	p(x)
60	20	7	3	0.840527	0.269764
60	20	7	4	0.964472	0.123945
60	20	7	5	0.995785	0.031313
60	20	7	6	0.999799	0.004014
60	20	7	7	1.000000	0.000201
60	20	8	0	0.030057	0.030057
60	20	8	1	0.175788	0.145731
60	20	8	2	0.460822	0.285033
60	20	8	3	0.753999	0.293177
60	20	8	4	0.927055	0.173056
60	20	8	5	0.986923	0.059868
60	20	8	6	0.998739	0.011816
60	20	8	7	0.999951	0.001212
60	20	8	8	1.000000	0.000049
60	20	9	0	0.018497	0.018497
60	20	9	1	0.122540	0.104044
60	20	9	2	0.362156	0.239616
60	20	9	3	0.658152	0.295996
60	20	9	4	0.873807	0.215654
60	20	9	5	0.969653	0.095846
60	20	9	6	0.995558	0.025904
60	20	9	7	0.999648	0.004090
60	20	9	8	0.999989	0.000341
60	20	9	9	1.000000	0.000011
60	20	10	0	0.011243	0.011243
60	20	10	1	0.083779	0.072536
60	20	10	2	0.277586	0.193807
60	20	10	3	0.559487	0.281901
60	20	10	4	0.806150	0.246663
60	20	10	5	0.941463	0.135312
60	20	10	6	0.988446	0.046984
60	20	10	7	0.998605	0.010159
60	20	10	8	0.999908	0.001303
60	20	10	9	0.999997	0.000089
60	20	10	10	1.000000	0.000002
60	20	11	0	0.006746	0.006746
60	20	11	1	0.056215	0.049470
60	20	11	2	0.207815	0.151600
60	20	11	3	0.463641	0.255825
60	20	11	4	0.727218	0.263578
60	20	11	5	0.900869	0.173651
60	20	11	6	0.975291	0.074422
60	20	11	7	0.995964	0.020673
60	20	11	8	0.999596	0.003632
60	20	11	9	0.999978	0.000382
60	20	11	10	0.999999	0.000022
60	20	11	11	1.000000	0.000000
60	20	12	0	0.003992	0.003992
60	20	12	1	0.037033	0.033041
60	20	12	2	0.152126	0.115092

Table for $N = 2$, $n = 1$, through $N = 100$, $n = 50$

N	n	k	x	P(x)	p(x)
60	20	12	3	0.374885	0.222759
60	20	12	4	0.641152	0.266267
60	20	12	5	0.847711	0.206559
60	20	12	6	0.954028	0.106317
60	20	12	7	0.990479	0.036452
60	20	12	8	0.998706	0.008227
60	20	12	9	0.999892	0.001186
60	20	12	10	0.999995	0.000103
60	20	12	11	1.000000	0.000005
60	20	12	12	1.000000	0.000000
60	20	13	0	0.002329	0.002329
60	20	13	1	0.023955	0.021626
60	20	13	2	0.108966	0.085011
60	20	13	3	0.295991	0.187025
60	20	13	4	0.552396	0.256405
60	20	13	5	0.783161	0.230765
60	20	13	6	0.923019	0.139857
60	20	13	7	0.980607	0.057588
60	20	13	8	0.996649	0.016042
60	20	13	9	0.999620	0.002971
60	20	13	10	0.999974	0.000353
60	20	13	11	0.999999	0.000025
60	20	13	12	1.000000	0.000001
60	20	13	13	1.000000	0.000000
60	20	14	0	0.001338	0.001338
60	20	14	1	0.015212	0.013874
60	20	14	2	0.076408	0.061196
60	20	14	3	0.228344	0.151935
60	20	14	4	0.465109	0.236766
60	20	14	5	0.709513	0.244403
60	20	14	6	0.881359	0.171846
60	20	14	7	0.964678	0.083319
60	20	14	8	0.992553	0.027875
60	20	14	9	0.998925	0.006371
60	20	14	10	0.999898	0.000973
60	20	14	11	0.999994	0.000096
60	20	14	12	1.000000	0.000006
60	20	14	13	1.000000	0.000000
60	20	14	14	1.000000	0.000000
60	20	15	0	0.000756	0.000756
60	20	15	1	0.009482	0.008725
60	20	15	2	0.052462	0.042981
60	20	15	3	0.172194	0.119732
60	20	15	4	0.382756	0.210562
60	20	15	5	0.629816	0.247060
60	20	15	6	0.829058	0.199242
60	20	15	7	0.941132	0.112074
60	20	15	8	0.985282	0.044150
60	20	15	9	0.997401	0.012120
60	20	15	10	0.999687	0.002285

N	n	k	x	P(x)	p(x)
60	20	15	11	0.999975	0.000289
60	20	15	12	0.999999	0.000023
60	20	15	13	1.000000	0.000001
60	20	15	14	1.000000	0.000000
60	20	15	15	1.000000	0.000000
60	20	16	0	0.000420	0.000420
60	20	16	1	0.005798	0.005377
60	20	16	2	0.035270	0.029472
60	20	16	3	0.126962	0.091692
60	20	16	4	0.307890	0.180928
60	20	16	5	0.547463	0.239573
60	20	16	6	0.767072	0.219609
60	20	16	7	0.908755	0.141683
60	20	16	8	0.973508	0.064754
60	20	16	9	0.994439	0.020930
60	20	16	10	0.999179	0.004740
60	20	16	11	0.999918	0.000739
60	20	16	12	0.999995	0.000077
60	20	16	13	1.000000	0.000005
60	20	16	14	1.000000	0.000000
60	20	16	15	1.000000	0.000000
60	20	16	16	1.000000	0.000000
60	20	17	0	0.000229	0.000229
60	20	17	1	0.003475	0.003246
60	20	17	2	0.023213	0.019738
60	20	17	3	0.091535	0.068322
60	20	17	4	0.242098	0.150562
60	20	17	5	0.465790	0.223692
60	20	17	6	0.697196	0.231406
60	20	17	7	0.866894	0.169698
60	20	17	8	0.955848	0.088954
60	20	17	9	0.989206	0.033358
60	20	17	10	0.998102	0.008895
60	20	17	11	0.999766	0.001665
60	20	17	12	0.999981	0.000214
60	20	17	13	0.999999	0.000018
60	20	17	14	1.000000	0.000001
60	20	17	15	1.000000	0.000000
60	20	17	16	1.000000	0.000000
60	20	17	17	1.000000	0.000000
60	20	18	0	0.000123	0.000123
60	20	18	1	0.002041	0.001918
60	20	18	2	0.014951	0.012910
60	20	18	3	0.064524	0.049573
60	20	18	4	0.186074	0.121550
60	20	18	5	0.387758	0.201683
60	20	18	6	0.621855	0.234097
60	20	18	7	0.815590	0.193735
60	20	18	8	0.931024	0.115434
60	20	18	9	0.980673	0.049649

N	n	k	x	P(x)	p(x)
60	20	18	10	0.996033	0.015360
60	20	18	11	0.999418	0.003385
60	20	18	12	0.999941	0.000523
60	20	18	13	0.999996	0.000055
60	20	18	14	1.000000	0.000004
60	20	18	15	1.000000	0.000000
60	20	18	16	1.000000	0.000000
60	20	18	17	1.000000	0.000000
60	20	18	18	1.000000	0.000000
60	20	19	0	0.000064	0.000064
60	20	19	1	0.001173	0.001109
60	20	19	2	0.009418	0.008245
60	20	19	3	0.044459	0.035041
60	20	19	4	0.139770	0.095311
60	20	19	5	0.315728	0.175958
60	20	19	6	0.543822	0.228094
60	20	19	7	0.755624	0.211802
60	20	19	8	0.898043	0.142418
60	20	19	9	0.967669	0.069627
60	20	19	10	0.992376	0.024706
60	20	19	11	0.998693	0.006317
60	20	19	12	0.999841	0.001149
60	20	19	13	0.999987	0.000146
60	20	19	14	0.999999	0.000012
60	20	19	15	1.000000	0.000001
60	20	19	16	1.000000	0.000000
60	20	19	17	1.000000	0.000000
60	20	19	18	1.000000	0.000000
60	20	19	19	1.000000	0.000000
60	20	20	0	0.000033	0.000033
60	20	20	1	0.000659	0.000626
60	20	20	2	0.005798	0.005139
60	20	20	3	0.029930	0.024131
60	20	20	4	0.102575	0.072645
60	20	20	5	0.251353	0.148778
60	20	20	6	0.465936	0.214584
60	20	20	7	0.688467	0.222531
60	20	20	8	0.856359	0.167892
60	20	20	9	0.948989	0.092630
60	20	20	10	0.986350	0.037361
60	20	20	11	0.997306	0.010956
60	20	20	12	0.999617	0.002311
60	20	20	13	0.999962	0.000345
60	20	20	14	0.999997	0.000035
60	20	20	15	1.000000	0.000002
60	20	20	16	1.000000	0.000000
60	20	20	17	1.000000	0.000000
60	20	20	18	1.000000	0.000000
60	20	20	19	1.000000	0.000000
60	20	20	20	1.000000	0.000000

N	n	k	x	P(x)	p(x)
60	21	1	0	0.650000	0.650000
60	21	1	1	1.000000	0.350000
60	21	2	0	0.418644	0.418644
60	21	2	1	0.881356	0.462712
60	21	2	2	1.000000	0.118644
60	21	3	0	0.267066	0.267066
60	21	3	1	0.721800	0.454734
60	21	3	2	0.961134	0.239334
60	21	3	3	1.000000	0.038866
60	21	4	0	0.168673	0.168673
60	21	4	1	0.562244	0.393571
60	21	4	2	0.881356	0.319112
60	21	4	3	0.987726	0.106371
60	21	4	4	1.000000	0.012274
60	21	5	0	0.105421	0.105421
60	21	5	1	0.421683	0.316262
60	21	5	2	0.773086	0.351403
60	21	5	3	0.953536	0.180450
60	21	5	4	0.996274	0.042738
60	21	5	5	1.000000	0.003726
60	21	6	0	0.065169	0.065169
60	21	6	1	0.306679	0.241509
60	21	6	2	0.651692	0.345014
60	21	6	3	0.894480	0.242787
60	21	6	4	0.983064	0.088585
60	21	6	5	0.998916	0.015852
60	21	6	6	1.000000	0.001084
60	21	7	0	0.039826	0.039826
60	21	7	1	0.217231	0.177405
60	21	7	2	0.530299	0.313068
60	21	7	3	0.813550	0.283252
60	21	7	4	0.955176	0.141626
60	21	7	5	0.994219	0.039043
60	21	7	6	0.999699	0.005480
60	21	7	7	1.000000	0.000301
60	21	8	0	0.024046	0.024046
60	21	8	1	0.150285	0.126240
60	21	8	2	0.418067	0.267781
60	21	8	3	0.717352	0.299285
60	21	8	4	0.909749	0.192397
60	21	8	5	0.982433	0.072683
60	21	8	6	0.998148	0.015715
60	21	8	7	0.999920	0.001772
60	21	8	8	1.000000	0.000080
60	21	9	0	0.014335	0.014335
60	21	9	1	0.101732	0.087397
60	21	9	2	0.320224	0.218492
60	21	9	3	0.613753	0.293529
60	21	9	4	0.846850	0.233097
60	21	9	5	0.960068	0.113219

Table for $N = 2$, $n = 1$, through $N = 100$, $n = 50$

N	n	k	x	P(x)	p(x)
60	21	9	6	0.993615	0.033546
60	21	9	7	0.999443	0.005828
60	21	9	8	0.999980	0.000537
60	21	9	9	1.000000	0.000020
60	21	10	0	0.008432	0.008432
60	21	10	1	0.067458	0.059026
60	21	10	2	0.238825	0.171366
60	21	10	3	0.510154	0.271330
60	21	10	4	0.769151	0.258997
60	21	10	5	0.924549	0.155398
60	21	10	6	0.983748	0.059199
60	21	10	7	0.997843	0.014095
60	21	10	8	0.999843	0.002000
60	21	10	9	0.999995	0.000152
60	21	10	10	1.000000	0.000005
60	21	11	0	0.004891	0.004891
60	21	11	1	0.043848	0.038957
60	21	11	2	0.173705	0.129857
60	21	11	3	0.412476	0.238770
60	21	11	4	0.681092	0.268616
60	21	11	5	0.874822	0.193729
60	21	11	6	0.965988	0.091167
60	21	11	7	0.993897	0.027908
60	21	11	8	0.999323	0.005427
60	21	11	9	0.999959	0.000636
60	21	11	10	0.999999	0.000040
60	21	11	11	1.000000	0.000001
60	21	12	0	0.002795	0.002795
60	21	12	1	0.027947	0.025152
60	21	12	2	0.123353	0.095405
60	21	12	3	0.324764	0.201412
60	21	12	4	0.587899	0.263135
60	21	12	5	0.811563	0.223664
60	21	12	6	0.938080	0.126517
60	21	12	7	0.985923	0.047843
60	21	12	8	0.997883	0.011961
60	21	12	9	0.999803	0.001920
60	21	12	10	0.999990	0.000187
60	21	12	11	1.000000	0.000010
60	21	12	12	1.000000	0.000000
60	21	13	0	0.001572	0.001572
60	21	13	1	0.017467	0.015895
60	21	13	2	0.085588	0.068121
60	21	13	3	0.249235	0.163647
60	21	13	4	0.494705	0.245470
60	21	13	5	0.737008	0.242303
60	21	13	6	0.898544	0.161535
60	21	13	7	0.971969	0.073425
60	21	13	8	0.994644	0.022675
60	21	13	9	0.999323	0.004679
60	21	13	10	0.999947	0.000624
60	21	13	11	0.999998	0.000051
60	21	13	12	1.000000	0.000002
60	21	13	13	1.000000	0.000000
60	21	14	0	0.000870	0.000870
60	21	14	1	0.010703	0.009833
60	21	14	2	0.058050	0.047346
60	21	14	3	0.186562	0.128512
60	21	14	4	0.405918	0.219356
60	21	14	5	0.654522	0.248604
60	21	14	6	0.846990	0.192468
60	21	14	7	0.950097	0.103108
60	21	14	8	0.988372	0.038275
60	21	14	9	0.998129	0.009756
60	21	14	10	0.999801	0.001673
60	21	14	11	0.999987	0.000186
60	21	14	12	0.999999	0.000013
60	21	14	13	1.000000	0.000000
60	21	14	14	1.000000	0.000000
60	21	15	0	0.000473	0.000473
60	21	15	1	0.006428	0.005955
60	21	15	2	0.038493	0.032066
60	21	15	3	0.136274	0.097781
60	21	15	4	0.324851	0.188577
60	21	15	5	0.568051	0.243200
60	21	15	6	0.784229	0.216177
60	21	15	7	0.918717	0.134488
60	21	15	8	0.977555	0.058839
60	21	15	9	0.995583	0.018028
60	21	15	10	0.999401	0.003818
60	21	15	11	0.999946	0.000545
60	21	15	12	0.999997	0.000050
60	21	15	13	1.000000	0.000003
60	21	15	14	1.000000	0.000000
60	21	15	15	1.000000	0.000000
60	21	16	0	0.000252	0.000252
60	21	16	1	0.003781	0.003529
60	21	16	2	0.024955	0.021174
60	21	16	3	0.097162	0.072207
60	21	16	4	0.253611	0.156449
60	21	16	5	0.481580	0.227969
60	21	16	6	0.712169	0.230589
60	21	16	7	0.876876	0.164707
60	21	16	8	0.960558	0.083682
60	21	16	9	0.990776	0.030218
60	21	16	10	0.998468	0.007692
60	21	16	11	0.999825	0.001357
60	21	16	12	0.999987	0.000162
60	21	16	13	0.999999	0.000012
60	21	16	14	1.000000	0.000001
60	21	16	15	1.000000	0.000000
60	21	16	16	1.000000	0.000000
60	21	17	0	0.000132	0.000132
60	21	17	1	0.002177	0.002045
60	21	17	2	0.015811	0.013634
60	21	17	3	0.067623	0.051811
60	21	17	4	0.193165	0.125542
60	21	17	5	0.398682	0.205517
60	21	17	6	0.633559	0.234877
60	21	17	7	0.824469	0.190910
60	21	17	8	0.935833	0.111364
60	21	17	9	0.982535	0.046701
60	21	17	10	0.996545	0.014010
60	21	17	11	0.999517	0.002972
60	21	17	12	0.999954	0.000437
60	21	17	13	0.999997	0.000043
60	21	17	14	1.000000	0.000003
60	21	17	15	1.000000	0.000000
60	21	17	16	1.000000	0.000000
60	21	17	17	1.000000	0.000000
60	21	18	0	0.000067	0.000067
60	21	18	1	0.001226	0.001158
60	21	18	2	0.009787	0.008561
60	21	18	3	0.045934	0.036147
60	21	18	4	0.143532	0.097598
60	21	18	5	0.322211	0.178679
60	21	18	6	0.551626	0.229415
60	21	18	7	0.762313	0.210687
60	21	18	8	0.902165	0.139853
60	21	18	9	0.969502	0.067336
60	21	18	10	0.992961	0.023459
60	21	18	11	0.998826	0.005865
60	21	18	12	0.999862	0.001037
60	21	18	13	0.999989	0.000127
60	21	18	14	0.999999	0.000010
60	21	18	15	1.000000	0.000001
60	21	18	16	1.000000	0.000000
60	21	18	17	1.000000	0.000000
60	21	18	18	1.000000	0.000000
60	21	19	0	0.000034	0.000034
60	21	19	1	0.000674	0.000640
60	21	19	2	0.005914	0.005240
60	21	19	3	0.030442	0.024529
60	21	19	4	0.104028	0.073586
60	21	19	5	0.254143	0.150115
60	21	19	6	0.469692	0.215549
60	21	19	7	0.692084	0.222392
60	21	19	8	0.858878	0.166794
60	21	19	9	0.950263	0.091385
60	21	19	10	0.986817	0.036554
60	21	19	11	0.997429	0.010612
60	21	19	12	0.999640	0.002211
60	21	19	13	0.999965	0.000325
60	21	19	14	0.999998	0.000033
60	21	19	15	1.000000	0.000002
60	21	19	16	1.000000	0.000000
60	21	19	17	1.000000	0.000000
60	21	19	18	1.000000	0.000000
60	21	19	19	1.000000	0.000000
60	21	20	0	0.000016	0.000016
60	21	20	1	0.000362	0.000345
60	21	20	2	0.003486	0.003124
60	21	20	3	0.019674	0.016188
60	21	20	4	0.073517	0.053843
60	21	20	5	0.195561	0.122044
60	21	20	6	0.390832	0.195271
60	21	20	7	0.616145	0.225313
60	21	20	8	0.805992	0.189847
60	21	20	9	0.923516	0.117524
60	21	20	10	0.977010	0.053494
60	21	20	11	0.994841	0.017831
60	21	20	12	0.999155	0.004314
60	21	20	13	0.999901	0.000747
60	21	20	14	0.999992	0.000091
60	21	20	15	0.999999	0.000007
60	21	20	16	1.000000	0.000000
60	21	20	17	1.000000	0.000000
60	21	20	18	1.000000	0.000000
60	21	20	19	1.000000	0.000000
60	21	20	20	1.000000	0.000000
60	21	21	0	0.000008	0.000008
60	21	21	1	0.000189	0.000181
60	21	21	2	0.002002	0.001813
60	21	21	3	0.012389	0.010387
60	21	21	4	0.050634	0.038244
60	21	21	5	0.146744	0.096110
60	21	21	6	0.317606	0.170862
60	21	21	7	0.537285	0.219680
60	21	21	8	0.744291	0.207006
60	21	21	9	0.888259	0.143967
60	21	21	10	0.962299	0.074040
60	21	21	11	0.990383	0.028084
60	21	21	12	0.998184	0.007801
60	21	21	13	0.999752	0.001568
60	21	21	14	0.999976	0.000224
60	21	21	15	0.999998	0.000022
60	21	21	16	1.000000	0.000001
60	21	21	17	1.000000	0.000000
60	21	21	18	1.000000	0.000000
60	21	21	19	1.000000	0.000000

Table for $N=2$, $n=1$, through $N=100$, $n=50$

N	n	k	x	P(x)	p(x)
60	21	21	20	1.000000	0.000000
60	21	21	21	1.000000	0.000000
60	22	1	0	0.633333	0.633333
60	22	1	1	1.000000	0.366667
60	22	2	0	0.397175	0.397175
60	22	2	1	0.869492	0.472316
60	22	2	2	1.000000	0.130508
60	22	3	0	0.246522	0.246522
60	22	3	1	0.698480	0.451958
60	22	3	2	0.954997	0.256517
60	22	3	3	1.000000	0.045003
60	22	4	0	0.151373	0.151373
60	22	4	1	0.531970	0.380596
60	22	4	2	0.864991	0.333022
60	22	4	3	0.984999	0.120008
60	22	4	4	1.000000	0.015001
60	22	5	0	0.091905	0.091905
60	22	5	1	0.389246	0.297341
60	22	5	2	0.746055	0.356809
60	22	5	3	0.944282	0.198227
60	22	5	4	0.995178	0.050896
60	22	5	5	1.000000	0.004822
60	22	6	0	0.055143	0.055143
60	22	6	1	0.275716	0.220573
60	22	6	2	0.616306	0.340590
60	22	6	3	0.875804	0.259497
60	22	6	4	0.978521	0.102718
60	22	6	5	0.998510	0.019988
60	22	6	6	1.000000	0.001490
60	22	7	0	0.032677	0.032677
60	22	7	1	0.189938	0.157260
60	22	7	2	0.490162	0.300224
60	22	7	3	0.784499	0.294337
60	22	7	4	0.944282	0.159783
60	22	7	5	0.992217	0.047935
60	22	7	6	0.999558	0.007341
60	22	7	7	1.000000	0.000442
60	22	8	0	0.019113	0.019113
60	22	8	1	0.127627	0.108514
60	22	8	2	0.376870	0.249243
60	22	8	3	0.678982	0.302112
60	22	8	4	0.890016	0.211034
60	22	8	5	0.976842	0.086826
60	22	8	6	0.997342	0.020500
60	22	8	7	0.999875	0.002533
60	22	8	8	1.000000	0.000125
60	22	9	0	0.011027	0.011027
60	22	9	1	0.083804	0.072777
60	22	9	2	0.281007	0.197203
60	22	9	3	0.568595	0.287588
60	22	9	4	0.816966	0.248371
60	22	9	5	0.948456	0.131491
60	22	9	6	0.991034	0.042578
60	22	9	7	0.999144	0.008110
60	22	9	8	0.999966	0.000822
60	22	9	9	1.000000	0.000034
60	22	10	0	0.006270	0.006270
60	22	10	1	0.053837	0.047567
60	22	10	2	0.203673	0.149836
60	22	10	3	0.461454	0.257782
60	22	10	4	0.729305	0.267851
60	22	10	5	0.904626	0.175321
60	22	10	6	0.977676	0.073050
60	22	10	7	0.996759	0.019083
60	22	10	8	0.999741	0.002982
60	22	10	9	0.999991	0.000251
60	22	10	10	1.000000	0.000009
60	22	11	0	0.003511	0.003511
60	22	11	1	0.033859	0.030348
60	22	11	2	0.143738	0.109879
60	22	11	3	0.363497	0.219759
60	22	11	4	0.632879	0.269382
60	22	11	5	0.845017	0.212138
60	22	11	6	0.954300	0.109283
60	22	11	7	0.991034	0.036734
60	22	11	8	0.998906	0.007872
60	22	11	9	0.999926	0.001020
60	22	11	10	0.999998	0.000072
60	22	11	11	1.000000	0.000002
60	22	12	0	0.001935	0.001935
60	22	12	1	0.020853	0.018918
60	22	12	2	0.098890	0.078037
60	22	12	3	0.278285	0.179395
60	22	12	4	0.533922	0.255638
60	22	12	5	0.771418	0.237496
60	22	12	6	0.918616	0.147198
60	22	12	7	0.979789	0.061173
60	22	12	8	0.996657	0.016868
60	22	12	9	0.999655	0.002999
60	22	12	10	0.999980	0.000325
60	22	12	11	0.999999	0.000019
60	22	12	12	1.000000	0.000000
60	22	13	0	0.001048	0.001048
60	22	13	1	0.012576	0.011528
60	22	13	2	0.066374	0.053798
60	22	13	3	0.207274	0.140900
60	22	13	4	0.438058	0.230784
60	22	13	5	0.687305	0.249247
60	22	13	6	0.869550	0.182245
60	22	13	7	0.960673	0.091123
60	22	13	8	0.991737	0.031064
60	22	13	9	0.998843	0.007106
60	22	13	10	0.999899	0.001056
60	22	13	11	0.999995	0.000096
60	22	13	12	1.000000	0.000005
60	22	13	13	1.000000	0.000000
60	22	14	0	0.000557	0.000557
60	22	14	1	0.007425	0.006868
60	22	14	2	0.043481	0.036056
60	22	14	3	0.150315	0.106833
60	22	14	4	0.349673	0.199358
60	22	14	5	0.597152	0.247479
60	22	14	6	0.807509	0.210357
60	22	14	7	0.931591	0.124082
60	22	14	8	0.982484	0.050893
60	22	14	9	0.996878	0.014394
60	22	14	10	0.999629	0.002752
60	22	14	11	0.999973	0.000343
60	22	14	12	0.999999	0.000026
60	22	14	13	1.000000	0.000001
60	22	14	14	1.000000	0.000000
60	22	15	0	0.000291	0.000291
60	22	15	1	0.004290	0.003999
60	22	15	2	0.027805	0.023515
60	22	15	3	0.106188	0.078383
60	22	15	4	0.271663	0.165475
60	22	15	5	0.505692	0.234029
60	22	15	6	0.734341	0.228649
60	22	15	7	0.891129	0.156788
60	22	15	8	0.966995	0.075865
60	22	15	9	0.992810	0.025815
60	22	15	10	0.998912	0.006102
60	22	15	11	0.999891	0.000979
60	22	15	12	0.999993	0.000103
60	22	15	13	1.000000	0.000007
60	22	15	14	1.000000	0.000000
60	22	15	15	1.000000	0.000000
60	22	16	0	0.000149	0.000149
60	22	16	1	0.002424	0.002275
60	22	16	2	0.017354	0.014930
60	22	16	3	0.073093	0.055739
60	22	16	4	0.205473	0.132380
60	22	16	5	0.417281	0.211808
60	22	16	6	0.653044	0.235763
60	22	16	7	0.838867	0.185823
60	22	16	8	0.943392	0.104525
60	22	16	9	0.985352	0.041960
60	22	16	10	0.997284	0.011932
60	22	16	11	0.999651	0.002367
60	22	16	12	0.999970	0.000319
60	22	16	13	0.999998	0.000028
60	22	16	14	1.000000	0.000001
60	22	16	15	1.000000	0.000000
60	22	16	16	1.000000	0.000000
60	22	17	0	0.000074	0.000074
60	22	17	1	0.001338	0.001264
60	22	17	2	0.010567	0.009230
60	22	17	3	0.049024	0.038456
60	22	17	4	0.151317	0.102294
60	22	17	5	0.335446	0.184129
60	22	17	6	0.567312	0.231866
60	22	17	7	0.775518	0.208206
60	22	17	8	0.910134	0.134616
60	22	17	9	0.972955	0.062821
60	22	17	10	0.994030	0.021075
60	22	17	11	0.999060	0.005029
60	22	17	12	0.999898	0.000838
60	22	17	13	0.999993	0.000095
60	22	17	14	1.000000	0.000007
60	22	17	15	1.000000	0.000000
60	22	17	16	1.000000	0.000000
60	22	17	17	1.000000	0.000000
60	22	18	0	0.000036	0.000036
60	22	18	1	0.000721	0.000684
60	22	18	2	0.006275	0.005554
60	22	18	3	0.032031	0.025757
60	22	18	4	0.108497	0.076465
60	22	18	5	0.262651	0.154154
60	22	18	6	0.481036	0.218385
60	22	18	7	0.702888	0.221852
60	22	18	8	0.866306	0.163418
60	22	18	9	0.953963	0.087657
60	22	18	10	0.988149	0.034186
60	22	18	11	0.997773	0.009624
60	22	18	12	0.999703	0.001930
60	22	18	13	0.999973	0.000270
60	22	18	14	0.999998	0.000026
60	22	18	15	1.000000	0.000002
60	22	18	16	1.000000	0.000000
60	22	18	17	1.000000	0.000000
60	22	18	18	1.000000	0.000000
60	22	19	0	0.000017	0.000017
60	22	19	1	0.000379	0.000361
60	22	19	2	0.003630	0.003251
60	22	19	3	0.020379	0.016750
60	22	19	4	0.075726	0.055346
60	22	19	5	0.200255	0.124529
60	22	19	6	0.397842	0.197587
60	22	19	7	0.623655	0.225813
60	22	19	8	0.811833	0.188178

N	n	k	x	P(x)	p(x)
60	22	19	9	0.926831	0.114998
60	22	19	10	0.978381	0.051551
60	22	19	11	0.995252	0.016871
60	22	19	12	0.999243	0.003991
60	22	19	13	0.999915	0.000672
60	22	19	14	0.999993	0.000078
60	22	19	15	1.000000	0.000006
60	22	19	16	1.000000	0.000000
60	22	19	17	1.000000	0.000000
60	22	19	18	1.000000	0.000000
60	22	19	19	1.000000	0.000000
60	22	20	0	0.000008	0.000008
60	22	20	1	0.000194	0.000185
60	22	20	2	0.002044	0.001850
60	22	20	3	0.012618	0.010574
60	22	20	4	0.051427	0.038810
60	22	20	5	0.148621	0.097194
60	22	20	6	0.320735	0.172114
60	22	20	7	0.541041	0.220306
60	22	20	8	0.747577	0.206537
60	22	20	9	0.890368	0.142791
60	22	20	10	0.963293	0.072925
60	22	20	11	0.990726	0.027433
60	22	20	12	0.998270	0.007544
60	22	20	13	0.999767	0.001498
60	22	20	14	0.999978	0.000211
60	22	20	15	0.999999	0.000020
60	22	20	16	1.000000	0.000001
60	22	20	17	1.000000	0.000000
60	22	20	18	1.000000	0.000000
60	22	20	19	1.000000	0.000000
60	22	20	20	1.000000	0.000000
60	22	21	0	0.000004	0.000004
60	22	21	1	0.000096	0.000093
60	22	21	2	0.001119	0.001023
60	22	21	3	0.007595	0.006476
60	22	21	4	0.033963	0.026368
60	22	21	5	0.107314	0.073351
60	22	21	6	0.251889	0.144576
60	22	21	7	0.458426	0.206537
60	22	21	8	0.675289	0.216863
60	22	21	9	0.843961	0.168672
60	22	21	10	0.941416	0.097455
60	22	21	11	0.983182	0.041766
60	22	21	12	0.996384	0.013202
60	22	21	13	0.999431	0.003047
60	22	21	14	0.999936	0.000505
60	22	21	15	0.999995	0.000059
60	22	21	16	1.000000	0.000005
60	22	21	17	1.000000	0 000000

N	n	k	x	P(x)	p(x)
60	22	21	18	1.000000	0.000000
60	22	21	19	1.000000	0.000000
60	22	21	20	1.000000	0.000000
60	22	21	21	1.000000	0.000000
60	22	22	0	0.000002	0.000002
60	22	22	1	0.000046	0.000045
60	22	22	2	0.000594	0.000548
60	22	22	3	0.004440	0.003846
60	22	22	4	0.021793	0.017353
60	22	22	5	0.075340	0.053547
60	22	22	6	0.192576	0.117236
60	22	22	7	0.378989	0.186413
60	22	22	8	0.597441	0.218452
60	22	22	9	0.787737	0.190296
60	22	22	10	0.911430	0.123692
60	22	22	11	0.971402	0.059972
60	22	22	12	0.992999	0.021597
60	22	22	13	0.998727	0.005729
60	22	22	14	0.999832	0.001105
60	22	22	15	0.999984	0.000152
60	22	22	16	0.999999	0.000015
60	22	22	17	1.000000	0.000001
60	22	22	18	1.000000	0.000000
60	22	22	19	1.000000	0.000000
60	22	22	20	1.000000	0.000000
60	22	22	21	1.000000	0.000000
60	22	22	22	1.000000	0.000000
60	23	1	0	0.616667	0.616667
60	23	1	1	1.000000	0.383333
60	23	2	0	0.376271	0.376271
60	23	2	1	0.857062	0.480791
60	23	2	2	1.000000	0.142938
60	23	3	0	0.227060	0.227060
60	23	3	1	0.674693	0.447633
60	23	3	2	0.948247	0.273553
60	23	3	3	1.000000	0.051753
60	23	4	0	0.135439	0.135439
60	23	4	1	0.501923	0.366483
60	23	4	2	0.847464	0.345541
60	23	4	3	0.981841	0.134377
60	23	4	4	1.000000	0.018159
60	23	5	0	0.079813	0.079813
60	23	5	1	0.357947	0.278134
60	23	5	2	0.717886	0.359939
60	23	5	3	0.933849	0.215963
60	23	5	4	0.993839	0.059990
60	23	5	5	1.000000	0.006161
60	23	6	0	0.046436	0.046436
60	23	6	1	0.246693	0.200257
60	23	6	2	0.580455	0.333761

N	n	k	x	P(x)	p(x)
60	23	6	3	0.855317	0.274862
60	23	6	4	0.973115	0.117798
60	23	6	5	0.997984	0.024868
60	23	6	6	1.000000	0.002016
60	23	7	0	0.026658	0.026658
60	23	7	1	0.165107	0.138449
60	23	7	2	0.450659	0.285551
60	23	7	3	0.753516	0.302858
60	23	7	4	0.931668	0.178152
60	23	7	5	0.989694	0.058026
60	23	7	6	0.999365	0.009671
60	23	7	7	1.000000	0.000635
60	23	8	0	0.015089	0.015089
60	23	8	1	0.107638	0.092548
60	23	8	2	0.337515	0.229878
60	23	8	3	0.639230	0.301715
60	23	8	4	0.867802	0.228572
60	23	8	5	0.969987	0.102185
60	23	8	6	0.996263	0.026276
60	23	8	7	0.999808	0.003545
60	23	8	8	1.000000	0.000192
60	23	9	0	0.008415	0.008415
60	23	9	1	0.068483	0.060067
60	23	9	2	0.244680	0.176198
60	23	9	3	0.523186	0.278506
60	23	9	4	0.784285	0.261099
60	23	9	5	0.934615	0.150330
60	23	9	6	0.987673	0.053058
60	23	9	7	0.998717	0.011045
60	23	9	8	0.999945	0.001227
60	23	9	9	1.000000	0.000055
60	23	10	0	0.004620	0.004620
60	23	10	1	0.042571	0.037951
60	23	10	2	0.172128	0.129557
60	23	10	3	0.413968	0.241840
60	23	10	4	0.687013	0.273045
60	23	10	5	0.881558	0.194545
60	23	10	6	0.969987	0.088429
60	23	10	7	0.995252	0.025266
60	23	10	8	0.999584	0.004331
60	23	10	9	0.999985	0.000401
60	23	10	10	1.000000	0.000015
60	23	11	0	0.002495	0.002495
60	23	11	1	0.025873	0.023378
60	23	11	2	0.117714	0.091842
60	23	11	3	0.317232	0.199518
60	23	11	4	0.583256	0.266024
60	23	11	5	0.811522	0.228266
60	23	11	6	0.939921	0.128399
60	23	11	7	0.987168	0.047247

N	n	k	x	P(x)	p(x)
60	23	11	8	0.998284	0.011117
60	23	11	9	0.999872	0.001588
60	23	11	10	0.999996	0.000124
60	23	11	11	1.000000	0.000004
60	23	12	0	0.001324	0.001324
60	23	12	1	0.015377	0.014053
60	23	12	2	0.078354	0.062977
60	23	12	3	0.235796	0.157443
60	23	12	4	0.480104	0.244308
60	23	12	5	0.727669	0.247565
60	23	12	6	0.895374	0.167705
60	23	12	7	0.971740	0.076366
60	23	12	8	0.994881	0.023141
60	23	12	9	0.999419	0.004537
60	23	12	10	0.999963	0.000544
60	23	12	11	0.999999	0.000036
60	23	12	12	1.000000	0.000001
60	23	13	0	0.000689	0.000689
60	23	13	1	0.008936	0.008246
60	23	13	2	0.050801	0.041865
60	23	13	3	0.170195	0.119394
60	23	13	4	0.383399	0.213204
60	23	13	5	0.634832	0.251433
60	23	13	6	0.835979	0.201147
60	23	13	7	0.946285	0.110306
60	23	13	8	0.987650	0.041365
60	23	13	9	0.998095	0.010446
60	23	13	10	0.999816	0.001720
60	23	13	11	0.999990	0.000174
60	23	13	12	1.000000	0.000010
60	23	13	13	1.000000	0.000000
60	23	14	0	0.000352	0.000352
60	23	14	1	0.005076	0.004724
60	23	14	2	0.032095	0.027020
60	23	14	3	0.119389	0.087294
60	23	14	4	0.297210	0.177821
60	23	14	5	0.538538	0.241328
60	23	14	6	0.763223	0.224685
60	23	14	7	0.908734	0.145510
60	23	14	8	0.974448	0.065714
60	23	14	9	0.994984	0.020536
60	23	14	10	0.999340	0.004356
60	23	14	11	0.999946	0.000606
60	23	14	12	0.999997	0.000052
60	23	14	13	1.000000	0.000002
60	23	14	14	1.000000	0.000000
60	23	15	0	0.000176	0.000176
60	23	15	1	0.002817	0.002641
60	23	15	2	0.019760	0.016944
60	23	15	3	0.081435	0.061675

Table for $N = 2$, $n = 1$, through $N = 100$, $n = 50$

N	n	k	x	P(x)	p(x)
60	23	15	4	0.223762	0.142327
60	23	15	5	0.444106	0.220343
60	23	15	6	0.680188	0.236082
60	23	15	7	0.858121	0.177934
60	23	15	8	0.953019	0.094898
60	23	15	9	0.988734	0.035714
60	23	15	10	0.998109	0.009375
60	23	15	11	0.999787	0.001679
60	23	15	12	0.999985	0.000197
60	23	15	13	0.999999	0.000014
60	23	15	14	1.000000	0.000001
60	23	15	15	1.000000	0.000000
60	23	16	0	0.000086	0.000086
60	23	16	1	0.001526	0.001440
60	23	16	2	0.011853	0.010328
60	23	16	3	0.054024	0.042171
60	23	16	4	0.163669	0.109645
60	23	16	5	0.355968	0.192300
60	23	16	6	0.591001	0.235033
60	23	16	7	0.794856	0.203855
60	23	16	8	0.921387	0.126531
60	23	16	9	0.977623	0.056236
60	23	16	10	0.995400	0.017778
60	23	16	11	0.999340	0.003939
60	23	16	12	0.999937	0.000597
60	23	16	13	0.999996	0.000059
60	23	16	14	1.000000	0.000004
60	23	16	15	1.000000	0.000000
60	23	16	16	1.000000	0.000000
60	23	17	0	0.000041	0.000041
60	23	17	1	0.000806	0.000765
60	23	17	2	0.006924	0.006118
60	23	17	3	0.034856	0.027931
60	23	17	4	0.116322	0.081467
60	23	17	5	0.277300	0.160978
60	23	17	6	0.500193	0.222893
60	23	17	7	0.720727	0.220534
60	23	17	8	0.878251	0.157524
60	23	17	9	0.959729	0.081478
60	23	17	10	0.990148	0.030418
60	23	17	11	0.998265	0.008118
60	23	17	12	0.999787	0.001522
60	23	17	13	0.999983	0.000195
60	23	17	14	0.999999	0.000016
60	23	17	15	1.000000	0.000001
60	23	17	16	1.000000	0.000000
60	23	17	17	1.000000	0.000000
60	23	18	0	0.000019	0.000019
60	23	18	1	0.000415	0.000395
60	23	18	2	0.003936	0.003522

N	n	k	x	P(x)	p(x)
60	23	18	3	0.021864	0.017928
60	23	18	4	0.080325	0.058461
60	23	18	5	0.209914	0.129589
60	23	18	6	0.412073	0.202158
60	23	18	7	0.638668	0.226595
60	23	18	8	0.823301	0.184633
60	23	18	9	0.933202	0.109901
60	23	18	10	0.980952	0.047750
60	23	18	11	0.996000	0.015048
60	23	18	12	0.999398	0.003398
60	23	18	13	0.999937	0.000539
60	23	18	14	0.999996	0.000058
60	23	18	15	1.000000	0.000004
60	23	18	16	1.000000	0.000000
60	23	18	17	1.000000	0.000000
60	23	18	18	1.000000	0.000000
60	23	19	0	0.000009	0.000009
60	23	19	1	0.000207	0.000199
60	23	19	2	0.002175	0.001968
60	23	19	3	0.013327	0.011152
60	23	19	4	0.053879	0.040552
60	23	19	5	0.154376	0.100497
60	23	19	6	0.330247	0.175870
60	23	19	7	0.552346	0.222099
60	23	19	8	0.757361	0.205015
60	23	19	9	0.896568	0.139208
60	23	19	10	0.966172	0.069604
60	23	19	11	0.991701	0.025529
60	23	19	12	0.998508	0.006808
60	23	19	13	0.999809	0.001301
60	23	19	14	0.999983	0.000174
60	23	19	15	0.999999	0.000016
60	23	19	16	1.000000	0.000001
60	23	19	17	1.000000	0.000000
60	23	19	18	1.000000	0.000000
60	23	19	19	1.000000	0.000000
60	23	20	0	0.000004	0.000004
60	23	20	1	0.000101	0.000097
60	23	20	2	0.001167	0.001067
60	23	20	3	0.007887	0.006720
60	23	20	4	0.035086	0.027199
60	23	20	5	0.110255	0.075169
60	23	20	6	0.257325	0.147069
60	23	20	7	0.465673	0.208348
60	23	20	8	0.682355	0.216682
60	23	20	9	0.849034	0.166679
60	23	20	10	0.944102	0.095069
60	23	20	11	0.984229	0.040126
60	23	20	12	0.996682	0.012453
60	23	20	13	0.999492	0.002810

N	n	k	x	P(x)	p(x)
60	23	20	14	0.999945	0.000453
60	23	20	15	0.999996	0.000051
60	23	20	16	1.000000	0.000004
60	23	20	17	1.000000	0.000000
60	23	20	18	1.000000	0.000000
60	23	20	19	1.000000	0.000000
60	23	20	20	1.000000	0.000000
60	23	21	0	0.000002	0.000002
60	23	21	1	0.000047	0.000046
60	23	21	2	0.000607	0.000560
60	23	21	3	0.004527	0.003920
60	23	21	4	0.022167	0.017639
60	23	21	5	0.076429	0.054262
60	23	21	6	0.194820	0.118391
60	23	21	7	0.382334	0.187513
60	23	21	8	0.601099	0.218766
60	23	21	9	0.790696	0.189597
60	23	21	10	0.913205	0.122509
60	23	21	11	0.972191	0.058986
60	23	21	12	0.993257	0.021066
60	23	21	13	0.998789	0.005532
60	23	21	14	0.999843	0.001054
60	23	21	15	0.999986	0.000143
60	23	21	16	0.999999	0.000013
60	23	21	17	1.000000	0.000001
60	23	21	18	1.000000	0.000000
60	23	21	19	1.000000	0.000000
60	23	21	20	1.000000	0.000000
60	23	21	21	1.000000	0.000000
60	23	22	0	0.000001	0.000001
60	23	22	1	0.000022	0.000021
60	23	22	2	0.000306	0.000284
60	23	22	3	0.002517	0.002211
60	23	22	4	0.013573	0.011056
60	23	22	5	0.051385	0.037812
60	23	22	6	0.143214	0.091829
60	23	22	7	0.305405	0.162191
60	23	22	8	0.516959	0.211554
60	23	22	9	0.722636	0.205677
60	23	22	10	0.872369	0.149733
60	23	22	11	0.954041	0.081673
60	23	22	12	0.987315	0.033274
60	23	22	13	0.997371	0.010055
60	23	22	14	0.999600	0.002229
60	23	22	15	0.999956	0.000357
60	23	22	16	0.999997	0.000040
60	23	22	17	1.000000	0.000003
60	23	22	18	1.000000	0.000000
60	23	22	19	1.000000	0.000000
60	23	22	20	1.000000	0.000000

N	n	k	x	P(x)	p(x)
60	23	22	21	1.000000	0.000000
60	23	22	22	1.000000	0.000000
60	23	23	0	0.000000	0.000000
60	23	23	1	0.000009	0.000009
60	23	23	2	0.000149	0.000139
60	23	23	3	0.001353	0.001205
60	23	23	4	0.008045	0.006692
60	23	23	5	0.033474	0.025429
60	23	23	6	0.102133	0.068658
60	23	23	7	0.237114	0.134981
60	23	23	8	0.433451	0.196337
60	23	23	9	0.646860	0.213409
60	23	23	10	0.821144	0.174284
60	23	23	11	0.928250	0.107106
60	23	23	12	0.977683	0.049433
60	23	23	13	0.994725	0.017041
60	23	23	14	0.999072	0.004347
60	23	23	15	0.999881	0.000809
60	23	23	16	0.999989	0.000108
60	23	23	17	0.999999	0.000010
60	23	23	18	1.000000	0.000001
60	23	23	19	1.000000	0.000000
60	23	23	20	1.000000	0.000000
60	23	23	21	1.000000	0.000000
60	23	23	22	1.000000	0.000000
60	23	23	23	1.000000	0.000000
60	24	1	0	0.600000	0.600000
60	24	1	1	1.000000	0.400000
60	24	2	0	0.355932	0.355932
60	24	2	1	0.844068	0.488136
60	24	2	2	1.000000	0.155932
60	24	3	0	0.208650	0.208650
60	24	3	1	0.650497	0.441847
60	24	3	2	0.940853	0.290356
60	24	3	3	1.000000	0.059147
60	24	4	0	0.120797	0.120797
60	24	4	1	0.472208	0.351410
60	24	4	2	0.828786	0.356578
60	24	4	3	0.978209	0.149423
60	24	4	4	1.000000	0.021791
60	24	5	0	0.069027	0.069027
60	24	5	1	0.327878	0.258851
60	24	5	2	0.688702	0.360823
60	24	5	3	0.922175	0.233474
60	24	5	4	0.992217	0.070042
60	24	5	5	1.000000	0.007782
60	24	6	0	0.038906	0.038906
60	24	6	1	0.219631	0.180725
60	24	6	2	0.544372	0.324741
60	24	6	3	0.833031	0.288658

Table for $N = 2$, $n = 1$, through $N = 100$, $n = 50$

N	n	k	x	P(x)	p(x)
60	24	6	4	0.966748	0.133717
60	24	6	5	0.997311	0.030564
60	24	6	6	1.000000	0.002688
60	24	7	0	0.021615	0.021615
60	24	7	1	0.142656	0.121041
60	24	7	2	0.412070	0.269415
60	24	7	3	0.720775	0.308704
60	24	7	4	0.917223	0.196448
60	24	7	5	0.986557	0.069335
60	24	7	6	0.999104	0.012546
60	24	7	7	1.000000	0.000896
60	24	8	0	0.011827	0.011827
60	24	8	1	0.090128	0.078302
60	24	8	2	0.300238	0.210110
60	24	8	3	0.598458	0.298220
60	24	8	4	0.843091	0.244634
60	24	8	5	0.961702	0.118610
60	24	8	6	0.994843	0.033141
60	24	8	7	0.999712	0.004870
60	24	8	8	1.000000	0.000287
60	24	9	0	0.006368	0.006368
60	24	9	1	0.055495	0.049127
60	24	9	2	0.211345	0.155850
60	24	9	3	0.478023	0.266677
60	24	9	4	0.749002	0.270979
60	24	9	5	0.918363	0.169362
60	24	9	6	0.983371	0.065007
60	24	9	7	0.998120	0.014750
60	24	9	8	0.999911	0.001791
60	24	9	9	1.000000	0.000088
60	24	10	0	0.003371	0.003371
60	24	10	1	0.033340	0.029968
60	24	10	2	0.144116	0.110776
60	24	10	3	0.368215	0.224099
60	24	10	4	0.642735	0.274521
60	24	10	5	0.855268	0.212532
60	24	10	6	0.960427	0.105159
60	24	10	7	0.993204	0.032777
60	24	10	8	0.999350	0.006146
60	24	10	9	0.999974	0.000624
60	24	10	10	1.000000	0.000026
60	24	11	0	0.001753	0.001753
60	24	11	1	0.019554	0.017801
60	24	11	2	0.095374	0.075820
60	24	11	3	0.274093	0.178719
60	24	11	4	0.532927	0.258834
60	24	11	5	0.774506	0.241578
60	24	11	6	0.922570	0.148064
60	24	11	7	0.982060	0.059490
60	24	11	8	0.997383	0.015323
60	24	11	9	0.999787	0.002404
60	24	11	10	0.999993	0.000206
60	24	11	11	1.000000	0.000007
60	24	12	0	0.000894	0.000894
60	24	12	1	0.011199	0.010304
60	24	12	2	0.061333	0.050134
60	24	12	3	0.197499	0.136167
60	24	12	4	0.427281	0.229781
60	24	12	5	0.680832	0.253552
60	24	12	6	0.868179	0.187346
60	24	12	7	0.961420	0.093242
60	24	12	8	0.992379	0.030959
60	24	12	9	0.999051	0.006671
60	24	12	10	0.999934	0.000883
60	24	12	11	0.999998	0.000064
60	24	12	12	1.000000	0.000002
60	24	13	0	0.000447	0.000447
60	24	13	1	0.006261	0.005814
60	24	13	2	0.038355	0.032093
60	24	13	3	0.137927	0.099572
60	24	13	4	0.331538	0.193612
60	24	13	5	0.580468	0.248930
60	24	13	6	0.797924	0.217456
60	24	13	7	0.928397	0.130473
60	24	13	8	0.982060	0.053662
60	24	13	9	0.996966	0.014906
60	24	13	10	0.999676	0.002710
60	24	13	11	0.999981	0.000304
60	24	13	12	0.999999	0.000019
60	24	13	13	1.000000	0.000000
60	24	14	0	0.000219	0.000219
60	24	14	1	0.003416	0.003197
60	24	14	2	0.023332	0.019916
60	24	14	3	0.093437	0.070105
60	24	14	4	0.249150	0.155713
60	24	14	5	0.479837	0.230687
60	24	14	6	0.714643	0.234806
60	24	14	7	0.881205	0.166562
60	24	14	8	0.963792	0.082587
60	24	14	9	0.992209	0.028417
60	24	14	10	0.998869	0.006660
60	24	14	11	0.999896	0.001027
60	24	14	12	0.999995	0.000098
60	24	14	13	1.000000	0.000005
60	24	14	14	1.000000	0.000000
60	24	15	0	0.000105	0.000105
60	24	15	1	0.001817	0.001713
60	24	15	2	0.013807	0.011990
60	24	15	3	0.061433	0.047625
60	24	15	4	0.181449	0.120016
60	24	15	5	0.384553	0.203104
60	24	15	6	0.622762	0.238209
60	24	15	7	0.819649	0.196887
60	24	15	8	0.935066	0.115417
60	24	15	9	0.982942	0.047876
60	24	15	10	0.996842	0.013900
60	24	15	11	0.999606	0.002764
60	24	15	12	0.999969	0.000363
60	24	15	13	0.999998	0.000030
60	24	15	14	1.000000	0.000001
60	24	15	15	1.000000	0.000000
60	24	16	0	0.000049	0.000049
60	24	16	1	0.000942	0.000893
60	24	16	2	0.007946	0.007003
60	24	16	3	0.039207	0.031262
60	24	16	4	0.128108	0.088901
60	24	16	5	0.298798	0.170690
60	24	16	6	0.527479	0.228681
60	24	16	7	0.745270	0.217791
60	24	16	8	0.894029	0.148759
60	24	16	9	0.966983	0.072955
60	24	16	10	0.992518	0.025534
60	24	16	11	0.998807	0.006290
60	24	16	12	0.999872	0.001065
60	24	16	13	0.999991	0.000119
60	24	16	14	1.000000	0.000008
60	24	16	15	1.000000	0.000000
60	24	16	16	1.000000	0.000000
60	24	17	0	0.000022	0.000022
60	24	17	1	0.000475	0.000453
60	24	17	2	0.004444	0.003969
60	24	17	3	0.024287	0.019843
60	24	17	4	0.087699	0.063412
60	24	17	5	0.225091	0.137392
60	24	17	6	0.433928	0.208836
60	24	17	7	0.661123	0.227196
60	24	17	8	0.839935	0.178811
60	24	17	9	0.942113	0.102178
60	24	17	10	0.984393	0.042281
60	24	17	11	0.996949	0.012556
60	24	17	12	0.999582	0.002633
60	24	17	13	0.999961	0.000380
60	24	17	14	0.999998	0.000036
60	24	17	15	1.000000	0.000002
60	24	17	16	1.000000	0.000000
60	24	17	17	1.000000	0.000000
60	24	18	0	0.000010	0.000010
60	24	18	1	0.000233	0.000223
60	24	18	2	0.002413	0.002180
60	24	18	3	0.014596	0.012183
60	24	18	4	0.058205	0.043609
60	24	18	5	0.164383	0.106178
60	24	18	6	0.346508	0.182125
60	24	18	7	0.571302	0.224794
60	24	18	8	0.773400	0.202098
60	24	18	9	0.906469	0.133069
60	24	18	10	0.970627	0.064158
60	24	18	11	0.993153	0.022526
60	24	18	12	0.998847	0.005694
60	24	18	13	0.999864	0.001017
60	24	18	14	0.999989	0.000125
60	24	18	15	0.999999	0.000010
60	24	18	16	1.000000	0.000000
60	24	18	17	1.000000	0.000000
60	24	18	18	1.000000	0.000000
60	24	19	0	0.000004	0.000004
60	24	19	1	0.000111	0.000107
60	24	19	2	0.001271	0.001160
60	24	19	3	0.008505	0.007234
60	24	19	4	0.037439	0.028934
60	24	19	5	0.116350	0.078911
60	24	19	6	0.268454	0.152104
60	24	19	7	0.480314	0.211859
60	24	19	8	0.696410	0.216097
60	24	19	9	0.858944	0.162534
60	24	19	10	0.949241	0.090297
60	24	19	11	0.986181	0.036940
60	24	19	12	0.997220	0.011039
60	24	19	13	0.999598	0.002378
60	24	19	14	0.999960	0.000362
60	24	19	15	0.999997	0.000038
60	24	19	16	1.000000	0.000003
60	24	19	17	1.000000	0.000000
60	24	19	18	1.000000	0.000000
60	24	19	19	1.000000	0.000000
60	24	20	0	0.000002	0.000002
60	24	20	1	0.000051	0.000049
60	24	20	2	0.000648	0.000598
60	24	20	3	0.004800	0.004151
60	24	20	4	0.023325	0.018525
60	24	20	5	0.079781	0.056457
60	24	20	6	0.201677	0.121895
60	24	20	7	0.392470	0.190793
60	24	20	8	0.612080	0.219610
60	24	20	9	0.799481	0.187401
60	24	20	10	0.918408	0.118927
60	24	20	11	0.974468	0.056060
60	24	20	12	0.993989	0.019521
60	24	20	13	0.998960	0.004971
60	24	20	14	0.999871	0.000911

$N = 60$ $n = 24$

Table for $N = 2$, $n = 1$, through $N = 100$, $n = 50$

N	n	k	x	P(x)	p(x)
60	24	20	15	0.999989	0.000118
60	24	20	16	0.999999	0.000010
60	24	20	17	1.000000	0.000001
60	24	20	18	1.000000	0.000000
60	24	20	19	1.000000	0.000000
60	24	20	20	1.000000	0.000000
60	24	21	0	0.000001	0.000001
60	24	21	1	0.000023	0.000022
60	24	21	2	0.000320	0.000297
60	24	21	3	0.002620	0.002300
60	24	21	4	0.014062	0.011442
60	24	21	5	0.052964	0.038902
60	24	21	6	0.146824	0.093859
60	24	21	7	0.311383	0.164559
60	24	21	8	0.524236	0.212853
60	24	21	9	0.729205	0.204970
60	24	21	10	0.876784	0.147578
60	24	21	11	0.956249	0.079465
60	24	21	12	0.988133	0.031884
60	24	21	13	0.997593	0.009460
60	24	21	14	0.999644	0.002050
60	24	21	15	0.999962	0.000319
60	24	21	16	0.999997	0.000035
60	24	21	17	1.000000	0.000003
60	24	21	18	1.000000	0.000000
60	24	21	19	1.000000	0.000000
60	24	21	20	1.000000	0.000000
60	24	21	21	1.000000	0.000000
60	24	22	0	0.000000	0.000000
60	24	22	1	0.000010	0.000009
60	24	22	2	0.000152	0.000142
60	24	22	3	0.001382	0.001229
60	24	22	4	0.008195	0.006813
60	24	22	5	0.034012	0.025817
60	24	22	6	0.103504	0.069492
60	24	22	7	0.239652	0.136148
60	24	22	8	0.436911	0.197260
60	24	22	9	0.650371	0.213460
60	24	22	10	0.823807	0.173436
60	24	22	11	0.929760	0.105954
60	24	22	12	0.978322	0.048562
60	24	22	13	0.994925	0.016602
60	24	22	14	0.999118	0.004193
60	24	22	15	0.999889	0.000771
60	24	22	16	0.999990	0.000101
60	24	22	17	0.999999	0.000009
60	24	22	18	1.000000	0.000001
60	24	22	19	1.000000	0.000000
60	24	22	20	1.000000	0.000000
60	24	22	21	1.000000	0.000000

N	n	k	x	P(x)	p(x)
60	24	22	22	1.000000	0.000000
60	24	23	0	0.000000	0.000000
60	24	23	1	0.000004	0.000004
60	24	23	2	0.000070	0.000066
60	24	23	3	0.000702	0.000632
60	24	23	4	0.004609	0.003907
60	24	23	5	0.021103	0.016495
60	24	23	6	0.070587	0.049484
60	24	23	7	0.178744	0.108157
60	24	23	8	0.353855	0.175111
60	24	23	9	0.566111	0.212256
60	24	23	10	0.759909	0.193799
60	24	23	11	0.893513	0.133604
60	24	23	12	0.962987	0.069474
60	24	23	13	0.990119	0.027132
60	24	23	14	0.998014	0.007895
60	24	23	15	0.999706	0.001692
60	24	23	16	0.999969	0.000263
60	24	23	17	0.999998	0.000029
60	24	23	18	1.000000	0.000002
60	24	23	19	1.000000	0.000000
60	24	23	20	1.000000	0.000000
60	24	23	21	1.000000	0.000000
60	24	23	22	1.000000	0.000000
60	24	23	23	1.000000	0.000000
60	24	24	0	0.000000	0.000000
60	24	24	1	0.000002	0.000002
60	24	24	2	0.000031	0.000029
60	24	24	3	0.000343	0.000313
60	24	24	4	0.002497	0.002154
60	24	24	5	0.012633	0.010136
60	24	24	6	0.046514	0.033881
60	24	24	7	0.129050	0.082536
60	24	24	8	0.278131	0.149081
60	24	24	9	0.480061	0.201930
60	24	24	10	0.686580	0.206519
60	24	24	11	0.846571	0.159991
60	24	24	12	0.940455	0.093884
60	24	24	13	0.982053	0.041598
60	24	24	14	0.995880	0.013828
60	24	24	15	0.999295	0.003414
60	24	24	16	0.999912	0.000617
60	24	24	17	0.999992	0.000080
60	24	24	18	0.999999	0.000007
60	24	24	19	1.000000	0.000000
60	24	24	20	1.000000	0.000000
60	24	24	21	1.000000	0.000000
60	24	24	22	1.000000	0.000000
60	24	24	23	1.000000	0.000000
60	24	24	24	1.000000	0.000000

N	n	k	x	P(x)	p(x)
60	25	1	0	0.583333	0.583333
60	25	1	1	1.000000	0.416667
60	25	2	0	0.336158	0.336158
60	25	2	1	0.830508	0.494350
60	25	2	2	1.000000	0.169492
60	25	3	0	0.191262	0.191262
60	25	3	1	0.625950	0.434687
60	25	3	2	0.932788	0.306838
60	25	3	3	1.000000	0.067212
60	25	4	0	0.107375	0.107375
60	25	4	1	0.442923	0.335548
60	25	4	2	0.808976	0.366052
60	25	4	3	0.974058	0.165082
60	25	4	4	1.000000	0.025942
60	25	5	0	0.059440	0.059440
60	25	5	1	0.299117	0.239677
60	25	5	2	0.658633	0.359516
60	25	5	3	0.909205	0.250572
60	25	5	4	0.990272	0.081067
60	25	5	5	1.000000	0.009728
60	25	6	0	0.032422	0.032422
60	25	6	1	0.194531	0.162109
60	25	6	2	0.508290	0.313759
60	25	6	3	0.808976	0.300686
60	25	6	4	0.959319	0.150343
60	25	6	5	0.996462	0.037144
60	25	6	6	1.000000	0.003537
60	25	7	0	0.017412	0.017412
60	25	7	1	0.122482	0.105071
60	25	7	2	0.374652	0.252169
60	25	7	3	0.686474	0.311822
60	25	7	4	0.900852	0.214378
60	25	7	5	0.982706	0.081853
60	25	7	6	0.998755	0.016050
60	25	7	7	1.000000	0.001245
60	25	8	0	0.009199	0.009199
60	25	8	1	0.074903	0.065705
60	25	8	2	0.265220	0.190317
60	25	8	3	0.557039	0.291819
60	25	8	4	0.815910	0.258871
60	25	8	5	0.951818	0.135908
60	25	8	6	0.993002	0.041184
60	25	8	7	0.999577	0.006576
60	25	8	8	1.000000	0.000423
60	25	9	0	0.004776	0.004776
60	25	9	1	0.044578	0.039802
60	25	9	2	0.181041	0.136463
60	25	9	3	0.433577	0.252535
60	25	9	4	0.711366	0.277789
60	25	9	5	0.899545	0.188180

N	n	k	x	P(x)	p(x)
60	25	9	6	0.977954	0.078408
60	25	9	7	0.997301	0.019347
60	25	9	8	0.999862	0.002561
60	25	9	9	1.000000	0.000138
60	25	10	0	0.002435	0.002435
60	25	10	1	0.025848	0.023413
60	25	10	2	0.119499	0.093651
60	25	10	3	0.324640	0.205141
60	25	10	4	0.596982	0.272342
60	25	10	5	0.825749	0.228767
60	25	10	6	0.948743	0.122993
60	25	10	7	0.990473	0.041730
60	25	10	8	0.999008	0.008536
60	25	10	9	0.999957	0.000948
60	25	10	10	1.000000	0.000043
60	25	11	0	0.001217	0.001217
60	25	11	1	0.014610	0.013392
60	25	11	2	0.076419	0.061810
60	25	11	3	0.234378	0.157958
60	25	11	4	0.482598	0.248220
60	25	11	5	0.734242	0.251644
60	25	11	6	0.902005	0.167763
60	25	11	7	0.975450	0.073445
60	25	11	8	0.996106	0.020656
60	25	11	9	0.999653	0.003547
60	25	11	10	0.999987	0.000334
60	25	11	11	1.000000	0.000013
60	25	12	0	0.000596	0.000596
60	25	12	1	0.008050	0.007454
60	25	12	2	0.047407	0.039356
60	25	12	3	0.163458	0.116051
60	25	12	4	0.376218	0.212760
60	25	12	5	0.631531	0.255312
60	25	12	6	0.836954	0.205424
60	25	12	7	0.948470	0.111516
60	25	12	8	0.988940	0.040469
60	25	12	9	0.998495	0.009555
60	25	12	10	0.999885	0.001390
60	25	12	11	0.999996	0.000111
60	25	12	12	1.000000	0.000004
60	25	13	0	0.000286	0.000286
60	25	13	1	0.004323	0.004038
60	25	13	2	0.028548	0.024225
60	25	13	3	0.110268	0.081719
60	25	13	4	0.283135	0.172868
60	25	13	5	0.525150	0.242015
60	25	13	6	0.755641	0.230490
60	25	13	7	0.906652	0.151011
60	25	13	8	0.974607	0.067955
60	25	13	9	0.995310	0.020703

Table for $N = 2$, $n = 1$, through $N = 100$, $n = 50$

N	n	k	x	P(x)	p(x)
60	25	13	10	0.999450	0.004141
60	25	13	11	0.999964	0.000513
60	25	13	12	0.999999	0.000035
60	25	13	13	1.000000	0.000001
60	25	14	0	0.000134	0.000134
60	25	14	1	0.002262	0.002128
60	25	14	2	0.016694	0.014432
60	25	14	3	0.072016	0.055323
60	25	14	4	0.205897	0.133881
60	25	14	5	0.422165	0.216269
60	25	14	6	0.662464	0.240299
60	25	14	7	0.848818	0.186354
60	25	14	8	0.950027	0.101209
60	25	14	9	0.988262	0.038235
60	25	14	10	0.998129	0.009867
60	25	14	11	0.999811	0.001682
60	25	14	12	0.999989	0.000178
60	25	14	13	1.000000	0.000010
60	25	14	14	1.000000	0.000000
60	25	15	0	0.000061	0.000061
60	25	15	1	0.001151	0.001090
60	25	15	2	0.009478	0.008326
60	25	15	3	0.045557	0.036080
60	25	15	4	0.144777	0.099220
60	25	15	5	0.328135	0.183358
60	25	15	6	0.563210	0.235075
60	25	15	7	0.775897	0.212687
60	25	15	8	0.912624	0.136727
60	25	15	9	0.974963	0.062339
60	25	15	10	0.994911	0.019949
60	25	15	11	0.999299	0.004387
60	25	15	12	0.999939	0.000640
60	25	15	13	0.999997	0.000058
60	25	15	14	1.000000	0.000003
60	25	15	15	1.000000	0.000000
60	25	16	0	0.000027	0.000027
60	25	16	1	0.000570	0.000543
60	25	16	2	0.005222	0.004652
60	25	16	3	0.027918	0.022696
60	25	16	4	0.098475	0.070556
60	25	16	5	0.246643	0.148168
60	25	16	6	0.463956	0.217313
60	25	16	7	0.690822	0.226866
60	25	16	8	0.860971	0.170149
60	25	16	9	0.952798	0.091827
60	25	16	10	0.988262	0.035464
60	25	16	11	0.997934	0.009672
60	25	16	12	0.999754	0.001820
60	25	16	13	0.999981	0.000227
60	25	16	14	0.999999	0.000018

N	n	k	x	P(x)	p(x)
60	25	16	15	1.000000	0.000001
60	25	16	16	1.000000	0.000000
60	25	17	0	0.000012	0.000012
60	25	17	1	0.000274	0.000262
60	25	17	2	0.002790	0.002516
60	25	17	3	0.016570	0.013780
60	25	17	4	0.064800	0.048230
60	25	17	5	0.179294	0.114494
60	25	17	6	0.370116	0.190823
60	25	17	7	0.598013	0.227897
60	25	17	8	0.795232	0.197218
60	25	17	9	0.919406	0.124175
60	25	17	10	0.976172	0.056766
60	25	17	11	0.994856	0.018685
60	25	17	12	0.999216	0.004360
60	25	17	13	0.999919	0.000703
60	25	17	14	0.999995	0.000075
60	25	17	15	1.000000	0.000005
60	25	17	16	1.000000	0.000000
60	25	17	17	1.000000	0.000000
60	25	18	0	0.000005	0.000005
60	25	18	1	0.000128	0.000123
60	25	18	2	0.001444	0.001317
60	25	18	3	0.009520	0.008076
60	25	18	4	0.041246	0.031726
60	25	18	5	0.126041	0.084795
60	25	18	6	0.285799	0.159759
60	25	18	7	0.502615	0.216815
60	25	18	8	0.717262	0.214647
60	25	18	9	0.873202	0.155940
60	25	18	10	0.956370	0.083168
60	25	18	11	0.988773	0.032403
60	25	18	12	0.997898	0.009125
60	25	18	13	0.999723	0.001825
60	25	18	14	0.999975	0.000252
60	25	18	15	0.999999	0.000023
60	25	18	16	1.000000	0.000001
60	25	18	17	1.000000	0.000000
60	25	18	18	1.000000	0.000000
60	25	19	0	0.000002	0.000002
60	25	19	1	0.000057	0.000055
60	25	19	2	0.000723	0.000666
60	25	19	3	0.005290	0.004567
60	25	19	4	0.025383	0.020093
60	25	19	5	0.085662	0.060279
60	25	19	6	0.213528	0.127865
60	25	19	7	0.409694	0.196166
60	25	19	8	0.630381	0.220687
60	25	19	9	0.813796	0.183415
60	25	19	10	0.926667	0.112871

N	n	k	x	P(x)	p(x)
60	25	19	11	0.977972	0.051305
60	25	19	12	0.995074	0.017102
60	25	19	13	0.999202	0.004128
60	25	19	14	0.999909	0.000708
60	25	19	15	0.999993	0.000084
60	25	19	16	1.000000	0.000007
60	25	19	17	1.000000	0.000000
60	25	19	18	1.000000	0.000000
60	25	19	19	1.000000	0.000000
60	25	20	0	0.000001	0.000001
60	25	20	1	0.000025	0.000024
60	25	20	2	0.000350	0.000325
60	25	20	3	0.002839	0.002490
60	25	20	4	0.015091	0.012252
60	25	20	5	0.056258	0.041166
60	25	20	6	0.154273	0.098015
60	25	20	7	0.323572	0.169299
60	25	20	8	0.538876	0.215304
60	25	20	9	0.742219	0.203343
60	25	20	10	0.885373	0.143153
60	25	20	11	0.960453	0.075080
60	25	20	12	0.989651	0.029198
60	25	20	13	0.997993	0.008342
60	25	20	14	0.999719	0.001726
60	25	20	15	0.999973	0.000253
60	25	20	16	0.999998	0.000026
60	25	20	17	1.000000	0.000002
60	25	20	18	1.000000	0.000000
60	25	20	19	1.000000	0.000000
60	25	20	20	1.000000	0.000000
60	25	21	0	0.000000	0.000000
60	25	21	1	0.000010	0.000010
60	25	21	2	0.000163	0.000153
60	25	21	3	0.001470	0.001307
60	25	21	4	0.008659	0.007189
50	25	21	5	0.035675	0.027015
60	25	21	6	0.107716	0.072041
60	25	21	7	0.247388	0.139672
60	25	21	8	0.447372	0.199985
60	25	21	9	0.660882	0.213510
60	25	21	10	0.831690	0.170808
60	25	21	11	0.934175	0.102485
60	25	21	12	0.980162	0.045987
60	25	21	13	0.995491	0.015329
60	25	21	14	0.999245	0.003754
60	25	21	15	0.999909	0.000665
60	25	21	16	0.999992	0.000083
60	25	21	17	1.000000	0.000007
60	25	21	18	1.000000	0.000000
60	25	21	19	1.000000	0.000000

N	n	k	x	P(x)	p(x)
60	25	21	20	1.000000	0.000000
60	25	21	21	1.000000	0.000000
60	25	22	0	0.000000	0.000000
60	25	22	1	0.000004	0.000004
60	25	22	2	0.000073	0.000069
60	25	22	3	0.000733	0.000660
60	25	22	4	0.004788	0.004055
60	25	22	5	0.021820	0.017032
60	25	22	6	0.072619	0.050798
60	25	22	7	0.182924	0.110305
60	25	22	8	0.360199	0.177276
60	25	22	9	0.573288	0.213089
60	25	22	10	0.765995	0.192707
60	25	22	11	0.897386	0.131391
60	25	22	12	0.964833	0.067447
60	25	22	13	0.990774	0.025941
60	25	22	14	0.998186	0.007412
60	25	22	15	0.999739	0.001553
60	25	22	16	0.999973	0.000234
60	25	22	17	0.999998	0.000025
60	25	22	18	1.000000	0.000002
60	25	22	19	1.000000	0.000000
60	25	22	20	1.000000	0.000000
60	25	22	21	1.000000	0.000000
60	25	22	22	1.000000	0.000000
60	25	23	0	0.000000	0.000000
60	25	23	1	0.000002	0.000002
60	25	23	2	0.000031	0.000030
60	25	23	3	0.000351	0.000319
60	25	23	4	0.002547	0.002196
60	25	23	5	0.012856	0.010309
60	25	23	6	0.047220	0.034364
60	25	23	7	0.130674	0.083454
60	25	23	8	0.280892	0.150218
60	25	23	9	0.483567	0.202675
60	25	23	10	0.689926	0.206360
60	25	23	11	0.848978	0.159052
60	25	23	12	0.941759	0.092780
60	25	23	13	0.982582	0.040823
60	25	23	14	0.996040	0.013458
60	25	23	15	0.999330	0.003290
60	25	23	16	0.999918	0.000587
60	25	23	17	0.999993	0.000075
60	25	23	18	0.999999	0.000007
60	25	23	19	1.000000	0.000000
60	25	23	20	1.000000	0.000000
60	25	23	21	1.000000	0.000000
60	25	23	22	1.000000	0.000000
60	25	23	23	1.000000	0.000000
60	25	24	0	0.000000	0.000000

Table for $N = 2$, $n = 1$, through $N = 100$, $n = 50$

N	n	k	x	P(x)	p(x)
60	25	24	1	0.000001	0.000001
60	25	24	2	0.000013	0.000012
60	25	24	3	0.000161	0.000148
60	25	24	4	0.001301	0.001140
60	25	24	5	0.007284	0.005983
60	25	24	6	0.029573	0.022290
60	25	24	7	0.090075	0.060501
60	25	24	8	0.211873	0.121798
60	25	24	9	0.395923	0.184051
60	25	24	10	0.606267	0.210344
60	25	24	11	0.788797	0.182530
60	25	24	12	0.909160	0.120364
60	25	24	13	0.969342	0.060182
60	25	24	14	0.992039	0.022697
60	25	24	15	0.998441	0.006402
60	25	24	16	0.999775	0.001334
60	25	24	17	0.999977	0.000202
60	25	24	18	0.999998	0.000022
60	25	24	19	1.000000	0.000002
60	25	24	20	1.000000	0.000000
60	25	24	21	1.000000	0.000000
60	25	24	22	1.000000	0.000000
60	25	24	23	1.000000	0.000000
60	25	24	24	1.000000	0.000000
60	25	25	0	0.000000	0.000000
60	25	25	1	0.000000	0.000000
60	25	25	2	0.000005	0.000005
60	25	25	3	0.000070	0.000065
60	25	25	4	0.000636	0.000565
60	25	25	5	0.003960	0.003324
60	25	25	6	0.017809	0.013850
60	25	25	7	0.059824	0.042015
60	25	25	8	0.154357	0.094533
60	25	25	9	0.314123	0.159766
60	25	25	10	0.518624	0.204501
60	25	25	11	0.717813	0.199189
60	25	25	12	0.865696	0.147883
60	25	25	13	0.949282	0.083586
60	25	25	14	0.985104	0.035823
60	25	25	15	0.996663	0.011559
60	25	25	16	0.999441	0.002779
60	25	25	17	0.999932	0.000490
60	25	25	18	0.999994	0.000062
60	25	25	19	0.999999	0.000006
60	25	25	20	1.000000	0.000000
60	25	25	21	1.000000	0.000000
60	25	25	22	1.000000	0.000000
60	25	25	23	1.000000	0.000000
60	25	25	24	1.000000	0.000000
60	25	25	25	1.000000	0.000000

N	n	k	x	P(x)	p(x)
60	26	1	0	0.566667	0.566667
60	26	1	1	1.000000	0.433333
60	26	2	0	0.316949	0.316949
60	26	2	1	0.816384	0.499435
60	26	2	2	1.000000	0.183616
60	26	3	0	0.174868	0.174868
60	26	3	1	0.601110	0.426242
60	26	3	2	0.924021	0.322911
60	26	3	3	1.000000	0.075979
60	26	4	0	0.095104	0.095104
60	26	4	1	0.414162	0.319058
60	26	4	2	0.788059	0.373896
60	26	4	3	0.969342	0.181283
60	26	4	4	1.000000	0.030658
60	26	5	0	0.050949	0.050949
60	26	5	1	0.271725	0.220777
60	26	5	2	0.627817	0.356092
60	26	5	3	0.894886	0.267069
60	26	5	4	0.987956	0.093069
60	26	5	5	1.000000	0.012044
60	26	6	0	0.026864	0.026864
60	26	6	1	0.171372	0.144509
60	26	6	2	0.472432	0.301059
60	26	6	3	0.783203	0.310771
60	26	6	4	0.950728	0.167525
60	26	6	5	0.995401	0.044673
60	26	6	6	1.000000	0.004599
60	26	7	0	0.013929	0.013929
60	26	7	1	0.104470	0.090541
60	26	7	2	0.338628	0.234157
60	26	7	3	0.650837	0.312210
60	26	7	4	0.882477	0.231640
60	26	7	5	0.978028	0.095551
60	26	7	6	0.998297	0.020268
60	26	7	7	1.000000	0.001703
60	26	8	0	0.007096	0.007096
60	26	8	1	0.061762	0.054666
60	26	8	2	0.232594	0.170832
60	26	8	3	0.515350	0.282756
60	26	8	4	0.786325	0.270975
60	26	8	5	0.940168	0.153844
60	26	8	6	0.990648	0.050480
60	26	8	7	0.999389	0.008741
60	26	8	8	1.000000	0.000611
60	26	9	0	0.003548	0.003548
60	26	9	1	0.035480	0.031932
60	26	9	2	0.153749	0.118268
60	26	9	3	0.390285	0.236536
60	26	9	4	0.671682	0.281397
60	26	9	5	0.878039	0.206358

N	n	k	x	P(x)	p(x)
60	26	9	6	0.971233	0.093194
60	26	9	7	0.996195	0.024963
60	26	9	8	0.999789	0.003593
60	26	9	9	1.000000	0.000211
60	26	10	0	0.001739	0.001739
60	26	10	1	0.019827	0.018088
60	26	10	2	0.098093	0.078266
60	26	10	3	0.283612	0.185519
60	26	10	4	0.550295	0.266683
60	26	10	5	0.793068	0.242774
60	26	10	6	0.934686	0.141618
60	26	10	7	0.986896	0.052209
60	26	10	8	0.998520	0.011625
60	26	10	9	0.999929	0.001409
60	26	10	10	1.000000	0.000070
60	26	11	0	0.000835	0.000835
60	26	11	1	0.010783	0.009948
60	26	11	2	0.060525	0.049742
60	26	11	3	0.198273	0.137748
60	26	11	4	0.432954	0.234681
60	26	11	5	0.691103	0.258149
60	26	11	6	0.878039	0.186936
60	26	11	7	0.967056	0.089017
60	26	11	8	0.994335	0.027279
60	26	11	9	0.999450	0.005115
60	26	11	10	0.999977	0.000527
60	26	11	11	1.000000	0.000023
60	26	12	0	0.000392	0.000392
60	26	12	1	0.005708	0.005316
60	26	12	2	0.036162	0.030454
60	26	12	3	0.133616	0.097454
60	26	12	4	0.327587	0.193971
60	26	12	5	0.580468	0.252881
60	26	12	6	0.801739	0.221271
60	26	12	7	0.932539	0.130800
60	26	12	8	0.984314	0.051775
60	26	12	9	0.997676	0.013361
60	26	12	10	0.999805	0.002129
60	26	12	11	0.999993	0.000188
60	26	12	12	1.000000	0.000007
60	26	13	0	0.000180	0.000180
60	26	13	1	0.002939	0.002759
60	26	13	2	0.020935	0.017996
60	26	13	3	0.086919	0.065985
60	26	13	4	0.238684	0.151764
60	26	13	5	0.469833	0.231149
60	26	13	6	0.709543	0.239710
60	26	13	7	0.880764	0.171221
60	26	13	8	0.964899	0.084135
60	26	13	9	0.992944	0.028045

N	n	k	x	P(x)	p(x)
60	26	13	10	0.999095	0.006152
60	26	13	11	0.999934	0.000839
60	26	13	12	0.999998	0.000064
60	26	13	13	1.000000	0.000002
60	26	14	0	0.000080	0.000080
60	26	14	1	0.001471	0.001391
60	26	14	2	0.011745	0.010274
60	26	14	3	0.054629	0.042884
60	26	14	4	0.167645	0.113016
60	26	14	5	0.366553	0.198908
60	26	14	6	0.607538	0.240985
60	26	14	7	0.811547	0.204009
60	26	14	8	0.932677	0.121130
60	26	14	9	0.982800	0.050123
60	26	14	10	0.997001	0.014201
60	26	14	11	0.999667	0.002665
60	26	14	12	0.999979	0.000312
60	26	14	13	0.999999	0.000020
60	26	14	14	1.000000	0.000001
60	26	15	0	0.000035	0.000035
60	26	15	1	0.000715	0.000680
60	26	15	2	0.006385	0.005670
60	26	15	3	0.033187	0.026802
60	26	15	4	0.113594	0.080407
60	26	15	5	0.275747	0.162154
60	26	15	6	0.502762	0.227015
60	26	15	7	0.727283	0.224520
60	26	15	8	0.885278	0.157996
60	26	15	9	0.964276	0.078998
60	26	15	10	0.992062	0.027785
60	26	15	11	0.998797	0.006736
60	26	15	12	0.999884	0.001086
60	26	15	13	0.999994	0.000110
60	26	15	14	1.000000	0.000006
60	26	15	15	1.000000	0.000000
60	26	16	0	0.000015	0.000015
60	26	16	1	0.000337	0.000323
60	26	16	2	0.003361	0.003024
60	26	16	3	0.019488	0.016127
60	26	16	4	0.074284	0.054796
60	26	16	5	0.200076	0.125792
60	26	16	6	0.401867	0.201791
60	26	16	7	0.632485	0.230618
60	26	16	8	0.822080	0.189595
60	26	16	9	0.934432	0.112353
60	26	16	10	0.982182	0.047750
60	26	16	11	0.996552	0.014370
60	26	16	12	0.999546	0.002994
60	26	16	13	0.999962	0.000416
60	26	16	14	0.999998	0.000036

Table for $N = 2$, $n = 1$, through $N = 100$, $n = 50$

N	n	k	x	P(x)	p(x)
60	26	16	15	1.000000	0.000002
60	26	16	16	1.000000	0.000000
60	26	17	0	0.000006	0.000006
60	26	17	1	0.000154	0.000148
60	26	17	2	0.001712	0.001558
60	26	17	3	0.011058	0.009346
60	26	17	4	0.046886	0.035828
60	26	17	5	0.140039	0.093153
60	26	17	6	0.310144	0.170105
60	26	17	7	0.532900	0.222756
60	26	17	8	0.744518	0.211619
60	26	17	9	0.891024	0.146505
60	26	17	10	0.964819	0.073795
60	26	17	11	0.991653	0.026835
60	26	17	12	0.998593	0.006940
60	26	17	13	0.999839	0.001246
60	26	17	14	0.999988	0.000149
60	26	17	15	0.999999	0.000011
60	26	17	16	1.000000	0.000000
60	26	17	17	1.000000	0.000000
60	26	18	0	0.000002	0.000002
60	26	18	1	0.000068	0.000066
60	26	18	2	0.000842	0.000774
60	26	18	3	0.006059	0.005217
60	26	18	4	0.028556	0.022497
60	26	18	5	0.094546	0.065990
60	26	18	6	0.231025	0.136479
60	26	18	7	0.434473	0.203448
60	26	18	8	0.655934	0.221461
60	26	18	9	0.833103	0.177169
60	26	18	10	0.937360	0.104257
60	26	18	11	0.982292	0.044932
60	26	18	12	0.996334	0.014041
60	26	18	13	0.999462	0.003129
60	26	18	14	0.999947	0.000484
60	26	18	15	0.999997	0.000050
60	26	18	16	1.000000	0.000003
60	26	18	17	1.000000	0.000000
60	26	18	18	1.000000	0.000000
60	26	19	0	0.000001	0.000001
60	26	19	1	0.000029	0.000028
60	26	19	2	0.000400	0.000371
60	26	19	3	0.003202	0.002802
60	26	19	4	0.016772	0.013569
60	26	19	5	0.061551	0.044779
60	26	19	6	0.166035	0.104484
60	26	19	7	0.342437	0.176402
60	26	19	8	0.561022	0.218585
60	26	19	9	0.761392	0.200370
60	26	19	10	0.897643	0.136251
60	26	19	11	0.966245	0.068602
60	26	19	12	0.991653	0.025408
60	26	19	13	0.998494	0.006841
60	26	19	14	0.999808	0.001314
60	26	19	15	0.999983	0.000175
60	26	19	16	0.999999	0.000016
60	26	19	17	1.000000	0.000001
60	26	19	18	1.000000	0.000000
60	26	19	19	1.000000	0.000000
60	26	20	0	0.000000	0.000000
60	26	20	1	0.000012	0.000012
60	26	20	2	0.000183	0.000171
60	26	20	3	0.001630	0.001447
60	26	20	4	0.009490	0.007860
60	26	20	5	0.038615	0.029125
60	26	20	6	0.115067	0.076452
60	26	20	7	0.260690	0.145623
60	26	20	8	0.465058	0.204368
60	26	20	9	0.678312	0.213254
60	26	20	10	0.844472	0.166160
60	26	20	11	0.941147	0.096675
60	26	20	12	0.982977	0.041831
60	26	20	13	0.996325	0.013348
60	26	20	14	0.999424	0.003099
60	26	20	15	0.999936	0.000513
60	26	20	16	0.999995	0.000059
60	26	20	17	1.000000	0.000004
60	26	20	18	1.000000	0.000000
60	26	20	19	1.000000	0.000000
60	26	20	20	1.000000	0.000000
60	26	21	0	0.000000	0.000000
60	26	21	1	0.000005	0.000005
60	26	21	2	0.000080	0.000076
60	26	21	3	0.000798	0.000718
60	26	21	4	0.005167	0.004369
60	26	21	5	0.023325	0.018157
60	26	21	6	0.076841	0.053516
60	26	21	7	0.191519	0.114678
60	26	21	8	0.373092	0.181573
60	26	21	9	0.587679	0.214587
60	26	21	10	0.778008	0.190329
60	26	21	11	0.904894	0.126886
60	26	21	12	0.968337	0.063443
60	26	21	13	0.991987	0.023650
60	26	21	14	0.998494	0.006507
60	26	21	15	0.999795	0.001301
60	26	21	16	0.999981	0.000185
60	26	21	17	0.999999	0.000018
60	26	21	18	1.000000	0.000001
60	26	21	19	1.000000	0.000000
60	26	21	20	1.000000	0.000000
60	26	21	21	1.000000	0.000000
60	26	22	0	0.000000	0.000000
60	26	22	1	0.000002	0.000002
60	26	22	2	0.000034	0.000032
60	26	22	3	0.000375	0.000341
60	26	22	4	0.002702	0.002328
60	26	22	5	0.013548	0.010845
60	26	22	6	0.049397	0.035849
60	26	22	7	0.135650	0.086253
60	26	22	8	0.289289	0.153639
60	26	22	9	0.494141	0.204852
60	26	22	10	0.699924	0.205783
60	26	22	11	0.856091	0.156167
60	26	22	12	0.945562	0.089471
60	26	22	13	0.984104	0.038541
60	26	22	14	0.996492	0.012388
60	26	22	15	0.999428	0.002936
60	26	22	16	0.999933	0.000505
60	26	22	17	0.999994	0.000061
60	26	22	18	1.000000	0.000005
60	26	22	19	1.000000	0.000000
60	26	22	20	1.000000	0.000000
60	26	22	21	1.000000	0.000000
60	26	22	22	1.000000	0.000000
60	26	23	0	0.000000	0.000000
60	26	23	1	0.000001	0.000001
60	26	23	2	0.000014	0.000013
60	26	23	3	0.000168	0.000155
60	26	23	4	0.001355	0.001186
60	26	23	5	0.007554	0.006199
60	26	23	6	0.030529	0.022974
60	26	23	7	0.092523	0.061995
60	26	23	8	0.216513	0.123989
60	26	23	9	0.402497	0.185984
60	26	23	10	0.613279	0.210782
60	26	23	11	0.794446	0.181168
60	26	23	12	0.912599	0.118153
60	26	23	13	0.970918	0.058319
60	26	23	14	0.992580	0.021661
60	26	23	15	0.998578	0.005999
60	26	23	16	0.999800	0.001222
60	26	23	17	0.999980	0.000180
60	26	23	18	0.999998	0.000019
60	26	23	19	1.000000	0.000001
60	26	23	20	1.000000	0.000000
60	26	23	21	1.000000	0.000000
60	26	23	22	1.000000	0.000000
60	26	23	23	1.000000	0.000000
60	26	24	0	0.000000	0.000000
60	26	24	1	0.000000	0.000000
60	26	24	2	0.000005	0.000005
60	26	24	3	0.000072	0.000067
60	26	24	4	0.000649	0.000577
60	26	24	5	0.004036	0.003386
60	26	24	6	0.018110	0.014074
60	26	24	7	0.060688	0.042578
60	26	24	8	0.156194	0.095505
60	26	24	9	0.317045	0.160851
60	26	24	10	0.522130	0.205085
60	26	24	11	0.721000	0.198870
60	26	24	12	0.867893	0.146893
60	26	24	13	0.950428	0.082535
60	26	24	14	0.985554	0.035127
60	26	24	15	0.996795	0.011240
60	26	24	16	0.999470	0.002675
60	26	24	17	0.999936	0.000466
60	26	24	18	0.999994	0.000058
60	26	24	19	1.000000	0.000005
60	26	24	20	1.000000	0.000000
60	26	24	21	1.000000	0.000000
60	26	24	22	1.000000	0.000000
60	26	24	23	1.000000	0.000000
60	26	24	24	1.000000	0.000000
60	26	25	0	0.000000	0.000000
60	26	25	1	0.000000	0.000000
60	26	25	2	0.000002	0.000002
60	26	25	3	0.000029	0.000027
60	26	25	4	0.000297	0.000267
60	26	25	5	0.002060	0.001764
60	26	25	6	0.010291	0.008231
60	26	25	7	0.038216	0.027926
60	26	25	8	0.108441	0.070224
60	26	25	9	0.241087	0.132646
60	26	25	10	0.430981	0.189894
60	26	25	11	0.638137	0.207157
60	26	25	12	0.810768	0.172630
60	26	25	13	0.920624	0.109856
60	26	25	14	0.973846	0.053222
60	26	25	15	0.993360	0.019515
60	26	25	16	0.998727	0.005367
60	26	25	17	0.999820	0.001093
60	26	25	18	0.999982	0.000162
60	26	25	19	0.999999	0.000017
60	26	25	20	1.000000	0.000001
60	26	25	21	1.000000	0.000000
60	26	25	22	1.000000	0.000000
60	26	25	23	1.000000	0.000000
60	26	25	24	1.000000	0.000000
60	26	25	25	1.000000	0.000000

N	n	k	x	P(x)	p(x)
60	26	26	0	0.000000	0.000000
60	26	26	1	0.000000	0.000000
60	26	26	2	0.000001	0.000001
60	26	26	3	0.000011	0.000011
60	26	26	4	0.000129	0.000117
60	26	26	5	0.001002	0.000873
60	26	26	6	0.005588	0.004586
60	26	26	7	0.023057	0.017469
60	26	26	8	0.072325	0.049269
60	26	26	9	0.176659	0.104333
60	26	26	10	0.344172	0.167513
60	26	26	11	0.549356	0.205184
60	26	26	12	0.741716	0.192360
60	26	26	13	0.879820	0.138104
60	26	26	14	0.955598	0.075778
60	26	26	15	0.987227	0.031629
60	26	26	16	0.997194	0.009966
60	26	26	17	0.999539	0.002345
60	26	26	18	0.999945	0.000406
60	26	26	19	0.999995	0.000051
60	26	26	20	1.000000	0.000004
60	26	26	21	1.000000	0.000000
60	26	26	22	1.000000	0.000000
60	26	26	23	1.000000	0.000000
60	26	26	24	1.000000	0.000000
60	26	26	25	1.000000	0.000000
60	26	26	26	1.000000	0.000000
60	27	1	0	0.550000	0.550000
60	27	1	1	1.000000	0.450000
60	27	2	0	0.298305	0.298305
60	27	2	1	0.801695	0.503390
60	27	2	2	1.000000	0.198305
60	27	3	0	0.159439	0.159439
60	27	3	1	0.576037	0.416598
60	27	3	2	0.914524	0.338486
60	27	3	3	1.000000	0.085476
60	27	4	0	0.083915	0.083915
60	27	4	1	0.386010	0.302095
60	27	4	2	0.766065	0.380055
60	27	4	3	0.964010	0.197945
60	27	4	4	1.000000	0.035990
60	27	5	0	0.043456	0.043456
60	27	5	1	0.245752	0.202296
60	27	5	2	0.596397	0.350646
60	27	5	3	0.879176	0.282779
60	27	5	4	0.985218	0.106042
60	27	5	5	1.000000	0.014782
60	27	6	0	0.022123	0.022123
60	27	6	1	0.150121	0.127998
60	27	6	2	0.437013	0.286892
60	27	6	3	0.755782	0.318769
60	27	6	4	0.940873	0.185092
60	27	6	5	0.994087	0.053214
60	27	6	6	1.000000	0.005913
60	27	7	0	0.011062	0.011062
60	27	7	1	0.088492	0.077431
60	27	7	2	0.304193	0.215700
60	27	7	3	0.614107	0.309914
60	27	7	4	0.862038	0.247931
60	27	7	5	0.972408	0.110369
60	27	7	6	0.997701	0.025293
60	27	7	7	1.000000	0.002299
60	27	8	0	0.005426	0.005426
60	27	8	1	0.050507	0.045081
60	27	8	2	0.202447	0.151940
60	27	8	3	0.473768	0.271321
60	27	8	4	0.754445	0.280677
60	27	8	5	0.926594	0.172149
60	27	8	6	0.987679	0.061085
60	27	8	7	0.999132	0.011453
60	27	8	8	1.000000	0.000868
60	27	9	0	0.002609	0.002609
60	27	9	1	0.027967	0.025358
60	27	9	2	0.129399	0.101432
60	27	9	3	0.348543	0.219144
60	27	9	4	0.630300	0.281756
60	27	9	5	0.853762	0.223462
60	27	9	6	0.963010	0.109248
60	27	9	7	0.994727	0.031717
60	27	9	8	0.999683	0.004956
60	27	9	9	1.000000	0.000317
60	27	10	0	0.001228	0.001228
60	27	10	1	0.015039	0.013812
60	27	10	2	0.079678	0.064638
60	27	10	3	0.245417	0.165739
60	27	10	4	0.503233	0.257816
60	27	10	5	0.757366	0.254133
60	27	10	6	0.918025	0.160659
60	27	10	7	0.982289	0.064264
60	27	10	8	0.997837	0.015548
60	27	10	9	0.999888	0.002051
60	27	10	10	1.000000	0.000112
60	27	11	0	0.000565	0.000565
60	27	11	1	0.007857	0.007293
60	27	11	2	0.047358	0.039501
60	27	11	3	0.165862	0.118503
60	27	11	4	0.384638	0.218776
60	27	11	5	0.645548	0.260910
60	27	11	6	0.850549	0.205001
60	27	11	7	0.956584	0.106035
60	27	11	8	0.991929	0.035345
60	27	11	9	0.999150	0.007221
60	27	11	10	0.999962	0.000812
60	27	11	11	1.000000	0.000038
60	27	12	0	0.000254	0.000254
60	27	12	1	0.003988	0.003734
60	27	12	2	0.027205	0.023217
60	27	12	3	0.107819	0.080615
60	27	12	4	0.281947	0.174128
60	27	12	5	0.528404	0.246457
60	27	12	6	0.762691	0.234287
60	27	12	7	0.913304	0.150613
60	27	12	8	0.978223	0.064919
60	27	12	9	0.996497	0.018274
60	27	12	10	0.999680	0.003183
60	27	12	11	0.999987	0.000307
60	27	12	12	1.000000	0.000012
60	27	13	0	0.000111	0.000111
60	27	13	1	0.001965	0.001854
60	27	13	2	0.015113	0.013148
60	27	13	3	0.067512	0.052399
60	27	13	4	0.198511	0.130999
60	27	13	5	0.415445	0.216934
60	27	13	6	0.660191	0.244746
60	27	13	7	0.850549	0.190358
60	27	13	8	0.952526	0.101977
60	27	13	9	0.989644	0.037118
60	27	13	10	0.998553	0.008908
60	27	13	11	0.999885	0.001332
60	27	13	12	0.999996	0.000111
60	27	13	13	1.000000	0.000004
60	27	14	0	0.000047	0.000047
60	27	14	1	0.000939	0.000892
60	27	14	2	0.008119	0.007180
60	27	14	3	0.040755	0.032636
60	27	14	4	0.134405	0.093650
60	27	14	5	0.313901	0.179496
60	27	14	6	0.550836	0.236935
60	27	14	7	0.769545	0.218709
60	27	14	8	0.911301	0.141756
60	27	14	9	0.975429	0.064128
60	27	14	10	0.995330	0.019902
60	27	14	11	0.999431	0.004101
60	27	14	12	0.999961	0.000529
60	27	14	13	0.999999	0.000038
60	27	14	14	1.000000	0.000001
60	27	15	0	0.000019	0.000019
60	27	15	1	0.000435	0.000416
60	27	15	2	0.004217	0.003782
60	27	15	3	0.023728	0.019510
60	27	15	4	0.087580	0.063852
60	27	15	5	0.228055	0.140475
60	27	15	6	0.442670	0.214615
60	27	15	7	0.674454	0.231784
60	27	15	8	0.852750	0.178295
60	27	15	9	0.950335	0.097586
60	27	15	10	0.987976	0.037640
60	27	15	11	0.998005	0.010030
60	27	15	12	0.999788	0.001783
60	27	15	13	0.999987	0.000199
60	27	15	14	1.000000	0.000012
60	27	15	15	1.000000	0.000000
60	27	16	0	0.000008	0.000008
60	27	16	1	0.000195	0.000187
60	27	16	2	0.002116	0.001921
60	27	16	3	0.013322	0.011206
60	27	16	4	0.054944	0.041622
60	27	16	5	0.159378	0.104434
60	27	16	6	0.342516	0.183138
60	27	16	7	0.571439	0.228923
60	27	16	8	0.777469	0.206030
60	27	16	9	0.911301	0.133832
60	27	16	10	0.973756	0.062455
60	27	16	11	0.994439	0.020683
60	27	16	12	0.999194	0.004755
60	27	16	13	0.999925	0.000731
60	27	16	14	0.999996	0.000071
60	27	16	15	1.000000	0.000004
60	27	16	16	1.000000	0.000000
60	27	17	0	0.000003	0.000003
60	27	17	1	0.000084	0.000081
60	27	17	2	0.001025	0.000940
60	27	17	3	0.007210	0.006185
60	27	17	4	0.033187	0.025978
60	27	17	5	0.107161	0.073974
60	27	17	6	0.255110	0.147948
60	27	17	7	0.467383	0.212274
60	27	17	8	0.688502	0.221118
60	27	17	9	0.856552	0.168050
60	27	17	10	0.949626	0.093074
60	27	17	11	0.986918	0.037292
60	27	17	12	0.997573	0.010655
60	27	17	13	0.999692	0.002120
60	27	17	14	0.999975	0.000283
60	27	17	15	0.999999	0.000024
60	27	17	16	1.000000	0.000001
60	27	17	17	1.000000	0.000000
60	27	18	0	0.000001	0.000001
60	27	18	1	0.000035	0.000034
60	27	18	2	0.000478	0.000443

N	n	k	x	P(x)	p(x)
60	27	18	3	0.003757	0.003280
60	27	18	4	0.019292	0.015535
60	27	18	5	0.069314	0.050022
60	27	18	6	0.182856	0.113542
60	27	18	7	0.368651	0.185796
60	27	18	8	0.590798	0.222147
60	27	18	9	0.786205	0.195407
60	27	18	10	0.912829	0.126624
60	27	18	11	0.973042	0.060213
60	27	18	12	0.993856	0.020814
60	27	18	13	0.999002	0.005146
60	27	18	14	0.999890	0.000887
60	27	18	15	0.999992	0.000103
60	27	18	16	1.000000	0.000007
60	27	18	17	1.000000	0.000000
60	27	18	18	1.000000	0.000000
60	27	19	0	0.000000	0.000000
60	27	19	1	0.000014	0.000014
60	27	19	2	0.000214	0.000200
60	27	19	3	0.001883	0.001669
60	27	19	4	0.010785	0.008902
60	27	19	5	0.043112	0.032327
60	27	19	6	0.126085	0.082973
60	27	19	7	0.280177	0.154092
60	27	19	8	0.490303	0.210126
60	27	19	9	0.702459	0.212156
60	27	19	10	0.861577	0.159117
60	27	19	11	0.950104	0.088527
60	27	19	12	0.986422	0.036319
60	27	19	13	0.997287	0.010865
60	27	19	14	0.999615	0.002328
60	27	19	15	0.999963	0.000348
60	27	19	16	0.999998	0.000035
60	27	19	17	1.000000	0.000002
60	27	19	18	1.000000	0.000000
60	27	19	19	1.000000	0.000000
60	27	20	0	0.000000	0.000000
60	27	20	1	0.000005	0.000005
60	27	20	2	0.000092	0.000087
60	27	20	3	0.000906	0.000814
60	27	20	4	0.005791	0.004885
60	27	20	5	0.025766	0.019974
60	27	20	6	0.083587	0.057821
60	27	20	7	0.205010	0.121424
60	27	20	8	0.392928	0.187918
60	27	20	9	0.609318	0.216390
60	27	20	10	0.795601	0.186283
60	27	20	11	0.915556	0.119955
60	27	20	12	0.973135	0.057579
60	27	20	13	0.993577	0.020442

N	n	k	x	P(x)	p(x)
60	27	20	14	0.998877	0.005300
60	27	20	15	0.999861	0.000984
60	27	20	16	0.999988	0.000127
60	27	20	17	0.999999	0.000011
60	27	20	18	1.000000	0.000001
60	27	20	19	1.000000	0.000000
60	27	20	20	1.000000	0.000000
60	27	21	0	0.000000	0.000000
60	27	21	1	0.000002	0.000002
60	27	21	2	0.000038	0.000036
60	27	21	3	0.000418	0.000380
60	27	21	4	0.002983	0.002565
60	27	21	5	0.014780	0.011797
60	27	21	6	0.053231	0.038451
60	27	21	7	0.144299	0.091068
60	27	21	8	0.303667	0.159369
60	27	21	9	0.511942	0.208275
60	27	21	10	0.716431	0.204488
60	27	21	11	0.867574	0.151144
60	27	21	12	0.951543	0.083969
60	27	21	13	0.986422	0.034879
60	27	21	14	0.997154	0.010732
60	27	21	15	0.999566	0.002411
60	27	21	16	0.999953	0.000388
60	27	21	17	0.999996	0.000043
60	27	21	18	1.000000	0.000003
60	27	21	19	1.000000	0.000000
60	27	21	20	1.000000	0.000000
60	27	21	21	1.000000	0.000000
60	27	22	0	0.000000	0.000000
60	27	22	1	0.000001	0.000001
60	27	22	2	0.000015	0.000014
60	27	22	3	0.000184	0.000169
60	27	22	4	0.001470	0.001286
60	27	22	5	0.008125	0.006655
60	27	22	6	0.032527	0.024401
60	27	22	7	0.097597	0.065071
60	27	22	8	0.226026	0.128429
60	27	22	9	0.415815	0.189789
60	27	22	10	0.627295	0.211479
60	27	22	11	0.805567	0.178272
60	27	22	12	0.919247	0.113681
60	27	22	13	0.973901	0.054654
60	27	22	14	0.993577	0.019675
60	27	22	15	0.998824	0.005247
60	27	22	16	0.999844	0.001020
60	27	22	17	0.999985	0.000141
60	27	22	18	0.999999	0.000014
60	27	22	19	1.000000	0.000001
60	27	22	20	1.000000	0.000000

N	n	k	x	P(x)	p(x)
60	27	22	21	1.000000	0.000000
60	27	22	22	1.000000	0.000000
60	27	23	0	0.000000	0.000000
60	27	23	1	0.000000	0.000000
60	27	23	2	0.000006	0.000005
60	27	23	3	0.000077	0.000072
60	27	23	4	0.000692	0.000614
60	27	23	5	0.004272	0.003580
60	27	23	6	0.019041	0.014769
60	27	23	7	0.063349	0.044308
60	27	23	8	0.161812	0.098462
60	27	23	9	0.325915	0.164104
60	27	23	10	0.532685	0.206770
60	27	23	11	0.730505	0.197819
60	27	23	12	0.874373	0.143869
60	27	23	13	0.953766	0.079392
60	27	23	14	0.986846	0.033080
60	27	23	15	0.997167	0.010321
60	27	23	16	0.999549	0.002382
60	27	23	17	0.999948	0.000400
60	27	23	18	0.999996	0.000048
60	27	23	19	1.000000	0.000004
60	27	23	20	1.000000	0.000000
60	27	23	21	1.000000	0.000000
60	27	23	22	1.000000	0.000000
60	27	23	23	1.000000	0.000000
60	27	24	0	0.000000	0.000000
60	27	24	1	0.000000	0.000000
60	27	24	2	0.000002	0.000002
60	27	24	3	0.000031	0.000029
60	27	24	4	0.000310	0.000279
60	27	24	5	0.002143	0.001834
60	27	24	6	0.010659	0.008516
60	27	24	7	0.039399	0.028740
60	27	24	8	0.111250	0.071851
60	27	24	9	0.246081	0.134831
60	27	24	10	0.437683	0.191602
60	27	24	11	0.644961	0.207278
60	27	24	12	0.816048	0.171087
60	27	24	13	0.923725	0.107677
60	27	24	14	0.975223	0.051498
60	27	24	15	0.993820	0.018596
60	27	24	16	0.998841	0.005021
60	27	24	17	0.999840	0.001000
60	27	24	18	0.999984	0.000144
60	27	24	19	0.999999	0.000015
60	27	24	20	1.000000	0.000001
60	27	24	21	1.000000	0.000000
60	27	24	22	1.000000	0.000000
60	27	24	23	1.000000	0.000000

N	n	k	x	P(x)	p(x)
60	27	24	24	1.000000	0.000000
60	27	25	0	0.000000	0.000000
60	27	25	1	0.000000	0.000000
60	27	25	2	0.000001	0.000001
60	27	25	3	0.000012	0.000011
60	27	25	4	0.000131	0.000120
60	27	25	5	0.001023	0.000891
60	27	25	6	0.005691	0.004669
60	27	25	7	0.023432	0.017741
60	27	25	8	0.073329	0.049896
60	27	25	9	0.178665	0.105337
60	27	25	10	0.347204	0.168539
60	27	25	11	0.552838	0.205633
60	27	25	12	0.744762	0.191925
60	27	25	13	0.881851	0.137089
60	27	25	14	0.956627	0.074776
60	27	25	15	0.987621	0.030994
60	27	25	16	0.997306	0.009686
60	27	25	17	0.999563	0.002256
60	27	25	18	0.999948	0.000386
60	27	25	19	0.999996	0.000047
60	27	25	20	1.000000	0.000004
60	27	25	21	1.000000	0.000000
60	27	25	22	1.000000	0.000000
60	27	25	23	1.000000	0.000000
60	27	25	24	1.000000	0.000000
60	27	25	25	1.000000	0.000000
60	27	26	0	0.000000	0.000000
60	27	26	1	0.000000	0.000000
60	27	26	2	0.000000	0.000000
60	27	26	3	0.000004	0.000004
60	27	26	4	0.000053	0.000049
60	27	26	5	0.000463	0.000410
60	27	26	6	0.002890	0.002428
60	27	26	7	0.013295	0.010404
60	27	26	8	0.046242	0.032947
60	27	26	9	0.124492	0.078250
60	27	26	10	0.265342	0.140850
60	27	26	11	0.458834	0.193491
60	27	26	12	0.662509	0.203675
60	27	26	13	0.827015	0.164507
60	27	26	14	0.928853	0.101837
60	27	26	15	0.976994	0.048141
60	27	26	16	0.994262	0.017268
60	27	26	17	0.998918	0.004656
60	27	26	18	0.999849	0.000931
60	27	26	19	0.999985	0.000136
60	27	26	20	0.999999	0.000014
60	27	26	21	1.000000	0.000001
60	27	26	22	1.000000	0.000000

Table for $N = 2$, $n = 1$, through $N = 100$, $n = 50$

N	n	k	x	P(x)	p(x)
60	27	26	23	1.000000	0.000000
60	27	26	24	1.000000	0.000000
60	27	26	25	1.000000	0.000000
60	27	26	26	1.000000	0.000000
60	27	27	0	0.000000	0.000000
60	27	27	1	0.000000	0.000000
60	27	27	2	0.000000	0.000000
60	27	27	3	0.000001	0.000001
60	27	27	4	0.000020	0.000018
60	27	27	5	0.000197	0.000178
60	27	27	6	0.001391	0.001193
60	27	27	7	0.007174	0.005784
60	27	27	8	0.027830	0.020656
60	27	27	9	0.083066	0.055235
60	27	27	10	0.194917	0.111852
60	27	27	11	0.367779	0.172862
60	27	27	12	0.572652	0.204873
60	27	27	13	0.759277	0.186625
60	27	27	14	0.889915	0.130638
60	27	27	15	0.960003	0.070088
60	27	27	16	0.988676	0.028672
60	27	27	17	0.997549	0.008873
60	27	27	18	0.999603	0.002054
60	27	27	19	0.999953	0.000350
60	27	27	20	0.999996	0.000043
60	27	27	21	1.000000	0.000004
60	27	27	22	1.000000	0.000000
60	27	27	23	1.000000	0.000000
60	27	27	24	1.000000	0.000000
60	27	27	25	1.000000	0.000000
60	27	27	26	1.000000	0.000000
60	27	27	27	1.000000	0.000000
60	28	1	0	0.533333	0.533333
60	28	1	1	1.000000	0.466667
60	28	2	0	0.280226	0.280226
60	28	2	1	0.786441	0.506215
60	28	2	2	1.000000	0.213559
60	28	3	0	0.144944	0.144944
60	28	3	1	0.550789	0.405845
60	28	3	2	0.904266	0.353477
60	28	3	3	1.000000	0.095733
60	28	4	0	0.073744	0.073744
60	28	4	1	0.358547	0.284803
60	28	4	2	0.743031	0.384484
60	28	4	3	0.958012	0.214980
60	28	4	4	1.000000	0.041988
60	28	5	0	0.036872	0.036872
60	28	5	1	0.221231	0.184359
60	28	5	2	0.564521	0.343290
60	28	5	3	0.862038	0.297518

N	n	k	x	P(x)	p(x)
60	28	5	4	0.982005	0.119967
60	28	5	5	1.000000	0.017995
60	28	6	0	0.018101	0.018101
60	28	6	1	0.130727	0.112627
60	28	6	2	0.402238	0.271511
60	28	6	3	0.726803	0.324565
60	28	6	4	0.929656	0.202853
60	28	6	5	0.992475	0.062819
60	28	6	6	1.000000	0.007525
60	28	7	0	0.008715	0.008715
60	28	7	1	0.074414	0.065699
60	28	7	2	0.271511	0.197097
60	28	7	3	0.576541	0.305031
60	28	7	4	0.839499	0.262957
60	28	7	5	0.965718	0.126220
60	28	7	6	0.996934	0.031216
60	28	7	7	1.000000	0.003066
60	28	8	0	0.004111	0.004111
60	28	8	1	0.040945	0.036834
60	28	8	2	0.174822	0.133877
60	28	8	3	0.432659	0.257837
60	28	8	4	0.720424	0.287765
60	28	8	5	0.910944	0.190520
60	28	8	6	0.983977	0.073033
60	28	8	7	0.998785	0.014808
60	28	8	8	1.000000	0.001215
60	28	9	0	0.001897	0.001897
60	28	9	1	0.021820	0.019922
60	28	9	2	0.107883	0.086064
60	28	9	3	0.308699	0.200816
60	28	9	4	0.587609	0.278910
60	28	9	5	0.826675	0.239066
60	28	9	6	0.953078	0.126403
60	28	9	7	0.992805	0.039727
60	28	9	8	0.999533	0.006728
60	28	9	9	1.000000	0.000467
60	28	10	0	0.000856	0.000856
60	28	10	1	0.011272	0.010417
60	28	10	2	0.064008	0.052735
60	28	10	3	0.210260	0.146252
60	28	10	4	0.456357	0.246097
60	28	10	5	0.718861	0.262504
60	28	10	6	0.898552	0.179690
60	28	10	7	0.976447	0.077895
60	28	10	8	0.996894	0.020448
60	28	10	9	0.999826	0.002932
60	28	10	10	1.000000	0.000174
60	28	11	0	0.000376	0.000376
60	28	11	1	0.005647	0.005271
60	28	11	2	0.036585	0.030938

N	n	k	x	P(x)	p(x)
60	28	11	3	0.137134	0.100548
60	28	11	4	0.338231	0.201097
60	28	11	5	0.598110	0.259879
60	28	11	6	0.819488	0.221378
60	28	11	7	0.943731	0.124243
60	28	11	8	0.988715	0.044985
60	28	11	9	0.998712	0.009997
60	28	11	10	0.999937	0.001225
60	28	11	11	1.000000	0.000063
60	28	12	0	0.000161	0.000161
60	28	12	1	0.002743	0.002582
60	28	12	2	0.020169	0.017426
60	28	12	3	0.085834	0.065664
60	28	12	4	0.239734	0.153901
60	28	12	5	0.476126	0.236391
60	28	12	6	0.720093	0.243968
60	28	12	7	0.890484	0.170390
60	28	12	8	0.970354	0.079870
60	28	12	9	0.994836	0.024481
60	28	12	10	0.999487	0.004651
60	28	12	11	0.999978	0.000491
60	28	12	12	1.000000	0.000022
60	28	13	0	0.000067	0.000067
60	28	13	1	0.001291	0.001224
60	28	13	2	0.010730	0.009439
60	28	13	3	0.051633	0.040903
60	28	13	4	0.162784	0.111150
60	28	13	5	0.362855	0.200071
60	28	13	6	0.608275	0.245420
60	28	13	7	0.815938	0.207663
60	28	13	8	0.937075	0.121137
60	28	13	9	0.985145	0.048070
60	28	13	10	0.997743	0.012598
60	28	13	11	0.999804	0.002061
60	28	13	12	0.999993	0.000188
60	28	13	13	1.000000	0.000007
60	28	14	0	0.000027	0.000027
60	28	14	1	0.000588	0.000561
60	28	14	2	0.005508	0.004920
60	28	14	3	0.029876	0.024368
60	28	14	4	0.106026	0.076150
60	28	14	5	0.264948	0.158921
60	28	14	6	0.493397	0.228450
60	28	14	7	0.723152	0.229755
60	28	14	8	0.885527	0.162375
60	28	14	9	0.965712	0.080185
60	28	14	10	0.992918	0.027206
60	28	14	11	0.999059	0.006140
60	28	14	12	0.999929	0.000870
60	28	14	13	0.999998	0.000069

N	n	k	x	P(x)	p(x)
60	28	14	14	1.000000	0.000002
60	28	15	0	0.000011	0.000011
60	28	15	1	0.000259	0.000248
60	28	15	2	0.002727	0.002468
60	28	15	3	0.016633	0.013906
60	28	15	4	0.066296	0.049663
60	28	15	5	0.185487	0.119191
60	28	15	6	0.384139	0.198652
60	28	15	7	0.618264	0.234125
60	28	15	8	0.814930	0.196665
60	28	15	9	0.932593	0.117663
60	28	15	10	0.982272	0.049680
60	28	15	11	0.996789	0.014517
60	28	15	12	0.999626	0.002837
60	28	15	13	0.999975	0.000349
60	28	15	14	0.999999	0.000024
60	28	15	15	1.000000	0.000001
60	28	16	0	0.000004	0.000004
60	28	16	1	0.000110	0.000106
60	28	16	2	0.001301	0.001191
60	28	16	3	0.008908	0.007606
60	28	16	4	0.039809	0.030901
60	28	16	5	0.124567	0.084758
60	28	16	6	0.287020	0.162453
60	28	16	7	0.509006	0.221986
60	28	16	8	0.727523	0.218517
60	28	16	9	0.882913	0.155390
60	28	16	10	0.962400	0.079488
60	28	16	11	0.991305	0.028905
60	28	16	12	0.998617	0.007312
60	28	16	13	0.999859	0.001241
60	28	16	14	0.999992	0.000133
60	28	16	15	1.000000	0.000008
60	28	16	16	1.000000	0.000000
60	28	17	0	0.000001	0.000001
60	28	17	1	0.000045	0.000043
60	28	17	2	0.000597	0.000552
60	28	17	3	0.004586	0.003988
60	28	17	4	0.022954	0.018368
60	28	17	5	0.080262	0.057308
60	28	17	6	0.205794	0.125532
60	28	17	7	0.403058	0.197264
60	28	17	8	0.628197	0.225139
60	28	17	9	0.815812	0.187616
60	28	17	10	0.929883	0.114070
60	28	17	11	0.980137	0.050255
60	28	17	12	0.995958	0.015821
60	28	17	13	0.999435	0.003477
60	28	17	14	0.999949	0.000514
60	28	17	15	0.999997	0.000048

N	n	k	x	P(x)	p(x)
60	28	17	16	1.000000	0.000003
60	28	17	17	1.000000	0.000000
60	28	18	0	0.000001	0.000001
60	28	18	1	0.000018	0.000017
60	28	18	2	0.000263	0.000246
60	28	18	3	0.002267	0.002004
60	28	18	4	0.012702	0.010435
60	28	18	5	0.049609	0.036907
60	28	18	6	0.141568	0.091959
60	28	18	7	0.306720	0.165152
60	28	18	8	0.523481	0.216762
60	28	18	9	0.732912	0.209431
60	28	18	10	0.882132	0.149220
60	28	18	11	0.960269	0.078137
60	28	18	12	0.990071	0.029802
60	28	18	13	0.998223	0.008151
60	28	18	14	0.999782	0.001560
60	28	18	15	0.999983	0.000201
60	28	18	16	0.999999	0.000016
60	28	18	17	1.000000	0.000001
60	28	18	18	1.000000	0.000000
60	28	19	0	0.000000	0.000000
60	28	19	1	0.000007	0.000006
60	28	19	2	0.000111	0.000105
60	28	19	3	0.001074	0.000963
60	28	19	4	0.006739	0.005665
60	28	19	5	0.029398	0.022659
60	28	19	6	0.093399	0.064001
60	28	19	7	0.224144	0.130745
60	28	19	8	0.420261	0.196118
60	28	19	9	0.638170	0.217908
60	28	19	10	0.818181	0.180011
60	28	19	11	0.928642	0.110461
60	28	19	12	0.978718	0.050076
60	28	19	13	0.995311	0.016593
60	28	19	14	0.999262	0.003951
60	28	19	15	0.999921	0.000658
60	28	19	16	0.999994	0.000074
60	28	19	17	1.000000	0.000005
60	28	19	18	1.000000	0.000000
60	28	19	19	1.000000	0.000000
60	28	20	0	0.000000	0.000000
60	28	20	1	0.000002	0.000002
60	28	20	2	0.000045	0.000043
60	28	20	3	0.000487	0.000442
60	28	20	4	0.003423	0.002936
60	28	20	5	0.016687	0.013264
60	28	20	6	0.059057	0.042370
60	28	20	7	0.157177	0.098120
60	28	20	8	0.324594	0.167417

N	n	k	x	P(x)	p(x)
60	28	20	9	0.537188	0.212594
60	28	20	10	0.739152	0.201964
60	28	20	11	0.882841	0.143690
60	28	20	12	0.959177	0.076335
60	28	20	13	0.989241	0.030064
60	28	20	14	0.997913	0.008672
60	28	20	15	0.999712	0.001799
60	28	20	16	0.999973	0.000261
60	28	20	17	0.999998	0.000025
60	28	20	18	1.000000	0.000002
60	28	20	19	1.000000	0.000000
60	28	20	20	1.000000	0.000000
60	28	21	0	0.000000	0.000000
60	28	21	1	0.000001	0.000001
60	28	21	2	0.000017	0.000016
60	28	21	3	0.000211	0.000193
60	28	21	4	0.001661	0.001451
60	28	21	5	0.009060	0.007399
60	28	21	6	0.035753	0.026693
60	28	21	7	0.105664	0.069911
60	28	21	8	0.240885	0.135222
60	28	21	9	0.436206	0.195320
60	28	21	10	0.648268	0.212062
60	28	21	11	0.821773	0.173505
60	28	21	12	0.928642	0.106869
60	28	21	13	0.977967	0.049324
60	28	21	14	0.994878	0.016911
60	28	21	15	0.999127	0.004249
60	28	21	16	0.999895	0.000767
60	28	21	17	0.999991	0.000097
60	28	21	18	0.999999	0.000008
60	28	21	19	1.000000	0.000000
60	28	21	20	1.000000	0.000000
60	28	21	21	1.000000	0.000000
60	28	22	0	0.000000	0.000000
60	28	22	1	0.000000	0.000000
60	28	22	2	0.000006	0.000006
60	28	22	3	0.000087	0.000080
60	28	22	4	0.000769	0.000682
60	28	22	5	0.004697	0.003928
60	28	22	6	0.020695	0.015999
60	28	22	7	0.068020	0.047324
60	28	22	8	0.171541	0.103521
60	28	22	9	0.341050	0.169509
60	28	22	10	0.550393	0.209343
60	28	22	11	0.746143	0.195750
60	28	22	12	0.884799	0.138656
60	28	22	13	0.958996	0.074197
60	28	22	14	0.988807	0.029811
60	28	22	15	0.997711	0.008904

N	n	k	x	P(x)	p(x)
60	28	22	16	0.999659	0.001948
60	28	22	17	0.999964	0.000306
60	28	22	18	0.999997	0.000033
60	28	22	19	1.000000	0.000002
60	28	22	20	1.000000	0.000000
60	28	22	21	1.000000	0.000000
60	28	22	22	1.000000	0.000000
60	28	23	0	0.000000	0.000000
60	28	23	1	0.000000	0.000000
60	28	23	2	0.000002	0.000002
60	28	23	3	0.000034	0.000032
60	28	23	4	0.000338	0.000304
60	28	23	5	0.002319	0.001981
60	28	23	6	0.011433	0.009114
60	28	23	7	0.041867	0.030434
60	28	23	8	0.117056	0.075189
60	28	23	9	0.256295	0.139239
60	28	23	10	0.451230	0.194935
60	28	23	11	0.658570	0.207340
60	28	23	12	0.826417	0.167847
60	28	23	13	0.929707	0.103290
60	28	23	14	0.977824	0.048117
60	28	23	15	0.994665	0.016841
60	28	23	16	0.999043	0.004379
60	28	23	17	0.999876	0.000832
60	28	23	18	0.999989	0.000113
60	28	23	19	0.999999	0.000011
60	28	23	20	1.000000	0.000001
60	28	23	21	1.000000	0.000000
60	28	23	22	1.000000	0.000000
60	28	23	23	1.000000	0.000000
60	28	24	0	0.000000	0.000000
60	28	24	1	0.000000	0.000000
60	28	24	2	0.000001	0.000001
60	28	24	3	0.000012	0.000012
60	28	24	4	0.000141	0.000128
60	28	24	5	0.001088	0.000947
60	28	24	6	0.006014	0.004926
60	28	24	7	0.024594	0.018580
60	28	24	8	0.076413	0.051820
60	28	24	9	0.184794	0.108381
60	28	24	10	0.356397	0.171603
60	28	24	11	0.563306	0.206909
60	28	24	12	0.753835	0.190529
60	28	24	13	0.887833	0.133998
60	28	24	14	0.959618	0.071785
60	28	24	15	0.988748	0.029130
60	28	24	16	0.997623	0.008876
60	28	24	17	0.999628	0.002005
60	28	24	18	0.999958	0.000330

N	n	k	x	P(x)	p(x)
60	28	24	19	0.999997	0.000039
60	28	24	20	1.000000	0.000003
60	28	24	21	1.000000	0.000000
60	28	24	22	1.000000	0.000000
60	28	24	23	1.000000	0.000000
60	28	24	24	1.000000	0.000000
60	28	25	0	0.000000	0.000000
60	28	25	1	0.000000	0.000000
60	28	25	2	0.000000	0.000000
60	28	25	3	0.000004	0.000004
60	28	25	4	0.000055	0.000051
60	28	25	5	0.000483	0.000427
60	28	25	6	0.003003	0.002521
60	28	25	7	0.013756	0.010752
60	28	25	8	0.047625	0.033869
60	28	25	9	0.127593	0.079969
60	28	25	10	0.270596	0.143003
60	28	25	11	0.465599	0.195004
60	28	25	12	0.669155	0.203556
60	28	25	13	0.832000	0.162845
60	28	25	14	0.931701	0.099701
60	28	25	15	0.978229	0.046527
60	28	25	16	0.994665	0.016436
60	28	25	17	0.999016	0.004351
60	28	25	18	0.999866	0.000851
60	28	25	19	0.999987	0.000121
60	28	25	20	0.999999	0.000012
60	28	25	21	1.000000	0.000001
60	28	25	22	1.000000	0.000000
60	28	25	23	1.000000	0.000000
60	28	25	24	1.000000	0.000000
60	28	25	25	1.000000	0.000000
60	28	26	0	0.000000	0.000000
60	28	26	1	0.000000	0.000000
60	28	26	2	0.000000	0.000000
60	28	26	3	0.000001	0.000001
60	28	26	4	0.000020	0.000019
60	28	26	5	0.000202	0.000181
60	28	26	6	0.001419	0.001217
60	28	26	7	0.007304	0.005885
60	28	26	8	0.028271	0.020967
60	28	26	9	0.084182	0.055911
60	28	26	10	0.197051	0.112870
60	28	26	11	0.370883	0.173832
60	28	26	12	0.576101	0.205218
60	28	26	13	0.762210	0.186109
60	28	26	14	0.891821	0.129611
60	28	26	15	0.960947	0.069126
60	28	26	16	0.989030	0.028082
60	28	26	17	0.997648	0.008619

Table for $N = 2$, $n = 1$, through $N = 100$, $n = 50$

N	n	k	x	P(x)	p(x)
60	28	26	18	0.999623	0.001975
60	28	26	19	0.999956	0.000333
60	28	26	20	0.999996	0.000040
60	28	26	21	1.000000	0.000003
60	28	26	22	1.000000	0.000000
60	28	26	23	1.000000	0.000000
60	28	26	24	1.000000	0.000000
60	28	26	25	1.000000	0.000000
60	28	26	26	1.000000	0.000000
60	28	27	0	0.000000	0.000000
60	28	27	1	0.000000	0.000000
60	28	27	2	0.000000	0.000000
60	28	27	3	0.000000	0.000000
60	28	27	4	0.000007	0.000007
60	28	27	5	0.000079	0.000072
60	28	27	6	0.000631	0.000552
60	28	27	7	0.003669	0.003038
60	28	27	8	0.015938	0.012268
60	28	27	9	0.052937	0.037000
60	28	27	10	0.137297	0.084359
60	28	27	11	0.283967	0.146670
60	28	27	12	0.479528	0.195561
60	28	27	13	0.680103	0.200575
60	28	27	14	0.838452	0.158349
60	28	27	15	0.934517	0.096065
60	28	27	16	0.979118	0.044602
60	28	27	17	0.994860	0.015742
60	28	27	18	0.999042	0.004183
60	28	27	19	0.999868	0.000826
60	28	27	20	0.999987	0.000119
60	28	27	21	0.999999	0.000012
60	28	27	22	1.000000	0.000001
60	28	27	23	1.000000	0.000000
60	28	27	24	1.000000	0.000000
60	28	27	25	1.000000	0.000000
60	28	27	26	1.000000	0.000000
60	28	27	27	1.000000	0.000000
60	28	28	0	0.000000	0.000000
60	28	28	1	0.000000	0.000000
60	28	28	2	0.000000	0.000000
60	28	28	3	0.000000	0.000000
60	28	28	4	0.000002	0.000002
60	28	28	5	0.000029	0.000027
60	28	28	6	0.000263	0.000234
60	28	28	7	0.001736	0.001473
60	28	28	8	0.008502	0.006766
60	28	28	9	0.031634	0.023132
60	28	28	10	0.091283	0.059648
60	28	28	11	0.208410	0.117127
60	28	28	12	0.384711	0.176301

N	n	k	x	P(x)	p(x)
60	28	28	13	0.588933	0.204222
60	28	28	14	0.771273	0.182341
60	28	28	15	0.896673	0.125399
60	28	28	16	0.962899	0.066227
60	28	28	17	0.989613	0.026713
60	28	28	18	0.997775	0.008162
60	28	28	19	0.999643	0.001868
60	28	28	20	0.999958	0.000315
60	28	28	21	0.999996	0.000038
60	28	28	22	1.000000	0.000003
60	28	28	23	1.000000	0.000000
60	28	28	24	1.000000	0.000000
60	28	28	25	1.000000	0.000000
60	28	28	26	1.000000	0.000000
60	28	28	27	1.000000	0.000000
60	28	28	28	1.000000	0.000000
60	29	1	0	0.516667	0.516667
60	29	1	1	1.000000	0.483333
60	29	2	0	0.262712	0.262712
60	29	2	1	0.770621	0.507910
60	29	2	2	1.000000	0.229379
60	29	3	0	0.131356	0.131356
60	29	3	1	0.525424	0.394068
60	29	3	2	0.893220	0.367797
60	29	3	3	1.000000	0.106780
60	29	4	0	0.064526	0.064526
60	29	4	1	0.331847	0.267321
60	29	4	2	0.719001	0.387154
60	29	4	3	0.951293	0.232293
60	29	4	4	1.000000	0.048707
60	29	5	0	0.031111	0.031111
60	29	5	1	0.198186	0.167075
60	29	5	2	0.532337	0.334151
60	29	5	3	0.843443	0.311106
60	29	5	4	0.978256	0.134813
60	29	5	5	1.000000	0.021744
60	29	6	0	0.014707	0.014707
60	29	6	1	0.113130	0.098423
60	29	6	2	0.368299	0.255170
60	29	6	3	0.696375	0.328076
60	29	6	4	0.916977	0.220603
60	29	6	5	0.990512	0.073534
60	29	6	6	1.000000	0.009488
60	29	7	0	0.006809	0.006809
60	29	7	1	0.062096	0.055287
60	29	7	2	0.240714	0.178619
60	29	7	3	0.538413	0.297698
60	29	7	4	0.814847	0.276434
60	29	7	5	0.957830	0.142983
60	29	7	6	0.995959	0.038129

N	n	k	x	P(x)	p(x)
60	29	7	7	1.000000	0.004041
60	29	8	0	0.003083	0.003083
60	29	8	1	0.032887	0.029804
60	29	8	2	0.149720	0.116832
60	29	8	3	0.392372	0.242652
60	29	8	4	0.684453	0.292081
60	29	8	5	0.893083	0.208629
60	29	8	6	0.979412	0.086329
60	29	8	7	0.998322	0.018910
60	29	8	8	1.000000	0.001678
60	29	9	0	0.001364	0.001364
60	29	9	1	0.016839	0.015475
60	29	9	2	0.089057	0.072218
60	29	9	3	0.271046	0.181989
60	29	9	4	0.544030	0.272984
60	29	9	5	0.796792	0.252763
60	29	9	6	0.941228	0.144436
60	29	9	7	0.990322	0.049094
60	29	9	8	0.999322	0.009001
60	29	9	9	1.000000	0.000677
60	29	10	0	0.000588	0.000588
60	29	10	1	0.008343	0.007754
60	29	10	2	0.050824	0.042481
60	29	10	3	0.178267	0.127443
60	29	10	4	0.410214	0.231947
60	29	10	5	0.677845	0.267631
60	29	10	6	0.876090	0.198245
60	29	10	7	0.969144	0.093054
60	29	10	8	0.995616	0.026472
60	29	10	9	0.999734	0.004118
60	29	10	10	1.000000	0.000266
60	29	11	0	0.000247	0.000247
60	29	11	1	0.004000	0.003753
60	29	11	2	0.027884	0.023884
60	29	11	3	0.111997	0.084113
60	29	11	4	0.294241	0.182244
60	29	11	5	0.549382	0.255142
60	29	11	6	0.784897	0.235515
60	29	11	7	0.928200	0.143303
60	29	11	8	0.984498	0.056298
60	29	11	9	0.998087	0.013589
60	29	11	10	0.999899	0.001812
60	29	11	11	1.000000	0.000101
60	29	12	0	0.000101	0.000101
60	29	12	1	0.001856	0.001755
60	29	12	2	0.014724	0.012868
60	29	12	3	0.067365	0.052642
60	29	12	4	0.201259	0.133894
60	29	12	5	0.424415	0.223156
60	29	12	6	0.674349	0.249935

N	n	k	x	P(x)	p(x)
60	29	12	7	0.863860	0.189511
60	29	12	8	0.960370	0.096510
60	29	12	9	0.992541	0.032170
60	29	12	10	0.999196	0.006656
60	29	12	11	0.999963	0.000766
60	29	12	12	1.000000	0.000037
60	29	13	0	0.000040	0.000040
60	29	13	1	0.000832	0.000792
60	29	13	2	0.007485	0.006653
60	29	13	3	0.038851	0.031366
60	29	13	4	0.131523	0.092672
60	29	13	5	0.312837	0.181314
60	29	13	6	0.554589	0.241752
60	29	13	7	0.777001	0.222412
60	29	13	8	0.918147	0.141146
60	29	13	9	0.979136	0.060989
60	29	13	10	0.996562	0.017425
60	29	13	11	0.999675	0.003114
60	29	13	12	0.999987	0.000311
60	29	13	13	1.000000	0.000013
60	29	14	0	0.000015	0.000015
60	29	14	1	0.000360	0.000345
60	29	14	2	0.003663	0.003303
60	29	14	3	0.021500	0.017837
60	29	14	4	0.082229	0.060730
60	29	14	5	0.220251	0.138022
60	29	14	6	0.436285	0.216034
60	29	14	7	0.672893	0.236609
60	29	14	8	0.855082	0.182189
60	29	14	9	0.953184	0.098102
60	29	14	10	0.989517	0.036334
60	29	14	11	0.998483	0.008965
60	29	14	12	0.999874	0.001391
60	29	14	13	0.999995	0.000121
60	29	14	14	1.000000	0.000004
60	29	15	0	0.000006	0.000006
60	29	15	1	0.000150	0.000145
60	29	15	2	0.001724	0.001574
60	29	15	3	0.011418	0.009694
60	29	15	4	0.049224	0.037806
60	29	15	5	0.148240	0.099016
60	29	15	6	0.328268	0.180028
60	29	15	7	0.559733	0.231465
60	29	15	8	0.771909	0.212176
60	29	15	9	0.910531	0.138622
60	29	15	10	0.974510	0.063979
60	29	15	11	0.994975	0.020465
60	29	15	12	0.999360	0.004385
60	29	15	13	0.999953	0.000593
60	29	15	14	0.999998	0.000045

N	n	k	x	P(x)	p(x)
60	29	15	15	1.000000	0.000001
60	29	16	0	0.000002	0.000002
60	29	16	1	0.000060	0.000058
60	29	16	2	0.000780	0.000720
60	29	16	3	0.005817	0.005037
60	29	16	4	0.028221	0.022404
60	29	16	5	0.095431	0.067211
60	29	16	6	0.236254	0.140822
60	29	16	7	0.446572	0.210319
60	29	16	8	0.672893	0.226321
60	29	16	9	0.848921	0.176028
60	29	16	10	0.947497	0.098575
60	29	16	11	0.986789	0.039292
60	29	16	12	0.997703	0.010915
60	29	16	13	0.999742	0.002039
60	29	16	14	0.999983	0.000241
60	29	16	15	0.999999	0.000016
60	29	16	16	1.000000	0.000000
60	29	17	0	0.000001	0.000001
60	29	17	1	0.000023	0.000023
60	29	17	2	0.000338	0.000315
60	29	17	3	0.002841	0.002502
60	29	17	4	0.015492	0.012651
60	29	17	5	0.058771	0.043279
60	29	17	6	0.162642	0.103871
60	29	17	7	0.341413	0.178771
60	29	17	8	0.564876	0.223464
60	29	17	9	0.768908	0.204032
60	29	17	10	0.904930	0.136021
60	29	17	11	0.970715	0.065785
60	29	17	12	0.993486	0.022772
60	29	17	13	0.999001	0.005515
60	29	17	14	0.999901	0.000900
60	29	17	15	0.999994	0.000093
60	29	17	16	1.000000	0.000005
60	29	17	17	1.000000	0.000000
60	29	18	0	0.000000	0.000000
60	29	18	1	0.000009	0.000008
60	29	18	2	0.000140	0.000132
60	29	18	3	0.001328	0.001187
60	29	18	4	0.008136	0.006809
60	29	18	5	0.034615	0.026479
60	29	18	6	0.107083	0.072468
60	29	18	7	0.249949	0.142866
60	29	18	8	0.455743	0.205794
60	29	18	9	0.674010	0.218267
60	29	18	10	0.844827	0.170817
60	29	18	11	0.943177	0.098349
60	29	18	12	0.984484	0.041307
60	29	18	13	0.996949	0.012465

N	n	k	x	P(x)	p(x)
60	29	18	14	0.999587	0.002638
60	29	18	15	0.999964	0.000377
60	29	18	16	0.999998	0.000034
60	29	18	17	1.000000	0.000002
60	29	18	18	1.000000	0.000000
60	29	19	0	0.000000	0.000000
60	29	19	1	0.000003	0.000003
60	29	19	2	0.000056	0.000053
60	29	19	3	0.000593	0.000537
60	29	19	4	0.004084	0.003491
60	29	19	5	0.019484	0.015401
60	29	19	6	0.067398	0.047914
60	29	19	7	0.175114	0.107716
60	29	19	8	0.352846	0.177732
60	29	19	9	0.570073	0.217227
60	29	19	10	0.767553	0.197479
60	29	19	11	0.901027	0.133474
60	29	19	12	0.967764	0.066737
60	29	19	13	0.992200	0.024436
60	29	19	14	0.998645	0.006445
60	29	19	15	0.999838	0.001193
60	29	19	16	0.999987	0.000149
60	29	19	17	0.999999	0.000012
60	29	19	18	1.000000	0.000001
60	29	19	19	1.000000	0.000000
60	29	20	0	0.000000	0.000000
60	29	20	1	0.000001	0.000001
60	29	20	2	0.000021	0.000020
60	29	20	3	0.000252	0.000231
60	29	20	4	0.001955	0.001703
60	29	20	5	0.010469	0.008514
60	29	20	6	0.040520	0.030050
60	29	20	7	0.117315	0.076796
60	29	20	8	0.261813	0.144497
60	29	20	9	0.464109	0.202296
60	29	20	10	0.676038	0.211929
60	29	20	11	0.842429	0.166391
60	29	20	12	0.940093	0.097664
60	29	20	13	0.982664	0.042572
60	29	20	14	0.996287	0.013623
60	29	20	15	0.999431	0.003144
60	29	20	16	0.999940	0.000509
60	29	20	17	0.999996	0.000056
60	29	20	18	1.000000	0.000004
60	29	20	19	1.000000	0.000000
60	29	20	20	1.000000	0.000000
60	29	21	0	0.000000	0.000000
60	29	21	1	0.000000	0.000000
60	29	21	2	0.000007	0.000007
60	29	21	3	0.000102	0.000094

N	n	k	x	P(x)	p(x)
60	29	21	4	0.000891	0.000789
60	29	21	5	0.005361	0.004470
60	29	21	6	0.023241	0.017880
60	29	21	7	0.075078	0.051837
60	29	21	8	0.185952	0.110874
60	29	21	9	0.362961	0.177009
60	29	21	10	0.575372	0.212411
60	29	21	11	0.767553	0.192181
60	29	21	12	0.898585	0.131033
60	29	21	13	0.965636	0.067050
60	29	21	14	0.991179	0.025543
60	29	21	15	0.998331	0.007152
60	29	21	16	0.999775	0.001444
60	29	21	17	0.999979	0.000205
60	29	21	18	0.999999	0.000019
60	29	21	19	1.000000	0.000001
60	29	21	20	1.000000	0.000000
60	29	21	21	1.000000	0.000000
60	29	22	0	0.000000	0.000000
60	29	22	1	0.000000	0.000000
60	29	22	2	0.000003	0.000002
60	29	22	3	0.000039	0.000036
60	29	22	4	0.000385	0.000346
60	29	22	5	0.002610	0.002225
60	29	22	6	0.012696	0.010086
60	29	22	7	0.045836	0.033140
60	29	22	8	0.126250	0.080414
60	29	22	9	0.272187	0.145936
60	29	22	10	0.471889	0.199703
60	29	22	11	0.678854	0.206964
60	29	22	12	0.841469	0.162615
60	29	22	13	0.938128	0.096659
60	29	22	14	0.981354	0.043226
60	29	22	15	0.995763	0.014409
60	29	22	16	0.999293	0.003530
60	29	22	17	0.999916	0.000623
60	29	22	18	0.999993	0.000077
60	29	22	19	1.000000	0.000006
60	29	22	20	1.000000	0.000000
60	29	22	21	1.000000	0.000000
60	29	22	22	1.000000	0.000000
60	29	23	0	0.000000	0.000000
60	29	23	1	0.000000	0.000000
60	29	23	2	0.000001	0.000001
60	29	23	3	0.000014	0.000013
60	29	23	4	0.000157	0.000143
60	29	23	5	0.001205	0.001047
60	29	23	6	0.006591	0.005387
60	29	23	7	0.026650	0.020059
60	29	23	8	0.081811	0.055161

N	n	k	x	P(x)	p(x)
60	29	23	9	0.195378	0.113567
60	29	23	10	0.372038	0.176660
60	29	23	11	0.580818	0.208780
60	29	23	12	0.768720	0.187902
60	29	23	13	0.897429	0.128709
60	29	23	14	0.964291	0.066862
60	29	23	15	0.990455	0.026163
60	29	23	16	0.998086	0.007631
60	29	23	17	0.999720	0.001634
60	29	23	18	0.999971	0.000251
60	29	23	19	0.999998	0.000027
60	29	23	20	1.000000	0.000002
60	29	23	21	1.000000	0.000000
60	29	23	22	1.000000	0.000000
60	29	23	23	1.000000	0.000000
60	29	24	0	0.000000	0.000000
60	29	24	1	0.000000	0.000000
60	29	24	2	0.000000	0.000000
60	29	24	3	0.000005	0.000004
60	29	24	4	0.000061	0.000056
60	29	24	5	0.000525	0.000465
60	29	24	6	0.003243	0.002718
60	29	24	7	0.014723	0.011480
60	29	24	8	0.050503	0.035780
60	29	24	9	0.133991	0.083487
60	29	24	10	0.281321	0.147330
60	29	24	11	0.479249	0.197929
60	29	24	12	0.682387	0.203137
60	29	24	13	0.841771	0.159385
60	29	24	14	0.937185	0.095414
60	29	24	15	0.980555	0.043370
60	29	24	16	0.995404	0.014849
60	29	24	17	0.999190	0.003785
60	29	24	18	0.999896	0.000707
60	29	24	19	0.999991	0.000094
60	29	24	20	0.999999	0.000009
60	29	24	21	1.000000	0.000001
60	29	24	22	1.000000	0.000000
60	29	24	23	1.000000	0.000000
60	29	24	24	1.000000	0.000000
60	29	25	0	0.000000	0.000000
60	29	25	1	0.000000	0.000000
60	29	25	2	0.000000	0.000000
60	29	25	3	0.000001	0.000001
60	29	25	4	0.000022	0.000020
60	29	25	5	0.000215	0.000194
60	29	25	6	0.001507	0.001291
60	29	25	7	0.007707	0.006201
60	29	25	8	0.029632	0.021924
60	29	25	9	0.087609	0.057977

Table for $N = 2$, $n = 1$, through $N = 100$, $n = 50$

N	n	k	x	P(x)	p(x)
60	29	25	10	0.203563	0.115954
60	29	25	11	0.380285	0.176722
60	29	25	12	0.586461	0.206176
60	29	25	13	0.770934	0.184473
60	29	25	14	0.897429	0.126496
60	29	25	15	0.963689	0.066260
60	29	25	16	0.990042	0.026353
60	29	25	17	0.997928	0.007886
60	29	25	18	0.999680	0.001752
60	29	25	19	0.999964	0.000284
60	29	25	20	0.999997	0.000033
60	29	25	21	1.000000	0.000003
60	29	25	22	1.000000	0.000000
60	29	25	23	1.000000	0.000000
60	29	25	24	1.000000	0.000000
60	29	25	25	1.000000	0.000000
60	29	26	0	0.000000	0.000000
60	29	26	1	0.000000	0.000000
60	29	26	2	0.000000	0.000000
60	29	26	3	0.000000	0.000000
60	29	26	4	0.000007	0.000007
60	29	26	5	0.000083	0.000075
60	29	26	6	0.000658	0.000576
60	29	26	7	0.003810	0.003152
60	29	26	8	0.016477	0.012667
60	29	26	9	0.054479	0.038002
60	29	26	10	0.140617	0.086138
60	29	26	11	0.289400	0.148783
60	29	26	12	0.486318	0.196919
60	29	26	13	0.686603	0.200285
60	29	26	14	0.843217	0.156614
60	29	26	15	0.937185	0.093968
60	29	26	16	0.980254	0.043069
60	29	26	17	0.995224	0.014970
60	29	26	18	0.999130	0.003905
60	29	26	19	0.999883	0.000754
60	29	26	20	0.999989	0.000106
60	29	26	21	0.999999	0.000010
60	29	26	22	1.000000	0.000001
60	29	26	23	1.000000	0.000000
60	29	26	24	1.000000	0.000000
60	29	26	25	1.000000	0.000000
60	29	26	26	1.000000	0.000000
60	29	27	0	0.000000	0.000000
60	29	27	1	0.000000	0.000000
60	29	27	2	0.000000	0.000000
60	29	27	3	0.000000	0.000000
60	29	27	4	0.000002	0.000002
60	29	27	5	0.000029	0.000027
60	29	27	6	0.000269	0.000239

N	n	k	x	P(x)	p(x)
60	29	27	7	0.001771	0.001502
60	29	27	8	0.008653	0.006883
60	29	27	9	0.032125	0.023472
60	29	27	10	0.092481	0.060356
60	29	27	11	0.210632	0.118151
60	29	27	12	0.387859	0.177227
60	29	27	13	0.592351	0.204492
60	29	27	14	0.774122	0.181771
60	29	27	15	0.898492	0.124370
60	29	27	16	0.963786	0.065294
60	29	27	17	0.989941	0.026154
60	29	27	18	0.997866	0.007926
60	29	27	19	0.999662	0.001795
60	29	27	20	0.999961	0.000299
60	29	27	21	0.999997	0.000036
60	29	27	22	1.000000	0.000003
60	29	27	23	1.000000	0.000000
60	29	27	24	1.000000	0.000000
60	29	27	25	1.000000	0.000000
60	29	27	26	1.000000	0.000000
60	29	27	27	1.000000	0.000000
60	29	28	0	0.000000	0.000000
60	29	28	1	0.000000	0.000000
60	29	28	2	0.000000	0.000000
60	29	28	3	0.000000	0.000000
60	29	28	4	0.000001	0.000001
60	29	28	5	0.000010	0.000009
60	29	28	6	0.000102	0.000092
60	29	28	7	0.000769	0.000667
60	29	28	8	0.004273	0.003504
60	29	28	9	0.017900	0.013626
60	29	28	10	0.057730	0.039831
60	29	28	11	0.146186	0.088455
60	29	28	12	0.296560	0.150374
60	29	28	13	0.493204	0.196643
60	29	28	14	0.691499	0.198296
60	29	28	15	0.845729	0.154230
60	29	28	16	0.938064	0.092335
60	29	28	17	0.980430	0.042366
60	29	28	18	0.995224	0.014794
60	29	28	19	0.999117	0.003893
60	29	28	20	0.999879	0.000762
60	29	28	21	0.999988	0.000109
60	29	28	22	0.999999	0.000011
60	29	28	23	1.000000	0.000001
60	29	28	24	1.000000	0.000000
60	29	28	25	1.000000	0.000000
60	29	28	26	1.000000	0.000000
60	29	28	27	1.000000	0.000000
60	29	28	28	1.000000	0.000000

N	n	k	x	P(x)	p(x)
60	29	29	0	0.000000	0.000000
60	29	29	1	0.000000	0.000000
60	29	29	2	0.000000	0.000000
60	29	29	3	0.000000	0.000000
60	29	29	4	0.000000	0.000000
60	29	29	5	0.000003	0.000003
60	29	29	6	0.000036	0.000033
60	29	29	7	0.000311	0.000275
60	29	29	8	0.001974	0.001663
60	29	29	9	0.009383	0.007409
60	29	29	10	0.034081	0.024698
60	29	29	11	0.096430	0.062349
60	29	29	12	0.216674	0.120244
60	29	29	13	0.394882	0.178208
60	29	29	14	0.598548	0.203666
60	29	29	15	0.778254	0.179706
60	29	29	16	0.900553	0.122300
60	29	29	17	0.964543	0.063990
60	29	29	18	0.990139	0.025596
60	29	29	19	0.997901	0.007762
60	29	29	20	0.999665	0.001764
60	29	29	21	0.999961	0.000296
60	29	29	22	0.999997	0.000036
60	29	29	23	1.000000	0.000003
60	29	29	24	1.000000	0.000000
60	29	29	25	1.000000	0.000000
60	29	29	26	1.000000	0.000000
60	29	29	27	1.000000	0.000000
60	29	29	28	1.000000	0.000000
60	29	29	29	1.000000	0.000000
60	30	1	0	0.500000	0.500000
60	30	1	1	1.000000	0.500000
60	30	2	0	0.245763	0.245763
60	30	2	1	0.754237	0.508475
60	30	2	2	1.000000	0.245763
60	30	3	0	0.118644	0.118644
60	30	3	1	0.500000	0.381356
60	30	3	2	0.881356	0.381356
60	30	3	3	1.000000	0.118644
60	30	4	0	0.056200	0.056200
60	30	4	1	0.305977	0.249777
60	30	4	2	0.694023	0.388046
60	30	4	3	0.943800	0.249777
60	30	4	4	1.000000	0.056200
60	30	5	0	0.026093	0.026093
60	30	5	1	0.176628	0.150535
60	30	5	2	0.500000	0.323372
60	30	5	3	0.823372	0.323372
60	30	5	4	0.973907	0.150535
60	30	5	5	1.000000	0.026093

N	n	k	x	P(x)	p(x)
60	30	6	0	0.011860	0.011860
60	30	6	1	0.097255	0.085395
60	30	6	2	0.335374	0.238119
60	30	6	3	0.664626	0.329251
60	30	6	4	0.902745	0.238119
60	30	6	5	0.988140	0.085395
60	30	6	6	1.000000	0.011860
60	30	7	0	0.005271	0.005271
60	30	7	1	0.051395	0.046124
60	30	7	2	0.211905	0.160510
60	30	7	3	0.500000	0.288095
60	30	7	4	0.788095	0.288095
60	30	7	5	0.948605	0.160510
60	30	7	6	0.994729	0.046124
60	30	7	7	1.000000	0.005271
60	30	8	0	0.002288	0.002288
60	30	8	1	0.026157	0.023870
60	30	8	2	0.127107	0.100950
60	30	8	3	0.353235	0.226127
60	30	8	4	0.646765	0.293531
60	30	8	5	0.872893	0.226127
60	30	8	6	0.973842	0.100950
60	30	8	7	0.997712	0.023870
60	30	8	8	1.000000	0.002288
60	30	9	0	0.000968	0.000968
60	30	9	1	0.012845	0.011878
60	30	9	2	0.072750	0.059904
60	30	9	3	0.235822	0.163073
60	30	9	4	0.500000	0.264178
60	30	9	5	0.764178	0.264178
60	30	9	6	0.927250	0.163073
60	30	9	7	0.987155	0.059904
60	30	9	8	0.999032	0.011878
60	30	9	9	1.000000	0.000968
60	30	10	0	0.000399	0.000399
60	30	10	1	0.006091	0.005693
60	30	10	2	0.039861	0.033770
60	30	10	3	0.149490	0.109629
60	30	10	4	0.365321	0.215831
60	30	10	5	0.634679	0.269358
60	30	10	6	0.850510	0.215831
60	30	10	7	0.960139	0.109629
60	30	10	8	0.993908	0.033770
60	30	10	9	0.999601	0.005693
60	30	10	10	1.000000	0.000399
60	30	11	0	0.000159	0.000159
60	30	11	1	0.002790	0.002630
60	30	11	2	0.020950	0.018160
60	30	11	3	0.090290	0.069340
60	30	11	4	0.253089	0.162799

N	n	k	x	P(x)	p(x)
60	30	11	5	0.500000	0.246911
60	30	11	6	0.746911	0.246911
60	30	11	7	0.909710	0.162799
60	30	11	8	0.979050	0.069340
60	30	11	9	0.997210	0.018160
60	30	11	10	0.999840	0.002630
60	30	11	11	1.000000	0.000159
60	30	12	0	0.000062	0.000062
60	30	12	1	0.001233	0.001171
60	30	12	2	0.010573	0.009340
60	30	12	3	0.052082	0.041510
60	30	12	4	0.166706	0.114623
60	30	12	5	0.374025	0.207319
60	30	12	6	0.625975	0.251950
60	30	12	7	0.833294	0.207319
60	30	12	8	0.947918	0.114623
60	30	12	9	0.989427	0.041510
60	30	12	10	0.998767	0.009340
60	30	12	11	0.999938	0.001171
60	30	12	12	1.000000	0.000062
60	30	13	0	0.000023	0.000023
60	30	13	1	0.000525	0.000502
60	30	13	2	0.005124	0.004599
60	30	13	3	0.028733	0.023609
60	30	13	4	0.104618	0.075885
60	30	13	5	0.266046	0.161428
60	30	13	6	0.500000	0.233954
60	30	13	7	0.733954	0.233954
60	30	13	8	0.895382	0.161428
60	30	13	9	0.971267	0.075885
60	30	13	10	0.994875	0.023609
60	30	13	11	0.999474	0.004599
60	30	13	12	0.999977	0.000502
60	30	13	13	1.000000	0.000023
60	30	14	0	0.000008	0.000008
60	30	14	1	0.000216	0.000207
60	30	14	2	0.002385	0.002169
60	30	14	3	0.015171	0.012786
60	30	14	4	0.062639	0.047468
60	30	14	5	0.180180	0.117541
60	30	14	6	0.380534	0.200354
60	30	14	7	0.619466	0.238931
60	30	14	8	0.819820	0.200354
60	30	14	9	0.937361	0.117541
60	30	14	10	0.984829	0.047468
60	30	14	11	0.997615	0.012786
60	30	14	12	0.999784	0.002169
60	30	14	13	0.999992	0.000207
60	30	14	14	1.000000	0.000008
60	30	15	0	0.000003	0.000003
60	30	15	1	0.000085	0.000082
60	30	15	2	0.001064	0.000979
60	30	15	3	0.007666	0.006602
60	30	15	4	0.035809	0.028143
60	30	15	5	0.116299	0.080490
60	30	15	6	0.276002	0.159702
60	30	15	7	0.500000	0.223998
60	30	15	8	0.723998	0.223998
60	30	15	9	0.883701	0.159702
60	30	15	10	0.964191	0.080490
60	30	15	11	0.992334	0.028143
60	30	15	12	0.998936	0.006602
60	30	15	13	0.999915	0.000979
60	30	15	14	0.999997	0.000082
60	30	15	15	1.000000	0.000003
60	30	16	0	0.000001	0.000001
60	30	16	1	0.000032	0.000031
60	30	16	2	0.000455	0.000423
60	30	16	3	0.003705	0.003250
60	30	16	4	0.019549	0.015844
60	30	16	5	0.071583	0.052034
60	30	16	6	0.190827	0.119244
60	30	16	7	0.385512	0.194685
60	30	16	8	0.614488	0.228976
60	30	16	9	0.809173	0.194685
60	30	16	10	0.928417	0.119244
60	30	16	11	0.980451	0.052034
60	30	16	12	0.996295	0.015844
60	30	16	13	0.999545	0.003250
60	30	16	14	0.999968	0.000423
60	30	16	15	0.999999	0.000031
60	30	16	16	1.000000	0.000001
60	30	17	0	0.000000	0.000000
60	30	17	1	0.000012	0.000011
60	30	17	2	0.000186	0.000174
60	30	17	3	0.001711	0.001525
60	30	17	4	0.010186	0.008476
60	30	17	5	0.042018	0.031831
60	30	17	6	0.125785	0.083767
60	30	17	7	0.283745	0.157960
60	30	17	8	0.500000	0.216255
60	30	17	9	0.716255	0.216255
60	30	17	10	0.874215	0.157960
60	30	17	11	0.957982	0.083767
60	30	17	12	0.989814	0.031831
60	30	17	13	0.998289	0.008476
60	30	17	14	0.999814	0.001525
60	30	17	15	0.999988	0.000174
60	30	17	16	1.000000	0.000011
60	30	17	17	1.000000	0.000000
60	30	18	0	0.000000	0.000000
60	30	18	1	0.000004	0.000004
60	30	18	2	0.000072	0.000068
60	30	18	3	0.000753	0.000681
60	30	18	4	0.005061	0.004308
60	30	18	5	0.023511	0.018450
60	30	18	6	0.079031	0.055520
60	30	18	7	0.199254	0.120223
60	30	18	8	0.389358	0.190103
60	30	18	9	0.610642	0.221284
60	30	18	10	0.800745	0.190103
60	30	18	11	0.920969	0.120223
60	30	18	12	0.976489	0.055520
60	30	18	13	0.994938	0.018450
60	30	18	14	0.999247	0.004308
60	30	18	15	0.999928	0.000681
60	30	18	16	0.999996	0.000068
60	30	18	17	1.000000	0.000004
60	30	18	18	1.000000	0.000000
60	30	19	0	0.000000	0.000000
60	30	19	1	0.000001	0.000001
60	30	19	2	0.000027	0.000025
60	30	19	3	0.000316	0.000289
60	30	19	4	0.002394	0.002079
60	30	19	5	0.012529	0.010135
60	30	19	6	0.047305	0.034776
60	30	19	7	0.133418	0.086112
60	30	19	8	0.289780	0.156362
60	30	19	9	0.500000	0.210220
60	30	19	10	0.710220	0.210220
60	30	19	11	0.866582	0.156362
60	30	19	12	0.952695	0.086112
60	30	19	13	0.987471	0.034776
60	30	19	14	0.997605	0.010135
60	30	19	15	0.999684	0.002079
60	30	19	16	0.999973	0.000289
60	30	19	17	0.999999	0.000025
60	30	19	18	1.000000	0.000001
60	30	19	19	1.000000	0.000000
60	30	20	0	0.000000	0.000000
60	30	20	1	0.000000	0.000000
60	30	20	2	0.000009	0.000009
60	30	20	3	0.000125	0.000116
60	30	20	4	0.001076	0.000951
60	30	20	5	0.006349	0.005273
60	30	20	6	0.026949	0.020599
60	30	20	7	0.085111	0.058162
60	30	20	8	0.205878	0.120767
60	30	20	9	0.392326	0.186448
60	30	20	10	0.607674	0.215347
60	30	20	11	0.794122	0.186448
60	30	20	12	0.914889	0.120767
60	30	20	13	0.973051	0.058162
60	30	20	14	0.993650	0.020599
60	30	20	15	0.998924	0.005273
60	30	20	16	0.999875	0.000951
60	30	20	17	0.999991	0.000116
60	30	20	18	1.000000	0.000009
60	30	20	19	1.000000	0.000000
60	30	20	20	1.000000	0.000000
60	30	21	0	0.000000	0.000000
60	30	21	1	0.000000	0.000000
60	30	21	2	0.000003	0.000003
60	30	21	3	0.000047	0.000044
60	30	21	4	0.000458	0.000411
60	30	21	5	0.003054	0.002595
60	30	21	6	0.014589	0.011536
60	30	21	7	0.051668	0.037078
60	30	21	8	0.139456	0.087789
60	30	21	9	0.294441	0.154985
60	30	21	10	0.500000	0.205559
60	30	21	11	0.705559	0.205559
60	30	21	12	0.860544	0.154985
60	30	21	13	0.948332	0.087789
60	30	21	14	0.985411	0.037078
60	30	21	15	0.996946	0.011536
60	30	21	16	0.999542	0.002595
60	30	21	17	0.999953	0.000411
60	30	21	18	0.999997	0.000044
60	30	21	19	1.000000	0.000003
60	30	21	20	1.000000	0.000000
60	30	21	21	1.000000	0.000000
60	30	22	0	0.000000	0.000000
60	30	22	1	0.000000	0.000000
60	30	22	2	0.000001	0.000001
60	30	22	3	0.000017	0.000016
60	30	22	4	0.000184	0.000167
60	30	22	5	0.001390	0.001206
60	30	22	6	0.007490	0.006101
60	30	22	7	0.029801	0.022310
60	30	22	8	0.089934	0.060134
60	30	22	9	0.210988	0.121053
60	30	22	10	0.394585	0.183597
60	30	22	11	0.605415	0.210830
60	30	22	12	0.789012	0.183597
60	30	22	13	0.910066	0.121053
60	30	22	14	0.970199	0.060134
60	30	22	15	0.992510	0.022310
60	30	22	16	0.998610	0.006101
60	30	22	17	0.999816	0.001206

Table for $N = 2$, $n = 1$, through $N = 100$, $n = 50$

N	n	k	x	P(x)	p(x)
60	30	22	18	0.999983	0.000167
60	30	22	19	0.999999	0.000016
60	30	22	20	1.000000	0.000001
60	30	22	21	1.000000	0.000000
60	30	22	22	1.000000	0.000000
60	30	23	0	0.000000	0.000000
60	30	23	1	0.000000	0.000000
60	30	23	2	0.000000	0.000000
60	30	23	3	0.000005	0.000005
60	30	23	4	0.000070	0.000064
60	30	23	5	0.000597	0.000527
60	30	23	6	0.003637	0.003041
60	30	23	7	0.016297	0.012660
60	30	23	8	0.055120	0.038823
60	30	23	9	0.144090	0.088970
60	30	23	10	0.297955	0.153865
60	30	23	11	0.500000	0.202045
60	30	23	12	0.702045	0.202045
60	30	23	13	0.855910	0.153865
60	30	23	14	0.944880	0.088970
60	30	23	15	0.983703	0.038823
60	30	23	16	0.996363	0.012660
60	30	23	17	0.999403	0.003041
60	30	23	18	0.999930	0.000527
60	30	23	19	0.999994	0.000064
60	30	23	20	1.000000	0.000005
60	30	23	21	1.000000	0.000000
60	30	23	22	1.000000	0.000000
60	30	23	23	1.000000	0.000000
60	30	24	0	0.000000	0.000000
60	30	24	1	0.000000	0.000000
60	30	24	2	0.000000	0.000000
60	30	24	3	0.000002	0.000002
60	30	24	4	0.000025	0.000023
60	30	24	5	0.000240	0.000216
60	30	24	6	0.001665	0.001425
60	30	24	7	0.008428	0.006763
60	30	24	8	0.032036	0.023609
60	30	24	9	0.093594	0.061557
60	30	24	10	0.214785	0.121191
60	30	24	11	0.396247	0.181463
60	30	24	12	0.603753	0.207506
60	30	24	13	0.785215	0.181463
60	30	24	14	0.906406	0.121191
60	30	24	15	0.967964	0.061557
60	30	24	16	0.991572	0.023609
60	30	24	17	0.998335	0.006763
60	30	24	18	0.999759	0.001425
60	30	24	19	0.999975	0.000216
60	30	24	20	0.999998	0.000023

N	n	k	x	P(x)	p(x)
60	30	24	21	1.000000	0.000002
60	30	24	22	1.000000	0.000000
60	30	24	23	1.000000	0.000000
60	30	24	24	1.000000	0.000000
60	30	25	0	0.000000	0.000000
60	30	25	1	0.000000	0.000000
60	30	25	2	0.000000	0.000000
60	30	25	3	0.000000	0.000000
60	30	25	4	0.000008	0.000008
60	30	25	5	0.000090	0.000082
60	30	25	6	0.000715	0.000625
60	30	25	7	0.004107	0.003392
60	30	25	8	0.017609	0.013502
60	30	25	9	0.057685	0.040076
60	30	25	10	0.147456	0.089771
60	30	25	11	0.300475	0.153019
60	30	25	12	0.500000	0.199525
60	30	25	13	0.699525	0.199525
60	30	25	14	0.852544	0.153019
60	30	25	15	0.942315	0.089771
60	30	25	16	0.982391	0.040076
60	30	25	17	0.995893	0.013502
60	30	25	18	0.999285	0.003392
60	30	25	19	0.999909	0.000625
60	30	25	20	0.999992	0.000082
60	30	25	21	0.999999	0.000008
60	30	25	22	1.000000	0.000000
60	30	25	23	1.000000	0.000000
60	30	25	24	1.000000	0.000000
60	30	25	25	1.000000	0.000000
60	30	26	0	0.000000	0.000000
60	30	26	1	0.000000	0.000000
60	30	26	2	0.000000	0.000000
60	30	26	3	0.000000	0.000000
60	30	26	4	0.000002	0.000002
60	30	26	5	0.000032	0.000029
60	30	26	6	0.000287	0.000255
60	30	26	7	0.001878	0.001591
60	30	26	8	0.009122	0.007244
60	30	26	9	0.033639	0.024517
60	30	26	10	0.096158	0.062519
60	30	26	11	0.217408	0.121249
60	30	26	12	0.397387	0.179979
60	30	26	13	0.602613	0.205225
60	30	26	14	0.782592	0.179979
60	30	26	15	0.903841	0.121249
60	30	26	16	0.966361	0.062519
60	30	26	17	0.990878	0.024517
60	30	26	18	0.998122	0.007244
60	30	26	19	0.999713	0.001591

N	n	k	x	P(x)	p(x)
60	30	26	20	0.999968	0.000255
60	30	26	21	0.999997	0.000029
60	30	26	22	1.000000	0.000002
60	30	26	23	1.000000	0.000000
60	30	26	24	1.000000	0.000000
60	30	26	25	1.000000	0.000000
60	30	26	26	1.000000	0.000000
60	30	27	0	0.000000	0.000000
60	30	27	1	0.000000	0.000000
60	30	27	2	0.000000	0.000000
60	30	27	3	0.000000	0.000000
60	30	27	4	0.000001	0.000001
60	30	27	5	0.000010	0.000009
60	30	27	6	0.000107	0.000097
60	30	27	7	0.000802	0.000695
60	30	27	8	0.004435	0.003633
60	30	27	9	0.018496	0.014061
60	30	27	10	0.059382	0.040886
60	30	27	11	0.149651	0.090268
60	30	27	12	0.302104	0.152453
60	30	27	13	0.500000	0.197896
60	30	27	14	0.697896	0.197896
60	30	27	15	0.850349	0.152453
60	30	27	16	0.940617	0.090268
60	30	27	17	0.981504	0.040886
60	30	27	18	0.995565	0.014061
60	30	27	19	0.999198	0.003633
60	30	27	20	0.999893	0.000695
60	30	27	21	0.999990	0.000097
60	30	27	22	0.999999	0.000009
60	30	27	23	1.000000	0.000001
60	30	27	24	1.000000	0.000000
60	30	27	25	1.000000	0.000000
60	30	27	26	1.000000	0.000000
60	30	27	27	1.000000	0.000000
60	30	28	0	0.000000	0.000000
60	30	28	1	0.000000	0.000000
60	30	28	2	0.000000	0.000000
60	30	28	3	0.000000	0.000000
60	30	28	4	0.000000	0.000000
60	30	28	5	0.000003	0.000003
60	30	28	6	0.000036	0.000034
60	30	28	7	0.000317	0.000281
60	30	28	8	0.002013	0.001695
60	30	28	9	0.009548	0.007535
60	30	28	10	0.034603	0.025055
60	30	28	11	0.097678	0.063075
60	30	28	12	0.218948	0.121270
60	30	28	13	0.398054	0.179106
60	30	28	14	0.601946	0.203893

N	n	k	x	P(x)	p(x)
60	30	28	15	0.781052	0.179106
60	30	28	16	0.902322	0.121270
60	30	28	17	0.965397	0.063075
60	30	28	18	0.990452	0.025055
60	30	28	19	0.997987	0.007535
60	30	28	20	0.999683	0.001695
60	30	28	21	0.999963	0.000281
60	30	28	22	0.999997	0.000034
60	30	28	23	1.000000	0.000003
60	30	28	24	1.000000	0.000000
60	30	28	25	1.000000	0.000000
60	30	28	26	1.000000	0.000000
60	30	28	27	1.000000	0.000000
60	30	28	28	1.000000	0.000000
60	30	29	0	0.000000	0.000000
60	30	29	1	0.000000	0.000000
60	30	29	2	0.000000	0.000000
60	30	29	3	0.000000	0.000000
60	30	29	4	0.000000	0.000000
60	30	29	5	0.000001	0.000001
60	30	29	6	0.000011	0.000011
60	30	29	7	0.000115	0.000104
60	30	29	8	0.000847	0.000732
60	30	29	9	0.004603	0.003756
60	30	29	10	0.018944	0.014341
60	30	29	11	0.060227	0.041284
60	30	29	12	0.150734	0.090506
60	30	29	13	0.302904	0.152170
60	30	29	14	0.500000	0.197096
60	30	29	15	0.697096	0.197096
60	30	29	16	0.849266	0.152170
60	30	29	17	0.939773	0.090506
60	30	29	18	0.981056	0.041284
60	30	29	19	0.995397	0.014341
60	30	29	20	0.999153	0.003756
60	30	29	21	0.999884	0.000732
60	30	29	22	0.999989	0.000104
60	30	29	23	0.999999	0.000011
60	30	29	24	1.000000	0.000001
60	30	29	25	1.000000	0.000000
60	30	29	26	1.000000	0.000000
60	30	29	27	1.000000	0.000000
60	30	29	28	1.000000	0.000000
60	30	29	29	1.000000	0.000000
60	30	30	0	0.000000	0.000000
60	30	30	1	0.000000	0.000000
60	30	30	2	0.000000	0.000000
60	30	30	3	0.000000	0.000000
60	30	30	4	0.000000	0.000000
60	30	30	5	0.000000	0.000000

N	n	k	x	P(x)	p(x)
60	30	30	6	0.000003	0.000003
60	30	30	7	0.000038	0.000035
60	30	30	8	0.000328	0.000290
60	30	30	9	0.002059	0.001731
60	30	30	10	0.009692	0.007633
60	30	30	11	0.034924	0.025233
60	30	30	12	0.098181	0.063257
60	30	30	13	0.219455	0.121274
60	30	30	14	0.398273	0.178817
60	30	30	15	0.601727	0.203454
60	30	30	16	0.780544	0.178817
60	30	30	17	0.901818	0.121274
60	30	30	18	0.965076	0.063257
60	30	30	19	0.990308	0.025233
60	30	30	20	0.997941	0.007633
60	30	30	21	0.999672	0.001731
60	30	30	22	0.999962	0.000290
60	30	30	23	0.999997	0.000035
60	30	30	24	1.000000	0.000003
60	30	30	25	1.000000	0.000000
60	30	30	26	1.000000	0.000000
60	30	30	27	1.000000	0.000000
60	30	30	28	1.000000	0.000000
60	30	30	29	1.000000	0.000000
60	30	30	30	1.000000	0.000000
70	1	1	0	0.985714	0.985714
70	1	1	1	1.000000	0.014286
70	2	1	0	0.971429	0.971429
70	2	1	1	1.000000	0.028571
70	2	2	0	0.943271	0.943271
70	2	2	1	0.999586	0.056315
70	2	2	2	1.000000	0.000414
70	3	1	0	0.957143	0.957143
70	3	1	1	1.000000	0.042857
70	3	2	0	0.915528	0.915528
70	3	2	1	0.998758	0.083230
70	3	2	2	1.000000	0.001242
70	3	3	0	0.875137	0.875137
70	3	3	1	0.996310	0.121173
70	3	3	2	0.999982	0.003672
70	3	3	3	1.000000	0.000018
70	4	1	0	0.942857	0.942857
70	4	1	1	1.000000	0.057143
70	4	2	0	0.888199	0.888199
70	4	2	1	0.997516	0.109317
70	4	2	2	1.000000	0.002484
70	4	3	0	0.835952	0.835952
70	4	3	1	0.992693	0.156741
70	4	3	2	0.999927	0.007234
70	4	3	3	1.000000	0.000073

N	n	k	x	P(x)	p(x)
70	4	4	0	0.786044	0.786044
70	4	4	1	0.985674	0.199630
70	4	4	2	0.999711	0.014036
70	4	4	3	0.999999	0.000288
70	4	4	4	1.000000	0.000001
70	5	1	0	0.928571	0.928571
70	5	1	1	1.000000	0.071429
70	5	2	0	0.861284	0.861284
70	5	2	1	0.995859	0.134576
70	5	2	2	1.000000	0.004141
70	5	3	0	0.797954	0.797954
70	5	3	1	0.987943	0.189989
70	5	3	2	0.999817	0.011874
70	5	3	3	1.000000	0.000183
70	5	4	0	0.738405	0.738405
70	5	4	1	0.976600	0.238195
70	5	4	2	0.999286	0.022685
70	5	4	3	0.999995	0.000709
70	5	4	4	1.000000	0.000005
70	5	5	0	0.682465	0.682465
70	5	5	1	0.962164	0.279699
70	5	5	2	0.998254	0.036090
70	5	5	3	0.999973	0.001719
70	5	5	4	1.000000	0.000027
70	5	5	5	1.000000	0.000000
70	6	1	0	0.914286	0.914286
70	6	1	1	1.000000	0.085714
70	6	2	0	0.834783	0.834783
70	6	2	1	0.993789	0.159006
70	6	2	2	1.000000	0.006211
70	6	3	0	0.761125	0.761125
70	6	3	1	0.982097	0.220972
70	6	3	2	0.999635	0.017537
70	6	3	3	1.000000	0.000365
70	6	4	0	0.692965	0.692965
70	6	4	1	0.965607	0.272642
70	6	4	2	0.998588	0.032981
70	6	4	3	0.999984	0.001396
70	6	4	4	1.000000	0.000016
70	6	5	0	0.629968	0.629968
70	6	5	1	0.944952	0.314984
70	6	5	2	0.996589	0.051637
70	6	5	3	0.999920	0.003331
70	6	5	4	0.999999	0.000079
70	6	5	5	1.000000	0.000000
70	6	6	0	0.571817	0.571817
70	6	6	1	0.920722	0.348905
70	6	6	2	0.993411	0.072689
70	6	6	3	0.999766	0.006355
70	6	6	4	0.999997	0.000231

N	n	k	x	P(x)	p(x)
70	6	6	5	1.000000	0.000003
70	6	6	6	1.000000	0.000000
70	7	1	0	0.900000	0.900000
70	7	1	1	1.000000	0.100000
70	7	2	0	0.808696	0.808696
70	7	2	1	0.991304	0.182609
70	7	2	2	1.000000	0.008696
70	7	3	0	0.725448	0.725448
70	7	3	1	0.975192	0.249744
70	7	3	2	0.999361	0.024169
70	7	3	3	1.000000	0.000639
70	7	4	0	0.649655	0.649655
70	7	4	1	0.952827	0.303172
70	7	4	2	0.997557	0.044730
70	7	4	3	0.999962	0.002405
70	7	4	4	1.000000	0.000038
70	7	5	0	0.580752	0.580752
70	7	5	1	0.925266	0.344514
70	7	5	2	0.994168	0.068903
70	7	5	3	0.999816	0.005648
70	7	5	4	0.999998	0.000182
70	7	5	5	1.000000	0.000002
70	7	6	0	0.518209	0.518209
70	7	6	1	0.893464	0.375255
70	7	6	2	0.988868	0.095404
70	7	6	3	0.999468	0.010600
70	7	6	4	0.999990	0.000521
70	7	6	5	1.000000	0.000010
70	7	6	6	1.000000	0.000000
70	7	7	0	0.461530	0.461530
70	7	7	1	0.858284	0.396754
70	7	7	2	0.981415	0.123131
70	7	7	3	0.998806	0.017391
70	7	7	4	0.999965	0.001159
70	7	7	5	1.000000	0.000034
70	7	7	6	1.000000	0.000000
70	7	7	7	1.000000	0.000000
70	8	1	0	0.885714	0.885714
70	8	1	1	1.000000	0.114286
70	8	2	0	0.783023	0.783023
70	8	2	1	0.988406	0.205383
70	8	2	2	1.000000	0.011594
70	8	3	0	0.690902	0.690902
70	8	3	1	0.967263	0.276361
70	8	3	2	0.998977	0.031714
70	8	3	3	1.000000	0.001023
70	8	4	0	0.608407	0.608407
70	8	4	1	0.938390	0.329983
70	8	4	2	0.996137	0.057747
70	8	4	3	0.999924	0.003787

N	n	k	x	P(x)	p(x)
70	8	4	4	1.000000	0.000076
70	8	5	0	0.534660	0.534660
70	8	5	1	0.903392	0.368731
70	8	5	2	0.990887	0.087496
70	8	5	3	0.999637	0.008750
70	8	5	4	0.999995	0.000359
70	8	5	5	1.000000	0.000005
70	8	6	0	0.468856	0.468856
70	8	6	1	0.863682	0.394826
70	8	6	2	0.982811	0.119129
70	8	6	3	0.998964	0.016153
70	8	6	4	0.999973	0.001010
70	8	6	5	1.000000	0.000026
70	8	6	6	1.000000	0.000000
70	8	7	0	0.410249	0.410249
70	8	7	1	0.820498	0.410249
70	8	7	2	0.971642	0.151144
70	8	7	3	0.997702	0.026059
70	8	7	4	0.999910	0.002208
70	8	7	5	0.999998	0.000088
70	8	7	6	1.000000	0.000001
70	8	7	7	1.000000	0.000000
70	8	8	0	0.358154	0.358154
70	8	8	1	0.774915	0.416761
70	8	8	2	0.957248	0.182333
70	8	8	3	0.995634	0.038386
70	8	8	4	0.999770	0.004136
70	8	8	5	0.999994	0.000224
70	8	8	6	1.000000	0.000006
70	8	8	7	1.000000	0.000000
70	8	8	8	1.000000	0.000000
70	9	1	0	0.871429	0.871429
70	9	1	1	1.000000	0.128571
70	9	2	0	0.757764	0.757764
70	9	2	1	0.985093	0.227329
70	9	2	2	1.000000	0.014907
70	9	3	0	0.657472	0.657472
70	9	3	1	0.958348	0.300877
70	9	3	2	0.998465	0.040117
70	9	3	3	1.000000	0.001535
70	9	4	0	0.569155	0.569155
70	9	4	1	0.922423	0.353268
70	9	4	2	0.994274	0.071851
70	9	4	3	0.999862	0.005588
70	9	4	4	1.000000	0.000137
70	9	5	0	0.491543	0.491543
70	9	5	1	0.879603	0.388060
70	9	5	2	0.986654	0.107051
70	9	5	3	0.999354	0.012701
70	9	5	4	0.999990	0.000635

Table for $N = 2$, $n = 1$, through $N = 100$, $n = 50$

N	n	k	x	P(x)	p(x)
70	9	5	5	1.000000	0.000010
70	9	6	0	0.423483	0.423483
70	9	6	1	0.831841	0.408358
70	9	6	2	0.975125	0.143284
70	9	6	3	0.998182	0.023057
70	9	6	4	0.999941	0.001759
70	9	6	5	0.999999	0.000059
70	9	6	6	1.000000	0.000001
70	9	7	0	0.363931	0.363931
70	9	7	1	0.780796	0.416866
70	9	7	2	0.959453	0.178657
70	9	7	3	0.996020	0.036567
70	9	7	4	0.999803	0.003783
70	9	7	5	0.999996	0.000192
70	9	7	6	1.000000	0.000004
70	9	7	7	1.000000	0.000000
70	9	8	0	0.311940	0.311940
70	9	8	1	0.727861	0.415921
70	9	8	2	0.939603	0.211741
70	9	8	3	0.992538	0.052935
70	9	8	4	0.999503	0.006965
70	9	8	5	0.999983	0.000480
70	9	8	6	1.000000	0.000016
70	9	8	7	1.000000	0.000000
70	9	8	8	1.000000	0.000000
70	9	9	0	0.266659	0.266659
70	9	9	1	0.674194	0.407535
70	9	9	2	0.915696	0.241502
70	9	9	3	0.987415	0.071719
70	9	9	4	0.998941	0.011526
70	9	9	5	0.999952	0.001011
70	9	9	6	0.999999	0.000046
70	9	9	7	1.000000	0.000001
70	9	9	8	1.000000	0.000000
70	9	9	9	1.000000	0.000000
70	10	1	0	0.857143	0.857143
70	10	1	1	1.000000	0.142857
70	10	2	0	0.732919	0.732919
70	10	2	1	0.981366	0.248447
70	10	2	2	1.000000	0.018634
70	10	3	0	0.625137	0.625137
70	10	3	1	0.948484	0.323347
70	10	3	2	0.997808	0.049324
70	10	3	3	1.000000	0.002192
70	10	4	0	0.531833	0.531833
70	10	4	1	0.905049	0.373216
70	10	4	2	0.991918	0.086869
70	10	4	3	0.999771	0.007853
70	10	4	4	1.000000	0.000229
70	10	5	0	0.451252	0.451252

N	n	k	x	P(x)	p(x)
70	10	5	1	0.854156	0.402904
70	10	5	2	0.981389	0.127233
70	10	5	3	0.998938	0.017549
70	10	5	4	0.999979	0.001041
70	10	5	5	1.000000	0.000021
70	10	6	0	0.381829	0.381829
70	10	6	1	0.798369	0.416540
70	10	6	2	0.965729	0.167360
70	10	6	3	0.997048	0.031319
70	10	6	4	0.999883	0.002835
70	10	6	5	0.999998	0.000115
70	10	6	6	1.000000	0.000002
70	10	7	0	0.322168	0.322168
70	10	7	1	0.739793	0.417625
70	10	7	2	0.944809	0.205016
70	10	7	3	0.993623	0.048813
70	10	7	4	0.999617	0.005995
70	10	7	5	0.999989	0.000372
70	10	7	6	1.000000	0.000011
70	10	7	7	1.000000	0.000000
70	10	8	0	0.271030	0.271030
70	10	8	1	0.680133	0.409102
70	10	8	2	0.918776	0.238643
70	10	8	3	0.988199	0.069423
70	10	8	4	0.999046	0.010847
70	10	8	5	0.999960	0.000913
70	10	8	6	0.999999	0.000039
70	10	8	7	1.000000	0.000001
70	10	8	8	1.000000	0.000000
70	10	9	0	0.227316	0.227316
70	10	9	1	0.620747	0.393431
70	10	9	2	0.887983	0.267236
70	10	9	3	0.980361	0.092378
70	10	9	4	0.997997	0.017636
70	10	9	5	0.999886	0.001890
70	10	9	6	0.999997	0.000110
70	10	9	7	1.000000	0.000003
70	10	9	8	1.000000	0.000000
70	10	9	9	1.000000	0.000000
70	10	10	0	0.190051	0.190051
70	10	10	1	0.562700	0.372649
70	10	10	2	0.852935	0.290236
70	10	10	3	0.969760	0.116825
70	10	10	4	0.996262	0.026502
70	10	10	5	0.999731	0.003469
70	10	10	6	0.999989	0.000258
70	10	10	7	1.000000	0.000010
70	10	10	8	1.000000	0.000000
70	10	10	9	1.000000	0.000000
70	10	10	10	1.000000	0.000000

N	n	k	x	P(x)	p(x)
70	11	1	0	0.842857	0.842857
70	11	1	1	1.000000	0.157143
70	11	2	0	0.708489	0.708489
70	11	2	1	0.977226	0.268737
70	11	2	2	1.000000	0.022774
70	11	3	0	0.593880	0.593880
70	11	3	1	0.937706	0.343825
70	11	3	2	0.996986	0.059280
70	11	3	3	1.000000	0.003014
70	11	4	0	0.496377	0.496377
70	11	4	1	0.886388	0.390011
70	11	4	2	0.989023	0.102634
70	11	4	3	0.999640	0.010617
70	11	4	4	1.000000	0.000360
70	11	5	0	0.413648	0.413648
70	11	5	1	0.827296	0.413648
70	11	5	2	0.975027	0.147731
70	11	5	3	0.998353	0.023326
70	11	5	4	0.999962	0.001609
70	11	5	5	1.000000	0.000038
70	11	6	0	0.343646	0.343646
70	11	6	1	0.763658	0.420012
70	11	6	2	0.954572	0.190914
70	11	6	3	0.995482	0.040910
70	11	6	4	0.999789	0.004306
70	11	6	5	0.999996	0.000208
70	11	6	6	1.000000	0.000004
70	11	7	0	0.284582	0.284582
70	11	7	1	0.698031	0.413449
70	11	7	2	0.927725	0.229694
70	11	7	3	0.990368	0.062644
70	11	7	4	0.999317	0.008949
70	11	7	5	0.999977	0.000659
70	11	7	6	1.000000	0.000023
70	11	7	7	1.000000	0.000000
70	11	8	0	0.234893	0.234893
70	11	8	1	0.632404	0.397511
70	11	8	2	0.894911	0.262507
70	11	8	3	0.982414	0.087502
70	11	8	4	0.998323	0.015910
70	11	8	5	0.999914	0.001591
70	11	8	6	0.999998	0.000084
70	11	8	7	1.000000	0.000002
70	11	8	8	1.000000	0.000000
70	11	9	0	0.193218	0.193218
70	11	9	1	0.568289	0.375071
70	11	9	2	0.856805	0.288516
70	11	9	3	0.971123	0.114318
70	11	9	4	0.996527	0.025404
70	11	9	5	0.999760	0.003233

N	n	k	x	P(x)	p(x)
70	11	9	6	0.999991	0.000231
70	11	9	7	1.000000	0.000009
70	11	9	8	1.000000	0.000000
70	11	9	9	1.000000	0.000000
70	11	10	0	0.158376	0.158376
70	11	10	1	0.506802	0.348426
70	11	10	2	0.814237	0.307435
70	11	10	3	0.956130	0.141893
70	11	10	4	0.993612	0.037481
70	11	10	5	0.999442	0.005830
70	11	10	6	0.999972	0.000530
70	11	10	7	0.999999	0.000027
70	11	10	8	1.000000	0.000001
70	11	10	9	1.000000	0.000000
70	11	10	10	1.000000	0.000000
70	11	11	0	0.129340	0.129340
70	11	11	1	0.448731	0.319391
70	11	11	2	0.768122	0.319391
70	11	11	3	0.937211	0.169089
70	11	11	4	0.989239	0.052027
70	11	11	5	0.998859	0.009620
70	11	11	6	0.999928	0.001069
70	11	11	7	0.999997	0.000069
70	11	11	8	1.000000	0.000002
70	11	11	9	1.000000	0.000000
70	11	11	10	1.000000	0.000000
70	11	11	11	1.000000	0.000000
70	12	1	0	0.828571	0.828571
70	12	1	1	1.000000	0.171429
70	12	2	0	0.684472	0.684472
70	12	2	1	0.972671	0.288199
70	12	2	2	1.000000	0.027329
70	12	3	0	0.563683	0.563683
70	12	3	1	0.926050	0.362368
70	12	3	2	0.995981	0.069931
70	12	3	3	1.000000	0.004019
70	12	4	0	0.462725	0.462725
70	12	4	1	0.866557	0.403832
70	12	4	2	0.985544	0.118986
70	12	4	3	0.999460	0.013917
70	12	4	4	1.000000	0.000540
70	12	5	0	0.378593	0.378593
70	12	5	1	0.799252	0.420659
70	12	5	2	0.967515	0.168264
70	12	5	3	0.997562	0.030047
70	12	5	4	0.999934	0.002372
70	12	5	5	1.000000	0.000065
70	12	6	0	0.308699	0.308699
70	12	6	1	0.728063	0.419364
70	12	6	2	0.941629	0.213565

Table for $N = 2$, $n = 1$, through $N = 100$, $n = 50$

N	n	k	x	P(x)	p(x)
70	12	6	3	0.993402	0.051773
70	12	6	4	0.999643	0.006241
70	12	6	5	0.999993	0.000350
70	12	6	6	1.000000	0.000007
70	12	7	0	0.250818	0.250818
70	12	7	1	0.655985	0.405167
70	12	7	2	0.908259	0.252274
70	12	7	3	0.986121	0.077862
70	12	7	4	0.998862	0.012741
70	12	7	5	0.999955	0.001092
70	12	7	6	0.999999	0.000045
70	12	7	7	1.000000	0.000001
70	12	8	0	0.203043	0.203043
70	12	8	1	0.585242	0.382199
70	12	8	2	0.868216	0.282974
70	12	8	3	0.974998	0.106783
70	12	8	4	0.997245	0.022246
70	12	8	5	0.999833	0.002589
70	12	8	6	0.999995	0.000162
70	12	8	7	1.000000	0.000005
70	12	8	8	1.000000	0.000000
70	12	9	0	0.163744	0.163744
70	12	9	1	0.517432	0.353688
70	12	9	2	0.822575	0.305142
70	12	9	3	0.959497	0.136923
70	12	9	4	0.994374	0.034877
70	12	9	5	0.999541	0.005167
70	12	9	6	0.999979	0.000438
70	12	9	7	0.999999	0.000020
70	12	9	8	1.000000	0.000000
70	12	9	9	1.000000	0.000000
70	12	10	0	0.131532	0.131532
70	12	10	1	0.453652	0.322120
70	12	10	2	0.772551	0.318899
70	12	10	3	0.939296	0.166745
70	12	10	4	0.989800	0.050504
70	12	10	5	0.998948	0.009148
70	12	10	6	0.999936	0.000988
70	12	10	7	0.999998	0.000062
70	12	10	8	1.000000	0.000002
70	12	10	9	1.000000	0.000000
70	12	10	10	1.000000	0.000000
70	12	11	0	0.105226	0.105226
70	12	11	1	0.394597	0.289371
70	12	11	2	0.719401	0.324804
70	12	11	3	0.914284	0.194883
70	12	11	4	0.983066	0.068782
70	12	11	5	0.997881	0.014815
70	12	11	6	0.999837	0.001957
70	12	11	7	0.999993	0.000155

N	n	k	x	P(x)	p(x)
70	12	11	8	1.000000	0.000007
70	12	11	9	1.000000	0.000000
70	12	11	10	1.000000	0.000000
70	12	11	11	1.000000	0.000000
70	12	12	0	0.083824	0.083824
70	12	12	1	0.340647	0.256823
70	12	12	2	0.664350	0.323703
70	12	12	3	0.884556	0.220206
70	12	12	4	0.973740	0.089184
70	12	12	5	0.996123	0.022383
70	12	12	6	0.999638	0.003515
70	12	12	7	0.999980	0.000341
70	12	12	8	0.999999	0.000020
70	12	12	9	1.000000	0.000001
70	12	12	10	1.000000	0.000000
70	12	12	11	1.000000	0.000000
70	12	12	12	1.000000	0.000000
70	13	1	0	0.814286	0.814286
70	13	1	1	1.000000	0.185714
70	13	2	0	0.660870	0.660870
70	13	2	1	0.967702	0.306832
70	13	2	2	1.000000	0.032298
70	13	3	0	0.534527	0.534527
70	13	3	1	0.913555	0.379028
70	13	3	2	0.994775	0.081220
70	13	3	3	1.000000	0.005225
70	13	4	0	0.430813	0.430813
70	13	4	1	0.845669	0.414857
70	13	4	2	0.981441	0.135771
70	13	4	3	0.999220	0.017780
70	13	4	4	1.000000	0.000780
70	13	5	0	0.345956	0.345956
70	13	5	1	0.770241	0.424285
70	13	5	2	0.958812	0.188571
70	13	5	3	0.996526	0.037714
70	13	5	4	0.999894	0.003367
70	13	5	5	1.000000	0.000106
70	13	6	0	0.276764	0.276764
70	13	6	1	0.691911	0.415147
70	13	6	2	0.926900	0.234989
70	13	6	3	0.990724	0.063824
70	13	6	4	0.999427	0.008703
70	13	6	5	0.999987	0.000559
70	13	6	6	1.000000	0.000013
70	13	7	0	0.220547	0.220547
70	13	7	1	0.614071	0.393525
70	13	7	2	0.886511	0.272440
70	13	7	3	0.980751	0.094240
70	13	7	4	0.998203	0.017452
70	13	7	5	0.999917	0.001713

N	n	k	x	P(x)	p(x)
70	13	7	6	0.999998	0.000082
70	13	7	7	1.000000	0.000001
70	13	8	0	0.175037	0.175037
70	13	8	1	0.539114	0.364077
70	13	8	2	0.838942	0.299828
70	13	8	3	0.965793	0.126850
70	13	8	4	0.995710	0.029918
70	13	8	5	0.999699	0.003989
70	13	8	6	0.999989	0.000290
70	13	8	7	1.000000	0.000010
70	13	8	8	1.000000	0.000000
70	13	9	0	0.138336	0.138336
70	13	9	1	0.468648	0.330312
70	13	9	2	0.785747	0.317099
70	13	9	3	0.945333	0.159586
70	13	9	4	0.991367	0.046034
70	13	9	5	0.999185	0.007817
70	13	9	6	0.999957	0.000772
70	13	9	7	0.999999	0.000042
70	13	9	8	1.000000	0.000001
70	13	9	9	1.000000	0.000000
70	13	10	0	0.108854	0.108854
70	13	10	1	0.403668	0.294814
70	13	10	2	0.728565	0.324897
70	13	10	3	0.919171	0.190606
70	13	10	4	0.984575	0.065404
70	13	10	5	0.998159	0.013584
70	13	10	6	0.999868	0.001709
70	13	10	7	0.999995	0.000127
70	13	10	8	1.000000	0.000005
70	13	10	9	1.000000	0.000000
70	13	10	10	1.000000	0.000000
70	13	11	0	0.085269	0.085269
70	13	11	1	0.344705	0.259436
70	13	11	2	0.669001	0.324295
70	13	11	3	0.887404	0.218403
70	13	11	4	0.974765	0.087361
70	13	11	5	0.996348	0.021583
70	13	11	6	0.999669	0.003321
70	13	11	7	0.999982	0.000313
70	13	11	8	0.999999	0.000017
70	13	11	9	1.000000	0.000001
70	13	11	10	1.000000	0.000000
70	13	11	11	1.000000	0.000000
70	13	12	0	0.066481	0.066481
70	13	12	1	0.291939	0.225458
70	13	12	2	0.608539	0.316600
70	13	12	3	0.850386	0.241847
70	13	12	4	0.961438	0.111052
70	13	12	5	0.993422	0.031983

N	n	k	x	P(x)	p(x)
70	13	12	6	0.999275	0.005853
70	13	12	7	0.999950	0.000675
70	13	12	8	0.999998	0.000048
70	13	12	9	1.000000	0.000002
70	13	12	10	1.000000	0.000000
70	13	12	11	1.000000	0.000000
70	13	12	12	1.000000	0.000000
70	13	13	0	0.051580	0.051580
70	13	13	1	0.245292	0.193712
70	13	13	2	0.548494	0.303202
70	13	13	3	0.808688	0.260194
70	13	13	4	0.944206	0.135518
70	13	13	5	0.989010	0.044804
70	13	13	6	0.998568	0.009558
70	13	13	7	0.999880	0.001312
70	13	13	8	0.999994	0.000114
70	13	13	9	1.000000	0.000006
70	13	13	10	1.000000	0.000000
70	13	13	11	1.000000	0.000000
70	13	13	12	1.000000	0.000000
70	13	13	13	1.000000	0.000000
70	14	1	0	0.800000	0.800000
70	14	1	1	1.000000	0.200000
70	14	2	0	0.637681	0.637681
70	14	2	1	0.962319	0.324638
70	14	2	2	1.000000	0.037681
70	14	3	0	0.506394	0.506394
70	14	3	1	0.900256	0.393862
70	14	3	2	0.993350	0.093095
70	14	3	3	1.000000	0.006650
70	14	4	0	0.400580	0.400580
70	14	4	1	0.823835	0.423255
70	14	4	2	0.976677	0.152842
70	14	4	3	0.998908	0.022232
70	14	4	4	1.000000	0.001092
70	14	5	0	0.315609	0.315609
70	14	5	1	0.740466	0.424858
70	14	5	2	0.948887	0.208421
70	14	5	3	0.995203	0.046316
70	14	5	4	0.999835	0.004632
70	14	5	5	1.000000	0.000165
70	14	6	0	0.247631	0.247631
70	14	6	1	0.655495	0.407863
70	14	6	2	0.910410	0.254915
70	14	6	3	0.987365	0.076955
70	14	6	4	0.999122	0.011757
70	14	6	5	0.999977	0.000855
70	14	6	6	1.000000	0.000023
70	14	7	0	0.193462	0.193462
70	14	7	1	0.572648	0.379186

N	n	k	x	P(x)	p(x)
70	14	7	2	0.862613	0.289965
70	14	7	3	0.974138	0.111525
70	14	7	4	0.997285	0.023147
70	14	7	5	0.999857	0.002572
70	14	7	6	0.999997	0.000140
70	14	7	7	1.000000	0.000003
70	14	8	0	0.150470	0.150470
70	14	8	1	0.494403	0.343932
70	14	8	2	0.807381	0.312979
70	14	8	3	0.954666	0.147284
70	14	8	4	0.993611	0.038945
70	14	8	5	0.999489	0.005879
70	14	8	6	0.999979	0.000490
70	14	8	7	1.000000	0.000020
70	14	8	8	1.000000	0.000000
70	14	9	0	0.116493	0.116493
70	14	9	1	0.422288	0.305795
70	14	9	2	0.746805	0.324517
70	14	9	3	0.928535	0.181730
70	14	9	4	0.987329	0.058795
70	14	9	5	0.998636	0.011307
70	14	9	6	0.999916	0.001280
70	14	9	7	0.999997	0.000081
70	14	9	8	1.000000	0.000003
70	14	9	9	1.000000	0.000000
70	14	10	0	0.089757	0.089757
70	14	10	1	0.357119	0.267362
70	14	10	2	0.682966	0.325847
70	14	10	3	0.895764	0.212798
70	14	10	4	0.977691	0.081927
70	14	10	5	0.996968	0.019277
70	14	10	6	0.999748	0.002780
70	14	10	7	0.999988	0.000240
70	14	10	8	1.000000	0.000012
70	14	10	9	1.000000	0.000000
70	14	10	10	1.000000	0.000000
70	14	11	0	0.068814	0.068814
70	14	11	1	0.299190	0.230377
70	14	11	2	0.617796	0.318606
70	14	11	3	0.856751	0.238954
70	14	11	4	0.964036	0.107286
70	14	11	5	0.994076	0.030040
70	14	11	6	0.999377	0.005301
70	14	11	7	0.999960	0.000583
70	14	11	8	0.999998	0.000038
70	14	11	9	1.000000	0.000001
70	14	11	10	1.000000	0.000000
70	14	11	11	1.000000	0.000000
70	14	12	0	0.052485	0.052485
70	14	12	1	0.248429	0.195944

N	n	k	x	P(x)	p(x)
70	14	12	2	0.552995	0.304566
70	14	12	3	0.812200	0.259205
70	14	12	4	0.945852	0.133652
70	14	12	5	0.989494	0.043642
70	14	12	6	0.998659	0.009165
70	14	12	7	0.999891	0.001232
70	14	12	8	0.999995	0.000104
70	14	12	9	1.000000	0.000005
70	14	12	10	1.000000	0.000000
70	14	12	11	1.000000	0.000000
70	14	12	12	1.000000	0.000000
70	14	13	0	0.039816	0.039816
70	14	13	1	0.204511	0.164695
70	14	13	2	0.489981	0.285471
70	14	13	3	0.763040	0.273059
70	14	13	4	0.922809	0.159768
70	14	13	5	0.982722	0.059913
70	14	13	6	0.997395	0.014673
70	14	13	7	0.999742	0.002348
70	14	13	8	0.999984	0.000242
70	14	13	9	0.999999	0.000015
70	14	13	10	1.000000	0.000001
70	14	13	11	1.000000	0.000000
70	14	13	12	1.000000	0.000000
70	14	13	13	1.000000	0.000000
70	14	14	0	0.030037	0.030037
70	14	14	1	0.166949	0.136912
70	14	14	2	0.429882	0.262933
70	14	14	3	0.710345	0.280462
70	14	14	4	0.894779	0.184434
70	14	14	5	0.973262	0.078483
70	14	14	6	0.995335	0.022073
70	14	14	7	0.999454	0.004119
70	14	14	8	0.999958	0.000505
70	14	14	9	0.999998	0.000040
70	14	14	10	1.000000	0.000002
70	14	14	11	1.000000	0.000000
70	14	14	12	1.000000	0.000000
70	14	14	13	1.000000	0.000000
70	14	14	14	1.000000	0.000000
70	15	1	0	0.785714	0.785714
70	15	1	1	1.000000	0.214286
70	15	2	0	0.614907	0.614907
70	15	2	1	0.956522	0.341615
70	15	2	2	1.000000	0.043478
70	15	3	0	0.479266	0.479266
70	15	3	1	0.886189	0.406924
70	15	3	2	0.991688	0.105499
70	15	3	3	1.000000	0.008312
70	15	4	0	0.371967	0.371967

N	n	k	x	P(x)	p(x)
70	15	4	1	0.801160	0.429193
70	15	4	2	0.971218	0.170058
70	15	4	3	0.998511	0.027293
70	15	4	4	1.000000	0.001489
70	15	5	0	0.287429	0.287429
70	15	5	1	0.710119	0.422690
70	15	5	2	0.937722	0.227602
70	15	5	3	0.993549	0.055827
70	15	5	4	0.999752	0.006203
70	15	5	5	1.000000	0.000248
70	15	6	0	0.221099	0.221099
70	15	6	1	0.619078	0.397979
70	15	6	2	0.892201	0.273123
70	15	6	3	0.983242	0.091041
70	15	6	4	0.998702	0.015460
70	15	6	5	0.999962	0.001260
70	15	6	6	1.000000	0.000038
70	15	7	0	0.169279	0.169279
70	15	7	1	0.532021	0.362741
70	15	7	2	0.836723	0.304703
70	15	7	3	0.966172	0.129449
70	15	7	4	0.996045	0.029873
70	15	7	5	0.999765	0.003720
70	15	7	6	0.999995	0.000230
70	15	7	7	1.000000	0.000005
70	15	8	0	0.128975	0.128975
70	15	8	1	0.451411	0.322437
70	15	8	2	0.773848	0.322437
70	15	8	3	0.941515	0.167667
70	15	8	4	0.990829	0.049314
70	15	8	5	0.999174	0.008345
70	15	8	6	0.999962	0.000787
70	15	8	7	0.999999	0.000037
70	15	8	8	1.000000	0.000001
70	15	9	0	0.097771	0.097771
70	15	9	1	0.378603	0.280832
70	15	9	2	0.706240	0.327637
70	15	9	3	0.909063	0.202823
70	15	9	4	0.982080	0.073016
70	15	9	5	0.997828	0.015749
70	15	9	6	0.999847	0.002019
70	15	9	7	0.999994	0.000147
70	15	9	8	1.000000	0.000005
70	15	9	9	1.000000	0.000000
70	15	10	0	0.073729	0.073729
70	15	10	1	0.314150	0.240421
70	15	10	2	0.636416	0.322266
70	15	10	3	0.869164	0.232748
70	15	10	4	0.968913	0.099749
70	15	10	5	0.995247	0.026334

N	n	k	x	P(x)	p(x)
70	15	10	6	0.999550	0.004303
70	15	10	7	0.999975	0.000426
70	15	10	8	0.999999	0.000024
70	15	10	9	1.000000	0.000001
70	15	10	10	1.000000	0.000000
70	15	11	0	0.055297	0.055297
70	15	11	1	0.258052	0.202755
70	15	11	2	0.566592	0.308540
70	15	11	3	0.822614	0.256023
70	15	11	4	0.950626	0.128011
70	15	11	5	0.990858	0.040232
70	15	11	6	0.998904	0.008046
70	15	11	7	0.999918	0.001014
70	15	11	8	0.999996	0.000078
70	15	11	9	1.000000	0.000003
70	15	11	10	1.000000	0.000000
70	15	11	11	1.000000	0.000000
70	15	12	0	0.041238	0.041238
70	15	12	1	0.209940	0.168702
70	15	12	2	0.498608	0.288668
70	15	12	3	0.770542	0.271934
70	15	12	4	0.926759	0.156217
70	15	12	5	0.984039	0.057280
70	15	12	6	0.997677	0.013638
70	15	12	7	0.999781	0.002104
70	15	12	8	0.999987	0.000206
70	15	12	9	0.999999	0.000012
70	15	12	10	1.000000	0.000000
70	15	12	11	1.000000	0.000000
70	15	12	12	1.000000	0.000000
70	15	13	0	0.030573	0.030573
70	15	13	1	0.169219	0.138646
70	15	13	2	0.433907	0.264688
70	15	13	3	0.714280	0.280373
70	15	13	4	0.897132	0.182852
70	15	13	5	0.974163	0.077031
70	15	13	6	0.995560	0.021398
70	15	13	7	0.999491	0.003930
70	15	13	8	0.999962	0.000472
70	15	13	9	0.999998	0.000036
70	15	13	10	1.000000	0.000002
70	15	13	11	1.000000	0.000000
70	15	13	12	1.000000	0.000000
70	15	13	13	1.000000	0.000000
70	15	14	0	0.022528	0.022528
70	15	14	1	0.135166	0.112638
70	15	14	2	0.373539	0.238374
70	15	14	3	0.655254	0.281714
70	15	14	4	0.861844	0.206591
70	15	14	5	0.960649	0.098804

N	n	k	x	P(x)	p(x)
70	15	14	6	0.992182	0.031533
70	15	14	7	0.998939	0.006757
70	15	14	8	0.999904	0.000965
70	15	14	9	0.999994	0.000090
70	15	14	10	1.000000	0.000005
70	15	14	11	1.000000	0.000000
70	15	14	12	1.000000	0.000000
70	15	14	13	1.000000	0.000000
70	15	14	14	1.000000	0.000000
70	15	15	0	0.016493	0.016493
70	15	15	1	0.107006	0.090513
70	15	15	2	0.318203	0.211196
70	15	15	3	0.594886	0.276684
70	15	15	4	0.821264	0.226378
70	15	15	5	0.943005	0.121741
70	15	15	6	0.987114	0.044109
70	15	15	7	0.997974	0.010860
70	15	15	8	0.999784	0.001810
70	15	15	9	0.999985	0.000201
70	15	15	10	0.999999	0.000014
70	15	15	11	1.000000	0.000001
70	15	15	12	1.000000	0.000000
70	15	15	13	1.000000	0.000000
70	15	15	14	1.000000	0.000000
70	15	15	15	1.000000	0.000000
70	16	1	0	0.771429	0.771429
70	16	1	1	1.000000	0.228571
70	16	2	0	0.592547	0.592547
70	16	2	1	0.950310	0.357764
70	16	2	2	1.000000	0.049689
70	16	3	0	0.453124	0.453124
70	16	3	1	0.871392	0.418268
70	16	3	2	0.989770	0.118378
70	16	3	3	1.000000	0.010230
70	16	4	0	0.344915	0.344915
70	16	4	1	0.777750	0.432835
70	16	4	2	0.965034	0.187284
70	16	4	3	0.998015	0.032981
70	16	4	4	1.000000	0.001985
70	16	5	0	0.261299	0.261299
70	16	5	1	0.679378	0.418079
70	16	5	2	0.925307	0.245929
70	16	5	3	0.991519	0.066212
70	16	5	4	0.999639	0.008120
70	16	5	5	1.000000	0.000361
70	16	6	0	0.196980	0.196980
70	16	6	1	0.582899	0.385919
70	16	6	2	0.872338	0.289439
70	16	6	3	0.978276	0.105939
70	16	6	4	0.998140	0.019863

N	n	k	x	P(x)	p(x)
70	16	6	5	0.999939	0.001799
70	16	6	6	1.000000	0.000061
70	16	7	0	0.147735	0.147735
70	16	7	1	0.492449	0.344714
70	16	7	2	0.809023	0.316574
70	16	7	3	0.956758	0.147735
70	16	7	4	0.994415	0.037658
70	16	7	5	0.999630	0.005214
70	16	7	6	0.999990	0.000361
70	16	7	7	1.000000	0.000010
70	16	8	0	0.110215	0.110215
70	16	8	1	0.410374	0.300159
70	16	8	2	0.738673	0.328299
70	16	8	3	0.926273	0.187600
70	16	8	4	0.987242	0.060970
70	16	8	5	0.998719	0.011477
70	16	8	6	0.999933	0.001214
70	16	8	7	0.999998	0.000065
70	16	8	8	1.000000	0.000001
70	16	9	0	0.081772	0.081772
70	16	9	1	0.337755	0.255983
70	16	9	2	0.664541	0.326786
70	16	9	3	0.886937	0.222396
70	16	9	4	0.975442	0.088505
70	16	9	5	0.996683	0.021241
70	16	9	6	0.999737	0.003054
70	16	9	7	0.999989	0.000252
70	16	9	8	1.000000	0.000011
70	16	9	9	1.000000	0.000000
70	16	10	0	0.060324	0.060324
70	16	10	1	0.274808	0.214485
70	16	10	2	0.589541	0.314733
70	16	10	3	0.839541	0.250000
70	16	10	4	0.958031	0.118490
70	16	10	5	0.992853	0.034821
70	16	10	6	0.999237	0.006384
70	16	10	7	0.999952	0.000715
70	16	10	8	0.999998	0.000046
70	16	10	9	1.000000	0.000002
70	16	10	10	1.000000	0.000000
70	16	11	0	0.044237	0.044237
70	16	11	1	0.221187	0.176950
70	16	11	2	0.516103	0.294916
70	16	11	3	0.785375	0.269271
70	16	11	4	0.934333	0.148959
70	16	11	5	0.986469	0.052136
70	16	11	6	0.998173	0.011704
70	16	11	7	0.999845	0.001672
70	16	11	8	0.999992	0.000148
70	16	11	9	1.000000	0.000008

N	n	k	x	P(x)	p(x)
70	16	11	10	1.000000	0.000000
70	16	11	11	1.000000	0.000000
70	16	12	0	0.032241	0.032241
70	16	12	1	0.176200	0.143959
70	16	12	2	0.446123	0.269923
70	16	12	3	0.726044	0.279920
70	16	12	4	0.904037	0.177993
70	16	12	5	0.976749	0.072712
70	16	12	6	0.996189	0.019440
70	16	12	7	0.999590	0.003401
70	16	12	8	0.999972	0.000383
70	16	12	9	0.999999	0.000027
70	16	12	10	1.000000	0.000001
70	16	12	11	1.000000	0.000000
70	16	12	12	1.000000	0.000000
70	16	13	0	0.023347	0.023347
70	16	13	1	0.138969	0.115622
70	16	13	2	0.380969	0.242000
70	16	13	3	0.663303	0.282334
70	16	13	4	0.867210	0.203908
70	16	13	5	0.962958	0.095748
70	16	13	6	0.992837	0.029879
70	16	13	7	0.999062	0.006225
70	16	13	8	0.999919	0.000857
70	16	13	9	0.999996	0.000076
70	16	13	10	1.000000	0.000004
70	16	13	11	1.000000	0.000000
70	16	13	12	1.000000	0.000000
70	16	13	13	1.000000	0.000000
70	16	14	0	0.016793	0.016793
70	16	14	1	0.108542	0.091749
70	16	14	2	0.321531	0.212989
70	16	14	3	0.598911	0.277380
70	16	14	4	0.824283	0.225372
70	16	14	5	0.944481	0.120198
70	16	14	6	0.987595	0.043115
70	16	14	7	0.998079	0.010484
70	16	14	8	0.999799	0.001720
70	16	14	9	0.999986	0.000187
70	16	14	10	0.999999	0.000013
70	16	14	11	1.000000	0.000001
70	16	14	12	1.000000	0.000000
70	16	14	13	1.000000	0.000000
70	16	14	14	1.000000	0.000000
70	16	15	0	0.011995	0.011995
70	16	15	1	0.083967	0.071971
70	16	15	2	0.268284	0.184317
70	16	15	3	0.534519	0.266236
70	16	15	4	0.775989	0.241469
70	16	15	5	0.920870	0.144882

N	n	k	x	P(x)	p(x)
70	16	15	6	0.979896	0.059026
70	16	15	7	0.996394	0.016498
70	16	15	8	0.999553	0.003159
70	16	15	9	0.999963	0.000410
70	16	15	10	0.999998	0.000035
70	16	15	11	1.000000	0.000002
70	16	15	12	1.000000	0.000000
70	16	15	13	1.000000	0.000000
70	16	15	14	1.000000	0.000000
70	16	15	15	1.000000	0.000000
70	16	16	0	0.008506	0.008506
70	16	16	1	0.064338	0.055832
70	16	16	2	0.221367	0.157028
70	16	16	3	0.471591	0.250224
70	16	16	4	0.723304	0.251714
70	16	16	5	0.891894	0.168590
70	16	16	6	0.969164	0.077270
70	16	16	7	0.993694	0.024530
70	16	16	8	0.999094	0.005399
70	16	16	9	0.999911	0.000817
70	16	16	10	0.999994	0.000083
70	16	16	11	1.000000	0.000006
70	16	16	12	1.000000	0.000000
70	16	16	13	1.000000	0.000000
70	16	16	14	1.000000	0.000000
70	16	16	15	1.000000	0.000000
70	16	16	16	1.000000	0.000000
70	17	1	0	0.757143	0.757143
70	17	1	1	1.000000	0.242857
70	17	2	0	0.570600	0.570600
70	17	2	1	0.943685	0.373085
70	17	2	2	1.000000	0.056315
70	17	3	0	0.427950	0.427950
70	17	3	1	0.855901	0.427950
70	17	3	2	0.987578	0.131677
70	17	3	3	1.000000	0.012422
70	17	4	0	0.319366	0.319366
70	17	4	1	0.753704	0.434338
70	17	4	2	0.958098	0.204394
70	17	4	3	0.997404	0.039307
70	17	4	4	1.000000	0.002596
70	17	5	0	0.237105	0.237105
70	17	5	1	0.648410	0.411305
70	17	5	2	0.911644	0.263235
70	17	5	3	0.989066	0.077422
70	17	5	4	0.999489	0.010422
70	17	5	5	1.000000	0.000511
70	17	6	0	0.175093	0.175093
70	17	6	1	0.547165	0.372072
70	17	6	2	0.850898	0.303733

Table for $N = 2$, $n = 1$, through $N = 100$, $n = 50$

N	n	k	x	P(x)	p(x)
70	17	6	3	0.972391	0.121493
70	17	6	4	0.997404	0.025013
70	17	6	5	0.999905	0.002501
70	17	6	6	1.000000	0.000094
70	17	7	0	0.128584	0.128584
70	17	7	1	0.454147	0.325563
70	17	7	2	0.779711	0.325563
70	17	7	3	0.945814	0.166104
70	17	7	4	0.992323	0.046509
70	17	7	5	0.999436	0.007113
70	17	7	6	0.999984	0.000547
70	17	7	7	1.000000	0.000016
70	17	8	0	0.093887	0.093887
70	17	8	1	0.371464	0.277578
70	17	8	2	0.702196	0.330731
70	17	8	3	0.908902	0.206707
70	17	8	4	0.982726	0.073824
70	17	8	5	0.998082	0.015355
70	17	8	6	0.999888	0.001807
70	17	8	7	0.999997	0.000109
70	17	8	8	1.000000	0.000003
70	17	9	0	0.068143	0.068143
70	17	9	1	0.299831	0.231688
70	17	9	2	0.622180	0.322348
70	17	9	3	0.862227	0.240047
70	17	9	4	0.967247	0.105020
70	17	9	5	0.995110	0.027863
70	17	9	6	0.999568	0.004458
70	17	9	7	0.999980	0.000412
70	17	9	8	0.999999	0.000020
70	17	9	9	1.000000	0.000000
70	17	10	0	0.049153	0.049153
70	17	10	1	0.239061	0.189908
70	17	10	2	0.542914	0.303853
70	17	10	3	0.807134	0.264220
70	17	10	4	0.944866	0.137732
70	17	10	5	0.989628	0.044763
70	17	10	6	0.998764	0.009135
70	17	10	7	0.999912	0.001148
70	17	10	8	0.999997	0.000084
70	17	10	9	1.000000	0.000003
70	17	10	10	1.000000	0.000000
70	17	11	0	0.035226	0.035226
70	17	11	1	0.188419	0.153193
70	17	11	2	0.466951	0.278532
70	17	11	3	0.745483	0.278532
70	17	11	4	0.915024	0.169541
70	17	11	5	0.980676	0.065652
70	17	11	6	0.997089	0.016413
70	17	11	7	0.999721	0.002632

N	n	k	x	P(x)	p(x)
70	17	11	8	0.999984	0.000263
70	17	11	9	0.999999	0.000015
70	17	11	10	1.000000	0.000000
70	17	11	11	1.000000	0.000000
70	17	12	0	0.025076	0.025076
70	17	12	1	0.146875	0.121799
70	17	12	2	0.396137	0.249262
70	17	12	3	0.679390	0.283253
70	17	12	4	0.877667	0.198277
70	17	12	5	0.967323	0.089656
70	17	12	6	0.994029	0.026706
70	17	12	7	0.999275	0.005246
70	17	12	8	0.999944	0.000669
70	17	12	9	0.999997	0.000054
70	17	12	10	1.000000	0.000003
70	17	12	11	1.000000	0.000000
70	17	12	12	1.000000	0.000000
70	17	13	0	0.017726	0.017726
70	17	13	1	0.113275	0.095549
70	17	13	2	0.331673	0.218398
70	17	13	3	0.611019	0.279346
70	17	13	4	0.833226	0.222207
70	17	13	5	0.948773	0.115548
70	17	13	6	0.988964	0.040190
70	17	13	7	0.998370	0.009406
70	17	13	8	0.999840	0.001470
70	17	13	9	0.999990	0.000150
70	17	13	10	0.999999	0.000010
70	17	13	11	1.000000	0.000000
70	17	13	12	1.000000	0.000000
70	17	13	13	1.000000	0.000000
70	17	14	0	0.012439	0.012439
70	17	14	1	0.086454	0.074015
70	17	14	2	0.274200	0.187745
70	17	14	3	0.542408	0.268208
70	17	14	4	0.782547	0.240139
70	17	14	5	0.924448	0.141901
70	17	14	6	0.981208	0.056760
70	17	14	7	0.996720	0.015512
70	17	14	8	0.999608	0.002888
70	17	14	9	0.999969	0.000361
70	17	14	10	0.999998	0.000029
70	17	14	11	1.000000	0.000001
70	17	14	12	1.000000	0.000000
70	17	14	13	1.000000	0.000000
70	17	14	14	1.000000	0.000000
70	17	15	0	0.008663	0.008663
70	17	15	1	0.065307	0.056644
70	17	15	2	0.223911	0.158604
70	17	15	3	0.475356	0.251445

N	n	k	x	P(x)	p(x)
70	17	15	4	0.726800	0.251445
70	17	15	5	0.894040	0.167240
70	17	15	6	0.970059	0.076018
70	17	15	7	0.993950	0.023891
70	17	15	8	0.999144	0.005194
70	17	15	9	0.999917	0.000774
70	17	15	10	0.999995	0.000077
70	17	15	11	1.000000	0.000005
70	17	15	12	1.000000	0.000000
70	17	15	13	1.000000	0.000000
70	17	15	14	1.000000	0.000000
70	17	15	15	1.000000	0.000000
70	17	16	0	0.005985	0.005985
70	17	16	1	0.048829	0.042844
70	17	16	2	0.180655	0.131826
70	17	16	3	0.411352	0.230696
70	17	16	4	0.667368	0.256016
70	17	16	5	0.857552	0.190184
70	17	16	6	0.954855	0.097303
70	17	16	7	0.989606	0.034751
70	17	16	8	0.998294	0.008688
70	17	16	9	0.999805	0.001511
70	17	16	10	0.999985	0.000180
70	17	16	11	0.999999	0.000014
70	17	16	12	1.000000	0.000001
70	17	16	13	1.000000	0.000000
70	17	16	14	1.000000	0.000000
70	17	16	15	1.000000	0.000000
70	17	16	16	1.000000	0.000000
70	17	17	0	0.004101	0.004101
70	17	17	1	0.036135	0.032033
70	17	17	2	0.144037	0.107902
70	17	17	3	0.351541	0.207504
70	17	17	4	0.605734	0.254193
70	17	17	5	0.815289	0.209554
70	17	17	6	0.935034	0.119745
70	17	17	7	0.983171	0.048137
70	17	17	8	0.996846	0.013675
70	17	17	9	0.999581	0.002735
70	17	17	10	0.999961	0.000381
70	17	17	11	0.999998	0.000036
70	17	17	12	1.000000	0.000002
70	17	17	13	1.000000	0.000000
70	17	17	14	1.000000	0.000000
70	17	17	15	1.000000	0.000000
70	17	17	16	1.000000	0.000000
70	17	17	17	1.000000	0.000000
70	18	1	0	0.742857	0.742857
70	18	1	1	1.000000	0.257143
70	18	2	0	0.549068	0.549068

N	n	k	x	P(x)	p(x)
70	18	2	1	0.936646	0.387578
70	18	2	2	1.000000	0.063354
70	18	3	0	0.403727	0.403727
70	18	3	1	0.839752	0.436025
70	18	3	2	0.985093	0.145342
70	18	3	3	1.000000	0.014907
70	18	4	0	0.295263	0.295263
70	18	4	1	0.729118	0.433856
70	18	4	2	0.950385	0.221266
70	18	4	3	0.996663	0.046278
70	18	4	4	1.000000	0.003337
70	18	5	0	0.214737	0.214737
70	18	5	1	0.617368	0.402631
70	18	5	2	0.896744	0.279377
70	18	5	3	0.986145	0.089401
70	18	5	4	0.999292	0.013147
70	18	5	5	1.000000	0.000708
70	18	6	0	0.155271	0.155271
70	18	6	1	0.512064	0.356793
70	18	6	2	0.827975	0.315911
70	18	6	3	0.965514	0.137539
70	18	6	4	0.996460	0.030946
70	18	6	5	0.999858	0.003398
70	18	6	6	1.000000	0.000142
70	18	7	0	0.111601	0.111601
70	18	7	1	0.417291	0.305690
70	18	7	2	0.748997	0.331706
70	18	7	3	0.933278	0.184281
70	18	7	4	0.989691	0.056413
70	18	7	5	0.999168	0.009477
70	18	7	6	0.999973	0.000805
70	18	7	7	1.000000	0.000027
70	18	8	0	0.079715	0.079715
70	18	8	1	0.334803	0.255088
70	18	8	2	0.664754	0.329951
70	18	8	3	0.889402	0.224648
70	18	8	4	0.977155	0.087753
70	18	8	5	0.997212	0.020058
70	18	8	6	0.999820	0.002608
70	18	8	7	0.999995	0.000175
70	18	8	8	1.000000	0.000005
70	18	9	0	0.056572	0.056572
70	18	9	1	0.264860	0.208288
70	18	9	2	0.579606	0.314746
70	18	9	3	0.835052	0.255446
70	18	9	4	0.957339	0.122288
70	18	9	5	0.993007	0.035667
70	18	9	6	0.999315	0.006309
70	18	9	7	0.999964	0.000649
70	18	9	8	0.999999	0.000035

N	n	k	x	P(x)	p(x)
70	18	9	9	1.000000	0.000001
70	18	10	0	0.039879	0.039879
70	18	10	1	0.206812	0.166934
70	18	10	2	0.497049	0.290237
70	18	10	3	0.772237	0.275188
70	18	10	4	0.929273	0.157036
70	18	10	5	0.985406	0.056132
70	18	10	6	0.998074	0.012669
70	18	10	7	0.999847	0.001773
70	18	10	8	0.999993	0.000146
70	18	10	9	1.000000	0.000006
70	18	10	10	1.000000	0.000000
70	18	11	0	0.027915	0.027915
70	18	11	1	0.159514	0.131599
70	18	11	2	0.419653	0.260138
70	18	11	3	0.703440	0.283787
70	18	11	4	0.892632	0.189192
70	18	11	5	0.973244	0.080612
70	18	11	6	0.995540	0.022297
70	18	11	7	0.999522	0.003982
70	18	11	8	0.999969	0.000447
70	18	11	9	0.999999	0.000030
70	18	11	10	1.000000	0.000001
70	18	11	11	1.000000	0.000000
70	18	12	0	0.019399	0.019399
70	18	12	1	0.121596	0.102197
70	18	12	2	0.349107	0.227511
70	18	12	3	0.631291	0.282184
70	18	12	4	0.847739	0.216448
70	18	12	5	0.955482	0.107743
70	18	12	6	0.991006	0.035524
70	18	12	7	0.998780	0.007774
70	18	12	8	0.999893	0.001113
70	18	12	9	0.999994	0.000101
70	18	12	10	1.000000	0.000005
70	18	12	11	1.000000	0.000000
70	18	12	12	1.000000	0.000000
70	18	13	0	0.013378	0.013378
70	18	13	1	0.091642	0.078263
70	18	13	2	0.286345	0.194704
70	18	13	3	0.558312	0.271967
70	18	13	4	0.795493	0.237180
70	18	13	5	0.931332	0.135840
70	18	13	6	0.983656	0.052323
70	18	13	7	0.997305	0.013650
70	18	13	8	0.999701	0.002396
70	18	13	9	0.999979	0.000277
70	18	13	10	0.999999	0.000020
70	18	13	11	1.000000	0.000001
70	18	13	12	1.000000	0.000000

N	n	k	x	P(x)	p(x)
70	18	13	13	1.000000	0.000000
70	18	14	0	0.009154	0.009154
70	18	14	1	0.068300	0.059146
70	18	14	2	0.231691	0.163392
70	18	14	3	0.486742	0.255050
70	18	14	4	0.737238	0.250496
70	18	14	5	0.900351	0.163114
70	18	14	6	0.972640	0.072289
70	18	14	7	0.994671	0.022031
70	18	14	8	0.999281	0.004610
70	18	14	9	0.999935	0.000654
70	18	14	10	0.999996	0.000061
70	18	14	11	1.000000	0.000004
70	18	14	12	1.000000	0.000000
70	18	14	13	1.000000	0.000000
70	18	14	14	1.000000	0.000000
70	18	15	0	0.006211	0.006211
70	18	15	1	0.050345	0.044133
70	18	15	2	0.185008	0.134663
70	18	15	3	0.418425	0.233417
70	18	15	4	0.674614	0.256189
70	18	15	5	0.862486	0.187872
70	18	15	6	0.957150	0.094664
70	18	15	7	0.990344	0.033194
70	18	15	8	0.998458	0.008114
70	18	15	9	0.999830	0.001372
70	18	15	10	0.999987	0.000158
70	18	15	11	0.999999	0.000012
70	18	15	12	1.000000	0.000001
70	18	15	13	1.000000	0.000000
70	18	15	14	1.000000	0.000000
70	18	15	15	1.000000	0.000000
70	18	16	0	0.004179	0.004179
70	18	16	1	0.036704	0.032525
70	18	16	2	0.145833	0.109130
70	18	16	3	0.354766	0.208932
70	18	16	4	0.609402	0.254636
70	18	16	5	0.818080	0.208678
70	18	16	6	0.936496	0.118416
70	18	16	7	0.983705	0.047209
70	18	16	8	0.996982	0.013278
70	18	16	9	0.999605	0.002623
70	18	16	10	0.999964	0.000359
70	18	16	11	0.999998	0.000033
70	18	16	12	1.000000	0.000002
70	18	16	13	1.000000	0.000000
70	18	16	14	1.000000	0.000000
70	18	16	15	1.000000	0.000000
70	18	16	16	1.000000	0.000000
70	18	17	0	0.002786	0.002786

N	n	k	x	P(x)	p(x)
70	18	17	1	0.026464	0.023678
70	18	17	2	0.113499	0.087034
70	18	17	3	0.296729	0.183230
70	18	17	4	0.543385	0.246656
70	18	17	5	0.767842	0.224457
70	18	17	6	0.910181	0.142339
70	18	17	7	0.974088	0.063907
70	18	17	8	0.994524	0.020435
70	18	17	9	0.999168	0.004644
70	18	17	10	0.999911	0.000743
70	18	17	11	0.999994	0.000082
70	18	17	12	1.000000	0.000006
70	18	17	13	1.000000	0.000000
70	18	17	14	1.000000	0.000000
70	18	17	15	1.000000	0.000000
70	18	17	16	1.000000	0.000000
70	18	17	17	1.000000	0.000000
70	18	18	0	0.001840	0.001840
70	18	18	1	0.018869	0.017030
70	18	18	2	0.087224	0.068355
70	18	18	3	0.244871	0.157647
70	18	18	4	0.478231	0.233359
70	18	18	5	0.712787	0.234556
70	18	18	6	0.877953	0.165167
70	18	18	7	0.960825	0.082871
70	18	18	8	0.990668	0.029843
70	18	18	9	0.998379	0.007711
70	18	18	10	0.999799	0.001420
70	18	18	11	0.999983	0.000184
70	18	18	12	0.999999	0.000016
70	18	18	13	1.000000	0.000001
70	18	18	14	1.000000	0.000000
70	18	18	15	1.000000	0.000000
70	18	18	16	1.000000	0.000000
70	18	18	17	1.000000	0.000000
70	18	18	18	1.000000	0.000000
70	19	1	0	0.728571	0.728571
70	19	1	1	1.000000	0.271429
70	19	2	0	0.527950	0.527950
70	19	2	1	0.929192	0.401242
70	19	2	2	1.000000	0.070807
70	19	3	0	0.380435	0.380435
70	19	3	1	0.822981	0.442547
70	19	3	2	0.982298	0.159317
70	19	3	3	1.000000	0.017702
70	19	4	0	0.272550	0.272550
70	19	4	1	0.704088	0.431538
70	19	4	2	0.941874	0.237786
70	19	4	3	0.995773	0.053898
70	19	4	4	1.000000	0.004227

N	n	k	x	P(x)	p(x)
70	19	5	0	0.194089	0.194089
70	19	5	1	0.586396	0.392307
70	19	5	2	0.880626	0.294230
70	19	5	3	0.982706	0.102080
70	19	5	4	0.999039	0.016333
70	19	5	5	1.000000	0.000961
70	19	6	0	0.137355	0.137355
70	19	6	1	0.477757	0.340402
70	19	6	2	0.803674	0.325917
70	19	6	3	0.957579	0.153905
70	19	6	4	0.995270	0.037691
70	19	6	5	0.999793	0.004523
70	19	6	6	1.000000	0.000207
70	19	7	0	0.096578	0.096578
70	19	7	1	0.382019	0.285441
70	19	7	2	0.717102	0.335083
70	19	7	3	0.919103	0.202000
70	19	7	4	0.986436	0.067333
70	19	7	5	0.998804	0.012367
70	19	7	6	0.999958	0.001154
70	19	7	7	1.000000	0.000042
70	19	8	0	0.067451	0.067451
70	19	8	1	0.300464	0.233013
70	19	8	2	0.626683	0.326219
70	19	8	3	0.867801	0.241118
70	19	8	4	0.970404	0.102603
70	19	8	5	0.996055	0.025651
70	19	8	6	0.999720	0.003664
70	19	8	7	0.999992	0.000272
70	19	8	8	1.000000	0.000008
70	19	9	0	0.046781	0.046781
70	19	9	1	0.232815	0.186035
70	19	9	2	0.537236	0.304421
70	19	9	3	0.805577	0.268341
70	19	9	4	0.945581	0.140004
70	19	9	5	0.990263	0.044682
70	19	9	6	0.998951	0.008688
70	19	9	7	0.999939	0.000988
70	19	9	8	0.999998	0.000059
70	19	9	9	1.000000	0.000001
70	19	10	0	0.032210	0.032210
70	19	10	1	0.177920	0.145710
70	19	10	2	0.452397	0.274478
70	19	10	3	0.735192	0.282795
70	19	10	4	0.911154	0.175961
70	19	10	5	0.980008	0.068854
70	19	10	6	0.997100	0.017092
70	19	10	7	0.999745	0.002645
70	19	10	8	0.999988	0.000243
70	19	10	9	1.000000	0.000012

N	n	k	x	P(x)	p(x)
70	19	10	10	1.000000	0.000000
70	19	11	0	0.022010	0.022010
70	19	11	1	0.134207	0.112197
70	19	11	2	0.374629	0.240422
70	19	11	3	0.659780	0.285152
70	19	11	4	0.867163	0.207383
70	19	11	5	0.963942	0.096779
70	19	11	6	0.993397	0.029454
70	19	11	7	0.999216	0.005819
70	19	11	8	0.999943	0.000727
70	19	11	9	0.999998	0.000054
70	19	11	10	1.000000	0.000002
70	19	11	11	1.000000	0.000000
70	19	12	0	0.014922	0.014922
70	19	12	1	0.099977	0.085055
70	19	12	2	0.305355	0.205377
70	19	12	3	0.582451	0.277097
70	19	12	4	0.814439	0.231988
70	19	12	5	0.940978	0.126539
70	19	12	6	0.986907	0.045929
70	19	12	7	0.998032	0.011126
70	19	12	8	0.999808	0.001775
70	19	12	9	0.999988	0.000181
70	19	12	10	1.000000	0.000011
70	19	12	11	1.000000	0.000000
70	19	12	12	1.000000	0.000000
70	19	13	0	0.010034	0.010034
70	19	13	1	0.073581	0.063547
70	19	13	2	0.245158	0.171577
70	19	13	3	0.506011	0.260853
70	19	13	4	0.754442	0.248431
70	19	13	5	0.910434	0.155992
70	19	13	6	0.976612	0.066178
70	19	13	7	0.995730	0.019118
70	19	13	8	0.999471	0.003741
70	19	13	9	0.999957	0.000486
70	19	13	10	0.999998	0.000041
70	19	13	11	1.000000	0.000002
70	19	13	12	1.000000	0.000000
70	19	13	13	1.000000	0.000000
70	19	14	0	0.006689	0.006689
70	19	14	1	0.053513	0.046824
70	19	14	2	0.193986	0.140472
70	19	14	3	0.432789	0.238803
70	19	14	4	0.689065	0.256277
70	19	14	5	0.872120	0.183055
70	19	14	6	0.961519	0.089399
70	19	14	7	0.991706	0.030187
70	19	14	8	0.998749	0.007044
70	19	14	9	0.999872	0.001123

N	n	k	x	P(x)	p(x)
70	19	14	10	0.999991	0.000119
70	19	14	11	1.000000	0.000008
70	19	14	12	1.000000	0.000000
70	19	14	13	1.000000	0.000000
70	19	14	14	1.000000	0.000000
70	19	15	0	0.004420	0.004420
70	19	15	1	0.038463	0.034043
70	19	15	2	0.151342	0.112880
70	19	15	3	0.364559	0.213217
70	19	15	4	0.620420	0.255861
70	19	15	5	0.826356	0.205937
70	19	15	6	0.940766	0.114409
70	19	15	7	0.985237	0.044471
70	19	15	8	0.997366	0.012129
70	19	15	9	0.999671	0.002306
70	19	15	10	0.999972	0.000301
70	19	15	11	0.999998	0.000026
70	19	15	12	1.000000	0.000001
70	19	15	13	1.000000	0.000000
70	19	15	14	1.000000	0.000000
70	19	15	15	1.000000	0.000000
70	19	16	0	0.002893	0.002893
70	19	16	1	0.027321	0.024428
70	19	16	2	0.116452	0.089131
70	19	16	3	0.302533	0.186080
70	19	16	4	0.550640	0.248107
70	19	16	5	0.773936	0.223296
70	19	16	6	0.913723	0.139787
70	19	16	7	0.975534	0.061811
70	19	16	8	0.994940	0.019406
70	19	16	9	0.999252	0.004312
70	19	16	10	0.999923	0.000671
70	19	16	11	0.999995	0.000072
70	19	16	12	1.000000	0.000005
70	19	16	13	1.000000	0.000000
70	19	16	14	1.000000	0.000000
70	19	16	15	1.000000	0.000000
70	19	16	16	1.000000	0.000000
70	19	17	0	0.001875	0.001875
70	19	17	1	0.019178	0.017304
70	19	17	2	0.088393	0.069214
70	19	17	3	0.247398	0.159005
70	19	17	4	0.481721	0.234323
70	19	17	5	0.716045	0.234323
70	19	17	6	0.880071	0.164026
70	19	17	7	0.961798	0.081727
70	19	17	8	0.990987	0.029188
70	19	17	9	0.998454	0.007467
70	19	17	10	0.999811	0.001358
70	19	17	11	0.999984	0.000173

N	n	k	x	P(x)	p(x)
70	19	17	12	0.999999	0.000015
70	19	17	13	1.000000	0.000001
70	19	17	14	1.000000	0.000000
70	19	17	15	1.000000	0.000000
70	19	17	16	1.000000	0.000000
70	19	17	17	1.000000	0.000000
70	19	18	0	0.001203	0.001203
70	19	18	1	0.013302	0.012099
70	19	18	2	0.066192	0.052890
70	19	18	3	0.199396	0.133204
70	19	18	4	0.415403	0.216007
70	19	18	5	0.654148	0.238745
70	19	18	6	0.839838	0.185690
70	19	18	7	0.943294	0.103456
70	19	18	8	0.984929	0.041635
70	19	18	9	0.997045	0.012116
70	19	18	10	0.999581	0.002536
70	19	18	11	0.999958	0.000377
70	19	18	12	0.999997	0.000039
70	19	18	13	1.000000	0.000003
70	19	18	14	1.000000	0.000000
70	19	18	15	1.000000	0.000000
70	19	18	16	1.000000	0.000000
70	19	18	17	1.000000	0.000000
70	19	18	18	1.000000	0.000000
70	19	19	0	0.000763	0.000763
70	19	19	1	0.009114	0.008350
70	19	19	2	0.048901	0.039787
70	19	19	3	0.158410	0.109509
70	19	19	4	0.353093	0.194683
70	19	19	5	0.589870	0.236777
70	19	19	6	0.793415	0.203545
70	19	19	7	0.919420	0.126004
70	19	19	8	0.976121	0.056702
70	19	19	9	0.994715	0.018593
70	19	19	10	0.999142	0.004427
70	19	19	11	0.999900	0.000758
70	19	19	12	0.999992	0.000092
70	19	19	13	0.999999	0.000008
70	19	19	14	1.000000	0.000000
70	19	19	15	1.000000	0.000000
70	19	19	16	1.000000	0.000000
70	19	19	17	1.000000	0.000000
70	19	19	18	1.000000	0.000000
70	19	19	19	1.000000	0.000000
70	20	1	0	0.714286	0.714286
70	20	1	1	1.000000	0.285714
70	20	2	0	0.507246	0.507246
70	20	2	1	0.921325	0.414079
70	20	2	2	1.000000	0.078675

N	n	k	x	P(x)	p(x)
70	20	3	0	0.358056	0.358056
70	20	3	1	0.805627	0.447570
70	20	3	2	0.979174	0.173548
70	20	3	3	1.000000	0.020826
70	20	4	0	0.251174	0.251174
70	20	4	1	0.678704	0.427530
70	20	4	2	0.932550	0.253846
70	20	4	3	0.994716	0.062166
70	20	4	4	1.000000	0.005284
70	20	5	0	0.175061	0.175061
70	20	5	1	0.555627	0.380566
70	20	5	2	0.863319	0.307692
70	20	5	3	0.978703	0.115384
70	20	5	4	0.998719	0.020016
70	20	5	5	1.000000	0.001281
70	20	6	0	0.121196	0.121196
70	20	6	1	0.444384	0.323189
70	20	6	2	0.778112	0.333727
70	20	6	3	0.948526	0.170414
70	20	6	4	0.993792	0.045266
70	20	6	5	0.999704	0.005912
70	20	6	6	1.000000	0.000296
70	20	7	0	0.083322	0.083322
70	20	7	1	0.348438	0.265116
70	20	7	2	0.684251	0.335813
70	20	7	3	0.903259	0.219009
70	20	7	4	0.982475	0.079216
70	20	7	5	0.998318	0.015843
70	20	7	6	0.999935	0.001617
70	20	7	7	1.000000	0.000065
70	20	8	0	0.056871	0.056871
70	20	8	1	0.268482	0.211612
70	20	8	2	0.588304	0.319822
70	20	8	3	0.844162	0.255858
70	20	8	4	0.962357	0.118195
70	20	8	5	0.994546	0.032189
70	20	8	6	0.999576	0.005030
70	20	8	7	0.999987	0.000411
70	20	8	8	1.000000	0.000013
70	20	9	0	0.038525	0.038525
70	20	9	1	0.203634	0.165108
70	20	9	2	0.495453	0.291819
70	20	9	3	0.774007	0.278555
70	20	9	4	0.931855	0.157848
70	20	9	5	0.986759	0.054904
70	20	9	6	0.998440	0.011682
70	20	9	7	0.999900	0.001460
70	20	9	8	0.999997	0.000097
70	20	9	9	1.000000	0.000003
70	20	10	0	0.025894	0.025894

Table for $N = 2$, $n = 1$, through $N = 100$, $n = 50$

N	n	k	x	P(x)	p(x)
70	20	10	1	0.152206	0.126312
70	20	10	2	0.409342	0.257136
70	20	10	3	0.696377	0.287035
70	20	10	4	0.890452	0.194075
70	20	10	5	0.973258	0.082805
70	20	10	6	0.995759	0.022501
70	20	10	7	0.999589	0.003830
70	20	10	8	0.999978	0.000389
70	20	10	9	0.999999	0.000021
70	20	10	10	1.000000	0.000000
70	20	11	0	0.017263	0.017263
70	20	11	1	0.112207	0.094945
70	20	11	2	0.332201	0.219994
70	20	11	3	0.615051	0.282849
70	20	11	4	0.838699	0.223648
70	20	11	5	0.952556	0.113857
70	20	11	6	0.990509	0.037952
70	20	11	7	0.998759	0.008251
70	20	11	8	0.999900	0.001141
70	20	11	9	0.999995	0.000095
70	20	11	10	1.000000	0.000004
70	20	11	11	1.000000	0.000000
70	20	12	0	0.011411	0.011411
70	20	12	1	0.081632	0.070221
70	20	12	2	0.265085	0.183453
70	20	12	3	0.533552	0.268467
70	20	12	4	0.778049	0.244497
70	20	12	5	0.923610	0.145561
70	20	12	6	0.981503	0.057894
70	20	12	7	0.996941	0.015438
70	20	12	8	0.999668	0.002727
70	20	12	9	0.999978	0.000309
70	20	12	10	0.999999	0.000021
70	20	12	11	1.000000	0.000001
70	20	12	12	1.000000	0.000000
70	20	13	0	0.007476	0.007476
70	20	13	1	0.058629	0.051152
70	20	13	2	0.208151	0.149522
70	20	13	3	0.454863	0.246712
70	20	13	4	0.710601	0.255738
70	20	13	5	0.885965	0.175363
70	20	13	6	0.967529	0.081564
70	20	13	7	0.993481	0.025952
70	20	13	8	0.999104	0.005623
70	20	13	9	0.999919	0.000815
70	20	13	10	0.999995	0.000076
70	20	13	11	1.000000	0.000004
70	20	13	12	1.000000	0.000000
70	20	13	13	1.000000	0.000000
70	20	14	0	0.004853	0.004853

N	n	k	x	P(x)	p(x)
70	20	14	1	0.041578	0.036725
70	20	14	2	0.160933	0.119356
70	20	14	3	0.381282	0.220349
70	20	14	4	0.638815	0.257533
70	20	14	5	0.839816	0.201001
70	20	14	6	0.947496	0.107679
70	20	14	7	0.987562	0.040067
70	20	14	8	0.997920	0.010358
70	20	14	9	0.999762	0.001841
70	20	14	10	0.999982	0.000220
70	20	14	11	0.999999	0.000017
70	20	14	12	1.000000	0.000001
70	20	14	13	1.000000	0.000000
70	20	14	14	1.000000	0.000000
70	20	15	0	0.003120	0.003120
70	20	15	1	0.029118	0.025998
70	20	15	2	0.122569	0.093452
70	20	15	3	0.314391	0.191822
70	20	15	4	0.565234	0.250844
70	20	15	5	0.785977	0.220742
70	20	15	6	0.920576	0.134599
70	20	15	7	0.978261	0.057685
70	20	15	8	0.995701	0.017440
70	20	15	9	0.999400	0.003699
70	20	15	10	0.999943	0.000543
70	20	15	11	0.999996	0.000054
70	20	15	12	1.000000	0.000003
70	20	15	13	1.000000	0.000000
70	20	15	14	1.000000	0.000000
70	20	15	15	1.000000	0.000000
70	20	16	0	0.001985	0.001985
70	20	16	1	0.020136	0.018151
70	20	16	2	0.091985	0.071848
70	20	16	3	0.255100	0.163115
70	20	16	4	0.492262	0.237161
70	20	16	5	0.725774	0.233513
70	20	16	6	0.886314	0.160540
70	20	16	7	0.964626	0.078312
70	20	16	8	0.991896	0.027269
70	20	16	9	0.998660	0.006765
70	20	16	10	0.999844	0.001184
70	20	16	11	0.999988	0.000143
70	20	16	12	0.999999	0.000012
70	20	16	13	1.000000	0.000001
70	20	16	14	1.000000	0.000000
70	20	16	15	1.000000	0.000000
70	20	16	16	1.000000	0.000000
70	20	17	0	0.001250	0.001250
70	20	17	1	0.013750	0.012500
70	20	17	2	0.068035	0.054286

N	n	k	x	P(x)	p(x)
70	20	17	3	0.203749	0.135714
70	20	17	4	0.421992	0.218242
70	20	17	5	0.660910	0.238918
70	20	17	6	0.844693	0.183783
70	20	17	7	0.945773	0.101081
70	20	17	8	0.985836	0.040062
70	20	17	9	0.997282	0.011446
70	20	17	10	0.999625	0.002343
70	20	17	11	0.999964	0.000339
70	20	17	12	0.999997	0.000034
70	20	17	13	1.000000	0.000002
70	20	17	14	1.000000	0.000000
70	20	17	15	1.000000	0.000000
70	20	17	16	1.000000	0.000000
70	20	17	17	1.000000	0.000000
70	20	18	0	0.000778	0.000778
70	20	18	1	0.009269	0.008491
70	20	18	2	0.049599	0.040330
70	20	18	3	0.160218	0.110620
70	20	18	4	0.356107	0.195889
70	20	18	5	0.593291	0.237184
70	20	18	6	0.796146	0.202855
70	20	18	7	0.920980	0.124834
70	20	18	8	0.976765	0.055785
70	20	18	9	0.994907	0.018141
70	20	18	10	0.999183	0.004276
70	20	18	11	0.999906	0.000723
70	20	18	12	0.999992	0.000086
70	20	18	13	0.999999	0.000007
70	20	18	14	1.000000	0.000000
70	20	18	15	1.000000	0.000000
70	20	18	16	1.000000	0.000000
70	20	18	17	1.000000	0.000000
70	20	18	18	1.000000	0.000000
70	20	19	0	0.000479	0.000479
70	20	19	1	0.006167	0.005688
70	20	19	2	0.035638	0.029472
70	20	19	3	0.124054	0.088416
70	20	19	4	0.295834	0.171779
70	20	19	5	0.524873	0.229039
70	20	19	6	0.741531	0.216659
70	20	19	7	0.889772	0.148240
70	20	19	8	0.963892	0.074120
70	20	19	9	0.991069	0.027177
70	20	19	10	0.998361	0.007291
70	20	19	11	0.999781	0.001420
70	20	19	12	0.999979	0.000198
70	20	19	13	0.999999	0.000019
70	20	19	14	1.000000	0.000001
70	20	19	15	1.000000	0.000000

N	n	k	x	P(x)	p(x)
70	20	19	16	1.000000	0.000000
70	20	19	17	1.000000	0.000000
70	20	19	18	1.000000	0.000000
70	20	19	19	1.000000	0.000000
70	20	20	0	0.000291	0.000291
70	20	20	1	0.004048	0.003756
70	20	20	2	0.025237	0.021189
70	20	20	3	0.094582	0.069346
70	20	20	4	0.241942	0.147360
70	20	20	5	0.457508	0.215566
70	20	20	6	0.682057	0.224548
70	20	20	7	0.851985	0.169928
70	20	20	8	0.946452	0.094467
70	20	20	9	0.985207	0.038756
70	20	20	10	0.996931	0.011724
70	20	20	11	0.999530	0.002599
70	20	20	12	0.999948	0.000418
70	20	20	13	0.999996	0.000048
70	20	20	14	1.000000	0.000004
70	20	20	15	1.000000	0.000000
70	20	20	16	1.000000	0.000000
70	20	20	17	1.000000	0.000000
70	20	20	18	1.000000	0.000000
70	20	20	19	1.000000	0.000000
70	20	20	20	1.000000	0.000000
70	21	1	0	0.700000	0.700000
70	21	1	1	1.000000	0.300000
70	21	2	0	0.486956	0.486956
70	21	2	1	0.913043	0.426087
70	21	2	2	1.000000	0.086957
70	21	3	0	0.336573	0.336573
70	21	3	1	0.787724	0.451151
70	21	3	2	0.975703	0.187980
70	21	3	3	1.000000	0.024297
70	21	4	0	0.231080	0.231080
70	21	4	1	0.653052	0.421972
70	21	4	2	0.922396	0.269344
70	21	4	3	0.993472	0.071077
70	21	4	4	1.000000	0.006527
70	21	5	0	0.157554	0.157554
70	21	5	1	0.525182	0.367627
70	21	5	2	0.844857	0.319676
70	21	5	3	0.974088	0.129231
70	21	5	4	0.998319	0.024231
70	21	5	5	1.000000	0.001681
70	21	6	0	0.106652	0.106652
70	21	6	1	0.412065	0.305413
70	21	6	2	0.751414	0.339348
70	21	6	3	0.938301	0.186887
70	21	6	4	0.991981	0.053680

Table for $N = 2$, $n = 1$, through $N = 100$, $n = 50$

N	n	k	x	P(x)	p(x)
70	21	6	5	0.999586	0.007605
70	21	6	6	1.000000	0.000414
70	21	7	0	0.071657	0.071657
70	21	7	1	0.316624	0.244967
70	21	7	2	0.650670	0.334046
70	21	7	3	0.885739	0.235069
70	21	7	4	0.977723	0.091984
70	21	7	5	0.997685	0.019962
70	21	7	6	0.999903	0.002218
70	21	7	7	1.000000	0.000097
70	21	8	0	0.047771	0.047771
70	21	8	1	0.238857	0.191085
70	21	8	2	0.549926	0.311069
70	21	8	3	0.818576	0.268651
70	21	8	4	0.952901	0.134325
70	21	8	5	0.992615	0.039714
70	21	8	6	0.999375	0.006760
70	21	8	7	0.999978	0.000604
70	21	8	8	1.000000	0.000022
70	21	9	0	0.031591	0.031591
70	21	9	1	0.177216	0.145625
70	21	9	2	0.454598	0.277382
70	21	9	3	0.740581	0.285983
70	21	9	4	0.916070	0.175489
70	21	9	5	0.982366	0.066296
70	21	9	6	0.997739	0.015373
70	21	9	7	0.999842	0.002103
70	21	9	8	0.999995	0.000153
70	21	9	9	1.000000	0.000005
70	21	10	0	0.020715	0.020715
70	21	10	1	0.129470	0.108755
70	21	10	2	0.368200	0.238730
70	21	10	3	0.656193	0.287992
70	21	10	4	0.867163	0.210971
70	21	10	5	0.964977	0.097814
70	21	10	6	0.993959	0.028982
70	21	10	7	0.999359	0.005400
70	21	10	8	0.999963	0.000603
70	21	10	9	0.999999	0.000036
70	21	10	10	1.000000	0.000001
70	21	11	0	0.013465	0.013465
70	21	11	1	0.093218	0.079754
70	21	11	2	0.292602	0.199384
70	21	11	3	0.569795	0.277192
70	21	11	4	0.807388	0.237593
70	21	11	5	0.938894	0.131505
70	21	11	6	0.986714	0.047820
70	21	11	7	0.998099	0.011386
70	21	11	8	0.999832	0.001733
70	21	11	9	0.999992	0.000160

N	n	k	x	P(x)	p(x)
70	21	11	10	1.000000	0.000008
70	21	11	11	1.000000	0.000000
70	21	12	0	0.008672	0.008672
70	21	12	1	0.066183	0.057511
70	21	12	2	0.228394	0.162211
70	21	12	3	0.485228	0.256834
70	21	12	4	0.738929	0.253702
70	21	12	5	0.903231	0.164302
70	21	12	6	0.974556	0.071325
70	21	12	7	0.995398	0.020842
70	21	12	8	0.999450	0.004053
70	21	12	9	0.999959	0.000509
70	21	12	10	0.999998	0.000039
70	21	12	11	1.000000	0.000002
70	21	12	12	1.000000	0.000000
70	21	13	0	0.005532	0.005532
70	21	13	1	0.046352	0.040820
70	21	13	2	0.175256	0.128904
70	21	13	3	0.405521	0.230265
70	21	13	4	0.664568	0.259048
70	21	13	5	0.857907	0.193338
70	21	13	6	0.956110	0.098203
70	21	13	7	0.990367	0.034257
70	21	13	8	0.998542	0.008175
70	21	13	9	0.999854	0.001312
70	21	13	10	0.999991	0.000137
70	21	13	11	1.000000	0.000009
70	21	13	12	1.000000	0.000000
70	21	13	13	1.000000	0.000000
70	21	14	0	0.003494	0.003494
70	21	14	1	0.032029	0.028535
70	21	14	2	0.132288	0.100259
70	21	14	3	0.332806	0.200518
70	21	14	4	0.587309	0.254503
70	21	14	5	0.803636	0.216328
70	21	14	6	0.930267	0.126631
70	21	14	7	0.981953	0.051686
70	21	14	8	0.996677	0.014724
70	21	14	9	0.999578	0.002900
70	21	14	10	0.999964	0.000387
70	21	14	11	0.999998	0.000034
70	21	14	12	1.000000	0.000002
70	21	14	13	1.000000	0.000000
70	21	14	14	1.000000	0.000000
70	21	15	0	0.002184	0.002184
70	21	15	1	0.021838	0.019654
70	21	15	2	0.098272	0.076434
70	21	15	3	0.268354	0.170082
70	21	15	4	0.510049	0.241695
70	21	15	5	0.741828	0.231780

N	n	k	x	P(x)	p(x)
70	21	15	6	0.896348	0.154520
70	21	15	7	0.969031	0.072683
70	21	15	8	0.993259	0.024228
70	21	15	9	0.998956	0.005697
70	21	15	10	0.999888	0.000932
70	21	15	11	0.999992	0.000104
70	21	15	12	1.000000	0.000008
70	21	15	13	1.000000	0.000000
70	21	15	14	1.000000	0.000000
70	21	15	15	1.000000	0.000000
70	21	16	0	0.001350	0.001350
70	21	16	1	0.014691	0.013341
70	21	16	2	0.071867	0.057176
70	21	16	3	0.212690	0.140823
70	21	16	4	0.435343	0.222653
70	21	16	5	0.674401	0.239059
70	21	16	6	0.854206	0.179805
70	21	16	7	0.950530	0.096324
70	21	16	8	0.987533	0.037002
70	21	16	9	0.997713	0.010181
70	21	16	10	0.999702	0.001989
70	21	16	11	0.999973	0.000271
70	21	16	12	0.999998	0.000025
70	21	16	13	1.000000	0.000002
70	21	16	14	1.000000	0.000000
70	21	16	15	1.000000	0.000000
70	21	16	16	1.000000	0.000000
70	21	17	0	0.000825	0.000825
70	21	17	1	0.009750	0.008925
70	21	17	2	0.051750	0.042000
70	21	17	3	0.165749	0.114000
70	21	17	4	0.365249	0.199499
70	21	17	5	0.603569	0.238321
70	21	17	6	0.804260	0.200691
70	21	17	7	0.925557	0.121297
70	21	17	8	0.978625	0.053067
70	21	17	9	0.995451	0.016826
70	21	17	10	0.999297	0.003846
70	21	17	11	0.999923	0.000626
70	21	17	12	0.999994	0.000071
70	21	17	13	1.000000	0.000005
70	21	17	14	1.000000	0.000000
70	21	17	15	1.000000	0.000000
70	21	17	16	1.000000	0.000000
70	21	17	17	1.000000	0.000000
70	21	18	0	0.000498	0.000498
70	21	18	1	0.006382	0.005884
70	21	18	2	0.036693	0.030311
70	21	18	3	0.127033	0.090339
70	21	18	4	0.301258	0.174226

N	n	k	x	P(x)	p(x)
70	21	18	5	0.531623	0.230365
70	21	18	6	0.747461	0.215838
70	21	18	7	0.893517	0.146056
70	21	18	8	0.965608	0.072092
70	21	18	9	0.991641	0.026033
70	21	18	10	0.998499	0.006857
70	21	18	11	0.999805	0.001306
70	21	18	12	0.999982	0.000177
70	21	18	13	0.999999	0.000017
70	21	18	14	1.000000	0.000001
70	21	18	15	1.000000	0.000000
70	21	18	16	1.000000	0.000000
70	21	18	17	1.000000	0.000000
70	21	18	18	1.000000	0.000000
70	21	19	0	0.000297	0.000297
70	21	19	1	0.004119	0.003822
70	21	19	2	0.025618	0.021499
70	21	19	3	0.095761	0.070143
70	21	19	4	0.244300	0.148539
70	21	19	5	0.460742	0.216442
70	21	19	6	0.685200	0.224458
70	21	19	7	0.854194	0.168994
70	21	19	8	0.947585	0.093391
70	21	19	9	0.985634	0.038048
70	21	19	10	0.997048	0.011414
70	21	19	11	0.999554	0.002506
70	21	19	12	0.999951	0.000398
70	21	19	13	0.999996	0.000045
70	21	19	14	1.000000	0.000003
70	21	19	15	1.000000	0.000000
70	21	19	16	1.000000	0.000000
70	21	19	17	1.000000	0.000000
70	21	19	18	1.000000	0.000000
70	21	19	19	1.000000	0.000000
70	21	20	0	0.000175	0.000175
70	21	20	1	0.002620	0.002445
70	21	20	2	0.017609	0.014988
70	21	20	3	0.071005	0.053396
70	21	20	4	0.194787	0.123782
70	21	20	5	0.392838	0.198052
70	21	20	6	0.619183	0.226345
70	21	20	7	0.807804	0.188620
70	21	20	8	0.923780	0.115976
70	21	20	9	0.976681	0.052901
70	21	20	10	0.994586	0.017905
70	21	20	11	0.999062	0.004476
70	21	20	12	0.999881	0.000819
70	21	20	13	0.999989	0.000108
70	21	20	14	0.999999	0.000010
70	21	20	15	1.000000	0.000001

N	n	k	x	P(x)	p(x)
70	21	20	16	1.000000	0.000000
70	21	20	17	1.000000	0.000000
70	21	20	18	1.000000	0.000000
70	21	20	19	1.000000	0.000000
70	21	20	20	1.000000	0.000000
70	21	21	0	0.000101	0.000101
70	21	21	1	0.001642	0.001541
70	21	21	2	0.011913	0.010271
70	21	21	3	0.051782	0.039869
70	21	21	4	0.152701	0.100919
70	21	21	5	0.329462	0.176761
70	21	21	6	0.551280	0.221818
70	21	21	7	0.754990	0.203710
70	21	21	8	0.893626	0.138636
70	21	21	9	0.963985	0.070359
70	21	21	10	0.990647	0.026662
70	21	21	11	0.998167	0.007520
70	21	21	12	0.999734	0.001567
70	21	21	13	0.999972	0.000238
70	21	21	14	0.999998	0.000026
70	21	21	15	1.000000	0.000002
70	21	21	16	1.000000	0.000000
70	21	21	17	1.000000	0.000000
70	21	21	18	1.000000	0.000000
70	21	21	19	1.000000	0.000000
70	21	21	20	1.000000	0.000000
70	21	21	21	1.000000	0.000000
70	22	1	0	0.685714	0.685714
70	22	1	1	1.000000	0.314286
70	22	2	0	0.467081	0.467081
70	22	2	1	0.904348	0.437267
70	22	2	2	1.000000	0.095652
70	22	3	0	0.315966	0.315966
70	22	3	1	0.769309	0.453343
70	22	3	2	0.971867	0.202558
70	22	3	3	1.000000	0.028133
70	22	4	0	0.212216	0.212216
70	22	4	1	0.627217	0.415001
70	22	4	2	0.911402	0.284185
70	22	4	3	0.992022	0.080620
70	22	4	4	1.000000	0.007978
70	22	5	0	0.141477	0.141477
70	22	5	1	0.495171	0.353694
70	22	5	2	0.825285	0.330114
70	22	5	3	0.968813	0.143528
70	22	5	4	0.997824	0.029011
70	22	5	5	1.000000	0.002176
70	22	6	0	0.093593	0.093593
70	22	6	1	0.380901	0.287308
70	22	6	2	0.723712	0.342811

N	n	k	x	P(x)	p(x)
70	22	6	3	0.926859	0.203147
70	22	6	4	0.989790	0.062931
70	22	6	5	0.999431	0.009641
70	22	6	6	1.000000	0.000569
70	22	7	0	0.061420	0.061420
70	22	7	1	0.286628	0.225208
70	22	7	2	0.616583	0.329955
70	22	7	3	0.866550	0.249966
70	22	7	4	0.972091	0.105541
70	22	7	5	0.996870	0.024779
70	22	7	6	0.999858	0.002988
70	22	7	7	1.000000	0.000142
70	22	8	0	0.039972	0.039972
70	22	8	1	0.211559	0.171587
70	22	8	2	0.511836	0.300277
70	22	8	3	0.791163	0.279327
70	22	8	4	0.941936	0.150773
70	22	8	5	0.990184	0.048247
70	22	8	6	0.999099	0.008915
70	22	8	7	0.999966	0.000867
70	22	8	8	1.000000	0.000034
70	22	9	0	0.025788	0.025788
70	22	9	1	0.153441	0.127652
70	22	9	2	0.414972	0.261531
70	22	9	3	0.705563	0.290591
70	22	9	4	0.898163	0.192601
70	22	9	5	0.976955	0.078791
70	22	9	6	0.996798	0.019844
70	22	9	7	0.999756	0.002958
70	22	9	8	0.999992	0.000236
70	22	9	9	1.000000	0.000008
70	22	10	0	0.016488	0.016488
70	22	10	1	0.109495	0.093007
70	22	10	2	0.329224	0.219729
70	22	10	3	0.615051	0.285827
70	22	10	4	0.841330	0.226279
70	22	10	5	0.954996	0.113666
70	22	10	6	0.991593	0.036597
70	22	10	7	0.999029	0.007436
70	22	10	8	0.999938	0.000909
70	22	10	9	0.999998	0.000060
70	22	10	10	1.000000	0.000002
70	22	11	0	0.010442	0.010442
70	22	11	1	0.076942	0.066500
70	22	11	2	0.255981	0.179039
70	22	11	3	0.524539	0.268558
70	22	11	4	0.773446	0.248907
70	22	11	5	0.922791	0.149344
70	22	11	6	0.981834	0.059043
70	22	11	7	0.997170	0.015336

N	n	k	x	P(x)	p(x)
70	22	11	8	0.999726	0.002556
70	22	11	9	0.999985	0.000259
70	22	11	10	1.000000	0.000014
70	22	11	11	1.000000	0.000000
70	22	12	0	0.006548	0.006548
70	22	12	1	0.053273	0.046724
70	22	12	2	0.195290	0.142017
70	22	12	3	0.438054	0.242764
70	22	12	4	0.697509	0.259454
70	22	12	5	0.879759	0.182251
70	22	12	6	0.965822	0.086063
70	22	12	7	0.993271	0.027449
70	22	12	8	0.999119	0.005848
70	22	12	9	0.999928	0.000809
70	22	12	10	0.999997	0.000069
70	22	12	11	1.000000	0.000003
70	22	12	12	1.000000	0.000000
70	22	13	0	0.004065	0.004065
70	22	13	1	0.036355	0.032291
70	22	13	2	0.146318	0.109963
70	22	13	3	0.358528	0.212210
70	22	13	4	0.616988	0.258460
70	22	13	5	0.826341	0.209353
70	22	13	6	0.942081	0.115740
70	22	13	7	0.986172	0.044091
70	22	13	8	0.997708	0.011536
70	22	13	9	0.999747	0.002039
70	22	13	10	0.999982	0.000236
70	22	13	11	0.999999	0.000017
70	22	13	12	1.000000	0.000001
70	22	13	13	1.000000	0.000000
70	22	14	0	0.002496	0.002496
70	22	14	1	0.024459	0.021963
70	22	14	2	0.107735	0.083276
70	22	14	3	0.287791	0.180057
70	22	14	4	0.535369	0.247578
70	22	14	5	0.763902	0.228533
70	22	14	6	0.909592	0.145690
70	22	14	7	0.974569	0.064977
70	22	14	8	0.994874	0.020305
70	22	14	9	0.999282	0.004407
70	22	14	10	0.999933	0.000651
70	22	14	11	0.999996	0.000063
70	22	14	12	1.000000	0.000004
70	22	14	13	1.000000	0.000000
70	22	14	14	1.000000	0.000000
70	22	15	0	0.001515	0.001515
70	22	15	1	0.016223	0.014707
70	22	15	2	0.077993	0.061771
70	22	15	3	0.226701	0.148707

N	n	k	x	P(x)	p(x)
70	22	15	4	0.455791	0.229090
70	22	15	5	0.694526	0.238736
70	22	15	6	0.867967	0.173440
70	22	15	7	0.957165	0.089198
70	22	15	8	0.989798	0.032633
70	22	15	9	0.998259	0.008460
70	22	15	10	0.999793	0.001535
70	22	15	11	0.999984	0.000190
70	22	15	12	0.999999	0.000015
70	22	15	13	1.000000	0.000001
70	22	15	14	1.000000	0.000000
70	22	15	15	1.000000	0.000000
70	22	16	0	0.000909	0.000909
70	22	16	1	0.010607	0.009698
70	22	16	2	0.055531	0.044924
70	22	16	3	0.175329	0.119798
70	22	16	4	0.380816	0.205487
70	22	16	5	0.620735	0.239920
70	22	16	6	0.817511	0.196776
70	22	16	7	0.932838	0.115327
70	22	16	8	0.981491	0.048653
70	22	16	9	0.996259	0.014767
70	22	16	10	0.999458	0.003200
70	22	16	11	0.999945	0.000487
70	22	16	12	0.999996	0.000051
70	22	16	13	1.000000	0.000003
70	22	16	14	1.000000	0.000000
70	22	16	15	1.000000	0.000000
70	22	16	16	1.000000	0.000000
70	22	17	0	0.000539	0.000539
70	22	17	1	0.006836	0.006297
70	22	17	2	0.038893	0.032057
70	22	17	3	0.133178	0.094285
70	22	17	4	0.312320	0.179142
70	22	17	5	0.545205	0.232885
70	22	17	6	0.759207	0.214002
70	22	17	7	0.900803	0.141595
70	22	17	8	0.968878	0.068075
70	22	17	9	0.992704	0.023826
70	22	17	10	0.998747	0.006044
70	22	17	11	0.999846	0.001099
70	22	17	12	0.999987	0.000141
70	22	17	13	0.999999	0.000012
70	22	17	14	1.000000	0.000001
70	22	17	15	1.000000	0.000000
70	22	17	16	1.000000	0.000000
70	22	17	17	1.000000	0.000000
70	22	18	0	0.000315	0.000315
70	22	18	1	0.004341	0.004026
70	22	18	2	0.026796	0.022455

Table for $N=2$, $n=1$, through $N=100$, $n=50$

N	n	k	x	P(x)	p(x)
70	22	18	3	0.099378	0.072582
70	22	18	4	0.251479	0.152102
70	22	18	5	0.470506	0.219027
70	22	18	6	0.694603	0.224097
70	22	18	7	0.860729	0.166126
70	22	18	8	0.950896	0.090167
70	22	18	9	0.986860	0.035964
70	22	18	10	0.997379	0.010519
70	22	18	11	0.999618	0.002239
70	22	18	12	0.999960	0.000342
70	22	18	13	0.999997	0.000037
70	22	18	14	1.000000	0.000003
70	22	18	15	1.000000	0.000000
70	22	18	16	1.000000	0.000000
70	22	18	17	1.000000	0.000000
70	22	18	18	1.000000	0.000000
70	22	19	0	0.000182	0.000182
70	22	19	1	0.002715	0.002533
70	22	19	2	0.018159	0.015444
70	22	19	3	0.072857	0.054698
70	22	19	4	0.198829	0.125972
70	22	19	5	0.398901	0.200072
70	22	19	6	0.625650	0.226749
70	22	19	7	0.812808	0.187158
70	22	19	8	0.926620	0.113812
70	22	19	9	0.977869	0.051249
70	22	19	10	0.994951	0.017083
70	22	19	11	0.999145	0.004193
70	22	19	12	0.999895	0.000750
70	22	19	13	0.999991	0.000096
70	22	19	14	0.999999	0.000009
70	22	19	15	1.000000	0.000001
70	22	19	16	1.000000	0.000000
70	22	19	17	1.000000	0.000000
70	22	19	18	1.000000	0.000000
70	22	19	19	1.000000	0.000000
70	22	20	0	0.000103	0.000103
70	22	20	1	0.001672	0.001569
70	22	20	2	0.012103	0.010431
70	22	20	3	0.052480	0.040377
70	22	20	4	0.154368	0.101889
70	22	20	5	0.332211	0.177842
70	22	20	6	0.554513	0.222303
70	22	20	7	0.757761	0.203248
70	22	20	8	0.895377	0.137616
70	22	20	9	0.964805	0.069428
70	22	20	10	0.990932	0.026127
70	22	20	11	0.998240	0.007308
70	22	20	12	0.999748	0.001507
70	22	20	13	0.999974	0.000226

N	n	k	x	P(x)	p(x)
70	22	20	14	0.999998	0.000024
70	22	20	15	1.000000	0.000002
70	22	20	16	1.000000	0.000000
70	22	20	17	1.000000	0.000000
70	22	20	18	1.000000	0.000000
70	22	20	19	1.000000	0.000000
70	22	20	20	1.000000	0.000000
70	22	21	0	0.000058	0.000058
70	22	21	1	0.001013	0.000955
70	22	21	2	0.007930	0.006917
70	22	21	3	0.037136	0.029206
70	22	21	4	0.117688	0.080552
70	22	21	5	0.271744	0.154056
70	22	21	6	0.483376	0.211632
70	22	21	7	0.696787	0.213411
70	22	21	8	0.856845	0.160058
70	22	21	9	0.946754	0.089909
70	22	21	10	0.984662	0.037908
70	22	21	11	0.996632	0.011971
70	22	21	12	0.999446	0.002814
70	22	21	13	0.999933	0.000487
70	22	21	14	0.999994	0.000061
70	22	21	15	1.000000	0.000005
70	22	21	16	1.000000	0.000000
70	22	21	17	1.000000	0.000000
70	22	21	18	1.000000	0.000000
70	22	21	19	1.000000	0.000000
70	22	21	20	1.000000	0.000000
70	22	21	21	1.000000	0.000000
70	22	22	0	0.000032	0.000032
70	22	22	1	0.000604	0.000572
70	22	22	2	0.005107	0.004503
70	22	22	3	0.025812	0.020705
70	22	22	4	0.088098	0.062286
70	22	22	5	0.218296	0.130198
70	22	22	6	0.414272	0.195976
70	22	22	7	0.631457	0.217185
70	22	22	8	0.811114	0.179657
70	22	22	9	0.922901	0.111786
70	22	22	10	0.975378	0.052478
70	22	22	11	0.993945	0.018567
70	22	22	12	0.998872	0.004927
70	22	22	13	0.999844	0.000972
70	22	22	14	0.999984	0.000141
70	22	22	15	0.999999	0.000015
70	22	22	16	1.000000	0.000001
70	22	22	17	1.000000	0.000000
70	22	22	18	1.000000	0.000000
70	22	22	19	1.000000	0.000000
70	22	22	20	1.000000	0.000000

N	n	k	x	P(x)	p(x)
70	22	22	21	1.000000	0.000000
70	22	22	22	1.000000	0.000000
70	23	1	0	0.671429	0.671429
70	23	1	1	1.000000	0.328571
70	23	2	0	0.447619	0.447619
70	23	2	1	0.895238	0.447619
70	23	2	2	1.000000	0.104762
70	23	3	0	0.296218	0.296218
70	23	3	1	0.750420	0.454202
70	23	3	2	0.967647	0.217227
70	23	3	3	1.000000	0.032353
70	23	4	0	0.194532	0.194532
70	23	4	1	0.601279	0.406748
70	23	4	2	0.899561	0.298282
70	23	4	3	0.990342	0.090781
70	23	4	4	1.000000	0.009658
70	23	5	0	0.126740	0.126740
70	23	5	1	0.465697	0.338956
70	23	5	2	0.804653	0.338956
70	23	5	3	0.962833	0.158180
70	23	5	4	0.997220	0.034387
70	23	5	5	1.000000	0.002780
70	23	6	0	0.081894	0.081894
70	23	6	1	0.350973	0.269079
70	23	6	2	0.695144	0.344171
70	23	6	3	0.914162	0.219018
70	23	6	4	0.987168	0.073006
70	23	6	5	0.999230	0.012062
70	23	6	6	1.000000	0.000770
70	23	7	0	0.052463	0.052463
70	23	7	1	0.258477	0.206014
70	23	7	2	0.582213	0.323736
70	23	7	3	0.845719	0.263506
70	23	7	4	0.965494	0.119775
70	23	7	5	0.995838	0.030343
70	23	7	6	0.999795	0.003958
70	23	7	7	1.000000	0.000205
70	23	8	0	0.033310	0.033310
70	23	8	1	0.186536	0.153226
70	23	8	2	0.474301	0.287765
70	23	8	3	0.762066	0.287765
70	23	8	4	0.929372	0.167305
70	23	8	5	0.987168	0.057796
70	23	8	6	0.998727	0.011559
70	23	8	7	0.999948	0.001221
70	23	8	8	1.000000	0.000052
70	23	9	0	0.020953	0.020953
70	23	9	1	0.132165	0.111212
70	23	9	2	0.376832	0.244667
70	23	9	3	0.669239	0.292407

N	n	k	x	P(x)	p(x)
70	23	9	4	0.878101	0.208862
70	23	9	5	0.970388	0.092288
70	23	9	6	0.995558	0.025169
70	23	9	7	0.999633	0.004075
70	23	9	8	0.999987	0.000354
70	23	9	9	1.000000	0.000013
70	23	10	0	0.013053	0.013053
70	23	10	1	0.092056	0.079003
70	23	10	2	0.292603	0.200547
70	23	10	3	0.573368	0.280765
70	23	10	4	0.813045	0.239678
70	23	10	5	0.943156	0.130111
70	23	10	6	0.988543	0.045387
70	23	10	7	0.998564	0.010021
70	23	10	8	0.999900	0.001336
70	23	10	9	0.999997	0.000097
70	23	10	10	1.000000	0.000003
70	23	11	0	0.008049	0.008049
70	23	11	1	0.063088	0.055039
70	23	11	2	0.222411	0.159323
70	23	11	3	0.479779	0.257368
70	23	11	4	0.737147	0.257368
70	23	11	5	0.904123	0.166975
70	23	11	6	0.975684	0.071561
70	23	11	7	0.995892	0.020208
70	23	11	8	0.999566	0.003674
70	23	11	9	0.999974	0.000408
70	23	11	10	0.999999	0.000025
70	23	11	11	1.000000	0.000001
70	23	12	0	0.004911	0.004911
70	23	12	1	0.042565	0.037654
70	23	12	2	0.165703	0.123138
70	23	12	3	0.392536	0.226833
70	23	12	4	0.654266	0.261730
70	23	12	5	0.853181	0.198915
70	23	12	6	0.955064	0.101883
70	23	12	7	0.990412	0.035347
70	23	12	8	0.998632	0.008220
70	23	12	9	0.999878	0.001245
70	23	12	10	0.999994	0.000116
70	23	12	11	1.000000	0.000006
70	23	12	12	1.000000	0.000000
70	23	13	0	0.002964	0.002964
70	23	13	1	0.028283	0.025319
70	23	13	2	0.121119	0.092836
70	23	13	3	0.314318	0.193199
70	23	13	4	0.568527	0.254209
70	23	13	5	0.791449	0.222922
70	23	13	6	0.925202	0.133753
70	23	13	7	0.980661	0.055459

N	n	k	x	P(x)	p(x)
70	23	13	8	0.996506	0.015845
70	23	13	9	0.999577	0.003071
70	23	13	10	0.999968	0.000391
70	23	13	11	0.999998	0.000031
70	23	13	12	1.000000	0.000001
70	23	13	13	1.000000	0.000000
70	23	14	0	0.001768	0.001768
70	23	14	1	0.018510	0.016743
70	23	14	2	0.086916	0.068405
70	23	14	3	0.246528	0.159613
70	23	14	4	0.483790	0.237262
70	23	14	5	0.721053	0.237262
70	23	14	6	0.885311	0.164258
70	23	14	7	0.965093	0.079783
70	23	14	8	0.992336	0.027243
70	23	14	9	0.998823	0.006486
70	23	14	10	0.999879	0.001056
70	23	14	11	0.999992	0.000113
70	23	14	12	1.000000	0.000008
70	23	14	13	1.000000	0.000000
70	23	14	14	1.000000	0.000000
70	23	15	0	0.001042	0.001042
70	23	15	1	0.011933	0.010891
70	23	15	2	0.061264	0.049331
70	23	15	3	0.189524	0.128260
70	23	15	4	0.403291	0.213767
70	23	15	5	0.644790	0.241499
70	23	15	6	0.835447	0.190657
70	23	15	7	0.942298	0.106852
70	23	15	8	0.985039	0.042741
70	23	15	9	0.997201	0.012162
70	23	15	10	0.999633	0.002432
70	23	15	11	0.999968	0.000334
70	23	15	12	0.999998	0.000030
70	23	15	13	1.000000	0.000002
70	23	15	14	1.000000	0.000000
70	23	15	15	1.000000	0.000000
70	23	16	0	0.000606	0.000606
70	23	16	1	0.007576	0.006970
70	23	16	2	0.042428	0.034852
70	23	16	3	0.142884	0.100455
70	23	16	4	0.329444	0.186560
70	23	16	5	0.565754	0.236310
70	23	16	6	0.776516	0.210763
70	23	16	7	0.911214	0.134698
70	23	16	8	0.973383	0.062168
70	23	16	9	0.994105	0.020723
70	23	16	10	0.999058	0.004953
70	23	16	11	0.999895	0.000836
70	23	16	12	0.999992	0.000097

N	n	k	x	P(x)	p(x)
70	23	16	13	0.999999	0.000007
70	23	16	14	1.000000	0.000000
70	23	16	15	1.000000	0.000000
70	23	16	16	1.000000	0.000000
70	23	17	0	0.000348	0.000348
70	23	17	1	0.004737	0.004389
70	23	17	2	0.028875	0.024138
70	23	17	3	0.105678	0.076803
70	23	17	4	0.263803	0.158124
70	23	17	5	0.486984	0.223181
70	23	17	6	0.710165	0.223181
70	23	17	7	0.871304	0.161139
70	23	17	8	0.956114	0.084810
70	23	17	9	0.988733	0.032619
70	23	17	10	0.997866	0.009133
70	23	17	11	0.999709	0.001843
70	23	17	12	0.999972	0.000263
70	23	17	13	0.999998	0.000026
70	23	17	14	1.000000	0.000002
70	23	17	15	1.000000	0.000000
70	23	17	16	1.000000	0.000000
70	23	17	17	1.000000	0.000000
70	23	18	0	0.000197	0.000197
70	23	18	1	0.002915	0.002718
70	23	18	2	0.019311	0.016396
70	23	18	3	0.076696	0.057385
70	23	18	4	0.207116	0.130421
70	23	18	5	0.411186	0.204070
70	23	18	6	0.638579	0.227392
70	23	18	7	0.822658	0.184079
70	23	18	8	0.932111	0.109453
70	23	18	9	0.980116	0.048006
70	23	18	10	0.995626	0.015509
70	23	18	11	0.999292	0.003666
70	23	18	12	0.999918	0.000626
70	23	18	13	0.999993	0.000076
70	23	18	14	1.000000	0.000006
70	23	18	15	1.000000	0.000000
70	23	18	16	1.000000	0.000000
70	23	18	17	1.000000	0.000000
70	23	18	18	1.000000	0.000000
70	23	19	0	0.000110	0.000110
70	23	19	1	0.001765	0.001655
70	23	19	2	0.012689	0.010924
70	23	19	3	0.054625	0.041935
70	23	19	4	0.159463	0.104838
70	23	19	5	0.340547	0.181084
70	23	19	6	0.564239	0.223692
70	23	19	7	0.766018	0.201779
70	23	19	8	0.900538	0.134520

N	n	k	x	P(x)	p(x)
70	23	19	9	0.967192	0.066654
70	23	19	10	0.991748	0.024557
70	23	19	11	0.998446	0.006697
70	23	19	12	0.999785	0.001339
70	23	19	13	0.999979	0.000193
70	23	19	14	0.999998	0.000020
70	23	19	15	1.000000	0.000001
70	23	19	16	1.000000	0.000000
70	23	19	17	1.000000	0.000000
70	23	19	18	1.000000	0.000000
70	23	19	19	1.000000	0.000000
70	23	20	0	0.000060	0.000060
70	23	20	1	0.001051	0.000991
70	23	20	2	0.008191	0.007140
70	23	20	3	0.038179	0.029988
70	23	20	4	0.120405	0.082226
70	23	20	5	0.276635	0.156229
70	23	20	6	0.489675	0.213040
70	23	20	7	0.702715	0.213040
70	23	20	8	0.860973	0.158258
70	23	20	9	0.948895	0.087921
70	23	20	10	0.985489	0.036594
70	23	20	11	0.996870	0.011381
70	23	20	12	0.999496	0.002626
70	23	20	13	0.999941	0.000444
70	23	20	14	0.999995	0.000054
70	23	20	15	1.000000	0.000005
70	23	20	16	1.000000	0.000000
70	23	20	17	1.000000	0.000000
70	23	20	18	1.000000	0.000000
70	23	20	19	1.000000	0.000000
70	23	20	20	1.000000	0.000000
70	23	21	0	0.000033	0.000033
70	23	21	1	0.000615	0.000583
70	23	21	2	0.005192	0.004577
70	23	21	3	0.026184	0.020992
70	23	21	4	0.089160	0.062975
70	23	21	5	0.220392	0.131233
70	23	21	6	0.417241	0.196849
70	23	21	7	0.634542	0.217301
70	23	21	8	0.813496	0.178954
70	23	21	9	0.924277	0.110781
70	23	21	10	0.975974	0.051698
70	23	21	11	0.994138	0.018164
70	23	21	12	0.998918	0.004780
70	23	21	13	0.999852	0.000933
70	23	21	14	0.999985	0.000133
70	23	21	15	0.999999	0.000014
70	23	21	16	1.000000	0.000001
70	23	21	17	1.000000	0.000000

N	n	k	x	P(x)	p(x)
70	23	21	18	1.000000	0.000000
70	23	21	19	1.000000	0.000000
70	23	21	20	1.000000	0.000000
70	23	21	21	1.000000	0.000000
70	23	22	0	0.000017	0.000017
70	23	22	1	0.000354	0.000336
70	23	22	2	0.003231	0.002877
70	23	22	3	0.017616	0.014385
70	23	22	4	0.064740	0.047124
70	23	22	5	0.172184	0.107444
70	23	22	6	0.348947	0.176762
70	23	22	7	0.563587	0.214640
70	23	22	8	0.758714	0.195127
70	23	22	9	0.892625	0.133911
70	23	22	10	0.962259	0.069634
70	23	22	11	0.989690	0.027431
70	23	22	12	0.997845	0.008155
70	23	22	13	0.999661	0.001816
70	23	22	14	0.999961	0.000299
70	23	22	15	0.999997	0.000036
70	23	22	16	1.000000	0.000003
70	23	22	17	1.000000	0.000000
70	23	22	18	1.000000	0.000000
70	23	22	19	1.000000	0.000000
70	23	22	20	1.000000	0.000000
70	23	22	21	1.000000	0.000000
70	23	22	22	1.000000	0.000000
70	23	23	0	0.000009	0.000009
70	23	23	1	0.000199	0.000190
70	23	23	2	0.001972	0.001772
70	23	23	3	0.011622	0.009650
70	23	23	4	0.046087	0.034465
70	23	23	5	0.131893	0.085806
70	23	23	6	0.286343	0.154450
70	23	23	7	0.492040	0.205697
70	23	23	8	0.697737	0.205697
70	23	23	9	0.853568	0.155831
70	23	23	10	0.943400	0.089832
70	23	23	11	0.982832	0.039433
70	23	23	12	0.995976	0.013144
70	23	23	13	0.999283	0.003307
70	23	23	14	0.999905	0.000622
70	23	23	15	0.999991	0.000086
70	23	23	16	0.999999	0.000009
70	23	23	17	1.000000	0.000001
70	23	23	18	1.000000	0.000000
70	23	23	19	1.000000	0.000000
70	23	23	20	1.000000	0.000000
70	23	23	21	1.000000	0.000000
70	23	23	22	1.000000	0.000000

Table for $N = 2$, $n = 1$, through $N = 100$, $n = 50$

N	n	k	x	P(x)	p(x)
70	23	23	23	1.000000	0.000000
70	24	1	0	0.657143	0.657143
70	24	1	1	1.000000	0.342857
70	24	2	0	0.428571	0.428571
70	24	2	1	0.885714	0.457143
70	24	2	2	1.000000	0.114286
70	24	3	0	0.277311	0.277311
70	24	3	1	0.731092	0.453781
70	24	3	2	0.963025	0.231933
70	24	3	3	1.000000	0.036975
70	24	4	0	0.177976	0.177976
70	24	4	1	0.575317	0.397341
70	24	4	2	0.886868	0.311551
70	24	4	3	0.988411	0.101543
70	24	4	4	1.000000	0.011589
70	24	5	0	0.113257	0.113257
70	24	5	1	0.436849	0.323592
70	24	5	2	0.783018	0.346168
70	24	5	3	0.956102	0.173084
70	24	5	4	0.996488	0.040386
70	24	5	5	1.000000	0.003512
70	24	6	0	0.071439	0.071439
70	24	6	1	0.322348	0.250908
70	24	6	2	0.665853	0.343505
70	24	6	3	0.900182	0.234329
70	24	6	4	0.984062	0.083879
70	24	6	5	0.998973	0.014912
70	24	6	6	1.000000	0.001027
70	24	7	0	0.044649	0.044649
70	24	7	1	0.232177	0.187528
70	24	7	2	0.547773	0.315596
70	24	7	3	0.823293	0.275520
70	24	7	4	0.957849	0.134556
70	24	7	5	0.994546	0.036697
70	24	7	6	0.999711	0.005165
70	24	7	7	1.000000	0.000289
70	24	8	0	0.027640	0.027640
70	24	8	1	0.163715	0.136075
70	24	8	2	0.437565	0.273850
70	24	8	3	0.731453	0.293888
70	24	8	4	0.915133	0.183680
70	24	8	5	0.983479	0.068346
70	24	8	6	0.998236	0.014757
70	24	8	7	0.999922	0.001686
70	24	8	8	1.000000	0.000078
70	24	9	0	0.016941	0.016941
70	24	9	1	0.113235	0.096295
70	24	9	2	0.340392	0.227157
70	24	9	3	0.631910	0.291518
70	24	9	4	0.855881	0.223971

N	n	k	x	P(x)	p(x)
70	24	9	5	0.962534	0.106653
70	24	9	6	0.993951	0.031417
70	24	9	7	0.999460	0.005508
70	24	9	8	0.999980	0.000520
70	24	9	9	1.000000	0.000020
70	24	10	0	0.010276	0.010276
70	24	10	1	0.076928	0.066652
70	24	10	2	0.258467	0.181539
70	24	10	3	0.531552	0.273085
70	24	10	4	0.782448	0.250897
70	24	10	5	0.929315	0.146866
70	24	10	6	0.984681	0.055366
70	24	10	7	0.997925	0.013244
70	24	10	8	0.999843	0.001919
70	24	10	9	0.999995	0.000152
70	24	10	10	1.000000	0.000005
70	24	11	0	0.006165	0.006165
70	24	11	1	0.051378	0.045212
70	24	11	2	0.191902	0.140525
70	24	11	3	0.435972	0.244069
70	24	11	4	0.698816	0.262844
70	24	11	5	0.882807	0.183991
70	24	11	6	0.968071	0.085264
70	24	11	7	0.994172	0.026101
70	24	11	8	0.999332	0.005160
70	24	11	9	0.999957	0.000625
70	24	11	10	0.999999	0.000042
70	24	11	11	1.000000	0.000001
70	24	12	0	0.003657	0.003657
70	24	12	1	0.033753	0.030095
70	24	12	2	0.139503	0.105751
70	24	12	3	0.349100	0.209596
70	24	12	4	0.609716	0.260617
70	24	12	5	0.823556	0.213839
70	24	12	6	0.942058	0.118503
70	24	12	7	0.986651	0.044593
70	24	12	8	0.997932	0.011281
70	24	12	9	0.999798	0.001866
70	24	12	10	0.999989	0.000191
70	24	12	11	1.000000	0.000011
70	24	12	12	1.000000	0.000000
70	24	13	0	0.002144	0.002144
70	24	13	1	0.021818	0.019674
70	24	13	2	0.099391	0.077573
70	24	13	3	0.273211	0.173820
70	24	13	4	0.519849	0.246637
70	24	13	5	0.753505	0.233656
70	24	13	6	0.905281	0.151777
70	24	13	7	0.973581	0.068300
70	24	13	8	0.994820	0.021239

N	n	k	x	P(x)	p(x)
70	24	13	9	0.999316	0.004495
70	24	13	10	0.999943	0.000627
70	24	13	11	0.999997	0.000054
70	24	13	12	1.000000	0.000003
70	24	13	13	1.000000	0.000000
70	24	14	0	0.001241	0.001241
70	24	14	1	0.013880	0.012638
70	24	14	2	0.069451	0.055571
70	24	14	3	0.209172	0.139722
70	24	14	4	0.493309	0.224137
70	24	14	5	0.675619	0.242310
70	24	14	6	0.857352	0.181733
70	24	14	7	0.953211	0.095859
70	24	14	8	0.988858	0.035648
70	24	14	9	0.998133	0.009274
70	24	14	10	0.999789	0.001656
70	24	14	11	0.999985	0.000196
70	24	14	12	0.999999	0.000014
70	24	14	13	1.000000	0.000001
70	24	14	14	1.000000	0.000000
70	24	15	0	0.000709	0.000709
70	24	15	1	0.008689	0.007979
70	24	15	2	0.047619	0.038930
70	24	15	3	0.156777	0.109158
70	24	15	4	0.353260	0.196484
70	24	15	5	0.593407	0.240147
70	24	15	6	0.798938	0.205531
70	24	15	7	0.924111	0.125173
70	24	15	8	0.978673	0.054563
70	24	15	9	0.995648	0.016975
70	24	15	10	0.999375	0.003726
70	24	15	11	0.999939	0.000565
70	24	15	12	0.999996	0.000057
70	24	15	13	1.000000	0.000004
70	24	15	14	1.000000	0.000000
70	24	15	15	1.000000	0.000000
70	24	16	0	0.000400	0.000400
70	24	16	1	0.005352	0.004952
70	24	16	2	0.032047	0.026695
70	24	16	3	0.115098	0.083051
70	24	16	4	0.281812	0.166713
70	24	16	5	0.510447	0.228636
70	24	16	6	0.731673	0.221226
70	24	16	7	0.885421	0.153748
70	24	16	8	0.962801	0.077380
70	24	16	9	0.991019	0.028218
70	24	16	10	0.998426	0.007407
70	24	16	11	0.999806	0.001380
70	24	16	12	0.999984	0.000178
70	24	16	13	0.999999	0.000015

N	n	k	x	P(x)	p(x)
70	24	16	14	1.000000	0.000001
70	24	16	15	1.000000	0.000000
70	24	16	16	1.000000	0.000000
70	24	17	0	0.000222	0.000222
70	24	17	1	0.003243	0.003021
70	24	17	2	0.021171	0.017929
70	24	17	3	0.082801	0.061629
70	24	17	4	0.220066	0.137265
70	24	17	5	0.430001	0.209935
70	24	17	6	0.657931	0.227930
70	24	17	7	0.837019	0.179088
70	24	17	8	0.939873	0.102854
70	24	17	9	0.983181	0.043307
70	24	17	10	0.996506	0.013325
70	24	17	11	0.999474	0.002968
70	24	17	12	0.999944	0.000471
70	24	17	13	0.999996	0.000052
70	24	17	14	1.000000	0.000004
70	24	17	15	1.000000	0.000000
70	24	17	16	1.000000	0.000000
70	24	17	17	1.000000	0.000000
70	24	18	0	0.000122	0.000122
70	24	18	1	0.001932	0.001810
70	24	18	2	0.013729	0.011797
70	24	18	3	0.058381	0.044652
70	24	18	4	0.168268	0.109886
70	24	18	5	0.354742	0.186474
70	24	18	6	0.580521	0.225780
70	24	18	7	0.779576	0.199055
70	24	18	8	0.908823	0.129247
70	24	18	9	0.970924	0.062101
70	24	18	10	0.992986	0.022062
70	24	18	11	0.998746	0.005760
70	24	18	12	0.999838	0.001092
70	24	18	13	0.999985	0.000148
70	24	18	14	0.999999	0.000014
70	24	18	15	1.000000	0.000001
70	24	18	16	1.000000	0.000000
70	24	18	17	1.000000	0.000000
70	24	18	18	1.000000	0.000000
70	24	19	0	0.000065	0.000065
70	24	19	1	0.001131	0.001066
70	24	19	2	0.008738	0.007607
70	24	19	3	0.040349	0.031611
70	24	19	4	0.126004	0.085655
70	24	19	5	0.286607	0.160603
70	24	19	6	0.502367	0.215760
70	24	19	7	0.714500	0.212134
70	24	19	8	0.869055	0.154554
70	24	19	9	0.953010	0.083955

N	n	k	x	P(x)	p(x)
70	24	19	10	0.987046	0.034036
70	24	19	11	0.997306	0.010260
70	24	19	12	0.999586	0.002280
70	24	19	13	0.999954	0.000368
70	24	19	14	0.999996	0.000042
70	24	19	15	1.000000	0.000003
70	24	19	16	1.000000	0.000000
70	24	19	17	1.000000	0.000000
70	24	19	18	1.000000	0.000000
70	24	19	19	1.000000	0.000000
70	24	20	0	0.000035	0.000035
70	24	20	1	0.000651	0.000616
70	24	20	2	0.005457	0.004806
70	24	20	3	0.027333	0.021876
70	24	20	4	0.092414	0.065081
70	24	20	5	0.226774	0.134361
70	24	20	6	0.426216	0.199442
70	24	20	7	0.643789	0.217573
70	24	20	8	0.820567	0.176778
70	24	20	9	0.928317	0.107750
70	24	20	10	0.977703	0.049386
70	24	20	11	0.994691	0.016988
70	24	20	12	0.999049	0.004359
70	24	20	13	0.999875	0.000825
70	24	20	14	0.999988	0.000113
70	24	20	15	0.999999	0.000011
70	24	20	16	1.000000	0.000001
70	24	20	17	1.000000	0.000000
70	24	20	18	1.000000	0.000000
70	24	20	19	1.000000	0.000000
70	24	20	20	1.000000	0.000000
70	24	21	0	0.000018	0.000018
70	24	21	1	0.000367	0.000349
70	24	21	2	0.003342	0.002975
70	24	21	3	0.018145	0.014803
70	24	21	4	0.066381	0.048237
70	24	21	5	0.175717	0.109336
70	24	21	6	0.354417	0.178700
70	24	21	7	0.569814	0.215397
70	24	21	8	0.763998	0.194184
70	24	21	9	0.895992	0.131994
70	24	21	10	0.963875	0.067883
70	24	21	11	0.990274	0.026399
70	24	21	12	0.998003	0.007729
70	24	21	13	0.999693	0.001690
70	24	21	14	0.999965	0.000272
70	24	21	15	0.999997	0.000032
70	24	21	16	1.000000	0.000003
70	24	21	17	1.000000	0.000000
70	24	21	18	1.000000	0.000000
70	24	21	19	1.000000	0.000000
70	24	21	20	1.000000	0.000000
70	24	21	21	1.000000	0.000000
70	24	22	0	0.000009	0.000009
70	24	22	1	0.000203	0.000194
70	24	22	2	0.002006	0.001803
70	24	22	3	0.011801	0.009794
70	24	22	4	0.046693	0.034892
70	24	22	5	0.133322	0.086629
70	24	22	6	0.288772	0.155451
70	24	22	7	0.495085	0.206312
70	24	22	8	0.700591	0.205506
70	24	22	9	0.855586	0.154995
70	24	22	10	0.944480	0.088894
70	24	22	11	0.983270	0.038790
70	24	22	12	0.996110	0.012840
70	24	22	13	0.999314	0.003203
70	24	22	14	0.999910	0.000596
70	24	22	15	0.999991	0.000082
70	24	22	16	0.999999	0.000008
70	24	22	17	1.000000	0.000001
70	24	22	18	1.000000	0.000000
70	24	22	19	1.000000	0.000000
70	24	22	20	1.000000	0.000000
70	24	22	21	1.000000	0.000000
70	24	22	22	1.000000	0.000000
70	24	23	0	0.000005	0.000005
70	24	23	1	0.000110	0.000106
70	24	23	2	0.001180	0.001070
70	24	23	3	0.007516	0.006336
70	24	23	4	0.032154	0.024639
70	24	23	5	0.099031	0.066877
70	24	23	6	0.230478	0.131447
70	24	23	7	0.422016	0.191537
70	24	23	8	0.632089	0.210073
70	24	23	9	0.807150	0.175061
70	24	23	10	0.918552	0.111402
70	24	23	11	0.972764	0.054212
70	24	23	12	0.992900	0.020136
70	24	23	13	0.998580	0.005679
70	24	23	14	0.999786	0.001206
70	24	23	15	0.999976	0.000190
70	24	23	16	0.999998	0.000022
70	24	23	17	1.000000	0.000002
70	24	23	18	1.000000	0.000000
70	24	23	19	1.000000	0.000000
70	24	23	20	1.000000	0.000000
70	24	23	21	1.000000	0.000000
70	24	23	22	1.000000	0.000000
70	24	23	23	1.000000	0.000000
70	24	24	0	0.000002	0.000002
70	24	24	1	0.000059	0.000056
70	24	24	2	0.000679	0.000621
70	24	24	3	0.004685	0.004006
70	24	24	4	0.021670	0.016985
70	24	24	5	0.071996	0.050326
70	24	24	6	0.180137	0.108141
70	24	24	7	0.352736	0.172599
70	24	24	8	0.560575	0.207838
70	24	24	9	0.751279	0.190705
70	24	24	10	0.885369	0.134089
70	24	24	11	0.957770	0.072401
70	24	24	12	0.987759	0.029990
70	24	24	13	0.997250	0.009491
70	24	24	14	0.999529	0.002279
70	24	24	15	0.999939	0.000411
70	24	24	16	0.999994	0.000055
70	24	24	17	0.999999	0.000005
70	24	24	18	1.000000	0.000000
70	24	24	19	1.000000	0.000000
70	24	24	20	1.000000	0.000000
70	24	24	21	1.000000	0.000000
70	24	24	22	1.000000	0.000000
70	24	24	23	1.000000	0.000000
70	24	24	24	1.000000	0.000000
70	25	1	0	0.642857	0.642857
70	25	1	1	1.000000	0.357143
70	25	2	0	0.409938	0.409938
70	25	2	1	0.875776	0.465838
70	25	2	2	1.000000	0.124224
70	25	3	0	0.259225	0.259225
70	25	3	1	0.711363	0.452137
70	25	3	2	0.957983	0.246620
70	25	3	3	1.000000	0.042017
70	25	4	0	0.162500	0.162500
70	25	4	1	0.549403	0.386904
70	25	4	2	0.873322	0.323919
70	25	4	3	0.986203	0.112881
70	25	4	4	1.000000	0.013797
70	25	5	0	0.100947	0.100947
70	25	5	1	0.408711	0.307764
70	25	5	2	0.760441	0.351731
70	25	5	3	0.948576	0.188135
70	25	5	4	0.995610	0.047034
70	25	5	5	1.000000	0.004390
70	25	6	0	0.062121	0.062121
70	25	6	1	0.295075	0.232954
70	25	6	2	0.635983	0.340908
70	25	6	3	0.884900	0.248917
70	25	6	4	0.980415	0.095515
70	25	6	5	0.998649	0.018235
70	25	6	6	1.000000	0.001351
70	25	7	0	0.037855	0.037855
70	25	7	1	0.207717	0.169862
70	25	7	2	0.513469	0.305752
70	25	7	3	0.799335	0.285866
70	25	7	4	0.949074	0.149739
70	25	7	5	0.992951	0.043877
70	25	7	6	0.999599	0.006648
70	25	7	7	1.000000	0.000401
70	25	8	0	0.022833	0.022833
70	25	8	1	0.143008	0.120175
70	25	8	2	0.401845	0.258838
70	25	8	3	0.699509	0.297663
70	25	8	4	0.899161	0.199652
70	25	8	5	0.979022	0.079861
70	25	8	6	0.997594	0.018572
70	25	8	7	0.999885	0.002291
70	25	8	8	1.000000	0.000115
70	25	9	0	0.013626	0.013626
70	25	9	1	0.096489	0.082862
70	25	9	2	0.305825	0.209336
70	25	9	3	0.593886	0.288061
70	25	9	4	0.831537	0.237650
70	25	9	5	0.953260	0.121723
70	25	9	6	0.991902	0.038642
70	25	9	7	0.999220	0.007318
70	25	9	8	0.999968	0.000748
70	25	9	9	1.000000	0.000031
70	25	10	0	0.008042	0.008042
70	25	10	1	0.063887	0.055845
70	25	10	2	0.226895	0.163008
70	25	10	3	0.489995	0.263100
70	25	10	4	0.749723	0.259727
70	25	10	5	0.913351	0.163628
70	25	10	6	0.979866	0.066516
70	25	10	7	0.997061	0.017194
70	25	10	8	0.999760	0.002699
70	25	10	9	0.999992	0.000232
70	25	10	10	1.000000	0.000008
70	25	11	0	0.004691	0.004691
70	25	11	1	0.041549	0.036858
70	25	11	2	0.164408	0.122860
70	25	11	3	0.393525	0.229117
70	25	11	4	0.658818	0.265293
70	25	11	5	0.858808	0.199990
70	25	11	6	0.958803	0.099995
70	25	11	7	0.991902	0.033099
70	25	11	8	0.998995	0.007093
70	25	11	9	0.999930	0.000935

Table for $N = 2$, $n = 1$, through $N = 100$, $n = 50$

N	n	k	x	P(x)	p(x)
70	25	11	10	0.999998	0.000068
70	25	11	11	1.000000	0.000002
70	25	12	0	0.002703	0.002703
70	25	12	1	0.026556	0.023853
70	25	12	2	0.116514	0.089958
70	25	12	3	0.308092	0.191578
70	25	12	4	0.564392	0.256300
70	25	12	5	0.791015	0.226623
70	25	12	6	0.926601	0.135586
70	25	12	7	0.981804	0.055203
70	25	12	8	0.996951	0.015147
70	25	12	9	0.999676	0.002725
70	25	12	10	0.999981	0.000304
70	25	12	11	0.999999	0.000019
70	25	12	12	1.000000	0.000000
70	25	13	0	0.001538	0.001538
70	25	13	1	0.016686	0.015148
70	25	13	2	0.080841	0.064155
70	25	13	3	0.235424	0.154583
70	25	13	4	0.471593	0.236169
70	25	13	5	0.712869	0.241275
70	25	13	6	0.882185	0.169316
70	25	13	7	0.964672	0.082487
70	25	13	8	0.992512	0.027839
70	25	13	9	0.998925	0.006413
70	25	13	10	0.999902	0.000977
70	25	13	11	0.999995	0.000093
70	25	13	12	1.000000	0.000005
70	25	13	13	1.000000	0.000000
70	25	14	0	0.000863	0.000863
70	25	14	1	0.010308	0.009444
70	25	14	2	0.054954	0.044646
70	25	14	3	0.175761	0.120807
70	25	14	4	0.384584	0.208823
70	25	14	5	0.628211	0.243627
70	25	14	6	0.825746	0.197535
70	25	14	7	0.938624	0.112877
70	25	14	8	0.984209	0.045585
70	25	14	9	0.997125	0.012916
70	25	14	10	0.999645	0.002520
70	25	14	11	0.999972	0.000327
70	25	14	12	0.999999	0.000027
70	25	14	13	1.000000	0.000001
70	25	14	14	1.000000	0.000000
70	25	15	0	0.000478	0.000478
70	25	15	1	0.006260	0.005782
70	25	15	2	0.036617	0.030357
70	25	15	3	0.128301	0.091684
70	25	15	4	0.306275	0.177974
70	25	15	5	0.541201	0.234926

N	n	k	x	P(x)	p(x)
70	25	15	6	0.758725	0.217524
70	25	15	7	0.902342	0.143616
70	25	15	8	0.970370	0.068029
70	25	15	9	0.993434	0.023064
70	25	15	10	0.998970	0.005535
70	25	15	11	0.999890	0.000921
70	25	15	12	0.999992	0.000102
70	25	15	13	1.000000	0.000007
70	25	15	14	1.000000	0.000000
70	25	15	15	1.000000	0.000000
70	25	16	0	0.000261	0.000261
70	25	16	1	0.003737	0.003476
70	25	16	2	0.023922	0.020185
70	25	16	3	0.091627	0.067705
70	25	16	4	0.238321	0.146694
70	25	16	5	0.455774	0.217452
70	25	16	6	0.683581	0.227807
70	25	16	7	0.855340	0.171759
70	25	16	8	0.949343	0.094003
70	25	16	9	0.986725	0.037381
70	25	16	10	0.997460	0.010735
70	25	16	11	0.999656	0.002196
70	25	16	12	0.999968	0.000312
70	25	16	13	0.999998	0.000030
70	25	16	14	1.000000	0.000002
70	25	16	15	1.000000	0.000000
70	25	16	16	1.000000	0.000000
70	25	17	0	0.000140	0.000140
70	25	17	1	0.002192	0.002052
70	25	17	2	0.015325	0.013133
70	25	17	3	0.064044	0.048719
70	25	17	4	0.181274	0.117230
70	25	17	5	0.375236	0.193962
70	25	17	6	0.603426	0.228191
70	25	17	7	0.798087	0.194661
70	25	17	8	0.919750	0.121663
70	25	17	9	0.975649	0.055899
70	25	17	10	0.994478	0.018829
70	25	17	11	0.999087	0.004609
70	25	17	12	0.999893	0.000806
70	25	17	13	0.999991	0.000098
70	25	17	14	0.999999	0.000008
70	25	17	15	1.000000	0.000000
70	25	17	16	1.000000	0.000000
70	25	17	17	1.000000	0.000000
70	25	18	0	0.000074	0.000074
70	25	18	1	0.001263	0.001189
70	25	18	2	0.009626	0.008363
70	25	18	3	0.043821	0.034195
70	25	18	4	0.134824	0.091003

N	n	k	x	P(x)	p(x)
70	25	18	5	0.302042	0.167218
70	25	18	6	0.521622	0.219580
70	25	18	7	0.731976	0.210354
70	25	18	8	0.880726	0.148750
70	25	18	9	0.958774	0.078048
70	25	18	10	0.989149	0.030375
70	25	18	11	0.997869	0.008720
70	25	18	12	0.999695	0.001826
70	25	18	13	0.999969	0.000274
70	25	18	14	0.999998	0.000029
70	25	18	15	1.000000	0.000002
70	25	18	16	1.000000	0.000000
70	25	18	17	1.000000	0.000000
70	25	18	18	1.000000	0.000000
70	25	19	0	0.000038	0.000038
70	25	19	1	0.000714	0.000676
70	25	19	2	0.005927	0.005213
70	25	19	3	0.029354	0.023427
70	25	19	4	0.098073	0.068719
70	25	19	5	0.237728	0.139655
70	25	19	6	0.441391	0.203663
70	25	19	7	0.659161	0.217770
70	25	19	8	0.832096	0.172935
70	25	19	9	0.934759	0.102663
70	25	19	10	0.980387	0.045628
70	25	19	11	0.995521	0.015135
70	25	19	12	0.999239	0.003717
70	25	19	13	0.999906	0.000667
70	25	19	14	0.999992	0.000086
70	25	19	15	0.999999	0.000008
70	25	19	16	1.000000	0.000000
70	25	19	17	1.000000	0.000000
70	25	19	18	1.000000	0.000000
70	25	19	19	1.000000	0.000000
70	25	20	0	0.000020	0.000020
70	25	20	1	0.000396	0.000377
70	25	20	2	0.003576	0.003180
70	25	20	3	0.019248	0.015672
70	25	20	4	0.069777	0.050529
70	25	20	5	0.182961	0.113184
70	25	20	6	0.365516	0.182555
70	25	20	7	0.582301	0.216785
70	25	20	8	0.774451	0.192150
70	25	20	9	0.902551	0.128100
70	25	20	10	0.966967	0.064416
70	25	20	11	0.991367	0.024400
70	25	20	12	0.998291	0.006924
70	25	20	13	0.999749	0.001458
70	25	20	14	0.999973	0.000224
70	25	20	15	0.999998	0.000025

N	n	k	x	P(x)	p(x)
70	25	20	16	1.000000	0.000002
70	25	20	17	1.000000	0.000000
70	25	20	18	1.000000	0.000000
70	25	20	19	1.000000	0.000000
70	25	20	20	1.000000	0.000000
70	25	21	0	0.000010	0.000010
70	25	21	1	0.000215	0.000206
70	25	21	2	0.002113	0.001898
70	25	21	3	0.012352	0.010239
70	25	21	4	0.048555	0.036202
70	25	21	5	0.137687	0.089133
70	25	21	6	0.296145	0.158458
70	25	21	7	0.504259	0.208113
70	25	21	8	0.709120	0.204861
70	25	21	9	0.861559	0.152439
70	25	21	10	0.947642	0.086083
70	25	21	11	0.984535	0.036893
70	25	21	12	0.996491	0.011956
70	25	21	13	0.999399	0.002908
70	25	21	14	0.999924	0.000525
70	25	21	15	0.999993	0.000069
70	25	21	16	0.999999	0.000006
70	25	21	17	1.000000	0.000000
70	25	21	18	1.000000	0.000000
70	25	21	19	1.000000	0.000000
70	25	21	20	1.000000	0.000000
70	25	21	21	1.000000	0.000000
70	25	22	0	0.000005	0.000005
70	25	22	1	0.000115	0.000110
70	25	22	2	0.001222	0.001108
70	25	22	3	0.007755	0.006533
70	25	22	4	0.033039	0.025284
70	25	22	5	0.101307	0.068267
70	25	22	6	0.234702	0.133396
70	25	22	7	0.427809	0.193106
70	25	22	8	0.638046	0.210237
70	25	22	9	0.811783	0.173737
70	25	22	10	0.921290	0.109507
70	25	22	11	0.973994	0.052704
70	25	22	12	0.993319	0.019325
70	25	22	13	0.998687	0.005368
70	25	22	14	0.999806	0.001119
70	25	22	15	0.999979	0.000173
70	25	22	16	0.999998	0.000019
70	25	22	17	1.000000	0.000002
70	25	22	18	1.000000	0.000000
70	25	22	19	1.000000	0.000000
70	25	22	20	1.000000	0.000000
70	25	22	21	1.000000	0.000000
70	25	22	22	1.000000	0.000000

N	n	k	x	P(x)	p(x)
70	25	23	0	0.000002	0.000002
70	25	23	1	0.000060	0.000057
70	25	23	2	0.000692	0.000632
70	25	23	3	0.004761	0.004069
70	25	23	4	0.021978	0.017217
70	25	23	5	0.072862	0.050884
70	25	23	6	0.181900	0.109038
70	25	23	7	0.355394	0.173494
70	25	23	8	0.563587	0.208193
70	25	23	9	0.753871	0.190284
70	25	23	10	0.887069	0.133199
70	25	23	11	0.958622	0.071553
70	25	23	12	0.988085	0.029463
70	25	23	13	0.997345	0.009260
70	25	23	14	0.999550	0.002205
70	25	23	15	0.999943	0.000393
70	25	23	16	0.999995	0.000052
70	25	23	17	1.000000	0.000005
70	25	23	18	1.000000	0.000000
70	25	23	19	1.000000	0.000000
70	25	23	20	1.000000	0.000000
70	25	23	21	1.000000	0.000000
70	25	23	22	1.000000	0.000000
70	25	23	23	1.000000	0.000000
70	25	24	0	0.000001	0.000001
70	25	24	1	0.000030	0.000029
70	25	24	2	0.000382	0.000352
70	25	24	3	0.002856	0.002474
70	25	24	4	0.014285	0.011429
70	25	24	5	0.051209	0.036924
70	25	24	6	0.137821	0.086612
70	25	24	7	0.288949	0.151129
70	25	24	8	0.488283	0.199334
70	25	24	9	0.689093	0.200810
70	25	24	10	0.844559	0.155466
70	25	24	11	0.937309	0.092750
70	25	24	12	0.979936	0.042627
70	25	24	13	0.994981	0.015045
70	25	24	14	0.999034	0.004053
70	25	24	15	0.999859	0.000826
70	25	24	16	0.999985	0.000126
70	25	24	17	0.999999	0.000014
70	25	24	18	1.000000	0.000001
70	25	24	19	1.000000	0.000000
70	25	24	20	1.000000	0.000000
70	25	24	21	1.000000	0.000000
70	25	24	22	1.000000	0.000000
70	25	24	23	1.000000	0.000000
70	25	24	24	1.000000	0.000000
70	25	25	0	0.000000	0.000000

N	n	k	x	P(x)	p(x)
70	25	25	1	0.000015	0.000015
70	25	25	2	0.000206	0.000191
70	25	25	3	0.001673	0.001467
70	25	25	4	0.009068	0.007394
70	25	25	5	0.035155	0.026088
70	25	25	6	0.102046	0.066891
70	25	25	7	0.229812	0.127766
70	25	25	8	0.414616	0.184804
70	25	25	9	0.619246	0.204630
70	25	25	10	0.793864	0.174617
70	25	25	11	0.909080	0.115217
70	25	25	12	0.967889	0.058809
70	25	25	13	0.991056	0.023167
70	25	25	14	0.998064	0.007009
70	25	25	15	0.999680	0.001615
70	25	25	16	0.999960	0.000280
70	25	25	17	0.999996	0.000036
70	25	25	18	1.000000	0.000003
70	25	25	19	1.000000	0.000000
70	25	25	20	1.000000	0.000000
70	25	25	21	1.000000	0.000000
70	25	25	22	1.000000	0.000000
70	25	25	23	1.000000	0.000000
70	25	25	24	1.000000	0.000000
70	25	25	25	1.000000	0.000000
70	26	1	0	0.628571	0.628571
70	26	1	1	1.000000	0.371429
70	26	2	0	0.391718	0.391718
70	26	2	1	0.865424	0.473706
70	26	2	2	1.000000	0.134576
70	26	3	0	0.241944	0.241944
70	26	3	1	0.691268	0.449324
70	26	3	2	0.952503	0.261235
70	26	3	3	1.000000	0.047497
70	26	4	0	0.148055	0.148055
70	26	4	1	0.523610	0.375554
70	26	4	2	0.858926	0.335316
70	26	4	3	0.983695	0.124769
70	26	4	4	1.000000	0.016305
70	26	5	0	0.089730	0.089730
70	26	5	1	0.381354	0.291624
70	26	5	2	0.736993	0.355639
70	26	5	3	0.940215	0.203222
70	26	5	4	0.994565	0.054350
70	26	5	5	1.000000	0.005435
70	26	6	0	0.053838	0.053838
70	26	6	1	0.269191	0.215353
70	26	6	2	0.605680	0.336489
70	26	6	3	0.868305	0.262625
70	26	6	4	0.976169	0.107864

N	n	k	x	P(x)	p(x)
70	26	6	5	0.998244	0.022074
70	26	6	6	1.000000	0.001756
70	26	7	0	0.031966	0.031966
70	26	7	1	0.185069	0.153102
70	26	7	2	0.479497	0.294428
70	26	7	3	0.773924	0.294428
70	26	7	4	0.939091	0.165167
70	26	7	5	0.991001	0.051910
70	26	7	6	0.999451	0.008450
70	26	7	7	1.000000	0.000549
70	26	8	0	0.018774	0.018774
70	26	8	1	0.124314	0.105540
70	26	8	2	0.367334	0.243020
70	26	8	3	0.666435	0.299101
70	26	8	4	0.881414	0.214979
70	26	8	5	0.973698	0.092284
70	26	8	6	0.996768	0.023071
70	26	8	7	0.999834	0.003066
70	26	8	8	1.000000	0.000165
70	26	9	0	0.010901	0.010901
70	26	9	1	0.081757	0.070856
70	26	9	2	0.273262	0.191504
70	26	9	3	0.555478	0.282216
70	26	9	4	0.805131	0.249653
70	26	9	5	0.942440	0.137309
70	26	9	6	0.989326	0.046886
70	26	9	7	0.998895	0.009569
70	26	9	8	0.999952	0.001057
70	26	9	9	1.000000	0.000048
70	26	10	0	0.006255	0.006255
70	26	10	1	0.052718	0.046463
70	26	10	2	0.197916	0.145198
70	26	10	3	0.449068	0.251153
70	26	10	4	0.715092	0.266024
70	26	10	5	0.895170	0.180078
70	26	10	6	0.973954	0.078784
70	26	10	7	0.995914	0.021961
70	26	10	8	0.999640	0.003725
70	26	10	9	0.999986	0.000347
70	26	10	10	1.000000	0.000013
70	26	11	0	0.003544	0.003544
70	26	11	1	0.033358	0.029814
70	26	11	2	0.139837	0.106478
70	26	11	3	0.352793	0.212957
70	26	11	4	0.617550	0.264757
70	26	11	5	0.832143	0.214592
70	26	11	6	0.947692	0.115550
70	26	11	7	0.988960	0.041268
70	26	11	8	0.998522	0.009562
70	26	11	9	0.999888	0.001366

N	n	k	x	P(x)	p(x)
70	26	11	10	0.999996	0.000108
70	26	11	11	1.000000	0.000004
70	26	12	0	0.001982	0.001982
70	26	12	1	0.020725	0.018743
70	26	12	2	0.096523	0.075798
70	26	12	3	0.269776	0.173253
70	26	12	4	0.518827	0.249051
70	26	12	5	0.755762	0.236935
70	26	12	6	0.908523	0.152761
70	26	12	7	0.975671	0.067148
70	26	12	8	0.995605	0.019934
70	26	12	9	0.999495	0.003890
70	26	12	10	0.999967	0.000472
70	26	12	11	0.999999	0.000032
70	26	12	12	1.000000	0.000001
70	26	13	0	0.001094	0.001094
70	26	13	1	0.012646	0.011553
70	26	13	2	0.065159	0.052512
70	26	13	3	0.201073	0.135914
70	26	13	4	0.424360	0.223287
70	26	13	5	0.669975	0.245616
70	26	13	6	0.855847	0.185871
70	26	13	7	0.953674	0.097827
70	26	13	8	0.989418	0.035745
70	26	13	9	0.998355	0.008936
70	26	13	10	0.999837	0.001482
70	26	13	11	0.999991	0.000154
70	26	13	12	1.000000	0.000009
70	26	13	13	1.000000	0.000000
70	26	14	0	0.000595	0.000595
70	26	14	1	0.007579	0.006985
70	26	14	2	0.043048	0.035469
70	26	14	3	0.146230	0.103182
70	26	14	4	0.338179	0.191949
70	26	14	5	0.579485	0.241307
70	26	14	6	0.790629	0.211143
70	26	14	7	0.921065	0.130436
70	26	14	8	0.978131	0.057066
70	26	14	9	0.995689	0.017559
70	26	14	10	0.999421	0.003731
70	26	14	11	0.999950	0.000529
70	26	14	12	0.999997	0.000047
70	26	14	13	1.000000	0.000002
70	26	14	14	1.000000	0.000000
70	26	15	0	0.000319	0.000319
70	26	15	1	0.004461	0.004143
70	26	15	2	0.027847	0.023386
70	26	15	3	0.103852	0.076004
70	26	15	4	0.262770	0.158918
70	26	15	5	0.488995	0.226225

Table for $N = 2$, $n = 1$, through $N = 100$, $n = 50$

N	n	k	x	P(x)	p(x)
70	26	15	6	0.715221	0.226225
70	26	15	7	0.876810	0.161589
70	26	15	8	0.959788	0.082978
70	26	15	9	0.990359	0.030571
70	26	15	10	0.998355	0.007995
70	26	15	11	0.999808	0.001454
70	26	15	12	0.999986	0.000177
70	26	15	13	0.999999	0.000014
70	26	15	14	1.000000	0.000001
70	26	15	15	1.000000	0.000000
70	26	16	0	0.000168	0.000168
70	26	16	1	0.002578	0.002410
70	26	16	2	0.017643	0.015064
70	26	16	3	0.072068	0.054426
70	26	16	4	0.199203	0.127135
70	26	16	5	0.402619	0.203416
70	26	16	6	0.632957	0.230338
70	26	16	7	0.820988	0.188031
70	26	16	8	0.932632	0.111644
70	26	16	9	0.980910	0.048278
70	26	16	10	0.996029	0.015119
70	26	16	11	0.999412	0.003383
70	26	16	12	0.999941	0.000529
70	26	16	13	0.999996	0.000056
70	26	16	14	1.000000	0.000004
70	26	16	15	1.000000	0.000000
70	26	16	16	1.000000	0.000000
70	26	17	0	0.000087	0.000087
70	26	17	1	0.001462	0.001375
70	26	17	2	0.010947	0.009485
70	26	17	3	0.048887	0.037940
70	26	17	4	0.147407	0.098520
70	26	17	5	0.323512	0.176105
70	26	17	6	0.547646	0.224134
70	26	17	7	0.754829	0.207183
70	26	17	8	0.895417	0.140588
70	26	17	9	0.965711	0.070294
70	26	17	10	0.991549	0.025838
70	26	17	11	0.998472	0.006923
70	26	17	12	0.999803	0.001331
70	26	17	13	0.999983	0.000179
70	26	17	14	0.999999	0.000016
70	26	17	15	1.000000	0.000001
70	26	17	16	1.000000	0.000000
70	26	17	17	1.000000	0.000000
70	26	18	0	0.000044	0.000044
70	26	18	1	0.000814	0.000769
70	26	18	2	0.006652	0.005839
70	26	18	3	0.032423	0.025770
70	26	18	4	0.106512	0.074089
70	26	18	5	0.253735	0.147223
70	26	18	6	0.463067	0.209333
70	26	18	7	0.680556	0.217488
70	26	18	8	0.847670	0.167114
70	26	18	9	0.943164	0.095494
70	26	18	10	0.983749	0.040585
70	26	18	11	0.996513	0.012764
70	26	18	12	0.999452	0.002939
70	26	18	13	0.999939	0.000487
70	26	18	14	0.999995	0.000057
70	26	18	15	1.000000	0.000004
70	26	18	16	1.000000	0.000000
70	26	18	17	1.000000	0.000000
70	26	18	18	1.000000	0.000000
70	26	19	0	0.000022	0.000022
70	26	19	1	0.000444	0.000422
70	26	19	2	0.003958	0.003514
70	26	19	3	0.021024	0.017067
70	26	19	4	0.075166	0.054142
70	26	19	5	0.194279	0.119113
70	26	19	6	0.382555	0.188275
70	26	19	7	0.601089	0.218534
70	26	19	8	0.789823	0.188734
70	26	19	9	0.911945	0.122122
70	26	19	10	0.971261	0.059316
70	26	19	11	0.992831	0.021570
70	26	19	12	0.998660	0.005830
70	26	19	13	0.999817	0.001156
70	26	19	14	0.999982	0.000165
70	26	19	15	0.999999	0.000017
70	26	19	16	1.000000	0.000001
70	26	19	17	1.000000	0.000000
70	26	19	18	1.000000	0.000000
70	26	19	19	1.000000	0.000000
70	26	20	0	0.000011	0.000011
70	26	20	1	0.000237	0.000226
70	26	20	2	0.002304	0.002067
70	26	20	3	0.013327	0.011023
70	26	20	4	0.051811	0.038483
70	26	20	5	0.145233	0.093422
70	26	20	6	0.308721	0.163489
70	26	20	7	0.519674	0.210953
70	26	20	8	0.723211	0.203537
70	26	20	9	0.871238	0.148027
70	26	20	10	0.952652	0.081415
70	26	20	11	0.986487	0.033835
70	26	20	12	0.997060	0.010573
70	26	20	13	0.999522	0.002462
70	26	20	14	0.999943	0.000421
70	26	20	15	0.999995	0.000052
70	26	20	16	1.000000	0.000004
70	26	20	17	1.000000	0.000000
70	26	20	18	1.000000	0.000000
70	26	20	19	1.000000	0.000000
70	26	20	20	1.000000	0.000000
70	26	21	0	0.000005	0.000005
70	26	21	1	0.000124	0.000119
70	26	21	2	0.001312	0.001188
70	26	21	3	0.008257	0.006945
70	26	21	4	0.034878	0.026622
70	26	21	5	0.105996	0.071117
70	26	21	6	0.243326	0.137330
70	26	21	7	0.439512	0.196186
70	26	21	8	0.649938	0.210426
70	26	21	9	0.820909	0.170971
70	26	21	10	0.926600	0.105691
70	26	21	11	0.976336	0.049737
70	26	21	12	0.994100	0.017763
70	26	21	13	0.998882	0.004782
70	26	21	14	0.999842	0.000960
70	26	21	15	0.999984	0.000141
70	26	21	16	0.999999	0.000015
70	26	21	17	1.000000	0.000001
70	26	21	18	1.000000	0.000000
70	26	21	19	1.000000	0.000000
70	26	21	20	1.000000	0.000000
70	26	21	21	1.000000	0.000000
70	26	22	0	0.000002	0.000002
70	26	22	1	0.000063	0.000061
70	26	22	2	0.000730	0.000667
70	26	22	3	0.004997	0.004267
70	26	22	4	0.022926	0.017929
70	26	22	5	0.075517	0.052591
70	26	22	6	0.187273	0.111756
70	26	22	7	0.363440	0.176167
70	26	22	8	0.572639	0.209199
70	26	22	9	0.761592	0.188954
70	26	22	10	0.892088	0.130496
70	26	22	11	0.961111	0.069023
70	26	22	12	0.989024	0.027914
70	26	22	13	0.997613	0.008589
70	26	22	14	0.999607	0.001994
70	26	22	15	0.999952	0.000345
70	26	22	16	0.999996	0.000044
70	26	22	17	1.000000	0.000004
70	26	22	18	1.000000	0.000000
70	26	22	19	1.000000	0.000000
70	26	22	20	1.000000	0.000000
70	26	22	21	1.000000	0.000000
70	26	22	22	1.000000	0.000000
70	26	23	0	0.000001	0.000001
70	26	23	1	0.000032	0.000031
70	26	23	2	0.000397	0.000365
70	26	23	3	0.002952	0.002556
70	26	23	4	0.014708	0.011756
70	26	23	5	0.052508	0.037800
70	26	23	6	0.140708	0.088200
70	26	23	7	0.293707	0.152999
70	26	23	8	0.494189	0.200482
70	26	23	9	0.694671	0.200482
70	26	23	10	0.848589	0.153918
70	26	23	11	0.939541	0.090952
70	26	23	12	0.980883	0.041342
70	26	23	13	0.995287	0.014404
70	26	23	14	0.999109	0.003821
70	26	23	15	0.999873	0.000764
70	26	23	16	0.999987	0.000114
70	26	23	17	0.999999	0.000012
70	26	23	18	1.000000	0.000001
70	26	23	19	1.000000	0.000000
70	26	23	20	1.000000	0.000000
70	26	23	21	1.000000	0.000000
70	26	23	22	1.000000	0.000000
70	26	23	23	1.000000	0.000000
70	26	24	0	0.000000	0.000000
70	26	24	1	0.000015	0.000015
70	26	24	2	0.000210	0.000195
70	26	24	3	0.001702	0.001491
70	26	24	4	0.009206	0.007504
70	26	24	5	0.035619	0.026413
70	26	24	6	0.103176	0.067557
70	26	24	7	0.231856	0.128680
70	26	24	8	0.417409	0.185552
70	26	24	9	0.622156	0.204748
70	26	24	10	0.796192	0.174035
70	26	24	11	0.910514	0.114322
70	26	24	12	0.968568	0.058054
70	26	24	13	0.991303	0.022735
70	26	24	14	0.998133	0.006830
70	26	24	15	0.999694	0.001561
70	26	24	16	0.999962	0.000268
70	26	24	17	0.999997	0.000034
70	26	24	18	1.000000	0.000003
70	26	24	19	1.000000	0.000000
70	26	24	20	1.000000	0.000000
70	26	24	21	1.000000	0.000000
70	26	24	22	1.000000	0.000000
70	26	24	23	1.000000	0.000000
70	26	24	24	1.000000	0.000000
70	26	25	0	0.000000	0.000000

N	n	k	x	P(x)	p(x)
70	26	25	1	0.000007	0.000007
70	26	25	2	0.000109	0.000101
70	26	25	3	0.000956	0.000847
70	26	25	4	0.005617	0.004661
70	26	25	5	0.023561	0.017944
70	26	25	6	0.073803	0.050243
70	26	25	7	0.178706	0.104902
70	26	25	8	0.344801	0.166096
70	26	25	9	0.546489	0.201687
70	26	25	10	0.735658	0.189169
70	26	25	11	0.873235	0.137577
70	26	25	12	0.950900	0.077665
70	26	25	13	0.984878	0.033978
70	26	25	14	0.996351	0.011473
70	26	25	15	0.999321	0.002970
70	26	25	16	0.999904	0.000583
70	26	25	17	0.999990	0.000086
70	26	25	18	0.999999	0.000009
70	26	25	19	1.000000	0.000001
70	26	25	20	1.000000	0.000000
70	26	25	21	1.000000	0.000000
70	26	25	22	1.000000	0.000000
70	26	25	23	1.000000	0.000000
70	26	25	24	1.000000	0.000000
70	26	25	25	1.000000	0.000000
70	26	26	0	0.000000	0.000000
70	26	26	1	0.000003	0.000003
70	26	26	2	0.000055	0.000051
70	26	26	3	0.000523	0.000468
70	26	26	4	0.003338	0.002815
70	26	26	5	0.015187	0.011849
70	26	26	6	0.051473	0.036286
70	26	26	7	0.134414	0.082940
70	26	26	8	0.278363	0.143949
70	26	26	9	0.470296	0.191933
70	26	26	10	0.668398	0.198102
70	26	26	11	0.827376	0.158978
70	26	26	12	0.926737	0.099361
70	26	26	13	0.975062	0.048325
70	26	26	14	0.993292	0.018230
70	26	26	15	0.998595	0.005303
70	26	26	16	0.999774	0.001180
70	26	26	17	0.999973	0.000198
70	26	26	18	0.999997	0.000025
70	26	26	19	1.000000	0.000002
70	26	26	20	1.000000	0.000000
70	26	26	21	1.000000	0.000000
70	26	26	22	1.000000	0.000000
70	26	26	23	1.000000	0.000000
70	26	26	24	1.000000	0.000000

N	n	k	x	P(x)	p(x)
70	26	26	25	1.000000	0.000000
70	26	26	26	1.000000	0.000000
70	27	1	0	0.614286	0.614286
70	27	1	1	1.000000	0.385714
70	27	2	0	0.373913	0.373913
70	27	2	1	0.854658	0.480745
70	27	2	2	1.000000	0.145342
70	27	3	0	0.225448	0.225448
70	27	3	1	0.670844	0.445396
70	27	3	2	0.946566	0.275722
70	27	3	3	1.000000	0.053434
70	27	4	0	0.134596	0.134596
70	27	4	1	0.498004	0.363408
70	27	4	2	0.843684	0.345681
70	27	4	3	0.980859	0.137175
70	27	4	4	1.000000	0.019141
70	27	5	0	0.079534	0.079534
70	27	5	1	0.354843	0.275309
70	27	5	2	0.712745	0.357902
70	27	5	3	0.930977	0.218233
70	27	5	4	0.993330	0.062352
70	27	5	5	1.000000	0.006670
70	27	6	0	0.046497	0.046497
70	27	6	1	0.244719	0.198223
70	27	6	2	0.575090	0.330371
70	27	6	3	0.850399	0.275309
70	27	6	4	0.971267	0.120867
70	27	6	5	0.997742	0.026476
70	27	6	6	1.000000	0.002258
70	27	7	0	0.026881	0.026881
70	27	7	1	0.164191	0.137310
70	27	7	2	0.446039	0.281848
70	27	7	3	0.747158	0.301119
70	27	7	4	0.927830	0.180672
70	27	7	5	0.988641	0.060811
70	27	7	6	0.999259	0.010618
70	27	7	7	1.000000	0.000741
70	27	8	0	0.015360	0.015360
70	27	8	1	0.107523	0.092163
70	27	8	2	0.334195	0.226671
70	27	8	3	0.632446	0.298251
70	27	8	4	0.861870	0.229424
70	27	8	5	0.967406	0.105535
70	27	8	6	0.995720	0.028314
70	27	8	7	0.999765	0.004045
70	27	8	8	1.000000	0.000235
70	27	9	0	0.008671	0.008671
70	27	9	1	0.068874	0.060203
70	27	9	2	0.242795	0.173921
70	27	9	3	0.516994	0.274199

N	n	k	x	P(x)	p(x)
70	27	9	4	0.776761	0.259767
70	27	9	5	0.929958	0.153196
70	27	9	6	0.986130	0.056172
70	27	9	7	0.998460	0.012330
70	27	9	8	0.999928	0.001468
70	27	9	9	1.000000	0.000072
70	27	10	0	0.004833	0.004833
70	27	10	1	0.043214	0.038381
70	27	10	2	0.171516	0.128302
70	27	10	3	0.409112	0.237596
70	27	10	4	0.678816	0.269704
70	27	10	5	0.874706	0.195890
70	27	10	6	0.966792	0.092085
70	27	10	7	0.994417	0.027626
70	27	10	8	0.999471	0.005053
70	27	10	9	0.999979	0.000508
70	27	10	10	1.000000	0.000021
70	27	11	0	0.002658	0.002658
70	27	11	1	0.026582	0.023924
70	27	11	2	0.118057	0.091475
70	27	11	3	0.314074	0.196017
70	27	11	4	0.575430	0.261356
70	27	11	5	0.802880	0.227450
70	27	11	6	0.934562	0.131682
70	27	11	7	0.985209	0.050647
70	27	11	8	0.997870	0.012662
70	27	11	9	0.999826	0.001956
70	27	11	10	0.999994	0.000168
70	27	11	11	1.000000	0.000006
70	27	12	0	0.001442	0.001442
70	27	12	1	0.016040	0.014598
70	27	12	2	0.079297	0.063257
70	27	12	3	0.234338	0.155042
70	27	12	4	0.473545	0.239207
70	27	12	5	0.718068	0.244523
70	27	12	6	0.887692	0.169624
70	27	12	7	0.968040	0.080348
70	27	12	8	0.993793	0.025753
70	27	12	9	0.999229	0.005437
70	27	12	10	0.999945	0.000716
70	27	12	11	0.999998	0.000053
70	27	12	12	1.000000	0.000002
70	27	13	0	0.000771	0.000771
70	27	13	1	0.009496	0.008725
70	27	13	2	0.052031	0.042535
70	27	13	3	0.170183	0.118152
70	27	13	4	0.378687	0.208504
70	27	13	5	0.625318	0.246631
70	27	13	6	0.826276	0.200958
70	27	13	7	0.940334	0.114057

N	n	k	x	P(x)	p(x)
70	27	13	8	0.985357	0.045023
70	27	13	9	0.997542	0.012186
70	27	13	10	0.999736	0.002193
70	27	13	11	0.999984	0.000248
70	27	13	12	0.999999	0.000016
70	27	13	13	1.000000	0.000000
70	27	14	0	0.000406	0.000406
70	27	14	1	0.005516	0.005110
70	27	14	2	0.033375	0.027859
70	27	14	3	0.120435	0.087060
70	27	14	4	0.294554	0.174119
70	27	14	5	0.530127	0.235573
70	27	14	6	0.752239	0.222112
70	27	14	7	0.900314	0.148075
70	27	14	8	0.970349	0.070035
70	27	14	9	0.993694	0.023345
70	27	14	10	0.999081	0.005387
70	27	14	11	0.999914	0.000833
70	27	14	12	0.999995	0.000081
70	27	14	13	1.000000	0.000004
70	27	14	14	1.000000	0.000000
70	27	15	0	0.000210	0.000210
70	27	15	1	0.003143	0.002933
70	27	15	2	0.020938	0.017795
70	27	15	3	0.083123	0.062185
70	27	15	4	0.223041	0.139917
70	27	15	5	0.437581	0.214540
70	27	15	6	0.668947	0.231367
70	27	15	7	0.847430	0.178483
70	27	15	8	0.946587	0.099157
70	27	15	9	0.986190	0.039603
70	27	15	10	0.997446	0.011256
70	27	15	11	0.999676	0.002230
70	27	15	12	0.999974	0.000297
70	27	15	13	0.999999	0.000025
70	27	15	14	1.000000	0.000001
70	27	15	15	1.000000	0.000000
70	27	16	0	0.000107	0.000107
70	27	16	1	0.001757	0.001650
70	27	16	2	0.012849	0.011093
70	27	16	3	0.055988	0.043138
70	27	16	4	0.164530	0.108542
70	27	16	5	0.351765	0.187235
70	27	16	6	0.580607	0.228843
70	27	16	7	0.782527	0.201920
70	27	16	8	0.912333	0.129806
70	27	16	9	0.973229	0.060896
70	27	16	10	0.993967	0.020738
70	27	16	11	0.999027	0.005060
70	27	16	12	0.999892	0.000865

Table for $N = 2$, $n = 1$, through $N = 100$, $n = 50$

N	n	k	x	P(x)	p(x)
70	27	16	13	0.999992	0.000100
70	27	16	14	1.000000	0.000007
70	27	16	15	1.000000	0.000000
70	27	16	16	1.000000	0.000000
70	27	17	0	0.000053	0.000053
70	27	17	1	0.000962	0.000909
70	27	17	2	0.007714	0.006752
70	27	17	3	0.036815	0.029101
70	27	17	4	0.118299	0.081484
70	27	17	5	0.275484	0.157185
70	27	17	6	0.491613	0.216129
70	27	17	7	0.707742	0.216129
70	27	17	8	0.866660	0.158918
70	27	17	9	0.952930	0.086270
70	27	17	10	0.987438	0.034508
70	27	17	11	0.997528	0.010090
70	27	17	12	0.999652	0.002124
70	27	17	13	0.999966	0.000314
70	27	17	14	0.999998	0.000031
70	27	17	15	1.000000	0.000002
70	27	17	16	1.000000	0.000000
70	27	17	17	1.000000	0.000000
70	27	18	0	0.000026	0.000026
70	27	18	1	0.000516	0.000490
70	27	18	2	0.004529	0.004013
70	27	18	3	0.023637	0.019108
70	27	18	4	0.082938	0.059301
70	27	18	5	0.210237	0.127299
70	27	18	6	0.405977	0.195740
70	27	18	7	0.626184	0.220207
70	27	18	8	0.809690	0.183506
70	27	18	9	0.923631	0.113942
70	27	18	10	0.976370	0.052739
70	27	18	11	0.994482	0.018112
70	27	18	12	0.999051	0.004569
70	27	18	13	0.999883	0.000832
70	27	18	14	0.999990	0.000107
70	27	18	15	0.999999	0.000009
70	27	18	16	1.000000	0.000001
70	27	18	17	1.000000	0.000000
70	27	18	18	1.000000	0.000000
70	27	19	0	0.000013	0.000013
70	27	19	1	0.000271	0.000259
70	27	19	2	0.002600	0.002329
70	27	19	3	0.014818	0.012218
70	27	19	4	0.056709	0.041891
70	27	19	5	0.156380	0.099671
70	27	19	6	0.326928	0.170548
70	27	19	7	0.541489	0.214561
70	27	19	8	0.742639	0.201151

N	n	k	x	P(x)	p(x)
70	27	19	9	0.884190	0.141550
70	27	19	10	0.959128	0.074938
70	27	19	11	0.988909	0.029781
70	27	19	12	0.997733	0.008824
70	27	19	13	0.999659	0.001926
70	27	19	14	0.999963	0.000304
70	27	19	15	0.999997	0.000034
70	27	19	16	1.000000	0.000003
70	27	19	17	1.000000	0.000000
70	27	19	18	1.000000	0.000000
70	27	19	19	1.000000	0.000000
70	27	20	0	0.000006	0.000006
70	27	20	1	0.000139	0.000134
70	27	20	2	0.001458	0.001319
70	27	20	3	0.009068	0.007610
70	27	20	4	0.037817	0.028749
70	27	20	5	0.113385	0.075568
70	27	20	6	0.256702	0.143318
70	27	20	7	0.457347	0.200645
70	27	20	8	0.667701	0.210354
70	27	20	9	0.834231	0.166530
70	27	20	10	0.934149	0.099918
70	27	20	11	0.979566	0.045417
70	27	20	12	0.995138	0.015572
70	27	20	13	0.999131	0.003993
70	27	20	14	0.999886	0.000755
70	27	20	15	0.999989	0.000103
70	27	20	16	0.999999	0.000010
70	27	20	17	1.000000	0.000001
70	27	20	18	1.000000	0.000000
70	27	20	19	1.000000	0.000000
70	27	20	20	1.000000	0.000000
70	27	21	0	0.000003	0.000003
70	27	21	1	0.000070	0.000067
70	27	21	2	0.000799	0.000729
70	27	21	3	0.005416	0.004617
70	27	21	4	0.024593	0.019177
70	27	21	5	0.080135	0.055542
70	27	21	6	0.196509	0.116374
70	27	21	7	0.377089	0.180580
70	27	21	8	0.587767	0.210677
70	27	21	9	0.774280	0.186514
70	27	21	10	0.900177	0.125897
70	27	21	11	0.965033	0.064856
70	27	21	12	0.990466	0.025434
70	27	21	13	0.998013	0.007546
70	27	21	14	0.999690	0.001677
70	27	21	15	0.999964	0.000275
70	27	21	16	0.999997	0.000033
70	27	21	17	1.000000	0.000003

N	n	k	x	P(x)	p(x)
70	27	21	18	1.000000	0.000000
70	27	21	19	1.000000	0.000000
70	27	21	20	1.000000	0.000000
70	27	21	21	1.000000	0.000000
70	27	22	0	0.000001	0.000001
70	27	22	1	0.000034	0.000033
70	27	22	2	0.000427	0.000393
70	27	22	3	0.003154	0.002727
70	27	22	4	0.015591	0.012437
70	27	22	5	0.055197	0.039606
70	27	22	6	0.146634	0.091437
70	27	22	7	0.303383	0.156749
70	27	22	8	0.506075	0.202692
70	27	22	9	0.705765	0.199690
70	27	22	10	0.856498	0.150733
70	27	22	11	0.943855	0.087357
70	27	22	12	0.982681	0.038825
70	27	22	13	0.995857	0.013176
70	27	22	14	0.999245	0.003388
70	27	22	15	0.999897	0.000653
70	27	22	16	0.999990	0.000093
70	27	22	17	0.999999	0.000009
70	27	22	18	1.000000	0.000001
70	27	22	19	1.000000	0.000000
70	27	22	20	1.000000	0.000000
70	27	22	21	1.000000	0.000000
70	27	22	22	1.000000	0.000000
70	27	23	0	0.000001	0.000001
70	27	23	1	0.000016	0.000016
70	27	23	2	0.000222	0.000206
70	27	23	3	0.001791	0.001568
70	27	23	4	0.009632	0.007841
70	27	23	5	0.037045	0.027413
70	27	23	6	0.106631	0.069586
70	27	23	7	0.238071	0.131440
70	27	23	8	0.425843	0.187772
70	27	23	9	0.630881	0.205038
70	27	23	10	0.803114	0.172232
70	27	23	11	0.914736	0.111623
70	27	23	12	0.970548	0.055811
70	27	23	13	0.992013	0.021466
70	27	23	14	0.998327	0.006313
70	27	23	15	0.999734	0.001407
70	27	23	16	0.999968	0.000234
70	27	23	17	0.999997	0.000029
70	27	23	18	1.000000	0.000003
70	27	23	19	1.000000	0.000000
70	27	23	20	1.000000	0.000000
70	27	23	21	1.000000	0.000000
70	27	23	22	1.000000	0.000000

N	n	k	x	P(x)	p(x)
70	27	23	23	1.000000	0.000000
70	27	24	0	0.000000	0.000000
70	27	24	1	0.000008	0.000007
70	27	24	2	0.000113	0.000105
70	27	24	3	0.000990	0.000877
70	27	24	4	0.005795	0.004805
70	27	24	5	0.024213	0.018418
70	27	24	6	0.075539	0.051326
70	27	24	7	0.182139	0.106600
70	27	24	8	0.349935	0.167796
70	27	24	9	0.552356	0.202421
70	27	24	10	0.740817	0.188461
70	27	24	11	0.876737	0.135920
70	27	24	12	0.952736	0.075998
70	27	24	13	0.985619	0.032884
70	27	24	14	0.996581	0.010961
70	27	24	15	0.999375	0.002794
70	27	24	16	0.999914	0.000539
70	27	24	17	0.999991	0.000077
70	27	24	18	0.999999	0.000008
70	27	24	19	1.000000	0.000001
70	27	24	20	1.000000	0.000000
70	27	24	21	1.000000	0.000000
70	27	24	22	1.000000	0.000000
70	27	24	23	1.000000	0.000000
70	27	24	24	1.000000	0.000000
70	27	25	0	0.000000	0.000000
70	27	25	1	0.000003	0.000003
70	27	25	2	0.000056	0.000052
70	27	25	3	0.000532	0.000477
70	27	25	4	0.003392	0.002860
70	27	25	5	0.015404	0.012012
70	27	25	6	0.052108	0.036703
70	27	25	7	0.135791	0.083684
70	27	25	8	0.280628	0.144837
70	27	25	9	0.473148	0.192520
70	27	25	10	0.671168	0.198020
70	27	25	11	0.829461	0.158292
70	27	25	12	0.927953	0.098493
70	27	25	13	0.975611	0.047658
70	27	25	14	0.993483	0.017872
70	27	25	15	0.998646	0.005163
70	27	25	16	0.999785	0.001139
70	27	25	17	0.999974	0.000189
70	27	25	18	0.999998	0.000023
70	27	25	19	1.000000	0.000002
70	27	25	20	1.000000	0.000000
70	27	25	21	1.000000	0.000000
70	27	25	22	1.000000	0.000000
70	27	25	23	1.000000	0.000000

Table for $N = 2$, $n = 1$, through $N = 100$, $n = 50$

N	n	k	x	P(x)	p(x)
70	27	25	24	1.000000	0.000000
70	27	25	25	1.000000	0.000000
70	27	26	0	0.000000	0.000000
70	27	26	1	0.000002	0.000001
70	27	26	2	0.000027	0.000025
70	27	26	3	0.000278	0.000251
70	27	26	4	0.001931	0.001652
70	27	26	5	0.009532	0.007601
70	27	26	6	0.034979	0.025448
70	27	26	7	0.098598	0.063619
70	27	26	8	0.219475	0.120876
70	27	26	9	0.396140	0.176665
70	27	26	10	0.596361	0.200221
70	27	26	11	0.773179	0.176818
70	27	26	12	0.895122	0.121944
70	27	26	13	0.960784	0.065662
70	27	26	14	0.988320	0.027536
70	27	26	15	0.997269	0.008949
70	27	26	16	0.999506	0.002237
70	27	26	17	0.999932	0.000426
70	27	26	18	0.999993	0.000061
70	27	26	19	0.999999	0.000006
70	27	26	20	1.000000	0.000000
70	27	26	21	1.000000	0.000000
70	27	26	22	1.000000	0.000000
70	27	26	23	1.000000	0.000000
70	27	26	24	1.000000	0.000000
70	27	26	25	1.000000	0.000000
70	27	26	26	1.000000	0.000000
70	27	27	0	0.000000	0.000000
70	27	27	1	0.000001	0.000001
70	27	27	2	0.000012	0.000012
70	27	27	3	0.000141	0.000129
70	27	27	4	0.001067	0.000926
70	27	27	5	0.005731	0.004664
70	27	27	6	0.022834	0.017103
70	27	27	7	0.069681	0.046847
70	27	27	8	0.167278	0.097597
70	27	27	9	0.323868	0.156590
70	27	27	10	0.519003	0.195135
70	27	27	11	0.708881	0.189879
70	27	27	12	0.853551	0.144669
70	27	27	13	0.939892	0.086341
70	27	27	14	0.980184	0.040293
70	27	27	15	0.994828	0.014644
70	27	27	16	0.998947	0.004119
70	27	27	17	0.999835	0.000888
70	27	27	18	0.999981	0.000145
70	27	27	19	0.999998	0.000018
70	27	27	20	1.000000	0.000002

N	n	k	x	P(x)	p(x)
70	27	27	21	1.000000	0.000000
70	27	27	22	1.000000	0.000000
70	27	27	23	1.000000	0.000000
70	27	27	24	1.000000	0.000000
70	27	27	25	1.000000	0.000000
70	27	27	26	1.000000	0.000000
70	27	27	27	1.000000	0.000000
70	28	1	0	0.600000	0.600000
70	28	1	1	1.000000	0.400000
70	28	2	0	0.356522	0.356522
70	28	2	1	0.843478	0.486956
70	28	2	2	1.000000	0.156522
70	28	3	0	0.209719	0.209719
70	28	3	1	0.650128	0.440409
70	28	3	2	0.940153	0.290026
70	28	3	3	1.000000	0.059847
70	28	4	0	0.122075	0.122075
70	28	4	1	0.472650	0.350574
70	28	4	2	0.827606	0.354957
70	28	4	3	0.977669	0.150063
70	28	4	4	1.000000	0.022331
70	28	5	0	0.070286	0.070286
70	28	5	1	0.329233	0.258947
70	28	5	2	0.687775	0.358542
70	28	5	3	0.920827	0.233052
70	28	5	4	0.991880	0.071053
70	28	5	5	1.000000	0.008120
70	28	6	0	0.040009	0.040009
70	28	6	1	0.221670	0.181661
70	28	6	2	0.544358	0.322688
70	28	6	3	0.831192	0.286834
70	28	6	4	0.965645	0.134453
70	28	6	5	0.997127	0.031482
70	28	6	6	1.000000	0.002873
70	28	7	0	0.022505	0.022505
70	28	7	1	0.145032	0.122527
70	28	7	2	0.413266	0.268234
70	28	7	3	0.719147	0.305881
70	28	7	4	0.915225	0.196078
70	28	7	5	0.985813	0.070588
70	28	7	6	0.999012	0.013199
70	28	7	7	1.000000	0.000988
70	28	8	0	0.012503	0.012503
70	28	8	1	0.092520	0.080017
70	28	8	2	0.302566	0.210046
70	28	8	3	0.597766	0.295200
70	28	8	4	0.840529	0.242763
70	28	8	5	0.960043	0.119514
70	28	8	6	0.994403	0.034360
70	28	8	7	0.999671	0.005268

N	n	k	x	P(x)	p(x)
70	28	8	8	1.000000	0.000329
70	28	9	0	0.006856	0.006856
70	28	9	1	0.057674	0.050818
70	28	9	2	0.214482	0.156808
70	28	9	3	0.478734	0.264251
70	28	9	4	0.746556	0.267822
70	28	9	5	0.915707	0.169151
70	28	9	6	0.982211	0.066504
70	28	9	7	0.997886	0.015676
70	28	9	8	0.999894	0.002007
70	28	9	9	1.000000	0.000106
70	28	10	0	0.003709	0.003709
70	28	10	1	0.035181	0.031472
70	28	10	2	0.147646	0.112465
70	28	10	3	0.370434	0.222788
70	28	10	4	0.641183	0.270749
70	28	10	5	0.851929	0.210745
70	28	10	6	0.958226	0.106297
70	28	10	7	0.992490	0.034264
70	28	10	8	0.999236	0.006746
70	28	10	9	0.999967	0.000731
70	28	10	10	1.000000	0.000033
70	28	11	0	0.001978	0.001978
70	28	11	1	0.021019	0.019040
70	28	11	2	0.098911	0.077892
70	28	11	3	0.277606	0.178695
70	28	11	4	0.532883	0.255278
70	28	11	5	0.771143	0.238259
70	28	11	6	0.919250	0.148107
70	28	11	7	0.980497	0.061247
70	28	11	8	0.996987	0.016490
70	28	11	9	0.999735	0.002748
70	28	11	10	0.999990	0.000255
70	28	11	11	1.000000	0.000010
70	28	12	0	0.001039	0.001039
70	28	12	1	0.012305	0.011266
70	28	12	2	0.064586	0.052280
70	28	12	3	0.201888	0.137302
70	28	12	4	0.429042	0.227154
70	28	12	5	0.678262	0.249220
70	28	12	6	0.864024	0.185762
70	28	12	7	0.958697	0.094674
70	28	12	8	0.991397	0.032700
70	28	12	9	0.998850	0.007453
70	28	12	10	0.999912	0.001062
70	28	12	11	0.999997	0.000085
70	28	12	12	1.000000	0.000003
70	28	13	0	0.000538	0.000538
70	28	13	1	0.007061	0.006523
70	28	13	2	0.041150	0.034089

N	n	k	x	P(x)	p(x)
70	28	13	3	0.142706	0.101556
70	28	13	4	0.335047	0.192341
70	28	13	5	0.579433	0.244386
70	28	13	6	0.793562	0.214129
70	28	13	7	0.924419	0.130857
70	28	13	8	0.980121	0.055702
70	28	13	9	0.996409	0.016287
70	28	13	10	0.999583	0.003174
70	28	13	11	0.999972	0.000390
70	28	13	12	0.999999	0.000027
70	28	13	13	1.000000	0.000001
70	28	14	0	0.000274	0.000274
70	28	14	1	0.003971	0.003697
70	28	14	2	0.025600	0.021629
70	28	14	3	0.098164	0.072563
70	28	14	4	0.254061	0.155898
70	28	14	5	0.480821	0.226760
70	28	14	6	0.710916	0.230095
70	28	14	7	0.876209	0.165293
70	28	14	8	0.960577	0.084368
70	28	14	9	0.990980	0.030403
70	28	14	10	0.998580	0.007601
70	28	14	11	0.999856	0.001276
70	28	14	12	0.999991	0.000136
70	28	14	13	1.000000	0.000008
70	28	14	14	1.000000	0.000000
70	28	15	0	0.000137	0.000137
70	28	15	1	0.002188	0.002051
70	28	15	2	0.015558	0.013370
70	28	15	3	0.065769	0.050211
70	28	15	4	0.187248	0.121479
70	28	15	5	0.387688	0.200440
70	28	15	6	0.620522	0.232834
70	28	15	7	0.814224	0.193702
70	28	15	8	0.930445	0.116221
70	28	15	9	0.980664	0.050219
70	28	15	10	0.996137	0.015473
70	28	15	11	0.999469	0.003331
70	28	15	12	0.999953	0.000484
70	28	15	13	0.999997	0.000045
70	28	15	14	1.000000	0.000002
70	28	15	15	1.000000	0.000000
70	28	16	0	0.000067	0.000067
70	28	16	1	0.001181	0.001114
70	28	16	2	0.009238	0.008057
70	28	16	3	0.042946	0.033708
70	28	16	4	0.134239	0.091293
70	28	16	5	0.303867	0.169628
70	28	16	6	0.527388	0.223521
70	28	16	7	0.740265	0.212877

Table for $N = 2$, $n = 1$, through $N = 100$, $n = 50$

N	n	k	x	P(x)	p(x)
70	28	16	8	0.888183	0.147918
70	28	16	9	0.963316	0.075133
70	28	16	10	0.991073	0.027757
70	28	16	11	0.998439	0.007366
70	28	16	12	0.999812	0.001373
70	28	16	13	0.999985	0.000173
70	28	16	14	0.999999	0.000014
70	28	16	15	1.000000	0.000001
70	28	16	16	1.000000	0.000000
70	28	17	0	0.000032	0.000032
70	28	17	1	0.000624	0.000592
70	28	17	2	0.005359	0.004735
70	28	17	3	0.027340	0.021982
70	28	17	4	0.093664	0.066324
70	28	17	5	0.231618	0.137954
70	28	17	6	0.436324	0.204706
70	28	17	7	0.657479	0.221155
70	28	17	8	0.833399	0.175919
70	28	17	9	0.936880	0.103482
70	28	17	10	0.981821	0.044941
70	28	17	11	0.996120	0.014299
70	28	17	12	0.999405	0.003285
70	28	17	13	0.999937	0.000532
70	28	17	14	0.999996	0.000058
70	28	17	15	1.000000	0.000004
70	28	17	16	1.000000	0.000000
70	28	17	17	1.000000	0.000000
70	28	18	0	0.000015	0.000015
70	28	18	1	0.000323	0.000307
70	28	18	2	0.003036	0.002713
70	28	18	3	0.016972	0.013936
70	28	18	4	0.063631	0.046659
70	28	18	5	0.171751	0.108121
70	28	18	6	0.351352	0.179600
70	28	18	7	0.569852	0.218500
70	28	18	8	0.767014	0.197162
70	28	18	9	0.899783	0.132769
70	28	18	10	0.966558	0.066775
70	28	18	11	0.991534	0.024976
70	28	18	12	0.998414	0.006880
70	28	18	13	0.999787	0.001373
70	28	18	14	0.999980	0.000194
70	28	18	15	0.999999	0.000019
70	28	18	16	1.000000	0.000001
70	28	18	17	1.000000	0.000000
70	28	18	18	1.000000	0.000000
70	28	19	0	0.000007	0.000007
70	28	19	1	0.000163	0.000156
70	28	19	2	0.001679	0.001516
70	28	19	3	0.010272	0.008592
70	28	19	4	0.042096	0.031824
70	28	19	5	0.123929	0.081833
70	28	19	6	0.275367	0.151438
70	28	19	7	0.481611	0.206244
70	28	19	8	0.691182	0.209571
70	28	19	9	0.851271	0.160089
70	28	19	10	0.943444	0.092172
70	28	19	11	0.983369	0.039925
70	28	19	12	0.996297	0.012928
70	28	19	13	0.999391	0.003094
70	28	19	14	0.999928	0.000538
70	28	19	15	0.999994	0.000066
70	28	19	16	1.000000	0.000005
70	28	19	17	1.000000	0.000000
70	28	19	18	1.000000	0.000000
70	28	19	19	1.000000	0.000000
70	28	20	0	0.000003	0.000003
70	28	20	1	0.000080	0.000077
70	28	20	2	0.000906	0.000826
70	28	20	3	0.006060	0.005153
70	28	20	4	0.027120	0.021060
70	28	20	5	0.087024	0.059904
70	28	20	6	0.210041	0.123017
70	28	20	7	0.396687	0.186646
70	28	20	8	0.608998	0.212310
70	28	20	9	0.791630	0.182633
70	28	20	10	0.910912	0.119282
70	28	20	11	0.970060	0.059148
70	28	20	12	0.992241	0.022181
70	28	20	13	0.998480	0.006240
70	28	20	14	0.999780	0.001300
70	28	20	15	0.999977	0.000197
70	28	20	16	0.999998	0.000021
70	28	20	17	1.000000	0.000002
70	28	20	18	1.000000	0.000000
70	28	20	19	1.000000	0.000000
70	28	20	20	1.000000	0.000000
70	28	21	0	0.000001	0.000001
70	28	21	1	0.000039	0.000037
70	28	21	2	0.000477	0.000438
70	28	21	3	0.003483	0.003006
70	28	21	4	0.017011	0.013528
70	28	21	5	0.059468	0.042457
70	28	21	6	0.155913	0.096445
70	28	21	7	0.318296	0.162382
70	28	21	8	0.524074	0.205778
70	28	21	9	0.722230	0.198156
70	28	21	10	0.867971	0.145741
70	28	21	11	0.949950	0.081979
70	28	21	12	0.985143	0.035193
70	28	21	13	0.996609	0.011466
70	28	21	14	0.999417	0.002808
70	28	21	15	0.999926	0.000510
70	28	21	16	0.999993	0.000067
70	28	21	17	0.999999	0.000006
70	28	21	18	1.000000	0.000000
70	28	21	19	1.000000	0.000000
70	28	21	20	1.000000	0.000000
70	28	21	21	1.000000	0.000000
70	28	22	0	0.000001	0.000001
70	28	22	1	0.000018	0.000018
70	28	22	2	0.000244	0.000226
70	28	22	3	0.001949	0.001705
70	28	22	4	0.010385	0.008436
70	28	22	5	0.039539	0.029154
70	28	22	6	0.112611	0.073072
70	28	22	7	0.248703	0.136092
70	28	22	8	0.440083	0.191379
70	28	22	9	0.645394	0.205311
70	28	22	10	0.814433	0.169040
70	28	22	11	0.921508	0.107075
70	28	22	12	0.973651	0.052143
70	28	22	13	0.993099	0.019447
70	28	22	14	0.998614	0.005516
70	28	22	15	0.999791	0.001177
70	28	22	16	0.999977	0.000186
70	28	22	17	0.999998	0.000021
70	28	22	18	1.000000	0.000002
70	28	22	19	1.000000	0.000000
70	28	22	20	1.000000	0.000000
70	28	22	21	1.000000	0.000000
70	28	22	22	1.000000	0.000000
70	28	23	0	0.000000	0.000000
70	28	23	1	0.000008	0.000008
70	28	23	2	0.000122	0.000114
70	28	23	3	0.001061	0.000939
70	28	23	4	0.006167	0.005106
70	28	23	5	0.025570	0.019402
70	28	23	6	0.079120	0.053550
70	28	23	7	0.189163	0.110043
70	28	23	8	0.360341	0.171178
70	28	23	9	0.564125	0.203784
70	28	23	10	0.751043	0.186919
70	28	23	11	0.883586	0.132542
70	28	23	12	0.956270	0.072685
70	28	23	13	0.987021	0.030751
70	28	23	14	0.997006	0.009984
70	28	23	15	0.999472	0.002467
70	28	23	16	0.999930	0.000458
70	28	23	17	0.999993	0.000063
70	28	23	18	0.999999	0.000006
70	28	23	19	1.000000	0.000000
70	28	23	20	1.000000	0.000000
70	28	23	21	1.000000	0.000000
70	28	23	22	1.000000	0.000000
70	28	23	23	1.000000	0.000000
70	28	24	0	0.000000	0.000000
70	28	24	1	0.000004	0.000004
70	28	24	2	0.000059	0.000055
70	28	24	3	0.000562	0.000503
70	28	24	4	0.003560	0.002998
70	28	24	5	0.016075	0.012515
70	28	24	6	0.054054	0.037979
70	28	24	7	0.139995	0.085941
70	28	24	8	0.287499	0.147505
70	28	24	9	0.481744	0.194245
70	28	24	10	0.679458	0.197713
70	28	24	11	0.835645	0.156187
70	28	24	12	0.931527	0.095882
70	28	24	13	0.977207	0.045681
70	28	24	14	0.994032	0.016824
70	28	24	15	0.998790	0.004758
70	28	24	16	0.999813	0.001023
70	28	24	17	0.999978	0.000165
70	28	24	18	0.999998	0.000020
70	28	24	19	1.000000	0.000002
70	28	24	20	1.000000	0.000000
70	28	24	21	1.000000	0.000000
70	28	24	22	1.000000	0.000000
70	28	24	23	1.000000	0.000000
70	28	24	24	1.000000	0.000000
70	28	25	0	0.000000	0.000000
70	28	25	1	0.000002	0.000002
70	28	25	2	0.000028	0.000026
70	28	25	3	0.000288	0.000261
70	28	25	4	0.001996	0.001707
70	28	25	5	0.009817	0.007822
70	28	25	6	0.035890	0.026073
70	28	25	7	0.100761	0.064871
70	28	25	8	0.223367	0.122606
70	28	25	9	0.401513	0.178146
70	28	25	10	0.602091	0.200579
70	28	25	11	0.777924	0.175832
70	28	25	12	0.898177	0.120253
70	28	25	13	0.962311	0.064135
70	28	25	14	0.988911	0.026600
70	28	25	15	0.997445	0.008534
70	28	25	16	0.999546	0.002101
70	28	25	17	0.999939	0.000393
70	28	25	18	0.999994	0.000055

Table for $N = 2$, $n = 1$, through $N = 100$, $n = 50$

N	n	k	x	P(x)	p(x)
70	28	25	19	1.000000	0.000006
70	28	25	20	1.000000	0.000000
70	28	25	21	1.000000	0.000000
70	28	25	22	1.000000	0.000000
70	28	25	23	1.000000	0.000000
70	28	25	24	1.000000	0.000000
70	28	25	25	1.000000	0.000000
70	28	26	0	0.000000	0.000000
70	28	26	1	0.000001	0.000001
70	28	26	2	0.000013	0.000012
70	28	26	3	0.000144	0.000131
70	28	26	4	0.001085	0.000942
70	28	26	5	0.005820	0.004734
70	28	26	6	0.023143	0.017324
70	28	26	7	0.070488	0.047344
70	28	26	8	0.168875	0.098388
70	28	26	9	0.326296	0.157420
70	28	26	10	0.521860	0.195564
70	28	26	11	0.711498	0.189638
70	28	26	12	0.855420	0.143922
70	28	26	13	0.940933	0.085513
70	28	26	14	0.980636	0.039703
70	28	26	15	0.994980	0.014344
70	28	26	16	0.998986	0.004006
70	28	26	17	0.999843	0.000857
70	28	26	18	0.999982	0.000139
70	28	26	19	0.999998	0.000017
70	28	26	20	1.000000	0.000001
70	28	26	21	1.000000	0.000000
70	28	26	22	1.000000	0.000000
70	28	26	23	1.000000	0.000000
70	28	26	24	1.000000	0.000000
70	28	26	25	1.000000	0.000000
70	28	26	26	1.000000	0.000000
70	28	27	0	0.000000	0.000000
70	28	27	1	0.000000	0.000000
70	28	27	2	0.000006	0.000005
70	28	27	3	0.000069	0.000064
70	28	27	4	0.000572	0.000502
70	28	27	5	0.003345	0.002773
70	28	27	6	0.014481	0.011137
70	28	27	7	0.047892	0.033410
70	28	27	8	0.124154	0.076262
70	28	27	9	0.258319	0.134165
70	28	27	10	0.441856	0.183538
70	28	27	11	0.638229	0.196372
70	28	27	12	0.803085	0.164856
70	28	27	13	0.911781	0.108696
70	28	27	14	0.968003	0.056222
70	28	27	15	0.990742	0.022739

N	n	k	x	P(x)	p(x)
70	28	27	16	0.997893	0.007152
70	28	27	17	0.999629	0.001735
70	28	27	18	0.999950	0.000321
70	28	27	19	0.999995	0.000045
70	28	27	20	1.000000	0.000005
70	28	27	21	1.000000	0.000000
70	28	27	22	1.000000	0.000000
70	28	27	23	1.000000	0.000000
70	28	27	24	1.000000	0.000000
70	28	27	25	1.000000	0.000000
70	28	27	26	1.000000	0.000000
70	28	27	27	1.000000	0.000000
70	28	28	0	0.000000	0.000000
70	28	28	1	0.000000	0.000000
70	28	28	2	0.000002	0.000002
70	28	28	3	0.000032	0.000030
70	28	28	4	0.000291	0.000259
70	28	28	5	0.001861	0.001570
70	28	28	6	0.008784	0.006922
70	28	28	7	0.031575	0.022791
70	28	28	8	0.088683	0.057108
70	28	28	9	0.199036	0.110354
70	28	28	10	0.365027	0.165990
70	28	28	11	0.560593	0.195567
70	28	28	12	0.741743	0.181150
70	28	28	13	0.873864	0.132121
70	28	28	14	0.949698	0.075835
70	28	28	15	0.983867	0.034169
70	28	28	16	0.995898	0.012030
70	28	28	17	0.999185	0.003287
70	28	28	18	0.999875	0.000691
70	28	28	19	0.999986	0.000110
70	28	28	20	0.999999	0.000013
70	28	28	21	1.000000	0.000001
70	28	28	22	1.000000	0.000000
70	28	28	23	1.000000	0.000000
70	28	28	24	1.000000	0.000000
70	28	28	25	1.000000	0.000000
70	28	28	26	1.000000	0.000000
70	28	28	27	1.000000	0.000000
70	28	28	28	1.000000	0.000000
70	29	1	0	0.585714	0.585714
70	29	1	1	1.000000	0.414286
70	29	2	0	0.339544	0.339544
70	29	2	1	0.831884	0.492340
70	29	2	2	1.000000	0.168116
70	29	3	0	0.194739	0.194739
70	29	3	1	0.629156	0.434417
70	29	3	2	0.933248	0.304092
70	29	3	3	1.000000	0.066752

N	n	k	x	P(x)	p(x)
70	29	4	0	0.110449	0.110449
70	29	4	1	0.447608	0.337160
70	29	4	2	0.810703	0.363095
70	29	4	3	0.974096	0.163393
70	29	4	4	1.000000	0.025904
70	29	5	0	0.061918	0.061918
70	29	5	1	0.304571	0.242653
70	29	5	2	0.662165	0.357594
70	29	5	3	0.909729	0.247565
70	29	5	4	0.990188	0.080459
70	29	5	5	1.000000	0.009812
70	29	6	0	0.034293	0.034293
70	29	6	1	0.200044	0.165750
70	29	6	2	0.513626	0.313582
70	29	6	3	0.810703	0.297078
70	29	6	4	0.959242	0.148539
70	29	6	5	0.996377	0.037135
70	29	6	6	1.000000	0.003623
70	29	7	0	0.018754	0.018754
70	29	7	1	0.127528	0.108774
70	29	7	2	0.381333	0.253805
70	29	7	3	0.690016	0.308682
70	29	7	4	0.901219	0.211204
70	29	7	5	0.982452	0.081232
70	29	7	6	0.998698	0.016246
70	29	7	7	1.000000	0.001302
70	29	8	0	0.010121	0.010121
70	29	8	1	0.079184	0.069063
70	29	8	2	0.272560	0.193376
70	29	8	3	0.562623	0.290063
70	29	8	4	0.817408	0.254785
70	29	8	5	0.951506	0.134098
70	29	8	6	0.992767	0.041261
70	29	8	7	0.999545	0.006779
70	29	8	8	1.000000	0.000455
70	29	9	0	0.005387	0.005387
70	29	9	1	0.047994	0.042607
70	29	9	2	0.188348	0.140353
70	29	9	3	0.440983	0.252636
70	29	9	4	0.714672	0.273689
70	29	9	5	0.899597	0.184925
70	29	9	6	0.977460	0.077863
70	29	9	7	0.997140	0.019680
70	29	9	8	0.999846	0.002706
70	29	9	9	1.000000	0.000154
70	29	10	0	0.002826	0.002826
70	29	10	1	0.028437	0.025611
70	29	10	2	0.126224	0.097787
70	29	10	3	0.333303	0.207079
70	29	10	4	0.602505	0.269202

N	n	k	x	P(x)	p(x)
70	29	10	5	0.826840	0.224335
70	29	10	6	0.948102	0.121262
70	29	10	7	0.990042	0.041940
70	29	10	8	0.998914	0.008872
70	29	10	9	0.999949	0.001035
70	29	10	10	1.000000	0.000050
70	29	11	0	0.001460	0.001460
70	29	11	1	0.016485	0.015025
70	29	11	2	0.082220	0.065735
70	29	11	3	0.243569	0.161349
70	29	11	4	0.490337	0.246769
70	29	11	5	0.737106	0.246769
70	29	11	6	0.901618	0.164512
70	29	11	7	0.974664	0.073046
70	29	11	8	0.995809	0.021145
70	29	11	9	0.999604	0.003795
70	29	11	10	0.999984	0.000380
70	29	11	11	1.000000	0.000016
70	29	12	0	0.000742	0.000742
70	29	12	1	0.009355	0.008612
70	29	12	2	0.052138	0.042783
70	29	12	3	0.172466	0.120328
70	29	12	4	0.385774	0.213308
70	29	12	5	0.636725	0.250951
70	29	12	6	0.837486	0.200761
70	29	12	7	0.947427	0.109941
70	29	12	8	0.988283	0.040856
70	29	12	9	0.998318	0.010035
70	29	12	10	0.999862	0.001544
70	29	12	11	0.999995	0.000133
70	29	12	12	1.000000	0.000005
70	29	13	0	0.000371	0.000371
70	29	13	1	0.005197	0.004826
70	29	13	2	0.032222	0.027025
70	29	13	3	0.118526	0.086304
70	29	13	4	0.293831	0.175305
70	29	13	5	0.532883	0.239053
70	29	13	6	0.757874	0.224991
70	29	13	7	0.905725	0.147851
70	29	13	8	0.973490	0.067765
70	29	13	9	0.994858	0.021367
70	29	13	10	0.999356	0.004498
70	29	13	11	0.999954	0.000598
70	29	13	12	0.999998	0.000045
70	29	13	13	1.000000	0.000001
70	29	14	0	0.000182	0.000182
70	29	14	1	0.002826	0.002644
70	29	14	2	0.019420	0.016594
70	29	14	3	0.079159	0.059738
70	29	14	4	0.216943	0.137784

Table for $N=2,\ n=1,$ through $N=100,\ n=50$

N	n	k	x	P(x)	p(x)
70	29	14	5	0.432230	0.215287
70	29	14	6	0.667088	0.234859
70	29	14	7	0.848660	0.181571
70	29	14	8	0.948524	0.099864
70	29	14	9	0.987360	0.038836
70	29	14	10	0.997856	0.010496
70	29	14	11	0.999765	0.001908
70	29	14	12	0.999985	0.000220
70	29	14	13	0.999999	0.000014
70	29	14	14	1.000000	0.000000
70	29	15	0	0.000088	0.000088
70	29	15	1	0.001504	0.001416
70	29	15	2	0.011420	0.009915
70	29	15	3	0.051423	0.040003
70	29	15	4	0.155432	0.104009
70	29	15	5	0.339964	0.184532
70	29	15	6	0.570629	0.230665
70	29	15	7	0.777328	0.206700
70	29	15	8	0.911075	0.133747
70	29	15	9	0.973490	0.062415
70	29	15	10	0.994295	0.020805
70	29	15	11	0.999151	0.004856
70	29	15	12	0.999918	0.000767
70	29	15	13	0.999995	0.000077
70	29	15	14	1.000000	0.000004
70	29	15	15	1.000000	0.000000
70	29	16	0	0.000042	0.000042
70	29	16	1	0.000783	0.000742
70	29	16	2	0.006552	0.005769
70	29	16	3	0.032512	0.025960
70	29	16	4	0.108155	0.075643
70	29	16	5	0.259441	0.151286
70	29	16	6	0.474169	0.214728
70	29	16	7	0.694648	0.220480
70	29	16	8	0.860008	0.165360
70	29	16	9	0.950794	0.090786
70	29	16	10	0.987108	0.036314
70	29	16	11	0.997562	0.010454
70	29	16	12	0.999681	0.002119
70	29	16	13	0.999973	0.000292
70	29	16	14	0.999999	0.000026
70	29	16	15	1.000000	0.000001
70	29	16	16	1.000000	0.000000
70	29	17	0	0.000019	0.000019
70	29	17	1	0.000399	0.000379
70	29	17	2	0.003668	0.003269
70	29	17	3	0.020013	0.016345
70	29	17	4	0.073135	0.053122
70	29	17	5	0.192203	0.119067
70	29	17	6	0.382711	0.190508
70	29	17	7	0.604823	0.222113
70	29	17	8	0.795702	0.190878
70	29	17	9	0.917169	0.121468
70	29	17	10	0.974331	0.057161
70	29	17	11	0.994077	0.019747
70	29	17	12	0.999014	0.004937
70	29	17	13	0.999886	0.000872
70	29	17	14	0.999991	0.000105
70	29	17	15	0.999999	0.000008
70	29	17	16	1.000000	0.000000
70	29	17	17	1.000000	0.000000
70	29	18	0	0.000009	0.000009
70	29	18	1	0.000198	0.000190
70	29	18	2	0.002002	0.001804
70	29	18	3	0.011995	0.009992
70	29	18	4	0.048078	0.036083
70	29	18	5	0.138286	0.090208
70	29	18	6	0.300037	0.161752
70	29	18	7	0.512626	0.212588
70	29	18	8	0.720071	0.207445
70	29	18	9	0.871332	0.151262
70	29	18	10	0.953839	0.082507
70	29	18	11	0.987371	0.033532
70	29	18	12	0.997431	0.010060
70	29	18	13	0.999623	0.002192
70	29	18	14	0.999962	0.000339
70	29	18	15	0.999997	0.000036
70	29	18	16	1.000000	0.000002
70	29	18	17	1.000000	0.000000
70	29	18	18	1.000000	0.000000
70	29	19	0	0.000004	0.000004
70	29	19	1	0.000096	0.000092
70	29	19	2	0.001066	0.000969
70	29	19	3	0.006998	0.005933
70	29	19	4	0.030730	0.023732
70	29	19	5	0.096651	0.065921
70	29	19	6	0.228493	0.131842
70	29	19	7	0.422684	0.194191
70	29	19	8	0.636295	0.213610
70	29	19	9	0.813155	0.176860
70	29	19	10	0.923692	0.110538
70	29	19	11	0.975764	0.052071
70	29	19	12	0.994142	0.018378
70	29	19	13	0.998949	0.004807
70	29	19	14	0.999864	0.000916
70	29	19	15	0.999988	0.000124
70	29	19	16	0.999999	0.000011
70	29	19	17	1.000000	0.000001
70	29	19	18	1.000000	0.000000
70	29	19	19	1.000000	0.000000
70	29	20	0	0.000002	0.000002
70	29	20	1	0.000045	0.000044
70	29	20	2	0.000552	0.000507
70	29	20	3	0.003974	0.003422
70	29	20	4	0.019097	0.015123
70	29	20	5	0.065629	0.046533
70	29	20	6	0.169035	0.103406
70	29	20	7	0.338916	0.169881
70	29	20	8	0.548338	0.209422
70	29	20	9	0.743798	0.195460
70	29	20	10	0.882512	0.138714
70	29	20	11	0.957386	0.074874
70	29	20	12	0.988016	0.030630
70	29	20	13	0.997441	0.009425
70	29	20	14	0.999595	0.002154
70	29	20	15	0.999954	0.000359
70	29	20	16	0.999996	0.000042
70	29	20	17	1.000000	0.000003
70	29	20	18	1.000000	0.000000
70	29	20	19	1.000000	0.000000
70	29	20	20	1.000000	0.000000
70	29	21	0	0.000001	0.000001
70	29	21	1	0.000021	0.000020
70	29	21	2	0.000279	0.000258
70	29	21	3	0.002195	0.001916
70	29	21	4	0.011535	0.009341
70	29	21	5	0.043294	0.031758
70	29	21	6	0.121468	0.078175
70	29	21	7	0.264168	0.142700
70	29	21	8	0.460380	0.196212
70	29	21	9	0.665614	0.205233
70	29	21	10	0.829800	0.164187
70	29	21	11	0.930431	0.100631
70	29	21	12	0.977602	0.047171
70	29	21	13	0.994425	0.016823
70	29	21	14	0.998949	0.004524
70	29	21	15	0.999853	0.000905
70	29	21	16	0.999985	0.000132
70	29	21	17	0.999999	0.000014
70	29	21	18	1.000000	0.000001
70	29	21	19	1.000000	0.000000
70	29	21	20	1.000000	0.000000
70	29	21	21	1.000000	0.000000
70	29	22	0	0.000000	0.000000
70	29	22	1	0.000009	0.000009
70	29	22	2	0.000137	0.000127
70	29	22	3	0.001178	0.001041
70	29	22	4	0.006770	0.005592
70	29	22	5	0.027739	0.020969
70	29	22	6	0.084774	0.057036
70	29	22	7	0.200099	0.115325
70	29	22	8	0.376289	0.176190
70	29	22	9	0.581845	0.205556
70	29	22	10	0.766136	0.184291
70	29	22	11	0.893465	0.127328
70	29	22	12	0.961236	0.067772
70	29	22	13	0.988931	0.027695
70	29	22	14	0.997564	0.008632
70	29	22	15	0.999595	0.002031
70	29	22	16	0.999950	0.000355
70	29	22	17	0.999996	0.000045
70	29	22	18	1.000000	0.000004
70	29	22	19	1.000000	0.000000
70	29	22	20	1.000000	0.000000
70	29	22	21	1.000000	0.000000
70	29	22	22	1.000000	0.000000
70	29	23	0	0.000000	0.000000
70	29	23	1	0.000004	0.000004
70	29	23	2	0.000065	0.000061
70	29	23	3	0.000614	0.000549
70	29	23	4	0.003857	0.003243
70	29	23	5	0.017254	0.013397
70	29	23	6	0.057445	0.040190
70	29	23	7	0.147242	0.089797
70	29	23	8	0.299206	0.151964
70	29	23	9	0.496197	0.196991
70	29	23	10	0.693188	0.196991
70	29	23	11	0.845717	0.152529
70	29	23	12	0.937234	0.091517
70	29	23	13	0.979700	0.042466
70	29	23	14	0.994866	0.015166
70	29	23	15	0.999002	0.004136
70	29	23	16	0.999854	0.000852
70	29	23	17	0.999984	0.000130
70	29	23	18	0.999999	0.000014
70	29	23	19	1.000000	0.000001
70	29	23	20	1.000000	0.000000
70	29	23	21	1.000000	0.000000
70	29	23	22	1.000000	0.000000
70	29	23	23	1.000000	0.000000
70	29	24	0	0.000000	0.000000
70	29	24	1	0.000002	0.000002
70	29	24	2	0.000030	0.000028
70	29	24	3	0.000310	0.000280
70	29	24	4	0.002132	0.001822
70	29	24	5	0.010413	0.008281
70	29	24	6	0.037777	0.027364
70	29	24	7	0.105209	0.067432
70	29	24	8	0.231307	0.126098
70	29	24	9	0.412371	0.181064

N	n	k	x	P(x)	p(x)
70	29	24	10	0.613553	0.201182
70	29	24	11	0.787301	0.173748
70	29	24	12	0.904132	0.116831
70	29	24	13	0.965243	0.061111
70	29	24	14	0.990026	0.024783
70	29	24	15	0.997770	0.007745
70	29	24	16	0.999618	0.001848
70	29	24	17	0.999951	0.000333
70	29	24	18	0.999995	0.000044
70	29	24	19	1.000000	0.000004
70	29	24	20	1.000000	0.000000
70	29	24	21	1.000000	0.000000
70	29	24	22	1.000000	0.000000
70	29	24	23	1.000000	0.000000
70	29	24	24	1.000000	0.000000
70	29	25	0	0.000000	0.000000
70	29	25	1	0.000001	0.000001
70	29	25	2	0.000013	0.000013
70	29	25	3	0.000152	0.000138
70	29	25	4	0.001142	0.000990
70	29	25	5	0.006093	0.004951
70	29	25	6	0.024095	0.018002
70	29	25	7	0.072959	0.048864
70	29	25	8	0.173741	0.100782
70	29	25	9	0.333648	0.159907
70	29	25	10	0.530456	0.196809
70	29	25	11	0.719313	0.188857
70	29	25	12	0.860955	0.141643
70	29	25	13	0.943987	0.083032
70	29	25	14	0.981945	0.037957
70	29	25	15	0.995413	0.013469
70	29	25	16	0.999096	0.003683
70	29	25	17	0.999864	0.000768
70	29	25	18	0.999985	0.000120
70	29	25	19	0.999999	0.000014
70	29	25	20	1.000000	0.000001
70	29	25	21	1.000000	0.000000
70	29	25	22	1.000000	0.000000
70	29	25	23	1.000000	0.000000
70	29	25	24	1.000000	0.000000
70	29	25	25	1.000000	0.000000
70	29	26	0	0.000000	0.000000
70	29	26	1	0.000000	0.000000
70	29	26	2	0.000006	0.000006
70	29	26	3	0.000072	0.000066
70	29	26	4	0.000592	0.000520
70	29	26	5	0.003452	0.002860
70	29	26	6	0.014894	0.011442
70	29	26	7	0.049070	0.034176
70	29	26	8	0.126709	0.077639
70	29	26	9	0.262578	0.135869
70	29	26	10	0.447359	0.184781
70	29	26	11	0.643770	0.196411
70	29	26	12	0.807446	0.163676
70	29	26	13	0.914465	0.107019
70	29	26	14	0.969292	0.054827
70	29	26	15	0.991223	0.021931
70	29	26	16	0.998032	0.006809
70	29	26	17	0.999659	0.001627
70	29	26	18	0.999955	0.000296
70	29	26	19	0.999995	0.000040
70	29	26	20	1.000000	0.000004
70	29	26	21	1.000000	0.000000
70	29	26	22	1.000000	0.000000
70	29	26	23	1.000000	0.000000
70	29	26	24	1.000000	0.000000
70	29	26	25	1.000000	0.000000
70	29	26	26	1.000000	0.000000
70	29	27	0	0.000000	0.000000
70	29	27	1	0.000000	0.000000
70	29	27	2	0.000002	0.000002
70	29	27	3	0.000033	0.000030
70	29	27	4	0.000296	0.000264
70	29	27	5	0.001892	0.001596
70	29	27	6	0.008913	0.007021
70	29	27	7	0.031982	0.023069
70	29	27	8	0.089654	0.057672
70	29	27	9	0.200819	0.111165
70	29	27	10	0.367567	0.166748
70	29	27	11	0.563420	0.195853
70	29	27	12	0.744208	0.180787
70	29	27	13	0.875549	0.131341
70	29	27	14	0.950601	0.075052
70	29	27	15	0.984245	0.033644
70	29	27	16	0.996020	0.011775
70	29	27	17	0.999216	0.003195
70	29	27	18	0.999881	0.000666
70	29	27	19	0.999986	0.000105
70	29	27	20	0.999999	0.000012
70	29	27	21	1.000000	0.000001
70	29	27	22	1.000000	0.000000
70	29	27	23	1.000000	0.000000
70	29	27	24	1.000000	0.000000
70	29	27	25	1.000000	0.000000
70	29	27	26	1.000000	0.000000
70	29	27	27	1.000000	0.000000
70	29	28	0	0.000000	0.000000
70	29	28	1	0.000000	0.000000
70	29	28	2	0.000001	0.000001
70	29	28	3	0.000014	0.000013
70	29	28	4	0.000143	0.000129
70	29	28	5	0.001002	0.000858
70	29	28	6	0.005158	0.004156
70	29	28	7	0.020179	0.015022
70	29	28	8	0.061489	0.041309
70	29	28	9	0.149115	0.087626
70	29	28	10	0.293888	0.144773
70	29	28	11	0.481435	0.187547
70	29	28	12	0.672734	0.191298
70	29	28	13	0.826678	0.153944
70	29	28	14	0.924420	0.097742
70	29	28	15	0.973291	0.048871
70	29	28	16	0.992460	0.019169
70	29	28	17	0.998324	0.005864
70	29	28	18	0.999711	0.001387
70	29	28	19	0.999962	0.000251
70	29	28	20	0.999996	0.000034
70	29	28	21	1.000000	0.000003
70	29	28	22	1.000000	0.000000
70	29	28	23	1.000000	0.000000
70	29	28	24	1.000000	0.000000
70	29	28	25	1.000000	0.000000
70	29	28	26	1.000000	0.000000
70	29	28	27	1.000000	0.000000
70	29	28	28	1.000000	0.000000
70	29	29	0	0.000000	0.000000
70	29	29	1	0.000000	0.000000
70	29	29	2	0.000000	0.000000
70	29	29	3	0.000006	0.000006
70	29	29	4	0.000067	0.000060
70	29	29	5	0.000511	0.000444
70	29	29	6	0.002882	0.002371
70	29	29	7	0.012311	0.009429
70	29	29	8	0.040834	0.028523
70	29	29	9	0.107388	0.066554
70	29	29	10	0.228395	0.121007
70	29	29	11	0.401058	0.172663
70	29	29	12	0.595303	0.194245
70	29	29	13	0.768032	0.172729
70	29	29	14	0.889512	0.121480
70	29	29	15	0.957001	0.067489
70	29	29	16	0.986527	0.029526
70	29	29	17	0.996649	0.010122
70	29	29	18	0.999348	0.002699
70	29	29	19	0.999902	0.000554
70	29	29	20	0.999989	0.000087
70	29	29	21	0.999999	0.000010
70	29	29	22	1.000000	0.000001
70	29	29	23	1.000000	0.000000
70	29	29	24	1.000000	0.000000
70	29	29	25	1.000000	0.000000
70	29	29	26	1.000000	0.000000
70	29	29	27	1.000000	0.000000
70	29	29	28	1.000000	0.000000
70	29	29	29	1.000000	0.000000
70	30	1	0	0.571429	0.571429
70	30	1	1	1.000000	0.428571
70	30	2	0	0.322981	0.322981
70	30	2	1	0.819876	0.496894
70	30	2	2	1.000000	0.180124
70	30	3	0	0.180490	0.180490
70	30	3	1	0.607965	0.427475
70	30	3	2	0.925831	0.317866
70	30	3	3	1.000000	0.074169
70	30	4	0	0.099673	0.099673
70	30	4	1	0.422938	0.323265
70	30	4	2	0.792992	0.370053
70	30	4	3	0.970111	0.177120
70	30	4	4	1.000000	0.029889
70	30	5	0	0.054367	0.054367
70	30	5	1	0.280898	0.226530
70	30	5	2	0.635999	0.355102
70	30	5	3	0.897653	0.261654
70	30	5	4	0.988225	0.090572
70	30	5	5	1.000000	0.011774
70	30	6	0	0.029275	0.029275
70	30	6	1	0.179830	0.150556
70	30	6	2	0.483032	0.303202
70	30	6	3	0.788966	0.305934
70	30	6	4	0.951997	0.163030
70	30	6	5	0.995471	0.043475
70	30	6	6	1.000000	0.004529
70	30	7	0	0.015552	0.015552
70	30	7	1	0.111610	0.096058
70	30	7	2	0.350381	0.238772
70	30	7	3	0.659900	0.309519
70	30	7	4	0.885765	0.225865
70	30	7	5	0.978489	0.092724
70	30	7	6	0.998302	0.019813
70	30	7	7	1.000000	0.001698
70	30	8	0	0.008146	0.008146
70	30	8	1	0.067393	0.059246
70	30	8	2	0.244261	0.176868
70	30	8	3	0.527249	0.282989
70	30	8	4	0.792551	0.265302
70	30	8	5	0.941694	0.149143
70	30	8	6	0.990754	0.049060
70	30	8	7	0.999380	0.008626
70	30	8	8	1.000000	0.000620
70	30	9	0	0.004205	0.004205

N	n	k	x	P(x)	p(x)
70	30	9	1	0.039681	0.035476
70	30	9	2	0.164385	0.124704
70	30	9	3	0.404012	0.239627
70	30	9	4	0.681296	0.277283
70	30	9	5	0.881556	0.200260
70	30	9	6	0.971763	0.090207
70	30	9	7	0.996180	0.024417
70	30	9	8	0.999780	0.003600
70	30	9	9	1.000000	0.000220
70	30	10	0	0.002137	0.002137
70	30	10	1	0.022815	0.020678
70	30	10	2	0.107144	0.084329
70	30	10	3	0.297948	0.190804
70	30	10	4	0.563109	0.265162
70	30	10	5	0.799482	0.236373
70	30	10	6	0.936272	0.136790
70	30	10	7	0.986974	0.050702
70	30	10	8	0.998482	0.011508
70	30	10	9	0.999924	0.001443
70	30	10	10	1.000000	0.000076
70	30	11	0	0.001068	0.001068
70	30	11	1	0.012821	0.011752
70	30	11	2	0.067790	0.054970
70	30	11	3	0.212086	0.144296
70	30	11	4	0.448206	0.236120
70	30	11	5	0.700993	0.252787
70	30	11	6	0.881556	0.180562
70	30	11	7	0.967538	0.085982
70	30	11	8	0.994262	0.026724
70	30	11	9	0.999419	0.005157
70	30	11	10	0.999975	0.000555
70	30	11	11	1.000000	0.000025
70	30	12	0	0.000525	0.000525
70	30	12	1	0.007044	0.006519
70	30	12	2	0.041703	0.034659
70	30	12	3	0.146052	0.104349
70	30	12	4	0.344153	0.198101
70	30	12	5	0.593880	0.249727
70	30	12	6	0.808107	0.214227
70	30	12	7	0.934019	0.125913
70	30	12	8	0.984297	0.050278
70	30	12	9	0.997584	0.013287
70	30	12	10	0.999786	0.002203
70	30	12	11	0.999992	0.000205
70	30	12	12	1.000000	0.000008
70	30	13	0	0.000254	0.000254
70	30	13	1	0.003785	0.003531
70	30	13	2	0.024971	0.021186
70	30	13	3	0.097476	0.072505
70	30	13	4	0.255349	0.157873
70	30	13	5	0.486239	0.230890
70	30	13	6	0.719461	0.233222
70	30	13	7	0.884088	0.164627
70	30	13	8	0.965226	0.081138
70	30	13	9	0.992773	0.027547
70	30	13	10	0.999027	0.006254
70	30	13	11	0.999924	0.000898
70	30	13	12	0.999997	0.000073
70	30	13	13	1.000000	0.000003
70	30	14	0	0.000120	0.000120
70	30	14	1	0.001988	0.001868
70	30	14	2	0.014564	0.012576
70	30	14	3	0.063131	0.048568
70	30	14	4	0.183337	0.120205
70	30	14	5	0.384972	0.201635
70	30	14	6	0.621262	0.236291
70	30	14	7	0.817660	0.196398
70	30	14	8	0.933910	0.116250
70	30	14	9	0.982624	0.048714
70	30	14	10	0.996832	0.014208
70	30	14	11	0.999625	0.002793
70	30	14	12	0.999974	0.000349
70	30	14	13	0.999999	0.000025
70	30	14	14	1.000000	0.000001
70	30	15	0	0.000056	0.000056
70	30	15	1	0.001021	0.000965
70	30	15	2	0.008276	0.007255
70	30	15	3	0.039715	0.031439
70	30	15	4	0.127527	0.087812
70	30	15	5	0.294956	0.167429
70	30	15	6	0.519995	0.225039
70	30	15	7	0.736997	0.217002
70	30	15	8	0.888240	0.151244
70	30	15	9	0.964356	0.076116
70	30	15	10	0.991758	0.027402
70	30	15	11	0.998678	0.006920
70	30	15	12	0.999862	0.001184
70	30	15	13	0.999992	0.000129
70	30	15	14	1.000000	0.000008
70	30	15	15	1.000000	0.000000
70	30	16	0	0.000025	0.000025
70	30	16	1	0.000512	0.000487
70	30	16	2	0.004582	0.004070
70	30	16	3	0.024281	0.019699
70	30	16	4	0.086016	0.061735
70	30	16	5	0.218852	0.132836
70	30	16	6	0.421796	0.202944
70	30	16	7	0.646250	0.224454
70	30	16	8	0.827743	0.181492
70	30	16	9	0.935294	0.107551
70	30	16	10	0.981794	0.046500
70	30	16	11	0.996287	0.014493
70	30	16	12	0.999474	0.003187
70	30	16	13	0.999952	0.000477
70	30	16	14	0.999997	0.000046
70	30	16	15	1.000000	0.000002
70	30	16	16	1.000000	0.000000
70	30	17	0	0.000011	0.000011
70	30	17	1	0.000251	0.000239
70	30	17	2	0.002472	0.002221
70	30	17	3	0.014432	0.011960
70	30	17	4	0.056292	0.041860
70	30	17	5	0.157354	0.101062
70	30	17	6	0.331599	0.174245
70	30	17	7	0.550650	0.219051
70	30	17	8	0.753801	0.203152
70	30	17	9	0.893468	0.139667
70	30	17	10	0.964572	0.071103
70	30	17	11	0.991188	0.026616
70	30	17	12	0.998412	0.007224
70	30	17	13	0.999801	0.001389
70	30	17	14	0.999984	0.000182
70	30	17	15	0.999999	0.000015
70	30	17	16	1.000000	0.000001
70	30	17	17	1.000000	0.000000
70	30	18	0	0.000005	0.000005
70	30	18	1	0.000120	0.000115
70	30	18	2	0.001298	0.001179
70	30	18	3	0.008339	0.007041
70	30	18	4	0.035757	0.027418
70	30	18	5	0.109683	0.073926
70	30	18	6	0.252695	0.143012
70	30	18	7	0.455590	0.202894
70	30	18	8	0.669474	0.213884
70	30	18	9	0.838129	0.168654
70	30	18	10	0.937740	0.099611
70	30	18	11	0.981646	0.043906
70	30	18	12	0.995959	0.014312
70	30	18	13	0.999356	0.003397
70	30	18	14	0.999929	0.000573
70	30	18	15	0.999995	0.000066
70	30	18	16	1.000000	0.000005
70	30	18	17	1.000000	0.000000
70	30	18	18	1.000000	0.000000
70	30	19	0	0.000002	0.000002
70	30	19	1	0.000056	0.000054
70	30	19	2	0.000664	0.000608
70	30	19	3	0.004683	0.004020
70	30	19	4	0.022048	0.017365
70	30	19	5	0.074142	0.052094
70	30	19	6	0.186690	0.112548
70	30	19	7	0.365848	0.179158
70	30	19	8	0.578985	0.213137
70	30	19	9	0.770018	0.191034
70	30	19	10	0.899428	0.129410
70	30	19	11	0.965603	0.066175
70	30	19	12	0.991004	0.025401
70	30	19	13	0.998245	0.007241
70	30	19	14	0.999752	0.001507
70	30	19	15	0.999976	0.000223
70	30	19	16	0.999998	0.000023
70	30	19	17	1.000000	0.000001
70	30	19	18	1.000000	0.000000
70	30	19	19	1.000000	0.000000
70	30	20	0	0.000001	0.000001
70	30	20	1	0.000025	0.000024
70	30	20	2	0.000330	0.000305
70	30	20	3	0.002555	0.002225
70	30	20	4	0.013195	0.010640
70	30	20	5	0.048605	0.035410
70	30	20	6	0.133726	0.085120
70	30	20	7	0.285051	0.151325
70	30	20	8	0.487043	0.201992
70	30	20	9	0.691357	0.204314
70	30	20	10	0.848679	0.157322
70	30	20	11	0.940950	0.092271
70	30	20	12	0.982039	0.041089
70	30	20	13	0.995831	0.013792
70	30	20	14	0.999280	0.003448
70	30	20	15	0.999910	0.000630
70	30	20	16	0.999992	0.000082
70	30	20	17	0.999999	0.000007
70	30	20	18	1.000000	0.000000
70	30	20	19	1.000000	0.000000
70	30	20	20	1.000000	0.000000
70	30	21	0	0.000000	0.000000
70	30	21	1	0.000011	0.000011
70	30	21	2	0.000159	0.000148
70	30	21	3	0.001354	0.001194
70	30	21	4	0.007662	0.006309
70	30	21	5	0.030900	0.023238
70	30	21	6	0.092868	0.061968
70	30	21	7	0.215441	0.122573
70	30	21	8	0.398167	0.182725
70	30	21	9	0.605545	0.207379
70	30	21	10	0.785750	0.180205
70	30	21	11	0.905887	0.120137
70	30	21	12	0.967247	0.061360
70	30	21	13	0.991142	0.023895
70	30	21	14	0.998176	0.007034

Table for $N = 2$, $n = 1$, through $N = 100$, $n = 50$

N	n	k	x	P(x)	p(x)
70	30	21	15	0.999721	0.001545
70	30	21	16	0.999969	0.000248
70	30	21	17	0.999998	0.000028
70	30	21	18	1.000000	0.000002
70	30	21	19	1.000000	0.000000
70	30	21	20	1.000000	0.000000
70	30	21	21	1.000000	0.000000
70	30	22	0	0.000000	0.000000
70	30	22	1	0.000005	0.000005
70	30	22	2	0.000075	0.000070
70	30	22	3	0.000695	0.000621
70	30	22	4	0.004315	0.003619
70	30	22	5	0.019044	0.014729
70	30	22	6	0.062516	0.043472
70	30	22	7	0.157907	0.095391
70	30	22	8	0.316127	0.158220
70	30	22	9	0.516669	0.200542
70	30	22	10	0.712197	0.195528
70	30	22	11	0.859303	0.147106
70	30	22	12	0.944707	0.085403
70	30	22	13	0.982852	0.038145
70	30	22	14	0.995879	0.013027
70	30	22	15	0.999248	0.003369
70	30	22	16	0.999898	0.000650
70	30	22	17	0.999990	0.000092
70	30	22	18	0.999999	0.000009
70	30	22	19	1.000000	0.000001
70	30	22	20	1.000000	0.000000
70	30	22	21	1.000000	0.000000
70	30	22	22	1.000000	0.000000
70	30	23	0	0.000000	0.000000
70	30	23	1	0.000002	0.000002
70	30	23	2	0.000034	0.000032
70	30	23	3	0.000346	0.000312
70	30	23	4	0.002354	0.002008
70	30	23	5	0.011373	0.009018
70	30	23	6	0.040780	0.029408
70	30	23	7	0.112199	0.071419
70	30	23	8	0.243609	0.131410
70	30	23	9	0.428932	0.185322
70	30	23	10	0.630727	0.201795
70	30	23	11	0.801074	0.170347
70	30	23	12	0.912680	0.111606
70	30	23	13	0.969342	0.056662
70	30	23	14	0.991537	0.022195
70	30	23	15	0.998195	0.006658
70	30	23	16	0.999709	0.001513
70	30	23	17	0.999965	0.000257
70	30	23	18	0.999997	0.000032
70	30	23	19	1.000000	0.000003
70	30	23	20	1.000000	0.000000
70	30	23	21	1.000000	0.000000
70	30	23	22	1.000000	0.000000
70	30	23	23	1.000000	0.000000
70	30	24	0	0.000000	0.000000
70	30	24	1	0.000001	0.000001
70	30	24	2	0.000015	0.000014
70	30	24	3	0.000167	0.000152
70	30	24	4	0.001243	0.001077
70	30	24	5	0.006576	0.005332
70	30	24	6	0.025764	0.019188
70	30	24	7	0.077249	0.051486
70	30	24	8	0.182098	0.104849
70	30	24	9	0.346128	0.164030
70	30	24	10	0.544857	0.198729
70	30	24	11	0.732210	0.187354
70	30	24	12	0.869938	0.137727
70	30	24	13	0.948848	0.078910
70	30	24	14	0.983981	0.035134
70	30	24	15	0.996070	0.012089
70	30	24	16	0.999258	0.003188
70	30	24	17	0.999894	0.000636
70	30	24	18	0.999989	0.000095
70	30	24	19	0.999999	0.000010
70	30	24	20	1.000000	0.000001
70	30	24	21	1.000000	0.000000
70	30	24	22	1.000000	0.000000
70	30	24	23	1.000000	0.000000
70	30	24	24	1.000000	0.000000
70	30	25	0	0.000000	0.000000
70	30	25	1	0.000000	0.000000
70	30	25	2	0.000006	0.000006
70	30	25	3	0.000078	0.000071
70	30	25	4	0.000635	0.000557
70	30	25	5	0.003678	0.003043
70	30	25	6	0.015753	0.012075
70	30	25	7	0.051507	0.035754
70	30	25	8	0.131953	0.080447
70	30	25	9	0.271245	0.139292
70	30	25	10	0.458453	0.187208
70	30	25	11	0.654825	0.196372
70	30	25	12	0.816044	0.161219
70	30	25	13	0.919685	0.103641
70	30	25	14	0.971761	0.052076
70	30	25	15	0.992128	0.020367
70	30	25	16	0.998288	0.006159
70	30	25	17	0.999714	0.001427
70	30	25	18	0.999964	0.000250
70	30	25	19	0.999997	0.000032
70	30	25	20	1.000000	0.000003
70	30	25	21	1.000000	0.000000
70	30	25	22	1.000000	0.000000
70	30	25	23	1.000000	0.000000
70	30	25	24	1.000000	0.000000
70	30	25	25	1.000000	0.000000
70	30	26	0	0.000000	0.000000
70	30	26	1	0.000000	0.000000
70	30	26	2	0.000003	0.000002
70	30	26	3	0.000035	0.000032
70	30	26	4	0.000313	0.000278
70	30	26	5	0.001987	0.001674
70	30	26	6	0.009313	0.007325
70	30	26	7	0.033232	0.023920
70	30	26	8	0.092624	0.059391
70	30	26	9	0.206242	0.113618
70	30	26	10	0.375249	0.169007
70	30	26	11	0.571912	0.196663
70	30	26	12	0.751557	0.179644
70	30	26	13	0.880532	0.128975
70	30	26	14	0.953245	0.072713
70	30	26	15	0.985339	0.032094
70	30	26	16	0.996371	0.011032
70	30	26	17	0.999302	0.002931
70	30	26	18	0.999897	0.000595
70	30	26	19	0.999989	0.000091
70	30	26	20	0.999999	0.000010
70	30	26	21	1.000000	0.000001
70	30	26	22	1.000000	0.000000
70	30	26	23	1.000000	0.000000
70	30	26	24	1.000000	0.000000
70	30	26	25	1.000000	0.000000
70	30	26	26	1.000000	0.000000
70	30	27	0	0.000000	0.000000
70	30	27	1	0.000000	0.000000
70	30	27	2	0.000001	0.000001
70	30	27	3	0.000015	0.000014
70	30	27	4	0.000149	0.000134
70	30	27	5	0.001036	0.000887
70	30	27	6	0.005317	0.004281
70	30	27	7	0.020729	0.015412
70	30	27	8	0.062928	0.042199
70	30	27	9	0.152015	0.089087
70	30	27	10	0.298428	0.146413
70	30	27	11	0.486990	0.188562
70	30	27	12	0.678066	0.191076
70	30	27	13	0.830701	0.152635
70	30	27	14	0.926804	0.096103
70	30	27	15	0.974398	0.047594
70	30	27	16	0.992861	0.018463
70	30	27	17	0.998436	0.005575
70	30	27	18	0.999735	0.001299
70	30	27	19	0.999966	0.000231
70	30	27	20	0.999997	0.000031
70	30	27	21	1.000000	0.000003
70	30	27	22	1.000000	0.000000
70	30	27	23	1.000000	0.000000
70	30	27	24	1.000000	0.000000
70	30	27	25	1.000000	0.000000
70	30	27	26	1.000000	0.000000
70	30	27	27	1.000000	0.000000
70	30	28	0	0.000000	0.000000
70	30	28	1	0.000000	0.000000
70	30	28	2	0.000000	0.000000
70	30	28	3	0.000006	0.000006
70	30	28	4	0.000068	0.000062
70	30	28	5	0.000520	0.000452
70	30	28	6	0.002928	0.002408
70	30	28	7	0.012485	0.009558
70	30	28	8	0.041338	0.028852
70	30	28	9	0.108507	0.067170
70	30	28	10	0.230329	0.121821
70	30	28	11	0.403672	0.173343
70	30	28	12	0.598081	0.194409
70	30	28	13	0.770357	0.172276
70	30	28	14	0.891045	0.120688
70	30	28	15	0.957795	0.066751
70	30	28	16	0.986850	0.029054
70	30	28	17	0.996751	0.009901
70	30	28	18	0.999373	0.002622
70	30	28	19	0.999907	0.000534
70	30	28	20	0.999989	0.000083
70	30	28	21	0.999999	0.000010
70	30	28	22	1.000000	0.000001
70	30	28	23	1.000000	0.000000
70	30	28	24	1.000000	0.000000
70	30	28	25	1.000000	0.000000
70	30	28	26	1.000000	0.000000
70	30	28	27	1.000000	0.000000
70	30	28	28	1.000000	0.000000
70	30	29	0	0.000000	0.000000
70	30	29	1	0.000000	0.000000
70	30	29	2	0.000000	0.000000
70	30	29	3	0.000002	0.000002
70	30	29	4	0.000030	0.000027
70	30	29	5	0.000251	0.000221
70	30	29	6	0.001552	0.001301
70	30	29	7	0.007251	0.005699
70	30	29	8	0.026225	0.018973
70	30	29	9	0.074923	0.048698
70	30	29	10	0.172319	0.097396

Table for $N = 2$, $n = 1$, through $N = 100$, $n = 50$

N	n	k	x	P(x)	p(x)
70	30	29	11	0.325255	0.152936
70	30	29	12	0.514762	0.189508
70	30	29	13	0.700626	0.185863
70	30	29	14	0.845068	0.144442
70	30	29	15	0.933956	0.088888
70	30	29	16	0.977165	0.043209
70	30	29	17	0.993686	0.016521
70	30	29	18	0.998624	0.004937
70	30	29	19	0.999767	0.001143
70	30	29	20	0.999970	0.000203
70	30	29	21	0.999997	0.000027
70	30	29	22	1.000000	0.000003
70	30	29	23	1.000000	0.000000
70	30	29	24	1.000000	0.000000
70	30	29	25	1.000000	0.000000
70	30	29	26	1.000000	0.000000
70	30	29	27	1.000000	0.000000
70	30	29	28	1.000000	0.000000
70	30	29	29	1.000000	0.000000
70	30	30	0	0.000000	0.000000
70	30	30	1	0.000000	0.000000
70	30	30	2	0.000000	0.000000
70	30	30	3	0.000001	0.000001
70	30	30	4	0.000012	0.000011
70	30	30	5	0.000116	0.000104
70	30	30	6	0.000790	0.000674
70	30	30	7	0.004054	0.003264
70	30	30	8	0.016044	0.011990
70	30	30	9	0.049980	0.033936
70	30	30	10	0.124808	0.074829
70	30	30	11	0.254382	0.129574
70	30	30	12	0.431564	0.177182
70	30	30	13	0.623561	0.191997
70	30	30	14	0.788700	0.165140
70	30	30	15	0.901436	0.112735
70	30	30	16	0.962411	0.060975
70	30	30	17	0.988448	0.026037
70	30	30	18	0.997178	0.008731
70	30	30	19	0.999460	0.002282
70	30	30	20	0.999920	0.000460
70	30	30	21	0.999991	0.000071
70	30	30	22	0.999999	0.000008
70	30	30	23	1.000000	0.000001
70	30	30	24	1.000000	0.000000
70	30	30	25	1.000000	0.000000
70	30	30	26	1.000000	0.000000
70	30	30	27	1.000000	0.000000
70	30	30	28	1.000000	0.000000
70	30	30	29	1.000000	0.000000
70	30	30	30	1.000000	0.000000

N	n	k	x	P(x)	p(x)
70	31	1	0	0.557143	0.557143
70	31	1	1	1.000000	0.442857
70	31	2	0	0.306832	0.306832
70	31	2	1	0.807453	0.500621
70	31	2	2	1.000000	0.192547
70	31	3	0	0.166953	0.166953
70	31	3	1	0.586591	0.419638
70	31	3	2	0.917884	0.331293
70	31	3	3	1.000000	0.082115
70	31	4	0	0.089706	0.089706
70	31	4	1	0.398693	0.308987
70	31	4	2	0.774489	0.375795
70	31	4	3	0.965683	0.191194
70	31	4	4	1.000000	0.034317
70	31	5	0	0.047571	0.047571
70	31	5	1	0.258245	0.210673
70	31	5	2	0.609367	0.351122
70	31	5	3	0.884570	0.275204
70	31	5	4	0.985961	0.101391
70	31	5	5	1.000000	0.014039
70	31	6	0	0.024883	0.024883
70	31	6	1	0.161011	0.136127
70	31	6	2	0.452712	0.291701
70	31	6	3	0.766021	0.313309
70	31	6	4	0.943845	0.177824
70	31	6	5	0.994384	0.050539
70	31	6	6	1.000000	0.005615
70	31	7	0	0.012831	0.012831
70	31	7	1	0.097201	0.084371
70	31	7	2	0.320535	0.223334
70	31	7	3	0.628948	0.308413
70	31	7	4	0.868826	0.239877
70	31	7	5	0.973853	0.105027
70	31	7	6	0.997806	0.023954
70	31	7	7	1.000000	0.002194
70	31	8	0	0.006517	0.006517
70	31	8	1	0.057025	0.050508
70	31	8	2	0.217730	0.160706
70	31	8	3	0.491876	0.274145
70	31	8	4	0.766021	0.274145
70	31	8	5	0.930508	0.164487
70	31	8	6	0.988301	0.057793
70	31	8	7	0.999164	0.010863
70	31	8	8	1.000000	0.000836
70	31	9	0	0.003259	0.003259
70	31	9	1	0.032586	0.029327
70	31	9	2	0.142562	0.109976
70	31	9	3	0.368068	0.225507
70	31	9	4	0.646635	0.278567
70	31	9	5	0.861530	0.214895

N	n	k	x	P(x)	p(x)
70	31	9	6	0.964997	0.103468
70	31	9	7	0.994959	0.029962
70	31	9	8	0.999690	0.004731
70	31	9	9	1.000000	0.000310
70	31	10	0	0.001603	0.001603
70	31	10	1	0.018162	0.016560
70	31	10	2	0.090278	0.072115
70	31	10	3	0.264557	0.174279
70	31	10	4	0.523335	0.258778
70	31	10	5	0.769935	0.246600
70	31	10	6	0.922593	0.152657
70	31	10	7	0.983171	0.060578
70	31	10	8	0.997906	0.014735
70	31	10	9	0.999888	0.001982
70	31	10	10	1.000000	0.000112
70	31	11	0	0.000775	0.000775
70	31	11	1	0.009882	0.009108
70	31	11	2	0.055422	0.045540
70	31	11	3	0.183227	0.127805
70	31	11	4	0.406885	0.223658
70	31	11	5	0.663075	0.256190
70	31	11	6	0.858985	0.195910
70	31	11	7	0.958940	0.099954
70	31	11	8	0.992258	0.033318
70	31	11	9	0.999161	0.006904
70	31	11	10	0.999961	0.000799
70	31	11	11	1.000000	0.000039
70	31	12	0	0.000368	0.000368
70	31	12	1	0.005251	0.004884
70	31	12	2	0.033038	0.027787
70	31	12	3	0.122574	0.089535
70	31	12	4	0.304533	0.181959
70	31	12	5	0.550178	0.245645
70	31	12	6	0.775973	0.225795
70	31	12	7	0.918280	0.142308
70	31	12	8	0.979269	0.060989
70	31	12	9	0.996587	0.017318
70	31	12	10	0.999676	0.003089
70	31	12	11	0.999987	0.000310
70	31	12	12	1.000000	0.000013
70	31	13	0	0.000171	0.000171
70	31	13	1	0.002725	0.002554
70	31	13	2	0.019145	0.016420
70	31	13	3	0.079350	0.060205
70	31	13	4	0.219828	0.140478
70	31	13	5	0.440061	0.220233
70	31	13	6	0.678647	0.238586
70	31	13	7	0.859394	0.180747
70	31	13	8	0.955084	0.095690
70	31	13	9	0.990018	0.034934

N	n	k	x	P(x)	p(x)
70	31	13	10	0.998558	0.008539
70	31	13	11	0.999880	0.001322
70	31	13	12	0.999996	0.000116
70	31	13	13	1.000000	0.000004
70	31	14	0	0.000078	0.000078
70	31	14	1	0.001381	0.001303
70	31	14	2	0.010791	0.009410
70	31	14	3	0.049775	0.038984
70	31	14	4	0.153286	0.103510
70	31	14	5	0.339604	0.186318
70	31	14	6	0.574004	0.234400
70	31	14	7	0.783290	0.209286
70	31	14	8	0.916472	0.133182
70	31	14	9	0.976535	0.060062
70	31	14	10	0.995412	0.018877
70	31	14	11	0.999416	0.004004
70	31	14	12	0.999957	0.000541
70	31	14	13	0.999999	0.000042
70	31	14	14	1.000000	0.000001
70	31	15	0	0.000035	0.000035
70	31	15	1	0.000683	0.000648
70	31	15	2	0.005918	0.005235
70	31	15	3	0.030283	0.024365
70	31	15	4	0.103379	0.073096
70	31	15	5	0.253099	0.149720
70	31	15	6	0.469361	0.216262
70	31	15	7	0.693596	0.224235
70	31	15	8	0.861773	0.168176
70	31	15	9	0.952939	0.091166
70	31	15	10	0.988333	0.035394
70	31	15	11	0.997986	0.009653
70	31	15	12	0.999773	0.001788
70	31	15	13	0.999985	0.000212
70	31	15	14	0.999999	0.000014
70	31	15	15	1.000000	0.000000
70	31	16	0	0.000015	0.000015
70	31	16	1	0.000329	0.000314
70	31	16	2	0.003158	0.002828
70	31	16	3	0.017879	0.014721
70	31	16	4	0.067495	0.049616
70	31	16	5	0.182322	0.114827
70	31	16	6	0.371060	0.188738
70	31	16	7	0.595748	0.224688
70	31	16	8	0.791444	0.195696
70	31	16	9	0.916472	0.125028
70	31	16	10	0.974819	0.058346
70	31	16	11	0.994476	0.019657
70	31	16	12	0.999156	0.004680
70	31	16	13	0.999916	0.000760
70	31	16	14	0.999995	0.000079

N	n	k	x	P(x)	p(x)
70	31	16	15	1.000000	0.000005
70	31	16	16	1.000000	0.000000
70	31	17	0	0.000006	0.000006
70	31	17	1	0.000155	0.000148
70	31	17	2	0.001639	0.001484
70	31	17	3	0.010246	0.008607
70	31	17	4	0.042687	0.032442
70	31	17	5	0.127035	0.084348
70	31	17	6	0.283681	0.156646
70	31	17	7	0.495887	0.212205
70	31	17	8	0.708092	0.212205
70	31	17	9	0.865535	0.157443
70	31	17	10	0.952128	0.086593
70	31	17	11	0.987195	0.035067
70	31	17	12	0.997509	0.010314
70	31	17	13	0.999662	0.002153
70	31	17	14	0.999970	0.000308
70	31	17	15	0.999998	0.000028
70	31	17	16	1.000000	0.000001
70	31	17	17	1.000000	0.000000
70	31	18	0	0.000003	0.000003
70	31	18	1	0.000071	0.000068
70	31	18	2	0.000827	0.000756
70	31	18	3	0.005699	0.004872
70	31	18	4	0.026160	0.020462
70	31	18	5	0.085657	0.059497
70	31	18	6	0.209792	0.124135
70	31	18	7	0.399794	0.190002
70	31	18	8	0.616003	0.216209
70	31	18	9	0.800181	0.184178
70	31	18	10	0.917818	0.117636
70	31	18	11	0.973963	0.056145
70	31	18	12	0.993812	0.019849
70	31	18	13	0.998931	0.005119
70	31	18	14	0.999871	0.000940
70	31	18	15	0.999990	0.000118
70	31	18	16	0.999999	0.000010
70	31	18	17	1.000000	0.000000
70	31	18	18	1.000000	0.000000
70	31	19	0	0.000001	0.000001
70	31	19	1	0.000032	0.000030
70	31	19	2	0.000405	0.000374
70	31	19	3	0.003075	0.002670
70	31	19	4	0.015536	0.012461
70	31	19	5	0.055909	0.040373
70	31	19	6	0.150112	0.094203
70	31	19	7	0.312100	0.161989
70	31	19	8	0.520372	0.208272
70	31	19	9	0.722260	0.201888
70	31	19	10	0.870311	0.148051

N	n	k	x	P(x)	p(x)
70	31	19	11	0.952368	0.082058
70	31	19	12	0.986559	0.034191
70	31	19	13	0.997159	0.010600
70	31	19	14	0.999564	0.002405
70	31	19	15	0.999953	0.000389
70	31	19	16	0.999997	0.000043
70	31	19	17	1.000000	0.000003
70	31	19	18	1.000000	0.000000
70	31	19	19	1.000000	0.000000
70	31	20	0	0.000000	0.000000
70	31	20	1	0.000014	0.000013
70	31	20	2	0.000193	0.000179
70	31	20	3	0.001609	0.001417
70	31	20	4	0.008939	0.007330
70	31	20	5	0.035327	0.026387
70	31	20	6	0.103934	0.068607
70	31	20	7	0.235870	0.131937
70	31	20	8	0.426446	0.190575
70	31	20	9	0.635171	0.208725
70	31	20	10	0.809349	0.174178
70	31	20	11	0.920189	0.110840
70	31	20	12	0.973821	0.053632
70	31	20	13	0.993418	0.019596
70	31	20	14	0.998762	0.005344
70	31	20	15	0.999831	0.001069
70	31	20	16	0.999984	0.000153
70	31	20	17	0.999999	0.000015
70	31	20	18	1.000000	0.000001
70	31	20	19	1.000000	0.000000
70	31	20	20	1.000000	0.000000
70	31	21	0	0.000000	0.000000
70	31	21	1	0.000006	0.000006
70	31	21	2	0.000089	0.000083
70	31	21	3	0.000816	0.000727
70	31	21	4	0.004981	0.004165
70	31	21	5	0.021605	0.016624
70	31	21	6	0.069630	0.048025
70	31	21	7	0.172541	0.102911
70	31	21	8	0.338781	0.166240
70	31	21	9	0.543332	0.204551
70	31	21	10	0.736194	0.192862
70	31	21	11	0.875853	0.139659
70	31	21	12	0.953441	0.077588
70	31	21	13	0.986363	0.032922
70	31	21	14	0.996945	0.010582
70	31	21	15	0.999489	0.002544
70	31	21	16	0.999938	0.000449
70	31	21	17	0.999995	0.000057
70	31	21	18	1.000000	0.000005
70	31	21	19	1.000000	0.000000

N	n	k	x	P(x)	p(x)
70	31	21	20	1.000000	0.000000
70	31	21	21	1.000000	0.000000
70	31	22	0	0.000000	0.000000
70	31	22	1	0.000002	0.000002
70	31	22	2	0.000040	0.000037
70	31	22	3	0.000401	0.000361
70	31	22	4	0.002686	0.002286
70	31	22	5	0.012784	0.010098
70	31	22	6	0.045128	0.032343
70	31	22	7	0.122136	0.077008
70	31	22	8	0.260750	0.138614
70	31	22	9	0.451493	0.190743
70	31	22	10	0.653539	0.202046
70	31	22	11	0.818849	0.165310
70	31	22	12	0.923356	0.104507
70	31	22	13	0.974269	0.050913
70	31	22	14	0.993274	0.019005
70	31	22	15	0.998658	0.005385
70	31	22	16	0.999801	0.001142
70	31	22	17	0.999979	0.000178
70	31	22	18	0.999998	0.000020
70	31	22	19	1.000000	0.000001
70	31	22	20	1.000000	0.000000
70	31	22	21	1.000000	0.000000
70	31	22	22	1.000000	0.000000
70	31	23	0	0.000000	0.000000
70	31	23	1	0.000001	0.000001
70	31	23	2	0.000017	0.000016
70	31	23	3	0.000190	0.000173
70	31	23	4	0.001400	0.001210
70	31	23	5	0.007314	0.005914
70	31	23	6	0.028282	0.020968
70	31	23	7	0.083632	0.055349
70	31	23	8	0.194331	0.110699
70	31	23	9	0.364069	0.169738
70	31	23	10	0.565144	0.201075
70	31	23	11	0.749970	0.184826
70	31	23	12	0.881989	0.132019
70	31	23	13	0.955177	0.073188
70	31	23	14	0.986543	0.031366
70	31	23	15	0.996864	0.010321
70	31	23	16	0.999444	0.002580
70	31	23	17	0.999927	0.000483
70	31	23	18	0.999993	0.000066
70	31	23	19	0.999999	0.000006
70	31	23	20	1.000000	0.000000
70	31	23	21	1.000000	0.000000
70	31	23	22	1.000000	0.000000
70	31	23	23	1.000000	0.000000
70	31	24	0	0.000000	0.000000

N	n	k	x	P(x)	p(x)
70	31	24	1	0.000000	0.000000
70	31	24	2	0.000007	0.000007
70	31	24	3	0.000087	0.000080
70	31	24	4	0.000705	0.000618
70	31	24	5	0.004043	0.003338
70	31	24	6	0.017129	0.013086
70	31	24	7	0.055368	0.038239
70	31	24	8	0.140159	0.084791
70	31	24	9	0.284617	0.144458
70	31	24	10	0.475302	0.190685
70	31	24	11	0.671320	0.196019
70	31	24	12	0.828619	0.157299
70	31	24	13	0.927147	0.098528
70	31	24	14	0.975198	0.048051
70	31	24	15	0.993350	0.018152
70	31	24	16	0.998620	0.005270
70	31	24	17	0.999783	0.001163
70	31	24	18	0.999975	0.000192
70	31	24	19	0.999998	0.000023
70	31	24	20	1.000000	0.000002
70	31	24	21	1.000000	0.000000
70	31	24	22	1.000000	0.000000
70	31	24	23	1.000000	0.000000
70	31	24	24	1.000000	0.000000
70	31	25	0	0.000000	0.000000
70	31	25	1	0.000000	0.000000
70	31	25	2	0.000003	0.000003
70	31	25	3	0.000038	0.000036
70	31	25	4	0.000342	0.000304
70	31	25	5	0.002156	0.001814
70	31	25	6	0.010017	0.007861
70	31	25	7	0.035417	0.025400
70	31	25	8	0.097763	0.062346
70	31	25	9	0.215528	0.117765
70	31	25	10	0.388250	0.172722
70	31	25	11	0.586095	0.197845
70	31	25	12	0.763648	0.177553
70	31	25	13	0.888593	0.124945
70	31	25	14	0.957440	0.068847
70	31	25	15	0.987036	0.029596
70	31	25	16	0.996902	0.009865
70	31	25	17	0.999429	0.002527
70	31	25	18	0.999920	0.000491
70	31	25	19	0.999992	0.000071
70	31	25	20	0.999999	0.000008
70	31	25	21	1.000000	0.000001
70	31	25	22	1.000000	0.000000
70	31	25	23	1.000000	0.000000
70	31	25	24	1.000000	0.000000
70	31	25	25	1.000000	0.000000

Table for $N = 2$, $n = 1$, through $N = 100$, $n = 50$

N	n	k	x	P(x)	p(x)
70	31	26	0	0.000000	0.000000
70	31	26	1	0.000000	0.000000
70	31	26	2	0.000001	0.000001
70	31	26	3	0.000016	0.000015
70	31	26	4	0.000160	0.000144
70	31	26	5	0.001108	0.000948
70	31	26	6	0.005650	0.004542
70	31	26	7	0.021870	0.016221
70	31	26	8	0.065898	0.044027
70	31	26	9	0.157954	0.092057
70	31	26	10	0.307646	0.149692
70	31	26	11	0.498164	0.190517
70	31	26	12	0.688681	0.190517
70	31	26	13	0.838615	0.149934
70	31	26	14	0.931431	0.092816
70	31	26	15	0.976513	0.045082
70	31	26	16	0.993613	0.017100
70	31	26	17	0.998643	0.005029
70	31	26	18	0.999778	0.001136
70	31	26	19	0.999973	0.000194
70	31	26	20	0.999997	0.000025
70	31	26	21	1.000000	0.000002
70	31	26	22	1.000000	0.000000
70	31	26	23	1.000000	0.000000
70	31	26	24	1.000000	0.000000
70	31	26	25	1.000000	0.000000
70	31	26	26	1.000000	0.000000
70	31	27	0	0.000000	0.000000
70	31	27	1	0.000000	0.000000
70	31	27	2	0.000000	0.000000
70	31	27	3	0.000007	0.000006
70	31	27	4	0.000072	0.000065
70	31	27	5	0.000548	0.000476
70	31	27	6	0.003069	0.002522
70	31	27	7	0.013023	0.009953
70	31	27	8	0.042883	0.029860
70	31	27	9	0.111926	0.069043
70	31	27	10	0.236202	0.124277
70	31	27	11	0.411565	0.175363
70	31	27	12	0.606412	0.194847
70	31	27	13	0.777278	0.170866
70	31	27	14	0.895570	0.118292
70	31	27	15	0.960120	0.064549
70	31	27	16	0.987784	0.027664
70	31	27	17	0.997042	0.009259
70	31	27	18	0.999443	0.002400
70	31	27	19	0.999920	0.000477
70	31	27	20	0.999991	0.000072
70	31	27	21	0.999999	0.000008
70	31	27	22	1.000000	0.000001

N	n	k	x	P(x)	p(x)
70	31	27	23	1.000000	0.000000
70	31	27	24	1.000000	0.000000
70	31	27	25	1.000000	0.000000
70	31	27	26	1.000000	0.000000
70	31	27	27	1.000000	0.000000
70	31	28	0	0.000000	0.000000
70	31	28	1	0.000000	0.000000
70	31	28	2	0.000000	0.000000
70	31	28	3	0.000003	0.000002
70	31	28	4	0.000031	0.000028
70	31	28	5	0.000260	0.000229
70	31	28	6	0.001603	0.001343
70	31	28	7	0.007467	0.005864
70	31	28	8	0.026912	0.019444
70	31	28	9	0.076602	0.049690
70	31	28	10	0.175509	0.098907
70	31	28	11	0.330001	0.154492
70	31	28	12	0.520317	0.190316
70	31	28	13	0.705753	0.185436
70	31	28	14	0.848804	0.143051
70	31	28	15	0.936101	0.087298
70	31	28	16	0.978133	0.042032
70	31	28	17	0.994028	0.015895
70	31	28	18	0.998717	0.004689
70	31	28	19	0.999787	0.001069
70	31	28	20	0.999973	0.000186
70	31	28	21	0.999997	0.000024
70	31	28	22	1.000000	0.000002
70	31	28	23	1.000000	0.000000
70	31	28	24	1.000000	0.000000
70	31	28	25	1.000000	0.000000
70	31	28	26	1.000000	0.000000
70	31	28	27	1.000000	0.000000
70	31	28	28	1.000000	0.000000
70	31	29	0	0.000000	0.000000
70	31	29	1	0.000000	0.000000
70	31	29	2	0.000000	0.000000
70	31	29	3	0.000001	0.000001
70	31	29	4	0.000013	0.000012
70	31	29	5	0.000118	0.000105
70	31	29	6	0.000804	0.000686
70	31	29	7	0.004117	0.003313
70	31	29	8	0.016264	0.012147
70	31	29	9	0.050574	0.034310
70	31	29	10	0.126056	0.075482
70	31	29	11	0.256434	0.130378
70	31	29	12	0.434222	0.177788
70	31	29	13	0.626281	0.192059
70	31	29	14	0.790902	0.164622
70	31	29	15	0.902845	0.111943

N	n	k	x	P(x)	p(x)
70	31	29	16	0.963122	0.060277
70	31	29	17	0.988730	0.025608
70	31	29	18	0.997266	0.008536
70	31	29	19	0.999481	0.002215
70	31	29	20	0.999924	0.000443
70	31	29	21	0.999992	0.000067
70	31	29	22	0.999999	0.000008
70	31	29	23	1.000000	0.000001
70	31	29	24	1.000000	0.000000
70	31	29	25	1.000000	0.000000
70	31	29	26	1.000000	0.000000
70	31	29	27	1.000000	0.000000
70	31	29	28	1.000000	0.000000
70	31	29	29	1.000000	0.000000
70	31	30	0	0.000000	0.000000
70	31	30	1	0.000000	0.000000
70	31	30	2	0.000000	0.000000
70	31	30	3	0.000000	0.000000
70	31	30	4	0.000005	0.000005
70	31	30	5	0.000051	0.000046
70	31	30	6	0.000386	0.000334
70	31	30	7	0.002177	0.001792
70	31	30	8	0.009449	0.007272
70	31	30	9	0.032163	0.022714
70	31	30	10	0.087394	0.055231
70	31	30	11	0.192835	0.105440
70	31	30	12	0.351832	0.158998
70	31	30	13	0.541962	0.190130
70	31	30	14	0.722644	0.180682
70	31	30	15	0.859160	0.136516
70	31	30	16	0.941069	0.081909
70	31	30	17	0.979986	0.038916
70	31	30	18	0.994559	0.014574
70	31	30	19	0.998833	0.004273
70	31	30	20	0.999805	0.000973
70	31	30	21	0.999975	0.000170
70	31	30	22	0.999997	0.000022
70	31	30	23	1.000000	0.000002
70	31	30	24	1.000000	0.000000
70	31	30	25	1.000000	0.000000
70	31	30	26	1.000000	0.000000
70	31	30	27	1.000000	0.000000
70	31	30	28	1.000000	0.000000
70	31	30	29	1.000000	0.000000
70	31	30	30	1.000000	0.000000
70	31	31	0	0.000000	0.000000
70	31	31	1	0.000000	0.000000
70	31	31	2	0.000000	0.000000
70	31	31	3	0.000000	0.000000
70	31	31	4	0.000002	0.000002

N	n	k	x	P(x)	p(x)
70	31	31	5	0.000021	0.000019
70	31	31	6	0.000177	0.000156
70	31	31	7	0.001102	0.000926
70	31	31	8	0.005268	0.004166
70	31	31	9	0.019671	0.014403
70	31	31	10	0.058398	0.038727
70	31	31	11	0.140114	0.081716
70	31	31	12	0.276308	0.136194
70	31	31	13	0.456404	0.180095
70	31	31	14	0.645854	0.189451
70	31	31	15	0.804554	0.158699
70	31	31	16	0.910353	0.105800
70	31	31	17	0.966365	0.056012
70	31	31	18	0.989823	0.023458
70	31	31	19	0.997551	0.007728
70	31	31	20	0.999538	0.001987
70	31	31	21	0.999933	0.000395
70	31	31	22	0.999992	0.000060
70	31	31	23	0.999999	0.000007
70	31	31	24	1.000000	0.000001
70	31	31	25	1.000000	0.000000
70	31	31	26	1.000000	0.000000
70	31	31	27	1.000000	0.000000
70	31	31	28	1.000000	0.000000
70	31	31	29	1.000000	0.000000
70	31	31	30	1.000000	0.000000
70	31	31	31	1.000000	0.000000
70	32	1	0	0.542857	0.542857
70	32	1	1	1.000000	0.457143
70	32	2	0	0.291097	0.291097
70	32	2	1	0.794617	0.503520
70	32	2	2	1.000000	0.205383
70	32	3	0	0.154110	0.154110
70	32	3	1	0.565071	0.410961
70	32	3	2	0.909390	0.344319
70	32	3	3	1.000000	0.090610
70	32	4	0	0.080505	0.080505
70	32	4	1	0.374925	0.294420
70	32	4	2	0.755217	0.380292
70	32	4	3	0.960781	0.205563
70	32	4	4	1.000000	0.039219
70	32	5	0	0.041472	0.041472
70	32	5	1	0.236637	0.195165
70	32	5	2	0.582357	0.345720
70	32	5	3	0.870457	0.288100
70	32	5	4	0.983361	0.112904
70	32	5	5	1.000000	0.016638
70	32	6	0	0.021055	0.021055
70	32	6	1	0.143559	0.122503
70	32	6	2	0.422794	0.279235

N	n	k	x	P(x)	p(x)
70	32	6	3	0.741920	0.319126
70	32	6	4	0.934726	0.192805
70	32	6	5	0.993088	0.058363
70	32	6	6	1.000000	0.006911
70	32	7	0	0.010528	0.010528
70	32	7	1	0.084221	0.073693
70	32	7	2	0.291902	0.207681
70	32	7	3	0.597316	0.305414
70	32	7	4	0.850373	0.253057
70	32	7	5	0.968467	0.118093
70	32	7	6	0.997192	0.028725
70	32	7	7	1.000000	0.002808
70	32	8	0	0.005180	0.005180
70	32	8	1	0.047959	0.042779
70	32	8	2	0.193006	0.145047
70	32	8	3	0.456729	0.263722
70	32	8	4	0.737904	0.281175
70	32	8	5	0.917855	0.179952
70	32	8	6	0.985337	0.067482
70	32	8	7	0.998886	0.013548
70	32	8	8	1.000000	0.001114
70	32	9	0	0.002507	0.002507
70	32	9	1	0.026570	0.024063
70	32	9	2	0.122822	0.096253
70	32	9	3	0.333375	0.210553
70	32	9	4	0.610921	0.277547
70	32	9	5	0.839489	0.228568
70	32	9	6	0.957038	0.117549
70	32	9	7	0.993423	0.036384
70	32	9	8	0.999569	0.006146
70	32	9	9	1.000000	0.000431
70	32	10	0	0.001192	0.001192
70	32	10	1	0.014341	0.013149
70	32	10	2	0.075485	0.061144
70	32	10	3	0.233276	0.157791
70	32	10	4	0.483523	0.250247
70	32	10	5	0.738320	0.254797
70	32	10	6	0.906935	0.168616
70	32	10	7	0.978511	0.071576
70	32	10	8	0.997151	0.018639
70	32	10	9	0.999837	0.002687
70	32	10	10	1.000000	0.000163
70	32	11	0	0.000556	0.000556
70	32	11	1	0.007547	0.006991
70	32	11	2	0.044913	0.037366
70	32	11	3	0.157010	0.112097
70	32	11	4	0.366741	0.209731
70	32	11	5	0.623661	0.256920
70	32	11	6	0.833869	0.210207
70	32	11	7	0.948688	0.114819

N	n	k	x	P(x)	p(x)
70	32	11	8	0.989695	0.041007
70	32	11	9	0.998807	0.009113
70	32	11	10	0.999940	0.001133
70	32	11	11	1.000000	0.000060
70	32	12	0	0.000254	0.000254
70	32	12	1	0.003874	0.003619
70	32	12	2	0.025913	0.022039
70	32	12	3	0.101912	0.075998
70	32	12	4	0.267208	0.165296
70	32	12	5	0.506088	0.238880
70	32	12	6	0.741235	0.235147
70	32	12	7	0.900036	0.158801
70	32	12	8	0.973014	0.072978
70	32	12	9	0.995255	0.022241
70	32	12	10	0.999518	0.004263
70	32	12	11	0.999979	0.000461
70	32	12	12	1.000000	0.000021
70	32	13	0	0.000114	0.000114
70	32	13	1	0.001939	0.001825
70	32	13	2	0.014514	0.012574
70	32	13	3	0.063913	0.049399
70	32	13	4	0.187410	0.123497
70	32	13	5	0.394885	0.207475
70	32	13	6	0.635824	0.240939
70	32	13	7	0.831587	0.195763
70	32	13	8	0.942816	0.111229
70	32	13	9	0.986435	0.043619
70	32	13	10	0.997901	0.011466
70	32	13	11	0.999812	0.001911
70	32	13	12	0.999993	0.000181
70	32	13	13	1.000000	0.000007
70	32	14	0	0.000050	0.000050
70	32	14	1	0.000947	0.000897
70	32	14	2	0.007896	0.006949
70	32	14	3	0.038780	0.030884
70	32	14	4	0.126744	0.087965
70	32	14	5	0.296607	0.169863
70	32	14	6	0.525922	0.229315
70	32	14	7	0.745726	0.219804
70	32	14	8	0.895983	0.150257
70	32	14	9	0.968835	0.072852
70	32	14	10	0.993476	0.024641
70	32	14	11	0.999108	0.005632
70	32	14	12	0.999929	0.000821
70	32	14	13	0.999997	0.000068
70	32	14	14	1.000000	0.000002
70	32	15	0	0.000021	0.000021
70	32	15	1	0.000450	0.000429
70	32	15	2	0.004173	0.003723
70	32	15	3	0.022786	0.018613

N	n	k	x	P(x)	p(x)
70	32	15	4	0.082762	0.059976
70	32	15	5	0.214709	0.131947
70	32	15	6	0.419454	0.204745
70	32	15	7	0.647599	0.228145
70	32	15	8	0.831587	0.183988
70	32	15	9	0.938913	0.107326
70	32	15	10	0.983795	0.044882
70	32	15	11	0.996996	0.013201
70	32	15	12	0.999636	0.002640
70	32	15	13	0.999974	0.000338
70	32	15	14	0.999999	0.000025
70	32	15	15	1.000000	0.000001
70	32	16	0	0.000009	0.000009
70	32	16	1	0.000209	0.000200
70	32	16	2	0.002142	0.001934
70	32	16	3	0.012972	0.010830
70	32	16	4	0.052229	0.039257
70	32	16	5	0.149935	0.097706
70	32	16	6	0.322666	0.172731
70	32	16	7	0.543897	0.221231
70	32	16	8	0.751301	0.207404
70	32	16	9	0.894031	0.142730
70	32	16	10	0.965842	0.071811
70	32	16	11	0.991955	0.026113
70	32	16	12	0.998676	0.006720
70	32	16	13	0.999857	0.001182
70	32	16	14	0.999991	0.000134
70	32	16	15	1.000000	0.000009
70	32	16	16	1.000000	0.000000
70	32	17	0	0.000004	0.000004
70	32	17	1	0.000094	0.000090
70	32	17	2	0.001068	0.000974
70	32	17	3	0.007156	0.006088
70	32	17	4	0.031873	0.024717
70	32	17	5	0.101082	0.069209
70	32	17	6	0.239499	0.138417
70	32	17	7	0.441475	0.201976
70	32	17	8	0.659122	0.217647
70	32	17	9	0.833239	0.174117
70	32	17	10	0.936586	0.103347
70	32	17	11	0.981800	0.045214
70	32	17	12	0.996187	0.014386
70	32	17	13	0.999442	0.003255
70	32	17	14	0.999946	0.000505
70	32	17	15	0.999997	0.000050
70	32	17	16	1.000000	0.000003
70	32	17	17	1.000000	0.000000
70	32	18	0	0.000001	0.000001
70	32	18	1	0.000041	0.000040
70	32	18	2	0.000517	0.000476

N	n	k	x	P(x)	p(x)
70	32	18	3	0.003825	0.003308
70	32	18	4	0.018815	0.014990
70	32	18	5	0.065825	0.047010
70	32	18	6	0.171596	0.105772
70	32	18	7	0.346203	0.174607
70	32	18	8	0.560565	0.214361
70	32	18	9	0.757679	0.197114
70	32	18	10	0.893687	0.136009
70	32	18	11	0.963885	0.070198
70	32	18	12	0.990758	0.026873
70	32	18	13	0.998275	0.007517
70	32	18	14	0.999775	0.001500
70	32	18	15	0.999981	0.000206
70	32	18	16	0.999999	0.000018
70	32	18	17	1.000000	0.000001
70	32	18	18	1.000000	0.000000
70	32	19	0	0.000001	0.000001
70	32	19	1	0.000017	0.000017
70	32	19	2	0.000242	0.000225
70	32	19	3	0.001980	0.001738
70	32	19	4	0.010744	0.008764
70	32	19	5	0.041416	0.030672
70	32	19	6	0.118711	0.077295
70	32	19	7	0.262258	0.143547
70	32	19	8	0.461629	0.199371
70	32	19	9	0.670494	0.208865
70	32	19	10	0.836145	0.165652
70	32	19	11	0.935536	0.099391
70	32	19	12	0.980422	0.044886
70	32	19	13	0.995528	0.015106
70	32	19	14	0.999256	0.003727
70	32	19	15	0.999913	0.000658
70	32	19	16	0.999993	0.000080
70	32	19	17	1.000000	0.000006
70	32	19	18	1.000000	0.000000
70	32	19	19	1.000000	0.000000
70	32	20	0	0.000000	0.000000
70	32	20	1	0.000007	0.000007
70	32	20	2	0.000110	0.000103
70	32	20	3	0.000992	0.000882
70	32	20	4	0.005932	0.004940
70	32	20	5	0.025178	0.019245
70	32	20	6	0.079306	0.054128
70	32	20	7	0.191891	0.112586
70	32	20	8	0.367807	0.175916
70	32	20	9	0.576300	0.208493
70	32	20	10	0.764688	0.188388
70	32	20	11	0.894610	0.129923
70	32	20	12	0.962820	0.068209
70	32	20	13	0.989901	0.027081

Table for $N = 2$, $n = 1$, through $N = 100$, $n = 50$

N	n	k	x	P(x)	p(x)
70	32	20	14	0.997940	0.008040
70	32	20	15	0.999694	0.001754
70	32	20	16	0.999968	0.000274
70	32	20	17	0.999998	0.000029
70	32	20	18	1.000000	0.000002
70	32	20	19	1.000000	0.000000
70	32	20	20	1.000000	0.000000
70	32	21	0	0.000000	0.000000
70	32	21	1	0.000003	0.000003
70	32	21	2	0.000048	0.000045
70	32	21	3	0.000480	0.000432
70	32	21	4	0.003166	0.002685
70	32	21	5	0.014785	0.011619
70	32	21	6	0.051159	0.036374
70	32	21	7	0.135598	0.084439
70	32	21	8	0.283368	0.147769
70	32	21	9	0.480393	0.197026
70	32	21	10	0.681797	0.201404
70	32	21	11	0.840043	0.158246
70	32	21	12	0.935536	0.095493
70	32	21	13	0.979610	0.044074
70	32	21	14	0.995046	0.015436
70	32	21	15	0.999098	0.004052
70	32	21	16	0.999881	0.000783
70	32	21	17	0.999989	0.000108
70	32	21	18	0.999999	0.000010
70	32	21	19	1.000000	0.000001
70	32	21	20	1.000000	0.000000
70	32	21	21	1.000000	0.000000
70	32	22	0	0.000000	0.000000
70	32	22	1	0.000001	0.000001
70	32	22	2	0.000020	0.000019
70	32	22	3	0.000225	0.000204
70	32	22	4	0.001631	0.001407
70	32	22	5	0.008382	0.006751
70	32	22	6	0.031859	0.023476
70	32	22	7	0.092517	0.060659
70	32	22	8	0.210991	0.118474
70	32	22	9	0.387912	0.176921
70	32	22	10	0.591371	0.203459
70	32	22	11	0.772223	0.180852
70	32	22	12	0.896559	0.124336
70	32	22	13	0.962520	0.065961
70	32	22	14	0.989375	0.026855
70	32	22	15	0.997692	0.008317
70	32	22	16	0.999625	0.001933
70	32	22	17	0.999956	0.000331
70	32	22	18	0.999996	0.000041
70	32	22	19	1.000000	0.000003
70	32	22	20	1.000000	0.000000

N	n	k	x	P(x)	p(x)
70	32	22	21	1.000000	0.000000
70	32	22	22	1.000000	0.000000
70	32	23	0	0.000000	0.000000
70	32	23	1	0.000000	0.000000
70	32	23	2	0.000008	0.000008
70	32	23	3	0.000101	0.000093
70	32	23	4	0.000811	0.000709
70	32	23	5	0.004585	0.003774
70	32	23	6	0.019142	0.014557
70	32	23	7	0.060924	0.041782
70	32	23	8	0.151754	0.090830
70	32	23	9	0.303137	0.151383
70	32	23	10	0.498119	0.194982
70	32	23	11	0.693100	0.194982
70	32	23	12	0.844753	0.151652
70	32	23	13	0.936411	0.091658
70	32	23	14	0.979305	0.042894
70	32	23	15	0.994747	0.015442
70	32	23	16	0.998981	0.004234
70	32	23	17	0.999852	0.000872
70	32	23	18	0.999984	0.000132
70	32	23	19	0.999999	0.000014
70	32	23	20	1.000000	0.000001
70	32	23	21	1.000000	0.000000
70	32	23	22	1.000000	0.000000
70	32	23	23	1.000000	0.000000
70	32	24	0	0.000000	0.000000
70	32	24	1	0.000000	0.000000
70	32	24	2	0.000003	0.000003
70	32	24	3	0.000044	0.000041
70	32	24	4	0.000388	0.000344
70	32	24	5	0.002417	0.002029
70	32	24	6	0.011089	0.008673
70	32	24	7	0.038700	0.027610
70	32	24	8	0.105373	0.066673
70	32	24	9	0.229056	0.123683
70	32	24	10	0.406851	0.177795
70	32	24	11	0.605981	0.199130
70	32	24	12	0.780220	0.174239
70	32	24	13	0.899357	0.119138
70	32	24	14	0.962877	0.063520
70	32	24	15	0.989161	0.026284
70	32	24	16	0.997539	0.008378
70	32	24	17	0.999574	0.002035
70	32	24	18	0.999945	0.000371
70	32	24	19	0.999995	0.000050
70	32	24	20	1.000000	0.000005
70	32	24	21	1.000000	0.000000
70	32	24	22	1.000000	0.000000
70	32	24	23	1.000000	0.000000

N	n	k	x	P(x)	p(x)
70	32	24	24	1.000000	0.000000
70	32	25	0	0.000000	0.000000
70	32	25	1	0.000000	0.000000
70	32	25	2	0.000001	0.000001
70	32	25	3	0.000018	0.000017
70	32	25	4	0.000179	0.000160
70	32	25	5	0.001226	0.001047
70	32	25	6	0.006187	0.004961
70	32	25	7	0.023694	0.017507
70	32	25	8	0.070587	0.046893
70	32	25	9	0.167214	0.096628
70	32	25	10	0.321819	0.154604
70	32	25	11	0.515074	0.193255
70	32	25	12	0.704464	0.189390
70	32	25	13	0.850148	0.145685
70	32	25	14	0.938022	0.087873
70	32	25	15	0.979448	0.041426
70	32	25	16	0.994625	0.015178
70	32	25	17	0.998911	0.004285
70	32	25	18	0.999832	0.000922
70	32	25	19	0.999981	0.000149
70	32	25	20	0.999998	0.000018
70	32	25	21	1.000000	0.000001
70	32	25	22	1.000000	0.000000
70	32	25	23	1.000000	0.000000
70	32	25	24	1.000000	0.000000
70	32	25	25	1.000000	0.000000
70	32	26	0	0.000000	0.000000
70	32	26	1	0.000000	0.000000
70	32	26	2	0.000000	0.000000
70	32	26	3	0.000007	0.000007
70	32	26	4	0.000079	0.000072
70	32	26	5	0.000598	0.000519
70	32	26	6	0.003321	0.002723
70	32	26	7	0.013968	0.010647
70	32	26	8	0.045577	0.031609
70	32	26	9	0.117827	0.072250
70	32	26	10	0.246234	0.128407
70	32	26	11	0.424888	0.178654
70	32	26	12	0.620290	0.195402
70	32	26	13	0.788637	0.168347
70	32	26	14	0.902872	0.114235
70	32	26	15	0.963798	0.060925
70	32	26	16	0.989229	0.025431
70	32	26	17	0.997482	0.008253
70	32	26	18	0.999545	0.002063
70	32	26	19	0.999938	0.000392
70	32	26	20	0.999994	0.000056
70	32	26	21	0.999999	0.000006
70	32	26	22	1.000000	0.000000

N	n	k	x	P(x)	p(x)
70	32	26	23	1.000000	0.000000
70	32	26	24	1.000000	0.000000
70	32	26	25	1.000000	0.000000
70	32	26	26	1.000000	0.000000
70	32	27	0	0.000000	0.000000
70	32	27	1	0.000000	0.000000
70	32	27	2	0.000000	0.000000
70	32	27	3	0.000003	0.000003
70	32	27	4	0.000033	0.000031
70	32	27	5	0.000279	0.000246
70	32	27	6	0.001712	0.001432
70	32	27	7	0.007919	0.006207
70	32	27	8	0.028336	0.020417
70	32	27	9	0.080060	0.051724
70	32	27	10	0.182031	0.101970
70	32	27	11	0.339621	0.157591
70	32	27	12	0.531471	0.191850
70	32	27	13	0.715942	0.184471
70	32	27	14	0.856140	0.140198
70	32	27	15	0.940258	0.084119
70	32	27	16	0.979981	0.039723
70	32	27	17	0.994668	0.014687
70	32	27	18	0.998889	0.004221
70	32	27	19	0.999822	0.000933
70	32	27	20	0.999978	0.000156
70	32	27	21	0.999998	0.000020
70	32	27	22	1.000000	0.000002
70	32	27	23	1.000000	0.000000
70	32	27	24	1.000000	0.000000
70	32	27	25	1.000000	0.000000
70	32	27	26	1.000000	0.000000
70	32	27	27	1.000000	0.000000
70	32	28	0	0.000000	0.000000
70	32	28	1	0.000000	0.000000
70	32	28	2	0.000000	0.000000
70	32	28	3	0.000001	0.000001
70	32	28	4	0.000013	0.000012
70	32	28	5	0.000125	0.000111
70	32	28	6	0.000846	0.000721
70	32	28	7	0.004310	0.003464
70	32	28	8	0.016940	0.012630
70	32	28	9	0.052394	0.035454
70	32	28	10	0.129860	0.077466
70	32	28	11	0.262658	0.132799
70	32	28	12	0.442239	0.179580
70	32	28	13	0.634432	0.192193
70	32	28	14	0.797452	0.163021
70	32	28	15	0.907002	0.109550
70	32	28	16	0.965201	0.058198
70	32	28	17	0.989545	0.024344

Table for $N = 2$, $n = 1$, through $N = 100$, $n = 50$

N	n	k	x	P(x)	p(x)
70	32	28	18	0.997515	0.007970
70	32	28	19	0.999540	0.002025
70	32	28	20	0.999935	0.000395
70	32	28	21	0.999993	0.000058
70	32	28	22	0.999999	0.000006
70	32	28	23	1.000000	0.000000
70	32	28	24	1.000000	0.000000
70	32	28	25	1.000000	0.000000
70	32	28	26	1.000000	0.000000
70	32	28	27	1.000000	0.000000
70	32	28	28	1.000000	0.000000
70	32	29	0	0.000000	0.000000
70	32	29	1	0.000000	0.000000
70	32	29	2	0.000000	0.000000
70	32	29	3	0.000000	0.000000
70	32	29	4	0.000005	0.000005
70	32	29	5	0.000053	0.000048
70	32	29	6	0.000399	0.000346
70	32	29	7	0.002248	0.001848
70	32	29	8	0.009723	0.007475
70	32	29	9	0.032979	0.023256
70	32	29	10	0.089282	0.056304
70	32	29	11	0.196259	0.106977
70	32	29	12	0.356724	0.160465
70	32	29	13	0.547487	0.190763
70	32	29	14	0.727586	0.180099
70	32	29	15	0.862661	0.135074
70	32	29	16	0.943030	0.080369
70	32	29	17	0.980851	0.037821
70	32	29	18	0.994858	0.014008
70	32	29	19	0.998913	0.004055
70	32	29	20	0.999822	0.000909
70	32	29	21	0.999978	0.000156
70	32	29	22	0.999998	0.000020
70	32	29	23	1.000000	0.000002
70	32	29	24	1.000000	0.000000
70	32	29	25	1.000000	0.000000
70	32	29	26	1.000000	0.000000
70	32	29	27	1.000000	0.000000
70	32	29	28	1.000000	0.000000
70	32	29	29	1.000000	0.000000
70	32	30	0	0.000000	0.000000
70	32	30	1	0.000000	0.000000
70	32	30	2	0.000000	0.000000
70	32	30	3	0.000000	0.000000
70	32	30	4	0.000002	0.000002
70	32	30	5	0.000022	0.000020
70	32	30	6	0.000180	0.000158
70	32	30	7	0.001121	0.000941
70	32	30	8	0.005347	0.004226

N	n	k	x	P(x)	p(x)
70	32	30	9	0.019933	0.014586
70	32	30	10	0.059071	0.039138
70	32	30	11	0.141466	0.082395
70	32	30	12	0.278449	0.136982
70	32	30	13	0.459085	0.180636
70	32	30	14	0.648518	0.189433
70	32	30	15	0.806654	0.158136
70	32	30	16	0.911666	0.105012
70	32	30	17	0.967014	0.055348
70	32	30	18	0.990075	0.023061
70	32	30	19	0.997627	0.007552
70	32	30	20	0.999556	0.001929
70	32	30	21	0.999936	0.000380
70	32	30	22	0.999993	0.000057
70	32	30	23	0.999999	0.000006
70	32	30	24	1.000000	0.000001
70	32	30	25	1.000000	0.000000
70	32	30	26	1.000000	0.000000
70	32	30	27	1.000000	0.000000
70	32	30	28	1.000000	0.000000
70	32	30	29	1.000000	0.000000
70	32	30	30	1.000000	0.000000
70	32	31	0	0.000000	0.000000
70	32	31	1	0.000000	0.000000
70	32	31	2	0.000000	0.000000
70	32	31	3	0.000000	0.000000
70	32	31	4	0.000001	0.000001
70	32	31	5	0.000008	0.000008
70	32	31	6	0.000077	0.000069
70	32	31	7	0.000533	0.000456
70	32	31	8	0.002811	0.002279
70	32	31	9	0.011546	0.008735
70	32	31	10	0.037545	0.025999
70	32	31	11	0.098208	0.060664
70	32	31	12	0.209957	0.111749
70	32	31	13	0.373283	0.163325
70	32	31	14	0.563273	0.189991
70	32	31	15	0.739446	0.176173
70	32	31	16	0.869661	0.130215
70	32	31	17	0.946258	0.076597
70	32	31	18	0.982004	0.035745
70	32	31	19	0.995173	0.013169
70	32	31	20	0.998977	0.003804
70	32	31	21	0.999831	0.000854
70	32	31	22	0.999979	0.000147
70	32	31	23	0.999998	0.000019
70	32	31	24	1.000000	0.000002
70	32	31	25	1.000000	0.000000
70	32	31	26	1.000000	0.000000
70	32	31	27	1.000000	0.000000

N	n	k	x	P(x)	p(x)
70	32	31	28	1.000000	0.000000
70	32	31	29	1.000000	0.000000
70	32	31	30	1.000000	0.000000
70	32	31	31	1.000000	0.000000
70	32	32	0	0.000000	0.000000
70	32	32	1	0.000000	0.000000
70	32	32	2	0.000000	0.000000
70	32	32	3	0.000000	0.000000
70	32	32	4	0.000000	0.000000
70	32	32	5	0.000003	0.000003
70	32	32	6	0.000031	0.000028
70	32	32	7	0.000241	0.000209
70	32	32	8	0.001409	0.001169
70	32	32	9	0.006395	0.004986
70	32	32	10	0.022879	0.016484
70	32	32	11	0.065543	0.042665
70	32	32	12	0.152650	0.087107
70	32	32	13	0.293714	0.141064
70	32	32	14	0.475585	0.181871
70	32	32	15	0.662653	0.187068
70	32	32	16	0.816240	0.153587
70	32	32	17	0.916798	0.100558
70	32	32	18	0.969172	0.052374
70	32	32	19	0.990783	0.021611
70	32	32	20	0.997807	0.007024
70	32	32	21	0.999591	0.001784
70	32	32	22	0.999941	0.000350
70	32	32	23	0.999993	0.000053
70	32	32	24	0.999999	0.000006
70	32	32	25	1.000000	0.000000
70	32	32	26	1.000000	0.000000
70	32	32	27	1.000000	0.000000
70	32	32	28	1.000000	0.000000
70	32	32	29	1.000000	0.000000
70	32	32	30	1.000000	0.000000
70	32	32	31	1.000000	0.000000
70	32	32	32	1.000000	0.000000
70	33	1	0	0.528571	0.528571
70	33	1	1	1.000000	0.471429
70	33	2	0	0.275776	0.275776
70	33	2	1	0.781366	0.505590
70	33	2	2	1.000000	0.218634
70	33	3	0	0.141944	0.141944
70	33	3	1	0.543442	0.401498
70	33	3	2	0.900329	0.356887
70	33	3	3	1.000000	0.099671
70	33	4	0	0.072031	0.072031
70	33	4	1	0.351681	0.279650
70	33	4	2	0.735202	0.383520
70	33	4	3	0.955371	0.220169

N	n	k	x	P(x)	p(x)
70	33	4	4	1.000000	0.044629
70	33	5	0	0.036016	0.036016
70	33	5	1	0.216093	0.180078
70	33	5	2	0.555064	0.338970
70	33	5	3	0.855294	0.300231
70	33	5	4	0.980390	0.125096
70	33	5	5	1.000000	0.019610
70	33	6	0	0.017731	0.017731
70	33	6	1	0.127440	0.109709
70	33	6	2	0.393401	0.265961
70	33	6	3	0.716726	0.323325
70	33	6	4	0.924578	0.207852
70	33	6	5	0.991553	0.066975
70	33	6	6	1.000000	0.008447
70	33	7	0	0.008588	0.008588
70	33	7	1	0.072585	0.063997
70	33	7	2	0.264576	0.191991
70	33	7	3	0.565167	0.300592
70	33	7	4	0.830395	0.265228
70	33	7	5	0.962251	0.131856
70	33	7	6	0.996436	0.034185
70	33	7	7	1.000000	0.003564
70	33	8	0	0.004090	0.004090
70	33	8	1	0.040079	0.035989
70	33	8	2	0.170104	0.130025
70	33	8	3	0.422029	0.251924
70	33	8	4	0.708306	0.286278
70	33	8	5	0.903649	0.195342
70	33	8	6	0.981785	0.078137
70	33	8	7	0.998529	0.016744
70	33	8	8	1.000000	0.001471
70	33	9	0	0.001913	0.001913
70	33	9	1	0.021504	0.019591
70	33	9	2	0.105092	0.083588
70	33	9	3	0.300130	0.195038
70	33	9	4	0.574402	0.274272
70	33	9	5	0.815429	0.241027
70	33	9	6	0.947758	0.132329
70	33	9	7	0.991508	0.043749
70	33	9	8	0.999407	0.007899
70	33	9	9	1.000000	0.000593
70	33	10	0	0.000878	0.000878
70	33	10	1	0.011227	0.010349
70	33	10	2	0.062613	0.051386
70	33	10	3	0.204209	0.141597
70	33	10	4	0.444010	0.239801
70	33	10	5	0.704794	0.260784
70	33	10	6	0.889186	0.184392
70	33	10	7	0.972860	0.083674
70	33	10	8	0.996169	0.023309

Table for $N = 2$, $n = 1$, through $N = 100$, $n = 50$

N	n	k	x	P(x)	p(x)
70	33	10	9	0.999767	0.003597
70	33	10	10	1.000000	0.000233
70	33	11	0	0.000395	0.000395
70	33	11	1	0.005707	0.005312
70	33	11	2	0.036063	0.030356
70	33	11	3	0.133411	0.097348
70	33	11	4	0.328107	0.194696
70	33	11	5	0.583095	0.254988
70	33	11	6	0.806210	0.223115
70	33	11	7	0.936602	0.130392
70	33	11	8	0.986457	0.049856
70	33	11	9	0.998328	0.011870
70	33	11	10	0.999910	0.001583
70	33	11	11	1.000000	0.000089
70	33	12	0	0.000174	0.000174
70	33	12	1	0.002826	0.002652
70	33	12	2	0.020114	0.017287
70	33	12	3	0.083912	0.063799
70	33	12	4	0.232409	0.148497
70	33	12	5	0.462084	0.229675
70	33	12	6	0.704106	0.242023
70	33	12	7	0.879141	0.175034
70	33	12	8	0.965332	0.086191
70	33	12	9	0.993499	0.028167
70	33	12	10	0.999293	0.005794
70	33	12	11	0.999967	0.000673
70	33	12	12	1.000000	0.000033
70	33	13	0	0.000075	0.000075
70	33	13	1	0.001363	0.001288
70	33	13	2	0.010874	0.009511
70	33	13	3	0.050913	0.040039
70	33	13	4	0.158160	0.107248
70	33	13	5	0.351206	0.193046
70	33	13	6	0.591441	0.240235
70	33	13	7	0.800677	0.209237
70	33	13	8	0.928181	0.127503
70	33	13	9	0.981844	0.053663
70	33	13	10	0.996996	0.015152
70	33	13	11	0.999711	0.002716
70	33	13	12	0.999988	0.000277
70	33	13	13	1.000000	0.000012
70	33	14	0	0.000032	0.000032
70	33	14	1	0.000640	0.000608
70	33	14	2	0.005701	0.005061
70	33	14	3	0.029840	0.024138
70	33	14	4	0.103596	0.073756
70	33	14	5	0.256377	0.152781
70	33	14	6	0.477645	0.221269
70	33	14	7	0.705236	0.227591
70	33	14	8	0.872258	0.167022

N	n	k	x	P(x)	p(x)
70	33	14	9	0.959249	0.086991
70	33	14	10	0.990882	0.031633
70	33	14	11	0.998663	0.007781
70	33	14	12	0.999886	0.001223
70	33	14	13	0.999996	0.000110
70	33	14	14	1.000000	0.000004
70	33	15	0	0.000013	0.000013
70	33	15	1	0.000292	0.000279
70	33	15	2	0.002899	0.002607
70	33	15	3	0.016908	0.014009
70	33	15	4	0.065401	0.048492
70	33	15	5	0.179986	0.114586
70	33	15	6	0.370962	0.190976
70	33	15	7	0.599569	0.228607
70	33	15	8	0.797695	0.198126
70	33	15	9	0.921967	0.124272
70	33	15	10	0.977890	0.055923
70	33	15	11	0.995606	0.017717
70	33	15	12	0.999427	0.003821
70	33	15	13	0.999956	0.000529
70	33	15	14	0.999998	0.000042
70	33	15	15	1.000000	0.000001
70	33	16	0	0.000005	0.000005
70	33	16	1	0.000130	0.000125
70	33	16	2	0.001430	0.001300
70	33	16	3	0.009267	0.007837
70	33	16	4	0.039832	0.030565
70	33	16	5	0.121652	0.081820
70	33	16	6	0.277210	0.155559
70	33	16	7	0.491500	0.214290
70	33	16	8	0.707637	0.216137
70	33	16	9	0.867739	0.160102
70	33	16	10	0.954504	0.086765
70	33	16	11	0.988520	0.034016
70	33	16	12	0.997968	0.009449
70	33	16	13	0.999764	0.001796
70	33	16	14	0.999984	0.000220
70	33	16	15	0.999999	0.000015
70	33	16	16	1.000000	0.000000
70	33	17	0	0.000002	0.000002
70	33	17	1	0.000056	0.000054
70	33	17	2	0.000684	0.000628
70	33	17	3	0.004913	0.004230
70	33	17	4	0.023417	0.018504
70	33	17	5	0.079227	0.055809
70	33	17	6	0.199431	0.120204
70	33	17	7	0.388324	0.188893
70	33	17	8	0.607574	0.219250
70	33	17	9	0.796583	0.189009
70	33	17	10	0.917548	0.120966

N	n	k	x	P(x)	p(x)
70	33	17	11	0.974661	0.057113
70	33	17	12	0.994294	0.019633
70	33	17	13	0.999099	0.004805
70	33	17	14	0.999907	0.000808
70	33	17	15	0.999994	0.000088
70	33	17	16	1.000000	0.000005
70	33	17	17	1.000000	0.000000
70	33	18	0	0.000001	0.000001
70	33	18	1	0.000023	0.000023
70	33	18	2	0.000316	0.000293
70	33	18	3	0.002519	0.002203
70	33	18	4	0.013292	0.010773
70	33	18	5	0.049743	0.036450
70	33	18	6	0.138195	0.088452
70	33	18	7	0.295659	0.157464
70	33	18	8	0.504154	0.208495
70	33	18	9	0.710994	0.206840
70	33	18	10	0.865054	0.154060
70	33	18	11	0.950954	0.085900
70	33	18	12	0.986515	0.035561
70	33	18	13	0.997286	0.010771
70	33	18	14	0.999617	0.002331
70	33	18	15	0.999965	0.000347
70	33	18	16	0.999998	0.000033
70	33	18	17	1.000000	0.000002
70	33	18	18	1.000000	0.000000
70	33	19	0	0.000000	0.000000
70	33	19	1	0.000009	0.000009
70	33	19	2	0.000142	0.000132
70	33	19	3	0.001248	0.001107
70	33	19	4	0.007284	0.006036
70	33	19	5	0.030116	0.022831
70	33	19	6	0.092268	0.062152
70	33	19	7	0.216927	0.124659
70	33	19	8	0.403916	0.186989
70	33	19	9	0.615529	0.211613
70	33	19	10	0.796912	0.181383
70	33	19	11	0.914612	0.117700
70	33	19	12	0.972154	0.057542
70	33	19	13	0.993143	0.020989
70	33	19	14	0.998765	0.005622
70	33	19	15	0.999844	0.001079
70	33	19	16	0.999987	0.000143
70	33	19	17	0.999999	0.000012
70	33	19	18	1.000000	0.000001
70	33	19	19	1.000000	0.000000
70	33	20	0	0.000000	0.000000
70	33	20	1	0.000004	0.000004
70	33	20	2	0.000061	0.000058
70	33	20	3	0.000597	0.000536

N	n	k	x	P(x)	p(x)
70	33	20	4	0.003852	0.003255
70	33	20	5	0.017581	0.013729
70	33	20	6	0.059364	0.041783
70	33	20	7	0.153375	0.094012
70	33	20	8	0.312255	0.158880
70	33	20	9	0.515947	0.203692
70	33	20	10	0.715112	0.199165
70	33	20	11	0.863839	0.148727
70	33	20	12	0.948460	0.084621
70	33	20	13	0.984912	0.036452
70	33	20	14	0.996671	0.011759
70	33	20	15	0.999463	0.002793
70	33	20	16	0.999939	0.000476
70	33	20	17	0.999995	0.000056
70	33	20	18	1.000000	0.000004
70	33	20	19	1.000000	0.000000
70	33	20	20	1.000000	0.000000
70	33	21	0	0.000000	0.000000
70	33	21	1	0.000001	0.000001
70	33	21	2	0.000026	0.000024
70	33	21	3	0.000276	0.000250
70	33	21	4	0.001964	0.001689
70	33	21	5	0.009893	0.007928
70	33	21	6	0.036801	0.026908
70	33	21	7	0.104489	0.067688
70	33	21	8	0.232815	0.128326
70	33	21	9	0.418175	0.185360
70	33	21	10	0.623496	0.205321
70	33	21	11	0.798399	0.174903
70	33	21	12	0.912919	0.114520
70	33	21	13	0.970331	0.057412
70	33	21	14	0.992202	0.021871
70	33	21	15	0.998458	0.006256
70	33	21	16	0.999778	0.001320
70	33	21	17	0.999977	0.000200
70	33	21	18	0.999998	0.000021
70	33	21	19	1.000000	0.000001
70	33	21	20	1.000000	0.000000
70	33	21	21	1.000000	0.000000
70	33	22	0	0.000000	0.000000
70	33	22	1	0.000000	0.000000
70	33	22	2	0.000010	0.000010
70	33	22	3	0.000123	0.000112
70	33	22	4	0.000965	0.000842
70	33	22	5	0.005362	0.004397
70	33	22	6	0.021974	0.016612
70	33	22	7	0.068573	0.046599
70	33	22	8	0.167343	0.098770
70	33	22	9	0.327386	0.160044
70	33	22	10	0.527121	0.199734

N	n	k	x	P(x)	p(x)
70	33	22	11	0.719871	0.192751
70	33	22	12	0.863839	0.143968
70	33	22	13	0.946898	0.083058
70	33	22	14	0.983722	0.036824
70	33	22	15	0.996160	0.012438
70	33	22	16	0.999320	0.003160
70	33	22	17	0.999912	0.000592
70	33	22	18	0.999992	0.000080
70	33	22	19	0.999999	0.000007
70	33	22	20	1.000000	0.000000
70	33	22	21	1.000000	0.000000
70	33	22	22	1.000000	0.000000
70	33	23	0	0.000000	0.000000
70	33	23	1	0.000000	0.000000
70	33	23	2	0.000004	0.000004
70	33	23	3	0.000052	0.000048
70	33	23	4	0.000456	0.000404
70	33	23	5	0.002797	0.002341
70	33	23	6	0.012630	0.009833
70	33	23	7	0.043332	0.030702
70	33	23	8	0.115900	0.072568
70	33	23	9	0.247364	0.131464
70	33	23	10	0.431415	0.184050
70	33	23	11	0.631527	0.200113
70	33	23	12	0.800853	0.169326
70	33	23	13	0.912290	0.111437
70	33	23	14	0.969146	0.056856
70	33	23	15	0.991496	0.022350
70	33	23	16	0.998201	0.006705
70	33	23	17	0.999715	0.001514
70	33	23	18	0.999967	0.000252
70	33	23	19	0.999997	0.000030
70	33	23	20	1.000000	0.000002
70	33	23	21	1.000000	0.000000
70	33	23	22	1.000000	0.000000
70	33	23	23	1.000000	0.000000
70	33	24	0	0.000000	0.000000
70	33	24	1	0.000000	0.000000
70	33	24	2	0.000001	0.000001
70	33	24	3	0.000021	0.000020
70	33	24	4	0.000207	0.000185
70	33	24	5	0.001402	0.001195
70	33	24	6	0.006981	0.005579
70	33	24	7	0.026348	0.019366
70	33	24	8	0.077300	0.050952
70	33	24	9	0.180234	0.102934
70	33	24	10	0.341347	0.161114
70	33	24	11	0.537857	0.196510
70	33	24	12	0.725197	0.187339
70	33	24	13	0.864870	0.139673

N	n	k	x	P(x)	p(x)
70	33	24	14	0.946161	0.081291
70	33	24	15	0.982936	0.036775
70	33	24	16	0.995775	0.012839
70	33	24	17	0.999199	0.003424
70	33	24	18	0.999887	0.000687
70	33	24	19	0.999988	0.000102
70	33	24	20	0.999999	0.000011
70	33	24	21	1.000000	0.000001
70	33	24	22	1.000000	0.000000
70	33	24	23	1.000000	0.000000
70	33	24	24	1.000000	0.000000
70	33	25	0	0.000000	0.000000
70	33	25	1	0.000000	0.000000
70	33	25	2	0.000001	0.000000
70	33	25	3	0.000008	0.000008
70	33	25	4	0.000090	0.000082
70	33	25	5	0.000675	0.000585
70	33	25	6	0.003707	0.003032
70	33	25	7	0.015401	0.011695
70	33	25	8	0.049609	0.034207
70	33	25	9	0.126529	0.076921
70	33	25	10	0.260791	0.134261
70	33	25	11	0.443874	0.183084
70	33	25	12	0.639672	0.195798
70	33	25	13	0.804143	0.164470
70	33	25	14	0.912585	0.108442
70	33	25	15	0.968546	0.055961
70	33	25	16	0.991030	0.022484
70	33	25	17	0.998008	0.006978
70	33	25	18	0.999662	0.001654
70	33	25	19	0.999957	0.000295
70	33	25	20	0.999996	0.000039
70	33	25	21	1.000000	0.000004
70	33	25	22	1.000000	0.000000
70	33	25	23	1.000000	0.000000
70	33	25	24	1.000000	0.000000
70	33	25	25	1.000000	0.000000
70	33	26	0	0.000000	0.000000
70	33	26	1	0.000000	0.000000
70	33	26	2	0.000000	0.000000
70	33	26	3	0.000003	0.000003
70	33	26	4	0.000037	0.000034
70	33	26	5	0.000311	0.000273
70	33	26	6	0.001888	0.001577
70	33	26	7	0.008645	0.006757
70	33	26	8	0.030605	0.021960
70	33	26	9	0.085505	0.054900
70	33	26	10	0.192168	0.106663
70	33	26	11	0.354367	0.162199
70	33	26	12	0.548300	0.193933

N	n	k	x	P(x)	p(x)
70	33	26	13	0.731045	0.182745
70	33	26	14	0.866798	0.135753
70	33	26	15	0.946161	0.079363
70	33	26	16	0.982536	0.036375
70	33	26	17	0.995527	0.012991
70	33	26	18	0.999111	0.003584
70	33	26	19	0.999866	0.000754
70	33	26	20	0.999985	0.000119
70	33	26	21	0.999999	0.000014
70	33	26	22	1.000000	0.000001
70	33	26	23	1.000000	0.000000
70	33	26	24	1.000000	0.000000
70	33	26	25	1.000000	0.000000
70	33	26	26	1.000000	0.000000
70	33	27	0	0.000000	0.000000
70	33	27	1	0.000000	0.000000
70	33	27	2	0.000000	0.000000
70	33	27	3	0.000001	0.000001
70	33	27	4	0.000015	0.000014
70	33	27	5	0.000137	0.000122
70	33	27	6	0.000920	0.000783
70	33	27	7	0.004652	0.003732
70	33	27	8	0.018127	0.013476
70	33	27	9	0.055559	0.037432
70	33	27	10	0.136412	0.080853
70	33	27	11	0.273267	0.136855
70	33	27	12	0.455741	0.182474
70	33	27	13	0.647979	0.192238
70	33	27	14	0.808177	0.160198
70	33	27	15	0.913695	0.105517
70	33	27	16	0.968482	0.054788
70	33	27	17	0.990803	0.022321
70	33	27	18	0.997889	0.007086
70	33	27	19	0.999626	0.001736
70	33	27	20	0.999950	0.000324
70	33	27	21	0.999995	0.000045
70	33	27	22	1.000000	0.000005
70	33	27	23	1.000000	0.000000
70	33	27	24	1.000000	0.000000
70	33	27	25	1.000000	0.000000
70	33	27	26	1.000000	0.000000
70	33	27	27	1.000000	0.000000
70	33	28	0	0.000000	0.000000
70	33	28	1	0.000000	0.000000
70	33	28	2	0.000000	0.000000
70	33	28	3	0.000000	0.000000
70	33	28	4	0.000006	0.000005
70	33	28	5	0.000057	0.000052
70	33	28	6	0.000428	0.000371
70	33	28	7	0.002395	0.001967

N	n	k	x	P(x)	p(x)
70	33	28	8	0.010293	0.007897
70	33	28	9	0.034667	0.024374
70	33	28	10	0.093165	0.058498
70	33	28	11	0.203249	0.110083
70	33	28	12	0.366626	0.163377
70	33	28	13	0.558566	0.191940
70	33	28	14	0.737392	0.178826
70	33	28	15	0.869525	0.132133
70	33	28	16	0.946822	0.077298
70	33	28	17	0.982498	0.035676
70	33	28	18	0.995417	0.012920
70	33	28	19	0.999060	0.003643
70	33	28	20	0.999852	0.000791
70	33	28	21	0.999982	0.000131
70	33	28	22	0.999998	0.000016
70	33	28	23	1.000000	0.000001
70	33	28	24	1.000000	0.000000
70	33	28	25	1.000000	0.000000
70	33	28	26	1.000000	0.000000
70	33	28	27	1.000000	0.000000
70	33	28	28	1.000000	0.000000
70	33	29	0	0.000000	0.000000
70	33	29	1	0.000000	0.000000
70	33	29	2	0.000000	0.000000
70	33	29	3	0.000000	0.000000
70	33	29	4	0.000002	0.000002
70	33	29	5	0.000023	0.000021
70	33	29	6	0.000190	0.000167
70	33	29	7	0.001178	0.000988
70	33	29	8	0.005592	0.004414
70	33	29	9	0.020739	0.015147
70	33	29	10	0.061131	0.040392
70	33	29	11	0.145586	0.084455
70	33	29	12	0.284937	0.139351
70	33	29	13	0.467166	0.182229
70	33	29	14	0.656494	0.189329
70	33	29	15	0.812896	0.156402
70	33	29	16	0.915535	0.102639
70	33	29	17	0.968907	0.053372
70	33	29	18	0.990803	0.021896
70	33	29	19	0.997846	0.007043
70	33	29	20	0.999607	0.001761
70	33	29	21	0.999945	0.000338
70	33	29	22	0.999994	0.000049
70	33	29	23	0.999999	0.000005
70	33	29	24	1.000000	0.000000
70	33	29	25	1.000000	0.000000
70	33	29	26	1.000000	0.000000
70	33	29	27	1.000000	0.000000
70	33	29	28	1.000000	0.000000

Table for $N = 2$, $n = 1$, through $N = 100$, $n = 50$

N	n	k	x	P(x)	p(x)
70	33	29	29	1.000000	0.000000
70	33	30	0	0.000000	0.000000
70	33	30	1	0.000000	0.000000
70	33	30	2	0.000000	0.000000
70	33	30	3	0.000000	0.000000
70	33	30	4	0.000001	0.000001
70	33	30	5	0.000009	0.000008
70	33	30	6	0.000080	0.000071
70	33	30	7	0.000551	0.000471
70	33	30	8	0.002900	0.002349
70	33	30	9	0.011872	0.008972
70	33	30	10	0.038472	0.026599
70	33	30	11	0.100268	0.061797
70	33	30	12	0.213562	0.113294
70	33	30	13	0.378274	0.164712
70	33	30	14	0.568757	0.190483
70	33	30	15	0.744232	0.175475
70	33	30	16	0.872977	0.128745
70	33	30	17	0.948079	0.075101
70	33	30	18	0.982792	0.034714
70	33	30	19	0.995441	0.012649
70	33	30	20	0.999048	0.003607
70	33	30	21	0.999846	0.000798
70	33	30	22	0.999981	0.000135
70	33	30	23	0.999998	0.000017
70	33	30	24	1.000000	0.000002
70	33	30	25	1.000000	0.000000
70	33	30	26	1.000000	0.000000
70	33	30	27	1.000000	0.000000
70	33	30	28	1.000000	0.000000
70	33	30	29	1.000000	0.000000
70	33	30	30	1.000000	0.000000
70	33	31	0	0.000000	0.000000
70	33	31	1	0.000000	0.000000
70	33	31	2	0.000000	0.000000
70	33	31	3	0.000000	0.000000
70	33	31	4	0.000000	0.000000
70	33	31	5	0.000003	0.000003
70	33	31	6	0.000032	0.000029
70	33	31	7	0.000245	0.000213
70	33	31	8	0.001432	0.001187
70	33	31	9	0.006489	0.005057
70	33	31	10	0.023177	0.016688
70	33	31	11	0.066280	0.043103
70	33	31	12	0.154083	0.087803
70	33	31	13	0.295918	0.141835
70	33	31	14	0.478277	0.182359
70	33	31	15	0.665268	0.186991
70	33	31	16	0.818261	0.152992
70	33	31	17	0.918038	0.099778

N	n	k	x	P(x)	p(x)
70	33	31	18	0.969775	0.051737
70	33	31	19	0.991014	0.021239
70	33	31	20	0.997876	0.006862
70	33	31	21	0.999607	0.001731
70	33	31	22	0.999944	0.000337
70	33	31	23	0.999994	0.000050
70	33	31	24	0.999999	0.000006
70	33	31	25	1.000000	0.000000
70	33	31	26	1.000000	0.000000
70	33	31	27	1.000000	0.000000
70	33	31	28	1.000000	0.000000
70	33	31	29	1.000000	0.000000
70	33	31	30	1.000000	0.000000
70	33	31	31	1.000000	0.000000
70	33	32	0	0.000000	0.000000
70	33	32	1	0.000000	0.000000
70	33	32	2	0.000000	0.000000
70	33	32	3	0.000000	0.000000
70	33	32	4	0.000000	0.000000
70	33	32	5	0.000001	0.000001
70	33	32	6	0.000012	0.000011
70	33	32	7	0.000103	0.000091
70	33	32	8	0.000671	0.000568
70	33	32	9	0.003377	0.002706
70	33	32	10	0.013335	0.009958
70	33	32	11	0.041966	0.028630
70	33	32	12	0.106805	0.064839
70	33	32	13	0.223182	0.116378
70	33	32	14	0.389436	0.166254
70	33	32	15	0.578965	0.189529
70	33	32	16	0.751572	0.172607
70	33	32	17	0.877104	0.125532
70	33	32	18	0.949876	0.072772
70	33	32	19	0.983390	0.033514
70	33	32	20	0.995589	0.012199
70	33	32	21	0.999074	0.003485
70	33	32	22	0.999849	0.000775
70	33	32	23	0.999981	0.000132
70	33	32	24	0.999998	0.000017
70	33	32	25	1.000000	0.000002
70	33	32	26	1.000000	0.000000
70	33	32	27	1.000000	0.000000
70	33	32	28	1.000000	0.000000
70	33	32	29	1.000000	0.000000
70	33	32	30	1.000000	0.000000
70	33	32	31	1.000000	0.000000
70	33	32	32	1.000000	0.000000
70	33	33	0	0.000000	0.000000
70	33	33	1	0.000000	0.000000
70	33	33	2	0.000000	0.000000

N	n	k	x	P(x)	p(x)
70	33	33	3	0.000000	0.000000
70	33	33	4	0.000000	0.000000
70	33	33	5	0.000000	0.000000
70	33	33	6	0.000004	0.000004
70	33	33	7	0.000041	0.000036
70	33	33	8	0.000297	0.000257
70	33	33	9	0.001668	0.001371
70	33	33	10	0.007308	0.005640
70	33	33	11	0.025390	0.018082
70	33	33	12	0.070973	0.045582
70	33	33	13	0.161931	0.090958
70	33	33	14	0.306309	0.144378
70	33	33	15	0.489188	0.182879
70	33	33	16	0.674353	0.185165
70	33	33	17	0.824248	0.149895
70	33	33	18	0.921150	0.096902
70	33	33	19	0.971043	0.049892
70	33	33	20	0.991415	0.020373
70	33	33	21	0.997973	0.006558
70	33	33	22	0.999624	0.001651
70	33	33	23	0.999946	0.000322
70	33	33	24	0.999994	0.000048
70	33	33	25	0.999999	0.000005
70	33	33	26	1.000000	0.000000
70	33	33	27	1.000000	0.000000
70	33	33	28	1.000000	0.000000
70	33	33	29	1.000000	0.000000
70	33	33	30	1.000000	0.000000
70	33	33	31	1.000000	0.000000
70	33	33	32	1.000000	0.000000
70	33	33	33	1.000000	0.000000
70	34	1	0	0.514286	0.514286
70	34	1	1	1.000000	0.485714
70	34	2	0	0.260870	0.260870
70	34	2	1	0.767702	0.506832
70	34	2	2	1.000000	0.232298
70	34	3	0	0.130435	0.130435
70	34	3	1	0.521739	0.391304
70	34	3	2	0.890683	0.368944
70	34	3	3	1.000000	0.109317
70	34	4	0	0.064244	0.064244
70	34	4	1	0.329007	0.264763
70	34	4	2	0.714471	0.385464
70	34	4	3	0.949421	0.234949
70	34	4	4	1.000000	0.050579
70	34	5	0	0.031149	0.031149
70	34	5	1	0.196626	0.165477
70	34	5	2	0.527579	0.330954
70	34	5	3	0.839065	0.311486
70	34	5	4	0.977009	0.137944

N	n	k	x	P(x)	p(x)
70	34	5	5	1.000000	0.022991
70	34	6	0	0.014855	0.014855
70	34	6	1	0.112614	0.097759
70	34	6	2	0.364648	0.252034
70	34	6	3	0.690511	0.325862
70	34	6	4	0.913343	0.222832
70	34	6	5	0.989743	0.076400
70	34	6	6	1.000000	0.010257
70	34	7	0	0.006964	0.006964
70	34	7	1	0.062207	0.055244
70	34	7	2	0.238631	0.176424
70	34	7	3	0.532671	0.294040
70	34	7	4	0.808890	0.276219
70	34	7	5	0.955124	0.146234
70	34	7	6	0.995512	0.040388
70	34	7	7	1.000000	0.004488
70	34	8	0	0.003205	0.003205
70	34	8	1	0.033270	0.030065
70	34	8	2	0.149019	0.115749
70	34	8	3	0.387985	0.238966
70	34	8	4	0.677357	0.289373
70	34	8	5	0.887810	0.210453
70	34	8	6	0.977562	0.089752
70	34	8	7	0.998077	0.020515
70	34	8	8	1.000000	0.001923
70	34	9	0	0.001448	0.001448
70	34	9	1	0.017268	0.015820
70	34	9	2	0.089278	0.072010
70	34	9	3	0.268502	0.179224
70	34	9	4	0.537338	0.268836
70	34	9	5	0.789372	0.252034
70	34	9	6	0.937029	0.147656
70	34	9	7	0.989143	0.052114
70	34	9	8	0.999193	0.010051
70	34	9	9	1.000000	0.000807
70	34	10	0	0.000641	0.000641
70	34	10	1	0.008709	0.008069
70	34	10	2	0.051502	0.042793
70	34	10	3	0.177421	0.125919
70	34	10	4	0.405124	0.227703
70	34	10	5	0.669553	0.264429
70	34	10	6	0.869252	0.199699
70	34	10	7	0.966076	0.096824
70	34	10	8	0.994910	0.028834
70	34	10	9	0.999669	0.004760
70	34	10	10	1.000000	0.000331
70	34	11	0	0.000278	0.000278
70	34	11	1	0.004272	0.003994
70	34	11	2	0.028679	0.024408
70	34	11	3	0.112363	0.083683

N	n	k	x	P(x)	p(x)
70	34	11	4	0.291272	0.178909
70	34	11	5	0.541745	0.250473
70	34	11	6	0.776059	0.234314
70	34	11	7	0.922505	0.146446
70	34	11	8	0.982415	0.059910
70	34	11	9	0.997686	0.015271
70	34	11	10	0.999868	0.002182
70	34	11	11	1.000000	0.000132
70	34	12	0	0.000118	0.000118
70	34	12	1	0.002038	0.001920
70	34	12	2	0.015441	0.013404
70	34	12	3	0.068393	0.052952
70	34	12	4	0.200301	0.131908
70	34	12	5	0.418631	0.218330
70	34	12	6	0.664859	0.246228
70	34	12	7	0.855488	0.190628
70	34	12	8	0.956014	0.100527
70	34	12	9	0.991215	0.035201
70	34	12	10	0.998980	0.007765
70	34	12	11	0.999948	0.000968
70	34	12	12	1.000000	0.000052
70	34	13	0	0.000049	0.000049
70	34	13	1	0.000945	0.000897
70	34	13	2	0.008046	0.007101
70	34	13	3	0.040091	0.032045
70	34	13	4	0.132073	0.091982
70	34	13	5	0.309466	0.177393
70	34	13	6	0.545991	0.236524
70	34	13	7	0.766747	0.220756
70	34	13	8	0.910950	0.144204
70	34	13	9	0.976042	0.065092
70	34	13	10	0.995767	0.019725
70	34	13	11	0.999564	0.003797
70	34	13	12	0.999980	0.000416
70	34	13	13	1.000000	0.000020
70	34	14	0	0.000020	0.000020
70	34	14	1	0.000426	0.000407
70	34	14	2	0.004060	0.003634
70	34	14	3	0.022663	0.018604
70	34	14	4	0.083662	0.060998
70	34	14	5	0.219214	0.135552
70	34	14	6	0.429803	0.210590
70	34	14	7	0.662178	0.232375
70	34	14	8	0.845173	0.182995
70	34	14	9	0.947493	0.102320
70	34	14	10	0.987462	0.039969
70	34	14	11	0.998032	0.010570
70	34	14	12	0.999820	0.001788
70	34	14	13	0.999993	0.000173
70	34	14	14	1.000000	0.000007

N	n	k	x	P(x)	p(x)
70	34	15	0	0.000008	0.000008
70	34	15	1	0.000187	0.000179
70	34	15	2	0.001983	0.001797
70	34	15	3	0.012365	0.010381
70	34	15	4	0.050984	0.038619
70	34	15	5	0.149017	0.098033
70	34	15	6	0.324509	0.175491
70	34	15	7	0.550140	0.225632
70	34	15	8	0.760211	0.210071
70	34	15	9	0.901815	0.141603
70	34	15	10	0.970333	0.068518
70	34	15	11	0.993691	0.023358
70	34	15	12	0.999118	0.005427
70	34	15	13	0.999928	0.000810
70	34	15	14	0.999997	0.000069
70	34	15	15	1.000000	0.000003
70	34	16	0	0.000003	0.000003
70	34	16	1	0.000079	0.000076
70	34	16	2	0.000938	0.000859
70	34	16	3	0.006514	0.005576
70	34	16	4	0.029919	0.023406
70	34	16	5	0.097327	0.067408
70	34	16	6	0.235167	0.137841
70	34	16	7	0.439376	0.204208
70	34	16	8	0.660905	0.221529
70	34	16	9	0.837450	0.176545
70	34	16	10	0.940434	0.102984
70	34	16	11	0.983923	0.043489
70	34	16	12	0.996947	0.013024
70	34	16	13	0.999618	0.002672
70	34	16	14	0.999972	0.000354
70	34	16	15	0.999999	0.000027
70	34	16	16	1.000000	0.000001
70	34	17	0	0.000001	0.000001
70	34	17	1	0.000033	0.000032
70	34	17	2	0.000429	0.000396
70	34	17	3	0.003313	0.002884
70	34	17	4	0.016916	0.013603
70	34	17	5	0.061126	0.044210
70	34	17	6	0.163695	0.102568
70	34	17	7	0.337272	0.173577
70	34	17	8	0.554243	0.216971
70	34	17	9	0.755716	0.201473
70	34	17	10	0.894663	0.138947
70	34	17	11	0.965400	0.070737
70	34	17	12	0.991641	0.026241
70	34	17	13	0.998580	0.006939
70	34	17	14	0.999841	0.001262
70	34	17	15	0.999990	0.000148
70	34	17	16	1.000000	0.000010

N	n	k	x	P(x)	p(x)
70	34	17	17	1.000000	0.000000
70	34	18	0	0.000000	0.000000
70	34	18	1	0.000013	0.000013
70	34	18	2	0.000190	0.000177
70	34	18	3	0.001626	0.001436
70	34	18	4	0.009216	0.007590
70	34	18	5	0.036936	0.027720
70	34	18	6	0.109508	0.072572
70	34	18	7	0.248846	0.139338
70	34	18	8	0.447804	0.198958
70	34	18	9	0.660681	0.212877
70	34	18	10	0.831744	0.171062
70	34	18	11	0.934703	0.102959
70	34	18	12	0.980748	0.046046
70	34	18	13	0.995830	0.015082
70	34	18	14	0.999365	0.003535
70	34	18	15	0.999936	0.000571
70	34	18	16	0.999996	0.000060
70	34	18	17	1.000000	0.000004
70	34	18	18	1.000000	0.000000
70	34	19	0	0.000000	0.000000
70	34	19	1	0.000005	0.000005
70	34	19	2	0.000081	0.000076
70	34	19	3	0.000770	0.000689
70	34	19	4	0.004837	0.004067
70	34	19	5	0.021477	0.016639
70	34	19	6	0.070430	0.048954
70	34	19	7	0.176497	0.106066
70	34	19	8	0.348325	0.171828
70	34	19	9	0.558336	0.210012
70	34	19	10	0.752792	0.194455
70	34	19	11	0.889163	0.136371
70	34	19	12	0.961267	0.072104
70	34	19	13	0.989739	0.028472
70	34	19	14	0.998005	0.008266
70	34	19	15	0.999727	0.001722
70	34	19	16	0.999975	0.000248
70	34	19	17	0.999999	0.000023
70	34	19	18	1.000000	0.000001
70	34	19	19	1.000000	0.000000
70	34	20	0	0.000000	0.000000
70	34	20	1	0.000002	0.000002
70	34	20	2	0.000033	0.000031
70	34	20	3	0.000351	0.000318
70	34	20	4	0.002445	0.002094
70	34	20	5	0.012015	0.009570
70	34	20	6	0.043554	0.031539
70	34	20	7	0.120344	0.076790
70	34	20	8	0.260726	0.140382
70	34	20	9	0.455390	0.194663

N	n	k	x	P(x)	p(x)
70	34	20	10	0.661283	0.205894
70	34	20	11	0.827662	0.166379
70	34	20	12	0.930164	0.102501
70	34	20	13	0.978016	0.047852
70	34	20	14	0.994764	0.016748
70	34	20	15	0.999086	0.004322
70	34	20	16	0.999888	0.000802
70	34	20	17	0.999991	0.000103
70	34	20	18	0.999999	0.000009
70	34	20	19	1.000000	0.000000
70	34	20	20	1.000000	0.000000
70	34	21	0	0.000000	0.000000
70	34	21	1	0.000001	0.000001
70	34	21	2	0.000013	0.000013
70	34	21	3	0.000154	0.000141
70	34	21	4	0.001189	0.001034
70	34	21	5	0.006464	0.005276
70	34	21	6	0.025892	0.019428
70	34	21	7	0.078877	0.052985
70	34	21	8	0.187728	0.108850
70	34	21	9	0.358058	0.170330
70	34	21	10	0.562454	0.204396
70	34	21	11	0.751128	0.188674
70	34	21	12	0.885063	0.133935
70	34	21	13	0.957918	0.072855
70	34	21	14	0.988065	0.030147
70	34	21	15	0.997444	0.009379
70	34	21	16	0.999599	0.002156
70	34	21	17	0.999956	0.000357
70	34	21	18	0.999997	0.000041
70	34	21	19	1.000000	0.000003
70	34	21	20	1.000000	0.000000
70	34	21	21	1.000000	0.000000
70	34	22	0	0.000000	0.000000
70	34	22	1	0.000000	0.000000
70	34	22	2	0.000005	0.000005
70	34	22	3	0.000065	0.000060
70	34	22	4	0.000555	0.000490
70	34	22	5	0.003342	0.002787
70	34	22	6	0.014790	0.011449
70	34	22	7	0.049681	0.034891
70	34	22	8	0.129970	0.080289
70	34	22	9	0.271155	0.141184
70	34	22	10	0.462342	0.191187
70	34	22	11	0.662567	0.200225
70	34	22	12	0.824929	0.162362
70	34	22	13	0.926694	0.101765
70	34	22	14	0.975760	0.049065
70	34	22	15	0.993807	0.018047
70	34	22	16	0.998807	0.005001

Table for $N = 2$, $n = 1$, through $N = 100$, $n = 50$

N	n	k	x	P(x)	p(x)
70	34	22	17	0.999832	0.001025
70	34	22	18	0.999983	0.000151
70	34	22	19	0.999999	0.000015
70	34	22	20	1.000000	0.000001
70	34	22	21	1.000000	0.000000
70	34	22	22	1.000000	0.000000
70	34	23	0	0.000000	0.000000
70	34	23	1	0.000000	0.000000
70	34	29	2	0.000002	0.000002
70	34	23	3	0.000026	0.000024
70	34	23	4	0.000249	0.000223
70	34	23	5	0.001658	0.001409
70	34	23	6	0.008112	0.006454
70	34	23	7	0.030055	0.021943
70	34	23	8	0.086481	0.056425
70	34	23	9	0.197621	0.111141
70	34	23	10	0.366748	0.169127
70	34	23	11	0.566626	0.199877
70	34	23	12	0.750513	0.183887
70	34	23	13	0.882172	0.131659
70	34	23	14	0.955316	0.073144
70	34	23	15	0.986663	0.031347
70	34	23	16	0.996932	0.010269
70	34	23	17	0.999469	0.002537
70	34	23	18	0.999933	0.000464
70	34	23	19	0.999994	0.000061
70	34	23	20	1.000000	0.000006
70	34	23	21	1.000000	0.000000
70	34	23	22	1.000000	0.000000
70	34	23	23	1.000000	0.000000
70	34	24	0	0.000000	0.000000
70	34	24	1	0.000000	0.000000
70	34	24	2	0.000001	0.000001
70	34	24	3	0.000010	0.000009
70	34	24	4	0.000107	0.000097
70	34	24	5	0.000789	0.000682
70	34	24	6	0.004267	0.003479
70	34	24	7	0.017450	0.013182
70	34	24	8	0.055267	0.037817
70	34	24	9	0.138504	0.083237
70	34	24	10	0.280386	0.141882
70	34	24	11	0.468813	0.188428
70	34	24	12	0.664438	0.195625
70	34	24	13	0.823346	0.158907
70	34	24	14	0.924191	0.100845
70	34	24	15	0.973991	0.049800
70	34	24	16	0.992999	0.019009
70	34	24	17	0.998552	0.005552
70	34	24	18	0.999775	0.001224
70	34	24	19	0.999974	0.000199

N	n	k	x	P(x)	p(x)
70	34	24	20	0.999998	0.000023
70	34	24	21	1.000000	0.000002
70	34	24	22	1.000000	0.000000
70	34	24	23	1.000000	0.000000
70	34	24	24	1.000000	0.000000
70	34	25	0	0.000000	0.000000
70	34	25	1	0.000000	0.000000
70	34	25	2	0.000000	0.000000
70	34	25	3	0.000004	0.000004
70	34	25	4	0.000044	0.000040
70	34	25	5	0.000359	0.000315
70	34	25	6	0.002150	0.001791
70	34	25	7	0.009712	0.007562
70	34	25	8	0.033892	0.024180
70	34	25	9	0.093266	0.059374
70	34	25	10	0.206360	0.113094
70	34	25	11	0.374599	0.168239
70	34	25	12	0.570879	0.196279
70	34	25	13	0.750801	0.179922
70	34	25	14	0.880345	0.129544
70	34	25	15	0.953421	0.073076
70	34	25	16	0.985561	0.032140
70	34	25	17	0.996500	0.010938
70	34	25	18	0.999350	0.002850
70	34	25	19	0.999910	0.000560
70	34	25	20	0.999991	0.000081
70	34	25	21	0.999999	0.000008
70	34	25	22	1.000000	0.000001
70	34	25	23	1.000000	0.000000
70	34	25	24	1.000000	0.000000
70	34	25	25	1.000000	0.000000
70	34	26	0	0.000000	0.000000
70	34	26	1	0.000000	0.000000
70	34	26	2	0.000000	0.000000
70	34	26	3	0.000001	0.000001
70	34	26	4	0.000017	0.000016
70	34	26	5	0.000156	0.000139
70	34	26	6	0.001035	0.000880
70	34	26	7	0.005175	0.004139
70	34	26	8	0.019921	0.014747
70	34	26	9	0.060280	0.040359
70	34	26	10	0.146043	0.085763
70	34	26	11	0.288611	0.142567
70	34	26	12	0.474920	0.186309
70	34	26	13	0.666837	0.191917
70	34	26	14	0.822770	0.155933
70	34	26	15	0.922567	0.099797
70	34	26	16	0.972705	0.050138
70	34	26	17	0.992367	0.019662
70	34	26	18	0.998336	0.005969

N	n	k	x	P(x)	p(x)
70	34	26	19	0.999723	0.001387
70	34	26	20	0.999965	0.000243
70	34	26	21	0.999997	0.000031
70	34	26	22	1.000000	0.000003
70	34	26	23	1.000000	0.000000
70	34	26	24	1.000000	0.000000
70	34	26	25	1.000000	0.000000
70	34	26	26	1.000000	0.000000
70	34	27	0	0.000000	0.000000
70	34	27	1	0.000000	0.000000
70	34	27	2	0.000000	0.000000
70	34	27	3	0.000000	0.000000
70	34	27	4	0.000006	0.000006
70	34	27	5	0.000064	0.000058
70	34	27	6	0.000476	0.000411
70	34	27	7	0.002635	0.002159
70	34	27	8	0.011207	0.008573
70	34	27	9	0.037349	0.026142
70	34	27	10	0.099264	0.061914
70	34	27	11	0.214087	0.114823
70	34	27	12	0.381765	0.167678
70	34	27	13	0.575240	0.193475
70	34	27	14	0.751891	0.176651
70	34	27	15	0.879473	0.127581
70	34	27	16	0.952194	0.072721
70	34	27	17	0.984771	0.032577
70	34	27	18	0.996166	0.011395
70	34	27	19	0.999250	0.003084
70	34	27	20	0.999888	0.000638
70	34	27	21	0.999988	0.000099
70	34	27	22	0.999999	0.000011
70	34	27	23	1.000000	0.000001
70	34	27	24	1.000000	0.000000
70	34	27	25	1.000000	0.000000
70	34	27	26	1.000000	0.000000
70	34	27	27	1.000000	0.000000
70	34	28	0	0.000000	0.000000
70	34	28	1	0.000000	0.000000
70	34	28	2	0.000000	0.000000
70	34	28	3	0.000000	0.000000
70	34	28	4	0.000002	0.000002
70	34	28	5	0.000025	0.000023
70	34	28	6	0.000208	0.000183
70	34	28	7	0.001279	0.001071
70	34	28	8	0.006024	0.004745
70	34	28	9	0.022150	0.016127
70	34	28	10	0.064707	0.042556
70	34	28	11	0.152670	0.087963
70	34	28	12	0.295976	0.143307
70	34	28	13	0.480752	0.184776

N	n	k	x	P(x)	p(x)
70	34	28	14	0.669728	0.188976
70	34	28	15	0.823100	0.153372
70	34	28	16	0.921753	0.098653
70	34	28	17	0.971892	0.050139
70	34	28	18	0.991926	0.020034
70	34	28	19	0.998174	0.006248
70	34	28	20	0.999680	0.001506
70	34	28	21	0.999958	0.000277
70	34	28	22	0.999996	0.000038
70	34	28	23	1.000000	0.000004
70	34	28	24	1.000000	0.000000
70	34	28	25	1.000000	0.000000
70	34	28	26	1.000000	0.000000
70	34	28	27	1.000000	0.000000
70	34	28	28	1.000000	0.000000
70	34	29	0	0.000000	0.000000
70	34	29	1	0.000000	0.000000
70	34	29	2	0.000000	0.000000
70	34	29	3	0.000000	0.000000
70	34	29	4	0.000001	0.000001
70	34	29	5	0.000009	0.000009
70	34	29	6	0.000086	0.000077
70	34	29	7	0.000590	0.000504
70	34	29	8	0.003087	0.002496
70	34	29	9	0.012551	0.009465
70	34	29	10	0.040389	0.027838
70	34	29	11	0.104499	0.064111
70	34	29	12	0.220911	0.116411
70	34	29	13	0.388364	0.167453
70	34	29	14	0.579740	0.191375
70	34	29	15	0.753717	0.173978
70	34	29	16	0.879473	0.125756
70	34	29	17	0.951597	0.072124
70	34	29	18	0.984294	0.032696
70	34	29	19	0.995943	0.011649
70	34	29	20	0.999178	0.003236
70	34	29	21	0.999872	0.000693
70	34	29	22	0.999985	0.000113
70	34	29	23	0.999999	0.000014
70	34	29	24	1.000000	0.000001
70	34	29	25	1.000000	0.000000
70	34	29	26	1.000000	0.000000
70	34	29	27	1.000000	0.000000
70	34	29	28	1.000000	0.000000
70	34	29	29	1.000000	0.000000
70	34	30	0	0.000000	0.000000
70	34	30	1	0.000000	0.000000
70	34	30	2	0.000000	0.000000
70	34	30	3	0.000000	0.000000
70	34	30	4	0.000000	0.000000

N	n	k	x	P(x)	p(x)
70	34	30	5	0.000003	0.000003
70	34	30	6	0.000034	0.000030
70	34	30	7	0.000258	0.000225
70	34	30	8	0.001504	0.001245
70	34	30	9	0.006780	0.005277
70	34	30	10	0.024094	0.017314
70	34	30	11	0.068535	0.044441
70	34	30	12	0.158446	0.089911
70	34	30	13	0.302596	0.144149
70	34	30	14	0.486386	0.183790
70	34	30	15	0.673094	0.186708
70	34	30	16	0.824263	0.151170
70	34	30	17	0.921692	0.097429
70	34	30	18	0.971534	0.049842
70	34	30	19	0.991681	0.020147
70	34	30	20	0.998074	0.006393
70	34	30	21	0.999652	0.001578
70	34	30	22	0.999952	0.000300
70	34	30	23	0.999995	0.000043
70	34	30	24	1.000000	0.000005
70	34	30	25	1.000000	0.000000
70	34	30	26	1.000000	0.000000
70	34	30	27	1.000000	0.000000
70	34	30	28	1.000000	0.000000
70	34	30	29	1.000000	0.000000
70	34	30	30	1.000000	0.000000
70	34	31	0	0.000000	0.000000
70	34	31	1	0.000000	0.000000
70	34	31	2	0.000000	0.000000
70	34	31	3	0.000000	0.000000
70	34	31	4	0.000000	0.000000
70	34	31	5	0.000001	0.000001
70	34	31	6	0.000012	0.000011
70	34	31	7	0.000107	0.000094
70	34	31	8	0.000694	0.000587
70	34	31	9	0.003482	0.002788
70	34	31	10	0.013706	0.010223
70	34	31	11	0.042981	0.029276
70	34	31	12	0.108995	0.066014
70	34	31	13	0.226917	0.117922
70	34	31	14	0.394491	0.167574
70	34	31	15	0.584407	0.189917
70	34	31	16	0.756237	0.171829
70	34	31	17	0.880285	0.124048
70	34	31	18	0.951597	0.071313
70	34	31	19	0.984126	0.032529
70	34	31	20	0.995836	0.011710
70	34	31	21	0.999139	0.003303
70	34	31	22	0.999862	0.000723
70	34	31	23	0.999983	0.000121
70	34	31	24	0.999998	0.000015
70	34	31	25	1.000000	0.000001
70	34	31	26	1.000000	0.000000
70	34	31	27	1.000000	0.000000
70	34	31	28	1.000000	0.000000
70	34	31	29	1.000000	0.000000
70	34	31	30	1.000000	0.000000
70	34	31	31	1.000000	0.000000
70	34	32	0	0.000000	0.000000
70	34	32	1	0.000000	0.000000
70	34	32	2	0.000000	0.000000
70	34	32	3	0.000000	0.000000
70	34	32	4	0.000000	0.000000
70	34	32	5	0.000000	0.000000
70	34	32	6	0.000004	0.000004
70	34	32	7	0.000041	0.000037
70	34	32	8	0.000302	0.000261
70	34	32	9	0.001695	0.001393
70	34	32	10	0.007414	0.005719
70	34	32	11	0.025716	0.018302
70	34	32	12	0.071756	0.046040
70	34	32	13	0.163421	0.091664
70	34	32	14	0.308556	0.145135
70	34	32	15	0.491884	0.183328
70	34	32	16	0.676931	0.185047
70	34	32	17	0.826212	0.149282
70	34	32	18	0.922341	0.096128
70	34	32	19	0.971615	0.049274
70	34	32	20	0.991632	0.020018
70	34	32	21	0.998038	0.006406
70	34	32	22	0.999639	0.001601
70	34	32	23	0.999949	0.000309
70	34	32	24	0.999994	0.000046
70	34	32	25	0.999999	0.000005
70	34	32	26	1.000000	0.000000
70	34	32	27	1.000000	0.000000
70	34	32	28	1.000000	0.000000
70	34	32	29	1.000000	0.000000
70	34	32	30	1.000000	0.000000
70	34	32	31	1.000000	0.000000
70	34	32	32	1.000000	0.000000
70	34	33	0	0.000000	0.000000
70	34	33	1	0.000000	0.000000
70	34	33	2	0.000000	0.000000
70	34	33	3	0.000000	0.000000
70	34	33	4	0.000000	0.000000
70	34	33	5	0.000000	0.000000
70	34	33	6	0.000001	0.000001
70	34	33	7	0.000015	0.000014
70	34	33	8	0.000124	0.000109
70	34	33	9	0.000779	0.000655
70	34	33	10	0.003802	0.003023
70	34	33	11	0.014639	0.010837
70	34	33	12	0.045102	0.030463
70	34	33	13	0.112764	0.067663
70	34	33	14	0.232169	0.119405
70	34	33	15	0.400220	0.168051
70	34	33	16	0.589277	0.189057
70	34	33	17	0.759429	0.170152
70	34	33	18	0.881866	0.122437
70	34	33	19	0.952165	0.070299
70	34	33	20	0.984257	0.032093
70	34	33	21	0.995847	0.011589
70	34	33	22	0.999134	0.003287
70	34	33	23	0.999859	0.000726
70	34	33	24	0.999982	0.000123
70	34	33	25	0.999998	0.000016
70	34	33	26	1.000000	0.000002
70	34	33	27	1.000000	0.000000
70	34	33	28	1.000000	0.000000
70	34	33	29	1.000000	0.000000
70	34	33	30	1.000000	0.000000
70	34	33	31	1.000000	0.000000
70	34	33	32	1.000000	0.000000
70	34	33	33	1.000000	0.000000
70	34	34	0	0.000000	0.000000
70	34	34	1	0.000000	0.000000
70	34	34	2	0.000000	0.000000
70	34	34	3	0.000000	0.000000
70	34	34	4	0.000000	0.000000
70	34	34	5	0.000000	0.000000
70	34	34	6	0.000000	0.000000
70	34	34	7	0.000005	0.000005
70	34	34	8	0.000047	0.000042
70	34	34	9	0.000336	0.000289
70	34	34	10	0.001841	0.001505
70	34	34	11	0.007902	0.006061
70	34	34	12	0.026988	0.019086
70	34	34	13	0.074361	0.047373
70	34	34	14	0.167626	0.093265
70	34	34	15	0.313923	0.146298
70	34	34	16	0.497303	0.183380
70	34	34	17	0.681251	0.183948
70	34	34	18	0.828920	0.147669
70	34	34	19	0.923665	0.094745
70	34	34	20	0.972114	0.048449
70	34	34	21	0.991775	0.019661
70	34	34	22	0.998068	0.006293
70	34	34	23	0.999644	0.001576
70	34	34	24	0.999949	0.000306
70	34	34	25	0.999994	0.000045
70	34	34	26	0.999999	0.000005
70	34	34	27	1.000000	0.000000
70	34	34	28	1.000000	0.000000
70	34	34	29	1.000000	0.000000
70	34	34	30	1.000000	0.000000
70	34	34	31	1.000000	0.000000
70	34	34	32	1.000000	0.000000
70	34	34	33	1.000000	0.000000
70	34	34	34	1.000000	0.000000
70	35	1	0	0.500000	0.500000
70	35	1	1	1.000000	0.500000
70	35	2	0	0.246377	0.246377
70	35	2	1	0.753623	0.507246
70	35	2	2	1.000000	0.246377
70	35	3	0	0.119565	0.119565
70	35	3	1	0.500000	0.380435
70	35	3	2	0.880435	0.380435
70	35	3	3	1.000000	0.119565
70	35	4	0	0.057106	0.057106
70	35	4	1	0.306944	0.249838
70	35	4	2	0.693056	0.386113
70	35	4	3	0.942894	0.249838
70	35	4	4	1.000000	0.057106
70	35	5	0	0.026822	0.026822
70	35	5	1	0.178239	0.151417
70	35	5	2	0.500000	0.321761
70	35	5	3	0.821761	0.321761
70	35	5	4	0.973178	0.151417
70	35	5	5	1.000000	0.026822
70	35	6	0	0.012380	0.012380
70	35	6	1	0.099037	0.086657
70	35	6	2	0.336644	0.237608
70	35	6	3	0.663355	0.326711
70	35	6	4	0.900963	0.237608
70	35	6	5	0.987620	0.086657
70	35	6	6	1.000000	0.012380
70	35	7	0	0.005609	0.005609
70	35	7	1	0.053000	0.047391
70	35	7	2	0.214128	0.161128
70	35	7	3	0.500000	0.285872
70	35	7	4	0.785872	0.285872
70	35	7	5	0.947000	0.161128
70	35	7	6	0.994390	0.047391
70	35	7	7	1.000000	0.005609
70	35	8	0	0.002493	0.002493
70	35	8	1	0.027424	0.024931
70	35	8	2	0.129728	0.102303
70	35	8	3	0.354795	0.225067
70	35	8	4	0.645205	0.290410

N	n	k	x	P(x)	p(x)
70	35	8	5	0.870272	0.225067
70	35	8	6	0.972576	0.102303
70	35	8	7	0.997507	0.024931
70	35	8	8	1.000000	0.002493
70	35	9	0	0.001086	0.001086
70	35	9	1	0.013752	0.012667
70	35	9	2	0.075276	0.061523
70	35	9	3	0.238631	0.163355
70	35	9	4	0.500000	0.261369
70	35	9	5	0.761369	0.261369
70	35	9	6	0.924724	0.163355
70	35	9	7	0.986248	0.061523
70	35	9	8	0.998914	0.012667
70	35	9	9	1.000000	0.001086
70	35	10	0	0.000463	0.000463
70	35	10	1	0.006692	0.006229
70	35	10	2	0.041993	0.035300
70	35	10	3	0.152937	0.110944
70	35	10	4	0.367173	0.214237
70	35	10	5	0.632827	0.265653
70	35	10	6	0.847063	0.214237
70	35	10	7	0.958007	0.110944
70	35	10	8	0.993308	0.035300
70	35	10	9	0.999537	0.006229
70	35	10	10	1.000000	0.000463
70	35	11	0	0.000193	0.000193
70	35	11	1	0.003162	0.002969
70	35	11	2	0.022577	0.019415
70	35	11	3	0.093766	0.071189
70	35	11	4	0.256484	0.162718
70	35	11	5	0.500000	0.243516
70	35	11	6	0.743516	0.243516
70	35	11	7	0.906233	0.162718
70	35	11	8	0.977422	0.071189
70	35	11	9	0.996838	0.019415
70	35	11	10	0.999807	0.002969
70	35	11	11	1.000000	0.000193
70	35	12	0	0.000078	0.000078
70	35	12	1	0.001451	0.001373
70	35	12	2	0.011718	0.010267
70	35	12	3	0.055155	0.043437
70	35	12	4	0.170988	0.115833
70	35	12	5	0.376178	0.205190
70	35	12	6	0.623821	0.247643
70	35	12	7	0.829011	0.205190
70	35	12	8	0.944844	0.115833
70	35	12	9	0.988282	0.043437
70	35	12	10	0.998549	0.010267
70	35	12	11	0.999921	0.001373
70	35	12	12	1.000000	0.000078

N	n	k	x	P(x)	p(x)
70	35	13	0	0.000031	0.000031
70	35	13	1	0.000646	0.000615
70	35	13	2	0.005876	0.005230
70	35	13	3	0.031190	0.025314
70	35	13	4	0.109078	0.077888
70	35	13	5	0.270046	0.160968
70	35	13	6	0.500000	0.229954
70	35	13	7	0.729954	0.229954
70	35	13	8	0.890922	0.160968
70	35	13	9	0.968810	0.077888
70	35	13	10	0.994123	0.025314
70	35	13	11	0.999353	0.005230
70	35	13	12	0.999969	0.000615
70	35	13	13	1.000000	0.000031
70	35	14	0	0.000012	0.000012
70	35	14	1	0.000279	0.000267
70	35	14	2	0.002849	0.002569
70	35	14	3	0.016979	0.014130
70	35	14	4	0.066718	0.049739
70	35	14	5	0.185326	0.118608
70	35	14	6	0.383006	0.197680
70	35	14	7	0.616994	0.233989
70	35	14	8	0.814674	0.197680
70	35	14	9	0.933282	0.118608
70	35	14	10	0.983021	0.049739
70	35	14	11	0.997151	0.014130
70	35	14	12	0.999720	0.002569
70	35	14	13	0.999988	0.000267
70	35	14	14	1.000000	0.000012
70	35	15	0	0.000005	0.000005
70	35	15	1	0.000117	0.000113
70	35	15	2	0.001335	0.001218
70	35	15	3	0.008904	0.007570
70	35	15	4	0.039184	0.030279
70	35	15	5	0.121786	0.082602
70	35	15	6	0.280636	0.158850
70	35	15	7	0.500000	0.219364
70	35	15	8	0.719364	0.219364
70	35	15	9	0.878214	0.158850
70	35	15	10	0.960816	0.082602
70	35	15	11	0.991096	0.030279
70	35	15	12	0.998665	0.007570
70	35	15	13	0.999883	0.001218
70	35	15	14	0.999995	0.000113
70	35	15	15	1.000000	0.000005
70	35	16	0	0.000002	0.000002
70	35	16	1	0.000047	0.000046
70	35	16	2	0.000604	0.000557
70	35	16	3	0.004500	0.003896
70	35	16	4	0.022117	0.017617

N	n	k	x	P(x)	p(x)
70	35	16	5	0.076730	0.054613
70	35	16	6	0.196878	0.120148
70	35	16	7	0.388324	0.191445
70	35	16	8	0.611676	0.223353
70	35	16	9	0.803121	0.191445
70	35	16	10	0.923270	0.120148
70	35	16	11	0.977883	0.054613
70	35	16	12	0.995500	0.017617
70	35	16	13	0.999396	0.003896
70	35	16	14	0.999952	0.000557
70	35	16	15	0.999998	0.000046
70	35	16	16	1.000000	0.000002
70	35	17	0	0.000001	0.000001
70	35	17	1	0.000019	0.000018
70	35	17	2	0.000264	0.000245
70	35	17	3	0.002191	0.001927
70	35	17	4	0.012004	0.009812
70	35	17	5	0.046390	0.034386
70	35	17	6	0.132354	0.085965
70	35	17	7	0.289056	0.156701
70	35	17	8	0.500000	0.210944
70	35	17	9	0.710944	0.210944
70	35	17	10	0.867646	0.156701
70	35	17	11	0.953610	0.085965
70	35	17	12	0.987996	0.034386
70	35	17	13	0.997808	0.009812
70	35	17	14	0.999736	0.001927
70	35	17	15	0.999981	0.000245
70	35	17	16	0.999999	0.000018
70	35	17	17	1.000000	0.000001
70	35	18	0	0.000000	0.000000
70	35	18	1	0.000007	0.000007
70	35	18	2	0.000111	0.000104
70	35	18	3	0.001028	0.000916
70	35	18	4	0.006264	0.005237
70	35	18	5	0.026926	0.020661
70	35	18	6	0.085317	0.058391
70	35	18	7	0.206270	0.120953
70	35	18	8	0.392538	0.186268
70	35	18	9	0.607462	0.214924
70	35	18	10	0.793730	0.186268
70	35	18	11	0.914683	0.120953
70	35	18	12	0.973074	0.058391
70	35	18	13	0.993735	0.020661
70	35	18	14	0.998972	0.005237
70	35	18	15	0.999889	0.000916
70	35	18	16	0.999993	0.000104
70	35	18	17	1.000000	0.000007
70	35	18	18	1.000000	0.000000
70	35	19	0	0.000000	0.000000

N	n	k	x	P(x)	p(x)
70	35	19	1	0.000003	0.000002
70	35	19	2	0.000045	0.000043
70	35	19	3	0.000464	0.000419
70	35	19	4	0.003142	0.002679
70	35	19	5	0.015006	0.011863
70	35	19	6	0.052753	0.037747
70	35	19	7	0.141142	0.088389
70	35	19	8	0.295822	0.154680
70	35	19	9	0.500000	0.204178
70	35	19	10	0.704178	0.204178
70	35	19	11	0.858858	0.154680
70	35	19	12	0.947247	0.088389
70	35	19	13	0.984994	0.037747
70	35	19	14	0.996857	0.011863
70	35	19	15	0.999536	0.002679
70	35	19	16	0.999955	0.000419
70	35	19	17	0.999997	0.000043
70	35	19	18	1.000000	0.000002
70	35	19	19	1.000000	0.000000
70	35	20	0	0.000000	0.000000
70	35	20	1	0.000001	0.000001
70	35	20	2	0.000018	0.000017
70	35	20	3	0.000201	0.000183
70	35	20	4	0.001514	0.001313
70	35	20	5	0.008027	0.006513
70	35	20	6	0.031289	0.023261
70	35	20	7	0.092614	0.061326
70	35	20	8	0.213932	0.121318
70	35	20	9	0.395909	0.181977
70	35	20	10	0.604091	0.208182
70	35	20	11	0.786068	0.181977
70	35	20	12	0.907386	0.121318
70	35	20	13	0.968711	0.061326
70	35	20	14	0.991972	0.023261
70	35	20	15	0.998486	0.006513
70	35	20	16	0.999799	0.001313
70	35	20	17	0.999982	0.000183
70	35	20	18	0.999999	0.000017
70	35	20	19	1.000000	0.000001
70	35	20	20	1.000000	0.000000
70	35	21	0	0.000000	0.000000
70	35	21	1	0.000000	0.000000
70	35	21	2	0.000007	0.000006
70	35	21	3	0.000084	0.000077
70	35	21	4	0.000700	0.000616
70	35	21	5	0.004119	0.003419
70	35	21	6	0.017797	0.013678
70	35	21	7	0.058272	0.040475
70	35	21	8	0.148421	0.090149
70	35	21	9	0.301281	0.152861

Table for $N = 2$, $n = 1$, through $N = 100$, $n = 50$

N	n	k	x	P(x)	p(x)
70	35	21	10	0.500000	0.198719
70	35	21	11	0.698719	0.198719
70	35	21	12	0.851579	0.152861
70	35	21	13	0.941728	0.090149
70	35	21	14	0.982203	0.040475
70	35	21	15	0.995880	0.013678
70	35	21	16	0.999300	0.003419
70	35	21	17	0.999916	0.000616
70	35	21	18	0.999993	0.000077
70	35	21	19	1.000000	0.000006
70	35	21	20	1.000000	0.000000
70	35	21	21	1.000000	0.000000
70	35	22	0	0.000000	0.000000
70	35	22	1	0.000000	0.000000
70	35	22	2	0.000002	0.000002
70	35	22	3	0.000033	0.000031
70	35	22	4	0.000310	0.000277
70	35	22	5	0.002026	0.001716
70	35	22	6	0.009702	0.007676
70	35	22	7	0.035144	0.025441
70	35	22	8	0.098747	0.063603
70	35	22	9	0.220171	0.121425
70	35	22	10	0.398613	0.178441
70	35	22	11	0.601387	0.202774
70	35	22	12	0.779828	0.178441
70	35	22	13	0.901253	0.121425
70	35	22	14	0.964856	0.063603
70	35	22	15	0.990298	0.025441
70	35	22	16	0.997974	0.007676
70	35	22	17	0.999690	0.001716
70	35	22	18	0.999967	0.000277
70	35	22	19	0.999998	0.000031
70	35	22	20	1.000000	0.000002
70	35	22	21	1.000000	0.000000
70	35	22	22	1.000000	0.000000
70	35	23	0	0.000000	0.000000
70	35	23	1	0.000000	0.000000
70	35	23	2	0.000001	0.000001
70	35	23	3	0.000013	0.000012
70	35	23	4	0.000131	0.000119
70	35	23	5	0.000954	0.000822
70	35	23	6	0.005064	0.004111
70	35	23	7	0.020303	0.015238
70	35	23	8	0.062970	0.042667
70	35	23	9	0.154400	0.091430
70	35	23	10	0.305675	0.151275
70	35	23	11	0.500000	0.194325
70	35	23	12	0.694325	0.194325
70	35	23	13	0.845600	0.151275
70	35	23	14	0.937030	0.091430

N	n	k	x	P(x)	p(x)
70	35	23	15	0.979697	0.042667
70	35	23	16	0.994935	0.015238
70	35	23	17	0.999046	0.004111
70	35	23	18	0.999869	0.000822
70	35	23	19	0.999987	0.000119
70	35	23	20	0.999999	0.000012
70	35	23	21	1.000000	0.000001
70	35	23	22	1.000000	0.000000
70	35	23	23	1.000000	0.000000
70	35	24	0	0.000000	0.000000
70	35	24	1	0.000000	0.000000
70	35	24	2	0.000000	0.000000
70	35	24	3	0.000005	0.000004
70	35	24	4	0.000053	0.000048
70	35	24	5	0.000429	0.000376
70	35	24	6	0.002528	0.002099
70	35	24	7	0.011225	0.008697
70	35	24	8	0.038459	0.027234
70	35	24	9	0.103822	0.065363
70	35	24	10	0.225209	0.121388
70	35	24	11	0.400770	0.175561
70	35	24	12	0.599230	0.198460
70	35	24	13	0.774791	0.175561
70	35	24	14	0.896178	0.121388
70	35	24	15	0.961541	0.065363
70	35	24	16	0.988775	0.027234
70	35	24	17	0.997472	0.008697
70	35	24	18	0.999571	0.002099
70	35	24	19	0.999947	0.000376
70	35	24	20	0.999995	0.000048
70	35	24	21	1.000000	0.000004
70	35	24	22	1.000000	0.000000
70	35	24	23	1.000000	0.000000
70	35	24	24	1.000000	0.000000
70	35	25	0	0.000000	0.000000
70	35	25	1	0.000000	0.000000
70	35	25	2	0.000000	0.000000
70	35	25	3	0.000002	0.000001
70	35	25	4	0.000020	0.000019
70	35	25	5	0.000184	0.000163
70	35	25	6	0.001204	0.001021
70	35	25	7	0.005931	0.004726
70	35	25	8	0.022474	0.016543
70	35	25	9	0.066878	0.044404
70	35	25	10	0.159238	0.092360
70	35	25	11	0.309173	0.149935
70	35	25	12	0.500000	0.190827
70	35	25	13	0.690827	0.190827
70	35	25	14	0.840762	0.149935
70	35	25	15	0.933122	0.092360

N	n	k	x	P(x)	p(x)
70	35	25	16	0.977526	0.044404
70	35	25	17	0.994069	0.016543
70	35	25	18	0.998795	0.004726
70	35	25	19	0.999816	0.001021
70	35	25	20	0.999979	0.000163
70	35	25	21	0.999998	0.000019
70	35	25	22	1.000000	0.000001
70	35	25	23	1.000000	0.000000
70	35	25	24	1.000000	0.000000
70	35	25	25	1.000000	0.000000
70	35	26	0	0.000000	0.000000
70	35	26	1	0.000000	0.000000
70	35	26	2	0.000000	0.000000
70	35	26	3	0.000001	0.000000
70	35	26	4	0.000007	0.000007
70	35	26	5	0.000075	0.000067
70	35	26	6	0.000547	0.000472
70	35	26	7	0.002990	0.002443
70	35	26	8	0.012548	0.009558
70	35	26	9	0.041222	0.028674
70	35	26	10	0.107927	0.066705
70	35	26	11	0.229208	0.121281
70	35	26	12	0.402466	0.173259
70	35	26	13	0.597534	0.195067
70	35	26	14	0.770792	0.173259
70	35	26	15	0.892073	0.121281
70	35	26	16	0.958778	0.066705
70	35	26	17	0.987452	0.028674
70	35	26	18	0.997010	0.009558
70	35	26	19	0.999453	0.002443
70	35	26	20	0.999925	0.000472
70	35	26	21	0.999992	0.000067
70	35	26	22	0.999999	0.000007
70	35	26	23	1.000000	0.000000
70	35	26	24	1.000000	0.000000
70	35	26	25	1.000000	0.000000
70	35	26	26	1.000000	0.000000
70	35	27	0	0.000000	0.000000
70	35	27	1	0.000000	0.000000
70	35	27	2	0.000000	0.000000
70	35	27	3	0.000000	0.000000
70	35	27	4	0.000003	0.000002
70	35	27	5	0.000029	0.000026
70	35	27	6	0.000236	0.000207
70	35	27	7	0.001435	0.001199
70	35	27	8	0.006683	0.005248
70	35	27	9	0.024278	0.017595
70	35	27	10	0.070026	0.045748
70	35	27	11	0.163054	0.093028
70	35	27	12	0.311899	0.148845

N	n	k	x	P(x)	p(x)
70	35	27	13	0.500000	0.188101
70	35	27	14	0.688101	0.188101
70	35	27	15	0.836946	0.148845
70	35	27	16	0.929974	0.093028
70	35	27	17	0.975722	0.045748
70	35	27	18	0.993317	0.017595
70	35	27	19	0.998565	0.005248
70	35	27	20	0.999764	0.001199
70	35	27	21	0.999971	0.000207
70	35	27	22	0.999997	0.000026
70	35	27	23	1.000000	0.000002
70	35	27	24	1.000000	0.000000
70	35	27	25	1.000000	0.000000
70	35	27	26	1.000000	0.000000
70	35	27	27	1.000000	0.000000
70	35	28	0	0.000000	0.000000
70	35	28	1	0.000000	0.000000
70	35	28	2	0.000000	0.000000
70	35	28	3	0.000000	0.000000
70	35	28	4	0.000001	0.000001
70	35	28	5	0.000011	0.000010
70	35	28	6	0.000096	0.000086
70	35	28	7	0.000654	0.000558
70	35	28	8	0.003388	0.002734
70	35	28	9	0.013639	0.010251
70	35	28	10	0.043429	0.029789
70	35	28	11	0.111132	0.067703
70	35	28	12	0.232284	0.121153
70	35	28	13	0.403762	0.171478
70	35	28	14	0.596238	0.192475
70	35	28	15	0.767715	0.171478
70	35	28	16	0.888868	0.121153
70	35	28	17	0.956571	0.067703
70	35	28	18	0.986361	0.029789
70	35	28	19	0.996612	0.010251
70	35	28	20	0.999346	0.002734
70	35	28	21	0.999904	0.000558
70	35	28	22	0.999989	0.000086
70	35	28	23	0.999999	0.000010
70	35	28	24	1.000000	0.000001
70	35	28	25	1.000000	0.000000
70	35	28	26	1.000000	0.000000
70	35	28	27	1.000000	0.000000
70	35	28	28	1.000000	0.000000
70	35	29	0	0.000000	0.000000
70	35	29	1	0.000000	0.000000
70	35	29	2	0.000000	0.000000
70	35	29	3	0.000000	0.000000
70	35	29	4	0.000000	0.000000
70	35	29	5	0.000004	0.000003

Table for $N = 2$, $n = 1$, through $N = 100$, $n = 50$

N	n	k	x	P(x)	p(x)
70	35	29	6	0.000037	0.000033
70	35	29	7	0.000282	0.000245
70	35	29	8	0.001630	0.001348
70	35	29	9	0.007293	0.005663
70	35	29	10	0.025697	0.018404
70	35	29	11	0.072444	0.046747
70	35	29	12	0.165939	0.093495
70	35	29	13	0.313941	0.148002
70	35	29	14	0.500000	0.186059
70	35	29	15	0.686059	0.186059
70	35	29	16	0.834061	0.148002
70	35	29	17	0.927556	0.093495
70	35	29	18	0.974303	0.046747
70	35	29	19	0.992707	0.018404
70	35	29	20	0.998369	0.005663
70	35	29	21	0.999718	0.001348
70	35	29	22	0.999963	0.000245
70	35	29	23	0.999996	0.000033
70	35	29	24	1.000000	0.000003
70	35	29	25	1.000000	0.000000
70	35	29	26	1.000000	0.000000
70	35	29	27	1.000000	0.000000
70	35	29	28	1.000000	0.000000
70	35	29	29	1.000000	0.000000
70	35	30	0	0.000000	0.000000
70	35	30	1	0.000000	0.000000
70	35	30	2	0.000000	0.000000
70	35	30	3	0.000000	0.000000
70	35	30	4	0.000000	0.000000
70	35	30	5	0.000001	0.000001
70	35	30	6	0.000013	0.000012
70	35	30	7	0.000115	0.000101
70	35	30	8	0.000743	0.000628
70	35	30	9	0.003702	0.002960
70	35	30	10	0.014475	0.010773
70	35	30	11	0.045080	0.030605
70	35	30	12	0.113491	0.068411
70	35	30	13	0.234525	0.121034
70	35	30	14	0.404701	0.170176
70	35	30	15	0.595299	0.190597
70	35	30	16	0.765475	0.170176
70	35	30	17	0.886509	0.121034
70	35	30	18	0.954920	0.068411
70	35	30	19	0.985525	0.030605
70	35	30	20	0.996298	0.010773
70	35	30	21	0.999257	0.002960
70	35	30	22	0.999885	0.000628
70	35	30	23	0.999986	0.000101
70	35	30	24	0.999999	0.000012
70	35	30	25	1.000000	0.000001
70	35	30	26	1.000000	0.000000
70	35	30	27	1.000000	0.000000
70	35	30	28	1.000000	0.000000
70	35	30	29	1.000000	0.000000
70	35	30	30	1.000000	0.000000
70	35	31	0	0.000000	0.000000
70	35	31	1	0.000000	0.000000
70	35	31	2	0.000000	0.000000
70	35	31	3	0.000000	0.000000
70	35	31	4	0.000000	0.000000
70	35	31	5	0.000000	0.000000
70	35	31	6	0.000005	0.000004
70	35	31	7	0.000044	0.000039
70	35	31	8	0.000319	0.000275
70	35	31	9	0.001778	0.001460
70	35	31	10	0.007742	0.005964
70	35	31	11	0.026717	0.018975
70	35	31	12	0.074154	0.047437
70	35	31	13	0.167956	0.093802
70	35	31	14	0.315359	0.147403
70	35	31	15	0.500000	0.184641
70	35	31	16	0.684641	0.184641
70	35	31	17	0.832044	0.147403
70	35	31	18	0.925845	0.093802
70	35	31	19	0.973283	0.047437
70	35	31	20	0.992258	0.018975
70	35	31	21	0.998221	0.005964
70	35	31	22	0.999681	0.001460
70	35	31	23	0.999956	0.000275
70	35	31	24	0.999995	0.000039
70	35	31	25	1.000000	0.000004
70	35	31	26	1.000000	0.000000
70	35	31	27	1.000000	0.000000
70	35	31	28	1.000000	0.000000
70	35	31	29	1.000000	0.000000
70	35	31	30	1.000000	0.000000
70	35	31	31	1.000000	0.000000
70	35	32	0	0.000000	0.000000
70	35	32	1	0.000000	0.000000
70	35	32	2	0.000000	0.000000
70	35	32	3	0.000000	0.000000
70	35	32	4	0.000000	0.000000
70	35	32	5	0.000000	0.000000
70	35	32	6	0.000001	0.000001
70	35	32	7	0.000016	0.000014
70	35	32	8	0.000128	0.000113
70	35	32	9	0.000805	0.000677
70	35	32	10	0.003919	0.003114
70	35	32	11	0.015040	0.011121
70	35	32	12	0.046178	0.031138
70	35	32	13	0.115042	0.068864
70	35	32	14	0.235988	0.120946
70	35	32	15	0.405312	0.169324
70	35	32	16	0.594688	0.189376
70	35	32	17	0.764012	0.169324
70	35	32	18	0.884958	0.120946
70	35	32	19	0.953821	0.068864
70	35	32	20	0.984960	0.031138
70	35	32	21	0.996081	0.011121
70	35	32	22	0.999195	0.003114
70	35	32	23	0.999872	0.000677
70	35	32	24	0.999984	0.000113
70	35	32	25	0.999999	0.000014
70	35	32	26	1.000000	0.000001
70	35	32	27	1.000000	0.000000
70	35	32	28	1.000000	0.000000
70	35	32	29	1.000000	0.000000
70	35	32	30	1.000000	0.000000
70	35	32	31	1.000000	0.000000
70	35	32	32	1.000000	0.000000
70	35	33	0	0.000000	0.000000
70	35	33	1	0.000000	0.000000
70	35	33	2	0.000000	0.000000
70	35	33	3	0.000000	0.000000
70	35	33	4	0.000000	0.000000
70	35	33	5	0.000000	0.000000
70	35	33	6	0.000000	0.000000
70	35	33	7	0.000005	0.000005
70	35	33	8	0.000048	0.000043
70	35	33	9	0.000342	0.000294
70	35	33	10	0.001871	0.001528
70	35	33	11	0.008016	0.006146
70	35	33	12	0.027332	0.019315
70	35	33	13	0.075174	0.047842
70	35	33	14	0.169150	0.093976
70	35	33	15	0.316194	0.147045
70	35	33	16	0.500000	0.183806
70	35	33	17	0.683806	0.183806
70	35	33	18	0.830850	0.147045
70	35	33	19	0.924826	0.093976
70	35	33	20	0.972668	0.047842
70	35	33	21	0.991984	0.019315
70	35	33	22	0.998129	0.006146
70	35	33	23	0.999658	0.001528
70	35	33	24	0.999952	0.000294
70	35	33	25	0.999995	0.000043
70	35	33	26	0.999999	0.000005
70	35	33	27	1.000000	0.000000
70	35	33	28	1.000000	0.000000
70	35	33	29	1.000000	0.000000
70	35	33	30	1.000000	0.000000
70	35	33	31	1.000000	0.000000
70	35	33	32	1.000000	0.000000
70	35	33	33	1.000000	0.000000
70	35	34	0	0.000000	0.000000
70	35	34	1	0.000000	0.000000
70	35	34	2	0.000000	0.000000
70	35	34	3	0.000000	0.000000
70	35	34	4	0.000000	0.000000
70	35	34	5	0.000000	0.000000
70	35	34	6	0.000000	0.000000
70	35	34	7	0.000002	0.000001
70	35	34	8	0.000017	0.000015
70	35	34	9	0.000136	0.000119
70	35	34	10	0.000838	0.000702
70	35	34	11	0.004030	0.003192
70	35	34	12	0.015325	0.011295
70	35	34	13	0.046727	0.031402
70	35	34	14	0.115812	0.069085
70	35	34	15	0.236711	0.120899
70	35	34	16	0.405613	0.168903
70	35	34	17	0.594387	0.188773
70	35	34	18	0.763289	0.168903
70	35	34	19	0.884188	0.120899
70	35	34	20	0.953273	0.069085
70	35	34	21	0.984675	0.031402
70	35	34	22	0.995970	0.011295
70	35	34	23	0.999162	0.003192
70	35	34	24	0.999864	0.000702
70	35	34	25	0.999983	0.000119
70	35	34	26	0.999998	0.000015
70	35	34	27	1.000000	0.000001
70	35	34	28	1.000000	0.000000
70	35	34	29	1.000000	0.000000
70	35	34	30	1.000000	0.000000
70	35	34	31	1.000000	0.000000
70	35	34	32	1.000000	0.000000
70	35	34	33	1.000000	0.000000
70	35	34	34	1.000000	0.000000
70	35	35	0	0.000000	0.000000
70	35	35	1	0.000000	0.000000
70	35	35	2	0.000000	0.000000
70	35	35	3	0.000000	0.000000
70	35	35	4	0.000000	0.000000
70	35	35	5	0.000000	0.000000
70	35	35	6	0.000000	0.000000
70	35	35	7	0.000000	0.000000
70	35	35	8	0.000005	0.000005
70	35	35	9	0.000050	0.000044
70	35	35	10	0.000350	0.000300

$N = 70$ $n = 35$

Table for $N = 2$, $n = 1$, through $N = 100$, $n = 50$

N	n	k	x	P(x)	p(x)
70	35	35	11	0.001902	0.001552
70	35	35	12	0.008109	0.006207
70	35	35	13	0.027537	0.019428
70	35	35	14	0.075512	0.047976
70	35	35	15	0.169545	0.094032
70	35	35	16	0.316470	0.146925
70	35	35	17	0.500000	0.183530
70	35	35	18	0.683530	0.183530
70	35	35	19	0.830455	0.146925
70	35	35	20	0.924487	0.094032
70	35	35	21	0.972463	0.047976
70	35	35	22	0.991891	0.019428
70	35	35	23	0.998098	0.006207
70	35	35	24	0.999650	0.001552
70	35	35	25	0.999950	0.000300
70	35	35	26	0.999995	0.000044
70	35	35	27	0.999999	0.000005
70	35	35	28	1.000000	0.000000
70	35	35	29	1.000000	0.000000
70	35	35	30	1.000000	0.000000
70	35	35	31	1.000000	0.000000
70	35	35	32	1.000000	0.000000
70	35	35	33	1.000000	0.000000
70	35	35	34	1.000000	0.000000
70	35	35	35	1.000000	0.000000
80	1	1	0	0.987500	0.987500
80	1	1	1	1.000000	0.012500
80	2	1	0	0.975000	0.975000
80	2	1	1	1.000000	0.025000
80	2	2	0	0.950316	0.950316
80	2	2	1	0.999684	0.049367
80	2	2	2	1.000000	0.000316
80	3	1	0	0.962500	0.962500
80	3	1	1	1.000000	0.037500
80	3	2	0	0.925949	0.925949
80	3	2	1	0.999051	0.073101
80	3	2	2	1.000000	0.000949
80	3	3	0	0.890336	0.890336
80	3	3	1	0.997176	0.106840
80	3	3	2	0.999988	0.002812
80	3	3	3	1.000000	0.000012
80	4	1	0	0.950000	0.950000
80	4	1	1	1.000000	0.050000
80	4	2	0	0.901899	0.901899
80	4	2	1	0.998101	0.096203
80	4	2	2	1.000000	0.001899
80	4	3	0	0.855648	0.855648
80	4	3	1	0.994401	0.138754
80	4	3	2	0.999951	0.005550
80	4	3	3	1.000000	0.000049

N	n	k	x	P(x)	p(x)
80	4	4	0	0.811198	0.811198
80	4	4	1	0.988995	0.177797
80	4	4	2	0.999807	0.010812
80	4	4	3	0.999999	0.000192
80	4	4	4	1.000000	0.000001
80	5	1	0	0.937500	0.937500
80	5	1	1	1.000000	0.062500
80	5	2	0	0.878165	0.878165
80	5	2	1	0.996835	0.118671
80	5	2	2	1.000000	0.003165
80	5	3	0	0.821872	0.821872
80	5	3	1	0.990750	0.168878
80	5	3	2	0.999878	0.009129
80	5	3	3	1.000000	0.000122
80	5	4	0	0.768504	0.768504
80	5	4	1	0.981977	0.213473
80	5	4	2	0.999523	0.017546
80	5	4	3	0.999997	0.000474
80	5	4	4	1.000000	0.000003
80	5	5	0	0.717944	0.717944
80	5	5	1	0.970741	0.252797
80	5	5	2	0.998830	0.028089
80	5	5	3	0.999984	0.001154
80	5	5	4	1.000000	0.000016
80	5	5	5	1.000000	0.000000
80	6	1	0	0.925000	0.925000
80	6	1	1	1.000000	0.075000
80	6	2	0	0.854747	0.854747
80	6	2	1	0.995253	0.140506
80	6	2	2	1.000000	0.004747
80	6	3	0	0.788997	0.788997
80	6	3	1	0.986246	0.197249
80	6	3	2	0.999757	0.013510
80	6	3	3	1.000000	0.000243
80	6	4	0	0.727517	0.727517
80	6	4	1	0.973438	0.245921
80	6	4	2	0.999055	0.025617
80	6	4	3	0.999990	0.000936
80	6	4	4	1.000000	0.000009
80	6	5	0	0.670081	0.670081
80	6	5	1	0.957259	0.287178
80	6	5	2	0.997706	0.040448
80	6	5	3	0.999954	0.002247
80	6	5	4	1.000000	0.000046
80	6	5	5	1.000000	0.000000
80	6	6	0	0.616475	0.616475
80	6	6	1	0.938114	0.321639
80	6	6	2	0.995549	0.057436
80	6	6	3	0.999864	0.004314
80	6	6	4	0.999998	0.000135

N	n	k	x	P(x)	p(x)
80	6	6	5	1.000000	0.000001
80	6	6	6	1.000000	0.000000
80	7	1	0	0.912500	0.912500
80	7	1	1	1.000000	0.087500
80	7	2	0	0.831646	0.831646
80	7	2	1	0.993354	0.161709
80	7	2	2	1.000000	0.006646
80	7	3	0	0.757011	0.757011
80	7	3	1	0.980915	0.223905
80	7	3	2	0.999574	0.018659
80	7	3	3	1.000000	0.000426
80	7	4	0	0.688192	0.688192
80	7	4	1	0.963468	0.275277
80	7	4	2	0.998362	0.034894
80	7	4	3	0.999978	0.001615
80	7	4	4	1.000000	0.000022
80	7	5	0	0.624805	0.624805
80	7	5	1	0.941736	0.316930
80	7	5	2	0.996067	0.054331
80	7	5	3	0.999893	0.003826
80	7	5	4	0.999999	0.000106
80	7	5	5	1.000000	0.000001
80	7	6	0	0.566490	0.566490
80	7	6	1	0.916381	0.349891
80	7	6	2	0.992445	0.076063
80	7	6	3	0.999689	0.007244
80	7	6	4	0.999995	0.000306
80	7	6	5	1.000000	0.000005
80	7	6	6	1.000000	0.000000
80	7	7	0	0.512903	0.512903
80	7	7	1	0.888012	0.375108
80	7	7	2	0.987305	0.099293
80	7	7	3	0.999297	0.011992
80	7	7	4	0.999982	0.000685
80	7	7	5	1.000000	0.000017
80	7	7	6	1.000000	0.000000
80	7	7	7	1.000000	0.000000
80	8	1	0	0.900000	0.900000
80	8	1	1	1.000000	0.100000
80	8	2	0	0.808861	0.808861
80	8	2	1	0.991139	0.182278
80	8	2	2	1.000000	0.008861
80	8	3	0	0.725901	0.725901
80	8	3	1	0.974781	0.248880
80	8	3	2	0.999318	0.024537
80	8	3	3	1.000000	0.000682
80	8	4	0	0.650482	0.650482
80	8	4	1	0.952155	0.301673
80	8	4	2	0.997406	0.045251
80	8	4	3	0.999956	0.002549

N	n	k	x	P(x)	p(x)
80	8	4	4	1.000000	0.000044
80	8	5	0	0.582011	0.582011
80	8	5	1	0.924370	0.342359
80	8	5	2	0.993834	0.069464
80	8	5	3	0.999788	0.005954
80	8	5	4	0.999998	0.000210
80	8	5	5	1.000000	0.000002
80	8	6	0	0.519929	0.519929
80	8	6	1	0.892416	0.372487
80	8	6	2	0.988277	0.095861
80	8	6	3	0.999391	0.011114
80	8	6	4	0.999986	0.000595
80	8	6	5	1.000000	0.000013
80	8	6	6	1.000000	0.000000
80	8	7	0	0.463721	0.463721
80	8	7	1	0.857181	0.393460
80	8	7	2	0.980504	0.123323
80	8	7	3	0.998640	0.018136
80	8	7	4	0.999954	0.001314
80	8	7	5	0.999999	0.000045
80	8	7	6	1.000000	0.000001
80	8	7	7	1.000000	0.000000
80	8	8	0	0.412902	0.412902
80	8	8	1	0.819452	0.406550
80	8	8	2	0.970368	0.150916
80	8	8	3	0.997398	0.027030
80	8	8	4	0.999882	0.002484
80	8	8	5	0.999997	0.000115
80	8	8	6	1.000000	0.000002
80	8	8	7	1.000000	0.000000
80	8	8	8	1.000000	0.000000
80	9	1	0	0.887500	0.887500
80	9	1	1	1.000000	0.112500
80	9	2	0	0.786392	0.786392
80	9	2	1	0.988608	0.202215
80	9	2	2	1.000000	0.011392
80	9	3	0	0.695655	0.695655
80	9	3	1	0.967868	0.272213
80	9	3	2	0.998978	0.031110
80	9	3	3	1.000000	0.001022
80	9	4	0	0.614344	0.614344
80	9	4	1	0.939586	0.325241
80	9	4	2	0.996149	0.056564
80	9	4	3	0.999920	0.003771
80	9	4	4	1.000000	0.000080
80	9	5	0	0.541593	0.541593
80	9	5	1	0.905350	0.363757
80	9	5	2	0.990940	0.085590
80	9	5	3	0.999623	0.008683
80	9	5	4	0.999995	0.000372

N	n	k	x	P(x)	p(x)
80	9	5	5	1.000000	0.000005
80	9	6	0	0.476602	0.476602
80	9	6	1	0.866549	0.389947
80	9	6	2	0.982951	0.116402
80	9	6	3	0.998928	0.015977
80	9	6	4	0.999970	0.001042
80	9	6	5	1.000000	0.000030
80	9	6	6	1.000000	0.000000
80	9	7	0	0.418637	0.418637
80	9	7	1	0.824393	0.405756
80	9	7	2	0.971940	0.147548
80	9	7	3	0.997633	0.025692
80	9	7	4	0.999899	0.002267
80	9	7	5	0.999998	0.000099
80	9	7	6	1.000000	0.000002
80	9	7	7	1.000000	0.000000
80	9	8	0	0.367024	0.367024
80	9	8	1	0.779926	0.412902
80	9	8	2	0.957792	0.177866
80	9	8	3	0.995521	0.037729
80	9	8	4	0.999744	0.004223
80	9	8	5	0.999993	0.000248
80	9	8	6	1.000000	0.000007
80	9	8	7	1.000000	0.000000
80	9	8	8	1.000000	0.000000
80	9	9	0	0.321146	0.321146
80	9	9	1	0.734048	0.412902
80	9	9	2	0.940499	0.206451
80	9	9	3	0.992377	0.051877
80	9	9	4	0.999451	0.007074
80	9	9	5	0.999979	0.000528
80	9	9	6	1.000000	0.000021
80	9	9	7	1.000000	0.000000
80	9	9	8	1.000000	0.000000
80	9	9	9	1.000000	0.000000
80	10	1	0	0.875000	0.875000
80	10	1	1	1.000000	0.125000
80	10	2	0	0.764241	0.764241
80	10	2	1	0.985759	0.221519
80	10	2	2	1.000000	0.014240
80	10	3	0	0.666261	0.666261
80	10	3	1	0.960200	0.293939
80	10	3	2	0.998539	0.038340
80	10	3	3	1.000000	0.001461
80	10	4	0	0.579734	0.579734
80	10	4	1	0.925843	0.346110
80	10	4	2	0.994556	0.068713
80	10	4	3	0.999867	0.005311
80	10	4	4	1.000000	0.000133
80	10	5	0	0.503453	0.503453

N	n	k	x	P(x)	p(x)
80	10	5	1	0.884856	0.381404
80	10	5	2	0.987323	0.102467
80	10	5	3	0.999378	0.012055
80	10	5	4	0.999989	0.000611
80	10	5	5	1.000000	0.000010
80	10	6	0	0.436326	0.436326
80	10	6	1	0.839088	0.402762
80	10	6	2	0.976393	0.137305
80	10	6	3	0.998253	0.021860
80	10	6	4	0.999941	0.001688
80	10	6	5	0.999999	0.000059
80	10	6	6	1.000000	0.000001
80	10	7	0	0.377363	0.377363
80	10	7	1	0.790103	0.412741
80	10	7	2	0.961550	0.171446
80	10	7	3	0.996185	0.034636
80	10	7	4	0.999804	0.003619
80	10	7	5	0.999995	0.000192
80	10	7	6	1.000000	0.000005
80	10	7	7	1.000000	0.000000
80	10	8	0	0.325669	0.325669
80	10	8	1	0.739218	0.413548
80	10	8	2	0.942761	0.203543
80	10	8	3	0.992864	0.050103
80	10	8	4	0.999506	0.006642
80	10	8	5	0.999982	0.000476
80	10	8	6	1.000000	0.000017
80	10	8	7	1.000000	0.000000
80	10	8	8	1.000000	0.000000
80	10	9	0	0.280437	0.280437
80	10	9	1	0.687524	0.407087
80	10	9	2	0.920145	0.232621
80	10	9	3	0.987993	0.067848
80	10	9	4	0.998953	0.010960
80	10	9	5	0.999949	0.000996
80	10	9	6	0.999999	0.000050
80	10	9	7	1.000000	0.000001
80	10	9	8	1.000000	0.000000
80	10	9	9	1.000000	0.000000
80	10	10	0	0.240939	0.240939
80	10	10	1	0.635922	0.394982
80	10	10	2	0.893934	0.258013
80	10	10	3	0.981304	0.087369
80	10	10	4	0.998027	0.016723
80	10	10	5	0.999879	0.001852
80	10	10	6	0.999996	0.000117
80	10	10	7	1.000000	0.000004
80	10	10	8	1.000000	0.000000
80	10	10	9	1.000000	0.000000
80	10	10	10	1.000000	0.000000

N	n	k	x	P(x)	p(x)
80	11	1	0	0.862500	0.862500
80	11	1	1	1.000000	0.137500
80	11	2	0	0.742405	0.742405
80	11	2	1	0.982595	0.240190
80	11	2	2	1.000000	0.017405
80	11	3	0	0.637707	0.637707
80	11	3	1	0.951801	0.314094
80	11	3	2	0.997992	0.046190
80	11	3	3	1.000000	0.002008
80	11	4	0	0.546606	0.546606
80	11	4	1	0.911010	0.364404
80	11	4	2	0.992593	0.081583
80	11	4	3	0.999791	0.007198
80	11	4	4	1.000000	0.000209
80	11	5	0	0.467492	0.467492
80	11	5	1	0.863062	0.395570
80	11	5	2	0.982932	0.119870
80	11	5	3	0.999034	0.016102
80	11	5	4	0.999981	0.000947
80	11	5	5	1.000000	0.000019
80	11	6	0	0.398926	0.398926
80	11	6	1	0.810319	0.411393
80	11	6	2	0.968547	0.158228
80	11	6	3	0.997316	0.028769
80	11	6	4	0.999892	0.002576
80	11	6	5	0.999998	0.000106
80	11	6	6	1.000000	0.000002
80	11	7	0	0.339627	0.339627
80	11	7	1	0.754726	0.415099
80	11	7	2	0.949303	0.194578
80	11	7	3	0.994206	0.044903
80	11	7	4	0.999649	0.005443
80	11	7	5	0.999990	0.000341
80	11	7	6	1.000000	0.000010
80	11	7	7	1.000000	0.000000
80	11	8	0	0.288450	0.288450
80	11	8	1	0.697863	0.409413
80	11	8	2	0.925314	0.227452
80	11	8	3	0.989285	0.063971
80	11	8	4	0.999127	0.009842
80	11	8	5	0.999962	0.000835
80	11	8	6	0.999999	0.000037
80	11	8	7	1.000000	0.000001
80	11	8	8	1.000000	0.000000
80	11	9	0	0.244381	0.244381
80	11	9	1	0.641000	0.396619
80	11	9	2	0.896883	0.255883
80	11	9	3	0.982177	0.085294
80	11	9	4	0.998170	0.015993
80	11	9	5	0.999892	0.001722

N	n	k	x	P(x)	p(x)
80	11	9	6	0.999996	0.000104
80	11	9	7	1.000000	0.000003
80	11	9	8	1.000000	0.000000
80	11	9	9	1.000000	0.000000
80	11	10	0	0.206519	0.206519
80	11	10	1	0.585138	0.378619
80	11	10	2	0.864447	0.279309
80	11	10	3	0.972567	0.108120
80	11	10	4	0.996593	0.024027
80	11	10	5	0.999747	0.003153
80	11	10	6	0.999989	0.000243
80	11	10	7	1.000000	0.000010
80	11	10	8	1.000000	0.000000
80	11	10	9	1.000000	0.000000
80	11	10	10	1.000000	0.000000
80	11	11	0	0.174066	0.174066
80	11	11	1	0.531050	0.356983
80	11	11	2	0.828536	0.297486
80	11	11	3	0.960210	0.131674
80	11	11	4	0.994190	0.033980
80	11	11	5	0.999476	0.005286
80	11	11	6	0.999972	0.000496
80	11	11	7	0.999999	0.000027
80	11	11	8	1.000000	0.000001
80	11	11	9	1.000000	0.000000
80	11	11	10	1.000000	0.000000
80	11	11	11	1.000000	0.000000
80	12	1	0	0.850000	0.850000
80	12	1	1	1.000000	0.150000
80	12	2	0	0.720886	0.720886
80	12	2	1	0.979114	0.258228
80	12	2	2	1.000000	0.020886
80	12	3	0	0.609981	0.609981
80	12	3	1	0.942697	0.332717
80	12	3	2	0.997322	0.054625
80	12	3	3	1.000000	0.002678
80	12	4	0	0.514919	0.514919
80	12	4	1	0.895166	0.380248
80	12	4	2	0.990228	0.095062
80	12	4	3	0.999687	0.009459
80	12	4	4	1.000000	0.000313
80	12	5	0	0.433616	0.433616
80	12	5	1	0.840130	0.406515
80	12	5	2	0.977720	0.137590
80	12	5	3	0.998567	0.020847
80	12	5	4	0.999967	0.001400
80	12	5	5	1.000000	0.000033
80	12	6	0	0.364237	0.364237
80	12	6	1	0.780508	0.416271
80	12	6	2	0.959375	0.178866

Table for $N = 2$, $n = 1$, through $N = 100$, $n = 50$

N	n	k	x	P(x)	p(x)
80	12	6	3	0.996065	0.036691
80	12	6	4	0.999818	0.003752
80	12	6	5	0.999997	0.000179
80	12	6	6	1.000000	0.000003
80	12	7	0	0.305172	0.305172
80	12	7	1	0.718630	0.413458
80	12	7	2	0.935203	0.216573
80	12	7	3	0.991603	0.056399
80	12	7	4	0.999412	0.007809
80	12	7	5	0.999980	0.000568
80	12	7	6	1.000000	0.000020
80	12	7	7	1.000000	0.000000
80	12	8	0	0.255006	0.255006
80	12	8	1	0.656328	0.401322
80	12	8	2	0.905536	0.249208
80	12	8	3	0.984649	0.079114
80	12	8	4	0.998556	0.013907
80	12	8	5	0.999925	0.001369
80	12	8	6	0.999998	0.000073
80	12	8	7	1.000000	0.000002
80	12	8	8	1.000000	0.000000
80	12	9	0	0.212505	0.212505
80	12	9	1	0.595015	0.382510
80	12	9	2	0.870924	0.275909
80	12	9	3	0.974760	0.103837
80	12	9	4	0.997011	0.022251
80	12	9	5	0.999792	0.002781
80	12	9	6	0.999992	0.000200
80	12	9	7	1.000000	0.000008
80	12	9	8	1.000000	0.000000
80	12	9	9	1.000000	0.000000
80	12	10	0	0.176589	0.176589
80	12	10	1	0.535753	0.359164
80	12	10	2	0.832063	0.296310
80	12	10	3	0.961598	0.129535
80	12	10	4	0.994504	0.032906
80	12	10	5	0.999518	0.005014
80	12	10	6	0.999975	0.000457
80	12	10	7	0.999999	0.000024
80	12	10	8	1.000000	0.000001
80	12	10	9	1.000000	0.000000
80	12	10	10	1.000000	0.000000
80	12	11	0	0.146317	0.146317
80	12	11	1	0.479313	0.332996
80	12	11	2	0.789733	0.310420
80	12	11	3	0.944943	0.155210
80	12	11	4	0.990743	0.045800
80	12	11	5	0.999017	0.008273
80	12	11	6	0.999936	0.000919
80	12	11	7	0.999998	0.000062

N	n	k	x	P(x)	p(x)
80	12	11	8	1.000000	0.000002
80	12	11	9	1.000000	0.000000
80	12	11	10	1.000000	0.000000
80	12	11	11	1.000000	0.000000
80	12	12	0	0.120870	0.120870
80	12	12	1	0.426227	0.305356
80	12	12	2	0.744745	0.318518
80	12	12	3	0.924699	0.179954
80	12	12	4	0.985433	0.060734
80	12	12	5	0.998177	0.012744
80	12	12	6	0.999856	0.001679
80	12	12	7	0.999993	0.000137
80	12	12	8	1.000000	0.000007
80	12	12	9	1.000000	0.000000
80	12	12	10	1.000000	0.000000
80	12	12	11	1.000000	0.000000
80	12	12	12	1.000000	0.000000
80	13	1	0	0.837500	0.837500
80	13	1	1	1.000000	0.162500
80	13	2	0	0.699684	0.699684
80	13	2	1	0.975316	0.275633
80	13	2	2	1.000000	0.024684
80	13	3	0	0.583070	0.583070
80	13	3	1	0.932911	0.349842
80	13	3	2	0.996519	0.063608
80	13	3	3	1.000000	0.003481
80	13	4	0	0.484629	0.484629
80	13	4	1	0.878391	0.393761
80	13	4	2	0.987432	0.109042
80	13	4	3	0.999548	0.012116
80	13	4	4	1.000000	0.000452
80	13	5	0	0.401732	0.401732
80	13	5	1	0.816218	0.414486
80	13	5	2	0.971650	0.155432
80	13	5	3	0.997954	0.026304
80	13	5	4	0.999946	0.001993
80	13	5	5	1.000000	0.000054
80	13	6	0	0.332099	0.332099
80	13	6	1	0.749900	0.417801
80	13	6	2	0.948853	0.198953
80	13	6	3	0.994446	0.045593
80	13	6	4	0.999707	0.005261
80	13	6	5	0.999994	0.000287
80	13	6	6	1.000000	0.000006
80	13	7	0	0.273757	0.273757
80	13	7	1	0.682148	0.408392
80	13	7	2	0.919279	0.237131
80	13	7	3	0.988285	0.069006
80	13	7	4	0.999067	0.010782
80	13	7	5	0.999963	0.000896

N	n	k	x	P(x)	p(x)
80	13	7	6	0.999999	0.000036
80	13	7	7	1.000000	0.000001
80	13	8	0	0.225006	0.225006
80	13	8	1	0.615016	0.390010
80	13	8	2	0.883547	0.268531
80	13	8	3	0.978832	0.095285
80	13	8	4	0.997738	0.018906
80	13	8	5	0.999865	0.002127
80	13	8	6	0.999996	0.000131
80	13	8	7	1.000000	0.000004
80	13	8	8	1.000000	0.000000
80	13	9	0	0.184380	0.184380
80	13	9	1	0.550014	0.365634
80	13	9	2	0.842521	0.292507
80	13	9	3	0.965598	0.123077
80	13	9	4	0.995375	0.029777
80	13	9	5	0.999629	0.004254
80	13	9	6	0.999983	0.000354
80	13	9	7	1.000000	0.000016
80	13	9	8	1.000000	0.000000
80	13	9	9	1.000000	0.000000
80	13	10	0	0.150620	0.150620
80	13	10	1	0.488217	0.337597
80	13	10	2	0.797203	0.308987
80	13	10	3	0.948263	0.151060
80	13	10	4	0.991600	0.043337
80	13	10	5	0.999149	0.007549
80	13	10	6	0.999948	0.000799
80	13	10	7	0.999998	0.000050
80	13	10	8	1.000000	0.000002
80	13	10	9	1.000000	0.000000
80	13	10	10	1.000000	0.000000
80	13	11	0	0.122648	0.122648
80	13	11	1	0.430343	0.307695
80	13	11	2	0.748648	0.318305
80	13	11	3	0.926683	0.178035
80	13	11	4	0.986028	0.059345
80	13	11	5	0.998287	0.012258
80	13	11	6	0.999868	0.001582
80	13	11	7	0.999994	0.000126
80	13	11	8	1.000000	0.000006
80	13	11	9	1.000000	0.000000
80	13	11	10	1.000000	0.000000
80	13	11	11	1.000000	0.000000
80	13	12	0	0.099540	0.099540
80	13	12	1	0.376831	0.277291
80	13	12	2	0.697904	0.321073
80	13	12	3	0.900881	0.202977
80	13	12	4	0.978288	0.077407
80	13	12	5	0.996865	0.018578

N	n	k	x	P(x)	p(x)
80	13	12	6	0.999708	0.002842
80	13	12	7	0.999983	0.000275
80	13	12	8	0.999999	0.000016
80	13	12	9	1.000000	0.000001
80	13	12	10	1.000000	0.000000
80	13	12	11	1.000000	0.000000
80	13	12	12	1.000000	0.000000
80	13	13	0	0.080510	0.080510
80	13	13	1	0.327897	0.247387
80	13	13	2	0.645966	0.318069
80	13	13	3	0.871032	0.225066
80	13	13	4	0.968043	0.097011
80	13	13	5	0.994680	0.026637
80	13	13	6	0.999415	0.004735
80	13	13	7	0.999959	0.000543
80	13	13	8	0.999998	0.000039
80	13	13	9	1.000000	0.000002
80	13	13	10	1.000000	0.000000
80	13	13	11	1.000000	0.000000
80	13	13	12	1.000000	0.000000
80	13	13	13	1.000000	0.000000
80	14	1	0	0.825000	0.825000
80	14	1	1	1.000000	0.175000
80	14	2	0	0.678797	0.678797
80	14	2	1	0.971203	0.292405
80	14	2	2	1.000000	0.028797
80	14	3	0	0.556962	0.556962
80	14	3	1	0.922468	0.365506
80	14	3	2	0.995570	0.073101
80	14	3	3	1.000000	0.004430
80	14	4	0	0.455696	0.455696
80	14	4	1	0.860759	0.405063
80	14	4	2	0.984177	0.123418
80	14	4	3	0.999367	0.015190
80	14	4	4	1.000000	0.000633
80	14	5	0	0.371752	0.371752
80	14	5	1	0.791472	0.419720
80	14	5	2	0.964690	0.173218
80	14	5	3	0.997168	0.032478
80	14	5	4	0.999917	0.002748
80	14	5	5	1.000000	0.000083
80	14	6	0	0.302358	0.302358
80	14	6	1	0.718721	0.416362
80	14	6	2	0.936975	0.218254
80	14	6	3	0.992405	0.055430
80	14	6	4	0.999550	0.007145
80	14	6	5	0.999990	0.000440
80	14	6	6	1.000000	0.000010
80	14	7	0	0.245155	0.245155
80	14	7	1	0.645576	0.400421

$N = 80$ $n = 12–14$

Table for $N = 2$, $n = 1$, through $N = 100$, $n = 50$

N	n	k	x	P(x)	p(x)
80	14	7	2	0.901583	0.256007
80	14	7	3	0.984165	0.082583
80	14	7	4	0.998585	0.014419
80	14	7	5	0.999936	0.001352
80	14	7	6	0.999999	0.000062
80	14	7	7	1.000000	0.000001
80	14	8	0	0.198139	0.198139
80	14	8	1	0.574268	0.376129
80	14	8	2	0.859499	0.285231
80	14	8	3	0.971721	0.112222
80	14	8	4	0.996609	0.024888
80	14	8	5	0.999770	0.003160
80	14	8	6	0.999992	0.000222
80	14	8	7	1.000000	0.000008
80	14	8	8	1.000000	0.000000
80	14	9	0	0.159612	0.159612
80	14	9	1	0.506356	0.346744
80	14	9	2	0.811961	0.305605
80	14	9	3	0.954576	0.142616
80	14	9	4	0.993153	0.038576
80	14	9	5	0.999375	0.006222
80	14	9	6	0.999967	0.000593
80	14	9	7	0.999999	0.000032
80	14	9	8	1.000000	0.000001
80	14	9	9	1.000000	0.000000
80	14	10	0	0.128139	0.128139
80	14	10	1	0.442868	0.314728
80	14	10	2	0.760309	0.317442
80	14	10	3	0.932481	0.172172
80	14	10	4	0.987719	0.055238
80	14	10	5	0.998586	0.010867
80	14	10	6	0.999901	0.001314
80	14	10	7	0.999996	0.000095
80	14	10	8	1.000000	0.000004
80	14	10	9	1.000000	0.000000
80	14	10	10	1.000000	0.000000
80	14	11	0	0.102512	0.102512
80	14	11	1	0.384418	0.281907
80	14	11	2	0.705891	0.321473
80	14	11	3	0.905426	0.199535
80	14	11	4	0.979828	0.074403
80	14	11	5	0.997189	0.017361
80	14	11	6	0.999750	0.002561
80	14	11	7	0.999986	0.000236
80	14	11	8	0.999999	0.000013
80	14	11	9	1.000000	0.000000
80	14	11	10	1.000000	0.000000
80	14	11	11	1.000000	0.000000
80	14	12	0	0.081712	0.081712
80	14	12	1	0.331305	0.249593
80	14	12	2	0.649983	0.318677
80	14	12	3	0.873616	0.223633
80	14	12	4	0.969045	0.095430
80	14	12	5	0.994924	0.025879
80	14	12	6	0.999453	0.004529
80	14	12	7	0.999962	0.000509
80	14	12	8	0.999998	0.000036
80	14	12	9	1.000000	0.000002
80	14	12	10	1.000000	0.000000
80	14	12	11	1.000000	0.000000
80	14	12	12	1.000000	0.000000
80	14	13	0	0.064889	0.064889
80	14	13	1	0.283589	0.218700
80	14	13	2	0.593745	0.310156
80	14	13	3	0.837440	0.243694
80	14	13	4	0.955012	0.117572
80	14	13	5	0.991499	0.036488
80	14	13	6	0.998921	0.007421
80	14	13	7	0.999910	0.000989
80	14	13	8	0.999995	0.000085
80	14	13	9	1.000000	0.000005
80	14	13	10	1.000000	0.000000
80	14	13	11	1.000000	0.000000
80	14	13	12	1.000000	0.000000
80	14	13	13	1.000000	0.000000
80	14	14	0	0.051330	0.051330
80	14	14	1	0.241155	0.189825
80	14	14	2	0.538195	0.297040
80	14	14	3	0.797430	0.259235
80	14	14	4	0.937463	0.140033
80	14	14	5	0.986598	0.049134
80	14	14	6	0.998034	0.011436
80	14	14	7	0.999807	0.001772
80	14	14	8	0.999988	0.000181
80	14	14	9	0.999999	0.000012
80	14	14	10	1.000000	0.000000
80	14	14	11	1.000000	0.000000
80	14	14	12	1.000000	0.000000
80	14	14	13	1.000000	0.000000
80	14	14	14	1.000000	0.000000
80	15	1	0	0.812500	0.812500
80	15	1	1	1.000000	0.187500
80	15	2	0	0.658228	0.658228
80	15	2	1	0.966772	0.308544
80	15	2	2	1.000000	0.033228
80	15	3	0	0.531646	0.531646
80	15	3	1	0.911392	0.379747
80	15	3	2	0.994462	0.083070
80	15	3	3	1.000000	0.005538
80	15	4	0	0.428078	0.428078
80	15	4	1	0.842348	0.414269
80	15	4	2	0.980437	0.138090
80	15	4	3	0.999137	0.018700
80	15	4	4	1.000000	0.000863
80	15	5	0	0.343589	0.343589
80	15	5	1	0.766035	0.422446
80	15	5	2	0.956817	0.190782
80	15	5	3	0.996184	0.039368
80	15	5	4	0.999875	0.003691
80	15	5	5	1.000000	0.000125
80	15	6	0	0.274871	0.274871
80	15	6	1	0.687178	0.412307
80	15	6	2	0.923748	0.236570
80	15	6	3	0.989885	0.066138
80	15	6	4	0.999334	0.009448
80	15	6	5	0.999983	0.000650
80	15	6	6	1.000000	0.000017
80	15	7	0	0.219154	0.219154
80	15	7	1	0.609174	0.390020
80	15	7	2	0.882188	0.273014
80	15	7	3	0.979160	0.096972
80	15	7	4	0.997929	0.018769
80	15	7	5	0.999895	0.001966
80	15	7	6	0.999998	0.000102
80	15	7	7	1.000000	0.000002
80	15	8	0	0.174122	0.174122
80	15	8	1	0.534376	0.360253
80	15	8	2	0.833569	0.299193
80	15	8	3	0.963220	0.129650
80	15	8	4	0.995101	0.031881
80	15	8	5	0.999626	0.004525
80	15	8	6	0.999985	0.000359
80	15	8	7	1.000000	0.000014
80	15	8	8	1.000000	0.000000
80	15	9	0	0.137847	0.137847
80	15	9	1	0.464327	0.326480
80	15	9	2	0.779548	0.315222
80	15	9	3	0.941611	0.162063
80	15	9	4	0.990230	0.048619
80	15	9	5	0.998998	0.008767
80	15	9	6	0.999940	0.000943
80	15	9	7	0.999998	0.000058
80	15	9	8	1.000000	0.000002
80	15	9	9	1.000000	0.000000
80	15	10	0	0.108724	0.108724
80	15	10	1	0.399950	0.291226
80	15	10	2	0.721832	0.321881
80	15	10	3	0.914220	0.192389
80	15	10	4	0.982698	0.068477
80	15	10	5	0.997763	0.015065
80	15	10	6	0.999821	0.002058
80	15	10	7	0.999992	0.000171
80	15	10	8	1.000000	0.000008
80	15	10	9	1.000000	0.000000
80	15	10	10	1.000000	0.000000
80	15	11	0	0.085426	0.085426
80	15	11	1	0.341705	0.256279
80	15	11	2	0.662054	0.320349
80	15	11	3	0.881239	0.219186
80	15	11	4	0.971937	0.090698
80	15	11	5	0.995611	0.023674
80	15	11	6	0.999556	0.003946
80	15	11	7	0.999972	0.000416
80	15	11	8	0.999999	0.000027
80	15	11	9	1.000000	0.000001
80	15	11	10	1.000000	0.000000
80	15	11	11	1.000000	0.000000
80	15	12	0	0.066855	0.066855
80	15	12	1	0.289706	0.222851
80	15	12	2	0.601698	0.311992
80	15	12	3	0.843120	0.241422
80	15	12	4	0.957478	0.114358
80	15	12	5	0.992180	0.034702
80	15	12	6	0.999042	0.006862
80	15	12	7	0.999924	0.000882
80	15	12	8	0.999996	0.000072
80	15	12	9	1.000000	0.000004
80	15	12	10	1.000000	0.000000
80	15	12	11	1.000000	0.000000
80	15	12	12	1.000000	0.000000
80	15	13	0	0.052108	0.052108
80	15	13	1	0.243825	0.191718
80	15	13	2	0.542053	0.298227
80	15	13	3	0.800516	0.258464
80	15	13	4	0.938979	0.138463
80	15	13	5	0.987076	0.048098
80	15	13	6	0.998133	0.011057
80	15	13	7	0.999820	0.001687
80	15	13	8	0.999989	0.000169
80	15	13	9	0.999999	0.000011
80	15	13	10	1.000000	0.000000
80	15	13	11	1.000000	0.000000
80	15	13	12	1.000000	0.000000
80	15	13	13	1.000000	0.000000
80	15	14	0	0.040442	0.040442
80	15	14	1	0.203765	0.163323
80	15	14	2	0.484188	0.280423
80	15	14	3	0.754224	0.270037
80	15	14	4	0.916246	0.162022
80	15	14	5	0.979898	0.063651

N	n	k	x	P(x)	p(x)
80	15	14	6	0.996648	0.016750
80	15	14	7	0.999619	0.002971
80	15	14	8	0.999971	0.000352
80	15	14	9	0.999998	0.000027
80	15	14	10	1.000000	0.000001
80	15	14	11	1.000000	0.000000
80	15	14	12	1.000000	0.000000
80	15	14	13	1.000000	0.000000
80	15	14	14	1.000000	0.000000
80	15	15	0	0.031251	0.031251
80	15	15	1	0.169121	0.137870
80	15	15	2	0.428953	0.259832
80	15	15	3	0.705127	0.276174
80	15	15	4	0.889243	0.184116
80	15	15	5	0.970254	0.081011
80	15	15	6	0.994364	0.024110
80	15	15	7	0.999259	0.004895
80	15	15	8	0.999934	0.000675
80	15	15	9	0.999996	0.000062
80	15	15	10	1.000000	0.000004
80	15	15	11	1.000000	0.000000
80	15	15	12	1.000000	0.000000
80	15	15	13	1.000000	0.000000
80	15	15	14	1.000000	0.000000
80	15	15	15	1.000000	0.000000
80	16	1	0	0.800000	0.800000
80	16	1	1	1.000000	0.200000
80	16	2	0	0.637975	0.637975
80	16	2	1	0.962025	0.324051
80	16	2	2	1.000000	0.037975
80	16	3	0	0.507108	0.507108
80	16	3	1	0.899708	0.392600
80	16	3	2	0.993184	0.093476
80	16	3	3	1.000000	0.006816
80	16	4	0	0.401735	0.401735
80	16	4	1	0.823227	0.421492
80	16	4	2	0.976188	0.152961
80	16	4	3	0.998849	0.022661
80	16	4	4	1.000000	0.001151
80	16	5	0	0.317159	0.317159
80	16	5	1	0.740038	0.422879
80	16	5	2	0.948011	0.207973
80	16	5	3	0.994973	0.046962
80	16	5	4	0.999818	0.004845
80	16	5	5	1.000000	0.000182
80	16	6	0	0.249499	0.249499
80	16	6	1	0.655462	0.405964
80	16	6	2	0.909190	0.253727
80	16	6	3	0.986833	0.077643
80	16	6	4	0.999043	0.012210
80	16	6	5	0.999973	0.000930
80	16	6	6	1.000000	0.000027
80	16	7	0	0.195553	0.195553
80	16	7	1	0.573172	0.377619
80	16	7	2	0.861187	0.288015
80	16	7	3	0.973193	0.112006
80	16	7	4	0.997063	0.023870
80	16	7	5	0.999835	0.002772
80	16	7	6	0.999996	0.000161
80	16	7	7	1.000000	0.000004
80	16	8	0	0.152692	0.152692
80	16	8	1	0.495579	0.342887
80	16	8	2	0.805951	0.310372
80	16	8	3	0.953247	0.147295
80	16	8	4	0.993139	0.039892
80	16	8	5	0.999417	0.006278
80	16	8	6	0.999974	0.000557
80	16	8	7	0.999999	0.000025
80	16	8	8	1.000000	0.000000
80	16	9	0	0.118760	0.118760
80	16	9	1	0.424144	0.305384
80	16	9	2	0.745601	0.321457
80	16	9	3	0.926652	0.181050
80	16	9	4	0.986490	0.059839
80	16	9	5	0.998458	0.011968
80	16	9	6	0.999897	0.001439
80	16	9	7	0.999996	0.000099
80	16	9	8	1.000000	0.000004
80	16	9	9	1.000000	0.000000
80	16	10	0	0.091998	0.091998
80	16	10	1	0.359627	0.267629
80	16	10	2	0.682215	0.322589
80	16	10	3	0.893502	0.211286
80	16	10	4	0.976377	0.082875
80	16	10	5	0.996604	0.020227
80	16	10	6	0.999694	0.003090
80	16	10	7	0.999984	0.000289
80	16	10	8	0.999999	0.000016
80	16	10	9	1.000000	0.000000
80	16	10	10	1.000000	0.000000
80	16	11	0	0.070970	0.070970
80	16	11	1	0.302278	0.231308
80	16	11	2	0.617698	0.315420
80	16	11	3	0.854263	0.236565
80	16	11	4	0.962170	0.107907
80	16	11	5	0.993425	0.031256
80	16	11	6	0.999253	0.005827
80	16	11	7	0.999947	0.000694
80	16	11	8	0.999998	0.000051
80	16	11	9	1.000000	0.000002
80	16	11	10	1.000000	0.000000
80	16	11	11	1.000000	0.000000
80	16	12	0	0.054513	0.054513
80	16	12	1	0.251993	0.197480
80	16	12	2	0.553699	0.301706
80	16	12	3	0.809693	0.255993
80	16	12	4	0.943403	0.133711
80	16	12	5	0.988443	0.045039
80	16	12	6	0.998408	0.009966
80	16	12	7	0.999856	0.001448
80	16	12	8	0.999992	0.000136
80	16	12	9	1.000000	0.000008
80	16	12	10	1.000000	0.000000
80	16	12	11	1.000000	0.000000
80	16	12	12	1.000000	0.000000
80	16	13	0	0.041686	0.041686
80	16	13	1	0.208431	0.166745
80	16	13	2	0.491583	0.283152
80	16	13	3	0.760753	0.269169
80	16	13	4	0.919807	0.159055
80	16	13	5	0.981157	0.061350
80	16	13	6	0.996943	0.015786
80	16	13	7	0.999664	0.002722
80	16	13	8	0.999976	0.000311
80	16	13	9	0.999999	0.000023
80	16	13	10	1.000000	0.000001
80	16	13	11	1.000000	0.000000
80	16	13	12	1.000000	0.000000
80	16	13	13	1.000000	0.000000
80	16	14	0	0.031731	0.031731
80	16	14	1	0.171100	0.139369
80	16	14	2	0.432417	0.261317
80	16	14	3	0.708526	0.276108
80	16	14	4	0.891320	0.182794
80	16	14	5	0.971084	0.079765
80	16	14	6	0.994587	0.023502
80	16	14	7	0.999299	0.004712
80	16	14	8	0.999939	0.000640
80	16	14	9	0.999996	0.000058
80	16	14	10	1.000000	0.000003
80	16	14	11	1.000000	0.000000
80	16	14	12	1.000000	0.000000
80	16	14	13	1.000000	0.000000
80	16	14	14	1.000000	0.000000
80	16	15	0	0.024039	0.024039
80	16	15	1	0.139426	0.115387
80	16	15	2	0.376986	0.237561
80	16	15	3	0.654141	0.277154
80	16	15	4	0.858085	0.203944
80	16	15	5	0.957790	0.099706
80	16	15	6	0.991026	0.033235
80	16	15	7	0.998656	0.007631
80	16	15	8	0.999861	0.001205
80	16	15	9	0.999990	0.000129
80	16	15	10	0.999999	0.000009
80	16	15	11	1.000000	0.000000
80	16	15	12	1.000000	0.000000
80	16	15	13	1.000000	0.000000
80	16	15	14	1.000000	0.000000
80	16	15	15	1.000000	0.000000
80	16	16	0	0.018122	0.018122
80	16	16	1	0.112798	0.094676
80	16	16	2	0.325819	0.213022
80	16	16	3	0.598710	0.272890
80	16	16	4	0.820433	0.221723
80	16	16	5	0.940917	0.120484
80	16	16	6	0.985912	0.044995
80	16	16	7	0.997600	0.011687
80	16	16	8	0.999713	0.002113
80	16	16	9	0.999976	0.000264
80	16	16	10	0.999999	0.000022
80	16	16	11	1.000000	0.000001
80	16	16	12	1.000000	0.000000
80	16	16	13	1.000000	0.000000
80	16	16	14	1.000000	0.000000
80	16	16	15	1.000000	0.000000
80	16	16	16	1.000000	0.000000
80	17	1	0	0.787500	0.787500
80	17	1	1	1.000000	0.212500
80	17	2	0	0.618038	0.618038
80	17	2	1	0.956962	0.338924
80	17	2	2	1.000000	0.043038
80	17	3	0	0.483337	0.483337
80	17	3	1	0.887439	0.404102
80	17	3	2	0.991723	0.104284
80	17	3	3	1.000000	0.008277
80	17	4	0	0.376627	0.376627
80	17	4	1	0.803470	0.426843
80	17	4	2	0.971408	0.167938
80	17	4	3	0.998495	0.027087
80	17	4	4	1.000000	0.001505
80	17	5	0	0.292381	0.292381
80	17	5	1	0.713608	0.421227
80	17	5	2	0.938263	0.224654
80	17	5	3	0.993505	0.055243
80	17	5	4	0.999743	0.006237
80	17	5	5	1.000000	0.000257
80	17	6	0	0.226108	0.226108
80	17	6	1	0.623746	0.397638
80	17	6	2	0.893332	0.269585

Table for $N = 2$, $n = 1$, through $N = 100$, $n = 50$

N	n	k	x	P(x)	p(x)
80	17	6	3	0.983193	0.089862
80	17	6	4	0.998661	0.015468
80	17	6	5	0.999959	0.001297
80	17	6	6	1.000000	0.000041
80	17	7	0	0.174164	0.174164
80	17	7	1	0.537771	0.363606
80	17	7	2	0.838686	0.300915
80	17	7	3	0.966193	0.127507
80	17	7	4	0.995944	0.029752
80	17	7	5	0.999748	0.003804
80	17	7	6	0.999994	0.000245
80	17	7	7	1.000000	0.000006
80	17	8	0	0.133606	0.133606
80	17	8	1	0.458076	0.324471
80	17	8	2	0.776854	0.318778
80	17	8	3	0.941739	0.164885
80	17	8	4	0.990646	0.048907
80	17	8	5	0.999123	0.008477
80	17	8	6	0.999957	0.000834
80	17	8	7	0.999999	0.000042
80	17	8	8	1.000000	0.000001
80	17	9	0	0.102060	0.102060
80	17	9	1	0.385971	0.283912
80	17	9	2	0.710442	0.324471
80	17	9	3	0.909678	0.199236
80	17	9	4	0.981816	0.072137
80	17	9	5	0.997710	0.015895
80	17	9	6	0.999829	0.002119
80	17	9	7	0.999993	0.000164
80	17	9	8	1.000000	0.000007
80	17	9	9	1.000000	0.000000
80	17	10	0	0.077623	0.077623
80	17	10	1	0.321991	0.244368
80	17	10	2	0.641892	0.319901
80	17	10	3	0.870392	0.228500
80	17	10	4	0.968607	0.098215
80	17	10	5	0.995024	0.026416
80	17	10	6	0.999501	0.004477
80	17	10	7	0.999970	0.000469
80	17	10	8	0.999999	0.000029
80	17	10	9	1.000000	0.000001
80	17	10	10	1.000000	0.000000
80	17	11	0	0.058772	0.058772
80	17	11	1	0.266136	0.207364
80	17	11	2	0.573342	0.307206
80	17	11	3	0.824692	0.251350
80	17	11	4	0.950367	0.125675
80	17	11	5	0.990495	0.040128
80	17	11	6	0.998797	0.008302
80	17	11	7	0.999903	0.001106
80	17	11	8	0.999995	0.000092
80	17	11	9	1.000000	0.000005
80	17	11	10	1.000000	0.000000
80	17	11	11	1.000000	0.000000
80	17	12	0	0.044292	0.044292
80	17	12	1	0.218051	0.173760
80	17	12	2	0.506558	0.288507
80	17	12	3	0.773693	0.267136
80	17	12	4	0.926689	0.152996
80	17	12	5	0.983516	0.056827
80	17	12	6	0.997474	0.013958
80	17	12	7	0.999743	0.002269
80	17	12	8	0.999983	0.000240
80	17	12	9	0.999999	0.000016
80	17	12	10	1.000000	0.000001
80	17	12	11	1.000000	0.000000
80	17	12	12	1.000000	0.000000
80	17	13	0	0.033219	0.033219
80	17	13	1	0.177167	0.143948
80	17	13	2	0.442917	0.265750
80	17	13	3	0.718695	0.275778
80	17	13	4	0.897440	0.178745
80	17	13	5	0.973488	0.076048
80	17	13	6	0.995216	0.021728
80	17	13	7	0.999409	0.004193
80	17	13	8	0.999951	0.000542
80	17	13	9	0.999997	0.000046
80	17	13	10	1.000000	0.000002
80	17	13	11	1.000000	0.000000
80	17	13	12	1.000000	0.000000
80	17	13	13	1.000000	0.000000
80	17	14	0	0.024790	0.024790
80	17	14	1	0.142791	0.118001
80	17	14	2	0.383420	0.240629
80	17	14	3	0.661070	0.277649
80	17	14	4	0.862758	0.201689
80	17	14	5	0.959868	0.097109
80	17	14	6	0.991649	0.031781
80	17	14	7	0.998783	0.007135
80	17	14	8	0.999879	0.001095
80	17	14	9	0.999992	0.000113
80	17	14	10	1.000000	0.000008
80	17	14	11	1.000000	0.000000
80	17	14	12	1.000000	0.000000
80	17	14	13	1.000000	0.000000
80	17	14	14	1.000000	0.000000
80	17	15	0	0.018405	0.018405
80	17	15	1	0.114185	0.095780
80	17	15	2	0.328732	0.214547
80	17	15	3	0.602174	0.273442
80	17	15	4	0.823032	0.220857
80	17	15	5	0.942211	0.119180
80	17	15	6	0.986352	0.044141
80	17	15	7	0.997702	0.011350
80	17	15	8	0.999729	0.002027
80	17	15	9	0.999978	0.000249
80	17	15	10	0.999999	0.000021
80	17	15	11	1.000000	0.000001
80	17	15	12	1.000000	0.000000
80	17	15	13	1.000000	0.000000
80	17	15	14	1.000000	0.000000
80	17	15	15	1.000000	0.000000
80	17	16	0	0.013591	0.013591
80	17	16	1	0.090608	0.077017
80	17	16	2	0.279221	0.188613
80	17	16	3	0.543279	0.264058
80	17	16	4	0.778860	0.235581
80	17	16	5	0.920209	0.141349
80	17	16	6	0.978882	0.058673
80	17	16	7	0.995956	0.017074
80	17	16	8	0.999449	0.003492
80	17	16	9	0.999948	0.000499
80	17	16	10	0.999997	0.000049
80	17	16	11	1.000000	0.000003
80	17	16	12	1.000000	0.000000
80	17	16	13	1.000000	0.000000
80	17	16	14	1.000000	0.000000
80	17	16	15	1.000000	0.000000
80	17	16	16	1.000000	0.000000
80	17	17	0	0.009981	0.009981
80	17	17	1	0.071354	0.061373
80	17	17	2	0.235015	0.163661
80	17	17	3	0.485516	0.250501
80	17	17	4	0.731008	0.245491
80	17	17	5	0.893706	0.162698
80	17	17	6	0.968798	0.075091
80	17	17	7	0.993288	0.024491
80	17	17	8	0.998957	0.005669
80	17	17	9	0.999885	0.000928
80	17	17	10	0.999991	0.000106
80	17	17	11	0.999999	0.000008
80	17	17	12	1.000000	0.000000
80	17	17	13	1.000000	0.000000
80	17	17	14	1.000000	0.000000
80	17	17	15	1.000000	0.000000
80	17	17	16	1.000000	0.000000
80	17	17	17	1.000000	0.000000
80	18	1	0	0.775000	0.775000
80	18	1	1	1.000000	0.225000
80	18	2	0	0.598418	0.598418
80	18	2	1	0.951582	0.353165
80	18	2	2	1.000000	0.048418
80	18	3	0	0.460321	0.460321
80	18	3	1	0.874610	0.414289
80	18	3	2	0.990068	0.115458
80	18	3	3	1.000000	0.009932
80	18	4	0	0.352714	0.352714
80	18	4	1	0.783144	0.430430
80	18	4	2	0.966077	0.182933
80	18	4	3	0.998065	0.031988
80	18	4	4	1.000000	0.001935
80	18	5	0	0.269176	0.269176
80	18	5	1	0.686864	0.417687
80	18	5	2	0.927565	0.240701
80	18	5	3	0.991752	0.064187
80	18	5	4	0.999643	0.007892
80	18	5	5	1.000000	0.000356
80	18	6	0	0.204574	0.204574
80	18	6	1	0.592188	0.387614
80	18	6	2	0.876215	0.284027
80	18	6	3	0.978914	0.102699
80	18	6	4	0.998170	0.019256
80	18	6	5	0.999938	0.001768
80	18	6	6	1.000000	0.000062
80	18	7	0	0.154813	0.154813
80	18	7	1	0.503141	0.348329
80	18	7	2	0.814804	0.311662
80	18	7	3	0.958097	0.143293
80	18	7	4	0.994527	0.036430
80	18	7	5	0.999628	0.005100
80	18	7	6	0.999990	0.000362
80	18	7	7	1.000000	0.000010
80	18	8	0	0.116640	0.116640
80	18	8	1	0.422024	0.305384
80	18	8	2	0.746494	0.324471
80	18	8	3	0.928653	0.182159
80	18	8	4	0.987541	0.058888
80	18	8	5	0.998719	0.011179
80	18	8	6	0.999930	0.001211
80	18	8	7	0.999998	0.000068
80	18	8	8	1.000000	0.000002
80	18	9	0	0.087480	0.087480
80	18	9	1	0.349919	0.262439
80	18	9	2	0.674390	0.324471
80	18	9	3	0.890703	0.216314
80	18	9	4	0.976090	0.085387
80	18	9	5	0.996701	0.020611
80	18	9	6	0.999728	0.003028
80	18	9	7	0.999988	0.000260
80	18	9	8	1.000000	0.000012

Table for $N = 2$, $n = 1$, through $N = 100$, $n = 50$

N	n	k	x	P(x)	p(x)
80	18	9	9	1.000000	0.000000
80	18	10	0	0.065302	0.065302
80	18	10	1	0.287082	0.221780
80	18	10	2	0.601269	0.314188
80	18	10	3	0.845003	0.243734
80	18	10	4	0.959253	0.114250
80	18	10	5	0.992927	0.033674
80	18	10	6	0.999217	0.006290
80	18	10	7	0.999948	0.000731
80	18	10	8	0.999998	0.000050
80	18	10	9	1.000000	0.000002
80	18	10	10	1.000000	0.000000
80	18	11	0	0.048510	0.048510
80	18	11	1	0.233221	0.184711
80	18	11	2	0.529455	0.296234
80	18	11	3	0.792775	0.263319
80	18	11	4	0.936403	0.143629
80	18	11	5	0.986673	0.050270
80	18	11	6	0.998139	0.011465
80	18	11	7	0.999833	0.001694
80	18	11	8	0.999991	0.000158
80	18	11	9	1.000000	0.000009
80	18	11	10	1.000000	0.000000
80	18	11	11	1.000000	0.000000
80	18	12	0	0.035855	0.035855
80	18	12	1	0.187712	0.151857
80	18	12	2	0.460763	0.273051
80	18	12	3	0.735531	0.274768
80	18	12	4	0.907261	0.171730
80	18	12	5	0.977202	0.069941
80	18	12	6	0.996145	0.018942
80	18	12	7	0.999563	0.003418
80	18	12	8	0.999968	0.000405
80	18	12	9	0.999999	0.000031
80	18	12	10	1.000000	0.000001
80	18	12	11	1.000000	0.000000
80	18	12	12	1.000000	0.000000
80	18	13	0	0.026364	0.026364
80	18	13	1	0.149748	0.123384
80	18	13	2	0.396516	0.246768
80	18	13	3	0.674921	0.278405
80	18	13	4	0.871905	0.196984
80	18	13	5	0.963831	0.091926
80	18	13	6	0.992802	0.028971
80	18	13	7	0.999010	0.006208
80	18	13	8	0.999908	0.000899
80	18	13	9	0.999994	0.000086
80	18	13	10	1.000000	0.000005
80	18	13	11	1.000000	0.000000
80	18	13	12	1.000000	0.000000

N	n	k	x	P(x)	p(x)
80	18	13	13	1.000000	0.000000
80	18	14	0	0.019281	0.019281
80	18	14	1	0.118442	0.099160
80	18	14	2	0.337586	0.219145
80	18	14	3	0.612591	0.275005
80	18	14	4	0.830744	0.218153
80	18	14	5	0.945995	0.115251
80	18	14	6	0.987613	0.041618
80	18	14	7	0.997991	0.010378
80	18	14	8	0.999774	0.001784
80	18	14	9	0.999983	0.000209
80	18	14	10	0.999999	0.000016
80	18	14	11	1.000000	0.000001
80	18	14	12	1.000000	0.000000
80	18	14	13	1.000000	0.000000
80	18	14	14	1.000000	0.000000
80	18	15	0	0.014023	0.014023
80	18	15	1	0.092900	0.078878
80	18	15	2	0.284460	0.191560
80	18	15	3	0.550090	0.265630
80	18	15	4	0.784469	0.234379
80	18	15	5	0.923294	0.138825
80	18	15	6	0.980046	0.056752
80	18	15	7	0.996261	0.016215
80	18	15	8	0.999504	0.003243
80	18	15	9	0.999954	0.000450
80	18	15	10	0.999997	0.000043
80	18	15	11	1.000000	0.000003
80	18	15	12	1.000000	0.000000
80	18	15	13	1.000000	0.000000
80	18	15	14	1.000000	0.000000
80	18	15	15	1.000000	0.000000
80	18	16	0	0.010139	0.010139
80	18	16	1	0.072271	0.062131
80	18	16	2	0.237307	0.165036
80	18	16	3	0.488791	0.251484
80	18	16	4	0.733988	0.245197
80	18	16	5	0.895529	0.161541
80	18	16	6	0.969569	0.074040
80	18	16	7	0.993517	0.023948
80	18	16	8	0.999005	0.005488
80	18	16	9	0.999892	0.000887
80	18	16	10	0.999992	0.000100
80	18	16	11	1.000000	0.000008
80	18	16	12	1.000000	0.000000
80	18	16	13	1.000000	0.000000
80	18	16	14	1.000000	0.000000
80	18	16	15	1.000000	0.000000
80	18	16	16	1.000000	0.000000
80	18	17	0	0.007288	0.007288

N	n	k	x	P(x)	p(x)
80	18	17	1	0.055767	0.048479
80	18	17	2	0.196048	0.140281
80	18	17	3	0.429849	0.233801
80	18	17	4	0.680351	0.250501
80	18	17	5	0.862716	0.182365
80	18	17	6	0.955686	0.092970
80	18	17	7	0.989401	0.033715
80	18	17	8	0.998148	0.008747
80	18	17	9	0.999767	0.001620
80	18	17	10	0.999979	0.000212
80	18	17	11	0.999999	0.000019
80	18	17	12	1.000000	0.000001
80	18	17	13	1.000000	0.000000
80	18	17	14	1.000000	0.000000
80	18	17	15	1.000000	0.000000
80	18	17	16	1.000000	0.000000
80	18	17	17	1.000000	0.000000
80	18	18	0	0.005206	0.005206
80	18	18	1	0.042685	0.037480
80	18	18	2	0.160421	0.117736
80	18	18	3	0.374182	0.213761
80	18	18	4	0.624684	0.250501
80	18	18	5	0.825085	0.200401
80	18	18	6	0.937978	0.112893
80	18	18	7	0.983514	0.045537
80	18	18	8	0.996759	0.013245
80	18	18	9	0.999536	0.002777
80	18	18	10	0.999952	0.000417
80	18	18	11	0.999996	0.000044
80	18	18	12	1.000000	0.000003
80	18	18	13	1.000000	0.000000
80	18	18	14	1.000000	0.000000
80	18	18	15	1.000000	0.000000
80	18	18	16	1.000000	0.000000
80	18	18	17	1.000000	0.000000
80	18	18	18	1.000000	0.000000
80	19	1	0	0.762500	0.762500
80	19	1	1	1.000000	0.237500
80	19	2	0	0.579114	0.579114
80	19	2	1	0.945886	0.366772
80	19	2	2	1.000000	0.054114
80	19	3	0	0.438048	0.438048
80	19	3	1	0.861246	0.423199
80	19	3	2	0.988206	0.126960
80	19	3	3	1.000000	0.011794
80	19	4	0	0.329958	0.329958
80	19	4	1	0.762317	0.432359
80	19	4	2	0.960176	0.197859
80	19	4	3	0.997549	0.037373
80	19	4	4	1.000000	0.002451

N	n	k	x	P(x)	p(x)
80	19	5	0	0.247468	0.247468
80	19	5	1	0.659916	0.412447
80	19	5	2	0.915918	0.256002
80	19	5	3	0.989681	0.073763
80	19	5	4	0.999516	0.009835
80	19	5	5	1.000000	0.000484
80	19	6	0	0.184776	0.184776
80	19	6	1	0.560929	0.376152
80	19	6	2	0.857891	0.296962
80	19	6	3	0.973945	0.116054
80	19	6	4	0.997549	0.023604
80	19	6	5	0.999910	0.002360
80	19	6	6	1.000000	0.000090
80	19	7	0	0.137334	0.137334
80	19	7	1	0.469432	0.332098
80	19	7	2	0.789670	0.320238
80	19	7	3	0.948852	0.159182
80	19	7	4	0.992765	0.043912
80	19	7	5	0.999463	0.006698
80	19	7	6	0.999984	0.000521
80	19	7	7	1.000000	0.000016
80	19	8	0	0.101589	0.101589
80	19	8	1	0.387545	0.285955
80	19	8	2	0.715094	0.327549
80	19	8	3	0.913963	0.198869
80	19	8	4	0.983741	0.069779
80	19	8	5	0.998178	0.014437
80	19	8	6	0.999891	0.001713
80	19	8	7	0.999997	0.000106
80	19	8	8	1.000000	0.000003
80	19	9	0	0.074781	0.074781
80	19	9	1	0.316056	0.241275
80	19	9	2	0.637756	0.321700
80	19	9	3	0.869770	0.232014
80	19	9	4	0.969204	0.099434
80	19	9	5	0.995371	0.026167
80	19	9	6	0.999582	0.004211
80	19	9	7	0.999980	0.000398
80	19	9	8	0.999999	0.000020
80	19	9	9	1.000000	0.000000
80	19	10	0	0.054769	0.054769
80	19	10	1	0.254888	0.200118
80	19	10	2	0.560729	0.305841
80	19	10	3	0.817485	0.256756
80	19	10	4	0.948197	0.130712
80	19	10	5	0.990212	0.042015
80	19	10	6	0.998811	0.008599
80	19	10	7	0.999912	0.001101
80	19	10	8	0.999996	0.000084
80	19	10	9	1.000000	0.000003

N	n	k	x	P(x)	p(x)
80	19	10	10	1.000000	0.000000
80	19	11	0	0.039903	0.039903
80	19	11	1	0.203429	0.163525
80	19	11	2	0.486453	0.283025
80	19	11	3	0.758798	0.272344
80	19	11	4	0.920187	0.161389
80	19	11	5	0.981809	0.061621
80	19	11	6	0.997214	0.015405
80	19	11	7	0.999724	0.002510
80	19	11	8	0.999983	0.000260
80	19	11	9	0.999999	0.000016
80	19	11	10	1.000000	0.000001
80	19	11	11	1.000000	0.000000
80	19	12	0	0.028915	0.028915
80	19	12	1	0.160770	0.131854
80	19	12	2	0.416723	0.255953
80	19	12	3	0.695646	0.278923
80	19	12	4	0.885103	0.189457
80	19	12	5	0.969306	0.084203
80	19	12	6	0.994312	0.025006
80	19	12	7	0.999287	0.004976
80	19	12	8	0.999942	0.000655
80	19	12	9	0.999997	0.000055
80	19	12	10	1.000000	0.000003
80	19	12	11	1.000000	0.000000
80	19	12	12	1.000000	0.000000
80	19	13	0	0.020836	0.020836
80	19	13	1	0.125867	0.105031
80	19	13	2	0.352734	0.226867
80	19	13	3	0.630017	0.277282
80	19	13	4	0.843311	0.213294
80	19	13	5	0.951970	0.108659
80	19	13	6	0.989531	0.037561
80	19	13	7	0.998409	0.008878
80	19	13	8	0.999836	0.001427
80	19	13	9	0.999989	0.000153
80	19	13	10	0.999999	0.000011
80	19	13	11	1.000000	0.000000
80	19	13	12	1.000000	0.000000
80	19	13	13	1.000000	0.000000
80	19	14	0	0.014927	0.014927
80	19	14	1	0.097650	0.082723
80	19	14	2	0.295171	0.197521
80	19	14	3	0.563800	0.268629
80	19	14	4	0.795558	0.231758
80	19	14	5	0.929265	0.133707
80	19	14	6	0.982243	0.052978
80	19	14	7	0.996819	0.014576
80	19	14	8	0.999602	0.002783
80	19	14	9	0.999966	0.000364

N	n	k	x	P(x)	p(x)
80	19	14	10	0.999998	0.000032
80	19	14	11	1.000000	0.000002
80	19	14	12	1.000000	0.000000
80	19	14	13	1.000000	0.000000
80	19	14	14	1.000000	0.000000
80	19	15	0	0.010630	0.010630
80	19	15	1	0.075089	0.064459
80	19	15	2	0.244294	0.169205
80	19	15	3	0.498678	0.254383
80	19	15	4	0.742886	0.244208
80	19	15	5	0.900903	0.158017
80	19	15	6	0.971808	0.070905
80	19	15	7	0.994169	0.022361
80	19	15	8	0.999138	0.004969
80	19	15	9	0.999911	0.000773
80	19	15	10	0.999994	0.000083
80	19	15	11	1.000000	0.000006
80	19	15	12	1.000000	0.000000
80	19	15	13	1.000000	0.000000
80	19	15	14	1.000000	0.000000
80	19	15	15	1.000000	0.000000
80	19	16	0	0.007523	0.007523
80	19	16	1	0.057239	0.049716
80	19	16	2	0.200041	0.142802
80	19	16	3	0.436060	0.236020
80	19	16	4	0.686530	0.250470
80	19	16	5	0.866868	0.180338
80	19	16	6	0.957627	0.090758
80	19	16	7	0.990041	0.032414
80	19	16	8	0.998297	0.008256
80	19	16	9	0.999792	0.001495
80	19	16	10	0.999982	0.000190
80	19	16	11	0.999999	0.000017
80	19	16	12	1.000000	0.000001
80	19	16	13	1.000000	0.000000
80	19	16	14	1.000000	0.000000
80	19	16	15	1.000000	0.000000
80	19	16	16	1.000000	0.000000
80	19	17	0	0.005289	0.005289
80	19	17	1	0.043256	0.037967
80	19	17	2	0.162109	0.118853
80	19	17	3	0.377055	0.214946
80	19	17	4	0.627826	0.250771
80	19	17	5	0.827419	0.199593
80	19	17	6	0.939192	0.111772
80	19	17	7	0.983963	0.044771
80	19	17	8	0.996878	0.012915
80	19	17	9	0.999558	0.002680
80	19	17	10	0.999955	0.000397
80	19	17	11	0.999997	0.000041

N	n	k	x	P(x)	p(x)
80	19	17	12	1.000000	0.000003
80	19	17	13	1.000000	0.000000
80	19	17	14	1.000000	0.000000
80	19	17	15	1.000000	0.000000
80	19	17	16	1.000000	0.000000
80	19	17	17	1.000000	0.000000
80	19	18	0	0.003694	0.003694
80	19	18	1	0.032409	0.028714
80	19	18	2	0.130038	0.097629
80	19	18	3	0.322466	0.192428
80	19	18	4	0.568119	0.245653
80	19	18	5	0.783065	0.214946
80	19	18	6	0.916127	0.133062
80	19	18	7	0.975435	0.059308
80	19	18	8	0.994623	0.019188
80	19	18	9	0.999133	0.004510
80	19	18	10	0.999899	0.000766
80	19	18	11	0.999992	0.000093
80	19	18	12	0.999999	0.000008
80	19	18	13	1.000000	0.000000
80	19	18	14	1.000000	0.000000
80	19	18	15	1.000000	0.000000
80	19	18	16	1.000000	0.000000
80	19	18	17	1.000000	0.000000
80	19	18	18	1.000000	0.000000
80	19	19	0	0.002562	0.002562
80	19	19	1	0.024072	0.021510
80	19	19	2	0.103268	0.079196
80	19	19	3	0.272807	0.169539
80	19	19	4	0.508687	0.235880
80	19	19	5	0.734529	0.225842
80	19	19	6	0.888227	0.153698
80	19	19	7	0.963956	0.075729
80	19	19	8	0.991219	0.027262
80	19	19	9	0.998405	0.007187
80	19	19	10	0.999788	0.001382
80	19	19	11	0.999980	0.000192
80	19	19	12	0.999999	0.000019
80	19	19	13	1.000000	0.000001
80	19	19	14	1.000000	0.000000
80	19	19	15	1.000000	0.000000
80	19	19	16	1.000000	0.000000
80	19	19	17	1.000000	0.000000
80	19	19	18	1.000000	0.000000
80	19	19	19	1.000000	0.000000
80	20	1	0	0.750000	0.750000
80	20	1	1	1.000000	0.250000
80	20	2	0	0.560127	0.560127
80	20	2	1	0.939873	0.379747
80	20	2	2	1.000000	0.060127

N	n	k	x	P(x)	p(x)
80	20	3	0	0.416504	0.416504
80	20	3	1	0.847371	0.430867
80	20	3	2	0.986125	0.138754
80	20	3	3	1.000000	0.013875
80	20	4	0	0.308321	0.308321
80	20	4	1	0.741053	0.432732
80	20	4	2	0.953689	0.212635
80	20	4	3	0.996937	0.043248
80	20	4	4	1.000000	0.003063
80	20	5	0	0.227184	0.227184
80	20	5	1	0.632870	0.405686
80	20	5	2	0.903328	0.270457
80	20	5	3	0.987263	0.083935
80	20	5	4	0.999355	0.012092
80	20	5	5	1.000000	0.000645
80	20	6	0	0.166602	0.166602
80	20	6	1	0.530096	0.363495
80	20	6	2	0.838418	0.308321
80	20	6	3	0.968237	0.129820
80	20	6	4	0.996775	0.028538
80	20	6	5	0.999871	0.003096
80	20	6	6	1.000000	0.000129
80	20	7	0	0.121574	0.121574
80	20	7	1	0.436767	0.315192
80	20	7	2	0.763421	0.326654
80	20	7	3	0.938414	0.174993
80	20	7	4	0.990605	0.052191
80	20	7	5	0.999243	0.008639
80	20	7	6	0.999976	0.000732
80	20	7	7	1.000000	0.000024
80	20	8	0	0.088266	0.088266
80	20	8	1	0.354730	0.266464
80	20	8	2	0.682876	0.328146
80	20	8	3	0.897662	0.214786
80	20	8	4	0.979166	0.081504
80	20	8	5	0.997468	0.018303
80	20	8	6	0.999835	0.002367
80	20	8	7	0.999996	0.000160
80	20	8	8	1.000000	0.000004
80	20	9	0	0.063748	0.063748
80	20	9	1	0.284413	0.220666
80	20	9	2	0.600839	0.316426
80	20	9	3	0.846949	0.246109
80	20	9	4	0.961054	0.114105
80	20	9	5	0.993655	0.032601
80	20	9	6	0.999375	0.005720
80	20	9	7	0.999967	0.000592
80	20	9	8	0.999999	0.000033
80	20	9	9	1.000000	0.000001
80	20	10	0	0.045791	0.045791

N	n	k	x	P(x)	p(x)
80	20	10	1	0.225362	0.179571
80	20	10	2	0.520619	0.295257
80	20	10	3	0.788021	0.267402
80	20	10	4	0.935340	0.147319
80	20	10	5	0.986768	0.051428
80	20	10	6	0.998247	0.011479
80	20	10	7	0.999858	0.001611
80	20	10	8	0.999994	0.000135
80	20	10	9	1.000000	0.000006
80	20	10	10	1.000000	0.000000
80	20	11	0	0.032708	0.032708
80	20	11	1	0.176621	0.143914
80	20	11	2	0.444696	0.268074
80	20	11	3	0.723081	0.278385
80	20	11	4	0.901667	0.178587
80	20	11	5	0.975747	0.074080
80	20	11	6	0.995951	0.020204
80	20	11	7	0.999559	0.003608
80	20	11	8	0.999970	0.000411
80	20	11	9	0.999999	0.000028
80	20	11	10	1.000000	0.000001
80	20	11	11	1.000000	0.000000
80	20	12	0	0.023227	0.023227
80	20	12	1	0.136993	0.113766
80	20	12	2	0.374763	0.237770
80	20	12	3	0.654493	0.279730
80	20	12	4	0.860256	0.205763
80	20	12	5	0.959643	0.099387
80	20	12	6	0.991852	0.032209
80	20	12	7	0.998879	0.007027
80	20	12	8	0.999899	0.001020
80	20	12	9	0.999994	0.000095
80	20	12	10	1.000000	0.000005
80	20	12	11	1.000000	0.000000
80	20	12	12	1.000000	0.000000
80	20	13	0	0.016396	0.016396
80	20	13	1	0.105205	0.088810
80	20	13	2	0.311824	0.206619
80	20	13	3	0.584560	0.272737
80	20	13	4	0.811841	0.227280
80	20	13	5	0.937719	0.125878
80	20	13	6	0.985221	0.047501
80	20	13	7	0.997536	0.012315
80	20	13	8	0.999719	0.002183
80	20	13	9	0.999979	0.000260
80	20	13	10	0.999999	0.000020
80	20	13	11	1.000000	0.000001
80	20	13	12	1.000000	0.000000
80	20	13	13	1.000000	0.000000
80	20	14	0	0.011501	0.011501

N	n	k	x	P(x)	p(x)
80	20	14	1	0.080021	0.068519
80	20	14	2	0.256315	0.176294
80	20	14	3	0.515359	0.259044
80	20	14	4	0.757565	0.242206
80	20	14	5	0.909538	0.151973
80	20	14	6	0.975295	0.065757
80	20	14	7	0.995146	0.019851
80	20	14	8	0.999328	0.004182
80	20	14	9	0.999936	0.000608
80	20	14	10	0.999996	0.000060
80	20	14	11	1.000000	0.000004
80	20	14	12	1.000000	0.000000
80	20	14	13	1.000000	0.000000
80	20	14	14	1.000000	0.000000
80	20	15	0	0.008016	0.008016
80	20	15	1	0.060295	0.052279
80	20	15	2	0.208234	0.147939
80	20	15	3	0.448635	0.240401
80	20	15	4	0.698848	0.250213
80	20	15	5	0.874998	0.176150
80	20	15	6	0.961347	0.086348
80	20	15	7	0.991236	0.029890
80	20	15	8	0.998568	0.007331
80	20	15	9	0.999835	0.001267
80	20	15	10	0.999987	0.000152
80	20	15	11	0.999999	0.000012
80	20	15	12	1.000000	0.000001
80	20	15	13	1.000000	0.000000
80	20	15	14	1.000000	0.000000
80	20	15	15	1.000000	0.000000
80	20	16	0	0.005550	0.005550
80	20	16	1	0.045014	0.039464
80	20	16	2	0.167267	0.122253
80	20	16	3	0.385761	0.218495
80	20	16	4	0.637257	0.251496
80	20	16	5	0.834348	0.197091
80	20	16	6	0.942749	0.108400
80	20	16	7	0.985258	0.042510
80	20	16	8	0.997214	0.011956
80	20	16	9	0.999620	0.002406
80	20	16	10	0.999964	0.000343
80	20	16	11	0.999998	0.000034
80	20	16	12	1.000000	0.000002
80	20	16	13	1.000000	0.000000
80	20	16	14	1.000000	0.000000
80	20	16	15	1.000000	0.000000
80	20	16	16	1.000000	0.000000
80	20	17	0	0.003815	0.003815
80	20	17	1	0.033298	0.029482
80	20	17	2	0.132883	0.099585

N	n	k	x	P(x)	p(x)
80	20	17	3	0.327724	0.194841
80	20	17	4	0.574383	0.246660
80	20	17	5	0.788155	0.213772
80	20	17	6	0.919036	0.130881
80	20	17	7	0.976623	0.057588
80	20	17	8	0.994972	0.018349
80	20	17	9	0.999207	0.004234
80	20	17	10	0.999910	0.000703
80	20	17	11	0.999993	0.000083
80	20	17	12	0.999999	0.000007
80	20	17	13	1.000000	0.000000
80	20	17	14	1.000000	0.000000
80	20	17	15	1.000000	0.000000
80	20	17	16	1.000000	0.000000
80	20	17	17	1.000000	0.000000
80	20	18	0	0.002604	0.002604
80	20	18	1	0.024406	0.021802
80	20	18	2	0.104430	0.080024
80	20	18	3	0.275148	0.170717
80	20	18	4	0.511740	0.236592
80	20	18	5	0.737257	0.225518
80	20	18	6	0.889951	0.152694
80	20	18	7	0.964740	0.074789
80	20	18	8	0.991477	0.026737
80	20	18	9	0.998467	0.006990
80	20	18	10	0.999798	0.001331
80	20	18	11	0.999981	0.000183
80	20	18	12	0.999999	0.000018
80	20	18	13	1.000000	0.000001
80	20	18	14	1.000000	0.000000
80	20	18	15	1.000000	0.000000
80	20	18	16	1.000000	0.000000
80	20	18	17	1.000000	0.000000
80	20	18	18	1.000000	0.000000
80	20	19	0	0.001764	0.001764
80	20	19	1	0.017725	0.015961
80	20	19	2	0.081197	0.063472
80	20	19	3	0.228338	0.147141
80	20	19	4	0.450684	0.222346
80	20	19	5	0.682696	0.232013
80	20	19	6	0.855472	0.172776
80	20	19	7	0.949059	0.093587
80	20	19	8	0.986302	0.037244
80	20	19	9	0.997227	0.010925
80	20	19	10	0.999584	0.002356
80	20	19	11	0.999954	0.000371
80	20	19	12	0.999996	0.000042
80	20	19	13	1.000000	0.000003
80	20	19	14	1.000000	0.000000
80	20	19	15	1.000000	0.000000

N	n	k	x	P(x)	p(x)
80	20	19	16	1.000000	0.000000
80	20	19	17	1.000000	0.000000
80	20	19	18	1.000000	0.000000
80	20	19	19	1.000000	0.000000
80	20	20	0	0.001186	0.001186
80	20	20	1	0.012754	0.011568
80	20	20	2	0.062468	0.049714
80	20	20	3	0.187332	0.124864
80	20	20	4	0.392363	0.205032
80	20	20	5	0.625644	0.233281
80	20	20	6	0.815819	0.190175
80	20	20	7	0.929114	0.113295
80	20	20	8	0.978976	0.049862
80	20	20	9	0.995257	0.016281
80	20	20	10	0.999197	0.003940
80	20	20	11	0.999900	0.000702
80	20	20	12	0.999991	0.000091
80	20	20	13	0.999999	0.000008
80	20	20	14	1.000000	0.000001
80	20	20	15	1.000000	0.000000
80	20	20	16	1.000000	0.000000
80	20	20	17	1.000000	0.000000
80	20	20	18	1.000000	0.000000
80	20	20	19	1.000000	0.000000
80	20	20	20	1.000000	0.000000
80	21	1	0	0.737500	0.737500
80	21	1	1	1.000000	0.262500
80	21	2	0	0.541456	0.541456
80	21	2	1	0.933544	0.392089
80	21	2	2	1.000000	0.066456
80	21	3	0	0.395679	0.395679
80	21	3	1	0.833009	0.437330
80	21	3	2	0.983812	0.150803
80	21	3	3	1.000000	0.016188
80	21	4	0	0.287767	0.287767
80	21	4	1	0.719417	0.431650
80	21	4	2	0.946601	0.227184
80	21	4	3	0.996216	0.049615
80	21	4	4	1.000000	0.003784
80	21	5	0	0.208252	0.208252
80	21	5	1	0.605825	0.397572
80	21	5	2	0.889805	0.283980
80	21	5	3	0.984465	0.094660
80	21	5	4	0.999153	0.014689
80	21	5	5	1.000000	0.000846
80	21	6	0	0.149942	0.149942
80	21	6	1	0.499805	0.349864
80	21	6	2	0.817863	0.318058
80	21	6	3	0.961746	0.143883
80	21	6	4	0.995824	0.034078

Table for $N = 2$, $n = 1$, through $N = 100$, $n = 50$

N	n	k	x	P(x)	p(x)
80	21	6	5	0.999819	0.003995
80	21	6	6	1.000000	0.000181
80	21	7	0	0.107391	0.107391
80	21	7	1	0.405247	0.297857
80	21	7	2	0.736200	0.330952
80	21	7	3	0.926748	0.190548
80	21	7	4	0.987995	0.061248
80	21	7	5	0.998955	0.010960
80	21	7	6	0.999963	0.001008
80	21	7	7	1.000000	0.000037
80	21	8	0	0.076497	0.076497
80	21	8	1	0.323643	0.247145
80	21	8	2	0.650061	0.326418
80	21	8	3	0.879763	0.229702
80	21	8	4	0.973732	0.093969
80	21	8	5	0.996553	0.022821
80	21	8	6	0.999756	0.003203
80	21	8	7	0.999993	0.000237
80	21	8	8	1.000000	0.000007
80	21	9	0	0.054186	0.054186
80	21	9	1	0.254991	0.200806
80	21	9	2	0.563923	0.308932
80	21	9	3	0.822338	0.258415
80	21	9	4	0.951545	0.129207
80	21	9	5	0.991482	0.039937
80	21	9	6	0.999089	0.007607
80	21	9	7	0.999947	0.000858
80	21	9	8	0.999999	0.000052
80	21	9	9	1.000000	0.000001
80	21	10	0	0.038159	0.038159
80	21	10	1	0.198426	0.160267
80	21	10	2	0.481251	0.282825
80	21	10	3	0.756824	0.275573
80	21	10	4	0.920608	0.163784
80	21	10	5	0.982482	0.061874
80	21	10	6	0.997482	0.015000
80	21	10	7	0.999778	0.002296
80	21	10	8	0.999989	0.000211
80	21	10	9	1.000000	0.000011
80	21	10	10	1.000000	0.000000
80	21	11	0	0.026711	0.026711
80	21	11	1	0.152636	0.125924
80	21	11	2	0.404484	0.251849
80	21	11	3	0.685963	0.281478
80	21	11	4	0.880832	0.194869
80	21	11	5	0.968339	0.087507
80	21	11	6	0.994268	0.025928
80	21	11	7	0.999319	0.005051
80	21	11	8	0.999950	0.000631
80	21	11	9	0.999998	0.000048

N	n	k	x	P(x)	p(x)
80	21	11	10	1.000000	0.000002
80	21	11	11	1.000000	0.000000
80	21	12	0	0.018582	0.018582
80	21	12	1	0.116136	0.097554
80	21	12	2	0.335135	0.218999
80	21	12	3	0.612534	0.277399
80	21	12	4	0.832821	0.220287
80	21	12	5	0.948048	0.115227
80	21	12	6	0.988631	0.040583
80	21	12	7	0.998294	0.009663
80	21	12	8	0.999831	0.001537
80	21	12	9	0.999990	0.000159
80	21	12	10	1.000000	0.000010
80	21	12	11	1.000000	0.000000
80	21	12	12	1.000000	0.000000
80	21	13	0	0.012843	0.012843
80	21	13	1	0.087443	0.074600
80	21	13	2	0.273944	0.186500
80	21	13	3	0.539104	0.265161
80	21	13	4	0.777749	0.238645
80	21	13	5	0.920936	0.143187
80	21	13	6	0.979679	0.058743
80	21	13	7	0.996304	0.016625
80	21	13	8	0.999537	0.003233
80	21	13	9	0.999962	0.000424
80	21	13	10	0.999998	0.000036
80	21	13	11	1.000000	0.000002
80	21	13	12	1.000000	0.000000
80	21	13	13	1.000000	0.000000
80	21	14	0	0.008818	0.008818
80	21	14	1	0.065175	0.056357
80	21	14	2	0.221056	0.155881
80	21	14	3	0.467867	0.246812
80	21	14	4	0.717197	0.249330
80	21	14	5	0.886742	0.169544
80	21	14	6	0.966527	0.079786
80	21	14	7	0.992830	0.026303
80	21	14	8	0.998910	0.006079
80	21	14	9	0.999886	0.000976
80	21	14	10	0.999992	0.000106
80	21	14	11	1.000000	0.000008
80	21	14	12	1.000000	0.000000
80	21	14	13	1.000000	0.000000
80	21	14	14	1.000000	0.000000
80	21	15	0	0.006012	0.006012
80	21	15	1	0.048097	0.042085
80	21	15	2	0.176181	0.128084
80	21	15	3	0.400555	0.224374
80	21	15	4	0.652976	0.252421
80	21	15	5	0.845640	0.192664

N	n	k	x	P(x)	p(x)
80	21	15	6	0.948394	0.102754
80	21	15	7	0.987251	0.038857
80	21	15	8	0.997712	0.010461
80	21	15	9	0.999708	0.001996
80	21	15	10	0.999974	0.000266
80	21	15	11	0.999998	0.000024
80	21	15	12	1.000000	0.000001
80	21	15	13	1.000000	0.000000
80	21	15	14	1.000000	0.000000
80	21	15	15	1.000000	0.000000
80	21	16	0	0.004070	0.004070
80	21	16	1	0.035148	0.031078
80	21	16	2	0.138741	0.103593
80	21	16	3	0.338421	0.199680
80	21	16	4	0.586958	0.248538
80	21	16	5	0.798215	0.211257
80	21	16	6	0.924682	0.126467
80	21	16	7	0.978882	0.054200
80	21	16	8	0.995620	0.016738
80	21	16	9	0.999340	0.003720
80	21	16	10	0.999929	0.000590
80	21	16	11	0.999995	0.000065
80	21	16	12	1.000000	0.000005
80	21	16	13	1.000000	0.000000
80	21	16	14	1.000000	0.000000
80	21	16	15	1.000000	0.000000
80	21	16	16	1.000000	0.000000
80	21	17	0	0.002734	0.002734
80	21	17	1	0.025436	0.022701
80	21	17	2	0.107987	0.082551
80	21	17	3	0.282261	0.174274
80	21	17	4	0.520940	0.238680
80	21	17	5	0.745401	0.224460
80	21	17	6	0.895041	0.149640
80	21	17	7	0.967026	0.071984
80	21	17	8	0.992220	0.025195
80	21	17	9	0.998642	0.006422
80	21	17	10	0.999828	0.001186
80	21	17	11	0.999984	0.000157
80	21	17	12	0.999999	0.000014
80	21	17	13	1.000000	0.000001
80	21	17	14	1.000000	0.000000
80	21	17	15	1.000000	0.000000
80	21	17	16	1.000000	0.000000
80	21	17	17	1.000000	0.000000
80	21	18	0	0.001823	0.001823
80	21	18	1	0.018229	0.016406
80	21	18	2	0.083090	0.064861
80	21	18	3	0.232468	0.149378
80	21	18	4	0.456535	0.224067

N	n	k	x	P(x)	p(x)
80	21	18	5	0.688395	0.231860
80	21	18	6	0.859413	0.171017
80	21	18	7	0.951029	0.091616
80	21	18	8	0.987021	0.035992
80	21	18	9	0.997419	0.010398
80	21	18	10	0.999621	0.002202
80	21	18	11	0.999960	0.000339
80	21	18	12	0.999997	0.000037
80	21	18	13	1.000000	0.000003
80	21	18	14	1.000000	0.000000
80	21	18	15	1.000000	0.000000
80	21	18	16	1.000000	0.000000
80	21	18	17	1.000000	0.000000
80	21	18	18	1.000000	0.000000
80	21	19	0	0.001205	0.001205
80	21	19	1	0.012937	0.011731
80	21	19	2	0.063214	0.050277
80	21	19	3	0.189100	0.125887
80	21	19	4	0.395097	0.205997
80	21	19	5	0.628560	0.233463
80	21	19	6	0.818037	0.189477
80	21	19	7	0.930341	0.112304
80	21	19	8	0.979474	0.049133
80	21	19	9	0.995406	0.015932
80	21	19	10	0.999230	0.003824
80	21	19	11	0.999905	0.000675
80	21	19	12	0.999991	0.000087
80	21	19	13	0.999999	0.000008
80	21	19	14	1.000000	0.000000
80	21	19	15	1.000000	0.000000
80	21	19	16	1.000000	0.000000
80	21	19	17	1.000000	0.000000
80	21	19	18	1.000000	0.000000
80	21	19	19	1.000000	0.000000
80	21	20	0	0.000790	0.000790
80	21	20	1	0.009090	0.008300
80	21	20	2	0.047554	0.038463
80	21	20	3	0.151953	0.104400
80	21	20	4	0.337688	0.185735
80	21	20	5	0.567324	0.229636
80	21	20	6	0.771445	0.204121
80	21	20	7	0.904567	0.133122
80	21	20	8	0.969003	0.064437
80	21	20	9	0.992272	0.023269
80	21	20	10	0.998541	0.006268
80	21	20	11	0.999794	0.001254
80	21	20	12	0.999979	0.000184
80	21	20	13	0.999998	0.000020
80	21	20	14	1.000000	0.000001
80	21	20	15	1.000000	0.000000

Table for $N = 2$, $n = 1$, through $N = 100$, $n = 50$

N	n	k	x	P(x)	p(x)
80	21	20	16	1.000000	0.000000
80	21	20	17	1.000000	0.000000
80	21	20	18	1.000000	0.000000
80	21	20	19	1.000000	0.000000
80	21	20	20	1.000000	0.000000
80	21	21	0	0.000514	0.000514
80	21	21	1	0.006324	0.005810
80	21	21	2	0.035374	0.029050
80	21	21	3	0.120633	0.085260
80	21	21	4	0.285063	0.164430
80	21	21	5	0.506088	0.221024
80	21	21	6	0.720414	0.214327
80	21	21	7	0.873505	0.153090
80	21	21	8	0.955042	0.081537
80	21	21	9	0.987619	0.032576
80	21	21	10	0.997391	0.009773
80	21	21	11	0.999585	0.002194
80	21	21	12	0.999951	0.000366
80	21	21	13	0.999996	0.000045
80	21	21	14	1.000000	0.000004
80	21	21	15	1.000000	0.000000
80	21	21	16	1.000000	0.000000
80	21	21	17	1.000000	0.000000
80	21	21	18	1.000000	0.000000
80	21	21	19	1.000000	0.000000
80	21	21	20	1.000000	0.000000
80	21	21	21	1.000000	0.000000
80	22	1	0	0.725000	0.725000
80	22	1	1	1.000000	0.275000
80	22	2	0	0.523101	0.523101
80	22	2	1	0.926899	0.403797
80	22	2	2	1.000000	0.073101
80	22	3	0	0.375560	0.375560
80	22	3	1	0.818184	0.442624
80	22	3	2	0.981256	0.163072
80	22	3	3	1.000000	0.018744
80	22	4	0	0.268257	0.268257
80	22	4	1	0.697468	0.429211
80	22	4	2	0.938900	0.241431
80	22	4	3	0.995375	0.056475
80	22	4	4	1.000000	0.004625
80	22	5	0	0.190604	0.190604
80	22	5	1	0.578870	0.388267
80	22	5	2	0.875365	0.296495
80	22	5	3	0.981256	0.105891
80	22	5	4	0.998905	0.017648
80	22	5	5	1.000000	0.001095
80	22	6	0	0.134693	0.134693
80	22	6	1	0.470156	0.335462
80	22	6	2	0.796300	0.326144

N	n	k	x	P(x)	p(x)
80	22	6	3	0.954430	0.158130
80	22	6	4	0.994669	0.040239
80	22	6	5	0.999752	0.005083
80	22	6	6	1.000000	0.000248
80	22	7	0	0.094649	0.094649
80	22	7	1	0.374957	0.280308
80	22	7	2	0.708153	0.333196
80	22	7	3	0.913829	0.205676
80	22	7	4	0.984881	0.071052
80	22	7	5	0.998584	0.013703
80	22	7	6	0.999946	0.001362
80	22	7	7	1.000000	0.000054
80	22	8	0	0.066125	0.066125
80	22	8	1	0.294321	0.228196
80	22	8	2	0.616866	0.322546
80	22	8	3	0.860297	0.243431
80	22	8	4	0.967361	0.107064
80	22	8	5	0.995393	0.028031
80	22	8	6	0.999648	0.004255
80	22	8	7	0.999989	0.000341
80	22	8	8	1.000000	0.000011
80	22	9	0	0.045920	0.045920
80	22	9	1	0.227763	0.181843
80	22	9	2	0.527270	0.299507
80	22	9	3	0.796058	0.268788
80	22	9	4	0.940595	0.144537
80	22	9	5	0.988774	0.048179
80	22	9	6	0.998702	0.009928
80	22	9	7	0.999918	0.001216
80	22	9	8	0.999998	0.000080
80	22	9	9	1.000000	0.000002
80	22	10	0	0.031691	0.031691
80	22	10	1	0.173979	0.142287
80	22	10	2	0.442902	0.268923
80	22	10	3	0.724129	0.281227
80	22	10	4	0.903952	0.179823
80	22	10	5	0.977238	0.073286
80	22	10	6	0.996465	0.019226
80	22	10	7	0.999661	0.003196
80	22	10	8	0.999982	0.000321
80	22	10	9	0.999999	0.000018
80	22	10	10	1.000000	0.000000
80	22	11	0	0.021731	0.021731
80	22	11	1	0.131293	0.109561
80	22	11	2	0.366067	0.234774
80	22	11	3	0.647796	0.281729
80	22	11	4	0.857712	0.209916
80	22	11	5	0.959440	0.101728
80	22	11	6	0.992070	0.032630
80	22	11	7	0.998976	0.006906

N	n	k	x	P(x)	p(x)
80	22	11	8	0.999918	0.000942
80	22	11	9	0.999996	0.000078
80	22	11	10	1.000000	0.000004
80	22	11	11	1.000000	0.000000
80	22	12	0	0.014802	0.014802
80	22	12	1	0.097948	0.083145
80	22	12	2	0.298016	0.200069
80	22	12	3	0.570218	0.272202
80	22	12	4	0.802951	0.232733
80	22	12	5	0.934377	0.131426
80	22	12	6	0.984504	0.050127
80	22	12	7	0.997475	0.012971
80	22	12	8	0.999727	0.002252
80	22	12	9	0.999981	0.000255
80	22	12	10	0.999999	0.000018
80	22	12	11	1.000000	0.000001
80	22	12	12	1.000000	0.000000
80	22	13	0	0.010013	0.010013
80	22	13	1	0.072271	0.062257
80	22	13	2	0.239173	0.166902
80	22	13	3	0.494162	0.254989
80	22	13	4	0.741346	0.247184
80	22	13	5	0.901520	0.160175
80	22	13	6	0.972709	0.071189
80	22	13	7	0.994614	0.021904
80	22	13	8	0.999263	0.004649
80	22	13	9	0.999933	0.000670
80	22	13	10	0.999996	0.000063
80	22	13	11	1.000000	0.000004
80	22	13	12	1.000000	0.000000
80	22	13	13	1.000000	0.000000
80	22	14	0	0.006725	0.006725
80	22	14	1	0.052757	0.046032
80	22	14	2	0.189351	0.136594
80	22	14	3	0.421852	0.232500
80	22	14	4	0.674938	0.253086
80	22	14	5	0.860879	0.185941
80	22	14	6	0.955709	0.094830
80	22	14	7	0.989710	0.034001
80	22	14	8	0.998292	0.008582
80	22	14	9	0.999803	0.001511
80	22	14	10	0.999985	0.000182
80	22	14	11	0.999999	0.000014
80	22	14	12	1.000000	0.000001
80	22	14	13	1.000000	0.000000
80	22	14	14	1.000000	0.000000
80	22	15	0	0.004484	0.004484
80	22	15	1	0.038111	0.033627
80	22	15	2	0.147959	0.109848
80	22	15	3	0.354920	0.206961

N	n	k	x	P(x)	p(x)
80	22	15	4	0.605914	0.250995
80	22	15	5	0.812985	0.207071
80	22	15	6	0.932720	0.119735
80	22	15	7	0.981982	0.049262
80	22	15	8	0.996471	0.014489
80	22	15	9	0.999505	0.003034
80	22	15	10	0.999952	0.000447
80	22	15	11	0.999997	0.000045
80	22	15	12	1.000000	0.000003
80	22	15	13	1.000000	0.000000
80	22	15	14	1.000000	0.000000
80	22	15	15	1.000000	0.000000
80	22	16	0	0.002966	0.002966
80	22	16	1	0.027247	0.024280
80	22	16	2	0.114159	0.086913
80	22	16	3	0.294423	0.180264
80	22	16	4	0.536408	0.241985
80	22	16	5	0.758828	0.222420
80	22	16	6	0.903247	0.144419
80	22	16	7	0.970614	0.067367
80	22	16	8	0.993350	0.022736
80	22	16	9	0.998898	0.005548
80	22	16	10	0.999869	0.000971
80	22	16	11	0.999989	0.000120
80	22	16	12	0.999999	0.000010
80	22	16	13	1.000000	0.000001
80	22	16	14	1.000000	0.000000
80	22	16	15	1.000000	0.000000
80	22	16	16	1.000000	0.000000
80	22	17	0	0.001946	0.001946
80	22	17	1	0.019280	0.017333
80	22	17	2	0.086999	0.067720
80	22	17	3	0.240908	0.153908
80	22	17	4	0.468350	0.227442
80	22	17	5	0.699748	0.231398
80	22	17	6	0.867142	0.167394
80	22	17	7	0.954825	0.087683
80	22	17	8	0.988377	0.033552
80	22	17	9	0.997771	0.009395
80	22	17	10	0.999687	0.001916
80	22	17	11	0.999969	0.000281
80	22	17	12	0.999998	0.000029
80	22	17	13	1.000000	0.000002
80	22	17	14	1.000000	0.000000
80	22	17	15	1.000000	0.000000
80	22	17	16	1.000000	0.000000
80	22	17	17	1.000000	0.000000
80	22	18	0	0.001267	0.001267
80	22	18	1	0.013502	0.012235
80	22	18	2	0.065501	0.051999

Table for $N = 2$, $n = 1$, through $N = 100$, $n = 50$

N	n	k	x	P(x)	p(x)
80	22	18	3	0.194491	0.128990
80	22	18	4	0.403366	0.208876
80	22	18	5	0.637307	0.233941
80	22	18	6	0.824629	0.187322
80	22	18	7	0.933948	0.109319
80	22	18	8	0.980921	0.046973
80	22	18	9	0.995833	0.014912
80	22	18	10	0.999322	0.003489
80	22	18	11	0.999919	0.000597
80	22	18	12	0.999993	0.000074
80	22	18	13	0.999999	0.000006
80	22	18	14	1.000000	0.000000
80	22	18	15	1.000000	0.000000
80	22	18	16	1.000000	0.000000
80	22	18	17	1.000000	0.000000
80	22	18	18	1.000000	0.000000
80	22	19	0	0.000817	0.000817
80	22	19	1	0.009358	0.008540
80	22	19	2	0.048727	0.039369
80	22	19	3	0.154962	0.106235
80	22	19	4	0.342725	0.187764
80	22	19	5	0.573162	0.230437
80	22	19	6	0.776288	0.203126
80	22	19	7	0.907500	0.131212
80	22	19	8	0.970314	0.062814
80	22	19	9	0.992706	0.022392
80	22	19	10	0.998647	0.005941
80	22	19	11	0.999813	0.001167
80	22	19	12	0.999981	0.000168
80	22	19	13	0.999999	0.000017
80	22	19	14	1.000000	0.000001
80	22	19	15	1.000000	0.000000
80	22	19	16	1.000000	0.000000
80	22	19	17	1.000000	0.000000
80	22	19	18	1.000000	0.000000
30	22	19	19	1.000000	0.000000
80	22	20	0	0.000523	0.000523
80	22	20	1	0.006418	0.005895
80	22	20	2	0.035819	0.029401
80	22	20	3	0.121872	0.086053
80	22	20	4	0.287319	0.165447
80	22	20	5	0.508942	0.221622
80	22	20	6	0.723009	0.214067
80	22	20	7	0.875235	0.152226
80	22	20	8	0.955898	0.080663
80	22	20	9	0.987934	0.032036
80	22	20	10	0.997478	0.009544
80	22	20	11	0.999603	0.002125
80	22	20	12	0.999954	0.000351
80	22	20	13	0.999996	0.000042

N	n	k	x	P(x)	p(x)
80	22	20	14	1.000000	0.000004
80	22	20	15	1.000000	0.000000
80	22	20	16	1.000000	0.000000
80	22	20	17	1.000000	0.000000
80	22	20	18	1.000000	0.000000
80	22	20	19	1.000000	0.000000
80	22	20	20	1.000000	0.000000
80	22	21	0	0.000331	0.000331
80	22	21	1	0.004354	0.004023
80	22	21	2	0.026019	0.021664
80	22	21	3	0.094622	0.068603
80	22	21	4	0.237685	0.143063
80	22	21	5	0.446149	0.208464
80	22	21	6	0.665925	0.219776
80	22	21	7	0.837178	0.171254
80	22	21	8	0.937076	0.099898
80	22	21	9	0.980993	0.043917
80	22	21	10	0.995569	0.014577
80	22	21	11	0.999214	0.003644
80	22	21	12	0.999895	0.000682
80	22	21	13	0.999990	0.000094
80	22	21	14	0.999999	0.000010
80	22	21	15	1.000000	0.000001
80	22	21	16	1.000000	0.000000
80	22	21	17	1.000000	0.000000
80	22	21	18	1.000000	0.000000
80	22	21	19	1.000000	0.000000
80	22	21	20	1.000000	0.000000
80	22	21	21	1.000000	0.000000
80	22	22	0	0.000208	0.000208
80	22	22	1	0.002922	0.002715
80	22	22	2	0.018675	0.015752
80	22	22	3	0.072529	0.053855
80	22	22	4	0.194039	0.121509
80	22	22	5	0.386083	0.192044
80	22	22	6	0.606324	0.220241
80	22	22	7	0.793639	0.187315
80	22	22	8	0.913372	0.119732
80	22	22	9	0.971316	0.057945
80	22	22	10	0.992605	0.021288
80	22	22	11	0.998534	0.005929
80	22	22	12	0.999780	0.001246
80	22	22	13	0.999975	0.000196
80	22	22	14	0.999998	0.000023
80	22	22	15	1.000000	0.000002
80	22	22	16	1.000000	0.000000
80	22	22	17	1.000000	0.000000
80	22	22	18	1.000000	0.000000
80	22	22	19	1.000000	0.000000
80	22	22	20	1.000000	0.000000

N	n	k	x	P(x)	p(x)
80	22	22	21	1.000000	0.000000
80	22	22	22	1.000000	0.000000
80	23	1	0	0.712500	0.712500
80	23	1	1	1.000000	0.287500
80	23	2	0	0.505063	0.505063
80	23	2	1	0.919937	0.414873
80	23	2	2	1.000000	0.080063
80	23	3	0	0.356134	0.356134
80	23	3	1	0.802921	0.446787
80	23	3	2	0.978444	0.175523
80	23	3	3	1.000000	0.021555
80	23	4	0	0.249757	0.249757
80	23	4	1	0.675268	0.425511
80	23	4	2	0.930574	0.255307
80	23	4	3	0.994401	0.063827
80	23	4	4	1.000000	0.005599
80	23	5	0	0.174172	0.174172
80	23	5	1	0.552093	0.377921
80	23	5	2	0.860029	0.307936
80	23	5	3	0.977605	0.117575
80	23	5	4	0.998600	0.020996
80	23	5	5	1.000000	0.001400
80	23	6	0	0.120759	0.120759
80	23	6	1	0.441237	0.320477
80	23	6	2	0.773807	0.332571
80	23	6	3	0.946251	0.172444
80	23	6	4	0.993281	0.047030
80	23	6	5	0.999664	0.006383
80	23	6	6	1.000000	0.000336
80	23	7	0	0.083226	0.083226
80	23	7	1	0.345960	0.262733
80	23	7	2	0.679429	0.333469
80	23	7	3	0.899645	0.220216
80	23	7	4	0.981206	0.081561
80	23	7	5	0.998111	0.016905
80	23	7	6	0.999923	0.001811
80	23	7	7	1.000000	0.000077
80	23	8	0	0.057004	0.057004
80	23	8	1	0.266780	0.209775
80	23	8	2	0.583499	0.316720
80	23	8	3	0.839312	0.255812
80	23	8	4	0.959978	0.120666
80	23	8	5	0.993943	0.033965
80	23	8	6	0.999501	0.005558
80	23	8	7	0.999983	0.000482
80	23	8	8	1.000000	0.000017
80	23	9	0	0.038795	0.038795
80	23	9	1	0.202682	0.163887
80	23	9	2	0.491123	0.288441
80	23	9	3	0.768253	0.277130

N	n	k	x	P(x)	p(x)
80	23	9	4	0.928135	0.159883
80	23	9	5	0.985452	0.057316
80	23	9	6	0.998189	0.012737
80	23	9	7	0.999876	0.001687
80	23	9	8	0.999996	0.000121
80	23	9	9	1.000000	0.000004
80	23	10	0	0.026227	0.026227
80	23	10	1	0.151900	0.125672
80	23	10	2	0.405809	0.253910
80	23	10	3	0.690188	0.284379
80	23	10	4	0.885350	0.195162
80	23	10	5	0.970921	0.085571
80	23	10	6	0.995139	0.024218
80	23	10	7	0.999496	0.004357
80	23	10	8	0.999971	0.000475
80	23	10	9	0.999999	0.000028
80	23	10	10	1.000000	0.000001
80	23	11	0	0.017610	0.017610
80	23	11	1	0.112403	0.094793
80	23	11	2	0.329636	0.217234
80	23	11	3	0.608937	0.279301
80	23	11	4	0.832377	0.223440
80	23	11	5	0.948917	0.116540
80	23	11	6	0.989257	0.040341
80	23	11	7	0.998500	0.009242
80	23	11	8	0.999869	0.001369
80	23	11	9	0.999994	0.000124
80	23	11	10	1.000000	0.000006
80	23	11	11	1.000000	0.000000
80	23	12	0	0.011740	0.011740
80	23	12	1	0.082179	0.070439
80	23	12	2	0.263522	0.181343
80	23	12	3	0.527980	0.264458
80	23	12	4	0.770850	0.242870
80	23	12	5	0.918515	0.147665
80	23	12	6	0.979318	0.060803
80	23	12	7	0.996357	0.017038
80	23	12	8	0.999571	0.003215
80	23	12	9	0.999968	0.000397
80	23	12	10	0.999999	0.000030
80	23	12	11	1.000000	0.000001
80	23	12	12	1.000000	0.000000
80	23	13	0	0.007769	0.007769
80	23	13	1	0.059390	0.051621
80	23	13	2	0.207519	0.148129
80	23	13	3	0.450198	0.242680
80	23	13	4	0.702989	0.252791
80	23	13	5	0.879427	0.176438
80	23	13	6	0.964118	0.084690
80	23	13	7	0.992348	0.028230

N	n	k	x	P(x)	p(x)
80	23	13	8	0.998862	0.006515
80	23	13	9	0.999887	0.001024
80	23	13	10	0.999993	0.000106
80	23	13	11	1.000000	0.000007
80	23	13	12	1.000000	0.000000
80	23	13	13	1.000000	0.000000
80	23	14	0	0.005102	0.005102
80	23	14	1	0.042440	0.037338
80	23	14	2	0.161090	0.118651
80	23	14	3	0.377757	0.216666
80	23	14	4	0.631302	0.253546
80	23	14	5	0.832026	0.200724
80	23	14	6	0.942629	0.110603
80	23	14	7	0.985606	0.042977
80	23	14	8	0.997404	0.011798
80	23	14	9	0.999673	0.002269
80	23	14	10	0.999972	0.000300
80	23	14	11	0.999998	0.000026
80	23	14	12	1.000000	0.000001
80	23	14	13	1.000000	0.000000
80	23	14	14	1.000000	0.000000
80	23	15	0	0.003324	0.003324
80	23	15	1	0.029994	0.026670
80	23	15	2	0.123338	0.093344
80	23	15	3	0.312100	0.188762
80	23	15	4	0.558312	0.246212
80	23	15	5	0.777283	0.218971
80	23	15	6	0.914140	0.136857
80	23	15	7	0.975187	0.061047
80	23	15	8	0.994723	0.019535
80	23	15	9	0.999191	0.004469
80	23	15	10	0.999913	0.000722
80	23	15	11	0.999994	0.000080
80	23	15	12	1.000000	0.000006
80	23	15	13	1.000000	0.000000
80	23	15	14	1.000000	0.000000
80	23	15	15	1.000000	0.000000
80	23	16	0	0.002148	0.002148
80	23	16	1	0.020967	0.018819
80	23	16	2	0.093181	0.072213
80	23	16	3	0.254019	0.160839
80	23	16	4	0.486342	0.232323
80	23	16	5	0.716645	0.230303
80	23	16	6	0.878347	0.161702
80	23	16	7	0.960160	0.081813
80	23	16	8	0.990214	0.030054
80	23	16	9	0.998229	0.008014
80	23	16	10	0.999769	0.001540
80	23	16	11	0.999979	0.000210
80	23	16	12	0.999999	0.000020

N	n	k	x	P(x)	p(x)
80	23	16	13	1.000000	0.000001
80	23	16	14	1.000000	0.000000
80	23	16	15	1.000000	0.000000
80	23	16	16	1.000000	0.000000
80	23	17	0	0.001376	0.001376
80	23	17	1	0.014498	0.013122
80	23	17	2	0.069485	0.054987
80	23	17	3	0.203757	0.134272
80	23	17	4	0.417371	0.213614
80	23	17	5	0.651872	0.234501
80	23	17	6	0.835395	0.183522
80	23	17	7	0.939707	0.104312
80	23	17	8	0.983170	0.043463
80	23	17	9	0.996476	0.013305
80	23	17	10	0.999456	0.002980
80	23	17	11	0.999939	0.000483
80	23	17	12	0.999995	0.000056
80	23	17	13	1.000000	0.000004
80	23	17	14	1.000000	0.000000
80	23	17	15	1.000000	0.000000
80	23	17	16	1.000000	0.000000
80	23	17	17	1.000000	0.000000
80	23	18	0	0.000874	0.000874
80	23	18	1	0.009916	0.009042
80	23	18	2	0.051156	0.041241
80	23	18	3	0.161131	0.109975
80	23	18	4	0.352948	0.191817
80	23	18	5	0.584872	0.231924
80	23	18	6	0.785873	0.201001
80	23	18	7	0.913215	0.127342
80	23	18	8	0.972822	0.059607
80	23	18	9	0.993519	0.020697
80	23	18	10	0.998841	0.005322
80	23	18	11	0.999847	0.001006
80	23	18	12	0.999985	0.000138
80	23	18	13	0.999999	0.000013
80	23	18	14	1.000000	0.000001
80	23	18	15	1.000000	0.000000
80	23	18	16	1.000000	0.000000
80	23	18	17	1.000000	0.000000
80	23	18	18	1.000000	0.000000
80	23	19	0	0.000550	0.000550
80	23	19	1	0.006707	0.006158
80	23	19	2	0.037188	0.030480
80	23	19	3	0.125655	0.088468
80	23	19	4	0.294166	0.168510
80	23	19	5	0.517539	0.223374
80	23	19	6	0.730760	0.213220
80	23	19	7	0.880353	0.149593
80	23	19	8	0.958401	0.078048

N	n	k	x	P(x)	p(x)
80	23	19	9	0.988845	0.030444
80	23	19	10	0.997725	0.008880
80	23	19	11	0.999653	0.001927
80	23	19	12	0.999961	0.000308
80	23	19	13	0.999997	0.000036
80	23	19	14	1.000000	0.000003
80	23	19	15	1.000000	0.000000
80	23	19	16	1.000000	0.000000
80	23	19	17	1.000000	0.000000
80	23	19	18	1.000000	0.000000
80	23	19	19	1.000000	0.000000
80	23	20	0	0.000342	0.000342
80	23	20	1	0.004486	0.004144
80	23	20	2	0.026694	0.022208
80	23	20	3	0.096650	0.069955
80	23	20	4	0.241679	0.145029
80	23	20	5	0.451626	0.209947
80	23	20	6	0.671338	0.219712
80	23	20	7	0.841115	0.169777
80	23	20	8	0.939209	0.098094
80	23	20	9	0.981858	0.042649
80	23	20	10	0.995833	0.013974
80	23	20	11	0.999273	0.003441
80	23	20	12	0.999905	0.000632
80	23	20	13	0.999991	0.000086
80	23	20	14	0.999999	0.000008
80	23	20	15	1.000000	0.000001
80	23	20	16	1.000000	0.000000
80	23	20	17	1.000000	0.000000
80	23	20	18	1.000000	0.000000
80	23	20	19	1.000000	0.000000
80	23	20	20	1.000000	0.000000
80	23	21	0	0.000211	0.000211
80	23	21	1	0.002967	0.002756
80	23	21	2	0.018922	0.015955
80	23	21	3	0.073331	0.054410
80	23	21	4	0.195753	0.122422
80	23	21	5	0.388642	0.192889
80	23	21	6	0.609086	0.220444
80	23	21	7	0.795841	0.186755
80	23	21	8	0.914685	0.118844
80	23	21	9	0.971907	0.057221
80	23	21	10	0.992805	0.020898
80	23	21	11	0.998585	0.005780
80	23	21	12	0.999789	0.001204
80	23	21	13	0.999977	0.000187
80	23	21	14	0.999998	0.000021
80	23	21	15	1.000000	0.000002
80	23	21	16	1.000000	0.000000
80	23	21	17	1.000000	0.000000

N	n	k	x	P(x)	p(x)
80	23	21	18	1.000000	0.000000
80	23	21	19	1.000000	0.000000
80	23	21	20	1.000000	0.000000
80	23	21	21	1.000000	0.000000
80	23	22	0	0.000129	0.000129
80	23	22	1	0.001939	0.001811
80	23	22	2	0.013243	0.011303
80	23	22	3	0.054887	0.041644
80	23	22	4	0.156329	0.101442
80	23	22	5	0.329794	0.173465
80	23	22	6	0.545568	0.215774
80	23	22	7	0.745196	0.199628
80	23	22	8	0.884471	0.139275
80	23	22	9	0.958329	0.073858
80	23	22	10	0.988200	0.029871
80	23	22	11	0.997410	0.009209
80	23	22	12	0.999565	0.002155
80	23	22	13	0.999945	0.000380
80	23	22	14	0.999995	0.000050
80	23	22	15	1.000000	0.000005
80	23	22	16	1.000000	0.000000
80	23	22	17	1.000000	0.000000
80	23	22	18	1.000000	0.000000
80	23	22	19	1.000000	0.000000
80	23	22	20	1.000000	0.000000
80	23	22	21	1.000000	0.000000
80	23	22	22	1.000000	0.000000
80	23	23	0	0.000078	0.000078
80	23	23	1	0.001253	0.001175
80	23	23	2	0.009150	0.007898
80	23	23	3	0.040527	0.031377
80	23	23	4	0.123098	0.082571
80	23	23	5	0.275960	0.152862
80	23	23	6	0.482324	0.206364
80	23	23	7	0.690126	0.207802
80	23	23	8	0.848451	0.158325
80	23	23	9	0.940501	0.092050
80	23	23	10	0.981505	0.041004
80	23	23	11	0.995504	0.013999
80	23	23	12	0.999156	0.003652
80	23	23	13	0.999879	0.000723
80	23	23	14	0.999987	0.000108
80	23	23	15	0.999999	0.000012
80	23	23	16	1.000000	0.000001
80	23	23	17	1.000000	0.000000
80	23	23	18	1.000000	0.000000
80	23	23	19	1.000000	0.000000
80	23	23	20	1.000000	0.000000
80	23	23	21	1.000000	0.000000
80	23	23	22	1.000000	0.000000

Table for $N = 2$, $n = 1$, through $N = 100$, $n = 50$

N	n	k	x	P(x)	p(x)
80	23	23	23	1.000000	0.000000
80	24	1	0	0.700000	0.700000
80	24	1	1	1.000000	0.300000
80	24	2	0	0.487342	0.487342
80	24	2	1	0.912658	0.425316
80	24	2	2	1.000000	0.087342
80	24	3	0	0.337390	0.337390
80	24	3	1	0.787244	0.449854
80	24	3	2	0.975365	0.188121
80	24	3	3	1.000000	0.024635
80	24	4	0	0.232230	0.232230
80	24	4	1	0.652872	0.420643
80	24	4	2	0.921616	0.268744
80	24	4	3	0.993281	0.071665
80	24	4	4	1.000000	0.006719
80	24	5	0	0.158894	0.158894
80	24	5	1	0.525573	0.366679
80	24	5	2	0.843822	0.318249
80	24	5	3	0.973479	0.129657
80	24	5	4	0.998232	0.024753
80	24	5	5	1.000000	0.001768
80	24	6	0	0.108048	0.108048
80	24	6	1	0.413125	0.305077
80	24	6	2	0.750469	0.337344
80	24	6	3	0.937175	0.186706
80	24	6	4	0.991631	0.054456
80	24	6	5	0.999552	0.007921
80	24	6	6	1.000000	0.000448
80	24	7	0	0.073005	0.073005
80	24	7	1	0.318303	0.245298
80	24	7	2	0.650177	0.331874
80	24	7	3	0.884191	0.234014
80	24	7	4	0.976913	0.092722
80	24	7	5	0.997518	0.020605
80	24	7	6	0.999891	0.002373
80	24	7	7	1.000000	0.000109
80	24	8	0	0.049004	0.049004
80	24	8	1	0.241018	0.192014
80	24	8	2	0.550161	0.309143
80	24	8	3	0.816872	0.266711
80	24	8	4	0.951510	0.134638
80	24	8	5	0.992155	0.040645
80	24	8	6	0.999306	0.007151
80	24	8	7	0.999975	0.000669
80	24	8	8	1.000000	0.000025
80	24	9	0	0.032669	0.032669
80	24	9	1	0.179680	0.147011
80	24	9	2	0.455700	0.276020
80	24	9	3	0.739081	0.283381
80	24	9	4	0.914110	0.175029

N	n	k	x	P(x)	p(x)
80	24	9	5	0.981429	0.067319
80	24	9	6	0.997518	0.016089
80	24	9	7	0.999817	0.002298
80	24	9	8	0.999994	0.000178
80	24	9	9	1.000000	0.000006
80	24	10	0	0.021626	0.021626
80	24	10	1	0.132057	0.110431
80	24	10	2	0.370173	0.238116
80	24	10	3	0.655264	0.285091
80	24	10	4	0.864806	0.209542
80	24	10	5	0.963415	0.098608
80	24	10	6	0.993439	0.030025
80	24	10	7	0.999266	0.005827
80	24	10	8	0.999954	0.000688
80	24	10	9	0.999999	0.000044
80	24	10	10	1.000000	0.000001
80	24	11	0	0.014211	0.014211
80	24	11	1	0.095772	0.081561
80	24	11	2	0.295336	0.199564
80	24	11	3	0.569737	0.274401
80	24	11	4	0.804937	0.235200
80	24	11	5	0.936649	0.131712
80	24	11	6	0.985719	0.049069
80	24	11	7	0.997851	0.012133
80	24	11	8	0.999797	0.001946
80	24	11	9	0.999989	0.000192
80	24	11	10	1.000000	0.000010
80	24	11	11	1.000000	0.000000
80	24	12	0	0.009268	0.009268
80	24	12	1	0.068585	0.059317
80	24	12	2	0.231707	0.163122
80	24	12	3	0.486224	0.254516
80	24	12	4	0.736763	0.250540
80	24	12	5	0.900381	0.163618
80	24	12	6	0.972918	0.072537
80	24	12	7	0.994862	0.021944
80	24	12	8	0.999346	0.004484
80	24	12	9	0.999947	0.000602
80	24	12	10	0.999998	0.000050
80	24	12	11	1.000000	0.000002
80	24	12	12	1.000000	0.000000
80	24	13	0	0.005997	0.005997
80	24	13	1	0.048522	0.042525
80	24	13	2	0.178932	0.130410
80	24	13	3	0.407623	0.228690
80	24	13	4	0.663075	0.255452
80	24	13	5	0.854664	0.191589
80	24	13	6	0.953717	0.099053
80	24	13	7	0.989376	0.035659
80	24	13	8	0.998291	0.008915

N	n	k	x	P(x)	p(x)
80	24	13	9	0.999815	0.001524
80	24	13	10	0.999987	0.000173
80	24	13	11	0.999999	0.000012
80	24	13	12	1.000000	0.000000
80	24	13	13	1.000000	0.000000
80	24	14	0	0.003849	0.003849
80	24	14	1	0.033924	0.030075
80	24	14	2	0.136111	0.102187
80	24	14	3	0.335944	0.199833
80	24	14	4	0.586821	0.250877
80	24	14	5	0.800333	0.213512
80	24	14	6	0.927106	0.126773
80	24	14	7	0.980328	0.053222
80	24	14	8	0.996162	0.015834
80	24	14	9	0.999473	0.003312
80	24	14	10	0.999951	0.000478
80	24	14	11	0.999997	0.000046
80	24	14	12	1.000000	0.000003
80	24	14	13	1.000000	0.000000
80	24	14	14	1.000000	0.000000
80	24	15	0	0.002449	0.002449
80	24	15	1	0.023443	0.020994
80	24	15	2	0.102049	0.078605
80	24	15	3	0.272361	0.170312
80	24	15	4	0.510797	0.238437
80	24	15	5	0.738867	0.228070
80	24	15	6	0.892531	0.153664
80	24	15	7	0.966619	0.074088
80	24	15	8	0.992323	0.025704
80	24	15	9	0.998721	0.006397
80	24	15	10	0.999850	0.001129
80	24	15	11	0.999988	0.000138
80	24	15	12	0.999999	0.000011
80	24	15	13	1.000000	0.000001
80	24	15	14	1.000000	0.000000
80	24	15	15	1.000000	0.000000
80	24	16	0	0.001545	0.001545
80	24	16	1	0.016015	0.014470
80	24	16	2	0.075444	0.059429
80	24	16	3	0.217337	0.141893
80	24	16	4	0.437432	0.220095
80	24	16	5	0.672201	0.234768
80	24	16	6	0.849978	0.177778
80	24	16	7	0.947243	0.097264
80	24	16	8	0.985996	0.038754
80	24	16	9	0.997245	0.011248
80	24	16	10	0.999607	0.002362
80	24	16	11	0.999960	0.000354
80	24	16	12	0.999997	0.000037
80	24	16	13	1.000000	0.000003

N	n	k	x	P(x)	p(x)
80	24	16	14	1.000000	0.000000
80	24	16	15	1.000000	0.000000
80	24	16	16	1.000000	0.000000
80	24	17	0	0.000966	0.000966
80	24	17	1	0.010815	0.009849
80	24	17	2	0.055015	0.044200
80	24	17	3	0.170778	0.115763
80	24	17	4	0.368652	0.197874
80	24	17	5	0.602504	0.233851
80	24	17	6	0.799978	0.197474
80	24	17	7	0.921407	0.121428
80	24	17	8	0.976308	0.054901
80	24	17	9	0.994608	0.018300
80	24	17	10	0.999090	0.004482
80	24	17	11	0.999888	0.000799
80	24	17	12	0.999990	0.000102
80	24	17	13	0.999999	0.000009
80	24	17	14	1.000000	0.000001
80	24	17	15	1.000000	0.000000
80	24	17	16	1.000000	0.000000
80	24	17	17	1.000000	0.000000
80	24	18	0	0.000598	0.000598
80	24	18	1	0.007219	0.006621
80	24	18	2	0.039580	0.032361
80	24	18	3	0.132190	0.092610
80	24	18	4	0.305835	0.173645
80	24	18	5	0.531977	0.226142
80	24	18	6	0.743557	0.211580
80	24	18	7	0.888640	0.145083
80	24	18	8	0.962365	0.073724
80	24	18	9	0.990251	0.027886
80	24	18	10	0.998094	0.007843
80	24	18	11	0.999724	0.001630
80	24	18	12	0.999971	0.000247
80	24	18	13	0.999998	0.000027
80	24	18	14	1.000000	0.000002
80	24	18	15	1.000000	0.000000
80	24	18	16	1.000000	0.000000
80	24	18	17	1.000000	0.000000
80	24	18	18	1.000000	0.000000
80	24	19	0	0.000366	0.000366
80	24	19	1	0.004763	0.004396
80	24	19	2	0.028097	0.023334
80	24	19	3	0.100822	0.072725
80	24	19	4	0.249821	0.148998
80	24	19	5	0.462676	0.212855
80	24	19	6	0.682131	0.219455
80	24	19	7	0.848859	0.166729
80	24	19	8	0.943339	0.094480
80	24	19	9	0.983504	0.040165

Table for $N = 2$, $n = 1$, through $N = 100$, $n = 50$

N	n	k	x	P(x)	p(x)
80	24	19	10	0.996323	0.012819
80	24	19	11	0.999382	0.003059
80	24	19	12	0.999923	0.000541
80	24	19	13	0.999993	0.000070
80	24	19	14	0.999999	0.000006
80	24	19	15	1.000000	0.000000
80	24	19	16	1.000000	0.000000
80	24	19	17	1.000000	0.000000
80	24	19	18	1.000000	0.000000
80	24	19	19	1.000000	0.000000
80	24	20	0	0.000222	0.000222
80	24	20	1	0.003105	0.002883
80	24	20	2	0.019681	0.016576
80	24	20	3	0.075786	0.056104
80	24	20	4	0.200969	0.125183
80	24	20	5	0.396377	0.195408
80	24	20	6	0.617373	0.220997
80	24	20	7	0.802394	0.185021
80	24	20	8	0.918557	0.116163
80	24	20	9	0.973628	0.055070
80	24	20	10	0.993381	0.019753
80	24	20	11	0.998730	0.005349
80	24	20	12	0.999817	0.001087
80	24	20	13	0.999980	0.000164
80	24	20	14	0.999998	0.000018
80	24	20	15	1.000000	0.000001
80	24	20	16	1.000000	0.000000
80	24	20	17	1.000000	0.000000
80	24	20	18	1.000000	0.000000
80	24	20	19	1.000000	0.000000
80	24	20	20	1.000000	0.000000
80	24	21	0	0.000133	0.000133
80	24	21	1	0.002000	0.001867
80	24	21	2	0.013603	0.011603
80	24	21	3	0.056149	0.042546
80	24	21	4	0.159241	0.103092
80	24	21	5	0.334498	0.175256
80	24	21	6	0.551074	0.216577
80	24	21	7	0.749972	0.198897
80	24	21	8	0.887581	0.137609
80	24	21	9	0.959860	0.072279
80	24	21	10	0.988772	0.028912
80	24	21	11	0.997571	0.008799
80	24	21	12	0.999599	0.002028
80	24	21	13	0.999950	0.000351
80	24	21	14	0.999995	0.000045
80	24	21	15	1.000000	0.000004
80	24	21	16	1.000000	0.000000
80	24	21	17	1.000000	0.000000
80	24	21	18	1.000000	0.000000

N	n	k	x	P(x)	p(x)
80	24	21	19	1.000000	0.000000
80	24	21	20	1.000000	0.000000
80	24	21	21	1.000000	0.000000
80	24	22	0	0.000079	0.000079
80	24	22	1	0.001272	0.001193
80	24	22	2	0.009277	0.008004
80	24	22	3	0.041006	0.031729
80	24	22	4	0.124295	0.083289
80	24	22	5	0.278059	0.153764
80	24	22	6	0.485000	0.206941
80	24	22	7	0.692662	0.207662
80	24	22	8	0.850263	0.157601
80	24	22	9	0.941484	0.091221
80	24	22	10	0.981911	0.040428
80	24	22	11	0.995632	0.013721
80	24	22	12	0.999187	0.003554
80	24	22	13	0.999885	0.000698
80	24	22	14	0.999988	0.000103
80	24	22	15	0.999999	0.000011
80	24	22	16	1.000000	0.000001
80	24	22	17	1.000000	0.000000
80	24	22	18	1.000000	0.000000
80	24	22	19	1.000000	0.000000
80	24	22	20	1.000000	0.000000
80	24	22	21	1.000000	0.000000
80	24	22	22	1.000000	0.000000
80	24	23	0	0.000046	0.000046
80	24	23	1	0.000799	0.000753
80	24	23	2	0.006241	0.005441
80	24	23	3	0.029518	0.023277
80	24	23	4	0.095574	0.066057
80	24	23	5	0.227688	0.132114
80	24	23	6	0.420777	0.193089
80	24	23	7	0.631796	0.211019
80	24	23	8	0.806787	0.174991
80	24	23	9	0.917892	0.111105
80	24	23	10	0.972153	0.054261
80	24	23	11	0.992557	0.020404
80	24	23	12	0.998451	0.005894
80	24	23	13	0.999753	0.001301
80	24	23	14	0.999970	0.000218
80	24	23	15	0.999997	0.000027
80	24	23	16	1.000000	0.000002
80	24	23	17	1.000000	0.000000
80	24	23	18	1.000000	0.000000
80	24	23	19	1.000000	0.000000
80	24	23	20	1.000000	0.000000
80	24	23	21	1.000000	0.000000
80	24	23	22	1.000000	0.000000
80	24	23	23	1.000000	0.000000

N	n	k	x	P(x)	p(x)
80	24	24	0	0.000027	0.000027
80	24	24	1	0.000495	0.000469
80	24	24	2	0.004140	0.003645
80	24	24	3	0.020942	0.016802
80	24	24	4	0.072397	0.051455
80	24	24	5	0.183650	0.111254
80	24	24	6	0.359802	0.176151
80	24	24	7	0.568860	0.209059
80	24	24	8	0.757667	0.188806
80	24	24	9	0.888654	0.130987
80	24	24	10	0.958826	0.070172
80	24	24	11	0.987904	0.029078
80	24	24	12	0.997211	0.009307
80	24	24	13	0.999501	0.002291
80	24	24	14	0.999932	0.000430
80	24	24	15	0.999993	0.000061
80	24	24	16	0.999999	0.000006
80	24	24	17	1.000000	0.000000
80	24	24	18	1.000000	0.000000
80	24	24	19	1.000000	0.000000
80	24	24	20	1.000000	0.000000
80	24	24	21	1.000000	0.000000
80	24	24	22	1.000000	0.000000
80	24	24	23	1.000000	0.000000
80	24	24	24	1.000000	0.000000
80	25	1	0	0.687500	0.687500
80	25	1	1	1.000000	0.312500
80	25	2	0	0.469937	0.469937
80	25	2	1	0.905063	0.435127
80	25	2	2	1.000000	0.094937
80	25	3	0	0.319316	0.319316
80	25	3	1	0.771178	0.451862
80	25	3	2	0.972006	0.200828
80	25	3	3	1.000000	0.027994
80	25	4	0	0.215642	0.215642
80	25	4	1	0.630338	0.414696
80	25	4	2	0.912018	0.281680
80	25	4	3	0.992002	0.079983
80	25	4	4	1.000000	0.007998
80	25	5	0	0.144707	0.144707
80	25	5	1	0.499381	0.354674
80	25	5	2	0.826773	0.327392
80	25	5	3	0.968849	0.142076
80	25	5	4	0.997790	0.028941
80	25	5	5	1.000000	0.002210
80	25	6	0	0.096471	0.096471
80	25	6	1	0.385886	0.289414
80	25	6	2	0.726373	0.340487
80	25	6	3	0.927173	0.200800
80	25	6	4	0.989686	0.062513

N	n	k	x	P(x)	p(x)
80	25	6	5	0.999411	0.009724
80	25	6	6	1.000000	0.000589
80	25	7	0	0.063880	0.063880
80	25	7	1	0.292022	0.228142
80	25	7	2	0.620546	0.328524
80	25	7	3	0.867476	0.246930
80	25	7	4	0.971946	0.104470
80	25	7	5	0.996782	0.024836
80	25	7	6	0.999849	0.003066
80	25	7	7	1.000000	0.000151
80	25	8	0	0.042003	0.042003
80	25	8	1	0.217016	0.175013
80	25	8	2	0.517038	0.300022
80	25	8	3	0.793058	0.276020
80	25	8	4	0.941893	0.148834
80	25	8	5	0.989978	0.048085
80	25	8	6	0.999050	0.009073
80	25	8	7	0.999963	0.000912
80	25	8	8	1.000000	0.000037
80	25	9	0	0.027419	0.027419
80	25	9	1	0.158678	0.131260
80	25	9	2	0.421198	0.262519
80	25	9	3	0.708719	0.287521
80	25	9	4	0.898483	0.189764
80	25	9	5	0.976621	0.078138
80	25	9	6	0.996656	0.020035
80	25	9	7	0.999735	0.003078
80	25	9	8	0.999991	0.000257
80	25	9	9	1.000000	0.000009
80	25	10	0	0.017764	0.017764
80	25	10	1	0.114309	0.096545
80	25	10	2	0.336156	0.221847
80	25	10	3	0.619628	0.283472
80	25	10	4	0.842356	0.222728
80	25	10	5	0.954610	0.112255
80	25	10	6	0.991295	0.036685
80	25	10	7	0.998954	0.007659
80	25	10	8	0.999930	0.000975
80	25	10	9	0.999998	0.000068
80	25	10	10	1.000000	0.000002
80	25	11	0	0.011420	0.011420
80	25	11	1	0.081208	0.069788
80	25	11	2	0.263264	0.182056
80	25	11	3	0.530537	0.267273
80	25	11	4	0.775537	0.245000
80	25	11	5	0.922537	0.147000
80	25	11	6	0.981338	0.058800
80	25	11	7	0.996985	0.015647
80	25	11	8	0.999693	0.002708
80	25	11	9	0.999982	0.000290

$N = 80$ $n = 24\text{–}25$

Table for $N = 2$, $n = 1$, through $N = 100$, $n = 50$

N	n	k	x	P(x)	p(x)
80	25	11	10	0.999999	0.000017
80	25	11	11	1.000000	0.000000
80	25	12	0	0.007282	0.007282
80	25	12	1	0.056934	0.049652
80	25	12	2	0.202578	0.145645
80	25	12	3	0.445319	0.242741
80	25	12	4	0.700972	0.255653
80	25	12	5	0.879929	0.178957
80	25	12	6	0.965146	0.085218
80	25	12	7	0.992903	0.027757
80	25	12	8	0.999026	0.006123
80	25	12	9	0.999915	0.000890
80	25	12	10	0.999996	0.000081
80	25	12	11	1.000000	0.000004
80	25	12	12	1.000000	0.000000
80	25	13	0	0.004605	0.004605
80	25	13	1	0.039410	0.034805
80	25	13	2	0.153316	0.113906
80	25	13	3	0.366785	0.213469
80	25	13	4	0.622020	0.255235
80	25	13	5	0.827294	0.205274
80	25	13	6	0.941335	0.114041
80	25	13	7	0.985555	0.044220
80	25	13	8	0.997495	0.011939
80	25	13	9	0.999706	0.002211
80	25	13	10	0.999978	0.000272
80	25	13	11	0.999999	0.000021
80	25	13	12	1.000000	0.000001
80	25	13	13	1.000000	0.000000
80	25	14	0	0.002887	0.002887
80	25	14	1	0.026942	0.024056
80	25	14	2	0.114214	0.087272
80	25	14	3	0.296691	0.182477
80	25	14	4	0.542021	0.245330
80	25	14	5	0.766019	0.223997
80	25	14	6	0.908995	0.142977
80	25	14	7	0.973675	0.064680
80	25	14	8	0.994465	0.020790
80	25	14	9	0.999178	0.004712
80	25	14	10	0.999917	0.000739
80	25	14	11	0.999995	0.000078
80	25	14	12	1.000000	0.000005
80	25	14	13	1.000000	0.000000
80	25	14	14	1.000000	0.000000
80	25	15	0	0.001793	0.001793
80	25	15	1	0.018195	0.016402
80	25	15	2	0.083801	0.065606
80	25	15	3	0.235865	0.152064
80	25	15	4	0.463962	0.228096
80	25	15	5	0.698140	0.234179

N	n	k	x	P(x)	p(x)
80	25	15	6	0.867835	0.169695
80	25	15	7	0.956035	0.088200
80	25	15	8	0.989110	0.033075
80	25	15	9	0.998035	0.008925
80	25	15	10	0.999749	0.001714
80	25	15	11	0.999978	0.000229
80	25	15	12	0.999999	0.000021
80	25	15	13	1.000000	0.000001
80	25	15	14	1.000000	0.000000
80	25	15	15	1.000000	0.000000
80	25	16	0	0.001104	0.001104
80	25	16	1	0.012139	0.011035
80	25	16	2	0.060587	0.048448
80	25	16	3	0.184397	0.123811
80	25	16	4	0.390269	0.205872
80	25	16	5	0.626085	0.235817
80	25	16	6	0.818232	0.192147
80	25	16	7	0.931611	0.113379
80	25	16	8	0.980460	0.048849
80	25	16	9	0.995839	0.015378
80	25	16	10	0.999354	0.003515
80	25	16	11	0.999929	0.000575
80	25	16	12	0.999995	0.000066
80	25	16	13	1.000000	0.000005
80	25	16	14	1.000000	0.000000
80	25	16	15	1.000000	0.000000
80	25	16	16	1.000000	0.000000
80	25	17	0	0.000672	0.000672
80	25	17	1	0.008001	0.007328
80	25	17	2	0.043176	0.035175
80	25	17	3	0.141837	0.098662
80	25	17	4	0.322717	0.180880
80	25	17	5	0.552393	0.229675
80	25	17	6	0.761189	0.208796
80	25	17	7	0.899723	0.138534
80	25	17	8	0.967484	0.067761
80	25	17	9	0.991994	0.024509
80	25	17	10	0.998530	0.006536
80	25	17	11	0.999803	0.001273
80	25	17	12	0.999981	0.000178
80	25	17	13	0.999999	0.000017
80	25	17	14	1.000000	0.000001
80	25	17	15	1.000000	0.000000
80	25	17	16	1.000000	0.000000
80	25	17	17	1.000000	0.000000
80	25	18	0	0.000406	0.000406
80	25	18	1	0.005209	0.004803
80	25	18	2	0.030334	0.025125
80	25	18	3	0.107384	0.077050
80	25	18	4	0.262424	0.155040

N	n	k	x	P(x)	p(x)
80	25	18	5	0.479480	0.217056
80	25	18	6	0.698219	0.218739
80	25	18	7	0.860142	0.161923
80	25	18	8	0.949200	0.089058
80	25	18	9	0.985769	0.036570
80	25	18	10	0.996974	0.011204
80	25	18	11	0.999520	0.002546
80	25	18	12	0.999944	0.000424
80	25	18	13	0.999995	0.000051
80	25	18	14	1.000000	0.000004
80	25	18	15	1.000000	0.000000
80	25	18	16	1.000000	0.000000
80	25	18	17	1.000000	0.000000
80	25	18	18	1.000000	0.000000
80	25	19	0	0.000242	0.000242
80	25	19	1	0.003350	0.003108
80	25	19	2	0.021013	0.017664
80	25	19	3	0.080044	0.059030
80	25	19	4	0.209911	0.129867
80	25	19	5	0.409462	0.199551
80	25	19	6	0.631186	0.221724
80	25	19	7	0.813132	0.181946
80	25	19	8	0.924781	0.111649
80	25	19	9	0.976332	0.051551
80	25	19	10	0.994263	0.017931
80	25	19	11	0.998945	0.004682
80	25	19	12	0.999855	0.000910
80	25	19	13	0.999985	0.000130
80	25	19	14	0.999999	0.000013
80	25	19	15	1.000000	0.000001
80	25	19	16	1.000000	0.000000
80	25	19	17	1.000000	0.000000
80	25	19	18	1.000000	0.000000
80	25	19	19	1.000000	0.000000
80	25	20	0	0.000143	0.000143
80	25	20	1	0.002127	0.001984
80	25	20	2	0.014353	0.012226
80	25	20	3	0.058754	0.044401
80	25	20	4	0.165202	0.106448
80	25	20	5	0.344035	0.178833
80	25	20	6	0.562124	0.218089
80	25	20	7	0.759443	0.197319
80	25	20	8	0.893665	0.134223
80	25	20	9	0.962810	0.069145
80	25	20	10	0.989854	0.027043
80	25	20	11	0.997870	0.008017
80	25	20	12	0.999661	0.001791
80	25	20	13	0.999960	0.000298
80	25	20	14	0.999996	0.000037
80	25	20	15	1.000000	0.000003

N	n	k	x	P(x)	p(x)
80	25	20	16	1.000000	0.000000
80	25	20	17	1.000000	0.000000
80	25	20	18	1.000000	0.000000
80	25	20	19	1.000000	0.000000
80	25	20	20	1.000000	0.000000
80	25	21	0	0.000083	0.000083
80	25	21	1	0.001333	0.001250
80	25	21	2	0.009667	0.008333
80	25	21	3	0.042474	0.032807
80	25	21	4	0.127945	0.085472
80	25	21	5	0.284424	0.156479
80	25	21	6	0.493063	0.208638
80	25	21	7	0.700247	0.207185
80	25	21	8	0.855636	0.155388
80	25	21	9	0.944372	0.088736
80	25	21	10	0.983093	0.038721
80	25	21	11	0.996000	0.012907
80	25	21	12	0.999273	0.003274
80	25	21	13	0.999900	0.000627
80	25	21	14	0.999990	0.000090
80	25	21	15	0.999999	0.000009
80	25	21	16	1.000000	0.000001
80	25	21	17	1.000000	0.000000
80	25	21	18	1.000000	0.000000
80	25	21	19	1.000000	0.000000
80	25	21	20	1.000000	0.000000
80	25	21	21	1.000000	0.000000
80	25	22	0	0.000048	0.000048
80	25	22	1	0.000825	0.000777
80	25	22	2	0.006418	0.005593
80	25	22	3	0.030241	0.023823
80	25	22	4	0.097523	0.067283
80	25	22	5	0.231381	0.133857
80	25	22	6	0.425874	0.194493
80	25	22	7	0.637038	0.211164
80	25	22	8	0.810863	0.173824
80	25	22	9	0.920308	0.109445
80	25	22	10	0.973248	0.052941
80	25	22	11	0.992937	0.019689
80	25	22	12	0.998552	0.005615
80	25	22	13	0.999773	0.001221
80	25	22	14	0.999973	0.000200
80	25	22	15	0.999998	0.000024
80	25	22	16	1.000000	0.000002
80	25	22	17	1.000000	0.000000
80	25	22	18	1.000000	0.000000
80	25	22	19	1.000000	0.000000
80	25	22	20	1.000000	0.000000
80	25	22	21	1.000000	0.000000
80	25	22	22	1.000000	0.000000

N	n	k	x	P(x)	p(x)
80	25	23	0	0.000027	0.000027
80	25	23	1	0.000503	0.000476
80	25	23	2	0.004200	0.003697
80	25	23	3	0.021204	0.017004
80	25	23	4	0.073162	0.051958
80	25	23	5	0.185223	0.112061
80	25	23	6	0.362161	0.176938
80	25	23	7	0.571505	0.209344
80	25	23	8	0.759914	0.188410
80	25	23	9	0.890116	0.130202
80	25	23	10	0.959557	0.069441
80	25	23	11	0.988185	0.028628
80	25	23	12	0.997294	0.009109
80	25	23	13	0.999520	0.002227
80	25	23	14	0.999935	0.000415
80	25	23	15	0.999993	0.000058
80	25	23	16	0.999999	0.000006
80	25	23	17	1.000000	0.000000
80	25	23	18	1.000000	0.000000
80	25	23	19	1.000000	0.000000
80	25	23	20	1.000000	0.000000
80	25	23	21	1.000000	0.000000
80	25	23	22	1.000000	0.000000
80	25	23	23	1.000000	0.000000
80	25	24	0	0.000015	0.000015
80	25	24	1	0.000303	0.000288
80	25	24	2	0.002708	0.002405
80	25	24	3	0.014641	0.011933
80	25	24	4	0.054020	0.039379
80	25	24	5	0.145903	0.091883
80	25	24	6	0.303181	0.157278
80	25	24	7	0.505396	0.202215
80	25	24	8	0.703722	0.198326
80	25	24	9	0.853568	0.149846
80	25	24	10	0.941283	0.087715
80	25	24	11	0.981153	0.039870
80	25	24	12	0.995216	0.014063
80	25	24	13	0.999051	0.003835
80	25	24	14	0.999855	0.000804
80	25	24	15	0.999983	0.000128
80	25	24	16	0.999998	0.000015
80	25	24	17	1.000000	0.000001
80	25	24	18	1.000000	0.000000
80	25	24	19	1.000000	0.000000
80	25	24	20	1.000000	0.000000
80	25	24	21	1.000000	0.000000
80	25	24	22	1.000000	0.000000
80	25	24	23	1.000000	0.000000
80	25	24	24	1.000000	0.000000
80	25	25	0	0.000008	0.000008

N	n	k	x	P(x)	p(x)
80	25	25	1	0.000180	0.000171
80	25	25	2	0.001720	0.001541
80	25	25	3	0.009953	0.008233
80	25	25	4	0.039253	0.029300
80	25	25	5	0.113088	0.073835
80	25	25	6	0.249819	0.136731
80	25	25	7	0.440398	0.190579
80	25	25	8	0.643516	0.203117
80	25	25	9	0.810755	0.167239
80	25	25	10	0.917788	0.107033
80	25	25	11	0.971186	0.053398
80	25	25	12	0.991952	0.020766
80	25	25	13	0.998230	0.006278
80	25	25	14	0.999697	0.001468
80	25	25	15	0.999960	0.000263
80	25	25	16	0.999996	0.000036
80	25	25	17	1.000000	0.000004
80	25	25	18	1.000000	0.000000
80	25	25	19	1.000000	0.000000
80	25	25	20	1.000000	0.000000
80	25	25	21	1.000000	0.000000
80	25	25	22	1.000000	0.000000
80	25	25	23	1.000000	0.000000
80	25	25	24	1.000000	0.000000
80	25	25	25	1.000000	0.000000
80	26	1	0	0.675000	0.675000
80	26	1	1	1.000000	0.325000
80	26	2	0	0.452848	0.452848
80	26	2	1	0.897152	0.444304
80	26	2	2	1.000000	0.102848
80	26	3	0	0.301899	0.301899
80	26	3	1	0.754747	0.452848
80	26	3	2	0.968354	0.213608
80	26	3	3	1.000000	0.031646
80	26	4	0	0.199959	0.199959
80	26	4	1	0.607718	0.407759
80	26	4	2	0.901775	0.294057
80	26	4	3	0.990547	0.088772
80	26	4	4	1.000000	0.009453
80	26	5	0	0.131552	0.131552
80	26	5	1	0.473587	0.342035
80	26	5	2	0.808915	0.335328
80	26	5	3	0.963682	0.154767
80	26	5	4	0.997264	0.033582
80	26	5	5	1.000000	0.002736
80	26	6	0	0.085947	0.085947
80	26	6	1	0.359575	0.273628
80	26	6	2	0.701610	0.342035
80	26	6	3	0.916220	0.214610
80	26	6	4	0.987413	0.071193

N	n	k	x	P(x)	p(x)
80	26	6	5	0.999234	0.011821
80	26	6	6	1.000000	0.000766
80	26	7	0	0.055750	0.055750
80	26	7	1	0.267133	0.211384
80	26	7	2	0.590680	0.323547
80	26	7	3	0.849517	0.258837
80	26	7	4	0.966248	0.116731
80	26	7	5	0.995879	0.029632
80	26	7	6	0.999793	0.003914
80	26	7	7	1.000000	0.000207
80	26	8	0	0.035894	0.035894
80	26	8	1	0.194742	0.158848
80	26	8	2	0.484308	0.289567
80	26	8	3	0.767966	0.283657
80	26	8	4	0.931069	0.163103
80	26	8	5	0.987355	0.056286
80	26	8	6	0.998721	0.011366
80	26	8	7	0.999946	0.001225
80	26	8	8	1.000000	0.000054
80	26	9	0	0.022932	0.022932
80	26	9	1	0.139586	0.116654
80	26	9	2	0.387786	0.248200
80	26	9	3	0.677353	0.289567
80	26	9	4	0.881232	0.203879
80	26	9	5	0.970938	0.089707
80	26	9	6	0.995563	0.024625
80	26	9	7	0.999623	0.004059
80	26	9	8	0.999986	0.000364
80	26	9	9	1.000000	0.000013
80	26	10	0	0.014534	0.014534
80	26	10	1	0.098511	0.083976
80	26	10	2	0.303888	0.205377
80	26	10	3	0.583550	0.279662
80	26	10	4	0.818058	0.234508
80	26	10	5	0.944405	0.126347
80	26	10	6	0.988627	0.044222
80	26	10	7	0.998536	0.009910
80	26	10	8	0.999894	0.001358
80	26	10	9	0.999997	0.000102
80	26	10	10	1.000000	0.000003
80	26	11	0	0.009136	0.009136
80	26	11	1	0.068519	0.059383
80	26	11	2	0.233473	0.164953
80	26	11	3	0.491661	0.258188
80	26	11	4	0.744355	0.252695
80	26	11	5	0.906501	0.162146
80	26	11	6	0.975992	0.069491
80	26	11	7	0.995847	0.019855
80	26	11	8	0.999545	0.003698
80	26	11	9	0.999972	0.000427

N	n	k	x	P(x)	p(x)
80	26	11	10	0.999999	0.000027
80	26	11	11	1.000000	0.000001
80	26	12	0	0.005693	0.005693
80	26	12	1	0.047003	0.041310
80	26	12	2	0.176097	0.129094
80	26	12	3	0.405598	0.229500
80	26	12	4	0.663786	0.258188
80	26	12	5	0.857152	0.193366
80	26	12	6	0.955850	0.098697
80	26	12	7	0.990379	0.034530
80	26	12	8	0.998580	0.008201
80	26	12	9	0.999867	0.001286
80	26	12	10	0.999993	0.000126
80	26	12	11	1.000000	0.000007
80	26	12	12	1.000000	0.000000
80	26	13	0	0.003516	0.003516
80	26	13	1	0.031816	0.028299
80	26	13	2	0.130535	0.098719
80	26	13	3	0.327973	0.197438
80	26	13	4	0.580255	0.252282
80	26	13	5	0.797436	0.217182
80	26	13	6	0.926821	0.129385
80	26	13	7	0.980731	0.053910
80	26	13	8	0.996409	0.015678
80	26	13	9	0.999545	0.003136
80	26	13	10	0.999963	0.000418
80	26	13	11	0.999998	0.000035
80	26	13	12	1.000000	0.000002
80	26	13	13	1.000000	0.000000
80	26	14	0	0.002152	0.002152
80	26	14	1	0.021256	0.019105
80	26	14	2	0.095173	0.073916
80	26	14	3	0.260196	0.165023
80	26	14	4	0.497416	0.237220
80	26	14	5	0.729364	0.231949
80	26	14	6	0.888199	0.158834
80	26	14	7	0.965443	0.077245
80	26	14	8	0.992197	0.026754
80	26	14	9	0.998749	0.006552
80	26	14	10	0.999863	0.001114
80	26	14	11	0.999990	0.000127
80	26	14	12	1.000000	0.000009
80	26	14	13	1.000000	0.000000
80	26	14	14	1.000000	0.000000
80	26	15	0	0.001304	0.001304
80	26	15	1	0.014020	0.012716
80	26	15	2	0.068294	0.054274
80	26	15	3	0.202688	0.134394
80	26	15	4	0.418342	0.215655
80	26	15	5	0.655563	0.237220

N	n	k	x	P(x)	p(x)
80	26	15	6	0.840067	0.184505
80	26	15	7	0.943206	0.103139
80	26	15	8	0.984901	0.041695
80	26	15	9	0.997062	0.012161
80	26	15	10	0.999593	0.002531
80	26	15	11	0.999962	0.000368
80	26	15	12	0.999998	0.000036
80	26	15	13	1.000000	0.000002
80	26	15	14	1.000000	0.000000
80	26	15	15	1.000000	0.000000
80	26	16	0	0.000782	0.000782
80	26	16	1	0.009129	0.008347
80	26	16	2	0.048254	0.039125
80	26	16	3	0.155133	0.106879
80	26	16	4	0.345352	0.190219
80	26	16	5	0.578922	0.233571
80	26	16	6	0.783297	0.204374
80	26	16	7	0.913058	0.129761
80	26	16	8	0.973355	0.060297
80	26	16	9	0.993881	0.020527
80	26	16	10	0.998970	0.005089
80	26	16	11	0.999877	0.000906
80	26	16	12	0.999990	0.000113
80	26	16	13	0.999999	0.000010
80	26	16	14	1.000000	0.000001
80	26	16	15	1.000000	0.000000
80	26	16	16	1.000000	0.000000
80	26	17	0	0.000465	0.000465
80	26	17	1	0.005869	0.005404
80	26	17	2	0.033582	0.027714
80	26	17	3	0.116724	0.083141
80	26	17	4	0.279964	0.163240
80	26	17	5	0.502282	0.222318
80	26	17	6	0.719430	0.217148
80	26	17	7	0.874535	0.155105
80	26	17	8	0.956396	0.081861
80	26	17	9	0.988429	0.032033
80	26	17	10	0.997698	0.009269
80	26	17	11	0.999664	0.001966
80	26	17	12	0.999965	0.000301
80	26	17	13	0.999997	0.000032
80	26	17	14	1.000000	0.000002
80	26	17	15	1.000000	0.000000
80	26	17	16	1.000000	0.000000
80	26	17	17	1.000000	0.000000
80	26	18	0	0.000273	0.000273
80	26	18	1	0.003724	0.003451
80	26	18	2	0.023025	0.019301
80	26	18	3	0.086370	0.063346
80	26	18	4	0.222959	0.136589

N	n	k	x	P(x)	p(x)
80	26	18	5	0.428176	0.205216
80	26	18	6	0.650494	0.222318
80	26	18	7	0.827757	0.177263
80	26	18	8	0.933007	0.105250
80	26	18	9	0.979785	0.046778
80	26	18	10	0.995344	0.015559
80	26	18	11	0.999196	0.003852
80	26	18	12	0.999898	0.000702
80	26	18	13	0.999991	0.000093
80	26	18	14	0.999999	0.000009
80	26	18	15	1.000000	0.000001
80	26	18	16	1.000000	0.000000
80	26	18	17	1.000000	0.000000
80	26	18	18	1.000000	0.000000
80	26	19	0	0.000158	0.000158
80	26	19	1	0.002333	0.002174
80	26	19	2	0.015554	0.013221
80	26	19	3	0.062871	0.047318
80	26	19	4	0.174492	0.111621
80	26	19	5	0.358667	0.184175
80	26	19	6	0.578778	0.220111
80	26	19	7	0.773434	0.194656
80	26	19	8	0.902451	0.129016
80	26	19	9	0.966959	0.064508
80	26	19	10	0.991329	0.024370
80	26	19	11	0.998264	0.006935
80	26	19	12	0.999740	0.001476
80	26	19	13	0.999971	0.000232
80	26	19	14	0.999998	0.000026
80	26	19	15	1.000000	0.000002
80	26	19	16	1.000000	0.000000
80	26	19	17	1.000000	0.000000
80	26	19	18	1.000000	0.000000
80	26	19	19	1.000000	0.000000
80	26	20	0	0.000091	0.000091
80	26	20	1	0.001442	0.001351
80	26	20	2	0.010352	0.008910
80	26	20	3	0.045030	0.034678
80	26	20	4	0.134236	0.089205
80	26	20	5	0.295263	0.161027
80	26	20	6	0.506611	0.211348
80	26	20	7	0.712804	0.206193
80	26	20	8	0.864380	0.151577
80	26	20	9	0.948981	0.084601
80	26	20	10	0.984937	0.035955
80	26	20	11	0.996559	0.011622
80	26	20	12	0.999401	0.002842
80	26	20	13	0.999922	0.000521
80	26	20	14	0.999992	0.000071
80	26	20	15	0.999999	0.000007

N	n	k	x	P(x)	p(x)
80	26	20	16	1.000000	0.000000
80	26	20	17	1.000000	0.000000
80	26	20	18	1.000000	0.000000
80	26	20	19	1.000000	0.000000
80	26	20	20	1.000000	0.000000
80	26	21	0	0.000052	0.000052
80	26	21	1	0.000879	0.000827
80	26	21	2	0.006788	0.005909
80	26	21	3	0.031737	0.024949
80	26	21	4	0.101527	0.069790
80	26	21	5	0.238903	0.137376
80	26	21	6	0.436161	0.197258
80	26	21	7	0.647509	0.211348
80	26	21	8	0.818907	0.171398
80	26	21	9	0.925011	0.106104
80	26	21	10	0.975349	0.050338
80	26	21	11	0.993653	0.018305
80	26	21	12	0.998738	0.005085
80	26	21	13	0.999809	0.001071
80	26	21	14	0.999978	0.000169
80	26	21	15	0.999998	0.000020
80	26	21	16	1.000000	0.000002
80	26	21	17	1.000000	0.000000
80	26	21	18	1.000000	0.000000
80	26	21	19	1.000000	0.000000
80	26	21	20	1.000000	0.000000
80	26	21	21	1.000000	0.000000
80	26	22	0	0.000029	0.000029
80	26	22	1	0.000528	0.000499
80	26	22	2	0.004384	0.003856
80	26	22	3	0.022011	0.017627
80	26	22	4	0.075504	0.053493
80	26	22	5	0.190007	0.114503
80	26	22	6	0.369294	0.179288
80	26	22	7	0.579448	0.210154
80	26	22	8	0.766617	0.187168
80	26	22	9	0.894439	0.127822
80	26	22	10	0.961698	0.067259
80	26	22	11	0.988999	0.027302
80	26	22	12	0.997531	0.008532
80	26	22	13	0.999573	0.002042
80	26	22	14	0.999944	0.000371
80	26	22	15	0.999994	0.000051
80	26	22	16	0.999999	0.000005
80	26	22	17	1.000000	0.000000
80	26	22	18	1.000000	0.000000
80	26	22	19	1.000000	0.000000
80	26	22	20	1.000000	0.000000
80	26	22	21	1.000000	0.000000
80	26	22	22	1.000000	0.000000

N	n	k	x	P(x)	p(x)
80	26	23	0	0.000016	0.000016
80	26	23	1	0.000313	0.000297
80	26	23	2	0.002789	0.002476
80	26	23	3	0.015021	0.012232
80	26	23	4	0.055213	0.040192
80	26	23	5	0.148549	0.093335
80	26	23	6	0.307471	0.158922
80	26	23	7	0.510605	0.203134
80	26	23	8	0.708530	0.197925
80	26	23	9	0.856974	0.148444
80	26	23	10	0.943144	0.086170
80	26	23	11	0.981939	0.038795
80	26	23	12	0.995472	0.013533
80	26	23	13	0.999115	0.003644
80	26	23	14	0.999867	0.000752
80	26	23	15	0.999985	0.000118
80	26	23	16	0.999999	0.000014
80	26	23	17	1.000000	0.000001
80	26	23	18	1.000000	0.000000
80	26	23	19	1.000000	0.000000
80	26	23	20	1.000000	0.000000
80	26	23	21	1.000000	0.000000
80	26	23	22	1.000000	0.000000
80	26	23	23	1.000000	0.000000
80	26	24	0	0.000009	0.000009
80	26	24	1	0.000183	0.000174
80	26	24	2	0.001746	0.001564
80	26	24	3	0.010085	0.008339
80	26	24	4	0.039700	0.029615
80	26	24	5	0.114162	0.074461
80	26	24	6	0.251709	0.137547
80	26	24	7	0.442893	0.191185
80	26	24	8	0.646027	0.203134
80	26	24	9	0.812701	0.166674
80	26	24	10	0.918956	0.106255
80	26	24	11	0.971729	0.052774
80	26	24	12	0.992148	0.020418
80	26	24	13	0.998284	0.006136
80	26	24	14	0.999709	0.001425
80	26	24	15	0.999962	0.000253
80	26	24	16	0.999996	0.000034
80	26	24	17	1.000000	0.000003
80	26	24	18	1.000000	0.000000
80	26	24	19	1.000000	0.000000
80	26	24	20	1.000000	0.000000
80	26	24	21	1.000000	0.000000
80	26	24	22	1.000000	0.000000
80	26	24	23	1.000000	0.000000
80	26	24	24	1.000000	0.000000
80	26	25	0	0.000005	0.000005

N	n	k	x	P(x)	p(x)
80	26	25	1	0.000105	0.000100
80	26	25	2	0.001076	0.000971
80	26	25	3	0.006660	0.005584
80	26	25	4	0.028066	0.021406
80	26	25	5	0.086239	0.058173
80	26	25	6	0.202585	0.116346
80	26	25	7	0.378027	0.175442
80	26	25	8	0.580734	0.202707
80	26	25	9	0.762103	0.181369
80	26	25	10	0.888597	0.126494
80	26	25	11	0.957594	0.068996
80	26	25	12	0.987043	0.029450
80	26	25	13	0.996860	0.009817
80	26	25	14	0.999404	0.002544
80	26	25	15	0.999912	0.000509
80	26	25	16	0.999990	0.000078
80	26	25	17	0.999999	0.000009
80	26	25	18	1.000000	0.000001
80	26	25	19	1.000000	0.000000
80	26	25	20	1.000000	0.000000
80	26	25	21	1.000000	0.000000
80	26	25	22	1.000000	0.000000
80	26	25	23	1.000000	0.000000
80	26	25	24	1.000000	0.000000
80	26	25	25	1.000000	0.000000
80	26	26	0	0.000002	0.000002
80	26	26	1	0.000059	0.000057
80	26	26	2	0.000652	0.000593
80	26	26	3	0.004325	0.003673
80	26	26	4	0.019504	0.015179
80	26	26	5	0.064027	0.044524
80	26	26	6	0.160277	0.096250
80	26	26	7	0.317420	0.157143
80	26	26	8	0.514394	0.196974
80	26	26	9	0.706044	0.191650
80	26	26	10	0.851799	0.145755
80	26	26	11	0.938776	0.086977
80	26	26	12	0.979547	0.040771
80	26	26	13	0.994540	0.014993
80	26	26	14	0.998849	0.004309
80	26	26	15	0.999811	0.000962
80	26	26	16	0.999976	0.000165
80	26	26	17	0.999998	0.000022
80	26	26	18	1.000000	0.000002
80	26	26	19	1.000000	0.000000
80	26	26	20	1.000000	0.000000
80	26	26	21	1.000000	0.000000
80	26	26	22	1.000000	0.000000
80	26	26	23	1.000000	0.000000
80	26	26	24	1.000000	0.000000

N	n	k	x	P(x)	p(x)
80	26	26	25	1.000000	0.000000
80	26	26	26	1.000000	0.000000
80	27	1	0	0.662500	0.662500
80	27	1	1	1.000000	0.337500
80	27	2	0	0.436076	0.436076
80	27	2	1	0.888924	0.452848
80	27	2	2	1.000000	0.111076
80	27	3	0	0.285127	0.285127
80	27	3	1	0.737975	0.452848
80	27	3	2	0.964399	0.226424
80	27	3	3	1.000000	0.035601
80	27	4	0	0.185147	0.185147
80	27	4	1	0.585065	0.399918
80	27	4	2	0.890884	0.305819
80	27	4	3	0.988903	0.098019
80	27	4	4	1.000000	0.011096
80	27	5	0	0.119371	0.119371
80	27	5	1	0.448251	0.328880
80	27	5	2	0.790286	0.342035
80	27	5	3	0.957950	0.167664
80	27	5	4	0.996642	0.038692
80	27	5	5	1.000000	0.003358
80	27	6	0	0.076398	0.076398
80	27	6	1	0.334239	0.257842
80	27	6	2	0.676274	0.342035
80	27	6	3	0.904297	0.228023
80	27	6	4	0.984776	0.080479
80	27	6	5	0.999015	0.014239
80	27	6	6	1.000000	0.000985
80	27	7	0	0.048523	0.048523
80	27	7	1	0.243646	0.195123
80	27	7	2	0.560722	0.317076
80	27	7	3	0.830344	0.269622
80	27	7	4	0.959763	0.129419
80	27	7	5	0.994782	0.035019
80	27	7	6	0.999720	0.004939
80	27	7	7	1.000000	0.000280
80	27	8	0	0.030576	0.030576
80	27	8	1	0.174150	0.143574
80	27	8	2	0.452134	0.277984
80	27	8	3	0.741701	0.289567
80	27	8	4	0.918987	0.177286
80	27	8	5	0.984228	0.065241
80	27	8	6	0.998300	0.014072
80	27	8	7	0.999923	0.001624
80	27	8	8	1.000000	0.000077
80	27	9	0	0.019110	0.019110
80	27	9	1	0.122304	0.103194
80	27	9	2	0.355612	0.233308
80	27	9	3	0.645179	0.289567

N	n	k	x	P(x)	p(x)
80	27	9	4	0.862354	0.217175
80	27	9	5	0.964293	0.101939
80	27	9	6	0.994195	0.029902
80	27	9	7	0.999472	0.005277
80	27	9	8	0.999980	0.000507
80	27	9	9	1.000000	0.000020
80	27	10	0	0.011843	0.011843
80	27	10	1	0.084515	0.072672
80	27	10	2	0.273461	0.188947
80	27	10	3	0.547297	0.273836
80	27	10	4	0.792001	0.244704
80	27	10	5	0.932706	0.140705
80	27	10	6	0.985351	0.052645
80	27	10	7	0.997986	0.012635
80	27	10	8	0.999844	0.001858
80	27	10	9	0.999995	0.000151
80	27	10	10	1.000000	0.000005
80	27	11	0	0.007275	0.007275
80	27	11	1	0.057522	0.050247
80	27	11	2	0.205980	0.148458
80	27	11	3	0.453411	0.247430
80	27	11	4	0.711599	0.258188
80	27	11	5	0.888485	0.176886
80	27	11	6	0.969558	0.081073
80	27	11	7	0.994376	0.024818
80	27	11	8	0.999340	0.004964
80	27	11	9	0.999956	0.000616
80	27	11	10	0.999999	0.000043
80	27	11	11	1.000000	0.000001
80	27	12	0	0.004428	0.004428
80	27	12	1	0.038588	0.034160
80	27	12	2	0.152191	0.113603
80	27	12	3	0.367348	0.215157
80	27	12	4	0.625536	0.258188
80	27	12	5	0.832086	0.206550
80	27	12	6	0.944883	0.112797
80	27	12	7	0.987182	0.042299
80	27	12	8	0.997973	0.010791
80	27	12	9	0.999795	0.001822
80	27	12	10	0.999988	0.000193
80	27	12	11	1.000000	0.000011
80	27	12	12	1.000000	0.000000
80	27	13	0	0.002670	0.002670
80	27	13	1	0.025527	0.022857
80	27	13	2	0.110425	0.084898
80	27	13	3	0.291410	0.180985
80	27	13	4	0.538208	0.246797
80	27	13	5	0.765261	0.227054
80	27	13	6	0.910049	0.144788
80	27	13	7	0.974741	0.064692

N	n	k	x	P(x)	p(x)
80	27	13	8	0.994958	0.020216
80	27	13	9	0.999313	0.004355
80	27	13	10	0.999940	0.000627
80	27	13	11	0.999997	0.000057
80	27	13	12	1.000000	0.000003
80	27	13	13	1.000000	0.000000
80	27	14	0	0.001594	0.001594
80	27	14	1	0.016657	0.015063
80	27	14	2	0.078747	0.062090
80	27	14	3	0.226580	0.147833
80	27	14	4	0.453486	0.226906
80	27	14	5	0.690706	0.237220
80	27	14	6	0.864668	0.173961
80	27	14	7	0.955430	0.090762
80	27	14	8	0.989225	0.033795
80	27	14	9	0.998143	0.008918
80	27	14	10	0.999781	0.001638
80	27	14	11	0.999983	0.000203
80	27	14	12	0.999999	0.000016
80	27	14	13	1.000000	0.000001
80	27	14	14	1.000000	0.000000
80	27	15	0	0.000942	0.000942
80	27	15	1	0.010723	0.009781
80	27	15	2	0.055228	0.044505
80	27	15	3	0.172822	0.117594
80	27	15	4	0.374413	0.201590
80	27	15	5	0.611633	0.237220
80	27	15	6	0.809316	0.197683
80	27	15	7	0.927926	0.118610
80	27	15	8	0.979496	0.051570
80	27	15	9	0.995711	0.016215
80	27	15	10	0.999359	0.003648
80	27	15	11	0.999934	0.000575
80	27	15	12	0.999996	0.000061
80	27	15	13	1.000000	0.000004
80	27	15	14	1.000000	0.000000
80	27	15	15	1.000000	0.000000
80	27	16	0	0.000551	0.000551
80	27	16	1	0.006811	0.006260
80	27	16	2	0.038111	0.031300
80	27	16	3	0.129403	0.091292
80	27	16	4	0.303081	0.173678
80	27	16	5	0.531343	0.228262
80	27	16	6	0.745449	0.214106
80	27	16	7	0.891431	0.145982
80	27	16	8	0.964422	0.072991
80	27	16	9	0.991220	0.026799
80	27	16	10	0.998405	0.007184
80	27	16	11	0.999793	0.001388
80	27	16	12	0.999981	0.000189

Table for $N = 2$, $n = 1$, through $N = 100$, $n = 50$

N	n	k	x	P(x)	p(x)
80	27	16	13	0.999999	0.000017
80	27	16	14	1.000000	0.000001
80	27	16	15	1.000000	0.000000
80	27	16	16	1.000000	0.000000
80	27	17	0	0.000318	0.000318
80	27	17	1	0.004268	0.003949
80	27	17	2	0.025884	0.021617
80	27	17	3	0.095168	0.069284
80	27	17	4	0.240665	0.145497
80	27	17	5	0.452878	0.212213
80	27	17	6	0.675196	0.222318
80	27	17	7	0.845812	0.170616
80	27	17	8	0.942753	0.096941
80	27	17	9	0.983683	0.040931
80	27	17	10	0.996496	0.012813
80	27	17	11	0.999446	0.002949
80	27	17	12	0.999937	0.000492
80	27	17	13	0.999995	0.000058
80	27	17	14	1.000000	0.000005
80	27	17	15	1.000000	0.000000
80	27	17	16	1.000000	0.000000
80	27	17	17	1.000000	0.000000
80	27	18	0	0.000182	0.000182
80	27	18	1	0.002638	0.002456
80	27	18	2	0.017306	0.014668
80	27	18	3	0.068774	0.051468
80	27	18	4	0.187547	0.118773
80	27	18	5	0.378772	0.191224
80	27	18	6	0.601090	0.222318
80	27	18	7	0.791648	0.190558
80	27	18	8	0.913517	0.121869
80	27	18	9	0.971989	0.058472
80	27	18	10	0.993039	0.021050
80	27	18	11	0.998697	0.005658
80	27	18	12	0.999820	0.001124
80	27	18	13	0.999982	0.000162
80	27	18	14	0.999999	0.000017
80	27	18	15	1.000000	0.000001
80	27	18	16	1.000000	0.000000
80	27	18	17	1.000000	0.000000
80	27	18	18	1.000000	0.000000
80	27	19	0	0.000103	0.000103
80	27	19	1	0.001608	0.001505
80	27	19	2	0.011391	0.009784
80	27	19	3	0.048851	0.037460
80	27	19	4	0.143486	0.094635
80	27	19	5	0.310918	0.167432
80	27	19	6	0.525789	0.214871
80	27	19	7	0.730178	0.204389
80	27	19	8	0.876170	0.145992
80	27	19	9	0.955013	0.078843
80	27	19	10	0.987267	0.032254
80	27	19	11	0.997237	0.009969
80	27	19	12	0.999548	0.002312
80	27	19	13	0.999946	0.000397
80	27	19	14	0.999995	0.000050
80	27	19	15	1.000000	0.000004
80	27	19	16	1.000000	0.000000
80	27	19	17	1.000000	0.000000
80	27	19	18	1.000000	0.000000
80	27	19	19	1.000000	0.000000
80	27	20	0	0.000057	0.000057
80	27	20	1	0.000966	0.000909
80	27	20	2	0.007382	0.006415
80	27	20	3	0.034113	0.026731
80	27	20	4	0.107804	0.073691
80	27	20	5	0.250533	0.142729
80	27	20	6	0.451817	0.201284
80	27	20	7	0.663165	0.211348
80	27	20	8	0.830697	0.167532
80	27	20	9	0.931748	0.101051
80	27	20	10	0.978278	0.046530
80	27	20	11	0.994622	0.016343
80	27	20	12	0.998980	0.004358
80	27	20	13	0.999854	0.000875
80	27	20	14	0.999985	0.000130
80	27	20	15	0.999999	0.000014
80	27	20	16	1.000000	0.000001
80	27	20	17	1.000000	0.000000
80	27	20	18	1.000000	0.000000
80	27	20	19	1.000000	0.000000
80	27	20	20	1.000000	0.000000
80	27	21	0	0.000031	0.000031
80	27	21	1	0.000572	0.000541
80	27	21	2	0.004709	0.004136
80	27	21	3	0.023421	0.018712
80	27	21	4	0.079556	0.056135
80	27	21	5	0.198199	0.118643
80	27	21	6	0.381367	0.183168
80	27	21	7	0.592715	0.211348
80	27	21	8	0.777645	0.184930
80	27	21	9	0.901432	0.123787
80	27	21	10	0.965095	0.063662
80	27	21	11	0.990263	0.025169
80	27	21	12	0.997890	0.007627
80	27	21	13	0.999650	0.001760
80	27	21	14	0.999956	0.000306
80	27	21	15	0.999996	0.000040
80	27	21	16	1.000000	0.000004
80	27	21	17	1.000000	0.000000
80	27	21	18	1.000000	0.000000
80	27	21	19	1.000000	0.000000
80	27	21	20	1.000000	0.000000
80	27	21	21	1.000000	0.000000
80	27	22	0	0.000017	0.000017
80	27	22	1	0.000334	0.000317
80	27	22	2	0.002956	0.002622
80	27	22	3	0.015809	0.012853
80	27	22	4	0.057673	0.041864
80	27	22	5	0.153959	0.096287
80	27	22	6	0.316172	0.162213
80	27	22	7	0.521072	0.204900
80	27	22	8	0.718091	0.197019
80	27	22	9	0.863667	0.145575
80	27	22	10	0.946751	0.083085
80	27	22	11	0.983438	0.036687
80	27	22	12	0.995951	0.012513
80	27	22	13	0.999233	0.003281
80	27	22	14	0.999889	0.000656
80	27	22	15	0.999988	0.000099
80	27	22	16	0.999999	0.000011
80	27	22	17	1.000000	0.000001
80	27	22	18	1.000000	0.000000
80	27	22	19	1.000000	0.000000
80	27	22	20	1.000000	0.000000
80	27	22	21	1.000000	0.000000
80	27	22	22	1.000000	0.000000
80	27	23	0	0.000009	0.000009
80	27	23	1	0.000192	0.000183
80	27	23	2	0.001826	0.001634
80	27	23	3	0.010490	0.008665
80	27	23	4	0.041072	0.030581
80	27	23	5	0.117437	0.076365
80	27	23	6	0.257440	0.140003
80	27	23	7	0.450417	0.192977
80	27	23	8	0.653551	0.203134
80	27	23	9	0.818488	0.164938
80	27	23	10	0.922399	0.103911
80	27	23	11	0.973317	0.050919
80	27	23	12	0.992715	0.019398
80	27	23	13	0.998441	0.005726
80	27	23	14	0.999742	0.001301
80	27	23	15	0.999967	0.000226
80	27	23	16	0.999997	0.000029
80	27	23	17	1.000000	0.000003
80	27	23	18	1.000000	0.000000
80	27	23	19	1.000000	0.000000
80	27	23	20	1.000000	0.000000
80	27	23	21	1.000000	0.000000
80	27	23	22	1.000000	0.000000
80	27	23	23	1.000000	0.000000
80	27	24	0	0.000005	0.000005
80	27	24	1	0.000109	0.000104
80	27	24	2	0.001109	0.001001
80	27	24	3	0.006842	0.005733
80	27	24	4	0.028732	0.021890
80	27	24	5	0.087962	0.059231
80	27	24	6	0.205860	0.117897
80	27	24	7	0.382706	0.176846
80	27	24	8	0.585839	0.203134
80	27	24	9	0.766403	0.180563
80	27	24	10	0.891408	0.125005
80	27	24	11	0.959025	0.067617
80	27	24	12	0.987610	0.028586
80	27	24	13	0.997034	0.009424
80	27	24	14	0.999445	0.002411
80	27	24	15	0.999920	0.000475
80	27	24	16	0.999991	0.000071
80	27	24	17	0.999999	0.000008
80	27	24	18	1.000000	0.000001
80	27	24	19	1.000000	0.000000
80	27	24	20	1.000000	0.000000
80	27	24	21	1.000000	0.000000
80	27	24	22	1.000000	0.000000
80	27	24	23	1.000000	0.000000
80	27	24	24	1.000000	0.000000
80	27	25	0	0.000002	0.000002
80	27	25	1	0.000060	0.000058
80	27	25	2	0.000662	0.000602
80	27	25	3	0.004385	0.003723
80	27	25	4	0.019741	0.015356
80	27	25	5	0.064693	0.044952
80	27	25	6	0.161648	0.096955
80	27	25	7	0.319546	0.157898
80	27	25	8	0.516919	0.197373
80	27	25	9	0.708364	0.191445
80	27	25	10	0.853460	0.145096
80	27	25	11	0.939706	0.086246
80	27	25	12	0.979953	0.040248
80	27	25	13	0.994678	0.014725
80	27	25	14	0.998885	0.004207
80	27	25	15	0.999818	0.000933
80	27	25	16	0.999977	0.000159
80	27	25	17	0.999998	0.000021
80	27	25	18	1.000000	0.000002
80	27	25	19	1.000000	0.000000
80	27	25	20	1.000000	0.000000
80	27	25	21	1.000000	0.000000
80	27	25	22	1.000000	0.000000
80	27	25	23	1.000000	0.000000

N	n	k	x	$P(x)$	$p(x)$
80	27	25	24	1.000000	0.000000
80	27	25	25	1.000000	0.000000
80	27	26	0	0.000001	0.000001
80	27	26	1	0.000033	0.000032
80	27	26	2	0.000389	0.000356
80	27	26	3	0.002761	0.002372
80	27	26	4	0.013320	0.010559
80	27	26	5	0.046713	0.033393
80	27	26	6	0.124629	0.077917
80	27	26	7	0.262129	0.137500
80	27	26	8	0.448736	0.186607
80	27	26	9	0.645710	0.196974
80	27	26	10	0.808612	0.162903
80	27	26	11	0.914616	0.106004
80	27	26	12	0.968977	0.054361
80	27	26	13	0.990930	0.021953
80	27	26	14	0.997891	0.006961
80	27	26	15	0.999615	0.001724
80	27	26	16	0.999945	0.000331
80	27	26	17	0.999994	0.000049
80	27	26	18	0.999999	0.000005
80	27	26	19	1.000000	0.000000
80	27	26	20	1.000000	0.000000
80	27	26	21	1.000000	0.000000
80	27	26	22	1.000000	0.000000
80	27	26	23	1.000000	0.000000
80	27	26	24	1.000000	0.000000
80	27	26	25	1.000000	0.000000
80	27	26	26	1.000000	0.000000
80	27	27	0	0.000001	0.000001
80	27	27	1	0.000018	0.000017
80	27	27	2	0.000224	0.000206
80	27	27	3	0.001707	0.001482
80	27	27	4	0.008822	0.007116
80	27	27	5	0.033108	0.024286
80	27	27	6	0.094328	0.061220
80	27	27	7	0.211203	0.116875
80	27	27	8	0.383078	0.171875
80	27	27	9	0.580052	0.196974
80	27	27	10	0.757328	0.177276
80	27	27	11	0.883207	0.125879
80	27	27	12	0.953877	0.070669
80	27	27	13	0.985239	0.031362
80	27	27	14	0.996215	0.010977
80	27	27	15	0.999232	0.003016
80	27	27	16	0.999878	0.000646
80	27	27	17	0.999985	0.000107
80	27	27	18	0.999998	0.000014
80	27	27	19	1.000000	0.000001
80	27	27	20	1.000000	0.000000

N	n	k	x	$P(x)$	$p(x)$
80	27	27	21	1.000000	0.000000
80	27	27	22	1.000000	0.000000
80	27	27	23	1.000000	0.000000
80	27	27	24	1.000000	0.000000
80	27	27	25	1.000000	0.000000
80	27	27	26	1.000000	0.000000
80	27	27	27	1.000000	0.000000
80	28	1	0	0.650000	0.650000
80	28	1	1	1.000000	0.350000
80	28	2	0	0.419620	0.419620
80	28	2	1	0.880380	0.460759
80	28	2	2	1.000000	0.119620
80	28	3	0	0.268987	0.268987
80	28	3	1	0.720886	0.451899
80	28	3	2	0.960127	0.239240
80	28	3	3	1.000000	0.039873
80	28	4	0	0.171174	0.171174
80	28	4	1	0.562428	0.391254
80	28	4	2	0.879344	0.316916
80	28	4	3	0.987054	0.107710
80	28	4	4	1.000000	0.012946
80	28	5	0	0.108110	0.108110
80	28	5	1	0.423430	0.315320
80	28	5	2	0.770925	0.347496
80	28	5	3	0.951623	0.180698
80	28	5	4	0.995912	0.044289
80	28	5	5	1.000000	0.004088
80	28	6	0	0.067749	0.067749
80	28	6	1	0.309915	0.242166
80	28	6	2	0.650460	0.340546
80	28	6	3	0.891390	0.240930
80	28	6	4	0.981739	0.090349
80	28	6	5	0.998746	0.017007
80	28	6	6	1.000000	0.001254
80	28	7	0	0.042114	0.042114
80	28	7	1	0.221557	0.179443
80	28	7	2	0.530809	0.309252
80	28	7	3	0.809995	0.279186
80	28	7	4	0.952437	0.142442
80	28	7	5	0.993460	0.041023
80	28	7	6	0.999627	0.006167
80	28	7	7	1.000000	0.000373
80	28	8	0	0.025961	0.025961
80	28	8	1	0.155188	0.129227
80	28	8	2	0.420664	0.265477
80	28	8	3	0.714383	0.293719
80	28	8	4	0.905607	0.191223
80	28	8	5	0.980535	0.074928
80	28	8	6	0.997769	0.017234
80	28	8	7	0.999893	0.002124

N	n	k	x	$P(x)$	$p(x)$
80	28	8	8	1.000000	0.000107
80	28	9	0	0.015865	0.015865
80	28	9	1	0.106727	0.090863
80	28	9	2	0.324798	0.218070
80	28	9	3	0.612398	0.287600
80	28	9	4	0.841866	0.229468
80	28	9	5	0.956600	0.114734
80	28	9	6	0.992503	0.035903
80	28	9	7	0.999273	0.006770
80	28	9	8	0.999970	0.000697
80	28	9	9	1.000000	0.000030
80	28	10	0	0.009608	0.009608
80	28	10	1	0.072174	0.062566
80	28	10	2	0.244941	0.172767
80	28	10	3	0.511130	0.266189
80	28	10	4	0.764299	0.253169
80	28	10	5	0.919432	0.155133
80	28	10	6	0.981378	0.061946
80	28	10	7	0.997271	0.015893
80	28	10	8	0.999774	0.002503
80	28	10	9	0.999992	0.000218
80	28	10	10	1.000000	0.000008
80	28	11	0	0.005765	0.005765
80	28	11	1	0.048042	0.042277
80	28	11	2	0.180770	0.132729
80	28	11	3	0.416063	0.235292
80	28	11	4	0.677498	0.261436
80	28	11	5	0.868460	0.190962
80	28	11	6	0.961909	0.093449
80	28	11	7	0.992503	0.030594
80	28	11	8	0.999059	0.006556
80	28	11	9	0.999933	0.000874
80	28	11	10	0.999998	0.000065
80	28	11	11	1.000000	0.000002
80	28	12	0	0.003426	0.003426
80	28	12	1	0.031499	0.028073
80	28	12	2	0.130757	0.099258
80	28	12	3	0.330812	0.200055
80	28	12	4	0.586564	0.255752
80	28	12	5	0.804806	0.218242
80	28	12	6	0.932114	0.127308
80	28	12	7	0.983192	0.051078
80	28	12	8	0.997158	0.013967
80	28	12	9	0.999692	0.002534
80	28	12	10	0.999981	0.000289
80	28	12	11	0.999999	0.000019
80	28	12	12	1.000000	0.000000
80	28	13	0	0.002015	0.002015
80	28	13	1	0.020352	0.018337
80	28	13	2	0.092805	0.072453

N	n	k	x	$P(x)$	$p(x)$
80	28	13	3	0.257262	0.164457
80	28	13	4	0.496299	0.239036
80	28	13	5	0.730989	0.234690
80	28	13	6	0.890926	0.159937
80	28	13	7	0.967418	0.076492
80	28	13	8	0.993051	0.025633
80	28	13	9	0.998984	0.005934
80	28	13	10	0.999904	0.000920
80	28	13	11	0.999995	0.000090
80	28	13	12	1.000000	0.000005
80	28	13	13	1.000000	0.000000
80	28	14	0	0.001173	0.001173
80	28	14	1	0.012962	0.011790
80	28	14	2	0.064689	0.051726
80	28	14	3	0.195898	0.131209
80	28	14	4	0.410674	0.214776
80	28	14	5	0.650424	0.239750
80	28	14	6	0.838409	0.187986
80	28	14	7	0.943443	0.105033
80	28	14	8	0.985399	0.041956
80	28	14	9	0.997301	0.011902
80	28	14	10	0.999657	0.002356
80	28	14	11	0.999972	0.000315
80	28	14	12	0.999999	0.000027
80	28	14	13	1.000000	0.000001
80	28	14	14	1.000000	0.000000
80	28	15	0	0.000675	0.000675
80	28	15	1	0.008139	0.007464
80	28	15	2	0.044312	0.036172
80	28	15	3	0.146197	0.101885
80	28	15	4	0.332573	0.186376
80	28	15	5	0.566874	0.234301
80	28	15	6	0.775747	0.208873
80	28	15	7	0.910023	0.134276
80	28	15	8	0.972685	0.062662
80	28	15	9	0.993875	0.021190
80	28	15	10	0.999015	0.005140
80	28	15	11	0.999891	0.000876
80	28	15	12	0.999992	0.000101
80	28	15	13	1.000000	0.000007
80	28	15	14	1.000000	0.000000
80	28	15	15	1.000000	0.000000
80	28	16	0	0.000384	0.000384
80	28	16	1	0.005039	0.004655
80	28	16	2	0.029843	0.024804
80	28	16	3	0.107011	0.077168
80	28	16	4	0.263757	0.156747
80	28	16	5	0.483968	0.220210
80	28	16	6	0.705052	0.221084
80	28	16	7	0.866642	0.161590

Table for $N = 2$, $n = 1$, through $N = 100$, $n = 50$

N	n	k	x	P(x)	p(x)
80	28	16	8	0.953404	0.086763
80	28	16	9	0.987681	0.034277
80	28	16	10	0.997591	0.009910
80	28	16	11	0.999662	0.002070
80	28	16	12	0.999967	0.000306
80	28	16	13	0.999998	0.000031
80	28	16	14	1.000000	0.000002
80	28	16	15	1.000000	0.000000
80	28	16	16	1.000000	0.000000
80	28	17	0	0.000216	0.000216
80	28	17	1	0.003075	0.002859
80	28	17	2	0.019766	0.016691
80	28	17	3	0.076867	0.057101
80	28	17	4	0.204977	0.128110
80	28	17	5	0.404830	0.199852
80	28	17	6	0.629054	0.224225
80	28	17	7	0.813620	0.184566
80	28	17	8	0.926291	0.112671
80	28	17	9	0.977505	0.051214
80	28	17	10	0.994804	0.017299
80	28	17	11	0.999112	0.004308
80	28	17	12	0.999891	0.000779
80	28	17	13	0.999991	0.000100
80	28	17	14	0.999999	0.000009
80	28	17	15	1.000000	0.000000
80	28	17	16	1.000000	0.000000
80	28	17	17	1.000000	0.000000
80	28	18	0	0.000120	0.000120
80	28	18	1	0.001850	0.001730
80	28	18	2	0.012878	0.011028
80	28	18	3	0.054208	0.041330
80	28	18	4	0.156173	0.101965
80	28	18	5	0.331868	0.175694
80	28	18	6	0.550754	0.218886
80	28	18	7	0.752098	0.201344
80	28	18	8	0.890522	0.138424
80	28	18	9	0.962060	0.071537
80	28	18	10	0.989862	0.027802
80	28	18	11	0.997949	0.008088
80	28	18	12	0.999693	0.001744
80	28	18	13	0.999967	0.000274
80	28	18	14	0.999997	0.000031
80	28	18	15	1.000000	0.000002
80	28	18	16	1.000000	0.000000
80	28	18	17	1.000000	0.000000
80	28	18	18	1.000000	0.000000
80	28	19	0	0.000066	0.000066
80	28	19	1	0.001097	0.001031
80	28	19	2	0.008253	0.007157
80	28	19	3	0.037543	0.029289

N	n	k	x	P(x)	p(x)
80	28	19	4	0.116703	0.079160
80	28	19	5	0.266691	0.149988
80	28	19	6	0.473084	0.206394
80	28	19	7	0.683901	0.210816
80	28	19	8	0.845869	0.161969
80	28	19	9	0.940137	0.094267
80	28	19	10	0.981790	0.041653
80	28	19	11	0.995732	0.013942
80	28	19	12	0.999243	0.003511
80	28	19	13	0.999901	0.000658
80	28	19	14	0.999991	0.000090
80	28	19	15	0.999999	0.000009
80	28	19	16	1.000000	0.000001
80	28	19	17	1.000000	0.000000
80	28	19	18	1.000000	0.000000
80	28	19	19	1.000000	0.000000
80	28	20	0	0.000036	0.000036
80	28	20	1	0.000640	0.000605
80	28	20	2	0.005203	0.004563
80	28	20	3	0.025539	0.020336
80	28	20	4	0.085558	0.060019
80	28	20	5	0.210138	0.124580
80	28	20	6	0.398647	0.188509
80	28	20	7	0.611325	0.212677
80	28	20	8	0.792765	0.181440
80	28	20	9	0.910775	0.118010
80	28	20	10	0.969499	0.058724
80	28	20	11	0.991846	0.022347
80	28	20	12	0.998322	0.006476
80	28	20	13	0.999739	0.001417
80	28	20	14	0.999970	0.000231
80	28	20	15	0.999997	0.000028
80	28	20	16	1.000000	0.000002
80	28	20	17	1.000000	0.000000
80	28	20	18	1.000000	0.000000
80	28	20	19	1.000000	0.000000
80	28	20	20	1.000000	0.000000
80	28	21	0	0.000019	0.000019
80	28	21	1	0.000368	0.000349
80	28	21	2	0.003226	0.002858
80	28	21	3	0.017066	0.013840
80	28	21	4	0.061550	0.044485
80	28	21	5	0.162382	0.100832
80	28	21	6	0.329527	0.167145
80	28	21	7	0.536888	0.207360
80	28	21	8	0.732285	0.195397
80	28	21	9	0.873405	0.141120
80	28	21	10	0.951882	0.078477
80	28	21	11	0.985515	0.033633
80	28	21	12	0.996595	0.011081

N	n	k	x	P(x)	p(x)
80	28	21	13	0.999385	0.002790
80	28	21	14	0.999916	0.000531
80	28	21	15	0.999991	0.000075
80	28	21	16	0.999999	0.000008
80	28	21	17	1.000000	0.000001
80	28	21	18	1.000000	0.000000
80	28	21	19	1.000000	0.000000
80	28	21	20	1.000000	0.000000
80	28	21	21	1.000000	0.000000
80	28	22	0	0.000010	0.000010
80	28	22	1	0.000208	0.000198
80	28	22	2	0.001967	0.001758
80	28	22	3	0.011201	0.009235
80	28	22	4	0.043455	0.032254
80	28	22	5	0.123075	0.079620
80	28	22	6	0.267202	0.144127
80	28	22	7	0.463081	0.195879
80	28	22	8	0.666049	0.202967
80	28	22	9	0.827960	0.161911
80	28	22	10	0.927940	0.099980
80	28	22	11	0.975824	0.047884
80	28	22	12	0.993590	0.017766
80	28	22	13	0.998675	0.005085
80	28	22	14	0.999790	0.001114
80	28	22	15	0.999975	0.000185
80	28	22	16	0.999998	0.000023
80	28	22	17	1.000000	0.000002
80	28	22	18	1.000000	0.000000
80	28	22	19	1.000000	0.000000
80	28	22	20	1.000000	0.000000
80	28	22	21	1.000000	0.000000
80	28	22	22	1.000000	0.000000
80	28	23	0	0.000005	0.000005
80	28	23	1	0.000116	0.000111
80	28	23	2	0.001178	0.001062
80	28	23	3	0.007221	0.006042
80	28	23	4	0.030109	0.022888
80	28	23	5	0.091501	0.061393
80	28	23	6	0.212533	0.121032
80	28	23	7	0.392160	0.179626
80	28	23	8	0.596060	0.203900
80	28	23	9	0.774920	0.178860
80	28	23	10	0.896911	0.121992
80	28	23	11	0.961789	0.064877
80	28	23	12	0.988689	0.026900
80	28	23	13	0.997360	0.008671
80	28	23	14	0.999521	0.002161
80	28	23	15	0.999933	0.000412
80	28	23	16	0.999993	0.000060
80	28	23	17	0.999999	0.000006

N	n	k	x	P(x)	p(x)
80	28	23	18	1.000000	0.000000
80	28	23	19	1.000000	0.000000
80	28	23	20	1.000000	0.000000
80	28	23	21	1.000000	0.000000
80	28	23	22	1.000000	0.000000
80	28	23	23	1.000000	0.000000
80	28	24	0	0.000003	0.000003
80	28	24	1	0.000064	0.000061
80	28	24	2	0.000694	0.000630
80	28	24	3	0.004571	0.003877
80	28	24	4	0.020472	0.015901
80	28	24	5	0.066729	0.046257
80	28	24	6	0.165819	0.099090
80	28	24	7	0.325981	0.160162
80	28	24	8	0.524516	0.198534
80	28	24	9	0.715300	0.190784
80	28	24	10	0.858388	0.143088
80	28	24	11	0.942439	0.084052
80	28	24	12	0.981138	0.038699
80	28	24	13	0.995078	0.013940
80	28	24	14	0.998990	0.003912
80	28	24	15	0.999839	0.000849
80	28	24	16	0.999980	0.000141
80	28	24	17	0.999998	0.000018
80	28	24	18	1.000000	0.000002
80	28	24	19	1.000000	0.000000
80	28	24	20	1.000000	0.000000
80	28	24	21	1.000000	0.000000
80	28	24	22	1.000000	0.000000
80	28	24	23	1.000000	0.000000
80	28	24	24	1.000000	0.000000
80	28	25	0	0.000001	0.000001
80	28	25	1	0.000034	0.000033
80	28	25	2	0.000401	0.000367
80	28	25	3	0.002840	0.002439
80	28	25	4	0.013657	0.010817
80	28	25	5	0.047730	0.034073
80	28	25	6	0.126891	0.079160
80	28	25	7	0.265921	0.139030
80	28	25	8	0.453611	0.187690
80	28	25	9	0.650570	0.196959
80	28	25	10	0.812395	0.161826
80	28	25	11	0.916924	0.104529
80	28	25	12	0.970081	0.053158
80	28	25	13	0.991344	0.021263
80	28	25	14	0.998012	0.006668
80	28	25	15	0.999642	0.001630
80	28	25	16	0.999950	0.000308
80	28	25	17	0.999995	0.000044
80	28	25	18	0.999999	0.000005

N	n	k	x	P(x)	p(x)
80	28	25	19	1.000000	0.000000
80	28	25	20	1.000000	0.000000
80	28	25	21	1.000000	0.000000
80	28	25	22	1.000000	0.000000
80	28	25	23	1.000000	0.000000
80	28	25	24	1.000000	0.000000
80	28	25	25	1.000000	0.000000
80	28	26	0	0.000001	0.000001
80	28	26	1	0.000018	0.000017
80	28	26	2	0.000228	0.000210
80	28	26	3	0.001731	0.001504
80	28	26	4	0.008937	0.007205
80	28	26	5	0.033481	0.024545
80	28	26	6	0.095227	0.061745
80	28	26	7	0.212836	0.117610
80	28	26	8	0.385360	0.172523
80	28	26	9	0.582529	0.197169
80	28	26	10	0.759434	0.176905
80	28	26	11	0.884615	0.125181
80	28	26	12	0.954617	0.070002
80	28	26	13	0.985545	0.030928
80	28	26	14	0.996315	0.010770
80	28	26	15	0.999257	0.002942
80	28	26	16	0.999883	0.000626
80	28	26	17	0.999986	0.000103
80	28	26	18	0.999999	0.000013
80	28	26	19	1.000000	0.000001
80	28	26	20	1.000000	0.000000
80	28	26	21	1.000000	0.000000
80	28	26	22	1.000000	0.000000
80	28	26	23	1.000000	0.000000
80	28	26	24	1.000000	0.000000
80	28	26	25	1.000000	0.000000
80	28	26	26	1.000000	0.000000
80	28	27	0	0.000000	0.000000
80	28	27	1	0.000009	0.000009
80	28	27	2	0.000127	0.000117
80	28	27	3	0.001035	0.000908
80	28	27	4	0.005734	0.004699
80	28	27	5	0.023027	0.017293
80	28	27	6	0.070071	0.047044
80	28	27	7	0.167099	0.097028
80	28	27	8	0.321462	0.154363
80	28	27	9	0.513155	0.191693
80	28	27	10	0.700466	0.187311
80	28	27	11	0.845206	0.144740
80	28	27	12	0.933876	0.088670
80	28	27	13	0.976954	0.043078
80	28	27	14	0.993523	0.016569
80	28	27	15	0.998549	0.005026

N	n	k	x	P(x)	p(x)
80	28	27	16	0.999744	0.001195
80	28	27	17	0.999965	0.000221
80	28	27	18	0.999996	0.000031
80	28	27	19	1.000000	0.000003
80	28	27	20	1.000000	0.000000
80	28	27	21	1.000000	0.000000
80	28	27	22	1.000000	0.000000
80	28	27	23	1.000000	0.000000
80	28	27	24	1.000000	0.000000
80	28	27	25	1.000000	0.000000
80	28	27	26	1.000000	0.000000
80	28	27	27	1.000000	0.000000
80	28	28	0	0.000000	0.000000
80	28	28	1	0.000005	0.000005
80	28	28	2	0.000069	0.000064
80	28	28	3	0.000607	0.000538
80	28	28	4	0.003606	0.003000
80	28	28	5	0.015523	0.011916
80	28	28	6	0.050543	0.035021
80	28	28	7	0.128654	0.078111
80	28	28	8	0.263212	0.134558
80	28	28	9	0.444435	0.181223
80	28	28	10	0.636851	0.192416
80	28	28	11	0.798780	0.161929
80	28	28	12	0.907108	0.108328
80	28	28	13	0.964762	0.057655
80	28	28	14	0.989146	0.024384
80	28	28	15	0.997316	0.008170
80	28	28	16	0.999473	0.002157
80	28	28	17	0.999919	0.000446
80	28	28	18	0.999990	0.000071
80	28	28	19	0.999999	0.000009
80	28	28	20	1.000000	0.000001
80	28	28	21	1.000000	0.000000
80	28	28	22	1.000000	0.000000
80	28	28	23	1.000000	0.000000
80	28	28	24	1.000000	0.000000
80	28	28	25	1.000000	0.000000
80	28	28	26	1.000000	0.000000
80	28	28	27	1.000000	0.000000
80	28	28	28	1.000000	0.000000
80	29	1	0	0.637500	0.637500
80	29	1	1	1.000000	0.362500
80	29	2	0	0.403481	0.403481
80	29	2	1	0.871519	0.468038
80	29	2	2	1.000000	0.128481
80	29	3	0	0.253469	0.253469
80	29	3	1	0.703505	0.450037
80	29	3	2	0.955526	0.252020
80	29	3	3	1.000000	0.044474

N	n	k	x	P(x)	p(x)
80	29	4	0	0.158007	0.158007
80	29	4	1	0.539856	0.381849
80	29	4	2	0.867155	0.327299
80	29	4	3	0.984983	0.117828
80	29	4	4	1.000000	0.015017
80	29	5	0	0.097715	0.097715
80	29	5	1	0.399174	0.301460
80	29	5	2	0.750878	0.351703
80	29	5	3	0.944673	0.193796
80	29	5	4	0.995060	0.050387
80	29	5	5	1.000000	0.004940
80	29	6	0	0.059932	0.059932
80	29	6	1	0.286629	0.226698
80	29	6	2	0.624264	0.337635
80	29	6	3	0.877491	0.253226
80	29	6	4	0.978264	0.100774
80	29	6	5	0.998419	0.020155
80	29	6	6	1.000000	0.001581
80	29	7	0	0.036445	0.036445
80	29	7	1	0.200852	0.164407
80	29	7	2	0.501073	0.300221
80	29	7	3	0.788519	0.287446
80	29	7	4	0.944219	0.155700
80	29	7	5	0.991882	0.047663
80	29	7	6	0.999509	0.007626
80	29	7	7	1.000000	0.000491
80	29	8	0	0.021967	0.021967
80	29	8	1	0.137792	0.115825
80	29	8	2	0.390032	0.252241
80	29	8	3	0.686141	0.296109
80	29	8	4	0.890897	0.204756
80	29	8	5	0.976212	0.085315
80	29	8	6	0.997106	0.020893
80	29	8	7	0.999852	0.002746
80	29	8	8	1.000000	0.000148
80	29	9	0	0.013119	0.013119
80	29	9	1	0.092749	0.079630
80	29	9	2	0.295442	0.202694
80	29	9	3	0.579213	0.283771
80	29	9	4	0.819801	0.240588
80	29	9	5	0.947774	0.127973
80	29	9	6	0.990431	0.042658
80	29	9	7	0.999013	0.008581
80	29	9	8	0.999957	0.000944
80	29	9	9	1.000000	0.000043
80	29	10	0	0.007761	0.007761
80	29	10	1	0.061345	0.053585
80	29	10	2	0.218362	0.157016
80	29	10	3	0.475297	0.256935
80	29	10	4	0.735087	0.259790

N	n	k	x	P(x)	p(x)
80	29	10	5	0.904516	0.169428
80	29	10	6	0.976613	0.072097
80	29	10	7	0.996354	0.019741
80	29	10	8	0.999678	0.003324
80	29	10	9	0.999988	0.000310
80	29	10	10	1.000000	0.000012
80	29	11	0	0.004545	0.004545
80	29	11	1	0.039911	0.035366
80	29	11	2	0.157798	0.117887
80	29	11	3	0.379864	0.222066
80	29	11	4	0.642305	0.262441
80	29	11	5	0.846426	0.204121
80	29	11	6	0.952924	0.106498
80	29	11	7	0.990150	0.037226
80	29	11	8	0.998680	0.008531
80	29	11	9	0.999899	0.001219
80	29	11	10	0.999997	0.000097
80	29	11	11	1.000000	0.000003
80	29	12	0	0.002635	0.002635
80	29	12	1	0.025560	0.022925
80	29	12	2	0.111669	0.086109
80	29	12	3	0.296187	0.184518
80	29	12	4	0.547218	0.251031
80	29	12	5	0.775427	0.228210
80	29	12	6	0.917424	0.141997
80	29	12	7	0.978280	0.060856
80	29	12	8	0.996084	0.017804
80	29	12	9	0.999546	0.003462
80	29	12	10	0.999970	0.000424
80	29	12	11	0.999999	0.000029
80	29	12	12	1.000000	0.000001
80	29	13	0	0.001511	0.001511
80	29	13	1	0.016120	0.014609
80	29	13	2	0.077478	0.061358
80	29	13	3	0.225636	0.148157
80	29	13	4	0.454927	0.229291
80	29	13	5	0.694883	0.239956
80	29	13	6	0.869396	0.174513
80	29	13	7	0.958592	0.089196
80	29	13	8	0.990586	0.031994
80	29	13	9	0.998528	0.007942
80	29	13	10	0.999851	0.001324
80	29	13	11	0.999991	0.000140
80	29	13	12	1.000000	0.000008
80	29	13	13	1.000000	0.000000
80	29	14	0	0.000857	0.000857
80	29	14	1	0.010015	0.009158
80	29	14	2	0.052752	0.042737
80	29	14	3	0.168142	0.115390
80	29	14	4	0.369371	0.201229

Table for $N = 2$, $n = 1$, through $N = 100$, $n = 50$

N	n	k	x	P(x)	p(x)
80	29	14	5	0.608928	0.239558
80	29	14	6	0.809489	0.200560
80	29	14	7	0.929304	0.119815
80	29	14	8	0.980558	0.051254
80	29	14	9	0.996157	0.015599
80	29	14	10	0.999476	0.003319
80	29	14	11	0.999954	0.000478
80	29	14	12	0.999998	0.000044
80	29	14	13	1.000000	0.000002
80	29	14	14	1.000000	0.000000
80	29	15	0	0.000481	0.000481
80	29	15	1	0.006130	0.005649
80	29	15	2	0.035269	0.029139
80	29	15	3	0.122685	0.087417
80	29	15	4	0.293148	0.170462
80	29	15	5	0.521816	0.228669
80	29	15	6	0.739596	0.217780
80	29	15	7	0.889365	0.149769
80	29	15	8	0.964250	0.074884
80	29	15	9	0.991430	0.027180
80	29	15	10	0.998521	0.007091
80	29	15	11	0.999823	0.001303
80	29	15	12	0.999986	0.000163
80	29	15	13	0.999999	0.000013
80	29	15	14	1.000000	0.000001
80	29	15	15	1.000000	0.000000
80	29	16	0	0.000266	0.000266
80	29	16	1	0.003696	0.003430
80	29	16	2	0.023165	0.019469
80	29	16	3	0.087719	0.064554
80	29	16	4	0.227585	0.139866
80	29	16	5	0.437385	0.209800
80	29	16	6	0.662536	0.225151
80	29	16	7	0.838674	0.176138
80	29	16	8	0.940056	0.101382
80	29	16	9	0.983067	0.043011
80	29	16	10	0.996448	0.013381
80	29	16	11	0.999463	0.003015
80	29	16	12	0.999944	0.000481
80	29	16	13	0.999996	0.000052
80	29	16	14	1.000000	0.000004
80	29	16	15	1.000000	0.000000
80	29	16	16	1.000000	0.000000
80	29	17	0	0.000146	0.000146
80	29	17	1	0.002196	0.002050
80	29	17	2	0.014952	0.012756
80	29	17	3	0.061494	0.046542
80	29	17	4	0.172950	0.111456
80	29	17	5	0.358710	0.185760
80	29	17	6	0.581622	0.222912
80	29	17	7	0.778127	0.196505
80	29	17	8	0.906790	0.128664
80	29	17	9	0.969626	0.062836
80	29	17	10	0.992475	0.022849
80	29	17	11	0.998615	0.006139
80	29	17	12	0.999816	0.001201
80	29	17	13	0.999983	0.000167
80	29	17	14	0.999999	0.000016
80	29	17	15	1.000000	0.000001
80	29	17	16	1.000000	0.000000
80	29	17	17	1.000000	0.000000
80	29	18	0	0.000079	0.000079
80	29	18	1	0.001284	0.001206
80	29	18	2	0.009485	0.008200
80	29	18	3	0.042286	0.032801
80	29	18	4	0.128721	0.086435
80	29	18	5	0.287944	0.159223
80	29	18	6	0.500242	0.212297
80	29	18	7	0.709506	0.209265
80	29	18	8	0.863902	0.154396
80	29	18	9	0.949678	0.085776
80	29	18	10	0.985584	0.035906
80	29	18	11	0.996861	0.011276
80	29	18	12	0.999492	0.002631
80	29	18	13	0.999941	0.000449
80	29	18	14	0.999995	0.000055
80	29	18	15	1.000000	0.000005
80	29	18	16	1.000000	0.000000
80	29	18	17	1.000000	0.000000
80	29	18	18	1.000000	0.000000
80	29	19	0	0.000042	0.000042
80	29	19	1	0.000740	0.000698
80	29	19	2	0.005914	0.005174
80	29	19	3	0.028531	0.022617
80	29	19	4	0.093868	0.065338
80	29	19	5	0.226309	0.132441
80	29	19	6	0.421486	0.195177
80	29	19	7	0.635251	0.213765
80	29	19	8	0.811607	0.176356
80	29	19	9	0.922009	0.110402
80	29	19	10	0.974581	0.052572
80	29	19	11	0.993587	0.019006
80	29	19	12	0.998770	0.005183
80	29	19	13	0.999825	0.001054
80	29	19	14	0.999982	0.000157
80	29	19	15	0.999999	0.000017
80	29	19	16	1.000000	0.000001
80	29	19	17	1.000000	0.000000
80	29	19	18	1.000000	0.000000
80	29	19	19	1.000000	0.000000
80	29	20	0	0.000022	0.000022
80	29	20	1	0.000419	0.000398
80	29	20	2	0.003624	0.003204
80	29	20	3	0.018891	0.015267
80	29	20	4	0.067090	0.048200
80	29	20	5	0.174201	0.107111
80	29	20	6	0.347895	0.173693
80	29	20	7	0.558155	0.210260
80	29	20	8	0.750894	0.192739
80	29	20	9	0.885811	0.134917
80	29	20	10	0.958206	0.072395
80	29	20	11	0.987978	0.029773
80	29	20	12	0.997326	0.009347
80	29	20	13	0.999548	0.002222
80	29	20	14	0.999943	0.000395
80	29	20	15	0.999995	0.000052
80	29	20	16	1.000000	0.000005
80	29	20	17	1.000000	0.000000
80	29	20	18	1.000000	0.000000
80	29	20	19	1.000000	0.000000
80	29	20	20	1.000000	0.000000
80	29	21	0	0.000011	0.000011
80	29	21	1	0.000234	0.000223
80	29	21	2	0.002182	0.001948
80	29	21	3	0.012275	0.010093
80	29	21	4	0.047007	0.034732
80	29	21	5	0.131357	0.084350
80	29	21	6	0.281312	0.149955
80	29	21	7	0.481060	0.199747
80	29	21	8	0.683436	0.202376
80	29	21	9	0.840839	0.157403
80	29	21	10	0.935281	0.094442
80	29	21	11	0.979047	0.043766
80	29	21	12	0.994677	0.015631
80	29	21	13	0.998956	0.004278
80	29	21	14	0.999844	0.000889
80	29	21	15	0.999983	0.000138
80	29	21	16	0.999999	0.000016
80	29	21	17	1.000000	0.000001
80	29	21	18	1.000000	0.000000
80	29	21	19	1.000000	0.000000
80	29	21	20	1.000000	0.000000
80	29	21	21	1.000000	0.000000
80	29	22	0	0.000006	0.000006
80	29	22	1	0.000128	0.000123
80	29	22	2	0.001290	0.001162
80	29	22	3	0.007827	0.006537
80	29	22	4	0.032290	0.024463
80	29	22	5	0.097045	0.064755
80	29	22	6	0.222855	0.125810
80	29	22	7	0.406578	0.183723
80	29	22	8	0.611404	0.204826
80	29	22	9	0.787482	0.176078
80	29	22	10	0.904867	0.117386
80	29	22	11	0.965694	0.060827
80	29	22	12	0.990174	0.024479
80	29	22	13	0.997795	0.007622
80	29	22	14	0.999618	0.001823
80	29	22	15	0.999950	0.000331
80	29	22	16	0.999995	0.000045
80	29	22	17	1.000000	0.000005
80	29	22	18	1.000000	0.000000
80	29	22	19	1.000000	0.000000
80	29	22	20	1.000000	0.000000
80	29	22	21	1.000000	0.000000
80	29	22	22	1.000000	0.000000
80	29	23	0	0.000003	0.000003
80	29	23	1	0.000069	0.000066
80	29	23	2	0.000749	0.000680
80	29	23	3	0.004897	0.004147
80	29	23	4	0.021746	0.016849
80	29	23	5	0.070250	0.048504
80	29	23	6	0.172965	0.102715
80	29	23	7	0.336890	0.163925
80	29	23	8	0.537242	0.200353
80	29	23	9	0.726765	0.189523
80	29	23	10	0.866413	0.139648
80	29	23	11	0.946817	0.080404
80	29	23	12	0.982999	0.036182
80	29	23	13	0.995693	0.012694
80	29	23	14	0.999147	0.003454
80	29	23	15	0.999870	0.000723
80	29	23	16	0.999985	0.000115
80	29	23	17	0.999999	0.000014
80	29	23	18	1.000000	0.000001
80	29	23	19	1.000000	0.000000
80	29	23	20	1.000000	0.000000
80	29	23	21	1.000000	0.000000
80	29	23	22	1.000000	0.000000
80	29	23	23	1.000000	0.000000
80	29	24	0	0.000001	0.000001
80	29	24	1	0.000037	0.000035
80	29	24	2	0.000427	0.000391
80	29	24	3	0.003005	0.002578
80	29	24	4	0.014356	0.011351
80	29	24	5	0.049827	0.035471
80	29	24	6	0.131519	0.081691
80	29	24	7	0.273621	0.142102
80	29	24	8	0.463428	0.189808
80	29	24	9	0.660266	0.196838

N	n	k	x	P(x)	p(x)
80	29	24	10	0.819864	0.159598
80	29	24	11	0.921426	0.101562
80	29	24	12	0.972208	0.050781
80	29	24	13	0.992130	0.019922
80	29	24	14	0.998238	0.006108
80	29	24	15	0.999692	0.001454
80	29	24	16	0.999959	0.000266
80	29	24	17	0.999996	0.000037
80	29	24	18	1.000000	0.000004
80	29	24	19	1.000000	0.000000
80	29	24	20	1.000000	0.000000
80	29	24	21	1.000000	0.000000
80	29	24	22	1.000000	0.000000
80	29	24	23	1.000000	0.000000
80	29	24	24	1.000000	0.000000
80	29	25	0	0.000001	0.000001
80	29	25	1	0.000019	0.000018
80	29	25	2	0.000239	0.000220
80	29	25	3	0.001808	0.001569
80	29	25	4	0.009288	0.007480
80	29	25	5	0.034625	0.025337
80	29	25	6	0.097967	0.063342
80	29	25	7	0.217795	0.119828
80	29	25	8	0.392250	0.174456
80	29	25	9	0.589967	0.197716
80	29	25	10	0.765715	0.175748
80	29	25	11	0.888781	0.123067
80	29	25	12	0.956792	0.068011
80	29	25	13	0.986438	0.029646
80	29	25	14	0.996602	0.010164
80	29	25	15	0.999329	0.002727
80	29	25	16	0.999897	0.000568
80	29	25	17	0.999988	0.000091
80	29	25	18	0.999999	0.000011
80	29	25	19	1.000000	0.000001
80	29	25	20	1.000000	0.000000
80	29	25	21	1.000000	0.000000
80	29	25	22	1.000000	0.000000
80	29	25	23	1.000000	0.000000
80	29	25	24	1.000000	0.000000
80	29	25	25	1.000000	0.000000
80	29	26	0	0.000000	0.000000
80	29	26	1	0.000010	0.000009
80	29	26	2	0.000131	0.000121
80	29	26	3	0.001066	0.000935
80	29	26	4	0.005888	0.004822
80	29	26	5	0.023569	0.017681
80	29	26	6	0.071478	0.047909
80	29	26	7	0.169864	0.098385
80	29	26	8	0.325640	0.155777

N	n	k	x	P(x)	p(x)
80	29	26	9	0.518070	0.192430
80	29	26	10	0.705002	0.186932
80	29	26	11	0.848505	0.143503
80	29	26	12	0.935770	0.087265
80	29	26	13	0.977813	0.042043
80	29	26	14	0.993830	0.016016
80	29	26	15	0.998635	0.004805
80	29	26	16	0.999763	0.001128
80	29	26	17	0.999968	0.000205
80	29	26	18	0.999997	0.000029
80	29	26	19	1.000000	0.000003
80	29	26	20	1.000000	0.000000
80	29	26	21	1.000000	0.000000
80	29	26	22	1.000000	0.000000
80	29	26	23	1.000000	0.000000
80	29	26	24	1.000000	0.000000
80	29	26	25	1.000000	0.000000
80	29	26	26	1.000000	0.000000
80	29	27	0	0.000000	0.000000
80	29	27	1	0.000005	0.000005
80	29	27	2	0.000070	0.000065
80	29	27	3	0.000616	0.000546
80	29	27	4	0.003656	0.003040
80	29	27	5	0.015711	0.012055
80	29	27	6	0.051073	0.035362
80	29	27	7	0.129781	0.078708
80	29	27	8	0.265060	0.135280
80	29	27	9	0.446800	0.181739
80	29	27	10	0.639229	0.192430
80	29	27	11	0.800671	0.161441
80	29	27	12	0.908298	0.107627
80	29	27	13	0.965356	0.057058
80	29	27	14	0.989381	0.024024
80	29	27	15	0.997389	0.008008
80	29	27	16	0.999491	0.002102
80	29	27	17	0.999922	0.000431
80	29	27	18	0.999991	0.000068
80	29	27	19	0.999999	0.000008
80	29	27	20	1.000000	0.000001
80	29	27	21	1.000000	0.000000
80	29	27	22	1.000000	0.000000
80	29	27	23	1.000000	0.000000
80	29	27	24	1.000000	0.000000
80	29	27	25	1.000000	0.000000
80	29	27	26	1.000000	0.000000
80	29	27	27	1.000000	0.000000
80	29	28	0	0.000000	0.000000
80	29	28	1	0.000002	0.000002
80	29	28	2	0.000037	0.000035
80	29	28	3	0.000348	0.000311

N	n	k	x	P(x)	p(x)
80	29	28	4	0.002222	0.001874
80	29	28	5	0.010252	0.008030
80	29	28	6	0.035727	0.025475
80	29	28	7	0.097109	0.061382
80	29	28	8	0.211459	0.114350
80	29	28	9	0.378219	0.166760
80	29	28	10	0.570245	0.192026
80	29	28	11	0.745842	0.175596
80	29	28	12	0.873776	0.127934
80	29	28	13	0.948131	0.074355
80	29	28	14	0.982581	0.034450
80	29	28	15	0.995274	0.012692
80	29	28	16	0.998975	0.003702
80	29	28	17	0.999825	0.000849
80	29	28	18	0.999977	0.000152
80	29	28	19	0.999998	0.000021
80	29	28	20	1.000000	0.000002
80	29	28	21	1.000000	0.000000
80	29	28	22	1.000000	0.000000
80	29	28	23	1.000000	0.000000
80	29	28	24	1.000000	0.000000
80	29	28	25	1.000000	0.000000
80	29	28	26	1.000000	0.000000
80	29	28	27	1.000000	0.000000
80	29	28	28	1.000000	0.000000
80	29	29	0	0.000000	0.000000
80	29	29	1	0.000001	0.000001
80	29	29	2	0.000019	0.000018
80	29	29	3	0.000193	0.000174
80	29	29	4	0.001321	0.001129
80	29	29	5	0.006546	0.005225
80	29	29	6	0.024459	0.017913
80	29	29	7	0.071140	0.046681
80	29	29	8	0.165279	0.094139
80	29	29	9	0.314080	0.148801
80	29	29	10	0.500082	0.186001
80	29	29	11	0.685058	0.184976
80	29	29	12	0.831951	0.146893
80	29	29	13	0.925253	0.093301
80	29	29	14	0.972644	0.047391
80	29	29	15	0.991856	0.019213
80	29	29	16	0.998050	0.006194
80	29	29	17	0.999629	0.001579
80	29	29	18	0.999944	0.000316
80	29	29	19	0.999993	0.000049
80	29	29	20	0.999999	0.000006
80	29	29	21	1.000000	0.000001
80	29	29	22	1.000000	0.000000
80	29	29	23	1.000000	0.000000
80	29	29	24	1.000000	0.000000

N	n	k	x	P(x)	p(x)
80	29	29	25	1.000000	0.000000
80	29	29	26	1.000000	0.000000
80	29	29	27	1.000000	0.000000
80	29	29	28	1.000000	0.000000
80	29	29	29	1.000000	0.000000
80	30	1	0	0.625000	0.625000
80	30	1	1	1.000000	0.375000
80	30	2	0	0.387658	0.387658
80	30	2	1	0.862342	0.474684
80	30	2	2	1.000000	0.137658
80	30	3	0	0.238559	0.238559
80	30	3	1	0.685857	0.447298
80	30	3	2	0.950584	0.264727
80	30	3	3	1.000000	0.049416
80	30	4	0	0.145614	0.145614
80	30	4	1	0.517394	0.371780
80	30	4	2	0.854320	0.336926
80	30	4	3	0.982672	0.128353
80	30	4	4	1.000000	0.017328
80	30	5	0	0.088135	0.088135
80	30	5	1	0.375530	0.287396
80	30	5	2	0.730189	0.354659
80	30	5	3	0.937073	0.206884
80	30	5	4	0.994072	0.056999
80	30	5	5	1.000000	0.005928
80	30	6	0	0.052881	0.052881
80	30	6	1	0.264404	0.211523
80	30	6	2	0.597783	0.333379
80	30	6	3	0.862595	0.264812
80	30	6	4	0.974312	0.111717
80	30	6	5	0.998024	0.023711
80	30	6	6	1.000000	0.001976
80	30	7	0	0.031443	0.031443
80	30	7	1	0.181510	0.150067
80	30	7	2	0.471640	0.290130
80	30	7	3	0.765975	0.294335
80	30	7	4	0.935060	0.169086
80	30	7	5	0.990013	0.054953
80	30	7	6	0.999359	0.009346
80	30	7	7	1.000000	0.000641
80	30	8	0	0.018521	0.018521
80	30	8	1	0.121894	0.103373
80	30	8	2	0.360357	0.238463
80	30	8	3	0.657111	0.296754
80	30	8	4	0.874838	0.217727
80	30	8	5	0.971194	0.096356
80	30	8	6	0.996286	0.025093
80	30	8	7	0.999798	0.003512
80	30	8	8	1.000000	0.000202
80	30	9	0	0.010804	0.010804

Table for $N = 2$, $n = 1$, through $N = 100$, $n = 50$

N	n	k	x	P(x)	p(x)
80	30	9	1	0.080258	0.069454
80	30	9	2	0.267621	0.187364
80	30	9	3	0.545828	0.278207
80	30	9	4	0.796214	0.250386
80	30	9	5	0.937737	0.141523
80	30	9	6	0.987922	0.050185
80	30	9	7	0.998676	0.010754
80	30	9	8	0.999938	0.001262
80	30	9	9	1.000000	0.000062
80	30	10	0	0.006239	0.006239
80	30	10	1	0.051889	0.045650
80	30	10	2	0.193732	0.141842
80	30	10	3	0.440031	0.246300
80	30	10	4	0.704524	0.264492
80	30	10	5	0.887905	0.183381
80	30	10	6	0.970958	0.083053
80	30	10	7	0.995193	0.024234
80	30	10	8	0.999547	0.004355
80	30	10	9	0.999982	0.000434
80	30	10	10	1.000000	0.000018
80	30	11	0	0.003565	0.003565
80	30	11	1	0.032977	0.029412
80	30	11	2	0.136995	0.104018
80	30	11	3	0.345030	0.208035
80	30	11	4	0.606284	0.261254
80	30	11	5	0.822412	0.216128
80	30	11	6	0.942483	0.120071
80	30	11	7	0.987230	0.044747
80	30	11	8	0.998179	0.010949
80	30	11	9	0.999851	0.001673
80	30	11	10	0.999995	0.000143
80	30	11	11	1.000000	0.000005
80	30	12	0	0.002015	0.002015
80	30	12	1	0.020615	0.018600
80	30	12	2	0.094785	0.074169
80	30	12	3	0.263625	0.168840
80	30	12	4	0.507840	0.244215
80	30	12	5	0.744104	0.236264
80	30	12	6	0.900719	0.156615
80	30	12	7	0.972314	0.071595
80	30	12	8	0.994688	0.022373
80	30	12	9	0.999342	0.004655
80	30	12	10	0.999953	0.000611
80	30	12	11	0.999998	0.000045
80	30	12	12	1.000000	0.000001
80	30	13	0	0.001126	0.001126
80	30	13	1	0.012683	0.011557
80	30	13	2	0.064244	0.051561
80	30	13	3	0.196585	0.132341
80	30	13	4	0.414464	0.217878

N	n	k	x	P(x)	p(x)
80	30	13	5	0.657243	0.242779
80	30	13	6	0.845443	0.188201
80	30	13	7	0.948098	0.102655
80	30	13	8	0.987449	0.039351
80	30	13	9	0.997905	0.010456
80	30	13	10	0.999773	0.001869
80	30	13	11	0.999986	0.000212
80	30	13	12	0.999999	0.000014
80	30	13	13	1.000000	0.000000
80	30	14	0	0.000622	0.000622
80	30	14	1	0.007681	0.007059
80	30	14	2	0.042696	0.035016
80	30	14	3	0.143254	0.100558
80	30	14	4	0.329914	0.186660
80	30	14	5	0.566654	0.236740
80	30	14	6	0.778028	0.211375
80	30	14	7	0.912858	0.134830
80	30	14	8	0.974528	0.061670
80	30	14	9	0.994628	0.020100
80	30	14	10	0.999216	0.004588
80	30	14	11	0.999926	0.000710
80	30	14	12	0.999996	0.000070
80	30	14	13	1.000000	0.000004
80	30	14	14	1.000000	0.000000
80	30	15	0	0.000339	0.000339
80	30	15	1	0.004579	0.004240
80	30	15	2	0.027841	0.023262
80	30	15	3	0.102117	0.074276
80	30	15	4	0.256381	0.154264
80	30	15	5	0.476979	0.220598
80	30	15	6	0.701165	0.224185
80	30	15	7	0.865872	0.164707
80	30	15	8	0.953971	0.088099
80	30	15	9	0.988232	0.034261
80	30	15	10	0.997825	0.009593
80	30	15	11	0.999721	0.001896
80	30	15	12	0.999977	0.000255
80	30	15	13	0.999999	0.000022
80	30	15	14	1.000000	0.000001
80	30	15	15	1.000000	0.000000
80	30	16	0	0.000183	0.000183
80	30	16	1	0.002687	0.002505
80	30	16	2	0.017821	0.015133
80	30	16	3	0.071264	0.053443
80	30	16	4	0.194675	0.123412
80	30	16	5	0.392134	0.197459
80	30	16	6	0.618389	0.226255
80	30	16	7	0.807591	0.189202
80	30	16	8	0.924153	0.116562
80	30	16	9	0.977163	0.053010

N	n	k	x	P(x)	p(x)
80	30	16	10	0.994874	0.017710
80	30	16	11	0.999167	0.004293
80	30	16	12	0.999906	0.000739
80	30	16	13	0.999993	0.000087
80	30	16	14	1.000000	0.000007
80	30	16	15	1.000000	0.000000
80	30	16	16	1.000000	0.000000
80	30	17	0	0.000097	0.000097
80	30	17	1	0.001552	0.001455
80	30	17	2	0.011200	0.009647
80	30	17	3	0.048717	0.037518
80	30	17	4	0.144539	0.095822
80	30	17	5	0.315002	0.170462
80	30	17	6	0.533543	0.218541
80	30	17	7	0.739596	0.206053
80	30	17	8	0.884085	0.144489
80	30	17	9	0.959769	0.075684
80	30	17	10	0.989339	0.029570
80	30	17	11	0.997892	0.008553
80	30	17	12	0.999698	0.001806
80	30	17	13	0.999970	0.000272
80	30	17	14	0.999998	0.000028
80	30	17	15	1.000000	0.000002
80	30	17	16	1.000000	0.000000
80	30	17	17	1.000000	0.000000
80	30	18	0	0.000051	0.000051
80	30	18	1	0.000882	0.000832
80	30	18	2	0.006912	0.006030
80	30	18	3	0.032638	0.025726
80	30	18	4	0.104994	0.072355
80	30	18	5	0.247358	0.142364
80	30	18	6	0.450289	0.202931
80	30	18	7	0.664371	0.214081
80	30	18	8	0.833629	0.169258
80	30	18	9	0.934541	0.100913
80	30	18	10	0.979952	0.045411
80	30	18	11	0.995313	0.015361
80	30	18	12	0.999182	0.003869
80	30	18	13	0.999897	0.000714
80	30	18	14	0.999991	0.000094
80	30	18	15	0.999999	0.000009
80	30	18	16	1.000000	0.000000
80	30	18	17	1.000000	0.000000
80	30	18	18	1.000000	0.000000
80	30	19	0	0.000026	0.000026
80	30	19	1	0.000493	0.000467
80	30	19	2	0.004189	0.003696
80	30	19	3	0.021435	0.017246
80	30	19	4	0.074651	0.053216
80	30	19	5	0.189953	0.115302

N	n	k	x	P(x)	p(x)
80	30	19	6	0.371735	0.181782
80	30	19	7	0.584953	0.213218
80	30	19	8	0.773569	0.188616
80	30	19	9	0.900361	0.126792
80	30	19	10	0.965303	0.064942
80	30	19	11	0.990605	0.025302
80	30	19	12	0.998059	0.007453
80	30	19	13	0.999701	0.001642
80	30	19	14	0.999966	0.000266
80	30	19	15	0.999997	0.000031
80	30	19	16	1.000000	0.000002
80	30	19	17	1.000000	0.000000
80	30	19	18	1.000000	0.000000
80	30	19	19	1.000000	0.000000
80	30	20	0	0.000013	0.000013
80	30	20	1	0.000271	0.000258
80	30	20	2	0.002493	0.002221
80	30	20	3	0.013802	0.011309
80	30	20	4	0.051969	0.038167
80	30	20	5	0.142698	0.090729
80	30	20	6	0.300214	0.157516
80	30	20	7	0.504560	0.204345
80	30	20	8	0.705544	0.200984
80	30	20	9	0.856711	0.151168
80	30	20	10	0.944011	0.087299
80	30	20	11	0.982724	0.038714
80	30	20	12	0.995859	0.013135
80	30	20	13	0.999243	0.003384
80	30	20	14	0.999897	0.000654
80	30	20	15	0.999990	0.000093
80	30	20	16	0.999999	0.000009
80	30	20	17	1.000000	0.000001
80	30	20	18	1.000000	0.000000
80	30	20	19	1.000000	0.000000
80	30	20	20	1.000000	0.000000
80	30	21	0	0.000007	0.000007
80	30	21	1	0.000147	0.000140
80	30	21	2	0.001456	0.001309
80	30	21	3	0.008713	0.007257
80	30	21	4	0.035430	0.026717
80	30	21	5	0.104894	0.069465
80	30	21	6	0.237208	0.132314
80	30	21	7	0.426227	0.189019
80	30	21	8	0.631850	0.205622
80	30	21	9	0.803803	0.171953
80	30	21	10	0.914911	0.111108
80	30	21	11	0.970465	0.055554
80	30	21	12	0.991919	0.021454
80	30	21	13	0.998284	0.006365
80	30	21	14	0.999722	0.001438

N	n	k	x	P(x)	p(x)
80	30	21	15	0.999967	0.000244
80	30	21	16	0.999997	0.000031
80	30	21	17	1.000000	0.000003
80	30	21	18	1.000000	0.000000
80	30	21	19	1.000000	0.000000
80	30	21	20	1.000000	0.000000
80	30	21	21	1.000000	0.000000
80	30	22	0	0.000003	0.000003
80	30	22	1	0.000078	0.000075
80	30	22	2	0.000835	0.000757
80	30	22	3	0.005392	0.004557
80	30	22	4	0.023656	0.018264
80	30	22	5	0.075460	0.051804
80	30	22	6	0.183385	0.107925
80	30	22	7	0.352542	0.169156
80	30	22	8	0.555177	0.202635
80	30	22	9	0.742599	0.187422
80	30	22	10	0.877247	0.134648
80	30	22	11	0.952575	0.075328
80	30	22	12	0.985374	0.032799
80	30	22	13	0.996450	0.011077
80	30	22	14	0.999332	0.002882
80	30	22	15	0.999904	0.000572
80	30	22	16	0.999990	0.000085
80	30	22	17	0.999999	0.000009
80	30	22	18	1.000000	0.000001
80	30	22	19	1.000000	0.000000
80	30	22	20	1.000000	0.000000
80	30	22	21	1.000000	0.000000
80	30	22	22	1.000000	0.000000
80	30	23	0	0.000002	0.000002
80	30	23	1	0.000041	0.000039
80	30	23	2	0.000469	0.000429
80	30	23	3	0.003270	0.002801
80	30	23	4	0.015469	0.012198
80	30	23	5	0.053131	0.037662
80	30	23	6	0.138727	0.085596
80	30	23	7	0.285462	0.146736
80	30	23	8	0.478315	0.192853
80	30	23	9	0.674739	0.196424
80	30	23	10	0.830817	0.156078
80	30	23	11	0.927899	0.097082
80	30	23	12	0.975195	0.047296
80	30	23	13	0.993204	0.018009
80	30	23	14	0.998537	0.005334
80	30	23	15	0.999756	0.001219
80	30	23	16	0.999969	0.000213
80	30	23	17	0.999997	0.000028
80	30	23	18	1.000000	0.000003
80	30	23	19	1.000000	0.000000

N	n	k	x	P(x)	p(x)
80	30	23	20	1.000000	0.000000
80	30	23	21	1.000000	0.000000
80	30	23	22	1.000000	0.000000
80	30	23	23	1.000000	0.000000
80	30	24	0	0.000001	0.000001
80	30	24	1	0.000021	0.000020
80	30	24	2	0.000259	0.000238
80	30	24	3	0.001944	0.001685
80	30	24	4	0.009905	0.007961
80	30	24	5	0.036612	0.026708
80	30	24	6	0.102686	0.066074
80	30	24	7	0.226253	0.123567
80	30	24	8	0.403881	0.177628
80	30	24	9	0.602373	0.198492
80	30	24	10	0.776053	0.173680
80	30	24	11	0.895538	0.119485
80	30	24	12	0.960259	0.064721
80	30	24	13	0.987833	0.027573
80	30	24	14	0.997040	0.009208
80	30	24	15	0.999436	0.002395
80	30	24	16	0.999917	0.000481
80	30	24	17	0.999991	0.000074
80	30	24	18	0.999999	0.000008
80	30	24	19	1.000000	0.000001
80	30	24	20	1.000000	0.000000
80	30	24	21	1.000000	0.000000
80	30	24	22	1.000000	0.000000
80	30	24	23	1.000000	0.000000
80	30	24	24	1.000000	0.000000
80	30	25	0	0.000000	0.000000
80	30	25	1	0.000010	0.000010
80	30	25	2	0.000140	0.000129
80	30	25	3	0.001131	0.000991
80	30	25	4	0.006208	0.005077
80	30	25	5	0.024689	0.018481
80	30	25	6	0.074369	0.049680
80	30	25	7	0.175503	0.101134
80	30	25	8	0.334099	0.158596
80	30	25	9	0.527938	0.193840
80	30	25	10	0.714024	0.186086
80	30	25	11	0.854998	0.140974
80	30	25	12	0.939456	0.084458
80	30	25	13	0.979462	0.040006
80	30	25	14	0.994409	0.014947
80	30	25	15	0.998794	0.004385
80	30	25	16	0.999797	0.001003
80	30	25	17	0.999974	0.000177
80	30	25	18	0.999997	0.000024
80	30	25	19	1.000000	0.000002
80	30	25	20	1.000000	0.000000

N	n	k	x	P(x)	p(x)
80	30	25	21	1.000000	0.000000
80	30	25	22	1.000000	0.000000
80	30	25	23	1.000000	0.000000
80	30	25	24	1.000000	0.000000
80	30	25	25	1.000000	0.000000
80	30	26	0	0.000000	0.000000
80	30	26	1	0.000005	0.000005
80	30	26	2	0.000074	0.000069
80	30	26	3	0.000644	0.000571
80	30	26	4	0.003808	0.003164
80	30	26	5	0.016289	0.012481
80	30	26	6	0.052690	0.036402
80	30	26	7	0.133210	0.080520
80	30	26	8	0.270660	0.137450
80	30	26	9	0.453927	0.183267
80	30	26	10	0.646357	0.192430
80	30	26	11	0.806298	0.159942
80	30	26	12	0.911815	0.105517
80	30	26	13	0.967097	0.055281
80	30	26	14	0.990061	0.022965
80	30	26	15	0.997598	0.007537
80	30	26	16	0.999541	0.001943
80	30	26	17	0.999932	0.000390
80	30	26	18	0.999992	0.000060
80	30	26	19	0.999999	0.000007
80	30	26	20	1.000000	0.000001
80	30	26	21	1.000000	0.000000
80	30	26	22	1.000000	0.000000
80	30	26	23	1.000000	0.000000
80	30	26	24	1.000000	0.000000
80	30	26	25	1.000000	0.000000
80	30	26	26	1.000000	0.000000
80	30	27	0	0.000000	0.000000
80	30	27	1	0.000002	0.000002
80	30	27	2	0.000038	0.000036
80	30	27	3	0.000359	0.000321
80	30	27	4	0.002285	0.001926
80	30	27	5	0.010511	0.008226
80	30	27	6	0.036512	0.026001
80	30	27	7	0.098915	0.062403
80	30	27	8	0.214662	0.115747
80	30	27	9	0.382656	0.167994
80	30	27	10	0.575086	0.192430
80	30	27	11	0.750022	0.174936
80	30	27	12	0.876643	0.126620
80	30	27	13	0.949693	0.073050
80	30	27	14	0.983257	0.033564
80	30	27	15	0.995505	0.012248
80	30	27	16	0.999038	0.003533
80	30	27	17	0.999838	0.000800

N	n	k	x	P(x)	p(x)
80	30	27	18	0.999979	0.000141
80	30	27	19	0.999998	0.000019
80	30	27	20	1.000000	0.000002
80	30	27	21	1.000000	0.000000
80	30	27	22	1.000000	0.000000
80	30	27	23	1.000000	0.000000
80	30	27	24	1.000000	0.000000
80	30	27	25	1.000000	0.000000
80	30	27	26	1.000000	0.000000
80	30	27	27	1.000000	0.000000
80	30	28	0	0.000000	0.000000
80	30	28	1	0.000001	0.000001
80	30	28	2	0.000019	0.000018
80	30	28	3	0.000196	0.000176
80	30	28	4	0.001340	0.001145
80	30	28	5	0.006631	0.005290
80	30	28	6	0.024738	0.018107
80	30	28	7	0.071834	0.047096
80	30	28	8	0.166616	0.094782
80	30	28	9	0.316093	0.149477
80	30	28	10	0.502471	0.186379
80	30	28	11	0.687309	0.184838
80	30	28	12	0.833640	0.146330
80	30	28	13	0.926262	0.092622
80	30	28	14	0.973124	0.046862
80	30	28	15	0.992038	0.018914
80	30	28	16	0.998104	0.006066
80	30	28	17	0.999642	0.001537
80	30	28	18	0.999947	0.000305
80	30	28	19	0.999994	0.000047
80	30	28	20	0.999999	0.000006
80	30	28	21	1.000000	0.000000
80	30	28	22	1.000000	0.000000
80	30	28	23	1.000000	0.000000
80	30	28	24	1.000000	0.000000
80	30	28	25	1.000000	0.000000
80	30	28	26	1.000000	0.000000
80	30	28	27	1.000000	0.000000
80	30	28	28	1.000000	0.000000
80	30	29	0	0.000000	0.000000
80	30	29	1	0.000001	0.000001
80	30	29	2	0.000010	0.000009
80	30	29	3	0.000104	0.000095
80	30	29	4	0.000768	0.000664
80	30	29	5	0.004087	0.003319
80	30	29	6	0.016381	0.012293
80	30	29	7	0.051003	0.034623
80	30	29	8	0.126516	0.075513
80	30	29	9	0.255727	0.129211
80	30	29	10	0.430787	0.175060

Table for $N = 2$, $n = 1$, through $N = 100$, $n = 50$

N	n	k	x	P(x)	p(x)
80	30	29	11	0.619772	0.188985
80	30	29	12	0.782987	0.163215
80	30	29	13	0.895982	0.112995
80	30	29	14	0.958705	0.062724
80	30	29	15	0.986582	0.027877
80	30	29	16	0.996471	0.009889
80	30	29	17	0.999257	0.002786
80	30	29	18	0.999876	0.000619
80	30	29	19	0.999984	0.000108
80	30	29	20	0.999998	0.000014
80	30	29	21	1.000000	0.000001
80	30	29	22	1.000000	0.000000
80	30	29	23	1.000000	0.000000
80	30	29	24	1.000000	0.000000
80	30	29	25	1.000000	0.000000
80	30	29	26	1.000000	0.000000
80	30	29	27	1.000000	0.000000
80	30	29	28	1.000000	0.000000
80	30	29	29	1.000000	0.000000
80	30	30	0	0.000000	0.000000
80	30	30	1	0.000000	0.000000
80	30	30	2	0.000005	0.000004
80	30	30	3	0.000054	0.000049
80	30	30	4	0.000430	0.000375
80	30	30	5	0.002460	0.002031
80	30	30	6	0.010596	0.008135
80	30	30	7	0.035389	0.024794
80	30	30	8	0.093942	0.058553
80	30	30	9	0.202523	0.108581
80	30	30	10	0.362136	0.159614
80	30	30	11	0.549366	0.187230
80	30	30	12	0.725382	0.176016
80	30	30	13	0.858317	0.132935
80	30	30	14	0.939027	0.080710
80	30	30	15	0.978383	0.039356
80	30	30	16	0.993757	0.015373
80	30	30	17	0.998547	0.004790
80	30	30	18	0.999731	0.001184
80	30	30	19	0.999961	0.000230
80	30	30	20	0.999995	0.000035
80	30	30	21	1.000000	0.000004
80	30	30	22	1.000000	0.000000
80	30	30	23	1.000000	0.000000
80	30	30	24	1.000000	0.000000
80	30	30	25	1.000000	0.000000
80	30	30	26	1.000000	0.000000
80	30	30	27	1.000000	0.000000
80	30	30	28	1.000000	0.000000
80	30	30	29	1.000000	0.000000
80	30	30	30	1.000000	0.000000

N	n	k	x	P(x)	p(x)
80	31	1	0	0.612500	0.612500
80	31	1	1	1.000000	0.387500
80	31	2	0	0.372152	0.372152
80	31	2	1	0.852848	0.480696
80	31	2	2	1.000000	0.147152
80	31	3	0	0.224245	0.224245
80	31	3	1	0.667965	0.443720
80	31	3	2	0.945290	0.277325
80	31	3	3	1.000000	0.054710
80	31	4	0	0.133965	0.133965
80	31	4	1	0.495087	0.361122
80	31	4	2	0.840843	0.345755
80	31	4	3	0.980105	0.139263
80	31	4	4	1.000000	0.019895
80	31	5	0	0.079321	0.079321
80	31	5	1	0.352539	0.273218
80	31	5	2	0.708910	0.356371
80	31	5	3	0.928798	0.219888
80	31	5	4	0.992932	0.064134
80	31	5	5	1.000000	0.007068
80	31	6	0	0.046535	0.046535
80	31	6	1	0.243252	0.196717
80	31	6	2	0.571113	0.327861
80	31	6	3	0.846706	0.275593
80	31	6	4	0.969844	0.123137
80	31	6	5	0.997550	0.027706
80	31	6	6	1.000000	0.002450
80	31	7	0	0.027041	0.027041
80	31	7	1	0.163502	0.136461
80	31	7	2	0.442627	0.279125
80	31	7	3	0.742428	0.299801
80	31	7	4	0.924915	0.182488
80	31	7	5	0.987815	0.062900
80	31	7	6	0.999172	0.011357
80	31	7	7	1.000000	0.000828
80	31	8	0	0.015558	0.015558
80	31	8	1	0.107422	0.091864
80	31	8	2	0.331742	0.224320
80	31	8	3	0.627436	0.295694
80	31	8	4	0.857420	0.229984
80	31	8	5	0.965412	0.107993
80	31	8	6	0.995283	0.029870
80	31	8	7	0.999728	0.004445
80	31	8	8	1.000000	0.000272
80	31	9	0	0.008859	0.008859
80	31	9	1	0.069145	0.060286
80	31	9	2	0.241391	0.172245
80	31	9	3	0.512443	0.271053
80	31	9	4	0.771176	0.258732
80	31	9	5	0.926415	0.155239

N	n	k	x	P(x)	p(x)
80	31	9	6	0.984911	0.058496
80	31	9	7	0.998246	0.013335
80	31	9	8	0.999913	0.001667
80	31	9	9	1.000000	0.000087
80	31	10	0	0.004991	0.004991
80	31	10	1	0.043672	0.038681
80	31	10	2	0.171037	0.127365
80	31	10	3	0.405549	0.234513
80	31	10	4	0.672785	0.267235
80	31	10	5	0.869567	0.196782
80	31	10	6	0.964314	0.094747
80	31	10	7	0.993738	0.029425
80	31	10	8	0.999373	0.005634
80	31	10	9	0.999973	0.000600
80	31	10	10	1.000000	0.000027
80	31	11	0	0.002781	0.002781
80	31	11	1	0.027095	0.024314
80	31	11	2	0.118271	0.091177
80	31	11	3	0.311744	0.193473
80	31	11	4	0.569708	0.257964
80	31	11	5	0.796476	0.226768
80	31	11	6	0.930476	0.133999
80	31	11	7	0.983650	0.053174
80	31	11	8	0.997522	0.013872
80	31	11	9	0.999784	0.002263
80	31	11	10	0.999992	0.000207
80	31	11	11	1.000000	0.000008
80	31	12	0	0.001531	0.001531
80	31	12	1	0.016523	0.014992
80	31	12	2	0.079951	0.063427
80	31	12	3	0.233234	0.153283
80	31	12	4	0.468766	0.235532
80	31	12	5	0.711027	0.242262
80	31	12	6	0.881925	0.170898
80	31	12	7	0.965155	0.083229
80	31	12	8	0.992898	0.027743
80	31	12	9	0.999063	0.006165
80	31	12	10	0.999929	0.000866
80	31	12	11	0.999998	0.000069
80	31	12	12	1.000000	0.000002
80	31	13	0	0.000833	0.000833
80	31	13	1	0.009909	0.009076
80	31	13	2	0.052901	0.042992
80	31	13	3	0.170117	0.117216
80	31	13	4	0.375246	0.205129
80	31	13	5	0.618398	0.243152
80	31	13	6	0.819095	0.200697
80	31	13	7	0.935780	0.116684
80	31	13	8	0.983514	0.047735
80	31	13	9	0.997068	0.013554

N	n	k	x	P(x)	p(x)
80	31	13	10	0.999661	0.002593
80	31	13	11	0.999977	0.000316
80	31	13	12	0.999999	0.000022
80	31	13	13	1.000000	0.000001
80	31	14	0	0.000448	0.000448
80	31	14	1	0.005845	0.005398
80	31	14	2	0.034293	0.028447
80	31	14	3	0.121131	0.086839
80	31	14	4	0.292582	0.171451
80	31	14	5	0.524040	0.231458
80	31	14	6	0.744208	0.220168
80	31	14	7	0.893982	0.149774
80	31	14	8	0.967128	0.073145
80	31	14	9	0.992618	0.025490
80	31	14	10	0.998849	0.006231
80	31	14	11	0.999883	0.001034
80	31	14	12	0.999993	0.000110
80	31	14	13	1.000000	0.000007
80	31	14	14	1.000000	0.000000
80	31	15	0	0.000237	0.000237
80	31	15	1	0.003392	0.003154
80	31	15	2	0.021793	0.018401
80	31	15	3	0.084291	0.062498
80	31	15	4	0.222443	0.138152
80	31	15	5	0.432860	0.210417
80	31	15	6	0.660811	0.227952
80	31	15	7	0.839519	0.178708
80	31	15	8	0.941638	0.102119
80	31	15	9	0.984121	0.042483
80	31	15	10	0.996866	0.012745
80	31	15	11	0.999569	0.002703
80	31	15	12	0.999961	0.000392
80	31	15	13	0.999998	0.000037
80	31	15	14	1.000000	0.000002
80	31	15	15	1.000000	0.000000
80	31	16	0	0.000124	0.000124
80	31	16	1	0.001936	0.001812
80	31	16	2	0.013583	0.011647
80	31	16	3	0.057369	0.043785
80	31	16	4	0.165057	0.107688
80	31	16	5	0.348693	0.183636
80	31	16	6	0.573138	0.224445
80	31	16	7	0.773535	0.200397
80	31	16	8	0.905503	0.131969
80	31	16	9	0.969742	0.064239
80	31	16	10	0.992748	0.023006
80	31	16	11	0.998738	0.005989
80	31	16	12	0.999847	0.001109
80	31	16	13	0.999988	0.000141
80	31	16	14	0.999999	0.000012

Table for N = 2, n = 1, through N = 100, n = 50

N	n	k	x	P(x)	p(x)
80	31	16	15	1.000000	0.000001
80	31	16	16	1.000000	0.000000
80	31	17	0	0.000064	0.000064
80	31	17	1	0.001087	0.001023
80	31	17	2	0.008306	0.007219
80	31	17	3	0.038213	0.029907
80	31	17	4	0.119626	0.081413
80	31	17	5	0.274091	0.154465
80	31	17	6	0.485464	0.211373
80	31	17	7	0.698386	0.212922
80	31	17	8	0.858077	0.159691
80	31	17	9	0.947660	0.089583
80	31	17	10	0.985199	0.037540
80	31	17	11	0.996866	0.011667
80	31	17	12	0.999518	0.002651
80	31	17	13	0.999948	0.000431
80	31	17	14	0.999996	0.000048
80	31	17	15	1.000000	0.000003
80	31	17	16	1.000000	0.000000
80	31	17	17	1.000000	0.000000
80	31	18	0	0.000033	0.000033
80	31	18	1	0.000600	0.000567
80	31	18	2	0.004983	0.004383
80	31	18	3	0.024921	0.019938
80	31	18	4	0.084734	0.059814
80	31	18	5	0.210343	0.125609
80	31	18	6	0.401586	0.191242
80	31	18	7	0.617273	0.215687
80	31	18	8	0.799777	0.182504
80	31	18	9	0.916377	0.116600
80	31	18	10	0.972686	0.056309
80	31	18	11	0.993162	0.020476
80	31	18	12	0.998718	0.005556
80	31	18	13	0.999825	0.001107
80	31	18	14	0.999983	0.000158
80	31	18	15	0.999999	0.000016
80	31	18	16	1.000000	0.000001
80	31	18	17	1.000000	0.000000
80	31	18	18	1.000000	0.000000
80	31	19	0	0.000016	0.000016
80	31	19	1	0.000325	0.000309
80	31	19	2	0.002933	0.002607
80	31	19	3	0.015916	0.012984
80	31	19	4	0.058686	0.042770
80	31	19	5	0.157669	0.098982
80	31	19	6	0.324472	0.166803
80	31	19	7	0.533781	0.209309
80	31	19	8	0.732074	0.198293
80	31	19	9	0.875003	0.142929
80	31	19	10	0.953614	0.078611

N	n	k	x	P(x)	p(x)
80	31	19	11	0.986557	0.032943
80	31	19	12	0.997015	0.010458
80	31	19	13	0.999504	0.002488
80	31	19	14	0.999940	0.000436
80	31	19	15	0.999995	0.000055
80	31	19	16	1.000000	0.000005
80	31	19	17	1.000000	0.000000
80	31	19	18	1.000000	0.000000
80	31	19	19	1.000000	0.000000
80	31	20	0	0.000008	0.000008
80	31	20	1	0.000173	0.000165
80	31	20	2	0.001693	0.001520
80	31	20	3	0.009957	0.008264
80	31	20	4	0.039755	0.029799
80	31	20	5	0.115479	0.075724
80	31	20	6	0.256110	0.140630
80	31	20	7	0.451430	0.195320
80	31	20	8	0.657308	0.205878
80	31	20	9	0.823455	0.166147
80	31	20	10	0.926551	0.103096
80	31	20	11	0.975756	0.049205
80	31	20	12	0.993758	0.018002
80	31	20	13	0.998769	0.005011
80	31	20	14	0.999818	0.001049
80	31	20	15	0.999980	0.000162
80	31	20	16	0.999998	0.000018
80	31	20	17	1.000000	0.000001
80	31	20	18	1.000000	0.000000
80	31	20	19	1.000000	0.000000
80	31	20	20	1.000000	0.000000
80	31	21	0	0.000004	0.000004
80	31	21	1	0.000091	0.000087
80	31	21	2	0.000958	0.000868
80	31	21	3	0.006100	0.005142
80	31	21	4	0.026346	0.020246
80	31	21	5	0.082666	0.056320
80	31	21	6	0.197514	0.114848
80	31	21	7	0.373302	0.175788
80	31	21	8	0.578388	0.205086
80	31	21	9	0.762534	0.184146
80	31	21	10	0.890467	0.127933
80	31	21	11	0.959354	0.068887
80	31	21	12	0.988057	0.028703
80	31	21	13	0.997266	0.009209
80	31	21	14	0.999521	0.002255
80	31	21	15	0.999937	0.000416
80	31	21	16	0.999994	0.000057
80	31	21	17	0.999999	0.000006
80	31	21	18	1.000000	0.000000
80	31	21	19	1.000000	0.000000

N	n	k	x	P(x)	p(x)
80	31	21	20	1.000000	0.000000
80	31	21	21	1.000000	0.000000
80	31	22	0	0.000002	0.000002
80	31	22	1	0.000047	0.000045
80	31	22	2	0.000532	0.000485
80	31	22	3	0.003660	0.003128
80	31	22	4	0.017081	0.013421
80	31	22	5	0.057847	0.040766
80	31	22	6	0.148849	0.091003
80	31	22	7	0.301795	0.152945
80	31	22	8	0.498439	0.196644
80	31	22	9	0.693869	0.195430
80	31	22	10	0.844932	0.151062
80	31	22	11	0.936003	0.091071
80	31	22	12	0.978814	0.042811
80	31	22	13	0.994457	0.015643
80	31	22	14	0.998871	0.004415
80	31	22	15	0.999824	0.000953
80	31	22	16	0.999979	0.000155
80	31	22	17	0.999998	0.000019
80	31	22	18	1.000000	0.000002
80	31	22	19	1.000000	0.000000
80	31	22	20	1.000000	0.000000
80	31	22	21	1.000000	0.000000
80	31	22	22	1.000000	0.000000
80	31	23	0	0.000001	0.000001
80	31	23	1	0.000023	0.000023
80	31	23	2	0.000289	0.000266
80	31	23	3	0.002150	0.001861
80	31	23	4	0.010833	0.008683
80	31	23	5	0.039572	0.028739
80	31	23	6	0.109624	0.070052
80	31	23	7	0.238507	0.128883
80	31	23	8	0.420459	0.181952
80	31	23	9	0.619741	0.199281
80	31	23	10	0.790237	0.170496
80	31	23	11	0.904599	0.114362
80	31	23	12	0.964790	0.060191
80	31	23	13	0.989602	0.024812
80	31	23	14	0.997577	0.007975
80	31	23	15	0.999561	0.001984
80	31	23	16	0.999939	0.000378
80	31	23	17	0.999994	0.000054
80	31	23	18	0.999999	0.000006
80	31	23	19	1.000000	0.000000
80	31	23	20	1.000000	0.000000
80	31	23	21	1.000000	0.000000
80	31	23	22	1.000000	0.000000
80	31	23	23	1.000000	0.000000
80	31	24	0	0.000000	0.000000

N	n	k	x	P(x)	p(x)
80	31	24	1	0.000012	0.000011
80	31	24	2	0.000154	0.000142
80	31	24	3	0.001236	0.001082
80	31	24	4	0.006720	0.005484
80	31	24	5	0.026463	0.019743
80	31	24	6	0.078900	0.052436
80	31	24	7	0.184241	0.105341
80	31	24	8	0.347040	0.162799
80	31	24	9	0.542825	0.195785
80	31	24	10	0.727422	0.184597
80	31	24	11	0.864472	0.137050
80	31	24	12	0.944726	0.080254
80	31	24	13	0.981766	0.037040
80	31	24	14	0.995199	0.013432
80	31	24	15	0.999004	0.003806
80	31	24	16	0.999840	0.000835
80	31	24	17	0.999980	0.000140
80	31	24	18	0.999998	0.000018
80	31	24	19	1.000000	0.000002
80	31	24	20	1.000000	0.000000
80	31	24	21	1.000000	0.000000
80	31	24	22	1.000000	0.000000
80	31	24	23	1.000000	0.000000
80	31	24	24	1.000000	0.000000
80	31	25	0	0.000000	0.000000
80	31	25	1	0.000006	0.000005
80	31	25	2	0.000080	0.000075
80	31	25	3	0.000695	0.000615
80	31	25	4	0.004076	0.003381
80	31	25	5	0.017297	0.013221
80	31	25	6	0.055491	0.038194
80	31	25	7	0.139094	0.083604
80	31	25	8	0.280176	0.141082
80	31	25	9	0.465910	0.185734
80	31	25	10	0.658198	0.192289
80	31	25	11	0.815526	0.157327
80	31	25	12	0.917497	0.101971
80	31	25	13	0.969861	0.052364
80	31	25	14	0.991121	0.021260
80	31	25	15	0.997917	0.006796
80	31	25	16	0.999616	0.001699
80	31	25	17	0.999945	0.000329
80	31	25	18	0.999994	0.000049
80	31	25	19	0.999999	0.000005
80	31	25	20	1.000000	0.000000
80	31	25	21	1.000000	0.000000
80	31	25	22	1.000000	0.000000
80	31	25	23	1.000000	0.000000
80	31	25	24	1.000000	0.000000
80	31	25	25	1.000000	0.000000

Table for $N = 2$, $n = 1$, through $N = 100$, $n = 50$

N	n	k	x	P(x)	p(x)
80	31	26	0	0.000000	0.000000
80	31	26	1	0.000003	0.000003
80	31	26	2	0.000041	0.000038
80	31	26	3	0.000382	0.000341
80	31	26	4	0.002416	0.002034
80	31	26	5	0.011047	0.008631
80	31	26	6	0.038130	0.027083
80	31	26	7	0.102613	0.064483
80	31	26	8	0.221178	0.118565
80	31	26	9	0.391616	0.170438
80	31	26	10	0.584779	0.193163
80	31	26	11	0.758316	0.173537
80	31	26	12	0.882270	0.123955
80	31	26	13	0.952723	0.070453
80	31	26	14	0.984550	0.031826
80	31	26	15	0.995940	0.011390
80	31	26	16	0.999153	0.003213
80	31	26	17	0.999861	0.000709
80	31	26	18	0.999982	0.000121
80	31	26	19	0.999998	0.000016
80	31	26	20	1.000000	0.000002
80	31	26	21	1.000000	0.000000
80	31	26	22	1.000000	0.000000
80	31	26	23	1.000000	0.000000
80	31	26	24	1.000000	0.000000
80	31	26	25	1.000000	0.000000
80	31	26	26	1.000000	0.000000
80	31	27	0	0.000000	0.000000
80	31	27	1	0.000001	0.000001
80	31	27	2	0.000020	0.000019
80	31	27	3	0.000205	0.000185
80	31	27	4	0.001399	0.001194
80	31	27	5	0.006891	0.005492
80	31	27	6	0.025591	0.018700
80	31	27	7	0.073954	0.048362
80	31	27	8	0.170678	0.096724
80	31	27	9	0.322178	0.151500
80	31	27	10	0.509660	0.187482
80	31	27	11	0.694043	0.184383
80	31	27	12	0.838657	0.144614
80	31	27	13	0.929239	0.090582
80	31	27	14	0.974530	0.045291
80	31	27	15	0.992565	0.018035
80	31	27	16	0.998260	0.005695
80	31	27	17	0.999678	0.001417
80	31	27	18	0.999953	0.000276
80	31	27	19	0.999995	0.000041
80	31	27	20	0.999999	0.000005
80	31	27	21	1.000000	0.000000
80	31	27	22	1.000000	0.000000

N	n	k	x	P(x)	p(x)
80	31	27	23	1.000000	0.000000
80	31	27	24	1.000000	0.000000
80	31	27	25	1.000000	0.000000
80	31	27	26	1.000000	0.000000
80	31	27	27	1.000000	0.000000
80	31	28	0	0.000000	0.000000
80	31	28	1	0.000001	0.000001
80	31	28	2	0.000010	0.000009
80	31	28	3	0.000107	0.000098
80	31	28	4	0.000791	0.000683
80	31	28	5	0.004197	0.003406
80	31	28	6	0.016771	0.012574
80	31	28	7	0.052054	0.035283
80	31	28	8	0.128703	0.076650
80	31	28	9	0.259292	0.130588
80	31	28	10	0.435375	0.176083
80	31	28	11	0.624464	0.189090
80	31	28	12	0.786814	0.162350
80	31	28	13	0.898475	0.111661
80	31	28	14	0.960003	0.061528
80	31	28	15	0.987121	0.027118
80	31	28	16	0.996649	0.009528
80	31	28	17	0.999303	0.002655
80	31	28	18	0.999886	0.000582
80	31	28	19	0.999985	0.000100
80	31	28	20	0.999999	0.000013
80	31	28	21	1.000000	0.000001
80	31	28	22	1.000000	0.000000
80	31	28	23	1.000000	0.000000
80	31	28	24	1.000000	0.000000
80	31	28	25	1.000000	0.000000
80	31	28	26	1.000000	0.000000
80	31	28	27	1.000000	0.000000
80	31	28	28	1.000000	0.000000
80	31	29	0	0.000000	0.000000
80	31	29	1	0.000000	0.000000
80	31	29	2	0.000005	0.000004
80	31	29	3	0.000055	0.000050
80	31	29	4	0.000436	0.000381
80	31	29	5	0.002494	0.002058
80	31	29	6	0.010726	0.008232
80	31	29	7	0.035769	0.025044
80	31	29	8	0.094801	0.059031
80	31	29	9	0.204043	0.109242
80	31	29	10	0.364264	0.160222
80	31	29	11	0.551738	0.187473
80	31	29	12	0.727494	0.175756
80	31	29	13	0.859823	0.132329
80	31	29	14	0.939888	0.080065
80	31	29	15	0.978777	0.038889

N	n	k	x	P(x)	p(x)
80	31	29	16	0.993900	0.015123
80	31	29	17	0.998589	0.004688
80	31	29	18	0.999740	0.001152
80	31	29	19	0.999962	0.000222
80	31	29	20	0.999996	0.000033
80	31	29	21	1.000000	0.000004
80	31	29	22	1.000000	0.000000
80	31	29	23	1.000000	0.000000
80	31	29	24	1.000000	0.000000
80	31	29	25	1.000000	0.000000
80	31	29	26	1.000000	0.000000
80	31	29	27	1.000000	0.000000
80	31	29	28	1.000000	0.000000
80	31	29	29	1.000000	0.000000
80	31	30	0	0.000000	0.000000
80	31	30	1	0.000000	0.000000
80	31	30	2	0.000002	0.000002
80	31	30	3	0.000027	0.000025
80	31	30	4	0.000234	0.000207
80	31	30	5	0.001445	0.001211
80	31	30	6	0.006691	0.005246
80	31	30	7	0.023984	0.017294
80	31	30	8	0.068179	0.044195
80	31	30	9	0.156919	0.088740
80	31	30	10	0.298291	0.141372
80	31	30	11	0.478219	0.179928
80	31	30	12	0.662016	0.183797
80	31	30	13	0.813119	0.151103
80	31	30	14	0.913200	0.100081
80	31	30	15	0.966576	0.053377
80	31	30	16	0.989452	0.022876
80	31	30	17	0.997302	0.007849
80	31	30	18	0.999447	0.002145
80	31	30	19	0.999910	0.000463
80	31	30	20	0.999988	0.000078
80	31	30	21	0.999999	0.000010
80	31	30	22	1.000000	0.000001
80	31	30	23	1.000000	0.000000
80	31	30	24	1.000000	0.000000
80	31	30	25	1.000000	0.000000
80	31	30	26	1.000000	0.000000
80	31	30	27	1.000000	0.000000
80	31	30	28	1.000000	0.000000
80	31	30	29	1.000000	0.000000
80	31	30	30	1.000000	0.000000
80	31	31	0	0.000000	0.000000
80	31	31	1	0.000000	0.000000
80	31	31	2	0.000001	0.000001
80	31	31	3	0.000013	0.000012
80	31	31	4	0.000123	0.000109

N	n	k	x	P(x)	p(x)
80	31	31	5	0.000815	0.000693
80	31	31	6	0.004068	0.003252
80	31	31	7	0.015683	0.011615
80	31	31	8	0.047849	0.032166
80	31	31	9	0.117873	0.070024
80	31	31	10	0.238914	0.121041
80	31	31	11	0.406248	0.167333
80	31	31	12	0.592173	0.185926
80	31	31	13	0.758722	0.166549
80	31	31	14	0.879172	0.120450
80	31	31	15	0.949496	0.070324
80	31	31	16	0.982589	0.033093
80	31	31	17	0.995104	0.012514
80	31	31	18	0.998889	0.003785
80	31	31	19	0.999799	0.000910
80	31	31	20	0.999971	0.000172
80	31	31	21	0.999997	0.000025
80	31	31	22	1.000000	0.000003
80	31	31	23	1.000000	0.000000
80	31	31	24	1.000000	0.000000
80	31	31	25	1.000000	0.000000
80	31	31	26	1.000000	0.000000
80	31	31	27	1.000000	0.000000
80	31	31	28	1.000000	0.000000
80	31	31	29	1.000000	0.000000
80	31	31	30	1.000000	0.000000
80	31	31	31	1.000000	0.000000
80	32	1	0	0.600000	0.600000
80	32	1	1	1.000000	0.400000
80	32	2	0	0.356962	0.356962
80	32	2	1	0.843038	0.486076
80	32	2	2	1.000000	0.156962
80	32	3	0	0.210516	0.210516
80	32	3	1	0.649854	0.439338
80	32	3	2	0.939630	0.289776
80	32	3	3	1.000000	0.060370
80	32	4	0	0.123029	0.123029
80	32	4	1	0.472978	0.349949
80	32	4	2	0.826730	0.353753
80	32	4	3	0.977263	0.150533
80	32	4	4	1.000000	0.022737
80	32	5	0	0.071227	0.071227
80	32	5	1	0.330235	0.259008
80	32	5	2	0.687091	0.356856
80	32	5	3	0.919823	0.232732
80	32	5	4	0.991623	0.071800
80	32	5	5	1.000000	0.008377
80	32	6	0	0.040837	0.040837
80	32	6	1	0.223179	0.182342
80	32	6	2	0.544349	0.321170

Table for $N = 2$, $n = 1$, through $N = 100$, $n = 50$

N	n	k	x	P(x)	p(x)
80	32	6	3	0.829833	0.285485
80	32	6	4	0.964818	0.134985
80	32	6	5	0.996984	0.032167
80	32	6	6	1.000000	0.003016
80	32	7	0	0.023178	0.023178
80	32	7	1	0.146792	0.123615
80	32	7	2	0.414145	0.267352
80	32	7	3	0.717954	0.303810
80	32	7	4	0.913743	0.195788
80	32	7	5	0.985248	0.071505
80	32	7	6	0.998940	0.013693
80	32	7	7	1.000000	0.001060
80	32	8	0	0.013018	0.013018
80	32	8	1	0.094298	0.081281
80	32	8	2	0.304274	0.209975
80	32	8	3	0.597263	0.292989
80	32	8	4	0.838646	0.241383
80	32	8	5	0.958801	0.120155
80	32	8	6	0.994064	0.035263
80	32	8	7	0.999637	0.005573
80	32	8	8	1.000000	0.000363
80	32	9	0	0.007232	0.007232
80	32	9	1	0.059303	0.052071
80	32	9	2	0.216784	0.157482
80	32	9	3	0.479253	0.262469
80	32	9	4	0.744775	0.265521
80	32	9	5	0.913743	0.168968
80	32	9	6	0.981330	0.067587
80	32	9	7	0.997702	0.016372
80	32	9	8	0.999879	0.002177
80	32	9	9	1.000000	0.000121
80	32	10	0	0.003973	0.003973
80	32	10	1	0.036568	0.032595
80	32	10	2	0.150243	0.113675
80	32	10	3	0.372048	0.221805
80	32	10	4	0.640062	0.268014
80	32	10	5	0.849487	0.209425
80	32	10	6	0.956580	0.107092
80	32	10	7	0.991937	0.035357
80	32	10	8	0.999143	0.007206
80	32	10	9	0.999961	0.000818
80	32	10	10	1.000000	0.000039
80	32	11	0	0.002157	0.002157
80	32	11	1	0.022133	0.019976
80	32	11	2	0.101525	0.079392
80	32	11	3	0.280157	0.178632
80	32	11	4	0.532856	0.252699
80	32	11	5	0.768709	0.235853
80	32	11	6	0.916802	0.148094
80	32	11	7	0.979309	0.062507

N	n	k	x	P(x)	p(x)
80	32	11	8	0.996672	0.017363
80	32	11	9	0.999692	0.003020
80	32	11	10	0.999988	0.000296
80	32	11	11	1.000000	0.000012
80	32	12	0	0.001156	0.001156
80	32	12	1	0.013158	0.012001
80	32	12	2	0.067006	0.053849
80	32	12	3	0.205080	0.138073
80	32	12	4	0.430312	0.225232
80	32	12	5	0.676419	0.246107
80	32	12	6	0.860999	0.184580
80	32	12	7	0.956662	0.095663
80	32	12	8	0.990633	0.033971
80	32	12	9	0.998685	0.008052
80	32	12	10	0.999893	0.001208
80	32	12	11	0.999996	0.000103
80	32	12	12	1.000000	0.000004
80	32	13	0	0.000612	0.000612
80	32	13	1	0.007687	0.007074
80	32	13	2	0.043250	0.035563
80	32	13	3	0.146195	0.102946
80	32	13	4	0.337569	0.191374
80	32	13	5	0.578700	0.241131
80	32	13	6	0.790424	0.211724
80	32	13	7	0.921492	0.131068
80	32	13	8	0.978643	0.057152
80	32	13	9	0.995962	0.017319
80	32	13	10	0.999503	0.003541
80	32	13	11	0.999964	0.000462
80	32	13	12	0.999999	0.000034
80	32	13	13	1.000000	0.000001
80	32	14	0	0.000320	0.000320
80	32	14	1	0.004413	0.004094
80	32	14	2	0.027326	0.022913
80	32	14	3	0.101637	0.074311
80	32	14	4	0.257592	0.155955
80	32	14	5	0.481528	0.223936
80	32	14	6	0.708262	0.226735
80	32	14	7	0.872586	0.164323
80	32	14	8	0.958171	0.085585
80	32	14	9	0.990017	0.031846
80	32	14	10	0.998340	0.008323
80	32	14	11	0.999820	0.001480
80	32	14	12	0.999989	0.000169
80	32	14	13	1.000000	0.000011
80	32	14	14	1.000000	0.000000
80	32	15	0	0.000165	0.000165
80	32	15	1	0.002491	0.002326
80	32	15	2	0.016911	0.014420
80	32	15	3	0.068985	0.052074

N	n	k	x	P(x)	p(x)
80	32	15	4	0.191429	0.122444
80	32	15	5	0.389918	0.198488
80	32	15	6	0.618943	0.229025
80	32	15	7	0.810342	0.191400
80	32	15	8	0.927049	0.116707
80	32	15	9	0.978919	0.051870
80	32	15	10	0.995565	0.016647
80	32	15	11	0.999349	0.003783
80	32	15	12	0.999937	0.000589
80	32	15	13	0.999996	0.000059
80	32	15	14	1.000000	0.000003
80	32	15	15	1.000000	0.000000
80	32	16	0	0.000084	0.000084
80	32	16	1	0.001381	0.001298
80	32	16	2	0.010256	0.008874
80	32	16	3	0.045752	0.035497
80	32	16	4	0.138684	0.092932
80	32	16	5	0.307469	0.168785
80	32	16	6	0.527333	0.219864
80	32	16	7	0.736727	0.209394
80	32	16	8	0.883957	0.147230
80	32	16	9	0.960565	0.076608
80	32	16	10	0.989931	0.029366
80	32	16	11	0.998127	0.008195
80	32	16	12	0.999756	0.001630
80	32	16	13	0.999979	0.000223
80	32	16	14	0.999999	0.000020
80	32	16	15	1.000000	0.000001
80	32	16	16	1.000000	0.000000
80	32	17	0	0.000042	0.000042
80	32	17	1	0.000753	0.000711
80	32	17	2	0.006096	0.005343
80	32	17	3	0.029668	0.023572
80	32	17	4	0.098026	0.068359
80	32	17	5	0.236263	0.138236
80	32	17	6	0.438013	0.201750
80	32	17	7	0.654932	0.216919
80	32	17	8	0.828746	0.173814
80	32	17	9	0.933034	0.104288
80	32	17	10	0.979837	0.046803
80	32	17	11	0.995437	0.015601
80	32	17	12	0.999247	0.003810
80	32	17	13	0.999913	0.000666
80	32	17	14	0.999993	0.000080
80	32	17	15	1.000000	0.000006
80	32	17	16	1.000000	0.000000
80	32	17	17	1.000000	0.000000
80	32	18	0	0.000021	0.000021
80	32	18	1	0.000403	0.000382
80	32	18	2	0.003551	0.003149

N	n	k	x	P(x)	p(x)
80	32	18	3	0.018817	0.015266
80	32	18	4	0.067645	0.048828
80	32	18	5	0.177018	0.109374
80	32	18	6	0.354751	0.177732
80	32	18	7	0.568853	0.214102
80	32	18	8	0.762531	0.193678
80	32	18	9	0.894961	0.132429
80	32	18	10	0.963493	0.068532
80	32	18	11	0.990237	0.026744
80	32	18	12	0.998038	0.007800
80	32	18	13	0.999712	0.001675
80	32	18	14	0.999970	0.000258
80	32	18	15	0.999998	0.000028
80	32	18	16	1.000000	0.000002
80	32	18	17	1.000000	0.000000
80	32	18	18	1.000000	0.000000
80	32	19	0	0.000010	0.000010
80	32	19	1	0.000212	0.000202
80	32	19	2	0.002028	0.001816
80	32	19	3	0.011677	0.009649
80	32	19	4	0.045594	0.033917
80	32	19	5	0.129388	0.083794
80	32	19	6	0.280218	0.150830
80	32	19	7	0.482522	0.202304
80	32	19	8	0.687559	0.205038
80	32	19	9	0.845834	0.158275
80	32	19	10	0.939175	0.093341
80	32	19	11	0.981179	0.042004
80	32	19	12	0.995521	0.014343
80	32	19	13	0.999199	0.003678
80	32	19	14	0.999895	0.000696
80	32	19	15	0.999990	0.000095
80	32	19	16	0.999999	0.000009
80	32	19	17	1.000000	0.000001
80	32	19	18	1.000000	0.000000
80	32	19	19	1.000000	0.000000
80	32	20	0	0.000005	0.000005
80	32	20	1	0.000109	0.000104
80	32	20	2	0.001135	0.001026
80	32	20	3	0.007090	0.005955
80	32	20	4	0.030025	0.022936
80	32	20	5	0.092299	0.062273
80	32	20	6	0.215930	0.123631
80	32	20	7	0.399610	0.183680
80	32	20	8	0.606889	0.207278
80	32	20	9	0.786156	0.179268
80	32	20	10	0.905511	0.119355
80	32	20	11	0.966718	0.061207
80	32	20	12	0.990819	0.024100
80	32	20	13	0.998053	0.007235

$N = 80 \quad n = 32$

Table for $N = 2$, $n = 1$, through $N = 100$, $n = 50$

N	n	k	x	P(x)	p(x)
80	32	20	14	0.999690	0.001636
80	32	20	15	0.999964	0.000274
80	32	20	16	0.999997	0.000033
80	32	20	17	1.000000	0.000003
80	32	20	18	1.000000	0.000000
80	32	20	19	1.000000	0.000000
80	32	20	20	1.000000	0.000000
80	32	21	0	0.000002	0.000002
80	32	21	1	0.000055	0.000053
80	32	21	2	0.000622	0.000567
80	32	21	3	0.004211	0.003589
80	32	21	4	0.019322	0.015110
80	32	21	5	0.064276	0.044954
80	32	21	6	0.162356	0.098081
80	32	21	7	0.323077	0.160720
80	32	21	8	0.523977	0.200901
80	32	21	9	0.717437	0.193460
80	32	21	10	0.861748	0.144311
80	32	21	11	0.945296	0.083548
80	32	21	12	0.982785	0.037490
80	32	21	13	0.995762	0.012977
80	32	21	14	0.999199	0.003436
80	32	21	15	0.999886	0.000687
80	32	21	16	0.999988	0.000102
80	32	21	17	0.999999	0.000011
80	32	21	18	1.000000	0.000001
80	32	21	19	1.000000	0.000000
80	32	21	20	1.000000	0.000000
80	32	21	21	1.000000	0.000000
80	32	22	0	0.000001	0.000001
80	32	22	1	0.000027	0.000026
80	32	22	2	0.000334	0.000306
80	32	22	3	0.002447	0.002113
80	32	22	4	0.012151	0.009704
80	32	22	5	0.043704	0.031553
80	32	22	6	0.119134	0.075431
80	32	22	7	0.254975	0.135841
80	32	22	8	0.442255	0.187280
80	32	22	9	0.642021	0.199766
80	32	22	10	0.807937	0.165916
80	32	22	11	0.915558	0.107621
80	32	22	12	0.970077	0.054519
80	32	22	13	0.991583	0.021506
80	32	22	14	0.998151	0.006567
80	32	22	15	0.999688	0.001538
80	32	22	16	0.999961	0.000272
80	32	22	17	0.999996	0.000036
80	32	22	18	1.000000	0.000003
80	32	22	19	1.000000	0.000000
80	32	22	20	1.000000	0.000000

N	n	k	x	P(x)	p(x)
80	32	22	21	1.000000	0.000000
80	32	22	22	1.000000	0.000000
80	32	23	0	0.000000	0.000000
80	32	23	1	0.000013	0.000013
80	32	23	2	0.000175	0.000162
80	32	23	3	0.001390	0.001215
80	32	23	4	0.007466	0.006076
80	32	23	5	0.029015	0.021549
80	32	23	6	0.085320	0.056305
80	32	23	7	0.196423	0.111102
80	32	23	8	0.364760	0.168337
80	32	23	9	0.562803	0.198043
80	32	23	10	0.745003	0.182200
80	32	23	11	0.876592	0.131589
80	32	23	12	0.951277	0.074686
80	32	23	13	0.984538	0.033261
80	32	23	14	0.996112	0.011574
80	32	23	15	0.999238	0.003125
80	32	23	16	0.999885	0.000648
80	32	23	17	0.999987	0.000102
80	32	23	18	0.999999	0.000012
80	32	23	19	1.000000	0.000001
80	32	23	20	1.000000	0.000000
80	32	23	21	1.000000	0.000000
80	32	23	22	1.000000	0.000000
80	32	23	23	1.000000	0.000000
80	32	24	0	0.000000	0.000000
80	32	24	1	0.000006	0.000006
80	32	24	2	0.000090	0.000084
80	32	24	3	0.000772	0.000682
80	32	24	4	0.004482	0.003709
80	32	24	5	0.018808	0.014326
80	32	24	6	0.059637	0.040830
80	32	24	7	0.147694	0.088056
80	32	24	8	0.293881	0.146187
80	32	24	9	0.482891	0.189010
80	32	24	10	0.674680	0.191789
80	32	24	11	0.828112	0.153432
80	32	24	12	0.925072	0.096960
80	32	24	13	0.973451	0.048379
80	32	24	14	0.992457	0.019006
80	32	24	15	0.998306	0.005848
80	32	24	16	0.999704	0.001398
80	32	24	17	0.999960	0.000257
80	32	24	18	0.999996	0.000036
80	32	24	19	1.000000	0.000004
80	32	24	20	1.000000	0.000000
80	32	24	21	1.000000	0.000000
80	32	24	22	1.000000	0.000000
80	32	24	23	1.000000	0.000000

N	n	k	x	P(x)	p(x)
80	32	24	24	1.000000	0.000000
80	32	25	0	0.000000	0.000000
80	32	25	1	0.000003	0.000003
80	32	25	2	0.000045	0.000042
80	32	25	3	0.000419	0.000374
80	32	25	4	0.002627	0.002208
80	32	25	5	0.011901	0.009274
80	32	25	6	0.040681	0.028780
80	32	25	7	0.108383	0.067702
80	32	25	8	0.231229	0.122846
80	32	25	9	0.405262	0.174033
80	32	25	10	0.599334	0.194073
80	32	25	11	0.770575	0.171241
80	32	25	12	0.890443	0.119868
80	32	25	13	0.957037	0.066594
80	32	25	14	0.986348	0.029311
80	32	25	15	0.996530	0.010182
80	32	25	16	0.999304	0.002774
80	32	25	17	0.999891	0.000587
80	32	25	18	0.999987	0.000096
80	32	25	19	0.999999	0.000012
80	32	25	20	1.000000	0.000001
80	32	25	21	1.000000	0.000000
80	32	25	22	1.000000	0.000000
80	32	25	23	1.000000	0.000000
80	32	25	24	1.000000	0.000000
80	32	25	25	1.000000	0.000000
80	32	26	0	0.000000	0.000000
80	32	26	1	0.000001	0.000001
80	32	26	2	0.000022	0.000021
80	32	26	3	0.000222	0.000200
80	32	26	4	0.001503	0.001281
80	32	26	5	0.007348	0.005845
80	32	26	6	0.027076	0.019728
80	32	26	7	0.077609	0.050534
80	32	26	8	0.177624	0.100014
80	32	26	9	0.332485	0.154861
80	32	26	10	0.521705	0.189221
80	32	26	11	0.705192	0.183487
80	32	26	12	0.846855	0.141663
80	32	26	13	0.934032	0.087177
80	32	26	14	0.976755	0.042724
80	32	26	15	0.993383	0.016628
80	32	26	16	0.998497	0.005114
80	32	26	17	0.999731	0.001234
80	32	26	18	0.999963	0.000231
80	32	26	19	0.999996	0.000033
80	32	26	20	1.000000	0.000004
80	32	26	21	1.000000	0.000000
80	32	26	22	1.000000	0.000000

N	n	k	x	P(x)	p(x)
80	32	26	23	1.000000	0.000000
80	32	26	24	1.000000	0.000000
80	32	26	25	1.000000	0.000000
80	32	26	26	1.000000	0.000000
80	32	27	0	0.000000	0.000000
80	32	27	1	0.000001	0.000001
80	32	27	2	0.000011	0.000010
80	32	27	3	0.000115	0.000104
80	32	27	4	0.000839	0.000724
80	32	27	5	0.004425	0.003587
80	32	27	6	0.017577	0.013152
80	32	27	7	0.054214	0.036637
80	32	27	8	0.133173	0.078959
80	32	27	9	0.266525	0.133353
80	32	27	10	0.444615	0.178090
80	32	27	11	0.633836	0.189221
80	32	27	12	0.794387	0.160551
80	32	27	13	0.903358	0.108971
80	32	27	14	0.962514	0.059156
80	32	27	15	0.988148	0.025634
80	32	27	16	0.996982	0.008833
80	32	27	17	0.999388	0.002407
80	32	27	18	0.999903	0.000514
80	32	27	19	0.999988	0.000085
80	32	27	20	0.999999	0.000011
80	32	27	21	1.000000	0.000001
80	32	27	22	1.000000	0.000000
80	32	27	23	1.000000	0.000000
80	32	27	24	1.000000	0.000000
80	32	27	25	1.000000	0.000000
80	32	27	26	1.000000	0.000000
80	32	27	27	1.000000	0.000000
80	32	28	0	0.000000	0.000000
80	32	28	1	0.000000	0.000000
80	32	28	2	0.000005	0.000005
80	32	28	3	0.000058	0.000053
80	32	28	4	0.000456	0.000398
80	32	28	5	0.002598	0.002142
80	32	28	6	0.011125	0.008527
80	32	28	7	0.036932	0.025807
80	32	28	8	0.097418	0.060485
80	32	28	9	0.208655	0.111237
80	32	28	10	0.370691	0.162036
80	32	28	11	0.558862	0.188171
80	32	28	12	0.733802	0.174940
80	32	28	13	0.864293	0.130491
80	32	28	14	0.942424	0.078130
80	32	28	15	0.979926	0.037503
80	32	28	16	0.994315	0.014389
80	32	28	17	0.998707	0.004392

Table for $N = 2$, $n = 1$, through $N = 100$, $n = 50$

N	n	k	x	P(x)	p(x)
80	32	28	18	0.999767	0.001060
80	32	28	19	0.999967	0.000200
80	32	28	20	0.999996	0.000029
80	32	28	21	1.000000	0.000003
80	32	28	22	1.000000	0.000000
80	32	28	23	1.000000	0.000000
80	32	28	24	1.000000	0.000000
80	32	28	25	1.000000	0.000000
80	32	28	26	1.000000	0.000000
80	32	28	27	1.000000	0.000000
80	32	28	28	1.000000	0.000000
80	32	29	0	0.000000	0.000000
80	32	29	1	0.000000	0.000000
80	32	29	2	0.000002	0.000002
80	32	29	3	0.000028	0.000026
80	32	29	4	0.000242	0.000213
80	32	29	5	0.001486	0.001244
80	32	29	6	0.006862	0.005376
80	32	29	7	0.024525	0.017663
80	32	29	8	0.069502	0.044976
80	32	29	9	0.159454	0.089953
80	32	29	10	0.302138	0.142683
80	32	29	11	0.482870	0.180732
80	32	29	12	0.666517	0.183647
80	32	29	13	0.816614	0.150096
80	32	29	14	0.915378	0.098765
80	32	29	15	0.967666	0.052287
80	32	29	16	0.989888	0.022222
80	32	29	17	0.997440	0.007553
80	32	29	18	0.999482	0.002041
80	32	29	19	0.999917	0.000435
80	32	29	20	0.999990	0.000073
80	32	29	21	0.999999	0.000009
80	32	29	22	1.000000	0.000001
80	32	29	23	1.000000	0.000000
80	32	29	24	1.000000	0.000000
80	32	29	25	1.000000	0.000000
80	32	29	26	1.000000	0.000000
80	32	29	27	1.000000	0.000000
80	32	29	28	1.000000	0.000000
80	32	29	29	1.000000	0.000000
80	32	30	0	0.000000	0.000000
80	32	30	1	0.000000	0.000000
80	32	30	2	0.000001	0.000001
80	32	30	3	0.000013	0.000012
80	32	30	4	0.000124	0.000111
80	32	30	5	0.000827	0.000703
80	32	30	6	0.004121	0.003294
80	32	30	7	0.015867	0.011745
80	32	30	8	0.048336	0.032470
80	32	30	9	0.118887	0.070551
80	32	30	10	0.240588	0.121701
80	32	30	11	0.408451	0.167863
80	32	30	12	0.594499	0.186048
80	32	30	13	0.760695	0.166197
80	32	30	14	0.880520	0.119825
80	32	30	15	0.950237	0.069716
80	32	30	16	0.982916	0.032680
80	32	30	17	0.995219	0.012303
80	32	30	18	0.998921	0.003702
80	32	30	19	0.999806	0.000885
80	32	30	20	0.999972	0.000166
80	32	30	21	0.999997	0.000024
80	32	30	22	1.000000	0.000003
80	32	30	23	1.000000	0.000000
80	32	30	24	1.000000	0.000000
80	32	30	25	1.000000	0.000000
80	32	30	26	1.000000	0.000000
80	32	30	27	1.000000	0.000000
80	32	30	28	1.000000	0.000000
80	32	30	29	1.000000	0.000000
80	32	30	30	1.000000	0.000000
80	32	31	0	0.000000	0.000000
80	32	31	1	0.000000	0.000000
80	32	31	2	0.000000	0.000000
80	32	31	3	0.000006	0.000006
80	32	31	4	0.000062	0.000056
80	32	31	5	0.000448	0.000385
80	32	31	6	0.002408	0.001961
80	32	31	7	0.009994	0.007586
80	32	31	8	0.032751	0.022757
80	32	31	9	0.086434	0.053683
80	32	31	10	0.187040	0.100606
80	32	31	11	0.337948	0.150909
80	32	31	12	0.520080	0.182131
80	32	31	13	0.697541	0.177461
80	32	31	14	0.837383	0.139843
80	32	31	15	0.926533	0.089150
80	32	31	16	0.972459	0.045926
80	32	31	17	0.991528	0.019069
80	32	31	18	0.997885	0.006356
80	32	31	19	0.999576	0.001691
80	32	31	20	0.999933	0.000357
80	32	31	21	0.999991	0.000059
80	32	31	22	0.999999	0.000008
80	32	31	23	1.000000	0.000001
80	32	31	24	1.000000	0.000000
80	32	31	25	1.000000	0.000000
80	32	31	26	1.000000	0.000000
80	32	31	27	1.000000	0.000000
80	32	31	28	1.000000	0.000000
80	32	31	29	1.000000	0.000000
80	32	31	30	1.000000	0.000000
80	32	31	31	1.000000	0.000000
80	32	32	0	0.000000	0.000000
80	32	32	1	0.000000	0.000000
80	32	32	2	0.000000	0.000000
80	32	32	3	0.000003	0.000003
80	32	32	4	0.000030	0.000027
80	32	32	5	0.000235	0.000205
80	32	32	6	0.001368	0.001133
80	32	32	7	0.006124	0.004756
80	32	32	8	0.021605	0.015481
80	32	32	9	0.061236	0.039631
80	32	32	10	0.141870	0.080634
80	32	32	11	0.273273	0.131404
80	32	32	12	0.445740	0.172467
80	32	32	13	0.628729	0.182989
80	32	32	14	0.786013	0.157283
80	32	32	15	0.895604	0.109591
80	32	32	16	0.957463	0.061859
80	32	32	17	0.985691	0.028228
80	32	32	18	0.996069	0.010378
80	32	32	19	0.999127	0.003059
80	32	32	20	0.999845	0.000718
80	32	32	21	0.999978	0.000133
80	32	32	22	0.999998	0.000019
80	32	32	23	1.000000	0.000002
80	32	32	24	1.000000	0.000000
80	32	32	25	1.000000	0.000000
80	32	32	26	1.000000	0.000000
80	32	32	27	1.000000	0.000000
80	32	32	28	1.000000	0.000000
80	32	32	29	1.000000	0.000000
80	32	32	30	1.000000	0.000000
80	32	32	31	1.000000	0.000000
80	32	32	32	1.000000	0.000000
80	33	1	0	0.587500	0.587500
80	33	1	1	1.000000	0.412500
80	33	2	0	0.342089	0.342089
80	33	2	1	0.832911	0.490823
80	33	2	2	1.000000	0.167089
80	33	3	0	0.197359	0.197359
80	33	3	1	0.631548	0.434189
80	33	3	2	0.933593	0.302045
80	33	3	3	1.000000	0.066407
80	33	4	0	0.112776	0.112776
80	33	4	1	0.451106	0.338329
80	33	4	2	0.811991	0.360885
80	33	4	3	0.974127	0.162137
80	33	4	4	1.000000	0.025873
80	33	5	0	0.063808	0.063808
80	33	5	1	0.308651	0.244844
80	33	5	2	0.664788	0.356136
80	33	5	3	0.910126	0.245338
80	33	5	4	0.990127	0.080002
80	33	5	5	1.000000	0.009873
80	33	6	0	0.035732	0.035732
80	33	6	1	0.204185	0.168452
80	33	6	2	0.517585	0.313400
80	33	6	3	0.811990	0.294406
80	33	6	4	0.959193	0.147203
80	33	6	5	0.996314	0.037121
80	33	6	6	1.000000	0.003686
80	33	7	0	0.019798	0.019798
80	33	7	1	0.131340	0.111543
80	33	7	2	0.386295	0.254955
80	33	7	3	0.692637	0.306341
80	33	7	4	0.901506	0.208869
80	33	7	5	0.982268	0.080763
80	33	7	6	0.998655	0.016387
80	33	7	7	1.000000	0.001345
80	33	8	0	0.010848	0.010848
80	33	8	1	0.082445	0.071597
80	33	8	2	0.278027	0.195582
80	33	8	3	0.566743	0.288716
80	33	8	4	0.818530	0.251787
80	33	8	5	0.951291	0.132761
80	33	8	6	0.992594	0.041303
80	33	8	7	0.999521	0.006927
80	33	8	8	1.000000	0.000479
80	33	9	0	0.005876	0.005876
80	33	9	1	0.050624	0.044748
80	33	9	2	0.193818	0.143194
80	33	9	3	0.446445	0.252627
80	33	9	4	0.717116	0.270671
80	33	9	5	0.899662	0.182546
80	33	9	6	0.977106	0.077444
80	33	9	7	0.997020	0.019914
80	33	9	8	0.999834	0.002814
80	33	9	9	1.000000	0.000166
80	33	10	0	0.003145	0.003145
80	33	10	1	0.030456	0.027311
80	33	10	2	0.131297	0.100841
80	33	10	3	0.339701	0.208404
80	33	10	4	0.606560	0.266859
80	33	10	5	0.827672	0.221112
80	33	10	6	0.947655	0.119983
80	33	10	7	0.989727	0.042072
80	33	10	8	0.998843	0.009116

Table for $N = 2$, $n = 1$, through $N = 100$, $n = 50$

N	n	k	x	P(x)	p(x)
80	33	10	9	0.999944	0.001101
80	33	10	10	1.000000	0.000056
80	33	11	0	0.001662	0.001662
80	33	11	1	0.017971	0.016309
80	33	11	2	0.086639	0.068668
80	33	11	3	0.250385	0.163746
80	33	11	4	0.496004	0.245619
80	33	11	5	0.739227	0.243223
80	33	11	6	0.901376	0.162149
80	33	11	7	0.974100	0.072724
80	33	11	8	0.995587	0.021487
80	33	11	9	0.999566	0.003979
80	33	11	10	0.999981	0.000415
80	33	11	11	1.000000	0.000018
80	33	12	0	0.000867	0.000867
80	33	12	1	0.010407	0.009540
80	33	12	2	0.055788	0.045380
80	33	12	3	0.179191	0.123403
80	33	12	4	0.392773	0.213582
80	33	12	5	0.640528	0.247755
80	33	12	6	0.837926	0.197398
80	33	12	7	0.946697	0.108771
80	33	12	8	0.987802	0.041105
80	33	12	9	0.998182	0.010380
80	33	12	10	0.999843	0.001661
80	33	12	11	0.999994	0.000151
80	33	12	12	1.000000	0.000006
80	33	13	0	0.000446	0.000446
80	33	13	1	0.005918	0.005472
80	33	13	2	0.035100	0.029182
80	33	13	3	0.124748	0.089649
80	33	13	4	0.301686	0.176938
80	33	13	5	0.538511	0.236825
80	33	13	6	0.759548	0.221036
80	33	13	7	0.905108	0.145561
80	33	13	8	0.972690	0.067582
80	33	13	9	0.994519	0.021829
80	33	13	10	0.999281	0.004763
80	33	13	11	0.999945	0.000664
80	33	13	12	0.999998	0.000053
80	33	13	13	1.000000	0.000002
80	33	14	0	0.000227	0.000227
80	33	14	1	0.003305	0.003078
80	33	14	2	0.021598	0.018293
80	33	14	3	0.084607	0.063009
80	33	14	4	0.225101	0.140494
80	33	14	5	0.439540	0.214438
80	33	14	6	0.670473	0.230934
80	33	14	7	0.848622	0.178149
80	33	14	8	0.947473	0.098851

N	n	k	x	P(x)	p(x)
80	33	14	9	0.986699	0.039227
80	33	14	10	0.997646	0.010947
80	33	14	11	0.999727	0.002081
80	33	14	12	0.999981	0.000254
80	33	14	13	0.999999	0.000018
80	33	14	14	1.000000	0.000001
80	33	15	0	0.000113	0.000113
80	33	15	1	0.001812	0.001699
80	33	15	2	0.013006	0.011193
80	33	15	3	0.055967	0.042961
80	33	15	4	0.163369	0.107402
80	33	15	5	0.348566	0.185197
80	33	15	6	0.576000	0.227435
80	33	15	7	0.778442	0.202442
80	33	15	8	0.910029	0.131587
80	33	15	9	0.972435	0.062406
80	33	15	10	0.993831	0.021396
80	33	15	11	0.999034	0.005202
80	33	15	12	0.999901	0.000867
80	33	15	13	0.999994	0.000093
80	33	15	14	1.000000	0.000006
80	33	15	15	1.000000	0.000000
80	33	16	0	0.000056	0.000056
80	33	16	1	0.000976	0.000920
80	33	16	2	0.007667	0.006691
80	33	16	3	0.036138	0.028471
80	33	16	4	0.115451	0.079313
80	33	16	5	0.268789	0.153338
80	33	16	6	0.481528	0.212739
80	33	16	7	0.697466	0.215938
80	33	16	8	0.859419	0.161953
80	33	16	9	0.949393	0.089974
80	33	16	10	0.986260	0.036867
80	33	16	11	0.997273	0.011012
80	33	16	12	0.999620	0.002348
80	33	16	13	0.999965	0.000345
80	33	16	14	0.999998	0.000033
80	33	16	15	1.000000	0.000002
80	33	16	16	1.000000	0.000000
80	33	17	0	0.000027	0.000027
80	33	17	1	0.000516	0.000489
80	33	17	2	0.004426	0.003910
80	33	17	3	0.022793	0.018366
80	33	17	4	0.079512	0.056720
80	33	17	5	0.201703	0.122191
80	33	17	6	0.391778	0.190075
80	33	17	7	0.609741	0.217962
80	33	17	8	0.796156	0.186415
80	33	17	9	0.915653	0.119497
80	33	17	10	0.973011	0.057359

N	n	k	x	P(x)	p(x)
80	33	17	11	0.993487	0.020476
80	33	17	12	0.998850	0.005363
80	33	17	13	0.999857	0.001007
80	33	17	14	0.999988	0.000131
80	33	17	15	0.999999	0.000011
80	33	17	16	1.000000	0.000001
80	33	17	17	1.000000	0.000000
80	33	18	0	0.000013	0.000013
80	33	18	1	0.000268	0.000255
80	33	18	2	0.002502	0.002234
80	33	18	3	0.014047	0.011545
80	33	18	4	0.053403	0.039357
80	33	18	5	0.147396	0.093993
80	33	18	6	0.310318	0.162921
80	33	18	7	0.519788	0.209470
80	33	18	8	0.722182	0.202394
80	33	18	9	0.870130	0.147949
80	33	18	10	0.952071	0.081941
80	33	18	11	0.986337	0.034266
80	33	18	12	0.997063	0.010726
80	33	18	13	0.999538	0.002475
80	33	18	14	0.999949	0.000411
80	33	18	15	0.999996	0.000047
80	33	18	16	1.000000	0.000004
80	33	18	17	1.000000	0.000000
80	33	18	18	1.000000	0.000000
80	33	19	0	0.000006	0.000006
80	33	19	1	0.000136	0.000130
80	33	19	2	0.001385	0.001249
80	33	19	3	0.008460	0.007076
80	33	19	4	0.034995	0.026534
80	33	19	5	0.104948	0.069953
80	33	19	6	0.239368	0.134420
80	33	19	7	0.431946	0.192578
80	33	19	8	0.640571	0.208626
80	33	19	9	0.812860	0.172288
80	33	19	10	0.921673	0.108814
80	33	19	11	0.974178	0.052505
80	33	19	12	0.993430	0.019252
80	33	19	13	0.998739	0.005310
80	33	19	14	0.999823	0.001084
80	33	19	15	0.999982	0.000160
80	33	19	16	0.999999	0.000016
80	33	19	17	1.000000	0.000001
80	33	19	18	1.000000	0.000000
80	33	19	19	1.000000	0.000000
80	33	20	0	0.000003	0.000003
80	33	20	1	0.000068	0.000065
80	33	20	2	0.000750	0.000682
80	33	20	3	0.004981	0.004230

N	n	k	x	P(x)	p(x)
80	33	20	4	0.022380	0.017399
80	33	20	5	0.072838	0.050458
80	33	20	6	0.179871	0.107032
80	33	20	7	0.349863	0.169993
80	33	20	8	0.555069	0.205206
80	33	20	9	0.745074	0.190005
80	33	20	10	0.880645	0.135571
80	33	20	11	0.955242	0.074597
80	33	20	12	0.986802	0.031560
80	33	20	13	0.996998	0.010196
80	33	20	14	0.999485	0.002487
80	33	20	15	0.999935	0.000450
80	33	20	16	0.999994	0.000059
80	33	20	17	1.000000	0.000005
80	33	20	18	1.000000	0.000000
80	33	20	19	1.000000	0.000000
80	33	20	20	1.000000	0.000000
80	33	21	0	0.000001	0.000001
80	33	21	1	0.000033	0.000032
80	33	21	2	0.000398	0.000365
80	33	21	3	0.002865	0.002468
80	33	21	4	0.013970	0.011105
80	33	21	5	0.049291	0.035321
80	33	21	6	0.131706	0.082415
80	33	21	7	0.276200	0.144494
80	33	21	8	0.469567	0.193367
80	33	21	9	0.669072	0.199505
80	33	21	10	0.828676	0.159604
80	33	21	11	0.927890	0.099213
80	33	21	12	0.975756	0.047866
80	33	21	13	0.993600	0.017844
80	33	21	14	0.998698	0.005098
80	33	21	15	0.999800	0.001103
80	33	21	16	0.999977	0.000177
80	33	21	17	0.999998	0.000021
80	33	21	18	1.000000	0.000002
80	33	21	19	1.000000	0.000000
80	33	21	20	1.000000	0.000000
80	33	21	21	1.000000	0.000000
80	33	22	0	0.000001	0.000001
80	33	22	1	0.000016	0.000015
80	33	22	2	0.000206	0.000190
80	33	22	3	0.001611	0.001404
80	33	22	4	0.008512	0.006901
80	33	22	5	0.032529	0.024017
80	33	22	6	0.093991	0.061462
80	33	22	7	0.212525	0.118534
80	33	22	8	0.387632	0.175107
80	33	22	9	0.587917	0.200286
80	33	22	10	0.766458	0.178540

Table for $N = 2$, $n = 1$, through $N = 100$, $n = 50$

N	n	k	x	P(x)	p(x)
80	33	22	11	0.890895	0.124437
80	33	22	12	0.958719	0.067824
80	33	22	13	0.987551	0.028832
80	33	22	14	0.997056	0.009505
80	33	22	15	0.999464	0.002408
80	33	22	16	0.999926	0.000463
80	33	22	17	0.999992	0.000066
80	33	22	18	0.999999	0.000007
80	33	22	19	1.000000	0.000000
80	33	22	20	1.000000	0.000000
80	33	22	21	1.000000	0.000000
80	33	22	22	1.000000	0.000000
80	33	23	0	0.000000	0.000000
80	33	23	1	0.000007	0.000007
80	33	23	2	0.000104	0.000097
80	33	23	3	0.000884	0.000780
80	33	23	4	0.005061	0.004177
80	33	23	5	0.020934	0.015873
80	33	23	6	0.065379	0.044445
80	33	23	7	0.159389	0.094010
80	33	23	8	0.312154	0.152766
80	33	23	9	0.505040	0.192886
80	33	23	10	0.695657	0.190617
80	33	23	11	0.843695	0.148037
80	33	23	12	0.934162	0.090467
80	33	23	13	0.977609	0.043447
80	33	23	14	0.993942	0.016333
80	33	23	15	0.998717	0.004774
80	33	23	16	0.999791	0.001074
80	33	23	17	0.999974	0.000183
80	33	23	18	0.999998	0.000023
80	33	23	19	1.000000	0.000002
80	33	23	20	1.000000	0.000000
80	33	23	21	1.000000	0.000000
80	33	23	22	1.000000	0.000000
80	33	23	23	1.000000	0.000000
80	33	24	0	0.000000	0.000000
80	33	24	1	0.000003	0.000003
80	33	24	2	0.000052	0.000048
80	33	24	3	0.000474	0.000422
80	33	24	4	0.002936	0.002462
80	33	24	5	0.013137	0.010201
80	33	24	6	0.044326	0.031189
80	33	24	7	0.116507	0.072181
80	33	24	8	0.245152	0.128645
80	33	24	9	0.423825	0.178673
80	33	24	10	0.618742	0.194916
80	33	24	11	0.786557	0.167816
80	33	24	12	0.900832	0.114275
80	33	24	13	0.962364	0.061532

N	n	k	x	P(x)	p(x)
80	33	24	14	0.988498	0.026133
80	33	24	15	0.997209	0.008711
80	33	24	16	0.999470	0.002262
80	33	24	17	0.999923	0.000452
80	33	24	18	0.999991	0.000069
80	33	24	19	0.999999	0.000008
80	33	24	20	1.000000	0.000001
80	33	24	21	1.000000	0.000000
80	33	24	22	1.000000	0.000000
80	33	24	23	1.000000	0.000000
80	33	24	24	1.000000	0.000000
80	33	25	0	0.000000	0.000000
80	33	25	1	0.000001	0.000001
80	33	25	2	0.000025	0.000023
80	33	25	3	0.000248	0.000223
80	33	25	4	0.001661	0.001413
80	33	25	5	0.008037	0.006376
80	33	25	6	0.029289	0.021252
80	33	25	7	0.082995	0.053706
80	33	25	8	0.187721	0.104727
80	33	25	9	0.347251	0.159530
80	33	25	10	0.538687	0.191436
80	33	25	11	0.720630	0.181943
80	33	25	12	0.857979	0.137349
80	33	25	13	0.940389	0.082410
80	33	25	14	0.979631	0.039243
80	33	25	15	0.994409	0.014778
80	33	25	16	0.998784	0.004375
80	33	25	17	0.999794	0.001010
80	33	25	18	0.999973	0.000179
80	33	25	19	0.999997	0.000024
80	33	25	20	1.000000	0.000002
80	33	25	21	1.000000	0.000000
80	33	25	22	1.000000	0.000000
80	33	25	23	1.000000	0.000000
80	33	25	24	1.000000	0.000000
80	33	25	25	1.000000	0.000000
80	33	26	0	0.000000	0.000000
80	33	26	1	0.000001	0.000001
80	33	26	2	0.000012	0.000011
80	33	26	3	0.000126	0.000114
80	33	26	4	0.000916	0.000790
80	33	26	5	0.004791	0.003875
80	33	26	6	0.018856	0.014065
80	33	26	7	0.057606	0.038751
80	33	26	8	0.140118	0.082512
80	33	26	9	0.277638	0.137520
80	33	26	10	0.458632	0.180994
80	33	26	11	0.647853	0.189221
80	33	26	12	0.805537	0.157684

N	n	k	x	P(x)	p(x)
80	33	26	13	0.910421	0.104885
80	33	26	14	0.966075	0.055653
80	33	26	15	0.989573	0.023498
80	33	26	16	0.997432	0.007859
80	33	26	17	0.999500	0.002068
80	33	26	18	0.999924	0.000424
80	33	26	19	0.999991	0.000067
80	33	26	20	0.999999	0.000008
80	33	26	21	1.000000	0.000001
80	33	26	22	1.000000	0.000000
80	33	26	23	1.000000	0.000000
80	33	26	24	1.000000	0.000000
80	33	26	25	1.000000	0.000000
80	33	26	26	1.000000	0.000000
80	33	27	0	0.000000	0.000000
80	33	27	1	0.000000	0.000000
80	33	27	2	0.000005	0.000005
80	33	27	3	0.000063	0.000057
80	33	27	4	0.000492	0.000429
80	33	27	5	0.002782	0.002290
80	33	27	6	0.011823	0.009042
80	33	27	7	0.038949	0.027125
80	33	27	8	0.101918	0.062970
80	33	27	9	0.216518	0.114600
80	33	27	10	0.381542	0.165024
80	33	27	11	0.570763	0.189221
80	33	27	12	0.744215	0.173452
80	33	27	13	0.871575	0.127360
80	33	27	14	0.946493	0.074918
80	33	27	15	0.981740	0.035247
80	33	27	16	0.994958	0.013218
80	33	27	17	0.998887	0.003930
80	33	27	18	0.999806	0.000919
80	33	27	19	0.999974	0.000167
80	33	27	20	0.999997	0.000023
80	33	27	21	1.000000	0.000002
80	33	27	22	1.000000	0.000000
80	33	27	23	1.000000	0.000000
80	33	27	24	1.000000	0.000000
80	33	27	25	1.000000	0.000000
80	33	27	26	1.000000	0.000000
80	33	27	27	1.000000	0.000000
80	33	28	0	0.000000	0.000000
80	33	28	1	0.000000	0.000000
80	33	28	2	0.000002	0.000002
80	33	28	3	0.000030	0.000028
80	33	28	4	0.000257	0.000227
80	33	28	5	0.001572	0.001315
80	33	28	6	0.007217	0.005645
80	33	28	7	0.025642	0.018425

N	n	k	x	P(x)	p(x)
80	33	28	8	0.072216	0.046574
80	33	28	9	0.164624	0.092408
80	33	28	10	0.309928	0.145304
80	33	28	11	0.492218	0.182290
80	33	28	12	0.675489	0.183270
80	33	28	13	0.823515	0.148026
80	33	28	14	0.919636	0.096121
80	33	28	15	0.969769	0.050134
80	33	28	16	0.990718	0.020949
80	33	28	17	0.997701	0.006983
80	33	28	18	0.999546	0.001845
80	33	28	19	0.999930	0.000383
80	33	28	20	0.999991	0.000062
80	33	28	21	0.999999	0.000008
80	33	28	22	1.000000	0.000001
80	33	28	23	1.000000	0.000000
80	33	28	24	1.000000	0.000000
80	33	28	25	1.000000	0.000000
80	33	28	26	1.000000	0.000000
80	33	28	27	1.000000	0.000000
80	33	28	28	1.000000	0.000000
80	33	29	0	0.000000	0.000000
80	33	29	1	0.000000	0.000000
80	33	29	2	0.000001	0.000001
80	33	29	3	0.000014	0.000013
80	33	29	4	0.000130	0.000116
80	33	29	5	0.000864	0.000733
80	33	29	6	0.004286	0.003422
80	33	29	7	0.016429	0.012144
80	33	29	8	0.049824	0.033395
80	33	29	9	0.121974	0.072149
80	33	29	10	0.245659	0.123685
80	33	29	11	0.415095	0.169437
80	33	29	12	0.601475	0.186380
80	33	29	13	0.766582	0.165106
80	33	29	14	0.884514	0.117933
80	33	29	15	0.952415	0.067901
80	33	29	16	0.983869	0.031454
80	33	29	17	0.995552	0.011683
80	33	29	18	0.999014	0.003462
80	33	29	19	0.999826	0.000812
80	33	29	20	0.999976	0.000150
80	33	29	21	0.999997	0.000021
80	33	29	22	1.000000	0.000002
80	33	29	23	1.000000	0.000000
80	33	29	24	1.000000	0.000000
80	33	29	25	1.000000	0.000000
80	33	29	26	1.000000	0.000000
80	33	29	27	1.000000	0.000000
80	33	29	28	1.000000	0.000000

Table for $N = 2$, $n = 1$, through $N = 100$, $n = 50$

N	n	k	x	P(x)	p(x)
80	33	29	29	1.000000	0.000000
80	33	30	0	0.000000	0.000000
80	33	30	1	0.000000	0.000000
80	33	30	2	0.000000	0.000000
80	33	30	3	0.000006	0.000006
80	33	30	4	0.000064	0.000058
80	33	30	5	0.000461	0.000397
80	33	30	6	0.002474	0.002013
80	33	30	7	0.010239	0.007764
80	33	30	8	0.033454	0.023216
80	33	30	9	0.088021	0.054567
80	33	30	10	0.189879	0.101858
80	33	30	11	0.342005	0.152126
80	33	30	12	0.524731	0.182726
80	33	30	13	0.701834	0.177103
80	33	30	14	0.840579	0.138745
80	33	30	15	0.928450	0.087872
80	33	30	16	0.973385	0.044934
80	33	30	17	0.991887	0.018502
80	33	30	18	0.997996	0.006109
80	33	30	19	0.999603	0.001608
80	33	30	20	0.999938	0.000335
80	33	30	21	0.999992	0.000054
80	33	30	22	0.999999	0.000007
80	33	30	23	1.000000	0.000001
80	33	30	24	1.000000	0.000000
80	33	30	25	1.000000	0.000000
80	33	30	26	1.000000	0.000000
80	33	30	27	1.000000	0.000000
80	33	30	28	1.000000	0.000000
80	33	30	29	1.000000	0.000000
80	33	30	30	1.000000	0.000000
80	33	31	0	0.000000	0.000000
80	33	31	1	0.000000	0.000000
80	33	31	2	0.000000	0.000000
80	33	31	3	0.000003	0.000003
80	33	31	4	0.000031	0.000028
80	33	31	5	0.000239	0.000208
80	33	31	6	0.001387	0.001148
80	33	31	7	0.006201	0.004814
80	33	31	8	0.021846	0.015645
80	33	31	9	0.061829	0.039983
80	33	31	10	0.143025	0.081195
80	33	31	11	0.275070	0.132045
80	33	31	12	0.447986	0.172916
80	33	31	13	0.630993	0.183007
80	33	31	14	0.787856	0.156863
80	33	31	15	0.896817	0.108961
80	33	31	16	0.958107	0.061290
80	33	31	17	0.985966	0.027859

N	n	k	x	P(x)	p(x)
80	33	31	18	0.996163	0.010197
80	33	31	19	0.999153	0.002990
80	33	31	20	0.999851	0.000698
80	33	31	21	0.999979	0.000128
80	33	31	22	0.999998	0.000018
80	33	31	23	1.000000	0.000002
80	33	31	24	1.000000	0.000000
80	33	31	25	1.000000	0.000000
80	33	31	26	1.000000	0.000000
80	33	31	27	1.000000	0.000000
80	33	31	28	1.000000	0.000000
80	33	31	29	1.000000	0.000000
80	33	31	30	1.000000	0.000000
80	33	31	31	1.000000	0.000000
80	33	32	0	0.000000	0.000000
80	33	32	1	0.000000	0.000000
80	33	32	2	0.000000	0.000000
80	33	32	3	0.000001	0.000001
80	33	32	4	0.000014	0.000013
80	33	32	5	0.000120	0.000106
80	33	32	6	0.000754	0.000634
80	33	32	7	0.003647	0.002892
80	33	32	8	0.013864	0.010217
80	33	32	9	0.042246	0.028382
80	33	32	10	0.104912	0.062667
80	33	32	11	0.215784	0.110872
80	33	32	12	0.373879	0.158095
80	33	32	13	0.556296	0.182417
80	33	32	14	0.727031	0.170735
80	33	32	15	0.856790	0.129759
80	33	32	16	0.936843	0.080053
80	33	32	17	0.976869	0.040026
80	33	32	18	0.993042	0.016172
80	33	32	19	0.998299	0.005257
80	33	32	20	0.999666	0.001367
80	33	32	21	0.999948	0.000282
80	33	32	22	0.999994	0.000046
80	33	32	23	0.999999	0.000006
80	33	32	24	1.000000	0.000001
80	33	32	25	1.000000	0.000000
80	33	32	26	1.000000	0.000000
80	33	32	27	1.000000	0.000000
80	33	32	28	1.000000	0.000000
80	33	32	29	1.000000	0.000000
80	33	32	30	1.000000	0.000000
80	33	32	31	1.000000	0.000000
80	33	32	32	1.000000	0.000000
80	33	33	0	0.000000	0.000000
80	33	33	1	0.000000	0.000000
80	33	33	2	0.000000	0.000000

N	n	k	x	P(x)	p(x)
80	33	33	3	0.000000	0.000000
80	33	33	4	0.000006	0.000006
80	33	33	5	0.000058	0.000052
80	33	33	6	0.000398	0.000339
80	33	33	7	0.002080	0.001683
80	33	33	8	0.008543	0.006462
80	33	33	9	0.028055	0.019512
80	33	33	10	0.074885	0.046830
80	33	33	11	0.164968	0.090083
80	33	33	12	0.304712	0.139745
80	33	33	13	0.480289	0.175576
80	33	33	14	0.659449	0.179160
80	33	33	15	0.808131	0.148682
80	33	33	16	0.908491	0.100360
80	33	33	17	0.963527	0.055036
80	33	33	18	0.987988	0.024461
80	33	33	19	0.996765	0.008778
80	33	33	20	0.999296	0.002530
80	33	33	21	0.999877	0.000582
80	33	33	22	0.999983	0.000106
80	33	33	23	0.999998	0.000015
80	33	33	24	1.000000	0.000002
80	33	33	25	1.000000	0.000000
80	33	33	26	1.000000	0.000000
80	33	33	27	1.000000	0.000000
80	33	33	28	1.000000	0.000000
80	33	33	29	1.000000	0.000000
80	33	33	30	1.000000	0.000000
80	33	33	31	1.000000	0.000000
80	33	33	32	1.000000	0.000000
80	33	33	33	1.000000	0.000000
80	34	1	0	0.575000	0.575000
80	34	1	1	1.000000	0.425000
80	34	2	0	0.327532	0.327532
80	34	2	1	0.822468	0.494937
80	34	2	2	1.000000	0.177532
80	34	3	0	0.184761	0.184761
80	34	3	1	0.613072	0.428311
80	34	3	2	0.927166	0.314094
80	34	3	3	1.000000	0.072833
80	34	4	0	0.103178	0.103178
80	34	4	1	0.429510	0.326332
80	34	4	2	0.796634	0.367123
80	34	4	3	0.970677	0.174044
80	34	4	4	1.000000	0.029323
80	34	5	0	0.057020	0.057020
80	34	5	1	0.287814	0.230794
80	34	5	2	0.642055	0.354242
80	34	5	3	0.899686	0.257630
80	34	5	4	0.988425	0.088739

N	n	k	x	P(x)	p(x)
80	34	5	5	1.000000	0.011575
80	34	6	0	0.031171	0.031171
80	34	6	1	0.186264	0.155094
80	34	6	2	0.490912	0.304648
80	34	6	3	0.793199	0.302286
80	34	6	4	0.952929	0.159731
80	34	6	5	0.995524	0.042595
80	34	6	6	1.000000	0.004476
80	34	7	0	0.016849	0.016849
80	34	7	1	0.117101	0.100252
80	34	7	2	0.359173	0.242072
80	34	7	3	0.666565	0.307393
80	34	7	4	0.888174	0.221609
80	34	7	5	0.978832	0.090658
80	34	7	6	0.998306	0.019475
80	34	7	7	1.000000	0.001693
80	34	8	0	0.009002	0.009002
80	34	8	1	0.071782	0.062780
80	34	8	2	0.253059	0.181277
80	34	8	3	0.536029	0.282970
80	34	8	4	0.797102	0.261073
80	34	8	5	0.942817	0.145715
80	34	8	6	0.990837	0.048020
80	34	8	7	0.999374	0.008537
80	34	8	8	1.000000	0.000626
80	34	9	0	0.004751	0.004751
80	34	9	1	0.043007	0.038257
80	34	9	2	0.172491	0.129484
80	34	9	3	0.414194	0.241703
80	34	9	4	0.688321	0.274127
80	34	9	5	0.884126	0.195805
80	34	9	6	0.972162	0.088036
80	34	9	7	0.996172	0.024010
80	34	9	8	0.999774	0.003601
80	34	9	9	1.000000	0.000226
80	34	10	0	0.002476	0.002476
80	34	10	1	0.025226	0.022750
80	34	10	2	0.114132	0.088906
80	34	10	3	0.308662	0.194530
80	34	10	4	0.572493	0.263831
80	34	10	5	0.804149	0.231656
80	34	10	6	0.937444	0.133294
80	34	10	7	0.987042	0.049598
80	34	10	8	0.998455	0.011413
80	34	10	9	0.999920	0.001465
80	34	10	10	1.000000	0.000080
80	34	11	0	0.001273	0.001273
80	34	11	1	0.014501	0.013228
80	34	11	2	0.073490	0.058989
80	34	11	3	0.222513	0.149024

N	n	k	x	P(x)	p(x)
80	34	11	4	0.459423	0.236909
80	34	11	5	0.708177	0.248755
80	34	11	6	0.884126	0.175949
80	34	11	7	0.967911	0.083785
80	34	11	8	0.994216	0.026305
80	34	11	9	0.999397	0.005181
80	34	11	10	0.999973	0.000576
80	34	11	11	1.000000	0.000027
80	34	12	0	0.000646	0.000646
80	34	12	1	0.008175	0.007529
80	34	12	2	0.046133	0.037958
80	34	12	3	0.155561	0.109428
80	34	12	4	0.356418	0.200858
80	34	12	5	0.603628	0.247210
80	34	12	6	0.812727	0.209098
80	34	12	7	0.935126	0.122399
80	34	12	8	0.984304	0.049178
80	34	12	9	0.997520	0.013216
80	34	12	10	0.999772	0.002253
80	34	12	11	0.999991	0.000218
80	34	12	12	1.000000	0.000009
80	34	13	0	0.000323	0.000323
80	34	13	1	0.004521	0.004198
80	34	13	2	0.028270	0.023749
80	34	13	3	0.105674	0.077404
80	34	13	4	0.267805	0.162130
80	34	13	5	0.498201	0.230396
80	34	13	6	0.726627	0.228427
80	34	13	7	0.886526	0.159899
80	34	13	8	0.965500	0.078974
80	34	13	9	0.992661	0.027160
80	34	13	10	0.998977	0.006316
80	34	13	11	0.999917	0.000940
80	34	13	12	0.999997	0.000080
80	34	13	13	1.000000	0.000003
80	34	14	0	0.000159	0.000159
80	34	14	1	0.002453	0.002294
80	34	14	2	0.016927	0.014474
80	34	14	3	0.069860	0.052933
80	34	14	4	0.195209	0.125349
80	34	14	5	0.398477	0.203268
80	34	14	6	0.631165	0.232688
80	34	14	7	0.822089	0.190924
80	34	14	8	0.934854	0.112764
80	34	14	9	0.982526	0.047673
80	34	14	10	0.996715	0.014188
80	34	14	11	0.999594	0.002880
80	34	14	12	0.999971	0.000376
80	34	14	13	0.999999	0.000028
80	34	14	14	1.000000	0.000001

N	n	k	x	P(x)	p(x)
80	34	15	0	0.000077	0.000077
80	34	15	1	0.001306	0.001229
80	34	15	2	0.009910	0.008603
80	34	15	3	0.044998	0.035088
80	34	15	4	0.138232	0.093235
80	34	15	5	0.309162	0.170930
80	34	15	6	0.532449	0.223287
80	34	15	7	0.743984	0.211535
80	34	15	8	0.890431	0.146447
80	34	15	9	0.964469	0.074037
80	34	15	10	0.991555	0.027087
80	34	15	11	0.998591	0.007036
80	34	15	12	0.999845	0.001254
80	34	15	13	0.999990	0.000145
80	34	15	14	1.000000	0.000010
80	34	15	15	1.000000	0.000000
80	34	16	0	0.000037	0.000037
80	34	16	1	0.000682	0.000645
80	34	16	2	0.005674	0.004992
80	34	16	3	0.028263	0.022589
80	34	16	4	0.095201	0.066938
80	34	16	5	0.232901	0.137700
80	34	16	6	0.436264	0.203363
80	34	16	7	0.656116	0.219852
80	34	16	8	0.831852	0.175737
80	34	16	9	0.935993	0.104140
80	34	16	10	0.981554	0.045561
80	34	16	11	0.996101	0.014547
80	34	16	12	0.999421	0.003319
80	34	16	13	0.999943	0.000523
80	34	16	14	0.999997	0.000053
80	34	16	15	1.000000	0.000003
80	34	16	16	1.000000	0.000000
80	34	17	0	0.000017	0.000017
80	34	17	1	0.000349	0.000332
80	34	17	2	0.003178	0.002829
80	34	17	3	0.017322	0.014144
80	34	17	4	0.063824	0.046502
80	34	17	5	0.170506	0.106682
80	34	17	6	0.347293	0.176787
80	34	17	7	0.563366	0.216073
80	34	17	8	0.760459	0.197094
80	34	17	9	0.895313	0.134853
80	34	17	10	0.964468	0.069156
80	34	17	11	0.990873	0.026405
80	34	17	12	0.998280	0.007406
80	34	17	13	0.999772	0.001492
80	34	17	14	0.999980	0.000208
80	34	17	15	0.999999	0.000019
80	34	17	16	1.000000	0.000001

N	n	k	x	P(x)	p(x)
80	34	17	17	1.000000	0.000000
80	34	18	0	0.000008	0.000008
80	34	18	1	0.000175	0.000167
80	34	18	2	0.001741	0.001566
80	34	18	3	0.010362	0.008621
80	34	18	4	0.041680	0.031318
80	34	18	5	0.121398	0.079718
80	34	18	6	0.268721	0.147322
80	34	18	7	0.470763	0.202042
80	34	18	8	0.679119	0.208356
80	34	18	9	0.841800	0.162680
80	34	18	10	0.938123	0.096324
80	34	18	11	0.981233	0.043110
80	34	18	12	0.995693	0.014460
80	34	18	13	0.999274	0.003581
80	34	18	14	0.999914	0.000639
80	34	18	15	0.999993	0.000079
80	34	18	16	1.000000	0.000006
80	34	18	17	1.000000	0.000000
80	34	18	18	1.000000	0.000000
80	34	19	0	0.000004	0.000004
80	34	19	1	0.000086	0.000083
80	34	19	2	0.000933	0.000847
80	34	19	3	0.006052	0.005119
80	34	19	4	0.026526	0.020474
80	34	19	5	0.084111	0.057584
80	34	19	6	0.202188	0.118077
80	34	19	7	0.382777	0.180589
80	34	19	8	0.591744	0.208967
80	34	19	9	0.776203	0.184459
80	34	19	10	0.900837	0.124634
80	34	19	11	0.965241	0.064404
80	34	19	12	0.990562	0.025321
80	34	19	13	0.998061	0.007499
80	34	19	14	0.999708	0.001646
80	34	19	15	0.999969	0.000261
80	34	19	16	0.999998	0.000029
80	34	19	17	1.000000	0.000002
80	34	19	18	1.000000	0.000000
80	34	19	19	1.000000	0.000000
80	34	20	0	0.000002	0.000002
80	34	20	1	0.000042	0.000040
80	34	20	2	0.000489	0.000447
80	34	20	3	0.003450	0.002962
80	34	20	4	0.016457	0.013006
80	34	20	5	0.056734	0.040278
80	34	20	6	0.147989	0.091254
80	34	20	7	0.302844	0.154856
80	34	20	8	0.502676	0.199832
80	34	20	9	0.700605	0.197929

N	n	k	x	P(x)	p(x)
80	34	20	10	0.851800	0.151196
80	34	20	11	0.940957	0.089157
80	34	20	12	0.981430	0.040473
80	34	20	13	0.995480	0.014050
80	34	20	14	0.999168	0.003688
80	34	20	15	0.999887	0.000720
80	34	20	16	0.999989	0.000102
80	34	20	17	0.999999	0.000010
80	34	20	18	1.000000	0.000001
80	34	20	19	1.000000	0.000000
80	34	20	20	1.000000	0.000000
80	34	21	0	0.000001	0.000001
80	34	21	1	0.000020	0.000019
80	34	21	2	0.000250	0.000231
80	34	21	3	0.001920	0.001670
80	34	21	4	0.009954	0.008033
80	34	21	5	0.037267	0.027313
80	34	21	6	0.105403	0.068136
80	34	21	7	0.233159	0.127756
80	34	21	8	0.416082	0.182923
80	34	21	9	0.618135	0.202052
80	34	21	10	0.791322	0.173188
80	34	21	11	0.906781	0.115458
80	34	21	12	0.966590	0.059810
80	34	21	13	0.990562	0.023972
80	34	21	14	0.997939	0.007376
80	34	21	15	0.999660	0.001721
80	34	21	16	0.999959	0.000299
80	34	21	17	0.999996	0.000038
80	34	21	18	1.000000	0.000003
80	34	21	19	1.000000	0.000000
80	34	21	20	1.000000	0.000000
80	34	21	21	1.000000	0.000000
80	34	22	0	0.000000	0.000000
80	34	22	1	0.000009	0.000009
80	34	22	2	0.000125	0.000116
80	34	22	3	0.001043	0.000918
80	34	22	4	0.005869	0.004826
80	34	22	5	0.023842	0.017973
80	34	22	6	0.073067	0.049226
80	34	22	7	0.174695	0.101627
80	34	22	8	0.335472	0.160777
80	34	22	9	0.532519	0.197047
80	34	22	10	0.720873	0.188354
80	34	22	11	0.861771	0.140898
80	34	22	12	0.944288	0.082517
80	34	22	13	0.982030	0.037742
80	34	22	14	0.995438	0.013408
80	34	22	15	0.999105	0.003667
80	34	22	16	0.999867	0.000762

Table for $N = 2$, $n = 1$, through $N = 100$, $n = 50$

N	n	k	x	P(x)	p(x)
80	34	22	17	0.999986	0.000118
80	34	22	18	0.999999	0.000013
80	34	22	19	1.000000	0.000001
80	34	22	20	1.000000	0.000000
80	34	22	21	1.000000	0.000000
80	34	22	22	1.000000	0.000000
80	34	23	0	0.000000	0.000000
80	34	23	1	0.000004	0.000004
80	34	23	2	0.000061	0.000057
80	34	23	3	0.000552	0.000491
80	34	23	4	0.003373	0.002820
80	34	23	5	0.014855	0.011483
80	34	23	6	0.049303	0.034448
80	34	23	7	0.127385	0.078082
80	34	23	8	0.263399	0.136014
80	34	23	9	0.447585	0.184186
80	34	23	10	0.642933	0.195348
80	34	23	11	0.805898	0.162964
80	34	23	12	0.912989	0.107091
80	34	23	13	0.968365	0.055376
80	34	23	14	0.990815	0.022450
80	34	23	15	0.997904	0.007089
80	34	23	16	0.999631	0.001727
80	34	23	17	0.999951	0.000320
80	34	23	18	0.999995	0.000044
80	34	23	19	1.000000	0.000004
80	34	23	20	1.000000	0.000000
80	34	23	21	1.000000	0.000000
80	34	23	22	1.000000	0.000000
80	34	23	23	1.000000	0.000000
80	34	24	0	0.000000	0.000000
80	34	24	1	0.000002	0.000002
80	34	24	2	0.000029	0.000027
80	34	24	3	0.000285	0.000256
80	34	24	4	0.001888	0.001603
80	34	24	5	0.009013	0.007125
80	34	24	6	0.032381	0.023368
80	34	24	7	0.090399	0.058018
80	34	24	8	0.201358	0.110959
80	34	24	9	0.366802	0.165444
80	34	24	10	0.560681	0.193880
80	34	24	11	0.740140	0.179459
80	34	24	12	0.871655	0.131515
80	34	24	13	0.947963	0.076308
80	34	24	14	0.982938	0.034974
80	34	24	15	0.995541	0.012603
80	34	24	16	0.999086	0.003545
80	34	24	17	0.999855	0.000770
80	34	24	18	0.999983	0.000127
80	34	24	19	0.999998	0.000016

N	n	k	x	P(x)	p(x)
80	34	24	20	1.000000	0.000001
80	34	24	21	1.000000	0.000000
80	34	24	22	1.000000	0.000000
80	34	24	23	1.000000	0.000000
80	34	24	24	1.000000	0.000000
80	34	25	0	0.000000	0.000000
80	34	25	1	0.000001	0.000001
80	34	25	2	0.000013	0.000013
80	34	25	3	0.000143	0.000130
80	34	25	4	0.001029	0.000886
80	34	25	5	0.005324	0.004294
80	34	25	6	0.020697	0.015374
80	34	25	7	0.062426	0.041729
80	34	25	8	0.149841	0.087415
80	34	25	9	0.292943	0.143102
80	34	25	10	0.477590	0.184647
80	34	25	11	0.666434	0.188844
80	34	25	12	0.819989	0.153555
80	34	25	13	0.919348	0.099359
80	34	25	14	0.970447	0.051099
80	34	25	15	0.991265	0.020818
80	34	25	16	0.997946	0.006681
80	34	25	17	0.999622	0.001676
80	34	25	18	0.999946	0.000325
80	34	25	19	0.999994	0.000048
80	34	25	20	0.999999	0.000005
80	34	25	21	1.000000	0.000000
80	34	25	22	1.000000	0.000000
80	34	25	23	1.000000	0.000000
80	34	25	24	1.000000	0.000000
80	34	25	25	1.000000	0.000000
80	34	26	0	0.000000	0.000000
80	34	26	1	0.000000	0.000000
80	34	26	2	0.000006	0.000006
80	34	26	3	0.000070	0.000064
80	34	26	4	0.000546	0.000476
80	34	26	5	0.003059	0.002513
80	34	26	6	0.012871	0.009811
80	34	26	7	0.041941	0.029071
80	34	26	8	0.108518	0.066577
80	34	26	9	0.227897	0.119379
80	34	26	10	0.397017	0.169120
80	34	26	11	0.587463	0.190446
80	34	26	12	0.758567	0.171104
80	34	26	13	0.881411	0.122844
80	34	26	14	0.951865	0.070455
80	34	26	15	0.984073	0.032208
80	34	26	16	0.995759	0.011687
80	34	26	17	0.999104	0.003344
80	34	26	18	0.999852	0.000748

N	n	k	x	P(x)	p(x)
80	34	26	19	0.999981	0.000129
80	34	26	20	0.999998	0.000017
80	34	26	21	1.000000	0.000002
80	34	26	22	1.000000	0.000000
80	34	26	23	1.000000	0.000000
80	34	26	24	1.000000	0.000000
80	34	26	25	1.000000	0.000000
80	34	26	26	1.000000	0.000000
80	34	27	0	0.000000	0.000000
80	34	27	1	0.000000	0.000000
80	34	27	2	0.000003	0.000003
80	34	27	3	0.000033	0.000031
80	34	27	4	0.000282	0.000248
80	34	27	5	0.001710	0.001428
80	34	27	6	0.007783	0.006074
80	34	27	7	0.027406	0.019623
80	34	27	8	0.076462	0.049057
80	34	27	9	0.172629	0.096166
80	34	27	10	0.321852	0.149224
80	34	27	11	0.506347	0.184495
80	34	27	12	0.688858	0.182511
80	34	27	13	0.833638	0.144780
80	34	27	14	0.925771	0.092133
80	34	27	15	0.972741	0.046970
80	34	27	16	0.991864	0.019123
80	34	27	17	0.998051	0.006187
80	34	27	18	0.999630	0.001579
80	34	27	19	0.999945	0.000315
80	34	27	20	0.999994	0.000048
80	34	27	21	0.999999	0.000006
80	34	27	22	1.000000	0.000000
80	34	27	23	1.000000	0.000000
80	34	27	24	1.000000	0.000000
80	34	27	25	1.000000	0.000000
80	34	27	26	1.000000	0.000000
80	34	27	27	1.000000	0.000000
80	34	28	0	0.000000	0.000000
80	34	28	1	0.000000	0.000000
80	34	28	2	0.000001	0.000001
80	34	28	3	0.000015	0.000014
80	34	28	4	0.000141	0.000126
80	34	28	5	0.000928	0.000787
80	34	28	6	0.004575	0.003646
80	34	28	7	0.017410	0.012835
80	34	28	8	0.052397	0.034987
80	34	28	9	0.127267	0.074870
80	34	28	10	0.254279	0.127012
80	34	28	11	0.426283	0.172004
80	34	28	12	0.613099	0.186815
80	34	28	13	0.776272	0.163174

N	n	k	x	P(x)	p(x)
80	34	28	14	0.891004	0.114731
80	34	28	15	0.955902	0.064899
80	34	28	16	0.985369	0.029467
80	34	28	17	0.996066	0.010697
80	34	28	18	0.999153	0.003087
80	34	28	19	0.999856	0.000703
80	34	28	20	0.999981	0.000125
80	34	28	21	0.999998	0.000017
80	34	28	22	1.000000	0.000002
80	34	28	23	1.000000	0.000000
80	34	28	24	1.000000	0.000000
80	34	28	25	1.000000	0.000000
80	34	28	26	1.000000	0.000000
80	34	28	27	1.000000	0.000000
80	34	28	28	1.000000	0.000000
80	34	29	0	0.000000	0.000000
80	34	29	1	0.000000	0.000000
80	34	29	2	0.000000	0.000000
80	34	29	3	0.000007	0.000006
80	34	29	4	0.000069	0.000062
80	34	29	5	0.000489	0.000421
80	34	29	6	0.002611	0.002122
80	34	29	7	0.010745	0.008134
80	34	29	8	0.034903	0.024158
80	34	29	9	0.091272	0.056369
80	34	29	10	0.195658	0.104386
80	34	29	11	0.350205	0.154546
80	34	29	12	0.534061	0.183857
80	34	29	13	0.710375	0.176314
80	34	29	14	0.846876	0.136501
80	34	29	15	0.932189	0.085313
80	34	29	16	0.975169	0.042980
80	34	29	17	0.992569	0.017400
80	34	29	18	0.998204	0.005634
80	34	29	19	0.999653	0.001450
80	34	29	20	0.999947	0.000294
80	34	29	21	0.999994	0.000046
80	34	29	22	0.999999	0.000006
80	34	29	23	1.000000	0.000001
80	34	29	24	1.000000	0.000000
80	34	29	25	1.000000	0.000000
80	34	29	26	1.000000	0.000000
80	34	29	27	1.000000	0.000000
80	34	29	28	1.000000	0.000000
80	34	29	29	1.000000	0.000000
80	34	30	0	0.000000	0.000000
80	34	30	1	0.000000	0.000000
80	34	30	2	0.000000	0.000000
80	34	30	3	0.000003	0.000003
80	34	30	4	0.000032	0.000029

$N = 80$ $n = 34$

N	n	k	x	P(x)	p(x)
80	34	30	5	0.000250	0.000218
80	34	30	6	0.001446	0.001196
80	34	30	7	0.006439	0.004993
80	34	30	8	0.022587	0.016148
80	34	30	9	0.063640	0.041053
80	34	30	10	0.146535	0.082895
80	34	30	11	0.280507	0.133972
80	34	30	12	0.454751	0.174243
80	34	30	13	0.637775	0.183025
80	34	30	14	0.793347	0.155571
80	34	30	15	0.900406	0.107060
80	34	30	16	0.960000	0.059594
80	34	30	17	0.986769	0.026769
80	34	30	18	0.996436	0.009667
80	34	30	19	0.999227	0.002791
80	34	30	20	0.999867	0.000640
80	34	30	21	0.999982	0.000115
80	34	30	22	0.999998	0.000016
80	34	30	23	1.000000	0.000002
80	34	30	24	1.000000	0.000000
80	34	30	25	1.000000	0.000000
80	34	30	26	1.000000	0.000000
80	34	30	27	1.000000	0.000000
80	34	30	28	1.000000	0.000000
80	34	30	29	1.000000	0.000000
80	34	30	30	1.000000	0.000000
80	34	31	0	0.000000	0.000000
80	34	31	1	0.000000	0.000000
80	34	31	2	0.000000	0.000000
80	34	31	3	0.000001	0.000001
80	34	31	4	0.000015	0.000013
80	34	31	5	0.000124	0.000109
80	34	31	6	0.000776	0.000653
80	34	31	7	0.003743	0.002967
80	34	31	8	0.014190	0.010447
80	34	31	9	0.043114	0.028924
80	34	31	10	0.106746	0.063632
80	34	31	11	0.218880	0.112135
80	34	31	12	0.378084	0.159203
80	34	31	13	0.560905	0.182821
80	34	31	14	0.731118	0.170213
80	34	31	15	0.859723	0.128605
80	34	31	16	0.938546	0.078823
80	34	31	17	0.977668	0.039122
80	34	31	18	0.993343	0.015675
80	34	31	19	0.998390	0.005047
80	34	31	20	0.999688	0.001298
80	34	31	21	0.999952	0.000264
80	34	31	22	0.999994	0.000042
80	34	31	23	0.999999	0.000005

N	n	k	x	P(x)	p(x)
80	34	31	24	1.000000	0.000000
80	34	31	25	1.000000	0.000000
80	34	31	26	1.000000	0.000000
80	34	31	27	1.000000	0.000000
80	34	31	28	1.000000	0.000000
80	34	31	29	1.000000	0.000000
80	34	31	30	1.000000	0.000000
80	34	31	31	1.000000	0.000000
80	34	32	0	0.000000	0.000000
80	34	32	1	0.000000	0.000000
80	34	32	2	0.000000	0.000000
80	34	32	3	0.000000	0.000000
80	34	32	4	0.000006	0.000006
80	34	32	5	0.000059	0.000053
80	34	32	6	0.000403	0.000344
80	34	32	7	0.002108	0.001705
80	34	32	8	0.008647	0.006538
80	34	32	9	0.028357	0.019710
80	34	32	10	0.075579	0.047222
80	34	32	11	0.166246	0.090667
80	34	32	12	0.306605	0.140359
80	34	32	13	0.482553	0.175948
80	34	32	14	0.661643	0.179090
80	34	32	15	0.809856	0.148213
80	34	32	16	0.909591	0.099735
80	34	32	17	0.964095	0.054504
80	34	32	18	0.988224	0.024129
80	34	32	19	0.996845	0.008620
80	34	32	20	0.999317	0.002472
80	34	32	21	0.999882	0.000565
80	34	32	22	0.999984	0.000102
80	34	32	23	0.999998	0.000014
80	34	32	24	1.000000	0.000002
80	34	32	25	1.000000	0.000000
80	34	32	26	1.000000	0.000000
80	34	32	27	1.000000	0.000000
80	34	32	28	1.000000	0.000000
80	34	32	29	1.000000	0.000000
80	34	32	30	1.000000	0.000000
80	34	32	31	1.000000	0.000000
80	34	32	32	1.000000	0.000000
80	34	33	0	0.000000	0.000000
80	34	33	1	0.000000	0.000000
80	34	33	2	0.000000	0.000000
80	34	33	3	0.000000	0.000000
80	34	33	4	0.000003	0.000003
80	34	33	5	0.000027	0.000025
80	34	33	6	0.000203	0.000175
80	34	33	7	0.001149	0.000947
80	34	33	8	0.005105	0.003956

N	n	k	x	P(x)	p(x)
80	34	33	9	0.018091	0.012986
80	34	33	10	0.051968	0.033877
80	34	33	11	0.122801	0.070833
80	34	33	12	0.242273	0.119472
80	34	33	13	0.405576	0.163302
80	34	33	14	0.587022	0.181447
80	34	33	15	0.751189	0.164166
80	34	33	16	0.872190	0.121002
80	34	33	17	0.944791	0.072601
80	34	33	18	0.980181	0.035390
80	34	33	19	0.994151	0.013970
80	34	33	20	0.998596	0.004445
80	34	33	21	0.999729	0.001133
80	34	33	22	0.999958	0.000230
80	34	33	23	0.999995	0.000037
80	34	33	24	0.999999	0.000005
80	34	33	25	1.000000	0.000000
80	34	33	26	1.000000	0.000000
80	34	33	27	1.000000	0.000000
80	34	33	28	1.000000	0.000000
80	34	33	29	1.000000	0.000000
80	34	33	30	1.000000	0.000000
80	34	33	31	1.000000	0.000000
80	34	33	32	1.000000	0.000000
80	34	33	33	1.000000	0.000000
80	34	34	0	0.000000	0.000000
80	34	34	1	0.000000	0.000000
80	34	34	2	0.000000	0.000000
80	34	34	3	0.000000	0.000000
80	34	34	4	0.000001	0.000001
80	34	34	5	0.000012	0.000011
80	34	34	6	0.000098	0.000086
80	34	34	7	0.000606	0.000507
80	34	34	8	0.002917	0.002311
80	34	34	9	0.011184	0.008267
80	34	34	10	0.034669	0.023485
80	34	34	11	0.088138	0.053469
80	34	34	12	0.186350	0.098212
80	34	34	13	0.332611	0.146260
80	34	34	14	0.509811	0.177200
80	34	34	15	0.684824	0.175013
80	34	34	16	0.825849	0.141026
80	34	34	17	0.918532	0.092682
80	34	34	18	0.968134	0.049602
80	34	34	19	0.989693	0.021559
80	34	34	20	0.997272	0.007579
80	34	34	21	0.999415	0.002144
80	34	34	22	0.999900	0.000484
80	34	34	23	0.999986	0.000087
80	34	34	24	0.999998	0.000012

N	n	k	x	P(x)	p(x)
80	34	34	25	1.000000	0.000001
80	34	34	26	1.000000	0.000000
80	34	34	27	1.000000	0.000000
80	34	34	28	1.000000	0.000000
80	34	34	29	1.000000	0.000000
80	34	34	30	1.000000	0.000000
80	34	34	31	1.000000	0.000000
80	34	34	32	1.000000	0.000000
80	34	34	33	1.000000	0.000000
80	34	34	34	1.000000	0.000000
80	35	1	0	0.562500	0.562500
80	35	1	1	1.000000	0.437500
80	35	2	0	0.313291	0.313291
80	35	2	1	0.811709	0.498418
80	35	2	2	1.000000	0.188291
80	35	3	0	0.172712	0.172712
80	35	3	1	0.594450	0.421738
80	35	3	2	0.920338	0.325888
80	35	3	3	1.000000	0.079662
80	35	4	0	0.094206	0.094206
80	35	4	1	0.408228	0.314021
80	35	4	2	0.780672	0.372444
80	35	4	3	0.966894	0.186222
80	35	4	4	1.000000	0.033106
80	35	5	0	0.050822	0.050822
80	35	5	1	0.267745	0.216923
80	35	5	2	0.618953	0.351208
80	35	5	3	0.888485	0.269532
80	35	5	4	0.986496	0.098012
80	35	5	5	1.000000	0.013504
80	35	6	0	0.027105	0.027105
80	35	6	1	0.169406	0.142301
80	35	6	2	0.464421	0.295015
80	35	6	3	0.773484	0.309063
80	35	6	4	0.945985	0.172500
80	35	6	5	0.994598	0.048614
80	35	6	6	1.000000	0.005402
80	35	7	0	0.014285	0.014285
80	35	7	1	0.104025	0.089740
80	35	7	2	0.332860	0.228836
80	35	7	3	0.639835	0.306975
80	35	7	4	0.873721	0.233886
80	35	7	5	0.974890	0.101169
80	35	7	6	0.997883	0.022993
80	35	7	7	1.000000	0.002117
80	35	8	0	0.007436	0.007436
80	35	8	1	0.062228	0.054792
80	35	8	2	0.229414	0.167186
80	35	8	3	0.505271	0.275857
80	35	8	4	0.774400	0.269129

Table for $N = 2$, $n = 1$, through $N = 100$, $n = 50$

N	n	k	x	P(x)	p(x)
80	35	8	5	0.933314	0.158914
80	35	8	6	0.988749	0.055435
80	35	8	7	0.999188	0.010439
80	35	8	8	1.000000	0.000812
80	35	9	0	0.003821	0.003821
80	35	9	1	0.036354	0.032533
80	35	9	2	0.152787	0.116433
80	35	9	3	0.382668	0.229881
80	35	9	4	0.658525	0.275857
80	35	9	5	0.867100	0.208575
80	35	9	6	0.966421	0.099321
80	35	9	7	0.995128	0.028707
80	35	9	8	0.999695	0.004567
80	35	9	9	1.000000	0.000304
80	35	10	0	0.001938	0.001938
80	35	10	1	0.020775	0.018837
80	35	10	2	0.098670	0.077895
80	35	10	3	0.279060	0.180389
80	35	10	4	0.538080	0.259021
80	35	10	5	0.778969	0.240889
80	35	10	6	0.925853	0.146884
80	35	10	7	0.983807	0.057954
80	35	10	8	0.997959	0.014152
80	35	10	9	0.999888	0.001930
80	35	10	10	1.000000	0.000111
80	35	11	0	0.000969	0.000969
80	35	11	1	0.011625	0.010657
80	35	11	2	0.061948	0.050323
80	35	11	3	0.196596	0.134648
80	35	11	4	0.423371	0.226775
80	35	11	5	0.675731	0.252360
80	35	11	6	0.865001	0.189270
80	35	11	7	0.960625	0.095624
80	35	11	8	0.992500	0.031875
80	35	11	9	0.999172	0.006671
80	35	11	10	0.999960	0.000788
80	35	11	11	1.000000	0.000040
80	35	12	0	0.000477	0.000477
80	35	12	1	0.006374	0.005897
80	35	12	2	0.037881	0.031507
80	35	12	3	0.134151	0.096270
80	35	12	4	0.321487	0.187336
80	35	12	5	0.566009	0.244523
80	35	12	6	0.785453	0.219443
80	35	12	7	0.921821	0.136368
80	35	12	8	0.980027	0.058206
80	35	12	9	0.996658	0.016630
80	35	12	10	0.999674	0.003017
80	35	12	11	0.999986	0.000312
80	35	12	12	1.000000	0.000014
80	35	13	0	0.000232	0.000232
80	35	13	1	0.003426	0.003194
80	35	13	2	0.022591	0.019165
80	35	13	3	0.088847	0.066256
80	35	13	4	0.236084	0.147236
80	35	13	5	0.458132	0.222048
80	35	13	6	0.691867	0.233735
80	35	13	7	0.865670	0.173803
80	35	13	8	0.956916	0.091247
80	35	13	9	0.990299	0.033383
80	35	13	10	0.998565	0.008266
80	35	13	11	0.999876	0.001311
80	35	13	12	0.999995	0.000119
80	35	13	13	1.000000	0.000005
80	35	14	0	0.000111	0.000111
80	35	14	1	0.001805	0.001694
80	35	14	2	0.013151	0.011346
80	35	14	3	0.057202	0.044051
80	35	14	4	0.167959	0.110757
80	35	14	5	0.358707	0.190748
80	35	14	6	0.590698	0.231991
80	35	14	7	0.793036	0.202338
80	35	14	8	0.920145	0.127110
80	35	14	9	0.977345	0.057199
80	35	14	10	0.995481	0.018136
80	35	14	11	0.999407	0.003926
80	35	14	12	0.999954	0.000548
80	35	14	13	0.999998	0.000044
80	35	14	14	1.000000	0.000002
80	35	15	0	0.000052	0.000052
80	35	15	1	0.000932	0.000880
80	35	15	2	0.007478	0.006546
80	35	15	3	0.035844	0.028366
80	35	15	4	0.115937	0.080093
80	35	15	5	0.272004	0.156066
80	35	15	6	0.488763	0.216759
80	35	15	7	0.707195	0.218433
80	35	15	8	0.868146	0.160950
80	35	15	9	0.954811	0.086666
80	35	15	10	0.988611	0.033800
80	35	15	11	0.997979	0.009368
80	35	15	12	0.999763	0.001784
80	35	15	13	0.999984	0.000220
80	35	15	14	0.999999	0.000016
80	35	15	15	1.000000	0.000000
80	35	16	0	0.000024	0.000024
80	35	16	1	0.000472	0.000448
80	35	16	2	0.004155	0.003683
80	35	16	3	0.021879	0.017725
80	35	16	4	0.077739	0.055860
80	35	16	5	0.199973	0.122234
80	35	16	6	0.392055	0.192082
80	35	16	7	0.613101	0.221047
80	35	16	8	0.801289	0.188188
80	35	16	9	0.920145	0.118856
80	35	16	10	0.975611	0.055466
80	35	16	11	0.994520	0.018909
80	35	16	12	0.999132	0.004612
80	35	16	13	0.999909	0.000777
80	35	16	14	0.999994	0.000085
80	35	16	15	1.000000	0.000005
80	35	16	16	1.000000	0.000000
80	35	17	0	0.000011	0.000011
80	35	17	1	0.000234	0.000223
80	35	17	2	0.002256	0.002022
80	35	17	3	0.013017	0.010761
80	35	17	4	0.050682	0.037665
80	35	17	5	0.142676	0.091994
80	35	17	6	0.305018	0.162342
80	35	17	7	0.516393	0.211376
80	35	17	8	0.721898	0.205504
80	35	17	9	0.871860	0.149963
80	35	17	10	0.953945	0.082085
80	35	17	11	0.987429	0.033484
80	35	17	12	0.997475	0.010045
80	35	17	13	0.999642	0.002167
80	35	17	14	0.999966	0.000324
80	35	17	15	0.999998	0.000032
80	35	17	16	1.000000	0.000002
80	35	17	17	1.000000	0.000000
80	35	18	0	0.000005	0.000005
80	35	18	1	0.000114	0.000109
80	35	18	2	0.001197	0.001083
80	35	18	3	0.007551	0.006354
80	35	18	4	0.032149	0.024598
80	35	18	5	0.098869	0.066721
80	35	18	6	0.230289	0.131420
80	35	18	7	0.422449	0.192160
80	35	18	8	0.633824	0.211376
80	35	18	9	0.809971	0.176146
80	35	18	10	0.921372	0.111401
80	35	18	11	0.974673	0.053302
80	35	18	12	0.993807	0.019134
80	35	18	13	0.998885	0.005078
80	35	18	14	0.999858	0.000973
80	35	18	15	0.999988	0.000130
80	35	18	16	0.999999	0.000011
80	35	18	17	1.000000	0.000001
80	35	18	18	1.000000	0.000000
80	35	19	0	0.000002	0.000002
80	35	19	1	0.000054	0.000052
80	35	19	2	0.000620	0.000566
80	35	19	3	0.004271	0.003651
80	35	19	4	0.019850	0.015578
80	35	19	5	0.066585	0.046735
80	35	19	6	0.168818	0.102233
80	35	19	7	0.335667	0.166848
80	35	19	8	0.541774	0.206107
80	35	19	9	0.736103	0.194329
80	35	19	10	0.876452	0.140349
80	35	19	11	0.954040	0.077588
80	35	19	12	0.986709	0.032669
80	35	19	13	0.997083	0.010374
80	35	19	14	0.999529	0.002445
80	35	19	15	0.999946	0.000417
80	35	19	16	0.999996	0.000050
80	35	19	17	1.000000	0.000004
80	35	19	18	1.000000	0.000000
80	35	19	19	1.000000	0.000000
80	35	20	0	0.000001	0.000001
80	35	20	1	0.000025	0.000024
80	35	20	2	0.000314	0.000289
80	35	20	3	0.002356	0.002042
80	35	20	4	0.011933	0.009577
80	35	20	5	0.043600	0.031668
80	35	20	6	0.120216	0.076615
80	35	20	7	0.259081	0.138865
80	35	20	8	0.450546	0.191465
80	35	20	9	0.653274	0.202728
80	35	20	10	0.818932	0.165658
80	35	20	11	0.923514	0.104582
80	35	20	12	0.974391	0.050878
80	35	20	13	0.993342	0.018950
80	35	20	14	0.998687	0.005345
80	35	20	15	0.999809	0.001122
80	35	20	16	0.999980	0.000171
80	35	20	17	0.999999	0.000018
80	35	20	18	1.000000	0.000001
80	35	20	19	1.000000	0.000000
80	35	20	20	1.000000	0.000000
80	35	21	0	0.000000	0.000000
80	35	21	1	0.000011	0.000011
80	35	21	2	0.000155	0.000144
80	35	21	3	0.001267	0.001112
80	35	21	4	0.006985	0.005718
80	35	21	5	0.027767	0.020782
80	35	21	6	0.083185	0.055418
80	35	21	7	0.194277	0.111092
80	35	21	8	0.364387	0.170110
80	35	21	9	0.565425	0.201039

N	n	k	x	P(x)	p(x)
80	35	21	10	0.749908	0.184483
80	35	21	11	0.881681	0.131773
80	35	21	12	0.954888	0.073207
80	35	21	13	0.986393	0.031505
80	35	21	14	0.996816	0.010423
80	35	21	15	0.999435	0.002619
80	35	21	16	0.999926	0.000491
80	35	21	17	0.999993	0.000067
80	35	21	18	0.999999	0.000006
80	35	21	19	1.000000	0.000000
80	35	21	20	1.000000	0.000000
80	35	21	21	1.000000	0.000000
80	35	22	0	0.000000	0.000000
80	35	22	1	0.000005	0.000005
80	35	22	2	0.000075	0.000070
80	35	22	3	0.000664	0.000589
80	35	22	4	0.003980	0.003317
80	35	22	5	0.017200	0.013219
80	35	22	6	0.055945	0.038746
80	35	22	7	0.141555	0.085610
80	35	22	8	0.286540	0.144984
80	35	22	9	0.476832	0.190292
80	35	22	10	0.671737	0.194905
80	35	22	11	0.828078	0.156341
80	35	22	12	0.926350	0.098272
80	35	22	13	0.974646	0.048296
80	35	22	14	0.993106	0.018461
80	35	22	15	0.998547	0.005441
80	35	22	16	0.999768	0.001221
80	35	22	17	0.999973	0.000205
80	35	22	18	0.999998	0.000025
80	35	22	19	1.000000	0.000002
80	35	22	20	1.000000	0.000000
80	35	22	21	1.000000	0.000000
80	35	22	22	1.000000	0.000000
80	35	23	0	0.000000	0.000000
80	35	23	1	0.000002	0.000002
80	35	23	2	0.000035	0.000033
80	35	23	3	0.000339	0.000304
80	35	23	4	0.002208	0.001869
80	35	23	5	0.010362	0.008154
80	35	23	6	0.036573	0.026210
80	35	23	7	0.100226	0.063654
80	35	23	8	0.219047	0.118821
80	35	23	9	0.391529	0.172482
80	35	23	10	0.587726	0.196198
80	35	23	11	0.763385	0.175659
80	35	23	12	0.887380	0.123995
80	35	23	13	0.956327	0.068946
80	35	23	14	0.986422	0.030096
80	35	23	15	0.996671	0.010249
80	35	23	16	0.999368	0.002697
80	35	23	17	0.999909	0.000541
80	35	23	18	0.999990	0.000081
80	35	23	19	0.999999	0.000009
80	35	23	20	1.000000	0.000001
80	35	23	21	1.000000	0.000000
80	35	23	22	1.000000	0.000000
80	35	23	23	1.000000	0.000000
80	35	24	0	0.000000	0.000000
80	35	24	1	0.000001	0.000001
80	35	24	2	0.000016	0.000015
80	35	24	3	0.000168	0.000152
80	35	24	4	0.001191	0.001023
80	35	24	5	0.006070	0.004879
80	35	24	6	0.023237	0.017167
80	35	24	7	0.068958	0.045720
80	35	24	8	0.162764	0.093806
80	35	24	9	0.312853	0.150089
80	35	24	10	0.501675	0.188822
80	35	24	11	0.689424	0.187749
80	35	24	12	0.837347	0.147923
80	35	24	13	0.929716	0.092368
80	35	24	14	0.975334	0.045619
80	35	24	15	0.993075	0.017741
80	35	24	16	0.998469	0.005394
80	35	24	17	0.999738	0.001269
80	35	24	18	0.999966	0.000228
80	35	24	19	0.999997	0.000031
80	35	24	20	1.000000	0.000003
80	35	24	21	1.000000	0.000000
80	35	24	22	1.000000	0.000000
80	35	24	23	1.000000	0.000000
80	35	24	24	1.000000	0.000000
80	35	25	0	0.000000	0.000000
80	35	25	1	0.000000	0.000000
80	35	25	2	0.000007	0.000007
80	35	25	3	0.000081	0.000074
80	35	25	4	0.000625	0.000544
80	35	25	5	0.003457	0.002832
80	35	25	6	0.014347	0.010891
80	35	25	7	0.046098	0.031750
80	35	25	8	0.117536	0.071438
80	35	25	9	0.243169	0.125633
80	35	25	10	0.417379	0.174211
80	35	25	11	0.608960	0.191581
80	35	25	12	0.776593	0.167633
80	35	25	13	0.893428	0.116835
80	35	25	14	0.958227	0.064799
80	35	25	15	0.986739	0.028512
80	35	25	16	0.996639	0.009900
80	35	25	17	0.999330	0.002691
80	35	25	18	0.999897	0.000567
80	35	25	19	0.999988	0.000091
80	35	25	20	0.999999	0.000011
80	35	25	21	1.000000	0.000001
80	35	25	22	1.000000	0.000000
80	35	25	23	1.000000	0.000000
80	35	25	24	1.000000	0.000000
80	35	25	25	1.000000	0.000000
80	35	26	0	0.000000	0.000000
80	35	26	1	0.000000	0.000000
80	35	26	2	0.000003	0.000003
80	35	26	3	0.000038	0.000035
80	35	26	4	0.000318	0.000280
80	35	26	5	0.001912	0.001594
80	35	26	6	0.008605	0.006693
80	35	26	7	0.029934	0.021329
80	35	26	8	0.082466	0.052532
80	35	26	9	0.183778	0.101312
80	35	26	10	0.338193	0.154414
80	35	26	11	0.525361	0.187168
80	35	26	12	0.706492	0.181131
80	35	26	13	0.846694	0.140202
80	35	26	14	0.933486	0.086792
80	35	26	15	0.976371	0.042885
80	35	26	16	0.993219	0.016848
80	35	26	17	0.998450	0.005231
80	35	26	18	0.999722	0.001272
80	35	26	19	0.999961	0.000240
80	35	26	20	0.999996	0.000034
80	35	26	21	1.000000	0.000004
80	35	26	22	1.000000	0.000000
80	35	26	23	1.000000	0.000000
80	35	26	24	1.000000	0.000000
80	35	26	25	1.000000	0.000000
80	35	26	26	1.000000	0.000000
80	35	27	0	0.000000	0.000000
80	35	27	1	0.000000	0.000000
80	35	27	2	0.000001	0.000001
80	35	27	3	0.000017	0.000016
80	35	27	4	0.000158	0.000140
80	35	27	5	0.001027	0.000869
80	35	27	6	0.005011	0.003984
80	35	27	7	0.018874	0.013864
80	35	27	8	0.056200	0.037326
80	35	27	9	0.134998	0.078798
80	35	27	10	0.266705	0.131706
80	35	27	11	0.442175	0.175470
80	35	27	12	0.629343	0.187168
80	35	27	13	0.789575	0.160231
80	35	27	14	0.899733	0.110159
80	35	27	15	0.960488	0.060754
80	35	27	16	0.987291	0.026803
80	35	27	17	0.996706	0.009415
80	35	27	18	0.999321	0.002615
80	35	27	19	0.999890	0.000569
80	35	27	20	0.999986	0.000096
80	35	27	21	0.999999	0.000012
80	35	27	22	1.000000	0.000001
80	35	27	23	1.000000	0.000000
80	35	27	24	1.000000	0.000000
80	35	27	25	1.000000	0.000000
80	35	27	26	1.000000	0.000000
80	35	27	27	1.000000	0.000000
80	35	28	0	0.000000	0.000000
80	35	28	1	0.000000	0.000000
80	35	28	2	0.000001	0.000000
80	35	28	3	0.000008	0.000007
80	35	28	4	0.000076	0.000068
80	35	28	5	0.000535	0.000459
80	35	28	6	0.002831	0.002296
80	35	28	7	0.011550	0.008719
80	35	28	8	0.037185	0.025635
80	35	28	9	0.096343	0.059158
80	35	28	10	0.204579	0.108236
80	35	28	11	0.362717	0.158138
80	35	28	12	0.548119	0.185403
80	35	28	13	0.723064	0.174944
80	35	28	14	0.856086	0.133022
80	35	28	15	0.937562	0.081476
80	35	28	16	0.977682	0.040121
80	35	28	17	0.993509	0.015826
80	35	28	18	0.998482	0.004974
80	35	28	19	0.999719	0.001236
80	35	28	20	0.999959	0.000241
80	35	28	21	0.999995	0.000036
80	35	28	22	1.000000	0.000004
80	35	28	23	1.000000	0.000000
80	35	28	24	1.000000	0.000000
80	35	28	25	1.000000	0.000000
80	35	28	26	1.000000	0.000000
80	35	28	27	1.000000	0.000000
80	35	28	28	1.000000	0.000000
80	35	29	0	0.000000	0.000000
80	35	29	1	0.000000	0.000000
80	35	29	2	0.000000	0.000000
80	35	29	3	0.000003	0.000003
80	35	29	4	0.000035	0.000032
80	35	29	5	0.000270	0.000235

N	n	k	x	P(x)	p(x)
80	35	29	6	0.001550	0.001280
80	35	29	7	0.006855	0.005305
80	35	29	8	0.023875	0.017020
80	35	29	9	0.066764	0.042889
80	35	29	10	0.152542	0.085778
80	35	29	11	0.289730	0.137188
80	35	29	12	0.466114	0.176384
80	35	29	13	0.649049	0.182934
80	35	29	14	0.802365	0.153316
80	35	29	15	0.906225	0.103859
80	35	29	16	0.963023	0.056798
80	35	29	17	0.988030	0.025007
80	35	29	18	0.996856	0.008826
80	35	29	19	0.999338	0.002482
80	35	29	20	0.999890	0.000552
80	35	29	21	0.999986	0.000096
80	35	29	22	0.999998	0.000013
80	35	29	23	1.000000	0.000001
80	35	29	24	1.000000	0.000000
80	35	29	25	1.000000	0.000000
80	35	29	26	1.000000	0.000000
80	35	29	27	1.000000	0.000000
80	35	29	28	1.000000	0.000000
80	35	29	29	1.000000	0.000000
80	35	30	0	0.000000	0.000000
80	35	30	1	0.000000	0.000000
80	35	30	2	0.000000	0.000000
80	35	30	3	0.000001	0.000001
80	35	30	4	0.000016	0.000014
80	35	30	5	0.000132	0.000116
80	35	30	6	0.000822	0.000690
80	35	30	7	0.003943	0.003120
80	35	30	8	0.014864	0.010922
80	35	30	9	0.044899	0.030034
80	35	30	10	0.110494	0.065595
80	35	30	11	0.225171	0.114677
80	35	30	12	0.386569	0.161397
80	35	30	13	0.570136	0.183567
80	35	30	14	0.739235	0.169099
80	35	30	15	0.865495	0.126261
80	35	30	16	0.941863	0.076367
80	35	30	17	0.979204	0.037341
80	35	30	18	0.993914	0.014710
80	35	30	19	0.998560	0.004645
80	35	30	20	0.999728	0.001168
80	35	30	21	0.999959	0.000232
80	35	30	22	0.999995	0.000036
80	35	30	23	0.999999	0.000004
80	35	30	24	1.000000	0.000000
80	35	30	25	1.000000	0.000000

N	n	k	x	P(x)	p(x)
80	35	30	26	1.000000	0.000000
80	35	30	27	1.000000	0.000000
80	35	30	28	1.000000	0.000000
80	35	30	29	1.000000	0.000000
80	35	30	30	1.000000	0.000000
80	35	31	0	0.000000	0.000000
80	35	31	1	0.000000	0.000000
80	35	31	2	0.000000	0.000000
80	35	31	3	0.000001	0.000000
80	35	31	4	0.000007	0.000006
80	35	31	5	0.000062	0.000055
80	35	31	6	0.000422	0.000360
80	35	31	7	0.002195	0.001773
80	35	31	8	0.008967	0.006771
80	35	31	9	0.029281	0.020314
80	35	31	10	0.077696	0.048416
80	35	31	11	0.170126	0.092430
80	35	31	12	0.312326	0.142200
80	35	31	13	0.489366	0.177041
80	35	31	14	0.668213	0.178847
80	35	31	15	0.814991	0.146778
80	35	31	16	0.912843	0.097852
80	35	31	17	0.965761	0.052918
80	35	31	18	0.988913	0.023152
80	35	31	19	0.997073	0.008160
80	35	31	20	0.999377	0.002304
80	35	31	21	0.999894	0.000517
80	35	31	22	0.999986	0.000091
80	35	31	23	0.999998	0.000013
80	35	31	24	1.000000	0.000001
80	35	31	25	1.000000	0.000000
80	35	31	26	1.000000	0.000000
80	35	31	27	1.000000	0.000000
80	35	31	28	1.000000	0.000000
80	35	31	29	1.000000	0.000000
80	35	31	30	1.000000	0.000000
80	35	31	31	1.000000	0.000000
80	35	32	0	0.000000	0.000000
80	35	32	1	0.000000	0.000000
80	35	32	2	0.000000	0.000000
80	35	32	3	0.000000	0.000000
80	35	32	4	0.000003	0.000003
80	35	32	5	0.000028	0.000025
80	35	32	6	0.000209	0.000181
80	35	32	7	0.001182	0.000973
80	35	32	8	0.005235	0.004054
80	35	32	9	0.018502	0.013266
80	35	32	10	0.052995	0.034493
80	35	32	11	0.124854	0.071860
80	35	32	12	0.245579	0.120725

N	n	k	x	P(x)	p(x)
80	35	32	13	0.409879	0.164300
80	35	32	14	0.591565	0.181686
80	35	32	15	0.755082	0.163517
80	35	32	16	0.874901	0.119819
80	35	32	17	0.946322	0.071421
80	35	32	18	0.980881	0.034559
80	35	32	19	0.994408	0.013528
80	35	32	20	0.998672	0.004263
80	35	32	21	0.999747	0.001075
80	35	32	22	0.999962	0.000215
80	35	32	23	0.999995	0.000034
80	35	32	24	0.999999	0.000004
80	35	32	25	1.000000	0.000000
80	35	32	26	1.000000	0.000000
80	35	32	27	1.000000	0.000000
80	35	32	28	1.000000	0.000000
80	35	32	29	1.000000	0.000000
80	35	32	30	1.000000	0.000000
80	35	32	31	1.000000	0.000000
80	35	32	32	1.000000	0.000000
80	35	33	0	0.000000	0.000000
80	35	33	1	0.000000	0.000000
80	35	33	2	0.000000	0.000000
80	35	33	3	0.000000	0.000000
80	35	33	4	0.000001	0.000001
80	35	33	5	0.000012	0.000011
80	35	33	6	0.000100	0.000087
80	35	33	7	0.000614	0.000514
80	35	33	8	0.002955	0.002341
80	35	33	9	0.011316	0.008361
80	35	33	10	0.035030	0.023714
80	35	33	11	0.088924	0.053895
80	35	33	12	0.187732	0.098807
80	35	33	13	0.334575	0.146843
80	35	33	14	0.512077	0.177502
80	35	33	15	0.686950	0.174873
80	35	33	16	0.827472	0.140523
80	35	33	17	0.919539	0.092067
80	35	33	18	0.968641	0.049102
80	35	33	19	0.989899	0.021258
80	35	33	20	0.997340	0.007440
80	35	33	21	0.999433	0.002094
80	35	33	22	0.999903	0.000470
80	35	33	23	0.999987	0.000084
80	35	33	24	0.999999	0.000012
80	35	33	25	1.000000	0.000001
80	35	33	26	1.000000	0.000000
80	35	33	27	1.000000	0.000000
80	35	33	28	1.000000	0.000000
80	35	33	29	1.000000	0.000000

N	n	k	x	P(x)	p(x)
80	35	33	30	1.000000	0.000000
80	35	33	31	1.000000	0.000000
80	35	33	32	1.000000	0.000000
80	35	33	33	1.000000	0.000000
80	35	34	0	0.000000	0.000000
80	35	34	1	0.000000	0.000000
80	35	34	2	0.000000	0.000000
80	35	34	3	0.000000	0.000000
80	35	34	4	0.000000	0.000000
80	35	34	5	0.000005	0.000005
80	35	34	6	0.000046	0.000041
80	35	34	7	0.000308	0.000262
80	35	34	8	0.001610	0.001303
80	35	34	9	0.006691	0.005080
80	35	34	10	0.022416	0.015725
80	35	34	11	0.061404	0.038988
80	35	34	12	0.139379	0.077976
80	35	34	13	0.265840	0.126460
80	35	34	14	0.432767	0.166928
80	35	34	15	0.612536	0.179768
80	35	34	16	0.770665	0.158130
80	35	34	17	0.884279	0.113614
80	35	34	18	0.950881	0.066601
80	35	34	19	0.982662	0.031782
80	35	34	20	0.994965	0.012303
80	35	34	21	0.998810	0.003845
80	35	34	22	0.999773	0.000964
80	35	34	23	0.999966	0.000192
80	35	34	24	0.999996	0.000030
80	35	34	25	1.000000	0.000004
80	35	34	26	1.000000	0.000000
80	35	34	27	1.000000	0.000000
80	35	34	28	1.000000	0.000000
80	35	34	29	1.000000	0.000000
80	35	34	30	1.000000	0.000000
80	35	34	31	1.000000	0.000000
80	35	34	32	1.000000	0.000000
80	35	34	33	1.000000	0.000000
80	35	34	34	1.000000	0.000000
80	35	35	0	0.000000	0.000000
80	35	35	1	0.000000	0.000000
80	35	35	2	0.000000	0.000000
80	35	35	3	0.000000	0.000000
80	35	35	4	0.000000	0.000000
80	35	35	5	0.000002	0.000002
80	35	35	6	0.000020	0.000018
80	35	35	7	0.000148	0.000128
80	35	35	8	0.000846	0.000697
80	35	35	9	0.003819	0.002973
80	35	35	10	0.013870	0.010050

N	n	k	x	P(x)	p(x)
80	35	35	11	0.041062	0.027193
80	35	35	12	0.100391	0.059329
80	35	35	13	0.205359	0.104967
80	35	35	14	0.356561	0.151203
80	35	35	15	0.534376	0.177814
80	35	35	16	0.705351	0.170975
80	35	35	17	0.839822	0.134471
80	35	35	18	0.926267	0.086445
80	35	35	19	0.971608	0.045341
80	35	35	20	0.990953	0.019345
80	35	35	21	0.997639	0.006686
80	35	35	22	0.999501	0.001861
80	35	35	23	0.999915	0.000414
80	35	35	24	0.999989	0.000073
80	35	35	25	0.999999	0.000010
80	35	35	26	1.000000	0.000001
80	35	35	27	1.000000	0.000000
80	35	35	28	1.000000	0.000000
80	35	35	29	1.000000	0.000000
80	35	35	30	1.000000	0.000000
80	35	35	31	1.000000	0.000000
80	35	35	32	1.000000	0.000000
80	35	35	33	1.000000	0.000000
80	35	35	34	1.000000	0.000000
80	35	35	35	1.000000	0.000000
80	36	1	0	0.550000	0.550000
80	36	1	1	1.000000	0.450000
80	36	2	0	0.299367	0.299367
80	36	2	1	0.800633	0.501266
80	36	2	2	1.000000	0.199367
80	36	3	0	0.161198	0.161198
80	36	3	1	0.575706	0.414508
80	36	3	2	0.913096	0.337390
80	36	3	3	1.000000	0.086904
80	36	4	0	0.085833	0.085833
80	36	4	1	0.387293	0.301461
80	36	4	2	0.764119	0.376826
80	36	4	3	0.962756	0.198637
80	36	4	4	1.000000	0.037244
80	36	5	0	0.045175	0.045175
80	36	5	1	0.248463	0.203288
80	36	5	2	0.595539	0.347076
80	36	5	3	0.876505	0.280966
80	36	5	4	0.984318	0.107813
80	36	5	5	1.000000	0.015682
80	36	6	0	0.023491	0.023491
80	36	6	1	0.153595	0.130104
80	36	6	2	0.438198	0.284603
80	36	6	3	0.752880	0.314682
80	36	6	4	0.938318	0.185438
80	36	6	5	0.993518	0.055200
80	36	6	6	1.000000	0.006482
80	36	7	0	0.012063	0.012063
80	36	7	1	0.092059	0.079996
80	36	7	2	0.307434	0.215375
80	36	7	3	0.612549	0.305114
80	36	7	4	0.858129	0.245580
80	36	7	5	0.970394	0.112265
80	36	7	6	0.997372	0.026978
80	36	7	7	1.000000	0.002628
80	36	8	0	0.006114	0.006114
80	36	8	1	0.053705	0.047591
80	36	8	2	0.207123	0.153418
80	36	8	3	0.474620	0.267498
80	36	8	4	0.750477	0.275857
80	36	8	5	0.922719	0.172242
80	36	8	6	0.986285	0.063566
80	36	8	7	0.998956	0.012671
80	36	8	8	1.000000	0.001044
80	36	9	0	0.003057	0.003057
80	36	9	1	0.030570	0.027513
80	36	9	2	0.134675	0.104105
80	36	9	3	0.352017	0.217342
80	36	9	4	0.627874	0.275857
80	36	9	5	0.848560	0.220685
80	36	9	6	0.959799	0.111240
80	36	9	7	0.993852	0.034053
80	36	9	8	0.999594	0.005741
80	36	9	9	1.000000	0.000406
80	36	10	0	0.001507	0.001507
80	36	10	1	0.017008	0.015501
80	36	10	2	0.084822	0.067815
80	36	10	3	0.250999	0.166177
80	36	10	4	0.503544	0.252545
80	36	10	5	0.752204	0.248660
80	36	10	6	0.912797	0.160593
80	36	10	7	0.979943	0.067147
80	36	10	8	0.997330	0.017386
80	36	10	9	0.999845	0.002516
80	36	10	10	1.000000	0.000154
80	36	11	0	0.000732	0.000732
80	36	11	1	0.009257	0.008525
80	36	11	2	0.051884	0.042626
80	36	11	3	0.172659	0.120775
80	36	11	4	0.388095	0.215436
80	36	11	5	0.642083	0.253988
80	36	11	6	0.843971	0.201888
80	36	11	7	0.952125	0.108154
80	36	11	8	0.990375	0.038250
80	36	11	9	0.998875	0.008500
80	36	11	10	0.999943	0.001067
80	36	11	11	1.000000	0.000057
80	36	12	0	0.000350	0.000350
80	36	12	1	0.004933	0.004583
80	36	12	2	0.030879	0.025947
80	36	12	3	0.114897	0.084017
80	36	12	4	0.288182	0.173286
80	36	12	5	0.527973	0.239790
80	36	12	6	0.756194	0.228221
80	36	12	7	0.906669	0.150476
80	36	12	8	0.974854	0.068184
80	36	12	9	0.995549	0.020695
80	36	12	10	0.999540	0.003991
80	36	12	11	0.999979	0.000439
80	36	12	12	1.000000	0.000021
80	36	13	0	0.000165	0.000165
80	36	13	1	0.002574	0.002409
80	36	13	2	0.017906	0.015332
80	36	13	3	0.074124	0.056217
80	36	13	4	0.206636	0.132513
80	36	13	5	0.418656	0.212020
80	36	13	6	0.655508	0.236851
80	36	13	7	0.842496	0.186988
80	36	13	8	0.946778	0.104282
80	36	13	9	0.987332	0.040554
80	36	13	10	0.998014	0.010683
80	36	13	11	0.999818	0.001804
80	36	13	12	0.999993	0.000175
80	36	13	13	1.000000	0.000007
80	36	14	0	0.000076	0.000076
80	36	14	1	0.001315	0.001239
80	36	14	2	0.010126	0.008810
80	36	14	3	0.046434	0.036309
80	36	14	4	0.143347	0.096912
80	36	14	5	0.320558	0.177211
80	36	14	6	0.549455	0.228898
80	36	14	7	0.761561	0.212106
80	36	14	8	0.903197	0.141636
80	36	14	9	0.970989	0.067792
80	36	14	10	0.993869	0.022880
80	36	14	11	0.999145	0.005276
80	36	14	12	0.999930	0.000785
80	36	14	13	0.999997	0.000067
80	36	14	14	1.000000	0.000003
80	36	15	0	0.000035	0.000035
80	36	15	1	0.000658	0.000624
80	36	15	2	0.005587	0.004929
80	36	15	3	0.028280	0.022693
80	36	15	4	0.096359	0.068079
80	36	15	5	0.237322	0.140963
80	36	15	6	0.445411	0.208089
80	36	15	7	0.668363	0.222952
80	36	15	8	0.843109	0.174746
80	36	15	9	0.943256	0.100147
80	36	15	10	0.984856	0.041599
80	36	15	11	0.997146	0.012291
80	36	15	12	0.999644	0.002498
80	36	15	13	0.999974	0.000329
80	36	15	14	0.999999	0.000025
80	36	15	15	1.000000	0.000001
80	36	16	0	0.000015	0.000015
80	36	16	1	0.000322	0.000307
80	36	16	2	0.003009	0.002686
80	36	16	3	0.016759	0.013750
80	36	16	4	0.062843	0.046084
80	36	16	5	0.170093	0.107250
80	36	16	6	0.349370	0.179276
80	36	16	7	0.568892	0.219522
80	36	16	8	0.767834	0.198942
80	36	16	9	0.901657	0.133823
80	36	16	10	0.968216	0.066559
80	36	16	11	0.992419	0.024203
80	36	16	12	0.998722	0.006303
80	36	16	13	0.999857	0.001135
80	36	16	14	0.999991	0.000133
80	36	16	15	1.000000	0.000009
80	36	16	16	1.000000	0.000000
80	36	17	0	0.000007	0.000007
80	36	17	1	0.000155	0.000148
80	36	17	2	0.001582	0.001427
80	36	17	3	0.009669	0.008087
80	36	17	4	0.039801	0.030132
80	36	17	5	0.118144	0.078343
80	36	17	6	0.265334	0.147190
80	36	17	7	0.469421	0.204087
80	36	17	8	0.680797	0.211376
80	36	17	9	0.845200	0.164403
80	36	17	10	0.941176	0.095976
80	36	17	11	0.982965	0.041789
80	36	17	12	0.996358	0.013394
80	36	17	13	0.999449	0.003091
80	36	17	14	0.999945	0.000495
80	36	17	15	0.999997	0.000052
80	36	17	16	1.000000	0.000003
80	36	17	17	1.000000	0.000000
80	36	18	0	0.000003	0.000003
80	36	18	1	0.000072	0.000070
80	36	18	2	0.000812	0.000739
80	36	18	3	0.005433	0.004621
80	36	18	4	0.024496	0.019063

N	n	k	x	P(x)	p(x)
80	36	18	5	0.079594	0.055098
80	36	18	6	0.195244	0.115649
80	36	18	7	0.375476	0.180233
80	36	18	8	0.586852	0.211376
80	36	18	9	0.774742	0.187890
80	36	18	10	0.901567	0.126825
80	36	18	11	0.966382	0.064815
80	36	18	12	0.991256	0.024874
80	36	18	13	0.998321	0.007065
80	36	18	14	0.999772	0.001451
80	36	18	15	0.999979	0.000208
80	36	18	16	0.999999	0.000019
80	36	18	17	1.000000	0.000001
80	36	18	18	1.000000	0.000000
80	36	19	0	0.000001	0.000001
80	36	19	1	0.000033	0.000032
80	36	19	2	0.000406	0.000373
80	36	19	3	0.002973	0.002567
80	36	19	4	0.014657	0.011684
80	36	19	5	0.052045	0.037388
80	36	19	6	0.139284	0.087239
80	36	19	7	0.291174	0.151890
80	36	19	8	0.491392	0.200218
80	36	19	9	0.692919	0.201527
80	36	19	10	0.848382	0.155463
80	36	19	11	0.940247	0.091865
80	36	19	12	0.981627	0.041381
80	36	19	13	0.995700	0.014073
80	36	19	14	0.999257	0.003557
80	36	19	15	0.999909	0.000652
80	36	19	16	0.999992	0.000083
80	36	19	17	0.999999	0.000007
80	36	19	18	1.000000	0.000000
80	36	19	19	1.000000	0.000000
80	36	20	0	0.000000	0.000000
80	36	20	1	0.000015	0.000014
80	36	20	2	0.000198	0.000183
80	36	20	3	0.001584	0.001386
80	36	20	4	0.008528	0.006943
80	36	20	5	0.033045	0.024517
80	36	20	6	0.096380	0.063335
80	36	20	7	0.218964	0.122584
80	36	20	8	0.399489	0.180524
80	36	20	9	0.603718	0.204230
80	36	20	10	0.782119	0.178401
80	36	20	11	0.902597	0.120478
80	36	20	12	0.965346	0.062749
80	36	20	13	0.990394	0.025047
80	36	20	14	0.997974	0.007580
80	36	20	15	0.999684	0.001710

N	n	k	x	P(x)	p(x)
80	36	20	16	0.999965	0.000281
80	36	20	17	0.999997	0.000032
80	36	20	18	1.000000	0.000002
80	36	20	19	1.000000	0.000000
80	36	20	20	1.000000	0.000000
80	36	21	0	0.000000	0.000000
80	36	21	1	0.000006	0.000006
80	36	21	2	0.000094	0.000088
80	36	21	3	0.000822	0.000728
80	36	21	4	0.004825	0.004003
80	36	21	5	0.020378	0.015553
80	36	21	6	0.064712	0.044335
80	36	21	7	0.159715	0.095003
80	36	21	8	0.315244	0.155529
80	36	21	9	0.511815	0.196571
80	36	21	10	0.704812	0.192997
80	36	21	11	0.852398	0.147586
80	36	21	12	0.940247	0.087849
80	36	21	13	0.980792	0.040546
80	36	21	14	0.995195	0.014402
80	36	21	15	0.999086	0.003891
80	36	21	16	0.999872	0.000786
80	36	21	17	0.999987	0.000116
80	36	21	18	0.999999	0.000012
80	36	21	19	1.000000	0.000001
80	36	21	20	1.000000	0.000000
80	36	21	21	1.000000	0.000000
80	36	22	0	0.000000	0.000000
80	36	22	1	0.000003	0.000003
80	36	22	2	0.000044	0.000041
80	36	22	3	0.000415	0.000371
80	36	22	4	0.002654	0.002239
80	36	22	5	0.012206	0.009552
80	36	22	6	0.042169	0.029963
80	36	22	7	0.113019	0.070850
80	36	22	8	0.241433	0.128415
80	36	22	9	0.421859	0.180425
80	36	22	10	0.619762	0.197904
80	36	22	11	0.789862	0.170099
80	36	22	12	0.904512	0.114650
80	36	22	13	0.964986	0.060475
80	36	22	14	0.989824	0.024838
80	36	22	15	0.997701	0.007877
80	36	22	16	0.999605	0.001904
80	36	22	17	0.999950	0.000345
80	36	22	18	0.999995	0.000045
80	36	22	19	1.000000	0.000004
80	36	22	20	1.000000	0.000000
80	36	22	21	1.000000	0.000000
80	36	22	22	1.000000	0.000000

N	n	k	x	P(x)	p(x)
80	36	23	0	0.000000	0.000000
80	36	23	1	0.000001	0.000001
80	36	23	2	0.000020	0.000019
80	36	23	3	0.000204	0.000184
80	36	23	4	0.001419	0.001215
80	36	23	5	0.007100	0.005682
80	36	23	6	0.026671	0.019570
80	36	23	7	0.077594	0.050923
80	36	23	8	0.179440	0.101846
80	36	23	9	0.337868	0.158427
80	36	23	10	0.531047	0.193179
80	36	23	11	0.716543	0.185496
80	36	23	12	0.857070	0.140527
80	36	23	13	0.941005	0.083935
80	36	23	14	0.980403	0.039398
80	36	23	15	0.994849	0.014446
80	36	23	16	0.998949	0.004100
80	36	23	17	0.999837	0.000888
80	36	23	18	0.999981	0.000144
80	36	23	19	0.999998	0.000017
80	36	23	20	1.000000	0.000001
80	36	23	21	1.000000	0.000000
80	36	23	22	1.000000	0.000000
80	36	23	23	1.000000	0.000000
80	36	24	0	0.000000	0.000000
80	36	24	1	0.000000	0.000000
80	36	24	2	0.000009	0.000008
80	36	24	3	0.000097	0.000089
80	36	24	4	0.000737	0.000639
80	36	24	5	0.004010	0.003274
80	36	24	6	0.016371	0.012360
80	36	24	7	0.051686	0.035315
80	36	24	8	0.129410	0.077725
80	36	24	9	0.262823	0.133413
80	36	24	10	0.442930	0.180107
80	36	24	11	0.635185	0.192255
80	36	24	12	0.797901	0.162716
80	36	24	13	0.907137	0.109236
80	36	24	14	0.965197	0.058060
80	36	24	15	0.989527	0.024330
80	36	24	16	0.997510	0.007983
80	36	24	17	0.999541	0.002031
80	36	24	18	0.999936	0.000395
80	36	24	19	0.999993	0.000058
80	36	24	20	0.999999	0.000006
80	36	24	21	1.000000	0.000000
80	36	24	22	1.000000	0.000000
80	36	24	23	1.000000	0.000000
80	36	24	24	1.000000	0.000000
80	36	25	0	0.000000	0.000000

N	n	k	x	P(x)	p(x)
80	36	25	1	0.000000	0.000000
80	36	25	2	0.000004	0.000003
80	36	25	3	0.000045	0.000041
80	36	25	4	0.000371	0.000326
80	36	25	5	0.002198	0.001827
80	36	25	6	0.009749	0.007551
80	36	25	7	0.033397	0.023649
80	36	25	8	0.090548	0.057151
80	36	25	9	0.198499	0.107951
80	36	25	10	0.359309	0.160810
80	36	25	11	0.549357	0.190048
80	36	25	12	0.728166	0.178809
80	36	25	13	0.862272	0.134106
80	36	25	14	0.942388	0.080116
80	36	25	15	0.980403	0.038016
80	36	25	16	0.994659	0.014256
80	36	25	17	0.998852	0.004193
80	36	25	18	0.999809	0.000957
80	36	25	19	0.999976	0.000167
80	36	25	20	0.999998	0.000022
80	36	25	21	1.000000	0.000002
80	36	25	22	1.000000	0.000000
80	36	25	23	1.000000	0.000000
80	36	25	24	1.000000	0.000000
80	36	25	25	1.000000	0.000000
80	36	26	0	0.000000	0.000000
80	36	26	1	0.000000	0.000000
80	36	26	2	0.000002	0.000001
80	36	26	3	0.000020	0.000019
80	36	26	4	0.000181	0.000161
80	36	26	5	0.001168	0.000987
80	36	26	6	0.005630	0.004462
80	36	26	7	0.020928	0.015298
80	36	26	8	0.061453	0.040525
80	36	26	9	0.145505	0.084052
80	36	26	10	0.283290	0.137785
80	36	26	11	0.462972	0.179682
80	36	26	12	0.650140	0.187168
80	36	26	13	0.806191	0.156051
80	36	26	14	0.910341	0.104150
80	36	26	15	0.965888	0.055547
80	36	26	16	0.989475	0.023587
80	36	26	17	0.997403	0.007928
80	36	26	18	0.999496	0.002092
80	36	26	19	0.999924	0.000429
80	36	26	20	0.999991	0.000067
80	36	26	21	0.999999	0.000008
80	36	26	22	1.000000	0.000001
80	36	26	23	1.000000	0.000000
80	36	26	24	1.000000	0.000000

Table for $N = 2$, $n = 1$, through $N = 100$, $n = 50$

N	n	k	x	P(x)	p(x)
80	36	26	25	1.000000	0.000000
80	36	26	26	1.000000	0.000000
80	36	27	0	0.000000	0.000000
80	36	27	1	0.000000	0.000000
80	36	27	2	0.000001	0.000001
80	36	27	3	0.000009	0.000008
80	36	27	4	0.000086	0.000077
80	36	27	5	0.000602	0.000516
80	36	27	6	0.003151	0.002550
80	36	27	7	0.012713	0.009561
80	36	27	8	0.040440	0.027728
80	36	27	9	0.103479	0.063039
80	36	27	10	0.216949	0.113470
80	36	27	11	0.379785	0.162837
80	36	27	12	0.566954	0.187168
80	36	27	13	0.739725	0.172771
80	36	27	14	0.867910	0.128185
80	36	27	15	0.944287	0.076377
80	36	27	16	0.980739	0.036453
80	36	27	17	0.994614	0.013875
80	36	27	18	0.998798	0.004184
80	36	27	19	0.999789	0.000991
80	36	27	20	0.999971	0.000182
80	36	27	21	0.999997	0.000026
80	36	27	22	1.000000	0.000003
80	36	27	23	1.000000	0.000000
80	36	27	24	1.000000	0.000000
80	36	27	25	1.000000	0.000000
80	36	27	26	1.000000	0.000000
80	36	27	27	1.000000	0.000000
80	36	28	0	0.000000	0.000000
80	36	28	1	0.000000	0.000000
80	36	28	2	0.000000	0.000000
80	36	28	3	0.000004	0.000003
80	36	28	4	0.000039	0.000036
80	36	28	5	0.000300	0.000261
80	36	28	6	0.001708	0.001408
80	36	28	7	0.007481	0.005773
80	36	28	8	0.025792	0.018311
80	36	28	9	0.071365	0.045573
80	36	28	10	0.161284	0.089919
80	36	28	11	0.302976	0.141691
80	36	28	12	0.482198	0.179223
80	36	28	13	0.664749	0.182550
80	36	28	14	0.814701	0.149952
80	36	28	15	0.914024	0.099323
80	36	28	16	0.966983	0.052959
80	36	28	17	0.989640	0.022656
80	36	28	18	0.997377	0.007737
80	36	28	19	0.999471	0.002094

N	n	k	x	P(x)	p(x)
80	36	28	20	0.999916	0.000445
80	36	28	21	0.999990	0.000073
80	36	28	22	0.999999	0.000009
80	36	28	23	1.000000	0.000001
80	36	28	24	1.000000	0.000000
80	36	28	25	1.000000	0.000000
80	36	28	26	1.000000	0.000000
80	36	28	27	1.000000	0.000000
80	36	28	28	1.000000	0.000000
80	36	29	0	0.000000	0.000000
80	36	29	1	0.000000	0.000000
80	36	29	2	0.000000	0.000000
80	36	29	3	0.000001	0.000001
80	36	29	4	0.000017	0.000016
80	36	29	5	0.000145	0.000127
80	36	29	6	0.000896	0.000751
80	36	29	7	0.004262	0.003366
80	36	29	8	0.015932	0.011671
80	36	29	9	0.047702	0.031770
80	36	29	10	0.116325	0.068623
80	36	29	11	0.234855	0.118530
80	36	29	12	0.399480	0.164625
80	36	29	13	0.584005	0.184525
80	36	29	14	0.751260	0.167254
80	36	29	15	0.873913	0.122653
80	36	29	16	0.946614	0.072702
80	36	29	17	0.981362	0.034747
80	36	29	18	0.994699	0.013337
80	36	29	19	0.998787	0.004088
80	36	29	20	0.999779	0.000993
80	36	29	21	0.999969	0.000189
80	36	29	22	0.999996	0.000028
80	36	29	23	1.000000	0.000003
80	36	29	24	1.000000	0.000000
80	36	29	25	1.000000	0.000000
80	36	29	26	1.000000	0.000000
80	36	29	27	1.000000	0.000000
80	36	29	28	1.000000	0.000000
80	36	29	29	1.000000	0.000000
80	36	30	0	0.000000	0.000000
80	36	30	1	0.000000	0.000000
80	36	30	2	0.000000	0.000000
80	36	30	3	0.000001	0.000001
80	36	30	4	0.000007	0.000007
80	36	30	5	0.000067	0.000060
80	36	30	6	0.000454	0.000387
80	36	30	7	0.002348	0.001894
80	36	30	8	0.009525	0.007177
80	36	30	9	0.030883	0.021358
80	36	30	10	0.081341	0.050458

N	n	k	x	P(x)	p(x)
80	36	30	11	0.176752	0.095411
80	36	30	12	0.322010	0.145258
80	36	30	13	0.500788	0.178779
80	36	30	14	0.679111	0.178323
80	36	30	15	0.823409	0.144298
80	36	30	16	0.918104	0.094695
80	36	30	17	0.968417	0.050313
80	36	30	18	0.989992	0.021575
80	36	30	19	0.997424	0.007433
80	36	30	20	0.999468	0.002044
80	36	30	21	0.999913	0.000445
80	36	30	22	0.999989	0.000076
80	36	30	23	0.999999	0.000010
80	36	30	24	1.000000	0.000001
80	36	30	25	1.000000	0.000000
80	36	30	26	1.000000	0.000000
80	36	30	27	1.000000	0.000000
80	36	30	28	1.000000	0.000000
80	36	30	29	1.000000	0.000000
80	36	30	30	1.000000	0.000000
80	36	31	0	0.000000	0.000000
80	36	31	1	0.000000	0.000000
80	36	31	2	0.000000	0.000000
80	36	31	3	0.000000	0.000000
80	36	31	4	0.000003	0.000003
80	36	31	5	0.000030	0.000027
80	36	31	6	0.000222	0.000192
80	36	31	7	0.001249	0.001027
80	36	31	8	0.005506	0.004256
80	36	31	9	0.019349	0.013844
80	36	31	10	0.055102	0.035753
80	36	31	11	0.129046	0.073944
80	36	31	12	0.252286	0.123240
80	36	31	13	0.418550	0.166264
80	36	31	14	0.600649	0.182099
80	36	31	15	0.762803	0.162155
80	36	31	16	0.880226	0.117422
80	36	31	17	0.949298	0.069072
80	36	31	18	0.982225	0.032927
80	36	31	19	0.994897	0.012672
80	36	31	20	0.998814	0.003917
80	36	31	21	0.999779	0.000966
80	36	31	22	0.999968	0.000188
80	36	31	23	0.999996	0.000029
80	36	31	24	1.000000	0.000003
80	36	31	25	1.000000	0.000000
80	36	31	26	1.000000	0.000000
80	36	31	27	1.000000	0.000000
80	36	31	28	1.000000	0.000000
80	36	31	29	1.000000	0.000000

N	n	k	x	P(x)	p(x)
80	36	31	30	1.000000	0.000000
80	36	31	31	1.000000	0.000000
80	36	32	0	0.000000	0.000000
80	36	32	1	0.000000	0.000000
80	36	32	2	0.000000	0.000000
80	36	32	3	0.000000	0.000000
80	36	32	4	0.000001	0.000001
80	36	32	5	0.000013	0.000012
80	36	32	6	0.000105	0.000092
80	36	32	7	0.000641	0.000537
80	36	32	8	0.003073	0.002432
80	36	32	9	0.011721	0.008648
80	36	32	10	0.036131	0.024410
80	36	32	11	0.091320	0.055188
80	36	32	12	0.191924	0.100604
80	36	32	13	0.340508	0.148584
80	36	32	14	0.518890	0.178382
80	36	32	15	0.693309	0.174418
80	36	32	16	0.832298	0.138990
80	36	32	17	0.922515	0.090216
80	36	32	18	0.970129	0.047614
80	36	32	19	0.990500	0.020371
80	36	32	20	0.997535	0.007035
80	36	32	21	0.999484	0.001949
80	36	32	22	0.999914	0.000430
80	36	32	23	0.999989	0.000075
80	36	32	24	0.999999	0.000010
80	36	32	25	1.000000	0.000001
80	36	32	26	1.000000	0.000000
80	36	32	27	1.000000	0.000000
80	36	32	28	1.000000	0.000000
80	36	32	29	1.000000	0.000000
80	36	32	30	1.000000	0.000000
80	36	32	31	1.000000	0.000000
80	36	32	32	1.000000	0.000000
80	36	33	0	0.000000	0.000000
80	36	33	1	0.000000	0.000000
80	36	33	2	0.000000	0.000000
80	36	33	3	0.000000	0.000000
80	36	33	4	0.000000	0.000000
80	36	33	5	0.000005	0.000005
80	36	33	6	0.000047	0.000042
80	36	33	7	0.000317	0.000270
80	36	33	8	0.001655	0.001338
80	36	33	9	0.006857	0.005202
80	36	33	10	0.022909	0.016052
80	36	33	11	0.062576	0.039667
80	36	33	12	0.141622	0.079046
80	36	33	13	0.269311	0.127689
80	36	33	14	0.437132	0.167820

N	n	k	x	P(x)	p(x)
80	36	33	15	0.617001	0.179869
80	36	33	16	0.774386	0.157385
80	36	33	17	0.886804	0.112418
80	36	33	18	0.952274	0.065470
80	36	33	19	0.983286	0.031012
80	36	33	20	0.995190	0.011905
80	36	33	21	0.998875	0.003685
80	36	33	22	0.999788	0.000914
80	36	33	23	0.999968	0.000180
80	36	33	24	0.999996	0.000028
80	36	33	25	0.999999	0.000003
80	36	33	26	1.000000	0.000000
80	36	33	27	1.000000	0.000000
80	36	33	28	1.000000	0.000000
80	36	33	29	1.000000	0.000000
80	36	33	30	1.000000	0.000000
80	36	33	31	1.000000	0.000000
80	36	33	32	1.000000	0.000000
80	36	33	33	1.000000	0.000000
80	36	34	0	0.000000	0.000000
80	36	34	1	0.000000	0.000000
80	36	34	2	0.000000	0.000000
80	36	34	3	0.000000	0.000000
80	36	34	4	0.000000	0.000000
80	36	34	5	0.000002	0.000002
80	36	34	6	0.000021	0.000018
80	36	34	7	0.000151	0.000130
80	36	34	8	0.000858	0.000707
80	36	34	9	0.003868	0.003011
80	36	34	10	0.014029	0.010161
80	36	34	11	0.041477	0.027447
80	36	34	12	0.101258	0.059781
80	36	34	13	0.206825	0.105567
80	36	34	14	0.358577	0.151752
80	36	34	15	0.536634	0.178056
80	36	34	16	0.707413	0.170780
80	36	34	17	0.841358	0.133945
80	36	34	18	0.927200	0.085842
80	36	34	19	0.972068	0.044868
80	36	34	20	0.991138	0.019069
80	36	34	21	0.997699	0.006561
80	36	34	22	0.999516	0.001817
80	36	34	23	0.999919	0.000402
80	36	34	24	0.999989	0.000070
80	36	34	25	0.999999	0.000010
80	36	34	26	1.000000	0.000001
80	36	34	27	1.000000	0.000000
80	36	34	28	1.000000	0.000000
80	36	34	29	1.000000	0.000000
80	36	34	30	1.000000	0.000000

N	n	k	x	P(x)	p(x)
80	36	34	31	1.000000	0.000000
80	36	34	32	1.000000	0.000000
80	36	34	33	1.000000	0.000000
80	36	34	34	1.000000	0.000000
80	36	35	0	0.000000	0.000000
80	36	35	1	0.000000	0.000000
80	36	35	2	0.000000	0.000000
80	36	35	3	0.000000	0.000000
80	36	35	4	0.000000	0.000000
80	36	35	5	0.000001	0.000001
80	36	35	6	0.000009	0.000008
80	36	35	7	0.000069	0.000060
80	36	35	8	0.000427	0.000359
80	36	35	9	0.002101	0.001674
80	36	35	10	0.008286	0.006185
80	36	35	11	0.026560	0.018273
80	36	35	12	0.070068	0.043508
80	36	35	13	0.154041	0.083974
80	36	35	14	0.286000	0.131959
80	36	35	15	0.455347	0.169347
80	36	35	16	0.633161	0.177814
80	36	35	17	0.786033	0.152872
80	36	35	18	0.893610	0.107577
80	36	35	19	0.955487	0.061877
80	36	35	20	0.984505	0.029018
80	36	35	21	0.995559	0.011055
80	36	35	22	0.998963	0.003404
80	36	35	23	0.999805	0.000842
80	36	35	24	0.999971	0.000166
80	36	35	25	0.999996	0.000026
80	36	35	26	1.000000	0.000003
80	36	35	27	1.000000	0.000000
80	36	35	28	1.000000	0.000000
80	36	35	29	1.000000	0.000000
80	36	35	30	1.000000	0.000000
80	36	35	31	1.000000	0.000000
80	36	35	32	1.000000	0.000000
80	36	35	33	1.000000	0.000000
80	36	35	34	1.000000	0.000000
80	36	35	35	1.000000	0.000000
80	36	36	0	0.000000	0.000000
80	36	36	1	0.000000	0.000000
80	36	36	2	0.000000	0.000000
80	36	36	3	0.000000	0.000000
80	36	36	4	0.000000	0.000000
80	36	36	5	0.000000	0.000000
80	36	36	6	0.000003	0.000003
80	36	36	7	0.000030	0.000027
80	36	36	8	0.000204	0.000174
80	36	36	9	0.001097	0.000893

N	n	k	x	P(x)	p(x)
80	36	36	10	0.004713	0.003616
80	36	36	11	0.016408	0.011695
80	36	36	12	0.046863	0.030456
80	36	36	13	0.111121	0.064258
80	36	36	14	0.221487	0.110365
80	36	36	15	0.376318	0.154831
80	36	36	16	0.554133	0.177814
80	36	36	17	0.721487	0.167355
80	36	36	18	0.850579	0.129092
80	36	36	19	0.932111	0.081532
80	36	36	20	0.974187	0.042076
80	36	36	21	0.991874	0.017687
80	36	36	22	0.997904	0.006030
80	36	36	23	0.999562	0.001658
80	36	36	24	0.999926	0.000365
80	36	36	25	0.999990	0.000064
80	36	36	26	0.999999	0.000009
80	36	36	27	1.000000	0.000001
80	36	36	28	1.000000	0.000000
80	36	36	29	1.000000	0.000000
80	36	36	30	1.000000	0.000000
80	36	36	31	1.000000	0.000000
80	36	36	32	1.000000	0.000000
80	36	36	33	1.000000	0.000000
80	36	36	34	1.000000	0.000000
80	36	36	35	1.000000	0.000000
80	36	36	36	1.000000	0.000000
80	37	1	0	0.537500	0.537500
80	37	1	1	1.000000	0.462500
80	37	2	0	0.285759	0.285759
80	37	2	1	0.789240	0.503481
80	37	2	2	1.000000	0.210759
80	37	3	0	0.150207	0.150207
80	37	3	1	0.556865	0.406658
80	37	3	2	0.905428	0.348564
80	37	3	3	1.000000	0.094572
80	37	4	0	0.078030	0.078030
80	37	4	1	0.366739	0.288709
80	37	4	2	0.746990	0.380251
80	37	4	3	0.958241	0.211251
80	37	4	4	1.000000	0.041759
80	37	5	0	0.040041	0.040041
80	37	5	1	0.229982	0.189940
80	37	5	2	0.571875	0.341893
80	37	5	3	0.863734	0.291860
80	37	5	4	0.981868	0.118134
80	37	5	5	1.000000	0.018132
80	37	6	0	0.020288	0.020288
80	37	6	1	0.138810	0.118523
80	37	6	2	0.412325	0.273514

N	n	k	x	P(x)	p(x)
80	37	6	3	0.731424	0.319100
80	37	6	4	0.929889	0.198465
80	37	6	5	0.992264	0.062375
80	37	6	6	1.000000	0.007736
80	37	7	0	0.010144	0.010144
80	37	7	1	0.081151	0.071007
80	37	7	2	0.282960	0.201809
80	37	7	3	0.584811	0.301851
80	37	7	4	0.841384	0.256574
80	37	7	5	0.965291	0.123906
80	37	7	6	0.996759	0.031468
80	37	7	7	1.000000	0.003241
80	37	8	0	0.005002	0.005002
80	37	8	1	0.046134	0.041131
80	37	8	2	0.186202	0.140068
80	37	8	3	0.444223	0.258021
80	37	8	4	0.725399	0.281176
80	37	8	5	0.910976	0.185576
80	37	8	6	0.983396	0.072420
80	37	8	7	0.998668	0.015272
80	37	8	8	1.000000	0.001332
80	37	9	0	0.002432	0.002432
80	37	9	1	0.025568	0.023136
80	37	9	2	0.118113	0.092545
80	37	9	3	0.322380	0.204266
80	37	9	4	0.596527	0.274147
80	37	9	5	0.828497	0.231971
80	37	9	6	0.952215	0.123718
80	37	9	7	0.992305	0.040090
80	37	9	8	0.999463	0.007159
80	37	9	9	1.000000	0.000536
80	37	10	0	0.001164	0.001164
80	37	10	1	0.013837	0.012672
80	37	10	2	0.072492	0.058655
80	37	10	3	0.224562	0.152070
80	37	10	4	0.469106	0.244544
80	37	10	5	0.723947	0.254841
80	37	10	6	0.898197	0.174250
80	37	10	7	0.975365	0.077168
80	37	10	8	0.996539	0.021174
80	37	10	9	0.999788	0.003249
80	37	10	10	1.000000	0.000212
80	37	11	0	0.000549	0.000549
80	37	11	1	0.007320	0.006771
80	37	11	2	0.043165	0.035845
80	37	11	3	0.150700	0.107535
80	37	11	4	0.353821	0.203121
80	37	11	5	0.607448	0.253627
80	37	11	6	0.821029	0.213581
80	37	11	7	0.942293	0.121264

N	n	k	x	P(x)	p(x)
80	37	11	8	0.987767	0.045474
80	37	11	9	0.998489	0.010721
80	37	11	10	0.999918	0.001430
80	37	11	11	1.000000	0.000082
80	37	12	0	0.000255	0.000255
80	37	12	1	0.003787	0.003533
80	37	12	2	0.024982	0.021195
80	37	12	3	0.097711	0.072729
80	37	12	4	0.256676	0.158965
80	37	12	5	0.489824	0.233148
80	37	12	6	0.725073	0.235249
80	37	12	7	0.889570	0.164497
80	37	12	8	0.968655	0.079085
80	37	12	9	0.994138	0.025483
80	37	12	10	0.999359	0.005221
80	37	12	11	0.999969	0.000610
80	37	12	12	1.000000	0.000031
80	37	13	0	0.000116	0.000116
80	37	13	1	0.001917	0.001801
80	37	13	2	0.014073	0.012156
80	37	13	3	0.061347	0.047274
80	37	13	4	0.179531	0.118185
80	37	13	5	0.380107	0.200576
80	37	13	6	0.617827	0.237720
80	37	13	7	0.816998	0.199171
80	37	13	8	0.934928	0.117930
80	37	13	9	0.983645	0.048717
80	37	13	10	0.997286	0.013641
80	37	13	11	0.999736	0.002450
80	37	13	12	0.999989	0.000253
80	37	13	13	1.000000	0.000011
80	37	14	0	0.000052	0.000052
80	37	14	1	0.000949	0.000897
80	37	14	2	0.007723	0.006774
80	37	14	3	0.037357	0.029634
80	37	14	4	0.121321	0.083964
80	37	14	5	0.284310	0.162989
80	37	14	6	0.507837	0.223527
80	37	14	7	0.727817	0.219979
80	37	14	8	0.883883	0.156066
80	37	14	9	0.963286	0.079402
80	37	14	10	0.991789	0.028503
80	37	14	11	0.998785	0.006996
80	37	14	12	0.999894	0.001109
80	37	14	13	0.999996	0.000102
80	37	14	14	1.000000	0.000004
80	37	15	0	0.000023	0.000023
80	37	15	1	0.000460	0.000437
80	37	15	2	0.004131	0.003671
80	37	15	3	0.022091	0.017960

N	n	k	x	P(x)	p(x)
80	37	15	4	0.079339	0.057248
80	37	15	5	0.205285	0.125946
80	37	15	6	0.402847	0.197562
80	37	15	7	0.627826	0.224979
80	37	15	8	0.815309	0.187482
80	37	15	9	0.929600	0.114291
80	37	15	10	0.980128	0.050529
80	37	15	11	0.996029	0.015901
80	37	15	12	0.999474	0.003445
80	37	15	13	0.999959	0.000485
80	37	15	14	0.999999	0.000040
80	37	15	15	1.000000	0.000001
80	37	16	0	0.000010	0.000010
80	37	16	1	0.000218	0.000208
80	37	16	2	0.002154	0.001936
80	37	16	3	0.012696	0.010542
80	37	16	4	0.050275	0.037578
80	37	16	5	0.143281	0.093006
80	37	16	6	0.308625	0.165344
80	37	16	7	0.523990	0.215364
80	37	16	8	0.731663	0.207673
80	37	16	9	0.880367	0.148704
80	37	16	10	0.959140	0.078773
80	37	16	11	0.989669	0.030529
80	37	16	12	0.998149	0.008480
80	37	16	13	0.999780	0.001631
80	37	16	14	0.999984	0.000205
80	37	16	15	0.999999	0.000015
80	37	16	16	1.000000	0.000000
80	37	17	0	0.000004	0.000004
80	37	17	1	0.000101	0.000097
80	37	17	2	0.001095	0.000994
80	37	17	3	0.007096	0.006001
80	37	17	4	0.030898	0.023803
80	37	17	5	0.096778	0.065879
80	37	17	6	0.228537	0.131759
80	37	17	7	0.423038	0.194501
80	37	17	8	0.637561	0.214523
80	37	17	9	0.815309	0.177748
80	37	17	10	0.925907	0.110599
80	37	17	11	0.977266	0.051359
80	37	17	12	0.994836	0.017570
80	37	17	13	0.999168	0.004332
80	37	17	14	0.999911	0.000743
80	37	17	15	0.999994	0.000083
80	37	17	16	1.000000	0.000005
80	37	17	17	1.000000	0.000000
80	37	18	0	0.000002	0.000002
80	37	18	1	0.000046	0.000044
80	37	18	2	0.000543	0.000497

N	n	k	x	P(x)	p(x)
80	37	18	3	0.003857	0.003315
80	37	18	4	0.018430	0.014573
80	37	18	5	0.063315	0.044885
80	37	18	6	0.163703	0.100388
80	37	18	7	0.330418	0.166715
80	37	18	8	0.538812	0.208394
80	37	18	9	0.736310	0.197498
80	37	18	10	0.878508	0.142198
80	37	18	11	0.956070	0.077563
80	37	18	12	0.987864	0.031794
80	37	18	13	0.997518	0.009654
80	37	18	14	0.999640	0.002122
80	37	18	15	0.999965	0.000325
80	37	18	16	0.999998	0.000033
80	37	18	17	1.000000	0.000002
80	37	18	18	1.000000	0.000000
80	37	19	0	0.000001	0.000001
80	37	19	1	0.000020	0.000019
80	37	19	2	0.000262	0.000242
80	37	19	3	0.002040	0.001778
80	37	19	4	0.010674	0.008634
80	37	19	5	0.040149	0.029475
80	37	19	6	0.113509	0.073360
80	37	19	7	0.249749	0.136240
80	37	19	8	0.441337	0.191588
80	37	19	9	0.647117	0.205780
80	37	19	10	0.816583	0.169466
80	37	19	11	0.923544	0.106961
80	37	19	12	0.975044	0.051500
80	37	19	13	0.993781	0.018737
80	37	19	14	0.998853	0.005072
80	37	19	15	0.999850	0.000997
80	37	19	16	0.999987	0.000137
80	37	19	17	0.999999	0.000012
80	37	19	18	1.000000	0.000001
80	37	19	19	1.000000	0.000000
80	37	20	0	0.000000	0.000000
80	37	20	1	0.000009	0.000008
80	37	20	2	0.000123	0.000115
80	37	20	3	0.001049	0.000926
80	37	20	4	0.006003	0.004954
80	37	20	5	0.024687	0.018684
80	37	20	6	0.076228	0.051541
80	37	20	7	0.182746	0.106518
80	37	20	8	0.350255	0.167509
80	37	20	9	0.552661	0.202406
80	37	20	10	0.741573	0.188913
80	37	20	11	0.877954	0.136381
80	37	20	12	0.953938	0.075984
80	37	20	13	0.986409	0.032472

N	n	k	x	P(x)	p(x)
80	37	20	14	0.996940	0.010531
80	37	20	15	0.999490	0.002550
80	37	20	16	0.999940	0.000449
80	37	20	17	0.999995	0.000056
80	37	20	18	1.000000	0.000005
80	37	20	19	1.000000	0.000000
80	37	20	20	1.000000	0.000000
80	37	21	0	0.000000	0.000000
80	37	21	1	0.000004	0.000004
80	37	21	2	0.000056	0.000053
80	37	21	3	0.000524	0.000468
80	37	21	4	0.003278	0.002754
80	37	21	5	0.014722	0.011444
80	37	21	6	0.049598	0.034876
80	37	21	7	0.129487	0.079889
80	37	21	8	0.269292	0.139805
80	37	21	9	0.458205	0.188913
80	37	21	10	0.656563	0.198358
80	37	21	11	0.818856	0.162293
80	37	21	12	0.922278	0.103422
80	37	21	13	0.973421	0.051143
80	37	21	14	0.992903	0.019483
80	37	21	15	0.998555	0.005652
80	37	21	16	0.999782	0.001227
80	37	21	17	0.999977	0.000194
80	37	21	18	0.999998	0.000022
80	37	21	19	1.000000	0.000002
80	37	21	20	1.000000	0.000000
80	37	21	21	1.000000	0.000000
80	37	22	0	0.000000	0.000000
80	37	22	1	0.000001	0.000001
80	37	22	2	0.000025	0.000024
80	37	22	3	0.000255	0.000230
80	37	22	4	0.001738	0.001483
80	37	22	5	0.008515	0.006777
80	37	22	6	0.031273	0.022758
80	37	22	7	0.088865	0.057592
80	37	22	8	0.200574	0.111709
80	37	22	9	0.368551	0.167977
80	37	22	10	0.565789	0.197237
80	37	22	11	0.747337	0.181548
80	37	22	12	0.878455	0.131118
80	37	22	13	0.952617	0.074162
80	37	22	14	0.985308	0.032692
80	37	22	15	0.996448	0.011139
80	37	22	16	0.999346	0.002898
80	37	22	17	0.999911	0.000565
80	37	22	18	0.999991	0.000081
80	37	22	19	0.999999	0.000008
80	37	22	20	1.000000	0.000001

Table for $N=2$, $n=1$, through $N=100$, $n=50$

N	n	k	x	P(x)	p(x)
80	37	22	21	1.000000	0.000000
80	37	22	22	1.000000	0.000000
80	37	23	0	0.000000	0.000000
80	37	23	1	0.000001	0.000001
80	37	23	2	0.000011	0.000010
80	37	23	3	0.000120	0.000109
80	37	23	4	0.000894	0.000774
80	37	23	5	0.004776	0.003882
80	37	23	6	0.019109	0.014333
80	37	23	7	0.059076	0.039967
80	37	23	8	0.144720	0.085643
80	37	23	9	0.287459	0.142739
80	37	23	10	0.473971	0.186512
80	37	23	11	0.665953	0.191982
80	37	23	12	0.821938	0.155985
80	37	23	13	0.921929	0.099991
80	37	23	14	0.972344	0.050415
80	37	23	15	0.992223	0.019878
80	37	23	16	0.998296	0.006074
80	37	23	17	0.999716	0.001419
80	37	23	18	0.999965	0.000249
80	37	23	19	0.999997	0.000032
80	37	23	20	1.000000	0.000003
80	37	23	21	1.000000	0.000000
80	37	23	22	1.000000	0.000000
80	37	23	23	1.000000	0.000000
80	37	24	0	0.000000	0.000000
80	37	24	1	0.000000	0.000000
80	37	24	2	0.000005	0.000004
80	37	24	3	0.000055	0.000050
80	37	24	4	0.000446	0.000391
80	37	24	5	0.002597	0.002151
80	37	24	6	0.011314	0.008717
80	37	24	7	0.038041	0.026727
80	37	24	8	0.101147	0.063106
80	37	24	9	0.217341	0.116195
80	37	24	10	0.385623	0.168282
80	37	24	11	0.578382	0.192759
80	37	24	12	0.753524	0.175141
80	37	24	13	0.879828	0.126304
80	37	24	14	0.952001	0.072174
80	37	24	15	0.984550	0.032549
80	37	24	16	0.996059	0.011508
80	37	24	17	0.999218	0.003159
80	37	24	18	0.999882	0.000664
80	37	24	19	0.999987	0.000105
80	37	24	20	0.999999	0.000012
80	37	24	21	1.000000	0.000001
80	37	24	22	1.000000	0.000000
80	37	24	23	1.000000	0.000000

N	n	k	x	P(x)	p(x)
80	37	24	24	1.000000	0.000000
80	37	25	0	0.000000	0.000000
80	37	25	1	0.000000	0.000000
80	37	25	2	0.000002	0.000002
80	37	25	3	0.000024	0.000022
80	37	25	4	0.000216	0.000191
80	37	25	5	0.001368	0.001152
80	37	25	6	0.006488	0.005121
80	37	25	7	0.023723	0.017235
80	37	25	8	0.068467	0.044744
80	37	25	9	0.159244	0.090777
80	37	25	10	0.304487	0.145243
80	37	25	11	0.488887	0.184400
80	37	25	12	0.675336	0.186449
80	37	25	13	0.825698	0.150362
80	37	25	14	0.922359	0.096661
80	37	25	15	0.971763	0.049405
80	37	25	16	0.991743	0.019980
80	37	25	17	0.998090	0.006347
80	37	25	18	0.999657	0.001567
80	37	25	19	0.999953	0.000296
80	37	25	20	0.999995	0.000042
80	37	25	21	1.000000	0.000004
80	37	25	22	1.000000	0.000000
80	37	25	23	1.000000	0.000000
80	37	25	24	1.000000	0.000000
80	37	25	25	1.000000	0.000000
80	37	26	0	0.000000	0.000000
80	37	26	1	0.000000	0.000000
80	37	26	2	0.000001	0.000001
80	37	26	3	0.000010	0.000010
80	37	26	4	0.000101	0.000090
80	37	26	5	0.000697	0.000597
80	37	26	6	0.003602	0.002905
80	37	26	7	0.014322	0.010720
80	37	26	8	0.044875	0.030552
80	37	26	9	0.113030	0.068156
80	37	26	10	0.233186	0.120156
80	37	26	11	0.401716	0.168530
80	37	26	12	0.590586	0.188870
80	37	26	13	0.760085	0.169499
80	37	26	14	0.881937	0.121852
80	37	26	15	0.952001	0.070065
80	37	26	16	0.984114	0.032113
80	37	26	17	0.995782	0.011667
80	37	26	18	0.999115	0.003334
80	37	26	19	0.999856	0.000741
80	37	26	20	0.999982	0.000126
80	37	26	21	0.999998	0.000016
80	37	26	22	1.000000	0.000001

N	n	k	x	P(x)	p(x)
80	37	26	23	1.000000	0.000000
80	37	26	24	1.000000	0.000000
80	37	26	25	1.000000	0.000000
80	37	26	26	1.000000	0.000000
80	37	27	0	0.000000	0.000000
80	37	27	1	0.000000	0.000000
80	37	27	2	0.000000	0.000000
80	37	27	3	0.000004	0.000004
80	37	27	4	0.000046	0.000041
80	37	27	5	0.000344	0.000298
80	37	27	6	0.001935	0.001591
80	37	27	7	0.008367	0.006432
80	37	27	8	0.028467	0.020100
80	37	27	9	0.077690	0.049223
80	37	27	10	0.173108	0.095418
80	37	27	11	0.320572	0.147464
80	37	27	12	0.503147	0.182574
80	37	27	13	0.684752	0.181606
80	37	27	14	0.830037	0.145285
80	37	27	15	0.923457	0.093420
80	37	27	16	0.971626	0.048169
80	37	27	17	0.991460	0.019834
80	37	27	18	0.997942	0.006482
80	37	27	19	0.999609	0.001667
80	37	27	20	0.999942	0.000333
80	37	27	21	0.999993	0.000051
80	37	27	22	0.999999	0.000006
80	37	27	23	1.000000	0.000000
80	37	27	24	1.000000	0.000000
80	37	27	25	1.000000	0.000000
80	37	27	26	1.000000	0.000000
80	37	27	27	1.000000	0.000000
80	37	28	0	0.000000	0.000000
80	37	28	1	0.000000	0.000000
80	37	28	2	0.000000	0.000000
80	37	28	3	0.000002	0.000002
80	37	28	4	0.000020	0.000018
80	37	28	5	0.000164	0.000144
80	37	28	6	0.001004	0.000840
80	37	28	7	0.004726	0.003722
80	37	28	8	0.017469	0.012743
80	37	28	9	0.051686	0.034217
80	37	28	10	0.124499	0.072814
80	37	28	11	0.248231	0.123732
80	37	28	12	0.417027	0.168795
80	37	28	13	0.602516	0.185489
80	37	28	14	0.766989	0.164473
80	37	28	15	0.884679	0.117690
80	37	28	16	0.952540	0.067861
80	37	28	17	0.983976	0.031436

N	n	k	x	P(x)	p(x)
80	37	28	18	0.995619	0.011643
80	37	28	19	0.999043	0.003424
80	37	28	20	0.999835	0.000792
80	37	28	21	0.999978	0.000143
80	37	28	22	0.999998	0.000020
80	37	28	23	1.000000	0.000002
80	37	28	24	1.000000	0.000000
80	37	28	25	1.000000	0.000000
80	37	28	26	1.000000	0.000000
80	37	28	27	1.000000	0.000000
80	37	28	28	1.000000	0.000000
80	37	29	0	0.000000	0.000000
80	37	29	1	0.000000	0.000000
80	37	29	2	0.000000	0.000000
80	37	29	3	0.000001	0.000001
80	37	29	4	0.000008	0.000008
80	37	29	5	0.000075	0.000067
80	37	29	6	0.000503	0.000428
80	37	29	7	0.002579	0.002076
80	37	29	8	0.010362	0.007783
80	37	29	9	0.033261	0.022899
80	37	29	10	0.086692	0.053431
80	37	29	11	0.186365	0.099673
80	37	29	12	0.335875	0.149510
80	37	29	13	0.516905	0.181030
80	37	29	14	0.694241	0.177336
80	37	29	15	0.834887	0.140646
80	37	29	16	0.925134	0.090248
80	37	29	17	0.971885	0.046751
80	37	29	18	0.991364	0.019479
80	37	29	19	0.997858	0.006493
80	37	29	20	0.999576	0.001719
80	37	29	21	0.999934	0.000358
80	37	29	22	0.999992	0.000058
80	37	29	23	0.999999	0.000007
80	37	29	24	1.000000	0.000001
80	37	29	25	1.000000	0.000000
80	37	29	26	1.000000	0.000000
80	37	29	27	1.000000	0.000000
80	37	29	28	1.000000	0.000000
80	37	29	29	1.000000	0.000000
80	37	30	0	0.000000	0.000000
80	37	30	1	0.000000	0.000000
80	37	30	2	0.000000	0.000000
80	37	30	3	0.000000	0.000000
80	37	30	4	0.000003	0.000003
80	37	30	5	0.000033	0.000030
80	37	30	6	0.000243	0.000210
80	37	30	7	0.001358	0.001115
80	37	30	8	0.005936	0.004578

N	n	k	x	P(x)	p(x)
80	37	30	9	0.020689	0.014753
80	37	30	10	0.058405	0.037716
80	37	30	11	0.135552	0.077146
80	37	30	12	0.262586	0.127034
80	37	30	13	0.431715	0.169129
80	37	30	14	0.614266	0.182552
80	37	30	15	0.774216	0.159950
80	37	30	16	0.887974	0.113757
80	37	30	17	0.953551	0.065578
80	37	30	18	0.984107	0.030556
80	37	30	19	0.995566	0.011458
80	37	30	20	0.999003	0.003438
80	37	30	21	0.999822	0.000818
80	37	30	22	0.999975	0.000153
80	37	30	23	0.999997	0.000022
80	37	30	24	1.000000	0.000002
80	37	30	25	1.000000	0.000000
80	37	30	26	1.000000	0.000000
80	37	30	27	1.000000	0.000000
80	37	30	28	1.000000	0.000000
80	37	30	29	1.000000	0.000000
80	37	30	30	1.000000	0.000000
80	37	31	0	0.000000	0.000000
80	37	31	1	0.000000	0.000000
80	37	31	2	0.000000	0.000000
80	37	31	3	0.000000	0.000000
80	37	31	4	0.000001	0.000001
80	37	31	5	0.000014	0.000013
80	37	31	6	0.000113	0.000099
80	37	31	7	0.000689	0.000576
80	37	31	8	0.003281	0.002592
80	37	31	9	0.012428	0.009147
80	37	31	10	0.038038	0.025611
80	37	31	11	0.095435	0.057397
80	37	31	12	0.199069	0.103633
80	37	31	13	0.350533	0.151464
80	37	31	14	0.530292	0.179760
80	37	31	15	0.703838	0.173546
80	37	31	16	0.840195	0.136357
80	37	31	17	0.927320	0.087125
80	37	31	18	0.972496	0.045176
80	37	31	19	0.991441	0.018945
80	37	31	20	0.997835	0.006394
80	37	31	21	0.999560	0.001725
80	37	31	22	0.999929	0.000369
80	37	31	23	0.999991	0.000062
80	37	31	24	0.999999	0.000008
80	37	31	25	1.000000	0.000001
80	37	31	26	1.000000	0.000000
80	37	31	27	1.000000	0.000000

N	n	k	x	P(x)	p(x)
80	37	31	28	1.000000	0.000000
80	37	31	29	1.000000	0.000000
80	37	31	30	1.000000	0.000000
80	37	31	31	1.000000	0.000000
80	37	32	0	0.000000	0.000000
80	37	32	1	0.000000	0.000000
80	37	32	2	0.000000	0.000000
80	37	32	3	0.000000	0.000000
80	37	32	4	0.000000	0.000000
80	37	32	5	0.000006	0.000005
80	37	32	6	0.000050	0.000045
80	37	32	7	0.000336	0.000286
80	37	32	8	0.001747	0.001411
80	37	32	9	0.007201	0.005454
80	37	32	10	0.023926	0.016726
80	37	32	11	0.064980	0.041054
80	37	32	12	0.146195	0.081215
80	37	32	13	0.276346	0.130152
80	37	32	14	0.445916	0.169569
80	37	32	15	0.625920	0.180004
80	37	32	16	0.781757	0.155837
80	37	32	17	0.891759	0.110003
80	37	32	18	0.954979	0.063220
80	37	32	19	0.984482	0.029503
80	37	32	20	0.995616	0.011135
80	37	32	21	0.998997	0.003380
80	37	32	22	0.999816	0.000819
80	37	32	23	0.999973	0.000157
80	37	32	24	0.999997	0.000024
80	37	32	25	1.000000	0.000003
80	37	32	26	1.000000	0.000000
80	37	32	27	1.000000	0.000000
80	37	32	28	1.000000	0.000000
80	37	32	29	1.000000	0.000000
80	37	32	30	1.000000	0.000000
80	37	32	31	1.000000	0.000000
80	37	32	32	1.000000	0.000000
80	37	33	0	0.000000	0.000000
80	37	33	1	0.000000	0.000000
80	37	33	2	0.000000	0.000000
80	37	33	3	0.000000	0.000000
80	37	33	4	0.000000	0.000000
80	37	33	5	0.000002	0.000002
80	37	33	6	0.000022	0.000019
80	37	33	7	0.000158	0.000136
80	37	33	8	0.000895	0.000737
80	37	33	9	0.004019	0.003125
80	37	33	10	0.014518	0.010499
80	37	33	11	0.042743	0.028224
80	37	33	12	0.103895	0.061153

N	n	k	x	P(x)	p(x)
80	37	33	13	0.211270	0.107375
80	37	33	14	0.364664	0.153393
80	37	33	15	0.543418	0.178754
80	37	33	16	0.713578	0.170160
80	37	33	17	0.845925	0.132347
80	37	33	18	0.929954	0.084030
80	37	33	19	0.973418	0.043464
80	37	33	20	0.991673	0.018255
80	37	33	21	0.997870	0.006197
80	37	33	22	0.999560	0.001690
80	37	33	23	0.999927	0.000367
80	37	33	24	0.999990	0.000063
80	37	33	25	0.999999	0.000008
80	37	33	26	1.000000	0.000001
80	37	33	27	1.000000	0.000000
80	37	33	28	1.000000	0.000000
80	37	33	29	1.000000	0.000000
80	37	33	30	1.000000	0.000000
80	37	33	31	1.000000	0.000000
80	37	33	32	1.000000	0.000000
80	37	33	33	1.000000	0.000000
80	37	34	0	0.000000	0.000000
80	37	34	1	0.000000	0.000000
80	37	34	2	0.000000	0.000000
80	37	34	3	0.000000	0.000000
80	37	34	4	0.000000	0.000000
80	37	34	5	0.000001	0.000001
80	37	34	6	0.000009	0.000008
80	37	34	7	0.000071	0.000062
80	37	34	8	0.000440	0.000369
80	37	34	9	0.002158	0.001718
80	37	34	10	0.008487	0.006329
80	37	34	11	0.027129	0.018642
80	37	34	12	0.071367	0.044238
80	37	34	13	0.156441	0.085073
80	37	34	14	0.289599	0.133158
80	37	34	15	0.459746	0.170147
80	37	34	16	0.637549	0.177803
80	37	34	17	0.789607	0.152058
80	37	34	18	0.895985	0.106378
80	37	34	19	0.956772	0.060787
80	37	34	20	0.985070	0.028298
80	37	34	21	0.995760	0.010690
80	37	34	22	0.999021	0.003260
80	37	34	23	0.999818	0.000797
80	37	34	24	0.999973	0.000155
80	37	34	25	0.999997	0.000024
80	37	34	26	1.000000	0.000003
80	37	34	27	1.000000	0.000000
80	37	34	28	1.000000	0.000000

N	n	k	x	P(x)	p(x)
80	37	34	29	1.000000	0.000000
80	37	34	30	1.000000	0.000000
80	37	34	31	1.000000	0.000000
80	37	34	32	1.000000	0.000000
80	37	34	33	1.000000	0.000000
80	37	34	34	1.000000	0.000000
80	37	35	0	0.000000	0.000000
80	37	35	1	0.000000	0.000000
80	37	35	2	0.000000	0.000000
80	37	35	3	0.000000	0.000000
80	37	35	4	0.000000	0.000000
80	37	35	5	0.000000	0.000000
80	37	35	6	0.000003	0.000003
80	37	35	7	0.000030	0.000027
80	37	35	8	0.000207	0.000177
80	37	35	9	0.001112	0.000905
80	37	35	10	0.004772	0.003660
80	37	35	11	0.016592	0.011820
80	37	35	12	0.047325	0.030733
80	37	35	13	0.112054	0.064730
80	37	35	14	0.223020	0.110965
80	37	35	15	0.378371	0.155351
80	37	35	16	0.556378	0.178007
80	37	35	17	0.723495	0.167117
80	37	35	18	0.852046	0.128551
80	37	35	19	0.932986	0.080940
80	37	35	20	0.974612	0.041626
80	37	35	21	0.992042	0.017430
80	37	35	22	0.997958	0.005916
80	37	35	23	0.999575	0.001618
80	37	35	24	0.999929	0.000354
80	37	35	25	0.999991	0.000061
80	37	35	26	0.999999	0.000008
80	37	35	27	1.000000	0.000001
80	37	35	28	1.000000	0.000000
80	37	35	29	1.000000	0.000000
80	37	35	30	1.000000	0.000000
80	37	35	31	1.000000	0.000000
80	37	35	32	1.000000	0.000000
80	37	35	33	1.000000	0.000000
80	37	35	34	1.000000	0.000000
80	37	35	35	1.000000	0.000000
80	37	36	0	0.000000	0.000000
80	37	36	1	0.000000	0.000000
80	37	36	2	0.000000	0.000000
80	37	36	3	0.000000	0.000000
80	37	36	4	0.000000	0.000000
80	37	36	5	0.000000	0.000000
80	37	36	6	0.000001	0.000001
80	37	36	7	0.000012	0.000011

N	n	k	x	P(x)	p(x)
80	37	36	8	0.000093	0.000081
80	37	36	9	0.000549	0.000456
80	37	36	10	0.002576	0.002027
80	37	36	11	0.009763	0.007187
80	37	36	12	0.030251	0.020488
80	37	36	13	0.077532	0.047281
80	37	36	14	0.166304	0.088772
80	37	36	15	0.302422	0.136117
80	37	36	16	0.473308	0.170887
80	37	36	17	0.649221	0.175913
80	37	36	18	0.797769	0.148548
80	37	36	19	0.900610	0.102841
80	37	36	20	0.958887	0.058277
80	37	36	21	0.985845	0.026958
80	37	36	22	0.995986	0.010141
80	37	36	23	0.999072	0.003086
80	37	36	24	0.999827	0.000755
80	37	36	25	0.999974	0.000147
80	37	36	26	0.999997	0.000023
80	37	36	27	1.000000	0.000003
80	37	36	28	1.000000	0.000000
80	37	36	29	1.000000	0.000000
80	37	36	30	1.000000	0.000000
80	37	36	31	1.000000	0.000000
80	37	36	32	1.000000	0.000000
80	37	36	33	1.000000	0.000000
80	37	36	34	1.000000	0.000000
80	37	36	35	1.000000	0.000000
80	37	36	36	1.000000	0.000000
80	37	37	0	0.000000	0.000000
80	37	37	1	0.000000	0.000000
80	37	37	2	0.000000	0.000000
80	37	37	3	0.000000	0.000000
80	37	37	4	0.000000	0.000000
80	37	37	5	0.000000	0.000000
80	37	37	6	0.000000	0.000000
80	37	37	7	0.000005	0.000004
80	37	37	8	0.000040	0.000035
80	37	37	9	0.000259	0.000219
80	37	37	10	0.001332	0.001073
80	37	37	11	0.005516	0.004184
80	37	37	12	0.018610	0.013094
80	37	37	13	0.051742	0.033132
80	37	37	14	0.119901	0.068158
80	37	37	15	0.234363	0.114462
80	37	37	16	0.391749	0.157386
80	37	37	17	0.569260	0.177512
80	37	37	18	0.733623	0.164363
80	37	37	19	0.858539	0.124916
80	37	37	20	0.936371	0.077832

N	n	k	x	P(x)	p(x)
80	37	37	21	0.976042	0.039671
80	37	37	22	0.992529	0.016487
80	37	37	23	0.998090	0.005561
80	37	37	24	0.999604	0.001514
80	37	37	25	0.999934	0.000330
80	37	37	26	0.999991	0.000057
80	37	37	27	0.999999	0.000008
80	37	37	28	1.000000	0.000001
80	37	37	29	1.000000	0.000000
80	37	37	30	1.000000	0.000000
80	37	37	31	1.000000	0.000000
80	37	37	32	1.000000	0.000000
80	37	37	33	1.000000	0.000000
80	37	37	34	1.000000	0.000000
80	37	37	35	1.000000	0.000000
80	37	37	36	1.000000	0.000000
80	37	37	37	1.000000	0.000000
80	38	1	0	0.525000	0.525000
80	38	1	1	1.000000	0.475000
80	38	2	0	0.272468	0.272468
80	38	2	1	0.777532	0.505063
80	38	2	2	1.000000	0.222468
80	38	3	0	0.139727	0.139727
80	38	3	1	0.537950	0.398223
80	38	3	2	0.897322	0.359372
80	38	3	3	1.000000	0.102678
80	38	4	0	0.070771	0.070771
80	38	4	1	0.346596	0.275825
80	38	4	2	0.729304	0.382708
80	38	4	3	0.953328	0.224024
80	38	4	4	1.000000	0.046672
80	38	5	0	0.035385	0.035385
80	38	5	1	0.212313	0.176927
80	38	5	2	0.548022	0.335709
80	38	5	3	0.850159	0.302138
80	38	5	4	0.979120	0.128961
80	38	5	5	1.000000	0.020879
80	38	6	0	0.017457	0.017457
80	38	6	1	0.125029	0.107572
80	38	6	2	0.386881	0.261853
80	38	6	3	0.709162	0.322280
80	38	6	4	0.920658	0.211496
80	38	6	5	0.990813	0.070155
80	38	6	6	1.000000	0.009187
80	38	7	0	0.008493	0.008493
80	38	7	1	0.071243	0.062750
80	38	7	2	0.259494	0.188251
80	38	7	3	0.556732	0.297238
80	38	7	4	0.823484	0.266752
80	38	7	5	0.959528	0.136044

N	n	k	x	P(x)	p(x)
80	38	7	6	0.996027	0.036500
80	38	7	7	1.000000	0.003973
80	38	8	0	0.004072	0.004072
80	38	8	1	0.039438	0.035366
80	38	8	2	0.166658	0.127220
80	38	8	3	0.414220	0.247563
80	38	8	4	0.699243	0.285023
80	38	8	5	0.898028	0.198785
80	38	8	6	0.980027	0.081999
80	38	8	7	0.998313	0.018285
80	38	8	8	1.000000	0.001687
80	38	9	0	0.001923	0.001923
80	38	9	1	0.021264	0.019341
80	38	9	2	0.103048	0.081784
80	38	9	3	0.293877	0.190830
80	38	9	4	0.564649	0.270772
80	38	9	5	0.806919	0.242269
80	38	9	6	0.943583	0.136665
80	38	9	7	0.990440	0.046857
80	38	9	8	0.999297	0.008857
80	38	9	9	1.000000	0.000703
80	38	10	0	0.000894	0.000894
80	38	10	1	0.011185	0.010291
80	38	10	2	0.061580	0.050395
80	38	10	3	0.199806	0.138227
80	38	10	4	0.434984	0.235177
80	38	10	5	0.694314	0.259331
80	38	10	6	0.881988	0.187674
80	38	10	7	0.969981	0.087993
80	38	10	8	0.995555	0.025573
80	38	10	9	0.999713	0.004158
80	38	10	10	1.000000	0.000287
80	38	11	0	0.000409	0.000409
80	38	11	1	0.005745	0.005337
80	38	11	2	0.035662	0.029917
80	38	11	3	0.130693	0.095031
80	38	11	4	0.320755	0.190062
80	38	11	5	0.572059	0.251304
80	38	11	6	0.796194	0.224136
80	38	11	7	0.931013	0.134819
80	38	11	8	0.984595	0.053582
80	38	11	9	0.997990	0.013395
80	38	11	10	0.999885	0.001895
80	38	11	11	1.000000	0.000115
80	38	12	0	0.000184	0.000184
80	38	12	1	0.002883	0.002700
80	38	12	2	0.020053	0.017170
80	38	12	3	0.082489	0.062436
80	38	12	4	0.227101	0.144612
80	38	12	5	0.451870	0.224769

N	n	k	x	P(x)	p(x)
80	38	12	6	0.692247	0.240378
80	38	12	7	0.870442	0.178195
80	38	12	8	0.961298	0.090856
80	38	12	9	0.992360	0.031062
80	38	12	10	0.999116	0.006756
80	38	12	11	0.999955	0.000839
80	38	12	12	1.000000	0.000045
80	38	13	0	0.000081	0.000081
80	38	13	1	0.001414	0.001333
80	38	13	2	0.010963	0.009549
80	38	13	3	0.050353	0.039390
80	38	13	4	0.154795	0.104442
80	38	13	5	0.342791	0.187996
80	38	13	6	0.579128	0.236338
80	38	13	7	0.789206	0.210078
80	38	13	8	0.921215	0.132008
80	38	13	9	0.979113	0.057898
80	38	13	10	0.996334	0.017221
80	38	13	11	0.999622	0.003288
80	38	13	12	0.999983	0.000361
80	38	13	13	1.000000	0.000017
80	38	14	0	0.000035	0.000035
80	38	14	1	0.000678	0.000643
80	38	14	2	0.005833	0.005155
80	38	14	3	0.029776	0.023944
80	38	14	4	0.101795	0.072018
80	38	14	5	0.250196	0.148401
80	38	14	6	0.466251	0.216055
80	38	14	7	0.692006	0.225755
80	38	14	8	0.862107	0.170100
80	38	14	9	0.954053	0.091946
80	38	14	10	0.989137	0.035085
80	38	14	11	0.998297	0.009160
80	38	14	12	0.999843	0.001546
80	38	14	13	0.999993	0.000151
80	38	14	14	1.000000	0.000006
80	38	15	0	0.000015	0.000015
80	38	15	1	0.000318	0.000303
80	38	15	2	0.003021	0.002703
80	38	15	3	0.017079	0.014058
80	38	15	4	0.064694	0.047615
80	38	15	5	0.175995	0.111301
80	38	15	6	0.361497	0.185502
80	38	15	7	0.585970	0.224473
80	38	15	8	0.784788	0.198819
80	38	15	9	0.913652	0.128864
80	38	15	10	0.974253	0.060601
80	38	15	11	0.994550	0.020297
80	38	15	12	0.999234	0.004684
80	38	15	13	0.999936	0.000703

Table for $N = 2$, $n = 1$, through $N = 100$, $n = 50$

N	n	k	x	P(x)	p(x)
80	38	15	14	0.999998	0.000061
80	38	15	15	1.000000	0.000002
80	38	16	0	0.000006	0.000006
80	38	16	1	0.000145	0.000139
80	38	16	2	0.001524	0.001378
80	38	16	3	0.009509	0.007986
80	38	16	4	0.039788	0.030279
80	38	16	5	0.119489	0.079701
80	38	16	6	0.270173	0.150684
80	38	16	7	0.478913	0.208740
80	38	16	8	0.693026	0.214112
80	38	16	9	0.856159	0.163133
80	38	16	10	0.948148	0.091989
80	38	16	11	0.986119	0.037971
80	38	16	12	0.997360	0.011241
80	38	16	13	0.999666	0.002306
80	38	16	14	0.999975	0.000309
80	38	16	15	0.999999	0.000024
80	38	16	16	1.000000	0.000001
80	38	17	0	0.000003	0.000003
80	38	17	1	0.000065	0.000062
80	38	17	2	0.000748	0.000683
80	38	17	3	0.005142	0.004394
80	38	17	4	0.023702	0.018560
80	38	17	5	0.078393	0.054691
80	38	17	6	0.194831	0.116438
80	38	17	7	0.377805	0.182974
80	38	17	8	0.592661	0.214856
80	38	17	9	0.782239	0.189579
80	38	17	10	0.907903	0.125664
80	38	17	11	0.970100	0.062197
80	38	17	12	0.992793	0.022694
80	38	17	13	0.998765	0.005972
80	38	17	14	0.999859	0.001094
80	38	17	15	0.999990	0.000131
80	38	17	16	1.000000	0.000009
80	38	17	17	1.000000	0.000000
80	38	18	0	0.000001	0.000001
80	38	18	1	0.000028	0.000027
80	38	18	2	0.000358	0.000330
80	38	18	3	0.002701	0.002343
80	38	18	4	0.013686	0.010985
80	38	18	5	0.049745	0.036060
80	38	18	6	0.135688	0.085942
80	38	18	7	0.287770	0.152082
80	38	18	8	0.490348	0.202578
80	38	18	9	0.694973	0.204625
80	38	18	10	0.852052	0.157079
80	38	18	11	0.943444	0.091392
80	38	18	12	0.983428	0.039984

N	n	k	x	P(x)	p(x)
80	38	18	13	0.996396	0.012968
80	38	18	14	0.999442	0.003047
80	38	18	15	0.999942	0.000500
80	38	18	16	0.999996	0.000054
80	38	18	17	1.000000	0.000003
80	38	18	18	1.000000	0.000000
80	38	19	0	0.000000	0.000000
80	38	19	1	0.000012	0.000012
80	38	19	2	0.000166	0.000154
80	38	19	3	0.001378	0.001212
80	38	19	4	0.007662	0.006284
80	38	19	5	0.030552	0.022890
80	38	19	6	0.091330	0.060778
80	38	19	7	0.211729	0.120398
80	38	19	8	0.392327	0.180598
80	38	19	9	0.599261	0.206935
80	38	19	10	0.781113	0.181852
80	38	19	11	0.903644	0.122531
80	38	19	12	0.966660	0.063016
80	38	19	13	0.991167	0.024506
80	38	19	14	0.998263	0.007096
80	38	19	15	0.999757	0.001494
80	38	19	16	0.999977	0.000220
80	38	19	17	0.999999	0.000021
80	38	19	18	1.000000	0.000001
80	38	19	19	1.000000	0.000000
80	38	20	0	0.000000	0.000000
80	38	20	1	0.000005	0.000005
80	38	20	2	0.000075	0.000070
80	38	20	3	0.000683	0.000608
80	38	20	4	0.004160	0.003477
80	38	20	5	0.018169	0.014009
80	38	20	6	0.059447	0.041278
80	38	20	7	0.150543	0.091096
80	38	20	8	0.303508	0.152965
80	38	20	9	0.500883	0.197375
80	38	20	10	0.697640	0.196758
80	38	20	11	0.849409	0.151769
80	38	20	12	0.939801	0.090392
80	38	20	13	0.981123	0.041322
80	38	20	14	0.995471	0.014348
80	38	20	15	0.999194	0.003723
80	38	20	16	0.999898	0.000704
80	38	20	17	0.999991	0.000093
80	38	20	18	0.999999	0.000008
80	38	20	19	1.000000	0.000000
80	38	20	20	1.000000	0.000000
80	38	21	0	0.000000	0.000000
80	38	21	1	0.000002	0.000002
80	38	21	2	0.000033	0.000031

N	n	k	x	P(x)	p(x)
80	38	21	3	0.000328	0.000295
80	38	21	4	0.002189	0.001861
80	38	21	5	0.010464	0.008274
80	38	21	6	0.037432	0.026968
80	38	21	7	0.103477	0.066045
80	38	21	8	0.227025	0.123549
80	38	21	9	0.405485	0.178459
80	38	21	10	0.605820	0.200335
80	38	21	11	0.781113	0.175293
80	38	21	12	0.900631	0.119518
80	38	21	13	0.963906	0.063274
80	38	21	14	0.989732	0.025826
80	38	21	15	0.997767	0.008035
80	38	21	16	0.999640	0.001873
80	38	21	17	0.999959	0.000319
80	38	21	18	0.999997	0.000038
80	38	21	19	1.000000	0.000003
80	38	21	20	1.000000	0.000000
80	38	21	21	1.000000	0.000000
80	38	22	0	0.000000	0.000000
80	38	22	1	0.000001	0.000001
80	38	22	2	0.000014	0.000013
80	38	22	3	0.000153	0.000139
80	38	22	4	0.001117	0.000964
80	38	22	5	0.005836	0.004719
80	38	22	6	0.022805	0.016969
80	38	22	7	0.068775	0.045970
80	38	22	8	0.164204	0.095429
80	38	22	9	0.317767	0.153564
80	38	22	10	0.510746	0.192978
80	38	22	11	0.700894	0.190149
80	38	22	12	0.847962	0.147068
80	38	22	13	0.937094	0.089132
80	38	22	14	0.979226	0.042132
80	38	22	15	0.994634	0.015408
80	38	22	16	0.998941	0.004307
80	38	22	17	0.999845	0.000904
80	38	22	18	0.999984	0.000139
80	38	22	19	0.999999	0.000015
80	38	22	20	1.000000	0.000001
80	38	22	21	1.000000	0.000000
80	38	22	22	1.000000	0.000000
80	38	23	0	0.000000	0.000000
80	38	23	1	0.000000	0.000000
80	38	23	2	0.000006	0.000006
80	38	23	3	0.000069	0.000063
80	38	23	4	0.000552	0.000483
80	38	23	5	0.003151	0.002599
80	38	23	6	0.013443	0.010292
80	38	23	7	0.044205	0.030762

N	n	k	x	P(x)	p(x)
80	38	23	8	0.114844	0.070639
80	38	23	9	0.240986	0.126141
80	38	23	10	0.417584	0.176598
80	38	23	11	0.612377	0.194793
80	38	23	12	0.782035	0.169658
80	38	23	13	0.898675	0.116640
80	38	23	14	0.961792	0.063117
80	38	23	15	0.988524	0.026732
80	38	23	16	0.997308	0.008783
80	38	23	17	0.999518	0.002210
80	38	23	18	0.999936	0.000418
80	38	23	19	0.999994	0.000058
80	38	23	20	1.000000	0.000006
80	38	23	21	1.000000	0.000000
80	38	23	22	1.000000	0.000000
80	38	23	23	1.000000	0.000000
80	38	24	0	0.000000	0.000000
80	38	24	1	0.000000	0.000000
80	38	24	2	0.000002	0.000002
80	38	24	3	0.000030	0.000028
80	38	24	4	0.000264	0.000234
80	38	24	5	0.001646	0.001382
80	38	24	6	0.007665	0.006019
80	38	24	7	0.027475	0.019810
80	38	24	8	0.077666	0.050191
80	38	24	9	0.176808	0.099143
80	38	24	10	0.330834	0.154025
80	38	24	11	0.520106	0.189273
80	38	24	12	0.704647	0.184541
80	38	24	13	0.847517	0.142870
80	38	24	14	0.935217	0.087699
80	38	24	15	0.977738	0.042521
80	38	24	16	0.993918	0.016180
80	38	24	17	0.998704	0.004786
80	38	24	18	0.999789	0.001086
80	38	24	19	0.999975	0.000185
80	38	24	20	0.999998	0.000023
80	38	24	21	1.000000	0.000002
80	38	24	22	1.000000	0.000000
80	38	24	23	1.000000	0.000000
80	38	24	24	1.000000	0.000000
80	38	25	0	0.000000	0.000000
80	38	25	1	0.000000	0.000000
80	38	25	2	0.000001	0.000001
80	38	25	3	0.000013	0.000012
80	38	25	4	0.000122	0.000109
80	38	25	5	0.000832	0.000710
80	38	25	6	0.004226	0.003394
80	38	25	7	0.016509	0.012283
80	38	25	8	0.050778	0.034269

N	n	k	x	P(x)	p(x)
80	38	25	9	0.125467	0.074689
80	38	25	10	0.253821	0.128354
80	38	25	11	0.428850	0.175029
80	38	25	12	0.618968	0.190118
80	38	25	13	0.783736	0.164769
80	38	25	14	0.897631	0.113895
80	38	25	15	0.960274	0.062642
80	38	25	16	0.987561	0.027287
80	38	25	17	0.996909	0.009348
80	38	25	18	0.999401	0.002493
80	38	25	19	0.999912	0.000510
80	38	25	20	0.999990	0.000079
80	38	25	21	0.999999	0.000009
80	38	25	22	1.000000	0.000001
80	38	25	23	1.000000	0.000000
80	38	25	24	1.000000	0.000000
80	38	25	25	1.000000	0.000000
80	38	26	0	0.000000	0.000000
80	38	26	1	0.000000	0.000000
80	38	26	2	0.000000	0.000000
80	38	26	3	0.000005	0.000005
80	38	26	4	0.000055	0.000049
80	38	26	5	0.000406	0.000351
80	38	26	6	0.002251	0.001845
80	38	26	7	0.009586	0.007334
80	38	26	8	0.032085	0.022500
80	38	26	9	0.086085	0.054000
80	38	26	10	0.188477	0.102392
80	38	26	11	0.342927	0.154450
80	38	26	12	0.529094	0.186167
80	38	26	13	0.708841	0.179748
80	38	26	14	0.847932	0.139090
80	38	26	15	0.934078	0.086146
80	38	26	16	0.976646	0.042568
80	38	26	17	0.993340	0.016693
80	38	26	18	0.998495	0.005155
80	38	26	19	0.999735	0.001240
80	38	26	20	0.999965	0.000229
80	38	26	21	0.999996	0.000032
80	38	26	22	1.000000	0.000003
80	38	26	23	1.000000	0.000000
80	38	26	24	1.000000	0.000000
80	38	26	25	1.000000	0.000000
80	38	26	26	1.000000	0.000000
80	38	27	0	0.000000	0.000000
80	38	27	1	0.000000	0.000000
80	38	27	2	0.000000	0.000000
80	38	27	3	0.000002	0.000002
80	38	27	4	0.000024	0.000021
80	38	27	5	0.000191	0.000168

N	n	k	x	P(x)	p(x)
80	38	27	6	0.001158	0.000966
80	38	27	7	0.005375	0.004217
80	38	27	8	0.019585	0.014210
80	38	27	9	0.057085	0.037500
80	38	27	10	0.135385	0.078300
80	38	27	11	0.265702	0.130317
80	38	27	12	0.439458	0.173756
80	38	27	13	0.625625	0.186167
80	38	27	14	0.786114	0.160489
80	38	27	15	0.897386	0.111272
80	38	27	16	0.959304	0.061918
80	38	27	17	0.986848	0.027544
80	38	27	18	0.996586	0.009738
80	38	27	19	0.999299	0.002713
80	38	27	20	0.999888	0.000589
80	38	27	21	0.999986	0.000098
80	38	27	22	0.999999	0.000012
80	38	27	23	1.000000	0.000001
80	38	27	24	1.000000	0.000000
80	38	27	25	1.000000	0.000000
80	38	27	26	1.000000	0.000000
80	38	27	27	1.000000	0.000000
80	38	28	0	0.000000	0.000000
80	38	28	1	0.000000	0.000000
80	38	28	2	0.000000	0.000000
80	38	28	3	0.000001	0.000001
80	38	28	4	0.000010	0.000009
80	38	28	5	0.000087	0.000077
80	38	28	6	0.000574	0.000487
80	38	28	7	0.002908	0.002334
80	38	28	8	0.011542	0.008634
80	38	28	9	0.036567	0.025025
80	38	28	10	0.094019	0.057453
80	38	28	11	0.199314	0.105295
80	38	28	12	0.354219	0.154905
80	38	28	13	0.537810	0.183591
80	38	28	14	0.713440	0.175629
80	38	28	15	0.849098	0.135658
80	38	28	16	0.933602	0.084504
80	38	28	17	0.975934	0.042332
80	38	28	18	0.992911	0.016977
80	38	28	19	0.998326	0.005415
80	38	28	20	0.999688	0.001362
80	38	28	21	0.999955	0.000267
80	38	28	22	0.999995	0.000040
80	38	28	23	0.999999	0.000005
80	38	28	24	1.000000	0.000000
80	38	28	25	1.000000	0.000000
80	38	28	26	1.000000	0.000000
80	38	28	27	1.000000	0.000000

N	n	k	x	P(x)	p(x)
80	38	28	28	1.000000	0.000000
80	38	29	0	0.000000	0.000000
80	38	29	1	0.000000	0.000000
80	38	29	2	0.000000	0.000000
80	38	29	3	0.000000	0.000000
80	38	29	4	0.000004	0.000004
80	38	29	5	0.000038	0.000034
80	38	29	6	0.000274	0.000236
80	38	29	7	0.001517	0.001243
80	38	29	8	0.006561	0.005044
80	38	29	9	0.022610	0.016049
80	38	29	10	0.063083	0.040473
80	38	29	11	0.144642	0.081559
80	38	29	12	0.276767	0.132125
80	38	29	13	0.449545	0.172779
80	38	29	14	0.632380	0.182835
80	38	29	15	0.789095	0.156715
80	38	29	16	0.897850	0.108755
80	38	29	17	0.958838	0.060988
80	38	29	18	0.986381	0.027543
80	38	29	19	0.996348	0.009966
80	38	29	20	0.999217	0.002869
80	38	29	21	0.999868	0.000651
80	38	29	22	0.999983	0.000115
80	38	29	23	0.999998	0.000016
80	38	29	24	1.000000	0.000002
80	38	29	25	1.000000	0.000000
80	38	29	26	1.000000	0.000000
80	38	29	27	1.000000	0.000000
80	38	29	28	1.000000	0.000000
80	38	29	29	1.000000	0.000000
80	38	30	0	0.000000	0.000000
80	38	30	1	0.000000	0.000000
80	38	30	2	0.000000	0.000000
80	38	30	3	0.000000	0.000000
80	38	30	4	0.000001	0.000001
80	38	30	5	0.000016	0.000014
80	38	30	6	0.000126	0.000110
80	38	30	7	0.000762	0.000636
80	38	30	8	0.003594	0.002832
80	38	30	9	0.013484	0.009890
80	38	30	10	0.040863	0.027379
80	38	30	11	0.101464	0.060601
80	38	30	12	0.209409	0.107945
80	38	30	13	0.364850	0.155441
80	38	30	14	0.546340	0.181490
80	38	30	15	0.718420	0.172080
80	38	30	16	0.850936	0.132517
80	38	30	17	0.933726	0.082789
80	38	30	18	0.975580	0.041855

N	n	k	x	P(x)	p(x)
80	38	30	19	0.992635	0.017054
80	38	30	20	0.998204	0.005569
80	38	30	21	0.999651	0.001447
80	38	30	22	0.999947	0.000296
80	38	30	23	0.999994	0.000047
80	38	30	24	0.999999	0.000006
80	38	30	25	1.000000	0.000001
80	38	30	26	1.000000	0.000000
80	38	30	27	1.000000	0.000000
80	38	30	28	1.000000	0.000000
80	38	30	29	1.000000	0.000000
80	38	30	30	1.000000	0.000000
80	38	31	0	0.000000	0.000000
80	38	31	1	0.000000	0.000000
80	38	31	2	0.000000	0.000000
80	38	31	3	0.000000	0.000000
80	38	31	4	0.000001	0.000001
80	38	31	5	0.000006	0.000006
80	38	31	6	0.000056	0.000049
80	38	31	7	0.000367	0.000312
80	38	31	8	0.001894	0.001527
80	38	31	9	0.007748	0.005853
80	38	31	10	0.025531	0.017783
80	38	31	11	0.068739	0.043208
80	38	31	12	0.153277	0.084538
80	38	31	13	0.287129	0.133852
80	38	31	14	0.459225	0.172095
80	38	31	15	0.639263	0.180038
80	38	31	16	0.792629	0.153366
80	38	31	17	0.898954	0.106325
80	38	31	18	0.958838	0.059884
80	38	31	19	0.986154	0.027316
80	38	31	20	0.996199	0.010045
80	38	31	21	0.999159	0.002960
80	38	31	22	0.999852	0.000693
80	38	31	23	0.999979	0.000128
80	38	31	24	0.999998	0.000018
80	38	31	25	1.000000	0.000002
80	38	31	26	1.000000	0.000000
80	38	31	27	1.000000	0.000000
80	38	31	28	1.000000	0.000000
80	38	31	29	1.000000	0.000000
80	38	31	30	1.000000	0.000000
80	38	31	31	1.000000	0.000000
80	38	32	0	0.000000	0.000000
80	38	32	1	0.000000	0.000000
80	38	32	2	0.000000	0.000000
80	38	32	3	0.000000	0.000000
80	38	32	4	0.000000	0.000000
80	38	32	5	0.000002	0.000002

Table for $N = 2$, $n = 1$, through $N = 100$, $n = 50$

N	n	k	x	P(x)	p(x)
80	38	32	6	0.000023	0.000021
80	38	32	7	0.000170	0.000147
80	38	32	8	0.000960	0.000789
80	38	32	9	0.004284	0.003324
80	38	32	10	0.015369	0.011086
80	38	32	11	0.044930	0.029561
80	38	32	12	0.108420	0.063490
80	38	32	13	0.218837	0.110417
80	38	32	14	0.374933	0.156096
80	38	32	15	0.554755	0.179822
80	38	32	16	0.723771	0.169016
80	38	32	17	0.853386	0.129615
80	38	32	18	0.934396	0.081010
80	38	32	19	0.975562	0.041166
80	38	32	20	0.992509	0.016947
80	38	32	21	0.998132	0.005623
80	38	32	22	0.999626	0.001494
80	38	32	23	0.999940	0.000315
80	38	32	24	0.999992	0.000052
80	38	32	25	0.999999	0.000007
80	38	32	26	1.000000	0.000001
80	38	32	27	1.000000	0.000000
80	38	32	28	1.000000	0.000000
80	38	32	29	1.000000	0.000000
80	38	32	30	1.000000	0.000000
80	38	32	31	1.000000	0.000000
80	38	32	32	1.000000	0.000000
80	38	33	0	0.000000	0.000000
80	38	33	1	0.000000	0.000000
80	38	33	2	0.000000	0.000000
80	38	33	3	0.000000	0.000000
80	38	33	4	0.000000	0.000000
80	38	33	5	0.000001	0.000001
80	38	33	6	0.000009	0.000009
80	38	33	7	0.000075	0.000066
80	38	33	8	0.000466	0.000391
80	38	33	9	0.002275	0.001809
80	38	33	10	0.008903	0.006627
80	38	33	11	0.028302	0.019400
80	38	33	12	0.074030	0.045728
80	38	33	13	0.161329	0.087299
80	38	33	14	0.296885	0.135557
80	38	33	15	0.468590	0.171705
80	38	33	16	0.646305	0.177715
80	38	33	17	0.796679	0.150374
80	38	33	18	0.900642	0.103962
80	38	33	19	0.959267	0.058625
80	38	33	20	0.986154	0.026887
80	38	33	21	0.996141	0.009987
80	38	33	22	0.999128	0.002987

N	n	k	x	P(x)	p(x)
80	38	33	23	0.999842	0.000714
80	38	33	24	0.999977	0.000135
80	38	33	25	0.999997	0.000020
80	38	33	26	1.000000	0.000002
80	38	33	27	1.000000	0.000000
80	38	33	28	1.000000	0.000000
80	38	33	29	1.000000	0.000000
80	38	33	30	1.000000	0.000000
80	38	33	31	1.000000	0.000000
80	38	33	32	1.000000	0.000000
80	38	33	33	1.000000	0.000000
80	38	34	0	0.000000	0.000000
80	38	34	1	0.000000	0.000000
80	38	34	2	0.000000	0.000000
80	38	34	3	0.000000	0.000000
80	38	34	4	0.000000	0.000000
80	38	34	5	0.000000	0.000000
80	38	34	6	0.000004	0.000003
80	38	34	7	0.000032	0.000028
80	38	34	8	0.000217	0.000185
80	38	34	9	0.001159	0.000942
80	38	34	10	0.004954	0.003795
80	38	34	11	0.017158	0.012203
80	38	34	12	0.048734	0.031576
80	38	34	13	0.114893	0.066159
80	38	34	14	0.227665	0.112772
80	38	34	15	0.384565	0.156900
80	38	34	16	0.563120	0.178555
80	38	34	17	0.729491	0.166371
80	38	34	18	0.856403	0.126912
80	38	34	19	0.935568	0.079165
80	38	34	20	0.975857	0.040289
80	38	34	21	0.992528	0.016671
80	38	34	22	0.998111	0.005582
80	38	34	23	0.999614	0.001503
80	38	34	24	0.999937	0.000323
80	38	34	25	0.999992	0.000055
80	38	34	26	0.999999	0.000007
80	38	34	27	1.000000	0.000001
80	38	34	28	1.000000	0.000000
80	38	34	29	1.000000	0.000000
80	38	34	30	1.000000	0.000000
80	38	34	31	1.000000	0.000000
80	38	34	32	1.000000	0.000000
80	38	34	33	1.000000	0.000000
80	38	34	34	1.000000	0.000000
80	38	35	0	0.000000	0.000000
80	38	35	1	0.000000	0.000000
80	38	35	2	0.000000	0.000000
80	38	35	3	0.000000	0.000000

N	n	k	x	P(x)	p(x)
80	38	35	4	0.000000	0.000000
80	38	35	5	0.000000	0.000000
80	38	35	6	0.000001	0.000001
80	38	35	7	0.000013	0.000012
80	38	35	8	0.000096	0.000083
80	38	35	9	0.000565	0.000469
80	38	35	10	0.002644	0.002079
80	38	35	11	0.009995	0.007351
80	38	35	12	0.030886	0.020892
80	38	35	13	0.078937	0.048051
80	38	35	14	0.168827	0.089890
80	38	35	15	0.306115	0.137287
80	38	35	16	0.477724	0.171609
80	38	35	17	0.653539	0.175815
80	38	35	18	0.801224	0.147685
80	38	35	19	0.902869	0.101645
80	38	35	20	0.960092	0.057223
80	38	35	21	0.986367	0.026276
80	38	35	22	0.996169	0.009802
80	38	35	23	0.999124	0.002955
80	38	35	24	0.999839	0.000715
80	38	35	25	0.999976	0.000138
80	38	35	26	0.999997	0.000021
80	38	35	27	1.000000	0.000002
80	38	35	28	1.000000	0.000000
80	38	35	29	1.000000	0.000000
80	38	35	30	1.000000	0.000000
80	38	35	31	1.000000	0.000000
80	38	35	32	1.000000	0.000000
80	38	35	33	1.000000	0.000000
80	38	35	34	1.000000	0.000000
80	38	35	35	1.000000	0.000000
80	38	36	0	0.000000	0.000000
80	38	36	1	0.000000	0.000000
80	38	36	2	0.000000	0.000000
80	38	36	3	0.000000	0.000000
80	38	36	4	0.000000	0.000000
80	38	36	5	0.000000	0.000000
80	38	36	6	0.000000	0.000000
80	38	36	7	0.000005	0.000004
80	38	36	8	0.000041	0.000036
80	38	36	9	0.000263	0.000222
80	38	36	10	0.001350	0.001088
80	38	36	11	0.005584	0.004234
80	38	36	12	0.018816	0.013231
80	38	36	13	0.052242	0.033426
80	38	36	14	0.120886	0.068644
80	38	36	15	0.235946	0.115060
80	38	36	16	0.393826	0.157880
80	38	36	17	0.571492	0.177666

N	n	k	x	P(x)	p(x)
80	38	36	18	0.735586	0.164094
80	38	36	19	0.859952	0.124366
80	38	36	20	0.937202	0.077250
80	38	36	21	0.976441	0.039238
80	38	36	22	0.992684	0.016243
80	38	36	23	0.998139	0.005455
80	38	36	24	0.999616	0.001477
80	38	36	25	0.999937	0.000320
80	38	36	26	0.999992	0.000055
80	38	36	27	0.999999	0.000007
80	38	36	28	1.000000	0.000001
80	38	36	29	1.000000	0.000000
80	38	36	30	1.000000	0.000000
80	38	36	31	1.000000	0.000000
80	38	36	32	1.000000	0.000000
80	38	36	33	1.000000	0.000000
80	38	36	34	1.000000	0.000000
80	38	36	35	1.000000	0.000000
80	38	36	36	1.000000	0.000000
80	38	37	0	0.000000	0.000000
80	38	37	1	0.000000	0.000000
80	38	37	2	0.000000	0.000000
80	38	37	3	0.000000	0.000000
80	38	37	4	0.000000	0.000000
80	38	37	5	0.000000	0.000000
80	38	37	6	0.000000	0.000000
80	38	37	7	0.000002	0.000002
80	38	37	8	0.000016	0.000014
80	38	37	9	0.000116	0.000100
80	38	37	10	0.000658	0.000542
80	38	37	11	0.002986	0.002328
80	38	37	12	0.010997	0.008011
80	38	37	13	0.033250	0.022253
80	38	37	14	0.083444	0.050194
80	38	37	15	0.175801	0.092357
80	38	37	16	0.314886	0.139085
80	38	37	17	0.486697	0.171811
80	38	37	18	0.660998	0.174301
80	38	37	19	0.806248	0.145251
80	38	37	20	0.905600	0.099352
80	38	37	21	0.961281	0.055681
80	38	37	22	0.986777	0.025497
80	38	37	23	0.996279	0.009502
80	38	37	24	0.999146	0.002867
80	38	37	25	0.999842	0.000696
80	38	37	26	0.999977	0.000135
80	38	37	27	0.999997	0.000021
80	38	37	28	1.000000	0.000002
80	38	37	29	1.000000	0.000000
80	38	37	30	1.000000	0.000000

Table for $N = 2$, $n = 1$, through $N = 100$, $n = 50$

N	n	k	x	P(x)	p(x)
80	38	37	31	1.000000	0.000000
80	38	37	32	1.000000	0.000000
80	38	37	33	1.000000	0.000000
80	38	37	34	1.000000	0.000000
80	38	37	35	1.000000	0.000000
80	38	37	36	1.000000	0.000000
80	38	37	37	1.000000	0.000000
80	38	38	0	0.000000	0.000000
80	38	38	1	0.000000	0.000000
80	38	38	2	0.000000	0.000000
80	38	38	3	0.000000	0.000000
80	38	38	4	0.000000	0.000000
80	38	38	5	0.000000	0.000000
80	38	38	6	0.000000	0.000000
80	38	38	7	0.000001	0.000001
80	38	38	8	0.000006	0.000006
80	38	38	9	0.000049	0.000043
80	38	38	10	0.000305	0.000257
80	38	38	11	0.001525	0.001219
80	38	38	12	0.006153	0.004629
80	38	38	13	0.020312	0.014159
80	38	38	14	0.055429	0.035116
80	38	38	15	0.126401	0.070972
80	38	38	16	0.243726	0.117325
80	38	38	17	0.402789	0.159063
80	38	38	18	0.579927	0.177138
80	38	38	19	0.742068	0.162140
80	38	38	20	0.864011	0.121943
80	38	38	21	0.939267	0.075256
80	38	38	22	0.977290	0.038023
80	38	38	23	0.992965	0.015674
80	38	38	24	0.998213	0.005248
80	38	38	25	0.999632	0.001419
80	38	38	26	0.999939	0.000307
80	38	38	27	0.999992	0.000053
80	38	38	28	0.999999	0.000007
80	38	38	29	1.000000	0.000001
80	38	38	30	1.000000	0.000000
80	38	38	31	1.000000	0.000000
80	38	38	32	1.000000	0.000000
80	38	38	33	1.000000	0.000000
80	38	38	34	1.000000	0.000000
80	38	38	35	1.000000	0.000000
80	38	38	36	1.000000	0.000000
80	38	38	37	1.000000	0.000000
80	38	38	38	1.000000	0.000000
80	39	1	0	0.512500	0.512500
80	39	1	1	1.000000	0.487500
80	39	2	0	0.259494	0.259494
80	39	2	1	0.765506	0.506013

N	n	k	x	P(x)	p(x)
80	39	2	2	1.000000	0.234494
80	39	3	0	0.129747	0.129747
80	39	3	1	0.518987	0.389240
80	39	3	2	0.888766	0.369778
80	39	3	3	1.000000	0.111234
80	39	4	0	0.064031	0.064031
80	39	4	1	0.326895	0.262864
80	39	4	2	0.711080	0.384185
80	39	4	3	0.947994	0.236914
80	39	4	4	1.000000	0.052006
80	39	5	0	0.031173	0.031173
80	39	5	1	0.195463	0.164290
80	39	5	2	0.524042	0.328580
80	39	5	3	0.835772	0.311729
80	39	5	4	0.976050	0.140278
80	39	5	5	1.000000	0.023950
80	39	6	0	0.014963	0.014963
80	39	6	1	0.112223	0.097260
80	39	6	2	0.361943	0.249721
80	39	6	3	0.686142	0.324199
80	39	6	4	0.910587	0.224445
80	39	6	5	0.989143	0.078556
80	39	6	6	1.000000	0.010857
80	39	7	0	0.007077	0.007077
80	39	7	1	0.062278	0.055201
80	39	7	2	0.237083	0.174804
80	39	7	3	0.528423	0.291341
80	39	7	4	0.804430	0.276007
80	39	7	5	0.953049	0.148619
80	39	7	6	0.995158	0.042109
80	39	7	7	1.000000	0.004842
80	39	8	0	0.003296	0.003296
80	39	8	1	0.033544	0.030247
80	39	8	2	0.148483	0.114940
80	39	8	3	0.384749	0.236265
80	39	8	4	0.672098	0.287350
80	39	8	5	0.883830	0.211731
80	39	8	6	0.976123	0.092293
80	39	8	7	0.997877	0.021755
80	39	8	8	1.000000	0.002122
80	39	9	0	0.001511	0.001511
80	39	9	1	0.017580	0.016069
80	39	9	2	0.089417	0.071837
80	39	9	3	0.266616	0.177199
80	39	9	4	0.532414	0.265798
80	39	9	5	0.783845	0.251431
80	39	9	6	0.933822	0.149976
80	39	9	7	0.988209	0.054387
80	39	9	8	0.999086	0.010877
80	39	9	9	1.000000	0.000914

N	n	k	x	P(x)	p(x)
80	39	10	0	0.000681	0.000681
80	39	10	1	0.008979	0.008298
80	39	10	2	0.051981	0.043001
80	39	10	3	0.176769	0.124788
80	39	10	4	0.401387	0.224618
80	39	10	5	0.663442	0.262055
80	39	10	6	0.864114	0.200673
80	39	10	7	0.963696	0.099582
80	39	10	8	0.994337	0.030641
80	39	10	9	0.999614	0.005277
80	39	10	10	1.000000	0.000386
80	39	11	0	0.000302	0.000302
80	39	11	1	0.004474	0.004173
80	39	11	2	0.029251	0.024777
80	39	11	3	0.112592	0.083341
80	39	11	4	0.289078	0.176486
80	39	11	5	0.536158	0.247080
80	39	11	6	0.769512	0.233354
80	39	11	7	0.918173	0.148662
80	39	11	8	0.980767	0.062594
80	39	11	9	0.997352	0.016585
80	39	11	10	0.999840	0.002488
80	39	11	11	1.000000	0.000160
80	39	12	0	0.000131	0.000131
80	39	12	1	0.002176	0.002045
80	39	12	2	0.015965	0.013789
80	39	12	3	0.069110	0.053145
80	39	12	4	0.199556	0.130446
80	39	12	5	0.414408	0.214852
80	39	12	6	0.657908	0.243499
80	39	12	7	0.849229	0.191321
80	39	12	8	0.952645	0.103417
80	39	12	9	0.990141	0.037496
80	39	12	10	0.998794	0.008653
80	39	12	11	0.999935	0.001141
80	39	12	12	1.000000	0.000065
80	39	13	0	0.000056	0.000056
80	39	13	1	0.001033	0.000978
80	39	13	2	0.008462	0.007429
80	39	13	3	0.040974	0.032512
80	39	13	4	0.132415	0.091440
80	39	13	5	0.306982	0.174568
80	39	13	6	0.539739	0.232757
80	39	13	7	0.759195	0.219456
80	39	13	8	0.905499	0.146304
80	39	13	9	0.973599	0.068100
80	39	13	10	0.995104	0.021505
80	39	13	11	0.999465	0.004361
80	39	13	12	0.999974	0.000509
80	39	13	13	1.000000	0.000026

N	n	k	x	P(x)	p(x)
80	39	14	0	0.000023	0.000023
80	39	14	1	0.000479	0.000456
80	39	14	2	0.004360	0.003881
80	39	14	3	0.023505	0.019145
80	39	14	4	0.084647	0.061142
80	39	14	5	0.218396	0.133748
80	39	14	6	0.425098	0.206702
80	39	14	7	0.654380	0.229283
80	39	14	8	0.837807	0.183426
80	39	14	9	0.943107	0.105300
80	39	14	10	0.985796	0.042689
80	39	14	11	0.997643	0.011847
80	39	14	12	0.999769	0.002126
80	39	14	13	0.999990	0.000221
80	39	14	14	1.000000	0.000010
80	39	15	0	0.000010	0.000010
80	39	15	1	0.000217	0.000207
80	39	15	2	0.002184	0.001968
80	39	15	3	0.013062	0.010878
80	39	15	4	0.052224	0.039161
80	39	15	5	0.149495	0.097271
80	39	15	6	0.321747	0.172252
80	39	15	7	0.543213	0.221466
80	39	15	8	0.751652	0.208439
80	39	15	9	0.895243	0.143591
80	39	15	10	0.967039	0.071796
80	39	15	11	0.992617	0.025578
80	39	15	12	0.998899	0.006282
80	39	15	13	0.999903	0.001004
80	39	15	14	0.999996	0.000093
80	39	15	15	1.000000	0.000004
80	39	16	0	0.000004	0.000004
80	39	16	1	0.000096	0.000092
80	39	16	2	0.001064	0.000969
80	39	16	3	0.007038	0.005973
80	39	16	4	0.031137	0.024099
80	39	16	5	0.098615	0.067478
80	39	16	6	0.234296	0.135681
80	39	16	7	0.434183	0.199888
80	39	16	8	0.652242	0.218059
80	39	16	9	0.828970	0.176728
80	39	16	10	0.935007	0.106037
80	39	16	11	0.981598	0.046592
80	39	16	12	0.996290	0.014691
80	39	16	13	0.999501	0.003212
80	39	16	14	0.999960	0.000459
80	39	16	15	0.999998	0.000038
80	39	16	16	1.000000	0.000001
80	39	17	0	0.000001	0.000001
80	39	17	1	0.000041	0.000040

N	n	k	x	P(x)	p(x)
80	39	17	2	0.000504	0.000463
80	39	17	3	0.003678	0.003173
80	39	17	4	0.017958	0.014280
80	39	17	5	0.062767	0.044809
80	39	17	6	0.164335	0.101568
80	39	17	7	0.334240	0.169904
80	39	17	8	0.546620	0.212380
80	39	17	9	0.746129	0.199509
80	39	17	10	0.886959	0.140830
80	39	17	11	0.961215	0.074256
80	39	17	12	0.990092	0.028877
80	39	17	13	0.998197	0.008105
80	39	17	14	0.999781	0.001584
80	39	17	15	0.999984	0.000203
80	39	17	16	0.999999	0.000015
80	39	17	17	1.000000	0.000000
80	39	18	0	0.000001	0.000001
80	39	18	1	0.000017	0.000017
80	39	18	2	0.000232	0.000215
80	39	18	3	0.001864	0.001632
80	39	18	4	0.010024	0.008160
80	39	18	5	0.038584	0.028560
80	39	18	6	0.111133	0.072549
80	39	18	7	0.247939	0.136806
80	39	18	8	0.442115	0.194176
80	39	18	9	0.651125	0.209009
80	39	18	10	0.822132	0.171008
80	39	18	11	0.928212	0.106080
80	39	18	12	0.977716	0.049504
80	39	18	13	0.994852	0.017136
80	39	18	14	0.999152	0.004301
80	39	18	15	0.999907	0.000754
80	39	18	16	0.999994	0.000087
80	39	18	17	1.000000	0.000006
80	39	18	18	1.000000	0.000000
80	39	19	0	0.000000	0.000000
80	39	19	1	0.000007	0.000007
80	39	19	2	0.000104	0.000097
80	39	19	3	0.000917	0.000813
80	39	19	4	0.005418	0.004501
80	39	19	5	0.022922	0.017504
80	39	19	6	0.072518	0.049596
80	39	19	7	0.177329	0.104811
80	39	19	8	0.345027	0.167698
80	39	19	9	0.549991	0.204964
80	39	19	10	0.742145	0.192154
80	39	19	11	0.880305	0.138160
80	39	19	12	0.956158	0.075853
80	39	19	13	0.987666	0.031508
80	39	19	14	0.997418	0.009752

N	n	k	x	P(x)	p(x)
80	39	19	15	0.999615	0.002196
80	39	19	16	0.999962	0.000347
80	39	19	17	0.999998	0.000036
80	39	19	18	1.000000	0.000002
80	39	19	19	1.000000	0.000000
80	39	20	0	0.000000	0.000000
80	39	20	1	0.000003	0.000003
80	39	20	2	0.000045	0.000042
80	39	20	3	0.000437	0.000392
80	39	20	4	0.002835	0.002398
80	39	20	5	0.013166	0.010330
80	39	20	6	0.045688	0.032522
80	39	20	7	0.122346	0.076659
80	39	20	8	0.259804	0.137457
80	39	20	9	0.449189	0.189386
80	39	20	10	0.650793	0.201604
80	39	20	11	0.816887	0.166094
80	39	20	12	0.922584	0.105696
80	39	20	13	0.974236	0.051652
80	39	20	14	0.993421	0.019185
80	39	20	15	0.998751	0.005329
80	39	20	16	0.999831	0.001080
80	39	20	17	0.999985	0.000154
80	39	20	18	0.999999	0.000014
80	39	20	19	1.000000	0.000001
80	39	20	20	1.000000	0.000000
80	39	21	0	0.000000	0.000000
80	39	21	1	0.000001	0.000001
80	39	21	2	0.000019	0.000018
80	39	21	3	0.000202	0.000183
80	39	21	4	0.001436	0.001234
80	39	21	5	0.007312	0.005875
80	39	21	6	0.027801	0.020489
80	39	21	7	0.081462	0.053661
80	39	21	8	0.188784	0.107322
80	39	21	9	0.354496	0.165712
80	39	21	10	0.553351	0.198855
80	39	21	11	0.739377	0.186025
80	39	21	12	0.875020	0.135644
80	39	21	13	0.951853	0.076833
80	39	21	14	0.985427	0.033574
80	39	21	15	0.996619	0.011191
80	39	21	16	0.999417	0.002798
80	39	21	17	0.999928	0.000512
80	39	21	18	0.999994	0.000066
80	39	21	19	1.000000	0.000006
80	39	21	20	1.000000	0.000000
80	39	21	21	1.000000	0.000000
80	39	22	0	0.000000	0.000000
80	39	22	1	0.000000	0.000000

N	n	k	x	P(x)	p(x)
80	39	22	2	0.000008	0.000007
80	39	22	3	0.000090	0.000083
80	39	22	4	0.000704	0.000614
80	39	22	5	0.003926	0.003222
80	39	22	6	0.016341	0.012415
80	39	22	7	0.052357	0.036017
80	39	22	8	0.132394	0.080037
80	39	22	9	0.270236	0.137842
80	39	22	10	0.455609	0.185373
80	39	22	11	0.651093	0.195484
80	39	22	12	0.812946	0.161853
80	39	22	13	0.917995	0.105049
80	39	22	14	0.971201	0.053206
80	39	22	15	0.992066	0.020865
80	39	22	16	0.998326	0.006260
80	39	22	17	0.999737	0.001411
80	39	22	18	0.999971	0.000233
80	39	22	19	0.999998	0.000027
80	39	22	20	1.000000	0.000002
80	39	22	21	1.000000	0.000000
80	39	22	22	1.000000	0.000000
80	39	23	0	0.000000	0.000000
80	39	23	1	0.000000	0.000000
80	39	23	2	0.000003	0.000003
80	39	23	3	0.000039	0.000036
80	39	23	4	0.000334	0.000295
80	39	23	5	0.002037	0.001704
80	39	23	6	0.009277	0.007240
80	39	23	7	0.032486	0.023209
80	39	23	8	0.089616	0.057130
80	39	23	9	0.198939	0.109323
80	39	23	10	0.362922	0.163984
80	39	23	11	0.556722	0.193799
80	39	23	12	0.737601	0.180879
80	39	23	13	0.870904	0.133303
80	39	23	14	0.948267	0.077363
80	39	23	15	0.983432	0.035165
80	39	23	16	0.995844	0.012411
80	39	23	17	0.999202	0.003358
80	39	23	18	0.999886	0.000684
80	39	23	19	0.999988	0.000102
80	39	23	20	0.999999	0.000011
80	39	23	21	1.000000	0.000001
80	39	23	22	1.000000	0.000000
80	39	23	23	1.000000	0.000000
80	39	24	0	0.000000	0.000000
80	39	24	1	0.000000	0.000000
80	39	24	2	0.000001	0.000001
80	39	24	3	0.000016	0.000015
80	39	24	4	0.000153	0.000136

N	n	k	x	P(x)	p(x)
80	39	24	5	0.001021	0.000868
80	39	24	6	0.005086	0.004064
80	39	24	7	0.019456	0.014371
80	39	24	8	0.058545	0.039089
80	39	24	9	0.141400	0.082855
80	39	24	10	0.279492	0.138092
80	39	24	11	0.461522	0.182030
80	39	24	12	0.651921	0.190399
80	39	24	13	0.810099	0.158178
80	39	24	14	0.914336	0.104237
80	39	24	15	0.968626	0.054290
80	39	24	16	0.990836	0.022210
80	39	24	17	0.997906	0.007070
80	39	24	18	0.999634	0.001728
80	39	24	19	0.999952	0.000318
80	39	24	20	0.999995	0.000043
80	39	24	21	1.000000	0.000004
80	39	24	22	1.000000	0.000000
80	39	24	23	1.000000	0.000000
80	39	24	24	1.000000	0.000000
80	39	25	0	0.000000	0.000000
80	39	25	1	0.000000	0.000000
80	39	25	2	0.000000	0.000000
80	39	25	3	0.000007	0.000006
80	39	25	4	0.000067	0.000061
80	39	25	5	0.000494	0.000426
80	39	25	6	0.002690	0.002197
80	39	25	7	0.011244	0.008554
80	39	25	8	0.036907	0.025662
80	39	25	9	0.097014	0.060107
80	39	25	10	0.207980	0.110967
80	39	25	11	0.370507	0.162527
80	39	25	12	0.560122	0.189615
80	39	25	13	0.736659	0.176538
80	39	25	14	0.867802	0.131142
80	39	25	15	0.945359	0.077557
80	39	25	16	0.981714	0.036355
80	39	25	17	0.995128	0.013414
80	39	25	18	0.998986	0.003858
80	39	25	19	0.999839	0.000853
80	39	25	20	0.999981	0.000142
80	39	25	21	0.999998	0.000017
80	39	25	22	1.000000	0.000001
80	39	25	23	1.000000	0.000000
80	39	25	24	1.000000	0.000000
80	39	25	25	1.000000	0.000000
80	39	26	0	0.000000	0.000000
80	39	26	1	0.000000	0.000000
80	39	26	2	0.000000	0.000000
80	39	26	3	0.000003	0.000002

N	n	k	x	P(x)	p(x)
80	39	26	4	0.000029	0.000026
80	39	26	5	0.000230	0.000202
80	39	26	6	0.001372	0.001142
80	39	26	7	0.006268	0.004895
80	39	26	8	0.022443	0.016175
80	39	26	9	0.064228	0.041786
80	39	26	10	0.149471	0.085242
80	39	26	11	0.287766	0.138296
80	39	26	12	0.467038	0.179272
80	39	26	13	0.653205	0.186167
80	39	26	14	0.808192	0.154986
80	39	26	15	0.911516	0.103324
80	39	26	16	0.966511	0.054995
80	39	26	17	0.989763	0.023252
80	39	26	18	0.997513	0.007751
80	39	26	19	0.999529	0.002016
80	39	26	20	0.999932	0.000403
80	39	26	21	0.999993	0.000061
80	39	26	22	0.999999	0.000007
80	39	26	23	1.000000	0.000001
80	39	26	24	1.000000	0.000000
80	39	26	25	1.000000	0.000000
80	39	26	26	1.000000	0.000000
80	39	27	0	0.000000	0.000000
80	39	27	1	0.000000	0.000000
80	39	27	2	0.000000	0.000000
80	39	27	3	0.000001	0.000001
80	39	27	4	0.000012	0.000011
80	39	27	5	0.000103	0.000092
80	39	27	6	0.000674	0.000571
80	39	27	7	0.003367	0.002692
80	39	27	8	0.013157	0.009790
80	39	27	9	0.041014	0.027857
80	39	27	10	0.103692	0.062678
80	39	27	11	0.216057	0.112365
80	39	27	12	0.377402	0.161345
80	39	27	13	0.563569	0.186167
80	39	27	14	0.736439	0.172869
80	39	27	15	0.865594	0.129155
80	39	27	16	0.943087	0.077493
80	39	27	17	0.980290	0.037203
80	39	27	18	0.994499	0.014209
80	39	27	19	0.998782	0.004283
80	39	27	20	0.999790	0.001008
80	39	27	21	0.999972	0.000182
80	39	27	22	0.999997	0.000025
80	39	27	23	1.000000	0.000002
80	39	27	24	1.000000	0.000000
80	39	27	25	1.000000	0.000000
80	39	27	26	1.000000	0.000000

N	n	k	x	P(x)	p(x)
80	39	27	27	1.000000	0.000000
80	39	28	0	0.000000	0.000000
80	39	28	1	0.000000	0.000000
80	39	28	2	0.000000	0.000000
80	39	28	3	0.000000	0.000000
80	39	28	4	0.000005	0.000004
80	39	28	5	0.000045	0.000040
80	39	28	6	0.000319	0.000274
80	39	28	7	0.001741	0.001422
80	39	28	8	0.007431	0.005689
80	39	28	9	0.025246	0.017815
80	39	28	10	0.069397	0.044151
80	39	28	11	0.156695	0.087298
80	39	28	12	0.295208	0.138513
80	39	28	13	0.472242	0.177034
80	39	28	14	0.654896	0.182654
80	39	28	15	0.807108	0.152212
80	39	28	16	0.909458	0.102349
80	39	28	17	0.964847	0.055389
80	39	28	18	0.988869	0.024022
80	39	28	19	0.997166	0.008297
80	39	28	20	0.999429	0.002263
80	39	28	21	0.999910	0.000482
80	39	28	22	0.999989	0.000079
80	39	28	23	0.999999	0.000010
80	39	28	24	1.000000	0.000001
80	39	28	25	1.000000	0.000000
80	39	28	26	1.000000	0.000000
80	39	28	27	1.000000	0.000000
80	39	28	28	1.000000	0.000000
80	39	29	0	0.000000	0.000000
80	39	29	1	0.000000	0.000000
80	39	29	2	0.000000	0.000000
80	39	29	3	0.000000	0.000000
80	39	29	4	0.000002	0.000002
80	39	29	5	0.000018	0.000017
80	39	29	6	0.000145	0.000126
80	39	29	7	0.000866	0.000721
80	39	29	8	0.004039	0.003173
80	39	29	9	0.014968	0.010929
80	39	29	10	0.044774	0.029806
80	39	29	11	0.109688	0.064914
80	39	29	12	0.223287	0.113599
80	39	29	13	0.383725	0.160437
80	39	29	14	0.567082	0.183357
80	39	29	15	0.736857	0.169775
80	39	29	16	0.864188	0.127331
80	39	29	17	0.941413	0.077225
80	39	29	18	0.979168	0.037755
80	39	29	19	0.993975	0.014807

N	n	k	x	P(x)	p(x)
80	39	29	20	0.998602	0.004627
80	39	29	21	0.999744	0.001142
80	39	29	22	0.999963	0.000220
80	39	29	23	0.999996	0.000032
80	39	29	24	1.000000	0.000004
80	39	29	25	1.000000	0.000000
80	39	29	26	1.000000	0.000000
80	39	29	27	1.000000	0.000000
80	39	29	28	1.000000	0.000000
80	39	29	29	1.000000	0.000000
80	39	30	0	0.000000	0.000000
80	39	30	1	0.000000	0.000000
80	39	30	2	0.000000	0.000000
80	39	30	3	0.000000	0.000000
80	39	30	4	0.000001	0.000001
80	39	30	5	0.000007	0.000007
80	39	30	6	0.000063	0.000056
80	39	30	7	0.000413	0.000350
80	39	30	8	0.002110	0.001697
80	39	30	9	0.008539	0.006429
80	39	30	10	0.027825	0.019286
80	39	30	11	0.074049	0.046224
80	39	30	12	0.163147	0.089098
80	39	30	13	0.301933	0.138787
80	39	30	14	0.477201	0.175268
80	39	30	15	0.656963	0.179762
80	39	30	16	0.806764	0.149801
80	39	30	17	0.908100	0.101336
80	39	30	18	0.963622	0.055521
80	39	30	19	0.988168	0.024546
80	39	30	20	0.996878	0.008710
80	39	30	21	0.999341	0.002463
80	39	30	22	0.999890	0.000550
80	39	30	23	0.999986	0.000096
80	39	30	24	0.999998	0.000013
80	39	30	25	1.000000	0.000001
80	39	30	26	1.000000	0.000000
80	39	30	27	1.000000	0.000000
80	39	30	28	1.000000	0.000000
80	39	30	29	1.000000	0.000000
80	39	30	30	1.000000	0.000000
80	39	31	0	0.000000	0.000000
80	39	31	1	0.000000	0.000000
80	39	31	2	0.000000	0.000000
80	39	31	3	0.000000	0.000000
80	39	31	4	0.000000	0.000000
80	39	31	5	0.000003	0.000003
80	39	31	6	0.000026	0.000023
80	39	31	7	0.000189	0.000163
80	39	31	8	0.001058	0.000869

N	n	k	x	P(x)	p(x)
80	39	31	9	0.004682	0.003624
80	39	31	10	0.016639	0.011958
80	39	31	11	0.048164	0.031524
80	39	31	12	0.115034	0.066870
80	39	31	13	0.229764	0.114730
80	39	31	14	0.389567	0.159803
80	39	31	15	0.570677	0.181110
80	39	31	16	0.737855	0.167178
80	39	31	17	0.863512	0.125657
80	39	31	18	0.940303	0.076790
80	39	31	19	0.978350	0.038047
80	39	31	20	0.993568	0.015219
80	39	31	21	0.998454	0.004886
80	39	31	22	0.999703	0.001249
80	39	31	23	0.999955	0.000252
80	39	31	24	0.999995	0.000039
80	39	31	25	0.999999	0.000005
80	39	31	26	1.000000	0.000000
80	39	31	27	1.000000	0.000000
80	39	31	28	1.000000	0.000000
80	39	31	29	1.000000	0.000000
80	39	31	30	1.000000	0.000000
80	39	31	31	1.000000	0.000000
80	39	32	0	0.000000	0.000000
80	39	32	1	0.000000	0.000000
80	39	32	2	0.000000	0.000000
80	39	32	3	0.000000	0.000000
80	39	32	4	0.000000	0.000000
80	39	32	5	0.000001	0.000001
80	39	32	6	0.000010	0.000009
80	39	32	7	0.000083	0.000072
80	39	32	8	0.000508	0.000426
80	39	32	9	0.002463	0.001955
80	39	32	10	0.009562	0.007099
80	39	32	11	0.030150	0.020587
80	39	32	12	0.078187	0.048037
80	39	32	13	0.168887	0.090700
80	39	32	14	0.308035	0.139148
80	39	32	15	0.481970	0.173935
80	39	32	16	0.659384	0.177414
80	39	32	17	0.807095	0.147711
80	39	32	18	0.907393	0.100298
80	39	32	19	0.962820	0.055428
80	39	32	20	0.987667	0.024847
80	39	32	21	0.996659	0.008992
80	39	32	22	0.999270	0.002611
80	39	32	23	0.999873	0.000603
80	39	32	24	0.999983	0.000110
80	39	32	25	0.999998	0.000015
80	39	32	26	1.000000	0.000002

N	n	k	x	P(x)	p(x)
80	39	32	27	1.000000	0.000000
80	39	32	28	1.000000	0.000000
80	39	32	29	1.000000	0.000000
80	39	32	30	1.000000	0.000000
80	39	32	31	1.000000	0.000000
80	39	32	32	1.000000	0.000000
80	39	33	0	0.000000	0.000000
80	39	33	1	0.000000	0.000000
80	39	33	2	0.000000	0.000000
80	39	33	3	0.000000	0.000000
80	39	33	4	0.000000	0.000000
80	39	33	5	0.000000	0.000000
80	39	33	6	0.000004	0.000004
80	39	33	7	0.000035	0.000031
80	39	33	8	0.000234	0.000199
80	39	33	9	0.001242	0.001008
80	39	33	10	0.005273	0.004032
80	39	33	11	0.018140	0.012867
80	39	33	12	0.051166	0.033026
80	39	33	13	0.119758	0.068592
80	39	33	14	0.235562	0.115804
80	39	33	15	0.395003	0.159441
80	39	33	16	0.574373	0.179371
80	39	33	17	0.739394	0.165021
80	39	33	18	0.863512	0.124118
80	39	33	19	0.939725	0.076213
80	39	33	20	0.977832	0.038107
80	39	33	21	0.993287	0.015455
80	39	33	22	0.998345	0.005058
80	39	33	23	0.999672	0.001327
80	39	33	24	0.999948	0.000276
80	39	33	25	0.999993	0.000045
80	39	33	26	0.999999	0.000006
80	39	33	27	1.000000	0.000001
80	39	33	28	1.000000	0.000000
80	39	33	29	1.000000	0.000000
80	39	33	30	1.000000	0.000000
80	39	33	31	1.000000	0.000000
80	39	33	32	1.000000	0.000000
80	39	33	33	1.000000	0.000000
80	39	34	0	0.000000	0.000000
80	39	34	1	0.000000	0.000000
80	39	34	2	0.000000	0.000000
80	39	34	3	0.000000	0.000000
80	39	34	4	0.000000	0.000000
80	39	34	5	0.000000	0.000000
80	39	34	6	0.000001	0.000001
80	39	34	7	0.000014	0.000012
80	39	34	8	0.000102	0.000089
80	39	34	9	0.000598	0.000496

N	n	k	x	P(x)	p(x)
80	39	34	10	0.002786	0.002187
80	39	34	11	0.010475	0.007689
80	39	34	12	0.032194	0.021719
80	39	34	13	0.081813	0.049620
80	39	34	14	0.173964	0.092151
80	39	34	15	0.313586	0.139622
80	39	34	16	0.486596	0.173010
80	39	34	17	0.662150	0.175554
80	39	34	18	0.808055	0.145905
80	39	34	19	0.907294	0.099239
80	39	34	20	0.962427	0.055133
80	39	34	21	0.987368	0.024941
80	39	34	22	0.996516	0.009148
80	39	34	23	0.999220	0.002705
80	39	34	24	0.999860	0.000640
80	39	34	25	0.999980	0.000120
80	39	34	26	0.999998	0.000018
80	39	34	27	1.000000	0.000002
80	39	34	28	1.000000	0.000000
80	39	34	29	1.000000	0.000000
80	39	34	30	1.000000	0.000000
80	39	34	31	1.000000	0.000000
80	39	34	32	1.000000	0.000000
80	39	34	33	1.000000	0.000000
80	39	34	34	1.000000	0.000000
80	39	35	0	0.000000	0.000000
80	39	35	1	0.000000	0.000000
80	39	35	2	0.000000	0.000000
80	39	35	3	0.000000	0.000000
80	39	35	4	0.000000	0.000000
80	39	35	5	0.000000	0.000000
80	39	35	6	0.000000	0.000000
80	39	35	7	0.000005	0.000005
80	39	35	8	0.000043	0.000037
80	39	35	9	0.000275	0.000232
80	39	35	10	0.001407	0.001132
80	39	35	11	0.005794	0.004388
80	39	35	12	0.019446	0.013651
80	39	35	13	0.053768	0.034322
80	39	35	14	0.123882	0.070115
80	39	35	15	0.240740	0.116858
80	39	35	16	0.400091	0.159351
80	39	35	17	0.578189	0.178098
80	39	35	18	0.741446	0.163257
80	39	35	19	0.864147	0.122700
80	39	35	20	0.939655	0.075508
80	39	35	21	0.977609	0.037954
80	39	35	22	0.993135	0.015527
80	39	35	23	0.998280	0.005144
80	39	35	24	0.999651	0.001372

N	n	k	x	P(x)	p(x)
80	39	35	25	0.999943	0.000292
80	39	35	26	0.999993	0.000049
80	39	35	27	0.999999	0.000006
80	39	35	28	1.000000	0.000001
80	39	35	29	1.000000	0.000000
80	39	35	30	1.000000	0.000000
80	39	35	31	1.000000	0.000000
80	39	35	32	1.000000	0.000000
80	39	35	33	1.000000	0.000000
80	39	35	34	1.000000	0.000000
80	39	35	35	1.000000	0.000000
80	39	36	0	0.000000	0.000000
80	39	36	1	0.000000	0.000000
80	39	36	2	0.000000	0.000000
80	39	36	3	0.000000	0.000000
80	39	36	4	0.000000	0.000000
80	39	36	5	0.000000	0.000000
80	39	36	6	0.000000	0.000000
80	39	36	7	0.000002	0.000002
80	39	36	8	0.000017	0.000015
80	39	36	9	0.000120	0.000103
80	39	36	10	0.000677	0.000557
80	39	36	11	0.003064	0.002387
80	39	36	12	0.011255	0.008191
80	39	36	13	0.033937	0.022682
80	39	36	14	0.084930	0.050992
80	39	36	15	0.178416	0.093486
80	39	36	16	0.318645	0.140229
80	39	36	17	0.491119	0.172474
80	39	36	18	0.665260	0.174141
80	39	36	19	0.809613	0.144353
80	39	36	20	0.907774	0.098160
80	39	36	21	0.962427	0.054653
80	39	36	22	0.987270	0.024842
80	39	36	23	0.996450	0.009181
80	39	36	24	0.999194	0.002744
80	39	36	25	0.999853	0.000658
80	39	36	26	0.999979	0.000126
80	39	36	27	0.999997	0.000019
80	39	36	28	1.000000	0.000002
80	39	36	29	1.000000	0.000000
80	39	36	30	1.000000	0.000000
80	39	36	31	1.000000	0.000000
80	39	36	32	1.000000	0.000000
80	39	36	33	1.000000	0.000000
80	39	36	34	1.000000	0.000000
80	39	36	35	1.000000	0.000000
80	39	36	36	1.000000	0.000000
80	39	37	0	0.000000	0.000000
80	39	37	1	0.000000	0.000000

N	n	k	x	P(x)	p(x)
80	39	37	2	0.000000	0.000000
80	39	37	3	0.000000	0.000000
80	39	37	4	0.000000	0.000000
80	39	37	5	0.000000	0.000000
80	39	37	6	0.000000	0.000000
80	39	37	7	0.000001	0.000001
80	39	37	8	0.000006	0.000006
80	39	37	9	0.000050	0.000043
80	39	37	10	0.000310	0.000260
80	39	37	11	0.001545	0.001235
80	39	37	12	0.006229	0.004684
80	39	37	13	0.020534	0.014305
80	39	37	14	0.055957	0.035423
80	39	37	15	0.127423	0.071467
80	39	37	16	0.245343	0.117920
80	39	37	17	0.404882	0.159539
80	39	37	18	0.582147	0.177265
80	39	37	19	0.743998	0.161851
80	39	37	20	0.865386	0.121388
80	39	37	21	0.940069	0.074683
80	39	37	22	0.977671	0.037602
80	39	37	23	0.993112	0.015441
80	39	37	24	0.998259	0.005147
80	39	37	25	0.999643	0.001384
80	39	37	26	0.999941	0.000298
80	39	37	27	0.999992	0.000051
80	39	37	28	0.999999	0.000007
80	39	37	29	1.000000	0.000001
80	39	37	30	1.000000	0.000000
80	39	37	31	1.000000	0.000000
80	39	37	32	1.000000	0.000000
80	39	37	33	1.000000	0.000000
80	39	37	34	1.000000	0.000000
80	39	37	35	1.000000	0.000000
80	39	37	36	1.000000	0.000000
80	39	37	37	1.000000	0.000000
80	39	38	0	0.000000	0.000000
80	39	38	1	0.000000	0.000000
80	39	38	2	0.000000	0.000000
80	39	38	3	0.000000	0.000000
80	39	38	4	0.000000	0.000000
80	39	38	5	0.000000	0.000000
80	39	38	6	0.000000	0.000000
80	39	38	7	0.000000	0.000000
80	39	38	8	0.000002	0.000002
80	39	38	9	0.000019	0.000017
80	39	38	10	0.000134	0.000115
80	39	38	11	0.000741	0.000606
80	39	38	12	0.003288	0.002547
80	39	38	13	0.011884	0.008596

Table for $N = 2$, $n = 1$, through $N = 100$, $n = 50$

N	n	k	x	P(x)	p(x)
80	39	38	14	0.035362	0.023478
80	39	38	15	0.087535	0.052173
80	39	38	16	0.182270	0.094735
80	39	38	17	0.323258	0.140988
80	39	38	18	0.495576	0.172318
80	39	38	19	0.668719	0.173143
80	39	38	20	0.811750	0.143031
80	39	38	21	0.908806	0.097057
80	39	38	22	0.962805	0.053999
80	39	38	23	0.987367	0.024561
80	39	38	24	0.996463	0.009097
80	39	38	25	0.999192	0.002729
80	39	38	26	0.999851	0.000659
80	39	38	27	0.999978	0.000127
80	39	38	28	0.999997	0.000019
80	39	38	29	1.000000	0.000002
80	39	38	30	1.000000	0.000000
80	39	38	31	1.000000	0.000000
80	39	38	32	1.000000	0.000000
80	39	38	33	1.000000	0.000000
80	39	38	34	1.000000	0.000000
80	39	38	35	1.000000	0.000000
80	39	38	36	1.000000	0.000000
80	39	38	37	1.000000	0.000000
80	39	38	38	1.000000	0.000000
80	39	39	0	0.000000	0.000000
80	39	39	1	0.000000	0.000000
80	39	39	2	0.000000	0.000000
80	39	39	3	0.000000	0.000000
80	39	39	4	0.000000	0.000000
80	39	39	5	0.000000	0.000000
80	39	39	6	0.000000	0.000000
80	39	39	7	0.000000	0.000000
80	39	39	8	0.000001	0.000001
80	39	39	9	0.000007	0.000006
80	39	39	10	0.000055	0.000048
80	39	39	11	0.000337	0.000282
80	39	39	12	0.001651	0.001314
80	39	39	13	0.006563	0.004912
80	39	39	14	0.021387	0.014825
80	39	39	15	0.057722	0.036335
80	39	39	16	0.130391	0.072669
80	39	39	17	0.249407	0.119016
80	39	39	18	0.409417	0.160010
80	39	39	19	0.586270	0.176853
80	39	39	20	0.747045	0.160775
80	39	39	21	0.867211	0.120165
80	39	39	22	0.940949	0.073738
80	39	39	23	0.978010	0.037061
80	39	39	24	0.993214	0.015205

N	n	k	x	P(x)	p(x)
80	39	39	25	0.998283	0.005068
80	39	39	26	0.999647	0.001365
80	39	39	27	0.999942	0.000295
80	39	39	28	0.999992	0.000050
80	39	39	29	0.999999	0.000007
80	39	39	30	1.000000	0.000001
80	39	39	31	1.000000	0.000000
80	39	39	32	1.000000	0.000000
80	39	39	33	1.000000	0.000000
80	39	39	34	1.000000	0.000000
80	39	39	35	1.000000	0.000000
80	39	39	36	1.000000	0.000000
80	39	39	37	1.000000	0.000000
80	39	39	38	1.000000	0.000000
80	39	39	39	1.000000	0.000000
80	40	1	0	0.500000	0.500000
80	40	1	1	1.000000	0.500000
80	40	2	0	0.246835	0.246835
80	40	2	1	0.753165	0.506329
80	40	2	2	1.000000	0.246835
80	40	3	0	0.120253	0.120253
80	40	3	1	0.500000	0.379747
80	40	3	2	0.879747	0.379747
80	40	3	3	1.000000	0.120253
80	40	4	0	0.057784	0.057784
80	40	4	1	0.307661	0.249877
80	40	4	2	0.692339	0.384679
80	40	4	3	0.942216	0.249877
80	40	4	4	1.000000	0.057784
80	40	5	0	0.027371	0.027371
80	40	5	1	0.179434	0.152063
80	40	5	2	0.500000	0.320565
80	40	5	3	0.820565	0.320565
80	40	5	4	0.972629	0.152063
80	40	5	5	1.000000	0.027371
80	40	6	0	0.012773	0.012773
80	40	6	1	0.100362	0.087588
80	40	6	2	0.337580	0.237218
80	40	6	3	0.662420	0.324840
80	40	6	4	0.899638	0.237218
80	40	6	5	0.987227	0.087588
80	40	6	6	1.000000	0.012773
80	40	7	0	0.005869	0.005869
80	40	7	1	0.054200	0.048331
80	40	7	2	0.215765	0.161565
80	40	7	3	0.500000	0.284235
80	40	7	4	0.784235	0.284235
80	40	7	5	0.945800	0.161565
80	40	7	6	0.994131	0.048331
80	40	7	7	1.000000	0.005869

N	n	k	x	P(x)	p(x)
80	40	8	0	0.002653	0.002653
80	40	8	1	0.028379	0.025726
80	40	8	2	0.131663	0.103284
80	40	8	3	0.355936	0.224273
80	40	8	4	0.644064	0.288128
80	40	8	5	0.868337	0.224273
80	40	8	6	0.971621	0.103284
80	40	8	7	0.997347	0.025726
80	40	8	8	1.000000	0.002653
80	40	9	0	0.001179	0.001179
80	40	9	1	0.014444	0.013265
80	40	9	2	0.077152	0.062708
80	40	9	3	0.240684	0.163532
80	40	9	4	0.500000	0.259316
80	40	9	5	0.759315	0.259316
80	40	9	6	0.922848	0.163532
80	40	9	7	0.985556	0.062708
80	40	9	8	0.998821	0.013265
80	40	9	9	1.000000	0.001179
80	40	10	0	0.000515	0.000515
80	40	10	1	0.007158	0.006643
80	40	10	2	0.043590	0.036432
80	40	10	3	0.155463	0.111873
80	40	10	4	0.368516	0.213053
80	40	10	5	0.631484	0.262968
80	40	10	6	0.844537	0.213053
80	40	10	7	0.956410	0.111873
80	40	10	8	0.992842	0.036432
80	40	10	9	0.999485	0.006643
80	40	10	10	1.000000	0.000515
80	40	11	0	0.000221	0.000221
80	40	11	1	0.003457	0.003236
80	40	11	2	0.023813	0.020356
80	40	11	3	0.096330	0.072518
80	40	11	4	0.258946	0.162616
80	40	11	5	0.500000	0.241054
80	40	11	6	0.741054	0.241054
80	40	11	7	0.903669	0.162616
80	40	11	8	0.976187	0.072518
80	40	11	9	0.996543	0.020356
80	40	11	10	0.999779	0.003236
80	40	11	11	1.000000	0.000221
80	40	12	0	0.000093	0.000093
80	40	12	1	0.001628	0.001535
80	40	12	2	0.012602	0.010974
80	40	12	3	0.057444	0.044842
80	40	12	4	0.174103	0.116659
80	40	12	5	0.377726	0.203623
80	40	12	6	0.622274	0.244547
80	40	12	7	0.825897	0.203623

N	n	k	x	P(x)	p(x)
80	40	12	8	0.942556	0.116659
80	40	12	9	0.987398	0.044842
80	40	12	10	0.998372	0.010974
80	40	12	11	0.999907	0.001535
80	40	12	12	1.000000	0.000093
80	40	13	0	0.000038	0.000038
80	40	13	1	0.000747	0.000709
80	40	13	2	0.006469	0.005722
80	40	13	3	0.033045	0.026575
80	40	13	4	0.112342	0.079298
80	40	13	5	0.272920	0.160578
80	40	13	6	0.500000	0.227080
80	40	13	7	0.727080	0.227080
80	40	13	8	0.887657	0.160578
80	40	13	9	0.966955	0.079298
80	40	13	10	0.993531	0.026575
80	40	13	11	0.999253	0.005722
80	40	13	12	0.999962	0.000709
80	40	13	13	1.000000	0.000038
80	40	14	0	0.000015	0.000015
80	40	14	1	0.000335	0.000319
80	40	14	2	0.003224	0.002889
80	40	14	3	0.018369	0.015145
80	40	14	4	0.069735	0.051366
80	40	14	5	0.189036	0.119302
80	40	14	6	0.384765	0.195729
80	40	14	7	0.615234	0.230469
80	40	14	8	0.810964	0.195729
80	40	14	9	0.930265	0.119302
80	40	14	10	0.981631	0.051366
80	40	14	11	0.996776	0.015145
80	40	14	12	0.999665	0.002889
80	40	14	13	0.999985	0.000319
80	40	14	14	1.000000	0.000015
80	40	15	0	0.000006	0.000006
80	40	15	1	0.000146	0.000140
80	40	15	2	0.001560	0.001414
80	40	15	3	0.009879	0.008318
80	40	15	4	0.041717	0.031838
80	40	15	5	0.125770	0.084053
80	40	15	6	0.283935	0.158165
80	40	15	7	0.500000	0.216065
80	40	15	8	0.716065	0.216065
80	40	15	9	0.874230	0.158165
80	40	15	10	0.958283	0.084053
80	40	15	11	0.990121	0.031838
80	40	15	12	0.998440	0.008318
80	40	15	13	0.999854	0.001414
80	40	15	14	0.999994	0.000140
80	40	15	15	1.000000	0.000006

Table for $N = 2,\ n = 1,$ through $N = 100,\ n = 50$

N	n	k	x	P(x)	p(x)
80	40	16	0	0.000002	0.000002
80	40	16	1	0.000062	0.000060
80	40	16	2	0.000733	0.000671
80	40	16	3	0.005144	0.004410
80	40	16	4	0.024083	0.018940
80	40	16	5	0.080511	0.056427
80	40	16	6	0.201203	0.120692
80	40	16	7	0.390306	0.189103
80	40	16	8	0.609694	0.219389
80	40	16	9	0.798797	0.189103
80	40	16	10	0.919489	0.120692
80	40	16	11	0.975917	0.056427
80	40	16	12	0.994856	0.018940
80	40	16	13	0.999266	0.004410
80	40	16	14	0.999938	0.000671
80	40	16	15	0.999998	0.000060
80	40	16	16	1.000000	0.000002
80	40	17	0	0.000001	0.000001
80	40	17	1	0.000026	0.000025
80	40	17	2	0.000335	0.000309
80	40	17	3	0.002594	0.002259
80	40	17	4	0.013430	0.010836
80	40	17	5	0.049652	0.036222
80	40	17	6	0.137085	0.087433
80	40	17	7	0.292799	0.155714
80	40	17	8	0.500000	0.207200
80	40	17	9	0.707200	0.207200
80	40	17	10	0.862915	0.155714
80	40	17	11	0.950348	0.087433
80	40	17	12	0.986570	0.036222
80	40	17	13	0.997406	0.010836
80	40	17	14	0.999665	0.002259
80	40	17	15	0.999974	0.000309
80	40	17	16	0.999999	0.000025
80	40	17	17	1.000000	0.000001
80	40	18	0	0.000000	0.000000
80	40	18	1	0.000010	0.000010
80	40	18	2	0.000148	0.000138
80	40	18	3	0.001267	0.001119
80	40	18	4	0.007238	0.005971
80	40	18	5	0.029529	0.022291
80	40	18	6	0.089899	0.060370
80	40	18	7	0.211235	0.121336
80	40	18	8	0.394755	0.183520
80	40	18	9	0.605245	0.210489
80	40	18	10	0.788765	0.183520
80	40	18	11	0.910101	0.121336
80	40	18	12	0.970471	0.060370
80	40	18	13	0.992762	0.022291
80	40	18	14	0.998733	0.005971

N	n	k	x	P(x)	p(x)
80	40	18	15	0.999852	0.001119
80	40	18	16	0.999990	0.000138
80	40	18	17	1.000000	0.000010
80	40	18	18	1.000000	0.000000
80	40	19	0	0.000000	0.000000
80	40	19	1	0.000004	0.000004
80	40	19	2	0.000064	0.000060
80	40	19	3	0.000599	0.000536
80	40	19	4	0.003771	0.003172
80	40	19	5	0.016945	0.013174
80	40	19	6	0.056793	0.039848
80	40	19	7	0.146653	0.089860
80	40	19	8	0.300035	0.153382
80	40	19	9	0.500000	0.199965
80	40	19	10	0.699965	0.199965
80	40	19	11	0.853347	0.153382
80	40	19	12	0.943207	0.089860
80	40	19	13	0.983055	0.039848
80	40	19	14	0.996229	0.013174
80	40	19	15	0.999400	0.003172
80	40	19	16	0.999936	0.000536
80	40	19	17	0.999996	0.000060
80	40	19	18	1.000000	0.000004
80	40	19	19	1.000000	0.000000
80	40	20	0	0.000000	0.000000
80	40	20	1	0.000002	0.000001
80	40	20	2	0.000027	0.000025
80	40	20	3	0.000275	0.000248
80	40	20	4	0.001899	0.001625
80	40	20	5	0.009386	0.007487
80	40	20	6	0.034583	0.025196
80	40	20	7	0.098040	0.063457
80	40	20	8	0.219572	0.121532
80	40	20	9	0.398378	0.178806
80	40	20	10	0.601621	0.203243
80	40	20	11	0.780428	0.178806
80	40	20	12	0.901960	0.121532
80	40	20	13	0.965417	0.063457
80	40	20	14	0.990614	0.025196
80	40	20	15	0.998101	0.007487
80	40	20	16	0.999725	0.001625
80	40	20	17	0.999973	0.000248
80	40	20	18	0.999998	0.000025
80	40	20	19	1.000000	0.000001
80	40	20	20	1.000000	0.000000
80	40	21	0	0.000000	0.000000
80	40	21	1	0.000001	0.000001
80	40	21	2	0.000011	0.000010
80	40	21	3	0.000122	0.000111
80	40	21	4	0.000924	0.000803

N	n	k	x	P(x)	p(x)
80	40	21	5	0.005019	0.004094
80	40	21	6	0.020305	0.015286
80	40	21	7	0.063138	0.042834
80	40	21	8	0.154755	0.091617
80	40	21	9	0.305995	0.151240
80	40	21	10	0.500000	0.194005
80	40	21	11	0.694005	0.194005
80	40	21	12	0.845245	0.151240
80	40	21	13	0.936862	0.091617
80	40	21	14	0.979695	0.042834
80	40	21	15	0.994981	0.015286
80	40	21	16	0.999075	0.004094
80	40	21	17	0.999878	0.000803
80	40	21	18	0.999989	0.000111
80	40	21	19	0.999999	0.000010
80	40	21	20	1.000000	0.000001
80	40	21	21	1.000000	0.000000
80	40	22	0	0.000000	0.000000
80	40	22	1	0.000000	0.000000
80	40	22	2	0.000004	0.000004
80	40	22	3	0.000052	0.000048
80	40	22	4	0.000435	0.000383
80	40	22	5	0.002590	0.002155
80	40	22	6	0.011496	0.008906
80	40	22	7	0.039180	0.027685
80	40	22	8	0.105065	0.065884
80	40	22	9	0.226530	0.121465
80	40	22	10	0.401353	0.174823
80	40	22	11	0.598646	0.197293
80	40	22	12	0.773470	0.174823
80	40	22	13	0.894935	0.121465
80	40	22	14	0.960819	0.065884
80	40	22	15	0.988504	0.027685
80	40	22	16	0.997410	0.008906
80	40	22	17	0.999565	0.002155
80	40	22	18	0.999948	0.000383
80	40	22	19	0.999996	0.000048
80	40	22	20	1.000000	0.000004
80	40	22	21	1.000000	0.000000
80	40	22	22	1.000000	0.000000
80	40	23	0	0.000000	0.000000
80	40	23	1	0.000000	0.000000
80	40	23	2	0.000002	0.000001
80	40	23	3	0.000022	0.000020
80	40	23	4	0.000197	0.000176
80	40	23	5	0.001289	0.001092
80	40	23	6	0.006275	0.004986
80	40	23	7	0.023429	0.017154
80	40	23	8	0.068715	0.045286
80	40	23	9	0.161609	0.092894

N	n	k	x	P(x)	p(x)
80	40	23	10	0.310928	0.149319
80	40	23	11	0.500000	0.189072
80	40	23	12	0.689072	0.189072
80	40	23	13	0.838391	0.149319
80	40	23	14	0.931285	0.092894
80	40	23	15	0.976571	0.045286
80	40	23	16	0.993725	0.017154
80	40	23	17	0.998711	0.004986
80	40	23	18	0.999803	0.001092
80	40	23	19	0.999978	0.000176
80	40	23	20	0.999998	0.000020
80	40	23	21	1.000000	0.000001
80	40	23	22	1.000000	0.000000
80	40	23	23	1.000000	0.000000
80	40	24	0	0.000000	0.000000
80	40	24	1	0.000000	0.000000
80	40	24	2	0.000001	0.000001
80	40	24	3	0.000009	0.000008
80	40	24	4	0.000086	0.000078
80	40	24	5	0.000619	0.000532
80	40	24	6	0.003301	0.002682
80	40	24	7	0.013498	0.010197
80	40	24	8	0.043291	0.029793
80	40	24	9	0.111088	0.067796
80	40	24	10	0.232339	0.121251
80	40	24	11	0.403805	0.171466
80	40	24	12	0.596195	0.192389
80	40	24	13	0.767661	0.171466
80	40	24	14	0.888912	0.121251
80	40	24	15	0.956709	0.067796
80	40	24	16	0.986502	0.029793
80	40	24	17	0.996699	0.010197
80	40	24	18	0.999381	0.002682
80	40	24	19	0.999914	0.000532
80	40	24	20	0.999991	0.000078
80	40	24	21	0.999999	0.000008
80	40	24	22	1.000000	0.000001
80	40	24	23	1.000000	0.000000
80	40	24	24	1.000000	0.000000
80	40	25	0	0.000000	0.000000
80	40	25	1	0.000000	0.000000
80	40	25	2	0.000000	0.000000
80	40	25	3	0.000003	0.000003
80	40	25	4	0.000036	0.000033
80	40	25	5	0.000286	0.000250
80	40	25	6	0.001672	0.001387
80	40	25	7	0.007489	0.005817
80	40	25	8	0.026266	0.018777
80	40	25	9	0.073557	0.047291
80	40	25	10	0.167383	0.093825

Table for $N = 2$, $n = 1$, through $N = 100$, $n = 50$

N	n	k	x	P(x)	p(x)
80	40	25	11	0.315010	0.147627
80	40	25	12	0.500000	0.184990
80	40	25	13	0.684990	0.184990
80	40	25	14	0.832617	0.147627
80	40	25	15	0.926443	0.093825
80	40	25	16	0.973734	0.047291
80	40	25	17	0.992511	0.018777
80	40	25	18	0.998327	0.005817
80	40	25	19	0.999714	0.001387
80	40	25	20	0.999964	0.000250
80	40	25	21	0.999997	0.000033
80	40	25	22	1.000000	0.000003
80	40	25	23	1.000000	0.000000
80	40	25	24	1.000000	0.000000
80	40	25	25	1.000000	0.000000
80	40	26	0	0.000000	0.000000
80	40	26	1	0.000000	0.000000
80	40	26	2	0.000000	0.000000
80	40	26	3	0.000001	0.000001
80	40	26	4	0.000015	0.000013
80	40	26	5	0.000127	0.000112
80	40	26	6	0.000815	0.000688
80	40	26	7	0.003999	0.003184
80	40	26	8	0.015341	0.011342
80	40	26	9	0.046902	0.031561
80	40	26	10	0.116205	0.069303
80	40	26	11	0.237170	0.120965
80	40	26	12	0.405823	0.168653
80	40	26	13	0.594177	0.188353
80	40	26	14	0.762830	0.168653
80	40	26	15	0.883795	0.120965
80	40	26	16	0.953097	0.069303
80	40	26	17	0.984658	0.031561
80	40	26	18	0.996001	0.011342
80	40	26	19	0.999185	0.003184
80	40	26	20	0.999873	0.000688
80	40	26	21	0.999985	0.000112
80	40	26	22	0.999999	0.000013
80	40	26	23	1.000000	0.000001
80	40	26	24	1.000000	0.000000
80	40	26	25	1.000000	0.000000
80	40	26	26	1.000000	0.000000
80	40	27	0	0.000000	0.000000
80	40	27	1	0.000000	0.000000
80	40	27	2	0.000000	0.000000
80	40	27	3	0.000000	0.000000
80	40	27	4	0.000006	0.000005
80	40	27	5	0.000054	0.000049
80	40	27	6	0.000382	0.000328
80	40	27	7	0.002053	0.001671
80	40	27	8	0.008620	0.006567
80	40	27	9	0.028784	0.020164
80	40	27	10	0.077704	0.048920
80	40	27	11	0.172208	0.094504
80	40	27	12	0.318374	0.146166
80	40	27	13	0.500000	0.181626
80	40	27	14	0.681626	0.181626
80	40	27	15	0.827792	0.146166
80	40	27	16	0.922296	0.094504
80	40	27	17	0.971216	0.048920
80	40	27	18	0.991380	0.020164
80	40	27	19	0.997946	0.006567
80	40	27	20	0.999618	0.001671
80	40	27	21	0.999946	0.000328
80	40	27	22	0.999994	0.000049
80	40	27	23	0.999999	0.000005
80	40	27	24	1.000000	0.000000
80	40	27	25	1.000000	0.000000
80	40	27	26	1.000000	0.000000
80	40	27	27	1.000000	0.000000
80	40	28	0	0.000000	0.000000
80	40	28	1	0.000000	0.000000
80	40	28	2	0.000000	0.000000
80	40	28	3	0.000000	0.000000
80	40	28	4	0.000002	0.000002
80	40	28	5	0.000022	0.000020
80	40	28	6	0.000172	0.000150
80	40	28	7	0.001013	0.000841
80	40	28	8	0.004655	0.003643
80	40	28	9	0.016990	0.012335
80	40	28	10	0.050013	0.033023
80	40	28	11	0.120498	0.070485
80	40	28	12	0.241154	0.120656
80	40	28	13	0.407473	0.166320
80	40	28	14	0.592527	0.185053
80	40	28	15	0.758846	0.166320
80	40	28	16	0.879502	0.120656
80	40	28	17	0.949987	0.070485
80	40	28	18	0.983010	0.033023
80	40	28	19	0.995345	0.012335
80	40	28	20	0.998987	0.003643
80	40	28	21	0.999828	0.000841
80	40	28	22	0.999978	0.000150
80	40	28	23	0.999998	0.000020
80	40	28	24	1.000000	0.000002
80	40	28	25	1.000000	0.000000
80	40	28	26	1.000000	0.000000
80	40	28	27	1.000000	0.000000
80	40	28	28	1.000000	0.000000
80	40	29	0	0.000000	0.000000
80	40	29	1	0.000000	0.000000
80	40	29	2	0.000000	0.000000
80	40	29	3	0.000000	0.000000
80	40	29	4	0.000001	0.000001
80	40	29	5	0.000009	0.000008
80	40	29	6	0.000074	0.000065
80	40	29	7	0.000479	0.000405
80	40	29	8	0.002414	0.001935
80	40	29	9	0.009637	0.007223
80	40	29	10	0.030961	0.021325
80	40	29	11	0.081189	0.050228
80	40	29	12	0.176185	0.094996
80	40	29	13	0.321115	0.144930
80	40	29	14	0.500000	0.178885
80	40	29	15	0.678885	0.178885
80	40	29	16	0.823815	0.144930
80	40	29	17	0.918811	0.094996
80	40	29	18	0.969039	0.050228
80	40	29	19	0.990363	0.021325
80	40	29	20	0.997586	0.007223
80	40	29	21	0.999521	0.001935
80	40	29	22	0.999926	0.000405
80	40	29	23	0.999991	0.000065
80	40	29	24	0.999999	0.000008
80	40	29	25	1.000000	0.000001
80	40	29	26	1.000000	0.000000
80	40	29	27	1.000000	0.000000
80	40	29	28	1.000000	0.000000
80	40	29	29	1.000000	0.000000
80	40	30	0	0.000000	0.000000
80	40	30	1	0.000000	0.000000
80	40	30	2	0.000000	0.000000
80	40	30	3	0.000000	0.000000
80	40	30	4	0.000000	0.000000
80	40	30	5	0.000003	0.000003
80	40	30	6	0.000030	0.000027
80	40	30	7	0.000217	0.000186
80	40	30	8	0.001200	0.000983
80	40	30	9	0.005246	0.004046
80	40	30	10	0.018417	0.013171
80	40	30	11	0.052628	0.034211
80	40	30	12	0.124031	0.071402
80	40	30	13	0.244388	0.120357
80	40	30	14	0.408804	0.164416
80	40	30	15	0.591196	0.182392
80	40	30	16	0.755612	0.164416
80	40	30	17	0.875969	0.120357
80	40	30	18	0.947372	0.071402
80	40	30	19	0.981583	0.034211
80	40	30	20	0.994754	0.013171
80	40	30	21	0.998800	0.004046
80	40	30	22	0.999783	0.000983
80	40	30	23	0.999969	0.000186
80	40	30	24	0.999997	0.000027
80	40	30	25	1.000000	0.000003
80	40	30	26	1.000000	0.000000
80	40	30	27	1.000000	0.000000
80	40	30	28	1.000000	0.000000
80	40	30	29	1.000000	0.000000
80	40	30	30	1.000000	0.000000
80	40	31	0	0.000000	0.000000
80	40	31	1	0.000000	0.000000
80	40	31	2	0.000000	0.000000
80	40	31	3	0.000000	0.000000
80	40	31	4	0.000000	0.000000
80	40	31	5	0.000001	0.000001
80	40	31	6	0.000012	0.000011
80	40	31	7	0.000094	0.000082
80	40	31	8	0.000571	0.000477
80	40	31	9	0.002737	0.002167
80	40	31	10	0.010515	0.007777
80	40	31	11	0.032786	0.022271
80	40	31	12	0.084045	0.051259
80	40	31	13	0.179395	0.095349
80	40	31	14	0.323307	0.143913
80	40	31	15	0.500000	0.176693
80	40	31	16	0.676693	0.176693
80	40	31	17	0.820605	0.143913
80	40	31	18	0.915955	0.095349
80	40	31	19	0.967214	0.051259
80	40	31	20	0.989485	0.022271
80	40	31	21	0.997262	0.007777
80	40	31	22	0.999429	0.002167
80	40	31	23	0.999906	0.000477
80	40	31	24	0.999988	0.000082
80	40	31	25	0.999999	0.000011
80	40	31	26	1.000000	0.000001
80	40	31	27	1.000000	0.000000
80	40	31	28	1.000000	0.000000
80	40	31	29	1.000000	0.000000
80	40	31	30	1.000000	0.000000
80	40	31	31	1.000000	0.000000
80	40	32	0	0.000000	0.000000
80	40	32	1	0.000000	0.000000
80	40	32	2	0.000000	0.000000
80	40	32	3	0.000000	0.000000
80	40	32	4	0.000000	0.000000
80	40	32	5	0.000000	0.000000
80	40	32	6	0.000004	0.000004
80	40	32	7	0.000039	0.000034

Table for $N = 2$, $n = 1$, through $N = 100$, $n = 50$

N	n	k	x	P(x)	p(x)
80	40	32	8	0.000259	0.000221
80	40	32	9	0.001367	0.001107
80	40	32	10	0.005753	0.004386
80	40	32	11	0.019605	0.013852
80	40	32	12	0.054754	0.035149
80	40	32	13	0.126855	0.072101
80	40	32	14	0.246946	0.120090
80	40	32	15	0.409851	0.162905
80	40	32	16	0.590149	0.180299
80	40	32	17	0.753054	0.162905
80	40	32	18	0.873145	0.120090
80	40	32	19	0.945246	0.072101
80	40	32	20	0.980395	0.035149
80	40	32	21	0.994247	0.013852
80	40	32	22	0.998633	0.004386
80	40	32	23	0.999741	0.001107
80	40	32	24	0.999961	0.000221
80	40	32	25	0.999995	0.000034
80	40	32	26	0.999999	0.000004
80	40	32	27	1.000000	0.000000
80	40	32	28	1.000000	0.000000
80	40	32	29	1.000000	0.000000
80	40	32	30	1.000000	0.000000
80	40	32	31	1.000000	0.000000
80	40	32	32	1.000000	0.000000
80	40	33	0	0.000000	0.000000
80	40	33	1	0.000000	0.000000
80	40	33	2	0.000000	0.000000
80	40	33	3	0.000000	0.000000
80	40	33	4	0.000000	0.000000
80	40	33	5	0.000000	0.000000
80	40	33	6	0.000002	0.000001
80	40	33	7	0.000015	0.000014
80	40	33	8	0.000112	0.000097
80	40	33	9	0.000651	0.000539
80	40	33	10	0.003012	0.002360
80	40	33	11	0.011236	0.008225
80	40	33	12	0.034251	0.023014
80	40	33	13	0.086298	0.052048
80	40	33	14	0.181897	0.095598
80	40	33	15	0.325004	0.143108
80	40	33	16	0.500000	0.174996
80	40	33	17	0.674996	0.174996
80	40	33	18	0.818103	0.143108
80	40	33	19	0.913701	0.095598
80	40	33	20	0.965749	0.052048
80	40	33	21	0.988764	0.023014
80	40	33	22	0.996988	0.008225
80	40	33	23	0.999348	0.002360
80	40	33	24	0.999888	0.000539

N	n	k	x	P(x)	p(x)
80	40	33	25	0.999985	0.000097
80	40	33	26	0.999998	0.000014
80	40	33	27	1.000000	0.000001
80	40	33	28	1.000000	0.000000
80	40	33	29	1.000000	0.000000
80	40	33	30	1.000000	0.000000
80	40	33	31	1.000000	0.000000
80	40	33	32	1.000000	0.000000
80	40	33	33	1.000000	0.000000
80	40	34	0	0.000000	0.000000
80	40	34	1	0.000000	0.000000
80	40	34	2	0.000000	0.000000
80	40	34	3	0.000000	0.000000
80	40	34	4	0.000000	0.000000
80	40	34	5	0.000000	0.000000
80	40	34	6	0.000001	0.000000
80	40	34	7	0.000006	0.000005
80	40	34	8	0.000046	0.000041
80	40	34	9	0.000296	0.000250
80	40	34	10	0.001505	0.001209
80	40	34	11	0.006161	0.004656
80	40	34	12	0.020540	0.014378
80	40	34	13	0.056399	0.035859
80	40	34	14	0.129013	0.072614
80	40	34	15	0.248883	0.119871
80	40	34	16	0.410640	0.161757
80	40	34	17	0.589359	0.178719
80	40	34	18	0.751117	0.161757
80	40	34	19	0.870987	0.119871
80	40	34	20	0.943601	0.072614
80	40	34	21	0.979460	0.035859
80	40	34	22	0.993838	0.014378
80	40	34	23	0.998495	0.004656
80	40	34	24	0.999704	0.001209
80	40	34	25	0.999954	0.000250
80	40	34	26	0.999994	0.000041
80	40	34	27	0.999999	0.000005
80	40	34	28	1.000000	0.000000
80	40	34	29	1.000000	0.000000
80	40	34	30	1.000000	0.000000
80	40	34	31	1.000000	0.000000
80	40	34	32	1.000000	0.000000
80	40	34	33	1.000000	0.000000
80	40	34	34	1.000000	0.000000
80	40	35	0	0.000000	0.000000
80	40	35	1	0.000000	0.000000
80	40	35	2	0.000000	0.000000
80	40	35	3	0.000000	0.000000
80	40	35	4	0.000000	0.000000
80	40	35	5	0.000000	0.000000

N	n	k	x	P(x)	p(x)
80	40	35	6	0.000000	0.000000
80	40	35	7	0.000002	0.000002
80	40	35	8	0.000018	0.000016
80	40	35	9	0.000128	0.000110
80	40	35	10	0.000716	0.000589
80	40	35	11	0.003226	0.002509
80	40	35	12	0.011788	0.008562
80	40	35	13	0.035351	0.023563
80	40	35	14	0.087970	0.052619
80	40	35	15	0.183736	0.095766
80	40	35	16	0.326245	0.142509
80	40	35	17	0.500000	0.173755
80	40	35	18	0.673755	0.173755
80	40	35	19	0.816264	0.142509
80	40	35	20	0.912030	0.095766
80	40	35	21	0.964649	0.052619
80	40	35	22	0.988212	0.023563
80	40	35	23	0.996774	0.008562
80	40	35	24	0.999283	0.002509
80	40	35	25	0.999872	0.000589
80	40	35	26	0.999982	0.000110
80	40	35	27	0.999998	0.000016
80	40	35	28	1.000000	0.000002
80	40	35	29	1.000000	0.000000
80	40	35	30	1.000000	0.000000
80	40	35	31	1.000000	0.000000
80	40	35	32	1.000000	0.000000
80	40	35	33	1.000000	0.000000
80	40	35	34	1.000000	0.000000
80	40	35	35	1.000000	0.000000
80	40	36	0	0.000000	0.000000
80	40	36	1	0.000000	0.000000
80	40	36	2	0.000000	0.000000
80	40	36	3	0.000000	0.000000
80	40	36	4	0.000000	0.000000
80	40	36	5	0.000000	0.000000
80	40	36	6	0.000000	0.000000
80	40	36	7	0.000001	0.000001
80	40	36	8	0.000007	0.000006
80	40	36	9	0.000052	0.000045
80	40	36	10	0.000324	0.000272
80	40	36	11	0.001609	0.001285
80	40	36	12	0.006460	0.004852
80	40	36	13	0.021213	0.014753
80	40	36	14	0.057568	0.036355
80	40	36	15	0.130533	0.072965
80	40	36	16	0.250240	0.119708
80	40	36	17	0.411192	0.160952
80	40	36	18	0.588808	0.177616
80	40	36	19	0.749759	0.160952

N	n	k	x	P(x)	p(x)
80	40	36	20	0.869467	0.119708
80	40	36	21	0.942432	0.072965
80	40	36	22	0.978787	0.036355
80	40	36	23	0.993539	0.014753
80	40	36	24	0.998391	0.004852
80	40	36	25	0.999676	0.001285
80	40	36	26	0.999948	0.000272
80	40	36	27	0.999993	0.000045
80	40	36	28	0.999999	0.000006
80	40	36	29	1.000000	0.000001
80	40	36	30	1.000000	0.000000
80	40	36	31	1.000000	0.000000
80	40	36	32	1.000000	0.000000
80	40	36	33	1.000000	0.000000
80	40	36	34	1.000000	0.000000
80	40	36	35	1.000000	0.000000
80	40	36	36	1.000000	0.000000
80	40	37	0	0.000000	0.000000
80	40	37	1	0.000000	0.000000
80	40	37	2	0.000000	0.000000
80	40	37	3	0.000000	0.000000
80	40	37	4	0.000000	0.000000
80	40	37	5	0.000000	0.000000
80	40	37	6	0.000000	0.000000
80	40	37	7	0.000000	0.000000
80	40	37	8	0.000002	0.000002
80	40	37	9	0.000020	0.000018
80	40	37	10	0.000139	0.000119
80	40	37	11	0.000762	0.000623
80	40	37	12	0.003373	0.002611
80	40	37	13	0.012160	0.008787
80	40	37	14	0.036085	0.023925
80	40	37	15	0.089075	0.052990
80	40	37	16	0.184945	0.095870
80	40	37	17	0.327058	0.142113
80	40	37	18	0.500000	0.172942
80	40	37	19	0.672942	0.172942
80	40	37	20	0.815055	0.142113
80	40	37	21	0.910924	0.095870
80	40	37	22	0.963914	0.052990
80	40	37	23	0.987840	0.023925
80	40	37	24	0.996627	0.008787
80	40	37	25	0.999238	0.002611
80	40	37	26	0.999861	0.000623
80	40	37	27	0.999980	0.000119
80	40	37	28	0.999998	0.000018
80	40	37	29	1.000000	0.000002
80	40	37	30	1.000000	0.000000
80	40	37	31	1.000000	0.000000
80	40	37	32	1.000000	0.000000

$N = 80$ $n = 40$

Table for $N = 2$, $n = 1$, through $N = 100$, $n = 50$

N	n	k	x	P(x)	p(x)
80	40	37	33	1.000000	0.000000
80	40	37	34	1.000000	0.000000
80	40	37	35	1.000000	0.000000
80	40	37	36	1.000000	0.000000
80	40	37	37	1.000000	0.000000
80	40	38	0	0.000000	0.000000
80	40	38	1	0.000000	0.000000
80	40	38	2	0.000000	0.000000
80	40	38	3	0.000000	0.000000
80	40	38	4	0.000000	0.000000
80	40	38	5	0.000000	0.000000
80	40	38	6	0.000000	0.000000
80	40	38	7	0.000000	0.000000
80	40	38	8	0.000001	0.000001
80	40	38	9	0.000007	0.000006
80	40	38	10	0.000056	0.000049
80	40	38	11	0.000341	0.000286
80	40	38	12	0.001673	0.001331
80	40	38	13	0.006643	0.004970
80	40	38	14	0.021619	0.014976
80	40	38	15	0.058267	0.036648
80	40	38	16	0.131436	0.073169
80	40	38	17	0.251044	0.119608
80	40	38	18	0.411518	0.160474
80	40	38	19	0.588482	0.176964
80	40	38	20	0.748956	0.160474
80	40	38	21	0.868563	0.119608
80	40	38	22	0.941733	0.073169
80	40	38	23	0.978381	0.036648
80	40	38	24	0.993357	0.014976
80	40	38	25	0.998327	0.004970
80	40	38	26	0.999658	0.001331
80	40	38	27	0.999944	0.000286
80	40	38	28	0.999993	0.000049
80	40	38	29	0.999999	0.000006
80	40	38	30	1.000000	0.000001
80	40	38	31	1.000000	0.000000
80	40	38	32	1.000000	0.000000
80	40	38	33	1.000000	0.000000
80	40	38	34	1.000000	0.000000
80	40	38	35	1.000000	0.000000
80	40	38	36	1.000000	0.000000
80	40	38	37	1.000000	0.000000
80	40	38	38	1.000000	0.000000
80	40	39	0	0.000000	0.000000
80	40	39	1	0.000000	0.000000
80	40	39	2	0.000000	0.000000
80	40	39	3	0.000000	0.000000
80	40	39	4	0.000000	0.000000
80	40	39	5	0.000000	0.000000

N	n	k	x	P(x)	p(x)
80	40	39	6	0.000000	0.000000
80	40	39	7	0.000000	0.000000
80	40	39	8	0.000000	0.000000
80	40	39	9	0.000002	0.000002
80	40	39	10	0.000021	0.000019
80	40	39	11	0.000144	0.000123
80	40	39	12	0.000785	0.000641
80	40	39	13	0.003448	0.002662
80	40	39	14	0.012348	0.008900
80	40	39	15	0.036453	0.024105
80	40	39	16	0.089626	0.053173
80	40	39	17	0.185545	0.095919
80	40	39	18	0.327460	0.141916
80	40	39	19	0.500000	0.172540
80	40	39	20	0.672540	0.172540
80	40	39	21	0.814455	0.141916
80	40	39	22	0.910374	0.095919
80	40	39	23	0.963547	0.053173
80	40	39	24	0.987652	0.024105
80	40	39	25	0.996552	0.008900
80	40	39	26	0.999215	0.002662
80	40	39	27	0.999856	0.000641
80	40	39	28	0.999979	0.000123
80	40	39	29	0.999997	0.000019
80	40	39	30	1.000000	0.000002
80	40	39	31	1.000000	0.000000
80	40	39	32	1.000000	0.000000
80	40	39	33	1.000000	0.000000
80	40	39	34	1.000000	0.000000
80	40	39	35	1.000000	0.000000
80	40	39	36	1.000000	0.000000
80	40	39	37	1.000000	0.000000
80	40	39	38	1.000000	0.000000
80	40	39	39	1.000000	0.000000
80	40	40	0	0.000000	0.000000
80	40	40	1	0.000000	0.000000
80	40	40	2	0.000000	0.000000
80	40	40	3	0.000000	0.000000
80	40	40	4	0.000000	0.000000
80	40	40	5	0.000000	0.000000
80	40	40	6	0.000000	0.000000
80	40	40	7	0.000000	0.000000
80	40	40	8	0.000000	0.000000
80	40	40	9	0.000001	0.000001
80	40	40	10	0.000007	0.000007
80	40	40	11	0.000057	0.000050
80	40	40	12	0.000347	0.000290
80	40	40	13	0.001694	0.001347
80	40	40	14	0.006704	0.005010
80	40	40	15	0.021755	0.015051

N	n	k	x	P(x)	p(x)
80	40	40	16	0.058500	0.036745
80	40	40	17	0.131736	0.073236
80	40	40	18	0.251311	0.119574
80	40	40	19	0.411626	0.160315
80	40	40	20	0.588374	0.176748
80	40	40	21	0.748689	0.160315
80	40	40	22	0.868263	0.119574
80	40	40	23	0.941500	0.073236
80	40	40	24	0.978245	0.036745
80	40	40	25	0.993296	0.015051
80	40	40	26	0.998306	0.005010
80	40	40	27	0.999652	0.001347
80	40	40	28	0.999943	0.000290
80	40	40	29	0.999992	0.000050
80	40	40	30	0.999999	0.000007
80	40	40	31	1.000000	0.000001
80	40	40	32	1.000000	0.000000
80	40	40	33	1.000000	0.000000
80	40	40	34	1.000000	0.000000
80	40	40	35	1.000000	0.000000
80	40	40	36	1.000000	0.000000
80	40	40	37	1.000000	0.000000
80	40	40	38	1.000000	0.000000
80	40	40	39	1.000000	0.000000
80	40	40	40	1.000000	0.000000
90	1	1	0	0.988889	0.988889
90	1	1	1	1.000000	0.011111
90	2	1	0	0.977778	0.977778
90	2	1	1	1.000000	0.022222
90	2	2	0	0.955805	0.955805
90	2	2	1	0.999750	0.043945
90	2	2	2	1.000000	0.000250
90	3	1	0	0.966667	0.966667
90	3	1	1	1.000000	0.033333
90	3	2	0	0.934082	0.934082
90	3	2	1	0.999251	0.065169
90	3	2	2	1.000000	0.000749
90	3	3	0	0.902239	0.902239
90	3	3	1	0.997770	0.095531
90	3	3	2	0.999991	0.002222
90	3	3	3	1.000000	0.000009
90	4	1	0	0.955556	0.955556
90	4	1	1	1.000000	0.044444
90	4	2	0	0.912609	0.912609
90	4	2	1	0.998502	0.085893
90	4	2	2	1.000000	0.001498
90	4	3	0	0.871127	0.871127
90	4	3	1	0.995574	0.124447
90	4	3	2	0.999966	0.004392
90	4	3	3	1.000000	0.000034

N	n	k	x	P(x)	p(x)
90	4	4	0	0.831075	0.831075
90	4	4	1	0.991282	0.160207
90	4	4	2	0.999865	0.008583
90	4	4	3	1.000000	0.000135
90	4	4	4	1.000000	0.000000
90	5	1	0	0.944444	0.944444
90	5	1	1	1.000000	0.055556
90	5	2	0	0.891386	0.891386
90	5	2	1	0.997503	0.106117
90	5	2	2	1.000000	0.002497
90	5	3	0	0.840739	0.840739
90	5	3	1	0.992680	0.151941
90	5	3	2	0.999915	0.007235
90	5	3	3	1.000000	0.000085
90	5	4	0	0.792421	0.792421
90	5	4	1	0.985694	0.193273
90	5	4	2	0.999665	0.013972
90	5	4	3	0.999998	0.000333
90	5	4	4	1.000000	0.000002
90	5	5	0	0.746350	0.746350
90	5	5	1	0.976704	0.230355
90	5	5	2	0.999178	0.022474
90	5	5	3	0.999990	0.000812
90	5	5	4	1.000000	0.000010
90	5	5	5	1.000000	0.000000
90	6	1	0	0.933333	0.933333
90	6	1	1	1.000000	0.066667
90	6	2	0	0.870412	0.870412
90	6	2	1	0.996255	0.125843
90	6	2	2	1.000000	0.003745
90	6	3	0	0.811066	0.811066
90	6	3	1	0.989105	0.178039
90	6	3	2	0.999830	0.010725
90	6	3	3	1.000000	0.000170
90	6	4	0	0.755130	0.755130
90	6	4	1	0.978872	0.223742
90	6	4	2	0.999337	0.020464
90	6	4	3	0.999994	0.000657
90	6	4	4	1.000000	0.000006
90	6	5	0	0.702447	0.702447
90	6	5	1	0.965864	0.263417
90	6	5	2	0.998385	0.032521
90	6	5	3	0.999971	0.001586
90	6	5	4	1.000000	0.000029
90	6	5	5	1.000000	0.000000
90	6	6	0	0.652862	0.652862
90	6	6	1	0.950369	0.297507
90	6	6	2	0.996854	0.046485
90	6	6	3	0.999915	0.003061
90	6	6	4	0.999999	0.000084

N	n	k	x	P(x)	p(x)
90	6	6	5	1.000000	0.000001
90	6	6	6	1.000000	0.000000
90	7	1	0	0.922222	0.922222
90	7	1	1	1.000000	0.077778
90	7	2	0	0.849688	0.849688
90	7	2	1	0.994757	0.145069
90	7	2	2	1.000000	0.005243
90	7	3	0	0.782099	0.782099
90	7	3	1	0.984865	0.202766
90	7	3	2	0.999702	0.014837
90	7	3	3	1.000000	0.000298
90	7	4	0	0.719172	0.719172
90	7	4	1	0.970882	0.251710
90	7	4	2	0.998849	0.027968
90	7	4	3	0.999986	0.001137
90	7	4	4	1.000000	0.000014
90	7	5	0	0.660634	0.660634
90	7	5	1	0.953320	0.292686
90	7	5	2	0.997223	0.043903
90	7	5	3	0.999933	0.002710
90	7	5	4	0.999999	0.000066
90	7	5	5	1.000000	0.000000
90	7	6	0	0.606229	0.606229
90	7	6	1	0.932660	0.326431
90	7	6	2	0.994641	0.061981
90	7	6	3	0.999806	0.005165
90	7	6	4	0.999997	0.000191
90	7	6	5	1.000000	0.000003
90	7	6	6	1.000000	0.000000
90	7	7	0	0.555710	0.555710
90	7	7	1	0.909344	0.353634
90	7	7	2	0.990951	0.081608
90	7	7	3	0.999560	0.008608
90	7	7	4	0.999990	0.000430
90	7	7	5	1.000000	0.000010
90	7	7	6	1.000000	0.000000
90	7	7	7	1.000000	0.000000
90	8	1	0	0.911111	0.911111
90	8	1	1	1.000000	0.088889
90	8	2	0	0.829213	0.829213
90	8	2	1	0.993009	0.163795
90	8	2	2	1.000000	0.006991
90	8	3	0	0.753830	0.753830
90	8	3	1	0.979980	0.226149
90	8	3	2	0.999523	0.019544
90	8	3	3	1.000000	0.000477
90	8	4	0	0.684513	0.684513
90	8	4	1	0.961784	0.277271
90	8	4	2	0.998175	0.036392
90	8	4	3	0.999973	0.001797
90	8	4	4	1.000000	0.000027
90	8	5	0	0.620837	0.620837
90	8	5	1	0.939215	0.318378
90	8	5	2	0.995636	0.056421
90	8	5	3	0.999868	0.004232
90	8	5	4	0.999999	0.000131
90	8	5	5	1.000000	0.000001
90	8	6	0	0.562405	0.562405
90	8	6	1	0.912996	0.350590
90	8	6	2	0.991654	0.078658
90	8	6	3	0.999619	0.007965
90	8	6	4	0.999992	0.000373
90	8	6	5	1.000000	0.000007
90	8	6	6	1.000000	0.000000
90	8	7	0	0.508843	0.508843
90	8	7	1	0.883780	0.374937
90	8	7	2	0.986035	0.102256
90	8	7	3	0.999145	0.013110
90	8	7	4	0.999975	0.000830
90	8	7	5	1.000000	0.000025
90	8	7	6	1.000000	0.000000
90	8	7	7	1.000000	0.000000
90	8	8	0	0.459798	0.459798
90	8	8	1	0.852159	0.392361
90	8	8	2	0.978643	0.126485
90	8	8	3	0.998355	0.019712
90	8	8	4	0.999935	0.001579
90	8	8	5	0.999999	0.000064
90	8	8	6	1.000000	0.000001
90	8	8	7	1.000000	0.000000
90	8	8	8	1.000000	0.000000
90	9	1	0	0.900000	0.900000
90	9	1	1	1.000000	0.100000
90	9	2	0	0.808989	0.808989
90	9	2	1	0.991011	0.182022
90	9	2	2	1.000000	0.008989
90	9	3	0	0.726251	0.726251
90	9	3	1	0.974464	0.248212
90	9	3	2	0.999285	0.024821
90	9	3	3	1.000000	0.000715
90	9	4	0	0.651122	0.651122
90	9	4	1	0.951640	0.300518
90	9	4	2	0.997288	0.045648
90	9	4	3	0.999951	0.002663
90	9	4	4	1.000000	0.000049
90	9	5	0	0.582981	0.582981
90	9	5	1	0.923684	0.340703
90	9	5	2	0.993572	0.069888
90	9	5	3	0.999765	0.006193
90	9	5	4	0.999997	0.000232
90	9	5	5	1.000000	0.000003
90	9	6	0	0.521254	0.521254
90	9	6	1	0.891618	0.370365
90	9	6	2	0.987817	0.096199
90	9	6	3	0.999328	0.011511
90	9	6	4	0.999983	0.000656
90	9	6	5	1.000000	0.000016
90	9	6	6	1.000000	0.000000
90	9	7	0	0.465405	0.465405
90	9	7	1	0.856345	0.390940
90	9	7	2	0.979800	0.123455
90	9	7	3	0.998506	0.018705
90	9	7	4	0.999944	0.001439
90	9	7	5	0.999999	0.000055
90	9	7	6	1.000000	0.000001
90	9	7	7	1.000000	0.000000
90	9	8	0	0.414940	0.414940
90	9	8	1	0.818664	0.403725
90	9	8	2	0.969388	0.150724
90	9	8	3	0.997153	0.027765
90	9	8	4	0.999858	0.002704
90	9	8	5	0.999996	0.000139
90	9	8	6	1.000000	0.000004
90	9	8	7	1.000000	0.000000
90	9	8	8	1.000000	0.000000
90	9	9	0	0.369397	0.369397
90	9	9	1	0.779277	0.409879
90	9	9	2	0.956522	0.177245
90	9	9	3	0.995122	0.038600
90	9	9	4	0.999693	0.004571
90	9	9	5	0.999990	0.000297
90	9	9	6	1.000000	0.000010
90	9	9	7	1.000000	0.000000
90	9	9	8	1.000000	0.000000
90	9	9	9	1.000000	0.000000
90	10	1	0	0.888889	0.888889
90	10	1	1	1.000000	0.111111
90	10	2	0	0.789014	0.789014
90	10	2	1	0.988764	0.199750
90	10	2	2	1.000000	0.011236
90	10	3	0	0.699353	0.699353
90	10	3	1	0.968335	0.268982
90	10	3	2	0.998978	0.030644
90	10	3	3	1.000000	0.001021
90	10	4	0	0.618968	0.618968
90	10	4	1	0.940509	0.321542
90	10	4	2	0.996161	0.055651
90	10	4	3	0.999918	0.003757
90	10	4	4	1.000000	0.000082
90	10	5	0	0.546995	0.546995
90	10	5	1	0.906860	0.359865
90	10	5	2	0.990984	0.084124
90	10	5	3	0.999612	0.008628
90	10	5	4	0.999994	0.000382
90	10	5	5	1.000000	0.000006
90	10	6	0	0.482642	0.482642
90	10	6	1	0.868756	0.386114
90	10	6	2	0.983066	0.114310
90	10	6	3	0.998901	0.015835
90	10	6	4	0.999967	0.001066
90	10	6	5	1.000000	0.000032
90	10	6	6	1.000000	0.000000
90	10	7	0	0.425185	0.425185
90	10	7	1	0.827387	0.402202
90	10	7	2	0.972180	0.144793
90	10	7	3	0.997582	0.025402
90	10	7	4	0.999891	0.002309
90	10	7	5	0.999998	0.000107
90	10	7	6	1.000000	0.000002
90	10	7	7	1.000000	0.000000
90	10	8	0	0.373958	0.373958
90	10	8	1	0.783775	0.409817
90	10	8	2	0.958224	0.174449
90	10	8	3	0.995439	0.037216
90	10	8	4	0.999724	0.004285
90	10	8	5	0.999991	0.000267
90	10	8	6	1.000000	0.000009
90	10	8	7	1.000000	0.000000
90	10	8	8	1.000000	0.000000
90	10	9	0	0.328353	0.328353
90	10	9	1	0.738795	0.410442
90	10	9	2	0.941204	0.202410
90	10	9	3	0.992263	0.051058
90	10	9	4	0.999411	0.007148
90	10	9	5	0.999975	0.000564
90	10	9	6	0.999999	0.000024
90	10	9	7	1.000000	0.000001
90	10	9	8	1.000000	0.000000
90	10	9	9	1.000000	0.000000
90	10	10	0	0.287816	0.287816
90	10	10	1	0.693190	0.405374
90	10	10	2	0.921213	0.228023
90	10	10	3	0.987850	0.066637
90	10	10	4	0.998881	0.011031
90	10	10	5	0.999940	0.001059
90	10	10	6	0.999998	0.000058
90	10	10	7	1.000000	0.000002
90	10	10	8	1.000000	0.000000
90	10	10	9	1.000000	0.000000
90	10	10	10	1.000000	0.000000

Table for $N = 2$, $n = 1$, through $N = 100$, $n = 50$

N	n	k	x	P(x)	p(x)
90	11	1	0	0.877778	0.877778
90	11	1	1	1.000000	0.122222
90	11	2	0	0.769288	0.769288
90	11	2	1	0.986267	0.216979
90	11	2	2	1.000000	0.013733
90	11	3	0	0.673127	0.673127
90	11	3	1	0.961610	0.288483
90	11	3	2	0.998595	0.036985
90	11	3	3	1.000000	0.001404
90	11	4	0	0.588019	0.588019
90	11	4	1	0.928451	0.340432
90	11	4	2	0.994769	0.066318
90	11	4	3	0.999871	0.005101
90	11	4	4	1.000000	0.000129
90	11	5	0	0.512807	0.512807
90	11	5	1	0.888866	0.376059
90	11	5	2	0.987829	0.098963
90	11	5	3	0.999396	0.011567
90	11	5	4	0.999989	0.000593
90	11	5	5	1.000000	0.000011
90	11	6	0	0.446444	0.446444
90	11	6	1	0.844624	0.398180
90	11	6	2	0.977351	0.132727
90	11	6	3	0.998308	0.020957
90	11	6	4	0.999941	0.001633
90	11	6	5	0.999999	0.000059
90	11	6	6	1.000000	0.000001
90	11	7	0	0.387981	0.387981
90	11	7	1	0.797222	0.409240
90	11	7	2	0.963130	0.165908
90	11	7	3	0.996312	0.033182
90	11	7	4	0.999804	0.003493
90	11	7	5	0.999995	0.000191
90	11	7	6	1.000000	0.000005
90	11	7	7	1.000000	0.000000
90	11	8	0	0.336562	0.336562
90	11	8	1	0.747916	0.411354
90	11	8	2	0.945140	0.197224
90	11	8	3	0.993113	0.047973
90	11	8	4	0.999510	0.006396
90	11	8	5	0.999981	0.000471
90	11	8	6	1.000000	0.000018
90	11	8	7	1.000000	0.000000
90	11	8	8	1.000000	0.000000
90	11	9	0	0.291413	0.291413
90	11	9	1	0.697751	0.406337
90	11	9	2	0.923493	0.225743
90	11	9	3	0.988433	0.064940
90	11	9	4	0.998964	0.010531
90	11	9	5	0.999947	0.000983
90	11	9	6	0.999998	0.000052
90	11	9	7	1.000000	0.000001
90	11	9	8	1.000000	0.000000
90	11	9	9	1.000000	0.000000
90	11	10	0	0.251839	0.251839
90	11	10	1	0.647585	0.395747
90	11	10	2	0.898411	0.250825
90	11	10	3	0.982019	0.083608
90	11	10	4	0.998054	0.016034
90	11	10	5	0.999874	0.001820
90	11	10	6	0.999995	0.000121
90	11	10	7	1.000000	0.000005
90	11	10	8	1.000000	0.000000
90	11	10	9	1.000000	0.000000
90	11	10	10	1.000000	0.000000
90	11	11	0	0.217211	0.217211
90	11	11	1	0.598117	0.380906
90	11	11	2	0.870193	0.272076
90	11	11	3	0.973658	0.103465
90	11	11	4	0.996651	0.022992
90	11	11	5	0.999737	0.003087
90	11	11	6	0.999988	0.000250
90	11	11	7	1.000000	0.000012
90	11	11	8	1.000000	0.000000
90	11	11	9	1.000000	0.000000
90	11	11	10	1.000000	0.000000
90	11	11	11	1.000000	0.000000
90	12	1	0	0.866667	0.866667
90	12	1	1	1.000000	0.133333
90	12	2	0	0.749813	0.749813
90	12	2	1	0.983521	0.233708
90	12	2	2	1.000000	0.016479
90	12	3	0	0.647566	0.647566
90	12	3	1	0.954307	0.306742
90	12	3	2	0.998127	0.043820
90	12	3	3	1.000000	0.001873
90	12	4	0	0.558246	0.558246
90	12	4	1	0.915524	0.357278
90	12	4	2	0.993090	0.077567
90	12	4	3	0.999806	0.006716
90	12	4	4	1.000000	0.000194
90	12	5	0	0.480351	0.480351
90	12	5	1	0.869825	0.389474
90	12	5	2	0.984071	0.114246
90	12	5	3	0.999103	0.015032
90	12	5	4	0.999982	0.000879
90	12	5	5	1.000000	0.000018
90	12	6	0	0.412537	0.412537
90	12	6	1	0.819423	0.406886
90	12	6	2	0.970630	0.151208
90	12	6	3	0.997512	0.026881
90	12	6	4	0.999899	0.002387
90	12	6	5	0.999998	0.000099
90	12	6	6	1.000000	0.000001
90	12	7	0	0.353603	0.353603
90	12	7	1	0.766140	0.412537
90	12	7	2	0.952630	0.186489
90	12	7	3	0.994632	0.042002
90	12	7	4	0.999672	0.005040
90	12	7	5	0.999990	0.000318
90	12	7	6	1.000000	0.000010
90	12	7	7	1.000000	0.000000
90	12	8	0	0.302480	0.302480
90	12	8	1	0.711467	0.408987
90	12	8	2	0.930161	0.218694
90	12	8	3	0.990077	0.059916
90	12	8	4	0.999186	0.009109
90	12	8	5	0.999963	0.000777
90	12	8	6	0.999999	0.000036
90	12	8	7	1.000000	0.000001
90	12	8	8	1.000000	0.000000
90	12	9	0	0.258214	0.258214
90	12	9	1	0.656603	0.398388
90	12	9	2	0.903491	0.246888
90	12	9	3	0.983501	0.080010
90	12	9	4	0.998297	0.014796
90	12	9	5	0.999897	0.001600
90	12	9	6	0.999997	0.000100
90	12	9	7	1.000000	0.000003
90	12	9	8	1.000000	0.000000
90	12	9	9	1.000000	0.000000
90	12	10	0	0.219960	0.219960
90	12	10	1	0.602500	0.382540
90	12	10	2	0.873011	0.270510
90	12	10	3	0.974611	0.101600
90	12	10	4	0.996836	0.022225
90	12	10	5	0.999759	0.002923
90	12	10	6	0.999989	0.000230
90	12	10	7	1.000000	0.000011
90	12	10	8	1.000000	0.000000
90	12	10	9	1.000000	0.000000
90	12	10	10	1.000000	0.000000
90	12	11	0	0.186966	0.186966
90	12	11	1	0.549901	0.362935
90	12	11	2	0.839197	0.289296
90	12	11	3	0.963181	0.123984
90	12	11	4	0.994613	0.031433
90	12	11	5	0.999503	0.004890
90	12	11	6	0.999972	0.000469
90	12	11	7	0.999999	0.000027
90	12	11	8	1.000000	0.000001
90	12	11	9	1.000000	0.000000
90	12	11	10	1.000000	0.000000
90	12	11	11	1.000000	0.000000
90	12	12	0	0.158566	0.158566
90	12	12	1	0.499366	0.340800
90	12	12	2	0.802577	0.303211
90	12	12	3	0.949056	0.146479
90	12	12	4	0.991430	0.042374
90	12	12	5	0.999070	0.007639
90	12	12	6	0.999936	0.000866
90	12	12	7	0.999997	0.000061
90	12	12	8	1.000000	0.000003
90	12	12	9	1.000000	0.000000
90	12	12	10	1.000000	0.000000
90	12	12	11	1.000000	0.000000
90	12	12	12	1.000000	0.000000
90	13	1	0	0.855556	0.855556
90	13	1	1	1.000000	0.144444
90	13	2	0	0.730587	0.730587
90	13	2	1	0.980524	0.249938
90	13	2	2	1.000000	0.019476
90	13	3	0	0.622659	0.622659
90	13	3	1	0.946442	0.323783
90	13	3	2	0.997566	0.051124
90	13	3	3	1.000000	0.002434
90	13	4	0	0.529618	0.529618
90	13	4	1	0.901782	0.372164
90	13	4	2	0.991102	0.089319
90	13	4	3	0.999720	0.008619
90	13	4	4	1.000000	0.000280
90	13	5	0	0.449560	0.449560
90	13	5	1	0.849852	0.400293
90	13	5	2	0.979677	0.129825
90	13	5	3	0.998718	0.019041
90	13	5	4	0.999971	0.001253
90	13	5	5	1.000000	0.000029
90	13	6	0	0.380803	0.380803
90	13	6	1	0.793340	0.412537
90	13	6	2	0.962876	0.169536
90	13	6	3	0.996478	0.033602
90	13	6	4	0.999838	0.003360
90	13	6	5	0.999997	0.000159
90	13	6	6	1.000000	0.000003
90	13	7	0	0.321870	0.321870
90	13	7	1	0.734407	0.412537
90	13	7	2	0.940675	0.206269
90	13	7	3	0.992478	0.051803
90	13	7	4	0.999478	0.007000
90	13	7	5	0.999982	0.000504

N	n	k	x	P(x)	p(x)
90	13	7	6	1.000000	0.000018
90	13	7	7	1.000000	0.000000
90	13	8	0	0.271456	0.271456
90	13	8	1	0.674763	0.403306
90	13	8	2	0.913338	0.238576
90	13	8	3	0.986236	0.072898
90	13	8	4	0.998719	0.012483
90	13	8	5	0.999933	0.001215
90	13	8	6	0.999998	0.000065
90	13	8	7	1.000000	0.000002
90	13	8	8	1.000000	0.000000
90	13	9	0	0.228420	0.228420
90	13	9	1	0.615742	0.387322
90	13	9	2	0.881334	0.265592
90	13	9	3	0.977346	0.096012
90	13	9	4	0.997349	0.020003
90	13	9	5	0.999815	0.002466
90	13	9	6	0.999993	0.000178
90	13	9	7	1.000000	0.000007
90	13	9	8	1.000000	0.000000
90	13	9	9	1.000000	0.000000
90	13	10	0	0.191760	0.191760
90	13	10	1	0.558361	0.366601
90	13	10	2	0.845266	0.286905
90	13	10	3	0.965493	0.120227
90	13	10	4	0.995126	0.029633
90	13	10	5	0.999571	0.004445
90	13	10	6	0.999977	0.000406
90	13	10	7	0.999999	0.000022
90	13	10	8	1.000000	0.000001
90	13	10	9	1.000000	0.000000
90	13	10	10	1.000000	0.000000
90	13	11	0	0.160599	0.160599
90	13	11	1	0.503371	0.342772
90	13	11	2	0.805817	0.302446
90	13	11	3	0.950465	0.144648
90	13	11	4	0.991793	0.041328
90	13	11	5	0.999127	0.007334
90	13	11	6	0.999942	0.000815
90	13	11	7	0.999998	0.000056
90	13	11	8	1.000000	0.000002
90	13	11	9	1.000000	0.000000
90	13	11	10	1.000000	0.000000
90	13	11	11	1.000000	0.000000
90	13	12	0	0.134172	0.134172
90	13	12	1	0.451304	0.317133
90	13	12	2	0.763704	0.312400
90	13	12	3	0.932155	0.168451
90	13	12	4	0.987084	0.054930
90	13	12	5	0.998384	0.011300

N	n	k	x	P(x)	p(x)
90	13	12	6	0.999870	0.001485
90	13	12	7	0.999993	0.000124
90	13	12	8	1.000000	0.000006
90	13	12	9	1.000000	0.000000
90	13	12	10	1.000000	0.000000
90	13	12	11	1.000000	0.000000
90	13	12	12	1.000000	0.000000
90	13	13	0	0.111810	0.111810
90	13	13	1	0.402515	0.290705
90	13	13	2	0.719648	0.317133
90	13	13	3	0.910559	0.190911
90	13	13	4	0.980746	0.070188
90	13	13	5	0.997225	0.016479
90	13	13	6	0.999736	0.002511
90	13	13	7	0.999984	0.000248
90	13	13	8	0.999999	0.000015
90	13	13	9	1.000000	0.000001
90	13	13	10	1.000000	0.000000
90	13	13	11	1.000000	0.000000
90	13	13	12	1.000000	0.000000
90	13	13	13	1.000000	0.000000
90	14	1	0	0.844444	0.844444
90	14	1	1	1.000000	0.155556
90	14	2	0	0.711610	0.711610
90	14	2	1	0.977278	0.265668
90	14	2	2	1.000000	0.022722
90	14	3	0	0.598400	0.598400
90	14	3	1	0.938032	0.339632
90	14	3	2	0.996902	0.058870
90	14	3	3	1.000000	0.003098
90	14	4	0	0.502105	0.502105
90	14	4	1	0.887282	0.385177
90	14	4	2	0.988782	0.101499
90	14	4	3	0.999608	0.010827
90	14	4	4	1.000000	0.000392
90	14	5	0	0.420367	0.420367
90	14	5	1	0.829058	0.408691
90	14	5	2	0.974619	0.145561
90	14	5	3	0.998223	0.023604
90	14	5	4	0.999954	0.001731
90	14	5	5	1.000000	0.000046
90	14	6	0	0.351130	0.351130
90	14	6	1	0.766552	0.415422
90	14	6	2	0.954069	0.187517
90	14	6	3	0.995169	0.041100
90	14	6	4	0.999751	0.004582
90	14	6	5	0.999995	0.000244
90	14	6	6	1.000000	0.000005
90	14	7	0	0.292609	0.292609
90	14	7	1	0.702261	0.409652

N	n	k	x	P(x)	p(x)
90	14	7	2	0.927281	0.225020
90	14	7	3	0.989787	0.062506
90	14	7	4	0.999205	0.009419
90	14	7	5	0.999969	0.000764
90	14	7	6	0.999999	0.000031
90	14	7	7	1.000000	0.000000
90	14	8	0	0.243253	0.243253
90	14	8	1	0.638098	0.394845
90	14	8	2	0.894748	0.256650
90	14	8	3	0.981503	0.086755
90	14	8	4	0.998070	0.016568
90	14	8	5	0.999886	0.001816
90	14	8	6	0.999996	0.000110
90	14	8	7	1.000000	0.000003
90	14	8	8	1.000000	0.000000
90	14	9	0	0.201722	0.201722
90	14	9	1	0.575501	0.373779
90	14	9	2	0.857189	0.281689
90	14	9	3	0.969865	0.112675
90	14	9	4	0.996050	0.026185
90	14	9	5	0.999687	0.003637
90	14	9	6	0.999986	0.000299
90	14	9	7	1.000000	0.000014
90	14	9	8	1.000000	0.000000
90	14	9	9	1.000000	0.000000
90	14	10	0	0.166856	0.166856
90	14	10	1	0.515512	0.348655
90	14	10	2	0.815458	0.299946
90	14	10	3	0.954563	0.139105
90	14	10	4	0.992817	0.038254
90	14	10	5	0.999283	0.006465
90	14	10	6	0.999956	0.000673
90	14	10	7	0.999998	0.000042
90	14	10	8	1.000000	0.000001
90	14	10	9	1.000000	0.000000
90	14	10	10	1.000000	0.000000
90	14	11	0	0.137657	0.137657
90	14	11	1	0.458855	0.321199
90	14	11	2	0.770466	0.311611
90	14	11	3	0.935436	0.164970
90	14	11	4	0.988036	0.052599
90	14	11	5	0.998555	0.010520
90	14	11	6	0.999889	0.001333
90	14	11	7	0.999995	0.000106
90	14	11	8	1.000000	0.000005
90	14	11	9	1.000000	0.000000
90	14	11	10	1.000000	0.000000
90	14	11	11	1.000000	0.000000
90	14	12	0	0.113262	0.113262
90	14	12	1	0.406000	0.292738

N	n	k	x	P(x)	p(x)
90	14	12	2	0.723133	0.317133
90	14	12	3	0.912466	0.189333
90	14	12	4	0.981377	0.068912
90	14	12	5	0.997357	0.015980
90	14	12	6	0.999754	0.002397
90	14	12	7	0.999985	0.000231
90	14	12	8	0.999999	0.000014
90	14	12	9	1.000000	0.000001
90	14	12	10	1.000000	0.000000
90	14	12	11	1.000000	0.000000
90	14	12	12	1.000000	0.000000
90	14	13	0	0.092933	0.092933
90	14	13	1	0.357210	0.264277
90	14	13	2	0.674343	0.317133
90	14	13	3	0.885765	0.211422
90	14	13	4	0.972543	0.086778
90	14	13	5	0.995513	0.022971
90	14	13	6	0.999508	0.003995
90	14	13	7	0.999965	0.000457
90	14	13	8	0.999998	0.000034
90	14	13	9	1.000000	0.000002
90	14	13	10	1.000000	0.000000
90	14	13	11	1.000000	0.000000
90	14	13	12	1.000000	0.000000
90	14	13	13	1.000000	0.000000
90	14	14	0	0.076036	0.076036
90	14	14	1	0.312592	0.236556
90	14	14	2	0.624920	0.312328
90	14	14	3	0.855562	0.230642
90	14	14	4	0.961273	0.105711
90	14	14	5	0.992828	0.031556
90	14	14	6	0.999093	0.006265
90	14	14	7	0.999923	0.000830
90	14	14	8	0.999996	0.000073
90	14	14	9	1.000000	0.000004
90	14	14	10	1.000000	0.000000
90	14	14	11	1.000000	0.000000
90	14	14	12	1.000000	0.000000
90	14	14	13	1.000000	0.000000
90	14	14	14	1.000000	0.000000
90	15	1	0	0.833333	0.833333
90	15	1	1	1.000000	0.166667
90	15	2	0	0.692884	0.692884
90	15	2	1	0.973783	0.280899
90	15	2	2	1.000000	0.026217
90	15	3	0	0.574779	0.574779
90	15	3	1	0.929094	0.354316
90	15	3	2	0.996127	0.067033
90	15	3	3	1.000000	0.003873
90	15	4	0	0.475679	0.475679

Table for $N = 2$, $n = 1$, through $N = 100$, $n = 50$

N	n	k	x	P(x)	p(x)
90	15	4	1	0.872078	0.396399
90	15	4	2	0.986111	0.114033
90	15	4	3	0.999466	0.013355
90	15	4	4	1.000000	0.000534
90	15	5	0	0.392712	0.392712
90	15	5	1	0.807548	0.414836
90	15	5	2	0.968873	0.161325
90	15	5	3	0.997602	0.028729
90	15	5	4	0.999932	0.002329
90	15	5	5	1.000000	0.000068
90	15	6	0	0.323410	0.323410
90	15	6	1	0.739222	0.415812
90	15	6	2	0.944200	0.204978
90	15	6	3	0.993546	0.049347
90	15	6	4	0.999630	0.006084
90	15	6	5	0.999992	0.000362
90	15	6	6	1.000000	0.000008
90	15	7	0	0.265658	0.265658
90	15	7	1	0.669920	0.404262
90	15	7	2	0.912477	0.242557
90	15	7	3	0.986497	0.074020
90	15	7	4	0.998833	0.012337
90	15	7	5	0.999949	0.001115
90	15	7	6	0.999999	0.000050
90	15	7	7	1.000000	0.000001
90	15	8	0	0.217647	0.217647
90	15	8	1	0.601731	0.384084
90	15	8	2	0.874486	0.272755
90	15	8	3	0.975795	0.101309
90	15	8	4	0.997199	0.021403
90	15	8	5	0.999814	0.002616
90	15	8	6	0.999994	0.000179
90	15	8	7	1.000000	0.000006
90	15	8	8	1.000000	0.000000
90	15	9	0	0.177834	0.177834
90	15	9	1	0.536156	0.358322
90	15	9	2	0.831244	0.295089
90	15	9	3	0.960969	0.129725
90	15	9	4	0.994327	0.033358
90	15	9	5	0.999495	0.005168
90	15	9	6	0.999974	0.000479
90	15	9	7	0.999999	0.000025
90	15	9	8	1.000000	0.000001
90	15	9	9	1.000000	0.000000
90	15	10	0	0.144902	0.144902
90	15	10	1	0.474224	0.329322
90	15	10	2	0.783885	0.309661
90	15	10	3	0.941751	0.157866
90	15	10	4	0.989797	0.048046
90	15	10	5	0.998857	0.009060

N	n	k	x	P(x)	p(x)
90	15	10	6	0.999921	0.001063
90	15	10	7	0.999997	0.000076
90	15	10	8	1.000000	0.000003
90	15	10	9	1.000000	0.000000
90	15	10	10	1.000000	0.000000
90	15	11	0	0.117733	0.117733
90	15	11	1	0.416592	0.298860
90	15	11	2	0.733565	0.316972
90	15	11	3	0.918071	0.184506
90	15	11	4	0.983191	0.065120
90	15	11	5	0.997725	0.014534
90	15	11	6	0.999801	0.002076
90	15	11	7	0.999989	0.000188
90	15	11	8	1.000000	0.000010
90	15	11	9	1.000000	0.000000
90	15	11	10	1.000000	0.000000
90	15	11	11	1.000000	0.000000
90	15	12	0	0.095378	0.095378
90	15	12	1	0.363630	0.268251
90	15	12	2	0.681405	0.317775
90	15	12	3	0.890045	0.208640
90	15	12	4	0.974123	0.084079
90	15	12	5	0.995885	0.021762
90	15	12	6	0.999565	0.003679
90	15	12	7	0.999970	0.000405
90	15	12	8	0.999999	0.000029
90	15	12	9	1.000000	0.000001
90	15	12	10	1.000000	0.000000
90	15	12	11	1.000000	0.000000
90	15	12	12	1.000000	0.000000
90	15	13	0	0.077036	0.077036
90	15	13	1	0.315482	0.238446
90	15	13	2	0.628442	0.312960
90	15	13	3	0.857946	0.229504
90	15	13	4	0.962266	0.104320
90	15	13	5	0.993095	0.030829
90	15	13	6	0.999140	0.006045
90	15	13	7	0.999929	0.000788
90	15	13	8	0.999996	0.000068
90	15	13	9	1.000000	0.000004
90	15	13	10	1.000000	0.000000
90	15	13	11	1.000000	0.000000
90	15	13	12	1.000000	0.000000
90	15	13	13	1.000000	0.000000
90	15	14	0	0.062029	0.062029
90	15	14	1	0.272128	0.210099
90	15	14	2	0.575605	0.303476
90	15	14	3	0.822179	0.246575
90	15	14	4	0.947363	0.125184
90	15	14	5	0.989091	0.041728

N	n	k	x	P(x)	p(x)
90	15	14	6	0.998434	0.009342
90	15	14	7	0.999847	0.001413
90	15	14	8	0.999990	0.000143
90	15	14	9	1.000000	0.000010
90	15	14	10	1.000000	0.000000
90	15	14	11	1.000000	0.000000
90	15	14	12	1.000000	0.000000
90	15	14	13	1.000000	0.000000
90	15	14	14	1.000000	0.000000
90	15	15	0	0.049787	0.049787
90	15	15	1	0.233426	0.183639
90	15	15	2	0.523694	0.290268
90	15	15	3	0.783247	0.259552
90	15	15	4	0.929245	0.145998
90	15	15	5	0.983601	0.054356
90	15	15	6	0.997327	0.013726
90	15	15	7	0.999698	0.002371
90	15	15	8	0.999977	0.000279
90	15	15	9	0.999999	0.000022
90	15	15	10	1.000000	0.000001
90	15	15	11	1.000000	0.000000
90	15	15	12	1.000000	0.000000
90	15	15	13	1.000000	0.000000
90	15	15	14	1.000000	0.000000
90	15	15	15	1.000000	0.000000
90	16	1	0	0.822222	0.822222
90	16	1	1	1.000000	0.177778
90	16	2	0	0.674407	0.674407
90	16	2	1	0.970037	0.295630
90	16	2	2	1.000000	0.029963
90	16	3	0	0.551788	0.551788
90	16	3	1	0.919646	0.367858
90	16	3	2	0.995233	0.075587
90	16	3	3	1.000000	0.004767
90	16	4	0	0.450309	0.450309
90	16	4	1	0.856222	0.405913
90	16	4	2	0.983070	0.126848
90	16	4	3	0.999288	0.016218
90	16	4	4	1.000000	0.000712
90	16	5	0	0.366531	0.366531
90	16	5	1	0.785423	0.418892
90	16	5	2	0.962420	0.176997
90	16	5	3	0.996836	0.034416
90	16	5	4	0.999901	0.003064
90	16	5	5	1.000000	0.000099
90	16	6	0	0.297537	0.297537
90	16	6	1	0.711501	0.413964
90	16	6	2	0.933268	0.221767
90	16	6	3	0.991572	0.058305
90	16	6	4	0.999468	0.007895

N	n	k	x	P(x)	p(x)
90	16	6	5	0.999987	0.000519
90	16	6	6	1.000000	0.000013
90	16	7	0	0.240863	0.240863
90	16	7	1	0.637579	0.396716
90	16	7	2	0.896307	0.258728
90	16	7	3	0.982549	0.086243
90	16	7	4	0.998340	0.015791
90	16	7	5	0.999919	0.001579
90	16	7	6	0.999998	0.000079
90	16	7	7	1.000000	0.000002
90	16	8	0	0.194432	0.194432
90	16	8	1	0.565883	0.371452
90	16	8	2	0.852666	0.286782
90	16	8	3	0.969041	0.116375
90	16	8	4	0.996057	0.027016
90	16	8	5	0.999710	0.003653
90	16	8	6	0.999989	0.000279
90	16	8	7	1.000000	0.000011
90	16	8	8	1.000000	0.000000
90	16	9	0	0.156494	0.156494
90	16	9	1	0.497935	0.341441
90	16	9	2	0.803703	0.305768
90	16	9	3	0.950591	0.146889
90	16	9	4	0.992103	0.041512
90	16	9	5	0.999220	0.007116
90	16	9	6	0.999955	0.000735
90	16	9	7	0.999999	0.000044
90	16	9	8	1.000000	0.000001
90	16	9	9	1.000000	0.000000
90	16	10	0	0.125581	0.125581
90	16	10	1	0.434705	0.309124
90	16	10	2	0.750854	0.316149
90	16	10	3	0.927017	0.176163
90	16	10	4	0.985953	0.058937
90	16	10	5	0.998253	0.012300
90	16	10	6	0.999864	0.001611
90	16	10	7	0.999994	0.000130
90	16	10	8	1.000000	0.000006
90	16	10	9	1.000000	0.000000
90	16	10	10	1.000000	0.000000
90	16	11	0	0.100465	0.100465
90	16	11	1	0.376744	0.276279
90	16	11	2	0.695528	0.318784
90	16	11	3	0.898390	0.202862
90	16	11	4	0.977113	0.078723
90	16	11	5	0.996562	0.019449
90	16	11	6	0.999663	0.003101
90	16	11	7	0.999979	0.000316
90	16	11	8	0.999999	0.000020
90	16	11	9	1.000000	0.000001

N	n	k	x	P(x)	p(x)
90	16	11	10	1.000000	0.000000
90	16	11	11	1.000000	0.000000
90	16	12	0	0.080118	0.080118
90	16	12	1	0.324286	0.244168
90	16	12	2	0.639035	0.314748
90	16	12	3	0.865008	0.225973
90	16	12	4	0.965155	0.100147
90	16	12	5	0.993854	0.028699
90	16	12	6	0.999270	0.005416
90	16	12	7	0.999943	0.000673
90	16	12	8	0.999997	0.000054
90	16	12	9	1.000000	0.000003
90	16	12	10	1.000000	0.000000
90	16	12	11	1.000000	0.000000
90	16	12	12	1.000000	0.000000
90	16	13	0	0.063683	0.063683
90	16	13	1	0.277331	0.213647
90	16	13	2	0.582541	0.305211
90	16	13	3	0.827346	0.244804
90	16	13	4	0.949748	0.122402
90	16	13	5	0.989807	0.040059
90	16	13	6	0.998576	0.008769
90	16	13	7	0.999865	0.001290
90	16	13	8	0.999991	0.000126
90	16	13	9	1.000000	0.000008
90	16	13	10	1.000000	0.000000
90	16	13	11	1.000000	0.000000
90	16	13	12	1.000000	0.000000
90	16	13	13	1.000000	0.000000
90	16	14	0	0.050450	0.050450
90	16	14	1	0.235711	0.185261
90	16	14	2	0.527049	0.291337
90	16	14	3	0.786015	0.258967
90	16	14	4	0.930672	0.144657
90	16	14	5	0.984084	0.053412
90	16	14	6	0.997437	0.013353
90	16	14	7	0.999715	0.002278
90	16	14	8	0.999978	0.000264
90	16	14	9	0.999999	0.000020
90	16	14	10	1.000000	0.000001
90	16	14	11	1.000000	0.000000
90	16	14	12	1.000000	0.000000
90	16	14	13	1.000000	0.000000
90	16	14	14	1.000000	0.000000
90	16	15	0	0.039829	0.039829
90	16	15	1	0.199147	0.159317
90	16	15	2	0.473381	0.274235
90	16	15	3	0.741718	0.268337
90	16	15	4	0.907832	0.166113
90	16	15	5	0.976353	0.068522

N	n	k	x	P(x)	p(x)
90	16	15	6	0.995680	0.019327
90	16	15	7	0.999445	0.003765
90	16	15	8	0.999951	0.000506
90	16	15	9	0.999997	0.000046
90	16	15	10	1.000000	0.000003
90	16	15	11	1.000000	0.000000
90	16	15	12	1.000000	0.000000
90	16	15	13	1.000000	0.000000
90	16	15	14	1.000000	0.000000
90	16	15	15	1.000000	0.000000
90	16	16	0	0.031332	0.031332
90	16	16	1	0.167283	0.135951
90	16	16	2	0.422191	0.254908
90	16	16	3	0.695206	0.273016
90	16	16	4	0.881253	0.186047
90	16	16	5	0.966304	0.085050
90	16	16	6	0.993103	0.026800
90	16	16	7	0.998993	0.005890
90	16	16	8	0.999897	0.000904
90	16	16	9	0.999993	0.000096
90	16	16	10	1.000000	0.000007
90	16	16	11	1.000000	0.000000
90	16	16	12	1.000000	0.000000
90	16	16	13	1.000000	0.000000
90	16	16	14	1.000000	0.000000
90	16	16	15	1.000000	0.000000
90	16	16	16	1.000000	0.000000
90	17	1	0	0.811111	0.811111
90	17	1	1	1.000000	0.188889
90	17	2	0	0.656180	0.656180
90	17	2	1	0.966042	0.309863
90	17	2	2	1.000000	0.033958
90	17	3	0	0.529418	0.529418
90	17	3	1	0.909704	0.380286
90	17	3	2	0.994212	0.084508
90	17	3	3	1.000000	0.005788
90	17	4	0	0.425968	0.425968
90	17	4	1	0.839766	0.413798
90	17	4	2	0.979641	0.139875
90	17	4	3	0.999069	0.019427
90	17	4	4	1.000000	0.000931
90	17	5	0	0.341765	0.341765
90	17	5	1	0.762780	0.421015
90	17	5	2	0.955244	0.192464
90	17	5	3	0.995906	0.040661
90	17	5	4	0.999859	0.003953
90	17	5	5	1.000000	0.000141
90	17	6	0	0.273412	0.273412
90	17	6	1	0.683531	0.410118
90	17	6	2	0.921280	0.237750

N	n	k	x	P(x)	p(x)
90	17	6	3	0.989209	0.067928
90	17	6	4	0.999254	0.010046
90	17	6	5	0.999980	0.000726
90	17	6	6	1.000000	0.000020
90	17	7	0	0.218079	0.218079
90	17	7	1	0.605413	0.387334
90	17	7	2	0.878825	0.273412
90	17	7	3	0.977887	0.099062
90	17	7	4	0.997700	0.019812
90	17	7	5	0.999876	0.002177
90	17	7	6	0.999997	0.000121
90	17	7	7	1.000000	0.000003
90	17	8	0	0.173412	0.173412
90	17	8	1	0.530746	0.357334
90	17	8	2	0.829413	0.298667
90	17	8	3	0.961178	0.131765
90	17	8	4	0.994597	0.033419
90	17	8	5	0.999562	0.004965
90	17	8	6	0.999981	0.000420
90	17	8	7	1.000000	0.000018
90	17	8	8	1.000000	0.000000
90	17	9	0	0.137461	0.137461
90	17	9	1	0.461022	0.323561
90	17	9	2	0.774779	0.313757
90	17	9	3	0.938682	0.163903
90	17	9	4	0.989299	0.050617
90	17	9	5	0.998835	0.009537
90	17	9	6	0.999925	0.001090
90	17	9	7	0.999997	0.000072
90	17	9	8	1.000000	0.000003
90	17	9	9	1.000000	0.000000
90	17	10	0	0.108611	0.108611
90	17	10	1	0.397109	0.288498
90	17	10	2	0.716676	0.319567
90	17	10	3	0.910353	0.193677
90	17	10	4	0.981175	0.070822
90	17	10	5	0.997422	0.016247
90	17	10	6	0.999777	0.002355
90	17	10	7	0.999988	0.000211
90	17	10	8	1.000000	0.000011
90	17	10	9	1.000000	0.000000
90	17	10	10	1.000000	0.000000
90	17	11	0	0.085531	0.085531
90	17	11	1	0.339409	0.253878
90	17	11	2	0.656757	0.317348
90	17	11	3	0.876459	0.219702
90	17	11	4	0.969666	0.093207
90	17	11	5	0.994985	0.025319
90	17	11	6	0.999453	0.004468
90	17	11	7	0.999962	0.000509

N	n	k	x	P(x)	p(x)
90	17	11	8	0.999998	0.000036
90	17	11	9	1.000000	0.000002
90	17	11	10	1.000000	0.000000
90	17	11	11	1.000000	0.000000
90	17	12	0	0.067126	0.067126
90	17	12	1	0.287991	0.220865
90	17	12	2	0.596501	0.308510
90	17	12	3	0.837525	0.241024
90	17	12	4	0.954328	0.116804
90	17	12	5	0.991139	0.036811
90	17	12	6	0.998831	0.007692
90	17	12	7	0.999898	0.001067
90	17	12	8	0.999994	0.000097
90	17	12	9	1.000000	0.000006
90	17	12	10	1.000000	0.000000
90	17	12	11	1.000000	0.000000
90	17	12	12	1.000000	0.000000
90	17	13	0	0.052496	0.052496
90	17	13	1	0.242685	0.190190
90	17	13	2	0.537172	0.294487
90	17	13	3	0.794264	0.257092
90	17	13	4	0.934861	0.140597
90	17	13	5	0.985476	0.050615
90	17	13	6	0.997746	0.012270
90	17	13	7	0.999761	0.002015
90	17	13	8	0.999983	0.000222
90	17	13	9	0.999999	0.000016
90	17	13	10	1.000000	0.000001
90	17	13	11	1.000000	0.000000
90	17	13	12	1.000000	0.000000
90	17	13	13	1.000000	0.000000
90	17	14	0	0.040906	0.040906
90	17	14	1	0.203165	0.162260
90	17	14	2	0.479805	0.276639
90	17	14	3	0.747520	0.267715
90	17	14	4	0.911124	0.163604
90	17	14	5	0.977588	0.066464
90	17	14	6	0.995993	0.018405
90	17	14	7	0.999499	0.003506
90	17	14	8	0.999957	0.000458
90	17	14	9	0.999997	0.000040
90	17	14	10	1.000000	0.000002
90	17	14	11	1.000000	0.000000
90	17	14	12	1.000000	0.000000
90	17	14	13	1.000000	0.000000
90	17	14	14	1.000000	0.000000
90	17	15	0	0.031756	0.031756
90	17	15	1	0.169005	0.137250
90	17	15	2	0.425205	0.256199
90	17	15	3	0.698204	0.272999

N	n	k	x	P(x)	p(x)
90	17	15	4	0.883139	0.184935
90	17	15	5	0.967094	0.083955
90	17	15	6	0.993329	0.026236
90	17	15	7	0.999038	0.005708
90	17	15	8	0.999903	0.000865
90	17	15	9	0.999993	0.000090
90	17	15	10	1.000000	0.000006
90	17	15	11	1.000000	0.000000
90	17	15	12	1.000000	0.000000
90	17	15	13	1.000000	0.000000
90	17	15	14	1.000000	0.000000
90	17	15	15	1.000000	0.000000
90	17	16	0	0.024558	0.024558
90	17	16	1	0.139725	0.115168
90	17	16	2	0.373965	0.234239
90	17	16	3	0.647244	0.273279
90	17	16	4	0.851084	0.203839
90	17	16	5	0.953661	0.102577
90	17	16	6	0.989482	0.035821
90	17	16	7	0.998277	0.008795
90	17	16	8	0.999799	0.001522
90	17	16	9	0.999984	0.000185
90	17	16	10	0.999999	0.000015
90	17	16	11	1.000000	0.000001
90	17	16	12	1.000000	0.000000
90	17	16	13	1.000000	0.000000
90	17	16	14	1.000000	0.000000
90	17	16	15	1.000000	0.000000
90	17	16	16	1.000000	0.000000
90	17	17	0	0.018916	0.018916
90	17	17	1	0.114824	0.095908
90	17	17	2	0.326484	0.211660
90	17	17	3	0.595543	0.269059
90	17	17	4	0.815274	0.219731
90	17	17	5	0.937027	0.121753
90	17	17	6	0.984157	0.047130
90	17	17	7	0.997088	0.012931
90	17	17	8	0.999614	0.002526
90	17	17	9	0.999964	0.000350
90	17	17	10	0.999998	0.000034
90	17	17	11	1.000000	0.000002
90	17	17	12	1.000000	0.000000
90	17	17	13	1.000000	0.000000
90	17	17	14	1.000000	0.000000
90	17	17	15	1.000000	0.000000
90	17	17	16	1.000000	0.000000
90	17	17	17	1.000000	0.000000
90	18	1	0	0.800000	0.800000
90	18	1	1	1.000000	0.200000
90	18	2	0	0.638202	0.638202

N	n	k	x	P(x)	p(x)
90	18	2	1	0.961798	0.323595
90	18	2	2	1.000000	0.038202
90	18	3	0	0.507661	0.507661
90	18	3	1	0.899285	0.391624
90	18	3	2	0.993054	0.093769
90	18	3	3	1.000000	0.006946
90	18	4	0	0.402628	0.402628
90	18	4	1	0.822761	0.420133
90	18	4	2	0.975809	0.153048
90	18	4	3	0.998802	0.022993
90	18	4	4	1.000000	0.001198
90	18	5	0	0.318357	0.318357
90	18	5	1	0.739711	0.421354
90	18	5	2	0.947335	0.207624
90	18	5	3	0.994792	0.047457
90	18	5	4	0.999805	0.005013
90	18	5	5	1.000000	0.000195
90	18	6	0	0.250940	0.250940
90	18	6	1	0.655440	0.404500
90	18	6	2	0.908253	0.252813
90	18	6	3	0.986417	0.078164
90	18	6	4	0.998979	0.012562
90	18	6	5	0.999970	0.000991
90	18	6	6	1.000000	0.000030
90	18	7	0	0.197167	0.197167
90	18	7	1	0.573577	0.376410
90	18	7	2	0.860098	0.286521
90	18	7	3	0.972459	0.112361
90	18	7	4	0.996886	0.024426
90	18	7	5	0.999817	0.002931
90	18	7	6	0.999996	0.000179
90	18	7	7	1.000000	0.000004
90	18	8	0	0.154408	0.154408
90	18	8	1	0.496481	0.342073
90	18	8	2	0.804865	0.308384
90	18	8	3	0.952153	0.147288
90	18	8	4	0.992766	0.040612
90	18	8	5	0.999358	0.006592
90	18	8	6	0.999970	0.000612
90	18	8	7	0.999999	0.000030
90	18	8	8	1.000000	0.000001
90	18	9	0	0.120514	0.120514
90	18	9	1	0.425563	0.305050
90	18	9	2	0.744693	0.319129
90	18	9	3	0.925210	0.180517
90	18	9	4	0.985832	0.060622
90	18	9	5	0.998313	0.012481
90	18	9	6	0.999880	0.001568
90	18	9	7	0.999995	0.000115
90	18	9	8	1.000000	0.000004

N	n	k	x	P(x)	p(x)
90	18	9	9	1.000000	0.000000
90	18	10	0	0.093733	0.093733
90	18	10	1	0.361541	0.267808
90	18	10	2	0.681655	0.320114
90	18	10	3	0.891781	0.210126
90	18	10	4	0.975354	0.083573
90	18	10	5	0.996309	0.020956
90	18	10	6	0.999648	0.003339
90	18	10	7	0.999980	0.000332
90	18	10	8	0.999999	0.000020
90	18	10	9	1.000000	0.000001
90	18	10	10	1.000000	0.000000
90	18	11	0	0.072643	0.072643
90	18	11	1	0.304631	0.231989
90	18	11	2	0.617632	0.313000
90	18	11	3	0.852382	0.234750
90	18	11	4	0.960729	0.108346
90	18	11	5	0.992904	0.032176
90	18	11	6	0.999147	0.006243
90	18	11	7	0.999934	0.000787
90	18	11	8	0.999997	0.000063
90	18	11	9	1.000000	0.000003
90	18	11	10	1.000000	0.000000
90	18	11	11	1.000000	0.000000
90	18	12	0	0.056091	0.056091
90	18	12	1	0.254710	0.198619
90	18	12	2	0.554239	0.299530
90	18	12	3	0.807809	0.253570
90	18	12	4	0.941528	0.133719
90	18	12	5	0.987609	0.046081
90	18	12	6	0.998199	0.010589
90	18	12	7	0.999824	0.001626
90	18	12	8	0.999989	0.000164
90	18	12	9	0.999999	0.000011
90	18	12	10	1.000000	0.000000
90	18	12	11	1.000000	0.000000
90	18	12	12	1.000000	0.000000
90	18	13	0	0.043147	0.043147
90	18	13	1	0.211421	0.168274
90	18	13	2	0.492797	0.281376
90	18	13	3	0.759046	0.266248
90	18	13	4	0.917527	0.158481
90	18	13	5	0.979929	0.062402
90	18	13	6	0.996570	0.016641
90	18	13	7	0.999595	0.003026
90	18	13	8	0.999968	0.000373
90	18	13	9	0.999998	0.000030
90	18	13	10	1.000000	0.000002
90	18	13	11	1.000000	0.000000
90	18	13	12	1.000000	0.000000

N	n	k	x	P(x)	p(x)
90	18	13	13	1.000000	0.000000
90	18	14	0	0.033061	0.033061
90	18	14	1	0.174270	0.141209
90	18	14	2	0.434330	0.260060
90	18	14	3	0.707179	0.272850
90	18	14	4	0.888712	0.181533
90	18	14	5	0.969394	0.080681
90	18	14	6	0.993976	0.024583
90	18	14	7	0.999163	0.005187
90	18	14	8	0.999919	0.000756
90	18	14	9	0.999995	0.000075
90	18	14	10	1.000000	0.000005
90	18	14	11	1.000000	0.000000
90	18	14	12	1.000000	0.000000
90	18	14	13	1.000000	0.000000
90	18	14	14	1.000000	0.000000
90	18	15	0	0.025231	0.025231
90	18	15	1	0.142684	0.117453
90	18	15	2	0.379580	0.236897
90	18	15	3	0.653327	0.273747
90	18	15	4	0.855272	0.201945
90	18	15	5	0.955593	0.100321
90	18	15	6	0.990095	0.034502
90	18	15	7	0.998412	0.008317
90	18	15	8	0.999820	0.001408
90	18	15	9	0.999986	0.000166
90	18	15	10	0.999999	0.000013
90	18	15	11	1.000000	0.000001
90	18	15	12	1.000000	0.000000
90	18	15	13	1.000000	0.000000
90	18	15	14	1.000000	0.000000
90	18	15	15	1.000000	0.000000
90	18	16	0	0.019175	0.019175
90	18	16	1	0.116061	0.096886
90	18	16	2	0.329042	0.212981
90	18	16	3	0.598578	0.269536
90	18	16	4	0.817576	0.218998
90	18	16	5	0.938204	0.120628
90	18	16	6	0.984575	0.046371
90	18	16	7	0.997192	0.012618
90	18	16	8	0.999632	0.002440
90	18	16	9	0.999966	0.000334
90	18	16	10	0.999998	0.000032
90	18	16	11	1.000000	0.000002
90	18	16	12	1.000000	0.000000
90	18	16	13	1.000000	0.000000
90	18	16	14	1.000000	0.000000
90	18	16	15	1.000000	0.000000
90	18	16	16	1.000000	0.000000
90	18	17	0	0.014511	0.014511

N	n	k	x	P(x)	p(x)
90	18	17	1	0.093803	0.079292
90	18	17	2	0.282992	0.189189
90	18	17	3	0.543942	0.260950
90	18	17	4	0.776144	0.232201
90	18	17	5	0.917013	0.140869
90	18	17	6	0.977055	0.060042
90	18	17	7	0.995317	0.018262
90	18	17	8	0.999303	0.003986
90	18	17	9	0.999925	0.000623
90	18	17	10	0.999994	0.000069
90	18	17	11	1.000000	0.000005
90	18	17	12	1.000000	0.000000
90	18	17	13	1.000000	0.000000
90	18	17	14	1.000000	0.000000
90	18	17	15	1.000000	0.000000
90	18	17	16	1.000000	0.000000
90	18	17	17	1.000000	0.000000
90	18	18	0	0.010933	0.010933
90	18	18	1	0.075338	0.064405
90	18	18	2	0.241526	0.166188
90	18	18	3	0.490322	0.248796
90	18	18	4	0.731612	0.241290
90	18	18	5	0.891926	0.160314
90	18	18	6	0.967185	0.075259
90	18	18	7	0.992565	0.025380
90	18	18	8	0.998757	0.006191
90	18	18	9	0.999848	0.001092
90	18	18	10	0.999987	0.000138
90	18	18	11	0.999999	0.000012
90	18	18	12	1.000000	0.000001
90	18	18	13	1.000000	0.000000
90	18	18	14	1.000000	0.000000
90	18	18	15	1.000000	0.000000
90	18	18	16	1.000000	0.000000
90	18	18	17	1.000000	0.000000
90	18	18	18	1.000000	0.000000
90	19	1	0	0.788889	0.788889
90	19	1	1	1.000000	0.211111
90	19	2	0	0.620474	0.620474
90	19	2	1	0.957303	0.336829
90	19	2	2	1.000000	0.042697
90	19	3	0	0.486508	0.486508
90	19	3	1	0.888406	0.401898
90	19	3	2	0.991752	0.103345
90	19	3	3	1.000000	0.008248
90	19	4	0	0.380259	0.380259
90	19	4	1	0.805255	0.424996
90	19	4	2	0.971558	0.166303
90	19	4	3	0.998483	0.026925
90	19	4	4	1.000000	0.001517

N	n	k	x	P(x)	p(x)
90	19	5	0	0.296249	0.296249
90	19	5	1	0.716303	0.420054
90	19	5	2	0.938684	0.222382
90	19	5	3	0.993474	0.054790
90	19	5	4	0.999735	0.006262
90	19	5	5	1.000000	0.000265
90	19	6	0	0.230028	0.230028
90	19	6	1	0.627350	0.397322
90	19	6	2	0.894208	0.266858
90	19	6	3	0.983160	0.088953
90	19	6	4	0.998630	0.015470
90	19	6	5	0.999956	0.001326
90	19	6	6	1.000000	0.000044
90	19	7	0	0.177998	0.177998
90	19	7	1	0.542210	0.364211
90	19	7	2	0.840201	0.297991
90	19	7	3	0.966217	0.126016
90	19	7	4	0.995868	0.029651
90	19	7	5	0.999735	0.003867
90	19	7	6	0.999993	0.000258
90	19	7	7	1.000000	0.000007
90	19	8	0	0.137252	0.137252
90	19	8	1	0.463224	0.325972
90	19	8	2	0.779166	0.315942
90	19	8	3	0.941925	0.162758
90	19	8	4	0.990509	0.048585
90	19	8	5	0.999083	0.008574
90	19	8	6	0.999953	0.000870
90	19	8	7	0.999999	0.000046
90	19	8	8	1.000000	0.000001
90	19	9	0	0.105449	0.105449
90	19	9	1	0.391669	0.286220
90	19	9	2	0.713666	0.321997
90	19	9	3	0.910167	0.196501
90	19	9	4	0.981622	0.071455
90	19	9	5	0.997619	0.015997
90	19	9	6	0.999815	0.002196
90	19	9	7	0.999992	0.000177
90	19	9	8	1.000000	0.000008
90	19	9	9	1.000000	0.000000
90	19	10	0	0.080714	0.080714
90	19	10	1	0.328065	0.247350
90	19	10	2	0.646087	0.318022
90	19	10	3	0.871352	0.225265
90	19	10	4	0.968389	0.097037
90	19	10	5	0.994854	0.026465
90	19	10	6	0.999462	0.004608
90	19	10	7	0.999966	0.000503
90	19	10	8	0.999999	0.000033
90	19	10	9	1.000000	0.000001

N	n	k	x	P(x)	p(x)
90	19	10	10	1.000000	0.000000
90	19	11	0	0.061545	0.061545
90	19	11	1	0.272411	0.210866
90	19	11	2	0.578507	0.306096
90	19	11	3	0.826299	0.247792
90	19	11	4	0.950195	0.123896
90	19	11	5	0.990223	0.040028
90	19	11	6	0.998714	0.008491
90	19	11	7	0.999890	0.001177
90	19	11	8	0.999994	0.000104
90	19	11	9	1.000000	0.000006
90	19	11	10	1.000000	0.000000
90	19	11	11	1.000000	0.000000
90	19	12	0	0.046743	0.046743
90	19	12	1	0.224365	0.177623
90	19	12	2	0.512638	0.288273
90	19	12	3	0.776113	0.263475
90	19	12	4	0.926670	0.150557
90	19	12	5	0.983129	0.056459
90	19	12	6	0.997316	0.014187
90	19	12	7	0.999712	0.002395
90	19	12	8	0.999980	0.000268
90	19	12	9	0.999999	0.000019
90	19	12	10	1.000000	0.000001
90	19	12	11	1.000000	0.000000
90	19	12	12	1.000000	0.000000
90	19	13	0	0.035357	0.035357
90	19	13	1	0.183376	0.148019
90	19	13	2	0.449809	0.266434
90	19	13	3	0.722067	0.272258
90	19	13	4	0.897717	0.175650
90	19	13	5	0.972996	0.075279
90	19	13	6	0.994952	0.021956
90	19	13	7	0.999343	0.004391
90	19	13	8	0.999942	0.000599
90	19	13	9	0.999997	0.000055
90	19	13	10	1.000000	0.000003
90	19	13	11	1.000000	0.000000
90	19	13	12	1.000000	0.000000
90	19	13	13	1.000000	0.000000
90	19	14	0	0.026632	0.026632
90	19	14	1	0.148774	0.122141
90	19	14	2	0.390986	0.242213
90	19	14	3	0.665494	0.274508
90	19	14	4	0.863500	0.198006
90	19	14	5	0.959309	0.095809
90	19	14	6	0.991245	0.031936
90	19	14	7	0.998659	0.007414
90	19	14	8	0.999856	0.001198
90	19	14	9	0.999989	0.000133

N	n	k	x	P(x)	p(x)
90	19	14	10	0.999999	0.000010
90	19	14	11	1.000000	0.000000
90	19	14	12	1.000000	0.000000
90	19	14	13	1.000000	0.000000
90	19	14	14	1.000000	0.000000
90	19	15	0	0.019974	0.019974
90	19	15	1	0.119846	0.099871
90	19	15	2	0.336807	0.216962
90	19	15	3	0.607703	0.270896
90	19	15	4	0.824419	0.216717
90	19	15	5	0.941660	0.117240
90	19	15	6	0.985782	0.044123
90	19	15	7	0.997488	0.011706
90	19	15	8	0.999683	0.002195
90	19	15	9	0.999972	0.000289
90	19	15	10	0.999998	0.000026
90	19	15	11	1.000000	0.000002
90	19	15	12	1.000000	0.000000
90	19	15	13	1.000000	0.000000
90	19	15	14	1.000000	0.000000
90	19	15	15	1.000000	0.000000
90	19	16	0	0.014914	0.014914
90	19	16	1	0.095876	0.080962
90	19	16	2	0.287629	0.191753
90	19	16	3	0.549912	0.262283
90	19	16	4	0.781076	0.231164
90	19	16	5	0.919775	0.138699
90	19	16	6	0.978134	0.058360
90	19	16	7	0.995615	0.017481
90	19	16	8	0.999361	0.003746
90	19	16	9	0.999933	0.000572
90	19	16	10	0.999995	0.000062
90	19	16	11	1.000000	0.000005
90	19	16	12	1.000000	0.000000
90	19	16	13	1.000000	0.000000
90	19	16	14	1.000000	0.000000
90	19	16	15	1.000000	0.000000
90	19	16	16	1.000000	0.000000
90	19	17	0	0.011085	0.011085
90	19	17	1	0.076183	0.065098
90	19	17	2	0.243578	0.167395
90	19	17	3	0.493202	0.249624
90	19	17	4	0.734219	0.241016
90	19	17	5	0.893534	0.159316
90	19	17	6	0.967882	0.074347
90	19	17	7	0.992780	0.024899
90	19	17	8	0.998804	0.006024
90	19	17	9	0.999856	0.001052
90	19	17	10	0.999988	0.000131
90	19	17	11	0.999999	0.000012

N	n	k	x	P(x)	p(x)
90	19	17	12	1.000000	0.000001
90	19	17	13	1.000000	0.000000
90	19	17	14	1.000000	0.000000
90	19	17	15	1.000000	0.000000
90	19	17	16	1.000000	0.000000
90	19	17	17	1.000000	0.000000
90	19	18	0	0.008200	0.008200
90	19	18	1	0.060131	0.051932
90	19	18	2	0.204595	0.144464
90	19	18	3	0.438490	0.233894
90	19	18	4	0.684695	0.246205
90	19	18	5	0.862981	0.178286
90	19	18	6	0.954642	0.091661
90	19	18	7	0.988687	0.034046
90	19	18	8	0.997897	0.009209
90	19	18	9	0.999712	0.001815
90	19	18	10	0.999971	0.000259
90	19	18	11	0.999998	0.000027
90	19	18	12	1.000000	0.000002
90	19	18	13	1.000000	0.000000
90	19	18	14	1.000000	0.000000
90	19	18	15	1.000000	0.000000
90	19	18	16	1.000000	0.000000
90	19	18	17	1.000000	0.000000
90	19	18	18	1.000000	0.000000
90	19	19	0	0.006036	0.006036
90	19	19	1	0.047148	0.041112
90	19	19	2	0.170486	0.123337
90	19	19	3	0.386513	0.216027
90	19	19	4	0.633402	0.246889
90	19	19	5	0.828314	0.194912
90	19	19	6	0.938092	0.109778
90	19	19	7	0.983013	0.044921
90	19	19	8	0.996490	0.013476
90	19	19	9	0.999460	0.002970
90	19	19	10	0.999939	0.000479
90	19	19	11	0.999995	0.000056
90	19	19	12	1.000000	0.000005
90	19	19	13	1.000000	0.000000
90	19	19	14	1.000000	0.000000
90	19	19	15	1.000000	0.000000
90	19	19	16	1.000000	0.000000
90	19	19	17	1.000000	0.000000
90	19	19	18	1.000000	0.000000
90	19	19	19	1.000000	0.000000
90	20	1	0	0.777778	0.777778
90	20	1	1	1.000000	0.222222
90	20	2	0	0.602996	0.602996
90	20	2	1	0.952559	0.349563
90	20	2	2	1.000000	0.047441

N	n	k	x	P(x)	p(x)
90	20	3	0	0.465952	0.465952
90	20	3	1	0.877085	0.411134
90	20	3	2	0.990296	0.113211
90	20	3	3	1.000000	0.009704
90	20	4	0	0.358836	0.358836
90	20	4	1	0.787298	0.428461
90	20	4	2	0.966873	0.179576
90	20	4	3	0.998104	0.031231
90	20	4	4	1.000000	0.001896
90	20	5	0	0.275386	0.275386
90	20	5	1	0.692637	0.417251
90	20	5	2	0.929288	0.236650
90	20	5	3	0.991930	0.062643
90	20	5	4	0.999647	0.007717
90	20	5	5	1.000000	0.000353
90	20	6	0	0.210589	0.210589
90	20	6	1	0.599370	0.388780
90	20	6	2	0.879173	0.279804
90	20	6	3	0.979402	0.100228
90	20	6	4	0.998195	0.018793
90	20	6	5	0.999938	0.001743
90	20	6	6	1.000000	0.000062
90	20	7	0	0.160449	0.160449
90	20	7	1	0.511431	0.350982
90	20	7	2	0.819216	0.307784
90	20	7	3	0.959118	0.139902
90	20	7	4	0.994615	0.035498
90	20	7	5	0.999626	0.005011
90	20	7	6	0.999990	0.000363
90	20	7	7	1.000000	0.000010
90	20	8	0	0.121787	0.121787
90	20	8	1	0.431086	0.309299
90	20	8	2	0.752467	0.321381
90	20	8	3	0.930463	0.177996
90	20	8	4	0.987772	0.057309
90	20	8	5	0.998721	0.010949
90	20	8	6	0.999928	0.001208
90	20	8	7	0.999998	0.000070
90	20	8	8	1.000000	0.000002
90	20	9	0	0.092083	0.092083
90	20	9	1	0.359419	0.267336
90	20	9	2	0.681920	0.322501
90	20	9	3	0.893561	0.211641
90	20	9	4	0.976590	0.083028
90	20	9	5	0.996718	0.020128
90	20	9	6	0.999722	0.003004
90	20	9	7	0.999987	0.000265
90	20	9	8	1.000000	0.000012
90	20	9	9	1.000000	0.000000
90	20	10	0	0.069346	0.069346

N	n	k	x	P(x)	p(x)
90	20	10	1	0.296710	0.227364
90	20	10	2	0.610253	0.313543
90	20	10	3	0.849143	0.238890
90	20	10	4	0.960189	0.111046
90	20	10	5	0.992990	0.032801
90	20	10	6	0.999203	0.006212
90	20	10	7	0.999945	0.000742
90	20	10	8	0.999998	0.000053
90	20	10	9	1.000000	0.000002
90	20	10	10	1.000000	0.000000
90	20	11	0	0.052010	0.052010
90	20	11	1	0.242711	0.190702
90	20	11	2	0.539706	0.296995
90	20	11	3	0.798379	0.258673
90	20	11	4	0.937980	0.139601
90	20	11	5	0.986840	0.048860
90	20	11	6	0.998116	0.011275
90	20	11	7	0.999824	0.001708
90	20	11	8	0.999990	0.000166
90	20	11	9	1.000000	0.000010
90	20	11	10	1.000000	0.000000
90	20	11	11	1.000000	0.000000
90	20	12	0	0.038843	0.038843
90	20	12	1	0.196846	0.158004
90	20	12	2	0.472036	0.275190
90	20	12	3	0.742715	0.270679
90	20	12	4	0.909706	0.166991
90	20	12	5	0.977563	0.067857
90	20	12	6	0.996118	0.018555
90	20	12	7	0.999543	0.003425
90	20	12	8	0.999965	0.000422
90	20	12	9	0.999998	0.000034
90	20	12	10	1.000000	0.000002
90	20	12	11	1.000000	0.000000
90	20	12	12	1.000000	0.000000
90	20	13	0	0.028883	0.028883
90	20	13	1	0.158358	0.129475
90	20	13	2	0.408531	0.250173
90	20	13	3	0.683721	0.275190
90	20	13	4	0.875452	0.191731
90	20	13	5	0.964514	0.089062
90	20	13	6	0.992787	0.028274
90	20	13	7	0.998972	0.006185
90	20	13	8	0.999900	0.000928
90	20	13	9	0.999994	0.000094
90	20	13	10	1.000000	0.000006
90	20	13	11	1.000000	0.000000
90	20	13	12	1.000000	0.000000
90	20	13	13	1.000000	0.000000
90	20	14	0	0.021381	0.021381

N	n	k	x	P(x)	p(x)
90	20	14	1	0.126410	0.105029
90	20	14	2	0.350049	0.223639
90	20	14	3	0.622965	0.272916
90	20	14	4	0.835611	0.212647
90	20	14	5	0.947164	0.111552
90	20	14	6	0.987646	0.040483
90	20	14	7	0.997928	0.010281
90	20	14	8	0.999755	0.001827
90	20	14	9	0.999980	0.000225
90	20	14	10	0.999999	0.000019
90	20	14	11	1.000000	0.000001
90	20	14	12	1.000000	0.000000
90	20	14	13	1.000000	0.000000
90	20	14	14	1.000000	0.000000
90	20	15	0	0.015754	0.015754
90	20	15	1	0.100153	0.084398
90	20	15	2	0.297082	0.196929
90	20	15	3	0.561918	0.264836
90	20	15	4	0.790844	0.228926
90	20	15	5	0.925147	0.134303
90	20	15	6	0.980189	0.055042
90	20	15	7	0.996169	0.015980
90	20	15	8	0.999467	0.003297
90	20	15	9	0.999948	0.000481
90	20	15	10	0.999996	0.000049
90	20	15	11	1.000000	0.000003
90	20	15	12	1.000000	0.000000
90	20	15	13	1.000000	0.000000
90	20	15	14	1.000000	0.000000
90	20	15	15	1.000000	0.000000
90	20	16	0	0.011553	0.011553
90	20	16	1	0.078772	0.067219
90	20	16	2	0.249819	0.171047
90	20	16	3	0.501888	0.252069
90	20	16	4	0.742006	0.240118
90	20	16	5	0.898286	0.156280
90	20	16	6	0.969915	0.071628
90	20	16	7	0.993399	0.023485
90	20	16	8	0.998939	0.005540
90	20	16	9	0.999877	0.000938
90	20	16	10	0.999990	0.000113
90	20	16	11	0.999999	0.000009
90	20	16	12	1.000000	0.000001
90	20	16	13	1.000000	0.000000
90	20	16	14	1.000000	0.000000
90	20	16	15	1.000000	0.000000
90	20	16	16	1.000000	0.000000
90	20	17	0	0.008431	0.008431
90	20	17	1	0.061513	0.053082
90	20	17	2	0.208213	0.146700

Table for $N = 2$, $n = 1$, through $N = 100$, $n = 50$

N	n	k	x	$P(x)$	$p(x)$
90	20	17	3	0.443980	0.235768
90	20	17	4	0.690089	0.246108
90	20	17	5	0.866608	0.176519
90	20	17	6	0.956363	0.089755
90	20	17	7	0.989274	0.032910
90	20	17	8	0.998041	0.008767
90	20	17	9	0.999738	0.001697
90	20	17	10	0.999975	0.000237
90	20	17	11	0.999998	0.000024
90	20	17	12	1.000000	0.000002
90	20	17	13	1.000000	0.000000
90	20	17	14	1.000000	0.000000
90	20	17	15	1.000000	0.000000
90	20	17	16	1.000000	0.000000
90	20	17	17	1.000000	0.000000
90	20	18	0	0.006121	0.006121
90	20	18	1	0.047697	0.041576
90	20	18	2	0.172040	0.124343
90	20	18	3	0.389076	0.217035
90	20	18	4	0.636147	0.247072
90	20	18	5	0.830337	0.194190
90	20	18	6	0.939150	0.108813
90	20	18	7	0.983413	0.044263
90	20	18	8	0.996600	0.013187
90	20	18	9	0.999482	0.002882
90	20	18	10	0.999942	0.000460
90	20	18	11	0.999995	0.000053
90	20	18	12	1.000000	0.000004
90	20	18	13	1.000000	0.000000
90	20	18	14	1.000000	0.000000
90	20	18	15	1.000000	0.000000
90	20	18	16	1.000000	0.000000
90	20	18	17	1.000000	0.000000
90	20	18	18	1.000000	0.000000
90	20	19	0	0.004421	0.004421
90	20	19	1	0.036726	0.032305
90	20	19	2	0.140954	0.104229
90	20	19	3	0.337831	0.196877
90	20	19	4	0.581242	0.243411
90	20	19	5	0.789881	0.208638
90	20	19	6	0.917992	0.128111
90	20	19	7	0.975421	0.057429
90	20	19	8	0.994402	0.018981
90	20	19	9	0.999042	0.004640
90	20	19	10	0.999878	0.000837
90	20	19	11	0.999989	0.000110
90	20	19	12	0.999999	0.000011
90	20	19	13	1.000000	0.000001
90	20	19	14	1.000000	0.000000
90	20	19	15	1.000000	0.000000

N	n	k	x	$P(x)$	$p(x)$
90	20	19	16	1.000000	0.000000
90	20	19	17	1.000000	0.000000
90	20	19	18	1.000000	0.000000
90	20	19	19	1.000000	0.000000
90	20	20	0	0.003175	0.003175
90	20	20	1	0.028081	0.024905
90	20	20	2	0.114530	0.086450
90	20	20	3	0.290692	0.176161
90	20	20	4	0.526389	0.235698
90	20	20	5	0.745802	0.219413
90	20	20	6	0.892730	0.146928
90	20	20	7	0.964906	0.072175
90	20	20	8	0.991194	0.026288
90	20	20	9	0.998323	0.007129
90	20	20	10	0.999760	0.001438
90	20	20	11	0.999975	0.000214
90	20	20	12	0.999998	0.000023
90	20	20	13	1.000000	0.000002
90	20	20	14	1.000000	0.000000
90	20	20	15	1.000000	0.000000
90	20	20	16	1.000000	0.000000
90	20	20	17	1.000000	0.000000
90	20	20	18	1.000000	0.000000
90	20	20	19	1.000000	0.000000
90	20	20	20	1.000000	0.000000
90	21	1	0	0.766667	0.766667
90	21	1	1	1.000000	0.233333
90	21	2	0	0.585768	0.585768
90	21	2	1	0.947565	0.361798
90	21	2	2	1.000000	0.052434
90	21	3	0	0.445982	0.445982
90	21	3	1	0.865339	0.419356
90	21	3	2	0.988679	0.123340
90	21	3	3	1.000000	0.011321
90	21	4	0	0.338331	0.338331
90	21	4	1	0.768935	0.430604
90	21	4	2	0.961743	0.192808
90	21	4	3	0.997658	0.035915
90	21	4	4	1.000000	0.002342
90	21	5	0	0.255716	0.255716
90	21	5	1	0.668795	0.413079
90	21	5	2	0.919145	0.250351
90	21	5	3	0.990141	0.070995
90	21	5	4	0.999537	0.009396
90	21	5	5	1.000000	0.000463
90	21	6	0	0.192539	0.192539
90	21	6	1	0.571600	0.379061
90	21	6	2	0.863185	0.291585
90	21	6	3	0.975106	0.111922
90	21	6	4	0.997658	0.022551

N	n	k	x	$P(x)$	$p(x)$
90	21	6	5	0.999913	0.002255
90	21	6	6	1.000000	0.000087
90	21	7	0	0.144404	0.144404
90	21	7	1	0.481347	0.336943
90	21	7	2	0.797231	0.315884
90	21	7	3	0.951123	0.153892
90	21	7	4	0.993094	0.041971
90	21	7	5	0.999483	0.006390
90	21	7	6	0.999984	0.000501
90	21	7	7	1.000000	0.000016
90	21	8	0	0.107868	0.107868
90	21	8	1	0.400156	0.292288
90	21	8	2	0.724920	0.324764
90	21	8	3	0.917749	0.192829
90	21	8	4	0.984497	0.066748
90	21	8	5	0.998251	0.013754
90	21	8	6	0.999894	0.001642
90	21	8	7	0.999997	0.000104
90	21	8	8	1.000000	0.000003
90	21	9	0	0.080243	0.080243
90	21	9	1	0.328866	0.248623
90	21	9	2	0.649670	0.320804
90	21	9	3	0.875421	0.225751
90	21	9	4	0.970659	0.095239
90	21	9	5	0.995568	0.024909
90	21	9	6	0.999593	0.004026
90	21	9	7	0.999980	0.000386
90	21	9	8	1.000000	0.000020
90	21	9	9	1.000000	0.000000
90	21	10	0	0.059440	0.059440
90	21	10	1	0.267478	0.208038
90	21	10	2	0.574420	0.306942
90	21	10	3	0.825254	0.250834
90	21	10	4	0.950671	0.125417
90	21	10	5	0.990648	0.039977
90	21	10	6	0.998848	0.008200
90	21	10	7	0.999913	0.001065
90	21	10	8	0.999996	0.000083
90	21	10	9	1.000000	0.000004
90	21	10	10	1.000000	0.000000
90	21	11	0	0.043837	0.043837
90	21	11	1	0.215468	0.171632
90	21	11	2	0.501521	0.286053
90	21	11	3	0.768816	0.267295
90	21	11	4	0.924020	0.155204
90	21	11	5	0.982652	0.058632
90	21	11	6	0.997310	0.014658
90	21	11	7	0.999726	0.002416
90	21	11	8	0.999983	0.000256
90	21	11	9	0.999999	0.000017

N	n	k	x	$P(x)$	$p(x)$
90	21	11	10	1.000000	0.000001
90	21	11	11	1.000000	0.000000
90	21	12	0	0.032184	0.032184
90	21	12	1	0.172017	0.139833
90	21	12	2	0.432723	0.260706
90	21	12	3	0.707913	0.275190
90	21	12	4	0.890621	0.182708
90	21	12	5	0.970777	0.080156
90	21	12	6	0.994527	0.023750
90	21	12	7	0.999298	0.004771
90	21	12	8	0.999941	0.000642
90	21	12	9	0.999997	0.000056
90	21	12	10	1.000000	0.000003
90	21	12	11	1.000000	0.000000
90	21	12	12	1.000000	0.000000
90	21	13	0	0.023519	0.023519
90	21	13	1	0.136163	0.112644
90	21	13	2	0.369218	0.233056
90	21	13	3	0.644408	0.275190
90	21	13	4	0.850800	0.206392
90	21	13	5	0.954335	0.103535
90	21	13	6	0.989960	0.035625
90	21	13	7	0.998442	0.008482
90	21	13	8	0.999834	0.001392
90	21	13	9	0.999988	0.000155
90	21	13	10	0.999999	0.000011
90	21	13	11	1.000000	0.000000
90	21	13	12	1.000000	0.000000
90	21	13	13	1.000000	0.000000
90	21	14	0	0.017105	0.017105
90	21	14	1	0.106904	0.089800
90	21	14	2	0.311711	0.204806
90	21	14	3	0.580078	0.268367
90	21	14	4	0.805233	0.225155
90	21	14	5	0.932821	0.127588
90	21	14	6	0.983020	0.050199
90	21	14	7	0.996900	0.013880
90	21	14	8	0.999599	0.002699
90	21	14	9	0.999964	0.000365
90	21	14	10	0.999998	0.000034
90	21	14	11	1.000000	0.000002
90	21	14	12	1.000000	0.000000
90	21	14	13	1.000000	0.000000
90	21	14	14	1.000000	0.000000
90	21	15	0	0.012378	0.012378
90	21	15	1	0.083273	0.070895
90	21	15	2	0.260509	0.177236
90	21	15	3	0.516517	0.256008
90	21	15	4	0.754870	0.238352
90	21	15	5	0.905961	0.151091

Table for $N = 2$, $n = 1$, through $N = 100$, $n = 50$

N	n	k	x	P(x)	p(x)
90	21	15	6	0.973112	0.067152
90	21	15	7	0.994343	0.021231
90	21	15	8	0.999137	0.004794
90	21	15	9	0.999906	0.000769
90	21	15	10	0.999993	0.000087
90	21	15	11	1.000000	0.000007
90	21	15	12	1.000000	0.000000
90	21	15	13	1.000000	0.000000
90	21	15	14	1.000000	0.000000
90	21	15	15	1.000000	0.000000
90	21	16	0	0.008912	0.008912
90	21	16	1	0.064368	0.055455
90	21	16	2	0.215609	0.151242
90	21	16	3	0.455075	0.239466
90	21	16	4	0.700843	0.245768
90	21	16	5	0.873728	0.172885
90	21	16	6	0.959682	0.085954
90	21	16	7	0.990380	0.030698
90	21	16	8	0.998306	0.007926
90	21	16	9	0.999783	0.001477
90	21	16	10	0.999980	0.000197
90	21	16	11	0.999999	0.000018
90	21	16	12	1.000000	0.000001
90	21	16	13	1.000000	0.000000
90	21	16	14	1.000000	0.000000
90	21	16	15	1.000000	0.000000
90	21	16	16	1.000000	0.000000
90	21	17	0	0.006383	0.006383
90	21	17	1	0.049380	0.042997
90	21	17	2	0.176777	0.127397
90	21	17	3	0.396827	0.220050
90	21	17	4	0.644383	0.247556
90	21	17	5	0.836347	0.191965
90	21	17	6	0.942259	0.105911
90	21	17	7	0.984572	0.042313
90	21	17	8	0.996914	0.012341
90	21	17	9	0.999544	0.002630
90	21	17	10	0.999951	0.000407
90	21	17	11	0.999996	0.000045
90	21	17	12	1.000000	0.000004
90	21	17	13	1.000000	0.000000
90	21	17	14	1.000000	0.000000
90	21	17	15	1.000000	0.000000
90	21	17	16	1.000000	0.000000
90	21	17	17	1.000000	0.000000
90	21	18	0	0.004547	0.004547
90	21	18	1	0.037600	0.033053
90	21	18	2	0.143619	0.106019
90	21	18	3	0.342568	0.198949
90	21	18	4	0.586733	0.244165

N	n	k	x	P(x)	p(x)
90	21	18	5	0.794273	0.207540
90	21	18	6	0.920496	0.126223
90	21	18	7	0.976457	0.055961
90	21	18	8	0.994716	0.018258
90	21	18	9	0.999111	0.004396
90	21	18	10	0.999890	0.000778
90	21	18	11	0.999990	0.000100
90	21	18	12	0.999999	0.000009
90	21	18	13	1.000000	0.000001
90	21	18	14	1.000000	0.000000
90	21	18	15	1.000000	0.000000
90	21	18	16	1.000000	0.000000
90	21	18	17	1.000000	0.000000
90	21	18	18	1.000000	0.000000
90	21	19	0	0.003221	0.003221
90	21	19	1	0.028419	0.025198
90	21	19	2	0.115642	0.087223
90	21	19	3	0.292831	0.177189
90	21	19	4	0.529083	0.236252
90	21	19	5	0.748153	0.219070
90	21	19	6	0.894200	0.146047
90	21	19	7	0.965576	0.071376
90	21	19	8	0.991419	0.025843
90	21	19	9	0.998379	0.006960
90	21	19	10	0.999771	0.001392
90	21	19	11	0.999976	0.000205
90	21	19	12	0.999998	0.000022
90	21	19	13	1.000000	0.000002
90	21	19	14	1.000000	0.000000
90	21	19	15	1.000000	0.000000
90	21	19	16	1.000000	0.000000
90	21	19	17	1.000000	0.000000
90	21	19	18	1.000000	0.000000
90	21	19	19	1.000000	0.000000
90	21	20	0	0.002268	0.002268
90	21	20	1	0.021321	0.019052
90	21	20	2	0.092300	0.070980
90	21	20	3	0.247910	0.155609
90	21	20	4	0.472515	0.224606
90	21	20	5	0.698785	0.226270
90	21	20	6	0.863345	0.164560
90	21	20	7	0.951502	0.088157
90	21	20	8	0.986687	0.035185
90	21	20	9	0.997202	0.010515
90	21	20	10	0.999555	0.002353
90	21	20	11	0.999947	0.000392
90	21	20	12	0.999995	0.000048
90	21	20	13	1.000000	0.000004
90	21	20	14	1.000000	0.000000
90	21	20	15	1.000000	0.000000

N	n	k	x	P(x)	p(x)
90	21	20	16	1.000000	0.000000
90	21	20	17	1.000000	0.000000
90	21	20	18	1.000000	0.000000
90	21	20	19	1.000000	0.000000
90	21	20	20	1.000000	0.000000
90	21	21	0	0.001588	0.001588
90	21	21	1	0.015877	0.014289
90	21	21	2	0.073034	0.057157
90	21	21	3	0.207896	0.134861
90	21	21	4	0.417968	0.210073
90	21	21	5	0.647066	0.229098
90	21	21	6	0.828082	0.181016
90	21	21	7	0.933870	0.105788
90	21	21	8	0.980153	0.046282
90	21	21	9	0.995400	0.015247
90	21	21	10	0.999185	0.003785
90	21	21	11	0.999891	0.000706
90	21	21	12	0.999989	0.000098
90	21	21	13	0.999999	0.000010
90	21	21	14	1.000000	0.000001
90	21	21	15	1.000000	0.000000
90	21	21	16	1.000000	0.000000
90	21	21	17	1.000000	0.000000
90	21	21	18	1.000000	0.000000
90	21	21	19	1.000000	0.000000
90	21	21	20	1.000000	0.000000
90	21	21	21	1.000000	0.000000
90	22	1	0	0.755556	0.755556
90	22	1	1	1.000000	0.244444
90	22	2	0	0.568789	0.568789
90	22	2	1	0.942322	0.373533
90	22	2	2	1.000000	0.057678
90	22	3	0	0.426592	0.426592
90	22	3	1	0.853183	0.426592
90	22	3	2	0.986891	0.133708
90	22	3	3	1.000000	0.013109
90	22	4	0	0.318718	0.318718
90	22	4	1	0.750213	0.431495
90	22	4	2	0.956154	0.205941
90	22	4	3	0.997137	0.040983
90	22	4	4	1.000000	0.002863
90	22	5	0	0.237185	0.237185
90	22	5	1	0.644848	0.407663
90	22	5	2	0.908261	0.263413
90	22	5	3	0.988083	0.079822
90	22	5	4	0.999401	0.011318
90	22	5	5	1.000000	0.000599
90	22	6	0	0.175796	0.175796
90	22	6	1	0.544131	0.368335
90	22	6	2	0.846281	0.302150

N	n	k	x	P(x)	p(x)
90	22	6	3	0.970240	0.123959
90	22	6	4	0.997004	0.026764
90	22	6	5	0.999880	0.002876
90	22	6	6	1.000000	0.000120
90	22	7	0	0.129754	0.129754
90	22	7	1	0.452048	0.322293
90	22	7	2	0.774341	0.322293
90	22	7	3	0.942202	0.167861
90	22	7	4	0.991269	0.049067
90	22	7	5	0.999298	0.008029
90	22	7	6	0.999977	0.000679
90	22	7	7	1.000000	0.000023
90	22	8	0	0.095362	0.095362
90	22	8	1	0.370503	0.275142
90	22	8	2	0.696680	0.326176
90	22	8	3	0.903776	0.207096
90	22	8	4	0.980628	0.076852
90	22	8	5	0.997653	0.017026
90	22	8	6	0.999846	0.002193
90	22	8	7	0.999996	0.000150
90	22	8	8	1.000000	0.000004
90	22	9	0	0.069777	0.069777
90	22	9	1	0.300040	0.230264
90	22	9	2	0.617125	0.317084
90	22	9	3	0.855790	0.238666
90	22	9	4	0.963758	0.107968
90	22	9	5	0.994124	0.030366
90	22	9	6	0.999418	0.005295
90	22	9	7	0.999968	0.000550
90	22	9	8	0.999999	0.000031
90	22	9	9	1.000000	0.000001
90	22	10	0	0.050825	0.050825
90	22	10	1	0.240342	0.189517
90	22	10	2	0.538832	0.298490
90	22	10	3	0.799807	0.260975
90	22	10	4	0.939765	0.139958
90	22	10	5	0.987751	0.047986
90	22	10	6	0.998372	0.010622
90	22	10	7	0.999867	0.001494
90	22	10	8	0.999994	0.000127
90	22	10	9	1.000000	0.000006
90	22	10	10	1.000000	0.000000
90	22	11	0	0.036848	0.036848
90	22	11	1	0.190594	0.153746
90	22	11	2	0.464210	0.273616
90	22	11	3	0.737825	0.273616
90	22	11	4	0.908274	0.170449
90	22	11	5	0.977554	0.069279
90	22	11	6	0.996248	0.018694
90	22	11	7	0.999586	0.003338

Table for $N=2$, $n=1$, through $N=100$, $n=50$

N	n	k	x	P(x)	p(x)
90	22	11	8	0.999972	0.000385
90	22	11	9	0.999999	0.000027
90	22	11	10	1.000000	0.000001
90	22	11	11	1.000000	0.000000
90	22	12	0	0.026587	0.026587
90	22	12	1	0.149725	0.123138
90	22	12	2	0.394940	0.245215
90	22	12	3	0.672019	0.277079
90	22	12	4	0.869438	0.197419
90	22	12	5	0.962646	0.093208
90	22	12	6	0.992462	0.029816
90	22	12	7	0.998953	0.006491
90	22	12	8	0.999903	0.000951
90	22	12	9	0.999994	0.000091
90	22	12	10	1.000000	0.000005
90	22	12	11	1.000000	0.000000
90	22	12	12	1.000000	0.000000
90	22	13	0	0.019088	0.019088
90	22	13	1	0.116572	0.097484
90	22	13	2	0.332064	0.215492
90	22	13	3	0.604525	0.272461
90	22	13	4	0.823880	0.219354
90	22	13	5	0.942331	0.118451
90	22	13	6	0.986346	0.044015
90	22	13	7	0.997704	0.011359
90	22	13	8	0.999733	0.002028
90	22	13	9	0.999979	0.000246
90	22	13	10	0.999999	0.000020
90	22	13	11	1.000000	0.000001
90	22	13	12	1.000000	0.000000
90	22	13	13	1.000000	0.000000
90	22	14	0	0.013634	0.013634
90	22	14	1	0.089986	0.076351
90	22	14	2	0.276092	0.186107
90	22	14	3	0.537295	0.261202
90	22	14	4	0.772602	0.235307
90	22	14	5	0.916179	0.143577
90	22	14	6	0.977200	0.061020
90	22	14	7	0.995492	0.018292
90	22	14	8	0.999364	0.003872
90	22	14	9	0.999938	0.000574
90	22	14	10	0.999996	0.000058
90	22	14	11	1.000000	0.000004
90	22	14	12	1.000000	0.000000
90	22	14	13	1.000000	0.000000
90	22	14	14	1.000000	0.000000
90	22	15	0	0.009687	0.009687
90	22	15	1	0.068889	0.059201
90	22	15	2	0.227117	0.158228
90	22	15	3	0.471994	0.244877
90	22	15	4	0.716871	0.244877
90	22	15	5	0.884063	0.167192
90	22	15	6	0.964353	0.080290
90	22	15	7	0.991881	0.027528
90	22	15	8	0.998651	0.006769
90	22	15	9	0.999839	0.001189
90	22	15	10	0.999987	0.000147
90	22	15	11	0.999999	0.000013
90	22	15	12	1.000000	0.000001
90	22	15	13	1.000000	0.000000
90	22	15	14	1.000000	0.000000
90	22	15	15	1.000000	0.000000
90	22	16	0	0.006846	0.006846
90	22	16	1	0.052312	0.045466
90	22	16	2	0.184923	0.132610
90	22	16	3	0.409959	0.225036
90	22	16	4	0.658101	0.248142
90	22	16	5	0.846167	0.188066
90	22	16	6	0.947225	0.101058
90	22	16	7	0.986376	0.039151
90	22	16	8	0.997387	0.011011
90	22	16	9	0.999633	0.002246
90	22	16	10	0.999963	0.000330
90	22	16	11	0.999997	0.000034
90	22	16	12	1.000000	0.000002
90	22	16	13	1.000000	0.000000
90	22	16	14	1.000000	0.000000
90	22	16	15	1.000000	0.000000
90	22	16	16	1.000000	0.000000
90	22	17	0	0.004811	0.004811
90	22	17	1	0.039410	0.034599
90	22	17	2	0.149082	0.109672
90	22	17	3	0.352179	0.203097
90	22	17	4	0.597742	0.245563
90	22	17	5	0.802962	0.205220
90	22	17	6	0.925375	0.122412
90	22	17	7	0.978440	0.053065
90	22	17	8	0.995304	0.016864
90	22	17	9	0.999239	0.003935
90	22	17	10	0.999910	0.000671
90	22	17	11	0.999992	0.000083
90	22	17	12	1.000000	0.000007
90	22	17	13	1.000000	0.000000
90	22	17	14	1.000000	0.000000
90	22	17	15	1.000000	0.000000
90	22	17	16	1.000000	0.000000
90	22	17	17	1.000000	0.000000
90	22	18	0	0.003361	0.003361
90	22	18	1	0.029456	0.026096
90	22	18	2	0.119035	0.089578
90	22	18	3	0.299318	0.180283
90	22	18	4	0.537192	0.237874
90	22	18	5	0.755171	0.217979
90	22	18	6	0.898544	0.143373
90	22	18	7	0.967536	0.068992
90	22	18	8	0.992070	0.024534
90	22	18	9	0.998538	0.006468
90	22	18	10	0.999799	0.001261
90	22	18	11	0.999980	0.000180
90	22	18	12	0.999998	0.000019
90	22	18	13	1.000000	0.000001
90	22	18	14	1.000000	0.000000
90	22	18	15	1.000000	0.000000
90	22	18	16	1.000000	0.000000
90	22	18	17	1.000000	0.000000
90	22	18	18	1.000000	0.000000
90	22	19	0	0.002334	0.002334
90	22	19	1	0.021845	0.019511
90	22	19	2	0.094152	0.072307
90	22	19	3	0.251743	0.157592
90	22	19	4	0.477724	0.225980
90	22	19	5	0.703704	0.225980
90	22	19	6	0.866684	0.162980
90	22	19	7	0.953163	0.086479
90	22	19	8	0.987299	0.034136
90	22	19	9	0.997370	0.010071
90	22	19	10	0.999589	0.002219
90	22	19	11	0.999952	0.000363
90	22	19	12	0.999996	0.000044
90	22	19	13	1.000000	0.000004
90	22	19	14	1.000000	0.000000
90	22	19	15	1.000000	0.000000
90	22	19	16	1.000000	0.000000
90	22	19	17	1.000000	0.000000
90	22	19	18	1.000000	0.000000
90	22	19	19	1.000000	0.000000
90	22	20	0	0.001611	0.001611
90	22	20	1	0.016074	0.014464
90	22	20	2	0.073784	0.057710
90	22	20	3	0.209571	0.135787
90	22	20	4	0.420433	0.210862
90	22	20	5	0.649596	0.229163
90	22	20	6	0.829956	0.180360
90	22	20	7	0.934892	0.104937
90	22	20	8	0.980568	0.045676
90	22	20	9	0.995526	0.014958
90	22	20	10	0.999214	0.003688
90	22	20	11	0.999896	0.000682
90	22	20	12	0.999990	0.000094
90	22	20	13	0.999999	0.000009
90	22	20	14	1.000000	0.000001
90	22	20	15	1.000000	0.000000
90	22	20	16	1.000000	0.000000
90	22	20	17	1.000000	0.000000
90	22	20	18	1.000000	0.000000
90	22	20	19	1.000000	0.000000
90	22	20	20	1.000000	0.000000
90	22	21	0	0.001104	0.001104
90	22	21	1	0.011735	0.010631
90	22	21	2	0.057295	0.045560
90	22	21	3	0.172715	0.115419
90	22	21	4	0.366211	0.193497
90	22	21	5	0.593942	0.227731
90	22	21	6	0.788731	0.194789
90	22	21	7	0.912406	0.123675
90	22	21	8	0.971433	0.059027
90	22	21	9	0.992748	0.021315
90	22	21	10	0.998582	0.005834
90	22	21	11	0.999789	0.001207
90	22	21	12	0.999976	0.000188
90	22	21	13	0.999998	0.000022
90	22	21	14	1.000000	0.000002
90	22	21	15	1.000000	0.000000
90	22	21	16	1.000000	0.000000
90	22	21	17	1.000000	0.000000
90	22	21	18	1.000000	0.000000
90	22	21	19	1.000000	0.000000
90	22	21	20	1.000000	0.000000
90	22	21	21	1.000000	0.000000
90	22	22	0	0.000752	0.000752
90	22	22	1	0.008500	0.007747
90	22	22	2	0.044090	0.035590
90	22	22	3	0.140932	0.096843
90	22	22	4	0.315734	0.174801
90	22	22	5	0.537835	0.222101
90	22	22	6	0.743562	0.205728
90	22	22	7	0.885520	0.141958
90	22	22	8	0.959457	0.073936
90	22	22	9	0.988732	0.029276
90	22	22	10	0.997567	0.008835
90	22	22	11	0.999596	0.002029
90	22	22	12	0.999949	0.000353
90	22	22	13	0.999995	0.000046
90	22	22	14	1.000000	0.000004
90	22	22	15	1.000000	0.000000
90	22	22	16	1.000000	0.000000
90	22	22	17	1.000000	0.000000
90	22	22	18	1.000000	0.000000
90	22	22	19	1.000000	0.000000
90	22	22	20	1.000000	0.000000

N	n	k	x	$P(x)$	$p(x)$
90	22	22	21	1.000000	0.000000
90	22	22	22	1.000000	0.000000
90	23	1	0	0.744444	0.744444
90	23	1	1	1.000000	0.255556
90	23	2	0	0.552060	0.552060
90	23	2	1	0.936829	0.384769
90	23	2	2	1.000000	0.063171
90	23	3	0	0.407772	0.407772
90	23	3	1	0.840637	0.432865
90	23	3	2	0.984925	0.144288
90	23	3	3	1.000000	0.015075
90	23	4	0	0.299970	0.299970
90	23	4	1	0.731177	0.431207
90	23	4	2	0.950097	0.218920
90	23	4	3	0.996534	0.046438
90	23	4	4	1.000000	0.003465
90	23	5	0	0.219745	0.219745
90	23	5	1	0.620868	0.401122
90	23	5	2	0.896640	0.275772
90	23	5	3	0.985735	0.089095
90	23	5	4	0.999234	0.013499
90	23	5	5	1.000000	0.000766
90	23	6	0	0.160285	0.160285
90	23	6	1	0.517048	0.356763
90	23	6	2	0.828508	0.311460
90	23	6	3	0.964771	0.136264
90	23	6	4	0.996217	0.031445
90	23	6	5	0.999838	0.003621
90	23	6	6	1.000000	0.000162
90	23	7	0	0.116397	0.116397
90	23	7	1	0.423610	0.307213
90	23	7	2	0.750643	0.327033
90	23	7	3	0.932328	0.181685
90	23	7	4	0.989104	0.056777
90	23	7	5	0.999062	0.009958
90	23	7	6	0.999967	0.000905
90	23	7	7	1.000000	0.000033
90	23	8	0	0.084143	0.084143
90	23	8	1	0.342180	0.258037
90	23	8	2	0.667899	0.325719
90	23	8	3	0.888548	0.220649
90	23	8	4	0.976107	0.087559
90	23	8	5	0.996902	0.020795
90	23	8	6	0.999782	0.002879
90	23	8	7	0.999994	0.000212
90	23	8	8	1.000000	0.000006
90	23	9	0	0.060542	0.060542
90	23	9	1	0.272951	0.212409
90	23	9	2	0.584484	0.311533
90	23	9	3	0.834731	0.250248
90	23	9	4	0.955819	0.121088
90	23	9	5	0.992337	0.036518
90	23	9	6	0.999185	0.006847
90	23	9	7	0.999952	0.000767
90	23	9	8	0.999999	0.000047
90	23	9	9	1.000000	0.000001
90	23	10	0	0.043351	0.043351
90	23	10	1	0.215259	0.171908
90	23	10	2	0.503716	0.288456
90	23	10	3	0.772942	0.269226
90	23	10	4	0.927416	0.154474
90	23	10	5	0.984222	0.056807
90	23	10	6	0.997748	0.013525
90	23	10	7	0.999801	0.002053
90	23	10	8	0.999990	0.000189
90	23	10	9	1.000000	0.000010
90	23	10	10	1.000000	0.000000
90	23	11	0	0.030887	0.030887
90	23	11	1	0.167984	0.137097
90	23	11	2	0.427996	0.260011
90	23	11	3	0.705635	0.277639
90	23	11	4	0.890728	0.185093
90	23	11	5	0.971441	0.080713
90	23	11	6	0.994873	0.023433
90	23	11	7	0.999390	0.004517
90	23	11	8	0.999955	0.000565
90	23	11	9	0.999998	0.000043
90	23	11	10	1.000000	0.000002
90	23	11	11	1.000000	0.000000
90	23	12	0	0.021895	0.021895
90	23	12	1	0.129806	0.107911
90	23	12	2	0.358879	0.229073
90	23	12	3	0.635347	0.276468
90	23	12	4	0.846212	0.210865
90	23	12	5	0.953051	0.106838
90	23	12	6	0.989831	0.036780
90	23	12	7	0.998475	0.008644
90	23	12	8	0.999847	0.001372
90	23	12	9	0.999990	0.000143
90	23	12	10	1.000000	0.000009
90	23	12	11	1.000000	0.000000
90	23	12	12	1.000000	0.000000
90	23	13	0	0.015439	0.015439
90	23	13	1	0.099369	0.083930
90	23	13	2	0.297205	0.197836
90	23	13	3	0.564458	0.267252
90	23	13	4	0.794848	0.230390
90	23	13	5	0.928396	0.133548
90	23	13	6	0.981815	0.053419
90	23	13	7	0.996702	0.014887
90	23	13	8	0.999583	0.002881
90	23	13	9	0.999965	0.000381
90	23	13	10	0.999998	0.000033
90	23	13	11	1.000000	0.000002
90	23	13	12	1.000000	0.000000
90	23	13	13	1.000000	0.000000
90	23	14	0	0.010827	0.010827
90	23	14	1	0.075389	0.064562
90	23	14	2	0.243250	0.167861
90	23	14	3	0.495041	0.251791
90	23	14	4	0.737998	0.242957
90	23	14	5	0.897177	0.159178
90	23	14	6	0.970021	0.072844
90	23	14	7	0.993609	0.023588
90	23	14	8	0.999022	0.005414
90	23	14	9	0.999895	0.000873
90	23	14	10	0.999992	0.000097
90	23	14	11	1.000000	0.000007
90	23	14	12	1.000000	0.000000
90	23	14	13	1.000000	0.000000
90	23	14	14	1.000000	0.000000
90	23	15	0	0.007551	0.007551
90	23	15	1	0.056700	0.049150
90	23	15	2	0.196867	0.140167
90	23	15	3	0.428781	0.231913
90	23	15	4	0.677259	0.248478
90	23	15	5	0.859476	0.182217
90	23	15	6	0.953727	0.094250
90	23	15	7	0.988643	0.034916
90	23	15	8	0.997954	0.009311
90	23	15	9	0.999734	0.001781
90	23	15	10	0.999976	0.000241
90	23	15	11	0.999998	0.000023
90	23	15	12	1.000000	0.000001
90	23	15	13	1.000000	0.000000
90	23	15	14	1.000000	0.000000
90	23	15	15	1.000000	0.000000
90	23	16	0	0.005235	0.005235
90	23	16	1	0.042283	0.037048
90	23	16	2	0.157621	0.115338
90	23	16	3	0.366937	0.209316
90	23	16	4	0.614311	0.247374
90	23	16	5	0.815744	0.201433
90	23	16	6	0.932363	0.116619
90	23	16	7	0.981194	0.048831
90	23	16	8	0.996092	0.014897
90	23	16	9	0.999402	0.003311
90	23	16	10	0.999934	0.000532
90	23	16	11	0.999995	0.000061
90	23	16	12	1.000000	0.000005
90	23	16	13	1.000000	0.000000
90	23	16	14	1.000000	0.000000
90	23	16	15	1.000000	0.000000
90	23	16	16	1.000000	0.000000
90	23	17	0	0.003608	0.003608
90	23	17	1	0.031269	0.027661
90	23	17	2	0.124890	0.093621
90	23	17	3	0.310365	0.185475
90	23	17	4	0.550796	0.240431
90	23	17	5	0.766747	0.215951
90	23	17	6	0.905572	0.138826
90	23	17	7	0.970636	0.065064
90	23	17	8	0.993072	0.022436
90	23	17	9	0.998776	0.005704
90	23	17	10	0.999840	0.001065
90	23	17	11	0.999985	0.000144
90	23	17	12	0.999999	0.000014
90	23	17	13	1.000000	0.000001
90	23	17	14	1.000000	0.000000
90	23	17	15	1.000000	0.000000
90	23	17	16	1.000000	0.000000
90	23	17	17	1.000000	0.000000
90	23	18	0	0.002471	0.002471
90	23	18	1	0.022933	0.020461
90	23	18	2	0.097957	0.075025
90	23	18	3	0.259550	0.161592
90	23	18	4	0.488218	0.228668
90	23	18	5	0.713499	0.225281
90	23	18	6	0.873243	0.159744
90	23	18	7	0.956376	0.083132
90	23	18	8	0.988462	0.032086
90	23	18	9	0.997682	0.009220
90	23	18	10	0.999651	0.001969
90	23	18	11	0.999961	0.000310
90	23	18	12	0.999997	0.000036
90	23	18	13	1.000000	0.000003
90	23	18	14	1.000000	0.000000
90	23	18	15	1.000000	0.000000
90	23	18	16	1.000000	0.000000
90	23	18	17	1.000000	0.000000
90	23	18	18	1.000000	0.000000
90	23	19	0	0.001682	0.001682
90	23	19	1	0.016680	0.014999
90	23	19	2	0.076075	0.059395
90	23	19	3	0.214663	0.138588
90	23	19	4	0.427875	0.213212
90	23	19	5	0.657179	0.229304
90	23	19	6	0.835526	0.178347
90	23	19	7	0.937902	0.102376
90	23	19	8	0.981777	0.043875

N	n	k	x	P(x)	p(x)
90	23	19	9	0.995889	0.014112
90	23	19	10	0.999295	0.003406
90	23	19	11	0.999909	0.000614
90	23	19	12	0.999991	0.000082
90	23	19	13	0.999999	0.000008
90	23	19	14	1.000000	0.000001
90	23	19	15	1.000000	0.000000
90	23	19	16	1.000000	0.000000
90	23	19	17	1.000000	0.000000
90	23	19	18	1.000000	0.000000
90	23	19	19	1.000000	0.000000
90	23	20	0	0.001137	0.001137
90	23	20	1	0.012033	0.010896
90	23	20	2	0.058508	0.046475
90	23	20	3	0.175624	0.117116
90	23	20	4	0.370818	0.195194
90	23	20	5	0.599045	0.228227
90	23	20	6	0.792823	0.193778
90	23	20	7	0.914831	0.122008
90	23	20	8	0.972508	0.057677
90	23	20	9	0.993106	0.020599
90	23	20	10	0.998672	0.005565
90	23	20	11	0.999806	0.001134
90	23	20	12	0.999979	0.000173
90	23	20	13	0.999998	0.000020
90	23	20	14	1.000000	0.000002
90	23	20	15	1.000000	0.000000
90	23	20	16	1.000000	0.000000
90	23	20	17	1.000000	0.000000
90	23	20	18	1.000000	0.000000
90	23	20	19	1.000000	0.000000
90	23	20	20	1.000000	0.000000
90	23	21	0	0.000763	0.000763
90	23	21	1	0.008609	0.007845
90	23	21	2	0.044565	0.035957
90	23	21	3	0.142162	0.097597
90	23	21	4	0.317837	0.175675
90	23	21	5	0.540358	0.222521
90	23	21	6	0.745763	0.205404
90	23	21	7	0.886943	0.141181
90	23	21	8	0.960148	0.073205
90	23	21	9	0.988987	0.028838
90	23	21	10	0.997638	0.008651
90	23	21	11	0.999611	0.001973
90	23	21	12	0.999951	0.000340
90	23	21	13	0.999995	0.000044
90	23	21	14	1.000000	0.000004
90	23	21	15	1.000000	0.000000
90	23	21	16	1.000000	0.000000
90	23	21	17	1.000000	0.000000
90	23	21	18	1.000000	0.000000
90	23	21	19	1.000000	0.000000
90	23	21	20	1.000000	0.000000
90	23	21	21	1.000000	0.000000
90	23	22	0	0.000509	0.000509
90	23	22	1	0.006107	0.005598
90	23	22	2	0.033622	0.027515
90	23	22	3	0.113873	0.080251
90	23	22	4	0.269463	0.155590
90	23	22	5	0.482309	0.212846
90	23	22	6	0.695156	0.212846
90	23	22	7	0.854206	0.159050
90	23	22	8	0.944234	0.090028
90	23	22	9	0.983135	0.038901
90	23	22	10	0.996008	0.012873
90	23	22	11	0.999268	0.003260
90	23	22	12	0.999897	0.000629
90	23	22	13	0.999989	0.000092
90	23	22	14	0.999999	0.000010
90	23	22	15	1.000000	0.000001
90	23	22	16	1.000000	0.000000
90	23	22	17	1.000000	0.000000
90	23	22	18	1.000000	0.000000
90	23	22	19	1.000000	0.000000
90	23	22	20	1.000000	0.000000
90	23	22	21	1.000000	0.000000
90	23	22	22	1.000000	0.000000
90	23	23	0	0.000337	0.000337
90	23	23	1	0.004296	0.003959
90	23	23	2	0.025125	0.020829
90	23	23	3	0.090270	0.065145
90	23	23	4	0.225989	0.135719
90	23	23	5	0.425968	0.199978
90	23	23	6	0.641944	0.215977
90	23	23	7	0.816782	0.174838
90	23	23	8	0.924375	0.107593
90	23	23	9	0.975126	0.050751
90	23	23	10	0.993547	0.018421
90	23	23	11	0.998693	0.005146
90	23	23	12	0.999795	0.001103
90	23	23	13	0.999975	0.000180
90	23	23	14	0.999998	0.000022
90	23	23	15	1.000000	0.000002
90	23	23	16	1.000000	0.000000
90	23	23	17	1.000000	0.000000
90	23	23	18	1.000000	0.000000
90	23	23	19	1.000000	0.000000
90	23	23	20	1.000000	0.000000
90	23	23	21	1.000000	0.000000
90	23	23	22	1.000000	0.000000
90	23	23	23	1.000000	0.000000
90	24	1	0	0.733333	0.733333
90	24	1	1	1.000000	0.266667
90	24	2	0	0.535581	0.535581
90	24	2	1	0.931086	0.395506
90	24	2	2	1.000000	0.068914
90	24	3	0	0.389513	0.389513
90	24	3	1	0.827715	0.438202
90	24	3	2	0.982772	0.155056
90	24	3	3	1.000000	0.017228
90	24	4	0	0.282061	0.282061
90	24	4	1	0.711869	0.429808
90	24	4	2	0.943562	0.231693
90	24	4	3	0.995841	0.052279
90	24	4	4	1.000000	0.004159
90	24	5	0	0.203346	0.203346
90	24	5	1	0.596920	0.393574
90	24	5	2	0.884291	0.287371
90	24	5	3	0.983075	0.098784
90	24	5	4	0.999033	0.015957
90	24	5	5	1.000000	0.000967
90	24	6	0	0.145931	0.145931
90	24	6	1	0.490424	0.344493
90	24	6	2	0.809913	0.319489
90	24	6	3	0.958670	0.148757
90	24	6	4	0.995278	0.036608
90	24	6	5	0.999784	0.004506
90	24	6	6	1.000000	0.000216
90	24	7	0	0.104236	0.104236
90	24	7	1	0.396098	0.291862
90	24	7	2	0.726237	0.330139
90	24	7	3	0.921481	0.195243
90	24	7	4	0.986562	0.065081
90	24	7	5	0.998765	0.012203
90	24	7	6	0.999954	0.001189
90	24	7	7	1.000000	0.000046
90	24	8	0	0.074096	0.074096
90	24	8	1	0.315221	0.241125
90	24	8	2	0.638731	0.323510
90	24	8	3	0.872082	0.233351
90	24	8	4	0.970880	0.098798
90	24	8	5	0.995971	0.025092
90	24	8	6	0.999696	0.003725
90	24	8	7	0.999990	0.000295
90	24	8	8	1.000000	0.000009
90	24	9	0	0.052409	0.052409
90	24	9	1	0.247588	0.195179
90	24	9	2	0.551935	0.304347
90	24	9	3	0.812321	0.260386
90	24	9	4	0.946783	0.134462
90	24	9	5	0.990157	0.043375
90	24	9	6	0.998878	0.008721
90	24	9	7	0.999929	0.001051
90	24	9	8	0.999998	0.000069
90	24	9	9	1.000000	0.000002
90	24	10	0	0.036881	0.036881
90	24	10	1	0.192167	0.155287
90	24	10	2	0.469273	0.277106
90	24	10	3	0.744814	0.275541
90	24	10	4	0.913582	0.168769
90	24	10	5	0.979983	0.066401
90	24	10	6	0.996940	0.016957
90	24	10	7	0.999709	0.002769
90	24	10	8	0.999985	0.000276
90	24	10	9	1.000000	0.000015
90	24	10	10	1.000000	0.000000
90	24	11	0	0.025816	0.025816
90	24	11	1	0.147522	0.121706
90	24	11	2	0.393069	0.245547
90	24	11	3	0.672484	0.279415
90	24	11	4	0.871390	0.198906
90	24	11	5	0.964213	0.092823
90	24	11	6	0.993125	0.028912
90	24	11	7	0.999120	0.005996
90	24	11	8	0.999929	0.000809
90	24	11	9	0.999997	0.000067
90	24	11	10	1.000000	0.000003
90	24	11	11	1.000000	0.000000
90	24	12	0	0.017973	0.017973
90	24	12	1	0.112089	0.094115
90	24	12	2	0.324689	0.212600
90	24	12	3	0.598209	0.273521
90	24	12	4	0.821034	0.222825
90	24	12	5	0.941888	0.120854
90	24	12	6	0.986537	0.044649
90	24	12	7	0.997830	0.011293
90	24	12	8	0.999765	0.001935
90	24	12	9	0.999984	0.000218
90	24	12	10	0.999999	0.000015
90	24	12	11	1.000000	0.000001
90	24	12	12	1.000000	0.000000
90	24	13	0	0.012443	0.012443
90	24	13	1	0.084337	0.071894
90	24	13	2	0.264725	0.180388
90	24	13	3	0.524569	0.259844
90	24	13	4	0.763900	0.239330
90	24	13	5	0.912450	0.148550
90	24	13	6	0.976234	0.063784
90	24	13	7	0.995369	0.019135
90	24	13	8	0.999369	0.004000

N	n	k	x	P(x)	p(x)
90	24	13	9	0.999942	0.000573
90	24	13	10	0.999997	0.000055
90	24	13	11	1.000000	0.000003
90	24	13	12	1.000000	0.000000
90	24	13	13	1.000000	0.000000
90	24	14	0	0.008565	0.008565
90	24	14	1	0.062862	0.054297
90	24	14	2	0.213185	0.150323
90	24	14	3	0.453703	0.240517
90	24	14	4	0.701736	0.248033
90	24	14	5	0.875794	0.174058
90	24	14	6	0.961323	0.085529
90	24	14	7	0.991144	0.029821
90	24	14	8	0.998537	0.007393
90	24	14	9	0.999830	0.001293
90	24	14	10	0.999987	0.000156
90	24	14	11	0.999999	0.000013
90	24	14	12	1.000000	0.000001
90	24	14	13	1.000000	0.000000
90	24	14	14	1.000000	0.000000
90	24	15	0	0.005860	0.005860
90	24	15	1	0.046430	0.040570
90	24	15	2	0.169671	0.123241
90	24	15	3	0.387244	0.217573
90	24	15	4	0.636464	0.249220
90	24	15	5	0.832280	0.195816
90	24	15	6	0.941066	0.108787
90	24	15	7	0.984474	0.043407
90	24	15	8	0.996981	0.012507
90	24	15	9	0.999575	0.002594
90	24	15	10	0.999958	0.000383
90	24	15	11	0.999997	0.000039
90	24	15	12	1.000000	0.000003
90	24	15	13	1.000000	0.000000
90	24	15	14	1.000000	0.000000
90	24	15	15	1.000000	0.000000
90	24	16	0	0.003985	0.003985
90	24	16	1	0.033989	0.030004
90	24	16	2	0.133520	0.099532
90	24	16	3	0.326323	0.192803
90	24	16	4	0.570005	0.243682
90	24	16	5	0.782673	0.212668
90	24	16	6	0.914957	0.132284
90	24	16	7	0.974635	0.059677
90	24	16	8	0.994313	0.019678
90	24	16	9	0.999056	0.004743
90	24	16	10	0.999886	0.000830
90	24	16	11	0.999990	0.000104
90	24	16	12	0.999999	0.000009
90	24	16	13	1.000000	0.000001

N	n	k	x	P(x)	p(x)
90	24	16	14	1.000000	0.000000
90	24	16	15	1.000000	0.000000
90	24	16	16	1.000000	0.000000
90	24	17	0	0.002692	0.002692
90	24	17	1	0.024663	0.021971
90	24	17	2	0.103930	0.079267
90	24	17	3	0.271609	0.167679
90	24	17	4	0.504145	0.232536
90	24	17	5	0.728069	0.223924
90	24	17	6	0.882780	0.154711
90	24	17	7	0.960925	0.078145
90	24	17	8	0.990058	0.029133
90	24	17	9	0.998095	0.008037
90	24	17	10	0.999729	0.001635
90	24	17	11	0.999972	0.000243
90	24	17	12	0.999998	0.000026
90	24	17	13	1.000000	0.000002
90	24	17	14	1.000000	0.000000
90	24	17	15	1.000000	0.000000
90	24	17	16	1.000000	0.000000
90	24	17	17	1.000000	0.000000
90	24	18	0	0.001807	0.001807
90	24	18	1	0.017741	0.015934
90	24	18	2	0.080041	0.062300
90	24	18	3	0.223372	0.143331
90	24	18	4	0.440437	0.217064
90	24	18	5	0.669788	0.229351
90	24	18	6	0.844632	0.174845
90	24	18	7	0.942727	0.098095
90	24	18	8	0.983673	0.040946
90	24	18	9	0.996443	0.012771
90	24	18	10	0.999416	0.002972
90	24	18	11	0.999929	0.000513
90	24	18	12	0.999994	0.000065
90	24	18	13	0.999999	0.000006
90	24	18	14	1.000000	0.000000
90	24	18	15	1.000000	0.000000
90	24	18	16	1.000000	0.000000
90	24	18	17	1.000000	0.000000
90	24	18	18	1.000000	0.000000
90	24	19	0	0.001205	0.001205
90	24	19	1	0.012651	0.011446
90	24	19	2	0.061005	0.048354
90	24	19	3	0.181567	0.120563
90	24	19	4	0.380141	0.198574
90	24	19	5	0.609264	0.229123
90	24	19	6	0.800921	0.191657
90	24	19	7	0.919566	0.118645
90	24	19	8	0.974574	0.055008
90	24	19	9	0.993783	0.019209

N	n	k	x	P(x)	p(x)
90	24	19	10	0.998838	0.005055
90	24	19	11	0.999836	0.000998
90	24	19	12	0.999983	0.000147
90	24	19	13	0.999999	0.000016
90	24	19	14	1.000000	0.000001
90	24	19	15	1.000000	0.000000
90	24	19	16	1.000000	0.000000
90	24	19	17	1.000000	0.000000
90	24	19	18	1.000000	0.000000
90	24	19	19	1.000000	0.000000
90	24	20	0	0.000798	0.000798
90	24	20	1	0.008943	0.008145
90	24	20	2	0.046022	0.037079
90	24	20	3	0.145908	0.099886
90	24	20	4	0.324205	0.178297
90	24	20	5	0.547950	0.223745
90	24	20	6	0.752332	0.204382
90	24	20	7	0.891158	0.138826
90	24	20	8	0.962178	0.071020
90	24	20	9	0.989725	0.027547
90	24	20	10	0.997841	0.008117
90	24	20	11	0.999653	0.001812
90	24	20	12	0.999958	0.000305
90	24	20	13	0.999996	0.000038
90	24	20	14	1.000000	0.000003
90	24	20	15	1.000000	0.000000
90	24	20	16	1.000000	0.000000
90	24	20	17	1.000000	0.000000
90	24	20	18	1.000000	0.000000
90	24	20	19	1.000000	0.000000
90	24	20	20	1.000000	0.000000
90	24	21	0	0.000524	0.000524
90	24	21	1	0.006267	0.005743
90	24	21	2	0.034369	0.028102
90	24	21	3	0.115942	0.081574
90	24	21	4	0.273263	0.157321
90	24	21	5	0.487219	0.213956
90	24	21	6	0.699777	0.212558
90	24	21	7	0.857443	0.157666
90	24	21	8	0.945944	0.088501
90	24	21	9	0.983822	0.037877
90	24	21	10	0.996218	0.012396
90	24	21	11	0.999317	0.003099
90	24	21	12	0.999906	0.000589
90	24	21	13	0.999990	0.000084
90	24	21	14	0.999999	0.000009
90	24	21	15	1.000000	0.000001
90	24	21	16	1.000000	0.000000
90	24	21	17	1.000000	0.000000
90	24	21	18	1.000000	0.000000

N	n	k	x	P(x)	p(x)
90	24	21	19	1.000000	0.000000
90	24	21	20	1.000000	0.000000
90	24	21	21	1.000000	0.000000
90	24	22	0	0.000342	0.000342
90	24	22	1	0.004352	0.004011
90	24	22	2	0.025409	0.021056
90	24	22	3	0.091115	0.065707
90	24	22	4	0.227663	0.136547
90	24	22	5	0.428304	0.200641
90	24	22	6	0.644327	0.216023
90	24	22	7	0.818598	0.174271
90	24	22	8	0.925422	0.106825
90	24	22	9	0.975587	0.050165
90	24	22	10	0.993703	0.018115
90	24	22	11	0.998733	0.005030
90	24	22	12	0.999803	0.001070
90	24	22	13	0.999977	0.000173
90	24	22	14	0.999998	0.000021
90	24	22	15	1.000000	0.000002
90	24	22	16	1.000000	0.000000
90	24	22	17	1.000000	0.000000
90	24	22	18	1.000000	0.000000
90	24	22	19	1.000000	0.000000
90	24	22	20	1.000000	0.000000
90	24	22	21	1.000000	0.000000
90	24	22	22	1.000000	0.000000
90	24	23	0	0.000221	0.000221
90	24	23	1	0.002996	0.002775
90	24	23	2	0.018596	0.015600
90	24	23	3	0.070824	0.052227
90	24	23	4	0.187502	0.116678
90	24	23	5	0.372242	0.184740
90	24	23	6	0.587144	0.214902
90	24	23	7	0.775030	0.187886
90	24	23	8	0.900287	0.125257
90	24	23	9	0.964522	0.064234
90	24	23	10	0.989973	0.025451
90	24	23	11	0.997771	0.007798
90	24	23	12	0.999614	0.001843
90	24	23	13	0.999949	0.000334
90	24	23	14	0.999995	0.000046
90	24	23	15	0.999999	0.000005
90	24	23	16	1.000000	0.000000
90	24	23	17	1.000000	0.000000
90	24	23	18	1.000000	0.000000
90	24	23	19	1.000000	0.000000
90	24	23	20	1.000000	0.000000
90	24	23	21	1.000000	0.000000
90	24	23	22	1.000000	0.000000
90	24	23	23	1.000000	0.000000

N	n	k	x	P(x)	p(x)
90	24	24	0	0.000142	0.000142
90	24	24	1	0.002043	0.001901
90	24	24	2	0.013474	0.011430
90	24	24	3	0.054454	0.040980
90	24	24	4	0.152672	0.098219
90	24	24	5	0.319853	0.167181
90	24	24	6	0.529409	0.209556
90	24	24	7	0.727357	0.197948
90	24	24	8	0.870375	0.143018
90	24	24	9	0.950141	0.079766
90	24	24	10	0.984655	0.034514
90	24	24	11	0.996258	0.011603
90	24	24	12	0.999284	0.003026
90	24	24	13	0.999894	0.000609
90	24	24	14	0.999988	0.000094
90	24	24	15	0.999999	0.000011
90	24	24	16	1.000000	0.000001
90	24	24	17	1.000000	0.000000
90	24	24	18	1.000000	0.000000
90	24	24	19	1.000000	0.000000
90	24	24	20	1.000000	0.000000
90	24	24	21	1.000000	0.000000
90	24	24	22	1.000000	0.000000
90	24	24	23	1.000000	0.000000
90	24	24	24	1.000000	0.000000
90	25	1	0	0.722222	0.722222
90	25	1	1	1.000000	0.277778
90	25	2	0	0.519351	0.519351
90	25	2	1	0.925094	0.405743
90	25	2	2	1.000000	0.074906
90	25	3	0	0.371808	0.371808
90	25	3	1	0.814436	0.442629
90	25	3	2	0.980422	0.165986
90	25	3	3	1.000000	0.019578
90	25	4	0	0.264967	0.264967
90	25	4	1	0.692332	0.427365
90	25	4	2	0.936541	0.244209
90	25	4	3	0.995049	0.058508
90	25	4	4	1.000000	0.004951
90	25	5	0	0.187941	0.187941
90	25	5	1	0.573067	0.385126
90	25	5	2	0.871229	0.298162
90	25	5	3	0.980082	0.108853
90	25	5	4	0.998791	0.018709
90	25	5	5	1.000000	0.001209
90	25	6	0	0.132665	0.132665
90	25	6	1	0.464326	0.331661
90	25	6	2	0.790550	0.326224
90	25	6	3	0.951908	0.161358
90	25	6	4	0.994169	0.042260
90	25	6	5	0.999715	0.005547
90	25	6	6	1.000000	0.000284
90	25	7	0	0.093181	0.093181
90	25	7	1	0.369565	0.276384
90	25	7	2	0.701227	0.331661
90	25	7	3	0.909648	0.208421
90	25	7	4	0.983604	0.073956
90	25	7	5	0.998395	0.014791
90	25	7	6	0.999936	0.001541
90	25	7	7	1.000000	0.000064
90	25	8	0	0.065114	0.065114
90	25	8	1	0.289647	0.224533
90	25	8	2	0.609321	0.319674
90	25	8	3	0.854404	0.245083
90	25	8	4	0.964892	0.110488
90	25	8	5	0.994831	0.029939
90	25	8	6	0.999583	0.004752
90	25	8	7	0.999986	0.000403
90	25	8	8	1.000000	0.000014
90	25	9	0	0.045262	0.045262
90	25	9	1	0.223930	0.178668
90	25	9	2	0.519656	0.295726
90	25	9	3	0.788650	0.268994
90	25	9	4	0.936596	0.147946
90	25	9	5	0.987529	0.050932
90	25	9	6	0.998482	0.010953
90	25	9	7	0.999897	0.001416
90	25	9	8	0.999997	0.000100
90	25	9	9	1.000000	0.000003
90	25	10	0	0.031293	0.031293
90	25	10	1	0.170992	0.139699
90	25	10	2	0.435685	0.264693
90	25	10	3	0.715590	0.279905
90	25	10	4	0.898240	0.182650
90	25	10	5	0.974953	0.076713
90	25	10	6	0.995912	0.020960
90	25	10	7	0.999583	0.003670
90	25	10	8	0.999976	0.000393
90	25	10	9	0.999999	0.000023
90	25	10	10	1.000000	0.000001
90	25	11	0	0.021514	0.021514
90	25	11	1	0.129082	0.107568
90	25	11	2	0.359585	0.230503
90	25	11	3	0.638616	0.279030
90	25	11	4	0.850294	0.211678
90	25	11	5	0.955774	0.105480
90	25	11	6	0.990934	0.035160
90	25	11	7	0.998757	0.007823
90	25	11	8	0.999893	0.001136
90	25	11	9	0.999995	0.000102
90	25	11	10	1.000000	0.000005
90	25	11	11	1.000000	0.000000
90	25	12	0	0.014706	0.014706
90	25	12	1	0.096403	0.081697
90	25	12	2	0.292477	0.196074
90	25	12	3	0.560911	0.268434
90	25	12	4	0.794025	0.233114
90	25	12	5	0.929071	0.135045
90	25	12	6	0.982478	0.053408
90	25	12	7	0.996975	0.014496
90	25	12	8	0.999648	0.002674
90	25	12	9	0.999974	0.000326
90	25	12	10	0.999999	0.000025
90	25	12	11	1.000000	0.000001
90	25	12	12	1.000000	0.000000
90	25	13	0	0.009992	0.009992
90	25	13	1	0.071265	0.061273
90	25	13	2	0.234660	0.163395
90	25	13	3	0.485199	0.250539
90	25	13	4	0.731264	0.246065
90	25	13	5	0.894444	0.163180
90	25	13	6	0.969469	0.075025
90	25	13	7	0.993629	0.024161
90	25	13	8	0.999066	0.005436
90	25	13	9	0.999907	0.000842
90	25	13	10	0.999994	0.000087
90	25	13	11	1.000000	0.000006
90	25	13	12	1.000000	0.000000
90	25	13	13	1.000000	0.000000
90	25	14	0	0.006748	0.006748
90	25	14	1	0.052167	0.045419
90	25	14	2	0.185854	0.133687
90	25	14	3	0.413616	0.227762
90	25	14	4	0.664155	0.250539
90	25	14	5	0.852059	0.187904
90	25	14	6	0.950956	0.098897
90	25	14	7	0.987981	0.037025
90	25	14	8	0.997865	0.009884
90	25	14	9	0.999732	0.001867
90	25	14	10	0.999977	0.000245
90	25	14	11	0.999999	0.000022
90	25	14	12	1.000000	0.000001
90	25	14	13	1.000000	0.000000
90	25	14	14	1.000000	0.000000
90	25	15	0	0.004528	0.004528
90	25	15	1	0.037824	0.033296
90	25	15	2	0.145396	0.107572
90	25	15	3	0.347685	0.202289
90	25	15	4	0.594927	0.247242
90	25	15	5	0.802611	0.207683
90	25	15	6	0.926232	0.123621
90	25	15	7	0.979212	0.052980
90	25	15	8	0.995654	0.016442
90	25	15	9	0.999339	0.003685
90	25	15	10	0.999929	0.000590
90	25	15	11	0.999995	0.000066
90	25	15	12	1.000000	0.000005
90	25	15	13	1.000000	0.000000
90	25	15	14	1.000000	0.000000
90	25	15	15	1.000000	0.000000
90	25	16	0	0.003019	0.003019
90	25	16	1	0.027170	0.024151
90	25	16	2	0.112407	0.085238
90	25	16	3	0.288347	0.175940
90	25	16	4	0.525699	0.237352
90	25	16	5	0.747228	0.221529
90	25	16	6	0.894914	0.147686
90	25	16	7	0.966497	0.071582
90	25	16	8	0.991928	0.025431
90	25	16	9	0.998553	0.006626
90	25	16	10	0.999811	0.001258
90	25	16	11	0.999982	0.000172
90	25	16	12	0.999999	0.000016
90	25	16	13	1.000000	0.000001
90	25	16	14	1.000000	0.000000
90	25	16	15	1.000000	0.000000
90	25	16	16	1.000000	0.000000
90	25	17	0	0.001999	0.001999
90	25	17	1	0.019337	0.017338
90	25	17	2	0.085915	0.066578
90	25	17	3	0.236041	0.150126
90	25	17	4	0.458343	0.222302
90	25	17	5	0.687356	0.229013
90	25	17	6	0.856995	0.169639
90	25	17	7	0.949085	0.092090
90	25	17	8	0.986085	0.037000
90	25	17	9	0.997121	0.011035
90	25	17	10	0.999556	0.002435
90	25	17	11	0.999950	0.000394
90	25	17	12	0.999996	0.000046
90	25	17	13	1.000000	0.000004
90	25	17	14	1.000000	0.000000
90	25	17	15	1.000000	0.000000
90	25	17	16	1.000000	0.000000
90	25	17	17	1.000000	0.000000
90	25	18	0	0.001314	0.001314
90	25	18	1	0.013637	0.012322
90	25	18	2	0.064938	0.051301
90	25	18	3	0.190797	0.125859
90	25	18	4	0.394393	0.203596

N	n	k	x	P(x)	p(x)
90	25	18	5	0.624612	0.230220
90	25	18	6	0.812842	0.188230
90	25	18	7	0.926378	0.113536
90	25	18	8	0.977469	0.051091
90	25	18	9	0.994702	0.017233
90	25	18	10	0.999056	0.004354
90	25	18	11	0.999874	0.000819
90	25	18	12	0.999988	0.000113
90	25	18	13	0.999999	0.000011
90	25	18	14	1.000000	0.000001
90	25	18	15	1.000000	0.000000
90	25	18	16	1.000000	0.000000
90	25	18	17	1.000000	0.000000
90	25	18	18	1.000000	0.000000
90	25	19	0	0.000858	0.000858
90	25	19	1	0.009529	0.008671
90	25	19	2	0.048550	0.039021
90	25	19	3	0.152340	0.103790
90	25	19	4	0.335011	0.182671
90	25	19	5	0.560663	0.225652
90	25	19	6	0.763171	0.202508
90	25	19	7	0.897994	0.134823
90	25	19	8	0.965406	0.067412
90	25	19	9	0.990872	0.025467
90	25	19	10	0.998149	0.007276
90	25	19	11	0.999715	0.001567
90	25	19	12	0.999967	0.000252
90	25	19	13	0.999997	0.000030
90	25	19	14	1.000000	0.000003
90	25	19	15	1.000000	0.000000
90	25	19	16	1.000000	0.000000
90	25	19	17	1.000000	0.000000
90	25	19	18	1.000000	0.000000
90	25	19	19	1.000000	0.000000
90	25	20	0	0.000556	0.000556
90	25	20	1	0.006598	0.006042
90	25	20	2	0.035910	0.029311
90	25	20	3	0.120180	0.084270
90	25	20	4	0.280981	0.160802
90	25	20	5	0.497099	0.216117
90	25	20	6	0.708978	0.211880
90	25	20	7	0.863813	0.154835
90	25	20	8	0.949265	0.085451
90	25	20	9	0.985133	0.035869
90	25	20	10	0.996611	0.011478
90	25	20	11	0.999406	0.002795
90	25	20	12	0.999921	0.000515
90	25	20	13	0.999992	0.000071
90	25	20	14	0.999999	0.000007
90	25	20	15	1.000000	0.000001

N	n	k	x	P(x)	p(x)
90	25	20	16	1.000000	0.000000
90	25	20	17	1.000000	0.000000
90	25	20	18	1.000000	0.000000
90	25	20	19	1.000000	0.000000
90	25	20	20	1.000000	0.000000
90	25	21	0	0.000357	0.000357
90	25	21	1	0.004527	0.004169
90	25	21	2	0.026279	0.021752
90	25	21	3	0.093695	0.067416
90	25	21	4	0.232741	0.139046
90	25	21	5	0.435351	0.202610
90	25	21	6	0.651468	0.216117
90	25	21	7	0.823999	0.172531
90	25	21	8	0.928512	0.104514
90	25	21	9	0.976935	0.048422
90	25	21	10	0.994152	0.017217
90	25	21	11	0.998847	0.004696
90	25	21	12	0.999825	0.000978
90	25	21	13	0.999980	0.000154
90	25	21	14	0.999998	0.000018
90	25	21	15	1.000000	0.000002
90	25	21	16	1.000000	0.000000
90	25	21	17	1.000000	0.000000
90	25	21	18	1.000000	0.000000
90	25	21	19	1.000000	0.000000
90	25	21	20	1.000000	0.000000
90	25	21	21	1.000000	0.000000
90	25	22	0	0.000228	0.000228
90	25	22	1	0.003076	0.002848
90	25	22	2	0.019028	0.015952
90	25	22	3	0.072200	0.053172
90	25	22	4	0.190423	0.118223
90	25	22	5	0.376623	0.186201
90	25	22	6	0.591958	0.215334
90	25	22	7	0.778991	0.187033
90	25	22	8	0.902763	0.123772
90	25	22	9	0.965706	0.062944
90	25	22	10	0.990409	0.024702
90	25	22	11	0.997895	0.007486
90	25	22	12	0.999641	0.001747
90	25	22	13	0.999953	0.000312
90	25	22	14	0.999995	0.000042
90	25	22	15	1.000000	0.000004
90	25	22	16	1.000000	0.000000
90	25	22	17	1.000000	0.000000
90	25	22	18	1.000000	0.000000
90	25	22	19	1.000000	0.000000
90	25	22	20	1.000000	0.000000
90	25	22	21	1.000000	0.000000
90	25	22	22	1.000000	0.000000

N	n	k	x	P(x)	p(x)
90	25	23	0	0.000144	0.000144
90	25	23	1	0.002071	0.001927
90	25	23	2	0.013633	0.011562
90	25	23	3	0.054997	0.041365
90	25	23	4	0.153913	0.098915
90	25	23	5	0.321858	0.167946
90	25	23	6	0.531791	0.209932
90	25	23	7	0.729482	0.197691
90	25	23	8	0.871820	0.142338
90	25	23	9	0.950896	0.079077
90	25	23	10	0.984960	0.034064
90	25	23	11	0.996353	0.011394
90	25	23	12	0.999307	0.002954
90	25	23	13	0.999898	0.000591
90	25	23	14	0.999988	0.000090
90	25	23	15	0.999999	0.000010
90	25	23	16	1.000000	0.000001
90	25	23	17	1.000000	0.000000
90	25	23	18	1.000000	0.000000
90	25	23	19	1.000000	0.000000
90	25	23	20	1.000000	0.000000
90	25	23	21	1.000000	0.000000
90	25	23	22	1.000000	0.000000
90	25	23	23	1.000000	0.000000
90	25	24	0	0.000090	0.000090
90	25	24	1	0.001381	0.001290
90	25	24	2	0.009664	0.008283
90	25	24	3	0.041415	0.031751
90	25	24	4	0.122909	0.081494
90	25	24	5	0.271725	0.148816
90	25	24	6	0.472258	0.200532
90	25	24	7	0.676371	0.204113
90	25	24	8	0.835704	0.159333
90	25	24	9	0.932012	0.096308
90	25	24	10	0.977334	0.045321
90	25	24	11	0.993973	0.016639
90	25	24	12	0.998734	0.004761
90	25	24	13	0.999792	0.001058
90	25	24	14	0.999974	0.000181
90	25	24	15	0.999997	0.000024
90	25	24	16	1.000000	0.000002
90	25	24	17	1.000000	0.000000
90	25	24	18	1.000000	0.000000
90	25	24	19	1.000000	0.000000
90	25	24	20	1.000000	0.000000
90	25	24	21	1.000000	0.000000
90	25	24	22	1.000000	0.000000
90	25	24	23	1.000000	0.000000
90	25	24	24	1.000000	0.000000
90	25	25	0	0.000056	0.000056

N	n	k	x	P(x)	p(x)
90	25	25	1	0.000912	0.000855
90	25	25	2	0.006777	0.005866
90	25	25	3	0.030831	0.024054
90	25	25	4	0.096979	0.066148
90	25	25	5	0.226630	0.129650
90	25	25	6	0.414529	0.187899
90	25	25	7	0.620704	0.206175
90	25	25	8	0.794664	0.173960
90	25	25	9	0.908665	0.114001
90	25	25	10	0.967033	0.058369
90	25	25	11	0.990443	0.023410
90	25	25	12	0.997796	0.007353
90	25	25	13	0.999600	0.001804
90	25	25	14	0.999943	0.000344
90	25	25	15	0.999994	0.000050
90	25	25	16	0.999999	0.000006
90	25	25	17	1.000000	0.000000
90	25	25	18	1.000000	0.000000
90	25	25	19	1.000000	0.000000
90	25	25	20	1.000000	0.000000
90	25	25	21	1.000000	0.000000
90	25	25	22	1.000000	0.000000
90	25	25	23	1.000000	0.000000
90	25	25	24	1.000000	0.000000
90	25	25	25	1.000000	0.000000
90	26	1	0	0.711111	0.711111
90	26	1	1	1.000000	0.288889
90	26	2	0	0.503371	0.503371
90	26	2	1	0.918851	0.415481
90	26	2	2	1.000000	0.081149
90	26	3	0	0.354648	0.354648
90	26	3	1	0.800817	0.446170
90	26	3	2	0.977869	0.177051
90	26	3	3	1.000000	0.022131
90	26	4	0	0.248661	0.248661
90	26	4	1	0.672607	0.423947
90	26	4	2	0.929027	0.256419
90	26	4	3	0.994149	0.065122
90	26	4	4	1.000000	0.005851
90	26	5	0	0.173484	0.173484
90	26	5	1	0.549367	0.375883
90	26	5	2	0.857468	0.308101
90	26	5	3	0.976733	0.119265
90	26	5	4	0.998503	0.021771
90	26	5	5	1.000000	0.001497
90	26	6	0	0.120419	0.120419
90	26	6	1	0.438813	0.318395
90	26	6	2	0.770475	0.331661
90	26	6	3	0.944461	0.173986
90	26	6	4	0.992868	0.048407

Table for $N = 2$, $n = 1$, through $N = 100$, $n = 50$

N	n	k	x	P(x)	p(x)
90	26	6	5	0.999630	0.006762
90	26	6	6	1.000000	0.000370
90	26	7	0	0.083146	0.083146
90	26	7	1	0.344053	0.260907
90	26	7	2	0.675714	0.331661
90	26	7	3	0.896822	0.221108
90	26	7	4	0.980190	0.083368
90	26	7	5	0.997940	0.017749
90	26	7	6	0.999912	0.001972
90	26	7	7	1.000000	0.000088
90	26	8	0	0.057100	0.057100
90	26	8	1	0.265467	0.208366
90	26	8	2	0.579812	0.314346
90	26	8	3	0.835551	0.255739
90	26	8	4	0.958093	0.122542
90	26	8	5	0.993449	0.035356
90	26	8	6	0.999437	0.005988
90	26	8	7	0.999980	0.000543
90	26	8	8	1.000000	0.000020
90	26	9	0	0.038995	0.038995
90	26	9	1	0.201940	0.162945
90	26	9	2	0.487809	0.285868
90	26	9	3	0.763820	0.276011
90	26	9	4	0.925216	0.161396
90	26	9	5	0.984394	0.059179
90	26	9	6	0.997976	0.013582
90	26	9	7	0.999854	0.001878
90	26	9	8	0.999995	0.000142
90	26	9	9	1.000000	0.000004
90	26	10	0	0.026478	0.026478
90	26	10	1	0.151649	0.125170
90	26	10	2	0.403107	0.251458
90	26	10	3	0.685446	0.282339
90	26	10	4	0.881380	0.195934
90	26	10	5	0.969052	0.087672
90	26	10	6	0.994623	0.025571
90	26	10	7	0.999413	0.004791
90	26	10	8	0.999964	0.000551
90	26	10	9	0.999999	0.000035
90	26	10	10	1.000000	0.000001
90	26	11	0	0.017873	0.017873
90	26	11	1	0.112533	0.094660
90	26	11	2	0.327669	0.215137
90	26	11	3	0.604274	0.276604
90	26	11	4	0.827498	0.223224
90	26	11	5	0.946038	0.118540
90	26	11	6	0.988230	0.042192
90	26	11	7	0.998276	0.010046
90	26	11	8	0.999840	0.001564
90	26	11	9	0.999991	0.000151

N	n	k	x	P(x)	p(x)
90	26	11	10	1.000000	0.000008
90	26	11	11	1.000000	0.000000
90	26	12	0	0.011991	0.011991
90	26	12	1	0.082577	0.070587
90	26	12	2	0.262312	0.179734
90	26	12	3	0.523743	0.261432
90	26	12	4	0.765334	0.241591
90	26	12	5	0.914527	0.149193
90	26	12	6	0.977548	0.063021
90	26	12	7	0.995860	0.018311
90	26	12	8	0.999484	0.003624
90	26	12	9	0.999959	0.000475
90	26	12	10	0.999998	0.000039
90	26	12	11	1.000000	0.000002
90	26	12	12	1.000000	0.000000
90	26	13	0	0.007994	0.007994
90	26	13	1	0.059953	0.051960
90	26	13	2	0.207009	0.147055
90	26	13	3	0.446654	0.239646
90	26	13	4	0.697193	0.250539
90	26	13	5	0.874360	0.177167
90	26	13	6	0.961389	0.087029
90	26	13	7	0.991399	0.030010
90	26	13	8	0.998647	0.007248
90	26	13	9	0.999855	0.001208
90	26	13	10	0.999990	0.000135
90	26	13	11	0.999999	0.000009
90	26	13	12	1.000000	0.000000
90	26	13	13	1.000000	0.000000
90	26	14	0	0.005295	0.005295
90	26	14	1	0.043083	0.037789
90	26	14	2	0.161173	0.118090
90	26	14	3	0.375072	0.213899
90	26	14	4	0.625611	0.250539
90	26	14	5	0.826042	0.200431
90	26	14	6	0.938784	0.112742
90	26	14	7	0.983994	0.045210
90	26	14	8	0.996953	0.012959
90	26	14	9	0.999589	0.002636
90	26	14	10	0.999962	0.000373
90	26	14	11	0.999998	0.000036
90	26	14	12	1.000000	0.000002
90	26	14	13	1.000000	0.000000
90	26	14	14	1.000000	0.000000
90	26	15	0	0.003483	0.003483
90	26	15	1	0.030653	0.027170
90	26	15	2	0.123882	0.093229
90	26	15	3	0.310339	0.186458
90	26	15	4	0.553086	0.242747
90	26	15	5	0.770659	0.217573

N	n	k	x	P(x)	p(x)
90	26	15	6	0.909115	0.138456
90	26	15	7	0.972692	0.063577
90	26	15	8	0.993884	0.021192
90	26	15	9	0.998999	0.005115
90	26	15	10	0.999883	0.000884
90	26	15	11	0.999991	0.000107
90	26	15	12	0.999999	0.000009
90	26	15	13	1.000000	0.000000
90	26	15	14	1.000000	0.000000
90	26	15	15	1.000000	0.000000
90	26	16	0	0.002276	0.002276
90	26	16	1	0.021596	0.019321
90	26	16	2	0.094048	0.072452
90	26	16	3	0.253159	0.159111
90	26	16	4	0.481881	0.228721
90	26	16	5	0.709739	0.227858
90	26	16	6	0.872193	0.162455
90	26	16	7	0.956585	0.084392
90	26	16	8	0.988798	0.032212
90	26	16	9	0.997840	0.009042
90	26	16	10	0.999695	0.001855
90	26	16	11	0.999969	0.000274
90	26	16	12	0.999998	0.000029
90	26	16	13	1.000000	0.000002
90	26	16	14	1.000000	0.000000
90	26	16	15	1.000000	0.000000
90	26	16	16	1.000000	0.000000
90	26	17	0	0.001476	0.001476
90	26	17	1	0.015069	0.013593
90	26	17	2	0.070550	0.055481
90	26	17	3	0.203706	0.133155
90	26	17	4	0.413882	0.210177
90	26	17	5	0.645076	0.231194
90	26	17	6	0.828287	0.183210
90	26	17	7	0.934917	0.106630
90	26	17	8	0.980962	0.046045
90	26	17	9	0.995762	0.014800
90	26	17	10	0.999294	0.003531
90	26	17	11	0.999914	0.000620
90	26	17	12	0.999992	0.000079
90	26	17	13	0.999999	0.000007
90	26	17	14	1.000000	0.000000
90	26	17	15	1.000000	0.000000
90	26	17	16	1.000000	0.000000
90	26	17	17	1.000000	0.000000
90	26	18	0	0.000950	0.000950
90	26	18	1	0.010414	0.009464
90	26	18	2	0.052310	0.041896
90	26	18	3	0.161753	0.109443
90	26	18	4	0.350541	0.188789

N	n	k	x	P(x)	p(x)
90	26	18	5	0.578569	0.228027
90	26	18	6	0.778092	0.199524
90	26	18	7	0.907164	0.129072
90	26	18	8	0.969609	0.062445
90	26	18	9	0.992316	0.022707
90	26	18	10	0.998520	0.006204
90	26	18	11	0.999786	0.001267
90	26	18	12	0.999977	0.000191
90	26	18	13	0.999998	0.000021
90	26	18	14	1.000000	0.000002
90	26	18	15	1.000000	0.000000
90	26	18	16	1.000000	0.000000
90	26	18	17	1.000000	0.000000
90	26	18	18	1.000000	0.000000
90	26	19	0	0.000607	0.000607
90	26	19	1	0.007128	0.006521
90	26	19	2	0.038345	0.031217
90	26	19	3	0.126792	0.088447
90	26	19	4	0.292856	0.166064
90	26	19	5	0.512061	0.219205
90	26	19	6	0.722669	0.210608
90	26	19	7	0.873104	0.150435
90	26	19	8	0.953998	0.080894
90	26	19	9	0.986954	0.032957
90	26	19	10	0.997141	0.010187
90	26	19	11	0.999522	0.002381
90	26	19	12	0.999940	0.000418
90	26	19	13	0.999994	0.000054
90	26	19	14	1.000000	0.000005
90	26	19	15	1.000000	0.000000
90	26	19	16	1.000000	0.000000
90	26	19	17	1.000000	0.000000
90	26	19	18	1.000000	0.000000
90	26	19	19	1.000000	0.000000
90	26	20	0	0.000385	0.000385
90	26	20	1	0.004832	0.004447
90	26	20	2	0.027793	0.022961
90	26	20	3	0.098140	0.070347
90	26	20	4	0.241399	0.143260
90	26	20	5	0.447225	0.205826
90	26	20	6	0.663343	0.216117
90	26	20	7	0.832846	0.169504
90	26	20	8	0.933489	0.100643
90	26	20	9	0.979063	0.045574
90	26	20	10	0.994845	0.015782
90	26	20	11	0.999019	0.004174
90	26	20	12	0.999858	0.000838
90	26	20	13	0.999984	0.000127
90	26	20	14	0.999999	0.000014
90	26	20	15	1.000000	0.000001

N	n	k	x	P(x)	p(x)
90	26	20	16	1.000000	0.000000
90	26	20	17	1.000000	0.000000
90	26	20	18	1.000000	0.000000
90	26	20	19	1.000000	0.000000
90	26	20	20	1.000000	0.000000
90	26	21	0	0.000242	0.000242
90	26	21	1	0.003244	0.003002
90	26	21	2	0.019920	0.016677
90	26	21	3	0.075026	0.055105
90	26	21	4	0.196375	0.121349
90	26	21	5	0.385478	0.189103
90	26	21	6	0.601595	0.216117
90	26	21	7	0.786838	0.185243
90	26	21	8	0.907610	0.120771
90	26	21	9	0.967995	0.060386
90	26	21	10	0.991238	0.023243
90	26	21	11	0.998125	0.006887
90	26	21	12	0.999690	0.001565
90	26	21	13	0.999961	0.000271
90	26	21	14	0.999996	0.000035
90	26	21	15	1.000000	0.000003
90	26	21	16	1.000000	0.000000
90	26	21	17	1.000000	0.000000
90	26	21	18	1.000000	0.000000
90	26	21	19	1.000000	0.000000
90	26	21	20	1.000000	0.000000
90	26	21	21	1.000000	0.000000
90	26	22	0	0.000151	0.000151
90	26	22	1	0.002156	0.002005
90	26	22	2	0.014120	0.011964
90	26	22	3	0.056657	0.042538
90	26	22	4	0.157684	0.101027
90	26	22	5	0.327925	0.170241
90	26	22	6	0.538952	0.211028
90	26	22	7	0.735829	0.196877
90	26	22	8	0.876104	0.140275
90	26	22	9	0.953118	0.077014
90	26	22	10	0.985848	0.032731
90	26	22	11	0.996628	0.010779
90	26	22	12	0.999372	0.002745
90	26	22	13	0.999910	0.000537
90	26	22	14	0.999990	0.000080
90	26	22	15	0.999999	0.000009
90	26	22	16	1.000000	0.000001
90	26	22	17	1.000000	0.000000
90	26	22	18	1.000000	0.000000
90	26	22	19	1.000000	0.000000
90	26	22	20	1.000000	0.000000
90	26	22	21	1.000000	0.000000
90	26	22	22	1.000000	0.000000

N	n	k	x	P(x)	p(x)
90	26	23	0	0.000093	0.000093
90	26	23	1	0.001419	0.001326
90	26	23	2	0.009897	0.008478
90	26	23	3	0.042270	0.032372
90	26	23	4	0.124999	0.082729
90	26	23	5	0.275350	0.150351
90	26	23	6	0.476885	0.201535
90	26	23	7	0.680819	0.203934
90	26	23	8	0.838972	0.158153
90	26	23	9	0.933864	0.094892
90	26	23	10	0.978147	0.044283
90	26	23	11	0.994250	0.016103
90	26	23	12	0.998807	0.004557
90	26	23	13	0.999807	0.001000
90	26	23	14	0.999976	0.000169
90	26	23	15	0.999998	0.000022
90	26	23	16	1.000000	0.000002
90	26	23	17	1.000000	0.000000
90	26	23	18	1.000000	0.000000
90	26	23	19	1.000000	0.000000
90	26	23	20	1.000000	0.000000
90	26	23	21	1.000000	0.000000
90	26	23	22	1.000000	0.000000
90	26	23	23	1.000000	0.000000
90	26	24	0	0.000057	0.000057
90	26	24	1	0.000924	0.000867
90	26	24	2	0.006860	0.005936
90	26	24	3	0.031157	0.024296
90	26	24	4	0.097834	0.066677
90	26	24	5	0.228225	0.130391
90	26	24	6	0.416726	0.188500
90	26	24	7	0.622987	0.206262
90	26	24	8	0.796484	0.173496
90	26	24	9	0.909787	0.113304
90	26	24	10	0.967572	0.057785
90	26	24	11	0.990645	0.023073
90	26	24	12	0.997855	0.007210
90	26	24	13	0.999613	0.001758
90	26	24	14	0.999946	0.000333
90	26	24	15	0.999994	0.000048
90	26	24	16	0.999999	0.000005
90	26	24	17	1.000000	0.000000
90	26	24	18	1.000000	0.000000
90	26	24	19	1.000000	0.000000
90	26	24	20	1.000000	0.000000
90	26	24	21	1.000000	0.000000
90	26	24	22	1.000000	0.000000
90	26	24	23	1.000000	0.000000
90	26	24	24	1.000000	0.000000
90	26	25	0	0.000035	0.000035

N	n	k	x	P(x)	p(x)
90	26	25	1	0.000596	0.000561
90	26	25	2	0.004702	0.004106
90	26	25	3	0.022690	0.017988
90	26	25	4	0.075608	0.052919
90	26	25	5	0.186737	0.111129
90	26	25	6	0.359604	0.172867
90	26	25	7	0.563609	0.204005
90	26	25	8	0.749166	0.185557
90	26	25	9	0.880603	0.131437
90	26	25	10	0.953564	0.072961
90	26	25	11	0.985401	0.031837
90	26	25	12	0.996326	0.010925
90	26	25	13	0.999267	0.002941
90	26	25	14	0.999885	0.000618
90	26	25	15	0.999986	0.000101
90	26	25	16	0.999999	0.000013
90	26	25	17	1.000000	0.000001
90	26	25	18	1.000000	0.000000
90	26	25	19	1.000000	0.000000
90	26	25	20	1.000000	0.000000
90	26	25	21	1.000000	0.000000
90	26	25	22	1.000000	0.000000
90	26	25	23	1.000000	0.000000
90	26	25	24	1.000000	0.000000
90	26	25	25	1.000000	0.000000
90	26	26	0	0.000021	0.000021
90	26	26	1	0.000380	0.000359
90	26	26	2	0.003186	0.002806
90	26	26	3	0.016325	0.013139
90	26	26	4	0.057697	0.041373
90	26	26	5	0.150834	0.093137
90	26	26	6	0.306414	0.155580
90	26	26	7	0.503977	0.197562
90	26	26	8	0.697781	0.193804
90	26	26	9	0.846227	0.148446
90	26	26	10	0.935604	0.089377
90	26	26	11	0.978054	0.042450
90	26	26	12	0.993973	0.015919
90	26	26	13	0.998679	0.004706
90	26	26	14	0.999771	0.001092
90	26	26	15	0.999969	0.000198
90	26	26	16	0.999997	0.000028
90	26	26	17	1.000000	0.000003
90	26	26	18	1.000000	0.000000
90	26	26	19	1.000000	0.000000
90	26	26	20	1.000000	0.000000
90	26	26	21	1.000000	0.000000
90	26	26	22	1.000000	0.000000
90	26	26	23	1.000000	0.000000
90	26	26	24	1.000000	0.000000

N	n	k	x	P(x)	p(x)
90	26	26	25	1.000000	0.000000
90	26	26	26	1.000000	0.000000
90	27	1	0	0.700000	0.700000
90	27	1	1	1.000000	0.300000
90	27	2	0	0.487640	0.487640
90	27	2	1	0.912360	0.424719
90	27	2	2	1.000000	0.087640
90	27	3	0	0.338023	0.338023
90	27	3	1	0.786874	0.448851
90	27	3	2	0.975102	0.188228
90	27	3	3	1.000000	0.024898
90	27	4	0	0.233120	0.233120
90	27	4	1	0.652735	0.419615
90	27	4	2	0.921014	0.268279
90	27	4	3	0.993132	0.072118
90	27	4	4	1.000000	0.006868
90	27	5	0	0.159931	0.159931
90	27	5	1	0.525875	0.365944
90	27	5	2	0.843026	0.317151
90	27	5	3	0.973006	0.129980
90	27	5	4	0.998163	0.025157
90	27	5	5	1.000000	0.001837
90	27	6	0	0.109129	0.109129
90	27	6	1	0.413939	0.304809
90	27	6	2	0.749746	0.335807
90	27	6	3	0.936305	0.186559
90	27	6	4	0.991356	0.055050
90	27	6	5	0.999524	0.008169
90	27	6	6	1.000000	0.000475
90	27	7	0	0.074052	0.074052
90	27	7	1	0.319593	0.245541
90	27	7	2	0.649803	0.330210
90	27	7	3	0.883003	0.233199
90	27	7	4	0.976282	0.093280
90	27	7	5	0.997385	0.021103
90	27	7	6	0.999881	0.002496
90	27	7	7	1.000000	0.000119
90	27	8	0	0.049963	0.049963
90	27	8	1	0.242677	0.192714
90	27	8	2	0.550342	0.307666
90	27	8	3	0.815571	0.265229
90	27	8	4	0.950434	0.134862
90	27	8	5	0.991792	0.041358
90	27	8	6	0.999249	0.007458
90	27	8	7	0.999971	0.000722
90	27	8	8	1.000000	0.000029
90	27	9	0	0.033512	0.033512
90	27	9	1	0.181572	0.148061
90	27	9	2	0.456542	0.274970
90	27	9	3	0.737943	0.281402

N	n	k	x	P(x)	p(x)
90	27	9	4	0.912607	0.174663
90	27	9	5	0.980696	0.068089
90	27	9	6	0.997340	0.016644
90	27	9	7	0.999795	0.002456
90	27	9	8	0.999993	0.000198
90	27	9	9	1.000000	0.000007
90	27	10	0	0.022341	0.022341
90	27	10	1	0.134047	0.111705
90	27	10	2	0.371675	0.237628
90	27	10	3	0.654565	0.282891
90	27	10	4	0.863011	0.208446
90	27	10	5	0.962202	0.099191
90	27	10	6	0.993024	0.030822
90	27	10	7	0.999189	0.006164
90	27	10	8	0.999947	0.000758
90	27	10	9	0.999998	0.000052
90	27	10	10	1.000000	0.000001
90	27	11	0	0.014801	0.014801
90	27	11	1	0.097742	0.082941
90	27	11	2	0.297416	0.199674
90	27	11	3	0.569698	0.272282
90	27	11	4	0.803083	0.233385
90	27	11	5	0.934925	0.131842
90	27	11	6	0.984934	0.050009
90	27	11	7	0.997648	0.012714
90	27	11	8	0.999767	0.002119
90	27	11	9	0.999987	0.000220
90	27	11	10	1.000000	0.000013
90	27	11	11	1.000000	0.000000
90	27	12	0	0.009742	0.009742
90	27	12	1	0.070445	0.060703
90	27	12	2	0.234228	0.163783
90	27	12	3	0.486979	0.252751
90	27	12	4	0.735135	0.248156
90	27	12	5	0.898209	0.163074
90	27	12	6	0.971640	0.073431
90	27	12	7	0.994429	0.022789
90	27	12	8	0.999257	0.004828
90	27	12	9	0.999937	0.000680
90	27	12	10	0.999997	0.000060
90	27	12	11	1.000000	0.000003
90	27	12	12	1.000000	0.000000
90	27	13	0	0.006370	0.006370
90	27	13	1	0.050211	0.043841
90	27	13	2	0.181734	0.131523
90	27	13	3	0.409210	0.227476
90	27	13	4	0.661961	0.252751
90	27	13	5	0.852214	0.190253
90	27	13	6	0.951870	0.099656
90	27	13	7	0.988586	0.036715
90	27	13	8	0.998081	0.009495
90	27	13	9	0.999780	0.001699
90	27	13	10	0.999984	0.000204
90	27	13	11	0.999999	0.000015
90	27	13	12	1.000000	0.000001
90	27	13	13	1.000000	0.000000
90	27	14	0	0.004136	0.004136
90	27	14	1	0.035408	0.031271
90	27	14	2	0.139031	0.103624
90	27	14	3	0.338308	0.199277
90	27	14	4	0.586464	0.248156
90	27	14	5	0.797856	0.211392
90	27	14	6	0.924691	0.126835
90	27	14	7	0.979049	0.054358
90	27	14	8	0.995738	0.016689
90	27	14	9	0.999383	0.003645
90	27	14	10	0.999939	0.000556
90	27	14	11	0.999996	0.000057
90	27	14	12	1.000000	0.000004
90	27	14	13	1.000000	0.000000
90	27	14	14	1.000000	0.000000
90	27	15	0	0.002667	0.002667
90	27	15	1	0.024709	0.022043
90	27	15	2	0.104945	0.080235
90	27	15	3	0.275379	0.170434
90	27	15	4	0.511364	0.235986
90	27	15	5	0.736664	0.225299
90	27	15	6	0.889645	0.152981
90	27	15	7	0.964744	0.075100
90	27	15	8	0.991566	0.026821
90	27	15	9	0.998519	0.006954
90	27	15	10	0.999814	0.001295
90	27	15	11	0.999984	0.000170
90	27	15	12	0.999999	0.000015
90	27	15	13	1.000000	0.000001
90	27	15	14	1.000000	0.000000
90	27	15	15	1.000000	0.000000
90	27	16	0	0.001707	0.001707
90	27	16	1	0.017068	0.015361
90	27	16	2	0.078200	0.061131
90	27	16	3	0.220840	0.142640
90	27	16	4	0.438995	0.218156
90	27	16	5	0.670576	0.231580
90	27	16	6	0.846810	0.176234
90	27	16	7	0.944718	0.097908
90	27	16	8	0.984771	0.040053
90	27	16	9	0.996851	0.012080
90	27	16	10	0.999521	0.002670
90	27	16	11	0.999948	0.000427
90	27	16	12	0.999996	0.000048
90	27	16	13	1.000000	0.000004
90	27	16	14	1.000000	0.000000
90	27	16	15	1.000000	0.000000
90	27	16	16	1.000000	0.000000
90	27	17	0	0.001084	0.001084
90	27	17	1	0.011671	0.010587
90	27	17	2	0.057547	0.045876
90	27	17	3	0.174578	0.117031
90	27	17	4	0.371190	0.196612
90	27	17	5	0.601727	0.230537
90	27	17	6	0.796798	0.195070
90	27	17	7	0.918256	0.121459
90	27	17	8	0.974487	0.056231
90	27	17	9	0.993912	0.019425
90	27	17	10	0.998907	0.004995
90	27	17	11	0.999855	0.000948
90	27	17	12	0.999986	0.000131
90	27	17	13	0.999999	0.000013
90	27	17	14	1.000000	0.000001
90	27	17	15	1.000000	0.000000
90	27	17	16	1.000000	0.000000
90	27	17	17	1.000000	0.000000
90	27	18	0	0.000683	0.000683
90	27	18	1	0.007900	0.007217
90	27	18	2	0.041836	0.033936
90	27	18	3	0.136102	0.094266
90	27	18	4	0.309244	0.173142
90	27	18	5	0.532251	0.223007
90	27	18	6	0.740682	0.208431
90	27	18	7	0.884980	0.144298
90	27	18	8	0.959852	0.074872
90	27	18	9	0.989123	0.029271
90	27	18	10	0.997744	0.008622
90	27	18	11	0.999648	0.001903
90	27	18	12	0.999959	0.000312
90	27	18	13	0.999997	0.000037
90	27	18	14	1.000000	0.000003
90	27	18	15	1.000000	0.000000
90	27	18	16	1.000000	0.000000
90	27	18	17	1.000000	0.000000
90	27	18	18	1.000000	0.000000
90	27	19	0	0.000427	0.000427
90	27	19	1	0.005294	0.004867
90	27	19	2	0.030053	0.024759
90	27	19	3	0.104680	0.074627
90	27	19	4	0.253935	0.149255
90	27	19	5	0.464110	0.210175
90	27	19	6	0.679889	0.215780
90	27	19	7	0.844897	0.165008
90	27	19	8	0.940094	0.095197
90	27	19	9	0.981805	0.041711
90	27	19	10	0.995709	0.013904
90	27	19	11	0.999225	0.003516
90	27	19	12	0.999894	0.000670
90	27	19	13	0.999989	0.000095
90	27	19	14	0.999999	0.000010
90	27	19	15	1.000000	0.000001
90	27	19	16	1.000000	0.000000
90	27	19	17	1.000000	0.000000
90	27	19	18	1.000000	0.000000
90	27	19	19	1.000000	0.000000
90	27	20	0	0.000265	0.000265
90	27	20	1	0.003512	0.003247
90	27	20	2	0.021335	0.017823
90	27	20	3	0.079454	0.058119
90	27	20	4	0.205585	0.126131
90	27	20	5	0.398985	0.193400
90	27	20	6	0.616067	0.217082
90	27	20	7	0.798416	0.182349
90	27	20	8	0.914619	0.116203
90	27	20	9	0.971230	0.056612
90	27	20	10	0.992379	0.021149
90	27	20	11	0.998432	0.006053
90	27	20	12	0.999753	0.001321
90	27	20	13	0.999971	0.000218
90	27	20	14	0.999997	0.000027
90	27	20	15	1.000000	0.000002
90	27	20	16	1.000000	0.000000
90	27	20	17	1.000000	0.000000
90	27	20	18	1.000000	0.000000
90	27	20	19	1.000000	0.000000
90	27	20	20	1.000000	0.000000
90	27	21	0	0.000163	0.000163
90	27	21	1	0.002306	0.002143
90	27	21	2	0.014969	0.012664
90	27	21	3	0.059527	0.044558
90	27	21	4	0.164142	0.104614
90	27	21	5	0.338202	0.174060
90	27	21	6	0.550942	0.212740
90	27	21	7	0.746316	0.195374
90	27	21	8	0.883078	0.136762
90	27	21	9	0.956673	0.073595
90	27	21	10	0.987243	0.030570
90	27	21	11	0.997049	0.009806
90	27	21	12	0.999470	0.002421
90	27	21	13	0.999927	0.000457
90	27	21	14	0.999992	0.000065
90	27	21	15	0.999999	0.000007
90	27	21	16	1.000000	0.000001
90	27	21	17	1.000000	0.000000

Table for $N = 2$, $n = 1$, through $N = 100$, $n = 50$

N	n	k	x	P(x)	p(x)
90	27	21	18	1.000000	0.000000
90	27	21	19	1.000000	0.000000
90	27	21	20	1.000000	0.000000
90	27	21	21	1.000000	0.000000
90	27	22	0	0.000099	0.000099
90	27	22	1	0.001498	0.001399
90	27	22	2	0.010381	0.008883
90	27	22	3	0.044029	0.033648
90	27	22	4	0.129270	0.085241
90	27	22	5	0.282704	0.153434
90	27	22	6	0.486195	0.203491
90	27	22	7	0.689686	0.203491
90	27	22	8	0.845419	0.155733
90	27	22	9	0.937474	0.092055
90	27	22	10	0.979711	0.042237
90	27	22	11	0.994775	0.015064
90	27	22	12	0.998944	0.004169
90	27	22	13	0.999834	0.000891
90	27	22	14	0.999980	0.000146
90	27	22	15	0.999998	0.000018
90	27	22	16	1.000000	0.000002
90	27	22	17	1.000000	0.000000
90	27	22	18	1.000000	0.000000
90	27	22	19	1.000000	0.000000
90	27	22	20	1.000000	0.000000
90	27	22	21	1.000000	0.000000
90	27	22	22	1.000000	0.000000
90	27	23	0	0.000060	0.000060
90	27	23	1	0.000963	0.000903
90	27	23	2	0.007115	0.006152
90	27	23	3	0.032153	0.025038
90	27	23	4	0.100439	0.068285
90	27	23	5	0.233064	0.132625
90	27	23	6	0.423353	0.190289
90	27	23	7	0.629836	0.206483
90	27	23	8	0.801905	0.172069
90	27	23	9	0.913107	0.111201
90	27	23	10	0.969152	0.056045
90	27	23	11	0.991231	0.022079
90	27	23	12	0.998024	0.006793
90	27	23	13	0.999651	0.001627
90	27	23	14	0.999952	0.000301
90	27	23	15	0.999995	0.000043
90	27	23	16	0.999999	0.000005
90	27	23	17	1.000000	0.000000
90	27	23	18	1.000000	0.000000
90	27	23	19	1.000000	0.000000
90	27	23	20	1.000000	0.000000
90	27	23	21	1.000000	0.000000
90	27	23	22	1.000000	0.000000

N	n	k	x	P(x)	p(x)
90	27	23	23	1.000000	0.000000
90	27	24	0	0.000036	0.000036
90	27	24	1	0.000613	0.000577
90	27	24	2	0.004820	0.004207
90	27	24	3	0.023184	0.018365
90	27	24	4	0.076997	0.053813
90	27	24	5	0.189515	0.112518
90	27	24	6	0.363710	0.174195
90	27	24	7	0.568199	0.204489
90	27	24	8	0.753110	0.184910
90	27	24	9	0.883232	0.130122
90	27	24	10	0.954932	0.071700
90	27	24	11	0.985958	0.031027
90	27	24	12	0.996503	0.010545
90	27	24	13	0.999311	0.002808
90	27	24	14	0.999894	0.000583
90	27	24	15	0.999987	0.000094
90	27	24	16	0.999999	0.000011
90	27	24	17	1.000000	0.000001
90	27	24	18	1.000000	0.000000
90	27	24	19	1.000000	0.000000
90	27	24	20	1.000000	0.000000
90	27	24	21	1.000000	0.000000
90	27	24	22	1.000000	0.000000
90	27	24	23	1.000000	0.000000
90	27	24	24	1.000000	0.000000
90	27	25	0	0.000021	0.000021
90	27	25	1	0.000385	0.000364
90	27	25	2	0.003226	0.002841
90	27	25	3	0.016506	0.013280
90	27	25	4	0.058244	0.041738
90	27	25	5	0.152009	0.093765
90	27	25	6	0.308284	0.156275
90	27	25	7	0.506233	0.197948
90	27	25	8	0.699878	0.193645
90	27	25	9	0.847744	0.147866
90	27	25	10	0.936463	0.088720
90	27	25	11	0.978437	0.041973
90	27	25	12	0.994107	0.015670
90	27	25	13	0.998715	0.004609
90	27	25	14	0.999779	0.001064
90	27	25	15	0.999970	0.000191
90	27	25	16	0.999997	0.000027
90	27	25	17	1.000000	0.000003
90	27	25	18	1.000000	0.000000
90	27	25	19	1.000000	0.000000
90	27	25	20	1.000000	0.000000
90	27	25	21	1.000000	0.000000
90	27	25	22	1.000000	0.000000
90	27	25	23	1.000000	0.000000

N	n	k	x	P(x)	p(x)
90	27	25	24	1.000000	0.000000
90	27	25	25	1.000000	0.000000
90	27	26	0	0.000012	0.000012
90	27	26	1	0.000240	0.000227
90	27	26	2	0.002133	0.001894
90	27	26	3	0.011603	0.009469
90	27	26	4	0.043476	0.031873
90	27	26	5	0.120274	0.076798
90	27	26	6	0.257796	0.137522
90	27	26	7	0.445325	0.187530
90	27	26	8	0.643274	0.197948
90	27	26	9	0.806796	0.163523
90	27	26	10	0.913260	0.106464
90	27	26	11	0.968105	0.054845
90	27	26	12	0.990490	0.022386
90	27	26	13	0.997723	0.007232
90	27	26	14	0.999566	0.001844
90	27	26	15	0.999935	0.000369
90	27	26	16	0.999992	0.000057
90	27	26	17	0.999999	0.000007
90	27	26	18	1.000000	0.000001
90	27	26	19	1.000000	0.000000
90	27	26	20	1.000000	0.000000
90	27	26	21	1.000000	0.000000
90	27	26	22	1.000000	0.000000
90	27	26	23	1.000000	0.000000
90	27	26	24	1.000000	0.000000
90	27	26	25	1.000000	0.000000
90	27	26	26	1.000000	0.000000
90	27	27	0	0.000007	0.000007
90	27	27	1	0.000147	0.000140
90	27	27	2	0.001394	0.001246
90	27	27	3	0.008052	0.006658
90	27	27	4	0.032021	0.023969
90	27	27	5	0.093874	0.061853
90	27	27	6	0.212671	0.118797
90	27	27	7	0.386722	0.174051
90	27	27	8	0.584508	0.197786
90	27	27	9	0.760806	0.176298
90	27	27	10	0.884980	0.124175
90	27	27	11	0.954394	0.069413
90	27	27	12	0.985244	0.030850
90	27	27	13	0.996141	0.010897
90	27	27	14	0.999192	0.003051
90	27	27	15	0.999866	0.000674
90	27	27	16	0.999982	0.000117
90	27	27	17	0.999998	0.000016
90	27	27	18	1.000000	0.000002
90	27	27	19	1.000000	0.000000
90	27	27	20	1.000000	0.000000

N	n	k	x	P(x)	p(x)
90	27	27	21	1.000000	0.000000
90	27	27	22	1.000000	0.000000
90	27	27	23	1.000000	0.000000
90	27	27	24	1.000000	0.000000
90	27	27	25	1.000000	0.000000
90	27	27	26	1.000000	0.000000
90	27	27	27	1.000000	0.000000
90	28	1	0	0.688889	0.688889
90	28	1	1	1.000000	0.311111
90	28	2	0	0.472160	0.472160
90	28	2	1	0.905618	0.433458
90	28	2	2	1.000000	0.094382
90	28	3	0	0.321927	0.321927
90	28	3	1	0.772625	0.450698
90	28	3	2	0.972114	0.199489
90	28	3	3	1.000000	0.027886
90	28	4	0	0.218318	0.218318
90	28	4	1	0.632753	0.414435
90	28	4	2	0.912497	0.279744
90	28	4	3	0.991987	0.079490
90	28	4	4	1.000000	0.008013
90	28	5	0	0.147238	0.147238
90	28	5	1	0.502640	0.355402
90	28	5	2	0.827923	0.325283
90	28	5	3	0.968879	0.140956
90	28	5	4	0.997764	0.028884
90	28	5	5	1.000000	0.002236
90	28	6	0	0.098736	0.098736
90	28	6	1	0.389748	0.291012
90	28	6	2	0.728425	0.338677
90	28	6	3	0.927422	0.198997
90	28	6	4	0.989608	0.062186
90	28	6	5	0.999395	0.009787
90	28	6	6	1.000000	0.000605
90	28	7	0	0.065824	0.065824
90	28	7	1	0.296208	0.230384
90	28	7	2	0.623596	0.327388
90	28	7	3	0.868196	0.244600
90	28	7	4	0.971840	0.103644
90	28	7	5	0.996715	0.024875
90	28	7	6	0.999841	0.003126
90	28	7	7	1.000000	0.000158
90	28	8	0	0.043618	0.043618
90	28	8	1	0.221264	0.177646
90	28	8	2	0.521041	0.299777
90	28	8	3	0.794522	0.273481
90	28	8	4	0.941871	0.147350
90	28	8	5	0.989822	0.047951
90	28	8	6	0.999013	0.009191
90	28	8	7	0.999960	0.000947

Table for $N = 2$, $n = 1$, through $N = 100$, $n = 50$

N	n	k	x	P(x)	p(x)
90	28	8	8	1.000000	0.000040
90	28	9	0	0.028724	0.028724
90	28	9	1	0.162771	0.134047
90	28	9	2	0.425990	0.263219
90	28	9	3	0.711143	0.285154
90	28	9	4	0.898744	0.187601
90	28	9	5	0.976372	0.077628
90	28	9	6	0.996547	0.020175
90	28	9	7	0.999717	0.003170
90	28	9	8	0.999990	0.000273
90	28	9	9	1.000000	0.000010
90	28	10	0	0.018795	0.018795
90	28	10	1	0.118089	0.099294
90	28	10	2	0.341500	0.223411
90	28	10	3	0.623133	0.281633
90	28	10	4	0.843159	0.220026
90	28	10	5	0.954330	0.111171
90	28	10	6	0.991067	0.036738
90	28	10	7	0.998895	0.007828
90	28	10	8	0.999923	0.001027
90	28	10	9	0.999998	0.000075
90	28	10	10	1.000000	0.000002
90	28	11	0	0.012217	0.012217
90	28	11	1	0.084577	0.072360
90	28	11	2	0.268891	0.184314
90	28	11	3	0.535123	0.266231
90	28	11	4	0.777151	0.242029
90	28	11	5	0.922368	0.145217
90	28	11	6	0.980965	0.058596
90	28	11	7	0.996841	0.015876
90	28	11	8	0.999666	0.002825
90	28	11	9	0.999980	0.000314
90	28	11	10	0.999999	0.000020
90	28	11	11	1.000000	0.000001
90	28	12	0	0.007887	0.007887
90	28	12	1	0.059846	0.051960
90	28	12	2	0.208231	0.148384
90	28	12	3	0.450872	0.242641
90	28	12	4	0.703623	0.252751
90	28	12	5	0.880090	0.176466
90	28	12	6	0.964647	0.084557
90	28	12	7	0.992620	0.027974
90	28	12	8	0.998951	0.006330
90	28	12	9	0.999904	0.000954
90	28	12	10	0.999995	0.000091
90	28	12	11	1.000000	0.000005
90	28	12	12	1.000000	0.000000
90	28	13	0	0.005056	0.005056
90	28	13	1	0.041860	0.036805
90	28	13	2	0.158769	0.116909

N	n	k	x	P(x)	p(x)
90	28	13	3	0.373102	0.214333
90	28	13	4	0.625854	0.252751
90	28	13	5	0.828055	0.202201
90	28	13	6	0.940797	0.112742
90	28	13	7	0.985089	0.044292
90	28	13	8	0.997327	0.012238
90	28	13	9	0.999672	0.002345
90	28	13	10	0.999974	0.000302
90	28	13	11	0.999999	0.000025
90	28	13	12	1.000000	0.000001
90	28	13	13	1.000000	0.000000
90	28	14	0	0.003217	0.003217
90	28	14	1	0.028955	0.025738
90	28	14	2	0.119293	0.090339
90	28	14	3	0.303514	0.184220
90	28	14	4	0.547074	0.243560
90	28	14	5	0.767657	0.220583
90	28	14	6	0.908585	0.140928
90	28	14	7	0.973009	0.064424
90	28	14	8	0.994149	0.021139
90	28	14	9	0.999093	0.004945
90	28	14	10	0.999903	0.000810
90	28	14	11	0.999993	0.000090
90	28	14	12	1.000000	0.000006
90	28	14	13	1.000000	0.000000
90	28	14	14	1.000000	0.000000
90	28	15	0	0.002032	0.002032
90	28	15	1	0.019811	0.017779
90	28	15	2	0.088388	0.068577
90	28	15	3	0.242915	0.154527
90	28	15	4	0.470160	0.227245
90	28	15	5	0.700902	0.230741
90	28	15	6	0.867790	0.166888
90	28	15	7	0.955208	0.087418
90	28	15	8	0.988586	0.033378
90	28	15	9	0.997857	0.009272
90	28	15	10	0.999712	0.001854
90	28	15	11	0.999973	0.000262
90	28	15	12	0.999998	0.000025
90	28	15	13	1.000000	0.000002
90	28	15	14	1.000000	0.000000
90	28	15	15	1.000000	0.000000
90	28	16	0	0.001273	0.001273
90	28	16	1	0.013411	0.012137
90	28	16	2	0.064615	0.051204
90	28	16	3	0.191406	0.126791
90	28	16	4	0.397442	0.206036
90	28	16	5	0.630141	0.232699
90	28	16	6	0.818836	0.188695
90	28	16	7	0.930731	0.111895

N	n	k	x	P(x)	p(x)
90	28	16	8	0.979685	0.048954
90	28	16	9	0.995508	0.015823
90	28	16	10	0.999266	0.003758
90	28	16	11	0.999914	0.000647
90	28	16	12	0.999993	0.000079
90	28	16	13	0.999999	0.000007
90	28	16	14	1.000000	0.000000
90	28	16	15	1.000000	0.000000
90	28	16	16	1.000000	0.000000
90	28	17	0	0.000792	0.000792
90	28	17	1	0.008982	0.008191
90	28	17	2	0.046624	0.037642
90	28	17	3	0.148571	0.101947
90	28	17	4	0.330619	0.182048
90	28	17	5	0.557816	0.227196
90	28	17	6	0.762738	0.204922
90	28	17	7	0.898977	0.136239
90	28	17	8	0.966454	0.067477
90	28	17	9	0.991446	0.024992
90	28	17	10	0.998352	0.006907
90	28	17	11	0.999765	0.001413
90	28	17	12	0.999976	0.000211
90	28	17	13	0.999998	0.000022
90	28	17	14	1.000000	0.000002
90	28	17	15	1.000000	0.000000
90	28	17	16	1.000000	0.000000
90	28	17	17	1.000000	0.000000
90	28	18	0	0.000488	0.000488
90	28	18	1	0.005953	0.005465
90	28	18	2	0.033217	0.027265
90	28	18	3	0.113658	0.080440
90	28	18	4	0.270768	0.157110
90	28	18	5	0.486233	0.215465
90	28	18	6	0.700980	0.214747
90	28	18	7	0.859785	0.158805
90	28	18	8	0.947967	0.088182
90	28	18	9	0.984941	0.036974
90	28	18	10	0.996649	0.011708
90	28	18	11	0.999436	0.002787
90	28	18	12	0.999930	0.000493
90	28	18	13	0.999994	0.000064
90	28	18	14	0.999999	0.000006
90	28	18	15	1.000000	0.000000
90	28	18	16	1.000000	0.000000
90	28	18	17	1.000000	0.000000
90	28	18	18	1.000000	0.000000
90	28	19	0	0.000298	0.000298
90	28	19	1	0.003903	0.003605
90	28	19	2	0.023372	0.019468
90	28	19	3	0.085727	0.062355

N	n	k	x	P(x)	p(x)
90	28	19	4	0.218398	0.132671
90	28	19	5	0.417404	0.199006
90	28	19	6	0.635363	0.217959
90	28	19	7	0.813467	0.178104
90	28	19	8	0.923472	0.110005
90	28	19	9	0.975184	0.051712
90	28	19	10	0.993722	0.018538
90	28	19	11	0.998778	0.005056
90	28	19	12	0.999820	0.001042
90	28	19	13	0.999980	0.000160
90	28	19	14	0.999998	0.000018
90	28	19	15	1.000000	0.000001
90	28	19	16	1.000000	0.000000
90	28	19	17	1.000000	0.000000
90	28	19	18	1.000000	0.000000
90	28	19	19	1.000000	0.000000
90	28	20	0	0.000181	0.000181
90	28	20	1	0.002532	0.002352
90	28	20	2	0.016243	0.013710
90	28	20	3	0.063771	0.047528
90	28	20	4	0.173551	0.109780
90	28	20	5	0.352937	0.179386
90	28	20	6	0.567827	0.214889
90	28	20	7	0.760788	0.192962
90	28	20	8	0.892485	0.131696
90	28	20	9	0.961346	0.068861
90	28	20	10	0.989022	0.027677
90	28	20	11	0.997568	0.008545
90	28	20	12	0.999585	0.002018
90	28	20	13	0.999946	0.000361
90	28	20	14	0.999995	0.000048
90	28	20	15	0.999999	0.000005
90	28	20	16	1.000000	0.000000
90	28	20	17	1.000000	0.000000
90	28	20	18	1.000000	0.000000
90	28	20	19	1.000000	0.000000
90	28	20	20	1.000000	0.000000
90	28	21	0	0.000108	0.000108
90	28	21	1	0.001625	0.001517
90	28	21	2	0.011150	0.009525
90	28	21	3	0.046797	0.035646
90	28	21	4	0.135912	0.089116
90	28	21	5	0.293996	0.158084
90	28	21	6	0.500290	0.206294
90	28	21	7	0.702900	0.202610
90	28	21	8	0.854857	0.151957
90	28	21	9	0.942655	0.087798
90	28	21	10	0.981906	0.039251
90	28	21	11	0.995492	0.013587
90	28	21	12	0.999124	0.003632

Table for $N = 2,\ n = 1,$ through $N = 100,\ n = 50$

N	n	k	x	P(x)	p(x)
90	28	21	13	0.999869	0.000745
90	28	21	14	0.999985	0.000116
90	28	21	15	0.999999	0.000014
90	28	21	16	1.000000	0.000001
90	28	21	17	1.000000	0.000000
90	28	21	18	1.000000	0.000000
90	28	21	19	1.000000	0.000000
90	28	21	20	1.000000	0.000000
90	28	21	21	1.000000	0.000000
90	28	22	0	0.000064	0.000064
90	28	22	1	0.001032	0.000967
90	28	22	2	0.007561	0.006529
90	28	22	3	0.033881	0.026320
90	28	22	4	0.104916	0.071034
90	28	22	5	0.241302	0.136386
90	28	22	6	0.434515	0.193214
90	28	22	7	0.641236	0.206721
90	28	22	8	0.810812	0.169576
90	28	22	9	0.918479	0.107667
90	28	22	10	0.971666	0.053188
90	28	22	11	0.992145	0.020479
90	28	22	12	0.998282	0.006137
90	28	22	13	0.999707	0.001425
90	28	22	14	0.999962	0.000254
90	28	22	15	0.999996	0.000035
90	28	22	16	1.000000	0.000004
90	28	22	17	1.000000	0.000000
90	28	22	18	1.000000	0.000000
90	28	22	19	1.000000	0.000000
90	28	22	20	1.000000	0.000000
90	28	22	21	1.000000	0.000000
90	28	22	22	1.000000	0.000000
90	28	23	0	0.000038	0.000038
90	28	23	1	0.000648	0.000610
90	28	23	2	0.005065	0.004417
90	28	23	3	0.024205	0.019140
90	28	23	4	0.079845	0.055640
90	28	23	5	0.195171	0.115326
90	28	23	6	0.372005	0.176834
90	28	23	7	0.577396	0.205391
90	28	23	8	0.760936	0.183541
90	28	23	9	0.888395	0.127459
90	28	23	10	0.957587	0.069192
90	28	23	11	0.987025	0.029438
90	28	23	12	0.996838	0.009813
90	28	23	13	0.999393	0.002555
90	28	23	14	0.999909	0.000516
90	28	23	15	0.999989	0.000080
90	28	23	16	0.999999	0.000009
90	28	23	17	1.000000	0.000001

N	n	k	x	P(x)	p(x)
90	28	23	18	1.000000	0.000000
90	28	23	19	1.000000	0.000000
90	28	23	20	1.000000	0.000000
90	28	23	21	1.000000	0.000000
90	28	23	22	1.000000	0.000000
90	28	23	23	1.000000	0.000000
90	28	24	0	0.000022	0.000022
90	28	24	1	0.000402	0.000380
90	28	24	2	0.003351	0.002949
90	28	24	3	0.017063	0.013712
90	28	24	4	0.059914	0.042851
90	28	24	5	0.155581	0.095667
90	28	24	6	0.313940	0.158359
90	28	24	7	0.513019	0.199079
90	28	24	8	0.706148	0.193129
90	28	24	9	0.852250	0.146102
90	28	24	10	0.938998	0.086748
90	28	24	11	0.979556	0.040558
90	28	24	12	0.994495	0.014939
90	28	24	13	0.998821	0.004326
90	28	24	14	0.999801	0.000981
90	28	24	15	0.999974	0.000173
90	28	24	16	0.999997	0.000023
90	28	24	17	1.000000	0.000002
90	28	24	18	1.000000	0.000000
90	28	24	19	1.000000	0.000000
90	28	24	20	1.000000	0.000000
90	28	24	21	1.000000	0.000000
90	28	24	22	1.000000	0.000000
90	28	24	23	1.000000	0.000000
90	28	24	24	1.000000	0.000000
90	28	25	0	0.000013	0.000013
90	28	25	1	0.000247	0.000234
90	28	25	2	0.002189	0.001942
90	28	25	3	0.011869	0.009680
90	28	25	4	0.044332	0.032463
90	28	25	5	0.122243	0.077911
90	28	25	6	0.261154	0.138911
90	28	25	7	0.449676	0.188522
90	28	25	8	0.647624	0.197948
90	28	25	9	0.810191	0.162566
90	28	25	10	0.915340	0.105149
90	28	25	11	0.969109	0.053769
90	28	25	12	0.990873	0.021764
90	28	25	13	0.997838	0.006964
90	28	25	14	0.999593	0.001756
90	28	25	15	0.999940	0.000347
90	28	25	16	0.999993	0.000053
90	28	25	17	0.999999	0.000006
90	28	25	18	1.000000	0.000001

N	n	k	x	P(x)	p(x)
90	28	25	19	1.000000	0.000000
90	28	25	20	1.000000	0.000000
90	28	25	21	1.000000	0.000000
90	28	25	22	1.000000	0.000000
90	28	25	23	1.000000	0.000000
90	28	25	24	1.000000	0.000000
90	28	25	25	1.000000	0.000000
90	28	26	0	0.000007	0.000007
90	28	26	1	0.000149	0.000142
90	28	26	2	0.001412	0.001263
90	28	26	3	0.008146	0.006734
90	28	26	4	0.032345	0.024200
90	28	26	5	0.094674	0.062329
90	28	26	6	0.214138	0.119464
90	28	26	7	0.388769	0.174631
90	28	26	8	0.586717	0.197948
90	28	26	9	0.762671	0.175954
90	28	26	10	0.886222	0.123550
90	28	26	11	0.955046	0.068825
90	28	26	12	0.985516	0.030469
90	28	26	13	0.996230	0.010715
90	28	26	14	0.999215	0.002985
90	28	26	15	0.999871	0.000655
90	28	26	16	0.999983	0.000113
90	28	26	17	0.999998	0.000015
90	28	26	18	1.000000	0.000002
90	28	26	19	1.000000	0.000000
90	28	26	20	1.000000	0.000000
90	28	26	21	1.000000	0.000000
90	28	26	22	1.000000	0.000000
90	28	26	23	1.000000	0.000000
90	28	26	24	1.000000	0.000000
90	28	26	25	1.000000	0.000000
90	28	26	26	1.000000	0.000000
90	28	27	0	0.000004	0.000004
90	28	27	1	0.000089	0.000085
90	28	27	2	0.000899	0.000810
90	28	27	3	0.005515	0.004616
90	28	27	4	0.023271	0.017755
90	28	27	5	0.072275	0.049004
90	28	27	6	0.173072	0.100797
90	28	27	7	0.331468	0.158396
90	28	27	8	0.524858	0.193390
90	28	27	9	0.710435	0.185577
90	28	27	10	0.851473	0.141038
90	28	27	11	0.936765	0.085292
90	28	27	12	0.977899	0.041134
90	28	27	13	0.993719	0.015821
90	28	27	14	0.998562	0.004843
90	28	27	15	0.999737	0.001175

N	n	k	x	P(x)	p(x)
90	28	27	16	0.999962	0.000225
90	28	27	17	0.999996	0.000034
90	28	27	18	1.000000	0.000004
90	28	27	19	1.000000	0.000000
90	28	27	20	1.000000	0.000000
90	28	27	21	1.000000	0.000000
90	28	27	22	1.000000	0.000000
90	28	27	23	1.000000	0.000000
90	28	27	24	1.000000	0.000000
90	28	27	25	1.000000	0.000000
90	28	27	26	1.000000	0.000000
90	28	27	27	1.000000	0.000000
90	28	28	0	0.000002	0.000002
90	28	28	1	0.000053	0.000051
90	28	28	2	0.000565	0.000512
90	28	28	3	0.003683	0.003119
90	28	28	4	0.016507	0.012823
90	28	28	5	0.054384	0.037878
90	28	28	6	0.137873	0.083489
90	28	28	7	0.278669	0.140796
90	28	28	8	0.463465	0.184795
90	28	28	9	0.654467	0.191003
90	28	28	10	0.811176	0.156709
90	28	28	11	0.913750	0.102573
90	28	28	12	0.967452	0.053702
90	28	28	13	0.989952	0.022500
90	28	28	14	0.997486	0.007534
90	28	28	15	0.999495	0.002009
90	28	28	16	0.999919	0.000424
90	28	28	17	0.999990	0.000070
90	28	28	18	0.999999	0.000009
90	28	28	19	1.000000	0.000001
90	28	28	20	1.000000	0.000000
90	28	28	21	1.000000	0.000000
90	28	28	22	1.000000	0.000000
90	28	28	23	1.000000	0.000000
90	28	28	24	1.000000	0.000000
90	28	28	25	1.000000	0.000000
90	28	28	26	1.000000	0.000000
90	28	28	27	1.000000	0.000000
90	28	28	28	1.000000	0.000000
90	29	1	0	0.677778	0.677778
90	29	1	1	1.000000	0.322222
90	29	2	0	0.456929	0.456929
90	29	2	1	0.898627	0.441698
90	29	2	2	1.000000	0.101373
90	29	3	0	0.306350	0.306350
90	29	3	1	0.758086	0.451736
90	29	3	2	0.968897	0.210810
90	29	3	3	1.000000	0.031103

Table for $N = 2$, $n = 1$, through $N = 100$, $n = 50$

N	n	k	x	P(x)	p(x)
90	29	4	0	0.204233	0.204233
90	29	4	1	0.612700	0.408467
90	29	4	2	0.903473	0.290773
90	29	4	3	0.990705	0.087232
90	29	4	4	1.000000	0.009295
90	29	5	0	0.135364	0.135364
90	29	5	1	0.479711	0.344347
90	29	5	2	0.812184	0.332473
90	29	5	3	0.964332	0.152149
90	29	5	4	0.997298	0.032966
90	29	5	5	1.000000	0.002702
90	29	6	0	0.089181	0.089181
90	29	6	1	0.366279	0.277098
90	29	6	2	0.706575	0.340296
90	29	6	3	0.917793	0.211218
90	29	6	4	0.987602	0.069809
90	29	6	5	0.999237	0.011635
90	29	6	6	1.000000	0.000763
90	29	7	0	0.058392	0.058392
90	29	7	1	0.273913	0.215521
90	29	7	2	0.597194	0.323281
90	29	7	3	0.852416	0.255222
90	29	7	4	0.966825	0.114410
90	29	7	5	0.995913	0.029087
90	29	7	6	0.999791	0.003878
90	29	7	7	1.000000	0.000209
90	29	8	0	0.037990	0.037990
90	29	8	1	0.201207	0.163217
90	29	8	2	0.492030	0.290823
90	29	8	3	0.772467	0.280436
90	29	8	4	0.932365	0.159898
90	29	8	5	0.987502	0.055137
90	29	8	6	0.998716	0.011214
90	29	8	7	0.999945	0.001228
90	29	8	8	1.000000	0.000055
90	29	9	0	0.024555	0.024555
90	29	9	1	0.145475	0.120920
90	29	9	2	0.396271	0.250797
90	29	9	3	0.683548	0.287276
90	29	9	4	0.883615	0.200067
90	29	9	5	0.971364	0.087749
90	29	9	6	0.995571	0.024207
90	29	9	7	0.999615	0.004044
90	29	9	8	0.999986	0.000371
90	29	9	9	1.000000	0.000014
90	29	10	0	0.015763	0.015763
90	29	10	1	0.103675	0.087912
90	29	10	2	0.312672	0.208997
90	29	10	3	0.591336	0.278663
90	29	10	4	0.821866	0.230530

N	n	k	x	P(x)	p(x)
90	29	10	5	0.945364	0.123498
90	29	10	6	0.988697	0.043333
90	29	10	7	0.998516	0.009819
90	29	10	8	0.999890	0.001373
90	29	10	9	0.999996	0.000107
90	29	10	10	1.000000	0.000003
90	29	11	0	0.010049	0.010049
90	29	11	1	0.072906	0.062857
90	29	11	2	0.242136	0.169230
90	29	11	3	0.500770	0.258634
90	29	11	4	0.749825	0.249055
90	29	11	5	0.908315	0.158490
90	29	11	6	0.976239	0.067924
90	29	11	7	0.995816	0.019577
90	29	11	8	0.999529	0.003713
90	29	11	9	0.999970	0.000441
90	29	11	10	0.999999	0.000029
90	29	11	11	1.000000	0.000001
90	29	12	0	0.006360	0.006360
90	29	12	1	0.050628	0.044267
90	29	12	2	0.184298	0.133670
90	29	12	3	0.415650	0.231352
90	29	12	4	0.671010	0.255360
90	29	12	5	0.860166	0.189156
90	29	12	6	0.956464	0.096298
90	29	12	7	0.990364	0.033901
90	29	12	8	0.998542	0.008178
90	29	12	9	0.999858	0.001316
90	29	12	10	0.999992	0.000134
90	29	12	11	1.000000	0.000008
90	29	12	12	1.000000	0.000000
90	29	13	0	0.003996	0.003996
90	29	13	1	0.034737	0.030741
90	29	13	2	0.138027	0.103291
90	29	13	3	0.338533	0.200505
90	29	13	4	0.589164	0.250632
90	29	13	5	0.801964	0.212800
90	29	13	6	0.928068	0.126104
90	29	13	7	0.980803	0.052734
90	29	13	8	0.996341	0.015538
90	29	13	9	0.999521	0.003180
90	29	13	10	0.999959	0.000439
90	29	13	11	0.999998	0.000039
90	29	13	12	1.000000	0.000002
90	29	13	13	1.000000	0.000000
90	29	14	0	0.002491	0.002491
90	29	14	1	0.023558	0.021067
90	29	14	2	0.101809	0.078250
90	29	14	3	0.270829	0.169021
90	29	14	4	0.507790	0.236961

N	n	k	x	P(x)	p(x)
90	29	14	5	0.735637	0.227847
90	29	14	6	0.890401	0.154764
90	29	14	7	0.965736	0.075335
90	29	14	8	0.992103	0.026367
90	29	14	9	0.998695	0.006592
90	29	14	10	0.999851	0.001156
90	29	14	11	0.999989	0.000138
90	29	14	12	0.999999	0.000011
90	29	14	13	1.000000	0.000000
90	29	14	14	1.000000	0.000000
90	29	15	0	0.001540	0.001540
90	29	15	1	0.015796	0.014256
90	29	15	2	0.074009	0.058213
90	29	15	3	0.213007	0.138997
90	29	15	4	0.429843	0.216836
90	29	15	5	0.663685	0.233843
90	29	15	6	0.843564	0.179879
90	29	15	7	0.943928	0.100364
90	29	15	8	0.984817	0.040889
90	29	15	9	0.996960	0.012143
90	29	15	10	0.999562	0.002602
90	29	15	11	0.999956	0.000394
90	29	15	12	0.999997	0.000041
90	29	15	13	1.000000	0.000003
90	29	15	14	1.000000	0.000000
90	29	15	15	1.000000	0.000000
90	29	16	0	0.000945	0.000945
90	29	16	1	0.010474	0.009529
90	29	16	2	0.053053	0.042578
90	29	16	3	0.164821	0.111768
90	29	16	4	0.357564	0.192743
90	29	16	5	0.588856	0.231292
90	29	16	6	0.788402	0.199546
90	29	16	7	0.914488	0.126087
90	29	16	8	0.973368	0.058880
90	29	16	9	0.993722	0.020354
90	29	16	10	0.998903	0.005181
90	29	16	11	0.999862	0.000959
90	29	16	12	0.999988	0.000126
90	29	16	13	0.999999	0.000011
90	29	16	14	1.000000	0.000001
90	29	16	15	1.000000	0.000000
90	29	16	16	1.000000	0.000000
90	29	17	0	0.000574	0.000574
90	29	17	1	0.006868	0.006294
90	29	17	2	0.037517	0.030649
90	29	17	3	0.125551	0.088034
90	29	17	4	0.292448	0.166897
90	29	17	5	0.513842	0.221394
90	29	17	6	0.726381	0.212538

N	n	k	x	P(x)	p(x)
90	29	17	7	0.877003	0.150622
90	29	17	8	0.956659	0.079656
90	29	17	9	0.988221	0.031562
90	29	17	10	0.997573	0.009352
90	29	17	11	0.999628	0.002056
90	29	17	12	0.999959	0.000330
90	29	17	13	0.999997	0.000038
90	29	17	14	1.000000	0.000003
90	29	17	15	1.000000	0.000000
90	29	17	16	1.000000	0.000000
90	29	17	17	1.000000	0.000000
90	29	18	0	0.000346	0.000346
90	29	18	1	0.004454	0.004108
90	29	18	2	0.026181	0.021727
90	29	18	3	0.094196	0.068015
90	29	18	4	0.235291	0.141095
90	29	18	5	0.441055	0.205764
90	29	18	6	0.659416	0.218361
90	29	18	7	0.831610	0.172194
90	29	18	8	0.933744	0.102134
90	29	18	9	0.979574	0.045830
90	29	18	10	0.995139	0.015565
90	29	18	11	0.999122	0.003983
90	29	18	12	0.999882	0.000760
90	29	18	13	0.999989	0.000107
90	29	18	14	0.999999	0.000011
90	29	18	15	1.000000	0.000001
90	29	18	16	1.000000	0.000000
90	29	18	17	1.000000	0.000000
90	29	18	18	1.000000	0.000000
90	29	19	0	0.000207	0.000207
90	29	19	1	0.002857	0.002650
90	29	19	2	0.018034	0.015177
90	29	19	3	0.069635	0.051602
90	29	19	4	0.186300	0.116665
90	29	19	5	0.372467	0.186167
90	29	19	6	0.589662	0.217195
90	29	19	7	0.778995	0.189333
90	29	19	8	0.903955	0.124960
90	29	19	9	0.966844	0.062888
90	29	19	10	0.991031	0.024188
90	29	19	11	0.998126	0.007095
90	29	19	12	0.999702	0.001577
90	29	19	13	0.999965	0.000262
90	29	19	14	0.999997	0.000032
90	29	19	15	1.000000	0.000003
90	29	19	16	1.000000	0.000000
90	29	19	17	1.000000	0.000000
90	29	19	18	1.000000	0.000000
90	29	19	19	1.000000	0.000000

Table for $N = 2$, $n = 1$, through $N = 100$, $n = 50$

N	n	k	x	P(x)	p(x)
90	29	20	0	0.000122	0.000122
90	29	20	1	0.001812	0.001689
90	29	20	2	0.012262	0.010450
90	29	20	3	0.050739	0.038477
90	29	20	4	0.145221	0.094482
90	29	20	5	0.309537	0.164316
90	29	20	6	0.519303	0.209766
90	29	20	7	0.720329	0.201025
90	29	20	8	0.866995	0.146667
90	29	20	9	0.949128	0.082133
90	29	20	10	0.984558	0.035430
90	29	20	11	0.996327	0.011769
90	29	20	12	0.999325	0.002998
90	29	20	13	0.999906	0.000581
90	29	20	14	0.999990	0.000084
90	29	20	15	0.999999	0.000009
90	29	20	16	1.000000	0.000001
90	29	20	17	1.000000	0.000000
90	29	20	18	1.000000	0.000000
90	29	20	19	1.000000	0.000000
90	29	20	20	1.000000	0.000000
90	29	21	0	0.000072	0.000072
90	29	21	1	0.001136	0.001064
90	29	21	2	0.008231	0.007095
90	29	21	3	0.036448	0.028216
90	29	21	4	0.111477	0.075030
90	29	21	5	0.253200	0.141723
90	29	21	6	0.450380	0.197180
90	29	21	7	0.657149	0.206769
90	29	21	8	0.822995	0.165846
90	29	21	9	0.925662	0.102667
90	29	21	10	0.974942	0.049280
90	29	21	11	0.993301	0.018359
90	29	21	12	0.998597	0.005296
90	29	21	13	0.999773	0.001176
90	29	21	14	0.999972	0.000199
90	29	21	15	0.999997	0.000025
90	29	21	16	1.000000	0.000002
90	29	21	17	1.000000	0.000000
90	29	21	18	1.000000	0.000000
90	29	21	19	1.000000	0.000000
90	29	21	20	1.000000	0.000000
90	29	21	21	1.000000	0.000000
90	29	22	0	0.000042	0.000042
90	29	22	1	0.000704	0.000663
90	29	22	2	0.005455	0.004751
90	29	22	3	0.025815	0.020361
90	29	22	4	0.084293	0.058477
90	29	22	5	0.203905	0.119613
90	29	22	6	0.384654	0.180748
90	29	22	7	0.591223	0.206569
90	29	22	8	0.772520	0.181297
90	29	22	9	0.895903	0.123383
90	29	22	10	0.961372	0.065469
90	29	22	11	0.988512	0.027140
90	29	22	12	0.997292	0.008780
90	29	22	13	0.999500	0.002208
90	29	22	14	0.999929	0.000429
90	29	22	15	0.999992	0.000063
90	29	22	16	0.999999	0.000007
90	29	22	17	1.000000	0.000001
90	29	22	18	1.000000	0.000000
90	29	22	19	1.000000	0.000000
90	29	22	20	1.000000	0.000000
90	29	22	21	1.000000	0.000000
90	29	22	22	1.000000	0.000000
90	29	23	0	0.000024	0.000024
90	29	23	1	0.000431	0.000407
90	29	23	2	0.003569	0.003137
90	29	23	3	0.018030	0.014462
90	29	23	4	0.062794	0.044763
90	29	23	5	0.161689	0.098895
90	29	23	6	0.323518	0.161829
90	29	23	7	0.524392	0.200873
90	29	23	8	0.716531	0.192140
90	29	23	9	0.859614	0.143083
90	29	23	10	0.943079	0.083465
90	29	23	11	0.981328	0.038248
90	29	23	12	0.995097	0.013769
90	29	23	13	0.998981	0.003884
90	29	23	14	0.999834	0.000854
90	29	23	15	0.999979	0.000145
90	29	23	16	0.999998	0.000019
90	29	23	17	1.000000	0.000002
90	29	23	18	1.000000	0.000000
90	29	23	19	1.000000	0.000000
90	29	23	20	1.000000	0.000000
90	29	23	21	1.000000	0.000000
90	29	23	22	1.000000	0.000000
90	29	23	23	1.000000	0.000000
90	29	24	0	0.000014	0.000014
90	29	24	1	0.000261	0.000247
90	29	24	2	0.002304	0.002043
90	29	24	3	0.012418	0.010114
90	29	24	4	0.046091	0.033673
90	29	24	5	0.126264	0.080173
90	29	24	6	0.267965	0.141701
90	29	24	7	0.458433	0.190468
90	29	24	8	0.656308	0.197875
90	29	24	9	0.816903	0.160594
90	29	24	10	0.919410	0.102507
90	29	24	11	0.971052	0.051642
90	29	24	12	0.991603	0.020551
90	29	24	13	0.998053	0.006450
90	29	24	14	0.999643	0.001590
90	29	24	15	0.999949	0.000306
90	29	24	16	0.999994	0.000045
90	29	24	17	0.999999	0.000005
90	29	24	18	1.000000	0.000000
90	29	24	19	1.000000	0.000000
90	29	24	20	1.000000	0.000000
90	29	24	21	1.000000	0.000000
90	29	24	22	1.000000	0.000000
90	29	24	23	1.000000	0.000000
90	29	24	24	1.000000	0.000000
90	29	25	0	0.000008	0.000008
90	29	25	1	0.000156	0.000148
90	29	25	2	0.001468	0.001312
90	29	25	3	0.008434	0.006966
90	29	25	4	0.033336	0.024902
90	29	25	5	0.097110	0.063774
90	29	25	6	0.218584	0.121474
90	29	25	7	0.394944	0.176359
90	29	25	8	0.593348	0.198404
90	29	25	9	0.768238	0.174890
90	29	25	10	0.889900	0.121662
90	29	25	11	0.956968	0.067067
90	29	25	12	0.986310	0.029342
90	29	25	13	0.996490	0.010180
90	29	25	14	0.999282	0.002792
90	29	25	15	0.999884	0.000602
90	29	25	16	0.999985	0.000101
90	29	25	17	0.999998	0.000013
90	29	25	18	1.000000	0.000001
90	29	25	19	1.000000	0.000000
90	29	25	20	1.000000	0.000000
90	29	25	21	1.000000	0.000000
90	29	25	22	1.000000	0.000000
90	29	25	23	1.000000	0.000000
90	29	25	24	1.000000	0.000000
90	29	25	25	1.000000	0.000000
90	29	26	0	0.000004	0.000004
90	29	26	1	0.000092	0.000088
90	29	26	2	0.000923	0.000831
90	29	26	3	0.005648	0.004725
90	29	26	4	0.023758	0.018111
90	29	26	5	0.073563	0.049804
90	29	26	6	0.175601	0.102038
90	29	26	7	0.335253	0.159652
90	29	26	8	0.529248	0.193995
90	29	26	9	0.714426	0.185177
90	29	26	10	0.854338	0.139912
90	29	26	11	0.938395	0.084058
90	29	26	12	0.978636	0.040240
90	29	26	13	0.993984	0.015348
90	29	26	14	0.998637	0.004654
90	29	26	15	0.999754	0.001117
90	29	26	16	0.999965	0.000211
90	29	26	17	0.999996	0.000031
90	29	26	18	0.999999	0.000004
90	29	26	19	1.000000	0.000000
90	29	26	20	1.000000	0.000000
90	29	26	21	1.000000	0.000000
90	29	26	22	1.000000	0.000000
90	29	26	23	1.000000	0.000000
90	29	26	24	1.000000	0.000000
90	29	26	25	1.000000	0.000000
90	29	26	26	1.000000	0.000000
90	29	27	0	0.000002	0.000002
90	29	27	1	0.000054	0.000051
90	29	27	2	0.000573	0.000519
90	29	27	3	0.003728	0.003156
90	29	27	4	0.016684	0.012956
90	29	27	5	0.054886	0.038202
90	29	27	6	0.138931	0.084045
90	29	27	7	0.280373	0.141441
90	29	27	8	0.465594	0.185221
90	29	27	9	0.656558	0.190964
90	29	27	10	0.812801	0.156243
90	29	27	11	0.914754	0.101953
90	29	27	12	0.967947	0.053193
90	29	27	13	0.990147	0.022200
90	29	27	14	0.997547	0.007400
90	29	27	15	0.999510	0.001963
90	29	27	16	0.999922	0.000412
90	29	27	17	0.999990	0.000068
90	29	27	18	0.999999	0.000009
90	29	27	19	1.000000	0.000001
90	29	27	20	1.000000	0.000000
90	29	27	21	1.000000	0.000000
90	29	27	22	1.000000	0.000000
90	29	27	23	1.000000	0.000000
90	29	27	24	1.000000	0.000000
90	29	27	25	1.000000	0.000000
90	29	27	26	1.000000	0.000000
90	29	27	27	1.000000	0.000000
90	29	28	0	0.000001	0.000001
90	29	28	1	0.000031	0.000030
90	29	28	2	0.000350	0.000319
90	29	28	3	0.002426	0.002076

Table for $N = 2$, $n = 1$, through $N = 100$, $n = 50$

N	n	k	x	P(x)	p(x)
90	29	28	4	0.011543	0.009117
90	29	28	5	0.040333	0.028790
90	29	28	6	0.108248	0.067915
90	29	28	7	0.230980	0.122732
90	29	28	8	0.403853	0.172873
90	29	28	9	0.595934	0.192081
90	29	28	10	0.765680	0.169746
90	29	28	11	0.885625	0.119944
90	29	28	12	0.953593	0.067969
90	29	28	13	0.984509	0.030915
90	29	28	14	0.995785	0.011276
90	29	28	15	0.999074	0.003289
90	29	28	16	0.999837	0.000763
90	29	28	17	0.999977	0.000140
90	29	28	18	0.999997	0.000020
90	29	28	19	1.000000	0.000002
90	29	28	20	1.000000	0.000000
90	29	28	21	1.000000	0.000000
90	29	28	22	1.000000	0.000000
90	29	28	23	1.000000	0.000000
90	29	28	24	1.000000	0.000000
90	29	28	25	1.000000	0.000000
90	29	28	26	1.000000	0.000000
90	29	28	27	1.000000	0.000000
90	29	28	28	1.000000	0.000000
90	29	29	0	0.000001	0.000001
90	29	29	1	0.000017	0.000017
90	29	29	2	0.000211	0.000194
90	29	29	3	0.001555	0.001344
90	29	29	4	0.007867	0.006311
90	29	29	5	0.029188	0.021322
90	29	29	6	0.083054	0.053865
90	29	29	7	0.187430	0.104376
90	29	29	8	0.345299	0.157869
90	29	29	9	0.533973	0.188673
90	29	29	10	0.713661	0.179689
90	29	29	11	0.850802	0.137141
90	29	29	12	0.934957	0.084155
90	29	29	13	0.976530	0.041574
90	29	29	14	0.993057	0.016526
90	29	29	15	0.998331	0.005274
90	29	29	16	0.999677	0.001346
90	29	29	17	0.999950	0.000273
90	29	29	18	0.999994	0.000044
90	29	29	19	0.999999	0.000005
90	29	29	20	1.000000	0.000001
90	29	29	21	1.000000	0.000000
90	29	29	22	1.000000	0.000000
90	29	29	23	1.000000	0.000000
90	29	29	24	1.000000	0.000000

N	n	k	x	P(x)	p(x)
90	29	29	25	1.000000	0.000000
90	29	29	26	1.000000	0.000000
90	29	29	27	1.000000	0.000000
90	29	29	28	1.000000	0.000000
90	29	29	29	1.000000	0.000000
90	30	1	0	0.666667	0.666667
90	30	1	1	1.000000	0.333333
90	30	2	0	0.441948	0.441948
90	30	2	1	0.891386	0.449438
90	30	2	2	1.000000	0.108614
90	30	3	0	0.291284	0.291284
90	30	3	1	0.743275	0.451992
90	30	3	2	0.965441	0.222165
90	30	3	3	1.000000	0.034559
90	30	4	0	0.190841	0.190841
90	30	4	1	0.592611	0.401771
90	30	4	2	0.893939	0.301328
90	30	4	3	0.989275	0.095335
90	30	4	4	1.000000	0.010725
90	30	5	0	0.124269	0.124269
90	30	5	1	0.457131	0.332862
90	30	5	2	0.795833	0.338702
90	30	5	3	0.959344	0.163511
90	30	5	4	0.996757	0.037414
90	30	5	5	1.000000	0.003243
90	30	6	0	0.080409	0.080409
90	30	6	1	0.343566	0.263157
90	30	6	2	0.684260	0.340694
90	30	6	3	0.907405	0.223145
90	30	6	4	0.985313	0.077908
90	30	6	5	0.999046	0.013733
90	30	6	6	1.000000	0.000954
90	30	7	0	0.051692	0.051692
90	30	7	1	0.252714	0.201023
90	30	7	2	0.570695	0.317981
90	30	7	3	0.835680	0.264984
90	30	7	4	0.961199	0.125519
90	30	7	5	0.994959	0.033760
90	30	7	6	0.999727	0.004768
90	30	7	7	1.000000	0.000272
90	30	8	0	0.033008	0.033008
90	30	8	1	0.182477	0.149469
90	30	8	2	0.463425	0.280947
90	30	8	3	0.749480	0.286055
90	30	8	4	0.921880	0.172399
90	30	8	5	0.984790	0.062911
90	30	8	6	0.998349	0.013558
90	30	8	7	0.999924	0.001576
90	30	8	8	1.000000	0.000076
90	30	9	0	0.020932	0.020932

N	n	k	x	P(x)	p(x)
90	30	9	1	0.129616	0.108684
90	30	9	2	0.367491	0.237875
90	30	9	3	0.655291	0.287800
90	30	9	4	0.867216	0.211925
90	30	9	5	0.965610	0.098394
90	30	9	6	0.994380	0.028770
90	30	9	7	0.999482	0.005102
90	30	9	8	0.999980	0.000497
90	30	9	9	1.000000	0.000020
90	30	10	0	0.013179	0.013179
90	30	10	1	0.090704	0.077525
90	30	10	2	0.285263	0.194558
90	30	10	3	0.559358	0.274095
90	30	10	4	0.799191	0.239833
90	30	10	5	0.935242	0.136051
90	30	10	6	0.985856	0.050614
90	30	10	7	0.998034	0.012178
90	30	10	8	0.999845	0.001811
90	30	10	9	0.999995	0.000150
90	30	10	10	1.000000	0.000005
90	30	11	0	0.008237	0.008237
90	30	11	1	0.062602	0.054365
90	30	11	2	0.217167	0.154566
90	30	11	3	0.466851	0.249683
90	30	11	4	0.721245	0.254394
90	30	11	5	0.892726	0.171481
90	30	11	6	0.970672	0.077946
90	30	11	7	0.994533	0.023861
90	30	11	8	0.999347	0.004814
90	30	11	9	0.999955	0.000609
90	30	11	10	0.999999	0.000043
90	30	11	11	1.000000	0.000001
90	30	12	0	0.005109	0.005109
90	30	12	1	0.042645	0.037536
90	30	12	2	0.162385	0.119740
90	30	12	3	0.381516	0.219131
90	30	12	4	0.637520	0.256004
90	30	12	5	0.838460	0.200939
90	30	12	6	0.946992	0.108532
90	30	12	7	0.987586	0.040594
90	30	12	8	0.998006	0.010420
90	30	12	9	0.999793	0.001787
90	30	12	10	0.999988	0.000194
90	30	12	11	1.000000	0.000012
90	30	12	12	1.000000	0.000000
90	30	13	0	0.003144	0.003144
90	30	13	1	0.028689	0.025545
90	30	13	2	0.119401	0.090712
90	30	13	3	0.305663	0.186262
90	30	13	4	0.552186	0.246523

N	n	k	x	P(x)	p(x)
90	30	13	5	0.774056	0.221871
90	30	13	6	0.913597	0.139541
90	30	13	7	0.975616	0.062018
90	30	13	8	0.995067	0.019451
90	30	13	9	0.999312	0.004245
90	30	13	10	0.999938	0.000626
90	30	13	11	0.999997	0.000059
90	30	13	12	1.000000	0.000003
90	30	13	13	1.000000	0.000000
90	30	14	0	0.001919	0.001919
90	30	14	1	0.019068	0.017149
90	30	14	2	0.086415	0.067347
90	30	14	3	0.240350	0.153935
90	30	14	4	0.468944	0.228594
90	30	14	5	0.702020	0.233076
90	30	14	6	0.870104	0.168084
90	30	14	7	0.957091	0.086987
90	30	14	8	0.989509	0.032419
90	30	14	9	0.998154	0.008645
90	30	14	10	0.999775	0.001621
90	30	14	11	0.999982	0.000207
90	30	14	12	0.999999	0.000017
90	30	14	13	1.000000	0.000001
90	30	14	14	1.000000	0.000000
90	30	15	0	0.001162	0.001162
90	30	15	1	0.012525	0.011363
90	30	15	2	0.061603	0.049079
90	30	15	3	0.185663	0.124060
90	30	15	4	0.390741	0.205078
90	30	15	5	0.625351	0.234610
90	30	15	6	0.817025	0.191674
90	30	15	7	0.930766	0.113741
90	30	15	8	0.980125	0.049359
90	30	15	9	0.995766	0.015641
90	30	15	10	0.999349	0.003583
90	30	15	11	0.999930	0.000582
90	30	15	12	0.999995	0.000065
90	30	15	13	1.000000	0.000005
90	30	15	14	1.000000	0.000000
90	30	15	15	1.000000	0.000000
90	30	16	0	0.000697	0.000697
90	30	16	1	0.008131	0.007434
90	30	16	2	0.043280	0.035150
90	30	16	3	0.141001	0.097721
90	30	16	4	0.319647	0.178646
90	30	16	5	0.547147	0.227500
90	30	16	6	0.755689	0.208542
90	30	16	7	0.895885	0.140196
90	30	16	8	0.965646	0.069761
90	30	16	9	0.991386	0.025740

Table for $N = 2$, $n = 1$, through $N = 100$, $n = 50$

N	n	k	x	P(x)	p(x)
90	30	16	10	0.998393	0.007007
90	30	16	11	0.999783	0.001390
90	30	16	12	0.999980	0.000196
90	30	16	13	0.999999	0.000019
90	30	16	14	1.000000	0.000001
90	30	16	15	1.000000	0.000000
90	30	16	16	1.000000	0.000000
90	30	17	0	0.000414	0.000414
90	30	17	1	0.005218	0.004803
90	30	17	2	0.029981	0.024763
90	30	17	3	0.105346	0.075366
90	30	17	4	0.256880	0.151533
90	30	17	5	0.470289	0.213409
90	30	17	6	0.688054	0.217765
90	30	17	7	0.852311	0.164257
90	30	17	8	0.944907	0.092596
90	30	17	9	0.984082	0.039175
90	30	17	10	0.996499	0.012418
90	30	17	11	0.999426	0.002927
90	30	17	12	0.999932	0.000506
90	30	17	13	0.999994	0.000062
90	30	17	14	1.000000	0.000005
90	30	17	15	1.000000	0.000000
90	30	17	16	1.000000	0.000000
90	30	17	17	1.000000	0.000000
90	30	18	0	0.000244	0.000244
90	30	18	1	0.003309	0.003065
90	30	18	2	0.020482	0.017173
90	30	18	3	0.077471	0.056989
90	30	18	4	0.202909	0.125438
90	30	18	5	0.397204	0.194295
90	30	18	6	0.616460	0.219256
90	30	18	7	0.800559	0.184099
90	30	18	8	0.917001	0.116442
90	30	18	9	0.972812	0.055811
90	30	18	10	0.993097	0.020285
90	30	18	11	0.998664	0.005567
90	30	18	12	0.999807	0.001143
90	30	18	13	0.999980	0.000173
90	30	18	14	0.999998	0.000019
90	30	18	15	1.000000	0.000001
90	30	18	16	1.000000	0.000000
90	30	18	17	1.000000	0.000000
90	30	18	18	1.000000	0.000000
90	30	19	0	0.000142	0.000142
90	30	19	1	0.002075	0.001932
90	30	19	2	0.013804	0.011729
90	30	19	3	0.056101	0.042296
90	30	19	4	0.157612	0.101512
90	30	19	5	0.329740	0.172128
90	30	19	6	0.543375	0.213634
90	30	19	7	0.741749	0.198375
90	30	19	8	0.881421	0.139672
90	30	19	9	0.956534	0.075112
90	30	19	10	0.987463	0.030929
90	30	19	11	0.997195	0.009733
90	30	19	12	0.999521	0.002326
90	30	19	13	0.999939	0.000418
90	30	19	14	0.999994	0.000055
90	30	19	15	1.000000	0.000005
90	30	19	16	1.000000	0.000000
90	30	19	17	1.000000	0.000000
90	30	19	18	1.000000	0.000000
90	30	19	19	1.000000	0.000000
90	30	20	0	0.000082	0.000082
90	30	20	1	0.001286	0.001203
90	30	20	2	0.009178	0.007893
90	30	20	3	0.040016	0.030838
90	30	20	4	0.120439	0.080423
90	30	20	5	0.269132	0.148693
90	30	20	6	0.471160	0.202028
90	30	20	7	0.677487	0.206327
90	30	20	8	0.838143	0.160656
90	30	20	9	0.934317	0.096175
90	30	20	10	0.978750	0.044433
90	30	20	11	0.994591	0.015841
90	30	20	12	0.998932	0.004341
90	30	20	13	0.999839	0.000907
90	30	20	14	0.999982	0.000143
90	30	20	15	0.999998	0.000017
90	30	20	16	1.000000	0.000001
90	30	20	17	1.000000	0.000000
90	30	20	18	1.000000	0.000000
90	30	20	19	1.000000	0.000000
90	30	20	20	1.000000	0.000000
90	30	21	0	0.000047	0.000047
90	30	21	1	0.000787	0.000740
90	30	21	2	0.006021	0.005234
90	30	21	3	0.028121	0.022100
90	30	21	4	0.090567	0.062446
90	30	21	5	0.216027	0.125460
90	30	21	6	0.401893	0.185866
90	30	21	7	0.609694	0.207801
90	30	21	8	0.787651	0.177957
90	30	21	9	0.905465	0.117814
90	30	21	10	0.966055	0.060590
90	30	21	11	0.990291	0.024236
90	30	21	12	0.997815	0.007524
90	30	21	13	0.999619	0.001803
90	30	21	14	0.999949	0.000330
90	30	21	15	0.999995	0.000046
90	30	21	16	0.999999	0.000005
90	30	21	17	1.000000	0.000000
90	30	21	18	1.000000	0.000000
90	30	21	19	1.000000	0.000000
90	30	21	20	1.000000	0.000000
90	30	21	21	1.000000	0.000000
90	30	22	0	0.000027	0.000027
90	30	22	1	0.000476	0.000449
90	30	22	2	0.003897	0.003421
90	30	22	3	0.019474	0.015576
90	30	22	4	0.067037	0.047564
90	30	22	5	0.170571	0.103534
90	30	22	6	0.337244	0.166673
90	30	22	7	0.540427	0.203183
90	30	22	8	0.730911	0.190484
90	30	22	9	0.869609	0.138698
90	30	22	10	0.948493	0.078884
90	30	22	11	0.983618	0.035125
90	30	22	12	0.995853	0.012235
90	30	22	13	0.999174	0.003322
90	30	22	14	0.999873	0.000698
90	30	22	15	0.999985	0.000112
90	30	22	16	0.999999	0.000014
90	30	22	17	1.000000	0.000001
90	30	22	18	1.000000	0.000000
90	30	22	19	1.000000	0.000000
90	30	22	20	1.000000	0.000000
90	30	22	21	1.000000	0.000000
90	30	22	22	1.000000	0.000000
90	30	23	0	0.000015	0.000015
90	30	23	1	0.000284	0.000269
90	30	23	2	0.002488	0.002204
90	30	23	3	0.013289	0.010800
90	30	23	4	0.048851	0.035562
90	30	23	5	0.132507	0.083656
90	30	23	6	0.278418	0.145911
90	30	23	7	0.471703	0.193285
90	30	23	8	0.669284	0.197580
90	30	23	9	0.826776	0.157492
90	30	23	10	0.925292	0.098516
90	30	23	11	0.973803	0.048512
90	30	23	12	0.992614	0.018811
90	30	23	13	0.998344	0.005730
90	30	23	14	0.999708	0.001364
90	30	23	15	0.999960	0.000252
90	30	23	16	0.999996	0.000036
90	30	23	17	1.000000	0.000004
90	30	23	18	1.000000	0.000000
90	30	23	19	1.000000	0.000000
90	30	23	20	1.000000	0.000000
90	30	23	21	1.000000	0.000000
90	30	23	22	1.000000	0.000000
90	30	23	23	1.000000	0.000000
90	30	24	0	0.000008	0.000008
90	30	24	1	0.000168	0.000159
90	30	24	2	0.001567	0.001400
90	30	24	3	0.008936	0.007369
90	30	24	4	0.035051	0.026114
90	30	24	5	0.101292	0.066241
90	30	24	6	0.226151	0.124859
90	30	24	7	0.405352	0.179200
90	30	24	8	0.604407	0.199055
90	30	24	9	0.777412	0.173006
90	30	24	10	0.895884	0.118471
90	30	24	11	0.960046	0.064162
90	30	24	12	0.987560	0.027514
90	30	24	13	0.996890	0.009330
90	30	24	14	0.999382	0.002492
90	30	24	15	0.999904	0.000521
90	30	24	16	0.999988	0.000085
90	30	24	17	0.999999	0.000011
90	30	24	18	1.000000	0.000001
90	30	24	19	1.000000	0.000000
90	30	24	20	1.000000	0.000000
90	30	24	21	1.000000	0.000000
90	30	24	22	1.000000	0.000000
90	30	24	23	1.000000	0.000000
90	30	24	24	1.000000	0.000000
90	30	25	0	0.000004	0.000004
90	30	25	1	0.000098	0.000093
90	30	25	2	0.000974	0.000876
90	30	25	3	0.005922	0.004948
90	30	25	4	0.024763	0.018842
90	30	25	5	0.076201	0.051437
90	30	25	6	0.180748	0.104547
90	30	25	7	0.342903	0.162155
90	30	25	8	0.538055	0.195152
90	30	25	9	0.722365	0.184310
90	30	25	10	0.859983	0.137618
90	30	25	11	0.941575	0.081592
90	30	25	12	0.980057	0.038481
90	30	25	13	0.994487	0.014430
90	30	25	14	0.998778	0.004291
90	30	25	15	0.999785	0.001007
90	30	25	16	0.999970	0.000185
90	30	25	17	0.999997	0.000026
90	30	25	18	1.000000	0.000003
90	30	25	19	1.000000	0.000000
90	30	25	20	1.000000	0.000000

N	n	k	x	P(x)	p(x)
90	30	25	21	1.000000	0.000000
90	30	25	22	1.000000	0.000000
90	30	25	23	1.000000	0.000000
90	30	25	24	1.000000	0.000000
90	30	25	25	1.000000	0.000000
90	30	26	0	0.000002	0.000002
90	30	26	1	0.000056	0.000054
90	30	26	2	0.000596	0.000540
90	30	26	3	0.003866	0.003270
90	30	26	4	0.017227	0.013360
90	30	26	5	0.056417	0.039190
90	30	26	6	0.142146	0.085729
90	30	26	7	0.285525	0.143379
90	30	26	8	0.472004	0.186478
90	30	26	9	0.662819	0.190815
90	30	26	10	0.817639	0.154820
90	30	26	11	0.917725	0.100086
90	30	26	12	0.969400	0.051675
90	30	26	13	0.990713	0.021313
90	30	26	14	0.997722	0.007009
90	30	26	15	0.999553	0.001831
90	30	26	16	0.999930	0.000378
90	30	26	17	0.999991	0.000061
90	30	26	18	0.999999	0.000008
90	30	26	19	1.000000	0.000001
90	30	26	20	1.000000	0.000000
90	30	26	21	1.000000	0.000000
90	30	26	22	1.000000	0.000000
90	30	26	23	1.000000	0.000000
90	30	26	24	1.000000	0.000000
90	30	26	25	1.000000	0.000000
90	30	26	26	1.000000	0.000000
90	30	27	0	0.000001	0.000001
90	30	27	1	0.000032	0.000030
90	30	27	2	0.000360	0.000328
90	30	27	3	0.002487	0.002127
90	30	27	4	0.011799	0.009312
90	30	27	5	0.041108	0.029309
90	30	27	6	0.109998	0.068889
90	30	27	7	0.233998	0.124001
90	30	27	8	0.407902	0.173903
90	30	27	9	0.600208	0.192306
90	30	27	10	0.769258	0.169050
90	30	27	11	0.888012	0.118754
90	30	27	12	0.954867	0.066854
90	30	27	13	0.985052	0.030185
90	30	27	14	0.995970	0.010918
90	30	27	15	0.999124	0.003154
90	30	27	16	0.999848	0.000724
90	30	27	17	0.999979	0.000131

N	n	k	x	P(x)	p(x)
90	30	27	18	0.999998	0.000019
90	30	27	19	1.000000	0.000002
90	30	27	20	1.000000	0.000000
90	30	27	21	1.000000	0.000000
90	30	27	22	1.000000	0.000000
90	30	27	23	1.000000	0.000000
90	30	27	24	1.000000	0.000000
90	30	27	25	1.000000	0.000000
90	30	27	26	1.000000	0.000000
90	30	27	27	1.000000	0.000000
90	30	28	0	0.000001	0.000001
90	30	28	1	0.000018	0.000017
90	30	28	2	0.000214	0.000196
90	30	28	3	0.001575	0.001361
90	30	28	4	0.007956	0.006381
90	30	28	5	0.029478	0.021522
90	30	28	6	0.083754	0.054276
90	30	28	7	0.188728	0.104974
90	30	28	8	0.347174	0.158445
90	30	28	9	0.536106	0.188932
90	30	28	10	0.715591	0.179486
90	30	28	11	0.852198	0.136606
90	30	28	12	0.935765	0.083568
90	30	28	13	0.976906	0.041141
90	30	28	14	0.993197	0.016290
90	30	28	15	0.998373	0.005176
90	30	28	16	0.999687	0.001314
90	30	28	17	0.999952	0.000265
90	30	28	18	0.999994	0.000042
90	30	28	19	0.999999	0.000005
90	30	28	20	1.000000	0.000000
90	30	28	21	1.000000	0.000000
90	30	28	22	1.000000	0.000000
90	30	28	23	1.000000	0.000000
90	30	28	24	1.000000	0.000000
90	30	28	25	1.000000	0.000000
90	30	28	26	1.000000	0.000000
90	30	28	27	1.000000	0.000000
90	30	28	28	1.000000	0.000000
90	30	29	0	0.000000	0.000000
90	30	29	1	0.000010	0.000009
90	30	29	2	0.000125	0.000116
90	30	29	3	0.000982	0.000857
90	30	29	4	0.005280	0.004298
90	30	29	5	0.020800	0.015519
90	30	29	6	0.062744	0.041944
90	30	29	7	0.149786	0.087042
90	30	29	8	0.290951	0.141165
90	30	29	9	0.472113	0.181162
90	30	29	10	0.657693	0.185580

N	n	k	x	P(x)	p(x)
90	30	29	11	0.810334	0.152642
90	30	29	12	0.911504	0.101169
90	30	29	13	0.965626	0.054122
90	30	29	14	0.988993	0.023367
90	30	29	15	0.997121	0.008128
90	30	29	16	0.999390	0.002270
90	30	29	17	0.999896	0.000506
90	30	29	18	0.999986	0.000090
90	30	29	19	0.999998	0.000012
90	30	29	20	1.000000	0.000001
90	30	29	21	1.000000	0.000000
90	30	29	22	1.000000	0.000000
90	30	29	23	1.000000	0.000000
90	30	29	24	1.000000	0.000000
90	30	29	25	1.000000	0.000000
90	30	29	26	1.000000	0.000000
90	30	29	27	1.000000	0.000000
90	30	29	28	1.000000	0.000000
90	30	29	29	1.000000	0.000000
90	30	30	0	0.000000	0.000000
90	30	30	1	0.000005	0.000005
90	30	30	2	0.000072	0.000067
90	30	30	3	0.000603	0.000531
90	30	30	4	0.003448	0.002845
90	30	30	5	0.014439	0.010991
90	30	30	6	0.046241	0.031802
90	30	30	7	0.116967	0.070726
90	30	30	8	0.240039	0.123072
90	30	30	9	0.409745	0.169706
90	30	30	10	0.596847	0.187101
90	30	30	11	0.762790	0.165943
90	30	30	12	0.881650	0.118860
90	30	30	13	0.950543	0.068892
90	30	30	14	0.982864	0.032321
90	30	30	15	0.995122	0.012258
90	30	30	16	0.998869	0.003747
90	30	30	17	0.999789	0.000919
90	30	30	18	0.999968	0.000180
90	30	30	19	0.999996	0.000028
90	30	30	20	0.999999	0.000003
90	30	30	21	1.000000	0.000000
90	30	30	22	1.000000	0.000000
90	30	30	23	1.000000	0.000000
90	30	30	24	1.000000	0.000000
90	30	30	25	1.000000	0.000000
90	30	30	26	1.000000	0.000000
90	30	30	27	1.000000	0.000000
90	30	30	28	1.000000	0.000000
90	30	30	29	1.000000	0.000000
90	30	30	30	1.000000	0.000000

N	n	k	x	P(x)	p(x)
90	31	1	0	0.655556	0.655556
90	31	1	1	1.000000	0.344444
90	31	2	0	0.427216	0.427216
90	31	2	1	0.883895	0.456679
90	31	2	2	1.000000	0.116105
90	31	3	0	0.276719	0.276719
90	31	3	1	0.728209	0.451490
90	31	3	2	0.961738	0.233529
90	31	3	3	1.000000	0.038262
90	31	4	0	0.178118	0.178118
90	31	4	1	0.572523	0.394405
90	31	4	2	0.883895	0.311372
90	31	4	3	0.987686	0.103791
90	31	4	4	1.000000	0.012314
90	31	5	0	0.113913	0.113913
90	31	5	1	0.434940	0.321027
90	31	5	2	0.778898	0.343958
90	31	5	3	0.953893	0.174996
90	31	5	4	0.996134	0.042240
90	31	5	5	1.000000	0.003866
90	31	6	0	0.072368	0.072368
90	31	6	1	0.321636	0.249268
90	31	6	2	0.661547	0.339911
90	31	6	3	0.896248	0.234700
90	31	6	4	0.982716	0.086469
90	31	6	5	0.998817	0.016101
90	31	6	6	1.000000	0.001183
90	31	7	0	0.045661	0.045661
90	31	7	1	0.232612	0.186951
90	31	7	2	0.544197	0.311585
90	31	7	3	0.818014	0.273817
90	31	7	4	0.954923	0.136909
90	31	7	5	0.993834	0.038911
90	31	7	6	0.999648	0.005814
90	31	7	7	1.000000	0.000352
90	31	8	0	0.028607	0.028607
90	31	8	1	0.165039	0.136432
90	31	8	2	0.435330	0.270291
90	31	8	3	0.725642	0.290312
90	31	8	4	0.910386	0.184744
90	31	8	5	0.981645	0.071258
90	31	8	6	0.997897	0.016252
90	31	8	7	0.999898	0.002001
90	31	8	8	1.000000	0.000102
90	31	9	0	0.017792	0.017792
90	31	9	1	0.115125	0.097333
90	31	9	2	0.339739	0.224614
90	31	9	3	0.626511	0.286772
90	31	9	4	0.849556	0.223045
90	31	9	5	0.959051	0.109495

N	n	k	x	P(x)	p(x)
90	31	9	6	0.992942	0.033891
90	31	9	7	0.999312	0.006371
90	31	9	8	0.999971	0.000659
90	31	9	9	1.000000	0.000029
90	31	10	0	0.010983	0.010983
90	31	10	1	0.079076	0.068093
90	31	10	2	0.259322	0.180246
90	31	10	3	0.527380	0.268058
90	31	10	4	0.775208	0.247827
90	31	10	5	0.923904	0.148697
90	31	10	6	0.982482	0.058577
90	31	10	7	0.997425	0.014943
90	31	10	8	0.999784	0.002359
90	31	10	9	0.999992	0.000208
90	31	10	10	1.000000	0.000008
90	31	11	0	0.006727	0.006727
90	31	11	1	0.053541	0.046814
90	31	11	2	0.193983	0.140442
90	31	11	3	0.433560	0.239577
90	31	11	4	0.691566	0.258006
90	31	11	5	0.875578	0.184012
90	31	11	6	0.964176	0.088598
90	31	11	7	0.992942	0.028766
90	31	11	8	0.999106	0.006164
90	31	11	9	0.999935	0.000829
90	31	11	10	0.999998	0.000063
90	31	11	11	1.000000	0.000002
90	31	12	0	0.004087	0.004087
90	31	12	1	0.035763	0.031676
90	31	12	2	0.142428	0.106665
90	31	12	3	0.348646	0.206218
90	31	12	4	0.603387	0.254740
90	31	12	5	0.815017	0.211630
90	31	12	6	0.936139	0.121122
90	31	12	7	0.984203	0.048064
90	31	12	8	0.997311	0.013108
90	31	12	9	0.999704	0.002393
90	31	12	10	0.999981	0.000277
90	31	12	11	0.999999	0.000018
90	31	12	12	1.000000	0.000001
90	31	13	0	0.002463	0.002463
90	31	13	1	0.023580	0.021117
90	31	13	2	0.102771	0.079190
90	31	13	3	0.274619	0.171849
90	31	13	4	0.515207	0.240588
90	31	13	5	0.744473	0.229266
90	31	13	6	0.897318	0.152844
90	31	13	7	0.969414	0.072096
90	31	13	8	0.993446	0.024032
90	31	13	9	0.999029	0.005583
90	31	13	10	0.999907	0.000877
90	31	13	11	0.999995	0.000088
90	31	13	12	1.000000	0.000005
90	31	13	13	1.000000	0.000000
90	31	14	0	0.001471	0.001471
90	31	14	1	0.015353	0.013881
90	31	14	2	0.072946	0.057593
90	31	14	3	0.212129	0.139183
90	31	14	4	0.430845	0.218716
90	31	14	5	0.667059	0.236214
90	31	14	6	0.847693	0.180634
90	31	14	7	0.946942	0.099249
90	31	14	8	0.986268	0.039325
90	31	14	9	0.997434	0.011166
90	31	14	10	0.999667	0.002233
90	31	14	11	0.999972	0.000305
90	31	14	12	0.999999	0.000027
90	31	14	13	1.000000	0.000001
90	31	14	14	1.000000	0.000000
90	31	15	0	0.000871	0.000871
90	31	15	1	0.009873	0.009002
90	31	15	2	0.050969	0.041096
90	31	15	3	0.160851	0.109881
90	31	15	4	0.353143	0.192293
90	31	15	5	0.586249	0.233106
90	31	15	6	0.788274	0.202025
90	31	15	7	0.915600	0.127327
90	31	15	8	0.974366	0.058766
90	31	15	9	0.994202	0.019835
90	31	15	10	0.999050	0.004849
90	31	15	11	0.999892	0.000841
90	31	15	12	0.999992	0.000100
90	31	15	13	1.000000	0.000008
90	31	15	14	1.000000	0.000000
90	31	15	15	1.000000	0.000000
90	31	16	0	0.000511	0.000511
90	31	16	1	0.006272	0.005761
90	31	16	2	0.035079	0.028807
90	31	16	3	0.119828	0.084750
90	31	16	4	0.283918	0.164090
90	31	16	5	0.505439	0.221521
90	31	16	6	0.720932	0.215493
90	31	16	7	0.874856	0.153924
90	31	16	8	0.956345	0.081489
90	31	16	9	0.988383	0.032038
90	31	16	10	0.997693	0.009309
90	31	16	11	0.999667	0.001975
90	31	16	12	0.999966	0.000299
90	31	16	13	0.999998	0.000031
90	31	16	14	1.000000	0.000002
90	31	16	15	1.000000	0.000000
90	31	16	16	1.000000	0.000000
90	31	17	0	0.000297	0.000297
90	31	17	1	0.003937	0.003640
90	31	17	2	0.023790	0.019853
90	31	17	3	0.087761	0.063971
90	31	17	4	0.224048	0.136286
90	31	17	5	0.427607	0.203560
90	31	17	6	0.648130	0.220523
90	31	17	7	0.824935	0.176804
90	31	17	8	0.931017	0.106083
90	31	17	9	0.978858	0.047841
90	31	17	10	0.995051	0.016192
90	31	17	11	0.999134	0.004083
90	31	17	12	0.999890	0.000756
90	31	17	13	0.999990	0.000100
90	31	17	14	0.999999	0.000009
90	31	17	15	1.000000	0.000001
90	31	17	16	1.000000	0.000000
90	31	17	17	1.000000	0.000000
90	31	18	0	0.000171	0.000171
90	31	18	1	0.002441	0.002270
90	31	18	2	0.015903	0.013462
90	31	18	3	0.063224	0.047321
90	31	18	4	0.173640	0.110416
90	31	18	5	0.355107	0.181466
90	31	18	6	0.572609	0.217502
90	31	18	7	0.766807	0.194198
90	31	18	8	0.897594	0.130787
90	31	18	9	0.964440	0.066847
90	31	18	10	0.990393	0.025952
90	31	18	11	0.998015	0.007622
90	31	18	12	0.999693	0.001678
90	31	18	13	0.999965	0.000272
90	31	18	14	0.999997	0.000032
90	31	18	15	1.000000	0.000003
90	31	18	16	1.000000	0.000000
90	31	18	17	1.000000	0.000000
90	31	18	18	1.000000	0.000000
90	31	19	0	0.000097	0.000097
90	31	19	1	0.001495	0.001398
90	31	19	2	0.010481	0.008986
90	31	19	3	0.044821	0.034341
90	31	19	4	0.132234	0.087413
90	31	19	5	0.289577	0.157343
90	31	19	6	0.497087	0.207510
90	31	19	7	0.702074	0.204987
90	31	19	8	0.855815	0.153740
90	31	19	9	0.944015	0.088200
90	31	19	10	0.982823	0.038808
90	31	19	11	0.995898	0.013074
90	31	19	12	0.999250	0.003352
90	31	19	13	0.999897	0.000647
90	31	19	14	0.999990	0.000092
90	31	19	15	0.999999	0.000010
90	31	19	16	1.000000	0.000001
90	31	19	17	1.000000	0.000000
90	31	19	18	1.000000	0.000000
90	31	19	19	1.000000	0.000000
90	31	20	0	0.000055	0.000055
90	31	20	1	0.000904	0.000850
90	31	20	2	0.006811	0.005906
90	31	20	3	0.031279	0.024468
90	31	20	4	0.098993	0.067714
90	31	20	5	0.231959	0.132966
90	31	20	6	0.424020	0.192062
90	31	20	7	0.632783	0.208763
90	31	20	8	0.806012	0.173229
90	31	20	9	0.916686	0.110674
90	31	20	10	0.971345	0.054659
90	31	20	11	0.992215	0.020870
90	31	20	12	0.998353	0.006138
90	31	20	13	0.999733	0.001380
90	31	20	14	0.999968	0.000234
90	31	20	15	0.999997	0.000030
90	31	20	16	1.000000	0.000003
90	31	20	17	1.000000	0.000000
90	31	20	18	1.000000	0.000000
90	31	20	19	1.000000	0.000000
90	31	20	20	1.000000	0.000000
90	31	21	0	0.000031	0.000031
90	31	21	1	0.000540	0.000510
90	31	21	2	0.004364	0.003823
90	31	21	3	0.021491	0.017128
90	31	21	4	0.072874	0.051383
90	31	21	5	0.182571	0.109697
90	31	21	6	0.355427	0.172856
90	31	21	7	0.561207	0.205780
90	31	21	8	0.749094	0.187886
90	31	21	9	0.881902	0.132809
90	31	21	10	0.954947	0.073045
90	31	21	11	0.986252	0.031305
90	31	21	12	0.996687	0.010435
90	31	21	13	0.999378	0.002691
90	31	21	14	0.999911	0.000532
90	31	21	15	0.999990	0.000080
90	31	21	16	0.999999	0.000009
90	31	21	17	1.000000	0.000001
90	31	21	18	1.000000	0.000000
90	31	21	19	1.000000	0.000000

Table for $N = 2$, $n = 1$, through $N = 100$, $n = 50$

N	n	k	x	P(x)	p(x)
90	31	21	20	1.000000	0.000000
90	31	21	21	1.000000	0.000000
90	31	22	0	0.000017	0.000017
90	31	22	1	0.000319	0.000302
90	31	22	2	0.002757	0.002438
90	31	22	3	0.014541	0.011784
90	31	22	4	0.052768	0.038227
90	31	22	5	0.141236	0.088468
90	31	22	6	0.292798	0.151562
90	31	22	7	0.489631	0.196833
90	31	22	8	0.686465	0.196833
90	31	22	9	0.839558	0.153093
90	31	22	10	0.932716	0.093159
90	31	22	11	0.977178	0.044462
90	31	22	12	0.993813	0.016635
90	31	22	13	0.998676	0.004863
90	31	22	14	0.999779	0.001103
90	31	22	15	0.999972	0.000192
90	31	22	16	0.999997	0.000025
90	31	22	17	1.000000	0.000002
90	31	22	18	1.000000	0.000000
90	31	22	19	1.000000	0.000000
90	31	22	20	1.000000	0.000000
90	31	22	21	1.000000	0.000000
90	31	22	22	1.000000	0.000000
90	31	23	0	0.000009	0.000009
90	31	23	1	0.000186	0.000176
90	31	23	2	0.001717	0.001532
90	31	23	3	0.009689	0.007972
90	31	23	4	0.037590	0.027901
90	31	23	5	0.107410	0.069820
90	31	23	6	0.237077	0.129667
90	31	23	7	0.420161	0.183084
90	31	23	8	0.619889	0.199728
90	31	23	9	0.790027	0.170139
90	31	23	10	0.903946	0.113919
90	31	23	11	0.964101	0.060155
90	31	23	12	0.989165	0.025064
90	31	23	13	0.997389	0.008224
90	31	23	14	0.999504	0.002115
90	31	23	15	0.999927	0.000423
90	31	23	16	0.999992	0.000065
90	31	23	17	0.999999	0.000008
90	31	23	18	1.000000	0.000001
90	31	23	19	1.000000	0.000000
90	31	23	20	1.000000	0.000000
90	31	23	21	1.000000	0.000000
90	31	23	22	1.000000	0.000000
90	31	23	23	1.000000	0.000000
90	31	24	0	0.000005	0.000005

N	n	k	x	P(x)	p(x)
90	31	24	1	0.000107	0.000102
90	31	24	2	0.001054	0.000948
90	31	24	3	0.006357	0.005303
90	31	24	4	0.026346	0.019989
90	31	24	5	0.080316	0.053970
90	31	24	6	0.188694	0.108378
90	31	24	7	0.354578	0.165885
90	31	24	8	0.551325	0.196747
90	31	24	9	0.734161	0.182836
90	31	24	10	0.868241	0.134079
90	31	24	11	0.946144	0.077904
90	31	24	12	0.982058	0.035913
90	31	24	13	0.995180	0.013122
90	31	24	14	0.998967	0.003787
90	31	24	15	0.999826	0.000858
90	31	24	16	0.999977	0.000151
90	31	24	17	0.999998	0.000021
90	31	24	18	1.000000	0.000002
90	31	24	19	1.000000	0.000000
90	31	24	20	1.000000	0.000000
90	31	24	21	1.000000	0.000000
90	31	24	22	1.000000	0.000000
90	31	24	23	1.000000	0.000000
90	31	24	24	1.000000	0.000000
90	31	25	0	0.000003	0.000003
90	31	25	1	0.000060	0.000058
90	31	25	2	0.000638	0.000577
90	31	25	3	0.004107	0.003470
90	31	25	4	0.018169	0.014061
90	31	25	5	0.059055	0.040886
90	31	25	6	0.147641	0.088587
90	31	25	7	0.294257	0.146615
90	31	25	8	0.482762	0.188505
90	31	25	9	0.673216	0.190454
90	31	25	10	0.825579	0.152363
90	31	25	11	0.922537	0.096958
90	31	25	12	0.971719	0.049182
90	31	25	13	0.991601	0.019882
90	31	25	14	0.997992	0.006391
90	31	25	15	0.999617	0.001626
90	31	25	16	0.999943	0.000325
90	31	25	17	0.999993	0.000051
90	31	25	18	0.999999	0.000006
90	31	25	19	1.000000	0.000001
90	31	25	20	1.000000	0.000000
90	31	25	21	1.000000	0.000000
90	31	25	22	1.000000	0.000000
90	31	25	23	1.000000	0.000000
90	31	25	24	1.000000	0.000000
90	31	25	25	1.000000	0.000000

N	n	k	x	P(x)	p(x)
90	31	26	0	0.000001	0.000001
90	31	26	1	0.000034	0.000032
90	31	26	2	0.000380	0.000346
90	31	26	3	0.002613	0.002233
90	31	26	4	0.012328	0.009715
90	31	26	5	0.042700	0.030373
90	31	26	6	0.113570	0.070869
90	31	26	7	0.240122	0.126552
90	31	26	8	0.416060	0.175938
90	31	26	9	0.608755	0.192694
90	31	26	10	0.776354	0.167599
90	31	26	11	0.892704	0.116350
90	31	26	12	0.957343	0.064639
90	31	26	13	0.986095	0.028752
90	31	26	14	0.996320	0.010225
90	31	26	15	0.999217	0.002897
90	31	26	16	0.999868	0.000650
90	31	26	17	0.999982	0.000115
90	31	26	18	0.999998	0.000016
90	31	26	19	1.000000	0.000002
90	31	26	20	1.000000	0.000000
90	31	26	21	1.000000	0.000000
90	31	26	22	1.000000	0.000000
90	31	26	23	1.000000	0.000000
90	31	26	24	1.000000	0.000000
90	31	26	25	1.000000	0.000000
90	31	26	26	1.000000	0.000000
90	31	27	0	0.000001	0.000001
90	31	27	1	0.000019	0.000018
90	31	27	2	0.000223	0.000205
90	31	27	3	0.001636	0.001413
90	31	27	4	0.008229	0.006593
90	31	27	5	0.030362	0.022132
90	31	27	6	0.085886	0.055525
90	31	27	7	0.192665	0.106778
90	31	27	8	0.352832	0.160168
90	31	27	9	0.542516	0.189684
90	31	27	10	0.721360	0.178845
90	31	27	11	0.856344	0.134984
90	31	27	12	0.938153	0.081809
90	31	27	13	0.978008	0.039855
90	31	27	14	0.993604	0.015596
90	31	27	15	0.998493	0.004889
90	31	27	16	0.999715	0.001222
90	31	27	17	0.999957	0.000242
90	31	27	18	0.999995	0.000038
90	31	27	19	0.999999	0.000005
90	31	27	20	1.000000	0.000000
90	31	27	21	1.000000	0.000000
90	31	27	22	1.000000	0.000000

N	n	k	x	P(x)	p(x)
90	31	27	23	1.000000	0.000000
90	31	27	24	1.000000	0.000000
90	31	27	25	1.000000	0.000000
90	31	27	26	1.000000	0.000000
90	31	27	27	1.000000	0.000000
90	31	28	0	0.000000	0.000000
90	31	28	1	0.000010	0.000010
90	31	28	2	0.000129	0.000119
90	31	28	3	0.001008	0.000879
90	31	28	4	0.005404	0.004396
90	31	28	5	0.021228	0.015824
90	31	28	6	0.063853	0.042625
90	31	28	7	0.151987	0.088135
90	31	28	8	0.294358	0.142371
90	31	28	9	0.476277	0.181919
90	31	28	10	0.661746	0.185468
90	31	28	11	0.813492	0.151747
90	31	28	12	0.913481	0.099988
90	31	28	13	0.966621	0.053141
90	31	28	14	0.989396	0.022775
90	31	28	15	0.997251	0.007856
90	31	28	16	0.999424	0.002173
90	31	28	17	0.999903	0.000479
90	31	28	18	0.999987	0.000084
90	31	28	19	0.999999	0.000011
90	31	28	20	1.000000	0.000001
90	31	28	21	1.000000	0.000000
90	31	28	22	1.000000	0.000000
90	31	28	23	1.000000	0.000000
90	31	28	24	1.000000	0.000000
90	31	28	25	1.000000	0.000000
90	31	28	26	1.000000	0.000000
90	31	28	27	1.000000	0.000000
90	31	28	28	1.000000	0.000000
90	31	29	0	0.000000	0.000000
90	31	29	1	0.000005	0.000005
90	31	29	2	0.000073	0.000068
90	31	29	3	0.000611	0.000538
90	31	29	4	0.003489	0.002878
90	31	29	5	0.014592	0.011102
90	31	29	6	0.046665	0.032073
90	31	29	7	0.117871	0.071205
90	31	29	8	0.241543	0.123673
90	31	29	9	0.411725	0.170182
90	31	29	10	0.598926	0.187200
90	31	29	11	0.764542	0.165616
90	31	29	12	0.882839	0.118297
90	31	29	13	0.951193	0.068354
90	31	29	14	0.983151	0.031958
90	31	29	15	0.995224	0.012073

$N = 90$ $n = 31$

N	n	k	x	P(x)	p(x)
90	31	29	16	0.998898	0.003674
90	31	29	17	0.999795	0.000897
90	31	29	18	0.999969	0.000174
90	31	29	19	0.999996	0.000027
90	31	29	20	0.999999	0.000003
90	31	29	21	1.000000	0.000000
90	31	29	22	1.000000	0.000000
90	31	29	23	1.000000	0.000000
90	31	29	24	1.000000	0.000000
90	31	29	25	1.000000	0.000000
90	31	29	26	1.000000	0.000000
90	31	29	27	1.000000	0.000000
90	31	29	28	1.000000	0.000000
90	31	29	29	1.000000	0.000000
90	31	30	0	0.000000	0.000000
90	31	30	1	0.000003	0.000003
90	31	30	2	0.000041	0.000038
90	31	30	3	0.000364	0.000323
90	31	30	4	0.002215	0.001851
90	31	30	5	0.009860	0.007644
90	31	30	6	0.033520	0.023661
90	31	30	7	0.089855	0.056335
90	31	30	8	0.194913	0.105057
90	31	30	9	0.350348	0.155436
90	31	30	10	0.534480	0.184131
90	31	30	11	0.710241	0.175762
90	31	30	12	0.845992	0.135751
90	31	30	13	0.931023	0.085031
90	31	30	14	0.974245	0.043222
90	31	30	15	0.992057	0.017813
90	31	30	16	0.997995	0.005938
90	31	30	17	0.999589	0.001594
90	31	30	18	0.999932	0.000343
90	31	30	19	0.999991	0.000059
90	31	30	20	0.999999	0.000008
90	31	30	21	1.000000	0.000001
90	31	30	22	1.000000	0.000000
90	31	30	23	1.000000	0.000000
90	31	30	24	1.000000	0.00000C
90	31	30	25	1.000000	0.000000
90	31	30	26	1.000000	0.000000
90	31	30	27	1.000000	0.000000
90	31	30	28	1.000000	0.000000
90	31	30	29	1.000000	0.000000
90	31	30	30	1.000000	0.000000
90	31	31	0	0.000000	0.000000
90	31	31	1	0.000001	0.000001
90	31	31	2	0.000023	0.000021
90	31	31	3	0.000213	0.000191
90	31	31	4	0.001382	0.001169
90	31	31	5	0.006547	0.005165
90	31	31	6	0.023662	0.017115
90	31	31	7	0.067321	0.043660
90	31	31	8	0.154641	0.087319
90	31	31	9	0.293355	0.138714
90	31	31	10	0.470034	0.176678
90	31	31	11	0.651654	0.181621
90	31	31	12	0.803005	0.151350
90	31	31	13	0.905514	0.102509
90	31	31	14	0.961999	0.056485
90	31	31	15	0.987307	0.025309
90	31	31	16	0.996511	0.009203
90	31	31	17	0.999217	0.002707
90	31	31	18	0.999858	0.000641
90	31	31	19	0.999979	0.000121
90	31	31	20	0.999998	0.000018
90	31	31	21	1.000000	0.000002
90	31	31	22	1.000000	0.000000
90	31	31	23	1.000000	0.000000
90	31	31	24	1.000000	0.000000
90	31	31	25	1.000000	0.000000
90	31	31	26	1.000000	0.000000
90	31	31	27	1.000000	0.000000
90	31	31	28	1.000000	0.000000
90	31	31	29	1.000000	0.000000
90	31	31	30	1.000000	0.000000
90	31	31	31	1.000000	0.000000
90	32	1	0	0.644444	0.644444
90	32	1	1	1.000000	0.355556
90	32	2	0	0.412734	0.412734
90	32	2	1	0.876155	0.463421
90	32	2	2	1.000000	0.123845
90	32	3	0	0.262649	0.262649
90	32	3	1	0.712904	0.450255
90	32	3	2	0.957780	0.244876
90	32	3	3	1.000000	0.042220
90	32	4	0	0.166042	0.166042
90	32	4	1	0.552468	0.386426
90	32	4	2	0.873340	0.320872
90	32	4	3	0.985927	0.112587
90	32	4	4	1.000000	0.014073
90	32	5	0	0.104259	0.104259
90	32	5	1	0.413175	0.308916
90	32	5	2	0.761408	0.348233
90	32	5	3	0.947961	0.186553
90	32	5	4	0.995418	0.047457
90	32	5	5	1.000000	0.004582
90	32	6	0	0.065009	0.065009
90	32	6	1	0.300512	0.235503
90	32	6	2	0.638502	0.337991
90	32	6	3	0.884314	0.245811
90	32	6	4	0.979785	0.095471
90	32	6	5	0.998544	0.018759
90	32	6	6	1.000000	0.001455
90	32	7	0	0.040243	0.040243
90	32	7	1	0.213600	0.173356
90	32	7	2	0.517792	0.304192
90	32	7	3	0.799450	0.281659
90	32	7	4	0.947961	0.148511
90	32	7	5	0.992515	0.044553
90	32	7	6	0.999549	0.007035
90	32	7	7	1.000000	0.000450
90	32	8	0	0.024728	0.024728
90	32	8	1	0.148852	0.124124
90	32	8	2	0.407843	0.258990
90	32	8	3	0.701039	0.293197
90	32	8	4	0.897861	0.196822
90	32	8	5	0.978021	0.080160
90	32	8	6	0.997346	0.019324
90	32	8	7	0.999864	0.002518
90	32	8	8	1.000000	0.000136
90	32	9	0	0.015078	0.015078
90	32	9	1	0.101927	0.086849
90	32	9	2	0.313090	0.211163
90	32	9	3	0.597348	0.284258
90	32	9	4	0.830654	0.233306
90	32	9	5	0.951627	0.120973
90	32	9	6	0.991219	0.039591
90	32	9	7	0.999096	0.007878
90	32	9	8	0.999960	0.000864
90	32	9	9	1.000000	0.000040
90	32	10	0	0.009121	0.009121
90	32	10	1	0.068689	0.059567
90	32	10	2	0.234882	0.166193
90	32	10	3	0.495577	0.260695
90	32	10	4	0.750005	0.254428
90	32	10	5	0.911303	0.161298
90	32	10	6	0.978510	0.067207
90	32	10	7	0.996665	0.018155
90	32	10	8	0.999704	0.003039
90	32	10	9	0.999989	0.000284
90	32	10	10	1.000000	0.000011
90	32	11	0	0.005473	0.005473
90	32	11	1	0.045606	0.040134
90	32	11	2	0.172559	0.126953
90	32	11	3	0.401075	0.228515
90	32	11	4	0.660955	0.259880
90	32	11	5	0.856865	0.195910
90	32	11	6	0.956668	0.099803
90	32	11	7	0.990992	0.034324
90	32	11	8	0.998792	0.007801
90	32	11	9	0.999907	0.001114
90	32	11	10	0.999997	0.000090
90	32	11	11	1.000000	0.000003
90	32	12	0	0.003256	0.003256
90	32	12	1	0.029858	0.026602
90	32	12	2	0.124349	0.094492
90	32	12	3	0.317189	0.192840
90	32	12	4	0.568845	0.251656
90	32	12	5	0.789908	0.221063
90	32	12	6	0.923821	0.133913
90	32	12	7	0.980130	0.056308
90	32	12	8	0.996423	0.016293
90	32	12	9	0.999582	0.003160
90	32	12	10	0.999972	0.000389
90	32	12	11	0.999999	0.000027
90	32	12	12	1.000000	0.000001
90	32	13	0	0.001920	0.001920
90	32	13	1	0.019285	0.017365
90	32	13	2	0.088006	0.068721
90	32	13	3	0.245492	0.157486
90	32	13	4	0.478507	0.233015
90	32	13	5	0.713387	0.234879
90	32	13	6	0.879184	0.165797
90	32	13	7	0.962082	0.082899
90	32	13	8	0.991409	0.029327
90	32	13	9	0.998651	0.007241
90	32	13	10	0.999862	0.001211
90	32	13	11	0.999992	0.000130
90	32	13	12	1.000000	0.000008
90	32	13	13	1.000000	0.000000
90	32	14	0	0.001122	0.001122
90	32	14	1	0.012294	0.011172
90	32	14	2	0.061232	0.048938
90	32	14	3	0.186179	0.124948
90	32	14	4	0.393775	0.207595
90	32	14	5	0.631026	0.237252
90	32	14	6	0.823200	0.192174
90	32	14	7	0.935167	0.111967
90	32	14	8	0.982268	0.047101
90	32	14	9	0.996488	0.014219
90	32	14	10	0.999516	0.003028
90	32	14	11	0.999956	0.000440
90	32	14	12	0.999998	0.000041
90	32	14	13	1.000000	0.000002
90	32	14	14	1.000000	0.000000
90	32	15	0	0.000650	0.000650
90	32	15	1	0.007737	0.007087
90	32	15	2	0.041914	0.034177
90	32	15	3	0.138502	0.096588

N	n	k	x	P(x)	p(x)
90	32	15	4	0.317292	0.178790
90	32	15	5	0.546739	0.229447
90	32	15	6	0.757456	0.210717
90	32	15	7	0.898336	0.140879
90	32	15	8	0.967394	0.069058
90	32	15	9	0.992184	0.024790
90	32	15	10	0.998639	0.006455
90	32	15	11	0.999835	0.001195
90	32	15	12	0.999987	0.000152
90	32	15	13	0.999999	0.000013
90	32	15	14	1.000000	0.000001
90	32	15	15	1.000000	0.000000
90	32	16	0	0.000372	0.000372
90	32	16	1	0.004808	0.004435
90	32	16	2	0.028243	0.023436
90	32	16	3	0.101155	0.072911
90	32	16	4	0.250544	0.149389
90	32	16	5	0.464138	0.213595
90	32	16	6	0.684408	0.220269
90	32	16	7	0.851376	0.166968
90	32	16	8	0.945296	0.093920
90	32	16	9	0.984582	0.039287
90	32	16	10	0.996746	0.012164
90	32	16	11	0.999500	0.002754
90	32	16	12	0.999946	0.000446
90	32	16	13	0.999996	0.000050
90	32	16	14	1.000000	0.000004
90	32	16	15	1.000000	0.000000
90	32	16	16	1.000000	0.000000
90	32	17	0	0.000211	0.000211
90	32	17	1	0.002950	0.002738
90	32	17	2	0.018742	0.015793
90	32	17	3	0.072581	0.053839
90	32	17	4	0.194018	0.121437
90	32	17	5	0.386205	0.192187
90	32	17	6	0.607016	0.220811
90	32	17	7	0.794968	0.187952
90	32	17	8	0.914835	0.119867
90	32	17	9	0.972371	0.057536
90	32	17	10	0.993130	0.020758
90	32	17	11	0.998718	0.005589
90	32	17	12	0.999826	0.001107
90	32	17	13	0.999983	0.000158
90	32	17	14	0.999999	0.000016
90	32	17	15	1.000000	0.000001
90	32	17	16	1.000000	0.000000
90	32	17	17	1.000000	0.000000
90	32	18	0	0.000119	0.000119
90	32	18	1	0.001787	0.001668
90	32	18	2	0.012252	0.010465

N	n	k	x	P(x)	p(x)
90	32	18	3	0.051193	0.038941
90	32	18	4	0.147440	0.096246
90	32	18	5	0.315122	0.167683
90	32	18	6	0.528371	0.213249
90	32	18	7	0.730601	0.202230
90	32	18	8	0.875427	0.144826
90	32	18	9	0.954243	0.078817
90	32	18	10	0.986874	0.032630
90	32	18	11	0.997111	0.010237
90	32	18	12	0.999522	0.002412
90	32	18	13	0.999942	0.000420
90	32	18	14	0.999995	0.000053
90	32	18	15	1.000000	0.000005
90	32	18	16	1.000000	0.000000
90	32	18	17	1.000000	0.000000
90	32	18	18	1.000000	0.000000
90	32	19	0	0.000066	0.000066
90	32	19	1	0.001069	0.001003
90	32	19	2	0.007892	0.006823
90	32	19	3	0.035509	0.027617
90	32	19	4	0.110011	0.074502
90	32	19	5	0.252241	0.142231
90	32	19	6	0.451365	0.199123
90	32	19	7	0.660382	0.209017
90	32	19	8	0.827151	0.166769
90	32	19	9	0.929066	0.101915
90	32	19	10	0.976903	0.047837
90	32	19	11	0.994125	0.017221
90	32	19	12	0.998852	0.004727
90	32	19	13	0.999831	0.000979
90	32	19	14	0.999982	0.000150
90	32	19	15	0.999998	0.000017
90	32	19	16	1.000000	0.000001
90	32	19	17	1.000000	0.000000
90	32	19	18	1.000000	0.000000
90	32	19	19	1.000000	0.000000
90	32	20	0	0.000036	0.000036
90	32	20	1	0.000631	0.000595
90	32	20	2	0.005009	0.004378
90	32	20	3	0.024229	0.019220
90	32	20	4	0.080630	0.056401
90	32	20	5	0.198154	0.117524
90	32	20	6	0.378446	0.180293
90	32	20	7	0.586784	0.208338
90	32	20	8	0.770779	0.183994
90	32	20	9	0.896051	0.125273
90	32	20	10	0.962080	0.066029
90	32	20	11	0.989031	0.026951
90	32	20	12	0.997521	0.008489
90	32	20	13	0.999569	0.002049

N	n	k	x	P(x)	p(x)
90	32	20	14	0.999944	0.000374
90	32	20	15	0.999995	0.000051
90	32	20	16	1.000000	0.000005
90	32	20	17	1.000000	0.000000
90	32	20	18	1.000000	0.000000
90	32	20	19	1.000000	0.000000
90	32	20	20	1.000000	0.000000
90	32	21	0	0.000020	0.000020
90	32	21	1	0.000368	0.000348
90	32	21	2	0.003132	0.002765
90	32	21	3	0.016266	0.013134
90	32	21	4	0.058069	0.041803
90	32	21	5	0.152823	0.094754
90	32	21	6	0.311480	0.158657
90	32	21	7	0.512378	0.200897
90	32	21	8	0.707695	0.195317
90	32	21	9	0.854890	0.147195
90	32	21	10	0.941328	0.086438
90	32	21	11	0.980946	0.039617
90	32	21	12	0.995095	0.014149
90	32	21	13	0.999013	0.003918
90	32	21	14	0.999847	0.000834
90	32	21	15	0.999982	0.000135
90	32	21	16	0.999998	0.000016
90	32	21	17	1.000000	0.000001
90	32	21	18	1.000000	0.000000
90	32	21	19	1.000000	0.000000
90	32	21	20	1.000000	0.000000
90	32	21	21	1.000000	0.000000
90	32	22	0	0.000011	0.000011
90	32	22	1	0.000211	0.000201
90	32	22	2	0.001930	0.001719
90	32	22	3	0.010746	0.008816
90	32	22	4	0.041106	0.030359
90	32	22	5	0.115745	0.074640
90	32	22	6	0.251696	0.135951
90	32	22	7	0.439589	0.187893
90	32	22	8	0.639759	0.200170
90	32	22	9	0.805825	0.166067
90	32	22	10	0.913768	0.107943
90	32	22	11	0.968888	0.055120
90	32	22	12	0.990994	0.022105
90	32	22	13	0.997934	0.006940
90	32	22	14	0.999630	0.001695
90	32	22	15	0.999949	0.000319
90	32	22	16	0.999995	0.000046
90	32	22	17	0.999999	0.000005
90	32	22	18	1.000000	0.000000
90	32	22	19	1.000000	0.000000
90	32	22	20	1.000000	0.000000

N	n	k	x	P(x)	p(x)
90	32	22	21	1.000000	0.000000
90	32	22	22	1.000000	0.000000
90	32	23	0	0.000006	0.000006
90	32	23	1	0.000120	0.000114
90	32	23	2	0.001172	0.001052
90	32	23	3	0.006986	0.005815
90	32	23	4	0.028605	0.021618
90	32	23	5	0.086109	0.057504
90	32	23	6	0.199715	0.113606
90	32	23	7	0.370511	0.170796
90	32	23	8	0.569110	0.198600
90	32	23	9	0.749656	0.180545
90	32	23	10	0.878846	0.129190
90	32	23	11	0.951866	0.073020
90	32	23	12	0.984492	0.032626
90	32	23	13	0.995995	0.011503
90	32	23	14	0.999181	0.003186
90	32	23	15	0.999869	0.000688
90	32	23	16	0.999984	0.000115
90	32	23	17	0.999998	0.000015
90	32	23	18	1.000000	0.000001
90	32	23	19	1.000000	0.000000
90	32	23	20	1.000000	0.000000
90	32	23	21	1.000000	0.000000
90	32	23	22	1.000000	0.000000
90	32	23	23	1.000000	0.000000
90	32	24	0	0.000003	0.000003
90	32	24	1	0.000067	0.000064
90	32	24	2	0.000701	0.000634
90	32	24	3	0.004470	0.003769
90	32	24	4	0.019570	0.015100
90	32	24	5	0.062936	0.043365
90	32	24	6	0.155629	0.092694
90	32	24	7	0.306781	0.151152
90	32	24	8	0.497970	0.191189
90	32	24	9	0.687677	0.189707
90	32	24	10	0.836425	0.148748
90	32	24	11	0.928979	0.092554
90	32	24	12	0.974753	0.045774
90	32	24	13	0.992733	0.017980
90	32	24	14	0.998325	0.005592
90	32	24	15	0.999695	0.001369
90	32	24	16	0.999956	0.000262
90	32	24	17	0.999995	0.000039
90	32	24	18	0.999999	0.000004
90	32	24	19	1.000000	0.000000
90	32	24	20	1.000000	0.000000
90	32	24	21	1.000000	0.000000
90	32	24	22	1.000000	0.000000
90	32	24	23	1.000000	0.000000

Table for $N = 2$, $n = 1$, through $N = 100$, $n = 50$

N	n	k	x	P(x)	p(x)
90	32	24	24	1.000000	0.000000
90	32	25	0	0.000001	0.000001
90	32	25	1	0.000037	0.000035
90	32	25	2	0.000413	0.000376
90	32	25	3	0.002814	0.002401
90	32	25	4	0.013164	0.010350
90	32	25	5	0.045195	0.032031
90	32	25	6	0.119114	0.073918
90	32	25	7	0.249527	0.130413
90	32	25	8	0.428447	0.178920
90	32	25	9	0.621567	0.193120
90	32	25	10	0.786843	0.165275
90	32	25	11	0.899530	0.112688
90	32	25	12	0.960882	0.061352
90	32	25	13	0.987557	0.026675
90	32	25	14	0.996800	0.009243
90	32	25	15	0.999342	0.002542
90	32	25	16	0.999893	0.000551
90	32	25	17	0.999986	0.000093
90	32	25	18	0.999999	0.000012
90	32	25	19	1.000000	0.000001
90	32	25	20	1.000000	0.000000
90	32	25	21	1.000000	0.000000
90	32	25	22	1.000000	0.000000
90	32	25	23	1.000000	0.000000
90	32	25	24	1.000000	0.000000
90	32	25	25	1.000000	0.000000
90	32	26	0	0.000001	0.000001
90	32	26	1	0.000020	0.000019
90	32	26	2	0.000239	0.000219
90	32	26	3	0.001742	0.001503
90	32	26	4	0.008705	0.006963
90	32	26	5	0.031890	0.023185
90	32	26	6	0.089546	0.057656
90	32	26	7	0.199368	0.109822
90	32	26	8	0.362384	0.163016
90	32	26	9	0.553232	0.190848
90	32	26	10	0.730903	0.177671
90	32	26	11	0.863123	0.132220
90	32	26	12	0.942005	0.078881
90	32	26	13	0.979760	0.037755
90	32	26	14	0.994240	0.014481
90	32	26	15	0.998677	0.004437
90	32	26	16	0.999757	0.001080
90	32	26	17	0.999965	0.000207
90	32	26	18	0.999996	0.000031
90	32	26	19	1.000000	0.000004
90	32	26	20	1.000000	0.000000
90	32	26	21	1.000000	0.000000
90	32	26	22	1.000000	0.000000
90	32	26	23	1.000000	0.000000
90	32	26	24	1.000000	0.000000
90	32	26	25	1.000000	0.000000
90	32	26	26	1.000000	0.000000
90	32	27	0	0.000000	0.000000
90	32	27	1	0.000011	0.000010
90	32	27	2	0.000136	0.000126
90	32	27	3	0.001061	0.000925
90	32	27	4	0.005659	0.004598
90	32	27	5	0.022109	0.016450
90	32	27	6	0.066123	0.044014
90	32	27	7	0.156469	0.090345
90	32	27	8	0.301253	0.144784
90	32	27	9	0.484646	0.183393
90	32	27	10	0.669829	0.185183
90	32	27	11	0.819739	0.149910
90	32	27	12	0.917354	0.097616
90	32	27	13	0.968551	0.051197
90	32	27	14	0.990168	0.021617
90	32	27	15	0.997499	0.007331
90	32	27	16	0.999487	0.001989
90	32	27	17	0.999916	0.000429
90	32	27	18	0.999989	0.000073
90	32	27	19	0.999999	0.000010
90	32	27	20	1.000000	0.000001
90	32	27	21	1.000000	0.000000
90	32	27	22	1.000000	0.000000
90	32	27	23	1.000000	0.000000
90	32	27	24	1.000000	0.000000
90	32	27	25	1.000000	0.000000
90	32	27	26	1.000000	0.000000
90	32	27	27	1.000000	0.000000
90	32	28	0	0.000000	0.000000
90	32	28	1	0.000006	0.000005
90	32	28	2	0.000077	0.000071
90	32	28	3	0.000636	0.000559
90	32	28	4	0.003616	0.002980
90	32	28	5	0.015059	0.011443
90	32	28	6	0.047959	0.032900
90	32	28	7	0.120617	0.072659
90	32	28	8	0.246097	0.125480
90	32	28	9	0.417693	0.171596
90	32	28	10	0.605162	0.187469
90	32	28	11	0.769769	0.164607
90	32	28	12	0.886365	0.116596
90	32	28	13	0.953111	0.066746
90	32	28	14	0.983992	0.030881
90	32	28	15	0.995520	0.011529
90	32	28	16	0.998982	0.003462
90	32	28	17	0.999814	0.000832
90	32	28	18	0.999973	0.000159
90	32	28	19	0.999997	0.000024
90	32	28	20	1.000000	0.000003
90	32	28	21	1.000000	0.000000
90	32	28	22	1.000000	0.000000
90	32	28	23	1.000000	0.000000
90	32	28	24	1.000000	0.000000
90	32	28	25	1.000000	0.000000
90	32	28	26	1.000000	0.000000
90	32	28	27	1.000000	0.000000
90	32	28	28	1.000000	0.000000
90	32	29	0	0.000000	0.000000
90	32	29	1	0.000003	0.000003
90	32	29	2	0.000042	0.000039
90	32	29	3	0.000374	0.000332
90	32	29	4	0.002270	0.001896
90	32	29	5	0.010076	0.007806
90	32	29	6	0.034162	0.024086
90	32	29	7	0.091319	0.057157
90	32	29	8	0.197524	0.106205
90	32	29	9	0.354037	0.156512
90	32	29	10	0.538641	0.184604
90	32	29	11	0.714015	0.175374
90	32	29	12	0.848753	0.134739
90	32	29	13	0.932656	0.083903
90	32	29	14	0.975026	0.042370
90	32	29	15	0.992359	0.017333
90	32	29	16	0.998089	0.005730
90	32	29	17	0.999613	0.001524
90	32	29	18	0.999937	0.000324
90	32	29	19	0.999992	0.000055
90	32	29	20	0.999999	0.000007
90	32	29	21	1.000000	0.000001
90	32	29	22	1.000000	0.000000
90	32	29	23	1.000000	0.000000
90	32	29	24	1.000000	0.000000
90	32	29	25	1.000000	0.000000
90	32	29	26	1.000000	0.000000
90	32	29	27	1.000000	0.000000
90	32	29	28	1.000000	0.000000
90	32	29	29	1.000000	0.000000
90	32	30	0	0.000000	0.000000
90	32	30	1	0.000001	0.000001
90	32	30	2	0.000023	0.000021
90	32	30	3	0.000216	0.000193
90	32	30	4	0.001400	0.001183
90	32	30	5	0.006621	0.005221
90	32	30	6	0.023896	0.017275
90	32	30	7	0.067894	0.043999
90	32	30	8	0.155739	0.087844
90	32	30	9	0.295024	0.139285
90	32	30	10	0.472062	0.177039
90	32	30	11	0.653640	0.181578
90	32	30	12	0.804577	0.150937
90	32	30	13	0.906523	0.101946
90	32	30	14	0.962524	0.056001
90	32	30	15	0.987529	0.025005
90	32	30	16	0.996586	0.009057
90	32	30	17	0.999238	0.002652
90	32	30	18	0.999863	0.000625
90	32	30	19	0.999980	0.000117
90	32	30	20	0.999998	0.000018
90	32	30	21	1.000000	0.000002
90	32	30	22	1.000000	0.000000
90	32	30	23	1.000000	0.000000
90	32	30	24	1.000000	0.000000
90	32	30	25	1.000000	0.000000
90	32	30	26	1.000000	0.000000
90	32	30	27	1.000000	0.000000
90	32	30	28	1.000000	0.000000
90	32	30	29	1.000000	0.000000
90	32	30	30	1.000000	0.000000
90	32	31	0	0.000000	0.000000
90	32	31	1	0.000001	0.000001
90	32	31	2	0.000012	0.000011
90	32	31	3	0.000123	0.000111
90	32	31	4	0.000847	0.000725
90	32	31	5	0.004271	0.003424
90	32	31	6	0.016410	0.012139
90	32	31	7	0.049562	0.033152
90	32	31	8	0.120601	0.071039
90	32	31	9	0.241631	0.121030
90	32	31	10	0.407148	0.165517
90	32	31	11	0.590088	0.182940
90	32	31	12	0.754265	0.164177
90	32	31	13	0.874240	0.119975
90	32	31	14	0.945723	0.071484
90	32	31	15	0.980444	0.034721
90	32	31	16	0.994171	0.013727
90	32	31	17	0.998575	0.004404
90	32	31	18	0.999717	0.001142
90	32	31	19	0.999955	0.000238
90	32	31	20	0.999994	0.000039
90	32	31	21	0.999999	0.000005
90	32	31	22	1.000000	0.000001
90	32	31	23	1.000000	0.000000
90	32	31	24	1.000000	0.000000
90	32	31	25	1.000000	0.000000
90	32	31	26	1.000000	0.000000
90	32	31	27	1.000000	0.000000

N	n	k	x	P(x)	p(x)
90	32	31	28	1.000000	0.000000
90	32	31	29	1.000000	0.000000
90	32	31	30	1.000000	0.000000
90	32	31	31	1.000000	0.000000
90	32	32	0	0.000000	0.000000
90	32	32	1	0.000000	0.000000
90	32	32	2	0.000006	0.000006
90	32	32	3	0.000068	0.000062
90	32	32	4	0.000504	0.000435
90	32	32	5	0.002704	0.002201
90	32	32	6	0.011061	0.008356
90	32	32	7	0.035514	0.024454
90	32	32	8	0.091704	0.056189
90	32	32	9	0.194450	0.102746
90	32	32	10	0.345430	0.150980
90	32	32	11	0.524974	0.179544
90	32	32	12	0.698612	0.173638
90	32	32	13	0.835604	0.136992
90	32	32	14	0.923915	0.088311
90	32	32	15	0.970440	0.046525
90	32	32	16	0.990448	0.020008
90	32	32	17	0.997455	0.007007
90	32	32	18	0.999446	0.001991
90	32	32	19	0.999902	0.000456
90	32	32	20	0.999986	0.000084
90	32	32	21	0.999998	0.000012
90	32	32	22	1.000000	0.000001
90	32	32	23	1.000000	0.000000
90	32	32	24	1.000000	0.000000
90	32	32	25	1.000000	0.000000
90	32	32	26	1.000000	0.000000
90	32	32	27	1.000000	0.000000
90	32	32	28	1.000000	0.000000
90	32	32	29	1.000000	0.000000
90	32	32	30	1.000000	0.000000
90	32	32	31	1.000000	0.000000
90	32	32	32	1.000000	0.000000
90	33	1	0	0.633333	0.633333
90	33	1	1	1.000000	0.366667
90	33	2	0	0.398502	0.398502
90	33	2	1	0.868165	0.469663
90	33	2	2	1.000000	0.131835
90	33	3	0	0.249064	0.249064
90	33	3	1	0.697378	0.448315
90	33	3	2	0.953558	0.256180
90	33	3	3	1.000000	0.046442
90	33	4	0	0.154591	0.154591
90	33	4	1	0.532481	0.377890
90	33	4	2	0.862276	0.329795
90	33	4	3	0.983985	0.121710

N	n	k	x	P(x)	p(x)
90	33	4	4	1.000000	0.016014
90	33	5	0	0.095271	0.095271
90	33	5	1	0.391871	0.296599
90	33	5	2	0.743396	0.351525
90	33	5	3	0.941529	0.198132
90	33	5	4	0.994600	0.053071
90	33	5	5	1.000000	0.005400
90	33	6	0	0.058284	0.058284
90	33	6	1	0.280210	0.221926
90	33	6	2	0.615193	0.334983
90	33	6	3	0.871599	0.256407
90	33	6	4	0.976493	0.104894
90	33	6	5	0.998221	0.021728
90	33	6	6	1.000000	0.001779
90	33	7	0	0.035386	0.035386
90	33	7	1	0.195666	0.160280
90	33	7	2	0.491568	0.295902
90	33	7	3	0.780026	0.288457
90	33	7	4	0.940280	0.160254
90	33	7	5	0.990978	0.050699
90	33	7	6	0.999428	0.008450
90	33	7	7	1.000000	0.000572
90	33	8	0	0.021317	0.021317
90	33	8	1	0.133872	0.112555
90	33	8	2	0.381051	0.247179
90	33	8	3	0.675764	0.294713
90	33	8	4	0.884287	0.208523
90	33	8	5	0.973875	0.089588
90	33	8	6	0.996679	0.022804
90	33	8	7	0.999821	0.003141
90	33	8	8	1.000000	0.000179
90	33	9	0	0.012738	0.012738
90	33	9	1	0.089948	0.077210
90	33	9	2	0.287605	0.197657
90	33	9	3	0.567942	0.280337
90	33	9	4	0.810541	0.242599
90	33	9	5	0.943284	0.132743
90	33	9	6	0.989171	0.045886
90	33	9	7	0.998825	0.009654
90	33	9	8	0.999945	0.001121
90	33	9	9	1.000000	0.000055
90	33	10	0	0.007549	0.007549
90	33	10	1	0.059445	0.051897
90	33	10	2	0.211958	0.152513
90	33	10	3	0.464113	0.252155
90	33	10	4	0.723685	0.259571
90	33	10	5	0.897398	0.173713
90	33	10	6	0.973875	0.076477
90	33	10	7	0.995726	0.021851
90	33	10	8	0.999599	0.003874

N	n	k	x	P(x)	p(x)
90	33	10	9	0.999984	0.000384
90	33	10	10	1.000000	0.000016
90	33	11	0	0.004435	0.004435
90	33	11	1	0.038687	0.034252
90	33	11	2	0.152860	0.114173
90	33	11	3	0.369555	0.216696
90	33	11	4	0.629590	0.260035
90	33	11	5	0.836598	0.207008
90	33	11	6	0.948064	0.111466
90	33	11	7	0.988624	0.040560
90	33	11	8	0.998389	0.009765
90	33	11	9	0.999868	0.001479
90	33	11	10	0.999995	0.000127
90	33	11	11	1.000000	0.000005
90	33	12	0	0.002582	0.002582
90	33	12	1	0.024813	0.022230
90	33	12	2	0.108058	0.083245
90	33	12	3	0.287266	0.179208
90	33	12	4	0.534134	0.246868
90	33	12	5	0.763228	0.229094
90	33	12	6	0.909968	0.146740
90	33	12	7	0.975275	0.065307
90	33	12	8	0.995299	0.020023
90	33	12	9	0.999419	0.004120
90	33	12	10	0.999958	0.000539
90	33	12	11	0.999999	0.000040
90	33	12	12	1.000000	0.000001
90	33	13	0	0.001490	0.001490
90	33	13	1	0.015692	0.014203
90	33	13	2	0.074973	0.059281
90	33	13	3	0.218340	0.143367
90	33	13	4	0.442350	0.224010
90	33	13	5	0.680989	0.238640
90	33	13	6	0.859174	0.178184
90	33	13	7	0.953506	0.094333
90	33	13	8	0.988881	0.035375
90	33	13	9	0.998151	0.009270
90	33	13	10	0.999799	0.001648
90	33	13	11	0.999987	0.000188
90	33	13	12	1.000000	0.000012
90	33	13	13	1.000000	0.000000
90	33	14	0	0.000851	0.000851
90	33	14	1	0.009790	0.008939
90	33	14	2	0.051107	0.041317
90	33	14	3	0.162482	0.111376
90	33	14	4	0.357982	0.195500
90	33	14	5	0.594211	0.236229
90	33	14	6	0.796693	0.202482
90	33	14	7	0.921654	0.124960
90	33	14	8	0.977396	0.055742

N	n	k	x	P(x)	p(x)
90	33	14	9	0.995262	0.017866
90	33	14	10	0.999307	0.004045
90	33	14	11	0.999934	0.000627
90	33	14	12	0.999996	0.000063
90	33	14	13	1.000000	0.000004
90	33	14	14	1.000000	0.000000
90	33	15	0	0.000482	0.000482
90	33	15	1	0.006026	0.005545
90	33	15	2	0.034254	0.028228
90	33	15	3	0.118518	0.084264
90	33	15	4	0.283384	0.164865
90	33	15	5	0.507180	0.223796
90	33	15	6	0.724759	0.217579
90	33	15	7	0.878904	0.154145
90	33	15	8	0.959060	0.080155
90	33	15	9	0.989620	0.030560
90	33	15	10	0.998083	0.008463
90	33	15	11	0.999752	0.001669
90	33	15	12	0.999979	0.000227
90	33	15	13	0.999999	0.000020
90	33	15	14	1.000000	0.000001
90	33	15	15	1.000000	0.000000
90	33	16	0	0.000270	0.000270
90	33	16	1	0.003661	0.003391
90	33	16	2	0.022587	0.018926
90	33	16	3	0.084813	0.062226
90	33	16	4	0.219636	0.134823
90	33	16	5	0.423629	0.203993
90	33	16	6	0.646430	0.222801
90	33	16	7	0.825467	0.179037
90	33	16	8	0.932341	0.106874
90	33	16	9	0.979841	0.047500
90	33	16	10	0.995488	0.015647
90	33	16	11	0.999262	0.003775
90	33	16	12	0.999915	0.000653
90	33	16	13	0.999993	0.000078
90	33	16	14	1.000000	0.000006
90	33	16	15	1.000000	0.000000
90	33	16	16	1.000000	0.000000
90	33	17	0	0.000149	0.000149
90	33	17	1	0.002194	0.002045
90	33	17	2	0.014658	0.012464
90	33	17	3	0.059586	0.044928
90	33	17	4	0.166800	0.107214
90	33	17	5	0.346442	0.179643
90	33	17	6	0.565138	0.218696
90	33	17	7	0.762562	0.197424
90	33	17	8	0.896235	0.133673
90	33	17	9	0.964435	0.068200
90	33	17	10	0.990624	0.026189

Table for $N=2$, $n=1$, through $N=100$, $n=50$

N	n	k	x	P(x)	p(x)
90	33	17	11	0.998140	0.007516
90	33	17	12	0.999730	0.001590
90	33	17	13	0.999972	0.000242
90	33	17	14	0.999998	0.000026
90	33	17	15	1.000000	0.000002
90	33	17	16	1.000000	0.000000
90	33	17	17	1.000000	0.000000
90	33	18	0	0.000082	0.000082
90	33	18	1	0.001298	0.001216
90	33	18	2	0.009365	0.008067
90	33	18	3	0.041122	0.031757
90	33	18	4	0.124208	0.083085
90	33	18	5	0.277538	0.153331
90	33	18	6	0.484251	0.206712
90	33	18	7	0.692247	0.207996
90	33	18	8	0.850457	0.158210
90	33	18	9	0.942013	0.091557
90	33	18	10	0.982373	0.040360
90	33	18	11	0.995875	0.013502
90	33	18	12	0.999273	0.003398
90	33	18	13	0.999906	0.000633
90	33	18	14	0.999991	0.000085
90	33	18	15	0.999999	0.000008
90	33	18	16	1.000000	0.000000
90	33	18	17	1.000000	0.000000
90	33	18	18	1.000000	0.000000
90	33	19	0	0.000044	0.000044
90	33	19	1	0.000757	0.000713
90	33	19	2	0.005892	0.005134
90	33	19	3	0.027890	0.021998
90	33	19	4	0.090743	0.062853
90	33	19	5	0.217910	0.127167
90	33	19	6	0.406734	0.188824
90	33	19	7	0.617137	0.210404
90	33	19	8	0.795523	0.178386
90	33	19	9	0.911494	0.115972
90	33	19	10	0.969480	0.057986
90	33	19	11	0.991749	0.022269
90	33	19	12	0.998282	0.006532
90	33	19	13	0.999730	0.001448
90	33	19	14	0.999969	0.000239
90	33	19	15	0.999997	0.000029
90	33	19	16	1.000000	0.000002
90	33	19	17	1.000000	0.000000
90	33	19	18	1.000000	0.000000
90	33	19	19	1.000000	0.000000
90	33	20	0	0.000024	0.000024
90	33	20	1	0.000436	0.000412
90	33	20	2	0.003650	0.003214
90	33	20	3	0.018595	0.014945

N	n	k	x	P(x)	p(x)
90	33	20	4	0.065071	0.046476
90	33	20	5	0.167759	0.102689
90	33	20	6	0.334927	0.167168
90	33	20	7	0.540088	0.205161
90	33	20	8	0.732711	0.192623
90	33	20	9	0.872293	0.139582
90	33	20	10	0.950696	0.078403
90	33	20	11	0.984849	0.034153
90	33	20	12	0.996350	0.011501
90	33	20	13	0.999322	0.002972
90	33	20	14	0.999905	0.000583
90	33	20	15	0.999990	0.000085
90	33	20	16	0.999999	0.000009
90	33	20	17	1.000000	0.000001
90	33	20	18	1.000000	0.000000
90	33	20	19	1.000000	0.000000
90	33	20	20	1.000000	0.000000
90	33	21	0	0.000013	0.000013
90	33	21	1	0.000248	0.000235
90	33	21	2	0.002227	0.001979
90	33	21	3	0.012190	0.009963
90	33	21	4	0.045817	0.033626
90	33	21	5	0.126684	0.080867
90	33	21	6	0.270448	0.143764
90	33	21	7	0.463885	0.193437
90	33	21	8	0.663917	0.200032
90	33	21	9	0.824436	0.160519
90	33	21	10	0.924935	0.100499
90	33	21	11	0.974115	0.049180
90	33	21	12	0.992900	0.018784
90	33	21	13	0.998473	0.005573
90	33	21	14	0.999747	0.001274
90	33	21	15	0.999968	0.000221
90	33	21	16	0.999997	0.000029
90	33	21	17	1.000000	0.000003
90	33	21	18	1.000000	0.000000
90	33	21	19	1.000000	0.000000
90	33	21	20	1.000000	0.000000
90	33	21	21	1.000000	0.000000
90	33	22	0	0.000007	0.000007
90	33	22	1	0.000139	0.000132
90	33	22	2	0.001338	0.001199
90	33	22	3	0.007858	0.006521
90	33	22	4	0.031684	0.023825
90	33	22	5	0.093868	0.062185
90	33	22	6	0.214193	0.120325
90	33	22	7	0.390996	0.176803
90	33	22	8	0.591442	0.200446
90	33	22	9	0.768603	0.177162
90	33	22	10	0.891435	0.122832

N	n	k	x	P(x)	p(x)
90	33	22	11	0.958435	0.066999
90	33	22	12	0.987183	0.028748
90	33	22	13	0.996857	0.009675
90	33	22	14	0.999396	0.002539
90	33	22	15	0.999910	0.000514
90	33	22	16	0.999990	0.000079
90	33	22	17	0.999999	0.000009
90	33	22	18	1.000000	0.000001
90	33	22	19	1.000000	0.000000
90	33	22	20	1.000000	0.000000
90	33	22	21	1.000000	0.000000
90	33	22	22	1.000000	0.000000
90	33	23	0	0.000003	0.000003
90	33	23	1	0.000076	0.000073
90	33	23	2	0.000791	0.000715
90	33	23	3	0.004981	0.004190
90	33	23	4	0.021523	0.016541
90	33	23	5	0.068263	0.046740
90	33	23	6	0.166417	0.098154
90	33	23	7	0.323395	0.156978
90	33	23	8	0.517748	0.194354
90	33	23	9	0.706076	0.188327
90	33	23	10	0.849889	0.143814
90	33	23	11	0.936758	0.086869
90	33	23	12	0.978305	0.041546
90	33	23	13	0.994012	0.015707
90	33	23	14	0.998687	0.004675
90	33	23	15	0.999774	0.001088
90	33	23	16	0.999970	0.000196
90	33	23	17	0.999997	0.000027
90	33	23	18	1.000000	0.000003
90	33	23	19	1.000000	0.000000
90	33	23	20	1.000000	0.000000
90	33	23	21	1.000000	0.000000
90	33	23	22	1.000000	0.000000
90	33	23	23	1.000000	0.000000
90	33	24	0	0.000002	0.000002
90	33	24	1	0.000042	0.000040
90	33	24	2	0.000460	0.000419
90	33	24	3	0.003105	0.002645
90	33	24	4	0.014363	0.011258
90	33	24	5	0.048730	0.034367
90	33	24	6	0.126862	0.078133
90	33	24	7	0.262478	0.135616
90	33	24	8	0.445228	0.182751
90	33	24	9	0.638615	0.193387
90	33	24	10	0.800520	0.161905
90	33	24	11	0.908234	0.107714
90	33	24	12	0.965283	0.057048
90	33	24	13	0.989323	0.024040

N	n	k	x	P(x)	p(x)
90	33	24	14	0.997361	0.008038
90	33	24	15	0.999482	0.002121
90	33	24	16	0.999920	0.000438
90	33	24	17	0.999990	0.000070
90	33	24	18	0.999999	0.000009
90	33	24	19	1.000000	0.000001
90	33	24	20	1.000000	0.000000
90	33	24	21	1.000000	0.000000
90	33	24	22	1.000000	0.000000
90	33	24	23	1.000000	0.000000
90	33	24	24	1.000000	0.000000
90	33	25	0	0.000001	0.000001
90	33	25	1	0.000022	0.000021
90	33	25	2	0.000264	0.000241
90	33	25	3	0.001903	0.001639
90	33	25	4	0.009416	0.007513
90	33	25	5	0.034150	0.024734
90	33	25	6	0.094899	0.060749
90	33	25	7	0.209054	0.114155
90	33	25	8	0.376005	0.166951
90	33	25	9	0.568293	0.192288
90	33	25	10	0.744099	0.175806
90	33	25	11	0.872330	0.128231
90	33	25	12	0.947131	0.074801
90	33	25	13	0.982038	0.034907
90	33	25	14	0.995047	0.013009
90	33	25	15	0.998904	0.003857
90	33	25	16	0.999808	0.000904
90	33	25	17	0.999974	0.000166
90	33	25	18	0.999997	0.000024
90	33	25	19	1.000000	0.000003
90	33	25	20	1.000000	0.000000
90	33	25	21	1.000000	0.000000
90	33	25	22	1.000000	0.000000
90	33	25	23	1.000000	0.000000
90	33	25	24	1.000000	0.000000
90	33	25	25	1.000000	0.000000
90	33	26	0	0.000000	0.000000
90	33	26	1	0.000012	0.000011
90	33	26	2	0.000149	0.000137
90	33	26	3	0.001146	0.000998
90	33	26	4	0.006064	0.004918
90	33	26	5	0.023495	0.017431
90	33	26	6	0.069665	0.046169
90	33	26	7	0.163392	0.093727
90	33	26	8	0.311793	0.148401
90	33	26	9	0.497294	0.185501
90	33	26	10	0.681891	0.184596
90	33	26	11	0.828929	0.147038
90	33	26	12	0.922964	0.094036

Table for $N=2$, $n=1$, through $N=100$, $n=50$

N	n	k	x	P(x)	p(x)
90	33	26	13	0.971297	0.048333
90	33	26	14	0.991244	0.019947
90	33	26	15	0.997836	0.006591
90	33	26	16	0.999571	0.001735
90	33	26	17	0.999933	0.000362
90	33	26	18	0.999992	0.000059
90	33	26	19	0.999999	0.000007
90	33	26	20	1.000000	0.000001
90	33	26	21	1.000000	0.000000
90	33	26	22	1.000000	0.000000
90	33	26	23	1.000000	0.000000
90	33	26	24	1.000000	0.000000
90	33	26	25	1.000000	0.000000
90	33	26	26	1.000000	0.000000
90	33	27	0	0.000000	0.000000
90	33	27	1	0.000006	0.000006
90	33	27	2	0.000082	0.000076
90	33	27	3	0.000679	0.000596
90	33	27	4	0.003836	0.003157
90	33	27	5	0.015869	0.012033
90	33	27	6	0.050187	0.034318
90	33	27	7	0.125315	0.075128
90	33	27	8	0.253824	0.128508
90	33	27	9	0.427731	0.173908
90	33	27	10	0.615551	0.187820
90	33	27	11	0.778384	0.162833
90	33	27	12	0.892109	0.113725
90	33	27	13	0.956193	0.064085
90	33	27	14	0.985323	0.029129
90	33	27	15	0.995982	0.010659
90	33	27	16	0.999110	0.003128
90	33	27	17	0.999842	0.000732
90	33	27	18	0.999978	0.000136
90	33	27	19	0.999997	0.000020
90	33	27	20	1.000000	0.000002
90	33	27	21	1.000000	0.000000
90	33	27	22	1.000000	0.000000
90	33	27	23	1.000000	0.000000
90	33	27	24	1.000000	0.000000
90	33	27	25	1.000000	0.000000
90	33	27	26	1.000000	0.000000
90	33	27	27	1.000000	0.000000
90	33	28	0	0.000000	0.000000
90	33	28	1	0.000003	0.000003
90	33	28	2	0.000045	0.000042
90	33	28	3	0.000395	0.000350
90	33	28	4	0.002382	0.001988
90	33	28	5	0.010521	0.008139
90	33	28	6	0.035479	0.024958
90	33	28	7	0.094310	0.058830

N	n	k	x	P(x)	p(x)
90	33	28	8	0.202828	0.108518
90	33	28	9	0.361481	0.158652
90	33	28	10	0.546982	0.185501
90	33	28	11	0.721522	0.174540
90	33	28	12	0.854201	0.132679
90	33	28	13	0.935849	0.081648
90	33	28	14	0.976538	0.040689
90	33	28	15	0.992936	0.016399
90	33	28	16	0.998266	0.005330
90	33	28	17	0.999656	0.001390
90	33	28	18	0.999946	0.000289
90	33	28	19	0.999993	0.000048
90	33	28	20	0.999999	0.000006
90	33	28	21	1.000000	0.000001
90	33	28	22	1.000000	0.000000
90	33	28	23	1.000000	0.000000
90	33	28	24	1.000000	0.000000
90	33	28	25	1.000000	0.000000
90	33	28	26	1.000000	0.000000
90	33	28	27	1.000000	0.000000
90	33	28	28	1.000000	0.000000
90	33	29	0	0.000000	0.000000
90	33	29	1	0.000002	0.000001
90	33	29	2	0.000024	0.000022
90	33	29	3	0.000225	0.000201
90	33	29	4	0.001453	0.001227
90	33	29	5	0.006846	0.005393
90	33	29	6	0.024610	0.017765
90	33	29	7	0.069639	0.045029
90	33	29	8	0.159071	0.089432
90	33	29	9	0.300067	0.140996
90	33	29	10	0.478167	0.178100
90	33	29	11	0.659588	0.181421
90	33	29	12	0.809261	0.149673
90	33	29	13	0.909511	0.100250
90	33	29	14	0.964069	0.054558
90	33	29	15	0.988175	0.024107
90	33	29	16	0.996805	0.008629
90	33	29	17	0.999297	0.002493
90	33	29	18	0.999876	0.000578
90	33	29	19	0.999982	0.000107
90	33	29	20	0.999998	0.000016
90	33	29	21	1.000000	0.000002
90	33	29	22	1.000000	0.000000
90	33	29	23	1.000000	0.000000
90	33	29	24	1.000000	0.000000
90	33	29	25	1.000000	0.000000
90	33	29	26	1.000000	0.000000
90	33	29	27	1.000000	0.000000
90	33	29	28	1.000000	0.000000

N	n	k	x	P(x)	p(x)
90	33	29	29	1.000000	0.000000
90	33	30	0	0.000000	0.000000
90	33	30	1	0.000001	0.000001
90	33	30	2	0.000013	0.000012
90	33	30	3	0.000126	0.000114
90	33	30	4	0.000869	0.000743
90	33	30	5	0.004370	0.003501
90	33	30	6	0.016747	0.012377
90	33	30	7	0.050447	0.033699
90	33	30	8	0.122418	0.071972
90	33	30	9	0.244593	0.122175
90	33	30	10	0.411014	0.166422
90	33	30	11	0.594158	0.183143
90	33	30	12	0.757734	0.163577
90	33	30	13	0.876642	0.118908
90	33	30	14	0.947075	0.070433
90	33	30	15	0.981062	0.033987
90	33	30	16	0.994400	0.013338
90	33	30	17	0.998644	0.004244
90	33	30	18	0.999733	0.001090
90	33	30	19	0.999958	0.000224
90	33	30	20	0.999995	0.000037
90	33	30	21	0.999999	0.000005
90	33	30	22	1.000000	0.000000
90	33	30	23	1.000000	0.000000
90	33	30	24	1.000000	0.000000
90	33	30	25	1.000000	0.000000
90	33	30	26	1.000000	0.000000
90	33	30	27	1.000000	0.000000
90	33	30	28	1.000000	0.000000
90	33	30	29	1.000000	0.000000
90	33	30	30	1.000000	0.000000
90	33	31	0	0.000000	0.000000
90	33	31	1	0.000000	0.000000
90	33	31	2	0.000006	0.000006
90	33	31	3	0.000069	0.000063
90	33	31	4	0.000510	0.000441
90	33	31	5	0.002736	0.002226
90	33	31	6	0.011178	0.008441
90	33	31	7	0.035844	0.024666
90	33	31	8	0.092430	0.056587
90	33	31	9	0.195723	0.103293
90	33	31	10	0.347220	0.151496
90	33	31	11	0.527005	0.179786
90	33	31	12	0.700483	0.173477
90	33	31	13	0.837006	0.136524
90	33	31	14	0.924771	0.087765
90	33	31	15	0.970866	0.046095
90	33	31	16	0.990621	0.019755
90	33	31	17	0.997512	0.006891

N	n	k	x	P(x)	p(x)
90	33	31	18	0.999461	0.001949
90	33	31	19	0.999905	0.000445
90	33	31	20	0.999987	0.000081
90	33	31	21	0.999998	0.000012
90	33	31	22	1.000000	0.000001
90	33	31	23	1.000000	0.000000
90	33	31	24	1.000000	0.000000
90	33	31	25	1.000000	0.000000
90	33	31	26	1.000000	0.000000
90	33	31	27	1.000000	0.000000
90	33	31	28	1.000000	0.000000
90	33	31	29	1.000000	0.000000
90	33	31	30	1.000000	0.000000
90	33	31	31	1.000000	0.000000
90	33	32	0	0.000000	0.000000
90	33	32	1	0.000000	0.000000
90	33	32	2	0.000003	0.000003
90	33	32	3	0.000037	0.000034
90	33	32	4	0.000294	0.000256
90	33	32	5	0.001680	0.001386
90	33	32	6	0.007315	0.005635
90	33	32	7	0.024974	0.017659
90	33	32	8	0.068453	0.043479
90	33	32	9	0.153706	0.085253
90	33	32	10	0.288162	0.134456
90	33	32	11	0.459966	0.171805
90	33	32	12	0.638736	0.178770
90	33	32	13	0.790727	0.151991
90	33	32	14	0.896508	0.105781
90	33	32	15	0.956803	0.060295
90	33	32	16	0.984929	0.028125
90	33	32	17	0.995643	0.010714
90	33	32	18	0.998965	0.003322
90	33	32	19	0.999800	0.000835
90	33	32	20	0.999969	0.000169
90	33	32	21	0.999996	0.000027
90	33	32	22	0.999999	0.000003
90	33	32	23	1.000000	0.000000
90	33	32	24	1.000000	0.000000
90	33	32	25	1.000000	0.000000
90	33	32	26	1.000000	0.000000
90	33	32	27	1.000000	0.000000
90	33	32	28	1.000000	0.000000
90	33	32	29	1.000000	0.000000
90	33	32	30	1.000000	0.000000
90	33	32	31	1.000000	0.000000
90	33	32	32	1.000000	0.000000
90	33	33	0	0.000000	0.000000
90	33	33	1	0.000000	0.000000
90	33	33	2	0.000002	0.000002

Table for $N = 2$, $n = 1$, through $N = 100$, $n = 50$

N	n	k	x	P(x)	p(x)
90	33	33	3	0.000020	0.000018
90	33	33	4	0.000165	0.000146
90	33	33	5	0.001011	0.000845
90	33	33	6	0.004692	0.003681
90	33	33	7	0.017058	0.012366
90	33	33	8	0.049712	0.032654
90	33	33	9	0.118429	0.068717
90	33	33	10	0.234843	0.116414
90	33	33	11	0.394799	0.159956
90	33	33	12	0.574009	0.179210
90	33	33	13	0.738316	0.164307
90	33	33	14	0.861856	0.123539
90	33	33	15	0.938091	0.076235
90	33	33	16	0.976685	0.038594
90	33	33	17	0.992687	0.016002
90	33	33	18	0.998106	0.005419
90	33	33	19	0.999598	0.001492
90	33	33	20	0.999931	0.000332
90	33	33	21	0.999990	0.000059
90	33	33	22	0.999999	0.000008
90	33	33	23	1.000000	0.000001
90	33	33	24	1.000000	0.000000
90	33	33	25	1.000000	0.000000
90	33	33	26	1.000000	0.000000
90	33	33	27	1.000000	0.000000
90	33	33	28	1.000000	0.000000
90	33	33	29	1.000000	0.000000
90	33	33	30	1.000000	0.000000
90	33	33	31	1.000000	0.000000
90	33	33	32	1.000000	0.000000
90	33	33	33	1.000000	0.000000
90	34	1	0	0.622222	0.622222
90	34	1	1	1.000000	0.377778
90	34	2	0	0.384519	0.384519
90	34	2	1	0.859925	0.475406
90	34	2	2	1.000000	0.140075
90	34	3	0	0.235955	0.235955
90	34	3	1	0.681648	0.445693
90	34	3	2	0.949064	0.267416
90	34	3	3	1.000000	0.050936
90	34	4	0	0.143743	0.143743
90	34	4	1	0.512592	0.368849
90	34	4	2	0.850704	0.338112
90	34	4	3	0.981850	0.131146
90	34	4	4	1.000000	0.018150
90	34	5	0	0.086914	0.086914
90	34	5	1	0.371057	0.284143
90	34	5	2	0.724895	0.353838
90	34	5	3	0.934577	0.209682
90	34	5	4	0.993669	0.059092

N	n	k	x	P(x)	p(x)
90	34	5	5	1.000000	0.006331
90	34	6	0	0.052149	0.052149
90	34	6	1	0.260743	0.208594
90	34	6	2	0.591685	0.330943
90	34	6	3	0.858104	0.266419
90	34	6	4	0.972813	0.114708
90	34	6	5	0.997840	0.025027
90	34	6	6	1.000000	0.002160
90	34	7	0	0.031041	0.031041
90	34	7	1	0.178795	0.147754
90	34	7	2	0.465612	0.286817
90	34	7	3	0.759783	0.294171
90	34	7	4	0.931845	0.172062
90	34	7	5	0.989199	0.057354
90	34	7	6	0.999280	0.010080
90	34	7	7	1.000000	0.000720
90	34	8	0	0.018325	0.018325
90	34	8	1	0.120049	0.101724
90	34	8	2	0.355032	0.234983
90	34	8	3	0.649912	0.294880
90	34	8	4	0.869654	0.219742
90	34	8	5	0.969160	0.099506
90	34	8	6	0.995879	0.026719
90	34	8	7	0.999766	0.003886
90	34	8	8	1.000000	0.000234
90	34	9	0	0.010727	0.010727
90	34	9	1	0.079112	0.068385
90	34	9	2	0.263331	0.184220
90	34	9	3	0.538433	0.275101
90	34	9	4	0.789261	0.250828
90	34	9	5	0.933969	0.144708
90	34	9	6	0.986756	0.052787
90	34	9	7	0.998486	0.011730
90	34	9	8	0.999926	0.001440
90	34	9	9	1.000000	0.000074
90	34	10	0	0.006224	0.006224
90	34	10	1	0.051251	0.045027
90	34	10	2	0.190553	0.139302
90	34	10	3	0.433147	0.242594
90	34	10	4	0.696361	0.263214
90	34	10	5	0.882160	0.185798
90	34	10	6	0.968508	0.086349
90	34	10	7	0.994576	0.026067
90	34	10	8	0.999463	0.004888
90	34	10	9	0.999977	0.000513
90	34	10	10	1.000000	0.000023
90	34	11	0	0.003579	0.003579
90	34	11	1	0.032678	0.029099
90	34	11	2	0.134832	0.102155
90	34	11	3	0.339142	0.204310

N	n	k	x	P(x)	p(x)
90	34	11	4	0.597656	0.258514
90	34	11	5	0.814808	0.217152
90	34	11	6	0.938286	0.123478
90	34	11	7	0.985778	0.047492
90	34	11	8	0.997875	0.012097
90	34	11	9	0.999817	0.001941
90	34	11	10	0.999993	0.000176
90	34	11	11	1.000000	0.000007
90	34	12	0	0.002039	0.002039
90	34	12	1	0.020522	0.018484
90	34	12	2	0.093453	0.072931
90	34	12	3	0.258970	0.165517
90	34	12	4	0.499486	0.240516
90	34	12	5	0.735094	0.235608
90	34	12	6	0.894522	0.159428
90	34	12	7	0.969547	0.075025
90	34	12	8	0.993894	0.024347
90	34	12	9	0.999202	0.005308
90	34	12	10	0.999939	0.000737
90	34	12	11	0.999998	0.000058
90	34	12	12	1.000000	0.000002
90	34	13	0	0.001150	0.001150
90	34	13	1	0.012702	0.011552
90	34	13	2	0.063533	0.050831
90	34	13	3	0.193188	0.129655
90	34	13	4	0.406980	0.213792
90	34	13	5	0.647496	0.240516
90	34	13	6	0.837291	0.189795
90	34	13	7	0.943577	0.106285
90	34	13	8	0.985778	0.042202
90	34	13	9	0.997501	0.011723
90	34	13	10	0.999713	0.002212
90	34	13	11	0.999981	0.000268
90	34	13	12	0.999999	0.000019
90	34	13	13	1.000000	0.000001
90	34	14	0	0.000642	0.000642
90	34	14	1	0.007751	0.007109
90	34	14	2	0.042409	0.034657
90	34	14	3	0.140989	0.098580
90	34	14	4	0.323684	0.182695
90	34	14	5	0.556912	0.233228
90	34	14	6	0.768275	0.211363
90	34	14	7	0.906308	0.138033
90	34	14	8	0.971528	0.065221
90	34	14	9	0.993695	0.022166
90	34	14	10	0.999023	0.005328
90	34	14	11	0.999901	0.000877
90	34	14	12	0.999994	0.000093
90	34	14	13	1.000000	0.000006
90	34	14	14	1.000000	0.000000

N	n	k	x	P(x)	p(x)
90	34	15	0	0.000355	0.000355
90	34	15	1	0.004664	0.004310
90	34	15	2	0.027816	0.023152
90	34	15	3	0.100779	0.072962
90	34	15	4	0.251568	0.150789
90	34	15	5	0.467917	0.216350
90	34	15	6	0.690404	0.222487
90	34	15	7	0.857270	0.166865
90	34	15	8	0.949216	0.091946
90	34	15	9	0.986403	0.037187
90	34	15	10	0.997341	0.010937
90	34	15	11	0.999635	0.002295
90	34	15	12	0.999967	0.000332
90	34	15	13	0.999998	0.000031
90	34	15	14	1.000000	0.000002
90	34	15	15	1.000000	0.000000
90	34	16	0	0.000194	0.000194
90	34	16	1	0.002768	0.002574
90	34	16	2	0.017938	0.015170
90	34	16	3	0.070621	0.052683
90	34	16	4	0.191252	0.120631
90	34	16	5	0.384262	0.193010
90	34	16	6	0.607343	0.223080
90	34	16	7	0.797198	0.189856
90	34	16	8	0.917341	0.120143
90	34	16	9	0.974007	0.056666
90	34	16	10	0.993841	0.019833
90	34	16	11	0.998931	0.005091
90	34	16	12	0.999870	0.000938
90	34	16	13	0.999989	0.000120
90	34	16	14	0.999999	0.000010
90	34	16	15	1.000000	0.000000
90	34	16	16	1.000000	0.000000
90	34	17	0	0.000105	0.000105
90	34	17	1	0.001620	0.001515
90	34	17	2	0.011378	0.009758
90	34	17	3	0.048551	0.037173
90	34	17	4	0.142347	0.093796
90	34	17	5	0.308623	0.166276
90	34	17	6	0.522934	0.214311
90	34	17	7	0.727927	0.204993
90	34	17	8	0.875129	0.147202
90	34	17	9	0.954863	0.079735
90	34	17	10	0.987408	0.032545
90	34	17	11	0.997349	0.009941
90	34	17	12	0.999591	0.002242
90	34	17	13	0.999955	0.000365
90	34	17	14	0.999997	0.000041
90	34	17	15	1.000000	0.000003
90	34	17	16	1.000000	0.000000

N	n	k	x	P(x)	p(x)
90	34	17	17	1.000000	0.000000
90	34	18	0	0.000056	0.000056
90	34	18	1	0.000935	0.000879
90	34	18	2	0.007101	0.006165
90	34	18	3	0.032765	0.025664
90	34	18	4	0.103801	0.071036
90	34	18	5	0.242568	0.138767
90	34	18	6	0.440732	0.198164
90	34	18	7	0.652107	0.211375
90	34	18	8	0.822701	0.170593
90	34	18	9	0.927557	0.104856
90	34	18	10	0.976708	0.049151
90	34	18	11	0.994217	0.017508
90	34	18	12	0.998915	0.004698
90	34	18	13	0.999850	0.000935
90	34	18	14	0.999985	0.000135
90	34	18	15	0.999999	0.000014
90	34	18	16	1.000000	0.000001
90	34	18	17	1.000000	0.000000
90	34	18	18	1.000000	0.000000
90	34	19	0	0.000030	0.000030
90	34	19	1	0.000532	0.000503
90	34	19	2	0.004361	0.003828
90	34	19	3	0.021715	0.017355
90	34	19	4	0.074203	0.052487
90	34	19	5	0.186676	0.112473
90	34	19	6	0.363669	0.176993
90	34	19	7	0.572842	0.209173
90	34	19	8	0.761097	0.188256
90	34	19	9	0.891149	0.130051
90	34	19	10	0.960325	0.069176
90	34	19	11	0.988624	0.028299
90	34	19	12	0.997480	0.008856
90	34	19	13	0.999578	0.002098
90	34	19	14	0.999948	0.000370
90	34	19	15	0.999995	0.000047
90	34	19	16	1.000000	0.000004
90	34	19	17	1.000000	0.000000
90	34	19	18	1.000000	0.000000
90	34	19	19	1.000000	0.000000
90	34	20	0	0.000015	0.000015
90	34	20	1	0.000299	0.000283
90	34	20	2	0.002635	0.002336
90	34	20	3	0.014138	0.011503
90	34	20	4	0.052025	0.037887
90	34	20	5	0.140736	0.088711
90	34	20	6	0.293869	0.153132
90	34	20	7	0.493297	0.199428
90	34	20	8	0.692159	0.198862
90	34	20	9	0.845356	0.153197

N	n	k	x	P(x)	p(x)
90	34	20	10	0.936941	0.091585
90	34	20	11	0.979457	0.042515
90	34	20	12	0.994736	0.015279
90	34	20	13	0.998957	0.004222
90	34	20	14	0.999844	0.000887
90	34	20	15	0.999983	0.000139
90	34	20	16	0.999999	0.000016
90	34	20	17	1.000000	0.000001
90	34	20	18	1.000000	0.000000
90	34	20	19	1.000000	0.000000
90	34	20	20	1.000000	0.000000
90	34	21	0	0.000008	0.000008
90	34	21	1	0.000165	0.000157
90	34	21	2	0.001567	0.001402
90	34	21	3	0.009044	0.007477
90	34	21	4	0.035788	0.026744
90	34	21	5	0.103984	0.068197
90	34	21	6	0.232616	0.128631
90	34	21	7	0.416375	0.183759
90	34	21	8	0.618296	0.201921
90	34	21	9	0.790643	0.172347
90	34	21	10	0.905541	0.114898
90	34	21	11	0.965487	0.059947
90	34	21	12	0.989934	0.024446
90	34	21	13	0.997691	0.007757
90	34	21	14	0.999590	0.001900
90	34	21	15	0.999945	0.000355
90	34	21	16	0.999994	0.000050
90	34	21	17	1.000000	0.000005
90	34	21	18	1.000000	0.000000
90	34	21	19	1.000000	0.000000
90	34	21	20	1.000000	0.000000
90	34	21	21	1.000000	0.000000
90	34	22	0	0.000004	0.000004
90	34	22	1	0.000090	0.000086
90	34	22	2	0.000917	0.000827
90	34	22	3	0.005685	0.004768
90	34	22	4	0.024160	0.018475
90	34	22	5	0.075322	0.051162
90	34	22	6	0.180417	0.105095
90	34	22	7	0.344469	0.164051
90	34	22	8	0.542210	0.197741
90	34	22	9	0.728198	0.185989
90	34	22	10	0.865576	0.137378
90	34	22	11	0.945505	0.079929
90	34	22	12	0.982139	0.036634
90	34	22	13	0.995330	0.013191
90	34	22	14	0.999040	0.003710
90	34	22	15	0.999847	0.000808
90	34	22	16	0.999982	0.000134

N	n	k	x	P(x)	p(x)
90	34	22	17	0.999998	0.000017
90	34	22	18	1.000000	0.000002
90	34	22	19	1.000000	0.000000
90	34	22	20	1.000000	0.000000
90	34	22	21	1.000000	0.000000
90	34	22	22	1.000000	0.000000
90	34	23	0	0.000002	0.000002
90	34	23	1	0.000048	0.000046
90	34	23	2	0.000528	0.000479
90	34	23	3	0.003511	0.002983
90	34	23	4	0.016009	0.012498
90	34	23	5	0.053503	0.037494
90	34	23	6	0.137143	0.083640
90	34	23	7	0.279331	0.142188
90	34	23	8	0.466603	0.187272
90	34	23	9	0.659820	0.193217
90	34	23	10	0.817090	0.157270
90	34	23	11	0.918470	0.101380
90	34	23	12	0.970287	0.051817
90	34	23	13	0.991256	0.020969
90	34	23	14	0.997949	0.006692
90	34	23	15	0.999622	0.001673
90	34	23	16	0.999946	0.000324
90	34	23	17	0.999994	0.000048
90	34	23	18	0.999999	0.000005
90	34	23	19	1.000000	0.000000
90	34	23	20	1.000000	0.000000
90	34	23	21	1.000000	0.000000
90	34	23	22	1.000000	0.000000
90	34	23	23	1.000000	0.000000
90	34	24	0	0.000001	0.000001
90	34	24	1	0.000025	0.000024
90	34	24	2	0.000299	0.000273
90	34	24	3	0.002131	0.001832
90	34	24	4	0.010413	0.008282
90	34	24	5	0.037274	0.026861
90	34	24	6	0.102189	0.064915
90	34	24	7	0.222031	0.119842
90	34	24	8	0.393930	0.171899
90	34	24	9	0.587724	0.193794
90	34	24	10	0.760754	0.173030
90	34	24	11	0.883668	0.122914
90	34	24	12	0.953272	0.069605
90	34	24	13	0.984684	0.031411
90	34	24	14	0.995951	0.011267
90	34	24	15	0.999147	0.003196
90	34	24	16	0.999859	0.000712
90	34	24	17	0.999982	0.000123
90	34	24	18	0.999998	0.000016
90	34	24	19	1.000000	0.000002

N	n	k	x	P(x)	p(x)
90	34	24	20	1.000000	0.000000
90	34	24	21	1.000000	0.000000
90	34	24	22	1.000000	0.000000
90	34	24	23	1.000000	0.000000
90	34	24	24	1.000000	0.000000
90	34	25	0	0.000000	0.000000
90	34	25	1	0.000013	0.000013
90	34	25	2	0.000166	0.000153
90	34	25	3	0.001270	0.001104
90	34	25	4	0.006648	0.005378
90	34	25	5	0.025471	0.018823
90	34	25	6	0.074649	0.049178
90	34	25	7	0.173005	0.098355
90	34	25	8	0.326212	0.153208
90	34	25	9	0.514317	0.188105
90	34	25	10	0.697834	0.183517
90	34	25	11	0.840834	0.143000
90	34	25	12	0.930071	0.089237
90	34	25	13	0.974689	0.044618
90	34	25	14	0.992537	0.017847
90	34	25	15	0.998227	0.005690
90	34	25	16	0.999665	0.001438
90	34	25	17	0.999950	0.000285
90	34	25	18	0.999994	0.000044
90	34	25	19	0.999999	0.000005
90	34	25	20	1.000000	0.000000
90	34	25	21	1.000000	0.000000
90	34	25	22	1.000000	0.000000
90	34	25	23	1.000000	0.000000
90	34	25	24	1.000000	0.000000
90	34	25	25	1.000000	0.000000
90	34	26	0	0.000000	0.000000
90	34	26	1	0.000007	0.000007
90	34	26	2	0.000091	0.000084
90	34	26	3	0.000744	0.000653
90	34	26	4	0.004166	0.003422
90	34	26	5	0.017073	0.012907
90	34	26	6	0.053465	0.036392
90	34	26	7	0.132149	0.078684
90	34	26	8	0.264929	0.132780
90	34	26	9	0.441969	0.177040
90	34	26	10	0.630074	0.188105
90	34	26	11	0.790234	0.160160
90	34	26	12	0.899868	0.109633
90	34	26	13	0.960274	0.060406
90	34	26	14	0.987045	0.026771
90	34	26	15	0.996564	0.009519
90	34	26	16	0.999267	0.002703
90	34	26	17	0.999876	0.000609
90	34	26	18	0.999983	0.000108

Table for $N = 2$, $n = 1$, through $N = 100$, $n = 50$

N	n	k	x	P(x)	p(x)
90	34	26	19	0.999998	0.000015
90	34	26	20	1.000000	0.000002
90	34	26	21	1.000000	0.000000
90	34	26	22	1.000000	0.000000
90	34	26	23	1.000000	0.000000
90	34	26	24	1.000000	0.000000
90	34	26	25	1.000000	0.000000
90	34	26	26	1.000000	0.000000
90	34	27	0	0.000000	0.000000
90	34	27	1	0.000003	0.000003
90	34	27	2	0.000049	0.000045
90	34	27	3	0.000428	0.000379
90	34	27	4	0.002562	0.002134
90	34	27	5	0.011225	0.008663
90	34	27	6	0.037544	0.026319
90	34	27	7	0.098954	0.061411
90	34	27	8	0.210987	0.112033
90	34	27	9	0.372813	0.161826
90	34	27	10	0.559535	0.186722
90	34	27	11	0.732677	0.173142
90	34	27	12	0.862181	0.129505
90	34	27	13	0.940453	0.078272
90	34	27	14	0.978679	0.038226
90	34	27	15	0.993738	0.015059
90	34	27	16	0.998506	0.004769
90	34	27	17	0.999714	0.001207
90	34	27	18	0.999956	0.000243
90	34	27	19	0.999995	0.000038
90	34	27	20	0.999999	0.000005
90	34	27	21	1.000000	0.000000
90	34	27	22	1.000000	0.000000
90	34	27	23	1.000000	0.000000
90	34	27	24	1.000000	0.000000
90	34	27	25	1.000000	0.000000
90	34	27	26	1.000000	0.000000
90	34	27	27	1.000000	0.000000
90	34	28	0	0.000000	0.000000
90	34	28	1	0.000002	0.000002
90	34	28	2	0.000026	0.000024
90	34	28	3	0.000241	0.000215
90	34	28	4	0.001545	0.001304
90	34	28	5	0.007237	0.005692
90	34	28	6	0.025846	0.018609
90	34	28	7	0.072636	0.046789
90	34	28	8	0.164752	0.092116
90	34	28	9	0.308596	0.143845
90	34	28	10	0.488403	0.179806
90	34	28	11	0.669466	0.181064
90	34	28	12	0.816957	0.147491
90	34	28	13	0.914363	0.097405

N	n	k	x	P(x)	p(x)
90	34	28	14	0.966544	0.052181
90	34	28	15	0.989196	0.022652
90	34	28	16	0.997144	0.007948
90	34	28	17	0.999388	0.002244
90	34	28	18	0.999895	0.000507
90	34	28	19	0.999986	0.000091
90	34	28	20	0.999998	0.000013
90	34	28	21	1.000000	0.000001
90	34	28	22	1.000000	0.000000
90	34	28	23	1.000000	0.000000
90	34	28	24	1.000000	0.000000
90	34	28	25	1.000000	0.000000
90	34	28	26	1.000000	0.000000
90	34	28	27	1.000000	0.000000
90	34	28	28	1.000000	0.000000
90	34	29	0	0.000000	0.000000
90	34	29	1	0.000001	0.000001
90	34	29	2	0.000013	0.000013
90	34	29	3	0.000133	0.000120
90	34	29	4	0.000914	0.000781
90	34	29	5	0.004575	0.003661
90	34	29	6	0.017442	0.012867
90	34	29	7	0.052260	0.034817
90	34	29	8	0.126122	0.073863
90	34	29	9	0.250595	0.124472
90	34	29	10	0.418800	0.168206
90	34	29	11	0.602297	0.183497
90	34	29	12	0.764622	0.162324
90	34	29	13	0.881371	0.116749
90	34	29	14	0.949711	0.068341
90	34	29	15	0.982254	0.032543
90	34	29	16	0.994837	0.012582
90	34	29	17	0.998773	0.003936
90	34	29	18	0.999764	0.000991
90	34	29	19	0.999964	0.000200
90	34	29	20	0.999995	0.000032
90	34	29	21	0.999999	0.000004
90	34	29	22	1.000000	0.000000
90	34	29	23	1.000000	0.000000
90	34	29	24	1.000000	0.000000
90	34	29	25	1.000000	0.000000
90	34	29	26	1.000000	0.000000
90	34	29	27	1.000000	0.000000
90	34	29	28	1.000000	0.000000
90	34	29	29	1.000000	0.000000
90	34	30	0	0.000000	0.000000
90	34	30	1	0.000000	0.000000
90	34	30	2	0.000007	0.000006
90	34	30	3	0.000072	0.000066
90	34	30	4	0.000530	0.000458

N	n	k	x	P(x)	p(x)
90	34	30	5	0.002835	0.002304
90	34	30	6	0.011536	0.008701
90	34	30	7	0.036849	0.025313
90	34	30	8	0.094640	0.057791
90	34	30	9	0.199581	0.104941
90	34	30	10	0.352621	0.153040
90	34	30	11	0.533110	0.180489
90	34	30	12	0.706079	0.172969
90	34	30	13	0.841178	0.135100
90	34	30	14	0.927304	0.086126
90	34	30	15	0.972118	0.044814
90	34	30	16	0.991124	0.019006
90	34	30	17	0.997676	0.006552
90	34	30	18	0.999504	0.001828
90	34	30	19	0.999915	0.000411
90	34	30	20	0.999988	0.000074
90	34	30	21	0.999999	0.000010
90	34	30	22	1.000000	0.000001
90	34	30	23	1.000000	0.000000
90	34	30	24	1.000000	0.000000
90	34	30	25	1.000000	0.000000
90	34	30	26	1.000000	0.000000
90	34	30	27	1.000000	0.000000
90	34	30	28	1.000000	0.000000
90	34	30	29	1.000000	0.000000
90	34	30	30	1.000000	0.000000
90	34	31	0	0.000000	0.000000
90	34	31	1	0.000000	0.000000
90	34	31	2	0.000003	0.000003
90	34	31	3	0.000038	0.000035
90	34	31	4	0.000301	0.000263
90	34	31	5	0.001721	0.001420
90	34	31	6	0.007475	0.005754
90	34	31	7	0.025458	0.017983
90	34	31	8	0.069597	0.044139
90	34	31	9	0.155856	0.086259
90	34	31	10	0.291405	0.135549
90	34	31	11	0.463923	0.172517
90	34	31	12	0.642657	0.178734
90	34	31	13	0.793893	0.151237
90	34	31	14	0.898596	0.104702
90	34	31	15	0.957927	0.059331
90	34	31	16	0.985422	0.027495
90	34	31	17	0.995819	0.010397
90	34	31	18	0.999016	0.003197
90	34	31	19	0.999812	0.000795
90	34	31	20	0.999971	0.000159
90	34	31	21	0.999996	0.000025
90	34	31	22	0.999999	0.000003
90	34	31	23	1.000000	0.000000

N	n	k	x	P(x)	p(x)
90	34	31	24	1.000000	0.000000
90	34	31	25	1.000000	0.000000
90	34	31	26	1.000000	0.000000
90	34	31	27	1.000000	0.000000
90	34	31	28	1.000000	0.000000
90	34	31	29	1.000000	0.000000
90	34	31	30	1.000000	0.000000
90	34	31	31	1.000000	0.000000
90	34	32	0	0.000000	0.000000
90	34	32	1	0.000000	0.000000
90	34	32	2	0.000002	0.000002
90	34	32	3	0.000020	0.000018
90	34	32	4	0.000168	0.000148
90	34	32	5	0.001023	0.000855
90	34	32	6	0.004744	0.003721
90	34	32	7	0.017229	0.012484
90	34	32	8	0.050146	0.032917
90	34	32	9	0.119305	0.069160
90	34	32	10	0.236266	0.116961
90	34	32	11	0.396670	0.160404
90	34	32	12	0.576010	0.179340
90	34	32	13	0.740063	0.164053
90	34	32	14	0.863103	0.123040
90	34	32	15	0.938820	0.075717
90	34	32	16	0.977034	0.038213
90	34	32	17	0.992824	0.015790
90	34	32	18	0.998149	0.005326
90	34	32	19	0.999610	0.001460
90	34	32	20	0.999933	0.000324
90	34	32	21	0.999991	0.000058
90	34	32	22	0.999999	0.000008
90	34	32	23	1.000000	0.000001
90	34	32	24	1.000000	0.000000
90	34	32	25	1.000000	0.000000
90	34	32	26	1.000000	0.000000
90	34	32	27	1.000000	0.000000
90	34	32	28	1.000000	0.000000
90	34	32	29	1.000000	0.000000
90	34	32	30	1.000000	0.000000
90	34	32	31	1.000000	0.000000
90	34	32	32	1.000000	0.000000
90	34	33	0	0.000000	0.000000
90	34	33	1	0.000000	0.000000
90	34	33	2	0.000001	0.000001
90	34	33	3	0.000010	0.000009
90	34	33	4	0.000091	0.000081
90	34	33	5	0.000595	0.000504
90	34	33	6	0.002948	0.002353
90	34	33	7	0.011417	0.008469
90	34	33	8	0.035390	0.023973

N	n	k	x	P(x)	p(x)
90	34	33	9	0.089495	0.054105
90	34	33	10	0.187869	0.098373
90	34	33	11	0.333062	0.145193
90	34	33	12	0.507985	0.174923
90	34	33	13	0.680665	0.172680
90	34	33	14	0.820676	0.140011
90	34	33	15	0.914017	0.093341
90	34	33	16	0.965174	0.051158
90	34	33	17	0.988196	0.023021
90	34	33	18	0.996680	0.008485
90	34	33	19	0.999232	0.002552
90	34	33	20	0.999855	0.000623
90	34	33	21	0.999978	0.000123
90	34	33	22	0.999997	0.000019
90	34	33	23	1.000000	0.000002
90	34	33	24	1.000000	0.000000
90	34	33	25	1.000000	0.000000
90	34	33	26	1.000000	0.000000
90	34	33	27	1.000000	0.000000
90	34	33	28	1.000000	0.000000
90	34	33	29	1.000000	0.000000
90	34	33	30	1.000000	0.000000
90	34	33	31	1.000000	0.000000
90	34	33	32	1.000000	0.000000
90	34	33	33	1.000000	0.000000
90	34	34	0	0.000000	0.000000
90	34	34	1	0.000000	0.000000
90	34	34	2	0.000000	0.000000
90	34	34	3	0.000005	0.000005
90	34	34	4	0.000049	0.000044
90	34	34	5	0.000339	0.000290
90	34	34	6	0.001792	0.001453
90	34	34	7	0.007405	0.005613
90	34	34	8	0.024455	0.017050
90	34	34	9	0.065765	0.041310
90	34	34	10	0.146448	0.080683
90	34	34	11	0.274475	0.128027
90	34	34	12	0.440470	0.165995
90	34	34	13	0.617046	0.176575
90	34	34	14	0.771549	0.154503
90	34	34	15	0.882903	0.111354
90	34	34	16	0.949019	0.066116
90	34	34	17	0.981330	0.032310
90	34	34	18	0.994299	0.012969
90	34	34	19	0.998560	0.004262
90	34	34	20	0.999702	0.001142
90	34	34	21	0.999950	0.000248
90	34	34	22	0.999993	0.000043
90	34	34	23	0.999999	0.000006
90	34	34	24	1.000000	0.000001

N	n	k	x	P(x)	p(x)
90	34	34	25	1.000000	0.000000
90	34	34	26	1.000000	0.000000
90	34	34	27	1.000000	0.000000
90	34	34	28	1.000000	0.000000
90	34	34	29	1.000000	0.000000
90	34	34	30	1.000000	0.000000
90	34	34	31	1.000000	0.000000
90	34	34	32	1.000000	0.000000
90	34	34	33	1.000000	0.000000
90	34	34	34	1.000000	0.000000
90	35	1	0	0.611111	0.611111
90	35	1	1	1.000000	0.388889
90	35	2	0	0.370786	0.370786
90	35	2	1	0.851436	0.480649
90	35	2	2	1.000000	0.148564
90	35	3	0	0.223315	0.223315
90	35	3	1	0.665730	0.442416
90	35	3	2	0.944288	0.278558
90	35	3	3	1.000000	0.055712
90	35	4	0	0.133475	0.133475
90	35	4	1	0.492832	0.359357
90	35	4	2	0.838628	0.345796
90	35	4	3	0.979508	0.140880
90	35	4	4	1.000000	0.020492
90	35	5	0	0.079154	0.079154
90	35	5	1	0.350761	0.271607
90	35	5	2	0.705939	0.355178
90	35	5	3	0.927088	0.221149
90	35	5	4	0.992613	0.065526
90	35	5	5	1.000000	0.007387
90	35	6	0	0.046561	0.046561
90	35	6	1	0.242118	0.195557
90	35	6	2	0.568046	0.325928
90	35	6	3	0.843832	0.275785
90	35	6	4	0.968716	0.124884
90	35	6	5	0.997393	0.028677
90	35	6	6	1.000000	0.002607
90	35	7	0	0.027161	0.027161
90	35	7	1	0.162964	0.135803
90	35	7	2	0.440003	0.277039
90	35	7	3	0.738771	0.298768
90	35	7	4	0.922628	0.183857
90	35	7	5	0.987151	0.064523
90	35	7	6	0.999100	0.011949
90	35	7	7	1.000000	0.000900
90	35	8	0	0.015707	0.015707
90	35	8	1	0.107334	0.091626
90	35	8	2	0.329855	0.222521
90	35	8	3	0.623583	0.293728
90	35	8	4	0.853958	0.230375

N	n	k	x	P(x)	p(x)
90	35	8	5	0.963829	0.109871
90	35	8	6	0.994925	0.031096
90	35	8	7	0.999696	0.004771
90	35	8	8	1.000000	0.000304
90	35	9	0	0.009003	0.009003
90	35	9	1	0.069342	0.060339
90	35	9	2	0.240304	0.170961
90	35	9	3	0.508958	0.268654
90	35	9	4	0.766865	0.257908
90	35	9	5	0.923633	0.156767
90	35	9	6	0.983928	0.060295
90	35	9	7	0.998067	0.014139
90	35	9	8	0.999900	0.001833
90	35	9	9	1.000000	0.000100
90	35	10	0	0.005113	0.005113
90	35	10	1	0.044015	0.038902
90	35	10	2	0.170653	0.126638
90	35	10	3	0.402823	0.232170
90	35	10	4	0.668160	0.265337
90	35	10	5	0.865571	0.197411
90	35	10	6	0.962341	0.096770
90	35	10	7	0.993179	0.030839
90	35	10	8	0.999289	0.006110
90	35	10	9	0.999968	0.000679
90	35	10	10	1.000000	0.000032
90	35	11	0	0.002876	0.002876
90	35	11	1	0.027481	0.024605
90	35	11	2	0.118415	0.090933
90	35	11	3	0.309955	0.191540
90	35	11	4	0.565342	0.255387
90	35	11	5	0.791542	0.226200
90	35	11	6	0.927261	0.135720
90	35	11	7	0.982386	0.055124
90	35	11	8	0.997227	0.014841
90	35	11	9	0.999747	0.002520
90	35	11	10	0.999990	0.000243
90	35	11	11	1.000000	0.000010
90	35	12	0	0.001602	0.001602
90	35	12	1	0.016892	0.015290
90	35	12	2	0.080430	0.063538
90	35	12	3	0.232369	0.151939
90	35	12	4	0.465127	0.232758
90	35	12	5	0.705643	0.240516
90	35	12	6	0.877440	0.171797
90	35	12	7	0.962848	0.085408
90	35	12	8	0.992155	0.029307
90	35	12	9	0.998918	0.006763
90	35	12	10	0.999913	0.000995
90	35	12	11	0.999997	0.000084
90	35	12	12	1.000000	0.000003

N	n	k	x	P(x)	p(x)
90	35	13	0	0.000883	0.000883
90	35	13	1	0.010227	0.009344
90	35	13	2	0.053548	0.043321
90	35	13	3	0.170035	0.116487
90	35	13	4	0.372620	0.202585
90	35	13	5	0.613137	0.240516
90	35	13	6	0.813567	0.200430
90	35	13	7	0.932189	0.118622
90	35	13	8	0.982010	0.049821
90	35	13	9	0.996663	0.014653
90	35	13	10	0.999594	0.002931
90	35	13	11	0.999971	0.000377
90	35	13	12	0.999999	0.000028
90	35	13	13	1.000000	0.000001
90	35	14	0	0.000482	0.000482
90	35	14	1	0.006101	0.005619
90	35	14	2	0.034982	0.028881
90	35	14	3	0.121625	0.086643
90	35	14	4	0.291060	0.169435
90	35	14	5	0.519429	0.228369
90	35	14	6	0.738080	0.218651
90	35	14	7	0.889054	0.150973
90	35	14	8	0.964540	0.075487
90	35	14	9	0.991716	0.027175
90	35	14	10	0.998643	0.006927
90	35	14	11	0.999854	0.001211
90	35	14	12	0.999991	0.000137
90	35	14	13	1.000000	0.000009
90	35	14	14	1.000000	0.000000
90	35	15	0	0.000260	0.000260
90	35	15	1	0.003587	0.003327
90	35	15	2	0.022442	0.018854
90	35	15	3	0.085144	0.062702
90	35	15	4	0.221948	0.136805
90	35	15	5	0.429283	0.207335
90	35	15	6	0.654647	0.225364
90	35	15	7	0.833432	0.178784
90	35	15	8	0.937723	0.104291
90	35	15	9	0.982419	0.044696
90	35	15	10	0.996364	0.013945
90	35	15	11	0.999471	0.003107
90	35	15	12	0.999949	0.000478
90	35	15	13	0.999997	0.000048
90	35	15	14	1.000000	0.000003
90	35	15	15	1.000000	0.000000
90	35	16	0	0.000139	0.000139
90	35	16	1	0.002079	0.001940
90	35	16	2	0.014146	0.012067
90	35	16	3	0.058391	0.044245
90	35	16	4	0.165402	0.107012

Table for $N = 2$, $n = 1$, through $N = 100$, $n = 50$

N	n	k	x	P(x)	p(x)
90	35	16	5	0.346349	0.180947
90	35	16	6	0.567507	0.221157
90	35	16	7	0.766686	0.199179
90	35	16	8	0.900178	0.133492
90	35	16	9	0.966924	0.066746
90	35	16	10	0.991715	0.024791
90	35	16	11	0.998477	0.006761
90	35	16	12	0.999803	0.001326
90	35	16	13	0.999983	0.000180
90	35	16	14	0.999999	0.000016
90	35	16	15	1.000000	0.000001
90	35	16	16	1.000000	0.000000
90	35	17	0	0.000073	0.000073
90	35	17	1	0.001187	0.001114
90	35	17	2	0.008764	0.007577
90	35	17	3	0.039258	0.030493
90	35	17	4	0.120573	0.081315
90	35	17	5	0.272993	0.152419
90	35	17	6	0.480837	0.207844
90	35	17	7	0.691321	0.210484
90	35	17	8	0.851471	0.160151
90	35	17	9	0.943473	0.092001
90	35	17	10	0.983340	0.039867
90	35	17	11	0.996284	0.012944
90	35	17	12	0.999391	0.003107
90	35	17	13	0.999929	0.000539
90	35	17	14	0.999994	0.000065
90	35	17	15	1.000000	0.000005
90	35	17	16	1.000000	0.000000
90	35	17	17	1.000000	0.000000
90	35	18	0	0.000038	0.000038
90	35	18	1	0.000668	0.000630
90	35	18	2	0.005339	0.004671
90	35	18	3	0.025891	0.020552
90	35	18	4	0.086042	0.060151
90	35	18	5	0.210354	0.124312
90	35	18	6	0.398269	0.187914
90	35	18	7	0.610587	0.212319
90	35	18	8	0.792238	0.181650
90	35	18	9	0.910705	0.118468
90	35	18	10	0.969687	0.058982
90	35	18	11	0.992028	0.022342
90	35	18	12	0.998412	0.006383
90	35	18	13	0.999767	0.001355
90	35	18	14	0.999976	0.000209
90	35	18	15	0.999998	0.000022
90	35	18	16	1.000000	0.000002
90	35	18	17	1.000000	0.000000
90	35	18	18	1.000000	0.000000
90	35	19	0	0.000020	0.000020

N	n	k	x	P(x)	p(x)
90	35	19	1	0.000371	0.000351
90	35	19	2	0.003198	0.002828
90	35	19	3	0.016757	0.013558
90	35	19	4	0.060144	0.043387
90	35	19	5	0.158558	0.098414
90	35	19	6	0.322581	0.164023
90	35	19	7	0.528019	0.205438
90	35	19	8	0.724119	0.196100
90	35	19	9	0.867925	0.143807
90	35	19	10	0.949207	0.081282
90	35	19	11	0.984581	0.035374
90	35	19	12	0.996373	0.011791
90	35	19	13	0.999353	0.002980
90	35	19	14	0.999915	0.000562
90	35	19	15	0.999992	0.000077
90	35	19	16	0.999999	0.000007
90	35	19	17	1.000000	0.000000
90	35	19	18	1.000000	0.000000
90	35	19	19	1.000000	0.000000
90	35	20	0	0.000010	0.000010
90	35	20	1	0.000203	0.000193
90	35	20	2	0.001884	0.001682
90	35	20	3	0.010646	0.008762
90	35	20	4	0.041200	0.030554
90	35	20	5	0.116974	0.075774
90	35	20	6	0.255585	0.138611
90	35	20	7	0.447001	0.191416
90	35	20	8	0.649545	0.202544
90	35	20	9	0.815264	0.165718
90	35	20	10	0.920587	0.105323
90	35	20	11	0.972624	0.052037
90	35	20	12	0.992553	0.019929
90	35	20	13	0.998429	0.005877
90	35	20	14	0.999749	0.001319
90	35	20	15	0.999970	0.000222
90	35	20	16	0.999997	0.000027
90	35	20	17	1.000000	0.000002
90	35	20	18	1.000000	0.000000
90	35	20	19	1.000000	0.000000
90	35	20	20	1.000000	0.000000
90	35	21	0	0.000005	0.000005
90	35	21	1	0.000109	0.000104
90	35	21	2	0.001091	0.000982
90	35	21	3	0.006641	0.005549
90	35	21	4	0.027669	0.021028
90	35	21	5	0.084500	0.056831
90	35	21	6	0.198161	0.113661
90	35	21	7	0.370435	0.172274
90	35	21	8	0.571421	0.200986
90	35	21	9	0.753711	0.182290

N	n	k	x	P(x)	p(x)
90	35	21	10	0.882971	0.129260
90	35	21	11	0.954782	0.071811
90	35	21	12	0.986005	0.031222
90	35	21	13	0.996582	0.010578
90	35	21	14	0.999353	0.002770
90	35	21	15	0.999907	0.000554
90	35	21	16	0.999990	0.000083
90	35	21	17	0.999999	0.000009
90	35	21	18	1.000000	0.000001
90	35	21	19	1.000000	0.000000
90	35	21	20	1.000000	0.000000
90	35	21	21	1.000000	0.000000
90	35	22	0	0.000002	0.000002
90	35	22	1	0.000058	0.000055
90	35	22	2	0.000622	0.000564
90	35	22	3	0.004067	0.003445
90	35	22	4	0.018221	0.014154
90	35	22	5	0.059791	0.041569
90	35	22	6	0.150390	0.090600
90	35	22	7	0.300527	0.150136
90	35	22	8	0.492774	0.192248
90	35	22	9	0.685022	0.192248
90	35	22	10	0.836138	0.151116
90	35	22	11	0.929805	0.093667
90	35	22	12	0.975597	0.045793
90	35	22	13	0.993210	0.017613
90	35	22	14	0.998510	0.005300
90	35	22	15	0.999746	0.001237
90	35	22	16	0.999967	0.000221
90	35	22	17	0.999997	0.000030
90	35	22	18	1.000000	0.000003
90	35	22	19	1.000000	0.000000
90	35	22	20	1.000000	0.000000
90	35	22	21	1.000000	0.000000
90	35	22	22	1.000000	0.000000
90	35	23	0	0.000001	0.000001
90	35	23	1	0.000030	0.000029
90	35	23	2	0.000348	0.000318
90	35	23	3	0.002446	0.002098
90	35	23	4	0.011769	0.009323
90	35	23	5	0.041451	0.029683
90	35	23	6	0.111752	0.070301
90	35	23	7	0.238706	0.126954
90	35	23	8	0.416441	0.177735
90	35	23	9	0.611516	0.195075
90	35	23	10	0.780581	0.169065
90	35	23	11	0.896746	0.116165
90	35	23	12	0.960109	0.063363
90	35	23	13	0.987512	0.027403
90	35	23	14	0.996873	0.009361

N	n	k	x	P(x)	p(x)
90	35	23	15	0.999383	0.002510
90	35	23	16	0.999905	0.000523
90	35	23	17	0.999989	0.000083
90	35	23	18	0.999999	0.000010
90	35	23	19	1.000000	0.000001
90	35	23	20	1.000000	0.000000
90	35	23	21	1.000000	0.000000
90	35	23	22	1.000000	0.000000
90	35	23	23	1.000000	0.000000
90	35	24	0	0.000001	0.000001
90	35	24	1	0.000015	0.000015
90	35	24	2	0.000191	0.000176
90	35	24	3	0.001444	0.001252
90	35	24	4	0.007455	0.006011
90	35	24	5	0.028161	0.020706
90	35	24	6	0.081323	0.053163
90	35	24	7	0.185650	0.104327
90	35	24	8	0.344816	0.159166
90	35	24	9	0.535815	0.190999
90	35	24	10	0.717497	0.181682
90	35	24	11	0.855134	0.137638
90	35	24	12	0.938357	0.083223
90	35	24	13	0.978514	0.040156
90	35	24	14	0.993939	0.015425
90	35	24	15	0.998633	0.004695
90	35	24	16	0.999757	0.001124
90	35	24	17	0.999966	0.000209
90	35	24	18	0.999996	0.000030
90	35	24	19	1.000000	0.000003
90	35	24	20	1.000000	0.000000
90	35	24	21	1.000000	0.000000
90	35	24	22	1.000000	0.000000
90	35	24	23	1.000000	0.000000
90	35	24	24	1.000000	0.000000
90	35	25	0	0.000000	0.000000
90	35	25	1	0.000008	0.000007
90	35	25	2	0.000103	0.000096
90	35	25	3	0.000837	0.000733
90	35	25	4	0.004632	0.003795
90	35	25	5	0.018749	0.014117
90	35	25	6	0.057964	0.039215
90	35	25	7	0.141390	0.083427
90	35	25	8	0.279703	0.138312
90	35	25	9	0.460573	0.180870
90	35	25	10	0.648678	0.188105
90	35	25	11	0.805084	0.156406
90	35	25	12	0.909355	0.104271
90	35	25	13	0.965128	0.055773
90	35	25	14	0.989031	0.023903
90	35	25	15	0.997211	0.008180

N	n	k	x	P(x)	p(x)
90	35	25	16	0.999434	0.002223
90	35	25	17	0.999909	0.000476
90	35	25	18	0.999989	0.000079
90	35	25	19	0.999999	0.000010
90	35	25	20	1.000000	0.000001
90	35	25	21	1.000000	0.000000
90	35	25	22	1.000000	0.000000
90	35	25	23	1.000000	0.000000
90	35	25	24	1.000000	0.000000
90	35	25	25	1.000000	0.000000
90	35	26	0	0.000000	0.000000
90	35	26	1	0.000004	0.000004
90	35	26	2	0.000055	0.000051
90	35	26	3	0.000476	0.000421
90	35	26	4	0.002822	0.002346
90	35	26	5	0.012233	0.009412
90	35	26	6	0.040468	0.028235
90	35	26	7	0.105453	0.064985
90	35	26	8	0.222250	0.116797
90	35	26	9	0.388225	0.165975
90	35	26	10	0.576330	0.188105
90	35	26	11	0.747334	0.171004
90	35	26	12	0.872459	0.125125
90	35	26	13	0.946251	0.073792
90	35	26	14	0.981308	0.035057
90	35	26	15	0.994694	0.013385
90	35	26	16	0.998784	0.004090
90	35	26	17	0.999778	0.000994
90	35	26	18	0.999968	0.000190
90	35	26	19	0.999996	0.000028
90	35	26	20	1.000000	0.000003
90	35	26	21	1.000000	0.000000
90	35	26	22	1.000000	0.000000
90	35	26	23	1.000000	0.000000
90	35	26	24	1.000000	0.000000
90	35	26	25	1.000000	0.000000
90	35	26	26	1.000000	0.000000
90	35	27	0	0.000000	0.000000
90	35	27	1	0.000002	0.000002
90	35	27	2	0.000029	0.000027
90	35	27	3	0.000265	0.000237
90	35	27	4	0.001685	0.001420
90	35	27	5	0.007822	0.006136
90	35	27	6	0.027674	0.019853
90	35	27	7	0.077022	0.049348
90	35	27	8	0.172976	0.095954
90	35	27	9	0.320798	0.147821
90	35	27	10	0.502851	0.182054
90	35	27	11	0.683208	0.180356
90	35	27	12	0.827493	0.144285

N	n	k	x	P(x)	p(x)
90	35	27	13	0.920885	0.093393
90	35	27	14	0.969805	0.048920
90	35	27	15	0.990511	0.020706
90	35	27	16	0.997570	0.007059
90	35	27	17	0.999498	0.001928
90	35	27	18	0.999917	0.000419
90	35	27	19	0.999989	0.000072
90	35	27	20	0.999999	0.000010
90	35	27	21	1.000000	0.000001
90	35	27	22	1.000000	0.000000
90	35	27	23	1.000000	0.000000
90	35	27	24	1.000000	0.000000
90	35	27	25	1.000000	0.000000
90	35	27	26	1.000000	0.000000
90	35	27	27	1.000000	0.000000
90	35	28	0	0.000000	0.000000
90	35	28	1	0.000001	0.000001
90	35	28	2	0.000015	0.000014
90	35	28	3	0.000145	0.000130
90	35	28	4	0.000987	0.000841
90	35	28	5	0.004899	0.003913
90	35	28	6	0.018536	0.013636
90	35	28	7	0.055090	0.036554
90	35	28	8	0.131853	0.076763
90	35	28	9	0.259792	0.127939
90	35	28	10	0.430608	0.170816
90	35	28	11	0.614500	0.183893
90	35	28	12	0.774817	0.160317
90	35	28	13	0.888272	0.113455
90	35	28	14	0.953499	0.065227
90	35	28	15	0.983938	0.030439
90	35	28	16	0.995441	0.011503
90	35	28	17	0.998947	0.003506
90	35	28	18	0.999804	0.000857
90	35	28	19	0.999971	0.000167
90	35	28	20	0.999996	0.000026
90	35	28	21	1.000000	0.000003
90	35	28	22	1.000000	0.000000
90	35	28	23	1.000000	0.000000
90	35	28	24	1.000000	0.000000
90	35	28	25	1.000000	0.000000
90	35	28	26	1.000000	0.000000
90	35	28	27	1.000000	0.000000
90	35	28	28	1.000000	0.000000
90	35	29	0	0.000000	0.000000
90	35	29	1	0.000000	0.000000
90	35	29	2	0.000007	0.000007
90	35	29	3	0.000078	0.000070
90	35	29	4	0.000566	0.000488
90	35	29	5	0.003006	0.002440

N	n	k	x	P(x)	p(x)
90	35	29	6	0.012157	0.009151
90	35	29	7	0.038581	0.026424
90	35	29	8	0.098424	0.059842
90	35	29	9	0.206140	0.107716
90	35	29	10	0.361730	0.155590
90	35	29	11	0.543316	0.181586
90	35	29	12	0.715345	0.172029
90	35	29	13	0.848014	0.132669
90	35	29	14	0.931406	0.083392
90	35	29	15	0.974119	0.042713
90	35	29	16	0.991916	0.017797
90	35	29	17	0.997929	0.006014
90	35	29	18	0.999569	0.001640
90	35	29	19	0.999928	0.000359
90	35	29	20	0.999990	0.000062
90	35	29	21	0.999999	0.000009
90	35	29	22	1.000000	0.000001
90	35	29	23	1.000000	0.000000
90	35	29	24	1.000000	0.000000
90	35	29	25	1.000000	0.000000
90	35	29	26	1.000000	0.000000
90	35	29	27	1.000000	0.000000
90	35	29	28	1.000000	0.000000
90	35	29	29	1.000000	0.000000
90	35	30	0	0.000000	0.000000
90	35	30	1	0.000000	0.000000
90	35	30	2	0.000004	0.000003
90	35	30	3	0.000041	0.000037
90	35	30	4	0.000318	0.000277
90	35	30	5	0.001806	0.001488
90	35	30	6	0.007807	0.006001
90	35	30	7	0.026452	0.018646
90	35	30	8	0.071936	0.045484
90	35	30	9	0.160228	0.088292
90	35	30	10	0.297964	0.137736
90	35	30	11	0.471873	0.173909
90	35	30	12	0.650482	0.178609
90	35	30	13	0.800166	0.149684
90	35	30	14	0.902697	0.102531
90	35	30	15	0.960114	0.057417
90	35	30	16	0.986372	0.026258
90	35	30	17	0.996155	0.009782
90	35	30	18	0.999112	0.002957
90	35	30	19	0.999834	0.000722
90	35	30	20	0.999975	0.000141
90	35	30	21	0.999997	0.000022
90	35	30	22	1.000000	0.000003
90	35	30	23	1.000000	0.000000
90	35	30	24	1.000000	0.000000
90	35	30	25	1.000000	0.000000

N	n	k	x	P(x)	p(x)
90	35	30	26	1.000000	0.000000
90	35	30	27	1.000000	0.000000
90	35	30	28	1.000000	0.000000
90	35	30	29	1.000000	0.000000
90	35	30	30	1.000000	0.000000
90	35	31	0	0.000000	0.000000
90	35	31	1	0.000000	0.000000
90	35	31	2	0.000002	0.000002
90	35	31	3	0.000021	0.000019
90	35	31	4	0.000175	0.000154
90	35	31	5	0.001062	0.000887
90	35	31	6	0.004906	0.003845
90	35	31	7	0.017751	0.012845
90	35	31	8	0.051468	0.033717
90	35	31	9	0.121968	0.070500
90	35	31	10	0.240574	0.118606
90	35	31	11	0.402309	0.161735
90	35	31	12	0.582015	0.179706
90	35	31	13	0.745282	0.163267
90	35	31	14	0.866811	0.121529
90	35	31	15	0.940975	0.074164
90	35	31	16	0.978057	0.037082
90	35	31	17	0.993220	0.015163
90	35	31	18	0.998274	0.005054
90	35	31	19	0.999641	0.001367
90	35	31	20	0.999940	0.000298
90	35	31	21	0.999992	0.000052
90	35	31	22	0.999999	0.000007
90	35	31	23	1.000000	0.000001
90	35	31	24	1.000000	0.000000
90	35	31	25	1.000000	0.000000
90	35	31	26	1.000000	0.000000
90	35	31	27	1.000000	0.000000
90	35	31	28	1.000000	0.000000
90	35	31	29	1.000000	0.000000
90	35	31	30	1.000000	0.000000
90	35	31	31	1.000000	0.000000
90	35	32	0	0.000000	0.000000
90	35	32	1	0.000000	0.000000
90	35	32	2	0.000001	0.000001
90	35	32	3	0.000010	0.000010
90	35	32	4	0.000094	0.000083
90	35	32	5	0.000611	0.000517
90	35	32	6	0.003017	0.002406
90	35	32	7	0.011655	0.008639
90	35	32	8	0.036038	0.024383
90	35	32	9	0.090901	0.054862
90	35	32	10	0.190317	0.099417
90	35	32	11	0.336519	0.146201
90	35	32	12	0.511960	0.175441

Table for $N=2$, $n=1$, through $N=100$, $n=50$

N	n	k	x	P(x)	p(x)
90	35	32	13	0.684403	0.172442
90	35	32	14	0.823555	0.139152
90	35	32	15	0.915835	0.092280
90	35	32	16	0.966116	0.050281
90	35	32	17	0.988594	0.022478
90	35	32	18	0.996818	0.008224
90	35	32	19	0.999271	0.002453
90	35	32	20	0.999864	0.000593
90	35	32	21	0.999979	0.000116
90	35	32	22	0.999997	0.000018
90	35	32	23	1.000000	0.000002
90	35	32	24	1.000000	0.000000
90	35	32	25	1.000000	0.000000
90	35	32	26	1.000000	0.000000
90	35	32	27	1.000000	0.000000
90	35	32	28	1.000000	0.000000
90	35	32	29	1.000000	0.000000
90	35	32	30	1.000000	0.000000
90	35	32	31	1.000000	0.000000
90	35	32	32	1.000000	0.000000
90	35	33	0	0.000000	0.000000
90	35	33	1	0.000000	0.000000
90	35	33	2	0.000000	0.000000
90	35	33	3	0.000005	0.000005
90	35	33	4	0.000049	0.000044
90	35	33	5	0.000343	0.000294
90	35	33	6	0.001814	0.001470
90	35	33	7	0.007485	0.005671
90	35	33	8	0.024688	0.017203
90	35	33	9	0.066307	0.041620
90	35	33	10	0.147465	0.081158
90	35	33	11	0.276022	0.128556
90	35	33	12	0.442388	0.166367
90	35	33	13	0.618993	0.176605
90	35	33	14	0.773172	0.154179
90	35	33	15	0.884014	0.110842
90	35	33	16	0.949644	0.065630
90	35	33	17	0.981618	0.031974
90	35	33	18	0.994407	0.012789
90	35	33	19	0.998594	0.004187
90	35	33	20	0.999710	0.001116
90	35	33	21	0.999952	0.000241
90	35	33	22	0.999993	0.000042
90	35	33	23	0.999999	0.000006
90	35	33	24	1.000000	0.000001
90	35	33	25	1.000000	0.000000
90	35	33	26	1.000000	0.000000
90	35	33	27	1.000000	0.000000
90	35	33	28	1.000000	0.000000
90	35	33	29	1.000000	0.000000
90	35	33	30	1.000000	0.000000
90	35	33	31	1.000000	0.000000
90	35	33	32	1.000000	0.000000
90	35	33	33	1.000000	0.000000
90	35	34	0	0.000000	0.000000
90	35	34	1	0.000000	0.000000
90	35	34	2	0.000000	0.000000
90	35	34	3	0.000002	0.000002
90	35	34	4	0.000025	0.000023
90	35	34	5	0.000189	0.000163
90	35	34	6	0.001066	0.000877
90	35	34	7	0.004699	0.003633
90	35	34	8	0.016539	0.011840
90	35	34	9	0.047323	0.030784
90	35	34	10	0.111870	0.064547
90	35	34	11	0.221893	0.110023
90	35	34	12	0.375258	0.153365
90	35	34	13	0.550830	0.175572
90	35	34	14	0.716369	0.165539
90	35	34	15	0.845122	0.128753
90	35	34	16	0.927768	0.082645
90	35	34	17	0.971521	0.043753
90	35	34	18	0.990593	0.019072
90	35	34	19	0.997419	0.006826
90	35	34	20	0.999417	0.001998
90	35	34	21	0.999892	0.000476
90	35	34	22	0.999984	0.000092
90	35	34	23	0.999998	0.000014
90	35	34	24	1.000000	0.000002
90	35	34	25	1.000000	0.000000
90	35	34	26	1.000000	0.000000
90	35	34	27	1.000000	0.000000
90	35	34	28	1.000000	0.000000
90	35	34	29	1.000000	0.000000
90	35	34	30	1.000000	0.000000
90	35	34	31	1.000000	0.000000
90	35	34	32	1.000000	0.000000
90	35	34	33	1.000000	0.000000
90	35	34	34	1.000000	0.000000
90	35	35	0	0.000000	0.000000
90	35	35	1	0.000000	0.000000
90	35	35	2	0.000000	0.000000
90	35	35	3	0.000001	0.000001
90	35	35	4	0.000013	0.000012
90	35	35	5	0.000101	0.000088
90	35	35	6	0.000611	0.000510
90	35	35	7	0.002882	0.002271
90	35	35	8	0.010830	0.007948
90	35	35	9	0.033030	0.022200
90	35	35	10	0.083054	0.050024
90	35	35	11	0.174740	0.091686
90	35	35	12	0.312269	0.137529
90	35	35	13	0.481855	0.169587
90	35	35	14	0.654292	0.172437
90	35	35	15	0.799139	0.144847
90	35	35	16	0.899727	0.100588
90	35	35	17	0.957457	0.057730
90	35	35	18	0.984803	0.027346
90	35	35	19	0.995469	0.010665
90	35	35	20	0.998882	0.003413
90	35	35	21	0.999773	0.000892
90	35	35	22	0.999963	0.000189
90	35	35	23	0.999995	0.000032
90	35	35	24	0.999999	0.000004
90	35	35	25	1.000000	0.000000
90	35	35	26	1.000000	0.000000
90	35	35	27	1.000000	0.000000
90	35	35	28	1.000000	0.000000
90	35	35	29	1.000000	0.000000
90	35	35	30	1.000000	0.000000
90	35	35	31	1.000000	0.000000
90	35	35	32	1.000000	0.000000
90	35	35	33	1.000000	0.000000
90	35	35	34	1.000000	0.000000
90	35	35	35	1.000000	0.000000
90	36	1	0	0.600000	0.600000
90	36	1	1	1.000000	0.400000
90	36	2	0	0.357303	0.357303
90	36	2	1	0.842697	0.485393
90	36	2	2	1.000000	0.157303
90	36	3	0	0.211134	0.211134
90	36	3	1	0.649642	0.438509
90	36	3	2	0.939224	0.289581
90	36	3	3	1.000000	0.060776
90	36	4	0	0.123768	0.123768
90	36	4	1	0.473231	0.349463
90	36	4	2	0.826054	0.352823
90	36	4	3	0.976947	0.150893
90	36	4	4	1.000000	0.023053
90	36	5	0	0.071958	0.071958
90	36	5	1	0.331008	0.259049
90	36	5	2	0.686566	0.355558
90	36	5	3	0.919046	0.232480
90	36	5	4	0.991422	0.072376
90	36	5	5	1.000000	0.008578
90	36	6	0	0.041482	0.041482
90	36	6	1	0.224340	0.182858
90	36	6	2	0.544343	0.320002
90	36	6	3	0.828789	0.284446
90	36	6	4	0.964175	0.135386
90	36	6	5	0.996871	0.032697
90	36	6	6	1.000000	0.003128
90	36	7	0	0.023704	0.023704
90	36	7	1	0.148149	0.124445
90	36	7	2	0.414818	0.266669
90	36	7	3	0.717042	0.302224
90	36	7	4	0.912599	0.195557
90	36	7	5	0.984805	0.072206
90	36	7	6	0.998883	0.014078
90	36	7	7	1.000000	0.001117
90	36	8	0	0.013423	0.013423
90	36	8	1	0.095672	0.082250
90	36	8	2	0.305580	0.209908
90	36	8	3	0.596881	0.291301
90	36	8	4	0.837204	0.240323
90	36	8	5	0.957836	0.120633
90	36	8	6	0.993794	0.035958
90	36	8	7	0.999609	0.005815
90	36	8	8	1.000000	0.000390
90	36	9	0	0.007530	0.007530
90	36	9	1	0.060566	0.053036
90	36	9	2	0.218545	0.157979
90	36	9	3	0.479650	0.261105
90	36	9	4	0.743419	0.263769
90	36	9	5	0.912231	0.168812
90	36	9	6	0.980639	0.068408
90	36	9	7	0.997553	0.016914
90	36	9	8	0.999867	0.002314
90	36	9	9	1.000000	0.000133
90	36	10	0	0.004183	0.004183
90	36	10	1	0.037649	0.033466
90	36	10	2	0.152233	0.114584
90	36	10	3	0.373274	0.221041
90	36	10	4	0.639214	0.265940
90	36	10	5	0.847624	0.208410
90	36	10	6	0.955303	0.107679
90	36	10	7	0.991497	0.036194
90	36	10	8	0.999067	0.007570
90	36	10	9	0.999956	0.000889
90	36	10	10	1.000000	0.000044
90	36	11	0	0.002301	0.002301
90	36	11	1	0.023008	0.020707
90	36	11	2	0.103535	0.080527
90	36	11	3	0.282094	0.178560
90	36	11	4	0.532838	0.250744
90	36	11	5	0.766865	0.234027
90	36	11	6	0.914923	0.148058
90	36	11	7	0.978377	0.063453
90	36	11	8	0.996418	0.018041
90	36	11	9	0.999656	0.003238

Table for $N = 2$, $n = 1$, through $N = 100$, $n = 50$

N	n	k	x	P(x)	p(x)
90	36	11	10	0.999985	0.000330
90	36	11	11	1.000000	0.000014
90	36	12	0	0.001252	0.001252
90	36	12	1	0.013834	0.012581
90	36	12	2	0.068877	0.055044
90	36	12	3	0.207506	0.138629
90	36	12	4	0.431271	0.223765
90	36	12	5	0.675032	0.243761
90	36	12	6	0.858699	0.183667
90	36	12	7	0.955084	0.096385
90	36	12	8	0.990023	0.034940
90	36	12	9	0.998549	0.008526
90	36	12	10	0.999877	0.001328
90	36	12	11	0.999995	0.000118
90	36	12	12	1.000000	0.000005
90	36	13	0	0.000674	0.000674
90	36	13	1	0.008188	0.007514
90	36	13	2	0.044884	0.036696
90	36	13	3	0.148856	0.103972
90	36	13	4	0.339470	0.190614
90	36	13	5	0.578152	0.238682
90	36	13	6	0.788058	0.209905
90	36	13	7	0.919248	0.131191
90	36	13	8	0.977481	0.058233
90	36	13	9	0.995598	0.018117
90	36	13	10	0.999434	0.003836
90	36	13	11	0.999957	0.000523
90	36	13	12	0.999998	0.000041
90	36	13	13	1.000000	0.000001
90	36	14	0	0.000359	0.000359
90	36	14	1	0.004773	0.004414
90	36	14	2	0.028681	0.023908
90	36	14	3	0.104296	0.075616
90	36	14	4	0.260254	0.155957
90	36	14	5	0.482060	0.221806
90	36	14	6	0.706276	0.224217
90	36	14	7	0.869839	0.163562
90	36	14	8	0.956305	0.086467
90	36	14	9	0.989245	0.032940
90	36	14	10	0.998139	0.008894
90	36	14	11	0.999787	0.001649
90	36	14	12	0.999986	0.000198
90	36	14	13	0.999999	0.000014
90	36	14	14	1.000000	0.000000
90	36	15	0	0.000189	0.000189
90	36	15	1	0.002740	0.002551
90	36	15	2	0.017985	0.015245
90	36	15	3	0.071463	0.053478
90	36	15	4	0.194587	0.123124
90	36	15	5	0.391586	0.196999
90	36	15	6	0.617770	0.226184
90	36	15	7	0.807427	0.189657
90	36	15	8	0.924449	0.117022
90	36	15	9	0.977543	0.053094
90	36	15	10	0.995096	0.017553
90	36	15	11	0.999245	0.004149
90	36	15	12	0.999923	0.000678
90	36	15	13	0.999995	0.000072
90	36	15	14	1.000000	0.000004
90	36	15	15	1.000000	0.000000
90	36	16	0	0.000098	0.000098
90	36	16	1	0.001550	0.001451
90	36	16	2	0.011074	0.009524
90	36	16	3	0.047933	0.036859
90	36	16	4	0.142054	0.094122
90	36	16	5	0.310160	0.168106
90	36	16	6	0.527296	0.217136
90	36	16	7	0.734093	0.206796
90	36	16	8	0.880761	0.146668
90	36	16	9	0.958429	0.077668
90	36	16	10	0.989011	0.030582
90	36	16	11	0.997862	0.008851
90	36	16	12	0.999706	0.001844
90	36	16	13	0.999973	0.000267
90	36	16	14	0.999998	0.000025
90	36	16	15	1.000000	0.000001
90	36	16	16	1.000000	0.000000
90	36	17	0	0.000050	0.000050
90	36	17	1	0.000863	0.000813
90	36	17	2	0.006698	0.005835
90	36	17	3	0.031496	0.024798
90	36	17	4	0.101353	0.069857
90	36	17	5	0.239737	0.138384
90	36	17	6	0.439268	0.199531
90	36	17	7	0.653051	0.213783
90	36	17	8	0.825265	0.172214
90	36	17	9	0.930091	0.104826
90	36	17	10	0.978266	0.048175
90	36	17	11	0.994872	0.016606
90	36	17	12	0.999108	0.004236
90	36	17	13	0.999890	0.000782
90	36	17	14	0.999991	0.000101
90	36	17	15	0.999999	0.000009
90	36	17	16	1.000000	0.000000
90	36	17	17	1.000000	0.000000
90	36	18	0	0.000026	0.000026
90	36	18	1	0.000474	0.000448
90	36	18	2	0.003980	0.003507
90	36	18	3	0.020286	0.016305
90	36	18	4	0.070731	0.050445
90	36	18	5	0.180971	0.110241
90	36	18	6	0.357269	0.176298
90	36	18	7	0.568123	0.210854
90	36	18	8	0.759210	0.191087
90	36	18	9	0.891320	0.132109
90	36	18	10	0.961108	0.069788
90	36	18	11	0.989185	0.028077
90	36	18	12	0.997715	0.008530
90	36	18	13	0.999644	0.001928
90	36	18	14	0.999961	0.000317
90	36	18	15	0.999997	0.000036
90	36	18	16	1.000000	0.000003
90	36	18	17	1.000000	0.000000
90	36	18	18	1.000000	0.000000
90	36	19	0	0.000013	0.000013
90	36	19	1	0.000256	0.000243
90	36	19	2	0.002324	0.002069
90	36	19	3	0.012812	0.010488
90	36	19	4	0.048311	0.035498
90	36	19	5	0.133507	0.085196
90	36	19	6	0.283812	0.150305
90	36	19	7	0.483196	0.199384
90	36	19	8	0.684899	0.201703
90	36	19	9	0.841779	0.156880
90	36	19	10	0.935906	0.094128
90	36	19	11	0.979436	0.043529
90	36	19	12	0.994872	0.015436
90	36	19	13	0.999028	0.004156
90	36	19	14	0.999864	0.000836
90	36	19	15	0.999986	0.000123
90	36	19	16	0.999999	0.000013
90	36	19	17	1.000000	0.000001
90	36	19	18	1.000000	0.000000
90	36	19	19	1.000000	0.000000
90	36	20	0	0.000006	0.000006
90	36	20	1	0.000136	0.000130
90	36	20	2	0.001334	0.001198
90	36	20	3	0.007938	0.006604
90	36	20	4	0.032312	0.024374
90	36	20	5	0.096309	0.063997
90	36	20	6	0.220303	0.123994
90	36	20	7	0.401757	0.181455
90	36	20	8	0.605354	0.203597
90	36	20	9	0.782120	0.176766
90	36	20	10	0.901437	0.119317
90	36	20	11	0.964109	0.062672
90	36	20	12	0.989654	0.025545
90	36	20	13	0.997681	0.008027
90	36	20	14	0.999605	0.001923
90	36	20	15	0.999950	0.000345
90	36	20	16	0.999995	0.000045
90	36	20	17	1.000000	0.000004
90	36	20	18	1.000000	0.000000
90	36	20	19	1.000000	0.000000
90	36	20	20	1.000000	0.000000
90	36	21	0	0.000003	0.000003
90	36	21	1	0.000071	0.000068
90	36	21	2	0.000752	0.000681
90	36	21	3	0.004824	0.004072
90	36	21	4	0.021169	0.016345
90	36	21	5	0.067967	0.046798
90	36	21	6	0.167162	0.099195
90	36	21	7	0.326583	0.159421
90	36	21	8	0.523915	0.197332
90	36	21	9	0.713939	0.190023
90	36	21	10	0.857119	0.143180
90	36	21	11	0.941726	0.084607
90	36	21	12	0.980896	0.039170
90	36	21	13	0.995044	0.014148
90	36	21	14	0.999000	0.003956
90	36	21	15	0.999846	0.000846
90	36	21	16	0.999982	0.000136
90	36	21	17	0.999998	0.000016
90	36	21	18	1.000000	0.000001
90	36	21	19	1.000000	0.000000
90	36	21	20	1.000000	0.000000
90	36	21	21	1.000000	0.000000
90	36	22	0	0.000001	0.000001
90	36	22	1	0.000037	0.000035
90	36	22	2	0.000416	0.000380
90	36	22	3	0.002877	0.002460
90	36	22	4	0.013589	0.010712
90	36	22	5	0.046942	0.033353
90	36	22	6	0.124034	0.077092
90	36	22	7	0.259580	0.135546
90	36	22	8	0.443839	0.184258
90	36	22	9	0.639582	0.195743
90	36	22	10	0.803167	0.163585
90	36	22	11	0.911071	0.107904
90	36	22	12	0.967271	0.056200
90	36	22	13	0.990328	0.023056
90	36	22	14	0.997739	0.007411
90	36	22	15	0.999589	0.001850
90	36	22	16	0.999943	0.000354
90	36	22	17	0.999994	0.000051
90	36	22	18	0.999999	0.000005
90	36	22	19	1.000000	0.000000
90	36	22	20	1.000000	0.000000
90	36	22	21	1.000000	0.000000
90	36	22	22	1.000000	0.000000

Table for $N = 2$, $n = 1$, through $N = 100$, $n = 50$

N	n	k	x	P(x)	p(x)
90	36	23	0	0.000001	0.000001
90	36	23	1	0.000019	0.000018
90	36	23	2	0.000227	0.000208
90	36	23	3	0.001683	0.001456
90	36	23	4	0.008548	0.006865
90	36	23	5	0.031737	0.023189
90	36	23	6	0.090023	0.058286
90	36	23	7	0.201774	0.111751
90	36	23	8	0.367967	0.166194
90	36	23	9	0.561860	0.193893
90	36	23	10	0.740620	0.178760
90	36	23	11	0.871401	0.130781
90	36	23	12	0.947436	0.076035
90	36	23	13	0.982529	0.035093
90	36	23	14	0.995341	0.012812
90	36	23	15	0.999018	0.003676
90	36	23	16	0.999839	0.000821
90	36	23	17	0.999980	0.000141
90	36	23	18	0.999998	0.000018
90	36	23	19	1.000000	0.000002
90	36	23	20	1.000000	0.000000
90	36	23	21	1.000000	0.000000
90	36	23	22	1.000000	0.000000
90	36	23	23	1.000000	0.000000
90	36	24	0	0.000000	0.000000
90	36	24	1	0.000009	0.000009
90	36	24	2	0.000121	0.000112
90	36	24	3	0.000966	0.000845
90	36	24	4	0.005269	0.004304
90	36	24	5	0.021008	0.015739
90	36	24	6	0.063925	0.042917
90	36	24	7	0.153404	0.089479
90	36	24	8	0.298513	0.145109
90	36	24	9	0.483724	0.185211
90	36	24	10	0.671250	0.187526
90	36	24	11	0.822602	0.151351
90	36	24	12	0.920199	0.097598
90	36	24	13	0.970482	0.050283
90	36	24	14	0.991134	0.020652
90	36	24	15	0.997865	0.006731
90	36	24	16	0.999594	0.001728
90	36	24	17	0.999940	0.000346
90	36	24	18	0.999993	0.000053
90	36	24	19	0.999999	0.000006
90	36	24	20	1.000000	0.000001
90	36	24	21	1.000000	0.000000
90	36	24	22	1.000000	0.000000
90	36	24	23	1.000000	0.000000
90	36	24	24	1.000000	0.000000
90	36	25	0	0.000000	0.000000

N	n	k	x	P(x)	p(x)
90	36	25	1	0.000004	0.000004
90	36	25	2	0.000063	0.000059
90	36	25	3	0.000543	0.000480
90	36	25	4	0.003183	0.002639
90	36	25	5	0.013615	0.010433
90	36	25	6	0.044417	0.030802
90	36	25	7	0.114087	0.069670
90	36	25	8	0.236952	0.122865
90	36	25	9	0.407956	0.171004
90	36	25	10	0.597377	0.189420
90	36	25	11	0.765272	0.167895
90	36	25	12	0.884709	0.119438
90	36	25	13	0.952960	0.068250
90	36	25	14	0.984250	0.031291
90	36	25	15	0.995724	0.011473
90	36	25	16	0.999070	0.003346
90	36	25	17	0.999840	0.000770
90	36	25	18	0.999979	0.000138
90	36	25	19	0.999998	0.000019
90	36	25	20	1.000000	0.000002
90	36	25	21	1.000000	0.000000
90	36	25	22	1.000000	0.000000
90	36	25	23	1.000000	0.000000
90	36	25	24	1.000000	0.000000
90	36	25	25	1.000000	0.000000
90	36	26	0	0.000000	0.000000
90	36	26	1	0.000002	0.000002
90	36	26	2	0.000033	0.000030
90	36	26	3	0.000300	0.000267
90	36	26	4	0.001883	0.001584
90	36	26	5	0.008640	0.006756
90	36	26	6	0.030201	0.021561
90	36	26	7	0.083004	0.052803
90	36	26	8	0.184025	0.101022
90	36	26	9	0.336924	0.152898
90	36	26	10	0.521608	0.184685
90	36	26	11	0.700697	0.179088
90	36	26	12	0.840609	0.139913
90	36	26	13	0.928810	0.088200
90	36	26	14	0.973660	0.044850
90	36	26	15	0.992017	0.018357
90	36	26	16	0.998040	0.006023
90	36	26	17	0.999615	0.001575
90	36	26	18	0.999940	0.000325
90	36	26	19	0.999993	0.000052
90	36	26	20	0.999999	0.000006
90	36	26	21	1.000000	0.000001
90	36	26	22	1.000000	0.000000
90	36	26	23	1.000000	0.000000
90	36	26	24	1.000000	0.000000

N	n	k	x	P(x)	p(x)
90	36	26	25	1.000000	0.000000
90	36	26	26	1.000000	0.000000
90	36	27	0	0.000000	0.000000
90	36	27	1	0.000001	0.000001
90	36	27	2	0.000016	0.000015
90	36	27	3	0.000162	0.000146
90	36	27	4	0.001091	0.000929
90	36	27	5	0.005367	0.004276
90	36	27	6	0.020094	0.014727
90	36	27	7	0.059077	0.038983
90	36	27	8	0.139828	0.080751
90	36	27	9	0.272420	0.132591
90	36	27	10	0.446580	0.174160
90	36	27	11	0.630740	0.184160
90	36	27	12	0.788142	0.157402
90	36	27	13	0.897113	0.108971
90	36	27	14	0.958242	0.061130
90	36	27	15	0.985993	0.027751
90	36	27	16	0.996158	0.010165
90	36	27	17	0.999148	0.002990
90	36	27	18	0.999849	0.000701
90	36	27	19	0.999979	0.000130
90	36	27	20	0.999998	0.000019
90	36	27	21	1.000000	0.000002
90	36	27	22	1.000000	0.000000
90	36	27	23	1.000000	0.000000
90	36	27	24	1.000000	0.000000
90	36	27	25	1.000000	0.000000
90	36	27	26	1.000000	0.000000
90	36	27	27	1.000000	0.000000
90	36	28	0	0.000000	0.000000
90	36	28	1	0.000000	0.000000
90	36	28	2	0.000008	0.000008
90	36	28	3	0.000086	0.000078
90	36	28	4	0.000619	0.000534
90	36	28	5	0.003263	0.002644
90	36	28	6	0.013081	0.009818
90	36	28	7	0.041133	0.028051
90	36	28	8	0.103939	0.062806
90	36	28	9	0.215595	0.111656
90	36	28	10	0.374704	0.159110
90	36	28	11	0.557661	0.182956
90	36	28	12	0.728180	0.170519
90	36	28	13	0.857330	0.129150
90	36	28	14	0.936895	0.079566
90	36	28	15	0.976743	0.039848
90	36	28	16	0.992931	0.016188
90	36	28	17	0.998246	0.005315
90	36	28	18	0.999648	0.001403
90	36	28	19	0.999944	0.000295

N	n	k	x	P(x)	p(x)
90	36	28	20	0.999993	0.000049
90	36	28	21	0.999999	0.000006
90	36	28	22	1.000000	0.000001
90	36	28	23	1.000000	0.000000
90	36	28	24	1.000000	0.000000
90	36	28	25	1.000000	0.000000
90	36	28	26	1.000000	0.000000
90	36	28	27	1.000000	0.000000
90	36	28	28	1.000000	0.000000
90	36	29	0	0.000000	0.000000
90	36	29	1	0.000000	0.000000
90	36	29	2	0.000004	0.000004
90	36	29	3	0.000044	0.000040
90	36	29	4	0.000344	0.000299
90	36	29	5	0.001941	0.001597
90	36	29	6	0.008331	0.006389
90	36	29	7	0.028012	0.019681
90	36	29	8	0.075575	0.047563
90	36	29	9	0.166971	0.091396
90	36	29	10	0.307981	0.141011
90	36	29	11	0.483888	0.175907
90	36	29	12	0.662172	0.178284
90	36	29	13	0.809419	0.147247
90	36	29	14	0.908663	0.099243
90	36	29	15	0.963246	0.054584
90	36	29	16	0.987709	0.024463
90	36	29	17	0.996617	0.008908
90	36	29	18	0.999241	0.002624
90	36	29	19	0.999863	0.000621
90	36	29	20	0.999980	0.000117
90	36	29	21	0.999998	0.000017
90	36	29	22	1.000000	0.000002
90	36	29	23	1.000000	0.000000
90	36	29	24	1.000000	0.000000
90	36	29	25	1.000000	0.000000
90	36	29	26	1.000000	0.000000
90	36	29	27	1.000000	0.000000
90	36	29	28	1.000000	0.000000
90	36	29	29	1.000000	0.000000
90	36	30	0	0.000000	0.000000
90	36	30	1	0.000000	0.000000
90	36	30	2	0.000002	0.000002
90	36	30	3	0.000022	0.000021
90	36	30	4	0.000187	0.000164
90	36	30	5	0.001129	0.000943
90	36	30	5	0.005188	0.004059
90	36	30	7	0.018655	0.013467
90	36	30	8	0.053743	0.035088
90	36	30	9	0.126517	0.072774
90	36	30	10	0.247878	0.121362

N	n	k	x	P(x)	p(x)
90	36	30	11	0.411795	0.163917
90	36	30	12	0.592028	0.180233
90	36	30	13	0.753900	0.161872
90	36	30	14	0.872870	0.118970
90	36	30	15	0.944455	0.071585
90	36	30	16	0.979689	0.035233
90	36	30	17	0.993843	0.014154
90	36	30	18	0.998467	0.004624
90	36	30	19	0.999690	0.001223
90	36	30	20	0.999949	0.000260
90	36	30	21	0.999993	0.000044
90	36	30	22	0.999999	0.000006
90	36	30	23	1.000000	0.000001
90	36	30	24	1.000000	0.000000
90	36	30	25	1.000000	0.000000
90	36	30	26	1.000000	0.000000
90	36	30	27	1.000000	0.000000
90	36	30	28	1.000000	0.000000
90	36	30	29	1.000000	0.000000
90	36	30	30	1.000000	0.000000
90	36	31	0	0.000000	0.000000
90	36	31	1	0.000000	0.000000
90	36	31	2	0.000001	0.000001
90	36	31	3	0.000011	0.000010
90	36	31	4	0.000099	0.000088
90	36	31	5	0.000642	0.000543
90	36	31	6	0.003159	0.002516
90	36	31	7	0.012146	0.008987
90	36	31	8	0.037368	0.025222
90	36	31	9	0.093768	0.056400
90	36	31	10	0.195288	0.101520
90	36	31	11	0.343496	0.148208
90	36	31	12	0.519935	0.176438
90	36	31	13	0.691849	0.171914
90	36	31	14	0.829248	0.137399
90	36	31	15	0.919400	0.090153
90	36	31	16	0.967944	0.048544
90	36	31	17	0.989360	0.021416
90	36	31	18	0.997080	0.007719
90	36	31	19	0.999343	0.002264
90	36	31	20	0.999880	0.000537
90	36	31	21	0.999982	0.000102
90	36	31	22	0.999998	0.000015
90	36	31	23	1.000000	0.000002
90	36	31	24	1.000000	0.000000
90	36	31	25	1.000000	0.000000
90	36	31	26	1.000000	0.000000
90	36	31	27	1.000000	0.000000
90	36	31	28	1.000000	0.000000
90	36	31	29	1.000000	0.000000
90	36	31	30	1.000000	0.000000
90	36	31	31	1.000000	0.000000
90	36	32	0	0.000000	0.000000
90	36	32	1	0.000000	0.000000
90	36	32	2	0.000000	0.000000
90	36	32	3	0.000005	0.000005
90	36	32	4	0.000051	0.000046
90	36	32	5	0.000357	0.000306
90	36	32	6	0.001879	0.001522
90	36	32	7	0.007729	0.005849
90	36	32	8	0.025399	0.017670
90	36	32	9	0.067958	0.042560
90	36	32	10	0.150551	0.082592
90	36	32	11	0.280696	0.130146
90	36	32	12	0.448163	0.167467
90	36	32	13	0.624831	0.176668
90	36	32	14	0.778014	0.153183
90	36	32	15	0.887312	0.109298
90	36	32	16	0.951489	0.064177
90	36	32	17	0.982464	0.030975
90	36	32	18	0.994725	0.012261
90	36	32	19	0.998691	0.003966
90	36	32	20	0.999735	0.001044
90	36	32	21	0.999956	0.000222
90	36	32	22	0.999994	0.000038
90	36	32	23	0.999999	0.000005
90	36	32	24	1.000000	0.000001
90	36	32	25	1.000000	0.000000
90	36	32	26	1.000000	0.000000
90	36	32	27	1.000000	0.000000
90	36	32	28	1.000000	0.000000
90	36	32	29	1.000000	0.000000
90	36	32	30	1.000000	0.000000
90	36	32	31	1.000000	0.000000
90	36	32	32	1.000000	0.000000
90	36	33	0	0.000000	0.000000
90	36	33	1	0.000000	0.000000
90	36	33	2	0.000000	0.000000
90	36	33	3	0.000003	0.000002
90	36	33	4	0.000026	0.000023
90	36	33	5	0.000194	0.000168
90	36	33	6	0.001092	0.000898
90	36	33	7	0.004804	0.003712
90	36	33	8	0.016868	0.012064
90	36	33	9	0.048146	0.031278
90	36	33	10	0.113526	0.065381
90	36	33	11	0.224599	0.111073
90	36	33	12	0.378867	0.154267
90	36	33	13	0.554773	0.175907
90	36	33	14	0.719910	0.165137
90	36	33	15	0.847739	0.127828
90	36	33	16	0.929359	0.081620
90	36	33	17	0.972317	0.042958
90	36	33	18	0.990920	0.018603
90	36	33	19	0.997528	0.006609
90	36	33	20	0.999447	0.001918
90	36	33	21	0.999899	0.000452
90	36	33	22	0.999985	0.000086
90	36	33	23	0.999998	0.000013
90	36	33	24	1.000000	0.000002
90	36	33	25	1.000000	0.000000
90	36	33	26	1.000000	0.000000
90	36	33	27	1.000000	0.000000
90	36	33	28	1.000000	0.000000
90	36	33	29	1.000000	0.000000
90	36	33	30	1.000000	0.000000
90	36	33	31	1.000000	0.000000
90	36	33	32	1.000000	0.000000
90	36	33	33	1.000000	0.000000
90	36	34	0	0.000000	0.000000
90	36	34	1	0.000000	0.000000
90	36	34	2	0.000000	0.000000
90	36	34	3	0.000001	0.000001
90	36	34	4	0.000013	0.000012
90	36	34	5	0.000102	0.000090
90	36	34	6	0.000619	0.000517
90	36	34	7	0.002915	0.002296
90	36	34	8	0.010942	0.008027
90	36	34	9	0.033330	0.022388
90	36	34	10	0.083704	0.050374
90	36	34	11	0.175883	0.092179
90	36	34	12	0.313912	0.138029
90	36	34	13	0.483793	0.169882
90	36	34	14	0.656173	0.172380
90	36	34	15	0.800644	0.144471
90	36	34	16	0.900720	0.100076
90	36	34	17	0.957997	0.057277
90	36	34	18	0.985045	0.027048
90	36	34	19	0.995557	0.010512
90	36	34	20	0.998908	0.003351
90	36	34	21	0.999780	0.000872
90	36	34	22	0.999964	0.000184
90	36	34	23	0.999995	0.000031
90	36	34	24	0.999999	0.000004
90	36	34	25	1.000000	0.000000
90	36	34	26	1.000000	0.000000
90	36	34	27	1.000000	0.000000
90	36	34	28	1.000000	0.000000
90	36	34	29	1.000000	0.000000
90	36	34	30	1.000000	0.000000
90	36	34	31	1.000000	0.000000
90	36	34	32	1.000000	0.000000
90	36	34	33	1.000000	0.000000
90	36	34	34	1.000000	0.000000
90	36	35	0	0.000000	0.000000
90	36	35	1	0.000000	0.000000
90	36	35	2	0.000000	0.000000
90	36	35	3	0.000001	0.000000
90	36	35	4	0.000006	0.000006
90	36	35	5	0.000053	0.000047
90	36	35	6	0.000342	0.000289
90	36	35	7	0.001726	0.001384
90	36	35	8	0.006929	0.005202
90	36	35	9	0.022536	0.015607
90	36	35	10	0.060316	0.037780
90	36	35	11	0.134732	0.074416
90	36	35	12	0.254757	0.120025
90	36	35	13	0.414021	0.159264
90	36	35	14	0.588453	0.174432
90	36	35	15	0.746467	0.158015
90	36	35	16	0.864979	0.118511
90	36	35	17	0.938564	0.073585
90	36	35	18	0.976351	0.037787
90	36	35	19	0.992366	0.016015
90	36	35	20	0.997951	0.005585
90	36	35	21	0.999546	0.001596
90	36	35	22	0.999918	0.000371
90	36	35	23	0.999988	0.000070
90	36	35	24	0.999998	0.000011
90	36	35	25	1.000000	0.000001
90	36	35	26	1.000000	0.000000
90	36	35	27	1.000000	0.000000
90	36	35	28	1.000000	0.000000
90	36	35	29	1.000000	0.000000
90	36	35	30	1.000000	0.000000
90	36	35	31	1.000000	0.000000
90	36	35	32	1.000000	0.000000
90	36	35	33	1.000000	0.000000
90	36	35	34	1.000000	0.000000
90	36	35	35	1.000000	0.000000
90	36	36	0	0.000000	0.000000
90	36	36	1	0.000000	0.000000
90	36	36	2	0.000000	0.000000
90	36	36	3	0.000000	0.000000
90	36	36	4	0.000003	0.000003
90	36	36	5	0.000027	0.000024
90	36	36	6	0.000184	0.000158
90	36	36	7	0.000997	0.000812
90	36	36	8	0.004280	0.003284
90	36	36	9	0.014874	0.010594

N	n	k	x	P(x)	p(x)
90	36	36	10	0.042456	0.027582
90	36	36	11	0.100906	0.058450
90	36	36	12	0.202382	0.101476
90	36	36	13	0.347419	0.145037
90	36	36	14	0.518680	0.171260
90	36	36	15	0.686135	0.167455
90	36	36	16	0.821884	0.135749
90	36	36	17	0.913144	0.091260
90	36	36	18	0.963984	0.050841
90	36	36	19	0.987416	0.023432
90	36	36	20	0.996326	0.008910
90	36	36	21	0.999111	0.002785
90	36	36	22	0.999823	0.000712
90	36	36	23	0.999971	0.000148
90	36	36	24	0.999996	0.000025
90	36	36	25	1.000000	0.000003
90	36	36	26	1.000000	0.000000
90	36	36	27	1.000000	0.000000
90	36	36	28	1.000000	0.000000
90	36	36	29	1.000000	0.000000
90	36	36	30	1.000000	0.000000
90	36	36	31	1.000000	0.000000
90	36	36	32	1.000000	0.000000
90	36	36	33	1.000000	0.000000
90	36	36	34	1.000000	0.000000
90	36	36	35	1.000000	0.000000
90	36	36	36	1.000000	0.000000
90	37	1	0	0.588889	0.588889
90	37	1	1	1.000000	0.411111
90	37	2	0	0.344070	0.344070
90	37	2	1	0.833708	0.489638
90	37	2	2	1.000000	0.166292
90	37	3	0	0.199404	0.199404
90	37	3	1	0.633401	0.433997
90	37	3	2	0.933861	0.300460
90	37	3	3	1.000000	0.066139
90	37	4	0	0.114600	0.114600
90	37	4	1	0.453816	0.339216
90	37	4	2	0.812986	0.359170
90	37	4	3	0.974153	0.161166
90	37	4	4	1.000000	0.025847
90	37	5	0	0.065295	0.065295
90	37	5	1	0.311819	0.246523
90	37	5	2	0.666813	0.354994
90	37	5	3	0.910436	0.243623
90	37	5	4	0.990082	0.079646
90	37	5	5	1.000000	0.009918
90	37	6	0	0.036873	0.036873
90	37	6	1	0.207409	0.170536
90	37	6	2	0.520639	0.313230

N	n	k	x	P(x)	p(x)
90	37	6	3	0.812986	0.292348
90	37	6	4	0.959160	0.146174
90	37	6	5	0.996266	0.037106
90	37	6	6	1.000000	0.003734
90	37	7	0	0.020631	0.020631
90	37	7	1	0.134322	0.113691
90	37	7	2	0.390126	0.255804
90	37	7	3	0.694655	0.304529
90	37	7	4	0.901735	0.207080
90	37	7	5	0.982130	0.080396
90	37	7	6	0.998622	0.016491
90	37	7	7	1.000000	0.001378
90	37	8	0	0.011434	0.011434
90	37	8	1	0.085010	0.073576
90	37	8	2	0.282257	0.197247
90	37	8	3	0.569908	0.287651
90	37	8	4	0.819402	0.249494
90	37	8	5	0.951135	0.131733
90	37	8	6	0.992462	0.041328
90	37	8	7	0.999502	0.007039
90	37	8	8	1.000000	0.000498
90	37	9	0	0.006275	0.006275
90	37	9	1	0.052709	0.046434
90	37	9	2	0.198066	0.145358
90	37	9	3	0.450638	0.252572
90	37	9	4	0.718996	0.268358
90	37	9	5	0.899727	0.180731
90	37	9	6	0.976839	0.077112
90	37	9	7	0.996926	0.020088
90	37	9	8	0.999824	0.002897
90	37	9	9	1.000000	0.000176
90	37	10	0	0.003409	0.003409
90	37	10	1	0.032071	0.028663
90	37	10	2	0.135257	0.103186
90	37	10	3	0.344620	0.209363
90	37	10	4	0.609665	0.265045
90	37	10	5	0.828327	0.218662
90	37	10	6	0.947327	0.119000
90	37	10	7	0.989486	0.042160
90	37	10	8	0.998786	0.009300
90	37	10	9	0.999939	0.001153
90	37	10	10	1.000000	0.000061
90	37	11	0	0.001832	0.001832
90	37	11	1	0.019173	0.017341
90	37	11	2	0.090113	0.070940
90	37	11	3	0.255641	0.165528
90	37	11	4	0.500334	0.244693
90	37	11	5	0.740862	0.240528
90	37	11	6	0.901214	0.160352
90	37	11	7	0.973677	0.072462

N	n	k	x	P(x)	p(x)
90	37	11	8	0.995415	0.021739
90	37	11	9	0.999536	0.004120
90	37	11	10	0.999979	0.000444
90	37	11	11	1.000000	0.000021
90	37	12	0	0.000974	0.000974
90	37	12	1	0.011271	0.010297
90	37	12	2	0.058684	0.047413
90	37	12	3	0.184401	0.125717
90	37	12	4	0.398121	0.213719
90	37	12	5	0.643433	0.245313
90	37	12	6	0.838291	0.194858
90	37	12	7	0.946159	0.107868
90	37	12	8	0.987435	0.041276
90	37	12	9	0.998075	0.010640
90	37	12	10	0.999828	0.001752
90	37	12	11	0.999993	0.000165
90	37	12	12	1.000000	0.000007
90	37	13	0	0.000512	0.000512
90	37	13	1	0.006518	0.006006
90	37	13	2	0.037409	0.030891
90	37	13	3	0.129602	0.092193
90	37	13	4	0.307701	0.178099
90	37	13	5	0.542792	0.235091
90	37	13	6	0.760848	0.218056
90	37	13	7	0.904672	0.143824
90	37	13	8	0.972089	0.067417
90	37	13	9	0.994256	0.022167
90	37	13	10	0.999221	0.004965
90	37	13	11	0.999938	0.000717
90	37	13	12	0.999998	0.000060
90	37	13	13	1.000000	0.000002
90	37	14	0	0.000266	0.000266
90	37	14	1	0.003710	0.003444
90	37	14	2	0.023368	0.019658
90	37	14	3	0.088893	0.065525
90	37	14	4	0.231373	0.142480
90	37	14	5	0.445092	0.213719
90	37	14	6	0.673059	0.227967
90	37	14	7	0.848636	0.175577
90	37	14	8	0.946698	0.098062
90	37	14	9	0.986195	0.039497
90	37	14	10	0.997480	0.011285
90	37	14	11	0.999696	0.002216
90	37	14	12	0.999978	0.000282
90	37	14	13	0.999999	0.000021
90	37	14	14	1.000000	0.000001
90	37	15	0	0.000136	0.000136
90	37	15	1	0.002079	0.001942
90	37	15	2	0.014315	0.012236
90	37	15	3	0.059579	0.045264

N	n	k	x	P(x)	p(x)
90	37	15	4	0.169507	0.109927
90	37	15	5	0.355105	0.185598
90	37	15	6	0.580072	0.224968
90	37	15	7	0.779329	0.199257
90	37	15	8	0.909280	0.129950
90	37	15	9	0.971644	0.062364
90	37	15	10	0.993471	0.021827
90	37	15	11	0.998938	0.005467
90	37	15	12	0.999885	0.000948
90	37	15	13	0.999993	0.000107
90	37	15	14	1.000000	0.000007
90	37	15	15	1.000000	0.000000
90	37	16	0	0.000069	0.000069
90	37	16	1	0.001146	0.001077
90	37	16	2	0.008605	0.007458
90	37	16	3	0.039059	0.030455
90	37	16	4	0.121139	0.082079
90	37	16	5	0.275916	0.154778
90	37	16	6	0.487086	0.211170
90	37	16	7	0.699627	0.212541
90	37	16	8	0.859032	0.159406
90	37	16	9	0.948361	0.089329
90	37	16	10	0.985613	0.037252
90	37	16	11	0.997043	0.011430
90	37	16	12	0.999570	0.002527
90	37	16	13	0.999958	0.000389
90	37	16	14	0.999998	0.000039
90	37	16	15	1.000000	0.000002
90	37	16	16	1.000000	0.000000
90	37	17	0	0.000035	0.000035
90	37	17	1	0.000622	0.000588
90	37	17	2	0.005077	0.004455
90	37	17	3	0.025067	0.019990
90	37	17	4	0.084536	0.059469
90	37	17	5	0.208985	0.124450
90	37	17	6	0.398623	0.189638
90	37	17	7	0.613461	0.214839
90	37	17	8	0.796563	0.183101
90	37	17	9	0.914561	0.117998
90	37	17	10	0.972021	0.057460
90	37	17	11	0.993027	0.021006
90	37	17	12	0.998716	0.005689
90	37	17	13	0.999832	0.001116
90	37	17	14	0.999985	0.000153
90	37	17	15	0.999999	0.000014
90	37	17	16	1.000000	0.000001
90	37	17	17	1.000000	0.000000
90	37	18	0	0.000017	0.000017
90	37	18	1	0.000332	0.000315
90	37	18	2	0.002941	0.002609

Table for $N = 2$, $n = 1$, through $N = 100$, $n = 50$

N	n	k	x	P(x)	p(x)
90	37	18	3	0.015757	0.012815
90	37	18	4	0.057652	0.041896
90	37	18	5	0.154432	0.096780
90	37	18	6	0.318092	0.163660
90	37	18	7	0.525172	0.207080
90	37	18	8	0.723824	0.198652
90	37	18	9	0.869301	0.145478
90	37	18	10	0.950769	0.081467
90	37	18	11	0.985545	0.034777
90	37	18	12	0.996768	0.011222
90	37	18	13	0.999465	0.002698
90	37	18	14	0.999937	0.000472
90	37	18	15	0.999995	0.000058
90	37	18	16	1.000000	0.000005
90	37	18	17	1.000000	0.000000
90	37	18	18	1.000000	0.000000
90	37	19	0	0.000008	0.000008
90	37	19	1	0.000175	0.000166
90	37	19	2	0.001673	0.001498
90	37	19	3	0.009705	0.008032
90	37	19	4	0.038450	0.028745
90	37	19	5	0.111419	0.072969
90	37	19	6	0.247627	0.136208
90	37	19	7	0.438888	0.191261
90	37	19	8	0.643811	0.204923
90	37	19	9	0.812727	0.168916
90	37	19	10	0.920218	0.107492
90	37	19	11	0.972987	0.052769
90	37	19	12	0.992871	0.019884
90	37	19	13	0.998566	0.005695
90	37	19	14	0.999786	0.001220
90	37	19	15	0.999977	0.000191
90	37	19	16	0.999998	0.000021
90	37	19	17	1.000000	0.000002
90	37	19	18	1.000000	0.000000
90	37	19	19	1.000000	0.000000
90	37	20	0	0.000004	0.000004
90	37	20	1	0.000090	0.000086
90	37	20	2	0.000935	0.000844
90	37	20	3	0.005859	0.004924
90	37	20	4	0.025090	0.019231
90	37	20	5	0.078532	0.053442
90	37	20	6	0.188156	0.109624
90	37	20	7	0.358074	0.169918
90	37	20	8	0.560110	0.202036
90	37	20	9	0.746112	0.186002
90	37	20	10	0.879341	0.133229
90	37	20	11	0.953663	0.074322
90	37	20	12	0.985870	0.032206
90	37	20	13	0.996641	0.010771

N	n	k	x	P(x)	p(x)
90	37	20	14	0.999391	0.002750
90	37	20	15	0.999918	0.000527
90	37	20	16	0.999992	0.000074
90	37	20	17	0.999999	0.000007
90	37	20	18	1.000000	0.000000
90	37	20	19	1.000000	0.000000
90	37	20	20	1.000000	0.000000
90	37	21	0	0.000002	0.000002
90	37	21	1	0.000046	0.000044
90	37	21	2	0.000512	0.000466
90	37	21	3	0.003467	0.002955
90	37	21	4	0.016024	0.012557
90	37	21	5	0.054101	0.038077
90	37	21	6	0.139608	0.085507
90	37	21	7	0.285252	0.145644
90	37	21	8	0.476409	0.191158
90	37	21	9	0.671711	0.195302
90	37	21	10	0.827953	0.156242
90	37	21	11	0.926058	0.098105
90	37	21	12	0.974367	0.048309
90	37	21	13	0.992948	0.018581
90	37	21	14	0.998487	0.005540
90	37	21	15	0.999752	0.001265
90	37	21	16	0.999970	0.000217
90	37	21	17	0.999997	0.000027
90	37	21	18	1.000000	0.000002
90	37	21	19	1.000000	0.000000
90	37	21	20	1.000000	0.000000
90	37	21	21	1.000000	0.000000
90	37	22	0	0.000001	0.000001
90	37	22	1	0.000023	0.000022
90	37	22	2	0.000276	0.000253
90	37	22	3	0.002011	0.001735
90	37	22	4	0.010018	0.008007
90	37	22	5	0.036442	0.026424
90	37	22	6	0.101192	0.064750
90	37	22	7	0.221928	0.120737
90	37	22	8	0.396068	0.174139
90	37	22	9	0.592458	0.196391
90	37	22	10	0.766815	0.174356
90	37	22	11	0.889091	0.122276
90	37	22	12	0.956864	0.067773
90	37	22	13	0.986485	0.029621
90	37	22	14	0.996641	0.010156
90	37	22	15	0.999349	0.002708
90	37	22	16	0.999904	0.000555
90	37	22	17	0.999989	0.000086
90	37	22	18	0.999999	0.000010
90	37	22	19	1.000000	0.000001
90	37	22	20	1.000000	0.000000

N	n	k	x	P(x)	p(x)
90	37	22	21	1.000000	0.000000
90	37	22	22	1.000000	0.000000
90	37	23	0	0.000000	0.000000
90	37	23	1	0.000011	0.000011
90	37	23	2	0.000146	0.000134
90	37	23	3	0.001143	0.000998
90	37	23	4	0.006132	0.004989
90	37	23	5	0.024007	0.017875
90	37	23	6	0.071673	0.047666
90	37	23	7	0.168662	0.096989
90	37	23	8	0.321803	0.153140
90	37	23	9	0.511592	0.189789
90	37	23	10	0.697585	0.185993
90	37	23	11	0.842338	0.144753
90	37	23	12	0.931947	0.089609
90	37	23	13	0.976031	0.044083
90	37	23	14	0.993206	0.017175
90	37	23	15	0.998473	0.005267
90	37	23	16	0.999732	0.001260
90	37	23	17	0.999964	0.000232
90	37	23	18	0.999996	0.000032
90	37	23	19	1.000000	0.000003
90	37	23	20	1.000000	0.000000
90	37	23	21	1.000000	0.000000
90	37	23	22	1.000000	0.000000
90	37	23	23	1.000000	0.000000
90	37	24	0	0.000000	0.000000
90	37	24	1	0.000005	0.000005
90	37	24	2	0.000075	0.000070
90	37	24	3	0.000637	0.000562
90	37	24	4	0.003675	0.003038
90	37	24	5	0.015470	0.011795
90	37	24	6	0.049619	0.034149
90	37	24	7	0.125234	0.075616
90	37	24	8	0.255518	0.130283
90	37	24	9	0.432277	0.176759
90	37	24	10	0.622633	0.190356
90	37	24	11	0.786165	0.163533
90	37	24	12	0.898511	0.112346
90	37	24	13	0.960240	0.061728
90	37	24	14	0.987310	0.027070
90	37	24	15	0.996743	0.009434
90	37	24	16	0.999338	0.002594
90	37	24	17	0.999895	0.000557
90	37	24	18	0.999987	0.000092
90	37	24	19	0.999999	0.000012
90	37	24	20	1.000000	0.000001
90	37	24	21	1.000000	0.000000
90	37	24	22	1.000000	0.000000
90	37	24	23	1.000000	0.000000

N	n	k	x	P(x)	p(x)
90	37	24	24	1.000000	0.000000
90	37	25	0	0.000000	0.000000
90	37	25	1	0.000003	0.000002
90	37	25	2	0.000038	0.000036
90	37	25	3	0.000348	0.000309
90	37	25	4	0.002156	0.001808
90	37	25	5	0.009751	0.007595
90	37	25	6	0.033579	0.023828
90	37	25	7	0.090864	0.057284
90	37	25	8	0.198272	0.107408
90	37	25	9	0.357288	0.159016
90	37	25	10	0.544760	0.187472
90	37	25	11	0.721743	0.176984
90	37	25	12	0.855956	0.134213
90	37	25	13	0.937793	0.081837
90	37	25	14	0.977876	0.040083
90	37	25	15	0.993599	0.015723
90	37	25	16	0.998512	0.004913
90	37	25	17	0.999726	0.001214
90	37	25	18	0.999961	0.000235
90	37	25	19	0.999996	0.000035
90	37	25	20	1.000000	0.000004
90	37	25	21	1.000000	0.000000
90	37	25	22	1.000000	0.000000
90	37	25	23	1.000000	0.000000
90	37	25	24	1.000000	0.000000
90	37	25	25	1.000000	0.000000
90	37	26	0	0.000000	0.000000
90	37	26	1	0.000001	0.000001
90	37	26	2	0.000019	0.000018
90	37	26	3	0.000186	0.000167
90	37	26	4	0.001238	0.001052
90	37	26	5	0.006012	0.004774
90	37	26	6	0.022215	0.016203
90	37	26	7	0.064425	0.042210
90	37	26	8	0.150351	0.085927
90	37	26	9	0.288789	0.138437
90	37	26	10	0.466887	0.178098
90	37	26	11	0.650950	0.184063
90	37	26	12	0.804336	0.153386
90	37	26	13	0.907576	0.103240
90	37	26	14	0.963693	0.056117
90	37	26	15	0.988277	0.024584
90	37	26	16	0.996925	0.008647
90	37	26	17	0.999353	0.002428
90	37	26	18	0.999892	0.000539
90	37	26	19	0.999986	0.000094
90	37	26	20	0.999998	0.000013
90	37	26	21	1.000000	0.000001
90	37	26	22	1.000000	0.000000

$N = 90$ $n = 37$

Table for $N = 2$, $n = 1$, through $N = 100$, $n = 50$

N	n	k	x	P(x)	p(x)
90	37	26	23	1.000000	0.000000
90	37	26	24	1.000000	0.000000
90	37	26	25	1.000000	0.000000
90	37	26	26	1.000000	0.000000
90	37	27	0	0.000000	0.000000
90	37	27	1	0.000001	0.000001
90	37	27	2	0.000009	0.000009
90	37	27	3	0.000097	0.000088
90	37	27	4	0.000696	0.000598
90	37	27	5	0.003625	0.002930
90	37	27	6	0.014367	0.010742
90	37	27	7	0.044639	0.030272
90	37	27	8	0.111416	0.066777
90	37	27	9	0.228223	0.116807
90	37	27	10	0.391752	0.163529
90	37	27	11	0.576174	0.184422
90	37	27	12	0.744419	0.168245
90	37	27	13	0.868861	0.124442
90	37	27	14	0.943526	0.074665
90	37	27	15	0.979826	0.036301
90	37	27	16	0.994087	0.014261
90	37	27	17	0.998594	0.004507
90	37	27	18	0.999732	0.001138
90	37	27	19	0.999959	0.000228
90	37	27	20	0.999995	0.000036
90	37	27	21	0.999999	0.000004
90	37	27	22	1.000000	0.000000
90	37	27	23	1.000000	0.000000
90	37	27	24	1.000000	0.000000
90	37	27	25	1.000000	0.000000
90	37	27	26	1.000000	0.000000
90	37	27	27	1.000000	0.000000
90	37	28	0	0.000000	0.000000
90	37	28	1	0.000000	0.000000
90	37	28	2	0.000004	0.000004
90	37	28	3	0.000050	0.000045
90	37	28	4	0.000382	0.000332
90	37	28	5	0.002137	0.001755
90	37	28	6	0.009081	0.006944
90	37	28	7	0.030224	0.021142
90	37	28	8	0.080677	0.050454
90	37	28	9	0.176309	0.095631
90	37	28	10	0.321668	0.145359
90	37	28	11	0.500063	0.178396
90	37	28	12	0.677655	0.177592
90	37	28	13	0.821455	0.143799
90	37	28	14	0.916267	0.094813
90	37	28	15	0.967150	0.050883
90	37	28	16	0.989334	0.022184
90	37	28	17	0.997163	0.007830

N	n	k	x	P(x)	p(x)
90	37	28	18	0.999389	0.002225
90	37	28	19	0.999895	0.000506
90	37	28	20	0.999986	0.000091
90	37	28	21	0.999998	0.000013
90	37	28	22	1.000000	0.000001
90	37	28	23	1.000000	0.000000
90	37	28	24	1.000000	0.000000
90	37	28	25	1.000000	0.000000
90	37	28	26	1.000000	0.000000
90	37	28	27	1.000000	0.000000
90	37	28	28	1.000000	0.000000
90	37	29	0	0.000000	0.000000
90	37	29	1	0.000000	0.000000
90	37	29	2	0.000002	0.000002
90	37	29	3	0.000025	0.000023
90	37	29	4	0.000205	0.000180
90	37	29	5	0.001231	0.001026
90	37	29	6	0.005609	0.004378
90	37	29	7	0.019993	0.014384
90	37	29	8	0.057078	0.037085
90	37	29	9	0.133120	0.076042
90	37	29	10	0.258366	0.125246
90	37	29	11	0.425252	0.166886
90	37	29	12	0.606046	0.180793
90	37	29	13	0.765790	0.159745
90	37	29	14	0.881095	0.115304
90	37	29	15	0.949095	0.068000
90	37	29	16	0.981820	0.032725
90	37	29	17	0.994638	0.012818
90	37	29	18	0.998707	0.004069
90	37	29	19	0.999748	0.001041
90	37	29	20	0.999961	0.000213
90	37	29	21	0.999995	0.000034
90	37	29	22	0.999999	0.000004
90	37	29	23	1.000000	0.000000
90	37	29	24	1.000000	0.000000
90	37	29	25	1.000000	0.000000
90	37	29	26	1.000000	0.000000
90	37	29	27	1.000000	0.000000
90	37	29	28	1.000000	0.000000
90	37	29	29	1.000000	0.000000
90	37	30	0	0.000000	0.000000
90	37	30	1	0.000000	0.000000
90	37	30	2	0.000001	0.000001
90	37	30	3	0.000012	0.000011
90	37	30	4	0.000108	0.000096
90	37	30	5	0.000693	0.000585
90	37	30	6	0.003384	0.002691
90	37	30	7	0.012919	0.009535
90	37	30	8	0.039448	0.026528

N	n	k	x	P(x)	p(x)
90	37	30	9	0.098216	0.058768
90	37	30	10	0.202930	0.104714
90	37	30	11	0.354121	0.151191
90	37	30	12	0.531950	0.177830
90	37	30	13	0.702940	0.170990
90	37	30	14	0.837619	0.134679
90	37	30	15	0.924570	0.086951
90	37	30	16	0.970554	0.045984
90	37	30	17	0.990435	0.019881
90	37	30	18	0.997439	0.007004
90	37	30	19	0.999441	0.002001
90	37	30	20	0.999901	0.000461
90	37	30	21	0.999986	0.000085
90	37	30	22	0.999998	0.000012
90	37	30	23	1.000000	0.000001
90	37	30	24	1.000000	0.000000
90	37	30	25	1.000000	0.000000
90	37	30	26	1.000000	0.000000
90	37	30	27	1.000000	0.000000
90	37	30	28	1.000000	0.000000
90	37	30	29	1.000000	0.000000
90	37	30	30	1.000000	0.000000
90	37	31	0	0.000000	0.000000
90	37	31	1	0.000000	0.000000
90	37	31	2	0.000000	0.000000
90	37	31	3	0.000006	0.000005
90	37	31	4	0.000055	0.000049
90	37	31	5	0.000381	0.000326
90	37	31	6	0.001994	0.001613
90	37	31	7	0.008152	0.006158
90	37	31	8	0.026626	0.018474
90	37	31	9	0.070791	0.044165
90	37	31	10	0.155808	0.085018
90	37	31	11	0.288604	0.132796
90	37	31	12	0.457855	0.169250
90	37	31	13	0.634544	0.176690
90	37	31	14	0.785992	0.151448
90	37	31	15	0.892688	0.106696
90	37	31	16	0.954460	0.061771
90	37	31	17	0.983808	0.029348
90	37	31	18	0.995221	0.011413
90	37	31	19	0.998840	0.003619
90	37	31	20	0.999771	0.000931
90	37	31	21	0.999963	0.000193
90	37	31	22	0.999995	0.000032
90	37	31	23	0.999999	0.000004
90	37	31	24	1.000000	0.000000
90	37	31	25	1.000000	0.000000
90	37	31	26	1.000000	0.000000
90	37	31	27	1.000000	0.000000

N	n	k	x	P(x)	p(x)
90	37	31	28	1.000000	0.000000
90	37	31	29	1.000000	0.000000
90	37	31	30	1.000000	0.000000
90	37	31	31	1.000000	0.000000
90	37	32	0	0.000000	0.000000
90	37	32	1	0.000000	0.000000
90	37	32	2	0.000000	0.000000
90	37	32	3	0.000003	0.000003
90	37	32	4	0.000028	0.000025
90	37	32	5	0.000204	0.000177
90	37	32	6	0.001146	0.000942
90	37	32	7	0.005021	0.003874
90	37	32	8	0.017545	0.012525
90	37	32	9	0.049831	0.032286
90	37	32	10	0.116902	0.067071
90	37	32	11	0.230084	0.113182
90	37	32	12	0.386138	0.156054
90	37	32	13	0.562670	0.176532
90	37	32	14	0.726953	0.164283
90	37	32	15	0.852903	0.125950
90	37	32	16	0.932473	0.079570
90	37	32	17	0.973859	0.041386
90	37	32	18	0.991546	0.017686
90	37	32	19	0.997736	0.006190
90	37	32	20	0.999503	0.001766
90	37	32	21	0.999911	0.000409
90	37	32	22	0.999987	0.000076
90	37	32	23	0.999998	0.000011
90	37	32	24	1.000000	0.000001
90	37	32	25	1.000000	0.000000
90	37	32	26	1.000000	0.000000
90	37	32	27	1.000000	0.000000
90	37	32	28	1.000000	0.000000
90	37	32	29	1.000000	0.000000
90	37	32	30	1.000000	0.000000
90	37	32	31	1.000000	0.000000
90	37	32	32	1.000000	0.000000
90	37	33	0	0.000000	0.000000
90	37	33	1	0.000000	0.000000
90	37	33	2	0.000000	0.000000
90	37	33	3	0.000001	0.000001
90	37	33	4	0.000013	0.000012
90	37	33	5	0.000107	0.000093
90	37	33	6	0.000643	0.000536
90	37	33	7	0.003017	0.002374
90	37	33	8	0.011283	0.008266
90	37	33	9	0.034245	0.022962
90	37	33	10	0.085679	0.051434
90	37	33	11	0.179347	0.093668
90	37	33	12	0.318874	0.139526

Table for $N = 2$, $n = 1$, through $N = 100$, $n = 50$

N	n	k	x	P(x)	p(x)
90	37	33	13	0.489623	0.170749
90	37	33	14	0.661807	0.172184
90	37	33	15	0.805129	0.143323
90	37	33	16	0.903664	0.098534
90	37	33	17	0.959588	0.055925
90	37	33	18	0.985752	0.026164
90	37	33	19	0.995815	0.010063
90	37	33	20	0.998985	0.003170
90	37	33	21	0.999798	0.000814
90	37	33	22	0.999968	0.000169
90	37	33	23	0.999996	0.000028
90	37	33	24	0.999999	0.000004
90	37	33	25	1.000000	0.000000
90	37	33	26	1.000000	0.000000
90	37	33	27	1.000000	0.000000
90	37	33	28	1.000000	0.000000
90	37	33	29	1.000000	0.000000
90	37	33	30	1.000000	0.000000
90	37	33	31	1.000000	0.000000
90	37	33	32	1.000000	0.000000
90	37	33	33	1.000000	0.000000
90	37	34	0	0.000000	0.000000
90	37	34	1	0.000000	0.000000
90	37	34	2	0.000000	0.000000
90	37	34	3	0.000001	0.000001
90	37	34	4	0.000006	0.000006
90	37	34	5	0.000054	0.000048
90	37	34	6	0.000351	0.000297
90	37	34	7	0.001767	0.001416
90	37	34	8	0.007077	0.005310
90	37	34	9	0.022965	0.015888
90	37	34	10	0.061315	0.038350
90	37	34	11	0.136621	0.075306
90	37	34	12	0.257678	0.121056
90	37	34	13	0.417728	0.160051
90	37	34	14	0.592329	0.174601
90	37	34	15	0.749812	0.157483
90	37	34	16	0.867361	0.117550
90	37	34	17	0.939966	0.072604
90	37	34	18	0.977031	0.037065
90	37	34	19	0.992637	0.015606
90	37	34	20	0.998039	0.005402
90	37	34	21	0.999570	0.001531
90	37	34	22	0.999923	0.000353
90	37	34	23	0.999989	0.000066
90	37	34	24	0.999999	0.000010
90	37	34	25	1.000000	0.000001
90	37	34	26	1.000000	0.000000
90	37	34	27	1.000000	0.000000
90	37	34	28	1.000000	0.000000

N	n	k	x	P(x)	p(x)
90	37	34	29	1.000000	0.000000
90	37	34	30	1.000000	0.000000
90	37	34	31	1.000000	0.000000
90	37	34	32	1.000000	0.000000
90	37	34	33	1.000000	0.000000
90	37	34	34	1.000000	0.000000
90	37	35	0	0.000000	0.000000
90	37	35	1	0.000000	0.000000
90	37	35	2	0.000000	0.000000
90	37	35	3	0.000000	0.000000
90	37	35	4	0.000003	0.000003
90	37	35	5	0.000027	0.000024
90	37	35	6	0.000187	0.000160
90	37	35	7	0.001009	0.000822
90	37	35	8	0.004327	0.003319
90	37	35	9	0.015021	0.010694
90	37	35	10	0.042825	0.027804
90	37	35	11	0.101658	0.058833
90	37	35	12	0.203635	0.101977
90	37	35	13	0.349135	0.145500
90	37	35	14	0.520618	0.171483
90	37	35	15	0.687943	0.167326
90	37	35	16	0.823280	0.135337
90	37	35	17	0.914036	0.090755
90	37	35	18	0.964455	0.050420
90	37	35	19	0.987621	0.023166
90	37	35	20	0.996399	0.008779
90	37	35	21	0.999133	0.002733
90	37	35	22	0.999828	0.000696
90	37	35	23	0.999972	0.000144
90	37	35	24	0.999996	0.000024
90	37	35	25	1.000000	0.000003
90	37	35	26	1.000000	0.000000
90	37	35	27	1.000000	0.000000
90	37	35	28	1.000000	0.000000
90	37	35	29	1.000000	0.000000
90	37	35	30	1.000000	0.000000
90	37	35	31	1.000000	0.000000
90	37	35	32	1.000000	0.000000
90	37	35	33	1.000000	0.000000
90	37	35	34	1.000000	0.000000
90	37	35	35	1.000000	0.000000
90	37	36	0	0.000000	0.000000
90	37	36	1	0.000000	0.000000
90	37	36	2	0.000000	0.000000
90	37	36	3	0.000000	0.000000
90	37	36	4	0.000001	0.000001
90	37	36	5	0.000013	0.000012
90	37	36	6	0.000097	0.000084
90	37	36	7	0.000560	0.000464

N	n	k	x	P(x)	p(x)
90	37	36	8	0.002578	0.002017
90	37	36	9	0.009577	0.007000
90	37	36	10	0.029176	0.019599
90	37	36	11	0.073846	0.044670
90	37	36	12	0.157282	0.083436
90	37	36	13	0.285644	0.128362
90	37	36	14	0.448907	0.163263
90	37	36	15	0.621013	0.172106
90	37	36	16	0.771606	0.150593
90	37	36	17	0.881034	0.109427
90	37	36	18	0.947037	0.066004
90	37	36	19	0.980039	0.033002
90	37	36	20	0.993686	0.013647
90	37	36	21	0.998338	0.004652
90	37	36	22	0.999639	0.001301
90	37	36	23	0.999936	0.000297
90	37	36	24	0.999991	0.000055
90	37	36	25	0.999999	0.000008
90	37	36	26	1.000000	0.000001
90	37	36	27	1.000000	0.000000
90	37	36	28	1.000000	0.000000
90	37	36	29	1.000000	0.000000
90	37	36	30	1.000000	0.000000
90	37	36	31	1.000000	0.000000
90	37	36	32	1.000000	0.000000
90	37	36	33	1.000000	0.000000
90	37	36	34	1.000000	0.000000
90	37	36	35	1.000000	0.000000
90	37	36	36	1.000000	0.000000
90	37	37	0	0.000000	0.000000
90	37	37	1	0.000000	0.000000
90	37	37	2	0.000000	0.000000
90	37	37	3	0.000000	0.000000
90	37	37	4	0.000001	0.000001
90	37	37	5	0.000006	0.000005
90	37	37	6	0.000049	0.000043
90	37	37	7	0.000303	0.000254
90	37	37	8	0.001494	0.001191
90	37	37	9	0.005948	0.004453
90	37	37	10	0.019377	0.013429
90	37	37	11	0.052338	0.032962
90	37	37	12	0.118654	0.066316
90	37	37	13	0.228594	0.109940
90	37	37	14	0.379369	0.150775
90	37	37	15	0.550896	0.171527
90	37	37	16	0.713042	0.162146
90	37	37	17	0.840505	0.127463
90	37	37	18	0.923814	0.083309
90	37	37	19	0.969039	0.045225
90	37	37	20	0.989390	0.020351

N	n	k	x	P(x)	p(x)
90	37	37	21	0.996959	0.007569
90	37	37	22	0.999277	0.002318
90	37	37	23	0.999859	0.000581
90	37	37	24	0.999977	0.000119
90	37	37	25	0.999997	0.000020
90	37	37	26	1.000000	0.000003
90	37	37	27	1.000000	0.000000
90	37	37	28	1.000000	0.000000
90	37	37	29	1.000000	0.000000
90	37	37	30	1.000000	0.000000
90	37	37	31	1.000000	0.000000
90	37	37	32	1.000000	0.000000
90	37	37	33	1.000000	0.000000
90	37	37	34	1.000000	0.000000
90	37	37	35	1.000000	0.000000
90	37	37	36	1.000000	0.000000
90	37	37	37	1.000000	0.000000
90	38	1	0	0.577778	0.577778
90	38	1	1	1.000000	0.422222
90	38	2	0	0.331086	0.331086
90	38	2	1	0.824469	0.493383
90	38	2	2	1.000000	0.175531
90	38	3	0	0.188117	0.188117
90	38	3	1	0.617024	0.428907
90	38	3	2	0.928192	0.311168
90	38	3	3	1.000000	0.071808
90	38	4	0	0.105951	0.105951
90	38	4	1	0.434615	0.328664
90	38	4	2	0.799433	0.364817
90	38	4	3	0.971112	0.171679
90	38	4	4	1.000000	0.028888
90	38	5	0	0.059135	0.059135
90	38	5	1	0.293213	0.234078
90	38	5	2	0.646719	0.353505
90	38	5	3	0.901242	0.254524
90	38	5	4	0.988579	0.087337
90	38	5	5	1.000000	0.011421
90	38	6	0	0.032698	0.032698
90	38	6	1	0.191321	0.158622
90	38	6	2	0.496999	0.305678
90	38	6	3	0.796438	0.299440
90	38	6	4	0.953644	0.157206
90	38	6	5	0.995566	0.041922
90	38	6	6	1.000000	0.004434
90	38	7	0	0.017906	0.017906
90	38	7	1	0.121451	0.103545
90	38	7	2	0.365994	0.244542
90	38	7	3	0.671672	0.305678
90	38	7	4	0.890013	0.218342
90	38	7	5	0.979097	0.089083

N	n	k	x	P(x)	p(x)
90	38	7	6	0.998311	0.019214
90	38	7	7	1.000000	0.001689
90	38	8	0	0.009708	0.009708
90	38	8	1	0.075293	0.065584
90	38	8	2	0.259927	0.184634
90	38	8	3	0.542771	0.282844
90	38	8	4	0.800572	0.257801
90	38	8	5	0.943678	0.143106
90	38	8	6	0.990903	0.047225
90	38	8	7	0.999369	0.008466
90	38	8	8	1.000000	0.000631
90	38	9	0	0.005209	0.005209
90	38	9	1	0.045700	0.040490
90	38	9	2	0.178868	0.133168
90	38	9	3	0.422045	0.243177
90	38	9	4	0.693679	0.271634
90	38	9	5	0.886087	0.192407
90	38	9	6	0.972474	0.086387
90	38	9	7	0.996168	0.023695
90	38	9	8	0.999769	0.003601
90	38	9	9	1.000000	0.000231
90	38	10	0	0.002765	0.002765
90	38	10	1	0.027204	0.024439
90	38	10	2	0.119682	0.092478
90	38	10	3	0.316969	0.197287
90	38	10	4	0.579660	0.262691
90	38	10	5	0.807699	0.228038
90	38	10	6	0.938346	0.130647
90	38	10	7	0.987100	0.048755
90	38	10	8	0.998436	0.011335
90	38	10	9	0.999917	0.001482
90	38	10	10	1.000000	0.000083
90	38	11	0	0.001452	0.001452
90	38	11	1	0.015901	0.014449
90	38	11	2	0.078067	0.062166
90	38	11	3	0.230656	0.152589
90	38	11	4	0.468016	0.237360
90	38	11	5	0.713633	0.245616
90	38	11	6	0.886087	0.172454
90	38	11	7	0.968208	0.082121
90	38	11	8	0.994185	0.025977
90	38	11	9	0.999380	0.005195
90	38	11	10	0.999971	0.000591
90	38	11	11	1.000000	0.000029
90	38	12	0	0.000753	0.000753
90	38	12	1	0.009134	0.008380
90	38	12	2	0.049738	0.040605
90	38	12	3	0.163053	0.113315
90	38	12	4	0.365861	0.202808
90	38	12	5	0.611033	0.245172

N	n	k	x	P(x)	p(x)
90	38	12	6	0.816232	0.205199
90	38	12	7	0.935983	0.119751
90	38	12	8	0.984320	0.048337
90	38	12	9	0.997473	0.013153
90	38	12	10	0.999762	0.002289
90	38	12	11	0.999990	0.000228
90	38	12	12	1.000000	0.000010
90	38	13	0	0.000386	0.000386
90	38	13	1	0.005159	0.004772
90	38	13	2	0.030998	0.025839
90	38	13	3	0.112207	0.081209
90	38	13	4	0.277458	0.165251
90	38	13	5	0.507307	0.229849
90	38	13	6	0.732048	0.224741
90	38	13	7	0.888390	0.156342
90	38	13	8	0.965729	0.077339
90	38	13	9	0.992583	0.026854
90	38	13	10	0.998940	0.006357
90	38	13	11	0.999911	0.000971
90	38	13	12	0.999997	0.000086
90	38	13	13	1.000000	0.000003
90	38	14	0	0.000196	0.000196
90	38	14	1	0.002865	0.002670
90	38	14	2	0.018917	0.016052
90	38	14	3	0.075294	0.056376
90	38	14	4	0.204490	0.129196
90	38	14	5	0.408800	0.204310
90	38	14	6	0.638649	0.229849
90	38	14	7	0.825447	0.186798
90	38	14	8	0.935597	0.110150
90	38	14	9	0.982469	0.046872
90	38	14	10	0.996628	0.014159
90	38	14	11	0.999570	0.002942
90	38	14	12	0.999968	0.000397
90	38	14	13	0.999999	0.000031
90	38	14	14	1.000000	0.000001
90	38	15	0	0.000098	0.000098
90	38	15	1	0.001566	0.001468
90	38	15	2	0.011314	0.009748
90	38	15	3	0.049331	0.038017
90	38	15	4	0.146691	0.097361
90	38	15	5	0.320086	0.173395
90	38	15	6	0.541870	0.221784
90	38	15	7	0.749253	0.207383
90	38	15	8	0.892116	0.142864
90	38	15	9	0.964583	0.072467
90	38	15	10	0.991412	0.026828
90	38	15	11	0.998525	0.007114
90	38	15	12	0.999832	0.001307
90	38	15	13	0.999989	0.000157

N	n	k	x	P(x)	p(x)
90	38	15	14	0.999999	0.000011
90	38	15	15	1.000000	0.000000
90	38	16	0	0.000048	0.000048
90	38	16	1	0.000842	0.000793
90	38	16	2	0.006635	0.005793
90	38	16	3	0.031589	0.024955
90	38	16	4	0.102555	0.070965
90	38	16	5	0.243793	0.141238
90	38	16	6	0.447243	0.203450
90	38	16	7	0.663535	0.216292
90	38	16	8	0.834971	0.171436
90	38	16	9	0.936563	0.101592
90	38	16	10	0.981396	0.044833
90	38	16	11	0.995964	0.014569
90	38	16	12	0.999379	0.003414
90	38	16	13	0.999936	0.000557
90	38	16	14	0.999996	0.000060
90	38	16	15	1.000000	0.000004
90	38	16	16	1.000000	0.000000
90	38	17	0	0.000023	0.000023
90	38	17	1	0.000445	0.000421
90	38	17	2	0.003816	0.003371
90	38	17	3	0.019786	0.015970
90	38	17	4	0.069949	0.050162
90	38	17	5	0.180808	0.110859
90	38	17	6	0.359264	0.178456
90	38	17	7	0.572926	0.213662
90	38	17	8	0.765470	0.192544
90	38	17	9	0.896750	0.131280
90	38	17	10	0.964432	0.067682
90	38	17	11	0.990649	0.026217
90	38	17	12	0.998179	0.007530
90	38	17	13	0.999748	0.001569
90	38	17	14	0.999977	0.000229
90	38	17	15	0.999999	0.000022
90	38	17	16	1.000000	0.000001
90	38	17	17	1.000000	0.000000
90	38	18	0	0.000011	0.000011
90	38	18	1	0.000231	0.000220
90	38	18	2	0.002154	0.001922
90	38	18	3	0.012130	0.009976
90	38	18	4	0.046586	0.034456
90	38	18	5	0.130694	0.084108
90	38	18	6	0.281037	0.150343
90	38	18	7	0.482193	0.201156
90	38	18	8	0.686342	0.204149
90	38	18	9	0.844598	0.158255
90	38	18	10	0.938472	0.093874
90	38	18	11	0.980952	0.042480
90	38	18	12	0.995497	0.014545

N	n	k	x	P(x)	p(x)
90	38	18	13	0.999211	0.003714
90	38	18	14	0.999901	0.000691
90	38	18	15	0.999992	0.000090
90	38	18	16	0.999999	0.000008
90	38	18	17	1.000000	0.000000
90	38	18	18	1.000000	0.000000
90	38	19	0	0.000005	0.000005
90	38	19	1	0.000118	0.000113
90	38	19	2	0.001193	0.001074
90	38	19	3	0.007280	0.006088
90	38	19	4	0.030315	0.023034
90	38	19	5	0.092144	0.061829
90	38	19	6	0.214218	0.122073
90	38	19	7	0.395584	0.181366
90	38	19	8	0.601280	0.205696
90	38	19	9	0.780856	0.179576
90	38	19	10	0.901965	0.121109
90	38	19	11	0.965022	0.063057
90	38	19	12	0.990245	0.025223
90	38	19	13	0.997921	0.007676
90	38	19	14	0.999671	0.001750
90	38	19	15	0.999963	0.000292
90	38	19	16	0.999997	0.000034
90	38	19	17	1.000000	0.000003
90	38	19	18	1.000000	0.000000
90	38	19	19	1.000000	0.000000
90	38	20	0	0.000002	0.000002
90	38	20	1	0.000059	0.000057
90	38	20	2	0.000648	0.000588
90	38	20	3	0.004279	0.003631
90	38	20	4	0.019284	0.015005
90	38	20	5	0.063407	0.044122
90	38	20	6	0.159199	0.095792
90	38	20	7	0.316396	0.157197
90	38	20	8	0.514366	0.197970
90	38	20	9	0.707508	0.193142
90	38	20	10	0.854204	0.146696
90	38	20	11	0.941043	0.086839
90	38	20	12	0.981008	0.039966
90	38	20	13	0.995218	0.014210
90	38	20	14	0.999080	0.003861
90	38	20	15	0.999868	0.000789
90	38	20	16	0.999987	0.000118
90	38	20	17	0.999999	0.000012
90	38	20	18	1.000000	0.000001
90	38	20	19	1.000000	0.000000
90	38	20	20	1.000000	0.000000
90	38	21	0	0.000001	0.000001
90	38	21	1	0.000029	0.000028
90	38	21	2	0.000345	0.000316

Table for $N=2$, $n=1$, through $N=100$, $n=50$

N	n	k	x	P(x)	p(x)
90	38	21	3	0.002464	0.002118
90	38	21	4	0.011996	0.009532
90	38	21	5	0.042606	0.030610
90	38	21	6	0.115408	0.072802
90	38	21	7	0.246780	0.131372
90	38	21	8	0.429522	0.182742
90	38	21	9	0.627492	0.197970
90	38	21	10	0.795525	0.168033
90	38	21	11	0.907548	0.112022
90	38	21	12	0.966164	0.058616
90	38	21	13	0.990143	0.023979
90	38	21	14	0.997756	0.007612
90	38	21	15	0.999609	0.001853
90	38	21	16	0.999949	0.000340
90	38	21	17	0.999995	0.000046
90	38	21	18	1.000000	0.000004
90	38	21	19	1.000000	0.000000
90	38	21	20	1.000000	0.000000
90	38	21	21	1.000000	0.000000
90	38	22	0	0.000001	0.000001
90	38	22	1	0.000014	0.000014
90	38	22	2	0.000180	0.000166
90	38	22	3	0.001389	0.001209
90	38	22	4	0.007299	0.005910
90	38	22	5	0.027966	0.020668
90	38	22	6	0.081645	0.053678
90	38	22	7	0.187758	0.106113
90	38	22	8	0.350069	0.162311
90	38	22	9	0.544287	0.194218
90	38	22	10	0.727338	0.183051
90	38	22	11	0.863713	0.136375
90	38	22	12	0.944077	0.080364
90	38	22	13	0.981455	0.037378
90	38	22	14	0.995108	0.013653
90	38	22	15	0.998991	0.003883
90	38	22	16	0.999841	0.000850
90	38	22	17	0.999981	0.000140
90	38	22	18	0.999998	0.000017
90	38	22	19	1.000000	0.000001
90	38	22	20	1.000000	0.000000
90	38	22	21	1.000000	0.000000
90	38	22	22	1.000000	0.000000
90	38	23	0	0.000000	0.000000
90	38	23	1	0.000007	0.000007
90	38	23	2	0.000092	0.000086
90	38	23	3	0.000767	0.000675
90	38	23	4	0.004344	0.003577
90	38	23	5	0.017937	0.013593
90	38	23	6	0.056384	0.038448
90	38	23	7	0.139383	0.082998

N	n	k	x	P(x)	p(x)
90	38	23	8	0.278461	0.139078
90	38	23	9	0.461459	0.182998
90	38	23	10	0.651964	0.190505
90	38	23	11	0.809564	0.157600
90	38	23	12	0.913349	0.103785
90	38	23	13	0.967713	0.054364
90	38	23	14	0.990289	0.022576
90	38	23	15	0.997678	0.007389
90	38	23	16	0.999566	0.001888
90	38	23	17	0.999938	0.000372
90	38	23	18	0.999993	0.000055
90	38	23	19	0.999999	0.000006
90	38	23	20	1.000000	0.000000
90	38	23	21	1.000000	0.000000
90	38	23	22	1.000000	0.000000
90	38	23	23	1.000000	0.000000
90	38	24	0	0.000000	0.000000
90	38	24	1	0.000003	0.000003
90	38	24	2	0.000046	0.000043
90	38	24	3	0.000415	0.000368
90	38	24	4	0.002529	0.002114
90	38	24	5	0.011242	0.008713
90	38	24	6	0.038021	0.026780
90	38	24	7	0.100980	0.062959
90	38	24	8	0.216187	0.115207
90	38	24	9	0.382251	0.166064
90	38	24	10	0.572350	0.190099
90	38	24	11	0.746054	0.173704
90	38	24	12	0.873074	0.127021
90	38	24	13	0.947428	0.074354
90	38	24	14	0.982202	0.034774
90	38	24	15	0.995141	0.012939
90	38	24	16	0.998946	0.003805
90	38	24	17	0.999821	0.000875
90	38	24	18	0.999977	0.000155
90	38	24	19	0.999998	0.000021
90	38	24	20	1.000000	0.000002
90	38	24	21	1.000000	0.000000
90	38	24	22	1.000000	0.000000
90	38	24	23	1.000000	0.000000
90	38	24	24	1.000000	0.000000
90	38	25	0	0.000000	0.000000
90	38	25	1	0.000001	0.000001
90	38	25	2	0.000023	0.000021
90	38	25	3	0.000219	0.000197
90	38	25	4	0.001440	0.001220
90	38	25	5	0.006885	0.005446
90	38	25	6	0.025037	0.018152
90	38	25	7	0.071409	0.046371
90	38	25	8	0.163820	0.092412

N	n	k	x	P(x)	p(x)
90	38	25	9	0.309283	0.145463
90	38	25	10	0.491702	0.182418
90	38	25	11	0.674993	0.183291
90	38	25	12	0.823036	0.148043
90	38	25	13	0.919264	0.096228
90	38	25	14	0.969557	0.050293
90	38	25	15	0.990632	0.021075
90	38	25	16	0.997678	0.007046
90	38	25	17	0.999543	0.001865
90	38	25	18	0.999930	0.000387
90	38	25	19	0.999992	0.000062
90	38	25	20	0.999999	0.000008
90	38	25	21	1.000000	0.000001
90	38	25	22	1.000000	0.000000
90	38	25	23	1.000000	0.000000
90	38	25	24	1.000000	0.000000
90	38	25	25	1.000000	0.000000
90	38	26	0	0.000000	0.000000
90	38	26	1	0.000001	0.000001
90	38	26	2	0.000011	0.000010
90	38	26	3	0.000113	0.000103
90	38	26	4	0.000801	0.000688
90	38	26	5	0.004121	0.003319
90	38	26	6	0.016101	0.011980
90	38	26	7	0.049293	0.033192
90	38	26	8	0.121169	0.071876
90	38	26	9	0.244384	0.123216
90	38	26	10	0.413122	0.168737
90	38	26	11	0.598857	0.185735
90	38	26	12	0.763819	0.164962
90	38	26	13	0.882253	0.118434
90	38	26	14	0.950987	0.068734
90	38	26	15	0.983175	0.032188
90	38	26	16	0.995293	0.012118
90	38	26	17	0.998940	0.003647
90	38	26	18	0.999811	0.000870
90	38	26	19	0.999974	0.000163
90	38	26	20	0.999997	0.000024
90	38	26	21	1.000000	0.000003
90	38	26	22	1.000000	0.000000
90	38	26	23	1.000000	0.000000
90	38	26	24	1.000000	0.000000
90	38	26	25	1.000000	0.000000
90	38	26	26	1.000000	0.000000
90	38	27	0	0.000000	0.000000
90	38	27	1	0.000000	0.000000
90	38	27	2	0.000005	0.000005
90	38	27	3	0.000057	0.000052
90	38	27	4	0.000436	0.000378
90	38	27	5	0.002409	0.001973

N	n	k	x	P(x)	p(x)
90	38	27	6	0.010111	0.007702
90	38	27	7	0.033216	0.023105
90	38	27	8	0.087477	0.054261
90	38	27	9	0.188552	0.101075
90	38	27	10	0.339299	0.150747
90	38	27	11	0.520500	0.181201
90	38	27	12	0.696803	0.176303
90	38	27	13	0.835990	0.139187
90	38	27	14	0.925212	0.089222
90	38	27	15	0.971607	0.046396
90	38	27	16	0.991128	0.019520
90	38	27	17	0.997744	0.006616
90	38	27	18	0.999539	0.001795
90	38	27	19	0.999925	0.000386
90	38	27	20	0.999990	0.000065
90	38	27	21	0.999999	0.000009
90	38	27	22	1.000000	0.000001
90	38	27	23	1.000000	0.000000
90	38	27	24	1.000000	0.000000
90	38	27	25	1.000000	0.000000
90	38	27	26	1.000000	0.000000
90	38	27	27	1.000000	0.000000
90	38	28	0	0.000000	0.000000
90	38	28	1	0.000000	0.000000
90	38	28	2	0.000002	0.000002
90	38	28	3	0.000028	0.000026
90	38	28	4	0.000232	0.000203
90	38	28	5	0.001375	0.001144
90	38	28	6	0.006199	0.004823
90	38	28	7	0.021846	0.015648
90	38	28	8	0.061638	0.039792
90	38	28	9	0.142026	0.080387
90	38	28	10	0.272301	0.130275
90	38	28	11	0.442842	0.170542
90	38	28	12	0.624043	0.181201
90	38	28	13	0.780757	0.156714
90	38	28	14	0.891222	0.110466
90	38	28	15	0.954669	0.063447
90	38	28	16	0.984311	0.029642
90	38	28	17	0.995538	0.011227
90	38	28	18	0.998969	0.003431
90	38	28	19	0.999809	0.000840
90	38	28	20	0.999972	0.000163
90	38	28	21	0.999997	0.000025
90	38	28	22	1.000000	0.000003
90	38	28	23	1.000000	0.000000
90	38	28	24	1.000000	0.000000
90	38	28	25	1.000000	0.000000
90	38	28	26	1.000000	0.000000
90	38	28	27	1.000000	0.000000

Table for $N = 2$, $n = 1$, through $N = 100$, $n = 50$

N	n	k	x	P(x)	p(x)
90	38	28	28	1.000000	0.000000
90	38	29	0	0.000000	0.000000
90	38	29	1	0.000000	0.000000
90	38	29	2	0.000001	0.000001
90	38	29	3	0.000014	0.000013
90	38	29	4	0.000120	0.000106
90	38	29	5	0.000767	0.000646
90	38	29	6	0.003709	0.002943
90	38	29	7	0.014023	0.010313
90	38	29	8	0.042384	0.028362
90	38	29	9	0.104425	0.062041
90	38	29	10	0.213467	0.109042
90	38	29	11	0.368574	0.155107
90	38	29	12	0.548055	0.179481
90	38	29	13	0.717566	0.169510
90	38	29	14	0.848462	0.130896
90	38	29	15	0.931133	0.082671
90	38	29	16	0.973793	0.042660
90	38	29	17	0.991735	0.017942
90	38	29	18	0.997862	0.006127
90	38	29	19	0.999551	0.001689
90	38	29	20	0.999924	0.000373
90	38	29	21	0.999990	0.000065
90	38	29	22	0.999999	0.000009
90	38	29	23	1.000000	0.000001
90	38	29	24	1.000000	0.000000
90	38	29	25	1.000000	0.000000
90	38	29	26	1.000000	0.000000
90	38	29	27	1.000000	0.000000
90	38	29	28	1.000000	0.000000
90	38	29	29	1.000000	0.000000
90	38	30	0	0.000000	0.000000
90	38	30	1	0.000000	0.000000
90	38	30	2	0.000000	0.000000
90	38	30	3	0.000006	0.000006
90	38	30	4	0.000061	0.000054
90	38	30	5	0.000417	0.000356
90	38	30	6	0.002166	0.001749
90	38	30	7	0.008781	0.006616
90	38	30	8	0.028436	0.019654
90	38	30	9	0.074930	0.046494
90	38	30	10	0.163415	0.088485
90	38	30	11	0.299920	0.136505
90	38	30	12	0.471555	0.171635
90	38	30	13	0.648094	0.176539
90	38	30	14	0.796962	0.148867
90	38	30	15	0.899962	0.103000
90	38	30	16	0.958407	0.058446
90	38	30	17	0.985559	0.027151
90	38	30	18	0.995853	0.010295

N	n	k	x	P(x)	p(x)
90	38	30	19	0.999025	0.003172
90	38	30	20	0.999814	0.000789
90	38	30	21	0.999972	0.000157
90	38	30	22	0.999996	0.000025
90	38	30	23	1.000000	0.000003
90	38	30	24	1.000000	0.000000
90	38	30	25	1.000000	0.000000
90	38	30	26	1.000000	0.000000
90	38	30	27	1.000000	0.000000
90	38	30	28	1.000000	0.000000
90	38	30	29	1.000000	0.000000
90	38	30	30	1.000000	0.000000
90	38	31	0	0.000000	0.000000
90	38	31	1	0.000000	0.000000
90	38	31	2	0.000000	0.000000
90	38	31	3	0.000003	0.000003
90	38	31	4	0.000030	0.000027
90	38	31	5	0.000221	0.000191
90	38	31	6	0.001233	0.001012
90	38	31	7	0.005363	0.004130
90	38	31	8	0.018609	0.013245
90	38	31	9	0.052458	0.033849
90	38	31	10	0.122122	0.069664
90	38	31	11	0.238493	0.116371
90	38	31	12	0.397180	0.158687
90	38	31	13	0.574536	0.177356
90	38	31	14	0.737415	0.162878
90	38	31	15	0.860478	0.123064
90	38	31	16	0.936977	0.076499
90	38	31	17	0.976056	0.039078
90	38	31	18	0.992422	0.016366
90	38	31	19	0.998021	0.005599
90	38	31	20	0.999578	0.001557
90	38	31	21	0.999927	0.000349
90	38	31	22	0.999990	0.000063
90	38	31	23	0.999999	0.000009
90	38	31	24	1.000000	0.000001
90	38	31	25	1.000000	0.000000
90	38	31	26	1.000000	0.000000
90	38	31	27	1.000000	0.000000
90	38	31	28	1.000000	0.000000
90	38	31	29	1.000000	0.000000
90	38	31	30	1.000000	0.000000
90	38	31	31	1.000000	0.000000
90	38	32	0	0.000000	0.000000
90	38	32	1	0.000000	0.000000
90	38	32	2	0.000000	0.000000
90	38	32	3	0.000001	0.000001
90	38	32	4	0.000014	0.000013
90	38	32	5	0.000114	0.000100

N	n	k	x	P(x)	p(x)
90	38	32	6	0.000684	0.000570
90	38	32	7	0.003193	0.002509
90	38	32	8	0.011874	0.008681
90	38	32	9	0.035820	0.023946
90	38	32	10	0.089061	0.053241
90	38	32	11	0.185238	0.096177
90	38	32	12	0.327250	0.142012
90	38	32	13	0.499385	0.172135
90	38	32	14	0.671159	0.171774
90	38	32	15	0.812504	0.141345
90	38	32	16	0.908452	0.095948
90	38	32	17	0.962146	0.053694
90	38	32	18	0.986874	0.024728
90	38	32	19	0.996218	0.009344
90	38	32	20	0.999103	0.002885
90	38	32	21	0.999826	0.000724
90	38	32	22	0.999973	0.000146
90	38	32	23	0.999996	0.000024
90	38	32	24	0.999999	0.000003
90	38	32	25	1.000000	0.000000
90	38	32	26	1.000000	0.000000
90	38	32	27	1.000000	0.000000
90	38	32	28	1.000000	0.000000
90	38	32	29	1.000000	0.000000
90	38	32	30	1.000000	0.000000
90	38	32	31	1.000000	0.000000
90	38	32	32	1.000000	0.000000
90	38	33	0	0.000000	0.000000
90	38	33	1	0.000000	0.000000
90	38	33	2	0.000000	0.000000
90	38	33	3	0.000001	0.000001
90	38	33	4	0.000007	0.000006
90	38	33	5	0.000057	0.000051
90	38	33	6	0.000370	0.000312
90	38	33	7	0.001852	0.001482
90	38	33	8	0.007384	0.005532
90	38	33	9	0.023847	0.016463
90	38	33	10	0.063359	0.039512
90	38	33	11	0.140466	0.077108
90	38	33	12	0.263590	0.123123
90	38	33	13	0.425189	0.161599
90	38	33	14	0.600080	0.174891
90	38	33	15	0.756454	0.156373
90	38	33	16	0.872058	0.115605
90	38	33	17	0.942705	0.070647
90	38	33	18	0.978347	0.035642
90	38	33	19	0.993157	0.014810
90	38	33	20	0.998207	0.005050
90	38	33	21	0.999614	0.001407
90	38	33	22	0.999932	0.000318

N	n	k	x	P(x)	p(x)
90	38	33	23	0.999990	0.000058
90	38	33	24	0.999999	0.000008
90	38	33	25	1.000000	0.000001
90	38	33	26	1.000000	0.000000
90	38	33	27	1.000000	0.000000
90	38	33	28	1.000000	0.000000
90	38	33	29	1.000000	0.000000
90	38	33	30	1.000000	0.000000
90	38	33	31	1.000000	0.000000
90	38	33	32	1.000000	0.000000
90	38	33	33	1.000000	0.000000
90	38	34	0	0.000000	0.000000
90	38	34	1	0.000000	0.000000
90	38	34	2	0.000000	0.000000
90	38	34	3	0.000000	0.000000
90	38	34	4	0.000003	0.000003
90	38	34	5	0.000028	0.000025
90	38	34	6	0.000194	0.000166
90	38	34	7	0.001046	0.000851
90	38	34	8	0.004472	0.003426
90	38	34	9	0.015471	0.010999
90	38	34	10	0.043949	0.028478
90	38	34	11	0.103942	0.059992
90	38	34	12	0.207428	0.103486
90	38	34	13	0.354312	0.146884
90	38	34	14	0.526442	0.172130
90	38	34	15	0.693356	0.166914
90	38	34	16	0.827439	0.134083
90	38	34	17	0.916677	0.089239
90	38	34	18	0.965841	0.049164
90	38	34	19	0.988220	0.022379
90	38	34	20	0.996612	0.008392
90	38	34	21	0.999195	0.002582
90	38	34	22	0.999843	0.000648
90	38	34	23	0.999975	0.000132
90	38	34	24	0.999997	0.000022
90	38	34	25	1.000000	0.000003
90	38	34	26	1.000000	0.000000
90	38	34	27	1.000000	0.000000
90	38	34	28	1.000000	0.000000
90	38	34	29	1.000000	0.000000
90	38	34	30	1.000000	0.000000
90	38	34	31	1.000000	0.000000
90	38	34	32	1.000000	0.000000
90	38	34	33	1.000000	0.000000
90	38	34	34	1.000000	0.000000
90	38	35	0	0.000000	0.000000
90	38	35	1	0.000000	0.000000
90	38	35	2	0.000000	0.000000
90	38	35	3	0.000000	0.000000

Table for $N = 2$, $n = 1$, through $N = 100$, $n = 50$

N	n	k	x	P(x)	p(x)
90	38	35	4	0.000001	0.000001
90	38	35	5	0.000013	0.000012
90	38	35	6	0.000099	0.000086
90	38	35	7	0.000575	0.000475
90	38	35	8	0.002637	0.002062
90	38	35	9	0.009775	0.007138
90	38	35	10	0.029710	0.019935
90	38	35	11	0.075017	0.045307
90	38	35	12	0.159381	0.084364
90	38	35	13	0.288739	0.129358
90	38	35	14	0.452672	0.163933
90	38	35	15	0.624802	0.172130
90	38	35	16	0.774763	0.149962
90	38	35	17	0.883213	0.108450
90	38	35	18	0.948283	0.065070
90	38	35	19	0.980627	0.032345
90	38	35	20	0.993915	0.013288
90	38	35	21	0.998411	0.004496
90	38	35	22	0.999658	0.001247
90	38	35	23	0.999940	0.000282
90	38	35	24	0.999991	0.000052
90	38	35	25	0.999999	0.000008
90	38	35	26	1.000000	0.000001
90	38	35	27	1.000000	0.000000
90	38	35	28	1.000000	0.000000
90	38	35	29	1.000000	0.000000
90	38	35	30	1.000000	0.000000
90	38	35	31	1.000000	0.000000
90	38	35	32	1.000000	0.000000
90	38	35	33	1.000000	0.000000
90	38	35	34	1.000000	0.000000
90	38	35	35	1.000000	0.000000
90	38	36	0	0.000000	0.000000
90	38	36	1	0.000000	0.000000
90	38	36	2	0.000000	0.000000
90	38	36	3	0.000000	0.000000
90	38	36	4	0.000001	0.000001
90	38	36	5	0.000006	0.000006
90	38	36	6	0.000049	0.000043
90	38	36	7	0.000307	0.000257
90	38	36	8	0.001512	0.001205
90	38	36	9	0.006011	0.004499
90	38	36	10	0.019561	0.013550
90	38	36	11	0.052775	0.033214
90	38	36	12	0.119500	0.066724
90	38	36	13	0.229940	0.110440
90	38	36	14	0.381138	0.151198
90	38	36	15	0.552820	0.171683
90	38	36	16	0.714779	0.161958
90	38	36	17	0.841805	0.127026

N	n	k	x	P(x)	p(x)
90	38	36	18	0.924621	0.082816
90	38	36	19	0.969454	0.044833
90	38	36	20	0.989566	0.020112
90	38	36	21	0.997021	0.007455
90	38	36	22	0.999295	0.002274
90	38	36	23	0.999863	0.000568
90	38	36	24	0.999978	0.000115
90	38	36	25	0.999997	0.000019
90	38	36	26	1.000000	0.000002
90	38	36	27	1.000000	0.000000
90	38	36	28	1.000000	0.000000
90	38	36	29	1.000000	0.000000
90	38	36	30	1.000000	0.000000
90	38	36	31	1.000000	0.000000
90	38	36	32	1.000000	0.000000
90	38	36	33	1.000000	0.000000
90	38	36	34	1.000000	0.000000
90	38	36	35	1.000000	0.000000
90	38	36	36	1.000000	0.000000
90	38	37	0	0.000000	0.000000
90	38	37	1	0.000000	0.000000
90	38	37	2	0.000000	0.000000
90	38	37	3	0.000000	0.000000
90	38	37	4	0.000000	0.000000
90	38	37	5	0.000003	0.000003
90	38	37	6	0.000024	0.000021
90	38	37	7	0.000159	0.000135
90	38	37	8	0.000842	0.000683
90	38	37	9	0.003595	0.002753
90	38	37	10	0.012535	0.008940
90	38	37	11	0.036168	0.023633
90	38	37	12	0.087373	0.051205
90	38	37	13	0.178810	0.091437
90	38	37	14	0.313938	0.135128
90	38	37	15	0.479696	0.165758
90	38	37	16	0.648796	0.169100
90	38	37	17	0.792406	0.143610
90	38	37	18	0.893948	0.101543
90	38	37	19	0.953679	0.059731
90	38	37	20	0.982862	0.029183
90	38	37	21	0.994674	0.011812
90	38	37	22	0.998621	0.003947
90	38	37	23	0.999705	0.001084
90	38	37	24	0.999948	0.000243
90	38	37	25	0.999993	0.000044
90	38	37	26	0.999999	0.000006
90	38	37	27	1.000000	0.000001
90	38	37	28	1.000000	0.000000
90	38	37	29	1.000000	0.000000
90	38	37	30	1.000000	0.000000

N	n	k	x	P(x)	p(x)
90	38	37	31	1.000000	0.000000
90	38	37	32	1.000000	0.000000
90	38	37	33	1.000000	0.000000
90	38	37	34	1.000000	0.000000
90	38	37	35	1.000000	0.000000
90	38	37	36	1.000000	0.000000
90	38	37	37	1.000000	0.000000
90	38	38	0	0.000000	0.000000
90	38	38	1	0.000000	0.000000
90	38	38	2	0.000000	0.000000
90	38	38	3	0.000000	0.000000
90	38	38	4	0.000000	0.000000
90	38	38	5	0.000001	0.000001
90	38	38	6	0.000011	0.000010
90	38	38	7	0.000080	0.000069
90	38	38	8	0.000456	0.000376
90	38	38	9	0.002089	0.001633
90	38	38	10	0.007812	0.005723
90	38	38	11	0.024129	0.016317
90	38	38	12	0.062254	0.038125
90	38	38	13	0.135679	0.073426
90	38	38	14	0.252748	0.117069
90	38	38	15	0.407764	0.155015
90	38	38	16	0.578603	0.170840
90	38	38	17	0.735504	0.156900
90	38	38	18	0.855630	0.120127
90	38	38	19	0.932266	0.076636
90	38	38	20	0.972951	0.040685
90	38	38	21	0.990885	0.017934
90	38	38	22	0.997430	0.006544
90	38	38	23	0.999398	0.001969
90	38	38	24	0.999884	0.000486
90	38	38	25	0.999982	0.000098
90	38	38	26	0.999998	0.000016
90	38	38	27	1.000000	0.000002
90	38	38	28	1.000000	0.000000
90	38	38	29	1.000000	0.000000
90	38	38	30	1.000000	0.000000
90	38	38	31	1.000000	0.000000
90	38	38	32	1.000000	0.000000
90	38	38	33	1.000000	0.000000
90	38	38	34	1.000000	0.000000
90	38	38	35	1.000000	0.000000
90	38	38	36	1.000000	0.000000
90	38	38	37	1.000000	0.000000
90	38	38	38	1.000000	0.000000
90	39	1	0	0.566667	0.566667
90	39	1	1	1.000000	0.433333
90	39	2	0	0.318352	0.318352
90	39	2	1	0.814981	0.496629

N	n	k	x	P(x)	p(x)
90	39	2	2	1.000000	0.185019
90	39	3	0	0.177264	0.177264
90	39	3	1	0.600528	0.423264
90	39	3	2	0.922208	0.321680
90	39	3	3	1.000000	0.077792
90	39	4	0	0.097801	0.097801
90	39	4	1	0.415654	0.317853
90	39	4	2	0.785401	0.369747
90	39	4	3	0.967810	0.182409
90	39	4	4	1.000000	0.032190
90	39	5	0	0.053449	0.053449
90	39	5	1	0.275207	0.221758
90	39	5	2	0.626324	0.351117
90	39	5	3	0.891453	0.265129
90	39	5	4	0.986899	0.095446
90	39	5	5	1.000000	0.013100
90	39	6	0	0.028926	0.028926
90	39	6	1	0.176068	0.147143
90	39	6	2	0.473485	0.297417
90	39	6	3	0.779163	0.305678
90	39	6	4	0.947598	0.168435
90	39	6	5	0.994760	0.047162
90	39	6	6	1.000000	0.005240
90	39	7	0	0.015496	0.015496
90	39	7	1	0.109504	0.094008
90	39	7	2	0.342480	0.232976
90	39	7	3	0.648158	0.305678
90	39	7	4	0.877417	0.229259
90	39	7	5	0.975670	0.098254
90	39	7	6	0.997941	0.022271
90	39	7	7	1.000000	0.002059
90	39	8	0	0.008215	0.008215
90	39	8	1	0.066464	0.058249
90	39	8	2	0.238623	0.172159
90	39	8	3	0.515575	0.276952
90	39	8	4	0.780741	0.265167
90	39	8	5	0.935422	0.154680
90	39	8	6	0.989087	0.053665
90	39	8	7	0.999206	0.010120
90	39	8	8	1.000000	0.000794
90	39	9	0	0.004308	0.004308
90	39	9	1	0.039470	0.035163
90	39	9	2	0.160942	0.121471
90	39	9	3	0.393986	0.233045
90	39	9	4	0.667561	0.273574
90	39	9	5	0.871286	0.203726
90	39	9	6	0.967490	0.096204
90	39	9	7	0.995257	0.027767
90	39	9	8	0.999700	0.004443
90	39	9	9	1.000000	0.000300

N	n	k	x	P(x)	p(x)
90	39	10	0	0.002234	0.002234
90	39	10	1	0.022974	0.020741
90	39	10	2	0.105455	0.082480
90	39	10	3	0.290411	0.184956
90	39	10	4	0.549350	0.258939
90	39	10	5	0.785772	0.236422
90	39	10	6	0.928296	0.142524
90	39	10	7	0.984287	0.055992
90	39	10	8	0.997999	0.013712
90	39	10	9	0.999889	0.001889
90	39	10	10	1.000000	0.000111
90	39	11	0	0.001145	0.001145
90	39	11	1	0.013122	0.011978
90	39	11	2	0.067308	0.054185
90	39	11	3	0.207181	0.139873
90	39	11	4	0.436064	0.228883
90	39	11	5	0.685292	0.249228
90	39	11	6	0.869505	0.184212
90	39	11	7	0.961891	0.092386
90	39	11	8	0.992686	0.030795
90	39	11	9	0.999180	0.006494
90	39	11	10	0.999960	0.000779
90	39	11	11	1.000000	0.000040
90	39	12	0	0.000580	0.000580
90	39	12	1	0.007361	0.006781
90	39	12	2	0.041930	0.034569
90	39	12	3	0.143441	0.101511
90	39	12	4	0.334660	0.191219
90	39	12	5	0.578029	0.243370
90	39	12	6	0.792555	0.214526
90	39	12	7	0.924468	0.131913
90	39	12	8	0.980602	0.056133
90	39	12	9	0.996714	0.016112
90	39	12	10	0.999674	0.002959
90	39	12	11	0.999986	0.000312
90	39	12	12	1.000000	0.000014
90	39	13	0	0.000290	0.000290
90	39	13	1	0.004057	0.003767
90	39	13	2	0.025532	0.021474
90	39	13	3	0.096590	0.071058
90	39	13	4	0.248857	0.152267
90	39	13	5	0.471945	0.223089
90	39	13	6	0.701794	0.229849
90	39	13	7	0.870350	0.168556
90	39	13	8	0.958292	0.087942
90	39	13	9	0.990517	0.032225
90	39	13	10	0.998573	0.008056
90	39	13	11	0.999874	0.001300
90	39	13	12	0.999995	0.000121
90	39	13	13	1.000000	0.000005

N	n	k	x	P(x)	p(x)
90	39	14	0	0.000143	0.000143
90	39	14	1	0.002198	0.002055
90	39	14	2	0.015213	0.013015
90	39	14	3	0.063368	0.048155
90	39	14	4	0.179644	0.116277
90	39	14	5	0.373439	0.193794
90	39	14	6	0.603288	0.229849
90	39	14	7	0.800301	0.197013
90	39	14	8	0.922887	0.122586
90	39	14	9	0.977962	0.055075
90	39	14	10	0.995539	0.017577
90	39	14	11	0.999401	0.003862
90	39	14	12	0.999952	0.000552
90	39	14	13	0.999998	0.000046
90	39	14	14	1.000000	0.000002
90	39	15	0	0.000070	0.000070
90	39	15	1	0.001171	0.001101
90	39	15	2	0.008877	0.007706
90	39	15	3	0.040558	0.031681
90	39	15	4	0.126096	0.085538
90	39	15	5	0.286741	0.160645
90	39	15	6	0.503485	0.216744
90	39	15	7	0.717348	0.213863
90	39	15	8	0.872885	0.155537
90	39	15	9	0.956222	0.083337
90	39	15	10	0.988832	0.032610
90	39	15	11	0.997978	0.009146
90	39	15	12	0.999756	0.001778
90	39	15	13	0.999983	0.000226
90	39	15	14	0.999999	0.000017
90	39	15	15	1.000000	0.000001
90	39	16	0	0.000033	0.000033
90	39	16	1	0.000613	0.000579
90	39	16	2	0.005075	0.004462
90	39	16	3	0.025351	0.020276
90	39	16	4	0.086178	0.060827
90	39	16	5	0.213915	0.127737
90	39	16	6	0.408118	0.194202
90	39	16	7	0.626100	0.217982
90	39	16	8	0.808596	0.182497
90	39	16	9	0.922887	0.114291
90	39	16	10	0.976223	0.053336
90	39	16	11	0.994563	0.018341
90	39	16	12	0.999116	0.004553
90	39	16	13	0.999904	0.000788
90	39	16	14	0.999994	0.000090
90	39	16	15	1.000000	0.000006
90	39	16	16	1.000000	0.000000
90	39	17	0	0.000016	0.000016
90	39	17	1	0.000315	0.000299

N	n	k	x	P(x)	p(x)
90	39	17	2	0.002844	0.002529
90	39	17	3	0.015487	0.012643
90	39	17	4	0.057408	0.041922
90	39	17	5	0.155225	0.097817
90	39	17	6	0.321514	0.166289
90	39	17	7	0.531837	0.210323
90	39	17	8	0.732145	0.200308
90	39	17	9	0.876553	0.144408
90	39	17	10	0.955321	0.078768
90	39	17	11	0.987624	0.032303
90	39	17	12	0.997455	0.009831
90	39	17	13	0.999627	0.002172
90	39	17	14	0.999963	0.000336
90	39	17	15	0.999998	0.000034
90	39	17	16	1.000000	0.000002
90	39	17	17	1.000000	0.000000
90	39	18	0	0.000007	0.000007
90	39	18	1	0.000159	0.000152
90	39	18	2	0.001562	0.001403
90	39	18	3	0.009252	0.007690
90	39	18	4	0.037309	0.028057
90	39	18	5	0.109667	0.072358
90	39	18	6	0.246342	0.136676
90	39	18	7	0.439641	0.193298
90	39	18	8	0.647083	0.207442
90	39	18	9	0.817207	0.170125
90	39	18	10	0.924030	0.106822
90	39	18	11	0.975234	0.051204
90	39	18	12	0.993819	0.018585
90	39	18	13	0.998854	0.005035
90	39	18	14	0.999848	0.000995
90	39	18	15	0.999986	0.000138
90	39	18	16	0.999999	0.000013
90	39	18	17	1.000000	0.000001
90	39	18	18	1.000000	0.000000
90	39	19	0	0.000003	0.000003
90	39	19	1	0.000079	0.000076
90	39	19	2	0.000841	0.000762
90	39	19	3	0.005407	0.004566
90	39	19	4	0.023670	0.018263
90	39	19	5	0.075498	0.051828
90	39	19	6	0.183699	0.108201
90	39	19	7	0.353730	0.170031
90	39	19	8	0.557767	0.204037
90	39	19	9	0.746322	0.188555
90	39	19	10	0.881004	0.134682
90	39	19	11	0.955321	0.074317
90	39	19	12	0.986849	0.031528
90	39	19	13	0.997035	0.010186
90	39	19	14	0.999503	0.002467

N	n	k	x	P(x)	p(x)
90	39	19	15	0.999940	0.000437
90	39	19	16	0.999995	0.000055
90	39	19	17	1.000000	0.000005
90	39	19	18	1.000000	0.000000
90	39	19	19	1.000000	0.000000
90	39	20	0	0.000002	0.000002
90	39	20	1	0.000039	0.000037
90	39	20	2	0.000444	0.000406
90	39	20	3	0.003092	0.002648
90	39	20	4	0.014667	0.011575
90	39	20	5	0.050679	0.036012
90	39	20	6	0.133408	0.082730
90	39	20	7	0.277097	0.143688
90	39	20	8	0.468681	0.191584
90	39	20	9	0.666651	0.197970
90	39	20	10	0.825993	0.159342
90	39	20	11	0.926013	0.100020
90	39	20	12	0.974860	0.048847
90	39	20	13	0.993305	0.018446
90	39	20	14	0.998634	0.005329
90	39	20	15	0.999793	0.001158
90	39	20	16	0.999977	0.000185
90	39	20	17	0.999998	0.000021
90	39	20	18	1.000000	0.000002
90	39	20	19	1.000000	0.000000
90	39	20	20	1.000000	0.000000
90	39	21	0	0.000001	0.000001
90	39	21	1	0.000018	0.000018
90	39	21	2	0.000230	0.000211
90	39	21	3	0.001730	0.001500
90	39	21	4	0.008880	0.007149
90	39	21	5	0.033188	0.024308
90	39	21	6	0.094407	0.061220
90	39	21	7	0.211411	0.117003
90	39	21	8	0.383836	0.172426
90	39	21	9	0.581807	0.197970
90	39	21	10	0.759980	0.178173
90	39	21	11	0.886005	0.126025
90	39	21	12	0.956019	0.070014
90	39	21	13	0.986454	0.030435
90	39	21	14	0.996731	0.010277
90	39	21	15	0.999395	0.002664
90	39	21	16	0.999917	0.000521
90	39	21	17	0.999992	0.000075
90	39	21	18	0.999999	0.000008
90	39	21	19	1.000000	0.000001
90	39	21	20	1.000000	0.000000
90	39	21	21	1.000000	0.000000
90	39	22	0	0.000000	0.000000
90	39	22	1	0.000009	0.000008

N	n	k	x	P(x)	p(x)
90	39	22	2	0.000116	0.000108
90	39	22	3	0.000947	0.000831
90	39	22	4	0.005253	0.004306
90	39	22	5	0.021210	0.015957
90	39	22	6	0.065128	0.043919
90	39	22	7	0.157148	0.092020
90	39	22	8	0.306370	0.149221
90	39	22	9	0.495733	0.189363
90	39	22	10	0.685095	0.189363
90	39	22	11	0.834864	0.149769
90	39	22	12	0.928622	0.093758
90	39	22	13	0.974986	0.046364
90	39	22	14	0.993007	0.018022
90	39	22	15	0.998469	0.005461
90	39	22	16	0.999743	0.001274
90	39	22	17	0.999968	0.000225
90	39	22	18	0.999997	0.000029
90	39	22	19	1.000000	0.000003
90	39	22	20	1.000000	0.000000
90	39	22	21	1.000000	0.000000
90	39	22	22	1.000000	0.000000
90	39	23	0	0.000000	0.000000
90	39	23	1	0.000004	0.000004
90	39	23	2	0.000058	0.000054
90	39	23	3	0.000507	0.000450
90	39	23	4	0.003037	0.002529
90	39	23	5	0.013231	0.010194
90	39	23	6	0.043815	0.030583
90	39	23	7	0.113845	0.070030
90	39	23	8	0.238342	0.124497
90	39	23	9	0.412190	0.173848
90	39	23	10	0.604338	0.192148
90	39	23	11	0.773195	0.168857
90	39	23	12	0.891395	0.118200
90	39	23	13	0.957258	0.065864
90	39	23	14	0.986382	0.029123
90	39	23	15	0.996541	0.010159
90	39	23	16	0.999312	0.002771
90	39	23	17	0.999895	0.000583
90	39	23	18	0.999988	0.000093
90	39	23	19	0.999999	0.000011
90	39	23	20	1.000000	0.000001
90	39	23	21	1.000000	0.000000
90	39	23	22	1.000000	0.000000
90	39	23	23	1.000000	0.000000
90	39	24	0	0.000000	0.000000
90	39	24	1	0.000002	0.000002
90	39	24	2	0.000028	0.000026
90	39	24	3	0.000266	0.000238
90	39	24	4	0.001716	0.001450
90	39	24	5	0.008058	0.006343
90	39	24	6	0.028751	0.020693
90	39	24	7	0.080398	0.051646
90	39	24	8	0.180739	0.100341
90	39	24	9	0.334348	0.153609
90	39	24	10	0.521169	0.186822
90	39	24	11	0.702628	0.181458
90	39	24	12	0.843762	0.141134
90	39	24	13	0.931699	0.087937
90	39	24	14	0.975515	0.043816
90	39	24	15	0.992902	0.017387
90	39	24	16	0.998361	0.005459
90	39	24	17	0.999704	0.001343
90	39	24	18	0.999959	0.000255
90	39	24	19	0.999996	0.000037
90	39	24	20	1.000000	0.000004
90	39	24	21	1.000000	0.000000
90	39	24	22	1.000000	0.000000
90	39	24	23	1.000000	0.000000
90	39	24	24	1.000000	0.000000
90	39	25	0	0.000000	0.000000
90	39	25	1	0.000001	0.000001
90	39	25	2	0.000013	0.000013
90	39	25	3	0.000136	0.000123
90	39	25	4	0.000947	0.000811
90	39	25	5	0.004791	0.003844
90	39	25	6	0.018405	0.013614
90	39	25	7	0.055357	0.036952
90	39	25	8	0.133609	0.078252
90	39	25	9	0.264526	0.130917
90	39	25	10	0.439081	0.174556
90	39	25	11	0.625645	0.186564
90	39	25	12	0.786025	0.160380
90	39	25	13	0.897057	0.111032
90	39	25	14	0.958918	0.061861
90	39	25	15	0.986579	0.027661
90	39	25	16	0.996458	0.009879
90	39	25	17	0.999256	0.002797
90	39	25	18	0.999878	0.000622
90	39	25	19	0.999984	0.000107
90	39	25	20	0.999998	0.000014
90	39	25	21	1.000000	0.000001
90	39	25	22	1.000000	0.000000
90	39	25	23	1.000000	0.000000
90	39	25	24	1.000000	0.000000
90	39	25	25	1.000000	0.000000
90	39	26	0	0.000000	0.000000
90	39	26	1	0.000000	0.000000
90	39	26	2	0.000006	0.000006
90	39	26	3	0.000068	0.000062
90	39	26	4	0.000510	0.000442
90	39	26	5	0.002780	0.002270
90	39	26	6	0.011493	0.008713
90	39	26	7	0.037165	0.025672
90	39	26	8	0.096289	0.059124
90	39	26	9	0.204102	0.107814
90	39	26	10	0.361202	0.157100
90	39	26	11	0.545279	0.184077
90	39	26	12	0.719406	0.174127
90	39	26	13	0.852644	0.133239
90	39	26	14	0.935125	0.082481
90	39	26	15	0.976366	0.041241
90	39	26	16	0.992963	0.016597
90	39	26	17	0.998309	0.005346
90	39	26	18	0.999677	0.001368
90	39	26	19	0.999952	0.000275
90	39	26	20	0.999994	0.000043
90	39	26	21	0.999999	0.000005
90	39	26	22	1.000000	0.000000
90	39	26	23	1.000000	0.000000
90	39	26	24	1.000000	0.000000
90	39	26	25	1.000000	0.000000
90	39	26	26	1.000000	0.000000
90	39	27	0	0.000000	0.000000
90	39	27	1	0.000000	0.000000
90	39	27	2	0.000003	0.000003
90	39	27	3	0.000033	0.000030
90	39	27	4	0.000268	0.000235
90	39	27	5	0.001574	0.001306
90	39	27	6	0.007000	0.005426
90	39	27	7	0.024329	0.017329
90	39	27	8	0.067651	0.043322
90	39	27	9	0.153565	0.085914
90	39	27	10	0.290017	0.136452
90	39	27	11	0.464746	0.174729
90	39	27	12	0.645946	0.181201
90	39	27	13	0.798516	0.152570
90	39	27	14	0.902906	0.104390
90	39	27	15	0.960901	0.057994
90	39	27	16	0.986998	0.026098
90	39	27	17	0.996471	0.009473
90	39	27	18	0.999228	0.002757
90	39	27	19	0.999866	0.000638
90	39	27	20	0.999982	0.000116
90	39	27	21	0.999998	0.000016
90	39	27	22	1.000000	0.000002
90	39	27	23	1.000000	0.000000
90	39	27	24	1.000000	0.000000
90	39	27	25	1.000000	0.000000
90	39	27	26	1.000000	0.000000
90	39	27	27	1.000000	0.000000
90	39	28	0	0.000000	0.000000
90	39	28	1	0.000000	0.000000
90	39	28	2	0.000001	0.000001
90	39	28	3	0.000016	0.000015
90	39	28	4	0.000138	0.000122
90	39	28	5	0.000870	0.000732
90	39	28	6	0.004158	0.003289
90	39	28	7	0.015527	0.011369
90	39	28	8	0.046334	0.030807
90	39	28	9	0.112653	0.066320
90	39	28	10	0.227205	0.114552
90	39	28	11	0.387088	0.159883
90	39	28	12	0.568289	0.181201
90	39	28	13	0.735551	0.167262
90	39	28	14	0.861482	0.125931
90	39	28	15	0.938808	0.077326
90	39	28	16	0.977471	0.038663
90	39	28	17	0.993163	0.015693
90	39	28	18	0.998309	0.005146
90	39	28	19	0.999663	0.001354
90	39	28	20	0.999947	0.000283
90	39	28	21	0.999993	0.000047
90	39	28	22	0.999999	0.000006
90	39	28	23	1.000000	0.000001
90	39	28	24	1.000000	0.000000
90	39	28	25	1.000000	0.000000
90	39	28	26	1.000000	0.000000
90	39	28	27	1.000000	0.000000
90	39	28	28	1.000000	0.000000
90	39	29	0	0.000000	0.000000
90	39	29	1	0.000000	0.000000
90	39	29	2	0.000001	0.000001
90	39	29	3	0.000007	0.000007
90	39	29	4	0.000069	0.000062
90	39	29	5	0.000468	0.000399
90	39	29	6	0.002408	0.001939
90	39	29	7	0.009659	0.007252
90	39	29	8	0.030930	0.021271
90	39	29	9	0.080563	0.049633
90	39	29	10	0.173625	0.093061
90	39	29	11	0.314883	0.141258
90	39	29	12	0.489379	0.174496
90	39	29	13	0.665409	0.176030
90	39	29	14	0.810703	0.145294
90	39	29	15	0.908875	0.098172
90	39	29	16	0.963128	0.054253
90	39	29	17	0.987595	0.024467
90	39	29	18	0.996566	0.008971
90	39	29	19	0.999226	0.002660

N	n	k	x	P(x)	p(x)
90	39	29	20	0.999860	0.000633
90	39	29	21	0.999980	0.000120
90	39	29	22	0.999998	0.000018
90	39	29	23	1.000000	0.000002
90	39	29	24	1.000000	0.000000
90	39	29	25	1.000000	0.000000
90	39	29	26	1.000000	0.000000
90	39	29	27	1.000000	0.000000
90	39	29	28	1.000000	0.000000
90	39	29	29	1.000000	0.000000
90	39	30	0	0.000000	0.000000
90	39	30	1	0.000000	0.000000
90	39	30	2	0.000000	0.000000
90	39	30	3	0.000003	0.000003
90	39	30	4	0.000034	0.000030
90	39	30	5	0.000246	0.000212
90	39	30	6	0.001359	0.001113
90	39	30	7	0.005855	0.004497
90	39	30	8	0.020121	0.014265
90	39	30	9	0.056154	0.036033
90	39	30	10	0.129382	0.073229
90	39	30	11	0.250043	0.120661
90	39	30	12	0.412143	0.162100
90	39	30	13	0.590380	0.178237
90	39	30	14	0.751156	0.160777
90	39	30	15	0.870250	0.119094
90	39	30	16	0.942672	0.072422
90	39	30	17	0.978771	0.036099
90	39	30	18	0.993478	0.014707
90	39	30	19	0.998354	0.004877
90	39	30	20	0.999663	0.001308
90	39	30	21	0.999944	0.000282
90	39	30	22	0.999993	0.000048
90	39	30	23	0.999999	0.000006
90	39	30	24	1.000000	0.000001
90	39	30	25	1.000000	0.000000
90	39	30	26	1.000000	0.000000
90	39	30	27	1.000000	0.000000
90	39	30	28	1.000000	0.000000
90	39	30	29	1.000000	0.000000
90	39	30	30	1.000000	0.000000
90	39	31	0	0.000000	0.000000
90	39	31	1	0.000000	0.000000
90	39	31	2	0.000000	0.000000
90	39	31	3	0.000001	0.000001
90	39	31	4	0.000016	0.000014
90	39	31	5	0.000126	0.000110
90	39	31	6	0.000746	0.000621
90	39	31	7	0.003457	0.002710
90	39	31	8	0.012750	0.009293

N	n	k	x	P(x)	p(x)
90	39	31	9	0.038137	0.025387
90	39	31	10	0.093988	0.055851
90	39	31	11	0.193735	0.099746
90	39	31	12	0.339198	0.145463
90	39	31	13	0.513144	0.173946
90	39	31	14	0.684166	0.171022
90	39	31	15	0.822612	0.138447
90	39	31	16	0.914910	0.092298
90	39	31	17	0.965535	0.050624
90	39	31	18	0.988330	0.022796
90	39	31	19	0.996729	0.008398
90	39	31	20	0.999248	0.002520
90	39	31	21	0.999860	0.000612
90	39	31	22	0.999979	0.000119
90	39	31	23	0.999997	0.000018
90	39	31	24	1.000000	0.000002
90	39	31	25	1.000000	0.000000
90	39	31	26	1.000000	0.000000
90	39	31	27	1.000000	0.000000
90	39	31	28	1.000000	0.000000
90	39	31	29	1.000000	0.000000
90	39	31	30	1.000000	0.000000
90	39	31	31	1.000000	0.000000
90	39	32	0	0.000000	0.000000
90	39	32	1	0.000000	0.000000
90	39	32	2	0.000000	0.000000
90	39	32	3	0.000001	0.000001
90	39	32	4	0.000007	0.000007
90	39	32	5	0.000062	0.000055
90	39	32	6	0.000399	0.000337
90	39	32	7	0.001987	0.001588
90	39	32	8	0.007867	0.005880
90	39	32	9	0.025228	0.017361
90	39	32	10	0.066536	0.041308
90	39	32	11	0.146398	0.079861
90	39	32	12	0.272630	0.126233
90	39	32	13	0.436490	0.163860
90	39	32	14	0.611699	0.175209
90	39	32	15	0.766295	0.154596
90	39	32	16	0.878930	0.112634
90	39	32	17	0.946658	0.067728
90	39	32	18	0.980217	0.033559
90	39	32	19	0.993882	0.013665
90	39	32	20	0.998437	0.004555
90	39	32	21	0.999673	0.001236
90	39	32	22	0.999945	0.000271
90	39	32	23	0.999992	0.000048
90	39	32	24	0.999999	0.000007
90	39	32	25	1.000000	0.000001
90	39	32	26	1.000000	0.000000

N	n	k	x	P(x)	p(x)
90	39	32	27	1.000000	0.000000
90	39	32	28	1.000000	0.000000
90	39	32	29	1.000000	0.000000
90	39	32	30	1.000000	0.000000
90	39	32	31	1.000000	0.000000
90	39	32	32	1.000000	0.000000
90	39	33	0	0.000000	0.000000
90	39	33	1	0.000000	0.000000
90	39	33	2	0.000000	0.000000
90	39	33	3	0.000000	0.000000
90	39	33	4	0.000003	0.000003
90	39	33	5	0.000030	0.000027
90	39	33	6	0.000208	0.000177
90	39	33	7	0.001111	0.000903
90	39	33	8	0.004724	0.003613
90	39	33	9	0.016249	0.011524
90	39	33	10	0.045882	0.029634
90	39	33	11	0.107844	0.061961
90	39	33	12	0.213867	0.106023
90	39	33	13	0.363035	0.149169
90	39	33	14	0.536178	0.173142
90	39	33	15	0.702324	0.166147
90	39	33	16	0.834264	0.131940
90	39	33	17	0.920968	0.086703
90	39	33	18	0.968066	0.047098
90	39	33	19	0.989170	0.021104
90	39	33	20	0.996945	0.007775
90	39	33	21	0.999290	0.002345
90	39	33	22	0.999865	0.000576
90	39	33	23	0.999979	0.000114
90	39	33	24	0.999997	0.000018
90	39	33	25	1.000000	0.000002
90	39	33	26	1.000000	0.000000
90	39	33	27	1.000000	0.000000
90	39	33	28	1.000000	0.000000
90	39	33	29	1.000000	0.000000
90	39	33	30	1.000000	0.000000
90	39	33	31	1.000000	0.000000
90	39	33	32	1.000000	0.000000
90	39	33	33	1.000000	0.000000
90	39	34	0	0.000000	0.000000
90	39	34	1	0.000000	0.000000
90	39	34	2	0.000000	0.000000
90	39	34	3	0.000000	0.000000
90	39	34	4	0.000001	0.000001
90	39	34	5	0.000014	0.000013
90	39	34	6	0.000105	0.000091
90	39	34	7	0.000604	0.000499
90	39	34	8	0.002759	0.002155
90	39	34	9	0.010183	0.007424

N	n	k	x	P(x)	p(x)
90	39	34	10	0.030805	0.020622
90	39	34	11	0.077407	0.046601
90	39	34	12	0.163645	0.086239
90	39	34	13	0.294994	0.131348
90	39	34	14	0.460238	0.165245
90	39	34	15	0.632368	0.172130
90	39	34	16	0.781025	0.148658
90	39	34	17	0.887503	0.106478
90	39	34	18	0.950714	0.063211
90	39	34	19	0.981765	0.031051
90	39	34	20	0.994353	0.012588
90	39	34	21	0.998549	0.004196
90	39	34	22	0.999693	0.001144
90	39	34	23	0.999947	0.000254
90	39	34	24	0.999993	0.000045
90	39	34	25	0.999999	0.000006
90	39	34	26	1.000000	0.000001
90	39	34	27	1.000000	0.000000
90	39	34	28	1.000000	0.000000
90	39	34	29	1.000000	0.000000
90	39	34	30	1.000000	0.000000
90	39	34	31	1.000000	0.000000
90	39	34	32	1.000000	0.000000
90	39	34	33	1.000000	0.000000
90	39	34	34	1.000000	0.000000
90	39	35	0	0.000000	0.000000
90	39	35	1	0.000000	0.000000
90	39	35	2	0.000000	0.000000
90	39	35	3	0.000000	0.000000
90	39	35	4	0.000001	0.000001
90	39	35	5	0.000006	0.000006
90	39	35	6	0.000051	0.000045
90	39	35	7	0.000319	0.000267
90	39	35	8	0.001566	0.001247
90	39	35	9	0.006206	0.004640
90	39	35	10	0.020126	0.013920
90	39	35	11	0.054106	0.033980
90	39	35	12	0.122066	0.067960
90	39	35	13	0.234010	0.111944
90	39	35	14	0.386468	0.152458
90	39	35	15	0.558598	0.172130
90	39	35	16	0.719970	0.161372
90	39	35	17	0.845673	0.125703
90	39	35	18	0.927010	0.081337
90	39	35	19	0.970675	0.043665
90	39	35	20	0.990082	0.019407
90	39	35	21	0.997200	0.007118
90	39	35	22	0.999346	0.002146
90	39	35	23	0.999875	0.000529
90	39	35	24	0.999980	0.000106

Table for $N = 2$, $n = 1$, through $N = 100$, $n = 50$

N	n	k	x	P(x)	p(x)
90	39	35	25	0.999997	0.000017
90	39	35	26	1.000000	0.000002
90	39	35	27	1.000000	0.000000
90	39	35	28	1.000000	0.000000
90	39	35	29	1.000000	0.000000
90	39	35	30	1.000000	0.000000
90	39	35	31	1.000000	0.000000
90	39	35	32	1.000000	0.000000
90	39	35	33	1.000000	0.000000
90	39	35	34	1.000000	0.000000
90	39	35	35	1.000000	0.000000
90	39	36	0	0.000000	0.000000
90	39	36	1	0.000000	0.000000
90	39	36	2	0.000000	0.000000
90	39	36	3	0.000000	0.000000
90	39	36	4	0.000000	0.000000
90	39	36	5	0.000003	0.000003
90	39	36	6	0.000024	0.000022
90	39	36	7	0.000163	0.000139
90	39	36	8	0.000863	0.000700
90	39	36	9	0.003675	0.002812
90	39	36	10	0.012786	0.009111
90	39	36	11	0.036807	0.024021
90	39	36	12	0.088704	0.051897
90	39	36	13	0.181091	0.092387
90	39	36	14	0.317169	0.136078
90	39	36	15	0.483487	0.166318
90	39	36	16	0.652487	0.169000
90	39	36	17	0.795391	0.142905
90	39	36	18	0.895954	0.100562
90	39	36	19	0.954797	0.058843
90	39	36	20	0.983378	0.028581
90	39	36	21	0.994871	0.011493
90	39	36	22	0.998683	0.003812
90	39	36	23	0.999721	0.001038
90	39	36	24	0.999952	0.000231
90	39	36	25	0.999993	0.000042
90	39	36	26	0.999999	0.000006
90	39	36	27	1.000000	0.000001
90	39	36	28	1.000000	0.000000
90	39	36	29	1.000000	0.000000
90	39	36	30	1.000000	0.000000
90	39	36	31	1.000000	0.000000
90	39	36	32	1.000000	0.000000
90	39	36	33	1.000000	0.000000
90	39	36	34	1.000000	0.000000
90	39	36	35	1.000000	0.000000
90	39	36	36	1.000000	0.000000
90	39	37	0	0.000000	0.000000
90	39	37	1	0.000000	0.000000
90	39	37	2	0.000000	0.000000
90	39	37	3	0.000000	0.000000
90	39	37	4	0.000000	0.000000
90	39	37	5	0.000001	0.000001
90	39	37	6	0.000011	0.000010
90	39	37	7	0.000081	0.000070
90	39	37	8	0.000461	0.000380
90	39	37	9	0.002113	0.001652
90	39	37	10	0.007893	0.005780
90	39	37	11	0.024352	0.016459
90	39	37	12	0.062755	0.038404
90	39	37	13	0.136608	0.073853
90	39	37	14	0.254170	0.117562
90	39	37	15	0.409568	0.155398
90	39	37	16	0.580505	0.170938
90	39	37	17	0.737171	0.156666
90	39	37	18	0.856846	0.119675
90	39	37	19	0.933003	0.076157
90	39	37	20	0.973322	0.040318
90	39	37	21	0.991040	0.017718
90	39	37	22	0.997483	0.006443
90	39	37	23	0.999413	0.001931
90	39	37	24	0.999888	0.000474
90	39	37	25	0.999982	0.000095
90	39	37	26	0.999998	0.000015
90	39	37	27	1.000000	0.000002
90	39	37	28	1.000000	0.000000
90	39	37	29	1.000000	0.000000
90	39	37	30	1.000000	0.000000
90	39	37	31	1.000000	0.000000
90	39	37	32	1.000000	0.000000
90	39	37	33	1.000000	0.000000
90	39	37	34	1.000000	0.000000
90	39	37	35	1.000000	0.000000
90	39	37	36	1.000000	0.000000
90	39	37	37	1.000000	0.000000
90	39	38	0	0.000000	0.000000
90	39	38	1	0.000000	0.000000
90	39	38	2	0.000000	0.000000
90	39	38	3	0.000000	0.000000
90	39	38	4	0.000000	0.000000
90	39	38	5	0.000000	0.000000
90	39	38	6	0.000005	0.000004
90	39	38	7	0.000039	0.000034
90	39	38	8	0.000239	0.000200
90	39	38	9	0.001178	0.000939
90	39	38	10	0.004730	0.003552
90	39	38	11	0.015657	0.010926
90	39	38	12	0.043191	0.027535
90	39	38	13	0.100379	0.057187
90	39	38	14	0.198716	0.098338
90	39	38	15	0.339199	0.140483
90	39	38	16	0.506325	0.167126
90	39	38	17	0.672140	0.165815
90	39	38	18	0.809428	0.137288
90	39	38	19	0.904265	0.094837
90	39	38	20	0.958868	0.054603
90	39	38	21	0.985022	0.026154
90	39	38	22	0.995416	0.010394
90	39	38	23	0.998830	0.003414
90	39	38	24	0.999753	0.000923
90	39	38	25	0.999957	0.000204
90	39	38	26	0.999994	0.000037
90	39	38	27	0.999999	0.000005
90	39	38	28	1.000000	0.000001
90	39	38	29	1.000000	0.000000
90	39	38	30	1.000000	0.000000
90	39	38	31	1.000000	0.000000
90	39	38	32	1.000000	0.000000
90	39	38	33	1.000000	0.000000
90	39	38	34	1.000000	0.000000
90	39	38	35	1.000000	0.000000
90	39	38	36	1.000000	0.000000
90	39	38	37	1.000000	0.000000
90	39	38	38	1.000000	0.000000
90	39	39	0	0.000000	0.000000
90	39	39	1	0.000000	0.000000
90	39	39	2	0.000000	0.000000
90	39	39	3	0.000000	0.000000
90	39	39	4	0.000000	0.000000
90	39	39	5	0.000000	0.000000
90	39	39	6	0.000002	0.000002
90	39	39	7	0.000018	0.000016
90	39	39	8	0.000120	0.000102
90	39	39	9	0.000636	0.000517
90	39	39	10	0.002749	0.002113
90	39	39	11	0.009773	0.007024
90	39	39	12	0.028895	0.019121
90	39	39	13	0.071785	0.042890
90	39	39	14	0.151439	0.079654
90	39	39	15	0.274361	0.122922
90	39	39	16	0.432404	0.158043
90	39	39	17	0.601988	0.169584
90	39	39	18	0.753985	0.151997
90	39	39	19	0.867789	0.113804
90	39	39	20	0.938917	0.071128
90	39	39	21	0.975969	0.037052
90	39	39	22	0.992018	0.016049
90	39	39	23	0.997780	0.005762
90	39	39	24	0.999487	0.001707
90	39	39	25	0.999902	0.000415
90	39	39	26	0.999985	0.000082
90	39	39	27	0.999998	0.000013
90	39	39	28	1.000000	0.000002
90	39	39	29	1.000000	0.000000
90	39	39	30	1.000000	0.000000
90	39	39	31	1.000000	0.000000
90	39	39	32	1.000000	0.000000
90	39	39	33	1.000000	0.000000
90	39	39	34	1.000000	0.000000
90	39	39	35	1.000000	0.000000
90	39	39	36	1.000000	0.000000
90	39	39	37	1.000000	0.000000
90	39	39	38	1.000000	0.000000
90	39	39	39	1.000000	0.000000
90	40	1	0	0.555556	0.555556
90	40	1	1	1.000000	0.444444
90	40	2	0	0.305868	0.305868
90	40	2	1	0.805243	0.499376
90	40	2	2	1.000000	0.194757
90	40	3	0	0.166837	0.166837
90	40	3	1	0.583929	0.417092
90	40	3	2	0.915901	0.331971
90	40	3	3	1.000000	0.084099
90	40	4	0	0.090130	0.090130
90	40	4	1	0.396957	0.306826
90	40	4	2	0.770902	0.373945
90	40	4	3	0.964233	0.193332
90	40	4	4	1.000000	0.035766
90	40	5	0	0.048209	0.048209
90	40	5	1	0.257815	0.209605
90	40	5	2	0.605670	0.347856
90	40	5	3	0.881056	0.275386
90	40	5	4	0.985028	0.103972
90	40	5	5	1.000000	0.014972
90	40	6	0	0.025523	0.025523
90	40	6	1	0.161643	0.136120
90	40	6	2	0.450158	0.288516
90	40	6	3	0.761182	0.311024
90	40	6	4	0.940993	0.179811
90	40	6	5	0.993835	0.052842
90	40	6	6	1.000000	0.006165
90	40	7	0	0.013369	0.013369
90	40	7	1	0.098444	0.085075
90	40	7	2	0.319639	0.221195
90	40	7	3	0.624183	0.304544
90	40	7	4	0.863931	0.239748
90	40	7	5	0.971817	0.107886
90	40	7	6	0.997505	0.025687
90	40	7	7	1.000000	0.002495

N	n	k	x	P(x)	p(x)
90	40	8	0	0.006926	0.006926
90	40	8	1	0.058469	0.051543
90	40	8	2	0.218369	0.159900
90	40	8	3	0.488423	0.270054
90	40	8	4	0.759944	0.271521
90	40	8	5	0.926323	0.166379
90	40	8	6	0.986982	0.060659
90	40	8	7	0.999008	0.012026
90	40	8	8	1.000000	0.000992
90	40	9	0	0.003547	0.003547
90	40	9	1	0.033955	0.030407
90	40	9	2	0.144269	0.110314
90	40	9	3	0.366569	0.222300
90	40	9	4	0.640740	0.274170
90	40	9	5	0.855308	0.214568
90	40	9	6	0.961831	0.106523
90	40	9	7	0.994168	0.032337
90	40	9	8	0.999613	0.005445
90	40	9	9	1.000000	0.000387
90	40	10	0	0.001796	0.001796
90	40	10	1	0.019314	0.017519
90	40	10	2	0.092517	0.073202
90	40	10	3	0.265025	0.172508
90	40	10	4	0.518886	0.253861
90	40	10	5	0.762593	0.243707
90	40	10	6	0.917117	0.154524
90	40	10	7	0.980994	0.063876
90	40	10	8	0.997462	0.016468
90	40	10	9	0.999852	0.002390
90	40	10	10	1.000000	0.000148
90	40	11	0	0.000898	0.000898
90	40	11	1	0.010774	0.009876
90	40	11	2	0.057745	0.046972
90	40	11	3	0.185240	0.127494
90	40	11	4	0.404648	0.219409
90	40	11	5	0.655971	0.251323
90	40	11	6	0.851445	0.195473
90	40	11	7	0.954645	0.103200
90	40	11	8	0.990875	0.036230
90	40	11	9	0.998926	0.008051
90	40	11	10	0.999944	0.001019
90	40	11	11	1.000000	0.000056
90	40	12	0	0.000443	0.000443
90	40	12	1	0.005898	0.005455
90	40	12	2	0.035152	0.029253
90	40	12	3	0.125527	0.090376
90	40	12	4	0.304665	0.179138
90	40	12	5	0.544626	0.239961
90	40	12	6	0.767317	0.222691
90	40	12	7	0.911536	0.144219

N	n	k	x	P(x)	p(x)
90	40	12	8	0.976199	0.064663
90	40	12	9	0.995766	0.019567
90	40	12	10	0.999557	0.003791
90	40	12	11	0.999979	0.000422
90	40	12	12	1.000000	0.000020
90	40	13	0	0.000216	0.000216
90	40	13	1	0.003171	0.002955
90	40	13	2	0.020900	0.017729
90	40	13	3	0.082657	0.061757
90	40	13	4	0.221986	0.139329
90	40	13	5	0.436951	0.214965
90	40	13	6	0.670246	0.233295
90	40	13	7	0.850520	0.180274
90	40	13	8	0.949671	0.099151
90	40	13	9	0.987990	0.038319
90	40	13	10	0.998099	0.010110
90	40	13	11	0.999823	0.001723
90	40	13	12	0.999993	0.000170
90	40	13	13	1.000000	0.000007
90	40	14	0	0.000104	0.000104
90	40	14	1	0.001674	0.001570
90	40	14	2	0.012151	0.010476
90	40	14	3	0.052981	0.040831
90	40	14	4	0.156845	0.103864
90	40	14	5	0.339240	0.182395
90	40	14	6	0.567233	0.227993
90	40	14	7	0.773260	0.206027
90	40	14	8	0.908465	0.135205
90	40	14	9	0.972562	0.064097
90	40	14	10	0.994160	0.021598
90	40	14	11	0.999174	0.005013
90	40	14	12	0.999931	0.000757
90	40	14	13	0.999997	0.000067
90	40	14	14	1.000000	0.000003
90	40	15	0	0.000049	0.000049
90	40	15	1	0.000868	0.000819
90	40	15	2	0.006912	0.006044
90	40	15	3	0.033103	0.026191
90	40	15	4	0.107646	0.074543
90	40	15	5	0.255242	0.147596
90	40	15	6	0.465236	0.209994
90	40	15	7	0.683801	0.218565
90	40	15	8	0.851537	0.167736
90	40	15	9	0.946418	0.094881
90	40	15	10	0.985635	0.039217
90	40	15	11	0.997261	0.011626
90	40	15	12	0.999652	0.002391
90	40	15	13	0.999974	0.000322
90	40	15	14	0.999999	0.000025
90	40	15	15	1.000000	0.000001

N	n	k	x	P(x)	p(x)
90	40	16	0	0.000023	0.000023
90	40	16	1	0.000442	0.000419
90	40	16	2	0.003850	0.003408
90	40	16	3	0.020182	0.016332
90	40	16	4	0.071866	0.051683
90	40	16	5	0.186364	0.114498
90	40	16	6	0.370039	0.183675
90	40	16	7	0.587632	0.217594
90	40	16	8	0.779969	0.192337
90	40	16	9	0.907200	0.127231
90	40	16	10	0.969948	0.062748
90	40	16	11	0.992765	0.022817
90	40	16	12	0.998759	0.005994
90	40	16	13	0.999858	0.001099
90	40	16	14	0.999990	0.000132
90	40	16	15	1.000000	0.000009
90	40	16	16	1.000000	0.000000
90	40	17	0	0.000011	0.000011
90	40	17	1	0.000221	0.000211
90	40	17	2	0.002100	0.001879
90	40	17	3	0.012016	0.009916
90	40	17	4	0.046722	0.034706
90	40	17	5	0.132209	0.085487
90	40	17	6	0.285648	0.153438
90	40	17	7	0.490597	0.204950
90	40	17	8	0.696797	0.206199
90	40	17	9	0.853901	0.157104
90	40	17	10	0.944510	0.090609
90	40	17	11	0.983823	0.039314
90	40	17	12	0.996491	0.012668
90	40	17	13	0.999457	0.002966
90	40	17	14	0.999944	0.000487
90	40	17	15	0.999996	0.000053
90	40	17	16	1.000000	0.000003
90	40	17	17	1.000000	0.000000
90	40	18	0	0.000005	0.000005
90	40	18	1	0.000109	0.000104
90	40	18	2	0.001122	0.001013
90	40	18	3	0.006990	0.005868
90	40	18	4	0.029607	0.022617
90	40	18	5	0.091223	0.061615
90	40	18	6	0.214183	0.122961
90	40	18	7	0.397949	0.183766
90	40	18	8	0.606408	0.208459
90	40	18	9	0.787185	0.180777
90	40	18	10	0.907273	0.120088
90	40	18	11	0.968206	0.060933
90	40	18	12	0.991632	0.023427
90	40	18	13	0.998360	0.006728
90	40	18	14	0.999770	0.001410

N	n	k	x	P(x)	p(x)
90	40	18	15	0.999978	0.000208
90	40	18	16	0.999999	0.000020
90	40	18	17	1.000000	0.000001
90	40	18	18	1.000000	0.000000
90	40	19	0	0.000002	0.000002
90	40	19	1	0.000052	0.000050
90	40	19	2	0.000587	0.000535
90	40	19	3	0.003975	0.003387
90	40	19	4	0.018299	0.014324
90	40	19	5	0.061271	0.042972
90	40	19	6	0.156118	0.094848
90	40	19	7	0.313723	0.157604
90	40	19	8	0.513759	0.200036
90	40	19	9	0.709351	0.195591
90	40	19	10	0.857237	0.147886
90	40	19	11	0.943663	0.086427
90	40	19	12	0.982522	0.038859
90	40	19	13	0.995837	0.013315
90	40	19	14	0.999261	0.003424
90	40	19	15	0.999906	0.000645
90	40	19	16	0.999992	0.000086
90	40	19	17	0.999999	0.000008
90	40	19	18	1.000000	0.000000
90	40	19	19	1.000000	0.000000
90	40	20	0	0.000001	0.000001
90	40	20	1	0.000025	0.000024
90	40	20	2	0.000301	0.000276
90	40	20	3	0.002209	0.001908
90	40	20	4	0.011036	0.008826
90	40	20	5	0.040087	0.029051
90	40	20	6	0.110698	0.070611
90	40	20	7	0.240470	0.129772
90	40	20	8	0.423602	0.183132
90	40	20	9	0.623952	0.200349
90	40	20	10	0.794749	0.170798
90	40	20	11	0.908362	0.113613
90	40	20	12	0.967197	0.058835
90	40	20	13	0.990774	0.023576
90	40	20	14	0.998007	0.007234
90	40	20	15	0.999679	0.001672
90	40	20	16	0.999963	0.000284
90	40	20	17	0.999997	0.000034
90	40	20	18	1.000000	0.000003
90	40	20	19	1.000000	0.000000
90	40	20	20	1.000000	0.000000
90	40	21	0	0.000000	0.000000
90	40	21	1	0.000011	0.000011
90	40	21	2	0.000151	0.000140
90	40	21	3	0.001201	0.001050
90	40	21	4	0.006497	0.005296

Table for $N = 2$, $n = 1$, through $N = 100$, $n = 50$

N	n	k	x	P(x)	p(x)
90	40	21	5	0.025562	0.019065
90	40	21	6	0.076402	0.050840
90	40	21	7	0.179292	0.102891
90	40	21	8	0.339885	0.160593
90	40	21	9	0.535225	0.195341
90	40	21	10	0.721550	0.186325
90	40	21	11	0.861294	0.139744
90	40	21	12	0.943663	0.082369
90	40	21	13	0.981680	0.038017
90	40	21	14	0.995320	0.013641
90	40	21	15	0.999082	0.003761
90	40	21	16	0.999866	0.000784
90	40	21	17	0.999986	0.000120
90	40	21	18	0.999999	0.000013
90	40	21	19	1.000000	0.000001
90	40	21	20	1.000000	0.000000
90	40	21	21	1.000000	0.000000
90	40	22	0	0.000000	0.000000
90	40	22	1	0.000005	0.000005
90	40	22	2	0.000074	0.000069
90	40	22	3	0.000638	0.000564
90	40	22	4	0.003733	0.003096
90	40	22	5	0.015891	0.012157
90	40	22	6	0.051350	0.035459
90	40	22	7	0.130084	0.078734
90	40	22	8	0.265407	0.135323
90	40	22	9	0.447464	0.182057
90	40	22	10	0.640539	0.193076
90	40	22	11	0.802561	0.162022
90	40	22	12	0.910238	0.107677
90	40	22	13	0.966804	0.056566
90	40	22	14	0.990181	0.023377
90	40	22	15	0.997719	0.007539
90	40	22	16	0.999593	0.001874
90	40	22	17	0.999946	0.000353
90	40	22	18	0.999995	0.000049
90	40	22	19	1.000000	0.000005
90	40	22	20	1.000000	0.000000
90	40	22	21	1.000000	0.000000
90	40	22	22	1.000000	0.000000
90	40	23	0	0.000000	0.000000
90	40	23	1	0.000002	0.000002
90	40	23	2	0.000036	0.000033
90	40	23	3	0.000331	0.000295
90	40	23	4	0.002095	0.001763
90	40	23	5	0.009633	0.007539
90	40	23	6	0.033620	0.023987
90	40	23	7	0.091875	0.058254
90	40	23	8	0.201725	0.109851
90	40	23	9	0.364467	0.162742

N	n	k	x	P(x)	p(x)
90	40	23	10	0.555359	0.190892
90	40	23	11	0.733464	0.178105
90	40	23	12	0.865901	0.132437
90	40	23	13	0.944344	0.078443
90	40	23	14	0.981242	0.036898
90	40	23	15	0.994948	0.013705
90	40	23	16	0.998932	0.003984
90	40	23	17	0.999826	0.000895
90	40	23	18	0.999979	0.000152
90	40	23	19	0.999998	0.000019
90	40	23	20	1.000000	0.000002
90	40	23	21	1.000000	0.000000
90	40	23	22	1.000000	0.000000
90	40	23	23	1.000000	0.000000
90	40	24	0	0.000000	0.000000
90	40	24	1	0.000001	0.000001
90	40	24	2	0.000017	0.000016
90	40	24	3	0.000168	0.000151
90	40	24	4	0.001147	0.000979
90	40	24	5	0.005695	0.004548
90	40	24	6	0.021448	0.015753
90	40	24	7	0.063182	0.041734
90	40	24	8	0.149259	0.086077
90	40	24	9	0.289169	0.139909
90	40	24	10	0.469885	0.180716
90	40	24	11	0.656374	0.186489
90	40	24	12	0.810554	0.154180
90	40	24	13	0.912732	0.102179
90	40	24	14	0.966924	0.054191
90	40	24	15	0.989834	0.022910
90	40	24	16	0.997504	0.007671
90	40	24	17	0.999519	0.002015
90	40	24	18	0.999929	0.000410
90	40	24	19	0.999992	0.000063
90	40	24	20	0.999999	0.000007
90	40	24	21	1.000000	0.000001
90	40	24	22	1.000000	0.000000
90	40	24	23	1.000000	0.000000
90	40	24	24	1.000000	0.000000
90	40	25	0	0.000000	0.000000
90	40	25	1	0.000000	0.000000
90	40	25	2	0.000008	0.000007
90	40	25	3	0.000083	0.000075
90	40	25	4	0.000613	0.000530
90	40	25	5	0.003283	0.002670
90	40	25	6	0.013333	0.010050
90	40	25	7	0.042315	0.028982
90	40	25	8	0.107525	0.065210
90	40	25	9	0.223454	0.115929
90	40	25	10	0.387741	0.164288

N	n	k	x	P(x)	p(x)
90	40	25	11	0.574432	0.186690
90	40	25	12	0.745144	0.170712
90	40	25	13	0.870932	0.125788
90	40	25	14	0.945576	0.074643
90	40	25	15	0.981156	0.035580
90	40	25	16	0.994715	0.013559
90	40	25	17	0.998817	0.004102
90	40	25	18	0.999792	0.000975
90	40	25	19	0.999972	0.000180
90	40	25	20	0.999997	0.000025
90	40	25	21	1.000000	0.000003
90	40	25	22	1.000000	0.000000
90	40	25	23	1.000000	0.000000
90	40	25	24	1.000000	0.000000
90	40	25	25	1.000000	0.000000
90	40	26	0	0.000000	0.000000
90	40	26	1	0.000000	0.000000
90	40	26	2	0.000003	0.000003
90	40	26	3	0.000040	0.000037
90	40	26	4	0.000320	0.000279
90	40	26	5	0.001845	0.001526
90	40	26	6	0.008076	0.006231
90	40	26	7	0.027601	0.019525
90	40	26	8	0.075422	0.047821
90	40	26	9	0.168165	0.092743
90	40	26	10	0.311916	0.143752
90	40	26	11	0.491139	0.179223
90	40	26	12	0.671606	0.180467
90	40	26	13	0.818682	0.147075
90	40	26	14	0.915718	0.097037
90	40	26	15	0.967471	0.051753
90	40	26	16	0.989709	0.022238
90	40	26	17	0.997366	0.007657
90	40	26	18	0.999462	0.002097
90	40	26	19	0.999914	0.000452
90	40	26	20	0.999989	0.000075
90	40	26	21	0.999999	0.000010
90	40	26	22	1.000000	0.000001
90	40	26	23	1.000000	0.000000
90	40	26	24	1.000000	0.000000
90	40	26	25	1.000000	0.000000
90	40	26	26	1.000000	0.000000
90	40	27	0	0.000000	0.000000
90	40	27	1	0.000000	0.000000
90	40	27	2	0.000001	0.000001
90	40	27	3	0.000019	0.000017
90	40	27	4	0.000162	0.000143
90	40	27	5	0.001011	0.000849
90	40	27	6	0.004766	0.003755
90	40	27	7	0.017534	0.012767

N	n	k	x	P(x)	p(x)
90	40	27	8	0.051511	0.033978
90	40	27	9	0.123242	0.071731
90	40	27	10	0.244533	0.121290
90	40	27	11	0.409929	0.165396
90	40	27	12	0.592652	0.182723
90	40	27	13	0.756634	0.163982
90	40	27	14	0.876297	0.119663
90	40	27	15	0.947255	0.070958
90	40	27	16	0.981369	0.034114
90	40	27	17	0.994614	0.013244
90	40	27	18	0.998741	0.004128
90	40	27	19	0.999766	0.001024
90	40	27	20	0.999966	0.000200
90	40	27	21	0.999996	0.000030
90	40	27	22	1.000000	0.000003
90	40	27	23	1.000000	0.000000
90	40	27	24	1.000000	0.000000
90	40	27	25	1.000000	0.000000
90	40	27	26	1.000000	0.000000
90	40	27	27	1.000000	0.000000
90	40	28	0	0.000000	0.000000
90	40	28	1	0.000000	0.000000
90	40	28	2	0.000001	0.000001
90	40	28	3	0.000009	0.000008
90	40	28	4	0.000080	0.000072
90	40	28	5	0.000540	0.000459
90	40	28	6	0.002740	0.002200
90	40	28	7	0.010846	0.008106
90	40	28	8	0.034253	0.023407
90	40	28	9	0.087946	0.053693
90	40	28	10	0.186775	0.098829
90	40	28	11	0.333794	0.147019
90	40	28	12	0.511442	0.177648
90	40	28	13	0.686356	0.174915
90	40	28	14	0.826912	0.140556
90	40	28	15	0.919097	0.092185
90	40	28	16	0.968374	0.049276
90	40	28	17	0.989779	0.021405
90	40	28	18	0.997300	0.007522
90	40	28	19	0.999424	0.002124
90	40	28	20	0.999902	0.000478
90	40	28	21	0.999987	0.000085
90	40	28	22	0.999999	0.000012
90	40	28	23	1.000000	0.000001
90	40	28	24	1.000000	0.000000
90	40	28	25	1.000000	0.000000
90	40	28	26	1.000000	0.000000
90	40	28	27	1.000000	0.000000
90	40	28	28	1.000000	0.000000
90	40	29	0	0.000000	0.000000

N	n	k	x	P(x)	p(x)
90	40	29	1	0.000000	0.000000
90	40	29	2	0.000000	0.000000
90	40	29	3	0.000004	0.000004
90	40	29	4	0.000039	0.000035
90	40	29	5	0.000280	0.000242
90	40	29	6	0.001533	0.001253
90	40	29	7	0.006531	0.004998
90	40	29	8	0.022172	0.015641
90	40	29	9	0.061099	0.038928
90	40	29	10	0.138955	0.077855
90	40	29	11	0.265027	0.126072
90	40	29	12	0.431214	0.166186
90	40	29	13	0.610184	0.178970
90	40	29	14	0.767969	0.157786
90	40	29	15	0.881926	0.113956
90	40	29	16	0.949299	0.067373
90	40	29	17	0.981838	0.032539
90	40	29	18	0.994631	0.012793
90	40	29	19	0.998705	0.004074
90	40	29	20	0.999748	0.001043
90	40	29	21	0.999961	0.000213
90	40	29	22	0.999995	0.000034
90	40	29	23	0.999999	0.000004
90	40	29	24	1.000000	0.000000
90	40	29	25	1.000000	0.000000
90	40	29	26	1.000000	0.000000
90	40	29	27	1.000000	0.000000
90	40	29	28	1.000000	0.000000
90	40	29	29	1.000000	0.000000
90	40	30	0	0.000000	0.000000
90	40	30	1	0.000000	0.000000
90	40	30	2	0.000000	0.000000
90	40	30	3	0.000002	0.000002
90	40	30	4	0.000018	0.000016
90	40	30	5	0.000142	0.000124
90	40	30	6	0.000835	0.000693
90	40	30	7	0.003827	0.002992
90	40	30	8	0.013967	0.010140
90	40	30	9	0.041316	0.027350
90	40	30	10	0.100665	0.059349
90	40	30	11	0.205091	0.104426
90	40	30	12	0.354931	0.149840
90	40	30	13	0.530967	0.176036
90	40	30	14	0.700717	0.169749
90	40	30	15	0.835222	0.134506
90	40	30	16	0.922791	0.087569
90	40	30	17	0.969569	0.046778
90	40	30	18	0.990017	0.020448
90	40	30	19	0.997302	0.007285
90	40	30	20	0.999406	0.002104

N	n	k	x	P(x)	p(x)
90	40	30	21	0.999895	0.000489
90	40	30	22	0.999985	0.000090
90	40	30	23	0.999998	0.000013
90	40	30	24	1.000000	0.000001
90	40	30	25	1.000000	0.000000
90	40	30	26	1.000000	0.000000
90	40	30	27	1.000000	0.000000
90	40	30	28	1.000000	0.000000
90	40	30	29	1.000000	0.000000
90	40	30	30	1.000000	0.000000
90	40	31	0	0.000000	0.000000
90	40	31	1	0.000000	0.000000
90	40	31	2	0.000000	0.000000
90	40	31	3	0.000001	0.000001
90	40	31	4	0.000008	0.000008
90	40	31	5	0.000070	0.000061
90	40	31	6	0.000442	0.000372
90	40	31	7	0.002181	0.001739
90	40	31	8	0.008559	0.006378
90	40	31	9	0.027186	0.018627
90	40	31	10	0.070991	0.043805
90	40	31	11	0.154619	0.083628
90	40	31	12	0.285006	0.130387
90	40	31	13	0.451751	0.166745
90	40	31	14	0.627159	0.175407
90	40	31	15	0.779178	0.152020
90	40	31	16	0.887764	0.108586
90	40	31	17	0.951638	0.063874
90	40	31	18	0.982520	0.030882
90	40	31	19	0.994753	0.012233
90	40	31	20	0.998705	0.003952
90	40	31	21	0.999740	0.001035
90	40	31	22	0.999958	0.000218
90	40	31	23	0.999995	0.000037
90	40	31	24	0.999999	0.000005
90	40	31	25	1.000000	0.000000
90	40	31	26	1.000000	0.000000
90	40	31	27	1.000000	0.000000
90	40	31	28	1.000000	0.000000
90	40	31	29	1.000000	0.000000
90	40	31	30	1.000000	0.000000
90	40	31	31	1.000000	0.000000
90	40	32	0	0.000000	0.000000
90	40	32	1	0.000000	0.000000
90	40	32	2	0.000000	0.000000
90	40	32	3	0.000000	0.000000
90	40	32	4	0.000004	0.000003
90	40	32	5	0.000033	0.000030
90	40	32	6	0.000227	0.000194
90	40	32	7	0.001209	0.000981

N	n	k	x	P(x)	p(x)
90	40	32	8	0.005100	0.003891
90	40	32	9	0.017399	0.012299
90	40	32	10	0.048717	0.031318
90	40	32	11	0.113513	0.064796
90	40	32	12	0.223127	0.109614
90	40	32	13	0.375444	0.152317
90	40	32	14	0.549860	0.174416
90	40	32	15	0.714763	0.164903
90	40	32	16	0.843593	0.128830
90	40	32	17	0.926738	0.083144
90	40	32	18	0.971004	0.044267
90	40	32	19	0.990398	0.019394
90	40	32	20	0.997365	0.006967
90	40	32	21	0.999407	0.002041
90	40	32	22	0.999891	0.000485
90	40	32	23	0.999984	0.000093
90	40	32	24	0.999998	0.000014
90	40	32	25	1.000000	0.000002
90	40	32	26	1.000000	0.000000
90	40	32	27	1.000000	0.000000
90	40	32	28	1.000000	0.000000
90	40	32	29	1.000000	0.000000
90	40	32	30	1.000000	0.000000
90	40	32	31	1.000000	0.000000
90	40	32	32	1.000000	0.000000
90	40	33	0	0.000000	0.000000
90	40	33	1	0.000000	0.000000
90	40	33	2	0.000000	0.000000
90	40	33	3	0.000000	0.000000
90	40	33	4	0.000002	0.000001
90	40	33	5	0.000015	0.000014
90	40	33	6	0.000114	0.000098
90	40	33	7	0.000650	0.000537
90	40	33	8	0.002953	0.002303
90	40	33	9	0.010825	0.007872
90	40	33	10	0.032518	0.021693
90	40	33	11	0.081115	0.048597
90	40	33	12	0.170210	0.089095
90	40	33	13	0.304538	0.134328
90	40	33	14	0.471674	0.167136
90	40	33	15	0.643684	0.172011
90	40	33	16	0.790284	0.146600
90	40	33	17	0.893767	0.103482
90	40	33	18	0.954214	0.060447
90	40	33	19	0.983376	0.029163
90	40	33	20	0.994963	0.011586
90	40	33	21	0.998738	0.003775
90	40	33	22	0.999741	0.001003
90	40	33	23	0.999957	0.000216
90	40	33	24	0.999994	0.000037

N	n	k	x	P(x)	p(x)
90	40	33	25	0.999999	0.000005
90	40	33	26	1.000000	0.000001
90	40	33	27	1.000000	0.000000
90	40	33	28	1.000000	0.000000
90	40	33	29	1.000000	0.000000
90	40	33	30	1.000000	0.000000
90	40	33	31	1.000000	0.000000
90	40	33	32	1.000000	0.000000
90	40	33	33	1.000000	0.000000
90	40	34	0	0.000000	0.000000
90	40	34	1	0.000000	0.000000
90	40	34	2	0.000000	0.000000
90	40	34	3	0.000000	0.000000
90	40	34	4	0.000001	0.000001
90	40	34	5	0.000007	0.000006
90	40	34	6	0.000055	0.000048
90	40	34	7	0.000340	0.000285
90	40	34	8	0.001660	0.001321
90	40	34	9	0.006544	0.004884
90	40	34	10	0.021101	0.014557
90	40	34	11	0.056390	0.035290
90	40	34	12	0.126444	0.070054
90	40	34	13	0.240909	0.114465
90	40	34	14	0.395436	0.154527
90	40	34	15	0.568241	0.172805
90	40	34	16	0.728558	0.160317
90	40	34	17	0.852011	0.123453
90	40	34	18	0.930883	0.078872
90	40	34	19	0.972632	0.041749
90	40	34	20	0.990897	0.018265
90	40	34	21	0.997479	0.006582
90	40	34	22	0.999424	0.001945
90	40	34	23	0.999892	0.000468
90	40	34	24	0.999984	0.000091
90	40	34	25	0.999998	0.000014
90	40	34	26	1.000000	0.000002
90	40	34	27	1.000000	0.000000
90	40	34	28	1.000000	0.000000
90	40	34	29	1.000000	0.000000
90	40	34	30	1.000000	0.000000
90	40	34	31	1.000000	0.000000
90	40	34	32	1.000000	0.000000
90	40	34	33	1.000000	0.000000
90	40	34	34	1.000000	0.000000
90	40	35	0	0.000000	0.000000
90	40	35	1	0.000000	0.000000
90	40	35	2	0.000000	0.000000
90	40	35	3	0.000000	0.000000
90	40	35	4	0.000000	0.000000
90	40	35	5	0.000003	0.000003

Table for $N = 2$, $n = 1$, through $N = 100$, $n = 50$

N	n	k	x	P(x)	p(x)
90	40	35	6	0.000026	0.000023
90	40	35	7	0.000172	0.000146
90	40	35	8	0.000906	0.000734
90	40	35	9	0.003841	0.002935
90	40	35	10	0.013303	0.009462
90	40	35	11	0.038115	0.024813
90	40	35	12	0.091417	0.053302
90	40	35	13	0.185721	0.094303
90	40	35	14	0.323691	0.137971
90	40	35	15	0.491096	0.167405
90	40	35	16	0.659851	0.168755
90	40	35	17	0.801307	0.141456
90	40	35	18	0.899898	0.098591
90	40	35	19	0.956976	0.057079
90	40	35	20	0.984374	0.027398
90	40	35	21	0.995246	0.010872
90	40	35	22	0.998799	0.003553
90	40	35	23	0.999750	0.000951
90	40	35	24	0.999958	0.000207
90	40	35	25	0.999994	0.000036
90	40	35	26	0.999999	0.000005
90	40	35	27	1.000000	0.000001
90	40	35	28	1.000000	0.000000
90	40	35	29	1.000000	0.000000
90	40	35	30	1.000000	0.000000
90	40	35	31	1.000000	0.000000
90	40	35	32	1.000000	0.000000
90	40	35	33	1.000000	0.000000
90	40	35	34	1.000000	0.000000
90	40	35	35	1.000000	0.000000
90	40	36	0	0.000000	0.000000
90	40	36	1	0.000000	0.000000
90	40	36	2	0.000000	0.000000
90	40	36	3	0.000000	0.000000
90	40	36	4	0.000000	0.000000
90	40	36	5	0.000001	0.000001
90	40	36	6	0.000012	0.000010
90	40	36	7	0.000084	0.000073
90	40	36	8	0.000479	0.000394
90	40	36	9	0.002186	0.001708
90	40	36	10	0.008141	0.005955
90	40	36	11	0.025032	0.016891
90	40	36	12	0.064282	0.039250
90	40	36	13	0.139426	0.075144
90	40	36	14	0.258469	0.119043
90	40	36	15	0.415003	0.156534
90	40	36	16	0.586212	0.171209
90	40	36	17	0.742153	0.155940
90	40	36	18	0.860461	0.118309
90	40	36	19	0.935183	0.074721

N	n	k	x	P(x)	p(x)
90	40	36	20	0.974411	0.039229
90	40	36	21	0.991490	0.017079
90	40	36	22	0.997636	0.006146
90	40	36	23	0.999456	0.001820
90	40	36	24	0.999897	0.000441
90	40	36	25	0.999984	0.000087
90	40	36	26	0.999998	0.000014
90	40	36	27	1.000000	0.000002
90	40	36	28	1.000000	0.000000
90	40	36	29	1.000000	0.000000
90	40	36	30	1.000000	0.000000
90	40	36	31	1.000000	0.000000
90	40	36	32	1.000000	0.000000
90	40	36	33	1.000000	0.000000
90	40	36	34	1.000000	0.000000
90	40	36	35	1.000000	0.000000
90	40	36	36	1.000000	0.000000
90	40	37	0	0.000000	0.000000
90	40	37	1	0.000000	0.000000
90	40	37	2	0.000000	0.000000
90	40	37	3	0.000000	0.000000
90	40	37	4	0.000000	0.000000
90	40	37	5	0.000001	0.000001
90	40	37	6	0.000005	0.000005
90	40	37	7	0.000040	0.000035
90	40	37	8	0.000245	0.000205
90	40	37	9	0.001206	0.000961
90	40	37	10	0.004833	0.003627
90	40	37	11	0.015961	0.011128
90	40	37	12	0.043930	0.027969
90	40	37	13	0.101854	0.057924
90	40	37	14	0.201152	0.099298
90	40	37	15	0.342534	0.141382
90	40	37	16	0.510119	0.167586
90	40	37	17	0.675734	0.165614
90	40	37	18	0.812261	0.136528
90	40	37	19	0.906124	0.093863
90	40	37	20	0.959882	0.053758
90	40	37	21	0.985481	0.025599
90	40	37	22	0.995588	0.010107
90	40	37	23	0.998883	0.003296
90	40	37	24	0.999767	0.000883
90	40	37	25	0.999960	0.000193
90	40	37	26	0.999994	0.000034
90	40	37	27	0.999999	0.000005
90	40	37	28	1.000000	0.000001
90	40	37	29	1.000000	0.000000
90	40	37	30	1.000000	0.000000
90	40	37	31	1.000000	0.000000
90	40	37	32	1.000000	0.000000

N	n	k	x	P(x)	p(x)
90	40	37	33	1.000000	0.000000
90	40	37	34	1.000000	0.000000
90	40	37	35	1.000000	0.000000
90	40	37	36	1.000000	0.000000
90	40	37	37	1.000000	0.000000
90	40	38	0	0.000000	0.000000
90	40	38	1	0.000000	0.000000
90	40	38	2	0.000000	0.000000
90	40	38	3	0.000000	0.000000
90	40	38	4	0.000000	0.000000
90	40	38	5	0.000000	0.000000
90	40	38	6	0.000002	0.000002
90	40	38	7	0.000018	0.000016
90	40	38	8	0.000121	0.000103
90	40	38	9	0.000644	0.000523
90	40	38	10	0.002780	0.002136
90	40	38	11	0.009872	0.007092
90	40	38	12	0.029154	0.019282
90	40	38	13	0.072346	0.043192
90	40	38	14	0.152440	0.080094
90	40	38	15	0.275844	0.123404
90	40	38	16	0.434231	0.158387
90	40	38	17	0.603863	0.169632
90	40	38	18	0.755589	0.151726
90	40	38	19	0.868933	0.113344
90	40	38	20	0.939596	0.070663
90	40	38	21	0.976304	0.036708
90	40	38	22	0.992155	0.015851
90	40	38	23	0.997826	0.005671
90	40	38	24	0.999500	0.001674
90	40	38	25	0.999905	0.000405
90	40	38	26	0.999985	0.000080
90	40	38	27	0.999998	0.000013
90	40	38	28	1.000000	0.000002
90	40	38	29	1.000000	0.000000
90	40	38	30	1.000000	0.000000
90	40	38	31	1.000000	0.000000
90	40	38	32	1.000000	0.000000
90	40	38	33	1.000000	0.000000
90	40	38	34	1.000000	0.000000
90	40	38	35	1.000000	0.000000
90	40	38	36	1.000000	0.000000
90	40	38	37	1.000000	0.000000
90	40	38	38	1.000000	0.000000
90	40	39	0	0.000000	0.000000
90	40	39	1	0.000000	0.000000
90	40	39	2	0.000000	0.000000
90	40	39	3	0.000000	0.000000
90	40	39	4	0.000000	0.000000
90	40	39	5	0.000000	0.000000

N	n	k	x	P(x)	p(x)
90	40	39	6	0.000001	0.000001
90	40	39	7	0.000008	0.000007
90	40	39	8	0.000058	0.000050
90	40	39	9	0.000332	0.000274
90	40	39	10	0.001548	0.001215
90	40	39	11	0.005917	0.004369
90	40	39	12	0.018772	0.012855
90	40	39	13	0.049919	0.031148
90	40	39	14	0.112393	0.062473
90	40	39	15	0.216515	0.104122
90	40	39	16	0.361130	0.144614
90	40	39	17	0.528834	0.167704
90	40	39	18	0.691398	0.162564
90	40	39	19	0.823160	0.131762
90	40	39	20	0.912418	0.089258
90	40	39	21	0.962892	0.050473
90	40	39	22	0.986668	0.023777
90	40	39	23	0.995972	0.009304
90	40	39	24	0.998985	0.003013
90	40	39	25	0.999788	0.000803
90	40	39	26	0.999964	0.000175
90	40	39	27	0.999995	0.000031
90	40	39	28	0.999999	0.000004
90	40	39	29	1.000000	0.000000
90	40	39	30	1.000000	0.000000
90	40	39	31	1.000000	0.000000
90	40	39	32	1.000000	0.000000
90	40	39	33	1.000000	0.000000
90	40	39	34	1.000000	0.000000
90	40	39	35	1.000000	0.000000
90	40	39	36	1.000000	0.000000
90	40	39	37	1.000000	0.000000
90	40	39	38	1.000000	0.000000
90	40	39	39	1.000000	0.000000
90	40	40	0	0.000000	0.000000
90	40	40	1	0.000000	0.000000
90	40	40	2	0.000000	0.000000
90	40	40	3	0.000000	0.000000
90	40	40	4	0.000000	0.000000
90	40	40	5	0.000000	0.000000
90	40	40	6	0.000000	0.000000
90	40	40	7	0.000003	0.000003
90	40	40	8	0.000027	0.000023
90	40	40	9	0.000165	0.000139
90	40	40	10	0.000833	0.000667
90	40	40	11	0.003432	0.002600
90	40	40	12	0.011714	0.008282
90	40	40	13	0.033429	0.021715
90	40	40	14	0.080544	0.047114
90	40	40	15	0.165475	0.084931

N	n	k	x	P(x)	p(x)
90	40	40	16	0.293076	0.127601
90	40	40	17	0.453203	0.160127
90	40	40	18	0.621272	0.168070
90	40	40	19	0.768905	0.147633
90	40	40	20	0.877415	0.108510
90	40	40	21	0.944088	0.066673
90	40	40	22	0.978277	0.034189
90	40	40	23	0.992871	0.014594
90	40	40	24	0.998040	0.005169
90	40	40	25	0.999552	0.001512
90	40	40	26	0.999916	0.000364
90	40	40	27	0.999987	0.000071
90	40	40	28	0.999998	0.000011
90	40	40	29	1.000000	0.000001
90	40	40	30	1.000000	0.000000
90	40	40	31	1.000000	0.000000
90	40	40	32	1.000000	0.000000
90	40	40	33	1.000000	0.000000
90	40	40	34	1.000000	0.000000
90	40	40	35	1.000000	0.000000
90	40	40	36	1.000000	0.000000
90	40	40	37	1.000000	0.000000
90	40	40	38	1.000000	0.000000
90	40	40	39	1.000000	0.000000
90	40	40	40	1.000000	0.000000
90	41	1	0	0.544444	0.544444
90	41	1	1	1.000000	0.455556
90	41	2	0	0.293633	0.293633
90	41	2	1	0.795256	0.501623
90	41	2	2	1.000000	0.204744
90	41	3	0	0.156827	0.156827
90	41	3	1	0.567245	0.410419
90	41	3	2	0.909261	0.342016
90	41	3	3	1.000000	0.090739
90	41	4	0	0.082920	0.082920
90	41	4	1	0.378547	0.295627
90	41	4	2	0.755944	0.377397
90	41	4	3	0.960367	0.204423
90	41	4	4	1.000000	0.039633
90	41	5	0	0.043388	0.043388
90	41	5	1	0.241046	0.197658
90	41	5	2	0.584799	0.343753
90	41	5	3	0.870040	0.285242
90	41	5	4	0.982948	0.112908
90	41	5	5	1.000000	0.017051
90	41	6	0	0.022460	0.022460
90	41	6	1	0.148031	0.125571
90	41	6	2	0.427077	0.279046
90	41	6	3	0.742521	0.315444
90	41	6	4	0.933800	0.191280

N	n	k	x	P(x)	p(x)
90	41	6	5	0.992778	0.058978
90	41	6	6	1.000000	0.007222
90	41	7	0	0.011497	0.011497
90	41	7	1	0.088235	0.076738
90	41	7	2	0.297520	0.209285
90	41	7	3	0.599820	0.302300
90	41	7	4	0.849546	0.249726
90	41	7	5	0.967502	0.117956
90	41	7	6	0.996991	0.029489
90	41	7	7	1.000000	0.003009
90	41	8	0	0.005818	0.005818
90	41	8	1	0.051253	0.045435
90	41	8	2	0.199181	0.147928
90	41	8	3	0.461417	0.262236
90	41	8	4	0.738222	0.276805
90	41	8	5	0.916340	0.178118
90	41	8	6	0.984556	0.068215
90	41	8	7	0.998767	0.014212
90	41	8	8	1.000000	0.001233
90	41	9	0	0.002909	0.002909
90	41	9	1	0.029090	0.026181
90	41	9	2	0.128825	0.099736
90	41	9	3	0.339893	0.211068
90	41	9	4	0.613323	0.273429
90	41	9	5	0.838142	0.224820
90	41	9	6	0.955439	0.117297
90	41	9	7	0.992875	0.037435
90	41	9	8	0.999504	0.006629
90	41	9	9	1.000000	0.000496
90	41	10	0	0.001437	0.001437
90	41	10	1	0.016161	0.014724
90	41	10	2	0.080804	0.064643
90	41	10	3	0.240874	0.160069
90	41	10	4	0.488423	0.247549
90	41	10	5	0.738222	0.249800
90	41	10	6	0.904755	0.166533
90	41	10	7	0.977161	0.072406
90	41	10	8	0.996803	0.019642
90	41	10	9	0.999804	0.003001
90	41	10	10	1.000000	0.000196
90	41	11	0	0.000700	0.000700
90	41	11	1	0.008799	0.008098
90	41	11	2	0.049291	0.040492
90	41	11	3	0.164841	0.115550
90	41	11	4	0.373931	0.209091
90	41	11	5	0.625813	0.251881
90	41	11	6	0.831897	0.206085
90	41	11	7	0.946389	0.114491
90	41	11	8	0.988701	0.042312
90	41	11	9	0.998604	0.009903

N	n	k	x	P(x)	p(x)
90	41	11	10	0.999924	0.001320
90	41	11	11	1.000000	0.000076
90	41	12	0	0.000337	0.000337
90	41	12	1	0.004698	0.004361
90	41	12	2	0.029301	0.024603
90	41	12	3	0.109260	0.079959
90	41	12	4	0.276003	0.166743
90	41	12	5	0.511031	0.235028
90	41	12	6	0.740594	0.229563
90	41	12	7	0.897114	0.156520
90	41	12	8	0.971026	0.073912
90	41	12	9	0.994592	0.023566
90	41	12	10	0.999406	0.004814
90	41	12	11	0.999971	0.000565
90	41	12	12	1.000000	0.000029
90	41	13	0	0.000160	0.000160
90	41	13	1	0.002462	0.002302
90	41	13	2	0.017000	0.014538
90	41	13	3	0.070305	0.053306
90	41	13	4	0.196907	0.126601
90	41	13	5	0.402557	0.205650
90	41	13	6	0.637585	0.235028
90	41	13	7	0.828887	0.191302
90	41	13	8	0.939756	0.110868
90	41	13	9	0.984924	0.045169
90	41	13	10	0.997493	0.012569
90	41	13	11	0.999754	0.002261
90	41	13	12	0.999989	0.000236
90	41	13	13	1.000000	0.000011
90	41	14	0	0.000075	0.000075
90	41	14	1	0.001266	0.001191
90	41	14	2	0.009636	0.008370
90	41	14	3	0.043999	0.034362
90	41	14	4	0.136072	0.092074
90	41	14	5	0.306409	0.170336
90	41	14	6	0.530754	0.224345
90	41	14	7	0.744416	0.213662
90	41	14	8	0.892241	0.147824
90	41	14	9	0.966153	0.073912
90	41	14	10	0.992433	0.026280
90	41	14	11	0.998873	0.006440
90	41	14	12	0.999900	0.001028
90	41	14	13	0.999996	0.000096
90	41	14	14	1.000000	0.000004
90	41	15	0	0.000034	0.000034
90	41	15	1	0.000639	0.000605
90	41	15	2	0.005341	0.004702
90	41	15	3	0.026817	0.021477
90	41	15	4	0.091247	0.064430
90	41	15	5	0.225723	0.134476

N	n	k	x	P(x)	p(x)
90	41	15	6	0.427437	0.201714
90	41	15	7	0.648830	0.221393
90	41	15	8	0.828054	0.179223
90	41	15	9	0.935032	0.106978
90	41	15	10	0.981713	0.046681
90	41	15	11	0.996331	0.014617
90	41	15	12	0.999508	0.003178
90	41	15	13	0.999961	0.000452
90	41	15	14	0.999999	0.000038
90	41	15	15	1.000000	0.000001
90	41	16	0	0.000016	0.000016
90	41	16	1	0.000317	0.000301
90	41	16	2	0.002896	0.002579
90	41	16	3	0.015936	0.013040
90	41	16	4	0.059462	0.043526
90	41	16	5	0.161174	0.101713
90	41	16	6	0.333304	0.172129
90	41	16	7	0.548465	0.215162
90	41	16	8	0.749195	0.200730
90	41	16	9	0.889388	0.140192
90	41	16	10	0.962418	0.073030
90	41	16	11	0.990484	0.028065
90	41	16	12	0.998280	0.007796
90	41	16	13	0.999792	0.001512
90	41	16	14	0.999985	0.000193
90	41	16	15	0.999999	0.000014
90	41	16	16	1.000000	0.000000
90	41	17	0	0.000007	0.000007
90	41	17	1	0.000154	0.000147
90	41	17	2	0.001536	0.001383
90	41	17	3	0.009240	0.007703
90	41	17	4	0.037699	0.028459
90	41	17	5	0.111693	0.073994
90	41	17	6	0.251891	0.140199
90	41	17	7	0.449607	0.197716
90	41	17	8	0.659681	0.210073
90	41	17	9	0.828764	0.169083
90	41	17	10	0.931824	0.103060
90	41	17	11	0.979106	0.047281
90	41	17	12	0.995224	0.016119
90	41	17	13	0.999220	0.003995
90	41	17	14	0.999915	0.000695
90	41	17	15	0.999994	0.000080
90	41	17	16	1.000000	0.000005
90	41	17	17	1.000000	0.000000
90	41	18	0	0.000003	0.000003
90	41	18	1	0.000073	0.000070
90	41	18	2	0.000798	0.000724
90	41	18	3	0.005230	0.004432
90	41	18	4	0.023274	0.018045

Table for $N = 2$, $n = 1$, through $N = 100$, $n = 50$

N	n	k	x	P(x)	p(x)
90	41	18	5	0.075203	0.051928
90	41	18	6	0.184673	0.109470
90	41	18	7	0.357520	0.172848
90	41	18	8	0.564716	0.207196
90	41	18	9	0.754645	0.189929
90	41	18	10	0.888059	0.133414
90	41	18	11	0.959675	0.071616
90	41	18	12	0.988821	0.029146
90	41	18	13	0.997687	0.008866
90	41	18	14	0.999658	0.001970
90	41	18	15	0.999966	0.000308
90	41	18	16	0.999998	0.000032
90	41	18	17	1.000000	0.000002
90	41	18	18	1.000000	0.000000
90	41	19	0	0.000001	0.000001
90	41	19	1	0.000034	0.000033
90	41	19	2	0.000405	0.000371
90	41	19	3	0.002891	0.002485
90	41	19	4	0.014001	0.011111
90	41	19	5	0.049238	0.035237
90	41	19	6	0.131458	0.082220
90	41	19	7	0.275898	0.144440
90	41	19	8	0.469751	0.193853
90	41	19	9	0.670232	0.200481
90	41	19	10	0.830617	0.160385
90	41	19	11	0.929835	0.099218
90	41	19	12	0.977082	0.047247
90	41	19	13	0.994239	0.017158
90	41	19	14	0.998919	0.004679
90	41	19	15	0.999855	0.000936
90	41	19	16	0.999987	0.000132
90	41	19	17	0.999999	0.000012
90	41	19	18	1.000000	0.000001
90	41	19	19	1.000000	0.000000
90	41	20	0	0.000001	0.000001
90	41	20	1	0.000016	0.000015
90	41	20	2	0.000202	0.000186
90	41	20	3	0.001561	0.001359
90	41	20	4	0.008211	0.006651
90	41	20	5	0.031372	0.023160
90	41	20	6	0.090927	0.059555
90	41	20	7	0.206730	0.115802
90	41	20	8	0.379650	0.172921
90	41	20	9	0.579875	0.200224
90	41	20	10	0.760590	0.180715
90	41	20	11	0.887912	0.127322
90	41	20	12	0.957784	0.069872
90	41	20	13	0.987473	0.029689
90	41	20	14	0.997139	0.009666
90	41	20	15	0.999512	0.002373
90	41	20	16	0.999940	0.000428
90	41	20	17	0.999995	0.000055
90	41	20	18	1.000000	0.000005
90	41	20	19	1.000000	0.000000
90	41	20	20	1.000000	0.000000
90	41	21	0	0.000000	0.000000
90	41	21	1	0.000007	0.000007
90	41	21	2	0.000098	0.000091
90	41	21	3	0.000823	0.000725
90	41	21	4	0.004696	0.003873
90	41	21	5	0.019461	0.014765
90	41	21	6	0.061150	0.041689
90	41	21	7	0.150483	0.089333
90	41	21	8	0.298131	0.147648
90	41	21	9	0.488344	0.190213
90	41	21	10	0.680559	0.192215
90	41	21	11	0.833345	0.152786
90	41	21	12	0.928837	0.095492
90	41	21	13	0.975597	0.046760
90	41	21	14	0.993411	0.017813
90	41	21	15	0.998631	0.005220
90	41	21	16	0.999787	0.001157
90	41	21	17	0.999976	0.000189
90	41	21	18	0.999998	0.000022
90	41	21	19	1.000000	0.000002
90	41	21	20	1.000000	0.000000
90	41	21	21	1.000000	0.000000
90	41	22	0	0.000000	0.000000
90	41	22	1	0.000003	0.000003
90	41	22	2	0.000047	0.000044
90	41	22	3	0.000424	0.000377
90	41	22	4	0.002619	0.002195
90	41	22	5	0.011757	0.009138
90	41	22	6	0.040003	0.028246
90	41	22	7	0.106463	0.066460
90	41	22	8	0.227516	0.121053
90	41	22	9	0.400129	0.172612
90	41	22	10	0.594201	0.194072
90	41	22	11	0.766916	0.172715
90	41	22	12	0.888703	0.121786
90	41	22	13	0.956622	0.067919
90	41	22	14	0.986440	0.029818
90	41	22	15	0.996664	0.010223
90	41	22	16	0.999368	0.002704
90	41	22	17	0.999910	0.000542
90	41	22	18	0.999991	0.000080
90	41	22	19	0.999999	0.000008
90	41	22	20	1.000000	0.000001
90	41	22	21	1.000000	0.000000
90	41	22	22	1.000000	0.000000
90	41	23	0	0.000000	0.000000
90	41	23	1	0.000001	0.000001
90	41	23	2	0.000022	0.000020
90	41	23	3	0.000213	0.000191
90	41	23	4	0.001424	0.001212
90	41	23	5	0.006919	0.005495
90	41	23	6	0.025465	0.018545
90	41	23	7	0.073233	0.047768
90	41	23	8	0.168770	0.095537
90	41	23	9	0.318899	0.150129
90	41	23	10	0.505727	0.186828
90	41	23	11	0.690719	0.184991
90	41	23	12	0.836764	0.146046
90	41	23	13	0.928655	0.091891
90	41	23	14	0.974601	0.045945
90	41	23	15	0.992755	0.018154
90	41	23	16	0.998374	0.005619
90	41	23	17	0.999719	0.001345
90	41	23	18	0.999964	0.000245
90	41	23	19	0.999996	0.000033
90	41	23	20	1.000000	0.000003
90	41	23	21	1.000000	0.000000
90	41	23	22	1.000000	0.000000
90	41	23	23	1.000000	0.000000
90	41	24	0	0.000000	0.000000
90	41	24	1	0.000001	0.000001
90	41	24	2	0.000010	0.000009
90	41	24	3	0.000104	0.000095
90	41	24	4	0.000755	0.000651
90	41	24	5	0.003967	0.003211
90	41	24	6	0.015777	0.011810
90	41	24	7	0.048993	0.033216
90	41	24	8	0.121715	0.072722
90	41	24	9	0.247196	0.125481
90	41	24	10	0.419284	0.172089
90	41	24	11	0.607887	0.188602
90	41	24	12	0.773551	0.165664
90	41	24	13	0.890253	0.116703
90	41	24	14	0.956085	0.065832
90	41	24	15	0.985710	0.029624
90	41	24	16	0.996277	0.010567
90	41	24	17	0.999237	0.002960
90	41	24	18	0.999880	0.000642
90	41	24	19	0.999986	0.000106
90	41	24	20	0.999999	0.000013
90	41	24	21	1.000000	0.000001
90	41	24	22	1.000000	0.000000
90	41	24	23	1.000000	0.000000
90	41	24	24	1.000000	0.000000
90	41	25	0	0.000000	0.000000
90	41	25	1	0.000000	0.000000
90	41	25	2	0.000004	0.000004
90	41	25	3	0.000050	0.000046
90	41	25	4	0.000390	0.000341
90	41	25	5	0.002215	0.001825
90	41	25	6	0.009514	0.007299
90	41	25	7	0.031881	0.022367
90	41	25	8	0.085354	0.053472
90	41	25	9	0.186356	0.101003
90	41	25	10	0.338455	0.152098
90	41	25	11	0.522158	0.183703
90	41	25	12	0.700759	0.178600
90	41	25	13	0.840743	0.139984
90	41	25	14	0.929154	0.088411
90	41	25	15	0.974040	0.044886
90	41	25	16	0.992274	0.018235
90	41	25	17	0.998161	0.005886
90	41	25	18	0.999656	0.001495
90	41	25	19	0.999950	0.000295
90	41	25	20	0.999995	0.000044
90	41	25	21	0.999999	0.000005
90	41	25	22	1.000000	0.000000
90	41	25	23	1.000000	0.000000
90	41	25	24	1.000000	0.000000
90	41	25	25	1.000000	0.000000
90	41	26	0	0.000000	0.000000
90	41	26	1	0.000000	0.000000
90	41	26	2	0.000002	0.000002
90	41	26	3	0.000023	0.000021
90	41	26	4	0.000197	0.000173
90	41	26	5	0.001205	0.001008
90	41	26	6	0.005584	0.004379
90	41	26	7	0.020182	0.014598
90	41	26	8	0.058206	0.038025
90	41	26	9	0.136632	0.078426
90	41	26	10	0.265916	0.129284
90	41	26	11	0.437372	0.171456
90	41	26	12	0.621075	0.183703
90	41	26	13	0.780442	0.159367
90	41	26	14	0.892429	0.111987
90	41	26	15	0.956085	0.063656
90	41	26	16	0.985261	0.029176
90	41	26	17	0.995987	0.010726
90	41	26	18	0.999127	0.003139
90	41	26	19	0.999851	0.000724
90	41	26	20	0.999980	0.000130
90	41	26	21	0.999998	0.000018
90	41	26	22	1.000000	0.000002
90	41	26	23	1.000000	0.000000
90	41	26	24	1.000000	0.000000

Table for $N = 2,\ n = 1,$ through $N = 100,\ n = 50$

N	n	k	x	P(x)	p(x)
90	41	26	25	1.000000	0.000000
90	41	26	26	1.000000	0.000000
90	41	27	0	0.000000	0.000000
90	41	27	1	0.000000	0.000000
90	41	27	2	0.000001	0.000001
90	41	27	3	0.000011	0.000010
90	41	27	4	0.000096	0.000086
90	41	27	5	0.000638	0.000541
90	41	27	6	0.003189	0.002551
90	41	27	7	0.012427	0.009238
90	41	27	8	0.038600	0.026173
90	41	27	9	0.097419	0.058819
90	41	27	10	0.203294	0.105875
90	41	27	11	0.357002	0.153708
90	41	27	12	0.537835	0.180833
90	41	27	13	0.710719	0.172884
90	41	27	14	0.845185	0.134466
90	41	27	15	0.930225	0.085040
90	41	27	16	0.973864	0.043639
90	41	27	17	0.991965	0.018101
90	41	27	18	0.997999	0.006034
90	41	27	19	0.999602	0.001603
90	41	27	20	0.999938	0.000336
90	41	27	21	0.999992	0.000055
90	41	27	22	0.999999	0.000007
90	41	27	23	1.000000	0.000001
90	41	27	24	1.000000	0.000000
90	41	27	25	1.000000	0.000000
90	41	27	26	1.000000	0.000000
90	41	27	27	1.000000	0.000000
90	41	28	0	0.000000	0.000000
90	41	28	1	0.000000	0.000000
90	41	28	2	0.000000	0.000000
90	41	28	3	0.000005	0.000004
90	41	28	4	0.000046	0.000041
90	41	28	5	0.000328	0.000282
90	41	28	6	0.001772	0.001443
90	41	28	7	0.007441	0.005670
90	41	28	8	0.024890	0.017449
90	41	28	9	0.067543	0.042653
90	41	28	10	0.151197	0.083654
90	41	28	11	0.283808	0.132611
90	41	28	12	0.454594	0.170787
90	41	28	13	0.633882	0.179287
90	41	28	14	0.787557	0.153675
90	41	28	15	0.895129	0.107572
90	41	28	16	0.956547	0.061418
90	41	28	17	0.985069	0.028522
90	41	28	18	0.995796	0.010726
90	41	28	19	0.999042	0.003246

N	n	k	x	P(x)	p(x)
90	41	28	20	0.999826	0.000784
90	41	28	21	0.999975	0.000149
90	41	28	22	0.999997	0.000022
90	41	28	23	1.000000	0.000002
90	41	28	24	1.000000	0.000000
90	41	28	25	1.000000	0.000000
90	41	28	26	1.000000	0.000000
90	41	28	27	1.000000	0.000000
90	41	28	28	1.000000	0.000000
90	41	29	0	0.000000	0.000000
90	41	29	1	0.000000	0.000000
90	41	29	2	0.000000	0.000000
90	41	29	3	0.000002	0.000002
90	41	29	4	0.000021	0.000019
90	41	29	5	0.000164	0.000143
90	41	29	6	0.000957	0.000792
90	41	29	7	0.004332	0.003375
90	41	29	8	0.015603	0.011271
90	41	29	9	0.045528	0.029926
90	41	29	10	0.109370	0.063841
90	41	29	11	0.219641	0.110271
90	41	29	12	0.374710	0.155069
90	41	29	13	0.552913	0.178203
90	41	29	14	0.720634	0.167720
90	41	29	15	0.850018	0.129384
90	41	29	16	0.931782	0.081764
90	41	29	17	0.974029	0.042247
90	41	29	18	0.991817	0.017788
90	41	29	19	0.997890	0.006073
90	41	29	20	0.999560	0.001670
90	41	29	21	0.999927	0.000367
90	41	29	22	0.999990	0.000063
90	41	29	23	0.999999	0.000009
90	41	29	24	1.000000	0.000001
90	41	29	25	1.000000	0.000000
90	41	29	26	1.000000	0.000000
90	41	29	27	1.000000	0.000000
90	41	29	28	1.000000	0.000000
90	41	29	29	1.000000	0.000000
90	41	30	0	0.000000	0.000000
90	41	30	1	0.000000	0.000000
90	41	30	2	0.000000	0.000000
90	41	30	3	0.000001	0.000001
90	41	30	4	0.000010	0.000009
90	41	30	5	0.000080	0.000070
90	41	30	6	0.000502	0.000422
90	41	30	7	0.002451	0.001949
90	41	30	8	0.009505	0.007055
90	41	30	9	0.029830	0.020324
90	41	30	10	0.076926	0.047096

N	n	k	x	P(x)	p(x)
90	41	30	11	0.165409	0.088484
90	41	30	12	0.300989	0.135580
90	41	30	13	0.471115	0.170126
90	41	30	14	0.646397	0.175282
90	41	30	15	0.794871	0.148474
90	41	30	16	0.898272	0.103401
90	41	30	17	0.957407	0.059135
90	41	30	18	0.985110	0.027703
90	41	30	19	0.995700	0.010590
90	41	30	20	0.998985	0.003286
90	41	30	21	0.999807	0.000821
90	41	30	22	0.999971	0.000164
90	41	30	23	0.999996	0.000026
90	41	30	24	1.000000	0.000003
90	41	30	25	1.000000	0.000000
90	41	30	26	1.000000	0.000000
90	41	30	27	1.000000	0.000000
90	41	30	28	1.000000	0.000000
90	41	30	29	1.000000	0.000000
90	41	30	30	1.000000	0.000000
90	41	31	0	0.000000	0.000000
90	41	31	1	0.000000	0.000000
90	41	31	2	0.000000	0.000000
90	41	31	3	0.000000	0.000000
90	41	31	4	0.000004	0.000004
90	41	31	5	0.000038	0.000034
90	41	31	6	0.000256	0.000218
90	41	31	7	0.001347	0.001091
90	41	31	8	0.005625	0.004279
90	41	31	9	0.018990	0.013365
90	41	31	10	0.052593	0.033603
90	41	31	11	0.121167	0.068575
90	41	31	12	0.235459	0.114291
90	41	31	13	0.391723	0.156264
90	41	31	14	0.567520	0.175797
90	41	31	15	0.730532	0.163012
90	41	31	16	0.855188	0.124656
90	41	31	17	0.933753	0.078565
90	41	31	18	0.974490	0.040737
90	41	31	19	0.991817	0.017326
90	41	31	20	0.997835	0.006019
90	41	31	21	0.999533	0.001698
90	41	31	22	0.999919	0.000386
90	41	31	23	0.999989	0.000070
90	41	31	24	0.999999	0.000010
90	41	31	25	1.000000	0.000001
90	41	31	26	1.000000	0.000000
90	41	31	27	1.000000	0.000000
90	41	31	28	1.000000	0.000000
90	41	31	29	1.000000	0.000000

N	n	k	x	P(x)	p(x)
90	41	31	30	1.000000	0.000000
90	41	31	31	1.000000	0.000000
90	41	32	0	0.000000	0.000000
90	41	32	1	0.000000	0.000000
90	41	32	2	0.000000	0.000000
90	41	32	3	0.000000	0.000000
90	41	32	4	0.000002	0.000002
90	41	32	5	0.000017	0.000016
90	41	32	6	0.000126	0.000109
90	41	32	7	0.000718	0.000592
90	41	32	8	0.003232	0.002514
90	41	32	9	0.011741	0.008509
90	41	32	10	0.034937	0.023196
90	41	32	11	0.086299	0.051362
90	41	32	12	0.179282	0.092983
90	41	32	13	0.317564	0.138282
90	41	32	14	0.487070	0.169507
90	41	32	15	0.658696	0.171626
90	41	32	16	0.802368	0.143672
90	41	32	17	0.901794	0.099427
90	41	32	18	0.958610	0.056815
90	41	32	19	0.985356	0.026746
90	41	32	20	0.995693	0.010337
90	41	32	21	0.998957	0.003264
90	41	32	22	0.999794	0.000837
90	41	32	23	0.999967	0.000173
90	41	32	24	0.999996	0.000028
90	41	32	25	0.999999	0.000004
90	41	32	26	1.000000	0.000000
90	41	32	27	1.000000	0.000000
90	41	32	28	1.000000	0.000000
90	41	32	29	1.000000	0.000000
90	41	32	30	1.000000	0.000000
90	41	32	31	1.000000	0.000000
90	41	32	32	1.000000	0.000000
90	41	33	0	0.000000	0.000000
90	41	33	1	0.000000	0.000000
90	41	33	2	0.000000	0.000000
90	41	33	3	0.000000	0.000000
90	41	33	4	0.000001	0.000001
90	41	33	5	0.000008	0.000007
90	41	33	6	0.000061	0.000053
90	41	33	7	0.000371	0.000311
90	41	33	8	0.001802	0.001430
90	41	33	9	0.007047	0.005245
90	41	33	10	0.022539	0.015493
90	41	33	11	0.059732	0.037193
90	41	33	12	0.132790	0.073058
90	41	33	13	0.250807	0.118016
90	41	33	14	0.408162	0.157355

N	n	k	x	P(x)	p(x)
90	41	33	15	0.581761	0.173598
90	41	33	16	0.740440	0.158680
90	41	33	17	0.860652	0.120212
90	41	33	18	0.936079	0.075427
90	41	33	19	0.975211	0.039131
90	41	33	20	0.991950	0.016740
90	41	33	21	0.997832	0.005881
90	41	33	22	0.999520	0.001688
90	41	33	23	0.999914	0.000393
90	41	33	24	0.999987	0.000074
90	41	33	25	0.999998	0.000011
90	41	33	26	1.000000	0.000001
90	41	33	27	1.000000	0.000000
90	41	33	28	1.000000	0.000000
90	41	33	29	1.000000	0.000000
90	41	33	30	1.000000	0.000000
90	41	33	31	1.000000	0.000000
90	41	33	32	1.000000	0.000000
90	41	33	33	1.000000	0.000000
90	41	34	0	0.000000	0.000000
90	41	34	1	0.000000	0.000000
90	41	34	2	0.000000	0.000000
90	41	34	3	0.000000	0.000000
90	41	34	4	0.000000	0.000000
90	41	34	5	0.000003	0.000003
90	41	34	6	0.000028	0.000025
90	41	34	7	0.000186	0.000158
90	41	34	8	0.000974	0.000788
90	41	34	9	0.004102	0.003129
90	41	34	10	0.014114	0.010011
90	41	34	11	0.040157	0.026044
90	41	34	12	0.095621	0.055463
90	41	34	13	0.192834	0.097213
90	41	34	14	0.333625	0.140792
90	41	34	15	0.502575	0.168950
90	41	34	16	0.670844	0.168269
90	41	34	17	0.810037	0.139193
90	41	34	18	0.905644	0.095607
90	41	34	19	0.960108	0.054464
90	41	34	20	0.985783	0.025676
90	41	34	21	0.995768	0.009985
90	41	34	22	0.998957	0.003189
90	41	34	23	0.999789	0.000832
90	41	34	24	0.999965	0.000176
90	41	34	25	0.999995	0.000030
90	41	34	26	0.999999	0.000004
90	41	34	27	1.000000	0.000000
90	41	34	28	1.000000	0.000000
90	41	34	29	1.000000	0.000000
90	41	34	30	1.000000	0.000000
90	41	34	31	1.000000	0.000000
90	41	34	32	1.000000	0.000000
90	41	34	33	1.000000	0.000000
90	41	34	34	1.000000	0.000000
90	41	35	0	0.000000	0.000000
90	41	35	1	0.000000	0.000000
90	41	35	2	0.000000	0.000000
90	41	35	3	0.000000	0.000000
90	41	35	4	0.000000	0.000000
90	41	35	5	0.000001	0.000001
90	41	35	6	0.000013	0.000011
90	41	35	7	0.000090	0.000078
90	41	35	8	0.000509	0.000419
90	41	35	9	0.002314	0.001805
90	41	35	10	0.008572	0.006257
90	41	35	11	0.026205	0.017634
90	41	35	12	0.066898	0.040693
90	41	35	13	0.144227	0.077329
90	41	35	14	0.265744	0.121517
90	41	35	15	0.424134	0.158391
90	41	35	16	0.595724	0.171590
90	41	35	17	0.750383	0.154659
90	41	35	18	0.866377	0.115994
90	41	35	19	0.938711	0.072334
90	41	35	20	0.976155	0.037444
90	41	35	21	0.992202	0.016047
90	41	35	22	0.997875	0.005673
90	41	35	23	0.999522	0.001647
90	41	35	24	0.999912	0.000390
90	41	35	25	0.999987	0.000075
90	41	35	26	0.999998	0.000012
90	41	35	27	1.000000	0.000001
90	41	35	28	1.000000	0.000000
90	41	35	29	1.000000	0.000000
90	41	35	30	1.000000	0.000000
90	41	35	31	1.000000	0.000000
90	41	35	32	1.000000	0.000000
90	41	35	33	1.000000	0.000000
90	41	35	34	1.000000	0.000000
90	41	35	35	1.000000	0.000000
90	41	36	0	0.000000	0.000000
90	41	36	1	0.000000	0.000000
90	41	36	2	0.000000	0.000000
90	41	36	3	0.000000	0.000000
90	41	36	4	0.000000	0.000000
90	41	36	5	0.000001	0.000000
90	41	36	6	0.000005	0.000005
90	41	36	7	0.000042	0.000037
90	41	36	8	0.000258	0.000216
90	41	36	9	0.001264	0.001006
90	41	36	10	0.005045	0.003781
90	41	36	11	0.016587	0.011542
90	41	36	12	0.045442	0.028855
90	41	36	13	0.104860	0.059418
90	41	36	14	0.206090	0.101230
90	41	36	15	0.349259	0.143169
90	41	36	16	0.517729	0.168470
90	41	36	17	0.682895	0.165167
90	41	36	18	0.817870	0.134975
90	41	36	19	0.909777	0.091907
90	41	36	20	0.961858	0.052081
90	41	36	21	0.986367	0.024509
90	41	36	22	0.995916	0.009549
90	41	36	23	0.998983	0.003068
90	41	36	24	0.999791	0.000808
90	41	36	25	0.999965	0.000174
90	41	36	26	0.999995	0.000030
90	41	36	27	0.999999	0.000004
90	41	36	28	1.000000	0.000000
90	41	36	29	1.000000	0.000000
90	41	36	30	1.000000	0.000000
90	41	36	31	1.000000	0.000000
90	41	36	32	1.000000	0.000000
90	41	36	33	1.000000	0.000000
90	41	36	34	1.000000	0.000000
90	41	36	35	1.000000	0.000000
90	41	36	36	1.000000	0.000000
90	41	37	0	0.000000	0.000000
90	41	37	1	0.000000	0.000000
90	41	37	2	0.000000	0.000000
90	41	37	3	0.000000	0.000000
90	41	37	4	0.000000	0.000000
90	41	37	5	0.000000	0.000000
90	41	37	6	0.000002	0.000002
90	41	37	7	0.000019	0.000017
90	41	37	8	0.000126	0.000107
90	41	37	9	0.000668	0.000542
90	41	37	10	0.002874	0.002207
90	41	37	11	0.010175	0.007300
90	41	37	12	0.029946	0.019771
90	41	37	13	0.074051	0.044105
90	41	37	14	0.155475	0.081424
90	41	37	15	0.280326	0.124851
90	41	37	16	0.439733	0.159408
90	41	37	17	0.609488	0.169755
90	41	37	18	0.760381	0.150893
90	41	37	19	0.872334	0.111953
90	41	37	20	0.941604	0.069271
90	41	37	21	0.977289	0.035685
90	41	37	22	0.992556	0.015266
90	41	37	23	0.997961	0.005405
90	41	37	24	0.999537	0.001576
90	41	37	25	0.999914	0.000377
90	41	37	26	0.999987	0.000073
90	41	37	27	0.999998	0.000011
90	41	37	28	1.000000	0.000001
90	41	37	29	1.000000	0.000000
90	41	37	30	1.000000	0.000000
90	41	37	31	1.000000	0.000000
90	41	37	32	1.000000	0.000000
90	41	37	33	1.000000	0.000000
90	41	37	34	1.000000	0.000000
90	41	37	35	1.000000	0.000000
90	41	37	36	1.000000	0.000000
90	41	37	37	1.000000	0.000000
90	41	38	0	0.000000	0.000000
90	41	38	1	0.000000	0.000000
90	41	38	2	0.000000	0.000000
90	41	38	3	0.000000	0.000000
90	41	38	4	0.000000	0.000000
90	41	38	5	0.000000	0.000000
90	41	38	6	0.000001	0.000001
90	41	38	7	0.000008	0.000007
90	41	38	8	0.000059	0.000051
90	41	38	9	0.000341	0.000281
90	41	38	10	0.001584	0.001243
90	41	38	11	0.006042	0.004459
90	41	38	12	0.019128	0.013085
90	41	38	13	0.050750	0.031622
90	41	38	14	0.113995	0.063245
90	41	38	15	0.219078	0.105084
90	41	38	16	0.364541	0.145463
90	41	38	17	0.532618	0.168077
90	41	38	18	0.694899	0.162281
90	41	38	19	0.825863	0.130964
90	41	38	20	0.914158	0.088295
90	41	38	21	0.963823	0.049666
90	41	38	22	0.987083	0.023259
90	41	38	23	0.996125	0.009042
90	41	38	24	0.999031	0.002906
90	41	38	25	0.999800	0.000769
90	41	38	26	0.999966	0.000166
90	41	38	27	0.999995	0.000029
90	41	38	28	0.999999	0.000004
90	41	38	29	1.000000	0.000000
90	41	38	30	1.000000	0.000000
90	41	38	31	1.000000	0.000000
90	41	38	32	1.000000	0.000000
90	41	38	33	1.000000	0.000000
90	41	38	34	1.000000	0.000000

Table for $N = 2$, $n = 1$, through $N = 100$, $n = 50$

N	n	k	x	P(x)	p(x)
90	41	38	35	1.000000	0.000000
90	41	38	36	1.000000	0.000000
90	41	38	37	1.000000	0.000000
90	41	38	38	1.000000	0.000000
90	41	39	0	0.000000	0.000000
90	41	39	1	0.000000	0.000000
90	41	39	2	0.000000	0.000000
90	41	39	3	0.000000	0.000000
90	41	39	4	0.000000	0.000000
90	41	39	5	0.000000	0.000000
90	41	39	6	0.000000	0.000000
90	41	39	7	0.000003	0.000003
90	41	39	8	0.000027	0.000024
90	41	39	9	0.000168	0.000141
90	41	39	10	0.000843	0.000675
90	41	39	11	0.003470	0.002627
90	41	39	12	0.011830	0.008360
90	41	39	13	0.033722	0.021892
90	41	39	14	0.081156	0.047434
90	41	39	15	0.166536	0.085380
90	41	39	16	0.294607	0.128071
90	41	39	17	0.455044	0.160437
90	41	39	18	0.623121	0.168077
90	41	39	19	0.770455	0.147334
90	41	39	20	0.878500	0.108045
90	41	39	21	0.944721	0.066221
90	41	39	22	0.978584	0.033863
90	41	39	23	0.992995	0.014411
90	41	39	24	0.998081	0.005086
90	41	39	25	0.999563	0.001482
90	41	39	26	0.999918	0.000355
90	41	39	27	0.999987	0.000069
90	41	39	28	0.999998	0.000011
90	41	39	29	1.000000	0.000001
90	41	39	30	1.000000	0.000000
90	41	39	31	1.000000	0.000000
90	41	39	32	1.000000	0.000000
90	41	39	33	1.000000	0.000000
90	41	39	34	1.000000	0.000000
90	41	39	35	1.000000	0.000000
90	41	39	36	1.000000	0.000000
90	41	39	37	1.000000	0.000000
90	41	39	38	1.000000	0.000000
90	41	39	39	1.000000	0.000000
90	41	40	0	0.000000	0.000000
90	41	40	1	0.000000	0.000000
90	41	40	2	0.000000	0.000000
90	41	40	3	0.000000	0.000000
90	41	40	4	0.000000	0.000000
90	41	40	5	0.000000	0.000000
90	41	40	6	0.000000	0.000000
90	41	40	7	0.000001	0.000001
90	41	40	8	0.000012	0.000010
90	41	40	9	0.000079	0.000068
90	41	40	10	0.000432	0.000353
90	41	40	11	0.001925	0.001492
90	41	40	12	0.007076	0.005152
90	41	40	13	0.021703	0.014627
90	41	40	14	0.056044	0.034341
90	41	40	15	0.123009	0.066965
90	41	40	16	0.231827	0.108818
90	41	40	17	0.379544	0.147717
90	41	40	18	0.547321	0.167777
90	41	40	19	0.706899	0.159578
90	41	40	20	0.834011	0.127112
90	41	40	21	0.918752	0.084741
90	41	40	22	0.965969	0.047217
90	41	40	23	0.987909	0.021940
90	41	40	24	0.996386	0.008477
90	41	40	25	0.999098	0.002713
90	41	40	26	0.999814	0.000715
90	41	40	27	0.999968	0.000155
90	41	40	28	0.999995	0.000027
90	41	40	29	0.999999	0.000004
90	41	40	30	1.000000	0.000000
90	41	40	31	1.000000	0.000000
90	41	40	32	1.000000	0.000000
90	41	40	33	1.000000	0.000000
90	41	40	34	1.000000	0.000000
90	41	40	35	1.000000	0.000000
90	41	40	36	1.000000	0.000000
90	41	40	37	1.000000	0.000000
90	41	40	38	1.000000	0.000000
90	41	40	39	1.000000	0.000000
90	41	40	40	1.000000	0.000000
90	41	41	0	0.000000	0.000000
90	41	41	1	0.000000	0.000000
90	41	41	2	0.000000	0.000000
90	41	41	3	0.000000	0.000000
90	41	41	4	0.000000	0.000000
90	41	41	5	0.000000	0.000000
90	41	41	6	0.000000	0.000000
90	41	41	7	0.000001	0.000000
90	41	41	8	0.000005	0.000004
90	41	41	9	0.000036	0.000031
90	41	41	10	0.000214	0.000177
90	41	41	11	0.001029	0.000816
90	41	41	12	0.004088	0.003059
90	41	41	13	0.013512	0.009424
90	41	41	14	0.037500	0.023988
90	41	41	15	0.088187	0.050687
90	41	41	16	0.177418	0.089231
90	41	41	17	0.308640	0.131222
90	41	41	18	0.470144	0.161504
90	41	41	19	0.636685	0.166541
90	41	41	20	0.780624	0.143939
90	41	41	21	0.884856	0.104232
90	41	41	22	0.948026	0.063171
90	41	41	23	0.980010	0.031984
90	41	41	24	0.993504	0.013493
90	41	41	25	0.998230	0.004727
90	41	41	26	0.999599	0.001369
90	41	41	27	0.999925	0.000326
90	41	41	28	0.999988	0.000063
90	41	41	29	0.999998	0.000010
90	41	41	30	1.000000	0.000001
90	41	41	31	1.000000	0.000000
90	41	41	32	1.000000	0.000000
90	41	41	33	1.000000	0.000000
90	41	41	34	1.000000	0.000000
90	41	41	35	1.000000	0.000000
90	41	41	36	1.000000	0.000000
90	41	41	37	1.000000	0.000000
90	41	41	38	1.000000	0.000000
90	41	41	39	1.000000	0.000000
90	41	41	40	1.000000	0.000000
90	41	41	41	1.000000	0.000000
90	42	1	0	0.533333	0.533333
90	42	1	1	1.000000	0.466667
90	42	2	0	0.281648	0.281648
90	42	2	1	0.785019	0.503371
90	42	2	2	1.000000	0.214981
90	42	3	0	0.147225	0.147225
90	42	3	1	0.550494	0.403269
90	42	3	2	0.902281	0.351788
90	42	3	3	1.000000	0.097719
90	42	4	0	0.076151	0.076151
90	42	4	1	0.360448	0.284297
90	42	4	2	0.740540	0.380092
90	42	4	3	0.956195	0.215655
90	42	4	4	1.000000	0.043805
90	42	5	0	0.038961	0.038961
90	42	5	1	0.224911	0.185950
90	42	5	2	0.563753	0.338842
90	42	5	3	0.858398	0.294645
90	42	5	4	0.980644	0.122246
90	42	5	5	1.000000	0.019356
90	42	6	0	0.019710	0.019710
90	42	6	1	0.135217	0.115508
90	42	6	2	0.404298	0.269080
90	42	6	3	0.723208	0.318910
90	42	6	4	0.925993	0.202785
90	42	6	5	0.991574	0.065582
90	42	6	6	1.000000	0.008425
90	42	7	0	0.009855	0.009855
90	42	7	1	0.078839	0.068984
90	42	7	2	0.276164	0.197326
90	42	7	3	0.575142	0.298978
90	42	7	4	0.834257	0.259114
90	42	7	5	0.962687	0.128431
90	42	7	6	0.996389	0.033702
90	42	7	7	1.000000	0.003611
90	42	8	0	0.004868	0.004868
90	42	8	1	0.044762	0.039894
90	42	8	2	0.181067	0.136305
90	42	8	3	0.434659	0.253591
90	42	8	4	0.715626	0.280967
90	42	8	5	0.905435	0.189809
90	42	8	6	0.981771	0.076336
90	42	8	7	0.998477	0.016706
90	42	8	8	1.000000	0.001523
90	42	9	0	0.002375	0.002375
90	42	9	1	0.024815	0.022440
90	42	9	2	0.114577	0.089762
90	42	9	3	0.314048	0.199471
90	42	9	4	0.585422	0.271373
90	42	9	5	0.819790	0.234368
90	42	9	6	0.948258	0.128468
90	42	9	7	0.991347	0.043089
90	42	9	8	0.999369	0.008022
90	42	9	9	1.000000	0.000631
90	42	10	0	0.001143	0.001143
90	42	10	1	0.013456	0.012313
90	42	10	2	0.070250	0.056794
90	42	10	3	0.218007	0.147756
90	42	10	4	0.458111	0.240104
90	42	10	5	0.712733	0.254622
90	42	10	6	0.891161	0.178428
90	42	10	7	0.972728	0.081567
90	42	10	8	0.996001	0.023273
90	42	10	9	0.999743	0.003741
90	42	10	10	1.000000	0.000257
90	42	11	0	0.000543	0.000543
90	42	11	1	0.007146	0.006603
90	42	11	2	0.041853	0.034707
90	42	11	3	0.145975	0.104122
90	42	11	4	0.344061	0.198086
90	42	11	5	0.594970	0.250909
90	42	11	6	0.810868	0.215898
90	42	11	7	0.937042	0.126174

Table for $N = 2$, $n = 1$, through $N = 100$, $n = 50$

N	n	k	x	P(x)	p(x)
90	42	11	8	0.986110	0.049068
90	42	11	9	0.998199	0.012089
90	42	11	10	0.999897	0.001698
90	42	11	11	1.000000	0.000103
90	42	12	0	0.000254	0.000254
90	42	12	1	0.003719	0.003465
90	42	12	2	0.024280	0.020561
90	42	12	3	0.094573	0.070293
90	42	12	4	0.248779	0.154206
90	42	12	5	0.477456	0.228676
90	42	12	6	0.712484	0.235028
90	42	12	7	0.881142	0.168658
90	42	12	8	0.964992	0.083850
90	42	12	9	0.993150	0.028157
90	42	12	10	0.999209	0.006060
90	42	12	11	0.999960	0.000750
90	42	12	12	1.000000	0.000040
90	42	13	0	0.000117	0.000117
90	42	13	1	0.001898	0.001781
90	42	13	2	0.013736	0.011838
90	42	13	3	0.059427	0.045691
90	42	13	4	0.173653	0.114227
90	42	13	5	0.368981	0.195328
90	42	13	6	0.604010	0.235028
90	42	13	7	0.805462	0.201453
90	42	13	8	0.928442	0.122980
90	42	13	9	0.981237	0.052794
90	42	13	10	0.996723	0.015486
90	42	13	11	0.999661	0.002938
90	42	13	12	0.999984	0.000323
90	42	13	13	1.000000	0.000016
90	42	14	0	0.000053	0.000053
90	42	14	1	0.000950	0.000896
90	42	14	2	0.007586	0.006636
90	42	14	3	0.036285	0.028698
90	42	14	4	0.117282	0.080997
90	42	14	5	0.275122	0.157841
90	42	14	6	0.494126	0.219004
90	42	14	7	0.713893	0.219767
90	42	14	8	0.874140	0.160247
90	42	14	9	0.958611	0.084471
90	42	14	10	0.990287	0.031677
90	42	14	11	0.998479	0.008191
90	42	14	12	0.999859	0.001380
90	42	14	13	0.999994	0.000136
90	42	14	14	1.000000	0.000006
90	42	15	0	0.000024	0.000024
90	42	15	1	0.000466	0.000442
90	42	15	2	0.004093	0.003627
90	42	15	3	0.021558	0.017464
90	42	15	4	0.076783	0.055225
90	42	15	5	0.198279	0.121496
90	42	15	6	0.390387	0.192109
90	42	15	7	0.612684	0.222297
90	42	15	8	0.802450	0.189766
90	42	15	9	0.921932	0.119482
90	42	15	10	0.976950	0.055017
90	42	15	11	0.995137	0.018188
90	42	15	12	0.999314	0.004176
90	42	15	13	0.999942	0.000629
90	42	15	14	0.999998	0.000055
90	42	15	15	1.000000	0.000002
90	42	16	0	0.000010	0.000010
90	42	16	1	0.000224	0.000214
90	42	16	2	0.002159	0.001935
90	42	16	3	0.012476	0.010317
90	42	16	4	0.048802	0.036326
90	42	16	5	0.138341	0.089539
90	42	16	6	0.298175	0.159834
90	42	16	7	0.508946	0.210771
90	42	16	8	0.716423	0.207477
90	42	16	9	0.869360	0.152937
90	42	16	10	0.953476	0.084115
90	42	16	11	0.987620	0.034144
90	42	16	12	0.997643	0.010023
90	42	16	13	0.999699	0.002056
90	42	16	14	0.999977	0.000278
90	42	16	15	0.999999	0.000022
90	42	16	16	1.000000	0.000001
90	42	17	0	0.000005	0.000005
90	42	17	1	0.000106	0.000101
90	42	17	2	0.001113	0.001007
90	42	17	3	0.007039	0.005926
90	42	17	4	0.030148	0.023110
90	42	17	5	0.093572	0.063423
90	42	17	6	0.220418	0.126846
90	42	17	7	0.409257	0.188839
90	42	17	8	0.621096	0.211839
90	42	17	9	0.801159	0.180063
90	42	17	10	0.917101	0.115943
90	42	17	11	0.973316	0.056215
90	42	17	12	0.993580	0.020263
90	42	17	13	0.998893	0.005314
90	42	17	14	0.999872	0.000978
90	42	17	15	0.999991	0.000119
90	42	17	16	1.000000	0.000009
90	42	17	17	1.000000	0.000000
90	42	18	0	0.000002	0.000002
90	42	18	1	0.000049	0.000047
90	42	18	2	0.000561	0.000512
90	42	18	3	0.003873	0.003312
90	42	18	4	0.018119	0.014246
90	42	18	5	0.061426	0.043307
90	42	18	6	0.157864	0.096438
90	42	18	7	0.318718	0.160854
90	42	18	8	0.522431	0.203713
90	42	18	9	0.719760	0.197329
90	42	18	10	0.866277	0.146517
90	42	18	11	0.949444	0.083167
90	42	18	12	0.985252	0.035808
90	42	18	13	0.996783	0.011530
90	42	18	14	0.999497	0.002714
90	42	18	15	0.999947	0.000450
90	42	18	16	0.999997	0.000050
90	42	18	17	1.000000	0.000003
90	42	18	18	1.000000	0.000000
90	42	19	0	0.000001	0.000001
90	42	19	1	0.000022	0.000021
90	42	19	2	0.000277	0.000254
90	42	19	3	0.002079	0.001803
90	42	19	4	0.010600	0.008521
90	42	19	5	0.039171	0.028571
90	42	19	6	0.109645	0.070474
90	42	19	7	0.240525	0.130880
90	42	19	8	0.426233	0.185708
90	42	19	9	0.629318	0.203085
90	42	19	10	0.801159	0.171841
90	42	19	11	0.913636	0.112478
90	42	19	12	0.970332	0.056696
90	42	19	13	0.992138	0.021806
90	42	19	14	0.998441	0.006303
90	42	19	15	0.999778	0.001337
90	42	19	16	0.999979	0.000201
90	42	19	17	0.999999	0.000020
90	42	19	18	1.000000	0.000001
90	42	19	19	1.000000	0.000000
90	42	20	0	0.000000	0.000000
90	42	20	1	0.000010	0.000010
90	42	20	2	0.000133	0.000123
90	42	20	3	0.001089	0.000956
90	42	20	4	0.006040	0.004951
90	42	20	5	0.024282	0.018242
90	42	20	6	0.073911	0.049630
90	42	20	7	0.176006	0.102095
90	42	20	8	0.337302	0.161296
90	42	20	9	0.534926	0.197624
90	42	20	10	0.723709	0.188783
90	42	20	11	0.864526	0.140817
90	42	20	12	0.946376	0.081850
90	42	20	13	0.983232	0.036855
90	42	20	14	0.995956	0.012724
90	42	20	15	0.999270	0.003314
90	42	20	16	0.999905	0.000636
90	42	20	17	0.999992	0.000086
90	42	20	18	0.999999	0.000008
90	42	20	19	1.000000	0.000000
90	42	20	20	1.000000	0.000000
90	42	21	0	0.000000	0.000000
90	42	21	1	0.000004	0.000004
90	42	21	2	0.000063	0.000058
90	42	21	3	0.000557	0.000494
90	42	21	4	0.003352	0.002796
90	42	21	5	0.014640	0.011287
90	42	21	6	0.048388	0.033748
90	42	21	7	0.124959	0.076571
90	42	21	8	0.258959	0.134000
90	42	21	9	0.441761	0.182802
90	42	21	10	0.637409	0.195648
90	42	21	11	0.802164	0.164756
90	42	21	12	0.911298	0.109133
90	42	21	13	0.967963	0.056665
90	42	21	14	0.990866	0.022903
90	42	21	15	0.997991	0.007125
90	42	21	16	0.999669	0.001678
90	42	21	17	0.999961	0.000292
90	42	21	18	0.999997	0.000036
90	42	21	19	1.000000	0.000003
90	42	21	20	1.000000	0.000000
90	42	21	21	1.000000	0.000000
90	42	22	0	0.000000	0.000000
90	42	22	1	0.000002	0.000002
90	42	22	2	0.000029	0.000027
90	42	22	3	0.000277	0.000249
90	42	22	4	0.001813	0.001535
90	42	22	5	0.008587	0.006774
90	42	22	6	0.030780	0.022193
90	42	22	7	0.086118	0.055338
90	42	22	8	0.192930	0.106811
90	42	22	9	0.354334	0.161404
90	42	22	10	0.546673	0.192340
90	42	22	11	0.728144	0.181470
90	42	22	12	0.863848	0.135705
90	42	22	13	0.944147	0.080299
90	42	22	14	0.981572	0.037425
90	42	22	15	0.995203	0.013631
90	42	22	16	0.999037	0.003834
90	42	22	17	0.999855	0.000818
90	42	22	18	0.999984	0.000129
90	42	22	19	0.999999	0.000014
90	42	22	20	1.000000	0.000001

Table for $N = 2$, $n = 1$, through $N = 100$, $n = 50$

N	n	k	x	P(x)	p(x)
90	42	22	21	1.000000	0.000000
90	42	22	22	1.000000	0.000000
90	42	23	0	0.000000	0.000000
90	42	23	1	0.000001	0.000001
90	42	23	2	0.000013	0.000012
90	42	23	3	0.000135	0.000122
90	42	23	4	0.000955	0.000820
90	42	23	5	0.004901	0.003946
90	42	23	6	0.019031	0.014130
90	42	23	7	0.057635	0.038605
90	42	23	8	0.139524	0.081889
90	42	23	9	0.276005	0.136481
90	42	23	10	0.456161	0.180155
90	42	23	11	0.645415	0.189254
90	42	23	12	0.803979	0.158564
90	42	23	13	0.909902	0.105923
90	42	23	14	0.966162	0.056260
90	42	23	15	0.989791	0.023629
90	42	23	16	0.997571	0.007780
90	42	23	17	0.999554	0.001983
90	42	23	18	0.999939	0.000384
90	42	23	19	0.999994	0.000055
90	42	23	20	0.999999	0.000006
90	42	23	21	1.000000	0.000000
90	42	23	22	1.000000	0.000000
90	42	23	23	1.000000	0.000000
90	42	24	0	0.000000	0.000000
90	42	24	1	0.000000	0.000000
90	42	24	2	0.000006	0.000005
90	42	24	3	0.000064	0.000058
90	42	24	4	0.000490	0.000426
90	42	24	5	0.002722	0.002232
90	42	24	6	0.011439	0.008717
90	42	24	7	0.037469	0.026030
90	42	24	8	0.097969	0.060500
90	42	24	9	0.208783	0.110815
90	42	24	10	0.370116	0.161333
90	42	24	11	0.557849	0.187733
90	42	24	12	0.732980	0.175131
90	42	24	13	0.864055	0.131075
90	42	24	14	0.942650	0.078596
90	42	24	15	0.980269	0.037618
90	42	24	16	0.994552	0.014283
90	42	24	17	0.998814	0.004262
90	42	24	18	0.999801	0.000987
90	42	24	19	0.999975	0.000174
90	42	24	20	0.999998	0.000023
90	42	24	21	1.000000	0.000002
90	42	24	22	1.000000	0.000000
90	42	24	23	1.000000	0.000000

N	n	k	x	P(x)	p(x)
90	42	24	24	1.000000	0.000000
90	42	25	0	0.000000	0.000000
90	42	25	1	0.000000	0.000000
90	42	25	2	0.000002	0.000002
90	42	25	3	0.000029	0.000027
90	42	25	4	0.000245	0.000215
90	42	25	5	0.001470	0.001226
90	42	25	6	0.006684	0.005213
90	42	25	7	0.023665	0.016981
90	42	25	8	0.066802	0.043137
90	42	25	9	0.153376	0.086574
90	42	25	10	0.291894	0.138518
90	42	25	11	0.469672	0.177777
90	42	25	12	0.653375	0.183703
90	42	25	13	0.806461	0.153086
90	42	25	14	0.909307	0.102846
90	42	25	15	0.964879	0.055573
90	42	25	16	0.988925	0.024046
90	42	25	17	0.997200	0.008275
90	42	25	18	0.999442	0.002242
90	42	25	19	0.999914	0.000472
90	42	25	20	0.999990	0.000076
90	42	25	21	0.999999	0.000009
90	42	25	22	1.000000	0.000001
90	42	25	23	1.000000	0.000000
90	42	25	24	1.000000	0.000000
90	42	25	25	1.000000	0.000000
90	42	26	0	0.000000	0.000000
90	42	26	1	0.000000	0.000000
90	42	26	2	0.000001	0.000001
90	42	26	3	0.000013	0.000012
90	42	26	4	0.000119	0.000106
90	42	26	5	0.000773	0.000654
90	42	26	6	0.003796	0.003024
90	42	26	7	0.014521	0.010725
90	42	26	8	0.044238	0.029717
90	42	26	9	0.109423	0.065185
90	42	26	10	0.223701	0.114278
90	42	26	11	0.384885	0.161185
90	42	26	12	0.568589	0.183703
90	42	26	13	0.738161	0.169572
90	42	26	14	0.865004	0.126843
90	42	26	15	0.941795	0.076791
90	42	26	16	0.979307	0.037512
90	42	26	17	0.994017	0.014710
90	42	26	18	0.998614	0.004597
90	42	26	19	0.999747	0.001133
90	42	26	20	0.999964	0.000217
90	42	26	21	0.999996	0.000032
90	42	26	22	1.000000	0.000003

N	n	k	x	P(x)	p(x)
90	42	26	23	1.000000	0.000000
90	42	26	24	1.000000	0.000000
90	42	26	25	1.000000	0.000000
90	42	26	26	1.000000	0.000000
90	42	27	0	0.000000	0.000000
90	42	27	1	0.000000	0.000000
90	42	27	2	0.000000	0.000000
90	42	27	3	0.000006	0.000005
90	42	27	4	0.000056	0.000050
90	42	27	5	0.000395	0.000339
90	42	27	6	0.002096	0.001701
90	42	27	7	0.008656	0.006561
90	42	27	8	0.028451	0.019795
90	42	27	9	0.075812	0.047361
90	42	27	10	0.166562	0.090750
90	42	27	11	0.306812	0.140250
90	42	27	12	0.482478	0.175666
90	42	27	13	0.661324	0.178846
90	42	27	14	0.809510	0.148187
90	42	27	15	0.909399	0.099889
90	42	27	16	0.964068	0.054669
90	42	27	17	0.988271	0.024203
90	42	27	18	0.996890	0.008619
90	42	27	19	0.999340	0.002450
90	42	27	20	0.999890	0.000550
90	42	27	21	0.999986	0.000096
90	42	27	22	0.999998	0.000013
90	42	27	23	1.000000	0.000001
90	42	27	24	1.000000	0.000000
90	42	27	25	1.000000	0.000000
90	42	27	26	1.000000	0.000000
90	42	27	27	1.000000	0.000000
90	42	28	0	0.000000	0.000000
90	42	28	1	0.000000	0.000000
90	42	28	2	0.000000	0.000000
90	42	28	3	0.000002	0.000002
90	42	28	4	0.000026	0.000023
90	42	28	5	0.000196	0.000170
90	42	28	6	0.001124	0.000928
90	42	28	7	0.005011	0.003888
90	42	28	8	0.017768	0.012757
90	42	28	9	0.051004	0.033236
90	42	28	10	0.120467	0.069463
90	42	28	11	0.237800	0.117333
90	42	28	12	0.398827	0.161027
90	42	28	13	0.578998	0.180171
90	42	28	14	0.743650	0.164652
90	42	28	15	0.866589	0.122940
90	42	28	16	0.941506	0.074917
90	42	28	17	0.978666	0.037161

N	n	k	x	P(x)	p(x)
90	42	28	18	0.993607	0.014940
90	42	28	19	0.998446	0.004839
90	42	28	20	0.999698	0.001252
90	42	28	21	0.999954	0.000256
90	42	28	22	0.999994	0.000041
90	42	28	23	0.999999	0.000005
90	42	28	24	1.000000	0.000000
90	42	28	25	1.000000	0.000000
90	42	28	26	1.000000	0.000000
90	42	28	27	1.000000	0.000000
90	42	28	28	1.000000	0.000000
90	42	29	0	0.000000	0.000000
90	42	29	1	0.000000	0.000000
90	42	29	2	0.000000	0.000000
90	42	29	3	0.000001	0.000001
90	42	29	4	0.000011	0.000010
90	42	29	5	0.000094	0.000083
90	42	29	6	0.000585	0.000491
90	42	29	7	0.002817	0.002232
90	42	29	8	0.010772	0.007956
90	42	29	9	0.033314	0.022541
90	42	29	10	0.084615	0.051301
90	42	29	11	0.179133	0.094518
90	42	29	12	0.320911	0.141778
90	42	29	13	0.494725	0.173814
90	42	29	14	0.669291	0.174566
90	42	29	15	0.813051	0.143760
90	42	29	16	0.910089	0.097038
90	42	29	17	0.963682	0.053593
90	42	29	18	0.987823	0.024141
90	42	29	19	0.996651	0.008827
90	42	29	20	0.999254	0.002603
90	42	29	21	0.999867	0.000614
90	42	29	22	0.999981	0.000114
90	42	29	23	0.999998	0.000017
90	42	29	24	1.000000	0.000002
90	42	29	25	1.000000	0.000000
90	42	29	26	1.000000	0.000000
90	42	29	27	1.000000	0.000000
90	42	29	28	1.000000	0.000000
90	42	29	29	1.000000	0.000000
90	42	30	0	0.000000	0.000000
90	42	30	1	0.000000	0.000000
90	42	30	2	0.000000	0.000000
90	42	30	3	0.000000	0.000000
90	42	30	4	0.000005	0.000005
90	42	30	5	0.000044	0.000039
90	42	30	6	0.000295	0.000251
90	42	30	7	0.001536	0.001241
90	42	30	8	0.006338	0.004802

N	n	k	x	P(x)	p(x)
90	42	30	9	0.021119	0.014781
90	42	30	10	0.057703	0.036584
90	42	30	11	0.131099	0.073396
90	42	30	12	0.251184	0.120085
90	42	30	13	0.412092	0.160908
90	42	30	14	0.589162	0.177070
90	42	30	15	0.749419	0.160258
90	42	30	16	0.868729	0.119309
90	42	30	17	0.941718	0.072989
90	42	30	18	0.978325	0.036607
90	42	30	19	0.993322	0.014997
90	42	30	20	0.998315	0.004992
90	42	30	21	0.999656	0.001341
90	42	30	22	0.999944	0.000288
90	42	30	23	0.999993	0.000049
90	42	30	24	0.999999	0.000006
90	42	30	25	1.000000	0.000001
90	42	30	26	1.000000	0.000000
90	42	30	27	1.000000	0.000000
90	42	30	28	1.000000	0.000000
90	42	30	29	1.000000	0.000000
90	42	30	30	1.000000	0.000000
90	42	31	0	0.000000	0.000000
90	42	31	1	0.000000	0.000000
90	42	31	2	0.000000	0.000000
90	42	31	3	0.000000	0.000000
90	42	31	4	0.000002	0.000002
90	42	31	5	0.000020	0.000018
90	42	31	6	0.000145	0.000125
90	42	31	7	0.000812	0.000668
90	42	31	8	0.003617	0.002805
90	42	31	9	0.012990	0.009373
90	42	31	10	0.038192	0.025202
90	42	31	11	0.093178	0.054986
90	42	31	12	0.191142	0.097964
90	42	31	13	0.334320	0.143178
90	42	31	14	0.506529	0.172210
90	42	31	15	0.677304	0.170774
90	42	31	16	0.817028	0.139725
90	42	31	17	0.911306	0.094278
90	42	31	18	0.963682	0.052377
90	42	31	19	0.987574	0.023891
90	42	31	20	0.996484	0.008911
90	42	31	21	0.999186	0.002702
90	42	31	22	0.999848	0.000661
90	42	31	23	0.999977	0.000129
90	42	31	24	0.999997	0.000020
90	42	31	25	1.000000	0.000002
90	42	31	26	1.000000	0.000000
90	42	31	27	1.000000	0.000000
90	42	31	28	1.000000	0.000000
90	42	31	29	1.000000	0.000000
90	42	31	30	1.000000	0.000000
90	42	31	31	1.000000	0.000000
90	42	32	0	0.000000	0.000000
90	42	32	1	0.000000	0.000000
90	42	32	2	0.000000	0.000000
90	42	32	3	0.000000	0.000000
90	42	32	4	0.000001	0.000001
90	42	32	5	0.000009	0.000008
90	42	32	6	0.000069	0.000060
90	42	32	7	0.000416	0.000348
90	42	32	8	0.002001	0.001585
90	42	32	9	0.007747	0.005747
90	42	32	10	0.024523	0.016775
90	42	32	11	0.064287	0.039764
90	42	32	12	0.141330	0.077043
90	42	32	13	0.263944	0.122615
90	42	32	14	0.424803	0.160859
90	42	32	15	0.599153	0.174350
90	42	32	16	0.755455	0.156302
90	42	32	17	0.871358	0.115903
90	42	32	18	0.942377	0.071019
90	42	32	19	0.978260	0.035883
90	42	32	20	0.993162	0.014902
90	42	32	21	0.998225	0.005063
90	42	32	22	0.999624	0.001399
90	42	32	23	0.999936	0.000312
90	42	32	24	0.999991	0.000056
90	42	32	25	0.999999	0.000008
90	42	32	26	1.000000	0.000001
90	42	32	27	1.000000	0.000000
90	42	32	28	1.000000	0.000000
90	42	32	29	1.000000	0.000000
90	42	32	30	1.000000	0.000000
90	42	32	31	1.000000	0.000000
90	42	32	32	1.000000	0.000000
90	42	33	0	0.000000	0.000000
90	42	33	1	0.000000	0.000000
90	42	33	2	0.000000	0.000000
90	42	33	3	0.000000	0.000000
90	42	33	4	0.000000	0.000000
90	42	33	5	0.000004	0.000003
90	42	33	6	0.000031	0.000028
90	42	33	7	0.000206	0.000175
90	42	33	8	0.001072	0.000865
90	42	33	9	0.004478	0.003406
90	42	33	10	0.015267	0.010790
90	42	33	11	0.043034	0.027766
90	42	33	12	0.101480	0.058446
90	42	33	13	0.202637	0.101157
90	42	33	14	0.347147	0.144510
90	42	33	15	0.517990	0.170843
90	42	33	16	0.685388	0.167398
90	42	33	17	0.821399	0.136011
90	42	33	18	0.912990	0.091590
90	42	33	19	0.964030	0.051041
90	42	33	20	0.987509	0.023479
90	42	33	21	0.996391	0.008882
90	42	33	22	0.999141	0.002750
90	42	33	23	0.999833	0.000692
90	42	33	24	0.999974	0.000140
90	42	33	25	0.999997	0.000023
90	42	33	26	1.000000	0.000003
90	42	33	27	1.000000	0.000000
90	42	33	28	1.000000	0.000000
90	42	33	29	1.000000	0.000000
90	42	33	30	1.000000	0.000000
90	42	33	31	1.000000	0.000000
90	42	33	32	1.000000	0.000000
90	42	33	33	1.000000	0.000000
90	42	34	0	0.000000	0.000000
90	42	34	1	0.000000	0.000000
90	42	34	2	0.000000	0.000000
90	42	34	3	0.000000	0.000000
90	42	34	4	0.000000	0.000000
90	42	34	5	0.000002	0.000001
90	42	34	6	0.000014	0.000012
90	42	34	7	0.000099	0.000085
90	42	34	8	0.000556	0.000457
90	42	34	9	0.002506	0.001950
90	42	34	10	0.009210	0.006704
90	42	34	11	0.027933	0.018723
90	42	34	12	0.070719	0.042786
90	42	34	13	0.151171	0.080452
90	42	34	14	0.276159	0.124988
90	42	34	15	0.437064	0.160905
90	42	34	16	0.609031	0.171967
90	42	34	17	0.761745	0.152714
90	42	34	18	0.874425	0.112680
90	42	34	19	0.943435	0.069010
90	42	34	20	0.978447	0.035012
90	42	34	21	0.993119	0.014672
90	42	34	22	0.998176	0.005057
90	42	34	23	0.999603	0.001426
90	42	34	24	0.999930	0.000327
90	42	34	25	0.999990	0.000060
90	42	34	26	0.999999	0.000009
90	42	34	27	1.000000	0.000001
90	42	34	28	1.000000	0.000000
90	42	34	29	1.000000	0.000000
90	42	34	30	1.000000	0.000000
90	42	34	31	1.000000	0.000000
90	42	34	32	1.000000	0.000000
90	42	34	33	1.000000	0.000000
90	42	34	34	1.000000	0.000000
90	42	35	0	0.000000	0.000000
90	42	35	1	0.000000	0.000000
90	42	35	2	0.000000	0.000000
90	42	35	3	0.000000	0.000000
90	42	35	4	0.000000	0.000000
90	42	35	5	0.000001	0.000001
90	42	35	6	0.000006	0.000005
90	42	35	7	0.000046	0.000040
90	42	35	8	0.000278	0.000233
90	42	35	9	0.001357	0.001078
90	42	35	10	0.005379	0.004022
90	42	35	11	0.017568	0.012189
90	42	35	12	0.047798	0.030229
90	42	35	13	0.109508	0.061711
90	42	35	14	0.213665	0.104157
90	42	35	15	0.359485	0.145820
90	42	35	16	0.529189	0.169704
90	42	35	17	0.693569	0.164380
90	42	35	18	0.826134	0.132565
90	42	35	19	0.915092	0.088958
90	42	35	20	0.964692	0.049601
90	42	35	21	0.987617	0.022925
90	42	35	22	0.996370	0.008753
90	42	35	23	0.999119	0.002749
90	42	35	24	0.999824	0.000706
90	42	35	25	0.999972	0.000147
90	42	35	26	0.999996	0.000025
90	42	35	27	0.999999	0.000003
90	42	35	28	1.000000	0.000000
90	42	35	29	1.000000	0.000000
90	42	35	30	1.000000	0.000000
90	42	35	31	1.000000	0.000000
90	42	35	32	1.000000	0.000000
90	42	35	33	1.000000	0.000000
90	42	35	34	1.000000	0.000000
90	42	35	35	1.000000	0.000000
90	42	36	0	0.000000	0.000000
90	42	36	1	0.000000	0.000000
90	42	36	2	0.000000	0.000000
90	42	36	3	0.000000	0.000000
90	42	36	4	0.000000	0.000000
90	42	36	5	0.000000	0.000000
90	42	36	6	0.000002	0.000002
90	42	36	7	0.000020	0.000018

Table for $N = 2$, $n = 1$, through $N = 100$, $n = 50$

N	n	k	x	P(x)	p(x)
90	42	36	8	0.000135	0.000114
90	42	36	9	0.000710	0.000575
90	42	36	10	0.003039	0.002329
90	42	36	11	0.010698	0.007659
90	42	36	12	0.031309	0.020611
90	42	36	13	0.076970	0.045661
90	42	36	14	0.160640	0.083670
90	42	36	15	0.287901	0.127261
90	42	36	16	0.448965	0.161065
90	42	36	17	0.618851	0.169886
90	42	36	18	0.768288	0.149436
90	42	36	19	0.877891	0.109604
90	42	36	20	0.944852	0.066961
90	42	36	21	0.978864	0.034012
90	42	36	22	0.993187	0.014323
90	42	36	23	0.998169	0.004982
90	42	36	24	0.999594	0.001424
90	42	36	25	0.999926	0.000333
90	42	36	26	0.999989	0.000063
90	42	36	27	0.999999	0.000010
90	42	36	28	1.000000	0.000001
90	42	36	29	1.000000	0.000000
90	42	36	30	1.000000	0.000000
90	42	36	31	1.000000	0.000000
90	42	36	32	1.000000	0.000000
90	42	36	33	1.000000	0.000000
90	42	36	34	1.000000	0.000000
90	42	36	35	1.000000	0.000000
90	42	36	36	1.000000	0.000000
90	42	37	0	0.000000	0.000000
90	42	37	1	0.000000	0.000000
90	42	37	2	0.000000	0.000000
90	42	37	3	0.000000	0.000000
90	42	37	4	0.000000	0.000000
90	42	37	5	0.000000	0.000000
90	42	37	6	0.000001	0.000001
90	42	37	7	0.000009	0.000008
90	42	37	8	0.000063	0.000054
90	42	37	9	0.000358	0.000296
90	42	37	10	0.001659	0.001300
90	42	37	11	0.006301	0.004642
90	42	37	12	0.019858	0.013557
90	42	37	13	0.052448	0.032590
90	42	37	14	0.117255	0.064807
90	42	37	15	0.224270	0.107015
90	42	37	16	0.371416	0.147146
90	42	37	17	0.540200	0.168785
90	42	37	18	0.701871	0.161671
90	42	37	19	0.831208	0.129337
90	42	37	20	0.917572	0.086364

N	n	k	x	P(x)	p(x)
90	42	37	21	0.965637	0.048065
90	42	37	22	0.987882	0.022245
90	42	37	23	0.996416	0.008534
90	42	37	24	0.999119	0.002702
90	42	37	25	0.999821	0.000703
90	42	37	26	0.999970	0.000149
90	42	37	27	0.999996	0.000026
90	42	37	28	0.999999	0.000004
90	42	37	29	1.000000	0.000000
90	42	37	30	1.000000	0.000000
90	42	37	31	1.000000	0.000000
90	42	37	32	1.000000	0.000000
90	42	37	33	1.000000	0.000000
90	42	37	34	1.000000	0.000000
90	42	37	35	1.000000	0.000000
90	42	37	36	1.000000	0.000000
90	42	37	37	1.000000	0.000000
90	42	38	0	0.000000	0.000000
90	42	38	1	0.000000	0.000000
90	42	38	2	0.000000	0.000000
90	42	38	3	0.000000	0.000000
90	42	38	4	0.000000	0.000000
90	42	38	5	0.000000	0.000000
90	42	38	6	0.000000	0.000000
90	42	38	7	0.000004	0.000003
90	42	38	8	0.000028	0.000024
90	42	38	9	0.000174	0.000146
90	42	38	10	0.000873	0.000699
90	42	38	11	0.003586	0.002712
90	42	38	12	0.012184	0.008599
90	42	38	13	0.034616	0.022432
90	42	38	14	0.083018	0.048402
90	42	38	15	0.169753	0.086736
90	42	38	16	0.299231	0.129478
90	42	38	17	0.460585	0.161354
90	42	38	18	0.628662	0.168077
90	42	38	19	0.775081	0.146419
90	42	38	20	0.881723	0.106642
90	42	38	21	0.946592	0.064870
90	42	38	22	0.979488	0.032896
90	42	38	23	0.993357	0.013869
90	42	38	24	0.998201	0.004844
90	42	38	25	0.999596	0.001395
90	42	38	26	0.999925	0.000329
90	42	38	27	0.999989	0.000063
90	42	38	28	0.999999	0.000010
90	42	38	29	1.000000	0.000001
90	42	38	30	1.000000	0.000000
90	42	38	31	1.000000	0.000000
90	42	38	32	1.000000	0.000000

N	n	k	x	P(x)	p(x)
90	42	38	33	1.000000	0.000000
90	42	38	34	1.000000	0.000000
90	42	38	35	1.000000	0.000000
90	42	38	36	1.000000	0.000000
90	42	38	37	1.000000	0.000000
90	42	38	38	1.000000	0.000000
90	42	39	0	0.000000	0.000000
90	42	39	1	0.000000	0.000000
90	42	39	2	0.000000	0.000000
90	42	39	3	0.000000	0.000000
90	42	39	4	0.000000	0.000000
90	42	39	5	0.000000	0.000000
90	42	39	6	0.000000	0.000000
90	42	39	7	0.000001	0.000001
90	42	39	8	0.000012	0.000011
90	42	39	9	0.000082	0.000069
90	42	39	10	0.000443	0.000362
90	42	39	11	0.001969	0.001526
90	42	39	12	0.007224	0.005255
90	42	39	13	0.022106	0.014883
90	42	39	14	0.056955	0.034849
90	42	39	15	0.124717	0.067762
90	42	39	16	0.234492	0.109775
90	42	39	17	0.383011	0.148519
90	42	39	18	0.551088	0.168077
90	42	39	19	0.710319	0.159231
90	42	39	20	0.836605	0.126286
90	42	39	21	0.920395	0.083790
90	42	39	22	0.966836	0.046441
90	42	39	23	0.988289	0.021454
90	42	39	24	0.996524	0.008235
90	42	39	25	0.999140	0.002616
90	42	39	26	0.999824	0.000684
90	42	39	27	0.999970	0.000146
90	42	39	28	0.999996	0.000025
90	42	39	29	0.999999	0.000004
90	42	39	30	1.000000	0.000000
90	42	39	31	1.000000	0.000000
90	42	39	32	1.000000	0.000000
90	42	39	33	1.000000	0.000000
90	42	39	34	1.000000	0.000000
90	42	39	35	1.000000	0.000000
90	42	39	36	1.000000	0.000000
90	42	39	37	1.000000	0.000000
90	42	39	38	1.000000	0.000000
90	42	39	39	1.000000	0.000000
90	42	40	0	0.000000	0.000000
90	42	40	1	0.000000	0.000000
90	42	40	2	0.000000	0.000000
90	42	40	3	0.000000	0.000000

N	n	k	x	P(x)	p(x)
90	42	40	4	0.000000	0.000000
90	42	40	5	0.000000	0.000000
90	42	40	6	0.000000	0.000000
90	42	40	7	0.000001	0.000000
90	42	40	8	0.000005	0.000004
90	42	40	9	0.000037	0.000032
90	42	40	10	0.000216	0.000180
90	42	40	11	0.001041	0.000825
90	42	40	12	0.004133	0.003091
90	42	40	13	0.013643	0.009511
90	42	40	14	0.037822	0.024179
90	42	40	15	0.088843	0.051021
90	42	40	16	0.178529	0.089685
90	42	40	17	0.310208	0.131679
90	42	40	18	0.471993	0.161785
90	42	40	19	0.638509	0.166516
90	42	40	20	0.782129	0.143620
90	42	40	21	0.885893	0.103765
90	42	40	22	0.948624	0.062730
90	42	40	23	0.980297	0.031673
90	42	40	24	0.993618	0.013321
90	42	40	25	0.998268	0.004650
90	42	40	26	0.999609	0.001341
90	42	40	27	0.999927	0.000318
90	42	40	28	0.999989	0.000062
90	42	40	29	0.999999	0.000010
90	42	40	30	1.000000	0.000001
90	42	40	31	1.000000	0.000000
90	42	40	32	1.000000	0.000000
90	42	40	33	1.000000	0.000000
90	42	40	34	1.000000	0.000000
90	42	40	35	1.000000	0.000000
90	42	40	36	1.000000	0.000000
90	42	40	37	1.000000	0.000000
90	42	40	38	1.000000	0.000000
90	42	40	39	1.000000	0.000000
90	42	40	40	1.000000	0.000000
90	42	41	0	0.000000	0.000000
90	42	41	1	0.000000	0.000000
90	42	41	2	0.000000	0.000000
90	42	41	3	0.000000	0.000000
90	42	41	4	0.000000	0.000000
90	42	41	5	0.000000	0.000000
90	42	41	6	0.000000	0.000000
90	42	41	7	0.000000	0.000000
90	42	41	8	0.000002	0.000002
90	42	41	9	0.000016	0.000014
90	42	41	10	0.000101	0.000086
90	42	41	11	0.000530	0.000429
90	42	41	12	0.002278	0.001748

N	n	k	x	P(x)	p(x)
90	42	41	13	0.008127	0.005849
90	42	41	14	0.024282	0.016155
90	42	41	15	0.061292	0.037010
90	42	41	16	0.131892	0.070600
90	42	41	17	0.244368	0.112476
90	42	41	18	0.394336	0.149968
90	42	41	19	0.561911	0.167575
90	42	41	20	0.718936	0.157024
90	42	41	21	0.842312	0.123376
90	42	41	22	0.923532	0.081220
90	42	41	23	0.968261	0.044730
90	42	41	24	0.988822	0.020561
90	42	41	25	0.996687	0.007865
90	42	41	26	0.999180	0.002493
90	42	41	27	0.999832	0.000652
90	42	41	28	0.999972	0.000140
90	42	41	29	0.999996	0.000024
90	42	41	30	0.999999	0.000003
90	42	41	31	1.000000	0.000000
90	42	41	32	1.000000	0.000000
90	42	41	33	1.000000	0.000000
90	42	41	34	1.000000	0.000000
90	42	41	35	1.000000	0.000000
90	42	41	36	1.000000	0.000000
90	42	41	37	1.000000	0.000000
90	42	41	38	1.000000	0.000000
90	42	41	39	1.000000	0.000000
90	42	41	40	1.000000	0.000000
90	42	41	41	1.000000	0.000000
90	42	42	0	0.000000	0.000000
90	42	42	1	0.000000	0.000000
90	42	42	2	0.000000	0.000000
90	42	42	3	0.000000	0.000000
90	42	42	4	0.000000	0.000000
90	42	42	5	0.000000	0.000000
90	42	42	6	0.000000	0.000000
90	42	42	7	0.000000	0.000000
90	42	42	8	0.000001	0.000001
90	42	42	9	0.000006	0.000006
90	42	42	10	0.000045	0.000039
90	42	42	11	0.000259	0.000213
90	42	42	12	0.001208	0.000949
90	42	42	13	0.004665	0.003458
90	42	42	14	0.015051	0.010385
90	42	42	15	0.040899	0.025848
90	42	42	16	0.094431	0.053532
90	42	42	17	0.186982	0.092552
90	42	42	18	0.320882	0.133900
90	42	42	19	0.483254	0.162371
90	42	42	20	0.648435	0.165181

N	n	k	x	P(x)	p(x)
90	42	42	21	0.789437	0.141001
90	42	42	22	0.890381	0.100944
90	42	42	23	0.950917	0.060536
90	42	42	24	0.981269	0.030352
90	42	42	25	0.993958	0.012689
90	42	42	26	0.998366	0.004408
90	42	42	27	0.999632	0.001266
90	42	42	28	0.999932	0.000299
90	42	42	29	0.999990	0.000058
90	42	42	30	0.999999	0.000009
90	42	42	31	1.000000	0.000001
90	42	42	32	1.000000	0.000000
90	42	42	33	1.000000	0.000000
90	42	42	34	1.000000	0.000000
90	42	42	35	1.000000	0.000000
90	42	42	36	1.000000	0.000000
90	42	42	37	1.000000	0.000000
90	42	42	38	1.000000	0.000000
90	42	42	39	1.000000	0.000000
90	42	42	40	1.000000	0.000000
90	42	42	41	1.000000	0.000000
90	42	42	42	1.000000	0.000000
90	43	1	0	0.522222	0.522222
90	43	1	1	1.000000	0.477778
90	43	2	0	0.269913	0.269913
90	43	2	1	0.774532	0.504619
90	43	2	2	1.000000	0.225468
90	43	3	0	0.138023	0.138023
90	43	3	1	0.533691	0.395667
90	43	3	2	0.894952	0.361261
90	43	3	3	1.000000	0.105048
90	43	4	0	0.069805	0.069805
90	43	4	1	0.342679	0.272874
90	43	4	2	0.724703	0.382024
90	43	4	3	0.951702	0.227000
90	43	4	4	1.000000	0.048298
90	43	5	0	0.034902	0.034902
90	43	5	1	0.209415	0.174512
90	43	5	2	0.542575	0.333160
90	43	5	3	0.846121	0.303546
90	43	5	4	0.978097	0.131976
90	43	5	5	1.000000	0.021902
90	43	6	0	0.017246	0.017246
90	43	6	1	0.123185	0.105939
90	43	6	2	0.381874	0.258689
90	43	6	3	0.703276	0.321402
90	43	6	4	0.917543	0.214268
90	43	6	5	0.990208	0.072665
90	43	6	6	1.000000	0.009792
90	43	7	0	0.008418	0.008418

N	n	k	x	P(x)	p(x)
90	43	7	1	0.070216	0.061798
90	43	7	2	0.255609	0.185394
90	43	7	3	0.550228	0.294618
90	43	7	4	0.818062	0.267835
90	43	7	5	0.957336	0.139274
90	43	7	6	0.995687	0.038351
90	43	7	7	1.000000	0.004313
90	43	8	0	0.004057	0.004057
90	43	8	1	0.038944	0.034888
90	43	8	2	0.164029	0.125085
90	43	8	3	0.408243	0.244214
90	43	8	4	0.692212	0.283969
90	43	8	5	0.893572	0.201360
90	43	8	6	0.978591	0.085019
90	43	8	7	0.998129	0.019538
90	43	8	8	1.000000	0.001871
90	43	9	0	0.001929	0.001929
90	43	9	1	0.021075	0.019146
90	43	9	2	0.101487	0.080412
90	43	9	3	0.289114	0.187627
90	43	9	4	0.557154	0.268039
90	43	9	5	0.800259	0.243105
90	43	9	6	0.940229	0.139970
90	43	9	7	0.989551	0.049323
90	43	9	8	0.999201	0.009650
90	43	9	9	1.000000	0.000798
90	43	10	0	0.000905	0.000905
90	43	10	1	0.011148	0.010243
90	43	10	2	0.060785	0.049637
90	43	10	3	0.196459	0.135674
90	43	10	4	0.428098	0.231639
90	43	10	5	0.686210	0.258112
90	43	10	6	0.876292	0.190082
90	43	10	7	0.967630	0.091338
90	43	10	8	0.995032	0.027401
90	43	10	9	0.999665	0.004633
90	43	10	10	1.000000	0.000335
90	43	11	0	0.000419	0.000419
90	43	11	1	0.005770	0.005352
90	43	11	2	0.035346	0.029575
90	43	11	3	0.128622	0.093276
90	43	11	4	0.315174	0.186552
90	43	11	5	0.563606	0.248433
90	43	11	6	0.788379	0.224772
90	43	11	7	0.926528	0.138149
90	43	11	8	0.983043	0.056516
90	43	11	9	0.997696	0.014652
90	43	11	10	0.999862	0.002166
90	43	11	11	1.000000	0.000138
90	43	12	0	0.000191	0.000191

N	n	k	x	P(x)	p(x)
90	43	12	1	0.002925	0.002734
90	43	12	2	0.019996	0.017071
90	43	12	3	0.081393	0.061397
90	43	12	4	0.223078	0.141685
90	43	12	5	0.444107	0.221029
90	43	12	6	0.683106	0.238999
90	43	12	7	0.863574	0.180468
90	43	12	8	0.958005	0.094431
90	43	12	9	0.991390	0.033385
90	43	12	10	0.998957	0.007567
90	43	12	11	0.999944	0.000987
90	43	12	12	1.000000	0.000056
90	43	13	0	0.000086	0.000086
90	43	13	1	0.001453	0.001367
90	43	13	2	0.011023	0.009570
90	43	13	3	0.049908	0.038885
90	43	13	4	0.152236	0.102328
90	43	13	5	0.336426	0.184191
90	43	13	6	0.569735	0.233308
90	43	13	7	0.780281	0.210546
90	43	13	8	0.915632	0.135351
90	43	13	9	0.976837	0.061205
90	43	13	10	0.995755	0.018918
90	43	13	11	0.999539	0.003784
90	43	13	12	0.999978	0.000439
90	43	13	13	1.000000	0.000022
90	43	14	0	0.000038	0.000038
90	43	14	1	0.000707	0.000669
90	43	14	2	0.005927	0.005220
90	43	14	3	0.029708	0.023781
90	43	14	4	0.100407	0.070699
90	43	14	5	0.245527	0.145120
90	43	14	6	0.457625	0.212098
90	43	14	7	0.681844	0.224218
90	43	14	8	0.854109	0.172265
90	43	14	9	0.949812	0.095703
90	43	14	10	0.987648	0.037836
90	43	14	11	0.997967	0.010319
90	43	14	12	0.999801	0.001834
90	43	14	13	0.999991	0.000190
90	43	14	14	1.000000	0.000009
90	43	15	0	0.000016	0.000016
90	43	15	1	0.000337	0.000321
90	43	15	2	0.003111	0.002774
90	43	15	3	0.017192	0.014081
90	43	15	4	0.064127	0.046936
90	43	15	5	0.172967	0.108840
90	43	15	6	0.354367	0.181400
90	43	15	7	0.575635	0.221268
90	43	15	8	0.774776	0.199141

Table for $N = 2$, $n = 1$, through $N = 100$, $n = 50$

N	n	k	x	P(x)	p(x)
90	43	15	9	0.906997	0.132221
90	43	15	10	0.971219	0.064222
90	43	15	11	0.993622	0.022403
90	43	15	12	0.999053	0.005431
90	43	15	13	0.999916	0.000863
90	43	15	14	0.999997	0.000080
90	43	15	15	1.000000	0.000003
90	43	16	0	0.000007	0.000007
90	43	16	1	0.000158	0.000151
90	43	16	2	0.001595	0.001437
90	43	16	3	0.009682	0.008087
90	43	16	4	0.039721	0.030039
90	43	16	5	0.117822	0.078101
90	43	16	6	0.264877	0.147055
90	43	16	7	0.469426	0.204550
90	43	16	8	0.681844	0.212417
90	43	16	9	0.847057	0.165213
90	43	16	10	0.942961	0.095904
90	43	16	11	0.984063	0.041102
90	43	16	12	0.996808	0.012745
90	43	16	13	0.999571	0.002763
90	43	16	14	0.999965	0.000395
90	43	16	15	0.999999	0.000033
90	43	16	16	1.000000	0.000001
90	43	17	0	0.000003	0.000003
90	43	17	1	0.000072	0.000069
90	43	17	2	0.000798	0.000726
90	43	17	3	0.005310	0.004512
90	43	17	4	0.023890	0.018579
90	43	17	5	0.077716	0.053826
90	43	17	6	0.191349	0.113633
90	43	17	7	0.369916	0.178567
90	43	17	8	0.581376	0.211460
90	43	17	9	0.771148	0.189772
90	43	17	10	0.900193	0.129045
90	43	17	11	0.966289	0.066096
90	43	17	12	0.991469	0.025180
90	43	17	13	0.998451	0.006982
90	43	17	14	0.999811	0.001360
90	43	17	15	0.999986	0.000175
90	43	17	16	0.999999	0.000013
90	43	17	17	1.000000	0.000000
90	43	18	0	0.000001	0.000001
90	43	18	1	0.000032	0.000031
90	43	18	2	0.000390	0.000358
90	43	18	3	0.002838	0.002448
90	43	18	4	0.013964	0.011126
90	43	18	5	0.049697	0.035733
90	43	18	6	0.133754	0.084057
90	43	18	7	0.281855	0.148101
90	43	18	8	0.479991	0.198135
90	43	18	9	0.682761	0.202770
90	43	18	10	0.841858	0.159097
90	43	18	11	0.937316	0.095458
90	43	18	12	0.980776	0.043461
90	43	18	13	0.995581	0.014805
90	43	18	14	0.999270	0.003689
90	43	18	15	0.999919	0.000648
90	43	18	16	0.999994	0.000076
90	43	18	17	1.000000	0.000005
90	43	18	18	1.000000	0.000000
90	43	19	0	0.000000	0.000000
90	43	19	1	0.000014	0.000014
90	43	19	2	0.000187	0.000172
90	43	19	3	0.001478	0.001292
90	43	19	4	0.007937	0.006459
90	43	19	5	0.030838	0.022900
90	43	19	6	0.090558	0.059720
90	43	19	7	0.207805	0.117247
90	43	19	8	0.383675	0.175870
90	43	19	9	0.587009	0.203333
90	43	19	10	0.768938	0.181930
90	43	19	11	0.894890	0.125951
90	43	19	12	0.962064	0.067174
90	43	19	13	0.989413	0.027349
90	43	19	14	0.997785	0.008372
90	43	19	15	0.999667	0.001882
90	43	19	16	0.999966	0.000299
90	43	19	17	0.999998	0.000032
90	43	19	18	1.000000	0.000002
90	43	19	19	1.000000	0.000000
90	43	20	0	0.000000	0.000000
90	43	20	1	0.000006	0.000006
90	43	20	2	0.000087	0.000081
90	43	20	3	0.000751	0.000664
90	43	20	4	0.004389	0.003639
90	43	20	5	0.018581	0.014192
90	43	20	6	0.059436	0.040855
90	43	20	7	0.148356	0.088920
90	43	20	8	0.296978	0.148623
90	43	20	9	0.489638	0.192659
90	43	20	10	0.684379	0.194742
90	43	20	11	0.838123	0.153744
90	43	20	12	0.932735	0.094611
90	43	20	13	0.977857	0.045122
90	43	20	14	0.994365	0.016508
90	43	20	15	0.998924	0.004559
90	43	20	16	0.999852	0.000928
90	43	20	17	0.999986	0.000134
90	43	20	18	0.999999	0.000013
90	43	20	19	1.000000	0.000001
90	43	20	20	1.000000	0.000000
90	43	21	0	0.000000	0.000000
90	43	21	1	0.000003	0.000002
90	43	21	2	0.000040	0.000037
90	43	21	3	0.000371	0.000332
90	43	21	4	0.002362	0.001991
90	43	21	5	0.010877	0.008515
90	43	21	6	0.037841	0.026964
90	43	21	7	0.102626	0.064784
90	43	21	8	0.222667	0.120041
90	43	21	9	0.396060	0.173393
90	43	21	10	0.592573	0.196512
90	43	21	11	0.767840	0.175268
90	43	21	12	0.890835	0.122995
90	43	21	13	0.958519	0.067684
90	43	21	14	0.987526	0.029007
90	43	21	15	0.997101	0.009575
90	43	21	16	0.999494	0.002394
90	43	21	17	0.999936	0.000442
90	43	21	18	0.999994	0.000058
90	43	21	19	1.000000	0.000005
90	43	21	20	1.000000	0.000000
90	43	21	21	1.000000	0.000000
90	43	22	0	0.000000	0.000000
90	43	22	1	0.000001	0.000001
90	43	22	2	0.000018	0.000017
90	43	22	3	0.000179	0.000161
90	43	22	4	0.001237	0.001058
90	43	22	5	0.006188	0.004951
90	43	22	6	0.023382	0.017195
90	43	22	7	0.068825	0.045443
90	43	22	8	0.161776	0.092951
90	43	22	9	0.310620	0.148844
90	43	22	10	0.498588	0.187968
90	43	22	11	0.686557	0.187968
90	43	22	12	0.835577	0.149020
90	43	22	13	0.929091	0.093514
90	43	22	14	0.975335	0.046243
90	43	22	15	0.993215	0.017881
90	43	22	16	0.998558	0.005342
90	43	22	17	0.999770	0.001212
90	43	22	18	0.999974	0.000204
90	43	22	19	0.999998	0.000024
90	43	22	20	1.000000	0.000002
90	43	22	21	1.000000	0.000000
90	43	22	22	1.000000	0.000000
90	43	23	0	0.000000	0.000000
90	43	23	1	0.000000	0.000000
90	43	23	2	0.000008	0.000007
90	43	23	3	0.000084	0.000076
90	43	23	4	0.000630	0.000546
90	43	23	5	0.003421	0.002791
90	43	23	6	0.014026	0.010605
90	43	23	7	0.044767	0.030741
90	43	23	8	0.113934	0.069167
90	43	23	9	0.236198	0.122264
90	43	23	10	0.407368	0.171170
90	43	23	11	0.598101	0.190733
90	43	23	12	0.767641	0.169540
90	43	23	13	0.887835	0.120194
90	43	23	14	0.955613	0.067778
90	43	23	15	0.985853	0.030240
90	43	23	16	0.996436	0.010584
90	43	23	17	0.999306	0.002870
90	43	23	18	0.999899	0.000592
90	43	23	19	0.999989	0.000091
90	43	23	20	0.999999	0.000010
90	43	23	21	1.000000	0.000001
90	43	23	22	1.000000	0.000000
90	43	23	23	1.000000	0.000000
90	43	24	0	0.000000	0.000000
90	43	24	1	0.000000	0.000000
90	43	24	2	0.000003	0.000003
90	43	24	3	0.000038	0.000035
90	43	24	4	0.000312	0.000274
90	43	24	5	0.001838	0.001526
90	43	24	6	0.008170	0.006332
90	43	24	7	0.028250	0.020080
90	43	24	8	0.077802	0.049552
90	43	24	9	0.174154	0.096352
90	43	24	10	0.323061	0.148907
90	43	24	11	0.507005	0.183944
90	43	24	12	0.689197	0.182192
90	43	24	13	0.834017	0.144819
90	43	24	14	0.926276	0.092259
90	43	24	15	0.973215	0.046939
90	43	24	16	0.992171	0.018956
90	43	24	17	0.998193	0.006021
90	43	24	18	0.999678	0.001485
90	43	24	19	0.999957	0.000279
90	43	24	20	0.999996	0.000039
90	43	24	21	1.000000	0.000004
90	43	24	22	1.000000	0.000000
90	43	24	23	1.000000	0.000000
90	43	24	24	1.000000	0.000000
90	43	25	0	0.000000	0.000000
90	43	25	1	0.000000	0.000000
90	43	25	2	0.000001	0.000001
90	43	25	3	0.000017	0.000016

Table for $N=2$, $n=1$, through $N=100$, $n=50$

N	n	k	x	P(x)	p(x)
90	43	25	4	0.000150	0.000133
90	43	25	5	0.000960	0.000809
90	43	25	6	0.004620	0.003661
90	43	25	7	0.017297	0.012677
90	43	25	8	0.051524	0.034227
90	43	25	9	0.124518	0.072994
90	43	25	10	0.248607	0.124089
90	43	25	11	0.417820	0.169213
90	43	25	12	0.603622	0.185802
90	43	25	13	0.768190	0.164568
90	43	25	14	0.885738	0.117548
90	43	25	15	0.953302	0.067564
90	43	25	16	0.984417	0.031115
90	43	25	17	0.995821	0.011404
90	43	25	18	0.999115	0.003295
90	43	25	19	0.999855	0.000740
90	43	25	20	0.999982	0.000127
90	43	25	21	0.999998	0.000016
90	43	25	22	1.000000	0.000001
90	43	25	23	1.000000	0.000000
90	43	25	24	1.000000	0.000000
90	43	25	25	1.000000	0.000000
90	43	26	0	0.000000	0.000000
90	43	26	1	0.000000	0.000000
90	43	26	2	0.000001	0.000000
90	43	26	3	0.000007	0.000007
90	43	26	4	0.000070	0.000063
90	43	26	5	0.000487	0.000416
90	43	26	6	0.002536	0.002050
90	43	26	7	0.010276	0.007739
90	43	26	8	0.033094	0.022818
90	43	26	9	0.086337	0.053242
90	43	26	10	0.185608	0.099271
90	43	26	11	0.334515	0.148907
90	43	26	12	0.515009	0.180493
90	43	26	13	0.692235	0.177227
90	43	26	14	0.833293	0.141058
90	43	26	15	0.924197	0.090904
90	43	26	16	0.971492	0.047295
90	43	26	17	0.991259	0.019767
90	43	26	18	0.997848	0.006589
90	43	26	19	0.999582	0.001734
90	43	26	20	0.999937	0.000355
90	43	26	21	0.999993	0.000056
90	43	26	22	0.999999	0.000006
90	43	26	23	1.000000	0.000001
90	43	26	24	1.000000	0.000000
90	43	26	25	1.000000	0.000000
90	43	26	26	1.000000	0.000000
90	43	27	0	0.000000	0.000000

N	n	k	x	P(x)	p(x)
90	43	27	1	0.000000	0.000000
90	43	27	2	0.000000	0.000000
90	43	27	3	0.000003	0.000003
90	43	27	4	0.000032	0.000029
90	43	27	5	0.000239	0.000207
90	43	27	6	0.001351	0.001112
90	43	27	7	0.005923	0.004571
90	43	27	8	0.020615	0.014693
90	43	27	9	0.058052	0.037436
90	43	27	10	0.134421	0.076370
90	43	27	11	0.260062	0.125640
90	43	27	12	0.427582	0.167521
90	43	27	13	0.609160	0.181578
90	43	27	14	0.769376	0.160216
90	43	27	15	0.884427	0.115050
90	43	27	16	0.951539	0.067113
90	43	27	17	0.983229	0.031689
90	43	27	18	0.995274	0.012046
90	43	27	19	0.998932	0.003658
90	43	27	20	0.999810	0.000878
90	43	27	21	0.999974	0.000164
90	43	27	22	0.999997	0.000023
90	43	27	23	1.000000	0.000002
90	43	27	24	1.000000	0.000000
90	43	27	25	1.000000	0.000000
90	43	27	26	1.000000	0.000000
90	43	27	27	1.000000	0.000000
90	43	28	0	0.000000	0.000000
90	43	28	1	0.000000	0.000000
90	43	28	2	0.000000	0.000000
90	43	28	3	0.000001	0.000001
90	43	28	4	0.000014	0.000013
90	43	28	5	0.000114	0.000100
90	43	28	6	0.000698	0.000584
90	43	28	7	0.003310	0.002612
90	43	28	8	0.012453	0.009142
90	43	28	9	0.037848	0.025395
90	43	28	10	0.094418	0.056570
90	43	28	11	0.196244	0.101826
90	43	28	12	0.345151	0.148907
90	43	28	13	0.522695	0.177543
90	43	28	14	0.695626	0.172932
90	43	28	15	0.833293	0.137667
90	43	28	16	0.922777	0.089484
90	43	28	17	0.970151	0.047374
90	43	28	18	0.990494	0.020344
90	43	28	19	0.997538	0.007044
90	43	28	20	0.999489	0.001951
90	43	28	21	0.999916	0.000427
90	43	28	22	0.999989	0.000073

N	n	k	x	P(x)	p(x)
90	43	28	23	0.999999	0.000010
90	43	28	24	1.000000	0.000001
90	43	28	25	1.000000	0.000000
90	43	28	26	1.000000	0.000000
90	43	28	27	1.000000	0.000000
90	43	28	28	1.000000	0.000000
90	43	29	0	0.000000	0.000000
90	43	29	1	0.000000	0.000000
90	43	29	2	0.000000	0.000000
90	43	29	3	0.000000	0.000000
90	43	29	4	0.000006	0.000006
90	43	29	5	0.000053	0.000047
90	43	29	6	0.000350	0.000297
90	43	29	7	0.001794	0.001444
90	43	29	8	0.007292	0.005498
90	43	29	9	0.023922	0.016630
90	43	29	10	0.064308	0.040387
90	43	29	11	0.143689	0.079381
90	43	29	12	0.270698	0.127009
90	43	29	13	0.436787	0.166089
90	43	29	14	0.614739	0.177952
90	43	29	15	0.771121	0.156382
90	43	29	16	0.883808	0.112687
90	43	29	17	0.950284	0.066476
90	43	29	18	0.982291	0.032007
90	43	29	19	0.994812	0.012520
90	43	29	20	0.998765	0.003954
90	43	29	21	0.999765	0.000999
90	43	29	22	0.999965	0.000200
90	43	29	23	0.999996	0.000031
90	43	29	24	1.000000	0.000004
90	43	29	25	1.000000	0.000000
90	43	29	26	1.000000	0.000000
90	43	29	27	1.000000	0.000000
90	43	29	28	1.000000	0.000000
90	43	29	29	1.000000	0.000000
90	43	30	0	0.000000	0.000000
90	43	30	1	0.000000	0.000000
90	43	30	2	0.000000	0.000000
90	43	30	3	0.000000	0.000000
90	43	30	4	0.000002	0.000002
90	43	30	5	0.000024	0.000021
90	43	30	6	0.000170	0.000146
90	43	30	7	0.000942	0.000772
90	43	30	8	0.004137	0.003196
90	43	30	9	0.014652	0.010515
90	43	30	10	0.042460	0.027807
90	43	30	11	0.102047	0.059587
90	43	30	12	0.206152	0.104106
90	43	30	13	0.355104	0.148951

N	n	k	x	P(x)	p(x)
90	43	30	14	0.530139	0.175035
90	43	30	15	0.699339	0.169200
90	43	30	16	0.833930	0.134591
90	43	30	17	0.921950	0.088020
90	43	30	18	0.969174	0.047223
90	43	30	19	0.989886	0.020712
90	43	30	20	0.997275	0.007389
90	43	30	21	0.999404	0.002130
90	43	30	22	0.999896	0.000491
90	43	30	23	0.999986	0.000090
90	43	30	24	0.999998	0.000013
90	43	30	25	1.000000	0.000001
90	43	30	26	1.000000	0.000000
90	43	30	27	1.000000	0.000000
90	43	30	28	1.000000	0.000000
90	43	30	29	1.000000	0.000000
90	43	30	30	1.000000	0.000000
90	43	31	0	0.000000	0.000000
90	43	31	1	0.000000	0.000000
90	43	31	2	0.000000	0.000000
90	43	31	3	0.000000	0.000000
90	43	31	4	0.000001	0.000001
90	43	31	5	0.000010	0.000009
90	43	31	6	0.000080	0.000069
90	43	31	7	0.000478	0.000399
90	43	31	8	0.002273	0.001795
90	43	31	9	0.008694	0.006421
90	43	31	10	0.027166	0.018472
90	43	31	11	0.070267	0.043101
90	43	31	12	0.152364	0.082097
90	43	31	13	0.280628	0.128264
90	43	31	14	0.445538	0.164910
90	43	31	15	0.620379	0.174840
90	43	31	16	0.773364	0.152985
90	43	31	17	0.883808	0.110444
90	43	31	18	0.949497	0.065689
90	43	31	19	0.981601	0.032104
90	43	31	20	0.994442	0.012841
90	43	31	21	0.998623	0.004181
90	43	31	22	0.999724	0.001100
90	43	31	23	0.999956	0.000232
90	43	31	24	0.999994	0.000039
90	43	31	25	0.999999	0.000005
90	43	31	26	1.000000	0.000000
90	43	31	27	1.000000	0.000000
90	43	31	28	1.000000	0.000000
90	43	31	29	1.000000	0.000000
90	43	31	30	1.000000	0.000000
90	43	31	31	1.000000	0.000000
90	43	32	0	0.000000	0.000000

Table for $N = 2,\ n = 1$, through $N = 100,\ n = 50$

N	n	k	x	P(x)	p(x)
90	43	32	1	0.000000	0.000000
90	43	32	2	0.000000	0.000000
90	43	32	3	0.000000	0.000000
90	43	32	4	0.000000	0.000000
90	43	32	5	0.000004	0.000004
90	43	32	6	0.000036	0.000032
90	43	32	7	0.000235	0.000199
90	43	32	8	0.001208	0.000973
90	43	32	9	0.004994	0.003785
90	43	32	10	0.016834	0.011840
90	43	32	11	0.046890	0.030056
90	43	32	12	0.109228	0.062338
90	43	32	13	0.215409	0.106181
90	43	32	14	0.364481	0.149072
90	43	32	15	0.537404	0.172923
90	43	32	16	0.703354	0.165950
90	43	32	17	0.835138	0.131784
90	43	32	18	0.921663	0.086525
90	43	32	19	0.968542	0.046879
90	43	32	20	0.989436	0.020895
90	43	32	21	0.997064	0.007628
90	43	32	22	0.999332	0.002268
90	43	32	23	0.999877	0.000545
90	43	32	24	0.999982	0.000105
90	43	32	25	0.999998	0.000016
90	43	32	26	1.000000	0.000002
90	43	32	27	1.000000	0.000000
90	43	32	28	1.000000	0.000000
90	43	32	29	1.000000	0.000000
90	43	32	30	1.000000	0.000000
90	43	32	31	1.000000	0.000000
90	43	32	32	1.000000	0.000000
90	43	33	0	0.000000	0.000000
90	43	33	1	0.000000	0.000000
90	43	33	2	0.000000	0.000000
90	43	33	3	0.000000	0.000000
90	43	33	4	0.000000	0.000000
90	43	33	5	0.000002	0.000002
90	43	33	6	0.000016	0.000014
90	43	33	7	0.000112	0.000096
90	43	33	8	0.000621	0.000509
90	43	33	9	0.002775	0.002154
90	43	33	10	0.010097	0.007322
90	43	33	11	0.030307	0.020210
90	43	33	12	0.075910	0.045602
90	43	33	13	0.160488	0.084578
90	43	33	14	0.289945	0.129457
90	43	33	15	0.453924	0.163979
90	43	33	16	0.626101	0.172178
90	43	33	17	0.776062	0.149961
90	43	33	18	0.884368	0.108305
90	43	33	19	0.949144	0.064776
90	43	33	20	0.981150	0.032007
90	43	33	21	0.994171	0.013020
90	43	33	22	0.998511	0.004340
90	43	33	23	0.999689	0.001178
90	43	33	24	0.999948	0.000258
90	43	33	25	0.999993	0.000045
90	43	33	26	0.999999	0.000006
90	43	33	27	1.000000	0.000001
90	43	33	28	1.000000	0.000000
90	43	33	29	1.000000	0.000000
90	43	33	30	1.000000	0.000000
90	43	33	31	1.000000	0.000000
90	43	33	32	1.000000	0.000000
90	43	33	33	1.000000	0.000000
90	43	34	0	0.000000	0.000000
90	43	34	1	0.000000	0.000000
90	43	34	2	0.000000	0.000000
90	43	34	3	0.000000	0.000000
90	43	34	4	0.000000	0.000000
90	43	34	5	0.000001	0.000001
90	43	34	6	0.000007	0.000006
90	43	34	7	0.000051	0.000044
90	43	34	8	0.000308	0.000257
90	43	34	9	0.001490	0.001182
90	43	34	10	0.005858	0.004368
90	43	34	11	0.018961	0.013103
90	43	34	12	0.051108	0.032147
90	43	34	13	0.115973	0.064865
90	43	34	14	0.224081	0.108108
90	43	34	15	0.373373	0.149292
90	43	34	16	0.544543	0.171171
90	43	34	17	0.707659	0.163116
90	43	34	18	0.836865	0.129206
90	43	34	19	0.921870	0.085004
90	43	34	20	0.968235	0.046366
90	43	34	21	0.989146	0.020910
90	43	34	22	0.996912	0.007767
90	43	34	23	0.999276	0.002364
90	43	34	24	0.999861	0.000586
90	43	34	25	0.999979	0.000117
90	43	34	26	0.999997	0.000019
90	43	34	27	1.000000	0.000002
90	43	34	28	1.000000	0.000000
90	43	34	29	1.000000	0.000000
90	43	34	30	1.000000	0.000000
90	43	34	31	1.000000	0.000000
90	43	34	32	1.000000	0.000000
90	43	34	33	1.000000	0.000000
90	43	34	34	1.000000	0.000000
90	43	35	0	0.000000	0.000000
90	43	35	1	0.000000	0.000000
90	43	35	2	0.000000	0.000000
90	43	35	3	0.000000	0.000000
90	43	35	4	0.000000	0.000000
90	43	35	5	0.000000	0.000000
90	43	35	6	0.000003	0.000002
90	43	35	7	0.000023	0.000020
90	43	35	8	0.000148	0.000125
90	43	35	9	0.000773	0.000625
90	43	35	10	0.003284	0.002511
90	43	35	11	0.011474	0.008190
90	43	35	12	0.033313	0.021839
90	43	35	13	0.081224	0.047911
90	43	35	14	0.168096	0.086872
90	43	35	15	0.298727	0.130630
90	43	35	16	0.462015	0.163288
90	43	35	17	0.631927	0.169912
90	43	35	18	0.779184	0.147257
90	43	35	19	0.885439	0.106255
90	43	35	20	0.949192	0.063753
90	43	35	21	0.980931	0.031739
90	43	35	22	0.994000	0.013069
90	43	35	23	0.998432	0.004432
90	43	35	24	0.999663	0.001231
90	43	35	25	0.999941	0.000278
90	43	35	26	0.999992	0.000051
90	43	35	27	0.999999	0.000007
90	43	35	28	1.000000	0.000001
90	43	35	29	1.000000	0.000000
90	43	35	30	1.000000	0.000000
90	43	35	31	1.000000	0.000000
90	43	35	32	1.000000	0.000000
90	43	35	33	1.000000	0.000000
90	43	35	34	1.000000	0.000000
90	43	35	35	1.000000	0.000000
90	43	36	0	0.000000	0.000000
90	43	36	1	0.000000	0.000000
90	43	36	2	0.000000	0.000000
90	43	36	3	0.000000	0.000000
90	43	36	4	0.000000	0.000000
90	43	36	5	0.000000	0.000000
90	43	36	6	0.000001	0.000001
90	43	36	7	0.000010	0.000009
90	43	36	8	0.000068	0.000058
90	43	36	9	0.000386	0.000318
90	43	36	10	0.001777	0.001391
90	43	36	11	0.006709	0.004932
90	43	36	12	0.021003	0.014295
90	43	36	13	0.055091	0.034087
90	43	36	14	0.122291	0.067200
90	43	36	15	0.232224	0.109933
90	43	36	16	0.381855	0.149631
90	43	36	17	0.551605	0.169750
90	43	36	18	0.712249	0.160644
90	43	36	19	0.839073	0.126824
90	43	36	20	0.922532	0.083459
90	43	36	21	0.968235	0.045704
90	43	36	22	0.989010	0.020774
90	43	36	23	0.996820	0.007810
90	43	36	24	0.999237	0.002417
90	43	36	25	0.999850	0.000612
90	43	36	26	0.999976	0.000126
90	43	36	27	0.999997	0.000021
90	43	36	28	1.000000	0.000003
90	43	36	29	1.000000	0.000000
90	43	36	30	1.000000	0.000000
90	43	36	31	1.000000	0.000000
90	43	36	32	1.000000	0.000000
90	43	36	33	1.000000	0.000000
90	43	36	34	1.000000	0.000000
90	43	36	35	1.000000	0.000000
90	43	36	36	1.000000	0.000000
90	43	37	0	0.000000	0.000000
90	43	37	1	0.000000	0.000000
90	43	37	2	0.000000	0.000000
90	43	37	3	0.000000	0.000000
90	43	37	4	0.000000	0.000000
90	43	37	5	0.000000	0.000000
90	43	37	6	0.000000	0.000000
90	43	37	7	0.000004	0.000003
90	43	37	8	0.000030	0.000026
90	43	37	9	0.000186	0.000156
90	43	37	10	0.000927	0.000741
90	43	37	11	0.003786	0.002859
90	43	37	12	0.012797	0.009011
90	43	37	13	0.036153	0.023356
90	43	37	14	0.086202	0.050049
90	43	37	15	0.175222	0.089020
90	43	37	16	0.307040	0.131818
90	43	37	17	0.469873	0.162834
90	43	37	18	0.637877	0.168003
90	43	37	19	0.782707	0.144830
90	43	37	20	0.886985	0.104278
90	43	37	21	0.949615	0.062631
90	43	37	22	0.980931	0.031315
90	43	37	23	0.993927	0.012996
90	43	37	24	0.998387	0.004460
90	43	37	25	0.999646	0.001259

Table for $N = 2$, $n = 1$, through $N = 100$, $n = 50$

N	n	k	x	P(x)	p(x)
90	43	37	26	0.999936	0.000291
90	43	37	27	0.999991	0.000054
90	43	37	28	0.999999	0.000008
90	43	37	29	1.000000	0.000001
90	43	37	30	1.000000	0.000000
90	43	37	31	1.000000	0.000000
90	43	37	32	1.000000	0.000000
90	43	37	33	1.000000	0.000000
90	43	37	34	1.000000	0.000000
90	43	37	35	1.000000	0.000000
90	43	37	36	1.000000	0.000000
90	43	37	37	1.000000	0.000000
90	43	38	0	0.000000	0.000000
90	43	38	1	0.000000	0.000000
90	43	38	2	0.000000	0.000000
90	43	38	3	0.000000	0.000000
90	43	38	4	0.000000	0.000000
90	43	38	5	0.000000	0.000000
90	43	38	6	0.000000	0.000000
90	43	38	7	0.000002	0.000001
90	43	38	8	0.000013	0.000011
90	43	38	9	0.000086	0.000073
90	43	38	10	0.000466	0.000380
90	43	38	11	0.002060	0.001594
90	43	38	12	0.007527	0.005467
90	43	38	13	0.022933	0.015406
90	43	38	14	0.058817	0.035884
90	43	38	15	0.128192	0.069376
90	43	38	16	0.239887	0.111695
90	43	38	17	0.389993	0.150106
90	43	38	18	0.558630	0.168637
90	43	38	19	0.717123	0.158494
90	43	38	20	0.841732	0.124609
90	43	38	21	0.923618	0.081886
90	43	38	22	0.968523	0.044905
90	43	38	23	0.989023	0.020500
90	43	38	24	0.996788	0.007765
90	43	38	25	0.999218	0.002430
90	43	38	26	0.999843	0.000625
90	43	38	27	0.999974	0.000131
90	43	38	28	0.999996	0.000022
90	43	38	29	1.000000	0.000003
90	43	38	30	1.000000	0.000000
90	43	38	31	1.000000	0.000000
90	43	38	32	1.000000	0.000000
90	43	38	33	1.000000	0.000000
90	43	38	34	1.000000	0.000000
90	43	38	35	1.000000	0.000000
90	43	38	36	1.000000	0.000000
90	43	38	37	1.000000	0.000000
90	43	38	38	1.000000	0.000000
90	43	39	0	0.000000	0.000000
90	43	39	1	0.000000	0.000000
90	43	39	2	0.000000	0.000000
90	43	39	3	0.000000	0.000000
90	43	39	4	0.000000	0.000000
90	43	39	5	0.000000	0.000000
90	43	39	6	0.000000	0.000000
90	43	39	7	0.000001	0.000001
90	43	39	8	0.000005	0.000005
90	43	39	9	0.000038	0.000033
90	43	39	10	0.000225	0.000187
90	43	39	11	0.001079	0.000854
90	43	39	12	0.004268	0.003189
90	43	39	13	0.014045	0.009777
90	43	39	14	0.038805	0.024760
90	43	39	15	0.090836	0.052032
90	43	39	16	0.181892	0.091055
90	43	39	17	0.314940	0.133048
90	43	39	18	0.477554	0.162614
90	43	39	19	0.643972	0.166418
90	43	39	20	0.786617	0.142644
90	43	39	21	0.888974	0.102357
90	43	39	22	0.950388	0.061414
90	43	39	23	0.981138	0.030750
90	43	39	24	0.993951	0.012813
90	43	39	25	0.998377	0.004426
90	43	39	26	0.999639	0.001262
90	43	39	27	0.999934	0.000295
90	43	39	28	0.999990	0.000056
90	43	39	29	0.999999	0.000009
90	43	39	30	1.000000	0.000001
90	43	39	31	1.000000	0.000000
90	43	39	32	1.000000	0.000000
90	43	39	33	1.000000	0.000000
90	43	39	34	1.000000	0.000000
90	43	39	35	1.000000	0.000000
90	43	39	36	1.000000	0.000000
90	43	39	37	1.000000	0.000000
90	43	39	38	1.000000	0.000000
90	43	39	39	1.000000	0.000000
90	43	40	0	0.000000	0.000000
90	43	40	1	0.000000	0.000000
90	43	40	2	0.000000	0.000000
90	43	40	3	0.000000	0.000000
90	43	40	4	0.000000	0.000000
90	43	40	5	0.000000	0.000000
90	43	40	6	0.000000	0.000000
90	43	40	7	0.000000	0.000000
90	43	40	8	0.000002	0.000002
90	43	40	9	0.000016	0.000014
90	43	40	10	0.000104	0.000088
90	43	40	11	0.000543	0.000439
90	43	40	12	0.002329	0.001787
90	43	40	13	0.008294	0.005964
90	43	40	14	0.024725	0.016432
90	43	40	15	0.062270	0.037544
90	43	40	16	0.133686	0.071416
90	43	40	17	0.247111	0.113425
90	43	40	18	0.397841	0.150730
90	43	40	19	0.565658	0.167817
90	43	40	20	0.722287	0.156629
90	43	40	21	0.844820	0.122533
90	43	40	22	0.925100	0.080280
90	43	40	23	0.969079	0.043980
90	43	40	24	0.989178	0.020098
90	43	40	25	0.996815	0.007637
90	43	40	26	0.999218	0.002403
90	43	40	27	0.999841	0.000623
90	43	40	28	0.999974	0.000132
90	43	40	29	0.999996	0.000023
90	43	40	30	0.999999	0.000003
90	43	40	31	1.000000	0.000000
90	43	40	32	1.000000	0.000000
90	43	40	33	1.000000	0.000000
90	43	40	34	1.000000	0.000000
90	43	40	35	1.000000	0.000000
90	43	40	36	1.000000	0.000000
90	43	40	37	1.000000	0.000000
90	43	40	38	1.000000	0.000000
90	43	40	39	1.000000	0.000000
90	43	40	40	1.000000	0.000000
90	43	41	0	0.000000	0.000000
90	43	41	1	0.000000	0.000000
90	43	41	2	0.000000	0.000000
90	43	41	3	0.000000	0.000000
90	43	41	4	0.000000	0.000000
90	43	41	5	0.000000	0.000000
90	43	41	6	0.000000	0.000000
90	43	41	7	0.000000	0.000000
90	43	41	8	0.000001	0.000001
90	43	41	9	0.000007	0.000006
90	43	41	10	0.000046	0.000039
90	43	41	11	0.000262	0.000216
90	43	41	12	0.001222	0.000960
90	43	41	13	0.004715	0.003493
90	43	41	14	0.015195	0.010480
90	43	41	15	0.041245	0.026050
90	43	41	16	0.095121	0.053876
90	43	41	17	0.188130	0.093009
90	43	41	18	0.322476	0.134346
90	43	41	19	0.485106	0.162630
90	43	41	20	0.650238	0.165132
90	43	41	21	0.790905	0.140668
90	43	41	22	0.891382	0.100477
90	43	41	23	0.951488	0.060105
90	43	41	24	0.981540	0.030053
90	43	41	25	0.994065	0.012525
90	43	41	26	0.998401	0.004336
90	43	41	27	0.999642	0.001241
90	43	41	28	0.999934	0.000292
90	43	41	29	0.999990	0.000056
90	43	41	30	0.999999	0.000009
90	43	41	31	1.000000	0.000001
90	43	41	32	1.000000	0.000000
90	43	41	33	1.000000	0.000000
90	43	41	34	1.000000	0.000000
90	43	41	35	1.000000	0.000000
90	43	41	36	1.000000	0.000000
90	43	41	37	1.000000	0.000000
90	43	41	38	1.000000	0.000000
90	43	41	39	1.000000	0.000000
90	43	41	40	1.000000	0.000000
90	43	41	41	1.000000	0.000000
90	43	42	0	0.000000	0.000000
90	43	42	1	0.000000	0.000000
90	43	42	2	0.000000	0.000000
90	43	42	3	0.000000	0.000000
90	43	42	4	0.000000	0.000000
90	43	42	5	0.000000	0.000000
90	43	42	6	0.000000	0.000000
90	43	42	7	0.000000	0.000000
90	43	42	8	0.000000	0.000000
90	43	42	9	0.000003	0.000002
90	43	42	10	0.000019	0.000017
90	43	42	11	0.000121	0.000102
90	43	42	12	0.000615	0.000494
90	43	42	13	0.002576	0.001962
90	43	42	14	0.008993	0.006416
90	43	42	15	0.026359	0.017367
90	43	42	16	0.065434	0.039075
90	43	42	17	0.138779	0.073344
90	43	42	18	0.253932	0.115154
90	43	42	19	0.405451	0.151518
90	43	42	20	0.572727	0.167276
90	43	42	21	0.727748	0.155022
90	43	42	22	0.848321	0.120572
90	43	42	23	0.926955	0.078634
90	43	42	24	0.969887	0.042932
90	43	42	25	0.989464	0.019577

Table for $N = 2$, $n = 1$, through $N = 100$, $n = 50$

N	n	k	x	P(x)	p(x)
90	43	42	26	0.996897	0.007433
90	43	42	27	0.999237	0.002340
90	43	42	28	0.999844	0.000608
90	43	42	29	0.999974	0.000129
90	43	42	30	0.999996	0.000022
90	43	42	31	0.999999	0.000003
90	43	42	32	1.000000	0.000000
90	43	42	33	1.000000	0.000000
90	43	42	34	1.000000	0.000000
90	43	42	35	1.000000	0.000000
90	43	42	36	1.000000	0.000000
90	43	42	37	1.000000	0.000000
90	43	42	38	1.000000	0.000000
90	43	42	39	1.000000	0.000000
90	43	42	40	1.000000	0.000000
90	43	42	41	1.000000	0.000000
90	43	42	42	1.000000	0.000000
90	43	43	0	0.000000	0.000000
90	43	43	1	0.000000	0.000000
90	43	43	2	0.000000	0.000000
90	43	43	3	0.000000	0.000000
90	43	43	4	0.000000	0.000000
90	43	43	5	0.000000	0.000000
90	43	43	6	0.000000	0.000000
90	43	43	7	0.000000	0.000000
90	43	43	8	0.000000	0.000000
90	43	43	9	0.000001	0.000001
90	43	43	10	0.000008	0.000007
90	43	43	11	0.000053	0.000045
90	43	43	12	0.000296	0.000242
90	43	43	13	0.001350	0.001054
90	43	43	14	0.005116	0.003766
90	43	43	15	0.016229	0.011113
90	43	43	16	0.043455	0.027226
90	43	43	17	0.099050	0.055596
90	43	43	18	0.193956	0.094906
90	43	43	19	0.329692	0.135735
90	43	43	20	0.492574	0.162882
90	43	43	21	0.656697	0.164123
90	43	43	22	0.795570	0.138873
90	43	43	23	0.894191	0.098620
90	43	43	24	0.952893	0.058703
90	43	43	25	0.982123	0.029230
90	43	43	26	0.994264	0.012142
90	43	43	27	0.998457	0.004192
90	43	43	28	0.999654	0.001198
90	43	43	29	0.999936	0.000282
90	43	43	30	0.999990	0.000054
90	43	43	31	0.999999	0.000008
90	43	43	32	1.000000	0.000001

N	n	k	x	P(x)	p(x)
90	43	43	33	1.000000	0.000000
90	43	43	34	1.000000	0.000000
90	43	43	35	1.000000	0.000000
90	43	43	36	1.000000	0.000000
90	43	43	37	1.000000	0.000000
90	43	43	38	1.000000	0.000000
90	43	43	39	1.000000	0.000000
90	43	43	40	1.000000	0.000000
90	43	43	41	1.000000	0.000000
90	43	43	42	1.000000	0.000000
90	43	43	43	1.000000	0.000000
90	44	1	0	0.511111	0.511111
90	44	1	1	1.000000	0.488889
90	44	2	0	0.258427	0.258427
90	44	2	1	0.763795	0.505368
90	44	2	2	1.000000	0.236205
90	44	3	0	0.129213	0.129213
90	44	3	1	0.516854	0.387640
90	44	3	2	0.887266	0.370412
90	44	3	3	1.000000	0.112734
90	44	4	0	0.063864	0.063864
90	44	4	1	0.325261	0.261397
90	44	4	2	0.708446	0.383185
90	44	4	3	0.946872	0.238426
90	44	4	4	1.000000	0.053128
90	44	5	0	0.031189	0.031189
90	44	5	1	0.194563	0.163373
90	44	5	2	0.521310	0.326747
90	44	5	3	0.833204	0.311895
90	44	5	4	0.975289	0.142085
90	44	5	5	1.000000	0.024710
90	44	6	0	0.015044	0.015044
90	44	6	1	0.111915	0.096871
90	44	6	2	0.359858	0.247943
90	44	6	3	0.682761	0.322903
90	44	6	4	0.908426	0.225665
90	44	6	5	0.988662	0.080236
90	44	6	6	1.000000	0.011338
90	44	7	0	0.007164	0.007164
90	44	7	1	0.062326	0.055163
90	44	7	2	0.235887	0.173560
90	44	7	3	0.525154	0.289267
90	44	7	4	0.800966	0.275813
90	44	7	5	0.951410	0.150443
90	44	7	6	0.994871	0.043461
90	44	7	7	1.000000	0.005129
90	44	8	0	0.003366	0.003366
90	44	8	1	0.033748	0.030382
90	44	8	2	0.148061	0.114313
90	44	8	3	0.382263	0.234202

N	n	k	x	P(x)	p(x)
90	44	8	4	0.668045	0.285782
90	44	8	5	0.880719	0.212675
90	44	8	6	0.974973	0.094254
90	44	8	7	0.997713	0.022741
90	44	8	8	1.000000	0.002286
90	44	9	0	0.001560	0.001560
90	44	9	1	0.017816	0.016256
90	44	9	2	0.089511	0.071694
90	44	9	3	0.265162	0.175651
90	44	9	4	0.528639	0.263477
90	44	9	5	0.779569	0.250930
90	44	9	6	0.931294	0.151725
90	44	9	7	0.987452	0.056158
90	44	9	8	0.998996	0.011544
90	44	9	9	1.000000	0.001004
90	44	10	0	0.000713	0.000713
90	44	10	1	0.009186	0.008474
90	44	10	2	0.052336	0.043149
90	44	10	3	0.176252	0.123916
90	44	10	4	0.398527	0.222275
90	44	10	5	0.658751	0.260224
90	44	10	6	0.860115	0.201364
90	44	10	7	0.961800	0.101685
90	44	10	8	0.993866	0.032066
90	44	10	9	0.999566	0.005701
90	44	10	10	1.000000	0.000434
90	44	11	0	0.000321	0.000321
90	44	11	1	0.004632	0.004311
90	44	11	2	0.029682	0.025051
90	44	11	3	0.112745	0.083063
90	44	11	4	0.287389	0.174644
90	44	11	5	0.531892	0.244502
90	44	11	6	0.764467	0.232575
90	44	11	7	0.914771	0.150304
90	44	11	8	0.979436	0.064666
90	44	11	9	0.997072	0.017636
90	44	11	10	0.999816	0.002743
90	44	11	11	1.000000	0.000184
90	44	12	0	0.000142	0.000142
90	44	12	1	0.002285	0.002143
90	44	12	2	0.016364	0.014079
90	44	12	3	0.069636	0.053272
90	44	12	4	0.198962	0.129325
90	44	12	5	0.411188	0.212226
90	44	12	6	0.652595	0.241407
90	44	12	7	0.844375	0.191780
90	44	12	8	0.949968	0.105593
90	44	12	9	0.989259	0.039290
90	44	12	10	0.998635	0.009376
90	44	12	11	0.999923	0.001288

N	n	k	x	P(x)	p(x)
90	44	12	12	1.000000	0.000077
90	44	13	0	0.000062	0.000062
90	44	13	1	0.001104	0.001042
90	44	13	2	0.008783	0.007679
90	44	13	3	0.041634	0.032851
90	44	13	4	0.132641	0.091007
90	44	13	5	0.305075	0.172434
90	44	13	6	0.534987	0.229912
90	44	13	7	0.753403	0.218416
90	44	13	8	0.901233	0.147830
90	44	13	9	0.971628	0.070395
90	44	13	10	0.994548	0.022919
90	44	13	11	0.999378	0.004830
90	44	13	12	0.999968	0.000590
90	44	13	13	1.000000	0.000032
90	44	14	0	0.000027	0.000027
90	44	14	1	0.000522	0.000495
90	44	14	2	0.004594	0.004072
90	44	14	3	0.024142	0.019548
90	44	14	4	0.085365	0.061223
90	44	14	5	0.217738	0.132373
90	44	14	6	0.421524	0.203785
90	44	14	7	0.648449	0.226926
90	44	14	8	0.832117	0.183668
90	44	14	9	0.939630	0.107513
90	44	14	10	0.984428	0.044797
90	44	14	11	0.997308	0.012880
90	44	14	12	0.999723	0.002415
90	44	14	13	0.999987	0.000264
90	44	14	14	1.000000	0.000013
90	44	15	0	0.000011	0.000011
90	44	15	1	0.000242	0.000230
90	44	15	2	0.002344	0.002102
90	44	15	3	0.013597	0.011253
90	44	15	4	0.053142	0.039546
90	44	15	5	0.149810	0.096667
90	44	15	6	0.319631	0.169821
90	44	15	7	0.537972	0.218341
90	44	15	8	0.745117	0.207144
90	44	15	9	0.890118	0.145001
90	44	15	10	0.964387	0.074269
90	44	15	11	0.991715	0.027328
90	44	15	12	0.998706	0.006991
90	44	15	13	0.999879	0.001173
90	44	15	14	0.999995	0.000115
90	44	15	15	1.000000	0.000005
90	44	16	0	0.000005	0.000005
90	44	16	1	0.000110	0.000105
90	44	16	2	0.001167	0.001057
90	44	16	3	0.007445	0.006279

Table for $N = 2$, $n = 1$, through $N = 100$, $n = 50$

N	n	k	x	P(x)	p(x)
90	44	16	4	0.032051	0.024606
90	44	16	5	0.099543	0.067491
90	44	16	6	0.233588	0.134045
90	44	16	7	0.430257	0.196669
90	44	16	8	0.645687	0.215430
90	44	16	9	0.822451	0.176763
90	44	16	10	0.930718	0.108268
90	44	16	11	0.979691	0.048972
90	44	16	12	0.995723	0.016033
90	44	16	13	0.999394	0.003671
90	44	16	14	0.999949	0.000554
90	44	16	15	0.999998	0.000049
90	44	16	16	1.000000	0.000002
90	44	17	0	0.000002	0.000002
90	44	17	1	0.000049	0.000047
90	44	17	2	0.000567	0.000518
90	44	17	3	0.003966	0.003400
90	44	17	4	0.018751	0.014784
90	44	17	5	0.063973	0.045222
90	44	17	6	0.164754	0.100781
90	44	17	7	0.331923	0.167169
90	44	17	8	0.540884	0.208961
90	44	17	9	0.738846	0.197963
90	44	17	10	0.880974	0.142127
90	44	17	11	0.957852	0.076878
90	44	17	12	0.988790	0.030939
90	44	17	13	0.997856	0.009066
90	44	17	14	0.999724	0.001867
90	44	17	15	0.999979	0.000255
90	44	17	16	0.999999	0.000021
90	44	17	17	1.000000	0.000001
90	44	18	0	0.000001	0.000001
90	44	18	1	0.000021	0.000020
90	44	18	2	0.000269	0.000248
90	44	18	3	0.002057	0.001788
90	44	18	4	0.010650	0.008593
90	44	18	5	0.039813	0.029163
90	44	18	6	0.112293	0.072480
90	44	18	7	0.247193	0.134901
90	44	18	8	0.437834	0.190641
90	44	18	9	0.643933	0.206098
90	44	18	10	0.814777	0.170845
90	44	18	11	0.923099	0.108321
90	44	18	12	0.975228	0.052130
90	44	18	13	0.994006	0.018778
90	44	18	14	0.998956	0.004950
90	44	18	15	0.999877	0.000921
90	44	18	16	0.999991	0.000114
90	44	18	17	1.000000	0.000008
90	44	18	18	1.000000	0.000000

N	n	k	x	P(x)	p(x)
90	44	19	0	0.000000	0.000000
90	44	19	1	0.000009	0.000009
90	44	19	2	0.000124	0.000115
90	44	19	3	0.001039	0.000914
90	44	19	4	0.005876	0.004837
90	44	19	5	0.024016	0.018140
90	44	19	6	0.074040	0.050023
90	44	19	7	0.177869	0.103830
90	44	19	8	0.342514	0.164644
90	44	19	9	0.543746	0.201232
90	44	19	10	0.734101	0.190355
90	44	19	11	0.873451	0.139351
90	44	19	12	0.952059	0.078608
90	44	19	13	0.985921	0.033862
90	44	19	14	0.996894	0.010973
90	44	19	15	0.999506	0.002613
90	44	19	16	0.999947	0.000440
90	44	19	17	0.999996	0.000049
90	44	19	18	1.000000	0.000003
90	44	19	19	1.000000	0.000000
90	44	20	0	0.000000	0.000000
90	44	20	1	0.000004	0.000004
90	44	20	2	0.000056	0.000052
90	44	20	3	0.000511	0.000455
90	44	20	4	0.003151	0.002640
90	44	20	5	0.014052	0.010901
90	44	20	6	0.047267	0.033215
90	44	20	7	0.123761	0.076494
90	44	20	8	0.259032	0.135271
90	44	20	9	0.444547	0.185515
90	44	20	10	0.642945	0.198398
90	44	20	11	0.808683	0.165738
90	44	20	12	0.916630	0.107948
90	44	20	13	0.971137	0.054506
90	44	20	14	0.992258	0.021121
90	44	20	15	0.998439	0.006182
90	44	20	16	0.999773	0.001334
90	44	20	17	0.999978	0.000204
90	44	20	18	0.999999	0.000021
90	44	20	19	1.000000	0.000001
90	44	20	20	1.000000	0.000000
90	44	21	0	0.000000	0.000000
90	44	21	1	0.000001	0.000001
90	44	21	2	0.000025	0.000023
90	44	21	3	0.000244	0.000220
90	44	21	4	0.001642	0.001398
90	44	21	5	0.007978	0.006336
90	44	21	6	0.029236	0.021257
90	44	21	7	0.083328	0.054092
90	44	21	8	0.189464	0.106136

N	n	k	x	P(x)	p(x)
90	44	21	9	0.351790	0.162326
90	44	21	10	0.546580	0.194791
90	44	21	11	0.730549	0.183965
90	44	21	12	0.867283	0.136734
90	44	21	13	0.946998	0.079715
90	44	21	14	0.983206	0.036208
90	44	21	15	0.995878	0.012673
90	44	21	16	0.999240	0.003361
90	44	21	17	0.999899	0.000659
90	44	21	18	0.999991	0.000092
90	44	21	19	0.999999	0.000009
90	44	21	20	1.000000	0.000000
90	44	21	21	1.000000	0.000000
90	44	22	0	0.000000	0.000000
90	44	22	1	0.000001	0.000001
90	44	22	2	0.000011	0.000010
90	44	22	3	0.000114	0.000103
90	44	22	4	0.000832	0.000718
90	44	22	5	0.004397	0.003565
90	44	22	6	0.017529	0.013132
90	44	22	7	0.054322	0.036793
90	44	22	8	0.134089	0.079767
90	44	22	9	0.269451	0.135362
90	44	22	10	0.450597	0.181146
90	44	22	11	0.642564	0.191968
90	44	22	12	0.803870	0.161306
90	44	22	13	0.911184	0.107314
90	44	22	14	0.967463	0.056279
90	44	22	15	0.990552	0.023089
90	44	22	16	0.997876	0.007324
90	44	22	17	0.999641	0.001765
90	44	22	18	0.999956	0.000315
90	44	22	19	0.999996	0.000040
90	44	22	20	1.000000	0.000003
90	44	22	21	1.000000	0.000000
90	44	22	22	1.000000	0.000000
90	44	23	0	0.000000	0.000000
90	44	23	1	0.000000	0.000000
90	44	23	2	0.000004	0.000004
90	44	23	3	0.000052	0.000047
90	44	23	4	0.000409	0.000358
90	44	23	5	0.002352	0.001943
90	44	23	6	0.010190	0.007838
90	44	23	7	0.034302	0.024112
90	44	23	8	0.091859	0.057557
90	44	23	9	0.199779	0.107920
90	44	23	10	0.360024	0.160244
90	44	23	11	0.549403	0.189380
90	44	23	12	0.727961	0.178558
90	44	23	13	0.862262	0.134300

N	n	k	x	P(x)	p(x)
90	44	23	14	0.942634	0.080373
90	44	23	15	0.980705	0.038071
90	44	23	16	0.994860	0.014155
90	44	23	17	0.998940	0.004080
90	44	23	18	0.999836	0.000896
90	44	23	19	0.999981	0.000146
90	44	23	20	0.999998	0.000017
90	44	23	21	1.000000	0.000001
90	44	23	22	1.000000	0.000000
90	44	23	23	1.000000	0.000000
90	44	24	0	0.000000	0.000000
90	44	24	1	0.000000	0.000000
90	44	24	2	0.000002	0.000002
90	44	24	3	0.000023	0.000021
90	44	24	4	0.000196	0.000173
90	44	24	5	0.001221	0.001026
90	44	24	6	0.005745	0.004524
90	44	24	7	0.020987	0.015242
90	44	24	8	0.060933	0.039946
90	44	24	9	0.143403	0.082470
90	44	24	10	0.278705	0.135302
90	44	24	11	0.456127	0.177421
90	44	24	12	0.642680	0.186553
90	44	24	13	0.800123	0.157443
90	44	24	14	0.906646	0.106524
90	44	24	15	0.964227	0.057580
90	44	24	16	0.988945	0.024718
90	44	24	17	0.997296	0.008351
90	44	24	18	0.999488	0.002192
90	44	24	19	0.999927	0.000439
90	44	24	20	0.999992	0.000065
90	44	24	21	0.999999	0.000007
90	44	24	22	1.000000	0.000000
90	44	24	23	1.000000	0.000000
90	44	24	24	1.000000	0.000000
90	44	25	0	0.000000	0.000000
90	44	25	1	0.000000	0.000000
90	44	25	2	0.000001	0.000001
90	44	25	3	0.000010	0.000009
90	44	25	4	0.000091	0.000081
90	44	25	5	0.000615	0.000524
90	44	25	6	0.003140	0.002525
90	44	25	7	0.012442	0.009302
90	44	25	8	0.039144	0.026702
90	44	25	9	0.099669	0.060525
90	44	25	10	0.209004	0.109335
90	44	25	11	0.367416	0.158412
90	44	25	12	0.552230	0.184814
90	44	25	13	0.726172	0.173942
90	44	25	14	0.858227	0.132054

$N = 90$ $n = 44$

Table for $N = 2$, $n = 1$, through $N = 100$, $n = 50$

N	n	k	x	P(x)	p(x)
90	44	25	15	0.938926	0.080700
90	44	25	16	0.978458	0.039532
90	44	25	17	0.993880	0.015421
90	44	25	18	0.998624	0.004745
90	44	25	19	0.999761	0.001136
90	44	25	20	0.999969	0.000208
90	44	25	21	0.999997	0.000028
90	44	25	22	1.000000	0.000003
90	44	25	23	1.000000	0.000000
90	44	25	24	1.000000	0.000000
90	44	25	25	1.000000	0.000000
90	44	26	0	0.000000	0.000000
90	44	26	1	0.000000	0.000000
90	44	26	2	0.000000	0.000000
90	44	26	3	0.000004	0.000004
90	44	26	4	0.000041	0.000037
90	44	26	5	0.000301	0.000260
90	44	26	6	0.001664	0.001364
90	44	26	7	0.007147	0.005483
90	44	26	8	0.024355	0.017208
90	44	26	9	0.067079	0.042723
90	44	26	10	0.151814	0.084735
90	44	26	11	0.286992	0.135178
90	44	26	12	0.461245	0.174253
90	44	26	13	0.643215	0.181971
90	44	26	14	0.797278	0.154063
90	44	26	15	0.902922	0.105643
90	44	26	16	0.961429	0.058507
90	44	26	17	0.987474	0.026045
90	44	26	18	0.996727	0.009253
90	44	26	19	0.999324	0.002597
90	44	26	20	0.999892	0.000568
90	44	26	21	0.999987	0.000095
90	44	26	22	0.999999	0.000012
90	44	26	23	1.000000	0.000001
90	44	26	24	1.000000	0.000000
90	44	26	25	1.000000	0.000000
90	44	26	26	1.000000	0.000000
90	44	27	0	0.000000	0.000000
90	44	27	1	0.000000	0.000000
90	44	27	2	0.000000	0.000000
90	44	27	3	0.000002	0.000002
90	44	27	4	0.000018	0.000016
90	44	27	5	0.000142	0.000125
90	44	27	6	0.000855	0.000712
90	44	27	7	0.003977	0.003123
90	44	27	8	0.014676	0.010698
90	44	27	9	0.043714	0.029039
90	44	27	10	0.106798	0.063084
90	44	27	11	0.217290	0.110492

N	n	k	x	P(x)	p(x)
90	44	27	12	0.374118	0.156828
90	44	27	13	0.555073	0.180955
90	44	27	14	0.725061	0.169988
90	44	27	15	0.855052	0.129991
90	44	27	16	0.935832	0.080780
90	44	27	17	0.976486	0.040654
90	44	27	18	0.992968	0.016481
90	44	27	19	0.998309	0.005342
90	44	27	20	0.999679	0.001370
90	44	27	21	0.999953	0.000274
90	44	27	22	0.999995	0.000042
90	44	27	23	0.999999	0.000005
90	44	27	24	1.000000	0.000000
90	44	27	25	1.000000	0.000000
90	44	27	26	1.000000	0.000000
90	44	27	27	1.000000	0.000000
90	44	28	0	0.000000	0.000000
90	44	28	1	0.000000	0.000000
90	44	28	2	0.000000	0.000000
90	44	28	3	0.000001	0.000001
90	44	28	4	0.000008	0.000007
90	44	28	5	0.000065	0.000058
90	44	28	6	0.000425	0.000360
90	44	28	7	0.002143	0.001718
90	44	28	8	0.008562	0.006419
90	44	28	9	0.027582	0.019019
90	44	28	10	0.072753	0.045171
90	44	28	11	0.159414	0.086661
90	44	28	12	0.294460	0.135046
90	44	28	13	0.466032	0.171572
90	44	28	14	0.644115	0.178083
90	44	28	15	0.795215	0.151101
90	44	28	16	0.899930	0.104715
90	44	28	17	0.959063	0.059133
90	44	28	18	0.986166	0.027103
90	44	28	19	0.996190	0.010024
90	44	28	20	0.999157	0.002968
90	44	28	21	0.999853	0.000696
90	44	28	22	0.999980	0.000127
90	44	28	23	0.999998	0.000018
90	44	28	24	1.000000	0.000002
90	44	28	25	1.000000	0.000000
90	44	28	26	1.000000	0.000000
90	44	28	27	1.000000	0.000000
90	44	28	28	1.000000	0.000000
90	44	29	0	0.000000	0.000000
90	44	29	1	0.000000	0.000000
90	44	29	2	0.000000	0.000000
90	44	29	3	0.000000	0.000000
90	44	29	4	0.000003	0.000003

N	n	k	x	P(x)	p(x)
90	44	29	5	0.000029	0.000026
90	44	29	6	0.000205	0.000176
90	44	29	7	0.001118	0.000913
90	44	29	8	0.004835	0.003717
90	44	29	9	0.016845	0.012010
90	44	29	10	0.047982	0.031137
90	44	29	11	0.113288	0.065306
90	44	29	12	0.224758	0.111471
90	44	29	13	0.380246	0.155487
90	44	29	14	0.557946	0.177700
90	44	29	15	0.724539	0.166594
90	44	29	16	0.852639	0.128100
90	44	29	17	0.933312	0.080672
90	44	29	18	0.974800	0.041489
90	44	29	19	0.992148	0.017348
90	44	29	20	0.998008	0.005861
90	44	29	21	0.999595	0.001586
90	44	29	22	0.999935	0.000340
90	44	29	23	0.999992	0.000057
90	44	29	24	0.999999	0.000007
90	44	29	25	1.000000	0.000001
90	44	29	26	1.000000	0.000000
90	44	29	27	1.000000	0.000000
90	44	29	28	1.000000	0.000000
90	44	29	29	1.000000	0.000000
90	44	30	0	0.000000	0.000000
90	44	30	1	0.000000	0.000000
90	44	30	2	0.000000	0.000000
90	44	30	3	0.000000	0.000000
90	44	30	4	0.000001	0.000001
90	44	30	5	0.000012	0.000011
90	44	30	6	0.000095	0.000083
90	44	30	7	0.000564	0.000469
90	44	30	8	0.002641	0.002077
90	44	30	9	0.009954	0.007313
90	44	30	10	0.030627	0.020673
90	44	30	11	0.077958	0.047331
90	44	30	12	0.166282	0.088324
90	44	30	13	0.301228	0.134946
90	44	30	14	0.470552	0.169325
90	44	30	15	0.645339	0.174787
90	44	30	16	0.793839	0.148500
90	44	30	17	0.897604	0.103765
90	44	30	18	0.957116	0.059512
90	44	30	19	0.985038	0.027922
90	44	30	20	0.995702	0.010664
90	44	30	21	0.998997	0.003294
90	44	30	22	0.999812	0.000816
90	44	30	23	0.999972	0.000160
90	44	30	24	0.999997	0.000025

N	n	k	x	P(x)	p(x)
90	44	30	25	1.000000	0.000003
90	44	30	26	1.000000	0.000000
90	44	30	27	1.000000	0.000000
90	44	30	28	1.000000	0.000000
90	44	30	29	1.000000	0.000000
90	44	30	30	1.000000	0.000000
90	44	31	0	0.000000	0.000000
90	44	31	1	0.000000	0.000000
90	44	31	2	0.000000	0.000000
90	44	31	3	0.000000	0.000000
90	44	31	4	0.000000	0.000000
90	44	31	5	0.000005	0.000005
90	44	31	6	0.000043	0.000038
90	44	31	7	0.000275	0.000232
90	44	31	8	0.001395	0.001120
90	44	31	9	0.005688	0.004293
90	44	31	10	0.018912	0.013224
90	44	31	11	0.051926	0.033014
90	44	31	12	0.119176	0.067250
90	44	31	13	0.231506	0.112330
90	44	31	14	0.385890	0.154384
90	44	31	15	0.560859	0.174969
90	44	31	16	0.724539	0.163680
90	44	31	17	0.850910	0.126371
90	44	31	18	0.931328	0.080418
90	44	31	19	0.973404	0.042076
90	44	31	20	0.991437	0.018033
90	44	31	21	0.997734	0.006297
90	44	31	22	0.999513	0.001779
90	44	31	23	0.999916	0.000403
90	44	31	24	0.999988	0.000072
90	44	31	25	0.999999	0.000010
90	44	31	26	1.000000	0.000001
90	44	31	27	1.000000	0.000000
90	44	31	28	1.000000	0.000000
90	44	31	29	1.000000	0.000000
90	44	31	30	1.000000	0.000000
90	44	31	31	1.000000	0.000000
90	44	32	0	0.000000	0.000000
90	44	32	1	0.000000	0.000000
90	44	32	2	0.000000	0.000000
90	44	32	3	0.000000	0.000000
90	44	32	4	0.000000	0.000000
90	44	32	5	0.000002	0.000002
90	44	32	6	0.000019	0.000016
90	44	32	7	0.000129	0.000111
90	44	32	8	0.000711	0.000582
90	44	32	9	0.003141	0.002430
90	44	32	10	0.011292	0.008150
90	44	32	11	0.033461	0.022169

Table for $N = 2$, $n = 1$, through $N = 100$, $n = 50$

N	n	k	x	P(x)	p(x)
90	44	32	12	0.082701	0.049241
90	44	32	13	0.172485	0.089784
90	44	32	14	0.307389	0.134904
90	44	32	15	0.474857	0.167468
90	44	32	16	0.646860	0.172003
90	44	32	17	0.793079	0.146219
90	44	32	18	0.895889	0.102810
90	44	32	19	0.955575	0.059686
90	44	32	20	0.984101	0.028526
90	44	32	21	0.995279	0.011178
90	44	32	22	0.998850	0.003571
90	44	32	23	0.999773	0.000923
90	44	32	24	0.999964	0.000191
90	44	32	25	0.999995	0.000031
90	44	32	26	0.999999	0.000004
90	44	32	27	1.000000	0.000000
90	44	32	28	1.000000	0.000000
90	44	32	29	1.000000	0.000000
90	44	32	30	1.000000	0.000000
90	44	32	31	1.000000	0.000000
90	44	32	32	1.000000	0.000000
90	44	33	0	0.000000	0.000000
90	44	33	1	0.000000	0.000000
90	44	33	2	0.000000	0.000000
90	44	33	3	0.000000	0.000000
90	44	33	4	0.000000	0.000000
90	44	33	5	0.000001	0.000001
90	44	33	6	0.000008	0.000007
90	44	33	7	0.000059	0.000051
90	44	33	8	0.000350	0.000291
90	44	33	9	0.001675	0.001325
90	44	33	10	0.006514	0.004839
90	44	33	11	0.020847	0.014333
90	44	33	12	0.055534	0.034687
90	44	33	13	0.124497	0.068963
90	44	33	14	0.237611	0.113114
90	44	33	15	0.391123	0.153512
90	44	33	16	0.563824	0.172701
90	44	33	17	0.725012	0.161188
90	44	33	18	0.849802	0.124790
90	44	33	19	0.929849	0.080046
90	44	33	20	0.972297	0.042449
90	44	33	21	0.990847	0.018549
90	44	33	22	0.997495	0.006649
90	44	33	23	0.999439	0.001943
90	44	33	24	0.999898	0.000460
90	44	33	25	0.999985	0.000087
90	44	33	26	0.999998	0.000013
90	44	33	27	1.000000	0.000002
90	44	33	28	1.000000	0.000000

N	n	k	x	P(x)	p(x)
90	44	33	29	1.000000	0.000000
90	44	33	30	1.000000	0.000000
90	44	33	31	1.000000	0.000000
90	44	33	32	1.000000	0.000000
90	44	33	33	1.000000	0.000000
90	44	34	0	0.000000	0.000000
90	44	34	1	0.000000	0.000000
90	44	34	2	0.000000	0.000000
90	44	34	3	0.000000	0.000000
90	44	34	4	0.000000	0.000000
90	44	34	5	0.000000	0.000000
90	44	34	6	0.000003	0.000003
90	44	34	7	0.000026	0.000022
90	44	34	8	0.000166	0.000140
90	44	34	9	0.000861	0.000695
90	44	34	10	0.003628	0.002766
90	44	34	11	0.012549	0.008921
90	44	34	12	0.036061	0.023512
90	44	34	13	0.086991	0.050930
90	44	34	14	0.178078	0.091087
90	44	34	15	0.313021	0.134943
90	44	34	16	0.478989	0.165968
90	44	34	17	0.648660	0.169671
90	44	34	18	0.792880	0.144220
90	44	34	19	0.894740	0.101860
90	44	34	20	0.954424	0.059684
90	44	34	21	0.983362	0.028938
90	44	34	22	0.994929	0.011567
90	44	34	23	0.998723	0.003793
90	44	34	24	0.999737	0.001014
90	44	34	25	0.999956	0.000219
90	44	34	26	0.999994	0.000038
90	44	34	27	0.999999	0.000005
90	44	34	28	1.000000	0.000001
90	44	34	29	1.000000	0.000000
90	44	34	30	1.000000	0.000000
90	44	34	31	1.000000	0.000000
90	44	34	32	1.000000	0.000000
90	44	34	33	1.000000	0.000000
90	44	34	34	1.000000	0.000000
90	44	35	0	0.000000	0.000000
90	44	35	1	0.000000	0.000000
90	44	35	2	0.000000	0.000000
90	44	35	3	0.000000	0.000000
90	44	35	4	0.000000	0.000000
90	44	35	5	0.000000	0.000000
90	44	35	6	0.000001	0.000001
90	44	35	7	0.000011	0.000010
90	44	35	8	0.000076	0.000065
90	44	35	9	0.000427	0.000351

N	n	k	x	P(x)	p(x)
90	44	35	10	0.001948	0.001521
90	44	35	11	0.007292	0.005344
90	44	35	12	0.022626	0.015334
90	44	35	13	0.058798	0.036172
90	44	35	14	0.129281	0.070484
90	44	35	15	0.243139	0.113858
90	44	35	16	0.396005	0.152865
90	44	35	17	0.566854	0.170849
90	44	35	18	0.725921	0.159067
90	44	35	19	0.849267	0.123346
90	44	35	20	0.928845	0.079578
90	44	35	21	0.971477	0.042631
90	44	35	22	0.990385	0.018908
90	44	35	23	0.997300	0.006915
90	44	35	24	0.999375	0.002075
90	44	35	25	0.999882	0.000507
90	44	35	26	0.999982	0.000100
90	44	35	27	0.999998	0.000016
90	44	35	28	1.000000	0.000002
90	44	35	29	1.000000	0.000000
90	44	35	30	1.000000	0.000000
90	44	35	31	1.000000	0.000000
90	44	35	32	1.000000	0.000000
90	44	35	33	1.000000	0.000000
90	44	35	34	1.000000	0.000000
90	44	35	35	1.000000	0.000000
90	44	36	0	0.000000	0.000000
90	44	36	1	0.000000	0.000000
90	44	36	2	0.000000	0.000000
90	44	36	3	0.000000	0.000000
90	44	36	4	0.000000	0.000000
90	44	36	5	0.000000	0.000000
90	44	36	6	0.000000	0.000000
90	44	36	7	0.000004	0.000004
90	44	36	8	0.000033	0.000029
90	44	36	9	0.000203	0.000170
90	44	36	10	0.001008	0.000804
90	44	36	11	0.004086	0.003078
90	44	36	12	0.013704	0.009618
90	44	36	13	0.038410	0.024706
90	44	36	14	0.090836	0.052426
90	44	36	15	0.183105	0.092270
90	44	36	16	0.318182	0.135077
90	44	36	17	0.482983	0.164800
90	44	36	18	0.650726	0.167743
90	44	36	19	0.793201	0.142475
90	44	36	20	0.894120	0.100920
90	44	36	21	0.953649	0.059529
90	44	36	22	0.982822	0.029172
90	44	36	23	0.994660	0.011838

N	n	k	x	P(x)	p(x)
90	44	36	24	0.998620	0.003961
90	44	36	25	0.999707	0.001086
90	44	36	26	0.999949	0.000243
90	44	36	27	0.999993	0.000044
90	44	36	28	0.999999	0.000006
90	44	36	29	1.000000	0.000001
90	44	36	30	1.000000	0.000000
90	44	36	31	1.000000	0.000000
90	44	36	32	1.000000	0.000000
90	44	36	33	1.000000	0.000000
90	44	36	34	1.000000	0.000000
90	44	36	35	1.000000	0.000000
90	44	36	36	1.000000	0.000000
90	44	37	0	0.000000	0.000000
90	44	37	1	0.000000	0.000000
90	44	37	2	0.000000	0.000000
90	44	37	3	0.000000	0.000000
90	44	37	4	0.000000	0.000000
90	44	37	5	0.000000	0.000000
90	44	37	6	0.000000	0.000000
90	44	37	7	0.000002	0.000001
90	44	37	8	0.000014	0.000012
90	44	37	9	0.000093	0.000079
90	44	37	10	0.000501	0.000408
90	44	37	11	0.002205	0.001703
90	44	37	12	0.008004	0.005800
90	44	37	13	0.024227	0.016223
90	44	37	14	0.061710	0.037483
90	44	37	15	0.133553	0.071843
90	44	37	16	0.248142	0.114589
90	44	37	17	0.400582	0.152440
90	44	37	18	0.569960	0.169378
90	44	37	19	0.727240	0.157280
90	44	37	20	0.849267	0.122027
90	44	37	21	0.928294	0.079027
90	44	37	22	0.970937	0.042642
90	44	37	23	0.990056	0.019119
90	44	37	24	0.997153	0.007097
90	44	37	25	0.999324	0.002171
90	44	37	26	0.999868	0.000544
90	44	37	27	0.999979	0.000111
90	44	37	28	0.999997	0.000018
90	44	37	29	1.000000	0.000002
90	44	37	30	1.000000	0.000000
90	44	37	31	1.000000	0.000000
90	44	37	32	1.000000	0.000000
90	44	37	33	1.000000	0.000000
90	44	37	34	1.000000	0.000000
90	44	37	35	1.000000	0.000000
90	44	37	36	1.000000	0.000000

Table for $N = 2$, $n = 1$, through $N = 100$, $n = 50$

N	n	k	x	P(x)	p(x)
90	44	37	37	1.000000	0.000000
90	44	38	0	0.000000	0.000000
90	44	38	1	0.000000	0.000000
90	44	38	2	0.000000	0.000000
90	44	38	3	0.000000	0.000000
90	44	38	4	0.000000	0.000000
90	44	38	5	0.000000	0.000000
90	44	38	6	0.000000	0.000000
90	44	38	7	0.000001	0.000001
90	44	38	8	0.000006	0.000005
90	44	38	9	0.000041	0.000035
90	44	38	10	0.000239	0.000199
90	44	38	11	0.001144	0.000905
90	44	38	12	0.004503	0.003359
90	44	38	13	0.014738	0.010236
90	44	38	14	0.040493	0.025755
90	44	38	15	0.094243	0.053750
90	44	38	16	0.187605	0.093362
90	44	38	17	0.322924	0.135320
90	44	38	18	0.486869	0.163945
90	44	38	19	0.653051	0.166182
90	44	38	20	0.794010	0.140958
90	44	38	21	0.893999	0.099990
90	44	38	22	0.953236	0.059236
90	44	38	23	0.982480	0.029244
90	44	38	24	0.994475	0.011995
90	44	38	25	0.998546	0.004071
90	44	38	26	0.999684	0.001137
90	44	38	27	0.999943	0.000260
90	44	38	28	0.999992	0.000048
90	44	38	29	0.999999	0.000007
90	44	38	30	1.000000	0.000001
90	44	38	31	1.000000	0.000000
90	44	38	32	1.000000	0.000000
90	44	38	33	1.000000	0.000000
90	44	38	34	1.000000	0.000000
90	44	38	35	1.000000	0.000000
90	44	38	36	1.000000	0.000000
90	44	38	37	1.000000	0.000000
90	44	38	38	1.000000	0.000000
90	44	39	0	0.000000	0.000000
90	44	39	1	0.000000	0.000000
90	44	39	2	0.000000	0.000000
90	44	39	3	0.000000	0.000000
90	44	39	4	0.000000	0.000000
90	44	39	5	0.000000	0.000000
90	44	39	6	0.000000	0.000000
90	44	39	7	0.000000	0.000000
90	44	39	8	0.000002	0.000002
90	44	39	9	0.000017	0.000015
90	44	39	10	0.000110	0.000092
90	44	39	11	0.000570	0.000460
90	44	39	12	0.002436	0.001866
90	44	39	13	0.008636	0.006200
90	44	39	14	0.025634	0.016998
90	44	39	15	0.064267	0.038632
90	44	39	16	0.137333	0.073066
90	44	39	17	0.252662	0.115329
90	44	39	18	0.404897	0.152235
90	44	39	19	0.573156	0.168259
90	44	39	20	0.728952	0.155796
90	44	39	21	0.849773	0.120821
90	44	39	22	0.928174	0.078401
90	44	39	23	0.970670	0.042496
90	44	39	24	0.989862	0.019192
90	44	39	25	0.997059	0.007197
90	44	39	26	0.999290	0.002231
90	44	39	27	0.999858	0.000569
90	44	39	28	0.999977	0.000118
90	44	39	29	0.999997	0.000020
90	44	39	30	0.999999	0.000003
90	44	39	31	1.000000	0.000000
90	44	39	32	1.000000	0.000000
90	44	39	33	1.000000	0.000000
90	44	39	34	1.000000	0.000000
90	44	39	35	1.000000	0.000000
90	44	39	36	1.000000	0.000000
90	44	39	37	1.000000	0.000000
90	44	39	38	1.000000	0.000000
90	44	39	39	1.000000	0.000000
90	44	40	0	0.000000	0.000000
90	44	40	1	0.000000	0.000000
90	44	40	2	0.000000	0.000000
90	44	40	3	0.000000	0.000000
90	44	40	4	0.000000	0.000000
90	44	40	5	0.000000	0.000000
90	44	40	6	0.000000	0.000000
90	44	40	7	0.000000	0.000000
90	44	40	8	0.000001	0.000001
90	44	40	9	0.000007	0.000006
90	44	40	10	0.000048	0.000041
90	44	40	11	0.000272	0.000224
90	44	40	12	0.001265	0.000993
90	44	40	13	0.004867	0.003602
90	44	40	14	0.015635	0.010768
90	44	40	15	0.042299	0.026664
90	44	40	16	0.097218	0.054919
90	44	40	17	0.191605	0.094387
90	44	40	18	0.327287	0.135682
90	44	40	19	0.490676	0.163389
90	44	40	20	0.655636	0.164960
90	44	40	21	0.795285	0.139649
90	44	40	22	0.894354	0.099069
90	44	40	23	0.953172	0.058818
90	44	40	24	0.982336	0.029164
90	44	40	25	0.994377	0.012042
90	44	40	26	0.998502	0.004125
90	44	40	27	0.999669	0.001167
90	44	40	28	0.999940	0.000271
90	44	40	29	0.999991	0.000051
90	44	40	30	0.999999	0.000008
90	44	40	31	1.000000	0.000001
90	44	40	32	1.000000	0.000000
90	44	40	33	1.000000	0.000000
90	44	40	34	1.000000	0.000000
90	44	40	35	1.000000	0.000000
90	44	40	36	1.000000	0.000000
90	44	40	37	1.000000	0.000000
90	44	40	38	1.000000	0.000000
90	44	40	39	1.000000	0.000000
90	44	40	40	1.000000	0.000000
90	44	41	0	0.000000	0.000000
90	44	41	1	0.000000	0.000000
90	44	41	2	0.000000	0.000000
90	44	41	3	0.000000	0.000000
90	44	41	4	0.000000	0.000000
90	44	41	5	0.000000	0.000000
90	44	41	6	0.000000	0.000000
90	44	41	7	0.000000	0.000000
90	44	41	8	0.000000	0.000000
90	44	41	9	0.000003	0.000002
90	44	41	10	0.000020	0.000017
90	44	41	11	0.000124	0.000104
90	44	41	12	0.000630	0.000505
90	44	41	13	0.002634	0.002004
90	44	41	14	0.009175	0.006541
90	44	41	15	0.026834	0.017660
90	44	41	16	0.066464	0.039629
90	44	41	17	0.140636	0.074173
90	44	41	18	0.256733	0.116096
90	44	41	19	0.408982	0.152249
90	44	41	20	0.576455	0.167474
90	44	41	21	0.731047	0.154591
90	44	41	22	0.850764	0.119717
90	44	41	23	0.928468	0.077705
90	44	41	24	0.970670	0.042202
90	44	41	25	0.989802	0.019131
90	44	41	26	0.997017	0.007216
90	44	41	27	0.999272	0.002255
90	44	41	28	0.999853	0.000581
90	44	41	29	0.999976	0.000123
90	44	41	30	0.999997	0.000021
90	44	41	31	1.000000	0.000003
90	44	41	32	1.000000	0.000000
90	44	41	33	1.000000	0.000000
90	44	41	34	1.000000	0.000000
90	44	41	35	1.000000	0.000000
90	44	41	36	1.000000	0.000000
90	44	41	37	1.000000	0.000000
90	44	41	38	1.000000	0.000000
90	44	41	39	1.000000	0.000000
90	44	41	40	1.000000	0.000000
90	44	41	41	1.000000	0.000000
90	44	42	0	0.000000	0.000000
90	44	42	1	0.000000	0.000000
90	44	42	2	0.000000	0.000000
90	44	42	3	0.000000	0.000000
90	44	42	4	0.000000	0.000000
90	44	42	5	0.000000	0.000000
90	44	42	6	0.000000	0.000000
90	44	42	7	0.000000	0.000000
90	44	42	8	0.000000	0.000000
90	44	42	9	0.000001	0.000001
90	44	42	10	0.000008	0.000007
90	44	42	11	0.000054	0.000046
90	44	42	12	0.000299	0.000245
90	44	42	13	0.001366	0.001066
90	44	42	14	0.005170	0.003804
90	44	42	15	0.016383	0.011213
90	44	42	16	0.043818	0.027436
90	44	42	17	0.099766	0.055947
90	44	42	18	0.195130	0.095365
90	44	42	19	0.331303	0.136173
90	44	42	20	0.494427	0.163124
90	44	42	21	0.658483	0.164056
90	44	42	22	0.797013	0.138530
90	44	42	23	0.895167	0.098153
90	44	42	24	0.953445	0.058278
90	44	42	25	0.982383	0.028938
90	44	42	26	0.994367	0.011983
90	44	42	27	0.998490	0.004123
90	44	42	28	0.999663	0.001173
90	44	42	29	0.999938	0.000275
90	44	42	30	0.999991	0.000053
90	44	42	31	0.999999	0.000008
90	44	42	32	1.000000	0.000001
90	44	42	33	1.000000	0.000000
90	44	42	34	1.000000	0.000000
90	44	42	35	1.000000	0.000000
90	44	42	36	1.000000	0.000000

$N = 90$ $n = 44$

Table for $N = 2$, $n = 1$, through $N = 100$, $n = 50$

N	n	k	x	P(x)	p(x)
90	44	42	37	1.000000	0.000000
90	44	42	38	1.000000	0.000000
90	44	42	39	1.000000	0.000000
90	44	42	40	1.000000	0.000000
90	44	42	41	1.000000	0.000000
90	44	42	42	1.000000	0.000000
90	44	43	0	0.000000	0.000000
90	44	43	1	0.000000	0.000000
90	44	43	2	0.000000	0.000000
90	44	43	3	0.000000	0.000000
90	44	43	4	0.000000	0.000000
90	44	43	5	0.000000	0.000000
90	44	43	6	0.000000	0.000000
90	44	43	7	0.000000	0.000000
90	44	43	8	0.000000	0.000000
90	44	43	9	0.000000	0.000000
90	44	43	10	0.000003	0.000003
90	44	43	11	0.000022	0.000019
90	44	43	12	0.000136	0.000114
90	44	43	13	0.000677	0.000541
90	44	43	14	0.002792	0.002115
90	44	43	15	0.009608	0.006816
90	44	43	16	0.027814	0.018206
90	44	43	17	0.068295	0.040481
90	44	43	18	0.143475	0.075179
90	44	43	19	0.260380	0.116905
90	44	43	20	0.412865	0.152485
90	44	43	21	0.579873	0.167008
90	44	43	22	0.733520	0.153647
90	44	43	23	0.852224	0.118704
90	44	43	24	0.929162	0.076938
90	44	43	25	0.970928	0.041766
90	44	43	26	0.989873	0.018944
90	44	43	27	0.997030	0.007157
90	44	43	28	0.999272	0.002243
90	44	43	29	0.999852	0.000580
90	44	43	30	0.999975	0.000123
90	44	43	31	0.999996	0.000021
90	44	43	32	0.999999	0.000003
90	44	43	33	1.000000	0.000000
90	44	43	34	1.000000	0.000000
90	44	43	35	1.000000	0.000000
90	44	43	36	1.000000	0.000000
90	44	43	37	1.000000	0.000000
90	44	43	38	1.000000	0.000000
90	44	43	39	1.000000	0.000000
90	44	43	40	1.000000	0.000000
90	44	43	41	1.000000	0.000000
90	44	43	42	1.000000	0.000000
90	44	43	43	1.000000	0.000000

N	n	k	x	P(x)	p(x)
90	44	44	0	0.000000	0.000000
90	44	44	1	0.000000	0.000000
90	44	44	2	0.000000	0.000000
90	44	44	3	0.000000	0.000000
90	44	44	4	0.000000	0.000000
90	44	44	5	0.000000	0.000000
90	44	44	6	0.000000	0.000000
90	44	44	7	0.000000	0.000000
90	44	44	8	0.000000	0.000000
90	44	44	9	0.000000	0.000000
90	44	44	10	0.000001	0.000001
90	44	44	11	0.000009	0.000008
90	44	44	12	0.000059	0.000050
90	44	44	13	0.000320	0.000262
90	44	44	14	0.001442	0.001122
90	44	44	15	0.005403	0.003961
90	44	44	16	0.016968	0.011565
90	44	44	17	0.045040	0.028072
90	44	44	18	0.101886	0.056846
90	44	44	19	0.198196	0.096310
90	44	44	20	0.335001	0.136804
90	44	44	21	0.498146	0.163145
90	44	44	22	0.661600	0.163454
90	44	44	23	0.799187	0.137586
90	44	44	24	0.896423	0.097236
90	44	44	25	0.954044	0.057621
90	44	44	26	0.982618	0.028573
90	44	44	27	0.994441	0.011823
90	44	44	28	0.998509	0.004068
90	44	44	29	0.999667	0.001158
90	44	44	30	0.999939	0.000271
90	44	44	31	0.999991	0.000052
90	44	44	32	0.999999	0.000008
90	44	44	33	1.000000	0.000001
90	44	44	34	1.000000	0.000000
90	44	44	35	1.000000	0.000000
90	44	44	36	1.000000	0.000000
90	44	44	37	1.000000	0.000000
90	44	44	38	1.000000	0.000000
90	44	44	39	1.000000	0.000000
90	44	44	40	1.000000	0.000000
90	44	44	41	1.000000	0.000000
90	44	44	42	1.000000	0.000000
90	44	44	43	1.000000	0.000000
90	44	44	44	1.000000	0.000000
90	45	1	0	0.500000	0.500000
90	45	1	1	1.000000	0.500000
90	45	2	0	0.247191	0.247191
90	45	2	1	0.752809	0.505618
90	45	2	2	1.000000	0.247191

N	n	k	x	P(x)	p(x)
90	45	3	0	0.120787	0.120787
90	45	3	1	0.500000	0.379213
90	45	3	2	0.879213	0.379213
90	45	3	3	1.000000	0.120787
90	45	4	0	0.058311	0.058311
90	45	4	1	0.308214	0.249903
90	45	4	2	0.691786	0.383572
90	45	4	3	0.941689	0.249903
90	45	4	4	1.000000	0.058311
90	45	5	0	0.027799	0.027799
90	45	5	1	0.180356	0.152557
90	45	5	2	0.500000	0.319644
90	45	5	3	0.819643	0.319644
90	45	5	4	0.972201	0.152557
90	45	5	5	1.000000	0.027799
90	45	6	0	0.013082	0.013082
90	45	6	1	0.101386	0.088304
90	45	6	2	0.338298	0.236912
90	45	6	3	0.661702	0.323404
90	45	6	4	0.898614	0.236912
90	45	6	5	0.986918	0.088304
90	45	6	6	1.000000	0.013082
90	45	7	0	0.006074	0.006074
90	45	7	1	0.055131	0.049058
90	45	7	2	0.217021	0.161890
90	45	7	3	0.500000	0.282979
90	45	7	4	0.782978	0.282979
90	45	7	5	0.944869	0.161890
90	45	7	6	0.993926	0.049058
90	45	7	7	1.000000	0.006074
90	45	8	0	0.002781	0.002781
90	45	8	1	0.029125	0.026344
90	45	8	2	0.133151	0.104026
90	45	8	3	0.356806	0.223655
90	45	8	4	0.643194	0.286388
90	45	8	5	0.866849	0.223655
90	45	8	6	0.970875	0.104026
90	45	8	7	0.997219	0.026344
90	45	8	8	1.000000	0.002781
90	45	9	0	0.001255	0.001255
90	45	9	1	0.014989	0.013734
90	45	9	2	0.078601	0.063612
90	45	9	3	0.242251	0.163650
90	45	9	4	0.500000	0.257749
90	45	9	5	0.757749	0.257749
90	45	9	6	0.921399	0.163650
90	45	9	7	0.985011	0.063612
90	45	9	8	0.998745	0.013734
90	45	9	9	1.000000	0.001255
90	45	10	0	0.000558	0.000558

N	n	k	x	P(x)	p(x)
90	45	10	1	0.007528	0.006971
90	45	10	2	0.044832	0.037303
90	45	10	3	0.157395	0.112564
90	45	10	4	0.369534	0.212139
90	45	10	5	0.630466	0.260931
90	45	10	6	0.842605	0.212139
90	45	10	7	0.955168	0.112564
90	45	10	8	0.992471	0.037303
90	45	10	9	0.999442	0.006971
90	45	10	10	1.000000	0.000558
90	45	11	0	0.000244	0.000244
90	45	11	1	0.003695	0.003451
90	45	11	2	0.024781	0.021087
90	45	11	3	0.098299	0.073518
90	45	11	4	0.260813	0.162514
90	45	11	5	0.500000	0.239187
90	45	11	6	0.739187	0.239187
90	45	11	7	0.901701	0.162514
90	45	11	8	0.975219	0.073518
90	45	11	9	0.996305	0.021087
90	45	11	10	0.999756	0.003451
90	45	11	11	1.000000	0.000244
90	45	12	0	0.000105	0.000105
90	45	12	1	0.001773	0.001668
90	45	12	2	0.013304	0.011531
90	45	12	3	0.059214	0.045910
90	45	12	4	0.176470	0.117257
90	45	12	5	0.378893	0.202422
90	45	12	6	0.621107	0.242215
90	45	12	7	0.823529	0.202422
90	45	12	8	0.940786	0.117257
90	45	12	9	0.986696	0.045910
90	45	12	10	0.998227	0.011531
90	45	12	11	0.999895	0.001668
90	45	12	12	1.000000	0.000105
90	45	13	0	0.000044	0.000044
90	45	13	1	0.000832	0.000788
90	45	13	2	0.006947	0.006115
90	45	13	3	0.034493	0.027546
90	45	13	4	0.114835	0.080343
90	45	13	5	0.275086	0.160251
90	45	13	6	0.500000	0.224914
90	45	13	7	0.724914	0.224914
90	45	13	8	0.885164	0.160251
90	45	13	9	0.965507	0.080343
90	45	13	10	0.993053	0.027546
90	45	13	11	0.999168	0.006115
90	45	13	12	0.999956	0.000788
90	45	13	13	1.000000	0.000044
90	45	14	0	0.000018	0.000018

Table for $N = 2$, $n = 1$, through $N = 100$, $n = 50$

N	n	k	x	P(x)	p(x)
90	45	14	1	0.000382	0.000363
90	45	14	2	0.003532	0.003150
90	45	14	3	0.019468	0.015936
90	45	14	4	0.072056	0.052588
90	45	14	5	0.191839	0.119784
90	45	14	6	0.386083	0.194244
90	45	14	7	0.613917	0.227835
90	45	14	8	0.808161	0.194244
90	45	14	9	0.927944	0.119784
90	45	14	10	0.980532	0.052588
90	45	14	11	0.996468	0.015936
90	45	14	12	0.999618	0.003150
90	45	14	13	0.999981	0.000363
90	45	14	14	1.000000	0.000018
90	45	15	0	0.000008	0.000008
90	45	15	1	0.000171	0.000164
90	45	15	2	0.001750	0.001578
90	45	15	3	0.010661	0.008911
90	45	15	4	0.043686	0.033025
90	45	15	5	0.128795	0.085109
90	45	15	6	0.286405	0.157610
90	45	15	7	0.500000	0.213595
90	45	15	8	0.713595	0.213595
90	45	15	9	0.871205	0.157610
90	45	15	10	0.956314	0.085109
90	45	15	11	0.989339	0.033025
90	45	15	12	0.998250	0.008911
90	45	15	13	0.999828	0.001578
90	45	15	14	0.999992	0.000164
90	45	15	15	1.000000	0.000008
90	45	16	0	0.000003	0.000003
90	45	16	1	0.000075	0.000072
90	45	16	2	0.000845	0.000770
90	45	16	3	0.005671	0.004826
90	45	16	4	0.025632	0.019962
90	45	16	5	0.083404	0.057771
90	45	16	6	0.204448	0.121044
90	45	16	7	0.391779	0.187331
90	45	16	8	0.608221	0.216443
90	45	16	9	0.795552	0.187331
90	45	16	10	0.916596	0.121044
90	45	16	11	0.974368	0.057771
90	45	16	12	0.994329	0.019962
90	45	16	13	0.999155	0.004826
90	45	16	14	0.999925	0.000770
90	45	16	15	0.999997	0.000072
90	45	16	16	1.000000	0.000003
90	45	17	0	0.000001	0.000001
90	45	17	1	0.000032	0.000031
90	45	17	2	0.000398	0.000365

N	n	k	x	P(x)	p(x)
90	45	17	3	0.002932	0.002534
90	45	17	4	0.014573	0.011641
90	45	17	5	0.052176	0.037603
90	45	17	6	0.140654	0.088478
90	45	17	7	0.295582	0.154927
90	45	17	8	0.500000	0.204418
90	45	17	9	0.704418	0.204418
90	45	17	10	0.859346	0.154927
90	45	17	11	0.947824	0.088478
90	45	17	12	0.985427	0.037603
90	45	17	13	0.997068	0.011641
90	45	17	14	0.999602	0.002534
90	45	17	15	0.999968	0.000365
90	45	17	16	0.999999	0.000031
90	45	17	17	1.000000	0.000001
90	45	18	0	0.000000	0.000000
90	45	18	1	0.000014	0.000013
90	45	18	2	0.000182	0.000169
90	45	18	3	0.001474	0.001291
90	45	18	4	0.008035	0.006561
90	45	18	5	0.031571	0.023537
90	45	18	6	0.093385	0.061814
90	45	18	7	0.214935	0.121550
90	45	18	8	0.396391	0.181456
90	45	18	9	0.603609	0.207218
90	45	18	10	0.785065	0.181456
90	45	18	11	0.906615	0.121550
90	45	18	12	0.968429	0.061814
90	45	18	13	0.991965	0.023537
90	45	18	14	0.998526	0.006561
90	45	18	15	0.999817	0.001291
90	45	18	16	0.999986	0.000169
90	45	18	17	0.999999	0.000013
90	45	18	18	1.000000	0.000000
90	45	19	0	0.000000	0.000000
90	45	19	1	0.000006	0.000005
90	45	19	2	0.000082	0.000076
90	45	19	3	0.000721	0.000639
90	45	19	4	0.004299	0.003578
90	45	19	5	0.018495	0.014197
90	45	19	6	0.059903	0.041407
90	45	19	7	0.150783	0.090881
90	45	19	8	0.303142	0.152359
90	45	19	9	0.500000	0.196858
90	45	19	10	0.696858	0.196858
90	45	19	11	0.849217	0.152359
90	45	19	12	0.940097	0.090881
90	45	19	13	0.981505	0.041407
90	45	19	14	0.995701	0.014197
90	45	19	15	0.999279	0.003578

N	n	k	x	P(x)	p(x)
90	45	19	16	0.999918	0.000639
90	45	19	17	0.999994	0.000076
90	45	19	18	1.000000	0.000005
90	45	19	19	1.000000	0.000000
90	45	20	0	0.000000	0.000000
90	45	20	1	0.000002	0.000002
90	45	20	2	0.000036	0.000033
90	45	20	3	0.000343	0.000307
90	45	20	4	0.002232	0.001890
90	45	20	5	0.010497	0.008265
90	45	20	6	0.037158	0.026661
90	45	20	7	0.102143	0.064985
90	45	20	8	0.223744	0.121601
90	45	20	9	0.400185	0.176441
90	45	20	10	0.599815	0.199630
90	45	20	11	0.776256	0.176441
90	45	20	12	0.897857	0.121601
90	45	20	13	0.962842	0.064985
90	45	20	14	0.989503	0.026661
90	45	20	15	0.997768	0.008265
90	45	20	16	0.999657	0.001890
90	45	20	17	0.999964	0.000307
90	45	20	18	0.999998	0.000033
90	45	20	19	1.000000	0.000002
90	45	20	20	1.000000	0.000000
90	45	21	0	0.000000	0.000000
90	45	21	1	0.000001	0.000001
90	45	21	2	0.000015	0.000014
90	45	21	3	0.000158	0.000143
90	45	21	4	0.001125	0.000967
90	45	21	5	0.005774	0.004649
90	45	21	6	0.022304	0.016530
90	45	21	7	0.066865	0.044561
90	45	21	8	0.159469	0.092604
90	45	21	9	0.309444	0.149975
90	45	21	10	0.500000	0.190556
90	45	21	11	0.690556	0.190556
90	45	21	12	0.840531	0.149975
90	45	21	13	0.933135	0.092604
90	45	21	14	0.977696	0.044561
90	45	21	15	0.994225	0.016530
90	45	21	16	0.998874	0.004649
90	45	21	17	0.999842	0.000967
90	45	21	18	0.999985	0.000143
90	45	21	19	0.999999	0.000014
90	45	21	20	1.000000	0.000001
90	45	21	21	1.000000	0.000000
90	45	22	0	0.000000	0.000000
90	45	22	1	0.000000	0.000000
90	45	22	2	0.000006	0.000006

N	n	k	x	P(x)	p(x)
90	45	22	3	0.000071	0.000065
90	45	22	4	0.000551	0.000480
90	45	22	5	0.003079	0.002529
90	45	22	6	0.012961	0.009882
90	45	22	7	0.042324	0.029363
90	45	22	8	0.109812	0.067488
90	45	22	9	0.231196	0.121384
90	45	22	10	0.403341	0.172145
90	45	22	11	0.596659	0.193318
90	45	22	12	0.768804	0.172145
90	45	22	13	0.890188	0.121384
90	45	22	14	0.957676	0.067488
90	45	22	15	0.987039	0.029363
90	45	22	16	0.996920	0.009882
90	45	22	17	0.999449	0.002529
90	45	22	18	0.999929	0.000480
90	45	22	19	0.999994	0.000065
90	45	22	20	1.000000	0.000006
90	45	22	21	1.000000	0.000000
90	45	22	22	1.000000	0.000000
90	45	23	0	0.000000	0.000000
90	45	23	1	0.000000	0.000000
90	45	23	2	0.000002	0.000002
90	45	23	3	0.000031	0.000029
90	45	23	4	0.000262	0.000231
90	45	23	5	0.001592	0.001330
90	45	23	6	0.007294	0.005702
90	45	23	7	0.025915	0.018622
90	45	23	8	0.073091	0.047175
90	45	23	9	0.166934	0.093843
90	45	23	10	0.314737	0.147803
90	45	23	11	0.500000	0.185263
90	45	23	12	0.685263	0.185263
90	45	23	13	0.833066	0.147803
90	45	23	14	0.926909	0.093843
90	45	23	15	0.974084	0.047175
90	45	23	16	0.992706	0.018622
90	45	23	17	0.998408	0.005702
90	45	23	18	0.999738	0.001330
90	45	23	19	0.999969	0.000231
90	45	23	20	0.999997	0.000029
90	45	23	21	1.000000	0.000002
90	45	23	22	1.000000	0.000000
90	45	23	23	1.000000	0.000000
90	45	24	0	0.000000	0.000000
90	45	24	1	0.000000	0.000000
90	45	24	2	0.000001	0.000001
90	45	24	3	0.000013	0.000012
90	45	24	4	0.000121	0.000107
90	45	24	5	0.000798	0.000677

N	n	k	x	P(x)	p(x)
90	45	24	6	0.003975	0.003177
90	45	24	7	0.015354	0.011379
90	45	24	8	0.047039	0.031685
90	45	24	9	0.116511	0.069472
90	45	24	10	0.237526	0.121016
90	45	24	11	0.405986	0.168459
90	45	24	12	0.594014	0.188028
90	45	24	13	0.762473	0.168459
90	45	24	14	0.883489	0.121016
90	45	24	15	0.952961	0.069472
90	45	24	16	0.984646	0.031685
90	45	24	17	0.996025	0.011379
90	45	24	18	0.999202	0.003177
90	45	24	19	0.999879	0.000677
90	45	24	20	0.999987	0.000107
90	45	24	21	0.999999	0.000012
90	45	24	22	1.000000	0.000001
90	45	24	23	1.000000	0.000000
90	45	24	24	1.000000	0.000000
90	45	25	0	0.000000	0.000000
90	45	25	1	0.000000	0.000000
90	45	25	2	0.000000	0.000000
90	45	25	3	0.000005	0.000005
90	45	25	4	0.000054	0.000048
90	45	25	5	0.000387	0.000333
90	45	25	6	0.002097	0.001710
90	45	25	7	0.008802	0.006705
90	45	25	8	0.029276	0.020474
90	45	25	9	0.078617	0.049341
90	45	25	10	0.173351	0.094735
90	45	25	11	0.319204	0.145852
90	45	25	12	0.500000	0.180796
90	45	25	13	0.680796	0.180796
90	45	25	14	0.826648	0.145852
90	45	25	15	0.921383	0.094735
90	45	25	16	0.970724	0.049341
90	45	25	17	0.991198	0.020474
90	45	25	18	0.997902	0.006705
90	45	25	19	0.999613	0.001710
90	45	25	20	0.999946	0.000333
90	45	25	21	0.999995	0.000048
90	45	25	22	1.000000	0.000005
90	45	25	23	1.000000	0.000000
90	45	25	24	1.000000	0.000000
90	45	25	25	1.000000	0.000000
90	45	26	0	0.000000	0.000000
90	45	26	1	0.000000	0.000000
90	45	26	2	0.000000	0.000000
90	45	26	3	0.000002	0.000002
90	45	26	4	0.000023	0.000021
90	45	26	5	0.000182	0.000159
90	45	26	6	0.001071	0.000889
90	45	26	7	0.004883	0.003811
90	45	26	8	0.017622	0.012739
90	45	26	9	0.051290	0.033668
90	45	26	10	0.122341	0.071051
90	45	26	11	0.242912	0.120571
90	45	26	12	0.408211	0.165299
90	45	26	13	0.591789	0.183578
90	45	26	14	0.757088	0.165299
90	45	26	15	0.877659	0.120571
90	45	26	16	0.948710	0.071051
90	45	26	17	0.982378	0.033668
90	45	26	18	0.995117	0.012739
90	45	26	19	0.998929	0.003811
90	45	26	20	0.999818	0.000889
90	45	26	21	0.999977	0.000159
90	45	26	22	0.999998	0.000021
90	45	26	23	1.000000	0.000002
90	45	26	24	1.000000	0.000000
90	45	26	25	1.000000	0.000000
90	45	26	26	1.000000	0.000000
90	45	27	0	0.000000	0.000000
90	45	27	1	0.000000	0.000000
90	45	27	2	0.000000	0.000000
90	45	27	3	0.000001	0.000001
90	45	27	4	0.000010	0.000009
90	45	27	5	0.000083	0.000073
90	45	27	6	0.000529	0.000447
90	45	27	7	0.002620	0.002090
90	45	27	8	0.010257	0.007637
90	45	27	9	0.032351	0.022095
90	45	27	10	0.083485	0.051133
90	45	27	11	0.178858	0.095374
90	45	27	12	0.322979	0.144120
90	45	27	13	0.500000	0.177021
90	45	27	14	0.677021	0.177021
90	45	27	15	0.821142	0.144120
90	45	27	16	0.916515	0.095374
90	45	27	17	0.967648	0.051133
90	45	27	18	0.989743	0.022095
90	45	27	19	0.997380	0.007637
90	45	27	20	0.999470	0.002090
90	45	27	21	0.999917	0.000447
90	45	27	22	0.999990	0.000073
90	45	27	23	0.999999	0.000009
90	45	27	24	1.000000	0.000001
90	45	27	25	1.000000	0.000000
90	45	27	26	1.000000	0.000000
90	45	27	27	1.000000	0.000000
90	45	28	0	0.000000	0.000000
90	45	28	1	0.000000	0.000000
90	45	28	2	0.000000	0.000000
90	45	28	3	0.000000	0.000000
90	45	28	4	0.000004	0.000004
90	45	28	5	0.000036	0.000032
90	45	28	6	0.000253	0.000217
90	45	28	7	0.001359	0.001106
90	45	28	8	0.005772	0.004413
90	45	28	9	0.019726	0.013954
90	45	28	10	0.055077	0.035351
90	45	28	11	0.127387	0.072310
90	45	28	12	0.247487	0.120100
90	45	28	13	0.410084	0.162597
90	45	28	14	0.589915	0.179831
90	45	28	15	0.752513	0.162597
90	45	28	16	0.872613	0.120100
90	45	28	17	0.944923	0.072310
90	45	28	18	0.980274	0.035351
90	45	28	19	0.994228	0.013954
90	45	28	20	0.998641	0.004413
90	45	28	21	0.999747	0.001106
90	45	28	22	0.999963	0.000217
90	45	28	23	0.999996	0.000032
90	45	28	24	1.000000	0.000004
90	45	28	25	1.000000	0.000000
90	45	28	26	1.000000	0.000000
90	45	28	27	1.000000	0.000000
90	45	28	28	1.000000	0.000000
90	45	29	0	0.000000	0.000000
90	45	29	1	0.000000	0.000000
90	45	29	2	0.000000	0.000000
90	45	29	3	0.000000	0.000000
90	45	29	4	0.000002	0.000001
90	45	29	5	0.000015	0.000014
90	45	29	6	0.000117	0.000101
90	45	29	7	0.000681	0.000564
90	45	29	8	0.003138	0.002457
90	45	29	9	0.011623	0.008485
90	45	29	10	0.035121	0.023498
90	45	29	11	0.087733	0.052612
90	45	29	12	0.183563	0.095830
90	45	29	13	0.326163	0.142600
90	45	29	14	0.500000	0.173837
90	45	29	15	0.673837	0.173837
90	45	29	16	0.816437	0.142600
90	45	29	17	0.912267	0.095830
90	45	29	18	0.964879	0.052612
90	45	29	19	0.988377	0.023498
90	45	29	20	0.996862	0.008485
90	45	29	21	0.999319	0.002457
90	45	29	22	0.999883	0.000564
90	45	29	23	0.999984	0.000101
90	45	29	24	0.999998	0.000014
90	45	29	25	1.000000	0.000001
90	45	29	26	1.000000	0.000000
90	45	29	27	1.000000	0.000000
90	45	29	28	1.000000	0.000000
90	45	29	29	1.000000	0.000000
90	45	30	0	0.000000	0.000000
90	45	30	1	0.000000	0.000000
90	45	30	2	0.000000	0.000000
90	45	30	3	0.000000	0.000000
90	45	30	4	0.000001	0.000001
90	45	30	5	0.000006	0.000006
90	45	30	6	0.000052	0.000046
90	45	30	7	0.000330	0.000278
90	45	30	8	0.001648	0.001318
90	45	30	9	0.006616	0.004968
90	45	30	10	0.021639	0.015023
90	45	30	11	0.058408	0.036770
90	45	30	12	0.131721	0.073312
90	45	30	13	0.251356	0.119636
90	45	30	14	0.411657	0.160300
90	45	30	15	0.588343	0.176687
90	45	30	16	0.748644	0.160300
90	45	30	17	0.868279	0.119636
90	45	30	18	0.941592	0.073312
90	45	30	19	0.978361	0.036770
90	45	30	20	0.993384	0.015023
90	45	30	21	0.998352	0.004968
90	45	30	22	0.999670	0.001318
90	45	30	23	0.999948	0.000278
90	45	30	24	0.999994	0.000046
90	45	30	25	0.999999	0.000006
90	45	30	26	1.000000	0.000001
90	45	30	27	1.000000	0.000000
90	45	30	28	1.000000	0.000000
90	45	30	29	1.000000	0.000000
90	45	30	30	1.000000	0.000000
90	45	31	0	0.000000	0.000000
90	45	31	1	0.000000	0.000000
90	45	31	2	0.000000	0.000000
90	45	31	3	0.000000	0.000000
90	45	31	4	0.000000	0.000000
90	45	31	5	0.000002	0.000002
90	45	31	6	0.000022	0.000020
90	45	31	7	0.000154	0.000131
90	45	31	8	0.000835	0.000681
90	45	31	9	0.003635	0.002800

N	n	k	x	P(x)	p(x)
90	45	31	10	0.012875	0.009240
90	45	31	11	0.037572	0.024697
90	45	31	12	0.091399	0.053827
90	45	31	13	0.187551	0.096152
90	45	31	14	0.328835	0.141284
90	45	31	15	0.500000	0.171165
90	45	31	16	0.671165	0.171165
90	45	31	17	0.812449	0.141284
90	45	31	18	0.908601	0.096152
90	45	31	19	0.962428	0.053827
90	45	31	20	0.987125	0.024697
90	45	31	21	0.996365	0.009240
90	45	31	22	0.999165	0.002800
90	45	31	23	0.999846	0.000681
90	45	31	24	0.999978	0.000131
90	45	31	25	0.999997	0.000020
90	45	31	26	1.000000	0.000002
90	45	31	27	1.000000	0.000000
90	45	31	28	1.000000	0.000000
90	45	31	29	1.000000	0.000000
90	45	31	30	1.000000	0.000000
90	45	31	31	1.000000	0.000000
90	45	32	0	0.000000	0.000000
90	45	32	1	0.000000	0.000000
90	45	32	2	0.000000	0.000000
90	45	32	3	0.000000	0.000000
90	45	32	4	0.000000	0.000000
90	45	32	5	0.000001	0.000001
90	45	32	6	0.000009	0.000008
90	45	32	7	0.000069	0.000060
90	45	32	8	0.000408	0.000339
90	45	32	9	0.001926	0.001519
90	45	32	10	0.007394	0.005467
90	45	32	11	0.023340	0.015946
90	45	32	12	0.061292	0.037952
90	45	32	13	0.135400	0.074108
90	45	32	14	0.254601	0.119200
90	45	32	15	0.412967	0.158366
90	45	32	16	0.587033	0.174066
90	45	32	17	0.745399	0.158366
90	45	32	18	0.864599	0.119200
90	45	32	19	0.938707	0.074108
90	45	32	20	0.976660	0.037952
90	45	32	21	0.992606	0.015946
90	45	32	22	0.998073	0.005467
90	45	32	23	0.999592	0.001519
90	45	32	24	0.999931	0.000339
90	45	32	25	0.999991	0.000060
90	45	32	26	0.999999	0.000008
90	45	32	27	1.000000	0.000001
90	45	32	28	1.000000	0.000000
90	45	32	29	1.000000	0.000000
90	45	32	30	1.000000	0.000000
90	45	32	31	1.000000	0.000000
90	45	32	32	1.000000	0.000000
90	45	33	0	0.000000	0.000000
90	45	33	1	0.000000	0.000000
90	45	33	2	0.000000	0.000000
90	45	33	3	0.000000	0.000000
90	45	33	4	0.000000	0.000000
90	45	33	5	0.000000	0.000000
90	45	33	6	0.000004	0.000003
90	45	33	7	0.000030	0.000026
90	45	33	8	0.000192	0.000162
90	45	33	9	0.000984	0.000792
90	45	33	10	0.004094	0.003111
90	45	33	11	0.013992	0.009898
90	45	33	12	0.039699	0.025707
90	45	33	13	0.094513	0.054814
90	45	33	14	0.190890	0.096377
90	45	33	15	0.331053	0.140163
90	45	33	16	0.500000	0.168947
90	45	33	17	0.668947	0.168947
90	45	33	18	0.809110	0.140163
90	45	33	19	0.905487	0.096377
90	45	33	20	0.960301	0.054814
90	45	33	21	0.986008	0.025707
90	45	33	22	0.995905	0.009898
90	45	33	23	0.999016	0.003111
90	45	33	24	0.999808	0.000792
90	45	33	25	0.999970	0.000162
90	45	33	26	0.999996	0.000026
90	45	33	27	0.999999	0.000003
90	45	33	28	1.000000	0.000000
90	45	33	29	1.000000	0.000000
90	45	33	30	1.000000	0.000000
90	45	33	31	1.000000	0.000000
90	45	33	32	1.000000	0.000000
90	45	33	33	1.000000	0.000000
90	45	34	0	0.000000	0.000000
90	45	34	1	0.000000	0.000000
90	45	34	2	0.000000	0.000000
90	45	34	3	0.000000	0.000000
90	45	34	4	0.000000	0.000000
90	45	34	5	0.000000	0.000000
90	45	34	6	0.000001	0.000001
90	45	34	7	0.000012	0.000011
90	45	34	8	0.000087	0.000074
90	45	34	9	0.000484	0.000397
90	45	34	10	0.002184	0.001701
90	45	34	11	0.008088	0.005904
90	45	34	12	0.024816	0.016728
90	45	34	13	0.063740	0.038924
90	45	34	14	0.138475	0.074734
90	45	34	15	0.257283	0.118809
90	45	34	16	0.414045	0.156761
90	45	34	17	0.585955	0.171911
90	45	34	18	0.742717	0.156761
90	45	34	19	0.861525	0.118809
90	45	34	20	0.936260	0.074734
90	45	34	21	0.975184	0.038924
90	45	34	22	0.991912	0.016728
90	45	34	23	0.997815	0.005904
90	45	34	24	0.999516	0.001701
90	45	34	25	0.999913	0.000397
90	45	34	26	0.999987	0.000074
90	45	34	27	0.999998	0.000011
90	45	34	28	1.000000	0.000001
90	45	34	29	1.000000	0.000000
90	45	34	30	1.000000	0.000000
90	45	34	31	1.000000	0.000000
90	45	34	32	1.000000	0.000000
90	45	34	33	1.000000	0.000000
90	45	34	34	1.000000	0.000000
90	45	35	0	0.000000	0.000000
90	45	35	1	0.000000	0.000000
90	45	35	2	0.000000	0.000000
90	45	35	3	0.000000	0.000000
90	45	35	4	0.000000	0.000000
90	45	35	5	0.000000	0.000000
90	45	35	6	0.000000	0.000000
90	45	35	7	0.000005	0.000004
90	45	35	8	0.000038	0.000033
90	45	35	9	0.000228	0.000191
90	45	35	10	0.001121	0.000893
90	45	35	11	0.004504	0.003382
90	45	35	12	0.014959	0.010455
90	45	35	13	0.041498	0.026539
90	45	35	14	0.097104	0.055606
90	45	35	15	0.193636	0.096532
90	45	35	16	0.332865	0.139229
90	45	35	17	0.500000	0.167135
90	45	35	18	0.667135	0.167135
90	45	35	19	0.806364	0.139229
90	45	35	20	0.902896	0.096532
90	45	35	21	0.958502	0.055606
90	45	35	22	0.985041	0.026539
90	45	35	23	0.995496	0.010455
90	45	35	24	0.998879	0.003382
90	45	35	25	0.999771	0.000893
90	45	35	26	0.999962	0.000191
90	45	35	27	0.999995	0.000033
90	45	35	28	0.999999	0.000004
90	45	35	29	1.000000	0.000000
90	45	35	30	1.000000	0.000000
90	45	35	31	1.000000	0.000000
90	45	35	32	1.000000	0.000000
90	45	35	33	1.000000	0.000000
90	45	35	34	1.000000	0.000000
90	45	35	35	1.000000	0.000000
90	45	36	0	0.000000	0.000000
90	45	36	1	0.000000	0.000000
90	45	36	2	0.000000	0.000000
90	45	36	3	0.000000	0.000000
90	45	36	4	0.000000	0.000000
90	45	36	5	0.000000	0.000000
90	45	36	6	0.000000	0.000000
90	45	36	7	0.000002	0.000002
90	45	36	8	0.000016	0.000014
90	45	36	9	0.000103	0.000088
90	45	36	10	0.000553	0.000450
90	45	36	11	0.002413	0.001860
90	45	36	12	0.008686	0.006273
90	45	36	13	0.026057	0.017371
90	45	36	14	0.065762	0.039705
90	45	36	15	0.140982	0.075220
90	45	36	16	0.259453	0.118471
90	45	36	17	0.414913	0.155460
90	45	36	18	0.585087	0.170174
90	45	36	19	0.740547	0.155460
90	45	36	20	0.859018	0.118471
90	45	36	21	0.934238	0.075220
90	45	36	22	0.973943	0.039705
90	45	36	23	0.991314	0.017371
90	45	36	24	0.997587	0.006273
90	45	36	25	0.999447	0.001860
90	45	36	26	0.999896	0.000450
90	45	36	27	0.999984	0.000088
90	45	36	28	0.999998	0.000014
90	45	36	29	1.000000	0.000002
90	45	36	30	1.000000	0.000000
90	45	36	31	1.000000	0.000000
90	45	36	32	1.000000	0.000000
90	45	36	33	1.000000	0.000000
90	45	36	34	1.000000	0.000000
90	45	36	35	1.000000	0.000000
90	45	36	36	1.000000	0.000000
90	45	37	0	0.000000	0.000000
90	45	37	1	0.000000	0.000000
90	45	37	2	0.000000	0.000000

N	n	k	x	P(x)	p(x)
90	45	37	3	0.000000	0.000000
90	45	37	4	0.000000	0.000000
90	45	37	5	0.000000	0.000000
90	45	37	6	0.000000	0.000000
90	45	37	7	0.000001	0.000001
90	45	37	8	0.000006	0.000006
90	45	37	9	0.000045	0.000039
90	45	37	10	0.000262	0.000217
90	45	37	11	0.001242	0.000980
90	45	37	12	0.004852	0.003610
90	45	37	13	0.015763	0.010911
90	45	37	14	0.042968	0.027206
90	45	37	15	0.099193	0.056225
90	45	37	16	0.195830	0.096636
90	45	37	17	0.334304	0.138474
90	45	37	18	0.500000	0.165696
90	45	37	19	0.665696	0.165696
90	45	37	20	0.804170	0.138474
90	45	37	21	0.900807	0.096636
90	45	37	22	0.957031	0.056225
90	45	37	23	0.984237	0.027206
90	45	37	24	0.995148	0.010911
90	45	37	25	0.998758	0.003610
90	45	37	26	0.999738	0.000980
90	45	37	27	0.999955	0.000217
90	45	37	28	0.999994	0.000039
90	45	37	29	0.999999	0.000006
90	45	37	30	1.000000	0.000001
90	45	37	31	1.000000	0.000000
90	45	37	32	1.000000	0.000000
90	45	37	33	1.000000	0.000000
90	45	37	34	1.000000	0.000000
90	45	37	35	1.000000	0.000000
90	45	37	36	1.000000	0.000000
90	45	37	37	1.000000	0.000000
90	45	38	0	0.000000	0.000000
90	45	38	1	0.000000	0.000000
90	45	38	2	0.000000	0.000000
90	45	38	3	0.000000	0.000000
90	45	38	4	0.000000	0.000000
90	45	38	5	0.000000	0.000000
90	45	38	6	0.000000	0.000000
90	45	38	7	0.000000	0.000000
90	45	38	8	0.000002	0.000002
90	45	38	9	0.000019	0.000016
90	45	38	10	0.000119	0.000100
90	45	38	11	0.000613	0.000495
90	45	38	12	0.002604	0.001991
90	45	38	13	0.009175	0.006571
90	45	38	14	0.027056	0.017880

N	n	k	x	P(x)	p(x)
90	45	38	15	0.067368	0.040312
90	45	38	16	0.142953	0.075585
90	45	38	17	0.261148	0.118195
90	45	38	18	0.415589	0.154441
90	45	38	19	0.584411	0.168822
90	45	38	20	0.738852	0.154441
90	45	38	21	0.857047	0.118195
90	45	38	22	0.932632	0.075585
90	45	38	23	0.972944	0.040312
90	45	38	24	0.990824	0.017880
90	45	38	25	0.997396	0.006571
90	45	38	26	0.999387	0.001991
90	45	38	27	0.999881	0.000495
90	45	38	28	0.999981	0.000100
90	45	38	29	0.999997	0.000016
90	45	38	30	1.000000	0.000002
90	45	38	31	1.000000	0.000000
90	45	38	32	1.000000	0.000000
90	45	38	33	1.000000	0.000000
90	45	38	34	1.000000	0.000000
90	45	38	35	1.000000	0.000000
90	45	38	36	1.000000	0.000000
90	45	38	37	1.000000	0.000000
90	45	38	38	1.000000	0.000000
90	45	39	0	0.000000	0.000000
90	45	39	1	0.000000	0.000000
90	45	39	2	0.000000	0.000000
90	45	39	3	0.000000	0.000000
90	45	39	4	0.000000	0.000000
90	45	39	5	0.000000	0.000000
90	45	39	6	0.000000	0.000000
90	45	39	7	0.000000	0.000000
90	45	39	8	0.000001	0.000001
90	45	39	9	0.000007	0.000007
90	45	39	10	0.000051	0.000044
90	45	39	11	0.000290	0.000238
90	45	39	12	0.001341	0.001051
90	45	39	13	0.005132	0.003791
90	45	39	14	0.016396	0.011265
90	45	39	15	0.044111	0.027715
90	45	39	16	0.100800	0.056689
90	45	39	17	0.197505	0.096705
90	45	39	18	0.335398	0.137894
90	45	39	19	0.500000	0.164602
90	45	39	20	0.664602	0.164602
90	45	39	21	0.802495	0.137894
90	45	39	22	0.899200	0.096705
90	45	39	23	0.955889	0.056689
90	45	39	24	0.983604	0.027715
90	45	39	25	0.994868	0.011265

N	n	k	x	P(x)	p(x)
90	45	39	26	0.998659	0.003791
90	45	39	27	0.999710	0.001051
90	45	39	28	0.999949	0.000238
90	45	39	29	0.999993	0.000044
90	45	39	30	0.999999	0.000007
90	45	39	31	1.000000	0.000001
90	45	39	32	1.000000	0.000000
90	45	39	33	1.000000	0.000000
90	45	39	34	1.000000	0.000000
90	45	39	35	1.000000	0.000000
90	45	39	36	1.000000	0.000000
90	45	39	37	1.000000	0.000000
90	45	39	38	1.000000	0.000000
90	45	39	39	1.000000	0.000000
90	45	40	0	0.000000	0.000000
90	45	40	1	0.000000	0.000000
90	45	40	2	0.000000	0.000000
90	45	40	3	0.000000	0.000000
90	45	40	4	0.000000	0.000000
90	45	40	5	0.000000	0.000000
90	45	40	6	0.000000	0.000000
90	45	40	7	0.000000	0.000000
90	45	40	8	0.000000	0.000000
90	45	40	9	0.000003	0.000002
90	45	40	10	0.000021	0.000018
90	45	40	11	0.000131	0.000110
90	45	40	12	0.000661	0.000530
90	45	40	13	0.002753	0.002092
90	45	40	14	0.009549	0.006796
90	45	40	15	0.027808	0.018259
90	45	40	16	0.068565	0.040757
90	45	40	17	0.144412	0.075847
90	45	40	18	0.262396	0.117984
90	45	40	19	0.416085	0.153690
90	45	40	20	0.583914	0.167829
90	45	40	21	0.737604	0.153690
90	45	40	22	0.855588	0.117984
90	45	40	23	0.931435	0.075847
90	45	40	24	0.972192	0.040757
90	45	40	25	0.990451	0.018259
90	45	40	26	0.997247	0.006796
90	45	40	27	0.999339	0.002092
90	45	40	28	0.999869	0.000530
90	45	40	29	0.999979	0.000110
90	45	40	30	0.999997	0.000018
90	45	40	31	1.000000	0.000002
90	45	40	32	1.000000	0.000000
90	45	40	33	1.000000	0.000000
90	45	40	34	1.000000	0.000000
90	45	40	35	1.000000	0.000000

N	n	k	x	P(x)	p(x)
90	45	40	36	1.000000	0.000000
90	45	40	37	1.000000	0.000000
90	45	40	38	1.000000	0.000000
90	45	40	39	1.000000	0.000000
90	45	40	40	1.000000	0.000000
90	45	41	0	0.000000	0.000000
90	45	41	1	0.000000	0.000000
90	45	41	2	0.000000	0.000000
90	45	41	3	0.000000	0.000000
90	45	41	4	0.000000	0.000000
90	45	41	5	0.000000	0.000000
90	45	41	6	0.000000	0.000000
90	45	41	7	0.000000	0.000000
90	45	41	8	0.000000	0.000000
90	45	41	9	0.000001	0.000001
90	45	41	10	0.000008	0.000007
90	45	41	11	0.000056	0.000048
90	45	41	12	0.000311	0.000255
90	45	41	13	0.001414	0.001103
90	45	41	14	0.005336	0.003922
90	45	41	15	0.016853	0.011517
90	45	41	16	0.044926	0.028073
90	45	41	17	0.101938	0.057012
90	45	41	18	0.198684	0.096747
90	45	41	19	0.336167	0.137482
90	45	41	20	0.500000	0.163833
90	45	41	21	0.663833	0.163833
90	45	41	22	0.801316	0.137482
90	45	41	23	0.898062	0.096747
90	45	41	24	0.955074	0.057012
90	45	41	25	0.983147	0.028073
90	45	41	26	0.994664	0.011517
90	45	41	27	0.998586	0.003922
90	45	41	28	0.999689	0.001103
90	45	41	29	0.999944	0.000255
90	45	41	30	0.999992	0.000048
90	45	41	31	0.999999	0.000007
90	45	41	32	1.000000	0.000001
90	45	41	33	1.000000	0.000000
90	45	41	34	1.000000	0.000000
90	45	41	35	1.000000	0.000000
90	45	41	36	1.000000	0.000000
90	45	41	37	1.000000	0.000000
90	45	41	38	1.000000	0.000000
90	45	41	39	1.000000	0.000000
90	45	41	40	1.000000	0.000000
90	45	41	41	1.000000	0.000000
90	45	42	0	0.000000	0.000000
90	45	42	1	0.000000	0.000000
90	45	42	2	0.000000	0.000000

Table for $N = 2$, $n = 1$, through $N = 100$, $n = 50$

N	n	k	x	P(x)	p(x)
90	45	42	3	0.000000	0.000000
90	45	42	4	0.000000	0.000000
90	45	42	5	0.000000	0.000000
90	45	42	6	0.000000	0.000000
90	45	42	7	0.000000	0.000000
90	45	42	8	0.000000	0.000000
90	45	42	9	0.000000	0.000000
90	45	42	10	0.000003	0.000003
90	45	42	11	0.000023	0.000020
90	45	42	12	0.000139	0.000116
90	45	42	13	0.000694	0.000554
90	45	42	14	0.002854	0.002161
90	45	42	15	0.009801	0.006947
90	45	42	16	0.028311	0.018510
90	45	42	17	0.069360	0.041048
90	45	42	18	0.145375	0.076015
90	45	42	19	0.263217	0.117842
90	45	42	20	0.416412	0.153195
90	45	42	21	0.583588	0.167177
90	45	42	22	0.736783	0.153195
90	45	42	23	0.854625	0.117842
90	45	42	24	0.930640	0.076015
90	45	42	25	0.971689	0.041048
90	45	42	26	0.990198	0.018510
90	45	42	27	0.997145	0.006947
90	45	42	28	0.999306	0.002161
90	45	42	29	0.999860	0.000554
90	45	42	30	0.999977	0.000116
90	45	42	31	0.999997	0.000020
90	45	42	32	1.000000	0.000003
90	45	42	33	1.000000	0.000000
90	45	42	34	1.000000	0.000000
90	45	42	35	1.000000	0.000000
90	45	42	36	1.000000	0.000000
90	45	42	37	1.000000	0.000000
90	45	42	38	1.000000	0.000000
90	45	42	39	1.000000	0.000000
90	45	42	40	1.000000	0.000000
90	45	42	41	1.000000	0.000000
90	45	42	42	1.000000	0.000000
90	45	43	0	0.000000	0.000000
90	45	43	1	0.000000	0.000000
90	45	43	2	0.000000	0.000000
90	45	43	3	0.000000	0.000000
90	45	43	4	0.000000	0.000000
90	45	43	5	0.000000	0.000000
90	45	43	6	0.000000	0.000000
90	45	43	7	0.000000	0.000000
90	45	43	8	0.000000	0.000000
90	45	43	9	0.000000	0.000000
90	45	43	10	0.000001	0.000001
90	45	43	11	0.000009	0.000008
90	45	43	12	0.000059	0.000050
90	45	43	13	0.000324	0.000265
90	45	43	14	0.001459	0.001135
90	45	43	15	0.005460	0.004001
90	45	43	16	0.017128	0.011669
90	45	43	17	0.045415	0.028286
90	45	43	18	0.102616	0.057202
90	45	43	19	0.199386	0.096770
90	45	43	20	0.336623	0.137237
90	45	43	21	0.500000	0.163377
90	45	43	22	0.663377	0.163377
90	45	43	23	0.800614	0.137237
90	45	43	24	0.897384	0.096770
90	45	43	25	0.954585	0.057202
90	45	43	26	0.982872	0.028286
90	45	43	27	0.994540	0.011669
90	45	43	28	0.998541	0.004001
90	45	43	29	0.999676	0.001135
90	45	43	30	0.999941	0.000265
90	45	43	31	0.999991	0.000050
90	45	43	32	0.999999	0.000008
90	45	43	33	1.000000	0.000001
90	45	43	34	1.000000	0.000000
90	45	43	35	1.000000	0.000000
90	45	43	36	1.000000	0.000000
90	45	43	37	1.000000	0.000000
90	45	43	38	1.000000	0.000000
90	45	43	39	1.000000	0.000000
90	45	43	40	1.000000	0.000000
90	45	43	41	1.000000	0.000000
90	45	43	42	1.000000	0.000000
90	45	43	43	1.000000	0.000000
90	45	44	0	0.000000	0.000000
90	45	44	1	0.000000	0.000000
90	45	44	2	0.000000	0.000000
90	45	44	3	0.000000	0.000000
90	45	44	4	0.000000	0.000000
90	45	44	5	0.000000	0.000000
90	45	44	6	0.000000	0.000000
90	45	44	7	0.000000	0.000000
90	45	44	8	0.000000	0.000000
90	45	44	9	0.000000	0.000000
90	45	44	10	0.000000	0.000000
90	45	44	11	0.000003	0.000003
90	45	44	12	0.000024	0.000021
90	45	44	13	0.000144	0.000120
90	45	44	14	0.000710	0.000567
90	45	44	15	0.002906	0.002196
90	45	44	16	0.009928	0.007022
90	45	44	17	0.028563	0.018635
90	45	44	18	0.069756	0.041193
90	45	44	19	0.145854	0.076098
90	45	44	20	0.263624	0.117771
90	45	44	21	0.416573	0.152949
90	45	44	22	0.583427	0.166853
90	45	44	23	0.736376	0.152949
90	45	44	24	0.854146	0.117771
90	45	44	25	0.930244	0.076098
90	45	44	26	0.971437	0.041193
90	45	44	27	0.990071	0.018635
90	45	44	28	0.997094	0.007022
90	45	44	29	0.999289	0.002196
90	45	44	30	0.999856	0.000567
90	45	44	31	0.999976	0.000120
90	45	44	32	0.999997	0.000021
90	45	44	33	1.000000	0.000003
90	45	44	34	1.000000	0.000000
90	45	44	35	1.000000	0.000000
90	45	44	36	1.000000	0.000000
90	45	44	37	1.000000	0.000000
90	45	44	38	1.000000	0.000000
90	45	44	39	1.000000	0.000000
90	45	44	40	1.000000	0.000000
90	45	44	41	1.000000	0.000000
90	45	44	42	1.000000	0.000000
90	45	44	43	1.000000	0.000000
90	45	44	44	1.000000	0.000000
90	45	45	0	0.000000	0.000000
90	45	45	1	0.000000	0.000000
90	45	45	2	0.000000	0.000000
90	45	45	3	0.000000	0.000000
90	45	45	4	0.000000	0.000000
90	45	45	5	0.000000	0.000000
90	45	45	6	0.000000	0.000000
90	45	45	7	0.000000	0.000000
90	45	45	8	0.000000	0.000000
90	45	45	9	0.000000	0.000000
90	45	45	10	0.000000	0.000000
90	45	45	11	0.000001	0.000001
90	45	45	12	0.000009	0.000008
90	45	45	13	0.000060	0.000051
90	45	45	14	0.000329	0.000268
90	45	45	15	0.001474	0.001145
90	45	45	16	0.005501	0.004027
90	45	45	17	0.017220	0.011719
90	45	45	18	0.045577	0.028357
90	45	45	19	0.102842	0.057264
90	45	45	20	0.199619	0.096777
90	45	45	21	0.336774	0.137155
90	45	45	22	0.500000	0.163226
90	45	45	23	0.663226	0.163226
90	45	45	24	0.800381	0.137155
90	45	45	25	0.897158	0.096777
90	45	45	26	0.954422	0.057264
90	45	45	27	0.982780	0.028357
90	45	45	28	0.994499	0.011719
90	45	45	29	0.998526	0.004027
90	45	45	30	0.999671	0.001145
90	45	45	31	0.999939	0.000268
90	45	45	32	0.999991	0.000051
90	45	45	33	0.999999	0.000008
90	45	45	34	1.000000	0.000001
90	45	45	35	1.000000	0.000000
90	45	45	36	1.000000	0.000000
90	45	45	37	1.000000	0.000000
90	45	45	38	1.000000	0.000000
90	45	45	39	1.000000	0.000000
90	45	45	40	1.000000	0.000000
90	45	45	41	1.000000	0.000000
90	45	45	42	1.000000	0.000000
90	45	45	43	1.000000	0.000000
90	45	45	44	1.000000	0.000000
90	45	45	45	1.000000	0.000000
100	1	1	0	0.990000	0.990000
100	1	1	1	1.000000	0.010000
100	2	1	0	0.980000	0.980000
100	2	1	1	1.000000	0.020000
100	2	2	0	0.960202	0.960202
100	2	2	1	0.999798	0.039596
100	2	2	2	1.000000	0.000202
100	3	1	0	0.970000	0.970000
100	3	1	1	1.000000	0.030000
100	3	2	0	0.940606	0.940606
100	3	2	1	0.999394	0.058788
100	3	2	2	1.000000	0.000606
100	3	3	0	0.911812	0.911812
100	3	3	1	0.998194	0.086382
100	3	3	2	0.999994	0.001800
100	3	3	3	1.000000	0.000006
100	4	1	0	0.960000	0.960000
100	4	1	1	1.000000	0.040000
100	4	2	0	0.921212	0.921212
100	4	2	1	0.998788	0.077576
100	4	2	2	1.000000	0.001212
100	4	3	0	0.883612	0.883612
100	4	3	1	0.996413	0.112801
100	4	3	2	0.999975	0.003562
100	4	3	3	1.000000	0.000025

Table for $N = 2$, $n = 1$, through $N = 100$, $n = 50$

N	n	k	x	P(x)	p(x)
100	4	4	0	0.847174	0.847174
100	4	4	1	0.992924	0.145750
100	4	4	2	0.999902	0.006977
100	4	4	3	1.000000	0.000098
100	4	4	4	1.000000	0.000000
100	5	1	0	0.950000	0.950000
100	5	1	1	1.000000	0.050000
100	5	2	0	0.902020	0.902020
100	5	2	1	0.997980	0.095960
100	5	2	2	1.000000	0.002020
100	5	3	0	0.855999	0.855999
100	5	3	1	0.994063	0.138064
100	5	3	2	0.999938	0.005875
100	5	3	3	1.000000	0.000062
100	5	4	0	0.811875	0.811875
100	5	4	1	0.988370	0.176495
100	5	4	2	0.999756	0.011387
100	5	4	3	0.999999	0.000242
100	5	4	4	1.000000	0.000001
100	5	5	0	0.769590	0.769590
100	5	5	1	0.981016	0.211426
100	5	5	2	0.999401	0.018385
100	5	5	3	0.999994	0.000593
100	5	5	4	1.000000	0.000006
100	5	5	5	1.000000	0.000000
100	6	1	0	0.940000	0.940000
100	6	1	1	1.000000	0.060000
100	6	2	0	0.883030	0.883030
100	6	2	1	0.996970	0.113939
100	6	2	2	1.000000	0.003030
100	6	3	0	0.828967	0.828967
100	6	3	1	0.991156	0.162189
100	6	3	2	0.999876	0.008720
100	6	3	3	1.000000	0.000124
100	6	4	0	0.777691	0.777691
100	6	4	1	0.982796	0.205105
100	6	4	2	0.999517	0.016721
100	6	4	3	0.999996	0.000479
100	6	4	4	1.000000	0.000004
100	6	5	0	0.729085	0.729085
100	6	5	1	0.972114	0.243028
100	6	5	2	0.998820	0.026706
100	6	5	3	0.999981	0.001161
100	6	5	4	1.000000	0.000019
100	6	5	5	1.000000	0.000000
100	6	6	0	0.683038	0.683038
100	6	6	1	0.959323	0.276285
100	6	6	2	0.997695	0.038373
100	6	6	3	0.999944	0.002249
100	6	6	4	0.999999	0.000055
100	6	6	5	1.000000	0.000000
100	6	6	6	1.000000	0.000000
100	7	1	0	0.930000	0.930000
100	7	1	1	1.000000	0.070000
100	7	2	0	0.864242	0.864242
100	7	2	1	0.995758	0.131515
100	7	2	2	1.000000	0.004242
100	7	3	0	0.802511	0.802511
100	7	3	1	0.987706	0.185195
100	7	3	2	0.999784	0.012078
100	7	3	3	1.000000	0.000216
100	7	4	0	0.744598	0.744598
100	7	4	1	0.976250	0.231653
100	7	4	2	0.999161	0.022911
100	7	4	3	0.999991	0.000830
100	7	4	4	1.000000	0.000009
100	7	5	0	0.690304	0.690304
100	7	5	1	0.961772	0.271468
100	7	5	2	0.997968	0.036196
100	7	5	3	0.999956	0.001989
100	7	5	4	1.000000	0.000043
100	7	5	5	1.000000	0.000000
100	7	6	0	0.639440	0.639440
100	7	6	1	0.944627	0.305187
100	7	6	2	0.996063	0.051436
100	7	6	3	0.999873	0.003810
100	7	6	4	0.999998	0.000126
100	7	6	5	1.000000	0.000002
100	7	6	6	1.000000	0.000000
100	7	7	0	0.591822	0.591822
100	7	7	1	0.925147	0.333325
100	7	7	2	0.993327	0.068180
100	7	7	3	0.999711	0.006384
100	7	7	4	0.999994	0.000284
100	7	7	5	1.000000	0.000006
100	7	7	6	1.000000	0.000000
100	7	7	7	1.000000	0.000000
100	8	1	0	0.920000	0.920000
100	8	1	1	1.000000	0.080000
100	8	2	0	0.845657	0.845657
100	8	2	1	0.994343	0.148687
100	8	2	2	1.000000	0.005657
100	8	3	0	0.776623	0.776623
100	8	3	1	0.983723	0.207100
100	8	3	2	0.999654	0.015931
100	8	3	3	1.000000	0.000346
100	8	4	0	0.712572	0.712572
100	8	4	1	0.968778	0.256206
100	8	4	2	0.998668	0.029891
100	8	4	3	0.999982	0.001314
100	8	4	4	1.000000	0.000018
100	8	5	0	0.653191	0.653191
100	8	5	1	0.950096	0.296905
100	8	5	2	0.996800	0.046704
100	8	5	3	0.999914	0.003114
100	8	5	4	0.999999	0.000086
100	8	5	5	1.000000	0.000001
100	8	6	0	0.598185	0.598185
100	8	6	1	0.928219	0.330033
100	8	6	2	0.993850	0.065632
100	8	6	3	0.999750	0.005899
100	8	6	4	0.999996	0.000246
100	8	6	5	1.000000	0.000004
100	8	6	6	1.000000	0.000000
100	8	7	0	0.547276	0.547276
100	8	7	1	0.903642	0.356366
100	8	7	2	0.989661	0.086019
100	8	7	3	0.999436	0.009775
100	8	7	4	0.999985	0.000549
100	8	7	5	1.000000	0.000015
100	8	7	6	1.000000	0.000000
100	8	7	7	1.000000	0.000000
100	8	8	0	0.500198	0.500198
100	8	8	1	0.876819	0.376620
100	8	8	2	0.984111	0.107293
100	8	8	3	0.998910	0.014799
100	8	8	4	0.999961	0.001051
100	8	8	5	0.999999	0.000038
100	8	8	6	1.000000	0.000001
100	8	8	7	1.000000	0.000000
100	8	8	8	1.000000	0.000000
100	9	1	0	0.910000	0.910000
100	9	1	1	1.000000	0.090000
100	9	2	0	0.827273	0.827273
100	9	2	1	0.992727	0.165455
100	9	2	2	1.000000	0.007273
100	9	3	0	0.751299	0.751299
100	9	3	1	0.979221	0.227922
100	9	3	2	0.999480	0.020260
100	9	3	3	1.000000	0.000519
100	9	4	0	0.681591	0.681591
100	9	4	1	0.960423	0.278832
100	9	4	2	0.998018	0.037595
100	9	4	3	0.999968	0.001949
100	9	4	4	1.000000	0.000032
100	9	5	0	0.617691	0.617691
100	9	5	1	0.937187	0.319496
100	9	5	2	0.995277	0.058090
100	9	5	3	0.999846	0.004569
100	9	5	4	0.999998	0.000152
100	9	5	5	1.000000	0.000002
100	9	6	0	0.559173	0.559173
100	9	6	1	0.910282	0.351109
100	9	6	2	0.990997	0.080715
100	9	6	3	0.999557	0.008561
100	9	6	4	0.999990	0.000433
100	9	6	5	1.000000	0.000010
100	9	6	6	1.000000	0.000000
100	9	7	0	0.505635	0.505635
100	9	7	1	0.880401	0.374765
100	9	7	2	0.984986	0.104586
100	9	7	3	0.999011	0.014025
100	9	7	4	0.999967	0.000956
100	9	7	5	0.999999	0.000032
100	9	7	6	1.000000	0.000000
100	9	7	7	1.000000	0.000000
100	9	8	0	0.456703	0.456703
100	9	8	1	0.848163	0.391460
100	9	8	2	0.977114	0.128951
100	9	8	3	0.998106	0.020992
100	9	8	4	0.999916	0.001810
100	9	8	5	0.999998	0.000082
100	9	8	6	1.000000	0.000002
100	9	8	7	1.000000	0.000000
100	9	8	8	1.000000	0.000000
100	9	9	0	0.412025	0.412025
100	9	9	1	0.814123	0.402097
100	9	9	2	0.967303	0.153180
100	9	9	3	0.996737	0.029435
100	9	9	4	0.999817	0.003080
100	9	9	5	0.999994	0.000177
100	9	9	6	1.000000	0.000005
100	9	9	7	1.000000	0.000000
100	9	9	8	1.000000	0.000000
100	9	9	9	1.000000	0.000000
100	10	1	0	0.900000	0.900000
100	10	1	1	1.000000	0.100000
100	10	2	0	0.809091	0.809091
100	10	2	1	0.990909	0.181818
100	10	2	2	1.000000	0.009091
100	10	3	0	0.726531	0.726531
100	10	3	1	0.974211	0.247681
100	10	3	2	0.999258	0.025046
100	10	3	3	1.000000	0.000742
100	10	4	0	0.651631	0.651631
100	10	4	1	0.951231	0.299600
100	10	4	2	0.997192	0.045961
100	10	4	3	0.999946	0.002754
100	10	4	4	1.000000	0.000054
100	10	5	0	0.583752	0.583752

Table for $N = 2$, $n = 1$, through $N = 100$, $n = 50$

N	n	k	x	P(x)	p(x)
100	10	5	1	0.923143	0.339391
100	10	5	2	0.993362	0.070219
100	10	5	3	0.999746	0.006384
100	10	5	4	0.999997	0.000251
100	10	5	5	1.000000	0.000003
100	10	6	0	0.522305	0.522305
100	10	6	1	0.890990	0.368686
100	10	6	2	0.987449	0.096458
100	10	6	3	0.999275	0.011826
100	10	6	4	0.999981	0.000706
100	10	6	5	1.000000	0.000019
100	10	6	6	1.000000	0.000000
100	10	7	0	0.466740	0.466740
100	10	7	1	0.855691	0.388950
100	10	7	2	0.979240	0.123549
100	10	7	3	0.998394	0.019155
100	10	7	4	0.999936	0.001541
100	10	7	5	0.999999	0.000063
100	10	7	6	1.000000	0.000001
100	10	7	7	1.000000	0.000000
100	10	8	0	0.416553	0.416553
100	10	8	1	0.818050	0.401497
100	10	8	2	0.968612	0.150561
100	10	8	3	0.996953	0.028341
100	10	8	4	0.999836	0.002884
100	10	8	5	0.999995	0.000159
100	10	8	6	1.000000	0.000005
100	10	8	7	1.000000	0.000000
100	10	8	8	1.000000	0.000000
100	10	9	0	0.371276	0.371276
100	10	9	1	0.778773	0.407498
100	10	9	2	0.955519	0.176746
100	10	9	3	0.994796	0.039277
100	10	9	4	0.999648	0.004852
100	10	9	5	0.999987	0.000338
100	10	9	6	1.000000	0.000013
100	10	9	7	1.000000	0.000000
100	10	9	8	1.000000	0.000000
100	10	9	9	1.000000	0.000000
100	10	10	0	0.330476	0.330476
100	10	10	1	0.738472	0.407995
100	10	10	2	0.939981	0.201510
100	10	10	3	0.991775	0.051794
100	10	10	4	0.999328	0.007553
100	10	10	5	0.999968	0.000640
100	10	10	6	0.999999	0.000031
100	10	10	7	1.000000	0.000001
100	10	10	8	1.000000	0.000000
100	10	10	9	1.000000	0.000000
100	10	10	10	1.000000	0.000000

N	n	k	x	P(x)	p(x)
100	11	1	0	0.890000	0.890000
100	11	1	1	1.000000	0.110000
100	11	2	0	0.791111	0.791111
100	11	2	1	0.988889	0.197778
100	11	2	2	1.000000	0.011111
100	11	3	0	0.702313	0.702313
100	11	3	1	0.968707	0.266395
100	11	3	2	0.998980	0.030272
100	11	3	3	1.000000	0.001020
100	11	4	0	0.622669	0.622669
100	11	4	1	0.941244	0.318575
100	11	4	2	0.996171	0.054927
100	11	4	3	0.999916	0.003745
100	11	4	4	1.000000	0.000084
100	11	5	0	0.551322	0.551322
100	11	5	1	0.908059	0.356738
100	11	5	2	0.991021	0.082962
100	11	5	3	0.999604	0.008582
100	11	5	4	0.999994	0.000390
100	11	5	5	1.000000	0.000006
100	11	6	0	0.487484	0.487484
100	11	6	1	0.870508	0.383023
100	11	6	2	0.983162	0.112654
100	11	6	3	0.998881	0.015719
100	11	6	4	0.999965	0.001084
100	11	6	5	1.000000	0.000034
100	11	6	6	1.000000	0.000000
100	11	7	0	0.430438	0.430438
100	11	7	1	0.829761	0.399322
100	11	7	2	0.972376	0.142615
100	11	7	3	0.997543	0.025167
100	11	7	4	0.999884	0.002341
100	11	7	5	0.999997	0.000113
100	11	7	6	1.000000	0.000003
100	11	7	7	1.000000	0.000000
100	11	8	0	0.379526	0.379526
100	11	8	1	0.786823	0.407297
100	11	8	2	0.958574	0.171752
100	11	8	3	0.995378	0.036804
100	11	8	4	0.999708	0.004330
100	11	8	5	0.999990	0.000282
100	11	8	6	1.000000	0.000010
100	11	8	7	1.000000	0.000000
100	11	8	8	1.000000	0.000000
100	11	9	0	0.334148	0.334148
100	11	9	1	0.742551	0.408403
100	11	9	2	0.941773	0.199221
100	11	9	3	0.992178	0.050405
100	11	9	4	0.999379	0.007201
100	11	9	5	0.999972	0.000593

N	n	k	x	P(x)	p(x)
100	11	9	6	0.999999	0.000028
100	11	9	7	1.000000	0.000001
100	11	9	8	1.000000	0.000000
100	11	9	9	1.000000	0.000000
100	11	10	0	0.293757	0.293757
100	11	10	1	0.697672	0.403915
100	11	10	2	0.922069	0.224397
100	11	10	3	0.987747	0.065677
100	11	10	4	0.998825	0.011078
100	11	10	5	0.999933	0.001108
100	11	10	6	0.999998	0.000065
100	11	10	7	1.000000	0.000002
100	11	10	8	1.000000	0.000000
100	11	10	9	1.000000	0.000000
100	11	10	10	1.000000	0.000000
100	11	11	0	0.257853	0.257853
100	11	11	1	0.652792	0.394939
100	11	11	2	0.899630	0.246837
100	11	11	3	0.981909	0.082279
100	11	11	4	0.997963	0.016054
100	11	11	5	0.999859	0.001896
100	11	11	6	0.999994	0.000135
100	11	11	7	1.000000	0.000006
100	11	11	8	1.000000	0.000000
100	11	11	9	1.000000	0.000000
100	11	11	10	1.000000	0.000000
100	11	11	11	1.000000	0.000000
100	12	1	0	0.880000	0.880000
100	12	1	1	1.000000	0.120000
100	12	2	0	0.773333	0.773333
100	12	2	1	0.986667	0.213333
100	12	2	2	1.000000	0.013333
100	12	3	0	0.678639	0.678639
100	12	3	1	0.962721	0.284082
100	12	3	2	0.998639	0.035918
100	12	3	3	1.000000	0.001361
100	12	4	0	0.594684	0.594684
100	12	4	1	0.930506	0.335822
100	12	4	2	0.994936	0.064431
100	12	4	3	0.999874	0.004937
100	12	4	4	1.000000	0.000126
100	12	5	0	0.520349	0.520349
100	12	5	1	0.892026	0.371678
100	12	5	2	0.988225	0.096199
100	12	5	3	0.999411	0.011186
100	12	5	4	0.999989	0.000579
100	12	5	5	1.000000	0.000011
100	12	6	0	0.454620	0.454620
100	12	6	1	0.848990	0.394369
100	12	6	2	0.978099	0.129109

N	n	k	x	P(x)	p(x)
100	12	6	3	0.998351	0.020252
100	12	6	4	0.999941	0.001590
100	12	6	5	0.999999	0.000058
100	12	6	6	1.000000	0.000001
100	12	7	0	0.396584	0.396584
100	12	7	1	0.802840	0.406256
100	12	7	2	0.964364	0.161524
100	12	7	3	0.996412	0.032048
100	12	7	4	0.999805	0.003393
100	12	7	5	0.999995	0.000189
100	12	7	6	1.000000	0.000005
100	12	7	7	1.000000	0.000000
100	12	8	0	0.345412	0.345412
100	12	8	1	0.754788	0.409377
100	12	8	2	0.946996	0.192207
100	12	8	3	0.993311	0.046315
100	12	8	4	0.999514	0.006203
100	12	8	5	0.999981	0.000467
100	12	8	6	1.000000	0.000019
100	12	8	7	1.000000	0.000000
100	12	8	8	1.000000	0.000000
100	12	9	0	0.300358	0.300358
100	12	9	1	0.705841	0.405483
100	12	9	2	0.926103	0.220262
100	12	9	3	0.988780	0.062676
100	12	9	4	0.998974	0.010194
100	12	9	5	0.999945	0.000971
100	12	9	6	0.999998	0.000053
100	12	9	7	1.000000	0.000002
100	12	9	8	1.000000	0.000000
100	12	9	9	1.000000	0.000000
100	12	10	0	0.260750	0.260750
100	12	10	1	0.656827	0.396076
100	12	10	2	0.901899	0.245072
100	12	10	3	0.982581	0.080682
100	12	10	4	0.998078	0.015497
100	12	10	5	0.999870	0.001792
100	12	10	6	0.999995	0.000124
100	12	10	7	1.000000	0.000005
100	12	10	8	1.000000	0.000000
100	12	10	9	1.000000	0.000000
100	12	10	10	1.000000	0.000000
100	12	11	0	0.225984	0.225984
100	12	11	1	0.608417	0.382434
100	12	11	2	0.874669	0.266251
100	12	11	3	0.974513	0.099844
100	12	11	4	0.996700	0.022188
100	12	11	5	0.999731	0.003031
100	12	11	6	0.999987	0.000256
100	12	11	7	1.000000	0.000013

Table for $N = 2$, $n = 1$, through $N = 100$, $n = 50$

N	n	k	x	P(x)	p(x)
100	12	11	8	1.000000	0.000000
100	12	11	9	1.000000	0.000000
100	12	11	10	1.000000	0.000000
100	12	11	11	1.000000	0.000000
100	12	12	0	0.195514	0.195514
100	12	12	1	0.561150	0.365636
100	12	12	2	0.844753	0.283603
100	12	12	3	0.964416	0.119663
100	12	12	4	0.994706	0.030290
100	12	12	5	0.999493	0.004787
100	12	12	6	0.999969	0.000477
100	12	12	7	0.999999	0.000030
100	12	12	8	1.000000	0.000001
100	12	12	9	1.000000	0.000000
100	12	12	10	1.000000	0.000000
100	12	12	11	1.000000	0.000000
100	12	12	12	1.000000	0.000000
100	13	1	0	0.870000	0.870000
100	13	1	1	1.000000	0.130000
100	13	2	0	0.755758	0.755758
100	13	2	1	0.984242	0.228485
100	13	2	2	1.000000	0.015758
100	13	3	0	0.655504	0.655504
100	13	3	1	0.956265	0.300761
100	13	3	2	0.998231	0.041967
100	13	3	3	1.000000	0.001769
100	13	4	0	0.567653	0.567653
100	13	4	1	0.919057	0.351404
100	13	4	2	0.993472	0.074415
100	13	4	3	0.999818	0.006345
100	13	4	4	1.000000	0.000182
100	13	5	0	0.490783	0.490783
100	13	5	1	0.875132	0.384348
100	13	5	2	0.984945	0.109814
100	13	5	3	0.999157	0.014211
100	13	5	4	0.999983	0.000826
100	13	5	5	1.000000	0.000017
100	13	6	0	0.423623	0.423623
100	13	6	1	0.826582	0.402959
100	13	6	2	0.972230	0.145648
100	13	6	3	0.997661	0.025431
100	13	6	4	0.999905	0.002244
100	13	6	5	0.999998	0.000094
100	13	6	6	1.000000	0.000001
100	13	7	0	0.365037	0.365037
100	13	7	1	0.775141	0.410104
100	13	7	2	0.955186	0.180045
100	13	7	3	0.994955	0.039769
100	13	7	4	0.999690	0.004734
100	13	7	5	0.999990	0.000301

N	n	k	x	P(x)	p(x)
100	13	7	6	1.000000	0.000009
100	13	7	7	1.000000	0.000000
100	13	8	0	0.314011	0.314011
100	13	8	1	0.722224	0.408214
100	13	8	2	0.933891	0.211666
100	13	8	3	0.990679	0.056789
100	13	8	4	0.999232	0.008552
100	13	8	5	0.999965	0.000733
100	13	8	6	0.999999	0.000034
100	13	8	7	1.000000	0.000001
100	13	8	8	1.000000	0.000000
100	13	9	0	0.269639	0.269639
100	13	9	1	0.668979	0.399339
100	13	9	2	0.908583	0.239604
100	13	9	3	0.984506	0.075924
100	13	9	4	0.998395	0.013888
100	13	9	5	0.999901	0.001506
100	13	9	6	0.999996	0.000096
100	13	9	7	1.000000	0.000003
100	13	9	8	1.000000	0.000000
100	13	9	9	1.000000	0.000000
100	13	10	0	0.231120	0.231120
100	13	10	1	0.616319	0.385199
100	13	10	2	0.879620	0.263301
100	13	10	3	0.976163	0.096544
100	13	10	4	0.997021	0.020858
100	13	10	5	0.999768	0.002747
100	13	10	6	0.999989	0.000221
100	13	10	7	1.000000	0.000011
100	13	10	8	1.000000	0.000000
100	13	10	9	1.000000	0.000000
100	13	10	10	1.000000	0.000000
100	13	11	0	0.197736	0.197736
100	13	11	1	0.564959	0.367223
100	13	11	2	0.847438	0.282479
100	13	11	3	0.965436	0.117998
100	13	11	4	0.994935	0.029499
100	13	11	5	0.999524	0.004589
100	13	11	6	0.999972	0.000448
100	13	11	7	0.999999	0.000027
100	13	11	8	1.000000	0.000001
100	13	11	9	1.000000	0.000000
100	13	11	10	1.000000	0.000000
100	13	11	11	1.000000	0.000000
100	13	12	0	0.168853	0.168853
100	13	12	1	0.515446	0.346593
100	13	12	2	0.812525	0.297080
100	13	12	3	0.952178	0.139653
100	13	12	4	0.991952	0.039775
100	13	12	5	0.999112	0.007159

N	n	k	x	P(x)	p(x)
100	13	12	6	0.999937	0.000825
100	13	12	7	0.999997	0.000060
100	13	12	8	1.000000	0.000003
100	13	12	9	1.000000	0.000000
100	13	12	10	1.000000	0.000000
100	13	12	11	1.000000	0.000000
100	13	12	12	1.000000	0.000000
100	13	13	0	0.143909	0.143909
100	13	13	1	0.468183	0.324274
100	13	13	2	0.775390	0.307207
100	13	13	3	0.936308	0.160918
100	13	13	4	0.987885	0.051576
100	13	13	5	0.998461	0.010576
100	13	13	6	0.999871	0.001410
100	13	13	7	0.999993	0.000122
100	13	13	8	1.000000	0.000007
100	13	13	9	1.000000	0.000000
100	13	13	10	1.000000	0.000000
100	13	13	11	1.000000	0.000000
100	13	13	12	1.000000	0.000000
100	13	13	13	1.000000	0.000000
100	14	1	0	0.860000	0.860000
100	14	1	1	1.000000	0.140000
100	14	2	0	0.738384	0.738384
100	14	2	1	0.981616	0.243232
100	14	2	2	1.000000	0.018384
100	14	3	0	0.632900	0.632900
100	14	3	1	0.949351	0.316450
100	14	3	2	0.997749	0.048398
100	14	3	3	1.000000	0.002251
100	14	4	0	0.541554	0.541554
100	14	4	1	0.906940	0.365386
100	14	4	2	0.991761	0.084822
100	14	4	3	0.999745	0.007983
100	14	4	4	1.000000	0.000255
100	14	5	0	0.462577	0.462577
100	14	5	1	0.857460	0.394883
100	14	5	2	0.981159	0.123698
100	14	5	3	0.998830	0.017671
100	14	5	4	0.999973	0.001143
100	14	5	5	1.000000	0.000027
100	14	6	0	0.394408	0.394408
100	14	6	1	0.803424	0.409016
100	14	6	2	0.965534	0.162110
100	14	6	3	0.996784	0.031250
100	14	6	4	0.999853	0.003069
100	14	6	5	0.999997	0.000144
100	14	6	6	1.000000	0.000003
100	14	7	0	0.335666	0.335666
100	14	7	1	0.746858	0.411191

N	n	k	x	P(x)	p(x)
100	14	7	2	0.944839	0.197981
100	14	7	3	0.993127	0.048288
100	14	7	4	0.999526	0.006400
100	14	7	5	0.999984	0.000457
100	14	7	6	1.000000	0.000016
100	14	7	7	1.000000	0.000000
100	14	8	0	0.285136	0.285136
100	14	8	1	0.689379	0.404243
100	14	8	2	0.919293	0.229913
100	14	8	3	0.987415	0.068122
100	14	8	4	0.998838	0.011423
100	14	8	5	0.999939	0.001101
100	14	8	6	0.999998	0.000059
100	14	8	7	1.000000	0.000002
100	14	8	8	1.000000	0.000000
100	14	9	0	0.241746	0.241746
100	14	9	1	0.632258	0.390512
100	14	9	2	0.889304	0.257046
100	14	9	3	0.979270	0.089966
100	14	9	4	0.997597	0.018326
100	14	9	5	0.999832	0.002235
100	14	9	6	0.999993	0.000162
100	14	9	7	1.000000	0.000007
100	14	9	8	1.000000	0.000000
100	14	9	9	1.000000	0.000000
100	14	10	0	0.204554	0.204554
100	14	10	1	0.576471	0.371917
100	14	10	2	0.855408	0.278937
100	14	10	3	0.968395	0.112987
100	14	10	4	0.995583	0.027188
100	14	10	5	0.999611	0.004028
100	14	10	6	0.999979	0.000368
100	14	10	7	0.999999	0.000020
100	14	10	8	1.000000	0.000001
100	14	10	9	1.000000	0.000000
100	14	10	10	1.000000	0.000000
100	14	11	0	0.172735	0.172735
100	14	11	1	0.522749	0.350015
100	14	11	2	0.818216	0.295467
100	14	11	3	0.954586	0.136369
100	14	11	4	0.992562	0.037976
100	14	11	5	0.999208	0.006646
100	14	11	6	0.999946	0.000738
100	14	11	7	0.999998	0.000051
100	14	11	8	1.000000	0.000002
100	14	11	9	1.000000	0.000000
100	14	11	10	1.000000	0.000000
100	14	11	11	1.000000	0.000000
100	14	12	0	0.145563	0.145563
100	14	12	1	0.471624	0.326061

N	n	k	x	P(x)	p(x)
100	14	12	2	0.778378	0.306755
100	14	12	3	0.937731	0.159353
100	14	12	4	0.988295	0.050564
100	14	12	5	0.998536	0.010241
100	14	12	6	0.999880	0.001344
100	14	12	7	0.999994	0.000114
100	14	12	8	1.000000	0.000006
100	14	12	9	1.000000	0.000000
100	14	12	10	1.000000	0.000000
100	14	12	11	1.000000	0.000000
100	14	12	12	1.000000	0.000000
100	14	13	0	0.122405	0.122405
100	14	13	1	0.423455	0.301050
100	14	13	2	0.736548	0.313092
100	14	13	3	0.917812	0.181264
100	14	13	4	0.982549	0.064737
100	14	13	5	0.997488	0.014939
100	14	13	6	0.999758	0.002269
100	14	13	7	0.999985	0.000227
100	14	13	8	0.999999	0.000015
100	14	13	9	1.000000	0.000001
100	14	13	10	1.000000	0.000000
100	14	13	11	1.000000	0.000000
100	14	13	12	1.000000	0.000000
100	14	13	13	1.000000	0.000000
100	14	14	0	0.102708	0.102708
100	14	14	1	0.378471	0.275763
100	14	14	2	0.693363	0.314892
100	14	14	3	0.894893	0.201531
100	14	14	4	0.975108	0.080215
100	14	14	5	0.995943	0.020835
100	14	14	6	0.999549	0.003606
100	14	14	7	0.999966	0.000417
100	14	14	8	0.999998	0.000032
100	14	14	9	1.000000	0.000002
100	14	14	10	1.000000	0.000000
100	14	14	11	1.000000	0.000000
100	14	14	12	1.000000	0.000000
100	14	14	13	1.000000	0.000000
100	14	14	14	1.000000	0.000000
100	15	1	0	0.850000	0.850000
100	15	1	1	1.000000	0.150000
100	15	2	0	0.721212	0.721212
100	15	2	1	0.978788	0.257576
100	15	2	2	1.000000	0.021212
100	15	3	0	0.610823	0.610823
100	15	3	1	0.941991	0.331169
100	15	3	2	0.997186	0.055195
100	15	3	3	1.000000	0.002814
100	15	4	0	0.516365	0.516365

N	n	k	x	P(x)	p(x)
100	15	4	1	0.894194	0.377828
100	15	4	2	0.989789	0.095595
100	15	4	3	0.999652	0.009863
100	15	4	4	1.000000	0.000348
100	15	5	0	0.435683	0.435683
100	15	5	1	0.839094	0.403410
100	15	5	2	0.976844	0.137750
100	15	5	3	0.998419	0.021575
100	15	5	4	0.999960	0.001541
100	15	5	5	1.000000	0.000040
100	15	6	0	0.366891	0.366891
100	15	6	1	0.779644	0.412753
100	15	6	2	0.957994	0.178350
100	15	6	3	0.995694	0.037700
100	15	6	4	0.999782	0.004088
100	15	6	5	0.999996	0.000214
100	15	6	6	1.000000	0.000004
100	15	7	0	0.308345	0.308345
100	15	7	1	0.718170	0.409825
100	15	7	2	0.933328	0.215158
100	15	7	3	0.990881	0.057553
100	15	7	4	0.999303	0.008422
100	15	7	5	0.999973	0.000670
100	15	7	6	0.999999	0.000027
100	15	7	7	1.000000	0.000000
100	15	8	0	0.258612	0.258612
100	15	8	1	0.656476	0.397864
100	15	8	2	0.903252	0.246777
100	15	8	3	0.983455	0.080202
100	15	8	4	0.998307	0.014852
100	15	8	5	0.999901	0.001594
100	15	8	6	0.999997	0.000096
100	15	8	7	1.000000	0.000003
100	15	8	8	1.000000	0.000000
100	15	9	0	0.216447	0.216447
100	15	9	1	0.595931	0.379485
100	15	9	2	0.868382	0.272450
100	15	9	3	0.972994	0.104612
100	15	9	4	0.996531	0.023538
100	15	9	5	0.999728	0.003196
100	15	9	6	0.999987	0.000260
100	15	9	7	1.000000	0.000012
100	15	9	8	1.000000	0.000000
100	15	9	9	1.000000	0.000000
100	15	10	0	0.180769	0.180769
100	15	10	1	0.537549	0.356780
100	15	10	2	0.829460	0.291911
100	15	10	3	0.959199	0.129738
100	15	10	4	0.993686	0.034487
100	15	10	5	0.999376	0.005690

N	n	k	x	P(x)	p(x)
100	15	10	6	0.999962	0.000585
100	15	10	7	0.999999	0.000037
100	15	10	8	1.000000	0.000001
100	15	10	9	1.000000	0.000000
100	15	10	10	1.000000	0.000000
100	15	11	0	0.150641	0.150641
100	15	11	1	0.482050	0.331409
100	15	11	2	0.787295	0.305245
100	15	11	3	0.941900	0.154605
100	15	11	4	0.989471	0.047571
100	15	11	5	0.998744	0.009273
100	15	11	6	0.999903	0.001159
100	15	11	7	0.999995	0.000092
100	15	11	8	1.000000	0.000004
100	15	11	9	1.000000	0.000000
100	15	11	10	1.000000	0.000000
100	15	11	11	1.000000	0.000000
100	15	12	0	0.125252	0.125252
100	15	12	1	0.429918	0.304666
100	15	12	2	0.742709	0.312791
100	15	12	3	0.921055	0.178346
100	15	12	4	0.983591	0.062537
100	15	12	5	0.997702	0.014111
100	15	12	6	0.999786	0.002084
100	15	12	7	0.999987	0.000201
100	15	12	8	0.999999	0.000012
100	15	12	9	1.000000	0.000000
100	15	12	10	1.000000	0.000000
100	15	12	11	1.000000	0.000000
100	15	12	12	1.000000	0.000000
100	15	13	0	0.103902	0.103902
100	15	13	1	0.381448	0.277546
100	15	13	2	0.696501	0.315053
100	15	13	3	0.896735	0.200233
100	15	13	4	0.975774	0.079040
100	15	13	5	0.996099	0.020324
100	15	13	6	0.999573	0.003474
100	15	13	7	0.999969	0.000396
100	15	13	8	0.999998	0.000030
100	15	13	9	1.000000	0.000001
100	15	13	10	1.000000	0.000000
100	15	13	11	1.000000	0.000000
100	15	13	12	1.000000	0.000000
100	15	13	13	1.000000	0.000000
100	15	14	0	0.085988	0.085988
100	15	14	1	0.336786	0.250798
100	15	14	2	0.649424	0.312639
100	15	14	3	0.869116	0.219692
100	15	14	4	0.965781	0.096664
100	15	14	5	0.993762	0.027982

N	n	k	x	P(x)	p(x)
100	15	14	6	0.999213	0.005451
100	15	14	7	0.999932	0.000719
100	15	14	8	0.999996	0.000064
100	15	14	9	1.000000	0.000004
100	15	14	10	1.000000	0.000000
100	15	14	11	1.000000	0.000000
100	15	14	12	1.000000	0.000000
100	15	14	13	1.000000	0.000000
100	15	14	14	1.000000	0.000000
100	15	15	0	0.070990	0.070990
100	15	15	1	0.295958	0.224968
100	15	15	2	0.602165	0.306207
100	15	15	3	0.838462	0.236297
100	15	15	4	0.953417	0.114955
100	15	15	5	0.990509	0.037092
100	15	15	6	0.998643	0.008134
100	15	15	7	0.999865	0.001222
100	15	15	8	0.999991	0.000125
100	15	15	9	1.000000	0.000009
100	15	15	10	1.000000	0.000000
100	15	15	11	1.000000	0.000000
100	15	15	12	1.000000	0.000000
100	15	15	13	1.000000	0.000000
100	15	15	14	1.000000	0.000000
100	15	15	15	1.000000	0.000000
100	16	1	0	0.840000	0.840000
100	16	1	1	1.000000	0.160000
100	16	2	0	0.704242	0.704242
100	16	2	1	0.975758	0.271515
100	16	2	2	1.000000	0.024242
100	16	3	0	0.589264	0.589264
100	16	3	1	0.934199	0.344935
100	16	3	2	0.996537	0.062338
100	16	3	3	1.000000	0.003463
100	16	4	0	0.492066	0.492066
100	16	4	1	0.880859	0.388793
100	16	4	2	0.987540	0.106681
100	16	4	3	0.999536	0.011996
100	16	4	4	1.000000	0.000464
100	16	5	0	0.410055	0.410055
100	16	5	1	0.820110	0.410055
100	16	5	2	0.971982	0.151872
100	16	5	3	0.997911	0.025929
100	16	5	4	0.999942	0.002031
100	16	5	5	1.000000	0.000058
100	16	6	0	0.340993	0.340993
100	16	6	1	0.755364	0.414371
100	16	6	2	0.949601	0.194237
100	16	6	3	0.994363	0.044762
100	16	6	4	0.999685	0.005322

Table for $N = 2$, $n = 1$, through $N = 100$, $n = 50$

N	n	k	x	P(x)	p(x)
100	16	6	5	0.999993	0.000308
100	16	6	6	1.000000	0.000007
100	16	7	0	0.282952	0.282952
100	16	7	1	0.689241	0.406290
100	16	7	2	0.920672	0.231431
100	16	7	3	0.988173	0.067501
100	16	7	4	0.999006	0.010833
100	16	7	5	0.999957	0.000951
100	16	7	6	0.999999	0.000042
100	16	7	7	1.000000	0.000001
100	16	8	0	0.234272	0.234272
100	16	8	1	0.623711	0.389439
100	16	8	2	0.885833	0.262122
100	16	8	3	0.978737	0.092904
100	16	8	4	0.997608	0.018871
100	16	8	5	0.999845	0.002237
100	16	8	6	0.999995	0.000150
100	16	8	7	1.000000	0.000005
100	16	8	8	1.000000	0.000000
100	16	9	0	0.193529	0.193529
100	16	9	1	0.560215	0.366686
100	16	9	2	0.845945	0.285730
100	16	9	3	0.965609	0.119664
100	16	9	4	0.995147	0.029537
100	16	9	5	0.999577	0.004431
100	16	9	6	0.999978	0.000401
100	16	9	7	0.999999	0.000021
100	16	9	8	1.000000	0.000001
100	16	9	9	1.000000	0.000000
100	16	10	0	0.159502	0.159502
100	16	10	1	0.499772	0.340271
100	16	10	2	0.801986	0.302214
100	16	10	3	0.948514	0.146528
100	16	10	4	0.991252	0.042737
100	16	10	5	0.999042	0.007790
100	16	10	6	0.999934	0.000893
100	16	10	7	0.999997	0.000063
100	16	10	8	1.000000	0.000003
100	16	10	9	1.000000	0.000000
100	16	10	10	1.000000	0.000000
100	16	11	0	0.131146	0.131146
100	16	11	1	0.443061	0.311915
100	16	11	2	0.754975	0.311915
100	16	11	3	0.927349	0.172374
100	16	11	4	0.985553	0.058204
100	16	11	5	0.998090	0.012536
100	16	11	6	0.999835	0.001746
100	16	11	7	0.999991	0.000156
100	16	11	8	1.000000	0.000009
100	16	11	9	1.000000	0.000000

N	n	k	x	P(x)	p(x)
100	16	11	10	1.000000	0.000000
100	16	11	11	1.000000	0.000000
100	16	12	0	0.107569	0.107569
100	16	12	1	0.390491	0.282922
100	16	12	2	0.705910	0.315419
100	16	12	3	0.902171	0.196261
100	16	12	4	0.977706	0.075535
100	16	12	5	0.996540	0.018835
100	16	12	6	0.999639	0.003099
100	16	12	7	0.999975	0.000336
100	16	12	8	0.999999	0.000024
100	16	12	9	1.000000	0.000001
100	16	12	10	1.000000	0.000000
100	16	12	11	1.000000	0.000000
100	16	12	12	1.000000	0.000000
100	16	13	0	0.088011	0.088011
100	16	13	1	0.342265	0.254254
100	16	13	2	0.655730	0.313464
100	16	13	3	0.873178	0.217448
100	16	13	4	0.967405	0.094228
100	16	13	5	0.994186	0.026780
100	16	13	6	0.999287	0.005101
100	16	13	7	0.999941	0.000654
100	16	13	8	0.999997	0.000056
100	16	13	9	1.000000	0.000003
100	16	13	10	1.000000	0.000000
100	16	13	11	1.000000	0.000000
100	16	13	12	1.000000	0.000000
100	16	13	13	1.000000	0.000000
100	16	14	0	0.071825	0.071825
100	16	14	1	0.298428	0.226603
100	16	14	2	0.605287	0.306859
100	16	14	3	0.840686	0.235398
100	16	14	4	0.954408	0.113723
100	16	14	5	0.990800	0.036391
100	16	14	6	0.998700	0.007901
100	16	14	7	0.999873	0.001173
100	16	14	8	0.999992	0.000118
100	16	14	9	1.000000	0.000008
100	16	14	10	1.000000	0.000000
100	16	14	11	1.000000	0.000000
100	16	14	12	1.000000	0.000000
100	16	14	13	1.000000	0.000000
100	16	14	14	1.000000	0.000000
100	16	15	0	0.058462	0.058462
100	16	15	1	0.258905	0.200442
100	16	15	2	0.555333	0.296429
100	16	15	3	0.805102	0.249769
100	16	15	4	0.938540	0.133438
100	16	15	5	0.986145	0.047605

N	n	k	x	P(x)	p(x)
100	16	15	6	0.997782	0.011637
100	16	15	7	0.999750	0.001969
100	16	15	8	0.999981	0.000230
100	16	15	9	0.999999	0.000018
100	16	15	10	1.000000	0.000001
100	16	15	11	1.000000	0.000000
100	16	15	12	1.000000	0.000000
100	16	15	13	1.000000	0.000000
100	16	15	14	1.000000	0.000000
100	16	15	15	1.000000	0.000000
100	16	16	0	0.047458	0.047458
100	16	16	1	0.223532	0.176075
100	16	16	2	0.506510	0.282977
100	16	16	3	0.766902	0.260392
100	16	16	4	0.919702	0.152800
100	16	16	5	0.979984	0.060283
100	16	16	6	0.996413	0.016428
100	16	16	7	0.999542	0.003129
100	16	16	8	0.999959	0.000417
100	16	16	9	0.999997	0.000038
100	16	16	10	1.000000	0.000002
100	16	16	11	1.000000	0.000000
100	16	16	12	1.000000	0.000000
100	16	16	13	1.000000	0.000000
100	16	16	14	1.000000	0.000000
100	16	16	15	1.000000	0.000000
100	16	16	16	1.000000	0.000000
100	17	1	0	0.830000	0.830000
100	17	1	1	1.000000	0.170000
100	17	2	0	0.687475	0.687475
100	17	2	1	0.972525	0.285050
100	17	2	2	1.000000	0.027475
100	17	3	0	0.568219	0.568219
100	17	3	1	0.925986	0.357767
100	17	3	2	0.995795	0.069808
100	17	3	3	1.000000	0.004205
100	17	4	0	0.468634	0.468634
100	17	4	1	0.866973	0.398339
100	17	4	2	0.985000	0.118026
100	17	4	3	0.999393	0.014393
100	17	4	4	1.000000	0.000607
100	17	5	0	0.385647	0.385647
100	17	5	1	0.800583	0.414936
100	17	5	2	0.966558	0.165975
100	17	5	3	0.997294	0.030736
100	17	5	4	0.999918	0.002624
100	17	5	5	1.000000	0.000082
100	17	6	0	0.316636	0.316636
100	17	6	1	0.730699	0.414063
100	17	6	2	0.940351	0.209652

N	n	k	x	P(x)	p(x)
100	17	6	3	0.992764	0.052413
100	17	6	4	0.999559	0.006794
100	17	6	5	0.999990	0.000431
100	17	6	6	1.000000	0.000010
100	17	7	0	0.259372	0.259372
100	17	7	1	0.660221	0.400848
100	17	7	2	0.906896	0.246676
100	17	7	3	0.984958	0.078062
100	17	7	4	0.998619	0.013661
100	17	7	5	0.999935	0.001315
100	17	7	6	0.999999	0.000064
100	17	7	7	1.000000	0.000001
100	17	8	0	0.211960	0.211960
100	17	8	1	0.591257	0.379297
100	17	8	2	0.867110	0.275852
100	17	8	3	0.973207	0.106097
100	17	8	4	0.996709	0.023503
100	17	8	5	0.999765	0.003055
100	17	8	6	0.999991	0.000226
100	17	8	7	1.000000	0.000009
100	17	8	8	1.000000	0.000000
100	17	9	0	0.172794	0.172794
100	17	9	1	0.525293	0.352499
100	17	9	2	0.822134	0.296841
100	17	9	3	0.957062	0.134928
100	17	9	4	0.993388	0.036327
100	17	9	5	0.999366	0.005978
100	17	9	6	0.999964	0.000598
100	17	9	7	0.999999	0.000035
100	17	9	8	1.000000	0.000001
100	17	9	9	1.000000	0.000000
100	17	10	0	0.140513	0.140513
100	17	10	1	0.463315	0.322801
100	17	10	2	0.773204	0.309889
100	17	10	3	0.936304	0.163100
100	17	10	4	0.988199	0.051895
100	17	10	5	0.998578	0.010379
100	17	10	6	0.999892	0.001314
100	17	10	7	0.999995	0.000103
100	17	10	8	1.000000	0.000005
100	17	10	9	1.000000	0.000000
100	17	10	10	1.000000	0.000000
100	17	11	0	0.113972	0.113972
100	17	11	1	0.405928	0.291956
100	17	11	2	0.721556	0.315628
100	17	11	3	0.910933	0.189377
100	17	11	4	0.980703	0.069770
100	17	11	5	0.997194	0.016491
100	17	11	6	0.999731	0.002537
100	17	11	7	0.999983	0.000252

N	n	k	x	P(x)	p(x)
100	17	11	8	0.999999	0.000016
100	17	11	9	1.000000	0.000001
100	17	11	10	1.000000	0.000000
100	17	11	11	1.000000	0.000000
100	17	12	0	0.092202	0.092202
100	17	12	1	0.353441	0.261239
100	17	12	2	0.668360	0.314919
100	17	12	3	0.881143	0.212783
100	17	12	4	0.970512	0.089369
100	17	12	5	0.994971	0.024459
100	17	12	6	0.999418	0.004447
100	17	12	7	0.999955	0.000538
100	17	12	8	0.999998	0.000043
100	17	12	9	1.000000	0.000002
100	17	12	10	1.000000	0.000000
100	17	12	11	1.000000	0.000000
100	17	12	12	1.000000	0.000000
100	17	13	0	0.074390	0.074390
100	17	13	1	0.305943	0.231553
100	17	13	2	0.614681	0.308737
100	17	13	3	0.847291	0.232610
100	17	13	4	0.957310	0.110018
100	17	13	5	0.991635	0.034326
100	17	13	6	0.998862	0.007226
100	17	13	7	0.999894	0.001032
100	17	13	8	0.999993	0.000099
100	17	13	9	1.000000	0.000006
100	17	13	10	1.000000	0.000000
100	17	13	11	1.000000	0.000000
100	17	13	12	1.000000	0.000000
100	17	13	13	1.000000	0.000000
100	17	14	0	0.059854	0.059854
100	17	14	1	0.263359	0.203505
100	17	14	2	0.561450	0.298091
100	17	14	3	0.809860	0.248409
100	17	14	4	0.940870	0.131010
100	17	14	5	0.986901	0.046031
100	17	14	6	0.997948	0.011047
100	17	14	7	0.999775	0.001827
100	17	14	8	0.999983	0.000208
100	17	14	9	0.999999	0.000016
100	17	14	10	1.000000	0.000001
100	17	14	11	1.000000	0.000000
100	17	14	12	1.000000	0.000000
100	17	14	13	1.000000	0.000000
100	17	14	14	1.000000	0.000000
100	17	15	0	0.048023	0.048023
100	17	15	1	0.225498	0.177475
100	17	15	2	0.509458	0.283960
100	17	15	3	0.769421	0.259963
100	17	15	4	0.921066	0.151645
100	17	15	5	0.980478	0.059412
100	17	15	6	0.996535	0.016057
100	17	15	7	0.999563	0.003028
100	17	15	8	0.999961	0.000398
100	17	15	9	0.999998	0.000036
100	17	15	10	1.000000	0.000002
100	17	15	11	1.000000	0.000000
100	17	15	12	1.000000	0.000000
100	17	15	13	1.000000	0.000000
100	17	15	14	1.000000	0.000000
100	17	15	15	1.000000	0.000000
100	17	16	0	0.038418	0.038418
100	17	16	1	0.192091	0.153672
100	17	16	2	0.459347	0.267256
100	17	16	3	0.726603	0.267256
100	17	16	4	0.897873	0.171270
100	17	16	5	0.972090	0.074217
100	17	16	6	0.994457	0.022367
100	17	16	7	0.999207	0.004750
100	17	16	8	0.999919	0.000712
100	17	16	9	0.999994	0.000075
100	17	16	10	1.000000	0.000005
100	17	16	11	1.000000	0.000000
100	17	16	12	1.000000	0.000000
100	17	16	13	1.000000	0.000000
100	17	16	14	1.000000	0.000000
100	17	16	15	1.000000	0.000000
100	17	16	16	1.000000	0.000000
100	17	17	0	0.030643	0.030643
100	17	17	1	0.162820	0.132177
100	17	17	2	0.411623	0.248803
100	17	17	3	0.682061	0.270438
100	17	17	4	0.871367	0.189307
100	17	17	5	0.961488	0.090121
100	17	17	6	0.991528	0.030040
100	17	17	7	0.998641	0.007113
100	17	17	8	0.999843	0.001202
100	17	17	9	0.999987	0.000144
100	17	17	10	0.999999	0.000012
100	17	17	11	1.000000	0.000001
100	17	17	12	1.000000	0.000000
100	17	17	13	1.000000	0.000000
100	17	17	14	1.000000	0.000000
100	17	17	15	1.000000	0.000000
100	17	17	16	1.000000	0.000000
100	17	17	17	1.000000	0.000000
100	18	1	0	0.820000	0.820000
100	18	1	1	1.000000	0.180000
100	18	2	0	0.670909	0.670909
100	18	2	1	0.969091	0.298182
100	18	2	2	1.000000	0.030909
100	18	3	0	0.547681	0.547681
100	18	3	1	0.917365	0.369685
100	18	3	2	0.994954	0.077588
100	18	3	3	1.000000	0.005046
100	18	4	0	0.446049	0.446049
100	18	4	1	0.852575	0.406526
100	18	4	2	0.982156	0.129580
100	18	4	3	0.999220	0.017064
100	18	4	4	1.000000	0.000780
100	18	5	0	0.362415	0.362415
100	18	5	1	0.780586	0.418171
100	18	5	2	0.960559	0.179972
100	18	5	3	0.996553	0.035994
100	18	5	4	0.999886	0.003333
100	18	5	5	1.000000	0.000114
100	18	6	0	0.293747	0.293747
100	18	6	1	0.705756	0.412009
100	18	6	2	0.930248	0.224492
100	18	6	3	0.990870	0.060622
100	18	6	4	0.999395	0.008525
100	18	6	5	0.999984	0.000589
100	18	6	6	1.000000	0.000016
100	18	7	0	0.237498	0.237498
100	18	7	1	0.631243	0.393746
100	18	7	2	0.892036	0.260793
100	18	7	3	0.981196	0.089160
100	18	7	4	0.998125	0.016929
100	18	7	5	0.999903	0.001778
100	18	7	6	0.999998	0.000095
100	18	7	7	1.000000	0.000002
100	18	8	0	0.191530	0.191530
100	18	8	1	0.559268	0.367738
100	18	8	2	0.847169	0.287900
100	18	8	3	0.966816	0.119647
100	18	8	4	0.995577	0.028761
100	18	8	5	0.999654	0.004078
100	18	8	6	0.999986	0.000331
100	18	8	7	1.000000	0.000014
100	18	8	8	1.000000	0.000000
100	18	9	0	0.154057	0.154057
100	18	9	1	0.491317	0.337260
100	18	9	2	0.797099	0.305782
100	18	9	3	0.947308	0.150209
100	18	9	4	0.991200	0.043892
100	18	9	5	0.999078	0.007878
100	18	9	6	0.999942	0.000864
100	18	9	7	0.999998	0.000056
100	18	9	8	1.000000	0.000002
100	18	9	9	1.000000	0.000000
100	18	10	0	0.123584	0.123584
100	18	10	1	0.428312	0.304728
100	18	10	2	0.743335	0.315023
100	18	10	3	0.922548	0.179213
100	18	10	4	0.984447	0.061899
100	18	10	5	0.997953	0.013505
100	18	10	6	0.999828	0.001876
100	18	10	7	0.999991	0.000163
100	18	10	8	1.000000	0.000008
100	18	10	9	1.000000	0.000000
100	18	10	10	1.000000	0.000000
100	18	11	0	0.098867	0.098867
100	18	11	1	0.370752	0.271885
100	18	11	2	0.687331	0.316579
100	18	11	3	0.892679	0.205348
100	18	11	4	0.974819	0.082139
100	18	11	5	0.996002	0.021183
100	18	11	6	0.999578	0.003576
100	18	11	7	0.999971	0.000393
100	18	11	8	0.999999	0.000027
100	18	11	9	1.000000	0.000001
100	18	11	10	1.000000	0.000000
100	18	11	11	1.000000	0.000000
100	18	12	0	0.078872	0.078872
100	18	12	1	0.318819	0.239948
100	18	12	2	0.630418	0.311599
100	18	12	3	0.858070	0.227652
100	18	12	4	0.961898	0.103828
100	18	12	5	0.992908	0.031010
100	18	12	6	0.999096	0.006188
100	18	12	7	0.999923	0.000827
100	18	12	8	0.999996	0.000073
100	18	12	9	1.000000	0.000004
100	18	12	10	1.000000	0.000000
100	18	12	11	1.000000	0.000000
100	18	12	12	1.000000	0.000000
100	18	13	0	0.062739	0.062739
100	18	13	1	0.272466	0.209727
100	18	13	2	0.573764	0.301298
100	18	13	3	0.819266	0.245502
100	18	13	4	0.945380	0.126114
100	18	13	5	0.988327	0.042947
100	18	13	6	0.998252	0.009926
100	18	13	7	0.999820	0.001567
100	18	13	8	0.999987	0.000168
100	18	13	9	0.999999	0.000012
100	18	13	10	1.000000	0.000001
100	18	13	11	1.000000	0.000000
100	18	13	12	1.000000	0.000000

Table for $N = 2$, $n = 1$, through $N = 100$, $n = 50$

N	n	k	x	P(x)	p(x)
100	18	13	13	1.000000	0.000000
100	18	14	0	0.049758	0.049758
100	18	14	1	0.231485	0.181726
100	18	14	2	0.518353	0.286868
100	18	14	3	0.776938	0.258585
100	18	14	4	0.925086	0.148148
100	18	14	5	0.981909	0.056824
100	18	14	6	0.996883	0.014974
100	18	14	7	0.999621	0.002738
100	18	14	8	0.999968	0.000347
100	18	14	9	0.999998	0.000030
100	18	14	10	1.000000	0.000002
100	18	14	11	1.000000	0.000000
100	18	14	12	1.000000	0.000000
100	18	14	13	1.000000	0.000000
100	18	14	14	1.000000	0.000000
100	18	15	0	0.039344	0.039344
100	18	15	1	0.195562	0.156218
100	18	15	2	0.464982	0.269420
100	18	15	3	0.731836	0.266854
100	18	15	4	0.900969	0.169133
100	18	15	5	0.973320	0.072351
100	18	15	6	0.994794	0.021474
100	18	15	7	0.999271	0.004477
100	18	15	8	0.999928	0.000657
100	18	15	9	0.999995	0.000067
100	18	15	10	1.000000	0.000005
100	18	15	11	1.000000	0.000000
100	18	15	12	1.000000	0.000000
100	18	15	13	1.000000	0.000000
100	18	15	14	1.000000	0.000000
100	18	15	15	1.000000	0.000000
100	18	16	0	0.031012	0.031012
100	18	16	1	0.164318	0.133306
100	18	16	2	0.414268	0.249949
100	18	16	3	0.684744	0.270476
100	18	16	4	0.873111	0.188367
100	18	16	5	0.962254	0.089143
100	18	16	6	0.991762	0.029508
100	18	16	7	0.998692	0.006929
100	18	16	8	0.999851	0.001159
100	18	16	9	0.999988	0.000137
100	18	16	10	0.999999	0.000011
100	18	16	11	1.000000	0.000001
100	18	16	12	1.000000	0.000000
100	18	16	13	1.000000	0.000000
100	18	16	14	1.000000	0.000000
100	18	16	15	1.000000	0.000000
100	18	16	16	1.000000	0.000000
100	18	17	0	0.024367	0.024367
100	18	17	1	0.137340	0.112973
100	18	17	2	0.366658	0.229318
100	18	17	3	0.636445	0.269786
100	18	17	4	0.841717	0.205272
100	18	17	5	0.948458	0.106742
100	18	17	6	0.987547	0.039088
100	18	17	7	0.997784	0.010237
100	18	17	8	0.999713	0.001928
100	18	17	9	0.999973	0.000261
100	18	17	10	0.999998	0.000025
100	18	17	11	1.000000	0.000002
100	18	17	12	1.000000	0.000000
100	18	17	13	1.000000	0.000000
100	18	17	14	1.000000	0.000000
100	18	17	15	1.000000	0.000000
100	18	17	16	1.000000	0.000000
100	18	17	17	1.000000	0.000000
100	18	18	0	0.019082	0.019082
100	18	18	1	0.114201	0.095118
100	18	18	2	0.322452	0.208252
100	18	18	3	0.587688	0.265236
100	18	18	4	0.807093	0.219405
100	18	18	5	0.931740	0.124647
100	18	18	6	0.981895	0.050156
100	18	18	7	0.996427	0.014532
100	18	18	8	0.999480	0.003053
100	18	18	9	0.999945	0.000465
100	18	18	10	0.999996	0.000051
100	18	18	11	1.000000	0.000004
100	18	18	12	1.000000	0.000000
100	18	18	13	1.000000	0.000000
100	18	18	14	1.000000	0.000000
100	18	18	15	1.000000	0.000000
100	18	18	16	1.000000	0.000000
100	18	18	17	1.000000	0.000000
100	18	18	18	1.000000	0.000000
100	19	1	0	0.810000	0.810000
100	19	1	1	1.000000	0.190000
100	19	2	0	0.654545	0.654545
100	19	2	1	0.965455	0.310909
100	19	2	2	1.000000	0.034545
100	19	3	0	0.527644	0.527644
100	19	3	1	0.908349	0.380705
100	19	3	2	0.994007	0.085659
100	19	3	3	1.000000	0.005993
100	19	4	0	0.424291	0.424291
100	19	4	1	0.837702	0.413412
100	19	4	2	0.978995	0.141293
100	19	4	3	0.999011	0.020016
100	19	4	4	1.000000	0.000988
100	19	5	0	0.340317	0.340317
100	19	5	1	0.760188	0.419871
100	19	5	2	0.953974	0.193787
100	19	5	3	0.995675	0.041701
100	19	5	4	0.999845	0.004170
100	19	5	5	1.000000	0.000154
100	19	6	0	0.272253	0.272253
100	19	6	1	0.680633	0.408380
100	19	6	2	0.919297	0.238664
100	19	6	3	0.988652	0.069355
100	19	6	4	0.999187	0.010535
100	19	6	5	0.999977	0.000790
100	19	6	6	1.000000	0.000023
100	19	7	0	0.217223	0.217223
100	19	7	1	0.602433	0.385209
100	19	7	2	0.876134	0.273701
100	19	7	3	0.976847	0.100713
100	19	7	4	0.997506	0.020659
100	19	7	5	0.999859	0.002354
100	19	7	6	0.999997	0.000137
100	19	7	7	1.000000	0.000003
100	19	8	0	0.172844	0.172844
100	19	8	1	0.527876	0.355032
100	19	8	2	0.826103	0.298227
100	19	8	3	0.959520	0.133417
100	19	8	4	0.994174	0.034654
100	19	8	5	0.999505	0.005331
100	19	8	6	0.999977	0.000472
100	19	8	7	0.999999	0.000022
100	19	8	8	1.000000	0.000000
100	19	9	0	0.137148	0.137148
100	19	9	1	0.458413	0.321265
100	19	9	2	0.770996	0.312582
100	19	9	3	0.936317	0.165321
100	19	9	4	0.988524	0.052207
100	19	9	5	0.998694	0.010170
100	19	9	6	0.999911	0.001217
100	19	9	7	0.999997	0.000086
100	19	9	8	1.000000	0.000003
100	19	9	9	1.000000	0.000000
100	19	10	0	0.108513	0.108513
100	19	10	1	0.394866	0.286354
100	19	10	2	0.712601	0.317735
100	19	10	3	0.907250	0.194648
100	19	10	4	0.979918	0.072669
100	19	10	5	0.997129	0.017211
100	19	10	6	0.999737	0.002608
100	19	10	7	0.999985	0.000248
100	19	10	8	0.999999	0.000014
100	19	10	9	1.000000	0.000000
100	19	10	10	1.000000	0.000000
100	19	11	0	0.085605	0.085605
100	19	11	1	0.337596	0.251991
100	19	11	2	0.652585	0.314989
100	19	11	3	0.872645	0.220061
100	19	11	4	0.967807	0.095161
100	19	11	5	0.994452	0.026645
100	19	11	6	0.999360	0.004908
100	19	11	7	0.999952	0.000592
100	19	11	8	0.999998	0.000046
100	19	11	9	1.000000	0.000002
100	19	11	10	1.000000	0.000000
100	19	11	11	1.000000	0.000000
100	19	12	0	0.067329	0.067329
100	19	12	1	0.286631	0.219302
100	19	12	2	0.592418	0.305787
100	19	12	3	0.833084	0.240666
100	19	12	4	0.951768	0.118684
100	19	12	5	0.990261	0.038492
100	19	12	6	0.998643	0.008383
100	19	12	7	0.999872	0.001229
100	19	12	8	0.999992	0.000120
100	19	12	9	1.000000	0.000008
100	19	12	10	1.000000	0.000000
100	19	12	11	1.000000	0.000000
100	19	12	12	1.000000	0.000000
100	19	13	0	0.052792	0.052792
100	19	13	1	0.241774	0.188982
100	19	13	2	0.533346	0.291572
100	19	13	3	0.789327	0.255981
100	19	13	4	0.931538	0.142212
100	19	13	5	0.984137	0.052599
100	19	13	6	0.997405	0.013268
100	19	13	7	0.999705	0.002300
100	19	13	8	0.999977	0.000272
100	19	13	9	0.999999	0.000022
100	19	13	10	1.000000	0.000001
100	19	13	11	1.000000	0.000000
100	19	13	12	1.000000	0.000000
100	19	13	13	1.000000	0.000000
100	19	14	0	0.041263	0.041263
100	19	14	1	0.202674	0.161411
100	19	14	2	0.476372	0.273698
100	19	14	3	0.742250	0.265878
100	19	14	4	0.907019	0.164769
100	19	14	5	0.975673	0.068654
100	19	14	6	0.995422	0.019750
100	19	14	7	0.999388	0.003965
100	19	14	8	0.999943	0.000555
100	19	14	9	0.999996	0.000054

Table for $N = 2$, $n = 1$, through $N = 100$, $n = 50$

N	n	k	x	P(x)	p(x)
100	19	14	10	1.000000	0.000003
100	19	14	11	1.000000	0.000000
100	19	14	12	1.000000	0.000000
100	19	14	13	1.000000	0.000000
100	19	14	14	1.000000	0.000000
100	19	15	0	0.032147	0.032147
100	19	15	1	0.168891	0.136744
100	19	15	2	0.422269	0.253378
100	19	15	3	0.692784	0.270515
100	19	15	4	0.878280	0.185496
100	19	15	5	0.964497	0.086216
100	19	15	6	0.992437	0.027941
100	19	15	7	0.998834	0.006397
100	19	15	8	0.999872	0.001037
100	19	15	9	0.999990	0.000118
100	19	15	10	0.999999	0.000009
100	19	15	11	1.000000	0.000000
100	19	15	12	1.000000	0.000000
100	19	15	13	1.000000	0.000000
100	19	15	14	1.000000	0.000000
100	19	15	15	1.000000	0.000000
100	19	16	0	0.024961	0.024961
100	19	16	1	0.139933	0.114972
100	19	16	2	0.371593	0.231660
100	19	16	3	0.641863	0.270270
100	19	16	4	0.845545	0.203682
100	19	16	5	0.950296	0.104751
100	19	16	6	0.988164	0.037868
100	19	16	7	0.997931	0.009767
100	19	16	8	0.999737	0.001806
100	19	16	9	0.999976	0.000239
100	19	16	10	0.999998	0.000022
100	19	16	11	1.000000	0.000001
100	19	16	12	1.000000	0.000000
100	19	16	13	1.000000	0.000000
100	19	16	14	1.000000	0.000000
100	19	16	15	1.000000	0.000000
100	19	16	16	1.000000	0.000000
100	19	17	0	0.019315	0.019315
100	19	17	1	0.115296	0.095981
100	19	17	2	0.324710	0.209413
100	19	17	3	0.590383	0.265674
100	19	17	4	0.809174	0.218790
100	19	17	5	0.932838	0.123664
100	19	17	6	0.982303	0.049466
100	19	17	7	0.996536	0.014233
100	19	17	8	0.999501	0.002965
100	19	17	9	0.999948	0.000447
100	19	17	10	0.999996	0.000048
100	19	17	11	1.000000	0.000004

N	n	k	x	P(x)	p(x)
100	19	17	12	1.000000	0.000000
100	19	17	13	1.000000	0.000000
100	19	17	14	1.000000	0.000000
100	19	17	15	1.000000	0.000000
100	19	17	16	1.000000	0.000000
100	19	17	17	1.000000	0.000000
100	19	18	0	0.014894	0.014894
100	19	18	1	0.094481	0.079587
100	19	18	2	0.281818	0.187337
100	19	18	3	0.539169	0.257351
100	19	18	4	0.769633	0.230464
100	19	18	5	0.911979	0.142345
100	19	18	6	0.974556	0.062577
100	19	18	7	0.994478	0.019922
100	19	18	8	0.999108	0.004630
100	19	18	9	0.999894	0.000786
100	19	18	10	0.999991	0.000097
100	19	18	11	0.999999	0.000009
100	19	18	12	1.000000	0.000001
100	19	18	13	1.000000	0.000000
100	19	18	14	1.000000	0.000000
100	19	18	15	1.000000	0.000000
100	19	18	16	1.000000	0.000000
100	19	18	17	1.000000	0.000000
100	19	18	18	1.000000	0.000000
100	19	19	0	0.011443	0.011443
100	19	19	1	0.077011	0.065568
100	19	19	2	0.242980	0.165969
100	19	19	3	0.488954	0.245975
100	19	19	4	0.727475	0.238521
100	19	19	5	0.887676	0.160201
100	19	19	6	0.964635	0.076959
100	19	19	7	0.991563	0.026928
100	19	19	8	0.998487	0.006924
100	19	19	9	0.999798	0.001311
100	19	19	10	0.999980	0.000182
100	19	19	11	0.999998	0.000018
100	19	19	12	1.000000	0.000001
100	19	19	13	1.000000	0.000000
100	19	19	14	1.000000	0.000000
100	19	19	15	1.000000	0.000000
100	19	19	16	1.000000	0.000000
100	19	19	17	1.000000	0.000000
100	19	19	18	1.000000	0.000000
100	19	19	19	1.000000	0.000000
100	20	1	0	0.800000	0.800000
100	20	1	1	1.000000	0.200000
100	20	2	0	0.638384	0.638384
100	20	2	1	0.961616	0.323232
100	20	2	2	1.000000	0.038384

N	n	k	x	P(x)	p(x)
100	20	3	0	0.508101	0.508101
100	20	3	1	0.898949	0.390847
100	20	3	2	0.992950	0.094001
100	20	3	3	1.000000	0.007050
100	20	4	0	0.403338	0.403338
100	20	4	1	0.822391	0.419053
100	20	4	2	0.975506	0.153115
100	20	4	3	0.998764	0.023258
100	20	4	4	1.000000	0.001236
100	20	5	0	0.319309	0.319309
100	20	5	1	0.739453	0.420144
100	20	5	2	0.946797	0.207344
100	20	5	3	0.994646	0.047849
100	20	5	4	0.999794	0.005148
100	20	5	5	1.000000	0.000206
100	20	6	0	0.252086	0.252086
100	20	6	1	0.655425	0.403338
100	20	6	2	0.907511	0.252086
100	20	6	3	0.986083	0.078572
100	20	6	4	0.998927	0.012844
100	20	6	5	0.999967	0.001040
100	20	6	6	1.000000	0.000033
100	20	7	0	0.198451	0.198451
100	20	7	1	0.573899	0.375448
100	20	7	2	0.859239	0.285340
100	20	7	3	0.971873	0.112634
100	20	7	4	0.996741	0.024867
100	20	7	5	0.999801	0.003061
100	20	7	6	0.999995	0.000194
100	20	7	7	1.000000	0.000005
100	20	8	0	0.155773	0.155773
100	20	8	1	0.497194	0.341421
100	20	8	2	0.804012	0.306818
100	20	8	3	0.951284	0.147272
100	20	8	4	0.992462	0.041178
100	20	8	5	0.999308	0.006845
100	20	8	6	0.999966	0.000658
100	20	8	7	0.999999	0.000033
100	20	8	8	1.000000	0.000001
100	20	9	0	0.121910	0.121910
100	20	9	1	0.426684	0.304774
100	20	9	2	0.743982	0.317299
100	20	9	3	0.924071	0.180089
100	20	9	4	0.985301	0.061230
100	20	9	5	0.998192	0.012891
100	20	9	6	0.999866	0.001674
100	20	9	7	0.999995	0.000129
100	20	9	8	1.000000	0.000005
100	20	9	9	1.000000	0.000000
100	20	10	0	0.095116	0.095116

N	n	k	x	P(x)	p(x)
100	20	10	1	0.363049	0.267933
100	20	10	2	0.681220	0.318171
100	20	10	3	0.890428	0.209208
100	20	10	4	0.974535	0.084107
100	20	10	5	0.996067	0.021531
100	20	10	6	0.999608	0.003541
100	20	10	7	0.999976	0.000368
100	20	10	8	0.999999	0.000023
100	20	10	9	1.000000	0.000001
100	20	10	10	1.000000	0.000000
100	20	11	0	0.073979	0.073979
100	20	11	1	0.306486	0.232506
100	20	11	2	0.617586	0.311100
100	20	11	3	0.850911	0.233325
100	20	11	4	0.959583	0.108672
100	20	11	5	0.992478	0.032895
100	20	11	6	0.999057	0.006579
100	20	11	7	0.999923	0.000866
100	20	11	8	0.999996	0.000073
100	20	11	9	1.000000	0.000004
100	20	11	10	1.000000	0.000000
100	20	11	11	1.000000	0.000000
100	20	12	0	0.057355	0.057355
100	20	12	1	0.256850	0.199495
100	20	12	2	0.554667	0.297817
100	20	12	3	0.806343	0.251677
100	20	12	4	0.940046	0.133703
100	20	12	5	0.986934	0.046888
100	20	12	6	0.998022	0.011088
100	20	12	7	0.999797	0.001774
100	20	12	8	0.999986	0.000190
100	20	12	9	0.999999	0.000013
100	20	12	10	1.000000	0.000001
100	20	12	11	1.000000	0.000000
100	20	12	12	1.000000	0.000000
100	20	13	0	0.044320	0.044320
100	20	13	1	0.213777	0.169457
100	20	13	2	0.493750	0.279973
100	20	13	3	0.757724	0.263974
100	20	13	4	0.915737	0.158013
100	20	13	5	0.978942	0.063205
100	20	13	6	0.996258	0.017316
100	20	13	7	0.999534	0.003276
100	20	13	8	0.999960	0.000426
100	20	13	9	0.999998	0.000037
100	20	13	10	1.000000	0.000002
100	20	13	11	1.000000	0.000000
100	20	13	12	1.000000	0.000000
100	20	13	13	1.000000	0.000000
100	20	14	0	0.034131	0.034131

Table for $N = 2$, $n = 1$, through $N = 100$, $n = 50$

N	n	k	x	P(x)	p(x)
100	20	14	1	0.176769	0.142638
100	20	14	2	0.435824	0.259055
100	20	14	3	0.706143	0.270319
100	20	14	4	0.886677	0.180534
100	20	14	5	0.968044	0.081368
100	20	14	6	0.993472	0.025427
100	20	14	7	0.999045	0.005573
100	20	14	8	0.999902	0.000857
100	20	14	9	0.999993	0.000091
100	20	14	10	1.000000	0.000007
100	20	14	11	1.000000	0.000000
100	20	14	12	1.000000	0.000000
100	20	14	13	1.000000	0.000000
100	20	14	14	1.000000	0.000000
100	20	15	0	0.026194	0.026194
100	20	15	1	0.145256	0.119062
100	20	15	2	0.381603	0.236347
100	20	15	3	0.652708	0.271104
100	20	15	4	0.853089	0.200381
100	20	15	5	0.953852	0.100763
100	20	15	6	0.989332	0.035480
100	20	15	7	0.998202	0.008870
100	20	15	8	0.999782	0.001580
100	20	15	9	0.999981	0.000199
100	20	15	10	0.999999	0.000018
100	20	15	11	1.000000	0.000001
100	20	15	12	1.000000	0.000000
100	20	15	13	1.000000	0.000000
100	20	15	14	1.000000	0.000000
100	20	15	15	1.000000	0.000000
100	20	16	0	0.020030	0.020030
100	20	16	1	0.118642	0.098612
100	20	16	2	0.331553	0.212911
100	20	16	3	0.598487	0.266934
100	20	16	4	0.815370	0.216883
100	20	16	5	0.936071	0.120700
100	20	16	6	0.983489	0.047418
100	20	16	7	0.996846	0.013357
100	20	16	8	0.999559	0.002713
100	20	16	9	0.999956	0.000396
100	20	16	10	0.999997	0.000041
100	20	16	11	1.000000	0.000003
100	20	16	12	1.000000	0.000000
100	20	16	13	1.000000	0.000000
100	20	16	14	1.000000	0.000000
100	20	16	15	1.000000	0.000000
100	20	16	16	1.000000	0.000000
100	20	17	0	0.015261	0.015261
100	20	17	1	0.096337	0.081076
100	20	17	2	0.285929	0.189592

N	n	k	x	P(x)	p(x)
100	20	17	3	0.544465	0.258535
100	20	17	4	0.774059	0.229595
100	20	17	5	0.914517	0.140458
100	20	17	6	0.975586	0.061069
100	20	17	7	0.994779	0.019193
100	20	17	8	0.999171	0.004393
100	20	17	9	0.999904	0.000732
100	20	17	10	0.999992	0.000088
100	20	17	11	0.999999	0.000008
100	20	17	12	1.000000	0.000000
100	20	17	13	1.000000	0.000000
100	20	17	14	1.000000	0.000000
100	20	17	15	1.000000	0.000000
100	20	17	16	1.000000	0.000000
100	20	17	17	1.000000	0.000000
100	20	18	0	0.011584	0.011584
100	20	18	1	0.077778	0.066194
100	20	18	2	0.244813	0.167035
100	20	18	3	0.491512	0.246699
100	20	18	4	0.729800	0.238288
100	20	18	5	0.889133	0.159333
100	20	18	6	0.965285	0.076152
100	20	18	7	0.991773	0.026488
100	20	18	8	0.998536	0.006764
100	20	18	9	0.999807	0.001270
100	20	18	10	0.999981	0.000175
100	20	18	11	0.999999	0.000017
100	20	18	12	1.000000	0.000001
100	20	18	13	1.000000	0.000000
100	20	18	14	1.000000	0.000000
100	20	18	15	1.000000	0.000000
100	20	18	16	1.000000	0.000000
100	20	18	17	1.000000	0.000000
100	20	18	18	1.000000	0.000000
100	20	19	0	0.008759	0.008759
100	20	19	1	0.062440	0.053681
100	20	19	2	0.208147	0.145707
100	20	19	3	0.440367	0.232220
100	20	19	4	0.683305	0.242938
100	20	19	5	0.859987	0.176682
100	20	19	6	0.952283	0.092297
100	20	19	7	0.987573	0.035290
100	20	19	8	0.997547	0.009973
100	20	19	9	0.999636	0.002090
100	20	19	10	0.999960	0.000324
100	20	19	11	0.999997	0.000037
100	20	19	12	1.000000	0.000003
100	20	19	13	1.000000	0.000000
100	20	19	14	1.000000	0.000000
100	20	19	15	1.000000	0.000000

N	n	k	x	P(x)	p(x)
100	20	19	16	1.000000	0.000000
100	20	19	17	1.000000	0.000000
100	20	19	18	1.000000	0.000000
100	20	19	19	1.000000	0.000000
100	20	20	0	0.006596	0.006596
100	20	20	1	0.049848	0.043252
100	20	20	2	0.175767	0.125919
100	20	20	3	0.391629	0.215862
100	20	20	4	0.635317	0.243688
100	20	20	5	0.827268	0.191951
100	20	20	6	0.936331	0.109063
100	20	20	7	0.981910	0.045579
100	20	20	8	0.996069	0.014160
100	20	20	9	0.999352	0.003283
100	20	20	10	0.999920	0.000568
100	20	20	11	0.999993	0.000073
100	20	20	12	0.999999	0.000007
100	20	20	13	1.000000	0.000000
100	20	20	14	1.000000	0.000000
100	20	20	15	1.000000	0.000000
100	20	20	16	1.000000	0.000000
100	20	20	17	1.000000	0.000000
100	20	20	18	1.000000	0.000000
100	20	20	19	1.000000	0.000000
100	20	20	20	1.000000	0.000000
100	21	1	0	0.790000	0.790000
100	21	1	1	1.000000	0.210000
100	21	2	0	0.622424	0.622424
100	21	2	1	0.957576	0.335151
100	21	2	2	1.000000	0.042424
100	21	3	0	0.489048	0.489048
100	21	3	1	0.889177	0.400130
100	21	3	2	0.991775	0.102597
100	21	3	3	1.000000	0.008225
100	21	4	0	0.383171	0.383171
100	21	4	1	0.806676	0.423505
100	21	4	2	0.971678	0.165002
100	21	4	3	0.998474	0.026795
100	21	4	4	1.000000	0.001526
100	21	5	0	0.299353	0.299353
100	21	5	1	0.718446	0.419094
100	21	5	2	0.939022	0.220576
100	21	5	3	0.993450	0.054428
100	21	5	4	0.999730	0.006280
100	21	5	5	1.000000	0.000270
100	21	6	0	0.233180	0.233180
100	21	6	1	0.630216	0.397036
100	21	6	2	0.894907	0.264691
100	21	6	3	0.983137	0.088230
100	21	6	4	0.998606	0.015469

N	n	k	x	P(x)	p(x)
100	21	6	5	0.999954	0.001349
100	21	6	6	1.000000	0.000046
100	21	7	0	0.181087	0.181087
100	21	7	1	0.545740	0.364654
100	21	7	2	0.841405	0.295665
100	21	7	3	0.966242	0.124836
100	21	7	4	0.995808	0.029567
100	21	7	5	0.999725	0.003917
100	21	7	6	0.999993	0.000268
100	21	7	7	1.000000	0.000007
100	21	8	0	0.140196	0.140196
100	21	8	1	0.467320	0.327124
100	21	8	2	0.781001	0.313681
100	21	8	3	0.942080	0.161079
100	21	8	4	0.990404	0.048324
100	21	8	5	0.999051	0.008647
100	21	8	6	0.999949	0.000898
100	21	8	7	0.999999	0.000049
100	21	8	8	1.000000	0.000001
100	21	9	0	0.108195	0.108195
100	21	9	1	0.396206	0.288011
100	21	9	2	0.716219	0.320013
100	21	9	3	0.910564	0.194346
100	21	9	4	0.981474	0.070910
100	21	9	5	0.997547	0.016073
100	21	9	6	0.999803	0.002256
100	21	9	7	0.999991	0.000188
100	21	9	8	1.000000	0.000008
100	21	9	9	1.000000	0.000000
100	21	10	0	0.083227	0.083227
100	21	10	1	0.332907	0.249680
100	21	10	2	0.649403	0.316496
100	21	10	3	0.872122	0.222719
100	21	10	4	0.968227	0.096105
100	21	10	5	0.994721	0.026494
100	21	10	6	0.999431	0.004710
100	21	10	7	0.999962	0.000531
100	21	10	8	0.999998	0.000036
100	21	10	9	1.000000	0.000001
100	21	10	10	1.000000	0.000000
100	21	11	0	0.063807	0.063807
100	21	11	1	0.277422	0.213615
100	21	11	2	0.582587	0.305165
100	21	11	3	0.827578	0.244991
100	21	11	4	0.950074	0.122496
100	21	11	5	0.990011	0.039937
100	21	11	6	0.998646	0.008635
100	21	11	7	0.999880	0.001234
100	21	11	8	0.999993	0.000114
100	21	11	9	1.000000	0.000006

Table for $N = 2$, $n = 1$, through $N = 100$, $n = 50$

N	n	k	x	P(x)	p(x)
100	21	11	10	1.000000	0.000000
100	21	11	11	1.000000	0.000000
100	21	12	0	0.048752	0.048752
100	21	12	1	0.229419	0.180667
100	21	12	2	0.517440	0.288021
100	21	12	3	0.778030	0.260590
100	21	12	4	0.926676	0.148646
100	21	12	5	0.982831	0.056155
100	21	12	6	0.997191	0.014359
100	21	12	7	0.999686	0.002495
100	21	12	8	0.999977	0.000291
100	21	12	9	0.999999	0.000022
100	21	12	10	1.000000	0.000001
100	21	12	11	1.000000	0.000000
100	21	12	12	1.000000	0.000000
100	21	13	0	0.037118	0.037118
100	21	13	1	0.188358	0.151241
100	21	13	2	0.455253	0.266895
100	21	13	3	0.724727	0.269474
100	21	13	4	0.897960	0.173233
100	21	13	5	0.972621	0.074661
100	21	13	6	0.994743	0.022122
100	21	13	7	0.999289	0.004546
100	21	13	8	0.999934	0.000645
100	21	13	9	0.999996	0.000062
100	21	13	10	1.000000	0.000004
100	21	13	11	1.000000	0.000000
100	21	13	12	1.000000	0.000000
100	21	13	13	1.000000	0.000000
100	21	14	0	0.028158	0.028158
100	21	14	1	0.153590	0.125432
100	21	14	2	0.396966	0.243376
100	21	14	3	0.668974	0.272008
100	21	14	4	0.864110	0.195136
100	21	14	5	0.958891	0.094780
100	21	14	6	0.990929	0.032038
100	21	14	7	0.998557	0.007628
100	21	14	8	0.999837	0.001280
100	21	14	9	0.999987	0.000150
100	21	14	10	0.999999	0.000012
100	21	14	11	1.000000	0.000001
100	21	14	12	1.000000	0.000000
100	21	14	13	1.000000	0.000000
100	21	14	14	1.000000	0.000000
100	21	15	0	0.021282	0.021282
100	21	15	1	0.124420	0.103138
100	21	15	2	0.343197	0.218777
100	21	15	3	0.612042	0.268845
100	21	15	4	0.825537	0.213495
100	21	15	5	0.941257	0.115720

N	n	k	x	P(x)	p(x)
100	21	15	6	0.985341	0.044084
100	21	15	7	0.997315	0.011974
100	21	15	8	0.999644	0.002328
100	21	15	9	0.999966	0.000322
100	21	15	10	0.999998	0.000031
100	21	15	11	1.000000	0.000002
100	21	15	12	1.000000	0.000000
100	21	15	13	1.000000	0.000000
100	21	15	14	1.000000	0.000000
100	21	15	15	1.000000	0.000000
100	21	16	0	0.016024	0.016024
100	21	16	1	0.100152	0.084128
100	21	16	2	0.294294	0.194141
100	21	16	3	0.555110	0.260816
100	21	16	4	0.782838	0.227728
100	21	16	5	0.919474	0.136637
100	21	16	6	0.977561	0.058087
100	21	16	7	0.995343	0.017782
100	21	16	8	0.999288	0.003945
100	21	16	9	0.999921	0.000633
100	21	16	10	0.999994	0.000073
100	21	16	11	0.999999	0.000006
100	21	16	12	1.000000	0.000000
100	21	16	13	1.000000	0.000000
100	21	16	14	1.000000	0.000000
100	21	16	15	1.000000	0.000000
100	21	16	16	1.000000	0.000000
100	21	17	0	0.012018	0.012018
100	21	17	1	0.080122	0.068104
100	21	17	2	0.250381	0.170259
100	21	17	3	0.499221	0.248840
100	21	17	4	0.736750	0.237529
100	21	17	5	0.893448	0.156698
100	21	17	6	0.967189	0.073740
100	21	17	7	0.992380	0.025191
100	21	17	8	0.998677	0.006298
100	21	17	9	0.999830	0.001153
100	21	17	10	0.999984	0.000154
100	21	17	11	0.999999	0.000015
100	21	17	12	1.000000	0.000001
100	21	17	13	1.000000	0.000000
100	21	17	14	1.000000	0.000000
100	21	17	15	1.000000	0.000000
100	21	17	16	1.000000	0.000000
100	21	17	17	1.000000	0.000000
100	21	18	0	0.008978	0.008978
100	21	18	1	0.063711	0.054734
100	21	18	2	0.211406	0.147695
100	21	18	3	0.445256	0.233850
100	21	18	4	0.688099	0.242844

N	n	k	x	P(x)	p(x)
100	21	18	5	0.863241	0.175142
100	21	18	6	0.953862	0.090621
100	21	18	7	0.988130	0.034268
100	21	18	8	0.997691	0.009560
100	21	18	9	0.999664	0.001973
100	21	18	10	0.999964	0.000300
100	21	18	11	0.999997	0.000033
100	21	18	12	1.000000	0.000003
100	21	18	13	1.000000	0.000000
100	21	18	14	1.000000	0.000000
100	21	18	15	1.000000	0.000000
100	21	18	16	1.000000	0.000000
100	21	18	17	1.000000	0.000000
100	21	18	18	1.000000	0.000000
100	21	19	0	0.006678	0.006678
100	21	19	1	0.050362	0.043683
100	21	19	2	0.177184	0.126822
100	21	19	3	0.393923	0.216739
100	21	19	4	0.637754	0.243831
100	21	19	5	0.829067	0.191314
100	21	19	6	0.937285	0.108218
100	21	19	7	0.982280	0.044995
100	21	19	8	0.996175	0.013895
100	21	19	9	0.999375	0.003200
100	21	19	10	0.999923	0.000549
100	21	19	11	0.999993	0.000070
100	21	19	12	0.999999	0.000006
100	21	19	13	1.000000	0.000000
100	21	19	14	1.000000	0.000000
100	21	19	15	1.000000	0.000000
100	21	19	16	1.000000	0.000000
100	21	19	17	1.000000	0.000000
100	21	19	18	1.000000	0.000000
100	21	19	19	1.000000	0.000000
100	21	20	0	0.004947	0.004947
100	21	20	1	0.039576	0.034629
100	21	20	2	0.147436	0.107860
100	21	20	3	0.345759	0.198323
100	21	20	4	0.586579	0.240821
100	21	20	5	0.791277	0.204698
100	21	20	6	0.917245	0.125968
100	21	20	7	0.974503	0.057258
100	21	20	8	0.993945	0.019442
100	21	20	9	0.998901	0.004956
100	21	20	10	0.999849	0.000948
100	21	20	11	0.999984	0.000135
100	21	20	12	0.999999	0.000014
100	21	20	13	1.000000	0.000001
100	21	20	14	1.000000	0.000000
100	21	20	15	1.000000	0.000000

N	n	k	x	P(x)	p(x)
100	21	20	16	1.000000	0.000000
100	21	20	17	1.000000	0.000000
100	21	20	18	1.000000	0.000000
100	21	20	19	1.000000	0.000000
100	21	20	20	1.000000	0.000000
100	21	21	0	0.003648	0.003648
100	21	21	1	0.030918	0.027270
100	21	21	2	0.121819	0.090900
100	21	21	3	0.301136	0.179317
100	21	21	4	0.535405	0.234269
100	21	21	5	0.750338	0.214933
100	21	21	6	0.893626	0.143288
100	21	21	7	0.964483	0.070857
100	21	21	8	0.990786	0.026303
100	21	21	9	0.998158	0.007372
100	21	21	10	0.999719	0.001561
100	21	21	11	0.999968	0.000249
100	21	21	12	0.999997	0.000030
100	21	21	13	1.000000	0.000003
100	21	21	14	1.000000	0.000000
100	21	21	15	1.000000	0.000000
100	21	21	16	1.000000	0.000000
100	21	21	17	1.000000	0.000000
100	21	21	18	1.000000	0.000000
100	21	21	19	1.000000	0.000000
100	21	21	20	1.000000	0.000000
100	21	21	21	1.000000	0.000000
100	22	1	0	0.780000	0.780000
100	22	1	1	1.000000	0.220000
100	22	2	0	0.606667	0.606667
100	22	2	1	0.953333	0.346667
100	22	2	2	1.000000	0.046667
100	22	3	0	0.470476	0.470476
100	22	3	1	0.879048	0.408571
100	22	3	2	0.990476	0.111429
100	22	3	3	1.000000	0.009524
100	22	4	0	0.363770	0.363770
100	22	4	1	0.790594	0.426824
100	22	4	2	0.967501	0.176907
100	22	4	3	0.998134	0.030633
100	22	4	4	1.000000	0.001865
100	22	5	0	0.280406	0.280406
100	22	5	1	0.697226	0.416820
100	22	5	2	0.930646	0.233419
100	22	5	3	0.992072	0.061426
100	22	5	4	0.999650	0.007579
100	22	5	5	1.000000	0.000350
100	22	6	0	0.215470	0.215470
100	22	6	1	0.605087	0.389617
100	22	6	2	0.881505	0.276417

Table for $N = 2$, $n = 1$, through $N = 100$, $n = 50$

N	n	k	x	P(x)	p(x)
100	22	6	3	0.979786	0.098282
100	22	6	4	0.998214	0.018428
100	22	6	5	0.999937	0.001723
100	22	6	6	1.000000	0.000063
100	22	7	0	0.165041	0.165041
100	22	7	1	0.518045	0.353004
100	22	7	2	0.822692	0.304647
100	22	7	3	0.959921	0.137229
100	22	7	4	0.994685	0.034765
100	22	7	5	0.999626	0.004940
100	22	7	6	0.999989	0.000364
100	22	7	7	1.000000	0.000011
100	22	8	0	0.125999	0.125999
100	22	8	1	0.438334	0.312335
100	22	8	2	0.757177	0.318842
100	22	8	3	0.931885	0.174708
100	22	8	4	0.987957	0.056072
100	22	8	5	0.998723	0.010766
100	22	8	6	0.999927	0.001204
100	22	8	7	0.999998	0.000071
100	22	8	8	1.000000	0.000002
100	22	9	0	0.095869	0.095869
100	22	9	1	0.367040	0.271172
100	22	9	2	0.687863	0.320823
100	22	9	3	0.895804	0.207941
100	22	9	4	0.976986	0.081182
100	22	9	5	0.996733	0.019747
100	22	9	6	0.999717	0.002984
100	22	9	7	0.999987	0.000269
100	22	9	8	1.000000	0.000013
100	22	9	9	1.000000	0.000000
100	22	10	0	0.072692	0.072692
100	22	10	1	0.304462	0.231771
100	22	10	2	0.617353	0.312890
100	22	10	3	0.852388	0.235035
100	22	10	4	0.960928	0.108540
100	22	10	5	0.993044	0.032116
100	22	10	6	0.999193	0.006148
100	22	10	7	0.999942	0.000750
100	22	10	8	0.999998	0.000055
100	22	10	9	1.000000	0.000002
100	22	10	10	1.000000	0.000000
100	22	11	0	0.054923	0.054923
100	22	11	1	0.250383	0.195460
100	22	11	2	0.547822	0.297439
100	22	11	3	0.802769	0.254948
100	22	11	4	0.939220	0.136451
100	22	11	5	0.986978	0.047758
100	22	11	6	0.998100	0.011122
100	22	11	7	0.999817	0.001718

N	n	k	x	P(x)	p(x)
100	22	11	8	0.999989	0.000172
100	22	11	9	1.000000	0.000011
100	22	11	10	1.000000	0.000000
100	22	11	11	1.000000	0.000000
100	22	12	0	0.041346	0.041346
100	22	12	1	0.204263	0.162917
100	22	12	2	0.480981	0.276719
100	22	12	3	0.748342	0.267361
100	22	12	4	0.911623	0.163281
100	22	12	5	0.977856	0.066232
100	22	12	6	0.996100	0.018245
100	22	12	7	0.999528	0.003428
100	22	12	8	0.999962	0.000434
100	22	12	9	0.999998	0.000036
100	22	12	10	1.000000	0.000002
100	22	12	11	1.000000	0.000000
100	22	12	12	1.000000	0.000000
100	22	13	0	0.031010	0.031010
100	22	13	1	0.165385	0.134375
100	22	13	2	0.418091	0.252706
100	22	13	3	0.690617	0.272526
100	22	13	4	0.878225	0.187608
100	22	13	5	0.965061	0.086836
100	22	13	6	0.992783	0.027722
100	22	13	7	0.998944	0.006160
100	22	13	8	0.999893	0.000949
100	22	13	9	0.999993	0.000100
100	22	13	10	1.000000	0.000007
100	22	13	11	1.000000	0.000000
100	22	13	12	1.000000	0.000000
100	22	13	13	1.000000	0.000000
100	22	14	0	0.023168	0.023168
100	22	14	1	0.132950	0.109781
100	22	14	2	0.359997	0.227048
100	22	14	3	0.631099	0.271102
100	22	14	4	0.839409	0.208310
100	22	14	5	0.948093	0.108683
100	22	14	6	0.987685	0.039592
100	22	14	7	0.997881	0.010197
100	22	14	8	0.999740	0.001859
100	22	14	9	0.999978	0.000238
100	22	14	10	0.999999	0.000021
100	22	14	11	1.000000	0.000001
100	22	14	12	1.000000	0.000000
100	22	14	13	1.000000	0.000000
100	22	14	14	1.000000	0.000000
100	22	15	0	0.017241	0.017241
100	22	15	1	0.106142	0.088901
100	22	15	2	0.307196	0.201053
100	22	15	3	0.571205	0.264009

N	n	k	x	P(x)	p(x)
100	22	15	4	0.795810	0.224605
100	22	15	5	0.926609	0.130799
100	22	15	6	0.980319	0.053710
100	22	15	7	0.996103	0.015784
100	22	15	8	0.999438	0.003335
100	22	15	9	0.999942	0.000504
100	22	15	10	0.999996	0.000054
100	22	15	11	1.000000	0.000004
100	22	15	12	1.000000	0.000000
100	22	15	13	1.000000	0.000000
100	22	15	14	1.000000	0.000000
100	22	15	15	1.000000	0.000000
100	22	16	0	0.012779	0.012779
100	22	16	1	0.084179	0.071400
100	22	16	2	0.259889	0.175710
100	22	16	3	0.512191	0.252302
100	22	16	4	0.748246	0.236055
100	22	16	5	0.900449	0.152203
100	22	16	6	0.970209	0.069760
100	22	16	7	0.993317	0.023109
100	22	16	8	0.998888	0.005571
100	22	16	9	0.999865	0.000976
100	22	16	10	0.999988	0.000123
100	22	16	11	0.999999	0.000011
100	22	16	12	1.000000	0.000001
100	22	16	13	1.000000	0.000000
100	22	16	14	1.000000	0.000000
100	22	16	15	1.000000	0.000000
100	22	16	16	1.000000	0.000000
100	22	17	0	0.009432	0.009432
100	22	17	1	0.066329	0.056897
100	22	17	2	0.218053	0.151724
100	22	17	3	0.455123	0.237069
100	22	17	4	0.697663	0.242540
100	22	17	5	0.869646	0.171983
100	22	17	6	0.956921	0.087275
100	22	17	7	0.989191	0.032270
100	22	17	8	0.997960	0.008769
100	22	17	9	0.999714	0.001754
100	22	17	10	0.999970	0.000257
100	22	17	11	0.999998	0.000027
100	22	17	12	1.000000	0.000002
100	22	17	13	1.000000	0.000000
100	22	17	14	1.000000	0.000000
100	22	17	15	1.000000	0.000000
100	22	17	16	1.000000	0.000000
100	22	17	17	1.000000	0.000000
100	22	18	0	0.006932	0.006932
100	22	18	1	0.051933	0.045001
100	22	18	2	0.181493	0.129560

N	n	k	x	P(x)	p(x)
100	22	18	3	0.400854	0.219361
100	22	18	4	0.645064	0.244210
100	22	18	5	0.834421	0.189357
100	22	18	6	0.940097	0.105676
100	22	18	7	0.983359	0.043262
100	22	18	8	0.996481	0.013122
100	22	18	9	0.999439	0.002958
100	22	18	10	0.999933	0.000494
100	22	18	11	0.999994	0.000061
100	22	18	12	1.000000	0.000005
100	22	18	13	1.000000	0.000000
100	22	18	14	1.000000	0.000000
100	22	18	15	1.000000	0.000000
100	22	18	16	1.000000	0.000000
100	22	18	17	1.000000	0.000000
100	22	18	18	1.000000	0.000000
100	22	19	0	0.005072	0.005072
100	22	19	1	0.040408	0.035336
100	22	19	2	0.149893	0.109485
100	22	19	3	0.350026	0.200133
100	22	19	4	0.591457	0.241430
100	22	19	5	0.795164	0.203707
100	22	19	6	0.919477	0.124313
100	22	19	7	0.975445	0.055968
100	22	19	8	0.994240	0.018795
100	22	19	9	0.998970	0.004730
100	22	19	10	0.999861	0.000891
100	22	19	11	0.999986	0.000125
100	22	19	12	0.999999	0.000013
100	22	19	13	1.000000	0.000001
100	22	19	14	1.000000	0.000000
100	22	19	15	1.000000	0.000000
100	22	19	16	1.000000	0.000000
100	22	19	17	1.000000	0.000000
100	22	19	18	1.000000	0.000000
100	22	19	19	1.000000	0.000000
100	22	20	0	0.003695	0.003695
100	22	20	1	0.031247	0.027553
100	22	20	2	0.122860	0.091613
100	22	20	3	0.303081	0.180222
100	22	20	4	0.537806	0.234724
100	22	20	5	0.752410	0.214605
100	22	20	6	0.894922	0.142511
100	22	20	7	0.965081	0.070159
100	22	20	8	0.990992	0.025911
100	22	20	9	0.998211	0.007219
100	22	20	10	0.999729	0.001518
100	22	20	11	0.999969	0.000240
100	22	20	12	0.999997	0.000028
100	22	20	13	1.000000	0.000002

Table for $N = 2$, $n = 1$, through $N = 100$, $n = 50$

N	n	k	x	P(x)	p(x)
100	22	20	14	1.000000	0.000000
100	22	20	15	1.000000	0.000000
100	22	20	16	1.000000	0.000000
100	22	20	17	1.000000	0.000000
100	22	20	18	1.000000	0.000000
100	22	20	19	1.000000	0.000000
100	22	20	20	1.000000	0.000000
100	22	21	0	0.002679	0.002679
100	22	21	1	0.024015	0.021336
100	22	21	2	0.099957	0.075942
100	22	21	3	0.260279	0.160322
100	22	21	4	0.484993	0.224714
100	22	21	5	0.706807	0.221814
100	22	21	6	0.866419	0.159612
100	22	21	7	0.951926	0.085507
100	22	21	8	0.986457	0.034532
100	22	21	9	0.997038	0.010580
100	22	21	10	0.999501	0.002463
100	22	21	11	0.999936	0.000435
100	22	21	12	0.999994	0.000058
100	22	21	13	0.999999	0.000006
100	22	21	14	1.000000	0.000000
100	22	21	15	1.000000	0.000000
100	22	21	16	1.000000	0.000000
100	22	21	17	1.000000	0.000000
100	22	21	18	1.000000	0.000000
100	22	21	19	1.000000	0.000000
100	22	21	20	1.000000	0.000000
100	22	21	21	1.000000	0.000000
100	22	22	0	0.001933	0.001933
100	22	22	1	0.018343	0.016410
100	22	22	2	0.080731	0.062388
100	22	22	3	0.221720	0.140989
100	22	22	4	0.433792	0.212072
100	22	22	5	0.659075	0.225283
100	22	22	6	0.834093	0.175018
100	22	22	7	0.935690	0.101598
100	22	22	8	0.980338	0.044647
100	22	22	9	0.995297	0.014959
100	22	22	10	0.999127	0.003830
100	22	22	11	0.999875	0.000748
100	22	22	12	0.999986	0.000111
100	22	22	13	0.999999	0.000012
100	22	22	14	1.000000	0.000001
100	22	22	15	1.000000	0.000000
100	22	22	16	1.000000	0.000000
100	22	22	17	1.000000	0.000000
100	22	22	18	1.000000	0.000000
100	22	22	19	1.000000	0.000000
100	22	22	20	1.000000	0.000000

N	n	k	x	P(x)	p(x)
100	22	22	21	1.000000	0.000000
100	22	22	22	1.000000	0.000000
100	23	1	0	0.770000	0.770000
100	23	1	1	1.000000	0.230000
100	23	2	0	0.591111	0.591111
100	23	2	1	0.948889	0.357778
100	23	2	2	1.000000	0.051111
100	23	3	0	0.452381	0.452381
100	23	3	1	0.868571	0.416190
100	23	3	2	0.989048	0.120476
100	23	3	3	1.000000	0.010952
100	23	4	0	0.345115	0.345115
100	23	4	1	0.774178	0.429062
100	23	4	2	0.962965	0.188787
100	23	4	3	0.997742	0.034777
100	23	4	4	1.000000	0.002258
100	23	5	0	0.262431	0.262431
100	23	5	1	0.675851	0.413419
100	23	5	2	0.921668	0.245817
100	23	5	3	0.990497	0.068829
100	23	5	4	0.999553	0.009056
100	23	5	5	1.000000	0.000447
100	23	6	0	0.198895	0.198895
100	23	6	1	0.580112	0.381216
100	23	6	2	0.867329	0.287218
100	23	6	3	0.976006	0.108677
100	23	6	4	0.997742	0.021735
100	23	6	5	0.999915	0.002174
100	23	6	6	1.000000	0.000085
100	23	7	0	0.150229	0.150229
100	23	7	1	0.490891	0.340661
100	23	7	2	0.803164	0.312273
100	23	7	3	0.952884	0.149720
100	23	7	4	0.993348	0.040465
100	23	7	5	0.999499	0.006151
100	23	7	6	0.999985	0.000486
100	23	7	7	1.000000	0.000015
100	23	8	0	0.113076	0.113076
100	23	8	1	0.410304	0.297228
100	23	8	2	0.732650	0.322346
100	23	8	3	0.920686	0.188035
100	23	8	4	0.985081	0.064396
100	23	8	5	0.998309	0.013227
100	23	8	6	0.999896	0.001587
100	23	8	7	0.999997	0.000101
100	23	8	8	1.000000	0.000003
100	23	9	0	0.084807	0.084807
100	23	9	1	0.339228	0.254421
100	23	9	2	0.659071	0.319843
100	23	9	3	0.879808	0.220737

N	n	k	x	P(x)	p(x)
100	23	9	4	0.971782	0.091974
100	23	9	5	0.995721	0.023938
100	23	9	6	0.999602	0.003882
100	23	9	7	0.999980	0.000377
100	23	9	8	0.999999	0.000020
100	23	9	9	1.000000	0.000000
100	23	10	0	0.063372	0.063372
100	23	10	1	0.277720	0.214347
100	23	10	2	0.585261	0.307542
100	23	10	3	0.831295	0.246033
100	23	10	4	0.952579	0.121284
100	23	10	5	0.990986	0.038407
100	23	10	6	0.998877	0.007892
100	23	10	7	0.999913	0.001036
100	23	10	8	0.999996	0.000083
100	23	10	9	1.000000	0.000004
100	23	10	10	1.000000	0.000000
100	23	11	0	0.047177	0.047177
100	23	11	1	0.225324	0.178146
100	23	11	2	0.513502	0.288178
100	23	11	3	0.776621	0.263119
100	23	11	4	0.926975	0.150354
100	23	11	5	0.983304	0.056330
100	23	11	6	0.997387	0.014082
100	23	11	7	0.999729	0.002342
100	23	11	8	0.999982	0.000253
100	23	11	9	0.999999	0.000017
100	23	11	10	1.000000	0.000001
100	23	11	11	1.000000	0.000000
100	23	12	0	0.034985	0.034985
100	23	12	1	0.181287	0.146302
100	23	12	2	0.445505	0.264217
100	23	12	3	0.717493	0.271988
100	23	12	4	0.894877	0.177384
100	23	12	5	0.971912	0.077035
100	23	12	6	0.994697	0.022785
100	23	12	7	0.999308	0.004611
100	23	12	8	0.999940	0.000632
100	23	12	9	0.999997	0.000057
100	23	12	10	1.000000	0.000003
100	23	12	11	1.000000	0.000000
100	23	12	12	1.000000	0.000000
100	23	13	0	0.025841	0.025841
100	23	13	1	0.144712	0.118870
100	23	13	2	0.382453	0.237741
100	23	13	3	0.655677	0.273225
100	23	13	4	0.856578	0.200900
100	23	13	5	0.956155	0.099577
100	23	13	6	0.990295	0.034141
100	23	13	7	0.998470	0.008175

N	n	k	x	P(x)	p(x)
100	23	13	8	0.999832	0.001362
100	23	13	9	0.999988	0.000156
100	23	13	10	0.999999	0.000012
100	23	13	11	1.000000	0.000001
100	23	13	12	1.000000	0.000000
100	23	13	13	1.000000	0.000000
100	23	14	0	0.019010	0.019010
100	23	14	1	0.114653	0.095643
100	23	14	2	0.325067	0.210414
100	23	14	3	0.592867	0.267800
100	23	14	4	0.812703	0.219836
100	23	14	5	0.935552	0.122849
100	23	14	6	0.983624	0.048072
100	23	14	7	0.996966	0.013342
100	23	14	8	0.999597	0.002631
100	23	14	9	0.999962	0.000365
100	23	14	10	0.999997	0.000035
100	23	14	11	1.000000	0.000002
100	23	14	12	1.000000	0.000000
100	23	14	13	1.000000	0.000000
100	23	14	14	1.000000	0.000000
100	23	15	0	0.013926	0.013926
100	23	15	1	0.090186	0.076260
100	23	15	2	0.273687	0.183501
100	23	15	3	0.530588	0.256901
100	23	15	4	0.764135	0.233547
100	23	15	5	0.909840	0.145705
100	23	15	6	0.974121	0.064282
100	23	15	7	0.994484	0.020363
100	23	15	8	0.999138	0.004654
100	23	15	9	0.999903	0.000765
100	23	15	10	0.999992	0.000089
100	23	15	11	0.999999	0.000007
100	23	15	12	1.000000	0.000000
100	23	15	13	1.000000	0.000000
100	23	15	14	1.000000	0.000000
100	23	15	15	1.000000	0.000000
100	23	16	0	0.010158	0.010158
100	23	16	1	0.070448	0.060290
100	23	16	2	0.228351	0.157903
100	23	16	3	0.470141	0.241789
100	23	16	4	0.711930	0.241789
100	23	16	5	0.878985	0.167055
100	23	16	6	0.961265	0.082281
100	23	16	7	0.990651	0.029386
100	23	16	8	0.998317	0.007666
100	23	16	9	0.999777	0.001460
100	23	16	10	0.999979	0.000202
100	23	16	11	0.999998	0.000020
100	23	16	12	1.000000	0.000001

Table for $N = 2$, $n = 1$, through $N = 100$, $n = 50$

N	n	k	x	P(x)	p(x)
100	23	16	13	1.000000	0.000000
100	23	16	14	1.000000	0.000000
100	23	16	15	1.000000	0.000000
100	23	16	16	1.000000	0.000000
100	23	17	0	0.007376	0.007376
100	23	17	1	0.054658	0.047281
100	23	17	2	0.188875	0.134218
100	23	17	3	0.412572	0.223696
100	23	17	4	0.657240	0.244668
100	23	17	5	0.843187	0.185948
100	23	17	6	0.944613	0.101426
100	23	17	7	0.985054	0.040441
100	23	17	8	0.996948	0.011894
100	23	17	9	0.999534	0.002586
100	23	17	10	0.999947	0.000414
100	23	17	11	0.999996	0.000048
100	23	17	12	1.000000	0.000004
100	23	17	13	1.000000	0.000000
100	23	17	14	1.000000	0.000000
100	23	17	15	1.000000	0.000000
100	23	17	16	1.000000	0.000000
100	23	17	17	1.000000	0.000000
100	23	18	0	0.005332	0.005332
100	23	18	1	0.042125	0.036793
100	23	18	2	0.154917	0.112791
100	23	18	3	0.358669	0.203752
100	23	18	4	0.601231	0.242562
100	23	18	5	0.802861	0.201630
100	23	18	6	0.923839	0.120978
100	23	18	7	0.977258	0.053419
100	23	18	8	0.994798	0.017540
100	23	18	9	0.999098	0.004299
100	23	18	10	0.999883	0.000785
100	23	18	11	0.999989	0.000106
100	23	18	12	0.999999	0.000010
100	23	18	13	1.000000	0.000001
100	23	18	14	1.000000	0.000000
100	23	18	15	1.000000	0.000000
100	23	18	16	1.000000	0.000000
100	23	18	17	1.000000	0.000000
100	23	18	18	1.000000	0.000000
100	23	19	0	0.003837	0.003837
100	23	19	1	0.032254	0.028417
100	23	19	2	0.126031	0.093777
100	23	19	3	0.308973	0.182942
100	23	19	4	0.545028	0.236055
100	23	19	5	0.758601	0.213573
100	23	19	6	0.898758	0.140157
100	23	19	7	0.966835	0.068076
100	23	19	8	0.991590	0.024755
100	23	19	9	0.998364	0.006774
100	23	19	10	0.999758	0.001395
100	23	19	11	0.999973	0.000215
100	23	19	12	0.999998	0.000025
100	23	19	13	1.000000	0.000002
100	23	19	14	1.000000	0.000000
100	23	19	15	1.000000	0.000000
100	23	19	16	1.000000	0.000000
100	23	19	17	1.000000	0.000000
100	23	19	18	1.000000	0.000000
100	23	19	19	1.000000	0.000000
100	23	20	0	0.002747	0.002747
100	23	20	1	0.024536	0.021788
100	23	20	2	0.101718	0.077183
100	23	20	3	0.263802	0.162084
100	23	20	4	0.489657	0.225855
100	23	20	5	0.711140	0.221483
100	23	20	6	0.869343	0.158202
100	23	20	7	0.953388	0.084045
100	23	20	8	0.987006	0.033618
100	23	20	9	0.997193	0.010187
100	23	20	10	0.999534	0.002342
100	23	20	11	0.999941	0.000407
100	23	20	12	0.999994	0.000053
100	23	20	13	1.000000	0.000005
100	23	20	14	1.000000	0.000000
100	23	20	15	1.000000	0.000000
100	23	20	16	1.000000	0.000000
100	23	20	17	1.000000	0.000000
100	23	20	18	1.000000	0.000000
100	23	20	19	1.000000	0.000000
100	23	20	20	1.000000	0.000000
100	23	21	0	0.001957	0.001957
100	23	21	1	0.018544	0.016586
100	23	21	2	0.081458	0.062914
100	23	21	3	0.223281	0.141823
100	23	21	4	0.436017	0.212735
100	23	21	5	0.661307	0.225290
100	23	21	6	0.835725	0.174418
100	23	21	7	0.936579	0.100854
100	23	21	8	0.980702	0.044124
100	23	21	9	0.995410	0.014708
100	23	21	10	0.999154	0.003744
100	23	21	11	0.999880	0.000726
100	23	21	12	0.999987	0.000107
100	23	21	13	0.999999	0.000012
100	23	21	14	1.000000	0.000001
100	23	21	15	1.000000	0.000000
100	23	21	16	1.000000	0.000000
100	23	21	17	1.000000	0.000000
100	23	21	18	1.000000	0.000000
100	23	21	19	1.000000	0.000000
100	23	21	20	1.000000	0.000000
100	23	21	21	1.000000	0.000000
100	23	22	0	0.001388	0.001388
100	23	22	1	0.013925	0.012537
100	23	22	2	0.064734	0.050809
100	23	22	3	0.187377	0.122643
100	23	22	4	0.384852	0.197476
100	23	22	5	0.609975	0.225122
100	23	22	6	0.798192	0.188217
100	23	22	7	0.916152	0.117961
100	23	22	8	0.972324	0.056172
100	23	22	9	0.992804	0.020479
100	23	22	10	0.998538	0.005734
100	23	22	11	0.999770	0.001232
100	23	22	12	0.999972	0.000202
100	23	22	13	0.999997	0.000025
100	23	22	14	1.000000	0.000002
100	23	22	15	1.000000	0.000000
100	23	22	16	1.000000	0.000000
100	23	22	17	1.000000	0.000000
100	23	22	18	1.000000	0.000000
100	23	22	19	1.000000	0.000000
100	23	22	20	1.000000	0.000000
100	23	22	21	1.000000	0.000000
100	23	22	22	1.000000	0.000000
100	23	23	0	0.000978	0.000978
100	23	23	1	0.010389	0.009410
100	23	23	2	0.051055	0.040666
100	23	23	3	0.155930	0.104875
100	23	23	4	0.336749	0.180819
100	23	23	5	0.558023	0.221274
100	23	23	6	0.757170	0.199147
100	23	23	7	0.891955	0.134785
100	23	23	8	0.961522	0.069567
100	23	23	9	0.989128	0.027606
100	23	23	10	0.997582	0.008454
100	23	23	11	0.999580	0.001998
100	23	23	12	0.999944	0.000363
100	23	23	13	0.999994	0.000050
100	23	23	14	0.999999	0.000005
100	23	23	15	1.000000	0.000000
100	23	23	16	1.000000	0.000000
100	23	23	17	1.000000	0.000000
100	23	23	18	1.000000	0.000000
100	23	23	19	1.000000	0.000000
100	23	23	20	1.000000	0.000000
100	23	23	21	1.000000	0.000000
100	23	23	22	1.000000	0.000000
100	23	23	23	1.000000	0.000000
100	24	1	0	0.760000	0.760000
100	24	1	1	1.000000	0.240000
100	24	2	0	0.575758	0.575758
100	24	2	1	0.944242	0.368485
100	24	2	2	1.000000	0.055758
100	24	3	0	0.434756	0.434756
100	24	3	1	0.857761	0.423006
100	24	3	2	0.987483	0.129722
100	24	3	3	1.000000	0.012517
100	24	4	0	0.327187	0.327187
100	24	4	1	0.757461	0.430274
100	24	4	2	0.958061	0.200601
100	24	4	3	0.997290	0.039229
100	24	4	4	1.000000	0.002710
100	24	5	0	0.245390	0.245390
100	24	5	1	0.654375	0.408984
100	24	5	2	0.912091	0.257716
100	24	5	3	0.988709	0.076618
100	24	5	4	0.999435	0.010727
100	24	5	5	1.000000	0.000565
100	24	6	0	0.183397	0.183397
100	24	6	1	0.555357	0.371960
100	24	6	2	0.852409	0.297052
100	24	6	3	0.971772	0.119363
100	24	6	4	0.997177	0.025405
100	24	6	5	0.999887	0.002710
100	24	6	6	1.000000	0.000113
100	24	7	0	0.136572	0.136572
100	24	7	1	0.464346	0.327774
100	24	7	2	0.782886	0.318540
100	24	7	3	0.945106	0.162220
100	24	7	4	0.991772	0.046666
100	24	7	5	0.999339	0.007567
100	24	7	6	0.999978	0.000639
100	24	7	7	1.000000	0.000022
100	24	8	0	0.101328	0.101328
100	24	8	1	0.383284	0.281956
100	24	8	2	0.707533	0.324249
100	24	8	3	0.908476	0.200943
100	24	8	4	0.981736	0.073260
100	24	8	5	0.997793	0.016057
100	24	8	6	0.999855	0.002061
100	24	8	7	0.999996	0.000141
100	24	8	8	1.000000	0.000004
100	24	9	0	0.074894	0.074894
100	24	9	1	0.312795	0.237900
100	24	9	2	0.629995	0.317200
100	24	9	3	0.862608	0.232613
100	24	9	4	0.965810	0.103202

Table for $N = 2$, $n = 1$, through $N = 100$, $n = 50$

N	n	k	x	P(x)	p(x)
100	24	9	5	0.994477	0.028667
100	24	9	6	0.999451	0.004974
100	24	9	7	0.999970	0.000519
100	24	9	8	0.999999	0.000029
100	24	9	9	1.000000	0.000001
100	24	10	0	0.055142	0.055142
100	24	10	1	0.252666	0.197524
100	24	10	2	0.553309	0.300643
100	24	10	3	0.808928	0.255619
100	24	10	4	0.943128	0.134200
100	24	10	5	0.988492	0.045363
100	24	10	6	0.998467	0.009976
100	24	10	7	0.999873	0.001406
100	24	10	8	0.999994	0.000121
100	24	10	9	1.000000	0.000006
100	24	10	10	1.000000	0.000000
100	24	11	0	0.040438	0.040438
100	24	11	1	0.202188	0.161750
100	24	11	2	0.479818	0.277631
100	24	11	3	0.749284	0.269465
100	24	11	4	0.913306	0.164022
100	24	11	5	0.978915	0.065609
100	24	11	6	0.996472	0.017557
100	24	11	7	0.999607	0.003135
100	24	11	8	0.999972	0.000365
100	24	11	9	0.999999	0.000026
100	24	11	10	1.000000	0.000001
100	24	11	11	1.000000	0.000000
100	24	12	0	0.029533	0.029533
100	24	12	1	0.160387	0.130854
100	24	12	2	0.411191	0.250804
100	24	12	3	0.685702	0.274511
100	24	12	4	0.876447	0.190745
100	24	12	5	0.964909	0.088461
100	24	12	6	0.992921	0.028013
100	24	12	7	0.999009	0.006087
100	24	12	8	0.999907	0.000898
100	24	12	9	0.999994	0.000087
100	24	12	10	1.000000	0.000005
100	24	12	11	1.000000	0.000000
100	24	12	12	1.000000	0.000000
100	24	13	0	0.021479	0.021479
100	24	13	1	0.126187	0.104708
100	24	13	2	0.348490	0.222303
100	24	13	3	0.620194	0.271704
100	24	13	4	0.833096	0.212902
100	24	13	5	0.945809	0.112713
100	24	13	6	0.987191	0.041383
100	24	13	7	0.997833	0.010641
100	24	13	8	0.999744	0.001911

N	n	k	x	P(x)	p(x)
100	24	13	9	0.999980	0.000236
100	24	13	10	0.999999	0.000019
100	24	13	11	1.000000	0.000001
100	24	13	12	1.000000	0.000000
100	24	13	13	1.000000	0.000000
100	24	14	0	0.015553	0.015553
100	24	14	1	0.098505	0.082952
100	24	14	2	0.292275	0.193770
100	24	14	3	0.554610	0.262335
100	24	14	4	0.784153	0.229543
100	24	14	5	0.921193	0.137041
100	24	14	6	0.978630	0.057436
100	24	14	7	0.995753	0.017124
100	24	14	8	0.999392	0.003639
100	24	14	9	0.999939	0.000547
100	24	14	10	0.999996	0.000057
100	24	14	11	1.000000	0.000004
100	24	14	12	1.000000	0.000000
100	24	14	13	1.000000	0.000000
100	24	14	14	1.000000	0.000000
100	24	15	0	0.011213	0.011213
100	24	15	1	0.076320	0.065107
100	24	15	2	0.242706	0.166386
100	24	15	3	0.490551	0.247845
100	24	15	4	0.730771	0.240219
100	24	15	5	0.890917	0.160146
100	24	15	6	0.966608	0.075691
100	24	15	7	0.992368	0.025760
100	24	15	8	0.998715	0.006347
100	24	15	9	0.999843	0.001128
100	24	15	10	0.999986	0.000143
100	24	15	11	0.999999	0.000013
100	24	15	12	1.000000	0.000001
100	24	15	13	1.000000	0.000000
100	24	15	14	1.000000	0.000000
100	24	15	15	1.000000	0.000000
100	24	16	0	0.008047	0.008047
100	24	16	1	0.058703	0.050656
100	24	16	2	0.199642	0.140938
100	24	16	3	0.429319	0.229677
100	24	16	4	0.674249	0.244929
100	24	16	5	0.855120	0.180871
100	24	16	6	0.950579	0.095460
100	24	16	7	0.987216	0.036637
100	24	16	8	0.997520	0.010304
100	24	16	9	0.999644	0.002124
100	24	16	10	0.999963	0.000319
100	24	16	11	0.999997	0.000034
100	24	16	12	1.000000	0.000003
100	24	16	13	1.000000	0.000000

N	n	k	x	P(x)	p(x)
100	24	16	14	1.000000	0.000000
100	24	16	15	1.000000	0.000000
100	24	16	16	1.000000	0.000000
100	24	17	0	0.005748	0.005748
100	24	17	1	0.044833	0.039085
100	24	17	2	0.162729	0.117896
100	24	17	3	0.371900	0.209171
100	24	17	4	0.615932	0.244032
100	24	17	5	0.814208	0.198276
100	24	17	6	0.930124	0.115915
100	24	17	7	0.979802	0.049678
100	24	17	8	0.995558	0.015756
100	24	17	9	0.999265	0.003707
100	24	17	10	0.999910	0.000645
100	24	17	11	0.999992	0.000082
100	24	17	12	0.999999	0.000008
100	24	17	13	1.000000	0.000000
100	24	17	14	1.000000	0.000000
100	24	17	15	1.000000	0.000000
100	24	17	16	1.000000	0.000000
100	24	17	17	1.000000	0.000000
100	24	18	0	0.004086	0.004086
100	24	18	1	0.034002	0.029916
100	24	18	2	0.131480	0.097477
100	24	18	3	0.318977	0.187497
100	24	18	4	0.557129	0.238152
100	24	18	5	0.768820	0.211691
100	24	18	6	0.904985	0.136166
100	24	18	7	0.969627	0.064641
100	24	18	8	0.992520	0.022894
100	24	18	9	0.998595	0.006075
100	24	18	10	0.999801	0.001206
100	24	18	11	0.999979	0.000178
100	24	18	12	0.999998	0.000019
100	24	18	13	1.000000	0.000001
100	24	18	14	1.000000	0.000000
100	24	18	15	1.000000	0.000000
100	24	18	16	1.000000	0.000000
100	24	18	17	1.000000	0.000000
100	24	18	18	1.000000	0.000000
100	24	19	0	0.002890	0.002890
100	24	19	1	0.025611	0.022721
100	24	19	2	0.105327	0.079716
100	24	19	3	0.270959	0.165632
100	24	19	4	0.499043	0.228084
100	24	19	5	0.719769	0.220726
100	24	19	6	0.875095	0.155326
100	24	19	7	0.956225	0.081130
100	24	19	8	0.988053	0.031828
100	24	19	9	0.997484	0.009430

N	n	k	x	P(x)	p(x)
100	24	19	10	0.999595	0.002111
100	24	19	11	0.999951	0.000356
100	24	19	12	0.999995	0.000045
100	24	19	13	1.000000	0.000004
100	24	19	14	1.000000	0.000000
100	24	19	15	1.000000	0.000000
100	24	19	16	1.000000	0.000000
100	24	19	17	1.000000	0.000000
100	24	19	18	1.000000	0.000000
100	24	19	19	1.000000	0.000000
100	24	20	0	0.002034	0.002034
100	24	20	1	0.019159	0.017126
100	24	20	2	0.083676	0.064516
100	24	20	3	0.228018	0.144342
100	24	20	4	0.442726	0.214709
100	24	20	5	0.667994	0.225268
100	24	20	6	0.840578	0.172584
100	24	20	7	0.939198	0.098620
100	24	20	8	0.981766	0.042568
100	24	20	9	0.995737	0.013971
100	24	20	10	0.999230	0.003493
100	24	20	11	0.999894	0.000663
100	24	20	12	0.999989	0.000095
100	24	20	13	0.999999	0.000010
100	24	20	14	1.000000	0.000001
100	24	20	15	1.000000	0.000000
100	24	20	16	1.000000	0.000000
100	24	20	17	1.000000	0.000000
100	24	20	18	1.000000	0.000000
100	24	20	19	1.000000	0.000000
100	24	20	20	1.000000	0.000000
100	24	21	0	0.001424	0.001424
100	24	21	1	0.014236	0.012812
100	24	21	2	0.065934	0.051698
100	24	21	3	0.190128	0.124194
100	24	21	4	0.389049	0.198921
100	24	21	5	0.614493	0.225444
100	24	21	6	0.801747	0.187254
100	24	21	7	0.918241	0.116494
100	24	21	8	0.973253	0.055011
100	24	21	9	0.993118	0.019865
100	24	21	10	0.998619	0.005501
100	24	21	11	0.999786	0.001167
100	24	21	12	0.999975	0.000189
100	24	21	13	0.999998	0.000023
100	24	21	14	1.000000	0.000002
100	24	21	15	1.000000	0.000000
100	24	21	16	1.000000	0.000000
100	24	21	17	1.000000	0.000000
100	24	21	18	1.000000	0.000000

N	n	k	x	P(x)	p(x)
100	24	21	19	1.000000	0.000000
100	24	21	20	1.000000	0.000000
100	24	21	21	1.000000	0.000000
100	24	22	0	0.000991	0.000991
100	24	22	1	0.010506	0.009514
100	24	22	2	0.051537	0.041031
100	24	22	3	0.157114	0.105577
100	24	22	4	0.338689	0.181575
100	24	22	5	0.560272	0.221583
100	24	22	6	0.759082	0.198809
100	24	22	7	0.893173	0.134091
100	24	22	8	0.962111	0.068938
100	24	22	9	0.989346	0.027235
100	24	22	10	0.997644	0.008298
100	24	22	11	0.999594	0.001950
100	24	22	12	0.999946	0.000352
100	24	22	13	0.999994	0.000048
100	24	22	14	0.999999	0.000005
100	24	22	15	1.000000	0.000000
100	24	22	16	1.000000	0.000000
100	24	22	17	1.000000	0.000000
100	24	22	18	1.000000	0.000000
100	24	22	19	1.000000	0.000000
100	24	22	20	1.000000	0.000000
100	24	22	21	1.000000	0.000000
100	24	22	22	1.000000	0.000000
100	24	23	0	0.000686	0.000686
100	24	23	1	0.007700	0.007014
100	24	23	2	0.039964	0.032264
100	24	23	3	0.128690	0.088726
100	24	23	4	0.292132	0.163442
100	24	23	5	0.506297	0.214165
100	24	23	6	0.713203	0.206906
100	24	23	7	0.863948	0.150745
100	24	23	8	0.947970	0.084022
100	24	23	9	0.984109	0.036139
100	24	23	10	0.996155	0.012046
100	24	23	11	0.999269	0.003114
100	24	23	12	0.999892	0.000623
100	24	23	13	0.999988	0.000096
100	24	23	14	0.999999	0.000011
100	24	23	15	1.000000	0.000001
100	24	23	16	1.000000	0.000000
100	24	23	17	1.000000	0.000000
100	24	23	18	1.000000	0.000000
100	24	23	19	1.000000	0.000000
100	24	23	20	1.000000	0.000000
100	24	23	21	1.000000	0.000000
100	24	23	22	1.000000	0.000000
100	24	23	23	1.000000	0.000000

N	n	k	x	P(x)	p(x)
100	24	24	0	0.000472	0.000472
100	24	24	1	0.005605	0.005133
100	24	24	2	0.030746	0.025141
100	24	24	3	0.104492	0.073746
100	24	24	4	0.249679	0.145187
100	24	24	5	0.453451	0.203772
100	24	24	6	0.664835	0.211384
100	24	24	7	0.830667	0.165832
100	24	24	8	0.930511	0.099844
100	24	24	9	0.977069	0.046558
100	24	24	10	0.993965	0.016896
100	24	24	11	0.998743	0.004779
100	24	24	12	0.999795	0.001052
100	24	24	13	0.999974	0.000179
100	24	24	14	0.999997	0.000023
100	24	24	15	1.000000	0.000002
100	24	24	16	1.000000	0.000000
100	24	24	17	1.000000	0.000000
100	24	24	18	1.000000	0.000000
100	24	24	19	1.000000	0.000000
100	24	24	20	1.000000	0.000000
100	24	24	21	1.000000	0.000000
100	24	24	22	1.000000	0.000000
100	24	24	23	1.000000	0.000000
100	24	24	24	1.000000	0.000000
100	25	1	0	0.750000	0.750000
100	25	1	1	1.000000	0.250000
100	25	2	0	0.560606	0.560606
100	25	2	1	0.939394	0.378788
100	25	2	2	1.000000	0.060606
100	25	3	0	0.417594	0.417594
100	25	3	1	0.846630	0.429035
100	25	3	2	0.985776	0.139147
100	25	3	3	1.000000	0.014224
100	25	4	0	0.309967	0.309967
100	25	4	1	0.740476	0.430510
100	25	4	2	0.952783	0.212306
100	25	4	3	0.996774	0.043991
100	25	4	4	1.000000	0.003226
100	25	5	0	0.229246	0.229246
100	25	5	1	0.632849	0.403603
100	25	5	2	0.901918	0.269068
100	25	5	3	0.986693	0.084775
100	25	5	4	0.999294	0.012602
100	25	5	5	1.000000	0.000706
100	25	6	0	0.168918	0.168918
100	25	6	1	0.530886	0.361968
100	25	6	2	0.836775	0.305888
100	25	6	3	0.967060	0.130286
100	25	6	4	0.996509	0.029448

N	n	k	x	P(x)	p(x)
100	25	6	5	0.999851	0.003343
100	25	6	6	1.000000	0.000149
100	25	7	0	0.123993	0.123993
100	25	7	1	0.438469	0.314476
100	25	7	2	0.761930	0.323461
100	25	7	3	0.936568	0.174638
100	25	7	4	0.989930	0.053362
100	25	7	5	0.999140	0.009210
100	25	7	6	0.999970	0.000830
100	25	7	7	1.000000	0.000030
100	25	8	0	0.090662	0.090662
100	25	8	1	0.357314	0.266652
100	25	8	2	0.681934	0.324620
100	25	8	3	0.895256	0.213322
100	25	8	4	0.977880	0.082625
100	25	8	5	0.997159	0.019279
100	25	8	6	0.999800	0.002641
100	25	8	7	0.999994	0.000194
100	25	8	8	1.000000	0.000006
100	25	9	0	0.066025	0.066025
100	25	9	1	0.287752	0.221727
100	25	9	2	0.600779	0.313026
100	25	9	3	0.844244	0.243465
100	25	9	4	0.959020	0.114776
100	25	9	5	0.992968	0.033948
100	25	9	6	0.999255	0.006287
100	25	9	7	0.999956	0.000701
100	25	9	8	0.999999	0.000043
100	25	9	9	1.000000	0.000001
100	25	10	0	0.047887	0.047887
100	25	10	1	0.229275	0.181388
100	25	10	2	0.521662	0.292387
100	25	10	3	0.785384	0.263722
100	25	10	4	0.932533	0.147149
100	25	10	5	0.985507	0.052974
100	25	10	6	0.997942	0.012435
100	25	10	7	0.999817	0.001875
100	25	10	8	0.999991	0.000173
100	25	10	9	1.000000	0.000009
100	25	10	10	1.000000	0.000000
100	25	11	0	0.034585	0.034585
100	25	11	1	0.180905	0.146320
100	25	11	2	0.446941	0.266036
100	25	11	3	0.720919	0.273978
100	25	11	4	0.898199	0.177280
100	25	11	5	0.973735	0.075537
100	25	11	6	0.995317	0.021582
100	25	11	7	0.999442	0.004125
100	25	11	8	0.999958	0.000516
100	25	11	9	0.999998	0.000040

N	n	k	x	P(x)	p(x)
100	25	11	10	1.000000	0.000002
100	25	11	11	1.000000	0.000000
100	25	12	0	0.024870	0.024870
100	25	12	1	0.141448	0.116578
100	25	12	2	0.378190	0.236743
100	25	12	3	0.653194	0.275004
100	25	12	4	0.856369	0.203175
100	25	12	5	0.956761	0.100392
100	25	12	6	0.990710	0.033949
100	25	12	7	0.998608	0.007898
100	25	12	8	0.999860	0.001251
100	25	12	9	0.999991	0.000131
100	25	12	10	1.000000	0.000009
100	25	12	11	1.000000	0.000000
100	25	12	12	1.000000	0.000000
100	25	13	0	0.017805	0.017805
100	25	13	1	0.109654	0.091849
100	25	13	2	0.316314	0.206661
100	25	13	3	0.584443	0.268129
100	25	13	4	0.807884	0.223441
100	25	13	5	0.933944	0.126061
100	25	13	6	0.983380	0.049436
100	25	13	7	0.996993	0.013613
100	25	13	8	0.999618	0.002625
100	25	13	9	0.999967	0.000349
100	25	13	10	0.999998	0.000031
100	25	13	11	1.000000	0.000002
100	25	13	12	1.000000	0.000000
100	25	13	13	1.000000	0.000000
100	25	14	0	0.012688	0.012688
100	25	14	1	0.084316	0.071628
100	25	14	2	0.261680	0.177364
100	25	14	3	0.516640	0.254961
100	25	14	4	0.753950	0.237309
100	25	14	5	0.904965	0.151015
100	25	14	6	0.972584	0.067619
100	25	14	7	0.994176	0.021593
100	25	14	8	0.999105	0.004929
100	25	14	9	0.999903	0.000798
100	25	14	10	0.999993	0.000090
100	25	14	11	1.000000	0.000007
100	25	14	12	1.000000	0.000000
100	25	14	13	1.000000	0.000000
100	25	14	14	1.000000	0.000000
100	25	15	0	0.009000	0.009000
100	25	15	1	0.064327	0.055327
100	25	15	2	0.214245	0.149918
100	25	15	3	0.451418	0.237173
100	25	15	4	0.696002	0.244584
100	25	15	5	0.869845	0.173843

Table for $N = 2$, $n = 1$, through $N = 100$, $n = 50$

N	n	k	x	P(x)	p(x)
100	25	15	6	0.957645	0.087799
100	25	15	7	0.989657	0.032012
100	25	15	8	0.998131	0.008474
100	25	15	9	0.999754	0.001624
100	25	15	10	0.999977	0.000223
100	25	15	11	0.999998	0.000021
100	25	15	12	1.000000	0.000001
100	25	15	13	1.000000	0.000000
100	25	15	14	1.000000	0.000000
100	25	15	15	1.000000	0.000000
100	25	16	0	0.006353	0.006353
100	25	16	1	0.048705	0.042352
100	25	16	2	0.173679	0.124974
100	25	16	3	0.390032	0.216353
100	25	16	4	0.635575	0.245543
100	25	16	5	0.828941	0.193365
100	25	16	6	0.938019	0.109078
100	25	16	7	0.982878	0.044859
100	25	16	8	0.996436	0.013558
100	25	16	9	0.999449	0.003013
100	25	16	10	0.999938	0.000489
100	25	16	11	0.999995	0.000057
100	25	16	12	1.000000	0.000005
100	25	16	13	1.000000	0.000000
100	25	16	14	1.000000	0.000000
100	25	16	15	1.000000	0.000000
100	25	16	16	1.000000	0.000000
100	25	17	0	0.004462	0.004462
100	25	17	1	0.036604	0.032142
100	25	17	2	0.139460	0.102856
100	25	17	3	0.333368	0.193908
100	25	17	4	0.574190	0.240821
100	25	17	5	0.782902	0.208712
100	25	17	6	0.913346	0.130445
100	25	17	7	0.973265	0.059919
100	25	17	8	0.993692	0.020427
100	25	17	9	0.998875	0.005183
100	25	17	10	0.999850	0.000976
100	25	17	11	0.999985	0.000135
100	25	17	12	0.999999	0.000013
100	25	17	13	1.000000	0.000001
100	25	17	14	1.000000	0.000000
100	25	17	15	1.000000	0.000000
100	25	17	16	1.000000	0.000000
100	25	17	17	1.000000	0.000000
100	25	18	0	0.003118	0.003118
100	25	18	1	0.027310	0.024192
100	25	18	2	0.110958	0.083648
100	25	18	3	0.281971	0.171013
100	25	18	4	0.513259	0.231288

N	n	k	x	P(x)	p(x)
100	25	18	5	0.732609	0.219351
100	25	18	6	0.883486	0.150876
100	25	18	7	0.960271	0.076785
100	25	18	8	0.989508	0.029237
100	25	18	9	0.997876	0.008368
100	25	18	10	0.999674	0.001798
100	25	18	11	0.999963	0.000289
100	25	18	12	0.999997	0.000034
100	25	18	13	1.000000	0.000003
100	25	18	14	1.000000	0.000000
100	25	18	15	1.000000	0.000000
100	25	18	16	1.000000	0.000000
100	25	18	17	1.000000	0.000000
100	25	18	18	1.000000	0.000000
100	25	19	0	0.002167	0.002167
100	25	19	1	0.020230	0.018062
100	25	19	2	0.087496	0.067266
100	25	19	3	0.236089	0.148594
100	25	19	4	0.454027	0.217937
100	25	19	5	0.679109	0.225083
100	25	19	6	0.848526	0.169417
100	25	19	7	0.943415	0.094889
100	25	19	8	0.983447	0.040031
100	25	19	9	0.996243	0.012796
100	25	19	10	0.999345	0.003102
100	25	19	11	0.999913	0.000568
100	25	19	12	0.999991	0.000078
100	25	19	13	0.999999	0.000008
100	25	19	14	1.000000	0.000001
100	25	19	15	1.000000	0.000000
100	25	19	16	1.000000	0.000000
100	25	19	17	1.000000	0.000000
100	25	19	18	1.000000	0.000000
100	25	19	19	1.000000	0.000000
100	25	20	0	0.001498	0.001498
100	25	20	1	0.014878	0.013379
100	25	20	2	0.068396	0.053518
100	25	20	3	0.195731	0.127335
100	25	20	4	0.397524	0.201794
100	25	20	5	0.623533	0.226009
100	25	20	6	0.808787	0.185253
100	25	20	7	0.922329	0.113542
100	25	20	8	0.975045	0.052716
100	25	20	9	0.993715	0.018670
100	25	20	10	0.998771	0.005055
100	25	20	11	0.999815	0.001044
100	25	20	12	0.999979	0.000164
100	25	20	13	0.999998	0.000019
100	25	20	14	1.000000	0.000002
100	25	20	15	1.000000	0.000000

N	n	k	x	P(x)	p(x)
100	25	20	16	1.000000	0.000000
100	25	20	17	1.000000	0.000000
100	25	20	18	1.000000	0.000000
100	25	20	19	1.000000	0.000000
100	25	20	20	1.000000	0.000000
100	25	21	0	0.001030	0.001030
100	25	21	1	0.010864	0.009834
100	25	21	2	0.053009	0.042145
100	25	21	3	0.160713	0.107704
100	25	21	4	0.344554	0.183840
100	25	21	5	0.567031	0.222478
100	25	21	6	0.764789	0.197758
100	25	21	7	0.896782	0.131993
100	25	21	8	0.963843	0.067061
100	25	21	9	0.989981	0.026138
100	25	21	10	0.997823	0.007842
100	25	21	11	0.999632	0.001810
100	25	21	12	0.999952	0.000320
100	25	21	13	0.999995	0.000043
100	25	21	14	1.000000	0.000004
100	25	21	15	1.000000	0.000000
100	25	21	16	1.000000	0.000000
100	25	21	17	1.000000	0.000000
100	25	21	18	1.000000	0.000000
100	25	21	19	1.000000	0.000000
100	25	21	20	1.000000	0.000000
100	25	21	21	1.000000	0.000000
100	25	22	0	0.000704	0.000704
100	25	22	1	0.007877	0.007172
100	25	22	2	0.040739	0.032863
100	25	22	3	0.130720	0.089981
100	25	22	4	0.295685	0.164965
100	25	22	5	0.510708	0.215023
100	25	22	6	0.717227	0.206519
100	25	22	7	0.866708	0.149481
100	25	22	8	0.949412	0.082704
100	25	22	9	0.984687	0.035275
100	25	22	10	0.996334	0.011646
100	25	22	11	0.999312	0.002978
100	25	22	12	0.999900	0.000588
100	25	22	13	0.999989	0.000089
100	25	22	14	0.999999	0.000010
100	25	22	15	1.000000	0.000001
100	25	22	16	1.000000	0.000000
100	25	22	17	1.000000	0.000000
100	25	22	18	1.000000	0.000000
100	25	22	19	1.000000	0.000000
100	25	22	20	1.000000	0.000000
100	25	22	21	1.000000	0.000000
100	25	22	22	1.000000	0.000000

N	n	k	x	P(x)	p(x)
100	25	23	0	0.000478	0.000478
100	25	23	1	0.005670	0.005191
100	25	23	2	0.031049	0.025379
100	25	23	3	0.105341	0.074292
100	25	23	4	0.251271	0.145930
100	25	23	5	0.455574	0.204302
100	25	23	6	0.666921	0.211347
100	25	23	7	0.832212	0.165291
100	25	23	8	0.931387	0.099175
100	25	23	9	0.977451	0.046065
100	25	23	10	0.994094	0.016643
100	25	23	11	0.998777	0.004683
100	25	23	12	0.999802	0.001024
100	25	23	13	0.999975	0.000173
100	25	23	14	0.999998	0.000023
100	25	23	15	1.000000	0.000002
100	25	23	16	1.000000	0.000000
100	25	23	17	1.000000	0.000000
100	25	23	18	1.000000	0.000000
100	25	23	19	1.000000	0.000000
100	25	23	20	1.000000	0.000000
100	25	23	21	1.000000	0.000000
100	25	23	22	1.000000	0.000000
100	25	23	23	1.000000	0.000000
100	25	24	0	0.000323	0.000323
100	25	24	1	0.004052	0.003729
100	25	24	2	0.023468	0.019416
100	25	24	3	0.084114	0.060646
100	25	24	4	0.211472	0.127357
100	25	24	5	0.402508	0.191036
100	25	24	6	0.614770	0.212262
100	25	24	7	0.793573	0.178802
100	25	24	8	0.909491	0.115918
100	25	24	9	0.967880	0.058389
100	25	24	10	0.990852	0.022973
100	25	24	11	0.997926	0.007074
100	25	24	12	0.999629	0.001703
100	25	24	13	0.999948	0.000319
100	25	24	14	0.999994	0.000046
100	25	24	15	0.999999	0.000005
100	25	24	16	1.000000	0.000000
100	25	24	17	1.000000	0.000000
100	25	24	18	1.000000	0.000000
100	25	24	19	1.000000	0.000000
100	25	24	20	1.000000	0.000000
100	25	24	21	1.000000	0.000000
100	25	24	22	1.000000	0.000000
100	25	24	23	1.000000	0.000000
100	25	24	24	1.000000	0.000000
100	25	25	0	0.000217	0.000217

Table for $N = 2$, $n = 1$, through $N = 100$, $n = 50$

N	n	k	x	P(x)	p(x)
100	25	25	1	0.002874	0.002657
100	25	25	2	0.017592	0.014718
100	25	25	3	0.066559	0.048967
100	25	25	4	0.176281	0.109722
100	25	25	5	0.352235	0.175954
100	25	25	6	0.561705	0.209469
100	25	25	7	0.751225	0.189520
100	25	25	8	0.883562	0.132337
100	25	25	9	0.955587	0.072025
100	25	25	10	0.986318	0.030731
100	25	25	11	0.996623	0.010305
100	25	25	12	0.999337	0.002715
100	25	25	13	0.999898	0.000560
100	25	25	14	0.999988	0.000090
100	25	25	15	0.999999	0.000011
100	25	25	16	1.000000	0.000001
100	25	25	17	1.000000	0.000000
100	25	25	18	1.000000	0.000000
100	25	25	19	1.000000	0.000000
100	25	25	20	1.000000	0.000000
100	25	25	21	1.000000	0.000000
100	25	25	22	1.000000	0.000000
100	25	25	23	1.000000	0.000000
100	25	25	24	1.000000	0.000000
100	25	25	25	1.000000	0.000000
100	26	1	0	0.740000	0.740000
100	26	1	1	1.000000	0.260000
100	26	2	0	0.545657	0.545657
100	26	2	1	0.934343	0.388687
100	26	2	2	1.000000	0.065657
100	26	3	0	0.400891	0.400891
100	26	3	1	0.835189	0.434298
100	26	3	2	0.983921	0.148732
100	26	3	3	1.000000	0.016079
100	26	4	0	0.293435	0.293435
100	26	4	1	0.723256	0.429821
100	26	4	2	0.947121	0.223865
100	26	4	3	0.996187	0.049066
100	26	4	4	1.000000	0.003813
100	26	5	0	0.213963	0.213963
100	26	5	1	0.611324	0.397360
100	26	5	2	0.891155	0.279831
100	26	5	3	0.984432	0.093277
100	26	5	4	0.999126	0.014694
100	26	5	5	1.000000	0.000874
100	26	6	0	0.155405	0.155405
100	26	6	1	0.506755	0.351350
100	26	6	2	0.820461	0.313706
100	26	6	3	0.961849	0.141388
100	26	6	4	0.995723	0.033874

N	n	k	x	P(x)	p(x)
100	26	6	5	0.999807	0.004083
100	26	6	6	1.000000	0.000193
100	26	7	0	0.112421	0.112421
100	26	7	1	0.413311	0.300890
100	26	7	2	0.740366	0.327055
100	26	7	3	0.927254	0.186888
100	26	7	4	0.987795	0.060541
100	26	7	5	0.998894	0.011099
100	26	7	6	0.999959	0.001064
100	26	7	7	1.000000	0.000041
100	26	8	0	0.080991	0.080991
100	26	8	1	0.332426	0.251435
100	26	8	2	0.655964	0.323538
100	26	8	3	0.881034	0.225070
100	26	8	4	0.973474	0.092439
100	26	8	5	0.996388	0.022915
100	26	8	6	0.999730	0.003342
100	26	8	7	0.999992	0.000262
100	26	8	8	1.000000	0.000008
100	26	9	0	0.058102	0.058102
100	26	9	1	0.264102	0.205999
100	26	9	2	0.571563	0.307462
100	26	9	3	0.824767	0.253204
100	26	9	4	0.951369	0.126602
100	26	9	5	0.991158	0.039789
100	26	9	6	0.999003	0.007846
100	26	9	7	0.999937	0.000934
100	26	9	8	0.999998	0.000061
100	26	9	9	1.000000	0.000002
100	26	10	0	0.041502	0.041502
100	26	10	1	0.207508	0.166007
100	26	10	2	0.490474	0.282966
100	26	10	3	0.760770	0.270296
100	26	10	4	0.920762	0.159991
100	26	10	5	0.981976	0.061214
100	26	10	6	0.997279	0.015304
100	26	10	7	0.999742	0.002463
100	26	10	8	0.999986	0.000244
100	26	10	9	1.000000	0.000013
100	26	10	10	1.000000	0.000000
100	26	11	0	0.029512	0.029512
100	26	11	1	0.161395	0.131883
100	26	11	2	0.415017	0.253621
100	26	11	3	0.691695	0.276678
100	26	11	4	0.881653	0.189958
100	26	11	5	0.967692	0.086040
100	26	11	6	0.993878	0.026186
100	26	11	7	0.999222	0.005344
100	26	11	8	0.999937	0.000715
100	26	11	9	0.999997	0.000060

N	n	k	x	P(x)	p(x)
100	26	11	10	1.000000	0.000003
100	26	11	11	1.000000	0.000000
100	26	12	0	0.020891	0.020891
100	26	12	1	0.124350	0.103459
100	26	12	2	0.346625	0.222275
100	26	12	3	0.620194	0.273569
100	26	12	4	0.834697	0.214503
100	26	12	5	0.947391	0.112694
100	26	12	6	0.987994	0.040603
100	26	12	7	0.998082	0.010088
100	26	12	8	0.999793	0.001711
100	26	12	9	0.999986	0.000193
100	26	12	10	0.999999	0.000014
100	26	12	11	1.000000	0.000001
100	26	12	12	1.000000	0.000000
100	26	13	0	0.014718	0.014718
100	26	13	1	0.094958	0.080239
100	26	13	2	0.286004	0.191046
100	26	13	3	0.548693	0.262689
100	26	13	4	0.781071	0.232378
100	26	13	5	0.920498	0.139427
100	26	13	6	0.978766	0.058268
100	26	13	7	0.995904	0.017138
100	26	13	8	0.999443	0.003539
100	26	13	9	0.999948	0.000506
100	26	13	10	0.999997	0.000048
100	26	13	11	1.000000	0.000003
100	26	13	12	1.000000	0.000000
100	26	13	13	1.000000	0.000000
100	26	14	0	0.010320	0.010320
100	26	14	1	0.071901	0.061581
100	26	14	2	0.233302	0.161401
100	26	14	3	0.479246	0.245945
100	26	14	4	0.722309	0.243062
100	26	14	5	0.886843	0.164535
100	26	14	6	0.965371	0.078528
100	26	14	7	0.992161	0.026790
100	26	14	8	0.998711	0.006550
100	26	14	9	0.999850	0.001139
100	26	14	10	0.999988	0.000138
100	26	14	11	0.999999	0.000011
100	26	14	12	1.000000	0.000001
100	26	14	13	1.000000	0.000000
100	26	14	14	1.000000	0.000000
100	26	15	0	0.007200	0.007200
100	26	15	1	0.053999	0.046799
100	26	15	2	0.188259	0.134260
100	26	15	3	0.413470	0.225211
100	26	15	4	0.660130	0.246660
100	26	15	5	0.846666	0.186536

N	n	k	x	P(x)	p(x)
100	26	15	6	0.947109	0.100443
100	26	15	7	0.986242	0.039133
100	26	15	8	0.997340	0.011098
100	26	15	9	0.999624	0.002285
100	26	15	10	0.999962	0.000338
100	26	15	11	0.999997	0.000035
100	26	15	12	1.000000	0.000002
100	26	15	13	1.000000	0.000000
100	26	15	14	1.000000	0.000000
100	26	15	15	1.000000	0.000000
100	26	16	0	0.004998	0.004998
100	26	16	1	0.040235	0.035237
100	26	16	2	0.150351	0.110116
100	26	16	3	0.352531	0.202180
100	26	16	4	0.596289	0.243758
100	26	16	5	0.800581	0.204292
100	26	16	6	0.923475	0.122894
100	26	16	7	0.977495	0.054020
100	26	16	8	0.994990	0.017495
100	26	16	9	0.999167	0.004178
100	26	16	10	0.999899	0.000731
100	26	16	11	0.999991	0.000092
100	26	16	12	0.999999	0.000008
100	26	16	13	1.000000	0.000000
100	26	16	14	1.000000	0.000000
100	26	16	15	1.000000	0.000000
100	26	16	16	1.000000	0.000000
100	26	17	0	0.003451	0.003451
100	26	17	1	0.029747	0.026297
100	26	17	2	0.118889	0.089142
100	26	17	3	0.297172	0.178283
100	26	17	4	0.532447	0.235275
100	26	17	5	0.749508	0.217060
100	26	17	6	0.894215	0.144707
100	26	17	7	0.965276	0.071061
100	26	17	8	0.991241	0.025965
100	26	17	9	0.998322	0.007081
100	26	17	10	0.999759	0.001437
100	26	17	11	0.999975	0.000215
100	26	17	12	0.999998	0.000023
100	26	17	13	1.000000	0.000002
100	26	17	14	1.000000	0.000000
100	26	17	15	1.000000	0.000000
100	26	17	16	1.000000	0.000000
100	26	17	17	1.000000	0.000000
100	26	18	0	0.002370	0.002370
100	26	18	1	0.021827	0.019457
100	26	18	2	0.093113	0.071286
100	26	18	3	0.247768	0.154655
100	26	18	4	0.470085	0.222317

Table for $N=2$, $n=1$, through $N=100$, $n=50$

N	n	k	x	P(x)	p(x)
100	26	18	5	0.694589	0.224504
100	26	18	6	0.859345	0.164757
100	26	18	7	0.949009	0.089663
100	26	18	8	0.985610	0.036601
100	26	18	9	0.996872	0.011262
100	26	18	10	0.999482	0.002611
100	26	18	11	0.999936	0.000453
100	26	18	12	0.999994	0.000058
100	26	18	13	1.000000	0.000005
100	26	18	14	1.000000	0.000000
100	26	18	15	1.000000	0.000000
100	26	18	16	1.000000	0.000000
100	26	18	17	1.000000	0.000000
100	26	18	18	1.000000	0.000000
100	26	19	0	0.001618	0.001618
100	26	19	1	0.015895	0.014276
100	26	19	2	0.072249	0.056354
100	26	19	3	0.204389	0.132141
100	26	19	4	0.410439	0.206050
100	26	19	5	0.637094	0.226655
100	26	19	6	0.819161	0.182067
100	26	19	7	0.928233	0.109072
100	26	19	8	0.977575	0.049342
100	26	19	9	0.994537	0.016961
100	26	19	10	0.998973	0.004436
100	26	19	11	0.999853	0.000880
100	26	19	12	0.999984	0.000131
100	26	19	13	0.999999	0.000015
100	26	19	14	1.000000	0.000001
100	26	19	15	1.000000	0.000000
100	26	19	16	1.000000	0.000000
100	26	19	17	1.000000	0.000000
100	26	19	18	1.000000	0.000000
100	26	19	19	1.000000	0.000000
100	26	20	0	0.001099	0.001099
100	26	20	1	0.011488	0.010390
100	26	20	2	0.055551	0.044063
100	26	20	3	0.166868	0.111317
100	26	20	4	0.354475	0.187607
100	26	20	5	0.578332	0.223857
100	26	20	6	0.774206	0.195874
100	26	20	7	0.902648	0.128442
100	26	20	8	0.966611	0.063962
100	26	20	9	0.990977	0.024367
100	26	20	10	0.998097	0.007120
100	26	20	11	0.999690	0.001593
100	26	20	12	0.999961	0.000272
100	26	20	13	0.999996	0.000035
100	26	20	14	1.000000	0.000003
100	26	20	15	1.000000	0.000000

N	n	k	x	P(x)	p(x)
100	26	20	16	1.000000	0.000000
100	26	20	17	1.000000	0.000000
100	26	20	18	1.000000	0.000000
100	26	20	19	1.000000	0.000000
100	26	20	20	1.000000	0.000000
100	26	21	0	0.000742	0.000742
100	26	21	1	0.008242	0.007500
100	26	21	2	0.042332	0.034091
100	26	21	3	0.134864	0.092532
100	26	21	4	0.302883	0.168019
100	26	21	5	0.519569	0.216686
100	26	21	6	0.725237	0.205668
100	26	21	7	0.872143	0.146906
100	26	21	8	0.952219	0.080076
100	26	21	9	0.985799	0.033580
100	26	21	10	0.996673	0.010874
100	26	21	11	0.999391	0.002718
100	26	21	12	0.999914	0.000523
100	26	21	13	0.999991	0.000077
100	26	21	14	0.999999	0.000009
100	26	21	15	1.000000	0.000001
100	26	21	16	1.000000	0.000000
100	26	21	17	1.000000	0.000000
100	26	21	18	1.000000	0.000000
100	26	21	19	1.000000	0.000000
100	26	21	20	1.000000	0.000000
100	26	21	21	1.000000	0.000000
100	26	22	0	0.000498	0.000498
100	26	22	1	0.005868	0.005371
100	26	22	2	0.031976	0.026107
100	26	22	3	0.107925	0.075949
100	26	22	4	0.256093	0.148168
100	26	22	5	0.461969	0.205876
100	26	22	6	0.673170	0.211200
100	26	22	7	0.836812	0.163642
100	26	22	8	0.933974	0.097162
100	26	22	9	0.978573	0.044599
100	26	22	10	0.994470	0.015897
100	26	22	11	0.998875	0.004404
100	26	22	12	0.999821	0.000946
100	26	22	13	0.999978	0.000157
100	26	22	14	0.999998	0.000020
100	26	22	15	1.000000	0.000002
100	26	22	16	1.000000	0.000000
100	26	22	17	1.000000	0.000000
100	26	22	18	1.000000	0.000000
100	26	22	19	1.000000	0.000000
100	26	22	20	1.000000	0.000000
100	26	22	21	1.000000	0.000000
100	26	22	22	1.000000	0.000000

N	n	k	x	P(x)	p(x)
100	26	23	0	0.000332	0.000332
100	26	23	1	0.004147	0.003815
100	26	23	2	0.023943	0.019796
100	26	23	3	0.085530	0.061587
100	26	23	4	0.214302	0.128772
100	26	23	5	0.406541	0.192239
100	26	23	6	0.619016	0.212475
100	26	23	7	0.796950	0.177934
100	26	23	8	0.911552	0.114602
100	26	23	9	0.968853	0.057301
100	26	23	10	0.991209	0.022357
100	26	23	11	0.998028	0.006818
100	26	23	12	0.999651	0.001623
100	26	23	13	0.999952	0.000300
100	26	23	14	0.999995	0.000043
100	26	23	15	0.999999	0.000005
100	26	23	16	1.000000	0.000000
100	26	23	17	1.000000	0.000000
100	26	23	18	1.000000	0.000000
100	26	23	19	1.000000	0.000000
100	26	23	20	1.000000	0.000000
100	26	23	21	1.000000	0.000000
100	26	23	22	1.000000	0.000000
100	26	23	23	1.000000	0.000000
100	26	24	0	0.000220	0.000220
100	26	24	1	0.002908	0.002689
100	26	24	2	0.017773	0.014864
100	26	24	3	0.067133	0.049361
100	26	24	4	0.177510	0.110376
100	26	24	5	0.354112	0.176602
100	26	24	6	0.563827	0.209715
100	26	24	7	0.753044	0.189217
100	26	24	8	0.884762	0.131718
100	26	24	9	0.956202	0.071440
100	26	24	10	0.986564	0.030362
100	26	24	11	0.996700	0.010136
100	26	24	12	0.999356	0.002657
100	26	24	13	0.999901	0.000545
100	26	24	14	0.999988	0.000087
100	26	24	15	0.999999	0.000011
100	26	24	16	1.000000	0.000001
100	26	24	17	1.000000	0.000000
100	26	24	18	1.000000	0.000000
100	26	24	19	1.000000	0.000000
100	26	24	20	1.000000	0.000000
100	26	24	21	1.000000	0.000000
100	26	24	22	1.000000	0.000000
100	26	24	23	1.000000	0.000000
100	26	24	24	1.000000	0.000000
100	26	25	0	0.000145	0.000145

N	n	k	x	P(x)	p(x)
100	26	25	1	0.002024	0.001879
100	26	25	2	0.013079	0.011055
100	26	25	3	0.052195	0.039117
100	26	25	4	0.145559	0.093363
100	26	25	5	0.305314	0.159755
100	26	25	6	0.508639	0.203325
100	26	25	7	0.705740	0.197101
100	26	25	8	0.853565	0.147826
100	26	25	9	0.940222	0.086656
100	26	25	10	0.980172	0.039950
100	26	25	11	0.994699	0.014527
100	26	25	12	0.998867	0.004168
100	26	25	13	0.999808	0.000941
100	26	25	14	0.999974	0.000166
100	26	25	15	0.999997	0.000023
100	26	25	16	1.000000	0.000002
100	26	25	17	1.000000	0.000000
100	26	25	18	1.000000	0.000000
100	26	25	19	1.000000	0.000000
100	26	25	20	1.000000	0.000000
100	26	25	21	1.000000	0.000000
100	26	25	22	1.000000	0.000000
100	26	25	23	1.000000	0.000000
100	26	25	24	1.000000	0.000000
100	26	25	25	1.000000	0.000000
100	26	26	0	0.000094	0.000094
100	26	26	1	0.001397	0.001303
100	26	26	2	0.009541	0.008144
100	26	26	3	0.040200	0.030658
100	26	26	4	0.118172	0.077973
100	26	26	5	0.260583	0.142410
100	26	26	6	0.454419	0.193836
100	26	26	7	0.655808	0.201389
100	26	26	8	0.818087	0.162280
100	26	26	9	0.920580	0.102492
100	26	26	10	0.971649	0.051069
100	26	26	11	0.991794	0.020145
100	26	26	12	0.998089	0.006295
100	26	26	13	0.999645	0.001556
100	26	26	14	0.999948	0.000303
100	26	26	15	0.999994	0.000046
100	26	26	16	0.999999	0.000005
100	26	26	17	1.000000	0.000000
100	26	26	18	1.000000	0.000000
100	26	26	19	1.000000	0.000000
100	26	26	20	1.000000	0.000000
100	26	26	21	1.000000	0.000000
100	26	26	22	1.000000	0.000000
100	26	26	23	1.000000	0.000000
100	26	26	24	1.000000	0.000000

Table for $N = 2$, $n = 1$, through $N = 100$, $n = 50$

N	n	k	x	P(x)	p(x)
100	26	26	25	1.000000	0.000000
100	26	26	26	1.000000	0.000000
100	27	1	0	0.730000	0.730000
100	27	1	1	1.000000	0.270000
100	27	2	0	0.530909	0.530909
100	27	2	1	0.929091	0.398182
100	27	2	2	1.000000	0.070909
100	27	3	0	0.384638	0.384638
100	27	3	1	0.823451	0.438813
100	27	3	2	0.981911	0.158460
100	27	3	3	1.000000	0.018089
100	27	4	0	0.277574	0.277574
100	27	4	1	0.705831	0.428257
100	27	4	2	0.941071	0.235240
100	27	4	3	0.995524	0.054454
100	27	4	4	1.000000	0.004476
100	27	5	0	0.199506	0.199506
100	27	5	1	0.589845	0.390338
100	27	5	2	0.879810	0.289966
100	27	5	3	0.981911	0.102101
100	27	5	4	0.998928	0.017017
100	27	5	5	1.000000	0.001072
100	27	6	0	0.142804	0.142804
100	27	6	1	0.483015	0.340211
100	27	6	2	0.803504	0.320488
100	27	6	3	0.956117	0.152613
100	27	6	4	0.994808	0.038691
100	27	6	5	0.999752	0.004944
100	27	6	6	1.000000	0.000248
100	27	7	0	0.101786	0.101786
100	27	7	1	0.388914	0.287128
100	27	7	2	0.718267	0.329353
100	27	7	3	0.917152	0.198885
100	27	7	4	0.985341	0.068189
100	27	7	5	0.998595	0.013254
100	27	7	6	0.999944	0.001350
100	27	7	7	1.000000	0.000055
100	27	8	0	0.072235	0.072235
100	27	8	1	0.308642	0.236407
100	27	8	2	0.629732	0.321090
100	27	8	3	0.865827	0.236095
100	27	8	4	0.968477	0.102650
100	27	8	5	0.995459	0.026982
100	27	8	6	0.999640	0.004180
100	27	8	7	0.999988	0.000348
100	27	8	8	1.000000	0.000012
100	27	9	0	0.051036	0.051036
100	27	9	1	0.241831	0.190796
100	27	9	2	0.542479	0.300648
100	27	9	3	0.804237	0.261758

N	n	k	x	P(x)	p(x)
100	27	9	4	0.942814	0.138578
100	27	9	5	0.989007	0.046193
100	27	9	6	0.998685	0.009678
100	27	9	7	0.999912	0.001227
100	27	9	8	0.999997	0.000085
100	27	9	9	1.000000	0.000002
100	27	10	0	0.035893	0.035893
100	27	10	1	0.187318	0.151425
100	27	10	2	0.459883	0.272565
100	27	10	3	0.735202	0.275318
100	27	10	4	0.907789	0.172588
100	27	10	5	0.977840	0.070050
100	27	10	6	0.996452	0.018612
100	27	10	7	0.999643	0.003191
100	27	10	8	0.999980	0.000337
100	27	10	9	0.999999	0.000020
100	27	10	10	1.000000	0.000000
100	27	11	0	0.025125	0.025125
100	27	11	1	0.143573	0.118448
100	27	11	2	0.384171	0.240598
100	27	11	3	0.661784	0.277613
100	27	11	4	0.863684	0.201900
100	27	11	5	0.960716	0.097033
100	27	11	6	0.992109	0.031393
100	27	11	7	0.998934	0.006825
100	27	11	8	0.999908	0.000975
100	27	11	9	0.999995	0.000087
100	27	11	10	1.000000	0.000004
100	27	11	11	1.000000	0.000000
100	27	12	0	0.017503	0.017503
100	27	12	1	0.108971	0.091468
100	27	12	2	0.316587	0.207617
100	27	12	3	0.586922	0.270334
100	27	12	4	0.811507	0.224585
100	27	12	5	0.936731	0.125223
100	27	12	6	0.984702	0.047971
100	27	12	7	0.997400	0.012698
100	27	12	8	0.999700	0.002300
100	27	12	9	0.999978	0.000278
100	27	12	10	0.999999	0.000021
100	27	12	11	1.000000	0.000001
100	27	12	12	1.000000	0.000000
100	27	13	0	0.012133	0.012133
100	27	13	1	0.081946	0.069813
100	27	13	2	0.257605	0.175659
100	27	13	3	0.513194	0.255589
100	27	13	4	0.752809	0.239615
100	27	13	5	0.905425	0.152616
100	27	13	6	0.973254	0.067829
100	27	13	7	0.994514	0.021260

N	n	k	x	P(x)	p(x)
100	27	13	8	0.999204	0.004690
100	27	13	9	0.999921	0.000717
100	27	13	10	0.999995	0.000074
100	27	13	11	1.000000	0.000005
100	27	13	12	1.000000	0.000000
100	27	13	13	1.000000	0.000000
100	27	14	0	0.008367	0.008367
100	27	14	1	0.061082	0.052715
100	27	14	2	0.207129	0.146046
100	27	14	3	0.442687	0.235558
100	27	14	4	0.689462	0.246775
100	27	14	5	0.866832	0.177370
100	27	14	6	0.956881	0.090049
100	27	14	7	0.989627	0.032745
100	27	14	8	0.998179	0.008553
100	27	14	9	0.999773	0.001593
100	27	14	10	0.999980	0.000208
100	27	14	11	0.999999	0.000018
100	27	14	12	1.000000	0.000001
100	27	14	13	1.000000	0.000000
100	27	14	14	1.000000	0.000000
100	27	15	0	0.005740	0.005740
100	27	15	1	0.045145	0.039405
100	27	15	2	0.164673	0.119528
100	27	15	3	0.376950	0.212276
100	27	15	4	0.623464	0.246515
100	27	15	5	0.821459	0.197994
100	27	15	6	0.934893	0.113434
100	27	15	7	0.982012	0.047119
100	27	15	8	0.996290	0.014278
100	27	15	9	0.999439	0.003149
100	27	15	10	0.999939	0.000500
100	27	15	11	0.999995	0.000056
100	27	15	12	1.000000	0.000004
100	27	15	13	1.000000	0.000000
100	27	15	14	1.000000	0.000000
100	27	15	15	1.000000	0.000000
100	27	16	0	0.003917	0.003917
100	27	16	1	0.033092	0.029175
100	27	16	2	0.129518	0.096426
100	27	16	3	0.317013	0.187495
100	27	16	4	0.556760	0.239747
100	27	16	5	0.770213	0.213453
100	27	16	6	0.906868	0.136655
100	27	16	7	0.970925	0.064057
100	27	16	8	0.993098	0.022174
100	27	16	9	0.998772	0.005674
100	27	16	10	0.999839	0.001067
100	27	16	11	0.999985	0.000146
100	27	16	12	0.999999	0.000014

N	n	k	x	P(x)	p(x)
100	27	16	13	1.000000	0.000001
100	27	16	14	1.000000	0.000000
100	27	16	15	1.000000	0.000000
100	27	16	16	1.000000	0.000000
100	27	17	0	0.002658	0.002658
100	27	17	1	0.024062	0.021404
100	27	17	2	0.100820	0.076758
100	27	17	3	0.263443	0.162623
100	27	17	4	0.491115	0.227672
100	27	17	5	0.714309	0.223193
100	27	17	6	0.872704	0.158395
100	27	17	7	0.955673	0.082969
100	27	17	8	0.988083	0.032410
100	27	17	9	0.997556	0.009474
100	27	17	10	0.999623	0.002067
100	27	17	11	0.999957	0.000334
100	27	17	12	0.999996	0.000039
100	27	17	13	1.000000	0.000003
100	27	17	14	1.000000	0.000000
100	27	17	15	1.000000	0.000000
100	27	17	16	1.000000	0.000000
100	27	17	17	1.000000	0.000000
100	27	18	0	0.001793	0.001793
100	27	18	1	0.017357	0.015564
100	27	18	2	0.077700	0.060343
100	27	18	3	0.216419	0.138719
100	27	18	4	0.428025	0.211606
100	27	18	5	0.655149	0.227124
100	27	18	6	0.832628	0.177479
100	27	18	7	0.935681	0.103052
100	27	18	8	0.980664	0.044983
100	27	18	9	0.995502	0.014838
100	27	18	10	0.999200	0.003698
100	27	18	11	0.999893	0.000693
100	27	18	12	0.999989	0.000096
100	27	18	13	0.999999	0.000010
100	27	18	14	1.000000	0.000001
100	27	18	15	1.000000	0.000000
100	27	18	16	1.000000	0.000000
100	27	18	17	1.000000	0.000000
100	27	18	18	1.000000	0.000000
100	27	19	0	0.001203	0.001203
100	27	19	1	0.012422	0.011219
100	27	19	2	0.059303	0.046881
100	27	19	3	0.175818	0.116516
100	27	19	4	0.368672	0.192854
100	27	19	5	0.594213	0.225541
100	27	19	6	0.787176	0.192963
100	27	19	7	0.910546	0.123370
100	27	19	8	0.970241	0.059695

Table for $N = 2$, $n = 1$, through $N = 100$, $n = 50$

N	n	k	x	P(x)	p(x)
100	27	19	9	0.992245	0.022004
100	27	19	10	0.998433	0.006189
100	27	19	11	0.999758	0.001324
100	27	19	12	0.999972	0.000214
100	27	19	13	0.999997	0.000026
100	27	19	14	1.000000	0.000002
100	27	19	15	1.000000	0.000000
100	27	19	16	1.000000	0.000000
100	27	19	17	1.000000	0.000000
100	27	19	18	1.000000	0.000000
100	27	19	19	1.000000	0.000000
100	27	20	0	0.000802	0.000802
100	27	20	1	0.008821	0.008019
100	27	20	2	0.044833	0.036012
100	27	20	3	0.141295	0.096462
100	27	20	4	0.313911	0.172616
100	27	20	5	0.532955	0.219044
100	27	20	6	0.737149	0.204193
100	27	20	7	0.880084	0.142935
100	27	20	8	0.956238	0.076154
100	27	20	9	0.987355	0.031117
100	27	20	10	0.997134	0.009780
100	27	20	11	0.999496	0.002362
100	27	20	12	0.999932	0.000436
100	27	20	13	0.999993	0.000061
100	27	20	14	0.999999	0.000006
100	27	20	15	1.000000	0.000000
100	27	20	16	1.000000	0.000000
100	27	20	17	1.000000	0.000000
100	27	20	18	1.000000	0.000000
100	27	20	19	1.000000	0.000000
100	27	20	20	1.000000	0.000000
100	27	21	0	0.000531	0.000531
100	27	21	1	0.006215	0.005683
100	27	21	2	0.033579	0.027365
100	27	21	3	0.112357	0.078777
100	27	21	4	0.264284	0.151928
100	27	21	5	0.472718	0.208434
100	27	21	6	0.683548	0.210830
100	27	21	7	0.844350	0.160802
100	27	21	8	0.938152	0.093801
100	27	21	9	0.980354	0.042202
100	27	21	10	0.995056	0.014703
100	27	21	11	0.999024	0.003967
100	27	21	12	0.999850	0.000827
100	27	21	13	0.999982	0.000132
100	27	21	14	0.999998	0.000016
100	27	21	15	1.000000	0.000001
100	27	21	16	1.000000	0.000000
100	27	21	17	1.000000	0.000000

N	n	k	x	P(x)	p(x)
100	27	21	18	1.000000	0.000000
100	27	21	19	1.000000	0.000000
100	27	21	20	1.000000	0.000000
100	27	21	21	1.000000	0.000000
100	27	22	0	0.000350	0.000350
100	27	22	1	0.004344	0.003995
100	27	22	2	0.024920	0.020576
100	27	22	3	0.088424	0.063505
100	27	22	4	0.220052	0.131628
100	27	22	5	0.414673	0.194621
100	27	22	6	0.627505	0.212831
100	27	22	7	0.803641	0.176136
100	27	22	8	0.915592	0.111951
100	27	22	9	0.970738	0.055146
100	27	22	10	0.991892	0.021154
100	27	22	11	0.998220	0.006328
100	27	22	12	0.999693	0.001473
100	27	22	13	0.999959	0.000266
100	27	22	14	0.999996	0.000037
100	27	22	15	1.000000	0.000004
100	27	22	16	1.000000	0.000000
100	27	22	17	1.000000	0.000000
100	27	22	18	1.000000	0.000000
100	27	22	19	1.000000	0.000000
100	27	22	20	1.000000	0.000000
100	27	22	21	1.000000	0.000000
100	27	22	22	1.000000	0.000000
100	27	23	0	0.000229	0.000229
100	27	23	1	0.003013	0.002784
100	27	23	2	0.018325	0.015312
100	27	23	3	0.068884	0.050559
100	27	23	4	0.181239	0.112354
100	27	23	5	0.359780	0.178541
100	27	23	6	0.570204	0.210424
100	27	23	7	0.758478	0.188274
100	27	23	8	0.888322	0.129844
100	27	23	9	0.958012	0.069690
100	27	23	10	0.987282	0.029270
100	27	23	11	0.996922	0.009640
100	27	23	12	0.999410	0.002488
100	27	23	13	0.999911	0.000501
100	27	23	14	0.999990	0.000078
100	27	23	15	0.999999	0.000009
100	27	23	16	1.000000	0.000001
100	27	23	17	1.000000	0.000000
100	27	23	18	1.000000	0.000000
100	27	23	19	1.000000	0.000000
100	27	23	20	1.000000	0.000000
100	27	23	21	1.000000	0.000000
100	27	23	22	1.000000	0.000000

N	n	k	x	P(x)	p(x)
100	27	23	23	1.000000	0.000000
100	27	24	0	0.000148	0.000148
100	27	24	1	0.002073	0.001924
100	27	24	2	0.013353	0.011281
100	27	24	3	0.053126	0.039772
100	27	24	4	0.147678	0.094553
100	27	24	5	0.308768	0.161090
100	27	24	6	0.512816	0.204047
100	27	24	7	0.709575	0.196760
100	27	24	8	0.856282	0.146707
100	27	24	9	0.941721	0.085438
100	27	24	10	0.980820	0.039099
100	27	24	11	0.994919	0.014099
100	27	24	12	0.998925	0.004006
100	27	24	13	0.999820	0.000895
100	27	24	14	0.999976	0.000156
100	27	24	15	0.999998	0.000021
100	27	24	16	1.000000	0.000002
100	27	24	17	1.000000	0.000000
100	27	24	18	1.000000	0.000000
100	27	24	19	1.000000	0.000000
100	27	24	20	1.000000	0.000000
100	27	24	21	1.000000	0.000000
100	27	24	22	1.000000	0.000000
100	27	24	23	1.000000	0.000000
100	27	24	24	1.000000	0.000000
100	27	25	0	0.000096	0.000096
100	27	25	1	0.001414	0.001319
100	27	25	2	0.009643	0.008228
100	27	25	3	0.040566	0.030923
100	27	25	4	0.119064	0.078498
100	27	25	5	0.262137	0.143073
100	27	25	6	0.456434	0.194297
100	27	25	7	0.657797	0.201362
100	27	25	8	0.819606	0.161809
100	27	25	9	0.921485	0.101880
100	27	25	10	0.972074	0.050589
100	27	25	11	0.991951	0.019877
100	27	25	12	0.998135	0.006184
100	27	25	13	0.999655	0.001521
100	27	25	14	0.999950	0.000294
100	27	25	15	0.999994	0.000045
100	27	25	16	0.999999	0.000005
100	27	25	17	1.000000	0.000000
100	27	25	18	1.000000	0.000000
100	27	25	19	1.000000	0.000000
100	27	25	20	1.000000	0.000000
100	27	25	21	1.000000	0.000000
100	27	25	22	1.000000	0.000000
100	27	25	23	1.000000	0.000000

N	n	k	x	P(x)	p(x)
100	27	25	24	1.000000	0.000000
100	27	25	25	1.000000	0.000000
100	27	26	0	0.000061	0.000061
100	27	26	1	0.000957	0.000896
100	27	26	2	0.006900	0.005943
100	27	26	3	0.030671	0.023771
100	27	26	4	0.094991	0.064321
100	27	26	5	0.220169	0.125178
100	27	26	6	0.402031	0.181862
100	27	26	7	0.604100	0.202069
100	27	26	8	0.778614	0.174514
100	27	26	9	0.897034	0.118420
100	27	26	10	0.960607	0.063573
100	27	26	11	0.987710	0.027103
100	27	26	12	0.996898	0.009188
100	27	26	13	0.999371	0.002474
100	27	26	14	0.999899	0.000527
100	27	26	15	0.999987	0.000088
100	27	26	16	0.999999	0.000012
100	27	26	17	1.000000	0.000001
100	27	26	18	1.000000	0.000000
100	27	26	19	1.000000	0.000000
100	27	26	20	1.000000	0.000000
100	27	26	21	1.000000	0.000000
100	27	26	22	1.000000	0.000000
100	27	26	23	1.000000	0.000000
100	27	26	24	1.000000	0.000000
100	27	26	25	1.000000	0.000000
100	27	26	26	1.000000	0.000000
100	27	27	0	0.000039	0.000039
100	27	27	1	0.000642	0.000604
100	27	27	2	0.004892	0.004250
100	27	27	3	0.022961	0.018069
100	27	27	4	0.075000	0.052038
100	27	27	5	0.182954	0.107954
100	27	27	6	0.350421	0.167468
100	27	27	7	0.549487	0.199065
100	27	27	8	0.733806	0.184320
100	27	27	9	0.868229	0.134423
100	27	27	10	0.946003	0.077773
100	27	27	11	0.981850	0.035848
100	27	27	12	0.995036	0.013185
100	27	27	13	0.998903	0.003868
100	27	27	14	0.999806	0.000903
100	27	27	15	0.999973	0.000167
100	27	27	16	0.999997	0.000024
100	27	27	17	1.000000	0.000003
100	27	27	18	1.000000	0.000000
100	27	27	19	1.000000	0.000000
100	27	27	20	1.000000	0.000000

Table for $N = 2$, $n = 1$, through $N = 100$, $n = 50$

N	n	k	x	P(x)	p(x)
100	27	27	21	1.000000	0.000000
100	27	27	22	1.000000	0.000000
100	27	27	23	1.000000	0.000000
100	27	27	24	1.000000	0.000000
100	27	27	25	1.000000	0.000000
100	27	27	26	1.000000	0.000000
100	27	27	27	1.000000	0.000000
100	28	1	0	0.720000	0.720000
100	28	1	1	1.000000	0.280000
100	28	2	0	0.516364	0.516364
100	28	2	1	0.923636	0.407273
100	28	2	2	1.000000	0.076364
100	28	3	0	0.368831	0.368831
100	28	3	1	0.811428	0.442597
100	28	3	2	0.979740	0.168312
100	28	3	3	1.000000	0.020260
100	28	4	0	0.262364	0.262364
100	28	4	1	0.688231	0.425867
100	28	4	2	0.934626	0.246394
100	28	4	3	0.994778	0.060153
100	28	4	4	1.000000	0.005222
100	28	5	0	0.185841	0.185841
100	28	5	1	0.568456	0.382615
100	28	5	2	0.867894	0.299438
100	28	5	3	0.979114	0.111220
100	28	5	4	0.998695	0.019581
100	28	5	5	1.000000	0.001305
100	28	6	0	0.131067	0.131067
100	28	6	1	0.459713	0.328646
100	28	6	2	0.785943	0.326229
100	28	6	3	0.949845	0.163903
100	28	6	4	0.993748	0.043903
100	28	6	5	0.999684	0.005936
100	28	6	6	1.000000	0.000316
100	28	7	0	0.092026	0.092026
100	28	7	1	0.365315	0.273289
100	28	7	2	0.695709	0.330394
100	28	7	3	0.906254	0.210545
100	28	7	4	0.982539	0.076284
100	28	7	5	0.998231	0.015693
100	28	7	6	0.999926	0.001695
100	28	7	7	1.000000	0.000074
100	28	8	0	0.064319	0.064319
100	28	8	1	0.285973	0.221654
100	28	8	2	0.603341	0.317368
100	28	8	3	0.849656	0.246315
100	28	8	4	0.962852	0.113196
100	28	8	5	0.994350	0.031498
100	28	8	6	0.999525	0.005175
100	28	8	7	0.999983	0.000458
100	28	8	8	1.000000	0.000017
100	28	9	0	0.044744	0.044744
100	28	9	1	0.220922	0.176179
100	28	9	2	0.513650	0.292727
100	28	9	3	0.782722	0.269073
100	28	9	4	0.933323	0.150600
100	28	9	5	0.986476	0.053153
100	28	9	6	0.998288	0.011812
100	28	9	7	0.999879	0.001591
100	28	9	8	0.999996	0.000118
100	28	9	9	1.000000	0.000004
100	28	10	0	0.030976	0.030976
100	28	10	1	0.168650	0.137673
100	28	10	2	0.430013	0.261364
100	28	10	3	0.708801	0.278788
100	28	10	4	0.893604	0.184803
100	28	10	5	0.973042	0.079438
100	28	10	6	0.995432	0.022390
100	28	10	7	0.999511	0.004079
100	28	10	8	0.999970	0.000459
100	28	10	9	0.999999	0.000029
100	28	10	10	1.000000	0.000001
100	28	11	0	0.021339	0.021339
100	28	11	1	0.127348	0.106008
100	28	11	2	0.354508	0.227161
100	28	11	3	0.631360	0.276852
100	28	11	4	0.844323	0.212963
100	28	11	5	0.952741	0.108418
100	28	11	6	0.989959	0.037218
100	28	11	7	0.998560	0.008601
100	28	11	8	0.999868	0.001309
100	28	11	9	0.999993	0.000125
100	28	11	10	1.000000	0.000007
100	28	11	11	1.000000	0.000000
100	28	12	0	0.014626	0.014626
100	28	12	1	0.095188	0.080562
100	28	12	2	0.288147	0.192959
100	28	12	3	0.553593	0.265446
100	28	12	4	0.786895	0.233302
100	28	12	5	0.924723	0.137828
100	28	12	6	0.980759	0.056036
100	28	12	7	0.996530	0.015771
100	28	12	8	0.999574	0.003044
100	28	12	9	0.999966	0.000392
100	28	12	10	0.999998	0.000032
100	28	12	11	1.000000	0.000001
100	28	12	12	1.000000	0.000000
100	28	13	0	0.009972	0.009972
100	28	13	1	0.070470	0.060498
100	28	13	2	0.231136	0.160666
100	28	13	3	0.478182	0.247046
100	28	13	4	0.723267	0.245085
100	28	13	5	0.888700	0.165432
100	28	13	6	0.966750	0.078050
100	28	13	7	0.992767	0.026017
100	28	13	8	0.998882	0.006116
100	28	13	9	0.999882	0.000999
100	28	13	10	0.999992	0.000110
100	28	13	11	1.000000	0.000008
100	28	13	12	1.000000	0.000000
100	28	13	13	1.000000	0.000000
100	28	14	0	0.006763	0.006763
100	28	14	1	0.051695	0.044932
100	28	14	2	0.183121	0.131426
100	28	14	3	0.407192	0.224071
100	28	14	4	0.655657	0.248466
100	28	14	5	0.844965	0.189307
100	28	14	6	0.947013	0.102048
100	28	14	7	0.986487	0.039474
100	28	14	8	0.997477	0.010990
100	28	14	9	0.999664	0.002187
100	28	14	10	0.999969	0.000306
100	28	14	11	0.999998	0.000029
100	28	14	12	1.000000	0.000002
100	28	14	13	1.000000	0.000000
100	28	14	14	1.000000	0.000000
100	28	15	0	0.004561	0.004561
100	28	15	1	0.037588	0.033027
100	28	15	2	0.143387	0.105799
100	28	15	3	0.342055	0.198667
100	28	15	4	0.586318	0.244263
100	28	15	5	0.794336	0.208018
100	28	15	6	0.920908	0.126572
100	28	15	7	0.976848	0.055940
100	28	15	8	0.994921	0.018073
100	28	15	9	0.999180	0.004260
100	28	15	10	0.999905	0.000725
100	28	15	11	0.999992	0.000087
100	28	15	12	0.999999	0.000007
100	28	15	13	1.000000	0.000000
100	28	15	14	1.000000	0.000000
100	28	15	15	1.000000	0.000000
100	28	16	0	0.003058	0.003058
100	28	16	1	0.027097	0.024039
100	28	16	2	0.111025	0.083928
100	28	16	3	0.283623	0.172598
100	28	16	4	0.517350	0.233726
100	28	16	5	0.738049	0.220699
100	28	16	6	0.888148	0.150099
100	28	16	7	0.963027	0.074879
100	28	16	8	0.990668	0.027641
100	28	16	9	0.998228	0.007560
100	28	16	10	0.999752	0.001523
100	28	16	11	0.999975	0.000223
100	28	16	12	0.999998	0.000023
100	28	16	13	1.000000	0.000002
100	28	16	14	1.000000	0.000000
100	28	16	15	1.000000	0.000000
100	28	16	16	1.000000	0.000000
100	28	17	0	0.002039	0.002039
100	28	17	1	0.019370	0.017331
100	28	17	2	0.085048	0.065677
100	28	17	3	0.232255	0.147207
100	28	17	4	0.450571	0.218316
100	28	17	5	0.677619	0.227049
100	28	17	6	0.848836	0.171217
100	28	17	7	0.944307	0.095471
100	28	17	8	0.984087	0.039780
100	28	17	9	0.996518	0.012431
100	28	17	10	0.999425	0.002907
100	28	17	11	0.999930	0.000505
100	28	17	12	0.999994	0.000064
100	28	17	13	0.999999	0.000006
100	28	17	14	1.000000	0.000000
100	28	17	15	1.000000	0.000000
100	28	17	16	1.000000	0.000000
100	28	17	17	1.000000	0.000000
100	28	18	0	0.001351	0.001351
100	28	18	1	0.013733	0.012381
100	28	18	2	0.064474	0.050742
100	28	18	3	0.187915	0.123441
100	28	18	4	0.387443	0.199528
100	28	18	5	0.614702	0.227259
100	28	18	6	0.803453	0.188751
100	28	18	7	0.920152	0.116699
100	28	18	8	0.974502	0.054350
100	28	18	9	0.993673	0.019171
100	28	18	10	0.998795	0.005122
100	28	18	11	0.999826	0.001032
100	28	18	12	0.999981	0.000155
100	28	18	13	0.999998	0.000017
100	28	18	14	1.000000	0.000001
100	28	18	15	1.000000	0.000000
100	28	18	16	1.000000	0.000000
100	28	18	17	1.000000	0.000000
100	28	18	18	1.000000	0.000000
100	28	19	0	0.000890	0.000890
100	28	19	1	0.009656	0.008766
100	28	19	2	0.048385	0.038730
100	28	19	3	0.150281	0.101896

Table for $N = 2$, $n = 1$, through $N = 100$, $n = 50$

N	n	k	x	P(x)	p(x)
100	28	19	4	0.329045	0.178764
100	28	19	5	0.550959	0.221914
100	28	19	6	0.752813	0.201854
100	28	19	7	0.890266	0.137453
100	28	19	8	0.961246	0.070980
100	28	19	9	0.989231	0.027985
100	28	19	10	0.997670	0.008440
100	28	19	11	0.999613	0.001942
100	28	19	12	0.999951	0.000339
100	28	19	13	0.999995	0.000044
100	28	19	14	1.000000	0.000004
100	28	19	15	1.000000	0.000000
100	28	19	16	1.000000	0.000000
100	28	19	17	1.000000	0.000000
100	28	19	18	1.000000	0.000000
100	28	19	19	1.000000	0.000000
100	28	20	0	0.000582	0.000582
100	28	20	1	0.006734	0.006152
100	28	20	2	0.035954	0.029220
100	28	20	3	0.118832	0.082878
100	28	20	4	0.276078	0.157246
100	28	20	5	0.487946	0.211869
100	28	20	6	0.697988	0.210042
100	28	20	7	0.854630	0.156642
100	28	20	8	0.943720	0.089090
100	28	20	9	0.982666	0.038946
100	28	20	10	0.995795	0.013129
100	28	20	11	0.999205	0.003410
100	28	20	12	0.999884	0.000679
100	28	20	13	0.999987	0.000103
100	28	20	14	0.999999	0.000012
100	28	20	15	1.000000	0.000001
100	28	20	16	1.000000	0.000000
100	28	20	17	1.000000	0.000000
100	28	20	18	1.000000	0.000000
100	28	20	19	1.000000	0.000000
100	28	20	20	1.000000	0.000000
100	28	21	0	0.000378	0.000378
100	28	21	1	0.004658	0.004279
100	28	21	2	0.026457	0.021800
100	28	21	3	0.092932	0.066475
100	28	21	4	0.228904	0.135972
100	28	21	5	0.427034	0.198130
100	28	21	6	0.640227	0.213193
100	28	21	7	0.813511	0.173285
100	28	21	8	0.921447	0.107936
100	28	21	9	0.973416	0.051969
100	28	21	10	0.992841	0.019425
100	28	21	11	0.998480	0.005639
100	28	21	12	0.999748	0.001268

N	n	k	x	P(x)	p(x)
100	28	21	13	0.999968	0.000219
100	28	21	14	0.999997	0.000029
100	28	21	15	1.000000	0.000003
100	28	21	16	1.000000	0.000000
100	28	21	17	1.000000	0.000000
100	28	21	18	1.000000	0.000000
100	28	21	19	1.000000	0.000000
100	28	21	20	1.000000	0.000000
100	28	21	21	1.000000	0.000000
100	28	22	0	0.000244	0.000244
100	28	22	1	0.003195	0.002951
100	28	22	2	0.019283	0.016087
100	28	22	3	0.071896	0.052613
100	28	22	4	0.187596	0.115700
100	28	22	5	0.369351	0.181755
100	28	22	6	0.580857	0.211506
100	28	22	7	0.767448	0.186592
100	28	22	8	0.894122	0.126673
100	28	22	9	0.960917	0.066796
100	28	22	10	0.988415	0.027498
100	28	22	11	0.997267	0.008852
100	28	22	12	0.999492	0.002225
100	28	22	13	0.999926	0.000435
100	28	22	14	0.999992	0.000065
100	28	22	15	0.999999	0.000008
100	28	22	16	1.000000	0.000001
100	28	22	17	1.000000	0.000000
100	28	22	18	1.000000	0.000000
100	28	22	19	1.000000	0.000000
100	28	22	20	1.000000	0.000000
100	28	22	21	1.000000	0.000000
100	28	22	22	1.000000	0.000000
100	28	23	0	0.000157	0.000157
100	28	23	1	0.002174	0.002017
100	28	23	2	0.013920	0.011746
100	28	23	3	0.055033	0.041112
100	28	23	4	0.151996	0.096963
100	28	23	5	0.315756	0.163760
100	28	23	6	0.521201	0.205445
100	28	23	7	0.717212	0.196011
100	28	23	8	0.861641	0.144429
100	28	23	9	0.944647	0.083005
100	28	23	10	0.982069	0.037423
100	28	23	11	0.995337	0.013268
100	28	23	12	0.999035	0.003698
100	28	23	13	0.999843	0.000807
100	28	23	14	0.999980	0.000137
100	28	23	15	0.999998	0.000018
100	28	23	16	1.000000	0.000002
100	28	23	17	1.000000	0.000000

N	n	k	x	P(x)	p(x)
100	28	23	18	1.000000	0.000000
100	28	23	19	1.000000	0.000000
100	28	23	20	1.000000	0.000000
100	28	23	21	1.000000	0.000000
100	28	23	22	1.000000	0.000000
100	28	23	23	1.000000	0.000000
100	28	24	0	0.000100	0.000100
100	28	24	1	0.001466	0.001367
100	28	24	2	0.009954	0.008487
100	28	24	3	0.041684	0.031731
100	28	24	4	0.121774	0.080089
100	28	24	5	0.266841	0.145067
100	28	24	6	0.462503	0.195662
100	28	24	7	0.663755	0.201252
100	28	24	8	0.824127	0.160373
100	28	24	9	0.924165	0.100038
100	28	24	10	0.973321	0.049156
100	28	24	11	0.992408	0.019087
100	28	24	12	0.998267	0.005859
100	28	24	13	0.999685	0.001418
100	28	24	14	0.999955	0.000270
100	28	24	15	0.999995	0.000040
100	28	24	16	0.999999	0.000005
100	28	24	17	1.000000	0.000000
100	28	24	18	1.000000	0.000000
100	28	24	19	1.000000	0.000000
100	28	24	20	1.000000	0.000000
100	28	24	21	1.000000	0.000000
100	28	24	22	1.000000	0.000000
100	28	24	23	1.000000	0.000000
100	28	24	24	1.000000	0.000000
100	28	25	0	0.000063	0.000063
100	28	25	1	0.000981	0.000918
100	28	25	2	0.007050	0.006069
100	28	25	3	0.031247	0.024196
100	28	25	4	0.096482	0.065236
100	28	25	5	0.222939	0.126457
100	28	25	6	0.405864	0.182925
100	28	25	7	0.608145	0.202282
100	28	25	8	0.781924	0.173778
100	28	25	9	0.899155	0.117232
100	28	25	10	0.961679	0.062523
100	28	25	11	0.988139	0.026460
100	28	25	12	0.997033	0.008895
100	28	25	13	0.999405	0.002372
100	28	25	14	0.999905	0.000500
100	28	25	15	0.999988	0.000083
100	28	25	16	0.999999	0.000011
100	28	25	17	1.000000	0.000001
100	28	25	18	1.000000	0.000000

N	n	k	x	P(x)	p(x)
100	28	25	19	1.000000	0.000000
100	28	25	20	1.000000	0.000000
100	28	25	21	1.000000	0.000000
100	28	25	22	1.000000	0.000000
100	28	25	23	1.000000	0.000000
100	28	25	24	1.000000	0.000000
100	28	25	25	1.000000	0.000000
100	28	26	0	0.000039	0.000039
100	28	26	1	0.000650	0.000611
100	28	26	2	0.004946	0.004296
100	28	26	3	0.023181	0.018235
100	28	26	4	0.075607	0.052426
100	28	26	5	0.184159	0.108552
100	28	26	6	0.352206	0.168047
100	28	26	7	0.551507	0.199301
100	28	26	8	0.735583	0.184076
100	28	26	9	0.869457	0.133874
100	28	26	10	0.946673	0.077216
100	28	26	11	0.982141	0.035468
100	28	26	12	0.995136	0.012995
100	28	26	13	0.998931	0.003795
100	28	26	14	0.999812	0.000881
100	28	26	15	0.999974	0.000162
100	28	26	16	0.999997	0.000023
100	28	26	17	1.000000	0.000003
100	28	26	18	1.000000	0.000000
100	28	26	19	1.000000	0.000000
100	28	26	20	1.000000	0.000000
100	28	26	21	1.000000	0.000000
100	28	26	22	1.000000	0.000000
100	28	26	23	1.000000	0.000000
100	28	26	24	1.000000	0.000000
100	28	26	25	1.000000	0.000000
100	28	26	26	1.000000	0.000000
100	28	27	0	0.000025	0.000025
100	28	27	1	0.000427	0.000403
100	28	27	2	0.003437	0.003009
100	28	27	3	0.017021	0.013584
100	28	27	4	0.058604	0.041583
100	28	27	5	0.150420	0.091816
100	28	27	6	0.302246	0.151826
100	28	27	7	0.494948	0.192702
100	28	27	8	0.685833	0.190884
100	28	27	9	0.835084	0.149251
100	28	27	10	0.927891	0.092807
100	28	27	11	0.973993	0.046102
100	28	27	12	0.992326	0.018333
100	28	27	13	0.998162	0.005835
100	28	27	14	0.999645	0.001484
100	28	27	15	0.999945	0.000300

Table for $N = 2$, $n = 1$, through $N = 100$, $n = 50$

N	n	k	x	P(x)	p(x)
100	28	27	16	0.999993	0.000048
100	28	27	17	0.999999	0.000006
100	28	27	18	1.000000	0.000001
100	28	27	19	1.000000	0.000000
100	28	27	20	1.000000	0.000000
100	28	27	21	1.000000	0.000000
100	28	27	22	1.000000	0.000000
100	28	27	23	1.000000	0.000000
100	28	27	24	1.000000	0.000000
100	28	27	25	1.000000	0.000000
100	28	27	26	1.000000	0.000000
100	28	27	27	1.000000	0.000000
100	28	28	0	0.000015	0.000015
100	28	28	1	0.000278	0.000263
100	28	28	2	0.002365	0.002087
100	28	28	3	0.012369	0.010004
100	28	28	4	0.044933	0.032564
100	28	28	5	0.121491	0.076559
100	28	28	6	0.256490	0.134999
100	28	28	7	0.439513	0.183023
100	28	28	8	0.633536	0.194022
100	28	28	9	0.796238	0.162702
100	28	28	10	0.905007	0.108769
100	28	28	11	0.963257	0.058250
100	28	28	12	0.988308	0.025051
100	28	28	13	0.996963	0.008655
100	28	28	14	0.999361	0.002398
100	28	28	15	0.999892	0.000531
100	28	28	16	0.999985	0.000093
100	28	28	17	0.999998	0.000013
100	28	28	18	1.000000	0.000001
100	28	28	19	1.000000	0.000000
100	28	28	20	1.000000	0.000000
100	28	28	21	1.000000	0.000000
100	28	28	22	1.000000	0.000000
100	28	28	23	1.000000	0.000000
100	28	28	24	1.000000	0.000000
100	28	28	25	1.000000	0.000000
100	28	28	26	1.000000	0.000000
100	28	28	27	1.000000	0.000000
100	28	28	28	1.000000	0.000000
100	29	1	0	0.710000	0.710000
100	29	1	1	1.000000	0.290000
100	29	2	0	0.502020	0.502020
100	29	2	1	0.917980	0.415960
100	29	2	2	1.000000	0.082020
100	29	3	0	0.353463	0.353463
100	29	3	1	0.799134	0.445671
100	29	3	2	0.977403	0.178268
100	29	3	3	1.000000	0.022597
100	29	4	0	0.247789	0.247789
100	29	4	1	0.670487	0.422698
100	29	4	2	0.927781	0.257295
100	29	4	3	0.993943	0.066161
100	29	4	4	1.000000	0.006057
100	29	5	0	0.172936	0.172936
100	29	5	1	0.547200	0.374264
100	29	5	2	0.855417	0.308217
100	29	5	3	0.976024	0.120607
100	29	5	4	0.998423	0.022398
100	29	5	5	1.000000	0.001577
100	29	6	0	0.120145	0.120145
100	29	6	1	0.436890	0.316746
100	29	6	2	0.767819	0.330928
100	29	6	3	0.943016	0.175197
100	29	6	4	0.992528	0.049512
100	29	6	5	0.999601	0.007073
100	29	6	6	1.000000	0.000398
100	29	7	0	0.083079	0.083079
100	29	7	1	0.342541	0.259462
100	29	7	2	0.672765	0.330224
100	29	7	3	0.894557	0.221792
100	29	7	4	0.979360	0.084803
100	29	7	5	0.997795	0.018435
100	29	7	6	0.999902	0.002107
100	29	7	7	1.000000	0.000097
100	29	8	0	0.057173	0.057173
100	29	8	1	0.264423	0.207251
100	29	8	2	0.576893	0.312470
100	29	8	3	0.832551	0.255657
100	29	8	4	0.956564	0.124013
100	29	8	5	0.993038	0.036474
100	29	8	6	0.999381	0.006343
100	29	8	7	0.999977	0.000595
100	29	8	8	1.000000	0.000023
100	29	9	0	0.039151	0.039151
100	29	9	1	0.201347	0.162196
100	29	9	2	0.485190	0.283843
100	29	9	3	0.760300	0.275110
100	29	9	4	0.922864	0.162565
100	29	9	5	0.983523	0.060658
100	29	9	6	0.997796	0.014273
100	29	9	7	0.999834	0.002039
100	29	9	8	0.999995	0.000160
100	29	9	9	1.000000	0.000005
100	29	10	0	0.026674	0.026674
100	29	10	1	0.151440	0.124766
100	29	10	2	0.400973	0.249533
100	29	10	3	0.681697	0.280724
100	29	10	4	0.878204	0.196507
100	29	10	5	0.967525	0.089321
100	29	10	6	0.994188	0.026663
100	29	10	7	0.999342	0.005153
100	29	10	8	0.999958	0.000616
100	29	10	9	0.999999	0.000041
100	29	10	10	1.000000	0.000001
100	29	11	0	0.018079	0.018079
100	29	11	1	0.112624	0.094545
100	29	11	2	0.326113	0.213489
100	29	11	3	0.600599	0.274486
100	29	11	4	0.823619	0.223020
100	29	11	5	0.943706	0.120088
100	29	11	6	0.987374	0.043668
100	29	11	7	0.998082	0.010708
100	29	11	8	0.999814	0.001732
100	29	11	9	0.999990	0.000176
100	29	11	10	1.000000	0.000010
100	29	11	11	1.000000	0.000000
100	29	12	0	0.012188	0.012188
100	29	12	1	0.082880	0.070692
100	29	12	2	0.261347	0.178467
100	29	12	3	0.520412	0.259065
100	29	12	4	0.760973	0.240561
100	29	12	5	0.911323	0.150350
100	29	12	6	0.976089	0.064766
100	29	12	7	0.995435	0.019346
100	29	12	8	0.999405	0.003970
100	29	12	9	0.999950	0.000545
100	29	12	10	0.999997	0.000047
100	29	12	11	1.000000	0.000002
100	29	12	12	1.000000	0.000000
100	29	13	0	0.008172	0.008172
100	29	13	1	0.060387	0.052215
100	29	13	2	0.206590	0.146203
100	29	13	3	0.443870	0.237280
100	29	13	4	0.692631	0.248761
100	29	13	5	0.870318	0.177687
100	29	13	6	0.959162	0.088843
100	29	13	7	0.990598	0.031437
100	29	13	8	0.998458	0.007859
100	29	13	9	0.999826	0.001369
100	29	13	10	0.999987	0.000161
100	29	13	11	0.999999	0.000012
100	29	13	12	1.000000	0.000001
100	29	13	13	1.000000	0.000000
100	29	14	0	0.005448	0.005448
100	29	14	1	0.043582	0.038134
100	29	14	2	0.161217	0.117635
100	29	14	3	0.372959	0.211742
100	29	14	4	0.621148	0.248190
100	29	14	5	0.821301	0.200153
100	29	14	6	0.935674	0.114373
100	29	14	7	0.982649	0.046975
100	29	14	8	0.996561	0.013912
100	29	14	9	0.999512	0.002951
100	29	14	10	0.999952	0.000440
100	29	14	11	0.999997	0.000045
100	29	14	12	1.000000	0.000003
100	29	14	13	1.000000	0.000000
100	29	14	14	1.000000	0.000000
100	29	15	0	0.003611	0.003611
100	29	15	1	0.031166	0.027555
100	29	15	2	0.124285	0.093119
100	29	15	3	0.308944	0.184659
100	29	15	4	0.549000	0.240057
100	29	15	5	0.765445	0.216444
100	29	15	6	0.905086	0.139642
100	29	15	7	0.970632	0.065546
100	29	15	8	0.993164	0.022531
100	29	15	9	0.998825	0.005662
100	29	15	10	0.999855	0.001029
100	29	15	11	0.999987	0.000133
100	29	15	12	0.999999	0.000012
100	29	15	13	1.000000	0.000001
100	29	15	14	1.000000	0.000000
100	29	15	15	1.000000	0.000000
100	29	16	0	0.002379	0.002379
100	29	16	1	0.022089	0.019710
100	29	16	2	0.094706	0.072617
100	29	16	3	0.252460	0.157754
100	29	16	4	0.478395	0.225936
100	29	16	5	0.704331	0.225936
100	29	16	6	0.867301	0.162970
100	29	16	7	0.953667	0.086367
100	29	16	8	0.987597	0.033930
100	29	16	9	0.997493	0.009896
100	29	16	10	0.999625	0.002131
100	29	16	11	0.999959	0.000335
100	29	16	12	0.999997	0.000037
100	29	16	13	1.000000	0.000003
100	29	16	14	1.000000	0.000000
100	29	16	15	1.000000	0.000000
100	29	16	16	1.000000	0.000000
100	29	17	0	0.001558	0.001558
100	29	17	1	0.015519	0.013961
100	29	17	2	0.071365	0.055846
100	29	17	3	0.203631	0.132266
100	29	17	4	0.411153	0.207521
100	29	17	5	0.639778	0.228625
100	29	17	6	0.822678	0.182900

Table for $N = 2$, $n = 1$, through $N = 100$, $n = 50$

N	n	k	x	P(x)	p(x)
100	29	17	7	0.931048	0.108369
100	29	17	8	0.979115	0.048067
100	29	17	9	0.995137	0.016022
100	29	17	10	0.999143	0.004006
100	29	17	11	0.999888	0.000745
100	29	17	12	0.999989	0.000102
100	29	17	13	0.999999	0.000010
100	29	17	14	1.000000	0.000001
100	29	17	15	1.000000	0.000000
100	29	17	16	1.000000	0.000000
100	29	17	17	1.000000	0.000000
100	29	18	0	0.001013	0.001013
100	29	18	1	0.010809	0.009796
100	29	18	2	0.053198	0.042389
100	29	18	3	0.162198	0.109000
100	29	18	4	0.348646	0.186448
100	29	18	5	0.573669	0.225023
100	29	18	6	0.771995	0.198326
100	29	18	7	0.902323	0.130328
100	29	18	8	0.966953	0.064630
100	29	18	9	0.991276	0.024323
100	29	18	10	0.998226	0.006949
100	29	18	11	0.999726	0.001500
100	29	18	12	0.999968	0.000242
100	29	18	13	0.999997	0.000029
100	29	18	14	1.000000	0.000002
100	29	18	15	1.000000	0.000000
100	29	18	16	1.000000	0.000000
100	29	18	17	1.000000	0.000000
100	29	18	18	1.000000	0.000000
100	29	19	0	0.000655	0.000655
100	29	19	1	0.007464	0.006809
100	29	19	2	0.039241	0.031777
100	29	19	3	0.127637	0.088397
100	29	19	4	0.291802	0.164165
100	29	19	5	0.507809	0.216007
100	29	19	6	0.716367	0.208558
100	29	19	7	0.867357	0.150990
100	29	19	8	0.950402	0.083044
100	29	19	9	0.985344	0.034942
100	29	19	10	0.996615	0.011272
100	29	19	11	0.999397	0.002781
100	29	19	12	0.999918	0.000521
100	29	19	13	0.999992	0.000073
100	29	19	14	0.999999	0.000008
100	29	19	15	1.000000	0.000001
100	29	19	16	1.000000	0.000000
100	29	19	17	1.000000	0.000000
100	29	19	18	1.000000	0.000000
100	29	19	19	1.000000	0.000000

N	n	k	x	P(x)	p(x)
100	29	20	0	0.000420	0.000420
100	29	20	1	0.005110	0.004690
100	29	20	2	0.028649	0.023538
100	29	20	3	0.099263	0.070615
100	29	20	4	0.241134	0.141871
100	29	20	5	0.443807	0.202673
100	29	20	6	0.657147	0.213340
100	29	20	7	0.826347	0.169201
100	29	20	8	0.928871	0.102524
100	29	20	9	0.976716	0.047845
100	29	20	10	0.993971	0.017255
100	29	20	11	0.998779	0.004807
100	29	20	12	0.999809	0.001030
100	29	20	13	0.999977	0.000168
100	29	20	14	0.999998	0.000021
100	29	20	15	1.000000	0.000002
100	29	20	16	1.000000	0.000000
100	29	20	17	1.000000	0.000000
100	29	20	18	1.000000	0.000000
100	29	20	19	1.000000	0.000000
100	29	20	20	1.000000	0.000000
100	29	21	0	0.000268	0.000268
100	29	21	1	0.003469	0.003201
100	29	21	2	0.020704	0.017236
100	29	21	3	0.076313	0.055609
100	29	21	4	0.196799	0.120486
100	29	21	5	0.383005	0.186206
100	29	21	6	0.595812	0.212806
100	29	21	7	0.779817	0.184006
100	29	21	8	0.901959	0.122142
100	29	21	9	0.964755	0.062796
100	29	21	10	0.989873	0.025118
100	29	21	11	0.997697	0.007824
100	29	21	12	0.999590	0.001893
100	29	21	13	0.999943	0.000354
100	29	21	14	0.999994	0.000051
100	29	21	15	0.999999	0.000005
100	29	21	16	1.000000	0.000000
100	29	21	17	1.000000	0.000000
100	29	21	18	1.000000	0.000000
100	29	21	19	1.000000	0.000000
100	29	21	20	1.000000	0.000000
100	29	21	21	1.000000	0.000000
100	29	22	0	0.000170	0.000170
100	29	22	1	0.002334	0.002165
100	29	22	2	0.014814	0.012479
100	29	22	3	0.058012	0.043198
100	29	22	4	0.158671	0.100659
100	29	22	5	0.326436	0.167765
100	29	22	6	0.533855	0.207419

N	n	k	x	P(x)	p(x)
100	29	22	7	0.728575	0.194720
100	29	22	8	0.869491	0.140916
100	29	22	9	0.948857	0.079366
100	29	22	10	0.983832	0.034975
100	29	22	11	0.995914	0.012082
100	29	22	12	0.999183	0.003268
100	29	22	13	0.999872	0.000689
100	29	22	14	0.999984	0.000113
100	29	22	15	0.999998	0.000014
100	29	22	16	1.000000	0.000001
100	29	22	17	1.000000	0.000000
100	29	22	18	1.000000	0.000000
100	29	22	19	1.000000	0.000000
100	29	22	20	1.000000	0.000000
100	29	22	21	1.000000	0.000000
100	29	22	22	1.000000	0.000000
100	29	23	0	0.000107	0.000107
100	29	23	1	0.001557	0.001451
100	29	23	2	0.010494	0.008937
100	29	23	3	0.043612	0.033118
100	29	23	4	0.126408	0.082796
100	29	23	5	0.274816	0.148408
100	29	23	6	0.472693	0.197877
100	29	23	7	0.673654	0.200961
100	29	23	8	0.831552	0.157898
100	29	23	9	0.928507	0.096955
100	29	23	10	0.975313	0.046806
100	29	23	11	0.993126	0.017814
100	29	23	12	0.998470	0.005344
100	29	23	13	0.999730	0.001260
100	29	23	14	0.999963	0.000232
100	29	23	15	0.999996	0.000033
100	29	23	16	1.000000	0.000004
100	29	23	17	1.000000	0.000000
100	29	23	18	1.000000	0.000000
100	29	23	19	1.000000	0.000000
100	29	23	20	1.000000	0.000000
100	29	23	21	1.000000	0.000000
100	29	23	22	1.000000	0.000000
100	29	23	23	1.000000	0.000000
100	29	24	0	0.000066	0.000066
100	29	24	1	0.001030	0.000963
100	29	24	2	0.007360	0.006331
100	29	24	3	0.032430	0.025069
100	29	24	4	0.099527	0.067097
100	29	24	5	0.228559	0.129033
100	29	24	6	0.413587	0.185028
100	29	24	7	0.616237	0.202650
100	29	24	8	0.788489	0.172252
100	29	24	9	0.903324	0.114835

N	n	k	x	P(x)	p(x)
100	29	24	10	0.963763	0.060439
100	29	24	11	0.988962	0.025199
100	29	24	12	0.997290	0.008328
100	29	24	13	0.999469	0.002178
100	29	24	14	0.999917	0.000449
100	29	24	15	0.999990	0.000072
100	29	24	16	0.999999	0.000009
100	29	24	17	1.000000	0.000001
100	29	24	18	1.000000	0.000000
100	29	24	19	1.000000	0.000000
100	29	24	20	1.000000	0.000000
100	29	24	21	1.000000	0.000000
100	29	24	22	1.000000	0.000000
100	29	24	23	1.000000	0.000000
100	29	24	24	1.000000	0.000000
100	29	25	0	0.000041	0.000041
100	29	25	1	0.000675	0.000634
100	29	25	2	0.005111	0.004436
100	29	25	3	0.023853	0.018742
100	29	25	4	0.077455	0.053602
100	29	25	5	0.187812	0.110357
100	29	25	6	0.357592	0.169780
100	29	25	7	0.557575	0.199983
100	29	25	8	0.740893	0.183318
100	29	25	9	0.873104	0.132211
100	29	25	10	0.948653	0.075549
100	29	25	11	0.982994	0.034341
100	29	25	12	0.995427	0.012434
100	29	25	13	0.999010	0.003583
100	29	25	14	0.999829	0.000819
100	29	25	15	0.999977	0.000148
100	29	25	16	0.999997	0.000021
100	29	25	17	1.000000	0.000002
100	29	25	18	1.000000	0.000000
100	29	25	19	1.000000	0.000000
100	29	25	20	1.000000	0.000000
100	29	25	21	1.000000	0.000000
100	29	25	22	1.000000	0.000000
100	29	25	23	1.000000	0.000000
100	29	25	24	1.000000	0.000000
100	29	25	25	1.000000	0.000000
100	29	26	0	0.000025	0.000025
100	29	26	1	0.000438	0.000413
100	29	26	2	0.003514	0.003076
100	29	26	3	0.017356	0.013842
100	29	26	4	0.059588	0.042232
100	29	26	5	0.152498	0.092910
100	29	26	6	0.305526	0.153028
100	29	26	7	0.498913	0.193387
100	29	26	8	0.689564	0.190651

Table for $N = 2$, $n = 1$, through $N = 100$, $n = 50$

N	n	k	x	P(x)	p(x)
100	29	26	9	0.837848	0.148284
100	29	26	10	0.929514	0.091666
100	29	26	11	0.974752	0.045238
100	29	26	12	0.992609	0.017857
100	29	26	13	0.998246	0.005637
100	29	26	14	0.999665	0.001419
100	29	26	15	0.999949	0.000284
100	29	26	16	0.999994	0.000045
100	29	26	17	0.999999	0.000006
100	29	26	18	1.000000	0.000001
100	29	26	19	1.000000	0.000000
100	29	26	20	1.000000	0.000000
100	29	26	21	1.000000	0.000000
100	29	26	22	1.000000	0.000000
100	29	26	23	1.000000	0.000000
100	29	26	24	1.000000	0.000000
100	29	26	25	1.000000	0.000000
100	29	26	26	1.000000	0.000000
100	29	27	0	0.000015	0.000015
100	29	27	1	0.000282	0.000267
100	29	27	2	0.002392	0.002110
100	29	27	3	0.012493	0.010101
100	29	27	4	0.045320	0.032828
100	29	27	5	0.122365	0.077044
100	29	27	6	0.257963	0.135598
100	29	27	7	0.441420	0.183457
100	29	27	8	0.635460	0.194041
100	29	27	9	0.797771	0.162311
100	29	27	10	0.905978	0.108207
100	29	27	11	0.963748	0.057770
100	29	27	12	0.988507	0.024759
100	29	27	13	0.997027	0.008520
100	29	27	14	0.999377	0.002350
100	29	27	15	0.999895	0.000518
100	29	27	16	0.999986	0.000091
100	29	27	17	0.999998	0.000012
100	29	27	18	1.000000	0.000001
100	29	27	19	1.000000	0.000000
100	29	27	20	1.000000	0.000000
100	29	27	21	1.000000	0.000000
100	29	27	22	1.000000	0.000000
100	29	27	23	1.000000	0.000000
100	29	27	24	1.000000	0.000000
100	29	27	25	1.000000	0.000000
100	29	27	26	1.000000	0.000000
100	29	27	27	1.000000	0.000000
100	29	28	0	0.000009	0.000009
100	29	28	1	0.000180	0.000170
100	29	28	2	0.001612	0.001432
100	29	28	3	0.008895	0.007284

N	n	k	x	P(x)	p(x)
100	29	28	4	0.034078	0.025183
100	29	28	5	0.097035	0.062957
100	29	28	6	0.215240	0.118205
100	29	28	7	0.386131	0.170891
100	29	28	8	0.579640	0.193509
100	29	28	9	0.753302	0.173662
100	29	28	10	0.877815	0.124512
100	29	28	11	0.949504	0.071689
100	29	28	12	0.982741	0.033238
100	29	28	13	0.995160	0.012418
100	29	28	14	0.998894	0.003735
100	29	28	15	0.999796	0.000902
100	29	28	16	0.999970	0.000174
100	29	28	17	0.999996	0.000027
100	29	28	18	1.000000	0.000003
100	29	28	19	1.000000	0.000000
100	29	28	20	1.000000	0.000000
100	29	28	21	1.000000	0.000000
100	29	28	22	1.000000	0.000000
100	29	28	23	1.000000	0.000000
100	29	28	24	1.000000	0.000000
100	29	28	25	1.000000	0.000000
100	29	28	26	1.000000	0.000000
100	29	28	27	1.000000	0.000000
100	29	28	28	1.000000	0.000000
100	29	29	0	0.000006	0.000006
100	29	29	1	0.000113	0.000108
100	29	29	2	0.001075	0.000961
100	29	29	3	0.006265	0.005190
100	29	29	4	0.025334	0.019069
100	29	29	5	0.076049	0.050715
100	29	29	6	0.177480	0.101431
100	29	29	7	0.333915	0.156434
100	29	29	8	0.523200	0.189286
100	29	29	9	0.705063	0.181863
100	29	29	10	0.844957	0.139894
100	29	29	11	0.931581	0.086624
100	29	29	12	0.974893	0.043312
100	29	29	13	0.992400	0.017507
100	29	29	14	0.998116	0.005716
100	29	29	15	0.999621	0.001504
100	29	29	16	0.999938	0.000318
100	29	29	17	0.999992	0.000054
100	29	29	18	0.999999	0.000007
100	29	29	19	1.000000	0.000001
100	29	29	20	1.000000	0.000000
100	29	29	21	1.000000	0.000000
100	29	29	22	1.000000	0.000000
100	29	29	23	1.000000	0.000000
100	29	29	24	1.000000	0.000000

N	n	k	x	P(x)	p(x)
100	29	29	25	1.000000	0.000000
100	29	29	26	1.000000	0.000000
100	29	29	27	1.000000	0.000000
100	29	29	28	1.000000	0.000000
100	29	29	29	1.000000	0.000000
100	30	1	0	0.700000	0.700000
100	30	1	1	1.000000	0.300000
100	30	2	0	0.487879	0.487879
100	30	2	1	0.912121	0.424242
100	30	2	2	1.000000	0.087879
100	30	3	0	0.338528	0.338528
100	30	3	1	0.786580	0.448052
100	30	3	2	0.974892	0.188312
100	30	3	3	1.000000	0.025108
100	30	4	0	0.233829	0.233829
100	30	4	1	0.652626	0.418798
100	30	4	2	0.920534	0.267907
100	30	4	3	0.993011	0.072477
100	30	4	4	1.000000	0.006989
100	30	5	0	0.160757	0.160757
100	30	5	1	0.526115	0.365357
100	30	5	2	0.842394	0.316279
100	30	5	3	0.972627	0.130233
100	30	5	4	0.998107	0.025480
100	30	5	5	1.000000	0.001893
100	30	6	0	0.109992	0.109992
100	30	6	1	0.414584	0.304593
100	30	6	2	0.749175	0.334590
100	30	6	3	0.935613	0.186438
100	30	6	4	0.991134	0.055520
100	30	6	5	0.999502	0.008368
100	30	6	6	1.000000	0.000498
100	30	7	0	0.074888	0.074888
100	30	7	1	0.320614	0.245726
100	30	7	2	0.649510	0.328895
100	30	7	3	0.882062	0.232552
100	30	7	4	0.975777	0.093715
100	30	7	5	0.997276	0.021499
100	30	7	6	0.999873	0.002597
100	30	7	7	1.000000	0.000127
100	30	8	0	0.050731	0.050731
100	30	8	1	0.243990	0.193259
100	30	8	2	0.550487	0.306497
100	30	8	3	0.814547	0.264059
100	30	8	4	0.949577	0.135030
100	30	8	5	0.991497	0.041920
100	30	8	6	0.999203	0.007706
100	30	8	7	0.999968	0.000766
100	30	8	8	1.000000	0.000031
100	30	9	0	0.034188	0.034188

N	n	k	x	P(x)	p(x)
100	30	9	1	0.183071	0.148883
100	30	9	2	0.457206	0.274134
100	30	9	3	0.737051	0.279845
100	30	9	4	0.911416	0.174365
100	30	9	5	0.980106	0.068689
100	30	9	6	0.997192	0.017087
100	30	9	7	0.999777	0.002585
100	30	9	8	0.999992	0.000215
100	30	9	9	1.000000	0.000008
100	30	10	0	0.022917	0.022917
100	30	10	1	0.135625	0.112708
100	30	10	2	0.372857	0.237232
100	30	10	3	0.654020	0.281163
100	30	10	4	0.861598	0.207578
100	30	10	5	0.961235	0.099637
100	30	10	6	0.992686	0.031451
100	30	10	7	0.999124	0.006438
100	30	10	8	0.999940	0.000817
100	30	10	9	0.999998	0.000058
100	30	10	10	1.000000	0.000002
100	30	11	0	0.015278	0.015278
100	30	11	1	0.099308	0.084030
100	30	11	2	0.299051	0.199743
100	30	11	3	0.569671	0.270620
100	30	11	4	0.801631	0.231960
100	30	11	5	0.933558	0.131927
100	30	11	6	0.984299	0.050741
100	30	11	7	0.997479	0.013180
100	30	11	8	0.999741	0.002262
100	30	11	9	0.999985	0.000244
100	30	11	10	0.999999	0.000015
100	30	11	11	1.000000	0.000000
100	30	12	0	0.010128	0.010128
100	30	12	1	0.071928	0.061799
100	30	12	2	0.236211	0.164283
100	30	12	3	0.487573	0.251362
100	30	12	4	0.733867	0.246294
100	30	12	5	0.896500	0.162632
100	30	12	6	0.970616	0.074116
100	30	12	7	0.994073	0.023457
100	30	12	8	0.999182	0.005109
100	30	12	9	0.999927	0.000746
100	30	12	10	0.999996	0.000069
100	30	12	11	1.000000	0.000004
100	30	12	12	1.000000	0.000000
100	30	13	0	0.006675	0.006675
100	30	13	1	0.051562	0.044886
100	30	13	2	0.183939	0.132377
100	30	13	3	0.410450	0.226512
100	30	13	4	0.661098	0.250648

N	n	k	x	P(x)	p(x)
100	30	13	5	0.850297	0.189199
100	30	13	6	0.950402	0.100105
100	30	13	7	0.987942	0.037539
100	30	13	8	0.997904	0.009962
100	30	13	9	0.999749	0.001845
100	30	13	10	0.999981	0.000231
100	30	13	11	0.999999	0.000019
100	30	13	12	1.000000	0.000001
100	30	13	13	1.000000	0.000000
100	30	14	0	0.004374	0.004374
100	30	14	1	0.036600	0.032226
100	30	14	2	0.141335	0.104735
100	30	14	3	0.340154	0.198819
100	30	14	4	0.586192	0.246038
100	30	14	5	0.795930	0.209738
100	30	14	6	0.922787	0.126857
100	30	14	7	0.978018	0.055230
100	30	14	8	0.995385	0.017367
100	30	14	9	0.999304	0.003919
100	30	14	10	0.999927	0.000623
100	30	14	11	0.999995	0.000068
100	30	14	12	1.000000	0.000005
100	30	14	13	1.000000	0.000000
100	30	14	14	1.000000	0.000000
100	30	15	0	0.002848	0.002848
100	30	15	1	0.025733	0.022885
100	30	15	2	0.107235	0.081502
100	30	15	3	0.277734	0.170499
100	30	15	4	0.511808	0.234075
100	30	15	5	0.734959	0.223151
100	30	15	6	0.887385	0.152426
100	30	15	7	0.963247	0.075862
100	30	15	8	0.990942	0.027695
100	30	15	9	0.998347	0.007405
100	30	15	10	0.999782	0.001435
100	30	15	11	0.999980	0.000198
100	30	15	12	0.999999	0.000019
100	30	15	13	1.000000	0.000001
100	30	15	14	1.000000	0.000000
100	30	15	15	1.000000	0.000000
100	30	16	0	0.001843	0.001843
100	30	16	1	0.017925	0.016082
100	30	16	2	0.080387	0.062462
100	30	16	3	0.223575	0.143188
100	30	16	4	0.440209	0.216634
100	30	16	5	0.669327	0.229118
100	30	16	6	0.844347	0.175021
100	30	16	7	0.942719	0.098372
100	30	16	8	0.983774	0.041055
100	30	16	9	0.996518	0.012744
100	30	16	10	0.999445	0.002927
100	30	16	11	0.999936	0.000491
100	30	16	12	0.999995	0.000059
100	30	16	13	1.000000	0.000005
100	30	16	14	1.000000	0.000000
100	30	16	15	1.000000	0.000000
100	30	16	16	1.000000	0.000000
100	30	17	0	0.001185	0.001185
100	30	17	1	0.012373	0.011188
100	30	17	2	0.059566	0.047194
100	30	17	3	0.177551	0.117984
100	30	17	4	0.373156	0.195605
100	30	17	5	0.601137	0.227981
100	30	17	6	0.794341	0.193204
100	30	17	7	0.915784	0.121443
100	30	17	8	0.973022	0.057237
100	30	17	9	0.993332	0.020310
100	30	17	10	0.998748	0.005416
100	30	17	11	0.999825	0.001077
100	30	17	12	0.999982	0.000157
100	30	17	13	0.999999	0.000017
100	30	17	14	1.000000	0.000001
100	30	17	15	1.000000	0.000000
100	30	17	16	1.000000	0.000000
100	30	17	17	1.000000	0.000000
100	30	18	0	0.000756	0.000756
100	30	18	1	0.008464	0.007707
100	30	18	2	0.043646	0.035182
100	30	18	3	0.139170	0.095524
100	30	18	4	0.311882	0.172712
100	30	18	5	0.532468	0.220586
100	30	18	6	0.738475	0.206007
100	30	18	7	0.882131	0.143656
100	30	18	8	0.957850	0.075719
100	30	18	9	0.988193	0.030343
100	30	18	10	0.997443	0.009250
100	30	18	11	0.999578	0.002136
100	30	18	12	0.999948	0.000370
100	30	18	13	0.999995	0.000047
100	30	18	14	1.000000	0.000004
100	30	18	15	1.000000	0.000000
100	30	18	16	1.000000	0.000000
100	30	18	17	1.000000	0.000000
100	30	18	18	1.000000	0.000000
100	30	19	0	0.000480	0.000480
100	30	19	1	0.005738	0.005258
100	30	19	2	0.031632	0.025894
100	30	19	3	0.107717	0.076085
100	30	19	4	0.257120	0.149403
100	30	19	5	0.465216	0.208096
100	30	19	6	0.678180	0.212964
100	30	19	7	0.841838	0.163657
100	30	19	8	0.937535	0.095698
100	30	19	9	0.980422	0.042887
100	30	19	10	0.995187	0.014764
100	30	19	11	0.999083	0.003897
100	30	19	12	0.999867	0.000783
100	30	19	13	0.999985	0.000119
100	30	19	14	0.999999	0.000013
100	30	19	15	1.000000	0.000001
100	30	19	16	1.000000	0.000000
100	30	19	17	1.000000	0.000000
100	30	19	18	1.000000	0.000000
100	30	19	19	1.000000	0.000000
100	30	20	0	0.000302	0.000302
100	30	20	1	0.003855	0.003553
100	30	20	2	0.022681	0.018826
100	30	20	3	0.082355	0.059674
100	30	20	4	0.209163	0.126808
100	30	20	5	0.400989	0.191826
100	30	20	6	0.615080	0.214091
100	30	20	7	0.795367	0.180287
100	30	20	8	0.911543	0.116176
100	30	20	9	0.969303	0.057760
100	30	20	10	0.991541	0.022238
100	30	20	11	0.998169	0.006628
100	30	20	12	0.999693	0.001523
100	30	20	13	0.999961	0.000268
100	30	20	14	0.999996	0.000036
100	30	20	15	1.000000	0.000004
100	30	20	16	1.000000	0.000000
100	30	20	17	1.000000	0.000000
100	30	20	18	1.000000	0.000000
100	30	20	19	1.000000	0.000000
100	30	20	20	1.000000	0.000000
100	30	21	0	0.000189	0.000189
100	30	21	1	0.002567	0.002379
100	30	21	2	0.016092	0.013525
100	30	21	3	0.062215	0.046123
100	30	21	4	0.167951	0.105735
100	30	21	5	0.341043	0.173093
100	30	21	6	0.550853	0.209809
100	30	21	7	0.743534	0.192682
100	30	21	8	0.879595	0.136060
100	30	21	9	0.954142	0.074547
100	30	21	10	0.985982	0.031840
100	30	21	11	0.996595	0.010613
100	30	21	12	0.999350	0.002755
100	30	21	13	0.999904	0.000554
100	30	21	14	0.999989	0.000085
100	30	21	15	0.999999	0.000010
100	30	21	16	1.000000	0.000001
100	30	21	17	1.000000	0.000000
100	30	21	18	1.000000	0.000000
100	30	21	19	1.000000	0.000000
100	30	21	20	1.000000	0.000000
100	30	21	21	1.000000	0.000000
100	30	22	0	0.000117	0.000117
100	30	22	1	0.001694	0.001577
100	30	22	2	0.011299	0.009604
100	30	22	3	0.046452	0.035153
100	30	22	4	0.133152	0.086700
100	30	22	5	0.286267	0.153115
100	30	22	6	0.487113	0.200846
100	30	22	7	0.687437	0.200324
100	30	22	8	0.841705	0.154268
100	30	22	9	0.934325	0.092621
100	30	22	10	0.977921	0.043596
100	30	22	11	0.994042	0.016122
100	30	22	12	0.998722	0.004680
100	30	22	13	0.999784	0.001062
100	30	22	14	0.999972	0.000187
100	30	22	15	0.999997	0.000025
100	30	22	16	1.000000	0.000003
100	30	22	17	1.000000	0.000000
100	30	22	18	1.000000	0.000000
100	30	22	19	1.000000	0.000000
100	30	22	20	1.000000	0.000000
100	30	22	21	1.000000	0.000000
100	30	22	22	1.000000	0.000000
100	30	23	0	0.000072	0.000072
100	30	23	1	0.001108	0.001036
100	30	23	2	0.007851	0.006743
100	30	23	3	0.034283	0.026433
100	30	23	4	0.104252	0.069968
100	30	23	5	0.237192	0.132940
100	30	23	6	0.425314	0.188123
100	30	23	7	0.628367	0.203053
100	30	23	8	0.798193	0.169826
100	30	23	9	0.909389	0.111196
100	30	23	10	0.966742	0.057354
100	30	23	11	0.990115	0.023373
100	30	23	12	0.997642	0.007527
100	30	23	13	0.999553	0.001911
100	30	23	14	0.999933	0.000380
100	30	23	15	0.999992	0.000059
100	30	23	16	0.999999	0.000007
100	30	23	17	1.000000	0.000001
100	30	23	18	1.000000	0.000000
100	30	23	19	1.000000	0.000000

Table for $N = 2$, $n = 1$, through $N = 100$, $n = 50$

N	n	k	x	P(x)	p(x)
100	30	23	20	1.000000	0.000000
100	30	23	21	1.000000	0.000000
100	30	23	22	1.000000	0.000000
100	30	23	23	1.000000	0.000000
100	30	24	0	0.000044	0.000044
100	30	24	1	0.000718	0.000674
100	30	24	2	0.005399	0.004681
100	30	24	3	0.025015	0.019616
100	30	24	4	0.080626	0.055611
100	30	24	5	0.194029	0.113403
100	30	24	6	0.366679	0.172649
100	30	24	7	0.567715	0.201036
100	30	24	8	0.749672	0.181957
100	30	24	9	0.879063	0.129391
100	30	24	10	0.951845	0.072783
100	30	24	11	0.984348	0.032503
100	30	24	12	0.995883	0.011535
100	30	24	13	0.999131	0.003248
100	30	24	14	0.999854	0.000723
100	30	24	15	0.999981	0.000126
100	30	24	16	0.999998	0.000017
100	30	24	17	1.000000	0.000002
100	30	24	18	1.000000	0.000000
100	30	24	19	1.000000	0.000000
100	30	24	20	1.000000	0.000000
100	30	24	21	1.000000	0.000000
100	30	24	22	1.000000	0.000000
100	30	24	23	1.000000	0.000000
100	30	24	24	1.000000	0.000000
100	30	25	0	0.000027	0.000027
100	30	25	1	0.000461	0.000434
100	30	25	2	0.003674	0.003214
100	30	25	3	0.018046	0.014372
100	30	25	4	0.061601	0.043555
100	30	25	5	0.156726	0.095125
100	30	25	6	0.312158	0.155432
100	30	25	7	0.506875	0.194717
100	30	25	8	0.697000	0.190125
100	30	25	9	0.843310	0.146310
100	30	25	10	0.932692	0.089382
100	30	25	11	0.976222	0.043530
100	30	25	12	0.993151	0.016928
100	30	25	13	0.998405	0.005254
100	30	25	14	0.999702	0.001298
100	30	25	15	0.999956	0.000254
100	30	25	16	0.999995	0.000039
100	30	25	17	0.999999	0.000005
100	30	25	18	1.000000	0.000000
100	30	25	19	1.000000	0.000000
100	30	25	20	1.000000	0.000000

N	n	k	x	P(x)	p(x)
100	30	25	21	1.000000	0.000000
100	30	25	22	1.000000	0.000000
100	30	25	23	1.000000	0.000000
100	30	25	24	1.000000	0.000000
100	30	25	25	1.000000	0.000000
100	30	26	0	0.000016	0.000016
100	30	26	1	0.000293	0.000277
100	30	26	2	0.002474	0.002182
100	30	26	3	0.012872	0.010398
100	30	26	4	0.046502	0.033630
100	30	26	5	0.125017	0.078515
100	30	26	6	0.262420	0.137402
100	30	26	7	0.447162	0.184742
100	30	26	8	0.641230	0.194068
100	30	26	9	0.802343	0.161113
100	30	26	10	0.908857	0.106514
100	30	26	11	0.965195	0.056338
100	30	26	12	0.989088	0.023893
100	30	26	13	0.997214	0.008126
100	30	26	14	0.999425	0.002212
100	30	26	15	0.999905	0.000480
100	30	26	16	0.999988	0.000082
100	30	26	17	0.999999	0.000011
100	30	26	18	1.000000	0.000001
100	30	26	19	1.000000	0.000000
100	30	26	20	1.000000	0.000000
100	30	26	21	1.000000	0.000000
100	30	26	22	1.000000	0.000000
100	30	26	23	1.000000	0.000000
100	30	26	24	1.000000	0.000000
100	30	26	25	1.000000	0.000000
100	30	26	26	1.000000	0.000000
100	30	27	0	0.000009	0.000009
100	30	27	1	0.000184	0.000175
100	30	27	2	0.001649	0.001465
100	30	27	3	0.009078	0.007429
100	30	27	4	0.034686	0.025608
100	30	27	5	0.098492	0.063806
100	30	27	6	0.217857	0.119365
100	30	27	7	0.389742	0.171885
100	30	27	8	0.583534	0.193792
100	30	27	9	0.756622	0.173088
100	30	27	10	0.880069	0.123448
100	30	27	11	0.950730	0.070660
100	30	27	12	0.983276	0.032547
100	30	27	13	0.995347	0.012071
100	30	27	14	0.998947	0.003600
100	30	27	15	0.999808	0.000861
100	30	27	16	0.999972	0.000164
100	30	27	17	0.999997	0.000025

N	n	k	x	P(x)	p(x)
100	30	27	18	1.000000	0.000003
100	30	27	19	1.000000	0.000000
100	30	27	20	1.000000	0.000000
100	30	27	21	1.000000	0.000000
100	30	27	22	1.000000	0.000000
100	30	27	23	1.000000	0.000000
100	30	27	24	1.000000	0.000000
100	30	27	25	1.000000	0.000000
100	30	27	26	1.000000	0.000000
100	30	27	27	1.000000	0.000000
100	30	28	0	0.000006	0.000006
100	30	28	1	0.000115	0.000109
100	30	28	2	0.001087	0.000972
100	30	28	3	0.006331	0.005243
100	30	28	4	0.025566	0.019235
100	30	28	5	0.076641	0.051075
100	30	28	6	0.178613	0.101973
100	30	28	7	0.335586	0.156973
100	30	28	8	0.525131	0.189545
100	30	28	9	0.706829	0.181699
100	30	28	10	0.846248	0.139419
100	30	28	11	0.932339	0.086090
100	30	28	12	0.975251	0.042912
100	30	28	13	0.992536	0.017285
100	30	28	14	0.998158	0.005622
100	30	28	15	0.999631	0.001473
100	30	28	16	0.999940	0.000309
100	30	28	17	0.999992	0.000052
100	30	28	18	0.999999	0.000007
100	30	28	19	1.000000	0.000001
100	30	28	20	1.000000	0.000000
100	30	28	21	1.000000	0.000000
100	30	28	22	1.000000	0.000000
100	30	28	23	1.000000	0.000000
100	30	28	24	1.000000	0.000000
100	30	28	25	1.000000	0.000000
100	30	28	26	1.000000	0.000000
100	30	28	27	1.000000	0.000000
100	30	28	28	1.000000	0.000000
100	30	29	0	0.000003	0.000003
100	30	29	1	0.000071	0.000068
100	30	29	2	0.000709	0.000638
100	30	29	3	0.004364	0.003655
100	30	29	4	0.018620	0.014255
100	30	29	5	0.058906	0.040287
100	30	29	6	0.144623	0.085716
100	30	29	7	0.285442	0.140820
100	30	29	8	0.467215	0.181772
100	30	29	9	0.653834	0.186620
100	30	29	10	0.807521	0.153687

N	n	k	x	P(x)	p(x)
100	30	29	11	0.909620	0.102100
100	30	29	12	0.964523	0.054903
100	30	29	13	0.988455	0.023932
100	30	29	14	0.996909	0.008454
100	30	29	15	0.999324	0.002415
100	30	29	16	0.999880	0.000556
100	30	29	17	0.999983	0.000103
100	30	29	18	0.999998	0.000015
100	30	29	19	1.000000	0.000002
100	30	29	20	1.000000	0.000000
100	30	29	21	1.000000	0.000000
100	30	29	22	1.000000	0.000000
100	30	29	23	1.000000	0.000000
100	30	29	24	1.000000	0.000000
100	30	29	25	1.000000	0.000000
100	30	29	26	1.000000	0.000000
100	30	29	27	1.000000	0.000000
100	30	29	28	1.000000	0.000000
100	30	29	29	1.000000	0.000000
100	30	30	0	0.000002	0.000002
100	30	30	1	0.000043	0.000041
100	30	30	2	0.000457	0.000414
100	30	30	3	0.002974	0.002517
100	30	30	4	0.013399	0.010425
100	30	30	5	0.044721	0.031321
100	30	30	6	0.115648	0.070927
100	30	30	7	0.239825	0.124177
100	30	30	8	0.410891	0.171066
100	30	30	9	0.598637	0.187746
100	30	30	10	0.764229	0.165592
100	30	30	11	0.882298	0.118069
100	30	30	12	0.950604	0.068306
100	30	30	13	0.982725	0.032121
100	30	30	14	0.995004	0.012279
100	30	30	15	0.998814	0.003810
100	30	30	16	0.999771	0.000957
100	30	30	17	0.999964	0.000194
100	30	30	18	0.999995	0.000031
100	30	30	19	0.999999	0.000004
100	30	30	20	1.000000	0.000000
100	30	30	21	1.000000	0.000000
100	30	30	22	1.000000	0.000000
100	30	30	23	1.000000	0.000000
100	30	30	24	1.000000	0.000000
100	30	30	25	1.000000	0.000000
100	30	30	26	1.000000	0.000000
100	30	30	27	1.000000	0.000000
100	30	30	28	1.000000	0.000000
100	30	30	29	1.000000	0.000000
100	30	30	30	1.000000	0.000000

N	n	k	x	P(x)	p(x)
100	31	1	0	0.690000	0.690000
100	31	1	1	1.000000	0.310000
100	31	2	0	0.473939	0.473939
100	31	2	1	0.906061	0.432121
100	31	2	2	1.000000	0.093939
100	31	3	0	0.324020	0.324020
100	31	3	1	0.773779	0.449759
100	31	3	2	0.972202	0.198423
100	31	3	3	1.000000	0.027798
100	31	4	0	0.220467	0.220467
100	31	4	1	0.634678	0.414211
100	31	4	2	0.912879	0.278201
100	31	4	3	0.991976	0.079096
100	31	4	4	1.000000	0.008024
100	31	5	0	0.149275	0.149275
100	31	5	1	0.505237	0.355962
100	31	5	2	0.828839	0.323602
100	31	5	3	0.968906	0.140067
100	31	5	4	0.997743	0.028837
100	31	5	5	1.000000	0.002257
100	31	6	0	0.100564	0.100564
100	31	6	1	0.392828	0.292264
100	31	6	2	0.730055	0.337228
100	31	6	3	0.927623	0.197568
100	31	6	4	0.989547	0.061924
100	31	6	5	0.999382	0.009835
100	31	6	6	1.000000	0.000618
100	31	7	0	0.067399	0.067399
100	31	7	1	0.299552	0.232153
100	31	7	2	0.626017	0.326465
100	31	7	3	0.868773	0.242756
100	31	7	4	0.971760	0.102987
100	31	7	5	0.996662	0.024901
100	31	7	6	0.999836	0.003174
100	31	7	7	1.000000	0.000164
100	31	8	0	0.044933	0.044933
100	31	8	1	0.224664	0.179731
100	31	8	2	0.524216	0.299552
100	31	8	3	0.795685	0.271469
100	31	8	4	0.941861	0.146176
100	31	8	5	0.989700	0.047839
100	31	8	6	0.998983	0.009282
100	31	8	7	0.999958	0.000975
100	31	8	8	1.000000	0.000042
100	31	9	0	0.029792	0.029792
100	31	9	1	0.166056	0.136264
100	31	9	2	0.429792	0.263736
100	31	9	3	0.713064	0.283272
100	31	9	4	0.898962	0.185897
100	31	9	5	0.976181	0.077219

N	n	k	x	P(x)	p(x)
100	31	9	6	0.996460	0.020280
100	31	9	7	0.999703	0.003243
100	31	9	8	0.999989	0.000286
100	31	9	9	1.000000	0.000011
100	31	10	0	0.019643	0.019643
100	31	10	1	0.121134	0.101491
100	31	10	2	0.345744	0.224610
100	31	10	3	0.625904	0.280159
100	31	10	4	0.843805	0.217902
100	31	10	5	0.954118	0.110313
100	31	10	6	0.990889	0.036771
100	31	10	7	0.998848	0.007959
100	31	10	8	0.999917	0.001069
100	31	10	9	0.999997	0.000080
100	31	10	10	1.000000	0.000003
100	31	11	0	0.012877	0.012877
100	31	11	1	0.087304	0.074426
100	31	11	2	0.273370	0.186066
100	31	11	3	0.538743	0.265373
100	31	11	4	0.778435	0.239692
100	31	11	5	0.922250	0.143815
100	31	11	6	0.980675	0.058425
100	31	11	7	0.996725	0.016051
100	31	11	8	0.999644	0.002918
100	31	11	9	0.999978	0.000334
100	31	11	10	0.999999	0.000022
100	31	11	11	1.000000	0.000001
100	31	12	0	0.008392	0.008392
100	31	12	1	0.062216	0.053824
100	31	12	2	0.212742	0.150525
100	31	12	3	0.455255	0.242513
100	31	12	4	0.705719	0.250464
100	31	12	5	0.880236	0.174517
100	31	12	6	0.964263	0.084027
100	31	12	7	0.992397	0.028134
100	31	12	8	0.998890	0.006492
100	31	12	9	0.999895	0.001006
100	31	12	10	0.999994	0.000099
100	31	12	11	1.000000	0.000006
100	31	12	12	1.000000	0.000000
100	31	13	0	0.005436	0.005436
100	31	13	1	0.043867	0.038431
100	31	13	2	0.163137	0.119270
100	31	13	3	0.378092	0.214955
100	31	13	4	0.628872	0.250781
100	31	13	5	0.828675	0.199802
100	31	13	6	0.940392	0.111717
100	31	13	7	0.984724	0.044332
100	31	13	8	0.997193	0.012468
100	31	13	9	0.999644	0.002451

N	n	k	x	P(x)	p(x)
100	31	13	10	0.999971	0.000327
100	31	13	11	0.999998	0.000028
100	31	13	12	1.000000	0.000001
100	31	13	13	1.000000	0.000000
100	31	14	0	0.003499	0.003499
100	31	14	1	0.030615	0.027116
100	31	14	2	0.123380	0.092765
100	31	14	3	0.308911	0.185531
100	31	14	4	0.551044	0.242133
100	31	14	5	0.768964	0.217920
100	31	14	6	0.908289	0.139326
100	31	14	7	0.972495	0.064205
100	31	14	8	0.993896	0.021402
100	31	14	9	0.999024	0.005128
100	31	14	10	0.999892	0.000868
100	31	14	11	0.999992	0.000100
100	31	14	12	1.000000	0.000007
100	31	14	13	1.000000	0.000000
100	31	14	14	1.000000	0.000000
100	31	15	0	0.002238	0.002238
100	31	15	1	0.021156	0.018918
100	31	15	2	0.092099	0.070943
100	31	15	3	0.248505	0.156407
100	31	15	4	0.475025	0.226520
100	31	15	5	0.703081	0.228056
100	31	15	6	0.867788	0.164707
100	31	15	7	0.954577	0.086789
100	31	15	8	0.988173	0.033596
100	31	15	9	0.997712	0.009540
100	31	15	10	0.999680	0.001968
100	31	15	11	0.999969	0.000289
100	31	15	12	0.999998	0.000029
100	31	15	13	1.000000	0.000002
100	31	15	14	1.000000	0.000000
100	31	15	15	1.000000	0.000000
100	31	16	0	0.001422	0.001422
100	31	16	1	0.014479	0.013057
100	31	16	2	0.067895	0.053416
100	31	16	3	0.196983	0.129088
100	31	16	4	0.403072	0.206089
100	31	16	5	0.633323	0.230251
100	31	16	6	0.819344	0.186022
100	31	16	7	0.930072	0.110727
100	31	16	8	0.979082	0.049010
100	31	16	9	0.995243	0.016161
100	31	16	10	0.999194	0.003950
100	31	16	11	0.999901	0.000707
100	31	16	12	0.999991	0.000091
100	31	16	13	0.999999	0.000008
100	31	16	14	1.000000	0.000000

N	n	k	x	P(x)	p(x)
100	31	16	15	1.000000	0.000000
100	31	16	16	1.000000	0.000000
100	31	17	0	0.000897	0.000897
100	31	17	1	0.009815	0.008919
100	31	17	2	0.049453	0.039638
100	31	17	3	0.153954	0.104500
100	31	17	4	0.336829	0.182875
100	31	17	5	0.562054	0.225225
100	31	17	6	0.763981	0.201926
100	31	17	7	0.898435	0.134455
100	31	17	8	0.965663	0.067227
100	31	17	9	0.991010	0.025348
100	31	17	10	0.998206	0.007196
100	31	17	11	0.999732	0.001526
100	31	17	12	0.999971	0.000238
100	31	17	13	0.999998	0.000027
100	31	17	14	1.000000	0.000002
100	31	17	15	1.000000	0.000000
100	31	17	16	1.000000	0.000000
100	31	17	17	1.000000	0.000000
100	31	18	0	0.000562	0.000562
100	31	18	1	0.006592	0.006030
100	31	18	2	0.035604	0.029012
100	31	18	3	0.118701	0.083097
100	31	18	4	0.277339	0.158639
100	31	18	5	0.491502	0.214162
100	31	18	6	0.703160	0.211658
100	31	18	7	0.859557	0.156397
100	31	18	8	0.947033	0.087476
100	31	18	9	0.984292	0.037258
100	31	18	10	0.996385	0.012094
100	31	18	11	0.999365	0.002979
100	31	18	12	0.999916	0.000552
100	31	18	13	0.999992	0.000076
100	31	18	14	0.999999	0.000007
100	31	18	15	1.000000	0.000001
100	31	18	16	1.000000	0.000000
100	31	18	17	1.000000	0.000000
100	31	18	18	1.000000	0.000000
100	31	19	0	0.000349	0.000349
100	31	19	1	0.004386	0.004036
100	31	19	2	0.025344	0.020958
100	31	19	3	0.090326	0.064982
100	31	19	4	0.225105	0.134779
100	31	19	5	0.423597	0.198492
100	31	19	6	0.638630	0.215033
100	31	19	7	0.813782	0.175152
100	31	19	8	0.922497	0.108715
100	31	19	9	0.974296	0.051798
100	31	19	10	0.993288	0.018993

Table for $N = 2$, $n = 1$, through $N = 100$, $n = 50$

N	n	k	x	P(x)	p(x)
100	31	19	11	0.998638	0.005350
100	31	19	12	0.999788	0.001150
100	31	19	13	0.999975	0.000187
100	31	19	14	0.999998	0.000023
100	31	19	15	1.000000	0.000002
100	31	19	16	1.000000	0.000000
100	31	19	17	1.000000	0.000000
100	31	19	18	1.000000	0.000000
100	31	19	19	1.000000	0.000000
100	31	20	0	0.000216	0.000216
100	31	20	1	0.002891	0.002675
100	31	20	2	0.017840	0.014949
100	31	20	3	0.067863	0.050023
100	31	20	4	0.180179	0.112315
100	31	20	5	0.359883	0.179705
100	31	20	6	0.572262	0.212378
100	31	20	7	0.761885	0.189623
100	31	20	8	0.891627	0.129742
100	31	20	9	0.960227	0.068599
100	31	20	10	0.988364	0.028137
100	31	20	11	0.997317	0.008953
100	31	20	12	0.999519	0.002202
100	31	20	13	0.999934	0.000415
100	31	20	14	0.999993	0.000059
100	31	20	15	0.999999	0.000006
100	31	20	16	1.000000	0.000000
100	31	20	17	1.000000	0.000000
100	31	20	18	1.000000	0.000000
100	31	20	19	1.000000	0.000000
100	31	20	20	1.000000	0.000000
100	31	21	0	0.000132	0.000132
100	31	21	1	0.001888	0.001756
100	31	21	2	0.012421	0.010533
100	31	21	3	0.050355	0.037934
100	31	21	4	0.142272	0.091917
100	31	21	5	0.301479	0.159207
100	31	21	6	0.505893	0.204414
100	31	21	7	0.704998	0.199105
100	31	21	8	0.854327	0.149329
100	31	21	9	0.941362	0.087036
100	31	21	10	0.980978	0.039616
100	31	21	11	0.995079	0.014101
100	31	21	12	0.998996	0.003917
100	31	21	13	0.999840	0.000845
100	31	21	14	0.999980	0.000140
100	31	21	15	0.999998	0.000018
100	31	21	16	1.000000	0.000002
100	31	21	17	1.000000	0.000000
100	31	21	18	1.000000	0.000000
100	31	21	19	1.000000	0.000000

N	n	k	x	P(x)	p(x)
100	31	21	20	1.000000	0.000000
100	31	21	21	1.000000	0.000000
100	31	22	0	0.000080	0.000080
100	31	22	1	0.001221	0.001141
100	31	22	2	0.008554	0.007333
100	31	22	3	0.036910	0.028356
100	31	22	4	0.110857	0.073947
100	31	22	5	0.249082	0.138225
100	31	22	6	0.441205	0.192123
100	31	22	7	0.644511	0.203305
100	31	22	8	0.810851	0.166341
100	31	22	9	0.917124	0.106273
100	31	22	10	0.970447	0.053323
100	31	22	11	0.991509	0.021062
100	31	22	12	0.998054	0.006545
100	31	22	13	0.999648	0.001594
100	31	22	14	0.999950	0.000302
100	31	22	15	0.999995	0.000044
100	31	22	16	0.999999	0.000005
100	31	22	17	1.000000	0.000000
100	31	22	18	1.000000	0.000000
100	31	22	19	1.000000	0.000000
100	31	22	20	1.000000	0.000000
100	31	22	21	1.000000	0.000000
100	31	22	22	1.000000	0.000000
100	31	23	0	0.000048	0.000048
100	31	23	1	0.000782	0.000734
100	31	23	2	0.005828	0.005046
100	31	23	3	0.026731	0.020903
100	31	23	4	0.085260	0.058529
100	31	23	5	0.203007	0.117747
100	31	23	6	0.379627	0.176620
100	31	23	7	0.581955	0.202328
100	31	23	8	0.761802	0.179847
100	31	23	9	0.887150	0.125348
100	31	23	10	0.956091	0.068941
100	31	23	11	0.986109	0.030017
100	31	23	12	0.996459	0.010351
100	31	23	13	0.999280	0.002821
100	31	23	14	0.999884	0.000604
100	31	23	15	0.999985	0.000101
100	31	23	16	0.999998	0.000013
100	31	23	17	1.000000	0.000001
100	31	23	18	1.000000	0.000000
100	31	23	19	1.000000	0.000000
100	31	23	20	1.000000	0.000000
100	31	23	21	1.000000	0.000000
100	31	23	22	1.000000	0.000000
100	31	23	23	1.000000	0.000000
100	31	24	0	0.000029	0.000029

N	n	k	x	P(x)	p(x)
100	31	24	1	0.000496	0.000467
100	31	24	2	0.003928	0.003431
100	31	24	3	0.019130	0.015202
100	31	24	4	0.064737	0.045607
100	31	24	5	0.163248	0.098511
100	31	24	6	0.322283	0.159035
100	31	24	7	0.518892	0.196609
100	31	24	8	0.708081	0.189190
100	31	24	9	0.851336	0.143255
100	31	24	10	0.937289	0.085953
100	31	24	11	0.978312	0.041023
100	31	24	12	0.993905	0.015593
100	31	24	13	0.998621	0.004715
100	31	24	14	0.999751	0.001130
100	31	24	15	0.999964	0.000213
100	31	24	16	0.999996	0.000031
100	31	24	17	1.000000	0.000004
100	31	24	18	1.000000	0.000000
100	31	24	19	1.000000	0.000000
100	31	24	20	1.000000	0.000000
100	31	24	21	1.000000	0.000000
100	31	24	22	1.000000	0.000000
100	31	24	23	1.000000	0.000000
100	31	24	24	1.000000	0.000000
100	31	25	0	0.000017	0.000017
100	31	25	1	0.000312	0.000295
100	31	25	2	0.002618	0.002306
100	31	25	3	0.013529	0.010911
100	31	25	4	0.048535	0.035005
100	31	25	5	0.129547	0.081013
100	31	25	6	0.269969	0.140422
100	31	25	7	0.456805	0.186836
100	31	25	8	0.650827	0.194022
100	31	25	9	0.809868	0.159041
100	31	25	10	0.913539	0.103671
100	31	25	11	0.967516	0.053978
100	31	25	12	0.990007	0.022491
100	31	25	13	0.997504	0.007497
100	31	25	14	0.999498	0.001994
100	31	25	15	0.999919	0.000421
100	31	25	16	0.999990	0.000070
100	31	25	17	0.999999	0.000009
100	31	25	18	1.000000	0.000001
100	31	25	19	1.000000	0.000000
100	31	25	20	1.000000	0.000000
100	31	25	21	1.000000	0.000000
100	31	25	22	1.000000	0.000000
100	31	25	23	1.000000	0.000000
100	31	25	24	1.000000	0.000000
100	31	25	25	1.000000	0.000000

N	n	k	x	P(x)	p(x)
100	31	26	0	0.000010	0.000010
100	31	26	1	0.000194	0.000184
100	31	26	2	0.001726	0.001533
100	31	26	3	0.009456	0.007729
100	31	26	4	0.035933	0.026477
100	31	26	5	0.101463	0.065530
100	31	26	6	0.223162	0.121699
100	31	26	7	0.397017	0.173856
100	31	26	8	0.591327	0.194309
100	31	26	9	0.763216	0.171889
100	31	26	10	0.884511	0.121295
100	31	26	11	0.953122	0.068611
100	31	26	12	0.984309	0.031187
100	31	26	13	0.995705	0.011395
100	31	26	14	0.999046	0.003341
100	31	26	15	0.999830	0.000784
100	31	26	16	0.999976	0.000146
100	31	26	17	0.999997	0.000021
100	31	26	18	1.000000	0.000002
100	31	26	19	1.000000	0.000000
100	31	26	20	1.000000	0.000000
100	31	26	21	1.000000	0.000000
100	31	26	22	1.000000	0.000000
100	31	26	23	1.000000	0.000000
100	31	26	24	1.000000	0.000000
100	31	26	25	1.000000	0.000000
100	31	26	26	1.000000	0.000000
100	31	27	0	0.000006	0.000006
100	31	27	1	0.000119	0.000114
100	31	27	2	0.001126	0.001006
100	31	27	3	0.006531	0.005405
100	31	27	4	0.026272	0.019741
100	31	27	5	0.078439	0.052167
100	31	27	6	0.182047	0.103609
100	31	27	7	0.340632	0.158585
100	31	27	8	0.530933	0.190301
100	31	27	9	0.712114	0.181180
100	31	27	10	0.850089	0.137976
100	31	27	11	0.934579	0.084489
100	31	27	12	0.976302	0.041723
100	31	27	13	0.992933	0.016631
100	31	27	14	0.998279	0.005346
100	31	27	15	0.999660	0.001382
100	31	27	16	0.999946	0.000286
100	31	27	17	0.999993	0.000047
100	31	27	18	0.999999	0.000006
100	31	27	19	1.000000	0.000001
100	31	27	20	1.000000	0.000000
100	31	27	21	1.000000	0.000000
100	31	27	22	1.000000	0.000000

N	n	k	x	P(x)	p(x)
100	31	27	23	1.000000	0.000000
100	31	27	24	1.000000	0.000000
100	31	27	25	1.000000	0.000000
100	31	27	26	1.000000	0.000000
100	31	27	27	1.000000	0.000000
100	31	28	0	0.000003	0.000003
100	31	28	1	0.000073	0.000069
100	31	28	2	0.000726	0.000653
100	31	28	3	0.004458	0.003732
100	31	28	4	0.018971	0.014513
100	31	28	5	0.059859	0.040888
100	31	28	6	0.146565	0.086706
100	31	28	7	0.288494	0.141930
100	31	28	8	0.470975	0.182481
100	31	28	9	0.657511	0.186536
100	31	28	10	0.810398	0.152886
100	31	28	11	0.911431	0.101033
100	31	28	12	0.965442	0.054011
100	31	28	13	0.988832	0.023390
100	31	28	14	0.997034	0.008202
100	31	28	15	0.999357	0.002324
100	31	28	16	0.999887	0.000530
100	31	28	17	0.999984	0.000097
100	31	28	18	0.999998	0.000014
100	31	28	19	1.000000	0.000002
100	31	28	20	1.000000	0.000000
100	31	28	21	1.000000	0.000000
100	31	28	22	1.000000	0.000000
100	31	28	23	1.000000	0.000000
100	31	28	24	1.000000	0.000000
100	31	28	25	1.000000	0.000000
100	31	28	26	1.000000	0.000000
100	31	28	27	1.000000	0.000000
100	31	28	28	1.000000	0.000000
100	31	29	0	0.000002	0.000002
100	31	29	1	0.000044	0.000042
100	31	29	2	0.000463	0.000419
100	31	29	3	0.003007	0.002544
100	31	29	4	0.013528	0.010522
100	31	29	5	0.045094	0.031565
100	31	29	6	0.116459	0.071365
100	31	29	7	0.241185	0.124726
100	31	29	8	0.412683	0.171498
100	31	29	9	0.600514	0.187831
100	31	29	10	0.765806	0.165292
100	31	29	11	0.883366	0.117560
100	31	29	12	0.951189	0.067823
100	31	29	13	0.982985	0.031795
100	31	29	14	0.995097	0.012112
100	31	29	15	0.998841	0.003744

N	n	k	x	P(x)	p(x)
100	31	29	16	0.999777	0.000936
100	31	29	17	0.999965	0.000188
100	31	29	18	0.999996	0.000030
100	31	29	19	0.999999	0.000004
100	31	29	20	1.000000	0.000000
100	31	29	21	1.000000	0.000000
100	31	29	22	1.000000	0.000000
100	31	29	23	1.000000	0.000000
100	31	29	24	1.000000	0.000000
100	31	29	25	1.000000	0.000000
100	31	29	26	1.000000	0.000000
100	31	29	27	1.000000	0.000000
100	31	29	28	1.000000	0.000000
100	31	29	29	1.000000	0.000000
100	31	30	0	0.000001	0.000001
100	31	30	1	0.000026	0.000025
100	31	30	2	0.000292	0.000266
100	31	30	3	0.002003	0.001712
100	31	30	4	0.009527	0.007524
100	31	30	5	0.033535	0.024007
100	31	30	6	0.091330	0.057796
100	31	30	7	0.199024	0.107694
100	31	30	8	0.357127	0.158103
100	31	30	9	0.542313	0.185186
100	31	30	10	0.716917	0.174604
100	31	30	11	0.850251	0.133334
100	31	30	12	0.933039	0.082789
100	31	30	13	0.974924	0.041884
100	31	30	14	0.992197	0.017273
100	31	30	15	0.997997	0.005800
100	31	30	16	0.999579	0.001582
100	31	30	17	0.999928	0.000349
100	31	30	18	0.999990	0.000062
100	31	30	19	0.999999	0.000009
100	31	30	20	1.000000	0.000001
100	31	30	21	1.000000	0.000000
100	31	30	22	1.000000	0.000000
100	31	30	23	1.000000	0.000000
100	31	30	24	1.000000	0.000000
100	31	30	25	1.000000	0.000000
100	31	30	26	1.000000	0.000000
100	31	30	27	1.000000	0.000000
100	31	30	28	1.000000	0.000000
100	31	30	29	1.000000	0.000000
100	31	30	30	1.000000	0.000000
100	31	31	0	0.000001	0.000001
100	31	31	1	0.000015	0.000015
100	31	31	2	0.000182	0.000166
100	31	31	3	0.001319	0.001137
100	31	31	4	0.006625	0.005306

N	n	k	x	P(x)	p(x)
100	31	31	5	0.024617	0.017992
100	31	31	6	0.070689	0.046071
100	31	31	7	0.162100	0.091411
100	31	31	8	0.305179	0.143079
100	31	31	9	0.484112	0.178933
100	31	31	10	0.664536	0.180424
100	31	31	11	0.812155	0.147620
100	31	31	12	0.910568	0.098413
100	31	31	13	0.964154	0.053585
100	31	31	14	0.988002	0.023848
100	31	31	15	0.996671	0.008669
100	31	31	16	0.999240	0.002569
100	31	31	17	0.999858	0.000618
100	31	31	18	0.999979	0.000120
100	31	31	19	0.999997	0.000019
100	31	31	20	1.000000	0.000002
100	31	31	21	1.000000	0.000000
100	31	31	22	1.000000	0.000000
100	31	31	23	1.000000	0.000000
100	31	31	24	1.000000	0.000000
100	31	31	25	1.000000	0.000000
100	31	31	26	1.000000	0.000000
100	31	31	27	1.000000	0.000000
100	31	31	28	1.000000	0.000000
100	31	31	29	1.000000	0.000000
100	31	31	30	1.000000	0.000000
100	31	31	31	1.000000	0.000000
100	32	1	0	0.680000	0.680000
100	32	1	1	1.000000	0.320000
100	32	2	0	0.460202	0.460202
100	32	2	1	0.899798	0.439596
100	32	2	2	1.000000	0.100202
100	32	3	0	0.309932	0.309932
100	32	3	1	0.760742	0.450810
100	32	3	2	0.969326	0.208584
100	32	3	3	1.000000	0.030674
100	32	4	0	0.207686	0.207686
100	32	4	1	0.616669	0.408982
100	32	4	2	0.904815	0.288147
100	32	4	3	0.990829	0.086014
100	32	4	4	1.000000	0.009171
100	32	5	0	0.138458	0.138458
100	32	5	1	0.484602	0.346144
100	32	5	2	0.814770	0.330168
100	32	5	3	0.964846	0.150076
100	32	5	4	0.997325	0.032479
100	32	5	5	1.000000	0.002675
100	32	6	0	0.091819	0.091819
100	32	6	1	0.371649	0.279830
100	32	6	2	0.710506	0.338857

N	n	k	x	P(x)	p(x)
100	32	6	3	0.919033	0.208527
100	32	6	4	0.987752	0.068719
100	32	6	5	0.999240	0.011487
100	32	6	6	1.000000	0.000760
100	32	7	0	0.060562	0.060562
100	32	7	1	0.279365	0.218803
100	32	7	2	0.602360	0.322995
100	32	7	3	0.854700	0.252340
100	32	7	4	0.967283	0.112583
100	32	7	5	0.995940	0.028657
100	32	7	6	0.999790	0.003849
100	32	7	7	1.000000	0.000210
100	32	8	0	0.039723	0.039723
100	32	8	1	0.206430	0.166707
100	32	8	2	0.498168	0.291738
100	32	8	3	0.776014	0.277845
100	32	8	4	0.933387	0.157373
100	32	8	5	0.987620	0.054233
100	32	8	6	0.998713	0.011093
100	32	8	7	0.999943	0.001230
100	32	8	8	1.000000	0.000057
100	32	9	0	0.025906	0.025906
100	32	9	1	0.150257	0.124351
100	32	9	2	0.403036	0.252779
100	32	9	3	0.688432	0.285396
100	32	9	4	0.885491	0.197059
100	32	9	5	0.971704	0.086213
100	32	9	6	0.995578	0.023874
100	32	9	7	0.999609	0.004031
100	32	9	8	0.999985	0.000376
100	32	9	9	1.000000	0.000015
100	32	10	0	0.016796	0.016796
100	32	10	1	0.107896	0.091100
100	32	10	2	0.319703	0.211807
100	32	10	3	0.597482	0.277779
100	32	10	4	0.824857	0.227376
100	32	10	5	0.946124	0.121267
100	32	10	6	0.988757	0.042633
100	32	10	7	0.998502	0.009745
100	32	10	8	0.999886	0.001384
100	32	10	9	0.999996	0.000110
100	32	10	10	1.000000	0.000004
100	32	11	0	0.010824	0.010824
100	32	11	1	0.076517	0.065693
100	32	11	2	0.249100	0.172583
100	32	11	3	0.507975	0.258875
100	32	11	4	0.754118	0.246143
100	32	11	5	0.909744	0.155626
100	32	11	6	0.976441	0.066697
100	32	11	7	0.995795	0.019354

Table for $N = 2$, $n = 1$, through $N = 100$, $n = 50$

N	n	k	x	P(x)	p(x)
100	32	11	8	0.999517	0.003722
100	32	11	9	0.999968	0.000451
100	32	11	10	0.999999	0.000031
100	32	11	11	1.000000	0.000001
100	32	12	0	0.006932	0.006932
100	32	12	1	0.053636	0.046703
100	32	12	2	0.190926	0.137291
100	32	12	3	0.423623	0.232696
100	32	12	4	0.676680	0.253057
100	32	12	5	0.862532	0.185852
100	32	12	6	0.956956	0.094425
100	32	12	7	0.990358	0.033402
100	32	12	8	0.998513	0.008155
100	32	12	9	0.999851	0.001338
100	32	12	10	0.999991	0.000140
100	32	12	11	1.000000	0.000008
100	32	12	12	1.000000	0.000000
100	32	13	0	0.004412	0.004412
100	32	13	1	0.037183	0.032772
100	32	13	2	0.144123	0.106939
100	32	13	3	0.346939	0.202816
100	32	13	4	0.596162	0.249223
100	32	13	5	0.805509	0.209347
100	32	13	6	0.929058	0.123549
100	32	13	7	0.980869	0.051811
100	32	13	8	0.996289	0.015420
100	32	13	9	0.999502	0.003212
100	32	13	10	0.999956	0.000455
100	32	13	11	0.999998	0.000041
100	32	13	12	1.000000	0.000002
100	32	13	13	1.000000	0.000000
100	32	14	0	0.002789	0.002789
100	32	14	1	0.025506	0.022717
100	32	14	2	0.107247	0.081741
100	32	14	3	0.279333	0.172086
100	32	14	4	0.515952	0.236619
100	32	14	5	0.740539	0.224587
100	32	14	6	0.892136	0.151596
100	32	14	7	0.965981	0.073846
100	32	14	8	0.992035	0.026054
100	32	14	9	0.998652	0.006617
100	32	14	10	0.999841	0.001189
100	32	14	11	0.999988	0.000146
100	32	14	12	0.999999	0.000012
100	32	14	13	1.000000	0.000001
100	32	14	14	1.000000	0.000000
100	32	15	0	0.001751	0.001751
100	32	15	1	0.017317	0.015566
100	32	15	2	0.078733	0.061415
100	32	15	3	0.221304	0.142571

N	n	k	x	P(x)	p(x)
100	32	15	4	0.438913	0.217609
100	32	15	5	0.670029	0.231116
100	32	15	6	0.846304	0.176275
100	32	15	7	0.944514	0.098210
100	32	15	8	0.984764	0.040250
100	32	15	9	0.996883	0.012118
100	32	15	10	0.999537	0.002654
100	32	15	11	0.999952	0.000415
100	32	15	12	0.999997	0.000045
100	32	15	13	1.000000	0.000003
100	32	15	14	1.000000	0.000000
100	32	15	15	1.000000	0.000000
100	32	16	0	0.001092	0.001092
100	32	16	1	0.011640	0.010548
100	32	16	2	0.057057	0.045416
100	32	16	3	0.172662	0.115605
100	32	16	4	0.367230	0.194568
100	32	16	5	0.596616	0.229386
100	32	16	6	0.792385	0.195769
100	32	16	7	0.915629	0.123244
100	32	16	8	0.973400	0.057771
100	32	16	9	0.993604	0.020204
100	32	16	10	0.998850	0.005247
100	32	16	11	0.999849	0.000999
100	32	16	12	0.999986	0.000137
100	32	16	13	0.999999	0.000013
100	32	16	14	1.000000	0.000001
100	32	16	15	1.000000	0.000000
100	32	16	16	1.000000	0.000000
100	32	17	0	0.000676	0.000676
100	32	17	1	0.007747	0.007071
100	32	17	2	0.040837	0.033089
100	32	17	3	0.132751	0.091914
100	32	17	4	0.302374	0.169624
100	32	17	5	0.522885	0.220511
100	32	17	6	0.731790	0.208905
100	32	17	7	0.878949	0.147159
100	32	17	8	0.956894	0.077945
100	32	17	9	0.988072	0.031178
100	32	17	10	0.997476	0.009404
100	32	17	11	0.999600	0.002124
100	32	17	12	0.999954	0.000354
100	32	17	13	0.999996	0.000043
100	32	17	14	1.000000	0.000004
100	32	17	15	1.000000	0.000000
100	32	17	16	1.000000	0.000000
100	32	17	17	1.000000	0.000000
100	32	18	0	0.000415	0.000415
100	32	18	1	0.005106	0.004691
100	32	18	2	0.028877	0.023770

N	n	k	x	P(x)	p(x)
100	32	18	3	0.100636	0.071760
100	32	18	4	0.245152	0.144516
100	32	18	5	0.451152	0.206001
100	32	18	6	0.666350	0.215197
100	32	18	7	0.834624	0.168274
100	32	18	8	0.934356	0.099732
100	32	18	9	0.979432	0.045076
100	32	18	10	0.994983	0.015551
100	32	18	11	0.999062	0.004079
100	32	18	12	0.999868	0.000806
100	32	18	13	0.999986	0.000118
100	32	18	14	0.999999	0.000013
100	32	18	15	1.000000	0.000001
100	32	18	16	1.000000	0.000000
100	32	18	17	1.000000	0.000000
100	32	18	18	1.000000	0.000000
100	32	19	0	0.000253	0.000253
100	32	19	1	0.003333	0.003080
100	32	19	2	0.020180	0.016847
100	32	19	3	0.075258	0.055078
100	32	19	4	0.195805	0.120547
100	32	19	5	0.383323	0.187518
100	32	19	6	0.598116	0.214793
100	32	19	7	0.783321	0.185204
100	32	19	8	0.905166	0.121845
100	32	19	9	0.966789	0.061623
100	32	19	10	0.990811	0.024022
100	32	19	11	0.998018	0.007207
100	32	19	12	0.999672	0.001654
100	32	19	13	0.999959	0.000287
100	32	19	14	0.999996	0.000037
100	32	19	15	1.000000	0.000003
100	32	19	16	1.000000	0.000000
100	32	19	17	1.000000	0.000000
100	32	19	18	1.000000	0.000000
100	32	19	19	1.000000	0.000000
100	32	20	0	0.000153	0.000153
100	32	20	1	0.002154	0.002001
100	32	20	2	0.013940	0.011786
100	32	20	3	0.055539	0.041598
100	32	20	4	0.154134	0.098596
100	32	20	5	0.320817	0.166683
100	32	20	6	0.529170	0.208353
100	32	20	7	0.726159	0.196989
100	32	20	8	0.869064	0.142905
100	32	20	9	0.949291	0.080227
100	32	20	10	0.984286	0.034996
100	32	20	11	0.996149	0.011863
100	32	20	12	0.999263	0.003114
100	32	20	13	0.999892	0.000628

N	n	k	x	P(x)	p(x)
100	32	20	14	0.999988	0.000096
100	32	20	15	0.999999	0.000011
100	32	20	16	1.000000	0.000001
100	32	20	17	1.000000	0.000000
100	32	20	18	1.000000	0.000000
100	32	20	19	1.000000	0.000000
100	32	20	20	1.000000	0.000000
100	32	21	0	0.000092	0.000092
100	32	21	1	0.001379	0.001287
100	32	21	2	0.009521	0.008142
100	32	21	3	0.040459	0.030939
100	32	21	4	0.119626	0.079167
100	32	21	5	0.264562	0.144936
100	32	21	6	0.461456	0.196894
100	32	21	7	0.664600	0.203144
100	32	21	8	0.826192	0.161592
100	32	21	9	0.926225	0.100033
100	32	21	10	0.974663	0.048437
100	32	21	11	0.993035	0.018373
100	32	21	12	0.998485	0.005450
100	32	21	13	0.999742	0.001258
100	32	21	14	0.999966	0.000224
100	32	21	15	0.999997	0.000030
100	32	21	16	1.000000	0.000003
100	32	21	17	1.000000	0.000000
100	32	21	18	1.000000	0.000000
100	32	21	19	1.000000	0.000000
100	32	21	20	1.000000	0.000000
100	32	21	21	1.000000	0.000000
100	32	22	0	0.000055	0.000055
100	32	22	1	0.000874	0.000819
100	32	22	2	0.006429	0.005555
100	32	22	3	0.029102	0.022673
100	32	22	4	0.091567	0.062465
100	32	22	5	0.215027	0.123460
100	32	22	6	0.396655	0.181628
100	32	22	7	0.600314	0.203659
100	32	22	8	0.777101	0.176787
100	32	22	9	0.897102	0.120001
100	32	22	10	0.961174	0.064072
100	32	22	11	0.988151	0.026978
100	32	22	12	0.997105	0.008954
100	32	22	13	0.999440	0.002335
100	32	22	14	0.999915	0.000475
100	32	22	15	0.999990	0.000075
100	32	22	16	0.999999	0.000009
100	32	22	17	1.000000	0.000001
100	32	22	18	1.000000	0.000000
100	32	22	19	1.000000	0.000000
100	32	22	20	1.000000	0.000000

$N = 100$ $n = 32$

N	n	k	x	P(x)	p(x)
100	32	22	21	1.000000	0.000000
100	32	22	22	1.000000	0.000000
100	32	23	0	0.000032	0.000032
100	32	23	1	0.000548	0.000516
100	32	23	2	0.004292	0.003744
100	32	23	3	0.020672	0.016380
100	32	23	4	0.069144	0.048471
100	32	23	5	0.172291	0.103147
100	32	23	6	0.336112	0.163822
100	32	23	7	0.535038	0.198926
100	32	23	8	0.722705	0.187666
100	32	23	9	0.861717	0.139012
100	32	23	10	0.943102	0.081385
100	32	23	11	0.980888	0.037786
100	32	23	12	0.994809	0.013921
100	32	23	13	0.998871	0.004062
100	32	23	14	0.999806	0.000934
100	32	23	15	0.999974	0.000168
100	32	23	16	0.999997	0.000023
100	32	23	17	1.000000	0.000002
100	32	23	18	1.000000	0.000000
100	32	23	19	1.000000	0.000000
100	32	23	20	1.000000	0.000000
100	32	23	21	1.000000	0.000000
100	32	23	22	1.000000	0.000000
100	32	23	23	1.000000	0.000000
100	32	24	0	0.000019	0.000019
100	32	24	1	0.000341	0.000322
100	32	24	2	0.002834	0.002493
100	32	24	3	0.014503	0.011670
100	32	24	4	0.051518	0.037014
100	32	24	5	0.136122	0.084605
100	32	24	6	0.280796	0.144674
100	32	24	7	0.470452	0.189656
100	32	24	8	0.664211	0.193759
100	32	24	9	0.820194	0.155982
100	32	24	10	0.919849	0.099655
100	32	24	11	0.970583	0.050734
100	32	24	12	0.991193	0.020611
100	32	24	13	0.997869	0.006675
100	32	24	14	0.999587	0.001718
100	32	24	15	0.999937	0.000349
100	32	24	16	0.999992	0.000056
100	32	24	17	0.999999	0.000007
100	32	24	18	1.000000	0.000001
100	32	24	19	1.000000	0.000000
100	32	24	20	1.000000	0.000000
100	32	24	21	1.000000	0.000000
100	32	24	22	1.000000	0.000000
100	32	24	23	1.000000	0.000000

N	n	k	x	P(x)	p(x)
100	32	24	24	1.000000	0.000000
100	32	25	0	0.000011	0.000011
100	32	25	1	0.000209	0.000198
100	32	25	2	0.001849	0.001640
100	32	25	3	0.010050	0.008201
100	32	25	4	0.037881	0.027830
100	32	25	5	0.106065	0.068185
100	32	25	6	0.231302	0.125237
100	32	25	7	0.408065	0.176763
100	32	25	8	0.603024	0.194959
100	32	25	9	0.772989	0.169964
100	32	25	10	0.891002	0.118013
100	32	25	11	0.956564	0.065563
100	32	25	12	0.985770	0.029205
100	32	25	13	0.996200	0.010430
100	32	25	14	0.999180	0.002980
100	32	25	15	0.999858	0.000678
100	32	25	16	0.999981	0.000122
100	32	25	17	0.999998	0.000017
100	32	25	18	1.000000	0.000002
100	32	25	19	1.000000	0.000000
100	32	25	20	1.000000	0.000000
100	32	25	21	1.000000	0.000000
100	32	25	22	1.000000	0.000000
100	32	25	23	1.000000	0.000000
100	32	25	24	1.000000	0.000000
100	32	25	25	1.000000	0.000000
100	32	26	0	0.000006	0.000006
100	32	26	1	0.000127	0.000121
100	32	26	2	0.001193	0.001066
100	32	26	3	0.006879	0.005686
100	32	26	4	0.027491	0.020611
100	32	26	5	0.081519	0.054028
100	32	26	6	0.187887	0.106368
100	32	26	7	0.349144	0.161257
100	32	26	8	0.540637	0.191493
100	32	26	9	0.720866	0.180229
100	32	26	10	0.856385	0.135518
100	32	26	11	0.938207	0.081822
100	32	26	12	0.977982	0.039775
100	32	26	13	0.993558	0.015576
100	32	26	14	0.998465	0.004907
100	32	26	15	0.999705	0.001240
100	32	26	16	0.999955	0.000250
100	32	26	17	0.999994	0.000040
100	32	26	18	0.999999	0.000005
100	32	26	19	1.000000	0.000000
100	32	26	20	1.000000	0.000000
100	32	26	21	1.000000	0.000000
100	32	26	22	1.000000	0.000000

N	n	k	x	P(x)	p(x)
100	32	26	23	1.000000	0.000000
100	32	26	24	1.000000	0.000000
100	32	26	25	1.000000	0.000000
100	32	26	26	1.000000	0.000000
100	32	27	0	0.000004	0.000004
100	32	27	1	0.000077	0.000073
100	32	27	2	0.000761	0.000685
100	32	27	3	0.004651	0.003890
100	32	27	4	0.019692	0.015041
100	32	27	5	0.061806	0.042114
100	32	27	6	0.150514	0.088708
100	32	27	7	0.294665	0.144151
100	32	27	8	0.478531	0.183866
100	32	27	9	0.664849	0.186318
100	32	27	10	0.816095	0.151246
100	32	27	11	0.914987	0.098892
100	32	27	12	0.967232	0.052245
100	32	27	13	0.989558	0.022327
100	32	27	14	0.997271	0.007713
100	32	27	15	0.999420	0.002149
100	32	27	16	0.999900	0.000481
100	32	27	17	0.999986	0.000086
100	32	27	18	0.999998	0.000012
100	32	27	19	1.000000	0.000001
100	32	27	20	1.000000	0.000000
100	32	27	21	1.000000	0.000000
100	32	27	22	1.000000	0.000000
100	32	27	23	1.000000	0.000000
100	32	27	24	1.000000	0.000000
100	32	27	25	1.000000	0.000000
100	32	27	26	1.000000	0.000000
100	32	27	27	1.000000	0.000000
100	32	28	0	0.000002	0.000002
100	32	28	1	0.000046	0.000044
100	32	28	2	0.000480	0.000434
100	32	28	3	0.003106	0.002626
100	32	28	4	0.013923	0.010817
100	32	28	5	0.046229	0.032307
100	32	28	6	0.118919	0.072690
100	32	28	7	0.245298	0.126379
100	32	28	8	0.418082	0.172784
100	32	28	9	0.606146	0.188064
100	32	28	10	0.770514	0.164368
100	32	28	11	0.886539	0.116024
100	32	28	12	0.952918	0.066379
100	32	28	13	0.983747	0.030829
100	32	28	14	0.995369	0.011622
100	32	28	15	0.998920	0.003550
100	32	28	16	0.999795	0.000876
100	32	28	17	0.999969	0.000173

N	n	k	x	P(x)	p(x)
100	32	28	18	0.999996	0.000027
100	32	28	19	0.999999	0.000003
100	32	28	20	1.000000	0.000000
100	32	28	21	1.000000	0.000000
100	32	28	22	1.000000	0.000000
100	32	28	23	1.000000	0.000000
100	32	28	24	1.000000	0.000000
100	32	28	25	1.000000	0.000000
100	32	28	26	1.000000	0.000000
100	32	28	27	1.000000	0.000000
100	32	28	28	1.000000	0.000000
100	32	29	0	0.000001	0.000001
100	32	29	1	0.000027	0.000026
100	32	29	2	0.000299	0.000272
100	32	29	3	0.002048	0.001749
100	32	29	4	0.009716	0.007668
100	32	29	5	0.034114	0.024398
100	32	29	6	0.092670	0.058556
100	32	29	7	0.201417	0.108747
100	32	29	8	0.360488	0.159071
100	32	29	9	0.546070	0.185583
100	32	29	10	0.720291	0.174220
100	32	29	11	0.852698	0.132408
100	32	29	12	0.934479	0.081781
100	32	29	13	0.975612	0.041133
100	32	29	14	0.992464	0.016852
100	32	29	15	0.998081	0.005617
100	32	29	16	0.999601	0.001519
100	32	29	17	0.999933	0.000332
100	32	29	18	0.999991	0.000058
100	32	29	19	0.999999	0.000008
100	32	29	20	1.000000	0.000001
100	32	29	21	1.000000	0.000000
100	32	29	22	1.000000	0.000000
100	32	29	23	1.000000	0.000000
100	32	29	24	1.000000	0.000000
100	32	29	25	1.000000	0.000000
100	32	29	26	1.000000	0.000000
100	32	29	27	1.000000	0.000000
100	32	29	28	1.000000	0.000000
100	32	29	29	1.000000	0.000000
100	32	30	0	0.000001	0.000001
100	32	30	1	0.000016	0.000015
100	32	30	2	0.000184	0.000168
100	32	30	3	0.001334	0.001150
100	32	30	4	0.006692	0.005358
100	32	30	5	0.024836	0.018144
100	32	30	6	0.071227	0.046391
100	32	30	7	0.163126	0.091899
100	32	30	8	0.306717	0.143591

Table for $N = 2$, $n = 1$, through $N = 100$, $n = 50$

N	n	k	x	P(x)	p(x)
100	32	30	9	0.485952	0.179235
100	32	30	10	0.666307	0.180355
100	32	30	11	0.813535	0.147229
100	32	30	12	0.911442	0.097907
100	32	30	13	0.964605	0.053162
100	32	30	14	0.988192	0.023587
100	32	30	15	0.996736	0.008545
100	32	30	16	0.999258	0.002522
100	32	30	17	0.999862	0.000604
100	32	30	18	0.999979	0.000117
100	32	30	19	0.999997	0.000018
100	32	30	20	1.000000	0.000002
100	32	30	21	1.000000	0.000000
100	32	30	22	1.000000	0.000000
100	32	30	23	1.000000	0.000000
100	32	30	24	1.000000	0.000000
100	32	30	25	1.000000	0.000000
100	32	30	26	1.000000	0.000000
100	32	30	27	1.000000	0.000000
100	32	30	28	1.000000	0.000000
100	32	30	29	1.000000	0.000000
100	32	30	30	1.000000	0.000000
100	32	31	0	0.000000	0.000000
100	32	31	1	0.000009	0.000009
100	32	31	2	0.000112	0.000103
100	32	31	3	0.000857	0.000746
100	32	31	4	0.004549	0.003691
100	32	31	5	0.017838	0.013289
100	32	31	6	0.053996	0.036159
100	32	31	7	0.130305	0.076309
100	32	31	8	0.257486	0.127181
100	32	31	9	0.427061	0.169575
100	32	31	10	0.609624	0.182563
100	32	31	11	0.769367	0.159743
100	32	31	12	0.883469	0.114102
100	32	31	13	0.950175	0.066706
100	32	31	14	0.982126	0.031952
100	32	31	15	0.994661	0.012535
100	32	31	16	0.998682	0.004021
100	32	31	17	0.999733	0.001051
100	32	31	18	0.999956	0.000223
100	32	31	19	0.999994	0.000038
100	32	31	20	0.999999	0.000005
100	32	31	21	1.000000	0.000001
100	32	31	22	1.000000	0.000000
100	32	31	23	1.000000	0.000000
100	32	31	24	1.000000	0.000000
100	32	31	25	1.000000	0.000000
100	32	31	26	1.000000	0.000000
100	32	31	27	1.000000	0.000000

N	n	k	x	P(x)	p(x)
100	32	31	28	1.000000	0.000000
100	32	31	29	1.000000	0.000000
100	32	31	30	1.000000	0.000000
100	32	31	31	1.000000	0.000000
100	32	32	0	0.000000	0.000000
100	32	32	1	0.000005	0.000005
100	32	32	2	0.000067	0.000062
100	32	32	3	0.000544	0.000477
100	32	32	4	0.003051	0.002507
100	32	32	5	0.012638	0.009587
100	32	32	6	0.040371	0.027734
100	32	32	7	0.102657	0.062286
100	32	32	8	0.213249	0.110592
100	32	32	9	0.370536	0.157287
100	32	32	10	0.551415	0.180880
100	32	32	11	0.720749	0.169334
100	32	32	12	0.850396	0.129646
100	32	32	13	0.931807	0.081411
100	32	32	14	0.973791	0.041985
100	32	32	15	0.991573	0.017782
100	32	32	16	0.997749	0.006177
100	32	32	17	0.999504	0.001755
100	32	32	18	0.999911	0.000406
100	32	32	19	0.999987	0.000076
100	32	32	20	0.999998	0.000011
100	32	32	21	1.000000	0.000001
100	32	32	22	1.000000	0.000000
100	32	32	23	1.000000	0.000000
100	32	32	24	1.000000	0.000000
100	32	32	25	1.000000	0.000000
100	32	32	26	1.000000	0.000000
100	32	32	27	1.000000	0.000000
100	32	32	28	1.000000	0.000000
100	32	32	29	1.000000	0.000000
100	32	32	30	1.000000	0.000000
100	32	32	31	1.000000	0.000000
100	32	32	32	1.000000	0.000000
100	33	1	0	0.670000	0.670000
100	33	1	1	1.000000	0.330000
100	33	2	0	0.446667	0.446667
100	33	2	1	0.893333	0.446667
100	33	2	2	1.000000	0.106667
100	33	3	0	0.296258	0.296258
100	33	3	1	0.747483	0.451224
100	33	3	2	0.966258	0.218775
100	33	3	3	1.000000	0.033741
100	33	4	0	0.195470	0.195470
100	33	4	1	0.598625	0.403156
100	33	4	2	0.896341	0.297715
100	33	4	3	0.989564	0.093224

N	n	k	x	P(x)	p(x)
100	33	4	4	1.000000	0.010436
100	33	5	0	0.128277	0.128277
100	33	5	1	0.464240	0.335963
100	33	5	2	0.800203	0.335963
100	33	5	3	0.960432	0.160229
100	33	5	4	0.996847	0.036416
100	33	5	5	1.000000	0.003152
100	33	6	0	0.083718	0.083718
100	33	6	1	0.351074	0.267356
100	33	6	2	0.690573	0.339500
100	33	6	3	0.909833	0.219260
100	33	6	4	0.985731	0.075898
100	33	6	5	0.999071	0.013340
100	33	6	6	1.000000	0.000929
100	33	7	0	0.054327	0.054327
100	33	7	1	0.260059	0.205731
100	33	7	2	0.578611	0.318552
100	33	7	3	0.839857	0.261246
100	33	7	4	0.962316	0.122459
100	33	7	5	0.995097	0.032781
100	33	7	6	0.999733	0.004636
100	33	7	7	1.000000	0.000267
100	33	8	0	0.035050	0.035050
100	33	8	1	0.189269	0.154220
100	33	8	2	0.472427	0.283157
100	33	8	3	0.755584	0.283157
100	33	8	4	0.924130	0.168546
100	33	8	5	0.985228	0.061098
100	33	8	6	0.998387	0.013160
100	33	8	7	0.999925	0.001538
100	33	8	8	1.000000	0.000075
100	33	9	0	0.022478	0.022478
100	33	9	1	0.135628	0.113150
100	33	9	2	0.377015	0.241387
100	33	9	3	0.663250	0.286235
100	33	9	4	0.871001	0.207751
100	33	9	5	0.966633	0.095631
100	33	9	6	0.994525	0.027893
100	33	9	7	0.999491	0.004965
100	33	9	8	0.999980	0.000489
100	33	9	9	1.000000	0.000020
100	33	10	0	0.014326	0.014326
100	33	10	1	0.095839	0.081512
100	33	10	2	0.294784	0.198945
100	33	10	3	0.568887	0.274103
100	33	10	4	0.804795	0.235908
100	33	10	5	0.937208	0.132413
100	33	10	6	0.986249	0.049042
100	33	10	7	0.998072	0.011823
100	33	10	8	0.999845	0.001773

N	n	k	x	P(x)	p(x)
100	33	10	9	0.999995	0.000149
100	33	10	10	1.000000	0.000005
100	33	11	0	0.009073	0.009073
100	33	11	1	0.066857	0.057783
100	33	11	2	0.226259	0.159402
100	33	11	3	0.477519	0.251261
100	33	11	4	0.728780	0.251261
100	33	11	5	0.896012	0.167233
100	33	11	6	0.971537	0.075524
100	33	11	7	0.994657	0.023120
100	33	11	8	0.999353	0.004696
100	33	11	9	0.999955	0.000602
100	33	11	10	0.999999	0.000044
100	33	11	11	1.000000	0.000001
100	33	12	0	0.005709	0.005709
100	33	12	1	0.046081	0.040372
100	33	12	2	0.170736	0.124656
100	33	12	3	0.392825	0.222088
100	33	12	4	0.646908	0.254084
100	33	12	5	0.843400	0.196492
100	33	12	6	0.948625	0.105225
100	33	12	7	0.987902	0.039278
100	33	12	8	0.998034	0.010131
100	33	12	9	0.999792	0.001759
100	33	12	10	0.999987	0.000195
100	33	12	11	1.000000	0.000012
100	33	12	12	1.000000	0.000000
100	33	13	0	0.003568	0.003568
100	33	13	1	0.031400	0.027832
100	33	13	2	0.126824	0.095424
100	33	13	3	0.317113	0.190289
100	33	13	4	0.563176	0.246064
100	33	13	5	0.780880	0.217704
100	33	13	6	0.916340	0.135460
100	33	13	7	0.976298	0.059958
100	33	13	8	0.995155	0.018858
100	33	13	9	0.999313	0.004157
100	33	13	10	0.999936	0.000624
100	33	13	11	0.999996	0.000060
100	33	13	12	1.000000	0.000003
100	33	13	13	1.000000	0.000000
100	33	14	0	0.002215	0.002215
100	33	14	1	0.021163	0.018948
100	33	14	2	0.092822	0.071659
100	33	14	3	0.251496	0.158674
100	33	14	4	0.481155	0.229659
100	33	14	5	0.710814	0.229659
100	33	14	6	0.874301	0.163486
100	33	14	7	0.958379	0.084079
100	33	14	8	0.989737	0.031357

Table for $N = 2$, $n = 1$, through $N = 100$, $n = 50$

N	n	k	x	P(x)	p(x)
100	33	14	9	0.998166	0.008429
100	33	14	10	0.999771	0.001606
100	33	14	11	0.999981	0.000210
100	33	14	12	0.999999	0.000018
100	33	14	13	1.000000	0.000001
100	33	14	14	1.000000	0.000000
100	33	15	0	0.001365	0.001365
100	33	15	1	0.014113	0.012748
100	33	15	2	0.066992	0.052879
100	33	15	3	0.196145	0.129153
100	33	15	4	0.403712	0.207567
100	33	15	5	0.636042	0.232330
100	33	15	6	0.822974	0.186932
100	33	15	7	0.932960	0.109987
100	33	15	8	0.980621	0.047661
100	33	15	9	0.995813	0.015192
100	33	15	10	0.999342	0.003529
100	33	15	11	0.999928	0.000586
100	33	15	12	0.999995	0.000067
100	33	15	13	1.000000	0.000005
100	33	15	14	1.000000	0.000000
100	33	15	15	1.000000	0.000000
100	33	16	0	0.000835	0.000835
100	33	16	1	0.009313	0.008478
100	33	16	2	0.047706	0.038393
100	33	16	3	0.150561	0.102855
100	33	16	4	0.332895	0.182334
100	33	16	5	0.559509	0.226615
100	33	16	6	0.763595	0.204086
100	33	16	7	0.899317	0.135722
100	33	16	8	0.966603	0.067286
100	33	16	9	0.991524	0.024921
100	33	16	10	0.998387	0.006863
100	33	16	11	0.999776	0.001389
100	33	16	12	0.999978	0.000202
100	33	16	13	0.999998	0.000020
100	33	16	14	1.000000	0.000001
100	33	16	15	1.000000	0.000000
100	33	16	16	1.000000	0.000000
100	33	17	0	0.000507	0.000507
100	33	17	1	0.006084	0.005577
100	33	17	2	0.033537	0.027454
100	33	17	3	0.113827	0.080290
100	33	17	4	0.269946	0.156119
100	33	17	5	0.483971	0.214025
100	33	17	6	0.697996	0.214025
100	33	17	7	0.857308	0.159312
100	33	17	8	0.946578	0.089270
100	33	17	9	0.984404	0.037826
100	33	17	10	0.996508	0.012104

N	n	k	x	P(x)	p(x)
100	33	17	11	0.999412	0.002904
100	33	17	12	0.999928	0.000515
100	33	17	13	0.999994	0.000066
100	33	17	14	0.999999	0.000006
100	33	17	15	1.000000	0.000000
100	33	17	16	1.000000	0.000000
100	33	17	17	1.000000	0.000000
100	33	18	0	0.000305	0.000305
100	33	18	1	0.003934	0.003628
100	33	18	2	0.023284	0.019350
100	33	18	3	0.084807	0.061523
100	33	18	4	0.215399	0.130592
100	33	18	5	0.411770	0.196371
100	33	18	6	0.628374	0.216604
100	33	18	7	0.807403	0.179029
100	33	18	8	0.919689	0.112286
100	33	18	9	0.973466	0.053777
100	33	18	10	0.993154	0.019688
100	33	18	11	0.998642	0.005489
100	33	18	12	0.999797	0.001155
100	33	18	13	0.999978	0.000181
100	33	18	14	0.999998	0.000020
100	33	18	15	1.000000	0.000002
100	33	18	16	1.000000	0.000000
100	33	18	17	1.000000	0.000000
100	33	18	18	1.000000	0.000000
100	33	19	0	0.000182	0.000182
100	33	19	1	0.002518	0.002335
100	33	19	2	0.015968	0.013451
100	33	19	3	0.062298	0.046330
100	33	19	4	0.169214	0.106915
100	33	19	5	0.344716	0.175503
100	33	19	6	0.557053	0.212337
100	33	19	7	0.750638	0.193585
100	33	19	8	0.885456	0.134818
100	33	19	9	0.957726	0.072271
100	33	19	10	0.987632	0.029905
100	33	19	11	0.997170	0.009538
100	33	19	12	0.999501	0.002332
100	33	19	13	0.999934	0.000432
100	33	19	14	0.999993	0.000060
100	33	19	15	0.999999	0.000006
100	33	19	16	1.000000	0.000000
100	33	19	17	1.000000	0.000000
100	33	19	18	1.000000	0.000000
100	33	19	19	1.000000	0.000000
100	33	20	0	0.000108	0.000108
100	33	20	1	0.001595	0.001487
100	33	20	2	0.010821	0.009225
100	33	20	3	0.045139	0.034319

N	n	k	x	P(x)	p(x)
100	33	20	4	0.130935	0.085796
100	33	20	5	0.284049	0.153113
100	33	20	6	0.486274	0.202225
100	33	20	7	0.688499	0.202225
100	33	20	8	0.843845	0.155346
100	33	20	9	0.936313	0.092468
100	33	20	10	0.979140	0.042827
100	33	20	11	0.994579	0.015439
100	33	20	12	0.998897	0.004318
100	33	20	13	0.999827	0.000930
100	33	20	14	0.999979	0.000152
100	33	20	15	0.999998	0.000019
100	33	20	16	1.000000	0.000002
100	33	20	17	1.000000	0.000000
100	33	20	18	1.000000	0.000000
100	33	20	19	1.000000	0.000000
100	33	20	20	1.000000	0.000000
100	33	21	0	0.000064	0.000064
100	33	21	1	0.001000	0.000937
100	33	21	2	0.007246	0.006245
100	33	21	3	0.032270	0.025024
100	33	21	4	0.099834	0.067565
100	33	21	5	0.230459	0.130625
100	33	21	6	0.418023	0.187564
100	33	21	7	0.622776	0.204753
100	33	21	8	0.795300	0.172523
100	33	21	9	0.908572	0.113273
100	33	21	10	0.966827	0.058255
100	33	21	11	0.990333	0.023506
100	33	21	12	0.997764	0.007430
100	33	21	13	0.999594	0.001831
100	33	21	14	0.999943	0.000349
100	33	21	15	0.999994	0.000051
100	33	21	16	0.999999	0.000006
100	33	21	17	1.000000	0.000000
100	33	21	18	1.000000	0.000000
100	33	21	19	1.000000	0.000000
100	33	21	20	1.000000	0.000000
100	33	21	21	1.000000	0.000000
100	33	22	0	0.000037	0.000037
100	33	22	1	0.000621	0.000584
100	33	22	2	0.004795	0.004174
100	33	22	3	0.022767	0.017972
100	33	22	4	0.075032	0.052265
100	33	22	5	0.184162	0.109130
100	33	22	6	0.353919	0.169757
100	33	22	7	0.555389	0.201470
100	33	22	8	0.740704	0.185314
100	33	22	9	0.874160	0.133457
100	33	22	10	0.949867	0.075706

N	n	k	x	P(x)	p(x)
100	33	22	11	0.983787	0.033920
100	33	22	12	0.995788	0.012001
100	33	22	13	0.999131	0.003342
100	33	22	14	0.999859	0.000728
100	33	22	15	0.999982	0.000123
100	33	22	16	0.999998	0.000016
100	33	22	17	1.000000	0.000002
100	33	22	18	1.000000	0.000000
100	33	22	19	1.000000	0.000000
100	33	22	20	1.000000	0.000000
100	33	22	21	1.000000	0.000000
100	33	22	22	1.000000	0.000000
100	33	23	0	0.000021	0.000021
100	33	23	1	0.000381	0.000360
100	33	23	2	0.003136	0.002755
100	33	23	3	0.015855	0.012719
100	33	23	4	0.055600	0.039745
100	33	23	5	0.144987	0.089387
100	33	23	6	0.295157	0.150170
100	33	23	7	0.488232	0.193076
100	33	23	8	0.681308	0.193076
100	33	23	9	0.833097	0.151789
100	33	23	10	0.927543	0.094446
100	33	23	11	0.974220	0.046677
100	33	23	12	0.992557	0.018337
100	33	23	13	0.998274	0.005716
100	33	23	14	0.999682	0.001408
100	33	23	15	0.999954	0.000272
100	33	23	16	0.999995	0.000041
100	33	23	17	0.999999	0.000005
100	33	23	18	1.000000	0.000000
100	33	23	19	1.000000	0.000000
100	33	23	20	1.000000	0.000000
100	33	23	21	1.000000	0.000000
100	33	23	22	1.000000	0.000000
100	33	23	23	1.000000	0.000000
100	33	24	0	0.000012	0.000012
100	33	24	1	0.000232	0.000220
100	33	24	2	0.002027	0.001795
100	33	24	3	0.010899	0.008872
100	33	24	4	0.040631	0.029732
100	33	24	5	0.112483	0.071852
100	33	24	6	0.242500	0.130017
100	33	24	7	0.423038	0.180538
100	33	24	8	0.618621	0.195583
100	33	24	9	0.785786	0.167165
100	33	24	10	0.899332	0.113546
100	33	24	11	0.960884	0.061552
100	33	24	12	0.987556	0.026672
100	33	24	13	0.996789	0.009233

N	n	k	x	P(x)	p(x)
100	33	24	14	0.999334	0.002545
100	33	24	15	0.999890	0.000556
100	33	24	16	0.999986	0.000095
100	33	24	17	0.999998	0.000013
100	33	24	18	1.000000	0.000001
100	33	24	19	1.000000	0.000000
100	33	24	20	1.000000	0.000000
100	33	24	21	1.000000	0.000000
100	33	24	22	1.000000	0.000000
100	33	24	23	1.000000	0.000000
100	33	24	24	1.000000	0.000000
100	33	25	0	0.000007	0.000007
100	33	25	1	0.000139	0.000132
100	33	25	2	0.001295	0.001155
100	33	25	3	0.007397	0.006102
100	33	25	4	0.029286	0.021889
100	33	25	5	0.086011	0.056725
100	33	25	6	0.196309	0.110299
100	33	25	7	0.361275	0.164965
100	33	25	8	0.554284	0.193010
100	33	25	9	0.732997	0.178713
100	33	25	10	0.864969	0.131972
100	33	25	11	0.943066	0.078097
100	33	25	12	0.980186	0.037120
100	33	25	13	0.994359	0.014173
100	33	25	14	0.998698	0.004339
100	33	25	15	0.999759	0.001061
100	33	25	16	0.999964	0.000206
100	33	25	17	0.999996	0.000031
100	33	25	18	0.999999	0.000004
100	33	25	19	1.000000	0.000000
100	33	25	20	1.000000	0.000000
100	33	25	21	1.000000	0.000000
100	33	25	22	1.000000	0.000000
100	33	25	23	1.000000	0.000000
100	33	25	24	1.000000	0.000000
100	33	25	25	1.000000	0.000000
100	33	26	0	0.000004	0.000004
100	33	26	1	0.000083	0.000079
100	33	26	2	0.000817	0.000734
100	33	26	3	0.004956	0.004139
100	33	26	4	0.020822	0.015866
100	33	26	5	0.064834	0.044011
100	33	26	6	0.156602	0.091768
100	33	26	7	0.304087	0.147485
100	33	26	8	0.489948	0.185861
100	33	26	9	0.675809	0.185861
100	33	26	10	0.824498	0.148689
100	33	26	11	0.920158	0.095660
100	33	26	12	0.969793	0.049635

N	n	k	x	P(x)	p(x)
100	33	26	13	0.990580	0.020787
100	33	26	14	0.997599	0.007019
100	33	26	15	0.999504	0.001905
100	33	26	16	0.999918	0.000414
100	33	26	17	0.999989	0.000071
100	33	26	18	0.999999	0.000010
100	33	26	19	1.000000	0.000001
100	33	26	20	1.000000	0.000000
100	33	26	21	1.000000	0.000000
100	33	26	22	1.000000	0.000000
100	33	26	23	1.000000	0.000000
100	33	26	24	1.000000	0.000000
100	33	26	25	1.000000	0.000000
100	33	26	26	1.000000	0.000000
100	33	27	0	0.000002	0.000002
100	33	27	1	0.000049	0.000047
100	33	27	2	0.000510	0.000461
100	33	27	3	0.003278	0.002769
100	33	27	4	0.014604	0.011326
100	33	27	5	0.048181	0.033576
100	33	27	6	0.123119	0.074938
100	33	27	7	0.252268	0.129149
100	33	27	8	0.427157	0.174889
100	33	27	9	0.615530	0.188373
100	33	27	10	0.778284	0.162754
100	33	27	11	0.891718	0.113435
100	33	27	12	0.955707	0.063989
100	33	27	13	0.984962	0.029255
100	33	27	14	0.995797	0.010835
100	33	27	15	0.999041	0.003244
100	33	27	16	0.999823	0.000782
100	33	27	17	0.999974	0.000151
100	33	27	18	0.999997	0.000023
100	33	27	19	1.000000	0.000003
100	33	27	20	1.000000	0.000000
100	33	27	21	1.000000	0.000000
100	33	27	22	1.000000	0.000000
100	33	27	23	1.000000	0.000000
100	33	27	24	1.000000	0.000000
100	33	27	25	1.000000	0.000000
100	33	27	26	1.000000	0.000000
100	33	27	27	1.000000	0.000000
100	33	28	0	0.000001	0.000001
100	33	28	1	0.000028	0.000027
100	33	28	2	0.000314	0.000286
100	33	28	3	0.002140	0.001827
100	33	28	4	0.010105	0.007965
100	33	28	5	0.035302	0.025197
100	33	28	6	0.095402	0.060100
100	33	28	7	0.206270	0.110868

N	n	k	x	P(x)	p(x)
100	33	28	8	0.367264	0.160994
100	33	28	9	0.553599	0.186335
100	33	28	10	0.727005	0.173406
100	33	28	11	0.857533	0.130527
100	33	28	12	0.937299	0.079767
100	33	28	13	0.976947	0.039647
100	33	28	14	0.992977	0.016030
100	33	28	15	0.998241	0.005264
100	33	28	16	0.999641	0.001400
100	33	28	17	0.999941	0.000300
100	33	28	18	0.999992	0.000051
100	33	28	19	0.999999	0.000007
100	33	28	20	1.000000	0.000001
100	33	28	21	1.000000	0.000000
100	33	28	22	1.000000	0.000000
100	33	28	23	1.000000	0.000000
100	33	28	24	1.000000	0.000000
100	33	28	25	1.000000	0.000000
100	33	28	26	1.000000	0.000000
100	33	28	27	1.000000	0.000000
100	33	28	28	1.000000	0.000000
100	33	29	0	0.000001	0.000001
100	33	29	1	0.000016	0.000016
100	33	29	2	0.000191	0.000175
100	33	29	3	0.001379	0.001188
100	33	29	4	0.006897	0.005518
100	33	29	5	0.025503	0.018606
100	33	29	6	0.072865	0.047361
100	33	29	7	0.166234	0.093369
100	33	29	8	0.311363	0.145129
100	33	29	9	0.491487	0.180124
100	33	29	10	0.671611	0.180124
100	33	29	11	0.817649	0.146038
100	33	29	12	0.914034	0.096385
100	33	29	13	0.965934	0.051900
100	33	29	14	0.988747	0.022813
100	33	29	15	0.996925	0.008178
100	33	29	16	0.999310	0.002385
100	33	29	17	0.999874	0.000564
100	33	29	18	0.999981	0.000107
100	33	29	19	0.999998	0.000016
100	33	29	20	1.000000	0.000002
100	33	29	21	1.000000	0.000000
100	33	29	22	1.000000	0.000000
100	33	29	23	1.000000	0.000000
100	33	29	24	1.000000	0.000000
100	33	29	25	1.000000	0.000000
100	33	29	26	1.000000	0.000000
100	33	29	27	1.000000	0.000000
100	33	29	28	1.000000	0.000000

N	n	k	x	P(x)	p(x)
100	33	29	29	1.000000	0.000000
100	33	30	0	0.000000	0.000000
100	33	30	1	0.000009	0.000009
100	33	30	2	0.000115	0.000105
100	33	30	3	0.000877	0.000763
100	33	30	4	0.004643	0.003766
100	33	30	5	0.018166	0.013522
100	33	30	6	0.054854	0.036688
100	33	30	7	0.132042	0.077189
100	33	30	8	0.260261	0.128219
100	33	30	9	0.430600	0.170339
100	33	30	10	0.613261	0.182661
100	33	30	11	0.772398	0.159137
100	33	30	12	0.885526	0.113128
100	33	30	13	0.951314	0.065788
100	33	30	14	0.982642	0.031328
100	33	30	15	0.994851	0.012210
100	33	30	16	0.998739	0.003888
100	33	30	17	0.999747	0.001008
100	33	30	18	0.999959	0.000212
100	33	30	19	0.999994	0.000036
100	33	30	20	0.999999	0.000005
100	33	30	21	1.000000	0.000001
100	33	30	22	1.000000	0.000000
100	33	30	23	1.000000	0.000000
100	33	30	24	1.000000	0.000000
100	33	30	25	1.000000	0.000000
100	33	30	26	1.000000	0.000000
100	33	30	27	1.000000	0.000000
100	33	30	28	1.000000	0.000000
100	33	30	29	1.000000	0.000000
100	33	30	30	1.000000	0.000000
100	33	31	0	0.000000	0.000000
100	33	31	1	0.000005	0.000005
100	33	31	2	0.000068	0.000063
100	33	31	3	0.000550	0.000482
100	33	31	4	0.003083	0.002533
100	33	31	5	0.012757	0.009674
100	33	31	6	0.040703	0.027946
100	33	31	7	0.103372	0.062670
100	33	31	8	0.214469	0.111096
100	33	31	9	0.372198	0.157729
100	33	31	10	0.553244	0.181046
100	33	31	11	0.722384	0.169140
100	33	31	12	0.851588	0.129204
100	33	31	13	0.932517	0.080930
100	33	31	14	0.974138	0.041621
100	33	31	15	0.991712	0.017573
100	33	31	16	0.997795	0.006083
100	33	31	17	0.999517	0.001722

N	n	k	x	P(x)	p(x)
100	33	31	18	0.999913	0.000397
100	33	31	19	0.999987	0.000074
100	33	31	20	0.999998	0.000011
100	33	31	21	1.000000	0.000001
100	33	31	22	1.000000	0.000000
100	33	31	23	1.000000	0.000000
100	33	31	24	1.000000	0.000000
100	33	31	25	1.000000	0.000000
100	33	31	26	1.000000	0.000000
100	33	31	27	1.000000	0.000000
100	33	31	28	1.000000	0.000000
100	33	31	29	1.000000	0.000000
100	33	31	30	1.000000	0.000000
100	33	31	31	1.000000	0.000000
100	33	32	0	0.000000	0.000000
100	33	32	1	0.000003	0.000003
100	33	32	2	0.000040	0.000037
100	33	32	3	0.000341	0.000301
100	33	32	4	0.002019	0.001678
100	33	32	5	0.008831	0.006813
100	33	32	6	0.029767	0.020936
100	33	32	7	0.079758	0.049990
100	33	32	8	0.174217	0.094459
100	33	32	9	0.317336	0.143119
100	33	32	10	0.492895	0.175560
100	33	32	11	0.668455	0.175560
100	33	32	12	0.812265	0.143809
100	33	32	13	0.909059	0.096795
100	33	32	14	0.962678	0.053618
100	33	32	15	0.987127	0.024450
100	33	32	16	0.996296	0.009169
100	33	32	17	0.999117	0.002821
100	33	32	18	0.999827	0.000710
100	33	32	19	0.999972	0.000145
100	33	32	20	0.999996	0.000024
100	33	32	21	0.999999	0.000003
100	33	32	22	1.000000	0.000000
100	33	32	23	1.000000	0.000000
100	33	32	24	1.000000	0.000000
100	33	32	25	1.000000	0.000000
100	33	32	26	1.000000	0.000000
100	33	32	27	1.000000	0.000000
100	33	32	28	1.000000	0.000000
100	33	32	29	1.000000	0.000000
100	33	32	30	1.000000	0.000000
100	33	32	31	1.000000	0.000000
100	33	32	32	1.000000	0.000000
100	33	33	0	0.000000	0.000000
100	33	33	1	0.000002	0.000001
100	33	33	2	0.000023	0.000021

N	n	k	x	P(x)	p(x)
100	33	33	3	0.000208	0.000185
100	33	33	4	0.001303	0.001095
100	33	33	5	0.006026	0.004723
100	33	33	6	0.021454	0.015428
100	33	33	7	0.060644	0.039189
100	33	33	8	0.139489	0.078845
100	33	33	9	0.266823	0.127334
100	33	33	10	0.433515	0.166692
100	33	33	11	0.611656	0.178141
100	33	33	12	0.767853	0.156196
100	33	33	13	0.880590	0.112738
100	33	33	14	0.947696	0.067106
100	33	33	15	0.980655	0.032959
100	33	33	16	0.994004	0.013349
100	33	33	17	0.998453	0.004450
100	33	33	18	0.999670	0.001217
100	33	33	19	0.999942	0.000272
100	33	33	20	0.999992	0.000049
100	33	33	21	0.999999	0.000007
100	33	33	22	1.000000	0.000001
100	33	33	23	1.000000	0.000000
100	33	33	24	1.000000	0.000000
100	33	33	25	1.000000	0.000000
100	33	33	26	1.000000	0.000000
100	33	33	27	1.000000	0.000000
100	33	33	28	1.000000	0.000000
100	33	33	29	1.000000	0.000000
100	33	33	30	1.000000	0.000000
100	33	33	31	1.000000	0.000000
100	33	33	32	1.000000	0.000000
100	33	33	33	1.000000	0.000000
100	34	1	0	0.660000	0.660000
100	34	1	1	1.000000	0.340000
100	34	2	0	0.433333	0.433333
100	34	2	1	0.886667	0.453333
100	34	2	2	1.000000	0.113333
100	34	3	0	0.282993	0.282993
100	34	3	1	0.734014	0.451020
100	34	3	2	0.962993	0.228980
100	34	3	3	1.000000	0.037007
100	34	4	0	0.183800	0.183800
100	34	4	1	0.580574	0.396774
100	34	4	2	0.887454	0.30688C
100	34	4	3	0.988173	0.10072C
100	34	4	4	1.000000	0.011827
100	34	5	0	0.118704	0.118704
100	34	5	1	0.444183	0.325479
100	34	5	2	0.785160	0.340978
100	34	5	3	0.955649	0.170489
100	34	5	4	0.996304	0.040655

N	n	k	x	P(x)	p(x)
100	34	5	5	1.000000	0.003696
100	34	6	0	0.076220	0.076220
100	34	6	1	0.331122	0.254901
100	34	6	2	0.670305	0.339183
100	34	6	3	0.900016	0.229711
100	34	6	4	0.983466	0.083450
100	34	6	5	0.998872	0.015406
100	34	6	6	1.000000	0.001128
100	34	7	0	0.048651	0.048651
100	34	7	1	0.241635	0.192984
100	34	7	2	0.554838	0.313203
100	34	7	3	0.824260	0.269422
100	34	7	4	0.956833	0.132573
100	34	7	5	0.994119	0.037286
100	34	7	6	0.999664	0.005545
100	34	7	7	1.000000	0.000336
100	34	8	0	0.030865	0.030865
100	34	8	1	0.173157	0.142292
100	34	8	2	0.447069	0.273912
100	34	8	3	0.734453	0.287383
100	34	8	4	0.914067	0.179615
100	34	8	5	0.982492	0.068425
100	34	8	6	0.997994	0.015502
100	34	8	7	0.999902	0.001908
100	34	8	8	1.000000	0.000098
100	34	9	0	0.019458	0.019458
100	34	9	1	0.122117	0.102659
100	34	9	2	0.351795	0.229678
100	34	9	3	0.637617	0.285822
100	34	9	4	0.855497	0.217880
100	34	9	5	0.960923	0.105426
100	34	9	6	0.993276	0.032353
100	34	9	7	0.999342	0.006066
100	34	9	8	0.999972	0.000630
100	34	9	9	1.000000	0.000028
100	34	10	0	0.012188	0.012188
100	34	10	1	0.084889	0.072701
100	34	10	2	0.271030	0.186140
100	34	10	3	0.540249	0.269220
100	34	10	4	0.783669	0.243419
100	34	10	5	0.927326	0.143657
100	34	10	6	0.983321	0.055995
100	34	10	7	0.997543	0.014221
100	34	10	8	0.999792	0.002250
100	34	10	9	0.999992	0.000200
100	34	10	10	1.000000	0.000008
100	34	11	0	0.007584	0.007584
100	34	11	1	0.058232	0.050648
100	34	11	2	0.204846	0.146614
100	34	11	3	0.447518	0.242672

N	n	k	x	P(x)	p(x)
100	34	11	4	0.702529	0.255011
100	34	11	5	0.881036	0.178508
100	34	11	6	0.965901	0.084864
100	34	11	7	0.993276	0.027376
100	34	11	8	0.999142	0.005866
100	34	11	9	0.999937	0.000794
100	34	11	10	0.999998	0.000061
100	34	11	11	1.000000	0.000002
100	34	12	0	0.004687	0.004687
100	34	12	1	0.039452	0.034766
100	34	12	2	0.152131	0.112679
100	34	12	3	0.362992	0.210861
100	34	12	4	0.616570	0.253578
100	34	12	5	0.822871	0.206301
100	34	12	6	0.939202	0.116331
100	34	12	7	0.984971	0.045769
100	34	12	8	0.997429	0.012457
100	34	12	9	0.999714	0.002285
100	34	12	10	0.999981	0.000268
100	34	12	11	0.999999	0.000018
100	34	12	12	1.000000	0.000001
100	34	13	0	0.002876	0.002876
100	34	13	1	0.026415	0.023539
100	34	13	2	0.111157	0.084742
100	34	13	3	0.288711	0.177554
100	34	13	4	0.530123	0.241412
100	34	13	5	0.754885	0.224762
100	34	13	6	0.902187	0.147302
100	34	13	7	0.970928	0.068741
100	34	13	8	0.993748	0.022820
100	34	13	9	0.999065	0.005316
100	34	13	10	0.999908	0.000844
100	34	13	11	0.999995	0.000086
100	34	13	12	1.000000	0.000005
100	34	13	13	1.000000	0.000000
100	34	14	0	0.001752	0.001752
100	34	14	1	0.017487	0.015735
100	34	14	2	0.079988	0.062501
100	34	14	3	0.225445	0.145457
100	34	14	4	0.446878	0.221433
100	34	14	5	0.679965	0.233087
100	34	14	6	0.854780	0.174815
100	34	14	7	0.949595	0.094815
100	34	14	8	0.986928	0.037333
100	34	14	9	0.997537	0.010608
100	34	14	10	0.999676	0.002139
100	34	14	11	0.999972	0.000296
100	34	14	12	0.999998	0.000027
100	34	14	13	1.000000	0.000001
100	34	14	14	1.000000	0.000000

N	n	k	x	P(x)	p(x)
100	34	15	0	0.001059	0.001059
100	34	15	1	0.011449	0.010390
100	34	15	2	0.056731	0.045283
100	34	15	3	0.173013	0.116281
100	34	15	4	0.369634	0.196621
100	34	15	5	0.601365	0.231732
100	34	15	6	0.797863	0.196498
100	34	15	7	0.919827	0.121964
100	34	15	8	0.975642	0.055814
100	34	15	9	0.994453	0.018811
100	34	15	10	0.999079	0.004626
100	34	15	11	0.999893	0.000814
100	34	15	12	0.999992	0.000099
100	34	15	13	0.999999	0.000008
100	34	15	14	1.000000	0.000000
100	34	15	15	1.000000	0.000000
100	34	16	0	0.000636	0.000636
100	34	16	1	0.007415	0.006780
100	34	16	2	0.039684	0.032269
100	34	16	3	0.130604	0.090920
100	34	16	4	0.300238	0.169634
100	34	16	5	0.522304	0.222066
100	34	16	6	0.733135	0.210831
100	34	16	7	0.881086	0.147951
100	34	16	8	0.958569	0.077483
100	34	16	9	0.988920	0.030351
100	34	16	10	0.997773	0.008852
100	34	16	11	0.999672	0.001900
100	34	16	12	0.999966	0.000294
100	34	16	13	0.999998	0.000032
100	34	16	14	1.000000	0.000002
100	34	16	15	1.000000	0.000000
100	34	16	16	1.000000	0.000000
100	34	17	0	0.000378	0.000378
100	34	17	1	0.004752	0.004373
100	34	17	2	0.027391	0.022639
100	34	17	3	0.097050	0.069659
100	34	17	4	0.239654	0.142604
100	34	17	5	0.445638	0.205984
100	34	17	6	0.662858	0.217219
100	34	17	7	0.833530	0.170672
100	34	17	8	0.934586	0.101056
100	34	17	9	0.979887	0.045301
100	34	17	10	0.995243	0.015356
100	34	17	11	0.999152	0.003909
100	34	17	12	0.999889	0.000737
100	34	17	13	0.999990	0.000101
100	34	17	14	0.999999	0.000010
100	34	17	15	1.000000	0.000001
100	34	17	16	1.000000	0.000000

N	n	k	x	P(x)	p(x)
100	34	17	17	1.000000	0.000000
100	34	18	0	0.000223	0.000223
100	34	18	1	0.003013	0.002790
100	34	18	2	0.018663	0.015650
100	34	18	3	0.071033	0.052370
100	34	18	4	0.188111	0.117078
100	34	18	5	0.373668	0.185557
100	34	18	6	0.589579	0.215911
100	34	18	7	0.778010	0.188431
100	34	18	8	0.902930	0.124920
100	34	18	9	0.966242	0.063312
100	34	18	10	0.990803	0.024561
100	34	18	11	0.998069	0.007266
100	34	18	12	0.999694	0.001625
100	34	18	13	0.999964	0.000270
100	34	18	14	0.999997	0.000033
100	34	18	15	1.000000	0.000003
100	34	18	16	1.000000	0.000000
100	34	18	17	1.000000	0.000000
100	34	18	18	1.000000	0.000000
100	34	19	0	0.000131	0.000131
100	34	19	1	0.001890	0.001760
100	34	19	2	0.012555	0.010665
100	34	19	3	0.051234	0.038679
100	34	19	4	0.145277	0.094043
100	34	19	5	0.308044	0.162767
100	34	19	6	0.515853	0.207809
100	34	19	7	0.715966	0.200112
100	34	19	8	0.863321	0.147356
100	34	19	9	0.946940	0.083618
100	34	19	10	0.983614	0.036675
100	34	19	11	0.996031	0.012417
100	34	19	12	0.999258	0.003227
100	34	19	13	0.999895	0.000637
100	34	19	14	0.999989	0.000094
100	34	19	15	0.999999	0.000010
100	34	19	16	1.000000	0.000001
100	34	19	17	1.000000	0.000000
100	34	19	18	1.000000	0.000000
100	34	19	19	1.000000	0.000000
100	34	20	0	0.000076	0.000076
100	34	20	1	0.001173	0.001098
100	34	20	2	0.008342	0.007169
100	34	20	3	0.036431	0.028089
100	34	20	4	0.110447	0.074015
100	34	20	5	0.249770	0.139323
100	34	20	6	0.444018	0.194248
100	34	20	7	0.649262	0.205244
100	34	20	8	0.816022	0.166760
100	34	20	9	0.921132	0.105110

N	n	k	x	P(x)	p(x)
100	34	20	10	0.972748	0.051616
100	34	20	11	0.992505	0.019757
100	34	20	12	0.998381	0.005876
100	34	20	13	0.999730	0.001348
100	34	20	14	0.999966	0.000236
100	34	20	15	0.999997	0.000031
100	34	20	16	1.000000	0.000003
100	34	20	17	1.000000	0.000000
100	34	20	18	1.000000	0.000000
100	34	20	19	1.000000	0.000000
100	34	20	20	1.000000	0.000000
100	34	21	0	0.000044	0.000044
100	34	21	1	0.000721	0.000677
100	34	21	2	0.005475	0.004754
100	34	21	3	0.025547	0.020072
100	34	21	4	0.082691	0.057144
100	34	21	5	0.199265	0.116574
100	34	21	6	0.376031	0.176766
100	34	21	7	0.579992	0.203961
100	34	21	8	0.761825	0.181833
100	34	21	9	0.888285	0.126460
100	34	21	10	0.957263	0.068978
100	34	21	11	0.986825	0.029562
100	34	21	12	0.996765	0.009940
100	34	21	13	0.999376	0.002610
100	34	21	14	0.999907	0.000531
100	34	21	15	0.999989	0.000083
100	34	21	16	0.999999	0.000010
100	34	21	17	1.000000	0.000001
100	34	21	18	1.000000	0.000000
100	34	21	19	1.000000	0.000000
100	34	21	20	1.000000	0.000000
100	34	21	21	1.000000	0.000000
100	34	22	0	0.000025	0.000025
100	34	22	1	0.000438	0.000413
100	34	22	2	0.003549	0.003111
100	34	22	3	0.017670	0.014121
100	34	22	4	0.060991	0.043320
100	34	22	5	0.156472	0.095481
100	34	22	6	0.313380	0.156908
100	34	22	7	0.510284	0.196904
100	34	22	8	0.701981	0.191697
100	34	22	9	0.848266	0.146285
100	34	22	10	0.936308	0.088042
100	34	22	11	0.978218	0.041911
100	34	22	12	0.993997	0.015779
100	34	22	13	0.998682	0.004685
100	34	22	14	0.999772	0.001090
100	34	22	15	0.999969	0.000197
100	34	22	16	0.999997	0.000027

N	n	k	x	P(x)	p(x)
100	34	22	17	1.000000	0.000003
100	34	22	18	1.000000	0.000000
100	34	22	19	1.000000	0.000000
100	34	22	20	1.000000	0.000000
100	34	22	21	1.000000	0.000000
100	34	22	22	1.000000	0.000000
100	34	23	0	0.000014	0.000014
100	34	23	1	0.000263	0.000249
100	34	23	2	0.002273	0.002009
100	34	23	3	0.012058	0.009785
100	34	23	4	0.044329	0.032271
100	34	23	5	0.120973	0.076644
100	34	23	6	0.257054	0.136081
100	34	23	7	0.442125	0.185071
100	34	23	8	0.638082	0.195957
100	34	23	9	0.801380	0.163298
100	34	23	10	0.909218	0.107838
100	34	23	11	0.965860	0.056642
100	34	23	12	0.989547	0.023687
100	34	23	13	0.997421	0.007874
100	34	23	14	0.999493	0.002072
100	34	23	15	0.999921	0.000429
100	34	23	16	0.999990	0.000069
100	34	23	17	0.999999	0.000009
100	34	23	18	1.000000	0.000001
100	34	23	19	1.000000	0.000000
100	34	23	20	1.000000	0.000000
100	34	23	21	1.000000	0.000000
100	34	23	22	1.000000	0.000000
100	34	23	23	1.000000	0.000000
100	34	24	0	0.000008	0.000008
100	34	24	1	0.000156	0.000149
100	34	24	2	0.001438	0.001281
100	34	24	3	0.008118	0.006681
100	34	24	4	0.031756	0.023637
100	34	24	5	0.092107	0.060351
100	34	24	6	0.207570	0.115463
100	34	24	7	0.377230	0.169660
100	34	24	8	0.571915	0.194685
100	34	24	9	0.748361	0.176446
100	34	24	10	0.875606	0.127245
100	34	24	11	0.948941	0.073335
100	34	24	12	0.982779	0.033838
100	34	24	13	0.995273	0.012494
100	34	24	14	0.998955	0.003681
100	34	24	15	0.999816	0.000861
100	34	24	16	0.999974	0.000159
100	34	24	17	0.999997	0.000023
100	34	24	18	1.000000	0.000003
100	34	24	19	1.000000	0.000000

Table for $N = 2$, $n = 1$, through $N = 100$, $n = 50$

N	n	k	x	P(x)	p(x)
100	34	24	20	1.000000	0.000000
100	34	24	21	1.000000	0.000000
100	34	24	22	1.000000	0.000000
100	34	24	23	1.000000	0.000000
100	34	24	24	1.000000	0.000000
100	34	25	0	0.000004	0.000004
100	34	25	1	0.000092	0.000088
100	34	25	2	0.000898	0.000806
100	34	25	3	0.005393	0.004495
100	34	25	4	0.022425	0.017032
100	34	25	5	0.069078	0.046653
100	34	25	6	0.165031	0.095953
100	34	25	7	0.316956	0.151925
100	34	25	8	0.505312	0.188356
100	34	25	9	0.690319	0.185007
100	34	25	10	0.835423	0.145104
100	34	25	11	0.926747	0.091324
100	34	25	12	0.972984	0.046236
100	34	25	13	0.991821	0.018837
100	34	25	14	0.997986	0.006165
100	34	25	15	0.999600	0.001615
100	34	25	16	0.999937	0.000336
100	34	25	17	0.999992	0.000055
100	34	25	18	0.999999	0.000007
100	34	25	19	1.000000	0.000001
100	34	25	20	1.000000	0.000000
100	34	25	21	1.000000	0.000000
100	34	25	22	1.000000	0.000000
100	34	25	23	1.000000	0.000000
100	34	25	24	1.000000	0.000000
100	34	25	25	1.000000	0.000000
100	34	26	0	0.000002	0.000002
100	34	26	1	0.000053	0.000051
100	34	26	2	0.000554	0.000501
100	34	26	3	0.003535	0.002981
100	34	26	4	0.015613	0.012077
100	34	26	5	0.051039	0.035427
100	34	26	6	0.129208	0.078169
100	34	26	7	0.262263	0.133054
100	34	26	8	0.440015	0.177752
100	34	26	9	0.628650	0.188635
100	34	26	10	0.788990	0.160340
100	34	26	11	0.898741	0.109751
100	34	26	12	0.959421	0.060680
100	34	26	13	0.986547	0.027125
100	34	26	14	0.996342	0.009795
100	34	26	15	0.999191	0.002850
100	34	26	16	0.999856	0.000665
100	34	26	17	0.999980	0.000123
100	34	26	18	0.999998	0.000018

N	n	k	x	P(x)	p(x)
100	34	26	19	1.000000	0.000002
100	34	26	20	1.000000	0.000000
100	34	26	21	1.000000	0.000000
100	34	26	22	1.000000	0.000000
100	34	26	23	1.000000	0.000000
100	34	26	24	1.000000	0.000000
100	34	26	25	1.000000	0.000000
100	34	26	26	1.000000	0.000000
100	34	27	0	0.000001	0.000001
100	34	27	1	0.000031	0.000029
100	34	27	2	0.000338	0.000307
100	34	27	3	0.002286	0.001949
100	34	27	4	0.010716	0.008430
100	34	27	5	0.037156	0.026439
100	34	27	6	0.099631	0.062475
100	34	27	7	0.213716	0.114085
100	34	27	8	0.377562	0.163846
100	34	27	9	0.564922	0.187361
100	34	27	10	0.736988	0.172066
100	34	27	11	0.864630	0.127642
100	34	27	12	0.941381	0.076752
100	34	27	13	0.978849	0.037468
100	34	27	14	0.993694	0.014846
100	34	27	15	0.998460	0.004765
100	34	27	16	0.999694	0.001235
100	34	27	17	0.999951	0.000257
100	34	27	18	0.999994	0.000043
100	34	27	19	0.999999	0.000006
100	34	27	20	1.000000	0.000001
100	34	27	21	1.000000	0.000000
100	34	27	22	1.000000	0.000000
100	34	27	23	1.000000	0.000000
100	34	27	24	1.000000	0.000000
100	34	27	25	1.000000	0.000000
100	34	27	26	1.000000	0.000000
100	34	27	27	1.000000	0.000000
100	34	28	0	0.000001	0.000001
100	34	28	1	0.000017	0.000017
100	34	28	2	0.000203	0.000186
100	34	28	3	0.001459	0.001256
100	34	28	4	0.007252	0.005793
100	34	28	5	0.026652	0.019400
100	34	28	6	0.075668	0.049015
100	34	28	7	0.171520	0.095852
100	34	28	8	0.319206	0.147686
100	34	28	9	0.500758	0.181552
100	34	28	10	0.680418	0.179661
100	34	28	11	0.824414	0.143995
100	34	28	12	0.918251	0.093837
100	34	28	13	0.968070	0.049820

N	n	k	x	P(x)	p(x)
100	34	28	14	0.989627	0.021557
100	34	28	15	0.997219	0.007592
100	34	28	16	0.999390	0.002170
100	34	28	17	0.999891	0.000501
100	34	28	18	0.999984	0.000093
100	34	28	19	0.999998	0.000014
100	34	28	20	1.000000	0.000002
100	34	28	21	1.000000	0.000000
100	34	28	22	1.000000	0.000000
100	34	28	23	1.000000	0.000000
100	34	28	24	1.000000	0.000000
100	34	28	25	1.000000	0.000000
100	34	28	26	1.000000	0.000000
100	34	28	27	1.000000	0.000000
100	34	28	28	1.000000	0.000000
100	34	29	0	0.000000	0.000000
100	34	29	1	0.000010	0.000009
100	34	29	2	0.000120	0.000111
100	34	29	3	0.000918	0.000798
100	34	29	4	0.004838	0.003920
100	34	29	5	0.018838	0.014000
100	34	29	6	0.056606	0.037768
100	34	29	7	0.135575	0.078969
100	34	29	8	0.265875	0.130299
100	34	29	9	0.437719	0.171844
100	34	29	10	0.620531	0.182813
100	34	29	11	0.778415	0.157884
100	34	29	12	0.889578	0.111163
100	34	29	13	0.953540	0.063961
100	34	29	14	0.983639	0.030100
100	34	29	15	0.995216	0.011577
100	34	29	16	0.998847	0.003631
100	34	29	17	0.999773	0.000926
100	34	29	18	0.999964	0.000191
100	34	29	19	0.999995	0.000032
100	34	29	20	0.999999	0.000004
100	34	29	21	1.000000	0.000000
100	34	29	22	1.000000	0.000000
100	34	29	23	1.000000	0.000000
100	34	29	24	1.000000	0.000000
100	34	29	25	1.000000	0.000000
100	34	29	26	1.000000	0.000000
100	34	29	27	1.000000	0.000000
100	34	29	28	1.000000	0.000000
100	34	29	29	1.000000	0.000000
100	34	30	0	0.000000	0.000000
100	34	30	1	0.000005	0.000005
100	34	30	2	0.000071	0.000065
100	34	30	3	0.000570	0.000499
100	34	30	4	0.003182	0.002612

N	n	k	x	P(x)	p(x)
100	34	30	5	0.013120	0.009938
100	34	30	6	0.041712	0.028592
100	34	30	7	0.105545	0.063833
100	34	30	8	0.218160	0.112615
100	34	30	9	0.377210	0.159051
100	34	30	10	0.558736	0.181525
100	34	30	11	0.727270	0.168534
100	34	30	12	0.855133	0.127864
100	34	30	13	0.934621	0.079488
100	34	30	14	0.975160	0.040539
100	34	30	15	0.992118	0.016957
100	34	30	16	0.997927	0.005809
100	34	30	17	0.999551	0.001625
100	34	30	18	0.999921	0.000369
100	34	30	19	0.999989	0.000068
100	34	30	20	0.999999	0.000010
100	34	30	21	1.000000	0.000001
100	34	30	22	1.000000	0.000000
100	34	30	23	1.000000	0.000000
100	34	30	24	1.000000	0.000000
100	34	30	25	1.000000	0.000000
100	34	30	26	1.000000	0.000000
100	34	30	27	1.000000	0.000000
100	34	30	28	1.000000	0.000000
100	34	30	29	1.000000	0.000000
100	34	30	30	1.000000	0.000000
100	34	31	0	0.000000	0.000000
100	34	31	1	0.000003	0.000003
100	34	31	2	0.000041	0.000038
100	34	31	3	0.000349	0.000308
100	34	31	4	0.002062	0.001714
100	34	31	5	0.009003	0.006940
100	34	31	6	0.030275	0.021272
100	34	31	7	0.080924	0.050648
100	34	31	8	0.176331	0.095408
100	34	31	9	0.320406	0.144075
100	34	31	10	0.496498	0.176092
100	34	31	11	0.671894	0.175396
100	34	31	12	0.814948	0.143053
100	34	31	13	0.910775	0.095827
100	34	31	14	0.963578	0.052803
100	34	31	15	0.987515	0.023937
100	34	31	16	0.996433	0.008918
100	34	31	17	0.999157	0.002724
100	34	31	18	0.999836	0.000680
100	34	31	19	0.999974	0.000138
100	34	31	20	0.999997	0.000023
100	34	31	21	1.000000	0.000003
100	34	31	22	1.000000	0.000000
100	34	31	23	1.000000	0.000000

Table for $N = 2$, $n = 1$, through $N = 100$, $n = 50$

N	n	k	x	P(x)	p(x)
100	34	31	24	1.000000	0.000000
100	34	31	25	1.000000	0.000000
100	34	31	26	1.000000	0.000000
100	34	31	27	1.000000	0.000000
100	34	31	28	1.000000	0.000000
100	34	31	29	1.000000	0.000000
100	34	31	30	1.000000	0.000000
100	34	31	31	1.000000	0.000000
100	34	32	0	0.000000	0.000000
100	34	32	1	0.000002	0.000002
100	34	32	2	0.000023	0.000022
100	34	32	3	0.000210	0.000187
100	34	32	4	0.001317	0.001107
100	34	32	5	0.006086	0.004768
100	34	32	6	0.021643	0.015557
100	34	32	7	0.061105	0.039462
100	34	32	8	0.140381	0.079276
100	34	32	9	0.268205	0.127825
100	34	32	10	0.435249	0.167044
100	34	32	11	0.613429	0.178180
100	34	32	12	0.769336	0.155907
100	34	32	13	0.881610	0.112274
100	34	32	14	0.948273	0.066663
100	34	32	15	0.980924	0.032651
100	34	32	16	0.994107	0.013183
100	34	32	17	0.998486	0.004379
100	34	32	18	0.999679	0.001193
100	34	32	19	0.999944	0.000265
100	34	32	20	0.999992	0.000048
100	34	32	21	0.999999	0.000007
100	34	32	22	1.000000	0.000001
100	34	32	23	1.000000	0.000000
100	34	32	24	1.000000	0.000000
100	34	32	25	1.000000	0.000000
100	34	32	26	1.000000	0.000000
100	34	32	27	1.000000	0.000000
100	34	32	28	1.000000	0.000000
100	34	32	29	1.000000	0.000000
100	34	32	30	1.000000	0.000000
100	34	32	31	1.000000	0.000000
100	34	32	32	1.000000	0.000000
100	34	33	0	0.000000	0.000000
100	34	33	1	0.000001	0.000001
100	34	33	2	0.000013	0.000012
100	34	33	3	0.000125	0.000112
100	34	33	4	0.000829	0.000704
100	34	33	5	0.004052	0.003223
100	34	33	6	0.015237	0.011185
100	34	33	7	0.045436	0.030199
100	34	33	8	0.110069	0.064633

N	n	k	x	P(x)	p(x)
100	34	33	9	0.221211	0.111142
100	34	33	10	0.376292	0.155081
100	34	33	11	0.553162	0.176870
100	34	33	12	0.718896	0.165734
100	34	33	13	0.846937	0.128042
100	34	33	14	0.928666	0.081729
100	34	33	15	0.971801	0.043135
100	34	33	16	0.990617	0.018816
100	34	33	17	0.997391	0.006774
100	34	33	18	0.999398	0.002007
100	34	33	19	0.999886	0.000488
100	34	33	20	0.999982	0.000097
100	34	33	21	0.999998	0.000015
100	34	33	22	1.000000	0.000002
100	34	33	23	1.000000	0.000000
100	34	33	24	1.000000	0.000000
100	34	33	25	1.000000	0.000000
100	34	33	26	1.000000	0.000000
100	34	33	27	1.000000	0.000000
100	34	33	28	1.000000	0.000000
100	34	33	29	1.000000	0.000000
100	34	33	30	1.000000	0.000000
100	34	33	31	1.000000	0.000000
100	34	33	32	1.000000	0.000000
100	34	33	33	1.000000	0.000000
100	34	34	0	0.000000	0.000000
100	34	34	1	0.000000	0.000000
100	34	34	2	0.000007	0.000007
100	34	34	3	0.000073	0.000066
100	34	34	4	0.000514	0.000441
100	34	34	5	0.002657	0.002143
100	34	34	6	0.010563	0.007906
100	34	34	7	0.033266	0.022703
100	34	34	8	0.084988	0.051721
100	34	34	9	0.179740	0.094752
100	34	34	10	0.320741	0.141001
100	34	34	11	0.492446	0.171705
100	34	34	12	0.664476	0.172030
100	34	34	13	0.806805	0.142329
100	34	34	14	0.904269	0.097465
100	34	34	15	0.959568	0.055299
100	34	34	16	0.985562	0.025993
100	34	34	17	0.995672	0.010110
100	34	34	18	0.998919	0.003247
100	34	34	19	0.999776	0.000858
100	34	34	20	0.999962	0.000186
100	34	34	21	0.999995	0.000033
100	34	34	22	0.999999	0.000005
100	34	34	23	1.000000	0.000001
100	34	34	24	1.000000	0.000000

N	n	k	x	P(x)	p(x)
100	34	34	25	1.000000	0.000000
100	34	34	26	1.000000	0.000000
100	34	34	27	1.000000	0.000000
100	34	34	28	1.000000	0.000000
100	34	34	29	1.000000	0.000000
100	34	34	30	1.000000	0.000000
100	34	34	31	1.000000	0.000000
100	34	34	32	1.000000	0.000000
100	34	34	33	1.000000	0.000000
100	34	34	34	1.000000	0.000000
100	35	1	0	0.650000	0.650000
100	35	1	1	1.000000	0.350000
100	35	2	0	0.420202	0.420202
100	35	2	1	0.879798	0.459596
100	35	2	2	1.000000	0.120202
100	35	3	0	0.270130	0.270130
100	35	3	1	0.720346	0.450216
100	35	3	2	0.959524	0.239177
100	35	3	3	1.000000	0.040476
100	35	4	0	0.172660	0.172660
100	35	4	1	0.562538	0.389878
100	35	4	2	0.878154	0.315616
100	35	4	3	0.986647	0.108493
100	35	4	4	1.000000	0.013353
100	35	5	0	0.109711	0.109711
100	35	5	1	0.424457	0.314745
100	35	5	2	0.769661	0.345205
100	35	5	3	0.950483	0.180821
100	35	5	4	0.995688	0.045205
100	35	5	5	1.000000	0.004312
100	35	6	0	0.069291	0.069291
100	35	6	1	0.311811	0.242520
100	35	6	2	0.649748	0.337937
100	35	6	3	0.889574	0.239826
100	35	6	4	0.980937	0.091362
100	35	6	5	0.998638	0.017701
100	35	6	6	1.000000	0.001362
100	35	7	0	0.043491	0.043491
100	35	7	1	0.224091	0.180600
100	35	7	2	0.531111	0.307019
100	35	7	3	0.807931	0.276821
100	35	7	4	0.950807	0.142875
100	35	7	5	0.992989	0.042182
100	35	7	6	0.999580	0.006591
100	35	7	7	1.000000	0.000420
100	35	8	0	0.027124	0.027124
100	35	8	1	0.158065	0.130942
100	35	8	2	0.422168	0.264103
100	35	8	3	0.712681	0.290513
100	35	8	4	0.903182	0.190500

N	n	k	x	P(x)	p(x)
100	35	8	5	0.979382	0.076200
100	35	8	6	0.997525	0.018143
100	35	8	7	0.999873	0.002349
100	35	8	8	1.000000	0.000126
100	35	9	0	0.016805	0.016805
100	35	9	1	0.109674	0.092869
100	35	9	2	0.327436	0.217762
100	35	9	3	0.611633	0.284198
100	35	9	4	0.838991	0.227358
100	35	9	5	0.954534	0.115543
100	35	9	6	0.991806	0.037272
100	35	9	7	0.999159	0.007353
100	35	9	8	0.999963	0.000804
100	35	9	9	1.000000	0.000037
100	35	10	0	0.010341	0.010341
100	35	10	1	0.074976	0.064634
100	35	10	2	0.248467	0.173492
100	35	10	3	0.511696	0.263228
100	35	10	4	0.761540	0.249844
100	35	10	5	0.916443	0.154903
100	35	10	6	0.979928	0.063485
100	35	10	7	0.996896	0.016968
100	35	10	8	0.999724	0.002828
100	35	10	9	0.999989	0.000265
100	35	10	10	1.000000	0.000011
100	35	11	0	0.006320	0.006320
100	35	11	1	0.050558	0.044238
100	35	11	2	0.184853	0.134295
100	35	11	3	0.418103	0.233250
100	35	11	4	0.675482	0.257379
100	35	11	5	0.864808	0.189326
100	35	11	6	0.959472	0.094663
100	35	11	7	0.991617	0.032146
100	35	11	8	0.998876	0.007259
100	35	11	9	0.999913	0.001037
100	35	11	10	0.999997	0.000084
100	35	11	11	1.000000	0.000003
100	35	12	0	0.003834	0.003834
100	35	12	1	0.033658	0.029824
100	35	12	2	0.135059	0.101400
100	35	12	3	0.334238	0.199180
100	35	12	4	0.585833	0.251595
100	35	12	5	0.800991	0.215157
100	35	12	6	0.928626	0.127636
100	35	12	7	0.981504	0.052878
100	35	12	8	0.996674	0.015170
100	35	12	9	0.999610	0.002936
100	35	12	10	0.999973	0.000364
100	35	12	11	0.999999	0.000026
100	35	12	12	1.000000	0.000001

Table for $N = 2$, $n = 1$, through $N = 100$, $n = 50$

N	n	k	x	P(x)	p(x)
100	35	13	0	0.002309	0.002309
100	35	13	1	0.022135	0.019826
100	35	13	2	0.097033	0.074898
100	35	13	3	0.261809	0.164776
100	35	13	4	0.497203	0.235394
100	35	13	5	0.727642	0.230438
100	35	13	6	0.886565	0.158923
100	35	13	7	0.964679	0.078115
100	35	13	8	0.992019	0.027340
100	35	13	9	0.998742	0.006723
100	35	13	10	0.999870	0.001128
100	35	13	11	0.999992	0.000122
100	35	13	12	1.000000	0.000008
100	35	13	13	1.000000	0.000000
100	35	14	0	0.001380	0.001380
100	35	14	1	0.014387	0.013007
100	35	14	2	0.068624	0.054237
100	35	14	3	0.201202	0.132578
100	35	14	4	0.413327	0.212125
100	35	14	5	0.648180	0.234853
100	35	14	6	0.833590	0.185410
100	35	14	7	0.939539	0.105949
100	35	14	8	0.983534	0.043996
100	35	14	9	0.996733	0.013199
100	35	14	10	0.999546	0.002813
100	35	14	11	0.999958	0.000412
100	35	14	12	0.999998	0.000039
100	35	14	13	1.000000	0.000002
100	35	14	14	1.000000	0.000000
100	35	15	0	0.000819	0.000819
100	35	15	1	0.009245	0.008426
100	35	15	2	0.047812	0.038567
100	35	15	3	0.151871	0.104058
100	35	15	4	0.336863	0.184993
100	35	15	5	0.566255	0.229391
100	35	15	6	0.771068	0.204814
100	35	15	7	0.905044	0.133976
100	35	15	8	0.969722	0.064678
100	35	15	9	0.992743	0.023021
100	35	15	10	0.998728	0.005985
100	35	15	11	0.999843	0.001115
100	35	15	12	0.999987	0.000144
100	35	15	13	0.999999	0.000012
100	35	15	14	1.000000	0.000001
100	35	15	15	1.000000	0.000000
100	35	16	0	0.000482	0.000482
100	35	16	1	0.005874	0.005393
100	35	16	2	0.032839	0.026965
100	35	16	3	0.112696	0.079857
100	35	16	4	0.269395	0.156700

N	n	k	x	P(x)	p(x)
100	35	16	5	0.485293	0.215898
100	35	16	6	0.701191	0.215898
100	35	16	7	0.860911	0.159720
100	35	16	8	0.949177	0.088266
100	35	16	9	0.985701	0.036524
100	35	16	10	0.996968	0.011267
100	35	16	11	0.999529	0.002561
100	35	16	12	0.999948	0.000420
100	35	16	13	0.999996	0.000048
100	35	16	14	1.000000	0.000004
100	35	16	15	1.000000	0.000000
100	35	16	16	1.000000	0.000000
100	35	17	0	0.000281	0.000281
100	35	17	1	0.003692	0.003411
100	35	17	2	0.022246	0.018554
100	35	17	3	0.082274	0.060028
100	35	17	4	0.211566	0.129292
100	35	17	5	0.408187	0.196621
100	35	17	6	0.626655	0.218468
100	35	17	7	0.807671	0.181016
100	35	17	8	0.920806	0.113135
100	35	17	9	0.974396	0.053590
100	35	17	10	0.993615	0.019219
100	35	17	11	0.998797	0.005182
100	35	17	12	0.999833	0.001036
100	35	17	13	0.999984	0.000150
100	35	17	14	0.999999	0.000015
100	35	17	15	1.000000	0.000001
100	35	17	16	1.000000	0.000000
100	35	17	17	1.000000	0.000000
100	35	18	0	0.000162	0.000162
100	35	18	1	0.002294	0.002132
100	35	18	2	0.014869	0.012574
100	35	18	3	0.059131	0.044262
100	35	18	4	0.163276	0.104145
100	35	18	5	0.337119	0.173843
100	35	18	6	0.550322	0.213203
100	35	18	7	0.746605	0.196283
100	35	18	8	0.884003	0.137398
100	35	18	9	0.957609	0.073606
100	35	18	10	0.987826	0.030217
100	35	18	11	0.997298	0.009472
100	35	18	12	0.999546	0.002248
100	35	18	13	0.999944	0.000398
100	35	18	14	0.999995	0.000051
100	35	18	15	1.000000	0.000005
100	35	18	16	1.000000	0.000000
100	35	18	17	1.000000	0.000000
100	35	18	18	1.000000	0.000000
100	35	19	0	0.000093	0.000093

N	n	k	x	P(x)	p(x)
100	35	19	1	0.001410	0.001317
100	35	19	2	0.009808	0.008398
100	35	19	3	0.041858	0.032049
100	35	19	4	0.123904	0.082046
100	35	19	5	0.273518	0.149614
100	35	19	6	0.474921	0.201403
100	35	19	7	0.679582	0.204660
100	35	19	8	0.838762	0.159180
100	35	19	9	0.934270	0.095508
100	35	19	10	0.978613	0.044343
100	35	19	11	0.994526	0.015913
100	35	19	12	0.998916	0.004390
100	35	19	13	0.999837	0.000921
100	35	19	14	0.999982	0.000145
100	35	19	15	0.999998	0.000017
100	35	19	16	1.000000	0.000001
100	35	19	17	1.000000	0.000000
100	35	19	18	1.000000	0.000000
100	35	19	19	1.000000	0.000000
100	35	20	0	0.000053	0.000053
100	35	20	1	0.000857	0.000805
100	35	20	2	0.006387	0.005530
100	35	20	3	0.029196	0.022809
100	35	20	4	0.092504	0.063307
100	35	20	5	0.218105	0.125602
100	35	20	6	0.402814	0.184708
100	35	20	7	0.608835	0.206021
100	35	20	8	0.785702	0.176867
100	35	20	9	0.903613	0.117911
100	35	20	10	0.964927	0.061314
100	35	20	11	0.989811	0.024884
100	35	20	12	0.997669	0.007858
100	35	20	13	0.999587	0.001918
100	35	20	14	0.999944	0.000358
100	35	20	15	0.999994	0.000050
100	35	20	16	0.999999	0.000005
100	35	20	17	1.000000	0.000000
100	35	20	18	1.000000	0.000000
100	35	20	19	1.000000	0.000000
100	35	20	20	1.000000	0.000000
100	35	21	0	0.000030	0.000030
100	35	21	1	0.000516	0.000486
100	35	21	2	0.004106	0.003591
100	35	21	3	0.020073	0.015966
100	35	21	4	0.067972	0.047899
100	35	21	5	0.171005	0.103033
100	35	21	6	0.335857	0.164852
100	35	21	7	0.536728	0.200870
100	35	21	8	0.726009	0.189282
100	35	21	9	0.865292	0.139283

N	n	k	x	P(x)	p(x)
100	35	21	10	0.945767	0.080474
100	35	21	11	0.982346	0.036579
100	35	21	12	0.995410	0.013064
100	35	21	13	0.999059	0.003649
100	35	21	14	0.999850	0.000791
100	35	21	15	0.999982	0.000131
100	35	21	16	0.999998	0.000016
100	35	21	17	1.000000	0.000001
100	35	21	18	1.000000	0.000000
100	35	21	19	1.000000	0.000000
100	35	21	20	1.000000	0.000000
100	35	21	21	1.000000	0.000000
100	35	22	0	0.000017	0.000017
100	35	22	1	0.000306	0.000290
100	35	22	2	0.002606	0.002300
100	35	22	3	0.013605	0.010999
100	35	22	4	0.049176	0.035571
100	35	22	5	0.131878	0.082702
100	35	22	6	0.275342	0.143463
100	35	22	7	0.465533	0.190191
100	35	22	8	0.661318	0.195785
100	35	22	9	0.819452	0.158134
100	35	22	10	0.920300	0.100848
100	35	22	11	0.971233	0.050933
100	35	22	12	0.991606	0.020373
100	35	22	13	0.998043	0.006437
100	35	22	14	0.999640	0.001597
100	35	22	15	0.999949	0.000308
100	35	22	16	0.999994	0.000046
100	35	22	17	0.999999	0.000005
100	35	22	18	1.000000	0.000000
100	35	22	19	1.000000	0.000000
100	35	22	20	1.000000	0.000000
100	35	22	21	1.000000	0.000000
100	35	22	22	1.000000	0.000000
100	35	23	0	0.000009	0.000009
100	35	23	1	0.000180	0.000171
100	35	23	2	0.001633	0.001453
100	35	23	3	0.009093	0.007460
100	35	23	4	0.035039	0.025946
100	35	23	5	0.100070	0.065031
100	35	23	6	0.222003	0.121933
100	35	23	7	0.397259	0.175256
100	35	23	8	0.593546	0.196287
100	35	23	9	0.766741	0.173195
100	35	23	10	0.887977	0.121236
100	35	23	11	0.955561	0.067584
100	35	23	12	0.985599	0.030038
100	35	23	13	0.996228	0.010629
100	35	23	14	0.999210	0.002983

N	n	k	x	P(x)	p(x)
100	35	23	15	0.999870	0.000659
100	35	23	16	0.999983	0.000114
100	35	23	17	0.999998	0.000015
100	35	23	18	1.000000	0.000002
100	35	23	19	1.000000	0.000000
100	35	23	20	1.000000	0.000000
100	35	23	21	1.000000	0.000000
100	35	23	22	1.000000	0.000000
100	35	23	23	1.000000	0.000000
100	35	24	0	0.000005	0.000005
100	35	24	1	0.000105	0.000100
100	35	24	2	0.001010	0.000906
100	35	24	3	0.005993	0.004982
100	35	24	4	0.024593	0.018600
100	35	24	5	0.074733	0.050140
100	35	24	6	0.176080	0.101347
100	35	24	7	0.333530	0.157450
100	35	24	8	0.524718	0.191189
100	35	24	9	0.708260	0.183541
100	35	24	10	0.848615	0.140355
100	35	24	11	0.934496	0.085882
100	35	24	12	0.976627	0.042131
100	35	24	13	0.993191	0.016564
100	35	24	14	0.998397	0.005206
100	35	24	15	0.999698	0.001301
100	35	24	16	0.999955	0.000257
100	35	24	17	0.999995	0.000040
100	35	24	18	0.999999	0.000005
100	35	24	19	1.000000	0.000000
100	35	24	20	1.000000	0.000000
100	35	24	21	1.000000	0.000000
100	35	24	22	1.000000	0.000000
100	35	24	23	1.000000	0.000000
100	35	24	24	1.000000	0.000000
100	35	25	0	0.000003	0.000003
100	35	25	1	0.000060	0.000057
100	35	25	2	0.000617	0.000557
100	35	25	3	0.003895	0.003278
100	35	25	4	0.017006	0.013111
100	35	25	5	0.054941	0.037935
100	35	25	6	0.137408	0.082467
100	35	25	7	0.275522	0.138114
100	35	25	8	0.456796	0.181274
100	35	25	9	0.645469	0.188673
100	35	25	10	0.802445	0.156976
100	35	25	11	0.907376	0.104931
100	35	25	12	0.963877	0.056501
100	35	25	13	0.988396	0.024519
100	35	25	14	0.996958	0.008562
100	35	25	15	0.999356	0.002397

N	n	k	x	P(x)	p(x)
100	35	25	16	0.999891	0.000535
100	35	25	17	0.999985	0.000094
100	35	25	18	0.999998	0.000013
100	35	25	19	1.000000	0.000001
100	35	25	20	1.000000	0.000000
100	35	25	21	1.000000	0.000000
100	35	25	22	1.000000	0.000000
100	35	25	23	1.000000	0.000000
100	35	25	24	1.000000	0.000000
100	35	25	25	1.000000	0.000000
100	35	26	0	0.000001	0.000001
100	35	26	1	0.000034	0.000033
100	35	26	2	0.000372	0.000338
100	35	26	3	0.002496	0.002124
100	35	26	4	0.011587	0.009090
100	35	26	5	0.039767	0.028180
100	35	26	6	0.105521	0.065754
100	35	26	7	0.223959	0.118439
100	35	26	8	0.391537	0.167578
100	35	26	9	0.580062	0.188525
100	35	26	10	0.750120	0.170057
100	35	26	11	0.873798	0.123678
100	35	26	12	0.946550	0.072752
100	35	26	13	0.981204	0.034654
100	35	26	14	0.994561	0.013357
100	35	26	15	0.998716	0.004156
100	35	26	16	0.999755	0.001039
100	35	26	17	0.999963	0.000207
100	35	26	18	0.999995	0.000033
100	35	26	19	0.999999	0.000004
100	35	26	20	1.000000	0.000000
100	35	26	21	1.000000	0.000000
100	35	26	22	1.000000	0.000000
100	35	26	23	1.000000	0.000000
100	35	26	24	1.000000	0.000000
100	35	26	25	1.000000	0.000000
100	35	26	26	1.000000	0.000000
100	35	27	0	0.000001	0.000001
100	35	27	1	0.000019	0.000018
100	35	27	2	0.000221	0.000202
100	35	27	3	0.001578	0.001356
100	35	27	4	0.007779	0.006201
100	35	27	5	0.028343	0.020564
100	35	27	6	0.079752	0.051410
100	35	27	7	0.179145	0.099392
100	35	27	8	0.330394	0.151249
100	35	27	9	0.513824	0.183430
100	35	27	10	0.692668	0.178844
100	35	27	11	0.833686	0.141018
100	35	27	12	0.923938	0.090252

N	n	k	x	P(x)	p(x)
100	35	27	13	0.970901	0.046965
100	35	27	14	0.990770	0.019869
100	35	27	15	0.997593	0.006823
100	35	27	16	0.999489	0.001895
100	35	27	17	0.999912	0.000424
100	35	27	18	0.999988	0.000076
100	35	27	19	0.999999	0.000011
100	35	27	20	1.000000	0.000001
100	35	27	21	1.000000	0.000000
100	35	27	22	1.000000	0.000000
100	35	27	23	1.000000	0.000000
100	35	27	24	1.000000	0.000000
100	35	27	25	1.000000	0.000000
100	35	27	26	1.000000	0.000000
100	35	27	27	1.000000	0.000000
100	35	28	0	0.000000	0.000000
100	35	28	1	0.000011	0.000010
100	35	28	2	0.000130	0.000119
100	35	28	3	0.000983	0.000853
100	35	28	4	0.005145	0.004162
100	35	28	5	0.019892	0.014746
100	35	28	6	0.059329	0.039438
100	35	28	7	0.141022	0.081692
100	35	28	8	0.274452	0.133431
100	35	28	9	0.448493	0.174040
100	35	28	10	0.631420	0.182927
100	35	28	11	0.787324	0.155904
100	35	28	12	0.895502	0.108178
100	35	28	13	0.956748	0.061246
100	35	28	14	0.985054	0.028307
100	35	28	15	0.995724	0.010669
100	35	28	16	0.998995	0.003271
100	35	28	17	0.999808	0.000812
100	35	28	18	0.999970	0.000162
100	35	28	19	0.999996	0.000026
100	35	28	20	0.999999	0.000003
100	35	28	21	1.000000	0.000000
100	35	28	22	1.000000	0.000000
100	35	28	23	1.000000	0.000000
100	35	28	24	1.000000	0.000000
100	35	28	25	1.000000	0.000000
100	35	28	26	1.000000	0.000000
100	35	28	27	1.000000	0.000000
100	35	28	28	1.000000	0.000000
100	35	29	0	0.000000	0.000000
100	35	29	1	0.000006	0.000006
100	35	29	2	0.000075	0.000069
100	35	29	3	0.000604	0.000529
100	35	29	4	0.003353	0.002749
100	35	29	5	0.013747	0.010394

N	n	k	x	P(x)	p(x)
100	35	29	6	0.043445	0.029697
100	35	29	7	0.109252	0.065808
100	35	29	8	0.224416	0.115163
100	35	29	9	0.385645	0.161229
100	35	29	10	0.567904	0.182259
100	35	29	11	0.735356	0.167452
100	35	29	12	0.860945	0.125589
100	35	29	13	0.938034	0.077089
100	35	29	14	0.976798	0.038765
100	35	29	15	0.992760	0.015962
100	35	29	16	0.998132	0.005372
100	35	29	17	0.999605	0.001473
100	35	29	18	0.999932	0.000327
100	35	29	19	0.999990	0.000059
100	35	29	20	0.999999	0.000008
100	35	29	21	1.000000	0.000001
100	35	29	22	1.000000	0.000000
100	35	29	23	1.000000	0.000000
100	35	29	24	1.000000	0.000000
100	35	29	25	1.000000	0.000000
100	35	29	26	1.000000	0.000000
100	35	29	27	1.000000	0.000000
100	35	29	28	1.000000	0.000000
100	35	29	29	1.000000	0.000000
100	35	30	0	0.000000	0.000000
100	35	30	1	0.000003	0.000003
100	35	30	2	0.000043	0.000040
100	35	30	3	0.000366	0.000323
100	35	30	4	0.002153	0.001787
100	35	30	5	0.009356	0.007203
100	35	30	6	0.031315	0.021959
100	35	30	7	0.083300	0.051985
100	35	30	8	0.180621	0.097321
100	35	30	9	0.326603	0.145982
100	35	30	10	0.503728	0.177125
100	35	30	11	0.678752	0.175024
100	35	30	12	0.820261	0.141509
100	35	30	13	0.914147	0.093886
100	35	30	14	0.965333	0.051186
100	35	30	15	0.988264	0.022931
100	35	30	16	0.996695	0.008431
100	35	30	17	0.999231	0.002537
100	35	30	18	0.999854	0.000622
100	35	30	19	0.999977	0.000124
100	35	30	20	0.999997	0.000020
100	35	30	21	1.000000	0.000003
100	35	30	22	1.000000	0.000000
100	35	30	23	1.000000	0.000000
100	35	30	24	1.000000	0.000000
100	35	30	25	1.000000	0.000000

Table for $N = 2$, $n = 1$, through $N = 100$, $n = 50$

N	n	k	x	P(x)	p(x)
100	35	30	26	1.000000	0.000000
100	35	30	27	1.000000	0.000000
100	35	30	28	1.000000	0.000000
100	35	30	29	1.000000	0.000000
100	35	30	30	1.000000	0.000000
100	35	31	0	0.000000	0.000000
100	35	31	1	0.000002	0.000002
100	35	31	2	0.000024	0.000022
100	35	31	3	0.000218	0.000194
100	35	31	4	0.001361	0.001143
100	35	31	5	0.006269	0.004907
100	35	31	6	0.022217	0.015949
100	35	31	7	0.062506	0.040289
100	35	31	8	0.143083	0.080577
100	35	31	9	0.272381	0.129298
100	35	31	10	0.440469	0.168088
100	35	31	11	0.618744	0.178275
100	35	31	12	0.773766	0.155022
100	35	31	13	0.884640	0.110875
100	35	31	14	0.949977	0.065337
100	35	31	15	0.981712	0.031735
100	35	31	16	0.994406	0.012694
100	35	31	17	0.998579	0.004173
100	35	31	18	0.999702	0.001123
100	35	31	19	0.999949	0.000247
100	35	31	20	0.999993	0.000044
100	35	31	21	0.999999	0.000006
100	35	31	22	1.000000	0.000001
100	35	31	23	1.000000	0.000000
100	35	31	24	1.000000	0.000000
100	35	31	25	1.000000	0.000000
100	35	31	26	1.000000	0.000000
100	35	31	27	1.000000	0.000000
100	35	31	28	1.000000	0.000000
100	35	31	29	1.000000	0.000000
100	35	31	30	1.000000	0.000000
100	35	31	31	1.000000	0.000000
100	35	32	0	0.000000	0.000000
100	35	32	1	0.000001	0.000001
100	35	32	2	0.000013	0.000013
100	35	32	3	0.000128	0.000115
100	35	32	4	0.000848	0.000720
100	35	32	5	0.004135	0.003287
100	35	32	6	0.015514	0.011379
100	35	32	7	0.046157	0.030643
100	35	32	8	0.111553	0.065396
100	35	32	9	0.223660	0.112107
100	35	32	10	0.379568	0.155907
100	35	32	11	0.556735	0.177168
100	35	32	12	0.722092	0.165356

N	n	k	x	P(x)	p(x)
100	35	32	13	0.849289	0.127197
100	35	32	14	0.930092	0.080803
100	35	32	15	0.972514	0.042422
100	35	32	16	0.990911	0.018397
100	35	32	17	0.997490	0.006580
100	35	32	18	0.999426	0.001935
100	35	32	19	0.999892	0.000466
100	35	32	20	0.999983	0.000091
100	35	32	21	0.999998	0.000015
100	35	32	22	1.000000	0.000002
100	35	32	23	1.000000	0.000000
100	35	32	24	1.000000	0.000000
100	35	32	25	1.000000	0.000000
100	35	32	26	1.000000	0.000000
100	35	32	27	1.000000	0.000000
100	35	32	28	1.000000	0.000000
100	35	32	29	1.000000	0.000000
100	35	32	30	1.000000	0.000000
100	35	32	31	1.000000	0.000000
100	35	32	32	1.000000	0.000000
100	35	33	0	0.000000	0.000000
100	35	33	1	0.000000	0.000000
100	35	33	2	0.000007	0.000007
100	35	33	3	0.000074	0.000067
100	35	33	4	0.000520	0.000446
100	35	33	5	0.002685	0.002165
100	35	33	6	0.010661	0.007977
100	35	33	7	0.033539	0.022878
100	35	33	8	0.085587	0.052047
100	35	33	9	0.180796	0.095209
100	35	33	10	0.322249	0.141453
100	35	33	11	0.494206	0.171957
100	35	33	12	0.666162	0.171957
100	35	33	13	0.808137	0.141975
100	35	33	14	0.905138	0.097001
100	35	33	15	0.960037	0.054899
100	35	33	16	0.985770	0.025734
100	35	33	17	0.995749	0.009978
100	35	33	18	0.998942	0.003193
100	35	33	19	0.999782	0.000840
100	35	33	20	0.999963	0.000181
100	35	33	21	0.999995	0.000032
100	35	33	22	0.999999	0.000004
100	35	33	23	1.000000	0.000001
100	35	33	24	1.000000	0.000000
100	35	33	25	1.000000	0.000000
100	35	33	26	1.000000	0.000000
100	35	33	27	1.000000	0.000000
100	35	33	28	1.000000	0.000000
100	35	33	29	1.000000	0.000000

N	n	k	x	P(x)	p(x)
100	35	33	30	1.000000	0.000000
100	35	33	31	1.000000	0.000000
100	35	33	32	1.000000	0.000000
100	35	33	33	1.000000	0.000000
100	35	34	0	0.000000	0.000000
100	35	34	1	0.000000	0.000000
100	35	34	2	0.000004	0.000004
100	35	34	3	0.000042	0.000038
100	35	34	4	0.000314	0.000271
100	35	34	5	0.001715	0.001402
100	35	34	6	0.007209	0.005493
100	35	34	7	0.023978	0.016770
100	35	34	8	0.064613	0.040634
100	35	34	9	0.143849	0.079236
100	35	34	10	0.269468	0.125619
100	35	34	11	0.432609	0.163141
100	35	34	12	0.607132	0.174523
100	35	34	13	0.761518	0.154386
100	35	34	14	0.874735	0.113216
100	35	34	15	0.943649	0.068914
100	35	34	16	0.978473	0.034824
100	35	34	17	0.993068	0.014595
100	35	34	18	0.998132	0.005064
100	35	34	19	0.999581	0.001450
100	35	34	20	0.999923	0.000341
100	35	34	21	0.999988	0.000066
100	35	34	22	0.999998	0.000010
100	35	34	23	1.000000	0.000001
100	35	34	24	1.000000	0.000000
100	35	34	25	1.000000	0.000000
100	35	34	26	1.000000	0.000000
100	35	34	27	1.000000	0.000000
100	35	34	28	1.000000	0.000000
100	35	34	29	1.000000	0.000000
100	35	34	30	1.000000	0.000000
100	35	34	31	1.000000	0.000000
100	35	34	32	1.000000	0.000000
100	35	34	33	1.000000	0.000000
100	35	34	34	1.000000	0.000000
100	35	35	0	0.000000	0.000000
100	35	35	1	0.000000	0.000000
100	35	35	2	0.000002	0.000002
100	35	35	3	0.000024	0.000022
100	35	35	4	0.000186	0.000162
100	35	35	5	0.001078	0.000892
100	35	35	6	0.004795	0.003717
100	35	35	7	0.016864	0.012069
100	35	35	8	0.047989	0.031125
100	35	35	9	0.112635	0.064645
100	35	35	10	0.221885	0.109250

N	n	k	x	P(x)	p(x)
100	35	35	11	0.373285	0.151400
100	35	35	12	0.546314	0.173029
100	35	35	13	0.710056	0.163743
100	35	35	14	0.838711	0.128655
100	35	35	15	0.922766	0.084055
100	35	35	16	0.968448	0.045682
100	35	35	17	0.989087	0.020640
100	35	35	18	0.996827	0.007740
100	35	35	19	0.999230	0.002403
100	35	35	20	0.999845	0.000615
100	35	35	21	0.999974	0.000129
100	35	35	22	0.999996	0.000022
100	35	35	23	0.999999	0.000003
100	35	35	24	1.000000	0.000000
100	35	35	25	1.000000	0.000000
100	35	35	26	1.000000	0.000000
100	35	35	27	1.000000	0.000000
100	35	35	28	1.000000	0.000000
100	35	35	29	1.000000	0.000000
100	35	35	30	1.000000	0.000000
100	35	35	31	1.000000	0.000000
100	35	35	32	1.000000	0.000000
100	35	35	33	1.000000	0.000000
100	35	35	34	1.000000	0.000000
100	35	35	35	1.000000	0.000000
100	36	1	0	0.640000	0.640000
100	36	1	1	1.000000	0.360000
100	36	2	0	0.407273	0.407273
100	36	2	1	0.872727	0.465455
100	36	2	2	1.000000	0.127273
100	36	3	0	0.257662	0.257662
100	36	3	1	0.706494	0.448831
100	36	3	2	0.955844	0.249351
100	36	3	3	1.000000	0.044156
100	36	4	0	0.162035	0.162035
100	36	4	1	0.544544	0.382509
100	36	4	2	0.868443	0.323899
100	36	4	3	0.984978	0.116535
100	36	4	4	1.000000	0.015022
100	36	5	0	0.101272	0.101272
100	36	5	1	0.405088	0.303816
100	36	5	2	0.753729	0.348641
100	36	5	3	0.944919	0.191190
100	36	5	4	0.994993	0.050074
100	36	5	5	1.000000	0.005007
100	36	6	0	0.062895	0.062895
100	36	6	1	0.293156	0.230260
100	36	6	2	0.628952	0.335796
100	36	6	3	0.878505	0.249554
100	36	6	4	0.978126	0.099620

Table for $N = 2$, $n = 1$, through $N = 100$, $n = 50$

N	n	k	x	P(x)	p(x)
100	36	6	5	0.998366	0.020240
100	36	6	6	1.000000	0.001634
100	36	7	0	0.038808	0.038808
100	36	7	1	0.207420	0.168613
100	36	7	2	0.507494	0.300073
100	36	7	3	0.790896	0.283403
100	36	7	4	0.944212	0.153316
100	36	7	5	0.991691	0.047479
100	36	7	6	0.999478	0.007787
100	36	7	7	1.000000	0.000521
100	36	8	0	0.023785	0.023785
100	36	8	1	0.143964	0.120179
100	36	8	2	0.397789	0.253825
100	36	8	3	0.690334	0.292545
100	36	8	4	0.891458	0.201124
100	36	8	5	0.975865	0.084406
100	36	8	6	0.996966	0.021102
100	36	8	7	0.999837	0.002871
100	36	8	8	1.000000	0.000163
100	36	9	0	0.014478	0.014478
100	36	9	1	0.098244	0.083766
100	36	9	2	0.303984	0.205741
100	36	9	3	0.585400	0.281415
100	36	9	4	0.821502	0.236103
100	36	9	5	0.947424	0.125921
100	36	9	6	0.990085	0.042662
100	36	9	7	0.998932	0.008847
100	36	9	8	0.999950	0.001018
100	36	9	9	1.000000	0.000049
100	36	10	0	0.008750	0.008750
100	36	10	1	0.066026	0.057276
100	36	10	2	0.227114	0.161088
100	36	10	3	0.483348	0.256234
100	36	10	4	0.738477	0.255129
100	36	10	5	0.904527	0.166050
100	36	10	6	0.976021	0.071494
100	36	10	7	0.996113	0.020092
100	36	10	8	0.999637	0.003524
100	36	10	9	0.999985	0.000348
100	36	10	10	1.000000	0.000015
100	36	11	0	0.005250	0.005250
100	36	11	1	0.043752	0.038502
100	36	11	2	0.166259	0.122506
100	36	11	3	0.389396	0.223137
100	36	11	4	0.647764	0.258369
100	36	11	5	0.847332	0.199568
100	36	11	6	0.952190	0.104858
100	36	11	7	0.989639	0.037449
100	36	11	8	0.998541	0.008902
100	36	11	9	0.999881	0.001340

N	n	k	x	P(x)	p(x)
100	36	11	10	0.999996	0.000115
100	36	11	11	1.000000	0.000004
100	36	12	0	0.003127	0.003127
100	36	12	1	0.028611	0.025484
100	36	12	2	0.119459	0.090847
100	36	12	3	0.306659	0.187201
100	36	12	4	0.554868	0.248208
100	36	12	5	0.777820	0.222952
100	36	12	6	0.916845	0.139025
100	36	12	7	0.977436	0.060592
100	36	12	8	0.995740	0.018304
100	36	12	9	0.999474	0.003734
100	36	12	10	0.999962	0.000488
100	36	12	11	0.999999	0.000037
100	36	12	12	1.000000	0.000001
100	36	13	0	0.001848	0.001848
100	36	13	1	0.018475	0.016628
100	36	13	2	0.084358	0.065883
100	36	13	3	0.236459	0.152101
100	36	13	4	0.464610	0.228151
100	36	13	5	0.699280	0.234670
100	36	13	6	0.869450	0.170170
100	36	13	7	0.957469	0.088019
100	36	13	8	0.989916	0.032448
100	36	13	9	0.998329	0.008412
100	36	13	10	0.999818	0.001489
100	36	13	11	0.999988	0.000170
100	36	13	12	1.000000	0.000011
100	36	13	13	1.000000	0.000000
100	36	14	0	0.001083	0.001083
100	36	14	1	0.011786	0.010703
100	36	14	2	0.058611	0.046825
100	36	14	3	0.178766	0.120155
100	36	14	4	0.380693	0.201927
100	36	14	5	0.615662	0.234969
100	36	14	6	0.810770	0.195109
100	36	14	7	0.928129	0.117359
100	36	14	8	0.979473	0.051344
100	36	14	9	0.995718	0.016245
100	36	14	10	0.999373	0.003655
100	36	14	11	0.999939	0.000566
100	36	14	12	0.999996	0.000057
100	36	14	13	1.000000	0.000003
100	36	14	14	1.000000	0.000000
100	36	15	0	0.000630	0.000630
100	36	15	1	0.007430	0.006800
100	36	15	2	0.040099	0.032669
100	36	15	3	0.132660	0.092561
100	36	15	4	0.305557	0.172897
100	36	15	5	0.530964	0.225407

N	n	k	x	P(x)	p(x)
100	36	15	6	0.742709	0.211746
100	36	15	7	0.888555	0.145845
100	36	15	8	0.962757	0.074202
100	36	15	9	0.990618	0.027861
100	36	15	10	0.998268	0.007650
100	36	15	11	0.999775	0.001507
100	36	15	12	0.999981	0.000206
100	36	15	13	0.999999	0.000018
100	36	15	14	1.000000	0.000001
100	36	15	15	1.000000	0.000000
100	36	16	0	0.000363	0.000363
100	36	16	1	0.004630	0.004267
100	36	16	2	0.027031	0.022401
100	36	16	3	0.096724	0.069693
100	36	16	4	0.240466	0.143742
100	36	16	5	0.448757	0.208290
100	36	16	6	0.667976	0.219219
100	36	16	7	0.838796	0.170820
100	36	16	8	0.938314	0.099518
100	36	16	9	0.981768	0.043454
100	36	16	10	0.995928	0.014160
100	36	16	11	0.999331	0.003404
100	36	16	12	0.999922	0.000591
100	36	16	13	0.999994	0.000072
100	36	16	14	1.000000	0.000006
100	36	16	15	1.000000	0.000000
100	36	16	16	1.000000	0.000000
100	36	17	0	0.000207	0.000207
100	36	17	1	0.002852	0.002645
100	36	17	2	0.017964	0.015112
100	36	17	3	0.069345	0.051381
100	36	17	4	0.185707	0.116363
100	36	17	5	0.371887	0.186180
100	36	17	6	0.589683	0.217796
100	36	17	7	0.779822	0.190139
100	36	17	8	0.905141	0.125319
100	36	17	9	0.967800	0.062659
100	36	17	10	0.991545	0.023745
100	36	17	11	0.998319	0.006774
100	36	17	12	0.999754	0.001435
100	36	17	13	0.999974	0.000221
100	36	17	14	0.999998	0.000024
100	36	17	15	1.000000	0.000002
100	36	17	16	1.000000	0.000000
100	36	17	17	1.000000	0.000000
100	36	18	0	0.000117	0.000117
100	36	18	1	0.001737	0.001619
100	36	18	2	0.011774	0.010037
100	36	18	3	0.048916	0.037143
100	36	18	4	0.140845	0.091928

N	n	k	x	P(x)	p(x)
100	36	18	5	0.302350	0.161506
100	36	18	6	0.510962	0.208611
100	36	18	7	0.713388	0.202426
100	36	18	8	0.862865	0.149477
100	36	18	9	0.947417	0.084552
100	36	18	10	0.984107	0.036690
100	36	18	11	0.996278	0.012171
100	36	18	12	0.999339	0.003060
100	36	18	13	0.999913	0.000575
100	36	18	14	0.999992	0.000079
100	36	18	15	0.999999	0.000008
100	36	18	16	1.000000	0.000000
100	36	18	17	1.000000	0.000000
100	36	18	18	1.000000	0.000000
100	36	19	0	0.000066	0.000066
100	36	19	1	0.001046	0.000980
100	36	19	2	0.007612	0.006566
100	36	19	3	0.033969	0.026357
100	36	19	4	0.104970	0.071002
100	36	19	5	0.241293	0.136323
100	36	19	6	0.434640	0.193347
100	36	19	7	0.641798	0.207158
100	36	19	8	0.811824	0.170026
100	36	19	9	0.919577	0.107753
100	36	19	10	0.972474	0.052897
100	36	19	11	0.992567	0.020094
100	36	19	12	0.998443	0.005875
100	36	19	13	0.999752	0.001309
100	36	19	14	0.999971	0.000219
100	36	19	15	0.999997	0.000027
100	36	19	16	1.000000	0.000002
100	36	19	17	1.000000	0.000000
100	36	19	18	1.000000	0.000000
100	36	19	19	1.000000	0.000000
100	36	20	0	0.000037	0.000037
100	36	20	1	0.000622	0.000586
100	36	20	2	0.004856	0.004233
100	36	20	3	0.023231	0.018375
100	36	20	4	0.076920	0.053689
100	36	20	5	0.189120	0.112200
100	36	20	6	0.363030	0.173910
100	36	20	7	0.567631	0.204600
100	36	20	8	0.753050	0.185419
100	36	20	9	0.883659	0.130610
100	36	20	10	0.955494	0.071835
100	36	20	11	0.986366	0.030871
100	36	20	12	0.996702	0.010336
100	36	20	13	0.999380	0.002678
100	36	20	14	0.999911	0.000531
100	36	20	15	0.999991	0.000079

Table for $N = 2$, $n = 1$, through $N = 100$, $n = 50$

N	n	k	x	P(x)	p(x)
100	36	20	16	0.999999	0.000009
100	36	20	17	1.000000	0.000001
100	36	20	18	1.000000	0.000000
100	36	20	19	1.000000	0.000000
100	36	20	20	1.000000	0.000000
100	36	21	0	0.000020	0.000020
100	36	21	1	0.000366	0.000346
100	36	21	2	0.003057	0.002690
100	36	21	3	0.015651	0.012595
100	36	21	4	0.055444	0.039793
100	36	21	5	0.145643	0.090198
100	36	21	6	0.297814	0.152171
100	36	21	7	0.493463	0.195649
100	36	21	8	0.688153	0.194690
100	36	21	9	0.839578	0.151425
100	36	21	10	0.932148	0.092569
100	36	21	11	0.976718	0.044570
100	36	21	12	0.993601	0.016883
100	36	21	13	0.998610	0.005009
100	36	21	14	0.999765	0.001155
100	36	21	15	0.999970	0.000204
100	36	21	16	0.999997	0.000027
100	36	21	17	1.000000	0.000003
100	36	21	18	1.000000	0.000000
100	36	21	19	1.000000	0.000000
100	36	21	20	1.000000	0.000000
100	36	21	21	1.000000	0.000000
100	36	22	0	0.000011	0.000011
100	36	22	1	0.000213	0.000202
100	36	22	2	0.001899	0.001686
100	36	22	3	0.010390	0.008491
100	36	22	4	0.039326	0.028936
100	36	22	5	0.110249	0.070923
100	36	22	6	0.240028	0.129779
100	36	22	7	0.421643	0.181615
100	36	22	8	0.619149	0.197506
100	36	22	9	0.787825	0.168676
100	36	22	10	0.901682	0.113857
100	36	22	11	0.962614	0.060932
100	36	22	12	0.988472	0.025858
100	36	22	13	0.997152	0.008680
100	36	22	14	0.999444	0.002292
100	36	22	15	0.999915	0.000472
100	36	22	16	0.999990	0.000075
100	36	22	17	0.999999	0.000009
100	36	22	18	1.000000	0.000001
100	36	22	19	1.000000	0.000000
100	36	22	20	1.000000	0.000000
100	36	22	21	1.000000	0.000000
100	36	22	22	1.000000	0.000000

N	n	k	x	P(x)	p(x)
100	36	23	0	0.000006	0.000006
100	36	23	1	0.000122	0.000116
100	36	23	2	0.001164	0.001042
100	36	23	3	0.006798	0.005634
100	36	23	4	0.027455	0.020657
100	36	23	5	0.082061	0.054607
100	36	23	6	0.190113	0.108051
100	36	23	7	0.354119	0.164006
100	36	23	8	0.548249	0.194130
100	36	23	9	0.729437	0.181188
100	36	23	10	0.863730	0.134292
100	36	23	11	0.943084	0.079355
100	36	23	12	0.980516	0.037431
100	36	23	13	0.994592	0.014077
100	36	23	14	0.998797	0.004205
100	36	23	15	0.999788	0.000991
100	36	23	16	0.999971	0.000183
100	36	23	17	0.999997	0.000026
100	36	23	18	1.000000	0.000003
100	36	23	19	1.000000	0.000000
100	36	23	20	1.000000	0.000000
100	36	23	21	1.000000	0.000000
100	36	23	22	1.000000	0.000000
100	36	23	23	1.000000	0.000000
100	36	24	0	0.000003	0.000003
100	36	24	1	0.000069	0.000066
100	36	24	2	0.000704	0.000635
100	36	24	3	0.004383	0.003679
100	36	24	4	0.018870	0.014487
100	36	24	5	0.060077	0.041207
100	36	24	6	0.148015	0.087938
100	36	24	7	0.292350	0.144336
100	36	24	8	0.477656	0.185306
100	36	24	9	0.665904	0.188247
100	36	24	10	0.818384	0.152480
100	36	24	11	0.917320	0.098936
100	36	24	12	0.968849	0.051529
100	36	24	13	0.990388	0.021539
100	36	24	14	0.997596	0.007208
100	36	24	15	0.999518	0.001922
100	36	24	16	0.999923	0.000405
100	36	24	17	0.999990	0.000067
100	36	24	18	0.999999	0.000009
100	36	24	19	1.000000	0.000001
100	36	24	20	1.000000	0.000000
100	36	24	21	1.000000	0.000000
100	36	24	22	1.000000	0.000000
100	36	24	23	1.000000	0.000000
100	36	24	24	1.000000	0.000000
100	36	25	0	0.000002	0.000002

N	n	k	x	P(x)	p(x)
100	36	25	1	0.000039	0.000037
100	36	25	2	0.000420	0.000381
100	36	25	3	0.002786	0.002365
100	36	25	4	0.012770	0.009985
100	36	25	5	0.043269	0.030499
100	36	25	6	0.113302	0.070034
100	36	25	7	0.237275	0.123973
100	36	25	8	0.409386	0.172111
100	36	25	9	0.599026	0.189641
100	36	25	10	0.766220	0.167193
100	36	25	11	0.884775	0.118555
100	36	25	12	0.952576	0.067801
100	36	25	13	0.983869	0.031293
100	36	25	14	0.995509	0.011640
100	36	25	15	0.998987	0.003478
100	36	25	16	0.999817	0.000830
100	36	25	17	0.999974	0.000157
100	36	25	18	0.999997	0.000023
100	36	25	19	1.000000	0.000003
100	36	25	20	1.000000	0.000000
100	36	25	21	1.000000	0.000000
100	36	25	22	1.000000	0.000000
100	36	25	23	1.000000	0.000000
100	36	25	24	1.000000	0.000000
100	36	25	25	1.000000	0.000000
100	36	26	0	0.000001	0.000001
100	36	26	1	0.000021	0.000021
100	36	26	2	0.000247	0.000226
100	36	26	3	0.001745	0.001497
100	36	26	4	0.008510	0.006765
100	36	26	5	0.030663	0.022153
100	36	26	6	0.085289	0.054626
100	36	26	7	0.189339	0.104050
100	36	26	8	0.345131	0.155792
100	36	26	9	0.530756	0.185625
100	36	26	10	0.708260	0.177504
100	36	26	11	0.845257	0.136997
100	36	26	12	0.930880	0.085623
100	36	26	13	0.974273	0.043393
100	36	26	14	0.992095	0.017822
100	36	26	15	0.998013	0.005918
100	36	26	16	0.999596	0.001582
100	36	26	17	0.999934	0.000338
100	36	26	18	0.999991	0.000057
100	36	26	19	0.999999	0.000008
100	36	26	20	1.000000	0.000001
100	36	26	21	1.000000	0.000000
100	36	26	22	1.000000	0.000000
100	36	26	23	1.000000	0.000000
100	36	26	24	1.000000	0.000000

N	n	k	x	P(x)	p(x)
100	36	26	25	1.000000	0.000000
100	36	26	26	1.000000	0.000000
100	36	27	0	0.000000	0.000000
100	36	27	1	0.000012	0.000011
100	36	27	2	0.000144	0.000132
100	36	27	3	0.001077	0.000933
100	36	27	4	0.005584	0.004508
100	36	27	5	0.021382	0.015798
100	36	27	6	0.063143	0.041761
100	36	27	7	0.148562	0.085419
100	36	27	8	0.286183	0.137620
100	36	27	9	0.463028	0.176845
100	36	27	10	0.645893	0.182865
100	36	27	11	0.798974	0.153081
100	36	27	12	0.903110	0.104136
100	36	27	13	0.960786	0.057676
100	36	27	14	0.986797	0.026011
100	36	27	15	0.996334	0.009537
100	36	27	16	0.999168	0.002834
100	36	27	17	0.999847	0.000679
100	36	27	18	0.999977	0.000130
100	36	27	19	0.999997	0.000020
100	36	27	20	1.000000	0.000002
100	36	27	21	1.000000	0.000000
100	36	27	22	1.000000	0.000000
100	36	27	23	1.000000	0.000000
100	36	27	24	1.000000	0.000000
100	36	27	25	1.000000	0.000000
100	36	27	26	1.000000	0.000000
100	36	27	27	1.000000	0.000000
100	36	28	0	0.000000	0.000000
100	36	28	1	0.000006	0.000006
100	36	28	2	0.000082	0.000076
100	36	28	3	0.000655	0.000573
100	36	28	4	0.003609	0.002954
100	36	28	5	0.014674	0.011065
100	36	28	6	0.045981	0.031307
100	36	28	7	0.114629	0.068648
100	36	28	8	0.233397	0.118768
100	36	28	9	0.397619	0.164223
100	36	28	10	0.580763	0.183144
100	36	28	11	0.746549	0.165786
100	36	28	12	0.868874	0.122325
100	36	28	13	0.942614	0.073740
100	36	28	14	0.978958	0.036344
100	36	28	15	0.993590	0.014632
100	36	28	16	0.998391	0.004801
100	36	28	17	0.999670	0.001279
100	36	28	18	0.999945	0.000275
100	36	28	19	0.999993	0.000047

Table for $N=2$, $n=1$, through $N=100$, $n=50$

N	n	k	x	P(x)	p(x)
100	36	28	20	0.999999	0.000006
100	36	28	21	1.000000	0.000001
100	36	28	22	1.000000	0.000000
100	36	28	23	1.000000	0.000000
100	36	28	24	1.000000	0.000000
100	36	28	25	1.000000	0.000000
100	36	28	26	1.000000	0.000000
100	36	28	27	1.000000	0.000000
100	36	28	28	1.000000	0.000000
100	36	29	0	0.000000	0.000000
100	36	29	1	0.000003	0.000003
100	36	29	2	0.000046	0.000043
100	36	29	3	0.000392	0.000346
100	36	29	4	0.002296	0.001903
100	36	29	5	0.009910	0.007614
100	36	29	6	0.032936	0.023027
100	36	29	7	0.086979	0.054043
100	36	29	8	0.187209	0.100230
100	36	29	9	0.336036	0.148827
100	36	29	10	0.514628	0.178592
100	36	29	11	0.688985	0.174357
100	36	29	12	0.828099	0.139114
100	36	29	13	0.919058	0.090959
100	36	29	14	0.967853	0.048795
100	36	29	15	0.989322	0.021470
100	36	29	16	0.997058	0.007735
100	36	29	17	0.999333	0.002275
100	36	29	18	0.999877	0.000544
100	36	29	19	0.999981	0.000105
100	36	29	20	0.999998	0.000016
100	36	29	21	1.000000	0.000002
100	36	29	22	1.000000	0.000000
100	36	29	23	1.000000	0.000000
100	36	29	24	1.000000	0.000000
100	36	29	25	1.000000	0.000000
100	36	29	26	1.000000	0.000000
100	36	29	27	1.000000	0.000000
100	36	29	28	1.000000	0.000000
100	36	29	29	1.000000	0.000000
100	36	30	0	0.000000	0.000000
100	36	30	1	0.000002	0.000002
100	36	30	2	0.000026	0.000024
100	36	30	3	0.000232	0.000206
100	36	30	4	0.001438	0.001206
100	36	30	5	0.006585	0.005147
100	36	30	6	0.023207	0.016621
100	36	30	7	0.064905	0.041699
100	36	30	8	0.147682	0.082777
100	36	30	9	0.279440	0.131758
100	36	30	10	0.449228	0.169788
100	36	30	11	0.627591	0.178363
100	36	30	12	0.781074	0.153483
100	36	30	13	0.889593	0.108518
100	36	30	14	0.952733	0.063141
100	36	30	15	0.982972	0.030239
100	36	30	16	0.994879	0.011907
100	36	30	17	0.998724	0.003845
100	36	30	18	0.999739	0.001015
100	36	30	19	0.999956	0.000218
100	36	30	20	0.999994	0.000038
100	36	30	21	0.999999	0.000005
100	36	30	22	1.000000	0.000001
100	36	30	23	1.000000	0.000000
100	36	30	24	1.000000	0.000000
100	36	30	25	1.000000	0.000000
100	36	30	26	1.000000	0.000000
100	36	30	27	1.000000	0.000000
100	36	30	28	1.000000	0.000000
100	36	30	29	1.000000	0.000000
100	36	30	30	1.000000	0.000000
100	36	31	0	0.000000	0.000000
100	36	31	1	0.000001	0.000001
100	36	31	2	0.000014	0.000013
100	36	31	3	0.000135	0.000120
100	36	31	4	0.000886	0.000752
100	36	31	5	0.004306	0.003419
100	36	31	6	0.016083	0.011778
100	36	31	7	0.047630	0.031547
100	36	31	8	0.114571	0.066941
100	36	31	9	0.228619	0.114048
100	36	31	10	0.386164	0.157545
100	36	31	11	0.563890	0.177726
100	36	31	12	0.728452	0.164561
100	36	31	13	0.853937	0.125485
100	36	31	14	0.932889	0.078953
100	36	31	15	0.973901	0.041011
100	36	31	16	0.991477	0.017576
100	36	31	17	0.997680	0.006203
100	36	31	18	0.999478	0.001797
100	36	31	19	0.999904	0.000426
100	36	31	20	0.999985	0.000082
100	36	31	21	0.999998	0.000013
100	36	31	22	1.000000	0.000002
100	36	31	23	1.000000	0.000000
100	36	31	24	1.000000	0.000000
100	36	31	25	1.000000	0.000000
100	36	31	26	1.000000	0.000000
100	36	31	27	1.000000	0.000000
100	36	31	28	1.000000	0.000000
100	36	31	29	1.000000	0.000000
100	36	31	30	1.000000	0.000000
100	36	31	31	1.000000	0.000000
100	36	32	0	0.000000	0.000000
100	36	32	1	0.000000	0.000000
100	36	32	2	0.000008	0.000007
100	36	32	3	0.000077	0.000069
100	36	32	4	0.000538	0.000461
100	36	32	5	0.002770	0.002232
100	36	32	6	0.010963	0.008193
100	36	32	7	0.034371	0.023409
100	36	32	8	0.087407	0.053035
100	36	32	9	0.183992	0.096585
100	36	32	10	0.326799	0.142808
100	36	32	11	0.499497	0.172697
100	36	32	12	0.671213	0.171716
100	36	32	13	0.812108	0.140895
100	36	32	14	0.907716	0.095608
100	36	32	15	0.961419	0.053703
100	36	32	16	0.986382	0.024963
100	36	32	17	0.995972	0.009590
100	36	32	18	0.999009	0.003037
100	36	32	19	0.999799	0.000790
100	36	32	20	0.999966	0.000168
100	36	32	21	0.999995	0.000029
100	36	32	22	0.999999	0.000004
100	36	32	23	1.000000	0.000000
100	36	32	24	1.000000	0.000000
100	36	32	25	1.000000	0.000000
100	36	32	26	1.000000	0.000000
100	36	32	27	1.000000	0.000000
100	36	32	28	1.000000	0.000000
100	36	32	29	1.000000	0.000000
100	36	32	30	1.000000	0.000000
100	36	32	31	1.000000	0.000000
100	36	32	32	1.000000	0.000000
100	36	33	0	0.000000	0.000000
100	36	33	1	0.000000	0.000000
100	36	33	2	0.000004	0.000004
100	36	33	3	0.000043	0.000039
100	36	33	4	0.000321	0.000278
100	36	33	5	0.001752	0.001431
100	36	33	6	0.007348	0.005596
100	36	33	7	0.024388	0.017040
100	36	33	8	0.065569	0.041180
100	36	33	9	0.145642	0.080073
100	36	33	10	0.272196	0.126554
100	36	33	11	0.436005	0.163809
100	36	33	12	0.610607	0.174602
100	36	33	13	0.764452	0.153845
100	36	33	14	0.876784	0.112332
100	36	33	15	0.944834	0.068050
100	36	33	16	0.979040	0.034206
100	36	33	17	0.993293	0.014253
100	36	33	18	0.998205	0.004912
100	36	33	19	0.999601	0.001396
100	36	33	20	0.999927	0.000326
100	36	33	21	0.999989	0.000062
100	36	33	22	0.999999	0.000010
100	36	33	23	1.000000	0.000001
100	36	33	24	1.000000	0.000000
100	36	33	25	1.000000	0.000000
100	36	33	26	1.000000	0.000000
100	36	33	27	1.000000	0.000000
100	36	33	28	1.000000	0.000000
100	36	33	29	1.000000	0.000000
100	36	33	30	1.000000	0.000000
100	36	33	31	1.000000	0.000000
100	36	33	32	1.000000	0.000000
100	36	33	33	1.000000	0.000000
100	36	34	0	0.000000	0.000000
100	36	34	1	0.000000	0.000000
100	36	34	2	0.000002	0.000002
100	36	34	3	0.000024	0.000022
100	36	34	4	0.000188	0.000164
100	36	34	5	0.001090	0.000902
100	36	34	6	0.004842	0.003752
100	36	34	7	0.017013	0.012170
100	36	34	8	0.048359	0.031346
100	36	34	9	0.113373	0.065015
100	36	34	10	0.223085	0.109712
100	36	34	11	0.374882	0.151797
100	36	34	12	0.548063	0.173181
100	36	34	13	0.711640	0.163577
100	36	34	14	0.839899	0.128259
100	36	34	15	0.923505	0.083606
100	36	34	16	0.968829	0.045324
100	36	34	17	0.989251	0.020422
100	36	34	18	0.996885	0.007634
100	36	34	19	0.999247	0.002362
100	36	34	20	0.999849	0.000602
100	36	34	21	0.999975	0.000126
100	36	34	22	0.999997	0.000021
100	36	34	23	0.999999	0.000003
100	36	34	24	1.000000	0.000000
100	36	34	25	1.000000	0.000000
100	36	34	26	1.000000	0.000000
100	36	34	27	1.000000	0.000000
100	36	34	28	1.000000	0.000000
100	36	34	29	1.000000	0.000000
100	36	34	30	1.000000	0.000000

Table for $N = 2$, $n = 1$, through $N = 100$, $n = 50$

N	n	k	x	P(x)	p(x)
100	36	34	31	1.000000	0.000000
100	36	34	32	1.000000	0.000000
100	36	34	33	1.000000	0.000000
100	36	34	34	1.000000	0.000000
100	36	35	0	0.000000	0.000000
100	36	35	1	0.000000	0.000000
100	36	35	2	0.000001	0.000001
100	36	35	3	0.000013	0.000012
100	36	35	4	0.000109	0.000096
100	36	35	5	0.000666	0.000558
100	36	35	6	0.003137	0.002470
100	36	35	7	0.011665	0.008528
100	36	35	8	0.035060	0.023395
100	36	35	9	0.086777	0.051716
100	36	35	10	0.179866	0.093089
100	36	35	11	0.317383	0.137518
100	36	35	12	0.485088	0.167705
100	36	35	13	0.654636	0.169548
100	36	35	14	0.797146	0.142510
100	36	35	15	0.896903	0.099757
100	36	35	16	0.955095	0.058192
100	36	35	17	0.983372	0.028277
100	36	35	18	0.994803	0.011431
100	36	35	19	0.998639	0.003835
100	36	35	20	0.999703	0.001065
100	36	35	21	0.999946	0.000243
100	36	35	22	0.999992	0.000046
100	36	35	23	0.999999	0.000007
100	36	35	24	1.000000	0.000001
100	36	35	25	1.000000	0.000000
100	36	35	26	1.000000	0.000000
100	36	35	27	1.000000	0.000000
100	36	35	28	1.000000	0.000000
100	36	35	29	1.000000	0.000000
100	36	35	30	1.000000	0.000000
100	36	35	31	1.000000	0.000000
100	36	35	32	1.000000	0.000000
100	36	35	33	1.000000	0.000000
100	36	35	34	1.000000	0.000000
100	36	35	35	1.000000	0.000000
100	36	36	0	0.000000	0.000000
100	36	36	1	0.000000	0.000000
100	36	36	2	0.000001	0.000001
100	36	36	3	0.000007	0.000006
100	36	36	4	0.000062	0.000055
100	36	36	5	0.000400	0.000339
100	36	36	6	0.001997	0.001596
100	36	36	7	0.007860	0.005864
100	36	36	8	0.024982	0.017122
100	36	36	9	0.065294	0.040312

N	n	k	x	P(x)	p(x)
100	36	36	10	0.142630	0.077335
100	36	36	11	0.264492	0.121862
100	36	36	12	0.423166	0.158674
100	36	36	13	0.594642	0.171476
100	36	36	14	0.748912	0.154270
100	36	36	15	0.864674	0.115762
100	36	36	16	0.937189	0.072516
100	36	36	17	0.975106	0.037917
100	36	36	18	0.991637	0.016531
100	36	36	19	0.997635	0.005998
100	36	36	20	0.999441	0.001806
100	36	36	21	0.999890	0.000449
100	36	36	22	0.999982	0.000092
100	36	36	23	0.999997	0.000015
100	36	36	24	1.000000	0.000002
100	36	36	25	1.000000	0.000000
100	36	36	26	1.000000	0.000000
100	36	36	27	1.000000	0.000000
100	36	36	28	1.000000	0.000000
100	36	36	29	1.000000	0.000000
100	36	36	30	1.000000	0.000000
100	36	36	31	1.000000	0.000000
100	36	36	32	1.000000	0.000000
100	36	36	33	1.000000	0.000000
100	36	36	34	1.000000	0.000000
100	36	36	35	1.000000	0.000000
100	36	36	36	1.000000	0.000000
100	37	1	0	0.630000	0.630000
100	37	1	1	1.000000	0.370000
100	37	2	0	0.394545	0.394545
100	37	2	1	0.865455	0.470909
100	37	2	2	1.000000	0.134545
100	37	3	0	0.245584	0.245584
100	37	3	1	0.692467	0.446883
100	37	3	2	0.951948	0.259481
100	37	3	3	1.000000	0.048052
100	37	4	0	0.151908	0.151908
100	37	4	1	0.526614	0.374706
100	37	4	2	0.858321	0.331707
100	37	4	3	0.983157	0.124836
100	37	4	4	1.000000	0.016843
100	37	5	0	0.093360	0.093360
100	37	5	1	0.386099	0.292739
100	37	5	2	0.737386	0.351287
100	37	5	3	0.938944	0.201558
100	37	5	4	0.994210	0.055266
100	37	5	5	1.000000	0.005790
100	37	6	0	0.056999	0.056999
100	37	6	1	0.275166	0.218168
100	37	6	2	0.607965	0.332798

N	n	k	x	P(x)	p(x)
100	37	6	3	0.866808	0.258843
100	37	6	4	0.975013	0.108205
100	37	6	5	0.998050	0.023037
100	37	6	6	1.000000	0.001950
100	37	7	0	0.034563	0.034563
100	37	7	1	0.191613	0.157050
100	37	7	2	0.484050	0.292438
100	37	7	3	0.773184	0.289133
100	37	7	4	0.937026	0.163842
100	37	7	5	0.990207	0.053182
100	37	7	6	0.999357	0.009150
100	37	7	7	1.000000	0.000643
100	37	8	0	0.020812	0.020812
100	37	8	1	0.130819	0.110007
100	37	8	2	0.373993	0.243174
100	37	8	3	0.667479	0.293486
100	37	8	4	0.878888	0.211409
100	37	8	5	0.971908	0.093020
100	37	8	6	0.996307	0.024399
100	37	8	7	0.999792	0.003486
100	37	8	8	1.000000	0.000207
100	37	9	0	0.012442	0.012442
100	37	9	1	0.087773	0.075331
100	37	9	2	0.281481	0.193708
100	37	9	3	0.559017	0.277535
100	37	9	4	0.803057	0.244040
100	37	9	5	0.939553	0.136497
100	37	9	6	0.988086	0.048532
100	37	9	7	0.998656	0.010570
100	37	9	8	0.999934	0.001279
100	37	9	9	1.000000	0.000065
100	37	10	0	0.007383	0.007383
100	37	10	1	0.057972	0.050589
100	37	10	2	0.206978	0.149006
100	37	10	3	0.455322	0.248344
100	37	10	4	0.714559	0.259236
100	37	10	5	0.891555	0.176996
100	37	10	6	0.971553	0.079998
100	37	10	7	0.995171	0.023618
100	37	10	8	0.999527	0.004356
100	37	10	9	0.999980	0.000453
100	37	10	10	1.000000	0.000020
100	37	11	0	0.004348	0.004348
100	37	11	1	0.037736	0.033388
100	37	11	2	0.149031	0.111295
100	37	11	3	0.361503	0.212472
100	37	11	4	0.619505	0.258002
100	37	11	5	0.828623	0.209117
100	37	11	6	0.943998	0.115375
100	37	11	7	0.987298	0.043301

N	n	k	x	P(x)	p(x)
100	37	11	8	0.998123	0.010825
100	37	11	9	0.999839	0.001715
100	37	11	10	0.999994	0.000155
100	37	11	11	1.000000	0.000006
100	37	12	0	0.002540	0.002540
100	37	12	1	0.024231	0.021691
100	37	12	2	0.105264	0.081033
100	37	12	3	0.280334	0.175071
100	37	12	4	0.523842	0.243507
100	37	12	5	0.753434	0.229593
100	37	12	6	0.903811	0.150377
100	37	12	7	0.972703	0.068892
100	37	12	8	0.994596	0.021894
100	37	12	9	0.999299	0.004703
100	37	12	10	0.999947	0.000648
100	37	12	11	0.999998	0.000051
100	37	12	12	1.000000	0.000002
100	37	13	0	0.001472	0.001472
100	37	13	1	0.015357	0.013885
100	37	13	2	0.073035	0.057677
100	37	13	3	0.212693	0.139659
100	37	13	4	0.432526	0.219833
100	37	13	5	0.669946	0.237420
100	37	13	6	0.850837	0.180891
100	37	13	7	0.949217	0.098379
100	37	13	8	0.987381	0.038164
100	37	13	9	0.997803	0.010422
100	37	13	10	0.999748	0.001945
100	37	13	11	0.999983	0.000235
100	37	13	12	0.999999	0.000016
100	37	13	13	1.000000	0.000000
100	37	14	0	0.000846	0.000846
100	37	14	1	0.009612	0.008766
100	37	14	2	0.049831	0.040219
100	37	14	3	0.158114	0.108283
100	37	14	4	0.349141	0.191027
100	37	14	5	0.582619	0.233478
100	37	14	6	0.786382	0.203763
100	37	14	7	0.915293	0.128911
100	37	14	8	0.974660	0.059367
100	37	14	9	0.994449	0.019789
100	37	14	10	0.999144	0.004696
100	37	14	11	0.999913	0.000768
100	37	14	12	0.999995	0.000082
100	37	14	13	1.000000	0.000005
100	37	14	14	1.000000	0.000000
100	37	15	0	0.000482	0.000482
100	37	15	1	0.005942	0.005460
100	37	15	2	0.033463	0.027520
100	37	15	3	0.115305	0.081842

Table for $N = 2$, $n = 1$, through $N = 100$, $n = 50$

N	n	k	x	P(x)	p(x)
100	37	15	4	0.275840	0.160536
100	37	15	5	0.495744	0.219904
100	37	15	6	0.712933	0.217189
100	37	15	7	0.870324	0.157391
100	37	15	8	0.954641	0.084317
100	37	15	9	0.988006	0.033365
100	37	15	10	0.997670	0.009664
100	37	15	11	0.999681	0.002010
100	37	15	12	0.999971	0.000290
100	37	15	13	0.999998	0.000027
100	37	15	14	1.000000	0.000002
100	37	15	15	1.000000	0.000000
100	37	16	0	0.000272	0.000272
100	37	16	1	0.003630	0.003358
100	37	16	2	0.022131	0.018501
100	37	16	3	0.082568	0.060437
100	37	16	4	0.213515	0.130947
100	37	16	5	0.412957	0.199442
100	37	16	6	0.633723	0.220766
100	37	16	7	0.814774	0.181051
100	37	16	8	0.925874	0.111100
100	37	16	9	0.977015	0.051141
100	37	16	10	0.994600	0.017585
100	37	16	11	0.999066	0.004465
100	37	16	12	0.999885	0.000820
100	37	16	13	0.999991	0.000105
100	37	16	14	0.999999	0.000009
100	37	16	15	1.000000	0.000000
100	37	16	16	1.000000	0.000000
100	37	17	0	0.000152	0.000152
100	37	17	1	0.002191	0.002039
100	37	17	2	0.014422	0.012231
100	37	17	3	0.058105	0.043683
100	37	17	4	0.162071	0.103966
100	37	17	5	0.336979	0.174907
100	37	17	6	0.552249	0.215271
100	37	17	7	0.750113	0.197863
100	37	17	8	0.887518	0.137405
100	37	17	9	0.959968	0.072450
100	37	17	10	0.988948	0.028980
100	37	17	11	0.997683	0.008736
100	37	17	12	0.999641	0.001958
100	37	17	13	0.999961	0.000319
100	37	17	14	0.999997	0.000036
100	37	17	15	1.000000	0.000003
100	37	17	16	1.000000	0.000000
100	37	17	17	1.000000	0.000000
100	37	18	0	0.000084	0.000084
100	37	18	1	0.001307	0.001222
100	37	18	2	0.009264	0.007958
100	37	18	3	0.040211	0.030947
100	37	18	4	0.120735	0.080524
100	37	18	5	0.269545	0.148809
100	37	18	6	0.471847	0.202303
100	37	18	7	0.678596	0.206749
100	37	18	8	0.839509	0.160913
100	37	18	9	0.935527	0.096018
100	37	18	10	0.979521	0.043994
100	37	18	11	0.994947	0.015426
100	37	18	12	0.999052	0.004105
100	37	18	13	0.999868	0.000817
100	37	18	14	0.999987	0.000119
100	37	18	15	0.999999	0.000012
100	37	18	16	1.000000	0.000001
100	37	18	17	1.000000	0.000000
100	37	18	18	1.000000	0.000000
100	37	19	0	0.000046	0.000046
100	37	19	1	0.000770	0.000724
100	37	19	2	0.005868	0.005098
100	37	19	3	0.027379	0.021512
100	37	19	4	0.088329	0.060950
100	37	19	5	0.211473	0.123143
100	37	19	6	0.395367	0.183894
100	37	19	7	0.602956	0.207589
100	37	19	8	0.782601	0.179645
100	37	19	9	0.902740	0.120140
100	37	19	10	0.965035	0.062295
100	37	19	11	0.990056	0.025021
100	37	19	12	0.997800	0.007745
100	37	19	13	0.999629	0.001829
100	37	19	14	0.999954	0.000324
100	37	19	15	0.999996	0.000042
100	37	19	16	1.000000	0.000004
100	37	19	17	1.000000	0.000000
100	37	19	18	1.000000	0.000000
100	37	19	19	1.000000	0.000000
100	37	20	0	0.000025	0.000025
100	37	20	1	0.000448	0.000423
100	37	20	2	0.003665	0.003217
100	37	20	3	0.018350	0.014685
100	37	20	4	0.063498	0.045148
100	37	20	5	0.162823	0.099326
100	37	20	6	0.324988	0.162164
100	37	20	7	0.526071	0.201084
100	37	20	8	0.718284	0.192212
100	37	20	9	0.861211	0.142927
100	37	20	10	0.944270	0.083059
100	37	20	11	0.982024	0.037754
100	37	20	12	0.995410	0.013386
100	37	20	13	0.999087	0.003677
100	37	20	14	0.999862	0.000774
100	37	20	15	0.999984	0.000123
100	37	20	16	0.999999	0.000014
100	37	20	17	1.000000	0.000001
100	37	20	18	1.000000	0.000000
100	37	20	19	1.000000	0.000000
100	37	20	20	1.000000	0.000000
100	37	21	0	0.000014	0.000014
100	37	21	1	0.000258	0.000244
100	37	21	2	0.002258	0.002000
100	37	21	3	0.012109	0.009851
100	37	21	4	0.044874	0.032765
100	37	21	5	0.123093	0.078219
100	37	21	6	0.262149	0.139056
100	37	21	7	0.450665	0.188516
100	37	21	8	0.648607	0.197942
100	37	21	9	0.811186	0.162580
100	37	21	10	0.916238	0.105051
100	37	21	11	0.969754	0.053517
100	37	21	12	0.991227	0.021473
100	37	21	13	0.997984	0.006757
100	37	21	14	0.999639	0.001655
100	37	21	15	0.999951	0.000312
100	37	21	16	0.999995	0.000044
100	37	21	17	1.000000	0.000005
100	37	21	18	1.000000	0.000000
100	37	21	19	1.000000	0.000000
100	37	21	20	1.000000	0.000000
100	37	21	21	1.000000	0.000000
100	37	22	0	0.000007	0.000007
100	37	22	1	0.000147	0.000139
100	37	22	2	0.001372	0.001225
100	37	22	3	0.007869	0.006497
100	37	22	4	0.031187	0.023318
100	37	22	5	0.091410	0.060222
100	37	22	6	0.207583	0.116173
100	37	22	7	0.379077	0.171494
100	37	22	8	0.575944	0.196868
100	37	22	9	0.753563	0.177619
100	37	22	10	0.880334	0.126771
100	37	22	11	0.952141	0.071807
100	37	22	12	0.984432	0.032291
100	37	22	13	0.995931	0.011500
100	37	22	14	0.999157	0.003226
100	37	22	15	0.999864	0.000707
100	37	22	16	0.999983	0.000119
100	37	22	17	0.999998	0.000015
100	37	22	18	1.000000	0.000001
100	37	22	19	1.000000	0.000000
100	37	22	20	1.000000	0.000000
100	37	22	21	1.000000	0.000000
100	37	22	22	1.000000	0.000000
100	37	23	0	0.000004	0.000004
100	37	23	1	0.000082	0.000078
100	37	23	2	0.000822	0.000740
100	37	23	3	0.005037	0.004215
100	37	23	4	0.021322	0.016285
100	37	23	5	0.066703	0.045381
100	37	23	6	0.161411	0.094708
100	37	23	7	0.313117	0.151706
100	37	23	8	0.502750	0.189633
100	37	23	9	0.689802	0.187052
100	37	23	10	0.836452	0.146649
100	37	23	11	0.928205	0.091754
100	37	23	12	0.974082	0.045877
100	37	23	13	0.992393	0.018311
100	37	23	14	0.998206	0.005813
100	37	23	15	0.999664	0.001459
100	37	23	16	0.999951	0.000286
100	37	23	17	0.999994	0.000043
100	37	23	18	0.999999	0.000005
100	37	23	19	1.000000	0.000000
100	37	23	20	1.000000	0.000000
100	37	23	21	1.000000	0.000000
100	37	23	22	1.000000	0.000000
100	37	23	23	1.000000	0.000000
100	37	24	0	0.000002	0.000002
100	37	24	1	0.000046	0.000044
100	37	24	2	0.000486	0.000440
100	37	24	3	0.003176	0.002690
100	37	24	4	0.014343	0.011167
100	37	24	5	0.047843	0.033501
100	37	24	6	0.123282	0.075439
100	37	24	7	0.254011	0.130729
100	37	24	8	0.431330	0.177319
100	37	24	9	0.621783	0.190453
100	37	24	10	0.785029	0.163246
100	37	24	11	0.897224	0.112194
100	37	24	12	0.959187	0.061964
100	37	24	13	0.986686	0.027499
100	37	24	14	0.996470	0.009784
100	37	24	15	0.999248	0.002778
100	37	24	16	0.999873	0.000625
100	37	24	17	0.999983	0.000110
100	37	24	18	0.999998	0.000015
100	37	24	19	1.000000	0.000002
100	37	24	20	1.000000	0.000000
100	37	24	21	1.000000	0.000000
100	37	24	22	1.000000	0.000000
100	37	24	23	1.000000	0.000000

Table for $N = 2$, $n = 1$, through $N = 100$, $n = 50$

N	n	k	x	P(x)	p(x)
100	37	24	24	1.000000	0.000000
100	37	25	0	0.000001	0.000001
100	37	25	1	0.000025	0.000024
100	37	25	2	0.000283	0.000258
100	37	25	3	0.001972	0.001689
100	37	25	4	0.009494	0.007522
100	37	25	5	0.033738	0.024244
100	37	25	6	0.092511	0.058773
100	37	25	7	0.202408	0.109897
100	37	25	8	0.363669	0.161261
100	37	25	9	0.551616	0.187947
100	37	25	10	0.727034	0.175418
100	37	25	11	0.858841	0.131807
100	37	25	12	0.938804	0.079963
100	37	25	13	0.978002	0.039198
100	37	25	14	0.993509	0.015507
100	37	25	15	0.998444	0.004935
100	37	25	16	0.999700	0.001257
100	37	25	17	0.999954	0.000254
100	37	25	18	0.999994	0.000040
100	37	25	19	0.999999	0.000005
100	37	25	20	1.000000	0.000000
100	37	25	21	1.000000	0.000000
100	37	25	22	1.000000	0.000000
100	37	25	23	1.000000	0.000000
100	37	25	24	1.000000	0.000000
100	37	25	25	1.000000	0.000000
100	37	26	0	0.000001	0.000001
100	37	26	1	0.000013	0.000013
100	37	26	2	0.000163	0.000149
100	37	26	3	0.001207	0.001044
100	37	26	4	0.006184	0.004978
100	37	26	5	0.023394	0.017209
1,00	37	26	6	0.068218	0.044824
100	37	26	7	0.158449	0.090231
100	37	26	8	0.301314	0.142865
100	37	26	9	0.481449	0.180135
100	37	26	10	0.663884	0.182434
100	37	26	11	0.813148	0.149264
100	37	26	12	0.912150	0.099002
100	37	26	13	0.965459	0.053309
100	37	26	14	0.988753	0.023295
100	37	26	15	0.996996	0.008243
100	37	26	16	0.999348	0.002352
100	37	26	17	0.999886	0.000538
100	37	26	18	0.999984	0.000098
100	37	26	19	0.999998	0.000014
100	37	26	20	1.000000	0.000002
100	37	26	21	1.000000	0.000000
100	37	26	22	1.000000	0.000000

N	n	k	x	P(x)	p(x)
100	37	26	23	1.000000	0.000000
100	37	26	24	1.000000	0.000000
100	37	26	25	1.000000	0.000000
100	37	26	26	1.000000	0.000000
100	37	27	0	0.000000	0.000000
100	37	27	1	0.000007	0.000007
100	37	27	2	0.000092	0.000085
100	37	27	3	0.000727	0.000635
100	37	27	4	0.003965	0.003238
100	37	27	5	0.015952	0.011987
100	37	27	6	0.049440	0.033488
100	37	27	7	0.121869	0.072429
100	37	27	8	0.245327	0.123458
100	37	27	9	0.413290	0.167963
100	37	27	10	0.597320	0.184030
100	37	27	11	0.760704	0.163384
100	37	27	12	0.878703	0.118000
100	37	27	13	0.948169	0.069466
100	37	27	14	0.981513	0.033344
100	37	27	15	0.994545	0.013032
100	37	27	16	0.998681	0.004135
100	37	27	17	0.999741	0.001060
100	37	27	18	0.999959	0.000218
100	37	27	19	0.999995	0.000036
100	37	27	20	0.999999	0.000005
100	37	27	21	1.000000	0.000000
100	37	27	22	1.000000	0.000000
100	37	27	23	1.000000	0.000000
100	37	27	24	1.000000	0.000000
100	37	27	25	1.000000	0.000000
100	37	27	26	1.000000	0.000000
100	37	27	27	1.000000	0.000000
100	37	28	0	0.000000	0.000000
100	37	28	1	0.000004	0.000004
100	37	28	2	0.000051	0.000048
100	37	28	3	0.000431	0.000380
100	37	28	4	0.002501	0.002070
100	37	28	5	0.010697	0.008196
100	37	28	6	0.035219	0.024522
100	37	28	7	0.092104	0.056885
100	37	28	8	0.196282	0.104178
100	37	28	9	0.348866	0.152584
100	37	28	10	0.529254	0.180388
100	37	28	11	0.702512	0.173258
100	37	28	12	0.838293	0.135780
100	37	28	13	0.925331	0.087039
100	37	28	14	0.971008	0.045676
100	37	28	15	0.990618	0.019610
100	37	28	16	0.997491	0.006873
100	37	28	17	0.999450	0.001959

N	n	k	x	P(x)	p(x)
100	37	28	18	0.999902	0.000452
100	37	28	19	0.999986	0.000084
100	37	28	20	0.999998	0.000012
100	37	28	21	1.000000	0.000001
100	37	28	22	1.000000	0.000000
100	37	28	23	1.000000	0.000000
100	37	28	24	1.000000	0.000000
100	37	28	25	1.000000	0.000000
100	37	28	26	1.000000	0.000000
100	37	28	27	1.000000	0.000000
100	37	28	28	1.000000	0.000000
100	37	29	0	0.000000	0.000000
100	37	29	1	0.000002	0.000002
100	37	29	2	0.000028	0.000026
100	37	29	3	0.000252	0.000224
100	37	29	4	0.001552	0.001300
100	37	29	5	0.007054	0.005502
100	37	29	6	0.024661	0.017607
100	37	29	7	0.068402	0.043741
100	37	29	8	0.154321	0.085919
100	37	29	9	0.289528	0.135206
100	37	29	10	0.461608	0.172081
100	37	29	11	0.639947	0.178338
100	37	29	12	0.791147	0.151200
100	37	29	13	0.896318	0.105172
100	37	29	14	0.956417	0.060098
100	37	29	15	0.984626	0.028209
100	37	29	16	0.995486	0.010861
100	37	29	17	0.998906	0.003420
100	37	29	18	0.999783	0.000877
100	37	29	19	0.999965	0.000182
100	37	29	20	0.999995	0.000030
100	37	29	21	0.999999	0.000004
100	37	29	22	1.000000	0.000000
100	37	29	23	1.000000	0.000000
100	37	29	24	1.000000	0.000000
100	37	29	25	1.000000	0.000000
100	37	29	26	1.000000	0.000000
100	37	29	27	1.000000	0.000000
100	37	29	28	1.000000	0.000000
100	37	29	29	1.000000	0.000000
100	37	30	0	0.000000	0.000000
100	37	30	1	0.000001	0.000001
100	37	30	2	0.000015	0.000014
100	37	30	3	0.000145	0.000129
100	37	30	4	0.000948	0.000803
100	37	30	5	0.004575	0.003627
100	37	30	6	0.016974	0.012399
100	37	30	7	0.049920	0.032946
100	37	30	8	0.119227	0.069308

N	n	k	x	P(x)	p(x)
100	37	30	9	0.236207	0.116979
100	37	30	10	0.396169	0.159962
100	37	30	11	0.574640	0.178471
100	37	30	12	0.737907	0.163267
100	37	30	13	0.860767	0.122860
100	37	30	14	0.936948	0.076181
100	37	30	15	0.975885	0.038937
100	37	30	16	0.992274	0.016389
100	37	30	17	0.997943	0.005669
100	37	30	18	0.999548	0.001606
100	37	30	19	0.999919	0.000371
100	37	30	20	0.999988	0.000069
100	37	30	21	0.999998	0.000010
100	37	30	22	1.000000	0.000001
100	37	30	23	1.000000	0.000000
100	37	30	24	1.000000	0.000000
100	37	30	25	1.000000	0.000000
100	37	30	26	1.000000	0.000000
100	37	30	27	1.000000	0.000000
100	37	30	28	1.000000	0.000000
100	37	30	29	1.000000	0.000000
100	37	30	30	1.000000	0.000000
100	37	31	0	0.000000	0.000000
100	37	31	1	0.000000	0.000000
100	37	31	2	0.000008	0.000008
100	37	31	3	0.000082	0.000074
100	37	31	4	0.000569	0.000487
100	37	31	5	0.002917	0.002347
100	37	31	6	0.011483	0.008566
100	37	31	7	0.035800	0.024317
100	37	31	8	0.090514	0.054714
100	37	31	9	0.189415	0.098901
100	37	31	10	0.334469	0.145054
100	37	31	11	0.508351	0.173881
100	37	31	12	0.679597	0.171247
100	37	31	13	0.818644	0.139046
100	37	31	14	0.911917	0.093273
100	37	31	15	0.963648	0.051730
100	37	31	16	0.987357	0.023710
100	37	31	17	0.996323	0.008966
100	37	31	18	0.999113	0.002789
100	37	31	19	0.999824	0.000711
100	37	31	20	0.999971	0.000148
100	37	31	21	0.999996	0.000025
100	37	31	22	0.999999	0.000003
100	37	31	23	1.000000	0.000000
100	37	31	24	1.000000	0.000000
100	37	31	25	1.000000	0.000000
100	37	31	26	1.000000	0.000000
100	37	31	27	1.000000	0.000000

Table for $N = 2$, $n = 1$, through $N = 100$, $n = 50$

N	n	k	x	P(x)	p(x)
100	37	31	28	1.000000	0.000000
100	37	31	29	1.000000	0.000000
100	37	31	30	1.000000	0.000000
100	37	31	31	1.000000	0.000000
100	37	32	0	0.000000	0.000000
100	37	32	1	0.000000	0.000000
100	37	32	2	0.000004	0.000004
100	37	32	3	0.000046	0.000041
100	37	32	4	0.000336	0.000291
100	37	32	5	0.001828	0.001492
100	37	32	6	0.007634	0.005806
100	37	32	7	0.025227	0.017593
100	37	32	8	0.067518	0.042291
100	37	32	9	0.149281	0.081763
100	37	32	10	0.277709	0.128428
100	37	32	11	0.442830	0.165121
100	37	32	12	0.617551	0.174721
100	37	32	13	0.770280	0.152728
100	37	32	14	0.880826	0.110546
100	37	32	15	0.947154	0.066328
100	37	32	16	0.980141	0.032987
100	37	32	17	0.993724	0.013583
100	37	32	18	0.998345	0.004620
100	37	32	19	0.999638	0.001294
100	37	32	20	0.999935	0.000297
100	37	32	21	0.999990	0.000055
100	37	32	22	0.999999	0.000008
100	37	32	23	1.000000	0.000001
100	37	32	24	1.000000	0.000000
100	37	32	25	1.000000	0.000000
100	37	32	26	1.000000	0.000000
100	37	32	27	1.000000	0.000000
100	37	32	28	1.000000	0.000000
100	37	32	29	1.000000	0.000000
100	37	32	30	1.000000	0.000000
100	37	32	31	1.000000	0.000000
100	37	32	32	1.000000	0.000000
100	37	33	0	0.000000	0.000000
100	37	33	1	0.000000	0.000000
100	37	33	2	0.000002	0.000002
100	37	33	3	0.000025	0.000023
100	37	33	4	0.000195	0.000170
100	37	33	5	0.001126	0.000931
100	37	33	6	0.004987	0.003861
100	37	33	7	0.017466	0.012478
100	37	33	8	0.049483	0.032017
100	37	33	9	0.115614	0.066132
100	37	33	10	0.226716	0.111101
100	37	33	11	0.379696	0.152980
100	37	33	12	0.553316	0.173620

N	n	k	x	P(x)	p(x)
100	37	33	13	0.716376	0.163060
100	37	33	14	0.843436	0.127060
100	37	33	15	0.925695	0.082259
100	37	33	16	0.969954	0.044259
100	37	33	17	0.989729	0.019775
100	37	33	18	0.997054	0.007324
100	37	33	19	0.999296	0.002242
100	37	33	20	0.999861	0.000565
100	37	33	21	0.999977	0.000117
100	37	33	22	0.999997	0.000020
100	37	33	23	1.000000	0.000003
100	37	33	24	1.000000	0.000000
100	37	33	25	1.000000	0.000000
100	37	33	26	1.000000	0.000000
100	37	33	27	1.000000	0.000000
100	37	33	28	1.000000	0.000000
100	37	33	29	1.000000	0.000000
100	37	33	30	1.000000	0.000000
100	37	33	31	1.000000	0.000000
100	37	33	32	1.000000	0.000000
100	37	33	33	1.000000	0.000000
100	37	34	0	0.000000	0.000000
100	37	34	1	0.000000	0.000000
100	37	34	2	0.000001	0.000001
100	37	34	3	0.000013	0.000012
100	37	34	4	0.000111	0.000098
100	37	34	5	0.000681	0.000570
100	37	34	6	0.003201	0.002519
100	37	34	7	0.011878	0.008678
100	37	34	8	0.035624	0.023746
100	37	34	9	0.087977	0.052353
100	37	34	10	0.181943	0.093966
100	37	34	11	0.320330	0.138387
100	37	34	12	0.488532	0.168202
100	37	34	13	0.657966	0.169434
100	37	34	14	0.799818	0.141852
100	37	34	15	0.898684	0.098866
100	37	34	16	0.956082	0.057397
100	37	34	17	0.983826	0.027745
100	37	34	18	0.994977	0.011150
100	37	34	19	0.998693	0.003717
100	37	34	20	0.999717	0.001024
100	37	34	21	0.999949	0.000232
100	37	34	22	0.999993	0.000043
100	37	34	23	0.999999	0.000006
100	37	34	24	1.000000	0.000001
100	37	34	25	1.000000	0.000000
100	37	34	26	1.000000	0.000000
100	37	34	27	1.000000	0.000000
100	37	34	28	1.000000	0.000000

N	n	k	x	P(x)	p(x)
100	37	34	29	1.000000	0.000000
100	37	34	30	1.000000	0.000000
100	37	34	31	1.000000	0.000000
100	37	34	32	1.000000	0.000000
100	37	34	33	1.000000	0.000000
100	37	34	34	1.000000	0.000000
100	37	35	0	0.000000	0.000000
100	37	35	1	0.000000	0.000000
100	37	35	2	0.000001	0.000001
100	37	35	3	0.000007	0.000006
100	37	35	4	0.000062	0.000055
100	37	35	5	0.000405	0.000343
100	37	35	6	0.002017	0.001612
100	37	35	7	0.007934	0.005917
100	37	35	8	0.025191	0.017257
100	37	35	9	0.065767	0.040576
100	37	35	10	0.143503	0.077736
100	37	35	11	0.265814	0.122312
100	37	35	12	0.424819	0.159005
100	37	35	13	0.596354	0.171535
100	37	35	14	0.750385	0.154031
100	37	35	15	0.865729	0.115344
100	37	35	16	0.937819	0.072090
100	37	35	17	0.975419	0.037600
100	37	35	18	0.991767	0.016348
100	37	35	19	0.997680	0.005913
100	37	35	20	0.999454	0.001774
100	37	35	21	0.999893	0.000440
100	37	35	22	0.999983	0.000090
100	37	35	23	0.999998	0.000015
100	37	35	24	1.000000	0.000002
100	37	35	25	1.000000	0.000000
100	37	35	26	1.000000	0.000000
100	37	35	27	1.000000	0.000000
100	37	35	28	1.000000	0.000000
100	37	35	29	1.000000	0.000000
100	37	35	30	1.000000	0.000000
100	37	35	31	1.000000	0.000000
100	37	35	32	1.000000	0.000000
100	37	35	33	1.000000	0.000000
100	37	35	34	1.000000	0.000000
100	37	35	35	1.000000	0.000000
100	37	36	0	0.000000	0.000000
100	37	36	1	0.000000	0.000000
100	37	36	2	0.000000	0.000000
100	37	36	3	0.000004	0.000003
100	37	36	4	0.000034	0.000031
100	37	36	5	0.000236	0.000202
100	37	36	6	0.001248	0.001012
100	37	36	7	0.005203	0.003955

N	n	k	x	P(x)	p(x)
100	37	36	8	0.017491	0.012288
100	37	36	9	0.048288	0.030796
100	37	36	10	0.111212	0.062925
100	37	36	11	0.216890	0.105677
100	37	36	12	0.363664	0.146774
100	37	36	13	0.533018	0.169354
100	37	36	14	0.695882	0.162864
100	37	36	15	0.826689	0.130808
100	37	36	16	0.914528	0.087839
100	37	36	17	0.963850	0.049321
100	37	36	18	0.986988	0.023138
100	37	36	19	0.996042	0.009054
100	37	36	20	0.998990	0.002947
100	37	36	21	0.999785	0.000795
100	37	36	22	0.999962	0.000177
100	37	36	23	0.999994	0.000032
100	37	36	24	0.999999	0.000005
100	37	36	25	1.000000	0.000001
100	37	36	26	1.000000	0.000000
100	37	36	27	1.000000	0.000000
100	37	36	28	1.000000	0.000000
100	37	36	29	1.000000	0.000000
100	37	36	30	1.000000	0.000000
100	37	36	31	1.000000	0.000000
100	37	36	32	1.000000	0.000000
100	37	36	33	1.000000	0.000000
100	37	36	34	1.000000	0.000000
100	37	36	35	1.000000	0.000000
100	37	36	36	1.000000	0.000000
100	37	37	0	0.000000	0.000000
100	37	37	1	0.000000	0.000000
100	37	37	2	0.000000	0.000000
100	37	37	3	0.000002	0.000002
100	37	37	4	0.000018	0.000017
100	37	37	5	0.000135	0.000117
100	37	37	6	0.000758	0.000623
100	37	37	7	0.003349	0.002591
100	37	37	8	0.011923	0.008574
100	37	37	9	0.034814	0.022891
100	37	37	10	0.084666	0.049852
100	37	37	11	0.173958	0.089292
100	37	37	12	0.306330	0.132372
100	37	37	13	0.469510	0.163180
100	37	37	14	0.637352	0.167842
100	37	37	15	0.781724	0.144372
100	37	37	16	0.885706	0.103982
100	37	37	17	0.948437	0.062731
100	37	37	18	0.980119	0.031682
100	37	37	19	0.993496	0.013377
100	37	37	20	0.998207	0.004711

Table for $N = 2$, $n = 1$, through $N = 100$, $n = 50$

N	n	k	x	P(x)	p(x)
100	37	37	21	0.999586	0.001379
100	37	37	22	0.999920	0.000334
100	37	37	23	0.999987	0.000067
100	37	37	24	0.999998	0.000011
100	37	37	25	1.000000	0.000001
100	37	37	26	1.000000	0.000000
100	37	37	27	1.000000	0.000000
100	37	37	28	1.000000	0.000000
100	37	37	29	1.000000	0.000000
100	37	37	30	1.000000	0.000000
100	37	37	31	1.000000	0.000000
100	37	37	32	1.000000	0.000000
100	37	37	33	1.000000	0.000000
100	37	37	34	1.000000	0.000000
100	37	37	35	1.000000	0.000000
100	37	37	36	1.000000	0.000000
100	37	37	37	1.000000	0.000000
100	38	1	0	0.620000	0.620000
100	38	1	1	1.000000	0.380000
100	38	2	0	0.382020	0.382020
100	38	2	1	0.857980	0.475960
100	38	2	2	1.000000	0.142020
100	38	3	0	0.233890	0.233890
100	38	3	1	0.678281	0.444391
100	38	3	2	0.947829	0.269549
100	38	3	3	1.000000	0.052171
100	38	4	0	0.142263	0.142263
100	38	4	1	0.508771	0.366508
100	38	4	2	0.847791	0.339020
100	38	4	3	0.981175	0.133385
100	38	4	4	1.000000	0.018824
100	38	5	0	0.085951	0.085951
100	38	5	1	0.367513	0.281562
100	38	5	2	0.720658	0.353146
100	38	5	3	0.932546	0.211887
100	38	5	4	0.993333	0.060787
100	38	5	5	1.000000	0.006667
100	38	6	0	0.051570	0.051570
100	38	6	1	0.257852	0.206281
100	38	6	2	0.586835	0.328983
100	38	6	3	0.854482	0.267647
100	38	6	4	0.971577	0.117096
100	38	6	5	0.997684	0.026107
100	38	6	6	1.000000	0.002316
100	38	7	0	0.030723	0.030723
100	38	7	1	0.176656	0.145933
100	38	7	2	0.460841	0.284185
100	38	7	3	0.754826	0.293985
100	38	7	4	0.929224	0.174398
100	38	7	5	0.988519	0.059295

N	n	k	x	P(x)	p(x)
100	38	7	6	0.999211	0.010693
100	38	7	7	1.000000	0.000788
100	38	8	0	0.018169	0.018169
100	38	8	1	0.118596	0.100427
100	38	8	2	0.350834	0.232237
100	38	8	3	0.644186	0.293353
100	38	8	4	0.865465	0.221279
100	38	8	5	0.967479	0.102013
100	38	8	6	0.995532	0.028054
100	38	8	7	0.999737	0.004205
100	38	8	8	1.000000	0.000263
100	38	9	0	0.010665	0.010665
100	38	9	1	0.078207	0.067543
100	38	9	2	0.259958	0.181751
100	38	9	3	0.532585	0.272627
100	38	9	4	0.783688	0.251103
100	38	9	5	0.930887	0.147199
100	38	9	6	0.985775	0.054888
100	38	9	7	0.998320	0.012546
100	38	9	8	0.999914	0.001594
100	38	9	9	1.000000	0.000086
100	38	10	0	0.006211	0.006211
100	38	10	1	0.050745	0.044534
100	38	10	2	0.188057	0.137312
100	38	10	3	0.427729	0.239672
100	38	10	4	0.689870	0.262141
100	38	10	5	0.877507	0.187638
100	38	10	6	0.966473	0.088966
100	38	10	7	0.994046	0.027573
100	38	10	8	0.999389	0.005342
100	38	10	9	0.999973	0.000584
100	38	10	10	1.000000	0.000027
100	38	11	0	0.003589	0.003589
100	38	11	1	0.032437	0.028848
100	38	11	2	0.133132	0.100695
100	38	11	3	0.334523	0.201391
100	38	11	4	0.590838	0.256316
100	38	11	5	0.808707	0.217868
100	38	11	6	0.934841	0.126134
100	38	11	7	0.984549	0.049708
100	38	11	8	0.997608	0.013059
100	38	11	9	0.999784	0.002176
100	38	11	10	0.999991	0.000207
100	38	11	11	1.000000	0.000008
100	38	12	0	0.002056	0.002056
100	38	12	1	0.020444	0.018387
100	38	12	2	0.092401	0.071958
100	38	12	3	0.255324	0.162923
100	38	12	4	0.492920	0.237596
100	38	12	5	0.727924	0.235004

N	n	k	x	P(x)	p(x)
100	38	12	6	0.889489	0.161565
100	38	12	7	0.967235	0.077746
100	38	12	8	0.993206	0.025971
100	38	12	9	0.999075	0.005869
100	38	12	10	0.999926	0.000851
100	38	12	11	0.999997	0.000071
100	38	12	12	1.000000	0.000003
100	38	13	0	0.001168	0.001168
100	38	13	1	0.012713	0.011544
100	38	13	2	0.062964	0.050251
100	38	13	3	0.190525	0.127561
100	38	13	4	0.401122	0.210596
100	38	13	5	0.639798	0.238676
100	38	13	6	0.830738	0.190941
100	38	13	7	0.939847	0.109109
100	38	13	8	0.984352	0.044505
100	38	13	9	0.997141	0.012789
100	38	13	10	0.999655	0.002514
100	38	13	11	0.999975	0.000320
100	38	13	12	0.999999	0.000024
100	38	13	13	1.000000	0.000001
100	38	14	0	0.000658	0.000658
100	38	14	1	0.007803	0.007145
100	38	14	2	0.042170	0.034367
100	38	14	3	0.139208	0.097037
100	38	14	4	0.318820	0.179612
100	38	14	5	0.549265	0.230446
100	38	14	6	0.760507	0.211242
100	38	14	7	0.900969	0.140462
100	38	14	8	0.969006	0.068036
100	38	14	9	0.992878	0.023872
100	38	14	10	0.998846	0.005968
100	38	14	11	0.999876	0.001030
100	38	14	12	0.999992	0.000116
100	38	14	13	1.000000	0.000008
100	38	14	14	1.000000	0.000000
100	38	15	0	0.000367	0.000367
100	38	15	1	0.004729	0.004362
100	38	15	2	0.027784	0.023055
100	38	15	3	0.099716	0.071932
100	38	15	4	0.247810	0.148095
100	38	15	5	0.460839	0.213028
100	38	15	6	0.681906	0.221067
100	38	15	7	0.850338	0.168432
100	38	15	8	0.945272	0.094934
100	38	15	9	0.984828	0.039556
100	38	15	10	0.996903	0.012075
100	38	15	11	0.999553	0.002650
100	38	15	12	0.999957	0.000404
100	38	15	13	0.999997	0.000040

N	n	k	x	P(x)	p(x)
100	38	15	14	1.000000	0.000002
100	38	15	15	1.000000	0.000000
100	38	16	0	0.000203	0.000203
100	38	16	1	0.002830	0.002627
100	38	16	2	0.018020	0.015189
100	38	16	3	0.070097	0.052077
100	38	16	4	0.188572	0.118476
100	38	16	5	0.378133	0.189561
100	38	16	6	0.598680	0.220547
100	38	16	7	0.788909	0.190229
100	38	16	8	0.911766	0.122856
100	38	16	9	0.971333	0.059567
100	38	16	10	0.992925	0.021593
100	38	16	11	0.998711	0.005786
100	38	16	12	0.999833	0.001122
100	38	16	13	0.999985	0.000152
100	38	16	14	0.999999	0.000014
100	38	16	15	1.000000	0.000001
100	38	16	16	1.000000	0.000000
100	38	17	0	0.000111	0.000111
100	38	17	1	0.001673	0.001562
100	38	17	2	0.011510	0.009837
100	38	17	3	0.048398	0.036888
100	38	17	4	0.140618	0.092220
100	38	17	5	0.303663	0.163045
100	38	17	6	0.514663	0.210999
100	38	17	7	0.718706	0.204043
100	38	17	8	0.867889	0.149183
100	38	17	9	0.950768	0.082879
100	38	17	10	0.985728	0.034960
100	38	17	11	0.996852	0.011124
100	38	17	12	0.999486	0.002635
100	38	17	13	0.999940	0.000454
100	38	17	14	0.999995	0.000055
100	38	17	15	1.000000	0.000004
100	38	17	16	1.000000	0.000000
100	38	17	17	1.000000	0.000000
100	38	18	0	0.000060	0.000060
100	38	18	1	0.000977	0.000917
100	38	18	2	0.007243	0.006267
100	38	18	3	0.032843	0.025599
100	38	18	4	0.102841	0.069998
100	38	18	5	0.238838	0.135997
100	38	18	6	0.433313	0.194475
100	38	18	7	0.642497	0.209184
100	38	18	8	0.813967	0.171470
100	38	18	9	0.921810	0.107843
100	38	18	10	0.973934	0.052124
100	38	18	11	0.993233	0.019299
100	38	18	12	0.998661	0.005428

Table for $N = 2$, $n = 1$, through $N = 100$, $n = 50$

N	n	k	x	P(x)	p(x)
100	38	18	13	0.999803	0.001143
100	38	18	14	0.999979	0.000176
100	38	18	15	0.999998	0.000019
100	38	18	16	1.000000	0.000001
100	38	18	17	1.000000	0.000000
100	38	18	18	1.000000	0.000000
100	38	19	0	0.000032	0.000032
100	38	19	1	0.000563	0.000531
100	38	19	2	0.004492	0.003929
100	38	19	3	0.021916	0.017424
100	38	19	4	0.073817	0.051901
100	38	19	5	0.184107	0.110290
100	38	19	6	0.357420	0.173313
100	38	19	7	0.563415	0.205995
100	38	19	8	0.751234	0.187819
100	38	19	9	0.883671	0.132436
100	38	19	10	0.956136	0.072465
100	38	19	11	0.986879	0.030743
100	38	19	12	0.996940	0.010061
100	38	19	13	0.999455	0.002515
100	38	19	14	0.999928	0.000473
100	38	19	15	0.999993	0.000065
100	38	19	16	0.999999	0.000006
100	38	19	17	1.000000	0.000000
100	38	19	18	1.000000	0.000000
100	38	19	19	1.000000	0.000000
100	38	20	0	0.000017	0.000017
100	38	20	1	0.000321	0.000304
100	38	20	2	0.002746	0.002425
100	38	20	3	0.014387	0.011641
100	38	20	4	0.052032	0.037644
100	38	20	5	0.139174	0.087143
100	38	20	6	0.288951	0.149777
100	38	20	7	0.484578	0.195627
100	38	20	8	0.681672	0.197094
100	38	20	9	0.836255	0.154583
100	38	20	10	0.931086	0.094831
100	38	20	11	0.976631	0.045545
100	38	20	12	0.993710	0.017079
100	38	20	13	0.998679	0.004969
100	38	20	14	0.999788	0.001109
100	38	20	15	0.999975	0.000187
100	38	20	16	0.999998	0.000023
100	38	20	17	1.000000	0.000002
100	38	20	18	1.000000	0.000000
100	38	20	19	1.000000	0.000000
100	38	20	20	1.000000	0.000000
100	38	21	0	0.000009	0.000009
100	38	21	1	0.000180	0.000171
100	38	21	2	0.001655	0.001474

N	n	k	x	P(x)	p(x)
100	38	21	3	0.009294	0.007640
100	38	21	4	0.036033	0.026739
100	38	21	5	0.103228	0.067195
100	38	21	6	0.229040	0.125812
100	38	21	7	0.408772	0.179732
100	38	21	8	0.607761	0.198989
100	38	21	9	0.780219	0.172457
100	38	21	10	0.897895	0.117677
100	38	21	11	0.961260	0.063364
100	38	21	12	0.988159	0.026900
100	38	21	13	0.997126	0.008967
100	38	21	14	0.999455	0.002329
100	38	21	15	0.999921	0.000466
100	38	21	16	0.999991	0.000070
100	38	21	17	0.999999	0.000008
100	38	21	18	1.000000	0.000001
100	38	21	19	1.000000	0.000000
100	38	21	20	1.000000	0.000000
100	38	21	21	1.000000	0.000000
100	38	22	0	0.000005	0.000005
100	38	22	1	0.000100	0.000095
100	38	22	2	0.000983	0.000883
100	38	22	3	0.005910	0.004927
100	38	22	4	0.024525	0.018615
100	38	22	5	0.075159	0.050634
100	38	22	6	0.178078	0.102919
100	38	22	7	0.338245	0.160166
100	38	22	8	0.532196	0.193951
100	38	22	9	0.716911	0.184716
100	38	22	10	0.856187	0.139275
100	38	22	11	0.939603	0.083416
100	38	22	12	0.979306	0.039703
100	38	22	13	0.994289	0.014982
100	38	22	14	0.998748	0.004459
100	38	22	15	0.999785	0.001038
100	38	22	16	0.999972	0.000186
100	38	22	17	0.999997	0.000025
100	38	22	18	1.000000	0.000003
100	38	22	19	1.000000	0.000000
100	38	22	20	1.000000	0.000000
100	38	22	21	1.000000	0.000000
100	38	22	22	1.000000	0.000000
100	38	23	0	0.000002	0.000002
100	38	23	1	0.000055	0.000052
100	38	23	2	0.000575	0.000521
100	38	23	3	0.003699	0.003123
100	38	23	4	0.016411	0.012712
100	38	23	5	0.053737	0.037326
100	38	23	6	0.135855	0.082118
100	38	23	7	0.274589	0.138734

N	n	k	x	P(x)	p(x)
100	38	23	8	0.457599	0.183010
100	38	23	9	0.648235	0.190636
100	38	23	10	0.806191	0.157955
100	38	23	11	0.910728	0.104538
100	38	23	12	0.966072	0.055344
100	38	23	13	0.989487	0.023415
100	38	23	14	0.997376	0.007889
100	38	23	15	0.999479	0.002104
100	38	23	16	0.999919	0.000440
100	38	23	17	0.999990	0.000071
100	38	23	18	0.999999	0.000009
100	38	23	19	1.000000	0.000001
100	38	23	20	1.000000	0.000000
100	38	23	21	1.000000	0.000000
100	38	23	22	1.000000	0.000000
100	38	23	23	1.000000	0.000000
100	38	24	0	0.000001	0.000001
100	38	24	1	0.000030	0.000028
100	38	24	2	0.000332	0.000302
100	38	24	3	0.002279	0.001947
100	38	24	4	0.010798	0.008519
100	38	24	5	0.037740	0.026942
100	38	24	6	0.101728	0.063988
100	38	24	7	0.218735	0.117007
100	38	24	8	0.386297	0.167562
100	38	24	9	0.576437	0.190141
100	38	24	10	0.748752	0.172315
100	38	24	11	0.874072	0.125320
100	38	24	12	0.947385	0.073312
100	38	24	13	0.981884	0.034500
100	38	24	14	0.994917	0.013032
100	38	24	15	0.998851	0.003934
100	38	24	16	0.999793	0.000943
100	38	24	17	0.999971	0.000177
100	38	24	18	0.999997	0.000026
100	38	24	19	1.000000	0.000003
100	38	24	20	1.000000	0.000000
100	38	24	21	1.000000	0.000000
100	38	24	22	1.000000	0.000000
100	38	24	23	1.000000	0.000000
100	38	24	24	1.000000	0.000000
100	38	25	0	0.000001	0.000001
100	38	25	1	0.000016	0.000015
100	38	25	2	0.000189	0.000173
100	38	25	3	0.001382	0.001194
100	38	25	4	0.006987	0.005604
100	38	25	5	0.026041	0.019055
100	38	25	6	0.074786	0.048744
100	38	25	7	0.171009	0.096223
100	38	25	8	0.320154	0.149145

N	n	k	x	P(x)	p(x)
100	38	25	9	0.503884	0.183730
100	38	25	10	0.685268	0.181384
100	38	25	11	0.829551	0.144283
100	38	25	12	0.922304	0.092753
100	38	25	13	0.970536	0.048232
100	38	25	14	0.990801	0.020265
100	38	25	15	0.997660	0.006859
100	38	25	16	0.999521	0.001860
100	38	25	17	0.999922	0.000401
100	38	25	18	0.999990	0.000068
100	38	25	19	0.999999	0.000009
100	38	25	20	1.000000	0.000001
100	38	25	21	1.000000	0.000000
100	38	25	22	1.000000	0.000000
100	38	25	23	1.000000	0.000000
100	38	25	24	1.000000	0.000000
100	38	25	25	1.000000	0.000000
100	38	26	0	0.000000	0.000000
100	38	26	1	0.000008	0.000008
100	38	26	2	0.000106	0.000097
100	38	26	3	0.000825	0.000720
100	38	26	4	0.004446	0.003621
100	38	26	5	0.017657	0.013211
100	38	26	6	0.053988	0.036331
100	38	26	7	0.131236	0.077248
100	38	26	8	0.260496	0.129259
100	38	26	9	0.432841	0.172346
100	38	26	10	0.617551	0.184710
100	38	26	11	0.777609	0.160058
100	38	26	12	0.890150	0.112541
100	38	26	13	0.954458	0.064309
100	38	26	14	0.984316	0.029858
100	38	26	15	0.995557	0.011241
100	38	26	16	0.998975	0.003418
100	38	26	17	0.999809	0.000835
100	38	26	18	0.999972	0.000162
100	38	26	19	0.999997	0.000025
100	38	26	20	1.000000	0.000003
100	38	26	21	1.000000	0.000000
100	38	26	22	1.000000	0.000000
100	38	26	23	1.000000	0.000000
100	38	26	24	1.000000	0.000000
100	38	26	25	1.000000	0.000000
100	38	26	26	1.000000	0.000000
100	38	27	0	0.000000	0.000000
100	38	27	1	0.000004	0.000004
100	38	27	2	0.000058	0.000054
100	38	27	3	0.000485	0.000427
100	38	27	4	0.002783	0.002298
100	38	27	5	0.011766	0.008983

Table for $N = 2$, $n = 1$, through $N = 100$, $n = 50$

N	n	k	x	P(x)	p(x)
100	38	27	6	0.038278	0.026512
100	38	27	7	0.098876	0.060598
100	38	27	8	0.208093	0.109218
100	38	27	9	0.365301	0.157207
100	38	27	10	0.547661	0.182360
100	38	27	11	0.719209	0.171548
100	38	27	12	0.850608	0.131399
100	38	27	13	0.932732	0.082124
100	38	27	14	0.974633	0.041900
100	38	27	15	0.992063	0.017430
100	38	27	16	0.997959	0.005896
100	38	27	17	0.999573	0.001614
100	38	27	18	0.999928	0.000355
100	38	27	19	0.999990	0.000062
100	38	27	20	0.999999	0.000009
100	38	27	21	1.000000	0.000001
100	38	27	22	1.000000	0.000000
100	38	27	23	1.000000	0.000000
100	38	27	24	1.000000	0.000000
100	38	27	25	1.000000	0.000000
100	38	27	26	1.000000	0.000000
100	38	27	27	1.000000	0.000000
100	38	28	0	0.000000	0.000000
100	38	28	1	0.000002	0.000002
100	38	28	2	0.000032	0.000029
100	38	28	3	0.000280	0.000249
100	38	28	4	0.001712	0.001432
100	38	28	5	0.007705	0.005992
100	38	28	6	0.026656	0.018951
100	38	28	7	0.073142	0.046486
100	38	28	8	0.163209	0.090067
100	38	28	9	0.302848	0.139639
100	38	28	10	0.477714	0.174866
100	38	28	11	0.655760	0.178045
100	38	28	12	0.803809	0.148049
100	38	28	13	0.904608	0.100799
100	38	28	14	0.960857	0.056249
100	38	28	15	0.986571	0.025714
100	38	28	16	0.996182	0.009611
100	38	28	17	0.999108	0.002926
100	38	28	18	0.999830	0.000722
100	38	28	19	0.999974	0.000143
100	38	28	20	0.999997	0.000023
100	38	28	21	0.999999	0.000003
100	38	28	22	1.000000	0.000000
100	38	28	23	1.000000	0.000000
100	38	28	24	1.000000	0.000000
100	38	28	25	1.000000	0.000000
100	38	28	26	1.000000	0.000000
100	38	28	27	1.000000	0.000000

N	n	k	x	P(x)	p(x)
100	38	28	28	1.000000	0.000000
100	38	29	0	0.000000	0.000000
100	38	29	1	0.000001	0.000001
100	38	29	2	0.000017	0.000016
100	38	29	3	0.000160	0.000143
100	38	29	4	0.001036	0.000877
100	38	29	5	0.004958	0.003922
100	38	29	6	0.018233	0.013275
100	38	29	7	0.053127	0.034894
100	38	29	8	0.125681	0.072554
100	38	29	9	0.246605	0.120923
100	38	29	10	0.409711	0.163106
100	38	29	11	0.588993	0.179282
100	38	29	12	0.750347	0.161354
100	38	29	13	0.869608	0.119261
100	38	29	14	0.942107	0.072499
100	38	29	15	0.978357	0.036250
100	38	29	16	0.993245	0.014888
100	38	29	17	0.998255	0.005009
100	38	29	18	0.999630	0.001375
100	38	29	19	0.999936	0.000306
100	38	29	20	0.999991	0.000055
100	38	29	21	0.999999	0.000008
100	38	29	22	1.000000	0.000001
100	38	29	23	1.000000	0.000000
100	38	29	24	1.000000	0.000000
100	38	29	25	1.000000	0.000000
100	38	29	26	1.000000	0.000000
100	38	29	27	1.000000	0.000000
100	38	29	28	1.000000	0.000000
100	38	29	29	1.000000	0.000000
100	38	30	0	0.000000	0.000000
100	38	30	1	0.000001	0.000001
100	38	30	2	0.000009	0.000008
100	38	30	3	0.000089	0.000080
100	38	30	4	0.000616	0.000527
100	38	30	5	0.003135	0.002519
100	38	30	6	0.012250	0.009115
100	38	30	7	0.037892	0.025642
100	38	30	8	0.095025	0.057133
100	38	30	9	0.197214	0.102189
100	38	30	10	0.345387	0.148174
100	38	30	11	0.520815	0.175428
100	38	30	12	0.691259	0.170444
100	38	30	13	0.827615	0.136355
100	38	30	14	0.917601	0.089986
100	38	30	15	0.966614	0.049014
100	38	30	16	0.988632	0.022018
100	38	30	17	0.996773	0.008141
100	38	30	18	0.999243	0.002469

N	n	k	x	P(x)	p(x)
100	38	30	19	0.999854	0.000612
100	38	30	20	0.999977	0.000123
100	38	30	21	0.999997	0.000020
100	38	30	22	1.000000	0.000003
100	38	30	23	1.000000	0.000000
100	38	30	24	1.000000	0.000000
100	38	30	25	1.000000	0.000000
100	38	30	26	1.000000	0.000000
100	38	30	27	1.000000	0.000000
100	38	30	28	1.000000	0.000000
100	38	30	29	1.000000	0.000000
100	38	30	30	1.000000	0.000000
100	38	31	0	0.000000	0.000000
100	38	31	1	0.000000	0.000000
100	38	31	2	0.000005	0.000004
100	38	31	3	0.000049	0.000044
100	38	31	4	0.000360	0.000311
100	38	31	5	0.001948	0.001588
100	38	31	6	0.008083	0.006136
100	38	31	7	0.026536	0.018453
100	38	31	8	0.070539	0.044003
100	38	31	9	0.154878	0.084339
100	38	31	10	0.286118	0.131240
100	38	31	11	0.453150	0.167032
100	38	31	12	0.627952	0.174801
100	38	31	13	0.778916	0.150965
100	38	31	14	0.886748	0.107832
100	38	31	15	0.950510	0.063761
100	38	31	16	0.981712	0.031202
100	38	31	17	0.994331	0.012619
100	38	31	18	0.998537	0.004206
100	38	31	19	0.999688	0.001151
100	38	31	20	0.999946	0.000257
100	38	31	21	0.999992	0.000047
100	38	31	22	0.999999	0.000007
100	38	31	23	1.000000	0.000001
100	38	31	24	1.000000	0.000000
100	38	31	25	1.000000	0.000000
100	38	31	26	1.000000	0.000000
100	38	31	27	1.000000	0.000000
100	38	31	28	1.000000	0.000000
100	38	31	29	1.000000	0.000000
100	38	31	30	1.000000	0.000000
100	38	31	31	1.000000	0.000000
100	38	32	0	0.000000	0.000000
100	38	32	1	0.000000	0.000000
100	38	32	2	0.000002	0.000002
100	38	32	3	0.000027	0.000024
100	38	32	4	0.000207	0.000180
100	38	32	5	0.001189	0.000982

N	n	k	x	P(x)	p(x)
100	38	32	6	0.005238	0.004049
100	38	32	7	0.018246	0.013008
100	38	32	8	0.051408	0.033162
100	38	32	9	0.119431	0.068024
100	38	32	10	0.232861	0.113430
100	38	32	11	0.387790	0.154929
100	38	32	12	0.562084	0.174295
100	38	32	13	0.724219	0.162135
100	38	32	14	0.849241	0.125023
100	38	32	15	0.929256	0.080014
100	38	32	16	0.971764	0.042508
100	38	32	17	0.990490	0.018727
100	38	32	18	0.997318	0.006827
100	38	32	19	0.999371	0.002053
100	38	32	20	0.999878	0.000507
100	38	32	21	0.999981	0.000102
100	38	32	22	0.999997	0.000017
100	38	32	23	1.000000	0.000002
100	38	32	24	1.000000	0.000000
100	38	32	25	1.000000	0.000000
100	38	32	26	1.000000	0.000000
100	38	32	27	1.000000	0.000000
100	38	32	28	1.000000	0.000000
100	38	32	29	1.000000	0.000000
100	38	32	30	1.000000	0.000000
100	38	32	31	1.000000	0.000000
100	38	32	32	1.000000	0.000000
100	38	33	0	0.000000	0.000000
100	38	33	1	0.000000	0.000000
100	38	33	2	0.000001	0.000001
100	38	33	3	0.000014	0.000013
100	38	33	4	0.000117	0.000103
100	38	33	5	0.000712	0.000595
100	38	33	6	0.003332	0.002620
100	38	33	7	0.012316	0.008983
100	38	33	8	0.036777	0.024462
100	38	33	9	0.090421	0.053644
100	38	33	10	0.186155	0.095734
100	38	33	11	0.326274	0.140119
100	38	33	12	0.495442	0.169168
100	38	33	13	0.664611	0.169168
100	38	33	14	0.805116	0.140505
100	38	33	15	0.902192	0.097076
100	38	33	16	0.958011	0.055819
100	38	33	17	0.984707	0.026696
100	38	33	18	0.995310	0.010603
100	38	33	19	0.998798	0.003488
100	38	33	20	0.999744	0.000947
100	38	33	21	0.999955	0.000211
100	38	33	22	0.999994	0.000038

Table for $N = 2$, $n = 1$, through $N = 100$, $n = 50$

N	n	k	x	P(x)	p(x)
100	38	33	23	0.999999	0.000006
100	38	33	24	1.000000	0.000001
100	38	33	25	1.000000	0.000000
100	38	33	26	1.000000	0.000000
100	38	33	27	1.000000	0.000000
100	38	33	28	1.000000	0.000000
100	38	33	29	1.000000	0.000000
100	38	33	30	1.000000	0.000000
100	38	33	31	1.000000	0.000000
100	38	33	32	1.000000	0.000000
100	38	33	33	1.000000	0.000000
100	38	34	0	0.000000	0.000000
100	38	34	1	0.000000	0.000000
100	38	34	2	0.000001	0.000001
100	38	34	3	0.000007	0.000007
100	38	34	4	0.000065	0.000057
100	38	34	5	0.000419	0.000354
100	38	34	6	0.002081	0.001662
100	38	34	7	0.008159	0.006078
100	38	34	8	0.025824	0.017665
100	38	34	9	0.067202	0.041378
100	38	34	10	0.146147	0.078944
100	38	34	11	0.269808	0.123661
100	38	34	12	0.429795	0.159987
100	38	34	13	0.601488	0.171693
100	38	34	14	0.754786	0.153298
100	38	34	15	0.868868	0.114082
100	38	34	16	0.939683	0.070815
100	38	34	17	0.976340	0.036657
100	38	34	18	0.992145	0.015805
100	38	34	19	0.997808	0.005664
100	38	34	20	0.999490	0.001681
100	38	34	21	0.999902	0.000412
100	38	34	22	0.999984	0.000083
100	38	34	23	0.999998	0.000014
100	38	34	24	1.000000	0.000002
100	38	34	25	1.000000	0.000000
100	38	34	26	1.000000	0.000000
100	38	34	27	1.000000	0.000000
100	38	34	28	1.000000	0.000000
100	38	34	29	1.000000	0.000000
100	38	34	30	1.000000	0.000000
100	38	34	31	1.000000	0.000000
100	38	34	32	1.000000	0.000000
100	38	34	33	1.000000	0.000000
100	38	34	34	1.000000	0.000000
100	38	35	0	0.000000	0.000000
100	38	35	1	0.000000	0.000000
100	38	35	2	0.000000	0.000000
100	38	35	3	0.000004	0.000003
100	38	35	4	0.000035	0.000031
100	38	35	5	0.000242	0.000207
100	38	35	6	0.001275	0.001033
100	38	35	7	0.005304	0.004029
100	38	35	8	0.017795	0.012490
100	38	35	9	0.049021	0.031226
100	38	35	10	0.112655	0.063634
100	38	35	11	0.219219	0.106564
100	38	35	12	0.366770	0.147550
100	38	35	13	0.536453	0.169683
100	38	35	14	0.699041	0.162588
100	38	35	15	0.829112	0.130071
100	38	35	16	0.916077	0.086966
100	38	35	17	0.964676	0.048599
100	38	35	18	0.987355	0.022679
100	38	35	19	0.996178	0.008823
100	38	35	20	0.999031	0.002853
100	38	35	21	0.999795	0.000764
100	38	35	22	0.999964	0.000169
100	38	35	23	0.999995	0.000031
100	38	35	24	0.999999	0.000004
100	38	35	25	1.000000	0.000001
100	38	35	26	1.000000	0.000000
100	38	35	27	1.000000	0.000000
100	38	35	28	1.000000	0.000000
100	38	35	29	1.000000	0.000000
100	38	35	30	1.000000	0.000000
100	38	35	31	1.000000	0.000000
100	38	35	32	1.000000	0.000000
100	38	35	33	1.000000	0.000000
100	38	35	34	1.000000	0.000000
100	38	35	35	1.000000	0.000000
100	38	36	0	0.000000	0.000000
100	38	36	1	0.000000	0.000000
100	38	36	2	0.000000	0.000000
100	38	36	3	0.000002	0.000002
100	38	36	4	0.000019	0.000017
100	38	36	5	0.000137	0.000118
100	38	36	6	0.000766	0.000630
100	38	36	7	0.003383	0.002616
100	38	36	8	0.012030	0.008647
100	38	36	9	0.035089	0.023059
100	38	36	10	0.085244	0.050154
100	38	36	11	0.174954	0.089711
100	38	36	12	0.307750	0.132795
100	38	36	13	0.471190	0.163441
100	38	36	14	0.639009	0.167818
100	38	36	15	0.783087	0.144078
100	38	36	16	0.886643	0.103556
100	38	36	17	0.948975	0.062332
100	38	36	18	0.980377	0.031402
100	38	36	19	0.993599	0.013222
100	38	36	20	0.998241	0.004642
100	38	36	21	0.999596	0.001355
100	38	36	22	0.999923	0.000327
100	38	36	23	0.999988	0.000065
100	38	36	24	0.999998	0.000011
100	38	36	25	1.000000	0.000001
100	38	36	26	1.000000	0.000000
100	38	36	27	1.000000	0.000000
100	38	36	28	1.000000	0.000000
100	38	36	29	1.000000	0.000000
100	38	36	30	1.000000	0.000000
100	38	36	31	1.000000	0.000000
100	38	36	32	1.000000	0.000000
100	38	36	33	1.000000	0.000000
100	38	36	34	1.000000	0.000000
100	38	36	35	1.000000	0.000000
100	38	36	36	1.000000	0.000000
100	38	37	0	0.000000	0.000000
100	38	37	1	0.000000	0.000000
100	38	37	2	0.000000	0.000000
100	38	37	3	0.000001	0.000001
100	38	37	4	0.000010	0.000009
100	38	37	5	0.000076	0.000066
100	38	37	6	0.000452	0.000376
100	38	37	7	0.002115	0.001664
100	38	37	8	0.007977	0.005861
100	38	37	9	0.024641	0.016664
100	38	37	10	0.063301	0.038661
100	38	37	11	0.137108	0.073806
100	38	37	12	0.253802	0.116694
100	38	37	13	0.407346	0.153545
100	38	37	14	0.576077	0.168730
100	38	37	15	0.731309	0.155232
100	38	37	16	0.851046	0.119737
100	38	37	17	0.928522	0.077477
100	38	37	18	0.970564	0.042042
100	38	37	19	0.989674	0.019110
100	38	37	20	0.996936	0.007262
100	38	37	21	0.999236	0.002300
100	38	37	22	0.999841	0.000605
100	38	37	23	0.999973	0.000132
100	38	37	24	0.999996	0.000023
100	38	37	25	0.999999	0.000003
100	38	37	26	1.000000	0.000000
100	38	37	27	1.000000	0.000000
100	38	37	28	1.000000	0.000000
100	38	37	29	1.000000	0.000000
100	38	37	30	1.000000	0.000000
100	38	37	31	1.000000	0.000000
100	38	37	32	1.000000	0.000000
100	38	37	33	1.000000	0.000000
100	38	37	34	1.000000	0.000000
100	38	37	35	1.000000	0.000000
100	38	37	36	1.000000	0.000000
100	38	37	37	1.000000	0.000000
100	38	38	0	0.000000	0.000000
100	38	38	1	0.000000	0.000000
100	38	38	2	0.000000	0.000000
100	38	38	3	0.000000	0.000000
100	38	38	4	0.000005	0.000005
100	38	38	5	0.000041	0.000036
100	38	38	6	0.000261	0.000220
100	38	38	7	0.001297	0.001036
100	38	38	8	0.005186	0.003889
100	38	38	9	0.016970	0.011784
100	38	38	10	0.046119	0.029149
100	38	38	11	0.105476	0.059358
100	38	38	12	0.205642	0.100166
100	38	38	13	0.346416	0.140774
100	38	38	14	0.511798	0.165383
100	38	38	15	0.674637	0.162838
100	38	38	16	0.809233	0.134596
100	38	38	17	0.902697	0.093464
100	38	38	18	0.957217	0.054521
100	38	38	19	0.983910	0.026693
100	38	38	20	0.994861	0.010950
100	38	38	21	0.998615	0.003754
100	38	38	22	0.999687	0.001072
100	38	38	23	0.999941	0.000254
100	38	38	24	0.999991	0.000050
100	38	38	25	0.999999	0.000008
100	38	38	26	1.000000	0.000001
100	38	38	27	1.000000	0.000000
100	38	38	28	1.000000	0.000000
100	38	38	29	1.000000	0.000000
100	38	38	30	1.000000	0.000000
100	38	38	31	1.000000	0.000000
100	38	38	32	1.000000	0.000000
100	38	38	33	1.000000	0.000000
100	38	38	34	1.000000	0.000000
100	38	38	35	1.000000	0.000000
100	38	38	36	1.000000	0.000000
100	38	38	37	1.000000	0.000000
100	38	38	38	1.000000	0.000000
100	39	1	0	0.610000	0.610000
100	39	1	1	1.000000	0.390000
100	39	2	0	0.369697	0.369697
100	39	2	1	0.850303	0.480606

Table for $N = 2$, $n = 1$, through $N = 100$, $n = 50$

N	n	k	x	P(x)	p(x)
100	39	2	2	1.000000	0.149697
100	39	3	0	0.222573	0.222573
100	39	3	1	0.663946	0.441373
100	39	3	2	0.943482	0.279536
100	39	3	3	1.000000	0.056518
100	39	4	0	0.133085	0.133085
100	39	4	1	0.491037	0.357952
100	39	4	2	0.836855	0.345818
100	39	4	3	0.979024	0.142170
100	39	4	4	1.000000	0.020976
100	39	5	0	0.079019	0.079019
100	39	5	1	0.349347	0.270328
100	39	5	2	0.703570	0.354223
100	39	5	3	0.925710	0.222140
100	39	5	4	0.992352	0.066642
100	39	5	5	1.000000	0.007647
100	39	6	0	0.046580	0.046580
100	39	6	1	0.241216	0.194636
100	39	6	2	0.565610	0.324394
100	39	6	3	0.841531	0.275921
100	39	6	4	0.967800	0.126269
100	39	6	5	0.997263	0.029463
100	39	6	6	1.000000	0.002737
100	39	7	0	0.027254	0.027254
100	39	7	1	0.162533	0.135279
100	39	7	2	0.437923	0.275390
100	39	7	3	0.735859	0.297936
100	39	7	4	0.920785	0.184926
100	39	7	5	0.986606	0.065821
100	39	7	6	0.999039	0.012433
100	39	7	7	1.000000	0.000961
100	39	8	0	0.015825	0.015825
100	39	8	1	0.107258	0.091433
100	39	8	2	0.328359	0.221101
100	39	8	3	0.620529	0.292170
100	39	8	4	0.851189	0.230660
100	39	8	5	0.962543	0.111353
100	39	8	6	0.994627	0.032085
100	39	8	7	0.999669	0.005042
100	39	8	8	1.000000	0.000331
100	39	9	0	0.009117	0.009117
100	39	9	1	0.069492	0.060376
100	39	9	2	0.239438	0.169946
100	39	9	3	0.506202	0.266764
100	39	9	4	0.763438	0.257236
100	39	9	5	0.921390	0.157952
100	39	9	6	0.983119	0.061728
100	39	9	7	0.997916	0.014797
100	39	9	8	0.999888	0.001973
100	39	9	9	1.000000	0.000111

N	n	k	x	P(x)	p(x)
100	39	10	0	0.005209	0.005209
100	39	10	1	0.044280	0.039071
100	39	10	2	0.170339	0.126059
100	39	10	3	0.400669	0.230330
100	39	10	4	0.664501	0.263832
100	39	10	5	0.862375	0.197874
100	39	10	6	0.960734	0.098358
100	39	10	7	0.992712	0.031979
100	39	10	8	0.999216	0.006504
100	39	10	9	0.999963	0.000747
100	39	10	10	1.000000	0.000037
100	39	11	0	0.002952	0.002952
100	39	11	1	0.027784	0.024832
100	39	11	2	0.118515	0.090731
100	39	11	3	0.308537	0.190022
100	39	11	4	0.561900	0.253363
100	39	11	5	0.787623	0.225723
100	39	11	6	0.924669	0.137046
100	39	11	7	0.981342	0.056673
100	39	11	8	0.996976	0.015634
100	39	11	9	0.999714	0.002738
100	39	11	10	0.999988	0.000274
100	39	11	11	1.000000	0.000012
100	39	12	0	0.001658	0.001658
100	39	12	1	0.017181	0.015523
100	39	12	2	0.080795	0.063614
100	39	12	3	0.231674	0.150879
100	39	12	4	0.462263	0.230589
100	39	12	5	0.701391	0.239129
100	39	12	6	0.873854	0.172463
100	39	12	7	0.960965	0.087111
100	39	12	8	0.991531	0.030565
100	39	12	9	0.998791	0.007261
100	39	12	10	0.999899	0.001108
100	39	12	11	0.999996	0.000097
100	39	12	12	1.000000	0.000004
100	39	13	0	0.000923	0.000923
100	39	13	1	0.010478	0.009555
100	39	13	2	0.054049	0.043570
100	39	13	3	0.169951	0.115902
100	39	13	4	0.370551	0.200600
100	39	13	5	0.609001	0.238449
100	39	13	6	0.809181	0.200180
100	39	13	7	0.929288	0.120108
100	39	13	8	0.980763	0.051475
100	39	13	9	0.996316	0.015553
100	39	13	10	0.999534	0.003218
100	39	13	11	0.999965	0.000431
100	39	13	12	0.999999	0.000034
100	39	13	13	1.000000	0.000001

N	n	k	x	P(x)	p(x)
100	39	14	0	0.000509	0.000509
100	39	14	1	0.006305	0.005795
100	39	14	2	0.035519	0.029214
100	39	14	3	0.121991	0.086473
100	39	14	4	0.289850	0.167859
100	39	14	5	0.515814	0.225964
100	39	14	6	0.733250	0.217437
100	39	14	7	0.885111	0.151861
100	39	14	8	0.962422	0.077311
100	39	14	9	0.990953	0.028531
100	39	14	10	0.998461	0.007508
100	39	14	11	0.999826	0.001365
100	39	14	12	0.999988	0.000162
100	39	14	13	1.000000	0.000011
100	39	14	14	1.000000	0.000000
100	39	15	0	0.000278	0.000278
100	39	15	1	0.003744	0.003466
100	39	15	2	0.022950	0.019206
100	39	15	3	0.085793	0.062843
100	39	15	4	0.221535	0.135742
100	39	15	5	0.426479	0.204944
100	39	15	6	0.649815	0.223336
100	39	15	7	0.828605	0.178789
100	39	15	8	0.934554	0.105949
100	39	15	9	0.981000	0.046446
100	39	15	10	0.995929	0.014929
100	39	15	11	0.999382	0.003453
100	39	15	12	0.999937	0.000556
100	39	15	13	0.999996	0.000059
100	39	15	14	1.000000	0.000004
100	39	15	15	1.000000	0.000000
100	39	16	0	0.000151	0.000151
100	39	16	1	0.002195	0.002044
100	39	16	2	0.014590	0.012395
100	39	16	3	0.059177	0.044588
100	39	16	4	0.165642	0.106464
100	39	16	5	0.344502	0.178860
100	39	16	6	0.563108	0.218607
100	39	16	7	0.761296	0.198187
100	39	16	8	0.895913	0.134618
100	39	16	9	0.964607	0.068694
100	39	16	10	0.990836	0.026229
100	39	16	11	0.998245	0.007409
100	39	16	12	0.999761	0.001516
100	39	16	13	0.999978	0.000217
100	39	16	14	0.999999	0.000021
100	39	16	15	1.000000	0.000001
100	39	16	16	1.000000	0.000000
100	39	17	0	0.000081	0.000081
100	39	17	1	0.001270	0.001189

N	n	k	x	P(x)	p(x)
100	39	17	2	0.009130	0.007860
100	39	17	3	0.040068	0.030938
100	39	17	4	0.121282	0.081213
100	39	17	5	0.272106	0.150824
100	39	17	6	0.477227	0.205121
100	39	17	7	0.685796	0.208569
100	39	17	8	0.846233	0.160437
100	39	17	9	0.940074	0.093841
100	39	17	10	0.981781	0.041707
100	39	17	11	0.995775	0.013994
100	39	17	12	0.999274	0.003499
100	39	17	13	0.999911	0.000637
100	39	17	14	0.999993	0.000082
100	39	17	15	1.000000	0.000007
100	39	17	16	1.000000	0.000000
100	39	17	17	1.000000	0.000000
100	39	18	0	0.000043	0.000043
100	39	18	1	0.000726	0.000683
100	39	18	2	0.005626	0.004901
100	39	18	3	0.026649	0.021023
100	39	18	4	0.087035	0.060386
100	39	18	5	0.210322	0.123287
100	39	18	6	0.395673	0.185350
100	39	18	7	0.605384	0.209711
100	39	18	8	0.786311	0.180927
100	39	18	9	0.906155	0.119845
100	39	18	10	0.967208	0.061053
100	39	18	11	0.991054	0.023846
100	39	18	12	0.998135	0.007081
100	39	18	13	0.999711	0.001576
100	39	18	14	0.999968	0.000257
100	39	18	15	0.999998	0.000030
100	39	18	16	1.000000	0.000002
100	39	18	17	1.000000	0.000000
100	39	18	18	1.000000	0.000000
100	39	19	0	0.000022	0.000022
100	39	19	1	0.000409	0.000387
100	39	19	2	0.003415	0.003006
100	39	19	3	0.017420	0.014005
100	39	19	4	0.061261	0.043841
100	39	19	5	0.159203	0.097943
100	39	19	6	0.321081	0.161877
100	39	19	7	0.523545	0.202465
100	39	19	8	0.717911	0.194366
100	39	19	9	0.862310	0.144398
100	39	19	10	0.945617	0.083307
100	39	19	11	0.982912	0.037295
100	39	19	12	0.995804	0.012892
100	39	19	13	0.999212	0.003408
100	39	19	14	0.999890	0.000678

Table for $N = 2$, $n = 1$, through $N = 100$, $n = 50$

N	n	k	x	P(x)	p(x)
100	39	19	15	0.999989	0.000099
100	39	19	16	0.999999	0.000010
100	39	19	17	1.000000	0.000001
100	39	19	18	1.000000	0.000000
100	39	19	19	1.000000	0.000000
100	39	20	0	0.000012	0.000012
100	39	20	1	0.000228	0.000216
100	39	20	2	0.002042	0.001814
100	39	20	3	0.011195	0.009153
100	39	20	4	0.042317	0.031122
100	39	20	5	0.118092	0.075775
100	39	20	6	0.255131	0.137039
100	39	20	7	0.443559	0.188429
100	39	20	8	0.643524	0.199965
100	39	20	9	0.808829	0.165305
100	39	20	10	0.915791	0.106962
100	39	20	11	0.970020	0.054229
100	39	20	12	0.991506	0.021487
100	39	20	13	0.998118	0.006611
100	39	20	14	0.999680	0.001563
100	39	20	15	0.999960	0.000279
100	39	20	16	0.999996	0.000037
100	39	20	17	1.000000	0.000003
100	39	20	18	1.000000	0.000000
100	39	20	19	1.000000	0.000000
100	39	20	20	1.000000	0.000000
100	39	21	0	0.000006	0.000006
100	39	21	1	0.000125	0.000119
100	39	21	2	0.001203	0.001078
100	39	21	3	0.007076	0.005873
100	39	21	4	0.028701	0.021625
100	39	21	5	0.085887	0.057186
100	39	21	6	0.198602	0.112715
100	39	21	7	0.368188	0.169586
100	39	21	8	0.566038	0.197850
100	39	21	9	0.746840	0.180802
100	39	21	10	0.877017	0.130177
100	39	21	11	0.951039	0.074022
100	39	21	12	0.984255	0.033215
100	39	21	13	0.995969	0.011714
100	39	21	14	0.999192	0.003223
100	39	21	15	0.999876	0.000684
100	39	21	16	0.999986	0.000110
100	39	21	17	0.999999	0.000013
100	39	21	18	1.000000	0.000001
100	39	21	19	1.000000	0.000000
100	39	21	20	1.000000	0.000000
100	39	21	21	1.000000	0.000000
100	39	22	0	0.000003	0.000003
100	39	22	1	0.000068	0.000065

N	n	k	x	P(x)	p(x)
100	39	22	2	0.000698	0.000630
100	39	22	3	0.004400	0.003702
100	39	22	4	0.019121	0.014721
100	39	22	5	0.061276	0.042155
100	39	22	6	0.151519	0.090243
100	39	22	7	0.299495	0.147976
100	39	22	8	0.488401	0.188906
100	39	22	9	0.678181	0.189780
100	39	22	10	0.829230	0.151050
100	39	22	11	0.924804	0.095573
100	39	22	12	0.972903	0.048099
100	39	22	13	0.992114	0.019211
100	39	22	14	0.998172	0.006058
100	39	22	15	0.999668	0.001496
100	39	22	16	0.999954	0.000286
100	39	22	17	0.999995	0.000041
100	39	22	18	1.000000	0.000004
100	39	22	19	1.000000	0.000000
100	39	22	20	1.000000	0.000000
100	39	22	21	1.000000	0.000000
100	39	22	22	1.000000	0.000000
100	39	23	0	0.000002	0.000002
100	39	23	1	0.000036	0.000035
100	39	23	2	0.000399	0.000363
100	39	23	3	0.002691	0.002292
100	39	23	4	0.012515	0.009824
100	39	23	5	0.042900	0.030385
100	39	23	6	0.113339	0.070439
100	39	23	7	0.238787	0.125448
100	39	23	8	0.413323	0.174536
100	39	23	9	0.605188	0.191866
100	39	23	10	0.773071	0.167883
100	39	23	11	0.890495	0.117424
100	39	23	12	0.956253	0.065758
100	39	23	13	0.985710	0.029457
100	39	23	14	0.996230	0.010520
100	39	23	15	0.999208	0.002977
100	39	23	16	0.999869	0.000662
100	39	23	17	0.999983	0.000114
100	39	23	18	0.999998	0.000015
100	39	23	19	1.000000	0.000001
100	39	23	20	1.000000	0.000000
100	39	23	21	1.000000	0.000000
100	39	23	22	1.000000	0.000000
100	39	23	23	1.000000	0.000000
100	39	24	0	0.000001	0.000001
100	39	24	1	0.000019	0.000018
100	39	24	2	0.000225	0.000206
100	39	24	3	0.001620	0.001395
100	39	24	4	0.008050	0.006430

N	n	k	x	P(x)	p(x)
100	39	24	5	0.029484	0.021434
100	39	24	6	0.083151	0.053667
100	39	24	7	0.186653	0.103502
100	39	24	8	0.343055	0.156402
100	39	24	9	0.530435	0.187381
100	39	24	10	0.709842	0.179407
100	39	24	11	0.847796	0.137953
100	39	24	12	0.933195	0.085400
100	39	24	13	0.975763	0.042568
100	39	24	14	0.992815	0.017051
100	39	24	15	0.998280	0.005465
100	39	24	16	0.999672	0.001392
100	39	24	17	0.999951	0.000279
100	39	24	18	0.999994	0.000043
100	39	24	19	0.999999	0.000005
100	39	24	20	1.000000	0.000000
100	39	24	21	1.000000	0.000000
100	39	24	22	1.000000	0.000000
100	39	24	23	1.000000	0.000000
100	39	24	24	1.000000	0.000000
100	39	25	0	0.000000	0.000000
100	39	25	1	0.000010	0.000010
100	39	25	2	0.000125	0.000115
100	39	25	3	0.000959	0.000834
100	39	25	4	0.005089	0.004130
100	39	25	5	0.019895	0.014806
100	39	25	6	0.059848	0.039953
100	39	25	7	0.143073	0.083225
100	39	25	8	0.279259	0.136186
100	39	25	9	0.456470	0.177210
100	39	25	10	0.641385	0.184915
100	39	25	11	0.796971	0.155586
100	39	25	12	0.902856	0.105885
100	39	25	13	0.961201	0.058345
100	39	25	14	0.987206	0.026005
100	39	25	15	0.996554	0.009348
100	39	25	16	0.999251	0.002697
100	39	25	17	0.999870	0.000620
100	39	25	18	0.999982	0.000112
100	39	25	19	0.999998	0.000016
100	39	25	20	1.000000	0.000002
100	39	25	21	1.000000	0.000000
100	39	25	22	1.000000	0.000000
100	39	25	23	1.000000	0.000000
100	39	25	24	1.000000	0.000000
100	39	25	25	1.000000	0.000000
100	39	26	0	0.000000	0.000000
100	39	26	1	0.000005	0.000005
100	39	26	2	0.000068	0.000063
100	39	26	3	0.000558	0.000490

N	n	k	x	P(x)	p(x)
100	39	26	4	0.003161	0.002603
100	39	26	5	0.013183	0.010021
100	39	26	6	0.042269	0.029086
100	39	26	7	0.107564	0.065295
100	39	26	8	0.222969	0.115405
100	39	26	9	0.385585	0.162616
100	39	26	10	0.569884	0.184299
100	39	26	11	0.738885	0.169001
100	39	26	12	0.864737	0.125852
100	39	26	13	0.940974	0.076237
100	39	26	14	0.978537	0.037563
100	39	26	15	0.993563	0.015025
100	39	26	16	0.998424	0.004861
100	39	26	17	0.999688	0.001265
100	39	26	18	0.999951	0.000262
100	39	26	19	0.999994	0.000043
100	39	26	20	0.999999	0.000005
100	39	26	21	1.000000	0.000001
100	39	26	22	1.000000	0.000000
100	39	26	23	1.000000	0.000000
100	39	26	24	1.000000	0.000000
100	39	26	25	1.000000	0.000000
100	39	26	26	1.000000	0.000000
100	39	27	0	0.000000	0.000000
100	39	27	1	0.000003	0.000002
100	39	27	2	0.000037	0.000034
100	39	27	3	0.000320	0.000283
100	39	27	4	0.001930	0.001610
100	39	27	5	0.008578	0.006648
100	39	27	6	0.029298	0.020720
100	39	27	7	0.079328	0.050030
100	39	27	8	0.174623	0.095295
100	39	27	9	0.319660	0.145036
100	39	27	10	0.497659	0.177999
100	39	27	11	0.674939	0.177280
100	39	27	12	0.818818	0.143879
100	39	27	13	0.914188	0.095370
100	39	27	14	0.965847	0.051659
100	39	27	15	0.988689	0.022842
100	39	27	16	0.996913	0.008223
100	39	27	17	0.999312	0.002400
100	39	27	18	0.999876	0.000564
100	39	27	19	0.999982	0.000106
100	39	27	20	0.999998	0.000016
100	39	27	21	1.000000	0.000002
100	39	27	22	1.000000	0.000000
100	39	27	23	1.000000	0.000000
100	39	27	24	1.000000	0.000000
100	39	27	25	1.000000	0.000000
100	39	27	26	1.000000	0.000000

Table for $N = 2$, $n = 1$, through $N = 100$, $n = 50$

N	n	k	x	P(x)	p(x)
100	39	27	27	1.000000	0.000000
100	39	28	0	0.000000	0.000000
100	39	28	1	0.000001	0.000001
100	39	28	2	0.000019	0.000018
100	39	28	3	0.000180	0.000161
100	39	28	4	0.001158	0.000978
100	39	28	5	0.005482	0.004324
100	39	28	6	0.019931	0.014449
100	39	28	7	0.057397	0.037465
100	39	28	8	0.134155	0.076759
100	39	28	9	0.260056	0.125900
100	39	28	10	0.426947	0.166891
100	39	28	11	0.606941	0.179994
100	39	28	12	0.765603	0.158661
100	39	28	13	0.880221	0.114618
100	39	28	14	0.948156	0.067935
100	39	28	15	0.981180	0.033024
100	39	28	16	0.994322	0.013142
100	39	28	17	0.998589	0.004267
100	39	28	18	0.999714	0.001125
100	39	28	19	0.999953	0.000239
100	39	28	20	0.999994	0.000041
100	39	28	21	0.999999	0.000005
100	39	28	22	1.000000	0.000001
100	39	28	23	1.000000	0.000000
100	39	28	24	1.000000	0.000000
100	39	28	25	1.000000	0.000000
100	39	28	26	1.000000	0.000000
100	39	28	27	1.000000	0.000000
100	39	28	28	1.000000	0.000000
100	39	29	0	0.000000	0.000000
100	39	29	1	0.000001	0.000001
100	39	29	2	0.000010	0.000009
100	39	29	3	0.000100	0.000090
100	39	29	4	0.000683	0.000583
100	39	29	5	0.003440	0.002757
100	39	29	6	0.013309	0.009869
100	39	29	7	0.040746	0.027437
100	39	29	8	0.101107	0.060361
100	39	29	9	0.207597	0.106491
100	39	29	10	0.359727	0.152130
100	39	29	11	0.536943	0.177216
100	39	29	12	0.706104	0.169161
100	39	29	13	0.838831	0.132726
100	39	29	14	0.924567	0.085736
100	39	29	15	0.970172	0.045604
100	39	29	16	0.990124	0.019952
100	39	29	17	0.997285	0.007162
100	39	29	18	0.999386	0.002101
100	39	29	19	0.999887	0.000501
100	39	29	20	0.999983	0.000096
100	39	29	21	0.999998	0.000015
100	39	29	22	1.000000	0.000002
100	39	29	23	1.000000	0.000000
100	39	29	24	1.000000	0.000000
100	39	29	25	1.000000	0.000000
100	39	29	26	1.000000	0.000000
100	39	29	27	1.000000	0.000000
100	39	29	28	1.000000	0.000000
100	39	29	29	1.000000	0.000000
100	39	30	0	0.000000	0.000000
100	39	30	1	0.000000	0.000000
100	39	30	2	0.000005	0.000005
100	39	30	3	0.000054	0.000049
100	39	30	4	0.000395	0.000341
100	39	30	5	0.002120	0.001724
100	39	30	6	0.008722	0.006602
100	39	30	7	0.028380	0.019658
100	39	30	8	0.074752	0.046372
100	39	30	9	0.162601	0.087849
100	39	30	10	0.297589	0.134988
100	39	30	11	0.467055	0.169466
100	39	30	12	0.641776	0.174720
100	39	30	13	0.790227	0.148451
100	39	30	14	0.894378	0.104152
100	39	30	15	0.954756	0.060378
100	39	30	16	0.983660	0.028904
100	39	30	17	0.995066	0.011406
100	39	30	18	0.998765	0.003698
100	39	30	19	0.999746	0.000981
100	39	30	20	0.999957	0.000212
100	39	30	21	0.999994	0.000037
100	39	30	22	0.999999	0.000005
100	39	30	23	1.000000	0.000001
100	39	30	24	1.000000	0.000000
100	39	30	25	1.000000	0.000000
100	39	30	26	1.000000	0.000000
100	39	30	27	1.000000	0.000000
100	39	30	28	1.000000	0.000000
100	39	30	29	1.000000	0.000000
100	39	30	30	1.000000	0.000000
100	39	31	0	0.000000	0.000000
100	39	31	1	0.000000	0.000000
100	39	31	2	0.000003	0.000002
100	39	31	3	0.000029	0.000026
100	39	31	4	0.000225	0.000196
100	39	31	5	0.001282	0.001057
100	39	31	6	0.005609	0.004327
100	39	31	7	0.019393	0.013784
100	39	31	8	0.054216	0.034822
100	39	31	9	0.124951	0.070736
100	39	31	10	0.241666	0.116714
100	39	31	11	0.399269	0.157603
100	39	31	12	0.574383	0.175115
100	39	31	13	0.735088	0.160704
100	39	31	14	0.857181	0.122094
100	39	31	15	0.934055	0.076874
100	39	31	16	0.974163	0.040108
100	39	31	17	0.991481	0.017318
100	39	31	18	0.997655	0.006174
100	39	31	19	0.999465	0.001810
100	39	31	20	0.999900	0.000434
100	39	31	21	0.999985	0.000085
100	39	31	22	0.999998	0.000013
100	39	31	23	1.000000	0.000002
100	39	31	24	1.000000	0.000000
100	39	31	25	1.000000	0.000000
100	39	31	26	1.000000	0.000000
100	39	31	27	1.000000	0.000000
100	39	31	28	1.000000	0.000000
100	39	31	29	1.000000	0.000000
100	39	31	30	1.000000	0.000000
100	39	31	31	1.000000	0.000000
100	39	32	0	0.000000	0.000000
100	39	32	1	0.000000	0.000000
100	39	32	2	0.000001	0.000001
100	39	32	3	0.000015	0.000014
100	39	32	4	0.000125	0.000110
100	39	32	5	0.000761	0.000636
100	39	32	6	0.003540	0.002779
100	39	32	7	0.013001	0.009461
100	39	32	8	0.038571	0.025570
100	39	32	9	0.094197	0.055626
100	39	32	10	0.192612	0.098415
100	39	32	11	0.335314	0.142702
100	39	32	12	0.505860	0.170546
100	39	32	13	0.674533	0.168672
100	39	32	14	0.812945	0.138412
100	39	32	15	0.907316	0.094372
100	39	32	16	0.960794	0.053477
100	39	32	17	0.985960	0.025166
100	39	32	18	0.995776	0.009816
100	39	32	19	0.998941	0.003165
100	39	32	20	0.999780	0.000840
100	39	32	21	0.999963	0.000182
100	39	32	22	0.999995	0.000032
100	39	32	23	0.999999	0.000005
100	39	32	24	1.000000	0.000001
100	39	32	25	1.000000	0.000000
100	39	32	26	1.000000	0.000000
100	39	32	27	1.000000	0.000000
100	39	32	28	1.000000	0.000000
100	39	32	29	1.000000	0.000000
100	39	32	30	1.000000	0.000000
100	39	32	31	1.000000	0.000000
100	39	32	32	1.000000	0.000000
100	39	33	0	0.000000	0.000000
100	39	33	1	0.000000	0.000000
100	39	33	2	0.000001	0.000001
100	39	33	3	0.000008	0.000007
100	39	33	4	0.000069	0.000061
100	39	33	5	0.000443	0.000375
100	39	33	6	0.002191	0.001748
100	39	33	7	0.008549	0.006357
100	39	33	8	0.026914	0.018365
100	39	33	9	0.069656	0.042742
100	39	33	10	0.150641	0.080985
100	39	33	11	0.276554	0.125914
100	39	33	12	0.438143	0.161589
100	39	33	13	0.610040	0.171897
100	39	33	14	0.762058	0.152018
100	39	33	15	0.874009	0.111951
100	39	33	16	0.942706	0.068697
100	39	33	17	0.977818	0.035112
100	39	33	18	0.992745	0.014927
100	39	33	19	0.998010	0.005265
100	39	33	20	0.999546	0.001536
100	39	33	21	0.999914	0.000369
100	39	33	22	0.999987	0.000072
100	39	33	23	0.999998	0.000012
100	39	33	24	1.000000	0.000001
100	39	33	25	1.000000	0.000000
100	39	33	26	1.000000	0.000000
100	39	33	27	1.000000	0.000000
100	39	33	28	1.000000	0.000000
100	39	33	29	1.000000	0.000000
100	39	33	30	1.000000	0.000000
100	39	33	31	1.000000	0.000000
100	39	33	32	1.000000	0.000000
100	39	33	33	1.000000	0.000000
100	39	34	0	0.000000	0.000000
100	39	34	1	0.000000	0.000000
100	39	34	2	0.000000	0.000000
100	39	34	3	0.000004	0.000004
100	39	34	4	0.000037	0.000033
100	39	34	5	0.000253	0.000216
100	39	34	6	0.001330	0.001077
100	39	34	7	0.005512	0.004182
100	39	34	8	0.018416	0.012904
100	39	34	9	0.050518	0.032101

$N = 100$ $n = 39$

N	n	k	x	P(x)	p(x)
100	39	34	10	0.115588	0.065070
100	39	34	11	0.223934	0.108346
100	39	34	12	0.373025	0.149092
100	39	34	13	0.543334	0.170309
100	39	34	14	0.705335	0.162001
100	39	34	15	0.833907	0.128572
100	39	34	16	0.919123	0.085216
100	39	34	17	0.966289	0.047165
100	39	34	18	0.988066	0.021778
100	39	34	19	0.996438	0.008372
100	39	34	20	0.999110	0.002672
100	39	34	21	0.999815	0.000705
100	39	34	22	0.999968	0.000153
100	39	34	23	0.999995	0.000027
100	39	34	24	0.999999	0.000004
100	39	34	25	1.000000	0.000000
100	39	34	26	1.000000	0.000000
100	39	34	27	1.000000	0.000000
100	39	34	28	1.000000	0.000000
100	39	34	29	1.000000	0.000000
100	39	34	30	1.000000	0.000000
100	39	34	31	1.000000	0.000000
100	39	34	32	1.000000	0.000000
100	39	34	33	1.000000	0.000000
100	39	34	34	1.000000	0.000000
100	39	35	0	0.000000	0.000000
100	39	35	1	0.000000	0.000000
100	39	35	2	0.000000	0.000000
100	39	35	3	0.000002	0.000002
100	39	35	4	0.000019	0.000017
100	39	35	5	0.000142	0.000122
100	39	35	6	0.000792	0.000650
100	39	35	7	0.003485	0.002693
100	39	35	8	0.012355	0.008871
100	39	35	9	0.035926	0.023571
100	39	35	10	0.086996	0.051070
100	39	35	11	0.177969	0.090972
100	39	35	12	0.312033	0.134065
100	39	35	13	0.476243	0.164209
100	39	35	14	0.643971	0.167728
100	39	35	15	0.787153	0.143183
100	39	35	16	0.889427	0.102273
100	39	35	17	0.950567	0.061140
100	39	35	18	0.981137	0.030570
100	39	35	19	0.993901	0.012764
100	39	35	20	0.998341	0.004440
100	39	35	21	0.999623	0.001282
100	39	35	22	0.999929	0.000306
100	39	35	23	0.999989	0.000060
100	39	35	24	0.999998	0.000010

N	n	k	x	P(x)	p(x)
100	39	35	25	1.000000	0.000001
100	39	35	26	1.000000	0.000000
100	39	35	27	1.000000	0.000000
100	39	35	28	1.000000	0.000000
100	39	35	29	1.000000	0.000000
100	39	35	30	1.000000	0.000000
100	39	35	31	1.000000	0.000000
100	39	35	32	1.000000	0.000000
100	39	35	33	1.000000	0.000000
100	39	35	34	1.000000	0.000000
100	39	35	35	1.000000	0.000000
100	39	36	0	0.000000	0.000000
100	39	36	1	0.000000	0.000000
100	39	36	2	0.000000	0.000000
100	39	36	3	0.000001	0.000001
100	39	36	4	0.000010	0.000009
100	39	36	5	0.000078	0.000068
100	39	36	6	0.000462	0.000384
100	39	36	7	0.002159	0.001697
100	39	36	8	0.008125	0.005966
100	39	36	9	0.025047	0.016923
100	39	36	10	0.064211	0.039164
100	39	36	11	0.138781	0.074569
100	39	36	12	0.256345	0.117564
100	39	36	13	0.410559	0.154214
100	39	36	14	0.579460	0.168901
100	39	36	15	0.734286	0.154826
100	39	36	16	0.853238	0.118952
100	39	36	17	0.929873	0.076636
100	39	36	18	0.971260	0.041387
100	39	36	19	0.989974	0.018713
100	39	36	20	0.997043	0.007069
100	39	36	21	0.999268	0.002225
100	39	36	22	0.999849	0.000581
100	39	36	23	0.999974	0.000125
100	39	36	24	0.999996	0.000022
100	39	36	25	0.999999	0.000003
100	39	36	26	1.000000	0.000000
100	39	36	27	1.000000	0.000000
100	39	36	28	1.000000	0.000000
100	39	36	29	1.000000	0.000000
100	39	36	30	1.000000	0.000000
100	39	36	31	1.000000	0.000000
100	39	36	32	1.000000	0.000000
100	39	36	33	1.000000	0.000000
100	39	36	34	1.000000	0.000000
100	39	36	35	1.000000	0.000000
100	39	36	36	1.000000	0.000000
100	39	37	0	0.000000	0.000000
100	39	37	1	0.000000	0.000000

N	n	k	x	P(x)	p(x)
100	39	37	2	0.000000	0.000000
100	39	37	3	0.000000	0.000000
100	39	37	4	0.000005	0.000005
100	39	37	5	0.000042	0.000037
100	39	37	6	0.000264	0.000222
100	39	37	7	0.001310	0.001047
100	39	37	8	0.005235	0.003925
100	39	37	9	0.017115	0.011880
100	39	37	10	0.046465	0.029350
100	39	37	11	0.106156	0.059691
100	39	37	12	0.206747	0.100591
100	39	37	13	0.347910	0.141162
100	39	37	14	0.513483	0.165574
100	39	37	15	0.676226	0.162743
100	39	37	16	0.810489	0.134263
100	39	37	17	0.903530	0.093040
100	39	37	18	0.957680	0.054150
100	39	37	19	0.984126	0.026446
100	39	37	20	0.994944	0.010819
100	39	37	21	0.998642	0.003698
100	39	37	22	0.999695	0.001052
100	39	37	23	0.999943	0.000248
100	39	37	24	0.999991	0.000048
100	39	37	25	0.999999	0.000008
100	39	37	26	1.000000	0.000001
100	39	37	27	1.000000	0.000000
100	39	37	28	1.000000	0.000000
100	39	37	29	1.000000	0.000000
100	39	37	30	1.000000	0.000000
100	39	37	31	1.000000	0.000000
100	39	37	32	1.000000	0.000000
100	39	37	33	1.000000	0.000000
100	39	37	34	1.000000	0.000000
100	39	37	35	1.000000	0.000000
100	39	37	36	1.000000	0.000000
100	39	37	37	1.000000	0.000000
100	39	38	0	0.000000	0.000000
100	39	38	1	0.000000	0.000000
100	39	38	2	0.000000	0.000000
100	39	38	3	0.000000	0.000000
100	39	38	4	0.000002	0.000002
100	39	38	5	0.000022	0.000019
100	39	38	6	0.000147	0.000126
100	39	38	7	0.000779	0.000631
100	39	38	8	0.003304	0.002525
100	39	38	9	0.011458	0.008154
100	39	38	10	0.032955	0.021497
100	39	38	11	0.079627	0.046672
100	39	38	12	0.163637	0.084010
100	39	38	13	0.289652	0.126015

N	n	k	x	P(x)	p(x)
100	39	38	14	0.447779	0.158127
100	39	38	15	0.614229	0.166450
100	39	38	16	0.761473	0.147244
100	39	38	17	0.871040	0.109567
100	39	38	18	0.939630	0.068591
100	39	38	19	0.975730	0.036100
100	39	38	20	0.991682	0.015951
100	39	38	21	0.997586	0.005904
100	39	38	22	0.999411	0.001825
100	39	38	23	0.999880	0.000469
100	39	38	24	0.999980	0.000100
100	39	38	25	0.999997	0.000017
100	39	38	26	1.000000	0.000002
100	39	38	27	1.000000	0.000000
100	39	38	28	1.000000	0.000000
100	39	38	29	1.000000	0.000000
100	39	38	30	1.000000	0.000000
100	39	38	31	1.000000	0.000000
100	39	38	32	1.000000	0.000000
100	39	38	33	1.000000	0.000000
100	39	38	34	1.000000	0.000000
100	39	38	35	1.000000	0.000000
100	39	38	36	1.000000	0.000000
100	39	38	37	1.000000	0.000000
100	39	38	38	1.000000	0.000000
100	39	39	0	0.000000	0.000000
100	39	39	1	0.000000	0.000000
100	39	39	2	0.000000	0.000000
100	39	39	3	0.000000	0.000000
100	39	39	4	0.000001	0.000001
100	39	39	5	0.000011	0.000010
100	39	39	6	0.000081	0.000069
100	39	39	7	0.000453	0.000372
100	39	39	8	0.002041	0.001588
100	39	39	9	0.007512	0.005471
100	39	39	10	0.022900	0.015387
100	39	39	11	0.058549	0.035649
100	39	39	12	0.127052	0.068503
100	39	39	13	0.236807	0.109755
100	39	39	14	0.384018	0.147211
100	39	39	15	0.549797	0.165778
100	39	39	16	0.706850	0.157053
100	39	39	17	0.832161	0.125311
100	39	39	18	0.916398	0.084237
100	39	39	19	0.964085	0.047687
100	39	39	20	0.986793	0.022708
100	39	39	21	0.995872	0.009078
100	39	39	22	0.998910	0.003039
100	39	39	23	0.999759	0.000848
100	39	39	24	0.999955	0.000197

Table for $N = 2$, $n = 1$, through $N = 100$, $n = 50$

N	n	k	x	P(x)	p(x)
100	39	39	25	0.999993	0.000038
100	39	39	26	0.999999	0.000006
100	39	39	27	1.000000	0.000001
100	39	39	28	1.000000	0.000000
100	39	39	29	1.000000	0.000000
100	39	39	30	1.000000	0.000000
100	39	39	31	1.000000	0.000000
100	39	39	32	1.000000	0.000000
100	39	39	33	1.000000	0.000000
100	39	39	34	1.000000	0.000000
100	39	39	35	1.000000	0.000000
100	39	39	36	1.000000	0.000000
100	39	39	37	1.000000	0.000000
100	39	39	38	1.000000	0.000000
100	39	39	39	1.000000	0.000000
100	40	1	0	0.600000	0.600000
100	40	1	1	1.000000	0.400000
100	40	2	0	0.357576	0.357576
100	40	2	1	0.842424	0.484848
100	40	2	2	1.000000	0.157576
100	40	3	0	0.211626	0.211626
100	40	3	1	0.649474	0.437848
100	40	3	2	0.938899	0.289425
100	40	3	3	1.000000	0.061101
100	40	4	0	0.124358	0.124358
100	40	4	1	0.473432	0.349075
100	40	4	2	0.825516	0.352084
100	40	4	3	0.976693	0.151177
100	40	4	4	1.000000	0.023306
100	40	5	0	0.072542	0.072542
100	40	5	1	0.331621	0.259079
100	40	5	2	0.686150	0.354529
100	40	5	3	0.918427	0.232278
100	40	5	4	0.991260	0.072833
100	40	5	5	1.000000	0.008740
100	40	6	0	0.041998	0.041998
100	40	6	1	0.225262	0.183264
100	40	6	2	0.544338	0.319076
100	40	6	3	0.827961	0.283623
100	40	6	4	0.963660	0.135699
100	40	6	5	0.996780	0.033120
100	40	6	6	1.000000	0.003220
100	40	7	0	0.024127	0.024127
100	40	7	1	0.149227	0.125101
100	40	7	2	0.415350	0.266123
100	40	7	3	0.716322	0.300972
100	40	7	4	0.911690	0.195368
100	40	7	5	0.984448	0.072758
100	40	7	6	0.998835	0.014387
100	40	7	7	1.000000	0.001165

N	n	k	x	P(x)	p(x)
100	40	8	0	0.013750	0.013750
100	40	8	1	0.096766	0.083016
100	40	8	2	0.306612	0.209846
100	40	8	3	0.596581	0.289969
100	40	8	4	0.836064	0.239483
100	40	8	5	0.957066	0.121002
100	40	8	6	0.993575	0.036509
100	40	8	7	0.999587	0.006011
100	40	8	8	1.000000	0.000413
100	40	9	0	0.007771	0.007771
100	40	9	1	0.061574	0.053802
100	40	9	2	0.219936	0.158362
100	40	9	3	0.479963	0.260027
100	40	9	4	0.742353	0.262390
100	40	9	5	0.911033	0.168680
100	40	9	6	0.980083	0.069050
100	40	9	7	0.997430	0.017348
100	40	9	8	0.999856	0.002426
100	40	9	9	1.000000	0.000144
100	40	10	0	0.004355	0.004355
100	40	10	1	0.038516	0.034160
100	40	10	2	0.153807	0.115291
100	40	10	3	0.374238	0.220431
100	40	10	4	0.638550	0.264313
100	40	10	5	0.846156	0.207606
100	40	10	6	0.954284	0.108128
100	40	10	7	0.991140	0.036856
100	40	10	8	0.999003	0.007864
100	40	10	9	0.999951	0.000948
100	40	10	10	1.000000	0.000049
100	40	11	0	0.002420	0.002420
100	40	11	1	0.023713	0.021293
100	40	11	2	0.105128	0.081415
100	40	11	3	0.283616	0.178488
100	40	11	4	0.532825	0.249209
100	40	11	5	0.765420	0.232595
100	40	11	6	0.913436	0.148015
100	40	11	7	0.977626	0.064190
100	40	11	8	0.996207	0.018581
100	40	11	9	0.999625	0.003417
100	40	11	10	0.999984	0.000359
100	40	11	11	1.000000	0.000016
100	40	12	0	0.001332	0.001332
100	40	12	1	0.014382	0.013050
100	40	12	2	0.070367	0.055985
100	40	12	3	0.209413	0.139047
100	40	12	4	0.432021	0.222608
100	40	12	5	0.673950	0.241929
100	40	12	6	0.856891	0.182940
100	40	12	7	0.953825	0.096934

N	n	k	x	P(x)	p(x)
100	40	12	8	0.989526	0.035701
100	40	12	9	0.998434	0.008908
100	40	12	10	0.999863	0.001428
100	40	12	11	0.999995	0.000132
100	40	12	12	1.000000	0.000005
100	40	13	0	0.000727	0.000727
100	40	13	1	0.008599	0.007872
100	40	13	2	0.046192	0.037593
100	40	13	3	0.150951	0.104759
100	40	13	4	0.340955	0.190004
100	40	13	5	0.577729	0.236774
100	40	13	6	0.786209	0.208480
100	40	13	7	0.917475	0.131265
100	40	13	8	0.976544	0.059069
100	40	13	9	0.995296	0.018752
100	40	13	10	0.999376	0.004079
100	40	13	11	0.999951	0.000575
100	40	13	12	0.999998	0.000047
100	40	13	13	1.000000	0.000002
100	40	14	0	0.000393	0.000393
100	40	14	1	0.005070	0.004677
100	40	14	2	0.029772	0.024702
100	40	14	3	0.106398	0.076626
100	40	14	4	0.262332	0.155934
100	40	14	5	0.482475	0.220143
100	40	14	6	0.704734	0.222259
100	40	14	7	0.867684	0.162950
100	40	14	8	0.954817	0.087133
100	40	14	9	0.988614	0.033797
100	40	14	10	0.997969	0.009355
100	40	14	11	0.999759	0.001790
100	40	14	12	0.999983	0.000224
100	40	14	13	0.999999	0.000016
100	40	14	14	1.000000	0.000001
100	40	15	0	0.000210	0.000210
100	40	15	1	0.002949	0.002739
100	40	15	2	0.018857	0.015908
100	40	15	3	0.073431	0.054574
100	40	15	4	0.197057	0.123627
100	40	15	5	0.392882	0.195824
100	40	15	6	0.616864	0.223982
100	40	15	7	0.805157	0.188293
100	40	15	8	0.922396	0.117239
100	40	15	9	0.976432	0.054036
100	40	15	10	0.994706	0.018274
100	40	15	11	0.999156	0.004450
100	40	15	12	0.999910	0.000755
100	40	15	13	0.999994	0.000084
100	40	15	14	1.000000	0.000005
100	40	15	15	1.000000	0.000000

N	n	k	x	P(x)	p(x)
100	40	16	0	0.000111	0.000111
100	40	16	1	0.001692	0.001581
100	40	16	2	0.011745	0.010053
100	40	16	3	0.049675	0.037930
100	40	16	4	0.144698	0.095023
100	40	16	5	0.312248	0.167550
100	40	16	6	0.527271	0.215023
100	40	16	7	0.732055	0.204784
100	40	16	8	0.878259	0.146204
100	40	16	9	0.956724	0.078466
100	40	16	10	0.988256	0.031532
100	40	16	11	0.997637	0.009381
100	40	16	12	0.999662	0.002024
100	40	16	13	0.999968	0.000306
100	40	16	14	0.999998	0.000031
100	40	16	15	1.000000	0.000002
100	40	16	16	1.000000	0.000000
100	40	17	0	0.000058	0.000058
100	40	17	1	0.000958	0.000900
100	40	17	2	0.007197	0.006239
100	40	17	3	0.032968	0.025771
100	40	17	4	0.103974	0.071006
100	40	17	5	0.242436	0.138462
100	40	17	6	0.440238	0.197802
100	40	17	7	0.651604	0.211366
100	40	17	8	0.822562	0.170958
100	40	17	9	0.927767	0.105205
100	40	17	10	0.976995	0.049228
100	40	17	11	0.994398	0.017404
100	40	17	12	0.998987	0.004588
100	40	17	13	0.999869	0.000882
100	40	17	14	0.999989	0.000119
100	40	17	15	0.999999	0.000011
100	40	17	16	1.000000	0.000001
100	40	17	17	1.000000	0.000000
100	40	18	0	0.000030	0.000030
100	40	18	1	0.000535	0.000505
100	40	18	2	0.004341	0.003806
100	40	18	3	0.021480	0.017139
100	40	18	4	0.073176	0.051696
100	40	18	5	0.184048	0.110872
100	40	18	6	0.359211	0.175162
100	40	18	7	0.567567	0.208356
100	40	18	8	0.756651	0.189083
100	40	18	9	0.888473	0.131823
100	40	18	10	0.959201	0.070728
100	40	18	11	0.988318	0.029116
100	40	18	12	0.997439	0.009121
100	40	18	13	0.999582	0.002143
100	40	18	14	0.999951	0.000369

Table for $N = 2$, $n = 1$, through $N = 100$, $n = 50$

N	n	k	x	P(x)	p(x)
100	40	18	15	0.999996	0.000045
100	40	18	16	1.000000	0.000004
100	40	18	17	1.000000	0.000000
100	40	18	18	1.000000	0.000000
100	40	19	0	0.000015	0.000015
100	40	19	1	0.000295	0.000280
100	40	19	2	0.002577	0.002282
100	40	19	3	0.013746	0.011169
100	40	19	4	0.050480	0.036734
100	40	19	5	0.136725	0.086245
100	40	19	6	0.286582	0.149858
100	40	19	7	0.483716	0.197134
100	40	19	8	0.682862	0.199146
100	40	19	9	0.838638	0.155776
100	40	19	10	0.933325	0.094687
100	40	19	11	0.978021	0.044695
100	40	19	12	0.994324	0.016304
100	40	19	13	0.998876	0.004552
100	40	19	14	0.999834	0.000958
100	40	19	15	0.999982	0.000148
100	40	19	16	0.999999	0.000016
100	40	19	17	1.000000	0.000001
100	40	19	18	1.000000	0.000000
100	40	19	19	1.000000	0.000000
100	40	20	0	0.000008	0.000008
100	40	20	1	0.000160	0.000153
100	40	20	2	0.001507	0.001346
100	40	20	3	0.008644	0.007138
100	40	20	4	0.034154	0.025510
100	40	20	5	0.099458	0.065304
100	40	20	6	0.223679	0.124221
100	40	20	7	0.403402	0.179723
100	40	20	8	0.604187	0.200785
100	40	20	9	0.779020	0.174833
100	40	20	10	0.898256	0.119236
100	40	20	11	0.962019	0.063763
100	40	20	12	0.988688	0.026670
100	40	20	13	0.997359	0.008671
100	40	20	14	0.999527	0.002168
100	40	20	15	0.999937	0.000410
100	40	20	16	0.999994	0.000057
100	40	20	17	0.999999	0.000006
100	40	20	18	1.000000	0.000000
100	40	20	19	1.000000	0.000000
100	40	20	20	1.000000	0.000000
100	40	21	0	0.000004	0.000004
100	40	21	1	0.000086	0.000082
100	40	21	2	0.000867	0.000781
100	40	21	3	0.005343	0.004476
100	40	21	4	0.022675	0.017332

N	n	k	x	P(x)	p(x)
100	40	21	5	0.070888	0.048213
100	40	21	6	0.170885	0.099998
100	40	21	7	0.329266	0.158381
100	40	21	8	0.523873	0.194607
100	40	21	9	0.711272	0.187399
100	40	21	10	0.853542	0.142270
100	40	21	11	0.938905	0.085362
100	40	21	12	0.979354	0.040449
100	40	21	13	0.994433	0.015079
100	40	21	14	0.998822	0.004390
100	40	21	15	0.999809	0.000986
100	40	21	16	0.999977	0.000168
100	40	21	17	0.999998	0.000021
100	40	21	18	1.000000	0.000002
100	40	21	19	1.000000	0.000000
100	40	21	20	1.000000	0.000000
100	40	21	21	1.000000	0.000000
100	40	22	0	0.000002	0.000002
100	40	22	1	0.000045	0.000044
100	40	22	2	0.000491	0.000446
100	40	22	3	0.003247	0.002755
100	40	22	4	0.014777	0.011530
100	40	22	5	0.049528	0.034751
100	40	22	6	0.127848	0.078321
100	40	22	7	0.263107	0.135259
100	40	22	8	0.445045	0.181938
100	40	22	9	0.637736	0.192691
100	40	22	10	0.799516	0.161780
100	40	22	11	0.907569	0.108053
100	40	22	12	0.965017	0.057448
100	40	22	13	0.989279	0.024262
100	40	22	14	0.997378	0.008098
100	40	22	15	0.999496	0.002119
100	40	22	16	0.999926	0.000429
100	40	22	17	0.999992	0.000066
100	40	22	18	0.999999	0.000008
100	40	22	19	1.000000	0.000001
100	40	22	20	1.000000	0.000000
100	40	22	21	1.000000	0.000000
100	40	22	22	1.000000	0.000000
100	40	23	0	0.000001	0.000001
100	40	23	1	0.000024	0.000023
100	40	23	2	0.000274	0.000250
100	40	23	3	0.001940	0.001666
100	40	23	4	0.009455	0.007515
100	40	23	5	0.033934	0.024479
100	40	23	6	0.093709	0.059774
100	40	23	7	0.205882	0.112174
100	40	23	8	0.370404	0.164522
100	40	23	9	0.561154	0.190750

N	n	k	x	P(x)	p(x)
100	40	23	10	0.737293	0.176139
100	40	23	11	0.867395	0.130103
100	40	23	12	0.944395	0.077000
100	40	23	13	0.980881	0.036486
100	40	23	14	0.994678	0.013797
100	40	23	15	0.998817	0.004139
100	40	23	16	0.999794	0.000976
100	40	23	17	0.999972	0.000179
100	40	23	18	0.999997	0.000025
100	40	23	19	1.000000	0.000003
100	40	23	20	1.000000	0.000000
100	40	23	21	1.000000	0.000000
100	40	23	22	1.000000	0.000000
100	40	23	23	1.000000	0.000000
100	40	24	0	0.000000	0.000000
100	40	24	1	0.000012	0.000012
100	40	24	2	0.000151	0.000138
100	40	24	3	0.001139	0.000989
100	40	24	4	0.005942	0.004802
100	40	24	5	0.022807	0.016866
100	40	24	6	0.067315	0.044507
100	40	24	7	0.157808	0.090493
100	40	24	8	0.302031	0.144223
100	40	24	9	0.484358	0.182327
100	40	24	10	0.668667	0.184309
100	40	24	11	0.818396	0.149729
100	40	24	12	0.916395	0.097999
100	40	24	13	0.968087	0.051692
100	40	24	14	0.990019	0.021932
100	40	24	15	0.997473	0.007454
100	40	24	16	0.999489	0.002016
100	40	24	17	0.999919	0.000430
100	40	24	18	0.999990	0.000071
100	40	24	19	0.999999	0.000009
100	40	24	20	1.000000	0.000001
100	40	24	21	1.000000	0.000000
100	40	24	22	1.000000	0.000000
100	40	24	23	1.000000	0.000000
100	40	24	24	1.000000	0.000000
100	40	25	0	0.000000	0.000000
100	40	25	1	0.000006	0.000006
100	40	25	2	0.000081	0.000075
100	40	25	3	0.000658	0.000577
100	40	25	4	0.003667	0.003009
100	40	25	5	0.015040	0.011373
100	40	25	6	0.047404	0.032363
100	40	25	7	0.118515	0.071111
100	40	25	8	0.241306	0.122791
100	40	25	9	0.409988	0.168682
100	40	25	10	0.595914	0.185926

N	n	k	x	P(x)	p(x)
100	40	25	11	0.761262	0.165349
100	40	25	12	0.880290	0.119028
100	40	25	13	0.949723	0.069433
100	40	25	14	0.982516	0.032793
100	40	25	15	0.995021	0.012505
100	40	25	16	0.998853	0.003831
100	40	25	17	0.999789	0.000936
100	40	25	18	0.999969	0.000181
100	40	25	19	0.999996	0.000027
100	40	25	20	1.000000	0.000003
100	40	25	21	1.000000	0.000000
100	40	25	22	1.000000	0.000000
100	40	25	23	1.000000	0.000000
100	40	25	24	1.000000	0.000000
100	40	25	25	1.000000	0.000000
100	40	26	0	0.000000	0.000000
100	40	26	1	0.000003	0.000003
100	40	26	2	0.000043	0.000040
100	40	26	3	0.000374	0.000330
100	40	26	4	0.002223	0.001849
100	40	26	5	0.009733	0.007510
100	40	26	6	0.032732	0.023000
100	40	26	7	0.087226	0.054494
100	40	26	8	0.188915	0.101689
100	40	26	9	0.340266	0.151351
100	40	26	10	0.521543	0.181277
100	40	26	11	0.697328	0.175784
100	40	26	12	0.835853	0.138526
100	40	26	13	0.924727	0.088874
100	40	26	14	0.971148	0.046421
100	40	26	15	0.990853	0.019705
100	40	26	16	0.997627	0.006774
100	40	26	17	0.999502	0.001875
100	40	26	18	0.999916	0.000415
100	40	26	19	0.999989	0.000072
100	40	26	20	0.999999	0.000010
100	40	26	21	1.000000	0.000001
100	40	26	22	1.000000	0.000000
100	40	26	23	1.000000	0.000000
100	40	26	24	1.000000	0.000000
100	40	26	25	1.000000	0.000000
100	40	26	26	1.000000	0.000000
100	40	27	0	0.000000	0.000000
100	40	27	1	0.000001	0.000001
100	40	27	2	0.000023	0.000021
100	40	27	3	0.000208	0.000186
100	40	27	4	0.001323	0.001115
100	40	27	5	0.006181	0.004858
100	40	27	6	0.022165	0.015984
100	40	27	7	0.062925	0.040760

Table for $N = 2$, $n = 1$, through $N = 100$, $n = 50$

N	n	k	x	P(x)	p(x)
100	40	27	8	0.144941	0.082017
100	40	27	9	0.276862	0.131921
100	40	27	10	0.448053	0.171190
100	40	27	11	0.628439	0.180387
100	40	27	12	0.783438	0.154999
100	40	27	13	0.892300	0.108862
100	40	27	14	0.954838	0.062538
100	40	27	15	0.984196	0.029358
100	40	27	16	0.995430	0.011234
100	40	27	17	0.998919	0.003489
100	40	27	18	0.999793	0.000874
100	40	27	19	0.999968	0.000175
100	40	27	20	0.999996	0.000028
100	40	27	21	0.999999	0.000003
100	40	27	22	1.000000	0.000000
100	40	27	23	1.000000	0.000000
100	40	27	24	1.000000	0.000000
100	40	27	25	1.000000	0.000000
100	40	27	26	1.000000	0.000000
100	40	27	27	1.000000	0.000000
100	40	28	0	0.000000	0.000000
100	40	28	1	0.000001	0.000001
100	40	28	2	0.000012	0.000011
100	40	28	3	0.000114	0.000103
100	40	28	4	0.000773	0.000659
100	40	28	5	0.003852	0.003078
100	40	28	6	0.014720	0.010868
100	40	28	7	0.044499	0.029779
100	40	28	8	0.108989	0.064490
100	40	28	9	0.220841	0.111852
100	40	28	10	0.377700	0.156859
100	40	28	11	0.556779	0.179078
100	40	28	12	0.723986	0.167208
100	40	28	13	0.852036	0.128050
100	40	28	14	0.932564	0.080528
100	40	28	15	0.974142	0.041578
100	40	28	16	0.991736	0.017595
100	40	28	17	0.997820	0.006083
100	40	28	18	0.999530	0.001710
100	40	28	19	0.999918	0.000388
100	40	28	20	0.999988	0.000071
100	40	28	21	0.999999	0.000010
100	40	28	22	1.000000	0.000001
100	40	28	23	1.000000	0.000000
100	40	28	24	1.000000	0.000000
100	40	28	25	1.000000	0.000000
100	40	28	26	1.000000	0.000000
100	40	28	27	1.000000	0.000000
100	40	28	28	1.000000	0.000000
100	40	29	0	0.000000	0.000000

N	n	k	x	P(x)	p(x)
100	40	29	1	0.000000	0.000000
100	40	29	2	0.000006	0.000006
100	40	29	3	0.000062	0.000056
100	40	29	4	0.000444	0.000382
100	40	29	5	0.002355	0.001911
100	40	29	6	0.009588	0.007233
100	40	29	7	0.030850	0.021263
100	40	29	8	0.080327	0.049476
100	40	29	9	0.172682	0.092356
100	40	29	10	0.312342	0.139660
100	40	29	11	0.484650	0.172308
100	40	29	12	0.658961	0.174311
100	40	29	13	0.804017	0.145056
100	40	29	14	0.903485	0.099467
100	40	29	15	0.959705	0.056221
100	40	29	16	0.985872	0.026166
100	40	29	17	0.995876	0.010005
100	40	29	18	0.999007	0.003131
100	40	29	19	0.999805	0.000798
100	40	29	20	0.999969	0.000164
100	40	29	21	0.999996	0.000027
100	40	29	22	0.999999	0.000004
100	40	29	23	1.000000	0.000000
100	40	29	24	1.000000	0.000000
100	40	29	25	1.000000	0.000000
100	40	29	26	1.000000	0.000000
100	40	29	27	1.000000	0.000000
100	40	29	28	1.000000	0.000000
100	40	29	29	1.000000	0.000000
100	40	30	0	0.000000	0.000000
100	40	30	1	0.000000	0.000000
100	40	30	2	0.000003	0.000003
100	40	30	3	0.000033	0.000030
100	40	30	4	0.000250	0.000217
100	40	30	5	0.001413	0.001163
100	40	30	6	0.006124	0.004711
100	40	30	7	0.020968	0.014843
100	40	30	8	0.058027	0.037060
100	40	30	9	0.132358	0.074331
100	40	30	10	0.253331	0.120973
100	40	30	11	0.414271	0.160940
100	40	30	12	0.590219	0.175948
100	40	30	13	0.748855	0.158636
100	40	30	14	0.867060	0.118205
100	40	30	15	0.939909	0.072849
100	40	30	16	0.977027	0.037117
100	40	30	17	0.992635	0.015609
100	40	30	18	0.998037	0.005402
100	40	30	19	0.999569	0.001532
100	40	30	20	0.999923	0.000354

N	n	k	x	P(x)	p(x)
100	40	30	21	0.999989	0.000066
100	40	30	22	0.999999	0.000010
100	40	30	23	1.000000	0.000001
100	40	30	24	1.000000	0.000000
100	40	30	25	1.000000	0.000000
100	40	30	26	1.000000	0.000000
100	40	30	27	1.000000	0.000000
100	40	30	28	1.000000	0.000000
100	40	30	29	1.000000	0.000000
100	40	30	30	1.000000	0.000000
100	40	31	0	0.000000	0.000000
100	40	31	1	0.000000	0.000000
100	40	31	2	0.000001	0.000001
100	40	31	3	0.000017	0.000015
100	40	31	4	0.000138	0.000121
100	40	31	5	0.000831	0.000693
100	40	31	6	0.003836	0.003004
100	40	31	7	0.013970	0.010134
100	40	31	8	0.041086	0.027116
100	40	31	9	0.099440	0.058354
100	40	31	10	0.201485	0.102045
100	40	31	11	0.347596	0.146110
100	40	31	12	0.519840	0.172244
100	40	31	13	0.687667	0.167827
100	40	31	14	0.823155	0.135488
100	40	31	15	0.913892	0.090736
100	40	31	16	0.964301	0.050409
100	40	31	17	0.987507	0.023206
100	40	31	18	0.996339	0.008833
100	40	31	19	0.999109	0.002770
100	40	31	20	0.999822	0.000712
100	40	31	21	0.999971	0.000149
100	40	31	22	0.999996	0.000025
100	40	31	23	0.999999	0.000003
100	40	31	24	1.000000	0.000000
100	40	31	25	1.000000	0.000000
100	40	31	26	1.000000	0.000000
100	40	31	27	1.000000	0.000000
100	40	31	28	1.000000	0.000000
100	40	31	29	1.000000	0.000000
100	40	31	30	1.000000	0.000000
100	40	31	31	1.000000	0.000000
100	40	32	0	0.000000	0.000000
100	40	32	1	0.000000	0.000000
100	40	32	2	0.000001	0.000001
100	40	32	3	0.000009	0.000008
100	40	32	4	0.000075	0.000066
100	40	32	5	0.000480	0.000405
100	40	32	6	0.002355	0.001876
100	40	32	7	0.009123	0.006768

N	n	k	x	P(x)	p(x)
100	40	32	8	0.028510	0.019387
100	40	32	9	0.073223	0.044713
100	40	32	10	0.157118	0.083895
100	40	32	11	0.286187	0.129069
100	40	32	12	0.449943	0.163756
100	40	32	13	0.621995	0.172052
100	40	32	14	0.772102	0.150106
100	40	32	15	0.881016	0.108914
100	40	32	16	0.946767	0.065751
100	40	32	17	0.979771	0.033004
100	40	32	18	0.993523	0.013752
100	40	32	19	0.998266	0.004743
100	40	32	20	0.999615	0.001349
100	40	32	21	0.999930	0.000315
100	40	32	22	0.999989	0.000060
100	40	32	23	0.999999	0.000009
100	40	32	24	1.000000	0.000001
100	40	32	25	1.000000	0.000000
100	40	32	26	1.000000	0.000000
100	40	32	27	1.000000	0.000000
100	40	32	28	1.000000	0.000000
100	40	32	29	1.000000	0.000000
100	40	32	30	1.000000	0.000000
100	40	32	31	1.000000	0.000000
100	40	32	32	1.000000	0.000000
100	40	33	0	0.000000	0.000000
100	40	33	1	0.000000	0.000000
100	40	33	2	0.000000	0.000000
100	40	33	3	0.000004	0.000004
100	40	33	4	0.000040	0.000035
100	40	33	5	0.000271	0.000232
100	40	33	6	0.001418	0.001146
100	40	33	7	0.005839	0.004421
100	40	33	8	0.019387	0.013548
100	40	33	9	0.052839	0.033452
100	40	33	10	0.120105	0.067266
100	40	33	11	0.231143	0.111037
100	40	33	12	0.382514	0.151372
100	40	33	13	0.553681	0.171166
100	40	33	14	0.714708	0.161027
100	40	33	15	0.840974	0.126266
100	40	33	16	0.923561	0.082587
100	40	33	17	0.968608	0.045047
100	40	33	18	0.989074	0.020466
100	40	33	19	0.996802	0.007727
100	40	33	20	0.999218	0.002417
100	40	33	21	0.999842	0.000623
100	40	33	22	0.999974	0.000132
100	40	33	23	0.999996	0.000023
100	40	33	24	0.999999	0.000003

Table for $N = 2$, $n = 1$, through $N = 100$, $n = 50$

N	n	k	x	P(x)	p(x)
100	40	33	25	1.000000	0.000000
100	40	33	26	1.000000	0.000000
100	40	33	27	1.000000	0.000000
100	40	33	28	1.000000	0.000000
100	40	33	29	1.000000	0.000000
100	40	33	30	1.000000	0.000000
100	40	33	31	1.000000	0.000000
100	40	33	32	1.000000	0.000000
100	40	33	33	1.000000	0.000000
100	40	34	0	0.000000	0.000000
100	40	34	1	0.000000	0.000000
100	40	34	2	0.000000	0.000000
100	40	34	3	0.000002	0.000002
100	40	34	4	0.000021	0.000019
100	40	34	5	0.000150	0.000130
100	40	34	6	0.000836	0.000686
100	40	34	7	0.003661	0.002825
100	40	34	8	0.012916	0.009255
100	40	34	9	0.037361	0.024445
100	40	34	10	0.089986	0.052625
100	40	34	11	0.183082	0.093096
100	40	34	12	0.319255	0.136173
100	40	34	13	0.484703	0.165449
100	40	34	14	0.652220	0.167517
100	40	34	15	0.793860	0.141640
100	40	34	16	0.893977	0.100118
100	40	34	17	0.953144	0.059167
100	40	34	18	0.982354	0.029210
100	40	34	19	0.994379	0.012026
100	40	34	20	0.998497	0.004117
100	40	34	21	0.999665	0.001168
100	40	34	22	0.999938	0.000273
100	40	34	23	0.999991	0.000052
100	40	34	24	0.999999	0.000008
100	40	34	25	1.000000	0.000001
100	40	34	26	1.000000	0.000000
100	40	34	27	1.000000	0.000000
100	40	34	28	1.000000	0.000000
100	40	34	29	1.000000	0.000000
100	40	34	30	1.000000	0.000000
100	40	34	31	1.000000	0.000000
100	40	34	32	1.000000	0.000000
100	40	34	33	1.000000	0.000000
100	40	34	34	1.000000	0.000000
100	40	35	0	0.000000	0.000000
100	40	35	1	0.000000	0.000000
100	40	35	2	0.000000	0.000000
100	40	35	3	0.000001	0.000001
100	40	35	4	0.000011	0.000010
100	40	35	5	0.000082	0.000071

N	n	k	x	P(x)	p(x)
100	40	35	6	0.000483	0.000401
100	40	35	7	0.002249	0.001766
100	40	35	8	0.008429	0.006180
100	40	35	9	0.025880	0.017451
100	40	35	10	0.066066	0.040186
100	40	35	11	0.142176	0.076110
100	40	35	12	0.261484	0.119308
100	40	35	13	0.417020	0.155535
100	40	35	14	0.586228	0.169209
100	40	35	15	0.740208	0.153980
100	40	35	16	0.857571	0.117363
100	40	35	17	0.932525	0.074954
100	40	35	18	0.972617	0.040092
100	40	35	19	0.990553	0.017936
100	40	35	20	0.997249	0.006696
100	40	35	21	0.999329	0.002080
100	40	35	22	0.999864	0.000535
100	40	35	23	0.999977	0.000113
100	40	35	24	0.999997	0.000020
100	40	35	25	1.000000	0.000003
100	40	35	26	1.000000	0.000000
100	40	35	27	1.000000	0.000000
100	40	35	28	1.000000	0.000000
100	40	35	29	1.000000	0.000000
100	40	35	30	1.000000	0.000000
100	40	35	31	1.000000	0.000000
100	40	35	32	1.000000	0.000000
100	40	35	33	1.000000	0.000000
100	40	35	34	1.000000	0.000000
100	40	35	35	1.000000	0.000000
100	40	36	0	0.000000	0.000000
100	40	36	1	0.000000	0.000000
100	40	36	2	0.000000	0.000000
100	40	36	3	0.000000	0.000000
100	40	36	4	0.000005	0.000005
100	40	36	5	0.000043	0.000038
100	40	36	6	0.000273	0.000230
100	40	36	7	0.001352	0.001079
100	40	36	8	0.005386	0.004034
100	40	36	9	0.017557	0.012171
100	40	36	10	0.047518	0.029961
100	40	36	11	0.108219	0.060701
100	40	36	12	0.210090	0.101871
100	40	36	13	0.352413	0.142323
100	40	36	14	0.518545	0.166132
100	40	36	15	0.680985	0.162440
100	40	36	16	0.814237	0.133252
100	40	36	17	0.906003	0.091766
100	40	36	18	0.959048	0.053045
100	40	36	19	0.984759	0.025711

N	n	k	x	P(x)	p(x)
100	40	36	20	0.995189	0.010430
100	40	36	21	0.998721	0.003532
100	40	36	22	0.999716	0.000995
100	40	36	23	0.999947	0.000232
100	40	36	24	0.999992	0.000044
100	40	36	25	0.999999	0.000007
100	40	36	26	1.000000	0.000001
100	40	36	27	1.000000	0.000000
100	40	36	28	1.000000	0.000000
100	40	36	29	1.000000	0.000000
100	40	36	30	1.000000	0.000000
100	40	36	31	1.000000	0.000000
100	40	36	32	1.000000	0.000000
100	40	36	33	1.000000	0.000000
100	40	36	34	1.000000	0.000000
100	40	36	35	1.000000	0.000000
100	40	36	36	1.000000	0.000000
100	40	37	0	0.000000	0.000000
100	40	37	1	0.000000	0.000000
100	40	37	2	0.000000	0.000000
100	40	37	3	0.000000	0.000000
100	40	37	4	0.000003	0.000002
100	40	37	5	0.000023	0.000020
100	40	37	6	0.000151	0.000128
100	40	37	7	0.000796	0.000645
100	40	37	8	0.003369	0.002574
100	40	37	9	0.011662	0.008293
100	40	37	10	0.033474	0.021812
100	40	37	11	0.080714	0.047240
100	40	37	12	0.165522	0.084807
100	40	37	13	0.292370	0.126849
100	40	37	14	0.451054	0.158683
100	40	37	15	0.617532	0.166478
100	40	37	16	0.764268	0.146736
100	40	37	17	0.873025	0.108757
100	40	37	18	0.940814	0.067789
100	40	37	19	0.976322	0.035508
100	40	37	20	0.991929	0.015607
100	40	37	21	0.997672	0.005743
100	40	37	22	0.999436	0.001763
100	40	37	23	0.999886	0.000450
100	40	37	24	0.999981	0.000095
100	40	37	25	0.999997	0.000016
100	40	37	26	1.000000	0.000002
100	40	37	27	1.000000	0.000000
100	40	37	28	1.000000	0.000000
100	40	37	29	1.000000	0.000000
100	40	37	30	1.000000	0.000000
100	40	37	31	1.000000	0.000000
100	40	37	32	1.000000	0.000000

N	n	k	x	P(x)	p(x)
100	40	37	33	1.000000	0.000000
100	40	37	34	1.000000	0.000000
100	40	37	35	1.000000	0.000000
100	40	37	36	1.000000	0.000000
100	40	37	37	1.000000	0.000000
100	40	38	0	0.000000	0.000000
100	40	38	1	0.000000	0.000000
100	40	38	2	0.000000	0.000000
100	40	38	3	0.000000	0.000000
100	40	38	4	0.000001	0.000001
100	40	38	5	0.000011	0.000010
100	40	38	6	0.000082	0.000070
100	40	38	7	0.000458	0.000376
100	40	38	8	0.002062	0.001604
100	40	38	9	0.007581	0.005519
100	40	38	10	0.023087	0.015506
100	40	38	11	0.058969	0.035881
100	40	38	12	0.127829	0.068861
100	40	38	13	0.238007	0.110177
100	40	38	14	0.385565	0.147559
100	40	38	15	0.551469	0.165904
100	40	38	16	0.708368	0.156899
100	40	38	17	0.833320	0.124951
100	40	38	18	0.917141	0.083822
100	40	38	19	0.964486	0.047345
100	40	38	20	0.986975	0.022489
100	40	38	21	0.995940	0.008966
100	40	38	22	0.998932	0.002992
100	40	38	23	0.999764	0.000832
100	40	38	24	0.999957	0.000192
100	40	38	25	0.999993	0.000037
100	40	38	26	0.999999	0.000006
100	40	38	27	1.000000	0.000001
100	40	38	28	1.000000	0.000000
100	40	38	29	1.000000	0.000000
100	40	38	30	1.000000	0.000000
100	40	38	31	1.000000	0.000000
100	40	38	32	1.000000	0.000000
100	40	38	33	1.000000	0.000000
100	40	38	34	1.000000	0.000000
100	40	38	35	1.000000	0.000000
100	40	38	36	1.000000	0.000000
100	40	38	37	1.000000	0.000000
100	40	38	38	1.000000	0.000000
100	40	39	0	0.000000	0.000000
100	40	39	1	0.000000	0.000000
100	40	39	2	0.000000	0.000000
100	40	39	3	0.000000	0.000000
100	40	39	4	0.000001	0.000001
100	40	39	5	0.000006	0.000005

N	n	k	x	P(x)	p(x)
100	40	39	6	0.000043	0.000037
100	40	39	7	0.000258	0.000215
100	40	39	8	0.001234	0.000976
100	40	39	9	0.004822	0.003588
100	40	39	10	0.015584	0.010763
100	40	39	11	0.042185	0.026601
100	40	39	12	0.096731	0.054546
100	40	39	13	0.190026	0.093295
100	40	39	14	0.323686	0.133660
100	40	39	15	0.484573	0.160887
100	40	39	16	0.647633	0.163061
100	40	39	17	0.786967	0.139334
100	40	39	18	0.887398	0.100431
100	40	39	19	0.948450	0.061052
100	40	39	20	0.979720	0.031270
100	40	39	21	0.993193	0.013472
100	40	39	22	0.998063	0.004871
100	40	39	23	0.999536	0.001473
100	40	39	24	0.999907	0.000371
100	40	39	25	0.999984	0.000077
100	40	39	26	0.999998	0.000013
100	40	39	27	1.000000	0.000002
100	40	39	28	1.000000	0.000000
100	40	39	29	1.000000	0.000000
100	40	39	30	1.000000	0.000000
100	40	39	31	1.000000	0.000000
100	40	39	32	1.000000	0.000000
100	40	39	33	1.000000	0.000000
100	40	39	34	1.000000	0.000000
100	40	39	35	1.000000	0.000000
100	40	39	36	1.000000	0.000000
100	40	39	37	1.000000	0.000000
100	40	39	38	1.000000	0.000000
100	40	39	39	1.000000	0.000000
100	40	40	0	0.000000	0.000000
100	40	40	1	0.000000	0.000000
100	40	40	2	0.000000	0.000000
100	40	40	3	0.000000	0.000000
100	40	40	4	0.000000	0.000000
100	40	40	5	0.000003	0.000002
100	40	40	6	0.000022	0.000020
100	40	40	7	0.000142	0.000119
100	40	40	8	0.000722	0.000580
100	40	40	9	0.002999	0.002277
100	40	40	10	0.010291	0.007293
100	40	40	11	0.029539	0.019248
100	40	40	12	0.071694	0.042155
100	40	40	13	0.148732	0.077038
100	40	40	14	0.266716	0.117985
100	40	40	15	0.418635	0.151919

N	n	k	x	P(x)	p(x)
100	40	40	16	0.583478	0.164843
100	40	40	17	0.734431	0.150953
100	40	40	18	0.851177	0.116746
100	40	40	19	0.927432	0.076255
100	40	40	20	0.969468	0.042036
100	40	40	21	0.988996	0.019529
100	40	40	22	0.996626	0.007630
100	40	40	23	0.999126	0.002500
100	40	40	24	0.999810	0.000684
100	40	40	25	0.999965	0.000156
100	40	40	26	0.999995	0.000029
100	40	40	27	0.999999	0.000005
100	40	40	28	1.000000	0.000001
100	40	40	29	1.000000	0.000000
100	40	40	30	1.000000	0.000000
100	40	40	31	1.000000	0.000000
100	40	40	32	1.000000	0.000000
100	40	40	33	1.000000	0.000000
100	40	40	34	1.000000	0.000000
100	40	40	35	1.000000	0.000000
100	40	40	36	1.000000	0.000000
100	40	40	37	1.000000	0.000000
100	40	40	38	1.000000	0.000000
100	40	40	39	1.000000	0.000000
100	40	40	40	1.000000	0.000000
100	41	1	0	0.590000	0.590000
100	41	1	1	1.000000	0.410000
100	41	2	0	0.345657	0.345657
100	41	2	1	0.834343	0.488687
100	41	2	2	1.000000	0.165657
100	41	3	0	0.201045	0.201045
100	41	3	1	0.634879	0.433834
100	41	3	2	0.934075	0.299196
100	41	3	3	1.000000	0.065925
100	41	4	0	0.116067	0.116067
100	41	4	1	0.455979	0.339911
100	41	4	2	0.813780	0.357801
100	41	4	3	0.974174	0.160394
100	41	4	4	1.000000	0.025826
100	41	5	0	0.066497	0.066497
100	41	5	1	0.314349	0.247852
100	41	5	2	0.668423	0.354074
100	41	5	3	0.910685	0.242261
100	41	5	4	0.990046	0.079361
100	41	5	5	1.000000	0.009954
100	41	6	0	0.037798	0.037798
100	41	6	1	0.209990	0.172192
100	41	6	2	0.523066	0.313076
100	41	6	3	0.813780	0.290714
100	41	6	4	0.959137	0.145357

N	n	k	x	P(x)	p(x)
100	41	6	5	0.996228	0.037091
100	41	6	6	1.000000	0.003772
100	41	7	0	0.021312	0.021312
100	41	7	1	0.136717	0.115405
100	41	7	2	0.393173	0.256456
100	41	7	3	0.696258	0.303084
100	41	7	4	0.901922	0.205664
100	41	7	5	0.982023	0.080101
100	41	7	6	0.998595	0.016573
100	41	7	7	1.000000	0.001404
100	41	8	0	0.011916	0.011916
100	41	8	1	0.087080	0.075164
100	41	8	2	0.285627	0.198547
100	41	8	3	0.572417	0.286790
100	41	8	4	0.820099	0.247682
100	41	8	5	0.951016	0.130918
100	41	8	6	0.992358	0.041342
100	41	8	7	0.999486	0.007128
100	41	8	8	1.000000	0.000513
100	41	9	0	0.006606	0.006606
100	41	9	1	0.054400	0.047795
100	41	9	2	0.201460	0.147060
100	41	9	3	0.453960	0.252500
100	41	9	4	0.720487	0.266527
100	41	9	5	0.899787	0.179300
100	41	9	6	0.976630	0.076843
100	41	9	7	0.996852	0.020222
100	41	9	8	0.999816	0.002964
100	41	9	9	1.000000	0.000184
100	41	10	0	0.003630	0.003630
100	41	10	1	0.033392	0.029762
100	41	10	2	0.138435	0.105043
100	41	10	3	0.348521	0.210086
100	41	10	4	0.612119	0.263598
100	41	10	5	0.828856	0.216736
100	41	10	6	0.947075	0.118220
100	41	10	7	0.989297	0.042221
100	41	10	8	0.998741	0.009444
100	41	10	9	0.999935	0.001194
100	41	10	10	1.000000	0.000065
100	41	11	0	0.001976	0.001976
100	41	11	1	0.020164	0.018188
100	41	11	2	0.092916	0.072752
100	41	11	3	0.259818	0.166902
100	41	11	4	0.503751	0.243933
100	41	11	5	0.742161	0.238410
100	41	11	6	0.901101	0.158940
100	41	11	7	0.973346	0.072245
100	41	11	8	0.995278	0.021932
100	41	11	9	0.999511	0.004232

N	n	k	x	P(x)	p(x)
100	41	11	10	0.999978	0.000467
100	41	11	11	1.000000	0.000022
100	41	12	0	0.001066	0.001066
100	41	12	1	0.011990	0.010924
100	41	12	2	0.061036	0.049046
100	41	12	3	0.188556	0.127520
100	41	12	4	0.402340	0.213784
100	41	12	5	0.645725	0.243385
100	41	12	6	0.838596	0.192871
100	41	12	7	0.945747	0.107151
100	41	12	8	0.987146	0.041399
100	41	12	9	0.997989	0.010843
100	41	12	10	0.999815	0.001826
100	41	12	11	0.999992	0.000177
100	41	12	12	1.000000	0.000008
100	41	13	0	0.000569	0.000569
100	41	13	1	0.007024	0.006455
100	41	13	2	0.039300	0.032275
100	41	13	3	0.133491	0.094191
100	41	13	4	0.312454	0.178963
100	41	13	5	0.546159	0.233705
100	41	13	6	0.761886	0.215728
100	41	13	7	0.904348	0.142462
100	41	13	8	0.971621	0.067274
100	41	13	9	0.994046	0.022425
100	41	13	10	0.999172	0.005126
100	41	13	11	0.999932	0.000760
100	41	13	12	0.999997	0.000066
100	41	13	13	1.000000	0.000002
100	41	14	0	0.000301	0.000301
100	41	14	1	0.004056	0.003755
100	41	14	2	0.024831	0.020775
100	41	14	3	0.092350	0.067518
100	41	14	4	0.236343	0.143993
100	41	14	5	0.449453	0.213110
100	41	14	6	0.675099	0.225646
100	41	14	7	0.848673	0.173574
100	41	14	8	0.946104	0.097431
100	41	14	9	0.985798	0.039694
100	41	14	10	0.997345	0.011547
100	41	14	11	0.999670	0.002324
100	41	14	12	0.999976	0.000306
100	41	14	13	0.999999	0.000024
100	41	14	14	1.000000	0.000001
100	41	15	0	0.000157	0.000157
100	41	15	1	0.002310	0.002152
100	41	15	2	0.015410	0.013100
100	41	15	3	0.062516	0.047106
100	41	15	4	0.174392	0.111876
100	41	15	5	0.360244	0.185852

Table for $N=2$, $n=1$, through $N=100$, $n=50$

N	n	k	x	P(x)	p(x)
100	41	15	6	0.583267	0.223022
100	41	15	7	0.780051	0.196784
100	41	15	8	0.908718	0.128667
100	41	15	9	0.971028	0.062310
100	41	15	10	0.993183	0.022155
100	41	15	11	0.998859	0.005676
100	41	15	12	0.999872	0.001014
100	41	15	13	0.999991	0.000119
100	41	15	14	1.000000	0.000008
100	41	15	15	1.000000	0.000000
100	41	16	0	0.000082	0.000082
100	41	16	1	0.001297	0.001215
100	41	16	2	0.009399	0.008102
100	41	16	3	0.041457	0.032058
100	41	16	4	0.125693	0.084236
100	41	16	5	0.281531	0.155837
100	41	16	6	0.491434	0.209903
100	41	16	7	0.701337	0.209903
100	41	16	8	0.858765	0.157428
100	41	16	9	0.947570	0.088805
100	41	16	10	0.985103	0.037533
100	41	16	11	0.996856	0.011753
100	41	16	12	0.999527	0.002671
100	41	16	13	0.999952	0.000426
100	41	16	14	0.999997	0.000045
100	41	16	15	1.000000	0.000003
100	41	16	16	1.000000	0.000000
100	41	17	0	0.000042	0.000042
100	41	17	1	0.000718	0.000676
100	41	17	2	0.005637	0.004919
100	41	17	3	0.026955	0.021317
100	41	17	4	0.088589	0.061635
100	41	17	5	0.214743	0.126154
100	41	17	6	0.403974	0.189231
100	41	17	7	0.616376	0.212402
100	41	17	8	0.796918	0.180542
100	41	17	9	0.913740	0.116821
100	41	17	10	0.971251	0.057512
100	41	17	11	0.992658	0.021407
100	41	17	12	0.998604	0.005946
100	41	17	13	0.999810	0.001206
100	41	17	14	0.999983	0.000172
100	41	17	15	0.999999	0.000016
100	41	17	16	1.000000	0.000001
100	41	17	17	1.000000	0.000000
100	41	18	0	0.000021	0.000021
100	41	18	1	0.000392	0.000371
100	41	18	2	0.003326	0.002934
100	41	18	3	0.017195	0.013869
100	41	18	4	0.061114	0.043919

N	n	k	x	P(x)	p(x)
100	41	18	5	0.160026	0.098912
100	41	18	6	0.324178	0.164152
100	41	18	7	0.529368	0.205190
100	41	18	8	0.725137	0.195768
100	41	18	9	0.868700	0.143563
100	41	18	10	0.949771	0.081071
100	41	18	11	0.984921	0.035150
100	41	18	12	0.996527	0.011606
100	41	18	13	0.999404	0.002877
100	41	18	14	0.999927	0.000523
100	41	18	15	0.999994	0.000067
100	41	18	16	1.000000	0.000006
100	41	18	17	1.000000	0.000000
100	41	18	18	1.000000	0.000000
100	41	19	0	0.000011	0.000011
100	41	19	1	0.000211	0.000201
100	41	19	2	0.001931	0.001719
100	41	19	3	0.010768	0.008837
100	41	19	4	0.041297	0.030529
100	41	19	5	0.116601	0.075304
100	41	19	6	0.254113	0.137512
100	41	19	7	0.444289	0.190176
100	41	19	8	0.646352	0.202062
100	41	19	9	0.812675	0.166323
100	41	19	10	0.919122	0.106447
100	41	19	11	0.972061	0.052939
100	41	19	12	0.992422	0.020361
100	41	19	13	0.998421	0.005999
100	41	19	14	0.999754	0.001333
100	41	19	15	0.999972	0.000218
100	41	19	16	0.999998	0.000025
100	41	19	17	1.000000	0.000002
100	41	19	18	1.000000	0.000000
100	41	19	19	1.000000	0.000000
100	41	20	0	0.000005	0.000005
100	41	20	1	0.000112	0.000107
100	41	20	2	0.001103	0.000991
100	41	20	3	0.006622	0.005519
100	41	20	4	0.027351	0.020729
100	41	20	5	0.083132	0.055781
100	41	20	6	0.194694	0.111562
100	41	20	7	0.364462	0.169768
100	41	20	8	0.564030	0.199568
100	41	20	9	0.746967	0.182937
100	41	20	10	0.878383	0.131416
100	41	20	11	0.952454	0.074071
100	41	20	12	0.985132	0.032678
100	41	20	13	0.996347	0.011215
100	41	20	14	0.999310	0.002962
100	41	20	15	0.999902	0.000592

N	n	k	x	P(x)	p(x)
100	41	20	16	0.999990	0.000088
100	41	20	17	0.999999	0.000009
100	41	20	18	1.000000	0.000001
100	41	20	19	1.000000	0.000000
100	41	20	20	1.000000	0.000000
100	41	21	0	0.000003	0.000003
100	41	21	1	0.000059	0.000056
100	41	21	2	0.000620	0.000561
100	41	21	3	0.004000	0.003381
100	41	21	4	0.017764	0.013764
100	41	21	5	0.058031	0.040267
100	41	21	6	0.145886	0.087855
100	41	21	7	0.292311	0.146425
100	41	21	8	0.481708	0.189398
100	41	21	9	0.673792	0.192084
100	41	21	10	0.827460	0.153667
100	41	21	11	0.924678	0.097218
100	41	21	12	0.973287	0.048609
100	41	21	13	0.992422	0.019136
100	41	21	14	0.998310	0.005888
100	41	21	15	0.999710	0.001400
100	41	21	16	0.999963	0.000253
100	41	21	17	0.999996	0.000034
100	41	21	18	1.000000	0.000003
100	41	21	19	1.000000	0.000000
100	41	21	20	1.000000	0.000000
100	41	21	21	1.000000	0.000000
100	41	22	0	0.000001	0.000001
100	41	22	1	0.000030	0.000029
100	41	22	2	0.000343	0.000313
100	41	22	3	0.002374	0.002031
100	41	22	4	0.011318	0.008944
100	41	22	5	0.039681	0.028364
100	41	22	6	0.106963	0.067281
100	41	22	7	0.229293	0.122330
100	41	22	8	0.402593	0.173300
100	41	22	9	0.595986	0.193393
100	41	22	10	0.767160	0.171174
100	41	22	11	0.887759	0.120600
100	41	22	12	0.955443	0.067683
100	41	22	13	0.985640	0.030197
100	41	22	14	0.996298	0.010658
100	41	22	15	0.999249	0.002951
100	41	22	16	0.999883	0.000633
100	41	22	17	0.999986	0.000103
100	41	22	18	0.999999	0.000013
100	41	22	19	1.000000	0.000001
100	41	22	20	1.000000	0.000000
100	41	22	21	1.000000	0.000000
100	41	22	22	1.000000	0.000000

N	n	k	x	P(x)	p(x)
100	41	23	0	0.000001	0.000001
100	41	23	1	0.000015	0.000015
100	41	23	2	0.000187	0.000171
100	41	23	3	0.001385	0.001198
100	41	23	4	0.007075	0.005691
100	41	23	5	0.026591	0.019515
100	41	23	6	0.076772	0.050182
100	41	23	7	0.175969	0.099197
100	41	23	8	0.329274	0.153304
100	41	23	9	0.516645	0.187372
100	41	23	10	0.699129	0.182484
100	41	23	11	0.841375	0.142246
100	41	23	12	0.930278	0.088904
100	41	23	13	0.974800	0.044522
100	41	23	14	0.992609	0.017809
100	41	23	15	0.998265	0.005657
100	41	23	16	0.999680	0.001414
100	41	23	17	0.999954	0.000275
100	41	23	18	0.999995	0.000041
100	41	23	19	1.000000	0.000004
100	41	23	20	1.000000	0.000000
100	41	23	21	1.000000	0.000000
100	41	23	22	1.000000	0.000000
100	41	23	23	1.000000	0.000000
100	41	24	0	0.000000	0.000000
100	41	24	1	0.000008	0.000007
100	41	24	2	0.000100	0.000092
100	41	24	3	0.000793	0.000693
100	41	24	4	0.004341	0.003547
100	41	24	5	0.017467	0.013126
100	41	24	6	0.053963	0.036496
100	41	24	7	0.132168	0.078206
100	41	24	8	0.263572	0.131404
100	41	24	9	0.438777	0.175205
100	41	24	10	0.625662	0.186885
100	41	24	11	0.785955	0.160293
100	41	24	12	0.896795	0.110841
100	41	24	13	0.958610	0.061815
100	41	24	14	0.986364	0.027754
100	41	24	15	0.996355	0.009991
100	41	24	16	0.999221	0.002865
100	41	24	17	0.999869	0.000648
100	41	24	18	0.999983	0.000114
100	41	24	19	0.999998	0.000015
100	41	24	20	1.000000	0.000002
100	41	24	21	1.000000	0.000000
100	41	24	22	1.000000	0.000000
100	41	24	23	1.000000	0.000000
100	41	24	24	1.000000	0.000000
100	41	25	0	0.000000	0.000000

Table for $N = 2$, $n = 1$, through $N = 100$, $n = 50$

N	n	k	x	P(x)	p(x)
100	41	25	1	0.000004	0.000004
100	41	25	2	0.000053	0.000049
100	41	25	3	0.000447	0.000394
100	41	25	4	0.002614	0.002167
100	41	25	5	0.011249	0.008635
100	41	25	6	0.037155	0.025906
100	41	25	7	0.097181	0.060026
100	41	25	8	0.206515	0.109334
100	41	25	9	0.365006	0.158491
100	41	25	10	0.549432	0.184426
100	41	25	11	0.722681	0.173249
100	41	25	12	0.854501	0.131820
100	41	25	13	0.935836	0.081336
100	41	25	14	0.976504	0.040668
100	41	25	15	0.992937	0.016433
100	41	25	16	0.998278	0.005341
100	41	25	17	0.999664	0.001386
100	41	25	18	0.999948	0.000284
100	41	25	19	0.999994	0.000045
100	41	25	20	0.999999	0.000006
100	41	25	21	1.000000	0.000000
100	41	25	22	1.000000	0.000000
100	41	25	23	1.000000	0.000000
100	41	25	24	1.000000	0.000000
100	41	25	25	1.000000	0.000000
100	41	26	0	0.000000	0.000000
100	41	26	1	0.000002	0.000002
100	41	26	2	0.000027	0.000025
100	41	26	3	0.000247	0.000220
100	41	26	4	0.001545	0.001298
100	41	26	5	0.007104	0.005560
100	41	26	6	0.025066	0.017962
100	41	26	7	0.069969	0.044904
100	41	26	8	0.158408	0.088439
100	41	26	9	0.297383	0.138975
100	41	26	10	0.473203	0.175820
100	41	26	11	0.653382	0.180179
100	41	26	12	0.803530	0.150149
100	41	26	13	0.905471	0.101941
100	41	26	14	0.961864	0.056393
100	41	26	15	0.987240	0.025377
100	41	26	16	0.996498	0.009257
100	41	26	17	0.999220	0.002723
100	41	26	18	0.999861	0.000641
100	41	26	19	0.999980	0.000119
100	41	26	20	0.999998	0.000017
100	41	26	21	1.000000	0.000002
100	41	26	22	1.000000	0.000000
100	41	26	23	1.000000	0.000000
100	41	26	24	1.000000	0.000000
100	41	26	25	1.000000	0.000000
100	41	26	26	1.000000	0.000000
100	41	27	0	0.000000	0.000000
100	41	27	1	0.000001	0.000001
100	41	27	2	0.000014	0.000013
100	41	27	3	0.000134	0.000120
100	41	27	4	0.000896	0.000762
100	41	27	5	0.004399	0.003504
100	41	27	6	0.016570	0.012171
100	41	27	7	0.049338	0.032768
100	41	27	8	0.118969	0.069631
100	41	27	9	0.237286	0.118317
100	41	27	10	0.399549	0.162263
100	41	27	11	0.580336	0.180787
100	41	27	12	0.744688	0.164352
100	41	27	13	0.866899	0.122211
100	41	27	14	0.941288	0.074389
100	41	27	15	0.978324	0.037036
100	41	27	16	0.993370	0.015046
100	41	27	17	0.998337	0.004967
100	41	27	18	0.999662	0.001325
100	41	27	19	0.999945	0.000283
100	41	27	20	0.999993	0.000048
100	41	27	21	0.999999	0.000006
100	41	27	22	1.000000	0.000001
100	41	27	23	1.000000	0.000000
100	41	27	24	1.000000	0.000000
100	41	27	25	1.000000	0.000000
100	41	27	26	1.000000	0.000000
100	41	27	27	1.000000	0.000000
100	41	28	0	0.000000	0.000000
100	41	28	1	0.000000	0.000000
100	41	28	2	0.000007	0.000006
100	41	28	3	0.000071	0.000065
100	41	28	4	0.000510	0.000438
100	41	28	5	0.002672	0.002162
100	41	28	6	0.010735	0.008063
100	41	28	7	0.034076	0.023341
100	41	28	8	0.087492	0.053416
100	41	28	9	0.185421	0.097929
100	41	28	10	0.330643	0.145221
100	41	28	11	0.506040	0.175397
100	41	28	12	0.679398	0.173358
100	41	28	13	0.820024	0.140626
100	41	28	14	0.913774	0.093751
100	41	28	15	0.965133	0.051359
100	41	28	16	0.988218	0.023084
100	41	28	17	0.996704	0.008487
100	41	28	18	0.999245	0.002540
100	41	28	19	0.999860	0.000615
100	41	28	20	0.999979	0.000119
100	41	28	21	0.999997	0.000018
100	41	28	22	1.000000	0.000002
100	41	28	23	1.000000	0.000000
100	41	28	24	1.000000	0.000000
100	41	28	25	1.000000	0.000000
100	41	28	26	1.000000	0.000000
100	41	28	27	1.000000	0.000000
100	41	28	28	1.000000	0.000000
100	41	29	0	0.000000	0.000000
100	41	29	1	0.000000	0.000000
100	41	29	2	0.000003	0.000003
100	41	29	3	0.000037	0.000034
100	41	29	4	0.000285	0.000247
100	41	29	5	0.001591	0.001306
100	41	29	6	0.006815	0.005225
100	41	29	7	0.023054	0.016239
100	41	29	8	0.063010	0.039956
100	41	29	9	0.141897	0.078887
100	41	29	10	0.268117	0.126220
100	41	29	11	0.432958	0.164841
100	41	29	12	0.609573	0.176615
100	41	29	13	0.765336	0.155763
100	41	29	14	0.878618	0.113282
100	41	29	15	0.946587	0.067969
100	41	29	16	0.980202	0.033615
100	41	29	17	0.993875	0.013673
100	41	29	18	0.998433	0.004558
100	41	29	19	0.999672	0.001239
100	41	29	20	0.999944	0.000272
100	41	29	21	0.999992	0.000048
100	41	29	22	0.999999	0.000007
100	41	29	23	1.000000	0.000001
100	41	29	24	1.000000	0.000000
100	41	29	25	1.000000	0.000000
100	41	29	26	1.000000	0.000000
100	41	29	27	1.000000	0.000000
100	41	29	28	1.000000	0.000000
100	41	29	29	1.000000	0.000000
100	41	30	0	0.000000	0.000000
100	41	30	1	0.000000	0.000000
100	41	30	2	0.000002	0.000002
100	41	30	3	0.000019	0.000018
100	41	30	4	0.000156	0.000137
100	41	30	5	0.000928	0.000773
100	41	30	6	0.004240	0.003311
100	41	30	7	0.015278	0.011038
100	41	30	8	0.044439	0.029161
100	41	30	9	0.106342	0.061903
100	41	30	10	0.213007	0.106664
100	41	30	11	0.363307	0.150300
100	41	30	12	0.537434	0.174128
100	41	30	13	0.703908	0.166474
100	41	30	14	0.835539	0.131630
100	41	30	15	0.921697	0.086158
100	41	30	16	0.968366	0.046669
100	41	30	17	0.989254	0.020888
100	41	30	18	0.996957	0.007703
100	41	30	19	0.999288	0.002331
100	41	30	20	0.999864	0.000576
100	41	30	21	0.999979	0.000115
100	41	30	22	0.999997	0.000018
100	41	30	23	1.000000	0.000002
100	41	30	24	1.000000	0.000000
100	41	30	25	1.000000	0.000000
100	41	30	26	1.000000	0.000000
100	41	30	27	1.000000	0.000000
100	41	30	28	1.000000	0.000000
100	41	30	29	1.000000	0.000000
100	41	30	30	1.000000	0.000000
100	41	31	0	0.000000	0.000000
100	41	31	1	0.000000	0.000000
100	41	31	2	0.000001	0.000001
100	41	31	3	0.000010	0.000009
100	41	31	4	0.000084	0.000074
100	41	31	5	0.000531	0.000447
100	41	31	6	0.002584	0.002053
100	41	31	7	0.009916	0.007332
100	41	31	8	0.030691	0.020775
100	41	31	9	0.078044	0.047352
100	41	31	10	0.165770	0.087726
100	41	31	11	0.298892	0.133123
100	41	31	12	0.465296	0.166403
100	41	31	13	0.637319	0.172023
100	41	31	14	0.784767	0.147448
100	41	31	15	0.889695	0.104928
100	41	31	16	0.951698	0.062003
100	41	31	17	0.982092	0.030394
100	41	31	18	0.994426	0.012334
100	41	31	19	0.998555	0.004130
100	41	31	20	0.999691	0.001136
100	41	31	21	0.999946	0.000255
100	41	31	22	0.999992	0.000046
100	41	31	23	0.999999	0.000007
100	41	31	24	1.000000	0.000001
100	41	31	25	1.000000	0.000000
100	41	31	26	1.000000	0.000000
100	41	31	27	1.000000	0.000000
100	41	31	28	1.000000	0.000000
100	41	31	29	1.000000	0.000000

N	n	k	x	P(x)	p(x)
100	41	31	30	1.000000	0.000000
100	41	31	31	1.000000	0.000000
100	41	32	0	0.000000	0.000000
100	41	32	1	0.000000	0.000000
100	41	32	2	0.000000	0.000000
100	41	32	3	0.000005	0.000004
100	41	32	4	0.000044	0.000039
100	41	32	5	0.000298	0.000254
100	41	32	6	0.001543	0.001245
100	41	32	7	0.006303	0.004761
100	41	32	8	0.020756	0.014452
100	41	32	9	0.056083	0.035328
100	41	32	10	0.126356	0.070273
100	41	32	11	0.241013	0.114656
100	41	32	12	0.395358	0.154345
100	41	32	13	0.567512	0.172154
100	41	32	14	0.727070	0.159558
100	41	32	15	0.850157	0.123087
100	41	32	16	0.929233	0.079076
100	41	32	17	0.971520	0.042287
100	41	32	18	0.990314	0.018794
100	41	32	19	0.997239	0.006924
100	41	32	20	0.999345	0.002107
100	41	32	21	0.999872	0.000527
100	41	32	22	0.999979	0.000107
100	41	32	23	0.999997	0.000018
100	41	32	24	1.000000	0.000002
100	41	32	25	1.000000	0.000000
100	41	32	26	1.000000	0.000000
100	41	32	27	1.000000	0.000000
100	41	32	28	1.000000	0.000000
100	41	32	29	1.000000	0.000000
100	41	32	30	1.000000	0.000000
100	41	32	31	1.000000	0.000000
100	41	32	32	1.000000	0.000000
100	41	33	0	0.000000	0.000000
100	41	33	1	0.000000	0.000000
100	41	33	2	0.000000	0.000000
100	41	33	3	0.000002	0.000002
100	41	33	4	0.000023	0.000020
100	41	33	5	0.000163	0.000141
100	41	33	6	0.000902	0.000739
100	41	33	7	0.003923	0.003021
100	41	33	8	0.013742	0.009819
100	41	33	9	0.039458	0.025716
100	41	33	10	0.094320	0.054862
100	41	33	11	0.190429	0.096109
100	41	33	12	0.329534	0.139105
100	41	33	13	0.496625	0.167091
100	41	33	14	0.663716	0.167091

N	n	k	x	P(x)	p(x)
100	41	33	15	0.803094	0.139378
100	41	33	16	0.900161	0.097067
100	41	33	17	0.956596	0.056434
100	41	33	18	0.983958	0.027362
100	41	33	19	0.994998	0.011041
100	41	33	20	0.998695	0.003696
100	41	33	21	0.999717	0.001022
100	41	33	22	0.999949	0.000232
100	41	33	23	0.999993	0.000043
100	41	33	24	0.999999	0.000006
100	41	33	25	1.000000	0.000001
100	41	33	26	1.000000	0.000000
100	41	33	27	1.000000	0.000000
100	41	33	28	1.000000	0.000000
100	41	33	29	1.000000	0.000000
100	41	33	30	1.000000	0.000000
100	41	33	31	1.000000	0.000000
100	41	33	32	1.000000	0.000000
100	41	33	33	1.000000	0.000000
100	41	34	0	0.000000	0.000000
100	41	34	1	0.000000	0.000000
100	41	34	2	0.000000	0.000000
100	41	34	3	0.000001	0.000001
100	41	34	4	0.000011	0.000010
100	41	34	5	0.000088	0.000076
100	41	34	6	0.000516	0.000428
100	41	34	7	0.002390	0.001874
100	41	34	8	0.008906	0.006516
100	41	34	9	0.027176	0.018270
100	41	34	10	0.068936	0.041760
100	41	34	11	0.147395	0.078459
100	41	34	12	0.269325	0.121929
100	41	34	13	0.426796	0.157472
100	41	34	14	0.596381	0.169585
100	41	34	15	0.749007	0.152626
100	41	34	16	0.863942	0.114935
100	41	34	17	0.936380	0.072438
100	41	34	18	0.974565	0.038184
100	41	34	19	0.991373	0.016808
100	41	34	20	0.997536	0.006163
100	41	34	21	0.999412	0.001876
100	41	34	22	0.999884	0.000472
100	41	34	23	0.999981	0.000097
100	41	34	24	0.999997	0.000016
100	41	34	25	1.000000	0.000002
100	41	34	26	1.000000	0.000000
100	41	34	27	1.000000	0.000000
100	41	34	28	1.000000	0.000000
100	41	34	29	1.000000	0.000000
100	41	34	30	1.000000	0.000000

N	n	k	x	P(x)	p(x)
100	41	34	31	1.000000	0.000000
100	41	34	32	1.000000	0.000000
100	41	34	33	1.000000	0.000000
100	41	34	34	1.000000	0.000000
100	41	35	0	0.000000	0.000000
100	41	35	1	0.000000	0.000000
100	41	35	2	0.000000	0.000000
100	41	35	3	0.000000	0.000000
100	41	35	4	0.000006	0.000005
100	41	35	5	0.000046	0.000040
100	41	35	6	0.000289	0.000243
100	41	35	7	0.001425	0.001136
100	41	35	8	0.005648	0.004223
100	41	35	9	0.018318	0.012670
100	41	35	10	0.049322	0.031004
100	41	35	11	0.111732	0.062410
100	41	35	12	0.215750	0.104017
100	41	35	13	0.359990	0.144240
100	41	35	14	0.527005	0.167015
100	41	35	15	0.688882	0.161876
100	41	35	16	0.820406	0.131525
100	41	35	17	0.910039	0.089633
100	41	35	18	0.961258	0.051219
100	41	35	19	0.985770	0.024512
100	41	35	20	0.995575	0.009805
100	41	35	21	0.998844	0.003268
100	41	35	22	0.999748	0.000904
100	41	35	23	0.999954	0.000207
100	41	35	24	0.999993	0.000039
100	41	35	25	0.999999	0.000006
100	41	35	26	1.000000	0.000001
100	41	35	27	1.000000	0.000000
100	41	35	28	1.000000	0.000000
100	41	35	29	1.000000	0.000000
100	41	35	30	1.000000	0.000000
100	41	35	31	1.000000	0.000000
100	41	35	32	1.000000	0.000000
100	41	35	33	1.000000	0.000000
100	41	35	34	1.000000	0.000000
100	41	35	35	1.000000	0.000000
100	41	36	0	0.000000	0.000000
100	41	36	1	0.000000	0.000000
100	41	36	2	0.000000	0.000000
100	41	36	3	0.000000	0.000000
100	41	36	4	0.000003	0.000002
100	41	36	5	0.000024	0.000021
100	41	36	6	0.000158	0.000134
100	41	36	7	0.000831	0.000672
100	41	36	8	0.003504	0.002673
100	41	36	9	0.012080	0.008577

N	n	k	x	P(x)	p(x)
100	41	36	10	0.034535	0.022455
100	41	36	11	0.082927	0.048392
100	41	36	12	0.169342	0.086415
100	41	36	13	0.297856	0.128514
100	41	36	14	0.457630	0.159774
100	41	36	15	0.624131	0.166501
100	41	36	16	0.769820	0.145689
100	41	36	17	0.876944	0.107124
100	41	36	18	0.943134	0.066190
100	41	36	19	0.977474	0.034339
100	41	36	20	0.992407	0.014934
100	41	36	21	0.997838	0.005430
100	41	36	22	0.999483	0.001646
100	41	36	23	0.999897	0.000414
100	41	36	24	0.999983	0.000086
100	41	36	25	0.999998	0.000015
100	41	36	26	1.000000	0.000002
100	41	36	27	1.000000	0.000000
100	41	36	28	1.000000	0.000000
100	41	36	29	1.000000	0.000000
100	41	36	30	1.000000	0.000000
100	41	36	31	1.000000	0.000000
100	41	36	32	1.000000	0.000000
100	41	36	33	1.000000	0.000000
100	41	36	34	1.000000	0.000000
100	41	36	35	1.000000	0.000000
100	41	36	36	1.000000	0.000000
100	41	37	0	0.000000	0.000000
100	41	37	1	0.000000	0.000000
100	41	37	2	0.000000	0.000000
100	41	37	3	0.000000	0.000000
100	41	37	4	0.000001	0.000001
100	41	37	5	0.000012	0.000011
100	41	37	6	0.000085	0.000073
100	41	37	7	0.000473	0.000389
100	41	37	8	0.002125	0.001652
100	41	37	9	0.007792	0.005667
100	41	37	10	0.023659	0.015867
100	41	37	11	0.060243	0.036585
100	41	37	12	0.130185	0.069942
100	41	37	13	0.241631	0.111446
100	41	37	14	0.390225	0.148594
100	41	37	15	0.556490	0.166265
100	41	37	16	0.712910	0.156420
100	41	37	17	0.836772	0.123862
100	41	37	18	0.919347	0.082575
100	41	37	19	0.965670	0.046322
100	41	37	20	0.987507	0.021838
100	41	37	21	0.996141	0.008634
100	41	37	22	0.998995	0.002854

Table for $N = 2$, $n = 1$, through $N = 100$, $n = 50$

N	n	k	x	P(x)	p(x)
100	41	37	23	0.999781	0.000786
100	41	37	24	0.999960	0.000179
100	41	37	25	0.999994	0.000034
100	41	37	26	0.999999	0.000005
100	41	37	27	1.000000	0.000001
100	41	37	28	1.000000	0.000000
100	41	37	29	1.000000	0.000000
100	41	37	30	1.000000	0.000000
100	41	37	31	1.000000	0.000000
100	41	37	32	1.000000	0.000000
100	41	37	33	1.000000	0.000000
100	41	37	34	1.000000	0.000000
100	41	37	35	1.000000	0.000000
100	41	37	36	1.000000	0.000000
100	41	37	37	1.000000	0.000000
100	41	38	0	0.000000	0.000000
100	41	38	1	0.000000	0.000000
100	41	38	2	0.000000	0.000000
100	41	38	3	0.000000	0.000000
100	41	38	4	0.000001	0.000001
100	41	38	5	0.000006	0.000005
100	41	38	6	0.000044	0.000038
100	41	38	7	0.000264	0.000219
100	41	38	8	0.001260	0.000996
100	41	38	9	0.004914	0.003654
100	41	38	10	0.015851	0.010937
100	41	38	11	0.042822	0.026971
100	41	38	12	0.097990	0.055168
100	41	38	13	0.192099	0.094110
100	41	38	14	0.326542	0.134442
100	41	38	15	0.487873	0.161331
100	41	38	16	0.650839	0.162966
100	41	38	17	0.789587	0.138748
100	41	38	18	0.889201	0.099614
100	41	38	19	0.949494	0.060293
100	41	38	20	0.980228	0.030735
100	41	38	21	0.993400	0.013172
100	41	38	22	0.998134	0.004734
100	41	38	23	0.999556	0.001422
100	41	38	24	0.999912	0.000356
100	41	38	25	0.999985	0.000074
100	41	38	26	0.999998	0.000013
100	41	38	27	1.000000	0.000002
100	41	38	28	1.000000	0.000000
100	41	38	29	1.000000	0.000000
100	41	38	30	1.000000	0.000000
100	41	38	31	1.000000	0.000000
100	41	38	32	1.000000	0.000000
100	41	38	33	1.000000	0.000000
100	41	38	34	1.000000	0.000000

N	n	k	x	P(x)	p(x)
100	41	38	35	1.000000	0.000000
100	41	38	36	1.000000	0.000000
100	41	38	37	1.000000	0.000000
100	41	38	38	1.000000	0.000000
100	41	39	0	0.000000	0.000000
100	41	39	1	0.000000	0.000000
100	41	39	2	0.000000	0.000000
100	41	39	3	0.000000	0.000000
100	41	39	4	0.000000	0.000000
100	41	39	5	0.000003	0.000003
100	41	39	6	0.000023	0.000020
100	41	39	7	0.000143	0.000121
100	41	39	8	0.000730	0.000586
100	41	39	9	0.003028	0.002298
100	41	39	10	0.010382	0.007355
100	41	39	11	0.029772	0.019389
100	41	39	12	0.072186	0.042414
100	41	39	13	0.149598	0.077413
100	41	39	14	0.267994	0.118396
100	41	39	15	0.420218	0.152223
100	41	39	16	0.585127	0.164909
100	41	39	17	0.735878	0.150751
100	41	39	18	0.852247	0.116369
100	41	39	19	0.928099	0.075852
100	41	39	20	0.969818	0.041719
100	41	39	21	0.989151	0.019333
100	41	39	22	0.996683	0.007532
100	41	39	23	0.999143	0.002460
100	41	39	24	0.999814	0.000671
100	41	39	25	0.999966	0.000152
100	41	39	26	0.999995	0.000028
100	41	39	27	0.999999	0.000004
100	41	39	28	1.000000	0.000001
100	41	39	29	1.000000	0.000000
100	41	39	30	1.000000	0.000000
100	41	39	31	1.000000	0.000000
100	41	39	32	1.000000	0.000000
100	41	39	33	1.000000	0.000000
100	41	39	34	1.000000	0.000000
100	41	39	35	1.000000	0.000000
100	41	39	36	1.000000	0.000000
100	41	39	37	1.000000	0.000000
100	41	39	38	1.000000	0.000000
100	41	39	39	1.000000	0.000000
100	41	40	0	0.000000	0.000000
100	41	40	1	0.000000	0.000000
100	41	40	2	0.000000	0.000000
100	41	40	3	0.000000	0.000000
100	41	40	4	0.000000	0.000000
100	41	40	5	0.000001	0.000001

N	n	k	x	P(x)	p(x)
100	41	40	6	0.000011	0.000010
100	41	40	7	0.000076	0.000065
100	41	40	8	0.000412	0.000336
100	41	40	9	0.001822	0.001410
100	41	40	10	0.006645	0.004823
100	41	40	11	0.020236	0.013591
100	41	40	12	0.052022	0.031786
100	41	40	13	0.114065	0.062043
100	41	40	14	0.215590	0.101525
100	41	40	15	0.355336	0.139746
100	41	40	16	0.517541	0.162205
100	41	40	17	0.676566	0.159025
100	41	40	18	0.808370	0.131804
100	41	40	19	0.900743	0.092373
100	41	40	20	0.955456	0.054713
100	41	40	21	0.982812	0.027356
100	41	40	22	0.994337	0.011525
100	41	40	23	0.998417	0.004080
100	41	40	24	0.999627	0.001210
100	41	40	25	0.999926	0.000299
100	41	40	26	0.999988	0.000061
100	41	40	27	0.999998	0.000010
100	41	40	28	1.000000	0.000001
100	41	40	29	1.000000	0.000000
100	41	40	30	1.000000	0.000000
100	41	40	31	1.000000	0.000000
100	41	40	32	1.000000	0.000000
100	41	40	33	1.000000	0.000000
100	41	40	34	1.000000	0.000000
100	41	40	35	1.000000	0.000000
100	41	40	36	1.000000	0.000000
100	41	40	37	1.000000	0.000000
100	41	40	38	1.000000	0.000000
100	41	40	39	1.000000	0.000000
100	41	40	40	1.000000	0.000000
100	41	41	0	0.000000	0.000000
100	41	41	1	0.000000	0.000000
100	41	41	2	0.000000	0.000000
100	41	41	3	0.000000	0.000000
100	41	41	4	0.000000	0.000000
100	41	41	5	0.000001	0.000001
100	41	41	6	0.000005	0.000005
100	41	41	7	0.000039	0.000034
100	41	41	8	0.000227	0.000188
100	41	41	9	0.001070	0.000843
100	41	41	10	0.004153	0.003083
100	41	41	11	0.013440	0.009287
100	41	41	12	0.036659	0.023218
100	41	41	13	0.085111	0.048453
100	41	41	14	0.169903	0.084792

N	n	k	x	P(x)	p(x)
100	41	41	15	0.294779	0.124876
100	41	41	16	0.449956	0.155176
100	41	41	17	0.612956	0.163000
100	41	41	18	0.757845	0.144889
100	41	41	19	0.866873	0.109028
100	41	41	20	0.936306	0.069433
100	41	41	21	0.973694	0.037387
100	41	41	22	0.990688	0.016994
100	41	41	23	0.997193	0.006506
100	41	41	24	0.999285	0.002091
100	41	41	25	0.999847	0.000562
100	41	41	26	0.999973	0.000126
100	41	41	27	0.999996	0.000023
100	41	41	28	0.999999	0.000004
100	41	41	29	1.000000	0.000000
100	41	41	30	1.000000	0.000000
100	41	41	31	1.000000	0.000000
100	41	41	32	1.000000	0.000000
100	41	41	33	1.000000	0.000000
100	41	41	34	1.000000	0.000000
100	41	41	35	1.000000	0.000000
100	41	41	36	1.000000	0.000000
100	41	41	37	1.000000	0.000000
100	41	41	38	1.000000	0.000000
100	41	41	39	1.000000	0.000000
100	41	41	40	1.000000	0.000000
100	41	41	41	1.000000	0.000000
100	42	1	0	0.580000	0.580000
100	42	1	1	1.000000	0.420000
100	42	2	0	0.333939	0.333939
100	42	2	1	0.826061	0.492121
100	42	2	2	1.000000	0.173939
100	42	3	0	0.190822	0.190822
100	42	3	1	0.620173	0.429351
100	42	3	2	0.929004	0.308831
100	42	3	3	1.000000	0.070996
100	42	4	0	0.108198	0.108198
100	42	4	1	0.438695	0.330497
100	42	4	2	0.801651	0.362956
100	42	4	3	0.971455	0.169804
100	42	4	4	1.000000	0.028545
100	42	5	0	0.060862	0.060862
100	42	5	1	0.297545	0.236684
100	42	5	2	0.650419	0.352874
100	42	5	3	0.902472	0.252053
100	42	5	4	0.988701	0.086229
100	42	5	5	1.000000	0.011299
100	42	6	0	0.033954	0.033954
100	42	6	1	0.195398	0.161443
100	42	6	2	0.501841	0.306443

Table for $N = 2$, $n = 1$, through $N = 100$, $n = 50$

N	n	k	x	P(x)	p(x)
100	42	6	3	0.798998	0.297157
100	42	6	4	0.954210	0.155212
100	42	6	5	0.995599	0.041390
100	42	6	6	1.000000	0.004401
100	42	7	0	0.018783	0.018783
100	42	7	1	0.124981	0.106198
100	42	7	2	0.371440	0.246459
100	42	7	3	0.675709	0.304270
100	42	7	4	0.891464	0.215755
100	42	7	5	0.979308	0.087843
100	42	7	6	0.998315	0.019007
100	42	7	7	1.000000	0.001685
100	42	8	0	0.010300	0.010300
100	42	8	1	0.078163	0.067862
100	42	8	2	0.265436	0.187273
100	42	8	3	0.548112	0.282677
100	42	8	4	0.803307	0.255194
100	42	8	5	0.944359	0.141053
100	42	8	6	0.990957	0.046598
100	42	8	7	0.999366	0.008409
100	42	8	8	1.000000	0.000634
100	42	9	0	0.005598	0.005598
100	42	9	1	0.047920	0.042322
100	42	9	2	0.184013	0.136093
100	42	9	3	0.428282	0.244269
100	42	9	4	0.697900	0.269618
100	42	9	5	0.887632	0.189731
100	42	9	6	0.972723	0.085092
100	42	9	7	0.996167	0.023444
100	42	9	8	0.999766	0.003599
100	42	9	9	1.000000	0.000234
100	42	10	0	0.003014	0.003014
100	42	10	1	0.028852	0.025837
100	42	10	2	0.124192	0.095340
100	42	10	3	0.323595	0.199404
100	42	10	4	0.585312	0.261717
100	42	10	5	0.810488	0.225176
100	42	10	6	0.939060	0.128572
100	42	10	7	0.987150	0.048089
100	42	10	8	0.998421	0.011271
100	42	10	9	0.999915	0.001494
100	42	10	10	1.000000	0.000085
100	42	11	0	0.001608	0.001608
100	42	11	1	0.017081	0.015474
100	42	11	2	0.081818	0.064737
100	42	11	3	0.237187	0.155369
100	42	11	4	0.474810	0.237623
100	42	11	5	0.717916	0.243106
100	42	11	6	0.887631	0.169716
100	42	11	7	0.968448	0.080817

N	n	k	x	P(x)	p(x)
100	42	11	8	0.994163	0.025714
100	42	11	9	0.999367	0.005204
100	42	11	10	0.999970	0.000603
100	42	11	11	1.000000	0.000030
100	42	12	0	0.000849	0.000849
100	42	12	1	0.009953	0.009104
100	42	12	2	0.052723	0.042770
100	42	12	3	0.169104	0.116381
100	42	12	4	0.373353	0.204249
100	42	12	5	0.616849	0.243496
100	42	12	6	0.818982	0.202133
100	42	12	7	0.936667	0.117684
100	42	12	8	0.984339	0.047673
100	42	12	9	0.997437	0.013098
100	42	12	10	0.999753	0.002316
100	42	12	11	0.999989	0.000236
100	42	12	12	1.000000	0.000011
100	42	13	0	0.000444	0.000444
100	42	13	1	0.005711	0.005268
100	42	13	2	0.033282	0.027571
100	42	13	3	0.117526	0.084244
100	42	13	4	0.285154	0.167628
100	42	13	5	0.514470	0.229316
100	42	13	6	0.736292	0.221822
100	42	13	7	0.889860	0.153569
100	42	13	8	0.965920	0.076060
100	42	13	9	0.992526	0.026605
100	42	13	10	0.998911	0.006385
100	42	13	11	0.999906	0.000995
100	42	13	12	0.999996	0.000090
100	42	13	13	1.000000	0.000004
100	42	14	0	0.000230	0.000230
100	42	14	1	0.003229	0.002999
100	42	14	2	0.020606	0.017377
100	42	14	3	0.079762	0.059156
100	42	14	4	0.211937	0.132176
100	42	14	5	0.416945	0.205008
100	42	14	6	0.644503	0.227558
100	42	14	7	0.828080	0.183576
100	42	14	8	0.936196	0.108116
100	42	14	9	0.982434	0.046238
100	42	14	10	0.996562	0.014128
100	42	14	11	0.999551	0.002989
100	42	14	12	0.999965	0.000414
100	42	14	13	0.999999	0.000033
100	42	14	14	1.000000	0.000001
100	42	15	0	0.000117	0.000117
100	42	15	1	0.001799	0.001682
100	42	15	2	0.012524	0.010725
100	42	15	3	0.052935	0.040412

N	n	k	x	P(x)	p(x)
100	42	15	4	0.153534	0.100599
100	42	15	5	0.328744	0.175210
100	42	15	6	0.549246	0.220502
100	42	15	7	0.753368	0.204122
100	42	15	8	0.893452	0.140084
100	42	15	9	0.964691	0.071239
100	42	15	10	0.991305	0.026614
100	42	15	11	0.998474	0.007169
100	42	15	12	0.999821	0.001347
100	42	15	13	0.999987	0.000167
100	42	15	14	0.999999	0.000012
100	42	15	15	1.000000	0.000000
100	42	16	0	0.000059	0.000059
100	42	16	1	0.000988	0.000928
100	42	16	2	0.007477	0.006489
100	42	16	3	0.034393	0.026917
100	42	16	4	0.108560	0.074167
100	42	16	5	0.252476	0.143916
100	42	16	6	0.455857	0.203381
100	42	16	7	0.669318	0.213461
100	42	16	8	0.837419	0.168101
100	42	16	9	0.937034	0.099615
100	42	16	10	0.981286	0.044252
100	42	16	11	0.995860	0.014574
100	42	16	12	0.999346	0.003486
100	42	16	13	0.999931	0.000585
100	42	16	14	0.999995	0.000065
100	42	16	15	1.000000	0.000004
100	42	16	16	1.000000	0.000000
100	42	17	0	0.000030	0.000030
100	42	17	1	0.000535	0.000505
100	42	17	2	0.004387	0.003852
100	42	17	3	0.021896	0.017510
100	42	17	4	0.075009	0.053112
100	42	17	5	0.189085	0.114076
100	42	17	6	0.368694	0.179609
100	42	17	7	0.580376	0.211682
100	42	17	8	0.769378	0.189002
100	42	17	9	0.897899	0.128521
100	42	17	10	0.964428	0.066529
100	42	17	11	0.990481	0.026053
100	42	17	12	0.998101	0.007619
100	42	17	13	0.999729	0.001628
100	42	17	14	0.999974	0.000245
100	42	17	15	0.999998	0.000025
100	42	17	16	1.000000	0.000001
100	42	17	17	1.000000	0.000000
100	42	18	0	0.000015	0.000015
100	42	18	1	0.000285	0.000271
100	42	18	2	0.002530	0.002245

N	n	k	x	P(x)	p(x)
100	42	18	3	0.013669	0.011139
100	42	18	4	0.050692	0.037023
100	42	18	5	0.138232	0.087539
100	42	18	6	0.290791	0.152560
100	42	18	7	0.491113	0.200321
100	42	18	8	0.691956	0.200843
100	42	18	9	0.846801	0.154845
100	42	18	10	0.938778	0.091978
100	42	18	11	0.980750	0.041972
100	42	18	12	0.995347	0.014596
100	42	18	13	0.999160	0.003813
100	42	18	14	0.999891	0.000731
100	42	18	15	0.999990	0.000099
100	42	18	16	0.999999	0.000009
100	42	18	17	1.000000	0.000000
100	42	18	18	1.000000	0.000000
100	42	19	0	0.000007	0.000007
100	42	19	1	0.000150	0.000143
100	42	19	2	0.001435	0.001285
100	42	19	3	0.008371	0.006936
100	42	19	4	0.033535	0.025164
100	42	19	5	0.098732	0.065197
100	42	19	6	0.223814	0.125082
100	42	19	7	0.405609	0.181796
100	42	19	8	0.608679	0.203070
100	42	19	9	0.784485	0.175806
100	42	19	10	0.902885	0.118400
100	42	19	11	0.964883	0.061998
100	42	19	12	0.990006	0.025124
100	42	19	13	0.997811	0.007805
100	42	19	14	0.999641	0.001830
100	42	19	15	0.999958	0.000316
100	42	19	16	0.999996	0.000039
100	42	19	17	1.000000	0.000003
100	42	19	18	1.000000	0.000000
100	42	19	19	1.000000	0.000000
100	42	20	0	0.000003	0.000003
100	42	20	1	0.000078	0.000074
100	42	20	2	0.000801	0.000723
100	42	20	3	0.005032	0.004231
100	42	20	4	0.021730	0.016698
100	42	20	5	0.068951	0.047221
100	42	20	6	0.168222	0.099271
100	42	20	7	0.327056	0.158834
100	42	20	8	0.523440	0.196384
100	42	20	9	0.712860	0.189420
100	42	20	10	0.856109	0.143249
100	42	20	11	0.941155	0.085046
100	42	20	12	0.980702	0.039546
100	42	20	13	0.995017	0.014315

Table for $N = 2$, $n = 1$, through $N = 100$, $n = 50$

N	n	k	x	P(x)	p(x)
100	42	20	14	0.999009	0.003992
100	42	20	15	0.999852	0.000844
100	42	20	16	0.999984	0.000132
100	42	20	17	0.999999	0.000015
100	42	20	18	1.000000	0.000001
100	42	20	19	1.000000	0.000000
100	42	20	20	1.000000	0.000000
100	42	21	0	0.000002	0.000002
100	42	21	1	0.000040	0.000038
100	42	21	2	0.000439	0.000399
100	42	21	3	0.002969	0.002530
100	42	21	4	0.013798	0.010829
100	42	21	5	0.047111	0.033313
100	42	21	6	0.123550	0.076439
100	42	21	7	0.257566	0.134016
100	42	21	8	0.439977	0.182411
100	42	21	9	0.634724	0.194748
100	42	21	10	0.798810	0.164085
100	42	21	11	0.908200	0.109390
100	42	21	12	0.965872	0.057672
100	42	21	13	0.989828	0.023956
100	42	21	14	0.997612	0.007784
100	42	21	15	0.999568	0.001956
100	42	21	16	0.999941	0.000374
100	42	21	17	0.999994	0.000053
100	42	21	18	0.999999	0.000005
100	42	21	19	1.000000	0.000000
100	42	21	20	1.000000	0.000000
100	42	21	21	1.000000	0.000000
100	42	22	0	0.000001	0.000001
100	42	22	1	0.000020	0.000019
100	42	22	2	0.000237	0.000217
100	42	22	3	0.001720	0.001483
100	42	22	4	0.008589	0.006869
100	42	22	5	0.031509	0.022920
100	42	22	6	0.088717	0.057208
100	42	22	7	0.198192	0.109475
100	42	22	8	0.361471	0.163279
100	42	22	9	0.553374	0.191903
100	42	22	10	0.732345	0.178971
100	42	22	11	0.865275	0.132930
100	42	22	12	0.943971	0.078696
100	42	22	13	0.981034	0.037063
100	42	22	14	0.994853	0.013819
100	42	22	15	0.998899	0.004046
100	42	22	16	0.999818	0.000919
100	42	22	17	0.999977	0.000159
100	42	22	18	0.999998	0.000020
100	42	22	19	1.000000	0.000002
100	42	22	20	1.000000	0.000000
100	42	22	21	1.000000	0.000000
100	42	22	22	1.000000	0.000000
100	42	23	0	0.000000	0.000000
100	42	23	1	0.000010	0.000009
100	42	23	2	0.000126	0.000116
100	42	23	3	0.000978	0.000853
100	42	23	4	0.005243	0.004264
100	42	23	5	0.020637	0.015394
100	42	23	6	0.062313	0.041677
100	42	23	7	0.149068	0.086755
100	42	23	8	0.290298	0.141229
100	42	23	9	0.472184	0.181886
100	42	23	10	0.658921	0.186737
100	42	23	11	0.812444	0.153523
100	42	23	12	0.913703	0.101260
100	42	23	13	0.967254	0.053551
100	42	23	14	0.989892	0.022638
100	42	23	15	0.997498	0.007606
100	42	23	16	0.999512	0.002013
100	42	23	17	0.999926	0.000415
100	42	23	18	0.999992	0.000065
100	42	23	19	0.999999	0.000008
100	42	23	20	1.000000	0.000001
100	42	23	21	1.000000	0.000000
100	42	23	22	1.000000	0.000000
100	42	23	23	1.000000	0.000000
100	42	24	0	0.000000	0.000000
100	42	24	1	0.000005	0.000005
100	42	24	2	0.000065	0.000061
100	42	24	3	0.000546	0.000481
100	42	24	4	0.003138	0.002592
100	42	24	5	0.013240	0.010101
100	42	24	6	0.042828	0.029589
100	42	24	7	0.109634	0.066806
100	42	24	8	0.227937	0.118303
100	42	24	9	0.394233	0.166296
100	42	24	10	0.581316	0.187083
100	42	24	11	0.750636	0.169320
100	42	24	12	0.874251	0.123616
100	42	24	13	0.947085	0.072834
100	42	24	14	0.981660	0.034575
100	42	24	15	0.994831	0.013171
100	42	24	16	0.998832	0.004001
100	42	24	17	0.999792	0.000960
100	42	24	18	0.999971	0.000179
100	42	24	19	0.999997	0.000026
100	42	24	20	1.000000	0.000003
100	42	24	21	1.000000	0.000000
100	42	24	22	1.000000	0.000000
100	42	24	23	1.000000	0.000000
100	42	24	24	1.000000	0.000000
100	42	25	0	0.000000	0.000000
100	42	25	1	0.000002	0.000002
100	42	25	2	0.000034	0.000031
100	42	25	3	0.000300	0.000266
100	42	25	4	0.001842	0.001543
100	42	25	5	0.008322	0.006479
100	42	25	6	0.028813	0.020491
100	42	25	7	0.078868	0.050056
100	42	25	8	0.175012	0.096144
100	42	25	9	0.322025	0.147013
100	42	25	10	0.502544	0.180519
100	42	25	11	0.681571	0.179027
100	42	25	12	0.825456	0.143885
100	42	25	13	0.919294	0.093838
100	42	25	14	0.968922	0.049628
100	42	25	15	0.990152	0.021230
100	42	25	16	0.997463	0.007311
100	42	25	17	0.999476	0.002013
100	42	25	18	0.999915	0.000439
100	42	25	19	0.999989	0.000075
100	42	25	20	0.999999	0.000010
100	42	25	21	1.000000	0.000001
100	42	25	22	1.000000	0.000000
100	42	25	23	1.000000	0.000000
100	42	25	24	1.000000	0.000000
100	42	25	25	1.000000	0.000000
100	42	26	0	0.000000	0.000000
100	42	26	1	0.000001	0.000001
100	42	26	2	0.000017	0.000016
100	42	26	3	0.000161	0.000144
100	42	26	4	0.001061	0.000899
100	42	26	5	0.005125	0.004065
100	42	26	6	0.018977	0.013852
100	42	26	7	0.055509	0.036532
100	42	26	8	0.131427	0.075918
100	42	26	9	0.257340	0.125913
100	42	26	10	0.425523	0.168183
100	42	26	11	0.607573	0.182050
100	42	26	12	0.767902	0.160329
100	42	26	13	0.883010	0.115108
100	42	26	14	0.950394	0.067384
100	42	26	15	0.982509	0.032115
100	42	26	16	0.994929	0.012420
100	42	26	17	0.998805	0.003876
100	42	26	18	0.999774	0.000969
100	42	26	19	0.999966	0.000192
100	42	26	20	0.999996	0.000030
100	42	26	21	1.000000	0.000004
100	42	26	22	1.000000	0.000000
100	42	26	23	1.000000	0.000000
100	42	26	24	1.000000	0.000000
100	42	26	25	1.000000	0.000000
100	42	26	26	1.000000	0.000000
100	42	27	0	0.000000	0.000000
100	42	27	1	0.000000	0.000000
100	42	27	2	0.000008	0.000008
100	42	27	3	0.000085	0.000077
100	42	27	4	0.000599	0.000514
100	42	27	5	0.003093	0.002494
100	42	27	6	0.012238	0.009145
100	42	27	7	0.038230	0.025992
100	42	27	8	0.096546	0.058315
100	42	27	9	0.201189	0.104644
100	42	27	10	0.352795	0.151606
100	42	27	11	0.531309	0.178514
100	42	27	12	0.702904	0.171595
100	42	27	13	0.837900	0.134996
100	42	27	14	0.924897	0.086997
100	42	27	15	0.970791	0.045894
100	42	27	16	0.990565	0.019774
100	42	27	17	0.997496	0.006930
100	42	27	18	0.999460	0.001964
100	42	27	19	0.999907	0.000447
100	42	27	20	0.999987	0.000081
100	42	27	21	0.999998	0.000011
100	42	27	22	1.000000	0.000001
100	42	27	23	1.000000	0.000000
100	42	27	24	1.000000	0.000000
100	42	27	25	1.000000	0.000000
100	42	27	26	1.000000	0.000000
100	42	27	27	1.000000	0.000000
100	42	28	0	0.000000	0.000000
100	42	28	1	0.000000	0.000000
100	42	28	2	0.000004	0.000004
100	42	28	3	0.000044	0.000040
100	42	28	4	0.000331	0.000287
100	42	28	5	0.001829	0.001497
100	42	28	6	0.007728	0.005899
100	42	28	7	0.025768	0.018040
100	42	28	8	0.069385	0.043617
100	42	28	9	0.153885	0.084500
100	42	28	10	0.286338	0.132453
100	42	28	11	0.455502	0.169164
100	42	28	12	0.632385	0.176884
100	42	28	13	0.784271	0.151886
100	42	28	14	0.891528	0.107257
100	42	28	15	0.953817	0.062289
100	42	28	16	0.983522	0.029706
100	42	28	17	0.995122	0.011600

Table for $N = 2$, $n = 1$, through $N = 100$, $n = 50$

N	n	k	x	P(x)	p(x)
100	42	28	18	0.998814	0.003692
100	42	28	19	0.999766	0.000952
100	42	28	20	0.999963	0.000197
100	42	28	21	0.999995	0.000032
100	42	28	22	0.999999	0.000004
100	42	28	23	1.000000	0.000000
100	42	28	24	1.000000	0.000000
100	42	28	25	1.000000	0.000000
100	42	28	26	1.000000	0.000000
100	42	28	27	1.000000	0.000000
100	42	28	28	1.000000	0.000000
100	42	29	0	0.000000	0.000000
100	42	29	1	0.000000	0.000000
100	42	29	2	0.000002	0.000002
100	42	29	3	0.000022	0.000020
100	42	29	4	0.000180	0.000157
100	42	29	5	0.001059	0.000880
100	42	29	6	0.004779	0.003719
100	42	29	7	0.016999	0.012220
100	42	29	8	0.048788	0.031789
100	42	29	9	0.115156	0.066367
100	42	29	10	0.227470	0.112314
100	42	29	11	0.382667	0.155198
100	42	29	12	0.558684	0.176017
100	42	29	13	0.723095	0.164411
100	42	29	14	0.849817	0.126722
100	42	29	15	0.930459	0.080641
100	42	29	16	0.972795	0.042337
100	42	29	17	0.991094	0.018299
100	42	29	18	0.997583	0.006489
100	42	29	19	0.999462	0.001878
100	42	29	20	0.999903	0.000441
100	42	29	21	0.999986	0.000083
100	42	29	22	0.999998	0.000012
100	42	29	23	1.000000	0.000001
100	42	29	24	1.000000	0.000000
100	42	29	25	1.000000	0.000000
100	42	29	26	1.000000	0.000000
100	42	29	27	1.000000	0.000000
100	42	29	28	1.000000	0.000000
100	42	29	29	1.000000	0.000000
100	42	30	0	0.000000	0.000000
100	42	30	1	0.000000	0.000000
100	42	30	2	0.000001	0.000001
100	42	30	3	0.000011	0.000010
100	42	30	4	0.000096	0.000084
100	42	30	5	0.000601	0.000505
100	42	30	6	0.002893	0.002292
100	42	30	7	0.010975	0.008082
100	42	30	8	0.033565	0.022590

N	n	k	x	P(x)	p(x)
100	42	30	9	0.084309	0.050744
100	42	30	10	0.176849	0.092540
100	42	30	11	0.314905	0.138056
100	42	30	12	0.484311	0.169406
100	42	30	13	0.655941	0.171631
100	42	30	14	0.799842	0.143901
100	42	30	15	0.899792	0.099950
100	42	30	16	0.957292	0.057500
100	42	30	17	0.984651	0.027359
100	42	30	18	0.995390	0.010739
100	42	30	19	0.998853	0.003463
100	42	30	20	0.999766	0.000913
100	42	30	21	0.999961	0.000195
100	42	30	22	0.999995	0.000034
100	42	30	23	0.999999	0.000005
100	42	30	24	1.000000	0.000000
100	42	30	25	1.000000	0.000000
100	42	30	26	1.000000	0.000000
100	42	30	27	1.000000	0.000000
100	42	30	28	1.000000	0.000000
100	42	30	29	1.000000	0.000000
100	42	30	30	1.000000	0.000000
100	42	31	0	0.000000	0.000000
100	42	31	1	0.000000	0.000000
100	42	31	2	0.000000	0.000000
100	42	31	3	0.000005	0.000005
100	42	31	4	0.000050	0.000044
100	42	31	5	0.000334	0.000284
100	42	31	6	0.001714	0.001380
100	42	31	7	0.006934	0.005220
100	42	31	8	0.022593	0.015659
100	42	31	9	0.060387	0.037794
100	42	31	10	0.134545	0.074158
100	42	31	11	0.253766	0.119221
100	42	31	12	0.411708	0.157942
100	42	31	13	0.584837	0.173129
100	42	31	14	0.742282	0.157445
100	42	31	15	0.861240	0.118958
100	42	31	16	0.935935	0.074695
100	42	31	17	0.974880	0.038945
100	42	31	18	0.991708	0.016828
100	42	31	19	0.997715	0.006007
100	42	31	20	0.999479	0.001764
100	42	31	21	0.999903	0.000423
100	42	31	22	0.999985	0.000082
100	42	31	23	0.999998	0.000013
100	42	31	24	1.000000	0.000002
100	42	31	25	1.000000	0.000000
100	42	31	26	1.000000	0.000000
100	42	31	27	1.000000	0.000000

N	n	k	x	P(x)	p(x)
100	42	31	28	1.000000	0.000000
100	42	31	29	1.000000	0.000000
100	42	31	30	1.000000	0.000000
100	42	31	31	1.000000	0.000000
100	42	32	0	0.000000	0.000000
100	42	32	1	0.000000	0.000000
100	42	32	2	0.000000	0.000000
100	42	32	3	0.000003	0.000002
100	42	32	4	0.000025	0.000023
100	42	32	5	0.000181	0.000156
100	42	32	6	0.000994	0.000812
100	42	32	7	0.004286	0.003292
100	42	32	8	0.014877	0.010591
100	42	32	9	0.042311	0.027435
100	42	32	10	0.100153	0.057841
100	42	32	11	0.200203	0.100050
100	42	32	12	0.343038	0.142835
100	42	32	13	0.512073	0.169035
100	42	32	14	0.678391	0.166318
100	42	32	15	0.814691	0.136300
100	42	32	16	0.907789	0.093098
100	42	32	17	0.960769	0.052980
100	42	32	18	0.985855	0.025085
100	42	32	19	0.995713	0.009858
100	42	32	20	0.998917	0.003204
100	42	32	21	0.999774	0.000857
100	42	32	22	0.999961	0.000187
100	42	32	23	0.999995	0.000033
100	42	32	24	0.999999	0.000005
100	42	32	25	1.000000	0.000001
100	42	32	26	1.000000	0.000000
100	42	32	27	1.000000	0.000000
100	42	32	28	1.000000	0.000000
100	42	32	29	1.000000	0.000000
100	42	32	30	1.000000	0.000000
100	42	32	31	1.000000	0.000000
100	42	32	32	1.000000	0.000000
100	42	33	0	0.000000	0.000000
100	42	33	1	0.000000	0.000000
100	42	33	2	0.000000	0.000000
100	42	33	3	0.000001	0.000001
100	42	33	4	0.000013	0.000011
100	42	33	5	0.000097	0.000084
100	42	33	6	0.000564	0.000467
100	42	33	7	0.002592	0.002028
100	42	33	8	0.009581	0.006990
100	42	33	9	0.028998	0.019416
100	42	33	10	0.072933	0.043936
100	42	33	11	0.154592	0.081659
100	42	33	12	0.280022	0.125430

N	n	k	x	P(x)	p(x)
100	42	33	13	0.439984	0.159962
100	42	33	14	0.609907	0.169923
100	42	33	15	0.760572	0.150665
100	42	33	16	0.872193	0.111621
100	42	33	17	0.941291	0.069099
100	42	33	18	0.977001	0.035710
100	42	33	19	0.992379	0.015377
100	42	33	20	0.997880	0.005502
100	42	33	21	0.999509	0.001629
100	42	33	22	0.999906	0.000397
100	42	33	23	0.999985	0.000079
100	42	33	24	0.999998	0.000013
100	42	33	25	1.000000	0.000002
100	42	33	26	1.000000	0.000000
100	42	33	27	1.000000	0.000000
100	42	33	28	1.000000	0.000000
100	42	33	29	1.000000	0.000000
100	42	33	30	1.000000	0.000000
100	42	33	31	1.000000	0.000000
100	42	33	32	1.000000	0.000000
100	42	33	33	1.000000	0.000000
100	42	34	0	0.000000	0.000000
100	42	34	1	0.000000	0.000000
100	42	34	2	0.000000	0.000000
100	42	34	3	0.000001	0.000001
100	42	34	4	0.000006	0.000006
100	42	34	5	0.000050	0.000044
100	42	34	6	0.000313	0.000263
100	42	34	7	0.001532	0.001220
100	42	34	8	0.006034	0.004502
100	42	34	9	0.019434	0.013400
100	42	34	10	0.051949	0.032515
100	42	34	11	0.116810	0.064861
100	42	34	12	0.223859	0.107050
100	42	34	13	0.370747	0.146888
100	42	34	14	0.538895	0.168148
100	42	34	15	0.699856	0.160962
100	42	34	16	0.828877	0.129021
100	42	34	17	0.915508	0.086631
100	42	34	18	0.964210	0.048701
100	42	34	19	0.987100	0.022890
100	42	34	20	0.996074	0.008974
100	42	34	21	0.998999	0.002925
100	42	34	22	0.999788	0.000789
100	42	34	23	0.999963	0.000175
100	42	34	24	0.999995	0.000032
100	42	34	25	0.999999	0.000005
100	42	34	26	1.000000	0.000001
100	42	34	27	1.000000	0.000000
100	42	34	28	1.000000	0.000000

Table for $N = 2$, $n = 1$, through $N = 100$, $n = 50$

N	n	k	x	P(x)	p(x)
100	42	34	29	1.000000	0.000000
100	42	34	30	1.000000	0.000000
100	42	34	31	1.000000	0.000000
100	42	34	32	1.000000	0.000000
100	42	34	33	1.000000	0.000000
100	42	34	34	1.000000	0.000000
100	42	35	0	0.000000	0.000000
100	42	35	1	0.000000	0.000000
100	42	35	2	0.000000	0.000000
100	42	35	3	0.000000	0.000000
100	42	35	4	0.000003	0.000003
100	42	35	5	0.000026	0.000023
100	42	35	6	0.000170	0.000144
100	42	35	7	0.000886	0.000716
100	42	35	8	0.003715	0.002830
100	42	35	9	0.012734	0.009019
100	42	35	10	0.036184	0.023450
100	42	35	11	0.086345	0.050160
100	42	35	12	0.175200	0.088856
100	42	35	13	0.306205	0.131005
100	42	35	14	0.467559	0.161354
100	42	35	15	0.634008	0.166449
100	42	35	16	0.778051	0.144043
100	42	35	17	0.882694	0.104643
100	42	35	18	0.946500	0.063806
100	42	35	19	0.979123	0.032623
100	42	35	20	0.993082	0.013960
100	42	35	21	0.998068	0.004986
100	42	35	22	0.999549	0.001481
100	42	35	23	0.999912	0.000364
100	42	35	24	0.999986	0.000074
100	42	35	25	0.999998	0.000012
100	42	35	26	1.000000	0.000002
100	42	35	27	1.000000	0.000000
100	42	35	28	1.000000	0.000000
100	42	35	29	1.000000	0.000000
100	42	35	30	1.000000	0.000000
100	42	35	31	1.000000	0.000000
100	42	35	32	1.000000	0.000000
100	42	35	33	1.000000	0.000000
100	42	35	34	1.000000	0.000000
100	42	35	35	1.000000	0.000000
100	42	36	0	0.000000	0.000000
100	42	36	1	0.000000	0.000000
100	42	36	2	0.000000	0.000000
100	42	36	3	0.000000	0.000000
100	42	36	4	0.000001	0.000001
100	42	36	5	0.000013	0.000011
100	42	36	6	0.000090	0.000077
100	42	36	7	0.000500	0.000410

N	n	k	x	P(x)	p(x)
100	42	36	8	0.002235	0.001735
100	42	36	9	0.008155	0.005920
100	42	36	10	0.024640	0.016484
100	42	36	11	0.062422	0.037782
100	42	36	12	0.134190	0.071768
100	42	36	13	0.247757	0.113567
100	42	36	14	0.398053	0.150296
100	42	36	15	0.564868	0.166815
100	42	36	16	0.720434	0.155566
100	42	36	17	0.842446	0.122012
100	42	36	18	0.922941	0.080494
100	42	36	19	0.967579	0.044639
100	42	36	20	0.988358	0.020778
100	42	36	21	0.996457	0.008100
100	42	36	22	0.999093	0.002636
100	42	36	23	0.999806	0.000713
100	42	36	24	0.999965	0.000160
100	42	36	25	0.999995	0.000029
100	42	36	26	0.999999	0.000004
100	42	36	27	1.000000	0.000001
100	42	36	28	1.000000	0.000000
100	42	36	29	1.000000	0.000000
100	42	36	30	1.000000	0.000000
100	42	36	31	1.000000	0.000000
100	42	36	32	1.000000	0.000000
100	42	36	33	1.000000	0.000000
100	42	36	34	1.000000	0.000000
100	42	36	35	1.000000	0.000000
100	42	36	36	1.000000	0.000000
100	42	37	0	0.000000	0.000000
100	42	37	1	0.000000	0.000000
100	42	37	2	0.000000	0.000000
100	42	37	3	0.000000	0.000000
100	42	37	4	0.000001	0.000001
100	42	37	5	0.000006	0.000006
100	42	37	6	0.000046	0.000040
100	42	37	7	0.000276	0.000229
100	42	37	8	0.001313	0.001038
100	42	37	9	0.005103	0.003789
100	42	37	10	0.016398	0.011295
100	42	37	11	0.044121	0.027724
100	42	37	12	0.100549	0.056428
100	42	37	13	0.196297	0.095748
100	42	37	14	0.332299	0.136001
100	42	37	15	0.494493	0.162194
100	42	37	16	0.657235	0.162742
100	42	37	17	0.794785	0.137550
100	42	37	18	0.892755	0.097970
100	42	37	19	0.951537	0.058782
100	42	37	20	0.981215	0.029678

N	n	k	x	P(x)	p(x)
100	42	37	21	0.993800	0.012584
100	42	37	22	0.998269	0.004470
100	42	37	23	0.999594	0.001325
100	42	37	24	0.999921	0.000326
100	42	37	25	0.999987	0.000066
100	42	37	26	0.999998	0.000011
100	42	37	27	1.000000	0.000002
100	42	37	28	1.000000	0.000000
100	42	37	29	1.000000	0.000000
100	42	37	30	1.000000	0.000000
100	42	37	31	1.000000	0.000000
100	42	37	32	1.000000	0.000000
100	42	37	33	1.000000	0.000000
100	42	37	34	1.000000	0.000000
100	42	37	35	1.000000	0.000000
100	42	37	36	1.000000	0.000000
100	42	37	37	1.000000	0.000000
100	42	38	0	0.000000	0.000000
100	42	38	1	0.000000	0.000000
100	42	38	2	0.000000	0.000000
100	42	38	3	0.000000	0.000000
100	42	38	4	0.000000	0.000000
100	42	38	5	0.000003	0.000003
100	42	38	6	0.000023	0.000020
100	42	38	7	0.000148	0.000125
100	42	38	8	0.000753	0.000605
100	42	38	9	0.003118	0.002364
100	42	38	10	0.010661	0.007543
100	42	38	11	0.030479	0.019819
100	42	38	12	0.073679	0.043199
100	42	38	13	0.152222	0.078544
100	42	38	14	0.271853	0.119631
100	42	38	15	0.424981	0.153128
100	42	38	16	0.590072	0.165091
100	42	38	17	0.740202	0.150130
100	42	38	18	0.855433	0.115231
100	42	38	19	0.930077	0.074644
100	42	38	20	0.970851	0.040774
100	42	38	21	0.989605	0.018753
100	42	38	22	0.996850	0.007246
100	42	38	23	0.999195	0.002344
100	42	38	24	0.999827	0.000633
100	42	38	25	0.999969	0.000142
100	42	38	26	0.999995	0.000026
100	42	38	27	0.999999	0.000004
100	42	38	28	1.000000	0.000000
100	42	38	29	1.000000	0.000000
100	42	38	30	1.000000	0.000000
100	42	38	31	1.000000	0.000000
100	42	38	32	1.000000	0.000000

N	n	k	x	P(x)	p(x)
100	42	38	33	1.000000	0.000000
100	42	38	34	1.000000	0.000000
100	42	38	35	1.000000	0.000000
100	42	38	36	1.000000	0.000000
100	42	38	37	1.000000	0.000000
100	42	38	38	1.000000	0.000000
100	42	39	0	0.000000	0.000000
100	42	39	1	0.000000	0.000000
100	42	39	2	0.000000	0.000000
100	42	39	3	0.000000	0.000000
100	42	39	4	0.000000	0.000000
100	42	39	5	0.000001	0.000001
100	42	39	6	0.000011	0.000010
100	42	39	7	0.000078	0.000066
100	42	39	8	0.000422	0.000344
100	42	39	9	0.001859	0.001438
100	42	39	10	0.006768	0.004908
100	42	39	11	0.020570	0.013802
100	42	39	12	0.052776	0.032206
100	42	39	13	0.115484	0.062708
100	42	39	14	0.217827	0.102342
100	42	39	15	0.358296	0.140470
100	42	39	16	0.520840	0.162544
100	42	39	17	0.679666	0.158825
100	42	39	18	0.810828	0.131162
100	42	39	19	0.902387	0.091559
100	42	39	20	0.956383	0.053996
100	42	39	21	0.983253	0.026870
100	42	39	22	0.994513	0.011260
100	42	39	23	0.998476	0.003963
100	42	39	24	0.999644	0.001167
100	42	39	25	0.999930	0.000287
100	42	39	26	0.999989	0.000058
100	42	39	27	0.999998	0.000010
100	42	39	28	1.000000	0.000001
100	42	39	29	1.000000	0.000000
100	42	39	30	1.000000	0.000000
100	42	39	31	1.000000	0.000000
100	42	39	32	1.000000	0.000000
100	42	39	33	1.000000	0.000000
100	42	39	34	1.000000	0.000000
100	42	39	35	1.000000	0.000000
100	42	39	36	1.000000	0.000000
100	42	39	37	1.000000	0.000000
100	42	39	38	1.000000	0.000000
100	42	39	39	1.000000	0.000000
100	42	40	0	0.000000	0.000000
100	42	40	1	0.000000	0.000000
100	42	40	2	0.000000	0.000000
100	42	40	3	0.000000	0.000000

N	n	k	x	P(x)	p(x)
100	42	40	4	0.000000	0.000000
100	42	40	5	0.000001	0.000001
100	42	40	6	0.000005	0.000005
100	42	40	7	0.000040	0.000034
100	42	40	8	0.000230	0.000190
100	42	40	9	0.001081	0.000852
100	42	40	10	0.004193	0.003111
100	42	40	11	0.013556	0.009363
100	42	40	12	0.036937	0.023381
100	42	40	13	0.085672	0.048735
100	42	40	14	0.170850	0.085178
100	42	40	15	0.296121	0.125271
100	42	40	16	0.451559	0.155438
100	42	40	17	0.614573	0.163014
100	42	40	18	0.759223	0.144650
100	42	40	19	0.867864	0.108642
100	42	40	20	0.936909	0.069045
100	42	40	21	0.974003	0.037094
100	42	40	22	0.990821	0.016819
100	42	40	23	0.997242	0.006421
100	42	40	24	0.999299	0.002057
100	42	40	25	0.999850	0.000551
100	42	40	26	0.999973	0.000123
100	42	40	27	0.999996	0.000023
100	42	40	28	0.999999	0.000003
100	42	40	29	1.000000	0.000000
100	42	40	30	1.000000	0.000000
100	42	40	31	1.000000	0.000000
100	42	40	32	1.000000	0.000000
100	42	40	33	1.000000	0.000000
100	42	40	34	1.000000	0.000000
100	42	40	35	1.000000	0.000000
100	42	40	36	1.000000	0.000000
100	42	40	37	1.000000	0.000000
100	42	40	38	1.000000	0.000000
100	42	40	39	1.000000	0.000000
100	42	40	40	1.000000	0.000000
100	42	41	0	0.000000	0.000000
100	42	41	1	0.000000	0.000000
100	42	41	2	0.000000	0.000000
100	42	41	3	0.000000	0.000000
100	42	41	4	0.000000	0.000000
100	42	41	5	0.000000	0.000000
100	42	41	6	0.000003	0.000002
100	42	41	7	0.000020	0.000017
100	42	41	8	0.000122	0.000102
100	42	41	9	0.000613	0.000491
100	42	41	10	0.002533	0.001920
100	42	41	11	0.008718	0.006185
100	42	41	12	0.025246	0.016528

N	n	k	x	P(x)	p(x)
100	42	41	13	0.062117	0.036870
100	42	41	14	0.131100	0.068983
100	42	41	15	0.239749	0.108649
100	42	41	16	0.384203	0.144454
100	42	41	17	0.546651	0.162448
100	42	41	18	0.701363	0.154712
100	42	41	19	0.826219	0.124856
100	42	41	20	0.911593	0.085374
100	42	41	21	0.961020	0.049427
100	42	41	22	0.985215	0.024195
100	42	41	23	0.995209	0.009994
100	42	41	24	0.998682	0.003473
100	42	41	25	0.999694	0.001012
100	42	41	26	0.999941	0.000246
100	42	41	27	0.999990	0.000050
100	42	41	28	0.999999	0.000008
100	42	41	29	1.000000	0.000001
100	42	41	30	1.000000	0.000000
100	42	41	31	1.000000	0.000000
100	42	41	32	1.000000	0.000000
100	42	41	33	1.000000	0.000000
100	42	41	34	1.000000	0.000000
100	42	41	35	1.000000	0.000000
100	42	41	36	1.000000	0.000000
100	42	41	37	1.000000	0.000000
100	42	41	38	1.000000	0.000000
100	42	41	39	1.000000	0.000000
100	42	41	40	1.000000	0.000000
100	42	41	41	1.000000	0.000000
100	42	42	0	0.000000	0.000000
100	42	42	1	0.000000	0.000000
100	42	42	2	0.000000	0.000000
100	42	42	3	0.000000	0.000000
100	42	42	4	0.000000	0.000000
100	42	42	5	0.000000	0.000000
100	42	42	6	0.000001	0.000001
100	42	42	7	0.000010	0.000008
100	42	42	8	0.000063	0.000054
100	42	42	9	0.000339	0.000275
100	42	42	10	0.001492	0.001153
100	42	42	11	0.005469	0.003977
100	42	42	12	0.016842	0.011374
100	42	42	13	0.043994	0.027152
100	42	42	14	0.098362	0.054368
100	42	42	15	0.190028	0.091666
100	42	42	16	0.320545	0.130517
100	42	42	17	0.477817	0.157272
100	42	42	18	0.638429	0.160612
100	42	42	19	0.777546	0.139117
100	42	42	20	0.879758	0.102212

N	n	k	x	P(x)	p(x)
100	42	42	21	0.943427	0.063669
100	42	42	22	0.977013	0.033586
100	42	42	23	0.991990	0.014977
100	42	42	24	0.997622	0.005632
100	42	42	25	0.999403	0.001780
100	42	42	26	0.999874	0.000471
100	42	42	27	0.999978	0.000104
100	42	42	28	0.999997	0.000019
100	42	42	29	0.999999	0.000003
100	42	42	30	1.000000	0.000000
100	42	42	31	1.000000	0.000000
100	42	42	32	1.000000	0.000000
100	42	42	33	1.000000	0.000000
100	42	42	34	1.000000	0.000000
100	42	42	35	1.000000	0.000000
100	42	42	36	1.000000	0.000000
100	42	42	37	1.000000	0.000000
100	42	42	38	1.000000	0.000000
100	42	42	39	1.000000	0.000000
100	42	42	40	1.000000	0.000000
100	42	42	41	1.000000	0.000000
100	42	42	42	1.000000	0.000000
100	43	1	0	0.570000	0.570000
100	43	1	1	1.000000	0.430000
100	43	2	0	0.322424	0.322424
100	43	2	1	0.817576	0.495151
100	43	2	2	1.000000	0.182424
100	43	3	0	0.180952	0.180952
100	43	3	1	0.605368	0.424416
100	43	3	2	0.923680	0.318312
100	43	3	3	1.000000	0.076320
100	43	4	0	0.100736	0.100736
100	43	4	1	0.421600	0.320864
100	43	4	2	0.789135	0.367535
100	43	4	3	0.968528	0.179392
100	43	4	4	1.000000	0.031472
100	43	5	0	0.055615	0.055615
100	43	5	1	0.281222	0.225608
100	43	5	2	0.632167	0.350945
100	43	5	3	0.893781	0.261614
100	43	5	4	0.987214	0.093433
100	43	5	5	1.000000	0.012786
100	43	6	0	0.030442	0.030442
100	43	6	1	0.181480	0.151038
100	43	6	2	0.480707	0.299227
100	43	6	3	0.783628	0.302921
100	43	6	4	0.948857	0.165230
100	43	6	5	0.994886	0.046028
100	43	6	6	1.000000	0.005114
100	43	7	0	0.016516	0.016516

N	n	k	x	P(x)	p(x)
100	43	7	1	0.113995	0.097479
100	43	7	2	0.350193	0.236198
100	43	7	3	0.654725	0.304532
100	43	7	4	0.880305	0.225579
100	43	7	5	0.976279	0.095974
100	43	7	6	0.997987	0.021708
100	43	7	7	1.000000	0.002013
100	43	8	0	0.008880	0.008880
100	43	8	1	0.069972	0.061093
100	43	8	2	0.246063	0.176090
100	43	8	3	0.523744	0.277681
100	43	8	4	0.785707	0.261963
100	43	8	5	0.937063	0.151357
100	43	8	6	0.989350	0.052287
100	43	8	7	0.999221	0.009870
100	43	8	8	1.000000	0.000779
100	43	9	0	0.004729	0.004729
100	43	9	1	0.042082	0.037353
100	43	9	2	0.167588	0.125505
100	43	9	3	0.403013	0.235425
100	43	9	4	0.674657	0.271644
100	43	9	5	0.874547	0.199889
100	43	9	6	0.968322	0.093775
100	43	9	7	0.995358	0.027036
100	43	9	8	0.999703	0.004345
100	43	9	9	1.000000	0.000296
100	43	10	0	0.002495	0.002495
100	43	10	1	0.024842	0.022348
100	43	10	2	0.111041	0.086199
100	43	10	3	0.299529	0.188488
100	43	10	4	0.558238	0.258709
100	43	10	5	0.791076	0.232838
100	43	10	6	0.930193	0.139117
100	43	10	7	0.984663	0.054469
100	43	10	8	0.998032	0.013370
100	43	10	9	0.999889	0.001857
100	43	10	10	1.000000	0.000111
100	43	11	0	0.001303	0.001303
100	43	11	1	0.014413	0.013111
100	43	11	2	0.071773	0.057359
100	43	11	3	0.215757	0.143984
100	43	11	4	0.446131	0.230374
100	43	11	5	0.692767	0.246636
100	43	11	6	0.873001	0.180234
100	43	11	7	0.962875	0.089874
100	43	11	8	0.992833	0.029958
100	43	11	9	0.999188	0.006355
100	43	11	10	0.999959	0.000772
100	43	11	11	1.000000	0.000041
100	43	12	0	0.000673	0.000673

N	n	k	x	P(x)	p(x)
100	43	12	1	0.008226	0.007553
100	43	12	2	0.045349	0.037122
100	43	12	3	0.151045	0.105696
100	43	12	4	0.345181	0.194136
100	43	12	5	0.587462	0.242281
100	43	12	6	0.798072	0.210610
100	43	12	7	0.926521	0.128449
100	43	12	8	0.981052	0.054530
100	43	12	9	0.996760	0.015708
100	43	12	10	0.999673	0.002913
100	43	12	11	0.999985	0.000312
100	43	12	12	1.000000	0.000015
100	43	13	0	0.000344	0.000344
100	43	13	1	0.004622	0.004277
100	43	13	2	0.028053	0.023432
100	43	13	3	0.103001	0.074948
100	43	13	4	0.259143	0.156142
100	43	13	5	0.482840	0.223697
100	43	13	6	0.709520	0.226680
100	43	13	7	0.873974	0.164454
100	43	13	8	0.959363	0.085390
100	43	13	9	0.990691	0.031327
100	43	13	10	0.998581	0.007890
100	43	13	11	0.999872	0.001291
100	43	13	12	0.999995	0.000123
100	43	13	13	1.000000	0.000005
100	43	14	0	0.000174	0.000174
100	43	14	1	0.002557	0.002383
100	43	14	2	0.017011	0.014454
100	43	14	3	0.068542	0.051532
100	43	14	4	0.189148	0.120606
100	43	14	5	0.385133	0.195985
100	43	14	6	0.613116	0.227982
100	43	14	7	0.805924	0.192808
100	43	14	8	0.925011	0.119087
100	43	14	9	0.978448	0.053437
100	43	14	10	0.995588	0.017140
100	43	14	11	0.999397	0.003809
100	43	14	12	0.999951	0.000554
100	43	14	13	0.999998	0.000047
100	43	14	14	1.000000	0.000002
100	43	15	0	0.000087	0.000087
100	43	15	1	0.001393	0.001306
100	43	15	2	0.010120	0.008727
100	43	15	3	0.044574	0.034454
100	43	15	4	0.134455	0.089881
100	43	15	5	0.298535	0.164080
100	43	15	6	0.515030	0.216495
100	43	15	7	0.725214	0.210183
100	43	15	8	0.876546	0.151332

N	n	k	x	P(x)	p(x)
100	43	15	9	0.957322	0.080776
100	43	15	10	0.989011	0.031689
100	43	15	11	0.997980	0.008969
100	43	15	12	0.999751	0.001772
100	43	15	13	0.999982	0.000230
100	43	15	14	0.999999	0.000018
100	43	15	15	1.000000	0.000001
100	43	16	0	0.000043	0.000043
100	43	16	1	0.000748	0.000705
100	43	16	2	0.005910	0.005163
100	43	16	3	0.028360	0.022450
100	43	16	4	0.093216	0.064855
100	43	16	5	0.225182	0.131966
100	43	16	6	0.420791	0.195610
100	43	16	7	0.636195	0.215403
100	43	16	8	0.814232	0.178038
100	43	16	9	0.925011	0.110779
100	43	16	10	0.976708	0.051697
100	43	16	11	0.994603	0.017895
100	43	16	12	0.999105	0.004502
100	43	16	13	0.999900	0.000795
100	43	16	14	0.999993	0.000093
100	43	16	15	1.000000	0.000006
100	43	16	16	1.000000	0.000000
100	43	17	0	0.000021	0.000021
100	43	17	1	0.000395	0.000374
100	43	17	2	0.003391	0.002995
100	43	17	3	0.017670	0.014279
100	43	17	4	0.063104	0.045434
100	43	17	5	0.165483	0.102379
100	43	17	6	0.334630	0.169147
100	43	17	7	0.543879	0.209249
100	43	17	8	0.740050	0.196171
100	43	17	9	0.880172	0.140122
100	43	17	10	0.956399	0.076226
100	43	17	11	0.987786	0.031387
100	43	17	12	0.997444	0.009658
100	43	17	13	0.999616	0.002173
100	43	17	14	0.999961	0.000345
100	43	17	15	0.999998	0.000036
100	43	17	16	1.000000	0.000002
100	43	17	17	1.000000	0.000000
100	43	18	0	0.000010	0.000010
100	43	18	1	0.000206	0.000196
100	43	18	2	0.001911	0.001705
100	43	18	3	0.010788	0.008877
100	43	18	4	0.041756	0.030967
100	43	18	5	0.118611	0.076855
100	43	18	6	0.259227	0.140616
100	43	18	7	0.453120	0.193893

N	n	k	x	P(x)	p(x)
100	43	18	8	0.657327	0.204207
100	43	18	9	0.822773	0.165445
100	43	18	10	0.926092	0.103319
100	43	18	11	0.975685	0.049593
100	43	18	12	0.993837	0.018152
100	43	18	13	0.998831	0.004994
100	43	18	14	0.999841	0.001010
100	43	18	15	0.999985	0.000145
100	43	18	16	0.999999	0.000014
100	43	18	17	1.000000	0.000001
100	43	18	18	1.000000	0.000000
100	43	19	0	0.000005	0.000005
100	43	19	1	0.000106	0.000101
100	43	19	2	0.001058	0.000953
100	43	19	3	0.006458	0.005399
100	43	19	4	0.027027	0.020569
100	43	19	5	0.082995	0.055968
100	43	19	6	0.195778	0.112783
100	43	19	7	0.367997	0.172218
100	43	19	8	0.570166	0.202169
100	43	19	9	0.754174	0.184008
100	43	19	10	0.884512	0.130339
100	43	19	11	0.956332	0.071819
100	43	19	12	0.986974	0.030643
100	43	19	13	0.997004	0.010029
100	43	19	14	0.999484	0.002480
100	43	19	15	0.999936	0.000452
100	43	19	16	0.999995	0.000059
100	43	19	17	1.000000	0.000005
100	43	19	18	1.000000	0.000000
100	43	19	19	1.000000	0.000000
100	43	20	0	0.000002	0.000002
100	43	20	1	0.000053	0.000051
100	43	20	2	0.000576	0.000523
100	43	20	3	0.003792	0.003215
100	43	20	4	0.017123	0.013332
100	43	20	5	0.056738	0.039615
100	43	20	6	0.144260	0.087521
100	43	20	7	0.291455	0.147195
100	43	20	8	0.482809	0.191354
100	43	20	9	0.676936	0.194127
100	43	20	10	0.831411	0.154476
100	43	20	11	0.927958	0.096547
100	43	20	12	0.975247	0.047288
100	43	20	13	0.993289	0.018042
100	43	20	14	0.998596	0.005307
100	43	20	15	0.999780	0.001184
100	43	20	16	0.999975	0.000195
100	43	20	17	0.999998	0.000023
100	43	20	18	1.000000	0.000002

N	n	k	x	P(x)	p(x)
100	43	20	19	1.000000	0.000000
100	43	20	20	1.000000	0.000000
100	43	21	0	0.000001	0.000001
100	43	21	1	0.000027	0.000025
100	43	21	2	0.000308	0.000282
100	43	21	3	0.002184	0.001876
100	43	21	4	0.010624	0.008440
100	43	21	5	0.037921	0.027297
100	43	21	6	0.103781	0.065860
100	43	21	7	0.225217	0.121436
100	43	21	8	0.399092	0.173874
100	43	21	9	0.594432	0.195340
100	43	21	10	0.767690	0.173258
100	43	21	11	0.889340	0.121649
100	43	21	12	0.956923	0.067583
100	43	21	13	0.986523	0.029601
100	43	21	14	0.996672	0.010149
100	43	21	15	0.999365	0.002693
100	43	21	16	0.999909	0.000544
100	43	21	17	0.999990	0.000081
100	43	21	18	0.999999	0.000009
100	43	21	19	1.000000	0.000001
100	43	21	20	1.000000	0.000000
100	43	21	21	1.000000	0.000000
100	43	22	0	0.000000	0.000000
100	43	22	1	0.000013	0.000013
100	43	22	2	0.000162	0.000149
100	43	22	3	0.001234	0.001072
100	43	22	4	0.006457	0.005223
100	43	22	5	0.024791	0.018334
100	43	22	6	0.072935	0.048144
100	43	22	7	0.169879	0.096944
100	43	22	8	0.322059	0.152179
100	43	22	9	0.510361	0.188303
100	43	22	10	0.695316	0.184955
100	43	22	11	0.840064	0.144747
100	43	22	12	0.930403	0.090339
100	43	22	13	0.975283	0.044880
100	43	22	14	0.992947	0.017664
100	43	22	15	0.998411	0.005464
100	43	22	16	0.999723	0.001312
100	43	22	17	0.999964	0.000241
100	43	22	18	0.999996	0.000033
100	43	22	19	1.000000	0.000003
100	43	22	20	1.000000	0.000000
100	43	22	21	1.000000	0.000000
100	43	22	22	1.000000	0.000000
100	43	23	0	0.000000	0.000000
100	43	23	1	0.000006	0.000006
100	43	23	2	0.000084	0.000077

Table for $N = 2$, $n = 1$, through $N = 100$, $n = 50$

N	n	k	x	P(x)	p(x)
100	43	23	3	0.000684	0.000601
100	43	23	4	0.003846	0.003161
100	43	23	5	0.015859	0.012013
100	43	23	6	0.050098	0.034238
100	43	23	7	0.125136	0.075038
100	43	23	8	0.253773	0.128637
100	43	23	9	0.428281	0.174508
100	43	23	10	0.617066	0.188786
100	43	23	11	0.780680	0.163614
100	43	23	12	0.894499	0.113819
100	43	23	13	0.958021	0.063522
100	43	23	14	0.986379	0.028358
100	43	23	15	0.996449	0.010070
100	43	23	16	0.999269	0.002820
100	43	23	17	0.999884	0.000615
100	43	23	18	0.999986	0.000102
100	43	23	19	0.999999	0.000013
100	43	23	20	1.000000	0.000001
100	43	23	21	1.000000	0.000000
100	43	23	22	1.000000	0.000000
100	43	23	23	1.000000	0.000000
100	43	24	0	0.000000	0.000000
100	43	24	1	0.000003	0.000003
100	43	24	2	0.000042	0.000039
100	43	24	3	0.000372	0.000330
100	43	24	4	0.002245	0.001872
100	43	24	5	0.009931	0.007686
100	43	24	6	0.033645	0.023715
100	43	24	7	0.090053	0.056408
100	43	24	8	0.195302	0.105248
100	43	24	9	0.351225	0.155924
100	43	24	10	0.536158	0.184933
100	43	24	11	0.712685	0.176527
100	43	24	12	0.848676	0.135991
100	43	24	13	0.933272	0.084596
100	43	24	14	0.975699	0.042427
100	43	24	15	0.992787	0.017089
100	43	24	16	0.998280	0.005493
100	43	24	17	0.999676	0.001396
100	43	24	18	0.999953	0.000277
100	43	24	19	0.999995	0.000042
100	43	24	20	0.999999	0.000005
100	43	24	21	1.000000	0.000000
100	43	24	22	1.000000	0.000000
100	43	24	23	1.000000	0.000000
100	43	24	24	1.000000	0.000000
100	43	25	0	0.000000	0.000000
100	43	25	1	0.000001	0.000001
100	43	25	2	0.000021	0.000020
100	43	25	3	0.000199	0.000178
100	43	25	4	0.001284	0.001085
100	43	25	5	0.006088	0.004804
100	43	25	6	0.022100	0.016013
100	43	25	7	0.063334	0.041234
100	43	25	8	0.146832	0.083498
100	43	25	9	0.281470	0.134638
100	43	25	10	0.455858	0.174388
100	43	25	11	0.638358	0.182499
100	43	25	12	0.793206	0.154848
100	43	25	13	0.899879	0.106673
100	43	25	14	0.959510	0.059631
100	43	25	15	0.986492	0.026982
100	43	25	16	0.996329	0.009837
100	43	25	17	0.999199	0.002870
100	43	25	18	0.999862	0.000663
100	43	25	19	0.999982	0.000120
100	43	25	20	0.999998	0.000017
100	43	25	21	1.000000	0.000002
100	43	25	22	1.000000	0.000000
100	43	25	23	1.000000	0.000000
100	43	25	24	1.000000	0.000000
100	43	25	25	1.000000	0.000000
100	43	26	0	0.000000	0.000000
100	43	26	1	0.000001	0.000001
100	43	26	2	0.000010	0.000010
100	43	26	3	0.000104	0.000094
100	43	26	4	0.000720	0.000616
100	43	26	5	0.003654	0.002934
100	43	26	6	0.014201	0.010547
100	43	26	7	0.043542	0.029341
100	43	26	8	0.107866	0.064325
100	43	26	9	0.220434	0.112568
100	43	26	10	0.379127	0.158693
100	43	26	11	0.560491	0.181364
100	43	26	12	0.729202	0.168711
100	43	26	13	0.857210	0.128008
100	43	26	14	0.936452	0.079243
100	43	26	15	0.976418	0.039966
100	43	26	16	0.992787	0.016369
100	43	26	17	0.998204	0.005416
100	43	26	18	0.999641	0.001437
100	43	26	19	0.999943	0.000303
100	43	26	20	0.999993	0.000050
100	43	26	21	0.999999	0.000006
100	43	26	22	1.000000	0.000001
100	43	26	23	1.000000	0.000000
100	43	26	24	1.000000	0.000000
100	43	26	25	1.000000	0.000000
100	43	26	26	1.000000	0.000000
100	43	27	0	0.000000	0.000000
100	43	27	1	0.000000	0.000000
100	43	27	2	0.000005	0.000005
100	43	27	3	0.000053	0.000048
100	43	27	4	0.000395	0.000342
100	43	27	5	0.002147	0.001752
100	43	27	6	0.008927	0.006780
100	43	27	7	0.029268	0.020340
100	43	27	8	0.077442	0.048175
100	43	27	9	0.168714	0.091271
100	43	27	10	0.308359	0.139645
100	43	27	11	0.482064	0.173705
100	43	27	12	0.658526	0.176462
100	43	27	13	0.805315	0.146789
100	43	27	14	0.905398	0.100083
100	43	27	15	0.961296	0.055898
100	43	27	16	0.986815	0.025519
100	43	27	17	0.996301	0.009486
100	43	27	18	0.999155	0.002854
100	43	27	19	0.999845	0.000690
100	43	27	20	0.999977	0.000132
100	43	27	21	0.999997	0.000020
100	43	27	22	1.000000	0.000002
100	43	27	23	1.000000	0.000000
100	43	27	24	1.000000	0.000000
100	43	27	25	1.000000	0.000000
100	43	27	26	1.000000	0.000000
100	43	27	27	1.000000	0.000000
100	43	28	0	0.000000	0.000000
100	43	28	1	0.000000	0.000000
100	43	28	2	0.000002	0.000002
100	43	28	3	0.000027	0.000025
100	43	28	4	0.000213	0.000186
100	43	28	5	0.001235	0.001022
100	43	28	6	0.005491	0.004256
100	43	28	7	0.019237	0.013746
100	43	28	8	0.054345	0.035108
100	43	28	9	0.126204	0.071859
100	43	28	10	0.245232	0.119028
100	43	28	11	0.405919	0.160688
100	43	28	12	0.583590	0.177671
100	43	28	13	0.744990	0.161401
100	43	28	14	0.865639	0.120648
100	43	28	15	0.939856	0.074217
100	43	28	16	0.977376	0.037521
100	43	28	17	0.992922	0.015546
100	43	28	18	0.998178	0.005255
100	43	28	19	0.999618	0.001441
100	43	28	20	0.999936	0.000318
100	43	28	21	0.999991	0.000056
100	43	28	22	0.999999	0.000008
100	43	28	23	1.000000	0.000001
100	43	28	24	1.000000	0.000000
100	43	28	25	1.000000	0.000000
100	43	28	26	1.000000	0.000000
100	43	28	27	1.000000	0.000000
100	43	28	28	1.000000	0.000000
100	43	29	0	0.000000	0.000000
100	43	29	1	0.000000	0.000000
100	43	29	2	0.000001	0.000001
100	43	29	3	0.000013	0.000012
100	43	29	4	0.000112	0.000099
100	43	29	5	0.000695	0.000583
100	43	29	6	0.003304	0.002608
100	43	29	7	0.012364	0.009060
100	43	29	8	0.037278	0.024915
100	43	29	9	0.092270	0.054992
100	43	29	10	0.190677	0.098407
100	43	29	11	0.334502	0.143825
100	43	29	12	0.507093	0.172590
100	43	29	13	0.677740	0.170647
100	43	29	14	0.817044	0.139304
100	43	29	15	0.910994	0.093949
100	43	29	16	0.963306	0.052313
100	43	29	17	0.987308	0.024002
100	43	29	18	0.996353	0.009044
100	43	29	19	0.999138	0.002785
100	43	29	20	0.999834	0.000696
100	43	29	21	0.999974	0.000140
100	43	29	22	0.999997	0.000022
100	43	29	23	1.000000	0.000003
100	43	29	24	1.000000	0.000000
100	43	29	25	1.000000	0.000000
100	43	29	26	1.000000	0.000000
100	43	29	27	1.000000	0.000000
100	43	29	28	1.000000	0.000000
100	43	29	29	1.000000	0.000000
100	43	30	0	0.000000	0.000000
100	43	30	1	0.000000	0.000000
100	43	30	2	0.000000	0.000000
100	43	30	3	0.000006	0.000006
100	43	30	4	0.000058	0.000051
100	43	30	5	0.000383	0.000325
100	43	30	6	0.001944	0.001561
100	43	30	7	0.007770	0.005825
100	43	30	8	0.024996	0.017227
100	43	30	9	0.065936	0.040940
100	43	30	10	0.144939	0.079003
100	43	30	11	0.269680	0.124741
100	43	30	12	0.431736	0.162057
100	43	30	13	0.605636	0.173899

Table for $N = 2$, $n = 1$, through $N = 100$, $n = 50$

N	n	k	x	P(x)	p(x)
100	43	30	14	0.760145	0.154510
100	43	30	15	0.873943	0.113798
100	43	30	16	0.943413	0.069469
100	43	30	17	0.978519	0.035106
100	43	30	18	0.993168	0.014649
100	43	30	19	0.998196	0.005028
100	43	30	20	0.999609	0.001412
100	43	30	21	0.999931	0.000322
100	43	30	22	0.999990	0.000059
100	43	30	23	0.999999	0.000009
100	43	30	24	1.000000	0.000001
100	43	30	25	1.000000	0.000000
100	43	30	26	1.000000	0.000000
100	43	30	27	1.000000	0.000000
100	43	30	28	1.000000	0.000000
100	43	30	29	1.000000	0.000000
100	43	30	30	1.000000	0.000000
100	43	31	0	0.000000	0.000000
100	43	31	1	0.000000	0.000000
100	43	31	2	0.000000	0.000000
100	43	31	3	0.000003	0.000003
100	43	31	4	0.000029	0.000026
100	43	31	5	0.000206	0.000177
100	43	31	6	0.001119	0.000913
100	43	31	7	0.004774	0.003655
100	43	31	8	0.016383	0.011609
100	43	31	9	0.046051	0.029668
100	43	31	10	0.107695	0.061644
100	43	31	11	0.212655	0.104961
100	43	31	12	0.359968	0.147313
100	43	31	13	0.531107	0.171139
100	43	31	14	0.696134	0.165027
100	43	31	15	0.828424	0.132290
100	43	31	16	0.916617	0.088193
100	43	31	17	0.965479	0.048862
100	43	31	18	0.987936	0.022457
100	43	31	19	0.996473	0.008536
100	43	31	20	0.999145	0.002672
100	43	31	21	0.999830	0.000685
100	43	31	22	0.999972	0.000143
100	43	31	23	0.999996	0.000024
100	43	31	24	1.000000	0.000003
100	43	31	25	1.000000	0.000000
100	43	31	26	1.000000	0.000000
100	43	31	27	1.000000	0.000000
100	43	31	28	1.000000	0.000000
100	43	31	29	1.000000	0.000000
100	43	31	30	1.000000	0.000000
100	43	31	31	1.000000	0.000000
100	43	32	0	0.000000	0.000000

N	n	k	x	P(x)	p(x)
100	43	32	1	0.000000	0.000000
100	43	32	2	0.000000	0.000000
100	43	32	3	0.000001	0.000001
100	43	32	4	0.000014	0.000013
100	43	32	5	0.000109	0.000094
100	43	32	6	0.000630	0.000521
100	43	32	7	0.002867	0.002237
100	43	32	8	0.010494	0.007627
100	43	32	9	0.031432	0.020938
100	43	32	10	0.078213	0.046781
100	43	32	11	0.163978	0.085765
100	43	32	12	0.293784	0.129806
100	43	32	13	0.456699	0.162915
100	43	32	14	0.626775	0.170076
100	43	32	15	0.774741	0.147966
100	43	32	16	0.882107	0.107366
100	43	32	17	0.947068	0.064961
100	43	32	18	0.979800	0.032732
100	43	32	19	0.993503	0.013704
100	43	32	20	0.998254	0.004751
100	43	32	21	0.999611	0.001357
100	43	32	22	0.999929	0.000318
100	43	32	23	0.999989	0.000060
100	43	32	24	0.999999	0.000009
100	43	32	25	1.000000	0.000001
100	43	32	26	1.000000	0.000000
100	43	32	27	1.000000	0.000000
100	43	32	28	1.000000	0.000000
100	43	32	29	1.000000	0.000000
100	43	32	30	1.000000	0.000000
100	43	32	31	1.000000	0.000000
100	43	32	32	1.000000	0.000000
100	43	33	0	0.000000	0.000000
100	43	33	1	0.000000	0.000000
100	43	33	2	0.000000	0.000000
100	43	33	3	0.000001	0.000001
100	43	33	4	0.000007	0.000006
100	43	33	5	0.000056	0.000049
100	43	33	6	0.000346	0.000290
100	43	33	7	0.001683	0.001336
100	43	33	8	0.006569	0.004886
100	43	33	9	0.020963	0.014395
100	43	33	10	0.055511	0.034547
100	43	33	11	0.123618	0.068107
100	43	33	12	0.234608	0.110990
100	43	33	13	0.384825	0.150217
100	43	33	14	0.554243	0.169418
100	43	33	15	0.713814	0.159571
100	43	33	16	0.839476	0.125662
100	43	33	17	0.922230	0.082753

N	n	k	x	P(x)	p(x)
100	43	33	18	0.967766	0.045536
100	43	33	19	0.988667	0.020901
100	43	33	20	0.996647	0.007980
100	43	33	21	0.999172	0.002525
100	43	33	22	0.999831	0.000659
100	43	33	23	0.999972	0.000141
100	43	33	24	0.999996	0.000024
100	43	33	25	0.999999	0.000003
100	43	33	26	1.000000	0.000000
100	43	33	27	1.000000	0.000000
100	43	33	28	1.000000	0.000000
100	43	33	29	1.000000	0.000000
100	43	33	30	1.000000	0.000000
100	43	33	31	1.000000	0.000000
100	43	33	32	1.000000	0.000000
100	43	33	33	1.000000	0.000000
100	43	34	0	0.000000	0.000000
100	43	34	1	0.000000	0.000000
100	43	34	2	0.000000	0.000000
100	43	34	3	0.000000	0.000000
100	43	34	4	0.000003	0.000003
100	43	34	5	0.000028	0.000025
100	43	34	6	0.000186	0.000158
100	43	34	7	0.000965	0.000779
100	43	34	8	0.004016	0.003052
100	43	34	9	0.013659	0.009642
100	43	34	10	0.038495	0.024836
100	43	34	11	0.091089	0.052594
100	43	34	12	0.183254	0.092165
100	43	34	13	0.317564	0.134309
100	43	34	14	0.480913	0.163349
100	43	34	15	0.647128	0.166215
100	43	34	16	0.788836	0.141709
100	43	34	17	0.890116	0.101280
100	43	34	18	0.950775	0.060658
100	43	34	19	0.981180	0.030405
100	43	34	20	0.993908	0.012728
100	43	34	21	0.998343	0.004435
100	43	34	22	0.999624	0.001281
100	43	34	23	0.999930	0.000305
100	43	34	24	0.999989	0.000060
100	43	34	25	0.999998	0.000009
100	43	34	26	1.000000	0.000001
100	43	34	27	1.000000	0.000000
100	43	34	28	1.000000	0.000000
100	43	34	29	1.000000	0.000000
100	43	34	30	1.000000	0.000000
100	43	34	31	1.000000	0.000000
100	43	34	32	1.000000	0.000000
100	43	34	33	1.000000	0.000000

N	n	k	x	P(x)	p(x)
100	43	34	34	1.000000	0.000000
100	43	35	0	0.000000	0.000000
100	43	35	1	0.000000	0.000000
100	43	35	2	0.000000	0.000000
100	43	35	3	0.000000	0.000000
100	43	35	4	0.000001	0.000001
100	43	35	5	0.000014	0.000012
100	43	35	6	0.000098	0.000084
100	43	35	7	0.000540	0.000442
100	43	35	8	0.002398	0.001858
100	43	35	9	0.008691	0.006293
100	43	35	10	0.026077	0.017385
100	43	35	11	0.065589	0.039512
100	43	35	12	0.139965	0.074376
100	43	35	13	0.256514	0.116549
100	43	35	14	0.409138	0.152624
100	43	35	15	0.576612	0.167474
100	43	35	16	0.730865	0.154252
100	43	35	17	0.850218	0.119354
100	43	35	18	0.927798	0.077580
100	43	35	19	0.970123	0.042325
100	43	35	20	0.989472	0.019349
100	43	35	21	0.996865	0.007392
100	43	35	22	0.999217	0.002352
100	43	35	23	0.999837	0.000620
100	43	35	24	0.999972	0.000135
100	43	35	25	0.999996	0.000024
100	43	35	26	0.999999	0.000003
100	43	35	27	1.000000	0.000000
100	43	35	28	1.000000	0.000000
100	43	35	29	1.000000	0.000000
100	43	35	30	1.000000	0.000000
100	43	35	31	1.000000	0.000000
100	43	35	32	1.000000	0.000000
100	43	35	33	1.000000	0.000000
100	43	35	34	1.000000	0.000000
100	43	35	35	1.000000	0.000000
100	43	36	0	0.000000	0.000000
100	43	36	1	0.000000	0.000000
100	43	36	2	0.000000	0.000000
100	43	36	3	0.000000	0.000000
100	43	36	4	0.000001	0.000001
100	43	36	5	0.000007	0.000006
100	43	36	6	0.000050	0.000043
100	43	36	7	0.000295	0.000245
100	43	36	8	0.001397	0.001103
100	43	36	9	0.005399	0.004002
100	43	36	10	0.017250	0.011851
100	43	36	11	0.046137	0.028886
100	43	36	12	0.104493	0.058356

N	n	k	x	P(x)	p(x)
100	43	36	13	0.202722	0.098229
100	43	36	14	0.341044	0.138322
100	43	36	15	0.504469	0.163425
100	43	36	16	0.666791	0.162321
100	43	36	17	0.802477	0.135686
100	43	36	18	0.897960	0.095483
100	43	36	19	0.954496	0.056536
100	43	36	20	0.982626	0.028130
100	43	36	21	0.994363	0.011737
100	43	36	22	0.998457	0.004094
100	43	36	23	0.999646	0.001189
100	43	36	24	0.999933	0.000286
100	43	36	25	0.999989	0.000057
100	43	36	26	0.999999	0.000009
100	43	36	27	1.000000	0.000001
100	43	36	28	1.000000	0.000000
100	43	36	29	1.000000	0.000000
100	43	36	30	1.000000	0.000000
100	43	36	31	1.000000	0.000000
100	43	36	32	1.000000	0.000000
100	43	36	33	1.000000	0.000000
100	43	36	34	1.000000	0.000000
100	43	36	35	1.000000	0.000000
100	43	36	36	1.000000	0.000000
100	43	37	0	0.000000	0.000000
100	43	37	1	0.000000	0.000000
100	43	37	2	0.000000	0.000000
100	43	37	3	0.000000	0.000000
100	43	37	4	0.000000	0.000000
100	43	37	5	0.000003	0.000003
100	43	37	6	0.000025	0.000022
100	43	37	7	0.000157	0.000132
100	43	37	8	0.000794	0.000637
100	43	37	9	0.003273	0.002479
100	43	37	10	0.011140	0.007866
100	43	37	11	0.031693	0.020554
100	43	37	12	0.076227	0.044533
100	43	37	13	0.156677	0.080450
100	43	37	14	0.278367	0.121690
100	43	37	15	0.432971	0.154604
100	43	37	16	0.598311	0.165340
100	43	37	17	0.747354	0.149043
100	43	37	18	0.860662	0.113307
100	43	37	19	0.933295	0.072633
100	43	37	20	0.972516	0.039222
100	43	37	21	0.990328	0.017812
100	43	37	22	0.997113	0.006785
100	43	37	23	0.999274	0.002161
100	43	37	24	0.999848	0.000573
100	43	37	25	0.999973	0.000126

N	n	k	x	P(x)	p(x)
100	43	37	26	0.999996	0.000023
100	43	37	27	0.999999	0.000003
100	43	37	28	1.000000	0.000000
100	43	37	29	1.000000	0.000000
100	43	37	30	1.000000	0.000000
100	43	37	31	1.000000	0.000000
100	43	37	32	1.000000	0.000000
100	43	37	33	1.000000	0.000000
100	43	37	34	1.000000	0.000000
100	43	37	35	1.000000	0.000000
100	43	37	36	1.000000	0.000000
100	43	37	37	1.000000	0.000000
100	43	38	0	0.000000	0.000000
100	43	38	1	0.000000	0.000000
100	43	38	2	0.000000	0.000000
100	43	38	3	0.000000	0.000000
100	43	38	4	0.000000	0.000000
100	43	38	5	0.000001	0.000001
100	43	38	6	0.000012	0.000011
100	43	38	7	0.000082	0.000069
100	43	38	8	0.000440	0.000359
100	43	38	9	0.001936	0.001495
100	43	38	10	0.007019	0.005084
100	43	38	11	0.021253	0.014234
100	43	38	12	0.054313	0.033060
100	43	38	13	0.118367	0.064054
100	43	38	14	0.222351	0.103984
100	43	38	15	0.364258	0.141907
100	43	38	16	0.527451	0.163193
100	43	38	17	0.685844	0.158393
100	43	38	18	0.815698	0.129854
100	43	38	19	0.905625	0.089927
100	43	38	20	0.958197	0.052572
100	43	38	21	0.984108	0.025911
100	43	38	22	0.994852	0.010743
100	43	38	23	0.998588	0.003737
100	43	38	24	0.999675	0.001086
100	43	38	25	0.999937	0.000263
100	43	38	26	0.999990	0.000053
100	43	38	27	0.999999	0.000009
100	43	38	28	1.000000	0.000001
100	43	38	29	1.000000	0.000000
100	43	38	30	1.000000	0.000000
100	43	38	31	1.000000	0.000000
100	43	38	32	1.000000	0.000000
100	43	38	33	1.000000	0.000000
100	43	38	34	1.000000	0.000000
100	43	38	35	1.000000	0.000000
100	43	38	36	1.000000	0.000000
100	43	38	37	1.000000	0.000000

N	n	k	x	P(x)	p(x)
100	43	38	38	1.000000	0.000000
100	43	39	0	0.000000	0.000000
100	43	39	1	0.000000	0.000000
100	43	39	2	0.000000	0.000000
100	43	39	3	0.000000	0.000000
100	43	39	4	0.000000	0.000000
100	43	39	5	0.000001	0.000001
100	43	39	6	0.000006	0.000005
100	43	39	7	0.000041	0.000035
100	43	39	8	0.000238	0.000197
100	43	39	9	0.001116	0.000878
100	43	39	10	0.004313	0.003198
100	43	39	11	0.013907	0.009593
100	43	39	12	0.037783	0.023877
100	43	39	13	0.087374	0.049590
100	43	39	14	0.173713	0.086340
100	43	39	15	0.300171	0.126457
100	43	39	16	0.456383	0.156212
100	43	39	17	0.619421	0.163038
100	43	39	18	0.763338	0.143917
100	43	39	19	0.870815	0.107477
100	43	39	20	0.938695	0.067880
100	43	39	21	0.974914	0.036219
100	43	39	22	0.991213	0.016299
100	43	39	23	0.997383	0.006170
100	43	39	24	0.999342	0.001959
100	43	39	25	0.999861	0.000519
100	43	39	26	0.999976	0.000114
100	43	39	27	0.999996	0.000021
100	43	39	28	0.999999	0.000003
100	43	39	29	1.000000	0.000000
100	43	39	30	1.000000	0.000000
100	43	39	31	1.000000	0.000000
100	43	39	32	1.000000	0.000000
100	43	39	33	1.000000	0.000000
100	43	39	34	1.000000	0.000000
100	43	39	35	1.000000	0.000000
100	43	39	36	1.000000	0.000000
100	43	39	37	1.000000	0.000000
100	43	39	38	1.000000	0.000000
100	43	39	39	1.000000	0.000000
100	43	40	0	0.000000	0.000000
100	43	40	1	0.000000	0.000000
100	43	40	2	0.000000	0.000000
100	43	40	3	0.000000	0.000000
100	43	40	4	0.000000	0.000000
100	43	40	5	0.000000	0.000000
100	43	40	6	0.000003	0.000002
100	43	40	7	0.000020	0.000018
100	43	40	8	0.000125	0.000105

N	n	k	x	P(x)	p(x)
100	43	40	9	0.000626	0.000501
100	43	40	10	0.002583	0.001957
100	43	40	11	0.008874	0.006291
100	43	40	12	0.025649	0.016775
100	43	40	13	0.062985	0.037336
100	43	40	14	0.132667	0.069682
100	43	40	15	0.242125	0.109458
100	43	40	16	0.387240	0.145115
100	43	40	17	0.549930	0.162690
100	43	40	18	0.704356	0.154426
100	43	40	19	0.828529	0.124173
100	43	40	20	0.913101	0.084572
100	43	40	21	0.961851	0.048751
100	43	40	22	0.985602	0.023750
100	43	40	23	0.995360	0.009758
100	43	40	24	0.998732	0.003372
100	43	40	25	0.999708	0.000976
100	43	40	26	0.999944	0.000236
100	43	40	27	0.999991	0.000047
100	43	40	28	0.999999	0.000008
100	43	40	29	1.000000	0.000001
100	43	40	30	1.000000	0.000000
100	43	40	31	1.000000	0.000000
100	43	40	32	1.000000	0.000000
100	43	40	33	1.000000	0.000000
100	43	40	34	1.000000	0.000000
100	43	40	35	1.000000	0.000000
100	43	40	36	1.000000	0.000000
100	43	40	37	1.000000	0.000000
100	43	40	38	1.000000	0.000000
100	43	40	39	1.000000	0.000000
100	43	40	40	1.000000	0.000000
100	43	41	0	0.000000	0.000000
100	43	41	1	0.000000	0.000000
100	43	41	2	0.000000	0.000000
100	43	41	3	0.000000	0.000000
100	43	41	4	0.000000	0.000000
100	43	41	5	0.000000	0.000000
100	43	41	6	0.000001	0.000001
100	43	41	7	0.000010	0.000009
100	43	41	8	0.000064	0.000054
100	43	41	9	0.000342	0.000278
100	43	41	10	0.001507	0.001165
100	43	41	11	0.005519	0.004012
100	43	41	12	0.016982	0.011463
100	43	41	13	0.044317	0.027335
100	43	41	14	0.098987	0.054670
100	43	41	15	0.191044	0.092057
100	43	41	16	0.321938	0.130894
100	43	41	17	0.479431	0.157493

Table for $N = 2$, $n = 1$, through $N = 100$, $n = 50$

N	n	k	x	P(x)	p(x)
100	43	41	18	0.640012	0.160581
100	43	41	19	0.778860	0.138848
100	43	41	20	0.880681	0.101822
100	43	41	21	0.943976	0.063295
100	43	41	22	0.977289	0.033313
100	43	41	23	0.992107	0.014818
100	43	41	24	0.997664	0.005557
100	43	41	25	0.999415	0.001751
100	43	41	26	0.999877	0.000462
100	43	41	27	0.999978	0.000101
100	43	41	28	0.999997	0.000018
100	43	41	29	0.999999	0.000003
100	43	41	30	1.000000	0.000000
100	43	41	31	1.000000	0.000000
100	43	41	32	1.000000	0.000000
100	43	41	33	1.000000	0.000000
100	43	41	34	1.000000	0.000000
100	43	41	35	1.000000	0.000000
100	43	41	36	1.000000	0.000000
100	43	41	37	1.000000	0.000000
100	43	41	38	1.000000	0.000000
100	43	41	39	1.000000	0.000000
100	43	41	40	1.000000	0.000000
100	43	41	41	1.000000	0.000000
100	43	42	0	0.000000	0.000000
100	43	42	1	0.000000	0.000000
100	43	42	2	0.000000	0.000000
100	43	42	3	0.000000	0.000000
100	43	42	4	0.000000	0.000000
100	43	42	5	0.000000	0.000000
100	43	42	6	0.000000	0.000000
100	43	42	7	0.000004	0.000004
100	43	42	8	0.000032	0.000027
100	43	42	9	0.000182	0.000150
100	43	42	10	0.000856	0.000674
100	43	42	11	0.003343	0.002488
100	43	42	12	0.010959	0.007616
100	43	42	13	0.030418	0.019459
100	43	42	14	0.072115	0.041697
100	43	42	15	0.147356	0.075241
100	43	42	16	0.262038	0.114681
100	43	42	17	0.410027	0.147990
100	43	42	18	0.571969	0.161942
100	43	42	19	0.722379	0.150410
100	43	42	20	0.840988	0.118609
100	43	42	21	0.920375	0.079387
100	43	42	22	0.965432	0.045057
100	43	42	23	0.987084	0.021652
100	43	42	24	0.995874	0.008790
100	43	42	25	0.998881	0.003006

N	n	k	x	P(x)	p(x)
100	43	42	26	0.999744	0.000863
100	43	42	27	0.999951	0.000207
100	43	42	28	0.999992	0.000041
100	43	42	29	0.999999	0.000007
100	43	42	30	1.000000	0.000001
100	43	42	31	1.000000	0.000000
100	43	42	32	1.000000	0.000000
100	43	42	33	1.000000	0.000000
100	43	42	34	1.000000	0.000000
100	43	42	35	1.000000	0.000000
100	43	42	36	1.000000	0.000000
100	43	42	37	1.000000	0.000000
100	43	42	38	1.000000	0.000000
100	43	42	39	1.000000	0.000000
100	43	42	40	1.000000	0.000000
100	43	42	41	1.000000	0.000000
100	43	42	42	1.000000	0.000000
100	43	43	0	0.000000	0.000000
100	43	43	1	0.000000	0.000000
100	43	43	2	0.000000	0.000000
100	43	43	3	0.000000	0.000000
100	43	43	4	0.000000	0.000000
100	43	43	5	0.000000	0.000000
100	43	43	6	0.000000	0.000000
100	43	43	7	0.000002	0.000002
100	43	43	8	0.000015	0.000013
100	43	43	9	0.000094	0.000079
100	43	43	10	0.000472	0.000378
100	43	43	11	0.001971	0.001498
100	43	43	12	0.006889	0.004918
100	43	43	13	0.020353	0.013465
100	43	43	14	0.051267	0.030914
100	43	43	15	0.111033	0.059766
100	43	43	16	0.208651	0.097618
100	43	43	17	0.343687	0.135036
100	43	43	18	0.502166	0.158479
100	43	43	19	0.660140	0.157974
100	43	43	20	0.793953	0.133813
100	43	43	21	0.890262	0.096309
100	43	43	22	0.949118	0.058856
100	43	43	23	0.979618	0.030500
100	43	43	24	0.992995	0.013377
100	43	43	25	0.997948	0.004953
100	43	43	26	0.999491	0.001543
100	43	43	27	0.999894	0.000403
100	43	43	28	0.999981	0.000088
100	43	43	29	0.999997	0.000016
100	43	43	30	0.999999	0.000002
100	43	43	31	1.000000	0.000000
100	43	43	32	1.000000	0.000000

N	n	k	x	P(x)	p(x)
100	43	43	33	1.000000	0.000000
100	43	43	34	1.000000	0.000000
100	43	43	35	1.000000	0.000000
100	43	43	36	1.000000	0.000000
100	43	43	37	1.000000	0.000000
100	43	43	38	1.000000	0.000000
100	43	43	39	1.000000	0.000000
100	43	43	40	1.000000	0.000000
100	43	43	41	1.000000	0.000000
100	43	43	42	1.000000	0.000000
100	43	43	43	1.000000	0.000000
100	44	1	0	0.560000	0.560000
100	44	1	1	1.000000	0.440000
100	44	2	0	0.311111	0.311111
100	44	2	1	0.808889	0.497778
100	44	2	2	1.000000	0.191111
100	44	3	0	0.171429	0.171429
100	44	3	1	0.590476	0.419048
100	44	3	2	0.918095	0.327619
100	44	3	3	1.000000	0.081905
100	44	4	0	0.093667	0.093667
100	44	4	1	0.404713	0.311046
100	44	4	2	0.776240	0.371527
100	44	4	3	0.965380	0.189141
100	44	4	4	1.000000	0.034620
100	44	5	0	0.050736	0.050736
100	44	5	1	0.265390	0.214654
100	44	5	2	0.613697	0.348306
100	44	5	3	0.884601	0.270905
100	44	5	4	0.985575	0.100974
100	44	5	5	1.000000	0.014425
100	44	6	0	0.027237	0.027237
100	44	6	1	0.168231	0.140994
100	44	6	2	0.459709	0.291477
100	44	6	3	0.767685	0.307976
100	44	6	4	0.943060	0.175375
100	44	6	5	0.994078	0.051018
100	44	6	6	1.000000	0.005922
100	44	7	0	0.014488	0.014488
100	44	7	1	0.103734	0.089246
100	44	7	2	0.329474	0.225740
100	44	7	3	0.633355	0.303881
100	44	7	4	0.868432	0.235077
100	44	7	5	0.972911	0.104479
100	44	7	6	0.997606	0.024695
100	44	7	7	1.000000	0.002394
100	44	8	0	0.007633	0.007633
100	44	8	1	0.062470	0.054836
100	44	8	2	0.227527	0.165057
100	44	8	3	0.499386	0.271859

N	n	k	x	P(x)	p(x)
100	44	8	4	0.767324	0.267938
100	44	8	5	0.929097	0.161774
100	44	8	6	0.987516	0.058418
100	44	8	7	0.999048	0.011532
100	44	8	8	1.000000	0.000952
100	44	9	0	0.003983	0.003983
100	44	9	1	0.036840	0.032857
100	44	9	2	0.152175	0.115335
100	44	9	3	0.378231	0.226057
100	44	9	4	0.650829	0.272598
100	44	9	5	0.860519	0.209690
100	44	9	6	0.963386	0.102867
100	44	9	7	0.994410	0.031023
100	44	9	8	0.999627	0.005218
100	44	9	9	1.000000	0.000373
100	44	10	0	0.002057	0.002057
100	44	10	1	0.021314	0.019257
100	44	10	2	0.098943	0.077629
100	44	10	3	0.276382	0.177438
100	44	10	4	0.531006	0.254624
100	44	10	5	0.770652	0.239646
100	44	10	6	0.920431	0.149779
100	44	10	7	0.981796	0.061365
100	44	10	8	0.997563	0.015767
100	44	10	9	0.999856	0.002293
100	44	10	10	1.000000	0.000143
100	44	11	0	0.001051	0.001051
100	44	11	1	0.012113	0.011062
100	44	11	2	0.062716	0.050603
100	44	11	3	0.195549	0.132832
100	44	11	4	0.417839	0.222291
100	44	11	5	0.666805	0.248966
100	44	11	6	0.857191	0.190386
100	44	11	7	0.956568	0.099377
100	44	11	8	0.991256	0.034688
100	44	11	9	0.998965	0.007708
100	44	11	10	0.999946	0.000981
100	44	11	11	1.000000	0.000054
100	44	12	0	0.000532	0.000532
100	44	12	1	0.006769	0.006237
100	44	12	2	0.038836	0.032067
100	44	12	3	0.134356	0.095520
100	44	12	4	0.317933	0.183577
100	44	12	5	0.557708	0.239774
100	44	12	6	0.775903	0.218195
100	44	12	7	0.915254	0.139351
100	44	12	8	0.977225	0.061971
100	44	12	9	0.995933	0.018708
100	44	12	10	0.999571	0.003638
100	44	12	11	0.999980	0.000409

Table for $N = 2$, $n = 1$, through $N = 100$, $n = 50$

N	n	k	x	P(x)	p(x)
100	44	12	12	1.000000	0.000020
100	44	13	0	0.000266	0.000266
100	44	13	1	0.003721	0.003455
100	44	13	2	0.023531	0.019810
100	44	13	3	0.089852	0.066321
100	44	13	4	0.234489	0.144637
100	44	13	5	0.451444	0.216955
100	44	13	6	0.681682	0.230238
100	44	13	7	0.856663	0.174981
100	44	13	8	0.951873	0.095210
100	44	13	9	0.988492	0.036619
100	44	13	10	0.998165	0.009673
100	44	13	11	0.999826	0.001661
100	44	13	12	0.999993	0.000166
100	44	13	13	1.000000	0.000007
100	44	14	0	0.000131	0.000131
100	44	14	1	0.002013	0.001882
100	44	14	2	0.013968	0.011954
100	44	14	3	0.058598	0.044630
100	44	14	4	0.167990	0.109392
100	44	14	5	0.354188	0.186199
100	44	14	6	0.581118	0.226930
100	44	14	7	0.782246	0.201127
100	44	14	8	0.912476	0.130230
100	44	14	9	0.973761	0.061285
100	44	14	10	0.994385	0.020625
100	44	14	11	0.999196	0.004811
100	44	14	12	0.999931	0.000735
100	44	14	13	0.999997	0.000066
100	44	14	14	1.000000	0.000003
100	44	15	0	0.000064	0.000064
100	44	15	1	0.001072	0.001008
100	44	15	2	0.008130	0.007057
100	44	15	3	0.037321	0.029191
100	44	15	4	0.117110	0.079789
100	44	15	5	0.269749	0.152640
100	44	15	6	0.480847	0.211098
100	44	15	7	0.695714	0.214867
100	44	15	8	0.857961	0.162247
100	44	15	9	0.948819	0.090858
100	44	15	10	0.986231	0.037412
100	44	15	11	0.997350	0.011119
100	44	15	12	0.999658	0.002308
100	44	15	13	0.999974	0.000316
100	44	15	14	0.999999	0.000025
100	44	15	15	1.000000	0.000001
100	44	16	0	0.000031	0.000031
100	44	16	1	0.000562	0.000531
100	44	16	2	0.004642	0.004080
100	44	16	3	0.023240	0.018598

N	n	k	x	P(x)	p(x)
100	44	16	4	0.079562	0.056322
100	44	16	5	0.199715	0.120153
100	44	16	6	0.386474	0.186759
100	44	16	7	0.602184	0.215710
100	44	16	8	0.789245	0.187061
100	44	16	9	0.911407	0.122162
100	44	16	10	0.971266	0.059859
100	44	16	11	0.993033	0.021767
100	44	16	12	0.998789	0.005756
100	44	16	13	0.999858	0.001069
100	44	16	14	0.999990	0.000132
100	44	16	15	1.000000	0.000010
100	44	16	16	1.000000	0.000000
100	44	17	0	0.000015	0.000015
100	44	17	1	0.000290	0.000276
100	44	17	2	0.002602	0.002312
100	44	17	3	0.014163	0.011560
100	44	17	4	0.052742	0.038579
100	44	17	5	0.143929	0.091187
100	44	17	6	0.301988	0.158058
100	44	17	7	0.507169	0.205181
100	44	17	8	0.709076	0.201907
100	44	17	9	0.860506	0.151430
100	44	17	10	0.947038	0.086532
100	44	17	11	0.984482	0.037445
100	44	17	12	0.996597	0.012114
100	44	17	13	0.999464	0.002867
100	44	17	14	0.999943	0.000479
100	44	17	15	0.999996	0.000053
100	44	17	16	1.000000	0.000004
100	44	17	17	1.000000	0.000000
100	44	18	0	0.000007	0.000007
100	44	18	1	0.000148	0.000141
100	44	18	2	0.001432	0.001285
100	44	18	3	0.008452	0.007020
100	44	18	4	0.034150	0.025697
100	44	18	5	0.101082	0.066933
100	44	18	6	0.229624	0.128541
100	44	18	7	0.415702	0.186079
100	44	18	8	0.621502	0.205799
100	44	18	9	0.796650	0.175148
100	44	18	10	0.911591	0.114941
100	44	18	11	0.969595	0.058004
100	44	18	12	0.991926	0.022331
100	44	18	13	0.998393	0.006467
100	44	18	14	0.999770	0.001377
100	44	18	15	0.999978	0.000208
100	44	18	16	0.999999	0.000021
100	44	18	17	1.000000	0.000001
100	44	18	18	1.000000	0.000000

N	n	k	x	P(x)	p(x)
100	44	19	0	0.000003	0.000003
100	44	19	1	0.000074	0.000071
100	44	19	2	0.000774	0.000700
100	44	19	3	0.004942	0.004168
100	44	19	4	0.021614	0.016672
100	44	19	5	0.069248	0.047634
100	44	19	6	0.170056	0.100807
100	44	19	7	0.331740	0.161684
100	44	19	8	0.531151	0.199411
100	44	19	9	0.721892	0.190741
100	44	19	10	0.863932	0.142041
100	44	19	11	0.946252	0.082319
100	44	19	12	0.983211	0.036960
100	44	19	13	0.995948	0.012737
100	44	19	14	0.999266	0.003318
100	44	19	15	0.999904	0.000638
100	44	19	16	0.999991	0.000087
100	44	19	17	0.999999	0.000008
100	44	19	18	1.000000	0.000000
100	44	19	19	1.000000	0.000000
100	44	20	0	0.000001	0.000001
100	44	20	1	0.000036	0.000035
100	44	20	2	0.000411	0.000375
100	44	20	3	0.002833	0.002421
100	44	20	4	0.013381	0.010549
100	44	20	5	0.046313	0.032932
100	44	20	6	0.122763	0.076450
100	44	20	7	0.257884	0.135121
100	44	20	8	0.442524	0.184640
100	44	20	9	0.639473	0.196949
100	44	20	10	0.804310	0.164838
100	44	20	11	0.912714	0.108404
100	44	20	12	0.968610	0.055896
100	44	20	13	0.991073	0.022464
100	44	20	14	0.998037	0.006964
100	44	20	15	0.999676	0.001639
100	44	20	16	0.999961	0.000286
100	44	20	17	0.999997	0.000035
100	44	20	18	1.000000	0.000003
100	44	20	19	1.000000	0.000000
100	44	20	20	1.000000	0.000000
100	44	21	0	0.000001	0.000001
100	44	21	1	0.000018	0.000017
100	44	21	2	0.000214	0.000197
100	44	21	3	0.001592	0.001377
100	44	21	4	0.008107	0.006515
100	44	21	5	0.030259	0.022152
100	44	21	6	0.086450	0.056191
100	44	21	7	0.195391	0.108941
100	44	21	8	0.359436	0.164045

N	n	k	x	P(x)	p(x)
100	44	21	9	0.553308	0.193872
100	44	21	10	0.734254	0.180947
100	44	21	11	0.867998	0.133743
100	44	21	12	0.946252	0.078254
100	44	21	13	0.982369	0.036117
100	44	21	14	0.995426	0.013057
100	44	21	15	0.999082	0.003656
100	44	21	16	0.999861	0.000780
100	44	21	17	0.999985	0.000123
100	44	21	18	0.999999	0.000014
100	44	21	19	1.000000	0.000001
100	44	21	20	1.000000	0.000000
100	44	21	21	1.000000	0.000000
100	44	22	0	0.000000	0.000000
100	44	22	1	0.000008	0.000008
100	44	22	2	0.000110	0.000101
100	44	22	3	0.000877	0.000767
100	44	22	4	0.004808	0.003931
100	44	22	5	0.019323	0.014515
100	44	22	6	0.059421	0.040098
100	44	22	7	0.144368	0.084947
100	44	22	8	0.284681	0.140313
100	44	22	9	0.467415	0.182734
100	44	22	10	0.656378	0.188963
100	44	22	11	0.812130	0.155752
100	44	22	12	0.914554	0.102424
100	44	22	13	0.968196	0.053642
100	44	22	14	0.990467	0.022271
100	44	22	15	0.997740	0.007272
100	44	22	16	0.999585	0.001845
100	44	22	17	0.999943	0.000358
100	44	22	18	0.999994	0.000052
100	44	22	19	0.999999	0.000005
100	44	22	20	1.000000	0.000000
100	44	22	21	1.000000	0.000000
100	44	22	22	1.000000	0.000000
100	44	23	0	0.000000	0.000000
100	44	23	1	0.000004	0.000004
100	44	23	2	0.000055	0.000051
100	44	23	3	0.000474	0.000418
100	44	23	4	0.002792	0.002318
100	44	23	5	0.012066	0.009274
100	44	23	6	0.039886	0.027821
100	44	23	7	0.104073	0.064186
100	44	23	8	0.219921	0.115849
100	44	23	9	0.385419	0.165498
100	44	23	10	0.574010	0.188591
100	44	23	11	0.746235	0.172225
100	44	23	12	0.872534	0.126299
100	44	23	13	0.946877	0.074343

Table for $N = 2$, $n = 1$, through $N = 100$, $n = 50$

N	n	k	x	P(x)	p(x)
100	44	23	14	0.981902	0.035025
100	44	23	15	0.995036	0.013134
100	44	23	16	0.998923	0.003887
100	44	23	17	0.999819	0.000896
100	44	23	18	0.999977	0.000158
100	44	23	19	0.999998	0.000021
100	44	23	20	1.000000	0.000002
100	44	23	21	1.000000	0.000000
100	44	23	22	1.000000	0.000000
100	44	23	23	1.000000	0.000000
100	44	24	0	0.000000	0.000000
100	44	24	1	0.000002	0.000002
100	44	24	2	0.000027	0.000025
100	44	24	3	0.000251	0.000224
100	44	24	4	0.001588	0.001337
100	44	24	5	0.007369	0.005781
100	44	24	6	0.026157	0.018788
100	44	24	7	0.073230	0.047073
100	44	24	8	0.165758	0.092528
100	44	24	9	0.310193	0.144435
100	44	24	10	0.490736	0.180543
100	44	24	11	0.672424	0.181688
100	44	24	12	0.820046	0.147622
100	44	24	13	0.916947	0.096900
100	44	24	14	0.968256	0.051309
100	44	24	15	0.990089	0.021834
100	44	24	16	0.997509	0.007420
100	44	24	17	0.999505	0.001995
100	44	24	18	0.999924	0.000419
100	44	24	19	0.999991	0.000067
100	44	24	20	0.999999	0.000008
100	44	24	21	1.000000	0.000001
100	44	24	22	1.000000	0.000000
100	44	24	23	1.000000	0.000000
100	44	24	24	1.000000	0.000000
100	44	25	0	0.000000	0.000000
100	44	25	1	0.000001	0.000001
100	44	25	2	0.000013	0.000012
100	44	25	3	0.000130	0.000117
100	44	25	4	0.000884	0.000754
100	44	25	5	0.004402	0.003518
100	44	25	6	0.016763	0.012361
100	44	25	7	0.050313	0.033550
100	44	25	8	0.121929	0.071616
100	44	25	9	0.243677	0.121748
100	44	25	10	0.409967	0.166290
100	44	25	11	0.593533	0.183567
100	44	25	12	0.757890	0.164356
100	44	25	13	0.877421	0.119532
100	44	25	14	0.948002	0.070581

N	n	k	x	P(x)	p(x)
100	44	25	15	0.981758	0.033756
100	44	25	16	0.994776	0.013018
100	44	25	17	0.998796	0.004020
100	44	25	18	0.999780	0.000985
100	44	25	19	0.999969	0.000189
100	44	25	20	0.999997	0.000028
100	44	25	21	1.000000	0.000003
100	44	25	22	1.000000	0.000000
100	44	25	23	1.000000	0.000000
100	44	25	24	1.000000	0.000000
100	44	25	25	1.000000	0.000000
100	44	26	0	0.000000	0.000000
100	44	26	1	0.000000	0.000000
100	44	26	2	0.000006	0.000006
100	44	26	3	0.000066	0.000060
100	44	26	4	0.000482	0.000416
100	44	26	5	0.002573	0.002091
100	44	26	6	0.010500	0.007927
100	44	26	7	0.033761	0.023261
100	44	26	8	0.087553	0.053792
100	44	26	9	0.186861	0.099308
100	44	26	10	0.334582	0.147721
100	44	26	11	0.512764	0.178182
100	44	26	12	0.687764	0.175000
100	44	26	13	0.828015	0.140251
100	44	26	14	0.919770	0.091755
100	44	26	15	0.968706	0.048936
100	44	26	16	0.989916	0.021210
100	44	26	17	0.997348	0.007433
100	44	26	18	0.999439	0.002090
100	44	26	19	0.999906	0.000467
100	44	26	20	0.999988	0.000082
100	44	26	21	0.999999	0.000011
100	44	26	22	1.000000	0.000001
100	44	26	23	1.000000	0.000000
100	44	26	24	1.000000	0.000000
100	44	26	25	1.000000	0.000000
100	44	26	26	1.000000	0.000000
100	44	27	0	0.000000	0.000000
100	44	27	1	0.000000	0.000000
100	44	27	2	0.000003	0.000003
100	44	27	3	0.000033	0.000030
100	44	27	4	0.000257	0.000224
100	44	27	5	0.001471	0.001214
100	44	27	6	0.006429	0.004958
100	44	27	7	0.022131	0.015701
100	44	27	8	0.061384	0.039254
100	44	27	9	0.139891	0.078507
100	44	27	10	0.266710	0.126819
100	44	27	11	0.433304	0.166594

N	n	k	x	P(x)	p(x)
100	44	27	12	0.612088	0.178784
100	44	27	13	0.769261	0.157173
100	44	27	14	0.882572	0.113311
100	44	27	15	0.949528	0.066956
100	44	27	16	0.981890	0.032362
100	44	27	17	0.994636	0.012746
100	44	27	18	0.998704	0.004068
100	44	27	19	0.999748	0.001044
100	44	27	20	0.999961	0.000213
100	44	27	21	0.999995	0.000034
100	44	27	22	0.999999	0.000004
100	44	27	23	1.000000	0.000000
100	44	27	24	1.000000	0.000000
100	44	27	25	1.000000	0.000000
100	44	27	26	1.000000	0.000000
100	44	27	27	1.000000	0.000000
100	44	28	0	0.000000	0.000000
100	44	28	1	0.000000	0.000000
100	44	28	2	0.000001	0.000001
100	44	28	3	0.000016	0.000015
100	44	28	4	0.000134	0.000118
100	44	28	5	0.000822	0.000688
100	44	28	6	0.003848	0.003026
100	44	28	7	0.014172	0.010324
100	44	28	8	0.042026	0.027854
100	44	28	9	0.102251	0.060225
100	44	28	10	0.207644	0.105393
100	44	28	11	0.357995	0.150351
100	44	28	12	0.533717	0.175723
100	44	28	13	0.702517	0.168799
100	44	28	14	0.836006	0.133489
100	44	28	15	0.922929	0.086923
100	44	28	16	0.969477	0.046548
100	44	28	17	0.989922	0.020445
100	44	28	18	0.997255	0.007333
100	44	28	19	0.999391	0.002135
100	44	28	20	0.999891	0.000500
100	44	28	21	0.999984	0.000093
100	44	28	22	0.999998	0.000014
100	44	28	23	1.000000	0.000002
100	44	28	24	1.000000	0.000000
100	44	28	25	1.000000	0.000000
100	44	28	26	1.000000	0.000000
100	44	28	27	1.000000	0.000000
100	44	28	28	1.000000	0.000000
100	44	29	0	0.000000	0.000000
100	44	29	1	0.000000	0.000000
100	44	29	2	0.000001	0.000001
100	44	29	3	0.000008	0.000007
100	44	29	4	0.000069	0.000061

N	n	k	x	P(x)	p(x)
100	44	29	5	0.000450	0.000381
100	44	29	6	0.002251	0.001801
100	44	29	7	0.008867	0.006616
100	44	29	8	0.028099	0.019232
100	44	29	9	0.072975	0.044876
100	44	29	10	0.157875	0.084900
100	44	29	11	0.289084	0.131209
100	44	29	12	0.455618	0.166535
100	44	29	13	0.629839	0.174221
100	44	29	14	0.780385	0.150546
100	44	29	15	0.887918	0.107533
100	44	29	16	0.951375	0.063457
100	44	29	17	0.982255	0.030880
100	44	29	18	0.994607	0.012352
100	44	29	19	0.998649	0.004042
100	44	29	20	0.999724	0.001075
100	44	29	21	0.999955	0.000230
100	44	29	22	0.999994	0.000039
100	44	29	23	0.999999	0.000005
100	44	29	24	1.000000	0.000001
100	44	29	25	1.000000	0.000000
100	44	29	26	1.000000	0.000000
100	44	29	27	1.000000	0.000000
100	44	29	28	1.000000	0.000000
100	44	29	29	1.000000	0.000000
100	44	30	0	0.000000	0.000000
100	44	30	1	0.000000	0.000000
100	44	30	2	0.000000	0.000000
100	44	30	3	0.000004	0.000003
100	44	30	4	0.000034	0.000031
100	44	30	5	0.000240	0.000206
100	44	30	6	0.001287	0.001047
100	44	30	7	0.005419	0.004132
100	44	30	8	0.018348	0.012928
100	44	30	9	0.050853	0.032506
100	44	30	10	0.117219	0.066365
100	44	30	11	0.228099	0.110881
100	44	30	12	0.380561	0.152461
100	44	30	13	0.553771	0.173210
100	44	30	14	0.716774	0.163003
100	44	30	15	0.843996	0.127222
100	44	30	16	0.926350	0.082354
100	44	30	17	0.970512	0.044162
100	44	30	18	0.990084	0.019572
100	44	30	19	0.997226	0.007142
100	44	30	20	0.999361	0.002135
100	44	30	21	0.999880	0.000519
100	44	30	22	0.999982	0.000102
100	44	30	23	0.999998	0.000016
100	44	30	24	1.000000	0.000002

Table for $N = 2$, $n = 1$, through $N = 100$, $n = 50$

N	n	k	x	P(x)	p(x)
100	44	30	25	1.000000	0.000000
100	44	30	26	1.000000	0.000000
100	44	30	27	1.000000	0.000000
100	44	30	28	1.000000	0.000000
100	44	30	29	1.000000	0.000000
100	44	30	30	1.000000	0.000000
100	44	31	0	0.000000	0.000000
100	44	31	1	0.000000	0.000000
100	44	31	2	0.000000	0.000000
100	44	31	3	0.000002	0.000002
100	44	31	4	0.000017	0.000015
100	44	31	5	0.000126	0.000109
100	44	31	6	0.000719	0.000593
100	44	31	7	0.003235	0.002516
100	44	31	8	0.011699	0.008464
100	44	31	9	0.034600	0.022902
100	44	31	10	0.084984	0.050384
100	44	31	11	0.175827	0.090843
100	44	31	12	0.310864	0.135037
100	44	31	13	0.477064	0.166199
100	44	31	14	0.646916	0.169852
100	44	31	15	0.791290	0.144374
100	44	31	16	0.893409	0.102118
100	44	31	17	0.953478	0.060070
100	44	31	18	0.982815	0.029336
100	44	31	19	0.994675	0.011861
100	44	31	20	0.998629	0.003954
100	44	31	21	0.999710	0.001080
100	44	31	22	0.999950	0.000240
100	44	31	23	0.999993	0.000043
100	44	31	24	0.999999	0.000006
100	44	31	25	1.000000	0.000001
100	44	31	26	1.000000	0.000000
100	44	31	27	1.000000	0.000000
100	44	31	28	1.000000	0.000000
100	44	31	29	1.000000	0.000000
100	44	31	30	1.000000	0.000000
100	44	31	31	1.000000	0.000000
100	44	32	0	0.000000	0.000000
100	44	32	1	0.000000	0.000000
100	44	32	2	0.000000	0.000000
100	44	32	3	0.000001	0.000001
100	44	32	4	0.000008	0.000007
100	44	32	5	0.000064	0.000056
100	44	32	6	0.000392	0.000328
100	44	32	7	0.001886	0.001494
100	44	32	8	0.007283	0.005397
100	44	32	9	0.022984	0.015701
100	44	32	10	0.060157	0.037174
100	44	32	11	0.132380	0.072223

N	n	k	x	P(x)	p(x)
100	44	32	12	0.248238	0.115858
100	44	32	13	0.402394	0.154156
100	44	32	14	0.573067	0.170673
100	44	32	15	0.730611	0.157544
100	44	32	16	0.851969	0.121358
100	44	32	17	0.929973	0.078004
100	44	32	18	0.971760	0.041788
100	44	32	19	0.990378	0.018618
100	44	32	20	0.997254	0.006876
100	44	32	21	0.999349	0.002095
100	44	32	22	0.999873	0.000524
100	44	32	23	0.999980	0.000107
100	44	32	24	0.999997	0.000017
100	44	32	25	1.000000	0.000002
100	44	32	26	1.000000	0.000000
100	44	32	27	1.000000	0.000000
100	44	32	28	1.000000	0.000000
100	44	32	29	1.000000	0.000000
100	44	32	30	1.000000	0.000000
100	44	32	31	1.000000	0.000000
100	44	32	32	1.000000	0.000000
100	44	33	0	0.000000	0.000000
100	44	33	1	0.000000	0.000000
100	44	33	2	0.000000	0.000000
100	44	33	3	0.000000	0.000000
100	44	33	4	0.000004	0.000003
100	44	33	5	0.000032	0.000028
100	44	33	6	0.000209	0.000177
100	44	33	7	0.001073	0.000864
100	44	33	8	0.004426	0.003353
100	44	33	9	0.014902	0.010477
100	44	33	10	0.041570	0.026668
100	44	33	11	0.097331	0.055760
100	44	33	12	0.193717	0.096386
100	44	33	13	0.332117	0.138400
100	44	33	14	0.497770	0.165653
100	44	33	15	0.663423	0.165653
100	44	33	16	0.801998	0.138575
100	44	33	17	0.899001	0.097003
100	44	33	18	0.955783	0.056782
100	44	33	19	0.983533	0.027751
100	44	33	20	0.994827	0.011294
100	44	33	21	0.998641	0.003813
100	44	33	22	0.999704	0.001063
100	44	33	23	0.999947	0.000243
100	44	33	24	0.999992	0.000045
100	44	33	25	0.999999	0.000007
100	44	33	26	1.000000	0.000001
100	44	33	27	1.000000	0.000000
100	44	33	28	1.000000	0.000000

N	n	k	x	P(x)	p(x)
100	44	33	29	1.000000	0.000000
100	44	33	30	1.000000	0.000000
100	44	33	31	1.000000	0.000000
100	44	33	32	1.000000	0.000000
100	44	33	33	1.000000	0.000000
100	44	34	0	0.000000	0.000000
100	44	34	1	0.000000	0.000000
100	44	34	2	0.000000	0.000000
100	44	34	3	0.000000	0.000000
100	44	34	4	0.000002	0.000002
100	44	34	5	0.000015	0.000014
100	44	34	6	0.000108	0.000093
100	44	34	7	0.000596	0.000487
100	44	34	8	0.002624	0.002028
100	44	34	9	0.009429	0.006805
100	44	34	10	0.028037	0.018608
100	44	34	11	0.069867	0.041829
100	44	34	12	0.147682	0.077815
100	44	34	13	0.268081	0.120399
100	44	34	14	0.423597	0.155516
100	44	34	15	0.591723	0.168125
100	44	34	16	0.744086	0.152364
100	44	34	17	0.859910	0.115824
100	44	34	18	0.933748	0.073838
100	44	34	19	0.973179	0.039431
100	44	34	20	0.990781	0.017603
100	44	34	21	0.997331	0.006550
100	44	34	22	0.999355	0.002023
100	44	34	23	0.999871	0.000516
100	44	34	24	0.999979	0.000108
100	44	34	25	0.999997	0.000018
100	44	34	26	1.000000	0.000003
100	44	34	27	1.000000	0.000000
100	44	34	28	1.000000	0.000000
100	44	34	29	1.000000	0.000000
100	44	34	30	1.000000	0.000000
100	44	34	31	1.000000	0.000000
100	44	34	32	1.000000	0.000000
100	44	34	33	1.000000	0.000000
100	44	34	34	1.000000	0.000000
100	44	35	0	0.000000	0.000000
100	44	35	1	0.000000	0.000000
100	44	35	2	0.000000	0.000000
100	44	35	3	0.000000	0.000000
100	44	35	4	0.000001	0.000001
100	44	35	5	0.000007	0.000007
100	44	35	6	0.000055	0.000048
100	44	35	7	0.000323	0.000268
100	44	35	8	0.001518	0.001195
100	44	35	9	0.005821	0.004303

N	n	k	x	P(x)	p(x)
100	44	35	10	0.018451	0.012631
100	44	35	11	0.048952	0.030501
100	44	35	12	0.109953	0.061001
100	44	35	13	0.211530	0.101577
100	44	35	14	0.352908	0.141378
100	44	35	15	0.517849	0.164941
100	44	35	16	0.679447	0.161598
100	44	35	17	0.812528	0.133081
100	44	35	18	0.904660	0.092133
100	44	35	19	0.958243	0.053582
100	44	35	20	0.984380	0.026138
100	44	35	21	0.995049	0.010668
100	44	35	22	0.998680	0.003631
100	44	35	23	0.999707	0.001026
100	44	35	24	0.999946	0.000239
100	44	35	25	0.999992	0.000046
100	44	35	26	0.999999	0.000007
100	44	35	27	1.000000	0.000001
100	44	35	28	1.000000	0.000000
100	44	35	29	1.000000	0.000000
100	44	35	30	1.000000	0.000000
100	44	35	31	1.000000	0.000000
100	44	35	32	1.000000	0.000000
100	44	35	33	1.000000	0.000000
100	44	35	34	1.000000	0.000000
100	44	35	35	1.000000	0.000000
100	44	36	0	0.000000	0.000000
100	44	36	1	0.000000	0.000000
100	44	36	2	0.000000	0.000000
100	44	36	3	0.000000	0.000000
100	44	36	4	0.000000	0.000000
100	44	36	5	0.000003	0.000003
100	44	36	6	0.000027	0.000024
100	44	36	7	0.000170	0.000143
100	44	36	8	0.000856	0.000686
100	44	36	9	0.003504	0.002648
100	44	36	10	0.011845	0.008341
100	44	36	11	0.033467	0.021623
100	44	36	12	0.079922	0.046455
100	44	36	13	0.163086	0.083164
100	44	36	14	0.287657	0.124571
100	44	36	15	0.444260	0.156604
100	44	36	16	0.609836	0.165576
100	44	36	17	0.757248	0.147412
100	44	36	18	0.867807	0.110559
100	44	36	19	0.937634	0.069827
100	44	36	20	0.974730	0.037096
100	44	36	21	0.991274	0.016544
100	44	36	22	0.997451	0.006177
100	44	36	23	0.999375	0.001924

Table for $N = 2$, $n = 1$, through $N = 100$, $n = 50$

N	n	k	x	P(x)	p(x)
100	44	36	24	0.999872	0.000497
100	44	36	25	0.999978	0.000106
100	44	36	26	0.999997	0.000019
100	44	36	27	0.999999	0.000003
100	44	36	28	1.000000	0.000000
100	44	36	29	1.000000	0.000000
100	44	36	30	1.000000	0.000000
100	44	36	31	1.000000	0.000000
100	44	36	32	1.000000	0.000000
100	44	36	33	1.000000	0.000000
100	44	36	34	1.000000	0.000000
100	44	36	35	1.000000	0.000000
100	44	36	36	1.000000	0.000000
100	44	37	0	0.000000	0.000000
100	44	37	1	0.000000	0.000000
100	44	37	2	0.000000	0.000000
100	44	37	3	0.000000	0.000000
100	44	37	4	0.000000	0.000000
100	44	37	5	0.000002	0.000001
100	44	37	6	0.000013	0.000012
100	44	37	7	0.000087	0.000074
100	44	37	8	0.000470	0.000383
100	44	37	9	0.002056	0.001585
100	44	37	10	0.007414	0.005358
100	44	37	11	0.022318	0.014905
100	44	37	12	0.056695	0.034377
100	44	37	13	0.122803	0.066109
100	44	37	14	0.229264	0.106461
100	44	37	15	0.373299	0.144035
100	44	37	16	0.537396	0.164097
100	44	37	17	0.695058	0.157662
100	44	37	18	0.822893	0.127834
100	44	37	19	0.910358	0.087465
100	44	37	20	0.960819	0.050461
100	44	37	21	0.985328	0.024510
100	44	37	22	0.995328	0.009999
100	44	37	23	0.998744	0.003416
100	44	37	24	0.999717	0.000973
100	44	37	25	0.999947	0.000230
100	44	37	26	0.999992	0.000045
100	44	37	27	0.999999	0.000007
100	44	37	28	1.000000	0.000001
100	44	37	29	1.000000	0.000000
100	44	37	30	1.000000	0.000000
100	44	37	31	1.000000	0.000000
100	44	37	32	1.000000	0.000000
100	44	37	33	1.000000	0.000000
100	44	37	34	1.000000	0.000000
100	44	37	35	1.000000	0.000000
100	44	37	36	1.000000	0.000000
100	44	37	37	1.000000	0.000000
100	44	38	0	0.000000	0.000000
100	44	38	1	0.000000	0.000000
100	44	38	2	0.000000	0.000000
100	44	38	3	0.000000	0.000000
100	44	38	4	0.000000	0.000000
100	44	38	5	0.000001	0.000001
100	44	38	6	0.000006	0.000005
100	44	38	7	0.000044	0.000038
100	44	38	8	0.000252	0.000208
100	44	38	9	0.001175	0.000923
100	44	38	10	0.004522	0.003347
100	44	38	11	0.014511	0.009989
100	44	38	12	0.039233	0.024723
100	44	38	13	0.090274	0.051040
100	44	38	14	0.178568	0.088295
100	44	38	15	0.306997	0.128429
100	44	38	16	0.464464	0.157467
100	44	38	17	0.627489	0.163025
100	44	38	18	0.770136	0.142647
100	44	38	19	0.875649	0.105514
100	44	38	20	0.941596	0.065946
100	44	38	21	0.976380	0.034785
100	44	38	22	0.991836	0.015455
100	44	38	23	0.997605	0.005769
100	44	38	24	0.999408	0.001803
100	44	38	25	0.999877	0.000470
100	44	38	26	0.999979	0.000101
100	44	38	27	0.999997	0.000018
100	44	38	28	0.999999	0.000003
100	44	38	29	1.000000	0.000000
100	44	38	30	1.000000	0.000000
100	44	38	31	1.000000	0.000000
100	44	38	32	1.000000	0.000000
100	44	38	33	1.000000	0.000000
100	44	38	34	1.000000	0.000000
100	44	38	35	1.000000	0.000000
100	44	38	36	1.000000	0.000000
100	44	38	37	1.000000	0.000000
100	44	38	38	1.000000	0.000000
100	44	39	0	0.000000	0.000000
100	44	39	1	0.000000	0.000000
100	44	39	2	0.000000	0.000000
100	44	39	3	0.000000	0.000000
100	44	39	4	0.000000	0.000000
100	44	39	5	0.000000	0.000000
100	44	39	6	0.000003	0.000002
100	44	39	7	0.000021	0.000019
100	44	39	8	0.000131	0.000110
100	44	39	9	0.000654	0.000523
100	44	39	10	0.002686	0.002033
100	44	39	11	0.009194	0.006508
100	44	39	12	0.026473	0.017279
100	44	39	13	0.064754	0.038280
100	44	39	14	0.135845	0.071092
100	44	39	15	0.246926	0.111081
100	44	39	16	0.393350	0.146424
100	44	39	17	0.556494	0.163144
100	44	39	18	0.710316	0.153822
100	44	39	19	0.833104	0.122787
100	44	39	20	0.916068	0.082965
100	44	39	21	0.963476	0.047408
100	44	39	22	0.986352	0.022875
100	44	39	23	0.995651	0.009299
100	44	39	24	0.998826	0.003175
100	44	39	25	0.999734	0.000907
100	44	39	26	0.999949	0.000216
100	44	39	27	0.999992	0.000043
100	44	39	28	0.999999	0.000007
100	44	39	29	1.000000	0.000001
100	44	39	30	1.000000	0.000000
100	44	39	31	1.000000	0.000000
100	44	39	32	1.000000	0.000000
100	44	39	33	1.000000	0.000000
100	44	39	34	1.000000	0.000000
100	44	39	35	1.000000	0.000000
100	44	39	36	1.000000	0.000000
100	44	39	37	1.000000	0.000000
100	44	39	38	1.000000	0.000000
100	44	39	39	1.000000	0.000000
100	44	40	0	0.000000	0.000000
100	44	40	1	0.000000	0.000000
100	44	40	2	0.000000	0.000000
100	44	40	3	0.000000	0.000000
100	44	40	4	0.000000	0.000000
100	44	40	5	0.000000	0.000000
100	44	40	6	0.000001	0.000001
100	44	40	7	0.000010	0.000009
100	44	40	8	0.000066	0.000056
100	44	40	9	0.000354	0.000287
100	44	40	10	0.001553	0.001200
100	44	40	11	0.005674	0.004120
100	44	40	12	0.017409	0.011735
100	44	40	13	0.045300	0.027891
100	44	40	14	0.100882	0.055582
100	44	40	15	0.194117	0.093235
100	44	40	16	0.326139	0.132022
100	44	40	17	0.484283	0.158144
100	44	40	18	0.644753	0.160470
100	44	40	19	0.782781	0.138028
100	44	40	20	0.883426	0.100645
100	44	40	21	0.945601	0.062175
100	44	40	22	0.978102	0.032500
100	44	40	23	0.992450	0.014348
100	44	40	24	0.997785	0.005336
100	44	40	25	0.999451	0.001666
100	44	40	26	0.999886	0.000435
100	44	40	27	0.999980	0.000094
100	44	40	28	0.999997	0.000017
100	44	40	29	0.999999	0.000002
100	44	40	30	1.000000	0.000000
100	44	40	31	1.000000	0.000000
100	44	40	32	1.000000	0.000000
100	44	40	33	1.000000	0.000000
100	44	40	34	1.000000	0.000000
100	44	40	35	1.000000	0.000000
100	44	40	36	1.000000	0.000000
100	44	40	37	1.000000	0.000000
100	44	40	38	1.000000	0.000000
100	44	40	39	1.000000	0.000000
100	44	40	40	1.000000	0.000000
100	44	41	0	0.000000	0.000000
100	44	41	1	0.000000	0.000000
100	44	41	2	0.000000	0.000000
100	44	41	3	0.000000	0.000000
100	44	41	4	0.000000	0.000000
100	44	41	5	0.000000	0.000000
100	44	41	6	0.000001	0.000000
100	44	41	7	0.000005	0.000004
100	44	41	8	0.000032	0.000028
100	44	41	9	0.000186	0.000153
100	44	41	10	0.000874	0.000688
100	44	41	11	0.003407	0.002534
100	44	41	12	0.011150	0.007743
100	44	41	13	0.030889	0.019739
100	44	41	14	0.073091	0.042201
100	44	41	15	0.149053	0.075963
100	44	41	16	0.264529	0.115475
100	44	41	17	0.413118	0.148590
100	44	41	18	0.575216	0.162098
100	44	41	19	0.725269	0.150053
100	44	41	20	0.843168	0.117899
100	44	41	21	0.921767	0.078599
100	44	41	22	0.966185	0.044417
100	44	41	23	0.987428	0.021243
100	44	41	24	0.996007	0.008579
100	44	41	25	0.998924	0.002917
100	44	41	26	0.999755	0.000832
100	44	41	27	0.999953	0.000198
100	44	41	28	0.999993	0.000039

Table for $N = 2$, $n = 1$, through $N = 100$, $n = 50$

N	n	k	x	P(x)	p(x)
100	44	41	29	0.999999	0.000006
100	44	41	30	1.000000	0.000001
100	44	41	31	1.000000	0.000000
100	44	41	32	1.000000	0.000000
100	44	41	33	1.000000	0.000000
100	44	41	34	1.000000	0.000000
100	44	41	35	1.000000	0.000000
100	44	41	36	1.000000	0.000000
100	44	41	37	1.000000	0.000000
100	44	41	38	1.000000	0.000000
100	44	41	39	1.000000	0.000000
100	44	41	40	1.000000	0.000000
100	44	41	41	1.000000	0.000000
100	44	42	0	0.000000	0.000000
100	44	42	1	0.000000	0.000000
100	44	42	2	0.000000	0.000000
100	44	42	3	0.000000	0.000000
100	44	42	4	0.000000	0.000000
100	44	42	5	0.000000	0.000000
100	44	42	6	0.000000	0.000000
100	44	42	7	0.000002	0.000002
100	44	42	8	0.000015	0.000013
100	44	42	9	0.000095	0.000079
100	44	42	10	0.000477	0.000382
100	44	42	11	0.001990	0.001513
100	44	42	12	0.006951	0.004961
100	44	42	13	0.020518	0.013567
100	44	42	14	0.051633	0.031115
100	44	42	15	0.111716	0.060083
100	44	42	16	0.209727	0.098011
100	44	42	17	0.345120	0.135393
100	44	42	18	0.503783	0.158663
100	44	42	19	0.661687	0.157904
100	44	42	20	0.795209	0.133522
100	44	42	21	0.891127	0.095918
100	44	42	22	0.949622	0.058495
100	44	42	23	0.979867	0.030244
100	44	42	24	0.993099	0.013232
100	44	42	25	0.997984	0.004886
100	44	42	26	0.999502	0.001517
100	44	42	27	0.999896	0.000395
100	44	42	28	0.999982	0.000086
100	44	42	29	0.999997	0.000015
100	44	42	30	0.999999	0.000002
100	44	42	31	1.000000	0.000000
100	44	42	32	1.000000	0.000000
100	44	42	33	1.000000	0.000000
100	44	42	34	1.000000	0.000000
100	44	42	35	1.000000	0.000000
100	44	42	36	1.000000	0.000000
100	44	42	37	1.000000	0.000000
100	44	42	38	1.000000	0.000000
100	44	42	39	1.000000	0.000000
100	44	42	40	1.000000	0.000000
100	44	42	41	1.000000	0.000000
100	44	42	42	1.000000	0.000000
100	44	43	0	0.000000	0.000000
100	44	43	1	0.000000	0.000000
100	44	43	2	0.000000	0.000000
100	44	43	3	0.000000	0.000000
100	44	43	4	0.000000	0.000000
100	44	43	5	0.000000	0.000000
100	44	43	6	0.000000	0.000000
100	44	43	7	0.000001	0.000001
100	44	43	8	0.000007	0.000006
100	44	43	9	0.000047	0.000040
100	44	43	10	0.000253	0.000206
100	44	43	11	0.001129	0.000876
100	44	43	12	0.004214	0.003084
100	44	43	13	0.013266	0.009053
100	44	43	14	0.035539	0.022272
100	44	43	15	0.081674	0.046135
100	44	43	16	0.162411	0.080737
100	44	43	17	0.282092	0.119681
100	44	43	18	0.432658	0.150566
100	44	43	19	0.593625	0.160967
100	44	43	20	0.739959	0.146334
100	44	43	21	0.853091	0.113132
100	44	43	22	0.927434	0.074344
100	44	43	23	0.968916	0.041482
100	44	43	24	0.988536	0.019620
100	44	43	25	0.996384	0.007848
100	44	43	26	0.999031	0.002647
100	44	43	27	0.999781	0.000750
100	44	43	28	0.999958	0.000178
100	44	43	29	0.999993	0.000035
100	44	43	30	0.999999	0.000006
100	44	43	31	1.000000	0.000001
100	44	43	32	1.000000	0.000000
100	44	43	33	1.000000	0.000000
100	44	43	34	1.000000	0.000000
100	44	43	35	1.000000	0.000000
100	44	43	36	1.000000	0.000000
100	44	43	37	1.000000	0.000000
100	44	43	38	1.000000	0.000000
100	44	43	39	1.000000	0.000000
100	44	43	40	1.000000	0.000000
100	44	43	41	1.000000	0.000000
100	44	43	42	1.000000	0.000000
100	44	43	43	1.000000	0.000000
100	44	44	0	0.000000	0.000000
100	44	44	1	0.000000	0.000000
100	44	44	2	0.000000	0.000000
100	44	44	3	0.000000	0.000000
100	44	44	4	0.000000	0.000000
100	44	44	5	0.000000	0.000000
100	44	44	6	0.000000	0.000000
100	44	44	7	0.000000	0.000000
100	44	44	8	0.000003	0.000003
100	44	44	9	0.000023	0.000019
100	44	44	10	0.000130	0.000108
100	44	44	11	0.000622	0.000492
100	44	44	12	0.002482	0.001860
100	44	44	13	0.008343	0.005861
100	44	44	14	0.023817	0.015473
100	44	44	15	0.058202	0.034385
100	44	44	16	0.122751	0.064549
100	44	44	17	0.225401	0.102650
100	44	44	18	0.363979	0.138578
100	44	44	19	0.523025	0.159047
100	44	44	20	0.678345	0.155319
100	44	44	21	0.807441	0.129096
100	44	44	22	0.898740	0.091299
100	44	44	23	0.953633	0.054893
100	44	44	24	0.981652	0.028018
100	44	44	25	0.993768	0.012116
100	44	44	26	0.998195	0.004427
100	44	44	27	0.999557	0.001362
100	44	44	28	0.999908	0.000351
100	44	44	29	0.999984	0.000076
100	44	44	30	0.999998	0.000014
100	44	44	31	1.000000	0.000002
100	44	44	32	1.000000	0.000000
100	44	44	33	1.000000	0.000000
100	44	44	34	1.000000	0.000000
100	44	44	35	1.000000	0.000000
100	44	44	36	1.000000	0.000000
100	44	44	37	1.000000	0.000000
100	44	44	38	1.000000	0.000000
100	44	44	39	1.000000	0.000000
100	44	44	40	1.000000	0.000000
100	44	44	41	1.000000	0.000000
100	44	44	42	1.000000	0.000000
100	44	44	43	1.000000	0.000000
100	44	44	44	1.000000	0.000000
100	45	1	0	0.550000	0.550000
100	45	1	1	1.000000	0.450000
100	45	2	0	0.300000	0.300000
100	45	2	1	0.800000	0.500000
100	45	2	2	1.000000	0.200000
100	45	3	0	0.162245	0.162245
100	45	3	1	0.575510	0.413265
100	45	3	2	0.912245	0.336735
100	45	3	3	1.000000	0.087755
100	45	4	0	0.086977	0.086977
100	45	4	1	0.388050	0.301073
100	45	4	2	0.762971	0.374921
100	45	4	3	0.962003	0.199032
100	45	4	4	1.000000	0.037997
100	45	5	0	0.046206	0.046206
100	45	5	1	0.250058	0.203851
100	45	5	2	0.595037	0.344979
100	45	5	3	0.874926	0.279889
100	45	5	4	0.983772	0.108846
100	45	5	5	1.000000	0.016228
100	45	6	0	0.024319	0.024319
100	45	6	1	0.155642	0.131323
100	45	6	2	0.438889	0.283246
100	45	6	3	0.751186	0.312297
100	45	6	4	0.936796	0.185611
100	45	6	5	0.993167	0.056371
100	45	6	6	1.000000	0.006833
100	45	7	0	0.012677	0.012677
100	45	7	1	0.094172	0.081495
100	45	7	2	0.309319	0.215147
100	45	7	3	0.611649	0.302330
100	45	7	4	0.855839	0.244190
100	45	7	5	0.969180	0.113341
100	45	7	6	0.997165	0.027985
100	45	7	7	1.000000	0.002835
100	45	8	0	0.006543	0.006543
100	45	8	1	0.055615	0.049072
100	45	8	2	0.209842	0.154227
100	45	8	3	0.475113	0.265270
100	45	8	4	0.748185	0.273072
100	45	8	5	0.920431	0.172246
100	45	8	6	0.985429	0.064998
100	45	8	7	0.998842	0.013412
100	45	8	8	1.000000	0.001158
100	45	9	0	0.003343	0.003343
100	45	9	1	0.032146	0.028803
100	45	9	2	0.137758	0.105612
100	45	9	3	0.354011	0.216253
100	45	9	4	0.626490	0.272479
100	45	9	5	0.845541	0.219052
100	45	9	6	0.957875	0.112334
100	45	9	7	0.993302	0.035426
100	45	9	8	0.999534	0.006232
100	45	9	9	1.000000	0.000466
100	45	10	0	0.001690	0.001690

$N = 100 \quad n = 44\text{–}45$

Table for $N = 2$, $n = 1$, through $N = 100$, $n = 50$

N	n	k	x	P(x)	p(x)
100	45	10	1	0.018219	0.016529
100	45	10	2	0.087853	0.069634
100	45	10	3	0.254202	0.166349
100	45	10	4	0.503725	0.249523
100	45	10	5	0.749255	0.245530
100	45	10	6	0.909732	0.160477
100	45	10	7	0.978508	0.068776
100	45	10	8	0.997000	0.018492
100	45	10	9	0.999816	0.002816
100	45	10	10	1.000000	0.000184
100	45	11	0	0.000845	0.000845
100	45	11	1	0.010138	0.009293
100	45	11	2	0.054584	0.044446
100	45	11	3	0.176572	0.121989
100	45	11	4	0.390053	0.213481
100	45	11	5	0.640130	0.250077
100	45	11	6	0.840192	0.200062
100	45	11	7	0.949470	0.109277
100	45	11	8	0.989398	0.039928
100	45	11	9	0.998689	0.009291
100	45	11	10	0.999928	0.001239
100	45	11	11	1.000000	0.000072
100	45	12	0	0.000418	0.000418
100	45	12	1	0.005544	0.005126
100	45	12	2	0.033110	0.027566
100	45	12	3	0.119005	0.085895
100	45	12	4	0.291708	0.172703
100	45	12	5	0.527736	0.236028
100	45	12	6	0.752525	0.224789
100	45	12	7	0.902812	0.150287
100	45	12	8	0.972799	0.069987
100	45	12	9	0.994931	0.022133
100	45	12	10	0.999441	0.004510
100	45	12	11	0.999973	0.000531
100	45	12	12	1.000000	0.000027
100	45	13	0	0.000204	0.000204
100	45	13	1	0.002981	0.002777
100	45	13	2	0.019640	0.016659
100	45	13	3	0.078009	0.058369
100	45	13	4	0.211244	0.133235
100	45	13	5	0.420451	0.209207
100	45	13	6	0.652902	0.232452
100	45	13	7	0.837915	0.185013
100	45	13	8	0.943372	0.105457
100	45	13	9	0.985877	0.042505
100	45	13	10	0.997647	0.011770
100	45	13	11	0.999767	0.002120
100	45	13	12	0.999990	0.000222
100	45	13	13	1.000000	0.000010
100	45	14	0	0.000099	0.000099

N	n	k	x	P(x)	p(x)
100	45	14	1	0.001576	0.001478
100	45	14	2	0.011406	0.009830
100	45	14	3	0.049831	0.038425
100	45	14	4	0.148455	0.098624
100	45	14	5	0.324264	0.175808
100	45	14	6	0.548700	0.224436
100	45	14	7	0.757105	0.208405
100	45	14	8	0.898523	0.141418
100	45	14	9	0.968289	0.069766
100	45	14	10	0.992912	0.024623
100	45	14	11	0.998939	0.006027
100	45	14	12	0.999905	0.000967
100	45	14	13	0.999996	0.000091
100	45	14	14	1.000000	0.000004
100	45	15	0	0.000047	0.000047
100	45	15	1	0.000820	0.000773
100	45	15	2	0.006491	0.005671
100	45	15	3	0.031065	0.024574
100	45	15	4	0.101437	0.070371
100	45	15	5	0.242492	0.141056
100	45	15	6	0.446921	0.204428
100	45	15	7	0.665019	0.218098
100	45	15	8	0.837680	0.172661
100	45	15	9	0.939084	0.101404
100	45	15	10	0.982891	0.043807
100	45	15	11	0.996556	0.013665
100	45	15	12	0.999534	0.002978
100	45	15	13	0.999962	0.000428
100	45	15	14	0.999999	0.000036
100	45	15	15	1.000000	0.000001
100	45	16	0	0.000022	0.000022
100	45	16	1	0.000420	0.000398
100	45	16	2	0.003622	0.003202
100	45	16	3	0.018923	0.015300
100	45	16	4	0.067493	0.048570
100	45	16	5	0.176113	0.108620
100	45	16	6	0.353124	0.177011
100	45	16	7	0.567516	0.214392
100	45	16	8	0.762522	0.195006
100	45	16	9	0.896137	0.133615
100	45	16	10	0.964853	0.068716
100	45	16	11	0.991090	0.026237
100	45	16	12	0.998378	0.007288
100	45	16	13	0.999801	0.001423
100	45	16	14	0.999985	0.000184
100	45	16	15	0.999999	0.000014
100	45	16	16	1.000000	0.000000
100	45	17	0	0.000010	0.000010
100	45	17	1	0.000212	0.000201
100	45	17	2	0.001983	0.001771

N	n	k	x	P(x)	p(x)
100	45	17	3	0.011273	0.009290
100	45	17	4	0.043786	0.032513
100	45	17	5	0.124389	0.080603
100	45	17	6	0.270940	0.146551
100	45	17	7	0.470529	0.199589
100	45	17	8	0.676626	0.206097
100	45	17	9	0.838873	0.162247
100	45	17	10	0.936221	0.097348
100	45	17	11	0.980470	0.044249
100	45	17	12	0.995515	0.015045
100	45	17	13	0.999259	0.003744
100	45	17	14	0.999917	0.000658
100	45	17	15	0.999994	0.000077
100	45	17	16	1.000000	0.000005
100	45	17	17	1.000000	0.000000
100	45	18	0	0.000005	0.000005
100	45	18	1	0.000105	0.000100
100	45	18	2	0.001065	0.000960
100	45	18	3	0.006572	0.005507
100	45	18	4	0.027725	0.021153
100	45	18	5	0.085544	0.057819
100	45	18	6	0.202079	0.116535
100	45	18	7	0.379151	0.177072
100	45	18	8	0.584752	0.205600
100	45	18	9	0.768501	0.183749
100	45	18	10	0.895171	0.126670
100	45	18	11	0.962344	0.067173
100	45	18	12	0.989533	0.027189
100	45	18	13	0.997815	0.008282
100	45	18	14	0.999671	0.001856
100	45	18	15	0.999966	0.000295
100	45	18	16	0.999998	0.000031
100	45	18	17	1.000000	0.000002
100	45	18	18	1.000000	0.000000
100	45	19	0	0.000002	0.000002
100	45	19	1	0.000051	0.000049
100	45	19	2	0.000562	0.000511
100	45	19	3	0.003751	0.003190
100	45	19	4	0.017149	0.013397
100	45	19	5	0.057340	0.040191
100	45	19	6	0.146654	0.089314
100	45	19	7	0.297093	0.150439
100	45	19	8	0.491981	0.194887
100	45	19	9	0.687831	0.195850
100	45	19	10	0.841104	0.153274
100	45	19	11	0.934492	0.093387
100	45	19	12	0.978591	0.044100
100	45	19	13	0.994583	0.015992
100	45	19	14	0.998970	0.004386
100	45	19	15	0.999859	0.000889

N	n	k	x	P(x)	p(x)
100	45	19	16	0.999987	0.000128
100	45	19	17	0.999999	0.000012
100	45	19	18	1.000000	0.000001
100	45	19	19	1.000000	0.000000
100	45	20	0	0.000001	0.000001
100	45	20	1	0.000024	0.000024
100	45	20	2	0.000291	0.000266
100	45	20	3	0.002097	0.001807
100	45	20	4	0.010367	0.008270
100	45	20	5	0.037492	0.027125
100	45	20	6	0.103651	0.066159
100	45	20	7	0.226517	0.122866
100	45	20	8	0.402958	0.176441
100	45	20	9	0.600786	0.197828
100	45	20	10	0.774875	0.174089
100	45	20	11	0.895292	0.120417
100	45	20	12	0.960625	0.065333
100	45	20	13	0.988265	0.027641
100	45	20	14	0.997291	0.009026
100	45	20	15	0.999529	0.002238
100	45	20	16	0.999941	0.000411
100	45	20	17	0.999995	0.000054
100	45	20	18	1.000000	0.000005
100	45	20	19	1.000000	0.000000
100	45	20	20	1.000000	0.000000
100	45	21	0	0.000000	0.000000
100	45	21	1	0.000012	0.000011
100	45	21	2	0.000148	0.000136
100	45	21	3	0.001149	0.001001
100	45	21	4	0.006129	0.004980
100	45	21	5	0.023930	0.017801
100	45	21	6	0.071399	0.047469
100	45	21	7	0.168156	0.096757
100	45	21	8	0.321354	0.153199
100	45	21	9	0.511764	0.190410
100	45	21	10	0.698711	0.186948
100	45	21	11	0.844115	0.145404
100	45	21	12	0.933675	0.089560
100	45	21	13	0.977209	0.043534
100	45	21	14	0.993794	0.016584
100	45	21	15	0.998690	0.004896
100	45	21	16	0.999792	0.001102
100	45	21	17	0.999976	0.000184
100	45	21	18	0.999998	0.000022
100	45	21	19	1.000000	0.000002
100	45	21	20	1.000000	0.000000
100	45	21	21	1.000000	0.000000
100	45	22	0	0.000000	0.000000
100	45	22	1	0.000005	0.000005
100	45	22	2	0.000074	0.000068

Table for $N = 2$, $n = 1$, through $N = 100$, $n = 50$

N	n	k	x	P(x)	p(x)
100	45	22	3	0.000617	0.000543
100	45	22	4	0.003544	0.002928
100	45	22	5	0.014917	0.011372
100	45	22	6	0.047965	0.033048
100	45	22	7	0.121614	0.073650
100	45	22	8	0.249603	0.127989
100	45	22	9	0.424995	0.175392
100	45	22	10	0.615886	0.190892
100	45	22	11	0.781536	0.165650
100	45	22	12	0.896264	0.114728
100	45	22	13	0.959575	0.063311
100	45	22	14	0.987286	0.027711
100	45	22	15	0.996831	0.009545
100	45	22	16	0.999387	0.002557
100	45	22	17	0.999911	0.000523
100	45	22	18	0.999990	0.000080
100	45	22	19	0.999999	0.000009
100	45	22	20	1.000000	0.000001
100	45	22	21	1.000000	0.000000
100	45	22	22	1.000000	0.000000
100	45	23	0	0.000000	0.000000
100	45	23	1	0.000002	0.000002
100	45	23	2	0.000036	0.000034
100	45	23	3	0.000324	0.000288
100	45	23	4	0.002005	0.001681
100	45	23	5	0.009085	0.007079
100	45	23	6	0.031441	0.022356
100	45	23	7	0.085734	0.054293
100	45	23	8	0.188890	0.103157
100	45	23	9	0.344045	0.155154
100	45	23	10	0.530230	0.186185
100	45	23	11	0.709330	0.179100
100	45	23	12	0.847725	0.138395
100	45	23	13	0.933601	0.085876
100	45	23	14	0.976273	0.042671
100	45	23	15	0.993160	0.016887
100	45	23	16	0.998437	0.005277
100	45	23	17	0.999723	0.001286
100	45	23	18	0.999963	0.000240
100	45	23	19	0.999996	0.000033
100	45	23	20	1.000000	0.000003
100	45	23	21	1.000000	0.000000
100	45	23	22	1.000000	0.000000
100	45	23	23	1.000000	0.000000
100	45	24	0	0.000000	0.000000
100	45	24	1	0.000001	0.000001
100	45	24	2	0.000017	0.000016
100	45	24	3	0.000167	0.000150
100	45	24	4	0.001110	0.000943
100	45	24	5	0.005407	0.004297

N	n	k	x	P(x)	p(x)
100	45	24	6	0.020118	0.014710
100	45	24	7	0.058940	0.038822
100	45	24	8	0.139322	0.080382
100	45	24	9	0.271505	0.132183
100	45	24	10	0.445600	0.174095
100	45	24	11	0.630247	0.184646
100	45	24	12	0.788413	0.158166
100	45	24	13	0.897912	0.109500
100	45	24	14	0.959093	0.061181
100	45	24	15	0.986580	0.027487
100	45	24	16	0.996449	0.009869
100	45	24	17	0.999255	0.002806
100	45	24	18	0.999879	0.000624
100	45	24	19	0.999985	0.000106
100	45	24	20	0.999998	0.000014
100	45	24	21	1.000000	0.000001
100	45	24	22	1.000000	0.000000
100	45	24	23	1.000000	0.000000
100	45	24	24	1.000000	0.000000
100	45	25	0	0.000000	0.000000
100	45	25	1	0.000000	0.000000
100	45	25	2	0.000008	0.000008
100	45	25	3	0.000084	0.000076
100	45	25	4	0.000601	0.000517
100	45	25	5	0.003146	0.002544
100	45	25	6	0.012569	0.009423
100	45	25	7	0.039529	0.026960
100	45	25	8	0.100188	0.060660
100	45	25	9	0.208892	0.108703
100	45	25	10	0.365425	0.156533
100	45	25	11	0.547642	0.182217
100	45	25	12	0.719735	0.172094
100	45	25	13	0.851807	0.132072
100	45	25	14	0.934138	0.082331
100	45	25	15	0.975730	0.041592
100	45	25	16	0.992683	0.016953
100	45	25	17	0.998221	0.005538
100	45	25	18	0.999657	0.001436
100	45	25	19	0.999949	0.000291
100	45	25	20	0.999994	0.000045
100	45	25	21	0.999999	0.000005
100	45	25	22	1.000000	0.000000
100	45	25	23	1.000000	0.000000
100	45	25	24	1.000000	0.000000
100	45	25	25	1.000000	0.000000
100	45	26	0	0.000000	0.000000
100	45	26	1	0.000000	0.000000
100	45	26	2	0.000004	0.000004
100	45	26	3	0.000042	0.000038
100	45	26	4	0.000319	0.000277

N	n	k	x	P(x)	p(x)
100	45	26	5	0.001789	0.001470
100	45	26	6	0.007669	0.005880
100	45	26	7	0.025869	0.018200
100	45	26	8	0.070263	0.044394
100	45	26	9	0.156714	0.086451
100	45	26	10	0.292376	0.135662
100	45	26	11	0.465037	0.172661
100	45	26	12	0.644014	0.178978
100	45	26	13	0.795457	0.151442
100	45	26	14	0.900108	0.104651
100	45	26	15	0.959093	0.058985
100	45	26	16	0.986128	0.027035
100	45	26	17	0.996154	0.010026
100	45	26	18	0.999140	0.002986
100	45	26	19	0.999848	0.000707
100	45	26	20	0.999979	0.000131
100	45	26	21	0.999998	0.000019
100	45	26	22	1.000000	0.000002
100	45	26	23	1.000000	0.000000
100	45	26	24	1.000000	0.000000
100	45	26	25	1.000000	0.000000
100	45	26	26	1.000000	0.000000
100	45	27	0	0.000000	0.000000
100	45	27	1	0.000000	0.000000
100	45	27	2	0.000002	0.000002
100	45	27	3	0.000020	0.000018
100	45	27	4	0.000165	0.000145
100	45	27	5	0.000994	0.000829
100	45	27	6	0.004570	0.003576
100	45	27	7	0.016523	0.011953
100	45	27	8	0.048066	0.031543
100	45	27	9	0.114657	0.066591
100	45	27	10	0.228212	0.113555
100	45	27	11	0.385706	0.157495
100	45	27	12	0.564200	0.178494
100	45	27	13	0.729968	0.165768
100	45	27	14	0.856268	0.126300
100	45	27	15	0.935180	0.078913
100	45	27	16	0.975533	0.040353
100	45	27	17	0.992360	0.016827
100	45	27	18	0.998051	0.005690
100	45	27	19	0.999599	0.001548
100	45	27	20	0.999935	0.000335
100	45	27	21	0.999992	0.000057
100	45	27	22	0.999999	0.000007
100	45	27	23	1.000000	0.000001
100	45	27	24	1.000000	0.000000
100	45	27	25	1.000000	0.000000
100	45	27	26	1.000000	0.000000
100	45	27	27	1.000000	0.000000

N	n	k	x	P(x)	p(x)
100	45	28	0	0.000000	0.000000
100	45	28	1	0.000000	0.000000
100	45	28	2	0.000001	0.000001
100	45	28	3	0.000009	0.000009
100	45	28	4	0.000084	0.000074
100	45	28	5	0.000540	0.000456
100	45	28	6	0.002659	0.002120
100	45	28	7	0.010301	0.007641
100	45	28	8	0.032078	0.021778
100	45	28	9	0.081817	0.049739
100	45	28	10	0.173767	0.091950
100	45	28	11	0.312353	0.138585
100	45	28	12	0.483511	0.171158
100	45	28	13	0.657303	0.173792
100	45	28	14	0.802634	0.145331
100	45	28	15	0.902750	0.100117
100	45	28	16	0.959503	0.056752
100	45	28	17	0.985906	0.026403
100	45	28	18	0.995946	0.010040
100	45	28	19	0.999048	0.003102
100	45	28	20	0.999820	0.000772
100	45	28	21	0.999973	0.000153
100	45	28	22	0.999997	0.000024
100	45	28	23	1.000000	0.000003
100	45	28	24	1.000000	0.000000
100	45	28	25	1.000000	0.000000
100	45	28	26	1.000000	0.000000
100	45	28	27	1.000000	0.000000
100	45	28	28	1.000000	0.000000
100	45	29	0	0.000000	0.000000
100	45	29	1	0.000000	0.000000
100	45	29	2	0.000000	0.000000
100	45	29	3	0.000004	0.000004
100	45	29	4	0.000041	0.000037
100	45	29	5	0.000286	0.000245
100	45	29	6	0.001511	0.001225
100	45	29	7	0.006268	0.004756
100	45	29	8	0.020887	0.014619
100	45	29	9	0.056948	0.036061
100	45	29	10	0.129070	0.072122
100	45	29	11	0.246910	0.117840
100	45	29	12	0.405063	0.158154
100	45	29	13	0.580062	0.174999
100	45	29	14	0.740060	0.159999
100	45	29	15	0.861035	0.120975
100	45	29	16	0.936644	0.075609
100	45	29	17	0.975638	0.038994
100	45	29	18	0.992181	0.016543
100	45	29	19	0.997928	0.005746
100	45	29	20	0.999552	0.001624

Table for $N = 2$, $n = 1$, through $N = 100$, $n = 50$

N	n	k	x	P(x)	p(x)
100	45	29	21	0.999922	0.000370
100	45	29	22	0.999989	0.000067
100	45	29	23	0.999999	0.000010
100	45	29	24	1.000000	0.000001
100	45	29	25	1.000000	0.000000
100	45	29	26	1.000000	0.000000
100	45	29	27	1.000000	0.000000
100	45	29	28	1.000000	0.000000
100	45	29	29	1.000000	0.000000
100	45	30	0	0.000000	0.000000
100	45	30	1	0.000000	0.000000
100	45	30	2	0.000000	0.000000
100	45	30	3	0.000002	0.000002
100	45	30	4	0.000020	0.000018
100	45	30	5	0.000148	0.000128
100	45	30	6	0.000838	0.000690
100	45	30	7	0.003722	0.002884
100	45	30	8	0.013269	0.009547
100	45	30	9	0.038664	0.025395
100	45	30	10	0.093517	0.054853
100	45	30	11	0.190479	0.096963
100	45	30	12	0.331555	0.141076
100	45	30	13	0.501189	0.169634
100	45	30	14	0.670202	0.169013
100	45	30	15	0.809919	0.139717
100	45	30	16	0.905762	0.095843
100	45	30	17	0.960260	0.054499
100	45	30	18	0.985890	0.025630
100	45	30	19	0.995823	0.009933
100	45	30	20	0.998980	0.003157
100	45	30	21	0.999797	0.000817
100	45	30	22	0.999967	0.000171
100	45	30	23	0.999996	0.000028
100	45	30	24	0.999999	0.000004
100	45	30	25	1.000000	0.000000
100	45	30	26	1.000000	0.000000
100	45	30	27	1.000000	0.000000
100	45	30	28	1.000000	0.000000
100	45	30	29	1.000000	0.000000
100	45	30	30	1.000000	0.000000
100	45	31	0	0.000000	0.000000
100	45	31	1	0.000000	0.000000
100	45	31	2	0.000000	0.000000
100	45	31	3	0.000001	0.000001
100	45	31	4	0.000009	0.000009
100	45	31	5	0.000075	0.000066
100	45	31	6	0.000454	0.000379
100	45	31	7	0.002157	0.001703
100	45	31	8	0.008223	0.006066
100	45	31	9	0.025603	0.017381

N	n	k	x	P(x)	p(x)
100	45	31	10	0.066090	0.040487
100	45	31	11	0.143383	0.077293
100	45	31	12	0.265048	0.121665
100	45	31	13	0.423642	0.158594
100	45	31	14	0.595353	0.171711
100	45	31	15	0.750040	0.154687
100	45	31	16	0.866055	0.116015
100	45	31	17	0.938461	0.072405
100	45	31	18	0.976004	0.037544
100	45	31	19	0.992134	0.016129
100	45	31	20	0.997852	0.005719
100	45	31	21	0.999517	0.001664
100	45	31	22	0.999911	0.000395
100	45	31	23	0.999987	0.000076
100	45	31	24	0.999998	0.000012
100	45	31	25	1.000000	0.000001
100	45	31	26	1.000000	0.000000
100	45	31	27	1.000000	0.000000
100	45	31	28	1.000000	0.000000
100	45	31	29	1.000000	0.000000
100	45	31	30	1.000000	0.000000
100	45	31	31	1.000000	0.000000
100	45	32	0	0.000000	0.000000
100	45	32	1	0.000000	0.000000
100	45	32	2	0.000000	0.000000
100	45	32	3	0.000000	0.000000
100	45	32	4	0.000004	0.000004
100	45	32	5	0.000037	0.000033
100	45	32	6	0.000240	0.000203
100	45	32	7	0.001219	0.000979
100	45	32	8	0.004970	0.003751
100	45	32	9	0.016535	0.011565
100	45	32	10	0.045553	0.029018
100	45	32	11	0.105297	0.059743
100	45	32	12	0.206860	0.101564
100	45	32	13	0.350091	0.143231
100	45	32	14	0.518208	0.168117
100	45	32	15	0.682785	0.164577
100	45	32	16	0.817295	0.134510
100	45	32	17	0.909079	0.091783
100	45	32	18	0.961313	0.052235
100	45	32	19	0.986056	0.024743
100	45	32	20	0.995780	0.009724
100	45	32	21	0.998938	0.003157
100	45	32	22	0.999780	0.000842
100	45	32	23	0.999963	0.000183
100	45	32	24	0.999995	0.000032
100	45	32	25	0.999999	0.000004
100	45	32	26	1.000000	0.000000
100	45	32	27	1.000000	0.000000

N	n	k	x	P(x)	p(x)
100	45	32	28	1.000000	0.000000
100	45	32	29	1.000000	0.000000
100	45	32	30	1.000000	0.000000
100	45	32	31	1.000000	0.000000
100	45	32	32	1.000000	0.000000
100	45	33	0	0.000000	0.000000
100	45	33	1	0.000000	0.000000
100	45	33	2	0.000000	0.000000
100	45	33	3	0.000000	0.000000
100	45	33	4	0.000002	0.000002
100	45	33	5	0.000018	0.000016
100	45	33	6	0.000124	0.000106
100	45	33	7	0.000672	0.000548
100	45	33	8	0.002929	0.002257
100	45	33	9	0.010412	0.007483
100	45	33	10	0.030617	0.020205
100	45	33	11	0.075425	0.044808
100	45	33	12	0.157572	0.082147
100	45	33	13	0.282689	0.125116
100	45	33	14	0.441566	0.158878
100	45	33	15	0.610178	0.168611
100	45	33	16	0.759931	0.149753
100	45	33	17	0.871286	0.111355
100	45	33	18	0.940573	0.069288
100	45	33	19	0.976596	0.036022
100	45	33	20	0.992205	0.015610
100	45	33	21	0.997823	0.005618
100	45	33	22	0.999495	0.001672
100	45	33	23	0.999903	0.000409
100	45	33	24	0.999985	0.000081
100	45	33	25	0.999998	0.000013
100	45	33	26	1.000000	0.000002
100	45	33	27	1.000000	0.000000
100	45	33	28	1.000000	0.000000
100	45	33	29	1.000000	0.000000
100	45	33	30	1.000000	0.000000
100	45	33	31	1.000000	0.000000
100	45	33	32	1.000000	0.000000
100	45	33	33	1.000000	0.000000
100	45	34	0	0.000000	0.000000
100	45	34	1	0.000000	0.000000
100	45	34	2	0.000000	0.000000
100	45	34	3	0.000000	0.000000
100	45	34	4	0.000001	0.000001
100	45	34	5	0.000008	0.000007
100	45	34	6	0.000062	0.000054
100	45	34	7	0.000361	0.000299
100	45	34	8	0.001682	0.001322
100	45	34	9	0.006391	0.004709
100	45	34	10	0.020063	0.013671

N	n	k	x	P(x)	p(x)
100	45	34	11	0.052687	0.032624
100	45	34	12	0.117112	0.064425
100	45	34	13	0.222931	0.105820
100	45	34	14	0.368056	0.145124
100	45	34	15	0.534680	0.166624
100	45	34	16	0.695112	0.160432
100	45	34	17	0.824749	0.129637
100	45	34	18	0.912651	0.087902
100	45	34	19	0.962617	0.049965
100	45	34	20	0.986381	0.023764
100	45	34	21	0.995811	0.009430
100	45	34	22	0.998921	0.003110
100	45	34	23	0.999769	0.000848
100	45	34	24	0.999959	0.000190
100	45	34	25	0.999994	0.000035
100	45	34	26	0.999999	0.000005
100	45	34	27	1.000000	0.000001
100	45	34	28	1.000000	0.000000
100	45	34	29	1.000000	0.000000
100	45	34	30	1.000000	0.000000
100	45	34	31	1.000000	0.000000
100	45	34	32	1.000000	0.000000
100	45	34	33	1.000000	0.000000
100	45	34	34	1.000000	0.000000
100	45	35	0	0.000000	0.000000
100	45	35	1	0.000000	0.000000
100	45	35	2	0.000000	0.000000
100	45	35	3	0.000000	0.000000
100	45	35	4	0.000000	0.000000
100	45	35	5	0.000004	0.000003
100	45	35	6	0.000030	0.000026
100	45	35	7	0.000189	0.000158
100	45	35	8	0.000942	0.000753
100	45	35	9	0.003823	0.002881
100	45	35	10	0.012813	0.008990
100	45	35	11	0.035880	0.023068
100	45	35	12	0.084899	0.049019
100	45	35	13	0.171625	0.086726
100	45	35	14	0.299891	0.128266
100	45	35	15	0.458942	0.159050
100	45	35	16	0.624619	0.165678
100	45	35	17	0.769752	0.145132
100	45	35	18	0.876691	0.106940
100	45	35	19	0.942933	0.066242
100	45	35	20	0.977379	0.034446
100	45	35	21	0.992382	0.015003
100	45	35	22	0.997837	0.005455
100	45	35	23	0.999487	0.001649
100	45	35	24	0.999899	0.000412
100	45	35	25	0.999984	0.000085

Table for $N = 2$, $n = 1$, through $N = 100$, $n = 50$

N	n	k	x	P(x)	p(x)
100	45	35	26	0.999998	0.000014
100	45	35	27	1.000000	0.000002
100	45	35	28	1.000000	0.000000
100	45	35	29	1.000000	0.000000
100	45	35	30	1.000000	0.000000
100	45	35	31	1.000000	0.000000
100	45	35	32	1.000000	0.000000
100	45	35	33	1.000000	0.000000
100	45	35	34	1.000000	0.000000
100	45	35	35	1.000000	0.000000
100	45	36	0	0.000000	0.000000
100	45	36	1	0.000000	0.000000
100	45	36	2	0.000000	0.000000
100	45	36	3	0.000000	0.000000
100	45	36	4	0.000000	0.000000
100	45	36	5	0.000002	0.000002
100	45	36	6	0.000014	0.000013
100	45	36	7	0.000096	0.000082
100	45	36	8	0.000513	0.000417
100	45	36	9	0.002227	0.001714
100	45	36	10	0.007972	0.005745
100	45	36	11	0.023814	0.015842
100	45	36	12	0.060013	0.036199
100	45	36	13	0.128929	0.068916
100	45	36	14	0.238718	0.109789
100	45	36	15	0.385534	0.146816
100	45	36	16	0.550702	0.165168
100	45	36	17	0.707233	0.156532
100	45	36	18	0.832270	0.125037
100	45	36	19	0.916437	0.084166
100	45	36	20	0.964131	0.047694
100	45	36	21	0.986842	0.022712
100	45	36	22	0.995907	0.009064
100	45	36	23	0.998928	0.003021
100	45	36	24	0.999766	0.000837
100	45	36	25	0.999958	0.000192
100	45	36	26	0.999994	0.000036
100	45	36	27	0.999999	0.000006
100	45	36	28	1.000000	0.000001
100	45	36	29	1.000000	0.000000
100	45	36	30	1.000000	0.000000
100	45	36	31	1.000000	0.000000
100	45	36	32	1.000000	0.000000
100	45	36	33	1.000000	0.000000
100	45	36	34	1.000000	0.000000
100	45	36	35	1.000000	0.000000
100	45	36	36	1.000000	0.000000
100	45	37	0	0.000000	0.000000
100	45	37	1	0.000000	0.000000
100	45	37	2	0.000000	0.000000

N	n	k	x	P(x)	p(x)
100	45	37	3	0.000000	0.000000
100	45	37	4	0.000000	0.000000
100	45	37	5	0.000001	0.000001
100	45	37	6	0.000007	0.000006
100	45	37	7	0.000048	0.000041
100	45	37	8	0.000272	0.000224
100	45	37	9	0.001263	0.000991
100	45	37	10	0.004830	0.003567
100	45	37	11	0.015398	0.010568
100	45	37	12	0.041348	0.025950
100	45	37	13	0.094471	0.053123
100	45	37	14	0.185539	0.091068
100	45	37	15	0.316714	0.131175
100	45	37	16	0.475860	0.159146
100	45	37	17	0.638751	0.162891
100	45	37	18	0.779520	0.140770
100	45	37	19	0.882244	0.102724
100	45	37	20	0.945500	0.063256
100	45	37	21	0.978326	0.032825
100	45	37	22	0.992649	0.014324
100	45	37	23	0.997890	0.005240
100	45	37	24	0.999491	0.001601
100	45	37	25	0.999898	0.000407
100	45	37	26	0.999983	0.000085
100	45	37	27	0.999998	0.000015
100	45	37	28	1.000000	0.000002
100	45	37	29	1.000000	0.000000
100	45	37	30	1.000000	0.000000
100	45	37	31	1.000000	0.000000
100	45	37	32	1.000000	0.000000
100	45	37	33	1.000000	0.000000
100	45	37	34	1.000000	0.000000
100	45	37	35	1.000000	0.000000
100	45	37	36	1.000000	0.000000
100	45	37	37	1.000000	0.000000
100	45	38	0	0.000000	0.000000
100	45	38	1	0.000000	0.000000
100	45	38	2	0.000000	0.000000
100	45	38	3	0.000000	0.000000
100	45	38	4	0.000000	0.000000
100	45	38	5	0.000000	0.000000
100	45	38	6	0.000003	0.000003
100	45	38	7	0.000023	0.000020
100	45	38	8	0.000140	0.000117
100	45	38	9	0.000697	0.000556
100	45	38	10	0.002848	0.002152
100	45	38	11	0.009695	0.006846
100	45	38	12	0.027755	0.018060
100	45	38	13	0.067488	0.039733
100	45	38	14	0.140728	0.073240

N	n	k	x	P(x)	p(x)
100	45	38	15	0.254250	0.113522
100	45	38	16	0.402602	0.148353
100	45	38	17	0.566355	0.163752
100	45	38	18	0.719190	0.152836
100	45	38	19	0.839850	0.120660
100	45	38	20	0.920399	0.080549
100	45	38	21	0.965821	0.045422
100	45	38	22	0.987420	0.021599
100	45	38	23	0.996060	0.008640
100	45	38	24	0.998957	0.002897
100	45	38	25	0.999769	0.000811
100	45	38	26	0.999957	0.000189
100	45	38	27	0.999993	0.000036
100	45	38	28	0.999999	0.000006
100	45	38	29	1.000000	0.000001
100	45	38	30	1.000000	0.000000
100	45	38	31	1.000000	0.000000
100	45	38	32	1.000000	0.000000
100	45	38	33	1.000000	0.000000
100	45	38	34	1.000000	0.000000
100	45	38	35	1.000000	0.000000
100	45	38	36	1.000000	0.000000
100	45	38	37	1.000000	0.000000
100	45	38	38	1.000000	0.000000
100	45	39	0	0.000000	0.000000
100	45	39	1	0.000000	0.000000
100	45	39	2	0.000000	0.000000
100	45	39	3	0.000000	0.000000
100	45	39	4	0.000000	0.000000
100	45	39	5	0.000000	0.000000
100	45	39	6	0.000001	0.000001
100	45	39	7	0.000011	0.000009
100	45	39	8	0.000070	0.000060
100	45	39	9	0.000374	0.000303
100	45	39	10	0.001634	0.001260
100	45	39	11	0.005940	0.004307
100	45	39	12	0.018142	0.012202
100	45	39	13	0.046981	0.028838
100	45	39	14	0.104108	0.057127
100	45	39	15	0.199320	0.095212
100	45	39	16	0.333212	0.133892
100	45	39	17	0.492402	0.159190
100	45	39	18	0.652633	0.160231
100	45	39	19	0.789251	0.136618
100	45	39	20	0.887919	0.098669
100	45	39	21	0.948238	0.060319
100	45	39	22	0.979407	0.031169
100	45	39	23	0.992994	0.013587
100	45	39	24	0.997976	0.004982
100	45	39	25	0.999507	0.001531

N	n	k	x	P(x)	p(x)
100	45	39	26	0.999899	0.000393
100	45	39	27	0.999983	0.000084
100	45	39	28	0.999997	0.000015
100	45	39	29	1.000000	0.000002
100	45	39	30	1.000000	0.000000
100	45	39	31	1.000000	0.000000
100	45	39	32	1.000000	0.000000
100	45	39	33	1.000000	0.000000
100	45	39	34	1.000000	0.000000
100	45	39	35	1.000000	0.000000
100	45	39	36	1.000000	0.000000
100	45	39	37	1.000000	0.000000
100	45	39	38	1.000000	0.000000
100	45	39	39	1.000000	0.000000
100	45	40	0	0.000000	0.000000
100	45	40	1	0.000000	0.000000
100	45	40	2	0.000000	0.000000
100	45	40	3	0.000000	0.000000
100	45	40	4	0.000000	0.000000
100	45	40	5	0.000000	0.000000
100	45	40	6	0.000001	0.000000
100	45	40	7	0.000005	0.000004
100	45	40	8	0.000034	0.000029
100	45	40	9	0.000195	0.000160
100	45	40	10	0.000911	0.000716
100	45	40	11	0.003540	0.002629
100	45	40	12	0.011541	0.008001
100	45	40	13	0.031852	0.020311
100	45	40	14	0.075076	0.043224
100	45	40	15	0.152494	0.077418
100	45	40	16	0.269558	0.117064
100	45	40	17	0.419331	0.149773
100	45	40	18	0.581711	0.162380
100	45	40	19	0.731020	0.149309
100	45	40	20	0.847481	0.116461
100	45	40	21	0.924506	0.077025
100	45	40	22	0.967655	0.043149
100	45	40	23	0.988094	0.020439
100	45	40	24	0.996261	0.008167
100	45	40	25	0.999005	0.002744
100	45	40	26	0.999777	0.000772
100	45	40	27	0.999958	0.000181
100	45	40	28	0.999993	0.000035
100	45	40	29	0.999999	0.000006
100	45	40	30	1.000000	0.000001
100	45	40	31	1.000000	0.000000
100	45	40	32	1.000000	0.000000
100	45	40	33	1.000000	0.000000
100	45	40	34	1.000000	0.000000
100	45	40	35	1.000000	0.000000

Table for $N = 2$, $n = 1$, through $N = 100$, $n = 50$

N	n	k	x	P(x)	p(x)
100	45	40	36	1.000000	0.000000
100	45	40	37	1.000000	0.000000
100	45	40	38	1.000000	0.000000
100	45	40	39	1.000000	0.000000
100	45	40	40	1.000000	0.000000
100	45	41	0	0.000000	0.000000
100	45	41	1	0.000000	0.000000
100	45	41	2	0.000000	0.000000
100	45	41	3	0.000000	0.000000
100	45	41	4	0.000000	0.000000
100	45	41	5	0.000000	0.000000
100	45	41	6	0.000000	0.000000
100	45	41	7	0.000002	0.000002
100	45	41	8	0.000016	0.000014
100	45	41	9	0.000098	0.000082
100	45	41	10	0.000493	0.000395
100	45	41	11	0.002050	0.001557
100	45	41	12	0.007141	0.005090
100	45	41	13	0.021020	0.013879
100	45	41	14	0.052744	0.031724
100	45	41	15	0.113785	0.061041
100	45	41	16	0.212977	0.099192
100	45	41	17	0.349437	0.136460
100	45	41	18	0.508640	0.159203
100	45	41	19	0.666320	0.157680
100	45	41	20	0.798956	0.132636
100	45	41	21	0.893696	0.094740
100	45	41	22	0.951114	0.057418
100	45	41	23	0.980600	0.029485
100	45	41	24	0.993402	0.012803
100	45	41	25	0.998090	0.004688
100	45	41	26	0.999532	0.001442
100	45	41	27	0.999904	0.000371
100	45	41	28	0.999983	0.000080
100	45	41	29	0.999997	0.000014
100	45	41	30	1.000000	0.000002
100	45	41	31	1.000000	0.000000
100	45	41	32	1.000000	0.000000
100	45	41	33	1.000000	0.000000
100	45	41	34	1.000000	0.000000
100	45	41	35	1.000000	0.000000
100	45	41	36	1.000000	0.000000
100	45	41	37	1.000000	0.000000
100	45	41	38	1.000000	0.000000
100	45	41	39	1.000000	0.000000
100	45	41	40	1.000000	0.000000
100	45	41	41	1.000000	0.000000
100	45	42	0	0.000000	0.000000
100	45	42	1	0.000000	0.000000
100	45	42	2	0.000000	0.000000

N	n	k	x	P(x)	p(x)
100	45	42	3	0.000000	0.000000
100	45	42	4	0.000000	0.000000
100	45	42	5	0.000000	0.000000
100	45	42	6	0.000000	0.000000
100	45	42	7	0.000001	0.000001
100	45	42	8	0.000007	0.000006
100	45	42	9	0.000048	0.000041
100	45	42	10	0.000259	0.000211
100	45	42	11	0.001153	0.000894
100	45	42	12	0.004293	0.003141
100	45	42	13	0.013492	0.009199
100	45	42	14	0.036075	0.022583
100	45	42	15	0.082747	0.046672
100	45	42	16	0.164222	0.081475
100	45	42	17	0.284676	0.120455
100	45	42	18	0.435784	0.151108
100	45	42	19	0.596834	0.161049
100	45	42	20	0.742754	0.145920
100	45	42	21	0.855158	0.112404
100	45	42	22	0.928731	0.073573
100	45	42	23	0.969605	0.040874
100	45	42	24	0.988845	0.019240
100	45	42	25	0.996501	0.007656
100	45	42	26	0.999068	0.002567
100	45	42	27	0.999791	0.000723
100	45	42	28	0.999961	0.000170
100	45	42	29	0.999994	0.000033
100	45	42	30	0.999999	0.000005
100	45	42	31	1.000000	0.000001
100	45	42	32	1.000000	0.000000
100	45	42	33	1.000000	0.000000
100	45	42	34	1.000000	0.000000
100	45	42	35	1.000000	0.000000
100	45	42	36	1.000000	0.000000
100	45	42	37	1.000000	0.000000
100	45	42	38	1.000000	0.000000
100	45	42	39	1.000000	0.000000
100	45	42	40	1.000000	0.000000
100	45	42	41	1.000000	0.000000
100	45	42	42	1.000000	0.000000
100	45	43	0	0.000000	0.000000
100	45	43	1	0.000000	0.000000
100	45	43	2	0.000000	0.000000
100	45	43	3	0.000000	0.000000
100	45	43	4	0.000000	0.000000
100	45	43	5	0.000000	0.000000
100	45	43	6	0.000000	0.000000
100	45	43	7	0.000000	0.000000
100	45	43	8	0.000003	0.000003
100	45	43	9	0.000023	0.000020

N	n	k	x	P(x)	p(x)
100	45	43	10	0.000132	0.000109
100	45	43	11	0.000629	0.000497
100	45	43	12	0.002506	0.001878
100	45	43	13	0.008417	0.005910
100	45	43	14	0.024005	0.015588
100	45	43	15	0.058606	0.034602
100	45	43	16	0.123484	0.064878
100	45	43	17	0.226526	0.103041
100	45	43	18	0.365441	0.138915
100	45	43	19	0.524639	0.159198
100	45	43	20	0.679857	0.155218
100	45	43	21	0.808646	0.128789
100	45	43	22	0.899556	0.090910
100	45	43	23	0.954101	0.054546
100	45	43	24	0.981879	0.027778
100	45	43	25	0.993861	0.011982
100	45	43	26	0.998227	0.004366
100	45	43	27	0.999566	0.001339
100	45	43	28	0.999911	0.000344
100	45	43	29	0.999985	0.000074
100	45	43	30	0.999998	0.000013
100	45	43	31	1.000000	0.000002
100	45	43	32	1.000000	0.000000
100	45	43	33	1.000000	0.000000
100	45	43	34	1.000000	0.000000
100	45	43	35	1.000000	0.000000
100	45	43	36	1.000000	0.000000
100	45	43	37	1.000000	0.000000
100	45	43	38	1.000000	0.000000
100	45	43	39	1.000000	0.000000
100	45	43	40	1.000000	0.000000
100	45	43	41	1.000000	0.000000
100	45	43	42	1.000000	0.000000
100	45	43	43	1.000000	0.000000
100	45	44	0	0.000000	0.000000
100	45	44	1	0.000000	0.000000
100	45	44	2	0.000000	0.000000
100	45	44	3	0.000000	0.000000
100	45	44	4	0.000000	0.000000
100	45	44	5	0.000000	0.000000
100	45	44	6	0.000000	0.000000
100	45	44	7	0.000000	0.000000
100	45	44	8	0.000001	0.000001
100	45	44	9	0.000010	0.000009
100	45	44	10	0.000065	0.000054
100	45	44	11	0.000332	0.000267
100	45	44	12	0.001419	0.001087
100	45	44	13	0.005099	0.003679
100	45	44	14	0.015527	0.010428
100	45	44	15	0.040395	0.024868

N	n	k	x	P(x)	p(x)
100	45	44	16	0.090476	0.050081
100	45	44	17	0.175909	0.085433
100	45	44	18	0.299639	0.123730
100	45	44	19	0.452022	0.152383
100	45	44	20	0.611779	0.159757
100	45	44	21	0.754419	0.142640
100	45	44	22	0.862873	0.108454
100	45	44	23	0.933048	0.070176
100	45	44	24	0.971645	0.038597
100	45	44	25	0.989657	0.018012
100	45	44	26	0.996772	0.007115
100	45	44	27	0.999143	0.002372
100	45	44	28	0.999808	0.000665
100	45	44	29	0.999964	0.000156
100	45	44	30	0.999994	0.000030
100	45	44	31	0.999999	0.000005
100	45	44	32	1.000000	0.000001
100	45	44	33	1.000000	0.000000
100	45	44	34	1.000000	0.000000
100	45	44	35	1.000000	0.000000
100	45	44	36	1.000000	0.000000
100	45	44	37	1.000000	0.000000
100	45	44	38	1.000000	0.000000
100	45	44	39	1.000000	0.000000
100	45	44	40	1.000000	0.000000
100	45	44	41	1.000000	0.000000
100	45	44	42	1.000000	0.000000
100	45	44	43	1.000000	0.000000
100	45	44	44	1.000000	0.000000
100	45	45	0	0.000000	0.000000
100	45	45	1	0.000000	0.000000
100	45	45	2	0.000000	0.000000
100	45	45	3	0.000000	0.000000
100	45	45	4	0.000000	0.000000
100	45	45	5	0.000000	0.000000
100	45	45	6	0.000000	0.000000
100	45	45	7	0.000000	0.000000
100	45	45	8	0.000001	0.000000
100	45	45	9	0.000005	0.000004
100	45	45	10	0.000031	0.000026
100	45	45	11	0.000170	0.000139
100	45	45	12	0.000779	0.000609
100	45	45	13	0.002996	0.002217
100	45	45	14	0.009754	0.006758
100	45	45	15	0.027073	0.017319
100	45	45	16	0.064541	0.037468
100	45	45	17	0.133193	0.068651
100	45	45	18	0.239984	0.106791
100	45	45	19	0.381273	0.141290
100	45	45	20	0.540459	0.159186

Table for $N = 2$, $n = 1$, through $N = 100$, $n = 50$

N	n	k	x	P(x)	p(x)
100	45	45	21	0.693288	0.152829
100	45	45	22	0.818329	0.125042
100	45	45	23	0.905479	0.087150
100	45	45	24	0.957171	0.051692
100	45	45	25	0.983224	0.026053
100	45	45	26	0.994358	0.011134
100	45	45	27	0.998381	0.004023
100	45	45	28	0.999606	0.001225
100	45	45	29	0.999919	0.000313
100	45	45	30	0.999986	0.000067
100	45	45	31	0.999998	0.000012
100	45	45	32	1.000000	0.000002
100	45	45	33	1.000000	0.000000
100	45	45	34	1.000000	0.000000
100	45	45	35	1.000000	0.000000
100	45	45	36	1.000000	0.000000
100	45	45	37	1.000000	0.000000
100	45	45	38	1.000000	0.000000
100	45	45	39	1.000000	0.000000
100	45	45	40	1.000000	0.000000
100	45	45	41	1.000000	0.000000
100	45	45	42	1.000000	0.000000
100	45	45	43	1.000000	0.000000
100	45	45	44	1.000000	0.000000
100	45	45	45	1.000000	0.000000
100	46	1	0	0.540000	0.540000
100	46	1	1	1.000000	0.460000
100	46	2	0	0.289091	0.289091
100	46	2	1	0.790909	0.501818
100	46	2	2	1.000000	0.209091
100	46	3	0	0.153395	0.153395
100	46	3	1	0.560482	0.407087
100	46	3	2	0.906122	0.345640
100	46	3	3	1.000000	0.093878
100	46	4	0	0.080651	0.080651
100	46	4	1	0.371627	0.290976
100	46	4	2	0.749337	0.377710
100	46	4	3	0.958384	0.209047
100	46	4	4	1.000000	0.041616
100	46	5	0	0.042006	0.042006
100	46	5	1	0.235232	0.193226
100	46	5	2	0.576220	0.340988
100	46	5	3	0.864749	0.288528
100	46	5	4	0.981793	0.117044
100	46	5	5	1.000000	0.018207
100	46	6	0	0.021666	0.021666
100	46	6	1	0.143704	0.122038
100	46	6	2	0.418289	0.274585
100	46	6	3	0.734151	0.315863
100	46	6	4	0.930047	0.195896

N	n	k	x	P(x)	p(x)
100	46	6	5	0.992142	0.062095
100	46	6	6	1.000000	0.007858
100	46	7	0	0.011064	0.011064
100	46	7	1	0.085282	0.074218
100	46	7	2	0.289760	0.204478
100	46	7	3	0.589661	0.299901
100	46	7	4	0.842519	0.252858
100	46	7	5	0.965058	0.122539
100	46	7	6	0.996656	0.031598
100	46	7	7	1.000000	0.003344
100	46	8	0	0.005591	0.005591
100	46	8	1	0.049370	0.043778
100	46	8	2	0.193017	0.143648
100	46	8	3	0.450997	0.257980
100	46	8	4	0.728325	0.277328
100	46	8	5	0.911036	0.182710
100	46	8	6	0.983066	0.072030
100	46	8	7	0.998598	0.015532
100	46	8	8	1.000000	0.001402
100	46	9	0	0.002796	0.002796
100	46	9	1	0.027956	0.025161
100	46	9	2	0.124316	0.096360
100	46	9	3	0.330420	0.206103
100	46	9	4	0.601719	0.271299
100	46	9	5	0.829610	0.227891
100	46	9	6	0.951748	0.122138
100	46	9	7	0.992013	0.040265
100	46	9	8	0.999421	0.007407
100	46	9	9	1.000000	0.000579
100	46	10	0	0.001382	0.001382
100	46	10	1	0.015514	0.014132
100	46	10	2	0.077725	0.062210
100	46	10	3	0.233030	0.155306
100	46	10	4	0.476504	0.243474
100	46	10	5	0.726934	0.250430
100	46	10	6	0.898061	0.171127
100	46	10	7	0.974757	0.076696
100	46	10	8	0.996328	0.021571
100	46	10	9	0.999764	0.003437
100	46	10	10	1.000000	0.000235
100	46	11	0	0.000676	0.000676
100	46	11	1	0.008448	0.007772
100	46	11	2	0.047311	0.038862
100	46	11	3	0.158829	0.111518
100	46	11	4	0.362883	0.204054
100	46	11	5	0.612849	0.249966
100	46	11	6	0.822005	0.209156
100	46	11	7	0.941522	0.119517
100	46	11	8	0.987220	0.045698
100	46	11	9	0.998351	0.011132

N	n	k	x	P(x)	p(x)
100	46	11	10	0.999906	0.001554
100	46	11	11	1.000000	0.000094
100	46	12	0	0.000327	0.000327
100	46	12	1	0.004518	0.004192
100	46	12	2	0.028098	0.023579
100	46	12	3	0.104949	0.076851
100	46	12	4	0.266588	0.161638
100	46	12	5	0.497696	0.231109
100	46	12	6	0.728002	0.230306
100	46	12	7	0.889149	0.161147
100	46	12	8	0.967709	0.078559
100	46	12	9	0.993724	0.026015
100	46	12	10	0.999277	0.005553
100	46	12	11	0.999963	0.000686
100	46	12	12	1.000000	0.000037
100	46	13	0	0.000156	0.000156
100	46	13	1	0.002375	0.002219
100	46	13	2	0.016308	0.013933
100	46	13	3	0.067397	0.051089
100	46	13	4	0.189442	0.122045
100	46	13	5	0.390021	0.200579
100	46	13	6	0.623318	0.233297
100	46	13	7	0.817732	0.194414
100	46	13	8	0.933785	0.116053
100	46	13	9	0.982786	0.049000
100	46	13	10	0.997005	0.014220
100	46	13	11	0.999690	0.002685
100	46	13	12	0.999986	0.000295
100	46	13	13	1.000000	0.000014
100	46	14	0	0.000073	0.000073
100	46	14	1	0.001227	0.001154
100	46	14	2	0.009261	0.008034
100	46	14	3	0.042146	0.032885
100	46	14	4	0.130524	0.088378
100	46	14	5	0.295495	0.164971
100	46	14	6	0.516055	0.220560
100	46	14	7	0.730581	0.214526
100	46	14	8	0.883095	0.152515
100	46	14	9	0.961946	0.078851
100	46	14	10	0.991121	0.029175
100	46	14	11	0.998610	0.007489
100	46	14	12	0.999870	0.001260
100	46	14	13	0.999994	0.000124
100	46	14	14	1.000000	0.000005
100	46	15	0	0.000034	0.000034
100	46	15	1	0.000623	0.000589
100	46	15	2	0.005151	0.004527
100	46	15	3	0.025704	0.020553
100	46	15	4	0.087362	0.061659
100	46	15	5	0.216846	0.129483

N	n	k	x	P(x)	p(x)
100	46	15	6	0.413469	0.196623
100	46	15	7	0.633296	0.219827
100	46	15	8	0.815705	0.182409
100	46	15	9	0.928022	0.112317
100	46	15	10	0.978908	0.050886
100	46	15	11	0.995562	0.016654
100	46	15	12	0.999372	0.003810
100	46	15	13	0.999947	0.000575
100	46	15	14	0.999998	0.000051
100	46	15	15	1.000000	0.000002
100	46	16	0	0.000016	0.000016
100	46	16	1	0.000311	0.000296
100	46	16	2	0.002807	0.002496
100	46	16	3	0.015306	0.012499
100	46	16	4	0.056896	0.041589
100	46	16	5	0.154389	0.097493
100	46	16	6	0.320940	0.166551
100	46	16	7	0.532434	0.211494
100	46	16	8	0.734157	0.201723
100	46	16	9	0.879131	0.144974
100	46	16	10	0.957357	0.078225
100	46	16	11	0.988705	0.031348
100	46	16	12	0.997848	0.009143
100	46	16	13	0.999724	0.001876
100	46	16	14	0.999979	0.000255
100	46	16	15	0.999999	0.000021
100	46	16	16	1.000000	0.000001
100	46	17	0	0.000007	0.000007
100	46	17	1	0.000153	0.000146
100	46	17	2	0.001500	0.001347
100	46	17	3	0.008908	0.007408
100	46	17	4	0.036101	0.027193
100	46	17	5	0.106803	0.070702
100	46	17	6	0.241630	0.134827
100	46	17	7	0.434240	0.192610
100	46	17	8	0.642901	0.208661
100	46	17	9	0.815274	0.172372
100	46	17	10	0.923831	0.108558
100	46	17	11	0.975643	0.051812
100	46	17	12	0.994147	0.018504
100	46	17	13	0.998987	0.004840
100	46	17	14	0.999881	0.000895
100	46	17	15	0.999992	0.000110
100	46	17	16	1.000000	0.000008
100	46	17	17	1.000000	0.000000
100	46	18	0	0.000003	0.000003
100	46	18	1	0.000074	0.000071
100	46	18	2	0.000786	0.000712
100	46	18	3	0.005070	0.004284
100	46	18	4	0.022341	0.017271

N	n	k	x	P(x)	p(x)
100	46	18	5	0.071878	0.049537
100	46	18	6	0.176653	0.104775
100	46	18	7	0.343737	0.167084
100	46	18	8	0.547370	0.203633
100	46	18	9	0.738433	0.191063
100	46	18	10	0.876746	0.138313
100	46	18	11	0.953795	0.077049
100	46	18	12	0.986567	0.032772
100	46	18	13	0.997063	0.010495
100	46	18	14	0.999536	0.002474
100	46	18	15	0.999950	0.000414
100	46	18	16	0.999997	0.000046
100	46	18	17	1.000000	0.000003
100	46	18	18	1.000000	0.000000
100	46	19	0	0.000001	0.000001
100	46	19	1	0.000035	0.000034
100	46	19	2	0.000404	0.000369
100	46	19	3	0.002824	0.002420
100	46	19	4	0.013495	0.010671
100	46	19	5	0.047109	0.033615
100	46	19	6	0.125543	0.078434
100	46	19	7	0.264270	0.138727
100	46	19	8	0.453003	0.188733
100	46	19	9	0.652222	0.199218
100	46	19	10	0.816023	0.163802
100	46	19	11	0.920908	0.104885
100	46	19	12	0.972979	0.052070
100	46	19	13	0.992839	0.019860
100	46	19	14	0.998571	0.005732
100	46	19	15	0.999794	0.001223
100	46	19	16	0.999980	0.000186
100	46	19	17	0.999999	0.000019
100	46	19	18	1.000000	0.000001
100	46	19	19	1.000000	0.000000
100	46	20	0	0.000001	0.000001
100	46	20	1	0.000016	0.000016
100	46	20	2	0.000204	0.000187
100	46	20	3	0.001539	0.001335
100	46	20	4	0.007962	0.006423
100	46	20	5	0.030095	0.022133
100	46	20	6	0.086811	0.056716
100	46	20	7	0.197476	0.110665
100	46	20	8	0.364462	0.166986
100	46	20	9	0.561221	0.196759
100	46	20	10	0.743222	0.182002
100	46	20	11	0.875587	0.132365
100	46	20	12	0.951122	0.075534
100	46	20	13	0.984748	0.033626
100	46	20	14	0.996306	0.011559
100	46	20	15	0.999326	0.003019

N	n	k	x	P(x)	p(x)
100	46	20	16	0.999911	0.000585
100	46	20	17	0.999992	0.000081
100	46	20	18	0.999999	0.000008
100	46	20	19	1.000000	0.000000
100	46	20	20	1.000000	0.000000
100	46	21	0	0.000000	0.000000
100	46	21	1	0.000007	0.000007
100	46	21	2	0.000101	0.000093
100	46	21	3	0.000821	0.000721
100	46	21	4	0.004590	0.003769
100	46	21	5	0.018751	0.014162
100	46	21	6	0.058453	0.039701
100	46	21	7	0.143527	0.085074
100	46	21	8	0.285144	0.141617
100	46	21	9	0.470220	0.185076
100	46	21	10	0.661322	0.191102
100	46	21	11	0.817678	0.156356
100	46	21	12	0.919020	0.101342
100	46	21	13	0.970877	0.051857
100	46	21	14	0.991683	0.020806
100	46	21	15	0.998156	0.006473
100	46	21	16	0.999692	0.001536
100	46	21	17	0.999963	0.000271
100	46	21	18	0.999997	0.000034
100	46	21	19	1.000000	0.000003
100	46	21	20	1.000000	0.000000
100	46	21	21	1.000000	0.000000
100	46	22	0	0.000000	0.000000
100	46	22	1	0.000003	0.000003
100	46	22	2	0.000049	0.000045
100	46	22	3	0.000429	0.000380
100	46	22	4	0.002586	0.002157
100	46	22	5	0.011402	0.008815
100	46	22	6	0.038351	0.026949
100	46	22	7	0.101528	0.063177
100	46	22	8	0.217024	0.115496
100	46	22	9	0.383538	0.166515
100	46	22	10	0.574237	0.190699
100	46	22	11	0.748406	0.174169
100	46	22	12	0.875404	0.126998
100	46	22	13	0.949215	0.073811
100	46	22	14	0.983255	0.034040
100	46	22	15	0.995616	0.012361
100	46	22	16	0.999108	0.003493
100	46	22	17	0.999863	0.000755
100	46	22	18	0.999985	0.000122
100	46	22	19	0.999999	0.000014
100	46	22	20	1.000000	0.000001
100	46	22	21	1.000000	0.000000
100	46	22	22	1.000000	0.000000

N	n	k	x	P(x)	p(x)
100	46	23	0	0.000000	0.000000
100	46	23	1	0.000001	0.000001
100	46	23	2	0.000023	0.000022
100	46	23	3	0.000219	0.000196
100	46	23	4	0.001425	0.001205
100	46	23	5	0.006768	0.005343
100	46	23	6	0.024531	0.017763
100	46	23	7	0.069939	0.045409
100	46	23	8	0.160757	0.090817
100	46	23	9	0.304551	0.143794
100	46	23	10	0.486222	0.181672
100	46	23	11	0.670253	0.184031
100	46	23	12	0.820046	0.149793
100	46	23	13	0.917987	0.097941
100	46	23	14	0.969290	0.051303
100	46	23	15	0.990703	0.021413
100	46	23	16	0.997765	0.007062
100	46	23	17	0.999582	0.001817
100	46	23	18	0.999941	0.000359
100	46	23	19	0.999994	0.000053
100	46	23	20	0.999999	0.000006
100	46	23	21	1.000000	0.000000
100	46	23	22	1.000000	0.000000
100	46	23	23	1.000000	0.000000
100	46	24	0	0.000000	0.000000
100	46	24	1	0.000001	0.000001
100	46	24	2	0.000011	0.000010
100	46	24	3	0.000110	0.000099
100	46	24	4	0.000767	0.000657
100	46	24	5	0.003923	0.003156
100	46	24	6	0.015303	0.011380
100	46	24	7	0.046940	0.031637
100	46	24	8	0.115938	0.068998
100	46	24	9	0.235455	0.119517
100	46	24	10	0.401285	0.165830
100	46	24	11	0.586603	0.185318
100	46	24	12	0.753904	0.167301
100	46	24	13	0.876013	0.122109
100	46	24	14	0.947969	0.071957
100	46	24	15	0.982082	0.034113
100	46	24	16	0.995014	0.012931
100	46	24	17	0.998898	0.003884
100	46	24	18	0.999811	0.000913
100	46	24	19	0.999975	0.000165
100	46	24	20	0.999998	0.000022
100	46	24	21	1.000000	0.000002
100	46	24	22	1.000000	0.000000
100	46	24	23	1.000000	0.000000
100	46	24	24	1.000000	0.000000
100	46	25	0	0.000000	0.000000

N	n	k	x	P(x)	p(x)
100	46	25	1	0.000000	0.000000
100	46	25	2	0.000005	0.000005
100	46	25	3	0.000054	0.000049
100	46	25	4	0.000404	0.000350
100	46	25	5	0.002220	0.001817
100	46	25	6	0.009314	0.007093
100	46	25	7	0.030705	0.021392
100	46	25	8	0.081439	0.050734
100	46	25	9	0.177269	0.095830
100	46	25	10	0.322734	0.145465
100	46	25	11	0.501259	0.178525
100	46	25	12	0.679059	0.177800
100	46	25	13	0.822992	0.143933
100	46	25	14	0.917672	0.094680
100	46	25	15	0.968168	0.050496
100	46	25	16	0.989909	0.021741
100	46	25	17	0.997416	0.007507
100	46	25	18	0.999474	0.002059
100	46	25	19	0.999917	0.000442
100	46	25	20	0.999990	0.000073
100	46	25	21	0.999999	0.000009
100	46	25	22	1.000000	0.000001
100	46	25	23	1.000000	0.000000
100	46	25	24	1.000000	0.000000
100	46	25	25	1.000000	0.000000
100	46	26	0	0.000000	0.000000
100	46	26	1	0.000000	0.000000
100	46	26	2	0.000002	0.000002
100	46	26	3	0.000026	0.000024
100	46	26	4	0.000208	0.000182
100	46	26	5	0.001227	0.001020
100	46	26	6	0.005531	0.004303
100	46	26	7	0.019582	0.014051
100	46	26	8	0.055734	0.036152
100	46	26	9	0.129993	0.074259
100	46	26	10	0.252911	0.122918
100	46	26	11	0.417947	0.165037
100	46	26	12	0.598456	0.180509
100	46	26	13	0.759661	0.161205
100	46	26	14	0.877275	0.117614
100	46	26	15	0.947296	0.070021
100	46	26	16	0.981213	0.033917
100	46	26	17	0.994513	0.013301
100	46	26	18	0.998706	0.004193
100	46	26	19	0.999758	0.001052
100	46	26	20	0.999965	0.000207
100	46	26	21	0.999996	0.000031
100	46	26	22	1.000000	0.000004
100	46	26	23	1.000000	0.000000
100	46	26	24	1.000000	0.000000

Table for $N = 2$, $n = 1$, through $N = 100$, $n = 50$

N	n	k	x	P(x)	p(x)
100	46	26	25	1.000000	0.000000
100	46	26	26	1.000000	0.000000
100	46	27	0	0.000000	0.000000
100	46	27	1	0.000000	0.000000
100	46	27	2	0.000001	0.000001
100	46	27	3	0.000012	0.000011
100	46	27	4	0.000104	0.000092
100	46	27	5	0.000662	0.000558
100	46	27	6	0.003204	0.002542
100	46	27	7	0.012176	0.008972
100	46	27	8	0.037169	0.024993
100	46	27	9	0.092863	0.055694
100	46	27	10	0.193113	0.100249
100	46	27	11	0.339890	0.146777
100	46	27	12	0.515520	0.175630
100	46	27	13	0.687773	0.172253
100	46	27	14	0.826415	0.138642
100	46	27	15	0.917963	0.091548
100	46	27	16	0.967463	0.049500
100	46	27	17	0.989301	0.021838
100	46	27	18	0.997119	0.007819
100	46	27	19	0.999374	0.002254
100	46	27	20	0.999892	0.000518
100	46	27	21	0.999985	0.000094
100	46	27	22	0.999998	0.000013
100	46	27	23	1.000000	0.000001
100	46	27	24	1.000000	0.000000
100	46	27	25	1.000000	0.000000
100	46	27	26	1.000000	0.000000
100	46	27	27	1.000000	0.000000
100	46	28	0	0.000000	0.000000
100	46	28	1	0.000000	0.000000
100	46	28	2	0.000000	0.000000
100	46	28	3	0.000006	0.000005
100	46	28	4	0.000051	0.000046
100	46	28	5	0.000349	0.000298
100	46	28	6	0.001812	0.001463
100	46	28	7	0.007383	0.005572
100	46	28	8	0.024159	0.016776
100	46	28	9	0.064635	0.040476
100	46	28	10	0.143675	0.079040
100	46	28	11	0.269517	0.125842
100	46	28	12	0.433719	0.164202
100	46	28	13	0.609905	0.176185
100	46	28	14	0.765640	0.155735
100	46	28	15	0.879086	0.113446
100	46	28	16	0.947120	0.068034
100	46	28	17	0.980626	0.033505
100	46	28	18	0.994121	0.013495
100	46	28	19	0.998540	0.004419

N	n	k	x	P(x)	p(x)
100	46	28	20	0.999707	0.001167
100	46	28	21	0.999953	0.000246
100	46	28	22	0.999994	0.000041
100	46	28	23	0.999999	0.000005
100	46	28	24	1.000000	0.000000
100	46	28	25	1.000000	0.000000
100	46	28	26	1.000000	0.000000
100	46	28	27	1.000000	0.000000
100	46	28	28	1.000000	0.000000
100	46	29	0	0.000000	0.000000
100	46	29	1	0.000000	0.000000
100	46	29	2	0.000000	0.000000
100	46	29	3	0.000002	0.000002
100	46	29	4	0.000025	0.000022
100	46	29	5	0.000180	0.000155
100	46	29	6	0.000999	0.000820
100	46	29	7	0.004365	0.003366
100	46	29	8	0.015305	0.010940
100	46	29	9	0.043835	0.028530
100	46	29	10	0.104155	0.060320
100	46	29	11	0.208344	0.104189
100	46	29	12	0.356179	0.147836
100	46	29	13	0.529153	0.172974
100	46	29	14	0.696424	0.167271
100	46	29	15	0.830241	0.133817
100	46	29	16	0.918773	0.088531
100	46	29	17	0.967130	0.048357
100	46	29	18	0.988872	0.021742
100	46	29	19	0.996883	0.008010
100	46	29	20	0.999286	0.002403
100	46	29	21	0.999868	0.000582
100	46	29	22	0.999980	0.000113
100	46	29	23	0.999998	0.000017
100	46	29	24	1.000000	0.000002
100	46	29	25	1.000000	0.000000
100	46	29	26	1.000000	0.000000
100	46	29	27	1.000000	0.000000
100	46	29	28	1.000000	0.000000
100	46	29	29	1.000000	0.000000
100	46	30	0	0.000000	0.000000
100	46	30	1	0.000000	0.000000
100	46	30	2	0.000000	0.000000
100	46	30	3	0.000001	0.000001
100	46	30	4	0.000012	0.000010
100	46	30	5	0.000090	0.000079
100	46	30	6	0.000537	0.000447
100	46	30	7	0.002516	0.001979
100	46	30	8	0.009450	0.006934
100	46	30	9	0.028967	0.019517
100	46	30	10	0.073570	0.044603

N	n	k	x	P(x)	p(x)
100	46	30	11	0.156983	0.083413
100	46	30	12	0.285385	0.128402
100	46	30	13	0.448757	0.163372
100	46	30	14	0.621035	0.172278
100	46	30	15	0.771814	0.150780
100	46	30	16	0.881365	0.109551
100	46	30	17	0.947379	0.066013
100	46	30	18	0.980298	0.032919
100	46	30	19	0.993837	0.013538
100	46	30	20	0.998406	0.004569
100	46	30	21	0.999663	0.001257
100	46	30	22	0.999943	0.000280
100	46	30	23	0.999992	0.000050
100	46	30	24	0.999999	0.000007
100	46	30	25	1.000000	0.000001
100	46	30	26	1.000000	0.000000
100	46	30	27	1.000000	0.000000
100	46	30	28	1.000000	0.000000
100	46	30	29	1.000000	0.000000
100	46	30	30	1.000000	0.000000
100	46	31	0	0.000000	0.000000
100	46	31	1	0.000000	0.000000
100	46	31	2	0.000000	0.000000
100	46	31	3	0.000000	0.000000
100	46	31	4	0.000005	0.000005
100	46	31	5	0.000044	0.000039
100	46	31	6	0.000282	0.000238
100	46	31	7	0.001414	0.001132
100	46	31	8	0.005686	0.004272
100	46	31	9	0.018651	0.012965
100	46	31	10	0.050631	0.031980
100	46	31	11	0.115277	0.064645
100	46	31	12	0.223018	0.107742
100	46	31	13	0.371739	0.148720
100	46	31	14	0.542279	0.170540
100	46	31	15	0.705040	0.162761
100	46	31	16	0.834415	0.129374
100	46	31	17	0.920030	0.085615
100	46	31	18	0.967130	0.047100
100	46	31	19	0.988615	0.021484
100	46	31	20	0.996709	0.008094
100	46	31	21	0.999214	0.002505
100	46	31	22	0.999847	0.000633
100	46	31	23	0.999976	0.000129
100	46	31	24	0.999997	0.000021
100	46	31	25	1.000000	0.000003
100	46	31	26	1.000000	0.000000
100	46	31	27	1.000000	0.000000
100	46	31	28	1.000000	0.000000
100	46	31	29	1.000000	0.000000

N	n	k	x	P(x)	p(x)
100	46	31	30	1.000000	0.000000
100	46	31	31	1.000000	0.000000
100	46	32	0	0.000000	0.000000
100	46	32	1	0.000000	0.000000
100	46	32	2	0.000000	0.000000
100	46	32	3	0.000000	0.000000
100	46	32	4	0.000002	0.000002
100	46	32	5	0.000021	0.000019
100	46	32	6	0.000144	0.000123
100	46	32	7	0.000774	0.000630
100	46	32	8	0.003333	0.002559
100	46	32	9	0.011699	0.008366
100	46	32	10	0.033946	0.022247
100	46	32	11	0.082486	0.048540
100	46	32	12	0.169928	0.087443
100	46	32	13	0.300612	0.130683
100	46	32	14	0.463188	0.162576
100	46	32	15	0.631916	0.168728
100	46	32	16	0.778165	0.146249
100	46	32	17	0.884047	0.105882
100	46	32	18	0.948017	0.063970
100	46	32	19	0.980208	0.032190
100	46	32	20	0.993659	0.013451
100	46	32	21	0.998306	0.004648
100	46	32	22	0.999627	0.001320
100	46	32	23	0.999933	0.000306
100	46	32	24	0.999990	0.000057
100	46	32	25	0.999999	0.000009
100	46	32	26	1.000000	0.000001
100	46	32	27	1.000000	0.000000
100	46	32	28	1.000000	0.000000
100	46	32	29	1.000000	0.000000
100	46	32	30	1.000000	0.000000
100	46	32	31	1.000000	0.000000
100	46	32	32	1.000000	0.000000
100	46	33	0	0.000000	0.000000
100	46	33	1	0.000000	0.000000
100	46	33	2	0.000000	0.000000
100	46	33	3	0.000000	0.000000
100	46	33	4	0.000001	0.000001
100	46	33	5	0.000010	0.000009
100	46	33	6	0.000072	0.000062
100	46	33	7	0.000413	0.000341
100	46	33	8	0.001903	0.001490
100	46	33	9	0.007147	0.005244
100	46	33	10	0.022168	0.015021
100	46	33	11	0.057502	0.035334
100	46	33	12	0.126207	0.068705
100	46	33	13	0.237192	0.110985
100	46	33	14	0.386681	0.149490

Table for $N = 2$, $n = 1$, through $N = 100$, $n = 50$

N	n	k	x	P(x)	p(x)
100	46	33	15	0.554996	0.168314
100	46	33	16	0.713643	0.158648
100	46	33	17	0.838891	0.125248
100	46	33	18	0.921677	0.082785
100	46	33	19	0.967426	0.045750
100	46	33	20	0.988516	0.021089
100	46	33	21	0.996598	0.008082
100	46	33	22	0.999161	0.002563
100	46	33	23	0.999829	0.000669
100	46	33	24	0.999972	0.000142
100	46	33	25	0.999996	0.000025
100	46	33	26	0.999999	0.000003
100	46	33	27	1.000000	0.000000
100	46	33	28	1.000000	0.000000
100	46	33	29	1.000000	0.000000
100	46	33	30	1.000000	0.000000
100	46	33	31	1.000000	0.000000
100	46	33	32	1.000000	0.000000
100	46	33	33	1.000000	0.000000
100	46	34	0	0.000000	0.000000
100	46	34	1	0.000000	0.000000
100	46	34	2	0.000000	0.000000
100	46	34	3	0.000000	0.000000
100	46	34	4	0.000000	0.000000
100	46	34	5	0.000004	0.000004
100	46	34	6	0.000035	0.000030
100	46	34	7	0.000214	0.000179
100	46	34	8	0.001058	0.000844
100	46	34	9	0.004251	0.003193
100	46	34	10	0.014097	0.009846
100	46	34	11	0.039044	0.024947
100	46	34	12	0.091342	0.052298
100	46	34	13	0.182528	0.091186
100	46	34	14	0.315283	0.132756
100	46	34	15	0.477119	0.161836
100	46	34	16	0.642607	0.165488
100	46	34	17	0.784680	0.142073
100	46	34	18	0.887080	0.102400
100	46	34	19	0.948990	0.061910
100	46	34	20	0.980332	0.031342
100	46	34	21	0.993582	0.013250
100	46	34	22	0.998243	0.004661
100	46	34	23	0.999600	0.001357
100	46	34	24	0.999925	0.000325
100	46	34	25	0.999988	0.000064
100	46	34	26	0.999999	0.000010
100	46	34	27	1.000000	0.000001
100	46	34	28	1.000000	0.000000
100	46	34	29	1.000000	0.000000
100	46	34	30	1.000000	0.000000

N	n	k	x	P(x)	p(x)
100	46	34	31	1.000000	0.000000
100	46	34	32	1.000000	0.000000
100	46	34	33	1.000000	0.000000
100	46	34	34	1.000000	0.000000
100	46	35	0	0.000000	0.000000
100	46	35	1	0.000000	0.000000
100	46	35	2	0.000000	0.000000
100	46	35	3	0.000000	0.000000
100	46	35	4	0.000000	0.000000
100	46	35	5	0.000002	0.000002
100	46	35	6	0.000016	0.000014
100	46	35	7	0.000108	0.000092
100	46	35	8	0.000572	0.000464
100	46	35	9	0.002461	0.001889
100	46	35	10	0.008726	0.006266
100	46	35	11	0.025814	0.017088
100	46	35	12	0.064400	0.038586
100	46	35	13	0.136935	0.072534
100	46	35	14	0.250917	0.113982
100	46	35	15	0.401105	0.150188
100	46	35	16	0.567385	0.166280
100	46	35	17	0.722254	0.154869
100	46	35	18	0.843637	0.121384
100	46	35	19	0.923663	0.080026
100	46	35	20	0.967985	0.044322
100	46	35	21	0.988563	0.020578
100	46	35	22	0.996548	0.007985
100	46	35	23	0.999127	0.002579
100	46	35	24	0.999816	0.000690
100	46	35	25	0.999968	0.000152
100	46	35	26	0.999995	0.000027
100	46	35	27	0.999999	0.000004
100	46	35	28	1.000000	0.000000
100	46	35	29	1.000000	0.000000
100	46	35	30	1.000000	0.000000
100	46	35	31	1.000000	0.000000
100	46	35	32	1.000000	0.000000
100	46	35	33	1.000000	0.000000
100	46	35	34	1.000000	0.000000
100	46	35	35	1.000000	0.000000
100	46	36	0	0.000000	0.000000
100	46	36	1	0.000000	0.000000
100	46	36	2	0.000000	0.000000
100	46	36	3	0.000000	0.000000
100	46	36	4	0.000000	0.000000
100	46	36	5	0.000001	0.000001
100	46	36	6	0.000007	0.000007
100	46	36	7	0.000053	0.000046
100	46	36	8	0.000301	0.000248
100	46	36	9	0.001386	0.001085

N	n	k	x	P(x)	p(x)
100	46	36	10	0.005256	0.003871
100	46	36	11	0.016613	0.011357
100	46	36	12	0.044217	0.027604
100	46	36	13	0.100110	0.055893
100	46	36	14	0.194802	0.094693
100	46	36	15	0.329477	0.134674
100	46	36	16	0.490641	0.161164
100	46	36	17	0.653159	0.162518
100	46	36	18	0.791349	0.138190
100	46	36	19	0.890422	0.099073
100	46	36	20	0.950256	0.059835
100	46	36	21	0.980648	0.030392
100	46	36	22	0.993600	0.012951
100	46	36	23	0.998214	0.004615
100	46	36	24	0.999583	0.001369
100	46	36	25	0.999919	0.000336
100	46	36	26	0.999987	0.000068
100	46	36	27	0.999998	0.000011
100	46	36	28	1.000000	0.000001
100	46	36	29	1.000000	0.000000
100	46	36	30	1.000000	0.000000
100	46	36	31	1.000000	0.000000
100	46	36	32	1.000000	0.000000
100	46	36	33	1.000000	0.000000
100	46	36	34	1.000000	0.000000
100	46	36	35	1.000000	0.000000
100	46	36	36	1.000000	0.000000
100	46	37	0	0.000000	0.000000
100	46	37	1	0.000000	0.000000
100	46	37	2	0.000000	0.000000
100	46	37	3	0.000000	0.000000
100	46	37	4	0.000000	0.000000
100	46	37	5	0.000000	0.000000
100	46	37	6	0.000003	0.000003
100	46	37	7	0.000025	0.000022
100	46	37	8	0.000154	0.000128
100	46	37	9	0.000758	0.000605
100	46	37	10	0.003079	0.002321
100	46	37	11	0.010402	0.007323
100	46	37	12	0.029553	0.019150
100	46	37	13	0.071290	0.041737
100	46	37	14	0.147456	0.076166
100	46	37	15	0.264244	0.116788
100	46	37	16	0.415095	0.150851
100	46	37	17	0.579518	0.164423
100	46	37	18	0.730891	0.151373
100	46	37	19	0.848625	0.117735
100	46	37	20	0.925948	0.077323
100	46	37	21	0.968776	0.042828
100	46	37	22	0.988743	0.019966

N	n	k	x	P(x)	p(x)
100	46	37	23	0.996556	0.007813
100	46	37	24	0.999112	0.002557
100	46	37	25	0.999809	0.000696
100	46	37	26	0.999966	0.000157
100	46	37	27	0.999995	0.000029
100	46	37	28	0.999999	0.000004
100	46	37	29	1.000000	0.000001
100	46	37	30	1.000000	0.000000
100	46	37	31	1.000000	0.000000
100	46	37	32	1.000000	0.000000
100	46	37	33	1.000000	0.000000
100	46	37	34	1.000000	0.000000
100	46	37	35	1.000000	0.000000
100	46	37	36	1.000000	0.000000
100	46	37	37	1.000000	0.000000
100	46	38	0	0.000000	0.000000
100	46	38	1	0.000000	0.000000
100	46	38	2	0.000000	0.000000
100	46	38	3	0.000000	0.000000
100	46	38	4	0.000000	0.000000
100	46	38	5	0.000000	0.000000
100	46	38	6	0.000001	0.000001
100	46	38	7	0.000012	0.000010
100	46	38	8	0.000076	0.000065
100	46	38	9	0.000403	0.000327
100	46	38	10	0.001753	0.001350
100	46	38	11	0.006334	0.004581
100	46	38	12	0.019218	0.012884
100	46	38	13	0.049427	0.030210
100	46	38	14	0.108769	0.059341
100	46	38	15	0.206777	0.098008
100	46	38	16	0.343261	0.136484
100	46	38	17	0.503831	0.160570
100	46	38	18	0.663614	0.159783
100	46	38	19	0.798168	0.134554
100	46	38	20	0.894037	0.095870
100	46	38	21	0.951781	0.057744
100	46	38	22	0.981136	0.029355
100	46	38	23	0.993703	0.012567
100	46	38	24	0.998220	0.004516
100	46	38	25	0.999577	0.001357
100	46	38	26	0.999916	0.000339
100	46	38	27	0.999986	0.000070
100	46	38	28	0.999998	0.000012
100	46	38	29	1.000000	0.000002
100	46	38	30	1.000000	0.000000
100	46	38	31	1.000000	0.000000
100	46	38	32	1.000000	0.000000
100	46	38	33	1.000000	0.000000
100	46	38	34	1.000000	0.000000

Table for $N = 2$, $n = 1$, through $N = 100$, $n = 50$

N	n	k	x	P(x)	p(x)
100	46	38	35	1.000000	0.000000
100	46	38	36	1.000000	0.000000
100	46	38	37	1.000000	0.000000
100	46	38	38	1.000000	0.000000
100	46	39	0	0.000000	0.000000
100	46	39	1	0.000000	0.000000
100	46	39	2	0.000000	0.000000
100	46	39	3	0.000000	0.000000
100	46	39	4	0.000000	0.000000
100	46	39	5	0.000000	0.000000
100	46	39	6	0.000001	0.000001
100	46	39	7	0.000005	0.000005
100	46	39	8	0.000037	0.000031
100	46	39	9	0.000208	0.000171
100	46	39	10	0.000969	0.000761
100	46	39	11	0.003748	0.002779
100	46	39	12	0.012152	0.008404
100	46	39	13	0.033348	0.021196
100	46	39	14	0.078141	0.044793
100	46	39	15	0.157773	0.079632
100	46	39	16	0.277220	0.119448
100	46	39	17	0.428726	0.151505
100	46	39	18	0.591454	0.162728
100	46	39	19	0.739572	0.148118
100	46	39	20	0.853834	0.114262
100	46	39	21	0.928497	0.074663
100	46	39	22	0.969773	0.041276
100	46	39	23	0.989042	0.019268
100	46	39	24	0.996617	0.007576
100	46	39	25	0.999117	0.002500
100	46	39	26	0.999807	0.000689
100	46	39	27	0.999965	0.000158
100	46	39	28	0.999995	0.000030
100	46	39	29	0.999999	0.000005
100	46	39	30	1.000000	0.000001
100	46	39	31	1.000000	0.000000
100	46	39	32	1.000000	0.000000
100	46	39	33	1.000000	0.000000
100	46	39	34	1.000000	0.000000
100	46	39	35	1.000000	0.000000
100	46	39	36	1.000000	0.000000
100	46	39	37	1.000000	0.000000
100	46	39	38	1.000000	0.000000
100	46	39	39	1.000000	0.000000
100	46	40	0	0.000000	0.000000
100	46	40	1	0.000000	0.000000
100	46	40	2	0.000000	0.000000
100	46	40	3	0.000000	0.000000
100	46	40	4	0.000000	0.000000
100	46	40	5	0.000000	0.000000

N	n	k	x	P(x)	p(x)
100	46	40	6	0.000000	0.000000
100	46	40	7	0.000002	0.000002
100	46	40	8	0.000017	0.000015
100	46	40	9	0.000104	0.000087
100	46	40	10	0.000520	0.000416
100	46	40	11	0.002154	0.001634
100	46	40	12	0.007468	0.005314
100	46	40	13	0.021881	0.014414
100	46	40	14	0.054643	0.032762
100	46	40	15	0.117304	0.062661
100	46	40	16	0.218476	0.101172
100	46	40	17	0.356699	0.138223
100	46	40	18	0.516759	0.160060
100	46	40	19	0.674011	0.157252
100	46	40	20	0.805132	0.131121
100	46	40	21	0.897898	0.092766
100	46	40	22	0.953534	0.055636
100	46	40	23	0.981776	0.028243
100	46	40	24	0.993885	0.012109
100	46	40	25	0.998256	0.004371
100	46	40	26	0.999581	0.001324
100	46	40	27	0.999915	0.000335
100	46	40	28	0.999986	0.000070
100	46	40	29	0.999998	0.000012
100	46	40	30	1.000000	0.000002
100	46	40	31	1.000000	0.000000
100	46	40	32	1.000000	0.000000
100	46	40	33	1.000000	0.000000
100	46	40	34	1.000000	0.000000
100	46	40	35	1.000000	0.000000
100	46	40	36	1.000000	0.000000
100	46	40	37	1.000000	0.000000
100	46	40	38	1.000000	0.000000
100	46	40	39	1.000000	0.000000
100	46	40	40	1.000000	0.000000
100	46	41	0	0.000000	0.000000
100	46	41	1	0.000000	0.000000
100	46	41	2	0.000000	0.000000
100	46	41	3	0.000000	0.000000
100	46	41	4	0.000000	0.000000
100	46	41	5	0.000000	0.000000
100	46	41	6	0.000000	0.000000
100	46	41	7	0.000001	0.000001
100	46	41	8	0.000008	0.000007
100	46	41	9	0.000050	0.000043
100	46	41	10	0.000270	0.000220
100	46	41	11	0.001201	0.000930
100	46	41	12	0.004456	0.003256
100	46	41	13	0.013954	0.009498
100	46	41	14	0.037170	0.023216

N	n	k	x	P(x)	p(x)
100	46	41	15	0.084929	0.047759
100	46	41	16	0.167890	0.082961
100	46	41	17	0.289891	0.122001
100	46	41	18	0.442064	0.152173
100	46	41	19	0.603248	0.161184
100	46	41	20	0.748313	0.145065
100	46	41	21	0.859245	0.110932
100	46	41	22	0.931279	0.072034
100	46	41	23	0.970950	0.039671
100	46	41	24	0.989445	0.018495
100	46	41	25	0.996726	0.007281
100	46	41	26	0.999139	0.002413
100	46	41	27	0.999809	0.000670
100	46	41	28	0.999965	0.000155
100	46	41	29	0.999994	0.000030
100	46	41	30	0.999999	0.000005
100	46	41	31	1.000000	0.000001
100	46	41	32	1.000000	0.000000
100	46	41	33	1.000000	0.000000
100	46	41	34	1.000000	0.000000
100	46	41	35	1.000000	0.000000
100	46	41	36	1.000000	0.000000
100	46	41	37	1.000000	0.000000
100	46	41	38	1.000000	0.000000
100	46	41	39	1.000000	0.000000
100	46	41	40	1.000000	0.000000
100	46	41	41	1.000000	0.000000
100	46	42	0	0.000000	0.000000
100	46	42	1	0.000000	0.000000
100	46	42	2	0.000000	0.000000
100	46	42	3	0.000000	0.000000
100	46	42	4	0.000000	0.000000
100	46	42	5	0.000000	0.000000
100	46	42	6	0.000000	0.000000
100	46	42	7	0.000000	0.000000
100	46	42	8	0.000003	0.000003
100	46	42	9	0.000024	0.000020
100	46	42	10	0.000136	0.000113
100	46	42	11	0.000649	0.000513
100	46	42	12	0.002580	0.001931
100	46	42	13	0.008642	0.006062
100	46	42	14	0.024578	0.015937
100	46	42	15	0.059835	0.035257
100	46	42	16	0.125706	0.065871
100	46	42	17	0.229924	0.104218
100	46	42	18	0.369846	0.139922
100	46	42	19	0.529486	0.159639
100	46	42	20	0.684386	0.154900
100	46	42	21	0.812240	0.127854
100	46	42	22	0.901977	0.089737

N	n	k	x	P(x)	p(x)
100	46	42	23	0.955485	0.053508
100	46	42	24	0.982549	0.027064
100	46	42	25	0.994135	0.011586
100	46	42	26	0.998321	0.004186
100	46	42	27	0.999593	0.001272
100	46	42	28	0.999917	0.000324
100	46	42	29	0.999986	0.000069
100	46	42	30	0.999998	0.000012
100	46	42	31	1.000000	0.000002
100	46	42	32	1.000000	0.000000
100	46	42	33	1.000000	0.000000
100	46	42	34	1.000000	0.000000
100	46	42	35	1.000000	0.000000
100	46	42	36	1.000000	0.000000
100	46	42	37	1.000000	0.000000
100	46	42	38	1.000000	0.000000
100	46	42	39	1.000000	0.000000
100	46	42	40	1.000000	0.000000
100	46	42	41	1.000000	0.000000
100	46	42	42	1.000000	0.000000
100	46	43	0	0.000000	0.000000
100	46	43	1	0.000000	0.000000
100	46	43	2	0.000000	0.000000
100	46	43	3	0.000000	0.000000
100	46	43	4	0.000000	0.000000
100	46	43	5	0.000000	0.000000
100	46	43	6	0.000000	0.000000
100	46	43	7	0.000000	0.000000
100	46	43	8	0.000001	0.000001
100	46	43	9	0.000011	0.000009
100	46	43	10	0.000066	0.000056
100	46	43	11	0.000340	0.000273
100	46	43	12	0.001448	0.001109
100	46	43	13	0.005193	0.003745
100	46	43	14	0.015786	0.010593
100	46	43	15	0.040991	0.025205
100	46	43	16	0.091635	0.050644
100	46	43	17	0.177815	0.086180
100	46	43	18	0.302298	0.124482
100	46	43	19	0.455171	0.152873
100	46	43	20	0.614948	0.159777
100	46	43	21	0.757130	0.142183
100	46	43	22	0.864845	0.107714
100	46	43	23	0.934266	0.069422
100	46	43	24	0.972283	0.038017
100	46	43	25	0.989940	0.017657
100	46	43	26	0.996878	0.006938
100	46	43	27	0.999177	0.002299
100	46	43	28	0.999817	0.000640
100	46	43	29	0.999966	0.000149

Table for $N = 2$, $n = 1$, through $N = 100$, $n = 50$

N	n	k	x	P(x)	p(x)
100	46	43	30	0.999995	0.000029
100	46	43	31	0.999999	0.000005
100	46	43	32	1.000000	0.000001
100	46	43	33	1.000000	0.000000
100	46	43	34	1.000000	0.000000
100	46	43	35	1.000000	0.000000
100	46	43	36	1.000000	0.000000
100	46	43	37	1.000000	0.000000
100	46	43	38	1.000000	0.000000
100	46	43	39	1.000000	0.000000
100	46	43	40	1.000000	0.000000
100	46	43	41	1.000000	0.000000
100	46	43	42	1.000000	0.000000
100	46	43	43	1.000000	0.000000
100	46	44	0	0.000000	0.000000
100	46	44	1	0.000000	0.000000
100	46	44	2	0.000000	0.000000
100	46	44	3	0.000000	0.000000
100	46	44	4	0.000000	0.000000
100	46	44	5	0.000000	0.000000
100	46	44	6	0.000000	0.000000
100	46	44	7	0.000000	0.000000
100	46	44	8	0.000001	0.000001
100	46	44	9	0.000005	0.000004
100	46	44	10	0.000031	0.000027
100	46	44	11	0.000172	0.000141
100	46	44	12	0.000787	0.000615
100	46	44	13	0.003025	0.002238
100	46	44	14	0.009839	0.006814
100	46	44	15	0.027283	0.017444
100	46	44	16	0.064980	0.037697
100	46	44	17	0.133969	0.068989
100	46	44	18	0.241149	0.107179
100	46	44	19	0.382757	0.141609
100	46	44	20	0.542067	0.159310
100	46	44	21	0.694770	0.152703
100	46	44	22	0.819491	0.124722
100	46	44	23	0.906254	0.086763
100	46	44	24	0.957610	0.051356
100	46	44	25	0.983435	0.025825
100	46	44	26	0.994443	0.011009
100	46	44	27	0.998410	0.003967
100	46	44	28	0.999615	0.001204
100	46	44	29	0.999921	0.000307
100	46	44	30	0.999986	0.000065
100	46	44	31	0.999998	0.000011
100	46	44	32	1.000000	0.000002
100	46	44	33	1.000000	0.000000
100	46	44	34	1.000000	0.000000
100	46	44	35	1.000000	0.000000

N	n	k	x	P(x)	p(x)
100	46	44	36	1.000000	0.000000
100	46	44	37	1.000000	0.000000
100	46	44	38	1.000000	0.000000
100	46	44	39	1.000000	0.000000
100	46	44	40	1.000000	0.000000
100	46	44	41	1.000000	0.000000
100	46	44	42	1.000000	0.000000
100	46	44	43	1.000000	0.000000
100	46	44	44	1.000000	0.000000
100	46	45	0	0.000000	0.000000
100	46	45	1	0.000000	0.000000
100	46	45	2	0.000000	0.000000
100	46	45	3	0.000000	0.000000
100	46	45	4	0.000000	0.000000
100	46	45	5	0.000000	0.000000
100	46	45	6	0.000000	0.000000
100	46	45	7	0.000000	0.000000
100	46	45	8	0.000000	0.000000
100	46	45	9	0.000002	0.000002
100	46	45	10	0.000014	0.000012
100	46	45	11	0.000084	0.000070
100	46	45	12	0.000413	0.000329
100	46	45	13	0.001706	0.001293
100	46	45	14	0.005945	0.004239
100	46	45	15	0.017626	0.011681
100	46	45	16	0.044785	0.027159
100	46	45	17	0.098243	0.053458
100	46	45	18	0.187559	0.089316
100	46	45	19	0.314482	0.126923
100	46	45	20	0.468102	0.153620
100	46	45	21	0.626599	0.158497
100	46	45	22	0.766039	0.139440
100	46	45	23	0.870619	0.104580
100	46	45	24	0.937434	0.066815
100	46	45	25	0.973750	0.036316
100	46	45	26	0.990512	0.016761
100	46	45	27	0.997064	0.006553
100	46	45	28	0.999228	0.002163
100	46	45	29	0.999828	0.000601
100	46	45	30	0.999968	0.000140
100	46	45	31	0.999995	0.000027
100	46	45	32	0.999999	0.000004
100	46	45	33	1.000000	0.000001
100	46	45	34	1.000000	0.000000
100	46	45	35	1.000000	0.000000
100	46	45	36	1.000000	0.000000
100	46	45	37	1.000000	0.000000
100	46	45	38	1.000000	0.000000
100	46	45	39	1.000000	0.000000
100	46	45	40	1.000000	0.000000

N	n	k	x	P(x)	p(x)
100	46	45	41	1.000000	0.000000
100	46	45	42	1.000000	0.000000
100	46	45	43	1.000000	0.000000
100	46	45	44	1.000000	0.000000
100	46	45	45	1.000000	0.000000
100	46	46	0	0.000000	0.000000
100	46	46	1	0.000000	0.000000
100	46	46	2	0.000000	0.000000
100	46	46	3	0.000000	0.000000
100	46	46	4	0.000000	0.000000
100	46	46	5	0.000000	0.000000
100	46	46	6	0.000000	0.000000
100	46	46	7	0.000000	0.000000
100	46	46	8	0.000000	0.000000
100	46	46	9	0.000001	0.000001
100	46	46	10	0.000006	0.000005
100	46	46	11	0.000040	0.000033
100	46	46	12	0.000210	0.000170
100	46	46	13	0.000930	0.000721
100	46	46	14	0.003479	0.002548
100	46	46	15	0.011042	0.007564
100	46	46	16	0.029971	0.018929
100	46	46	17	0.070056	0.040085
100	46	46	18	0.142089	0.072033
100	46	46	19	0.252174	0.110085
100	46	46	20	0.395481	0.143307
100	46	46	21	0.554555	0.159073
100	46	46	22	0.705193	0.150638
100	46	46	23	0.826886	0.121693
100	46	46	24	0.910708	0.083823
100	46	46	25	0.959884	0.049176
100	46	46	26	0.984417	0.024532
100	46	46	27	0.994801	0.010384
100	46	46	28	0.998520	0.003719
100	46	46	29	0.999643	0.001123
100	46	46	30	0.999927	0.000285
100	46	46	31	0.999988	0.000060
100	46	46	32	0.999998	0.000011
100	46	46	33	1.000000	0.000002
100	46	46	34	1.000000	0.000000
100	46	46	35	1.000000	0.000000
100	46	46	36	1.000000	0.000000
100	46	46	37	1.000000	0.000000
100	46	46	38	1.000000	0.000000
100	46	46	39	1.000000	0.000000
100	46	46	40	1.000000	0.000000
100	46	46	41	1.000000	0.000000
100	46	46	42	1.000000	0.000000
100	46	46	43	1.000000	0.000000
100	46	46	44	1.000000	0.000000

N	n	k	x	P(x)	p(x)
100	46	46	45	1.000000	0.000000
100	46	46	46	1.000000	0.000000
100	47	1	0	0.530000	0.530000
100	47	1	1	1.000000	0.470000
100	47	2	0	0.278384	0.278384
100	47	2	1	0.781616	0.503232
100	47	2	2	1.000000	0.218384
100	47	3	0	0.144873	0.144873
100	47	3	1	0.545405	0.400532
100	47	3	2	0.899722	0.354317
100	47	3	3	1.000000	0.100278
100	47	4	0	0.074677	0.074677
100	47	4	1	0.355462	0.280785
100	47	4	2	0.735348	0.379886
100	47	4	3	0.954513	0.219165
100	47	4	4	1.000000	0.045487
100	47	5	0	0.038116	0.038116
100	47	5	1	0.220919	0.182803
100	47	5	2	0.557277	0.336357
100	47	5	3	0.854062	0.296786
100	47	5	4	0.979626	0.125563
100	47	5	5	1.000000	0.020374
100	47	6	0	0.019259	0.019259
100	47	6	1	0.132404	0.113145
100	47	6	2	0.397949	0.265545
100	47	6	3	0.716604	0.318654
100	47	6	4	0.922792	0.206188
100	47	6	5	0.990992	0.068201
100	47	6	6	1.000000	0.009008
100	47	7	0	0.009629	0.009629
100	47	7	1	0.077035	0.067406
100	47	7	2	0.270827	0.193792
100	47	7	3	0.567446	0.296620
100	47	7	[illegible]	0.828472	0.261025
100	47	7	[illegible]	0.960520	0.132048
100	47	7	[illegible]	0.996071	0.035551
100	47	7	7	1.000000	0.003929
100	47	8	[illegible]	0.004763	0.004763
100	47	8	1	0.043695	0.038932
100	47	8	2	0.177057	0.133362
100	47	8	3	0.427110	0.250054
100	47	8	4	0.707782	0.280672
100	47	8	5	0.900885	0.193103
100	47	8	6	0.980398	0.079513
100	47	8	7	0.998310	0.017912
100	47	8	8	1.000000	0.001690
100	47	9	0	0.002330	0.002330
100	47	9	1	0.024229	0.021899
100	47	9	2	0.111825	0.087596
100	47	9	3	0.307519	0.195694

Table for $N = 2$, $n = 1$, through $N = 100$, $n = 50$

N	n	k	x	P(x)	p(x)
100	47	9	4	0.576599	0.269079
100	47	9	5	0.812729	0.236131
100	47	9	6	0.944963	0.132233
100	47	9	7	0.990522	0.045559
100	47	9	8	0.999284	0.008761
100	47	9	9	1.000000	0.000716
100	47	10	0	0.001126	0.001126
100	47	10	1	0.013159	0.012032
100	47	10	2	0.068508	0.055349
100	47	10	3	0.212898	0.144390
100	47	10	4	0.449451	0.236553
100	47	10	5	0.703746	0.254295
100	47	10	6	0.885385	0.181639
100	47	10	7	0.970496	0.085111
100	47	10	8	0.995529	0.025033
100	47	10	9	0.999701	0.004172
100	47	10	10	1.000000	0.000299
100	47	11	0	0.000538	0.000538
100	47	11	1	0.007009	0.006471
100	47	11	2	0.040834	0.033825
100	47	11	3	0.142307	0.101474
100	47	11	4	0.336431	0.194124
100	47	11	5	0.585075	0.248644
100	47	11	6	0.802638	0.217563
100	47	11	7	0.932669	0.130031
100	47	11	8	0.984681	0.052012
100	47	11	9	0.997939	0.013258
100	47	11	10	0.999877	0.001938
100	47	11	11	1.000000	0.000123
100	47	12	0	0.000254	0.000254
100	47	12	1	0.003665	0.003411
100	47	12	2	0.023731	0.020067
100	47	12	3	0.092141	0.068409
100	47	12	4	0.242641	0.150501
100	47	12	5	0.467738	0.225096
100	47	12	6	0.702413	0.234675
100	47	12	7	0.874228	0.171816
100	47	12	8	0.961889	0.087661
100	47	12	9	0.992278	0.030389
100	47	12	10	0.999071	0.006793
100	47	12	11	0.999950	0.000879
100	47	12	12	1.000000	0.000050
100	47	13	0	0.000118	0.000118
100	47	13	1	0.001882	0.001763
100	47	13	2	0.013470	0.011588
100	47	13	3	0.057936	0.044466
100	47	13	4	0.169101	0.111165
100	47	13	5	0.360305	0.191204
100	47	13	6	0.593075	0.232770
100	47	13	7	0.796130	0.203055

N	n	k	x	P(x)	p(x)
100	47	13	8	0.923040	0.126909
100	47	13	9	0.979156	0.056116
100	47	13	10	0.996215	0.017059
100	47	13	11	0.999591	0.003375
100	47	13	12	0.999980	0.000389
100	47	13	13	1.000000	0.000020
100	47	14	0	0.000054	0.000054
100	47	14	1	0.000949	0.000895
100	47	14	2	0.007476	0.006527
100	47	14	3	0.035447	0.027971
100	47	14	4	0.114157	0.078710
100	47	14	5	0.268000	0.153842
100	47	14	6	0.483379	0.215379
100	47	14	7	0.702772	0.219393
100	47	14	8	0.866149	0.163377
100	47	14	9	0.954645	0.088496
100	47	14	10	0.988960	0.034315
100	47	14	11	0.998194	0.009234
100	47	14	12	0.999823	0.001629
100	47	14	13	0.999992	0.000169
100	47	14	14	1.000000	0.000008
100	47	15	0	0.000025	0.000025
100	47	15	1	0.000471	0.000446
100	47	15	2	0.004061	0.003590
100	47	15	3	0.021136	0.017076
100	47	15	4	0.074802	0.053666
100	47	15	5	0.192867	0.118065
100	47	15	6	0.380698	0.187831
100	47	15	7	0.600729	0.220030
100	47	15	8	0.792059	0.191331
100	47	15	9	0.915542	0.123483
100	47	15	10	0.974197	0.058654
100	47	15	11	0.994329	0.020132
100	47	15	12	0.999160	0.004832
100	47	15	13	0.999925	0.000765
100	47	15	14	0.999997	0.000071
100	47	15	15	1.000000	0.000003
100	47	16	0	0.000011	0.000011
100	47	16	1	0.000229	0.000218
100	47	16	2	0.002160	0.001931
100	47	16	3	0.012297	0.010137
100	47	16	4	0.047654	0.035356
100	47	16	5	0.134529	0.086876
100	47	16	6	0.290097	0.155568
100	47	16	7	0.497185	0.207087
100	47	16	8	0.704272	0.207087
100	47	16	9	0.860338	0.156066
100	47	16	10	0.948665	0.088327
100	47	16	11	0.985802	0.037137
100	47	16	12	0.997171	0.011369

N	n	k	x	P(x)	p(x)
100	47	16	13	0.999619	0.002449
100	47	16	14	0.999969	0.000350
100	47	16	15	0.999999	0.000030
100	47	16	16	1.000000	0.000001
100	47	17	0	0.000005	0.000005
100	47	17	1	0.000110	0.000105
100	47	17	2	0.001126	0.001016
100	47	17	3	0.006987	0.005862
100	47	17	4	0.029555	0.022567
100	47	17	5	0.091092	0.061537
100	47	17	6	0.214165	0.123074
100	47	17	7	0.398572	0.184406
100	47	17	8	0.608125	0.209553
100	47	17	9	0.789737	0.181612
100	47	17	10	0.909759	0.120022
100	47	17	11	0.969886	0.060127
100	47	17	12	0.992434	0.022548
100	47	17	13	0.998628	0.006194
100	47	17	14	0.999832	0.001203
100	47	17	15	0.999987	0.000156
100	47	17	16	0.999999	0.000012
100	47	17	17	1.000000	0.000000
100	47	18	0	0.000002	0.000002
100	47	18	1	0.000052	0.000050
100	47	18	2	0.000575	0.000523
100	47	18	3	0.003880	0.003305
100	47	18	4	0.017863	0.013983
100	47	18	5	0.059952	0.042089
100	47	18	6	0.153370	0.093418
100	47	18	7	0.309701	0.156331
100	47	18	8	0.509660	0.199959
100	47	18	9	0.706589	0.196929
100	47	18	10	0.856255	0.149666
100	47	18	11	0.943807	0.087552
100	47	18	12	0.982926	0.039119
100	47	18	13	0.996091	0.013165
100	47	18	14	0.999353	0.003262
100	47	18	15	0.999927	0.000574
100	47	18	16	0.999995	0.000068
100	47	18	17	1.000000	0.000005
100	47	18	18	1.000000	0.000000
100	47	19	0	0.000001	0.000001
100	47	19	1	0.000024	0.000023
100	47	19	2	0.000288	0.000264
100	47	19	3	0.002107	0.001819
100	47	19	4	0.010531	0.008424
100	47	19	5	0.038394	0.027864
100	47	19	6	0.106661	0.068267
100	47	19	7	0.233442	0.126781
100	47	19	8	0.414558	0.181116

N	n	k	x	P(x)	p(x)
100	47	19	9	0.615329	0.200772
100	47	19	10	0.788723	0.173394
100	47	19	11	0.905370	0.116647
100	47	19	12	0.966229	0.060859
100	47	19	13	0.990632	0.024403
100	47	19	14	0.998040	0.007408
100	47	19	15	0.999703	0.001663
100	47	19	16	0.999969	0.000266
100	47	19	17	0.999998	0.000029
100	47	19	18	1.000000	0.000002
100	47	19	19	1.000000	0.000000
100	47	20	0	0.000000	0.000000
100	47	20	1	0.000011	0.000010
100	47	20	2	0.000141	0.000130
100	47	20	3	0.001119	0.000977
100	47	20	4	0.006059	0.004940
100	47	20	5	0.023947	0.017888
100	47	20	6	0.072106	0.048160
100	47	20	7	0.170834	0.098728
100	47	20	8	0.327354	0.156520
100	47	20	9	0.521140	0.193786
100	47	20	10	0.709519	0.188378
100	47	20	11	0.853527	0.144008
100	47	20	12	0.939932	0.086405
100	47	20	13	0.980389	0.040457
100	47	20	14	0.995022	0.014633
100	47	20	15	0.999046	0.004024
100	47	20	16	0.999868	0.000821
100	47	20	17	0.999987	0.000120
100	47	20	18	0.999999	0.000012
100	47	20	19	1.000000	0.000001
100	47	20	20	1.000000	0.000000
100	47	21	0	0.000000	0.000000
100	47	21	1	0.000005	0.000005
100	47	21	2	0.000068	0.000063
100	47	21	3	0.000581	0.000513
100	47	21	4	0.003403	0.002822
100	47	21	5	0.014555	0.011152
100	47	21	6	0.047424	0.032869
100	47	21	7	0.121470	0.074046
100	47	21	8	0.251051	0.129580
100	47	21	9	0.429092	0.178041
100	47	21	10	0.622394	0.193302
100	47	21	11	0.788723	0.166329
100	47	21	12	0.902129	0.113406
100	47	21	13	0.963194	0.061065
100	47	21	14	0.988986	0.025791
100	47	21	15	0.997437	0.008451
100	47	21	16	0.999549	0.002113
100	47	21	17	0.999943	0.000393

Table for $N = 2$, $n = 1$, through $N = 100$, $n = 50$

N	n	k	x	P(x)	p(x)
100	47	21	18	0.999995	0.000052
100	47	21	19	1.000000	0.000005
100	47	21	20	1.000000	0.000000
100	47	21	21	1.000000	0.000000
100	47	22	0	0.000000	0.000000
100	47	22	1	0.000002	0.000002
100	47	22	2	0.000032	0.000030
100	47	22	3	0.000295	0.000263
100	47	22	4	0.001867	0.001572
100	47	22	5	0.008626	0.006759
100	47	22	6	0.030366	0.021739
100	47	22	7	0.083979	0.053613
100	47	22	8	0.187081	0.103102
100	47	22	9	0.343452	0.156371
100	47	22	10	0.531860	0.188408
100	47	22	11	0.712927	0.181068
100	47	22	12	0.851886	0.138959
100	47	22	13	0.936913	0.085027
100	47	22	14	0.978212	0.041299
100	47	22	15	0.994014	0.015801
100	47	22	16	0.998720	0.004707
100	47	22	17	0.999793	0.001073
100	47	22	18	0.999976	0.000182
100	47	22	19	0.999998	0.000022
100	47	22	20	1.000000	0.000002
100	47	22	21	1.000000	0.000000
100	47	22	22	1.000000	0.000000
100	47	23	0	0.000000	0.000000
100	47	23	1	0.000001	0.000001
100	47	23	2	0.000015	0.000014
100	47	23	3	0.000147	0.000132
100	47	23	4	0.001001	0.000854
100	47	23	5	0.004987	0.003986
100	47	23	6	0.018939	0.013952
100	47	23	7	0.056485	0.037546
100	47	23	8	0.135530	0.079045
100	47	23	9	0.267271	0.131741
100	47	23	10	0.442487	0.175216
100	47	23	11	0.629358	0.186871
100	47	23	12	0.789533	0.160175
100	47	23	13	0.899850	0.110317
100	47	23	14	0.960740	0.060889
100	47	23	15	0.987531	0.026791
100	47	23	16	0.996850	0.009319
100	47	23	17	0.999381	0.002531
100	47	23	18	0.999908	0.000527
100	47	23	19	0.999990	0.000082
100	47	23	20	0.999999	0.000009
100	47	23	21	1.000000	0.000001
100	47	23	22	1.000000	0.000000

N	n	k	x	P(x)	p(x)
100	47	23	23	1.000000	0.000000
100	47	24	0	0.000000	0.000000
100	47	24	1	0.000000	0.000000
100	47	24	2	0.000007	0.000006
100	47	24	3	0.000071	0.000065
100	47	24	4	0.000524	0.000452
100	47	24	5	0.002812	0.002289
100	47	24	6	0.011510	0.008697
100	47	24	7	0.036980	0.025471
100	47	24	8	0.095494	0.058514
100	47	24	9	0.202256	0.106762
100	47	24	10	0.358292	0.156036
100	47	24	11	0.541989	0.183697
100	47	24	12	0.716726	0.174736
100	47	24	13	0.851139	0.134413
100	47	24	14	0.934644	0.083506
100	47	24	15	0.976397	0.041753
100	47	24	16	0.993098	0.016701
100	47	24	17	0.998394	0.005297
100	47	24	18	0.999709	0.001315
100	47	24	19	0.999960	0.000251
100	47	24	20	0.999996	0.000036
100	47	24	21	1.000000	0.000004
100	47	24	22	1.000000	0.000000
100	47	24	23	1.000000	0.000000
100	47	24	24	1.000000	0.000000
100	47	25	0	0.000000	0.000000
100	47	25	1	0.000000	0.000000
100	47	25	2	0.000003	0.000003
100	47	25	3	0.000034	0.000031
100	47	25	4	0.000268	0.000234
100	47	25	5	0.001548	0.001280
100	47	25	6	0.006818	0.005270
100	47	25	7	0.023575	0.016757
100	47	25	8	0.065467	0.041892
100	47	25	9	0.148875	0.083408
100	47	25	10	0.282327	0.133452
100	47	25	11	0.454975	0.172648
100	47	25	12	0.636255	0.181280
100	47	25	13	0.791007	0.154751
100	47	25	14	0.898385	0.107379
100	47	25	15	0.958817	0.060432
100	47	25	16	0.986286	0.027469
100	47	25	17	0.996304	0.010018
100	47	25	18	0.999208	0.002904
100	47	25	19	0.999868	0.000660
100	47	25	20	0.999983	0.000116
100	47	25	21	0.999998	0.000015
100	47	25	22	1.000000	0.000001
100	47	25	23	1.000000	0.000000

N	n	k	x	P(x)	p(x)
100	47	25	24	1.000000	0.000000
100	47	25	25	1.000000	0.000000
100	47	26	0	0.000000	0.000000
100	47	26	1	0.000000	0.000000
100	47	26	2	0.000001	0.000001
100	47	26	3	0.000016	0.000014
100	47	26	4	0.000134	0.000118
100	47	26	5	0.000831	0.000697
100	47	26	6	0.003937	0.003106
100	47	26	7	0.014638	0.010701
100	47	26	8	0.043683	0.029045
100	47	26	9	0.106615	0.062932
100	47	26	10	0.216491	0.109876
100	47	26	11	0.372104	0.155613
100	47	26	12	0.551658	0.179554
100	47	26	13	0.720853	0.169195
100	47	26	14	0.851139	0.130286
100	47	26	15	0.933033	0.081894
100	47	26	16	0.974932	0.041899
100	47	26	17	0.992297	0.017365
100	47	26	18	0.998085	0.005788
100	47	26	19	0.999621	0.001536
100	47	26	20	0.999942	0.000320
100	47	26	21	0.999993	0.000051
100	47	26	22	0.999999	0.000006
100	47	26	23	1.000000	0.000001
100	47	26	24	1.000000	0.000000
100	47	26	25	1.000000	0.000000
100	47	26	26	1.000000	0.000000
100	47	27	0	0.000000	0.000000
100	47	27	1	0.000000	0.000000
100	47	27	2	0.000001	0.000001
100	47	27	3	0.000007	0.000007
100	47	27	4	0.000065	0.000058
100	47	27	5	0.000435	0.000370
100	47	27	6	0.002216	0.001781
100	47	27	7	0.008853	0.006637
100	47	27	8	0.028375	0.019522
100	47	27	9	0.074299	0.045923
100	47	27	10	0.161553	0.087254
100	47	27	11	0.296400	0.134847
100	47	27	12	0.466734	0.170333
100	47	27	13	0.643114	0.176381
100	47	27	14	0.793038	0.149924
100	47	27	15	0.897619	0.104581
100	47	27	16	0.957379	0.059761
100	47	27	17	0.985257	0.027877
100	47	27	18	0.995816	0.010560
100	47	27	19	0.999040	0.003223
100	47	27	20	0.999825	0.000785

N	n	k	x	P(x)	p(x)
100	47	27	21	0.999975	0.000150
100	47	27	22	0.999997	0.000022
100	47	27	23	1.000000	0.000002
100	47	27	24	1.000000	0.000000
100	47	27	25	1.000000	0.000000
100	47	27	26	1.000000	0.000000
100	47	27	27	1.000000	0.000000
100	47	28	0	0.000000	0.000000
100	47	28	1	0.000000	0.000000
100	47	28	2	0.000000	0.000000
100	47	28	3	0.000003	0.000003
100	47	28	4	0.000031	0.000028
100	47	28	5	0.000222	0.000191
100	47	28	6	0.001216	0.000994
100	47	28	7	0.005216	0.004001
100	47	28	8	0.017946	0.012729
100	47	28	9	0.050393	0.032448
100	47	28	10	0.117328	0.066935
100	47	28	11	0.229900	0.112572
100	47	28	12	0.385067	0.155167
100	47	28	13	0.560964	0.175898
100	47	28	14	0.725264	0.164300
100	47	28	15	0.851775	0.126511
100	47	28	16	0.932002	0.080226
100	47	28	17	0.973800	0.041799
100	47	28	18	0.991622	0.017821
100	47	28	19	0.997803	0.006182
100	47	28	20	0.999534	0.001731
100	47	28	21	0.999921	0.000387
100	47	28	22	0.999990	0.000068
100	47	28	23	0.999999	0.000009
100	47	28	24	1.000000	0.000001
100	47	28	25	1.000000	0.000000
100	47	28	26	1.000000	0.000000
100	47	28	27	1.000000	0.000000
100	47	28	28	1.000000	0.000000
100	47	29	0	0.000000	0.000000
100	47	29	1	0.000000	0.000000
100	47	29	2	0.000000	0.000000
100	47	29	3	0.000001	0.000001
100	47	29	4	0.000014	0.000013
100	47	29	5	0.000111	0.000096
100	47	29	6	0.000650	0.000539
100	47	29	7	0.002994	0.002344
100	47	29	8	0.011051	0.008057
100	47	29	9	0.033268	0.022218
100	47	29	10	0.082931	0.049663
100	47	29	11	0.173614	0.090683
100	47	29	12	0.309639	0.136024
100	47	29	13	0.477902	0.168263

Table for $N=2$, $n=1$, through $N=100$, $n=50$

N	n	k	x	P(x)	p(x)
100	47	29	14	0.649960	0.172059
100	47	29	15	0.795548	0.145588
100	47	29	16	0.897460	0.101912
100	47	29	17	0.956384	0.058924
100	47	29	18	0.984444	0.028059
100	47	29	19	0.995399	0.010956
100	47	29	20	0.998885	0.003486
100	47	29	21	0.999782	0.000896
100	47	29	22	0.999966	0.000184
100	47	29	23	0.999996	0.000030
100	47	29	24	1.000000	0.000004
100	47	29	25	1.000000	0.000000
100	47	29	26	1.000000	0.000000
100	47	29	27	1.000000	0.000000
100	47	29	28	1.000000	0.000000
100	47	29	29	1.000000	0.000000
100	47	30	0	0.000000	0.000000
100	47	30	1	0.000000	0.000000
100	47	30	2	0.000000	0.000000
100	47	30	3	0.000001	0.000001
100	47	30	4	0.000006	0.000006
100	47	30	5	0.000054	0.000047
100	47	30	6	0.000339	0.000285
100	47	30	7	0.001673	0.001335
100	47	30	8	0.006625	0.004952
100	47	30	9	0.021377	0.014752
100	47	30	10	0.057051	0.035673
100	47	30	11	0.127634	0.070583
100	47	30	12	0.242584	0.114950
100	47	30	13	0.397325	0.154741
100	47	30	14	0.569989	0.172664
100	47	30	15	0.729931	0.159942
100	47	30	16	0.852963	0.123032
100	47	30	17	0.931487	0.078523
100	47	30	18	0.972983	0.041496
100	47	30	19	0.991079	0.018096
100	47	30	20	0.997560	0.006481
100	47	30	21	0.999453	0.001894
100	47	30	22	0.999901	0.000448
100	47	30	23	0.999986	0.000085
100	47	30	24	0.999998	0.000013
100	47	30	25	1.000000	0.000001
100	47	30	26	1.000000	0.000000
100	47	30	27	1.000000	0.000000
100	47	30	28	1.000000	0.000000
100	47	30	29	1.000000	0.000000
100	47	30	30	1.000000	0.000000
100	47	31	0	0.000000	0.000000
100	47	31	1	0.000000	0.000000
100	47	31	2	0.000000	0.000000
100	47	31	3	0.000000	0.000000
100	47	31	4	0.000003	0.000003
100	47	31	5	0.000025	0.000023
100	47	31	6	0.000172	0.000146
100	47	31	7	0.000911	0.000739
100	47	31	8	0.003866	0.002956
100	47	31	9	0.013369	0.009503
100	47	31	10	0.038195	0.024826
100	47	31	11	0.091334	0.053139
100	47	31	12	0.185109	0.093775
100	47	31	13	0.322165	0.137056
100	47	31	14	0.488590	0.166425
100	47	31	15	0.656815	0.168224
100	47	31	16	0.798478	0.141663
100	47	31	17	0.897834	0.099356
100	47	31	18	0.955791	0.057958
100	47	31	19	0.983840	0.028049
100	47	31	20	0.995060	0.011220
100	47	31	21	0.998750	0.003690
100	47	31	22	0.999741	0.000991
100	47	31	23	0.999957	0.000215
100	47	31	24	0.999994	0.000037
100	47	31	25	0.999999	0.000005
100	47	31	26	1.000000	0.000001
100	47	31	27	1.000000	0.000000
100	47	31	28	1.000000	0.000000
100	47	31	29	1.000000	0.000000
100	47	31	30	1.000000	0.000000
100	47	31	31	1.000000	0.000000
100	47	32	0	0.000000	0.000000
100	47	32	1	0.000000	0.000000
100	47	32	2	0.000000	0.000000
100	47	32	3	0.000000	0.000000
100	47	32	4	0.000001	0.000001
100	47	32	5	0.000012	0.000010
100	47	32	6	0.000085	0.000073
100	47	32	7	0.000482	0.000398
100	47	32	8	0.002196	0.001713
100	47	32	9	0.008136	0.005940
100	47	32	10	0.024882	0.016747
100	47	32	11	0.063609	0.038727
100	47	32	12	0.137542	0.073933
100	47	32	13	0.254630	0.117088
100	47	32	14	0.408996	0.154365
100	47	32	15	0.578798	0.169802
100	47	32	16	0.734832	0.156034
100	47	32	17	0.854635	0.119803
100	47	32	18	0.931432	0.076797
100	47	32	19	0.972458	0.041026
100	47	32	20	0.990670	0.018211
100	47	32	21	0.997360	0.006690
100	47	32	22	0.999382	0.002023
100	47	32	23	0.999882	0.000500
100	47	32	24	0.999982	0.000100
100	47	32	25	0.999998	0.000016
100	47	32	26	1.000000	0.000002
100	47	32	27	1.000000	0.000000
100	47	32	28	1.000000	0.000000
100	47	32	29	1.000000	0.000000
100	47	32	30	1.000000	0.000000
100	47	32	31	1.000000	0.000000
100	47	32	32	1.000000	0.000000
100	47	33	0	0.000000	0.000000
100	47	33	1	0.000000	0.000000
100	47	33	2	0.000000	0.000000
100	47	33	3	0.000000	0.000000
100	47	33	4	0.000001	0.000000
100	47	33	5	0.000005	0.000005
100	47	33	6	0.000041	0.000035
100	47	33	7	0.000248	0.000208
100	47	33	8	0.001213	0.000965
100	47	33	9	0.004816	0.003603
100	47	33	10	0.015770	0.010954
100	47	33	11	0.043107	0.027337
100	47	33	12	0.099488	0.056382
100	47	33	13	0.196086	0.096598
100	47	33	14	0.334083	0.137997
100	47	33	15	0.498891	0.164808
100	47	33	16	0.663699	0.164808
100	47	33	17	0.801781	0.138082
100	47	33	18	0.898681	0.096900
100	47	33	19	0.955565	0.056884
100	47	33	20	0.983439	0.027873
100	47	33	21	0.994802	0.011363
100	47	33	22	0.998639	0.003837
100	47	33	23	0.999705	0.001067
100	47	33	24	0.999948	0.000242
100	47	33	25	0.999992	0.000045
100	47	33	26	0.999999	0.000007
100	47	33	27	1.000000	0.000001
100	47	33	28	1.000000	0.000000
100	47	33	29	1.000000	0.000000
100	47	33	30	1.000000	0.000000
100	47	33	31	1.000000	0.000000
100	47	33	32	1.000000	0.000000
100	47	33	33	1.000000	0.000000
100	47	34	0	0.000000	0.000000
100	47	34	1	0.000000	0.000000
100	47	34	2	0.000000	0.000000
100	47	34	3	0.000000	0.000000
100	47	34	4	0.000000	0.000000
100	47	34	5	0.000002	0.000002
100	47	34	6	0.000019	0.000017
100	47	34	7	0.000124	0.000105
100	47	34	8	0.000652	0.000527
100	47	34	9	0.002773	0.002121
100	47	34	10	0.009721	0.006948
100	47	34	11	0.028418	0.018697
100	47	34	12	0.070035	0.041617
100	47	34	13	0.147066	0.077031
100	47	34	14	0.266114	0.119048
100	47	34	15	0.420177	0.154062
100	47	34	16	0.587444	0.167268
100	47	34	17	0.739953	0.152509
100	47	34	18	0.856739	0.116786
100	47	34	19	0.931793	0.075054
100	47	34	20	0.972206	0.040413
100	47	34	21	0.990392	0.018186
100	47	34	22	0.997207	0.006815
100	47	34	23	0.999323	0.002116
100	47	34	24	0.999865	0.000541
100	47	34	25	0.999978	0.000113
100	47	34	26	0.999997	0.000019
100	47	34	27	1.000000	0.000003
100	47	34	28	1.000000	0.000000
100	47	34	29	1.000000	0.000000
100	47	34	30	1.000000	0.000000
100	47	34	31	1.000000	0.000000
100	47	34	32	1.000000	0.000000
100	47	34	33	1.000000	0.000000
100	47	34	34	1.000000	0.000000
100	47	35	0	0.000000	0.000000
100	47	35	1	0.000000	0.000000
100	47	35	2	0.000000	0.000000
100	47	35	3	0.000000	0.000000
100	47	35	4	0.000000	0.000000
100	47	35	5	0.000001	0.000001
100	47	35	6	0.000009	0.000008
100	47	35	7	0.000061	0.000052
100	47	35	8	0.000340	0.000280
100	47	35	9	0.001551	0.001211
100	47	35	10	0.005826	0.004274
100	47	35	11	0.018220	0.012394
100	47	35	12	0.047966	0.029746
100	47	35	13	0.107384	0.059418
100	47	35	14	0.206590	0.099207
100	47	35	15	0.345480	0.138889
100	47	35	16	0.508879	0.163399
100	47	35	17	0.670631	0.161752
100	47	35	18	0.805424	0.134793

Table for $N = 2$, $n = 1$, through $N = 100$, $n = 50$

N	n	k	x	P(x)	p(x)
100	47	35	19	0.899952	0.094528
100	47	35	20	0.955673	0.055722
100	47	35	21	0.983228	0.027555
100	47	35	22	0.994626	0.011398
100	47	35	23	0.998554	0.003928
100	47	35	24	0.999676	0.001122
100	47	35	25	0.999940	0.000264
100	47	35	26	0.999991	0.000051
100	47	35	27	0.999999	0.000008
100	47	35	28	1.000000	0.000001
100	47	35	29	1.000000	0.000000
100	47	35	30	1.000000	0.000000
100	47	35	31	1.000000	0.000000
100	47	35	32	1.000000	0.000000
100	47	35	33	1.000000	0.000000
100	47	35	34	1.000000	0.000000
100	47	35	35	1.000000	0.000000
100	47	36	0	0.000000	0.000000
100	47	36	1	0.000000	0.000000
100	47	36	2	0.000000	0.000000
100	47	36	3	0.000000	0.000000
100	47	36	4	0.000000	0.000000
100	47	36	5	0.000000	0.000000
100	47	36	6	0.000004	0.000003
100	47	36	7	0.000029	0.000025
100	47	36	8	0.000172	0.000144
100	47	36	9	0.000843	0.000671
100	47	36	10	0.003393	0.002549
100	47	36	11	0.011355	0.007963
100	47	36	12	0.031949	0.020593
100	47	36	13	0.076303	0.044355
100	47	36	14	0.156224	0.079920
100	47	36	15	0.277104	0.120880
100	47	36	16	0.430950	0.153847
100	47	36	17	0.595976	0.165026
100	47	36	18	0.745286	0.149309
100	47	36	19	0.859232	0.113946
100	47	36	20	0.932527	0.073295
100	47	36	21	0.972206	0.039679
100	47	36	22	0.990242	0.018036
100	47	36	23	0.997103	0.006861
100	47	36	24	0.999279	0.002176
100	47	36	25	0.999851	0.000572
100	47	36	26	0.999974	0.000124
100	47	36	27	0.999996	0.000022
100	47	36	28	0.999999	0.000003
100	47	36	29	1.000000	0.000000
100	47	36	30	1.000000	0.000000
100	47	36	31	1.000000	0.000000
100	47	36	32	1.000000	0.000000

N	n	k	x	P(x)	p(x)
100	47	36	33	1.000000	0.000000
100	47	36	34	1.000000	0.000000
100	47	36	35	1.000000	0.000000
100	47	36	36	1.000000	0.000000
100	47	37	0	0.000000	0.000000
100	47	37	1	0.000000	0.000000
100	47	37	2	0.000000	0.000000
100	47	37	3	0.000000	0.000000
100	47	37	4	0.000000	0.000000
100	47	37	5	0.000000	0.000000
100	47	37	6	0.000002	0.000001
100	47	37	7	0.000013	0.000011
100	47	37	8	0.000085	0.000072
100	47	37	9	0.000445	0.000360
100	47	37	10	0.001919	0.001474
100	47	37	11	0.006876	0.004958
100	47	37	12	0.020687	0.013810
100	47	37	13	0.052740	0.032053
100	47	37	14	0.115015	0.062275
100	47	37	15	0.216664	0.101649
100	47	37	16	0.356431	0.139767
100	47	37	17	0.518620	0.162190
100	47	37	18	0.677630	0.159009
100	47	37	19	0.809380	0.131751
100	47	37	20	0.901606	0.092225
100	47	37	21	0.956087	0.054481
100	47	37	22	0.983197	0.027110
100	47	37	23	0.994530	0.011334
100	47	37	24	0.998497	0.003967
100	47	37	25	0.999654	0.001157
100	47	37	26	0.999934	0.000280
100	47	37	27	0.999989	0.000056
100	47	37	28	0.999999	0.000009
100	47	37	29	1.000000	0.000001
100	47	37	30	1.000000	0.000000
100	47	37	31	1.000000	0.000000
100	47	37	32	1.000000	0.000000
100	47	37	33	1.000000	0.000000
100	47	37	34	1.000000	0.000000
100	47	37	35	1.000000	0.000000
100	47	37	36	1.000000	0.000000
100	47	37	37	1.000000	0.000000
100	47	38	0	0.000000	0.000000
100	47	38	1	0.000000	0.000000
100	47	38	2	0.000000	0.000000
100	47	38	3	0.000000	0.000000
100	47	38	4	0.000000	0.000000
100	47	38	5	0.000000	0.000000
100	47	38	6	0.000001	0.000001
100	47	38	7	0.000006	0.000005

N	n	k	x	P(x)	p(x)
100	47	38	8	0.000040	0.000035
100	47	38	9	0.000228	0.000187
100	47	38	10	0.001053	0.000826
100	47	38	11	0.004043	0.002990
100	47	38	12	0.013014	0.008971
100	47	38	13	0.035441	0.022427
100	47	38	14	0.082395	0.046953
100	47	38	15	0.165033	0.082638
100	47	38	16	0.287657	0.122624
100	47	38	17	0.441387	0.153731
100	47	38	18	0.604435	0.163048
100	47	38	19	0.750825	0.146390
100	47	38	20	0.862081	0.111256
100	47	38	21	0.933602	0.071522
100	47	38	22	0.972439	0.038836
100	47	38	23	0.990213	0.017774
100	47	38	24	0.997049	0.006836
100	47	38	25	0.999250	0.002201
100	47	38	26	0.999841	0.000591
100	47	38	27	0.999972	0.000131
100	47	38	28	0.999996	0.000024
100	47	38	29	0.999999	0.000004
100	47	38	30	1.000000	0.000000
100	47	38	31	1.000000	0.000000
100	47	38	32	1.000000	0.000000
100	47	38	33	1.000000	0.000000
100	47	38	34	1.000000	0.000000
100	47	38	35	1.000000	0.000000
100	47	38	36	1.000000	0.000000
100	47	38	37	1.000000	0.000000
100	47	38	38	1.000000	0.000000
100	47	39	0	0.000000	0.000000
100	47	39	1	0.000000	0.000000
100	47	39	2	0.000000	0.000000
100	47	39	3	0.000000	0.000000
100	47	39	4	0.000000	0.000000
100	47	39	5	0.000000	0.000000
100	47	39	6	0.000000	0.000000
100	47	39	7	0.000002	0.000002
100	47	39	8	0.000019	0.000016
100	47	39	9	0.000113	0.000094
100	47	39	10	0.000560	0.000448
100	47	39	11	0.002307	0.001747
100	47	39	12	0.007950	0.005643
100	47	39	13	0.023143	0.015193
100	47	39	14	0.057403	0.034261
100	47	39	15	0.122381	0.064977
100	47	39	16	0.226345	0.103964
100	47	39	17	0.367001	0.140657
100	47	39	18	0.528171	0.161169

N	n	k	x	P(x)	p(x)
100	47	39	19	0.684713	0.156542
100	47	39	20	0.813631	0.128917
100	47	39	21	0.903610	0.089979
100	47	39	22	0.956779	0.053169
100	47	39	23	0.983332	0.026553
100	47	39	24	0.994513	0.011180
100	47	39	25	0.998469	0.003956
100	47	39	26	0.999641	0.001172
100	47	39	27	0.999930	0.000289
100	47	39	28	0.999988	0.000059
100	47	39	29	0.999998	0.000010
100	47	39	30	1.000000	0.000001
100	47	39	31	1.000000	0.000000
100	47	39	32	1.000000	0.000000
100	47	39	33	1.000000	0.000000
100	47	39	34	1.000000	0.000000
100	47	39	35	1.000000	0.000000
100	47	39	36	1.000000	0.000000
100	47	39	37	1.000000	0.000000
100	47	39	38	1.000000	0.000000
100	47	39	39	1.000000	0.000000
100	47	40	0	0.000000	0.000000
100	47	40	1	0.000000	0.000000
100	47	40	2	0.000000	0.000000
100	47	40	3	0.000000	0.000000
100	47	40	4	0.000000	0.000000
100	47	40	5	0.000000	0.000000
100	47	40	6	0.000000	0.000000
100	47	40	7	0.000001	0.000001
100	47	40	8	0.000008	0.000007
100	47	40	9	0.000054	0.000046
100	47	40	10	0.000289	0.000235
100	47	40	11	0.001276	0.000987
100	47	40	12	0.004712	0.003436
100	47	40	13	0.014675	0.009962
100	47	40	14	0.038869	0.024194
100	47	40	15	0.088294	0.049425
100	47	40	16	0.173510	0.085216
100	47	40	17	0.297826	0.124315
100	47	40	18	0.451549	0.153723
100	47	40	19	0.612858	0.161308
100	47	40	20	0.756569	0.143711
100	47	40	21	0.865258	0.108689
100	47	40	22	0.934988	0.069730
100	47	40	23	0.972885	0.037897
100	47	40	24	0.990297	0.017412
100	47	40	25	0.997042	0.006745
100	47	40	26	0.999237	0.002195
100	47	40	27	0.999835	0.000598
100	47	40	28	0.999970	0.000135

Table for $N = 2,\ n = 1,$ through $N = 100,\ n = 50$

N	n	k	x	P(x)	p(x)
100	47	40	29	0.999995	0.000025
100	47	40	30	0.999999	0.000004
100	47	40	31	1.000000	0.000000
100	47	40	32	1.000000	0.000000
100	47	40	33	1.000000	0.000000
100	47	40	34	1.000000	0.000000
100	47	40	35	1.000000	0.000000
100	47	40	36	1.000000	0.000000
100	47	40	37	1.000000	0.000000
100	47	40	38	1.000000	0.000000
100	47	40	39	1.000000	0.000000
100	47	40	40	1.000000	0.000000
100	47	41	0	0.000000	0.000000
100	47	41	1	0.000000	0.000000
100	47	41	2	0.000000	0.000000
100	47	41	3	0.000000	0.000000
100	47	41	4	0.000000	0.000000
100	47	41	5	0.000000	0.000000
100	47	41	6	0.000000	0.000000
100	47	41	7	0.000000	0.000000
100	47	41	8	0.000004	0.000003
100	47	41	9	0.000025	0.000022
100	47	41	10	0.000144	0.000119
100	47	41	11	0.000684	0.000540
100	47	41	12	0.002708	0.002024
100	47	41	13	0.009029	0.006321
100	47	41	14	0.025562	0.016533
100	47	41	15	0.061934	0.036372
100	47	41	16	0.129482	0.067548
100	47	41	17	0.235668	0.106186
100	47	41	18	0.377250	0.141581
100	47	41	19	0.537580	0.160331
100	47	41	20	0.691899	0.154318
100	47	41	21	0.818159	0.126260
100	47	41	22	0.905934	0.087775
100	47	41	23	0.957726	0.051792
100	47	41	24	0.983623	0.025896
100	47	41	25	0.994569	0.010946
100	47	41	26	0.998469	0.003900
100	47	41	27	0.999636	0.001167
100	47	41	28	0.999927	0.000292
100	47	41	29	0.999988	0.000061
100	47	41	30	0.999998	0.000010
100	47	41	31	1.000000	0.000001
100	47	41	32	1.000000	0.000000
100	47	41	33	1.000000	0.000000
100	47	41	34	1.000000	0.000000
100	47	41	35	1.000000	0.000000
100	47	41	36	1.000000	0.000000
100	47	41	37	1.000000	0.000000

N	n	k	x	P(x)	p(x)
100	47	41	38	1.000000	0.000000
100	47	41	39	1.000000	0.000000
100	47	41	40	1.000000	0.000000
100	47	41	41	1.000000	0.000000
100	47	42	0	0.000000	0.000000
100	47	42	1	0.000000	0.000000
100	47	42	2	0.000000	0.000000
100	47	42	3	0.000000	0.000000
100	47	42	4	0.000000	0.000000
100	47	42	5	0.000000	0.000000
100	47	42	6	0.000000	0.000000
100	47	42	7	0.000000	0.000000
100	47	42	8	0.000001	0.000001
100	47	42	9	0.000011	0.000010
100	47	42	10	0.000069	0.000058
100	47	42	11	0.000355	0.000285
100	47	42	12	0.001507	0.001153
100	47	42	13	0.005387	0.003879
100	47	42	14	0.016315	0.010928
100	47	42	15	0.042207	0.025892
100	47	42	16	0.093991	0.051784
100	47	42	17	0.181675	0.087684
100	47	42	18	0.307659	0.125983
100	47	42	19	0.461491	0.153832
100	47	42	20	0.621278	0.159787
100	47	42	21	0.762519	0.141241
100	47	42	22	0.868741	0.106222
100	47	42	23	0.936658	0.067917
100	47	42	24	0.973527	0.036869
100	47	42	25	0.990487	0.016960
100	47	42	26	0.997081	0.006594
100	47	42	27	0.999240	0.002159
100	47	42	28	0.999833	0.000593
100	47	42	29	0.999969	0.000136
100	47	42	30	0.999995	0.000026
100	47	42	31	0.999999	0.000004
100	47	42	32	1.000000	0.000001
100	47	42	33	1.000000	0.000000
100	47	42	34	1.000000	0.000000
100	47	42	35	1.000000	0.000000
100	47	42	36	1.000000	0.000000
100	47	42	37	1.000000	0.000000
100	47	42	38	1.000000	0.000000
100	47	42	39	1.000000	0.000000
100	47	42	40	1.000000	0.000000
100	47	42	41	1.000000	0.000000
100	47	42	42	1.000000	0.000000
100	47	43	0	0.000000	0.000000
100	47	43	1	0.000000	0.000000
100	47	43	2	0.000000	0.000000

N	n	k	x	P(x)	p(x)
100	47	43	3	0.000000	0.000000
100	47	43	4	0.000000	0.000000
100	47	43	5	0.000000	0.000000
100	47	43	6	0.000000	0.000000
100	47	43	7	0.000000	0.000000
100	47	43	8	0.000001	0.000001
100	47	43	9	0.000005	0.000004
100	47	43	10	0.000032	0.000027
100	47	43	11	0.000178	0.000145
100	47	43	12	0.000812	0.000634
100	47	43	13	0.003112	0.002301
100	47	43	14	0.010097	0.006985
100	47	43	15	0.027922	0.017825
100	47	43	16	0.066313	0.038392
100	47	43	17	0.136321	0.070008
100	47	43	18	0.244667	0.108346
100	47	43	19	0.387227	0.142560
100	47	43	20	0.546895	0.159667
100	47	43	21	0.699204	0.152310
100	47	43	22	0.822956	0.123751
100	47	43	23	0.908555	0.085599
100	47	43	24	0.958907	0.050352
100	47	43	25	0.984054	0.025147
100	47	43	26	0.994693	0.010639
100	47	43	27	0.998496	0.003802
100	47	43	28	0.999639	0.001143
100	47	43	29	0.999927	0.000288
100	47	43	30	0.999988	0.000061
100	47	43	31	0.999998	0.000011
100	47	43	32	1.000000	0.000001
100	47	43	33	1.000000	0.000000
100	47	43	34	1.000000	0.000000
100	47	43	35	1.000000	0.000000
100	47	43	36	1.000000	0.000000
100	47	43	37	1.000000	0.000000
100	47	43	38	1.000000	0.000000
100	47	43	39	1.000000	0.000000
100	47	43	40	1.000000	0.000000
100	47	43	41	1.000000	0.000000
100	47	43	42	1.000000	0.000000
100	47	43	43	1.000000	0.000000
100	47	44	0	0.000000	0.000000
100	47	44	1	0.000000	0.000000
100	47	44	2	0.000000	0.000000
100	47	44	3	0.000000	0.000000
100	47	44	4	0.000000	0.000000
100	47	44	5	0.000000	0.000000
100	47	44	6	0.000000	0.000000
100	47	44	7	0.000000	0.000000
100	47	44	8	0.000000	0.000000

N	n	k	x	P(x)	p(x)
100	47	44	9	0.000002	0.000002
100	47	44	10	0.000014	0.000012
100	47	44	11	0.000086	0.000071
100	47	44	12	0.000422	0.000336
100	47	44	13	0.001740	0.001318
100	47	44	14	0.006053	0.004313
100	47	44	15	0.017915	0.011862
100	47	44	16	0.045434	0.027519
100	47	44	17	0.099475	0.054041
100	47	44	18	0.189544	0.090069
100	47	44	19	0.317198	0.127654
100	47	44	20	0.471263	0.154065
100	47	44	21	0.629730	0.158467
100	47	44	22	0.768679	0.138949
100	47	44	23	0.872513	0.103834
100	47	44	24	0.938589	0.066076
100	47	44	25	0.974348	0.035759
100	47	44	26	0.990774	0.016426
100	47	44	27	0.997161	0.006388
100	47	44	28	0.999258	0.002096
100	47	44	29	0.999836	0.000578
100	47	44	30	0.999970	0.000133
100	47	44	31	0.999995	0.000026
100	47	44	32	0.999999	0.000004
100	47	44	33	1.000000	0.000001
100	47	44	34	1.000000	0.000000
100	47	44	35	1.000000	0.000000
100	47	44	36	1.000000	0.000000
100	47	44	37	1.000000	0.000000
100	47	44	38	1.000000	0.000000
100	47	44	39	1.000000	0.000000
100	47	44	40	1.000000	0.000000
100	47	44	41	1.000000	0.000000
100	47	44	42	1.000000	0.000000
100	47	44	43	1.000000	0.000000
100	47	44	44	1.000000	0.000000
100	47	45	0	0.000000	0.000000
100	47	45	1	0.000000	0.000000
100	47	45	2	0.000000	0.000000
100	47	45	3	0.000000	0.000000
100	47	45	4	0.000000	0.000000
100	47	45	5	0.000000	0.000000
100	47	45	6	0.000000	0.000000
100	47	45	7	0.000000	0.000000
100	47	45	8	0.000000	0.000000
100	47	45	9	0.000001	0.000001
100	47	45	10	0.000006	0.000005
100	47	45	11	0.000040	0.000034
100	47	45	12	0.000212	0.000172
100	47	45	13	0.000940	0.000728

N	n	k	x	P(x)	p(x)
100	47	45	14	0.003512	0.002572
100	47	45	15	0.011137	0.007625
100	47	45	16	0.030200	0.019063
100	47	45	17	0.070524	0.040324
100	47	45	18	0.142901	0.072377
100	47	45	19	0.253371	0.110470
100	47	45	20	0.396981	0.143611
100	47	45	21	0.556156	0.159174
100	47	45	22	0.706648	0.150492
100	47	45	23	0.828013	0.121365
100	47	45	24	0.911451	0.083438
100	47	45	25	0.960300	0.048849
100	47	45	26	0.984614	0.024314
100	47	45	27	0.994880	0.010266
100	47	45	28	0.998547	0.003666
100	47	45	29	0.999650	0.001104
100	47	45	30	0.999929	0.000279
100	47	45	31	0.999988	0.000059
100	47	45	32	0.999998	0.000010
100	47	45	33	1.000000	0.000001
100	47	45	34	1.000000	0.000000
100	47	45	35	1.000000	0.000000
100	47	45	36	1.000000	0.000000
100	47	45	37	1.000000	0.000000
100	47	45	38	1.000000	0.000000
100	47	45	39	1.000000	0.000000
100	47	45	40	1.000000	0.000000
100	47	45	41	1.000000	0.000000
100	47	45	42	1.000000	0.000000
100	47	45	43	1.000000	0.000000
100	47	45	44	1.000000	0.000000
100	47	45	45	1.000000	0.000000
100	47	46	0	0.000000	0.000000
100	47	46	1	0.000000	0.000000
100	47	46	2	0.000000	0.000000
100	47	46	3	0.000000	0.000000
100	47	46	4	0.000000	0.000000
100	47	46	5	0.000000	0.000000
100	47	46	6	0.000000	0.000000
100	47	46	7	0.000000	0.000000
100	47	46	8	0.000000	0.000000
100	47	46	9	0.000000	0.000000
100	47	46	10	0.000003	0.000002
100	47	46	11	0.000018	0.000015
100	47	46	12	0.000103	0.000085
100	47	46	13	0.000490	0.000387
100	47	46	14	0.001969	0.001479
100	47	46	15	0.006700	0.004732
100	47	46	16	0.019455	0.012755
100	47	46	17	0.048529	0.029074

N	n	k	x	P(x)	p(x)
100	47	46	18	0.104739	0.056209
100	47	46	19	0.197132	0.092393
100	47	46	20	0.326482	0.129350
100	47	46	21	0.480910	0.154428
100	47	46	22	0.638242	0.157333
100	47	46	23	0.775053	0.136811
100	47	46	24	0.876558	0.101505
100	47	46	25	0.940760	0.064202
100	47	46	26	0.975331	0.034570
100	47	46	27	0.991147	0.015816
100	47	46	28	0.997280	0.006133
100	47	46	29	0.999289	0.002009
100	47	46	30	0.999843	0.000554
100	47	46	31	0.999971	0.000128
100	47	46	32	0.999995	0.000025
100	47	46	33	0.999999	0.000004
100	47	46	34	1.000000	0.000001
100	47	46	35	1.000000	0.000000
100	47	46	36	1.000000	0.000000
100	47	46	37	1.000000	0.000000
100	47	46	38	1.000000	0.000000
100	47	46	39	1.000000	0.000000
100	47	46	40	1.000000	0.000000
100	47	46	41	1.000000	0.000000
100	47	46	42	1.000000	0.000000
100	47	46	43	1.000000	0.000000
100	47	46	44	1.000000	0.000000
100	47	46	45	1.000000	0.000000
100	47	46	46	1.000000	0.000000
100	47	47	0	0.000000	0.000000
100	47	47	1	0.000000	0.000000
100	47	47	2	0.000000	0.000000
100	47	47	3	0.000000	0.000000
100	47	47	4	0.000000	0.000000
100	47	47	5	0.000000	0.000000
100	47	47	6	0.000000	0.000000
100	47	47	7	0.000000	0.000000
100	47	47	8	0.000000	0.000000
100	47	47	9	0.000000	0.000000
100	47	47	10	0.000001	0.000001
100	47	47	11	0.000008	0.000007
100	47	47	12	0.000048	0.000040
100	47	47	13	0.000246	0.000198
100	47	47	14	0.001065	0.000819
100	47	47	15	0.003896	0.002831
100	47	47	16	0.012133	0.008237
100	47	47	17	0.032377	0.020244
100	47	47	18	0.074552	0.042175
100	47	47	19	0.149224	0.074672
100	47	47	20	0.261807	0.112582

N	n	k	x	P(x)	p(x)
100	47	47	21	0.406555	0.144749
100	47	47	22	0.565403	0.158848
100	47	47	23	0.714249	0.148845
100	47	47	24	0.833325	0.119076
100	47	47	25	0.914604	0.081279
100	47	47	26	0.961887	0.047283
100	47	47	27	0.985289	0.023402
100	47	47	28	0.995122	0.009833
100	47	47	29	0.998619	0.003497
100	47	47	30	0.999668	0.001049
100	47	47	31	0.999933	0.000264
100	47	47	32	0.999988	0.000056
100	47	47	33	0.999998	0.000010
100	47	47	34	1.000000	0.000001
100	47	47	35	1.000000	0.000000
100	47	47	36	1.000000	0.000000
100	47	47	37	1.000000	0.000000
100	47	47	38	1.000000	0.000000
100	47	47	39	1.000000	0.000000
100	47	47	40	1.000000	0.000000
100	47	47	41	1.000000	0.000000
100	47	47	42	1.000000	0.000000
100	47	47	43	1.000000	0.000000
100	47	47	44	1.000000	0.000000
100	47	47	45	1.000000	0.000000
100	47	47	46	1.000000	0.000000
100	47	47	47	1.000000	0.000000
100	48	1	0	0.520000	0.520000
100	48	1	1	1.000000	0.480000
100	48	2	0	0.267879	0.267879
100	48	2	1	0.772121	0.504242
100	48	2	2	1.000000	0.227879
100	48	3	0	0.136673	0.136673
100	48	3	1	0.530291	0.393618
100	48	3	2	0.893036	0.362746
100	48	3	3	1.000000	0.106963
100	48	4	0	0.069041	0.069041
100	48	4	1	0.339569	0.270528
100	48	4	2	0.721013	0.381444
100	48	4	3	0.950378	0.229365
100	48	4	4	1.000000	0.049622
100	48	5	0	0.034520	0.034520
100	48	5	1	0.207123	0.172602
100	48	5	2	0.538237	0.331115
100	48	5	3	0.842863	0.304625
100	48	5	4	0.977256	0.134394
100	48	5	5	1.000000	0.022744
100	48	6	0	0.017079	0.017079
100	48	6	1	0.121730	0.104652
100	48	6	2	0.377908	0.256178

N	n	k	x	P(x)	p(x)
100	48	6	3	0.698567	0.320658
100	48	6	4	0.915011	0.216444
100	48	6	5	0.989705	0.074695
100	48	6	6	1.000000	0.010294
100	48	7	0	0.008358	0.008358
100	48	7	1	0.069404	0.061047
100	48	7	2	0.252544	0.183140
100	48	7	3	0.545060	0.292515
100	48	7	4	0.813697	0.268637
100	48	7	5	0.955537	0.141840
100	48	7	6	0.995400	0.039864
100	48	7	7	1.000000	0.004600
100	48	8	0	0.004044	0.004044
100	48	8	1	0.038553	0.034509
100	48	8	2	0.161959	0.123406
100	48	8	3	0.403520	0.241561
100	48	8	4	0.686600	0.283079
100	48	8	5	0.889955	0.203355
100	48	8	6	0.977397	0.087443
100	48	8	7	0.997972	0.020575
100	48	8	8	1.000000	0.002028
100	48	9	0	0.001934	0.001934
100	48	9	1	0.020923	0.018989
100	48	9	2	0.100256	0.079333
100	48	9	3	0.285365	0.185109
100	48	9	4	0.551214	0.265849
100	48	9	5	0.794908	0.243695
100	48	9	6	0.937478	0.142570
100	48	9	7	0.988803	0.051325
100	48	9	8	0.999118	0.010315
100	48	9	9	1.000000	0.000882
100	48	10	0	0.000914	0.000914
100	48	10	1	0.011116	0.010202
100	48	10	2	0.060154	0.049038
100	48	10	3	0.193828	0.133674
100	48	10	4	0.422672	0.228844
100	48	10	5	0.679756	0.257084
100	48	10	6	0.871677	0.191921
100	48	10	7	0.965678	0.094002
100	48	10	8	0.994584	0.028906
100	48	10	9	0.999622	0.005038
100	48	10	10	1.000000	0.000378
100	48	11	0	0.000426	0.000426
100	48	11	1	0.005788	0.005362
100	48	11	2	0.035090	0.029302
100	48	11	3	0.126991	0.091901
100	48	11	4	0.310792	0.183802
100	48	11	5	0.556927	0.246134
100	48	11	6	0.782114	0.225187
100	48	11	7	0.922855	0.140742

N	n	k	x	P(x)	p(x)
100	48	11	8	0.981737	0.058882
100	48	11	9	0.997439	0.015702
100	48	11	10	0.999840	0.002401
100	48	11	11	1.000000	0.000160
100	48	12	0	0.000196	0.000196
100	48	12	1	0.002957	0.002760
100	48	12	2	0.019945	0.016988
100	48	12	3	0.080524	0.060579
100	48	12	4	0.219924	0.139400
100	48	12	5	0.438008	0.218084
100	48	12	6	0.675846	0.237838
100	48	12	7	0.858019	0.182174
100	48	12	8	0.955273	0.097254
100	48	12	9	0.990558	0.035285
100	48	12	10	0.998815	0.008257
100	48	12	11	0.999934	0.001119
100	48	12	12	1.000000	0.000066
100	48	13	0	0.000089	0.000089
100	48	13	1	0.001482	0.001393
100	48	13	2	0.011065	0.009582
100	48	13	3	0.049546	0.038481
100	48	13	4	0.150224	0.100678
100	48	13	5	0.331444	0.181220
100	48	13	6	0.562332	0.230888
100	48	13	7	0.773143	0.210811
100	48	13	8	0.911067	0.137924
100	48	13	9	0.974921	0.063854
100	48	13	10	0.995250	0.020329
100	48	13	11	0.999463	0.004214
100	48	13	12	0.999973	0.000509
100	48	13	13	1.000000	0.000027
100	48	14	0	0.000040	0.000040
100	48	14	1	0.000730	0.000690
100	48	14	2	0.005998	0.005268
100	48	14	3	0.029642	0.023644
100	48	14	4	0.099306	0.069664
100	48	14	5	0.241876	0.142569
100	48	14	6	0.450869	0.208993
100	48	14	7	0.673795	0.222926
100	48	14	8	0.847654	0.173858
100	48	14	9	0.946297	0.098643
100	48	14	10	0.986370	0.040074
100	48	14	11	0.997671	0.011301
100	48	14	12	0.999762	0.002091
100	48	14	13	0.999989	0.000227
100	48	14	14	1.000000	0.000011
100	48	15	0	0.000018	0.000018
100	48	15	1	0.000353	0.000335
100	48	15	2	0.003180	0.002827
100	48	15	3	0.017270	0.014090

N	n	k	x	P(x)	p(x)
100	48	15	4	0.063664	0.046394
100	48	15	5	0.170591	0.106927
100	48	15	6	0.348802	0.178211
100	48	15	7	0.567516	0.218714
100	48	15	8	0.766789	0.199273
100	48	15	9	0.901563	0.134774
100	48	15	10	0.968663	0.067100
100	48	15	11	0.992809	0.024146
100	48	15	12	0.998887	0.006078
100	48	15	13	0.999897	0.001010
100	48	15	14	0.999996	0.000099
100	48	15	15	1.000000	0.000004
100	48	16	0	0.000008	0.000008
100	48	16	1	0.000168	0.000160
100	48	16	2	0.001650	0.001483
100	48	16	3	0.009811	0.008161
100	48	16	4	0.039648	0.029838
100	48	16	5	0.116499	0.076850
100	48	16	6	0.260745	0.144246
100	48	16	7	0.462019	0.201274
100	48	16	8	0.673014	0.210995
100	48	16	9	0.839726	0.166712
100	48	16	10	0.938666	0.098940
100	48	16	11	0.982299	0.043633
100	48	16	12	0.996313	0.014014
100	48	16	13	0.999481	0.003168
100	48	16	14	0.999956	0.000475
100	48	16	15	0.999998	0.000042
100	48	16	16	1.000000	0.000002
100	48	17	0	0.000003	0.000003
100	48	17	1	0.000078	0.000075
100	48	17	2	0.000838	0.000760
100	48	17	3	0.005439	0.004601
100	48	17	4	0.024019	0.018580
100	48	17	5	0.077159	0.053139
100	48	17	6	0.188622	0.111463
100	48	17	7	0.363778	0.175156
100	48	17	8	0.572540	0.208762
100	48	17	9	0.762324	0.189784
100	48	17	10	0.893907	0.131583
100	48	17	11	0.963079	0.069172
100	48	17	12	0.990307	0.027227
100	48	17	13	0.998161	0.007854
100	48	17	14	0.999764	0.001603
100	48	17	15	0.999982	0.000218
100	48	17	16	0.999999	0.000018
100	48	17	17	1.000000	0.000001
100	48	18	0	0.000001	0.000001
100	48	18	1	0.000036	0.000034
100	48	18	2	0.000417	0.000381

N	n	k	x	P(x)	p(x)
100	48	18	3	0.002945	0.002528
100	48	18	4	0.014170	0.011225
100	48	18	5	0.049629	0.035459
100	48	18	6	0.132219	0.082590
100	48	18	7	0.277255	0.145036
100	48	18	8	0.471932	0.194677
100	48	18	9	0.673148	0.201216
100	48	18	10	0.833664	0.160516
100	48	18	11	0.932244	0.098579
100	48	18	12	0.978497	0.046254
100	48	18	13	0.994849	0.016352
100	48	18	14	0.999107	0.004258
100	48	18	15	0.999895	0.000788
100	48	18	16	0.999992	0.000097
100	48	18	17	1.000000	0.000007
100	48	18	18	1.000000	0.000000
100	48	19	0	0.000001	0.000001
100	48	19	1	0.000016	0.000015
100	48	19	2	0.000203	0.000187
100	48	19	3	0.001557	0.001354
100	48	19	4	0.008146	0.006589
100	48	19	5	0.031034	0.022888
100	48	19	6	0.089916	0.058882
100	48	19	7	0.204737	0.114820
100	48	19	8	0.376968	0.172231
100	48	19	9	0.577448	0.200480
100	48	19	10	0.759279	0.181831
100	48	19	11	0.887763	0.128484
100	48	19	12	0.958191	0.070428
100	48	19	13	0.987870	0.029679
100	48	19	14	0.997341	0.009472
100	48	19	15	0.999578	0.002236
100	48	19	16	0.999954	0.000377
100	48	19	17	0.999997	0.000043
100	48	19	18	1.000000	0.000003
100	48	19	19	1.000000	0.000000
100	48	20	0	0.000000	0.000000
100	48	20	1	0.000007	0.000007
100	48	20	2	0.000097	0.000090
100	48	20	3	0.000805	0.000708
100	48	20	4	0.004567	0.003762
100	48	20	5	0.018884	0.014317
100	48	20	6	0.059385	0.040501
100	48	20	7	0.146618	0.087233
100	48	20	8	0.291916	0.145298
100	48	20	9	0.480920	0.189005
100	48	20	10	0.673975	0.193055
100	48	20	11	0.829072	0.155097
100	48	20	12	0.926889	0.097817
100	48	20	13	0.975045	0.048156

N	n	k	x	P(x)	p(x)
100	48	20	14	0.993366	0.018320
100	48	20	15	0.998667	0.005301
100	48	20	16	0.999806	0.001139
100	48	20	17	0.999981	0.000175
100	48	20	18	0.999999	0.000018
100	48	20	19	1.000000	0.000001
100	48	20	20	1.000000	0.000000
100	48	21	0	0.000000	0.000000
100	48	21	1	0.000003	0.000003
100	48	21	2	0.000045	0.000042
100	48	21	3	0.000407	0.000361
100	48	21	4	0.002498	0.002091
100	48	21	5	0.011189	0.008691
100	48	21	6	0.038122	0.026933
100	48	21	7	0.101911	0.063789
100	48	21	8	0.219267	0.117356
100	48	21	9	0.388780	0.169514
100	48	21	10	0.582274	0.193494
100	48	21	11	0.757340	0.175066
100	48	21	12	0.882872	0.125532
100	48	21	13	0.953977	0.071105
100	48	21	14	0.985579	0.031602
100	48	21	15	0.996480	0.010901
100	48	21	16	0.999350	0.002870
100	48	21	17	0.999913	0.000563
100	48	21	18	0.999992	0.000079
100	48	21	19	0.999999	0.000007
100	48	21	20	1.000000	0.000000
100	48	21	21	1.000000	0.000000
100	48	22	0	0.000000	0.000000
100	48	22	1	0.000001	0.000001
100	48	22	2	0.000021	0.000019
100	48	22	3	0.000201	0.000180
100	48	22	4	0.001333	0.001132
100	48	22	5	0.006458	0.005125
100	48	22	6	0.023803	0.017345
100	48	22	7	0.068805	0.045002
100	48	22	8	0.159846	0.091041
100	48	22	9	0.305096	0.145250
100	48	22	10	0.489201	0.184105
100	48	22	11	0.675347	0.186146
100	48	22	12	0.825667	0.150320
100	48	22	13	0.922475	0.096807
100	48	22	14	0.971978	0.049504
100	48	22	15	0.991927	0.019948
100	48	22	16	0.998188	0.006261
100	48	22	17	0.999692	0.001504
100	48	22	18	0.999962	0.000270
100	48	22	19	0.999997	0.000035
100	48	22	20	1.000000	0.000003

Table for $N = 2$, $n = 1$, through $N = 100$, $n = 50$

N	n	k	x	P(x)	p(x)
100	48	22	21	1.000000	0.000000
100	48	22	22	1.000000	0.000000
100	48	23	0	0.000000	0.000000
100	48	23	1	0.000001	0.000001
100	48	23	2	0.000009	0.000009
100	48	23	3	0.000097	0.000088
100	48	23	4	0.000694	0.000598
100	48	23	5	0.003633	0.002939
100	48	23	6	0.014464	0.010831
100	48	23	7	0.045150	0.030687
100	48	23	8	0.113158	0.068008
100	48	23	9	0.232471	0.119313
100	48	23	10	0.399509	0.167038
100	48	23	11	0.587047	0.187538
100	48	23	12	0.756289	0.169242
100	48	23	13	0.879036	0.122747
100	48	23	14	0.950400	0.071364
100	48	23	15	0.983487	0.033087
100	48	23	16	0.995619	0.012132
100	48	23	17	0.999094	0.003475
100	48	23	18	0.999858	0.000764
100	48	23	19	0.999984	0.000126
100	48	23	20	0.999999	0.000015
100	48	23	21	1.000000	0.000001
100	48	23	22	1.000000	0.000000
100	48	23	23	1.000000	0.000000
100	48	24	0	0.000000	0.000000
100	48	24	1	0.000000	0.000000
100	48	24	2	0.000004	0.000004
100	48	24	3	0.000046	0.000042
100	48	24	4	0.000353	0.000307
100	48	24	5	0.001992	0.001639
100	48	24	6	0.008556	0.006564
100	48	24	7	0.028811	0.020255
100	48	24	8	0.077830	0.049019
100	48	24	9	0.172040	0.094211
100	48	24	10	0.317075	0.145035
100	48	24	11	0.496932	0.179857
100	48	24	12	0.677163	0.180231
100	48	24	13	0.823242	0.146079
100	48	24	14	0.918888	0.095647
100	48	24	15	0.969307	0.050418
100	48	24	16	0.990577	0.021270
100	48	24	17	0.997695	0.007118
100	48	24	18	0.999560	0.001865
100	48	24	19	0.999936	0.000376
100	48	24	20	0.999993	0.000057
100	48	24	21	0.999999	0.000006
100	48	24	22	1.000000	0.000000
100	48	24	23	1.000000	0.000000

N	n	k	x	P(x)	p(x)
100	48	24	24	1.000000	0.000000
100	48	25	0	0.000000	0.000000
100	48	25	1	0.000000	0.000000
100	48	25	2	0.000002	0.000002
100	48	25	3	0.000021	0.000019
100	48	25	4	0.000175	0.000154
100	48	25	5	0.001065	0.000890
100	48	25	6	0.004928	0.003864
100	48	25	7	0.017884	0.012955
100	48	25	8	0.052030	0.034146
100	48	25	9	0.123695	0.071665
100	48	25	10	0.244558	0.120862
100	48	25	11	0.409370	0.164812
100	48	25	12	0.591790	0.182420
100	48	25	13	0.755968	0.164178
100	48	25	14	0.876099	0.120130
100	48	25	15	0.947415	0.071316
100	48	25	16	0.981621	0.034207
100	48	25	17	0.994792	0.013170
100	48	25	18	0.998824	0.004032
100	48	25	19	0.999793	0.000969
100	48	25	20	0.999972	0.000179
100	48	25	21	0.999997	0.000025
100	48	25	22	1.000000	0.000002
100	48	25	23	1.000000	0.000000
100	48	25	24	1.000000	0.000000
100	48	25	25	1.000000	0.000000
100	48	26	0	0.000000	0.000000
100	48	26	1	0.000000	0.000000
100	48	26	2	0.000001	0.000001
100	48	26	3	0.000009	0.000009
100	48	26	4	0.000085	0.000075
100	48	26	5	0.000555	0.000470
100	48	26	6	0.002765	0.002210
100	48	26	7	0.010802	0.008037
100	48	26	8	0.033819	0.023017
100	48	26	9	0.086429	0.052611
100	48	26	10	0.183321	0.096891
100	48	26	11	0.328062	0.144742
100	48	26	12	0.504228	0.176166
100	48	26	13	0.679352	0.175124
100	48	26	14	0.821640	0.142288
100	48	26	15	0.916036	0.094396
100	48	26	16	0.967026	0.050991
100	48	26	17	0.989348	0.022321
100	48	26	18	0.997211	0.007863
100	48	26	19	0.999418	0.002207
100	48	26	20	0.999905	0.000487
100	48	26	21	0.999988	0.000083
100	48	26	22	0.999999	0.000011

N	n	k	x	P(x)	p(x)
100	48	26	23	1.000000	0.000001
100	48	26	24	1.000000	0.000000
100	48	26	25	1.000000	0.000000
100	48	26	26	1.000000	0.000000
100	48	27	0	0.000000	0.000000
100	48	27	1	0.000000	0.000000
100	48	27	2	0.000000	0.000000
100	48	27	3	0.000004	0.000004
100	48	27	4	0.000040	0.000036
100	48	27	5	0.000282	0.000242
100	48	27	6	0.001510	0.001229
100	48	27	7	0.006349	0.004838
100	48	27	8	0.021377	0.015028
100	48	27	9	0.058702	0.037325
100	48	27	10	0.133566	0.074864
100	48	27	11	0.255692	0.122126
100	48	27	12	0.418526	0.162834
100	48	27	13	0.596523	0.177997
100	48	27	14	0.756264	0.159741
100	48	27	15	0.873940	0.117676
100	48	27	16	0.944976	0.071036
100	48	27	17	0.979997	0.035021
100	48	27	18	0.994023	0.014026
100	48	27	19	0.998553	0.004530
100	48	27	20	0.999721	0.001168
100	48	27	21	0.999958	0.000237
100	48	27	22	0.999995	0.000037
100	48	27	23	0.999999	0.000004
100	48	27	24	1.000000	0.000000
100	48	27	25	1.000000	0.000000
100	48	27	26	1.000000	0.000000
100	48	27	27	1.000000	0.000000
100	48	28	0	0.000000	0.000000
100	48	28	1	0.000000	0.000000
100	48	28	2	0.000000	0.000000
100	48	28	3	0.000002	0.000002
100	48	28	4	0.000018	0.000017
100	48	28	5	0.000139	0.000121
100	48	28	6	0.000803	0.000664
100	48	28	7	0.003631	0.002828
100	48	28	8	0.013142	0.009511
100	48	28	9	0.038761	0.025619
100	48	28	10	0.094596	0.055834
100	48	28	11	0.193792	0.099197
100	48	28	12	0.338224	0.144432
100	48	28	13	0.511182	0.172958
100	48	28	14	0.681864	0.170682
100	48	28	15	0.820744	0.138880
100	48	28	16	0.913837	0.093093
100	48	28	17	0.965125	0.051288

N	n	k	x	P(x)	p(x)
100	48	28	18	0.988259	0.023134
100	48	28	19	0.996754	0.008495
100	48	28	20	0.999273	0.002519
100	48	28	21	0.999870	0.000597
100	48	28	22	0.999982	0.000112
100	48	28	23	0.999998	0.000016
100	48	28	24	1.000000	0.000002
100	48	28	25	1.000000	0.000000
100	48	28	26	1.000000	0.000000
100	48	28	27	1.000000	0.000000
100	48	28	28	1.000000	0.000000
100	48	29	0	0.000000	0.000000
100	48	29	1	0.000000	0.000000
100	48	29	2	0.000000	0.000000
100	48	29	3	0.000001	0.000001
100	48	29	4	0.000008	0.000007
100	48	29	5	0.000067	0.000059
100	48	29	6	0.000416	0.000349
100	48	29	7	0.002021	0.001605
100	48	29	8	0.007858	0.005837
100	48	29	9	0.024884	0.017026
100	48	29	10	0.065127	0.040243
100	48	29	11	0.142816	0.077689
100	48	29	12	0.266008	0.123192
100	48	29	13	0.427105	0.161097
100	48	29	14	0.601264	0.174159
100	48	29	15	0.757091	0.155827
100	48	29	16	0.872463	0.115372
100	48	29	17	0.943043	0.070580
100	48	29	18	0.978620	0.035577
100	48	29	19	0.993332	0.014712
100	48	29	20	0.998293	0.004961
100	48	29	21	0.999646	0.001353
100	48	29	22	0.999942	0.000295
100	48	29	23	0.999992	0.000051
100	48	29	24	0.999999	0.000007
100	48	29	25	1.000000	0.000001
100	48	29	26	1.000000	0.000000
100	48	29	27	1.000000	0.000000
100	48	29	28	1.000000	0.000000
100	48	29	29	1.000000	0.000000
100	48	30	0	0.000000	0.000000
100	48	30	1	0.000000	0.000000
100	48	30	2	0.000000	0.000000
100	48	30	3	0.000000	0.000000
100	48	30	4	0.000004	0.000003
100	48	30	5	0.000031	0.000028
100	48	30	6	0.000210	0.000178
100	48	30	7	0.001094	0.000885
100	48	30	8	0.004570	0.003476

Table for $N = 2$, $n = 1$, through $N = 100$, $n = 50$

N	n	k	x	P(x)	p(x)
100	48	30	9	0.015532	0.010962
100	48	30	10	0.043589	0.028057
100	48	30	11	0.102330	0.058742
100	48	30	12	0.203545	0.101214
100	48	30	13	0.347691	0.144147
100	48	30	14	0.517864	0.170173
100	48	30	15	0.684665	0.166800
100	48	30	16	0.820464	0.135800
100	48	30	17	0.912226	0.091762
100	48	30	18	0.963587	0.051361
100	48	30	19	0.987323	0.023736
100	48	30	20	0.996337	0.009014
100	48	30	21	0.999132	0.002795
100	48	30	22	0.999833	0.000702
100	48	30	23	0.999974	0.000141
100	48	30	24	0.999997	0.000022
100	48	30	25	1.000000	0.000003
100	48	30	26	1.000000	0.000000
100	48	30	27	1.000000	0.000000
100	48	30	28	1.000000	0.000000
100	48	30	29	1.000000	0.000000
100	48	30	30	1.000000	0.000000
100	48	31	0	0.000000	0.000000
100	48	31	1	0.000000	0.000000
100	48	31	2	0.000000	0.000000
100	48	31	3	0.000000	0.000000
100	48	31	4	0.000002	0.000001
100	48	31	5	0.000014	0.000013
100	48	31	6	0.000103	0.000088
100	48	31	7	0.000576	0.000473
100	48	31	8	0.002584	0.002008
100	48	31	9	0.009424	0.006841
100	48	31	10	0.028358	0.018934
100	48	31	11	0.071281	0.042923
100	48	31	12	0.151492	0.080210
100	48	31	13	0.275618	0.124126
100	48	31	14	0.435209	0.159591
100	48	31	15	0.606030	0.170821
100	48	31	16	0.758384	0.152354
100	48	31	17	0.871589	0.113204
100	48	31	18	0.941575	0.069987
100	48	31	19	0.977490	0.035914
100	48	31	20	0.992731	0.015242
100	48	31	21	0.998054	0.005322
100	48	31	22	0.999573	0.001519
100	48	31	23	0.999924	0.000351
100	48	31	24	0.999989	0.000065
100	48	31	25	0.999999	0.000009
100	48	31	26	1.000000	0.000001
100	48	31	27	1.000000	0.000000

N	n	k	x	P(x)	p(x)
100	48	31	28	1.000000	0.000000
100	48	31	29	1.000000	0.000000
100	48	31	30	1.000000	0.000000
100	48	31	31	1.000000	0.000000
100	48	32	0	0.000000	0.000000
100	48	32	1	0.000000	0.000000
100	48	32	2	0.000000	0.000000
100	48	32	3	0.000000	0.000000
100	48	32	4	0.000001	0.000001
100	48	32	5	0.000006	0.000006
100	48	32	6	0.000049	0.000043
100	48	32	7	0.000295	0.000246
100	48	32	8	0.001420	0.001125
100	48	32	9	0.005558	0.004138
100	48	32	10	0.017931	0.012373
100	48	32	11	0.048265	0.030334
100	48	32	12	0.109643	0.061378
100	48	32	13	0.212655	0.103012
100	48	32	14	0.356570	0.143914
100	48	32	15	0.524333	0.167763
100	48	32	16	0.687727	0.163394
100	48	32	17	0.820729	0.133001
100	48	32	18	0.911146	0.090418
100	48	32	19	0.962395	0.051249
100	48	32	20	0.986546	0.024151
100	48	32	21	0.995971	0.009425
100	48	32	22	0.999000	0.003029
100	48	32	23	0.999797	0.000796
100	48	32	24	0.999967	0.000170
100	48	32	25	0.999995	0.000029
100	48	32	26	0.999999	0.000004
100	48	32	27	1.000000	0.000000
100	48	32	28	1.000000	0.000000
100	48	32	29	1.000000	0.000000
100	48	32	30	1.000000	0.000000
100	48	32	31	1.000000	0.000000
100	48	32	32	1.000000	0.000000
100	48	33	0	0.000000	0.000000
100	48	33	1	0.000000	0.000000
100	48	33	2	0.000000	0.000000
100	48	33	3	0.000000	0.000000
100	48	33	4	0.000000	0.000000
100	48	33	5	0.000003	0.000002
100	48	33	6	0.000023	0.000020
100	48	33	7	0.000147	0.000124
100	48	33	8	0.000758	0.000611
100	48	33	9	0.003185	0.002427
100	48	33	10	0.011017	0.007832
100	48	33	11	0.031759	0.020743
100	48	33	12	0.077148	0.045389

N	n	k	x	P(x)	p(x)
100	48	33	13	0.159634	0.082486
100	48	33	14	0.284613	0.124978
100	48	33	15	0.442918	0.158306
100	48	33	16	0.610836	0.167917
100	48	33	17	0.760096	0.149260
100	48	33	18	0.871256	0.111161
100	48	33	19	0.940539	0.069283
100	48	33	20	0.976602	0.036063
100	48	33	21	0.992229	0.015627
100	48	33	22	0.997842	0.005613
100	48	33	23	0.999504	0.001662
100	48	33	24	0.999907	0.000403
100	48	33	25	0.999986	0.000079
100	48	33	26	0.999998	0.000012
100	48	33	27	1.000000	0.000002
100	48	33	28	1.000000	0.000000
100	48	33	29	1.000000	0.000000
100	48	33	30	1.000000	0.000000
100	48	33	31	1.000000	0.000000
100	48	33	32	1.000000	0.000000
100	48	33	33	1.000000	0.000000
100	48	34	0	0.000000	0.000000
100	48	34	1	0.000000	0.000000
100	48	34	2	0.000000	0.000000
100	48	34	3	0.000000	0.000000
100	48	34	4	0.000000	0.000000
100	48	34	5	0.000001	0.000001
100	48	34	6	0.000010	0.000009
100	48	34	7	0.000071	0.000061
100	48	34	8	0.000393	0.000322
100	48	34	9	0.001772	0.001379
100	48	34	10	0.006575	0.004802
100	48	34	11	0.020304	0.013730
100	48	34	12	0.052760	0.032456
100	48	34	13	0.116545	0.063784
100	48	34	14	0.221191	0.104646
100	48	34	15	0.364947	0.143756
100	48	34	16	0.530636	0.165690
100	48	34	17	0.691035	0.160399
100	48	34	18	0.821483	0.130448
100	48	34	19	0.910551	0.089068
100	48	34	20	0.961531	0.050980
100	48	34	21	0.985931	0.024401
100	48	34	22	0.995664	0.009732
100	48	34	23	0.998884	0.003220
100	48	34	24	0.999762	0.000878
100	48	34	25	0.999959	0.000196
100	48	34	26	0.999994	0.000035
100	48	34	27	0.999999	0.000005
100	48	34	28	1.000000	0.000001

N	n	k	x	P(x)	p(x)
100	48	34	29	1.000000	0.000000
100	48	34	30	1.000000	0.000000
100	48	34	31	1.000000	0.000000
100	48	34	32	1.000000	0.000000
100	48	34	33	1.000000	0.000000
100	48	34	34	1.000000	0.000000
100	48	35	0	0.000000	0.000000
100	48	35	1	0.000000	0.000000
100	48	35	2	0.000000	0.000000
100	48	35	3	0.000000	0.000000
100	48	35	4	0.000000	0.000000
100	48	35	5	0.000000	0.000000
100	48	35	6	0.000004	0.000004
100	48	35	7	0.000033	0.000029
100	48	35	8	0.000198	0.000165
100	48	35	9	0.000957	0.000759
100	48	35	10	0.003810	0.002852
100	48	35	11	0.012607	0.008798
100	48	35	12	0.035057	0.022450
100	48	35	13	0.082720	0.047662
100	48	35	14	0.167282	0.084562
100	48	35	15	0.293069	0.125787
100	48	35	16	0.450302	0.157233
100	48	35	17	0.615696	0.165394
100	48	35	18	0.762188	0.146492
100	48	35	19	0.871415	0.109227
100	48	35	20	0.939903	0.068488
100	48	35	21	0.975949	0.036046
100	48	35	22	0.991830	0.015881
100	48	35	23	0.997664	0.005834
100	48	35	24	0.999443	0.001779
100	48	35	25	0.999890	0.000447
100	48	35	26	0.999982	0.000092
100	48	35	27	0.999998	0.000015
100	48	35	28	1.000000	0.000002
100	48	35	29	1.000000	0.000000
100	48	35	30	1.000000	0.000000
100	48	35	31	1.000000	0.000000
100	48	35	32	1.000000	0.000000
100	48	35	33	1.000000	0.000000
100	48	35	34	1.000000	0.000000
100	48	35	35	1.000000	0.000000
100	48	36	0	0.000000	0.000000
100	48	36	1	0.000000	0.000000
100	48	36	2	0.000000	0.000000
100	48	36	3	0.000000	0.000000
100	48	36	4	0.000000	0.000000
100	48	36	5	0.000000	0.000000
100	48	36	6	0.000002	0.000002
100	48	36	7	0.000015	0.000013

Table for $N = 2$, $n = 1$, through $N = 100$, $n = 50$

N	n	k	x	P(x)	p(x)
100	48	36	8	0.000096	0.000081
100	48	36	9	0.000501	0.000405
100	48	36	10	0.002142	0.001641
100	48	36	11	0.007599	0.005457
100	48	36	12	0.022623	0.015024
100	48	36	13	0.057055	0.034432
100	48	36	14	0.123049	0.065994
100	48	36	15	0.229208	0.106158
100	48	36	16	0.372895	0.143687
100	48	36	17	0.536816	0.163921
100	48	36	18	0.694577	0.157761
100	48	36	19	0.822683	0.128106
100	48	36	20	0.910400	0.087717
100	48	36	21	0.960976	0.050576
100	48	36	22	0.985478	0.024501
100	48	36	23	0.995420	0.009943
100	48	36	24	0.998786	0.003366
100	48	36	25	0.999732	0.000946
100	48	36	26	0.999951	0.000219
100	48	36	27	0.999993	0.000042
100	48	36	28	0.999999	0.000006
100	48	36	29	1.000000	0.000001
100	48	36	30	1.000000	0.000000
100	48	36	31	1.000000	0.000000
100	48	36	32	1.000000	0.000000
100	48	36	33	1.000000	0.000000
100	48	36	34	1.000000	0.000000
100	48	36	35	1.000000	0.000000
100	48	36	36	1.000000	0.000000
100	48	37	0	0.000000	0.000000
100	48	37	1	0.000000	0.000000
100	48	37	2	0.000000	0.000000
100	48	37	3	0.000000	0.000000
100	48	37	4	0.000000	0.000000
100	48	37	5	0.000000	0.000000
100	48	37	6	0.000001	0.000001
100	48	37	7	0.000007	0.000006
100	48	37	8	0.000046	0.000039
100	48	37	9	0.000255	0.000209
100	48	37	10	0.001168	0.000913
100	48	37	11	0.004444	0.003276
100	48	37	12	0.014172	0.009728
100	48	37	13	0.038225	0.024053
100	48	37	14	0.087990	0.049765
100	48	37	15	0.174470	0.086480
100	48	37	16	0.301051	0.126582
100	48	37	17	0.457417	0.156365
100	48	37	18	0.620626	0.163210
100	48	37	19	0.764635	0.144009
100	48	37	20	0.872024	0.107389

N	n	k	x	P(x)	p(x)
100	48	37	21	0.939640	0.067615
100	48	37	22	0.975524	0.035884
100	48	37	23	0.991536	0.016012
100	48	37	24	0.997524	0.005988
100	48	37	25	0.999392	0.001868
100	48	37	26	0.999876	0.000484
100	48	37	27	0.999979	0.000103
100	48	37	28	0.999997	0.000018
100	48	37	29	1.000000	0.000003
100	48	37	30	1.000000	0.000000
100	48	37	31	1.000000	0.000000
100	48	37	32	1.000000	0.000000
100	48	37	33	1.000000	0.000000
100	48	37	34	1.000000	0.000000
100	48	37	35	1.000000	0.000000
100	48	37	36	1.000000	0.000000
100	48	37	37	1.000000	0.000000
100	48	38	0	0.000000	0.000000
100	48	38	1	0.000000	0.000000
100	48	38	2	0.000000	0.000000
100	48	38	3	0.000000	0.000000
100	48	38	4	0.000000	0.000000
100	48	38	5	0.000000	0.000000
100	48	38	6	0.000000	0.000000
100	48	38	7	0.000003	0.000002
100	48	38	8	0.000021	0.000018
100	48	38	9	0.000125	0.000104
100	48	38	10	0.000617	0.000492
100	48	38	11	0.002520	0.001903
100	48	38	12	0.008613	0.006093
100	48	38	13	0.024863	0.016249
100	48	38	14	0.061133	0.036270
100	48	38	15	0.129171	0.068038
100	48	38	16	0.236756	0.107585
100	48	38	17	0.380475	0.143719
100	48	38	18	0.542908	0.162432
100	48	38	19	0.698345	0.155438
100	48	38	20	0.824296	0.125950
100	48	38	21	0.910662	0.086366
100	48	38	22	0.960714	0.050053
100	48	38	23	0.985182	0.024468
100	48	38	24	0.995243	0.010061
100	48	38	25	0.998710	0.003467
100	48	38	26	0.999707	0.000997
100	48	38	27	0.999945	0.000238
100	48	38	28	0.999991	0.000047
100	48	38	29	0.999999	0.000007
100	48	38	30	1.000000	0.000001
100	48	38	31	1.000000	0.000000
100	48	38	32	1.000000	0.000000

N	n	k	x	P(x)	p(x)
100	48	38	33	1.000000	0.000000
100	48	38	34	1.000000	0.000000
100	48	38	35	1.000000	0.000000
100	48	38	36	1.000000	0.000000
100	48	38	37	1.000000	0.000000
100	48	38	38	1.000000	0.000000
100	48	39	0	0.000000	0.000000
100	48	39	1	0.000000	0.000000
100	48	39	2	0.000000	0.000000
100	48	39	3	0.000000	0.000000
100	48	39	4	0.000000	0.000000
100	48	39	5	0.000000	0.000000
100	48	39	6	0.000000	0.000000
100	48	39	7	0.000001	0.000001
100	48	39	8	0.000009	0.000008
100	48	39	9	0.000060	0.000050
100	48	39	10	0.000316	0.000256
100	48	39	11	0.001384	0.001069
100	48	39	12	0.005075	0.003691
100	48	39	13	0.015690	0.010614
100	48	39	14	0.041243	0.025553
100	48	39	15	0.092957	0.051714
100	48	39	16	0.181228	0.088271
100	48	39	17	0.308616	0.127387
100	48	39	18	0.464311	0.155696
100	48	39	19	0.625641	0.161329
100	48	39	20	0.767415	0.141774
100	48	39	21	0.873051	0.105636
100	48	39	22	0.939725	0.066674
100	48	39	23	0.975316	0.035592
100	48	39	24	0.991349	0.016032
100	48	39	25	0.997424	0.006075
100	48	39	26	0.999353	0.001929
100	48	39	27	0.999864	0.000511
100	48	39	28	0.999976	0.000112
100	48	39	29	0.999997	0.000020
100	48	39	30	0.999999	0.000003
100	48	39	31	1.000000	0.000000
100	48	39	32	1.000000	0.000000
100	48	39	33	1.000000	0.000000
100	48	39	34	1.000000	0.000000
100	48	39	35	1.000000	0.000000
100	48	39	36	1.000000	0.000000
100	48	39	37	1.000000	0.000000
100	48	39	38	1.000000	0.000000
100	48	39	39	1.000000	0.000000
100	48	40	0	0.000000	0.000000
100	48	40	1	0.000000	0.000000
100	48	40	2	0.000000	0.000000
100	48	40	3	0.000000	0.000000

N	n	k	x	P(x)	p(x)
100	48	40	4	0.000000	0.000000
100	48	40	5	0.000000	0.000000
100	48	40	6	0.000000	0.000000
100	48	40	7	0.000000	0.000000
100	48	40	8	0.000004	0.000003
100	48	40	9	0.000027	0.000023
100	48	40	10	0.000156	0.000129
100	48	40	11	0.000736	0.000580
100	48	40	12	0.002897	0.002161
100	48	40	13	0.009600	0.006702
100	48	40	14	0.027000	0.017401
100	48	40	15	0.064980	0.037980
100	48	40	16	0.134922	0.069942
100	48	40	17	0.243878	0.108956
100	48	40	18	0.387740	0.143862
100	48	40	19	0.548943	0.161203
100	48	40	20	0.702338	0.153395
100	48	40	21	0.826294	0.123956
100	48	40	22	0.911306	0.085012
100	48	40	23	0.960730	0.049423
100	48	40	24	0.985041	0.024311
100	48	40	25	0.995133	0.010092
100	48	40	26	0.998657	0.003524
100	48	40	27	0.999688	0.001031
100	48	40	28	0.999939	0.000251
100	48	40	29	0.999990	0.000051
100	48	40	30	0.999999	0.000008
100	48	40	31	1.000000	0.000001
100	48	40	32	1.000000	0.000000
100	48	40	33	1.000000	0.000000
100	48	40	34	1.000000	0.000000
100	48	40	35	1.000000	0.000000
100	48	40	36	1.000000	0.000000
100	48	40	37	1.000000	0.000000
100	48	40	38	1.000000	0.000000
100	48	40	39	1.000000	0.000000
100	48	40	40	1.000000	0.000000
100	48	41	0	0.000000	0.000000
100	48	41	1	0.000000	0.000000
100	48	41	2	0.000000	0.000000
100	48	41	3	0.000000	0.000000
100	48	41	4	0.000000	0.000000
100	48	41	5	0.000000	0.000000
100	48	41	6	0.000000	0.000000
100	48	41	7	0.000000	0.000000
100	48	41	8	0.000002	0.000001
100	48	41	9	0.000012	0.000010
100	48	41	10	0.000075	0.000062
100	48	41	11	0.000378	0.000304
100	48	41	12	0.001600	0.001222

N	n	k	x	P(x)	p(x)
100	48	41	13	0.005690	0.004089
100	48	41	14	0.017140	0.011450
100	48	41	15	0.044091	0.026951
100	48	41	16	0.097620	0.053529
100	48	41	17	0.187584	0.089964
100	48	41	18	0.315809	0.128225
100	48	41	19	0.471028	0.155219
100	48	41	20	0.630754	0.159726
100	48	41	21	0.770514	0.139760
100	48	41	22	0.874468	0.103954
100	48	41	23	0.940136	0.065669
100	48	41	24	0.975316	0.035180
100	48	41	25	0.991264	0.015948
100	48	41	26	0.997365	0.006101
100	48	41	27	0.999327	0.001962
100	48	41	28	0.999856	0.000528
100	48	41	29	0.999974	0.000118
100	48	41	30	0.999996	0.000022
100	48	41	31	0.999999	0.000003
100	48	41	32	1.000000	0.000000
100	48	41	33	1.000000	0.000000
100	48	41	34	1.000000	0.000000
100	48	41	35	1.000000	0.000000
100	48	41	36	1.000000	0.000000
100	48	41	37	1.000000	0.000000
100	48	41	38	1.000000	0.000000
100	48	41	39	1.000000	0.000000
100	48	41	40	1.000000	0.000000
100	48	41	41	1.000000	0.000000
100	48	42	0	0.000000	0.000000
100	48	42	1	0.000000	0.000000
100	48	42	2	0.000000	0.000000
100	48	42	3	0.000000	0.000000
100	48	42	4	0.000000	0.000000
100	48	42	5	0.000000	0.000000
100	48	42	6	0.000000	0.000000
100	48	42	7	0.000000	0.000000
100	48	42	8	0.000001	0.000001
100	48	42	9	0.000005	0.000005
100	48	42	10	0.000034	0.000029
100	48	42	11	0.000188	0.000154
100	48	42	12	0.000855	0.000667
100	48	42	13	0.003264	0.002409
100	48	42	14	0.010541	0.007278
100	48	42	15	0.029017	0.018475
100	48	42	16	0.068587	0.039571
100	48	42	17	0.140315	0.071727
100	48	42	18	0.250610	0.110295
100	48	42	19	0.394734	0.144124
100	48	42	20	0.554952	0.160218

N	n	k	x	P(x)	p(x)
100	48	42	21	0.706556	0.151604
100	48	42	22	0.828657	0.122102
100	48	42	23	0.912311	0.083653
100	48	42	24	0.961006	0.048695
100	48	42	25	0.985047	0.024042
100	48	42	26	0.995090	0.010043
100	48	42	27	0.998629	0.003539
100	48	42	28	0.999677	0.001048
100	48	42	29	0.999936	0.000259
100	48	42	30	0.999989	0.000053
100	48	42	31	0.999998	0.000009
100	48	42	32	1.000000	0.000001
100	48	42	33	1.000000	0.000000
100	48	42	34	1.000000	0.000000
100	48	42	35	1.000000	0.000000
100	48	42	36	1.000000	0.000000
100	48	42	37	1.000000	0.000000
100	48	42	38	1.000000	0.000000
100	48	42	39	1.000000	0.000000
100	48	42	40	1.000000	0.000000
100	48	42	41	1.000000	0.000000
100	48	42	42	1.000000	0.000000
100	48	43	0	0.000000	0.000000
100	48	43	1	0.000000	0.000000
100	48	43	2	0.000000	0.000000
100	48	43	3	0.000000	0.000000
100	48	43	4	0.000000	0.000000
100	48	43	5	0.000000	0.000000
100	48	43	6	0.000000	0.000000
100	48	43	7	0.000000	0.000000
100	48	43	8	0.000000	0.000000
100	48	43	9	0.000002	0.000002
100	48	43	10	0.000015	0.000013
100	48	43	11	0.000090	0.000075
100	48	43	12	0.000441	0.000351
100	48	43	13	0.001810	0.001369
100	48	43	14	0.006275	0.004465
100	48	43	15	0.018505	0.012230
100	48	43	16	0.046755	0.028250
100	48	43	17	0.101978	0.055222
100	48	43	18	0.193561	0.091583
100	48	43	19	0.322672	0.129111
100	48	43	20	0.477605	0.154933
100	48	43	21	0.635981	0.158376
100	48	43	22	0.773922	0.137941
100	48	43	23	0.876253	0.102331
100	48	43	24	0.940856	0.064603
100	48	43	25	0.975514	0.034658
100	48	43	26	0.991281	0.015767
100	48	43	27	0.997348	0.006067

N	n	k	x	P(x)	p(x)
100	48	43	28	0.999315	0.001968
100	48	43	29	0.999851	0.000536
100	48	43	30	0.999973	0.000122
100	48	43	31	0.999996	0.000023
100	48	43	32	0.999999	0.000004
100	48	43	33	1.000000	0.000000
100	48	43	34	1.000000	0.000000
100	48	43	35	1.000000	0.000000
100	48	43	36	1.000000	0.000000
100	48	43	37	1.000000	0.000000
100	48	43	38	1.000000	0.000000
100	48	43	39	1.000000	0.000000
100	48	43	40	1.000000	0.000000
100	48	43	41	1.000000	0.000000
100	48	43	42	1.000000	0.000000
100	48	43	43	1.000000	0.000000
100	48	44	0	0.000000	0.000000
100	48	44	1	0.000000	0.000000
100	48	44	2	0.000000	0.000000
100	48	44	3	0.000000	0.000000
100	48	44	4	0.000000	0.000000
100	48	44	5	0.000000	0.000000
100	48	44	6	0.000000	0.000000
100	48	44	7	0.000000	0.000000
100	48	44	8	0.000000	0.000000
100	48	44	9	0.000001	0.000001
100	48	44	10	0.000006	0.000006
100	48	44	11	0.000041	0.000035
100	48	44	12	0.000219	0.000178
100	48	44	13	0.000969	0.000750
100	48	44	14	0.003612	0.002643
100	48	44	15	0.011425	0.007813
100	48	44	16	0.030895	0.019471
100	48	44	17	0.071944	0.041049
100	48	44	18	0.145359	0.073415
100	48	44	19	0.256984	0.111625
100	48	44	20	0.401497	0.144514
100	48	44	21	0.560961	0.159464
100	48	44	22	0.711002	0.150041
100	48	44	23	0.831371	0.120369
100	48	44	24	0.913655	0.082284
100	48	44	25	0.961529	0.047874
100	48	44	26	0.985195	0.023666
100	48	44	27	0.995113	0.009917
100	48	44	28	0.998625	0.003512
100	48	44	29	0.999672	0.001047
100	48	44	30	0.999934	0.000262
100	48	44	31	0.999989	0.000055
100	48	44	32	0.999998	0.000009
100	48	44	33	1.000000	0.000001

N	n	k	x	P(x)	p(x)
100	48	44	34	1.000000	0.000000
100	48	44	35	1.000000	0.000000
100	48	44	36	1.000000	0.000000
100	48	44	37	1.000000	0.000000
100	48	44	38	1.000000	0.000000
100	48	44	39	1.000000	0.000000
100	48	44	40	1.000000	0.000000
100	48	44	41	1.000000	0.000000
100	48	44	42	1.000000	0.000000
100	48	44	43	1.000000	0.000000
100	48	44	44	1.000000	0.000000
100	48	45	0	0.000000	0.000000
100	48	45	1	0.000000	0.000000
100	48	45	2	0.000000	0.000000
100	48	45	3	0.000000	0.000000
100	48	45	4	0.000000	0.000000
100	48	45	5	0.000000	0.000000
100	48	45	6	0.000000	0.000000
100	48	45	7	0.000000	0.000000
100	48	45	8	0.000000	0.000000
100	48	45	9	0.000000	0.000000
100	48	45	10	0.000003	0.000002
100	48	45	11	0.000018	0.000016
100	48	45	12	0.000105	0.000087
100	48	45	13	0.000500	0.000396
100	48	45	14	0.002007	0.001507
100	48	45	15	0.006821	0.004813
100	48	45	16	0.019769	0.012949
100	48	45	17	0.049221	0.029452
100	48	45	18	0.106030	0.056809
100	48	45	19	0.199178	0.093148
100	48	45	20	0.329240	0.130062
100	48	45	21	0.484077	0.154836
100	48	45	22	0.641340	0.157263
100	48	45	23	0.777635	0.136295
100	48	45	24	0.878390	0.100756
100	48	45	25	0.941866	0.063476
100	48	45	26	0.975898	0.034031
100	48	45	27	0.991394	0.015496
100	48	45	28	0.997371	0.005977
100	48	45	29	0.999317	0.001947
100	48	45	30	0.999850	0.000533
100	48	45	31	0.999972	0.000122
100	48	45	32	0.999996	0.000023
100	48	45	33	0.999999	0.000004
100	48	45	34	1.000000	0.000000
100	48	45	35	1.000000	0.000000
100	48	45	36	1.000000	0.000000
100	48	45	37	1.000000	0.000000
100	48	45	38	1.000000	0.000000

Table for $N = 2$, $n = 1$, through $N = 100$, $n = 50$

N	n	k	x	P(x)	p(x)
100	48	45	39	1.000000	0.000000
100	48	45	40	1.000000	0.000000
100	48	45	41	1.000000	0.000000
100	48	45	42	1.000000	0.000000
100	48	45	43	1.000000	0.000000
100	48	45	44	1.000000	0.000000
100	48	45	45	1.000000	0.000000
100	48	46	0	0.000000	0.000000
100	48	46	1	0.000000	0.000000
100	48	46	2	0.000000	0.000000
100	48	46	3	0.000000	0.000000
100	48	46	4	0.000000	0.000000
100	48	46	5	0.000000	0.000000
100	48	46	6	0.000000	0.000000
100	48	46	7	0.000000	0.000000
100	48	46	8	0.000000	0.000000
100	48	46	9	0.000000	0.000000
100	48	46	10	0.000001	0.000001
100	48	46	11	0.000008	0.000007
100	48	46	12	0.000048	0.000040
100	48	46	13	0.000249	0.000201
100	48	46	14	0.001076	0.000827
100	48	46	15	0.003933	0.002857
100	48	46	16	0.012236	0.008303
100	48	46	17	0.032621	0.020385
100	48	46	18	0.075043	0.042422
100	48	46	19	0.150063	0.075020
100	48	46	20	0.263027	0.112963
100	48	46	21	0.408066	0.145039
100	48	46	22	0.566997	0.158931
100	48	46	23	0.715683	0.148685
100	48	46	24	0.834424	0.118742
100	48	46	25	0.915322	0.080898
100	48	46	26	0.962285	0.046963
100	48	46	27	0.985477	0.023192
100	48	46	28	0.995197	0.009720
100	48	46	29	0.998645	0.003448
100	48	46	30	0.999676	0.001031
100	48	46	31	0.999935	0.000259
100	48	46	32	0.999989	0.000054
100	48	46	33	0.999998	0.000009
100	48	46	34	1.000000	0.000001
100	48	46	35	1.000000	0.000000
100	48	46	36	1.000000	0.000000
100	48	46	37	1.000000	0.000000
100	48	46	38	1.000000	0.000000
100	48	46	39	1.000000	0.000000
100	48	46	40	1.000000	0.000000
100	48	46	41	1.000000	0.000000
100	48	46	42	1.000000	0.000000
100	48	46	43	1.000000	0.000000
100	48	46	44	1.000000	0.000000
100	48	46	45	1.000000	0.000000
100	48	46	46	1.000000	0.000000
100	48	47	0	0.000000	0.000000
100	48	47	1	0.000000	0.000000
100	48	47	2	0.000000	0.000000
100	48	47	3	0.000000	0.000000
100	48	47	4	0.000000	0.000000
100	48	47	5	0.000000	0.000000
100	48	47	6	0.000000	0.000000
100	48	47	7	0.000000	0.000000
100	48	47	8	0.000000	0.000000
100	48	47	9	0.000000	0.000000
100	48	47	10	0.000000	0.000000
100	48	47	11	0.000003	0.000003
100	48	47	12	0.000021	0.000018
100	48	47	13	0.000119	0.000098
100	48	47	14	0.000555	0.000436
100	48	47	15	0.002187	0.001632
100	48	47	16	0.007315	0.005128
100	48	47	17	0.020918	0.013603
100	48	47	18	0.051475	0.030557
100	48	47	19	0.109775	0.058300
100	48	47	20	0.204453	0.094678
100	48	47	21	0.335547	0.131093
100	48	47	22	0.490475	0.154928
100	48	47	23	0.646847	0.156372
100	48	47	24	0.781650	0.134803
100	48	47	25	0.880866	0.099215
100	48	47	26	0.943152	0.062286
100	48	47	27	0.976458	0.033306
100	48	47	28	0.991597	0.015139
100	48	47	29	0.997432	0.005835
100	48	47	30	0.999332	0.001900
100	48	47	31	0.999853	0.000521
100	48	47	32	0.999973	0.000120
100	48	47	33	0.999996	0.000023
100	48	47	34	0.999999	0.000004
100	48	47	35	1.000000	0.000000
100	48	47	36	1.000000	0.000000
100	48	47	37	1.000000	0.000000
100	48	47	38	1.000000	0.000000
100	48	47	39	1.000000	0.000000
100	48	47	40	1.000000	0.000000
100	48	47	41	1.000000	0.000000
100	48	47	42	1.000000	0.000000
100	48	47	43	1.000000	0.000000
100	48	47	44	1.000000	0.000000
100	48	47	45	1.000000	0.000000
100	48	47	46	1.000000	0.000000
100	48	47	47	1.000000	0.000000
100	48	48	0	0.000000	0.000000
100	48	48	1	0.000000	0.000000
100	48	48	2	0.000000	0.000000
100	48	48	3	0.000000	0.000000
100	48	48	4	0.000000	0.000000
100	48	48	5	0.000000	0.000000
100	48	48	6	0.000000	0.000000
100	48	48	7	0.000000	0.000000
100	48	48	8	0.000000	0.000000
100	48	48	9	0.000000	0.000000
100	48	48	10	0.000000	0.000000
100	48	48	11	0.000001	0.000001
100	48	48	12	0.000009	0.000008
100	48	48	13	0.000054	0.000045
100	48	48	14	0.000275	0.000221
100	48	48	15	0.001171	0.000896
100	48	48	16	0.004219	0.003048
100	48	48	17	0.012962	0.008743
100	48	48	18	0.034179	0.021217
100	48	48	19	0.077875	0.043696
100	48	48	20	0.154434	0.076559
100	48	48	21	0.268763	0.114329
100	48	48	22	0.414472	0.145709
100	48	48	23	0.573087	0.158614
100	48	48	24	0.720607	0.147521
100	48	48	25	0.837810	0.117203
100	48	48	26	0.917297	0.079487
100	48	48	27	0.963261	0.045964
100	48	48	28	0.985884	0.022623
100	48	48	29	0.995340	0.009456
100	48	48	30	0.998687	0.003347
100	48	48	31	0.999686	0.000999
100	48	48	32	0.999937	0.000251
100	48	48	33	0.999989	0.000053
100	48	48	34	0.999998	0.000009
100	48	48	35	1.000000	0.000001
100	48	48	36	1.000000	0.000000
100	48	48	37	1.000000	0.000000
100	48	48	38	1.000000	0.000000
100	48	48	39	1.000000	0.000000
100	48	48	40	1.000000	0.000000
100	48	48	41	1.000000	0.000000
100	48	48	42	1.000000	0.000000
100	48	48	43	1.000000	0.000000
100	48	48	44	1.000000	0.000000
100	48	48	45	1.000000	0.000000
100	48	48	46	1.000000	0.000000
100	48	48	47	1.000000	0.000000
100	48	48	48	1.000000	0.000000
100	49	1	0	0.510000	0.510000
100	49	1	1	1.000000	0.490000
100	49	2	0	0.257576	0.257576
100	49	2	1	0.762424	0.504848
100	49	2	2	1.000000	0.237576
100	49	3	0	0.128788	0.128788
100	49	3	1	0.515151	0.386364
100	49	3	2	0.886061	0.370909
100	49	3	3	1.000000	0.113939
100	49	4	0	0.063730	0.063730
100	49	4	1	0.323961	0.260231
100	49	4	2	0.706342	0.382380
100	49	4	3	0.945967	0.239625
100	49	4	4	1.000000	0.054033
100	49	5	0	0.031201	0.031201
100	49	5	1	0.193846	0.162644
100	49	5	2	0.519135	0.325289
100	49	5	3	0.831146	0.312012
100	49	5	4	0.974672	0.143525
100	49	5	5	1.000000	0.025328
100	49	6	0	0.015108	0.015108
100	49	6	1	0.111667	0.096559
100	49	6	2	0.358202	0.246535
100	49	6	3	0.680067	0.321865
100	49	6	4	0.906686	0.226619
100	49	6	5	0.988269	0.081583
100	49	6	6	1.000000	0.011731
100	49	7	0	0.007233	0.007233
100	49	7	1	0.062360	0.055128
100	49	7	2	0.234935	0.172574
100	49	7	3	0.522559	0.287624
100	49	7	4	0.798198	0.275640
100	49	7	5	0.950081	0.151883
100	49	7	6	0.994634	0.044552
100	49	7	7	1.000000	0.005366
100	49	8	0	0.003422	0.003422
100	49	8	1	0.033907	0.030485
100	49	8	2	0.147720	0.113812
100	49	8	3	0.380293	0.232573
100	49	8	4	0.664824	0.284531
100	49	8	5	0.878223	0.213398
100	49	8	6	0.974034	0.095812
100	49	8	7	0.997576	0.023542
100	49	8	8	1.000000	0.002423
100	49	9	0	0.001599	0.001599
100	49	9	1	0.018002	0.016403
100	49	9	2	0.089576	0.071575
100	49	9	3	0.264006	0.174430
100	49	9	4	0.525651	0.261645

Table for $N = 2$, $n = 1$, through $N = 100$, $n = 50$

N	n	k	x	P(x)	p(x)
100	49	9	5	0.776163	0.250511
100	49	9	6	0.929253	0.153090
100	49	9	7	0.986829	0.057576
100	49	9	8	0.998920	0.012091
100	49	9	9	1.000000	0.001080
100	49	10	0	0.000738	0.000738
100	49	10	1	0.009350	0.008612
100	49	10	2	0.052609	0.043259
100	49	10	3	0.175833	0.123224
100	49	10	4	0.396267	0.220434
100	49	10	5	0.655036	0.258770
100	49	10	6	0.856913	0.201877
100	49	10	7	0.960255	0.103342
100	49	10	8	0.993472	0.033217
100	49	10	9	0.999525	0.006053
100	49	10	10	1.000000	0.000475
100	49	11	0	0.000336	0.000336
100	49	11	1	0.004757	0.004421
100	49	11	2	0.030018	0.025261
100	49	11	3	0.112852	0.082834
100	49	11	4	0.286050	0.173198
100	49	11	5	0.528527	0.242477
100	49	11	6	0.760461	0.231934
100	49	11	7	0.912029	0.151568
100	49	11	8	0.978340	0.066311
100	49	11	9	0.996835	0.018495
100	49	11	10	0.999794	0.002959
100	49	11	11	1.000000	0.000206
100	49	12	0	0.000151	0.000151
100	49	12	1	0.002373	0.002222
100	49	12	2	0.016678	0.014305
100	49	12	3	0.070039	0.053361
100	49	12	4	0.198478	0.128439
100	49	12	5	0.408650	0.210173
100	49	12	6	0.648403	0.239752
100	49	12	7	0.840503	0.192100
100	49	12	8	0.947792	0.107290
100	49	12	9	0.988523	0.040730
100	49	12	10	0.998497	0.009975
100	49	12	11	0.999912	0.001415
100	49	12	12	1.000000	0.000088
100	49	13	0	0.000067	0.000067
100	49	13	1	0.001161	0.001094
100	49	13	2	0.009038	0.007877
100	49	13	3	0.042146	0.033108
100	49	13	4	0.132799	0.090653
100	49	13	5	0.303564	0.170765
100	49	13	6	0.531251	0.227687
100	49	13	7	0.748819	0.217568
100	49	13	8	0.897805	0.148986

N	n	k	x	P(x)	p(x)
100	49	13	9	0.970009	0.072204
100	49	13	10	0.994077	0.024068
100	49	13	11	0.999301	0.005224
100	49	13	12	0.999963	0.000662
100	49	13	13	1.000000	0.000037
100	49	14	0	0.000029	0.000029
100	49	14	1	0.000557	0.000528
100	49	14	2	0.004782	0.004225
100	49	14	3	0.024640	0.019858
100	49	14	4	0.085909	0.061269
100	49	14	5	0.217200	0.131291
100	49	14	6	0.418716	0.201516
100	49	14	7	0.643786	0.225070
100	49	14	8	0.827593	0.183807
100	49	14	9	0.936812	0.109219
100	49	14	10	0.983288	0.046476
100	49	14	11	0.997019	0.013732
100	49	14	12	0.999682	0.002662
100	49	14	13	0.999985	0.000303
100	49	14	14	1.000000	0.000015
100	49	15	0	0.000013	0.000013
100	49	15	1	0.000263	0.000250
100	49	15	2	0.002473	0.002211
100	49	15	3	0.014019	0.011545
100	49	15	4	0.053850	0.039831
100	49	15	5	0.150028	0.096178
100	49	15	6	0.317958	0.167930
100	49	15	7	0.533868	0.215910
100	49	15	8	0.739964	0.206096
100	49	15	9	0.886012	0.146048
100	49	15	10	0.962211	0.076199
100	49	15	11	0.990952	0.028740
100	49	15	12	0.998536	0.007584
100	49	15	13	0.999858	0.001322
100	49	15	14	0.999994	0.000136
100	49	15	15	1.000000	0.000006
100	49	16	0	0.000005	0.000005
100	49	16	1	0.000121	0.000116
100	49	16	2	0.001251	0.001130
100	49	16	3	0.007771	0.006520
100	49	16	4	0.032763	0.024992
100	49	16	5	0.100242	0.067479
100	49	16	6	0.233005	0.132763
100	49	16	7	0.427183	0.194178
100	49	16	8	0.640553	0.213370
100	49	16	9	0.817284	0.176731
100	49	16	10	0.927249	0.109966
100	49	16	11	0.978103	0.050854
100	49	16	12	0.995235	0.017132
100	49	16	13	0.999298	0.004063

N	n	k	x	P(x)	p(x)
100	49	16	14	0.999938	0.000640
100	49	16	15	0.999997	0.000060
100	49	16	16	1.000000	0.000002
100	49	17	0	0.000002	0.000002
100	49	17	1	0.000055	0.000053
100	49	17	2	0.000619	0.000564
100	49	17	3	0.004200	0.003581
100	49	17	4	0.019374	0.015174
100	49	17	5	0.064896	0.045522
100	49	17	6	0.165043	0.100147
100	49	17	7	0.330094	0.165051
100	49	17	8	0.536408	0.206314
100	49	17	9	0.733126	0.196718
100	49	17	10	0.876194	0.143068
100	49	17	11	0.955098	0.078904
100	49	17	12	0.987689	0.032591
100	49	17	13	0.997557	0.009868
100	49	17	14	0.999671	0.002115
100	49	17	15	0.999973	0.000302
100	49	17	16	0.999999	0.000026
100	49	17	17	1.000000	0.000001
100	49	18	0	0.000001	0.000001
100	49	18	1	0.000025	0.000024
100	49	18	2	0.000300	0.000275
100	49	18	3	0.002215	0.001916
100	49	18	4	0.011147	0.008932
100	49	18	5	0.040764	0.029616
100	49	18	6	0.113160	0.072396
100	49	18	7	0.246574	0.133415
100	49	18	8	0.434494	0.187920
100	49	18	9	0.638322	0.203828
100	49	18	10	0.808969	0.170647
100	49	18	11	0.918973	0.110004
100	49	18	12	0.973160	0.054187
100	49	18	13	0.993277	0.020116
100	49	18	14	0.998779	0.005503
100	49	18	15	0.999849	0.001070
100	49	18	16	0.999989	0.000139
100	49	18	17	0.999999	0.000011
100	49	18	18	1.000000	0.000000
100	49	19	0	0.000000	0.000000
100	49	19	1	0.000011	0.000010
100	49	19	2	0.000142	0.000131
100	49	19	3	0.001141	0.000999
100	49	19	4	0.006246	0.005105
100	49	19	5	0.024872	0.018626
100	49	19	6	0.075196	0.050324
100	49	19	7	0.178240	0.103044
100	49	19	8	0.340534	0.162294
100	49	19	9	0.538894	0.198360

N	n	k	x	P(x)	p(x)
100	49	19	10	0.727808	0.188914
100	49	19	11	0.867996	0.140188
100	49	19	12	0.948710	0.080714
100	49	19	13	0.984445	0.035735
100	49	19	14	0.996431	0.011986
100	49	19	15	0.999406	0.002975
100	49	19	16	0.999933	0.000527
100	49	19	17	0.999995	0.000063
100	49	19	18	1.000000	0.000004
100	49	19	19	1.000000	0.000000
100	49	20	0	0.000000	0.000000
100	49	20	1	0.000005	0.000004
100	49	20	2	0.000066	0.000061
100	49	20	3	0.000574	0.000508
100	49	20	4	0.003410	0.002836
100	49	20	5	0.014754	0.011344
100	49	20	6	0.048481	0.033727
100	49	20	7	0.124810	0.076329
100	49	20	8	0.258385	0.133576
100	49	20	9	0.440939	0.182553
100	49	20	10	0.636849	0.195911
100	49	20	11	0.802229	0.165379
100	49	20	12	0.911841	0.109612
100	49	20	13	0.968563	0.056722
100	49	20	14	0.991252	0.022689
100	49	20	15	0.998157	0.006905
100	49	20	16	0.999718	0.001561
100	49	20	17	0.999971	0.000253
100	49	20	18	0.999998	0.000027
100	49	20	19	1.000000	0.000002
100	49	20	20	1.000000	0.000000
100	49	21	0	0.000000	0.000000
100	49	21	1	0.000002	0.000002
100	49	21	2	0.000030	0.000028
100	49	21	3	0.000282	0.000252
100	49	21	4	0.001814	0.001533
100	49	21	5	0.008515	0.006700
100	49	21	6	0.030353	0.021838
100	49	21	7	0.084737	0.054384
100	49	21	8	0.189928	0.105191
100	49	21	9	0.349662	0.159734
100	49	21	10	0.541343	0.191681
100	49	21	11	0.723674	0.182331
100	49	21	12	0.861145	0.137472
100	49	21	13	0.943038	0.081893
100	49	21	14	0.981325	0.038287
100	49	21	15	0.995222	0.013897
100	49	21	16	0.999074	0.003852
100	49	21	17	0.999870	0.000795
100	49	21	18	0.999987	0.000118

Table for $N = 2$, $n = 1$, through $N = 100$, $n = 50$

N	n	k	x	P(x)	p(x)
100	49	21	19	0.999999	0.000012
100	49	21	20	1.000000	0.000001
100	49	21	21	1.000000	0.000000
100	49	22	0	0.000000	0.000000
100	49	22	1	0.000001	0.000001
100	49	22	2	0.000013	0.000012
100	49	22	3	0.000135	0.000122
100	49	22	4	0.000941	0.000806
100	49	22	5	0.004783	0.003842
100	49	22	6	0.018466	0.013683
100	49	22	7	0.055824	0.037358
100	49	22	8	0.135335	0.079511
100	49	22	9	0.268784	0.133449
100	49	22	10	0.446716	0.177932
100	49	22	11	0.635970	0.189255
100	49	22	12	0.796760	0.160789
100	49	22	13	0.905720	0.108960
100	49	22	14	0.964362	0.058643
100	49	22	15	0.989241	0.024879
100	49	22	16	0.997465	0.008224
100	49	22	17	0.999547	0.002082
100	49	22	18	0.999941	0.000394
100	49	22	19	0.999995	0.000054
100	49	22	20	1.000000	0.000005
100	49	22	21	1.000000	0.000000
100	49	22	22	1.000000	0.000000
100	49	23	0	0.000000	0.000000
100	49	23	1	0.000000	0.000000
100	49	23	2	0.000006	0.000005
100	49	23	3	0.000063	0.000057
100	49	23	4	0.000476	0.000413
100	49	23	5	0.002616	0.002140
100	49	23	6	0.010923	0.008307
100	49	23	7	0.035708	0.024785
100	49	23	8	0.093541	0.057833
100	49	23	9	0.200349	0.106808
100	49	23	10	0.357750	0.157401
100	49	23	11	0.543769	0.186019
100	49	23	12	0.720488	0.176719
100	49	23	13	0.855430	0.134943
100	49	23	14	0.938048	0.082618
100	49	23	15	0.978397	0.040348
100	49	23	16	0.993986	0.015589
100	49	23	17	0.998693	0.004707
100	49	23	18	0.999785	0.001092
100	49	23	19	0.999974	0.000189
100	49	23	20	0.999998	0.000024
100	49	23	21	1.000000	0.000002
100	49	23	22	1.000000	0.000000
100	49	23	23	1.000000	0.000000

N	n	k	x	P(x)	p(x)
100	49	24	0	0.000000	0.000000
100	49	24	1	0.000000	0.000000
100	49	24	2	0.000002	0.000002
100	49	24	3	0.000029	0.000026
100	49	24	4	0.000235	0.000206
100	49	24	5	0.001393	0.001158
100	49	24	6	0.006284	0.004891
100	49	24	7	0.022189	0.015905
100	49	24	8	0.062747	0.040558
100	49	24	9	0.144864	0.082117
100	49	24	10	0.278027	0.133163
100	49	24	11	0.451967	0.173940
100	49	24	12	0.635571	0.183604
100	49	24	13	0.792340	0.156769
100	49	24	14	0.900495	0.108154
100	49	24	15	0.960581	0.060086
100	49	24	16	0.987305	0.026724
100	49	24	17	0.996737	0.009432
100	49	24	18	0.999345	0.002608
100	49	24	19	0.999900	0.000555
100	49	24	20	0.999989	0.000089
100	49	24	21	0.999999	0.000010
100	49	24	22	1.000000	0.000001
100	49	24	23	1.000000	0.000000
100	49	24	24	1.000000	0.000000
100	49	25	0	0.000000	0.000000
100	49	25	1	0.000000	0.000000
100	49	25	2	0.000001	0.000001
100	49	25	3	0.000013	0.000012
100	49	25	4	0.000113	0.000100
100	49	25	5	0.000722	0.000610
100	49	25	6	0.003517	0.002794
100	49	25	7	0.013399	0.009883
100	49	25	8	0.040867	0.027468
100	49	25	9	0.101644	0.060778
100	49	25	10	0.209693	0.108049
100	49	25	11	0.364997	0.155304
100	49	25	12	0.546185	0.181188
100	49	25	13	0.718081	0.171896
100	49	25	14	0.850687	0.132606
100	49	25	15	0.933700	0.083013
100	49	25	16	0.975701	0.042001
100	49	25	17	0.992765	0.017065
100	49	25	18	0.998281	0.005516
100	49	25	19	0.999681	0.001400
100	49	25	20	0.999955	0.000274
100	49	25	21	0.999995	0.000040
100	49	25	22	1.000000	0.000004
100	49	25	23	1.000000	0.000000
100	49	25	24	1.000000	0.000000

N	n	k	x	P(x)	p(x)
100	49	25	25	1.000000	0.000000
100	49	26	0	0.000000	0.000000
100	49	26	1	0.000000	0.000000
100	49	26	2	0.000000	0.000000
100	49	26	3	0.000006	0.000005
100	49	26	4	0.000053	0.000047
100	49	26	5	0.000365	0.000312
100	49	26	6	0.001915	0.001550
100	49	26	7	0.007865	0.005950
100	49	26	8	0.025851	0.017986
100	49	26	9	0.069230	0.043378
100	49	26	10	0.153508	0.084278
100	49	26	11	0.286310	0.132802
100	49	26	12	0.456799	0.170489
100	49	26	13	0.635571	0.178772
100	49	26	14	0.788804	0.153233
100	49	26	15	0.896067	0.107263
100	49	26	16	0.957220	0.061153
100	49	26	17	0.985485	0.028264
100	49	26	18	0.996001	0.010517
100	49	26	19	0.999121	0.003120
100	49	26	20	0.999849	0.000728
100	49	26	21	0.999980	0.000131
100	49	26	22	0.999998	0.000018
100	49	26	23	1.000000	0.000002
100	49	26	24	1.000000	0.000000
100	49	26	25	1.000000	0.000000
100	49	26	26	1.000000	0.000000
100	49	27	0	0.000000	0.000000
100	49	27	1	0.000000	0.000000
100	49	27	2	0.000000	0.000000
100	49	27	3	0.000002	0.000002
100	49	27	4	0.000024	0.000022
100	49	27	5	0.000179	0.000155
100	49	27	6	0.001014	0.000835
100	49	27	7	0.004488	0.003474
100	49	27	8	0.015886	0.011398
100	49	27	9	0.045782	0.029896
100	49	27	10	0.109091	0.063309
100	49	27	11	0.218114	0.109023
100	49	27	12	0.371555	0.153440
100	49	27	13	0.548601	0.177046
100	49	27	14	0.716329	0.167728
100	49	27	15	0.846784	0.130455
100	49	27	16	0.929949	0.083165
100	49	27	17	0.973262	0.043313
100	49	27	18	0.991596	0.018333
100	49	27	19	0.997856	0.006261
100	49	27	20	0.999564	0.001707
100	49	27	21	0.999931	0.000367

N	n	k	x	P(x)	p(x)
100	49	27	22	0.999992	0.000061
100	49	27	23	0.999999	0.000008
100	49	27	24	1.000000	0.000001
100	49	27	25	1.000000	0.000000
100	49	27	26	1.000000	0.000000
100	49	27	27	1.000000	0.000000
100	49	28	0	0.000000	0.000000
100	49	28	1	0.000000	0.000000
100	49	28	2	0.000000	0.000000
100	49	28	3	0.000001	0.000001
100	49	28	4	0.000011	0.000010
100	49	28	5	0.000086	0.000075
100	49	28	6	0.000522	0.000437
100	49	28	7	0.002489	0.001967
100	49	28	8	0.009484	0.006995
100	49	28	9	0.029401	0.019916
100	49	28	10	0.075268	0.045868
100	49	28	11	0.161362	0.086094
100	49	28	12	0.293783	0.132421
100	49	28	13	0.461290	0.167507
100	49	28	14	0.635911	0.174621
100	49	28	15	0.786024	0.150113
100	49	28	16	0.892354	0.106330
100	49	28	17	0.954276	0.061922
100	49	28	18	0.983810	0.029534
100	49	28	19	0.995283	0.011473
100	49	28	20	0.998886	0.003602
100	49	28	21	0.999790	0.000904
100	49	28	22	0.999969	0.000179
100	49	28	23	0.999996	0.000027
100	49	28	24	1.000000	0.000003
100	49	28	25	1.000000	0.000000
100	49	28	26	1.000000	0.000000
100	49	28	27	1.000000	0.000000
100	49	28	28	1.000000	0.000000
100	49	29	0	0.000000	0.000000
100	49	29	1	0.000000	0.000000
100	49	29	2	0.000000	0.000000
100	49	29	3	0.000000	0.000000
100	49	29	4	0.000005	0.000004
100	49	29	5	0.000040	0.000035
100	49	29	6	0.000262	0.000222
100	49	29	7	0.001342	0.001080
100	49	29	8	0.005501	0.004159
100	49	29	9	0.018336	0.012835
100	49	29	10	0.050423	0.032087
100	49	29	11	0.115924	0.065501
100	49	29	12	0.225734	0.109810
100	49	29	13	0.377537	0.151803
100	49	29	14	0.551026	0.173489

Table for $N = 2$, $n = 1$, through $N = 100$, $n = 50$

N	n	k	x	P(x)	p(x)
100	49	29	15	0.715138	0.164111
100	49	29	16	0.843620	0.128482
100	49	29	17	0.926755	0.083135
100	49	29	18	0.971094	0.044339
100	49	29	19	0.990503	0.019409
100	49	29	20	0.997435	0.006932
100	49	29	21	0.999438	0.002004
100	49	29	22	0.999902	0.000464
100	49	29	23	0.999987	0.000085
100	49	29	24	0.999998	0.000012
100	49	29	25	1.000000	0.000001
100	49	29	26	1.000000	0.000000
100	49	29	27	1.000000	0.000000
100	49	29	28	1.000000	0.000000
100	49	29	29	1.000000	0.000000
100	49	30	0	0.000000	0.000000
100	49	30	1	0.000000	0.000000
100	49	30	2	0.000000	0.000000
100	49	30	3	0.000000	0.000000
100	49	30	4	0.000002	0.000002
100	49	30	5	0.000018	0.000016
100	49	30	6	0.000127	0.000109
100	49	30	7	0.000703	0.000576
100	49	30	8	0.003099	0.002396
100	49	30	9	0.011105	0.008006
100	49	30	10	0.032798	0.021693
100	49	30	11	0.080867	0.048069
100	49	30	12	0.168509	0.087642
100	49	30	13	0.300566	0.132057
100	49	30	14	0.465503	0.164937
100	49	30	15	0.636549	0.171046
100	49	30	16	0.783903	0.147354
100	49	30	17	0.889286	0.105383
100	49	30	18	0.951735	0.062449
100	49	30	19	0.982302	0.030567
100	49	30	20	0.994603	0.012301
100	49	30	21	0.998648	0.004045
100	49	30	22	0.999726	0.001077
100	49	30	23	0.999956	0.000230
100	49	30	24	0.999994	0.000039
100	49	30	25	0.999999	0.000005
100	49	30	26	1.000000	0.000000
100	49	30	27	1.000000	0.000000
100	49	30	28	1.000000	0.000000
100	49	30	29	1.000000	0.000000
100	49	30	30	1.000000	0.000000
100	49	31	0	0.000000	0.000000
100	49	31	1	0.000000	0.000000
100	49	31	2	0.000000	0.000000
100	49	31	3	0.000000	0.000000

N	n	k	x	P(x)	p(x)
100	49	31	4	0.000001	0.000001
100	49	31	5	0.000008	0.000007
100	49	31	6	0.000060	0.000052
100	49	31	7	0.000358	0.000297
100	49	31	8	0.001696	0.001338
100	49	31	9	0.006530	0.004835
100	49	31	10	0.020712	0.014182
100	49	31	11	0.054772	0.034061
100	49	31	12	0.122184	0.067412
100	49	31	13	0.232651	0.110467
100	49	31	14	0.383035	0.150384
100	49	31	15	0.553470	0.170435
100	49	31	16	0.714436	0.160966
100	49	31	17	0.841111	0.126675
100	49	31	18	0.924079	0.082968
100	49	31	19	0.969202	0.045123
100	49	31	20	0.989507	0.020305
100	49	31	21	0.997030	0.007523
100	49	31	22	0.999310	0.002280
100	49	31	23	0.999870	0.000560
100	49	31	24	0.999980	0.000110
100	49	31	25	0.999998	0.000017
100	49	31	26	1.000000	0.000002
100	49	31	27	1.000000	0.000000
100	49	31	28	1.000000	0.000000
100	49	31	29	1.000000	0.000000
100	49	31	30	1.000000	0.000000
100	49	31	31	1.000000	0.000000
100	49	32	0	0.000000	0.000000
100	49	32	1	0.000000	0.000000
100	49	32	2	0.000000	0.000000
100	49	32	3	0.000000	0.000000
100	49	32	4	0.000000	0.000000
100	49	32	5	0.000003	0.000003
100	49	32	6	0.000028	0.000024
100	49	32	7	0.000177	0.000149
100	49	32	8	0.000901	0.000724
100	49	32	9	0.003728	0.002827
100	49	32	10	0.012696	0.008969
100	49	32	11	0.036014	0.023318
100	49	32	12	0.086036	0.050021
100	49	32	13	0.175016	0.088980
100	49	32	14	0.306753	0.131737
100	49	32	15	0.469487	0.162734
100	49	32	16	0.637452	0.167965
100	49	32	17	0.782363	0.144911
100	49	32	18	0.886803	0.104440
100	49	32	19	0.949583	0.062780
100	49	32	20	0.980973	0.031390
100	49	32	21	0.993977	0.013004

N	n	k	x	P(x)	p(x)
100	49	32	22	0.998418	0.004441
100	49	32	23	0.999659	0.001241
100	49	32	24	0.999940	0.000281
100	49	32	25	0.999992	0.000051
100	49	32	26	0.999999	0.000007
100	49	32	27	1.000000	0.000001
100	49	32	28	1.000000	0.000000
100	49	32	29	1.000000	0.000000
100	49	32	30	1.000000	0.000000
100	49	32	31	1.000000	0.000000
100	49	32	32	1.000000	0.000000
100	49	33	0	0.000000	0.000000
100	49	33	1	0.000000	0.000000
100	49	33	2	0.000000	0.000000
100	49	33	3	0.000000	0.000000
100	49	33	4	0.000000	0.000000
100	49	33	5	0.000001	0.000001
100	49	33	6	0.000012	0.000011
100	49	33	7	0.000085	0.000072
100	49	33	8	0.000464	0.000379
100	49	33	9	0.002065	0.001601
100	49	33	10	0.007552	0.005488
100	49	33	11	0.022984	0.015431
100	49	33	12	0.058818	0.035835
100	49	33	13	0.127909	0.069091
100	49	33	14	0.238947	0.111038
100	49	33	15	0.388120	0.149173
100	49	33	16	0.555940	0.167820
100	49	33	17	0.714170	0.158230
100	49	33	18	0.839191	0.125021
100	49	33	19	0.921886	0.082695
100	49	33	20	0.967586	0.045700
100	49	33	21	0.988623	0.021037
100	49	33	22	0.996655	0.008032
100	49	33	23	0.999185	0.002530
100	49	33	24	0.999837	0.000653
100	49	33	25	0.999974	0.000137
100	49	33	26	0.999996	0.000023
100	49	33	27	1.000000	0.000003
100	49	33	28	1.000000	0.000000
100	49	33	29	1.000000	0.000000
100	49	33	30	1.000000	0.000000
100	49	33	31	1.000000	0.000000
100	49	33	32	1.000000	0.000000
100	49	33	33	1.000000	0.000000
100	49	34	0	0.000000	0.000000
100	49	34	1	0.000000	0.000000
100	49	34	2	0.000000	0.000000
100	49	34	3	0.000000	0.000000
100	49	34	4	0.000000	0.000000

N	n	k	x	P(x)	p(x)
100	49	34	5	0.000001	0.000001
100	49	34	6	0.000005	0.000005
100	49	34	7	0.000039	0.000034
100	49	34	8	0.000232	0.000193
100	49	34	9	0.001109	0.000877
100	49	34	10	0.004358	0.003249
100	49	34	11	0.014232	0.009874
100	49	34	12	0.039029	0.024797
100	49	34	13	0.090786	0.051757
100	49	34	14	0.180942	0.090157
100	49	34	15	0.312421	0.131478
100	49	34	16	0.473282	0.160862
100	49	34	17	0.638597	0.165315
100	49	34	18	0.781345	0.142748
100	49	34	19	0.884858	0.103513
100	49	34	20	0.947805	0.062947
100	49	34	21	0.979831	0.032026
100	49	34	22	0.993418	0.013587
100	49	34	23	0.998203	0.004785
100	49	34	24	0.999594	0.001391
100	49	34	25	0.999925	0.000331
100	49	34	26	0.999989	0.000064
100	49	34	27	0.999999	0.000010
100	49	34	28	1.000000	0.000001
100	49	34	29	1.000000	0.000000
100	49	34	30	1.000000	0.000000
100	49	34	31	1.000000	0.000000
100	49	34	32	1.000000	0.000000
100	49	34	33	1.000000	0.000000
100	49	34	34	1.000000	0.000000
100	49	35	0	0.000000	0.000000
100	49	35	1	0.000000	0.000000
100	49	35	2	0.000000	0.000000
100	49	35	3	0.000000	0.000000
100	49	35	4	0.000000	0.000000
100	49	35	5	0.000000	0.000000
100	49	35	6	0.000002	0.000002
100	49	35	7	0.000018	0.000015
100	49	35	8	0.000112	0.000095
100	49	35	9	0.000577	0.000465
100	49	35	10	0.002438	0.001861
100	49	35	11	0.008547	0.006109
100	49	35	12	0.025127	0.016581
100	49	35	13	0.062555	0.037427
100	49	35	14	0.133132	0.070577
100	49	35	15	0.244689	0.111557
100	49	35	16	0.392851	0.148162
100	49	35	17	0.558445	0.165593
100	49	35	18	0.714297	0.155852
100	49	35	19	0.837807	0.123510

Table for $N = 2$, $n = 1$, through $N = 100$, $n = 50$

N	n	k	x	P(x)	p(x)
100	49	35	20	0.920147	0.082340
100	49	35	21	0.966245	0.046098
100	49	35	22	0.987860	0.021615
100	49	35	23	0.996318	0.008458
100	49	35	24	0.999067	0.002749
100	49	35	25	0.999804	0.000738
100	49	35	26	0.999966	0.000162
100	49	35	27	0.999995	0.000029
100	49	35	28	0.999999	0.000004
100	49	35	29	1.000000	0.000000
100	49	35	30	1.000000	0.000000
100	49	35	31	1.000000	0.000000
100	49	35	32	1.000000	0.000000
100	49	35	33	1.000000	0.000000
100	49	35	34	1.000000	0.000000
100	49	35	35	1.000000	0.000000
100	49	36	0	0.000000	0.000000
100	49	36	1	0.000000	0.000000
100	49	36	2	0.000000	0.000000
100	49	36	3	0.000000	0.000000
100	49	36	4	0.000000	0.000000
100	49	36	5	0.000000	0.000000
100	49	36	6	0.000001	0.000001
100	49	36	7	0.000008	0.000007
100	49	36	8	0.000053	0.000045
100	49	36	9	0.000291	0.000239
100	49	36	10	0.001322	0.001031
100	49	36	11	0.004976	0.003654
100	49	36	12	0.015689	0.010714
100	49	36	13	0.041826	0.026137
100	49	36	14	0.095129	0.053303
100	49	36	15	0.186336	0.091207
100	49	36	16	0.317631	0.131295
100	49	36	17	0.476922	0.159291
100	49	36	18	0.639967	0.163045
100	49	36	19	0.780802	0.140835
100	49	36	20	0.883411	0.102608
100	49	36	21	0.946387	0.062977
100	49	36	22	0.978881	0.032494
100	49	36	23	0.992935	0.014053
100	49	36	24	0.998010	0.005075
100	49	36	25	0.999532	0.001522
100	49	36	26	0.999909	0.000377
100	49	36	27	0.999985	0.000076
100	49	36	28	0.999998	0.000013
100	49	36	29	1.000000	0.000002
100	49	36	30	1.000000	0.000000
100	49	36	31	1.000000	0.000000
100	49	36	32	1.000000	0.000000
100	49	36	33	1.000000	0.000000

N	n	k	x	P(x)	p(x)
100	49	36	34	1.000000	0.000000
100	49	36	35	1.000000	0.000000
100	49	36	36	1.000000	0.000000
100	49	37	0	0.000000	0.000000
100	49	37	1	0.000000	0.000000
100	49	37	2	0.000000	0.000000
100	49	37	3	0.000000	0.000000
100	49	37	4	0.000000	0.000000
100	49	37	5	0.000000	0.000000
100	49	37	6	0.000000	0.000000
100	49	37	7	0.000003	0.000003
100	49	37	8	0.000024	0.000021
100	49	37	9	0.000142	0.000118
100	49	37	10	0.000694	0.000552
100	49	37	11	0.002806	0.002112
100	49	37	12	0.009495	0.006689
100	49	37	13	0.027124	0.017629
100	49	37	14	0.065979	0.038855
100	49	37	15	0.137882	0.071904
100	49	37	16	0.249932	0.112050
100	49	37	17	0.397276	0.147344
100	49	37	18	0.560992	0.163716
100	49	37	19	0.714786	0.153794
100	49	37	20	0.836916	0.122130
100	49	37	21	0.918835	0.081919
100	49	37	22	0.965173	0.046338
100	49	37	23	0.987225	0.022053
100	49	37	24	0.996027	0.008802
100	49	37	25	0.998961	0.002934
100	49	37	26	0.999773	0.000812
100	49	37	27	0.999959	0.000186
100	49	37	28	0.999994	0.000035
100	49	37	29	0.999999	0.000005
100	49	37	30	1.000000	0.000001
100	49	37	31	1.000000	0.000000
100	49	37	32	1.000000	0.000000
100	49	37	33	1.000000	0.000000
100	49	37	34	1.000000	0.000000
100	49	37	35	1.000000	0.000000
100	49	37	36	1.000000	0.000000
100	49	37	37	1.000000	0.000000
100	49	38	0	0.000000	0.000000
100	49	38	1	0.000000	0.000000
100	49	38	2	0.000000	0.000000
100	49	38	3	0.000000	0.000000
100	49	38	4	0.000000	0.000000
100	49	38	5	0.000000	0.000000
100	49	38	6	0.000000	0.000000
100	49	38	7	0.000001	0.000001
100	49	38	8	0.000010	0.000009

N	n	k	x	P(x)	p(x)
100	49	38	9	0.000067	0.000057
100	49	38	10	0.000352	0.000285
100	49	38	11	0.001532	0.001180
100	49	38	12	0.005567	0.004035
100	49	38	13	0.017050	0.011484
100	49	38	14	0.044393	0.027342
100	49	38	15	0.099077	0.054684
100	49	38	16	0.191239	0.092162
100	49	38	17	0.322435	0.131195
100	49	38	18	0.480434	0.157999
100	49	38	19	0.641551	0.161117
100	49	38	20	0.780697	0.139147
100	49	38	21	0.882427	0.101729
100	49	38	22	0.945314	0.062887
100	49	38	23	0.978124	0.032811
100	49	38	24	0.992534	0.014410
100	49	38	25	0.997843	0.005309
100	49	38	26	0.999477	0.001634
100	49	38	27	0.999894	0.000417
100	49	38	28	0.999982	0.000088
100	49	38	29	0.999998	0.000015
100	49	38	30	1.000000	0.000002
100	49	38	31	1.000000	0.000000
100	49	38	32	1.000000	0.000000
100	49	38	33	1.000000	0.000000
100	49	38	34	1.000000	0.000000
100	49	38	35	1.000000	0.000000
100	49	38	36	1.000000	0.000000
100	49	38	37	1.000000	0.000000
100	49	38	38	1.000000	0.000000
100	49	39	0	0.000000	0.000000
100	49	39	1	0.000000	0.000000
100	49	39	2	0.000000	0.000000
100	49	39	3	0.000000	0.000000
100	49	39	4	0.000000	0.000000
100	49	39	5	0.000000	0.000000
100	49	39	6	0.000000	0.000000
100	49	39	7	0.000001	0.000000
100	49	39	8	0.000004	0.000004
100	49	39	9	0.000030	0.000026
100	49	39	10	0.000173	0.000142
100	49	39	11	0.000809	0.000636
100	49	39	12	0.003159	0.002350
100	49	39	13	0.010383	0.007224
100	49	39	14	0.028958	0.018575
100	49	39	15	0.069089	0.040131
100	49	39	16	0.142185	0.073096
100	49	39	17	0.254721	0.112536
100	49	39	18	0.401434	0.146713
100	49	39	19	0.563591	0.162157

N	n	k	x	P(x)	p(x)
100	49	39	20	0.715613	0.152022
100	49	39	21	0.836484	0.120872
100	49	39	22	0.917927	0.081443
100	49	39	23	0.964365	0.046438
100	49	39	24	0.986724	0.022359
100	49	39	25	0.995788	0.009064
100	49	39	26	0.998871	0.003083
100	49	39	27	0.999746	0.000875
100	49	39	28	0.999952	0.000206
100	49	39	29	0.999993	0.000040
100	49	39	30	0.999999	0.000006
100	49	39	31	1.000000	0.000001
100	49	39	32	1.000000	0.000000
100	49	39	33	1.000000	0.000000
100	49	39	34	1.000000	0.000000
100	49	39	35	1.000000	0.000000
100	49	39	36	1.000000	0.000000
100	49	39	37	1.000000	0.000000
100	49	39	38	1.000000	0.000000
100	49	39	39	1.000000	0.000000
100	49	40	0	0.000000	0.000000
100	49	40	1	0.000000	0.000000
100	49	40	2	0.000000	0.000000
100	49	40	3	0.000000	0.000000
100	49	40	4	0.000000	0.000000
100	49	40	5	0.000000	0.000000
100	49	40	6	0.000000	0.000000
100	49	40	7	0.000000	0.000000
100	49	40	8	0.000002	0.000002
100	49	40	9	0.000013	0.000012
100	49	40	10	0.000082	0.000068
100	49	40	11	0.000413	0.000331
100	49	40	12	0.001733	0.001321
100	49	40	13	0.006119	0.004386
100	49	40	14	0.018300	0.012180
100	49	40	15	0.046721	0.028421
100	49	40	16	0.102641	0.055921
100	49	40	17	0.195686	0.093044
100	49	40	18	0.326875	0.131189
100	49	40	19	0.483842	0.156967
100	49	40	20	0.643340	0.159498
100	49	40	21	0.781002	0.137662
100	49	40	22	0.881879	0.100877
100	49	40	23	0.944572	0.062693
100	49	40	24	0.977560	0.032988
100	49	40	25	0.992222	0.014662
100	49	40	26	0.997709	0.005487
100	49	40	27	0.999430	0.001722
100	49	40	28	0.999881	0.000451
100	49	40	29	0.999979	0.000098

$N = 100 \quad n = 49$

Table for $N = 2$, $n = 1$, through $N = 100$, $n = 50$

N	n	k	x	P(x)	p(x)
100	49	40	30	0.999997	0.000018
100	49	40	31	1.000000	0.000003
100	49	40	32	1.000000	0.000000
100	49	40	33	1.000000	0.000000
100	49	40	34	1.000000	0.000000
100	49	40	35	1.000000	0.000000
100	49	40	36	1.000000	0.000000
100	49	40	37	1.000000	0.000000
100	49	40	38	1.000000	0.000000
100	49	40	39	1.000000	0.000000
100	49	40	40	1.000000	0.000000
100	49	41	0	0.000000	0.000000
100	49	41	1	0.000000	0.000000
100	49	41	2	0.000000	0.000000
100	49	41	3	0.000000	0.000000
100	49	41	4	0.000000	0.000000
100	49	41	5	0.000000	0.000000
100	49	41	6	0.000000	0.000000
100	49	41	7	0.000000	0.000000
100	49	41	8	0.000001	0.000001
100	49	41	9	0.000006	0.000005
100	49	41	10	0.000037	0.000032
100	49	41	11	0.000203	0.000166
100	49	41	12	0.000919	0.000716
100	49	41	13	0.003488	0.002569
100	49	41	14	0.011195	0.007707
100	49	41	15	0.030615	0.019421
100	49	41	16	0.071885	0.041269
100	49	41	17	0.146062	0.074177
100	49	41	18	0.259094	0.113032
100	49	41	19	0.405358	0.146264
100	49	41	20	0.566249	0.160891
100	49	41	21	0.716760	0.150511
100	49	41	22	0.836484	0.119724
100	49	41	23	0.917405	0.080921
100	49	41	24	0.963815	0.046410
100	49	41	25	0.986357	0.022542
100	49	41	26	0.995605	0.009248
100	49	41	27	0.998799	0.003194
100	49	41	28	0.999724	0.000925
100	49	41	29	0.999947	0.000223
100	49	41	30	0.999991	0.000045
100	49	41	31	0.999999	0.000007
100	49	41	32	1.000000	0.000001
100	49	41	33	1.000000	0.000000
100	49	41	34	1.000000	0.000000
100	49	41	35	1.000000	0.000000
100	49	41	36	1.000000	0.000000
100	49	41	37	1.000000	0.000000
100	49	41	38	1.000000	0.000000

N	n	k	x	P(x)	p(x)
100	49	41	39	1.000000	0.000000
100	49	41	40	1.000000	0.000000
100	49	41	41	1.000000	0.000000
100	49	42	0	0.000000	0.000000
100	49	42	1	0.000000	0.000000
100	49	42	2	0.000000	0.000000
100	49	42	3	0.000000	0.000000
100	49	42	4	0.000000	0.000000
100	49	42	5	0.000000	0.000000
100	49	42	6	0.000000	0.000000
100	49	42	7	0.000000	0.000000
100	49	42	8	0.000000	0.000000
100	49	42	9	0.000002	0.000002
100	49	42	10	0.000016	0.000014
100	49	42	11	0.000096	0.000080
100	49	42	12	0.000470	0.000374
100	49	42	13	0.001920	0.001450
100	49	42	14	0.006623	0.004702
100	49	42	15	0.019424	0.012801
100	49	42	16	0.048802	0.029378
100	49	42	17	0.105830	0.057028
100	49	42	18	0.199704	0.093874
100	49	42	19	0.330987	0.131282
100	49	42	20	0.487167	0.156181
100	49	42	21	0.645331	0.158164
100	49	42	22	0.781695	0.136364
100	49	42	23	0.881745	0.100050
100	49	42	24	0.944150	0.062405
100	49	42	25	0.977188	0.033038
100	49	42	26	0.992000	0.014813
100	49	42	27	0.997608	0.005608
100	49	42	28	0.999395	0.001786
100	49	42	29	0.999871	0.000477
100	49	42	30	0.999977	0.000106
100	49	42	31	0.999996	0.000019
100	49	42	32	0.999999	0.000003
100	49	42	33	1.000000	0.000000
100	49	42	34	1.000000	0.000000
100	49	42	35	1.000000	0.000000
100	49	42	36	1.000000	0.000000
100	49	42	37	1.000000	0.000000
100	49	42	38	1.000000	0.000000
100	49	42	39	1.000000	0.000000
100	49	42	40	1.000000	0.000000
100	49	42	41	1.000000	0.000000
100	49	42	42	1.000000	0.000000
100	49	43	0	0.000000	0.000000
100	49	43	1	0.000000	0.000000
100	49	43	2	0.000000	0.000000
100	49	43	3	0.000000	0.000000

N	n	k	x	P(x)	p(x)
100	49	43	4	0.000000	0.000000
100	49	43	5	0.000000	0.000000
100	49	43	6	0.000000	0.000000
100	49	43	7	0.000000	0.000000
100	49	43	8	0.000000	0.000000
100	49	43	9	0.000001	0.000001
100	49	43	10	0.000007	0.000006
100	49	43	11	0.000044	0.000037
100	49	43	12	0.000232	0.000188
100	49	43	13	0.001020	0.000789
100	49	43	14	0.003785	0.002765
100	49	43	15	0.011920	0.008135
100	49	43	16	0.032087	0.020167
100	49	43	17	0.074366	0.042280
100	49	43	18	0.149530	0.075164
100	49	43	19	0.263082	0.113552
100	49	43	20	0.409077	0.145995
100	49	43	21	0.568976	0.159899
100	49	43	22	0.718215	0.149239
100	49	43	23	0.836895	0.118680
100	49	43	24	0.917251	0.080356
100	49	43	25	0.963517	0.046266
100	49	43	26	0.986126	0.022609
100	49	43	27	0.995481	0.009355
100	49	43	28	0.998748	0.003267
100	49	43	29	0.999707	0.000959
100	49	43	30	0.999942	0.000236
100	49	43	31	0.999990	0.000048
100	49	43	32	0.999999	0.000008
100	49	43	33	1.000000	0.000001
100	49	43	34	1.000000	0.000000
100	49	43	35	1.000000	0.000000
100	49	43	36	1.000000	0.000000
100	49	43	37	1.000000	0.000000
100	49	43	38	1.000000	0.000000
100	49	43	39	1.000000	0.000000
100	49	43	40	1.000000	0.000000
100	49	43	41	1.000000	0.000000
100	49	43	42	1.000000	0.000000
100	49	43	43	1.000000	0.000000
100	49	44	0	0.000000	0.000000
100	49	44	1	0.000000	0.000000
100	49	44	2	0.000000	0.000000
100	49	44	3	0.000000	0.000000
100	49	44	4	0.000000	0.000000
100	49	44	5	0.000000	0.000000
100	49	44	6	0.000000	0.000000
100	49	44	7	0.000000	0.000000
100	49	44	8	0.000000	0.000000
100	49	44	9	0.000000	0.000000

N	n	k	x	P(x)	p(x)
100	49	44	10	0.000003	0.000002
100	49	44	11	0.000019	0.000016
100	49	44	12	0.000110	0.000091
100	49	44	13	0.000522	0.000412
100	49	44	14	0.002087	0.001565
100	49	44	15	0.007068	0.004980
100	49	44	16	0.020411	0.013344
100	49	44	17	0.050631	0.030219
100	49	44	18	0.108652	0.058021
100	49	44	19	0.203318	0.094666
100	49	44	20	0.334799	0.131481
100	49	44	21	0.490429	0.155630
100	49	44	22	0.647523	0.157094
100	49	44	23	0.782760	0.135237
100	49	44	24	0.882007	0.099247
100	49	44	25	0.944037	0.062029
100	49	44	26	0.977003	0.032967
100	49	44	27	0.991870	0.014867
100	49	44	28	0.997544	0.005674
100	49	44	29	0.999370	0.001826
100	49	44	30	0.999864	0.000494
100	49	44	31	0.999975	0.000111
100	49	44	32	0.999996	0.000021
100	49	44	33	0.999999	0.000003
100	49	44	34	1.000000	0.000000
100	49	44	35	1.000000	0.000000
100	49	44	36	1.000000	0.000000
100	49	44	37	1.000000	0.000000
100	49	44	38	1.000000	0.000000
100	49	44	39	1.000000	0.000000
100	49	44	40	1.000000	0.000000
100	49	44	41	1.000000	0.000000
100	49	44	42	1.000000	0.000000
100	49	44	43	1.000000	0.000000
100	49	44	44	1.000000	0.000000
100	49	45	0	0.000000	0.000000
100	49	45	1	0.000000	0.000000
100	49	45	2	0.000000	0.000000
100	49	45	3	0.000000	0.000000
100	49	45	4	0.000000	0.000000
100	49	45	5	0.000000	0.000000
100	49	45	6	0.000000	0.000000
100	49	45	7	0.000000	0.000000
100	49	45	8	0.000000	0.000000
100	49	45	9	0.000000	0.000000
100	49	45	10	0.000001	0.000001
100	49	45	11	0.000008	0.000007
100	49	45	12	0.000050	0.000042
100	49	45	13	0.000257	0.000207
100	49	45	14	0.001109	0.000852

Table for $N = 2$, $n = 1$, through $N = 100$, $n = 50$

N	n	k	x	P(x)	p(x)
100	49	45	15	0.004044	0.002935
100	49	45	16	0.012548	0.008504
100	49	45	17	0.033362	0.020814
100	49	45	18	0.076533	0.043171
100	49	45	19	0.152604	0.076071
100	49	45	20	0.266710	0.114107
100	49	45	21	0.412614	0.145903
100	49	45	22	0.571781	0.159167
100	49	45	23	0.719972	0.148190
100	49	45	24	0.837701	0.117729
100	49	45	25	0.917453	0.079752
100	49	45	26	0.963463	0.046011
100	49	45	27	0.986030	0.022566
100	49	45	28	0.995417	0.009387
100	49	45	29	0.998718	0.003302
100	49	45	30	0.999696	0.000978
100	49	45	31	0.999939	0.000243
100	49	45	32	0.999990	0.000050
100	49	45	33	0.999998	0.000009
100	49	45	34	1.000000	0.000001
100	49	45	35	1.000000	0.000000
100	49	45	36	1.000000	0.000000
100	49	45	37	1.000000	0.000000
100	49	45	38	1.000000	0.000000
100	49	45	39	1.000000	0.000000
100	49	45	40	1.000000	0.000000
100	49	45	41	1.000000	0.000000
100	49	45	42	1.000000	0.000000
100	49	45	43	1.000000	0.000000
100	49	45	44	1.000000	0.000000
100	49	45	45	1.000000	0.000000
100	49	46	0	0.000000	0.000000
100	49	46	1	0.000000	0.000000
100	49	46	2	0.000000	0.000000
100	49	46	3	0.000000	0.000000
100	49	46	4	0.000000	0.000000
100	49	46	5	0.000000	0.000000
100	49	46	6	0.000000	0.000000
100	49	46	7	0.000000	0.000000
100	49	46	8	0.000000	0.000000
100	49	46	9	0.000000	0.000000
100	49	46	10	0.000000	0.000000
100	49	46	11	0.000003	0.000003
100	49	46	12	0.000022	0.000019
100	49	46	13	0.000122	0.000100
100	49	46	14	0.000567	0.000445
100	49	46	15	0.002230	0.001663
100	49	46	16	0.007445	0.005216
100	49	46	17	0.021252	0.013807
100	49	46	18	0.052200	0.030948

N	n	k	x	P(x)	p(x)
100	49	46	19	0.111111	0.058910
100	49	46	20	0.206545	0.095435
100	49	46	21	0.338336	0.131791
100	49	46	22	0.493645	0.155309
100	49	46	23	0.649918	0.156273
100	49	46	24	0.784188	0.134269
100	49	46	25	0.882652	0.098464
100	49	46	26	0.944222	0.061571
100	49	46	27	0.977003	0.032781
100	49	46	28	0.991833	0.014829
100	49	46	29	0.997518	0.005685
100	49	46	30	0.999358	0.001841
100	49	46	31	0.999860	0.000501
100	49	46	32	0.999974	0.000114
100	49	46	33	0.999996	0.000022
100	49	46	34	0.999999	0.000003
100	49	46	35	1.000000	0.000000
100	49	46	36	1.000000	0.000000
100	49	46	37	1.000000	0.000000
100	49	46	38	1.000000	0.000000
100	49	46	39	1.000000	0.000000
100	49	46	40	1.000000	0.000000
100	49	46	41	1.000000	0.000000
100	49	46	42	1.000000	0.000000
100	49	46	43	1.000000	0.000000
100	49	46	44	1.000000	0.000000
100	49	46	45	1.000000	0.000000
100	49	46	46	1.000000	0.000000
100	49	47	0	0.000000	0.000000
100	49	47	1	0.000000	0.000000
100	49	47	2	0.000000	0.000000
100	49	47	3	0.000000	0.000000
100	49	47	4	0.000000	0.000000
100	49	47	5	0.000000	0.000000
100	49	47	6	0.000000	0.000000
100	49	47	7	0.000000	0.000000
100	49	47	8	0.000000	0.000000
100	49	47	9	0.000000	0.000000
100	49	47	10	0.000000	0.000000
100	49	47	11	0.000001	0.000001
100	49	47	12	0.000009	0.000008
100	49	47	13	0.000055	0.000046
100	49	47	14	0.000278	0.000223
100	49	47	15	0.001183	0.000904
100	49	47	16	0.004258	0.003075
100	49	47	17	0.013070	0.008812
100	49	47	18	0.034434	0.021363
100	49	47	19	0.078383	0.043949
100	49	47	20	0.155293	0.076911
100	49	47	21	0.270000	0.114707

N	n	k	x	P(x)	p(x)
100	49	47	22	0.415990	0.145990
100	49	47	23	0.574675	0.158685
100	49	47	24	0.722026	0.147350
100	49	47	25	0.838890	0.116864
100	49	47	26	0.917998	0.079108
100	49	47	27	0.963648	0.045650
100	49	47	28	0.986065	0.022417
100	49	47	29	0.995412	0.009346
100	49	47	30	0.998711	0.003299
100	49	47	31	0.999693	0.000982
100	49	47	32	0.999938	0.000246
100	49	47	33	0.999989	0.000051
100	49	47	34	0.999998	0.000009
100	49	47	35	1.000000	0.000001
100	49	47	36	1.000000	0.000000
100	49	47	37	1.000000	0.000000
100	49	47	38	1.000000	0.000000
100	49	47	39	1.000000	0.000000
100	49	47	40	1.000000	0.000000
100	49	47	41	1.000000	0.000000
100	49	47	42	1.000000	0.000000
100	49	47	43	1.000000	0.000000
100	49	47	44	1.000000	0.000000
100	49	47	45	1.000000	0.000000
100	49	47	46	1.000000	0.000000
100	49	47	47	1.000000	0.000000
100	49	48	0	0.000000	0.000000
100	49	48	1	0.000000	0.000000
100	49	48	2	0.000000	0.000000
100	49	48	3	0.000000	0.000000
100	49	48	4	0.000000	0.000000
100	49	48	5	0.000000	0.000000
100	49	48	6	0.000000	0.000000
100	49	48	7	0.000000	0.000000
100	49	48	8	0.000000	0.000000
100	49	48	9	0.000000	0.000000
100	49	48	10	0.000000	0.000000
100	49	48	11	0.000000	0.000000
100	49	48	12	0.000004	0.000003
100	49	48	13	0.000024	0.000020
100	49	48	14	0.000131	0.000107
100	49	48	15	0.000602	0.000472
100	49	48	16	0.002343	0.001741
100	49	48	17	0.007750	0.005406
100	49	48	18	0.021938	0.014188
100	49	48	19	0.053506	0.031568
100	49	48	20	0.113210	0.059704
100	49	48	21	0.209400	0.096190
100	49	48	22	0.341618	0.132218
100	49	48	23	0.496830	0.155212

N	n	k	x	P(x)	p(x)
100	49	48	24	0.652521	0.155691
100	49	48	25	0.785970	0.133449
100	49	48	26	0.883668	0.097698
100	49	48	27	0.944699	0.061031
100	49	48	28	0.977183	0.032484
100	49	48	29	0.991885	0.014702
100	49	48	30	0.997528	0.005643
100	49	48	31	0.999359	0.001831
100	49	48	32	0.999859	0.000500
100	49	48	33	0.999974	0.000115
100	49	48	34	0.999996	0.000022
100	49	48	35	0.999999	0.000003
100	49	48	36	1.000000	0.000000
100	49	48	37	1.000000	0.000000
100	49	48	38	1.000000	0.000000
100	49	48	39	1.000000	0.000000
100	49	48	40	1.000000	0.000000
100	49	48	41	1.000000	0.000000
100	49	48	42	1.000000	0.000000
100	49	48	43	1.000000	0.000000
100	49	48	44	1.000000	0.000000
100	49	48	45	1.000000	0.000000
100	49	48	46	1.000000	0.000000
100	49	48	47	1.000000	0.000000
100	49	48	48	1.000000	0.000000
100	49	49	0	0.000000	0.000000
100	49	49	1	0.000000	0.000000
100	49	49	2	0.000000	0.000000
100	49	49	3	0.000000	0.000000
100	49	49	4	0.000000	0.000000
100	49	49	5	0.000000	0.000000
100	49	49	6	0.000000	0.000000
100	49	49	7	0.000000	0.000000
100	49	49	8	0.000000	0.000000
100	49	49	9	0.000000	0.000000
100	49	49	10	0.000000	0.000000
100	49	49	11	0.000000	0.000000
100	49	49	12	0.000001	0.000001
100	49	49	13	0.000010	0.000008
100	49	49	14	0.000059	0.000049
100	49	49	15	0.000294	0.000235
100	49	49	16	0.001239	0.000944
100	49	49	17	0.004423	0.003184
100	49	49	18	0.013480	0.009057
100	49	49	19	0.035294	0.021814
100	49	49	20	0.079914	0.044620
100	49	49	21	0.157606	0.077692
100	49	49	22	0.272966	0.115361
100	49	49	23	0.419224	0.146257
100	49	49	24	0.577669	0.158446

Table for $N = 2$, $n = 1$, through $N = 100$, $n = 50$

N	n	k	x	P(x)	p(x)
100	49	49	25	0.724378	0.146709
100	49	49	26	0.840456	0.116077
100	49	49	27	0.918878	0.078423
100	49	49	28	0.964064	0.045186
100	49	49	29	0.986230	0.022166
100	49	49	30	0.995466	0.009236
100	49	49	31	0.998725	0.003259
100	49	49	32	0.999696	0.000971
100	49	49	33	0.999939	0.000243
100	49	49	34	0.999990	0.000051
100	49	49	35	0.999998	0.000009
100	49	49	36	1.000000	0.000001
100	49	49	37	1.000000	0.000000
100	49	49	38	1.000000	0.000000
100	49	49	39	1.000000	0.000000
100	49	49	40	1.000000	0.000000
100	49	49	41	1.000000	0.000000
100	49	49	42	1.000000	0.000000
100	49	49	43	1.000000	0.000000
100	49	49	44	1.000000	0.000000
100	49	49	45	1.000000	0.000000
100	49	49	46	1.000000	0.000000
100	49	49	47	1.000000	0.000000
100	49	49	48	1.000000	0.000000
100	49	49	49	1.000000	0.000000
100	50	1	0	0.500000	0.500000
100	50	1	1	1.000000	0.500000
100	50	2	0	0.247475	0.247475
100	50	2	1	0.752525	0.505050
100	50	2	2	1.000000	0.247475
100	50	3	0	0.121212	0.121212
100	50	3	1	0.500000	0.378788
100	50	3	2	0.878788	0.378788
100	50	3	3	1.000000	0.121212
100	50	4	0	0.058732	0.058732
100	50	4	1	0.308654	0.249922
100	50	4	2	0.691346	0.382693
100	50	4	3	0.941268	0.249922
100	50	4	4	1.000000	0.058732
100	50	5	0	0.028142	0.028142
100	50	5	1	0.181089	0.152947
100	50	5	2	0.500000	0.318911
100	50	5	3	0.818911	0.318911
100	50	5	4	0.971858	0.152947
100	50	5	5	1.000000	0.028142
100	50	6	0	0.013331	0.013331
100	50	6	1	0.102201	0.088870
100	50	6	2	0.338866	0.236665
100	50	6	3	0.661134	0.322268
100	50	6	4	0.897799	0.236665

N	n	k	x	P(x)	p(x)
100	50	6	5	0.986669	0.088870
100	50	6	6	1.000000	0.013331
100	50	7	0	0.006240	0.006240
100	50	7	1	0.055875	0.049635
100	50	7	2	0.218016	0.162141
100	50	7	3	0.500000	0.281984
100	50	7	4	0.781984	0.281984
100	50	7	5	0.944125	0.162141
100	50	7	6	0.993760	0.049635
100	50	7	7	1.000000	0.006240
100	50	8	0	0.002885	0.002885
100	50	8	1	0.029723	0.026838
100	50	8	2	0.134330	0.104607
100	50	8	3	0.357492	0.223162
100	50	8	4	0.642508	0.285016
100	50	8	5	0.865670	0.223162
100	50	8	6	0.970277	0.104607
100	50	8	7	0.997115	0.026838
100	50	8	8	1.000000	0.002885
100	50	9	0	0.001317	0.001317
100	50	9	1	0.015429	0.014112
100	50	9	2	0.079752	0.064324
100	50	9	3	0.243485	0.163733
100	50	9	4	0.500000	0.256515
100	50	9	5	0.756515	0.256515
100	50	9	6	0.920247	0.163733
100	50	9	7	0.984571	0.064324
100	50	9	8	0.998683	0.014112
100	50	9	9	1.000000	0.001317
100	50	10	0	0.000593	0.000593
100	50	10	1	0.007830	0.007237
100	50	10	2	0.045824	0.037993
100	50	10	3	0.158920	0.113096
100	50	10	4	0.370333	0.211413
100	50	10	5	0.629667	0.259334
100	50	10	6	0.841080	0.211413
100	50	10	7	0.954176	0.113096
100	50	10	8	0.992170	0.037993
100	50	10	9	0.999406	0.007237
100	50	10	10	1.000000	0.000593
100	50	11	0	0.000264	0.000264
100	50	11	1	0.003890	0.003626
100	50	11	2	0.025560	0.021670
100	50	11	3	0.099859	0.074298
100	50	11	4	0.262278	0.162419
100	50	11	5	0.500000	0.237722
100	50	11	6	0.737722	0.237722
100	50	11	7	0.900141	0.162419
100	50	11	8	0.974439	0.074298
100	50	11	9	0.996110	0.021670

N	n	k	x	P(x)	p(x)
100	50	11	10	0.999736	0.003626
100	50	11	11	1.000000	0.000264
100	50	12	0	0.000116	0.000116
100	50	12	1	0.001894	0.001778
100	50	12	2	0.013873	0.011980
100	50	12	3	0.060622	0.046749
100	50	12	4	0.178331	0.117708
100	50	12	5	0.379803	0.201473
100	50	12	6	0.620197	0.240393
100	50	12	7	0.821669	0.201473
100	50	12	8	0.939377	0.117708
100	50	12	9	0.986127	0.046749
100	50	12	10	0.998106	0.011980
100	50	12	11	0.999884	0.001778
100	50	12	12	1.000000	0.000116
100	50	13	0	0.000050	0.000050
100	50	13	1	0.000904	0.000854
100	50	13	2	0.007339	0.006435
100	50	13	3	0.035654	0.028315
100	50	13	4	0.116801	0.081147
100	50	13	5	0.276777	0.159976
100	50	13	6	0.500000	0.223222
100	50	13	7	0.723222	0.223222
100	50	13	8	0.883199	0.159976
100	50	13	9	0.964346	0.081147
100	50	13	10	0.992661	0.028315
100	50	13	11	0.999096	0.006435
100	50	13	12	0.999950	0.000854
100	50	13	13	1.000000	0.000050
100	50	14	0	0.000021	0.000021
100	50	14	1	0.000423	0.000402
100	50	14	2	0.003788	0.003366
100	50	14	3	0.020357	0.016569
100	50	14	4	0.073896	0.053539
100	50	14	5	0.194031	0.120135
100	50	14	6	0.387106	0.193075
100	50	14	7	0.612894	0.225788
100	50	14	8	0.805969	0.193075
100	50	14	9	0.926104	0.120135
100	50	14	10	0.979643	0.053539
100	50	14	11	0.996212	0.016569
100	50	14	12	0.999577	0.003366
100	50	14	13	0.999979	0.000402
100	50	14	14	1.000000	0.000021
100	50	15	0	0.000009	0.000009
100	50	15	1	0.000194	0.000185
100	50	15	2	0.001910	0.001716
100	50	15	3	0.011302	0.009392
100	50	15	4	0.045259	0.033957
100	50	15	5	0.131170	0.085911

N	n	k	x	P(x)	p(x)
100	50	15	6	0.288323	0.157154
100	50	15	7	0.500000	0.211677
100	50	15	8	0.711676	0.211677
100	50	15	9	0.868830	0.157154
100	50	15	10	0.954741	0.085911
100	50	15	11	0.988698	0.033957
100	50	15	12	0.998090	0.009392
100	50	15	13	0.999806	0.001716
100	50	15	14	0.999991	0.000185
100	50	15	15	1.000000	0.000009
100	50	16	0	0.000004	0.000004
100	50	16	1	0.000087	0.000084
100	50	16	2	0.000941	0.000854
100	50	16	3	0.006109	0.005168
100	50	16	4	0.026882	0.020774
100	50	16	5	0.085688	0.058805
100	50	16	6	0.206973	0.121286
100	50	16	7	0.392917	0.185943
100	50	16	8	0.607083	0.214167
100	50	16	9	0.793027	0.185943
100	50	16	10	0.914312	0.121286
100	50	16	11	0.973117	0.058805
100	50	16	12	0.993891	0.020774
100	50	16	13	0.999059	0.005168
100	50	16	14	0.999913	0.000854
100	50	16	15	0.999996	0.000084
100	50	16	16	1.000000	0.000004
100	50	17	0	0.000001	0.000001
100	50	17	1	0.000038	0.000037
100	50	17	2	0.000453	0.000415
100	50	17	3	0.003217	0.002764
100	50	17	4	0.015506	0.012289
100	50	17	5	0.054185	0.038678
100	50	17	6	0.143443	0.089258
100	50	17	7	0.297731	0.154289
100	50	17	8	0.500000	0.202269
100	50	17	9	0.702269	0.202269
100	50	17	10	0.856557	0.154289
100	50	17	11	0.945815	0.089258
100	50	17	12	0.984494	0.038678
100	50	17	13	0.996783	0.012289
100	50	17	14	0.999547	0.002764
100	50	17	15	0.999961	0.000415
100	50	17	16	0.999998	0.000037
100	50	17	17	1.000000	0.000001
100	50	18	0	0.000001	0.000001
100	50	18	1	0.000017	0.000016
100	50	18	2	0.000213	0.000197
100	50	18	3	0.001652	0.001439
100	50	18	4	0.008696	0.007044

Table for $N = 2$, $n = 1$, through $N = 100$, $n = 50$

N	n	k	x	P(x)	p(x)
100	50	18	5	0.033215	0.024519
100	50	18	6	0.096125	0.062911
100	50	18	7	0.217799	0.121673
100	50	18	8	0.397647	0.179849
100	50	18	9	0.602353	0.204706
100	50	18	10	0.782201	0.179849
100	50	18	11	0.903875	0.121673
100	50	18	12	0.966785	0.062911
100	50	18	13	0.991304	0.024519
100	50	18	14	0.998348	0.007044
100	50	18	15	0.999787	0.001439
100	50	18	16	0.999983	0.000197
100	50	18	17	0.999999	0.000016
100	50	18	18	1.000000	0.000001
100	50	19	0	0.000000	0.000000
100	50	19	1	0.000007	0.000007
100	50	19	2	0.000098	0.000091
100	50	19	3	0.000827	0.000729
100	50	19	4	0.004744	0.003917
100	50	19	5	0.019759	0.015015
100	50	19	6	0.062368	0.042609
100	50	19	7	0.153994	0.091626
100	50	19	8	0.305530	0.151535
100	50	19	9	0.500000	0.194470
100	50	19	10	0.694470	0.194470
100	50	19	11	0.846006	0.151535
100	50	19	12	0.937632	0.091626
100	50	19	13	0.980241	0.042609
100	50	19	14	0.995256	0.015015
100	50	19	15	0.999172	0.003917
100	50	19	16	0.999902	0.000729
100	50	19	17	0.999993	0.000091
100	50	19	18	1.000000	0.000007
100	50	19	19	1.000000	0.000000
100	50	20	0	0.000000	0.000000
100	50	20	1	0.000003	0.000003
100	50	20	2	0.000044	0.000041
100	50	20	3	0.000404	0.000360
100	50	20	4	0.002520	0.002116
100	50	20	5	0.011417	0.008898
100	50	20	6	0.039222	0.027805
100	50	20	7	0.105353	0.066131
100	50	20	8	0.226956	0.121602
100	50	20	9	0.401564	0.174609
100	50	20	10	0.598436	0.196871
100	50	20	11	0.773044	0.174609
100	50	20	12	0.894647	0.121602
100	50	20	13	0.960777	0.066131
100	50	20	14	0.988582	0.027805
100	50	20	15	0.997480	0.008898
100	50	20	16	0.999596	0.002116
100	50	20	17	0.999956	0.000360
100	50	20	18	0.999997	0.000041
100	50	20	19	1.000000	0.000003
100	50	20	20	1.000000	0.000000
100	50	21	0	0.000000	0.000000
100	50	21	1	0.000001	0.000001
100	50	21	2	0.000019	0.000018
100	50	21	3	0.000193	0.000173
100	50	21	4	0.001303	0.001111
100	50	21	5	0.006413	0.005109
100	50	21	6	0.023930	0.017517
100	50	21	7	0.069808	0.045878
100	50	21	8	0.163114	0.093306
100	50	21	9	0.312077	0.148963
100	50	21	10	0.500000	0.187923
100	50	21	11	0.687922	0.187923
100	50	21	12	0.836885	0.148963
100	50	21	13	0.930192	0.093306
100	50	21	14	0.976070	0.045878
100	50	21	15	0.993587	0.017517
100	50	21	16	0.998696	0.005109
100	50	21	17	0.999807	0.001111
100	50	21	18	0.999980	0.000173
100	50	21	19	0.999999	0.000018
100	50	21	20	1.000000	0.000001
100	50	21	21	1.000000	0.000000
100	50	22	0	0.000000	0.000000
100	50	22	1	0.000000	0.000000
100	50	22	2	0.000008	0.000008
100	50	22	3	0.000090	0.000081
100	50	22	4	0.000657	0.000567
100	50	22	5	0.003502	0.002846
100	50	22	6	0.014173	0.010671
100	50	22	7	0.044836	0.030663
100	50	22	8	0.113509	0.068672
100	50	22	9	0.234768	0.121259
100	50	22	10	0.404849	0.170082
100	50	22	11	0.595151	0.190301
100	50	22	12	0.765232	0.170082
100	50	22	13	0.886491	0.121259
100	50	22	14	0.955164	0.068672
100	50	22	15	0.985827	0.030663
100	50	22	16	0.996498	0.010671
100	50	22	17	0.999343	0.002846
100	50	22	18	0.999910	0.000567
100	50	22	19	0.999992	0.000081
100	50	22	20	0.999999	0.000008
100	50	22	21	1.000000	0.000000
100	50	22	22	1.000000	0.000000
100	50	23	0	0.000000	0.000000
100	50	23	1	0.000000	0.000000
100	50	23	2	0.000003	0.000003
100	50	23	3	0.000041	0.000037
100	50	23	4	0.000322	0.000282
100	50	23	5	0.001861	0.001538
100	50	23	6	0.008154	0.006293
100	50	23	7	0.027932	0.019779
100	50	23	8	0.076531	0.048599
100	50	23	9	0.171029	0.094498
100	50	23	10	0.317628	0.146599
100	50	23	11	0.500000	0.182372
100	50	23	12	0.682372	0.182372
100	50	23	13	0.828971	0.146599
100	50	23	14	0.923469	0.094498
100	50	23	15	0.972067	0.048599
100	50	23	16	0.991846	0.019779
100	50	23	17	0.998139	0.006293
100	50	23	18	0.999678	0.001538
100	50	23	19	0.999959	0.000282
100	50	23	20	0.999996	0.000037
100	50	23	21	1.000000	0.000003
100	50	23	22	1.000000	0.000000
100	50	23	23	1.000000	0.000000
100	50	24	0	0.000000	0.000000
100	50	24	1	0.000000	0.000000
100	50	24	2	0.000001	0.000001
100	50	24	3	0.000018	0.000017
100	50	24	4	0.000154	0.000136
100	50	24	5	0.000962	0.000808
100	50	24	6	0.004558	0.003596
100	50	24	7	0.016887	0.012329
100	50	24	8	0.050023	0.033136
100	50	24	9	0.120712	0.070689
100	50	24	10	0.241472	0.120761
100	50	24	11	0.407630	0.166157
100	50	24	12	0.592370	0.184741
100	50	24	13	0.758527	0.166157
100	50	24	14	0.879288	0.120761
100	50	24	15	0.949977	0.070689
100	50	24	16	0.983113	0.033136
100	50	24	17	0.995442	0.012329
100	50	24	18	0.999038	0.003596
100	50	24	19	0.999846	0.000808
100	50	24	20	0.999982	0.000136
100	50	24	21	0.999998	0.000017
100	50	24	22	1.000000	0.000001
100	50	24	23	1.000000	0.000000
100	50	24	24	1.000000	0.000000
100	50	25	0	0.000000	0.000000
100	50	25	1	0.000000	0.000000
100	50	25	2	0.000001	0.000001
100	50	25	3	0.000008	0.000007
100	50	25	4	0.000072	0.000064
100	50	25	5	0.000483	0.000412
100	50	25	6	0.002476	0.001992
100	50	25	7	0.009911	0.007436
100	50	25	8	0.031711	0.021800
100	50	25	9	0.082577	0.050866
100	50	25	10	0.177914	0.095337
100	50	25	11	0.322365	0.144451
100	50	25	12	0.500000	0.177635
100	50	25	13	0.677635	0.177635
100	50	25	14	0.822086	0.144451
100	50	25	15	0.917423	0.095337
100	50	25	16	0.968289	0.050866
100	50	25	17	0.990089	0.021800
100	50	25	18	0.997524	0.007436
100	50	25	19	0.999516	0.001992
100	50	25	20	0.999928	0.000412
100	50	25	21	0.999992	0.000064
100	50	25	22	0.999999	0.000007
100	50	25	23	1.000000	0.000001
100	50	25	24	1.000000	0.000000
100	50	25	25	1.000000	0.000000
100	50	26	0	0.000000	0.000000
100	50	26	1	0.000000	0.000000
100	50	26	2	0.000000	0.000000
100	50	26	3	0.000003	0.000003
100	50	26	4	0.000032	0.000029
100	50	26	5	0.000236	0.000204
100	50	26	6	0.001307	0.001071
100	50	26	7	0.005648	0.004341
100	50	26	8	0.019503	0.013855
100	50	26	9	0.054770	0.035267
100	50	26	10	0.127068	0.072297
100	50	26	11	0.247251	0.120183
100	50	26	12	0.409998	0.162748
100	50	26	13	0.590002	0.180004
100	50	26	14	0.752749	0.162748
100	50	26	15	0.872932	0.120183
100	50	26	16	0.945230	0.072297
100	50	26	17	0.980497	0.035267
100	50	26	18	0.994352	0.013855
100	50	26	19	0.998693	0.004341
100	50	26	20	0.999764	0.001071
100	50	26	21	0.999967	0.000204
100	50	26	22	0.999997	0.000029
100	50	26	23	1.000000	0.000003
100	50	26	24	1.000000	0.000000

Table for $N = 2$, $n = 1$, through $N = 100$, $n = 50$

N	n	k	x	P(x)	p(x)
100	50	26	25	1.000000	0.000000
100	50	26	26	1.000000	0.000000
100	50	27	0	0.000000	0.000000
100	50	27	1	0.000000	0.000000
100	50	27	2	0.000000	0.000000
100	50	27	3	0.000001	0.000001
100	50	27	4	0.000014	0.000013
100	50	27	5	0.000112	0.000098
100	50	27	6	0.000670	0.000558
100	50	27	7	0.003126	0.002455
100	50	27	8	0.011640	0.008514
100	50	27	9	0.035230	0.023591
100	50	27	10	0.087988	0.052758
100	50	27	11	0.183911	0.095923
100	50	27	12	0.326425	0.142514
100	50	27	13	0.500000	0.173575
100	50	27	14	0.673575	0.173575
100	50	27	15	0.816089	0.142514
100	50	27	16	0.912012	0.095923
100	50	27	17	0.964769	0.052758
100	50	27	18	0.988360	0.023591
100	50	27	19	0.996874	0.008514
100	50	27	20	0.999329	0.002455
100	50	27	21	0.999887	0.000558
100	50	27	22	0.999986	0.000098
100	50	27	23	0.999998	0.000013
100	50	27	24	1.000000	0.000001
100	50	27	25	1.000000	0.000000
100	50	27	26	1.000000	0.000000
100	50	27	27	1.000000	0.000000
100	50	28	0	0.000000	0.000000
100	50	28	1	0.000000	0.000000
100	50	28	2	0.000000	0.000000
100	50	28	3	0.000001	0.000000
100	50	28	4	0.000006	0.000006
100	50	28	5	0.000052	0.000046
100	50	28	6	0.000334	0.000282
100	50	28	7	0.001679	0.001345
100	50	28	8	0.006741	0.005062
100	50	28	9	0.021981	0.015240
100	50	28	10	0.059080	0.037099
100	50	28	11	0.132664	0.073585
100	50	28	12	0.252240	0.119575
100	50	28	13	0.412024	0.159784
100	50	28	14	0.587976	0.175953
100	5C	28	15	0.747760	0.159784
100	50	28	16	0.867335	0.119575
100	50	28	17	0.940920	0.073585
100	50	28	18	0.978019	0.037099
100	50	28	19	0.993259	0.015240

N	n	k	x	P(x)	p(x)
100	50	28	20	0.998321	0.005062
100	50	28	21	0.999666	0.001345
100	50	28	22	0.999948	0.000282
100	50	28	23	0.999994	0.000046
100	50	28	24	0.999999	0.000006
100	50	28	25	1.000000	0.000000
100	50	28	26	1.000000	0.000000
100	50	28	27	1.000000	0.000000
100	50	28	28	1.000000	0.000000
100	50	29	0	0.000000	0.000000
100	50	29	1	0.000000	0.000000
100	50	29	2	0.000000	0.000000
100	50	29	3	0.000000	0.000000
100	50	29	4	0.000003	0.000002
100	50	29	5	0.000023	0.000021
100	50	29	6	0.000162	0.000138
100	50	29	7	0.000876	0.000714
100	50	29	8	0.003788	0.002912
100	50	29	9	0.013303	0.009514
100	50	29	10	0.038469	0.025167
100	50	29	11	0.092806	0.054337
100	50	29	12	0.189131	0.096324
100	50	29	13	0.329912	0.140782
100	50	29	14	0.500000	0.170088
100	50	29	15	0.670087	0.170088
100	50	29	16	0.810869	0.140782
100	50	29	17	0.907194	0.096324
100	50	29	18	0.961531	0.054337
100	50	29	19	0.986697	0.025167
100	50	29	20	0.996211	0.009514
100	50	29	21	0.999124	0.002912
100	50	29	22	0.999838	0.000714
100	50	29	23	0.999977	0.000138
100	50	29	24	0.999997	0.000021
100	50	29	25	1.000000	0.000002
100	50	29	26	1.000000	0.000000
100	50	29	27	1.000000	0.000000
100	50	29	28	1.000000	0.000000
100	50	29	29	1.000000	0.000000
100	50	30	0	0.000000	0.000000
100	50	30	1	0.000000	0.000000
100	50	30	2	0.000000	0.000000
100	50	30	3	0.000000	0.000000
100	50	30	4	0.000001	0.000001
100	50	30	5	0.000010	0.000009
100	50	30	6	0.000076	0.000066
100	50	30	7	0.000443	0.000367
100	50	30	8	0.002066	0.001622
100	50	30	9	0.007808	0.005743
100	50	30	10	0.024291	0.016482

N	n	k	x	P(x)	p(x)
100	50	30	11	0.062959	0.038668
100	50	30	12	0.137577	0.074618
100	50	30	13	0.256547	0.118971
100	50	30	14	0.413758	0.157211
100	50	30	15	0.586242	0.172483
100	50	30	16	0.743453	0.157211
100	50	30	17	0.862423	0.118971
100	50	30	18	0.937041	0.074618
100	50	30	19	0.975709	0.038668
100	50	30	20	0.992191	0.016482
100	50	30	21	0.997934	0.005743
100	50	30	22	0.999557	0.001622
100	50	30	23	0.999924	0.000367
100	50	30	24	0.999990	0.000066
100	50	30	25	0.999999	0.000009
100	50	30	26	1.000000	0.000001
100	50	30	27	1.000000	0.000000
100	50	30	28	1.000000	0.000000
100	50	30	29	1.000000	0.000000
100	50	30	30	1.000000	0.000000
100	50	31	0	0.000000	0.000000
100	50	31	1	0.000000	0.000000
100	50	31	2	0.000000	0.000000
100	50	31	3	0.000000	0.000000
100	50	31	4	0.000000	0.000000
100	50	31	5	0.000004	0.000004
100	50	31	6	0.000035	0.000030
100	50	31	7	0.000218	0.000183
100	50	31	8	0.001092	0.000875
100	50	31	9	0.004445	0.003353
100	50	31	10	0.014872	0.010428
100	50	31	11	0.041415	0.026543
100	50	31	12	0.097070	0.055655
100	50	31	13	0.193663	0.096593
100	50	31	14	0.332907	0.139244
100	50	31	15	0.500000	0.167093
100	50	31	16	0.667093	0.167093
100	50	31	17	0.806337	0.139244
100	50	31	18	0.902930	0.096593
100	50	31	19	0.958585	0.055655
100	50	31	20	0.985128	0.026543
100	50	31	21	0.995555	0.010428
100	50	31	22	0.998908	0.003353
100	50	31	23	0.999782	0.000875
100	50	31	24	0.999965	0.000183
100	50	31	25	0.999996	0.000030
100	50	31	26	0.999999	0.000004
100	50	31	27	1.000000	0.000000
100	50	31	28	1.000000	0.000000
100	50	31	29	1.000000	0.000000

N	n	k	x	P(x)	p(x)
100	50	31	30	1.000000	0.000000
100	50	31	31	1.000000	0.000000
100	50	32	0	0.000000	0.000000
100	50	32	1	0.000000	0.000000
100	50	32	2	0.000000	0.000000
100	50	32	3	0.000000	0.000000
100	50	32	4	0.000000	0.000000
100	50	32	5	0.000002	0.000002
100	50	32	6	0.000015	0.000014
100	50	32	7	0.000104	0.000088
100	50	32	8	0.000560	0.000456
100	50	32	9	0.002453	0.001893
100	50	32	10	0.008827	0.006375
100	50	32	11	0.026413	0.017585
100	50	32	12	0.066420	0.040007
100	50	32	13	0.141867	0.075447
100	50	32	14	0.260258	0.118391
100	50	32	15	0.415243	0.154985
100	50	32	16	0.584757	0.169515
100	50	32	17	0.739742	0.154985
100	50	32	18	0.858133	0.118391
100	50	32	19	0.933580	0.075447
100	50	32	20	0.973587	0.040007
100	50	32	21	0.991172	0.017585
100	50	32	22	0.997547	0.006375
100	50	32	23	0.999440	0.001893
100	50	32	24	0.999896	0.000456
100	50	32	25	0.999985	0.000088
100	50	32	26	0.999998	0.000014
100	50	32	27	1.000000	0.000002
100	50	32	28	1.000000	0.000000
100	50	32	29	1.000000	0.000000
100	50	32	30	1.000000	0.000000
100	50	32	31	1.000000	0.000000
100	50	32	32	1.000000	0.000000
100	50	33	0	0.000000	0.000000
100	50	33	1	0.000000	0.000000
100	50	33	2	0.000000	0.000000
100	50	33	3	0.000000	0.000000
100	50	33	4	0.000000	0.000000
100	50	33	5	0.000001	0.000001
100	50	33	6	0.000007	0.000006
100	50	33	7	0.000048	0.000041
100	50	33	8	0.000278	0.000230
100	50	33	9	0.001311	0.001033
100	50	33	10	0.005078	0.003766
100	50	33	11	0.016327	0.011249
100	50	33	12	0.044063	0.027736
100	50	33	13	0.100814	0.056752
100	50	33	14	0.197580	0.096766

Table for $N = 2,\ n = 1,$ through $N = 100,\ n = 50$

N	n	k	x	P(x)	p(x)
100	50	33	15	0.335471	0.137891
100	50	33	16	0.500000	0.164529
100	50	33	17	0.664529	0.164529
100	50	33	18	0.802420	0.137891
100	50	33	19	0.899185	0.096766
100	50	33	20	0.955937	0.056752
100	50	33	21	0.983673	0.027736
100	50	33	22	0.994922	0.011249
100	50	33	23	0.998688	0.003766
100	50	33	24	0.999722	0.001033
100	50	33	25	0.999952	0.000230
100	50	33	26	0.999993	0.000041
100	50	33	27	0.999999	0.000006
100	50	33	28	1.000000	0.000001
100	50	33	29	1.000000	0.000000
100	50	33	30	1.000000	0.000000
100	50	33	31	1.000000	0.000000
100	50	33	32	1.000000	0.000000
100	50	33	33	1.000000	0.000000
100	50	34	0	0.000000	0.000000
100	50	34	1	0.000000	0.000000
100	50	34	2	0.000000	0.000000
100	50	34	3	0.000000	0.000000
100	50	34	4	0.000000	0.000000
100	50	34	5	0.000000	0.000000
100	50	34	6	0.000003	0.000002
100	50	34	7	0.000021	0.000019
100	50	34	8	0.000134	0.000112
100	50	34	9	0.000679	0.000545
100	50	34	10	0.002829	0.002150
100	50	34	11	0.009779	0.006950
100	50	34	12	0.028332	0.018553
100	50	34	13	0.069474	0.041142
100	50	34	14	0.145587	0.076113
100	50	34	15	0.263438	0.117852
100	50	34	16	0.416508	0.153069
100	50	34	17	0.583492	0.166985
100	50	34	18	0.736561	0.153069
100	50	34	19	0.854413	0.117852
100	50	34	20	0.930526	0.076113
100	50	34	21	0.971668	0.041142
100	50	34	22	0.990221	0.018553
100	50	34	23	0.997171	0.006950
100	50	34	24	0.999321	0.002150
100	50	34	25	0.999866	0.000545
100	50	34	26	0.999979	0.000112
100	50	34	27	0.999997	0.000019
100	50	34	28	1.000000	0.000002
100	50	34	29	1.000000	0.000000
100	50	34	30	1.000000	0.000000

N	n	k	x	P(x)	p(x)
100	50	34	31	1.000000	0.000000
100	50	34	32	1.000000	0.000000
100	50	34	33	1.000000	0.000000
100	50	34	34	1.000000	0.000000
100	50	35	0	0.000000	0.000000
100	50	35	1	0.000000	0.000000
100	50	35	2	0.000000	0.000000
100	50	35	3	0.000000	0.000000
100	50	35	4	0.000000	0.000000
100	50	35	5	0.000000	0.000000
100	50	35	6	0.000001	0.000001
100	50	35	7	0.000009	0.000008
100	50	35	8	0.000062	0.000053
100	50	35	9	0.000340	0.000278
100	50	35	10	0.001526	0.001186
100	50	35	11	0.005672	0.004146
100	50	35	12	0.017650	0.011978
100	50	35	13	0.046410	0.028760
100	50	35	14	0.104071	0.057661
100	50	35	15	0.200941	0.096871
100	50	35	16	0.337654	0.136713
100	50	35	17	0.500000	0.162346
100	50	35	18	0.662346	0.162346
100	50	35	19	0.799059	0.136713
100	50	35	20	0.895929	0.096871
100	50	35	21	0.953590	0.057661
100	50	35	22	0.982350	0.028760
100	50	35	23	0.994328	0.011978
100	50	35	24	0.998474	0.004146
100	50	35	25	0.999660	0.001186
100	50	35	26	0.999938	0.000278
100	50	35	27	0.999991	0.000053
100	50	35	28	0.999999	0.000008
100	50	35	29	1.000000	0.000001
100	50	35	30	1.000000	0.000000
100	50	35	31	1.000000	0.000000
100	50	35	32	1.000000	0.000000
100	50	35	33	1.000000	0.000000
100	50	35	34	1.000000	0.000000
100	50	35	35	1.000000	0.000000
100	50	36	0	0.000000	0.000000
100	50	36	1	0.000000	0.000000
100	50	36	2	0.000000	0.000000
100	50	36	3	0.000000	0.000000
100	50	36	4	0.000000	0.000000
100	50	36	5	0.000000	0.000000
100	50	36	6	0.000000	0.000000
100	50	36	7	0.000004	0.000003
100	50	36	8	0.000028	0.000024
100	50	36	9	0.000165	0.000137

N	n	k	x	P(x)	p(x)
100	50	36	10	0.000796	0.000631
100	50	36	11	0.003184	0.002388
100	50	36	12	0.010648	0.007463
100	50	36	13	0.030039	0.019391
100	50	36	14	0.072135	0.042097
100	50	36	15	0.148780	0.076645
100	50	36	16	0.266143	0.117362
100	50	36	17	0.417578	0.151435
100	50	36	18	0.582422	0.164844
100	50	36	19	0.733857	0.151435
100	50	36	20	0.851220	0.117362
100	50	36	21	0.927865	0.076645
100	50	36	22	0.969961	0.042097
100	50	36	23	0.989352	0.019391
100	50	36	24	0.996815	0.007463
100	50	36	25	0.999204	0.002388
100	50	36	26	0.999835	0.000631
100	50	36	27	0.999972	0.000137
100	50	36	28	0.999996	0.000024
100	50	36	29	0.999999	0.000003
100	50	36	30	1.000000	0.000000
100	50	36	31	1.000000	0.000000
100	50	36	32	1.000000	0.000000
100	50	36	33	1.000000	0.000000
100	50	36	34	1.000000	0.000000
100	50	36	35	1.000000	0.000000
100	50	36	36	1.000000	0.000000
100	50	37	0	0.000000	0.000000
100	50	37	1	0.000000	0.000000
100	50	37	2	0.000000	0.000000
100	50	37	3	0.000000	0.000000
100	50	37	4	0.000000	0.000000
100	50	37	5	0.000000	0.000000
100	50	37	6	0.000000	0.000000
100	50	37	7	0.000002	0.000001
100	50	37	8	0.000012	0.000011
100	50	37	9	0.000077	0.000065
100	50	37	10	0.000402	0.000325
100	50	37	11	0.001729	0.001328
100	50	37	12	0.006216	0.004487
100	50	37	13	0.018828	0.012612
100	50	37	14	0.048456	0.029628
100	50	37	15	0.106865	0.058409
100	50	37	16	0.203794	0.096929
100	50	37	17	0.339494	0.135700
100	50	37	18	0.500000	0.160506
100	50	37	19	0.660506	0.160506
100	50	37	20	0.796206	0.135700
100	50	37	21	0.893135	0.096929
100	50	37	22	0.951544	0.058409

N	n	k	x	P(x)	p(x)
100	50	37	23	0.981172	0.029628
100	50	37	24	0.993784	0.012612
100	50	37	25	0.998271	0.004487
100	50	37	26	0.999598	0.001328
100	50	37	27	0.999923	0.000325
100	50	37	28	0.999988	0.000065
100	50	37	29	0.999998	0.000011
100	50	37	30	1.000000	0.000001
100	50	37	31	1.000000	0.000000
100	50	37	32	1.000000	0.000000
100	50	37	33	1.000000	0.000000
100	50	37	34	1.000000	0.000000
100	50	37	35	1.000000	0.000000
100	50	37	36	1.000000	0.000000
100	50	37	37	1.000000	0.000000
100	50	38	0	0.000000	0.000000
100	50	38	1	0.000000	0.000000
100	50	38	2	0.000000	0.000000
100	50	38	3	0.000000	0.000000
100	50	38	4	0.000000	0.000000
100	50	38	5	0.000000	0.000000
100	50	38	6	0.000000	0.000000
100	50	38	7	0.000001	0.000001
100	50	38	8	0.000005	0.000004
100	50	38	9	0.000035	0.000030
100	50	38	10	0.000196	0.000161
100	50	38	11	0.000907	0.000712
100	50	38	12	0.003510	0.002602
100	50	38	13	0.011421	0.007911
100	50	38	14	0.031526	0.020105
100	50	38	15	0.074416	0.042890
100	50	38	16	0.151483	0.077067
100	50	38	17	0.268413	0.116930
100	50	38	18	0.418473	0.150060
100	50	38	19	0.581527	0.163053
100	50	38	20	0.731587	0.150060
100	50	38	21	0.848517	0.116930
100	50	38	22	0.925584	0.077067
100	50	38	23	0.968474	0.042890
100	50	38	24	0.988579	0.020105
100	50	38	25	0.996490	0.007911
100	50	38	26	0.999092	0.002602
100	50	38	27	0.999804	0.000712
100	50	38	28	0.999965	0.000161
100	50	38	29	0.999995	0.000030
100	50	38	30	0.999999	0.000004
100	50	38	31	1.000000	0.000001
100	50	38	32	1.000000	0.000000
100	50	38	33	1.000000	0.000000
100	50	38	34	1.000000	0.000000

Table for $N = 2$, $n = 1$, through $N = 100$, $n = 50$

N	n	k	x	P(x)	p(x)
100	50	38	35	1.000000	0.000000
100	50	38	36	1.000000	0.000000
100	50	38	37	1.000000	0.000000
100	50	38	38	1.000000	0.000000
100	50	39	0	0.000000	0.000000
100	50	39	1	0.000000	0.000000
100	50	39	2	0.000000	0.000000
100	50	39	3	0.000000	0.000000
100	50	39	4	0.000000	0.000000
100	50	39	5	0.000000	0.000000
100	50	39	6	0.000000	0.000000
100	50	39	7	0.000000	0.000000
100	50	39	8	0.000002	0.000002
100	50	39	9	0.000015	0.000013
100	50	39	10	0.000092	0.000077
100	50	39	11	0.000460	0.000368
100	50	39	12	0.001915	0.001455
100	50	39	13	0.006700	0.004785
100	50	39	14	0.019852	0.013152
100	50	39	15	0.050204	0.030351
100	50	39	16	0.109220	0.059017
100	50	39	17	0.206176	0.096956
100	50	39	18	0.341023	0.134847
100	50	39	19	0.500000	0.158977
100	50	39	20	0.658977	0.158977
100	50	39	21	0.793824	0.134847
100	50	39	22	0.890780	0.096956
100	50	39	23	0.949796	0.059017
100	50	39	24	0.980148	0.030351
100	50	39	25	0.993300	0.013152
100	50	39	26	0.998085	0.004785
100	50	39	27	0.999540	0.001455
100	50	39	28	0.999908	0.000368
100	50	39	29	0.999985	0.000077
100	50	39	30	0.999998	0.000013
100	50	39	31	1.000000	0.000002
100	50	39	32	1.000000	0.000000
100	50	39	33	1.000000	0.000000
100	50	39	34	1.000000	0.000000
100	50	39	35	1.000000	0.000000
100	50	39	36	1.000000	0.000000
100	50	39	37	1.000000	0.000000
100	50	39	38	1.000000	0.000000
100	50	39	39	1.000000	0.000000
100	50	40	0	0.000000	0.000000
100	50	40	1	0.000000	0.000000
100	50	40	2	0.000000	0.000000
100	50	40	3	0.000000	0.000000
100	50	40	4	0.000000	0.000000
100	50	40	5	0.000000	0.000000
100	50	40	6	0.000000	0.000000
100	50	40	7	0.000000	0.000000
100	50	40	8	0.000001	0.000001
100	50	40	9	0.000006	0.000006
100	50	40	10	0.000042	0.000035
100	50	40	11	0.000224	0.000183
100	50	40	12	0.001008	0.000784
100	50	40	13	0.003797	0.002789
100	50	40	14	0.012090	0.008293
100	50	40	15	0.032789	0.020699
100	50	40	16	0.076326	0.043537
100	50	40	17	0.153725	0.077399
100	50	40	18	0.270284	0.116559
100	50	40	19	0.419208	0.148925
100	50	40	20	0.580792	0.161583
100	50	40	21	0.729716	0.148925
100	50	40	22	0.846275	0.116559
100	50	40	23	0.923674	0.077399
100	50	40	24	0.967211	0.043537
100	50	40	25	0.987910	0.020699
100	50	40	26	0.996202	0.008293
100	50	40	27	0.998992	0.002789
100	50	40	28	0.999775	0.000784
100	50	40	29	0.999958	0.000183
100	50	40	30	0.999993	0.000035
100	50	40	31	0.999999	0.000006
100	50	40	32	1.000000	0.000001
100	50	40	33	1.000000	0.000000
100	50	40	34	1.000000	0.000000
100	50	40	35	1.000000	0.000000
100	50	40	36	1.000000	0.000000
100	50	40	37	1.000000	0.000000
100	50	40	38	1.000000	0.000000
100	50	40	39	1.000000	0.000000
100	50	40	40	1.000000	0.000000
100	50	41	0	0.000000	0.000000
100	50	41	1	0.000000	0.000000
100	50	41	2	0.000000	0.000000
100	50	41	3	0.000000	0.000000
100	50	41	4	0.000000	0.000000
100	50	41	5	0.000000	0.000000
100	50	41	6	0.000000	0.000000
100	50	41	7	0.000000	0.000000
100	50	41	8	0.000000	0.000000
100	50	41	9	0.000003	0.000002
100	50	41	10	0.000018	0.000016
100	50	41	11	0.000106	0.000088
100	50	41	12	0.000512	0.000406
100	50	41	13	0.002077	0.001566
100	50	41	14	0.007115	0.005037
100	50	41	15	0.020715	0.013600
100	50	41	16	0.051655	0.030940
100	50	41	17	0.111155	0.059500
100	50	41	18	0.208119	0.096964
100	50	41	19	0.342264	0.134145
100	50	41	20	0.500000	0.157736
100	50	41	21	0.657736	0.157736
100	50	41	22	0.791881	0.134145
100	50	41	23	0.888845	0.096964
100	50	41	24	0.948345	0.059500
100	50	41	25	0.979285	0.030940
100	50	41	26	0.992885	0.013600
100	50	41	27	0.997922	0.005037
100	50	41	28	0.999488	0.001566
100	50	41	29	0.999894	0.000406
100	50	41	30	0.999982	0.000088
100	50	41	31	0.999997	0.000016
100	50	41	32	0.999999	0.000002
100	50	41	33	1.000000	0.000000
100	50	41	34	1.000000	0.000000
100	50	41	35	1.000000	0.000000
100	50	41	36	1.000000	0.000000
100	50	41	37	1.000000	0.000000
100	50	41	38	1.000000	0.000000
100	50	41	39	1.000000	0.000000
100	50	41	40	1.000000	0.000000
100	50	41	41	1.000000	0.000000
100	50	42	0	0.000000	0.000000
100	50	42	1	0.000000	0.000000
100	50	42	2	0.000000	0.000000
100	50	42	3	0.000000	0.000000
100	50	42	4	0.000000	0.000000
100	50	42	5	0.000000	0.000000
100	50	42	6	0.000000	0.000000
100	50	42	7	0.000000	0.000000
100	50	42	8	0.000000	0.000000
100	50	42	9	0.000001	0.000001
100	50	42	10	0.000008	0.000007
100	50	42	11	0.000048	0.000040
100	50	42	12	0.000250	0.000202
100	50	42	13	0.001096	0.000845
100	50	42	14	0.004041	0.002945
100	50	42	15	0.012647	0.008606
100	50	42	16	0.033825	0.021178
100	50	42	17	0.077875	0.044050
100	50	42	18	0.155528	0.077653
100	50	42	19	0.271781	0.116252
100	50	42	20	0.419795	0.148014
100	50	42	21	0.580205	0.160410
100	50	42	22	0.728219	0.148014
100	50	42	23	0.844472	0.116252
100	50	42	24	0.922125	0.077653
100	50	42	25	0.966175	0.044050
100	50	42	26	0.987353	0.021178
100	50	42	27	0.995959	0.008606
100	50	42	28	0.998904	0.002945
100	50	42	29	0.999750	0.000845
100	50	42	30	0.999952	0.000202
100	50	42	31	0.999992	0.000040
100	50	42	32	0.999999	0.000007
100	50	42	33	1.000000	0.000001
100	50	42	34	1.000000	0.000000
100	50	42	35	1.000000	0.000000
100	50	42	36	1.000000	0.000000
100	50	42	37	1.000000	0.000000
100	50	42	38	1.000000	0.000000
100	50	42	39	1.000000	0.000000
100	50	42	40	1.000000	0.000000
100	50	42	41	1.000000	0.000000
100	50	42	42	1.000000	0.000000
100	50	43	0	0.000000	0.000000
100	50	43	1	0.000000	0.000000
100	50	43	2	0.000000	0.000000
100	50	43	3	0.000000	0.000000
100	50	43	4	0.000000	0.000000
100	50	43	5	0.000000	0.000000
100	50	43	6	0.000000	0.000000
100	50	43	7	0.000000	0.000000
100	50	43	8	0.000000	0.000000
100	50	43	9	0.000000	0.000000
100	50	43	10	0.000003	0.000003
100	50	43	11	0.000021	0.000018
100	50	43	12	0.000118	0.000097
100	50	43	13	0.000556	0.000439
100	50	43	14	0.002213	0.001657
100	50	43	15	0.007454	0.005241
100	50	43	16	0.021410	0.013956
100	50	43	17	0.052812	0.031402
100	50	43	18	0.112685	0.059873
100	50	43	19	0.209646	0.096961
100	50	43	20	0.343236	0.133590
100	50	43	21	0.500000	0.156764
100	50	43	22	0.656764	0.156764
100	50	43	23	0.790354	0.133590
100	50	43	24	0.887315	0.096961
100	50	43	25	0.947188	0.059873
100	50	43	26	0.978590	0.031402
100	50	43	27	0.992546	0.013956
100	50	43	28	0.997787	0.005241
100	50	43	29	0.999444	0.001657

Table for $N = 2$, $n = 1$, through $N = 100$, $n = 50$

N	n	k	x	P(x)	p(x)
100	50	43	30	0.999882	0.000439
100	50	43	31	0.999979	0.000097
100	50	43	32	0.999997	0.000018
100	50	43	33	1.000000	0.000003
100	50	43	34	1.000000	0.000000
100	50	43	35	1.000000	0.000000
100	50	43	36	1.000000	0.000000
100	50	43	37	1.000000	0.000000
100	50	43	38	1.000000	0.000000
100	50	43	39	1.000000	0.000000
100	50	43	40	1.000000	0.000000
100	50	43	41	1.000000	0.000000
100	50	43	42	1.000000	0.000000
100	50	43	43	1.000000	0.000000
100	50	44	0	0.000000	0.000000
100	50	44	1	0.000000	0.000000
100	50	44	2	0.000000	0.000000
100	50	44	3	0.000000	0.000000
100	50	44	4	0.000000	0.000000
100	50	44	5	0.000000	0.000000
100	50	44	6	0.000000	0.000000
100	50	44	7	0.000000	0.000000
100	50	44	8	0.000000	0.000000
100	50	44	9	0.000000	0.000000
100	50	44	10	0.000001	0.000001
100	50	44	11	0.000009	0.000007
100	50	44	12	0.000053	0.000044
100	50	44	13	0.000271	0.000219
100	50	44	14	0.001167	0.000895
100	50	44	15	0.004236	0.003069
100	50	44	16	0.013085	0.008850
100	50	44	17	0.034632	0.021547
100	50	44	18	0.079072	0.044440
100	50	44	19	0.156913	0.077841
100	50	44	20	0.272925	0.116012
100	50	44	21	0.420243	0.147317
100	50	44	22	0.579757	0.159514
100	50	44	23	0.727074	0.147317
100	50	44	24	0.843087	0.116012
100	50	44	25	0.920928	0.077841
100	50	44	26	0.965368	0.044440
100	50	44	27	0.986915	0.021547
100	50	44	28	0.995764	0.008850
100	50	44	29	0.998833	0.003069
100	50	44	30	0.999728	0.000895
100	50	44	31	0.999947	0.000219
100	50	44	32	0.999991	0.000044
100	50	44	33	0.999999	0.000007
100	50	44	34	1.000000	0.000001
100	50	44	35	1.000000	0.000000
100	50	44	36	1.000000	0.000000
100	50	44	37	1.000000	0.000000
100	50	44	38	1.000000	0.000000
100	50	44	39	1.000000	0.000000
100	50	44	40	1.000000	0.000000
100	50	44	41	1.000000	0.000000
100	50	44	42	1.000000	0.000000
100	50	44	43	1.000000	0.000000
100	50	44	44	1.000000	0.000000
100	50	45	0	0.000000	0.000000
100	50	45	1	0.000000	0.000000
100	50	45	2	0.000000	0.000000
100	50	45	3	0.000000	0.000000
100	50	45	4	0.000000	0.000000
100	50	45	5	0.000000	0.000000
100	50	45	6	0.000000	0.000000
100	50	45	7	0.000000	0.000000
100	50	45	8	0.000000	0.000000
100	50	45	9	0.000000	0.000000
100	50	45	10	0.000000	0.000000
100	50	45	11	0.000003	0.000003
100	50	45	12	0.000023	0.000019
100	50	45	13	0.000127	0.000104
100	50	45	14	0.000591	0.000464
100	50	45	15	0.002318	0.001726
100	50	45	16	0.007712	0.005395
100	50	45	17	0.021935	0.014223
100	50	45	18	0.053678	0.031743
100	50	45	19	0.113823	0.060145
100	50	45	20	0.210776	0.096953
100	50	45	21	0.343953	0.133178
100	50	45	22	0.500000	0.156046
100	50	45	23	0.656046	0.156046
100	50	45	24	0.789224	0.133178
100	50	45	25	0.886177	0.096953
100	50	45	26	0.946322	0.060145
100	50	45	27	0.978065	0.031743
100	50	45	28	0.992288	0.014223
100	50	45	29	0.997682	0.005395
100	50	45	30	0.999409	0.001726
100	50	45	31	0.999873	0.000464
100	50	45	32	0.999977	0.000104
100	50	45	33	0.999996	0.000019
100	50	45	34	0.999999	0.000003
100	50	45	35	1.000000	0.000000
100	50	45	36	1.000000	0.000000
100	50	45	37	1.000000	0.000000
100	50	45	38	1.000000	0.000000
100	50	45	39	1.000000	0.000000
100	50	45	40	1.000000	0.000000
100	50	45	41	1.000000	0.000000
100	50	45	42	1.000000	0.000000
100	50	45	43	1.000000	0.000000
100	50	45	44	1.000000	0.000000
100	50	45	45	1.000000	0.000000
100	50	46	0	0.000000	0.000000
100	50	46	1	0.000000	0.000000
100	50	46	2	0.000000	0.000000
100	50	46	3	0.000000	0.000000
100	50	46	4	0.000000	0.000000
100	50	46	5	0.000000	0.000000
100	50	46	6	0.000000	0.000000
100	50	46	7	0.000000	0.000000
100	50	46	8	0.000000	0.000000
100	50	46	9	0.000000	0.000000
100	50	46	10	0.000000	0.000000
100	50	46	11	0.000001	0.000001
100	50	46	12	0.000009	0.000008
100	50	46	13	0.000057	0.000048
100	50	46	14	0.000287	0.000230
100	50	46	15	0.001219	0.000932
100	50	46	16	0.004377	0.003158
100	50	46	17	0.013401	0.009024
100	50	46	18	0.035209	0.021808
100	50	46	19	0.079923	0.044714
100	50	46	20	0.157892	0.077969
100	50	46	21	0.273733	0.115840
100	50	46	22	0.420558	0.146826
100	50	46	23	0.579442	0.158884
100	50	46	24	0.726267	0.146826
100	50	46	25	0.842108	0.115840
100	50	46	26	0.920077	0.077969
100	50	46	27	0.964791	0.044714
100	50	46	28	0.986599	0.021808
100	50	46	29	0.995623	0.009024
100	50	46	30	0.998781	0.003158
100	50	46	31	0.999712	0.000932
100	50	46	32	0.999943	0.000230
100	50	46	33	0.999990	0.000048
100	50	46	34	0.999999	0.000008
100	50	46	35	1.000000	0.000001
100	50	46	36	1.000000	0.000000
100	50	46	37	1.000000	0.000000
100	50	46	38	1.000000	0.000000
100	50	46	39	1.000000	0.000000
100	50	46	40	1.000000	0.000000
100	50	46	41	1.000000	0.000000
100	50	46	42	1.000000	0.000000
100	50	46	43	1.000000	0.000000
100	50	46	44	1.000000	0.000000
100	50	46	45	1.000000	0.000000
100	50	46	46	1.000000	0.000000
100	50	47	0	0.000000	0.000000
100	50	47	1	0.000000	0.000000
100	50	47	2	0.000000	0.000000
100	50	47	3	0.000000	0.000000
100	50	47	4	0.000000	0.000000
100	50	47	5	0.000000	0.000000
100	50	47	6	0.000000	0.000000
100	50	47	7	0.000000	0.000000
100	50	47	8	0.000000	0.000000
100	50	47	9	0.000000	0.000000
100	50	47	10	0.000000	0.000000
100	50	47	11	0.000000	0.000000
100	50	47	12	0.000004	0.000003
100	50	47	13	0.000024	0.000021
100	50	47	14	0.000134	0.000109
100	50	47	15	0.000615	0.000481
100	50	47	16	0.002389	0.001774
100	50	47	17	0.007887	0.005498
100	50	47	18	0.022286	0.014399
100	50	47	19	0.054254	0.031968
100	50	47	20	0.114576	0.060322
100	50	47	21	0.211522	0.096946
100	50	47	22	0.344426	0.132904
100	50	47	23	0.500000	0.155574
100	50	47	24	0.655574	0.155574
100	50	47	25	0.788478	0.132904
100	50	47	26	0.885424	0.096946
100	50	47	27	0.945746	0.060322
100	50	47	28	0.977714	0.031968
100	50	47	29	0.992113	0.014399
100	50	47	30	0.997611	0.005498
100	50	47	31	0.999385	0.001774
100	50	47	32	0.999866	0.000481
100	50	47	33	0.999975	0.000109
100	50	47	34	0.999996	0.000021
100	50	47	35	0.999999	0.000003
100	50	47	36	1.000000	0.000000
100	50	47	37	1.000000	0.000000
100	50	47	38	1.000000	0.000000
100	50	47	39	1.000000	0.000000
100	50	47	40	1.000000	0.000000
100	50	47	41	1.000000	0.000000
100	50	47	42	1.000000	0.000000
100	50	47	43	1.000000	0.000000
100	50	47	44	1.000000	0.000000
100	50	47	45	1.000000	0.000000
100	50	47	46	1.000000	0.000000
100	50	47	47	1.000000	0.000000

N	n	k	x	P(x)	p(x)
100	50	48	0	0.000000	0.000000
100	50	48	1	0.000000	0.000000
100	50	48	2	0.000000	0.000000
100	50	48	3	0.000000	0.000000
100	50	48	4	0.000000	0.000000
100	50	48	5	0.000000	0.000000
100	50	48	6	0.000000	0.000000
100	50	48	7	0.000000	0.000000
100	50	48	8	0.000000	0.000000
100	50	48	9	0.000000	0.000000
100	50	48	10	0.000000	0.000000
100	50	48	11	0.000000	0.000000
100	50	48	12	0.000001	0.000001
100	50	48	13	0.000010	0.000009
100	50	48	14	0.000059	0.000050
100	50	48	15	0.000297	0.000238
100	50	48	16	0.001251	0.000954
100	50	48	17	0.004463	0.003212
100	50	48	18	0.013592	0.009129
100	50	48	19	0.035556	0.021964
100	50	48	20	0.080431	0.044876
100	50	48	21	0.158476	0.078045
100	50	48	22	0.274213	0.115737
100	50	48	23	0.420745	0.146533
100	50	48	24	0.579254	0.158509
100	50	48	25	0.725787	0.146533
100	50	48	26	0.841524	0.115737
100	50	48	27	0.919568	0.078045
100	50	48	28	0.964444	0.044876
100	50	48	29	0.986408	0.021964
100	50	48	30	0.995536	0.009129
100	50	48	31	0.998749	0.003212
100	50	48	32	0.999702	0.000954
100	50	48	33	0.999940	0.000238
100	50	48	34	0.999990	0.000050
100	50	48	35	0.999998	0.000009
100	50	48	36	1.000000	0.000001
100	50	48	37	1.000000	0.000000
100	50	48	38	1.000000	0.000000
100	50	48	39	1.000000	0.000000
100	50	48	40	1.000000	0.000000
100	50	48	41	1.000000	0.000000
100	50	48	42	1.000000	0.000000
100	50	48	43	1.000000	0.000000
100	50	48	44	1.000000	0.000000
100	50	48	45	1.000000	0.000000
100	50	48	46	1.000000	0.000000
100	50	48	47	1.000000	0.000000
100	50	48	48	1.000000	0.000000
100	50	49	0	0.000000	0.000000

N	n	k	x	P(x)	p(x)
100	50	49	1	0.000000	0.000000
100	50	49	2	0.000000	0.000000
100	50	49	3	0.000000	0.000000
100	50	49	4	0.000000	0.000000
100	50	49	5	0.000000	0.000000
100	50	49	6	0.000000	0.000000
100	50	49	7	0.000000	0.000000
100	50	49	8	0.000000	0.000000
100	50	49	9	0.000000	0.000000
100	50	49	10	0.000000	0.000000
100	50	49	11	0.000000	0.000000
100	50	49	12	0.000000	0.000000
100	50	49	13	0.000004	0.000003
100	50	49	14	0.000025	0.000021
100	50	49	15	0.000137	0.000112
100	50	49	16	0.000627	0.000490
100	50	49	17	0.002425	0.001797
100	50	49	18	0.007974	0.005550
100	50	49	19	0.022462	0.014488
100	50	49	20	0.054542	0.032080
100	50	49	21	0.114951	0.060410
100	50	49	22	0.211893	0.096942
100	50	49	23	0.344661	0.132768
100	50	49	24	0.500000	0.155339
100	50	49	25	0.655339	0.155339
100	50	49	26	0.788107	0.132768
100	50	49	27	0.885049	0.096942
100	50	49	28	0.945458	0.060410
100	50	49	29	0.977538	0.032080
100	50	49	30	0.992025	0.014488
100	50	49	31	0.997575	0.005550
100	50	49	32	0.999372	0.001797
100	50	49	33	0.999863	0.000490
100	50	49	34	0.999975	0.000112
100	50	49	35	0.999996	0.000021
100	50	49	36	0.999999	0.000003
100	50	49	37	1.000000	0.000000
100	50	49	38	1.000000	0.000000
100	50	49	39	1.000000	0.000000
100	50	49	40	1.000000	0.000000
100	50	49	41	1.000000	0.000000
100	50	49	42	1.000000	0.000000
100	50	49	43	1.000000	0.000000
100	50	49	44	1.000000	0.000000
100	50	49	45	1.000000	0.000000
100	50	49	46	1.000000	0.000000
100	50	49	47	1.000000	0.000000
100	50	49	48	1.000000	0.000000
100	50	49	49	1.000000	0.000000
100	50	50	0	0.000000	0.000000

N	n	k	x	P(x)	p(x)
100	50	50	1	0.000000	0.000000
100	50	50	2	0.000000	0.000000
100	50	50	3	0.000000	0.000000
100	50	50	4	0.000000	0.000000
100	50	50	5	0.000000	0.000000
100	50	50	6	0.000000	0.000000
100	50	50	7	0.000000	0.000000
100	50	50	8	0.000000	0.000000
100	50	50	9	0.000000	0.000000
100	50	50	10	0.000000	0.000000
100	50	50	11	0.000000	0.000000
100	50	50	12	0.000000	0.000000
100	50	50	13	0.000001	0.000001
100	50	50	14	0.000010	0.000009
100	50	50	15	0.000060	0.000050
100	50	50	16	0.000301	0.000240
100	50	50	17	0.001262	0.000961
100	50	50	18	0.004492	0.003230
100	50	50	19	0.013656	0.009164
100	50	50	20	0.035671	0.022015
100	50	50	21	0.080601	0.044929
100	50	50	22	0.158670	0.078069
100	50	50	23	0.274372	0.115702
100	50	50	24	0.420808	0.146436
100	50	50	25	0.579192	0.158385
100	50	50	26	0.725628	0.146436
100	50	50	27	0.841330	0.115702
100	50	50	28	0.919399	0.078069
100	50	50	29	0.964329	0.044929
100	50	50	30	0.986344	0.022015
100	50	50	31	0.995508	0.009164
100	50	50	32	0.998738	0.003230
100	50	50	33	0.999699	0.000961
100	50	50	34	0.999940	0.000240
100	50	50	35	0.999990	0.000050
100	50	50	36	0.999998	0.000009
100	50	50	37	1.000000	0.000001
100	50	50	38	1.000000	0.000000
100	50	50	39	1.000000	0.000000
100	50	50	40	1.000000	0.000000
100	50	50	41	1.000000	0.000000
100	50	50	42	1.000000	0.000000
100	50	50	43	1.000000	0.000000
100	50	50	44	1.000000	0.000000
100	50	50	45	1.000000	0.000000
100	50	50	46	1.000000	0.000000
100	50	50	47	1.000000	0.000000
100	50	50	48	1.000000	0.000000
100	50	50	49	1.000000	0.000000
100	50	50	50	1.000000	0.000000

N	n	k	x	P(x)	p(x)

Table for $N = 1000$, $n = 500$

N	n	k	x	P(x)	p(x)
1000	500	1	0	0.500000	0.500000
1000	500	2	0	0.249750	0.249750
1000	500	2	1	0.750250	0.500500
1000	500	3	0	0.124625	0.124625
1000	500	3	1	0.500000	0.375375
1000	500	4	0	0.062125	0.062125
1000	500	4	1	0.312124	0.249999
1000	500	4	2	0.687876	0.375752
1000	500	5	0	0.030938	0.030938
1000	500	5	1	0.186873	0.155936
1000	500	5	2	0.500000	0.313127
1000	500	6	0	0.015391	0.015391
1000	500	6	1	0.108670	0.093279
1000	500	6	2	0.343279	0.234609
1000	500	6	3	0.656721	0.313441
1000	500	7	0	0.007649	0.007649
1000	500	7	1	0.061843	0.054194
1000	500	7	2	0.225739	0.163896
1000	500	7	3	0.500000	0.274261
1000	500	8	0	0.003798	0.003798
1000	500	8	1	0.034610	0.030812
1000	500	8	2	0.143543	0.108934
1000	500	8	3	0.362731	0.219188
1000	500	8	4	0.637269	0.274537
1000	500	9	0	0.001883	0.001883
1000	500	9	1	0.019110	0.017227
1000	500	9	2	0.088857	0.069746
1000	500	9	3	0.252916	0.164059
1000	500	9	4	0.500000	0.247084
1000	500	10	0	0.000933	0.000933
1000	500	10	1	0.010436	0.009503
1000	500	10	2	0.053808	0.043372
1000	500	10	3	0.170638	0.116831
1000	500	10	4	0.376333	0.205695
1000	500	10	5	0.623666	0.247333
1000	500	11	0	0.000462	0.000462
1000	500	11	1	0.005646	0.005184
1000	500	11	2	0.031990	0.026344
1000	500	11	3	0.111987	0.079997
1000	500	11	4	0.273278	0.161291
1000	500	11	5	0.500000	0.226722
1000	500	12	0	0.000228	0.000228
1000	500	12	1	0.003030	0.002802
1000	500	12	2	0.018725	0.015695
1000	500	12	3	0.071787	0.053061
1000	500	12	4	0.192388	0.120601
1000	500	12	5	0.386524	0.194136
1000	500	12	6	0.613475	0.226951
1000	500	13	0	0.000113	0.000113
1000	500	13	1	0.001615	0.001502
1000	500	13	2	0.010814	0.009199
1000	500	13	3	0.045095	0.034280
1000	500	13	4	0.131843	0.086748
1000	500	13	5	0.289260	0.157417
1000	500	13	6	0.500000	0.210740
1000	500	14	0	0.000056	0.000056
1000	500	14	1	0.000856	0.000800
1000	500	14	2	0.006173	0.005317
1000	500	14	3	0.027833	0.021660
1000	500	14	4	0.088249	0.060416
1000	500	14	5	0.210312	0.122063
1000	500	14	6	0.394523	0.184211
1000	500	14	7	0.605477	0.210954
1000	500	15	0	0.000027	0.000027
1000	500	15	1	0.000451	0.000423
1000	500	15	2	0.003487	0.003036
1000	500	15	3	0.016915	0.013428
1000	500	15	4	0.057857	0.040942
1000	500	15	5	0.149033	0.091176
1000	500	15	6	0.302231	0.153198
1000	500	15	7	0.500000	0.197769
1000	500	16	0	0.000014	0.000014
1000	500	16	1	0.000236	0.000223
1000	500	16	2	0.001952	0.001716
1000	500	16	3	0.010140	0.008188
1000	500	16	4	0.037241	0.027101
1000	500	16	5	0.103214	0.065973
1000	500	16	6	0.225399	0.122185
1000	500	16	7	0.401015	0.175616
1000	500	16	8	0.598985	0.197970
1000	500	17	0	0.000007	0.000007
1000	500	17	1	0.000123	0.000117
1000	500	17	2	0.001084	0.000960
1000	500	17	3	0.006004	0.004921
1000	500	17	4	0.023580	0.017576
1000	500	17	5	0.070026	0.046446
1000	500	17	6	0.164058	0.094032
1000	500	17	7	0.313028	0.148970
1000	500	17	8	0.500000	0.186972
1000	500	18	0	0.000003	0.000003
1000	500	18	1	0.000064	0.000061
1000	500	18	2	0.000597	0.000533
1000	500	18	3	0.003516	0.002919
1000	500	18	4	0.014712	0.011195
1000	500	18	5	0.046638	0.031926
1000	500	18	6	0.116803	0.070165
1000	500	18	7	0.238316	0.121513
1000	500	18	8	0.406419	0.168103
1000	500	18	9	0.593581	0.187162
1000	500	19	0	0.000002	0.000002
1000	500	19	1	0.000033	0.000032
1000	500	19	2	0.000327	0.000294
1000	500	19	3	0.002039	0.001712
1000	500	19	4	0.009057	0.007018
1000	500	19	5	0.030545	0.021488
1000	500	19	6	0.081506	0.050961
1000	500	19	7	0.177312	0.095806
1000	500	19	8	0.322196	0.144884
1000	500	19	9	0.500000	0.177804
1000	500	20	0	0.000001	0.000001
1000	500	20	1	0.000017	0.000016
1000	500	20	2	0.000178	0.000161
1000	500	20	3	0.001172	0.000994
1000	500	20	4	0.005509	0.004337
1000	500	20	5	0.019702	0.014194
1000	500	20	6	0.055843	0.036141
1000	500	20	7	0.129165	0.073321
1000	500	20	8	0.249532	0.120368
1000	500	20	9	0.411007	0.161475
1000	500	20	10	0.588992	0.177985
1000	500	21	1	0.000009	0.000008
1000	500	21	2	0.000096	0.000087
1000	500	21	3	0.000668	0.000571
1000	500	21	4	0.003314	0.002646
1000	500	21	5	0.012533	0.009220
1000	500	21	6	0.037625	0.025092
1000	500	21	7	0.092280	0.054654
1000	500	21	8	0.189103	0.096823
1000	500	21	9	0.330105	0.141002
1000	500	21	10	0.500000	0.169895
1000	500	22	1	0.000005	0.000004
1000	500	22	2	0.000052	0.000047
1000	500	22	3	0.000377	0.000326
1000	500	22	4	0.001973	0.001596
1000	500	22	5	0.007872	0.005899
1000	500	22	6	0.024964	0.017092
1000	500	22	7	0.064757	0.039793
1000	500	22	8	0.140444	0.075687
1000	500	22	9	0.259388	0.118944
1000	500	22	10	0.414966	0.155578
1000	500	22	11	0.585034	0.170068
1000	500	23	1	0.000002	0.000002
1000	500	23	2	0.000028	0.000025
1000	500	23	3	0.000212	0.000184
1000	500	23	4	0.001164	0.000952
1000	500	23	5	0.004886	0.003722
1000	500	23	6	0.016330	0.011444
1000	500	23	7	0.044698	0.028367
1000	500	23	8	0.102368	0.057671
1000	500	23	9	0.199673	0.097304
1000	500	23	10	0.337018	0.137345
1000	500	23	11	0.500000	0.162982
1000	500	24	1	0.000001	0.000001
1000	500	24	2	0.000015	0.000014
1000	500	24	3	0.000118	0.000103
1000	500	24	4	0.000681	0.000562
1000	500	24	5	0.003000	0.002320
1000	500	24	6	0.010544	0.007544
1000	500	24	7	0.030383	0.019840
1000	500	24	8	0.073326	0.042943
1000	500	24	9	0.150772	0.077445
1000	500	24	10	0.268134	0.117363
1000	500	24	11	0.418425	0.150291
1000	500	24	12	0.581574	0.163149
1000	500	25	1	0.000001	0.000001
1000	500	25	2	0.000008	0.000007
1000	500	25	3	0.000066	0.000058
1000	500	25	4	0.000395	0.000329
1000	500	25	5	0.001824	0.001429
1000	500	25	6	0.006726	0.004902
1000	500	25	7	0.020362	0.013636
1000	500	25	8	0.051679	0.031317
1000	500	25	9	0.111811	0.060132
1000	500	25	10	0.209213	0.097402
1000	500	25	11	0.343126	0.133913
1000	500	25	12	0.500000	0.156874
1000	500	26	2	0.000004	0.000004
1000	500	26	3	0.000036	0.000032
1000	500	26	4	0.000227	0.000191
1000	500	26	5	0.001098	0.000871
1000	500	26	6	0.004242	0.003144
1000	500	26	7	0.013467	0.009225
1000	500	26	8	0.035876	0.022409
1000	500	26	9	0.081529	0.045653
1000	500	26	10	0.160262	0.078733
1000	500	26	11	0.275963	0.115701
1000	500	26	12	0.421482	0.145519
1000	500	26	13	0.578517	0.157035
1000	500	27	2	0.000002	0.000002
1000	500	27	3	0.000020	0.000018
1000	500	27	4	0.000130	0.000110
1000	500	27	5	0.000656	0.000526
1000	500	27	6	0.002648	0.001992
1000	500	27	7	0.008798	0.006150
1000	500	27	8	0.024556	0.015759
1000	500	27	9	0.058515	0.033959
1000	500	27	10	0.120652	0.062138
1000	500	27	11	0.217875	0.097222
1000	500	27	12	0.348573	0.130698
1000	500	27	13	0.500000	0.151426

Table for $N = 1000$, $n = 500$

N	n	k	x	P(x)	p(x)
1000	500	28	2	0.000001	0.000001
1000	500	28	3	0.000011	0.000010
1000	500	28	4	0.000074	0.000063
1000	500	28	5	0.000388	0.000314
1000	500	28	6	0.001636	0.001248
1000	500	28	7	0.005681	0.004045
1000	500	28	8	0.016588	0.010906
1000	500	28	9	0.041379	0.024791
1000	500	28	10	0.089361	0.047982
1000	500	28	11	0.169014	0.079654
1000	500	28	12	0.283023	0.114009
1000	500	28	13	0.424209	0.141186
1000	500	28	14	0.575791	0.151582
1000	500	29	2	0.000001	0.000001
1000	500	29	3	0.000006	0.000005
1000	500	29	4	0.000042	0.000036
1000	500	29	5	0.000228	0.000186
1000	500	29	6	0.001002	0.000774
1000	500	29	7	0.003630	0.002628
1000	500	29	8	0.011067	0.007438
1000	500	29	9	0.028856	0.017788
1000	500	29	10	0.065172	0.036316
1000	500	29	11	0.128942	0.063769
1000	500	29	12	0.225784	0.096842
1000	500	29	13	0.353471	0.127687
1000	500	29	14	0.500000	0.146529
1000	500	30	3	0.000003	0.000003
1000	500	30	4	0.000023	0.000020
1000	500	30	5	0.000133	0.000110
1000	500	30	6	0.000608	0.000475
1000	500	30	7	0.002296	0.001687
1000	500	30	8	0.007299	0.005003
1000	500	30	9	0.019861	0.012562
1000	500	30	10	0.046846	0.026985
1000	500	30	11	0.096827	0.049981
1000	500	30	12	0.177113	0.080286
1000	500	30	13	0.289430	0.112317
1000	500	30	14	0.426660	0.137230
1000	500	30	15	0.573340	0.146680
1000	500	31	3	0.000002	0.000002
1000	500	31	4	0.000013	0.000011
1000	500	31	5	0.000077	0.000064
1000	500	31	6	0.000366	0.000289
1000	500	31	7	0.001438	0.001072
1000	500	31	8	0.004761	0.003323
1000	500	31	9	0.013502	0.008741
1000	500	31	10	0.033214	0.019712
1000	500	31	11	0.071630	0.038416
1000	500	31	12	0.136722	0.065092
1000	500	31	13	0.233040	0.096318

N	n	k	x	P(x)	p(x)
1000	500	31	14	0.357903	0.124863
1000	500	31	15	0.500000	0.142097
1000	500	32	3	0.000001	0.000001
1000	500	32	4	0.000007	0.000006
1000	500	32	5	0.000044	0.000037
1000	500	32	6	0.000219	0.000174
1000	500	32	7	0.000893	0.000674
1000	500	32	8	0.003074	0.002181
1000	500	32	9	0.009073	0.005999
1000	500	32	10	0.023246	0.014173
1000	500	32	11	0.052244	0.028998
1000	500	32	12	0.103941	0.051697
1000	500	32	13	0.184632	0.080692
1000	500	32	14	0.295278	0.110646
1000	500	32	15	0.428878	0.133600
1000	500	32	16	0.571122	0.142243
1000	500	33	4	0.000004	0.000004
1000	500	33	5	0.000025	0.000021
1000	500	33	6	0.000130	0.000104
1000	500	33	7	0.000549	0.000420
1000	500	33	8	0.001965	0.001416
1000	500	33	9	0.006030	0.004065
1000	500	33	10	0.016072	0.010042
1000	500	33	11	0.037595	0.021523
1000	500	33	12	0.077879	0.040284
1000	500	33	13	0.144037	0.066158
1000	500	33	14	0.239727	0.095690
1000	500	33	15	0.361940	0.122214
1000	500	33	16	0.500000	0.138060
1000	500	34	4	0.000002	0.000002
1000	500	34	5	0.000014	0.000012
1000	500	34	6	0.000076	0.000062
1000	500	34	7	0.000336	0.000259
1000	500	34	8	0.001245	0.000909
1000	500	34	9	0.003966	0.002721
1000	500	34	10	0.010983	0.007017
1000	500	34	11	0.026711	0.015728
1000	500	34	12	0.057549	0.030838
1000	500	34	13	0.110718	0.053169
1000	500	34	14	0.191634	0.080916
1000	500	34	15	0.300644	0.109010
1000	500	34	16	0.430899	0.130255
1000	500	34	17	0.569101	0.138202
1000	500	35	4	0.000001	0.000001
1000	500	35	5	0.000008	0.000007
1000	500	35	6	0.000045	0.000037
1000	500	35	7	0.000203	0.000159
1000	500	35	8	0.000782	0.000578
1000	500	35	9	0.002583	0.001801
1000	500	35	10	0.007424	0.004841

N	n	k	x	P(x)	p(x)
1000	500	35	11	0.018749	0.011326
1000	500	35	12	0.041971	0.023221
1000	500	35	13	0.083914	0.041943
1000	500	35	14	0.150925	0.067011
1000	500	35	15	0.245913	0.094989
1000	500	35	16	0.365637	0.119723
1000	500	35	17	0.500000	0.134363
1000	500	36	4	0.000001	0.000001
1000	500	36	5	0.000005	0.000004
1000	500	36	6	0.000026	0.000021
1000	500	36	7	0.000122	0.000096
1000	500	36	8	0.000487	0.000365
1000	500	36	9	0.001666	0.001180
1000	500	36	10	0.004966	0.003299
1000	500	36	11	0.013010	0.008045
1000	500	36	12	0.030228	0.017217
1000	500	36	13	0.062747	0.032519
1000	500	36	14	0.117176	0.054430
1000	500	36	15	0.198173	0.080997
1000	500	36	16	0.305589	0.107416
1000	500	36	17	0.432749	0.127160
1000	500	36	18	0.567251	0.134503
1000	500	37	5	0.000003	0.000002
1000	500	37	6	0.000015	0.000012
1000	500	37	7	0.000073	0.000058
1000	500	37	8	0.000301	0.000228
1000	500	37	9	0.001066	0.000765
1000	500	37	10	0.003289	0.002223
1000	500	37	11	0.008930	0.005641
1000	500	37	12	0.021512	0.012582
1000	500	37	13	0.046318	0.024807
1000	500	37	14	0.089736	0.043417
1000	500	37	15	0.157423	0.067687
1000	500	37	16	0.251658	0.094236
1000	500	37	17	0.369037	0.117379
1000	500	37	18	0.500000	0.130963
1000	500	38	5	0.000001	0.000001
1000	500	38	6	0.000009	0.000007
1000	500	38	7	0.000043	0.000035
1000	500	38	8	0.000184	0.000141
1000	500	38	9	0.000676	0.000491
1000	500	38	10	0.002158	0.001482
1000	500	38	11	0.006065	0.003908
1000	500	38	12	0.015136	0.009071
1000	500	38	13	0.033774	0.018638
1000	500	38	14	0.067824	0.034051
1000	500	38	15	0.123333	0.055509
1000	500	38	16	0.204296	0.080963
1000	500	38	17	0.310165	0.105869
1000	500	38	18	0.434450	0.124286

N	n	k	x	P(x)	p(x)
1000	500	38	19	0.565549	0.131099
1000	500	39	5	0.000001	0.000001
1000	500	39	6	0.000005	0.000004
1000	500	39	7	0.000026	0.000021
1000	500	39	8	0.000112	0.000087
1000	500	39	9	0.000425	0.000313
1000	500	39	10	0.001403	0.000978
1000	500	39	11	0.004079	0.002676
1000	500	39	12	0.010535	0.006456
1000	500	39	13	0.024338	0.013804
1000	500	39	14	0.050622	0.026283
1000	500	39	15	0.095348	0.044726
1000	500	39	16	0.163562	0.068214
1000	500	39	17	0.257010	0.093448
1000	500	39	18	0.372178	0.115168
1000	500	39	19	0.500000	0.127822
1000	500	40	6	0.000003	0.000002
1000	500	40	7	0.000015	0.000012
1000	500	40	8	0.000068	0.000053
1000	500	40	9	0.000265	0.000197
1000	500	40	10	0.000904	0.000639
1000	500	40	11	0.002717	0.001813
1000	500	40	12	0.007256	0.004539
1000	500	40	13	0.017343	0.010087
1000	500	40	14	0.037330	0.019987
1000	500	40	15	0.072775	0.035446
1000	500	40	16	0.129206	0.056431
1000	500	40	17	0.210043	0.080836
1000	500	40	18	0.314414	0.104372
1000	500	40	19	0.436023	0.121608
1000	500	40	20	0.563977	0.127955
1000	500	41	6	0.000002	0.000001
1000	500	41	7	0.000009	0.000007
1000	500	41	8	0.000041	0.000032
1000	500	41	9	0.000164	0.000124
1000	500	41	10	0.000578	0.000414
1000	500	41	11	0.001794	0.001216
1000	500	41	12	0.004949	0.003155
1000	500	41	13	0.012226	0.007277
1000	500	41	14	0.027212	0.014985
1000	500	41	15	0.054868	0.027656
1000	500	41	16	0.100756	0.045888
1000	500	41	17	0.169372	0.068616
1000	500	41	18	0.262011	0.092639
1000	500	41	19	0.375092	0.113081
1000	500	41	20	0.500000	0.124908
1000	500	42	6	0.000001	0.000001
1000	500	42	7	0.000005	0.000004
1000	500	42	8	0.000024	0.000019
1000	500	42	9	0.000101	0.000077

Table for $N = 1000$, $n = 500$

N	n	k	x	P(x)	p(x)
1000	500	42	10	0.000367	0.000266
1000	500	42	11	0.001174	0.000807
1000	500	42	12	0.003343	0.002170
1000	500	42	13	0.008531	0.005187
1000	500	42	14	0.019617	0.011087
1000	500	42	15	0.040881	0.021264
1000	500	42	16	0.077597	0.036715
1000	500	42	17	0.134814	0.057217
1000	500	42	18	0.215450	0.080637
1000	500	42	19	0.318375	0.102925
1000	500	42	20	0.437481	0.119106
1000	500	42	21	0.562519	0.125038
1000	500	43	6	0.000000	0.000000
1000	500	43	7	0.000003	0.000002
1000	500	43	8	0.000014	0.000012
1000	500	43	9	0.000062	0.000047
1000	500	43	10	0.000231	0.000169
1000	500	43	11	0.000762	0.000531
1000	500	43	12	0.002238	0.001476
1000	500	43	13	0.005894	0.003655
1000	500	43	14	0.013993	0.008099
1000	500	43	15	0.030116	0.016123
1000	500	43	16	0.059047	0.028931
1000	500	43	17	0.105966	0.046919
1000	500	43	18	0.174879	0.068913
1000	500	43	19	0.266698	0.091818
1000	500	43	20	0.377803	0.111106
1000	500	43	21	0.500000	0.122197
1000	500	44	7	0.000002	0.000001
1000	500	44	8	0.000009	0.000007
1000	500	44	9	0.000037	0.000029
1000	500	44	10	0.000144	0.000107
1000	500	44	11	0.000490	0.000346
1000	500	44	12	0.001485	0.000995
1000	500	44	13	0.004033	0.002548
1000	500	44	14	0.009880	0.005846
1000	500	44	15	0.021945	0.012065
1000	500	44	16	0.044415	0.022470
1000	500	44	17	0.082286	0.037871
1000	500	44	18	0.140171	0.057885
1000	500	44	19	0.220549	0.080378
1000	500	44	20	0.322076	0.101528
1000	500	44	21	0.438838	0.116761
1000	500	44	22	0.561162	0.122324
1000	500	45	7	0.000001	0.000001
1000	500	45	8	0.000005	0.000004
1000	500	45	9	0.000023	0.000018
1000	500	45	10	0.000089	0.000067
1000	500	45	11	0.000313	0.000224
1000	500	45	12	0.000977	0.000664
1000	500	45	13	0.002735	0.001758
1000	500	45	14	0.006908	0.004172
1000	500	45	15	0.015824	0.008916
1000	500	45	16	0.033039	0.017215
1000	500	45	17	0.063153	0.030113
1000	500	45	18	0.110986	0.047834
1000	500	45	19	0.180107	0.069121
1000	500	45	20	0.271100	0.090993
1000	500	45	21	0.380335	0.109235
1000	500	45	22	0.500000	0.119665
1000	500	46	7	0.000001	0.000000
1000	500	46	8	0.000003	0.000002
1000	500	46	9	0.000014	0.000011
1000	500	46	10	0.000055	0.000042
1000	500	46	11	0.000199	0.000144
1000	500	46	12	0.000638	0.000439
1000	500	46	13	0.001839	0.001201
1000	500	46	14	0.004784	0.002945
1000	500	46	15	0.011296	0.006511
1000	500	46	16	0.024314	0.013019
1000	500	46	17	0.047922	0.023608
1000	500	46	18	0.086844	0.038922
1000	500	46	19	0.145293	0.058449
1000	500	46	20	0.225366	0.080072
1000	500	46	21	0.325546	0.100180
1000	500	46	22	0.440105	0.114559
1000	500	46	23	0.559895	0.119790
1000	500	47	8	0.000002	0.000001
1000	500	47	9	0.000008	0.000006
1000	500	47	10	0.000034	0.000026
1000	500	47	11	0.000125	0.000091
1000	500	47	12	0.000413	0.000288
1000	500	47	13	0.001226	0.000813
1000	500	47	14	0.003284	0.002058
1000	500	47	15	0.007986	0.004702
1000	500	47	16	0.017710	0.009724
1000	500	47	17	0.035970	0.018260
1000	500	47	18	0.067179	0.031209
1000	500	47	19	0.115824	0.048644
1000	500	47	20	0.185078	0.069254
1000	500	47	21	0.275246	0.090168
1000	500	47	22	0.382705	0.107460
1000	500	47	23	0.500000	0.117295
1000	500	48	8	0.000001	0.000001
1000	500	48	9	0.000005	0.000004
1000	500	48	10	0.000021	0.000016
1000	500	48	11	0.000078	0.000058
1000	500	48	12	0.000266	0.000188
1000	500	48	13	0.000811	0.000545
1000	500	48	14	0.002234	0.001424
1000	500	48	15	0.005593	0.003358
1000	500	48	16	0.012771	0.007178
1000	500	48	17	0.026715	0.013944
1000	500	48	18	0.051394	0.024679
1000	500	48	19	0.091272	0.039877
1000	500	48	20	0.150196	0.058925
1000	500	48	21	0.229925	0.079729
1000	500	48	22	0.328806	0.098881
1000	500	48	23	0.441291	0.112485
1000	500	48	24	0.558709	0.117418
1000	500	49	8	0.000001	0.000000
1000	500	49	9	0.000003	0.000002
1000	500	49	10	0.000012	0.000010
1000	500	49	11	0.000049	0.000036
1000	500	49	12	0.000170	0.000121
1000	500	49	13	0.000532	0.000362
1000	500	49	14	0.001508	0.000976
1000	500	49	15	0.003882	0.002374
1000	500	49	16	0.009122	0.005240
1000	500	49	17	0.019641	0.010519
1000	500	49	18	0.038899	0.019259
1000	500	49	19	0.071124	0.032224
1000	500	49	20	0.120486	0.049363
1000	500	49	21	0.189810	0.069323
1000	500	49	22	0.279158	0.089349
1000	500	49	23	0.384931	0.105773
1000	500	49	24	0.500000	0.115069
1000	500	50	9	0.000002	0.000001
1000	500	50	10	0.000007	0.000006
1000	500	50	11	0.000030	0.000023
1000	500	50	12	0.000108	0.000078
1000	500	50	13	0.000346	0.000239
1000	500	50	14	0.001009	0.000663
1000	500	50	15	0.002671	0.001662
1000	500	50	16	0.006455	0.003784
1000	500	50	17	0.014298	0.007843
1000	500	50	18	0.029138	0.014840
1000	500	50	19	0.054825	0.025687
1000	500	50	20	0.095572	0.040746
1000	500	50	21	0.154893	0.059321
1000	500	50	22	0.234249	0.079356
1000	500	50	23	0.331877	0.097629
1000	500	50	24	0.442405	0.110527
1000	500	50	25	0.557595	0.115190
1000	500	51	9	0.000001	0.000001
1000	500	51	10	0.000004	0.000004
1000	500	51	11	0.000018	0.000014
1000	500	51	12	0.000068	0.000049
1000	500	51	13	0.000224	0.000156
1000	500	51	14	0.000670	0.000446
1000	500	51	15	0.001823	0.001152
1000	500	51	16	0.004527	0.002705
1000	500	51	17	0.010310	0.005783
1000	500	51	18	0.021609	0.011298
1000	500	51	19	0.041820	0.020211
1000	500	51	20	0.074984	0.033164
1000	500	51	21	0.124982	0.049999
1000	500	51	22	0.194320	0.069338
1000	500	51	23	0.282858	0.088537
1000	500	51	24	0.387025	0.104167
1000	500	51	25	0.500000	0.112975
1000	500	52	9	0.000001	0.000000
1000	500	52	10	0.000003	0.000002
1000	500	52	11	0.000011	0.000009
1000	500	52	12	0.000042	0.000031
1000	500	52	13	0.000144	0.000101
1000	500	52	14	0.000442	0.000298
1000	500	52	15	0.001234	0.000792
1000	500	52	16	0.003148	0.001914
1000	500	52	17	0.007367	0.004219
1000	500	52	18	0.015870	0.008503
1000	500	52	19	0.031576	0.015705
1000	500	52	20	0.058209	0.026634
1000	500	52	21	0.099746	0.041537
1000	500	52	22	0.159396	0.059650
1000	500	52	23	0.238356	0.078960
1000	500	52	24	0.334777	0.096421
1000	500	52	25	0.443453	0.108676
1000	500	52	26	0.556547	0.113094
1000	500	53	10	0.000002	0.000001
1000	500	53	11	0.000007	0.000005
1000	500	53	12	0.000026	0.000020
1000	500	53	13	0.000092	0.000065
1000	500	53	14	0.000289	0.000197
1000	500	53	15	0.000828	0.000539
1000	500	53	16	0.002170	0.001342
1000	500	53	17	0.005217	0.003047
1000	500	53	18	0.011547	0.006330
1000	500	53	19	0.023607	0.012060
1000	500	53	20	0.044724	0.021117
1000	500	53	21	0.078759	0.034035
1000	500	53	22	0.129319	0.050560
1000	500	53	23	0.198626	0.069307
1000	500	53	24	0.286363	0.087737
1000	500	53	25	0.389000	0.102637
1000	500	53	26	0.500000	0.111000
1000	500	54	10	0.000001	0.000001
1000	500	54	11	0.000004	0.000003
1000	500	54	12	0.000016	0.000012
1000	500	54	13	0.000058	0.000042

Table for $N = 1000$, $n = 500$

N	n	k	x	P(x)	p(x)
1000	500	54	14	0.000188	0.000130
1000	500	54	15	0.000552	0.000364
1000	500	54	16	0.001485	0.000932
1000	500	54	17	0.003663	0.002179
1000	500	54	18	0.008325	0.004662
1000	500	54	19	0.017482	0.009156
1000	500	54	20	0.034021	0.016539
1000	500	54	21	0.061544	0.027523
1000	500	54	22	0.103799	0.042255
1000	500	54	23	0.163716	0.059918
1000	500	54	24	0.242263	0.078546
1000	500	54	25	0.337519	0.095256
1000	500	54	26	0.444441	0.106923
1000	500	54	27	0.555558	0.111117
1000	500	55	10	0.000001	0.000000
1000	500	55	11	0.000002	0.000002
1000	500	55	12	0.000010	0.000008
1000	500	55	13	0.000037	0.000027
1000	500	55	14	0.000121	0.000085
1000	500	55	15	0.000365	0.000244
1000	500	55	16	0.001007	0.000642
1000	500	55	17	0.002551	0.001543
1000	500	55	18	0.005950	0.003399
1000	500	55	19	0.012826	0.006876
1000	500	55	20	0.025629	0.012803
1000	500	55	21	0.047608	0.021979
1000	500	55	22	0.082448	0.034840
1000	500	55	23	0.133504	0.051056
1000	500	55	24	0.202740	0.069236
1000	500	55	25	0.289689	0.086949
1000	500	55	26	0.390867	0.101178
1000	500	55	27	0.500000	0.109133
1000	500	56	11	0.000001	0.000001
1000	500	56	12	0.000006	0.000005
1000	500	56	13	0.000023	0.000017
1000	500	56	14	0.000078	0.000055
1000	500	56	15	0.000240	0.000162
1000	500	56	16	0.000679	0.000439
1000	500	56	17	0.001762	0.001083
1000	500	56	18	0.004216	0.002454
1000	500	56	19	0.009326	0.005110
1000	500	56	20	0.019126	0.009800
1000	500	56	21	0.036467	0.017341
1000	500	56	22	0.064825	0.028358
1000	500	56	23	0.107733	0.042908
1000	500	56	24	0.167866	0.060133
1000	500	56	25	0.245985	0.078119
1000	500	56	26	0.340118	0.094133
1000	500	56	27	0.445376	0.105258
1000	500	56	28	0.554624	0.109248
1000	500	57	11	0.000001	0.000001
1000	500	57	12	0.000004	0.000003
1000	500	57	13	0.000014	0.000011
1000	500	57	14	0.000049	0.000035
1000	500	57	15	0.000157	0.000107
1000	500	57	16	0.000454	0.000297
1000	500	57	17	0.001208	0.000754
1000	500	57	18	0.002963	0.001755
1000	500	57	19	0.006722	0.003759
1000	500	57	20	0.014143	0.007420
1000	500	57	21	0.027668	0.013525
1000	500	57	22	0.050466	0.022798
1000	500	57	23	0.086052	0.035586
1000	500	57	24	0.137545	0.051493
1000	500	57	25	0.206677	0.069133
1000	500	57	26	0.292852	0.086175
1000	500	57	27	0.392635	0.099783
1000	500	57	28	0.500000	0.107365
1000	500	58	11	0.000001	0.000000
1000	500	58	12	0.000002	0.000002
1000	500	58	13	0.000009	0.000007
1000	500	58	14	0.000031	0.000022
1000	500	58	15	0.000101	0.000070
1000	500	58	16	0.000301	0.000200
1000	500	58	17	0.000821	0.000520
1000	500	58	18	0.002066	0.001244
1000	500	58	19	0.004805	0.002739
1000	500	58	20	0.010366	0.005561
1000	500	58	21	0.020798	0.010432
1000	500	58	22	0.038910	0.018112
1000	500	58	23	0.068051	0.029142
1000	500	58	24	0.111552	0.043501
1000	500	58	25	0.171854	0.060302
1000	500	58	26	0.249536	0.077682
1000	500	58	27	0.342585	0.093049
1000	500	58	28	0.446261	0.103676
1000	500	58	29	0.553739	0.107479
1000	500	59	12	0.000001	0.000001
1000	500	59	13	0.000005	0.000004
1000	500	59	14	0.000020	0.000014
1000	500	59	15	0.000065	0.000046
1000	500	59	16	0.000199	0.000133
1000	500	59	17	0.000555	0.000356
1000	500	59	18	0.001429	0.000874
1000	500	59	19	0.003406	0.001977
1000	500	59	20	0.007532	0.004126
1000	500	59	21	0.015493	0.007961
1000	500	59	22	0.029719	0.014226
1000	500	59	23	0.053295	0.023576
1000	500	59	24	0.089571	0.036276
1000	500	59	25	0.141447	0.051876
1000	500	59	26	0.210448	0.069000
1000	500	59	27	0.295863	0.085415
1000	500	59	28	0.394313	0.098450
1000	500	59	29	0.500000	0.105687
1000	500	60	12	0.000001	0.000001
1000	500	60	13	0.000003	0.000002
1000	500	60	14	0.000012	0.000009
1000	500	60	15	0.000042	0.000029
1000	500	60	16	0.000130	0.000088
1000	500	60	17	0.000372	0.000242
1000	500	60	18	0.000981	0.000609
1000	500	60	19	0.002396	0.001414
1000	500	60	20	0.005428	0.003032
1000	500	60	21	0.011441	0.006013
1000	500	60	22	0.022493	0.011052
1000	500	60	23	0.041344	0.018851
1000	500	60	24	0.071221	0.029877
1000	500	60	25	0.115261	0.044040
1000	500	60	26	0.175691	0.060430
1000	500	60	27	0.252928	0.077237
1000	500	60	28	0.344931	0.092002
1000	500	60	29	0.447100	0.102169
1000	500	60	30	0.552900	0.105800
1000	500	61	13	0.000002	0.000002
1000	500	61	14	0.000008	0.000006
1000	500	61	15	0.000027	0.000019
1000	500	61	16	0.000085	0.000058
1000	500	61	17	0.000248	0.000163
1000	500	61	18	0.000669	0.000421
1000	500	61	19	0.001672	0.001003
1000	500	61	20	0.003879	0.002207
1000	500	61	21	0.008377	0.004497
1000	500	61	22	0.016873	0.008496
1000	500	61	23	0.031778	0.014905
1000	500	61	24	0.056092	0.024314
1000	500	61	25	0.093007	0.036915
1000	500	61	26	0.145219	0.052212
1000	500	61	27	0.214063	0.068844
1000	500	61	28	0.298734	0.084671
1000	500	61	29	0.395907	0.097173
1000	500	61	30	0.500000	0.104093
1000	500	62	13	0.000001	0.000001
1000	500	62	14	0.000005	0.000004
1000	500	62	15	0.000017	0.000012
1000	500	62	16	0.000055	0.000038
1000	500	62	17	0.000164	0.000109
1000	500	62	18	0.000453	0.000289
1000	500	62	19	0.001158	0.000705
1000	500	62	20	0.002751	0.001593
1000	500	62	21	0.006083	0.003332
1000	500	62	22	0.012548	0.006465
1000	500	62	23	0.024206	0.011659
1000	500	62	24	0.043767	0.019560
1000	500	62	25	0.074333	0.030566
1000	500	62	26	0.118862	0.044529
1000	500	62	27	0.179384	0.060522
1000	500	62	28	0.256173	0.076789
1000	500	62	29	0.347165	0.090992
1000	500	62	30	0.447898	0.100733
1000	500	62	31	0.552102	0.104204
1000	500	63	13	0.000001	0.000001
1000	500	63	14	0.000003	0.000002
1000	500	63	15	0.000011	0.000008
1000	500	63	16	0.000035	0.000025
1000	500	63	17	0.000108	0.000072
1000	500	63	18	0.000304	0.000197
1000	500	63	19	0.000796	0.000492
1000	500	63	20	0.001936	0.001140
1000	500	63	21	0.004381	0.002445
1000	500	63	22	0.009253	0.004872
1000	500	63	23	0.018278	0.009025
1000	500	63	24	0.033840	0.015563
1000	500	63	25	0.058854	0.025014
1000	500	63	26	0.096360	0.037506
1000	500	63	27	0.148865	0.052505
1000	500	63	28	0.217533	0.068668
1000	500	63	29	0.301475	0.083942
1000	500	63	30	0.397424	0.095949
1000	500	63	31	0.500000	0.102576
1000	500	64	14	0.000002	0.000001
1000	500	64	15	0.000007	0.000005
1000	500	64	16	0.000022	0.000016
1000	500	64	17	0.000070	0.000048
1000	500	64	18	0.000203	0.000133
1000	500	64	19	0.000544	0.000341
1000	500	64	20	0.001352	0.000808
1000	500	64	21	0.003131	0.001779
1000	500	64	22	0.006768	0.003637
1000	500	64	23	0.013683	0.006916
1000	500	64	24	0.025934	0.012251
1000	500	64	25	0.046174	0.020239
1000	500	64	26	0.077387	0.031213
1000	500	64	27	0.122361	0.044974
1000	500	64	28	0.182943	0.060582
1000	500	64	29	0.259280	0.076337
1000	500	64	30	0.349296	0.090016
1000	500	64	31	0.448657	0.099361
1000	500	64	32	0.551343	0.102685
1000	500	65	14	0.000001	0.000001

Table for $N = 1000$, $n = 500$

N	n	k	x	P(x)	p(x)
1000	500	65	15	0.000004	0.000003
1000	500	65	16	0.000014	0.000010
1000	500	65	17	0.000046	0.000031
1000	500	65	18	0.000135	0.000089
1000	500	65	19	0.000369	0.000234
1000	500	65	20	0.000938	0.000569
1000	500	65	21	0.002221	0.001283
1000	500	65	22	0.004911	0.002690
1000	500	65	23	0.010159	0.005249
1000	500	65	24	0.019704	0.009545
1000	500	65	25	0.035903	0.016199
1000	500	65	26	0.061580	0.025678
1000	500	65	27	0.099634	0.038053
1000	500	65	28	0.152393	0.052759
1000	500	65	29	0.220867	0.068474
1000	500	65	30	0.304096	0.083229
1000	500	65	31	0.398870	0.094775
1000	500	65	32	0.500000	0.101129
1000	500	66	14	0.000001	0.000000
1000	500	66	15	0.000003	0.000002
1000	500	66	16	0.000009	0.000006
1000	500	66	17	0.000029	0.000020
1000	500	66	18	0.000089	0.000059
1000	500	66	19	0.000248	0.000160
1000	500	66	20	0.000645	0.000397
1000	500	66	21	0.001563	0.000918
1000	500	66	22	0.003535	0.001972
1000	500	66	23	0.007482	0.003946
1000	500	66	24	0.014845	0.007363
1000	500	66	25	0.027674	0.012829
1000	500	66	26	0.048563	0.020889
1000	500	66	27	0.080383	0.031820
1000	500	66	28	0.125759	0.045376
1000	500	66	29	0.186373	0.060614
1000	500	66	30	0.262259	0.075885
1000	500	66	31	0.351331	0.089072
1000	500	66	32	0.449381	0.098050
1000	500	66	33	0.550619	0.101238
1000	500	67	15	0.000002	0.000001
1000	500	67	16	0.000006	0.000004
1000	500	67	17	0.000019	0.000013
1000	500	67	18	0.000058	0.000039
1000	500	67	19	0.000166	0.000108
1000	500	67	20	0.000441	0.000275
1000	500	67	21	0.001093	0.000651
1000	500	67	22	0.002526	0.001434
1000	500	67	23	0.005466	0.002940
1000	500	67	24	0.011092	0.005626
1000	500	67	25	0.021149	0.010057
1000	500	67	26	0.037962	0.016812

N	n	k	x	P(x)	p(x)
1000	500	67	27	0.064269	0.026307
1000	500	67	28	0.102828	0.038560
1000	500	67	29	0.155807	0.052978
1000	500	67	30	0.224072	0.068266
1000	500	67	31	0.306604	0.082532
1000	500	67	32	0.400251	0.093647
1000	500	67	33	0.500000	0.099749
1000	500	68	15	0.000001	0.000001
1000	500	68	16	0.000004	0.000003
1000	500	68	17	0.000012	0.000008
1000	500	68	18	0.000038	0.000026
1000	500	68	19	0.000110	0.000073
1000	500	68	20	0.000300	0.000189
1000	500	68	21	0.000758	0.000458
1000	500	68	22	0.001792	0.001033
1000	500	68	23	0.003963	0.002171
1000	500	68	24	0.008222	0.004259
1000	500	68	25	0.016029	0.007807
1000	500	68	26	0.029420	0.013391
1000	500	68	27	0.050932	0.021512
1000	500	68	28	0.083321	0.032389
1000	500	68	29	0.129062	0.045741
1000	500	68	30	0.189683	0.060621
1000	500	68	31	0.265117	0.075434
1000	500	68	32	0.353277	0.088160
1000	500	68	33	0.450072	0.096795
1000	500	68	34	0.549928	0.099856
1000	500	69	15	0.000001	0.000000
1000	500	69	16	0.000002	0.000002
1000	500	69	17	0.000008	0.000005
1000	500	69	18	0.000024	0.000017
1000	500	69	19	0.000073	0.000048
1000	500	69	20	0.000202	0.000129
1000	500	69	21	0.000523	0.000320
1000	500	69	22	0.001262	0.000739
1000	500	69	23	0.002852	0.001590
1000	500	69	24	0.006047	0.003195
1000	500	69	25	0.012050	0.006003
1000	500	69	26	0.022610	0.010560
1000	500	69	27	0.040015	0.017405
1000	500	69	28	0.066918	0.026903
1000	500	69	29	0.105946	0.039028
1000	500	69	30	0.159113	0.053166
1000	500	69	31	0.227157	0.068045
1000	500	69	32	0.309008	0.081850
1000	500	69	33	0.401571	0.092563
1000	500	69	34	0.500000	0.098429
1000	500	70	16	0.000001	0.000001
1000	500	70	17	0.000005	0.000003
1000	500	70	18	0.000016	0.000011

N	n	k	x	P(x)	p(x)
1000	500	70	19	0.000048	0.000032
1000	500	70	20	0.000136	0.000088
1000	500	70	21	0.000358	0.000222
1000	500	70	22	0.000882	0.000524
1000	500	70	23	0.002037	0.001155
1000	500	70	24	0.004413	0.002376
1000	500	70	25	0.008987	0.004574
1000	500	70	26	0.017234	0.008246
1000	500	70	27	0.031172	0.013938
1000	500	70	28	0.053279	0.022106
1000	500	70	29	0.086202	0.032923
1000	500	70	30	0.132272	0.046071
1000	500	70	31	0.192879	0.060607
1000	500	70	32	0.267863	0.074984
1000	500	70	33	0.355140	0.087277
1000	500	70	34	0.450732	0.095592
1000	500	70	35	0.549267	0.098535
1000	500	71	16	0.000001	0.000001
1000	500	71	17	0.000003	0.000002
1000	500	71	18	0.000010	0.000007
1000	500	71	19	0.000031	0.000021
1000	500	71	20	0.000090	0.000059
1000	500	71	21	0.000243	0.000153
1000	500	71	22	0.000613	0.000369
1000	500	71	23	0.001445	0.000832
1000	500	71	24	0.003197	0.001752
1000	500	71	25	0.006651	0.003454
1000	500	71	26	0.013031	0.006380
1000	500	71	27	0.024083	0.011052
1000	500	71	28	0.042059	0.017976
1000	500	71	29	0.069528	0.027469
1000	500	71	30	0.108989	0.039462
1000	500	71	31	0.162315	0.053326
1000	500	71	32	0.230129	0.067814
1000	500	71	33	0.311314	0.081185
1000	500	71	34	0.402834	0.091520
1000	500	71	35	0.500000	0.097166
1000	500	72	16	0.000001	0.000000
1000	500	72	17	0.000002	0.000001
1000	500	72	18	0.000006	0.000005
1000	500	72	19	0.000020	0.000014
1000	500	72	20	0.000060	0.000040
1000	500	72	21	0.000164	0.000105
1000	500	72	22	0.000423	0.000258
1000	500	72	23	0.001018	0.000595
1000	500	72	24	0.002299	0.001282
1000	500	72	25	0.004885	0.002586
1000	500	72	26	0.009776	0.004891
1000	500	72	27	0.018456	0.008680
1000	500	72	28	0.032926	0.014470

N	n	k	x	P(x)	p(x)
1000	500	72	29	0.055601	0.022675
1000	500	72	30	0.089025	0.033424
1000	500	72	31	0.135394	0.046369
1000	500	72	32	0.195967	0.060573
1000	500	72	33	0.270503	0.074536
1000	500	72	34	0.356926	0.086423
1000	500	72	35	0.451364	0.094439
1000	500	72	36	0.548635	0.097271
1000	500	73	17	0.000001	0.000001
1000	500	73	18	0.000004	0.000003
1000	500	73	19	0.000013	0.000009
1000	500	73	20	0.000039	0.000026
1000	500	73	21	0.000110	0.000071
1000	500	73	22	0.000290	0.000179
1000	500	73	23	0.000712	0.000422
1000	500	73	24	0.001642	0.000930
1000	500	73	25	0.003561	0.001920
1000	500	73	26	0.007278	0.003716
1000	500	73	27	0.014031	0.006754
1000	500	73	28	0.025566	0.011535
1000	500	73	29	0.044092	0.018526
1000	500	73	30	0.072097	0.028004
1000	500	73	31	0.111960	0.039863
1000	500	73	32	0.165419	0.053460
1000	500	73	33	0.232994	0.067575
1000	500	73	34	0.313528	0.080534
1000	500	73	35	0.404043	0.090515
1000	500	73	36	0.500000	0.095957
1000	500	74	17	0.000001	0.000001
1000	500	74	18	0.000003	0.000002
1000	500	74	19	0.000008	0.000006
1000	500	74	20	0.000026	0.000017
1000	500	74	21	0.000074	0.000048
1000	500	74	22	0.000197	0.000124
1000	500	74	23	0.000495	0.000297
1000	500	74	24	0.001164	0.000670
1000	500	74	25	0.002578	0.001414
1000	500	74	26	0.005377	0.002800
1000	500	74	27	0.010585	0.005208
1000	500	74	28	0.019693	0.009108
1000	500	74	29	0.034679	0.014986
1000	500	74	30	0.057898	0.023219
1000	500	74	31	0.091792	0.033893
1000	500	74	32	0.138430	0.046638
1000	500	74	33	0.198951	0.060522
1000	500	74	34	0.273044	0.074092
1000	500	74	35	0.358639	0.085595
1000	500	74	36	0.451970	0.093331
1000	500	74	37	0.548030	0.096060
1000	500	75	18	0.000002	0.000001

Table for $N = 1000$, $n = 500$

N	n	k	x	P(x)	p(x)
1000	500	75	19	0.000005	0.000004
1000	500	75	20	0.000017	0.000011
1000	500	75	21	0.000049	0.000032
1000	500	75	22	0.000133	0.000085
1000	500	75	23	0.000341	0.000208
1000	500	75	24	0.000820	0.000479
1000	500	75	25	0.001853	0.001033
1000	500	75	26	0.003944	0.002092
1000	500	75	27	0.007925	0.003981
1000	500	75	28	0.015051	0.007125
1000	500	75	29	0.027057	0.012006
1000	500	75	30	0.046113	0.019057
1000	500	75	31	0.074625	0.028512
1000	500	75	32	0.114859	0.040234
1000	500	75	33	0.168429	0.053570
1000	500	75	34	0.235758	0.067328
1000	500	75	35	0.315657	0.079899
1000	500	75	36	0.405204	0.089547
1000	500	75	37	0.500000	0.094796
1000	500	76	18	0.000001	0.000001
1000	500	76	19	0.000003	0.000002
1000	500	76	20	0.000011	0.000007
1000	500	76	21	0.000032	0.000021
1000	500	76	22	0.000090	0.000058
1000	500	76	23	0.000234	0.000144
1000	500	76	24	0.000574	0.000340
1000	500	76	25	0.001322	0.000749
1000	500	76	26	0.002873	0.001550
1000	500	76	27	0.005890	0.003017
1000	500	76	28	0.011415	0.005525
1000	500	76	29	0.020943	0.009529
1000	500	76	30	0.036430	0.015487
1000	500	76	31	0.060169	0.023739
1000	500	76	32	0.094503	0.034334
1000	500	76	33	0.141383	0.046881
1000	500	76	34	0.201839	0.060455
1000	500	76	35	0.275491	0.073653
1000	500	76	36	0.360285	0.084794
1000	500	76	37	0.452551	0.092266
1000	500	76	38	0.547449	0.094899
1000	500	77	18	0.000001	0.000000
1000	500	77	19	0.000002	0.000002
1000	500	77	20	0.000007	0.000005
1000	500	77	21	0.000021	0.000014
1000	500	77	22	0.000060	0.000039
1000	500	77	23	0.000160	0.000100
1000	500	77	24	0.000399	0.000239
1000	500	77	25	0.000938	0.000539
1000	500	77	26	0.002077	0.001140
1000	500	77	27	0.004345	0.002268

N	n	k	x	P(x)	p(x)
1000	500	77	28	0.008592	0.004247
1000	500	77	29	0.016086	0.007494
1000	500	77	30	0.028553	0.012467
1000	500	77	31	0.048120	0.019567
1000	500	77	32	0.077113	0.028993
1000	500	77	33	0.117690	0.040577
1000	500	77	34	0.171349	0.053660
1000	500	77	35	0.238426	0.067076
1000	500	77	36	0.317705	0.079279
1000	500	77	37	0.406318	0.088613
1000	500	77	38	0.500000	0.093682
1000	500	78	19	0.000001	0.000001
1000	500	78	20	0.000004	0.000003
1000	500	78	21	0.000014	0.000009
1000	500	78	22	0.000040	0.000026
1000	500	78	23	0.000108	0.000068
1000	500	78	24	0.000275	0.000167
1000	500	78	25	0.000660	0.000385
1000	500	78	26	0.001492	0.000832
1000	500	78	27	0.003183	0.001691
1000	500	78	28	0.006420	0.003237
1000	500	78	29	0.012262	0.005842
1000	500	78	30	0.022205	0.009943
1000	500	78	31	0.038177	0.015973
1000	500	78	32	0.062412	0.024235
1000	500	78	33	0.097159	0.034747
1000	500	78	34	0.144258	0.047098
1000	500	78	35	0.204633	0.060375
1000	500	78	36	0.277850	0.073217
1000	500	78	37	0.361867	0.084017
1000	500	78	38	0.453108	0.091241
1000	500	78	39	0.546892	0.093783
1000	500	79	19	0.000001	0.000001
1000	500	79	20	0.000003	0.000002
1000	500	79	21	0.000009	0.000006
1000	500	79	22	0.000026	0.000017
1000	500	79	23	0.000073	0.000046
1000	500	79	24	0.000189	0.000116
1000	500	79	25	0.000462	0.000273
1000	500	79	26	0.001064	0.000602
1000	500	79	27	0.002315	0.001251
1000	500	79	28	0.004763	0.002448
1000	500	79	29	0.009278	0.004515
1000	500	79	30	0.017136	0.007858
1000	500	79	31	0.030052	0.012916
1000	500	79	32	0.050111	0.020058
1000	500	79	33	0.079560	0.029449
1000	500	79	34	0.120453	0.040893
1000	500	79	35	0.174184	0.053730
1000	500	79	36	0.241004	0.066820

N	n	k	x	P(x)	p(x)
1000	500	79	37	0.319677	0.078673
1000	500	79	38	0.407389	0.087712
1000	500	79	39	0.500000	0.092611
1000	500	80	19	0.000001	0.000000
1000	500	80	20	0.000002	0.000001
1000	500	80	21	0.000006	0.000004
1000	500	80	22	0.000017	0.000012
1000	500	80	23	0.000049	0.000031
1000	500	80	24	0.000129	0.000080
1000	500	80	25	0.000321	0.000192
1000	500	80	26	0.000754	0.000433
1000	500	80	27	0.001673	0.000919
1000	500	80	28	0.003509	0.001836
1000	500	80	29	0.006969	0.003460
1000	500	80	30	0.013126	0.006157
1000	500	80	31	0.023475	0.010349
1000	500	80	32	0.039918	0.016443
1000	500	80	33	0.064627	0.024709
1000	500	80	34	0.099762	0.035135
1000	500	80	35	0.147056	0.047294
1000	500	80	36	0.207340	0.060284
1000	500	80	37	0.280127	0.072787
1000	500	80	38	0.363390	0.083263
1000	500	80	39	0.453644	0.090254
1000	500	80	40	0.546356	0.092712
1000	500	81	20	0.000001	0.000001
1000	500	81	21	0.000004	0.000003
1000	500	81	22	0.000011	0.000008
1000	500	81	23	0.000032	0.000021
1000	500	81	24	0.000088	0.000055
1000	500	81	25	0.000222	0.000134
1000	500	81	26	0.000531	0.000309
1000	500	81	27	0.001201	0.000669
1000	500	81	28	0.002567	0.001366
1000	500	81	29	0.005198	0.002631
1000	500	81	30	0.009981	0.004783
1000	500	81	31	0.018199	0.008219
1000	500	81	32	0.031554	0.013354
1000	500	81	33	0.052085	0.020531
1000	500	81	34	0.081966	0.029881
1000	500	81	35	0.123152	0.041186
1000	500	81	36	0.176936	0.053784
1000	500	81	37	0.243496	0.066560
1000	500	81	38	0.321577	0.078081
1000	500	81	39	0.408419	0.086842
1000	500	81	40	0.500000	0.091581
1000	500	82	20	0.000001	0.000001
1000	500	82	21	0.000002	0.000002
1000	500	82	22	0.000007	0.000005
1000	500	82	23	0.000022	0.000014

N	n	k	x	P(x)	p(x)
1000	500	82	24	0.000059	0.000037
1000	500	82	25	0.000152	0.000093
1000	500	82	26	0.000372	0.000219
1000	500	82	27	0.000856	0.000484
1000	500	82	28	0.001865	0.001009
1000	500	82	29	0.003849	0.001984
1000	500	82	30	0.007534	0.003685
1000	500	82	31	0.014005	0.006471
1000	500	82	32	0.024753	0.010748
1000	500	82	33	0.041652	0.016899
1000	500	82	34	0.066814	0.025162
1000	500	82	35	0.102312	0.035498
1000	500	82	36	0.149780	0.047468
1000	500	82	37	0.209962	0.060182
1000	500	82	38	0.282324	0.072362
1000	500	82	39	0.364856	0.082532
1000	500	82	40	0.454160	0.089303
1000	500	82	41	0.545840	0.091681
1000	500	83	21	0.000002	0.000001
1000	500	83	22	0.000005	0.000003
1000	500	83	23	0.000014	0.000009
1000	500	83	24	0.000040	0.000025
1000	500	83	25	0.000104	0.000065
1000	500	83	26	0.000259	0.000154
1000	500	83	27	0.000607	0.000348
1000	500	83	28	0.001346	0.000740
1000	500	83	29	0.002831	0.001485
1000	500	83	30	0.005648	0.002816
1000	500	83	31	0.010699	0.005052
1000	500	83	32	0.019274	0.008574
1000	500	83	33	0.033055	0.013781
1000	500	83	34	0.054041	0.020986
1000	500	83	35	0.084331	0.030290
1000	500	83	36	0.125788	0.041457
1000	500	83	37	0.179609	0.053821
1000	500	83	38	0.245907	0.066298
1000	500	83	39	0.323410	0.077504
1000	500	83	40	0.409411	0.086000
1000	500	83	41	0.500000	0.090589
1000	500	84	21	0.000001	0.000001
1000	500	84	22	0.000003	0.000002
1000	500	84	23	0.000009	0.000006
1000	500	84	24	0.000026	0.000017
1000	500	84	25	0.000071	0.000044
1000	500	84	26	0.000179	0.000108
1000	500	84	27	0.000427	0.000248
1000	500	84	28	0.000966	0.000539
1000	500	84	29	0.002069	0.001103
1000	500	84	30	0.004204	0.002136
1000	500	84	31	0.008116	0.003911

Table for $N = 1000$, $n = 500$

N	n	k	x	P(x)	p(x)
1000	500	84	32	0.014898	0.006782
1000	500	84	33	0.026037	0.011139
1000	500	84	34	0.043376	0.017340
1000	500	84	35	0.068971	0.025595
1000	500	84	36	0.104811	0.035840
1000	500	84	37	0.152434	0.047624
1000	500	84	38	0.212505	0.060070
1000	500	84	39	0.284447	0.071942
1000	500	84	40	0.366270	0.081823
1000	500	84	41	0.454656	0.088386
1000	500	84	42	0.545344	0.090688
1000	500	85	21	0.000001	0.000000
1000	500	85	22	0.000002	0.000001
1000	500	85	23	0.000006	0.000004
1000	500	85	24	0.000017	0.000011
1000	500	85	25	0.000048	0.000030
1000	500	85	26	0.000123	0.000075
1000	500	85	27	0.000299	0.000176
1000	500	85	28	0.000688	0.000389
1000	500	85	29	0.001501	0.000813
1000	500	85	30	0.003108	0.001607
1000	500	85	31	0.006113	0.003005
1000	500	85	32	0.011433	0.005320
1000	500	85	33	0.020358	0.008925
1000	500	85	34	0.034555	0.014197
1000	500	85	35	0.055978	0.021423
1000	500	85	36	0.086656	0.030678
1000	500	85	37	0.128363	0.041706
1000	500	85	38	0.182207	0.053845
1000	500	85	39	0.248240	0.066033
1000	500	85	40	0.325180	0.076939
1000	500	85	41	0.410366	0.085186
1000	500	85	42	0.500000	0.089634
1000	500	86	22	0.000001	0.000001
1000	500	86	23	0.000004	0.000003
1000	500	86	24	0.000012	0.000008
1000	500	86	25	0.000032	0.000020
1000	500	86	26	0.000084	0.000052
1000	500	86	27	0.000208	0.000124
1000	500	86	28	0.000487	0.000279
1000	500	86	29	0.001083	0.000595
1000	500	86	30	0.002283	0.001200
1000	500	86	31	0.004573	0.002290
1000	500	86	32	0.008712	0.004139
1000	500	86	33	0.015802	0.007091
1000	500	86	34	0.027325	0.011522
1000	500	86	35	0.045091	0.017767
1000	500	86	36	0.071099	0.026008
1000	500	86	37	0.107259	0.036159
1000	500	86	38	0.155020	0.047761

N	n	k	x	P(x)	p(x)
1000	500	86	39	0.214971	0.059951
1000	500	86	40	0.286500	0.071529
1000	500	86	41	0.367633	0.081134
1000	500	86	42	0.455134	0.087501
1000	500	86	43	0.544866	0.089732
1000	500	87	22	0.000001	0.000001
1000	500	87	23	0.000003	0.000002
1000	500	87	24	0.000008	0.000005
1000	500	87	25	0.000021	0.000014
1000	500	87	26	0.000057	0.000036
1000	500	87	27	0.000144	0.000087
1000	500	87	28	0.000343	0.000199
1000	500	87	29	0.000776	0.000433
1000	500	87	30	0.001666	0.000890
1000	500	87	31	0.003398	0.001732
1000	500	87	32	0.006593	0.003195
1000	500	87	33	0.012180	0.005587
1000	500	87	34	0.021450	0.009271
1000	500	87	35	0.036052	0.014602
1000	500	87	36	0.057896	0.021844
1000	500	87	37	0.088942	0.031046
1000	500	87	38	0.130878	0.041936
1000	500	87	39	0.184733	0.053855
1000	500	87	40	0.250501	0.065767
1000	500	87	41	0.326889	0.076388
1000	500	87	42	0.411288	0.084399
1000	500	87	43	0.500000	0.088712
1000	500	88	22	0.000000	0.000000
1000	500	88	23	0.000002	0.000001
1000	500	88	24	0.000005	0.000003
1000	500	88	25	0.000014	0.000009
1000	500	88	26	0.000038	0.000024
1000	500	88	27	0.000099	0.000060
1000	500	88	28	0.000240	0.000141
1000	500	88	29	0.000553	0.000313
1000	500	88	30	0.001208	0.000655
1000	500	88	31	0.002508	0.001300
1000	500	88	32	0.004955	0.002447
1000	500	88	33	0.009322	0.004367
1000	500	88	34	0.016718	0.007397
1000	500	88	35	0.028615	0.011897
1000	500	88	36	0.046795	0.018180
1000	500	88	37	0.073198	0.026403
1000	500	88	38	0.109657	0.036459
1000	500	88	39	0.157540	0.047883
1000	500	88	40	0.217365	0.059825
1000	500	88	41	0.288486	0.071121
1000	500	88	42	0.368950	0.080464
1000	500	88	43	0.455595	0.086645
1000	500	88	44	0.544404	0.088809

N	n	k	x	P(x)	p(x)
1000	500	89	23	0.000001	0.000001
1000	500	89	24	0.000003	0.000002
1000	500	89	25	0.000009	0.000006
1000	500	89	26	0.000026	0.000016
1000	500	89	27	0.000068	0.000042
1000	500	89	28	0.000167	0.000099
1000	500	89	29	0.000391	0.000224
1000	500	89	30	0.000870	0.000479
1000	500	89	31	0.001839	0.000969
1000	500	89	32	0.003699	0.001860
1000	500	89	33	0.007086	0.003386
1000	500	89	34	0.012939	0.005853
1000	500	89	35	0.022549	0.009610
1000	500	89	36	0.037546	0.014996
1000	500	89	37	0.059794	0.022248
1000	500	89	38	0.091188	0.031394
1000	500	89	39	0.133336	0.042149
1000	500	89	40	0.187190	0.053854
1000	500	89	41	0.252691	0.065501
1000	500	89	42	0.328541	0.075850
1000	500	89	43	0.412178	0.083636
1000	500	89	44	0.500000	0.087822
1000	500	90	23	0.000001	0.000000
1000	500	90	24	0.000002	0.000001
1000	500	90	25	0.000006	0.000004
1000	500	90	26	0.000017	0.000011
1000	500	90	27	0.000046	0.000029
1000	500	90	28	0.000115	0.000070
1000	500	90	29	0.000275	0.000160
1000	500	90	30	0.000623	0.000348
1000	500	90	31	0.001340	0.000717
1000	500	90	32	0.002744	0.001403
1000	500	90	33	0.005350	0.002606
1000	500	90	34	0.009945	0.004595
1000	500	90	35	0.017644	0.007699
1000	500	90	36	0.029908	0.012264
1000	500	90	37	0.048486	0.018579
1000	500	90	38	0.075267	0.026780
1000	500	90	39	0.112008	0.036741
1000	500	90	40	0.159997	0.047990
1000	500	90	41	0.219689	0.059692
1000	500	90	42	0.290408	0.070719
1000	500	90	43	0.370222	0.079814
1000	500	90	44	0.456041	0.085819
1000	500	90	45	0.543959	0.087919
1000	500	91	24	0.000001	0.000001
1000	500	91	25	0.000004	0.000003
1000	500	91	26	0.000011	0.000007
1000	500	91	27	0.000031	0.000020
1000	500	91	28	0.000079	0.000048

N	n	k	x	P(x)	p(x)
1000	500	91	29	0.000192	0.000113
1000	500	91	30	0.000443	0.000251
1000	500	91	31	0.000971	0.000527
1000	500	91	32	0.002022	0.001051
1000	500	91	33	0.004012	0.001990
1000	500	91	34	0.007592	0.003579
1000	500	91	35	0.013710	0.006118
1000	500	91	36	0.023654	0.009945
1000	500	91	37	0.039034	0.015379
1000	500	91	38	0.061671	0.022637
1000	500	91	39	0.093395	0.031724
1000	500	91	40	0.135739	0.042344
1000	500	91	41	0.189581	0.053842
1000	500	91	42	0.254816	0.065235
1000	500	91	43	0.330140	0.075324
1000	500	91	44	0.413037	0.082897
1000	500	91	45	0.500000	0.086963
1000	500	92	24	0.000001	0.000001
1000	500	92	25	0.000003	0.000002
1000	500	92	26	0.000008	0.000005
1000	500	92	27	0.000021	0.000013
1000	500	92	28	0.000054	0.000033
1000	500	92	29	0.000134	0.000080
1000	500	92	30	0.000314	0.000180
1000	500	92	31	0.000699	0.000385
1000	500	92	32	0.001481	0.000782
1000	500	92	33	0.002990	0.001509
1000	500	92	34	0.005757	0.002767
1000	500	92	35	0.010580	0.004823
1000	500	92	36	0.018578	0.007998
1000	500	92	37	0.031200	0.012622
1000	500	92	38	0.050165	0.018965
1000	500	92	39	0.077306	0.027141
1000	500	92	40	0.114311	0.037005
1000	500	92	41	0.162393	0.048083
1000	500	92	42	0.221947	0.059554
1000	500	92	43	0.292270	0.070323
1000	500	92	44	0.371452	0.079181
1000	500	92	45	0.456471	0.085019
1000	500	92	46	0.543529	0.087059
1000	500	93	24	0.000001	0.000000
1000	500	93	25	0.000002	0.000001
1000	500	93	26	0.000005	0.000003
1000	500	93	27	0.000014	0.000009
1000	500	93	28	0.000037	0.000023
1000	500	93	29	0.000093	0.000056
1000	500	93	30	0.000221	0.000128
1000	500	93	31	0.000500	0.000279
1000	500	93	32	0.001078	0.000578
1000	500	93	33	0.002214	0.001136

Table for $N = 1000$, $n = 500$

N	n	k	x	P(x)	p(x)
1000	500	93	34	0.004337	0.002123
1000	500	93	35	0.008110	0.003773
1000	500	93	36	0.014491	0.006381
1000	500	93	37	0.024764	0.010273
1000	500	93	38	0.040516	0.015752
1000	500	93	39	0.063526	0.023011
1000	500	93	40	0.095564	0.032037
1000	500	93	41	0.138087	0.042523
1000	500	93	42	0.191908	0.053821
1000	500	93	43	0.256876	0.064969
1000	500	93	44	0.331687	0.074810
1000	500	93	45	0.413867	0.082181
1000	500	93	46	0.500000	0.086133
1000	500	94	25	0.000001	0.000001
1000	500	94	26	0.000003	0.000002
1000	500	94	27	0.000009	0.000006
1000	500	94	28	0.000025	0.000016
1000	500	94	29	0.000064	0.000039
1000	500	94	30	0.000154	0.000091
1000	500	94	31	0.000355	0.000201
1000	500	94	32	0.000779	0.000424
1000	500	94	33	0.001629	0.000849
1000	500	94	34	0.003246	0.001617
1000	500	94	35	0.006176	0.002930
1000	500	94	36	0.011227	0.005051
1000	500	94	37	0.019520	0.008293
1000	500	94	38	0.032492	0.012972
1000	500	94	39	0.051831	0.019338
1000	500	94	40	0.079315	0.027485
1000	500	94	41	0.116568	0.037252
1000	500	94	42	0.164730	0.048163
1000	500	94	43	0.224141	0.059411
1000	500	94	44	0.294075	0.069934
1000	500	94	45	0.372641	0.078566
1000	500	94	46	0.456886	0.084245
1000	500	94	47	0.543114	0.086227
1000	500	95	25	0.000001	0.000000
1000	500	95	26	0.000002	0.000001
1000	500	95	27	0.000006	0.000004
1000	500	95	28	0.000017	0.000011
1000	500	95	29	0.000044	0.000027
1000	500	95	30	0.000107	0.000064
1000	500	95	31	0.000251	0.000144
1000	500	95	32	0.000560	0.000309
1000	500	95	33	0.001191	0.000631
1000	500	95	34	0.002414	0.001223
1000	500	95	35	0.004672	0.002258
1000	500	95	36	0.008640	0.003968
1000	500	95	37	0.015282	0.006642
1000	500	95	38	0.025877	0.010595

N	n	k	x	P(x)	p(x)
1000	500	95	39	0.041991	0.016113
1000	500	95	40	0.065361	0.023370
1000	500	95	41	0.097695	0.032334
1000	500	95	42	0.140383	0.042688
1000	500	95	43	0.194174	0.053790
1000	500	95	44	0.258877	0.064703
1000	500	95	45	0.333185	0.074308
1000	500	95	46	0.414671	0.081486
1000	500	95	47	0.500000	0.085329
1000	500	96	26	0.000001	0.000001
1000	500	96	27	0.000004	0.000003
1000	500	96	28	0.000011	0.000007
1000	500	96	29	0.000030	0.000018
1000	500	96	30	0.000074	0.000045
1000	500	96	31	0.000177	0.000102
1000	500	96	32	0.000401	0.000224
1000	500	96	33	0.000866	0.000465
1000	500	96	34	0.001784	0.000919
1000	500	96	35	0.003512	0.001728
1000	500	96	36	0.006605	0.003094
1000	500	96	37	0.011884	0.005278
1000	500	96	38	0.020468	0.008585
1000	500	96	39	0.033783	0.013314
1000	500	96	40	0.053482	0.019699
1000	500	96	41	0.081296	0.027813
1000	500	96	42	0.118780	0.037484
1000	500	96	43	0.167011	0.048231
1000	500	96	44	0.226275	0.059264
1000	500	96	45	0.295825	0.069551
1000	500	96	46	0.373793	0.077967
1000	500	96	47	0.457288	0.083495
1000	500	96	48	0.542712	0.085424
1000	500	97	26	0.000001	0.000001
1000	500	97	27	0.000003	0.000002
1000	500	97	28	0.000007	0.000005
1000	500	97	29	0.000020	0.000013
1000	500	97	30	0.000051	0.000031
1000	500	97	31	0.000123	0.000072
1000	500	97	32	0.000285	0.000161
1000	500	97	33	0.000625	0.000341
1000	500	97	34	0.001311	0.000685
1000	500	97	35	0.002623	0.001313
1000	500	97	36	0.005018	0.002394
1000	500	97	37	0.009180	0.004163
1000	500	97	38	0.016081	0.006901
1000	500	97	39	0.026993	0.010912
1000	500	97	40	0.043458	0.016465
1000	500	97	41	0.067173	0.023715
1000	500	97	42	0.099789	0.032615
1000	500	97	43	0.142628	0.042840

N	n	k	x	P(x)	p(x)
1000	500	97	44	0.196381	0.053752
1000	500	97	45	0.260819	0.064439
1000	500	97	46	0.334637	0.073817
1000	500	97	47	0.415448	0.080811
1000	500	97	48	0.500000	0.084552
1000	500	98	26	0.000001	0.000000
1000	500	98	27	0.000002	0.000001
1000	500	98	28	0.000005	0.000003
1000	500	98	29	0.000013	0.000009
1000	500	98	30	0.000035	0.000021
1000	500	98	31	0.000086	0.000051
1000	500	98	32	0.000201	0.000115
1000	500	98	33	0.000449	0.000248
1000	500	98	34	0.000957	0.000508
1000	500	98	35	0.001947	0.000990
1000	500	98	36	0.003787	0.001840
1000	500	98	37	0.007046	0.003259
1000	500	98	38	0.012550	0.005504
1000	500	98	39	0.021422	0.008872
1000	500	98	40	0.035070	0.013648
1000	500	98	41	0.055119	0.020048
1000	500	98	42	0.083246	0.028128
1000	500	98	43	0.120948	0.037701
1000	500	98	44	0.169237	0.048289
1000	500	98	45	0.228350	0.059114
1000	500	98	46	0.297524	0.069173
1000	500	98	47	0.374908	0.077384
1000	500	98	48	0.457677	0.082769
1000	500	98	49	0.542323	0.084646
1000	500	99	27	0.000001	0.000001
1000	500	99	28	0.000003	0.000002
1000	500	99	29	0.000009	0.000006
1000	500	99	30	0.000024	0.000015
1000	500	99	31	0.000059	0.000036
1000	500	99	32	0.000141	0.000082
1000	500	99	33	0.000321	0.000179
1000	500	99	34	0.000695	0.000374
1000	500	99	35	0.001437	0.000742
1000	500	99	36	0.002841	0.001404
1000	500	99	37	0.005373	0.002532
1000	500	99	38	0.009731	0.004358
1000	500	99	39	0.016888	0.007157
1000	500	99	40	0.028110	0.011222
1000	500	99	41	0.044917	0.016807
1000	500	99	42	0.068964	0.024047
1000	500	99	43	0.101846	0.032882
1000	500	99	44	0.144825	0.042978
1000	500	99	45	0.198531	0.053707
1000	500	99	46	0.262707	0.064176
1000	500	99	47	0.336044	0.073337

N	n	k	x	P(x)	p(x)
1000	500	99	48	0.416201	0.080157
1000	500	99	49	0.500000	0.083799
1000	500	100	27	0.000001	0.000000
1000	500	100	28	0.000002	0.000001
1000	500	100	29	0.000006	0.000004
1000	500	100	30	0.000016	0.000010
1000	500	100	31	0.000041	0.000025
1000	500	100	32	0.000099	0.000058
1000	500	100	33	0.000228	0.000129
1000	500	100	34	0.000501	0.000274
1000	500	100	35	0.001054	0.000553
1000	500	100	36	0.002118	0.001064
1000	500	100	37	0.004072	0.001954
1000	500	100	38	0.007497	0.003425
1000	500	100	39	0.013226	0.005730
1000	500	100	40	0.022381	0.009155
1000	500	100	41	0.036355	0.013974
1000	500	100	42	0.056740	0.020385
1000	500	100	43	0.085168	0.028428
1000	500	100	44	0.123073	0.037905
1000	500	100	45	0.171410	0.048337
1000	500	100	46	0.230370	0.058960
1000	500	100	47	0.299172	0.068802
1000	500	100	48	0.375989	0.076817
1000	500	100	49	0.458054	0.082065
1000	500	100	50	0.541946	0.083892
1000	500	101	28	0.000001	0.000001
1000	500	101	29	0.000004	0.000003
1000	500	101	30	0.000011	0.000007
1000	500	101	31	0.000028	0.000017
1000	500	101	32	0.000069	0.000041
1000	500	101	33	0.000161	0.000092
1000	500	101	34	0.000360	0.000199
1000	500	101	35	0.000768	0.000409
1000	500	101	36	0.001569	0.000801
1000	500	101	37	0.003067	0.001497
1000	500	101	38	0.005739	0.002672
1000	500	101	39	0.010291	0.004553
1000	500	101	40	0.017702	0.007410
1000	500	101	41	0.029228	0.011526
1000	500	101	42	0.046366	0.017138
1000	500	101	43	0.070733	0.024367
1000	500	101	44	0.103868	0.033135
1000	500	101	45	0.146973	0.043105
1000	500	101	46	0.200628	0.053655
1000	500	101	47	0.264542	0.063914
1000	500	101	48	0.337410	0.072868
1000	500	101	49	0.416930	0.079520
1000	500	101	50	0.500000	0.083070
1000	500	102	28	0.000001	0.000001

Table for $N = 1000$, $n = 500$

N	n	k	x	P(x)	p(x)
1000	500	102	29	0.000003	0.000002
1000	500	102	30	0.000007	0.000005
1000	500	102	31	0.000019	0.000012
1000	500	102	32	0.000047	0.000028
1000	500	102	33	0.000113	0.000065
1000	500	102	34	0.000257	0.000144
1000	500	102	35	0.000557	0.000300
1000	500	102	36	0.001156	0.000599
1000	500	102	37	0.002295	0.001140
1000	500	102	38	0.004365	0.002070
1000	500	102	39	0.007957	0.003591
1000	500	102	40	0.013910	0.005953
1000	500	102	41	0.023343	0.009433
1000	500	102	42	0.037635	0.014292
1000	500	102	43	0.058346	0.020711
1000	500	102	44	0.087061	0.028714
1000	500	102	45	0.125157	0.038096
1000	500	102	46	0.173532	0.048375
1000	500	102	47	0.232336	0.058804
1000	500	102	48	0.300773	0.068437
1000	500	102	49	0.377037	0.076264
1000	500	102	50	0.458419	0.081381
1000	500	102	51	0.541581	0.083162
1000	500	103	28	0.000001	0.000000
1000	500	103	29	0.000002	0.000001
1000	500	103	30	0.000005	0.000003
1000	500	103	31	0.000013	0.000008
1000	500	103	32	0.000033	0.000020
1000	500	103	33	0.000079	0.000046
1000	500	103	34	0.000182	0.000103
1000	500	103	35	0.000401	0.000219
1000	500	103	36	0.000846	0.000445
1000	500	103	37	0.001708	0.000861
1000	500	103	38	0.003300	0.001593
1000	500	103	39	0.006113	0.002813
1000	500	103	40	0.010860	0.004747
1000	500	103	41	0.018521	0.007661
1000	500	103	42	0.030346	0.011824
1000	500	103	43	0.047806	0.017460
1000	500	103	44	0.072480	0.024674
1000	500	103	45	0.105854	0.033374
1000	500	103	46	0.149075	0.043221
1000	500	103	47	0.202672	0.053597
1000	500	103	48	0.266326	0.063654
1000	500	103	49	0.338735	0.072410
1000	500	103	50	0.417638	0.078902
1000	500	103	51	0.500000	0.082362
1000	500	104	29	0.000001	0.000001
1000	500	104	30	0.000003	0.000002
1000	500	104	31	0.000009	0.000005
1000	500	104	32	0.000022	0.000014
1000	500	104	33	0.000055	0.000032
1000	500	104	34	0.000128	0.000074
1000	500	104	35	0.000288	0.000159
1000	500	104	36	0.000616	0.000329
1000	500	104	37	0.001263	0.000647
1000	500	104	38	0.002480	0.001217
1000	500	104	39	0.004668	0.002187
1000	500	104	40	0.008426	0.003758
1000	500	104	41	0.014601	0.006175
1000	500	104	42	0.024308	0.009707
1000	500	104	43	0.038911	0.014602
1000	500	104	44	0.059937	0.021026
1000	500	104	45	0.088925	0.028988
1000	500	104	46	0.127200	0.038274
1000	500	104	47	0.175604	0.048405
1000	500	104	48	0.234251	0.058646
1000	500	104	49	0.302329	0.068078
1000	500	104	50	0.378055	0.075726
1000	500	104	51	0.458773	0.080718
1000	500	104	52	0.541227	0.082454
1000	500	105	29	0.000001	0.000000
1000	500	105	30	0.000002	0.000001
1000	500	105	31	0.000006	0.000004
1000	500	105	32	0.000015	0.000009
1000	500	105	33	0.000038	0.000023
1000	500	105	34	0.000090	0.000052
1000	500	105	35	0.000205	0.000115
1000	500	105	36	0.000446	0.000241
1000	500	105	37	0.000929	0.000483
1000	500	105	38	0.001853	0.000924
1000	500	105	39	0.003542	0.001689
1000	500	105	40	0.006496	0.002954
1000	500	105	41	0.011438	0.004941
1000	500	105	42	0.019347	0.007909
1000	500	105	43	0.031463	0.012116
1000	500	105	44	0.049236	0.017773
1000	500	105	45	0.074205	0.024969
1000	500	105	46	0.107806	0.033602
1000	500	105	47	0.151132	0.043326
1000	500	105	48	0.204666	0.053534
1000	500	105	49	0.268062	0.063396
1000	500	105	50	0.340023	0.071961
1000	500	105	51	0.418324	0.078301
1000	500	105	52	0.500000	0.081676
1000	500	106	30	0.000001	0.000001
1000	500	106	31	0.000004	0.000002
1000	500	106	32	0.000010	0.000006
1000	500	106	33	0.000026	0.000016
1000	500	106	34	0.000063	0.000037
1000	500	106	35	0.000145	0.000082
1000	500	106	36	0.000321	0.000176
1000	500	106	37	0.000679	0.000358
1000	500	106	38	0.001376	0.000697
1000	500	106	39	0.002672	0.001296
1000	500	106	40	0.004978	0.002306
1000	500	106	41	0.008904	0.003926
1000	500	106	42	0.015299	0.006396
1000	500	106	43	0.025276	0.009977
1000	500	106	44	0.040181	0.014905
1000	500	106	45	0.061511	0.021330
1000	500	106	46	0.090761	0.029250
1000	500	106	47	0.129203	0.038441
1000	500	106	48	0.177630	0.048427
1000	500	106	49	0.236116	0.058486
1000	500	106	50	0.303841	0.067725
1000	500	106	51	0.379042	0.075201
1000	500	106	52	0.459116	0.080074
1000	500	106	53	0.540884	0.081768
1000	500	107	30	0.000001	0.000001
1000	500	107	31	0.000003	0.000002
1000	500	107	32	0.000007	0.000004
1000	500	107	33	0.000018	0.000011
1000	500	107	34	0.000044	0.000026
1000	500	107	35	0.000102	0.000059
1000	500	107	36	0.000230	0.000127
1000	500	107	37	0.000494	0.000264
1000	500	107	38	0.001016	0.000522
1000	500	107	39	0.002004	0.000988
1000	500	107	40	0.003791	0.001788
1000	500	107	41	0.006888	0.003097
1000	500	107	42	0.012023	0.005135
1000	500	107	43	0.020176	0.008153
1000	500	107	44	0.032578	0.012402
1000	500	107	45	0.050655	0.018077
1000	500	107	46	0.075907	0.025252
1000	500	107	47	0.109724	0.033817
1000	500	107	48	0.153145	0.043421
1000	500	107	49	0.206611	0.053466
1000	500	107	50	0.269751	0.063140
1000	500	107	51	0.341273	0.071522
1000	500	107	52	0.418989	0.077716
1000	500	107	53	0.500000	0.081010
1000	500	108	30	0.000001	0.000000
1000	500	108	31	0.000002	0.000001
1000	500	108	32	0.000005	0.000003
1000	500	108	33	0.000012	0.000007
1000	500	108	34	0.000030	0.000018
1000	500	108	35	0.000072	0.000042
1000	500	108	36	0.000164	0.000092
1000	500	108	37	0.000357	0.000193
1000	500	108	38	0.000746	0.000389
1000	500	108	39	0.001494	0.000748
1000	500	108	40	0.002870	0.001377
1000	500	108	41	0.005296	0.002426
1000	500	108	42	0.009389	0.004093
1000	500	108	43	0.016004	0.006614
1000	500	108	44	0.026245	0.010242
1000	500	108	45	0.041445	0.015199
1000	500	108	46	0.063069	0.021624
1000	500	108	47	0.092569	0.029500
1000	500	108	48	0.131167	0.038598
1000	500	108	49	0.179609	0.048442
1000	500	108	50	0.237934	0.058325
1000	500	108	51	0.305312	0.067378
1000	500	108	52	0.380001	0.074689
1000	500	108	53	0.459449	0.079449
1000	500	108	54	0.540551	0.081101
1000	500	109	31	0.000001	0.000001
1000	500	109	32	0.000003	0.000002
1000	500	109	33	0.000008	0.000005
1000	500	109	34	0.000021	0.000013
1000	500	109	35	0.000050	0.000029
1000	500	109	36	0.000116	0.000066
1000	500	109	37	0.000257	0.000141
1000	500	109	38	0.000544	0.000288
1000	500	109	39	0.001107	0.000563
1000	500	109	40	0.002160	0.001053
1000	500	109	41	0.004048	0.001888
1000	500	109	42	0.007288	0.003240
1000	500	109	43	0.012615	0.005327
1000	500	109	44	0.021010	0.008395
1000	500	109	45	0.033692	0.012682
1000	500	109	46	0.052063	0.018371
1000	500	109	47	0.077588	0.025525
1000	500	109	48	0.111609	0.034021
1000	500	109	49	0.155116	0.043508
1000	500	109	50	0.208510	0.053393
1000	500	109	51	0.271396	0.062887
1000	500	109	52	0.342489	0.071093
1000	500	109	53	0.419636	0.077147
1000	500	109	54	0.500000	0.080364
1000	500	110	31	0.000001	0.000000
1000	500	110	32	0.000002	0.000001
1000	500	110	33	0.000006	0.000003
1000	500	110	34	0.000014	0.000009
1000	500	110	35	0.000035	0.000021
1000	500	110	36	0.000082	0.000047
1000	500	110	37	0.000183	0.000102
1000	500	110	38	0.000395	0.000212

Table for $N = 1000$, $n = 500$

N	n	k	x	P(x)	p(x)
1000	500	110	39	0.000816	0.000421
1000	500	110	40	0.001617	0.000801
1000	500	110	41	0.003076	0.001459
1000	500	110	42	0.005622	0.002547
1000	500	110	43	0.009883	0.004260
1000	500	110	44	0.016713	0.006831
1000	500	110	45	0.027215	0.010502
1000	500	110	46	0.042702	0.015487
1000	500	110	47	0.064610	0.021908
1000	500	110	48	0.094350	0.029740
1000	500	110	49	0.133094	0.038744
1000	500	110	50	0.181543	0.048449
1000	500	110	51	0.239706	0.058163
1000	500	110	52	0.306743	0.067037
1000	500	110	53	0.380933	0.074189
1000	500	110	54	0.459773	0.078840
1000	500	110	55	0.540227	0.080454
1000	500	111	32	0.000001	0.000001
1000	500	111	33	0.000004	0.000002
1000	500	111	34	0.000010	0.000006
1000	500	111	35	0.000024	0.000014
1000	500	111	36	0.000057	0.000033
1000	500	111	37	0.000131	0.000073
1000	500	111	38	0.000285	0.000155
1000	500	111	39	0.000598	0.000313
1000	500	111	40	0.001203	0.000605
1000	500	111	41	0.002323	0.001120
1000	500	111	42	0.004312	0.001989
1000	500	111	43	0.007695	0.003383
1000	500	111	44	0.013214	0.005518
1000	500	111	45	0.021846	0.008633
1000	500	111	46	0.034802	0.012956
1000	500	111	47	0.053459	0.018657
1000	500	111	48	0.079246	0.025787
1000	500	111	49	0.113461	0.034215
1000	500	111	50	0.157047	0.043586
1000	500	111	51	0.210363	0.053317
1000	500	111	52	0.272999	0.062635
1000	500	111	53	0.343671	0.070673
1000	500	111	54	0.420264	0.076593
1000	500	111	55	0.500000	0.079736
1000	500	112	32	0.000001	0.000001
1000	500	112	33	0.000002	0.000002
1000	500	112	34	0.000007	0.000004
1000	500	112	35	0.000017	0.000010
1000	500	112	36	0.000040	0.000023
1000	500	112	37	0.000092	0.000052
1000	500	112	38	0.000205	0.000112
1000	500	112	39	0.000436	0.000231
1000	500	112	40	0.000890	0.000454

N	n	k	x	P(x)	p(x)
1000	500	112	41	0.001745	0.000855
1000	500	112	42	0.003287	0.001543
1000	500	112	43	0.005956	0.002669
1000	500	112	44	0.010383	0.004427
1000	500	112	45	0.017428	0.007045
1000	500	112	46	0.028186	0.010758
1000	500	112	47	0.043952	0.015767
1000	500	112	48	0.066135	0.022183
1000	500	112	49	0.096103	0.029968
1000	500	112	50	0.134984	0.038881
1000	500	112	51	0.183435	0.048451
1000	500	112	52	0.241434	0.057999
1000	500	112	53	0.308136	0.066702
1000	500	112	54	0.381838	0.073702
1000	500	112	55	0.460087	0.078249
1000	500	112	56	0.539913	0.079825
1000	500	113	32	0.000001	0.000000
1000	500	113	33	0.000002	0.000001
1000	500	113	34	0.000004	0.000003
1000	500	113	35	0.000011	0.000007
1000	500	113	36	0.000028	0.000016
1000	500	113	37	0.000065	0.000037
1000	500	113	38	0.000146	0.000081
1000	500	113	39	0.000316	0.000170
1000	500	113	40	0.000655	0.000339
1000	500	113	41	0.001303	0.000648
1000	500	113	42	0.002492	0.001189
1000	500	113	43	0.004583	0.002091
1000	500	113	44	0.008110	0.003527
1000	500	113	45	0.013818	0.005709
1000	500	113	46	0.022686	0.008868
1000	500	113	47	0.035909	0.013223
1000	500	113	48	0.054844	0.018935
1000	500	113	49	0.080883	0.026039
1000	500	113	50	0.115281	0.034398
1000	500	113	51	0.158937	0.043656
1000	500	113	52	0.212173	0.053237
1000	500	113	53	0.274560	0.062386
1000	500	113	54	0.344821	0.070261
1000	500	113	55	0.420875	0.076053
1000	500	113	56	0.500000	0.079125
1000	500	114	33	0.000001	0.000001
1000	500	114	34	0.000003	0.000002
1000	500	114	35	0.000008	0.000005
1000	500	114	36	0.000019	0.000011
1000	500	114	37	0.000046	0.000026
1000	500	114	38	0.000104	0.000058
1000	500	114	39	0.000228	0.000124
1000	500	114	40	0.000479	0.000251
1000	500	114	41	0.000967	0.000488

N	n	k	x	P(x)	p(x)
1000	500	114	42	0.001878	0.000910
1000	500	114	43	0.003505	0.001627
1000	500	114	44	0.006296	0.002791
1000	500	114	45	0.010890	0.004593
1000	500	114	46	0.018147	0.007257
1000	500	114	47	0.029156	0.011009
1000	500	114	48	0.045195	0.016039
1000	500	114	49	0.067643	0.022448
1000	500	114	50	0.097830	0.030187
1000	500	114	51	0.136838	0.039008
1000	500	114	52	0.185285	0.048447
1000	500	114	53	0.243120	0.057835
1000	500	114	54	0.309493	0.066372
1000	500	114	55	0.382719	0.073227
1000	500	114	56	0.460393	0.077673
1000	500	114	57	0.539607	0.079214
1000	500	115	33	0.000001	0.000000
1000	500	115	34	0.000002	0.000001
1000	500	115	35	0.000005	0.000003
1000	500	115	36	0.000013	0.000008
1000	500	115	37	0.000032	0.000019
1000	500	115	38	0.000073	0.000042
1000	500	115	39	0.000163	0.000090
1000	500	115	40	0.000349	0.000185
1000	500	115	41	0.000714	0.000366
1000	500	115	42	0.001407	0.000693
1000	500	115	43	0.002666	0.001259
1000	500	115	44	0.004860	0.002194
1000	500	115	45	0.008531	0.003671
1000	500	115	46	0.014428	0.005897
1000	500	115	47	0.023527	0.009099
1000	500	115	48	0.037013	0.013485
1000	500	115	49	0.056217	0.019204
1000	500	115	50	0.082498	0.026282
1000	500	115	51	0.117070	0.034572
1000	500	115	52	0.160788	0.043718
1000	500	115	53	0.213942	0.053153
1000	500	115	54	0.276082	0.062140
1000	500	115	55	0.345941	0.069859
1000	500	115	56	0.421468	0.075528
1000	500	115	57	0.500000	0.078532
1000	500	116	34	0.000001	0.000001
1000	500	116	35	0.000004	0.000002
1000	500	116	36	0.000009	0.000005
1000	500	116	37	0.000022	0.000013
1000	500	116	38	0.000052	0.000030
1000	500	116	39	0.000117	0.000065
1000	500	116	40	0.000252	0.000136
1000	500	116	41	0.000525	0.000272
1000	500	116	42	0.001049	0.000524

N	n	k	x	P(x)	p(x)
1000	500	116	43	0.002016	0.000967
1000	500	116	44	0.003729	0.001713
1000	500	116	45	0.006644	0.002914
1000	500	116	46	0.011403	0.004759
1000	500	116	47	0.018870	0.007467
1000	500	116	48	0.030125	0.011255
1000	500	116	49	0.046430	0.016305
1000	500	116	50	0.069135	0.022705
1000	500	116	51	0.099530	0.030395
1000	500	116	52	0.138657	0.039127
1000	500	116	53	0.187095	0.048437
1000	500	116	54	0.244766	0.057671
1000	500	116	55	0.310814	0.066048
1000	500	116	56	0.383576	0.072763
1000	500	116	57	0.460690	0.077113
1000	500	116	58	0.539310	0.078620
1000	500	117	34	0.000001	0.000001
1000	500	117	35	0.000002	0.000001
1000	500	117	36	0.000006	0.000004
1000	500	117	37	0.000015	0.000009
1000	500	117	38	0.000036	0.000021
1000	500	117	39	0.000083	0.000047
1000	500	117	40	0.000182	0.000099
1000	500	117	41	0.000383	0.000202
1000	500	117	42	0.000777	0.000394
1000	500	117	43	0.001516	0.000739
1000	500	117	44	0.002846	0.001330
1000	500	117	45	0.005144	0.002298
1000	500	117	46	0.008959	0.003815
1000	500	117	47	0.015043	0.006085
1000	500	117	48	0.024370	0.009327
1000	500	117	49	0.038112	0.013741
1000	500	117	50	0.057577	0.019465
1000	500	117	51	0.084092	0.026515
1000	500	117	52	0.118828	0.034736
1000	500	117	53	0.162602	0.043774
1000	500	117	54	0.215669	0.053067
1000	500	117	55	0.277566	0.061896
1000	500	117	56	0.347031	0.069465
1000	500	117	57	0.422046	0.075015
1000	500	117	58	0.500000	0.077954
1000	500	118	34	0.000001	0.000000
1000	500	118	35	0.000002	0.000001
1000	500	118	36	0.000004	0.000003
1000	500	118	37	0.000010	0.000006
1000	500	118	38	0.000025	0.000015
1000	500	118	39	0.000058	0.000033
1000	500	118	40	0.000130	0.000072
1000	500	118	41	0.000278	0.000148
1000	500	118	42	0.000573	0.000294

Table for $N = 1000,\ n = 500$

N	n	k	x	P(x)	p(x)
1000	500	118	43	0.001134	0.000561
1000	500	118	44	0.002159	0.001025
1000	500	118	45	0.003959	0.001801
1000	500	118	46	0.006997	0.003038
1000	500	118	47	0.011922	0.004925
1000	500	118	48	0.019596	0.007674
1000	500	118	49	0.031093	0.011497
1000	500	118	50	0.047657	0.016564
1000	500	118	51	0.070609	0.022952
1000	500	118	52	0.101204	0.030595
1000	500	118	53	0.140442	0.039238
1000	500	118	54	0.188865	0.048423
1000	500	118	55	0.246372	0.057507
1000	500	118	56	0.312101	0.065729
1000	500	118	57	0.384411	0.072309
1000	500	118	58	0.460979	0.076568
1000	500	118	59	0.539021	0.078042
1000	500	119	35	0.000001	0.000001
1000	500	119	36	0.000003	0.000002
1000	500	119	37	0.000007	0.000004
1000	500	119	38	0.000017	0.000010
1000	500	119	39	0.000041	0.000024
1000	500	119	40	0.000093	0.000052
1000	500	119	41	0.000201	0.000109
1000	500	119	42	0.000420	0.000219
1000	500	119	43	0.000843	0.000423
1000	500	119	44	0.001629	0.000786
1000	500	119	45	0.003031	0.001402
1000	500	119	46	0.005434	0.002403
1000	500	119	47	0.009392	0.003959
1000	500	119	48	0.015663	0.006271
1000	500	119	49	0.025214	0.009551
1000	500	119	50	0.039206	0.013992
1000	500	119	51	0.058925	0.019719
1000	500	119	52	0.085664	0.026739
1000	500	119	53	0.120556	0.034893
1000	500	119	54	0.164379	0.043823
1000	500	119	55	0.217358	0.052979
1000	500	119	56	0.279013	0.061655
1000	500	119	57	0.348092	0.069079
1000	500	119	58	0.422608	0.074516
1000	500	119	59	0.500000	0.077392
1000	500	120	35	0.000001	0.000000
1000	500	120	36	0.000002	0.000001
1000	500	120	37	0.000005	0.000003
1000	500	120	38	0.000012	0.000007
1000	500	120	39	0.000029	0.000017
1000	500	120	40	0.000066	0.000037
1000	500	120	41	0.000145	0.000079
1000	500	120	42	0.000306	0.000162
1000	500	120	43	0.000624	0.000317
1000	500	120	44	0.001222	0.000598
1000	500	120	45	0.002307	0.001085
1000	500	120	46	0.004195	0.001889
1000	500	120	47	0.007357	0.003162
1000	500	120	48	0.012446	0.005089
1000	500	120	49	0.020325	0.007879
1000	500	120	50	0.032059	0.011734
1000	500	120	51	0.048875	0.016816
1000	500	120	52	0.072067	0.023192
1000	500	120	53	0.102853	0.030786
1000	500	120	54	0.142194	0.039342
1000	500	120	55	0.190598	0.048404
1000	500	120	56	0.247941	0.057343
1000	500	120	57	0.313357	0.065416
1000	500	120	58	0.385223	0.071867
1000	500	120	59	0.461260	0.076037
1000	500	120	60	0.538740	0.077480
1000	500	121	36	0.000001	0.000001
1000	500	121	37	0.000003	0.000002
1000	500	121	38	0.000008	0.000005
1000	500	121	39	0.000020	0.000012
1000	500	121	40	0.000046	0.000026
1000	500	121	41	0.000104	0.000057
1000	500	121	42	0.000222	0.000119
1000	500	121	43	0.000459	0.000237
1000	500	121	44	0.000912	0.000453
1000	500	121	45	0.001746	0.000834
1000	500	121	46	0.003221	0.001475
1000	500	121	47	0.005729	0.002508
1000	500	121	48	0.009832	0.004103
1000	500	121	49	0.016287	0.006455
1000	500	121	50	0.026059	0.009772
1000	500	121	51	0.040295	0.014237
1000	500	121	52	0.060260	0.019965
1000	500	121	53	0.087215	0.026955
1000	500	121	54	0.122255	0.035040
1000	500	121	55	0.166121	0.043866
1000	500	121	56	0.219009	0.052888
1000	500	121	57	0.280426	0.061417
1000	500	121	58	0.349127	0.068701
1000	500	121	59	0.423155	0.074029
1000	500	121	60	0.500000	0.076845
1000	500	122	36	0.000001	0.000001
1000	500	122	37	0.000002	0.000001
1000	500	122	38	0.000006	0.000003
1000	500	122	39	0.000014	0.000008
1000	500	122	40	0.000033	0.000019
1000	500	122	41	0.000074	0.000041
1000	500	122	42	0.000160	0.000087
1000	500	122	43	0.000336	0.000175
1000	500	122	44	0.000677	0.000341
1000	500	122	45	0.001314	0.000637
1000	500	122	46	0.002459	0.001145
1000	500	122	47	0.004437	0.001978
1000	500	122	48	0.007722	0.003286
1000	500	122	49	0.012975	0.005253
1000	500	122	50	0.021056	0.008081
1000	500	122	51	0.033023	0.011967
1000	500	122	52	0.050085	0.017062
1000	500	122	53	0.073507	0.023423
1000	500	122	54	0.104476	0.030968
1000	500	122	55	0.143914	0.039438
1000	500	122	56	0.192294	0.048380
1000	500	122	57	0.249473	0.057178
1000	500	122	58	0.314581	0.065108
1000	500	122	59	0.386015	0.071434
1000	500	122	60	0.461534	0.075519
1000	500	122	61	0.538466	0.076932
1000	500	123	36	0.000001	0.000000
1000	500	123	37	0.000001	0.000001
1000	500	123	38	0.000004	0.000002
1000	500	123	39	0.000010	0.000006
1000	500	123	40	0.000023	0.000013
1000	500	123	41	0.000052	0.000029
1000	500	123	42	0.000115	0.000063
1000	500	123	43	0.000244	0.000129
1000	500	123	44	0.000500	0.000255
1000	500	123	45	0.000984	0.000484
1000	500	123	46	0.001867	0.000883
1000	500	123	47	0.003416	0.001549
1000	500	123	48	0.006031	0.002615
1000	500	123	49	0.010277	0.004246
1000	500	123	50	0.016914	0.006637
1000	500	123	51	0.026903	0.009989
1000	500	123	52	0.041379	0.014476
1000	500	123	53	0.061583	0.020204
1000	500	123	54	0.088745	0.027162
1000	500	123	55	0.123925	0.035181
1000	500	123	56	0.167828	0.043903
1000	500	123	57	0.220623	0.052794
1000	500	123	58	0.281804	0.061181
1000	500	123	59	0.350135	0.068331
1000	500	123	60	0.423688	0.073553
1000	500	123	61	0.500000	0.076312
1000	500	124	37	0.000001	0.000001
1000	500	124	38	0.000003	0.000002
1000	500	124	39	0.000007	0.000004
1000	500	124	40	0.000016	0.000009
1000	500	124	41	0.000037	0.000021
1000	500	124	42	0.000082	0.000045
1000	500	124	43	0.000177	0.000095
1000	500	124	44	0.000367	0.000190
1000	500	124	45	0.000733	0.000366
1000	500	124	46	0.001410	0.000677
1000	500	124	47	0.002616	0.001206
1000	500	124	48	0.004683	0.002067
1000	500	124	49	0.008093	0.003410
1000	500	124	50	0.013509	0.005415
1000	500	124	51	0.021789	0.008281
1000	500	124	52	0.033984	0.012195
1000	500	124	53	0.051285	0.017301
1000	500	124	54	0.074931	0.023646
1000	500	124	55	0.106074	0.031143
1000	500	124	56	0.145601	0.039527
1000	500	124	57	0.193955	0.048353
1000	500	124	58	0.250969	0.057015
1000	500	124	59	0.315775	0.064805
1000	500	124	60	0.386786	0.071011
1000	500	124	61	0.461801	0.075015
1000	500	124	62	0.538199	0.076399
1000	500	125	37	0.000001	0.000000
1000	500	125	38	0.000002	0.000001
1000	500	125	39	0.000004	0.000003
1000	500	125	40	0.000011	0.000006
1000	500	125	41	0.000026	0.000015
1000	500	125	42	0.000058	0.000033
1000	500	125	43	0.000128	0.000069
1000	500	125	44	0.000268	0.000141
1000	500	125	45	0.000543	0.000275
1000	500	125	46	0.001059	0.000516
1000	500	125	47	0.001992	0.000933
1000	500	125	48	0.003617	0.001624
1000	500	125	49	0.006338	0.002721
1000	500	125	50	0.010727	0.004389
1000	500	125	51	0.017545	0.006818
1000	500	125	52	0.027747	0.010203
1000	500	125	53	0.042457	0.014709
1000	500	125	54	0.062892	0.020435
1000	500	125	55	0.090254	0.027362
1000	500	125	56	0.125567	0.035313
1000	500	125	57	0.169502	0.043935
1000	500	125	58	0.222202	0.052700
1000	500	125	59	0.283150	0.060948
1000	500	125	60	0.351118	0.067968
1000	500	125	61	0.424208	0.073089
1000	500	125	62	0.500000	0.075792
1000	500	126	38	0.000001	0.000001
1000	500	126	39	0.000003	0.000002
1000	500	126	40	0.000008	0.000005

Table for $N = 1000,\ n = 500$

N	n	k	x	P(x)	p(x)
1000	500	126	41	0.000018	0.000010
1000	500	126	42	0.000041	0.000023
1000	500	126	43	0.000091	0.000050
1000	500	126	44	0.000195	0.000103
1000	500	126	45	0.000400	0.000205
1000	500	126	46	0.000791	0.000391
1000	500	126	47	0.001509	0.000718
1000	500	126	48	0.002778	0.001268
1000	500	126	49	0.004935	0.002158
1000	500	126	50	0.008470	0.003534
1000	500	126	51	0.014046	0.005577
1000	500	126	52	0.022524	0.008478
1000	500	126	53	0.034942	0.012419
1000	500	126	54	0.052476	0.017534
1000	500	126	55	0.076339	0.023863
1000	500	126	56	0.107648	0.031310
1000	500	126	57	0.147258	0.039610
1000	500	126	58	0.195581	0.048322
1000	500	126	59	0.252432	0.056851
1000	500	126	60	0.316940	0.064508
1000	500	126	61	0.387538	0.070598
1000	500	126	62	0.462060	0.074523
1000	500	126	63	0.537939	0.075879
1000	500	127	38	0.000001	0.000000
1000	500	127	39	0.000002	0.000001
1000	500	127	40	0.000005	0.000003
1000	500	127	41	0.000012	0.000007
1000	500	127	42	0.000029	0.000017
1000	500	127	43	0.000065	0.000036
1000	500	127	44	0.000141	0.000076
1000	500	127	45	0.000293	0.000152
1000	500	127	46	0.000588	0.000295
1000	500	127	47	0.001137	0.000549
1000	500	127	48	0.002122	0.000985
1000	500	127	49	0.003822	0.001700
1000	500	127	50	0.006650	0.002828
1000	500	127	51	0.011181	0.004532
1000	500	127	52	0.018178	0.006997
1000	500	127	53	0.028591	0.010413
1000	500	127	54	0.043529	0.014938
1000	500	127	55	0.064189	0.020660
1000	500	127	56	0.091743	0.027554
1000	500	127	57	0.127181	0.035439
1000	500	127	58	0.171143	0.043962
1000	500	127	59	0.223746	0.052603
1000	500	127	60	0.284465	0.060718
1000	500	127	61	0.352077	0.067613
1000	500	127	62	0.424714	0.072637
1000	500	127	63	0.500000	0.075286
1000	500	128	38	0.000001	0.000000

N	n	k	x	P(x)	p(x)
1000	500	128	39	0.000001	0.000001
1000	500	128	40	0.000004	0.000002
1000	500	128	41	0.000009	0.000005
1000	500	128	42	0.000020	0.000012
1000	500	128	43	0.000046	0.000026
1000	500	128	44	0.000101	0.000055
1000	500	128	45	0.000214	0.000112
1000	500	128	46	0.000435	0.000221
1000	500	128	47	0.000852	0.000417
1000	500	128	48	0.001612	0.000760
1000	500	128	49	0.002943	0.001332
1000	500	128	50	0.005192	0.002249
1000	500	128	51	0.008851	0.003659
1000	500	128	52	0.014588	0.005737
1000	500	128	53	0.023259	0.008672
1000	500	128	54	0.035897	0.012638
1000	500	128	55	0.053658	0.017760
1000	500	128	56	0.077729	0.024071
1000	500	128	57	0.109198	0.031469
1000	500	128	58	0.148885	0.039687
1000	500	128	59	0.197174	0.048288
1000	500	128	60	0.253862	0.056688
1000	500	128	61	0.318077	0.064215
1000	500	128	62	0.388271	0.070193
1000	500	128	63	0.462314	0.074043
1000	500	128	64	0.537686	0.075372
1000	500	129	39	0.000001	0.000001
1000	500	129	40	0.000002	0.000001
1000	500	129	41	0.000006	0.000004
1000	500	129	42	0.000014	0.000008
1000	500	129	43	0.000033	0.000019
1000	500	129	44	0.000073	0.000040
1000	500	129	45	0.000155	0.000083
1000	500	129	46	0.000320	0.000164
1000	500	129	47	0.000635	0.000316
1000	500	129	48	0.001218	0.000583
1000	500	129	49	0.002255	0.001037
1000	500	129	50	0.004032	0.001777
1000	500	129	51	0.006967	0.002935
1000	500	129	52	0.011640	0.004673
1000	500	129	53	0.018814	0.007174
1000	500	129	54	0.029433	0.010619
1000	500	129	55	0.044594	0.015161
1000	500	129	56	0.065472	0.020878
1000	500	129	57	0.093211	0.027738
1000	500	129	58	0.128769	0.035558
1000	500	129	59	0.172753	0.043984
1000	500	129	60	0.225258	0.052505
1000	500	129	61	0.285749	0.060491
1000	500	129	62	0.353013	0.067264

N	n	k	x	P(x)	p(x)
1000	500	129	63	0.425207	0.072194
1000	500	129	64	0.500000	0.074793
1000	500	130	39	0.000001	0.000000
1000	500	130	40	0.000002	0.000001
1000	500	130	41	0.000004	0.000002
1000	500	130	42	0.000010	0.000006
1000	500	130	43	0.000023	0.000013
1000	500	130	44	0.000052	0.000029
1000	500	130	45	0.000112	0.000060
1000	500	130	46	0.000234	0.000122
1000	500	130	47	0.000471	0.000237
1000	500	130	48	0.000916	0.000444
1000	500	130	49	0.001718	0.000802
1000	500	130	50	0.003114	0.001396
1000	500	130	51	0.005454	0.002340
1000	500	130	52	0.009236	0.003783
1000	500	130	53	0.015132	0.005896
1000	500	130	54	0.023996	0.008863
1000	500	130	55	0.036848	0.012853
1000	500	130	56	0.054830	0.017981
1000	500	130	57	0.079103	0.024273
1000	500	130	58	0.110724	0.031622
1000	500	130	59	0.150483	0.039758
1000	500	130	60	0.198734	0.048251
1000	500	130	61	0.255260	0.056526
1000	500	130	62	0.319188	0.063928
1000	500	130	63	0.388986	0.069798
1000	500	130	64	0.462561	0.073575
1000	500	130	65	0.537439	0.074879
1000	500	131	40	0.000001	0.000001
1000	500	131	41	0.000003	0.000002
1000	500	131	42	0.000007	0.000004
1000	500	131	43	0.000016	0.000009
1000	500	131	44	0.000037	0.000021
1000	500	131	45	0.000081	0.000044
1000	500	131	46	0.000170	0.000090
1000	500	131	47	0.000348	0.000177
1000	500	131	48	0.000685	0.000337
1000	500	131	49	0.001302	0.000617
1000	500	131	50	0.002392	0.001090
1000	500	131	51	0.004246	0.001854
1000	500	131	52	0.007289	0.003043
1000	500	131	53	0.012103	0.004815
1000	500	131	54	0.019452	0.007349
1000	500	131	55	0.030274	0.010822
1000	500	131	56	0.045653	0.015379
1000	500	131	57	0.066743	0.021090
1000	500	131	58	0.094659	0.027916
1000	500	131	59	0.130329	0.035670
1000	500	131	60	0.174331	0.044002

N	n	k	x	P(x)	p(x)
1000	500	131	61	0.226737	0.052406
1000	500	131	62	0.287004	0.060266
1000	500	131	63	0.353926	0.066923
1000	500	131	64	0.425689	0.071762
1000	500	131	65	0.500000	0.074311
1000	500	132	40	0.000001	0.000000
1000	500	132	41	0.000002	0.000001
1000	500	132	42	0.000005	0.000003
1000	500	132	43	0.000011	0.000007
1000	500	132	44	0.000026	0.000015
1000	500	132	45	0.000058	0.000032
1000	500	132	46	0.000123	0.000066
1000	500	132	47	0.000255	0.000132
1000	500	132	48	0.000509	0.000254
1000	500	132	49	0.000982	0.000472
1000	500	132	50	0.001827	0.000846
1000	500	132	51	0.003288	0.001461
1000	500	132	52	0.005720	0.002432
1000	500	132	53	0.009627	0.003907
1000	500	132	54	0.015680	0.006054
1000	500	132	55	0.024732	0.009052
1000	500	132	56	0.037796	0.013063
1000	500	132	57	0.055992	0.018196
1000	500	132	58	0.080460	0.024468
1000	500	132	59	0.112228	0.031767
1000	500	132	60	0.152052	0.039824
1000	500	132	61	0.200263	0.048211
1000	500	132	62	0.256628	0.056365
1000	500	132	63	0.320272	0.063645
1000	500	132	64	0.389684	0.069411
1000	500	132	65	0.462802	0.073118
1000	500	132	66	0.537198	0.074397
1000	500	133	41	0.000001	0.000001
1000	500	133	42	0.000003	0.000002
1000	500	133	43	0.000008	0.000005
1000	500	133	44	0.000018	0.000010
1000	500	133	45	0.000041	0.000023
1000	500	133	46	0.000089	0.000048
1000	500	133	47	0.000186	0.000097
1000	500	133	48	0.000377	0.000191
1000	500	133	49	0.000736	0.000359
1000	500	133	50	0.001389	0.000653
1000	500	133	51	0.002532	0.001144
1000	500	133	52	0.004465	0.001932
1000	500	133	53	0.007615	0.003150
1000	500	133	54	0.012570	0.004955
1000	500	133	55	0.020092	0.007522
1000	500	133	56	0.031113	0.011022
1000	500	133	57	0.046705	0.015591
1000	500	133	58	0.068000	0.021295

Table for $N = 1000$, $n = 500$

N	n	k	x	P(x)	p(x)
1000	500	133	59	0.096088	0.028087
1000	500	133	60	0.131864	0.035777
1000	500	133	61	0.175880	0.044015
1000	500	133	62	0.228185	0.052306
1000	500	133	63	0.288230	0.060045
1000	500	133	64	0.354818	0.066588
1000	500	133	65	0.426158	0.071340
1000	500	133	66	0.500000	0.073842
1000	500	134	41	0.000001	0.000001
1000	500	134	42	0.000002	0.000001
1000	500	134	43	0.000005	0.000003
1000	500	134	44	0.000013	0.000007
1000	500	134	45	0.000029	0.000016
1000	500	134	46	0.000064	0.000035
1000	500	134	47	0.000136	0.000072
1000	500	134	48	0.000278	0.000142
1000	500	134	49	0.000549	0.000272
1000	500	134	50	0.001050	0.000501
1000	500	134	51	0.001940	0.000890
1000	500	134	52	0.003466	0.001526
1000	500	134	53	0.005991	0.002524
1000	500	134	54	0.010021	0.004030
1000	500	134	55	0.016231	0.006210
1000	500	134	56	0.025469	0.009238
1000	500	134	57	0.038738	0.013269
1000	500	134	58	0.057144	0.018406
1000	500	134	59	0.081801	0.024657
1000	500	134	60	0.113708	0.031907
1000	500	134	61	0.153593	0.039885
1000	500	134	62	0.201761	0.048168
1000	500	134	63	0.257965	0.056204
1000	500	134	64	0.321332	0.063367
1000	500	134	65	0.390365	0.069033
1000	500	134	66	0.463037	0.072672
1000	500	134	67	0.536963	0.073927
1000	500	135	41	0.000001	0.000000
1000	500	135	42	0.000001	0.000001
1000	500	135	43	0.000004	0.000002
1000	500	135	44	0.000009	0.000005
1000	500	135	45	0.000020	0.000012
1000	500	135	46	0.000046	0.000025
1000	500	135	47	0.000098	0.000052
1000	500	135	48	0.000203	0.000105
1000	500	135	49	0.000408	0.000204
1000	500	135	50	0.000790	0.000382
1000	500	135	51	0.001479	0.000689
1000	500	135	52	0.002677	0.001198
1000	500	135	53	0.004688	0.002011
1000	500	135	54	0.007945	0.003258
1000	500	135	55	0.013040	0.005095

N	n	k	x	P(x)	p(x)
1000	500	135	56	0.020733	0.007693
1000	500	135	57	0.031951	0.011218
1000	500	135	58	0.047750	0.015799
1000	500	135	59	0.069245	0.021495
1000	500	135	60	0.097497	0.028252
1000	500	135	61	0.133374	0.035877
1000	500	135	62	0.177399	0.044025
1000	500	135	63	0.229604	0.052205
1000	500	135	64	0.289429	0.059826
1000	500	135	65	0.355689	0.066260
1000	500	135	66	0.426617	0.070928
1000	500	135	67	0.500000	0.073383
1000	500	136	42	0.000001	0.000001
1000	500	136	43	0.000003	0.000002
1000	500	136	44	0.000006	0.000004
1000	500	136	45	0.000014	0.000008
1000	500	136	46	0.000032	0.000018
1000	500	136	47	0.000071	0.000038
1000	500	136	48	0.000148	0.000078
1000	500	136	49	0.000301	0.000153
1000	500	136	50	0.000591	0.000290
1000	500	136	51	0.001121	0.000530
1000	500	136	52	0.002056	0.000935
1000	500	136	53	0.003649	0.001592
1000	500	136	54	0.006266	0.002617
1000	500	136	55	0.010419	0.004153
1000	500	136	56	0.016784	0.006365
1000	500	136	57	0.026205	0.009421
1000	500	136	58	0.039677	0.013471
1000	500	136	59	0.058286	0.018609
1000	500	136	60	0.083126	0.024840
1000	500	136	61	0.115166	0.032040
1000	500	136	62	0.155106	0.039941
1000	500	136	63	0.203230	0.048123
1000	500	136	64	0.259274	0.056045
1000	500	136	65	0.322368	0.063093
1000	500	136	66	0.391030	0.068662
1000	500	136	67	0.463266	0.072236
1000	500	136	68	0.536734	0.073468
1000	500	137	42	0.000001	0.000000
1000	500	137	43	0.000002	0.000001
1000	500	137	44	0.000004	0.000003
1000	500	137	45	0.000010	0.000006
1000	500	137	46	0.000023	0.000013
1000	500	137	47	0.000051	0.000028
1000	500	137	48	0.000108	0.000057
1000	500	137	49	0.000221	0.000114
1000	500	137	50	0.000440	0.000219
1000	500	137	51	0.000846	0.000406
1000	500	137	52	0.001571	0.000726

N	n	k	x	P(x)	p(x)
1000	500	137	53	0.002825	0.001253
1000	500	137	54	0.004915	0.002090
1000	500	137	55	0.008280	0.003365
1000	500	137	56	0.013513	0.005233
1000	500	137	57	0.021375	0.007862
1000	500	137	58	0.032785	0.011410
1000	500	137	59	0.048788	0.016002
1000	500	137	60	0.070476	0.021688
1000	500	137	61	0.098886	0.028410
1000	500	137	62	0.134859	0.035973
1000	500	137	63	0.178890	0.044031
1000	500	137	64	0.230993	0.052103
1000	500	137	65	0.290602	0.059609
1000	500	137	66	0.356540	0.065938
1000	500	137	67	0.427064	0.070524
1000	500	137	68	0.500000	0.072936
1000	500	138	43	0.000001	0.000001
1000	500	138	44	0.000003	0.000002
1000	500	138	45	0.000007	0.000004
1000	500	138	46	0.000016	0.000009
1000	500	138	47	0.000036	0.000020
1000	500	138	48	0.000078	0.000042
1000	500	138	49	0.000162	0.000084
1000	500	138	50	0.000326	0.000164
1000	500	138	51	0.000635	0.000309
1000	500	138	52	0.001195	0.000560
1000	500	138	53	0.002176	0.000981
1000	500	138	54	0.003835	0.001659
1000	500	138	55	0.006545	0.002710
1000	500	138	56	0.010821	0.004276
1000	500	138	57	0.017339	0.006519
1000	500	138	58	0.026941	0.009602
1000	500	138	59	0.040610	0.013669
1000	500	138	60	0.059418	0.018808
1000	500	138	61	0.084434	0.025016
1000	500	138	62	0.116601	0.032167
1000	500	138	63	0.156594	0.039992
1000	500	138	64	0.204669	0.048076
1000	500	138	65	0.260555	0.055886
1000	500	138	66	0.323380	0.062824
1000	500	138	67	0.391679	0.068300
1000	500	138	68	0.463490	0.071810
1000	500	138	69	0.536510	0.073020
1000	500	139	43	0.000001	0.000000
1000	500	139	44	0.000002	0.000001
1000	500	139	45	0.000005	0.000003
1000	500	139	46	0.000011	0.000006
1000	500	139	47	0.000026	0.000014
1000	500	139	48	0.000056	0.000030
1000	500	139	49	0.000118	0.000062

N	n	k	x	P(x)	p(x)
1000	500	139	50	0.000240	0.000123
1000	500	139	51	0.000474	0.000234
1000	500	139	52	0.000904	0.000430
1000	500	139	53	0.001667	0.000763
1000	500	139	54	0.002976	0.001309
1000	500	139	55	0.005146	0.002169
1000	500	139	56	0.008618	0.003473
1000	500	139	57	0.013989	0.005371
1000	500	139	58	0.022018	0.008028
1000	500	139	59	0.033617	0.011599
1000	500	139	60	0.049818	0.016201
1000	500	139	61	0.071694	0.021876
1000	500	139	62	0.100257	0.028563
1000	500	139	63	0.136319	0.036063
1000	500	139	64	0.180353	0.044033
1000	500	139	65	0.232353	0.052000
1000	500	139	66	0.291749	0.059396
1000	500	139	67	0.357371	0.065622
1000	500	139	68	0.427501	0.070130
1000	500	139	69	0.500000	0.072499
1000	500	140	43	0.000001	0.000000
1000	500	140	44	0.000001	0.000001
1000	500	140	45	0.000003	0.000002
1000	500	140	46	0.000008	0.000005
1000	500	140	47	0.000018	0.000010
1000	500	140	48	0.000040	0.000022
1000	500	140	49	0.000085	0.000045
1000	500	140	50	0.000176	0.000091
1000	500	140	51	0.000352	0.000176
1000	500	140	52	0.000680	0.000328
1000	500	140	53	0.001271	0.000591
1000	500	140	54	0.002298	0.001027
1000	500	140	55	0.004024	0.001726
1000	500	140	56	0.006827	0.002803
1000	500	140	57	0.011226	0.004398
1000	500	140	58	0.017896	0.006670
1000	500	140	59	0.027676	0.009780
1000	500	140	60	0.041539	0.013863
1000	500	140	61	0.060540	0.019001
1000	500	140	62	0.085726	0.025187
1000	500	140	63	0.118016	0.032289
1000	500	140	64	0.158055	0.040039
1000	500	140	65	0.206081	0.048026
1000	500	140	66	0.261810	0.055729
1000	500	140	67	0.324370	0.062560
1000	500	140	68	0.392314	0.067945
1000	500	140	69	0.463709	0.071394
1000	500	140	70	0.536291	0.072583
1000	500	141	44	0.000001	0.000001
1000	500	141	45	0.000002	0.000001

N	n	k	x	P(x)	p(x)
1000	500	141	46	0.000005	0.000003
1000	500	141	47	0.000013	0.000007
1000	500	141	48	0.000028	0.000016
1000	500	141	49	0.000062	0.000033
1000	500	141	50	0.000129	0.000067
1000	500	141	51	0.000260	0.000132
1000	500	141	52	0.000509	0.000249
1000	500	141	53	0.000964	0.000454
1000	500	141	54	0.001765	0.000802
1000	500	141	55	0.003131	0.001366
1000	500	141	56	0.005380	0.002249
1000	500	141	57	0.008960	0.003580
1000	500	141	58	0.014468	0.005508
1000	500	141	59	0.022661	0.008193
1000	500	141	60	0.034446	0.011785
1000	500	141	61	0.050840	0.016394
1000	500	141	62	0.072899	0.022058
1000	500	141	63	0.101608	0.028709
1000	500	141	64	0.137756	0.036148
1000	500	141	65	0.181789	0.044033
1000	500	141	66	0.233686	0.051898
1000	500	141	67	0.292871	0.059185
1000	500	141	68	0.358184	0.065312
1000	500	141	69	0.427928	0.069744
1000	500	141	70	0.500000	0.072072
1000	500	142	44	0.000001	0.000000
1000	500	142	45	0.000002	0.000001
1000	500	142	46	0.000004	0.000002
1000	500	142	47	0.000009	0.000005
1000	500	142	48	0.000020	0.000011
1000	500	142	49	0.000044	0.000024
1000	500	142	50	0.000094	0.000049
1000	500	142	51	0.000192	0.000098
1000	500	142	52	0.000379	0.000188
1000	500	142	53	0.000727	0.000348
1000	500	142	54	0.001349	0.000622
1000	500	142	55	0.002423	0.001074
1000	500	142	56	0.004218	0.001794
1000	500	142	57	0.007114	0.002896
1000	500	142	58	0.011634	0.004520
1000	500	142	59	0.018455	0.006821
1000	500	142	60	0.028409	0.009955
1000	500	142	61	0.042462	0.014052
1000	500	142	62	0.061651	0.019189
1000	500	142	63	0.087003	0.025351
1000	500	142	64	0.119408	0.032406
1000	500	142	65	0.159490	0.040082
1000	500	142	66	0.207465	0.047975
1000	500	142	67	0.263038	0.055572
1000	500	142	68	0.325337	0.062300

N	n	k	x	P(x)	p(x)
1000	500	142	69	0.392934	0.067597
1000	500	142	70	0.463922	0.070988
1000	500	142	71	0.536078	0.072156
1000	500	143	45	0.000001	0.000001
1000	500	143	46	0.000003	0.000002
1000	500	143	47	0.000006	0.000004
1000	500	143	48	0.000014	0.000008
1000	500	143	49	0.000032	0.000017
1000	500	143	50	0.000068	0.000036
1000	500	143	51	0.000140	0.000073
1000	500	143	52	0.000281	0.000141
1000	500	143	53	0.000546	0.000265
1000	500	143	54	0.001026	0.000480
1000	500	143	55	0.001866	0.000840
1000	500	143	56	0.003289	0.001423
1000	500	143	57	0.005619	0.002329
1000	500	143	58	0.009305	0.003687
1000	500	143	59	0.014949	0.005643
1000	500	143	60	0.023304	0.008356
1000	500	143	61	0.035272	0.011967
1000	500	143	62	0.051855	0.016583
1000	500	143	63	0.074091	0.022235
1000	500	143	64	0.102941	0.028851
1000	500	143	65	0.139169	0.036228
1000	500	143	66	0.183198	0.044029
1000	500	143	67	0.234992	0.051794
1000	500	143	68	0.293970	0.058977
1000	500	143	69	0.358978	0.065009
1000	500	143	70	0.428345	0.069367
1000	500	143	71	0.500000	0.071654
1000	500	144	45	0.000001	0.000000
1000	500	144	46	0.000002	0.000001
1000	500	144	47	0.000004	0.000003
1000	500	144	48	0.000010	0.000006
1000	500	144	49	0.000022	0.000012
1000	500	144	50	0.000049	0.000026
1000	500	144	51	0.000102	0.000054
1000	500	144	52	0.000208	0.000105
1000	500	144	53	0.000408	0.000200
1000	500	144	54	0.000776	0.000368
1000	500	144	55	0.001430	0.000654
1000	500	144	56	0.002552	0.001122
1000	500	144	57	0.004414	0.001862
1000	500	144	58	0.007404	0.002989
1000	500	144	59	0.012045	0.004641
1000	500	144	60	0.019015	0.006970
1000	500	144	61	0.029142	0.010127
1000	500	144	62	0.043380	0.014238
1000	500	144	63	0.062753	0.019373
1000	500	144	64	0.088264	0.025511

N	n	k	x	P(x)	p(x)
1000	500	144	65	0.120780	0.032517
1000	500	144	66	0.160901	0.040121
1000	500	144	67	0.208823	0.047922
1000	500	144	68	0.264240	0.055417
1000	500	144	69	0.326284	0.062044
1000	500	144	70	0.393540	0.067256
1000	500	144	71	0.464131	0.070590
1000	500	144	72	0.535869	0.071738
1000	500	145	46	0.000001	0.000001
1000	500	145	47	0.000003	0.000002
1000	500	145	48	0.000007	0.000004
1000	500	145	49	0.000016	0.000009
1000	500	145	50	0.000035	0.000019
1000	500	145	51	0.000074	0.000039
1000	500	145	52	0.000152	0.000078
1000	500	145	53	0.000303	0.000151
1000	500	145	54	0.000584	0.000281
1000	500	145	55	0.001090	0.000506
1000	500	145	56	0.001970	0.000880
1000	500	145	57	0.003451	0.001481
1000	500	145	58	0.005860	0.002410
1000	500	145	59	0.009654	0.003794
1000	500	145	60	0.015432	0.005778
1000	500	145	61	0.023948	0.008516
1000	500	145	62	0.036094	0.012146
1000	500	145	63	0.052862	0.016768
1000	500	145	64	0.075270	0.022407
1000	500	145	65	0.104256	0.028986
1000	500	145	66	0.140559	0.036303
1000	500	145	67	0.184582	0.044023
1000	500	145	68	0.236273	0.051691
1000	500	145	69	0.295045	0.058772
1000	500	145	70	0.359755	0.064710
1000	500	145	71	0.428753	0.068998
1000	500	145	72	0.500000	0.071247
1000	500	146	46	0.000001	0.000000
1000	500	146	47	0.000002	0.000001
1000	500	146	48	0.000005	0.000003
1000	500	146	49	0.000011	0.000006
1000	500	146	50	0.000025	0.000014
1000	500	146	51	0.000054	0.000029
1000	500	146	52	0.000111	0.000058
1000	500	146	53	0.000224	0.000113
1000	500	146	54	0.000438	0.000213
1000	500	146	55	0.000827	0.000389
1000	500	146	56	0.001513	0.000686
1000	500	146	57	0.002683	0.001170
1000	500	146	58	0.004615	0.001931
1000	500	146	59	0.007697	0.003083
1000	500	146	60	0.012458	0.004761

N	n	k	x	P(x)	p(x)
1000	500	146	61	0.019575	0.007117
1000	500	146	62	0.029872	0.010297
1000	500	146	63	0.044292	0.014420
1000	500	146	64	0.063843	0.019551
1000	500	146	65	0.089509	0.025665
1000	500	146	66	0.122132	0.032623
1000	500	146	67	0.162288	0.040156
1000	500	146	68	0.210156	0.047868
1000	500	146	69	0.265418	0.055263
1000	500	146	70	0.327211	0.061793
1000	500	146	71	0.394133	0.066922
1000	500	146	72	0.464335	0.070202
1000	500	146	73	0.535665	0.071330
1000	500	147	46	0.000001	0.000000
1000	500	147	47	0.000001	0.000001
1000	500	147	48	0.000003	0.000002
1000	500	147	49	0.000008	0.000004
1000	500	147	50	0.000018	0.000010
1000	500	147	51	0.000038	0.000021
1000	500	147	52	0.000081	0.000043
1000	500	147	53	0.000165	0.000084
1000	500	147	54	0.000326	0.000161
1000	500	147	55	0.000624	0.000298
1000	500	147	56	0.001156	0.000532
1000	500	147	57	0.002076	0.000920
1000	500	147	58	0.003615	0.001539
1000	500	147	59	0.006105	0.002490
1000	500	147	60	0.010005	0.003900
1000	500	147	61	0.015917	0.005912
1000	500	147	62	0.024591	0.008674
1000	500	147	63	0.036913	0.012322
1000	500	147	64	0.053862	0.016948
1000	500	147	65	0.076435	0.022574
1000	500	147	66	0.105553	0.029117
1000	500	147	67	0.141928	0.036375
1000	500	147	68	0.185941	0.044014
1000	500	147	69	0.237528	0.051587
1000	500	147	70	0.296097	0.058569
1000	500	147	71	0.360515	0.064418
1000	500	147	72	0.429152	0.068637
1000	500	147	73	0.500000	0.070848
1000	500	148	47	0.000001	0.000001
1000	500	148	48	0.000002	0.000001
1000	500	148	49	0.000005	0.000003
1000	500	148	50	0.000012	0.000007
1000	500	148	51	0.000028	0.000015
1000	500	148	52	0.000059	0.000031
1000	500	148	53	0.000121	0.000062
1000	500	148	54	0.000242	0.000121
1000	500	148	55	0.000469	0.000227

Table for $N = 1000,\ n = 500$

N	n	k	x	P(x)	p(x)
1000	500	148	56	0.000879	0.000410
1000	500	148	57	0.001598	0.000719
1000	500	148	58	0.002818	0.001219
1000	500	148	59	0.004818	0.002000
1000	500	148	60	0.007994	0.003176
1000	500	148	61	0.012874	0.004881
1000	500	148	62	0.020137	0.007263
1000	500	148	63	0.030600	0.010463
1000	500	148	64	0.045199	0.014598
1000	500	148	65	0.064924	0.019725
1000	500	148	66	0.090738	0.025814
1000	500	148	67	0.123463	0.032725
1000	500	148	68	0.163651	0.040188
1000	500	148	69	0.211463	0.047812
1000	500	148	70	0.266572	0.055110
1000	500	148	71	0.328118	0.061545
1000	500	148	72	0.394713	0.066595
1000	500	148	73	0.464534	0.069821
1000	500	148	74	0.535465	0.070931
1000	500	149	47	0.000001	0.000000
1000	500	149	48	0.000002	0.000001
1000	500	149	49	0.000004	0.000002
1000	500	149	50	0.000009	0.000005
1000	500	149	51	0.000020	0.000011
1000	500	149	52	0.000042	0.000023
1000	500	149	53	0.000088	0.000046
1000	500	149	54	0.000179	0.000090
1000	500	149	55	0.000350	0.000172
1000	500	149	56	0.000665	0.000315
1000	500	149	57	0.001225	0.000559
1000	500	149	58	0.002185	0.000961
1000	500	149	59	0.003783	0.001597
1000	500	149	60	0.006354	0.002571
1000	500	149	61	0.010360	0.004006
1000	500	149	62	0.016403	0.006044
1000	500	149	63	0.025234	0.008830
1000	500	149	64	0.037728	0.012495
1000	500	149	65	0.054853	0.017125
1000	500	149	66	0.077588	0.022736
1000	500	149	67	0.106832	0.029243
1000	500	149	68	0.143274	0.036442
1000	500	149	69	0.187276	0.044002
1000	500	149	70	0.238759	0.051483
1000	500	149	71	0.297128	0.058370
1000	500	149	72	0.361259	0.064131
1000	500	149	73	0.429542	0.068283
1000	500	149	74	0.500000	0.070458
1000	500	150	48	0.000001	0.000001
1000	500	150	49	0.000003	0.000002
1000	500	150	50	0.000006	0.000004

N	n	k	x	P(x)	p(x)
1000	500	150	51	0.000014	0.000008
1000	500	150	52	0.000030	0.000016
1000	500	150	53	0.000064	0.000034
1000	500	150	54	0.000131	0.000067
1000	500	150	55	0.000261	0.000129
1000	500	150	56	0.000501	0.000240
1000	500	150	57	0.000933	0.000432
1000	500	150	58	0.001686	0.000753
1000	500	150	59	0.002955	0.001269
1000	500	150	60	0.005024	0.002069
1000	500	150	61	0.008293	0.003269
1000	500	150	62	0.013293	0.005000
1000	500	150	63	0.020699	0.007406
1000	500	150	64	0.031327	0.010628
1000	500	150	65	0.046099	0.014773
1000	500	150	66	0.065994	0.019894
1000	500	150	67	0.091952	0.025959
1000	500	150	68	0.124774	0.032822
1000	500	150	69	0.164991	0.040216
1000	500	150	70	0.212745	0.047755
1000	500	150	71	0.267703	0.054958
1000	500	150	72	0.329005	0.061302
1000	500	150	73	0.395280	0.066275
1000	500	150	74	0.464729	0.069449
1000	500	150	75	0.535270	0.070541
1000	500	151	48	0.000001	0.000000
1000	500	151	49	0.000002	0.000001
1000	500	151	50	0.000004	0.000002
1000	500	151	51	0.000010	0.000006
1000	500	151	52	0.000022	0.000012
1000	500	151	53	0.000046	0.000025
1000	500	151	54	0.000096	0.000050
1000	500	151	55	0.000193	0.000097
1000	500	151	56	0.000376	0.000182
1000	500	151	57	0.000708	0.000333
1000	500	151	58	0.001295	0.000587
1000	500	151	59	0.002297	0.001002
1000	500	151	60	0.003953	0.001656
1000	500	151	61	0.006605	0.002652
1000	500	151	62	0.010716	0.004112
1000	500	151	63	0.016891	0.006175
1000	500	151	64	0.025875	0.008984
1000	500	151	65	0.038539	0.012664
1000	500	151	66	0.055836	0.017297
1000	500	151	67	0.078729	0.022893
1000	500	151	68	0.108093	0.029364
1000	500	151	69	0.144599	0.036506
1000	500	151	70	0.188587	0.043988
1000	500	151	71	0.239966	0.051379
1000	500	151	72	0.298138	0.058172

N	n	k	x	P(x)	p(x)
1000	500	151	73	0.361987	0.063849
1000	500	151	74	0.429923	0.067936
1000	500	151	75	0.500000	0.070077
1000	500	152	48	0.000000	0.000000
1000	500	152	49	0.000001	0.000001
1000	500	152	50	0.000003	0.000002
1000	500	152	51	0.000007	0.000004
1000	500	152	52	0.000015	0.000009
1000	500	152	53	0.000033	0.000018
1000	500	152	54	0.000070	0.000037
1000	500	152	55	0.000142	0.000072
1000	500	152	56	0.000280	0.000138
1000	500	152	57	0.000535	0.000254
1000	500	152	58	0.000989	0.000455
1000	500	152	59	0.001776	0.000787
1000	500	152	60	0.003095	0.001318
1000	500	152	61	0.005234	0.002139
1000	500	152	62	0.008595	0.003362
1000	500	152	63	0.013713	0.005118
1000	500	152	64	0.021262	0.007549
1000	500	152	65	0.032051	0.010789
1000	500	152	66	0.046994	0.014943
1000	500	152	67	0.067053	0.020059
1000	500	152	68	0.093151	0.026098
1000	500	152	69	0.126066	0.032915
1000	500	152	70	0.166308	0.040241
1000	500	152	71	0.214004	0.047696
1000	500	152	72	0.268812	0.054808
1000	500	152	73	0.329874	0.061062
1000	500	152	74	0.395835	0.065960
1000	500	152	75	0.464920	0.069085
1000	500	152	76	0.535080	0.070160
1000	500	153	49	0.000001	0.000001
1000	500	153	50	0.000002	0.000001
1000	500	153	51	0.000005	0.000003
1000	500	153	52	0.000011	0.000006
1000	500	153	53	0.000024	0.000013
1000	500	153	54	0.000051	0.000027
1000	500	153	55	0.000104	0.000054
1000	500	153	56	0.000208	0.000104
1000	500	153	57	0.000402	0.000194
1000	500	153	58	0.000752	0.000351
1000	500	153	59	0.001367	0.000615
1000	500	153	60	0.002411	0.001044
1000	500	153	61	0.004126	0.001716
1000	500	153	62	0.006859	0.002733
1000	500	153	63	0.011076	0.004217
1000	500	153	64	0.017381	0.006305
1000	500	153	65	0.026516	0.009136
1000	500	153	66	0.039346	0.012830

N	n	k	x	P(x)	p(x)
1000	500	153	67	0.056811	0.017464
1000	500	153	68	0.079856	0.023045
1000	500	153	69	0.109337	0.029481
1000	500	153	70	0.145902	0.036565
1000	500	153	71	0.189875	0.043972
1000	500	153	72	0.241149	0.051275
1000	500	153	73	0.299127	0.057978
1000	500	153	74	0.362699	0.063572
1000	500	153	75	0.430296	0.067597
1000	500	153	76	0.500000	0.069704
1000	500	154	49	0.000001	0.000000
1000	500	154	50	0.000001	0.000001
1000	500	154	51	0.000003	0.000002
1000	500	154	52	0.000008	0.000004
1000	500	154	53	0.000017	0.000009
1000	500	154	54	0.000037	0.000020
1000	500	154	55	0.000076	0.000040
1000	500	154	56	0.000154	0.000077
1000	500	154	57	0.000300	0.000147
1000	500	154	58	0.000569	0.000269
1000	500	154	59	0.001047	0.000478
1000	500	154	60	0.001868	0.000821
1000	500	154	61	0.003237	0.001369
1000	500	154	62	0.005446	0.002209
1000	500	154	63	0.008900	0.003454
1000	500	154	64	0.014135	0.005235
1000	500	154	65	0.021824	0.007689
1000	500	154	66	0.032772	0.010948
1000	500	154	67	0.047882	0.015110
1000	500	154	68	0.068102	0.020220
1000	500	154	69	0.094336	0.026234
1000	500	154	70	0.127339	0.033004
1000	500	154	71	0.167603	0.040264
1000	500	154	72	0.215240	0.047637
1000	500	154	73	0.269899	0.054659
1000	500	154	74	0.330725	0.060827
1000	500	154	75	0.396378	0.065652
1000	500	154	76	0.465107	0.068729
1000	500	154	77	0.534893	0.069786
1000	500	155	50	0.000001	0.000001
1000	500	155	51	0.000002	0.000001
1000	500	155	52	0.000005	0.000003
1000	500	155	53	0.000012	0.000007
1000	500	155	54	0.000026	0.000014
1000	500	155	55	0.000055	0.000029
1000	500	155	56	0.000113	0.000058
1000	500	155	57	0.000223	0.000111
1000	500	155	58	0.000429	0.000205
1000	500	155	59	0.000798	0.000369
1000	500	155	60	0.001441	0.000643

N	n	k	x	P(x)	p(x)
1000	500	155	61	0.002527	0.001086
1000	500	155	62	0.004302	0.001775
1000	500	155	63	0.007116	0.002814
1000	500	155	64	0.011437	0.004321
1000	500	155	65	0.017871	0.006434
1000	500	155	66	0.027156	0.009285
1000	500	155	67	0.040149	0.012993
1000	500	155	68	0.057777	0.017628
1000	500	155	69	0.080971	0.023194
1000	500	155	70	0.110564	0.029594
1000	500	155	71	0.147186	0.036621
1000	500	155	72	0.191140	0.043954
1000	500	155	73	0.242311	0.051171
1000	500	155	74	0.300096	0.057786
1000	500	155	75	0.363396	0.063300
1000	500	155	76	0.430661	0.067265
1000	500	155	77	0.500000	0.069339
1000	500	156	50	0.000001	0.000000
1000	500	156	51	0.000002	0.000001
1000	500	156	52	0.000004	0.000002
1000	500	156	53	0.000009	0.000005
1000	500	156	54	0.000019	0.000010
1000	500	156	55	0.000040	0.000021
1000	500	156	56	0.000083	0.000043
1000	500	156	57	0.000166	0.000083
1000	500	156	58	0.000321	0.000156
1000	500	156	59	0.000605	0.000284
1000	500	156	60	0.001106	0.000501
1000	500	156	61	0.001963	0.000857
1000	500	156	62	0.003382	0.001419
1000	500	156	63	0.005661	0.002279
1000	500	156	64	0.009207	0.003547
1000	500	156	65	0.014559	0.005351
1000	500	156	66	0.022387	0.007828
1000	500	156	67	0.033491	0.011104
1000	500	156	68	0.048765	0.015274
1000	500	156	69	0.069141	0.020376
1000	500	156	70	0.095505	0.026364
1000	500	156	71	0.128593	0.033088
1000	500	156	72	0.168877	0.040283
1000	500	156	73	0.216453	0.047576
1000	500	156	74	0.270964	0.054511
1000	500	156	75	0.331559	0.060595
1000	500	156	76	0.396909	0.065350
1000	500	156	77	0.465289	0.068380
1000	500	156	78	0.534710	0.069421
1000	500	157	51	0.000001	0.000001
1000	500	157	52	0.000003	0.000002
1000	500	157	53	0.000006	0.000003
1000	500	157	54	0.000013	0.000007

N	n	k	x	P(x)	p(x)
1000	500	157	55	0.000029	0.000015
1000	500	157	56	0.000060	0.000031
1000	500	157	57	0.000122	0.000062
1000	500	157	58	0.000240	0.000118
1000	500	157	59	0.000457	0.000217
1000	500	157	60	0.000845	0.000388
1000	500	157	61	0.001517	0.000672
1000	500	157	62	0.002645	0.001128
1000	500	157	63	0.004481	0.001835
1000	500	157	64	0.007375	0.002894
1000	500	157	65	0.011801	0.004425
1000	500	157	66	0.018361	0.006561
1000	500	157	67	0.027794	0.009433
1000	500	157	68	0.040947	0.013153
1000	500	157	69	0.058735	0.017789
1000	500	157	70	0.082073	0.023337
1000	500	157	71	0.111775	0.029702
1000	500	157	72	0.148448	0.036674
1000	500	157	73	0.192383	0.043934
1000	500	157	74	0.243450	0.051067
1000	500	157	75	0.301046	0.057596
1000	500	157	76	0.364079	0.063033
1000	500	157	77	0.431018	0.066939
1000	500	157	78	0.500000	0.068982
1000	500	158	51	0.000001	0.000000
1000	500	158	52	0.000002	0.000001
1000	500	158	53	0.000004	0.000002
1000	500	158	54	0.000009	0.000005
1000	500	158	55	0.000021	0.000011
1000	500	158	56	0.000044	0.000023
1000	500	158	57	0.000090	0.000046
1000	500	158	58	0.000178	0.000088
1000	500	158	59	0.000343	0.000165
1000	500	158	60	0.000642	0.000299
1000	500	158	61	0.001167	0.000525
1000	500	158	62	0.002059	0.000892
1000	500	158	63	0.003529	0.001470
1000	500	158	64	0.005878	0.002349
1000	500	158	65	0.009517	0.003639
1000	500	158	66	0.014984	0.005467
1000	500	158	67	0.022949	0.007965
1000	500	158	68	0.034207	0.011258
1000	500	158	69	0.049641	0.015434
1000	500	158	70	0.070169	0.020528
1000	500	158	71	0.096659	0.026491
1000	500	158	72	0.129829	0.033169
1000	500	158	73	0.170129	0.040300
1000	500	158	74	0.217644	0.047515
1000	500	158	75	0.272009	0.054365
1000	500	158	76	0.332376	0.060367

N	n	k	x	P(x)	p(x)
1000	500	158	77	0.397430	0.065054
1000	500	158	78	0.465468	0.068038
1000	500	158	79	0.534532	0.069063
1000	500	159	51	0.000001	0.000000
1000	500	159	52	0.000001	0.000001
1000	500	159	53	0.000003	0.000002
1000	500	159	54	0.000007	0.000004
1000	500	159	55	0.000015	0.000008
1000	500	159	56	0.000032	0.000017
1000	500	159	57	0.000065	0.000034
1000	500	159	58	0.000132	0.000066
1000	500	159	59	0.000257	0.000125
1000	500	159	60	0.000486	0.000229
1000	500	159	61	0.000894	0.000407
1000	500	159	62	0.001595	0.000702
1000	500	159	63	0.002766	0.001171
1000	500	159	64	0.004662	0.001896
1000	500	159	65	0.007637	0.002975
1000	500	159	66	0.012166	0.004529
1000	500	159	67	0.018853	0.006687
1000	500	159	68	0.028431	0.009578
1000	500	159	69	0.041740	0.013310
1000	500	159	70	0.059685	0.017945
1000	500	159	71	0.083162	0.023477
1000	500	159	72	0.112968	0.029806
1000	500	159	73	0.149692	0.036723
1000	500	159	74	0.193604	0.043913
1000	500	159	75	0.244568	0.050964
1000	500	159	76	0.301977	0.057409
1000	500	159	77	0.364748	0.062771
1000	500	159	78	0.431368	0.066620
1000	500	159	79	0.500000	0.068632
1000	500	160	52	0.000001	0.000001
1000	500	160	53	0.000002	0.000001
1000	500	160	54	0.000005	0.000003
1000	500	160	55	0.000010	0.000006
1000	500	160	56	0.000023	0.000012
1000	500	160	57	0.000048	0.000025
1000	500	160	58	0.000097	0.000049
1000	500	160	59	0.000191	0.000094
1000	500	160	60	0.000366	0.000175
1000	500	160	61	0.000681	0.000315
1000	500	160	62	0.001230	0.000549
1000	500	160	63	0.002158	0.000928
1000	500	160	64	0.003679	0.001521
1000	500	160	65	0.006098	0.002419
1000	500	160	66	0.009829	0.003731
1000	500	160	67	0.015410	0.005581
1000	500	160	68	0.023511	0.008101
1000	500	160	69	0.034919	0.011409

N	n	k	x	P(x)	p(x)
1000	500	160	70	0.050510	0.015591
1000	500	160	71	0.071187	0.020676
1000	500	160	72	0.097800	0.026613
1000	500	160	73	0.131046	0.033247
1000	500	160	74	0.171361	0.040315
1000	500	160	75	0.218814	0.047453
1000	500	160	76	0.273034	0.054220
1000	500	160	77	0.333176	0.060142
1000	500	160	78	0.397939	0.064763
1000	500	160	79	0.465643	0.067704
1000	500	160	80	0.534357	0.068713
1000	500	161	52	0.000001	0.000000
1000	500	161	53	0.000001	0.000001
1000	500	161	54	0.000003	0.000002
1000	500	161	55	0.000007	0.000004
1000	500	161	56	0.000016	0.000009
1000	500	161	57	0.000034	0.000018
1000	500	161	58	0.000071	0.000036
1000	500	161	59	0.000142	0.000071
1000	500	161	60	0.000274	0.000133
1000	500	161	61	0.000516	0.000242
1000	500	161	62	0.000944	0.000427
1000	500	161	63	0.001675	0.000731
1000	500	161	64	0.002890	0.001215
1000	500	161	65	0.004846	0.001956
1000	500	161	66	0.007901	0.003056
1000	500	161	67	0.012533	0.004632
1000	500	161	68	0.019345	0.006812
1000	500	161	69	0.029066	0.009721
1000	500	161	70	0.042529	0.013463
1000	500	161	71	0.060627	0.018098
1000	500	161	72	0.084240	0.023613
1000	500	161	73	0.114146	0.029906
1000	500	161	74	0.150916	0.036770
1000	500	161	75	0.194805	0.043889
1000	500	161	76	0.245666	0.050861
1000	500	161	77	0.302890	0.057225
1000	500	161	78	0.365404	0.062513
1000	500	161	79	0.431711	0.066307
1000	500	161	80	0.500000	0.068289
1000	500	162	53	0.000001	0.000001
1000	500	162	54	0.000002	0.000001
1000	500	162	55	0.000005	0.000003
1000	500	162	56	0.000012	0.000006
1000	500	162	57	0.000025	0.000013
1000	500	162	58	0.000052	0.000027
1000	500	162	59	0.000104	0.000053
1000	500	162	60	0.000205	0.000100
1000	500	162	61	0.000390	0.000185
1000	500	162	62	0.000721	0.000331

Table for $N = 1000$, $n = 500$

N	n	k	x	P(x)	p(x)
1000	500	162	63	0.001294	0.000573
1000	500	162	64	0.002258	0.000964
1000	500	162	65	0.003831	0.001573
1000	500	162	66	0.006321	0.002489
1000	500	162	67	0.010143	0.003822
1000	500	162	68	0.015837	0.005695
1000	500	162	69	0.024072	0.008235
1000	500	162	70	0.035629	0.011557
1000	500	162	71	0.051373	0.015744
1000	500	162	72	0.072194	0.020821
1000	500	162	73	0.098925	0.026731
1000	500	162	74	0.132246	0.033320
1000	500	162	75	0.172572	0.040327
1000	500	162	76	0.219962	0.047390
1000	500	162	77	0.274039	0.054077
1000	500	162	78	0.333961	0.059922
1000	500	162	79	0.398439	0.064478
1000	500	162	80	0.465815	0.067376
1000	500	162	81	0.534185	0.068371
1000	500	163	53	0.000001	0.000000
1000	500	163	54	0.000002	0.000001
1000	500	163	55	0.000004	0.000002
1000	500	163	56	0.000008	0.000005
1000	500	163	57	0.000018	0.000010
1000	500	163	58	0.000038	0.000020
1000	500	163	59	0.000077	0.000039
1000	500	163	60	0.000152	0.000075
1000	500	163	61	0.000293	0.000141
1000	500	163	62	0.000548	0.000255
1000	500	163	63	0.000995	0.000447
1000	500	163	64	0.001757	0.000762
1000	500	163	65	0.003015	0.001258
1000	500	163	66	0.005032	0.002017
1000	500	163	67	0.008168	0.003136
1000	500	163	68	0.012902	0.004734
1000	500	163	69	0.019837	0.006935
1000	500	163	70	0.029699	0.009862
1000	500	163	71	0.043313	0.013614
1000	500	163	72	0.061560	0.018247
1000	500	163	73	0.085304	0.023745
1000	500	163	74	0.115307	0.030003
1000	500	163	75	0.152120	0.036813
1000	500	163	76	0.195985	0.043864
1000	500	163	77	0.246743	0.050758
1000	500	163	78	0.303785	0.057043
1000	500	163	79	0.366046	0.062260
1000	500	163	80	0.432046	0.066000
1000	500	163	81	0.500000	0.067954
1000	500	164	54	0.000001	0.000001
1000	500	164	55	0.000003	0.000001

N	n	k	x	P(x)	p(x)
1000	500	164	56	0.000006	0.000003
1000	500	164	57	0.000013	0.000007
1000	500	164	58	0.000027	0.000014
1000	500	164	59	0.000056	0.000029
1000	500	164	60	0.000112	0.000056
1000	500	164	61	0.000219	0.000107
1000	500	164	62	0.000414	0.000195
1000	500	164	63	0.000761	0.000347
1000	500	164	64	0.001360	0.000598
1000	500	164	65	0.002361	0.001001
1000	500	164	66	0.003986	0.001625
1000	500	164	67	0.006545	0.002559
1000	500	164	68	0.010458	0.003913
1000	500	164	69	0.016266	0.005807
1000	500	164	70	0.024632	0.008367
1000	500	164	71	0.036335	0.011703
1000	500	164	72	0.052229	0.015894
1000	500	164	73	0.073191	0.020962
1000	500	164	74	0.100037	0.026846
1000	500	164	75	0.133428	0.033391
1000	500	164	76	0.173764	0.040336
1000	500	164	77	0.221091	0.047326
1000	500	164	78	0.275025	0.053935
1000	500	164	79	0.334730	0.059704
1000	500	164	80	0.398928	0.064198
1000	500	164	81	0.465982	0.067055
1000	500	164	82	0.534017	0.068035
1000	500	165	54	0.000001	0.000000
1000	500	165	55	0.000002	0.000001
1000	500	165	56	0.000004	0.000002
1000	500	165	57	0.000009	0.000005
1000	500	165	58	0.000020	0.000010
1000	500	165	59	0.000041	0.000021
1000	500	165	60	0.000083	0.000042
1000	500	165	61	0.000163	0.000080
1000	500	165	62	0.000312	0.000149
1000	500	165	63	0.000580	0.000268
1000	500	165	64	0.001048	0.000468
1000	500	165	65	0.001840	0.000792
1000	500	165	66	0.003142	0.001302
1000	500	165	67	0.005220	0.002077
1000	500	165	68	0.008436	0.003217
1000	500	165	69	0.013272	0.004835
1000	500	165	70	0.020329	0.007057
1000	500	165	71	0.030330	0.010001
1000	500	165	72	0.044092	0.013762
1000	500	165	73	0.062485	0.018392
1000	500	165	74	0.086357	0.023873
1000	500	165	75	0.116453	0.030096
1000	500	165	76	0.153307	0.036854

N	n	k	x	P(x)	p(x)
1000	500	165	77	0.197144	0.043838
1000	500	165	78	0.247800	0.050656
1000	500	165	79	0.304663	0.056863
1000	500	165	80	0.366675	0.062012
1000	500	165	81	0.432375	0.065700
1000	500	165	82	0.500000	0.067625
1000	500	166	54	0.000001	0.000000
1000	500	166	55	0.000001	0.000001
1000	500	166	56	0.000003	0.000002
1000	500	166	57	0.000006	0.000004
1000	500	166	58	0.000014	0.000008
1000	500	166	59	0.000030	0.000016
1000	500	166	60	0.000061	0.000031
1000	500	166	61	0.000121	0.000060
1000	500	166	62	0.000234	0.000113
1000	500	166	63	0.000440	0.000206
1000	500	166	64	0.000804	0.000364
1000	500	166	65	0.001427	0.000624
1000	500	166	66	0.002465	0.001038
1000	500	166	67	0.004142	0.001677
1000	500	166	68	0.006772	0.002630
1000	500	166	69	0.010776	0.004004
1000	500	166	70	0.016695	0.005919
1000	500	166	71	0.025192	0.008497
1000	500	166	72	0.037038	0.011846
1000	500	166	73	0.053079	0.016041
1000	500	166	74	0.074178	0.021099
1000	500	166	75	0.101135	0.026957
1000	500	166	76	0.134593	0.033458
1000	500	166	77	0.174937	0.040344
1000	500	166	78	0.222199	0.047262
1000	500	166	79	0.275993	0.053794
1000	500	166	80	0.335484	0.059490
1000	500	166	81	0.399407	0.063923
1000	500	166	82	0.466147	0.066740
1000	500	166	83	0.533853	0.067706
1000	500	167	55	0.000001	0.000000
1000	500	167	56	0.000002	0.000001
1000	500	167	57	0.000005	0.000003
1000	500	167	58	0.000010	0.000005
1000	500	167	59	0.000021	0.000011
1000	500	167	60	0.000044	0.000023
1000	500	167	61	0.000089	0.000045
1000	500	167	62	0.000175	0.000085
1000	500	167	63	0.000332	0.000157
1000	500	167	64	0.000613	0.000281
1000	500	167	65	0.001102	0.000489
1000	500	167	66	0.001925	0.000823
1000	500	167	67	0.003272	0.001347
1000	500	167	68	0.005410	0.002138

N	n	k	x	P(x)	p(x)
1000	500	167	69	0.008707	0.003297
1000	500	167	70	0.013643	0.004936
1000	500	167	71	0.020821	0.007178
1000	500	167	72	0.030959	0.010138
1000	500	167	73	0.044866	0.013908
1000	500	167	74	0.063401	0.018535
1000	500	167	75	0.087398	0.023997
1000	500	167	76	0.117583	0.030185
1000	500	167	77	0.154475	0.036892
1000	500	167	78	0.198285	0.043810
1000	500	167	79	0.248838	0.050554
1000	500	167	80	0.305524	0.056686
1000	500	167	81	0.367292	0.061768
1000	500	167	82	0.432697	0.065405
1000	500	167	83	0.500000	0.067303
1000	500	168	55	0.000001	0.000000
1000	500	168	56	0.000001	0.000001
1000	500	168	57	0.000003	0.000002
1000	500	168	58	0.000007	0.000004
1000	500	168	59	0.000015	0.000008
1000	500	168	60	0.000032	0.000017
1000	500	168	61	0.000066	0.000033
1000	500	168	62	0.000130	0.000064
1000	500	168	63	0.000249	0.000120
1000	500	168	64	0.000466	0.000217
1000	500	168	65	0.000847	0.000381
1000	500	168	66	0.001496	0.000649
1000	500	168	67	0.002572	0.001076
1000	500	168	68	0.004301	0.001729
1000	500	168	69	0.007001	0.002700
1000	500	168	70	0.011095	0.004094
1000	500	168	71	0.017124	0.006029
1000	500	168	72	0.025750	0.008625
1000	500	168	73	0.037737	0.011987
1000	500	168	74	0.053922	0.016185
1000	500	168	75	0.075154	0.021232
1000	500	168	76	0.102219	0.027064
1000	500	168	77	0.135741	0.033522
1000	500	168	78	0.176090	0.040349
1000	500	168	79	0.223288	0.047198
1000	500	168	80	0.276943	0.053655
1000	500	168	81	0.336223	0.059280
1000	500	168	82	0.399877	0.063654
1000	500	168	83	0.466308	0.066431
1000	500	168	84	0.533692	0.067384
1000	500	169	56	0.000001	0.000001
1000	500	169	57	0.000002	0.000001
1000	500	169	58	0.000005	0.000003
1000	500	169	59	0.000011	0.000006
1000	500	169	60	0.000023	0.000012

Table for $N = 1000$, $n = 500$

N	n	k	x	P(x)	p(x)
1000	500	169	61	0.000048	0.000025
1000	500	169	62	0.000096	0.000048
1000	500	169	63	0.000186	0.000091
1000	500	169	64	0.000353	0.000166
1000	500	169	65	0.000648	0.000295
1000	500	169	66	0.001158	0.000510
1000	500	169	67	0.002012	0.000855
1000	500	169	68	0.003403	0.001391
1000	500	169	69	0.005602	0.002199
1000	500	169	70	0.008979	0.003377
1000	500	169	71	0.014015	0.005036
1000	500	169	72	0.021313	0.007297
1000	500	169	73	0.031585	0.010272
1000	500	169	74	0.045635	0.014050
1000	500	169	75	0.064309	0.018674
1000	500	169	76	0.088426	0.024118
1000	500	169	77	0.118698	0.030271
1000	500	169	78	0.155625	0.036927
1000	500	169	79	0.199406	0.043781
1000	500	169	80	0.249858	0.050452
1000	500	169	81	0.306369	0.056511
1000	500	169	82	0.367897	0.061527
1000	500	169	83	0.433013	0.065116
1000	500	169	84	0.500000	0.066987
1000	500	170	56	0.000001	0.000000
1000	500	170	57	0.000002	0.000001
1000	500	170	58	0.000004	0.000002
1000	500	170	59	0.000008	0.000004
1000	500	170	60	0.000017	0.000009
1000	500	170	61	0.000035	0.000018
1000	500	170	62	0.000071	0.000036
1000	500	170	63	0.000139	0.000068
1000	500	170	64	0.000265	0.000127
1000	500	170	65	0.000493	0.000228
1000	500	170	66	0.000891	0.000398
1000	500	170	67	0.001567	0.000676
1000	500	170	68	0.002680	0.001113
1000	500	170	69	0.004462	0.001782
1000	500	170	70	0.007232	0.002770
1000	500	170	71	0.011415	0.004184
1000	500	170	72	0.017554	0.006139
1000	500	170	73	0.026307	0.008752
1000	500	170	74	0.038432	0.012126
1000	500	170	75	0.054758	0.016326
1000	500	170	76	0.076121	0.021362
1000	500	170	77	0.103289	0.027168
1000	500	170	78	0.136872	0.033584
1000	500	170	79	0.177225	0.040353
1000	500	170	80	0.224359	0.047133
1000	500	170	81	0.277876	0.053517
1000	500	170	82	0.336948	0.059072
1000	500	170	83	0.400337	0.063389
1000	500	170	84	0.466466	0.066129
1000	500	170	85	0.533534	0.067068
1000	500	171	57	0.000001	0.000001
1000	500	171	58	0.000002	0.000001
1000	500	171	59	0.000006	0.000003
1000	500	171	60	0.000012	0.000006
1000	500	171	61	0.000025	0.000013
1000	500	171	62	0.000052	0.000026
1000	500	171	63	0.000103	0.000051
1000	500	171	64	0.000199	0.000096
1000	500	171	65	0.000374	0.000175
1000	500	171	66	0.000683	0.000309
1000	500	171	67	0.001214	0.000531
1000	500	171	68	0.002101	0.000886
1000	500	171	69	0.003537	0.001436
1000	500	171	70	0.005797	0.002260
1000	500	171	71	0.009253	0.003456
1000	500	171	72	0.014389	0.005136
1000	500	171	73	0.021804	0.007415
1000	500	171	74	0.032209	0.010405
1000	500	171	75	0.046399	0.014190
1000	500	171	76	0.065208	0.018809
1000	500	171	77	0.089443	0.024235
1000	500	171	78	0.119797	0.030354
1000	500	171	79	0.156757	0.036960
1000	500	171	80	0.200508	0.043751
1000	500	171	81	0.250859	0.050351
1000	500	171	82	0.307198	0.056339
1000	500	171	83	0.368490	0.061292
1000	500	171	84	0.433322	0.064832
1000	500	171	85	0.500000	0.066678
1000	500	172	57	0.000001	0.000000
1000	500	172	58	0.000002	0.000001
1000	500	172	59	0.000004	0.000002
1000	500	172	60	0.000009	0.000005
1000	500	172	61	0.000018	0.000010
1000	500	172	62	0.000038	0.000020
1000	500	172	63	0.000076	0.000038
1000	500	172	64	0.000148	0.000072
1000	500	172	65	0.000282	0.000134
1000	500	172	66	0.000521	0.000239
1000	500	172	67	0.000937	0.000416
1000	500	172	68	0.001639	0.000702
1000	500	172	69	0.002790	0.001151
1000	500	172	70	0.004624	0.001834
1000	500	172	71	0.007464	0.002840
1000	500	172	72	0.011737	0.004273
1000	500	172	73	0.017984	0.006247
1000	500	172	74	0.026862	0.008878
1000	500	172	75	0.039124	0.012262
1000	500	172	76	0.055588	0.016464
1000	500	172	77	0.077077	0.021489
1000	500	172	78	0.104345	0.027269
1000	500	172	79	0.137987	0.033642
1000	500	172	80	0.178342	0.040355
1000	500	172	81	0.225410	0.047068
1000	500	172	82	0.278791	0.053381
1000	500	172	83	0.337659	0.058868
1000	500	172	84	0.400788	0.063129
1000	500	172	85	0.466621	0.065832
1000	500	172	86	0.533379	0.066759
1000	500	173	57	0.000001	0.000000
1000	500	173	58	0.000001	0.000001
1000	500	173	59	0.000003	0.000002
1000	500	173	60	0.000006	0.000003
1000	500	173	61	0.000013	0.000007
1000	500	173	62	0.000027	0.000014
1000	500	173	63	0.000056	0.000028
1000	500	173	64	0.000110	0.000054
1000	500	173	65	0.000212	0.000102
1000	500	173	66	0.000396	0.000184
1000	500	173	67	0.000720	0.000324
1000	500	173	68	0.001273	0.000553
1000	500	173	69	0.002191	0.000918
1000	500	173	70	0.003672	0.001481
1000	500	173	71	0.005993	0.002321
1000	500	173	72	0.009529	0.003536
1000	500	173	73	0.014763	0.005234
1000	500	173	74	0.022295	0.007532
1000	500	173	75	0.032830	0.010536
1000	500	173	76	0.047157	0.014327
1000	500	173	77	0.066099	0.018942
1000	500	173	78	0.090448	0.024349
1000	500	173	79	0.120882	0.030434
1000	500	173	80	0.157872	0.036990
1000	500	173	81	0.201592	0.043720
1000	500	173	82	0.251843	0.050251
1000	500	173	83	0.308012	0.056169
1000	500	173	84	0.369071	0.061060
1000	500	173	85	0.433625	0.064554
1000	500	173	86	0.500000	0.066375
1000	500	174	58	0.000001	0.000000
1000	500	174	59	0.000002	0.000001
1000	500	174	60	0.000004	0.000002
1000	500	174	61	0.000009	0.000005
1000	500	174	62	0.000020	0.000010
1000	500	174	63	0.000041	0.000021
1000	500	174	64	0.000082	0.000041
1000	500	174	65	0.000158	0.000077
1000	500	174	66	0.000299	0.000141
1000	500	174	67	0.000550	0.000251
1000	500	174	68	0.000984	0.000434
1000	500	174	69	0.001712	0.000729
1000	500	174	70	0.002902	0.001189
1000	500	174	71	0.004789	0.001887
1000	500	174	72	0.007699	0.002910
1000	500	174	73	0.012060	0.004361
1000	500	174	74	0.018415	0.006355
1000	500	174	75	0.027416	0.009001
1000	500	174	76	0.039812	0.012396
1000	500	174	77	0.056410	0.016599
1000	500	174	78	0.078023	0.021613
1000	500	174	79	0.105389	0.027366
1000	500	174	80	0.139087	0.033698
1000	500	174	81	0.179442	0.040355
1000	500	174	82	0.226444	0.047002
1000	500	174	83	0.279689	0.053245
1000	500	174	84	0.338357	0.058667
1000	500	174	85	0.401231	0.062874
1000	500	174	86	0.466772	0.065541
1000	500	174	87	0.533227	0.066455
1000	500	175	58	0.000001	0.000000
1000	500	175	59	0.000001	0.000001
1000	500	175	60	0.000003	0.000002
1000	500	175	61	0.000007	0.000004
1000	500	175	62	0.000014	0.000008
1000	500	175	63	0.000030	0.000015
1000	500	175	64	0.000060	0.000030
1000	500	175	65	0.000118	0.000058
1000	500	175	66	0.000225	0.000107
1000	500	175	67	0.000419	0.000193
1000	500	175	68	0.000757	0.000338
1000	500	175	69	0.001332	0.000575
1000	500	175	70	0.002283	0.000950
1000	500	175	71	0.003809	0.001526
1000	500	175	72	0.006191	0.002382
1000	500	175	73	0.009806	0.003615
1000	500	175	74	0.015138	0.005332
1000	500	175	75	0.022785	0.007647
1000	500	175	76	0.033449	0.010664
1000	500	175	77	0.047910	0.014461
1000	500	175	78	0.066981	0.019071
1000	500	175	79	0.091441	0.024460
1000	500	175	80	0.121952	0.030511
1000	500	175	81	0.158971	0.037019
1000	500	175	82	0.202658	0.043687
1000	500	175	83	0.252809	0.050151
1000	500	175	84	0.308810	0.056001

N	n	k	x	P(x)	p(x)
1000	500	175	85	0.369642	0.060832
1000	500	175	86	0.433922	0.064281
1000	500	175	87	0.500000	0.066078
1000	500	176	59	0.000001	0.000001
1000	500	176	60	0.000002	0.000001
1000	500	176	61	0.000005	0.000003
1000	500	176	62	0.000010	0.000006
1000	500	176	63	0.000022	0.000011
1000	500	176	64	0.000044	0.000022
1000	500	176	65	0.000087	0.000043
1000	500	176	66	0.000169	0.000081
1000	500	176	67	0.000317	0.000148
1000	500	176	68	0.000580	0.000263
1000	500	176	69	0.001032	0.000452
1000	500	176	70	0.001787	0.000756
1000	500	176	71	0.003015	0.001228
1000	500	176	72	0.004955	0.001940
1000	500	176	73	0.007935	0.002980
1000	500	176	74	0.012384	0.004450
1000	500	176	75	0.018845	0.006461
1000	500	176	76	0.027968	0.009123
1000	500	176	77	0.040495	0.012527
1000	500	176	78	0.057226	0.016731
1000	500	176	79	0.078959	0.021733
1000	500	176	80	0.106419	0.027460
1000	500	176	81	0.140170	0.033751
1000	500	176	82	0.180523	0.040353
1000	500	176	83	0.227460	0.046937
1000	500	176	84	0.280572	0.053112
1000	500	176	85	0.339042	0.058470
1000	500	176	86	0.401665	0.062624
1000	500	176	87	0.466921	0.065256
1000	500	176	88	0.533079	0.066158
1000	500	177	59	0.000001	0.000000
1000	500	177	60	0.000001	0.000001
1000	500	177	61	0.000003	0.000002
1000	500	177	62	0.000007	0.000004
1000	500	177	63	0.000016	0.000008
1000	500	177	64	0.000032	0.000017
1000	500	177	65	0.000065	0.000032
1000	500	177	66	0.000126	0.000061
1000	500	177	67	0.000239	0.000113
1000	500	177	68	0.000442	0.000203
1000	500	177	69	0.000795	0.000353
1000	500	177	70	0.001393	0.000598
1000	500	177	71	0.002376	0.000983
1000	500	177	72	0.003947	0.001572
1000	500	177	73	0.006391	0.002443
1000	500	177	74	0.010084	0.003693
1000	500	177	75	0.015513	0.005429

N	n	k	x	P(x)	p(x)
1000	500	177	76	0.023274	0.007761
1000	500	177	77	0.034064	0.010791
1000	500	177	78	0.048658	0.014593
1000	500	177	79	0.067855	0.019198
1000	500	177	80	0.092422	0.024567
1000	500	177	81	0.123008	0.030585
1000	500	177	82	0.160053	0.037045
1000	500	177	83	0.203707	0.043654
1000	500	177	84	0.253758	0.050051
1000	500	177	85	0.309594	0.055836
1000	500	177	86	0.370201	0.060608
1000	500	177	87	0.434214	0.064013
1000	500	177	88	0.500000	0.065786
1000	500	178	60	0.000001	0.000001
1000	500	178	61	0.000002	0.000001
1000	500	178	62	0.000005	0.000003
1000	500	178	63	0.000011	0.000006
1000	500	178	64	0.000023	0.000012
1000	500	178	65	0.000047	0.000024
1000	500	178	66	0.000094	0.000046
1000	500	178	67	0.000180	0.000086
1000	500	178	68	0.000336	0.000156
1000	500	178	69	0.000610	0.000275
1000	500	178	70	0.001081	0.000470
1000	500	178	71	0.001863	0.000783
1000	500	178	72	0.003130	0.001267
1000	500	178	73	0.005123	0.001993
1000	500	178	74	0.008172	0.003049
1000	500	178	75	0.012709	0.004537
1000	500	178	76	0.019275	0.006566
1000	500	178	77	0.028518	0.009243
1000	500	178	78	0.041175	0.012657
1000	500	178	79	0.058035	0.016860
1000	500	178	80	0.079885	0.021850
1000	500	178	81	0.107436	0.027551
1000	500	178	82	0.141238	0.033801
1000	500	178	83	0.181588	0.040350
1000	500	178	84	0.228459	0.046871
1000	500	178	85	0.281439	0.052980
1000	500	178	86	0.339714	0.058275
1000	500	178	87	0.402091	0.062377
1000	500	178	88	0.467067	0.064976
1000	500	178	89	0.532933	0.065866
1000	500	179	60	0.000001	0.000000
1000	500	179	61	0.000002	0.000001
1000	500	179	62	0.000004	0.000002
1000	500	179	63	0.000008	0.000004
1000	500	179	64	0.000017	0.000009
1000	500	179	65	0.000035	0.000018
1000	500	179	66	0.000069	0.000034

N	n	k	x	P(x)	p(x)
1000	500	179	67	0.000134	0.000065
1000	500	179	68	0.000254	0.000119
1000	500	179	69	0.000466	0.000213
1000	500	179	70	0.000835	0.000368
1000	500	179	71	0.001455	0.000620
1000	500	179	72	0.002471	0.001016
1000	500	179	73	0.004088	0.001617
1000	500	179	74	0.006592	0.002504
1000	500	179	75	0.010364	0.003771
1000	500	179	76	0.015889	0.005525
1000	500	179	77	0.023762	0.007873
1000	500	179	78	0.034677	0.010915
1000	500	179	79	0.049400	0.014722
1000	500	179	80	0.068721	0.019321
1000	500	179	81	0.093393	0.024672
1000	500	179	82	0.124049	0.030657
1000	500	179	83	0.161118	0.037069
1000	500	179	84	0.204738	0.043620
1000	500	179	85	0.254691	0.049953
1000	500	179	86	0.310363	0.055672
1000	500	179	87	0.370750	0.060387
1000	500	179	88	0.434500	0.063749
1000	500	179	89	0.500000	0.065500
1000	500	180	61	0.000001	0.000001
1000	500	180	62	0.000003	0.000001
1000	500	180	63	0.000006	0.000003
1000	500	180	64	0.000012	0.000006
1000	500	180	65	0.000025	0.000013
1000	500	180	66	0.000051	0.000026
1000	500	180	67	0.000100	0.000049
1000	500	180	68	0.000191	0.000091
1000	500	180	69	0.000355	0.000164
1000	500	180	70	0.000642	0.000287
1000	500	180	71	0.001131	0.000489
1000	500	180	72	0.001941	0.000810
1000	500	180	73	0.003247	0.001306
1000	500	180	74	0.005293	0.002046
1000	500	180	75	0.008411	0.003119
1000	500	180	76	0.013035	0.004624
1000	500	180	77	0.019705	0.006670
1000	500	180	78	0.029067	0.009361
1000	500	180	79	0.041850	0.012783
1000	500	180	80	0.058837	0.016986
1000	500	180	81	0.080801	0.021965
1000	500	180	82	0.108441	0.027640
1000	500	180	83	0.142290	0.033849
1000	500	180	84	0.182636	0.040346
1000	500	180	85	0.229441	0.046805
1000	500	180	86	0.282290	0.052849
1000	500	180	87	0.340373	0.058083

N	n	k	x	P(x)	p(x)
1000	500	180	88	0.402509	0.062136
1000	500	180	89	0.467210	0.064701
1000	500	180	90	0.532790	0.065580
1000	500	181	61	0.000001	0.000000
1000	500	181	62	0.000002	0.000001
1000	500	181	63	0.000004	0.000002
1000	500	181	64	0.000009	0.000005
1000	500	181	65	0.000018	0.000010
1000	500	181	66	0.000037	0.000019
1000	500	181	67	0.000074	0.000037
1000	500	181	68	0.000143	0.000069
1000	500	181	69	0.000269	0.000126
1000	500	181	70	0.000491	0.000223
1000	500	181	71	0.000875	0.000384
1000	500	181	72	0.001518	0.000643
1000	500	181	73	0.002567	0.001049
1000	500	181	74	0.004230	0.001663
1000	500	181	75	0.006795	0.002565
1000	500	181	76	0.010644	0.003849
1000	500	181	77	0.016265	0.005620
1000	500	181	78	0.024249	0.007984
1000	500	181	79	0.035287	0.011038
1000	500	181	80	0.050136	0.014849
1000	500	181	81	0.069578	0.019442
1000	500	181	82	0.094351	0.024774
1000	500	181	83	0.125077	0.030725
1000	500	181	84	0.162168	0.037091
1000	500	181	85	0.205753	0.043585
1000	500	181	86	0.255607	0.049854
1000	500	181	87	0.311119	0.055511
1000	500	181	88	0.371289	0.060170
1000	500	181	89	0.434780	0.063491
1000	500	181	90	0.500000	0.065219
1000	500	182	61	0.000001	0.000000
1000	500	182	62	0.000001	0.000001
1000	500	182	63	0.000003	0.000002
1000	500	182	64	0.000006	0.000003
1000	500	182	65	0.000013	0.000007
1000	500	182	66	0.000027	0.000014
1000	500	182	67	0.000055	0.000027
1000	500	182	68	0.000107	0.000052
1000	500	182	69	0.000202	0.000096
1000	500	182	70	0.000374	0.000172
1000	500	182	71	0.000674	0.000300
1000	500	182	72	0.001182	0.000508
1000	500	182	73	0.002020	0.000838
1000	500	182	74	0.003365	0.001345
1000	500	182	75	0.005464	0.002099
1000	500	182	76	0.008651	0.003188
1000	500	182	77	0.013362	0.004710

N	n	k	x	P(x)	p(x)
1000	500	182	78	0.020135	0.006773
1000	500	182	79	0.029613	0.009478
1000	500	182	80	0.042521	0.012908
1000	500	182	81	0.059632	0.017110
1000	500	182	82	0.081707	0.022076
1000	500	182	83	0.109433	0.027725
1000	500	182	84	0.143328	0.033895
1000	500	182	85	0.183668	0.040340
1000	500	182	86	0.230406	0.046739
1000	500	182	87	0.283126	0.052720
1000	500	182	88	0.341020	0.057894
1000	500	182	89	0.402919	0.061898
1000	500	182	90	0.467350	0.064432
1000	500	182	91	0.532649	0.065299
1000	500	183	62	0.000001	0.000001
1000	500	183	63	0.000002	0.000001
1000	500	183	64	0.000004	0.000002
1000	500	183	65	0.000010	0.000005
1000	500	183	66	0.000020	0.000010
1000	500	183	67	0.000040	0.000020
1000	500	183	68	0.000079	0.000039
1000	500	183	69	0.000152	0.000073
1000	500	183	70	0.000284	0.000132
1000	500	183	71	0.000517	0.000233
1000	500	183	72	0.000916	0.000399
1000	500	183	73	0.001583	0.000666
1000	500	183	74	0.002664	0.001082
1000	500	183	75	0.004373	0.001709
1000	500	183	76	0.006999	0.002626
1000	500	183	77	0.010926	0.003927
1000	500	183	78	0.016641	0.005715
1000	500	183	79	0.024735	0.008094
1000	500	183	80	0.035893	0.011159
1000	500	183	81	0.050867	0.014974
1000	500	183	82	0.070427	0.019559
1000	500	183	83	0.095299	0.024872
1000	500	183	84	0.126090	0.030791
1000	500	183	85	0.163202	0.037112
1000	500	183	86	0.206751	0.043549
1000	500	183	87	0.256508	0.049757
1000	500	183	88	0.311861	0.055353
1000	500	183	89	0.371818	0.059957
1000	500	183	90	0.435056	0.063238
1000	500	183	91	0.500000	0.064944
1000	500	184	62	0.000001	0.000000
1000	500	184	63	0.000001	0.000001
1000	500	184	64	0.000003	0.000002
1000	500	184	65	0.000007	0.000004
1000	500	184	66	0.000014	0.000008
1000	500	184	67	0.000029	0.000015

N	n	k	x	P(x)	p(x)
1000	500	184	68	0.000059	0.000029
1000	500	184	69	0.000114	0.000055
1000	500	184	70	0.000214	0.000101
1000	500	184	71	0.000395	0.000180
1000	500	184	72	0.000707	0.000312
1000	500	184	73	0.001234	0.000527
1000	500	184	74	0.002100	0.000866
1000	500	184	75	0.003484	0.001384
1000	500	184	76	0.005636	0.002152
1000	500	184	77	0.008893	0.003256
1000	500	184	78	0.013689	0.004796
1000	500	184	79	0.020564	0.006875
1000	500	184	80	0.030157	0.009593
1000	500	184	81	0.043188	0.013031
1000	500	184	82	0.060419	0.017232
1000	500	184	83	0.082604	0.022185
1000	500	184	84	0.110412	0.027808
1000	500	184	85	0.144351	0.033939
1000	500	184	86	0.184684	0.040333
1000	500	184	87	0.231356	0.046672
1000	500	184	88	0.283947	0.052592
1000	500	184	89	0.341656	0.057709
1000	500	184	90	0.403321	0.061665
1000	500	184	91	0.467488	0.064167
1000	500	184	92	0.532512	0.065024
1000	500	185	63	0.000001	0.000001
1000	500	185	64	0.000002	0.000001
1000	500	185	65	0.000005	0.000003
1000	500	185	66	0.000010	0.000005
1000	500	185	67	0.000021	0.000011
1000	500	185	68	0.000043	0.000022
1000	500	185	69	0.000084	0.000041
1000	500	185	70	0.000161	0.000077
1000	500	185	71	0.000300	0.000139
1000	500	185	72	0.000543	0.000243
1000	500	185	73	0.000958	0.000415
1000	500	185	74	0.001648	0.000690
1000	500	185	75	0.002764	0.001115
1000	500	185	76	0.004518	0.001755
1000	500	185	77	0.007205	0.002687
1000	500	185	78	0.011209	0.004004
1000	500	185	79	0.017017	0.005808
1000	500	185	80	0.025219	0.008203
1000	500	185	81	0.036497	0.011277
1000	500	185	82	0.051593	0.015096
1000	500	185	83	0.071267	0.019675
1000	500	185	84	0.096235	0.024968
1000	500	185	85	0.127090	0.030855
1000	500	185	86	0.164221	0.037130
1000	500	185	87	0.207734	0.043513

N	n	k	x	P(x)	p(x)
1000	500	185	88	0.257394	0.049660
1000	500	185	89	0.312590	0.055196
1000	500	185	90	0.372337	0.059748
1000	500	185	91	0.435326	0.062989
1000	500	185	92	0.500000	0.064674
1000	500	186	63	0.000001	0.000000
1000	500	186	64	0.000002	0.000001
1000	500	186	65	0.000003	0.000002
1000	500	186	66	0.000007	0.000004
1000	500	186	67	0.000016	0.000008
1000	500	186	68	0.000032	0.000016
1000	500	186	69	0.000063	0.000031
1000	500	186	70	0.000121	0.000058
1000	500	186	71	0.000227	0.000106
1000	500	186	72	0.000415	0.000188
1000	500	186	73	0.000741	0.000325
1000	500	186	74	0.001288	0.000547
1000	500	186	75	0.002182	0.000894
1000	500	186	76	0.003605	0.001423
1000	500	186	77	0.005810	0.002205
1000	500	186	78	0.009135	0.003325
1000	500	186	79	0.014017	0.004881
1000	500	186	80	0.020992	0.006976
1000	500	186	81	0.030699	0.009707
1000	500	186	82	0.043850	0.013151
1000	500	186	83	0.061200	0.017350
1000	500	186	84	0.083491	0.022290
1000	500	186	85	0.111379	0.027888
1000	500	186	86	0.145359	0.033980
1000	500	186	87	0.185684	0.040324
1000	500	186	88	0.232290	0.046606
1000	500	186	89	0.284755	0.052465
1000	500	186	90	0.342280	0.057525
1000	500	186	91	0.403716	0.061435
1000	500	186	92	0.467623	0.063908
1000	500	186	93	0.532377	0.064753
1000	500	187	64	0.000001	0.000001
1000	500	187	65	0.000002	0.000001
1000	500	187	66	0.000005	0.000003
1000	500	187	67	0.000011	0.000006
1000	500	187	68	0.000023	0.000012
1000	500	187	69	0.000046	0.000023
1000	500	187	70	0.000090	0.000044
1000	500	187	71	0.000171	0.000081
1000	500	187	72	0.000316	0.000145
1000	500	187	73	0.000570	0.000254
1000	500	187	74	0.001001	0.000431
1000	500	187	75	0.001715	0.000714
1000	500	187	76	0.002864	0.001149
1000	500	187	77	0.004664	0.001801

N	n	k	x	P(x)	p(x)
1000	500	187	78	0.007412	0.002747
1000	500	187	79	0.011492	0.004080
1000	500	187	80	0.017393	0.005901
1000	500	187	81	0.025703	0.008310
1000	500	187	82	0.037097	0.011394
1000	500	187	83	0.052312	0.015215
1000	500	187	84	0.072099	0.019787
1000	500	187	85	0.097161	0.025062
1000	500	187	86	0.128077	0.030916
1000	500	187	87	0.165224	0.037147
1000	500	187	88	0.208700	0.043476
1000	500	187	89	0.258264	0.049564
1000	500	187	90	0.313305	0.055041
1000	500	187	91	0.372847	0.059542
1000	500	187	92	0.435591	0.062744
1000	500	187	93	0.500000	0.064409
1000	500	188	64	0.000001	0.000000
1000	500	188	65	0.000002	0.000001
1000	500	188	66	0.000004	0.000002
1000	500	188	67	0.000008	0.000004
1000	500	188	68	0.000017	0.000009
1000	500	188	69	0.000034	0.000017
1000	500	188	70	0.000067	0.000033
1000	500	188	71	0.000128	0.000061
1000	500	188	72	0.000240	0.000111
1000	500	188	73	0.000437	0.000197
1000	500	188	74	0.000775	0.000339
1000	500	188	75	0.001342	0.000567
1000	500	188	76	0.002265	0.000923
1000	500	188	77	0.003728	0.001463
1000	500	188	78	0.005986	0.002258
1000	500	188	79	0.009379	0.003393
1000	500	188	80	0.014345	0.004965
1000	500	188	81	0.021420	0.007076
1000	500	188	82	0.031239	0.009819
1000	500	188	83	0.044508	0.013269
1000	500	188	84	0.061974	0.017466
1000	500	188	85	0.084368	0.022393
1000	500	188	86	0.112334	0.027966
1000	500	188	87	0.146353	0.034020
1000	500	188	88	0.186668	0.040315
1000	500	188	89	0.233208	0.046540
1000	500	188	90	0.285548	0.052340
1000	500	188	91	0.342893	0.057345
1000	500	188	92	0.404103	0.061210
1000	500	188	93	0.467756	0.063653
1000	500	188	94	0.532244	0.064488
1000	500	189	64	0.000001	0.000000
1000	500	189	65	0.000001	0.000001
1000	500	189	66	0.000003	0.000001

Table for $N = 1000$, $n = 500$

N	n	k	x	P(x)	p(x)
1000	500	189	67	0.000006	0.000003
1000	500	189	68	0.000012	0.000006
1000	500	189	69	0.000025	0.000013
1000	500	189	70	0.000049	0.000025
1000	500	189	71	0.000096	0.000046
1000	500	189	72	0.000181	0.000085
1000	500	189	73	0.000333	0.000152
1000	500	189	74	0.000598	0.000265
1000	500	189	75	0.001045	0.000448
1000	500	189	76	0.001783	0.000737
1000	500	189	77	0.002966	0.001183
1000	500	189	78	0.004812	0.001847
1000	500	189	79	0.007620	0.002807
1000	500	189	80	0.011776	0.004157
1000	500	189	81	0.017769	0.005993
1000	500	189	82	0.026184	0.008415
1000	500	189	83	0.037694	0.011509
1000	500	189	84	0.053026	0.015333
1000	500	189	85	0.072923	0.019897
1000	500	189	86	0.098075	0.025152
1000	500	189	87	0.129050	0.030975
1000	500	189	88	0.166213	0.037163
1000	500	189	89	0.209651	0.043439
1000	500	189	90	0.259120	0.049469
1000	500	189	91	0.314009	0.054889
1000	500	189	92	0.373347	0.059339
1000	500	189	93	0.435851	0.062504
1000	500	189	94	0.500000	0.064149
1000	500	190	65	0.000001	0.000000
1000	500	190	66	0.000002	0.000001
1000	500	190	67	0.000004	0.000002
1000	500	190	68	0.000009	0.000005
1000	500	190	69	0.000018	0.000009
1000	500	190	70	0.000036	0.000018
1000	500	190	71	0.000071	0.000035
1000	500	190	72	0.000136	0.000065
1000	500	190	73	0.000253	0.000117
1000	500	190	74	0.000459	0.000206
1000	500	190	75	0.000811	0.000352
1000	500	190	76	0.001397	0.000587
1000	500	190	77	0.002349	0.000951
1000	500	190	78	0.003851	0.001503
1000	500	190	79	0.006162	0.002311
1000	500	190	80	0.009624	0.003461
1000	500	190	81	0.014673	0.005049
1000	500	190	82	0.021847	0.007174
1000	500	190	83	0.031776	0.009929
1000	500	190	84	0.045161	0.013386
1000	500	190	85	0.062741	0.017580
1000	500	190	86	0.085235	0.022494

N	n	k	x	P(x)	p(x)
1000	500	190	87	0.113276	0.028041
1000	500	190	88	0.147334	0.034057
1000	500	190	89	0.187638	0.040304
1000	500	190	90	0.234111	0.046474
1000	500	190	91	0.286327	0.052216
1000	500	190	92	0.343495	0.057168
1000	500	190	93	0.404484	0.060989
1000	500	190	94	0.467886	0.063402
1000	500	190	95	0.532114	0.064228
1000	500	191	65	0.000001	0.000000
1000	500	191	66	0.000001	0.000001
1000	500	191	67	0.000003	0.000002
1000	500	191	68	0.000006	0.000003
1000	500	191	69	0.000013	0.000007
1000	500	191	70	0.000027	0.000014
1000	500	191	71	0.000053	0.000026
1000	500	191	72	0.000102	0.000049
1000	500	191	73	0.000191	0.000090
1000	500	191	74	0.000351	0.000159
1000	500	191	75	0.000626	0.000276
1000	500	191	76	0.001090	0.000464
1000	500	191	77	0.001852	0.000762
1000	500	191	78	0.003068	0.001217
1000	500	191	79	0.004961	0.001893
1000	500	191	80	0.007829	0.002868
1000	500	191	81	0.012061	0.004232
1000	500	191	82	0.018145	0.006084
1000	500	191	83	0.026664	0.008519
1000	500	191	84	0.038287	0.011623
1000	500	191	85	0.053734	0.015448
1000	500	191	86	0.073739	0.020004
1000	500	191	87	0.098979	0.025240
1000	500	191	88	0.130011	0.031032
1000	500	191	89	0.167187	0.037176
1000	500	191	90	0.210588	0.043400
1000	500	191	91	0.259962	0.049374
1000	500	191	92	0.314700	0.054738
1000	500	191	93	0.373839	0.059139
1000	500	191	94	0.436107	0.062268
1000	500	191	95	0.500000	0.063893
1000	500	192	66	0.000001	0.000001
1000	500	192	67	0.000002	0.000001
1000	500	192	68	0.000004	0.000002
1000	500	192	69	0.000009	0.000005
1000	500	192	70	0.000019	0.000010
1000	500	192	71	0.000039	0.000019
1000	500	192	72	0.000076	0.000037
1000	500	192	73	0.000144	0.000068
1000	500	192	74	0.000267	0.000123
1000	500	192	75	0.000481	0.000215

N	n	k	x	P(x)	p(x)
1000	500	192	76	0.000847	0.000366
1000	500	192	77	0.001454	0.000607
1000	500	192	78	0.002434	0.000980
1000	500	192	79	0.003976	0.001542
1000	500	192	80	0.006340	0.002364
1000	500	192	81	0.009869	0.003529
1000	500	192	82	0.015001	0.005132
1000	500	192	83	0.022273	0.007272
1000	500	192	84	0.032310	0.010037
1000	500	192	85	0.045810	0.013500
1000	500	192	86	0.063502	0.017692
1000	500	192	87	0.086094	0.022592
1000	500	192	88	0.114207	0.028114
1000	500	192	89	0.148300	0.034093
1000	500	192	90	0.188592	0.040292
1000	500	192	91	0.235000	0.046408
1000	500	192	92	0.287094	0.052094
1000	500	192	93	0.344086	0.056993
1000	500	192	94	0.404857	0.060771
1000	500	192	95	0.468014	0.063157
1000	500	192	96	0.531986	0.063972
1000	500	193	66	0.000001	0.000000
1000	500	193	67	0.000001	0.000001
1000	500	193	68	0.000003	0.000002
1000	500	193	69	0.000007	0.000004
1000	500	193	70	0.000014	0.000007
1000	500	193	71	0.000029	0.000014
1000	500	193	72	0.000056	0.000028
1000	500	193	73	0.000108	0.000052
1000	500	193	74	0.000202	0.000094
1000	500	193	75	0.000368	0.000166
1000	500	193	76	0.000655	0.000287
1000	500	193	77	0.001136	0.000481
1000	500	193	78	0.001922	0.000786
1000	500	193	79	0.003173	0.001251
1000	500	193	80	0.005111	0.001939
1000	500	193	81	0.008039	0.002928
1000	500	193	82	0.012346	0.004307
1000	500	193	83	0.018520	0.006174
1000	500	193	84	0.027142	0.008622
1000	500	193	85	0.038876	0.011734
1000	500	193	86	0.054437	0.015560
1000	500	193	87	0.074546	0.020109
1000	500	193	88	0.099872	0.025326
1000	500	193	89	0.130958	0.031086
1000	500	193	90	0.168147	0.037189
1000	500	193	91	0.211509	0.043362
1000	500	193	92	0.260789	0.049280
1000	500	193	93	0.315378	0.054590
1000	500	193	94	0.374321	0.058943

N	n	k	x	P(x)	p(x)
1000	500	193	95	0.436357	0.062036
1000	500	193	96	0.500000	0.063643
1000	500	194	67	0.000001	0.000001
1000	500	194	68	0.000002	0.000001
1000	500	194	69	0.000005	0.000003
1000	500	194	70	0.000010	0.000005
1000	500	194	71	0.000021	0.000011
1000	500	194	72	0.000042	0.000021
1000	500	194	73	0.000081	0.000039
1000	500	194	74	0.000152	0.000072
1000	500	194	75	0.000281	0.000128
1000	500	194	76	0.000505	0.000224
1000	500	194	77	0.000884	0.000379
1000	500	194	78	0.001511	0.000627
1000	500	194	79	0.002520	0.001009
1000	500	194	80	0.004102	0.001582
1000	500	194	81	0.006519	0.002417
1000	500	194	82	0.010115	0.003596
1000	500	194	83	0.015330	0.005215
1000	500	194	84	0.022698	0.007368
1000	500	194	85	0.032842	0.010144
1000	500	194	86	0.046454	0.013612
1000	500	194	87	0.064255	0.017801
1000	500	194	88	0.086942	0.022688
1000	500	194	89	0.115126	0.028184
1000	500	194	90	0.149253	0.034127
1000	500	194	91	0.189532	0.040279
1000	500	194	92	0.235874	0.046342
1000	500	194	93	0.287847	0.051973
1000	500	194	94	0.344667	0.056820
1000	500	194	95	0.405224	0.060557
1000	500	194	96	0.468139	0.062915
1000	500	194	97	0.531861	0.063721
1000	500	195	67	0.000001	0.000000
1000	500	195	68	0.000002	0.000001
1000	500	195	69	0.000003	0.000002
1000	500	195	70	0.000007	0.000004
1000	500	195	71	0.000015	0.000008
1000	500	195	72	0.000031	0.000015
1000	500	195	73	0.000060	0.000029
1000	500	195	74	0.000115	0.000055
1000	500	195	75	0.000213	0.000099
1000	500	195	76	0.000387	0.000174
1000	500	195	77	0.000685	0.000298
1000	500	195	78	0.001183	0.000498
1000	500	195	79	0.001993	0.000810
1000	500	195	80	0.003278	0.001285
1000	500	195	81	0.005263	0.001985
1000	500	195	82	0.008250	0.002988
1000	500	195	83	0.012632	0.004382

Table for $N = 1000$, $n = 500$

N	n	k	x	P(x)	p(x)
1000	500	195	84	0.018895	0.006263
1000	500	195	85	0.027619	0.008724
1000	500	195	86	0.039462	0.011843
1000	500	195	87	0.055134	0.015671
1000	500	195	88	0.075345	0.020212
1000	500	195	89	0.100754	0.025409
1000	500	195	90	0.131893	0.031139
1000	500	195	91	0.169093	0.037200
1000	500	195	92	0.212416	0.043323
1000	500	195	93	0.261603	0.049187
1000	500	195	94	0.316046	0.054443
1000	500	195	95	0.374795	0.058750
1000	500	195	96	0.436604	0.061808
1000	500	195	97	0.500000	0.063396
1000	500	196	67	0.000000	0.000000
1000	500	196	68	0.000001	0.000001
1000	500	196	69	0.000002	0.000001
1000	500	196	70	0.000005	0.000003
1000	500	196	71	0.000011	0.000006
1000	500	196	72	0.000022	0.000011
1000	500	196	73	0.000044	0.000022
1000	500	196	74	0.000086	0.000041
1000	500	196	75	0.000161	0.000075
1000	500	196	76	0.000295	0.000134
1000	500	196	77	0.000528	0.000233
1000	500	196	78	0.000922	0.000393
1000	500	196	79	0.001569	0.000648
1000	500	196	80	0.002607	0.001038
1000	500	196	81	0.004230	0.001622
1000	500	196	82	0.006699	0.002469
1000	500	196	83	0.010362	0.003663
1000	500	196	84	0.015658	0.005296
1000	500	196	85	0.023121	0.007463
1000	500	196	86	0.033371	0.010250
1000	500	196	87	0.047094	0.013722
1000	500	196	88	0.065001	0.017907
1000	500	196	89	0.087781	0.022781
1000	500	196	90	0.116034	0.028252
1000	500	196	91	0.150193	0.034159
1000	500	196	92	0.190458	0.040266
1000	500	196	93	0.236734	0.046276
1000	500	196	94	0.288587	0.051853
1000	500	196	95	0.345238	0.056651
1000	500	196	96	0.405584	0.060346
1000	500	196	97	0.468262	0.062678
1000	500	196	98	0.531737	0.063475
1000	500	197	68	0.000001	0.000000
1000	500	197	69	0.000002	0.000001
1000	500	197	70	0.000004	0.000002
1000	500	197	71	0.000008	0.000004

N	n	k	x	P(x)	p(x)
1000	500	197	72	0.000016	0.000008
1000	500	197	73	0.000033	0.000016
1000	500	197	74	0.000064	0.000031
1000	500	197	75	0.000121	0.000057
1000	500	197	76	0.000225	0.000103
1000	500	197	77	0.000406	0.000181
1000	500	197	78	0.000715	0.000309
1000	500	197	79	0.001230	0.000515
1000	500	197	80	0.002065	0.000835
1000	500	197	81	0.003384	0.001319
1000	500	197	82	0.005415	0.002031
1000	500	197	83	0.008462	0.003047
1000	500	197	84	0.012918	0.004456
1000	500	197	85	0.019269	0.006351
1000	500	197	86	0.028093	0.008824
1000	500	197	87	0.040045	0.011951
1000	500	197	88	0.055824	0.015780
1000	500	197	89	0.076136	0.020312
1000	500	197	90	0.101627	0.025490
1000	500	197	91	0.132816	0.031189
1000	500	197	92	0.170025	0.037209
1000	500	197	93	0.213309	0.043283
1000	500	197	94	0.262403	0.049094
1000	500	197	95	0.316701	0.054299
1000	500	197	96	0.375261	0.058560
1000	500	197	97	0.436845	0.061584
1000	500	197	98	0.500000	0.063155
1000	500	198	68	0.000001	0.000000
1000	500	198	69	0.000001	0.000001
1000	500	198	70	0.000003	0.000001
1000	500	198	71	0.000006	0.000003
1000	500	198	72	0.000012	0.000006
1000	500	198	73	0.000024	0.000012
1000	500	198	74	0.000047	0.000023
1000	500	198	75	0.000091	0.000043
1000	500	198	76	0.000170	0.000079
1000	500	198	77	0.000310	0.000140
1000	500	198	78	0.000553	0.000242
1000	500	198	79	0.000960	0.000408
1000	500	198	80	0.001628	0.000668
1000	500	198	81	0.002696	0.001068
1000	500	198	82	0.004358	0.001662
1000	500	198	83	0.006880	0.002522
1000	500	198	84	0.010610	0.003730
1000	500	198	85	0.015987	0.005377
1000	500	198	86	0.023544	0.007557
1000	500	198	87	0.033898	0.010354
1000	500	198	88	0.047728	0.013830
1000	500	198	89	0.065740	0.018012
1000	500	198	90	0.088611	0.022872

N	n	k	x	P(x)	p(x)
1000	500	198	91	0.116930	0.028318
1000	500	198	92	0.151119	0.034189
1000	500	198	93	0.191370	0.040251
1000	500	198	94	0.237581	0.046210
1000	500	198	95	0.289315	0.051735
1000	500	198	96	0.345799	0.056483
1000	500	198	97	0.405938	0.060139
1000	500	198	98	0.468383	0.062445
1000	500	198	99	0.531616	0.063233
1000	500	199	69	0.000001	0.000000
1000	500	199	70	0.000002	0.000001
1000	500	199	71	0.000004	0.000002
1000	500	199	72	0.000009	0.000005
1000	500	199	73	0.000018	0.000009
1000	500	199	74	0.000035	0.000017
1000	500	199	75	0.000068	0.000033
1000	500	199	76	0.000128	0.000060
1000	500	199	77	0.000236	0.000108
1000	500	199	78	0.000425	0.000189
1000	500	199	79	0.000746	0.000321
1000	500	199	80	0.001278	0.000532
1000	500	199	81	0.002138	0.000860
1000	500	199	82	0.003492	0.001353
1000	500	199	83	0.005569	0.002077
1000	500	199	84	0.008675	0.003106
1000	500	199	85	0.013204	0.004529
1000	500	199	86	0.019643	0.006439
1000	500	199	87	0.028566	0.008923
1000	500	199	88	0.040623	0.012057
1000	500	199	89	0.056510	0.015886
1000	500	199	90	0.076919	0.020410
1000	500	199	91	0.102488	0.025569
1000	500	199	92	0.133726	0.031238
1000	500	199	93	0.170944	0.037218
1000	500	199	94	0.214187	0.043244
1000	500	199	95	0.263190	0.049003
1000	500	199	96	0.317346	0.054156
1000	500	199	97	0.375719	0.058373
1000	500	199	98	0.437083	0.061364
1000	500	199	99	0.500000	0.062917
1000	500	200	69	0.000001	0.000000
1000	500	200	70	0.000001	0.000001
1000	500	200	71	0.000003	0.000002
1000	500	200	72	0.000006	0.000003
1000	500	200	73	0.000013	0.000007
1000	500	200	74	0.000026	0.000013
1000	500	200	75	0.000050	0.000025
1000	500	200	76	0.000096	0.000046
1000	500	200	77	0.000179	0.000083
1000	500	200	78	0.000326	0.000146

N	n	k	x	P(x)	p(x)
1000	500	200	79	0.000577	0.000252
1000	500	200	80	0.000999	0.000422
1000	500	200	81	0.001688	0.000689
1000	500	200	82	0.002785	0.001097
1000	500	200	83	0.004487	0.001702
1000	500	200	84	0.007062	0.002574
1000	500	200	85	0.010858	0.003796
1000	500	200	86	0.016315	0.005457
1000	500	200	87	0.023965	0.007650
1000	500	200	88	0.034422	0.010456
1000	500	200	89	0.048358	0.013937
1000	500	200	90	0.066472	0.018114
1000	500	200	91	0.089432	0.022960
1000	500	200	92	0.117815	0.028382
1000	500	200	93	0.152033	0.034218
1000	500	200	94	0.192269	0.040236
1000	500	200	95	0.238413	0.046145
1000	500	200	96	0.290031	0.051618
1000	500	200	97	0.346350	0.056319
1000	500	200	98	0.406286	0.059936
1000	500	200	99	0.468502	0.062216
1000	500	200	100	0.531498	0.062996
1000	500	201	70	0.000001	0.000001
1000	500	201	71	0.000002	0.000001
1000	500	201	72	0.000004	0.000002
1000	500	201	73	0.000009	0.000005
1000	500	201	74	0.000019	0.000010
1000	500	201	75	0.000037	0.000018
1000	500	201	76	0.000072	0.000035
1000	500	201	77	0.000135	0.000063
1000	500	201	78	0.000248	0.000113
1000	500	201	79	0.000445	0.000197
1000	500	201	80	0.000778	0.000333
1000	500	201	81	0.001328	0.000550
1000	500	201	82	0.002212	0.000885
1000	500	201	83	0.003600	0.001388
1000	500	201	84	0.005723	0.002123
1000	500	201	85	0.008888	0.003165
1000	500	201	86	0.013491	0.004602
1000	500	201	87	0.020016	0.006525
1000	500	201	88	0.029037	0.009021
1000	500	201	89	0.041198	0.012162
1000	500	201	90	0.057189	0.015991
1000	500	201	91	0.077694	0.020505
1000	500	201	92	0.103340	0.025646
1000	500	201	93	0.134624	0.031285
1000	500	201	94	0.171849	0.037225
1000	500	201	95	0.215053	0.043204
1000	500	201	96	0.263964	0.048912
1000	500	201	97	0.317979	0.054015

Table for $N = 1000$, $n = 500$

N	n	k	x	P(x)	p(x)
1000	500	201	98	0.376168	0.058189
1000	500	201	99	0.437316	0.061148
1000	500	201	100	0.500000	0.062684
1000	500	202	70	0.000001	0.000000
1000	500	202	71	0.000001	0.000001
1000	500	202	72	0.000003	0.000002
1000	500	202	73	0.000007	0.000004
1000	500	202	74	0.000014	0.000007
1000	500	202	75	0.000027	0.000014
1000	500	202	76	0.000054	0.000026
1000	500	202	77	0.000102	0.000048
1000	500	202	78	0.000189	0.000087
1000	500	202	79	0.000341	0.000153
1000	500	202	80	0.000603	0.000261
1000	500	202	81	0.001039	0.000436
1000	500	202	82	0.001749	0.000710
1000	500	202	83	0.002876	0.001126
1000	500	202	84	0.004618	0.001742
1000	500	202	85	0.007244	0.002627
1000	500	202	86	0.011106	0.003862
1000	500	202	87	0.016643	0.005537
1000	500	202	88	0.024385	0.007742
1000	500	202	89	0.034942	0.010557
1000	500	202	90	0.048983	0.014041
1000	500	202	91	0.067198	0.018214
1000	500	202	92	0.090244	0.023046
1000	500	202	93	0.118688	0.028444
1000	500	202	94	0.152934	0.034245
1000	500	202	95	0.193153	0.040220
1000	500	202	96	0.239233	0.046080
1000	500	202	97	0.290735	0.051502
1000	500	202	98	0.346892	0.056156
1000	500	202	99	0.406627	0.059736
1000	500	202	100	0.468619	0.061991
1000	500	202	101	0.531381	0.062762
1000	500	203	71	0.000001	0.000001
1000	500	203	72	0.000002	0.000001
1000	500	203	73	0.000005	0.000003
1000	500	203	74	0.000010	0.000005
1000	500	203	75	0.000020	0.000010
1000	500	203	76	0.000040	0.000020
1000	500	203	77	0.000076	0.000036
1000	500	203	78	0.000143	0.000067
1000	500	203	79	0.000261	0.000118
1000	500	203	80	0.000465	0.000204
1000	500	203	81	0.000810	0.000345
1000	500	203	82	0.001377	0.000567
1000	500	203	83	0.002287	0.000910
1000	500	203	84	0.003710	0.001422
1000	500	203	85	0.005878	0.002169

N	n	k	x	P(x)	p(x)
1000	500	203	86	0.009103	0.003224
1000	500	203	87	0.013778	0.004675
1000	500	203	88	0.020388	0.006610
1000	500	203	89	0.029505	0.009117
1000	500	203	90	0.041769	0.012264
1000	500	203	91	0.057862	0.016093
1000	500	203	92	0.078461	0.020599
1000	500	203	93	0.104181	0.025720
1000	500	203	94	0.135511	0.031330
1000	500	203	95	0.172741	0.037230
1000	500	203	96	0.215904	0.043163
1000	500	203	97	0.264726	0.048821
1000	500	203	98	0.318602	0.053876
1000	500	203	99	0.376610	0.058007
1000	500	203	100	0.437545	0.060936
1000	500	203	101	0.500000	0.062454
1000	500	204	71	0.000001	0.000000
1000	500	204	72	0.000002	0.000001
1000	500	204	73	0.000003	0.000002
1000	500	204	74	0.000007	0.000004
1000	500	204	75	0.000015	0.000008
1000	500	204	76	0.000029	0.000015
1000	500	204	77	0.000057	0.000028
1000	500	204	78	0.000107	0.000051
1000	500	204	79	0.000198	0.000091
1000	500	204	80	0.000357	0.000159
1000	500	204	81	0.000629	0.000271
1000	500	204	82	0.001080	0.000451
1000	500	204	83	0.001811	0.000731
1000	500	204	84	0.002967	0.001156
1000	500	204	85	0.004749	0.001782
1000	500	204	86	0.007428	0.002679
1000	500	204	87	0.011355	0.003927
1000	500	204	88	0.016971	0.005616
1000	500	204	89	0.024804	0.007833
1000	500	204	90	0.035460	0.010657
1000	500	204	91	0.049604	0.014144
1000	500	204	92	0.067916	0.018312
1000	500	204	93	0.091047	0.023131
1000	500	204	94	0.119551	0.028504
1000	500	204	95	0.153822	0.034271
1000	500	204	96	0.194025	0.040203
1000	500	204	97	0.240040	0.046015
1000	500	204	98	0.291428	0.051388
1000	500	204	99	0.347424	0.055997
1000	500	204	100	0.406963	0.059539
1000	500	204	101	0.468733	0.061771
1000	500	204	102	0.531266	0.062533
1000	500	205	71	0.000001	0.000000
1000	500	205	72	0.000001	0.000001

N	n	k	x	P(x)	p(x)
1000	500	205	73	0.000002	0.000001
1000	500	205	74	0.000005	0.000003
1000	500	205	75	0.000011	0.000006
1000	500	205	76	0.000022	0.000011
1000	500	205	77	0.000042	0.000021
1000	500	205	78	0.000081	0.000038
1000	500	205	79	0.000150	0.000070
1000	500	205	80	0.000274	0.000123
1000	500	205	81	0.000486	0.000212
1000	500	205	82	0.000843	0.000357
1000	500	205	83	0.001428	0.000585
1000	500	205	84	0.002363	0.000935
1000	500	205	85	0.003820	0.001457
1000	500	205	86	0.006035	0.002215
1000	500	205	87	0.009317	0.003283
1000	500	205	88	0.014064	0.004747
1000	500	205	89	0.020759	0.006695
1000	500	205	90	0.029971	0.009212
1000	500	205	91	0.042337	0.012365
1000	500	205	92	0.058530	0.016193
1000	500	205	93	0.079220	0.020690
1000	500	205	94	0.105012	0.025793
1000	500	205	95	0.136385	0.031373
1000	500	205	96	0.173621	0.037235
1000	500	205	97	0.216743	0.043123
1000	500	205	98	0.265475	0.048732
1000	500	205	99	0.319215	0.053739
1000	500	205	100	0.377044	0.057829
1000	500	205	101	0.437771	0.060727
1000	500	205	102	0.500000	0.062229
1000	500	206	72	0.000001	0.000000
1000	500	206	73	0.000002	0.000001
1000	500	206	74	0.000004	0.000002
1000	500	206	75	0.000008	0.000004
1000	500	206	76	0.000016	0.000008
1000	500	206	77	0.000031	0.000015
1000	500	206	78	0.000060	0.000029
1000	500	206	79	0.000113	0.000053
1000	500	206	80	0.000208	0.000095
1000	500	206	81	0.000374	0.000166
1000	500	206	82	0.000655	0.000281
1000	500	206	83	0.001121	0.000466
1000	500	206	84	0.001874	0.000753
1000	500	206	85	0.003060	0.001186
1000	500	206	86	0.004881	0.001822
1000	500	206	87	0.007612	0.002731
1000	500	206	88	0.011604	0.003992
1000	500	206	89	0.017298	0.005694
1000	500	206	90	0.025220	0.007922
1000	500	206	91	0.035975	0.010755

N	n	k	x	P(x)	p(x)
1000	500	206	92	0.050219	0.014244
1000	500	206	93	0.068628	0.018408
1000	500	206	94	0.091840	0.023213
1000	500	206	95	0.120403	0.028563
1000	500	206	96	0.154699	0.034296
1000	500	206	97	0.194884	0.040185
1000	500	206	98	0.240833	0.045950
1000	500	206	99	0.292108	0.051275
1000	500	206	100	0.347948	0.055839
1000	500	206	101	0.407293	0.059345
1000	500	206	102	0.468846	0.061553
1000	500	206	103	0.531154	0.062308
1000	500	207	72	0.000001	0.000000
1000	500	207	73	0.000001	0.000001
1000	500	207	74	0.000003	0.000001
1000	500	207	75	0.000006	0.000003
1000	500	207	76	0.000012	0.000006
1000	500	207	77	0.000023	0.000011
1000	500	207	78	0.000045	0.000022
1000	500	207	79	0.000085	0.000040
1000	500	207	80	0.000158	0.000073
1000	500	207	81	0.000287	0.000128
1000	500	207	82	0.000507	0.000221
1000	500	207	83	0.000877	0.000369
1000	500	207	84	0.001480	0.000603
1000	500	207	85	0.002440	0.000960
1000	500	207	86	0.003931	0.001492
1000	500	207	87	0.006192	0.002260
1000	500	207	88	0.009533	0.003341
1000	500	207	89	0.014351	0.004818
1000	500	207	90	0.021129	0.006779
1000	500	207	91	0.030435	0.009306
1000	500	207	92	0.042900	0.012465
1000	500	207	93	0.059192	0.016292
1000	500	207	94	0.079971	0.020779
1000	500	207	95	0.105834	0.025863
1000	500	207	96	0.137248	0.031415
1000	500	207	97	0.174487	0.037239
1000	500	207	98	0.217569	0.043082
1000	500	207	99	0.266213	0.048643
1000	500	207	100	0.319817	0.053604
1000	500	207	101	0.377471	0.057654
1000	500	207	102	0.437992	0.060521
1000	500	207	103	0.500000	0.062008
1000	500	208	73	0.000001	0.000000
1000	500	208	74	0.000002	0.000001
1000	500	208	75	0.000004	0.000002
1000	500	208	76	0.000008	0.000004
1000	500	208	77	0.000017	0.000009
1000	500	208	78	0.000033	0.000016

Table for $N = 1000,\ n = 500$

N	n	k	x	P(x)	p(x)
1000	500	208	79	0.000064	0.000031
1000	500	208	80	0.000120	0.000056
1000	500	208	81	0.000219	0.000099
1000	500	208	82	0.000391	0.000172
1000	500	208	83	0.000682	0.000291
1000	500	208	84	0.001163	0.000481
1000	500	208	85	0.001937	0.000774
1000	500	208	86	0.003153	0.001215
1000	500	208	87	0.005015	0.001862
1000	500	208	88	0.007797	0.002782
1000	500	208	89	0.011854	0.004057
1000	500	208	90	0.017625	0.005771
1000	500	208	91	0.025636	0.008011
1000	500	208	92	0.036487	0.010851
1000	500	208	93	0.050830	0.014343
1000	500	208	94	0.069332	0.018502
1000	500	208	95	0.092625	0.023293
1000	500	208	96	0.121244	0.028619
1000	500	208	97	0.155563	0.034319
1000	500	208	98	0.195729	0.040167
1000	500	208	99	0.241615	0.045885
1000	500	208	100	0.292778	0.051163
1000	500	208	101	0.348462	0.055684
1000	500	208	102	0.407617	0.059155
1000	500	208	103	0.468957	0.061340
1000	500	208	104	0.531043	0.062086
1000	500	209	73	0.000001	0.000000
1000	500	209	74	0.000001	0.000001
1000	500	209	75	0.000003	0.000002
1000	500	209	76	0.000006	0.000003
1000	500	209	77	0.000012	0.000006
1000	500	209	78	0.000024	0.000012
1000	500	209	79	0.000048	0.000023
1000	500	209	80	0.000090	0.000042
1000	500	209	81	0.000166	0.000076
1000	500	209	82	0.000300	0.000134
1000	500	209	83	0.000529	0.000229
1000	500	209	84	0.000911	0.000382
1000	500	209	85	0.001532	0.000621
1000	500	209	86	0.002517	0.000986
1000	500	209	87	0.004044	0.001526
1000	500	209	88	0.006350	0.002306
1000	500	209	89	0.009748	0.003399
1000	500	209	90	0.014637	0.004889
1000	500	209	91	0.021499	0.006862
1000	500	209	92	0.030897	0.009398
1000	500	209	93	0.043459	0.012562
1000	500	209	94	0.059848	0.016388
1000	500	209	95	0.080714	0.020866
1000	500	209	96	0.106645	0.025931

N	n	k	x	P(x)	p(x)
1000	500	209	97	0.138100	0.031455
1000	500	209	98	0.175342	0.037242
1000	500	209	99	0.218382	0.043041
1000	500	209	100	0.266938	0.048556
1000	500	209	101	0.320409	0.053471
1000	500	209	102	0.377890	0.057481
1000	500	209	103	0.438209	0.060319
1000	500	209	104	0.500000	0.061791
1000	500	210	74	0.000001	0.000001
1000	500	210	75	0.000002	0.000001
1000	500	210	76	0.000004	0.000002
1000	500	210	77	0.000009	0.000005
1000	500	210	78	0.000018	0.000009
1000	500	210	79	0.000035	0.000017
1000	500	210	80	0.000067	0.000032
1000	500	210	81	0.000126	0.000058
1000	500	210	82	0.000229	0.000104
1000	500	210	83	0.000408	0.000179
1000	500	210	84	0.000710	0.000302
1000	500	210	85	0.001206	0.000496
1000	500	210	86	0.002001	0.000796
1000	500	210	87	0.003247	0.001245
1000	500	210	88	0.005148	0.001902
1000	500	210	89	0.007982	0.002834
1000	500	210	90	0.012103	0.004121
1000	500	210	91	0.017951	0.005848
1000	500	210	92	0.026049	0.008098
1000	500	210	93	0.036995	0.010946
1000	500	210	94	0.051436	0.014440
1000	500	210	95	0.070030	0.018594
1000	500	210	96	0.093401	0.023371
1000	500	210	97	0.122074	0.028673
1000	500	210	98	0.156415	0.034340
1000	500	210	99	0.196563	0.040148
1000	500	210	100	0.242384	0.045821
1000	500	210	101	0.293437	0.051053
1000	500	210	102	0.348968	0.055531
1000	500	210	103	0.407935	0.058967
1000	500	210	104	0.469066	0.061130
1000	500	210	105	0.530934	0.061869
1000	500	211	74	0.000001	0.000000
1000	500	211	75	0.000001	0.000001
1000	500	211	76	0.000003	0.000002
1000	500	211	77	0.000006	0.000003
1000	500	211	78	0.000013	0.000007
1000	500	211	79	0.000026	0.000013
1000	500	211	80	0.000050	0.000024
1000	500	211	81	0.000095	0.000045
1000	500	211	82	0.000175	0.000080
1000	500	211	83	0.000314	0.000139

N	n	k	x	P(x)	p(x)
1000	500	211	84	0.000551	0.000237
1000	500	211	85	0.000945	0.000394
1000	500	211	86	0.001584	0.000639
1000	500	211	87	0.002596	0.001011
1000	500	211	88	0.004157	0.001561
1000	500	211	89	0.006508	0.002351
1000	500	211	90	0.009965	0.003456
1000	500	211	91	0.014923	0.004959
1000	500	211	92	0.021867	0.006943
1000	500	211	93	0.031356	0.009489
1000	500	211	94	0.044015	0.012658
1000	500	211	95	0.060498	0.016483
1000	500	211	96	0.081449	0.020952
1000	500	211	97	0.107447	0.025998
1000	500	211	98	0.138940	0.031493
1000	500	211	99	0.176184	0.037244
1000	500	211	100	0.219183	0.043000
1000	500	211	101	0.267652	0.048469
1000	500	211	102	0.320992	0.053340
1000	500	211	103	0.378302	0.057311
1000	500	211	104	0.438423	0.060121
1000	500	211	105	0.500000	0.061577
1000	500	212	75	0.000001	0.000001
1000	500	212	76	0.000002	0.000001
1000	500	212	77	0.000005	0.000002
1000	500	212	78	0.000010	0.000005
1000	500	212	79	0.000019	0.000010
1000	500	212	80	0.000037	0.000018
1000	500	212	81	0.000071	0.000034
1000	500	212	82	0.000132	0.000061
1000	500	212	83	0.000240	0.000108
1000	500	212	84	0.000426	0.000186
1000	500	212	85	0.000738	0.000312
1000	500	212	86	0.001249	0.000511
1000	500	212	87	0.002066	0.000817
1000	500	212	88	0.003342	0.001275
1000	500	212	89	0.005283	0.001942
1000	500	212	90	0.008168	0.002885
1000	500	212	91	0.012353	0.004184
1000	500	212	92	0.018276	0.005923
1000	500	212	93	0.026461	0.008185
1000	500	212	94	0.037501	0.011040
1000	500	212	95	0.052037	0.014536
1000	500	212	96	0.070721	0.018685
1000	500	212	97	0.094168	0.023447
1000	500	212	98	0.122895	0.028726
1000	500	212	99	0.157255	0.034361
1000	500	212	100	0.197384	0.040129
1000	500	212	101	0.243141	0.045757
1000	500	212	102	0.294085	0.050944

N	n	k	x	P(x)	p(x)
1000	500	212	103	0.349465	0.055380
1000	500	212	104	0.408248	0.058783
1000	500	212	105	0.469172	0.060924
1000	500	212	106	0.530827	0.061655
1000	500	213	75	0.000001	0.000000
1000	500	213	76	0.000002	0.000001
1000	500	213	77	0.000003	0.000002
1000	500	213	78	0.000007	0.000004
1000	500	213	79	0.000014	0.000007
1000	500	213	80	0.000028	0.000014
1000	500	213	81	0.000053	0.000026
1000	500	213	82	0.000100	0.000047
1000	500	213	83	0.000183	0.000083
1000	500	213	84	0.000328	0.000145
1000	500	213	85	0.000574	0.000246
1000	500	213	86	0.000981	0.000407
1000	500	213	87	0.001638	0.000657
1000	500	213	88	0.002675	0.001037
1000	500	213	89	0.004271	0.001596
1000	500	213	90	0.006667	0.002397
1000	500	213	91	0.010181	0.003514
1000	500	213	92	0.015209	0.005028
1000	500	213	93	0.022234	0.007024
1000	500	213	94	0.031813	0.009579
1000	500	213	95	0.044566	0.012753
1000	500	213	96	0.061142	0.016576
1000	500	213	97	0.082177	0.021035
1000	500	213	98	0.108240	0.026063
1000	500	213	99	0.139770	0.031530
1000	500	213	100	0.177014	0.037244
1000	500	213	101	0.219972	0.042958
1000	500	213	102	0.268355	0.048382
1000	500	213	103	0.321564	0.053210
1000	500	213	104	0.378708	0.057143
1000	500	213	105	0.438633	0.059925
1000	500	213	106	0.500000	0.061367
1000	500	214	75	0.000001	0.000000
1000	500	214	76	0.000001	0.000001
1000	500	214	77	0.000002	0.000001
1000	500	214	78	0.000005	0.000003
1000	500	214	79	0.000010	0.000005
1000	500	214	80	0.000020	0.000010
1000	500	214	81	0.000040	0.000019
1000	500	214	82	0.000075	0.000035
1000	500	214	83	0.000139	0.000064
1000	500	214	84	0.000251	0.000112
1000	500	214	85	0.000444	0.000193
1000	500	214	86	0.000766	0.000322
1000	500	214	87	0.001293	0.000527
1000	500	214	88	0.002132	0.000839

Table for $N = 1000,\ n = 500$

N	n	k	x	P(x)	p(x)
1000	500	214	89	0.003437	0.001305
1000	500	214	90	0.005419	0.001981
1000	500	214	91	0.008355	0.002936
1000	500	214	92	0.012602	0.004248
1000	500	214	93	0.018601	0.005998
1000	500	214	94	0.026871	0.008270
1000	500	214	95	0.038003	0.011132
1000	500	214	96	0.052633	0.014630
1000	500	214	97	0.071406	0.018773
1000	500	214	98	0.094927	0.023521
1000	500	214	99	0.123704	0.028777
1000	500	214	100	0.158084	0.034380
1000	500	214	101	0.198193	0.040109
1000	500	214	102	0.243887	0.045694
1000	500	214	103	0.294723	0.050836
1000	500	214	104	0.349955	0.055232
1000	500	214	105	0.408556	0.058602
1000	500	214	106	0.469278	0.060721
1000	500	214	107	0.530722	0.061445
1000	500	215	76	0.000001	0.000000
1000	500	215	77	0.000002	0.000001
1000	500	215	78	0.000004	0.000002
1000	500	215	79	0.000007	0.000004
1000	500	215	80	0.000015	0.000008
1000	500	215	81	0.000029	0.000014
1000	500	215	82	0.000056	0.000027
1000	500	215	83	0.000105	0.000049
1000	500	215	84	0.000192	0.000087
1000	500	215	85	0.000342	0.000150
1000	500	215	86	0.000597	0.000254
1000	500	215	87	0.001016	0.000420
1000	500	215	88	0.001692	0.000676
1000	500	215	89	0.002755	0.001062
1000	500	215	90	0.004385	0.001630
1000	500	215	91	0.006827	0.002442
1000	500	215	92	0.010398	0.003571
1000	500	215	93	0.015495	0.005097
1000	500	215	94	0.022599	0.007104
1000	500	215	95	0.032267	0.009668
1000	500	215	96	0.045113	0.012846
1000	500	215	97	0.061780	0.016667
1000	500	215	98	0.082897	0.021117
1000	500	215	99	0.109023	0.026126
1000	500	215	100	0.140588	0.031565
1000	500	215	101	0.177832	0.037244
1000	500	215	102	0.220749	0.042917
1000	500	215	103	0.269046	0.048297
1000	500	215	104	0.322128	0.053082
1000	500	215	105	0.379106	0.056978
1000	500	215	106	0.438840	0.059733

N	n	k	x	P(x)	p(x)
1000	500	215	107	0.500000	0.061160
1000	500	216	76	0.000001	0.000000
1000	500	216	77	0.000001	0.000001
1000	500	216	78	0.000003	0.000001
1000	500	216	79	0.000005	0.000003
1000	500	216	80	0.000011	0.000006
1000	500	216	81	0.000022	0.000011
1000	500	216	82	0.000042	0.000020
1000	500	216	83	0.000079	0.000037
1000	500	216	84	0.000146	0.000067
1000	500	216	85	0.000263	0.000117
1000	500	216	86	0.000463	0.000200
1000	500	216	87	0.000796	0.000333
1000	500	216	88	0.001338	0.000542
1000	500	216	89	0.002198	0.000861
1000	500	216	90	0.003534	0.001335
1000	500	216	91	0.005555	0.002021
1000	500	216	92	0.008542	0.002987
1000	500	216	93	0.012852	0.004310
1000	500	216	94	0.018925	0.006073
1000	500	216	95	0.027279	0.008354
1000	500	216	96	0.038502	0.011223
1000	500	216	97	0.053224	0.014722
1000	500	216	98	0.072083	0.018859
1000	500	216	99	0.095677	0.023594
1000	500	216	100	0.124504	0.028827
1000	500	216	101	0.158902	0.034398
1000	500	216	102	0.198990	0.040088
1000	500	216	103	0.244621	0.045631
1000	500	216	104	0.295350	0.050730
1000	500	216	105	0.350436	0.055085
1000	500	216	106	0.408859	0.058423
1000	500	216	107	0.469381	0.060522
1000	500	216	108	0.530619	0.061238
1000	500	217	77	0.000001	0.000000
1000	500	217	78	0.000002	0.000001
1000	500	217	79	0.000004	0.000002
1000	500	217	80	0.000008	0.000004
1000	500	217	81	0.000016	0.000008
1000	500	217	82	0.000031	0.000015
1000	500	217	83	0.000059	0.000028
1000	500	217	84	0.000110	0.000051
1000	500	217	85	0.000201	0.000090
1000	500	217	86	0.000357	0.000156
1000	500	217	87	0.000620	0.000263
1000	500	217	88	0.001053	0.000433
1000	500	217	89	0.001747	0.000694
1000	500	217	90	0.002835	0.001088
1000	500	217	91	0.004500	0.001665
1000	500	217	92	0.006987	0.002487

N	n	k	x	P(x)	p(x)
1000	500	217	93	0.010614	0.003627
1000	500	217	94	0.015780	0.005166
1000	500	217	95	0.022964	0.007183
1000	500	217	96	0.032719	0.009755
1000	500	217	97	0.045656	0.012937
1000	500	217	98	0.062413	0.016757
1000	500	217	99	0.083609	0.021196
1000	500	217	100	0.109796	0.026187
1000	500	217	101	0.141396	0.031600
1000	500	217	102	0.178639	0.037244
1000	500	217	103	0.221515	0.042875
1000	500	217	104	0.269727	0.048212
1000	500	217	105	0.322682	0.052955
1000	500	217	106	0.379498	0.056816
1000	500	217	107	0.439043	0.059544
1000	500	217	108	0.500000	0.060957
1000	500	218	77	0.000001	0.000000
1000	500	218	78	0.000001	0.000001
1000	500	218	79	0.000003	0.000001
1000	500	218	80	0.000006	0.000003
1000	500	218	81	0.000012	0.000006
1000	500	218	82	0.000023	0.000011
1000	500	218	83	0.000044	0.000021
1000	500	218	84	0.000083	0.000039
1000	500	218	85	0.000153	0.000070
1000	500	218	86	0.000275	0.000122
1000	500	218	87	0.000481	0.000207
1000	500	218	88	0.000825	0.000344
1000	500	218	89	0.001383	0.000558
1000	500	218	90	0.002266	0.000883
1000	500	218	91	0.003631	0.001365
1000	500	218	92	0.005691	0.002061
1000	500	218	93	0.008729	0.003038
1000	500	218	94	0.013102	0.004373
1000	500	218	95	0.019248	0.006146
1000	500	218	96	0.027685	0.008437
1000	500	218	97	0.038997	0.011312
1000	500	218	98	0.053810	0.014812
1000	500	218	99	0.072754	0.018944
1000	500	218	100	0.096418	0.023665
1000	500	218	101	0.125293	0.028875
1000	500	218	102	0.159708	0.034415
1000	500	218	103	0.199776	0.040067
1000	500	218	104	0.245344	0.045568
1000	500	218	105	0.295968	0.050624
1000	500	218	106	0.350909	0.054941
1000	500	218	107	0.409156	0.058247
1000	500	218	108	0.469482	0.060326
1000	500	218	109	0.530517	0.061035
1000	500	219	78	0.000001	0.000001

N	n	k	x	P(x)	p(x)
1000	500	219	79	0.000002	0.000001
1000	500	219	80	0.000004	0.000002
1000	500	219	81	0.000009	0.000004
1000	500	219	82	0.000017	0.000008
1000	500	219	83	0.000033	0.000016
1000	500	219	84	0.000063	0.000030
1000	500	219	85	0.000116	0.000053
1000	500	219	86	0.000210	0.000094
1000	500	219	87	0.000372	0.000162
1000	500	219	88	0.000644	0.000272
1000	500	219	89	0.001090	0.000446
1000	500	219	90	0.001803	0.000713
1000	500	219	91	0.002917	0.001114
1000	500	219	92	0.004616	0.001700
1000	500	219	93	0.007148	0.002532
1000	500	219	94	0.010831	0.003683
1000	500	219	95	0.016065	0.005233
1000	500	219	96	0.023326	0.007262
1000	500	219	97	0.033168	0.009841
1000	500	219	98	0.046195	0.013027
1000	500	219	99	0.063040	0.016844
1000	500	219	100	0.084314	0.021274
1000	500	219	101	0.110560	0.026247
1000	500	219	102	0.142193	0.031632
1000	500	219	103	0.179435	0.037242
1000	500	219	104	0.222268	0.042834
1000	500	219	105	0.270397	0.048128
1000	500	219	106	0.323228	0.052831
1000	500	219	107	0.379884	0.056656
1000	500	219	108	0.439242	0.059359
1000	500	219	109	0.500000	0.060758
1000	500	220	78	0.000001	0.000000
1000	500	220	79	0.000001	0.000001
1000	500	220	80	0.000003	0.000002
1000	500	220	81	0.000006	0.000003
1000	500	220	82	0.000012	0.000006
1000	500	220	83	0.000024	0.000012
1000	500	220	84	0.000047	0.000022
1000	500	220	85	0.000088	0.000041
1000	500	220	86	0.000160	0.000073
1000	500	220	87	0.000286	0.000126
1000	500	220	88	0.000501	0.000214
1000	500	220	89	0.000855	0.000355
1000	500	220	90	0.001428	0.000573
1000	500	220	91	0.002333	0.000905
1000	500	220	92	0.003728	0.001395
1000	500	220	93	0.005829	0.002100
1000	500	220	94	0.008917	0.003088
1000	500	220	95	0.013351	0.004434
1000	500	220	96	0.019570	0.006219

Table for $N = 1000$, $n = 500$

N	n	k	x	P(x)	p(x)
1000	500	220	97	0.028090	0.008520
1000	500	220	98	0.039490	0.011400
1000	500	220	99	0.054391	0.014901
1000	500	220	100	0.073418	0.019027
1000	500	220	101	0.097151	0.023734
1000	500	220	102	0.126073	0.028921
1000	500	220	103	0.160504	0.034431
1000	500	220	104	0.200550	0.040046
1000	500	220	105	0.246055	0.045505
1000	500	220	106	0.296575	0.050520
1000	500	220	107	0.351374	0.054799
1000	500	220	108	0.409449	0.058075
1000	500	220	109	0.469582	0.060133
1000	500	220	110	0.530418	0.060836
1000	500	221	79	0.000001	0.000001
1000	500	221	80	0.000002	0.000001
1000	500	221	81	0.000004	0.000002
1000	500	221	82	0.000009	0.000005
1000	500	221	83	0.000018	0.000009
1000	500	221	84	0.000035	0.000017
1000	500	221	85	0.000066	0.000031
1000	500	221	86	0.000122	0.000056
1000	500	221	87	0.000220	0.000098
1000	500	221	88	0.000387	0.000168
1000	500	221	89	0.000668	0.000281
1000	500	221	90	0.001127	0.000459
1000	500	221	91	0.001859	0.000732
1000	500	221	92	0.002999	0.001140
1000	500	221	93	0.004733	0.001734
1000	500	221	94	0.007309	0.002577
1000	500	221	95	0.011049	0.003739
1000	500	221	96	0.016349	0.005300
1000	500	221	97	0.023688	0.007339
1000	500	221	98	0.033614	0.009926
1000	500	221	99	0.046730	0.013116
1000	500	221	100	0.063660	0.016930
1000	500	221	101	0.085011	0.021350
1000	500	221	102	0.111315	0.026305
1000	500	221	103	0.142979	0.031664
1000	500	221	104	0.180219	0.037239
1000	500	221	105	0.223011	0.042792
1000	500	221	106	0.271056	0.048045
1000	500	221	107	0.323764	0.052708
1000	500	221	108	0.380263	0.056499
1000	500	221	109	0.439438	0.059176
1000	500	221	110	0.500000	0.060561
1000	500	222	79	0.000001	0.000000
1000	500	222	80	0.000002	0.000001
1000	500	222	81	0.000003	0.000002
1000	500	222	82	0.000007	0.000003

N	n	k	x	P(x)	p(x)
1000	500	222	83	0.000013	0.000007
1000	500	222	84	0.000026	0.000013
1000	500	222	85	0.000049	0.000023
1000	500	222	86	0.000092	0.000043
1000	500	222	87	0.000168	0.000076
1000	500	222	88	0.000299	0.000131
1000	500	222	89	0.000520	0.000221
1000	500	222	90	0.000886	0.000365
1000	500	222	91	0.001475	0.000589
1000	500	222	92	0.002402	0.000927
1000	500	222	93	0.003827	0.001425
1000	500	222	94	0.005966	0.002140
1000	500	222	95	0.009105	0.003138
1000	500	222	96	0.013600	0.004496
1000	500	222	97	0.019891	0.006291
1000	500	222	98	0.028492	0.008601
1000	500	222	99	0.039979	0.011487
1000	500	222	100	0.054967	0.014989
1000	500	222	101	0.074075	0.019108
1000	500	222	102	0.097876	0.023801
1000	500	222	103	0.126843	0.028966
1000	500	222	104	0.161288	0.034446
1000	500	222	105	0.201313	0.040024
1000	500	222	106	0.246756	0.045443
1000	500	222	107	0.297173	0.050417
1000	500	222	108	0.351832	0.054659
1000	500	222	109	0.409737	0.057905
1000	500	222	110	0.469680	0.059944
1000	500	222	111	0.530319	0.060639
1000	500	223	79	0.000001	0.000000
1000	500	223	80	0.000001	0.000001
1000	500	223	81	0.000002	0.000001
1000	500	223	82	0.000005	0.000002
1000	500	223	83	0.000010	0.000005
1000	500	223	84	0.000019	0.000009
1000	500	223	85	0.000037	0.000018
1000	500	223	86	0.000069	0.000032
1000	500	223	87	0.000128	0.000058
1000	500	223	88	0.000229	0.000102
1000	500	223	89	0.000403	0.000174
1000	500	223	90	0.000693	0.000290
1000	500	223	91	0.001165	0.000472
1000	500	223	92	0.001915	0.000751
1000	500	223	93	0.003081	0.001166
1000	500	223	94	0.004850	0.001769
1000	500	223	95	0.007471	0.002621
1000	500	223	96	0.011266	0.003795
1000	500	223	97	0.016632	0.005367
1000	500	223	98	0.024048	0.007415
1000	500	223	99	0.034058	0.010010

N	n	k	x	P(x)	p(x)
1000	500	223	100	0.047261	0.013203
1000	500	223	101	0.064276	0.017015
1000	500	223	102	0.085700	0.021425
1000	500	223	103	0.112061	0.026361
1000	500	223	104	0.143755	0.031694
1000	500	223	105	0.180992	0.037236
1000	500	223	106	0.223742	0.042750
1000	500	223	107	0.271705	0.047963
1000	500	223	108	0.324292	0.052586
1000	500	223	109	0.380635	0.056344
1000	500	223	110	0.439631	0.058996
1000	500	223	111	0.500000	0.060369
1000	500	224	80	0.000001	0.000000
1000	500	224	81	0.000002	0.000001
1000	500	224	82	0.000003	0.000002
1000	500	224	83	0.000007	0.000004
1000	500	224	84	0.000014	0.000007
1000	500	224	85	0.000027	0.000013
1000	500	224	86	0.000052	0.000025
1000	500	224	87	0.000097	0.000045
1000	500	224	88	0.000175	0.000079
1000	500	224	89	0.000311	0.000136
1000	500	224	90	0.000540	0.000229
1000	500	224	91	0.000917	0.000376
1000	500	224	92	0.001521	0.000605
1000	500	224	93	0.002471	0.000949
1000	500	224	94	0.003926	0.001455
1000	500	224	95	0.006105	0.002179
1000	500	224	96	0.009293	0.003188
1000	500	224	97	0.013849	0.004556
1000	500	224	98	0.020211	0.006362
1000	500	224	99	0.028892	0.008681
1000	500	224	100	0.040464	0.011572
1000	500	224	101	0.055538	0.015074
1000	500	224	102	0.074726	0.019188
1000	500	224	103	0.098593	0.023866
1000	500	224	104	0.127603	0.029010
1000	500	224	105	0.162062	0.034460
1000	500	224	106	0.202064	0.040002
1000	500	224	107	0.247446	0.045382
1000	500	224	108	0.297762	0.050316
1000	500	224	109	0.352282	0.054521
1000	500	224	110	0.410020	0.057737
1000	500	224	111	0.469777	0.059757
1000	500	224	112	0.530223	0.060446
1000	500	225	80	0.000001	0.000000
1000	500	225	81	0.000001	0.000001
1000	500	225	82	0.000002	0.000001
1000	500	225	83	0.000005	0.000003
1000	500	225	84	0.000010	0.000005

N	n	k	x	P(x)	p(x)
1000	500	225	85	0.000020	0.000010
1000	500	225	86	0.000039	0.000019
1000	500	225	87	0.000073	0.000034
1000	500	225	88	0.000134	0.000061
1000	500	225	89	0.000239	0.000106
1000	500	225	90	0.000419	0.000180
1000	500	225	91	0.000718	0.000299
1000	500	225	92	0.001203	0.000485
1000	500	225	93	0.001973	0.000769
1000	500	225	94	0.003164	0.001192
1000	500	225	95	0.004967	0.001803
1000	500	225	96	0.007633	0.002665
1000	500	225	97	0.011483	0.003850
1000	500	225	98	0.016915	0.005433
1000	500	225	99	0.024406	0.007491
1000	500	225	100	0.034499	0.010093
1000	500	225	101	0.047788	0.013289
1000	500	225	102	0.064885	0.017097
1000	500	225	103	0.086382	0.021498
1000	500	225	104	0.112798	0.026416
1000	500	225	105	0.144521	0.031723
1000	500	225	106	0.181754	0.037232
1000	500	225	107	0.224463	0.042709
1000	500	225	108	0.272345	0.047882
1000	500	225	109	0.324811	0.052467
1000	500	225	110	0.381002	0.056191
1000	500	225	111	0.439821	0.058819
1000	500	225	112	0.500000	0.060179
1000	500	226	81	0.000001	0.000000
1000	500	226	82	0.000002	0.000001
1000	500	226	83	0.000004	0.000002
1000	500	226	84	0.000008	0.000004
1000	500	226	85	0.000015	0.000007
1000	500	226	86	0.000029	0.000014
1000	500	226	87	0.000055	0.000026
1000	500	226	88	0.000101	0.000047
1000	500	226	89	0.000183	0.000082
1000	500	226	90	0.000324	0.000141
1000	500	226	91	0.000560	0.000236
1000	500	226	92	0.000948	0.000388
1000	500	226	93	0.001569	0.000621
1000	500	226	94	0.002540	0.000971
1000	500	226	95	0.004025	0.001485
1000	500	226	96	0.006243	0.002218
1000	500	226	97	0.009481	0.003237
1000	500	226	98	0.014097	0.004617
1000	500	226	99	0.020530	0.006433
1000	500	226	100	0.029290	0.008760
1000	500	226	101	0.040946	0.011656
1000	500	226	102	0.056105	0.015158

Table for $N = 1000$, $n = 500$

N	n	k	x	P(x)	p(x)
1000	500	226	103	0.075370	0.019266
1000	500	226	104	0.099301	0.023931
1000	500	226	105	0.128353	0.029052
1000	500	226	106	0.162825	0.034472
1000	500	226	107	0.202805	0.039980
1000	500	226	108	0.248126	0.045320
1000	500	226	109	0.298341	0.050215
1000	500	226	110	0.352725	0.054384
1000	500	226	111	0.410298	0.057572
1000	500	226	112	0.469872	0.059574
1000	500	226	113	0.530128	0.060256
1000	500	227	81	0.000001	0.000000
1000	500	227	82	0.000001	0.000001
1000	500	227	83	0.000003	0.000001
1000	500	227	84	0.000005	0.000003
1000	500	227	85	0.000011	0.000005
1000	500	227	86	0.000021	0.000010
1000	500	227	87	0.000041	0.000020
1000	500	227	88	0.000077	0.000036
1000	500	227	89	0.000140	0.000063
1000	500	227	90	0.000249	0.000110
1000	500	227	91	0.000435	0.000186
1000	500	227	92	0.000744	0.000308
1000	500	227	93	0.001242	0.000499
1000	500	227	94	0.002031	0.000788
1000	500	227	95	0.003248	0.001218
1000	500	227	96	0.005086	0.001837
1000	500	227	97	0.007795	0.002710
1000	500	227	98	0.011700	0.003904
1000	500	227	99	0.017197	0.005498
1000	500	227	100	0.024763	0.007565
1000	500	227	101	0.034937	0.010174
1000	500	227	102	0.048310	0.013373
1000	500	227	103	0.065488	0.017179
1000	500	227	104	0.087057	0.021569
1000	500	227	105	0.113527	0.026469
1000	500	227	106	0.145277	0.031751
1000	500	227	107	0.182505	0.037228
1000	500	227	108	0.225173	0.042667
1000	500	227	109	0.272974	0.047801
1000	500	227	110	0.325322	0.052349
1000	500	227	111	0.381363	0.056040
1000	500	227	112	0.440008	0.058645
1000	500	227	113	0.500000	0.059992
1000	500	228	82	0.000001	0.000000
1000	500	228	83	0.000002	0.000001
1000	500	228	84	0.000004	0.000002
1000	500	228	85	0.000008	0.000004
1000	500	228	86	0.000016	0.000008
1000	500	228	87	0.000031	0.000015

N	n	k	x	P(x)	p(x)
1000	500	228	88	0.000058	0.000027
1000	500	228	89	0.000106	0.000049
1000	500	228	90	0.000191	0.000085
1000	500	228	91	0.000337	0.000146
1000	500	228	92	0.000581	0.000244
1000	500	228	93	0.000980	0.000399
1000	500	228	94	0.001616	0.000637
1000	500	228	95	0.002610	0.000994
1000	500	228	96	0.004125	0.001515
1000	500	228	97	0.006383	0.002257
1000	500	228	98	0.009669	0.003287
1000	500	228	99	0.014345	0.004676
1000	500	228	100	0.020848	0.006503
1000	500	228	101	0.029686	0.008838
1000	500	228	102	0.041425	0.011739
1000	500	228	103	0.056666	0.015241
1000	500	228	104	0.076008	0.019342
1000	500	228	105	0.100001	0.023993
1000	500	228	106	0.129094	0.029093
1000	500	228	107	0.163578	0.034484
1000	500	228	108	0.203535	0.039957
1000	500	228	109	0.248795	0.045259
1000	500	228	110	0.298911	0.050116
1000	500	228	111	0.353161	0.054250
1000	500	228	112	0.410572	0.057410
1000	500	228	113	0.469965	0.059394
1000	500	228	114	0.530035	0.060070
1000	500	229	82	0.000001	0.000000
1000	500	229	83	0.000001	0.000001
1000	500	229	84	0.000003	0.000001
1000	500	229	85	0.000006	0.000003
1000	500	229	86	0.000012	0.000006
1000	500	229	87	0.000023	0.000011
1000	500	229	88	0.000043	0.000020
1000	500	229	89	0.000080	0.000037
1000	500	229	90	0.000146	0.000066
1000	500	229	91	0.000260	0.000114
1000	500	229	92	0.000452	0.000192
1000	500	229	93	0.000769	0.000317
1000	500	229	94	0.001282	0.000512
1000	500	229	95	0.002089	0.000807
1000	500	229	96	0.003332	0.001244
1000	500	229	97	0.005204	0.001872
1000	500	229	98	0.007958	0.002754
1000	500	229	99	0.011916	0.003959
1000	500	229	100	0.017479	0.005562
1000	500	229	101	0.025118	0.007639
1000	500	229	102	0.035373	0.010255
1000	500	229	103	0.048828	0.013455
1000	500	229	104	0.066086	0.017258

N	n	k	x	P(x)	p(x)
1000	500	229	105	0.087725	0.021638
1000	500	229	106	0.114246	0.026521
1000	500	229	107	0.146023	0.031778
1000	500	229	108	0.183246	0.037223
1000	500	229	109	0.225872	0.042626
1000	500	229	110	0.273593	0.047721
1000	500	229	111	0.325825	0.052232
1000	500	229	112	0.381718	0.055892
1000	500	229	113	0.440191	0.058474
1000	500	229	114	0.500000	0.059809
1000	500	230	83	0.000001	0.000001
1000	500	230	84	0.000002	0.000001
1000	500	230	85	0.000004	0.000002
1000	500	230	86	0.000009	0.000004
1000	500	230	87	0.000017	0.000008
1000	500	230	88	0.000032	0.000015
1000	500	230	89	0.000060	0.000028
1000	500	230	90	0.000111	0.000051
1000	500	230	91	0.000199	0.000088
1000	500	230	92	0.000350	0.000151
1000	500	230	93	0.000602	0.000252
1000	500	230	94	0.001012	0.000410
1000	500	230	95	0.001665	0.000653
1000	500	230	96	0.002681	0.001016
1000	500	230	97	0.004226	0.001545
1000	500	230	98	0.006522	0.002296
1000	500	230	99	0.009857	0.003335
1000	500	230	100	0.014593	0.004735
1000	500	230	101	0.021165	0.006572
1000	500	230	102	0.030079	0.008914
1000	500	230	103	0.041900	0.011821
1000	500	230	104	0.057222	0.015322
1000	500	230	105	0.076639	0.019417
1000	500	230	106	0.100693	0.024054
1000	500	230	107	0.129826	0.029133
1000	500	230	108	0.164321	0.034496
1000	500	230	109	0.204255	0.039934
1000	500	230	110	0.249454	0.045199
1000	500	230	111	0.299472	0.050018
1000	500	230	112	0.353590	0.054118
1000	500	230	113	0.410841	0.057250
1000	500	230	114	0.470057	0.059216
1000	500	230	115	0.529943	0.059886
1000	500	231	83	0.000001	0.000000
1000	500	231	84	0.000001	0.000001
1000	500	231	85	0.000003	0.000002
1000	500	231	86	0.000006	0.000003
1000	500	231	87	0.000012	0.000006
1000	500	231	88	0.000024	0.000012
1000	500	231	89	0.000045	0.000021

N	n	k	x	P(x)	p(x)
1000	500	231	90	0.000084	0.000039
1000	500	231	91	0.000153	0.000068
1000	500	231	92	0.000270	0.000118
1000	500	231	93	0.000469	0.000199
1000	500	231	94	0.000796	0.000327
1000	500	231	95	0.001321	0.000526
1000	500	231	96	0.002148	0.000826
1000	500	231	97	0.003417	0.001269
1000	500	231	98	0.005323	0.001906
1000	500	231	99	0.008120	0.002797
1000	500	231	100	0.012133	0.004013
1000	500	231	101	0.017759	0.005626
1000	500	231	102	0.025471	0.007712
1000	500	231	103	0.035805	0.010334
1000	500	231	104	0.049342	0.013537
1000	500	231	105	0.066678	0.017336
1000	500	231	106	0.088385	0.021706
1000	500	231	107	0.114957	0.026572
1000	500	231	108	0.146760	0.031803
1000	500	231	109	0.183977	0.037217
1000	500	231	110	0.226561	0.042584
1000	500	231	111	0.274203	0.047642
1000	500	231	112	0.326320	0.052117
1000	500	231	113	0.382067	0.055746
1000	500	231	114	0.440372	0.058305
1000	500	231	115	0.500000	0.059628
1000	500	232	84	0.000001	0.000001
1000	500	232	85	0.000002	0.000001
1000	500	232	86	0.000005	0.000002
1000	500	232	87	0.000009	0.000005
1000	500	232	88	0.000018	0.000009
1000	500	232	89	0.000034	0.000016
1000	500	232	90	0.000063	0.000030
1000	500	232	91	0.000116	0.000053
1000	500	232	92	0.000208	0.000092
1000	500	232	93	0.000364	0.000156
1000	500	232	94	0.000623	0.000259
1000	500	232	95	0.001045	0.000421
1000	500	232	96	0.001714	0.000669
1000	500	232	97	0.002752	0.001038
1000	500	232	98	0.004327	0.001575
1000	500	232	99	0.006662	0.002335
1000	500	232	100	0.010046	0.003384
1000	500	232	101	0.014840	0.004794
1000	500	232	102	0.021480	0.006640
1000	500	232	103	0.030470	0.008990
1000	500	232	104	0.042371	0.011901
1000	500	232	105	0.057773	0.015402
1000	500	232	106	0.077263	0.019490
1000	500	232	107	0.101377	0.024114

Table for $N = 1000,\ n = 500$

N	n	k	x	P(x)	p(x)
1000	500	232	108	0.130548	0.029171
1000	500	232	109	0.165054	0.034506
1000	500	232	110	0.204965	0.039911
1000	500	232	111	0.250104	0.045139
1000	500	232	112	0.300025	0.049921
1000	500	232	113	0.354012	0.053988
1000	500	232	114	0.411106	0.057093
1000	500	232	115	0.470147	0.059041
1000	500	232	116	0.529853	0.059706
1000	500	233	84	0.000001	0.000000
1000	500	233	85	0.000002	0.000001
1000	500	233	86	0.000003	0.000002
1000	500	233	87	0.000007	0.000003
1000	500	233	88	0.000013	0.000006
1000	500	233	89	0.000025	0.000012
1000	500	233	90	0.000048	0.000022
1000	500	233	91	0.000088	0.000040
1000	500	233	92	0.000159	0.000071
1000	500	233	93	0.000281	0.000122
1000	500	233	94	0.000486	0.000205
1000	500	233	95	0.000822	0.000336
1000	500	233	96	0.001362	0.000539
1000	500	233	97	0.002207	0.000845
1000	500	233	98	0.003502	0.001295
1000	500	233	99	0.005443	0.001940
1000	500	233	100	0.008283	0.002841
1000	500	233	101	0.012349	0.004066
1000	500	233	102	0.018039	0.005690
1000	500	233	103	0.025823	0.007784
1000	500	233	104	0.036235	0.010412
1000	500	233	105	0.049852	0.013617
1000	500	233	106	0.067265	0.017413
1000	500	233	107	0.089038	0.021773
1000	500	233	108	0.115659	0.026621
1000	500	233	109	0.147486	0.031828
1000	500	233	110	0.184697	0.037211
1000	500	233	111	0.227240	0.042543
1000	500	233	112	0.274804	0.047564
1000	500	233	113	0.326808	0.052004
1000	500	233	114	0.382410	0.055603
1000	500	233	115	0.440550	0.058139
1000	500	233	116	0.500000	0.059450
1000	500	234	84	0.000001	0.000000
1000	500	234	85	0.000001	0.000001
1000	500	234	86	0.000002	0.000001
1000	500	234	87	0.000005	0.000002
1000	500	234	88	0.000010	0.000005
1000	500	234	89	0.000019	0.000009
1000	500	234	90	0.000036	0.000017
1000	500	234	91	0.000067	0.000031

N	n	k	x	P(x)	p(x)
1000	500	234	92	0.000121	0.000055
1000	500	234	93	0.000216	0.000095
1000	500	234	94	0.000378	0.000161
1000	500	234	95	0.000645	0.000267
1000	500	234	96	0.001078	0.000433
1000	500	234	97	0.001763	0.000685
1000	500	234	98	0.002824	0.001061
1000	500	234	99	0.004428	0.001605
1000	500	234	100	0.006802	0.002374
1000	500	234	101	0.010234	0.003432
1000	500	234	102	0.015087	0.004852
1000	500	234	103	0.021794	0.006707
1000	500	234	104	0.030859	0.009065
1000	500	234	105	0.042839	0.011980
1000	500	234	106	0.058320	0.015481
1000	500	234	107	0.077881	0.019562
1000	500	234	108	0.102053	0.024172
1000	500	234	109	0.131261	0.029208
1000	500	234	110	0.165776	0.034515
1000	500	234	111	0.205664	0.039887
1000	500	234	112	0.250743	0.045079
1000	500	234	113	0.300569	0.049826
1000	500	234	114	0.354428	0.053859
1000	500	234	115	0.411366	0.056938
1000	500	234	116	0.470236	0.058870
1000	500	234	117	0.529764	0.059528
1000	500	235	85	0.000001	0.000000
1000	500	235	86	0.000002	0.000001
1000	500	235	87	0.000004	0.000002
1000	500	235	88	0.000007	0.000004
1000	500	235	89	0.000014	0.000007
1000	500	235	90	0.000027	0.000013
1000	500	235	91	0.000050	0.000023
1000	500	235	92	0.000092	0.000042
1000	500	235	93	0.000166	0.000074
1000	500	235	94	0.000292	0.000126
1000	500	235	95	0.000504	0.000211
1000	500	235	96	0.000849	0.000346
1000	500	235	97	0.001402	0.000553
1000	500	235	98	0.002267	0.000865
1000	500	235	99	0.003588	0.001321
1000	500	235	100	0.005562	0.001974
1000	500	235	101	0.008446	0.002884
1000	500	235	102	0.012565	0.004119
1000	500	235	103	0.018318	0.005752
1000	500	235	104	0.026173	0.007855
1000	500	235	105	0.036661	0.010489
1000	500	235	106	0.050357	0.013696
1000	500	235	107	0.067845	0.017488
1000	500	235	108	0.089683	0.021838

N	n	k	x	P(x)	p(x)
1000	500	235	109	0.116352	0.026669
1000	500	235	110	0.148203	0.031851
1000	500	235	111	0.185408	0.037204
1000	500	235	112	0.227909	0.042501
1000	500	235	113	0.275396	0.047487
1000	500	235	114	0.327287	0.051892
1000	500	235	115	0.382748	0.055461
1000	500	235	116	0.440724	0.057976
1000	500	235	117	0.500000	0.059276
1000	500	236	85	0.000001	0.000000
1000	500	236	86	0.000001	0.000001
1000	500	236	87	0.000003	0.000001
1000	500	236	88	0.000005	0.000003
1000	500	236	89	0.000010	0.000005
1000	500	236	90	0.000020	0.000010
1000	500	236	91	0.000038	0.000018
1000	500	236	92	0.000070	0.000032
1000	500	236	93	0.000127	0.000057
1000	500	236	94	0.000225	0.000098
1000	500	236	95	0.000392	0.000166
1000	500	236	96	0.000667	0.000275
1000	500	236	97	0.001111	0.000444
1000	500	236	98	0.001813	0.000702
1000	500	236	99	0.002896	0.001083
1000	500	236	100	0.004530	0.001635
1000	500	236	101	0.006942	0.002412
1000	500	236	102	0.010422	0.003480
1000	500	236	103	0.015332	0.004910
1000	500	236	104	0.022107	0.006774
1000	500	236	105	0.031246	0.009139
1000	500	236	106	0.043303	0.012058
1000	500	236	107	0.058861	0.015558
1000	500	236	108	0.078493	0.019632
1000	500	236	109	0.102721	0.024228
1000	500	236	110	0.131966	0.029244
1000	500	236	111	0.166489	0.034524
1000	500	236	112	0.206353	0.039863
1000	500	236	113	0.251373	0.045020
1000	500	236	114	0.301104	0.049731
1000	500	236	115	0.354837	0.053733
1000	500	236	116	0.411623	0.056786
1000	500	236	117	0.470323	0.058701
1000	500	236	118	0.529676	0.059353
1000	500	237	86	0.000001	0.000000
1000	500	237	87	0.000002	0.000001
1000	500	237	88	0.000004	0.000002
1000	500	237	89	0.000007	0.000004
1000	500	237	90	0.000015	0.000007
1000	500	237	91	0.000028	0.000013
1000	500	237	92	0.000053	0.000025

N	n	k	x	P(x)	p(x)
1000	500	237	93	0.000096	0.000044
1000	500	237	94	0.000173	0.000076
1000	500	237	95	0.000303	0.000131
1000	500	237	96	0.000521	0.000218
1000	500	237	97	0.000877	0.000355
1000	500	237	98	0.001443	0.000567
1000	500	237	99	0.002327	0.000884
1000	500	237	100	0.003674	0.001347
1000	500	237	101	0.005682	0.002008
1000	500	237	102	0.008609	0.002927
1000	500	237	103	0.012781	0.004172
1000	500	237	104	0.018595	0.005814
1000	500	237	105	0.026521	0.007925
1000	500	237	106	0.037085	0.010565
1000	500	237	107	0.050858	0.013773
1000	500	237	108	0.068420	0.017562
1000	500	237	109	0.090322	0.021902
1000	500	237	110	0.117037	0.026715
1000	500	237	111	0.148911	0.031874
1000	500	237	112	0.186108	0.037197
1000	500	237	113	0.228568	0.042460
1000	500	237	114	0.275978	0.047410
1000	500	237	115	0.327759	0.051781
1000	500	237	116	0.383081	0.055322
1000	500	237	117	0.440896	0.057815
1000	500	237	118	0.500000	0.059104
1000	500	238	86	0.000001	0.000000
1000	500	238	87	0.000001	0.000001
1000	500	238	88	0.000003	0.000001
1000	500	238	89	0.000005	0.000003
1000	500	238	90	0.000011	0.000005
1000	500	238	91	0.000021	0.000010
1000	500	238	92	0.000039	0.000019
1000	500	238	93	0.000073	0.000034
1000	500	238	94	0.000132	0.000059
1000	500	238	95	0.000234	0.000102
1000	500	238	96	0.000406	0.000172
1000	500	238	97	0.000689	0.000283
1000	500	238	98	0.001145	0.000456
1000	500	238	99	0.001863	0.000718
1000	500	238	100	0.002968	0.001105
1000	500	238	101	0.004632	0.001664
1000	500	238	102	0.007083	0.002450
1000	500	238	103	0.010610	0.003528
1000	500	238	104	0.015578	0.004967
1000	500	238	105	0.022418	0.006840
1000	500	238	106	0.031630	0.009212
1000	500	238	107	0.043764	0.012134
1000	500	238	108	0.059398	0.015633
1000	500	238	109	0.079098	0.019701

Table for $N = 1000$, $n = 500$

N	n	k	x	P(x)	p(x)
1000	500	238	110	0.103382	0.024284
1000	500	238	111	0.132661	0.029279
1000	500	238	112	0.167193	0.034532
1000	500	238	113	0.207032	0.039839
1000	500	238	114	0.251994	0.044962
1000	500	238	115	0.301631	0.049638
1000	500	238	116	0.355239	0.053608
1000	500	238	117	0.411875	0.056636
1000	500	238	118	0.470409	0.058534
1000	500	238	119	0.529591	0.059181
1000	500	239	87	0.000001	0.000000
1000	500	239	88	0.000002	0.000001
1000	500	239	89	0.000004	0.000002
1000	500	239	90	0.000008	0.000004
1000	500	239	91	0.000015	0.000008
1000	500	239	92	0.000029	0.000014
1000	500	239	93	0.000055	0.000026
1000	500	239	94	0.000101	0.000046
1000	500	239	95	0.000180	0.000079
1000	500	239	96	0.000315	0.000135
1000	500	239	97	0.000539	0.000224
1000	500	239	98	0.000904	0.000365
1000	500	239	99	0.001485	0.000581
1000	500	239	100	0.002388	0.000903
1000	500	239	101	0.003761	0.001373
1000	500	239	102	0.005803	0.002042
1000	500	239	103	0.008772	0.002970
1000	500	239	104	0.012996	0.004224
1000	500	239	105	0.018872	0.005876
1000	500	239	106	0.026867	0.007995
1000	500	239	107	0.037506	0.010639
1000	500	239	108	0.051355	0.013849
1000	500	239	109	0.068989	0.017634
1000	500	239	110	0.090953	0.021964
1000	500	239	111	0.117714	0.026761
1000	500	239	112	0.149610	0.031896
1000	500	239	113	0.186799	0.037189
1000	500	239	114	0.229218	0.042419
1000	500	239	115	0.276552	0.047334
1000	500	239	116	0.328224	0.051672
1000	500	239	117	0.383408	0.055184
1000	500	239	118	0.441065	0.057657
1000	500	239	119	0.500000	0.058935
1000	500	240	87	0.000001	0.000000
1000	500	240	88	0.000001	0.000001
1000	500	240	89	0.000003	0.000001
1000	500	240	90	0.000006	0.000003
1000	500	240	91	0.000011	0.000006
1000	500	240	92	0.000022	0.000011
1000	500	240	93	0.000041	0.000019

N	n	k	x	P(x)	p(x)
1000	500	240	94	0.000076	0.000035
1000	500	240	95	0.000138	0.000061
1000	500	240	96	0.000243	0.000105
1000	500	240	97	0.000420	0.000177
1000	500	240	98	0.000711	0.000291
1000	500	240	99	0.001179	0.000468
1000	500	240	100	0.001913	0.000734
1000	500	240	101	0.003041	0.001128
1000	500	240	102	0.004735	0.001694
1000	500	240	103	0.007223	0.002488
1000	500	240	104	0.010798	0.003575
1000	500	240	105	0.015822	0.005024
1000	500	240	106	0.022727	0.006905
1000	500	240	107	0.032011	0.009284
1000	500	240	108	0.044221	0.012210
1000	500	240	109	0.059929	0.015708
1000	500	240	110	0.079697	0.019768
1000	500	240	111	0.104035	0.024338
1000	500	240	112	0.133347	0.029313
1000	500	240	113	0.167887	0.034539
1000	500	240	114	0.207702	0.039815
1000	500	240	115	0.252605	0.044903
1000	500	240	116	0.302150	0.049545
1000	500	240	117	0.355635	0.053485
1000	500	240	118	0.412123	0.056488
1000	500	240	119	0.470494	0.058371
1000	500	240	120	0.529506	0.059012
1000	500	241	88	0.000001	0.000001
1000	500	241	89	0.000002	0.000001
1000	500	241	90	0.000004	0.000002
1000	500	241	91	0.000008	0.000004
1000	500	241	92	0.000016	0.000008
1000	500	241	93	0.000031	0.000015
1000	500	241	94	0.000058	0.000027
1000	500	241	95	0.000105	0.000047
1000	500	241	96	0.000187	0.000082
1000	500	241	97	0.000326	0.000139
1000	500	241	98	0.000557	0.000231
1000	500	241	99	0.000932	0.000375
1000	500	241	100	0.001527	0.000594
1000	500	241	101	0.002449	0.000922
1000	500	241	102	0.003848	0.001399
1000	500	241	103	0.005923	0.002076
1000	500	241	104	0.008936	0.003012
1000	500	241	105	0.013211	0.004276
1000	500	241	106	0.019148	0.005937
1000	500	241	107	0.027211	0.008063
1000	500	241	108	0.037924	0.010713
1000	500	241	109	0.051848	0.013924
1000	500	241	110	0.069553	0.017705

N	n	k	x	P(x)	p(x)
1000	500	241	111	0.091578	0.022025
1000	500	241	112	0.118383	0.026805
1000	500	241	113	0.150299	0.031916
1000	500	241	114	0.187480	0.037181
1000	500	241	115	0.229858	0.042378
1000	500	241	116	0.277117	0.047259
1000	500	241	117	0.328681	0.051565
1000	500	241	118	0.383730	0.055049
1000	500	241	119	0.441232	0.057501
1000	500	241	120	0.500000	0.058768
1000	500	242	88	0.000001	0.000000
1000	500	242	89	0.000001	0.000001
1000	500	242	90	0.000003	0.000002
1000	500	242	91	0.000006	0.000003
1000	500	242	92	0.000012	0.000006
1000	500	242	93	0.000023	0.000011
1000	500	242	94	0.000043	0.000020
1000	500	242	95	0.000080	0.000036
1000	500	242	96	0.000143	0.000064
1000	500	242	97	0.000252	0.000109
1000	500	242	98	0.000435	0.000183
1000	500	242	99	0.000734	0.000299
1000	500	242	100	0.001213	0.000479
1000	500	242	101	0.001964	0.000751
1000	500	242	102	0.003114	0.001150
1000	500	242	103	0.004838	0.001723
1000	500	242	104	0.007364	0.002526
1000	500	242	105	0.010986	0.003622
1000	500	242	106	0.016066	0.005080
1000	500	242	107	0.023036	0.006970
1000	500	242	108	0.032391	0.009355
1000	500	242	109	0.044675	0.012284
1000	500	242	110	0.060455	0.015781
1000	500	242	111	0.080290	0.019834
1000	500	242	112	0.104680	0.024390
1000	500	242	113	0.134025	0.029345
1000	500	242	114	0.168571	0.034546
1000	500	242	115	0.208362	0.039791
1000	500	242	116	0.253207	0.044845
1000	500	242	117	0.302661	0.049454
1000	500	242	118	0.356025	0.053363
1000	500	242	119	0.412367	0.056343
1000	500	242	120	0.470577	0.058210
1000	500	242	121	0.529423	0.058846
1000	500	243	88	0.000000	0.000000
1000	500	243	89	0.000001	0.000001
1000	500	243	90	0.000002	0.000001
1000	500	243	91	0.000004	0.000002
1000	500	243	92	0.000009	0.000004
1000	500	243	93	0.000017	0.000008

N	n	k	x	P(x)	p(x)
1000	500	243	94	0.000033	0.000015
1000	500	243	95	0.000060	0.000028
1000	500	243	96	0.000109	0.000049
1000	500	243	97	0.000194	0.000085
1000	500	243	98	0.000338	0.000144
1000	500	243	99	0.000576	0.000238
1000	500	243	100	0.000961	0.000385
1000	500	243	101	0.001569	0.000608
1000	500	243	102	0.002510	0.000941
1000	500	243	103	0.003935	0.001425
1000	500	243	104	0.006044	0.002109
1000	500	243	105	0.009099	0.003054
1000	500	243	106	0.013425	0.004327
1000	500	243	107	0.019422	0.005997
1000	500	243	108	0.027553	0.008131
1000	500	243	109	0.038338	0.010785
1000	500	243	110	0.052336	0.013998
1000	500	243	111	0.070111	0.017775
1000	500	243	112	0.092195	0.022084
1000	500	243	113	0.119043	0.026848
1000	500	243	114	0.150979	0.031936
1000	500	243	115	0.188151	0.037172
1000	500	243	116	0.230489	0.042337
1000	500	243	117	0.277673	0.047185
1000	500	243	118	0.329132	0.051458
1000	500	243	119	0.384047	0.054916
1000	500	243	120	0.441395	0.057348
1000	500	243	121	0.500000	0.058604
1000	500	244	89	0.000001	0.000000
1000	500	244	90	0.000002	0.000001
1000	500	244	91	0.000003	0.000002
1000	500	244	92	0.000006	0.000003
1000	500	244	93	0.000013	0.000006
1000	500	244	94	0.000024	0.000012
1000	500	244	95	0.000045	0.000021
1000	500	244	96	0.000083	0.000038
1000	500	244	97	0.000149	0.000066
1000	500	244	98	0.000262	0.000113
1000	500	244	99	0.000450	0.000188
1000	500	244	100	0.000757	0.000307
1000	500	244	101	0.001248	0.000491
1000	500	244	102	0.002015	0.000767
1000	500	244	103	0.003188	0.001172
1000	500	244	104	0.004941	0.001753
1000	500	244	105	0.007505	0.002564
1000	500	244	106	0.011173	0.003669
1000	500	244	107	0.016309	0.005136
1000	500	244	108	0.023342	0.007034
1000	500	244	109	0.032767	0.009425
1000	500	244	110	0.045124	0.012357

Table for $N = 1000$, $n = 500$

N	n	k	x	P(x)	p(x)
1000	500	244	111	0.060977	0.015852
1000	500	244	112	0.080876	0.019899
1000	500	244	113	0.105318	0.024442
1000	500	244	114	0.134695	0.029377
1000	500	244	115	0.169246	0.034551
1000	500	244	116	0.209012	0.039766
1000	500	244	117	0.253801	0.044788
1000	500	244	118	0.303165	0.049364
1000	500	244	119	0.356408	0.053244
1000	500	244	120	0.412608	0.056200
1000	500	244	121	0.470659	0.058051
1000	500	244	122	0.529341	0.058682
1000	500	245	89	0.000001	0.000000
1000	500	245	90	0.000001	0.000001
1000	500	245	91	0.000002	0.000001
1000	500	245	92	0.000005	0.000002
1000	500	245	93	0.000009	0.000005
1000	500	245	94	0.000018	0.000009
1000	500	245	95	0.000034	0.000016
1000	500	245	96	0.000063	0.000029
1000	500	245	97	0.000114	0.000051
1000	500	245	98	0.000202	0.000088
1000	500	245	99	0.000350	0.000148
1000	500	245	100	0.000595	0.000245
1000	500	245	101	0.000989	0.000394
1000	500	245	102	0.001612	0.000622
1000	500	245	103	0.002572	0.000961
1000	500	245	104	0.004023	0.001450
1000	500	245	105	0.006165	0.002143
1000	500	245	106	0.009261	0.003096
1000	500	245	107	0.013639	0.004378
1000	500	245	108	0.019695	0.006056
1000	500	245	109	0.027893	0.008197
1000	500	245	110	0.038750	0.010857
1000	500	245	111	0.052820	0.014070
1000	500	245	112	0.070663	0.017843
1000	500	245	113	0.092806	0.022143
1000	500	245	114	0.119695	0.026889
1000	500	245	115	0.151650	0.031955
1000	500	245	116	0.188814	0.037163
1000	500	245	117	0.231110	0.042297
1000	500	245	118	0.278221	0.047111
1000	500	245	119	0.329575	0.051354
1000	500	245	120	0.384359	0.054784
1000	500	245	121	0.441557	0.057197
1000	500	245	122	0.500000	0.058443
1000	500	246	90	0.000001	0.000000
1000	500	246	91	0.000002	0.000001
1000	500	246	92	0.000003	0.000002
1000	500	246	93	0.000007	0.000003

N	n	k	x	P(x)	p(x)
1000	500	246	94	0.000013	0.000007
1000	500	246	95	0.000026	0.000012
1000	500	246	96	0.000048	0.000022
1000	500	246	97	0.000087	0.000039
1000	500	246	98	0.000155	0.000068
1000	500	246	99	0.000271	0.000116
1000	500	246	100	0.000465	0.000194
1000	500	246	101	0.000781	0.000316
1000	500	246	102	0.001284	0.000503
1000	500	246	103	0.002067	0.000784
1000	500	246	104	0.003262	0.001195
1000	500	246	105	0.005044	0.001782
1000	500	246	106	0.007646	0.002601
1000	500	246	107	0.011360	0.003715
1000	500	246	108	0.016551	0.005191
1000	500	246	109	0.023648	0.007097
1000	500	246	110	0.033142	0.009494
1000	500	246	111	0.045571	0.012429
1000	500	246	112	0.061493	0.015923
1000	500	246	113	0.081456	0.019962
1000	500	246	114	0.105948	0.024492
1000	500	246	115	0.135355	0.029408
1000	500	246	116	0.169912	0.034557
1000	500	246	117	0.209654	0.039742
1000	500	246	118	0.254385	0.044731
1000	500	246	119	0.303660	0.049275
1000	500	246	120	0.356786	0.053126
1000	500	246	121	0.412844	0.056058
1000	500	246	122	0.470739	0.057895
1000	500	246	123	0.529260	0.058521
1000	500	247	90	0.000001	0.000000
1000	500	247	91	0.000001	0.000001
1000	500	247	92	0.000003	0.000001
1000	500	247	93	0.000005	0.000003
1000	500	247	94	0.000010	0.000005
1000	500	247	95	0.000019	0.000009
1000	500	247	96	0.000036	0.000017
1000	500	247	97	0.000066	0.000030
1000	500	247	98	0.000119	0.000053
1000	500	247	99	0.000210	0.000091
1000	500	247	100	0.000362	0.000153
1000	500	247	101	0.000614	0.000251
1000	500	247	102	0.001018	0.000404
1000	500	247	103	0.001655	0.000636
1000	500	247	104	0.002634	0.000980
1000	500	247	105	0.004111	0.001476
1000	500	247	106	0.006286	0.002176
1000	500	247	107	0.009424	0.003138
1000	500	247	108	0.013853	0.004428
1000	500	247	109	0.019968	0.006115

N	n	k	x	P(x)	p(x)
1000	500	247	110	0.028231	0.008263
1000	500	247	111	0.039158	0.010927
1000	500	247	112	0.053300	0.014141
1000	500	247	113	0.071210	0.017910
1000	500	247	114	0.093409	0.022200
1000	500	247	115	0.120340	0.026930
1000	500	247	116	0.152313	0.031973
1000	500	247	117	0.189467	0.037154
1000	500	247	118	0.231723	0.042256
1000	500	247	119	0.278761	0.047038
1000	500	247	120	0.330012	0.051250
1000	500	247	121	0.384666	0.054655
1000	500	247	122	0.441715	0.057049
1000	500	247	123	0.500000	0.058285
1000	500	248	91	0.000001	0.000000
1000	500	248	92	0.000002	0.000001
1000	500	248	93	0.000004	0.000002
1000	500	248	94	0.000007	0.000004
1000	500	248	95	0.000014	0.000007
1000	500	248	96	0.000027	0.000013
1000	500	248	97	0.000050	0.000023
1000	500	248	98	0.000090	0.000041
1000	500	248	99	0.000161	0.000071
1000	500	248	100	0.000281	0.000120
1000	500	248	101	0.000481	0.000199
1000	500	248	102	0.000804	0.000324
1000	500	248	103	0.001319	0.000515
1000	500	248	104	0.002119	0.000800
1000	500	248	105	0.003336	0.001217
1000	500	248	106	0.005148	0.001812
1000	500	248	107	0.007787	0.002639
1000	500	248	108	0.011547	0.003761
1000	500	248	109	0.016792	0.005245
1000	500	248	110	0.023951	0.007159
1000	500	248	111	0.033513	0.009562
1000	500	248	112	0.046013	0.012500
1000	500	248	113	0.062005	0.015992
1000	500	248	114	0.082030	0.020025
1000	500	248	115	0.106570	0.024541
1000	500	248	116	0.136008	0.029437
1000	500	248	117	0.170569	0.034561
1000	500	248	118	0.210286	0.039717
1000	500	248	119	0.254961	0.044675
1000	500	248	120	0.304148	0.049187
1000	500	248	121	0.357157	0.053009
1000	500	248	122	0.413077	0.055920
1000	500	248	123	0.470819	0.057742
1000	500	248	124	0.529181	0.058362
1000	500	249	91	0.000001	0.000000
1000	500	249	92	0.000001	0.000001

N	n	k	x	P(x)	p(x)
1000	500	249	93	0.000003	0.000001
1000	500	249	94	0.000005	0.000003
1000	500	249	95	0.000010	0.000005
1000	500	249	96	0.000020	0.000010
1000	500	249	97	0.000037	0.000017
1000	500	249	98	0.000069	0.000031
1000	500	249	99	0.000124	0.000055
1000	500	249	100	0.000217	0.000094
1000	500	249	101	0.000375	0.000157
1000	500	249	102	0.000633	0.000258
1000	500	249	103	0.001047	0.000414
1000	500	249	104	0.001698	0.000650
1000	500	249	105	0.002697	0.000999
1000	500	249	106	0.004199	0.001502
1000	500	249	107	0.006408	0.002209
1000	500	249	108	0.009587	0.003179
1000	500	249	109	0.014065	0.004478
1000	500	249	110	0.020238	0.006173
1000	500	249	111	0.028567	0.008328
1000	500	249	112	0.039564	0.010997
1000	500	249	113	0.053775	0.014211
1000	500	249	114	0.071751	0.017976
1000	500	249	115	0.094006	0.022255
1000	500	249	116	0.120976	0.026970
1000	500	249	117	0.152967	0.031991
1000	500	249	118	0.190111	0.037144
1000	500	249	119	0.232326	0.042216
1000	500	249	120	0.279293	0.046966
1000	500	249	121	0.330441	0.051148
1000	500	249	122	0.384969	0.054527
1000	500	249	123	0.441871	0.056903
1000	500	249	124	0.500000	0.058129
1000	500	250	92	0.000001	0.000000
1000	500	250	93	0.000002	0.000001
1000	500	250	94	0.000004	0.000002
1000	500	250	95	0.000008	0.000004
1000	500	250	96	0.000015	0.000007
1000	500	250	97	0.000028	0.000013
1000	500	250	98	0.000052	0.000024
1000	500	250	99	0.000094	0.000042
1000	500	250	100	0.000167	0.000073
1000	500	250	101	0.000291	0.000124
1000	500	250	102	0.000496	0.000205
1000	500	250	103	0.000828	0.000332
1000	500	250	104	0.001355	0.000526
1000	500	250	105	0.002171	0.000817
1000	500	250	106	0.003411	0.001239
1000	500	250	107	0.005252	0.001841
1000	500	250	108	0.007927	0.002676
1000	500	250	109	0.011734	0.003806

Table for $N = 1000$, $n = 500$

N	n	k	x	P(x)	p(x)
1000	500	250	110	0.017033	0.005299
1000	500	250	111	0.024253	0.007220
1000	500	250	112	0.033882	0.009629
1000	500	250	113	0.046452	0.012570
1000	500	250	114	0.062512	0.016060
1000	500	250	115	0.082597	0.020085
1000	500	250	116	0.107186	0.024589
1000	500	250	117	0.136652	0.029466
1000	500	250	118	0.171217	0.034565
1000	500	250	119	0.210909	0.039692
1000	500	250	120	0.255528	0.044619
1000	500	250	121	0.304629	0.049101
1000	500	250	122	0.357523	0.052895
1000	500	250	123	0.413306	0.055783
1000	500	250	124	0.470897	0.057591
1000	500	250	125	0.529103	0.058206
1000	500	251	92	0.000001	0.000000
1000	500	251	93	0.000001	0.000001
1000	500	251	94	0.000003	0.000001
1000	500	251	95	0.000006	0.000003
1000	500	251	96	0.000011	0.000005
1000	500	251	97	0.000021	0.000010
1000	500	251	98	0.000039	0.000018
1000	500	251	99	0.000072	0.000032
1000	500	251	100	0.000128	0.000057
1000	500	251	101	0.000225	0.000097
1000	500	251	102	0.000387	0.000162
1000	500	251	103	0.000653	0.000265
1000	500	251	104	0.001077	0.000424
1000	500	251	105	0.001741	0.000664
1000	500	251	106	0.002760	0.001018
1000	500	251	107	0.004287	0.001527
1000	500	251	108	0.006529	0.002242
1000	500	251	109	0.009749	0.003220
1000	500	251	110	0.014277	0.004528
1000	500	251	111	0.020508	0.006231
1000	500	251	112	0.028901	0.008393
1000	500	251	113	0.039966	0.011065
1000	500	251	114	0.054246	0.014280
1000	500	251	115	0.072287	0.018040
1000	500	251	116	0.094596	0.022310
1000	500	251	117	0.121605	0.027008
1000	500	251	118	0.153612	0.032007
1000	500	251	119	0.190746	0.037134
1000	500	251	120	0.232921	0.042176
1000	500	251	121	0.279817	0.046895
1000	500	251	122	0.330865	0.051048
1000	500	251	123	0.385266	0.054402
1000	500	251	124	0.442025	0.056759
1000	500	251	125	0.500000	0.057975

N	n	k	x	P(x)	p(x)
1000	500	252	93	0.000001	0.000001
1000	500	252	94	0.000002	0.000001
1000	500	252	95	0.000004	0.000002
1000	500	252	96	0.000008	0.000004
1000	500	252	97	0.000016	0.000007
1000	500	252	98	0.000029	0.000014
1000	500	252	99	0.000054	0.000025
1000	500	252	100	0.000098	0.000044
1000	500	252	101	0.000174	0.000076
1000	500	252	102	0.000301	0.000128
1000	500	252	103	0.000512	0.000211
1000	500	252	104	0.000852	0.000340
1000	500	252	105	0.001391	0.000538
1000	500	252	106	0.002224	0.000833
1000	500	252	107	0.003486	0.001262
1000	500	252	108	0.005356	0.001870
1000	500	252	109	0.008068	0.002713
1000	500	252	110	0.011919	0.003851
1000	500	252	111	0.017272	0.005352
1000	500	252	112	0.024553	0.007281
1000	500	252	113	0.034249	0.009696
1000	500	252	114	0.046887	0.012638
1000	500	252	115	0.063014	0.016127
1000	500	252	116	0.083159	0.020145
1000	500	252	117	0.107794	0.024635
1000	500	252	118	0.137288	0.029494
1000	500	252	119	0.171857	0.034569
1000	500	252	120	0.211524	0.039667
1000	500	252	121	0.256087	0.044563
1000	500	252	122	0.305102	0.049015
1000	500	252	123	0.357884	0.052782
1000	500	252	124	0.413532	0.055648
1000	500	252	125	0.470974	0.057442
1000	500	252	126	0.529026	0.058053
1000	500	253	93	0.000001	0.000000
1000	500	253	94	0.000001	0.000001
1000	500	253	95	0.000003	0.000002
1000	500	253	96	0.000006	0.000003
1000	500	253	97	0.000012	0.000006
1000	500	253	98	0.000022	0.000010
1000	500	253	99	0.000041	0.000019
1000	500	253	100	0.000075	0.000034
1000	500	253	101	0.000133	0.000059
1000	500	253	102	0.000233	0.000100
1000	500	253	103	0.000400	0.000167
1000	500	253	104	0.000672	0.000272
1000	500	253	105	0.001107	0.000434
1000	500	253	106	0.001785	0.000679
1000	500	253	107	0.002823	0.001038
1000	500	253	108	0.004376	0.001553

N	n	k	x	P(x)	p(x)
1000	500	253	109	0.006650	0.002275
1000	500	253	110	0.009911	0.003261
1000	500	253	111	0.014488	0.004577
1000	500	253	112	0.020776	0.006288
1000	500	253	113	0.029232	0.008456
1000	500	253	114	0.040365	0.011133
1000	500	253	115	0.054713	0.014348
1000	500	253	116	0.072817	0.018104
1000	500	253	117	0.095180	0.022363
1000	500	253	118	0.122225	0.027046
1000	500	253	119	0.154249	0.032023
1000	500	253	120	0.191372	0.037123
1000	500	253	121	0.233508	0.042136
1000	500	253	122	0.280333	0.046825
1000	500	253	123	0.331281	0.050949
1000	500	253	124	0.385559	0.054278
1000	500	253	125	0.442176	0.056617
1000	500	253	126	0.500000	0.057824
1000	500	254	93	0.000001	0.000000
1000	500	254	94	0.000001	0.000001
1000	500	254	95	0.000002	0.000001
1000	500	254	96	0.000004	0.000002
1000	500	254	97	0.000009	0.000004
1000	500	254	98	0.000016	0.000008
1000	500	254	99	0.000031	0.000014
1000	500	254	100	0.000057	0.000026
1000	500	254	101	0.000102	0.000045
1000	500	254	102	0.000180	0.000078
1000	500	254	103	0.000312	0.000131
1000	500	254	104	0.000528	0.000216
1000	500	254	105	0.000877	0.000349
1000	500	254	106	0.001427	0.000550
1000	500	254	107	0.002277	0.000850
1000	500	254	108	0.003561	0.001284
1000	500	254	109	0.005460	0.001899
1000	500	254	110	0.008209	0.002749
1000	500	254	111	0.012105	0.003896
1000	500	254	112	0.017510	0.005405
1000	500	254	113	0.024851	0.007341
1000	500	254	114	0.034612	0.009761
1000	500	254	115	0.047318	0.012706
1000	500	254	116	0.063510	0.016192
1000	500	254	117	0.083714	0.020204
1000	500	254	118	0.108395	0.024681
1000	500	254	119	0.137916	0.029521
1000	500	254	120	0.172487	0.034572
1000	500	254	121	0.212130	0.039642
1000	500	254	122	0.256638	0.044508
1000	500	254	123	0.305568	0.048930
1000	500	254	124	0.358238	0.052670

N	n	k	x	P(x)	p(x)
1000	500	254	125	0.413754	0.055515
1000	500	254	126	0.471049	0.057295
1000	500	254	127	0.528951	0.057901
1000	500	255	94	0.000001	0.000000
1000	500	255	95	0.000002	0.000001
1000	500	255	96	0.000003	0.000002
1000	500	255	97	0.000006	0.000003
1000	500	255	98	0.000012	0.000006
1000	500	255	99	0.000023	0.000011
1000	500	255	100	0.000043	0.000020
1000	500	255	101	0.000078	0.000035
1000	500	255	102	0.000138	0.000061
1000	500	255	103	0.000242	0.000103
1000	500	255	104	0.000413	0.000172
1000	500	255	105	0.000692	0.000279
1000	500	255	106	0.001137	0.000444
1000	500	255	107	0.001829	0.000693
1000	500	255	108	0.002886	0.001057
1000	500	255	109	0.004464	0.001578
1000	500	255	110	0.006772	0.002307
1000	500	255	111	0.010073	0.003302
1000	500	255	112	0.014699	0.004626
1000	500	255	113	0.021043	0.006344
1000	500	255	114	0.029562	0.008519
1000	500	255	115	0.040761	0.011199
1000	500	255	116	0.055175	0.014415
1000	500	255	117	0.073341	0.018166
1000	500	255	118	0.095757	0.022415
1000	500	255	119	0.122839	0.027082
1000	500	255	120	0.154877	0.032038
1000	500	255	121	0.191990	0.037113
1000	500	255	122	0.234086	0.042096
1000	500	255	123	0.280841	0.046755
1000	500	255	124	0.331691	0.050851
1000	500	255	125	0.385847	0.054156
1000	500	255	126	0.442325	0.056477
1000	500	255	127	0.500000	0.057675
1000	500	256	94	0.000001	0.000000
1000	500	256	95	0.000001	0.000001
1000	500	256	96	0.000002	0.000001
1000	500	256	97	0.000005	0.000002
1000	500	256	98	0.000009	0.000004
1000	500	256	99	0.000017	0.000008
1000	500	256	100	0.000032	0.000015
1000	500	256	101	0.000059	0.000027
1000	500	256	102	0.000106	0.000047
1000	500	256	103	0.000187	0.000081
1000	500	256	104	0.000322	0.000135
1000	500	256	105	0.000544	0.000222
1000	500	256	106	0.000902	0.000357

Table for $N = 1000,\ n = 500$

N	n	k	x	P(x)	p(x)
1000	500	256	107	0.001464	0.000562
1000	500	256	108	0.002330	0.000866
1000	500	256	109	0.003636	0.001306
1000	500	256	110	0.005564	0.001928
1000	500	256	111	0.008349	0.002786
1000	500	256	112	0.012290	0.003941
1000	500	256	113	0.017747	0.005458
1000	500	256	114	0.025148	0.007401
1000	500	256	115	0.034973	0.009825
1000	500	256	116	0.047746	0.012772
1000	500	256	117	0.064002	0.016257
1000	500	256	118	0.084263	0.020261
1000	500	256	119	0.108989	0.024726
1000	500	256	120	0.138535	0.029547
1000	500	256	121	0.173110	0.034574
1000	500	256	122	0.212727	0.039617
1000	500	256	123	0.257181	0.044454
1000	500	256	124	0.306027	0.048846
1000	500	256	125	0.358588	0.052560
1000	500	256	126	0.413972	0.055385
1000	500	256	127	0.471124	0.057151
1000	500	256	128	0.528876	0.057753
1000	500	257	95	0.000001	0.000000
1000	500	257	96	0.000002	0.000001
1000	500	257	97	0.000003	0.000002
1000	500	257	98	0.000007	0.000003
1000	500	257	99	0.000013	0.000006
1000	500	257	100	0.000024	0.000011
1000	500	257	101	0.000045	0.000020
1000	500	257	102	0.000081	0.000036
1000	500	257	103	0.000144	0.000063
1000	500	257	104	0.000250	0.000106
1000	500	257	105	0.000426	0.000176
1000	500	257	106	0.000712	0.000286
1000	500	257	107	0.001167	0.000455
1000	500	257	108	0.001874	0.000707
1000	500	257	109	0.002950	0.001076
1000	500	257	110	0.004553	0.001604
1000	500	257	111	0.006893	0.002340
1000	500	257	112	0.010235	0.003342
1000	500	257	113	0.014909	0.004674
1000	500	257	114	0.021308	0.006400
1000	500	257	115	0.029889	0.008581
1000	500	257	116	0.041153	0.011264
1000	500	257	117	0.055634	0.014480
1000	500	257	118	0.073860	0.018227
1000	500	257	119	0.096327	0.022466
1000	500	257	120	0.123444	0.027117
1000	500	257	121	0.155497	0.032053
1000	500	257	122	0.192599	0.037102

N	n	k	x	P(x)	p(x)
1000	500	257	123	0.234655	0.042056
1000	500	257	124	0.281341	0.046686
1000	500	257	125	0.332095	0.050754
1000	500	257	126	0.386131	0.054036
1000	500	257	127	0.442471	0.056340
1000	500	257	128	0.500000	0.057529
1000	500	258	95	0.000001	0.000000
1000	500	258	96	0.000001	0.000001
1000	500	258	97	0.000002	0.000001
1000	500	258	98	0.000005	0.000002
1000	500	258	99	0.000009	0.000005
1000	500	258	100	0.000018	0.000009
1000	500	258	101	0.000034	0.000016
1000	500	258	102	0.000061	0.000028
1000	500	258	103	0.000110	0.000049
1000	500	258	104	0.000193	0.000083
1000	500	258	105	0.000332	0.000139
1000	500	258	106	0.000561	0.000228
1000	500	258	107	0.000926	0.000366
1000	500	258	108	0.001501	0.000574
1000	500	258	109	0.002383	0.000883
1000	500	258	110	0.003712	0.001328
1000	500	258	111	0.005668	0.001956
1000	500	258	112	0.008490	0.002822
1000	500	258	113	0.012474	0.003985
1000	500	258	114	0.017984	0.005509
1000	500	258	115	0.025443	0.007459
1000	500	258	116	0.035332	0.009889
1000	500	258	117	0.048169	0.012838
1000	500	258	118	0.064489	0.016320
1000	500	258	119	0.084806	0.020317
1000	500	258	120	0.109575	0.024769
1000	500	258	121	0.139147	0.029572
1000	500	258	122	0.173723	0.034576
1000	500	258	123	0.213316	0.039592
1000	500	258	124	0.257715	0.044400
1000	500	258	125	0.306479	0.048764
1000	500	258	126	0.358931	0.052452
1000	500	258	127	0.414187	0.055256
1000	500	258	128	0.471197	0.057009
1000	500	258	129	0.528803	0.057606
1000	500	259	96	0.000001	0.000000
1000	500	259	97	0.000002	0.000001
1000	500	259	98	0.000004	0.000002
1000	500	259	99	0.000007	0.000003
1000	500	259	100	0.000013	0.000006
1000	500	259	101	0.000025	0.000012
1000	500	259	102	0.000046	0.000021
1000	500	259	103	0.000084	0.000038
1000	500	259	104	0.000149	0.000065

N	n	k	x	P(x)	p(x)
1000	500	259	105	0.000258	0.000110
1000	500	259	106	0.000439	0.000181
1000	500	259	107	0.000733	0.000293
1000	500	259	108	0.001197	0.000465
1000	500	259	109	0.001918	0.000721
1000	500	259	110	0.003013	0.001095
1000	500	259	111	0.004642	0.001629
1000	500	259	112	0.007014	0.002372
1000	500	259	113	0.010396	0.003382
1000	500	259	114	0.015118	0.004722
1000	500	259	115	0.021573	0.006455
1000	500	259	116	0.030214	0.008642
1000	500	259	117	0.041543	0.011329
1000	500	259	118	0.056088	0.014545
1000	500	259	119	0.074374	0.018287
1000	500	259	120	0.096890	0.022516
1000	500	259	121	0.124042	0.027152
1000	500	259	122	0.156109	0.032067
1000	500	259	123	0.193199	0.037090
1000	500	259	124	0.235216	0.042017
1000	500	259	125	0.281834	0.046618
1000	500	259	126	0.332493	0.050659
1000	500	259	127	0.386410	0.053917
1000	500	259	128	0.442615	0.056205
1000	500	259	129	0.500000	0.057385
1000	500	260	96	0.000001	0.000000
1000	500	260	97	0.000001	0.000001
1000	500	260	98	0.000003	0.000001
1000	500	260	99	0.000005	0.000003
1000	500	260	100	0.000010	0.000005
1000	500	260	101	0.000019	0.000009
1000	500	260	102	0.000035	0.000016
1000	500	260	103	0.000064	0.000029
1000	500	260	104	0.000114	0.000050
1000	500	260	105	0.000200	0.000086
1000	500	260	106	0.000343	0.000143
1000	500	260	107	0.000577	0.000234
1000	500	260	108	0.000952	0.000374
1000	500	260	109	0.001538	0.000586
1000	500	260	110	0.002437	0.000899
1000	500	260	111	0.003787	0.001350
1000	500	260	112	0.005772	0.001985
1000	500	260	113	0.008630	0.002858
1000	500	260	114	0.012658	0.004028
1000	500	260	115	0.018219	0.005561
1000	500	260	116	0.025736	0.007517
1000	500	260	117	0.035687	0.009952
1000	500	260	118	0.048590	0.012902
1000	500	260	119	0.064972	0.016382
1000	500	260	120	0.085344	0.020372

N	n	k	x	P(x)	p(x)
1000	500	260	121	0.110155	0.024811
1000	500	260	122	0.139751	0.029596
1000	500	260	123	0.174329	0.034578
1000	500	260	124	0.213896	0.039567
1000	500	260	125	0.258242	0.044346
1000	500	260	126	0.306925	0.048682
1000	500	260	127	0.359270	0.052345
1000	500	260	128	0.414399	0.055129
1000	500	260	129	0.471269	0.056870
1000	500	260	130	0.528731	0.057462
1000	500	261	97	0.000001	0.000000
1000	500	261	98	0.000002	0.000001
1000	500	261	99	0.000004	0.000002
1000	500	261	100	0.000007	0.000004
1000	500	261	101	0.000014	0.000007
1000	500	261	102	0.000026	0.000012
1000	500	261	103	0.000048	0.000022
1000	500	261	104	0.000087	0.000039
1000	500	261	105	0.000154	0.000067
1000	500	261	106	0.000267	0.000113
1000	500	261	107	0.000453	0.000186
1000	500	261	108	0.000753	0.000300
1000	500	261	109	0.001228	0.000475
1000	500	261	110	0.001963	0.000735
1000	500	261	111	0.003077	0.001114
1000	500	261	112	0.004731	0.001654
1000	500	261	113	0.007135	0.002404
1000	500	261	114	0.010557	0.003421
1000	500	261	115	0.015326	0.004769
1000	500	261	116	0.021835	0.006509
1000	500	261	117	0.030537	0.008702
1000	500	261	118	0.041929	0.011392
1000	500	261	119	0.056537	0.014608
1000	500	261	120	0.074882	0.018345
1000	500	261	121	0.097448	0.022565
1000	500	261	122	0.124633	0.027186
1000	500	261	123	0.156713	0.032080
1000	500	261	124	0.193792	0.037079
1000	500	261	125	0.235769	0.041978
1000	500	261	126	0.282320	0.046551
1000	500	261	127	0.332885	0.050565
1000	500	261	128	0.386686	0.053800
1000	500	261	129	0.442757	0.056072
1000	500	261	130	0.500000	0.057243
1000	500	262	97	0.000001	0.000000
1000	500	262	98	0.000001	0.000001
1000	500	262	99	0.000003	0.000001
1000	500	262	100	0.000005	0.000003
1000	500	262	101	0.000010	0.000005
1000	500	262	102	0.000020	0.000009

Table for $N = 1000$, $n = 500$

N	n	k	x	P(x)	p(x)
1000	500	262	103	0.000037	0.000017
1000	500	262	104	0.000066	0.000030
1000	500	262	105	0.000118	0.000052
1000	500	262	106	0.000207	0.000088
1000	500	262	107	0.000354	0.000147
1000	500	262	108	0.000594	0.000240
1000	500	262	109	0.000977	0.000383
1000	500	262	110	0.001575	0.000598
1000	500	262	111	0.002491	0.000916
1000	500	262	112	0.003863	0.001372
1000	500	262	113	0.005876	0.002013
1000	500	262	114	0.008770	0.002893
1000	500	262	115	0.012841	0.004072
1000	500	262	116	0.018453	0.005611
1000	500	262	117	0.026027	0.007574
1000	500	262	118	0.036040	0.010013
1000	500	262	119	0.049006	0.012966
1000	500	262	120	0.065449	0.016443
1000	500	262	121	0.085875	0.020426
1000	500	262	122	0.110728	0.024853
1000	500	262	123	0.140347	0.029620
1000	500	262	124	0.174926	0.034579
1000	500	262	125	0.214468	0.039542
1000	500	262	126	0.258761	0.044293
1000	500	262	127	0.307363	0.048602
1000	500	262	128	0.359603	0.052240
1000	500	262	129	0.414608	0.055004
1000	500	262	130	0.471340	0.056732
1000	500	262	131	0.528660	0.057320
1000	500	263	98	0.000001	0.000000
1000	500	263	99	0.000002	0.000001
1000	500	263	100	0.000004	0.000002
1000	500	263	101	0.000008	0.000004
1000	500	263	102	0.000015	0.000007
1000	500	263	103	0.000027	0.000013
1000	500	263	104	0.000050	0.000023
1000	500	263	105	0.000091	0.000040
1000	500	263	106	0.000160	0.000069
1000	500	263	107	0.000276	0.000116
1000	500	263	108	0.000466	0.000191
1000	500	263	109	0.000774	0.000307
1000	500	263	110	0.001259	0.000485
1000	500	263	111	0.002008	0.000749
1000	500	263	112	0.003142	0.001133
1000	500	263	113	0.004821	0.001679
1000	500	263	114	0.007256	0.002436
1000	500	263	115	0.010717	0.003461
1000	500	263	116	0.015533	0.004816
1000	500	263	117	0.022096	0.006563
1000	500	263	118	0.030857	0.008761

N	n	k	x	P(x)	p(x)
1000	500	263	119	0.042312	0.011455
1000	500	263	120	0.056982	0.014670
1000	500	263	121	0.075385	0.018403
1000	500	263	122	0.097998	0.022613
1000	500	263	123	0.125216	0.027218
1000	500	263	124	0.157309	0.032092
1000	500	263	125	0.194376	0.037067
1000	500	263	126	0.236315	0.041939
1000	500	263	127	0.282799	0.046484
1000	500	263	128	0.333271	0.050472
1000	500	263	129	0.386956	0.053685
1000	500	263	130	0.442897	0.055940
1000	500	263	131	0.500000	0.057103
1000	500	264	98	0.000001	0.000000
1000	500	264	99	0.000001	0.000001
1000	500	264	100	0.000003	0.000001
1000	500	264	101	0.000006	0.000003
1000	500	264	102	0.000011	0.000005
1000	500	264	103	0.000021	0.000010
1000	500	264	104	0.000038	0.000018
1000	500	264	105	0.000069	0.000031
1000	500	264	106	0.000123	0.000054
1000	500	264	107	0.000214	0.000091
1000	500	264	108	0.000365	0.000151
1000	500	264	109	0.000611	0.000246
1000	500	264	110	0.001002	0.000391
1000	500	264	111	0.001613	0.000610
1000	500	264	112	0.002545	0.000932
1000	500	264	113	0.003939	0.001394
1000	500	264	114	0.005981	0.002042
1000	500	264	115	0.008909	0.002929
1000	500	264	116	0.013024	0.004114
1000	500	264	117	0.018685	0.005662
1000	500	264	118	0.026316	0.007631
1000	500	264	119	0.036390	0.010074
1000	500	264	120	0.049418	0.013028
1000	500	264	121	0.065922	0.016503
1000	500	264	122	0.086400	0.020479
1000	500	264	123	0.111293	0.024893
1000	500	264	124	0.140936	0.029643
1000	500	264	125	0.175515	0.034579
1000	500	264	126	0.215032	0.039517
1000	500	264	127	0.259273	0.044241
1000	500	264	128	0.307795	0.048522
1000	500	264	129	0.359932	0.052136
1000	500	264	130	0.414813	0.054881
1000	500	264	131	0.471410	0.056597
1000	500	264	132	0.528590	0.057181
1000	500	265	99	0.000001	0.000001
1000	500	265	100	0.000002	0.000001

N	n	k	x	P(x)	p(x)
1000	500	265	101	0.000004	0.000002
1000	500	265	102	0.000008	0.000004
1000	500	265	103	0.000015	0.000007
1000	500	265	104	0.000029	0.000013
1000	500	265	105	0.000052	0.000024
1000	500	265	106	0.000094	0.000042
1000	500	265	107	0.000165	0.000071
1000	500	265	108	0.000284	0.000119
1000	500	265	109	0.000480	0.000196
1000	500	265	110	0.000795	0.000315
1000	500	265	111	0.001290	0.000495
1000	500	265	112	0.002054	0.000763
1000	500	265	113	0.003206	0.001152
1000	500	265	114	0.004910	0.001704
1000	500	265	115	0.007377	0.002467
1000	500	265	116	0.010877	0.003500
1000	500	265	117	0.015739	0.004862
1000	500	265	118	0.022356	0.006616
1000	500	265	119	0.031175	0.008820
1000	500	265	120	0.042692	0.011517
1000	500	265	121	0.057423	0.014731
1000	500	265	122	0.075883	0.018459
1000	500	265	123	0.098542	0.022660
1000	500	265	124	0.125793	0.027250
1000	500	265	125	0.157897	0.032104
1000	500	265	126	0.194952	0.037055
1000	500	265	127	0.236852	0.041900
1000	500	265	128	0.283270	0.046418
1000	500	265	129	0.333651	0.050380
1000	500	265	130	0.387223	0.053572
1000	500	265	131	0.443034	0.055811
1000	500	265	132	0.500000	0.056966
1000	500	266	99	0.000001	0.000000
1000	500	266	100	0.000002	0.000001
1000	500	266	101	0.000003	0.000002
1000	500	266	102	0.000006	0.000003
1000	500	266	103	0.000011	0.000005
1000	500	266	104	0.000021	0.000010
1000	500	266	105	0.000040	0.000018
1000	500	266	106	0.000072	0.000032
1000	500	266	107	0.000127	0.000055
1000	500	266	108	0.000221	0.000094
1000	500	266	109	0.000376	0.000155
1000	500	266	110	0.000628	0.000252
1000	500	266	111	0.001028	0.000400
1000	500	266	112	0.001651	0.000623
1000	500	266	113	0.002599	0.000949
1000	500	266	114	0.004015	0.001416
1000	500	266	115	0.006085	0.002070
1000	500	266	116	0.009049	0.002964

N	n	k	x	P(x)	p(x)
1000	500	266	117	0.013206	0.004157
1000	500	266	118	0.018917	0.005711
1000	500	266	119	0.026604	0.007687
1000	500	266	120	0.036738	0.010134
1000	500	266	121	0.049827	0.013089
1000	500	266	122	0.066389	0.016562
1000	500	266	123	0.086920	0.020530
1000	500	266	124	0.111852	0.024933
1000	500	266	125	0.141517	0.029665
1000	500	266	126	0.176096	0.034579
1000	500	266	127	0.215588	0.039492
1000	500	266	128	0.259777	0.044189
1000	500	266	129	0.308221	0.048444
1000	500	266	130	0.360255	0.052034
1000	500	266	131	0.415015	0.054760
1000	500	266	132	0.471478	0.056464
1000	500	266	133	0.528521	0.057043
1000	500	267	99	0.000001	0.000000
1000	500	267	100	0.000001	0.000001
1000	500	267	101	0.000002	0.000001
1000	500	267	102	0.000004	0.000002
1000	500	267	103	0.000008	0.000004
1000	500	267	104	0.000016	0.000008
1000	500	267	105	0.000030	0.000014
1000	500	267	106	0.000055	0.000025
1000	500	267	107	0.000097	0.000043
1000	500	267	108	0.000171	0.000073
1000	500	267	109	0.000293	0.000123
1000	500	267	110	0.000494	0.000201
1000	500	267	111	0.000816	0.000322
1000	500	267	112	0.001321	0.000505
1000	500	267	113	0.002099	0.000778
1000	500	267	114	0.003270	0.001171
1000	500	267	115	0.004999	0.001729
1000	500	267	116	0.007498	0.002499
1000	500	267	117	0.011037	0.003538
1000	500	267	118	0.015945	0.004908
1000	500	267	119	0.022614	0.006669
1000	500	267	120	0.031491	0.008878
1000	500	267	121	0.043069	0.011577
1000	500	267	122	0.057860	0.014791
1000	500	267	123	0.076375	0.018515
1000	500	267	124	0.099080	0.022705
1000	500	267	125	0.126361	0.027281
1000	500	267	126	0.158477	0.032116
1000	500	267	127	0.195520	0.037043
1000	500	267	128	0.237382	0.041862
1000	500	267	129	0.283735	0.046353
1000	500	267	130	0.334025	0.050290
1000	500	267	131	0.387485	0.053460

Table for $N = 1000$, $n = 500$

N	n	k	x	P(x)	p(x)
1000	500	267	132	0.443170	0.055684
1000	500	267	133	0.500000	0.056830
1000	500	268	100	0.000001	0.000000
1000	500	268	101	0.000002	0.000001
1000	500	268	102	0.000003	0.000002
1000	500	268	103	0.000006	0.000003
1000	500	268	104	0.000012	0.000006
1000	500	268	105	0.000022	0.000010
1000	500	268	106	0.000041	0.000019
1000	500	268	107	0.000074	0.000033
1000	500	268	108	0.000132	0.000057
1000	500	268	109	0.000228	0.000096
1000	500	268	110	0.000387	0.000159
1000	500	268	111	0.000645	0.000258
1000	500	268	112	0.001054	0.000409
1000	500	268	113	0.001688	0.000635
1000	500	268	114	0.002654	0.000965
1000	500	268	115	0.004091	0.001438
1000	500	268	116	0.006189	0.002098
1000	500	268	117	0.009188	0.002999
1000	500	268	118	0.013387	0.004199
1000	500	268	119	0.019147	0.005760
1000	500	268	120	0.026889	0.007742
1000	500	268	121	0.037082	0.010194
1000	500	268	122	0.050232	0.013150
1000	500	268	123	0.066852	0.016620
1000	500	268	124	0.087433	0.020581
1000	500	268	125	0.112404	0.024971
1000	500	268	126	0.142091	0.029686
1000	500	268	127	0.176670	0.034579
1000	500	268	128	0.216136	0.039467
1000	500	268	129	0.260274	0.044137
1000	500	268	130	0.308640	0.048366
1000	500	268	131	0.360573	0.051933
1000	500	268	132	0.415214	0.054641
1000	500	268	133	0.471546	0.056332
1000	500	268	134	0.528454	0.056908
1000	500	269	100	0.000001	0.000000
1000	500	269	101	0.000001	0.000001
1000	500	269	102	0.000002	0.000001
1000	500	269	103	0.000005	0.000002
1000	500	269	104	0.000009	0.000004
1000	500	269	105	0.000017	0.000008
1000	500	269	106	0.000031	0.000014
1000	500	269	107	0.000057	0.000025
1000	500	269	108	0.000101	0.000044
1000	500	269	109	0.000176	0.000076
1000	500	269	110	0.000302	0.000126
1000	500	269	111	0.000508	0.000206
1000	500	269	112	0.000837	0.000329

N	n	k	x	P(x)	p(x)
1000	500	269	113	0.001353	0.000516
1000	500	269	114	0.002145	0.000792
1000	500	269	115	0.003335	0.001190
1000	500	269	116	0.005089	0.001753
1000	500	269	117	0.007619	0.002530
1000	500	269	118	0.011195	0.003577
1000	500	269	119	0.016149	0.004953
1000	500	269	120	0.022870	0.006721
1000	500	269	121	0.031805	0.008935
1000	500	269	122	0.043442	0.011637
1000	500	269	123	0.058292	0.014851
1000	500	269	124	0.076862	0.018569
1000	500	269	125	0.099612	0.022750
1000	500	269	126	0.126923	0.027311
1000	500	269	127	0.159049	0.032126
1000	500	269	128	0.196080	0.037030
1000	500	269	129	0.237904	0.041824
1000	500	269	130	0.284192	0.046289
1000	500	269	131	0.334394	0.050201
1000	500	269	132	0.387744	0.053350
1000	500	269	133	0.443303	0.055559
1000	500	269	134	0.500000	0.056697
1000	500	270	101	0.000001	0.000000
1000	500	270	102	0.000002	0.000001
1000	500	270	103	0.000003	0.000002
1000	500	270	104	0.000007	0.000003
1000	500	270	105	0.000012	0.000006
1000	500	270	106	0.000023	0.000011
1000	500	270	107	0.000043	0.000020
1000	500	270	108	0.000077	0.000034
1000	500	270	109	0.000136	0.000059
1000	500	270	110	0.000235	0.000099
1000	500	270	111	0.000399	0.000163
1000	500	270	112	0.000663	0.000264
1000	500	270	113	0.001080	0.000417
1000	500	270	114	0.001727	0.000647
1000	500	270	115	0.002708	0.000981
1000	500	270	116	0.004167	0.001459
1000	500	270	117	0.006293	0.002126
1000	500	270	118	0.009326	0.003033
1000	500	270	119	0.013567	0.004241
1000	500	270	120	0.019376	0.005809
1000	500	270	121	0.027172	0.007796
1000	500	270	122	0.037424	0.010252
1000	500	270	123	0.050633	0.013209
1000	500	270	124	0.067310	0.016677
1000	500	270	125	0.087941	0.020631
1000	500	270	126	0.112950	0.025009
1000	500	270	127	0.142657	0.029707
1000	500	270	128	0.177235	0.034579

N	n	k	x	P(x)	p(x)
1000	500	270	129	0.216677	0.039442
1000	500	270	130	0.260763	0.044086
1000	500	270	131	0.309053	0.048290
1000	500	270	132	0.360886	0.051834
1000	500	270	133	0.415409	0.054523
1000	500	270	134	0.471613	0.056203
1000	500	270	135	0.528387	0.056775
1000	500	271	101	0.000001	0.000000
1000	500	271	102	0.000001	0.000001
1000	500	271	103	0.000002	0.000001
1000	500	271	104	0.000005	0.000002
1000	500	271	105	0.000009	0.000004
1000	500	271	106	0.000018	0.000008
1000	500	271	107	0.000032	0.000015
1000	500	271	108	0.000059	0.000026
1000	500	271	109	0.000105	0.000046
1000	500	271	110	0.000182	0.000078
1000	500	271	111	0.000312	0.000129
1000	500	271	112	0.000522	0.000211
1000	500	271	113	0.000859	0.000336
1000	500	271	114	0.001385	0.000526
1000	500	271	115	0.002190	0.000806
1000	500	271	116	0.003400	0.001209
1000	500	271	117	0.005178	0.001778
1000	500	271	118	0.007739	0.002561
1000	500	271	119	0.011354	0.003615
1000	500	271	120	0.016352	0.004998
1000	500	271	121	0.023125	0.006772
1000	500	271	122	0.032116	0.008991
1000	500	271	123	0.043812	0.011696
1000	500	271	124	0.058721	0.014909
1000	500	271	125	0.077343	0.018622
1000	500	271	126	0.100137	0.022794
1000	500	271	127	0.127478	0.027341
1000	500	271	128	0.159614	0.032137
1000	500	271	129	0.196632	0.037018
1000	500	271	130	0.238418	0.041786
1000	500	271	131	0.284643	0.046225
1000	500	271	132	0.334756	0.050113
1000	500	271	133	0.387998	0.053242
1000	500	271	134	0.443434	0.055436
1000	500	271	135	0.500000	0.056566
1000	500	272	102	0.000001	0.000000
1000	500	272	103	0.000002	0.000001
1000	500	272	104	0.000004	0.000002
1000	500	272	105	0.000007	0.000003
1000	500	272	106	0.000013	0.000006
1000	500	272	107	0.000024	0.000011
1000	500	272	108	0.000045	0.000020
1000	500	272	109	0.000080	0.000035

N	n	k	x	P(x)	p(x)
1000	500	272	110	0.000141	0.000061
1000	500	272	111	0.000243	0.000102
1000	500	272	112	0.000410	0.000168
1000	500	272	113	0.000680	0.000270
1000	500	272	114	0.001106	0.000426
1000	500	272	115	0.001765	0.000659
1000	500	272	116	0.002763	0.000998
1000	500	272	117	0.004244	0.001481
1000	500	272	118	0.006397	0.002153
1000	500	272	119	0.009465	0.003068
1000	500	272	120	0.013747	0.004282
1000	500	272	121	0.019604	0.005857
1000	500	272	122	0.027454	0.007850
1000	500	272	123	0.037763	0.010309
1000	500	272	124	0.051031	0.013268
1000	500	272	125	0.067764	0.016733
1000	500	272	126	0.088443	0.020679
1000	500	272	127	0.113489	0.025046
1000	500	272	128	0.143215	0.029727
1000	500	272	129	0.177793	0.034578
1000	500	272	130	0.217210	0.039417
1000	500	272	131	0.261245	0.044036
1000	500	272	132	0.309459	0.048214
1000	500	272	133	0.361195	0.051736
1000	500	272	134	0.415602	0.054407
1000	500	272	135	0.471678	0.056076
1000	500	272	136	0.528322	0.056644
1000	500	273	102	0.000001	0.000000
1000	500	273	103	0.000001	0.000001
1000	500	273	104	0.000003	0.000001
1000	500	273	105	0.000005	0.000002
1000	500	273	106	0.000010	0.000005
1000	500	273	107	0.000018	0.000009
1000	500	273	108	0.000034	0.000015
1000	500	273	109	0.000061	0.000027
1000	500	273	110	0.000108	0.000047
1000	500	273	111	0.000188	0.000080
1000	500	273	112	0.000321	0.000133
1000	500	273	113	0.000537	0.000216
1000	500	273	114	0.000880	0.000344
1000	500	273	115	0.001417	0.000536
1000	500	273	116	0.002236	0.000820
1000	500	273	117	0.003465	0.001228
1000	500	273	118	0.005267	0.001802
1000	500	273	119	0.007859	0.002592
1000	500	273	120	0.011512	0.003652
1000	500	273	121	0.016555	0.005043
1000	500	273	122	0.023378	0.006823
1000	500	273	123	0.032424	0.009047
1000	500	273	124	0.044178	0.011754

Table for $N = 1000$, $n = 500$

N	n	k	x	P(x)	p(x)
1000	500	273	125	0.059144	0.014966
1000	500	273	126	0.077819	0.018675
1000	500	273	127	0.100656	0.022837
1000	500	273	128	0.128025	0.027369
1000	500	273	129	0.160172	0.032146
1000	500	273	130	0.197177	0.037005
1000	500	273	131	0.238925	0.041749
1000	500	273	132	0.285087	0.046162
1000	500	273	133	0.335114	0.050026
1000	500	273	134	0.388249	0.053135
1000	500	273	135	0.443563	0.055314
1000	500	273	136	0.500000	0.056437
1000	500	274	103	0.000001	0.000000
1000	500	274	104	0.000002	0.000001
1000	500	274	105	0.000004	0.000002
1000	500	274	106	0.000007	0.000003
1000	500	274	107	0.000014	0.000006
1000	500	274	108	0.000025	0.000012
1000	500	274	109	0.000046	0.000021
1000	500	274	110	0.000083	0.000037
1000	500	274	111	0.000145	0.000062
1000	500	274	112	0.000250	0.000105
1000	500	274	113	0.000422	0.000172
1000	500	274	114	0.000698	0.000276
1000	500	274	115	0.001133	0.000435
1000	500	274	116	0.001803	0.000671
1000	500	274	117	0.002818	0.001014
1000	500	274	118	0.004320	0.001502
1000	500	274	119	0.006501	0.002181
1000	500	274	120	0.009602	0.003102
1000	500	274	121	0.013926	0.004323
1000	500	274	122	0.019830	0.005905
1000	500	274	123	0.027733	0.007903
1000	500	274	124	0.038099	0.010366
1000	500	274	125	0.051425	0.013325
1000	500	274	126	0.068212	0.016787
1000	500	274	127	0.088939	0.020727
1000	500	274	128	0.114021	0.025081
1000	500	274	129	0.143767	0.029746
1000	500	274	130	0.178343	0.034576
1000	500	274	131	0.217735	0.039392
1000	500	274	132	0.261720	0.043986
1000	500	274	133	0.309860	0.048139
1000	500	274	134	0.361499	0.051639
1000	500	274	135	0.415792	0.054293
1000	500	274	136	0.471743	0.055951
1000	500	274	137	0.528257	0.056514
1000	500	275	103	0.000001	0.000000
1000	500	275	104	0.000001	0.000001
1000	500	275	105	0.000003	0.000001
1000	500	275	106	0.000005	0.000003
1000	500	275	107	0.000010	0.000005
1000	500	275	108	0.000019	0.000009
1000	500	275	109	0.000035	0.000016
1000	500	275	110	0.000063	0.000028
1000	500	275	111	0.000112	0.000049
1000	500	275	112	0.000194	0.000082
1000	500	275	113	0.000330	0.000136
1000	500	275	114	0.000551	0.000221
1000	500	275	115	0.000902	0.000351
1000	500	275	116	0.001449	0.000547
1000	500	275	117	0.002283	0.000834
1000	500	275	118	0.003529	0.001247
1000	500	275	119	0.005356	0.001827
1000	500	275	120	0.007979	0.002623
1000	500	275	121	0.011669	0.003690
1000	500	275	122	0.016756	0.005087
1000	500	275	123	0.023629	0.006873
1000	500	275	124	0.032731	0.009101
1000	500	275	125	0.044542	0.011811
1000	500	275	126	0.059564	0.015022
1000	500	275	127	0.078290	0.018726
1000	500	275	128	0.101169	0.022879
1000	500	275	129	0.128566	0.027397
1000	500	275	130	0.160722	0.032156
1000	500	275	131	0.197714	0.036992
1000	500	275	132	0.239425	0.041711
1000	500	275	133	0.285525	0.046100
1000	500	275	134	0.335466	0.049941
1000	500	275	135	0.388496	0.053030
1000	500	275	136	0.443690	0.055194
1000	500	275	137	0.500000	0.056310
1000	500	276	104	0.000001	0.000000
1000	500	276	105	0.000002	0.000001
1000	500	276	106	0.000004	0.000002
1000	500	276	107	0.000008	0.000004
1000	500	276	108	0.000014	0.000007
1000	500	276	109	0.000026	0.000012
1000	500	276	110	0.000048	0.000022
1000	500	276	111	0.000086	0.000038
1000	500	276	112	0.000150	0.000064
1000	500	276	113	0.000258	0.000107
1000	500	276	114	0.000433	0.000176
1000	500	276	115	0.000716	0.000282
1000	500	276	116	0.001159	0.000443
1000	500	276	117	0.001842	0.000683
1000	500	276	118	0.002872	0.001030
1000	500	276	119	0.004396	0.001524
1000	500	276	120	0.006604	0.002208
1000	500	276	121	0.009740	0.003136
1000	500	276	122	0.014104	0.004364
1000	500	276	123	0.020055	0.005952
1000	500	276	124	0.028010	0.007955
1000	500	276	125	0.038432	0.010422
1000	500	276	126	0.051815	0.013382
1000	500	276	127	0.068656	0.016841
1000	500	276	128	0.089430	0.020774
1000	500	276	129	0.114546	0.025117
1000	500	276	130	0.144311	0.029765
1000	500	276	131	0.178886	0.034575
1000	500	276	132	0.218253	0.039367
1000	500	276	133	0.262189	0.043936
1000	500	276	134	0.310254	0.048066
1000	500	276	135	0.361798	0.051544
1000	500	276	136	0.415979	0.054181
1000	500	276	137	0.471806	0.055827
1000	500	276	138	0.528194	0.056387
1000	500	277	104	0.000001	0.000000
1000	500	277	105	0.000001	0.000001
1000	500	277	106	0.000003	0.000001
1000	500	277	107	0.000006	0.000003
1000	500	277	108	0.000011	0.000005
1000	500	277	109	0.000020	0.000009
1000	500	277	110	0.000036	0.000017
1000	500	277	111	0.000066	0.000029
1000	500	277	112	0.000116	0.000050
1000	500	277	113	0.000200	0.000084
1000	500	277	114	0.000340	0.000140
1000	500	277	115	0.000566	0.000226
1000	500	277	116	0.000924	0.000358
1000	500	277	117	0.001481	0.000557
1000	500	277	118	0.002329	0.000848
1000	500	277	119	0.003594	0.001266
1000	500	277	120	0.005445	0.001851
1000	500	277	121	0.008098	0.002653
1000	500	277	122	0.011825	0.003727
1000	500	277	123	0.016956	0.005131
1000	500	277	124	0.023879	0.006923
1000	500	277	125	0.033034	0.009155
1000	500	277	126	0.044902	0.011867
1000	500	277	127	0.059979	0.015078
1000	500	277	128	0.078756	0.018777
1000	500	277	129	0.101676	0.022920
1000	500	277	130	0.129100	0.027424
1000	500	277	131	0.161264	0.032164
1000	500	277	132	0.198243	0.036979
1000	500	277	133	0.239917	0.041674
1000	500	277	134	0.285956	0.046039
1000	500	277	135	0.335813	0.049857
1000	500	277	136	0.388739	0.052926
1000	500	277	137	0.443815	0.055077
1000	500	277	138	0.500000	0.056184
1000	500	278	104	0.000001	0.000000
1000	500	278	105	0.000001	0.000001
1000	500	278	106	0.000002	0.000001
1000	500	278	107	0.000004	0.000002
1000	500	278	108	0.000008	0.000004
1000	500	278	109	0.000015	0.000007
1000	500	278	110	0.000027	0.000013
1000	500	278	111	0.000050	0.000022
1000	500	278	112	0.000089	0.000039
1000	500	278	113	0.000155	0.000066
1000	500	278	114	0.000265	0.000110
1000	500	278	115	0.000445	0.000180
1000	500	278	116	0.000734	0.000288
1000	500	278	117	0.001186	0.000452
1000	500	278	118	0.001881	0.000695
1000	500	278	119	0.002927	0.001047
1000	500	278	120	0.004473	0.001545
1000	500	278	121	0.006708	0.002235
1000	500	278	122	0.009877	0.003169
1000	500	278	123	0.014281	0.004404
1000	500	278	124	0.020279	0.005998
1000	500	278	125	0.028286	0.008007
1000	500	278	126	0.038763	0.010477
1000	500	278	127	0.052201	0.013438
1000	500	278	128	0.069095	0.016894
1000	500	278	129	0.089915	0.020820
1000	500	278	130	0.115066	0.025151
1000	500	278	131	0.144848	0.029783
1000	500	278	132	0.179421	0.034573
1000	500	278	133	0.218763	0.039342
1000	500	278	134	0.262650	0.043887
1000	500	278	135	0.310643	0.047993
1000	500	278	136	0.362093	0.051450
1000	500	278	137	0.416163	0.054070
1000	500	278	138	0.471869	0.055706
1000	500	278	139	0.528131	0.056262
1000	500	279	105	0.000001	0.000000
1000	500	279	106	0.000002	0.000001
1000	500	279	107	0.000003	0.000001
1000	500	279	108	0.000006	0.000003
1000	500	279	109	0.000011	0.000005
1000	500	279	110	0.000021	0.000010
1000	500	279	111	0.000038	0.000017
1000	500	279	112	0.000068	0.000030
1000	500	279	113	0.000119	0.000052
1000	500	279	114	0.000206	0.000087
1000	500	279	115	0.000349	0.000143
1000	500	279	116	0.000580	0.000231

Table for $N = 1000$, $n = 500$

N	n	k	x	P(x)	p(x)
1000	500	279	117	0.000946	0.000366
1000	500	279	118	0.001513	0.000567
1000	500	279	119	0.002375	0.000862
1000	500	279	120	0.003659	0.001284
1000	500	279	121	0.005534	0.001875
1000	500	279	122	0.008218	0.002683
1000	500	279	123	0.011981	0.003763
1000	500	279	124	0.017155	0.005174
1000	500	279	125	0.024127	0.006972
1000	500	279	126	0.033336	0.009209
1000	500	279	127	0.045258	0.011923
1000	500	279	128	0.060390	0.015132
1000	500	279	129	0.079217	0.018826
1000	500	279	130	0.102177	0.022960
1000	500	279	131	0.129627	0.027450
1000	500	279	132	0.161800	0.032173
1000	500	279	133	0.198765	0.036966
1000	500	279	134	0.240403	0.041637
1000	500	279	135	0.286381	0.045978
1000	500	279	136	0.336154	0.049773
1000	500	279	137	0.388978	0.052824
1000	500	279	138	0.443939	0.054961
1000	500	279	139	0.500000	0.056061
1000	500	280	105	0.000001	0.000000
1000	500	280	106	0.000001	0.000001
1000	500	280	107	0.000002	0.000001
1000	500	280	108	0.000004	0.000002
1000	500	280	109	0.000008	0.000004
1000	500	280	110	0.000015	0.000007
1000	500	280	111	0.000029	0.000013
1000	500	280	112	0.000052	0.000023
1000	500	280	113	0.000092	0.000040
1000	500	280	114	0.000160	0.000068
1000	500	280	115	0.000273	0.000113
1000	500	280	116	0.000457	0.000184
1000	500	280	117	0.000752	0.000294
1000	500	280	118	0.001213	0.000461
1000	500	280	119	0.001920	0.000707
1000	500	280	120	0.002982	0.001063
1000	500	280	121	0.004549	0.001566
1000	500	280	122	0.006811	0.002262
1000	500	280	123	0.010013	0.003202
1000	500	280	124	0.014457	0.004444
1000	500	280	125	0.020501	0.006044
1000	500	280	126	0.028559	0.008058
1000	500	280	127	0.039090	0.010532
1000	500	280	128	0.052583	0.013493
1000	500	280	129	0.069529	0.016946
1000	500	280	130	0.090394	0.020865
1000	500	280	131	0.115578	0.025184
1000	500	280	132	0.145379	0.029800
1000	500	280	133	0.179949	0.034571
1000	500	280	134	0.219266	0.039317
1000	500	280	135	0.263105	0.043838
1000	500	280	136	0.311026	0.047921
1000	500	280	137	0.362383	0.051357
1000	500	280	138	0.416344	0.053961
1000	500	280	139	0.471930	0.055586
1000	500	280	140	0.528069	0.056139
1000	500	281	106	0.000001	0.000000
1000	500	281	107	0.000002	0.000001
1000	500	281	108	0.000003	0.000002
1000	500	281	109	0.000006	0.000003
1000	500	281	110	0.000012	0.000005
1000	500	281	111	0.000021	0.000010
1000	500	281	112	0.000039	0.000018
1000	500	281	113	0.000070	0.000031
1000	500	281	114	0.000123	0.000053
1000	500	281	115	0.000212	0.000089
1000	500	281	116	0.000359	0.000147
1000	500	281	117	0.000595	0.000236
1000	500	281	118	0.000968	0.000373
1000	500	281	119	0.001545	0.000577
1000	500	281	120	0.002421	0.000876
1000	500	281	121	0.003724	0.001303
1000	500	281	122	0.005623	0.001899
1000	500	281	123	0.008336	0.002713
1000	500	281	124	0.012136	0.003800
1000	500	281	125	0.017353	0.005217
1000	500	281	126	0.024373	0.007020
1000	500	281	127	0.033634	0.009261
1000	500	281	128	0.045612	0.011977
1000	500	281	129	0.060797	0.015185
1000	500	281	130	0.079672	0.018875
1000	500	281	131	0.102671	0.022999
1000	500	281	132	0.130147	0.027476
1000	500	281	133	0.162328	0.032181
1000	500	281	134	0.199280	0.036952
1000	500	281	135	0.240881	0.041601
1000	500	281	136	0.286799	0.045918
1000	500	281	137	0.336490	0.049691
1000	500	281	138	0.389213	0.052723
1000	500	281	139	0.444060	0.054847
1000	500	281	140	0.500000	0.055940
1000	500	282	106	0.000001	0.000000
1000	500	282	107	0.000001	0.000001
1000	500	282	108	0.000002	0.000001
1000	500	282	109	0.000004	0.000002
1000	500	282	110	0.000009	0.000004
1000	500	282	111	0.000016	0.000008
1000	500	282	112	0.000030	0.000014
1000	500	282	113	0.000054	0.000024
1000	500	282	114	0.000095	0.000041
1000	500	282	115	0.000165	0.000070
1000	500	282	116	0.000281	0.000116
1000	500	282	117	0.000469	0.000189
1000	500	282	118	0.000770	0.000301
1000	500	282	119	0.001239	0.000470
1000	500	282	120	0.001958	0.000719
1000	500	282	121	0.003038	0.001079
1000	500	282	122	0.004625	0.001587
1000	500	282	123	0.006914	0.002289
1000	500	282	124	0.010149	0.003235
1000	500	282	125	0.014632	0.004483
1000	500	282	126	0.020721	0.006089
1000	500	282	127	0.028830	0.008108
1000	500	282	128	0.039415	0.010585
1000	500	282	129	0.052962	0.013547
1000	500	282	130	0.069959	0.016997
1000	500	282	131	0.090868	0.020909
1000	500	282	132	0.116085	0.025217
1000	500	282	133	0.145902	0.029817
1000	500	282	134	0.180470	0.034568
1000	500	282	135	0.219762	0.039292
1000	500	282	136	0.263553	0.043790
1000	500	282	137	0.311403	0.047850
1000	500	282	138	0.362669	0.051266
1000	500	282	139	0.416522	0.053854
1000	500	282	140	0.471991	0.055469
1000	500	282	141	0.528009	0.056018
1000	500	283	107	0.000001	0.000000
1000	500	283	108	0.000002	0.000001
1000	500	283	109	0.000003	0.000002
1000	500	283	110	0.000006	0.000003
1000	500	283	111	0.000012	0.000006
1000	500	283	112	0.000022	0.000010
1000	500	283	113	0.000041	0.000018
1000	500	283	114	0.000073	0.000032
1000	500	283	115	0.000127	0.000055
1000	500	283	116	0.000219	0.000091
1000	500	283	117	0.000369	0.000150
1000	500	283	118	0.000610	0.000241
1000	500	283	119	0.000990	0.000380
1000	500	283	120	0.001578	0.000588
1000	500	283	121	0.002468	0.000890
1000	500	283	122	0.003789	0.001321
1000	500	283	123	0.005712	0.001923
1000	500	283	124	0.008455	0.002743
1000	500	283	125	0.012291	0.003836
1000	500	283	126	0.017550	0.005259
1000	500	283	127	0.024617	0.007068
1000	500	283	128	0.033931	0.009313
1000	500	283	129	0.045962	0.012031
1000	500	283	130	0.061200	0.015238
1000	500	283	131	0.080122	0.018922
1000	500	283	132	0.103160	0.023038
1000	500	283	133	0.130661	0.027501
1000	500	283	134	0.162849	0.032188
1000	500	283	135	0.199788	0.036939
1000	500	283	136	0.241353	0.041565
1000	500	283	137	0.287211	0.045859
1000	500	283	138	0.336821	0.049610
1000	500	283	139	0.389445	0.052624
1000	500	283	140	0.444179	0.054734
1000	500	283	141	0.500000	0.055821
1000	500	284	107	0.000001	0.000000
1000	500	284	108	0.000001	0.000001
1000	500	284	109	0.000002	0.000001
1000	500	284	110	0.000005	0.000002
1000	500	284	111	0.000009	0.000004
1000	500	284	112	0.000017	0.000008
1000	500	284	113	0.000031	0.000014
1000	500	284	114	0.000055	0.000025
1000	500	284	115	0.000098	0.000042
1000	500	284	116	0.000170	0.000072
1000	500	284	117	0.000288	0.000119
1000	500	284	118	0.000481	0.000193
1000	500	284	119	0.000788	0.000307
1000	500	284	120	0.001266	0.000478
1000	500	284	121	0.001997	0.000731
1000	500	284	122	0.003093	0.001095
1000	500	284	123	0.004701	0.001608
1000	500	284	124	0.007016	0.002316
1000	500	284	125	0.010285	0.003268
1000	500	284	126	0.014806	0.004522
1000	500	284	127	0.020941	0.006134
1000	500	284	128	0.029099	0.008158
1000	500	284	129	0.039737	0.010638
1000	500	284	130	0.053336	0.013600
1000	500	284	131	0.070384	0.017047
1000	500	284	132	0.091336	0.020952
1000	500	284	133	0.116584	0.025249
1000	500	284	134	0.146418	0.029834
1000	500	284	135	0.180983	0.034565
1000	500	284	136	0.220251	0.039268
1000	500	284	137	0.263994	0.043743
1000	500	284	138	0.311774	0.047780
1000	500	284	139	0.362950	0.051176
1000	500	284	140	0.416698	0.053748
1000	500	284	141	0.472051	0.055353

Table for $N = 1000$, $n = 500$

N	n	k	x	P(x)	p(x)
1000	500	284	142	0.527949	0.055898
1000	500	285	108	0.000001	0.000000
1000	500	285	109	0.000002	0.000001
1000	500	285	110	0.000003	0.000002
1000	500	285	111	0.000007	0.000003
1000	500	285	112	0.000013	0.000006
1000	500	285	113	0.000023	0.000011
1000	500	285	114	0.000042	0.000019
1000	500	285	115	0.000075	0.000033
1000	500	285	116	0.000131	0.000056
1000	500	285	117	0.000225	0.000094
1000	500	285	118	0.000378	0.000154
1000	500	285	119	0.000625	0.000246
1000	500	285	120	0.001013	0.000388
1000	500	285	121	0.001611	0.000598
1000	500	285	122	0.002514	0.000904
1000	500	285	123	0.003854	0.001340
1000	500	285	124	0.005801	0.001946
1000	500	285	125	0.008573	0.002772
1000	500	285	126	0.012444	0.003872
1000	500	285	127	0.017745	0.005301
1000	500	285	128	0.024860	0.007115
1000	500	285	129	0.034224	0.009364
1000	500	285	130	0.046309	0.012084
1000	500	285	131	0.061598	0.015290
1000	500	285	132	0.080567	0.018969
1000	500	285	133	0.103643	0.023076
1000	500	285	134	0.131168	0.027525
1000	500	285	135	0.163363	0.032195
1000	500	285	136	0.200288	0.036925
1000	500	285	137	0.241817	0.041529
1000	500	285	138	0.287617	0.045800
1000	500	285	139	0.337148	0.049530
1000	500	285	140	0.389674	0.052526
1000	500	285	141	0.444297	0.054623
1000	500	285	142	0.500000	0.055703
1000	500	286	108	0.000001	0.000000
1000	500	286	109	0.000001	0.000001
1000	500	286	110	0.000003	0.000001
1000	500	286	111	0.000005	0.000002
1000	500	286	112	0.000009	0.000004
1000	500	286	113	0.000017	0.000008
1000	500	286	114	0.000032	0.000014
1000	500	286	115	0.000057	0.000025
1000	500	286	116	0.000101	0.000044
1000	500	286	117	0.000175	0.000074
1000	500	286	118	0.000296	0.000122
1000	500	286	119	0.000494	0.000197
1000	500	286	120	0.000807	0.000313
1000	500	286	121	0.001294	0.000487

N	n	k	x	P(x)	p(x)
1000	500	286	122	0.002037	0.000743
1000	500	286	123	0.003148	0.001111
1000	500	286	124	0.004777	0.001629
1000	500	286	125	0.007119	0.002342
1000	500	286	126	0.010419	0.003300
1000	500	286	127	0.014980	0.004560
1000	500	286	128	0.021158	0.006178
1000	500	286	129	0.029366	0.008207
1000	500	286	130	0.040055	0.010690
1000	500	286	131	0.053708	0.013652
1000	500	286	132	0.070804	0.017097
1000	500	286	133	0.091798	0.020994
1000	500	286	134	0.117078	0.025280
1000	500	286	135	0.146928	0.029850
1000	500	286	136	0.181490	0.034562
1000	500	286	137	0.220733	0.039243
1000	500	286	138	0.264429	0.043696
1000	500	286	139	0.312140	0.047711
1000	500	286	140	0.363227	0.051087
1000	500	286	141	0.416871	0.053644
1000	500	286	142	0.472110	0.055239
1000	500	286	143	0.527890	0.055781
1000	500	287	109	0.000001	0.000000
1000	500	287	110	0.000002	0.000001
1000	500	287	111	0.000004	0.000002
1000	500	287	112	0.000007	0.000003
1000	500	287	113	0.000013	0.000006
1000	500	287	114	0.000024	0.000011
1000	500	287	115	0.000044	0.000020
1000	500	287	116	0.000078	0.000034
1000	500	287	117	0.000135	0.000058
1000	500	287	118	0.000231	0.000096
1000	500	287	119	0.000388	0.000157
1000	500	287	120	0.000640	0.000252
1000	500	287	121	0.001035	0.000395
1000	500	287	122	0.001643	0.000608
1000	500	287	123	0.002561	0.000918
1000	500	287	124	0.003919	0.001358
1000	500	287	125	0.005889	0.001970
1000	500	287	126	0.008691	0.002802
1000	500	287	127	0.012597	0.003907
1000	500	287	128	0.017939	0.005342
1000	500	287	129	0.025101	0.007162
1000	500	287	130	0.034516	0.009415
1000	500	287	131	0.046652	0.012136
1000	500	287	132	0.061992	0.015341
1000	500	287	133	0.081007	0.019015
1000	500	287	134	0.104120	0.023112
1000	500	287	135	0.131669	0.027549
1000	500	287	136	0.163870	0.032201

N	n	k	x	P(x)	p(x)
1000	500	287	137	0.200782	0.036912
1000	500	287	138	0.242275	0.041493
1000	500	287	139	0.288017	0.045742
1000	500	287	140	0.337469	0.049452
1000	500	287	141	0.389899	0.052430
1000	500	287	142	0.444413	0.054514
1000	500	287	143	0.500000	0.055587
1000	500	288	109	0.000001	0.000000
1000	500	288	110	0.000001	0.000001
1000	500	288	111	0.000003	0.000001
1000	500	288	112	0.000005	0.000002
1000	500	288	113	0.000010	0.000005
1000	500	288	114	0.000018	0.000008
1000	500	288	115	0.000033	0.000015
1000	500	288	116	0.000059	0.000026
1000	500	288	117	0.000104	0.000045
1000	500	288	118	0.000180	0.000076
1000	500	288	119	0.000304	0.000125
1000	500	288	120	0.000506	0.000201
1000	500	288	121	0.000825	0.000319
1000	500	288	122	0.001321	0.000496
1000	500	288	123	0.002076	0.000755
1000	500	288	124	0.003203	0.001127
1000	500	288	125	0.004853	0.001650
1000	500	288	126	0.007221	0.002368
1000	500	288	127	0.010554	0.003333
1000	500	288	128	0.015152	0.004599
1000	500	288	129	0.021374	0.006222
1000	500	288	130	0.029630	0.008256
1000	500	288	131	0.040371	0.010741
1000	500	288	132	0.054075	0.013704
1000	500	288	133	0.071220	0.017145
1000	500	288	134	0.092256	0.021036
1000	500	288	135	0.117566	0.025310
1000	500	288	136	0.147431	0.029865
1000	500	288	137	0.181989	0.034559
1000	500	288	138	0.221208	0.039219
1000	500	288	139	0.264857	0.043649
1000	500	288	140	0.312500	0.047643
1000	500	288	141	0.363500	0.051000
1000	500	288	142	0.417041	0.053541
1000	500	288	143	0.472167	0.055126
1000	500	288	144	0.527832	0.055665
1000	500	289	110	0.000001	0.000000
1000	500	289	111	0.000002	0.000001
1000	500	289	112	0.000004	0.000002
1000	500	289	113	0.000007	0.000003
1000	500	289	114	0.000014	0.000006
1000	500	289	115	0.000025	0.000011
1000	500	289	116	0.000045	0.000020

N	n	k	x	P(x)	p(x)
1000	500	289	117	0.000080	0.000035
1000	500	289	118	0.000139	0.000059
1000	500	289	119	0.000238	0.000099
1000	500	289	120	0.000398	0.000161
1000	500	289	121	0.000655	0.000257
1000	500	289	122	0.001058	0.000402
1000	500	289	123	0.001676	0.000618
1000	500	289	124	0.002608	0.000932
1000	500	289	125	0.003984	0.001376
1000	500	289	126	0.005977	0.001993
1000	500	289	127	0.008808	0.002831
1000	500	289	128	0.012750	0.003942
1000	500	289	129	0.018132	0.005383
1000	500	289	130	0.025340	0.007208
1000	500	289	131	0.034804	0.009465
1000	500	289	132	0.046992	0.012187
1000	500	289	133	0.062382	0.015390
1000	500	289	134	0.081442	0.019060
1000	500	289	135	0.104591	0.023148
1000	500	289	136	0.132163	0.027572
1000	500	289	137	0.164370	0.032208
1000	500	289	138	0.201268	0.036898
1000	500	289	139	0.242726	0.041458
1000	500	289	140	0.288411	0.045685
1000	500	289	141	0.337785	0.049374
1000	500	289	142	0.390120	0.052335
1000	500	289	143	0.444527	0.054407
1000	500	289	144	0.500000	0.055473
1000	500	290	110	0.000001	0.000000
1000	500	290	111	0.000001	0.000001
1000	500	290	112	0.000003	0.000001
1000	500	290	113	0.000005	0.000003
1000	500	290	114	0.000010	0.000005
1000	500	290	115	0.000019	0.000009
1000	500	290	116	0.000034	0.000015
1000	500	290	117	0.000061	0.000027
1000	500	290	118	0.000107	0.000046
1000	500	290	119	0.000185	0.000078
1000	500	290	120	0.000313	0.000128
1000	500	290	121	0.000518	0.000206
1000	500	290	122	0.000844	0.000325
1000	500	290	123	0.001348	0.000504
1000	500	290	124	0.002115	0.000767
1000	500	290	125	0.003258	0.001143
1000	500	290	126	0.004929	0.001671
1000	500	290	127	0.007323	0.002394
1000	500	290	128	0.010687	0.003364
1000	500	290	129	0.015324	0.004636
1000	500	290	130	0.021589	0.006266
1000	500	290	131	0.029892	0.008303

Table for $N = 1000$, $n = 500$

N	n	k	x	P(x)	p(x)
1000	500	290	132	0.040684	0.010791
1000	500	290	133	0.054438	0.013754
1000	500	290	134	0.071631	0.017193
1000	500	290	135	0.092707	0.021076
1000	500	290	136	0.118047	0.025340
1000	500	290	137	0.147927	0.029880
1000	500	290	138	0.182482	0.034555
1000	500	290	139	0.221677	0.039195
1000	500	290	140	0.265280	0.043603
1000	500	290	141	0.312855	0.047575
1000	500	290	142	0.363769	0.050913
1000	500	290	143	0.417209	0.053440
1000	500	290	144	0.472224	0.055016
1000	500	290	145	0.527775	0.055551
1000	500	291	110	0.000000	0.000000
1000	500	291	111	0.000001	0.000001
1000	500	291	112	0.000002	0.000001
1000	500	291	113	0.000004	0.000002
1000	500	291	114	0.000008	0.000004
1000	500	291	115	0.000014	0.000007
1000	500	291	116	0.000026	0.000012
1000	500	291	117	0.000047	0.000021
1000	500	291	118	0.000083	0.000036
1000	500	291	119	0.000143	0.000061
1000	500	291	120	0.000244	0.000101
1000	500	291	121	0.000408	0.000164
1000	500	291	122	0.000670	0.000262
1000	500	291	123	0.001080	0.000410
1000	500	291	124	0.001709	0.000629
1000	500	291	125	0.002654	0.000945
1000	500	291	126	0.004049	0.001394
1000	500	291	127	0.006065	0.002016
1000	500	291	128	0.008924	0.002859
1000	500	291	129	0.012901	0.003977
1000	500	291	130	0.018324	0.005423
1000	500	291	131	0.025577	0.007253
1000	500	291	132	0.035091	0.009514
1000	500	291	133	0.047329	0.012238
1000	500	291	134	0.062768	0.015440
1000	500	291	135	0.081872	0.019104
1000	500	291	136	0.105056	0.023184
1000	500	291	137	0.132650	0.027594
1000	500	291	138	0.164864	0.032214
1000	500	291	139	0.201748	0.036885
1000	500	291	140	0.243171	0.041423
1000	500	291	141	0.288799	0.045628
1000	500	291	142	0.338097	0.049297
1000	500	291	143	0.390338	0.052241
1000	500	291	144	0.444639	0.054301
1000	500	291	145	0.500000	0.055361

N	n	k	x	P(x)	p(x)
1000	500	292	111	0.000001	0.000000
1000	500	292	112	0.000001	0.000001
1000	500	292	113	0.000003	0.000001
1000	500	292	114	0.000006	0.000003
1000	500	292	115	0.000011	0.000005
1000	500	292	116	0.000019	0.000009
1000	500	292	117	0.000035	0.000016
1000	500	292	118	0.000063	0.000028
1000	500	292	119	0.000111	0.000047
1000	500	292	120	0.000190	0.000079
1000	500	292	121	0.000321	0.000130
1000	500	292	122	0.000531	0.000210
1000	500	292	123	0.000862	0.000332
1000	500	292	124	0.001375	0.000513
1000	500	292	125	0.002154	0.000779
1000	500	292	126	0.003313	0.001159
1000	500	292	127	0.005004	0.001691
1000	500	292	128	0.007424	0.002420
1000	500	292	129	0.010820	0.003396
1000	500	292	130	0.015494	0.004674
1000	500	292	131	0.021802	0.006308
1000	500	292	132	0.030153	0.008351
1000	500	292	133	0.040994	0.010841
1000	500	292	134	0.054798	0.013804
1000	500	292	135	0.072037	0.017239
1000	500	292	136	0.093153	0.021116
1000	500	292	137	0.118522	0.025369
1000	500	292	138	0.148416	0.029894
1000	500	292	139	0.182968	0.034552
1000	500	292	140	0.222138	0.039171
1000	500	292	141	0.265696	0.043557
1000	500	292	142	0.313205	0.047509
1000	500	292	143	0.364033	0.050829
1000	500	292	144	0.417374	0.053340
1000	500	292	145	0.472280	0.054907
1000	500	292	146	0.527719	0.055439
1000	500	293	111	0.000001	0.000000
1000	500	293	112	0.000001	0.000001
1000	500	293	113	0.000002	0.000001
1000	500	293	114	0.000004	0.000002
1000	500	293	115	0.000008	0.000004
1000	500	293	116	0.000015	0.000007
1000	500	293	117	0.000027	0.000012
1000	500	293	118	0.000048	0.000021
1000	500	293	119	0.000085	0.000037
1000	500	293	120	0.000148	0.000062
1000	500	293	121	0.000251	0.000103
1000	500	293	122	0.000419	0.000168
1000	500	293	123	0.000686	0.000267
1000	500	293	124	0.001103	0.000417

N	n	k	x	P(x)	p(x)
1000	500	293	125	0.001742	0.000639
1000	500	293	126	0.002701	0.000959
1000	500	293	127	0.004113	0.001412
1000	500	293	128	0.006153	0.002039
1000	500	293	129	0.009040	0.002888
1000	500	293	130	0.013051	0.004011
1000	500	293	131	0.018514	0.005463
1000	500	293	132	0.025812	0.007298
1000	500	293	133	0.035374	0.009562
1000	500	293	134	0.047662	0.012288
1000	500	293	135	0.063150	0.015488
1000	500	293	136	0.082297	0.019147
1000	500	293	137	0.105515	0.023218
1000	500	293	138	0.133131	0.027616
1000	500	293	139	0.165350	0.032219
1000	500	293	140	0.202221	0.036871
1000	500	293	141	0.243609	0.041388
1000	500	293	142	0.289182	0.045572
1000	500	293	143	0.338404	0.049222
1000	500	293	144	0.390553	0.052149
1000	500	293	145	0.444749	0.054197
1000	500	293	146	0.500000	0.055250
1000	500	294	112	0.000001	0.000000
1000	500	294	113	0.000002	0.000001
1000	500	294	114	0.000003	0.000001
1000	500	294	115	0.000006	0.000003
1000	500	294	116	0.000011	0.000005
1000	500	294	117	0.000020	0.000009
1000	500	294	118	0.000037	0.000016
1000	500	294	119	0.000065	0.000029
1000	500	294	120	0.000114	0.000049
1000	500	294	121	0.000195	0.000081
1000	500	294	122	0.000329	0.000133
1000	500	294	123	0.000543	0.000214
1000	500	294	124	0.000881	0.000338
1000	500	294	125	0.001403	0.000522
1000	500	294	126	0.002194	0.000791
1000	500	294	127	0.003368	0.001175
1000	500	294	128	0.005080	0.001711
1000	500	294	129	0.007525	0.002446
1000	500	294	130	0.010952	0.003427
1000	500	294	131	0.015663	0.004711
1000	500	294	132	0.022014	0.006351
1000	500	294	133	0.030411	0.008397
1000	500	294	134	0.041301	0.010890
1000	500	294	135	0.055154	0.013853
1000	500	294	136	0.072439	0.017285
1000	500	294	137	0.093594	0.021155
1000	500	294	138	0.118991	0.025397
1000	500	294	139	0.148899	0.029908

N	n	k	x	P(x)	p(x)
1000	500	294	140	0.183447	0.034548
1000	500	294	141	0.222593	0.039147
1000	500	294	142	0.266105	0.043512
1000	500	294	143	0.313549	0.047443
1000	500	294	144	0.364294	0.050745
1000	500	294	145	0.417536	0.053242
1000	500	294	146	0.472336	0.054799
1000	500	294	147	0.527664	0.055329
1000	500	295	112	0.000001	0.000000
1000	500	295	113	0.000001	0.000001
1000	500	295	114	0.000002	0.000001
1000	500	295	115	0.000004	0.000002
1000	500	295	116	0.000008	0.000004
1000	500	295	117	0.000015	0.000007
1000	500	295	118	0.000028	0.000013
1000	500	295	119	0.000050	0.000022
1000	500	295	120	0.000088	0.000038
1000	500	295	121	0.000152	0.000064
1000	500	295	122	0.000257	0.000106
1000	500	295	123	0.000429	0.000171
1000	500	295	124	0.000701	0.000272
1000	500	295	125	0.001126	0.000425
1000	500	295	126	0.001775	0.000649
1000	500	295	127	0.002748	0.000973
1000	500	295	128	0.004178	0.001430
1000	500	295	129	0.006240	0.002062
1000	500	295	130	0.009156	0.002916
1000	500	295	131	0.013201	0.004045
1000	500	295	132	0.018703	0.005503
1000	500	295	133	0.026046	0.007342
1000	500	295	134	0.035655	0.009610
1000	500	295	135	0.047992	0.012337
1000	500	295	136	0.063527	0.015535
1000	500	295	137	0.082717	0.019190
1000	500	295	138	0.105969	0.023252
1000	500	295	139	0.133606	0.027637
1000	500	295	140	0.165830	0.032224
1000	500	295	141	0.202687	0.036857
1000	500	295	142	0.244041	0.041354
1000	500	295	143	0.289558	0.045517
1000	500	295	144	0.338706	0.049147
1000	500	295	145	0.390764	0.052058
1000	500	295	146	0.444858	0.054094
1000	500	295	147	0.500000	0.055142
1000	500	296	113	0.000001	0.000000
1000	500	296	114	0.000002	0.000001
1000	500	296	115	0.000003	0.000002
1000	500	296	116	0.000006	0.000003
1000	500	296	117	0.000011	0.000005
1000	500	296	118	0.000021	0.000010

Table for $N = 1000$, $n = 500$

N	n	k	x	P(x)	p(x)
1000	500	296	119	0.000038	0.000017
1000	500	296	120	0.000067	0.000029
1000	500	296	121	0.000117	0.000050
1000	500	296	122	0.000201	0.000083
1000	500	296	123	0.000337	0.000136
1000	500	296	124	0.000556	0.000219
1000	500	296	125	0.000900	0.000344
1000	500	296	126	0.001430	0.000531
1000	500	296	127	0.002233	0.000802
1000	500	296	128	0.003423	0.001190
1000	500	296	129	0.005155	0.001732
1000	500	296	130	0.007626	0.002471
1000	500	296	131	0.011084	0.003458
1000	500	296	132	0.015831	0.004747
1000	500	296	133	0.022224	0.006392
1000	500	296	134	0.030666	0.008443
1000	500	296	135	0.041605	0.010938
1000	500	296	136	0.055506	0.013901
1000	500	296	137	0.072836	0.017330
1000	500	296	138	0.094030	0.021193
1000	500	296	139	0.119454	0.025425
1000	500	296	140	0.149376	0.029921
1000	500	296	141	0.183919	0.034543
1000	500	296	142	0.223042	0.039123
1000	500	296	143	0.266509	0.043468
1000	500	296	144	0.313888	0.047379
1000	500	296	145	0.364550	0.050662
1000	500	296	146	0.417696	0.053146
1000	500	296	147	0.472390	0.054694
1000	500	296	148	0.527610	0.055220
1000	500	297	113	0.000001	0.000000
1000	500	297	114	0.000001	0.000001
1000	500	297	115	0.000002	0.000001
1000	500	297	116	0.000004	0.000002
1000	500	297	117	0.000008	0.000004
1000	500	297	118	0.000016	0.000007
1000	500	297	119	0.000029	0.000013
1000	500	297	120	0.000051	0.000023
1000	500	297	121	0.000090	0.000039
1000	500	297	122	0.000156	0.000066
1000	500	297	123	0.000264	0.000108
1000	500	297	124	0.000439	0.000175
1000	500	297	125	0.000717	0.000278
1000	500	297	126	0.001149	0.000432
1000	500	297	127	0.001808	0.000659
1000	500	297	128	0.002794	0.000986
1000	500	297	129	0.004242	0.001448
1000	500	297	130	0.006327	0.002085
1000	500	297	131	0.009271	0.002944
1000	500	297	132	0.013350	0.004078

N	n	k	x	P(x)	p(x)
1000	500	297	133	0.018891	0.005542
1000	500	297	134	0.026277	0.007386
1000	500	297	135	0.035934	0.009657
1000	500	297	136	0.048319	0.012385
1000	500	297	137	0.063900	0.015582
1000	500	297	138	0.083132	0.019231
1000	500	297	139	0.106417	0.023285
1000	500	297	140	0.134075	0.027658
1000	500	297	141	0.166304	0.032229
1000	500	297	142	0.203147	0.036843
1000	500	297	143	0.244467	0.041320
1000	500	297	144	0.289929	0.045463
1000	500	297	145	0.339003	0.049074
1000	500	297	146	0.390972	0.051969
1000	500	297	147	0.444965	0.053993
1000	500	297	148	0.500000	0.055035
1000	500	298	114	0.000001	0.000000
1000	500	298	115	0.000002	0.000001
1000	500	298	116	0.000003	0.000002
1000	500	298	117	0.000006	0.000003
1000	500	298	118	0.000012	0.000005
1000	500	298	119	0.000022	0.000010
1000	500	298	120	0.000039	0.000017
1000	500	298	121	0.000069	0.000030
1000	500	298	122	0.000121	0.000051
1000	500	298	123	0.000206	0.000085
1000	500	298	124	0.000345	0.000139
1000	500	298	125	0.000569	0.000223
1000	500	298	126	0.000919	0.000350
1000	500	298	127	0.001458	0.000539
1000	500	298	128	0.002272	0.000814
1000	500	298	129	0.003478	0.001206
1000	500	298	130	0.005230	0.001752
1000	500	298	131	0.007726	0.002496
1000	500	298	132	0.011215	0.003489
1000	500	298	133	0.015998	0.004783
1000	500	298	134	0.022432	0.006434
1000	500	298	135	0.030920	0.008488
1000	500	298	136	0.041906	0.010986
1000	500	298	137	0.055855	0.013949
1000	500	298	138	0.073229	0.017374
1000	500	298	139	0.094460	0.021231
1000	500	298	140	0.119912	0.025452
1000	500	298	141	0.149846	0.029934
1000	500	298	142	0.184385	0.034539
1000	500	298	143	0.223484	0.039099
1000	500	298	144	0.266907	0.043424
1000	500	298	145	0.314222	0.047315
1000	500	298	146	0.364803	0.050581
1000	500	298	147	0.417854	0.053051

N	n	k	x	P(x)	p(x)
1000	500	298	148	0.472444	0.054590
1000	500	298	149	0.527556	0.055113
1000	500	299	114	0.000001	0.000000
1000	500	299	115	0.000001	0.000001
1000	500	299	116	0.000002	0.000001
1000	500	299	117	0.000005	0.000002
1000	500	299	118	0.000009	0.000004
1000	500	299	119	0.000016	0.000008
1000	500	299	120	0.000030	0.000013
1000	500	299	121	0.000053	0.000023
1000	500	299	122	0.000093	0.000040
1000	500	299	123	0.000160	0.000067
1000	500	299	124	0.000271	0.000111
1000	500	299	125	0.000449	0.000179
1000	500	299	126	0.000732	0.000283
1000	500	299	127	0.001171	0.000439
1000	500	299	128	0.001841	0.000669
1000	500	299	129	0.002841	0.001000
1000	500	299	130	0.004306	0.001466
1000	500	299	131	0.006414	0.002107
1000	500	299	132	0.009386	0.002972
1000	500	299	133	0.013497	0.004112
1000	500	299	134	0.019078	0.005580
1000	500	299	135	0.026507	0.007429
1000	500	299	136	0.036209	0.009703
1000	500	299	137	0.048642	0.012432
1000	500	299	138	0.064270	0.015628
1000	500	299	139	0.083542	0.019272
1000	500	299	140	0.106859	0.023317
1000	500	299	141	0.134537	0.027678
1000	500	299	142	0.166771	0.032233
1000	500	299	143	0.203600	0.036829
1000	500	299	144	0.244886	0.041286
1000	500	299	145	0.290295	0.045409
1000	500	299	146	0.339296	0.049001
1000	500	299	147	0.391177	0.051881
1000	500	299	148	0.445071	0.053894
1000	500	299	149	0.500000	0.054929
1000	500	300	115	0.000001	0.000000
1000	500	300	116	0.000002	0.000001
1000	500	300	117	0.000003	0.000002
1000	500	300	118	0.000007	0.000003
1000	500	300	119	0.000012	0.000006
1000	500	300	120	0.000022	0.000010
1000	500	300	121	0.000040	0.000018
1000	500	300	122	0.000071	0.000031
1000	500	300	123	0.000124	0.000053
1000	500	300	124	0.000212	0.000087
1000	500	300	125	0.000354	0.000142
1000	500	300	126	0.000581	0.000227

N	n	k	x	P(x)	p(x)
1000	500	300	127	0.000938	0.000356
1000	500	300	128	0.001486	0.000548
1000	500	300	129	0.002311	0.000826
1000	500	300	130	0.003533	0.001221
1000	500	300	131	0.005305	0.001772
1000	500	300	132	0.007826	0.002521
1000	500	300	133	0.011345	0.003519
1000	500	300	134	0.016164	0.004819
1000	500	300	135	0.022638	0.006474
1000	500	300	136	0.031171	0.008533
1000	500	300	137	0.042204	0.011033
1000	500	300	138	0.056199	0.013995
1000	500	300	139	0.073617	0.017418
1000	500	300	140	0.094885	0.021268
1000	500	300	141	0.120363	0.025478
1000	500	300	142	0.150309	0.029946
1000	500	300	143	0.184844	0.034535
1000	500	300	144	0.223919	0.039075
1000	500	300	145	0.267299	0.043380
1000	500	300	146	0.314551	0.047252
1000	500	300	147	0.365052	0.050501
1000	500	300	148	0.418009	0.052957
1000	500	300	149	0.472496	0.054488
1000	500	300	150	0.527504	0.055007
1000	500	301	115	0.000001	0.000000
1000	500	301	116	0.000001	0.000001
1000	500	301	117	0.000003	0.000001
1000	500	301	118	0.000005	0.000002
1000	500	301	119	0.000009	0.000004
1000	500	301	120	0.000017	0.000008
1000	500	301	121	0.000031	0.000014
1000	500	301	122	0.000055	0.000024
1000	500	301	123	0.000096	0.000041
1000	500	301	124	0.000165	0.000069
1000	500	301	125	0.000278	0.000113
1000	500	301	126	0.000460	0.000182
1000	500	301	127	0.000748	0.000288
1000	500	301	128	0.001194	0.000447
1000	500	301	129	0.001874	0.000679
1000	500	301	130	0.002887	0.001013
1000	500	301	131	0.004371	0.001483
1000	500	301	132	0.006500	0.002130
1000	500	301	133	0.009500	0.003000
1000	500	301	134	0.013644	0.004145
1000	500	301	135	0.019263	0.005618
1000	500	301	136	0.026734	0.007472
1000	500	301	137	0.036483	0.009749
1000	500	301	138	0.048962	0.012479
1000	500	301	139	0.064634	0.015673
1000	500	301	140	0.083947	0.019313

Table for $N = 1000,\ n = 500$

N	n	k	x	P(x)	p(x)
1000	500	301	141	0.107296	0.023349
1000	500	301	142	0.134994	0.027698
1000	500	301	143	0.167231	0.032237
1000	500	301	144	0.204047	0.036815
1000	500	301	145	0.245299	0.041253
1000	500	301	146	0.290655	0.045356
1000	500	301	147	0.339585	0.048930
1000	500	301	148	0.391379	0.051794
1000	500	301	149	0.445175	0.053796
1000	500	301	150	0.500000	0.054825
1000	500	302	116	0.000001	0.000000
1000	500	302	117	0.000002	0.000001
1000	500	302	118	0.000004	0.000002
1000	500	302	119	0.000007	0.000003
1000	500	302	120	0.000013	0.000006
1000	500	302	121	0.000023	0.000011
1000	500	302	122	0.000042	0.000019
1000	500	302	123	0.000074	0.000032
1000	500	302	124	0.000128	0.000054
1000	500	302	125	0.000217	0.000089
1000	500	302	126	0.000362	0.000145
1000	500	302	127	0.000594	0.000232
1000	500	302	128	0.000957	0.000363
1000	500	302	129	0.001513	0.000556
1000	500	302	130	0.002351	0.000838
1000	500	302	131	0.003588	0.001237
1000	500	302	132	0.005379	0.001791
1000	500	302	133	0.007925	0.002546
1000	500	302	134	0.011474	0.003549
1000	500	302	135	0.016329	0.004855
1000	500	302	136	0.022843	0.006514
1000	500	302	137	0.031420	0.008577
1000	500	302	138	0.042499	0.011079
1000	500	302	139	0.056540	0.014041
1000	500	302	140	0.074001	0.017460
1000	500	302	141	0.095304	0.021304
1000	500	302	142	0.120808	0.025504
1000	500	302	143	0.150767	0.029958
1000	500	302	144	0.185296	0.034530
1000	500	302	145	0.224348	0.039052
1000	500	302	146	0.267685	0.043337
1000	500	302	147	0.314875	0.047190
1000	500	302	148	0.365297	0.050422
1000	500	302	149	0.418161	0.052865
1000	500	302	150	0.472548	0.054387
1000	500	302	151	0.527452	0.054904
1000	500	303	116	0.000001	0.000000
1000	500	303	117	0.000001	0.000001
1000	500	303	118	0.000003	0.000001
1000	500	303	119	0.000005	0.000002
1000	500	303	120	0.000009	0.000004
1000	500	303	121	0.000017	0.000008
1000	500	303	122	0.000032	0.000014
1000	500	303	123	0.000056	0.000025
1000	500	303	124	0.000099	0.000042
1000	500	303	125	0.000169	0.000070
1000	500	303	126	0.000284	0.000115
1000	500	303	127	0.000470	0.000186
1000	500	303	128	0.000764	0.000293
1000	500	303	129	0.001217	0.000454
1000	500	303	130	0.001907	0.000689
1000	500	303	131	0.002934	0.001027
1000	500	303	132	0.004435	0.001501
1000	500	303	133	0.006586	0.002152
1000	500	303	134	0.009613	0.003027
1000	500	303	135	0.013790	0.004177
1000	500	303	136	0.019446	0.005656
1000	500	303	137	0.026960	0.007514
1000	500	303	138	0.036753	0.009794
1000	500	303	139	0.049278	0.012525
1000	500	303	140	0.064995	0.015717
1000	500	303	141	0.084347	0.019352
1000	500	303	142	0.107727	0.023380
1000	500	303	143	0.135444	0.027717
1000	500	303	144	0.167685	0.032241
1000	500	303	145	0.204487	0.036801
1000	500	303	146	0.245706	0.041220
1000	500	303	147	0.291010	0.045303
1000	500	303	148	0.339869	0.048859
1000	500	303	149	0.391577	0.051708
1000	500	303	150	0.445277	0.053699
1000	500	303	151	0.500000	0.054723
1000	500	304	117	0.000001	0.000000
1000	500	304	118	0.000002	0.000001
1000	500	304	119	0.000004	0.000002
1000	500	304	120	0.000007	0.000003
1000	500	304	121	0.000013	0.000006
1000	500	304	122	0.000024	0.000011
1000	500	304	123	0.000043	0.000019
1000	500	304	124	0.000076	0.000033
1000	500	304	125	0.000131	0.000055
1000	500	304	126	0.000222	0.000091
1000	500	304	127	0.000371	0.000148
1000	500	304	128	0.000607	0.000236
1000	500	304	129	0.000976	0.000369
1000	500	304	130	0.001541	0.000565
1000	500	304	131	0.002390	0.000849
1000	500	304	132	0.003642	0.001252
1000	500	304	133	0.005453	0.001811
1000	500	304	134	0.008024	0.002570
1000	500	304	135	0.011603	0.003579
1000	500	304	136	0.016492	0.004890
1000	500	304	137	0.023047	0.006554
1000	500	304	138	0.031667	0.008620
1000	500	304	139	0.042791	0.011124
1000	500	304	140	0.056877	0.014086
1000	500	304	141	0.074380	0.017502
1000	500	304	142	0.095719	0.021339
1000	500	304	143	0.121248	0.025529
1000	500	304	144	0.151218	0.029970
1000	500	304	145	0.185743	0.034525
1000	500	304	146	0.224771	0.039028
1000	500	304	147	0.268065	0.043294
1000	500	304	148	0.315194	0.047129
1000	500	304	149	0.365538	0.050344
1000	500	304	150	0.418312	0.052774
1000	500	304	151	0.472599	0.054288
1000	500	304	152	0.527401	0.054802
1000	500	305	117	0.000001	0.000000
1000	500	305	118	0.000001	0.000001
1000	500	305	119	0.000003	0.000001
1000	500	305	120	0.000005	0.000002
1000	500	305	121	0.000010	0.000005
1000	500	305	122	0.000018	0.000008
1000	500	305	123	0.000033	0.000015
1000	500	305	124	0.000058	0.000025
1000	500	305	125	0.000101	0.000043
1000	500	305	126	0.000173	0.000072
1000	500	305	127	0.000291	0.000118
1000	500	305	128	0.000481	0.000189
1000	500	305	129	0.000779	0.000298
1000	500	305	130	0.001240	0.000461
1000	500	305	131	0.001940	0.000699
1000	500	305	132	0.002980	0.001040
1000	500	305	133	0.004498	0.001518
1000	500	305	134	0.006672	0.002174
1000	500	305	135	0.009726	0.003054
1000	500	305	136	0.013935	0.004209
1000	500	305	137	0.019628	0.005693
1000	500	305	138	0.027183	0.007555
1000	500	305	139	0.037021	0.009838
1000	500	305	140	0.049591	0.012570
1000	500	305	141	0.065352	0.015761
1000	500	305	142	0.084743	0.019391
1000	500	305	143	0.108153	0.023410
1000	500	305	144	0.135888	0.027735
1000	500	305	145	0.168133	0.032245
1000	500	305	146	0.204921	0.036788
1000	500	305	147	0.246107	0.041187
1000	500	305	148	0.291359	0.045252
1000	500	305	149	0.340149	0.048790
1000	500	305	150	0.391773	0.051624
1000	500	305	151	0.445377	0.053604
1000	500	305	152	0.500000	0.054623
1000	500	306	117	0.000001	0.000000
1000	500	306	118	0.000001	0.000001
1000	500	306	119	0.000002	0.000001
1000	500	306	120	0.000004	0.000002
1000	500	306	121	0.000007	0.000003
1000	500	306	122	0.000014	0.000006
1000	500	306	123	0.000025	0.000011
1000	500	306	124	0.000044	0.000020
1000	500	306	125	0.000078	0.000034
1000	500	306	126	0.000135	0.000057
1000	500	306	127	0.000228	0.000093
1000	500	306	128	0.000379	0.000151
1000	500	306	129	0.000620	0.000241
1000	500	306	130	0.000995	0.000375
1000	500	306	131	0.001569	0.000574
1000	500	306	132	0.002429	0.000861
1000	500	306	133	0.003697	0.001267
1000	500	306	134	0.005527	0.001831
1000	500	306	135	0.008122	0.002595
1000	500	306	136	0.011731	0.003609
1000	500	306	137	0.016655	0.004924
1000	500	306	138	0.023248	0.006593
1000	500	306	139	0.031911	0.008663
1000	500	306	140	0.043080	0.011169
1000	500	306	141	0.057211	0.014131
1000	500	306	142	0.074754	0.017543
1000	500	306	143	0.096128	0.021373
1000	500	306	144	0.121681	0.025554
1000	500	306	145	0.151663	0.029981
1000	500	306	146	0.186183	0.034520
1000	500	306	147	0.225188	0.039005
1000	500	306	148	0.268440	0.043252
1000	500	306	149	0.315508	0.047068
1000	500	306	150	0.365775	0.050267
1000	500	306	151	0.418459	0.052684
1000	500	306	152	0.472649	0.054190
1000	500	306	153	0.527350	0.054701
1000	500	307	118	0.000001	0.000000
1000	500	307	119	0.000001	0.000001
1000	500	307	120	0.000003	0.000001
1000	500	307	121	0.000005	0.000003
1000	500	307	122	0.000010	0.000005
1000	500	307	123	0.000019	0.000009
1000	500	307	124	0.000034	0.000015
1000	500	307	125	0.000060	0.000026
1000	500	307	126	0.000104	0.000044

Table for $N = 1000$, $n = 500$

N	n	k	x	P(x)	p(x)
1000	500	307	127	0.000178	0.000074
1000	500	307	128	0.000298	0.000120
1000	500	307	129	0.000491	0.000193
1000	500	307	130	0.000795	0.000304
1000	500	307	131	0.001264	0.000469
1000	500	307	132	0.001973	0.000709
1000	500	307	133	0.003026	0.001054
1000	500	307	134	0.004562	0.001535
1000	500	307	135	0.006757	0.002196
1000	500	307	136	0.009838	0.003081
1000	500	307	137	0.014079	0.004241
1000	500	307	138	0.019809	0.005730
1000	500	307	139	0.027405	0.007596
1000	500	307	140	0.037286	0.009882
1000	500	307	141	0.049901	0.012615
1000	500	307	142	0.065705	0.015803
1000	500	307	143	0.085133	0.019429
1000	500	307	144	0.108573	0.023440
1000	500	307	145	0.136326	0.027753
1000	500	307	146	0.168574	0.032248
1000	500	307	147	0.205348	0.036774
1000	500	307	148	0.246502	0.041154
1000	500	307	149	0.291703	0.045200
1000	500	307	150	0.340424	0.048722
1000	500	307	151	0.391966	0.051541
1000	500	307	152	0.445476	0.053511
1000	500	307	153	0.500000	0.054524
1000	500	308	118	0.000001	0.000000
1000	500	308	119	0.000001	0.000001
1000	500	308	120	0.000002	0.000001
1000	500	308	121	0.000004	0.000002
1000	500	308	122	0.000008	0.000004
1000	500	308	123	0.000014	0.000006
1000	500	308	124	0.000026	0.000012
1000	500	308	125	0.000046	0.000020
1000	500	308	126	0.000080	0.000034
1000	500	308	127	0.000138	0.000058
1000	500	308	128	0.000234	0.000095
1000	500	308	129	0.000388	0.000154
1000	500	308	130	0.000633	0.000245
1000	500	308	131	0.001014	0.000381
1000	500	308	132	0.001596	0.000582
1000	500	308	133	0.002468	0.000872
1000	500	308	134	0.003751	0.001282
1000	500	308	135	0.005601	0.001850
1000	500	308	136	0.008220	0.002619
1000	500	308	137	0.011858	0.003638
1000	500	308	138	0.016816	0.004958
1000	500	308	139	0.023448	0.006632
1000	500	308	140	0.032153	0.008705

N	n	k	x	P(x)	p(x)
1000	500	308	141	0.043366	0.011213
1000	500	308	142	0.057541	0.014174
1000	500	308	143	0.075124	0.017584
1000	500	308	144	0.096532	0.021407
1000	500	308	145	0.122109	0.025578
1000	500	308	146	0.152101	0.029992
1000	500	308	147	0.186616	0.034515
1000	500	308	148	0.225599	0.038982
1000	500	308	149	0.268809	0.043210
1000	500	308	150	0.315818	0.047008
1000	500	308	151	0.366009	0.050191
1000	500	308	152	0.418605	0.052596
1000	500	308	153	0.472699	0.054094
1000	500	308	154	0.527301	0.054602
1000	500	309	119	0.000001	0.000000
1000	500	309	120	0.000002	0.000001
1000	500	309	121	0.000003	0.000001
1000	500	309	122	0.000006	0.000003
1000	500	309	123	0.000011	0.000005
1000	500	309	124	0.000019	0.000009
1000	500	309	125	0.000035	0.000015
1000	500	309	126	0.000062	0.000027
1000	500	309	127	0.000107	0.000045
1000	500	309	128	0.000182	0.000075
1000	500	309	129	0.000305	0.000123
1000	500	309	130	0.000502	0.000197
1000	500	309	131	0.000811	0.000309
1000	500	309	132	0.001287	0.000476
1000	500	309	133	0.002006	0.000719
1000	500	309	134	0.003073	0.001067
1000	500	309	135	0.004625	0.001553
1000	500	309	136	0.006842	0.002217
1000	500	309	137	0.009949	0.003107
1000	500	309	138	0.014222	0.004273
1000	500	309	139	0.019988	0.005766
1000	500	309	140	0.027624	0.007636
1000	500	309	141	0.037549	0.009925
1000	500	309	142	0.050208	0.012659
1000	500	309	143	0.066053	0.015845
1000	500	309	144	0.085519	0.019466
1000	500	309	145	0.108988	0.023469
1000	500	309	146	0.136759	0.027771
1000	500	309	147	0.169010	0.032251
1000	500	309	148	0.205769	0.036760
1000	500	309	149	0.246892	0.041122
1000	500	309	150	0.292042	0.045150
1000	500	309	151	0.340696	0.048654
1000	500	309	152	0.392155	0.051459
1000	500	309	153	0.445574	0.053419
1000	500	309	154	0.500000	0.054426

N	n	k	x	P(x)	p(x)
1000	500	310	119	0.000001	0.000000
1000	500	310	120	0.000001	0.000001
1000	500	310	121	0.000002	0.000001
1000	500	310	122	0.000004	0.000002
1000	500	310	123	0.000008	0.000004
1000	500	310	124	0.000015	0.000007
1000	500	310	125	0.000026	0.000012
1000	500	310	126	0.000047	0.000021
1000	500	310	127	0.000082	0.000035
1000	500	310	128	0.000142	0.000059
1000	500	310	129	0.000239	0.000097
1000	500	310	130	0.000397	0.000157
1000	500	310	131	0.000646	0.000249
1000	500	310	132	0.001033	0.000387
1000	500	310	133	0.001624	0.000591
1000	500	310	134	0.002508	0.000884
1000	500	310	135	0.003805	0.001297
1000	500	310	136	0.005674	0.001869
1000	500	310	137	0.008317	0.002643
1000	500	310	138	0.011984	0.003667
1000	500	310	139	0.016976	0.004992
1000	500	310	140	0.023646	0.006670
1000	500	310	141	0.032393	0.008747
1000	500	310	142	0.043649	0.011257
1000	500	310	143	0.057867	0.014217
1000	500	310	144	0.075490	0.017623
1000	500	310	145	0.096931	0.021441
1000	500	310	146	0.122532	0.025601
1000	500	310	147	0.152534	0.030002
1000	500	310	148	0.187044	0.034510
1000	500	310	149	0.226003	0.038959
1000	500	310	150	0.269173	0.043169
1000	500	310	151	0.316122	0.046950
1000	500	310	152	0.366239	0.050117
1000	500	310	153	0.418748	0.052509
1000	500	310	154	0.472747	0.053999
1000	500	310	155	0.527252	0.054505
1000	500	311	120	0.000001	0.000000
1000	500	311	121	0.000002	0.000001
1000	500	311	122	0.000003	0.000001
1000	500	311	123	0.000006	0.000003
1000	500	311	124	0.000011	0.000005
1000	500	311	125	0.000020	0.000009
1000	500	311	126	0.000036	0.000016
1000	500	311	127	0.000063	0.000027
1000	500	311	128	0.000110	0.000046
1000	500	311	129	0.000187	0.000077
1000	500	311	130	0.000312	0.000125
1000	500	311	131	0.000513	0.000200
1000	500	311	132	0.000827	0.000314

N	n	k	x	P(x)	p(x)
1000	500	311	133	0.001310	0.000483
1000	500	311	134	0.002039	0.000729
1000	500	311	135	0.003119	0.001080
1000	500	311	136	0.004688	0.001570
1000	500	311	137	0.006927	0.002239
1000	500	311	138	0.010060	0.003133
1000	500	311	139	0.014364	0.004304
1000	500	311	140	0.020166	0.005802
1000	500	311	141	0.027842	0.007676
1000	500	311	142	0.037809	0.009967
1000	500	311	143	0.050511	0.012702
1000	500	311	144	0.066397	0.015887
1000	500	311	145	0.085900	0.019502
1000	500	311	146	0.109397	0.023498
1000	500	311	147	0.137185	0.027788
1000	500	311	148	0.169439	0.032254
1000	500	311	149	0.206185	0.036746
1000	500	311	150	0.247275	0.041090
1000	500	311	151	0.292375	0.045100
1000	500	311	152	0.340963	0.048588
1000	500	311	153	0.392342	0.051379
1000	500	311	154	0.445670	0.053328
1000	500	311	155	0.500000	0.054330
1000	500	312	120	0.000001	0.000000
1000	500	312	121	0.000001	0.000001
1000	500	312	122	0.000002	0.000001
1000	500	312	123	0.000004	0.000002
1000	500	312	124	0.000008	0.000004
1000	500	312	125	0.000015	0.000007
1000	500	312	126	0.000027	0.000012
1000	500	312	127	0.000048	0.000021
1000	500	312	128	0.000085	0.000036
1000	500	312	129	0.000145	0.000061
1000	500	312	130	0.000245	0.000100
1000	500	312	131	0.000405	0.000160
1000	500	312	132	0.000659	0.000254
1000	500	312	133	0.001052	0.000394
1000	500	312	134	0.001652	0.000599
1000	500	312	135	0.002547	0.000895
1000	500	312	136	0.003859	0.001312
1000	500	312	137	0.005747	0.001888
1000	500	312	138	0.008414	0.002666
1000	500	312	139	0.012109	0.003695
1000	500	312	140	0.017134	0.005025
1000	500	312	141	0.023842	0.006708
1000	500	312	142	0.032630	0.008788
1000	500	312	143	0.043929	0.011299
1000	500	312	144	0.058189	0.014259
1000	500	312	145	0.075851	0.017662
1000	500	312	146	0.097324	0.021473

Table for $N = 1000$, $n = 500$

N	n	k	x	P(x)	p(x)
1000	500	312	147	0.122948	0.025624
1000	500	312	148	0.152961	0.030013
1000	500	312	149	0.187465	0.034504
1000	500	312	150	0.226402	0.038937
1000	500	312	151	0.269531	0.043129
1000	500	312	152	0.316422	0.046892
1000	500	312	153	0.366466	0.050044
1000	500	312	154	0.418890	0.052424
1000	500	312	155	0.472795	0.053906
1000	500	312	156	0.527204	0.054409
1000	500	313	121	0.000001	0.000000
1000	500	313	122	0.000002	0.000001
1000	500	313	123	0.000003	0.000002
1000	500	313	124	0.000006	0.000003
1000	500	313	125	0.000011	0.000005
1000	500	313	126	0.000021	0.000009
1000	500	313	127	0.000037	0.000016
1000	500	313	128	0.000065	0.000028
1000	500	313	129	0.000113	0.000048
1000	500	313	130	0.000191	0.000079
1000	500	313	131	0.000319	0.000128
1000	500	313	132	0.000523	0.000204
1000	500	313	133	0.000842	0.000319
1000	500	313	134	0.001333	0.000490
1000	500	313	135	0.002072	0.000739
1000	500	313	136	0.003165	0.001093
1000	500	313	137	0.004751	0.001586
1000	500	313	138	0.007011	0.002260
1000	500	313	139	0.010170	0.003159
1000	500	313	140	0.014505	0.004335
1000	500	313	141	0.020342	0.005837
1000	500	313	142	0.028057	0.007715
1000	500	313	143	0.038066	0.010009
1000	500	313	144	0.050810	0.012744
1000	500	313	145	0.066738	0.015927
1000	500	313	146	0.086276	0.019538
1000	500	313	147	0.109801	0.023525
1000	500	313	148	0.137606	0.027804
1000	500	313	149	0.169862	0.032256
1000	500	313	150	0.206594	0.036732
1000	500	313	151	0.247653	0.041059
1000	500	313	152	0.292704	0.045051
1000	500	313	153	0.341226	0.048522
1000	500	313	154	0.392525	0.051299
1000	500	313	155	0.445764	0.053239
1000	500	313	156	0.500000	0.054236
1000	500	314	121	0.000001	0.000000
1000	500	314	122	0.000001	0.000001
1000	500	314	123	0.000002	0.000001
1000	500	314	124	0.000004	0.000002

N	n	k	x	P(x)	p(x)
1000	500	314	125	0.000008	0.000004
1000	500	314	126	0.000016	0.000007
1000	500	314	127	0.000028	0.000013
1000	500	314	128	0.000050	0.000022
1000	500	314	129	0.000087	0.000037
1000	500	314	130	0.000149	0.000062
1000	500	314	131	0.000250	0.000102
1000	500	314	132	0.000414	0.000163
1000	500	314	133	0.000672	0.000258
1000	500	314	134	0.001072	0.000400
1000	500	314	135	0.001679	0.000608
1000	500	314	136	0.002586	0.000906
1000	500	314	137	0.003913	0.001327
1000	500	314	138	0.005820	0.001907
1000	500	314	139	0.008510	0.002690
1000	500	314	140	0.012233	0.003723
1000	500	314	141	0.017292	0.005058
1000	500	314	142	0.024037	0.006745
1000	500	314	143	0.032865	0.008828
1000	500	314	144	0.044207	0.011341
1000	500	314	145	0.058507	0.014301
1000	500	314	146	0.076208	0.017701
1000	500	314	147	0.097713	0.021505
1000	500	314	148	0.123359	0.025646
1000	500	314	149	0.153382	0.030022
1000	500	314	150	0.187880	0.034499
1000	500	314	151	0.226795	0.038914
1000	500	314	152	0.269883	0.043089
1000	500	314	153	0.316718	0.046834
1000	500	314	154	0.366689	0.049971
1000	500	314	155	0.419028	0.052340
1000	500	314	156	0.472843	0.053814
1000	500	314	157	0.527157	0.054315
1000	500	315	122	0.000001	0.000000
1000	500	315	123	0.000002	0.000001
1000	500	315	124	0.000003	0.000002
1000	500	315	125	0.000006	0.000003
1000	500	315	126	0.000012	0.000005
1000	500	315	127	0.000021	0.000010
1000	500	315	128	0.000038	0.000017
1000	500	315	129	0.000067	0.000029
1000	500	315	130	0.000115	0.000049
1000	500	315	131	0.000196	0.000080
1000	500	315	132	0.000326	0.000130
1000	500	315	133	0.000534	0.000208
1000	500	315	134	0.000858	0.000324
1000	500	315	135	0.001356	0.000498
1000	500	315	136	0.002105	0.000749
1000	500	315	137	0.003211	0.001106
1000	500	315	138	0.004814	0.001603

N	n	k	x	P(x)	p(x)
1000	500	315	139	0.007094	0.002281
1000	500	315	140	0.010279	0.003185
1000	500	315	141	0.014645	0.004365
1000	500	315	142	0.020517	0.005872
1000	500	315	143	0.028271	0.007754
1000	500	315	144	0.038321	0.010050
1000	500	315	145	0.051107	0.012786
1000	500	315	146	0.067074	0.015967
1000	500	315	147	0.086647	0.019573
1000	500	315	148	0.110200	0.023553
1000	500	315	149	0.138020	0.027821
1000	500	315	150	0.170279	0.032259
1000	500	315	151	0.206997	0.036718
1000	500	315	152	0.248025	0.041028
1000	500	315	153	0.293027	0.045002
1000	500	315	154	0.341485	0.048457
1000	500	315	155	0.392706	0.051221
1000	500	315	156	0.445857	0.053151
1000	500	315	157	0.500000	0.054143
1000	500	316	122	0.000001	0.000000
1000	500	316	123	0.000001	0.000001
1000	500	316	124	0.000002	0.000001
1000	500	316	125	0.000005	0.000002
1000	500	316	126	0.000009	0.000004
1000	500	316	127	0.000016	0.000007
1000	500	316	128	0.000029	0.000013
1000	500	316	129	0.000051	0.000022
1000	500	316	130	0.000089	0.000038
1000	500	316	131	0.000153	0.000063
1000	500	316	132	0.000256	0.000104
1000	500	316	133	0.000423	0.000166
1000	500	316	134	0.000685	0.000262
1000	500	316	135	0.001091	0.000406
1000	500	316	136	0.001707	0.000616
1000	500	316	137	0.002624	0.000918
1000	500	316	138	0.003966	0.001342
1000	500	316	139	0.005892	0.001926
1000	500	316	140	0.008606	0.002713
1000	500	316	141	0.012357	0.003751
1000	500	316	142	0.017448	0.005091
1000	500	316	143	0.024230	0.006782
1000	500	316	144	0.033098	0.008868
1000	500	316	145	0.044481	0.011383
1000	500	316	146	0.058822	0.014342
1000	500	316	147	0.076561	0.017738
1000	500	316	148	0.098097	0.021536
1000	500	316	149	0.123765	0.025668
1000	500	316	150	0.153797	0.030032
1000	500	316	151	0.188290	0.034493
1000	500	316	152	0.227182	0.038892

N	n	k	x	P(x)	p(x)
1000	500	316	153	0.270230	0.043049
1000	500	316	154	0.317008	0.046778
1000	500	316	155	0.366908	0.049900
1000	500	316	156	0.419165	0.052257
1000	500	316	157	0.472889	0.053724
1000	500	316	158	0.527111	0.054222
1000	500	317	123	0.000001	0.000000
1000	500	317	124	0.000002	0.000001
1000	500	317	125	0.000003	0.000002
1000	500	317	126	0.000006	0.000003
1000	500	317	127	0.000012	0.000006
1000	500	317	128	0.000022	0.000010
1000	500	317	129	0.000039	0.000017
1000	500	317	130	0.000069	0.000030
1000	500	317	131	0.000118	0.000050
1000	500	317	132	0.000200	0.000082
1000	500	317	133	0.000333	0.000133
1000	500	317	134	0.000545	0.000211
1000	500	317	135	0.000874	0.000330
1000	500	317	136	0.001379	0.000505
1000	500	317	137	0.002137	0.000758
1000	500	317	138	0.003256	0.001119
1000	500	317	139	0.004876	0.001620
1000	500	317	140	0.007178	0.002302
1000	500	317	141	0.010388	0.003210
1000	500	317	142	0.014783	0.004395
1000	500	317	143	0.020690	0.005907
1000	500	317	144	0.028482	0.007792
1000	500	317	145	0.038573	0.010091
1000	500	317	146	0.051400	0.012827
1000	500	317	147	0.067406	0.016006
1000	500	317	148	0.087014	0.019608
1000	500	317	149	0.110593	0.023579
1000	500	317	150	0.138430	0.027836
1000	500	317	151	0.170690	0.032261
1000	500	317	152	0.207395	0.036704
1000	500	317	153	0.248391	0.040997
1000	500	317	154	0.293346	0.044955
1000	500	317	155	0.341740	0.048394
1000	500	317	156	0.392884	0.051144
1000	500	317	157	0.445949	0.053065
1000	500	317	158	0.500000	0.054051
1000	500	318	123	0.000001	0.000000
1000	500	318	124	0.000001	0.000001
1000	500	318	125	0.000003	0.000001
1000	500	318	126	0.000005	0.000002
1000	500	318	127	0.000009	0.000004
1000	500	318	128	0.000017	0.000008
1000	500	318	129	0.000030	0.000013
1000	500	318	130	0.000053	0.000023

Table for $N = 1000$, $n = 500$

N	n	k	x	P(x)	p(x)
1000	500	318	131	0.000092	0.000039
1000	500	318	132	0.000156	0.000065
1000	500	318	133	0.000262	0.000106
1000	500	318	134	0.000431	0.000169
1000	500	318	135	0.000698	0.000267
1000	500	318	136	0.001110	0.000412
1000	500	318	137	0.001734	0.000624
1000	500	318	138	0.002663	0.000929
1000	500	318	139	0.004020	0.001357
1000	500	318	140	0.005964	0.001945
1000	500	318	141	0.008701	0.002736
1000	500	318	142	0.012479	0.003779
1000	500	318	143	0.017603	0.005123
1000	500	318	144	0.024421	0.006818
1000	500	318	145	0.033328	0.008907
1000	500	318	146	0.044752	0.011423
1000	500	318	147	0.059134	0.014382
1000	500	318	148	0.076909	0.017775
1000	500	318	149	0.098476	0.021567
1000	500	318	150	0.124165	0.025689
1000	500	318	151	0.154206	0.030041
1000	500	318	152	0.188693	0.034487
1000	500	318	153	0.227563	0.038870
1000	500	318	154	0.270572	0.043010
1000	500	318	155	0.317295	0.046722
1000	500	318	156	0.367125	0.049830
1000	500	318	157	0.419300	0.052175
1000	500	318	158	0.472935	0.053635
1000	500	318	159	0.527065	0.054131
1000	500	319	124	0.000001	0.000000
1000	500	319	125	0.000002	0.000001
1000	500	319	126	0.000004	0.000002
1000	500	319	127	0.000007	0.000003
1000	500	319	128	0.000012	0.000006
1000	500	319	129	0.000023	0.000010
1000	500	319	130	0.000040	0.000018
1000	500	319	131	0.000071	0.000030
1000	500	319	132	0.000121	0.000051
1000	500	319	133	0.000205	0.000084
1000	500	319	134	0.000340	0.000135
1000	500	319	135	0.000555	0.000215
1000	500	319	136	0.000890	0.000335
1000	500	319	137	0.001402	0.000512
1000	500	319	138	0.002170	0.000768
1000	500	319	139	0.003302	0.001132
1000	500	319	140	0.004938	0.001636
1000	500	319	141	0.007260	0.002322
1000	500	319	142	0.010496	0.003236
1000	500	319	143	0.014921	0.004425
1000	500	319	144	0.020862	0.005941

N	n	k	x	P(x)	p(x)
1000	500	319	145	0.028692	0.007830
1000	500	319	146	0.038822	0.010131
1000	500	319	147	0.051690	0.012867
1000	500	319	148	0.067734	0.016045
1000	500	319	149	0.087376	0.019641
1000	500	319	150	0.110981	0.023605
1000	500	319	151	0.138833	0.027852
1000	500	319	152	0.171095	0.032262
1000	500	319	153	0.207786	0.036691
1000	500	319	154	0.248752	0.040966
1000	500	319	155	0.293660	0.044907
1000	500	319	156	0.341990	0.048331
1000	500	319	157	0.393059	0.051069
1000	500	319	158	0.446039	0.052979
1000	500	319	159	0.500000	0.053961
1000	500	320	124	0.000001	0.000000
1000	500	320	125	0.000001	0.000001
1000	500	320	126	0.000003	0.000001
1000	500	320	127	0.000005	0.000002
1000	500	320	128	0.000009	0.000004
1000	500	320	129	0.000017	0.000008
1000	500	320	130	0.000031	0.000014
1000	500	320	131	0.000054	0.000023
1000	500	320	132	0.000094	0.000040
1000	500	320	133	0.000160	0.000066
1000	500	320	134	0.000268	0.000108
1000	500	320	135	0.000440	0.000172
1000	500	320	136	0.000711	0.000271
1000	500	320	137	0.001129	0.000418
1000	500	320	138	0.001762	0.000633
1000	500	320	139	0.002702	0.000940
1000	500	320	140	0.004073	0.001371
1000	500	320	141	0.006036	0.001963
1000	500	320	142	0.008795	0.002759
1000	500	320	143	0.012601	0.003806
1000	500	320	144	0.017756	0.005155
1000	500	320	145	0.024610	0.006854
1000	500	320	146	0.033556	0.008946
1000	500	320	147	0.045020	0.011464
1000	500	320	148	0.059441	0.014421
1000	500	320	149	0.077252	0.017811
1000	500	320	150	0.098849	0.021597
1000	500	320	151	0.124559	0.025710
1000	500	320	152	0.154609	0.030049
1000	500	320	153	0.189090	0.034482
1000	500	320	154	0.227938	0.038848
1000	500	320	155	0.270909	0.042971
1000	500	320	156	0.317577	0.046667
1000	500	320	157	0.367337	0.049761
1000	500	320	158	0.419432	0.052095

N	n	k	x	P(x)	p(x)
1000	500	320	159	0.472980	0.053547
1000	500	320	160	0.527020	0.054041
1000	500	321	125	0.000001	0.000000
1000	500	321	126	0.000002	0.000001
1000	500	321	127	0.000004	0.000002
1000	500	321	128	0.000007	0.000003
1000	500	321	129	0.000013	0.000006
1000	500	321	130	0.000023	0.000010
1000	500	321	131	0.000041	0.000018
1000	500	321	132	0.000072	0.000031
1000	500	321	133	0.000124	0.000052
1000	500	321	134	0.000210	0.000085
1000	500	321	135	0.000348	0.000138
1000	500	321	136	0.000566	0.000219
1000	500	321	137	0.000906	0.000340
1000	500	321	138	0.001425	0.000519
1000	500	321	139	0.002203	0.000778
1000	500	321	140	0.003347	0.001144
1000	500	321	141	0.005000	0.001653
1000	500	321	142	0.007342	0.002343
1000	500	321	143	0.010603	0.003260
1000	500	321	144	0.015057	0.004454
1000	500	321	145	0.021032	0.005975
1000	500	321	146	0.028899	0.007867
1000	500	321	147	0.039069	0.010170
1000	500	321	148	0.051976	0.012907
1000	500	321	149	0.068059	0.016083
1000	500	321	150	0.087733	0.019675
1000	500	321	151	0.111364	0.023631
1000	500	321	152	0.139231	0.027867
1000	500	321	153	0.171495	0.032264
1000	500	321	154	0.208172	0.036677
1000	500	321	155	0.249108	0.040936
1000	500	321	156	0.293968	0.044861
1000	500	321	157	0.342237	0.048269
1000	500	321	158	0.393232	0.050994
1000	500	321	159	0.446127	0.052896
1000	500	321	160	0.500000	0.053873
1000	500	322	125	0.000001	0.000000
1000	500	322	126	0.000001	0.000001
1000	500	322	127	0.000003	0.000001
1000	500	322	128	0.000005	0.000002
1000	500	322	129	0.000010	0.000004
1000	500	322	130	0.000018	0.000008
1000	500	322	131	0.000032	0.000014
1000	500	322	132	0.000056	0.000024
1000	500	322	133	0.000096	0.000041
1000	500	322	134	0.000164	0.000067
1000	500	322	135	0.000273	0.000110
1000	500	322	136	0.000449	0.000176

N	n	k	x	P(x)	p(x)
1000	500	322	137	0.000724	0.000275
1000	500	322	138	0.001148	0.000424
1000	500	322	139	0.001789	0.000641
1000	500	322	140	0.002740	0.000951
1000	500	322	141	0.004126	0.001386
1000	500	322	142	0.006107	0.001981
1000	500	322	143	0.008889	0.002781
1000	500	322	144	0.012722	0.003833
1000	500	322	145	0.017908	0.005186
1000	500	322	146	0.024797	0.006889
1000	500	322	147	0.033782	0.008984
1000	500	322	148	0.045285	0.011503
1000	500	322	149	0.059745	0.014460
1000	500	322	150	0.077597	0.017847
1000	500	322	151	0.099218	0.021626
1000	500	322	152	0.124948	0.025730
1000	500	322	153	0.155006	0.030058
1000	500	322	154	0.189482	0.034476
1000	500	322	155	0.228308	0.038826
1000	500	322	156	0.271241	0.042933
1000	500	322	157	0.317854	0.046613
1000	500	322	158	0.367547	0.049693
1000	500	322	159	0.419562	0.052016
1000	500	322	160	0.473024	0.053461
1000	500	322	161	0.526976	0.053952
1000	500	323	125	0.000001	0.000000
1000	500	323	126	0.000001	0.000001
1000	500	323	127	0.000002	0.000001
1000	500	323	128	0.000004	0.000002
1000	500	323	129	0.000007	0.000003
1000	500	323	130	0.000013	0.000006
1000	500	323	131	0.000024	0.000011
1000	500	323	132	0.000043	0.000019
1000	500	323	133	0.000074	0.000032
1000	500	323	134	0.000127	0.000053
1000	500	323	135	0.000214	0.000087
1000	500	323	136	0.000355	0.000140
1000	500	323	137	0.000577	0.000222
1000	500	323	138	0.000922	0.000345
1000	500	323	139	0.001448	0.000526
1000	500	323	140	0.002235	0.000787
1000	500	323	141	0.003392	0.001157
1000	500	323	142	0.005061	0.001669
1000	500	323	143	0.007424	0.002363
1000	500	323	144	0.010709	0.003285
1000	500	323	145	0.015193	0.004483
1000	500	323	146	0.021200	0.006008
1000	500	323	147	0.029104	0.007904
1000	500	323	148	0.039313	0.010209
1000	500	323	149	0.052259	0.012946

Table for $N = 1000$, $n = 500$

N	n	k	x	P(x)	p(x)
1000	500	323	150	0.068379	0.016120
1000	500	323	151	0.088086	0.019707
1000	500	323	152	0.111742	0.023656
1000	500	323	153	0.139623	0.027881
1000	500	323	154	0.171888	0.032265
1000	500	323	155	0.208551	0.036663
1000	500	323	156	0.249458	0.040906
1000	500	323	157	0.294272	0.044815
1000	500	323	158	0.342481	0.048208
1000	500	323	159	0.393401	0.050921
1000	500	323	160	0.446214	0.052813
1000	500	323	161	0.500000	0.053786
1000	500	324	126	0.000001	0.000000
1000	500	324	127	0.000001	0.000001
1000	500	324	128	0.000003	0.000001
1000	500	324	129	0.000005	0.000003
1000	500	324	130	0.000010	0.000005
1000	500	324	131	0.000018	0.000008
1000	500	324	132	0.000032	0.000014
1000	500	324	133	0.000057	0.000025
1000	500	324	134	0.000099	0.000042
1000	500	324	135	0.000167	0.000069
1000	500	324	136	0.000279	0.000112
1000	500	324	137	0.000458	0.000179
1000	500	324	138	0.000737	0.000280
1000	500	324	139	0.001167	0.000430
1000	500	324	140	0.001817	0.000649
1000	500	324	141	0.002779	0.000962
1000	500	324	142	0.004179	0.001400
1000	500	324	143	0.006178	0.001999
1000	500	324	144	0.008982	0.002804
1000	500	324	145	0.012842	0.003860
1000	500	324	146	0.018059	0.005217
1000	500	324	147	0.024983	0.006924
1000	500	324	148	0.034005	0.009022
1000	500	324	149	0.045547	0.011542
1000	500	324	150	0.060045	0.014498
1000	500	324	151	0.077927	0.017882
1000	500	324	152	0.099582	0.021655
1000	500	324	153	0.125332	0.025750
1000	500	324	154	0.155398	0.030066
1000	500	324	155	0.189868	0.034470
1000	500	324	156	0.228672	0.038805
1000	500	324	157	0.271567	0.042895
1000	500	324	158	0.318127	0.046560
1000	500	324	159	0.367753	0.049626
1000	500	324	160	0.419691	0.051938
1000	500	324	161	0.473067	0.053377
1000	500	324	162	0.526932	0.053865
1000	500	325	126	0.000001	0.000000
1000	500	325	127	0.000001	0.000001
1000	500	325	128	0.000002	0.000001
1000	500	325	129	0.000004	0.000002
1000	500	325	130	0.000007	0.000003
1000	500	325	131	0.000014	0.000006
1000	500	325	132	0.000025	0.000011
1000	500	325	133	0.000044	0.000019
1000	500	325	134	0.000076	0.000032
1000	500	325	135	0.000130	0.000054
1000	500	325	136	0.000219	0.000089
1000	500	325	137	0.000362	0.000143
1000	500	325	138	0.000588	0.000226
1000	500	325	139	0.000938	0.000350
1000	500	325	140	0.001471	0.000533
1000	500	325	141	0.002268	0.000797
1000	500	325	142	0.003437	0.001169
1000	500	325	143	0.005122	0.001685
1000	500	325	144	0.007505	0.002383
1000	500	325	145	0.010815	0.003309
1000	500	325	146	0.015327	0.004512
1000	500	325	147	0.021367	0.006040
1000	500	325	148	0.029307	0.007940
1000	500	325	149	0.039554	0.010247
1000	500	325	150	0.052538	0.012985
1000	500	325	151	0.068695	0.016156
1000	500	325	152	0.088434	0.019739
1000	500	325	153	0.112114	0.023680
1000	500	325	154	0.140009	0.027895
1000	500	325	155	0.172276	0.032267
1000	500	325	156	0.208926	0.036650
1000	500	325	157	0.249803	0.040877
1000	500	325	158	0.294572	0.044769
1000	500	325	159	0.342720	0.048148
1000	500	325	160	0.393568	0.050848
1000	500	325	161	0.446300	0.052732
1000	500	325	162	0.500000	0.053700
1000	500	326	127	0.000001	0.000000
1000	500	326	128	0.000002	0.000001
1000	500	326	129	0.000003	0.000001
1000	500	326	130	0.000006	0.000003
1000	500	326	131	0.000010	0.000005
1000	500	326	132	0.000019	0.000008
1000	500	326	133	0.000033	0.000015
1000	500	326	134	0.000059	0.000025
1000	500	326	135	0.000101	0.000042
1000	500	326	136	0.000171	0.000070
1000	500	326	137	0.000285	0.000114
1000	500	326	138	0.000467	0.000182
1000	500	326	139	0.000751	0.000284
1000	500	326	140	0.001187	0.000436
1000	500	326	141	0.001844	0.000657
1000	500	326	142	0.002817	0.000973
1000	500	326	143	0.004231	0.001414
1000	500	326	144	0.006248	0.002017
1000	500	326	145	0.009074	0.002826
1000	500	326	146	0.012960	0.003886
1000	500	326	147	0.018208	0.005248
1000	500	326	148	0.025167	0.006958
1000	500	326	149	0.034225	0.009059
1000	500	326	150	0.045806	0.011580
1000	500	326	151	0.060341	0.014536
1000	500	326	152	0.078257	0.017916
1000	500	326	153	0.099941	0.021683
1000	500	326	154	0.125710	0.025770
1000	500	326	155	0.155784	0.030074
1000	500	326	156	0.190248	0.034464
1000	500	326	157	0.229031	0.038783
1000	500	326	158	0.271889	0.042858
1000	500	326	159	0.318396	0.046507
1000	500	326	160	0.367956	0.049559
1000	500	326	161	0.419817	0.051861
1000	500	326	162	0.473110	0.053293
1000	500	326	163	0.526890	0.053779
1000	500	327	127	0.000001	0.000000
1000	500	327	128	0.000001	0.000001
1000	500	327	129	0.000002	0.000001
1000	500	327	130	0.000004	0.000002
1000	500	327	131	0.000008	0.000004
1000	500	327	132	0.000014	0.000006
1000	500	327	133	0.000025	0.000011
1000	500	327	134	0.000045	0.000020
1000	500	327	135	0.000078	0.000033
1000	500	327	136	0.000133	0.000055
1000	500	327	137	0.000224	0.000090
1000	500	327	138	0.000369	0.000145
1000	500	327	139	0.000598	0.000229
1000	500	327	140	0.000954	0.000355
1000	500	327	141	0.001494	0.000540
1000	500	327	142	0.002300	0.000806
1000	500	327	143	0.003482	0.001182
1000	500	327	144	0.005183	0.001701
1000	500	327	145	0.007586	0.002403
1000	500	327	146	0.010919	0.003333
1000	500	327	147	0.015460	0.004541
1000	500	327	148	0.021533	0.006073
1000	500	327	149	0.029508	0.007975
1000	500	327	150	0.039792	0.010284
1000	500	327	151	0.052815	0.013023
1000	500	327	152	0.069007	0.016192
1000	500	327	153	0.088777	0.019770
1000	500	327	154	0.112481	0.023704
1000	500	327	155	0.140390	0.027909
1000	500	327	156	0.172658	0.032268
1000	500	327	157	0.209294	0.036636
1000	500	327	158	0.250142	0.040848
1000	500	327	159	0.294866	0.044724
1000	500	327	160	0.342955	0.048089
1000	500	327	161	0.393732	0.050777
1000	500	327	162	0.446385	0.052652
1000	500	327	163	0.500000	0.053615
1000	500	328	128	0.000001	0.000000
1000	500	328	129	0.000002	0.000001
1000	500	328	130	0.000003	0.000001
1000	500	328	131	0.000006	0.000003
1000	500	328	132	0.000011	0.000005
1000	500	328	133	0.000019	0.000009
1000	500	328	134	0.000034	0.000015
1000	500	328	135	0.000060	0.000026
1000	500	328	136	0.000103	0.000043
1000	500	328	137	0.000175	0.000072
1000	500	328	138	0.000291	0.000116
1000	500	328	139	0.000475	0.000185
1000	500	328	140	0.000764	0.000288
1000	500	328	141	0.001206	0.000442
1000	500	328	142	0.001871	0.000666
1000	500	328	143	0.002855	0.000984
1000	500	328	144	0.004283	0.001428
1000	500	328	145	0.006318	0.002035
1000	500	328	146	0.009166	0.002848
1000	500	328	147	0.013078	0.003912
1000	500	328	148	0.018356	0.005278
1000	500	328	149	0.025348	0.006992
1000	500	328	150	0.034444	0.009095
1000	500	328	151	0.046062	0.011618
1000	500	328	152	0.060634	0.014573
1000	500	328	153	0.078584	0.017950
1000	500	328	154	0.100295	0.021711
1000	500	328	155	0.126083	0.025789
1000	500	328	156	0.156165	0.030081
1000	500	328	157	0.190622	0.034458
1000	500	328	158	0.229384	0.038762
1000	500	328	159	0.272205	0.042821
1000	500	328	160	0.318661	0.046456
1000	500	328	161	0.368155	0.049494
1000	500	328	162	0.419941	0.051786
1000	500	328	163	0.473152	0.053211
1000	500	328	164	0.526847	0.053695
1000	500	329	128	0.000001	0.000000
1000	500	329	129	0.000001	0.000001
1000	500	329	130	0.000002	0.000001

Table for $N = 1000$, $n = 500$

N	n	k	x	P(x)	p(x)
1000	500	329	131	0.000004	0.000002
1000	500	329	132	0.000008	0.000004
1000	500	329	133	0.000014	0.000007
1000	500	329	134	0.000026	0.000012
1000	500	329	135	0.000046	0.000020
1000	500	329	136	0.000080	0.000034
1000	500	329	137	0.000136	0.000056
1000	500	329	138	0.000228	0.000092
1000	500	329	139	0.000376	0.000148
1000	500	329	140	0.000609	0.000233
1000	500	329	141	0.000970	0.000360
1000	500	329	142	0.001517	0.000547
1000	500	329	143	0.002333	0.000816
1000	500	329	144	0.003527	0.001194
1000	500	329	145	0.005243	0.001716
1000	500	329	146	0.007666	0.002423
1000	500	329	147	0.011023	0.003357
1000	500	329	148	0.015592	0.004569
1000	500	329	149	0.021696	0.006105
1000	500	329	150	0.029707	0.008010
1000	500	329	151	0.040028	0.010321
1000	500	329	152	0.053088	0.013060
1000	500	329	153	0.069315	0.016227
1000	500	329	154	0.089116	0.019801
1000	500	329	155	0.112844	0.023728
1000	500	329	156	0.140766	0.027922
1000	500	329	157	0.173034	0.032269
1000	500	329	158	0.209657	0.036623
1000	500	329	159	0.250476	0.040819
1000	500	329	160	0.295156	0.044680
1000	500	329	161	0.343187	0.048031
1000	500	329	162	0.393894	0.050707
1000	500	329	163	0.446468	0.052574
1000	500	329	164	0.500000	0.053532
1000	500	330	129	0.000001	0.000000
1000	500	330	130	0.000002	0.000001
1000	500	330	131	0.000003	0.000001
1000	500	330	132	0.000006	0.000003
1000	500	330	133	0.000011	0.000005
1000	500	330	134	0.000020	0.000009
1000	500	330	135	0.000035	0.000015
1000	500	330	136	0.000061	0.000026
1000	500	330	137	0.000106	0.000044
1000	500	330	138	0.000179	0.000073
1000	500	330	139	0.000297	0.000118
1000	500	330	140	0.000484	0.000188
1000	500	330	141	0.000777	0.000293
1000	500	330	142	0.001225	0.000448
1000	500	330	143	0.001898	0.000674
1000	500	330	144	0.002893	0.000995

N	n	k	x	P(x)	p(x)
1000	500	330	145	0.004335	0.001442
1000	500	330	146	0.006388	0.002053
1000	500	330	147	0.009257	0.002869
1000	500	330	148	0.013195	0.003938
1000	500	330	149	0.018503	0.005308
1000	500	330	150	0.025528	0.007025
1000	500	330	151	0.034660	0.009131
1000	500	330	152	0.046314	0.011655
1000	500	330	153	0.060923	0.014609
1000	500	330	154	0.078906	0.017983
1000	500	330	155	0.100644	0.021738
1000	500	330	156	0.126451	0.025807
1000	500	330	157	0.156539	0.030088
1000	500	330	158	0.190991	0.034452
1000	500	330	159	0.229732	0.038741
1000	500	330	160	0.272517	0.042785
1000	500	330	161	0.318921	0.046405
1000	500	330	162	0.368352	0.049430
1000	500	330	163	0.420063	0.051712
1000	500	330	164	0.473194	0.053131
1000	500	330	165	0.526806	0.053612
1000	500	331	129	0.000001	0.000000
1000	500	331	130	0.000001	0.000001
1000	500	331	131	0.000002	0.000001
1000	500	331	132	0.000004	0.000002
1000	500	331	133	0.000008	0.000004
1000	500	331	134	0.000015	0.000007
1000	500	331	135	0.000027	0.000012
1000	500	331	136	0.000047	0.000020
1000	500	331	137	0.000082	0.000035
1000	500	331	138	0.000139	0.000058
1000	500	331	139	0.000233	0.000094
1000	500	331	140	0.000384	0.000150
1000	500	331	141	0.000620	0.000237
1000	500	331	142	0.000985	0.000365
1000	500	331	143	0.001540	0.000554
1000	500	331	144	0.002365	0.000825
1000	500	331	145	0.003571	0.001206
1000	500	331	146	0.005303	0.001732
1000	500	331	147	0.007745	0.002442
1000	500	331	148	0.011126	0.003381
1000	500	331	149	0.015722	0.004596
1000	500	331	150	0.021858	0.006136
1000	500	331	151	0.029903	0.008045
1000	500	331	152	0.040261	0.010358
1000	500	331	153	0.053357	0.013096
1000	500	331	154	0.069619	0.016262
1000	500	331	155	0.089450	0.019831
1000	500	331	156	0.113201	0.023750
1000	500	331	157	0.141136	0.027935

N	n	k	x	P(x)	p(x)
1000	500	331	158	0.173405	0.032269
1000	500	331	159	0.210015	0.036609
1000	500	331	160	0.250805	0.040790
1000	500	331	161	0.295442	0.044637
1000	500	331	162	0.343415	0.047973
1000	500	331	163	0.394053	0.050638
1000	500	331	164	0.446549	0.052496
1000	500	331	165	0.500000	0.053451
1000	500	332	130	0.000001	0.000000
1000	500	332	131	0.000002	0.000001
1000	500	332	132	0.000003	0.000002
1000	500	332	133	0.000006	0.000003
1000	500	332	134	0.000011	0.000005
1000	500	332	135	0.000020	0.000009
1000	500	332	136	0.000036	0.000016
1000	500	332	137	0.000063	0.000027
1000	500	332	138	0.000108	0.000045
1000	500	332	139	0.000182	0.000074
1000	500	332	140	0.000303	0.000120
1000	500	332	141	0.000493	0.000191
1000	500	332	142	0.000790	0.000297
1000	500	332	143	0.001244	0.000454
1000	500	332	144	0.001926	0.000682
1000	500	332	145	0.002931	0.001005
1000	500	332	146	0.004387	0.001456
1000	500	332	147	0.006457	0.002070
1000	500	332	148	0.009347	0.002890
1000	500	332	149	0.013311	0.003964
1000	500	332	150	0.018648	0.005337
1000	500	332	151	0.025706	0.007058
1000	500	332	152	0.034873	0.009167
1000	500	332	153	0.046564	0.011691
1000	500	332	154	0.061209	0.014644
1000	500	332	155	0.079224	0.018015
1000	500	332	156	0.100988	0.021765
1000	500	332	157	0.126813	0.025825
1000	500	332	158	0.156909	0.030095
1000	500	332	159	0.191354	0.034445
1000	500	332	160	0.230074	0.038720
1000	500	332	161	0.272823	0.042749
1000	500	332	162	0.319177	0.046354
1000	500	332	163	0.368545	0.049367
1000	500	332	164	0.420183	0.051639
1000	500	332	165	0.473235	0.053051
1000	500	332	166	0.526765	0.053531
1000	500	333	130	0.000001	0.000000
1000	500	333	131	0.000001	0.000001
1000	500	333	132	0.000002	0.000001
1000	500	333	133	0.000005	0.000002
1000	500	333	134	0.000008	0.000004

N	n	k	x	P(x)	p(x)
1000	500	333	135	0.000015	0.000007
1000	500	333	136	0.000027	0.000012
1000	500	333	137	0.000048	0.000021
1000	500	333	138	0.000084	0.000035
1000	500	333	139	0.000142	0.000059
1000	500	333	140	0.000238	0.000096
1000	500	333	141	0.000391	0.000153
1000	500	333	142	0.000631	0.000240
1000	500	333	143	0.001001	0.000370
1000	500	333	144	0.001562	0.000561
1000	500	333	145	0.002397	0.000834
1000	500	333	146	0.003615	0.001219
1000	500	333	147	0.005363	0.001748
1000	500	333	148	0.007824	0.002461
1000	500	333	149	0.011228	0.003404
1000	500	333	150	0.015852	0.004623
1000	500	333	151	0.022019	0.006167
1000	500	333	152	0.030098	0.008079
1000	500	333	153	0.040491	0.010394
1000	500	333	154	0.053623	0.013132
1000	500	333	155	0.069919	0.016296
1000	500	333	156	0.089780	0.019861
1000	500	333	157	0.113553	0.023773
1000	500	333	158	0.141501	0.027948
1000	500	333	159	0.173771	0.032270
1000	500	333	160	0.210367	0.036596
1000	500	333	161	0.251129	0.040762
1000	500	333	162	0.295723	0.044594
1000	500	333	163	0.343639	0.047917
1000	500	333	164	0.394210	0.050570
1000	500	333	165	0.446630	0.052420
1000	500	333	166	0.500000	0.053370
1000	500	334	131	0.000001	0.000000
1000	500	334	132	0.000002	0.000001
1000	500	334	133	0.000003	0.000002
1000	500	334	134	0.000006	0.000003
1000	500	334	135	0.000012	0.000005
1000	500	334	136	0.000021	0.000009
1000	500	334	137	0.000037	0.000016
1000	500	334	138	0.000065	0.000028
1000	500	334	139	0.000111	0.000046
1000	500	334	140	0.000186	0.000076
1000	500	334	141	0.000308	0.000122
1000	500	334	142	0.000502	0.000194
1000	500	334	143	0.000803	0.000301
1000	500	334	144	0.001263	0.000460
1000	500	334	145	0.001953	0.000690
1000	500	334	146	0.002968	0.001016
1000	500	334	147	0.004438	0.001469
1000	500	334	148	0.006525	0.002087

Table for $N = 1000$, $n = 500$

N	n	k	x	P(x)	p(x)
1000	500	334	149	0.009437	0.002912
1000	500	334	150	0.013425	0.003989
1000	500	334	151	0.018792	0.005366
1000	500	334	152	0.025882	0.007091
1000	500	334	153	0.035084	0.009201
1000	500	334	154	0.046811	0.011727
1000	500	334	155	0.061490	0.014679
1000	500	334	156	0.079537	0.018047
1000	500	334	157	0.101328	0.021791
1000	500	334	158	0.127171	0.025843
1000	500	334	159	0.157273	0.030102
1000	500	334	160	0.191712	0.034439
1000	500	334	161	0.230412	0.038700
1000	500	334	162	0.273125	0.042713
1000	500	334	163	0.319430	0.046305
1000	500	334	164	0.368735	0.049305
1000	500	334	165	0.420302	0.051567
1000	500	334	166	0.473275	0.052973
1000	500	334	167	0.526725	0.053450
1000	500	335	131	0.000001	0.000000
1000	500	335	132	0.000001	0.000001
1000	500	335	133	0.000002	0.000001
1000	500	335	134	0.000005	0.000002
1000	500	335	135	0.000009	0.000004
1000	500	335	136	0.000016	0.000007
1000	500	335	137	0.000028	0.000012
1000	500	335	138	0.000050	0.000021
1000	500	335	139	0.000086	0.000036
1000	500	335	140	0.000145	0.000060
1000	500	335	141	0.000243	0.000097
1000	500	335	142	0.000398	0.000155
1000	500	335	143	0.000642	0.000244
1000	500	335	144	0.001017	0.000375
1000	500	335	145	0.001585	0.000568
1000	500	335	146	0.002428	0.000844
1000	500	335	147	0.003659	0.001231
1000	500	335	148	0.005422	0.001763
1000	500	335	149	0.007902	0.002480
1000	500	335	150	0.011329	0.003427
1000	500	335	151	0.015980	0.004650
1000	500	335	152	0.022177	0.006198
1000	500	335	153	0.030290	0.008112
1000	500	335	154	0.040719	0.010429
1000	500	335	155	0.053886	0.013168
1000	500	335	156	0.070216	0.016329
1000	500	335	157	0.090105	0.019890
1000	500	335	158	0.113900	0.023795
1000	500	335	159	0.141860	0.027960
1000	500	335	160	0.174130	0.032270
1000	500	335	161	0.210713	0.036583
1000	500	335	162	0.251448	0.040734
1000	500	335	163	0.295999	0.044552
1000	500	335	164	0.343860	0.047861
1000	500	335	165	0.394363	0.050503
1000	500	335	166	0.446709	0.052345
1000	500	335	167	0.500000	0.053291
1000	500	336	132	0.000001	0.000000
1000	500	336	133	0.000002	0.000001
1000	500	336	134	0.000003	0.000002
1000	500	336	135	0.000006	0.000003
1000	500	336	136	0.000012	0.000005
1000	500	336	137	0.000021	0.000010
1000	500	336	138	0.000038	0.000017
1000	500	336	139	0.000066	0.000028
1000	500	336	140	0.000113	0.000047
1000	500	336	141	0.000190	0.000077
1000	500	336	142	0.000314	0.000124
1000	500	336	143	0.000511	0.000196
1000	500	336	144	0.000816	0.000305
1000	500	336	145	0.001282	0.000466
1000	500	336	146	0.001979	0.000698
1000	500	336	147	0.003006	0.001026
1000	500	336	148	0.004489	0.001483
1000	500	336	149	0.006593	0.002104
1000	500	336	150	0.009526	0.002932
1000	500	336	151	0.013539	0.004014
1000	500	336	152	0.018934	0.005395
1000	500	336	153	0.026057	0.007123
1000	500	336	154	0.035292	0.009236
1000	500	336	155	0.047055	0.011762
1000	500	336	156	0.061769	0.014714
1000	500	336	157	0.079846	0.018078
1000	500	336	158	0.101663	0.021816
1000	500	336	159	0.127523	0.025860
1000	500	336	160	0.157631	0.030109
1000	500	336	161	0.192064	0.034433
1000	500	336	162	0.230743	0.038679
1000	500	336	163	0.273422	0.042678
1000	500	336	164	0.319678	0.046256
1000	500	336	165	0.368922	0.049244
1000	500	336	166	0.420418	0.051496
1000	500	336	167	0.473314	0.052896
1000	500	336	168	0.526686	0.053371
1000	500	337	132	0.000001	0.000000
1000	500	337	133	0.000001	0.000001
1000	500	337	134	0.000003	0.000001
1000	500	337	135	0.000005	0.000002
1000	500	337	136	0.000009	0.000004
1000	500	337	137	0.000016	0.000007
1000	500	337	138	0.000029	0.000013
1000	500	337	139	0.000051	0.000022
1000	500	337	140	0.000088	0.000037
1000	500	337	141	0.000148	0.000061
1000	500	337	142	0.000247	0.000099
1000	500	337	143	0.000405	0.000158
1000	500	337	144	0.000652	0.000247
1000	500	337	145	0.001033	0.000380
1000	500	337	146	0.001607	0.000575
1000	500	337	147	0.002460	0.000853
1000	500	337	148	0.003703	0.001242
1000	500	337	149	0.005481	0.001778
1000	500	337	150	0.007980	0.002499
1000	500	337	151	0.011430	0.003450
1000	500	337	152	0.016107	0.004677
1000	500	337	153	0.022335	0.006228
1000	500	337	154	0.030480	0.008145
1000	500	337	155	0.040943	0.010464
1000	500	337	156	0.054146	0.013202
1000	500	337	157	0.070508	0.016362
1000	500	337	158	0.090426	0.019918
1000	500	337	159	0.114242	0.023816
1000	500	337	160	0.142214	0.027972
1000	500	337	161	0.174485	0.032270
1000	500	337	162	0.211054	0.036570
1000	500	337	163	0.251761	0.040707
1000	500	337	164	0.296271	0.044510
1000	500	337	165	0.344077	0.047806
1000	500	337	166	0.394515	0.050437
1000	500	337	167	0.446786	0.052272
1000	500	337	168	0.500000	0.053213
1000	500	338	133	0.000001	0.000000
1000	500	338	134	0.000002	0.000001
1000	500	338	135	0.000004	0.000002
1000	500	338	136	0.000007	0.000003
1000	500	338	137	0.000012	0.000006
1000	500	338	138	0.000022	0.000010
1000	500	338	139	0.000039	0.000017
1000	500	338	140	0.000068	0.000029
1000	500	338	141	0.000115	0.000048
1000	500	338	142	0.000194	0.000078
1000	500	338	143	0.000320	0.000126
1000	500	338	144	0.000520	0.000199
1000	500	338	145	0.000829	0.000309
1000	500	338	146	0.001301	0.000472
1000	500	338	147	0.002006	0.000706
1000	500	338	148	0.003043	0.001037
1000	500	338	149	0.004539	0.001496
1000	500	338	150	0.006661	0.002121
1000	500	338	151	0.009614	0.002953
1000	500	338	152	0.013652	0.004038
1000	500	338	153	0.019075	0.005423
1000	500	338	154	0.026229	0.007154
1000	500	338	155	0.035499	0.009270
1000	500	338	156	0.047296	0.011797
1000	500	338	157	0.062043	0.014747
1000	500	338	158	0.080152	0.018108
1000	500	338	159	0.101993	0.021841
1000	500	338	160	0.127870	0.025877
1000	500	338	161	0.157985	0.030115
1000	500	338	162	0.192411	0.034427
1000	500	338	163	0.231070	0.038659
1000	500	338	164	0.273714	0.042644
1000	500	338	165	0.319922	0.046208
1000	500	338	166	0.369106	0.049184
1000	500	338	167	0.420532	0.051426
1000	500	338	168	0.473353	0.052821
1000	500	338	169	0.526647	0.053294
1000	500	339	133	0.000001	0.000000
1000	500	339	134	0.000001	0.000001
1000	500	339	135	0.000003	0.000001
1000	500	339	136	0.000005	0.000002
1000	500	339	137	0.000009	0.000004
1000	500	339	138	0.000017	0.000007
1000	500	339	139	0.000030	0.000013
1000	500	339	140	0.000052	0.000022
1000	500	339	141	0.000089	0.000038
1000	500	339	142	0.000152	0.000062
1000	500	339	143	0.000252	0.000101
1000	500	339	144	0.000412	0.000160
1000	500	339	145	0.000663	0.000251
1000	500	339	146	0.001048	0.000385
1000	500	339	147	0.001630	0.000581
1000	500	339	148	0.002492	0.000862
1000	500	339	149	0.003746	0.001254
1000	500	339	150	0.005539	0.001793
1000	500	339	151	0.008057	0.002518
1000	500	339	152	0.011529	0.003472
1000	500	339	153	0.016232	0.004703
1000	500	339	154	0.022490	0.006258
1000	500	339	155	0.030668	0.008178
1000	500	339	156	0.041166	0.010498
1000	500	339	157	0.054402	0.013237
1000	500	339	158	0.070796	0.016394
1000	500	339	159	0.090742	0.019946
1000	500	339	160	0.114579	0.023837
1000	500	339	161	0.142563	0.027984
1000	500	339	162	0.174834	0.032271
1000	500	339	163	0.211390	0.036556
1000	500	339	164	0.252070	0.040680
1000	500	339	165	0.296539	0.044469

Table for $N = 1000$, $n = 500$

N	n	k	x	P(x)	p(x)
1000	500	339	166	0.344291	0.047752
1000	500	339	167	0.394664	0.050373
1000	500	339	168	0.446863	0.052199
1000	500	339	169	0.500000	0.053137
1000	500	340	133	0.000001	0.000000
1000	500	340	134	0.000001	0.000000
1000	500	340	135	0.000002	0.000001
1000	500	340	136	0.000004	0.000002
1000	500	340	137	0.000007	0.000003
1000	500	340	138	0.000013	0.000006
1000	500	340	139	0.000023	0.000010
1000	500	340	140	0.000040	0.000017
1000	500	340	141	0.000069	0.000029
1000	500	340	142	0.000118	0.000049
1000	500	340	143	0.000198	0.000080
1000	500	340	144	0.000326	0.000128
1000	500	340	145	0.000528	0.000202
1000	500	340	146	0.000842	0.000314
1000	500	340	147	0.001319	0.000477
1000	500	340	148	0.002033	0.000713
1000	500	340	149	0.003080	0.001047
1000	500	340	150	0.004590	0.001510
1000	500	340	151	0.006727	0.002138
1000	500	340	152	0.009701	0.002973
1000	500	340	153	0.013763	0.004062
1000	500	340	154	0.019214	0.005451
1000	500	340	155	0.026399	0.007185
1000	500	340	156	0.035702	0.009303
1000	500	340	157	0.047533	0.011831
1000	500	340	158	0.062314	0.014781
1000	500	340	159	0.080453	0.018138
1000	500	340	160	0.102318	0.021866
1000	500	340	161	0.128211	0.025893
1000	500	340	162	0.158332	0.030121
1000	500	340	163	0.192752	0.034420
1000	500	340	164	0.231392	0.038639
1000	500	340	165	0.274002	0.042610
1000	500	340	166	0.320162	0.046160
1000	500	340	167	0.369287	0.049125
1000	500	340	168	0.420645	0.051358
1000	500	340	169	0.473391	0.052746
1000	500	340	170	0.526609	0.053217
1000	500	341	134	0.000001	0.000000
1000	500	341	135	0.000001	0.000001
1000	500	341	136	0.000003	0.000001
1000	500	341	137	0.000005	0.000002
1000	500	341	138	0.000009	0.000004
1000	500	341	139	0.000017	0.000008
1000	500	341	140	0.000030	0.000013
1000	500	341	141	0.000053	0.000023

N	n	k	x	P(x)	p(x)
1000	500	341	142	0.000091	0.000038
1000	500	341	143	0.000155	0.000063
1000	500	341	144	0.000257	0.000102
1000	500	341	145	0.000420	0.000163
1000	500	341	146	0.000674	0.000254
1000	500	341	147	0.001064	0.000390
1000	500	341	148	0.001652	0.000588
1000	500	341	149	0.002523	0.000871
1000	500	341	150	0.003789	0.001266
1000	500	341	151	0.005597	0.001808
1000	500	341	152	0.008133	0.002536
1000	500	341	153	0.011627	0.003494
1000	500	341	154	0.016357	0.004729
1000	500	341	155	0.022644	0.006287
1000	500	341	156	0.030854	0.008210
1000	500	341	157	0.041385	0.010531
1000	500	341	158	0.054655	0.013270
1000	500	341	159	0.071081	0.016426
1000	500	341	160	0.091054	0.019973
1000	500	341	161	0.114912	0.023858
1000	500	341	162	0.142907	0.027995
1000	500	341	163	0.175177	0.032271
1000	500	341	164	0.211721	0.036543
1000	500	341	165	0.252374	0.040653
1000	500	341	166	0.296802	0.044428
1000	500	341	167	0.344501	0.047699
1000	500	341	168	0.394810	0.050309
1000	500	341	169	0.446938	0.052128
1000	500	341	170	0.500000	0.053062
1000	500	342	134	0.000001	0.000000
1000	500	342	135	0.000001	0.000001
1000	500	342	136	0.000002	0.000001
1000	500	342	137	0.000004	0.000002
1000	500	342	138	0.000007	0.000003
1000	500	342	139	0.000013	0.000006
1000	500	342	140	0.000023	0.000010
1000	500	342	141	0.000041	0.000018
1000	500	342	142	0.000071	0.000030
1000	500	342	143	0.000120	0.000050
1000	500	342	144	0.000202	0.000081
1000	500	342	145	0.000332	0.000130
1000	500	342	146	0.000537	0.000205
1000	500	342	147	0.000855	0.000318
1000	500	342	148	0.001338	0.000483
1000	500	342	149	0.002059	0.000721
1000	500	342	150	0.003117	0.001057
1000	500	342	151	0.004640	0.001523
1000	500	342	152	0.006794	0.002154
1000	500	342	153	0.009787	0.002994
1000	500	342	154	0.013874	0.004086

N	n	k	x	P(x)	p(x)
1000	500	342	155	0.019352	0.005479
1000	500	342	156	0.026568	0.007216
1000	500	342	157	0.035903	0.009336
1000	500	342	158	0.047768	0.011865
1000	500	342	159	0.062581	0.014813
1000	500	342	160	0.080749	0.018168
1000	500	342	161	0.102639	0.021890
1000	500	342	162	0.128548	0.025909
1000	500	342	163	0.158675	0.030127
1000	500	342	164	0.193089	0.034414
1000	500	342	165	0.231708	0.038619
1000	500	342	166	0.274285	0.042577
1000	500	342	167	0.320398	0.046114
1000	500	342	168	0.369465	0.049067
1000	500	342	169	0.420756	0.051291
1000	500	342	170	0.473429	0.052673
1000	500	342	171	0.526571	0.053142
1000	500	343	135	0.000001	0.000000
1000	500	343	136	0.000001	0.000001
1000	500	343	137	0.000003	0.000001
1000	500	343	138	0.000005	0.000002
1000	500	343	139	0.000010	0.000004
1000	500	343	140	0.000018	0.000008
1000	500	343	141	0.000031	0.000014
1000	500	343	142	0.000054	0.000023
1000	500	343	143	0.000093	0.000039
1000	500	343	144	0.000158	0.000064
1000	500	343	145	0.000262	0.000104
1000	500	343	146	0.000427	0.000165
1000	500	343	147	0.000685	0.000258
1000	500	343	148	0.001080	0.000395
1000	500	343	149	0.001675	0.000595
1000	500	343	150	0.002554	0.000880
1000	500	343	151	0.003832	0.001278
1000	500	343	152	0.005654	0.001823
1000	500	343	153	0.008209	0.002554
1000	500	343	154	0.011725	0.003516
1000	500	343	155	0.016480	0.004755
1000	500	343	156	0.022796	0.006316
1000	500	343	157	0.031037	0.008241
1000	500	343	158	0.041601	0.010564
1000	500	343	159	0.054905	0.013303
1000	500	343	160	0.071362	0.016457
1000	500	343	161	0.091362	0.020000
1000	500	343	162	0.115239	0.023878
1000	500	343	163	0.143245	0.028006
1000	500	343	164	0.175516	0.032270
1000	500	343	165	0.212046	0.036531
1000	500	343	166	0.252673	0.040627
1000	500	343	167	0.297061	0.044388

N	n	k	x	P(x)	p(x)
1000	500	343	168	0.344708	0.047647
1000	500	343	169	0.394954	0.050246
1000	500	343	170	0.447012	0.052058
1000	500	343	171	0.500000	0.052988
1000	500	344	135	0.000001	0.000000
1000	500	344	136	0.000001	0.000001
1000	500	344	137	0.000002	0.000001
1000	500	344	138	0.000004	0.000002
1000	500	344	139	0.000007	0.000003
1000	500	344	140	0.000013	0.000006
1000	500	344	141	0.000024	0.000010
1000	500	344	142	0.000042	0.000018
1000	500	344	143	0.000072	0.000030
1000	500	344	144	0.000123	0.000051
1000	500	344	145	0.000205	0.000083
1000	500	344	146	0.000338	0.000132
1000	500	344	147	0.000546	0.000208
1000	500	344	148	0.000868	0.000322
1000	500	344	149	0.001357	0.000489
1000	500	344	150	0.002086	0.000729
1000	500	344	151	0.003153	0.001068
1000	500	344	152	0.004689	0.001536
1000	500	344	153	0.006860	0.002171
1000	500	344	154	0.009873	0.003014
1000	500	344	155	0.013983	0.004110
1000	500	344	156	0.019489	0.005506
1000	500	344	157	0.026734	0.007246
1000	500	344	158	0.036102	0.009368
1000	500	344	159	0.048000	0.011898
1000	500	344	160	0.062845	0.014845
1000	500	344	161	0.081042	0.018197
1000	500	344	162	0.102955	0.021913
1000	500	344	163	0.128880	0.025925
1000	500	344	164	0.159012	0.030132
1000	500	344	165	0.193419	0.034407
1000	500	344	166	0.232019	0.038600
1000	500	344	167	0.274563	0.042544
1000	500	344	168	0.320631	0.046068
1000	500	344	169	0.369640	0.049009
1000	500	344	170	0.420864	0.051224
1000	500	344	171	0.473466	0.052601
1000	500	344	172	0.526534	0.053068
1000	500	345	136	0.000001	0.000000
1000	500	345	137	0.000002	0.000001
1000	500	345	138	0.000003	0.000001
1000	500	345	139	0.000005	0.000003
1000	500	345	140	0.000010	0.000005
1000	500	345	141	0.000018	0.000008
1000	500	345	142	0.000032	0.000014
1000	500	345	143	0.000056	0.000024

Table for $N = 1000,\ n = 500$

N	n	k	x	P(x)	p(x)
1000	500	345	144	0.000095	0.000040
1000	500	345	145	0.000161	0.000065
1000	500	345	146	0.000266	0.000106
1000	500	345	147	0.000434	0.000168
1000	500	345	148	0.000695	0.000261
1000	500	345	149	0.001095	0.000400
1000	500	345	150	0.001697	0.000601
1000	500	345	151	0.002585	0.000888
1000	500	345	152	0.003874	0.001289
1000	500	345	153	0.005711	0.001837
1000	500	345	154	0.008284	0.002572
1000	500	345	155	0.011822	0.003538
1000	500	345	156	0.016602	0.004780
1000	500	345	157	0.022946	0.006344
1000	500	345	158	0.031218	0.008273
1000	500	345	159	0.041815	0.010597
1000	500	345	160	0.055151	0.013336
1000	500	345	161	0.071638	0.016487
1000	500	345	162	0.091665	0.020026
1000	500	345	163	0.115562	0.023897
1000	500	345	164	0.143578	0.028017
1000	500	345	165	0.175849	0.032270
1000	500	345	166	0.212366	0.036518
1000	500	345	167	0.252967	0.040601
1000	500	345	168	0.297316	0.044349
1000	500	345	169	0.344911	0.047595
1000	500	345	170	0.395096	0.050185
1000	500	345	171	0.447085	0.051989
1000	500	345	172	0.500000	0.052915
1000	500	346	136	0.000001	0.000000
1000	500	346	137	0.000001	0.000001
1000	500	346	138	0.000002	0.000001
1000	500	346	139	0.000004	0.000002
1000	500	346	140	0.000007	0.000003
1000	500	346	141	0.000014	0.000006
1000	500	346	142	0.000024	0.000011
1000	500	346	143	0.000043	0.000018
1000	500	346	144	0.000074	0.000031
1000	500	346	145	0.000125	0.000052
1000	500	346	146	0.000209	0.000084
1000	500	346	147	0.000344	0.000134
1000	500	346	148	0.000555	0.000211
1000	500	346	149	0.000881	0.000326
1000	500	346	150	0.001375	0.000494
1000	500	346	151	0.002112	0.000736
1000	500	346	152	0.003189	0.001078
1000	500	346	153	0.004738	0.001549
1000	500	346	154	0.006925	0.002187
1000	500	346	155	0.009958	0.003033
1000	500	346	156	0.014091	0.004133

N	n	k	x	P(x)	p(x)
1000	500	346	157	0.019624	0.005532
1000	500	346	158	0.026899	0.007275
1000	500	346	159	0.036298	0.009399
1000	500	346	160	0.048229	0.011930
1000	500	346	161	0.063105	0.014877
1000	500	346	162	0.081330	0.018225
1000	500	346	163	0.103267	0.021936
1000	500	346	164	0.129207	0.025940
1000	500	346	165	0.159344	0.030137
1000	500	346	166	0.193745	0.034401
1000	500	346	167	0.232326	0.038581
1000	500	346	168	0.274837	0.042511
1000	500	346	169	0.320859	0.046023
1000	500	346	170	0.369812	0.048953
1000	500	346	171	0.420971	0.051159
1000	500	346	172	0.473502	0.052531
1000	500	346	173	0.526498	0.052996
1000	500	347	137	0.000001	0.000000
1000	500	347	138	0.000002	0.000001
1000	500	347	139	0.000003	0.000001
1000	500	347	140	0.000006	0.000003
1000	500	347	141	0.000010	0.000005
1000	500	347	142	0.000018	0.000008
1000	500	347	143	0.000033	0.000014
1000	500	347	144	0.000057	0.000024
1000	500	347	145	0.000097	0.000040
1000	500	347	146	0.000164	0.000067
1000	500	347	147	0.000271	0.000107
1000	500	347	148	0.000441	0.000170
1000	500	347	149	0.000706	0.000265
1000	500	347	150	0.001111	0.000405
1000	500	347	151	0.001719	0.000608
1000	500	347	152	0.002616	0.000897
1000	500	347	153	0.003916	0.001300
1000	500	347	154	0.005768	0.001852
1000	500	347	155	0.008358	0.002590
1000	500	347	156	0.011917	0.003559
1000	500	347	157	0.016722	0.004805
1000	500	347	158	0.023094	0.006372
1000	500	347	159	0.031397	0.008303
1000	500	347	160	0.042026	0.010629
1000	500	347	161	0.055394	0.013368
1000	500	347	162	0.071911	0.016517
1000	500	347	163	0.091963	0.020052
1000	500	347	164	0.115880	0.023917
1000	500	347	165	0.143907	0.028027
1000	500	347	166	0.176176	0.032270
1000	500	347	167	0.212681	0.036505
1000	500	347	168	0.253256	0.040575
1000	500	347	169	0.297567	0.044310

N	n	k	x	P(x)	p(x)
1000	500	347	170	0.345111	0.047545
1000	500	347	171	0.395235	0.050124
1000	500	347	172	0.447156	0.051921
1000	500	347	173	0.500000	0.052844
1000	500	348	137	0.000001	0.000000
1000	500	348	138	0.000001	0.000001
1000	500	348	139	0.000002	0.000001
1000	500	348	140	0.000004	0.000002
1000	500	348	141	0.000008	0.000004
1000	500	348	142	0.000014	0.000006
1000	500	348	143	0.000025	0.000011
1000	500	348	144	0.000044	0.000019
1000	500	348	145	0.000075	0.000032
1000	500	348	146	0.000128	0.000052
1000	500	348	147	0.000213	0.000085
1000	500	348	148	0.000350	0.000136
1000	500	348	149	0.000564	0.000214
1000	500	348	150	0.000894	0.000330
1000	500	348	151	0.001394	0.000500
1000	500	348	152	0.002138	0.000744
1000	500	348	153	0.003225	0.001088
1000	500	348	154	0.004787	0.001562
1000	500	348	155	0.006990	0.002203
1000	500	348	156	0.010042	0.003053
1000	500	348	157	0.014198	0.004156
1000	500	348	158	0.019757	0.005559
1000	500	348	159	0.027062	0.007305
1000	500	348	160	0.036492	0.009431
1000	500	348	161	0.048454	0.011962
1000	500	348	162	0.063362	0.014907
1000	500	348	163	0.081615	0.018253
1000	500	348	164	0.103573	0.021959
1000	500	348	165	0.129528	0.025955
1000	500	348	166	0.159671	0.030143
1000	500	348	167	0.194066	0.034395
1000	500	348	168	0.232627	0.038562
1000	500	348	169	0.275106	0.042479
1000	500	348	170	0.321084	0.045978
1000	500	348	171	0.369981	0.048897
1000	500	348	172	0.421077	0.051095
1000	500	348	173	0.473538	0.052461
1000	500	348	174	0.526462	0.052924
1000	500	349	138	0.000001	0.000000
1000	500	349	139	0.000002	0.000001
1000	500	349	140	0.000003	0.000001
1000	500	349	141	0.000006	0.000003
1000	500	349	142	0.000010	0.000005
1000	500	349	143	0.000019	0.000008
1000	500	349	144	0.000033	0.000015
1000	500	349	145	0.000058	0.000025

N	n	k	x	P(x)	p(x)
1000	500	349	146	0.000099	0.000041
1000	500	349	147	0.000167	0.000068
1000	500	349	148	0.000276	0.000109
1000	500	349	149	0.000448	0.000172
1000	500	349	150	0.000717	0.000268
1000	500	349	151	0.001126	0.000410
1000	500	349	152	0.001741	0.000615
1000	500	349	153	0.002647	0.000906
1000	500	349	154	0.003958	0.001312
1000	500	349	155	0.005824	0.001866
1000	500	349	156	0.008432	0.002608
1000	500	349	157	0.012012	0.003580
1000	500	349	158	0.016841	0.004829
1000	500	349	159	0.023241	0.006400
1000	500	349	160	0.031574	0.008333
1000	500	349	161	0.042235	0.010660
1000	500	349	162	0.055634	0.013399
1000	500	349	163	0.072180	0.016547
1000	500	349	164	0.092257	0.020077
1000	500	349	165	0.116193	0.023935
1000	500	349	166	0.144230	0.028037
1000	500	349	167	0.176499	0.032269
1000	500	349	168	0.212991	0.036493
1000	500	349	169	0.253541	0.040549
1000	500	349	170	0.297813	0.044272
1000	500	349	171	0.345308	0.047495
1000	500	349	172	0.395372	0.050065
1000	500	349	173	0.447227	0.051854
1000	500	349	174	0.500000	0.052773
1000	500	350	138	0.000001	0.000000
1000	500	350	139	0.000001	0.000001
1000	500	350	140	0.000002	0.000001
1000	500	350	141	0.000004	0.000002
1000	500	350	142	0.000008	0.000004
1000	500	350	143	0.000014	0.000006
1000	500	350	144	0.000025	0.000011
1000	500	350	145	0.000045	0.000019
1000	500	350	146	0.000077	0.000032
1000	500	350	147	0.000130	0.000053
1000	500	350	148	0.000217	0.000087
1000	500	350	149	0.000355	0.000138
1000	500	350	150	0.000572	0.000217
1000	500	350	151	0.000907	0.000334
1000	500	350	152	0.001412	0.000506
1000	500	350	153	0.002164	0.000752
1000	500	350	154	0.003261	0.001097
1000	500	350	155	0.004835	0.001574
1000	500	350	156	0.007054	0.002219
1000	500	350	157	0.010126	0.003072
1000	500	350	158	0.014304	0.004179

Table for $N = 1000$, $n = 500$

N	n	k	x	P(x)	p(x)
1000	500	350	159	0.019889	0.005585
1000	500	350	160	0.027222	0.007333
1000	500	350	161	0.036683	0.009461
1000	500	350	162	0.048677	0.011993
1000	500	350	163	0.063615	0.014938
1000	500	350	164	0.081895	0.018280
1000	500	350	165	0.103876	0.021981
1000	500	350	166	0.129845	0.025969
1000	500	350	167	0.159993	0.030147
1000	500	350	168	0.194381	0.034388
1000	500	350	169	0.232924	0.038543
1000	500	350	170	0.275371	0.042447
1000	500	350	171	0.321305	0.045934
1000	500	350	172	0.370148	0.048843
1000	500	350	173	0.421180	0.051032
1000	500	350	174	0.473573	0.052393
1000	500	350	175	0.526427	0.052854
1000	500	351	139	0.000001	0.000000
1000	500	351	140	0.000002	0.000001
1000	500	351	141	0.000003	0.000001
1000	500	351	142	0.000006	0.000003
1000	500	351	143	0.000011	0.000005
1000	500	351	144	0.000019	0.000009
1000	500	351	145	0.000034	0.000015
1000	500	351	146	0.000059	0.000025
1000	500	351	147	0.000101	0.000042
1000	500	351	148	0.000170	0.000069
1000	500	351	149	0.000281	0.000111
1000	500	351	150	0.000456	0.000175
1000	500	351	151	0.000727	0.000272
1000	500	351	152	0.001142	0.000414
1000	500	351	153	0.001763	0.000621
1000	500	351	154	0.002677	0.000914
1000	500	351	155	0.004000	0.001323
1000	500	351	156	0.005880	0.001880
1000	500	351	157	0.008505	0.002625
1000	500	351	158	0.012106	0.003601
1000	500	351	159	0.016959	0.004853
1000	500	351	160	0.023386	0.006427
1000	500	351	161	0.031749	0.008363
1000	500	351	162	0.042440	0.010691
1000	500	351	163	0.055870	0.013430
1000	500	351	164	0.072445	0.016575
1000	500	351	165	0.092547	0.020102
1000	500	351	166	0.116501	0.023954
1000	500	351	167	0.144548	0.028047
1000	500	351	168	0.176816	0.032269
1000	500	351	169	0.213296	0.036480
1000	500	351	170	0.253821	0.040524
1000	500	351	171	0.298055	0.044235

N	n	k	x	P(x)	p(x)
1000	500	351	172	0.345501	0.047446
1000	500	351	173	0.395507	0.050006
1000	500	351	174	0.447296	0.051789
1000	500	351	175	0.500000	0.052704
1000	500	352	139	0.000001	0.000000
1000	500	352	140	0.000001	0.000001
1000	500	352	141	0.000002	0.000001
1000	500	352	142	0.000004	0.000002
1000	500	352	143	0.000008	0.000004
1000	500	352	144	0.000015	0.000007
1000	500	352	145	0.000026	0.000011
1000	500	352	146	0.000046	0.000020
1000	500	352	147	0.000078	0.000033
1000	500	352	148	0.000133	0.000054
1000	500	352	149	0.000221	0.000088
1000	500	352	150	0.000361	0.000140
1000	500	352	151	0.000581	0.000220
1000	500	352	152	0.000919	0.000338
1000	500	352	153	0.001430	0.000511
1000	500	352	154	0.002189	0.000759
1000	500	352	155	0.003297	0.001107
1000	500	352	156	0.004883	0.001587
1000	500	352	157	0.007117	0.002234
1000	500	352	158	0.010208	0.003091
1000	500	352	159	0.014409	0.004201
1000	500	352	160	0.020019	0.005610
1000	500	352	161	0.027381	0.007362
1000	500	352	162	0.036872	0.009491
1000	500	352	163	0.048896	0.012024
1000	500	352	164	0.063864	0.014968
1000	500	352	165	0.082171	0.018307
1000	500	352	166	0.104173	0.022002
1000	500	352	167	0.130157	0.025984
1000	500	352	168	0.160309	0.030152
1000	500	352	169	0.194691	0.034382
1000	500	352	170	0.233215	0.038524
1000	500	352	171	0.275631	0.042416
1000	500	352	172	0.321522	0.045891
1000	500	352	173	0.370311	0.048789
1000	500	352	174	0.421282	0.050970
1000	500	352	175	0.473607	0.052326
1000	500	352	176	0.526393	0.052785
1000	500	353	140	0.000001	0.000000
1000	500	353	141	0.000002	0.000001
1000	500	353	142	0.000003	0.000002
1000	500	353	143	0.000006	0.000003
1000	500	353	144	0.000011	0.000005
1000	500	353	145	0.000020	0.000009
1000	500	353	146	0.000035	0.000015
1000	500	353	147	0.000061	0.000026

N	n	k	x	P(x)	p(x)
1000	500	353	148	0.000103	0.000043
1000	500	353	149	0.000173	0.000070
1000	500	353	150	0.000285	0.000112
1000	500	353	151	0.000463	0.000177
1000	500	353	152	0.000738	0.000275
1000	500	353	153	0.001157	0.000419
1000	500	353	154	0.001784	0.000627
1000	500	353	155	0.002707	0.000923
1000	500	353	156	0.004041	0.001334
1000	500	353	157	0.005935	0.001894
1000	500	353	158	0.008577	0.002642
1000	500	353	159	0.012198	0.003621
1000	500	353	160	0.017075	0.004877
1000	500	353	161	0.023530	0.006454
1000	500	353	162	0.031921	0.008392
1000	500	353	163	0.042643	0.010722
1000	500	353	164	0.056103	0.013460
1000	500	353	165	0.072707	0.016604
1000	500	353	166	0.092833	0.020126
1000	500	353	167	0.116804	0.023972
1000	500	353	168	0.144861	0.028057
1000	500	353	169	0.177129	0.032268
1000	500	353	170	0.213596	0.036468
1000	500	353	171	0.254096	0.040500
1000	500	353	172	0.298293	0.044198
1000	500	353	173	0.345691	0.047397
1000	500	353	174	0.395639	0.049948
1000	500	353	175	0.447364	0.051725
1000	500	353	176	0.500000	0.052636
1000	500	354	140	0.000001	0.000000
1000	500	354	141	0.000001	0.000001
1000	500	354	142	0.000002	0.000001
1000	500	354	143	0.000004	0.000002
1000	500	354	144	0.000008	0.000004
1000	500	354	145	0.000015	0.000007
1000	500	354	146	0.000027	0.000012
1000	500	354	147	0.000047	0.000020
1000	500	354	148	0.000080	0.000033
1000	500	354	149	0.000135	0.000055
1000	500	354	150	0.000225	0.000089
1000	500	354	151	0.000367	0.000142
1000	500	354	152	0.000590	0.000223
1000	500	354	153	0.000932	0.000342
1000	500	354	154	0.001449	0.000517
1000	500	354	155	0.002215	0.000766
1000	500	354	156	0.003332	0.001117
1000	500	354	157	0.004931	0.001599
1000	500	354	158	0.007180	0.002249
1000	500	354	159	0.010290	0.003109
1000	500	354	160	0.014512	0.004223

N	n	k	x	P(x)	p(x)
1000	500	354	161	0.020148	0.005635
1000	500	354	162	0.027537	0.007390
1000	500	354	163	0.037059	0.009521
1000	500	354	164	0.049113	0.012054
1000	500	354	165	0.064110	0.014997
1000	500	354	166	0.082443	0.018333
1000	500	354	167	0.104466	0.022023
1000	500	354	168	0.130464	0.025997
1000	500	354	169	0.160621	0.030157
1000	500	354	170	0.194996	0.034375
1000	500	354	171	0.233502	0.038506
1000	500	354	172	0.275887	0.042385
1000	500	354	173	0.321736	0.045848
1000	500	354	174	0.370472	0.048736
1000	500	354	175	0.421381	0.050909
1000	500	354	176	0.473641	0.052260
1000	500	354	177	0.526359	0.052718
1000	500	355	141	0.000001	0.000000
1000	500	355	142	0.000002	0.000001
1000	500	355	143	0.000003	0.000002
1000	500	355	144	0.000006	0.000003
1000	500	355	145	0.000011	0.000005
1000	500	355	146	0.000020	0.000009
1000	500	355	147	0.000036	0.000015
1000	500	355	148	0.000062	0.000026
1000	500	355	149	0.000105	0.000043
1000	500	355	150	0.000176	0.000071
1000	500	355	151	0.000290	0.000114
1000	500	355	152	0.000470	0.000180
1000	500	355	153	0.000748	0.000278
1000	500	355	154	0.001172	0.000424
1000	500	355	155	0.001806	0.000634
1000	500	355	156	0.002737	0.000931
1000	500	355	157	0.004082	0.001345
1000	500	355	158	0.005989	0.001908
1000	500	355	159	0.008648	0.002659
1000	500	355	160	0.012290	0.003642
1000	500	355	161	0.017191	0.004901
1000	500	355	162	0.023671	0.006480
1000	500	355	163	0.032092	0.008421
1000	500	355	164	0.042843	0.010752
1000	500	355	165	0.056333	0.013490
1000	500	355	166	0.072964	0.016631
1000	500	355	167	0.093114	0.020150
1000	500	355	168	0.117103	0.023989
1000	500	355	169	0.145169	0.028066
1000	500	355	170	0.177436	0.032267
1000	500	355	171	0.213891	0.036455
1000	500	355	172	0.254367	0.040475
1000	500	355	173	0.298528	0.044161

Table for $N = 1000$, $n = 500$

N	n	k	x	P(x)	p(x)
1000	500	355	174	0.345878	0.047350
1000	500	355	175	0.395769	0.049892
1000	500	355	176	0.447430	0.051661
1000	500	355	177	0.500000	0.052570
1000	500	356	141	0.000001	0.000000
1000	500	356	142	0.000001	0.000001
1000	500	356	143	0.000002	0.000001
1000	500	356	144	0.000005	0.000002
1000	500	356	145	0.000008	0.000004
1000	500	356	146	0.000015	0.000007
1000	500	356	147	0.000027	0.000012
1000	500	356	148	0.000048	0.000020
1000	500	356	149	0.000082	0.000034
1000	500	356	150	0.000138	0.000056
1000	500	356	151	0.000228	0.000091
1000	500	356	152	0.000373	0.000144
1000	500	356	153	0.000598	0.000226
1000	500	356	154	0.000945	0.000346
1000	500	356	155	0.001467	0.000522
1000	500	356	156	0.002240	0.000774
1000	500	356	157	0.003367	0.001126
1000	500	356	158	0.004978	0.001611
1000	500	356	159	0.007243	0.002265
1000	500	356	160	0.010370	0.003128
1000	500	356	161	0.014615	0.004245
1000	500	356	162	0.020275	0.005660
1000	500	356	163	0.027692	0.007417
1000	500	356	164	0.037242	0.009550
1000	500	356	165	0.049327	0.012084
1000	500	356	166	0.064352	0.015025
1000	500	356	167	0.082711	0.018359
1000	500	356	168	0.104755	0.022044
1000	500	356	169	0.130766	0.026011
1000	500	356	170	0.160927	0.030161
1000	500	356	171	0.195296	0.034369
1000	500	356	172	0.233784	0.038488
1000	500	356	173	0.276139	0.042355
1000	500	356	174	0.321946	0.045807
1000	500	356	175	0.370630	0.048684
1000	500	356	176	0.421480	0.050850
1000	500	356	177	0.473674	0.052195
1000	500	356	178	0.526325	0.052651
1000	500	357	142	0.000001	0.000000
1000	500	357	143	0.000002	0.000001
1000	500	357	144	0.000003	0.000002
1000	500	357	145	0.000006	0.000003
1000	500	357	146	0.000012	0.000005
1000	500	357	147	0.000021	0.000009
1000	500	357	148	0.000036	0.000016
1000	500	357	149	0.000063	0.000027
1000	500	357	150	0.000107	0.000044
1000	500	357	151	0.000179	0.000072
1000	500	357	152	0.000295	0.000116
1000	500	357	153	0.000477	0.000182
1000	500	357	154	0.000759	0.000282
1000	500	357	155	0.001187	0.000428
1000	500	357	156	0.001827	0.000640
1000	500	357	157	0.002767	0.000940
1000	500	357	158	0.004122	0.001355
1000	500	357	159	0.006043	0.001921
1000	500	357	160	0.008719	0.002676
1000	500	357	161	0.012381	0.003662
1000	500	357	162	0.017304	0.004924
1000	500	357	163	0.023811	0.006506
1000	500	357	164	0.032260	0.008449
1000	500	357	165	0.043041	0.010781
1000	500	357	166	0.056559	0.013519
1000	500	357	167	0.073218	0.016659
1000	500	357	168	0.093391	0.020173
1000	500	357	169	0.117397	0.024006
1000	500	357	170	0.145472	0.028075
1000	500	357	171	0.177738	0.032266
1000	500	357	172	0.214181	0.036443
1000	500	357	173	0.254633	0.040451
1000	500	357	174	0.298758	0.044125
1000	500	357	175	0.346061	0.047303
1000	500	357	176	0.395897	0.049836
1000	500	357	177	0.447496	0.051599
1000	500	357	178	0.500000	0.052504
1000	500	358	142	0.000001	0.000000
1000	500	358	143	0.000001	0.000001
1000	500	358	144	0.000003	0.000001
1000	500	358	145	0.000005	0.000002
1000	500	358	146	0.000009	0.000004
1000	500	358	147	0.000016	0.000007
1000	500	358	148	0.000028	0.000012
1000	500	358	149	0.000048	0.000021
1000	500	358	150	0.000083	0.000035
1000	500	358	151	0.000140	0.000057
1000	500	358	152	0.000232	0.000092
1000	500	358	153	0.000379	0.000146
1000	500	358	154	0.000607	0.000228
1000	500	358	155	0.000957	0.000350
1000	500	358	156	0.001485	0.000528
1000	500	358	157	0.002266	0.000781
1000	500	358	158	0.003401	0.001136
1000	500	358	159	0.005025	0.001623
1000	500	358	160	0.007304	0.002280
1000	500	358	161	0.010450	0.003146
1000	500	358	162	0.014716	0.004266
1000	500	358	163	0.020401	0.005685
1000	500	358	164	0.027845	0.007444
1000	500	358	165	0.037424	0.009579
1000	500	358	166	0.049537	0.012113
1000	500	358	167	0.064591	0.015054
1000	500	358	168	0.082975	0.018384
1000	500	358	169	0.105039	0.022064
1000	500	358	170	0.131063	0.026024
1000	500	358	171	0.161229	0.030165
1000	500	358	172	0.195591	0.034363
1000	500	358	173	0.234061	0.038470
1000	500	358	174	0.276386	0.042325
1000	500	358	175	0.322152	0.045766
1000	500	358	176	0.370785	0.048633
1000	500	358	177	0.421576	0.050791
1000	500	358	178	0.473707	0.052131
1000	500	358	179	0.526293	0.052586
1000	500	359	142	0.000000	0.000000
1000	500	359	143	0.000001	0.000000
1000	500	359	144	0.000002	0.000001
1000	500	359	145	0.000004	0.000002
1000	500	359	146	0.000007	0.000003
1000	500	359	147	0.000012	0.000005
1000	500	359	148	0.000021	0.000009
1000	500	359	149	0.000037	0.000016
1000	500	359	150	0.000064	0.000027
1000	500	359	151	0.000109	0.000045
1000	500	359	152	0.000182	0.000073
1000	500	359	153	0.000300	0.000117
1000	500	359	154	0.000484	0.000184
1000	500	359	155	0.000769	0.000285
1000	500	359	156	0.001202	0.000433
1000	500	359	157	0.001848	0.000646
1000	500	359	158	0.002796	0.000948
1000	500	359	159	0.004162	0.001366
1000	500	359	160	0.006097	0.001935
1000	500	359	161	0.008789	0.002692
1000	500	359	162	0.012470	0.003681
1000	500	359	163	0.017417	0.004947
1000	500	359	164	0.023949	0.006532
1000	500	359	165	0.032425	0.008477
1000	500	359	166	0.043235	0.010810
1000	500	359	167	0.056783	0.013547
1000	500	359	168	0.073468	0.016685
1000	500	359	169	0.093663	0.020196
1000	500	359	170	0.117687	0.024023
1000	500	359	171	0.145770	0.028084
1000	500	359	172	0.178035	0.032265
1000	500	359	173	0.214467	0.036431
1000	500	359	174	0.254894	0.040427
1000	500	359	175	0.298984	0.044090
1000	500	359	176	0.346241	0.047257
1000	500	359	177	0.396022	0.049781
1000	500	359	178	0.447560	0.051538
1000	500	359	179	0.500000	0.052440
1000	500	360	143	0.000001	0.000000
1000	500	360	144	0.000001	0.000001
1000	500	360	145	0.000003	0.000001
1000	500	360	146	0.000005	0.000002
1000	500	360	147	0.000009	0.000004
1000	500	360	148	0.000016	0.000007
1000	500	360	149	0.000028	0.000012
1000	500	360	150	0.000049	0.000021
1000	500	360	151	0.000085	0.000035
1000	500	360	152	0.000143	0.000058
1000	500	360	153	0.000236	0.000093
1000	500	360	154	0.000384	0.000148
1000	500	360	155	0.000616	0.000231
1000	500	360	156	0.000970	0.000354
1000	500	360	157	0.001502	0.000533
1000	500	360	158	0.002291	0.000788
1000	500	360	159	0.003436	0.001145
1000	500	360	160	0.005071	0.001635
1000	500	360	161	0.007365	0.002294
1000	500	360	162	0.010529	0.003164
1000	500	360	163	0.014816	0.004287
1000	500	360	164	0.020525	0.005709
1000	500	360	165	0.027995	0.007471
1000	500	360	166	0.037603	0.009607
1000	500	360	167	0.049745	0.012142
1000	500	360	168	0.064826	0.015081
1000	500	360	169	0.083235	0.018409
1000	500	360	170	0.105319	0.022084
1000	500	360	171	0.131356	0.026037
1000	500	360	172	0.161525	0.030169
1000	500	360	173	0.195882	0.034356
1000	500	360	174	0.234333	0.038452
1000	500	360	175	0.276629	0.042296
1000	500	360	176	0.322355	0.045725
1000	500	360	177	0.370938	0.048583
1000	500	360	178	0.421671	0.050733
1000	500	360	179	0.473739	0.052068
1000	500	360	180	0.526261	0.052521
1000	500	361	143	0.000001	0.000000
1000	500	361	144	0.000001	0.000000
1000	500	361	145	0.000002	0.000001
1000	500	361	146	0.000004	0.000002
1000	500	361	147	0.000007	0.000003
1000	500	361	148	0.000012	0.000005
1000	500	361	149	0.000022	0.000010

Table for $N = 1000$, $n = 500$

N	n	k	x	P(x)	p(x)
1000	500	361	150	0.000038	0.000016
1000	500	361	151	0.000065	0.000028
1000	500	361	152	0.000111	0.000046
1000	500	361	153	0.000185	0.000074
1000	500	361	154	0.000304	0.000119
1000	500	361	155	0.000491	0.000187
1000	500	361	156	0.000779	0.000288
1000	500	361	157	0.001217	0.000438
1000	500	361	158	0.001869	0.000653
1000	500	361	159	0.002826	0.000956
1000	500	361	160	0.004202	0.001377
1000	500	361	161	0.006150	0.001948
1000	500	361	162	0.008858	0.002708
1000	500	361	163	0.012559	0.003701
1000	500	361	164	0.017528	0.004969
1000	500	361	165	0.024085	0.006557
1000	500	361	166	0.032589	0.008504
1000	500	361	167	0.043427	0.010838
1000	500	361	168	0.057003	0.013575
1000	500	361	169	0.073714	0.016711
1000	500	361	170	0.093932	0.020218
1000	500	361	171	0.117971	0.024039
1000	500	361	172	0.146063	0.028092
1000	500	361	173	0.178328	0.032264
1000	500	361	174	0.214747	0.036419
1000	500	361	175	0.255151	0.040404
1000	500	361	176	0.299206	0.044055
1000	500	361	177	0.346418	0.047212
1000	500	361	178	0.396146	0.049727
1000	500	361	179	0.447624	0.051478
1000	500	361	180	0.500000	0.052376
1000	500	362	144	0.000001	0.000000
1000	500	362	145	0.000001	0.000001
1000	500	362	146	0.000003	0.000001
1000	500	362	147	0.000005	0.000002
1000	500	362	148	0.000009	0.000004
1000	500	362	149	0.000016	0.000007
1000	500	362	150	0.000029	0.000013
1000	500	362	151	0.000050	0.000021
1000	500	362	152	0.000086	0.000036
1000	500	362	153	0.000145	0.000059
1000	500	362	154	0.000240	0.000095
1000	500	362	155	0.000390	0.000150
1000	500	362	156	0.000624	0.000234
1000	500	362	157	0.000982	0.000358
1000	500	362	158	0.001520	0.000538
1000	500	362	159	0.002315	0.000795
1000	500	362	160	0.003470	0.001154
1000	500	362	161	0.005117	0.001647
1000	500	362	162	0.007426	0.002309

N	n	k	x	P(x)	p(x)
1000	500	362	163	0.010607	0.003181
1000	500	362	164	0.014915	0.004308
1000	500	362	165	0.020647	0.005732
1000	500	362	166	0.028144	0.007497
1000	500	362	167	0.037779	0.009635
1000	500	362	168	0.049949	0.012170
1000	500	362	169	0.065057	0.015108
1000	500	362	170	0.083491	0.018433
1000	500	362	171	0.105594	0.022103
1000	500	362	172	0.131644	0.026050
1000	500	362	173	0.161817	0.030173
1000	500	362	174	0.196167	0.034350
1000	500	362	175	0.234601	0.038434
1000	500	362	176	0.276868	0.042267
1000	500	362	177	0.322554	0.045685
1000	500	362	178	0.371087	0.048534
1000	500	362	179	0.421764	0.050676
1000	500	362	180	0.473771	0.052007
1000	500	362	181	0.526229	0.052458
1000	500	363	144	0.000001	0.000000
1000	500	363	145	0.000001	0.000000
1000	500	363	146	0.000002	0.000001
1000	500	363	147	0.000004	0.000002
1000	500	363	148	0.000007	0.000003
1000	500	363	149	0.000012	0.000006
1000	500	363	150	0.000022	0.000010
1000	500	363	151	0.000039	0.000017
1000	500	363	152	0.000067	0.000028
1000	500	363	153	0.000113	0.000046
1000	500	363	154	0.000188	0.000075
1000	500	363	155	0.000309	0.000120
1000	500	363	156	0.000498	0.000189
1000	500	363	157	0.000790	0.000292
1000	500	363	158	0.001232	0.000442
1000	500	363	159	0.001890	0.000659
1000	500	363	160	0.002854	0.000964
1000	500	363	161	0.004241	0.001387
1000	500	363	162	0.006202	0.001961
1000	500	363	163	0.008927	0.002724
1000	500	363	164	0.012646	0.003720
1000	500	363	165	0.017637	0.004991
1000	500	363	166	0.024219	0.006582
1000	500	363	167	0.032750	0.008531
1000	500	363	168	0.043616	0.010866
1000	500	363	169	0.057219	0.013603
1000	500	363	170	0.073956	0.016737
1000	500	363	171	0.094196	0.020240
1000	500	363	172	0.118251	0.024055
1000	500	363	173	0.146352	0.028100
1000	500	363	174	0.178615	0.032263

N	n	k	x	P(x)	p(x)
1000	500	363	175	0.215023	0.036408
1000	500	363	176	0.255404	0.040381
1000	500	363	177	0.299425	0.044021
1000	500	363	178	0.346592	0.047168
1000	500	363	179	0.396267	0.049674
1000	500	363	180	0.447686	0.051419
1000	500	363	181	0.500000	0.052314
1000	500	364	145	0.000001	0.000000
1000	500	364	146	0.000001	0.000001
1000	500	364	147	0.000003	0.000001
1000	500	364	148	0.000005	0.000002
1000	500	364	149	0.000009	0.000004
1000	500	364	150	0.000017	0.000007
1000	500	364	151	0.000030	0.000013
1000	500	364	152	0.000051	0.000022
1000	500	364	153	0.000088	0.000036
1000	500	364	154	0.000148	0.000060
1000	500	364	155	0.000244	0.000096
1000	500	364	156	0.000396	0.000152
1000	500	364	157	0.000633	0.000237
1000	500	364	158	0.000994	0.000362
1000	500	364	159	0.001538	0.000543
1000	500	364	160	0.002340	0.000802
1000	500	364	161	0.003503	0.001163
1000	500	364	162	0.005162	0.001659
1000	500	364	163	0.007485	0.002323
1000	500	364	164	0.010684	0.003199
1000	500	364	165	0.015013	0.004328
1000	500	364	166	0.020768	0.005756
1000	500	364	167	0.028291	0.007522
1000	500	364	168	0.037953	0.009662
1000	500	364	169	0.050151	0.012198
1000	500	364	170	0.065286	0.015135
1000	500	364	171	0.083742	0.018457
1000	500	364	172	0.105865	0.022122
1000	500	364	173	0.131927	0.026062
1000	500	364	174	0.162104	0.030177
1000	500	364	175	0.196447	0.034344
1000	500	364	176	0.234864	0.038417
1000	500	364	177	0.277103	0.042239
1000	500	364	178	0.322749	0.045646
1000	500	364	179	0.371235	0.048485
1000	500	364	180	0.421855	0.050621
1000	500	364	181	0.473802	0.051947
1000	500	364	182	0.526198	0.052396
1000	500	365	145	0.000001	0.000000
1000	500	365	146	0.000001	0.000001
1000	500	365	147	0.000002	0.000001
1000	500	365	148	0.000004	0.000002
1000	500	365	149	0.000007	0.000003

N	n	k	x	P(x)	p(x)
1000	500	365	150	0.000013	0.000006
1000	500	365	151	0.000023	0.000010
1000	500	365	152	0.000039	0.000017
1000	500	365	153	0.000068	0.000028
1000	500	365	154	0.000115	0.000047
1000	500	365	155	0.000192	0.000076
1000	500	365	156	0.000314	0.000122
1000	500	365	157	0.000505	0.000191
1000	500	365	158	0.000800	0.000295
1000	500	365	159	0.001246	0.000447
1000	500	365	160	0.001911	0.000665
1000	500	365	161	0.002883	0.000972
1000	500	365	162	0.004280	0.001397
1000	500	365	163	0.006254	0.001974
1000	500	365	164	0.008994	0.002740
1000	500	365	165	0.012733	0.003738
1000	500	365	166	0.017746	0.005013
1000	500	365	167	0.024352	0.006606
1000	500	365	168	0.032909	0.008557
1000	500	365	169	0.043802	0.010893
1000	500	365	170	0.057433	0.013630
1000	500	365	171	0.074195	0.016762
1000	500	365	172	0.094456	0.020261
1000	500	365	173	0.118527	0.024071
1000	500	365	174	0.146636	0.028109
1000	500	365	175	0.178897	0.032262
1000	500	365	176	0.215293	0.036396
1000	500	365	177	0.255652	0.040358
1000	500	365	178	0.299639	0.043987
1000	500	365	179	0.346763	0.047124
1000	500	365	180	0.396386	0.049622
1000	500	365	181	0.447747	0.051361
1000	500	365	182	0.500000	0.052253
1000	500	366	146	0.000001	0.000000
1000	500	366	147	0.000001	0.000001
1000	500	366	148	0.000003	0.000001
1000	500	366	149	0.000005	0.000002
1000	500	366	150	0.000010	0.000004
1000	500	366	151	0.000017	0.000008
1000	500	366	152	0.000030	0.000013
1000	500	366	153	0.000052	0.000022
1000	500	366	154	0.000089	0.000037
1000	500	366	155	0.000150	0.000061
1000	500	366	156	0.000247	0.000097
1000	500	366	157	0.000402	0.000154
1000	500	366	158	0.000641	0.000239
1000	500	366	159	0.001007	0.000366
1000	500	366	160	0.001555	0.000549
1000	500	366	161	0.002364	0.000809
1000	500	366	162	0.003537	0.001173

Table for $N = 1000$, $n = 500$

N	n	k	x	P(x)	p(x)
1000	500	366	163	0.005207	0.001670
1000	500	366	164	0.007544	0.002338
1000	500	366	165	0.010760	0.003216
1000	500	366	166	0.015109	0.004348
1000	500	366	167	0.020887	0.005779
1000	500	366	168	0.028435	0.007548
1000	500	366	169	0.038124	0.009689
1000	500	366	170	0.050349	0.012225
1000	500	366	171	0.065510	0.015161
1000	500	366	172	0.083990	0.018480
1000	500	366	173	0.106131	0.022141
1000	500	366	174	0.132205	0.026074
1000	500	366	175	0.162385	0.030180
1000	500	366	176	0.196723	0.034337
1000	500	366	177	0.235123	0.038400
1000	500	366	178	0.277334	0.042211
1000	500	366	179	0.322941	0.045608
1000	500	366	180	0.371379	0.048438
1000	500	366	181	0.421945	0.050566
1000	500	366	182	0.473832	0.051887
1000	500	366	183	0.526168	0.052335
1000	500	367	146	0.000001	0.000000
1000	500	367	147	0.000001	0.000001
1000	500	367	148	0.000002	0.000001
1000	500	367	149	0.000004	0.000002
1000	500	367	150	0.000007	0.000003
1000	500	367	151	0.000013	0.000006
1000	500	367	152	0.000023	0.000010
1000	500	367	153	0.000040	0.000017
1000	500	367	154	0.000069	0.000029
1000	500	367	155	0.000117	0.000048
1000	500	367	156	0.000195	0.000078
1000	500	367	157	0.000318	0.000124
1000	500	367	158	0.000512	0.000194
1000	500	367	159	0.000810	0.000298
1000	500	367	160	0.001261	0.000451
1000	500	367	161	0.001932	0.000671
1000	500	367	162	0.002912	0.000980
1000	500	367	163	0.004319	0.001407
1000	500	367	164	0.006306	0.001986
1000	500	367	165	0.009061	0.002756
1000	500	367	166	0.012818	0.003757
1000	500	367	167	0.017852	0.005034
1000	500	367	168	0.024483	0.006630
1000	500	367	169	0.033066	0.008583
1000	500	367	170	0.043986	0.010920
1000	500	367	171	0.057643	0.013657
1000	500	367	172	0.074430	0.016787
1000	500	367	173	0.094712	0.020282
1000	500	367	174	0.118798	0.024086

N	n	k	x	P(x)	p(x)
1000	500	367	175	0.146914	0.028116
1000	500	367	176	0.179175	0.032261
1000	500	367	177	0.215560	0.036384
1000	500	367	178	0.255895	0.040336
1000	500	367	179	0.299850	0.043954
1000	500	367	180	0.346931	0.047081
1000	500	367	181	0.396503	0.049571
1000	500	367	182	0.447807	0.051304
1000	500	367	183	0.500000	0.052193
1000	500	368	147	0.000001	0.000000
1000	500	368	148	0.000002	0.000001
1000	500	368	149	0.000003	0.000001
1000	500	368	150	0.000005	0.000002
1000	500	368	151	0.000010	0.000004
1000	500	368	152	0.000017	0.000008
1000	500	368	153	0.000031	0.000013
1000	500	368	154	0.000053	0.000023
1000	500	368	155	0.000091	0.000038
1000	500	368	156	0.000152	0.000061
1000	500	368	157	0.000251	0.000099
1000	500	368	158	0.000407	0.000156
1000	500	368	159	0.000649	0.000242
1000	500	368	160	0.001019	0.000369
1000	500	368	161	0.001572	0.000554
1000	500	368	162	0.002388	0.000816
1000	500	368	163	0.003570	0.001181
1000	500	368	164	0.005251	0.001681
1000	500	368	165	0.007603	0.002352
1000	500	368	166	0.010836	0.003233
1000	500	368	167	0.015204	0.004368
1000	500	368	168	0.021005	0.005801
1000	500	368	169	0.028578	0.007572
1000	500	368	170	0.038293	0.009715
1000	500	368	171	0.050545	0.012252
1000	500	368	172	0.065731	0.015187
1000	500	368	173	0.084234	0.018503
1000	500	368	174	0.106393	0.022159
1000	500	368	175	0.132479	0.026085
1000	500	368	176	0.162663	0.030184
1000	500	368	177	0.196994	0.034331
1000	500	368	178	0.235377	0.038383
1000	500	368	179	0.277560	0.042183
1000	500	368	180	0.323130	0.045570
1000	500	368	181	0.371521	0.048391
1000	500	368	182	0.422033	0.050512
1000	500	368	183	0.473862	0.051829
1000	500	368	184	0.526138	0.052276
1000	500	369	147	0.000001	0.000000
1000	500	369	148	0.000001	0.000001
1000	500	369	149	0.000002	0.000001

N	n	k	x	P(x)	p(x)
1000	500	369	150	0.000004	0.000002
1000	500	369	151	0.000007	0.000003
1000	500	369	152	0.000013	0.000006
1000	500	369	153	0.000023	0.000010
1000	500	369	154	0.000041	0.000018
1000	500	369	155	0.000070	0.000029
1000	500	369	156	0.000119	0.000049
1000	500	369	157	0.000198	0.000079
1000	500	369	158	0.000323	0.000125
1000	500	369	159	0.000519	0.000196
1000	500	369	160	0.000820	0.000301
1000	500	369	161	0.001275	0.000455
1000	500	369	162	0.001952	0.000677
1000	500	369	163	0.002940	0.000988
1000	500	369	164	0.004357	0.001417
1000	500	369	165	0.006356	0.001999
1000	500	369	166	0.009127	0.002771
1000	500	369	167	0.012902	0.003775
1000	500	369	168	0.017958	0.005055
1000	500	369	169	0.024612	0.006654
1000	500	369	170	0.033220	0.008608
1000	500	369	171	0.044167	0.010947
1000	500	369	172	0.057850	0.013683
1000	500	369	173	0.074661	0.016811
1000	500	369	174	0.094963	0.020303
1000	500	369	175	0.119065	0.024101
1000	500	369	176	0.147189	0.028124
1000	500	369	177	0.179448	0.032259
1000	500	369	178	0.215821	0.036373
1000	500	369	179	0.256135	0.040314
1000	500	369	180	0.300057	0.043922
1000	500	369	181	0.347096	0.047039
1000	500	369	182	0.396617	0.049521
1000	500	369	183	0.447866	0.051248
1000	500	369	184	0.500000	0.052134
1000	500	370	148	0.000001	0.000000
1000	500	370	149	0.000002	0.000001
1000	500	370	150	0.000003	0.000001
1000	500	370	151	0.000005	0.000003
1000	500	370	152	0.000010	0.000004
1000	500	370	153	0.000018	0.000008
1000	500	370	154	0.000031	0.000014
1000	500	370	155	0.000054	0.000023
1000	500	370	156	0.000093	0.000038
1000	500	370	157	0.000155	0.000062
1000	500	370	158	0.000255	0.000100
1000	500	370	159	0.000413	0.000158
1000	500	370	160	0.000658	0.000245
1000	500	370	161	0.001031	0.000373
1000	500	370	162	0.001590	0.000559

N	n	k	x	P(x)	p(x)
1000	500	370	163	0.002412	0.000823
1000	500	370	164	0.003602	0.001190
1000	500	370	165	0.005295	0.001692
1000	500	370	166	0.007660	0.002366
1000	500	370	167	0.010910	0.003250
1000	500	370	168	0.015298	0.004388
1000	500	370	169	0.021121	0.005823
1000	500	370	170	0.028718	0.007597
1000	500	370	171	0.038459	0.009741
1000	500	370	172	0.050737	0.012278
1000	500	370	173	0.065949	0.015212
1000	500	370	174	0.084474	0.018525
1000	500	370	175	0.106651	0.022177
1000	500	370	176	0.132748	0.026097
1000	500	370	177	0.162935	0.030187
1000	500	370	178	0.197260	0.034325
1000	500	370	179	0.235627	0.038367
1000	500	370	180	0.277783	0.042156
1000	500	370	181	0.323315	0.045533
1000	500	370	182	0.371660	0.048345
1000	500	370	183	0.422120	0.050459
1000	500	370	184	0.473891	0.051772
1000	500	370	185	0.526108	0.052217
1000	500	371	148	0.000001	0.000000
1000	500	371	149	0.000001	0.000001
1000	500	371	150	0.000002	0.000001
1000	500	371	151	0.000004	0.000002
1000	500	371	152	0.000007	0.000003
1000	500	371	153	0.000014	0.000006
1000	500	371	154	0.000024	0.000010
1000	500	371	155	0.000042	0.000018
1000	500	371	156	0.000072	0.000030
1000	500	371	157	0.000121	0.000049
1000	500	371	158	0.000201	0.000080
1000	500	371	159	0.000327	0.000127
1000	500	371	160	0.000525	0.000198
1000	500	371	161	0.000830	0.000304
1000	500	371	162	0.001290	0.000460
1000	500	371	163	0.001972	0.000682
1000	500	371	164	0.002968	0.000995
1000	500	371	165	0.004395	0.001427
1000	500	371	166	0.006406	0.002011
1000	500	371	167	0.009192	0.002786
1000	500	371	168	0.012986	0.003793
1000	500	371	169	0.018062	0.005076
1000	500	371	170	0.024739	0.006677
1000	500	371	171	0.033372	0.008633
1000	500	371	172	0.044345	0.010973
1000	500	371	173	0.058053	0.013708
1000	500	371	174	0.074888	0.016835

Table for $N = 1000,\ n = 500$

N	n	k	x	P(x)	p(x)
1000	500	371	175	0.095211	0.020323
1000	500	371	176	0.119327	0.024116
1000	500	371	177	0.147458	0.028131
1000	500	371	178	0.179716	0.032258
1000	500	371	179	0.216078	0.036362
1000	500	371	180	0.256370	0.040292
1000	500	371	181	0.300260	0.043890
1000	500	371	182	0.347258	0.046998
1000	500	371	183	0.396730	0.049472
1000	500	371	184	0.447923	0.051194
1000	500	371	185	0.500000	0.052077
1000	500	372	149	0.000001	0.000000
1000	500	372	150	0.000002	0.000001
1000	500	372	151	0.000003	0.000001
1000	500	372	152	0.000006	0.000003
1000	500	372	153	0.000010	0.000005
1000	500	372	154	0.000018	0.000008
1000	500	372	155	0.000032	0.000014
1000	500	372	156	0.000055	0.000023
1000	500	372	157	0.000094	0.000039
1000	500	372	158	0.000157	0.000063
1000	500	372	159	0.000259	0.000101
1000	500	372	160	0.000418	0.000160
1000	500	372	161	0.000666	0.000247
1000	500	372	162	0.001043	0.000377
1000	500	372	163	0.001607	0.000564
1000	500	372	164	0.002436	0.000829
1000	500	372	165	0.003635	0.001199
1000	500	372	166	0.005338	0.001704
1000	500	372	167	0.007717	0.002379
1000	500	372	168	0.010983	0.003266
1000	500	372	169	0.015390	0.004407
1000	500	372	170	0.021236	0.005845
1000	500	372	171	0.028856	0.007621
1000	500	372	172	0.038623	0.009767
1000	500	372	173	0.050927	0.012304
1000	500	372	174	0.066163	0.015236
1000	500	372	175	0.084710	0.018547
1000	500	372	176	0.106904	0.022194
1000	500	372	177	0.133012	0.026108
1000	500	372	178	0.163202	0.030190
1000	500	372	179	0.197521	0.034319
1000	500	372	180	0.235872	0.038351
1000	500	372	181	0.278001	0.042129
1000	500	372	182	0.323497	0.045496
1000	500	372	183	0.371797	0.048300
1000	500	372	184	0.422204	0.050407
1000	500	372	185	0.473920	0.051716
1000	500	372	186	0.526080	0.052159
1000	500	373	149	0.000001	0.000000

N	n	k	x	P(x)	p(x)
1000	500	373	150	0.000001	0.000001
1000	500	373	151	0.000002	0.000001
1000	500	373	152	0.000004	0.000002
1000	500	373	153	0.000008	0.000003
1000	500	373	154	0.000014	0.000006
1000	500	373	155	0.000024	0.000011
1000	500	373	156	0.000043	0.000018
1000	500	373	157	0.000073	0.000030
1000	500	373	158	0.000123	0.000050
1000	500	373	159	0.000204	0.000081
1000	500	373	160	0.000332	0.000128
1000	500	373	161	0.000532	0.000200
1000	500	373	162	0.000840	0.000308
1000	500	373	163	0.001304	0.000464
1000	500	373	164	0.001992	0.000688
1000	500	373	165	0.002995	0.001003
1000	500	373	166	0.004432	0.001437
1000	500	373	167	0.006456	0.002024
1000	500	373	168	0.009257	0.002801
1000	500	373	169	0.013068	0.003811
1000	500	373	170	0.018164	0.005096
1000	500	373	171	0.024864	0.006700
1000	500	373	172	0.033522	0.008658
1000	500	373	173	0.044520	0.010998
1000	500	373	174	0.058254	0.013733
1000	500	373	175	0.075111	0.016858
1000	500	373	176	0.095454	0.020342
1000	500	373	177	0.119584	0.024130
1000	500	373	178	0.147723	0.028139
1000	500	373	179	0.179979	0.032257
1000	500	373	180	0.216330	0.036351
1000	500	373	181	0.256601	0.040271
1000	500	373	182	0.300459	0.043858
1000	500	373	183	0.347416	0.046957
1000	500	373	184	0.396840	0.049424
1000	500	373	185	0.447980	0.051140
1000	500	373	186	0.500000	0.052020
1000	500	374	150	0.000001	0.000000
1000	500	374	151	0.000002	0.000001
1000	500	374	152	0.000003	0.000001
1000	500	374	153	0.000006	0.000003
1000	500	374	154	0.000010	0.000005
1000	500	374	155	0.000019	0.000008
1000	500	374	156	0.000033	0.000014
1000	500	374	157	0.000056	0.000024
1000	500	374	158	0.000096	0.000039
1000	500	374	159	0.000160	0.000064
1000	500	374	160	0.000262	0.000103
1000	500	374	161	0.000424	0.000162
1000	500	374	162	0.000674	0.000250

N	n	k	x	P(x)	p(x)
1000	500	374	163	0.001055	0.000380
1000	500	374	164	0.001623	0.000569
1000	500	374	165	0.002459	0.000836
1000	500	374	166	0.003667	0.001207
1000	500	374	167	0.005381	0.001714
1000	500	374	168	0.007774	0.002393
1000	500	374	169	0.011056	0.003282
1000	500	374	170	0.015482	0.004426
1000	500	374	171	0.021348	0.005867
1000	500	374	172	0.028993	0.007644
1000	500	374	173	0.038784	0.009791
1000	500	374	174	0.051113	0.012329
1000	500	374	175	0.066373	0.015260
1000	500	374	176	0.084942	0.018569
1000	500	374	177	0.107153	0.022211
1000	500	374	178	0.133272	0.026119
1000	500	374	179	0.163465	0.030193
1000	500	374	180	0.197778	0.034313
1000	500	374	181	0.236112	0.038334
1000	500	374	182	0.278215	0.042103
1000	500	374	183	0.323676	0.045460
1000	500	374	184	0.371931	0.048255
1000	500	374	185	0.422288	0.050356
1000	500	374	186	0.473948	0.051661
1000	500	374	187	0.526051	0.052103
1000	500	375	150	0.000001	0.000000
1000	500	375	151	0.000001	0.000001
1000	500	375	152	0.000002	0.000001
1000	500	375	153	0.000004	0.000002
1000	500	375	154	0.000008	0.000004
1000	500	375	155	0.000014	0.000006
1000	500	375	156	0.000025	0.000011
1000	500	375	157	0.000043	0.000018
1000	500	375	158	0.000074	0.000031
1000	500	375	159	0.000125	0.000051
1000	500	375	160	0.000207	0.000082
1000	500	375	161	0.000336	0.000130
1000	500	375	162	0.000539	0.000203
1000	500	375	163	0.000850	0.000311
1000	500	375	164	0.001318	0.000468
1000	500	375	165	0.002012	0.000694
1000	500	375	166	0.003023	0.001011
1000	500	375	167	0.004469	0.001447
1000	500	375	168	0.006505	0.002036
1000	500	375	169	0.009320	0.002816
1000	500	375	170	0.013149	0.003828
1000	500	375	171	0.018265	0.005117
1000	500	375	172	0.024987	0.006722
1000	500	375	173	0.033669	0.008682
1000	500	375	174	0.044693	0.011023

N	n	k	x	P(x)	p(x)
1000	500	375	175	0.058451	0.013758
1000	500	375	176	0.075331	0.016881
1000	500	375	177	0.095693	0.020361
1000	500	375	178	0.119837	0.024144
1000	500	375	179	0.147983	0.028146
1000	500	375	180	0.180238	0.032255
1000	500	375	181	0.216577	0.036340
1000	500	375	182	0.256827	0.040250
1000	500	375	183	0.300655	0.043828
1000	500	375	184	0.347572	0.046917
1000	500	375	185	0.396949	0.049376
1000	500	375	186	0.448036	0.051087
1000	500	375	187	0.500000	0.051964
1000	500	376	151	0.000001	0.000000
1000	500	376	152	0.000002	0.000001
1000	500	376	153	0.000003	0.000001
1000	500	376	154	0.000006	0.000003
1000	500	376	155	0.000011	0.000005
1000	500	376	156	0.000019	0.000008
1000	500	376	157	0.000033	0.000014
1000	500	376	158	0.000057	0.000024
1000	500	376	159	0.000097	0.000040
1000	500	376	160	0.000162	0.000065
1000	500	376	161	0.000266	0.000104
1000	500	376	162	0.000429	0.000163
1000	500	376	163	0.000682	0.000253
1000	500	376	164	0.001066	0.000384
1000	500	376	165	0.001640	0.000574
1000	500	376	166	0.002482	0.000842
1000	500	376	167	0.003698	0.001216
1000	500	376	168	0.005424	0.001725
1000	500	376	169	0.007829	0.002406
1000	500	376	170	0.011127	0.003298
1000	500	376	171	0.015572	0.004444
1000	500	376	172	0.021460	0.005888
1000	500	376	173	0.029127	0.007667
1000	500	376	174	0.038943	0.009816
1000	500	376	175	0.051296	0.012353
1000	500	376	176	0.066580	0.015284
1000	500	376	177	0.085170	0.018590
1000	500	376	178	0.107398	0.022228
1000	500	376	179	0.133527	0.026129
1000	500	376	180	0.163723	0.030196
1000	500	376	181	0.198030	0.034307
1000	500	376	182	0.236348	0.038319
1000	500	376	183	0.278426	0.042077
1000	500	376	184	0.323851	0.045425
1000	500	376	185	0.372063	0.048212
1000	500	376	186	0.422370	0.050307
1000	500	376	187	0.473976	0.051607

Table for $N = 1000$, $n = 500$

N	n	k	x	P(x)	p(x)
1000	500	376	188	0.526024	0.052047
1000	500	377	151	0.000001	0.000000
1000	500	377	152	0.000001	0.000001
1000	500	377	153	0.000002	0.000001
1000	500	377	154	0.000004	0.000002
1000	500	377	155	0.000008	0.000004
1000	500	377	156	0.000014	0.000006
1000	500	377	157	0.000025	0.000011
1000	500	377	158	0.000044	0.000019
1000	500	377	159	0.000075	0.000031
1000	500	377	160	0.000127	0.000051
1000	500	377	161	0.000210	0.000083
1000	500	377	162	0.000341	0.000131
1000	500	377	163	0.000546	0.000205
1000	500	377	164	0.000859	0.000314
1000	500	377	165	0.001332	0.000473
1000	500	377	166	0.002032	0.000700
1000	500	377	167	0.003050	0.001018
1000	500	377	168	0.004506	0.001456
1000	500	377	169	0.006553	0.002047
1000	500	377	170	0.009383	0.002830
1000	500	377	171	0.013228	0.003845
1000	500	377	172	0.018365	0.005136
1000	500	377	173	0.025109	0.006744
1000	500	377	174	0.033815	0.008706
1000	500	377	175	0.044862	0.011048
1000	500	377	176	0.058644	0.013782
1000	500	377	177	0.075547	0.016903
1000	500	377	178	0.095928	0.020380
1000	500	377	179	0.120086	0.024158
1000	500	377	180	0.148238	0.028152
1000	500	377	181	0.180491	0.032254
1000	500	377	182	0.216820	0.036329
1000	500	377	183	0.257050	0.040229
1000	500	377	184	0.300847	0.043797
1000	500	377	185	0.347725	0.046878
1000	500	377	186	0.397055	0.049330
1000	500	377	187	0.448090	0.051035
1000	500	377	188	0.500000	0.051910
1000	500	378	152	0.000001	0.000000
1000	500	378	153	0.000002	0.000001
1000	500	378	154	0.000003	0.000002
1000	500	378	155	0.000006	0.000003
1000	500	378	156	0.000011	0.000005
1000	500	378	157	0.000019	0.000008
1000	500	378	158	0.000034	0.000015
1000	500	378	159	0.000058	0.000024
1000	500	378	160	0.000099	0.000040
1000	500	378	161	0.000164	0.000066
1000	500	378	162	0.000270	0.000105

N	n	k	x	P(x)	p(x)
1000	500	378	163	0.000435	0.000165
1000	500	378	164	0.000690	0.000255
1000	500	378	165	0.001078	0.000388
1000	500	378	166	0.001657	0.000579
1000	500	378	167	0.002505	0.000849
1000	500	378	168	0.003730	0.001224
1000	500	378	169	0.005465	0.001736
1000	500	378	170	0.007884	0.002419
1000	500	378	171	0.011198	0.003314
1000	500	378	172	0.015661	0.004463
1000	500	378	173	0.021569	0.005908
1000	500	378	174	0.029259	0.007690
1000	500	378	175	0.039099	0.009840
1000	500	378	176	0.051477	0.012378
1000	500	378	177	0.066784	0.015307
1000	500	378	178	0.085394	0.018610
1000	500	378	179	0.107638	0.022244
1000	500	378	180	0.133778	0.026139
1000	500	378	181	0.163976	0.030199
1000	500	378	182	0.198277	0.034301
1000	500	378	183	0.236580	0.038303
1000	500	378	184	0.278632	0.042052
1000	500	378	185	0.324023	0.045391
1000	500	378	186	0.372192	0.048169
1000	500	378	187	0.422450	0.050258
1000	500	378	188	0.474003	0.051554
1000	500	378	189	0.525996	0.051993
1000	500	379	152	0.000001	0.000000
1000	500	379	153	0.000001	0.000001
1000	500	379	154	0.000002	0.000001
1000	500	379	155	0.000004	0.000002
1000	500	379	156	0.000008	0.000004
1000	500	379	157	0.000015	0.000006
1000	500	379	158	0.000026	0.000011
1000	500	379	159	0.000045	0.000019
1000	500	379	160	0.000077	0.000032
1000	500	379	161	0.000129	0.000052
1000	500	379	162	0.000212	0.000084
1000	500	379	163	0.000345	0.000133
1000	500	379	164	0.000552	0.000207
1000	500	379	165	0.000869	0.000317
1000	500	379	166	0.001346	0.000477
1000	500	379	167	0.002051	0.000705
1000	500	379	168	0.003076	0.001025
1000	500	379	169	0.004542	0.001465
1000	500	379	170	0.006601	0.002059
1000	500	379	171	0.009445	0.002844
1000	500	379	172	0.013307	0.003862
1000	500	379	173	0.018463	0.005156
1000	500	379	174	0.025229	0.006766

N	n	k	x	P(x)	p(x)
1000	500	379	175	0.033958	0.008729
1000	500	379	176	0.045029	0.011072
1000	500	379	177	0.058835	0.013806
1000	500	379	178	0.075760	0.016925
1000	500	379	179	0.096159	0.020399
1000	500	379	180	0.120330	0.024171
1000	500	379	181	0.148489	0.028159
1000	500	379	182	0.180741	0.032252
1000	500	379	183	0.217059	0.036318
1000	500	379	184	0.257268	0.040209
1000	500	379	185	0.301036	0.043768
1000	500	379	186	0.347875	0.046840
1000	500	379	187	0.397159	0.049284
1000	500	379	188	0.448144	0.050984
1000	500	379	189	0.500000	0.051856
1000	500	380	153	0.000001	0.000000
1000	500	380	154	0.000002	0.000001
1000	500	380	155	0.000003	0.000002
1000	500	380	156	0.000006	0.000003
1000	500	380	157	0.000011	0.000005
1000	500	380	158	0.000020	0.000009
1000	500	380	159	0.000034	0.000015
1000	500	380	160	0.000059	0.000025
1000	500	380	161	0.000100	0.000041
1000	500	380	162	0.000167	0.000067
1000	500	380	163	0.000273	0.000106
1000	500	380	164	0.000440	0.000167
1000	500	380	165	0.000698	0.000258
1000	500	380	166	0.001089	0.000391
1000	500	380	167	0.001673	0.000583
1000	500	380	168	0.002528	0.000855
1000	500	380	169	0.003761	0.001232
1000	500	380	170	0.005507	0.001746
1000	500	380	171	0.007938	0.002432
1000	500	380	172	0.011267	0.003329
1000	500	380	173	0.015748	0.004481
1000	500	380	174	0.021677	0.005929
1000	500	380	175	0.029389	0.007712
1000	500	380	176	0.039253	0.009863
1000	500	380	177	0.051654	0.012401
1000	500	380	178	0.066984	0.015330
1000	500	380	179	0.085614	0.018630
1000	500	380	180	0.107874	0.022260
1000	500	380	181	0.134024	0.026149
1000	500	380	182	0.164225	0.030201
1000	500	380	183	0.198520	0.034295
1000	500	380	184	0.236807	0.038288
1000	500	380	185	0.278835	0.042027
1000	500	380	186	0.324191	0.045357
1000	500	380	187	0.372319	0.048127

N	n	k	x	P(x)	p(x)
1000	500	380	188	0.422528	0.050210
1000	500	380	189	0.474030	0.051502
1000	500	380	190	0.525970	0.051940
1000	500	381	153	0.000001	0.000000
1000	500	381	154	0.000001	0.000001
1000	500	381	155	0.000002	0.000001
1000	500	381	156	0.000005	0.000002
1000	500	381	157	0.000008	0.000004
1000	500	381	158	0.000015	0.000007
1000	500	381	159	0.000026	0.000011
1000	500	381	160	0.000046	0.000019
1000	500	381	161	0.000078	0.000032
1000	500	381	162	0.000131	0.000053
1000	500	381	163	0.000215	0.000085
1000	500	381	164	0.000350	0.000134
1000	500	381	165	0.000559	0.000209
1000	500	381	166	0.000879	0.000320
1000	500	381	167	0.001359	0.000481
1000	500	381	168	0.002070	0.000711
1000	500	381	169	0.003103	0.001032
1000	500	381	170	0.004577	0.001475
1000	500	381	171	0.006648	0.002071
1000	500	381	172	0.009506	0.002858
1000	500	381	173	0.013385	0.003879
1000	500	381	174	0.018559	0.005175
1000	500	381	175	0.025347	0.006787
1000	500	381	176	0.034098	0.008752
1000	500	381	177	0.045194	0.011095
1000	500	381	178	0.059022	0.013829
1000	500	381	179	0.075969	0.016946
1000	500	381	180	0.096385	0.020417
1000	500	381	181	0.120569	0.024184
1000	500	381	182	0.148735	0.028165
1000	500	381	183	0.180985	0.032250
1000	500	381	184	0.217293	0.036308
1000	500	381	185	0.257482	0.040189
1000	500	381	186	0.301220	0.043738
1000	500	381	187	0.348022	0.046802
1000	500	381	188	0.397262	0.049239
1000	500	381	189	0.448196	0.050935
1000	500	381	190	0.500000	0.051804
1000	500	382	154	0.000001	0.000000
1000	500	382	155	0.000002	0.000001
1000	500	382	156	0.000003	0.000002
1000	500	382	157	0.000006	0.000003
1000	500	382	158	0.000011	0.000005
1000	500	382	159	0.000020	0.000009
1000	500	382	160	0.000035	0.000015
1000	500	382	161	0.000060	0.000025
1000	500	382	162	0.000102	0.000042

Table for $N = 1000$, $n = 500$

N	n	k	x	P(x)	p(x)
1000	500	382	163	0.000169	0.000067
1000	500	382	164	0.000277	0.000108
1000	500	382	165	0.000446	0.000169
1000	500	382	166	0.000706	0.000260
1000	500	382	167	0.001101	0.000395
1000	500	382	168	0.001689	0.000588
1000	500	382	169	0.002550	0.000862
1000	500	382	170	0.003791	0.001241
1000	500	382	171	0.005547	0.001756
1000	500	382	172	0.007992	0.002444
1000	500	382	173	0.011336	0.003344
1000	500	382	174	0.015834	0.004498
1000	500	382	175	0.021783	0.005949
1000	500	382	176	0.029517	0.007734
1000	500	382	177	0.039404	0.009887
1000	500	382	178	0.051829	0.012425
1000	500	382	179	0.067181	0.015352
1000	500	382	180	0.085831	0.018650
1000	500	382	181	0.108106	0.022276
1000	500	382	182	0.134265	0.026159
1000	500	382	183	0.164469	0.030204
1000	500	382	184	0.198758	0.034289
1000	500	382	185	0.237030	0.038273
1000	500	382	186	0.279033	0.042003
1000	500	382	187	0.324357	0.045323
1000	500	382	188	0.372443	0.048086
1000	500	382	189	0.422605	0.050162
1000	500	382	190	0.474056	0.051451
1000	500	382	191	0.525944	0.051887
1000	500	383	154	0.000001	0.000000
1000	500	383	155	0.000001	0.000001
1000	500	383	156	0.000003	0.000001
1000	500	383	157	0.000005	0.000002
1000	500	383	158	0.000008	0.000004
1000	500	383	159	0.000015	0.000007
1000	500	383	160	0.000027	0.000012
1000	500	383	161	0.000046	0.000020
1000	500	383	162	0.000079	0.000033
1000	500	383	163	0.000132	0.000053
1000	500	383	164	0.000218	0.000086
1000	500	383	165	0.000354	0.000136
1000	500	383	166	0.000565	0.000211
1000	500	383	167	0.000888	0.000323
1000	500	383	168	0.001373	0.000485
1000	500	383	169	0.002089	0.000716
1000	500	383	170	0.003129	0.001040
1000	500	383	171	0.004612	0.001484
1000	500	383	172	0.006694	0.002082
1000	500	383	173	0.009566	0.002872
1000	500	383	174	0.013461	0.003895

N	n	k	x	P(x)	p(x)
1000	500	383	175	0.018655	0.005193
1000	500	383	176	0.025463	0.006808
1000	500	383	177	0.034237	0.008774
1000	500	383	178	0.045355	0.011118
1000	500	383	179	0.059207	0.013852
1000	500	383	180	0.076174	0.016967
1000	500	383	181	0.096608	0.020434
1000	500	383	182	0.120805	0.024197
1000	500	383	183	0.148976	0.028171
1000	500	383	184	0.181225	0.032249
1000	500	383	185	0.217523	0.036297
1000	500	383	186	0.257692	0.040169
1000	500	383	187	0.301402	0.043710
1000	500	383	188	0.348166	0.046765
1000	500	383	189	0.397362	0.049195
1000	500	383	190	0.448248	0.050886
1000	500	383	191	0.500000	0.051752
1000	500	384	154	0.000001	0.000000
1000	500	384	155	0.000001	0.000000
1000	500	384	156	0.000002	0.000001
1000	500	384	157	0.000003	0.000002
1000	500	384	158	0.000006	0.000003
1000	500	384	159	0.000011	0.000005
1000	500	384	160	0.000020	0.000009
1000	500	384	161	0.000036	0.000015
1000	500	384	162	0.000061	0.000026
1000	500	384	163	0.000103	0.000042
1000	500	384	164	0.000172	0.000068
1000	500	384	165	0.000280	0.000109
1000	500	384	166	0.000451	0.000171
1000	500	384	167	0.000714	0.000263
1000	500	384	168	0.001112	0.000398
1000	500	384	169	0.001705	0.000593
1000	500	384	170	0.002573	0.000868
1000	500	384	171	0.003821	0.001249
1000	500	384	172	0.005588	0.001766
1000	500	384	173	0.008044	0.002457
1000	500	384	174	0.011403	0.003359
1000	500	384	175	0.015919	0.004516
1000	500	384	176	0.021888	0.005968
1000	500	384	177	0.029643	0.007756
1000	500	384	178	0.039553	0.009909
1000	500	384	179	0.052000	0.012448
1000	500	384	180	0.067374	0.015374
1000	500	384	181	0.086043	0.018669
1000	500	384	182	0.108334	0.022291
1000	500	384	183	0.134502	0.026169
1000	500	384	184	0.164709	0.030206
1000	500	384	185	0.198991	0.034283
1000	500	384	186	0.237249	0.038258

N	n	k	x	P(x)	p(x)
1000	500	384	187	0.279228	0.041979
1000	500	384	188	0.324519	0.045291
1000	500	384	189	0.372565	0.048046
1000	500	384	190	0.422681	0.050116
1000	500	384	191	0.474082	0.051401
1000	500	384	192	0.525918	0.051836
1000	500	385	155	0.000001	0.000000
1000	500	385	156	0.000001	0.000001
1000	500	385	157	0.000003	0.000001
1000	500	385	158	0.000005	0.000002
1000	500	385	159	0.000009	0.000004
1000	500	385	160	0.000015	0.000007
1000	500	385	161	0.000027	0.000012
1000	500	385	162	0.000047	0.000020
1000	500	385	163	0.000080	0.000033
1000	500	385	164	0.000134	0.000054
1000	500	385	165	0.000221	0.000087
1000	500	385	166	0.000359	0.000137
1000	500	385	167	0.000572	0.000213
1000	500	385	168	0.000898	0.000326
1000	500	385	169	0.001386	0.000489
1000	500	385	170	0.002108	0.000721
1000	500	385	171	0.003154	0.001047
1000	500	385	172	0.004647	0.001493
1000	500	385	173	0.006740	0.002093
1000	500	385	174	0.009626	0.002885
1000	500	385	175	0.013537	0.003911
1000	500	385	176	0.018748	0.005212
1000	500	385	177	0.025577	0.006828
1000	500	385	178	0.034373	0.008796
1000	500	385	179	0.045514	0.011141
1000	500	385	180	0.059388	0.013874
1000	500	385	181	0.076375	0.016988
1000	500	385	182	0.096827	0.020451
1000	500	385	183	0.121036	0.024209
1000	500	385	184	0.149213	0.028177
1000	500	385	185	0.181461	0.032247
1000	500	385	186	0.217748	0.036287
1000	500	385	187	0.257898	0.040150
1000	500	385	188	0.301579	0.043681
1000	500	385	189	0.348308	0.046728
1000	500	385	190	0.397460	0.049152
1000	500	385	191	0.448298	0.050838
1000	500	385	192	0.500000	0.051702
1000	500	386	155	0.000001	0.000000
1000	500	386	156	0.000001	0.000000
1000	500	386	157	0.000002	0.000001
1000	500	386	158	0.000004	0.000002
1000	500	386	159	0.000006	0.000003
1000	500	386	160	0.000012	0.000005

N	n	k	x	P(x)	p(x)
1000	500	386	161	0.000021	0.000009
1000	500	386	162	0.000036	0.000015
1000	500	386	163	0.000062	0.000026
1000	500	386	164	0.000105	0.000043
1000	500	386	165	0.000174	0.000069
1000	500	386	166	0.000284	0.000110
1000	500	386	167	0.000456	0.000172
1000	500	386	168	0.000722	0.000265
1000	500	386	169	0.001123	0.000402
1000	500	386	170	0.001721	0.000597
1000	500	386	171	0.002594	0.000874
1000	500	386	172	0.003851	0.001256
1000	500	386	173	0.005627	0.001776
1000	500	386	174	0.008096	0.002469
1000	500	386	175	0.011470	0.003374
1000	500	386	176	0.016003	0.004533
1000	500	386	177	0.021990	0.005988
1000	500	386	178	0.029767	0.007777
1000	500	386	179	0.039699	0.009931
1000	500	386	180	0.052169	0.012470
1000	500	386	181	0.067564	0.015395
1000	500	386	182	0.086252	0.018688
1000	500	386	183	0.108557	0.022306
1000	500	386	184	0.134735	0.026178
1000	500	386	185	0.164944	0.030209
1000	500	386	186	0.199221	0.034277
1000	500	386	187	0.237464	0.038243
1000	500	386	188	0.279419	0.041955
1000	500	386	189	0.324678	0.045259
1000	500	386	190	0.372684	0.048007
1000	500	386	191	0.422755	0.050071
1000	500	386	192	0.474107	0.051352
1000	500	386	193	0.525893	0.051786
1000	500	387	156	0.000001	0.000000
1000	500	387	157	0.000001	0.000001
1000	500	387	158	0.000003	0.000001
1000	500	387	159	0.000005	0.000002
1000	500	387	160	0.000009	0.000004
1000	500	387	161	0.000016	0.000007
1000	500	387	162	0.000028	0.000012
1000	500	387	163	0.000048	0.000020
1000	500	387	164	0.000081	0.000034
1000	500	387	165	0.000136	0.000055
1000	500	387	166	0.000224	0.000088
1000	500	387	167	0.000363	0.000139
1000	500	387	168	0.000578	0.000215
1000	500	387	169	0.000907	0.000329
1000	500	387	170	0.001400	0.000493
1000	500	387	171	0.002126	0.000727
1000	500	387	172	0.003180	0.001053

N	n	k	x	P(x)	p(x)
1000	500	387	173	0.004681	0.001501
1000	500	387	174	0.006785	0.002104
1000	500	387	175	0.009684	0.002899
1000	500	387	176	0.013611	0.003927
1000	500	387	177	0.018840	0.005230
1000	500	387	178	0.025689	0.006848
1000	500	387	179	0.034507	0.008818
1000	500	387	180	0.045670	0.011163
1000	500	387	181	0.059565	0.013896
1000	500	387	182	0.076573	0.017008
1000	500	387	183	0.097041	0.020468
1000	500	387	184	0.121262	0.024221
1000	500	387	185	0.149446	0.028183
1000	500	387	186	0.181691	0.032246
1000	500	387	187	0.217968	0.036277
1000	500	387	188	0.258100	0.040131
1000	500	387	189	0.301753	0.043654
1000	500	387	190	0.348446	0.046693
1000	500	387	191	0.397557	0.049110
1000	500	387	192	0.448347	0.050791
1000	500	387	193	0.500000	0.051653
1000	500	388	156	0.000001	0.000000
1000	500	388	157	0.000001	0.000000
1000	500	388	158	0.000002	0.000001
1000	500	388	159	0.000004	0.000002
1000	500	388	160	0.000007	0.000003
1000	500	388	161	0.000012	0.000005
1000	500	388	162	0.000021	0.000009
1000	500	388	163	0.000037	0.000016
1000	500	388	164	0.000063	0.000026
1000	500	388	165	0.000106	0.000043
1000	500	388	166	0.000176	0.000070
1000	500	388	167	0.000287	0.000111
1000	500	388	168	0.000462	0.000174
1000	500	388	169	0.000729	0.000268
1000	500	388	170	0.001134	0.000405
1000	500	388	171	0.001736	0.000602
1000	500	388	172	0.002616	0.000880
1000	500	388	173	0.003880	0.001264
1000	500	388	174	0.005666	0.001786
1000	500	388	175	0.008147	0.002481
1000	500	388	176	0.011535	0.003388
1000	500	388	177	0.016085	0.004550
1000	500	388	178	0.022091	0.006007
1000	500	388	179	0.029889	0.007798
1000	500	388	180	0.039842	0.009953
1000	500	388	181	0.052334	0.012492
1000	500	388	182	0.067750	0.015416
1000	500	388	183	0.086456	0.018706
1000	500	388	184	0.108776	0.022320

N	n	k	x	P(x)	p(x)
1000	500	388	185	0.134963	0.026187
1000	500	388	186	0.165174	0.030211
1000	500	388	187	0.199445	0.034271
1000	500	388	188	0.237674	0.038229
1000	500	388	189	0.279606	0.041932
1000	500	388	190	0.324833	0.045227
1000	500	388	191	0.372801	0.047968
1000	500	388	192	0.422828	0.050027
1000	500	388	193	0.474132	0.051304
1000	500	388	194	0.525868	0.051737
1000	500	389	157	0.000001	0.000000
1000	500	389	158	0.000001	0.000001
1000	500	389	159	0.000003	0.000001
1000	500	389	160	0.000005	0.000002
1000	500	389	161	0.000009	0.000004
1000	500	389	162	0.000016	0.000007
1000	500	389	163	0.000028	0.000012
1000	500	389	164	0.000049	0.000020
1000	500	389	165	0.000083	0.000034
1000	500	389	166	0.000138	0.000055
1000	500	389	167	0.000227	0.000089
1000	500	389	168	0.000367	0.000140
1000	500	389	169	0.000585	0.000217
1000	500	389	170	0.000916	0.000331
1000	500	389	171	0.001413	0.000497
1000	500	389	172	0.002145	0.000732
1000	500	389	173	0.003205	0.001060
1000	500	389	174	0.004715	0.001510
1000	500	389	175	0.006830	0.002115
1000	500	389	176	0.009742	0.002912
1000	500	389	177	0.013684	0.003942
1000	500	389	178	0.018931	0.005247
1000	500	389	179	0.025799	0.006868
1000	500	389	180	0.034638	0.008839
1000	500	389	181	0.045823	0.011185
1000	500	389	182	0.059740	0.013917
1000	500	389	183	0.076767	0.017027
1000	500	389	184	0.097252	0.020485
1000	500	389	185	0.121485	0.024233
1000	500	389	186	0.149674	0.028189
1000	500	389	187	0.181918	0.032244
1000	500	389	188	0.218185	0.036267
1000	500	389	189	0.258297	0.040113
1000	500	389	190	0.301924	0.043627
1000	500	389	191	0.348582	0.046658
1000	500	389	192	0.397651	0.049069
1000	500	389	193	0.448396	0.050745
1000	500	389	194	0.500000	0.051604
1000	500	390	157	0.000001	0.000000
1000	500	390	158	0.000001	0.000000

N	n	k	x	P(x)	p(x)
1000	500	390	159	0.000002	0.000001
1000	500	390	160	0.000004	0.000002
1000	500	390	161	0.000007	0.000003
1000	500	390	162	0.000012	0.000005
1000	500	390	163	0.000021	0.000009
1000	500	390	164	0.000037	0.000016
1000	500	390	165	0.000064	0.000027
1000	500	390	166	0.000108	0.000044
1000	500	390	167	0.000178	0.000071
1000	500	390	168	0.000291	0.000112
1000	500	390	169	0.000467	0.000176
1000	500	390	170	0.000737	0.000270
1000	500	390	171	0.001145	0.000408
1000	500	390	172	0.001752	0.000606
1000	500	390	173	0.002637	0.000886
1000	500	390	174	0.003909	0.001272
1000	500	390	175	0.005705	0.001796
1000	500	390	176	0.008197	0.002493
1000	500	390	177	0.011600	0.003402
1000	500	390	178	0.016166	0.004566
1000	500	390	179	0.022191	0.006025
1000	500	390	180	0.030009	0.007818
1000	500	390	181	0.039983	0.009974
1000	500	390	182	0.052497	0.012513
1000	500	390	183	0.067933	0.015436
1000	500	390	184	0.086657	0.018724
1000	500	390	185	0.108992	0.022334
1000	500	390	186	0.135187	0.026196
1000	500	390	187	0.165400	0.030213
1000	500	390	188	0.199665	0.034265
1000	500	390	189	0.237880	0.038215
1000	500	390	190	0.279789	0.041909
1000	500	390	191	0.324986	0.045196
1000	500	390	192	0.372916	0.047930
1000	500	390	193	0.422899	0.049983
1000	500	390	194	0.474156	0.051257
1000	500	390	195	0.525844	0.051689
1000	500	391	158	0.000001	0.000000
1000	500	391	159	0.000001	0.000001
1000	500	391	160	0.000003	0.000001
1000	500	391	161	0.000005	0.000002
1000	500	391	162	0.000009	0.000004
1000	500	391	163	0.000016	0.000007
1000	500	391	164	0.000029	0.000012
1000	500	391	165	0.000049	0.000021
1000	500	391	166	0.000084	0.000034
1000	500	391	167	0.000140	0.000056
1000	500	391	168	0.000230	0.000090
1000	500	391	169	0.000371	0.000142
1000	500	391	170	0.000591	0.000219

N	n	k	x	P(x)	p(x)
1000	500	391	171	0.000925	0.000334
1000	500	391	172	0.001426	0.000501
1000	500	391	173	0.002163	0.000737
1000	500	391	174	0.003229	0.001067
1000	500	391	175	0.004748	0.001519
1000	500	391	176	0.006873	0.002125
1000	500	391	177	0.009798	0.002925
1000	500	391	178	0.013755	0.003957
1000	500	391	179	0.019020	0.005265
1000	500	391	180	0.025908	0.006887
1000	500	391	181	0.034767	0.008860
1000	500	391	182	0.045973	0.011206
1000	500	391	183	0.059911	0.013938
1000	500	391	184	0.076958	0.017046
1000	500	391	185	0.097458	0.020501
1000	500	391	186	0.121703	0.024245
1000	500	391	187	0.149897	0.028194
1000	500	391	188	0.182139	0.032242
1000	500	391	189	0.218397	0.036257
1000	500	391	190	0.258491	0.040094
1000	500	391	191	0.302091	0.043600
1000	500	391	192	0.348715	0.046624
1000	500	391	193	0.397743	0.049028
1000	500	391	194	0.448443	0.050700
1000	500	391	195	0.500000	0.051557
1000	500	392	158	0.000001	0.000000
1000	500	392	159	0.000001	0.000001
1000	500	392	160	0.000002	0.000001
1000	500	392	161	0.000004	0.000002
1000	500	392	162	0.000007	0.000003
1000	500	392	163	0.000012	0.000005
1000	500	392	164	0.000022	0.000009
1000	500	392	165	0.000038	0.000016
1000	500	392	166	0.000065	0.000027
1000	500	392	167	0.000109	0.000044
1000	500	392	168	0.000181	0.000072
1000	500	392	169	0.000294	0.000114
1000	500	392	170	0.000472	0.000178
1000	500	392	171	0.000744	0.000273
1000	500	392	172	0.001156	0.000412
1000	500	392	173	0.001767	0.000611
1000	500	392	174	0.002658	0.000892
1000	500	392	175	0.003938	0.001279
1000	500	392	176	0.005743	0.001805
1000	500	392	177	0.008247	0.002504
1000	500	392	178	0.011663	0.003416
1000	500	392	179	0.016245	0.004582
1000	500	392	180	0.022289	0.006043
1000	500	392	181	0.030126	0.007838
1000	500	392	182	0.040122	0.009995

Table for $N = 1000$, $n = 500$

N	n	k	x	P(x)	p(x)
1000	500	392	183	0.052656	0.012534
1000	500	392	184	0.068112	0.015456
1000	500	392	185	0.086854	0.018742
1000	500	392	186	0.109202	0.022348
1000	500	392	187	0.135407	0.026204
1000	500	392	188	0.165621	0.030215
1000	500	392	189	0.199881	0.034260
1000	500	392	190	0.238082	0.038201
1000	500	392	191	0.279969	0.041887
1000	500	392	192	0.325135	0.045166
1000	500	392	193	0.373028	0.047893
1000	500	392	194	0.422968	0.049940
1000	500	392	195	0.474179	0.051211
1000	500	392	196	0.525821	0.051641
1000	500	393	159	0.000001	0.000000
1000	500	393	160	0.000001	0.000001
1000	500	393	161	0.000003	0.000001
1000	500	393	162	0.000005	0.000002
1000	500	393	163	0.000009	0.000004
1000	500	393	164	0.000017	0.000007
1000	500	393	165	0.000029	0.000012
1000	500	393	166	0.000050	0.000021
1000	500	393	167	0.000085	0.000035
1000	500	393	168	0.000142	0.000057
1000	500	393	169	0.000232	0.000091
1000	500	393	170	0.000376	0.000143
1000	500	393	171	0.000597	0.000221
1000	500	393	172	0.000934	0.000337
1000	500	393	173	0.001438	0.000504
1000	500	393	174	0.002180	0.000742
1000	500	393	175	0.003254	0.001073
1000	500	393	176	0.004781	0.001527
1000	500	393	177	0.006917	0.002136
1000	500	393	178	0.009854	0.002937
1000	500	393	179	0.013826	0.003972
1000	500	393	180	0.019108	0.005282
1000	500	393	181	0.026014	0.006906
1000	500	393	182	0.034894	0.008880
1000	500	393	183	0.046121	0.011227
1000	500	393	184	0.060079	0.013958
1000	500	393	185	0.077144	0.017065
1000	500	393	186	0.097661	0.020516
1000	500	393	187	0.121917	0.024256
1000	500	393	188	0.150116	0.028200
1000	500	393	189	0.182357	0.032240
1000	500	393	190	0.218605	0.036248
1000	500	393	191	0.258681	0.040076
1000	500	393	192	0.302255	0.043574
1000	500	393	193	0.348845	0.046590
1000	500	393	194	0.397834	0.048989
1000	500	393	195	0.448490	0.050656
1000	500	393	196	0.500000	0.051510
1000	500	394	159	0.000001	0.000000
1000	500	394	160	0.000001	0.000001
1000	500	394	161	0.000002	0.000001
1000	500	394	162	0.000004	0.000002
1000	500	394	163	0.000007	0.000003
1000	500	394	164	0.000013	0.000006
1000	500	394	165	0.000022	0.000010
1000	500	394	166	0.000039	0.000016
1000	500	394	167	0.000066	0.000027
1000	500	394	168	0.000111	0.000045
1000	500	394	169	0.000183	0.000072
1000	500	394	170	0.000298	0.000115
1000	500	394	171	0.000477	0.000179
1000	500	394	172	0.000752	0.000275
1000	500	394	173	0.001167	0.000415
1000	500	394	174	0.001782	0.000615
1000	500	394	175	0.002679	0.000897
1000	500	394	176	0.003966	0.001287
1000	500	394	177	0.005780	0.001814
1000	500	394	178	0.008296	0.002516
1000	500	394	179	0.011725	0.003430
1000	500	394	180	0.016323	0.004598
1000	500	394	181	0.022385	0.006061
1000	500	394	182	0.030242	0.007857
1000	500	394	183	0.040258	0.010016
1000	500	394	184	0.052813	0.012555
1000	500	394	185	0.068288	0.015476
1000	500	394	186	0.087048	0.018759
1000	500	394	187	0.109409	0.022361
1000	500	394	188	0.135622	0.026212
1000	500	394	189	0.165838	0.030217
1000	500	394	190	0.200093	0.034254
1000	500	394	191	0.238280	0.038187
1000	500	394	192	0.280145	0.041865
1000	500	394	193	0.325281	0.045136
1000	500	394	194	0.373138	0.047856
1000	500	394	195	0.423037	0.049899
1000	500	394	196	0.474202	0.051166
1000	500	394	197	0.525798	0.051595
1000	500	395	160	0.000001	0.000000
1000	500	395	161	0.000002	0.000001
1000	500	395	162	0.000003	0.000001
1000	500	395	163	0.000005	0.000002
1000	500	395	164	0.000009	0.000004
1000	500	395	165	0.000017	0.000007
1000	500	395	166	0.000030	0.000013
1000	500	395	167	0.000051	0.000021
1000	500	395	168	0.000086	0.000035
1000	500	395	169	0.000143	0.000057
1000	500	395	170	0.000235	0.000092
1000	500	395	171	0.000380	0.000144
1000	500	395	172	0.000603	0.000223
1000	500	395	173	0.000943	0.000340
1000	500	395	174	0.001451	0.000508
1000	500	395	175	0.002198	0.000747
1000	500	395	176	0.003278	0.001080
1000	500	395	177	0.004813	0.001535
1000	500	395	178	0.006959	0.002146
1000	500	395	179	0.009909	0.002950
1000	500	395	180	0.013895	0.003987
1000	500	395	181	0.019194	0.005299
1000	500	395	182	0.026119	0.006925
1000	500	395	183	0.035018	0.008900
1000	500	395	184	0.046266	0.011247
1000	500	395	185	0.060244	0.013978
1000	500	395	186	0.077328	0.017083
1000	500	395	187	0.097859	0.020532
1000	500	395	188	0.122126	0.024267
1000	500	395	189	0.150331	0.028205
1000	500	395	190	0.182570	0.032239
1000	500	395	191	0.218808	0.036238
1000	500	395	192	0.258867	0.040059
1000	500	395	193	0.302415	0.043548
1000	500	395	194	0.348973	0.046557
1000	500	395	195	0.397922	0.048950
1000	500	395	196	0.448535	0.050613
1000	500	395	197	0.500000	0.051465
1000	500	396	160	0.000001	0.000000
1000	500	396	161	0.000001	0.000001
1000	500	396	162	0.000002	0.000001
1000	500	396	163	0.000004	0.000002
1000	500	396	164	0.000007	0.000003
1000	500	396	165	0.000013	0.000006
1000	500	396	166	0.000023	0.000010
1000	500	396	167	0.000039	0.000017
1000	500	396	168	0.000067	0.000028
1000	500	396	169	0.000112	0.000045
1000	500	396	170	0.000185	0.000073
1000	500	396	171	0.000301	0.000116
1000	500	396	172	0.000482	0.000181
1000	500	396	173	0.000759	0.000277
1000	500	396	174	0.001177	0.000418
1000	500	396	175	0.001797	0.000619
1000	500	396	176	0.002699	0.000903
1000	500	396	177	0.003993	0.001294
1000	500	396	178	0.005817	0.001823
1000	500	396	179	0.008344	0.002527
1000	500	396	180	0.011787	0.003443
1000	500	396	181	0.016400	0.004613
1000	500	396	182	0.022479	0.006079
1000	500	396	183	0.030355	0.007876
1000	500	396	184	0.040391	0.010036
1000	500	396	185	0.052966	0.012575
1000	500	396	186	0.068461	0.015495
1000	500	396	187	0.087237	0.018776
1000	500	396	188	0.109612	0.022375
1000	500	396	189	0.135832	0.026221
1000	500	396	190	0.166051	0.030218
1000	500	396	191	0.200300	0.034249
1000	500	396	192	0.238473	0.038174
1000	500	396	193	0.280317	0.041844
1000	500	396	194	0.325425	0.045107
1000	500	396	195	0.373245	0.047821
1000	500	396	196	0.423103	0.049858
1000	500	396	197	0.474225	0.051122
1000	500	396	198	0.525775	0.051550
1000	500	397	161	0.000001	0.000000
1000	500	397	162	0.000002	0.000001
1000	500	397	163	0.000003	0.000001
1000	500	397	164	0.000005	0.000002
1000	500	397	165	0.000010	0.000004
1000	500	397	166	0.000017	0.000007
1000	500	397	167	0.000030	0.000013
1000	500	397	168	0.000052	0.000022
1000	500	397	169	0.000087	0.000036
1000	500	397	170	0.000145	0.000058
1000	500	397	171	0.000238	0.000093
1000	500	397	172	0.000384	0.000146
1000	500	397	173	0.000609	0.000225
1000	500	397	174	0.000952	0.000342
1000	500	397	175	0.001463	0.000512
1000	500	397	176	0.002215	0.000752
1000	500	397	177	0.003301	0.001086
1000	500	397	178	0.004845	0.001543
1000	500	397	179	0.007001	0.002156
1000	500	397	180	0.009963	0.002962
1000	500	397	181	0.013964	0.004001
1000	500	397	182	0.019279	0.005315
1000	500	397	183	0.026222	0.006943
1000	500	397	184	0.035141	0.008919
1000	500	397	185	0.046408	0.011267
1000	500	397	186	0.060406	0.013998
1000	500	397	187	0.077507	0.017101
1000	500	397	188	0.098054	0.020547
1000	500	397	189	0.122331	0.024277
1000	500	397	190	0.150541	0.028210
1000	500	397	191	0.182778	0.032237
1000	500	397	192	0.219007	0.036229

Table for $N = 1000$, $n = 500$

N	n	k	x	P(x)	p(x)
1000	500	397	193	0.259049	0.040042
1000	500	397	194	0.302572	0.043523
1000	500	397	195	0.349097	0.046525
1000	500	397	196	0.398009	0.048912
1000	500	397	197	0.448579	0.050570
1000	500	397	198	0.500000	0.051420
1000	500	398	161	0.000001	0.000000
1000	500	398	162	0.000001	0.000001
1000	500	398	163	0.000002	0.000001
1000	500	398	164	0.000004	0.000002
1000	500	398	165	0.000007	0.000003
1000	500	398	166	0.000013	0.000006
1000	500	398	167	0.000023	0.000010
1000	500	398	168	0.000040	0.000017
1000	500	398	169	0.000068	0.000028
1000	500	398	170	0.000113	0.000046
1000	500	398	171	0.000187	0.000074
1000	500	398	172	0.000304	0.000117
1000	500	398	173	0.000487	0.000182
1000	500	398	174	0.000766	0.000280
1000	500	398	175	0.001187	0.000421
1000	500	398	176	0.001811	0.000624
1000	500	398	177	0.002720	0.000908
1000	500	398	178	0.004021	0.001301
1000	500	398	179	0.005853	0.001832
1000	500	398	180	0.008391	0.002538
1000	500	398	181	0.011847	0.003456
1000	500	398	182	0.016476	0.004629
1000	500	398	183	0.022572	0.006096
1000	500	398	184	0.030467	0.007895
1000	500	398	185	0.040522	0.010055
1000	500	398	186	0.053117	0.012595
1000	500	398	187	0.068630	0.015514
1000	500	398	188	0.087423	0.018792
1000	500	398	189	0.109810	0.022387
1000	500	398	190	0.136039	0.026228
1000	500	398	191	0.166259	0.030220
1000	500	398	192	0.200502	0.034243
1000	500	398	193	0.238663	0.038161
1000	500	398	194	0.280486	0.041823
1000	500	398	195	0.325565	0.045079
1000	500	398	196	0.373351	0.047786
1000	500	398	197	0.423169	0.049818
1000	500	398	198	0.474247	0.051079
1000	500	398	199	0.525753	0.051506
1000	500	399	162	0.000001	0.000000
1000	500	399	163	0.000002	0.000001
1000	500	399	164	0.000003	0.000001
1000	500	399	165	0.000005	0.000002
1000	500	399	166	0.000010	0.000004

N	n	k	x	P(x)	p(x)
1000	500	399	167	0.000017	0.000008
1000	500	399	168	0.000030	0.000013
1000	500	399	169	0.000052	0.000022
1000	500	399	170	0.000088	0.000036
1000	500	399	171	0.000147	0.000059
1000	500	399	172	0.000241	0.000094
1000	500	399	173	0.000388	0.000147
1000	500	399	174	0.000615	0.000227
1000	500	399	175	0.000960	0.000345
1000	500	399	176	0.001475	0.000515
1000	500	399	177	0.002232	0.000757
1000	500	399	178	0.003325	0.001093
1000	500	399	179	0.004876	0.001551
1000	500	399	180	0.007042	0.002166
1000	500	399	181	0.010016	0.002974
1000	500	399	182	0.014031	0.004015
1000	500	399	183	0.019362	0.005331
1000	500	399	184	0.026322	0.006961
1000	500	399	185	0.035261	0.008938
1000	500	399	186	0.046547	0.011287
1000	500	399	187	0.060564	0.014017
1000	500	399	188	0.077683	0.017119
1000	500	399	189	0.098245	0.020561
1000	500	399	190	0.122532	0.024288
1000	500	399	191	0.150747	0.028215
1000	500	399	192	0.182982	0.032235
1000	500	399	193	0.219202	0.036220
1000	500	399	194	0.259227	0.040025
1000	500	399	195	0.302726	0.043499
1000	500	399	196	0.349220	0.046494
1000	500	399	197	0.398094	0.048874
1000	500	399	198	0.448623	0.050529
1000	500	399	199	0.500000	0.051377
1000	500	400	162	0.000001	0.000000
1000	500	400	163	0.000001	0.000001
1000	500	400	164	0.000002	0.000001
1000	500	400	165	0.000004	0.000002
1000	500	400	166	0.000007	0.000003
1000	500	400	167	0.000013	0.000006
1000	500	400	168	0.000023	0.000010
1000	500	400	169	0.000040	0.000017
1000	500	400	170	0.000069	0.000028
1000	500	400	171	0.000115	0.000046
1000	500	400	172	0.000190	0.000075
1000	500	400	173	0.000308	0.000118
1000	500	400	174	0.000492	0.000184
1000	500	400	175	0.000774	0.000282
1000	500	400	176	0.001198	0.000424
1000	500	400	177	0.001825	0.000628
1000	500	400	178	0.002739	0.000914

N	n	k	x	P(x)	p(x)
1000	500	400	179	0.004047	0.001308
1000	500	400	180	0.005889	0.001841
1000	500	400	181	0.008437	0.002549
1000	500	400	182	0.011906	0.003469
1000	500	400	183	0.016550	0.004644
1000	500	400	184	0.022662	0.006113
1000	500	400	185	0.030576	0.007913
1000	500	400	186	0.040650	0.010075
1000	500	400	187	0.053264	0.012614
1000	500	400	188	0.068796	0.015532
1000	500	400	189	0.087605	0.018808
1000	500	400	190	0.110004	0.022400
1000	500	400	191	0.136241	0.026236
1000	500	400	192	0.166462	0.030222
1000	500	400	193	0.200700	0.034238
1000	500	400	194	0.238848	0.038148
1000	500	400	195	0.280651	0.041802
1000	500	400	196	0.325702	0.045051
1000	500	400	197	0.373453	0.047752
1000	500	400	198	0.423232	0.049779
1000	500	400	199	0.474269	0.051036
1000	500	400	200	0.525731	0.051462
1000	500	401	163	0.000001	0.000000
1000	500	401	164	0.000002	0.000001
1000	500	401	165	0.000003	0.000001
1000	500	401	166	0.000006	0.000003
1000	500	401	167	0.000010	0.000004
1000	500	401	168	0.000018	0.000008
1000	500	401	169	0.000031	0.000013
1000	500	401	170	0.000053	0.000022
1000	500	401	171	0.000089	0.000037
1000	500	401	172	0.000149	0.000059
1000	500	401	173	0.000243	0.000095
1000	500	401	174	0.000392	0.000148
1000	500	401	175	0.000621	0.000229
1000	500	401	176	0.000969	0.000348
1000	500	401	177	0.001487	0.000519
1000	500	401	178	0.002249	0.000761
1000	500	401	179	0.003348	0.001099
1000	500	401	180	0.004907	0.001559
1000	500	401	181	0.007082	0.002176
1000	500	401	182	0.010068	0.002985
1000	500	401	183	0.014096	0.004029
1000	500	401	184	0.019443	0.005347
1000	500	401	185	0.026421	0.006978
1000	500	401	186	0.035378	0.008957
1000	500	401	187	0.046684	0.011306
1000	500	401	188	0.060720	0.014036
1000	500	401	189	0.077856	0.017136
1000	500	401	190	0.098431	0.020576

N	n	k	x	P(x)	p(x)
1000	500	401	191	0.122729	0.024298
1000	500	401	192	0.150949	0.028220
1000	500	401	193	0.183182	0.032233
1000	500	401	194	0.219393	0.036211
1000	500	401	195	0.259401	0.040008
1000	500	401	196	0.302876	0.043475
1000	500	401	197	0.349339	0.046463
1000	500	401	198	0.398177	0.048838
1000	500	401	199	0.448665	0.050488
1000	500	401	200	0.500000	0.051334
1000	500	402	163	0.000001	0.000000
1000	500	402	164	0.000001	0.000001
1000	500	402	165	0.000002	0.000001
1000	500	402	166	0.000004	0.000002
1000	500	402	167	0.000007	0.000003
1000	500	402	168	0.000013	0.000006
1000	500	402	169	0.000024	0.000010
1000	500	402	170	0.000041	0.000017
1000	500	402	171	0.000069	0.000029
1000	500	402	172	0.000116	0.000047
1000	500	402	173	0.000192	0.000075
1000	500	402	174	0.000311	0.000119
1000	500	402	175	0.000497	0.000186
1000	500	402	176	0.000781	0.000284
1000	500	402	177	0.001208	0.000427
1000	500	402	178	0.001840	0.000632
1000	500	402	179	0.002759	0.000919
1000	500	402	180	0.004074	0.001315
1000	500	402	181	0.005924	0.001850
1000	500	402	182	0.008483	0.002559
1000	500	402	183	0.011964	0.003482
1000	500	402	184	0.016622	0.004658
1000	500	402	185	0.022752	0.006129
1000	500	402	186	0.030683	0.007931
1000	500	402	187	0.040776	0.010093
1000	500	402	188	0.053409	0.012633
1000	500	402	189	0.068959	0.015550
1000	500	402	190	0.087783	0.018824
1000	500	402	191	0.110195	0.022412
1000	500	402	192	0.136438	0.026244
1000	500	402	193	0.166662	0.030223
1000	500	402	194	0.200895	0.034233
1000	500	402	195	0.239030	0.038135
1000	500	402	196	0.280812	0.041782
1000	500	402	197	0.325836	0.045024
1000	500	402	198	0.373554	0.047718
1000	500	402	199	0.423295	0.049741
1000	500	402	200	0.474290	0.050995
1000	500	402	201	0.525710	0.051420
1000	500	403	164	0.000001	0.000000

Table for $N = 1000$, $n = 500$

N	n	k	x	P(x)	p(x)
1000	500	403	165	0.000002	0.000001
1000	500	403	166	0.000003	0.000001
1000	500	403	167	0.000006	0.000003
1000	500	403	168	0.000010	0.000005
1000	500	403	169	0.000018	0.000008
1000	500	403	170	0.000031	0.000013
1000	500	403	171	0.000054	0.000022
1000	500	403	172	0.000091	0.000037
1000	500	403	173	0.000150	0.000060
1000	500	403	174	0.000246	0.000095
1000	500	403	175	0.000396	0.000150
1000	500	403	176	0.000627	0.000231
1000	500	403	177	0.000977	0.000350
1000	500	403	178	0.001499	0.000522
1000	500	403	179	0.002265	0.000766
1000	500	403	180	0.003370	0.001105
1000	500	403	181	0.004937	0.001567
1000	500	403	182	0.007122	0.002185
1000	500	403	183	0.010119	0.002997
1000	500	403	184	0.014161	0.004042
1000	500	403	185	0.019523	0.005362
1000	500	403	186	0.026518	0.006995
1000	500	403	187	0.035493	0.008975
1000	500	403	188	0.046818	0.011325
1000	500	403	189	0.060872	0.014054
1000	500	403	190	0.078024	0.017153
1000	500	403	191	0.098614	0.020590
1000	500	403	192	0.122922	0.024308
1000	500	403	193	0.151146	0.028224
1000	500	403	194	0.183377	0.032232
1000	500	403	195	0.219579	0.036202
1000	500	403	196	0.259571	0.039992
1000	500	403	197	0.303023	0.043451
1000	500	403	198	0.349456	0.046433
1000	500	403	199	0.398258	0.048802
1000	500	403	200	0.448707	0.050449
1000	500	403	201	0.500000	0.051293
1000	500	404	164	0.000001	0.000000
1000	500	404	165	0.000001	0.000001
1000	500	404	166	0.000002	0.000001
1000	500	404	167	0.000004	0.000002
1000	500	404	168	0.000008	0.000003
1000	500	404	169	0.000014	0.000006
1000	500	404	170	0.000024	0.000010
1000	500	404	171	0.000041	0.000017
1000	500	404	172	0.000070	0.000029
1000	500	404	173	0.000118	0.000047
1000	500	404	174	0.000194	0.000076
1000	500	404	175	0.000314	0.000120
1000	500	404	176	0.000501	0.000187

N	n	k	x	P(x)	p(x)
1000	500	404	177	0.000787	0.000286
1000	500	404	178	0.001218	0.000430
1000	500	404	179	0.001853	0.000636
1000	500	404	180	0.002778	0.000924
1000	500	404	181	0.004100	0.001322
1000	500	404	182	0.005958	0.001858
1000	500	404	183	0.008527	0.002569
1000	500	404	184	0.012021	0.003494
1000	500	404	185	0.016694	0.004672
1000	500	404	186	0.022839	0.006145
1000	500	404	187	0.030788	0.007949
1000	500	404	188	0.040900	0.010112
1000	500	404	189	0.053551	0.012651
1000	500	404	190	0.069118	0.015567
1000	500	404	191	0.087957	0.018839
1000	500	404	192	0.110381	0.022424
1000	500	404	193	0.136632	0.026251
1000	500	404	194	0.166857	0.030225
1000	500	404	195	0.201084	0.034228
1000	500	404	196	0.239207	0.038123
1000	500	404	197	0.280970	0.041763
1000	500	404	198	0.325967	0.044997
1000	500	404	199	0.373652	0.047686
1000	500	404	200	0.423356	0.049703
1000	500	404	201	0.474311	0.050955
1000	500	404	202	0.525689	0.051379
1000	500	405	165	0.000001	0.000000
1000	500	405	166	0.000002	0.000001
1000	500	405	167	0.000003	0.000001
1000	500	405	168	0.000006	0.000003
1000	500	405	169	0.000010	0.000005
1000	500	405	170	0.000018	0.000008
1000	500	405	171	0.000032	0.000014
1000	500	405	172	0.000054	0.000023
1000	500	405	173	0.000092	0.000037
1000	500	405	174	0.000152	0.000060
1000	500	405	175	0.000249	0.000096
1000	500	405	176	0.000400	0.000151
1000	500	405	177	0.000632	0.000233
1000	500	405	178	0.000985	0.000353
1000	500	405	179	0.001511	0.000526
1000	500	405	180	0.002282	0.000771
1000	500	405	181	0.003392	0.001111
1000	500	405	182	0.004966	0.001574
1000	500	405	183	0.007161	0.002194
1000	500	405	184	0.010169	0.003008
1000	500	405	185	0.014224	0.004055
1000	500	405	186	0.019601	0.005377
1000	500	405	187	0.026613	0.007012
1000	500	405	188	0.035606	0.008993

N	n	k	x	P(x)	p(x)
1000	500	405	189	0.046949	0.011343
1000	500	405	190	0.061021	0.014072
1000	500	405	191	0.078190	0.017169
1000	500	405	192	0.098793	0.020603
1000	500	405	193	0.123110	0.024317
1000	500	405	194	0.151338	0.028229
1000	500	405	195	0.183568	0.032230
1000	500	405	196	0.219762	0.036193
1000	500	405	197	0.259738	0.039976
1000	500	405	198	0.303166	0.043428
1000	500	405	199	0.349570	0.046403
1000	500	405	200	0.398337	0.048768
1000	500	405	201	0.448748	0.050410
1000	500	405	202	0.500000	0.051252
1000	500	406	165	0.000001	0.000000
1000	500	406	166	0.000001	0.000001
1000	500	406	167	0.000002	0.000001
1000	500	406	168	0.000004	0.000002
1000	500	406	169	0.000008	0.000003
1000	500	406	170	0.000014	0.000006
1000	500	406	171	0.000024	0.000010
1000	500	406	172	0.000042	0.000018
1000	500	406	173	0.000071	0.000029
1000	500	406	174	0.000119	0.000048
1000	500	406	175	0.000196	0.000077
1000	500	406	176	0.000317	0.000121
1000	500	406	177	0.000506	0.000189
1000	500	406	178	0.000794	0.000288
1000	500	406	179	0.001227	0.000433
1000	500	406	180	0.001867	0.000640
1000	500	406	181	0.002797	0.000930
1000	500	406	182	0.004125	0.001328
1000	500	406	183	0.005992	0.001867
1000	500	406	184	0.008571	0.002580
1000	500	406	185	0.012077	0.003506
1000	500	406	186	0.016764	0.004686
1000	500	406	187	0.022925	0.006161
1000	500	406	188	0.030891	0.007966
1000	500	406	189	0.041020	0.010130
1000	500	406	190	0.053689	0.012669
1000	500	406	191	0.069273	0.015584
1000	500	406	192	0.088128	0.018854
1000	500	406	193	0.110563	0.022435
1000	500	406	194	0.136821	0.026258
1000	500	406	195	0.167047	0.030226
1000	500	406	196	0.201270	0.034223
1000	500	406	197	0.239380	0.038111
1000	500	406	198	0.281124	0.041743
1000	500	406	199	0.326095	0.044971
1000	500	406	200	0.373749	0.047654

N	n	k	x	P(x)	p(x)
1000	500	406	201	0.423415	0.049667
1000	500	406	202	0.474331	0.050915
1000	500	406	203	0.525669	0.051338
1000	500	407	166	0.000001	0.000000
1000	500	407	167	0.000002	0.000001
1000	500	407	168	0.000003	0.000001
1000	500	407	169	0.000006	0.000003
1000	500	407	170	0.000010	0.000005
1000	500	407	171	0.000018	0.000008
1000	500	407	172	0.000032	0.000014
1000	500	407	173	0.000055	0.000023
1000	500	407	174	0.000093	0.000038
1000	500	407	175	0.000154	0.000061
1000	500	407	176	0.000251	0.000097
1000	500	407	177	0.000403	0.000152
1000	500	407	178	0.000638	0.000235
1000	500	407	179	0.000993	0.000355
1000	500	407	180	0.001522	0.000529
1000	500	407	181	0.002298	0.000775
1000	500	407	182	0.003414	0.001116
1000	500	407	183	0.004996	0.001582
1000	500	407	184	0.007199	0.002203
1000	500	407	185	0.010218	0.003019
1000	500	407	186	0.014286	0.004068
1000	500	407	187	0.019678	0.005392
1000	500	407	188	0.026707	0.007028
1000	500	407	189	0.035717	0.009010
1000	500	407	190	0.047078	0.011361
1000	500	407	191	0.061167	0.014089
1000	500	407	192	0.078351	0.017185
1000	500	407	193	0.098968	0.020616
1000	500	407	194	0.123294	0.024327
1000	500	407	195	0.151527	0.028233
1000	500	407	196	0.183755	0.032228
1000	500	407	197	0.219940	0.036185
1000	500	407	198	0.259901	0.039961
1000	500	407	199	0.303306	0.043406
1000	500	407	200	0.349681	0.046375
1000	500	407	201	0.398415	0.048734
1000	500	407	202	0.448787	0.050373
1000	500	407	203	0.500000	0.051213
1000	500	408	166	0.000001	0.000000
1000	500	408	167	0.000001	0.000001
1000	500	408	168	0.000002	0.000001
1000	500	408	169	0.000004	0.000002
1000	500	408	170	0.000008	0.000004
1000	500	408	171	0.000014	0.000006
1000	500	408	172	0.000025	0.000011
1000	500	408	173	0.000042	0.000018
1000	500	408	174	0.000072	0.000030

Table for $N = 1000$, $n = 500$

N	n	k	x	P(x)	p(x)
1000	500	408	175	0.000120	0.000048
1000	500	408	176	0.000198	0.000078
1000	500	408	177	0.000320	0.000122
1000	500	408	178	0.000511	0.000190
1000	500	408	179	0.000801	0.000290
1000	500	408	180	0.001237	0.000436
1000	500	408	181	0.001881	0.000644
1000	500	408	182	0.002815	0.000935
1000	500	408	183	0.004150	0.001335
1000	500	408	184	0.006025	0.001875
1000	500	408	185	0.008614	0.002590
1000	500	408	186	0.012132	0.003518
1000	500	408	187	0.016832	0.004700
1000	500	408	188	0.023009	0.006176
1000	500	408	189	0.030991	0.007983
1000	500	408	190	0.041138	0.010147
1000	500	408	191	0.053825	0.012687
1000	500	408	192	0.069426	0.015601
1000	500	408	193	0.088294	0.018869
1000	500	408	194	0.110741	0.022447
1000	500	408	195	0.137006	0.026265
1000	500	408	196	0.167233	0.030228
1000	500	408	197	0.201451	0.034218
1000	500	408	198	0.239550	0.038099
1000	500	408	199	0.281274	0.041724
1000	500	408	200	0.326220	0.044946
1000	500	408	201	0.373843	0.047622
1000	500	408	202	0.423474	0.049631
1000	500	408	203	0.474350	0.050877
1000	500	408	204	0.525649	0.051299
1000	500	409	167	0.000001	0.000000
1000	500	409	168	0.000002	0.000001
1000	500	409	169	0.000003	0.000001
1000	500	409	170	0.000006	0.000003
1000	500	409	171	0.000011	0.000005
1000	500	409	172	0.000019	0.000008
1000	500	409	173	0.000033	0.000014
1000	500	409	174	0.000056	0.000023
1000	500	409	175	0.000094	0.000038
1000	500	409	176	0.000155	0.000062
1000	500	409	177	0.000254	0.000098
1000	500	409	178	0.000407	0.000154
1000	500	409	179	0.000644	0.000236
1000	500	409	180	0.001001	0.000358
1000	500	409	181	0.001534	0.000532
1000	500	409	182	0.002313	0.000779
1000	500	409	183	0.003435	0.001122
1000	500	409	184	0.005024	0.001589
1000	500	409	185	0.007236	0.002212
1000	500	409	186	0.010266	0.003030

N	n	k	x	P(x)	p(x)
1000	500	409	187	0.014347	0.004081
1000	500	409	188	0.019754	0.005406
1000	500	409	189	0.026798	0.007044
1000	500	409	190	0.035825	0.009027
1000	500	409	191	0.047203	0.011378
1000	500	409	192	0.061309	0.014106
1000	500	409	193	0.078509	0.017200
1000	500	409	194	0.099139	0.020629
1000	500	409	195	0.123474	0.024336
1000	500	409	196	0.151711	0.028237
1000	500	409	197	0.183938	0.032226
1000	500	409	198	0.220114	0.036177
1000	500	409	199	0.260060	0.039945
1000	500	409	200	0.303443	0.043384
1000	500	409	201	0.349790	0.046347
1000	500	409	202	0.398490	0.048700
1000	500	409	203	0.448826	0.050336
1000	500	409	204	0.500000	0.051174
1000	500	410	167	0.000001	0.000000
1000	500	410	168	0.000001	0.000001
1000	500	410	169	0.000002	0.000001
1000	500	410	170	0.000004	0.000002
1000	500	410	171	0.000008	0.000004
1000	500	410	172	0.000014	0.000006
1000	500	410	173	0.000025	0.000011
1000	500	410	174	0.000043	0.000018
1000	500	410	175	0.000073	0.000030
1000	500	410	176	0.000122	0.000049
1000	500	410	177	0.000200	0.000078
1000	500	410	178	0.000323	0.000124
1000	500	410	179	0.000515	0.000192
1000	500	410	180	0.000808	0.000292
1000	500	410	181	0.001246	0.000439
1000	500	410	182	0.001894	0.000647
1000	500	410	183	0.002833	0.000940
1000	500	410	184	0.004175	0.001341
1000	500	410	185	0.006057	0.001883
1000	500	410	186	0.008657	0.002599
1000	500	410	187	0.012186	0.003529
1000	500	410	188	0.016900	0.004714
1000	500	410	189	0.023091	0.006191
1000	500	410	190	0.031090	0.007999
1000	500	410	191	0.041254	0.010164
1000	500	410	192	0.053958	0.012704
1000	500	410	193	0.069575	0.015617
1000	500	410	194	0.088458	0.018883
1000	500	410	195	0.110915	0.022458
1000	500	410	196	0.137187	0.026271
1000	500	410	197	0.167415	0.030229
1000	500	410	198	0.201628	0.034213

N	n	k	x	P(x)	p(x)
1000	500	410	199	0.239715	0.038087
1000	500	410	200	0.281421	0.041706
1000	500	410	201	0.326342	0.044921
1000	500	410	202	0.373934	0.047592
1000	500	410	203	0.423531	0.049596
1000	500	410	204	0.474370	0.050839
1000	500	410	205	0.525630	0.051260
1000	500	411	168	0.000001	0.000000
1000	500	411	169	0.000002	0.000001
1000	500	411	170	0.000003	0.000002
1000	500	411	171	0.000006	0.000003
1000	500	411	172	0.000011	0.000005
1000	500	411	173	0.000019	0.000008
1000	500	411	174	0.000033	0.000014
1000	500	411	175	0.000056	0.000023
1000	500	411	176	0.000095	0.000038
1000	500	411	177	0.000157	0.000062
1000	500	411	178	0.000256	0.000099
1000	500	411	179	0.000411	0.000155
1000	500	411	180	0.000649	0.000238
1000	500	411	181	0.001009	0.000360
1000	500	411	182	0.001545	0.000536
1000	500	411	183	0.002329	0.000784
1000	500	411	184	0.003456	0.001128
1000	500	411	185	0.005052	0.001596
1000	500	411	186	0.007273	0.002221
1000	500	411	187	0.010314	0.003040
1000	500	411	188	0.014407	0.004093
1000	500	411	189	0.019827	0.005420
1000	500	411	190	0.026887	0.007060
1000	500	411	191	0.035931	0.009044
1000	500	411	192	0.047326	0.011395
1000	500	411	193	0.061449	0.014123
1000	500	411	194	0.078664	0.017215
1000	500	411	195	0.099306	0.020642
1000	500	411	196	0.123650	0.024344
1000	500	411	197	0.151891	0.028241
1000	500	411	198	0.184116	0.032225
1000	500	411	199	0.220284	0.036168
1000	500	411	200	0.260215	0.039930
1000	500	411	201	0.303577	0.043362
1000	500	411	202	0.349896	0.046319
1000	500	411	203	0.398564	0.048668
1000	500	411	204	0.448864	0.050300
1000	500	411	205	0.500000	0.051136
1000	500	412	168	0.000001	0.000000
1000	500	412	169	0.000001	0.000001
1000	500	412	170	0.000002	0.000001
1000	500	412	171	0.000004	0.000002
1000	500	412	172	0.000008	0.000004

N	n	k	x	P(x)	p(x)
1000	500	412	173	0.000014	0.000006
1000	500	412	174	0.000025	0.000011
1000	500	412	175	0.000043	0.000018
1000	500	412	176	0.000074	0.000030
1000	500	412	177	0.000123	0.000049
1000	500	412	178	0.000202	0.000079
1000	500	412	179	0.000327	0.000125
1000	500	412	180	0.000520	0.000193
1000	500	412	181	0.000814	0.000294
1000	500	412	182	0.001256	0.000442
1000	500	412	183	0.001907	0.000651
1000	500	412	184	0.002851	0.000945
1000	500	412	185	0.004199	0.001347
1000	500	412	186	0.006089	0.001890
1000	500	412	187	0.008698	0.002609
1000	500	412	188	0.012239	0.003541
1000	500	412	189	0.016965	0.004727
1000	500	412	190	0.023171	0.006206
1000	500	412	191	0.031186	0.008015
1000	500	412	192	0.041367	0.010181
1000	500	412	193	0.054088	0.012720
1000	500	412	194	0.069720	0.015633
1000	500	412	195	0.088617	0.018897
1000	500	412	196	0.111085	0.022468
1000	500	412	197	0.137363	0.026278
1000	500	412	198	0.167593	0.030230
1000	500	412	199	0.201801	0.034208
1000	500	412	200	0.239877	0.038076
1000	500	412	201	0.281565	0.041688
1000	500	412	202	0.326461	0.044897
1000	500	412	203	0.374024	0.047562
1000	500	412	204	0.423586	0.049562
1000	500	412	205	0.474389	0.050803
1000	500	412	206	0.525611	0.051223
1000	500	413	169	0.000001	0.000000
1000	500	413	170	0.000002	0.000001
1000	500	413	171	0.000003	0.000002
1000	500	413	172	0.000006	0.000003
1000	500	413	173	0.000011	0.000005
1000	500	413	174	0.000019	0.000008
1000	500	413	175	0.000033	0.000014
1000	500	413	176	0.000057	0.000024
1000	500	413	177	0.000096	0.000039
1000	500	413	178	0.000159	0.000063
1000	500	413	179	0.000259	0.000100
1000	500	413	180	0.000415	0.000156
1000	500	413	181	0.000654	0.000240
1000	500	413	182	0.001017	0.000362
1000	500	413	183	0.001556	0.000539
1000	500	413	184	0.002344	0.000788

Table for $N = 1000$, $n = 500$

N	n	k	x	P(x)	p(x)
1000	500	413	185	0.003477	0.001133
1000	500	413	186	0.005080	0.001603
1000	500	413	187	0.007309	0.002230
1000	500	413	188	0.010360	0.003051
1000	500	413	189	0.014465	0.004105
1000	500	413	190	0.019899	0.005434
1000	500	413	191	0.026974	0.007075
1000	500	413	192	0.036034	0.009060
1000	500	413	193	0.047446	0.011412
1000	500	413	194	0.061585	0.014139
1000	500	413	195	0.078815	0.017230
1000	500	413	196	0.099469	0.020654
1000	500	413	197	0.123822	0.024353
1000	500	413	198	0.152067	0.028245
1000	500	413	199	0.184290	0.032223
1000	500	413	200	0.220450	0.036160
1000	500	413	201	0.260366	0.039916
1000	500	413	202	0.303708	0.043341
1000	500	413	203	0.350000	0.046292
1000	500	413	204	0.398636	0.048636
1000	500	413	205	0.448901	0.050265
1000	500	413	206	0.500000	0.051099
1000	500	414	169	0.000001	0.000000
1000	500	414	170	0.000001	0.000001
1000	500	414	171	0.000002	0.000001
1000	500	414	172	0.000005	0.000002
1000	500	414	173	0.000008	0.000004
1000	500	414	174	0.000015	0.000006
1000	500	414	175	0.000026	0.000011
1000	500	414	176	0.000044	0.000018
1000	500	414	177	0.000075	0.000031
1000	500	414	178	0.000124	0.000050
1000	500	414	179	0.000204	0.000080
1000	500	414	180	0.000329	0.000126
1000	500	414	181	0.000524	0.000195
1000	500	414	182	0.000820	0.000296
1000	500	414	183	0.001265	0.000444
1000	500	414	184	0.001919	0.000655
1000	500	414	185	0.002869	0.000949
1000	500	414	186	0.004222	0.001353
1000	500	414	187	0.006120	0.001898
1000	500	414	188	0.008738	0.002618
1000	500	414	189	0.012290	0.003552
1000	500	414	190	0.017030	0.004739
1000	500	414	191	0.023250	0.006220
1000	500	414	192	0.031281	0.008030
1000	500	414	193	0.041478	0.010197
1000	500	414	194	0.054214	0.012737
1000	500	414	195	0.069862	0.015648
1000	500	414	196	0.088773	0.018910

N	n	k	x	P(x)	p(x)
1000	500	414	197	0.111251	0.022479
1000	500	414	198	0.137535	0.026284
1000	500	414	199	0.167767	0.030231
1000	500	414	200	0.201970	0.034203
1000	500	414	201	0.240034	0.038065
1000	500	414	202	0.281705	0.041670
1000	500	414	203	0.326578	0.044873
1000	500	414	204	0.374111	0.047533
1000	500	414	205	0.423640	0.049529
1000	500	414	206	0.474407	0.050767
1000	500	414	207	0.525593	0.051186
1000	500	415	169	0.000001	0.000000
1000	500	415	170	0.000001	0.000000
1000	500	415	171	0.000002	0.000001
1000	500	415	172	0.000003	0.000002
1000	500	415	173	0.000006	0.000003
1000	500	415	174	0.000011	0.000005
1000	500	415	175	0.000019	0.000008
1000	500	415	176	0.000034	0.000014
1000	500	415	177	0.000058	0.000024
1000	500	415	178	0.000097	0.000039
1000	500	415	179	0.000160	0.000063
1000	500	415	180	0.000261	0.000101
1000	500	415	181	0.000418	0.000157
1000	500	415	182	0.000660	0.000241
1000	500	415	183	0.001024	0.000365
1000	500	415	184	0.001566	0.000542
1000	500	415	185	0.002358	0.000792
1000	500	415	186	0.003497	0.001139
1000	500	415	187	0.005107	0.001610
1000	500	415	188	0.007344	0.002238
1000	500	415	189	0.010405	0.003061
1000	500	415	190	0.014522	0.004117
1000	500	415	191	0.019970	0.005448
1000	500	415	192	0.027060	0.007090
1000	500	415	193	0.036136	0.009076
1000	500	415	194	0.047564	0.011428
1000	500	415	195	0.061718	0.014154
1000	500	415	196	0.078963	0.017245
1000	500	415	197	0.099628	0.020666
1000	500	415	198	0.123990	0.024361
1000	500	415	199	0.152238	0.028249
1000	500	415	200	0.184460	0.032221
1000	500	415	201	0.220612	0.036153
1000	500	415	202	0.260514	0.039902
1000	500	415	203	0.303835	0.043321
1000	500	415	204	0.350101	0.046266
1000	500	415	205	0.398706	0.048605
1000	500	415	206	0.448937	0.050230
1000	500	415	207	0.500000	0.051063

N	n	k	x	P(x)	p(x)
1000	500	416	170	0.000001	0.000000
1000	500	416	171	0.000001	0.000001
1000	500	416	172	0.000003	0.000001
1000	500	416	173	0.000005	0.000002
1000	500	416	174	0.000008	0.000004
1000	500	416	175	0.000015	0.000006
1000	500	416	176	0.000026	0.000011
1000	500	416	177	0.000044	0.000019
1000	500	416	178	0.000075	0.000031
1000	500	416	179	0.000126	0.000050
1000	500	416	180	0.000206	0.000080
1000	500	416	181	0.000332	0.000127
1000	500	416	182	0.000528	0.000196
1000	500	416	183	0.000827	0.000298
1000	500	416	184	0.001274	0.000447
1000	500	416	185	0.001932	0.000658
1000	500	416	186	0.002886	0.000954
1000	500	416	187	0.004245	0.001359
1000	500	416	188	0.006151	0.001906
1000	500	416	189	0.008778	0.002627
1000	500	416	190	0.012341	0.003563
1000	500	416	191	0.017093	0.004752
1000	500	416	192	0.023327	0.006234
1000	500	416	193	0.031373	0.008046
1000	500	416	194	0.041586	0.010213
1000	500	416	195	0.054338	0.012752
1000	500	416	196	0.070001	0.015663
1000	500	416	197	0.088925	0.018923
1000	500	416	198	0.111413	0.022489
1000	500	416	199	0.137704	0.026290
1000	500	416	200	0.167936	0.030232
1000	500	416	201	0.202134	0.034198
1000	500	416	202	0.240188	0.038054
1000	500	416	203	0.281841	0.041653
1000	500	416	204	0.326691	0.044850
1000	500	416	205	0.374196	0.047505
1000	500	416	206	0.423693	0.049497
1000	500	416	207	0.474425	0.050732
1000	500	416	208	0.525575	0.051150
1000	500	417	170	0.000001	0.000000
1000	500	417	171	0.000001	0.000000
1000	500	417	172	0.000002	0.000001
1000	500	417	173	0.000003	0.000002
1000	500	417	174	0.000006	0.000003
1000	500	417	175	0.000011	0.000005
1000	500	417	176	0.000020	0.000009
1000	500	417	177	0.000034	0.000014
1000	500	417	178	0.000058	0.000024
1000	500	417	179	0.000098	0.000040
1000	500	417	180	0.000162	0.000064

N	n	k	x	P(x)	p(x)
1000	500	417	181	0.000263	0.000101
1000	500	417	182	0.000422	0.000158
1000	500	417	183	0.000665	0.000243
1000	500	417	184	0.001032	0.000367
1000	500	417	185	0.001577	0.000545
1000	500	417	186	0.002373	0.000796
1000	500	417	187	0.003517	0.001144
1000	500	417	188	0.005133	0.001616
1000	500	417	189	0.007379	0.002246
1000	500	417	190	0.010450	0.003071
1000	500	417	191	0.014578	0.004129
1000	500	417	192	0.020039	0.005461
1000	500	417	193	0.027143	0.007104
1000	500	417	194	0.036234	0.009091
1000	500	417	195	0.047678	0.011444
1000	500	417	196	0.061848	0.014170
1000	500	417	197	0.079106	0.017259
1000	500	417	198	0.099784	0.020677
1000	500	417	199	0.124153	0.024369
1000	500	417	200	0.152406	0.028252
1000	500	417	201	0.184625	0.032220
1000	500	417	202	0.220770	0.036145
1000	500	417	203	0.260658	0.039888
1000	500	417	204	0.303959	0.043301
1000	500	417	205	0.350200	0.046241
1000	500	417	206	0.398775	0.048575
1000	500	417	207	0.448972	0.050197
1000	500	417	208	0.500000	0.051028
1000	500	418	171	0.000001	0.000000
1000	500	418	172	0.000001	0.000001
1000	500	418	173	0.000003	0.000001
1000	500	418	174	0.000005	0.000002
1000	500	418	175	0.000008	0.000004
1000	500	418	176	0.000015	0.000007
1000	500	418	177	0.000026	0.000011
1000	500	418	178	0.000045	0.000019
1000	500	418	179	0.000076	0.000031
1000	500	418	180	0.000127	0.000051
1000	500	418	181	0.000208	0.000081
1000	500	418	182	0.000335	0.000127
1000	500	418	183	0.000533	0.000197
1000	500	418	184	0.000833	0.000300
1000	500	418	185	0.001282	0.000450
1000	500	418	186	0.001944	0.000662
1000	500	418	187	0.002903	0.000958
1000	500	418	188	0.004268	0.001365
1000	500	418	189	0.006181	0.001913
1000	500	418	190	0.008817	0.002636
1000	500	418	191	0.012390	0.003573
1000	500	418	192	0.017154	0.004764

Table for $N = 1000$, $n = 500$

N	n	k	x	P(x)	p(x)
1000	500	418	193	0.023402	0.006248
1000	500	418	194	0.031463	0.008060
1000	500	418	195	0.041691	0.010229
1000	500	418	196	0.054459	0.012768
1000	500	418	197	0.070137	0.015677
1000	500	418	198	0.089073	0.018936
1000	500	418	199	0.111572	0.022499
1000	500	418	200	0.137868	0.026296
1000	500	418	201	0.168101	0.030233
1000	500	418	202	0.202295	0.034194
1000	500	418	203	0.240338	0.038043
1000	500	418	204	0.281974	0.041637
1000	500	418	205	0.326802	0.044827
1000	500	418	206	0.374279	0.047477
1000	500	418	207	0.423744	0.049465
1000	500	418	208	0.474442	0.050698
1000	500	418	209	0.525558	0.051115
1000	500	419	171	0.000001	0.000000
1000	500	419	172	0.000001	0.000000
1000	500	419	173	0.000002	0.000001
1000	500	419	174	0.000003	0.000002
1000	500	419	175	0.000006	0.000003
1000	500	419	176	0.000011	0.000005
1000	500	419	177	0.000020	0.000009
1000	500	419	178	0.000035	0.000015
1000	500	419	179	0.000059	0.000024
1000	500	419	180	0.000099	0.000040
1000	500	419	181	0.000163	0.000064
1000	500	419	182	0.000266	0.000102
1000	500	419	183	0.000425	0.000159
1000	500	419	184	0.000670	0.000245
1000	500	419	185	0.001039	0.000369
1000	500	419	186	0.001587	0.000548
1000	500	419	187	0.002387	0.000800
1000	500	419	188	0.003536	0.001149
1000	500	419	189	0.005159	0.001623
1000	500	419	190	0.007413	0.002254
1000	500	419	191	0.010493	0.003080
1000	500	419	192	0.014633	0.004140
1000	500	419	193	0.020106	0.005474
1000	500	419	194	0.027225	0.007118
1000	500	419	195	0.036331	0.009106
1000	500	419	196	0.047790	0.011459
1000	500	419	197	0.061975	0.014185
1000	500	419	198	0.079247	0.017272
1000	500	419	199	0.099936	0.020689
1000	500	419	200	0.124313	0.024377
1000	500	419	201	0.152569	0.028256
1000	500	419	202	0.184786	0.032218
1000	500	419	203	0.220924	0.036137

N	n	k	x	P(x)	p(x)
1000	500	419	204	0.260798	0.039874
1000	500	419	205	0.304080	0.043282
1000	500	419	206	0.350296	0.046216
1000	500	419	207	0.398842	0.048546
1000	500	419	208	0.449006	0.050164
1000	500	419	209	0.500000	0.050994
1000	500	420	172	0.000001	0.000000
1000	500	420	173	0.000001	0.000001
1000	500	420	174	0.000003	0.000001
1000	500	420	175	0.000005	0.000002
1000	500	420	176	0.000009	0.000004
1000	500	420	177	0.000015	0.000007
1000	500	420	178	0.000026	0.000011
1000	500	420	179	0.000045	0.000019
1000	500	420	180	0.000077	0.000031
1000	500	420	181	0.000128	0.000051
1000	500	420	182	0.000210	0.000082
1000	500	420	183	0.000338	0.000128
1000	500	420	184	0.000537	0.000199
1000	500	420	185	0.000839	0.000302
1000	500	420	186	0.001291	0.000452
1000	500	420	187	0.001956	0.000665
1000	500	420	188	0.002919	0.000963
1000	500	420	189	0.004290	0.001371
1000	500	420	190	0.006210	0.001920
1000	500	420	191	0.008855	0.002645
1000	500	420	192	0.012438	0.003583
1000	500	420	193	0.017214	0.004776
1000	500	420	194	0.023476	0.006261
1000	500	420	195	0.031551	0.008075
1000	500	420	196	0.041794	0.010244
1000	500	420	197	0.054577	0.012783
1000	500	420	198	0.070269	0.015692
1000	500	420	199	0.089217	0.018948
1000	500	420	200	0.111726	0.022508
1000	500	420	201	0.138027	0.026302
1000	500	420	202	0.168262	0.030234
1000	500	420	203	0.202451	0.034189
1000	500	420	204	0.240484	0.038033
1000	500	420	205	0.282104	0.041620
1000	500	420	206	0.326909	0.044805
1000	500	420	207	0.374360	0.047450
1000	500	420	208	0.423794	0.049435
1000	500	420	209	0.474459	0.050665
1000	500	420	210	0.525541	0.051082
1000	500	421	172	0.000001	0.000000
1000	500	421	173	0.000001	0.000000
1000	500	421	174	0.000002	0.000001
1000	500	421	175	0.000004	0.000002
1000	500	421	176	0.000006	0.000003

N	n	k	x	P(x)	p(x)
1000	500	421	177	0.000011	0.000005
1000	500	421	178	0.000020	0.000009
1000	500	421	179	0.000035	0.000015
1000	500	421	180	0.000060	0.000025
1000	500	421	181	0.000100	0.000040
1000	500	421	182	0.000165	0.000065
1000	500	421	183	0.000268	0.000103
1000	500	421	184	0.000428	0.000161
1000	500	421	185	0.000675	0.000246
1000	500	421	186	0.001046	0.000371
1000	500	421	187	0.001597	0.000551
1000	500	421	188	0.002401	0.000804
1000	500	421	189	0.003555	0.001154
1000	500	421	190	0.005184	0.001629
1000	500	421	191	0.007446	0.002262
1000	500	421	192	0.010535	0.003089
1000	500	421	193	0.014686	0.004151
1000	500	421	194	0.020172	0.005486
1000	500	421	195	0.027304	0.007132
1000	500	421	196	0.036425	0.009121
1000	500	421	197	0.047899	0.011474
1000	500	421	198	0.062098	0.014199
1000	500	421	199	0.079384	0.017285
1000	500	421	200	0.100083	0.020700
1000	500	421	201	0.124468	0.024385
1000	500	421	202	0.152727	0.028259
1000	500	421	203	0.184944	0.032216
1000	500	421	204	0.221074	0.036130
1000	500	421	205	0.260935	0.039861
1000	500	421	206	0.304198	0.043263
1000	500	421	207	0.350389	0.046192
1000	500	421	208	0.398907	0.048518
1000	500	421	209	0.449039	0.050133
1000	500	421	210	0.500000	0.050961
1000	500	422	173	0.000001	0.000000
1000	500	422	174	0.000001	0.000001
1000	500	422	175	0.000003	0.000001
1000	500	422	176	0.000005	0.000002
1000	500	422	177	0.000009	0.000004
1000	500	422	178	0.000015	0.000007
1000	500	422	179	0.000027	0.000011
1000	500	422	180	0.000046	0.000019
1000	500	422	181	0.000078	0.000032
1000	500	422	182	0.000129	0.000052
1000	500	422	183	0.000211	0.000082
1000	500	422	184	0.000341	0.000129
1000	500	422	185	0.000541	0.000200
1000	500	422	186	0.000845	0.000304
1000	500	422	187	0.001299	0.000455
1000	500	422	188	0.001968	0.000669

N	n	k	x	P(x)	p(x)
1000	500	422	189	0.002935	0.000967
1000	500	422	190	0.004312	0.001377
1000	500	422	191	0.006239	0.001927
1000	500	422	192	0.008892	0.002653
1000	500	422	193	0.012485	0.003593
1000	500	422	194	0.017273	0.004788
1000	500	422	195	0.023547	0.006274
1000	500	422	196	0.031636	0.008089
1000	500	422	197	0.041895	0.010258
1000	500	422	198	0.054692	0.012798
1000	500	422	199	0.070398	0.015705
1000	500	422	200	0.089358	0.018960
1000	500	422	201	0.111876	0.022518
1000	500	422	202	0.138183	0.026307
1000	500	422	203	0.168418	0.030235
1000	500	422	204	0.202603	0.034185
1000	500	422	205	0.240626	0.038023
1000	500	422	206	0.282230	0.041604
1000	500	422	207	0.327014	0.044784
1000	500	422	208	0.374438	0.047424
1000	500	422	209	0.423843	0.049405
1000	500	422	210	0.474476	0.050633
1000	500	422	211	0.525524	0.051049
1000	500	423	173	0.000001	0.000000
1000	500	423	174	0.000001	0.000000
1000	500	423	175	0.000002	0.000001
1000	500	423	176	0.000004	0.000002
1000	500	423	177	0.000007	0.000003
1000	500	423	178	0.000012	0.000005
1000	500	423	179	0.000020	0.000009
1000	500	423	180	0.000035	0.000015
1000	500	423	181	0.000060	0.000025
1000	500	423	182	0.000101	0.000041
1000	500	423	183	0.000166	0.000065
1000	500	423	184	0.000270	0.000104
1000	500	423	185	0.000432	0.000162
1000	500	423	186	0.000680	0.000248
1000	500	423	187	0.001053	0.000374
1000	500	423	188	0.001607	0.000554
1000	500	423	189	0.002415	0.000808
1000	500	423	190	0.003573	0.001159
1000	500	423	191	0.005209	0.001635
1000	500	423	192	0.007478	0.002269
1000	500	423	193	0.010577	0.003099
1000	500	423	194	0.014738	0.004162
1000	500	423	195	0.020236	0.005498
1000	500	423	196	0.027382	0.007145
1000	500	423	197	0.036517	0.009135
1000	500	423	198	0.048006	0.011489
1000	500	423	199	0.062219	0.014213

Table for $N = 1000$, $n = 500$

N	n	k	x	P(x)	p(x)
1000	500	423	200	0.079517	0.017298
1000	500	423	201	0.100227	0.020710
1000	500	423	202	0.124620	0.024392
1000	500	423	203	0.152882	0.028262
1000	500	423	204	0.185097	0.032214
1000	500	423	205	0.221220	0.036123
1000	500	423	206	0.261068	0.039848
1000	500	423	207	0.304312	0.043244
1000	500	423	208	0.350480	0.046168
1000	500	423	209	0.398970	0.048490
1000	500	423	210	0.449072	0.050102
1000	500	423	211	0.500000	0.050928
1000	500	424	174	0.000001	0.000000
1000	500	424	175	0.000001	0.000001
1000	500	424	176	0.000003	0.000001
1000	500	424	177	0.000005	0.000002
1000	500	424	178	0.000009	0.000004
1000	500	424	179	0.000016	0.000007
1000	500	424	180	0.000027	0.000012
1000	500	424	181	0.000046	0.000019
1000	500	424	182	0.000078	0.000032
1000	500	424	183	0.000130	0.000052
1000	500	424	184	0.000213	0.000083
1000	500	424	185	0.000344	0.000130
1000	500	424	186	0.000545	0.000201
1000	500	424	187	0.000851	0.000306
1000	500	424	188	0.001308	0.000457
1000	500	424	189	0.001979	0.000672
1000	500	424	190	0.002951	0.000972
1000	500	424	191	0.004333	0.001382
1000	500	424	192	0.006267	0.001934
1000	500	424	193	0.008928	0.002661
1000	500	424	194	0.012531	0.003603
1000	500	424	195	0.017330	0.004799
1000	500	424	196	0.023617	0.006287
1000	500	424	197	0.031720	0.008103
1000	500	424	198	0.041993	0.010273
1000	500	424	199	0.054804	0.012812
1000	500	424	200	0.070523	0.015719
1000	500	424	201	0.089495	0.018972
1000	500	424	202	0.112022	0.022527
1000	500	424	203	0.138335	0.026313
1000	500	424	204	0.168571	0.030236
1000	500	424	205	0.202751	0.034181
1000	500	424	206	0.240764	0.038013
1000	500	424	207	0.282353	0.041589
1000	500	424	208	0.327116	0.044763
1000	500	424	209	0.374515	0.047399
1000	500	424	210	0.423891	0.049376
1000	500	424	211	0.474492	0.050601

N	n	k	x	P(x)	p(x)
1000	500	424	212	0.525508	0.051016
1000	500	425	174	0.000001	0.000000
1000	500	425	175	0.000001	0.000000
1000	500	425	176	0.000002	0.000001
1000	500	425	177	0.000004	0.000002
1000	500	425	178	0.000007	0.000003
1000	500	425	179	0.000012	0.000005
1000	500	425	180	0.000021	0.000009
1000	500	425	181	0.000036	0.000015
1000	500	425	182	0.000061	0.000025
1000	500	425	183	0.000102	0.000041
1000	500	425	184	0.000168	0.000066
1000	500	425	185	0.000272	0.000105
1000	500	425	186	0.000435	0.000163
1000	500	425	187	0.000684	0.000249
1000	500	425	188	0.001060	0.000376
1000	500	425	189	0.001617	0.000557
1000	500	425	190	0.002428	0.000811
1000	500	425	191	0.003592	0.001163
1000	500	425	192	0.005233	0.001641
1000	500	425	193	0.007509	0.002277
1000	500	425	194	0.010617	0.003108
1000	500	425	195	0.014789	0.004172
1000	500	425	196	0.020299	0.005510
1000	500	425	197	0.027458	0.007159
1000	500	425	198	0.036607	0.009149
1000	500	425	199	0.048109	0.011503
1000	500	425	200	0.062336	0.014227
1000	500	425	201	0.079647	0.017311
1000	500	425	202	0.100368	0.020721
1000	500	425	203	0.124767	0.024400
1000	500	425	204	0.153033	0.028266
1000	500	425	205	0.185246	0.032213
1000	500	425	206	0.221362	0.036116
1000	500	425	207	0.261197	0.039836
1000	500	425	208	0.304424	0.043226
1000	500	425	209	0.350569	0.046145
1000	500	425	210	0.399031	0.048463
1000	500	425	211	0.449103	0.050072
1000	500	425	212	0.500000	0.050897
1000	500	426	175	0.000001	0.000000
1000	500	426	176	0.000001	0.000001
1000	500	426	177	0.000003	0.000001
1000	500	426	178	0.000005	0.000002
1000	500	426	179	0.000009	0.000004
1000	500	426	180	0.000016	0.000007
1000	500	426	181	0.000027	0.000012
1000	500	426	182	0.000047	0.000020
1000	500	426	183	0.000079	0.000032
1000	500	426	184	0.000132	0.000052

N	n	k	x	P(x)	p(x)
1000	500	426	185	0.000215	0.000084
1000	500	426	186	0.000346	0.000131
1000	500	426	187	0.000549	0.000202
1000	500	426	188	0.000856	0.000308
1000	500	426	189	0.001316	0.000459
1000	500	426	190	0.001991	0.000675
1000	500	426	191	0.002966	0.000976
1000	500	426	192	0.004354	0.001387
1000	500	426	193	0.006294	0.001940
1000	500	426	194	0.008963	0.002669
1000	500	426	195	0.012576	0.003613
1000	500	426	196	0.017386	0.004810
1000	500	426	197	0.023685	0.006299
1000	500	426	198	0.031801	0.008116
1000	500	426	199	0.042088	0.010287
1000	500	426	200	0.054914	0.012826
1000	500	426	201	0.070646	0.015732
1000	500	426	202	0.089629	0.018983
1000	500	426	203	0.112164	0.022535
1000	500	426	204	0.138482	0.026318
1000	500	426	205	0.168719	0.030237
1000	500	426	206	0.202895	0.034176
1000	500	426	207	0.240898	0.038003
1000	500	426	208	0.282472	0.041574
1000	500	426	209	0.327215	0.044743
1000	500	426	210	0.374589	0.047374
1000	500	426	211	0.423937	0.049347
1000	500	426	212	0.474507	0.050571
1000	500	426	213	0.525492	0.050985
1000	500	427	175	0.000001	0.000000
1000	500	427	176	0.000001	0.000000
1000	500	427	177	0.000002	0.000001
1000	500	427	178	0.000004	0.000002
1000	500	427	179	0.000007	0.000003
1000	500	427	180	0.000012	0.000005
1000	500	427	181	0.000021	0.000009
1000	500	427	182	0.000036	0.000015
1000	500	427	183	0.000061	0.000025
1000	500	427	184	0.000103	0.000041
1000	500	427	185	0.000169	0.000067
1000	500	427	186	0.000274	0.000105
1000	500	427	187	0.000438	0.000164
1000	500	427	188	0.000689	0.000251
1000	500	427	189	0.001067	0.000378
1000	500	427	190	0.001626	0.000559
1000	500	427	191	0.002441	0.000815
1000	500	427	192	0.003609	0.001168
1000	500	427	193	0.005256	0.001647
1000	500	427	194	0.007540	0.002284
1000	500	427	195	0.010656	0.003116

N	n	k	x	P(x)	p(x)
1000	500	427	196	0.014839	0.004182
1000	500	427	197	0.020360	0.005521
1000	500	427	198	0.027531	0.007171
1000	500	427	199	0.036694	0.009162
1000	500	427	200	0.048210	0.011517
1000	500	427	201	0.062451	0.014240
1000	500	427	202	0.079773	0.017323
1000	500	427	203	0.100504	0.020731
1000	500	427	204	0.124911	0.024407
1000	500	427	205	0.153179	0.028269
1000	500	427	206	0.185390	0.032211
1000	500	427	207	0.221500	0.036109
1000	500	427	208	0.261323	0.039824
1000	500	427	209	0.304532	0.043209
1000	500	427	210	0.350655	0.046123
1000	500	427	211	0.399091	0.048436
1000	500	427	212	0.449134	0.050043
1000	500	427	213	0.500000	0.050866
1000	500	428	176	0.000001	0.000000
1000	500	428	177	0.000001	0.000001
1000	500	428	178	0.000003	0.000001
1000	500	428	179	0.000005	0.000002
1000	500	428	180	0.000009	0.000004
1000	500	428	181	0.000016	0.000007
1000	500	428	182	0.000028	0.000012
1000	500	428	183	0.000047	0.000020
1000	500	428	184	0.000080	0.000033
1000	500	428	185	0.000133	0.000053
1000	500	428	186	0.000217	0.000084
1000	500	428	187	0.000349	0.000132
1000	500	428	188	0.000552	0.000204
1000	500	428	189	0.000862	0.000309
1000	500	428	190	0.001323	0.000462
1000	500	428	191	0.002002	0.000678
1000	500	428	192	0.002981	0.000980
1000	500	428	193	0.004374	0.001392
1000	500	428	194	0.006320	0.001947
1000	500	428	195	0.008998	0.002677
1000	500	428	196	0.012620	0.003622
1000	500	428	197	0.017440	0.004821
1000	500	428	198	0.023752	0.006311
1000	500	428	199	0.031881	0.008129
1000	500	428	200	0.042181	0.010300
1000	500	428	201	0.055020	0.012839
1000	500	428	202	0.070764	0.015745
1000	500	428	203	0.089759	0.018994
1000	500	428	204	0.112303	0.022544
1000	500	428	205	0.138625	0.026323
1000	500	428	206	0.168863	0.030238
1000	500	428	207	0.203035	0.034172

Table for $N = 1000$, $n = 500$

N	n	k	x	P(x)	p(x)
1000	500	428	208	0.241029	0.037994
1000	500	428	209	0.282588	0.041559
1000	500	428	210	0.327311	0.044723
1000	500	428	211	0.374661	0.047350
1000	500	428	212	0.423981	0.049320
1000	500	428	213	0.474523	0.050541
1000	500	428	214	0.525477	0.050955
1000	500	429	176	0.000001	0.000000
1000	500	429	177	0.000001	0.000001
1000	500	429	178	0.000002	0.000001
1000	500	429	179	0.000004	0.000002
1000	500	429	180	0.000007	0.000003
1000	500	429	181	0.000012	0.000005
1000	500	429	182	0.000021	0.000009
1000	500	429	183	0.000036	0.000015
1000	500	429	184	0.000062	0.000025
1000	500	429	185	0.000104	0.000042
1000	500	429	186	0.000171	0.000067
1000	500	429	187	0.000277	0.000106
1000	500	429	188	0.000441	0.000165
1000	500	429	189	0.000694	0.000252
1000	500	429	190	0.001073	0.000380
1000	500	429	191	0.001635	0.000562
1000	500	429	192	0.002454	0.000819
1000	500	429	193	0.003626	0.001173
1000	500	429	194	0.005279	0.001653
1000	500	429	195	0.007570	0.002291
1000	500	429	196	0.010695	0.003125
1000	500	429	197	0.014887	0.004192
1000	500	429	198	0.020419	0.005533
1000	500	429	199	0.027603	0.007184
1000	500	429	200	0.036779	0.009176
1000	500	429	201	0.048308	0.011530
1000	500	429	202	0.062562	0.014253
1000	500	429	203	0.079896	0.017335
1000	500	429	204	0.100637	0.020740
1000	500	429	205	0.125050	0.024413
1000	500	429	206	0.153321	0.028272
1000	500	429	207	0.185531	0.032210
1000	500	429	208	0.221634	0.036103
1000	500	429	209	0.261445	0.039812
1000	500	429	210	0.304637	0.043192
1000	500	429	211	0.350738	0.046101
1000	500	429	212	0.399149	0.048411
1000	500	429	213	0.449164	0.050015
1000	500	429	214	0.500000	0.050836
1000	500	430	177	0.000001	0.000000
1000	500	430	178	0.000001	0.000001
1000	500	430	179	0.000003	0.000001
1000	500	430	180	0.000005	0.000002
1000	500	430	181	0.000009	0.000004
1000	500	430	182	0.000016	0.000007
1000	500	430	183	0.000028	0.000012
1000	500	430	184	0.000048	0.000020
1000	500	430	185	0.000081	0.000033
1000	500	430	186	0.000134	0.000053
1000	500	430	187	0.000218	0.000085
1000	500	430	188	0.000351	0.000133
1000	500	430	189	0.000556	0.000205
1000	500	430	190	0.000867	0.000311
1000	500	430	191	0.001331	0.000464
1000	500	430	192	0.002012	0.000681
1000	500	430	193	0.002996	0.000984
1000	500	430	194	0.004393	0.001397
1000	500	430	195	0.006346	0.001953
1000	500	430	196	0.009031	0.002685
1000	500	430	197	0.012662	0.003631
1000	500	430	198	0.017493	0.004831
1000	500	430	199	0.023816	0.006323
1000	500	430	200	0.031958	0.008142
1000	500	430	201	0.042271	0.010313
1000	500	430	202	0.055123	0.012852
1000	500	430	203	0.070880	0.015757
1000	500	430	204	0.089885	0.019005
1000	500	430	205	0.112437	0.022552
1000	500	430	206	0.138765	0.026328
1000	500	430	207	0.169003	0.030238
1000	500	430	208	0.203171	0.034168
1000	500	430	209	0.241156	0.037985
1000	500	430	210	0.282701	0.041545
1000	500	430	211	0.327405	0.044704
1000	500	430	212	0.374732	0.047327
1000	500	430	213	0.424025	0.049293
1000	500	430	214	0.474537	0.050512
1000	500	430	215	0.525463	0.050925
1000	500	431	177	0.000001	0.000000
1000	500	431	178	0.000001	0.000001
1000	500	431	179	0.000002	0.000001
1000	500	431	180	0.000004	0.000002
1000	500	431	181	0.000007	0.000003
1000	500	431	182	0.000012	0.000005
1000	500	431	183	0.000021	0.000009
1000	500	431	184	0.000037	0.000015
1000	500	431	185	0.000062	0.000026
1000	500	431	186	0.000104	0.000042
1000	500	431	187	0.000172	0.000067
1000	500	431	188	0.000279	0.000107
1000	500	431	189	0.000444	0.000166
1000	500	431	190	0.000698	0.000254
1000	500	431	191	0.001080	0.000382
1000	500	431	192	0.001644	0.000565
1000	500	431	193	0.002466	0.000822
1000	500	431	194	0.003643	0.001177
1000	500	431	195	0.005301	0.001658
1000	500	431	196	0.007599	0.002298
1000	500	431	197	0.010732	0.003133
1000	500	431	198	0.014934	0.004202
1000	500	431	199	0.020477	0.005544
1000	500	431	200	0.027673	0.007196
1000	500	431	201	0.036861	0.009188
1000	500	431	202	0.048404	0.011543
1000	500	431	203	0.062670	0.014266
1000	500	431	204	0.080016	0.017346
1000	500	431	205	0.100765	0.020750
1000	500	431	206	0.125185	0.024420
1000	500	431	207	0.153460	0.028274
1000	500	431	208	0.185668	0.032208
1000	500	431	209	0.221764	0.036096
1000	500	431	210	0.261564	0.039800
1000	500	431	211	0.304739	0.043175
1000	500	431	212	0.350819	0.046080
1000	500	431	213	0.399205	0.048386
1000	500	431	214	0.449192	0.049987
1000	500	431	215	0.500000	0.050807
1000	500	432	178	0.000001	0.000000
1000	500	432	179	0.000002	0.000001
1000	500	432	180	0.000003	0.000001
1000	500	432	181	0.000005	0.000002
1000	500	432	182	0.000009	0.000004
1000	500	432	183	0.000016	0.000007
1000	500	432	184	0.000028	0.000012
1000	500	432	185	0.000048	0.000020
1000	500	432	186	0.000081	0.000033
1000	500	432	187	0.000135	0.000054
1000	500	432	188	0.000220	0.000085
1000	500	432	189	0.000354	0.000134
1000	500	432	190	0.000560	0.000206
1000	500	432	191	0.000872	0.000312
1000	500	432	192	0.001339	0.000466
1000	500	432	193	0.002023	0.000684
1000	500	432	194	0.003010	0.000988
1000	500	432	195	0.004412	0.001402
1000	500	432	196	0.006372	0.001959
1000	500	432	197	0.009064	0.002692
1000	500	432	198	0.012704	0.003640
1000	500	432	199	0.017545	0.004841
1000	500	432	200	0.023879	0.006334
1000	500	432	201	0.032033	0.008154
1000	500	432	202	0.042359	0.010326
1000	500	432	203	0.055223	0.012865
1000	500	432	204	0.070992	0.015769
1000	500	432	205	0.090008	0.019016
1000	500	432	206	0.112568	0.022560
1000	500	432	207	0.138900	0.026332
1000	500	432	208	0.169139	0.030239
1000	500	432	209	0.203303	0.034164
1000	500	432	210	0.241279	0.037976
1000	500	432	211	0.282810	0.041531
1000	500	432	212	0.327496	0.044686
1000	500	432	213	0.374800	0.047304
1000	500	432	214	0.424067	0.049267
1000	500	432	215	0.474552	0.050484
1000	500	432	216	0.525448	0.050897
1000	500	433	178	0.000001	0.000000
1000	500	433	179	0.000001	0.000001
1000	500	433	180	0.000002	0.000001
1000	500	433	181	0.000004	0.000002
1000	500	433	182	0.000007	0.000003
1000	500	433	183	0.000012	0.000005
1000	500	433	184	0.000022	0.000009
1000	500	433	185	0.000037	0.000016
1000	500	433	186	0.000063	0.000026
1000	500	433	187	0.000105	0.000042
1000	500	433	188	0.000173	0.000068
1000	500	433	189	0.000281	0.000107
1000	500	433	190	0.000447	0.000167
1000	500	433	191	0.000702	0.000255
1000	500	433	192	0.001086	0.000383
1000	500	433	193	0.001653	0.000567
1000	500	433	194	0.002478	0.000825
1000	500	433	195	0.003659	0.001181
1000	500	433	196	0.005323	0.001664
1000	500	433	197	0.007628	0.002305
1000	500	433	198	0.010768	0.003141
1000	500	433	199	0.014979	0.004211
1000	500	433	200	0.020533	0.005554
1000	500	433	201	0.027741	0.007207
1000	500	433	202	0.036941	0.009201
1000	500	433	203	0.048497	0.011555
1000	500	433	204	0.062775	0.014278
1000	500	433	205	0.080132	0.017357
1000	500	433	206	0.100890	0.020759
1000	500	433	207	0.125317	0.024426
1000	500	433	208	0.153594	0.028277
1000	500	433	209	0.185800	0.032207
1000	500	433	210	0.221890	0.036090
1000	500	433	211	0.261679	0.039789
1000	500	433	212	0.304838	0.043159
1000	500	433	213	0.350898	0.046060
1000	500	433	214	0.399260	0.048362

Table for $N = 1000$, $n = 500$

N	n	k	x	P(x)	p(x)
1000	500	433	215	0.449220	0.049961
1000	500	433	216	0.500000	0.050779
1000	500	434	179	0.000001	0.000000
1000	500	434	180	0.000002	0.000001
1000	500	434	181	0.000003	0.000001
1000	500	434	182	0.000005	0.000002
1000	500	434	183	0.000009	0.000004
1000	500	434	184	0.000016	0.000007
1000	500	434	185	0.000028	0.000012
1000	500	434	186	0.000049	0.000020
1000	500	434	187	0.000082	0.000033
1000	500	434	188	0.000136	0.000054
1000	500	434	189	0.000222	0.000086
1000	500	434	190	0.000356	0.000134
1000	500	434	191	0.000563	0.000207
1000	500	434	192	0.000877	0.000314
1000	500	434	193	0.001346	0.000468
1000	500	434	194	0.002033	0.000687
1000	500	434	195	0.003024	0.000991
1000	500	434	196	0.004431	0.001407
1000	500	434	197	0.006396	0.001965
1000	500	434	198	0.009096	0.002699
1000	500	434	199	0.012744	0.003648
1000	500	434	200	0.017595	0.004851
1000	500	434	201	0.023940	0.006345
1000	500	434	202	0.032106	0.008166
1000	500	434	203	0.042444	0.010338
1000	500	434	204	0.055321	0.012877
1000	500	434	205	0.071101	0.015780
1000	500	434	206	0.090127	0.019026
1000	500	434	207	0.112694	0.022568
1000	500	434	208	0.139031	0.026337
1000	500	434	209	0.169271	0.030240
1000	500	434	210	0.203432	0.034161
1000	500	434	211	0.241399	0.037967
1000	500	434	212	0.282916	0.041517
1000	500	434	213	0.327584	0.044668
1000	500	434	214	0.374866	0.047282
1000	500	434	215	0.424108	0.049242
1000	500	434	216	0.474565	0.050457
1000	500	434	217	0.525434	0.050869
1000	500	435	179	0.000001	0.000000
1000	500	435	180	0.000001	0.000001
1000	500	435	181	0.000002	0.000001
1000	500	435	182	0.000004	0.000002
1000	500	435	183	0.000007	0.000003
1000	500	435	184	0.000012	0.000005
1000	500	435	185	0.000022	0.000009
1000	500	435	186	0.000037	0.000016
1000	500	435	187	0.000064	0.000026

N	n	k	x	P(x)	p(x)
1000	500	435	188	0.000106	0.000043
1000	500	435	189	0.000175	0.000068
1000	500	435	190	0.000283	0.000108
1000	500	435	191	0.000450	0.000168
1000	500	435	192	0.000707	0.000256
1000	500	435	193	0.001092	0.000385
1000	500	435	194	0.001661	0.000570
1000	500	435	195	0.002490	0.000828
1000	500	435	196	0.003675	0.001185
1000	500	435	197	0.005344	0.001669
1000	500	435	198	0.007655	0.002311
1000	500	435	199	0.010804	0.003148
1000	500	435	200	0.015024	0.004220
1000	500	435	201	0.020588	0.005564
1000	500	435	202	0.027806	0.007218
1000	500	435	203	0.037019	0.009213
1000	500	435	204	0.048587	0.011568
1000	500	435	205	0.062876	0.014290
1000	500	435	206	0.080244	0.017368
1000	500	435	207	0.101012	0.020768
1000	500	435	208	0.125444	0.024432
1000	500	435	209	0.153724	0.028280
1000	500	435	210	0.185929	0.032205
1000	500	435	211	0.222013	0.036084
1000	500	435	212	0.261791	0.039778
1000	500	435	213	0.304934	0.043144
1000	500	435	214	0.350974	0.046040
1000	500	435	215	0.399313	0.048339
1000	500	435	216	0.449248	0.049935
1000	500	435	217	0.500000	0.050752
1000	500	436	180	0.000001	0.000000
1000	500	436	181	0.000002	0.000001
1000	500	436	182	0.000003	0.000001
1000	500	436	183	0.000005	0.000002
1000	500	436	184	0.000009	0.000004
1000	500	436	185	0.000017	0.000007
1000	500	436	186	0.000029	0.000012
1000	500	436	187	0.000049	0.000020
1000	500	436	188	0.000083	0.000034
1000	500	436	189	0.000137	0.000054
1000	500	436	190	0.000223	0.000086
1000	500	436	191	0.000359	0.000135
1000	500	436	192	0.000567	0.000208
1000	500	436	193	0.000882	0.000316
1000	500	436	194	0.001353	0.000470
1000	500	436	195	0.002043	0.000690
1000	500	436	196	0.003037	0.000995
1000	500	436	197	0.004449	0.001412
1000	500	436	198	0.006420	0.001971
1000	500	436	199	0.009126	0.002706

N	n	k	x	P(x)	p(x)
1000	500	436	200	0.012783	0.003657
1000	500	436	201	0.017643	0.004860
1000	500	436	202	0.023999	0.006356
1000	500	436	203	0.032176	0.008177
1000	500	436	204	0.042526	0.010350
1000	500	436	205	0.055415	0.012889
1000	500	436	206	0.071207	0.015791
1000	500	436	207	0.090242	0.019035
1000	500	436	208	0.112817	0.022575
1000	500	436	209	0.139158	0.026341
1000	500	436	210	0.169399	0.030240
1000	500	436	211	0.203556	0.034157
1000	500	436	212	0.241514	0.037959
1000	500	436	213	0.283019	0.041504
1000	500	436	214	0.327669	0.044650
1000	500	436	215	0.374930	0.047261
1000	500	436	216	0.424148	0.049218
1000	500	436	217	0.474579	0.050431
1000	500	436	218	0.525421	0.050842
1000	500	437	180	0.000001	0.000000
1000	500	437	181	0.000001	0.000001
1000	500	437	182	0.000002	0.000001
1000	500	437	183	0.000004	0.000002
1000	500	437	184	0.000007	0.000003
1000	500	437	185	0.000013	0.000005
1000	500	437	186	0.000022	0.000009
1000	500	437	187	0.000038	0.000016
1000	500	437	188	0.000064	0.000026
1000	500	437	189	0.000107	0.000043
1000	500	437	190	0.000176	0.000069
1000	500	437	191	0.000284	0.000109
1000	500	437	192	0.000453	0.000169
1000	500	437	193	0.000711	0.000258
1000	500	437	194	0.001098	0.000387
1000	500	437	195	0.001670	0.000572
1000	500	437	196	0.002501	0.000832
1000	500	437	197	0.003691	0.001190
1000	500	437	198	0.005365	0.001674
1000	500	437	199	0.007682	0.002317
1000	500	437	200	0.010838	0.003156
1000	500	437	201	0.015067	0.004229
1000	500	437	202	0.020641	0.005574
1000	500	437	203	0.027870	0.007229
1000	500	437	204	0.037094	0.009224
1000	500	437	205	0.048674	0.011579
1000	500	437	206	0.062975	0.014301
1000	500	437	207	0.080353	0.017378
1000	500	437	208	0.101129	0.020776
1000	500	437	209	0.125567	0.024438
1000	500	437	210	0.153850	0.028282

N	n	k	x	P(x)	p(x)
1000	500	437	211	0.186053	0.032204
1000	500	437	212	0.222131	0.036078
1000	500	437	213	0.261899	0.039768
1000	500	437	214	0.305027	0.043129
1000	500	437	215	0.351048	0.046021
1000	500	437	216	0.399364	0.048316
1000	500	437	217	0.449274	0.049910
1000	500	437	218	0.500000	0.050726
1000	500	438	181	0.000001	0.000000
1000	500	438	182	0.000002	0.000001
1000	500	438	183	0.000003	0.000001
1000	500	438	184	0.000005	0.000002
1000	500	438	185	0.000009	0.000004
1000	500	438	186	0.000017	0.000007
1000	500	438	187	0.000029	0.000012
1000	500	438	188	0.000050	0.000021
1000	500	438	189	0.000083	0.000034
1000	500	438	190	0.000138	0.000055
1000	500	438	191	0.000225	0.000087
1000	500	438	192	0.000361	0.000136
1000	500	438	193	0.000570	0.000209
1000	500	438	194	0.000887	0.000317
1000	500	438	195	0.001360	0.000472
1000	500	438	196	0.002052	0.000692
1000	500	438	197	0.003051	0.000998
1000	500	438	198	0.004467	0.001416
1000	500	438	199	0.006443	0.001976
1000	500	438	200	0.009156	0.002713
1000	500	438	201	0.012821	0.003665
1000	500	438	202	0.017690	0.004870
1000	500	438	203	0.024056	0.006366
1000	500	438	204	0.032245	0.008188
1000	500	438	205	0.042606	0.010362
1000	500	438	206	0.055507	0.012900
1000	500	438	207	0.071309	0.015802
1000	500	438	208	0.090354	0.019045
1000	500	438	209	0.112936	0.022582
1000	500	438	210	0.139282	0.026346
1000	500	438	211	0.169522	0.030241
1000	500	438	212	0.203676	0.034153
1000	500	438	213	0.241626	0.037951
1000	500	438	214	0.283118	0.041492
1000	500	438	215	0.327752	0.044633
1000	500	438	216	0.374991	0.047240
1000	500	438	217	0.424186	0.049195
1000	500	438	218	0.474592	0.050406
1000	500	438	219	0.525408	0.050816
1000	500	439	181	0.000001	0.000000
1000	500	439	182	0.000001	0.000001
1000	500	439	183	0.000002	0.000001

Table for $N = 1000$, $n = 500$

N	n	k	x	P(x)	p(x)
1000	500	439	184	0.000004	0.000002
1000	500	439	185	0.000007	0.000003
1000	500	439	186	0.000013	0.000006
1000	500	439	187	0.000022	0.000009
1000	500	439	188	0.000038	0.000016
1000	500	439	189	0.000065	0.000026
1000	500	439	190	0.000108	0.000043
1000	500	439	191	0.000177	0.000069
1000	500	439	192	0.000286	0.000109
1000	500	439	193	0.000456	0.000170
1000	500	439	194	0.000715	0.000259
1000	500	439	195	0.001103	0.000389
1000	500	439	196	0.001678	0.000574
1000	500	439	197	0.002512	0.000835
1000	500	439	198	0.003706	0.001193
1000	500	439	199	0.005385	0.001679
1000	500	439	200	0.007708	0.002323
1000	500	439	201	0.010871	0.003163
1000	500	439	202	0.015108	0.004237
1000	500	439	203	0.020692	0.005584
1000	500	439	204	0.027932	0.007240
1000	500	439	205	0.037167	0.009235
1000	500	439	206	0.048758	0.011591
1000	500	439	207	0.063071	0.014312
1000	500	439	208	0.080459	0.017388
1000	500	439	209	0.101243	0.020784
1000	500	439	210	0.125687	0.024444
1000	500	439	211	0.153971	0.028285
1000	500	439	212	0.186174	0.032202
1000	500	439	213	0.222246	0.036072
1000	500	439	214	0.262003	0.039757
1000	500	439	215	0.305117	0.043114
1000	500	439	216	0.351119	0.046002
1000	500	439	217	0.399414	0.048294
1000	500	439	218	0.449299	0.049886
1000	500	439	219	0.500000	0.050701
1000	500	440	182	0.000001	0.000000
1000	500	440	183	0.000002	0.000001
1000	500	440	184	0.000003	0.000001
1000	500	440	185	0.000005	0.000002
1000	500	440	186	0.000010	0.000004
1000	500	440	187	0.000017	0.000007
1000	500	440	188	0.000029	0.000012
1000	500	440	189	0.000050	0.000021
1000	500	440	190	0.000084	0.000034
1000	500	440	191	0.000139	0.000055
1000	500	440	192	0.000226	0.000087
1000	500	440	193	0.000363	0.000137
1000	500	440	194	0.000573	0.000210
1000	500	440	195	0.000892	0.000319

N	n	k	x	P(x)	p(x)
1000	500	440	196	0.001366	0.000474
1000	500	440	197	0.002061	0.000695
1000	500	440	198	0.003063	0.001002
1000	500	440	199	0.004484	0.001421
1000	500	440	200	0.006465	0.001982
1000	500	440	201	0.009185	0.002720
1000	500	440	202	0.012858	0.003672
1000	500	440	203	0.017736	0.004878
1000	500	440	204	0.024112	0.006376
1000	500	440	205	0.032311	0.008199
1000	500	440	206	0.042684	0.010373
1000	500	440	207	0.055595	0.012912
1000	500	440	208	0.071408	0.015813
1000	500	440	209	0.090462	0.019054
1000	500	440	210	0.113051	0.022589
1000	500	440	211	0.139401	0.026350
1000	500	440	212	0.169642	0.030241
1000	500	440	213	0.203792	0.034150
1000	500	440	214	0.241735	0.037943
1000	500	440	215	0.283214	0.041480
1000	500	440	216	0.327831	0.044617
1000	500	440	217	0.375051	0.047220
1000	500	440	218	0.424223	0.049172
1000	500	440	219	0.474604	0.050381
1000	500	440	220	0.525395	0.050791
1000	500	441	182	0.000001	0.000000
1000	500	441	183	0.000001	0.000001
1000	500	441	184	0.000002	0.000001
1000	500	441	185	0.000004	0.000002
1000	500	441	186	0.000007	0.000003
1000	500	441	187	0.000013	0.000006
1000	500	441	188	0.000022	0.000010
1000	500	441	189	0.000038	0.000016
1000	500	441	190	0.000065	0.000027
1000	500	441	191	0.000109	0.000043
1000	500	441	192	0.000178	0.000070
1000	500	441	193	0.000288	0.000110
1000	500	441	194	0.000459	0.000170
1000	500	441	195	0.000719	0.000260
1000	500	441	196	0.001109	0.000390
1000	500	441	197	0.001685	0.000577
1000	500	441	198	0.002523	0.000838
1000	500	441	199	0.003720	0.001197
1000	500	441	200	0.005404	0.001684
1000	500	441	201	0.007733	0.002329
1000	500	441	202	0.010903	0.003170
1000	500	441	203	0.015149	0.004245
1000	500	441	204	0.020742	0.005593
1000	500	441	205	0.027992	0.007250
1000	500	441	206	0.037238	0.009246

N	n	k	x	P(x)	p(x)
1000	500	441	207	0.048840	0.011602
1000	500	441	208	0.063163	0.014323
1000	500	441	209	0.080561	0.017398
1000	500	441	210	0.101353	0.020792
1000	500	441	211	0.125802	0.024449
1000	500	441	212	0.154089	0.028287
1000	500	441	213	0.186290	0.032201
1000	500	441	214	0.222357	0.036067
1000	500	441	215	0.262104	0.039747
1000	500	441	216	0.305204	0.043100
1000	500	441	217	0.351188	0.045984
1000	500	441	218	0.399462	0.048273
1000	500	441	219	0.449324	0.049862
1000	500	441	220	0.500000	0.050676
1000	500	442	183	0.000001	0.000000
1000	500	442	184	0.000002	0.000001
1000	500	442	185	0.000003	0.000001
1000	500	442	186	0.000005	0.000002
1000	500	442	187	0.000010	0.000004
1000	500	442	188	0.000017	0.000007
1000	500	442	189	0.000029	0.000012
1000	500	442	190	0.000050	0.000021
1000	500	442	191	0.000085	0.000034
1000	500	442	192	0.000140	0.000055
1000	500	442	193	0.000228	0.000088
1000	500	442	194	0.000365	0.000137
1000	500	442	195	0.000577	0.000211
1000	500	442	196	0.000897	0.000320
1000	500	442	197	0.001373	0.000476
1000	500	442	198	0.002071	0.000698
1000	500	442	199	0.003076	0.001005
1000	500	442	200	0.004500	0.001425
1000	500	442	201	0.006487	0.001987
1000	500	442	202	0.009213	0.002726
1000	500	442	203	0.012893	0.003680
1000	500	442	204	0.017780	0.004887
1000	500	442	205	0.024166	0.006386
1000	500	442	206	0.032375	0.008209
1000	500	442	207	0.042759	0.010384
1000	500	442	208	0.055681	0.012922
1000	500	442	209	0.071504	0.015823
1000	500	442	210	0.090566	0.019062
1000	500	442	211	0.113162	0.022596
1000	500	442	212	0.139516	0.026354
1000	500	442	213	0.169758	0.030242
1000	500	442	214	0.203904	0.034146
1000	500	442	215	0.241839	0.037935
1000	500	442	216	0.283307	0.041468
1000	500	442	217	0.327908	0.044601
1000	500	442	218	0.375109	0.047201

N	n	k	x	P(x)	p(x)
1000	500	442	219	0.424259	0.049150
1000	500	442	220	0.474616	0.050358
1000	500	442	221	0.525383	0.050767
1000	500	443	183	0.000001	0.000000
1000	500	443	184	0.000001	0.000001
1000	500	443	185	0.000002	0.000001
1000	500	443	186	0.000004	0.000002
1000	500	443	187	0.000007	0.000003
1000	500	443	188	0.000013	0.000006
1000	500	443	189	0.000023	0.000010
1000	500	443	190	0.000039	0.000016
1000	500	443	191	0.000066	0.000027
1000	500	443	192	0.000109	0.000044
1000	500	443	193	0.000179	0.000070
1000	500	443	194	0.000290	0.000110
1000	500	443	195	0.000461	0.000171
1000	500	443	196	0.000722	0.000261
1000	500	443	197	0.001114	0.000392
1000	500	443	198	0.001693	0.000579
1000	500	443	199	0.002533	0.000840
1000	500	443	200	0.003734	0.001201
1000	500	443	201	0.005423	0.001688
1000	500	443	202	0.007757	0.002335
1000	500	443	203	0.010934	0.003177
1000	500	443	204	0.015188	0.004253
1000	500	443	205	0.020790	0.005602
1000	500	443	206	0.028050	0.007260
1000	500	443	207	0.037307	0.009257
1000	500	443	208	0.048919	0.011612
1000	500	443	209	0.063252	0.014333
1000	500	443	210	0.080659	0.017407
1000	500	443	211	0.101459	0.020800
1000	500	443	212	0.125914	0.024455
1000	500	443	213	0.154203	0.028289
1000	500	443	214	0.186402	0.032199
1000	500	443	215	0.222464	0.036061
1000	500	443	216	0.262202	0.039738
1000	500	443	217	0.305288	0.043086
1000	500	443	218	0.351255	0.045967
1000	500	443	219	0.399508	0.048253
1000	500	443	220	0.449347	0.049840
1000	500	443	221	0.500000	0.050653
1000	500	444	184	0.000001	0.000000
1000	500	444	185	0.000002	0.000001
1000	500	444	186	0.000003	0.000001
1000	500	444	187	0.000005	0.000002
1000	500	444	188	0.000010	0.000004
1000	500	444	189	0.000017	0.000007
1000	500	444	190	0.000030	0.000013
1000	500	444	191	0.000051	0.000021

Table for $N = 1000,\ n = 500$

N	n	k	x	P(x)	p(x)
1000	500	444	192	0.000085	0.000034
1000	500	444	193	0.000141	0.000056
1000	500	444	194	0.000229	0.000088
1000	500	444	195	0.000367	0.000138
1000	500	444	196	0.000580	0.000212
1000	500	444	197	0.000901	0.000321
1000	500	444	198	0.001379	0.000478
1000	500	444	199	0.002079	0.000700
1000	500	444	200	0.003088	0.001008
1000	500	444	201	0.004516	0.001429
1000	500	444	202	0.006508	0.001992
1000	500	444	203	0.009240	0.002732
1000	500	444	204	0.012927	0.003687
1000	500	444	205	0.017823	0.004895
1000	500	444	206	0.024218	0.006395
1000	500	444	207	0.032437	0.008219
1000	500	444	208	0.042831	0.010394
1000	500	444	209	0.055764	0.012933
1000	500	444	210	0.071596	0.015832
1000	500	444	211	0.090667	0.019071
1000	500	444	212	0.113270	0.022602
1000	500	444	213	0.139627	0.026357
1000	500	444	214	0.169869	0.030242
1000	500	444	215	0.204012	0.034143
1000	500	444	216	0.241940	0.037928
1000	500	444	217	0.283397	0.041456
1000	500	444	218	0.327983	0.044586
1000	500	444	219	0.375165	0.047182
1000	500	444	220	0.424293	0.049129
1000	500	444	221	0.474628	0.050335
1000	500	444	222	0.525372	0.050743
1000	500	445	184	0.000001	0.000000
1000	500	445	185	0.000001	0.000001
1000	500	445	186	0.000002	0.000001
1000	500	445	187	0.000004	0.000002
1000	500	445	188	0.000007	0.000003
1000	500	445	189	0.000013	0.000006
1000	500	445	190	0.000023	0.000010
1000	500	445	191	0.000039	0.000016
1000	500	445	192	0.000066	0.000027
1000	500	445	193	0.000110	0.000044
1000	500	445	194	0.000181	0.000070
1000	500	445	195	0.000292	0.000111
1000	500	445	196	0.000464	0.000172
1000	500	445	197	0.000726	0.000262
1000	500	445	198	0.001119	0.000394
1000	500	445	199	0.001700	0.000581
1000	500	445	200	0.002543	0.000843
1000	500	445	201	0.003748	0.001204
1000	500	445	202	0.005441	0.001693
1000	500	445	203	0.007781	0.002340
1000	500	445	204	0.010965	0.003184
1000	500	445	205	0.015225	0.004261
1000	500	445	206	0.020836	0.005611
1000	500	445	207	0.028106	0.007269
1000	500	445	208	0.037373	0.009267
1000	500	445	209	0.048995	0.011623
1000	500	445	210	0.063338	0.014343
1000	500	445	211	0.080755	0.017416
1000	500	445	212	0.101562	0.020807
1000	500	445	213	0.126021	0.024460
1000	500	445	214	0.154313	0.028291
1000	500	445	215	0.186511	0.032198
1000	500	445	216	0.222567	0.036056
1000	500	445	217	0.262296	0.039729
1000	500	445	218	0.305369	0.043073
1000	500	445	219	0.351319	0.045950
1000	500	445	220	0.399552	0.048233
1000	500	445	221	0.449370	0.049818
1000	500	445	222	0.500000	0.050630
1000	500	446	185	0.000001	0.000000
1000	500	446	186	0.000002	0.000001
1000	500	446	187	0.000003	0.000001
1000	500	446	188	0.000005	0.000002
1000	500	446	189	0.000010	0.000004
1000	500	446	190	0.000017	0.000007
1000	500	446	191	0.000030	0.000013
1000	500	446	192	0.000051	0.000021
1000	500	446	193	0.000086	0.000035
1000	500	446	194	0.000142	0.000056
1000	500	446	195	0.000231	0.000089
1000	500	446	196	0.000369	0.000139
1000	500	446	197	0.000583	0.000213
1000	500	446	198	0.000905	0.000323
1000	500	446	199	0.001385	0.000480
1000	500	446	200	0.002088	0.000702
1000	500	446	201	0.003099	0.001011
1000	500	446	202	0.004532	0.001433
1000	500	446	203	0.006529	0.001997
1000	500	446	204	0.009267	0.002738
1000	500	446	205	0.012961	0.003694
1000	500	446	206	0.017864	0.004903
1000	500	446	207	0.024268	0.006404
1000	500	446	208	0.032497	0.008229
1000	500	446	209	0.042901	0.010404
1000	500	446	210	0.055844	0.012943
1000	500	446	211	0.071686	0.015842
1000	500	446	212	0.090765	0.019079
1000	500	446	213	0.113373	0.022608
1000	500	446	214	0.139734	0.026361
1000	500	446	215	0.169977	0.030243
1000	500	446	216	0.204117	0.034140
1000	500	446	217	0.242038	0.037921
1000	500	446	218	0.283483	0.041445
1000	500	446	219	0.328054	0.044571
1000	500	446	220	0.375218	0.047164
1000	500	446	221	0.424327	0.049108
1000	500	446	222	0.474639	0.050313
1000	500	446	223	0.525360	0.050721
1000	500	447	185	0.000001	0.000000
1000	500	447	186	0.000001	0.000001
1000	500	447	187	0.000002	0.000001
1000	500	447	188	0.000004	0.000002
1000	500	447	189	0.000007	0.000003
1000	500	447	190	0.000013	0.000006
1000	500	447	191	0.000023	0.000010
1000	500	447	192	0.000039	0.000016
1000	500	447	193	0.000067	0.000027
1000	500	447	194	0.000111	0.000044
1000	500	447	195	0.000182	0.000071
1000	500	447	196	0.000293	0.000112
1000	500	447	197	0.000466	0.000173
1000	500	447	198	0.000729	0.000263
1000	500	447	199	0.001125	0.000395
1000	500	447	200	0.001707	0.000583
1000	500	447	201	0.002553	0.000846
1000	500	447	202	0.003761	0.001208
1000	500	447	203	0.005458	0.001697
1000	500	447	204	0.007804	0.002346
1000	500	447	205	0.010994	0.003190
1000	500	447	206	0.015262	0.004268
1000	500	447	207	0.020881	0.005619
1000	500	447	208	0.028160	0.007279
1000	500	447	209	0.037436	0.009276
1000	500	447	210	0.049069	0.011633
1000	500	447	211	0.063422	0.014353
1000	500	447	212	0.080846	0.017425
1000	500	447	213	0.101661	0.020814
1000	500	447	214	0.126125	0.024465
1000	500	447	215	0.154419	0.028293
1000	500	447	216	0.186615	0.032197
1000	500	447	217	0.222667	0.036051
1000	500	447	218	0.262386	0.039720
1000	500	447	219	0.305447	0.043061
1000	500	447	220	0.351381	0.045934
1000	500	447	221	0.399595	0.048214
1000	500	447	222	0.449392	0.049797
1000	500	447	223	0.500000	0.050608
1000	500	448	186	0.000001	0.000000
1000	500	448	187	0.000002	0.000001
1000	500	448	188	0.000003	0.000001
1000	500	448	189	0.000006	0.000002
1000	500	448	190	0.000010	0.000004
1000	500	448	191	0.000017	0.000008
1000	500	448	192	0.000030	0.000013
1000	500	448	193	0.000051	0.000021
1000	500	448	194	0.000086	0.000035
1000	500	448	195	0.000143	0.000056
1000	500	448	196	0.000232	0.000089
1000	500	448	197	0.000371	0.000139
1000	500	448	198	0.000586	0.000214
1000	500	448	199	0.000909	0.000324
1000	500	448	200	0.001391	0.000482
1000	500	448	201	0.002096	0.000705
1000	500	448	202	0.003110	0.001014
1000	500	448	203	0.004547	0.001436
1000	500	448	204	0.006548	0.002001
1000	500	448	205	0.009292	0.002744
1000	500	448	206	0.012993	0.003701
1000	500	448	207	0.017904	0.004911
1000	500	448	208	0.024316	0.006413
1000	500	448	209	0.032555	0.008238
1000	500	448	210	0.042969	0.010414
1000	500	448	211	0.055921	0.012952
1000	500	448	212	0.071772	0.015851
1000	500	448	213	0.090858	0.019087
1000	500	448	214	0.113473	0.022614
1000	500	448	215	0.139837	0.026364
1000	500	448	216	0.170080	0.030243
1000	500	448	217	0.204217	0.034137
1000	500	448	218	0.242131	0.037914
1000	500	448	219	0.283566	0.041435
1000	500	448	220	0.328123	0.044557
1000	500	448	221	0.375270	0.047147
1000	500	448	222	0.424359	0.049089
1000	500	448	223	0.474650	0.050292
1000	500	448	224	0.525349	0.050699
1000	500	449	186	0.000001	0.000000
1000	500	449	187	0.000001	0.000001
1000	500	449	188	0.000002	0.000001
1000	500	449	189	0.000004	0.000002
1000	500	449	190	0.000007	0.000003
1000	500	449	191	0.000013	0.000006
1000	500	449	192	0.000023	0.000010
1000	500	449	193	0.000040	0.000017
1000	500	449	194	0.000067	0.000027
1000	500	449	195	0.000111	0.000044
1000	500	449	196	0.000183	0.000071
1000	500	449	197	0.000295	0.000112
1000	500	449	198	0.000468	0.000174

N	n	k	x	P(x)	p(x)
1000	500	449	199	0.000733	0.000265
1000	500	449	200	0.001129	0.000397
1000	500	449	201	0.001714	0.000585
1000	500	449	202	0.002563	0.000848
1000	500	449	203	0.003774	0.001211
1000	500	449	204	0.005475	0.001701
1000	500	449	205	0.007826	0.002351
1000	500	449	206	0.011022	0.003196
1000	500	449	207	0.015297	0.004275
1000	500	449	208	0.020924	0.005627
1000	500	449	209	0.028212	0.007287
1000	500	449	210	0.037498	0.009286
1000	500	449	211	0.049140	0.011642
1000	500	449	212	0.063502	0.014362
1000	500	449	213	0.080935	0.017433
1000	500	449	214	0.101756	0.020821
1000	500	449	215	0.126225	0.024469
1000	500	449	216	0.154520	0.028295
1000	500	449	217	0.186716	0.032196
1000	500	449	218	0.222762	0.036046
1000	500	449	219	0.262474	0.039711
1000	500	449	220	0.305522	0.043048
1000	500	449	221	0.351440	0.045919
1000	500	449	222	0.399637	0.048196
1000	500	449	223	0.449413	0.049777
1000	500	449	224	0.500000	0.050587
1000	500	450	187	0.000001	0.000000
1000	500	450	188	0.000002	0.000001
1000	500	450	189	0.000003	0.000001
1000	500	450	190	0.000006	0.000003
1000	500	450	191	0.000010	0.000004
1000	500	450	192	0.000018	0.000008
1000	500	450	193	0.000030	0.000013
1000	500	450	194	0.000052	0.000021
1000	500	450	195	0.000087	0.000035
1000	500	450	196	0.000143	0.000057
1000	500	450	197	0.000233	0.000090
1000	500	450	198	0.000373	0.000140
1000	500	450	199	0.000588	0.000215
1000	500	450	200	0.000914	0.000325
1000	500	450	201	0.001397	0.000483
1000	500	450	202	0.002104	0.000707
1000	500	450	203	0.003121	0.001017
1000	500	450	204	0.004561	0.001440
1000	500	450	205	0.006567	0.002006
1000	500	450	206	0.009316	0.002749
1000	500	450	207	0.013024	0.003707
1000	500	450	208	0.017942	0.004919
1000	500	450	209	0.024363	0.006421
1000	500	450	210	0.032610	0.008247
1000	500	450	211	0.043033	0.010423
1000	500	450	212	0.055995	0.012961
1000	500	450	213	0.071854	0.015859
1000	500	450	214	0.090948	0.019094
1000	500	450	215	0.113568	0.022620
1000	500	450	216	0.139936	0.026368
1000	500	450	217	0.170180	0.030244
1000	500	450	218	0.204314	0.034134
1000	500	450	219	0.242221	0.037908
1000	500	450	220	0.283646	0.041425
1000	500	450	221	0.328189	0.044543
1000	500	450	222	0.375320	0.047130
1000	500	450	223	0.424389	0.049070
1000	500	450	224	0.474661	0.050271
1000	500	450	225	0.525339	0.050678
1000	500	451	187	0.000001	0.000000
1000	500	451	188	0.000001	0.000001
1000	500	451	189	0.000002	0.000001
1000	500	451	190	0.000004	0.000002
1000	500	451	191	0.000008	0.000003
1000	500	451	192	0.000013	0.000006
1000	500	451	193	0.000023	0.000010
1000	500	451	194	0.000040	0.000017
1000	500	451	195	0.000067	0.000028
1000	500	451	196	0.000112	0.000045
1000	500	451	197	0.000184	0.000072
1000	500	451	198	0.000296	0.000113
1000	500	451	199	0.000471	0.000174
1000	500	451	200	0.000736	0.000266
1000	500	451	201	0.001134	0.000398
1000	500	451	202	0.001721	0.000587
1000	500	451	203	0.002572	0.000851
1000	500	451	204	0.003786	0.001214
1000	500	451	205	0.005491	0.001705
1000	500	451	206	0.007847	0.002356
1000	500	451	207	0.011049	0.003202
1000	500	451	208	0.015331	0.004282
1000	500	451	209	0.020966	0.005635
1000	500	451	210	0.028262	0.007296
1000	500	451	211	0.037556	0.009295
1000	500	451	212	0.049208	0.011651
1000	500	451	213	0.063578	0.014371
1000	500	451	214	0.081019	0.017441
1000	500	451	215	0.101847	0.020828
1000	500	451	216	0.126321	0.024474
1000	500	451	217	0.154618	0.028297
1000	500	451	218	0.186812	0.032194
1000	500	451	219	0.222854	0.036042
1000	500	451	220	0.262557	0.039703
1000	500	451	221	0.305594	0.043037
1000	500	451	222	0.351498	0.045904
1000	500	451	223	0.399676	0.048179
1000	500	451	224	0.449434	0.049757
1000	500	451	225	0.500000	0.050566
1000	500	452	188	0.000001	0.000000
1000	500	452	189	0.000002	0.000001
1000	500	452	190	0.000003	0.000001
1000	500	452	191	0.000006	0.000003
1000	500	452	192	0.000010	0.000004
1000	500	452	193	0.000018	0.000008
1000	500	452	194	0.000031	0.000013
1000	500	452	195	0.000052	0.000022
1000	500	452	196	0.000087	0.000035
1000	500	452	197	0.000144	0.000057
1000	500	452	198	0.000234	0.000090
1000	500	452	199	0.000375	0.000141
1000	500	452	200	0.000591	0.000216
1000	500	452	201	0.000917	0.000326
1000	500	452	202	0.001402	0.000485
1000	500	452	203	0.002111	0.000709
1000	500	452	204	0.003131	0.001020
1000	500	452	205	0.004575	0.001444
1000	500	452	206	0.006585	0.002010
1000	500	452	207	0.009340	0.002755
1000	500	452	208	0.013053	0.003713
1000	500	452	209	0.017979	0.004926
1000	500	452	210	0.024408	0.006429
1000	500	452	211	0.032664	0.008256
1000	500	452	212	0.043096	0.010432
1000	500	452	213	0.056066	0.012970
1000	500	452	214	0.071934	0.015868
1000	500	452	215	0.091035	0.019101
1000	500	452	216	0.113661	0.022626
1000	500	452	217	0.140031	0.026371
1000	500	452	218	0.170275	0.030244
1000	500	452	219	0.204407	0.034131
1000	500	452	220	0.242308	0.037901
1000	500	452	221	0.283723	0.041415
1000	500	452	222	0.328253	0.044530
1000	500	452	223	0.375367	0.047114
1000	500	452	224	0.424419	0.049052
1000	500	452	225	0.474671	0.050252
1000	500	452	226	0.525329	0.050658
1000	500	453	188	0.000001	0.000000
1000	500	453	189	0.000001	0.000001
1000	500	453	190	0.000002	0.000001
1000	500	453	191	0.000004	0.000002
1000	500	453	192	0.000008	0.000003
1000	500	453	193	0.000013	0.000006
1000	500	453	194	0.000023	0.000010
1000	500	453	195	0.000040	0.000017
1000	500	453	196	0.000068	0.000028
1000	500	453	197	0.000113	0.000045
1000	500	453	198	0.000185	0.000072
1000	500	453	199	0.000298	0.000113
1000	500	453	200	0.000473	0.000175
1000	500	453	201	0.000739	0.000267
1000	500	453	202	0.001139	0.000399
1000	500	453	203	0.001727	0.000588
1000	500	453	204	0.002580	0.000853
1000	500	453	205	0.003798	0.001217
1000	500	453	206	0.005507	0.001709
1000	500	453	207	0.007867	0.002360
1000	500	453	208	0.011075	0.003208
1000	500	453	209	0.015363	0.004289
1000	500	453	210	0.021006	0.005642
1000	500	453	211	0.028310	0.007304
1000	500	453	212	0.037613	0.009303
1000	500	453	213	0.049273	0.011660
1000	500	453	214	0.063652	0.014379
1000	500	453	215	0.081101	0.017449
1000	500	453	216	0.101935	0.020834
1000	500	453	217	0.126413	0.024478
1000	500	453	218	0.154712	0.028299
1000	500	453	219	0.186905	0.032193
1000	500	453	220	0.222942	0.036037
1000	500	453	221	0.262637	0.039695
1000	500	453	222	0.305663	0.043025
1000	500	453	223	0.351552	0.045889
1000	500	453	224	0.399714	0.048162
1000	500	453	225	0.449453	0.049739
1000	500	453	226	0.500000	0.050547
1000	500	454	189	0.000001	0.000000
1000	500	454	190	0.000002	0.000001
1000	500	454	191	0.000003	0.000001
1000	500	454	192	0.000006	0.000003
1000	500	454	193	0.000010	0.000004
1000	500	454	194	0.000018	0.000008
1000	500	454	195	0.000031	0.000013
1000	500	454	196	0.000052	0.000022
1000	500	454	197	0.000088	0.000035
1000	500	454	198	0.000145	0.000057
1000	500	454	199	0.000236	0.000091
1000	500	454	200	0.000377	0.000141
1000	500	454	201	0.000594	0.000217
1000	500	454	202	0.000921	0.000327
1000	500	454	203	0.001408	0.000486
1000	500	454	204	0.002119	0.000711
1000	500	454	205	0.003141	0.001023
1000	500	454	206	0.004588	0.001447

Table for $N = 1000$, $n = 500$

N	n	k	x	P(x)	p(x)
1000	500	454	207	0.006603	0.002015
1000	500	454	208	0.009362	0.002760
1000	500	454	209	0.013082	0.003719
1000	500	454	210	0.018014	0.004933
1000	500	454	211	0.024451	0.006436
1000	500	454	212	0.032715	0.008264
1000	500	454	213	0.043155	0.010441
1000	500	454	214	0.056134	0.012979
1000	500	454	215	0.072010	0.015876
1000	500	454	216	0.091118	0.019108
1000	500	454	217	0.113749	0.022631
1000	500	454	218	0.140123	0.026374
1000	500	454	219	0.170367	0.030244
1000	500	454	220	0.204495	0.034128
1000	500	454	221	0.242391	0.037895
1000	500	454	222	0.283796	0.041405
1000	500	454	223	0.328314	0.044518
1000	500	454	224	0.375413	0.047099
1000	500	454	225	0.424447	0.049034
1000	500	454	226	0.474680	0.050233
1000	500	454	227	0.525319	0.050639
1000	500	455	189	0.000001	0.000000
1000	500	455	190	0.000001	0.000001
1000	500	455	191	0.000002	0.000001
1000	500	455	192	0.000004	0.000002
1000	500	455	193	0.000008	0.000003
1000	500	455	194	0.000014	0.000006
1000	500	455	195	0.000024	0.000010
1000	500	455	196	0.000040	0.000017
1000	500	455	197	0.000068	0.000028
1000	500	455	198	0.000113	0.000045
1000	500	455	199	0.000186	0.000072
1000	500	455	200	0.000299	0.000114
1000	500	455	201	0.000475	0.000176
1000	500	455	202	0.000742	0.000267
1000	500	455	203	0.001143	0.000401
1000	500	455	204	0.001733	0.000590
1000	500	455	205	0.002589	0.000856
1000	500	455	206	0.003809	0.001220
1000	500	455	207	0.005522	0.001713
1000	500	455	208	0.007886	0.002365
1000	500	455	209	0.011099	0.003213
1000	500	455	210	0.015394	0.004295
1000	500	455	211	0.021044	0.005649
1000	500	455	212	0.028356	0.007312
1000	500	455	213	0.037667	0.009312
1000	500	455	214	0.049336	0.011668
1000	500	455	215	0.063723	0.014387
1000	500	455	216	0.081179	0.017456
1000	500	455	217	0.102019	0.020840

N	n	k	x	P(x)	p(x)
1000	500	455	218	0.126501	0.024482
1000	500	455	219	0.154802	0.028301
1000	500	455	220	0.186994	0.032192
1000	500	455	221	0.223027	0.036033
1000	500	455	222	0.262714	0.039688
1000	500	455	223	0.305729	0.043015
1000	500	455	224	0.351605	0.045876
1000	500	455	225	0.399751	0.048146
1000	500	455	226	0.449472	0.049721
1000	500	455	227	0.500000	0.050528
1000	500	456	190	0.000001	0.000000
1000	500	456	191	0.000002	0.000001
1000	500	456	192	0.000003	0.000001
1000	500	456	193	0.000006	0.000003
1000	500	456	194	0.000010	0.000004
1000	500	456	195	0.000018	0.000008
1000	500	456	196	0.000031	0.000013
1000	500	456	197	0.000053	0.000022
1000	500	456	198	0.000088	0.000036
1000	500	456	199	0.000146	0.000057
1000	500	456	200	0.000237	0.000091
1000	500	456	201	0.000378	0.000142
1000	500	456	202	0.000596	0.000218
1000	500	456	203	0.000925	0.000329
1000	500	456	204	0.001413	0.000488
1000	500	456	205	0.002126	0.000713
1000	500	456	206	0.003151	0.001025
1000	500	456	207	0.004601	0.001450
1000	500	456	208	0.006619	0.002019
1000	500	456	209	0.009384	0.002764
1000	500	456	210	0.013109	0.003725
1000	500	456	211	0.018048	0.004939
1000	500	456	212	0.024492	0.006444
1000	500	456	213	0.032764	0.008272
1000	500	456	214	0.043212	0.010449
1000	500	456	215	0.056199	0.012987
1000	500	456	216	0.072083	0.015883
1000	500	456	217	0.091197	0.019115
1000	500	456	218	0.113833	0.022636
1000	500	456	219	0.140210	0.026377
1000	500	456	220	0.170455	0.030245
1000	500	456	221	0.204580	0.034126
1000	500	456	222	0.242470	0.037890
1000	500	456	223	0.283866	0.041397
1000	500	456	224	0.328372	0.044506
1000	500	456	225	0.375457	0.047085
1000	500	456	226	0.424474	0.049018
1000	500	456	227	0.474689	0.050215
1000	500	456	228	0.525310	0.050621
1000	500	457	190	0.000001	0.000000

N	n	k	x	P(x)	p(x)
1000	500	457	191	0.000001	0.000001
1000	500	457	192	0.000002	0.000001
1000	500	457	193	0.000004	0.000002
1000	500	457	194	0.000008	0.000003
1000	500	457	195	0.000014	0.000006
1000	500	457	196	0.000024	0.000010
1000	500	457	197	0.000041	0.000017
1000	500	457	198	0.000069	0.000028
1000	500	457	199	0.000114	0.000045
1000	500	457	200	0.000186	0.000073
1000	500	457	201	0.000301	0.000114
1000	500	457	202	0.000477	0.000176
1000	500	457	203	0.000745	0.000268
1000	500	457	204	0.001147	0.000402
1000	500	457	205	0.001739	0.000592
1000	500	457	206	0.002597	0.000858
1000	500	457	207	0.003820	0.001223
1000	500	457	208	0.005536	0.001716
1000	500	457	209	0.007905	0.002369
1000	500	457	210	0.011123	0.003218
1000	500	457	211	0.015424	0.004301
1000	500	457	212	0.021080	0.005656
1000	500	457	213	0.028400	0.007319
1000	500	457	214	0.037719	0.009319
1000	500	457	215	0.049396	0.011676
1000	500	457	216	0.063791	0.014395
1000	500	457	217	0.081253	0.017463
1000	500	457	218	0.102099	0.020846
1000	500	457	219	0.126585	0.024486
1000	500	457	220	0.154888	0.028302
1000	500	457	221	0.187078	0.032191
1000	500	457	222	0.223107	0.036029
1000	500	457	223	0.262788	0.039680
1000	500	457	224	0.305792	0.043004
1000	500	457	225	0.351655	0.045863
1000	500	457	226	0.399785	0.048131
1000	500	457	227	0.449490	0.049704
1000	500	457	228	0.500000	0.050510
1000	500	458	191	0.000001	0.000000
1000	500	458	192	0.000002	0.000001
1000	500	458	193	0.000003	0.000001
1000	500	458	194	0.000006	0.000003
1000	500	458	195	0.000010	0.000005
1000	500	458	196	0.000018	0.000008
1000	500	458	197	0.000031	0.000013
1000	500	458	198	0.000053	0.000022
1000	500	458	199	0.000089	0.000036
1000	500	458	200	0.000146	0.000058
1000	500	458	201	0.000238	0.000091
1000	500	458	202	0.000380	0.000142

N	n	k	x	P(x)	p(x)
1000	500	458	203	0.000598	0.000218
1000	500	458	204	0.000928	0.000330
1000	500	458	205	0.001417	0.000489
1000	500	458	206	0.002132	0.000715
1000	500	458	207	0.003160	0.001028
1000	500	458	208	0.004613	0.001453
1000	500	458	209	0.006635	0.002022
1000	500	458	210	0.009405	0.002769
1000	500	458	211	0.013135	0.003731
1000	500	458	212	0.018080	0.004945
1000	500	458	213	0.024531	0.006451
1000	500	458	214	0.032811	0.008279
1000	500	458	215	0.043267	0.010456
1000	500	458	216	0.056262	0.012995
1000	500	458	217	0.072152	0.015890
1000	500	458	218	0.091273	0.019121
1000	500	458	219	0.113914	0.022641
1000	500	458	220	0.140293	0.026380
1000	500	458	221	0.170538	0.030245
1000	500	458	222	0.204662	0.034123
1000	500	458	223	0.242546	0.037884
1000	500	458	224	0.283934	0.041388
1000	500	458	225	0.328428	0.044494
1000	500	458	226	0.375499	0.047071
1000	500	458	227	0.424500	0.049002
1000	500	458	228	0.474698	0.050198
1000	500	458	229	0.525302	0.050603
1000	500	459	191	0.000001	0.000000
1000	500	459	192	0.000001	0.000001
1000	500	459	193	0.000002	0.000001
1000	500	459	194	0.000004	0.000002
1000	500	459	195	0.000008	0.000003
1000	500	459	196	0.000014	0.000006
1000	500	459	197	0.000024	0.000010
1000	500	459	198	0.000041	0.000017
1000	500	459	199	0.000069	0.000028
1000	500	459	200	0.000115	0.000046
1000	500	459	201	0.000187	0.000073
1000	500	459	202	0.000302	0.000114
1000	500	459	203	0.000479	0.000177
1000	500	459	204	0.000748	0.000269
1000	500	459	205	0.001151	0.000403
1000	500	459	206	0.001744	0.000593
1000	500	459	207	0.002604	0.000860
1000	500	459	208	0.003830	0.001226
1000	500	459	209	0.005550	0.001719
1000	500	459	210	0.007923	0.002373
1000	500	459	211	0.011146	0.003223
1000	500	459	212	0.015453	0.004307
1000	500	459	213	0.021115	0.005663

Table for $N = 1000$, $n = 500$

N	n	k	x	P(x)	p(x)
1000	500	459	214	0.028442	0.007326
1000	500	459	215	0.037769	0.009327
1000	500	459	216	0.049453	0.011684
1000	500	459	217	0.063855	0.014402
1000	500	459	218	0.081325	0.017470
1000	500	459	219	0.102176	0.020851
1000	500	459	220	0.126666	0.024490
1000	500	459	221	0.154969	0.028304
1000	500	459	222	0.187159	0.032190
1000	500	459	223	0.223184	0.036025
1000	500	459	224	0.262858	0.039674
1000	500	459	225	0.305852	0.042995
1000	500	459	226	0.351703	0.045850
1000	500	459	227	0.399819	0.048116
1000	500	459	228	0.449507	0.049688
1000	500	459	229	0.500000	0.050493
1000	500	460	192	0.000001	0.000000
1000	500	460	193	0.000002	0.000001
1000	500	460	194	0.000003	0.000001
1000	500	460	195	0.000006	0.000003
1000	500	460	196	0.000010	0.000005
1000	500	460	197	0.000018	0.000008
1000	500	460	198	0.000031	0.000013
1000	500	460	199	0.000053	0.000022
1000	500	460	200	0.000089	0.000036
1000	500	460	201	0.000147	0.000058
1000	500	460	202	0.000239	0.000092
1000	500	460	203	0.000382	0.000143
1000	500	460	204	0.000601	0.000219
1000	500	460	205	0.000931	0.000331
1000	500	460	206	0.001422	0.000491
1000	500	460	207	0.002139	0.000717
1000	500	460	208	0.003169	0.001030
1000	500	460	209	0.004625	0.001456
1000	500	460	210	0.006651	0.002026
1000	500	460	211	0.009424	0.002773
1000	500	460	212	0.013160	0.003736
1000	500	460	213	0.018111	0.004951
1000	500	460	214	0.024569	0.006457
1000	500	460	215	0.032855	0.008287
1000	500	460	216	0.043319	0.010464
1000	500	460	217	0.056321	0.013002
1000	500	460	218	0.072219	0.015897
1000	500	460	219	0.091346	0.019127
1000	500	460	220	0.113991	0.022645
1000	500	460	221	0.140373	0.026382
1000	500	460	222	0.170618	0.030245
1000	500	460	223	0.204739	0.034121
1000	500	460	224	0.242618	0.037879
1000	500	460	225	0.283997	0.041380

N	n	k	x	P(x)	p(x)
1000	500	460	226	0.328481	0.044483
1000	500	460	227	0.375538	0.047057
1000	500	460	228	0.424525	0.048987
1000	500	460	229	0.474707	0.050182
1000	500	460	230	0.525293	0.050587
1000	500	461	192	0.000001	0.000000
1000	500	461	193	0.000001	0.000001
1000	500	461	194	0.000002	0.000001
1000	500	461	195	0.000004	0.000002
1000	500	461	196	0.000008	0.000003
1000	500	461	197	0.000014	0.000006
1000	500	461	198	0.000024	0.000010
1000	500	461	199	0.000041	0.000017
1000	500	461	200	0.000069	0.000028
1000	500	461	201	0.000115	0.000046
1000	500	461	202	0.000188	0.000073
1000	500	461	203	0.000303	0.000115
1000	500	461	204	0.000481	0.000178
1000	500	461	205	0.000751	0.000270
1000	500	461	206	0.001155	0.000404
1000	500	461	207	0.001750	0.000595
1000	500	461	208	0.002612	0.000862
1000	500	461	209	0.003840	0.001228
1000	500	461	210	0.005563	0.001723
1000	500	461	211	0.007940	0.002377
1000	500	461	212	0.011168	0.003228
1000	500	461	213	0.015480	0.004312
1000	500	461	214	0.021149	0.005669
1000	500	461	215	0.028482	0.007333
1000	500	461	216	0.037816	0.009334
1000	500	461	217	0.049507	0.011691
1000	500	461	218	0.063917	0.014409
1000	500	461	219	0.081392	0.017476
1000	500	461	220	0.102249	0.020856
1000	500	461	221	0.126742	0.024493
1000	500	461	222	0.155047	0.028305
1000	500	461	223	0.187236	0.032189
1000	500	461	224	0.223257	0.036021
1000	500	461	225	0.262924	0.039667
1000	500	461	226	0.305910	0.042985
1000	500	461	227	0.351748	0.045839
1000	500	461	228	0.399850	0.048102
1000	500	461	229	0.449523	0.049673
1000	500	461	230	0.500000	0.050477
1000	500	462	193	0.000001	0.000000
1000	500	462	194	0.000002	0.000001
1000	500	462	195	0.000003	0.000001
1000	500	462	196	0.000006	0.000003
1000	500	462	197	0.000010	0.000005
1000	500	462	198	0.000018	0.000008

N	n	k	x	P(x)	p(x)
1000	500	462	199	0.000032	0.000013
1000	500	462	200	0.000054	0.000022
1000	500	462	201	0.000090	0.000036
1000	500	462	202	0.000148	0.000058
1000	500	462	203	0.000240	0.000092
1000	500	462	204	0.000383	0.000143
1000	500	462	205	0.000603	0.000220
1000	500	462	206	0.000934	0.000331
1000	500	462	207	0.001426	0.000492
1000	500	462	208	0.002145	0.000718
1000	500	462	209	0.003177	0.001032
1000	500	462	210	0.004636	0.001459
1000	500	462	211	0.006665	0.002029
1000	500	462	212	0.009443	0.002778
1000	500	462	213	0.013184	0.003741
1000	500	462	214	0.018141	0.004957
1000	500	462	215	0.024604	0.006464
1000	500	462	216	0.032898	0.008293
1000	500	462	217	0.043369	0.010471
1000	500	462	218	0.056378	0.013009
1000	500	462	219	0.072282	0.015904
1000	500	462	220	0.091414	0.019133
1000	500	462	221	0.114064	0.022649
1000	500	462	222	0.140448	0.026385
1000	500	462	223	0.170694	0.030246
1000	500	462	224	0.204812	0.034119
1000	500	462	225	0.242686	0.037874
1000	500	462	226	0.284058	0.041372
1000	500	462	227	0.328531	0.044473
1000	500	462	228	0.375576	0.047045
1000	500	462	229	0.424548	0.048972
1000	500	462	230	0.474715	0.050166
1000	500	462	231	0.525285	0.050571
1000	500	463	193	0.000001	0.000000
1000	500	463	194	0.000001	0.000001
1000	500	463	195	0.000002	0.000001
1000	500	463	196	0.000004	0.000002
1000	500	463	197	0.000008	0.000003
1000	500	463	198	0.000014	0.000006
1000	500	463	199	0.000024	0.000010
1000	500	463	200	0.000041	0.000017
1000	500	463	201	0.000070	0.000028
1000	500	463	202	0.000116	0.000046
1000	500	463	203	0.000189	0.000073
1000	500	463	204	0.000304	0.000115
1000	500	463	205	0.000482	0.000178
1000	500	463	206	0.000753	0.000271
1000	500	463	207	0.001158	0.000405
1000	500	463	208	0.001755	0.000596
1000	500	463	209	0.002619	0.000864

N	n	k	x	P(x)	p(x)
1000	500	463	210	0.003849	0.001231
1000	500	463	211	0.005575	0.001726
1000	500	463	212	0.007956	0.002381
1000	500	463	213	0.011188	0.003232
1000	500	463	214	0.015506	0.004317
1000	500	463	215	0.021180	0.005675
1000	500	463	216	0.028520	0.007340
1000	500	463	217	0.037861	0.009341
1000	500	463	218	0.049559	0.011698
1000	500	463	219	0.063975	0.014416
1000	500	463	220	0.081457	0.017482
1000	500	463	221	0.102318	0.020861
1000	500	463	222	0.126815	0.024497
1000	500	463	223	0.155121	0.028307
1000	500	463	224	0.187309	0.032188
1000	500	463	225	0.223327	0.036018
1000	500	463	226	0.262988	0.039661
1000	500	463	227	0.305964	0.042976
1000	500	463	228	0.351791	0.045827
1000	500	463	229	0.399880	0.048089
1000	500	463	230	0.449538	0.049658
1000	500	463	231	0.500000	0.050462
1000	500	464	194	0.000001	0.000000
1000	500	464	195	0.000002	0.000001
1000	500	464	196	0.000003	0.000001
1000	500	464	197	0.000006	0.000003
1000	500	464	198	0.000010	0.000005
1000	500	464	199	0.000018	0.000008
1000	500	464	200	0.000032	0.000013
1000	500	464	201	0.000054	0.000022
1000	500	464	202	0.000090	0.000036
1000	500	464	203	0.000148	0.000058
1000	500	464	204	0.000241	0.000092
1000	500	464	205	0.000385	0.000144
1000	500	464	206	0.000605	0.000220
1000	500	464	207	0.000937	0.000332
1000	500	464	208	0.001431	0.000493
1000	500	464	209	0.002151	0.000720
1000	500	464	210	0.003185	0.001034
1000	500	464	211	0.004646	0.001462
1000	500	464	212	0.006679	0.002033
1000	500	464	213	0.009461	0.002782
1000	500	464	214	0.013206	0.003745
1000	500	464	215	0.018169	0.004962
1000	500	464	216	0.024638	0.006470
1000	500	464	217	0.032938	0.008300
1000	500	464	218	0.043416	0.010478
1000	500	464	219	0.056431	0.013016
1000	500	464	220	0.072341	0.015910
1000	500	464	221	0.091479	0.019138

Table for $N = 1000,\ n = 500$

N	n	k	x	P(x)	p(x)
1000	500	464	222	0.114133	0.022653
1000	500	464	223	0.140520	0.026387
1000	500	464	224	0.170766	0.030246
1000	500	464	225	0.204882	0.034116
1000	500	464	226	0.242751	0.037869
1000	500	464	227	0.284116	0.041365
1000	500	464	228	0.328579	0.044463
1000	500	464	229	0.375612	0.047033
1000	500	464	230	0.424570	0.048959
1000	500	464	231	0.474722	0.050152
1000	500	464	232	0.525278	0.050556
1000	500	465	194	0.000001	0.000000
1000	500	465	195	0.000001	0.000001
1000	500	465	196	0.000002	0.000001
1000	500	465	197	0.000004	0.000002
1000	500	465	198	0.000008	0.000003
1000	500	465	199	0.000014	0.000006
1000	500	465	200	0.000024	0.000010
1000	500	465	201	0.000041	0.000017
1000	500	465	202	0.000070	0.000028
1000	500	465	203	0.000116	0.000046
1000	500	465	204	0.000190	0.000074
1000	500	465	205	0.000305	0.000116
1000	500	465	206	0.000484	0.000179
1000	500	465	207	0.000756	0.000272
1000	500	465	208	0.001162	0.000406
1000	500	465	209	0.001760	0.000598
1000	500	465	210	0.002625	0.000866
1000	500	465	211	0.003858	0.001233
1000	500	465	212	0.005587	0.001729
1000	500	465	213	0.007971	0.002384
1000	500	465	214	0.011208	0.003236
1000	500	465	215	0.015530	0.004322
1000	500	465	216	0.021210	0.005680
1000	500	465	217	0.028556	0.007346
1000	500	465	218	0.037903	0.009347
1000	500	465	219	0.049608	0.011705
1000	500	465	220	0.064030	0.014422
1000	500	465	221	0.081518	0.017488
1000	500	465	222	0.102384	0.020866
1000	500	465	223	0.126883	0.024500
1000	500	465	224	0.155191	0.028308
1000	500	465	225	0.187378	0.032187
1000	500	465	226	0.223393	0.036014
1000	500	465	227	0.263047	0.039655
1000	500	465	228	0.306016	0.042968
1000	500	465	229	0.351832	0.045817
1000	500	465	230	0.399908	0.048076
1000	500	465	231	0.449553	0.049644
1000	500	465	232	0.500000	0.050447
1000	500	466	194	0.000000	0.000000
1000	500	466	195	0.000001	0.000000
1000	500	466	196	0.000002	0.000001
1000	500	466	197	0.000003	0.000001
1000	500	466	198	0.000006	0.000003
1000	500	466	199	0.000011	0.000005
1000	500	466	200	0.000018	0.000008
1000	500	466	201	0.000032	0.000013
1000	500	466	202	0.000054	0.000022
1000	500	466	203	0.000090	0.000036
1000	500	466	204	0.000149	0.000059
1000	500	466	205	0.000242	0.000093
1000	500	466	206	0.000386	0.000144
1000	500	466	207	0.000607	0.000221
1000	500	466	208	0.000940	0.000333
1000	500	466	209	0.001435	0.000494
1000	500	466	210	0.002156	0.000721
1000	500	466	211	0.003192	0.001036
1000	500	466	212	0.004656	0.001464
1000	500	466	213	0.006692	0.002036
1000	500	466	214	0.009478	0.002785
1000	500	466	215	0.013227	0.003750
1000	500	466	216	0.018195	0.004967
1000	500	466	217	0.024670	0.006475
1000	500	466	218	0.032976	0.008306
1000	500	466	219	0.043460	0.010484
1000	500	466	220	0.056482	0.013022
1000	500	466	221	0.072398	0.015916
1000	500	466	222	0.091541	0.019143
1000	500	466	223	0.114198	0.022657
1000	500	466	224	0.140588	0.026389
1000	500	466	225	0.170834	0.030246
1000	500	466	226	0.204948	0.034114
1000	500	466	227	0.242812	0.037864
1000	500	466	228	0.284170	0.041358
1000	500	466	229	0.328624	0.044454
1000	500	466	230	0.375646	0.047022
1000	500	466	231	0.424591	0.048946
1000	500	466	232	0.474729	0.050138
1000	500	466	233	0.525271	0.050542
1000	500	467	195	0.000001	0.000000
1000	500	467	196	0.000001	0.000001
1000	500	467	197	0.000002	0.000001
1000	500	467	198	0.000004	0.000002
1000	500	467	199	0.000008	0.000004
1000	500	467	200	0.000014	0.000006
1000	500	467	201	0.000024	0.000010
1000	500	467	202	0.000042	0.000017
1000	500	467	203	0.000070	0.000029
1000	500	467	204	0.000117	0.000046
1000	500	467	205	0.000190	0.000074
1000	500	467	206	0.000306	0.000116
1000	500	467	207	0.000486	0.000179
1000	500	467	208	0.000758	0.000272
1000	500	467	209	0.001165	0.000407
1000	500	467	210	0.001764	0.000599
1000	500	467	211	0.002631	0.000867
1000	500	467	212	0.003867	0.001235
1000	500	467	213	0.005598	0.001731
1000	500	467	214	0.007986	0.002388
1000	500	467	215	0.011226	0.003240
1000	500	467	216	0.015553	0.004327
1000	500	467	217	0.021239	0.005686
1000	500	467	218	0.028590	0.007351
1000	500	467	219	0.037943	0.009353
1000	500	467	220	0.049654	0.011711
1000	500	467	221	0.064082	0.014428
1000	500	467	222	0.081575	0.017493
1000	500	467	223	0.102445	0.020870
1000	500	467	224	0.126948	0.024503
1000	500	467	225	0.155257	0.028309
1000	500	467	226	0.187444	0.032186
1000	500	467	227	0.223455	0.036011
1000	500	467	228	0.263104	0.039649
1000	500	467	229	0.306064	0.042960
1000	500	467	230	0.351871	0.045807
1000	500	467	231	0.399935	0.048064
1000	500	467	232	0.449566	0.049631
1000	500	467	233	0.500000	0.050434
1000	500	468	195	0.000000	0.000000
1000	500	468	196	0.000001	0.000000
1000	500	468	197	0.000002	0.000001
1000	500	468	198	0.000003	0.000001
1000	500	468	199	0.000006	0.000003
1000	500	468	200	0.000011	0.000005
1000	500	468	201	0.000019	0.000008
1000	500	468	202	0.000032	0.000013
1000	500	468	203	0.000054	0.000022
1000	500	468	204	0.000091	0.000037
1000	500	468	205	0.000150	0.000059
1000	500	468	206	0.000242	0.000093
1000	500	468	207	0.000387	0.000145
1000	500	468	208	0.000609	0.000222
1000	500	468	209	0.000943	0.000334
1000	500	468	210	0.001438	0.000495
1000	500	468	211	0.002161	0.000723
1000	500	468	212	0.003199	0.001038
1000	500	468	213	0.004666	0.001467
1000	500	468	214	0.006705	0.002039
1000	500	468	215	0.009493	0.002789
1000	500	468	216	0.013247	0.003754
1000	500	468	217	0.018220	0.004972
1000	500	468	218	0.024700	0.006481
1000	500	468	219	0.033012	0.008312
1000	500	468	220	0.043502	0.010490
1000	500	468	221	0.056530	0.013028
1000	500	468	222	0.072451	0.015922
1000	500	468	223	0.091599	0.019148
1000	500	468	224	0.114260	0.022661
1000	500	468	225	0.140651	0.026391
1000	500	468	226	0.170898	0.030246
1000	500	468	227	0.205010	0.034112
1000	500	468	228	0.242870	0.037860
1000	500	468	229	0.284221	0.041351
1000	500	468	230	0.328667	0.044445
1000	500	468	231	0.375677	0.047011
1000	500	468	232	0.424611	0.048934
1000	500	468	233	0.474736	0.050125
1000	500	468	234	0.525264	0.050528
1000	500	469	196	0.000001	0.000000
1000	500	469	197	0.000001	0.000001
1000	500	469	198	0.000002	0.000001
1000	500	469	199	0.000004	0.000002
1000	500	469	200	0.000008	0.000004
1000	500	469	201	0.000014	0.000006
1000	500	469	202	0.000024	0.000010
1000	500	469	203	0.000042	0.000017
1000	500	469	204	0.000070	0.000029
1000	500	469	205	0.000117	0.000046
1000	500	469	206	0.000191	0.000074
1000	500	469	207	0.000307	0.000116
1000	500	469	208	0.000487	0.000180
1000	500	469	209	0.000760	0.000273
1000	500	469	210	0.001168	0.000408
1000	500	469	211	0.001768	0.000600
1000	500	469	212	0.002637	0.000869
1000	500	469	213	0.003875	0.001237
1000	500	469	214	0.005609	0.001734
1000	500	469	215	0.007999	0.002391
1000	500	469	216	0.011243	0.003244
1000	500	469	217	0.015575	0.004331
1000	500	469	218	0.021265	0.005690
1000	500	469	219	0.028622	0.007357
1000	500	469	220	0.037981	0.009359
1000	500	469	221	0.049698	0.011717
1000	500	469	222	0.064131	0.014434
1000	500	469	223	0.081629	0.017498
1000	500	469	224	0.102504	0.020874
1000	500	469	225	0.127009	0.024506
1000	500	469	226	0.155320	0.028310

Table for $N = 1000$, $n = 500$

N	n	k	x	P(x)	p(x)
1000	500	469	227	0.187505	0.032185
1000	500	469	228	0.223513	0.036008
1000	500	469	229	0.263157	0.039644
1000	500	469	230	0.306110	0.042953
1000	500	469	231	0.351907	0.045797
1000	500	469	232	0.399960	0.048053
1000	500	469	233	0.449579	0.049619
1000	500	469	234	0.500000	0.050421
1000	500	470	196	0.000001	0.000000
1000	500	470	197	0.000001	0.000000
1000	500	470	198	0.000002	0.000001
1000	500	470	199	0.000003	0.000002
1000	500	470	200	0.000006	0.000003
1000	500	470	201	0.000011	0.000005
1000	500	470	202	0.000019	0.000008
1000	500	470	203	0.000032	0.000013
1000	500	470	204	0.000055	0.000022
1000	500	470	205	0.000091	0.000037
1000	500	470	206	0.000150	0.000059
1000	500	470	207	0.000243	0.000093
1000	500	470	208	0.000388	0.000145
1000	500	470	209	0.000610	0.000222
1000	500	470	210	0.000945	0.000335
1000	500	470	211	0.001442	0.000496
1000	500	470	212	0.002166	0.000724
1000	500	470	213	0.003206	0.001040
1000	500	470	214	0.004675	0.001469
1000	500	470	215	0.006716	0.002042
1000	500	470	216	0.009508	0.002792
1000	500	470	217	0.013266	0.003758
1000	500	470	218	0.018243	0.004977
1000	500	470	219	0.024729	0.006486
1000	500	470	220	0.033046	0.008317
1000	500	470	221	0.043541	0.010496
1000	500	470	222	0.056575	0.013033
1000	500	470	223	0.072501	0.015927
1000	500	470	224	0.091654	0.019152
1000	500	470	225	0.114318	0.022664
1000	500	470	226	0.140711	0.026393
1000	500	470	227	0.170958	0.030246
1000	500	470	228	0.205068	0.034111
1000	500	470	229	0.242924	0.037856
1000	500	470	230	0.284269	0.041345
1000	500	470	231	0.328707	0.044437
1000	500	470	232	0.375707	0.047001
1000	500	470	233	0.424630	0.048922
1000	500	470	234	0.474742	0.050112
1000	500	470	235	0.525258	0.050516
1000	500	471	197	0.000001	0.000000
1000	500	471	198	0.000001	0.000001

N	n	k	x	P(x)	p(x)
1000	500	471	199	0.000002	0.000001
1000	500	471	200	0.000004	0.000002
1000	500	471	201	0.000008	0.000004
1000	500	471	202	0.000014	0.000006
1000	500	471	203	0.000025	0.000010
1000	500	471	204	0.000042	0.000017
1000	500	471	205	0.000071	0.000029
1000	500	471	206	0.000117	0.000047
1000	500	471	207	0.000192	0.000074
1000	500	471	208	0.000308	0.000117
1000	500	471	209	0.000488	0.000180
1000	500	471	210	0.000762	0.000274
1000	500	471	211	0.001171	0.000409
1000	500	471	212	0.001772	0.000601
1000	500	471	213	0.002643	0.000870
1000	500	471	214	0.003882	0.001239
1000	500	471	215	0.005618	0.001736
1000	500	471	216	0.008012	0.002394
1000	500	471	217	0.011260	0.003248
1000	500	471	218	0.015595	0.004335
1000	500	471	219	0.021290	0.005695
1000	500	471	220	0.028652	0.007362
1000	500	471	221	0.038016	0.009364
1000	500	471	222	0.049738	0.011722
1000	500	471	223	0.064178	0.014439
1000	500	471	224	0.081680	0.017503
1000	500	471	225	0.102558	0.020878
1000	500	471	226	0.127067	0.024508
1000	500	471	227	0.155378	0.028311
1000	500	471	228	0.187562	0.032185
1000	500	471	229	0.223568	0.036005
1000	500	471	230	0.263207	0.039639
1000	500	471	231	0.306153	0.042946
1000	500	471	232	0.351941	0.045788
1000	500	471	233	0.399984	0.048043
1000	500	471	234	0.449591	0.049607
1000	500	471	235	0.500000	0.050409
1000	500	472	197	0.000001	0.000000
1000	500	472	198	0.000001	0.000000
1000	500	472	199	0.000002	0.000001
1000	500	472	200	0.000003	0.000002
1000	500	472	201	0.000006	0.000003
1000	500	472	202	0.000011	0.000005
1000	500	472	203	0.000019	0.000008
1000	500	472	204	0.000032	0.000014
1000	500	472	205	0.000055	0.000022
1000	500	472	206	0.000091	0.000037
1000	500	472	207	0.000151	0.000059
1000	500	472	208	0.000244	0.000093
1000	500	472	209	0.000389	0.000145

N	n	k	x	P(x)	p(x)
1000	500	472	210	0.000612	0.000223
1000	500	472	211	0.000948	0.000336
1000	500	472	212	0.001445	0.000497
1000	500	472	213	0.002171	0.000726
1000	500	472	214	0.003212	0.001041
1000	500	472	215	0.004683	0.001471
1000	500	472	216	0.006727	0.002044
1000	500	472	217	0.009522	0.002795
1000	500	472	218	0.013284	0.003762
1000	500	472	219	0.018265	0.004981
1000	500	472	220	0.024755	0.006490
1000	500	472	221	0.033078	0.008322
1000	500	472	222	0.043578	0.010501
1000	500	472	223	0.056617	0.013038
1000	500	472	224	0.072548	0.015932
1000	500	472	225	0.091705	0.019156
1000	500	472	226	0.114372	0.022667
1000	500	472	227	0.140767	0.026395
1000	500	472	228	0.171014	0.030247
1000	500	472	229	0.205123	0.034109
1000	500	472	230	0.242975	0.037853
1000	500	472	231	0.284314	0.041339
1000	500	472	232	0.328744	0.044429
1000	500	472	233	0.375735	0.046992
1000	500	472	234	0.424647	0.048912
1000	500	472	235	0.474748	0.050101
1000	500	472	236	0.525252	0.050504
1000	500	473	198	0.000001	0.000000
1000	500	473	199	0.000001	0.000001
1000	500	473	200	0.000002	0.000001
1000	500	473	201	0.000004	0.000002
1000	500	473	202	0.000008	0.000004
1000	500	473	203	0.000014	0.000006
1000	500	473	204	0.000025	0.000010
1000	500	473	205	0.000042	0.000018
1000	500	473	206	0.000071	0.000029
1000	500	473	207	0.000118	0.000047
1000	500	473	208	0.000192	0.000075
1000	500	473	209	0.000309	0.000117
1000	500	473	210	0.000490	0.000181
1000	500	473	211	0.000764	0.000274
1000	500	473	212	0.001174	0.000410
1000	500	473	213	0.001776	0.000602
1000	500	473	214	0.002648	0.000872
1000	500	473	215	0.003889	0.001241
1000	500	473	216	0.005628	0.001739
1000	500	473	217	0.008024	0.002397
1000	500	473	218	0.011275	0.003251
1000	500	473	219	0.015614	0.004339
1000	500	473	220	0.021314	0.005699

N	n	k	x	P(x)	p(x)
1000	500	473	221	0.028680	0.007367
1000	500	473	222	0.038050	0.009369
1000	500	473	223	0.049777	0.011727
1000	500	473	224	0.064220	0.014444
1000	500	473	225	0.081727	0.017507
1000	500	473	226	0.102609	0.020882
1000	500	473	227	0.127120	0.024511
1000	500	473	228	0.155432	0.028312
1000	500	473	229	0.187616	0.032184
1000	500	473	230	0.223619	0.036003
1000	500	473	231	0.263253	0.039635
1000	500	473	232	0.306193	0.042939
1000	500	473	233	0.351973	0.045780
1000	500	473	234	0.400006	0.048033
1000	500	473	235	0.449603	0.049597
1000	500	473	236	0.500000	0.050397
1000	500	474	198	0.000001	0.000000
1000	500	474	199	0.000001	0.000000
1000	500	474	200	0.000002	0.000001
1000	500	474	201	0.000003	0.000002
1000	500	474	202	0.000006	0.000003
1000	500	474	203	0.000011	0.000005
1000	500	474	204	0.000019	0.000008
1000	500	474	205	0.000032	0.000014
1000	500	474	206	0.000055	0.000023
1000	500	474	207	0.000092	0.000037
1000	500	474	208	0.000151	0.000059
1000	500	474	209	0.000245	0.000094
1000	500	474	210	0.000390	0.000146
1000	500	474	211	0.000614	0.000223
1000	500	474	212	0.000950	0.000336
1000	500	474	213	0.001448	0.000498
1000	500	474	214	0.002175	0.000727
1000	500	474	215	0.003218	0.001043
1000	500	474	216	0.004691	0.001473
1000	500	474	217	0.006737	0.002047
1000	500	474	218	0.009535	0.002798
1000	500	474	219	0.013300	0.003765
1000	500	474	220	0.018285	0.004985
1000	500	474	221	0.024780	0.006495
1000	500	474	222	0.033107	0.008327
1000	500	474	223	0.043613	0.010506
1000	500	474	224	0.056656	0.013043
1000	500	474	225	0.072592	0.015936
1000	500	474	226	0.091752	0.019160
1000	500	474	227	0.114423	0.022670
1000	500	474	228	0.140819	0.026397
1000	500	474	229	0.171066	0.030247
1000	500	474	230	0.205173	0.034107
1000	500	474	231	0.243022	0.037849

Table for $N = 1000,\ n = 500$

N	n	k	x	P(x)	p(x)
1000	500	474	232	0.284356	0.041334
1000	500	474	233	0.328779	0.044422
1000	500	474	234	0.375761	0.046983
1000	500	474	235	0.424663	0.048902
1000	500	474	236	0.474753	0.050090
1000	500	474	237	0.525246	0.050493
1000	500	475	199	0.000001	0.000000
1000	500	475	200	0.000001	0.000001
1000	500	475	201	0.000002	0.000001
1000	500	475	202	0.000005	0.000002
1000	500	475	203	0.000008	0.000004
1000	500	475	204	0.000014	0.000006
1000	500	475	205	0.000025	0.000010
1000	500	475	206	0.000042	0.000018
1000	500	475	207	0.000071	0.000029
1000	500	475	208	0.000118	0.000047
1000	500	475	209	0.000193	0.000075
1000	500	475	210	0.000310	0.000117
1000	500	475	211	0.000491	0.000181
1000	500	475	212	0.000766	0.000275
1000	500	475	213	0.001176	0.000410
1000	500	475	214	0.001780	0.000603
1000	500	475	215	0.002653	0.000873
1000	500	475	216	0.003895	0.001243
1000	500	475	217	0.005636	0.001741
1000	500	475	218	0.008035	0.002399
1000	500	475	219	0.011289	0.003254
1000	500	475	220	0.015632	0.004343
1000	500	475	221	0.021335	0.005703
1000	500	475	222	0.028706	0.007371
1000	500	475	223	0.038080	0.009374
1000	500	475	224	0.049812	0.011732
1000	500	475	225	0.064260	0.014448
1000	500	475	226	0.081771	0.017511
1000	500	475	227	0.102656	0.020885
1000	500	475	228	0.127169	0.024513
1000	500	475	229	0.155483	0.028313
1000	500	475	230	0.187666	0.032183
1000	500	475	231	0.223666	0.036000
1000	500	475	232	0.263296	0.039630
1000	500	475	233	0.306230	0.042933
1000	500	475	234	0.352002	0.045772
1000	500	475	235	0.400026	0.048024
1000	500	475	236	0.449613	0.049587
1000	500	475	237	0.500000	0.050387
1000	500	476	199	0.000001	0.000000
1000	500	476	200	0.000001	0.000000
1000	500	476	201	0.000002	0.000001
1000	500	476	202	0.000003	0.000002
1000	500	476	203	0.000006	0.000003

N	n	k	x	P(x)	p(x)
1000	500	476	204	0.000011	0.000005
1000	500	476	205	0.000019	0.000008
1000	500	476	206	0.000032	0.000014
1000	500	476	207	0.000055	0.000023
1000	500	476	208	0.000092	0.000037
1000	500	476	209	0.000151	0.000059
1000	500	476	210	0.000245	0.000094
1000	500	476	211	0.000391	0.000146
1000	500	476	212	0.000615	0.000224
1000	500	476	213	0.000952	0.000337
1000	500	476	214	0.001451	0.000499
1000	500	476	215	0.002179	0.000728
1000	500	476	216	0.003223	0.001044
1000	500	476	217	0.004698	0.001475
1000	500	476	218	0.006747	0.002049
1000	500	476	219	0.009548	0.002801
1000	500	476	220	0.013316	0.003768
1000	500	476	221	0.018304	0.004989
1000	500	476	222	0.024803	0.006499
1000	500	476	223	0.033134	0.008331
1000	500	476	224	0.043644	0.010510
1000	500	476	225	0.056692	0.013048
1000	500	476	226	0.072632	0.015940
1000	500	476	227	0.091796	0.019164
1000	500	476	228	0.114469	0.022673
1000	500	476	229	0.140868	0.026398
1000	500	476	230	0.171115	0.030247
1000	500	476	231	0.205220	0.034106
1000	500	476	232	0.243066	0.037846
1000	500	476	233	0.284395	0.041329
1000	500	476	234	0.328811	0.044416
1000	500	476	235	0.375785	0.046975
1000	500	476	236	0.424678	0.048893
1000	500	476	237	0.474758	0.050081
1000	500	476	238	0.525241	0.050483
1000	500	477	200	0.000001	0.000000
1000	500	477	201	0.000001	0.000001
1000	500	477	202	0.000002	0.000001
1000	500	477	203	0.000005	0.000002
1000	500	477	204	0.000008	0.000004
1000	500	477	205	0.000014	0.000006
1000	500	477	206	0.000025	0.000011
1000	500	477	207	0.000042	0.000018
1000	500	477	208	0.000071	0.000029
1000	500	477	209	0.000118	0.000047
1000	500	477	210	0.000193	0.000075
1000	500	477	211	0.000311	0.000117
1000	500	477	212	0.000492	0.000181
1000	500	477	213	0.000767	0.000275
1000	500	477	214	0.001179	0.000411

N	n	k	x	P(x)	p(x)
1000	500	477	215	0.001783	0.000604
1000	500	477	216	0.002657	0.000874
1000	500	477	217	0.003901	0.001244
1000	500	477	218	0.005644	0.001743
1000	500	477	219	0.008045	0.002401
1000	500	477	220	0.011302	0.003257
1000	500	477	221	0.015648	0.004346
1000	500	477	222	0.021355	0.005707
1000	500	477	223	0.028730	0.007375
1000	500	477	224	0.038108	0.009378
1000	500	477	225	0.049845	0.011736
1000	500	477	226	0.064297	0.014453
1000	500	477	227	0.081812	0.017515
1000	500	477	228	0.102700	0.020888
1000	500	477	229	0.127215	0.024515
1000	500	477	230	0.155529	0.028314
1000	500	477	231	0.187712	0.032182
1000	500	477	232	0.223710	0.035998
1000	500	477	233	0.263336	0.039626
1000	500	477	234	0.306264	0.042928
1000	500	477	235	0.352029	0.045765
1000	500	477	236	0.400045	0.048016
1000	500	477	237	0.449623	0.049578
1000	500	477	238	0.500000	0.050377
1000	500	478	200	0.000001	0.000000
1000	500	478	201	0.000001	0.000000
1000	500	478	202	0.000002	0.000001
1000	500	478	203	0.000003	0.000002
1000	500	478	204	0.000006	0.000003
1000	500	478	205	0.000011	0.000005
1000	500	478	206	0.000019	0.000008
1000	500	478	207	0.000033	0.000014
1000	500	478	208	0.000055	0.000023
1000	500	478	209	0.000092	0.000037
1000	500	478	210	0.000152	0.000060
1000	500	478	211	0.000246	0.000094
1000	500	478	212	0.000392	0.000146
1000	500	478	213	0.000616	0.000224
1000	500	478	214	0.000954	0.000337
1000	500	478	215	0.001454	0.000500
1000	500	478	216	0.002182	0.000729
1000	500	478	217	0.003228	0.001046
1000	500	478	218	0.004704	0.001476
1000	500	478	219	0.006755	0.002051
1000	500	478	220	0.009559	0.002803
1000	500	478	221	0.013330	0.003771
1000	500	478	222	0.018322	0.004992
1000	500	478	223	0.024824	0.006502
1000	500	478	224	0.033159	0.008335
1000	500	478	225	0.043674	0.010514

N	n	k	x	P(x)	p(x)
1000	500	478	226	0.056725	0.013052
1000	500	478	227	0.072669	0.015944
1000	500	478	228	0.091837	0.019167
1000	500	478	229	0.114512	0.022676
1000	500	478	230	0.140912	0.026400
1000	500	478	231	0.171159	0.030247
1000	500	478	232	0.205263	0.034104
1000	500	478	233	0.243106	0.037843
1000	500	478	234	0.284431	0.041324
1000	500	478	235	0.328840	0.044410
1000	500	478	236	0.375808	0.046967
1000	500	478	237	0.424692	0.048884
1000	500	478	238	0.474763	0.050071
1000	500	478	239	0.525237	0.050474
1000	500	479	201	0.000001	0.000000
1000	500	479	202	0.000001	0.000001
1000	500	479	203	0.000002	0.000001
1000	500	479	204	0.000005	0.000002
1000	500	479	205	0.000008	0.000004
1000	500	479	206	0.000014	0.000006
1000	500	479	207	0.000025	0.000011
1000	500	479	208	0.000043	0.000018
1000	500	479	209	0.000072	0.000029
1000	500	479	210	0.000119	0.000047
1000	500	479	211	0.000194	0.000075
1000	500	479	212	0.000312	0.000118
1000	500	479	213	0.000493	0.000182
1000	500	479	214	0.000769	0.000276
1000	500	479	215	0.001181	0.000412
1000	500	479	216	0.001786	0.000605
1000	500	479	217	0.002661	0.000875
1000	500	479	218	0.003907	0.001246
1000	500	479	219	0.005651	0.001744
1000	500	479	220	0.008055	0.002404
1000	500	479	221	0.011314	0.003259
1000	500	479	222	0.015663	0.004349
1000	500	479	223	0.021374	0.005710
1000	500	479	224	0.028752	0.007379
1000	500	479	225	0.038134	0.009382
1000	500	479	226	0.049874	0.011740
1000	500	479	227	0.064331	0.014456
1000	500	479	228	0.081849	0.017518
1000	500	479	229	0.102740	0.020891
1000	500	479	230	0.127257	0.024517
1000	500	479	231	0.155572	0.028315
1000	500	479	232	0.187754	0.032182
1000	500	479	233	0.223750	0.035996
1000	500	479	234	0.263373	0.039623
1000	500	479	235	0.306295	0.042922
1000	500	479	236	0.352054	0.045759

Table for $N = 1000$, $n = 500$

N	n	k	x	P(x)	p(x)
1000	500	479	237	0.400062	0.048008
1000	500	479	238	0.449631	0.049569
1000	500	479	239	0.500000	0.050368
1000	500	480	201	0.000001	0.000000
1000	500	480	202	0.000001	0.000000
1000	500	480	203	0.000002	0.000001
1000	500	480	204	0.000003	0.000002
1000	500	480	205	0.000006	0.000003
1000	500	480	206	0.000011	0.000005
1000	500	480	207	0.000019	0.000008
1000	500	480	208	0.000033	0.000014
1000	500	480	209	0.000055	0.000023
1000	500	480	210	0.000093	0.000037
1000	500	480	211	0.000152	0.000060
1000	500	480	212	0.000246	0.000094
1000	500	480	213	0.000393	0.000147
1000	500	480	214	0.000618	0.000224
1000	500	480	215	0.000955	0.000338
1000	500	480	216	0.001456	0.000501
1000	500	480	217	0.002186	0.000730
1000	500	480	218	0.003233	0.001047
1000	500	480	219	0.004710	0.001478
1000	500	480	220	0.006763	0.002053
1000	500	480	221	0.009569	0.002806
1000	500	480	222	0.013343	0.003774
1000	500	480	223	0.018337	0.004995
1000	500	480	224	0.024843	0.006506
1000	500	480	225	0.033182	0.008339
1000	500	480	226	0.043700	0.010518
1000	500	480	227	0.056756	0.013056
1000	500	480	228	0.072703	0.015948
1000	500	480	229	0.091874	0.019170
1000	500	480	230	0.114552	0.022678
1000	500	480	231	0.140953	0.026401
1000	500	480	232	0.171200	0.030247
1000	500	480	233	0.205303	0.034103
1000	500	480	234	0.243143	0.037840
1000	500	480	235	0.284463	0.041320
1000	500	480	236	0.328867	0.044404
1000	500	480	237	0.375828	0.046961
1000	500	480	238	0.424704	0.048876
1000	500	480	239	0.474767	0.050063
1000	500	480	240	0.525232	0.050465
1000	500	481	202	0.000001	0.000000
1000	500	481	203	0.000001	0.000001
1000	500	481	204	0.000002	0.000001
1000	500	481	205	0.000005	0.000002
1000	500	481	206	0.000008	0.000004
1000	500	481	207	0.000014	0.000006
1000	500	481	208	0.000025	0.000011
1000	500	481	209	0.000043	0.000018
1000	500	481	210	0.000072	0.000029
1000	500	481	211	0.000119	0.000047
1000	500	481	212	0.000194	0.000075
1000	500	481	213	0.000312	0.000118
1000	500	481	214	0.000494	0.000182
1000	500	481	215	0.000770	0.000276
1000	500	481	216	0.001183	0.000412
1000	500	481	217	0.001789	0.000606
1000	500	481	218	0.002665	0.000876
1000	500	481	219	0.003912	0.001247
1000	500	481	220	0.005658	0.001746
1000	500	481	221	0.008063	0.002406
1000	500	481	222	0.011325	0.003262
1000	500	481	223	0.015677	0.004352
1000	500	481	224	0.021390	0.005714
1000	500	481	225	0.028772	0.007382
1000	500	481	226	0.038158	0.009386
1000	500	481	227	0.049902	0.011744
1000	500	481	228	0.064362	0.014460
1000	500	481	229	0.081883	0.017521
1000	500	481	230	0.102776	0.020893
1000	500	481	231	0.127295	0.024519
1000	500	481	232	0.155611	0.028316
1000	500	481	233	0.187792	0.032181
1000	500	481	234	0.223786	0.035994
1000	500	481	235	0.263406	0.039619
1000	500	481	236	0.306324	0.042918
1000	500	481	237	0.352077	0.045753
1000	500	481	238	0.400078	0.048001
1000	500	481	239	0.449639	0.049561
1000	500	481	240	0.500000	0.050360
1000	500	482	202	0.000001	0.000000
1000	500	482	203	0.000001	0.000000
1000	500	482	204	0.000002	0.000001
1000	500	482	205	0.000003	0.000002
1000	500	482	206	0.000006	0.000003
1000	500	482	207	0.000011	0.000005
1000	500	482	208	0.000019	0.000008
1000	500	482	209	0.000033	0.000014
1000	500	482	210	0.000056	0.000023
1000	500	482	211	0.000093	0.000037
1000	500	482	212	0.000153	0.000060
1000	500	482	213	0.000247	0.000094
1000	500	482	214	0.000394	0.000147
1000	500	482	215	0.000619	0.000225
1000	500	482	216	0.000957	0.000338
1000	500	482	217	0.001458	0.000501
1000	500	482	218	0.002189	0.000731
1000	500	482	219	0.003237	0.001048
1000	500	482	220	0.004716	0.001479
1000	500	482	221	0.006770	0.002054
1000	500	482	222	0.009578	0.002808
1000	500	482	223	0.013354	0.003776
1000	500	482	224	0.018352	0.004998
1000	500	482	225	0.024861	0.006509
1000	500	482	226	0.033203	0.008342
1000	500	482	227	0.043724	0.010521
1000	500	482	228	0.056783	0.013059
1000	500	482	229	0.072734	0.015951
1000	500	482	230	0.091907	0.019173
1000	500	482	231	0.114587	0.022680
1000	500	482	232	0.140989	0.026402
1000	500	482	233	0.171236	0.030247
1000	500	482	234	0.205338	0.034102
1000	500	482	235	0.243176	0.037838
1000	500	482	236	0.284493	0.041316
1000	500	482	237	0.328892	0.044399
1000	500	482	238	0.375846	0.046955
1000	500	482	239	0.424715	0.048869
1000	500	482	240	0.474771	0.050056
1000	500	482	241	0.525229	0.050457
1000	500	483	203	0.000001	0.000000
1000	500	483	204	0.000001	0.000001
1000	500	483	205	0.000003	0.000001
1000	500	483	206	0.000005	0.000002
1000	500	483	207	0.000008	0.000004
1000	500	483	208	0.000014	0.000006
1000	500	483	209	0.000025	0.000011
1000	500	483	210	0.000043	0.000018
1000	500	483	211	0.000072	0.000029
1000	500	483	212	0.000119	0.000047
1000	500	483	213	0.000195	0.000075
1000	500	483	214	0.000313	0.000118
1000	500	483	215	0.000495	0.000182
1000	500	483	216	0.000771	0.000276
1000	500	483	217	0.001184	0.000413
1000	500	483	218	0.001791	0.000607
1000	500	483	219	0.002668	0.000877
1000	500	483	220	0.003916	0.001248
1000	500	483	221	0.005664	0.001747
1000	500	483	222	0.008071	0.002407
1000	500	483	223	0.011335	0.003264
1000	500	483	224	0.015689	0.004354
1000	500	483	225	0.021405	0.005716
1000	500	483	226	0.028790	0.007385
1000	500	483	227	0.038179	0.009389
1000	500	483	228	0.049926	0.011747
1000	500	483	229	0.064389	0.014463
1000	500	483	230	0.081913	0.017524
1000	500	483	231	0.102809	0.020896
1000	500	483	232	0.127329	0.024520
1000	500	483	233	0.155646	0.028316
1000	500	483	234	0.187827	0.032181
1000	500	483	235	0.223819	0.035992
1000	500	483	236	0.263436	0.039617
1000	500	483	237	0.306349	0.042914
1000	500	483	238	0.352097	0.045748
1000	500	483	239	0.400092	0.047995
1000	500	483	240	0.449647	0.049555
1000	500	483	241	0.500000	0.050353
1000	500	484	203	0.000001	0.000000
1000	500	484	204	0.000001	0.000000
1000	500	484	205	0.000002	0.000001
1000	500	484	206	0.000003	0.000002
1000	500	484	207	0.000006	0.000003
1000	500	484	208	0.000011	0.000005
1000	500	484	209	0.000019	0.000008
1000	500	484	210	0.000033	0.000014
1000	500	484	211	0.000056	0.000023
1000	500	484	212	0.000093	0.000037
1000	500	484	213	0.000153	0.000060
1000	500	484	214	0.000247	0.000095
1000	500	484	215	0.000395	0.000147
1000	500	484	216	0.000620	0.000225
1000	500	484	217	0.000958	0.000339
1000	500	484	218	0.001460	0.000502
1000	500	484	219	0.002191	0.000731
1000	500	484	220	0.003240	0.001049
1000	500	484	221	0.004721	0.001480
1000	500	484	222	0.006777	0.002056
1000	500	484	223	0.009586	0.002810
1000	500	484	224	0.013365	0.003778
1000	500	484	225	0.018365	0.005000
1000	500	484	226	0.024876	0.006512
1000	500	484	227	0.033221	0.008345
1000	500	484	228	0.043746	0.010525
1000	500	484	229	0.056808	0.013062
1000	500	484	230	0.072761	0.015954
1000	500	484	231	0.091937	0.019175
1000	500	484	232	0.114619	0.022682
1000	500	484	233	0.141022	0.026403
1000	500	484	234	0.171269	0.030247
1000	500	484	235	0.205370	0.034101
1000	500	484	236	0.243206	0.037836
1000	500	484	237	0.284519	0.041313
1000	500	484	238	0.328913	0.044394
1000	500	484	239	0.375862	0.046949
1000	500	484	240	0.424726	0.048863
1000	500	484	241	0.474775	0.050049

Table for $N = 1000$, $n = 500$

N	n	k	x	P(x)	p(x)
1000	500	484	242	0.525225	0.050451
1000	500	485	204	0.000001	0.000000
1000	500	485	205	0.000001	0.000001
1000	500	485	206	0.000003	0.000001
1000	500	485	207	0.000005	0.000002
1000	500	485	208	0.000008	0.000004
1000	500	485	209	0.000014	0.000006
1000	500	485	210	0.000025	0.000011
1000	500	485	211	0.000043	0.000018
1000	500	485	212	0.000072	0.000029
1000	500	485	213	0.000120	0.000047
1000	500	485	214	0.000195	0.000075
1000	500	485	215	0.000313	0.000118
1000	500	485	216	0.000496	0.000182
1000	500	485	217	0.000773	0.000277
1000	500	485	218	0.001186	0.000413
1000	500	485	219	0.001793	0.000607
1000	500	485	220	0.002671	0.000878
1000	500	485	221	0.003920	0.001249
1000	500	485	222	0.005669	0.001749
1000	500	485	223	0.008078	0.002409
1000	500	485	224	0.011344	0.003266
1000	500	485	225	0.015700	0.004356
1000	500	485	226	0.021419	0.005719
1000	500	485	227	0.028806	0.007388
1000	500	485	228	0.038198	0.009392
1000	500	485	229	0.049948	0.011750
1000	500	485	230	0.064414	0.014466
1000	500	485	231	0.081940	0.017527
1000	500	485	232	0.102838	0.020898
1000	500	485	233	0.127360	0.024522
1000	500	485	234	0.155677	0.028317
1000	500	485	235	0.187857	0.032181
1000	500	485	236	0.223848	0.035991
1000	500	485	237	0.263462	0.039614
1000	500	485	238	0.306372	0.042910
1000	500	485	239	0.352115	0.045743
1000	500	485	240	0.400105	0.047990
1000	500	485	241	0.449653	0.049549
1000	500	485	242	0.500000	0.050347
1000	500	486	204	0.000001	0.000000
1000	500	486	205	0.000001	0.000000
1000	500	486	206	0.000002	0.000001
1000	500	486	207	0.000003	0.000002
1000	500	486	208	0.000006	0.000003
1000	500	486	209	0.000011	0.000005
1000	500	486	210	0.000019	0.000008
1000	500	486	211	0.000033	0.000014
1000	500	486	212	0.000056	0.000023
1000	500	486	213	0.000093	0.000037

N	n	k	x	P(x)	p(x)
1000	500	486	214	0.000153	0.000060
1000	500	486	215	0.000248	0.000095
1000	500	486	216	0.000395	0.000147
1000	500	486	217	0.000620	0.000225
1000	500	486	218	0.000960	0.000339
1000	500	486	219	0.001462	0.000502
1000	500	486	220	0.002194	0.000732
1000	500	486	221	0.003244	0.001050
1000	500	486	222	0.004725	0.001481
1000	500	486	223	0.006782	0.002057
1000	500	486	224	0.009593	0.002811
1000	500	486	225	0.013374	0.003780
1000	500	486	226	0.018376	0.005002
1000	500	486	227	0.024890	0.006514
1000	500	486	228	0.033238	0.008348
1000	500	486	229	0.043765	0.010527
1000	500	486	230	0.056830	0.013065
1000	500	486	231	0.072786	0.015956
1000	500	486	232	0.091963	0.019178
1000	500	486	233	0.114647	0.022683
1000	500	486	234	0.141051	0.026404
1000	500	486	235	0.171298	0.030247
1000	500	486	236	0.205398	0.034100
1000	500	486	237	0.243232	0.037834
1000	500	486	238	0.284542	0.041310
1000	500	486	239	0.328933	0.044390
1000	500	486	240	0.375877	0.046944
1000	500	486	241	0.424735	0.048858
1000	500	486	242	0.474778	0.050043
1000	500	486	243	0.525222	0.050445
1000	500	487	205	0.000001	0.000000
1000	500	487	206	0.000001	0.000001
1000	500	487	207	0.000003	0.000001
1000	500	487	208	0.000005	0.000002
1000	500	487	209	0.000008	0.000004
1000	500	487	210	0.000014	0.000006
1000	500	487	211	0.000025	0.000011
1000	500	487	212	0.000043	0.000018
1000	500	487	213	0.000072	0.000029
1000	500	487	214	0.000120	0.000047
1000	500	487	215	0.000195	0.000076
1000	500	487	216	0.000314	0.000118
1000	500	487	217	0.000496	0.000183
1000	500	487	218	0.000773	0.000277
1000	500	487	219	0.001187	0.000414
1000	500	487	220	0.001795	0.000608
1000	500	487	221	0.002674	0.000879
1000	500	487	222	0.003924	0.001250
1000	500	487	223	0.005674	0.001750
1000	500	487	224	0.008084	0.002410

N	n	k	x	P(x)	p(x)
1000	500	487	225	0.011351	0.003267
1000	500	487	226	0.015709	0.004358
1000	500	487	227	0.021430	0.005721
1000	500	487	228	0.028820	0.007390
1000	500	487	229	0.038215	0.009394
1000	500	487	230	0.049967	0.011752
1000	500	487	231	0.064435	0.014468
1000	500	487	232	0.081964	0.017529
1000	500	487	233	0.102864	0.020900
1000	500	487	234	0.127386	0.024523
1000	500	487	235	0.155704	0.028317
1000	500	487	236	0.187884	0.032180
1000	500	487	237	0.223874	0.035990
1000	500	487	238	0.263485	0.039612
1000	500	487	239	0.306392	0.042907
1000	500	487	240	0.352131	0.045739
1000	500	487	241	0.400116	0.047985
1000	500	487	242	0.449659	0.049543
1000	500	487	243	0.500000	0.050341
1000	500	488	205	0.000001	0.000000
1000	500	488	206	0.000001	0.000000
1000	500	488	207	0.000002	0.000001
1000	500	488	208	0.000003	0.000002
1000	500	488	209	0.000006	0.000003
1000	500	488	210	0.000011	0.000005
1000	500	488	211	0.000019	0.000008
1000	500	488	212	0.000033	0.000014
1000	500	488	213	0.000056	0.000023
1000	500	488	214	0.000093	0.000037
1000	500	488	215	0.000153	0.000060
1000	500	488	216	0.000248	0.000095
1000	500	488	217	0.000396	0.000147
1000	500	488	218	0.000621	0.000226
1000	500	488	219	0.000961	0.000339
1000	500	488	220	0.001463	0.000503
1000	500	488	221	0.002196	0.000733
1000	500	488	222	0.003246	0.001051
1000	500	488	223	0.004729	0.001482
1000	500	488	224	0.006787	0.002058
1000	500	488	225	0.009600	0.002813
1000	500	488	226	0.013382	0.003782
1000	500	488	227	0.018386	0.005004
1000	500	488	228	0.024902	0.006516
1000	500	488	229	0.033252	0.008350
1000	500	488	230	0.043782	0.010530
1000	500	488	231	0.056849	0.013067
1000	500	488	232	0.072807	0.015958
1000	500	488	233	0.091986	0.019179
1000	500	488	234	0.114671	0.022685
1000	500	488	235	0.141076	0.026405

N	n	k	x	P(x)	p(x)
1000	500	488	236	0.171323	0.030247
1000	500	488	237	0.205423	0.034099
1000	500	488	238	0.243255	0.037832
1000	500	488	239	0.284562	0.041307
1000	500	488	240	0.328949	0.044387
1000	500	488	241	0.375889	0.046940
1000	500	488	242	0.424742	0.048853
1000	500	488	243	0.474780	0.050038
1000	500	488	244	0.525220	0.050439
1000	500	489	206	0.000001	0.000000
1000	500	489	207	0.000001	0.000001
1000	500	489	208	0.000003	0.000001
1000	500	489	209	0.000005	0.000002
1000	500	489	210	0.000008	0.000004
1000	500	489	211	0.000015	0.000006
1000	500	489	212	0.000025	0.000011
1000	500	489	213	0.000043	0.000018
1000	500	489	214	0.000072	0.000029
1000	500	489	215	0.000120	0.000048
1000	500	489	216	0.000196	0.000076
1000	500	489	217	0.000314	0.000119
1000	500	489	218	0.000497	0.000183
1000	500	489	219	0.000774	0.000277
1000	500	489	220	0.001188	0.000414
1000	500	489	221	0.001797	0.000608
1000	500	489	222	0.002676	0.000879
1000	500	489	223	0.003927	0.001251
1000	500	489	224	0.005678	0.001751
1000	500	489	225	0.008089	0.002412
1000	500	489	226	0.011358	0.003269
1000	500	489	227	0.015718	0.004360
1000	500	489	228	0.021440	0.005723
1000	500	489	229	0.028832	0.007392
1000	500	489	230	0.038229	0.009396
1000	500	489	231	0.049983	0.011754
1000	500	489	232	0.064454	0.014470
1000	500	489	233	0.081984	0.017531
1000	500	489	234	0.102885	0.020901
1000	500	489	235	0.127409	0.024524
1000	500	489	236	0.155727	0.028318
1000	500	489	237	0.187907	0.032180
1000	500	489	238	0.223895	0.035988
1000	500	489	239	0.263505	0.039610
1000	500	489	240	0.306409	0.042904
1000	500	489	241	0.352144	0.045735
1000	500	489	242	0.400125	0.047981
1000	500	489	243	0.449663	0.049539
1000	500	489	244	0.500000	0.050336
1000	500	490	206	0.000001	0.000000
1000	500	490	207	0.000001	0.000000

Table for $N = 1000$, $n = 500$

N	n	k	x	P(x)	p(x)
1000	500	490	208	0.000002	0.000001
1000	500	490	209	0.000003	0.000002
1000	500	490	210	0.000006	0.000003
1000	500	490	211	0.000011	0.000005
1000	500	490	212	0.000019	0.000008
1000	500	490	213	0.000033	0.000014
1000	500	490	214	0.000056	0.000023
1000	500	490	215	0.000093	0.000037
1000	500	490	216	0.000153	0.000060
1000	500	490	217	0.000248	0.000095
1000	500	490	218	0.000396	0.000148
1000	500	490	219	0.000622	0.000226
1000	500	490	220	0.000961	0.000340
1000	500	490	221	0.001465	0.000503
1000	500	490	222	0.002198	0.000733
1000	500	490	223	0.003249	0.001051
1000	500	490	224	0.004732	0.001483
1000	500	490	225	0.006791	0.002059
1000	500	490	226	0.009605	0.002814
1000	500	490	227	0.013388	0.003783
1000	500	490	228	0.018394	0.005006
1000	500	490	229	0.024912	0.006518
1000	500	490	230	0.033264	0.008352
1000	500	490	231	0.043796	0.010532
1000	500	490	232	0.056864	0.013069
1000	500	490	233	0.072824	0.015960
1000	500	490	234	0.092005	0.019181
1000	500	490	235	0.114691	0.022686
1000	500	490	236	0.141097	0.026406
1000	500	490	237	0.171345	0.030248
1000	500	490	238	0.205443	0.034099
1000	500	490	239	0.243274	0.037831
1000	500	490	240	0.284579	0.041305
1000	500	490	241	0.328963	0.044384
1000	500	490	242	0.375900	0.046937
1000	500	490	243	0.424749	0.048849
1000	500	490	244	0.474782	0.050034
1000	500	490	245	0.525217	0.050435
1000	500	491	207	0.000001	0.000000
1000	500	491	208	0.000001	0.000001
1000	500	491	209	0.000003	0.000001
1000	500	491	210	0.000005	0.000002
1000	500	491	211	0.000008	0.000004
1000	500	491	212	0.000015	0.000006
1000	500	491	213	0.000025	0.000011
1000	500	491	214	0.000043	0.000018
1000	500	491	215	0.000072	0.000029
1000	500	491	216	0.000120	0.000048
1000	500	491	217	0.000196	0.000076
1000	500	491	218	0.000314	0.000119
1000	500	491	219	0.000497	0.000183
1000	500	491	220	0.000775	0.000278
1000	500	491	221	0.001189	0.000414
1000	500	491	222	0.001798	0.000609
1000	500	491	223	0.002678	0.000880
1000	500	491	224	0.003929	0.001251
1000	500	491	225	0.005681	0.001752
1000	500	491	226	0.008093	0.002412
1000	500	491	227	0.011363	0.003270
1000	500	491	228	0.015724	0.004361
1000	500	491	229	0.021449	0.005724
1000	500	491	230	0.028842	0.007394
1000	500	491	231	0.038241	0.009398
1000	500	491	232	0.049997	0.011756
1000	500	491	233	0.064469	0.014472
1000	500	491	234	0.082001	0.017532
1000	500	491	235	0.102904	0.020902
1000	500	491	236	0.127428	0.024525
1000	500	491	237	0.155746	0.028318
1000	500	491	238	0.187926	0.032180
1000	500	491	239	0.223914	0.035988
1000	500	491	240	0.263522	0.039608
1000	500	491	241	0.306423	0.042902
1000	500	491	242	0.352155	0.045732
1000	500	491	243	0.400133	0.047977
1000	500	491	244	0.449668	0.049535
1000	500	491	245	0.500000	0.050332
1000	500	492	207	0.000001	0.000000
1000	500	492	208	0.000001	0.000000
1000	500	492	209	0.000002	0.000001
1000	500	492	210	0.000003	0.000002
1000	500	492	211	0.000006	0.000003
1000	500	492	212	0.000011	0.000005
1000	500	492	213	0.000019	0.000008
1000	500	492	214	0.000033	0.000014
1000	500	492	215	0.000056	0.000023
1000	500	492	216	0.000093	0.000037
1000	500	492	217	0.000154	0.000060
1000	500	492	218	0.000249	0.000095
1000	500	492	219	0.000396	0.000148
1000	500	492	220	0.000622	0.000226
1000	500	492	221	0.000962	0.000340
1000	500	492	222	0.001466	0.000503
1000	500	492	223	0.002199	0.000733
1000	500	492	224	0.003251	0.001052
1000	500	492	225	0.004734	0.001484
1000	500	492	226	0.006795	0.002060
1000	500	492	227	0.009609	0.002815
1000	500	492	228	0.013394	0.003784
1000	500	492	229	0.018401	0.005007
1000	500	492	230	0.024920	0.006519
1000	500	492	231	0.033274	0.008353
1000	500	492	232	0.043807	0.010533
1000	500	492	233	0.056877	0.013071
1000	500	492	234	0.072839	0.015961
1000	500	492	235	0.092021	0.019182
1000	500	492	236	0.114708	0.022687
1000	500	492	237	0.141114	0.026406
1000	500	492	238	0.171362	0.030248
1000	500	492	239	0.205460	0.034098
1000	500	492	240	0.243290	0.037829
1000	500	492	241	0.284593	0.041303
1000	500	492	242	0.328975	0.044382
1000	500	492	243	0.375908	0.046934
1000	500	492	244	0.424754	0.048846
1000	500	492	245	0.474784	0.050030
1000	500	492	246	0.525216	0.050431
1000	500	493	208	0.000001	0.000000
1000	500	493	209	0.000001	0.000001
1000	500	493	210	0.000003	0.000001
1000	500	493	211	0.000005	0.000002
1000	500	493	212	0.000008	0.000004
1000	500	493	213	0.000015	0.000006
1000	500	493	214	0.000025	0.000011
1000	500	493	215	0.000043	0.000018
1000	500	493	216	0.000073	0.000029
1000	500	493	217	0.000120	0.000048
1000	500	493	218	0.000196	0.000076
1000	500	493	219	0.000315	0.000119
1000	500	493	220	0.000498	0.000183
1000	500	493	221	0.000775	0.000278
1000	500	493	222	0.001190	0.000415
1000	500	493	223	0.001799	0.000609
1000	500	493	224	0.002679	0.000880
1000	500	493	225	0.003931	0.001252
1000	500	493	226	0.005683	0.001752
1000	500	493	227	0.008097	0.002413
1000	500	493	228	0.011368	0.003271
1000	500	493	229	0.015730	0.004362
1000	500	493	230	0.021455	0.005726
1000	500	493	231	0.028850	0.007395
1000	500	493	232	0.038250	0.009400
1000	500	493	233	0.050008	0.011758
1000	500	493	234	0.064481	0.014473
1000	500	493	235	0.082015	0.017534
1000	500	493	236	0.102918	0.020903
1000	500	493	237	0.127444	0.024525
1000	500	493	238	0.155762	0.028318
1000	500	493	239	0.187941	0.032179
1000	500	493	240	0.223928	0.035987
1000	500	493	241	0.263535	0.039607
1000	500	493	242	0.306434	0.042900
1000	500	493	243	0.352164	0.045730
1000	500	493	244	0.400139	0.047975
1000	500	493	245	0.449671	0.049532
1000	500	493	246	0.500000	0.050329
1000	500	494	208	0.000001	0.000000
1000	500	494	209	0.000001	0.000000
1000	500	494	210	0.000002	0.000001
1000	500	494	211	0.000003	0.000002
1000	500	494	212	0.000006	0.000003
1000	500	494	213	0.000011	0.000005
1000	500	494	214	0.000019	0.000008
1000	500	494	215	0.000033	0.000014
1000	500	494	216	0.000056	0.000023
1000	500	494	217	0.000094	0.000037
1000	500	494	218	0.000154	0.000060
1000	500	494	219	0.000249	0.000095
1000	500	494	220	0.000397	0.000148
1000	500	494	221	0.000623	0.000226
1000	500	494	222	0.000963	0.000340
1000	500	494	223	0.001466	0.000504
1000	500	494	224	0.002200	0.000734
1000	500	494	225	0.003252	0.001052
1000	500	494	226	0.004737	0.001484
1000	500	494	227	0.006797	0.002061
1000	500	494	228	0.009613	0.002815
1000	500	494	229	0.013398	0.003785
1000	500	494	230	0.018406	0.005008
1000	500	494	231	0.024927	0.006521
1000	500	494	232	0.033281	0.008355
1000	500	494	233	0.043816	0.010534
1000	500	494	234	0.056888	0.013072
1000	500	494	235	0.072850	0.015963
1000	500	494	236	0.092034	0.019183
1000	500	494	237	0.114721	0.022688
1000	500	494	238	0.141128	0.026407
1000	500	494	239	0.171376	0.030248
1000	500	494	240	0.205473	0.034098
1000	500	494	241	0.243302	0.037828
1000	500	494	242	0.284604	0.041302
1000	500	494	243	0.328984	0.044380
1000	500	494	244	0.375915	0.046931
1000	500	494	245	0.424758	0.048843
1000	500	494	246	0.474786	0.050027
1000	500	494	247	0.525214	0.050428
1000	500	495	209	0.000001	0.000000
1000	500	495	210	0.000001	0.000001
1000	500	495	211	0.000003	0.000001
1000	500	495	212	0.000005	0.000002

Table for $N = 1000$, $n = 500$

N	n	k	x	P(x)	p(x)
1000	500	495	213	0.000008	0.000004
1000	500	495	214	0.000015	0.000006
1000	500	495	215	0.000025	0.000011
1000	500	495	216	0.000043	0.000018
1000	500	495	217	0.000073	0.000029
1000	500	495	218	0.000120	0.000048
1000	500	495	219	0.000196	0.000076
1000	500	495	220	0.000315	0.000119
1000	500	495	221	0.000498	0.000183
1000	500	495	222	0.000776	0.000278
1000	500	495	223	0.001191	0.000415
1000	500	495	224	0.001800	0.000609
1000	500	495	225	0.002680	0.000881
1000	500	495	226	0.003933	0.001252
1000	500	495	227	0.005685	0.001753
1000	500	495	228	0.008099	0.002414
1000	500	495	229	0.011371	0.003271
1000	500	495	230	0.015734	0.004363
1000	500	495	231	0.021460	0.005726
1000	500	495	232	0.028856	0.007396
1000	500	495	233	0.038257	0.009401
1000	500	495	234	0.050016	0.011759
1000	500	495	235	0.064490	0.014474
1000	500	495	236	0.082025	0.017534
1000	500	495	237	0.102929	0.020904
1000	500	495	238	0.127455	0.024526
1000	500	495	239	0.155774	0.028319
1000	500	495	240	0.187953	0.032179
1000	500	495	241	0.223939	0.035986
1000	500	495	242	0.263545	0.039606
1000	500	495	243	0.306443	0.042898
1000	500	495	244	0.352171	0.045728
1000	500	495	245	0.400144	0.047972
1000	500	495	246	0.449673	0.049529
1000	500	495	247	0.500000	0.050327
1000	500	496	209	0.000001	0.000000
1000	500	496	210	0.000001	0.000000
1000	500	496	211	0.000002	0.000001
1000	500	496	212	0.000003	0.000002
1000	500	496	213	0.000006	0.000003
1000	500	496	214	0.000011	0.000005
1000	500	496	215	0.000019	0.000008
1000	500	496	216	0.000033	0.000014
1000	500	496	217	0.000056	0.000023
1000	500	496	218	0.000094	0.000038
1000	500	496	219	0.000154	0.000060
1000	500	496	220	0.000249	0.000095
1000	500	496	221	0.000397	0.000148
1000	500	496	222	0.000623	0.000226
1000	500	496	223	0.000963	0.000340

N	n	k	x	P(x)	p(x)
1000	500	496	224	0.001467	0.000504
1000	500	496	225	0.002201	0.000734
1000	500	496	226	0.003253	0.001052
1000	500	496	227	0.004738	0.001485
1000	500	496	228	0.006799	0.002061
1000	500	496	229	0.009615	0.002816
1000	500	496	230	0.013401	0.003786
1000	500	496	231	0.018410	0.005009
1000	500	496	232	0.024931	0.006521
1000	500	496	233	0.033287	0.008355
1000	500	496	234	0.043822	0.010535
1000	500	496	235	0.056895	0.013073
1000	500	496	236	0.072858	0.015963
1000	500	496	237	0.092042	0.019184
1000	500	496	238	0.114731	0.022688
1000	500	496	239	0.141138	0.026407
1000	500	496	240	0.171385	0.030248
1000	500	496	241	0.205483	0.034097
1000	500	496	242	0.243311	0.037828
1000	500	496	243	0.284612	0.041301
1000	500	496	244	0.328990	0.044379
1000	500	496	245	0.375920	0.046930
1000	500	496	246	0.424761	0.048841
1000	500	496	247	0.474787	0.050025
1000	500	496	248	0.525213	0.050426
1000	500	497	210	0.000001	0.000000
1000	500	497	211	0.000001	0.000001
1000	500	497	212	0.000003	0.000001
1000	500	497	213	0.000005	0.000002
1000	500	497	214	0.000008	0.000004
1000	500	497	215	0.000015	0.000006
1000	500	497	216	0.000025	0.000011
1000	500	497	217	0.000043	0.000018
1000	500	497	218	0.000073	0.000029
1000	500	497	219	0.000120	0.000048
1000	500	497	220	0.000196	0.000076
1000	500	497	221	0.000315	0.000119
1000	500	497	222	0.000498	0.000183
1000	500	497	223	0.000776	0.000278
1000	500	497	224	0.001191	0.000415
1000	500	497	225	0.001800	0.000609
1000	500	497	226	0.002681	0.000881
1000	500	497	227	0.003934	0.001253
1000	500	497	228	0.005687	0.001753
1000	500	497	229	0.008101	0.002414
1000	500	497	230	0.011373	0.003272
1000	500	497	231	0.015737	0.004364
1000	500	497	232	0.021464	0.005727
1000	500	497	233	0.028860	0.007397
1000	500	497	234	0.038262	0.009401

N	n	k	x	P(x)	p(x)
1000	500	497	235	0.050021	0.011759
1000	500	497	236	0.064496	0.014475
1000	500	497	237	0.082032	0.017535
1000	500	497	238	0.102936	0.020905
1000	500	497	239	0.127463	0.024526
1000	500	497	240	0.155781	0.028319
1000	500	497	241	0.187960	0.032179
1000	500	497	242	0.223946	0.035986
1000	500	497	243	0.263551	0.039605
1000	500	497	244	0.306449	0.042897
1000	500	497	245	0.352176	0.045727
1000	500	497	246	0.400147	0.047971
1000	500	497	247	0.449675	0.049528
1000	500	497	248	0.500000	0.050325
1000	500	498	210	0.000001	0.000000
1000	500	498	211	0.000001	0.000000
1000	500	498	212	0.000002	0.000001
1000	500	498	213	0.000003	0.000002
1000	500	498	214	0.000006	0.000003
1000	500	498	215	0.000011	0.000005
1000	500	498	216	0.000019	0.000008
1000	500	498	217	0.000033	0.000014
1000	500	498	218	0.000056	0.000023
1000	500	498	219	0.000094	0.000038
1000	500	498	220	0.000154	0.000060
1000	500	498	221	0.000249	0.000095
1000	500	498	222	0.000397	0.000148
1000	500	498	223	0.000623	0.000226
1000	500	498	224	0.000963	0.000340
1000	500	498	225	0.001467	0.000504
1000	500	498	226	0.002201	0.000734
1000	500	498	227	0.003254	0.001053
1000	500	498	228	0.004739	0.001485
1000	500	498	229	0.006800	0.002061
1000	500	498	230	0.009617	0.002816
1000	500	498	231	0.013403	0.003786
1000	500	498	232	0.018412	0.005009
1000	500	498	233	0.024934	0.006522
1000	500	498	234	0.033290	0.008356
1000	500	498	235	0.043826	0.010536
1000	500	498	236	0.056899	0.013073
1000	500	498	237	0.072863	0.015964
1000	500	498	238	0.092048	0.019184
1000	500	498	239	0.114736	0.022689
1000	500	498	240	0.141143	0.026407
1000	500	498	241	0.171391	0.030248
1000	500	498	242	0.205488	0.034097
1000	500	498	243	0.243316	0.037827
1000	500	498	244	0.284616	0.041300
1000	500	498	245	0.328994	0.044378

N	n	k	x	P(x)	p(x)
1000	500	498	246	0.375923	0.046929
1000	500	498	247	0.424763	0.048840
1000	500	498	248	0.474787	0.050024
1000	500	498	249	0.525213	0.050425
1000	500	499	211	0.000001	0.000000
1000	500	499	212	0.000001	0.000001
1000	500	499	213	0.000003	0.000001
1000	500	499	214	0.000005	0.000002
1000	500	499	215	0.000008	0.000004
1000	500	499	216	0.000015	0.000006
1000	500	499	217	0.000025	0.000011
1000	500	499	218	0.000043	0.000018
1000	500	499	219	0.000073	0.000029
1000	500	499	220	0.000120	0.000048
1000	500	499	221	0.000196	0.000076
1000	500	499	222	0.000315	0.000119
1000	500	499	223	0.000498	0.000183
1000	500	499	224	0.000776	0.000278
1000	500	499	225	0.001191	0.000415
1000	500	499	226	0.001801	0.000609
1000	500	499	227	0.002682	0.000881
1000	500	499	228	0.003934	0.001253
1000	500	499	229	0.005687	0.001753
1000	500	499	230	0.008102	0.002414
1000	500	499	231	0.011374	0.003272
1000	500	499	232	0.015738	0.004364
1000	500	499	233	0.021465	0.005727
1000	500	499	234	0.028862	0.007397
1000	500	499	235	0.038264	0.009402
1000	500	499	236	0.050024	0.011760
1000	500	499	237	0.064500	0.014476
1000	500	499	238	0.082035	0.017535
1000	500	499	239	0.102940	0.020905
1000	500	499	240	0.127466	0.024526
1000	500	499	241	0.155785	0.028319
1000	500	499	242	0.187964	0.032179
1000	500	499	243	0.223950	0.035986
1000	500	499	244	0.263555	0.039605
1000	500	499	245	0.306451	0.042897
1000	500	499	246	0.352178	0.045726
1000	500	499	247	0.400148	0.047970
1000	500	499	248	0.449676	0.049527
1000	500	499	249	0.500000	0.050324
1000	500	500	211	0.000001	0.000000
1000	500	500	212	0.000001	0.000000
1000	500	500	213	0.000002	0.000001
1000	500	500	214	0.000003	0.000002
1000	500	500	215	0.000006	0.000003
1000	500	500	216	0.000011	0.000005
1000	500	500	217	0.000019	0.000008

N	n	k	x	P(x)	p(x)
1000	500	500	218	0.000033	0.000014
1000	500	500	219	0.000056	0.000023
1000	500	500	220	0.000094	0.000038
1000	500	500	221	0.000154	0.000060
1000	500	500	222	0.000249	0.000095
1000	500	500	223	0.000397	0.000148
1000	500	500	224	0.000623	0.000226
1000	500	500	225	0.000963	0.000340
1000	500	500	226	0.001467	0.000504
1000	500	500	227	0.002202	0.000734
1000	500	500	228	0.003254	0.001053
1000	500	500	229	0.004739	0.001485
1000	500	500	230	0.006801	0.002062
1000	500	500	231	0.009617	0.002816
1000	500	500	232	0.013404	0.003786
1000	500	500	233	0.018413	0.005009
1000	500	500	234	0.024935	0.006522
1000	500	500	235	0.033291	0.008356
1000	500	500	236	0.043827	0.010536
1000	500	500	237	0.056901	0.013073
1000	500	500	238	0.072865	0.015964
1000	500	500	239	0.092049	0.019185
1000	500	500	240	0.114738	0.022689
1000	500	500	241	0.141145	0.026407
1000	500	500	242	0.171393	0.030248
1000	500	500	243	0.205490	0.034097
1000	500	500	244	0.243317	0.037827
1000	500	500	245	0.284618	0.041300
1000	500	500	246	0.328995	0.044378
1000	500	500	247	0.375924	0.046929
1000	500	500	248	0.424764	0.048840
1000	500	500	249	0.474787	0.050024
1000	500	500	250	0.525212	0.050425

N	n	k	x	P(x)	p(x)

N	n	k	x	P(x)	p(x)

N	n	k	x	P(x)	p(x)

Table for $k = n - 1,\ n;\ \ n = N/2,$

$N = 100,\ n = 50$ through $N = 2000,\ n = 1000$

N	n	k	x	P(x)	p(x)
100	50	49	13	0.000004	0.000003
100	50	49	14	0.000025	0.000021
100	50	49	15	0.000137	0.000112
100	50	49	16	0.000627	0.000490
100	50	49	17	0.002425	0.001797
100	50	49	18	0.007974	0.005550
100	50	49	19	0.022462	0.014488
100	50	49	20	0.054542	0.032080
100	50	49	21	0.114951	0.060410
100	50	49	22	0.211893	0.096942
100	50	49	23	0.344661	0.132768
100	50	49	24	0.500000	0.155339
100	50	50	13	0.000001	0.000001
100	50	50	14	0.000010	0.000009
100	50	50	15	0.000060	0.000050
100	50	50	16	0.000301	0.000240
100	50	50	17	0.001262	0.000961
100	50	50	18	0.004492	0.003230
100	50	50	19	0.013656	0.009164
100	50	50	20	0.035671	0.022015
100	50	50	21	0.080601	0.044929
100	50	50	22	0.158670	0.078069
100	50	50	23	0.274372	0.115702
100	50	50	24	0.420808	0.146436
100	50	50	25	0.579192	0.158385
200	100	99	32	0.000001	0.000000
200	100	99	33	0.000003	0.000002
200	100	99	34	0.000010	0.000007
200	100	99	35	0.000034	0.000024
200	100	99	36	0.000109	0.000075
200	100	99	37	0.000325	0.000216
200	100	99	38	0.000896	0.000570
200	100	99	39	0.002278	0.001382
200	100	99	40	0.005362	0.003084
200	100	99	41	0.011703	0.006341
200	100	99	42	0.023717	0.012015
200	100	99	43	0.044711	0.020994
200	100	99	44	0.078556	0.033844
200	100	99	45	0.128913	0.050356
200	100	99	46	0.198091	0.069178
200	100	99	47	0.285852	0.087760
200	100	99	48	0.388686	0.102835
200	100	99	49	0.500000	0.111313
200	100	100	33	0.000001	0.000001
200	100	100	34	0.000005	0.000004
200	100	100	35	0.000018	0.000013
200	100	100	36	0.000061	0.000043
200	100	100	37	0.000191	0.000129
200	100	100	38	0.000546	0.000355
200	100	100	39	0.001443	0.000897

N	n	k	x	P(x)	p(x)
200	100	100	40	0.003530	0.002087
200	100	100	41	0.007999	0.004469
200	100	100	42	0.016818	0.008819
200	100	100	43	0.032863	0.016045
200	100	100	44	0.059790	0.026927
200	100	100	45	0.101491	0.041700
200	100	100	46	0.161105	0.059614
200	100	100	47	0.239799	0.078694
200	100	100	48	0.335742	0.095943
200	100	100	49	0.443792	0.108050
200	100	100	50	0.556208	0.112416
300	150	149	53	0.000001	0.000000
300	150	149	54	0.000002	0.000001
300	150	149	55	0.000005	0.000003
300	150	149	56	0.000015	0.000010
300	150	149	57	0.000040	0.000026
300	150	149	58	0.000104	0.000064
300	150	149	59	0.000256	0.000151
300	150	149	60	0.000594	0.000338
300	150	149	61	0.001310	0.000717
300	150	149	62	0.002747	0.001437
300	150	149	63	0.005475	0.002728
300	150	149	64	0.010382	0.004907
300	150	149	65	0.018743	0.008361
300	150	149	66	0.032244	0.013500
300	150	149	67	0.052903	0.020659
300	150	149	68	0.082871	0.029968
300	150	149	69	0.124081	0.041210
300	150	149	70	0.177812	0.053731
300	150	149	71	0.244239	0.066428
300	150	149	72	0.322117	0.077878
300	150	149	73	0.408703	0.086586
300	150	149	74	0.500000	0.091297
300	150	150	54	0.000001	0.000001
300	150	150	55	0.000003	0.000002
300	150	150	56	0.000009	0.000006
300	150	150	57	0.000025	0.000016
300	150	150	58	0.000065	0.000041
300	150	150	59	0.000164	0.000099
300	150	150	60	0.000392	0.000228
300	150	150	61	0.000888	0.000496
300	150	150	62	0.001910	0.001022
300	150	150	63	0.003904	0.001994
300	150	150	64	0.007588	0.003684
300	150	150	65	0.014037	0.006449
300	150	150	66	0.024734	0.010697
300	150	150	67	0.041547	0.016814
300	150	150	68	0.066597	0.025050
300	150	150	69	0.101975	0.035378
300	150	150	70	0.149345	0.047370

N	n	k	x	P(x)	p(x)
300	150	150	71	0.209486	0.060141
300	150	150	72	0.281889	0.072403
300	150	150	73	0.364550	0.082661
300	150	150	74	0.454049	0.089499
300	150	150	75	0.545951	0.091902
400	200	199	75	0.000001	0.000000
400	200	199	76	0.000002	0.000001
400	200	199	77	0.000005	0.000003
400	200	199	78	0.000012	0.000007
400	200	199	79	0.000030	0.000018
400	200	199	80	0.000069	0.000039
400	200	199	81	0.000154	0.000084
400	200	199	82	0.000328	0.000174
400	200	199	83	0.000672	0.000345
400	200	199	84	0.001328	0.000655
400	200	199	85	0.002523	0.001196
400	200	199	86	0.004618	0.002095
400	200	199	87	0.008144	0.003525
400	200	199	88	0.013840	0.005697
400	200	199	89	0.022682	0.008841
400	200	199	90	0.035863	0.013181
400	200	199	91	0.054740	0.018877
400	200	199	92	0.080713	0.025973
400	200	199	93	0.115046	0.034333
400	200	199	94	0.158653	0.043607
400	200	199	95	0.211871	0.053218
400	200	199	96	0.274278	0.062407
400	200	199	97	0.344603	0.070325
400	200	199	98	0.420755	0.076152
400	200	199	99	0.500000	0.079245
400	200	200	76	0.000001	0.000001
400	200	200	77	0.000003	0.000002
400	200	200	78	0.000008	0.000005
400	200	200	79	0.000019	0.000012
400	200	200	80	0.000046	0.000026
400	200	200	81	0.000104	0.000058
400	200	200	82	0.000225	0.000122
400	200	200	83	0.000472	0.000246
400	200	200	84	0.000949	0.000478
400	200	200	85	0.001839	0.000890
400	200	200	86	0.003430	0.001591
400	200	200	87	0.006162	0.002732
400	200	200	88	0.010666	0.004504
400	200	200	89	0.017799	0.007133
400	200	200	90	0.028649	0.010850
400	200	200	91	0.044503	0.015854
400	200	200	92	0.066757	0.022254
400	200	200	93	0.096769	0.030012
400	200	200	94	0.135656	0.038887
400	200	200	95	0.184070	0.048414

N	n	k	x	P(x)	p(x)
400	200	200	96	0.241988	0.057917
400	200	200	97	0.308566	0.066578
400	200	200	98	0.382111	0.073545
400	200	200	99	0.460180	0.078070
400	200	200	100	0.539819	0.079639
500	250	249	97	0.000001	0.000000
500	250	249	98	0.000002	0.000001
500	250	249	99	0.000004	0.000002
500	250	249	100	0.000008	0.000005
500	250	249	101	0.000018	0.000010
500	250	249	102	0.000040	0.000021
500	250	249	103	0.000083	0.000043
500	250	249	104	0.000169	0.000085
500	250	249	105	0.000331	0.000162
500	250	249	106	0.000630	0.000299
500	250	249	107	0.001162	0.000533
500	250	249	108	0.002081	0.000919
500	250	249	109	0.003615	0.001534
500	250	249	110	0.006095	0.002480
500	250	249	111	0.009976	0.003882
500	250	249	112	0.015860	0.005883
500	250	249	113	0.024494	0.008634
500	250	249	114	0.036765	0.012271
500	250	249	115	0.053655	0.016889
500	250	249	116	0.076166	0.022511
500	250	249	117	0.105226	0.029060
500	250	249	118	0.141558	0.036332
500	250	249	119	0.185553	0.043995
500	250	249	120	0.237154	0.051601
500	250	249	121	0.295773	0.058620
500	250	249	122	0.360276	0.064503
500	250	249	123	0.429025	0.068749
500	250	249	124	0.500000	0.070975
500	250	250	98	0.000001	0.000001
500	250	250	99	0.000002	0.000001
500	250	250	100	0.000006	0.000003
500	250	250	101	0.000012	0.000007
500	250	250	102	0.000027	0.000015
500	250	250	103	0.000058	0.000031
500	250	250	104	0.000119	0.000061
500	250	250	105	0.000237	0.000118
500	250	250	106	0.000458	0.000221
500	250	250	107	0.000859	0.000401
500	250	250	108	0.001561	0.000702
500	250	250	109	0.002753	0.001192
500	250	250	110	0.004711	0.001958
500	250	250	111	0.007827	0.003115
500	250	250	112	0.012625	0.004798
500	250	250	113	0.019781	0.007156
500	250	250	114	0.030116	0.010335

Table for $k = n-1,\ n;\ \ n = N/2,$

$N = 100,\ n = 50$ through $N = 2000,\ n = 1000$

N	n	k	x	P(x)	p(x)
500	250	250	115	0.044571	0.014454
500	250	250	116	0.064148	0.019577
500	250	250	117	0.089828	0.025680
500	250	250	118	0.122451	0.032623
500	250	250	119	0.162592	0.040141
500	250	250	120	0.210428	0.047837
500	250	250	121	0.265646	0.055218
500	250	250	122	0.327382	0.061736
500	250	250	123	0.394239	0.066857
500	250	250	124	0.464371	0.070131
500	250	250	125	0.535629	0.071258
600	300	299	120	0.000001	0.000001
600	300	299	121	0.000002	0.000001
600	300	299	122	0.000005	0.000003
600	300	299	123	0.000010	0.000005
600	300	299	124	0.000021	0.000011
600	300	299	125	0.000043	0.000022
600	300	299	126	0.000084	0.000041
600	300	299	127	0.000160	0.000076
600	300	299	128	0.000297	0.000137
600	300	299	129	0.000537	0.000240
600	300	299	130	0.000947	0.000410
600	300	299	131	0.001628	0.000681
600	300	299	132	0.002729	0.001101
600	300	299	133	0.004462	0.001734
600	300	299	134	0.007119	0.002657
600	300	299	135	0.011082	0.003963
600	300	299	136	0.016838	0.005756
600	300	299	137	0.024976	0.008138
600	300	299	138	0.036179	0.011203
600	300	299	139	0.051195	0.015015
600	300	299	140	0.070789	0.019595
600	300	299	141	0.095687	0.024897
600	300	299	142	0.126488	0.030802
600	300	299	143	0.163594	0.037105
600	300	299	144	0.207118	0.043524
600	300	299	145	0.256830	0.049713
600	300	299	146	0.312121	0.055290
600	300	299	147	0.372001	0.059880
600	300	299	148	0.435150	0.063149
600	300	299	149	0.500000	0.064850
600	300	300	120	0.000001	0.000000
600	300	300	121	0.000002	0.000001
600	300	300	122	0.000003	0.000002
600	300	300	123	0.000007	0.000004
600	300	300	124	0.000015	0.000008
600	300	300	125	0.000030	0.000015
600	300	300	126	0.000060	0.000030
600	300	300	127	0.000116	0.000056
600	300	300	128	0.000219	0.000102

N	n	k	x	P(x)	p(x)
600	300	300	129	0.000401	0.000182
600	300	300	130	0.000715	0.000315
600	300	300	131	0.001245	0.000530
600	300	300	132	0.002114	0.000869
600	300	300	133	0.003501	0.001386
600	300	300	134	0.005654	0.002153
600	300	300	135	0.008910	0.003256
600	300	300	136	0.013702	0.004792
600	300	300	137	0.020569	0.006867
600	300	300	138	0.030150	0.009581
600	300	300	139	0.043163	0.013014
600	300	300	140	0.060374	0.017210
600	300	300	141	0.082535	0.022161
600	300	300	142	0.110320	0.027785
600	300	300	143	0.144240	0.033920
600	300	300	144	0.184561	0.040321
600	300	300	145	0.231231	0.046670
600	300	300	146	0.283833	0.052602
600	300	300	147	0.341563	0.057731
600	300	300	148	0.403260	0.061697
600	300	300	149	0.467467	0.064207
600	300	300	150	0.532533	0.065066
700	350	349	142	0.000001	0.000000
700	350	349	143	0.000001	0.000001
700	350	349	144	0.000003	0.000001
700	350	349	145	0.000006	0.000003
700	350	349	146	0.000011	0.000006
700	350	349	147	0.000022	0.000011
700	350	349	148	0.000041	0.000020
700	350	349	149	0.000077	0.000036
700	350	349	150	0.000140	0.000063
700	350	349	151	0.000249	0.000109
700	350	349	152	0.000434	0.000185
700	350	349	153	0.000741	0.000306
700	350	349	154	0.001236	0.000496
700	350	349	155	0.002020	0.000784
700	350	349	156	0.003230	0.001210
700	350	349	157	0.005057	0.001827
700	350	349	158	0.007752	0.002695
700	350	349	159	0.011637	0.003885
700	350	349	160	0.017109	0.005473
700	350	349	161	0.024644	0.007535
700	350	349	162	0.034783	0.010139
700	350	349	163	0.048116	0.013334
700	350	349	164	0.065255	0.017139
700	350	349	165	0.086786	0.021531
700	350	349	166	0.113225	0.026438
700	350	349	167	0.144955	0.031731
700	350	349	168	0.182177	0.037222
700	350	349	169	0.224857	0.042679

N	n	k	x	P(x)	p(x)
700	350	349	170	0.272689	0.047833
700	350	349	171	0.325089	0.052399
700	350	349	172	0.381196	0.056108
700	350	349	173	0.439921	0.058725
700	350	349	174	0.500000	0.060079
700	350	350	143	0.000001	0.000000
700	350	350	144	0.000002	0.000001
700	350	350	145	0.000004	0.000002
700	350	350	146	0.000008	0.000004
700	350	350	147	0.000016	0.000008
700	350	350	148	0.000030	0.000014
700	350	350	149	0.000056	0.000026
700	350	350	150	0.000104	0.000047
700	350	350	151	0.000187	0.000083
700	350	350	152	0.000330	0.000143
700	350	350	153	0.000569	0.000239
700	350	350	154	0.000959	0.000391
700	350	350	155	0.001585	0.000625
700	350	350	156	0.002561	0.000977
700	350	350	157	0.004053	0.001491
700	350	350	158	0.006278	0.002225
700	350	350	159	0.009523	0.003245
700	350	350	160	0.014147	0.004624
700	350	350	161	0.020587	0.006440
700	350	350	162	0.029352	0.008765
700	350	350	163	0.041013	0.011660
700	350	350	164	0.056173	0.015160
700	350	350	165	0.075438	0.019265
700	350	350	166	0.099365	0.023927
700	350	350	167	0.128412	0.029047
700	350	350	168	0.162877	0.034465
700	350	350	169	0.202848	0.039971
700	350	350	170	0.248160	0.045312
700	350	350	171	0.298367	0.050207
700	350	350	172	0.352743	0.054376
700	350	350	173	0.410308	0.057565
700	350	350	174	0.469875	0.059567
700	350	350	175	0.530125	0.060250
800	400	399	165	0.000001	0.000000
800	400	399	166	0.000001	0.000001
800	400	399	167	0.000003	0.000001
800	400	399	168	0.000006	0.000003
800	400	399	169	0.000011	0.000005
800	400	399	170	0.000020	0.000009
800	400	399	171	0.000037	0.000017
800	400	399	172	0.000066	0.000029
800	400	399	173	0.000116	0.000050
800	400	399	174	0.000200	0.000084
800	400	399	175	0.000340	0.000139
800	400	399	176	0.000565	0.000225

N	n	k	x	P(x)	p(x)
800	400	399	177	0.000923	0.000357
800	400	399	178	0.001478	0.000555
800	400	399	179	0.002324	0.000846
800	400	399	180	0.003586	0.001262
800	400	399	181	0.005432	0.001846
800	400	399	182	0.008078	0.002646
800	400	399	183	0.011796	0.003718
800	400	399	184	0.016915	0.005119
800	400	399	185	0.023823	0.006909
800	400	399	186	0.032962	0.009139
800	400	399	187	0.044811	0.011849
800	400	399	188	0.059870	0.015058
800	400	399	189	0.078627	0.018758
800	400	399	190	0.101531	0.022903
800	400	399	191	0.128942	0.027411
800	400	399	192	0.161099	0.032157
800	400	399	193	0.198078	0.036979
800	400	399	194	0.239761	0.041683
800	400	399	195	0.285817	0.046056
800	400	399	196	0.335699	0.049882
800	400	399	197	0.388659	0.052959
800	400	399	198	0.443774	0.055115
800	400	399	199	0.500000	0.056226
800	400	400	165	0.000000	0.000000
800	400	400	166	0.000001	0.000001
800	400	400	167	0.000002	0.000001
800	400	400	168	0.000004	0.000002
800	400	400	169	0.000008	0.000004
800	400	400	170	0.000015	0.000007
800	400	400	171	0.000027	0.000012
800	400	400	172	0.000049	0.000022
800	400	400	173	0.000087	0.000038
800	400	400	174	0.000153	0.000065
800	400	400	175	0.000261	0.000109
800	400	400	176	0.000439	0.000178
800	400	400	177	0.000724	0.000285
800	400	400	178	0.001170	0.000447
800	400	400	179	0.001858	0.000687
800	400	400	180	0.002893	0.001036
800	400	400	181	0.004424	0.001530
800	400	400	182	0.006640	0.002216
800	400	400	183	0.009784	0.003144
800	400	400	184	0.014157	0.004373
800	400	400	185	0.020119	0.005962
800	400	400	186	0.028085	0.007966
800	400	400	187	0.038517	0.010432
800	400	400	188	0.051909	0.013391
800	400	400	189	0.068757	0.016849
800	400	400	190	0.089537	0.020779
800	400	400	191	0.114655	0.025119

Table for $k = n - 1,\ n;\ \ n = N/2,$

$N = 100,\ n = 50$ through $N = 2000,\ n = 1000$

N	n	k	x	P(x)	p(x)
800	400	400	192	0.144419	0.029764
800	400	400	193	0.178989	0.034570
800	400	400	194	0.218348	0.039359
800	400	400	195	0.262272	0.043924
800	400	400	196	0.310323	0.048051
800	400	400	197	0.361849	0.051526
800	400	400	198	0.416010	0.054161
800	400	400	199	0.471817	0.055807
800	400	400	200	0.528183	0.056366
900	450	449	188	0.000001	0.000000
900	450	449	189	0.000001	0.000001
900	450	449	190	0.000003	0.000001
900	450	449	191	0.000005	0.000002
900	450	449	192	0.000010	0.000004
900	450	449	193	0.000017	0.000008
900	450	449	194	0.000031	0.000014
900	450	449	195	0.000054	0.000023
900	450	449	196	0.000093	0.000039
900	450	449	197	0.000157	0.000064
900	450	449	198	0.000260	0.000103
900	450	449	199	0.000424	0.000164
900	450	449	200	0.000681	0.000256
900	450	449	201	0.001074	0.000393
900	450	449	202	0.001666	0.000592
900	450	449	203	0.002541	0.000876
900	450	449	204	0.003813	0.001272
900	450	449	205	0.005629	0.001815
900	450	449	206	0.008174	0.002545
900	450	449	207	0.011679	0.003505
900	450	449	208	0.016420	0.004741
900	450	449	209	0.022720	0.006300
900	450	449	210	0.030944	0.008224
900	450	449	211	0.041489	0.010546
900	450	449	212	0.054773	0.013284
900	450	449	213	0.071212	0.016438
900	450	449	214	0.091195	0.019983
900	450	449	215	0.115060	0.023865
900	450	449	216	0.143058	0.027998
900	450	449	217	0.175326	0.032269
900	450	449	218	0.211863	0.036536
900	450	449	219	0.252503	0.040640
900	450	449	220	0.296913	0.044410
900	450	449	221	0.344589	0.047676
900	450	449	222	0.394871	0.050282
900	450	449	223	0.446969	0.052098
900	450	449	224	0.500000	0.053030
900	450	450	188	0.000001	0.000000
900	450	450	189	0.000001	0.000001
900	450	450	190	0.000002	0.000001
900	450	450	191	0.000004	0.000002
900	450	450	192	0.000007	0.000003
900	450	450	193	0.000013	0.000006
900	450	450	194	0.000023	0.000010
900	450	450	195	0.000041	0.000018
900	450	450	196	0.000071	0.000030
900	450	450	197	0.000121	0.000050
900	450	450	198	0.000202	0.000081
900	450	450	199	0.000333	0.000131
900	450	450	200	0.000538	0.000206
900	450	450	201	0.000857	0.000318
900	450	450	202	0.001340	0.000483
900	450	450	203	0.002062	0.000721
900	450	450	204	0.003119	0.001058
900	450	450	205	0.004642	0.001523
900	450	450	206	0.006797	0.002154
900	450	450	207	0.009790	0.002993
900	450	450	208	0.013876	0.004086
900	450	450	209	0.019353	0.005478
900	450	450	210	0.026567	0.007214
900	450	450	211	0.035901	0.009333
900	450	450	212	0.047763	0.011862
900	450	450	213	0.062573	0.014810
900	450	450	214	0.080738	0.018165
900	450	450	215	0.102625	0.021887
900	450	450	216	0.128531	0.025906
900	450	450	217	0.158655	0.030124
900	450	450	218	0.193068	0.034413
900	450	450	219	0.231687	0.038619
900	450	450	220	0.274265	0.042578
900	450	450	221	0.320381	0.046116
900	450	450	222	0.369452	0.049070
900	450	450	223	0.420747	0.051295
900	450	450	224	0.473426	0.052679
900	450	450	225	0.526574	0.053148
1000	500	499	211	0.000001	0.000000
1000	500	499	212	0.000001	0.000001
1000	500	499	213	0.000003	0.000001
1000	500	499	214	0.000005	0.000002
1000	500	499	215	0.000008	0.000004
1000	500	499	216	0.000015	0.000006
1000	500	499	217	0.000025	0.000011
1000	500	499	218	0.000043	0.000018
1000	500	499	219	0.000073	0.000029
1000	500	499	220	0.000120	0.000048
1000	500	499	221	0.000196	0.000076
1000	500	499	222	0.000315	0.000119
1000	500	499	223	0.000498	0.000183
1000	500	499	224	0.000776	0.000278
1000	500	499	225	0.001191	0.000415
1000	500	499	226	0.001801	0.000609
1000	500	499	227	0.002682	0.000881
1000	500	499	228	0.003934	0.001253
1000	500	499	229	0.005687	0.001753
1000	500	499	230	0.008102	0.002414
1000	500	499	231	0.011374	0.003272
1000	500	499	232	0.015738	0.004364
1000	500	499	233	0.021465	0.005727
1000	500	499	234	0.028862	0.007397
1000	500	499	235	0.038264	0.009402
1000	500	499	236	0.050024	0.011760
1000	500	499	237	0.064500	0.014476
1000	500	499	238	0.082035	0.017535
1000	500	499	239	0.102940	0.020905
1000	500	499	240	0.127466	0.024526
1000	500	499	241	0.155785	0.028319
1000	500	499	242	0.187964	0.032179
1000	500	499	243	0.223950	0.035986
1000	500	499	244	0.263555	0.039605
1000	500	499	245	0.306451	0.042897
1000	500	499	246	0.352178	0.045726
1000	500	499	247	0.400148	0.047970
1000	500	499	248	0.449676	0.049527
1000	500	499	249	0.500000	0.050324
1000	500	500	211	0.000001	0.000000
1000	500	500	212	0.000001	0.000000
1000	500	500	213	0.000002	0.000001
1000	500	500	214	0.000003	0.000002
1000	500	500	215	0.000006	0.000003
1000	500	500	216	0.000011	0.000005
1000	500	500	217	0.000019	0.000008
1000	500	500	218	0.000033	0.000014
1000	500	500	219	0.000056	0.000023
1000	500	500	220	0.000094	0.000038
1000	500	500	221	0.000154	0.000060
1000	500	500	222	0.000249	0.000095
1000	500	500	223	0.000397	0.000148
1000	500	500	224	0.000623	0.000226
1000	500	500	225	0.000963	0.000340
1000	500	500	226	0.001467	0.000504
1000	500	500	227	0.002202	0.000734
1000	500	500	228	0.003254	0.001053
1000	500	500	229	0.004739	0.001485
1000	500	500	230	0.006801	0.002062
1000	500	500	231	0.009617	0.002816
1000	500	500	232	0.013404	0.003786
1000	500	500	233	0.018413	0.005009
1000	500	500	234	0.024935	0.006522
1000	500	500	235	0.033291	0.008356
1000	500	500	236	0.043827	0.010536
1000	500	500	237	0.056901	0.013073
1000	500	500	238	0.072865	0.015964
1000	500	500	239	0.092049	0.019185
1000	500	500	240	0.114738	0.022689
1000	500	500	241	0.141145	0.026407
1000	500	500	242	0.171393	0.030248
1000	500	500	243	0.205490	0.034097
1000	500	500	244	0.243317	0.037827
1000	500	500	245	0.284618	0.041300
1000	500	500	246	0.328995	0.044378
1000	500	500	247	0.375924	0.046929
1000	500	500	248	0.424764	0.048840
1000	500	500	249	0.474787	0.050024
1000	500	500	250	0.525212	0.050425
1100	550	549	234	0.000001	0.000000
1100	550	549	235	0.000001	0.000001
1100	550	549	236	0.000002	0.000001
1100	550	549	237	0.000004	0.000002
1100	550	549	238	0.000007	0.000003
1100	550	549	239	0.000012	0.000005
1100	550	549	240	0.000020	0.000008
1100	550	549	241	0.000034	0.000014
1100	550	549	242	0.000056	0.000022
1100	550	549	243	0.000091	0.000035
1100	550	549	244	0.000146	0.000055
1100	550	549	245	0.000232	0.000086
1100	550	549	246	0.000363	0.000131
1100	550	549	247	0.000560	0.000197
1100	550	549	248	0.000851	0.000291
1100	550	549	249	0.001277	0.000426
1100	550	549	250	0.001889	0.000612
1100	550	549	251	0.002757	0.000868
1100	550	549	252	0.003971	0.001214
1100	550	549	253	0.005643	0.001672
1100	550	549	254	0.007911	0.002269
1100	550	549	255	0.010946	0.003035
1100	550	549	256	0.014947	0.004001
1100	550	549	257	0.020144	0.005197
1100	550	549	258	0.026799	0.006655
1100	550	549	259	0.035196	0.008397
1100	550	549	260	0.045638	0.010442
1100	550	549	261	0.058437	0.012798
1100	550	549	262	0.073896	0.015459
1100	550	549	263	0.092299	0.018404
1100	550	549	264	0.113892	0.021592
1100	550	549	265	0.138860	0.024968
1100	550	549	266	0.167314	0.028455
1100	550	549	267	0.199275	0.031960
1100	550	549	268	0.234655	0.035380
1100	550	549	269	0.273256	0.038601
1100	550	549	270	0.314764	0.041508

Table for $k = n - 1, n;\ n = N/2,$

$N = 100,\ n = 50$ through $N = 2000,\ n = 1000$

N	n	k	x	P(x)	p(x)
1100	550	549	271	0.358754	0.043990
1100	550	549	272	0.404703	0.045949
1100	550	549	273	0.452006	0.047303
1100	550	549	274	0.500000	0.047994
1100	550	550	235	0.000001	0.000000
1100	550	550	236	0.000002	0.000001
1100	550	550	237	0.000003	0.000001
1100	550	550	238	0.000005	0.000002
1100	550	550	239	0.000009	0.000004
1100	550	550	240	0.000015	0.000006
1100	550	550	241	0.000026	0.000011
1100	550	550	242	0.000044	0.000017
1100	550	550	243	0.000071	0.000028
1100	550	550	244	0.000116	0.000044
1100	550	550	245	0.000185	0.000069
1100	550	550	246	0.000291	0.000106
1100	550	550	247	0.000451	0.000161
1100	550	550	248	0.000691	0.000240
1100	550	550	249	0.001044	0.000353
1100	550	550	250	0.001556	0.000511
1100	550	550	251	0.002286	0.000731
1100	550	550	252	0.003315	0.001029
1100	550	550	253	0.004742	0.001427
1100	550	550	254	0.006693	0.001951
1100	550	550	255	0.009321	0.002629
1100	550	550	256	0.012812	0.003491
1100	550	550	257	0.017380	0.004568
1100	550	550	258	0.023272	0.005892
1100	550	550	259	0.030761	0.007489
1100	550	550	260	0.040142	0.009381
1100	550	550	261	0.051724	0.011582
1100	550	550	262	0.065815	0.014092
1100	550	550	263	0.082713	0.016898
1100	550	550	264	0.102684	0.019971
1100	550	550	265	0.125945	0.023261
1100	550	550	266	0.152648	0.026703
1100	550	550	267	0.182860	0.030211
1100	550	550	268	0.216548	0.033688
1100	550	550	269	0.253570	0.037023
1100	550	550	270	0.293671	0.040101
1100	550	550	271	0.336480	0.042809
1100	550	550	272	0.381520	0.045040
1100	550	550	273	0.428225	0.046705
1100	550	550	274	0.475959	0.047734
1100	550	550	275	0.524041	0.048081
1200	600	599	257	0.000001	0.000000
1200	600	599	258	0.000001	0.000000
1200	600	599	259	0.000002	0.000001
1200	600	599	260	0.000003	0.000001
1200	600	599	261	0.000006	0.000002

N	n	k	x	P(x)	p(x)
1200	600	599	262	0.000009	0.000004
1200	600	599	263	0.000016	0.000006
1200	600	599	264	0.000026	0.000010
1200	600	599	265	0.000042	0.000016
1200	600	599	266	0.000068	0.000026
1200	600	599	267	0.000109	0.000040
1200	600	599	268	0.000170	0.000062
1200	600	599	269	0.000263	0.000093
1200	600	599	270	0.000403	0.000139
1200	600	599	271	0.000608	0.000205
1200	600	599	272	0.000906	0.000298
1200	600	599	273	0.001333	0.000427
1200	600	599	274	0.001937	0.000604
1200	600	599	275	0.002781	0.000844
1200	600	599	276	0.003943	0.001162
1200	600	599	277	0.005522	0.001579
1200	600	599	278	0.007640	0.002118
1200	600	599	279	0.010441	0.002802
1200	600	599	280	0.014099	0.003658
1200	600	599	281	0.018811	0.004712
1200	600	599	282	0.024801	0.005990
1200	600	599	283	0.032313	0.007512
1200	600	599	284	0.041611	0.009297
1200	600	599	285	0.052965	0.011354
1200	600	599	286	0.066646	0.013682
1200	600	599	287	0.082915	0.016268
1200	600	599	288	0.102002	0.019087
1200	600	599	289	0.124101	0.022099
1200	600	599	290	0.149347	0.025246
1200	600	599	291	0.177808	0.028461
1200	600	599	292	0.209467	0.031659
1200	600	599	293	0.244219	0.034752
1200	600	599	294	0.281860	0.037641
1200	600	599	295	0.322092	0.040232
1200	600	599	296	0.364525	0.042433
1200	600	599	297	0.408686	0.044161
1200	600	599	298	0.454039	0.045353
1200	600	599	299	0.500000	0.045961
1200	600	600	258	0.000001	0.000000
1200	600	600	259	0.000001	0.000001
1200	600	600	260	0.000002	0.000001
1200	600	600	261	0.000004	0.000002
1200	600	600	262	0.000007	0.000003
1200	600	600	263	0.000012	0.000005
1200	600	600	264	0.000020	0.000008
1200	600	600	265	0.000033	0.000013
1200	600	600	266	0.000054	0.000021
1200	600	600	267	0.000086	0.000032
1200	600	600	268	0.000136	0.000050
1200	600	600	269	0.000212	0.000076

N	n	k	x	P(x)	p(x)
1200	600	600	270	0.000326	0.000114
1200	600	600	271	0.000495	0.000169
1200	600	600	272	0.000743	0.000248
1200	600	600	273	0.001100	0.000357
1200	600	600	274	0.001609	0.000509
1200	600	600	275	0.002325	0.000715
1200	600	600	276	0.003316	0.000992
1200	600	600	277	0.004673	0.001357
1200	600	600	278	0.006505	0.001832
1200	600	600	279	0.008945	0.002440
1200	600	600	280	0.012152	0.003207
1200	600	600	281	0.016310	0.004159
1200	600	600	282	0.021632	0.005321
1200	600	600	283	0.028351	0.006719
1200	600	600	284	0.036722	0.008371
1200	600	600	285	0.047014	0.010292
1200	600	600	286	0.059498	0.012484
1200	600	600	287	0.074442	0.014944
1200	600	600	288	0.092093	0.017651
1200	600	600	289	0.112665	0.020572
1200	600	600	290	0.136325	0.023660
1200	600	600	291	0.163175	0.026850
1200	600	600	292	0.193242	0.030067
1200	600	600	293	0.226467	0.033225
1200	600	600	294	0.262695	0.036228
1200	600	600	295	0.301675	0.038980
1200	600	600	296	0.343062	0.041387
1200	600	600	297	0.386422	0.043360
1200	600	600	298	0.431250	0.044828
1200	600	600	299	0.476981	0.045732
1200	600	600	300	0.523018	0.046037
1300	650	649	280	0.000001	0.000000
1300	650	649	281	0.000001	0.000000
1300	650	649	282	0.000002	0.000001
1300	650	649	283	0.000003	0.000001
1300	650	649	284	0.000004	0.000002
1300	650	649	285	0.000007	0.000003
1300	650	649	286	0.000012	0.000005
1300	650	649	287	0.000020	0.000008
1300	650	649	288	0.000032	0.000012
1300	650	649	289	0.000051	0.000019
1300	650	649	290	0.000080	0.000029
1300	650	649	291	0.000124	0.000044
1300	650	649	292	0.000191	0.000066
1300	650	649	293	0.000289	0.000099
1300	650	649	294	0.000434	0.000144
1300	650	649	295	0.000643	0.000209
1300	650	649	296	0.000942	0.000299
1300	650	649	297	0.001364	0.000422
1300	650	649	298	0.001952	0.000588

N	n	k	x	P(x)	p(x)
1300	650	649	299	0.002763	0.000811
1300	650	649	300	0.003865	0.001103
1300	650	649	301	0.005347	0.001482
1300	650	649	302	0.007314	0.001967
1300	650	649	303	0.009892	0.002578
1300	650	649	304	0.013231	0.003339
1300	650	649	305	0.017501	0.004270
1300	650	649	306	0.022896	0.005395
1300	650	649	307	0.029628	0.006732
1300	650	649	308	0.037925	0.008297
1300	650	649	309	0.048027	0.010102
1300	650	649	310	0.060175	0.012148
1300	650	649	311	0.074606	0.014430
1300	650	649	312	0.091537	0.016932
1300	650	649	313	0.111160	0.019623
1300	650	649	314	0.133625	0.022465
1300	650	649	315	0.159028	0.025403
1300	650	649	316	0.187402	0.028375
1300	650	649	317	0.218709	0.031306
1300	650	649	318	0.252828	0.034119
1300	650	649	319	0.289558	0.036730
1300	650	649	320	0.328616	0.039058
1300	650	649	321	0.369642	0.041026
1300	650	649	322	0.412209	0.042567
1300	650	649	323	0.455835	0.043626
1300	650	649	324	0.500000	0.044165
1300	650	650	281	0.000001	0.000000
1300	650	650	282	0.000001	0.000000
1300	650	650	283	0.000002	0.000001
1300	650	650	284	0.000003	0.000001
1300	650	650	285	0.000006	0.000002
1300	650	650	286	0.000010	0.000004
1300	650	650	287	0.000016	0.000006
1300	650	650	288	0.000025	0.000010
1300	650	650	289	0.000040	0.000015
1300	650	650	290	0.000064	0.000023
1300	650	650	291	0.000100	0.000036
1300	650	650	292	0.000154	0.000054
1300	650	650	293	0.000235	0.000081
1300	650	650	294	0.000355	0.000120
1300	650	650	295	0.000529	0.000174
1300	650	650	296	0.000779	0.000250
1300	650	650	297	0.001135	0.000356
1300	650	650	298	0.001634	0.000499
1300	650	650	299	0.002326	0.000692
1300	650	650	300	0.003273	0.000947
1300	650	650	301	0.004553	0.001280
1300	650	650	302	0.006263	0.001710
1300	650	650	303	0.008518	0.002255
1300	650	650	304	0.011457	0.002939

$N = 1100\text{–}1300$

$n = 550\text{–}650 \quad k = 549\text{–}650$

Table for $k = n - 1, n;\ n = N/2,$

$N = 100,\ n = 50$ through $N = 2000,\ n = 1000$

N	n	k	x	P(x)	p(x)
1300	650	650	305	0.015238	0.003782
1300	650	650	306	0.020045	0.004807
1300	650	650	307	0.026081	0.006036
1300	650	650	308	0.033566	0.007485
1300	650	650	309	0.042736	0.009169
1300	650	650	310	0.053831	0.011095
1300	650	650	311	0.067091	0.013261
1300	650	650	312	0.082746	0.015655
1300	650	650	313	0.101002	0.018256
1300	650	650	314	0.122030	0.021028
1300	650	650	315	0.145956	0.023925
1300	650	650	316	0.172845	0.026889
1300	650	650	317	0.202695	0.029850
1300	650	650	318	0.235428	0.032733
1300	650	650	319	0.270883	0.035455
1300	650	650	320	0.308817	0.037934
1300	650	650	321	0.348908	0.040091
1300	650	650	322	0.390762	0.041853
1300	650	650	323	0.433921	0.043159
1300	650	650	324	0.477883	0.043962
1300	650	650	325	0.522116	0.044233
1400	700	699	304	0.000001	0.000000
1400	700	699	305	0.000001	0.000001
1400	700	699	306	0.000002	0.000001
1400	700	699	307	0.000003	0.000001
1400	700	699	308	0.000006	0.000002
1400	700	699	309	0.000009	0.000004
1400	700	699	310	0.000015	0.000006
1400	700	699	311	0.000024	0.000009
1400	700	699	312	0.000038	0.000014
1400	700	699	313	0.000059	0.000021
1400	700	699	314	0.000090	0.000032
1400	700	699	315	0.000138	0.000047
1400	700	699	316	0.000208	0.000070
1400	700	699	317	0.000310	0.000102
1400	700	699	318	0.000457	0.000147
1400	700	699	319	0.000666	0.000210
1400	700	699	320	0.000962	0.000295
1400	700	699	321	0.001373	0.000412
1400	700	699	322	0.001940	0.000567
1400	700	699	323	0.002713	0.000772
1400	700	699	324	0.003752	0.001040
1400	700	699	325	0.005136	0.001383
1400	700	699	326	0.006956	0.001820
1400	700	699	327	0.009323	0.002367
1400	700	699	328	0.012367	0.003044
1400	700	699	329	0.016236	0.003869
1400	700	699	330	0.021099	0.004863
1400	700	699	331	0.027140	0.006041
1400	700	699	332	0.034561	0.007420

N	n	k	x	P(x)	p(x)
1400	700	699	333	0.043571	0.009010
1400	700	699	334	0.054388	0.010817
1400	700	699	335	0.067226	0.012838
1400	700	699	336	0.082289	0.015063
1400	700	699	337	0.099763	0.017474
1400	700	699	338	0.119802	0.020039
1400	700	699	339	0.142522	0.022720
1400	700	699	340	0.167990	0.025468
1400	700	699	341	0.196214	0.028223
1400	700	699	342	0.227136	0.030922
1400	700	699	343	0.260630	0.033494
1400	700	699	344	0.296498	0.035868
1400	700	699	345	0.334472	0.037974
1400	700	699	346	0.374220	0.039748
1400	700	699	347	0.415353	0.041133
1400	700	699	348	0.457435	0.042082
1400	700	699	349	0.500000	0.042565
1400	700	700	304	0.000001	0.000000
1400	700	700	305	0.000001	0.000000
1400	700	700	306	0.000002	0.000001
1400	700	700	307	0.000003	0.000001
1400	700	700	308	0.000004	0.000002
1400	700	700	309	0.000007	0.000003
1400	700	700	310	0.000012	0.000005
1400	700	700	311	0.000019	0.000007
1400	700	700	312	0.000030	0.000011
1400	700	700	313	0.000047	0.000017
1400	700	700	314	0.000073	0.000026
1400	700	700	315	0.000112	0.000039
1400	700	700	316	0.000169	0.000058
1400	700	700	317	0.000254	0.000085
1400	700	700	318	0.000376	0.000123
1400	700	700	319	0.000552	0.000176
1400	700	700	320	0.000801	0.000249
1400	700	700	321	0.001151	0.000349
1400	700	700	322	0.001635	0.000484
1400	700	700	323	0.002297	0.000663
1400	700	700	324	0.003195	0.000897
1400	700	700	325	0.004396	0.001201
1400	700	700	326	0.005985	0.001589
1400	700	700	327	0.008063	0.002079
1400	700	700	328	0.010752	0.002688
1400	700	700	329	0.014188	0.003437
1400	700	700	330	0.018532	0.004344
1400	700	700	331	0.023960	0.005428
1400	700	700	332	0.030665	0.006705
1400	700	700	333	0.038854	0.008189
1400	700	700	334	0.048740	0.009887
1400	700	700	335	0.060541	0.011801
1400	700	700	336	0.074467	0.013926

N	n	k	x	P(x)	p(x)
1400	700	700	337	0.090714	0.016247
1400	700	700	338	0.109454	0.018739
1400	700	700	339	0.130822	0.021368
1400	700	700	340	0.154911	0.024089
1400	700	700	341	0.181760	0.026849
1400	700	700	342	0.211344	0.029584
1400	700	700	343	0.243572	0.032228
1400	700	700	344	0.278282	0.034710
1400	700	700	345	0.315241	0.036959
1400	700	700	346	0.354148	0.038907
1400	700	700	347	0.394640	0.040492
1400	700	700	348	0.436304	0.041664
1400	700	700	349	0.478687	0.042383
1400	700	700	350	0.521313	0.042626
1500	750	749	327	0.000001	0.000000
1500	750	749	328	0.000001	0.000000
1500	750	749	329	0.000002	0.000001
1500	750	749	330	0.000003	0.000001
1500	750	749	331	0.000004	0.000002
1500	750	749	332	0.000007	0.000003
1500	750	749	333	0.000011	0.000004
1500	750	749	334	0.000018	0.000006
1500	750	749	335	0.000028	0.000010
1500	750	749	336	0.000043	0.000015
1500	750	749	337	0.000066	0.000023
1500	750	749	338	0.000099	0.000034
1500	750	749	339	0.000149	0.000050
1500	750	749	340	0.000221	0.000072
1500	750	749	341	0.000325	0.000104
1500	750	749	342	0.000472	0.000147
1500	750	749	343	0.000679	0.000207
1500	750	749	344	0.000968	0.000289
1500	750	749	345	0.001366	0.000398
1500	750	749	346	0.001908	0.000542
1500	750	749	347	0.002639	0.000731
1500	750	749	348	0.003614	0.000975
1500	750	749	349	0.004900	0.001287
1500	750	749	350	0.006580	0.001680
1500	750	749	351	0.008750	0.002170
1500	750	749	352	0.011523	0.002773
1500	750	749	353	0.015029	0.003506
1500	750	749	354	0.019416	0.004387
1500	750	749	355	0.024845	0.005429
1500	750	749	356	0.031493	0.006648
1500	750	749	357	0.039548	0.008055
1500	750	749	358	0.049203	0.009655
1500	750	749	359	0.060653	0.011450
1500	750	749	360	0.074088	0.013435
1500	750	749	361	0.089685	0.015597
1500	750	749	362	0.107600	0.017915

N	n	k	x	P(x)	p(x)
1500	750	749	363	0.127959	0.020358
1500	750	749	364	0.150849	0.022890
1500	750	749	365	0.176312	0.025464
1500	750	749	366	0.204339	0.028026
1500	750	749	367	0.234859	0.030520
1500	750	749	368	0.267742	0.032883
1500	750	749	369	0.302795	0.035054
1500	750	749	370	0.339767	0.036971
1500	750	749	371	0.378348	0.038581
1500	750	749	372	0.418181	0.039834
1500	750	749	373	0.458873	0.040691
1500	750	749	374	0.500000	0.041127
1500	750	750	328	0.000001	0.000000
1500	750	750	329	0.000001	0.000001
1500	750	750	330	0.000002	0.000001
1500	750	750	331	0.000003	0.000001
1500	750	750	332	0.000006	0.000002
1500	750	750	333	0.000009	0.000003
1500	750	750	334	0.000014	0.000005
1500	750	750	335	0.000022	0.000008
1500	750	750	336	0.000035	0.000012
1500	750	750	337	0.000053	0.000019
1500	750	750	338	0.000081	0.000028
1500	750	750	339	0.000122	0.000041
1500	750	750	340	0.000182	0.000060
1500	750	750	341	0.000268	0.000087
1500	750	750	342	0.000392	0.000124
1500	750	750	343	0.000567	0.000175
1500	750	750	344	0.000812	0.000245
1500	750	750	345	0.001151	0.000339
1500	750	750	346	0.001616	0.000465
1500	750	750	347	0.002246	0.000630
1500	750	750	348	0.003092	0.000845
1500	750	750	349	0.004213	0.001121
1500	750	750	350	0.005685	0.001472
1500	750	750	351	0.007597	0.001912
1500	750	750	352	0.010053	0.002456
1500	750	750	353	0.013176	0.003122
1500	750	750	354	0.017103	0.003927
1500	750	750	355	0.021989	0.004887
1500	750	750	356	0.028005	0.006016
1500	750	750	357	0.035333	0.007328
1500	750	750	358	0.044163	0.008830
1500	750	750	359	0.054692	0.010528
1500	750	750	360	0.067111	0.012420
1500	750	750	361	0.081606	0.014495
1500	750	750	362	0.098345	0.016738
1500	750	750	363	0.117468	0.019123
1500	750	750	364	0.139084	0.021616
1500	750	750	365	0.163258	0.024175

Table for $k = n - 1,\ n;\ \ n = N/2,$

$N = 100,\ n = 50$ through $N = 2000,\ n = 1000$

N	n	k	x	P(x)	p(x)
1500	750	750	366	0.190008	0.026750
1500	750	750	367	0.219294	0.029286
1500	750	750	368	0.251016	0.031722
1500	750	750	369	0.285012	0.033996
1500	750	750	370	0.321060	0.036048
1500	750	750	371	0.358877	0.037818
1500	750	750	372	0.398132	0.039254
1500	750	750	373	0.438446	0.040314
1500	750	750	374	0.479409	0.040963
1500	750	750	375	0.520591	0.041182
1600	800	799	351	0.000001	0.000000
1600	800	799	352	0.000001	0.000000
1600	800	799	353	0.000002	0.000001
1600	800	799	354	0.000003	0.000001
1600	800	799	355	0.000005	0.000002
1600	800	799	356	0.000008	0.000003
1600	800	799	357	0.000013	0.000005
1600	800	799	358	0.000020	0.000007
1600	800	799	359	0.000031	0.000011
1600	800	799	360	0.000048	0.000016
1600	800	799	361	0.000072	0.000024
1600	800	799	362	0.000107	0.000035
1600	800	799	363	0.000158	0.000051
1600	800	799	364	0.000231	0.000073
1600	800	799	365	0.000335	0.000104
1600	800	799	366	0.000480	0.000146
1600	800	799	367	0.000683	0.000203
1600	800	799	368	0.000963	0.000280
1600	800	799	369	0.001344	0.000381
1600	800	799	370	0.001859	0.000515
1600	800	799	371	0.002547	0.000688
1600	800	799	372	0.003458	0.000911
1600	800	799	373	0.004651	0.001193
1600	800	799	374	0.006198	0.001547
1600	800	799	375	0.008184	0.001987
1600	800	799	376	0.010710	0.002525
1600	800	799	377	0.013888	0.003178
1600	800	799	378	0.017848	0.003960
1600	800	799	379	0.022733	0.004885
1600	800	799	380	0.028700	0.005966
1600	800	799	381	0.035914	0.007214
1600	800	799	382	0.044550	0.008636
1600	800	799	383	0.054785	0.010235
1600	800	799	384	0.066794	0.012010
1600	800	799	385	0.080746	0.013952
1600	800	799	386	0.096792	0.016046
1600	800	799	387	0.115064	0.018272
1600	800	799	388	0.135663	0.020599
1600	800	799	389	0.158655	0.022992
1600	800	799	390	0.184062	0.025407

N	n	k	x	P(x)	p(x)
1600	800	799	391	0.211860	0.027797
1600	800	799	392	0.241969	0.030110
1600	800	799	393	0.274260	0.032290
1600	800	799	394	0.308544	0.034285
1600	800	799	395	0.344585	0.036040
1600	800	799	396	0.382094	0.037509
1600	800	799	397	0.420744	0.038650
1600	800	799	398	0.460174	0.039430
1600	800	799	399	0.500000	0.039826
1600	800	800	351	0.000001	0.000000
1600	800	800	352	0.000001	0.000000
1600	800	800	353	0.000002	0.000001
1600	800	800	354	0.000003	0.000001
1600	800	800	355	0.000004	0.000002
1600	800	800	356	0.000007	0.000002
1600	800	800	357	0.000011	0.000004
1600	800	800	358	0.000016	0.000006
1600	800	800	359	0.000025	0.000009
1600	800	800	360	0.000039	0.000013
1600	800	800	361	0.000058	0.000020
1600	800	800	362	0.000088	0.000029
1600	800	800	363	0.000130	0.000042
1600	800	800	364	0.000191	0.000061
1600	800	800	365	0.000278	0.000087
1600	800	800	366	0.000401	0.000123
1600	800	800	367	0.000574	0.000172
1600	800	800	368	0.000812	0.000239
1600	800	800	369	0.001139	0.000327
1600	800	800	370	0.001583	0.000444
1600	800	800	371	0.002179	0.000596
1600	800	800	372	0.002971	0.000793
1600	800	800	373	0.004015	0.001043
1600	800	800	374	0.005375	0.001360
1600	800	800	375	0.007130	0.001755
1600	800	800	376	0.009373	0.002243
1600	800	800	377	0.012209	0.002837
1600	800	800	378	0.015762	0.003552
1600	800	800	379	0.020166	0.004404
1600	800	800	380	0.025571	0.005406
1600	800	800	381	0.032140	0.006569
1600	800	800	382	0.040043	0.007903
1600	800	800	383	0.049456	0.009413
1600	800	800	384	0.060557	0.011101
1600	800	800	385	0.073518	0.012960
1600	800	800	386	0.088499	0.014981
1600	800	800	387	0.105643	0.017144
1600	800	800	388	0.125068	0.019425
1600	800	800	389	0.146858	0.021790
1600	800	800	390	0.171057	0.024200
1600	800	800	391	0.197666	0.026609

N	n	k	x	P(x)	p(x)
1600	800	800	392	0.226632	0.028967
1600	800	800	393	0.257852	0.031220
1600	800	800	394	0.291167	0.033314
1600	800	800	395	0.326362	0.035195
1600	800	800	396	0.363175	0.036813
1600	800	800	397	0.401298	0.038123
1600	800	800	398	0.440385	0.039087
1600	800	800	399	0.480062	0.039677
1600	800	800	400	0.519938	0.039876
1700	850	849	374	0.000001	0.000000
1700	850	849	375	0.000001	0.000000
1700	850	849	376	0.000002	0.000001
1700	850	849	377	0.000003	0.000001
1700	850	849	378	0.000004	0.000001
1700	850	849	379	0.000006	0.000002
1700	850	849	380	0.000010	0.000003
1700	850	849	381	0.000015	0.000005
1700	850	849	382	0.000023	0.000008
1700	850	849	383	0.000034	0.000012
1700	850	849	384	0.000052	0.000017
1700	850	849	385	0.000077	0.000025
1700	850	849	386	0.000113	0.000036
1700	850	849	387	0.000164	0.000052
1700	850	849	388	0.000238	0.000073
1700	850	849	389	0.000340	0.000103
1700	850	849	390	0.000483	0.000143
1700	850	849	391	0.000680	0.000197
1700	850	849	392	0.000949	0.000269
1700	850	849	393	0.001312	0.000363
1700	850	849	394	0.001799	0.000486
1700	850	849	395	0.002444	0.000645
1700	850	849	396	0.003291	0.000848
1700	850	849	397	0.004395	0.001103
1700	850	849	398	0.005817	0.001423
1700	850	849	399	0.007634	0.001817
1700	850	849	400	0.009934	0.002299
1700	850	849	401	0.012815	0.002882
1700	850	849	402	0.016393	0.003578
1700	850	849	403	0.020794	0.004401
1700	850	849	404	0.026156	0.005362
1700	850	849	405	0.032629	0.006472
1700	850	849	406	0.040368	0.007739
1700	850	849	407	0.049535	0.009167
1700	850	849	408	0.060291	0.010756
1700	850	849	409	0.072794	0.012503
1700	850	849	410	0.087191	0.014397
1700	850	849	411	0.103614	0.016423
1700	850	849	412	0.122173	0.018559
1700	850	849	413	0.142949	0.020776
1700	850	849	414	0.165988	0.023040

N	n	k	x	P(x)	p(x)
1700	850	849	415	0.191299	0.025311
1700	850	849	416	0.218845	0.027546
1700	850	849	417	0.248543	0.029698
1700	850	849	418	0.280261	0.031718
1700	850	849	419	0.313819	0.033558
1700	850	849	420	0.348993	0.035174
1700	850	849	421	0.385514	0.036521
1700	850	849	422	0.423080	0.037566
1700	850	849	423	0.461359	0.038279
1700	850	849	424	0.500000	0.038641
1700	850	850	375	0.000001	0.000000
1700	850	850	376	0.000001	0.000000
1700	850	850	377	0.000002	0.000001
1700	850	850	378	0.000003	0.000001
1700	850	850	379	0.000005	0.000002
1700	850	850	380	0.000008	0.000003
1700	850	850	381	0.000012	0.000004
1700	850	850	382	0.000018	0.000006
1700	850	850	383	0.000028	0.000010
1700	850	850	384	0.000042	0.000014
1700	850	850	385	0.000063	0.000021
1700	850	850	386	0.000093	0.000030
1700	850	850	387	0.000136	0.000043
1700	850	850	388	0.000198	0.000062
1700	850	850	389	0.000285	0.000087
1700	850	850	390	0.000406	0.000121
1700	850	850	391	0.000574	0.000168
1700	850	850	392	0.000804	0.000230
1700	850	850	393	0.001117	0.000313
1700	850	850	394	0.001538	0.000421
1700	850	850	395	0.002099	0.000561
1700	850	850	396	0.002839	0.000740
1700	850	850	397	0.003807	0.000968
1700	850	850	398	0.005062	0.001254
1700	850	850	399	0.006671	0.001610
1700	850	850	400	0.008718	0.002046
1700	850	850	401	0.011295	0.002577
1700	850	850	402	0.014510	0.003215
1700	850	850	403	0.018483	0.003973
1700	850	850	404	0.023346	0.004864
1700	850	850	405	0.029244	0.005898
1700	850	850	406	0.036330	0.007086
1700	850	850	407	0.044763	0.008433
1700	850	850	408	0.054704	0.009941
1700	850	850	409	0.066315	0.011610
1700	850	850	410	0.079747	0.013433
1700	850	850	411	0.095142	0.015395
1700	850	850	412	0.112621	0.017479
1700	850	850	413	0.132280	0.019659
1700	850	850	414	0.154184	0.021904

Table for $k = n - 1,\ n;\ \ n = N/2,$

$N = 100,\ n = 50$ through $N = 2000,\ n = 1000$

N	n	k	x	$P(x)$	$p(x)$
1700	850	850	415	0.178361	0.024177
1700	850	850	416	0.204797	0.026436
1700	850	850	417	0.233432	0.028635
1700	850	850	418	0.264160	0.030727
1700	850	850	419	0.296823	0.032664
1700	850	850	420	0.331220	0.034397
1700	850	850	421	0.367103	0.035883
1700	850	850	422	0.404187	0.037084
1700	850	850	423	0.442152	0.037965
1700	850	850	424	0.480657	0.038505
1700	850	850	425	0.519343	0.038686
1800	900	899	398	0.000001	0.000000
1800	900	899	399	0.000001	0.000000
1800	900	899	400	0.000002	0.000001
1800	900	899	401	0.000003	0.000001
1800	900	899	402	0.000005	0.000002
1800	900	899	403	0.000007	0.000003
1800	900	899	404	0.000011	0.000004
1800	900	899	405	0.000017	0.000006
1800	900	899	406	0.000025	0.000008
1800	900	899	407	0.000037	0.000012
1800	900	899	408	0.000055	0.000018
1800	900	899	409	0.000080	0.000026
1800	900	899	410	0.000117	0.000037
1800	900	899	411	0.000169	0.000052
1800	900	899	412	0.000241	0.000073
1800	900	899	413	0.000342	0.000101
1800	900	899	414	0.000481	0.000139
1800	900	899	415	0.000671	0.000190
1800	900	899	416	0.000927	0.000257
1800	900	899	417	0.001272	0.000344
1800	900	899	418	0.001729	0.000458
1800	900	899	419	0.002332	0.000603
1800	900	899	420	0.003119	0.000787
1800	900	899	421	0.004138	0.001019
1800	900	899	422	0.005445	0.001307
1800	900	899	423	0.007106	0.001661
1800	900	899	424	0.009199	0.002093
1800	900	899	425	0.011812	0.002614
1800	900	899	426	0.015048	0.003235
1800	900	899	427	0.019017	0.003969
1800	900	899	428	0.023842	0.004826
1800	900	899	429	0.029658	0.005816
1800	900	899	430	0.036605	0.006947
1800	900	899	431	0.044829	0.008224
1800	900	899	432	0.054479	0.009650
1800	900	899	433	0.065703	0.011223
1800	900	899	434	0.078640	0.012937
1800	900	899	435	0.093421	0.014781
1800	900	899	436	0.110159	0.016738
1800	900	899	437	0.128946	0.018787
1800	900	899	438	0.149846	0.020900
1800	900	899	439	0.172890	0.023044
1800	900	899	440	0.198075	0.025184
1800	900	899	441	0.225355	0.027280
1800	900	899	442	0.254643	0.029288
1800	900	899	443	0.285810	0.031167
1800	900	899	444	0.318682	0.032872
1800	900	899	445	0.353046	0.034364
1800	900	899	446	0.388653	0.035607
1800	900	899	447	0.425221	0.036568
1800	900	899	448	0.462445	0.037223
1800	900	899	449	0.500000	0.037555
1800	900	900	398	0.000001	0.000000
1800	900	900	399	0.000001	0.000000
1800	900	900	400	0.000001	0.000001
1800	900	900	401	0.000002	0.000001
1800	900	900	402	0.000004	0.000001
1800	900	900	403	0.000006	0.000002
1800	900	900	404	0.000009	0.000003
1800	900	900	405	0.000013	0.000005
1800	900	900	406	0.000020	0.000007
1800	900	900	407	0.000030	0.000010
1800	900	900	408	0.000045	0.000015
1800	900	900	409	0.000067	0.000021
1800	900	900	410	0.000097	0.000031
1800	900	900	411	0.000141	0.000044
1800	900	900	412	0.000202	0.000061
1800	900	900	413	0.000288	0.000086
1800	900	900	414	0.000406	0.000119
1800	900	900	415	0.000569	0.000163
1800	900	900	416	0.000790	0.000221
1800	900	900	417	0.001087	0.000298
1800	900	900	418	0.001485	0.000397
1800	900	900	419	0.002010	0.000526
1800	900	900	420	0.002700	0.000690
1800	900	900	421	0.003596	0.000896
1800	900	900	422	0.004751	0.001155
1800	900	900	423	0.006226	0.001475
1800	900	900	424	0.008093	0.001867
1800	900	900	425	0.010435	0.002342
1800	900	900	426	0.013346	0.002911
1800	900	900	427	0.016933	0.003587
1800	900	900	428	0.021314	0.004381
1800	900	900	429	0.026618	0.005304
1800	900	900	430	0.032981	0.006363
1800	900	900	431	0.040548	0.007567
1800	900	900	432	0.049467	0.008919
1800	900	900	433	0.059885	0.010419
1800	900	900	434	0.071949	0.012063
1800	900	900	435	0.085793	0.013844
1800	900	900	436	0.101539	0.015747
1800	900	900	437	0.119292	0.017753
1800	900	900	438	0.139129	0.019837
1800	900	900	439	0.161100	0.021970
1800	900	900	440	0.185217	0.024117
1800	900	900	441	0.211457	0.026240
1800	900	900	442	0.239755	0.028298
1800	900	900	443	0.270002	0.030246
1800	900	900	444	0.302045	0.032044
1800	900	900	445	0.335692	0.033647
1800	900	900	446	0.370711	0.035019
1800	900	900	447	0.406836	0.036124
1800	900	900	448	0.443771	0.036935
1800	900	900	449	0.481201	0.037430
1800	900	900	450	0.518798	0.037597
1900	950	949	421	0.000001	0.000000
1900	950	949	422	0.000001	0.000000
1900	950	949	423	0.000001	0.000001
1900	950	949	424	0.000002	0.000001
1900	950	949	425	0.000003	0.000001
1900	950	949	426	0.000005	0.000002
1900	950	949	427	0.000008	0.000003
1900	950	949	428	0.000012	0.000004
1900	950	949	429	0.000018	0.000006
1900	950	949	430	0.000027	0.000009
1900	950	949	431	0.000039	0.000013
1900	950	949	432	0.000057	0.000018
1900	950	949	433	0.000083	0.000026
1900	950	949	434	0.000120	0.000037
1900	950	949	435	0.000171	0.000051
1900	950	949	436	0.000243	0.000071
1900	950	949	437	0.000341	0.000098
1900	950	949	438	0.000475	0.000134
1900	950	949	439	0.000656	0.000182
1900	950	949	440	0.000900	0.000244
1900	950	949	441	0.001225	0.000325
1900	950	949	442	0.001654	0.000429
1900	950	949	443	0.002216	0.000562
1900	950	949	444	0.002946	0.000729
1900	950	949	445	0.003885	0.000939
1900	950	949	446	0.005084	0.001199
1900	950	949	447	0.006601	0.001518
1900	950	949	448	0.008507	0.001905
1900	950	949	449	0.010878	0.002372
1900	950	949	450	0.013806	0.002927
1900	950	949	451	0.017388	0.003583
1900	950	949	452	0.021737	0.004348
1900	950	949	453	0.026970	0.005233
1900	950	949	454	0.033215	0.006245
1900	950	949	455	0.040604	0.007390
1900	950	949	456	0.049275	0.008671
1900	950	949	457	0.059365	0.010090
1900	950	949	458	0.071006	0.011641
1900	950	949	459	0.084326	0.013319
1900	950	949	460	0.099437	0.015111
1900	950	949	461	0.116438	0.017001
1900	950	949	462	0.135403	0.018966
1900	950	949	463	0.156384	0.020981
1900	950	949	464	0.179399	0.023015
1900	950	949	465	0.204434	0.025035
1900	950	949	466	0.231439	0.027005
1900	950	949	467	0.260324	0.028885
1900	950	949	468	0.290960	0.030637
1900	950	949	469	0.323183	0.032223
1900	950	949	470	0.356790	0.033607
1900	950	949	471	0.391547	0.034757
1900	950	949	472	0.427193	0.035646
1900	950	949	473	0.463443	0.036250
1900	950	949	474	0.500000	0.036557
1900	950	950	422	0.000001	0.000000
1900	950	950	423	0.000001	0.000000
1900	950	950	424	0.000002	0.000001
1900	950	950	425	0.000003	0.000001
1900	950	950	426	0.000004	0.000001
1900	950	950	427	0.000006	0.000002
1900	950	950	428	0.000010	0.000003
1900	950	950	429	0.000015	0.000005
1900	950	950	430	0.000022	0.000007
1900	950	950	431	0.000032	0.000011
1900	950	950	432	0.000048	0.000015
1900	950	950	433	0.000069	0.000022
1900	950	950	434	0.000100	0.000031
1900	950	950	435	0.000144	0.000043
1900	950	950	436	0.000204	0.000061
1900	950	950	437	0.000288	0.000084
1900	950	950	438	0.000403	0.000115
1900	950	950	439	0.000559	0.000156
1900	950	950	440	0.000770	0.000211
1900	950	950	441	0.001051	0.000282
1900	950	950	442	0.001425	0.000374
1900	950	950	443	0.001917	0.000491
1900	950	950	444	0.002557	0.000641
1900	950	950	445	0.003386	0.000829
1900	950	950	446	0.004448	0.001062
1900	950	950	447	0.005799	0.001350
1900	950	950	448	0.007501	0.001702
1900	950	950	449	0.009629	0.002128
1900	950	950	450	0.012267	0.002638
1900	950	950	451	0.015509	0.003242

Table for $k = n - 1,\ n;\ \ n = N/2,$

$N = 100,\ n = 50$ through $N = 2000,\ n = 1000$

N	n	k	x	P(x)	p(x)
1900	950	950	452	0.019460	0.003951
1900	950	950	453	0.024235	0.004775
1900	950	950	454	0.029958	0.005723
1900	950	950	455	0.036758	0.006800
1900	950	950	456	0.044771	0.008013
1900	950	950	457	0.054135	0.009363
1900	950	950	458	0.064984	0.010849
1900	950	950	459	0.077449	0.012465
1900	950	950	460	0.091651	0.014202
1900	950	950	461	0.107696	0.016045
1900	950	950	462	0.125671	0.017975
1900	950	950	463	0.145640	0.019969
1900	950	950	464	0.167638	0.021997
1900	950	950	465	0.191667	0.024029
1900	950	950	466	0.217695	0.026029
1900	950	950	467	0.245654	0.027958
1900	950	950	468	0.275433	0.029779
1900	950	950	469	0.306885	0.031453
1900	950	950	470	0.339828	0.032942
1900	950	950	471	0.374041	0.034213
1900	950	950	472	0.409277	0.035236
1900	950	950	473	0.445261	0.035984
1900	950	950	474	0.481702	0.036441
1900	950	950	475	0.518297	0.036595
2000	1000	999	445	0.000001	0.000000
2000	1000	999	446	0.000001	0.000000
2000	1000	999	447	0.000002	0.000001
2000	1000	999	448	0.000002	0.000001
2000	1000	999	449	0.000004	0.000001
2000	1000	999	450	0.000006	0.000002
2000	1000	999	451	0.000009	0.000003
2000	1000	999	452	0.000013	0.000004
2000	1000	999	453	0.000019	0.000006
2000	1000	999	454	0.000028	0.000009
2000	1000	999	455	0.000041	0.000013
2000	1000	999	456	0.000059	0.000018
2000	1000	999	457	0.000085	0.000026
2000	1000	999	458	0.000122	0.000036
2000	1000	999	459	0.000172	0.000050
2000	1000	999	460	0.000242	0.000069
2000	1000	999	461	0.000337	0.000095
2000	1000	999	462	0.000465	0.000129
2000	1000	999	463	0.000638	0.000173
2000	1000	999	464	0.000869	0.000231
2000	1000	999	465	0.001175	0.000306
2000	1000	999	466	0.001576	0.000401
2000	1000	999	467	0.002098	0.000522
2000	1000	999	468	0.002773	0.000675
2000	1000	999	469	0.003638	0.000865
2000	1000	999	470	0.004737	0.001099

N	n	k	x	P(x)	p(x)
2000	1000	999	471	0.006123	0.001386
2000	1000	999	472	0.007858	0.001735
2000	1000	999	473	0.010011	0.002153
2000	1000	999	474	0.012661	0.002651
2000	1000	999	475	0.015899	0.003237
2000	1000	999	476	0.019821	0.003922
2000	1000	999	477	0.024535	0.004714
2000	1000	999	478	0.030157	0.005621
2000	1000	999	479	0.036806	0.006649
2000	1000	999	480	0.044608	0.007802
2000	1000	999	481	0.053691	0.009082
2000	1000	999	482	0.064179	0.010488
2000	1000	999	483	0.076194	0.012015
2000	1000	999	484	0.089849	0.013655
2000	1000	999	485	0.105243	0.015394
2000	1000	999	486	0.122461	0.017217
2000	1000	999	487	0.141564	0.019103
2000	1000	999	488	0.162590	0.021026
2000	1000	999	489	0.185549	0.022959
2000	1000	999	490	0.210418	0.024869
2000	1000	999	491	0.237142	0.026724
2000	1000	999	492	0.265630	0.028489
2000	1000	999	493	0.295758	0.030128
2000	1000	999	494	0.327366	0.031608
2000	1000	999	495	0.360262	0.032896
2000	1000	999	496	0.394227	0.033965
2000	1000	999	497	0.429017	0.034789
2000	1000	999	498	0.464366	0.035350
2000	1000	999	499	0.500000	0.035633
2000	1000	1000	445	0.000001	0.000000
2000	1000	1000	446	0.000001	0.000000
2000	1000	1000	447	0.000001	0.000000
2000	1000	1000	448	0.000002	0.000001
2000	1000	1000	449	0.000003	0.000001
2000	1000	1000	450	0.000005	0.000002
2000	1000	1000	451	0.000007	0.000002
2000	1000	1000	452	0.000011	0.000004
2000	1000	1000	453	0.000016	0.000005
2000	1000	1000	454	0.000023	0.000007
2000	1000	1000	455	0.000034	0.000011
2000	1000	1000	456	0.000049	0.000015
2000	1000	1000	457	0.000071	0.000022
2000	1000	1000	458	0.000102	0.000031
2000	1000	1000	459	0.000145	0.000043
2000	1000	1000	460	0.000204	0.000059
2000	1000	1000	461	0.000285	0.000081
2000	1000	1000	462	0.000396	0.000111
2000	1000	1000	463	0.000545	0.000149
2000	1000	1000	464	0.000746	0.000200
2000	1000	1000	465	0.001011	0.000266

N	n	k	x	P(x)	p(x)
2000	1000	1000	466	0.001362	0.000350
2000	1000	1000	467	0.001820	0.000458
2000	1000	1000	468	0.002414	0.000594
2000	1000	1000	469	0.003179	0.000765
2000	1000	1000	470	0.004155	0.000976
2000	1000	1000	471	0.005391	0.001236
2000	1000	1000	472	0.006943	0.001552
2000	1000	1000	473	0.008877	0.001934
2000	1000	1000	474	0.011269	0.002391
2000	1000	1000	475	0.014201	0.002932
2000	1000	1000	476	0.017768	0.003567
2000	1000	1000	477	0.022072	0.004304
2000	1000	1000	478	0.027225	0.005153
2000	1000	1000	479	0.033345	0.006120
2000	1000	1000	480	0.040555	0.007210
2000	1000	1000	481	0.048982	0.008426
2000	1000	1000	482	0.058751	0.009770
2000	1000	1000	483	0.069988	0.011237
2000	1000	1000	484	0.082810	0.012822
2000	1000	1000	485	0.097323	0.014513
2000	1000	1000	486	0.113620	0.016297
2000	1000	1000	487	0.131774	0.018154
2000	1000	1000	488	0.151835	0.020062
2000	1000	1000	489	0.173828	0.021993
2000	1000	1000	490	0.197747	0.023919
2000	1000	1000	491	0.223553	0.025806
2000	1000	1000	492	0.251172	0.027620
2000	1000	1000	493	0.280498	0.029326
2000	1000	1000	494	0.311388	0.030890
2000	1000	1000	495	0.343666	0.032278
2000	1000	1000	496	0.377126	0.033460
2000	1000	1000	497	0.411535	0.034409
2000	1000	1000	498	0.446639	0.035103
2000	1000	1000	499	0.482165	0.035527
2000	1000	1000	500	0.517834	0.035669

N	n	k	x	P(x)	p(x)

APPENDIX

LOGARITHMS OF FACTORIALS

N	Log N!	N	Log N!	N	Log N!	N	Log N!
1	0.00000 00000 00000	51	66.19064 50485 69972	101	159.97432 50284 98431	151	264.93587 03582 19209
2	0.30102 99956 63981	52	67.90664 83922 04771	102	161.98292 52002 60349	152	267.11771 39461 63981
3	0.77815 12503 83644	53	69.63092 42618 05560	103	163.99576 24249 65521	153	269.30240 53769 81580
4	1.38021 12417 11606	54	71.36331 80216 28528	104	166.01279 57642 64301	154	271.48992 60978 18043
5	2.07918 12460 47625	55	73.10368 07111 22772	105	168.03398 50633 34239	155	273.68025 77959 88335
6	2.85733 24964 31268	56	74.85186 87381 28973	106	170.05929 09285 99009	156	275.87338 23943 42796
7	3.70243 05364 45525	57	76.60774 35938 01464	107	172.08867 47062 84219	157	278.06928 20467 52030
8	4.60552 05234 37469	58	78.37117 15873 64401	108	174.12209 84617 71169	158	280.26793 91337 06453
9	5.55976 30328 76794	59	80.14202 35990 06546	109	176.15952 49597 11792	159	282.46933 62580 26904
10	6.55976 30328 76794	60	81.92017 48493 90189	110	178.20091 76448 70017	160	284.67345 62406 82829
11	7.60115 57180 35019	61	83.70550 46844 00956	111	180.24624 06236 56675	161	286.88028 21167 14679
12	8.68033 69640 82644	62	85.49789 63738 99210	112	182.29545 86463 26856	162	289.08979 71312 57310
13	9.79428 03163 89480	63	87.29723 69233 52792	113	184.34853 70898 10276	163	291.30198 47356 61267
14	10.94040 83520 67718	64	89.10341 68973 36679	114	186.40544 19411 46749	164	293.51682 85837 08965
15	12.11649 96111 23400	65	90.91633 02539 79535	115	188.46613 97815 00360	165	295.73431 25279 22871
16	13.32061 95937 79324	66	92.73587 41895 21403	116	190.53059 77707 27279	166	297.95442 06159 62927
17	14.55106 85151 57598	67	94.56194 89922 22230	117	192.59878 36324 73441	167	300.17713 70871 10510
18	15.80634 10202 60904	68	96.39445 79049 28466	118	194.67066 56397 79566	168	302.40244 63688 36373
19	17.08509 46212 13733	69	98.23330 69956 65721	119	196.74621 26011 72097	169	304.63033 30734 50046
20	18.38612 46168 77715	70	100.07840 50356 79978	120	198.82539 38472 19722	170	306.86078 19948 28320
21	19.70834 39116 11634	71	101.92966 33843 99053	121	200.90817 92175 36172	171	309.09377 81052 20474
22	21.05076 65924 33840	72	103.78699 58808 30322	122	202.99453 90482 10920	172	311.32930 65521 28023
23	22.41249 44284 51433	73	105.65031 87409 50778	123	205.08444 41596 50318	173	313.56735 26552 56818
24	23.79270 56701 63039	74	107.51955 04606 81754	124	207.17786 58448 12553	174	315.80790 19035 39418
25	25.19064 56788 35077	75	109.39461 17240 73454	125	209.27477 58578 20609	175	318.05093 99522 25713
26	26.60561 90268 05895	76	111.27542 53163 54245	126	211.37514 64029 38172	176	320.29645 26200 39862
27	28.03698 27909 64882	77	113.16191 60415 26727	127	213.47895 01238 94129	177	322.54442 58864 01669
28	29.48414 08223 07101	78	115.05401 06442 17208	128	215.58616 00935 41997	178	324.79484 58887 10563
29	30.94653 88202 06057	79	116.95163 77355 07649	129	217.69674 98038 41246	179	327.04769 89196 90456
30	32.42366 00749 25720	80	118.85472 77224 99593	130	219.81069 31561 48083	180	329.30297 14247 93762
31	33.91502 17687 59992	81	120.76321 27413 78242	131	221.92796 44518 03847	181	331.56064 99996 62947
32	35.42017 17470 79898	82	122.67702 65937 61959	132	224.04853 83830 09697	182	333.82072 13876 48021
33	36.93868 56869 57786	83	124.59610 46861 38033	133	226.17239 00239 76783	183	336.08317 24773 78451
34	38.47016 46040 00041	84	126.52038 39721 99915	134	228.29949 48223 41591	184	338.34799 03003 87987
35	40.01423 26483 50317	85	128.44980 28979 14207	135	230.42982 85908 36597	185	340.61516 20287 91001
36	41.57053 51491 17604	86	130.38430 13491 57775	136	232.56336 74992 06814	186	342.88467 49730 08918
37	43.13873 68731 84599	87	132.32382 06017 76394	137	234.70008 80663 63221	187	345.15651 65795 45417
38	44.71852 04698 01409	88	134.26830 32739 26562	138	236.83996 71527 64458	188	347.43067 44288 09096
39	46.30958 50768 27908	89	136.21769 32805 71475	139	238.98298 19530 18553	189	349.70713 62329 82341
40	47.91164 50681 55871	90	138.17193 57900 10800	140	241.12910 99886 96791	190	351.98588 98339 35169
41	49.52442 89248 75606	91	140.13097 71823 31894	141	243.27832 91013 52171	191	354.26692 32011 82897
42	51.14767 82152 73506	92	142.09476 50096 77449	142	245.43061 74457 35227	192	356.55022 44298 86447
43	52.78114 66708 53093	93	144.06324 79582 31384	143	247.58595 34832 00289	193	358.83578 17388 94220
44	54.42459 93473 39280	94	146.03637 58118 31083	144	249.74431 59752 95539	194	361.12358 34688 24446
45	56.07781 18611 14624	95	148.01409 94171 19930	145	251.90568 39775 30513	195	363.41361 80801 86964
46	57.74056 96927 96198	96	149.99637 06501 59499	146	254.07003 68333 14951	196	365.70587 41515 43440
47	59.41266 75507 31916	97	151.98314 23844 25744	147	256.23735 41680 63127	197	368.00034 03777 05033
48	61.09390 87881 07503	98	153.97436 84601 18238	148	258.40761 58834 58084	198	370.29700 55679 66565
49	62.78410 48681 36017	99	155.97000 36547 15788	149	260.58080 21518 70358	199	372.59585 86443 76271
50	64.48307 48724 72035	100	157.97000 36547 15788	150	262.75689 34109 26039	200	374.89688 86400 40252

N	Log N!	N	Log N!	N	Log N!	N	Log N!
201	377.20008 46974 60741	251	494.90926 01169 42653	301	616.96436 95393 67319	351	742.63728 13326 98287
202	379.50543 60669 07365	252	497.31066 06577 24198	302	619.44437 64823 24470	352	745.18382 39961 76418
203	381.81293 21048 20578	253	499.71378 11789 00015	303	621.92581 91108 26775	353	747.73159 87015 64241
204	384.12256 22722 46477	254	502.11861 48955 19954	304	624.40869 26944 35528	354	750.28060 19635 90029
205	386.43431 61333 02231	255	504.52515 50759 53909	305	626.89299 25337 82314	355	752.83083 03166 45123
206	388.74818 33536 71384	256	506.93339 50412 65758	306	629.37871 39602 63894	356	755.38228 03146 17998
207	391.06415 36991 28302	257	509.34332 81645 97053	307	631.86585 23357 41081	357	757.93494 85307 30191
208	393.38221 70340 91064	258	511.75494 78705 60283	308	634.35440 30522 41525	358	760.48883 15573 74066
209	395.70236 33202 02118	259	514.16824 76346 41535	309	636.84436 15316 66359	359	763.04392 60059 52385
210	398.02458 26149 36037	260	516.58322 09826 12353	310	639.33572 32255 00632	360	765.60022 85067 19672
211	400.34886 50702 33730	261	518.99986 14899 50634	311	641.82848 36145 27470	361	768.15773 57086 25330
212	402.67520 09311 62481	262	521.41816 27812 70379	312	644.32263 82085 45912	362	770.71644 42791 58496
213	405.00358 05346 01219	263	523.83811 85297 60137	313	646.81818 25460 92361	363	773.27635 09041 94608
214	407.33399 43079 50410	264	526.25972 24566 29968	314	649.31511 21941 65576	364	775.83745 22878 43664
215	409.66643 27678 66015	265	528.68296 83305 66776	315	651.81342 27479 55176	365	778.39974 51523 00139
216	412.00088 65190 16946	266	531.10784 99671 97843	316	654.31310 98305 73580	366	780.96322 62376 94550
217	414.33734 62528 65475	267	533.53436 12285 62418	317	656.81416 90927 91332	367	783.52789 23019 46639
218	416.67580 27464 70080	268	535.96249 60225 91207	318	659.31659 62127 75764	368	786.09374 01206 20157
219	419.01624 68613 10199	269	538.39224 83025 93615	319	661.82038 68958 32946	369	788.66076 64867 79217
220	421.35866 95421 32405	270	540.82361 20667 52602	320	664.32553 68741 52851	370	791.22896 82108 46212
221	423.70306 18158 17515	271	543.25658 13576 27008	321	666.83204 19065 57724	371	793.79834 21204 61258
222	426.04941 47902 68154	272	545.69115 02616 61207	322	669.33989 77782 53554	372	796.36888 50603 43155
223	428.39771 96533 16315	273	548.12731 29087 01963	323	671.84910 03005 84657	373	798.94059 38921 51843
224	430.74796 76716 50478	274	550.56506 34715 22351	324	674.35964 53107 91269	374	801.51346 54943 52323
225	433.10015 01897 61840	275	553.00439 61653 52613	325	676.87152 86717 70144	375	804.08749 67620 80042
226	435.45425 86289 09241	276	555.44530 52474 17831	326	679.38474 62718 38083	376	806.66268 46070 07703
227	437.81028 44861 02364	277	557.88778 50164 82280	327	681.89929 40244 98369	377	809.23902 59572 13496
228	440.16821 93331 02817	278	560.33182 98124 00356	328	684.41516 78682 10048	378	811.81651 77570 50721
229	442.52805 48154 42705	279	562.77743 40156 73953	329	686.93236 37661 60022	379	814.39515 69670 18794
230	444.88978 26514 60298	280	565.22459 20470 16173	330	689.45087 77060 37910	380	816.97494 05636 35604
231	447.25339 46313 52443	281	567.67329 83669 21253	331	691.97070 56998 13629	381	819.55586 55393 11223
232	449.61888 26162 43342	282	570.12354 74752 40614	332	694.49184 37835 17665	382	822.13792 89022 22932
233	451.98623 85372 69361	283	572.57533 39107 64904	333	697.01428 80170 23985	383	824.72112 76761 91554
234	454.35545 43946 79504	284	575.02865 22508 11942	334	699.53803 44838 35549	384	827.30545 89005 59085
235	456.72652 22569 51240	285	577.48349 71108 20452	335	702.06307 92908 72394	385	829.89091 96300 67586
236	459.09943 42599 21347	286	579.93986 31439 49495	336	704.58941 85682 62238	386	832.47750 69347 39341
237	461.47418 26059 31451	287	582.39774 50406 83487	337	707.11704 84691 33577	387	835.06521 78997 58252
238	463.85075 95629 87963	288	584.85713 75284 42718	338	709.64596 51694 11232	388	837.65404 96253 52460
239	466.22915 74639 36101	289	587.31803 53711 99266	339	712.17616 48676 14314	389	840.24399 92266 78167
240	468.60936 87056 47707	290	589.78043 33690 98222	340	714.70764 37846 56569	390	842.83506 38337 04667
241	470.99138 57482 22575	291	592.24432 63580 84129	341	717.24039 81636 49067	391	845.42724 05911 00533
242	473.37520 11142 03006	292	594.70970 92095 32548	342	719.77442 42697 05202	392	848.02052 66581 20991
243	475.76080 73878 01318	293	597.17657 68298 86657	343	722.30971 83897 47972	393	850.61491 92084 96417
244	478.14819 72141 40048	294	599.64492 41602 98814	344	724.84627 68323 19502	394	853.21041 54303 21991
245	480.53736 32985 04580	295	602.11474 61762 76977	345	727.38409 59273 92777	395	855.80701 25259 48452
246	482.92829 84056 07959	296	604.58603 78873 35916	346	729.92317 20261 85553	396	858.40470 77118 73964
247	485.32099 53588 67625	297	607.05879 43366 53128	347	732.46350 15009 76427	397	861.00349 82186 37079
248	487.71544 70396 93841	298	609.53301 06007 29383	348	735.00508 07449 23008	398	863.60338 12907 10767
249	490.11164 63867 89578	299	612.00868 17890 53813	349	737.54790 61718 82188	399	866.20435 41863 97515
250	492.50958 63954 61615	300	614.48580 30437 73476	350	740.09197 42162 32463	400	868.80641 41777 25477

N	Log N!	N	Log N!	N	Log N!	N	Log N!
401	871.40955 85503 45660	451	1002.89306 75002 76841	501	1136.78624 62610 01801	551	1272.84800 28550 10551
402	874.01378 46034 30130	452	1005.54820 59350 88223	502	1139.48694 99781 46820	552	1275.58994 19327 39750
403	876.61908 96495 71239	453	1008.20430 41371 01055	503	1142.18851 79632 02748	553	1278.33266 70640 44448
404	879.22547 10146 81844	454	1010.86135 99899 58159	504	1144.89094 84996 48273	554	1281.07617 68287 72876
405	881.83292 60378 96513	455	1013.51937 13866 15271	505	1147.59423 98777 66934	555	1283.82046 98118 95554
406	884.44145 20714 73707	456	1016.17833 62292 79706	506	1150.29839 03946 06733	556	1286.56554 46034 77612
407	887.05104 64806 98927	457	1018.83825 24293 49556	507	1153.00339 83539 40069	557	1289.31139 97986 51341
408	889.66170 66437 88807	458	1021.49911 79073 53425	508	1155.70926 20662 23989	558	1292.05803 39975 88919
409	892.27342 99517 96149	459	1024.16093 05928 90687	509	1158.41597 98485 60747	559	1294.80544 58054 75343
410	894.88621 38085 15884	460	1026.82368 84245 72261	510	1161.12355 00246 58684	560	1297.55363 38324 81543
411	897.50005 56303 91953	461	1029.48738 93499 61909	511	1163.83197 09247 93396	561	1300.30259 66937 37705
412	900.11495 28464 25088	462	1032.15203 13255 18034	512	1166.54124 08857 69227	562	1303.05233 30093 06766
413	902.73090 28980 81489	463	1034.81761 23165 35987	513	1169.25135 82508 81043	563	1305.80284 14041 58112
414	905.34790 32392 02388	464	1037.48413 02970 90868	514	1171.96232 13698 76319	564	1308.55412 05081 41454
415	907.96595 13359 14481	465	1040.15158 32499 80822	515	1174.67412 85989 17510	565	1311.30616 89559 60893
416	910.58504 46665 41223	466	1042.81996 91666 70822	516	1177.38677 83005 44721	566	1314.05898 53871 49164
417	913.20518 07215 14981	467	1045.48928 60472 36935	517	1180.10026 88436 38664	567	1316.81256 84460 42071
418	915.82635 70032 90016	468	1048.15953 19003 11059	518	1182.81459 86033 83897	568	1319.56691 67817 53090
419	918.44857 10262 56311	469	1050.83070 47430 26142	519	1185.52976 59612 32355	569	1322.32202 90481 48161
420	921.07182 03166 54212	470	1053.50280 26009 61859	520	1188.24576 93048 67154	570	1325.07790 39038 20652
421	923.69610 24124 89880	471	1056.17582 35080 90756	521	1190.96260 70281 66679	571	1327.83454 00120 66500
422	926.32141 48634 51554	472	1058.84976 55067 24843	522	1193.68027 75311 68941	572	1330.59193 60408 59524
423	928.94775 52308 26596	473	1061.52462 66474 62655	523	1196.39877 92200 36215	573	1333.35009 06628 26914
424	931.57512 10874 19329	474	1064.20040 49891 36740	524	1199.11811 05070 19942	574	1336.10900 25552 24888
425	934.20351 00174 69641	475	1066.87709 85987 61607	525	1201.83826 98104 25898	575	1338.86867 03999 14518
426	936.83291 96165 72360	476	1069.55470 55514 82100	526	1204.55925 55545 79637	576	1341.62909 28833 37730
427	939.46334 74915 97383	477	1072.23322 39305 22214	527	1207.28106 61697 92184	577	1344.39026 86964 93462
428	942.09479 12606 10555	478	1074.91265 18271 34332	528	1210.00370 00923 25996	578	1347.15219 65349 13991
429	944.72724 85527 95280	479	1077.59298 73405 48896	529	1212.72715 57643 61182	579	1349.91487 50986 41427
430	947.36071 70083 74866	480	1080.27422 85779 24483	530	1215.45143 16339 61971	580	1352.67830 30922 04364
431	949.99519 42785 35598	481	1082.95637 36542 98315	531	1218.17652 61550 43440	581	1355.44247 92245 94695
432	952.63067 80253 50510	482	1085.63942 06925 37164	532	1220.90243 77873 38488	582	1358.20740 22092 44584
433	955.26716 59217 03875	483	1088.32336 78232 88676	533	1223.62916 49963 65061	583	1360.97307 07640 03598
434	957.90465 56512 16386	484	1091.00821 31849 33089	534	1226.35670 62533 93617	584	1363.73948 36111 15997
435	960.54314 49081 71023	485	1093.69395 49235 35353	535	1229.08506 00354 14846	585	1366.50663 94771 98178
436	963.18263 13974 39609	486	1096.38059 11927 97646	536	1231.81422 48251 07616	586	1369.27453 70932 16268
437	965.82311 28344 10031	487	1099.06812 01540 12280	537	1234.54419 91108 07171	587	1372.04317 51944 63883
438	968.46458 69449 14131	488	1101.75653 99760 14991	538	1237.27498 13864 73560	588	1374.81255 25205 40021
439	971.10705 14651 56252	489	1104.44584 88351 38611	539	1240.00657 01516 60299	589	1377.58266 78153 27123
440	973.75050 41416 42440	490	1107.13604 49151 67125	540	1242.73896 39114 83268	590	1380.35351 98269 69267
441	976.39494 27311 10278	491	1109.82712 64072 90093	541	1245.47216 11765 89837	591	1383.12510 73078 50522
442	979.04036 50004 59370	492	1112.51909 15100 57454	542	1248.20616 04631 28224	592	1385.89742 90145 73442
443	981.68676 87266 82440	493	1115.21193 84293 34684	543	1250.94096 02927 17071	593	1388.67048 37079 37705
444	984.33415 16967 97059	494	1117.90566 53782 58331	544	1253.67655 91924 15251	594	1391.44427 01529 18898
445	986.98251 17077 77991	495	1120.60027 05771 91899	545	1256.41295 56946 91893	595	1394.21878 71186 47448
446	989.63184 65664 90133	496	1123.29575 22536 82097	546	1259.15014 83373 96630	596	1396.99403 33783 87684
447	992.28215 40896 22069	497	1125.99210 86424 15429	547	1261.88813 56637 30061	597	1399.77000 77095 17053
448	994.93343 21036 20213	498	1128.68933 79851 75146	548	1264.62691 62222 14430	598	1402.54670 88935 05464
449	997.58567 84446 23536	499	1131.38743 85307 98536	549	1267.36648 85666 64522	599	1405.32413 57158 94776
450	1000.23889 09583 98880	500	1134.08640 85351 34555	550	1270.10685 12561 58766	600	1408.10228 69662 78419

N	Log N!
601	1410.88116 14382 81159
602	1413.66075 79295 38983
603	1416.44107 52416 79135
604	1419.22211 21803 00267
605	1422.00386 75549 52735
606	1424.78634 01791 19022
607	1427.56952 88701 94279
608	1430.35343 24494 67014
609	1433.13804 97420 99889
610	1435.92337 95771 10657
611	1438.70942 07873 53211
612	1441.49617 22094 98772
613	1444.28363 26840 17187
614	1447.07180 10551 58355
615	1449.86067 61709 33771
616	1452.65025 68830 98197
617	1455.44054 20471 31439
618	1458.23153 05222 20254
619	1461.02322 11712 40372
620	1463.81561 28607 38626
621	1466.60870 44609 15206
622	1469.40249 48456 06025
623	1472.19698 28922 65195
624	1474.99216 74819 47619
625	1477.78804 74992 91694
626	1480.58462 18325 02124
627	1483.38188 93733 32840
628	1486.17984 90170 70036
629	1488.97849 96625 15305
630	1491.77784 02119 68887
631	1494.57786 95712 13021
632	1497.37858 66494 95406
633	1500.17999 03595 12761
634	1502.98207 96173 94494
635	1505.78485 33426 86470
636	1508.58831 04583 34883
637	1511.39244 98906 70234
638	1514.19727 05693 91396
639	1517.00277 14275 49796
640	1519.80895 14015 33684
641	1522.61580 94310 52501
642	1525.42334 44591 21354
643	1528.23155 54320 45576
644	1531.04044 12994 05388
645	1533.85000 10140 40656
646	1536.66023 35320 35740
647	1539.47113 78127 04441
648	1542.28271 28185 75034
649	1545.09495 75153 75403
650	1547.90787 08720 18259

N	Log N!
651	1550.72145 18605 86451
652	1553.53569 94563 18371
653	1556.35061 26375 93445
654	1559.16619 03859 17712
655	1561.98243 16859 09495
656	1564.79933 55252 85155
657	1567.61690 08948 44936
658	1570.43512 67884 58892
659	1573.25401 22030 52902
660	1576.07355 61385 94770
661	1578.89375 75980 80411
662	1581.71461 55875 20110
663	1584.53612 91159 24884
664	1587.35829 71952 92901
665	1590.18111 88405 96006
666	1593.00459 30697 66307
667	1595.82871 89036 82856
668	1598.65349 53661 58401
669	1601.47892 14839 26225
670	1604.30499 62866 27051
671	1607.13171 88067 96043
672	1609.95908 80798 49868
673	1612.78710 31440 73845
674	1615.61576 30406 09165
675	1618.44506 68134 40190
676	1621.27501 35093 81826
677	1624.10560 21780 66970
678	1626.93683 18719 34033
679	1629.76870 16462 14535
680	1632.60121 05589 20771
681	1635.43435 76708 33557
682	1638.26814 20454 90036
683	1641.10256 27491 71568
684	1643.93761 88508 91684
685	1646.77330 94223 84110
686	1649.60963 35380 90862
687	1652.44659 02751 50412
688	1655.28417 87133 85923
689	1658.12239 79352 93549
690	1660.96124 70260 30804
691	1663.80072 50734 05003
692	1666.64083 11678 61761
693	1669.48156 44024 73567
694	1672.32292 38729 28422
695	1675.16490 86775 18536
696	1678.00751 79171 29098
697	1680.85075 06952 27108
698	1683.69460 61178 50269
699	1686.53908 32935 95950
700	1689.38418 13336 10207

N	Log N!
701	1692.22989 93515 76866
702	1695.07623 64637 06671
703	1697.92319 17887 26495
704	1700.77076 44478 68607
705	1703.61895 35648 60006
706	1706.46775 82659 11810
707	1709.31717 76797 08709
708	1712.16721 09373 98478
709	1715.01785 71725 81545
710	1717.86911 55213 00620
711	1720.72098 51220 30386
712	1723.57346 51156 67243
713	1726.42655 46455 19108
714	1729.28025 28572 95283
715	1732.13455 88990 96363
716	1734.98947 19214 04219
717	1737.84499 10770 72019
718	1740.70111 55213 14319
719	1743.55784 44116 97202
720	1746.41517 69081 28470
721	1749.27311 21728 47899
722	1752.13164 93704 17538
723	1754.99078 76677 12069
724	1757.85052 62339 09216
725	1760.71086 42404 80210
726	1763.57180 08611 80304
727	1766.43333 52720 39341
728	1769.29546 66513 52379
729	1772.15819 41796 70353
730	1775.02151 70397 90809
731	1777.88543 44167 48670
732	1780.74994 54978 07061
733	1783.61504 94724 48189
734	1786.48074 55323 64260
735	1789.34703 28714 48455
736	1792.21391 06857 85954
737	1795.08137 81736 45005
738	1797.94943 45354 68047
739	1800.81807 89738 62872
740	1803.68731 06935 93849
741	1806.55712 89015 73177
742	1809.42753 28068 52204
743	1812.29852 16206 12779
744	1815.17009 45561 58658
745	1818.04225 08289 06951
746	1820.91498 96563 79619
747	1823.78831 02581 95018
748	1826.66221 18560 59480
749	1829.53669 36737 58946
750	1832.41175 49371 50646

N	Log N!
751	1835.28739 48741 54814
752	1838.16361 27147 46457
753	1841.04040 76909 47157
754	1843.91777 90368 16931
755	1846.79572 59884 46120
756	1849.67424 77839 47326
757	1852.55334 36634 47399
758	1855.43301 28690 79452
759	1858.31325 46449 74933
760	1861.19406 82372 55724
761	1864.07545 28940 26297
762	1866.95740 78653 65897
763	1869.83993 24033 20778
764	1872.72302 57618 96468
765	1875.60668 71970 50085
766	1878.49091 59666 82689
767	1881.37571 13306 31670
768	1884.26107 25506 63182
769	1887.14699 88904 64613
770	1890.03348 96156 37095
771	1892.92054 39936 88052
772	1895.80816 12940 23788
773	1898.69634 07879 42113
774	1901.58508 17486 25006
775	1904.47438 34511 31316
776	1907.36424 51723 89505
777	1910.25466 61911 90419
778	1913.14564 57881 80108
779	1916.03718 32458 52672
780	1918.92927 78485 43153
781	1921.82192 88824 20453
782	1924.71513 56354 80301
783	1927.60889 73975 38244
784	1930.50321 34602 22683
785	1933.39808 31169 67935
786	1936.29350 56630 07343
787	1939.18948 03953 66408
788	1942.08600 66128 55963
789	1944.98308 36160 65383
790	1947.88071 07073 55825
791	1950.77888 71908 53501
792	1953.67761 23724 42995
793	1956.57688 55597 60599
794	1959.47670 60621 87695
795	1962.37707 31908 44165
796	1965.27798 62585 81834
797	1968.17944 45799 77947
798	1971.08144 74713 28676
799	1973.98399 42506 42668
800	1976.88708 42376 34611

N	Log N!	N	Log N!	N	Log N!	N	Log N!
801	1979.79071 67537 18849	851	2125.64954 87743 93980	901	2272.78420 09747 93898	951	2421.12383 75660 82815
802	1982.69489 11220 03012	852	2128.57998 83691 60680	902	2275.73940 75123 35840	952	2424.10247 45144 67289
803	1985.59960 66672 81693	853	2131.51093 74003 28203	903	2278.69509 52626 49346	953	2427.08156 74151 05615
804	1988.50486 27160 30144	854	2134.44239 52710 17208	904	2281.65126 36931 24709	954	2430.06111 57898 09710
805	1991.41065 85963 98013	855	2137.37436 13857 45381	905	2284.60791 22723 29913	955	2433.04111 91613 93457
806	1994.31699 36382 03104	856	2140.30683 51504 22534	906	2287.56504 04700 06726	956	2436.02157 70536 69557
807	1997.22386 71729 25174	857	2143.23981 59723 45732	907	2290.52264 77570 66821	957	2439.00248 89914 46400
808	2000.13127 85336 99760	858	2146.17330 32601 94438	908	2293.48073 36055 87906	958	2441.98385 45005 24945
809	2003.03922 70553 12033	859	2149.10729 64240 25680	909	2296.43929 74888 09873	959	2444.96567 31076 95608
810	2005.94771 20741 90682	860	2152.04179 48752 69248	910	2299.39833 88811 30967	960	2447.94794 43407 35177
811	2008.85673 29284 01838	861	2154.97679 80267 22902	911	2302.35785 72581 03965	961	2450.93066 77284 03722
812	2011.76628 89576 43014	862	2157.91230 52925 47615	912	2305.31785 20964 32381	962	2453.91384 28004 41535
813	2014.67637 95032 37082	863	2160.84831 60882 62825	913	2308.27832 28739 66680	963	2456.89746 90875 66070
814	2017.58700 39081 26283	864	2163.78482 98307 41718	914	2311.23926 90697 00512	964	2459.88154 61214 68900
815	2020.49816 15168 66260	865	2166.72184 59382 06532	915	2314.20069 01637 66960	965	2462.86607 34348 12693
816	2023.40985 16756 20121	866	2169.65936 38302 23879	916	2317.16258 56374 34811	966	2465.85105 05612 28186
817	2026.32207 37321 52536	867	2172.59738 29277 00089	917	2320.12495 49731 04832	967	2468.83647 70353 11188
818	2029.23482 70358 23859	868	2175.53590 26528 76581	918	2323.08779 76543 06074	968	2471.82235 23926 19582
819	2032.14811 09375 84278	869	2178.47492 24293 25248	919	2326.05111 31656 92185	969	2474.80867 61696 70347
820	2035.06192 47899 67994	870	2181.41444 16819 43866	920	2329.01490 09930 37741	970	2477.79544 79039 36592
821	2037.97626 79470 87435	871	2184.35445 98369 51529	921	2331.97916 06232 34589	971	2480.78266 71338 44597
822	2040.89113 97646 27486	872	2187.29497 63218 84096	922	2334.94389 15442 88219	972	2483.77033 33987 70871
823	2043.80653 95998 39755	873	2190.23599 05655 89666	923	2337.90909 32453 14131	973	2486.75844 62390 39223
824	2046.72246 68115 36871	874	2193.17750 19982 24069	924	2340.87476 52165 34238	974	2489.74700 51959 17839
825	2049.63892 07600 86796	875	2196.11951 00512 46383	925	2343.84090 69492 73270	975	2492.73600 98116 16376
826	2052.55590 08074 07179	876	2199.06201 41574 14463	926	2346.80751 79359 55205	976	2495.72545 96292 83067
827	2055.47340 63169 59725	877	2202.00501 37507 80504	927	2349.77459 76700 99702	977	2498.71535 41930 01840
828	2058.39143 66537 44605	878	2204.94850 82666 86606	928	2352.74214 56463 18564	978	2501.70569 30477 89442
829	2061.30999 11842 94879	879	2207.89249 71417 60378	929	2355.71016 13603 12205	979	2504.69647 57395 92580
830	2064.22906 92766 70953	880	2210.83697 98139 10547	930	2358.67864 43088 66141	980	2507.68770 18152 85075
831	2067.14867 03004 55064	881	2213.78195 57223 22595	931	2361.64759 39898 47483	981	2510.67937 08226 65023
832	2070.06879 36267 45788	882	2216.72742 43074 54414	932	2364.61700 99022 01465	982	2513.67148 23104 51973
833	2072.98943 86281 52575	883	2219.67338 50110 31983	933	2367.58689 15459 47964	983	2516.66403 58282 84108
834	2075.91060 46787 90314	884	2222.61983 72760 45056	934	2370.55723 84221 78058	984	2519.65703 09267 15450
835	2078.83229 11542 73916	885	2225.56678 05467 42882	935	2373.52805 00330 50576	985	2522.65046 71572 13062
836	2081.75449 74317 12932	886	2228.51421 42686 29932	936	2376.49932 58817 88681	986	2525.64434 40721 54273
837	2084.67722 28897 06192	887	2231.46213 78884 61659	937	2379.47106 54726 76459	987	2528.63866 12248 23910
838	2087.60046 69083 36469	888	2234.41055 08542 40260	938	2382.44326 83110 55524	988	2531.63341 81694 11538
839	2090.52422 88691 65169	889	2237.35945 26152 10473	939	2385.41593 39033 21634	989	2534.62861 44610 08717
840	2093.44850 81552 27051	890	2240.30884 26218 55386	940	2388.38906 17569 21333	990	2537.62424 96556 06267
841	2096.37330 41510 24963	891	2243.25872 03258 92261	941	2391.36265 13803 48590	991	2540.62032 33100 91542
842	2099.29861 62425 24613	892	2246.20908 51802 68384	942	2394.33670 22831 41467	992	2543.61683 49822 45721
843	2102.22444 38171 49355	893	2249.15993 66391 56931	943	2397.31121 39758 78796	993	2546.61378 42307 41102
844	2105.15078 62637 75010	894	2252.11127 41579 52848	944	2400.28618 59701 76865	994	2549.61117 06151 38415
845	2108.07764 29727 24702	895	2255.06309 71932 68760	945	2403.26161 77786 86128	995	2552.60899 36958 84141
846	2111.00501 33357 63726	896	2258.01540 52029 30885	946	2406.23750 89150 87920	996	2555.60725 30343 07840
847	2113.93289 67460 94433	897	2260.96819 76459 74977	947	2409.21385 88940 91194	997	2558.60594 81926 19495
848	2116.86129 25983 51147	898	2263.92147 39826 42282	948	2412.19066 72314 29260	998	2561.60507 87339 06866
849	2119.79020 02885 95099	899	2266.87523 36743 75511	949	2415.16793 34438 56553	999	2564.60464 42221 32849
850	2122.71961 92143 09392	900	2269.82947 61838 14835	950	2418.14565 70491 45401	1000	2567.60464 42221 32849

N	Log N!
1001	2570.60507 82996 12167
1002	2573.60594 60211 43394
1003	2576.60724 69541 63812
1004	2579.60898 06669 72813
1005	2582.61114 67287 29321
1006	2585.61374 47094 49229
1007	2588.61677 41800 02847
1008	2591.62023 47121 12354
1009	2594.62412 58783 49264
1010	2597.62844 72521 31907
1011	2600.63319 84077 22908
1012	2603.63837 89202 26688
1013	2606.64398 83655 86969
1014	2609.65002 63205 84286
1015	2612.65649 23628 33518
1016	2615.66338 60707 81418
1017	2618.67070 70237 04163
1018	2621.67845 48017 04903
1019	2624.68662 89857 11329
1020	2627.69522 91574 73246
1021	2630.70425 48995 60157
1022	2633.71370 57953 58851
1023	2636.72358 14290 71011
1024	2639.73388 13857 10823
1025	2642.74460 52511 02596
1026	2645.75575 26118 78393
1027	2648.76732 30554 75672
1028	2651.77931 61701 34928
1029	2654.79173 15448 97361
1030	2657.80456 87696 02534
1031	2660.81782 74348 86050
1032	2663.83150 71321 77243
1033	2666.84560 74536 96863
1034	2669.86012 79924 54787
1035	2672.87506 83422 47724
1036	2675.89042 80976 56938
1037	2678.90620 68540 45979
1038	2681.92240 42075 58418
1039	2684.93901 97551 15595
1040	2687.95605 30944 14376
1041	2690.97350 38239 24912
1042	2693.99137 15428 88417
1043	2697.00965 58513 14948
1044	2700.02835 63499 81192
1045	2703.04747 26404 28264
1046	2706.06700 43249 59520
1047	2709.08695 10066 38362
1048	2712.10731 22892 86070
1049	2715.12808 77774 79628
1050	2718.14927 70765 49566

N	Log N!
1051	2721.17087 97925 77808
1052	2724.19289 55323 95528
1053	2727.21532 39035 81015
1054	2730.23816 45144 57543
1055	2733.26141 69740 91254
1056	2736.28508 08922 89048
1057	2739.30915 58795 96474
1058	2742.33364 15472 95641
1059	2745.35853 75074 03126
1060	2748.38384 33726 67896
1061	2751.40955 87565 69237
1062	2754.43568 32733 14687
1063	2757.46221 65378 37984
1064	2760.48915 81657 97013
1065	2763.51650 77735 71770
1066	2766.54426 49782 62323
1067	2769.57242 93976 86793
1068	2772.60100 06503 79331
1069	2775.62997 83555 88109
1070	2778.65936 21332 73318
1071	2781.68915 16041 05174
1072	2784.71934 63894 61925
1073	2787.74994 61114 27876
1074	2790.78095 03927 91413
1075	2793.81235 88570 43037
1076	2796.84417 11283 73408
1077	2799.87638 68316 71389
1078	2802.90900 55925 22109
1079	2805.94202 70372 05020
1080	2808.97545 07926 91969
1081	2812.00927 64866 45280
1082	2815.04350 37474 15830
1083	2818.07813 22040 41151
1084	2821.11316 14862 43519
1085	2824.14859 12244 28067
1086	2827.18442 10496 80895
1087	2830.22065 05937 67190
1088	2833.25727 94891 29351
1089	2836.29430 73688 85126
1090	2839.33173 38668 25750
1091	2842.36955 86174 14091
1092	2845.40778 12557 82810
1093	2848.44640 14177 32513
1094	2851.48541 87397 29925
1095	2854.52483 28589 06062
1096	2857.56464 34130 54412
1097	2860.60485 00406 29123
1098	2863.64545 23807 43196
1099	2866.68645 00731 66687
1100	2869.72784 27583 24912

N	Log N!
1101	2872.76963 00772 96664
1102	2875.81181 16718 12430
1103	2878.85438 71842 52621
1104	2881.89735 62576 45801
1105	2884.94071 85356 66930
1106	2887.98447 36626 35610
1107	2891.02862 12835 14332
1108	2894.07316 10439 06743
1109	2897.11809 25900 55903
1110	2900.16341 55688 42561
1111	2903.20912 96277 83429
1112	2906.25523 44150 29467
1113	2909.30172 95793 64175
1114	2912.34861 47702 01886
1115	2915.39588 96375 86065
1116	2918.44355 38321 87625
1117	2921.49160 70053 03234
1118	2924.54004 88088 53639
1119	2927.58887 88953 81989
1120	2930.63809 69180 52170
1121	2933.68770 25306 47143
1122	2936.73769 53875 67286
1123	2939.78807 51438 28744
1124	2942.83884 14550 61786
1125	2945.88999 39775 09167
1126	2948.94153 23680 24495
1127	2951.99345 62840 70601
1128	2955.04576 53837 17925
1129	2958.09845 93256 42893
1130	2961.15153 77691 26312
1131	2964.20500 03740 51768
1132	2967.25884 68009 04020
1133	2970.31307 67107 67418
1134	2973.36768 97653 24305
1135	2976.42268 56268 53447
1136	2979.47806 39582 28447
1137	2982.53382 44229 16182
1138	2985.58996 66849 75234
1139	2988.64649 04090 54334
1140	2991.70339 52603 90807
1141	2994.76068 09048 09022
1142	2997.81834 70087 18851
1143	3000.87639 32391 14133
1144	3003.93481 92635 71138
1145	3006.99362 47502 47045
1146	3010.05280 93678 78416
1147	3013.11237 27857 79684
1148	3016.17231 46738 41638
1149	3019.23263 47025 29924
1150	3022.29333 25428 83535

N	Log N!
1151	3025.35440 78665 13327
1152	3028.41586 03456 00520
1153	3031.47768 96528 95219
1154	3034.53989 54617 14932
1155	3037.60247 74459 43095
1156	3040.66543 52800 27605
1157	3043.72876 86389 79355
1158	3046.79247 71983 70772
1159	3049.85656 06343 34368
1160	3052.92101 86235 61287
1161	3055.98585 08432 99861
1162	3059.05105 69713 54172
1163	3062.11663 66860 82621
1164	3065.18258 96663 96491
1165	3068.24891 55917 58528
1166	3071.31561 41421 81524
1167	3074.38268 49982 26894
1168	3077.45012 78410 03274
1169	3080.51794 23521 65115
1170	3083.58612 82139 11276
1171	3086.65468 51089 83639
1172	3089.72361 27206 65711
1173	3092.79291 07327 81240
1174	3095.86257 88296 92836
1175	3098.93261 66963 00591
1176	3102.00302 40180 40711
1177	3105.07380 04808 84146
1178	3108.14494 57713 35228
1179	3111.21645 95764 30318
1180	3114.28834 15837 36443
1181	3117.36059 14813 49958
1182	3120.43320 89578 95194
1183	3123.50619 37025 23125
1184	3126.57954 54049 10026
1185	3129.65326 37552 56148
1186	3132.72734 84442 84392
1187	3135.80179 91632 38983
1188	3138.87661 56038 84158
1189	3141.95179 74585 02850
1190	3145.02734 44198 95380
1191	3148.10325 61813 78158
1192	3151.17953 24367 82376
1193	3154.25617 28804 52717
1194	3157.33317 72072 46068
1195	3160.41054 51125 30224
1196	3163.48827 62921 82616
1197	3166.56637 04425 89027
1198	3169.64482 72606 42319
1199	3172.72364 64437 41168
1200	3175.80282 76897 88793

N	Log N!	N	Log N!	N	Log N!	N	Log N!
1201	3178.88237 06971 91699	1251	3333.31424 50843 47082	1301	3488.61389 82709 43803	1351	3644.74797 58746 17743
1202	3181.96227 51648 58420	1252	3336.41184 94132 21493	1302	3491.72850 92551 75976	1352	3647.87895 25662 23360
1203	3185.04254 07921 98264	1253	3339.50980 04842 15643	1303	3494.84345 36708 88561	1353	3651.01025 03628 20983
1204	3188.12316 72791 20070	1254	3342.60809 80207 10340	1304	3497.95873 12622 84462	1354	3654.14186 90271 70109
1205	3191.20415 43260 30957	1255	3345.70674 17465 27397	1305	3501.07434 17739 58762	1355	3657.27380 83223 80533
1206	3194.28550 16338 35090	1256	3348.80573 13859 28574	1306	3504.19028 49508 97817	1356	3660.40606 80119 11578
1207	3197.36720 89039 32439	1257	3351.90506 66636 14532	1307	3507.30656 05384 78361	1357	3663.53864 78595 71315
1208	3200.44927 58382 17552	1258	3355.00474 73047 23782	1308	3510.42316 82824 66610	1358	3666.67154 76295 15798
1209	3203.53170 21390 78324	1259	3358.10477 30348 31645	1309	3513.54010 79290 17365	1359	3669.80476 70862 48292
1210	3206.61448 75093 94774	1260	3361.20514 35799 49208	1310	3516.65737 92246 73130	1360	3672.93830 59946 18510
1211	3209.69763 16525 37826	1261	3364.30585 86665 22289	1311	3519.77498 19163 63214	1361	3676.07216 41198 21844
1212	3212.78113 42723 68094	1262	3367.40691 80214 30405	1312	3522.89291 57514 02855	1362	3679.20634 12273 98611
1213	3215.86499 50732 34667	1263	3370.50832 13719 85736	1313	3526.01118 04774 92335	1363	3682.34083 70832 33284
1214	3218.94921 37599 73906	1264	3373.61006 84459 32102	1314	3529.12977 58427 16097	1364	3685.47565 14535 53744
1215	3222.03379 00379 08236	1265	3376.71215 89714 43939	1315	3532.24870 15955 41873	1365	3688.61078 41049 30519
1216	3225.11872 36128 44953	1266	3379.81459 26771 25275	1316	3535.36795 74848 19810	1366	3691.74623 48042 76033
1217	3228.20401 41910 75018	1267	3382.91736 92920 08716	1317	3538.48754 32597 81594	1367	3694.88700 33188 43855
1218	3231.28966 14793 71874	1268	3386.02048 85455 54430	1318	3541.60745 86700 39585	1368	3698.01808 94162 27953
1219	3234.37566 51849 90256	1269	3389.12395 01676 49135	1319	3544.72770 34655 85950	1369	3701.15449 28643 61943
1220	3237.46202 50156 65004	1270	3392.22775 38886 05092	1320	3547.84827 73967 91800	1370	3704.29121 34315 18349
1221	3240.54874 06796 09887	1271	3395.33189 94391 59100	1321	3550.96918 02144 06327	1371	3707.42825 08863 07862
1222	3243.63581 18855 16422	1272	3398.43638 65504 71495	1322	3554.09041 16695 55949	1372	3710.56560 49976 78595
1223	3246.72323 83425 52708	1273	3401.54121 49541 25150	1323	3557.21197 15137 43450	1373	3713.70327 55349 15350
1224	3249.81101 97603 62250	1274	3404.64638 43821 24482	1324	3560.33385 94988 47131	1374	3716.84126 22676 38882
1225	3252.89915 58490 62801	1275	3407.75189 45668 94456	1325	3563.45607 53771 19957	1375	3719.97956 49658 05163
1226	3255.98764 63192 45198	1276	3410.85774 52412 79599	1326	3566.57861 89011 88712	1376	3723.11818 33997 04656
1227	3259.07649 08819 72202	1277	3413.96393 61385 43015	1327	3569.70148 98240 53147	1377	3726.25711 73399 61579
1228	3262.16568 92487 77351	1278	3417.07046 69923 65396	1328	3572.82468 78990 85146	1378	3729.39636 65575 33186
1229	3265.25524 11316 63805	1279	3420.17733 75368 44050	1329	3575.94821 28800 27878	1379	3732.53593 08237 09036
1230	3268.34514 62431 03203	1280	3423.28454 75064 91918	1330	3579.07206 45209 94964	1380	3735.67580 99101 10273
1231	3271.43540 42960 34519	1281	3426.39209 66362 36605	1331	3582.19624 25764 69639	1381	3738.81600 35886 88904
1232	3274.52601 50038 62926	1282	3429.49998 46614 19403	1332	3585.32074 68013 03921	1382	3741.95651 16317 27083
1233	3277.61697 80804 58657	1283	3432.60821 13177 94332	1333	3588.44557 69507 17780	1383	3745.09733 38118 36394
1234	3280.70829 32401 55880	1284	3435.71677 63415 27166	1334	3591.57073 27802 98311	1384	3748.23846 99019 57133
1235	3283.79996 01977 51565	1285	3438.82567 94691 94480	1335	3594.69621 40459 98905	1385	3751.37991 96753 57600
1236	3286.89197 86685 04362	1286	3441.93492 04377 82683	1336	3597.82202 05041 38431	1386	3754.52168 29056 33388
1237	3289.98434 83681 33482	1287	3445.04449 89846 87070	1337	3600.94815 19114 00416	1387	3757.66375 93667 06673
1238	3293.07706 90128 17582	1288	3448.15441 48477 10863	1338	3604.07460 80248 32220	1388	3760.80614 88328 25509
1239	3296.17014 03191 93645	1289	3451.26466 77650 64266	1339	3607.20138 86018 44229	1389	3763.94885 10785 63125
1240	3299.26356 20043 55880	1290	3454.37525 74753 63515	1340	3610.32849 34002 09037	1390	3767.09186 58788 17220
1241	3302.35733 37858 54610	1291	3457.48618 37176 29935	1341	3613.45592 21780 60636	1391	3770.23519 30088 09266
1242	3305.45145 53816 95171	1292	3460.59744 62312 89000	1342	3616.58367 46938 93609	1392	3773.37883 22440 83810
1243	3308.54592 65103 36816	1293	3463.70904 47561 69395	1343	3619.71175 07065 62324	1393	3776.52278 33605 07773
1244	3311.64074 68906 91616	1294	3466.82097 90325 02076	1344	3622.84014 99752 80131	1394	3779.66704 61342 69764
1245	3314.73591 62421 23371	1295	3469.93324 88009 19347	1345	3625.96887 22596 18557	1395	3782.81162 03418 79380
1246	3317.83143 42844 46522	1296	3473.04585 38024 53921	1346	3629.09791 73195 06515	1396	3785.95650 57601 66522
1247	3320.92730 07379 25065	1297	3476.15879 37785 38001	1347	3632.22728 49152 29501	1397	3789.10170 21662 80704
1248	3324.02351 53232 71470	1298	3479.27206 84710 02352	1348	3635.35697 48074 28802	1398	3792.24720 93376 90367
1249	3327.12007 77616 45605	1299	3482.38567 76220 75380	1349	3638.48698 67571 00706	1399	3795.39302 70521 82195
1250	3330.21698 77746 53662	1300	3485.49962 09743 82216	1350	3641.61732 05255 95712	1400	3798.53915 50878 60433

N	Log N!
1401	3801.68559 32231 46207
1402	3804.83234 12367 76847
1403	3807.97939 89078 05207
1404	3811.12676 60155 98993
1405	3814.27444 23398 40092
1406	3817.42242 76605 23897
1407	3820.57072 17579 58643
1408	3823.71932 44127 64736
1409	3826.86823 54058 74093
1410	3830.01745 45185 29473
1411	3833.16698 15322 83820
1412	3836.31681 62289 99605
1413	3839.46695 83908 48164
1414	3842.61740 78003 09045
1415	3845.76816 42401 69354
1416	3848.91922 74935 23104
1417	3852.07059 73437 70564
1418	3855.22227 35746 17612
1419	3858.37425 59700 75086
1420	3861.52654 43144 58143
1421	3864.67913 83923 85612
1422	3867.83203 79887 79360
1423	3870.98524 28888 63644
1424	3874.13875 28781 64482
1425	3877.29256 77425 09011
1426	3880.44668 72680 24857
1427	3883.60111 12411 39504
1428	3886.75583 94485 79660
1429	3889.91087 16773 70630
1430	3893.06620 77148 35692
1431	3896.22184 73485 95468
1432	3899.37779 03665 67305
1433	3902.53403 65569 64650
1434	3905.69058 57082 96431
1435	3908.84743 76093 66442
1436	3912.00459 20492 72724
1437	3915.16204 88174 06949
1438	3918.31980 77034 53813
1439	3921.47786 84973 90418
1440	3924.63623 09894 85668
1441	3927.79489 49702 99657
1442	3930.95386 02306 83067
1443	3934.11312 65617 76562
1444	3937.27269 37550 10182
1445	3940.43256 16021 02749
1446	3943.59272 98950 61261
1447	3946.75319 84261 80298
1448	3949.91396 69880 41426
1449	3953.07503 53735 12601
1450	3956.23640 33757 47576

N	Log N!
1451	3959.39807 07881 85312
1452	3962.56003 74045 49386
1453	3965.72230 30188 47408
1454	3968.88486 74253 70427
1455	3972.04773 04186 92353
1456	3975.21089 17936 69372
1457	3978.37435 13454 39362
1458	3981.53810 88694 21317
1459	3984.70216 41613 14769
1460	3987.86651 70170 99206
1461	3991.03116 72330 33503
1462	3994.19611 46056 55345
1463	3997.36135 89317 80655
1464	4000.52690 00085 03028
1465	4003.69273 76331 93157
1466	4006.85887 16034 98266
1467	4010.02530 17173 41549
1468	4013.19202 77729 21600
1469	4016.35904 95687 11857
1470	4019.52636 69034 60033
1471	4022.69397 95761 87563
1472	4025.86188 73861 89043
1473	4029.03009 01330 31674
1474	4032.19858 76165 54707
1475	4035.36737 96368 68888
1476	4038.53646 59943 55911
1477	4041.70584 64896 67861
1478	4044.87552 09237 26668
1479	4048.04548 90977 23560
1480	4051.21575 08131 18517
1481	4054.38630 58716 39726
1482	4057.55715 40752 83035
1483	4060.72829 52263 11417
1484	4063.89972 91272 54426
1485	4067.07145 55809 07657
1486	4070.24347 43903 32213
1487	4073.41578 53588 54167
1488	4076.58838 82900 64027
1489	4079.76128 29878 16203
1490	4082.93446 92562 28478
1491	4086.10794 68996 81472
1492	4089.28171 57228 18122
1493	4092.45577 55305 43147
1494	4095.63012 61280 22527
1495	4098.80476 73206 82976
1496	4101.97969 89142 11418
1497	4105.15492 07145 54471
1498	4108.33043 25279 17918
1499	4111.50623 41607 66198
1500	4114.68232 54198 21879

N	Log N!
1501	4117.85870 61120 65150
1502	4121.03537 60447 33299
1503	4124.21233 50253 20207
1504	4127.38958 28615 75831
1505	4130.56711 93615 05693
1506	4133.74494 43333 70375
1507	4136.92305 75856 85007
1508	4140.10145 89272 18762
1509	4143.28014 81669 94352
1510	4146.45912 51142 87521
1511	4149.63838 95786 26546
1512	4152.81794 13697 91734
1513	4155.99778 02978 14921
1514	4159.17790 61729 78975
1515	4162.35831 88058 17299
1516	4165.53901 80071 13333
1517	4168.72000 35879 00064
1518	4171.90127 53594 59525
1519	4175.08283 31333 22312
1520	4178.26467 67212 67084
1521	4181.44680 59353 20083
1522	4184.62922 05877 54637
1523	4187.81192 04910 90679
1524	4190.99490 54580 94261
1525	4194.17817 53017 77066
1526	4197.36172 98353 95927
1527	4200.54556 88724 52348
1528	4203.72969 22266 92020
1529	4206.91409 97121 04340
1530	4210.09879 11429 21939
1531	4213.28376 63336 20200
1532	4216.46902 50989 16785
1533	4219.65456 72537 71160
1534	4222.84039 26133 84122
1535	4226.02650 09931 97327
1536	4229.21289 22088 92820
1537	4232.39956 60763 92566
1538	4235.58652 24118 57978
1539	4238.77376 10316 89457
1540	4241.96128 17525 25920
1541	4245.14908 43912 44339
1542	4248.33716 87649 59277
1543	4251.52553 46910 22425
1544	4254.71418 19870 22143
1545	4257.90311 04707 82996
1546	4261.09231 99603 65302
1547	4264.28181 02740 64670
1548	4267.47158 12304 11544
1549	4270.66163 26481 70750
1550	4273.85196 43463 41041

N	Log N!
1551	4277.04257 61441 54646
1552	4280.23346 78610 76816
1553	4283.42463 93168 05374
1554	4286.61609 03312 70270
1555	4289.80782 07246 33126
1556	4292.99983 03172 86796
1557	4296.19211 89298 54916
1558	4299.38468 63831 91462
1559	4302.57753 24983 80304
1560	4305.77065 70967 34765
1561	4308.96405 99997 97183
1562	4312.15774 10293 38464
1563	4315.35170 00073 57651
1564	4318.54593 67560 81480
1565	4321.74045 10979 63948
1566	4324.93524 28556 85872
1567	4328.13031 18521 54462
1568	4331.32565 79105 02882
1569	4334.52128 08540 89819
1570	4337.71718 05064 99052
1571	4340.91335 66915 39026
1572	4344.10980 92332 42415
1573	4347.30653 79558 65702
1574	4350.50354 26838 88748
1575	4353.70082 32420 14367
1576	4356.89837 94551 67903
1577	4360.09621 11484 96806
1578	4363.29431 81473 70208
1579	4366.49270 02773 78502
1580	4369.69135 73643 32925
1581	4372.89028 92342 65134
1582	4376.08949 57134 26791
1583	4379.28897 66282 89147
1584	4382.48873 18055 42622
1585	4385.68876 10720 96392
1586	4388.88906 42550 77977
1587	4392.08964 11818 32825
1588	4395.29049 16799 23903
1589	4398.49161 55771 31282
1590	4401.69301 27014 51734
1591	4404.89468 28810 98315
1592	4408.09662 59444 99966
1593	4411.29884 17203 01097
1594	4414.50133 00373 61191
1595	4417.70409 07247 54391
1596	4420.90712 36117 69101
1597	4424.11042 85279 07584
1598	4427.31400 53028 85557
1599	4430.51785 37666 31792
1600	4433.72197 37492 87716

N	Log N!
1601	4436.92636 50812 07016
1602	4440.13102 75929 55235
1603	4443.33596 11153 09380
1604	4446.54116 54792 57524
1605	4449.74664 05159 98415
1606	4452.95238 60569 41077
1607	4456.15840 19337 04422
1608	4459.36468 79781 16854
1609	4462.57124 40222 15884
1610	4465.77806 98982 47734
1611	4468.98516 54386 66952
1612	4472.19253 04761 36024
1613	4475.40016 48435 24985
1614	4478.60806 83739 11037
1615	4481.81624 09005 78158
1616	4485.02468 22570 16726
1617	4488.23339 22769 23127
1618	4491.44237 07941 99380
1619	4494.65161 76429 52754
1620	4497.86113 26574 95385
1621	4501.07091 56723 43900
1622	4504.28096 65222 19037
1623	4507.49128 50420 45269
1624	4510.70187 10669 50426
1625	4513.91272 44322 65319
1626	4517.12384 49735 23368
1627	4520.33523 25264 60227
1628	4523.54688 69270 13409
1629	4526.75880 80113 21919
1630	4529.97099 56157 25877
1631	4533.18344 95767 66152
1632	4536.39616 97311 83995
1633	4539.60915 59159 20663
1634	4542.82240 79681 17060
1635	4546.03592 57251 13364
1636	4549.24970 90244 48669
1637	4552.46375 77038 60610
1638	4555.67807 16012 85010
1639	4558.89265 05548 55509
1640	4562.10749 44029 03207
1641	4565.32260 29839 56300
1642	4568.53797 61367 39722
1643	4571.75361 37001 74784
1644	4574.96951 55133 78815
1645	4578.18568 14156 64808
1646	4581.40211 12465 41059
1647	4584.61880 48457 10814
1648	4587.83576 20530 71911
1649	4591.05298 27087 16429
1650	4594.27046 66529 30336

N	Log N!
1651	4597.48821 37261 93129
1652	4600.70622 37691 77493
1653	4603.92449 66227 48940
1654	4607.14303 21279 65468
1655	4610.36183 01260 77206
1656	4613.58089 04585 26067
1657	4616.80021 29669 45404
1658	4620.01979 74931 59659
1659	4623.23964 38791 84019
1660	4626.45975 19672 24074
1661	4629.68012 15996 75469
1662	4632.90075 26191 23561
1663	4636.12164 48683 43080
1664	4639.34279 81902 97785
1665	4642.56421 24281 40124
1666	4645.78588 74252 10893
1667	4649.00782 30250 38898
1668	4652.23001 90713 40618
1669	4655.45247 54080 19865
1670	4658.67519 18791 67448
1671	4661.89816 83290 60839
1672	4665.12140 46021 63837
1673	4668.34490 05431 26231
1674	4671.56865 59967 83473
1675	4674.79267 08081 56337
1676	4678.01694 48224 50594
1677	4681.24147 78850 56680
1678	4684.46626 98415 49362
1679	4687.69132 05376 87410
1680	4690.91662 98194 13273
1681	4694.14219 75328 52744
1682	4697.36802 35243 14638
1683	4700.59410 76402 90461
1684	4703.82044 97274 54092
1685	4707.04704 96326 61450
1686	4710.27390 72029 50173
1687	4713.50102 22855 39298
1688	4716.72839 47278 28935
1689	4719.95602 43773 99943
1690	4723.18391 10820 13617
1691	4726.41205 46896 11359
1692	4729.64045 50483 14363
1693	4732.86911 20064 23298
1694	4736.09802 54124 17987
1695	4739.32719 51149 57088
1696	4742.55662 09628 77783
1697	4745.78630 28051 95458
1698	4749.01624 04911 03392
1699	4752.24643 38699 72438
1700	4755.47688 27913 50712

N	Log N!
1701	4758.70758 71049 63281
1702	4761.93854 66607 11850
1703	4765.16976 13086 74451
1704	4768.40123 08991 05132
1705	4771.63295 52824 33649
1706	4774.86493 43092 65153
1707	4778.09716 78303 79887
1708	4781.32965 56967 32873
1709	4784.56239 77594 53610
1710	4787.79539 38698 45764
1711	4791.02864 38793 86864
1712	4794.26214 76397 27998
1713	4797.49590 50026 93509
1714	4800.72991 58202 80688
1715	4803.96417 99446 59477
1716	4807.19869 72281 72164
1717	4810.43346 75233 33080
1718	4813.66849 06828 28304
1719	4816.90376 65595 15356
1720	4820.13929 50064 22905
1721	4823.37507 58767 50466
1722	4826.61110 90238 68102
1723	4829.84739 43013 16130
1724	4833.08393 15628 04824
1725	4836.32072 06622 14117
1726	4839.55776 14535 93308
1727	4842.79505 37911 60767
1728	4846.03259 75293 03641
1729	4849.27039 25225 77564
1730	4852.50843 86257 06359
1731	4855.74673 56935 81753
1732	4858.98528 35812 63081
1733	4862.22408 21439 76998
1734	4865.46313 12371 17189
1735	4868.70243 07162 44082
1736	4871.94198 04370 84555
1737	4875.18178 02555 31653
1738	4878.42183 00276 44301
1739	4881.66212 96096 47014
1740	4884.90267 88579 29613
1741	4888.14347 76290 46944
1742	4891.38452 57797 18589
1743	4894.62582 31668 28582
1744	4897.86736 96474 25130
1745	4901.10916 50787 20329
1746	4904.35120 93180 89880
1747	4907.59350 22230 72811
1748	4910.83604 36513 71195
1749	4914.07883 34608 49872
1750	4917.32187 15095 36166

N	Log N!
1751	4920.56515 76556 19612
1752	4923.80869 17574 51674
1753	4927.05247 36735 45469
1754	4930.29650 32625 75491
1755	4933.54078 03833 77334
1756	4936.78530 48949 47418
1757	4940.03007 66564 42713
1758	4943.27509 55271 80466
1759	4946.52036 13666 37927
1760	4949.76587 40344 52077
1761	4953.01163 33904 19354
1762	4956.25763 92944 95383
1763	4959.50389 16067 94705
1764	4962.75039 01875 90506
1765	4965.99713 48973 14347
1766	4969.24412 55965 55897
1767	4972.49136 21460 62661
1768	4975.73884 44067 39715
1769	4978.98657 22396 49438
1770	4982.23454 55060 11245
1771	4985.48276 40672 01320
1772	4988.73122 77847 52352
1773	4991.97993 65203 53269
1774	4995.22889 01358 48977
1775	4998.47808 84932 40090
1776	5001.72753 14546 82672
1777	5004.97721 88824 87974
1778	5008.22715 06391 22169
1779	5011.47732 65872 06094
1780	5014.72774 65895 14988
1781	5017.97841 05089 78231
1782	5021.22931 82086 79087
1783	5024.48046 95518 54442
1784	5027.73186 44018 94546
1785	5030.98350 26223 42758
1786	5034.23538 40768 95286
1787	5037.48750 86294 00930
1788	5040.73987 61438 60829
1789	5043.99248 64844 28202
1790	5047.24533 95154 08095
1791	5050.49843 51012 57126
1792	5053.75177 31065 83233
1793	5057.00535 33961 45416
1794	5060.25917 58348 53489
1795	5063.51324 02877 67827
1796	5066.76754 66200 99112
1797	5070.02209 46972 08086
1798	5073.27688 43846 05296
1799	5076.53191 55479 50848
1800	5079.78718 80530 54154

N	Log N!	N	Log N!	N	Log N!	N	Log N!
1801	5083.04270 17658 73687	1851	5246.12300 78045 07430	1901	5409.78979 40937 20886	1951	5574.02763 55144 16974
1802	5086.29845 65525 16731	1852	5249.39064 87868 53345	1902	5413.06900 46063 22281	1952	5577.31811 53277 47647
1803	5089.55445 22792 39133	1853	5252.65852 42061 72243	1903	5416.34844 33946 09301	1953	5580.60881 75710 35501
1804	5092.81068 88124 45056	1854	5255.92663 39359 80721	1904	5419.62811 03386 57757	1954	5583.89974 21304 18255
1805	5096.06716 60186 86733	1855	5259.19497 78499 31786	1905	5422.90800 53186 69395	1955	5587.19088 88921 50141
1806	5099.32388 37646 64220	1856	5262.46355 58218 14629	1906	5426.18812 82149 71703	1956	5590.48225 77426 01724
1807	5102.58084 19172 25152	1857	5265.73236 77255 54410	1907	5429.46847 89080 17708	1957	5593.77384 85682 59725
1808	5105.83804 03433 64496	1858	5269.00141 34352 12033	1908	5432.74905 72783 85785	1958	5597.06566 12557 26844
1809	5109.09547 89102 24310	1859	5272.27069 28249 83931	1909	5436.02986 32067 79451	1959	5600.35769 56917 21580
1810	5112.35315 74850 93494	1860	5275.54020 57692 01848	1910	5439.31089 65740 27179	1960	5603.64995 17630 78056
1811	5115.61107 59354 07553	1861	5278.80995 21423 32615	1911	5442.59215 72610 82192	1961	5606.94242 93567 45840
1812	5118.86923 41287 48347	1862	5282.07993 18189 77938	1912	5445.87364 51490 22273	1962	5610.23512 83597 89770
1813	5122.12763 19328 43856	1863	5285.35014 46738 74181	1913	5449.15536 01190 49569	1963	5613.52804 86593 89776
1814	5125.38626 92155 67932	1864	5288.62059 05818 92144	1914	5452.43730 20524 90394	1964	5616.82119 01428 40707
1815	5128.64514 58449 40063	1865	5291.89126 94180 36850	1915	5455.71947 08307 95035	1965	5620.11455 26975 52153
1816	5131.90426 16891 25130	1866	5295.16218 10574 47331	1916	5459.00186 63355 37561	1966	5623.40813 62110 48269
1817	5135.16361 66164 33164	1867	5298.43332 53753 96409	1917	5462.28448 84484 15623	1967	5626.70194 05709 67606
1818	5138.42321 04953 19113	1868	5301.70470 22472 90484	1918	5465.56733 70512 50268	1968	5629.99596 56650 62929
1819	5141.68304 31943 82596	1869	5304.97631 15486 69316	1919	5468.85041 20259 85740	1969	5633.29021 13812 01047
1820	5144.94311 45823 67671	1870	5308.24815 31552 05815	1920	5472.13371 32546 89289	1970	5636.58467 76073 62640
1821	5148.20342 45281 62591	1871	5311.52022 69427 05825	1921	5475.41724 06195 50983	1971	5639.87936 42316 42083
1822	5151.46397 29007 99571	1872	5314.79253 27871 07911	1922	5478.70099 40028 83510	1972	5643.17427 11422 47276
1823	5154.72475 95694 54547	1873	5318.06507 05644 83149	1923	5481.98497 32871 21989	1973	5646.46939 82274 99467
1824	5157.98578 44034 46944	1874	5321.33784 01510 34908	1924	5485.26917 83548 23784	1974	5649.76474 53758 33085
1825	5161.24704 72722 39438	1875	5324.61084 14230 98646	1925	5488.55360 90886 68303	1975	5653.06031 24757 95564
1826	5164.50854 80454 37718	1876	5327.88407 42571 41691	1926	5491.83826 53714 56819	1976	5656.35609 94160 47173
1827	5167.77028 65927 90256	1877	5331.15753 85297 63038	1927	5495.12314 70861 12272	1977	5659.65210 60853 60845
1828	5171.03226 27841 88068	1878	5334.43123 41176 93130	1928	5498.40825 41156 79084	1978	5662.94833 23726 22006
1829	5174.29447 64896 64485	1879	5337.70516 08977 93655	1929	5501.69358 63433 22968	1979	5666.24477 81668 28402
1830	5177.55692 75793 94915	1880	5340.97931 87470 57335	1930	5504.97914 36523 30742	1980	5669.54144 33570 89933
1831	5180.81961 59236 96611	1881	5344.25370 75426 07714	1931	5508.26492 59261 10137	1981	5672.83832 78326 28480
1832	5184.08254 13930 28443	1882	5347.52832 71616 98952	1932	5511.55093 30481 89611	1982	5676.13543 14827 77737
1833	5187.34570 38579 90659	1883	5350.80317 74817 15617	1933	5514.83716 49022 18164	1983	5679.43275 41969 83040
1834	5190.60910 31893 24662	1884	5354.07825 83801 72476	1934	5518.12362 13719 65147	1984	5682.73029 58648 01199
1835	5193.87273 92579 12770	1885	5357.35356 97347 14287	1935	5521.41030 23413 20077	1985	5686.02805 63759 00333
1836	5197.13661 19347 77993	1886	5360.62911 14231 15597	1936	5524.69720 76942 92452	1986	5689.32603 56200 59696
1837	5200.40072 10910 83802	1887	5363.90488 33232 80528	1937	5527.98433 73150 11563	1987	5692.62423 34871 69511
1838	5203.66506 65981 33894	1888	5367.18088 53132 42578	1938	5531.27169 10877 26310	1988	5695.92264 98672 30805
1839	5206.92964 83273 71972	1889	5370.45711 72711 64412	1939	5534.55926 88968 05015	1989	5699.22128 46503 55241
1840	5210.19446 61503 81508	1890	5373.73357 90753 37656	1940	5537.84707 06267 35241	1990	5702.52013 77267 64948
1841	5213.45951 99388 85523	1891	5377.01027 06041 82696	1941	5541.13509 61621 23604	1991	5705.81920 89867 92357
1842	5216.72480 95647 46353	1892	5380.28719 17362 48470	1942	5544.42334 53876 95590	1992	5709.11849 83208 80037
1843	5219.99033 48999 65427	1893	5383.56434 23502 12267	1943	5547.71181 81882 95372	1993	5712.41800 56195 80525
1844	5223.25609 58166 83037	1894	5386.84172 23248 79521	1944	5551.00051 44488 85628	1994	5715.71773 07735 56162
1845	5226.52209 21871 78116	1895	5390.11933 15391 83612	1945	5554.28943 40545 47355	1995	5719.01767 36735 78929
1846	5229.78832 38838 68010	1896	5393.39716 98721 85660	1946	5557.57857 68904 79688	1996	5722.31783 42105 30281
1847	5233.05479 07793 08251	1897	5396.67523 72030 74322	1947	5560.86794 28419 99719	1997	5725.61821 22754 00983
1848	5236.32149 27461 92339	1898	5399.95353 34111 65596	1948	5564.15753 17945 42316	1998	5728.91880 77592 90947
1849	5239.58842 96573 51512	1899	5403.23205 83759 02614	1949	5567.44734 36336 59938	1999	5732.21962 05534 09064
1850	5242.85560 13857 54526	1900	5406.51081 19768 55443	1950	5570.73737 82450 22456	2000	5735.52065 05490 73045